Olmsted/Williams
TECHNOLOGY INTEGRATOR

  **ilw**

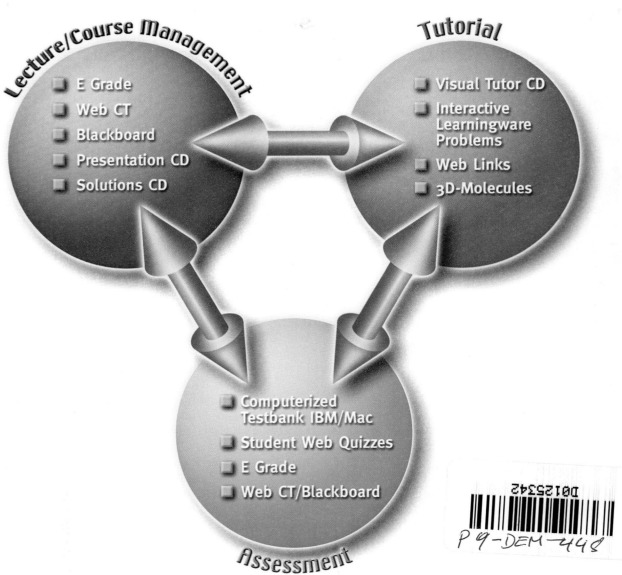

Lecture/Course Management
- E Grade
- Web CT
- Blackboard
- Presentation CD
- Solutions CD

Tutorial
- Visual Tutor CD
- Interactive Learningware Problems
- Web Links
- 3D-Molecules

Assessment
- Computerized Testbank IBM/Mac
- Student Web Quizzes
- E Grade
- Web CT/Blackboard

IMPORTANT NOTICE FOR FACULTY AND STUDENTS

The password card enclosed in the CD pouch on the opposite page contains your registration access code for the Student Companion Site for *Chemistry, 3rd edition*. Be sure to register for the website at www.wiley.com/college/olmsted to use the valuable features such as interactive problem tutorials, on-line quizzing, practice tests, molecular models, and links to other important resources.

About the Authors

John Olmsted III is currently Professor of Chemistry at California State University, Fullerton, where he was named the Outstanding Professor in 1997–1998 and was chairman of his department from 1998–2001. John has also taught at the American University of Beirut, UCLA, and the University of North Carolina at Chapel Hill. He has had visiting research appointments at the Max-Planck-Institut für Biophysikalische Chemie, Göttingen, University of California at San Diego and Sandia National Laboratory in Albuquerque.

John has a BS degree in chemistry from Carnegie Institute of Technology (now Carnegie-Mellon University), a PhD in physical chemistry from UC Berkeley, and postdoctoral work at the Lawrence Berkeley Laboratory.

His research in experimental physical chemistry has been supported by research grants from several sources and has led to more than 30 refereed publications. John has also published regularly on chemical education topics in the *Journal of Chemical Education*. In his spare time, John enjoys gardening, photography, and travelling with his wife Eileen.

(Photo of John Olmsted III courtesy of Patrick O'Donnell, staff photographer of California State University at Fullerton.)

Greg Williams is an Adjunct Professor of Chemistry at the University of Oregon. His teaching background includes introductory, general, organic, and inorganic chemistry. He writes textbooks and develops multimedia materials for chemistry students. Greg has also taught and conducted research at California State University, Fullerton, UCLA, and the University of California, Irvine. He earned his undergraduate degree from UCLA and his PhD in inorganic chemistry from Princeton University. His research interests are in synthetic and mechanistic inorganic and organometallic chemistry.

When he is not teaching or writing about chemistry, Greg can be found somewhere in western North America climbing, backpacking, river rafting, skiing, or fly fishing. He also sings with the Eugene Concert Choir. He is married to Trudy Cameron, a professor of economics at the University of Oregon. They have two daughters, Casey (13) and Perry (6). Greg absolutely insists on enjoying life.

*To all those students
whose determination to learn chemistry
inspired us to write this book*

CHEMISTRY

THIRD EDITION

John Olmsted III

California State University, Fullerton

Gregory M. Williams

University of Oregon

WILEY

John Wiley & Sons, Inc.

ASSOCIATE PUBLISHER	David Harris
DEVELOPMENTAL EDITOR	Ellen Ford
MARKETING MANAGER	Bob Smith
PRODUCTION EDITOR	Barbara Russiello
SENIOR DESIGNER	Harry Nolan
SENIOR ILLUSTRATION EDITOR	Sigmund Malinowski
PHOTO EDITORS	Hilary Newman Jennifer MacMillan
PHOTO RESEARCHER	Elyse Rieder
COVER DESIGN	Harry Nolan
INTERIOR DESIGN	Lisa Delgado/Delgado Design
COVER PHOTOGRAPH	© Roderick Chen/SUPERSTOCK
COVER ILLUSTRATION	Norm Christiansen
ELECTRONIC ILLUSTRATIONS	Imagineering, Inc.
MOLECULAR ART	Gregory M. Williams, Davi Erickson

This book was set in 10.5/12 Bembo by Progressive Information Technologies and printed and bound by Von Hoffman Press, Inc. The cover was printed by Von Hoffman Press, Inc.

This book is printed on acid free paper. ∞

Copyright © 2002 John Wiley & Sons, Inc. All rights reserved.

Previous editions copyrighted 1994 by Mosby-Year Book, Inc., 1997 by Wm. C. Brown

No part of this publication may be reproduced, stored in a retrieval system or transmitted in any form or by any means, electronic, mechanical, photocopying, recording, scanning or otherwise, except as permitted under Sections 107 or 108 of the 1976 United States Copyright Act, without either the prior written permission of the Publisher, or authorization through payment of the appropriate per-copy fee to the Copyright Clearance Center, 222 Rosewood Drive, Danvers, MA 01923, (508)750-8400, fax (508)750-4470. Requests to the Publisher for permission should be addressed to the Permissions Department, John Wiley & Sons, Inc., 605 Third Avenue, New York, NY 10158-0012, (212) 850-6011, fax (212) 850-6008, E-Mail: PERMREQ@WILEY.COM.
To order books or for customer service please call 1(800)225-5945.

Library of Congress Cataloging in Publication Data
Olmsted, John.
   Chemistry / John Olmsted III, Gregory M. Williams.—3rd ed.
      p.   cm.
   Includes index.
   ISBN 0-471-39071-2
   1. Chemistry.   I. Williams, Gregory M.   II. Title.
QD 2001
540—dc20   CIP
Printed in the United States of America
10 9 8 7 6 5 4 3 2

Preface

Why should anybody bother to read a preface? Why, for that matter, should any textbook author bother to write one? Our dictionary defines a preface as an introductory statement that explains the subject, plan, or purpose of the book. There are many general chemistry textbooks on the market, each claiming to be unique. Having compared many of these textbooks, we know that there is a tremendous amount of similarity among them. We believe that our text is different from all the rest in significant and important ways, which we describe in this preface.

The subject of this book is obvious from its title, and the general plan is clear from the table of contents. In those respects, our text does not deviate substantially from others, although there are significant differences which we describe below. Our goal in this preface is to explain our purpose and clarify the plan. We want to help our potential readers discover how the third edition of *Chemistry* can guide students to a confident mastery of the fundamentals of chemistry.

THE MOLECULAR PERSPECTIVE: INFORMING QUANTITATIVE REASONING

Our purpose in the third edition of our book is the same as in previous editions: to describe the fundamentals of chemistry as a chemist thinks about the subject. Matter is composed of atoms and molecules, and chemistry explains the behavior of matter by describing how atoms and molecules behave and interact. We believe that students will master chemistry more quickly and thoroughly when they grasp this molecular perspective; hence our theme song, "Think molecules."

When we published our first edition, no major general chemistry textbook emphasized the molecular approach. Now almost every textbook shows some molecular pictures. *Our text remains unique, however, in reinforcing each chemical topic with explanations of what the atoms and molecules are doing.* Open to any chapter and you will find phenomena illustrated at the molecular scale: reaction mixtures demonstrating stoichiometry, solutions showing the presence of dissociated ions or intact molecules, liquid/vapor equilibria showing the physical differences between the two phases, geometrical structures of both simple and more complex molecules. Beyond these illustrations, we have also shaped the text itself to focus on the molecular viewpoint.

Our molecular approach to concepts emerges particularly strongly in Chapter 4, where we emphasize the molecular underpinnings of stoichiometry; in Chapter 5, where we introduce the ideal gas equation by describing how molecular kinetic behavior varies with macroscopic variables; in Chapter 14, where we develop chemical kinetics starting with reaction mechanisms; and in Chapter 16, where we emphasize the importance of identifying the species in solution as the first step in understanding aqueous chemical equilibria.

Because chemists work quantitatively just as much as they think molecularly, our molecular approach is designed to facilitate successful quantitative reasoning, not to replace it. To see one example of this approach, explore our presentation (Section 15.6, pp. 748–751) of the species in aqueous solution as an essential prologue to equilibrium calculations involving solution phase reactions. Look, too, at how we introduce the ideal gas equation: a qualitative description of how gaseous molecules behave is used to develop the quantitative statement (Section 5.2, pp. 197–200).

■■■■ EMPHASIS ON PROBLEM SOLVING

Students usually want easy algorithms for problem-solving, but instructors usually want them to reason through problems rather than relying mindlessly on algorithms. We believe there is validity to both views, because whereas reasoning is at the heart of any science, every experienced scientist regularly employs algorithms as "short-cuts." Reflecting this view, we have developed step-wise approaches to problem-solving that are algorithmic but not mindlessly so. We use a seven-step problem-solving template consistently for quantitative problems, including a version designed specifically for attacking equilibrium problems. We also provide step-by-step procedures for the construction of Lewis structures and for the balancing of redox reactions. Beyond this, for each Example Problem we provide a "strategy" of attack, designed to encourage students to reason their way to solutions rather than search aimlessly for algorithms.

■■■■ NEW IN THE THIRD EDITION

In preparing this third edition, we wanted to improve upon the second edition while retaining what readers identified as its strengths. We made one major change in organization and major changes in the artwork. New also are many of the chapter introductions and problems designed for group learning. A major thrust has been the development of electronic materials—CD and web-based—not as supplements to the textbook, but rather integrated with the textbook. Key media components have been developed by the authors themselves, a practice unusual in this discipline. Our integrated media falls into three distinct categories: **Lecture/Course management, Tutorial, and Assessment.**

Expanded Coverage

This edition has one more chapter than previously, because we have expanded our coverage of equilibrium from two to three chapters. Two chapters simply are not enough to cover the basic features of this important topic. Chapter 15, "Principles of Chemical Equilibrium," retains much of the content of its counterpart in previous editions. Chapter 16, "Aqueous Acid-Base Equilibria," is devoted entirely to principles of aqueous acid-base equilibria, which play central roles in biological and geological systems as well as in chemistry. Chapter 17, "Applications of Aqueous Equilibria," provides expanded coverage of buffers and titrations as well as solubility and complexation.

Improved Art Program

Computer graphics have become versatile enough to make it possible to generate molecular art of outstanding quality. We have exploited this capability, adding molecular views and refining those that were part of the "signature" of the first two editions. To complement the molecular artwork, an entire set of new photographs was selected for this edition.

OTHER UNIQUE FEATURES

When we wrote the first edition of *Chemistry* we introduced features that were not present in other general chemistry textbooks but that we felt reflected how contemporary chemistry should be taught. These resonated well with our users, so we have retained them: molecular approaches to many traditional topics (described above); use of organic examples equally with inorganic ones; thermodynamics presented all together; a "rational" approach to Lewis structures; and a separate chapter on macromolecules.

Whereas most other general chemistry textbooks restrict their use of organic chemicals to a separate chapter, we feel that general chemistry is just that—a general introduction to principles that apply equally well to all categories of chemical compounds. Moreover, students encounter many organic (and biochemical) substances in their daily lives, so the use of organic examples can help to excite student interest and motivation. We introduce the "line structure" approach to organic structures in Chapter 3, allowing us to use "interesting" molecules as we introduce basic chemical concepts. As one reviewer of our second edition wrote, "Olmsted and Williams take advantage of this [line structures] by immediately showing students interesting molecules including cholesterol, malathion, various amino acids and hundreds of other examples. Seeing these molecules piques students' interests and, furthermore, makes them feel like scientists because they are learning the language used by practicing chemists."

Many years' teaching experience convinced us that the standard approach to thermodynamics, with thermochemistry presented in a stand-alone chapter in the first half of the course and other aspects introduced later, can intimidate students. We chose to introduce energy and its conservation—one of the central themes of chemistry—early and often in "just in time" fashion. We treat all the rest of thermodynamics in two linked chapters, 12 and 13, early in the second half of the textbook. Several advantages result from this organization. First, students learn about energy in the context of chemical processes. Second, the more sophisticated aspects of enthalpy and path independence are delayed until students are better ready to understand them. Third, a "seamless" connection between enthalpy, entropy, and free energy provides better continuity and greater efficiency. Fourth, presenting thermodynamics before kinetics, equilibrium, and electrochemistry allows us to apply thermodynamic reasoning to these topics.

We and others (see Packer and Woodgate, *Journal of Chemical Education* **68**, 456 (1991)) have developed a procedure for constructing Lewis structures that generates correct structures without invoking exceptions and special chemical knowledge. The procedure is both rational and general, allowing students to build correct structures for molecules and ions containing multiple bonds, resonance structures, and "expanded octets."

Macromolecules, both natural and synthetic, are everywhere in the contemporary world. In recognition of their importance, we devote Chapter 11 to describing macromolecule formation and structure. The chapter builds on the structure and bonding information of earlier chapters, but it can be omitted without loss of continuity. We placed this chapter at the natural break between semesters for a two-semester sequence, allowing instructors to cover it at the end of semester one, beginning of semester two, or not at all.

◼◼◼ PROVIDING WHAT STUDENTS NEED

Our discussions with students and colleagues, over the many years that we have taught general chemistry, have convinced us that visualization, relevance, timeliness, readability, and problem-solving strategies are all important for student success.

Students taking general chemistry generally tend not to major in chemistry. They are pursuing careers in the health sciences, biology, agriculture, engineering, geology, or physics. Many of these students need to be convinced that a sound understanding of chemical principles is essential to their field of interest. Our chapter introductions offer short descriptions of topics that have "real-world" importance and are related to material that appears in the chapter. Boxed material provides further examples that are grouped into five general themes: *Chemistry and Life, Chemistry and the Environment, Chemistry and Technology, Chemical Milestones,* and *Tools for Discovery*. In addition, some chapters end with a section that is of interest to a particular group of students but that can be omitted without loss of continuity. For example, see sections 13.6, Bioenergetics; 17.5, Complexation Equilibria; and 19.6, Transition Metals in Biology.

Students frequently do not see the importance of chemical topics that are presented outside a relevant context. For this reason, we have chosen to use "just-in-time" presentations of topics that the bulk of our audience needs to know. As an example, although we introduce energy and its conservation in Chapter 2, we defer detailed discussion of energetic concepts until they are needed, in the context of gas behavior, interactions of light and atoms/molecules, orbital and bond energies.

◼◼◼ LEARNING RESOURCES

Our third edition comes with an array of learning resources. Some of these were included in earlier editions and have been refined, while others are new to this edition. These resources can be divided into three categories: *features in the textbook, tools designed for the student,* and *ancillaries addressed to the instructor.*

Text Features

To provide students with early and continuous reinforcement, each section of the text ends with three *Section Exercises* designed to test comprehension of concepts introduced in that section. Answers to these exercises appear at the end of each chapter. Also, at the end of each chapter we provide a *Chapter Summary* including *Key Terms*, a list of *Skills to Master*, and a set of *Learning Exercises* to help students review and solidify their understanding.

Earlier editions received high marks for the variety and quality of *Chapter Problems*. We have sought to improve on this already-strong feature. After a set of *Paired Problems* specific to each section of the chapter, we present *Additional Paired Problems* that are not section-specific. These are followed by additional unpaired problems identified as *More Challenging Problems*, the last several of which—a new feature in this edition—are particularly suited for and designated as *Group Study Problems*.

Student Tools

Technology based resources are essential parts of the general chemistry toolkit, but for the most part these have not been well-integrated with the textbook. Our technology components for students fall into two distinct categories: *Tutorial and Assessment.*

Tutorial

To truly enhance visualization, molecular understanding, and problem solving, we believe that the computer-based materials for student use should closely mirror the molecular and problem-solving approach of the textbook, and be developed simultaneously with the textbook. For that reason we have worked actively with the software developers to ensure products that truly complement our text.

- **The *Visual Tutor CD* by Greg Williams,** included with the text, contains animations, 3-D molecules, and tutorials that we designed and edited. These address *key concepts* such as identification of molecular/ionic species in solution, construction of Lewis structures, properties of atomic orbitals, and other molecular visualizations. The CD-ROM also has a number of simple *molecular visualization* clips to aid students in learning how to "think molecules." The CD content is also linked to the on-line problem tutorials. A Web/CD icon in the text margin indicates where the media is relevant and will direct the reader to the appropriate figure or exercise.

Assessment

- The website for the text (*www.wiley.com/college/olmsted*) includes ***Interactive Learningware (ILW)* by John Olmsted,** a step-by-step problem solving tutorial program that guides students through selected problems from the book. Students are also able to access specific correlated text content on-line to assist in any difficulty understanding these ILW problems. When appropriate, students can also link to a molecular representation of a problem. The ILW problems are synthesis type problems that require students to use many chemical concepts. This, too, was designed and edited by us. ILW problems are indicated by this icon in the text.

- Other website features include on-line quizzing, links to other outside on-line resources and practice tests.
- ***E-grade*** is an on-line quizzing and homework management program that allows students to do practice tests and email homework assignments directly to the professor.
- **The *Student Study Guide*** *(0471-03511-4),* by Debbie Finocchio, University of San Diego, includes chapter overviews, learning objectives, sample exercises with worked out solutions, and self-test questions for each chapter.
- **The *Student Solutions Manual*** (0471-03512-2), by David Robichaud, California State University, Fullerton, provides worked-out solutions for the odd-numbered problems in the text.

Instructor Ancillaries

Instructor ancillaries are further integrated with the text by providing useful lecture and course management tools.

- A CD-ROM contains the full Solutions Manual and permits an instructor to select, collate, and print selected solutions to whatever set of chapter problems she or he has assigned.
- A presentation CD by William Zoller, University of Washington, with PowerPoint rendered text art is available for classroom projection. Molecular animations by Greg Williams based on art from the text are also included on this CD.

- A set of full color transparency acetates suitable for classroom presentation reproduces important text art. Illustrations have been rendered for projection clarity.
- For further course and classroom management, we offer instructors Web CT, Blackboard, and E-Grade options. These web-based programs allow instructors to set up complete on-line courses with chat rooms, bulletin boards, quizzing, student tracking, and homework management.
- A completely new print and web-based Instructor's Manual, prepared by Donna Friedman, St Louis Community College, provides chapter overviews, learning objectives and lecture outlines for each chapter.
- A thoroughly revised Test Bank, written by Wayne Tikkanen, California State University Los Angeles, includes a significant number of molecular reasoning problems.

■■■ ACKNOWLEDGMENTS

We have benefited immensely from the constructive criticisms made by early development editors, Dick Morel and David Chelton; from thoughtful and careful proofreading and indexing of pages by Gloria Hamilton; from the combination of enthusiasm and determination with which David Harris and Ellen Ford guided our work; from the steering through production by Barbara Russiello; from the efficient coordination of supplementary materials provided by Jennifer Yee; and from the willingness of Eileen Olmsted and Trudy Cameron to tolerate and even understand one more all-consuming labor of love by their author husbands.

The art for this edition was a major undertaking ably coordinated by Sigmund Malinowski at Wiley. For the many fine photographs that help convey the beauty of chemistry, we are indebted to Wiley's photo research team, Hilary Newman, Jennifer MacMillan, and Elyse Rieder, and to two excellent professional photographers, Stephen Frisch and Andy Washnik (John Olmsted also contributed his amateur skills). The many molecular views were created by Davi Erickson, a master's student in chemistry at California State University Fullerton, with assistance from David Rios, a postgraduate student at CSUF. The line art was faithfully rendered by Imagineering.

The electronic ancillary materials for this project were the work of a diverse and talented group, coordinated by Martin Batey at Wiley. Under Greg Williams' guidance, the animations for the Visual Tutor CD-ROM were developed by Virgil Carroll using molecular depictions that were designed and rendered by Davi Erickson. Professor Arnold Holland of the CSUF Art Department contributed the overall design for this element. Working with John Olmsted, Lenox Software (particularly, Rob Kestyn and Tom Lewis) produced the Interactive Learningware.

We are especially grateful to the many peer reviewers who offered their feedback, expertise and comments on the third and previous editions. Their comments helped us enormously in terms of accuracy, content and style. Reviewers of this edition include:

Robert Balahura
University of Guelph

David Ball
Cleveland State University

Gary Buckley
Cameron University

Larry Brown
Texas A&M University

Jim Byrd
California State University-Stanislaus

Michael Chetcuti
Universite Louis Pasteur, France

Stephen R. Daniel
Colorado School of Mines

James Falcone
West Chester University

Debbie Finocchio
University of San Diego

J. S. Francisco
Purdue University

L. Peter Gold
Pennsylvania State University

Thomas Greenbowe
Iowa State University

Hans Gunderson
Northern Arizona University

James Harrison
Michigan State University

James Hovick
*University of North Carolina,
Charlotte*

Ronald C. Johnson
Emory University

Philip Keller
University of Arizona

Pamela Kerrigan
*Manhattan College/College of Mt.
St Vincent*

Robley Light
Florida State University

Jeffry Madura
Duquesne University

Gary Mort
Lane Community College

Wyatt Murphy
Seton Hall University

George Reilly
University of Delaware

Dale Russell
Boise State University

Brian Sanctuary
McGill University

John R. Sowa
Union College

Stephen Summers
Seminole Community College

Larry Thompson
University of Minnesota, Duluth

Worth Vaughan
University of Wisconsin, Madison

John S. Winn
Dartmouth College

To those reviewers who tested and provided feedback on the *Visual Tutor CD*, we also send our thanks. Their feedback was practical, insightful, and helpful. These media reviewers include Mufeed Basti, North Carolina A&T University; Pamela Brown, New York City Technical College; John J. Dolhun, Norwalk Community College; Richard H. Langley, Stephen F. Austin State University; Jack McKenna, St. Cloud State University; Michael Louis Norton, Marshall University; Thomas Pentecost, Aims Community College; Michael Russell, Mt. Hood Community College; Richard S. Treptow, Chicago State University; In addition, thanks go to reviewers on the previous two editions who gave us the foundation on which to build. They include:

Reviewers of
the First Edition

Bruce Ault
University of Cincinnati

Caroline Ayers
East Carolina University

George Baldwin
University of Manitoba

Jon M. Bellama
University of Maryland

Allan R. Burkett
Dillard University

Donald Campbell
University of Wisconsin, Eau Claire

John F. Cannon
Brigham Young University

Grover W. Everett
University of Kansas

Michael D. Fryzuk
University of British Columbia

Steven D. Gammon
University of Idaho

Michael F. Golde
University of Pittsburgh

Paul Hunter
Michigan State University

Richard F. Jordan
University of Iowa

Paul J. Karol
Carnegie Mellon University

Robert Kiser
University of Kentucky

Joseph W. Kolis
Clemson University

David F. Koster
*Southern Illinois University
at Carbondale*

Glenn D. Kuehn
New Mexico State University

Richard S. Lumpkin
University of Alabama

John Luoma
Cleveland State University

Bruce E. Norcross
SUNY, Binghamton

Henry W. Offen
*University of California,
Santa Barbara*

M. Larry Peck
Texas A & M University

John V. Rund
University of Arizona

Martha E. Russell
Iowa State University

Sanford A. Safron
Florida State University

Caesar V. Senoff
University of Guelph

Joanne Stewart
Hope College

Dwight A. Sweigart
Brown University

Wayne Tikkanen
California State University, Los Angeles

Charles A. Trapp
University of Louisville

D. Rodney Truax
University of Calgary

Reviewers of the Second Edition

Lavoirs Banks
Elgin Community College

Paul Braterman
University of North Texas

Larry Brown
Texas A & M University

John F. Cannon
Brigham Young University

Terry S. Carlton
Oberlin College

T.A. George
University of Nebraska, Lincoln

John M. Halpin
New York University

Thomas A. Holme
University of Wisconsin, Milwaukee

James A. Ibers
Northwestern University

Virginia Indivero
Swarthmore College

Milton Johnston, Jr.
University of South Florida

Glenn D. Kuehn
New Mexico State University

Glenn Millhauser
University of California, Santa Cruz

Gholam Mirafzal
Drake University

Gary Mort
Dixie Community College

Melinda E. Oliver
Louisiana State University

John V. Rund
University of Arizona

Venkatesh M. Shanbhag
Mississippi State University

William Zoller
University of Washington

To the Student—Steps to Success in Chemistry

S uccess in chemistry requires a blend of ingredients. It requires a clearly-presented body of information; we hope you will find that in this text-book. It requires lucid instruction from a committed teacher; we hope that our text facilitates such instruction. Finally, success in chemistry requires commitment and hard work from the student. We have tried to structure our program into three categories: *Lecture/Text, Tutorial,* and *Assessment.*

STEP ONE: Lecture/Text

Attend Lectures, Read Your Assignment, and Visualize Chemistry!

- While there is no single formula that is guaranteed to work for every type of student, there are strategies that successful students consistently recommend. Foremost among these is a focus on understanding concepts, because memorization without understanding leads to frustration, not to success. We explain principles using logical underpinnings that can make them easier to understand.

- To truly understand chemical principles as a chemist does, you must 'think molecules.' **Visualizing** what molecules are and what they do is vital to this understanding. The illustration and photo program for the text was designed and created with that purpose in mind. Examine and think about the art as you study.

  Read the chapter introductions while considering the chapter opening photos. Pay close attention to the photos and diagrams in the text. Note the 3-dimensional look of the molecules, see and compare the different ways of representing molecules through molecular art, chemical formulas, photographs and drawings.

- Regardless of your major, a sound **understanding** of chemical principles and applications is essential to your field of interest. To make the study of chemistry more relevant, understandable and enjoyable, we provide boxed 'real-world' examples of how chemistry is all around you. Don't skip over these readings. Use them to enhance your understanding of *Chemistry and Life, Chemistry and the Environment, Chemistry and Technology, Chemical Milestones,* and *Tools for Discovery.*

  Other text sections will relate to such timely topics as atmospheric chemistry, bioenergetics, and industrial chemistry.

- Much of chemistry is concerned with the **applications of concepts** to practical problems. We have designed **problem-solving strategy steps** for typical quantitative problems, as well as for equilibrium problems, construction of Lewis struc-

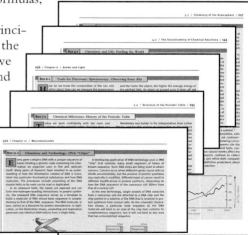

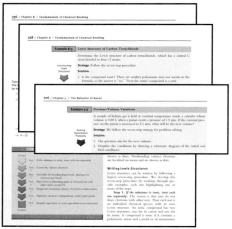

tures, and balancing of redox equations. These are indicated by problem solving logos throughout the text. Our text is laced with **Example Problems, Section Exercises**, and **Chapter Problems** designed to help you learn such applications.

- Each **Example Problem** will remind you of those steps by a margin icon and includes a brief explanation of the method, which outlines how the problem should be approached. Then it provides a step-by-step description of the solution.
- **Section Exercises** appear at the end of each section and are designed to give you immediate practice in applying the concepts presented in that section. So that you can know immediately whether or not you are reasoning correctly, we provide the answers to all Section Exercises at the end of each chapter.

STEP TWO: Tutorial

- To enhance your **visualization** skills, use the Visual Tutor CD by Greg Williams included with this text. *Key concept tutorials* and *molecular animations* are provided to enable you to further examine molecular illustrations in an interactive format.
  - Icons in the text margin will alert you when an appropriate animation or exercise is relevant to the text topic.
  - To build and reinforce your skills in **problem-solving**, the website (*www.wiley.com/college/olmsted*) offers *Interactive Learningware* **(ILW)** by John Olmsted, step-by step problem tutorials on key concepts through guided problem solutions to selected problems from the book, marked with the ilw icon.
- Additional web-based activities include practice quizzing as well as E-Grade, an on-line homework and quizzing program that affords your further practice and problem solving.

STEP THREE: Assessment: Are You Ready to Take the Test?

- At the end of each chapter, we provide additional material designed to engage you in **active learning**. For greatest effectiveness, use this material to guide the manner in which you study.

Each chapter ends with **Summary and Key Terms** to provide a brief overview of the major themes of the chapter and flag the terms with which you must be familiar. **Skills to Master** reminds you what problem solving techniques require your attention, while the **Learning Exercises** are qualitative questions designed to help you organize your ideas about the material in the chapter.

- Web-based quizzing offers additional problems to test yourself.
- **E-Grade,** an on-line homework and quizzing program, allows you to complete and deliver your homework to your professor over the internet.
- Proficiency in using chemical concepts comes only with practice. The **Problems** at the end of each chapter are designed to give you the opportunity for such practice. About half the chapter problems are grouped by section, and these problems are "paired." The full solution to each odd-numbered problem appears in the **Student Solutions Manual**, but

the solution to its companion even-numbered problem is not provided. Although you will know what concepts to use in working these problems, an equally important skill in problem solving is the ability to identify the concepts underlying the problem. For this reason, we have included **Additional Paired Problems** that are not identified by section. These problems are not necessarily more difficult than those identified by section; but by placing them randomly, we give you the opportunity to learn how to recognize problem types. These problems also are "paired," allowing you to use odd-numbered problems to guide your learning and then to test your abilities on even-numbered problems. The chapter problems also include **More Challenging Problems**, designed to stretch your thinking about chemistry, and the last several of these in each chapter have been designed to be particularly suitable for **Group Study Problems**. In working these problems, we think you will benefit from engaging in discussion with fellow students.

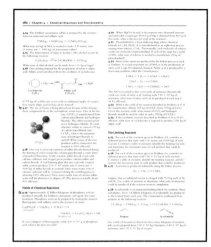

A FEW CLOSING THOUGHTS

Two axioms characterize successful students, in our experience. The first is an attitude: *Be an active learner*. *Ask* questions, *seek* help from many sources, *form* study groups, *work* extra problems, *prepare* chapter outlines. Try a combination of these and additional strategies until you find a set that works best for you. The second is a perspective: *Think molecules*. Every phenomenon in chemistry has an atomic/molecular basis that can make the phenomenon easier to understand. Ask yourself *what* the molecules are doing, and *why*. Imagine yourself to be the size of a molecule and ask what you would see. Learn how to draw pictures showing what goes on at the molecular level. When you have mastered this perspective, you will have learned how to think like a chemist and will appreciate the unity of chemistry. You may even decide that you *are* a chemist!

Contents in Brief

Contents

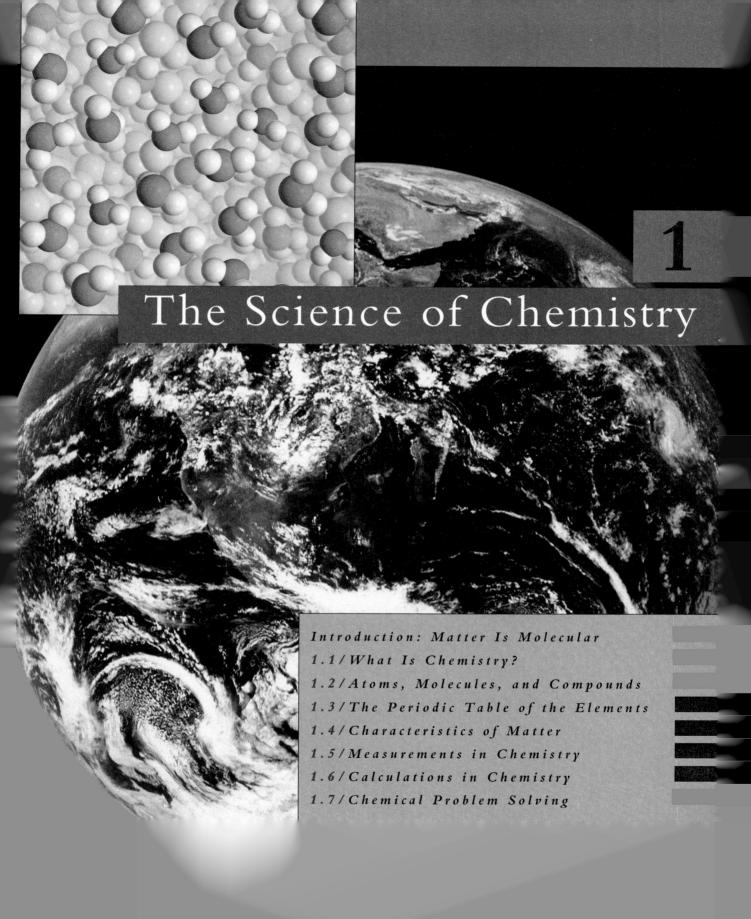

1

The Science of Chemistry

INTRODUCTION: MATTER IS MOLECULAR

A view of the Earth from space shows that our planet is an integrated whole. At the same time, we know that the Earth is almost mind-boggling in its diversity. Nevertheless, the stunning complexity of our world can be described using a small set of fundamental chemical principles. These principles of chemistry are the subject of this book.

The entire universe is made up of matter, from the diversity on Earth, to the vast reaches of the galaxies, to the contents of a simple glass of water. As we describe in the coming chapters, matter is composed of tiny particles called "atoms." On Earth there are around 100 different kinds of atoms, each kind with its own unique combination of properties. The complexity of our world arises from the unlimited number of ways that atoms can combine to form different molecules. The principles of modern chemistry are organized around the molecular nature of matter. Our book presents this perspective while at the same time emphasizing the quantitative aspects of chemistry.

The principles of chemistry focus on the structure and interactions of atoms and molecules. In keeping with this theme, each chapter of this book opens with an introduction that includes a molecular view of some aspect of chemistry. The view in Chapter 1 shows molecules of water. The smallest drop of water contains an unimaginable number of molecules, each one of identical composition. Water is essential to life as we know it. The simple yet unusual fact that solid water (ice) floats atop liquid water allows life to exist on our planet. Just as important is the fact that water dissolves an immense range of chemical compounds: Water is the solvent of life. In fact, water is so important to our perspective of life that the search for water is a key feature of our quest to discover life in other quarters of the galaxy.

Does the same chemistry that takes place on the Earth occur within the galaxies and nebulae in the far reaches of the universe? We have no way of knowing for certain, but observations made by astronomers are consistent with chemistry being the same throughout the universe. Moreover, from research on stars, chemists have learned that the various kinds of atoms probably form during stellar evolution and are dispersed throughout the universe by supernova explosions.

Chemists are interested in a huge range of problems, extending from the galactic scale to what takes place between individual atoms and molecules. Despite its immense perspective, much of chemistry is focused on practical issues, such as:

- Can we make new plastics that are both nontoxic and biodegradable?
- What causes the ozone hole in the Earth's upper atmosphere?
- What role does lithium play in the brain chemistry of schizophrenia?
- How small can we make computer chips?
- Can we find effective treatments for cancer, heart disease, and AIDS?

As this list suggests, chemistry is important to everything from consumer products, to brain functions, to the operation of a computer. We hope you enjoy your study of chemistry!

1.1 WHAT IS CHEMISTRY?

Science, in the broadest sense, can be viewed as an attempt to organize and understand our observations of nature. Because this is a vast undertaking, science is subdivided into various disciplines, including chemistry, biology, geology, and physics. Chemistry is the science that studies the properties and interactions of matter. Chemists seek to understand how chemical transformations occur by studying the properties of matter. Because of the broad scope of chemistry, the interests of chemists intertwine with those of physicists, biologists, and geologists.

Matter is anything that possesses mass and occupies space.

How Chemistry Advances

Chemists learn about chemical properties by performing experiments. They organize information about chemical properties using general principles and theories. The periodic table, for example, organizes the elements according to chemical properties. Chemists use general principles and theories to make predictions about yet-unknown substances. These predictions generate experiments whose results may extend the scope of the principles and theories.

Chemical research is driven by many goals, and it progresses in many different ways. The essential traits of a good researcher are curiosity, creativity, flexibility, and dedication. Some chemical advances come from a direct assault on a known problem. A classic example is the development of the Hall process for refining aluminum from its ores, which we describe in Chapter 19. As a contemporary example, many scientists around the world are working at a feverish pace to develop a vaccine against the AIDS virus.

Chemistry also advances when an imaginative researcher recognizes the potential of a lucky accident. Synthetic dye-making, the first major chemical industry, arose from one such lucky accident. While searching for a way to synthesize quinine, a drug for the prevention of malaria, English chemist William Perkin accidentally made a beautiful reddish-violet dye, which he called mauveine. Perkin had the imagination to realize the commercial potential of the new substance and switched his research interests from drugs to dyes. His insight made him rich and famous.

Chemical advances frequently are driven by technology. The discovery that atoms have inner structure was an outgrowth of the technology for working with radioactive materials. In Chapter 2 we describe a famous experiment in which the structure of atoms was studied by bombarding a thin gold foil with subatomic particles. A contemporary example is the use of lasers to study the details of chemical reactions. We introduce these ideas in Chapters 6 and 7.

Methods of Science

However a new chemical discovery arises, an essential component of science is to explain that discovery on the basis of general principles. When a new general principle is posed, it is termed a **hypothesis.** A hypothesis is tentative until it can be confirmed in two ways. First, additional observations must be consistent with the hypothesis; second, the hypothesis must predict new results that can be confirmed by experiments. If a hypothesis meets these tests, it is promoted to the status of a **theory.** A theory is a unifying principle that explains a collection of facts.

Experimental observations are at the heart of chemical research. Many experiments are designed specifically to answer some particular chemical question. Often, the results of these experiments are unexpected and lead to new hypotheses. New hypotheses, in turn, suggest additional experiments. The Chemistry and Life Box describes how the hypothesis of extraterrestrial life can be tested.

A typical example of the interaction between hypothesis and experiment is the story of the work that resulted in world-wide concern over the depletion of the

Box 1-1 Chemistry and Life: Is There Life on Other Planets?

Speculation about life on other planets probably began when humans discovered that the Earth is not unique. We know that several other planets of the solar system bear at least some resemblance to our own. Why, then, should there not be life on Mars, or Venus, or perhaps on undiscovered Earth-like planets orbiting some other star?

How can scientists collect experimental evidence about possible life on another planet? Sending astronauts to see for themselves is impractical at our current level of technology. Nevertheless, it is possible to search for life on other worlds without sending humans into space. In the late 1970s, NASA's *Viking* spacecraft lander collected a sample of dirt from Mars, the planet in our solar system most like Earth. The sample showed no signs of life.

Indirect evidence can be collected without actually visiting a planet. Recent photographs taken from fly-by spacecraft offer tantalizing hints. NASA's *Galileo* took the photographs, shown here, of Europa, one of Jupiter's moons. The close-up photo of the surface of Europa shows what appear to be huge broken chunks of ice, which suggests that there may be liquid water under the ice, warmed by tidal forces generated by Jupiter's huge mass and strong gravity. Because life seems to require the presence of water, this observation indicates that there could be life on Europa.

The photo above, taken by the *Viking* spacecraft, shows that the surface of Mars has been eroded, apparently by liquid water, indicating that Mars may once have been much warmer than it is today. Planetary scientists speculate that at one time the atmosphere of Mars may have contained large amounts of carbon dioxide, setting up a "greenhouse" effect that made the surface of that planet warmer and wetter. Might there, then, have been life on Mars at some earlier time? Molecular structures found in meteorites thought to come from Mars have been interpreted to show that there was once life there, but these results are controversial.

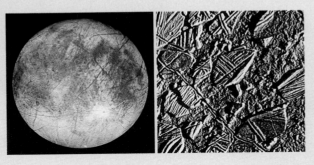

Conditions on other planets seem too hostile for life as we know it, but recent discoveries on our own planet indicate that life is much more robust than was once thought. Deep-sea explorers have discovered flourishing life around hydrothermal vents. Whereas life on the surface of the Earth relies on sunlight and photosynthesis for energy, these deep-sea life forms exploit energy-rich compounds spewed forth by volcanic vents. The warm water around hydrothermal vents teems with bacteria, which in turn support higher life forms such as worms and crustaceans. This terrestrial life thrives in an environment similar to one that might exist on Europa, reinforcing speculation that this moon of Jupiter could support some forms of life.

Outside our own solar system, might there be planetary environments where life flourishes? In recent years, astronomers have discovered planets orbiting stars other than our own. Whether or not these planets support life is still impossible to say. Nevertheless, the more we discover about the variety of the universe, the more likely it becomes that we are not alone.

ozone layer in the stratosphere. These studies led to the awarding of the 1995 Nobel Prize for Chemistry to Paul Crutzen, Mario Molina, and F. Sherwood Rowland. Figure 1-1 provides a schematic view of how this prize-winning research advanced. It began in 1971 when experiments revealed that chlorofluorocarbons, or CFCs, had appeared in the Earth's atmosphere. At the time, these CFCs were widely used as refrigerants and as aerosol propellants. Rowland wondered what eventually would happen to these gaseous compounds. He carried out a theoretical analysis, from which he concluded that CFCs are very durable and could persist in the atmosphere for many years.

Meanwhile, Crutzen had done experiments showing that ozone in the upper atmosphere can be destroyed easily by reactions with nitrogen oxides. This work demonstrated that the ozone layer is in a delicate balance that could be disturbed significantly by changes in atmospheric composition. In 1974, Molina and Rowland combined Crutzen's experimental work with their own theoretical analysis and published a prediction (hypothesis) that CFCs pose a serious threat to the ozone layer.

Following this interplay between observations and theory, many atmospheric scientists began studying chemical reactions of ozone in the upper atmosphere. Chemists duplicated atmospheric conditions in the laboratory and measured how

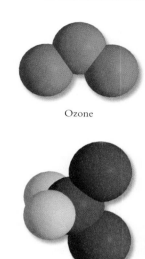

Ozone

Freon-12
An example of a CFC

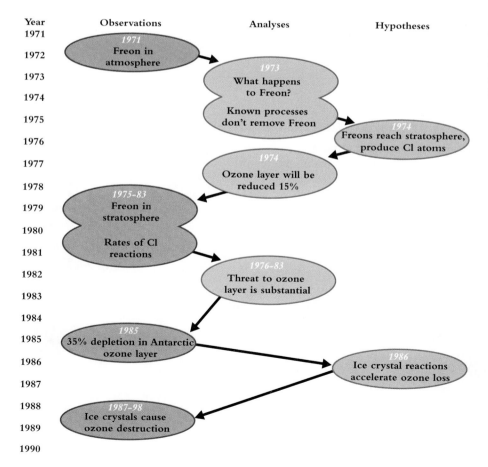

Figure 1-1
This flow chart illustrates how the scientific process led to worldwide concern over the effect of chlorofluorocarbons on the ozone layer.

fast various chemical reactions occur. The results of these experiments were used to create theoretical models of the upper atmosphere and predict how the ozone concentration would change as CFCs were introduced. Meanwhile, atmospheric scientists carried out measurements showing that ozone was being depleted in the upper atmosphere, at a rate even faster than had been predicted.

Today, scientists realize that the chemistry of the upper atmosphere is quite complex. In addition to gaseous molecules, solid particles such as tiny ice crystals play important roles in the chemistry that affects ozone. The original hypothesis of Rowland and Molina, that CFCs reach the upper atmosphere and deplete the ozone layer, has been fully confirmed. Exactly how this occurs, what other chemicals are involved, and how this process might be controlled, are still under intense study by chemists and other scientists, leading to yet more hypotheses and experiments.

CHAPTERS 6 and 14 →
More information on the ozone layer and the atmospheric chemistry of CFCs can be found in Chapters 6 and 14.

The story of research into the depletion of atmospheric ozone is just one example of how scientific understanding and theories develop. The fundamental theories and laws of chemistry that we present in this text all went through similar intense scrutiny and study.

Section Exercises

1.1.1 List four ways that chemistry applies to cooking.

1.1.2 Describe how chemistry applies to the automobile industry.

1.1.3 A planetary scientist announces a theory predicting that substances on Venus react differently than they do on Earth. Write a paragraph that describes ways to test this theory.

1.2 ATOMS, MOLECULES, AND COMPOUNDS

SECTION 1.4 →
We describe physical and chemical properties in Section 1.4.

Every substance has physical properties that we can measure and describe; these properties include shape, color, and texture. For example, the iron girders shown in Figure 1-2 have a lustrous silvery color and a smooth texture. Substances also have chemical properties, such as whether or not they will burn. Physical and chemical properties that can be observed with the eye are called **macroscopic.**

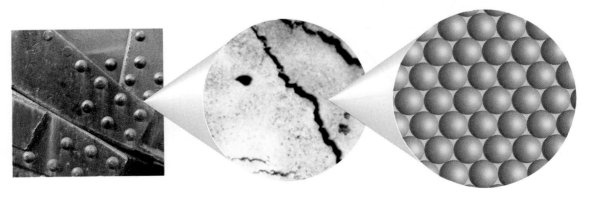

Figure 1-2
Iron appears different at the macroscopic, microscopic, and atomic levels.

The underlying structure of a chemical substance, which is called **microscopic,** can be explored using magnifying devices. The magnified view of iron shown in Figure 1–2 reveals fissures and pits that are not visible in the macroscopic view. Still further magnification eventually reveals the building blocks of matter. This, the **molecular** or **atomic view,** is an essential part of every chemist's thinking.

Atoms

The fundamental unit of a chemical substance is called an **atom.** The word is derived from the Greek *atomos,* meaning "uncuttable." An atom is the smallest possible particle of a substance.

Atoms are extremely small. Measurements show that the diameter of a single carbon atom is approximately 0.0000000003 meters (about 0.000000001 feet). To give you some idea of just how small that is, a sample of carbon the size of the period at the end of this sentence contains more atoms than the number of stars in the Milky Way. Any sample of matter large enough for us to see, feel, smell, or taste consists of an unfathomable number of atoms.

Molecules

Atoms combine to make all the substances in the world around us, but they do so in very orderly ways. Most substances that we encounter in day-to-day life are made up of small units called **molecules.**

/// *A molecule is a combination of two or more atoms held together in a specific shape by attractive forces.*

The simplest molecules contain just two atoms. For example, a molecule of hydrogen is made up of two hydrogen atoms. A molecule that contains two atoms is classified as a *diatomic* molecule. Figure 1–3 represents a diatomic hydrogen molecule as two spheres connected together.

Because chemistry deals mostly with the behavior of molecules, this book emphasizes chemistry's molecular foundation. Throughout this book you will see many figures that represent molecules, with each atom represented by a sphere. Although there are more than 100 different types of atoms, only about 20 are encountered frequently in our world. Many of the molecules described in this book are made up of just 10 different types of atoms: hydrogen, carbon, nitrogen, oxygen, phosphorus, sulfur, fluorine, chlorine, bromine, and iodine. Figure 1–4 shows the color scheme that we use to represent these atoms. We introduce other atoms as the need arises. Although most substances are colored, individual atoms do not have the colors shown in Figure 1–4. Figure 1–5 shows scale models of a few molecules whose names may be familiar to you.

Two hydrogen atoms

One hydrogen molecule

Figure 1-3
A hydrogen molecule can be represented by connecting two spheres together, with each sphere representing one hydrogen atom.

The name *hydrogen* refers to both atoms and molecules. To minimize confusion, we refer to *atomic hydrogen* when we mean hydrogen atoms and *molecular hydrogen* when we mean hydrogen molecules.

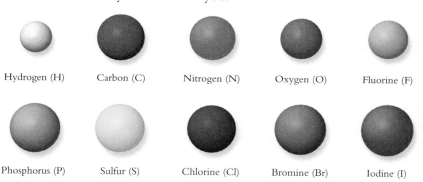

Hydrogen (H)   Carbon (C)   Nitrogen (N)   Oxygen (O)   Fluorine (F)

Phosphorus (P)   Sulfur (S)   Chlorine (Cl)   Bromine (Br)   Iodine (I)

Figure 1-4
Color-coded scale models of 10 types of atoms that appear frequently in this book.

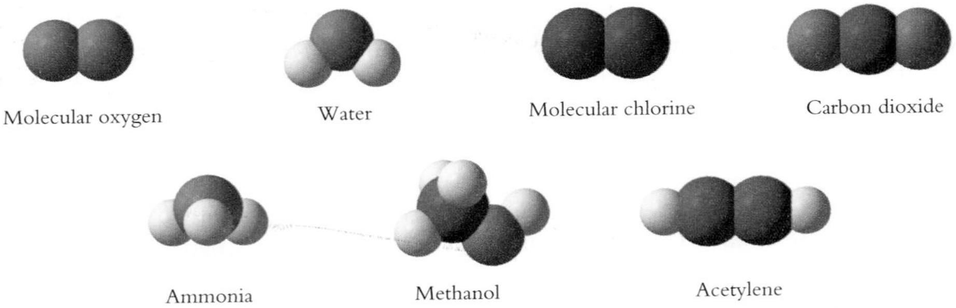

Molecular oxygen	Water	Molecular chlorine	Carbon dioxide

Ammonia	Methanol	Acetylene

Figure 1-5
Scale models of seven relatively simple molecules

Animation

CHAPTERS 8 and 9 →
We describe how atoms are held
together in Chapters 8 and 9.

Tutorial

With practice, simple molecules, such as carbon dioxide and water, can be identified just by looking at their models. The structures of larger molecules, however, are too complex to be recognized at a glance. Consequently, chemists have created a shorthand language of symbols, formulas, and equations that convey information about atoms and molecules in a simple manner. A specific letter or pair of letters designates each type of atom. These symbols, in turn, are combined into formulas that describe the compositions of more complicated chemical substances. Formulas can then be used to write chemical equations that describe how molecules change in chemical reactions.

The Elements

A substance that contains only one type of atom is called a chemical **element,** and each different element contains a different type of atom. Each chemical element has a unique name, such as hydrogen, carbon, oxygen, uranium, tantalum, and iron. The name of an element may refer to its history or to one of its properties. The Romans gave copper its name, *cuprum,* after the island of Cyprus, where copper was mined as early as 5000 BC. Bromine received its name from the Greek word *bromos* meaning "stench." If you ever work with bromine, you will understand the reason for its name. Other examples of evocative names include xenon (from the Greek *xenos,* "stranger"), rubidium (from the Latin "dark red") and neptunium (named for the planet Neptune).

Each element is represented by a unique one- or two-letter symbol. For example, the symbol for hydrogen is H, oxygen's symbol is O, and nitrogen's symbol is N. When more than one element begins with the same English letter, a second letter is added to the symbol. The second letter is always lower case. For example, carbon is C, cobalt is Co, and chromium is Cr. Chemists understand that the symbol for an element represents more than a letter or letters. Instead, a chemist sees the symbol Ni and immediately thinks of nickel *atoms*.

The elements listed in Table 1-1 have symbols derived from their names in other languages. Most of these elements were known in ancient times, so their symbols reflect the Latin language that was dominant when they were named.

Copper, one of the first elements to be purified by humans, is used for sculpture worldwide.

Table 1-1
Elemental Symbols with Non-English Roots

Name	Symbol	Root	Language	Name	Symbol	Root	Language
Antimony	Sb	stibium	Latin	Potassium	K	kalium	Latin
Copper	Cu	cuprum	Latin	Silver	Ag	argentum	Latin
Gold	Au	aurum	Latin	Sodium	Na	natrium	Latin
Iron	Fe	ferrum	Latin	Tin	Sn	stannum	Latin
Lead	Pb	plumbum	Latin	Tungsten	W	wolfram	German
Mercury	Hg	hydrargyrum	Latin				

Chemical Formulas

A chemical **compound** is a substance that contains more than one element. The relative amounts of the elements in a particular compound do not change: Every molecule of a particular chemical substance contains a characteristic number of atoms of its constituent elements. For example, every water molecule contains two hydrogen atoms and one oxygen atom. To describe this atomic composition, chemists write the **chemical formula** for water as H_2O.

The chemical formula for water shows how formulas are constructed. The formula lists the symbols of all elements found in the compound, in this case H (hydrogen) and O (oxygen). A subscript number after an element's symbol denotes how many atoms of that element are present in the molecule. The subscript 2 in the formula for water indicates that each molecule contains two hydrogen atoms. No subscript is used when only one atom is present, as is the case for the oxygen atom in a water molecule. Atoms are indivisible, so molecules always contain whole numbers of atoms. Consequently, the subscripts in chemical formulas of molecular substances are always integers. We explore chemical formulas in greater detail in Chapter 3.

Molecules vary considerably in complexity. Molecular oxygen is made up of two oxygen atoms, so its chemical formula is O_2. A carbon monoxide molecule contains one atom of carbon and one atom of oxygen, so its chemical formula is CO. Each molecule of methane, the major constituent of natural gas, contains one carbon atom and four hydrogen atoms, so its formula is CH_4. You will encounter still more complicated structures, such as methanol (three different elements, CH_4O) as you progress through this book.

Not all chemical compounds exist as molecules. Some compounds are made up of chemical species called ions. We introduce and describe ions in Chapter 2.

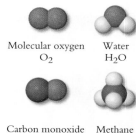

Molecular oxygen
O_2

Water
H_2O

Carbon monoxide
CO

Methane
CH_4

1.2.1 What are the elemental symbols for cerium, cesium, copper, calcium, and carbon?

1.2.2 What are the names of the elements represented by the symbols Zr, Ni, Sn, W, Se, Be, and Au?

1.2.3 Molecular pictures of some common molecules are shown here. What are their formulas?

Hydrogen peroxide (a common disinfectant and bleaching agent)

Sulfur dioxide (a common air pollutant)

Dinitrogen monoxide (laughing gas)

Acetic acid (vinegar)

Ethylene (used to make polyethylene plastic)

Box 1-2 | Tools for Discovery: Atomic–Level Microscopy

Atoms and molecules are much too tiny to see, even with the most powerful light microscopes. In recent years, however, scientists have developed a set of immensely powerful magnifying techniques that make it possible to visualize how individual atoms are arranged in solids. These techniques are scanning tunneling microscopy (STM) and atomic force microscopy (AFM).

Scanning tunneling microscopy and atomic force microscopy collect images of atoms and molecules using highly miniaturized lever arms with atomically sharp tips—like phonograph needles scaled down to the atomic level. These probes respond to the contours of individual atoms. In atomic force microscopy, the tip can be moved extremely precisely across a surface. The tip responds to individual atoms on that surface, moving up and down by tiny amounts as it passes over each atom. A laser beam (like the scanner at a supermarket checkout stand) focused on the edge of the lever arm detects this tiny motion, and an optical detector creates an image of the surface.

At their highest sensitivities, STM and AFM generate images that show how atoms are arranged on the surfaces that they probe. At first, scientists used these tools to explore how atoms are arranged on surfaces. The example below shows individual atoms on the surface of nickel metal.

Increasingly, scientists use STM and AFM to explore chemical reactions. For example, molecules often undergo chemical reactions when they "stick" to a metal surface. Scanning tunneling microscopy can be used to monitor the atomic changes that take place during such chemical reactions. It is even possible to use STM to manipulate individual atoms. The figures at the bottom of this box show how iron atoms can be arranged on the surface of copper metal to make an "atomic corral."

Whereas STM is best suited for imaging atoms, atomic force microscopy is more appropriate for larger structures. The following image shows a strand of DNA. The two blue regions of the figure are protein molecules bound to the DNA.

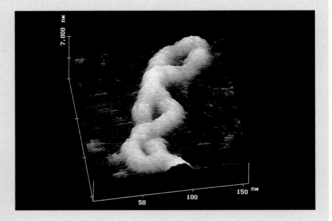

The next generation of molecular probes may be able to look within atoms to image their underlying structure. These probes combine magnetic resonance imaging (MRI) with the atomic force microscope. Physicians use MRI to do brain scans to pinpoint tumors or blood clots. By coupling MRI with an atomic force scanner, and cooling the sample to extremely low temperatures, it may be possible to create images of the interiors of individual atoms.

1.3 THE PERIODIC TABLE OF THE ELEMENTS

More chemical reactions exist than anyone can imagine. Nevertheless, certain patterns of chemical reactivity have been recognized for more than 100 years. These patterns remain valid even though new reactions are always being discovered. Each chemical element has characteristic chemical properties. Moreover, certain groups of elements display similar chemical properties. In 1869, the Russian chemist Dmitri Mendeleev and the German chemist Julius Lothar Meyer independently discovered how to arrange the chemical elements in a table so that elements with similar chemical properties were in the same columns. This arrangement, the **periodic table,** contains all the known chemical elements.

We show in Chapter 2 that the periodic table is based on the structure of atoms rather than on their masses. Elemental masses correlate closely with atomic structure, however, so ordering by mass is almost the same as ordering by structure. There are only three exceptions among more than 100 elements.

Arrangement

The **periodic table** lists all the known elements in numerical order, starting with the lightest (hydrogen) and proceeding to the heaviest (uranium, among naturally occurring elements). The list is broken into seven rows. Each row is placed below the previous row in a way that places elements with similar chemical properties in the same column of the table. Moving *across a row* of the periodic table, the elements generally increase in mass and change dramatically in their chemical properties. Moving *down a column,* mass also increases, but the elements have similar chemical properties.

Figure 1-6 shows the periodic table. Notice that rows 6 and 7 are quite long, which makes the table rather cumbersome. For convenience, 14 of the elements in the sixth and seventh rows usually are separated from the rest of the table and placed beneath the main portion, as shown in Figure 1-7 and on the front end papers of the book. This is the most common format for the periodic table.

Metals, Nonmetals, and Metalloids

The elements can be divided into categories: metals, nonmetals, and metalloids. Examples of each appear in Figure 1–8. All the elements in the left and central regions of the periodic table are **metals.** Metals display several characteristic properties. For example, they are good conductors of heat and electricity and usually appear shiny. Metals are malleable, meaning that they can be hammered into thin sheets, and ductile, meaning that they can be drawn into wires. Except for mercury, which is a liquid, all metals are solids at room temperature.

1	1 H																															2 He
2	3 Li	4 Be																									5 B	6 C	7 N	8 O	9 F	10 Ne
3	11 Na	12 Mg																									13 Al	14 Si	15 P	16 S	17 Cl	18 Ar
4	19 K	20 Ca															21 Sc	22 Ti	23 V	24 Cr	25 Mn	26 Fe	27 Co	28 Ni	29 Cu	30 Zn	31 Ga	32 Ge	33 As	34 Se	35 Br	36 Kr
5	37 Rb	38 Sr															39 Y	40 Zr	41 Nb	42 Mo	43 Tc	44 Ru	45 Rh	46 Pd	47 Ag	48 Cd	49 In	50 Sn	51 Sb	52 Te	53 I	54 Xe
6	55 Cs	56 Ba	57 La	58 Ce	59 Pr	60 Nd	61 Pm	62 Sm	63 Eu	64 Gd	65 Tb	66 Dy	67 Ho	68 Er	69 Tm	70 Yb	71 Lu	72 Hf	73 Ta	74 W	75 Re	76 Os	77 Ir	78 Pt	79 Au	80 Hg	81 Tl	82 Pb	83 Bi	84 Po	85 At	86 Rn
7	87 Fr	88 Ra	89 Ac	90 Th	91 Pa	92 U	93 Np	94 Pu	95 Am	96 Cm	97 Bk	98 Cf	99 Es	100 Fm	101 Md	102 No	103 Lr	104 Rf	105 Db	106 Sg	107 Bh	108 Hs	109 Mt									

Figure 1-6
Periodic table of the elements with the lanthanides and actinides included in their proper rows.

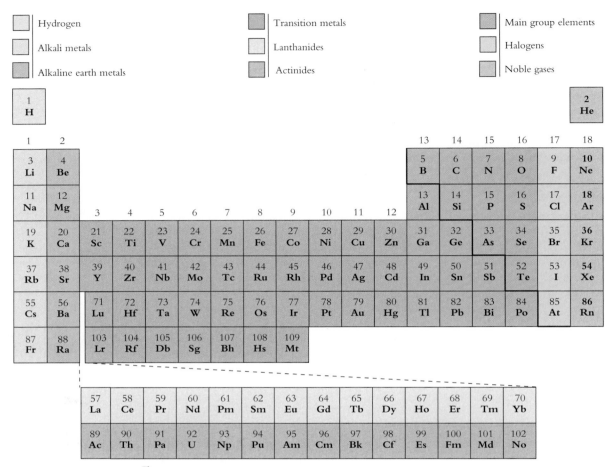

Figure 1-7
Periodic table of the elements as used in common practice.

The **nonmetals** are found in the upper right-hand corner of the periodic table. The properties of nonmetals are highly variable, but most nonmetals are poor conductors of electricity and heat. The six elements that are categorized as **metalloids** (B, Si, Ge, As, Sb, and Te) are dull-appearing, brittle solids at room temperature. The metalloids are sometimes called semiconductors because they conduct electricity better than nonmetals but not as well as metals. In fact, silicon and germanium are used in the manufacture of semiconductor chips in the electronics industry.

Periodic Properties

The first column of the periodic table, Group 1, contains elements that are soft, shiny solids. These are the **alkali metals,** and they include lithium, sodium, and potassium. On the other end of the table, fluorine, chlorine, bromine, and iodine appear in the next-to-last column. These are the **halogens,** or Group 17 elements. These four elements exist as diatomic molecules, so their formulas have the form X_2. Samples of three halogens are shown in Figure 1-9.

Each alkali metal combines with any of the halogens in a 1:1 ratio to form a white crystalline solid (Figure 1-10). The general formula of these compounds is $AX,$ where A represents the alkali metal and X represents the halogen. ($AX = $ NaCl, LiBr, CsBr, KI, etc.).

Figure 1-8
Pure elements show considerable variation in their appearance. Shown here from left to right are aluminum (metal), silicon (metalloid), sulfur (nonmetal), copper (metal), tellurium (metalloid), and mercury (metal).

Figure 1-9
The halogens all form diatomic molecules, X_2. Cl_2 is a gas under normal conditions. Br_2 is a dark red, volatile liquid, and I_2 is a purple-black solid. Astatine is highly radioactive and unstable.

Figure 1-10
The alkali metals are shiny, soft, reactive metals that combine with the halogens in 1:1 atomic ratios. The figure shows a tube of chlorine gas (Cl_2), a cube of sodium (Na), and a sample of sodium chloride (NaCl).

The elements in the second column of the table resemble the alkali metals in their appearance, but they have different chemical properties. For example, each of these metals combines with the halogens in a 1:2 ratio ($AX_2 = CaCl_2$, $MgBr_2$, BaF_2, etc.). Each also reacts with atmospheric oxygen, as shown in Figure 1-11, to form a solid with the formula AO (BaO, CaO, etc.). These are the **alkaline earth metals,** or Group 2 elements.

The last column of the periodic table contains the **noble gases,** or Group 18 elements, all of which occur in nature as gases. With a few exceptions, these elements do not undergo chemical reactions.

Elements also can be divided into three broad categories. The elements in Groups 3 through 12 are known as **transition metals.** The elements in rows 6 and 7 that are normally shown below the rest of the table are the rare earths, subdivided into **lanthanides** (row 6) and **actinides** (row 7). All other elements are **main group elements.**

The periodic table is a useful way to organize chemical properties. To help you see the patterns, the periodic table on the front endpapers of this book highlights the various groups of elements. As you learn more about chemical structure and behavior, you will discover the principles that account for similarities and differences in the chemical behavior of the elements.

Figure 1-11
Magnesium metal burns in air with a bright white flame to form a white solid, magnesium oxide.

← **CHAPTERS 6 and 7**
We describe atomic structure, which dictates periodic properties, in Chapters 6 and 7.

Section Exercises

■ **1.3.1** Predict the formulas of the compounds formed in the reactions between (a) calcium and chlorine; (b) cesium and iodine; (c) barium and oxygen; and (d) magnesium and fluorine.

■ **1.3.2** Boron and fluorine form a compound with the formula BF_3. Based on this, suggest formulas for compounds of aluminum with bromine and gallium with chlorine.

■ **1.3.3** Classify each of the following elements as alkali metal, alkaline earth metal, halogen, noble gas, main group, transition metal, lanthanide, or actinide: aluminum, fluorine, cobalt, phosphorus, krypton, europium, thorium, barium, and sodium.

Figure 1-12
Carbon *(left)*, gold *(center)*, and sulfur *(right)* are pure elements that occur naturally.

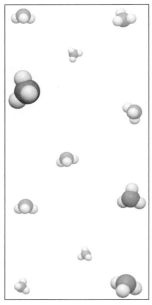

Figure 1-13
The composition of a pure substance is homogeneous and invariant. A sample of pure ammonia contains nothing but molecules made of nitrogen atoms and hydrogen atoms in a 1:3 ratio.

The gas of the atmosphere is held in place by gravity, not by the walls of a container.

CHAPTER 10 →
Describes how phases are affected by temperature.

1.4 CHARACTERISTICS OF MATTER

Matter is anything that has mass and occupies space. A sample of matter can contain a single substance or any number of different substances. As already described, the building blocks of most substances are molecules, which in turn are composed of atoms. It is convenient to classify samples of matter according to the complexity of their composition, both at the atomic level and at the macroscopic level.

The elements are the simplest form of matter. An element contains only one type of atom and cannot be decomposed into other chemical components. Of the more than 100 known chemical elements, only a few are found in our world in their pure form. Figure 1-12 shows three of these: Diamonds are pure carbon, nuggets of pure gold can be found by panning in the right stream bed, and sulfur is found in abundance in its elemental form.

When two or more different chemical elements combine, they form a chemical compound. Even though they are made of more than one type of element, pure chemical compounds are uniform in composition; that is, all samples of a particular chemical compound contain the same proportions of each element. For example, ammonia is a chemical compound that contains the elements nitrogen and hydrogen in a 1:3 atomic ratio. A sample of pure NH_3 always contains nothing but ammonia molecules, each one containing three hydrogen atoms and one nitrogen atom (Figure 1-13).

Although pure elements and pure compounds occur often, both in nature and in the laboratory, matter is usually encountered as a **mixture** of substances. A mixture contains two or more chemical substances. Unlike pure compounds, mixtures vary in composition because the proportions of the substances in a mixture can change. For example, dissolving sucrose, table sugar, in water forms a mixture that contains water molecules and sucrose molecules. A wide range of mixtures can be prepared by varying the relative amounts of sucrose and water.

A sample is **homogeneous** if it always has the same composition, no matter what part of the sample is examined. Pure elements and pure chemical compounds are homogeneous. Mixtures can be homogeneous, too; a homogeneous mixture usually is called a **solution.** As shown in Figure 1-14, the difference between a pure substance and a homogeneous mixture can be illustrated using hydrogen and chlorine. Under the right conditions, these two elements react in a 1:1 atomic ratio to give diatomic molecules of hydrogen chloride. Hydrogen chloride gas is a

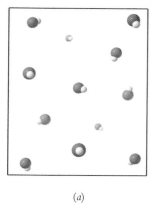  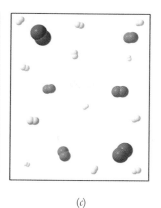

(a)   (b)   (c)

Figure 1-14

A sample of hydrogen chloride (*a*) is homogeneous and has constant composition because atoms of the two elements are always linked in a 1:1 atomic ratio. A mixture of hydrogen gas and chlorine gas is homogeneous but can have different compositions because the amount of either gas can be varied without changing the amount of the other (*b* and *c*).

homogeneous *pure substance* and always contains equal numbers of hydrogen atoms and chlorine atoms linked in HCl molecules. Under other conditions, molecular hydrogen and molecular chlorine do not react with each other. The two gases form a homogeneous *solution* whose composition can be changed by adding more of either substance.

When different portions of a mixture have different compositions, the mixture is said to be **heterogeneous.** For example, quartz is a pure chemical compound made from silicon and oxygen, and gold is a pure element, but the lump of quartz containing a vein of gold that appears in Figure 1-15 is a heterogeneous mixture because different parts of the lump have different compositions.

Phases of Matter

Matter can also be categorized into three distinct phases: solid, liquid, and gas. An object that is **solid** has a definite shape and volume that cannot be changed easily. Trees, automobiles, ice, and coffee mugs are all in the solid phase. Matter that is **liquid** has a definite volume but changes shape quite easily. A liquid flows to take on the shape of its container. Gasoline, water, and cooking oil are examples of common liquids. Solids and liquids are termed *condensed phases* because of their well-defined volumes. A **gas** has neither specific shape nor constant volume. A gas expands or contracts as the container confining it expands or contracts. Helium balloons are filled with helium gas, and the Earth's atmosphere is made up of gas that flows continually from place to place. Molecular pictures that illustrate the three phases of matter appear in Figure 1-16.

Figure 1-15

Quartz containing a vein of gold is a heterogeneous mixture because it contains two distinct substances that are not uniformly distributed through the sample.

  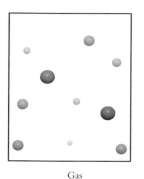

Solid   Liquid   Gas

Figure 1-16

The atoms or molecules in a solid are arranged in a specific shape and volume. In a liquid, the atoms or molecules fill a definite volume but are free to flow from place to place. A gas fills its entire container and changes shape and volume as the container expands or contracts.

| Example 1-1 | Characteristics of Matter |

Decide whether the following molecular pictures represent a pure substance, a homogeneous mixture, or a heterogeneous mixture. Tell whether the sample is a solid, a liquid, or a gas.

Strategy: Apply the characteristics described in the preceding paragraphs. Each circle in the figure represents an atom. Different colors distinguish one type of atom from another.

Solution: The sample on the left contains a collection of eight diatomic molecules. All the molecules have the same composition, so this is a pure substance. The molecules in the sample are distributed evenly through the entire volume of the container. This is the defining characteristic of a gas.

In the sample on the right, four atoms of one type are distributed evenly through a larger collection of atoms of a second type. This is a homogeneous mixture. Notice that the sample is spread across the bottom of the container, but it is also confined to a specific volume. These features identify the sample as a liquid.

Transformations of Matter

Although every substance exists in one particular phase under ordinary circumstances, under the right conditions most substances can be converted from one phase to another. Figure 1-17 shows water in all three phases. Water changes from the liquid phase into a solid (ice) when the temperature drops below 0 °C. When

Figure 1-17
Water can exist as a solid (ice and snow), as a liquid, and as a gas (steam).

the temperature rises above 100 °C, water changes from a liquid into a gas (steam). Similarly, lava is rock that has been heated sufficiently to convert it to the liquid phase, and "dry ice" is carbon dioxide that has been cooled enough to change it from the gas phase to the solid phase.

A process that changes the properties of a substance is a transformation. Transformations of matter are either physical or chemical. In a **physical transformation,** physical properties change but the substance's chemical nature remains the same. For example, when water freezes, it undergoes a physical transformation because the chemical makeup of ice is the same as that of liquid water. That is, both ice and liquid water contain molecules made up of two atoms of hydrogen and one atom of oxygen. Another physical transformation is the dissolving of a sugar cube in a cup of hot coffee. Although the sugar mixes uniformly with the coffee in this process, the chemical nature of each remains unchanged.

A **chemical transformation,** on the other hand, produces new substances. For example, when magnesium metal burns in air (Figure 1-11), elemental magnesium metal and molecular oxygen combine chemically to form magnesium oxide, a white solid containing Mg and O atoms in 1:1 ratio. This is a chemical transformation that rearranges the atoms in Mg and O_2 to yield MgO.

Section Exercises

■ **1.4.1** Decide whether each of the following molecular pictures represents a single substance or a mixture. Identify the mixtures as homogeneous or heterogeneous. If the picture represents a single substance, tell whether it is an element or a chemical compound.

 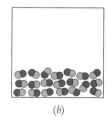 

(a)   (b)   (c)

■ **1.4.2** Classify each of the following as a pure substance, a solution, or a heterogeneous mixture: a cup of coffee, a lump of sugar, seasoned salt, a silver coin, and sea water.

■ **1.4.3** Classify each of the following as a physical or a chemical transformation: melting snow, burning coal, chopping wood, and digesting food.

1.5 MEASUREMENTS IN CHEMISTRY

The knowledge that allows chemists to describe, interpret, and predict the behavior of chemical substances is gained by making careful experimental measurements. The properties of a sample can be divided into *physical properties,* which can be measured without observing a chemical reaction, and *chemical properties,* which are displayed only during a chemical transformation.

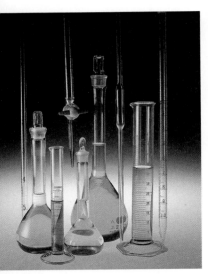

Figure 1-18
Some common laboratory glassware for measuring volume.

The process of determining mass is called *weighing*. Mass and weight are related, but they are not the same property. Mass is a fundamental characteristic of an object, whereas weight results from gravitational force acting on an object's mass. In outer space, objects have mass but no weight because there is no gravitational force.

CHAPTERS 2, 12, 13 →
We discuss energy changes in Chapter 2 and consider temperature and heat flow in detail in Chapters 12 and 13.

Physical Properties

Length, area, and volume measure the size of an object. **Length** refers to one dimension, **area** refers to two dimensions, and **volume** refers to three dimensions of space. Size measurements require standard measuring devices, such as rulers or measuring cups. Figure 1-18 shows some standard laboratory equipment for measuring volume.

In addition to its volume, every object possesses a certain quantity of matter, called its **mass (m)**. Mass measurements are particularly important in chemistry. Consequently, highly accurate mass-measuring machines, called analytical balances, are essential instruments in chemistry laboratories. Analytical balances work by comparing forces acting on masses. Modern balances, both in a delicatessen and in a chemistry laboratory (Figure 1-19), compare forces quickly and automatically, providing a digital readout of an object's mass.

Chemists measure **time (t)** because they want to know how long it takes for chemical transformations to occur. Some chemical reactions, such as the conversion of green plants into petroleum, may take millions of years. Other chemical processes, such as an explosion of dynamite, are incredibly fast. Whereas wristwatches typically measure time only to the nearest second, chemists have developed instruments that make it possible to study processes that occur in less than 0.00000000001 second.

Most of us associate **temperature (T)** with the concepts of hot and cold. More accurately, however, temperature is the property of an object that determines the direction of heat flow. Heat always flows from a warm object to a cool object, from higher temperature to lower temperature. Heat is a form of energy, and because energy changes in chemical systems have important consequences, chemists are interested in temperature changes that occur during chemical transformations.

All experimental sciences rely on quantitative measurements of properties. Every measurement gives a numerical result that has three aspects: a numerical *magnitude;* an indicator of scale, called a *unit;* and a *precision*. Each aspect is essential, and all three must be reported to make a measurement scientifically valuable.

(a)

(b)

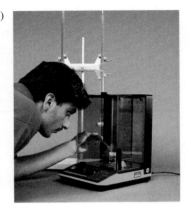

Figure 1-19
Masses are determined using balances that compare two forces. (*a*) Delicatessen balance. (*b*) Laboratory analytical balance.

Figure 1-20

Dimensions of objects known to humans span an immense range that covers 40 orders of magnitude, from the diameter of a hydrogen nucleus (about 10^{-15} meters) to the diameter of the known universe (about 10^{25} meters).

Magnitude

The magnitudes of experimental values in science range from infinitesimally small to astronomically large, as summarized in Figure 1-20. To simplify manipulating very large and very small numbers, we use powers of ten, also called **scientific notation.** For example, the diameter of a carbon atom is 0.0000000003 meters (symbol: m). This cumbersome number can be simplified by the use of scientific notation: $0.0000000003 \text{ m} = 3 \times 10^{-10} \text{ m}$.

Scientists routinely study objects whose sizes extend far beyond the narrow range encountered in daily life. Physicists, for example, study atomic nuclei measuring 10^{-15} m across, and astronomers study our universe, which spans about 10^{25} m. Chemists are most often interested in matter on the smaller side of this range. Length measurements in the laboratory vary from meters to sub-atomic sizes, 10^{-12} m. To further simplify the use of the very large and very small numbers that they encounter, scientists use prefixes that change the unit sizes by multiples of 10. For instance, the prefix *pico* means "10^{-12}". The symbol for picometer is *pm*. The diameter of a carbon atom is 3×10^{-10} m, which is 300 pm.

$$\text{Carbon diameter} = 3 \times 10^{-10} \text{ m} = 300 \times 10^{-12} \text{ m}$$
$$= 300 \text{ pm} \qquad \uparrow \text{pico}$$

The most common magnitude prefixes are listed in Table 1-2.

Units

The units associated with a numerical value are just as important as the value itself. If a recipe instruction reads, "Add 1 of sugar," that instruction is useless. The unit of measure, such as "teaspoon," must also be included. It is *essential* to include a unit with every experimental value.

Table 1-2
Frequently Used Scientific Prefixes for Magnitudes

Prefix	Symbol	Number	Exponential Notation
giga	*G*	1,000,000,000	10^9
mega	M	1,000,000	10^6
kilo	k	1000	10^3
—	—	1	10^0
centi	c	0.01	10^{-2}
milli	m	0.001	10^{-3}
micro	μ	0.000001	10^{-6}
nano	n	0.000000001	10^{-9}
pico	p	0.000000000001	10^{-12}

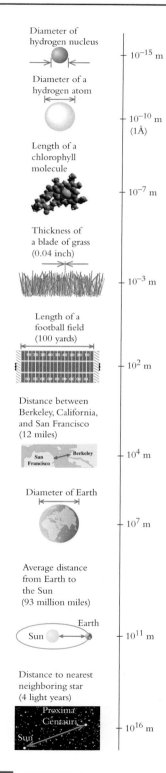

Diameter of hydrogen nucleus — 10^{-15} m

Diameter of a hydrogen atom — 10^{-10} m (1Å)

Length of a chlorophyll molecule — 10^{-7} m

Thickness of a blade of grass (0.04 inch) — 10^{-3} m

Length of a football field (100 yards) — 10^2 m

Distance between Berkeley, California, and San Francisco (12 miles) — 10^4 m

Diameter of Earth — 10^7 m

Average distance from Earth to the Sun (93 million miles) — 10^{11} m

Distance to nearest neighboring star (4 light years) — 10^{16} m

← APPENDIX A
See Appendix A for a review of scientific notation.

Table 1-3
Base SI Units

Quantity	Unit	Symbol
Mass	Kilogram	kg
Length	Meter	m
Time	Second	s
Temperature	Kelvin	K
Amount	Mole	mol
Electric current	Ampere	A
Luminous intensity	Candela	cd

A unit of measurement is an agreed-upon standard with which other values are compared. Scientists use the meter as the standard unit of length. The meter was originally chosen to be 10^{-7} times the length of a line from the North Pole to the equator. Volume can be measured in pints, quarts, and gallons, but the scientific standards are the cubic meter and the liter. Temperature can be measured in degrees Fahrenheit, degrees Celsius, or kelvins.

The international scientific community prefers to work exclusively with a single set of units, the Système International (SI), which expresses each fundamental physical quantity in decimally (power of 10) related units. The seven base units of the SI are listed in Table 1-3. The SI unit for volume is obtained from the base unit for length: A cube that measures 1 meter on a side has a volume of 1 cubic meter.

Unit Conversions

It is often necessary to convert measurements from one set of units to another. As an everyday example, travelers between the United States and Canada need to be able to convert between miles and kilometers. Chemists frequently need to convert volumes from one unit to another. The SI unit of volume is the cubic meter, but chemists usually work with much smaller volumes. Hence chemists often express volume using the liter (L), which is defined to be exactly 10^{-3} m^3. Another volume unit in common use is the milliliter (mL), or 10^{-3} L. The milliliter is the same as the cubic centimeter (cm^3). Figure 1-21 illustrates these volume measurements.

$$1 \text{ L} = 10^{-3} \text{ m}^3 \qquad 1 \text{ mL} = 10^{-3} \text{ L} = 1 \text{ cm}^3$$

A unit conversion starts with an equality between different units. For example, the equivalence between quarts and liters is:

$$1 \text{ quart} = 0.946353 \text{ L}$$

Each equality gives two conversion ratios. The equality between quarts and liters can be rearranged to give two ratios:

$$1 = \frac{1 \text{ quart}}{0.946353 \text{ L}} \quad and \quad 1 = \frac{0.946353 \text{ L}}{1 \text{ quart}}$$

To convert from one unit to another, we multiply by the ratio that leads to an appropriate cancellation of units. For example, a two-liter volume is expressed in quarts as follows.

$$2\cancel{L}\left(\frac{1 \text{ quart}}{0.946353 \cancel{L}}\right) = 2.11338 \text{ quarts}$$

/// The correct conversion ratio leads to cancellation of unwanted units.

A unit equality may link SI units and non-SI units (1 quart = 0.946353 L), different decimally related units (10^6 cm^3 = 1 m^3), or base units and derived units (1 L = 10^3 cm^3). Some of the more common unit equalities are given on the inside back cover of this text.

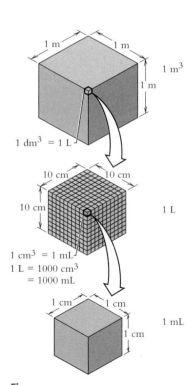

1 m **1 m**
1 m³
1 m

1 dm³ = 1 L

10 cm **10 cm**
10 cm
1 L

1 cm³ = 1 mL
1 L = 1000 cm³
= 1000 mL

1 cm **1 cm**
1 mL
1 cm

Figure 1-21
The defined unit of volume in SI is the cubic meter (m³). Chemists more commonly use the liter (L) or the milliliter (mL).

Unit Conversions

Example 1-2

A pair of rock climbers stand at the bottom of a rock face they estimate to be 155 feet high. Their rope is 65 m long. Is the rope long enough to reach the top of the cliff?

Strategy: The team must compare the English measure for the height of the cliff and the SI measure for the length of their rope. Is 65 m more or less than 155 feet? The equivalence between feet and meters can be found inside the back cover of the book: 1 ft = 0.3048 m.

Solution: The length of the rope is given in meters. To set up a ratio that converts meters to feet, we need to have feet in the numerator and meters in the denominator:

$$0.3048 \text{ m} = 1 \text{ ft} \qquad so \qquad 1 = \frac{1 \text{ ft}}{0.3048 \text{ m}}$$

Now multiply the length of the rope by the conversion ratio:

$$(65 \text{ m})\left(\frac{1 \text{ ft}}{0.3048 \text{ m}}\right) = 2.1 \times 10^2 \text{ ft}$$
$$\uparrow \text{ conversion ratio}$$

The two climbers have plenty of rope to climb the rock face.

Notice that we multiply the length of the rope in meters by feet/meter so that meters cancel. Multiplying the height of the cliff by the same conversion ratio would have given nonsensical units: (ft)(ft/m) = ft^2/m.

Multiple Unit Conversions

Example 1-3

The speed limit on many highways in the United States is 55 miles/hr. What is this speed limit in SI units?

Strategy: We are asked to make a unit conversion. The SI base unit of length is the meter, and the SI base unit of time is the second. It is necessary to convert from miles to meters and from hours to seconds. The appropriate unit equivalences are:

1 mile = 1.6093 km, 1 km = 10^3 m, 1 hr = 60 min, and 1 min = 60 s.

Solution: The speed limit in miles per hour is given. To cancel miles, multiply by a ratio that has miles in the denominator. To obtain meters, use a ratio that has meters in the numerator. We convert miles to kilometers and then convert kilometers to meters:

$$\left(\frac{55 \text{ miles}}{\text{hr}}\right)\left(\frac{1.6093 \text{ km}}{1 \text{ mile}}\right)\left(\frac{10^3 \text{ m}}{\text{km}}\right) = 8.9 \times 10^4 \text{ m/hr}$$

$$\text{conversion} \uparrow \qquad \uparrow \text{conversion}$$
$$\text{ratio} \qquad\qquad \text{ratio}$$

Notice that the conversion ratios are set up to cancel the unwanted units.

| Example 1-3 | Multiple Unit Conversions *(continued)* |

This completes the conversion into SI units of length. Now convert from hours to seconds, following a similar procedure. Because hours appear in the denominator, multiply by a ratio that has hours in the numerator:

$$\left(\frac{8.9 \times 10^4 \text{ m}}{\text{hr}}\right)\left(\frac{1 \text{ hr}}{60 \text{ min}}\right)\left(\frac{1 \text{ min}}{60 \text{ s}}\right) = 25 \text{ m/s}$$

Scientists use two units for temperature, the Celsius scale and the kelvin scale. These scales are shown schematically in Figure 1-22. Unlike other scientific units, the unit size of the Celsius and kelvin scales is the same, but their zero points differ. For both scales, the difference in temperature between the freezing and boiling points of water is defined to be 100 units. However, the temperature at which ice melts to liquid water is 0 °C and 273.15 K.

The conversion between kelvins and degrees Celsius is straightforward because their temperature (T) units are the same size. A temperature change of 1 °C is the same as a temperature change of 1 K. To convert from one scale to the other, add or subtract 273.15:

$$T \text{ (K)} = T \text{ (°C)} + 273.15 \qquad T \text{ (°C)} = T \text{ (K)} - 273.15 \qquad \text{(1-1)}$$

The Fahrenheit scale, in which water freezes at 32 °F and boils at 212 °F, is still in common use in the United States, but scientists rarely use the Fahrenheit scale. The formula for converting temperature from Fahrenheit to Celsius is:

$$T_C = \frac{5 \text{ °C}}{9 \text{ °F}} (T_F - 32 \text{ °F})$$

Precision and Accuracy

The exactness of a measurement is expressed by its **precision.** This concept can be explained with an example. Suppose three swimmers are discussing the temperature of a swimming pool. The first dips a finger in the water and says the temperature is "about 24 °C." The second examines an immersed pool thermometer and reports the temperature to be 26 °C. The third swimmer, who has been monitoring daily variations in the pool's temperature, uses a portable precision digital thermometer and reports, "According to my precision thermometer, the pool temperature is 25.8 °C."

The swimmers have measured the water temperature using different measuring devices with different levels of precision. The first swimmer's "finger test" is precise to about 3 °C: $T = 24 \pm 3$ °C (read "twenty-four plus or minus three degrees"). The pool thermometer gives a reading that is precise to the nearest degree: $T = 26 \pm 1$ °C. The precision thermometer measures to the nearest tenth of a degree: $T = 25.8 \pm 0.1$ °C. An actual temperature between 25.7 and 25.9 °C falls within the precision ranges of all three measurements, so all three are correct to within their stated limits of precision.

Whereas precision describes the exactness of a measurement, **accuracy** describes how close a measurement is to the true value. Figure 1-23 illustrates the difference between precision and accuracy. A rifle sharpshooter fires very precisely (*a* and *b*): Every bullet strikes the target close to the same spot. A novice shooter is not very precise (*c* and *d*): The bullets scatter over a rather large area. A sharpshooter with a well-adjusted sight also fires accurately (*a*): All the bullets strike the "bull's-eye." If the sight is poorly adjusted, however, the shooter may be

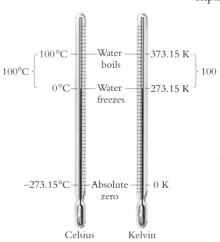

Figure 1-22
Celsius and kelvin temperature scales. Kelvins and degrees Celsius have the same unit size but different zero points. "Room temperature" is typically about 295 K (22 °C).

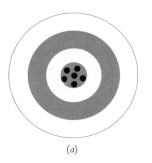
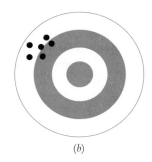
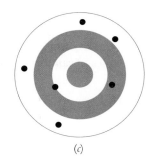
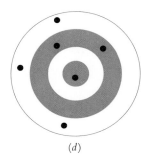

(a)   (b)   (c)   (d)

Figure 1-23
The patterns of bullets striking a target illustrate the notions of precision and accuracy. (a) A precise and accurate pattern; (b) A precise but inaccurate pattern; (c) an imprecise and inaccurate pattern; (d) an imprecise pattern with one accurate shot.

precise but inaccurate (b). The novice shooter is unlikely to be accurate (c). A low-precision result may occasionally be accurate (d).

The goal of any measurement in science is to be as accurate as possible. However, determining the accuracy of a measurement is much harder than determining precision. A shooter can examine the target to see if the bullets found their mark, but a scientific researcher studying a new phenomenon does not know the correct value. A scientist can assess precision by repeating measurements to find out how closely repeated measurements agree with one another. Assessing accuracy requires careful attention to the design of an experiment and the instruments used in that experiment.

Significant Figures

Scientific measurements should always include both magnitude and precision. Indicating the precision limits with a plus/minus (±) statement is cumbersome, particularly when many numerical values are reported at one time. Scientists have agreed to simplify the reporting of precision. Experimental measurements are written so that there is an uncertainty of up to one unit in the last reported digit. For example, a temperature reported as 26 °C is greater than 25 °C but less than 27 °C, or 26 ± 1 °C. A temperature reported as 25.8 °C is 25.8 ± 0.1 °C. The number of digits expressed in a numerical value is called the number of **significant figures**. The value 26 has two significant figures, whereas 25.8 has three.

Zeros can present a problem when determining the precision of a numerical value. This is because zeros are needed both to locate the decimal point and to express precision. From the statement that the sun is 93,000,000 miles from the Earth, we cannot tell whether the measurement is precise to eight significant figures or whether the zeros are there only to put the decimal point in the right place. Scientific notation eliminates this ambiguity because a power of 10 locates the decimal point, leaving us free to indicate the precision by the number of digits. For example, a distance of 9.3×10^7 miles means that the number is precise to $\pm 0.1 \times 10^7$ miles; the number has two significant figures. If this distance is known to a precision of $\pm 0.01 \times 10^7$ miles, it is written as 9.30×10^7 miles, with three significant figures. Zeros at the end of a number with a decimal point always indicate increased precision. Writing "0.010" in scientific notation clarifies that it has two significant figures rather than three or four: $0.010 = 1.0 \times 10^{-2}$.

To determine how many significant figures there are in a particular numerical value, read the number from left to right and count all the digits, starting with the first digit that is not zero. In this book we place a decimal point after a value when its trailing zeros are significant. For example, "110" has only two significant figures (110 ± 10), whereas "110." has three significant figures (110 ± 1). Here are some examples of significant figures:

500	1 significant figure
0.05	1 significant figure
55	2 significant figures
50.	2 significant figures
505	3 significant figures
5.00×10^3	3 significant figures
5.000	4 significant figures
505.0	4 significant figures

Section Exercises

■ **1.5.1** Convert the following measurements to scientific notation and express in base SI units: 0.000463 L, 17,935 km, and 260,000 hours (precise to three significant figures).

■ **1.5.2** One light-year is the distance light travels in exactly one year. The speed of light is 6.7×10^8 miles/hr. Express the speed of light and the length of one light-year in SI units.

■ **1.5.3** Convert each of these measurements to SI units: 155 pounds (mass of a typical person), 120.0 yards (full length of a football field), 38.5 °C (body temperature of someone with a slight fever), and 365.2422 days (length of 1 year).

1.6 CALCULATIONS IN CHEMISTRY

Chemical experiments are designed to provide information about a chemical system. During an experiment, a chemist may measure physical quantities such as mass, volume, and temperature. Usually the chemist seeks information that is related to the measured quantities but must be found by doing calculations. Many of our presentations in later chapters involve equations that relate measured physical quantities to important chemical properties. Calculations are an essential part of all of chemistry; therefore, they play important roles in much of general chemistry. The physical property of density illustrates this type of role.

Density

The symbol for density is the greek lower-case letter rho, ρ

The **density** (ρ) of an object is its mass (m) divided by its volume (V):

$$\text{Density} = \frac{\text{Mass}}{\text{Volume}} \quad \text{or in symbols} \quad \rho = \frac{m}{V} \qquad \textbf{(1-2)}$$

Every pure liquid or solid has a characteristic density that helps distinguish it from other substances. To give one example, the density of pure gold is 19.3 g/cm^3, whether the sample is a nugget in a miner's pan or an ingot in a bank vault. Pyrite, an iron compound that resembles gold, has a much lower density of

Table 1-4
Densities of Some Common Substances

Substance	Density (g cm^{-3})	Substance	Density (g cm^{-3})
Wood (balsa)	0.12	Aluminum	2.70
Cork	0.24	Diamond	3.51
Wood (white pine)	0.35 – 0.50	Silver	10.50
Alcohol (ethanol)	0.785	Lead	11.34
Water	1.00	Mercury	13.55
Quartz	2.65	Gold	19.3

5.0 g/cm^3. Table 1-4 lists the densities of several common substances. Most liquids and solids have densities in the range from 0.1 to 20 g/cm^3. This 200-fold variation is readily apparent to us and dictates how different materials are used in applications in which density is important. Fishermen use floats and sinkers made of low- and high-density materials, respectively, as Figure 1-24 illustrates. Low-density cork is used for floats, because low-density materials float in water. High-density lead is used for sinkers, because high-density materials sink.

Unlike mass and volume, density does not vary with the amount of a substance. Notice in Figure 1-24 that all the corks float, regardless of their sizes. Notice also that all the pieces of lead sink, regardless of their sizes. Dividing a sample into portions changes the mass and volume of each portion but leaves the density unchanged. A property that depends on amount is called **extensive.** Mass and volume are two examples of extensive properties. A property that is independent of amount is called **intensive.** Density is an example of an intensive property.

Mass and volume often can be measured easily, and density is then calculated using Equation 1-2. The equation can also be rearranged to find an object's volume or mass, as Example 1–4 illustrates.

Figure 1-24
Cork has a lower density than water, so corks of all sizes float on water. Lead has a higher density than water, so lead pieces of all sizes sink in water.

Using Density

Example 1-4

A diamond is a pure sample of the element carbon. The tabulated density of diamond is 3.51 g/cm^3. Jewelers use a unit called a carat to describe the mass of a diamond: 1 carat = 0.200 g. What is the volume of the stone in a 2.00-carat diamond engagement ring?

Strategy: Given the diamond's mass and density, we are asked to find its volume. Rearranging the density equation makes this possible:

$$\text{Density} = \frac{\text{Mass}}{\text{Volume}} \qquad so \qquad \text{Volume} = \frac{\text{Mass}}{\text{Density}}$$

Solution: A list of the information given in the problem allows us to determine what to substitute into the equation:

$$\text{Density}_{\text{diamond}} = 3.51 \text{ g/cm}^3$$

| Example 1-4 | *Using Density (continued)* |

$$\text{Mass} = 2.00 \text{ carat}$$

$$1 \text{ carat} = 0.200 \text{ g}$$

In the volume equation, density and mass must have consistent units. Thus mass must be converted from carats to grams using a conversion ratio that eliminates the unwanted unit.

$$2.00 \text{ carat} \left(\frac{0.200 \text{ g}}{1 \text{ carat}} \right) = 0.400 \text{ g}$$

↑ conversion ratio

Now substitute into the equation for volume:

$$\text{Volume} = \frac{\text{Mass}}{\text{Density}} = \frac{0.400 \text{ g}}{3.51 \text{ g}/\text{cm}^3} = 0.114 \text{ cm}^3$$

Precision of Calculations

The precision of a measuring instrument determines the precision of a single measurement. However, many scientific investigations require several measurements, often involving more than one instrument. Data obtained from multiple measurements are then used to calculate a quantity of interest. How precise is a value obtained by calculations? As a general rule, the least precise measurement determines the precision of a result. The following example, again using density, illustrates this guideline.

A chemical manufacturer prepares a silicone fluid for use as a lubricant. To determine the density of the fluid, a technician measures 25.0 cm³ into a graduated cylinder and determines the mass of this amount of fluid to be 39.086 g. Dividing mass by volume on a calculator, the technician reads the following result: 1.56344. What value does the technician report for the fluid's density? Are all six of these digits significant? A set of measurements is limited by the precision of the least-sensitive instrument used in the experiments. The mass is known to five significant figures, but there are only three significant figures in the volume. Thus only three significant figures are used to report the density of the fluid:

APPENDIX B →
Appendix B describes in more detail the way precision changes when numbers are combined.

$$\rho = \frac{39.0876 \text{ g}}{25.0 \text{ cm}^3} = 1.56 \text{ g}/\text{cm}^3$$

Three guidelines can be used to determine the precision of a sequence of mathematical operations.

1. **When *adding* or *subtracting*, the number of decimal places in the result is the *number of decimal places* in the number with the fewest places.**

0.0120	**4 decimal places**	3 significant figures
1.6	**1 decimal place**	2 significant figures
8.49026	**5 decimal places**	6 significant figures
10.1	**1 decimal place**	3 significant figures

The value with the *fewest decimal places* determines the number of decimal places in a sum.

2. When *multiplying* or *dividing*, the number of significant figures in the result is the same as in the quantity with the *fewest significant figures*.

	(0.0120)	×	(1.6)	×	(8.49026)	=	(0.16)
Significant figures	3		2		6		2
Decimal place(s)	4		1		5		2

The value with the *fewest significant figures* determines the number of significant figures in a product.

In a sequence of computations, adjusting the number of significant figures in intermediate results can lead to errors in the final value. Instead, wait until the computations are complete, and then express the final value with the appropriate number of significant figures.

3. **Postpone adjusting results to the correct number of significant figures until a calculation is complete.**

When exact numbers are used, their presence has no effect on the significant figures in the result. For example, we can convert 1.855 hours into seconds.

$$1.855 \ \text{hr} \left(\frac{60 \ \text{min}}{1 \ \text{hr}} \right) \left(\frac{60 \ s}{1 \ \text{min}} \right) = 6678 \ s$$

How many significant figures should be used? Although the conversion ratios contain only two digits, they are exact numbers. By definition, there are *exactly* 60 minutes in an hour and *exactly* 60 seconds in a minute. Thus the precision of this product is determined by the precision of the time in hours, which is four significant figures. We report the time in seconds to four significant figures.

Calculators usually give more significant figures than are justified. For example:

$$0.0120 + 1.6 + 8.49026 = \boxed{10.1}0226 \ \text{(calculator result)}$$

$$(0.0120)(1.6)(8.49026) = \boxed{0.16}3012992 \ \text{(calculator result)}$$

In each of these calculations, the calculator displays extra digits that are not significant. The significant digits are highlighted in brown. A calculator cannot distinguish the number of significant figures that is appropriate for a calculation. Many calculators can be set so that they display a predetermined number of digits; however, these also do not necessarily represent the correct number of significant figures.

/// Always adjust calculator results to the appropriate number of significant figures using the three guidelines given earlier.

In the process of dropping extra digits beyond the precision of the result, a conventional guideline is used for rounding off results. If the first digit that will be removed is 5 or greater, round the last remaining digit upward by one unit. If the first digit that will be removed is smaller than 5, leave the last remaining digit unchanged. Example 1-5 shows how to apply these guidelines.

When a number is exact (such as 60 min/hr), precision need not be specified.

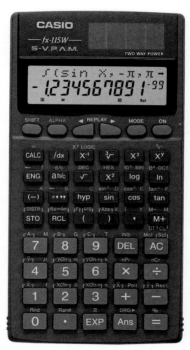

Calculators often display more significant figures than the data justify.

| Example 1-5 | Significant Figures |

A farmer owns a rectangular field that fronts along a road. State highway engineers surveyed the frontage and found that it measures 138.3 m in length. The farmer, who wants to build a fence around the field, paces off the field's width and estimates it to be 52 m. How many meters of fence will the farmer have to build? What mass of fertilizer will the farmer need to fertilize the field with 0.0050 kg of fertilizer for each square meter of field?

Strategy: First, calculate the field's perimeter to determine the amount of fencing needed. Then calculate the area of the field to determine how much fertilizer is needed. The precision of the results must also be determined.

Solution:

$$\text{Perimeter} = \text{Length} + \text{Width} + \text{Length} + \text{Width}$$

$$\text{Perimeter} = 138.3 \text{ m} + 52 \text{ m} + 138.3 \text{ m} + 52 \text{ m} = 380.6 \text{ m}$$

According to the guideline for adding or subtracting, this result must have the same number of decimal places as the least precise measurement. In this case, pacing off the width of the field limits the precision of the calculation to the nearest meter. The value should be rounded up because the "6" in "380.6" is larger than 5. The farmer requires 381 m of fence.

Calculate the area of the field to determine how much fertilizer is needed:

$$\text{Area} = (\text{Length})(\text{Width}) = (138.3\ )(52 \text{ m}) = 7191.6 \text{ m}^2$$

This result has too many significant figures, but we carry the extra digits until the calculation is complete. Each square meter of field requires 0.0050 kg of fertilizer. Find the required amount of fertilizer by multiplying:

$$\text{Mass}_{\text{fertilizer}} = (7191.6 \text{ m}^2)(0.0050 \text{ kg/m}^2) = 35.958 \text{ kg}$$

Now we are ready to round off. The multiplication steps include two numbers that have only two significant figures. The guideline for multiplying or dividing indicates that the result should also have two significant figures. The farmer requires 36 kg of fertilizer for this field.

Notice that the units work out correctly in this calculation: area multiplied by mass per unit area gives mass.

| Section Exercises |

1.6.1 Calculate the mass of a cylindrical cork of radius 1.00 cm and length 4.00 cm. (Volume of a cylinder is $V = \pi r^2 h$.)

1.6.2 A cylindrical jar has an inside diameter of 3.00 cm and a height of 8.00 cm. The empty jar weighs 185.65 g. Filled with gasoline, the same jar weighs 225.40 g. Find the density of gasoline.

1.6.3 Which of the following quantities can be determined exactly, and which must be measured with some degree of uncertainty? (a) mass of a gold nugget, (b) number of seconds in 1 day, (c) time it takes a sprinter to run the 100-m dash, (d) speed of light, (e) number of potatoes in a bushel, and (f) number of centimeters in exactly 1 mile.

1.7 CHEMICAL PROBLEM SOLVING

Many students have difficulty solving chemistry problems, but problem solving is a skill that you can learn and master. To help you, we devote extra attention to setting up and solving problems. Throughout this book, the Examples include information about methods for finding solutions.

Chemical problems that call for calculations often involve several tasks. You must identify what is to be calculated, be aware of the chemistry that is taking place, find equations that apply to the calculation, organize the data that are provided, do appropriate manipulations and substitutions, and carry out the calculation correctly. A stepwise procedure can help you master these types of calculations. We recommend the sequence of steps shown in the box. Example 1-6 provides a first illustration of this stepwise procedure.

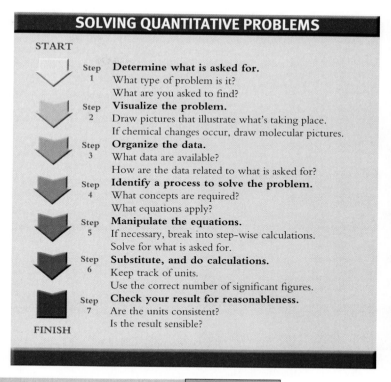

SOLVING QUANTITATIVE PROBLEMS

START

Step 1 **Determine what is asked for.**
What type of problem is it?
What are you asked to find?

Step 2 **Visualize the problem.**
Draw pictures that illustrate what's taking place.
If chemical changes occur, draw molecular pictures.

Step 3 **Organize the data.**
What data are available?
How are the data related to what is asked for?

Step 4 **Identify a process to solve the problem.**
What concepts are required?
What equations apply?

Step 5 **Manipulate the equations.**
If necessary, break into step-wise calculations.
Solve for what is asked for.

Step 6 **Substitute, and do calculations.**
Keep track of units.
Use the correct number of significant figures.

Step 7 **Check your result for reasonableness.**
Are the units consistent?
Is the result sensible?

FINISH

Problem-Solving Strategy   **Example 1-6**

Electroplating is a process in which a metal such as copper is coated with another metal, such as silver or chromium. The transfer of metal atoms is driven by an electrical current. In an electroplating process, a spoon is coated with silver from a silver rod. In the process, 1.0×10^{21} atoms are transferred from the rod to the spoon, and the rod loses 0.179 g of its mass. Use this information and the density of silver to estimate the volume of one silver atom.

Strategy: Apply the seven-step approach.

Solution:

1. Determine what is asked for. The problem asks for the volume of a silver atom.

2. Visualize the problem. Draw a picture that shows the setup and the information provided in the problem. A picture often helps interpret and summarize a problem. In this case, the figure should show the spoon, the silver wire, and some indication that atoms are transferred from the wire to the spoon.

Solving
Quantitative
Problems

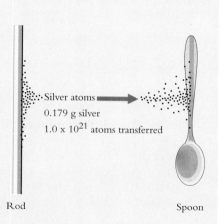

Silver atoms →
0.179 g silver
1.0×10^{21} atoms transferred

Rod   Spoon

Example 1-6 | **Problem-Solving Strategy** *(Continued)*

3. Organize the data. First, what data are given in the problem?

$$\text{Mass of silver} = 0.179 \text{ g}$$

$$\text{Number of atoms transferred} = 1.0 \times 10^{21} \text{ atoms}$$

Second, what information can be found in tables?

$$\text{Density of silver} = \rho = 10.50 \text{ g/cm}^3 \text{ (see Table 1-4)}$$

4. Identify a process to solve the problem. The question asks about the volume of one silver atom. Mass and volume are related through density: $\rho = m/V$. From this equation, we can calculate the total volume of the silver atoms. The problem also gives the total number of silver atoms transferred from the wire to the spoon. We can combine these data to find the volume of a single atom. Oftentimes, a flow chart helps summarize the process:

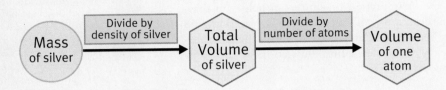

Spoons are electroplated with silver to give them an attractive finish.

5. Manipulate the equations as needed. We must use two equations. We can work through these step by step or combine them into a single equation. Generally, the stepwise approach is easier to follow.

To begin the calculations, we use the mass and density to calculate the volume:

$$\rho = m/V \qquad so \qquad V = m/\rho$$

This is V_{total}, which can be substituted into the second equation to find the volume of a single atom:

$$V_{\text{atom}} = \frac{V_{\text{total}}}{\text{number of atoms}}$$

6. Substitute and do the calculations.

$$V_{\text{total}} = \frac{m}{\rho} = \frac{0.179 \text{ g}}{10.50 \text{ g/cm}^3} = 1.70 \times 10^{-2} \text{cm}^3$$

$$V_{\text{atom}} = \frac{V_{\text{total}}}{\# \text{ of atoms}} = \frac{1.70 \times 10^{-2} \text{ cm}^3}{1.0 \times 10^{21} \text{ atoms}} = 1.7 \times 10^{-23} \text{ cm}^3/\text{atom}$$

Notice that units and significant figures are included in each step.

7. Check the result for reasonableness. Is this a "reasonable" value for the volume of an atom? At this stage of your study of chemistry, you cannot easily answer this question. Atoms are unimaginably small, however, and a volume of 10^{-23} cm^3 certainly is unimaginably small. Also, the units are right: cm^3 is a measure of volume, and the calculation yields units of volume per atom, as the question asked.

■ **1.7.1** The standard unit of length in the land of Ferdovia is the frud (2.000 frud $= 10^{-2}$ m). Fast-food restaurants in Ferdovia sell fruit drinks in rectangular cartons measuring 50.00 by 40.00 by 80.0 fruds. The Frod family left a freshly opened carton on their kitchen table, and their pet ferret (Fred) drank some of the fruit drink. Mrs. Frod found the height of the remaining liquid to be 61 fruds. How many cubic meters of fruit drink did Fred drink?

■ **1.7.2** A technician used a section of glass tubing to measure the density of a liquid that the laboratory needed to identify. The inside diameter of the tubing was known to be 0.87 mm. An empty piece of the tubing weighed 0.785 g. When liquid was drawn into the tubing to a height of 4.0 cm, the tubing and liquid weighed 0.816 g. Find the density of the liquid.

CHAPTER REVIEW

Summary and Key Terms

1. Chemistry, which deals with the properties and reactions of matter, shares topics and methods with many other disciplines. These include a reliance on experiments. A generalization from experiments is a **hypothesis,** and if a hypothesis is confirmed by many experiments, it becomes a **theory.**

2. **Macroscopic** properties can be observed directly, but **microscopic** observations require magnifying instruments. Chemists routinely visualize events at the **atomic/molecular level.** At any of these levels, **atoms** and **molecules** determine the properties of matter. Each **element** contains a different type of atom, represented by a chemical symbol. Each type of molecule has a specific **chemical formula.**

3. The **periodic table** organizes the elements into groups with similar properties, including the **alkali metals, alkaline earth metals, halogens,** and **noble gases;** and into broad categories, such as **transition metals, lanthanides, actinides,** and **main group elements.** Most of the elements are **metals;** some are **nonmetals,** and a few are **metalloids.**

4. Matter can consist of a pure substance or a **mixture** of substances. Whereas a pure substance is **homogeneous,** a mixture may be either a homogeneous **solution** or **heterogeneous.** The **phases** of matter are **solid, liquid,** and **gas.** A change of phase is a **physical transformation,** whereas a rearrangement that forms a new substance is a **chemical transformation.**

5. Measurements of physical properties such as **length, area, volume, mass, time,** and **temperature** are important in chemistry. Any measurement has a magnitude, units, and a precision. **Accuracy** describes how close a measurement is to the true value. **Precision** is expressed by using **significant figures.**

6. **Density** is an **intensive** property formed from the ratio of two **extensive** properties, mass and volume. In calculating density and other derived quantities, rules must be used to determine the correct number of significant figures.

7. A seven-step procedure provides a "blueprint" for solving chemical problems.

Skills to Master

▶ Using the periodic table of the elements

▶ Applying unit conversions

▶ Keeping track of significant figures

▶ Analyzing and solving problems

Learning Exercises

1.1 Make a list of all terms new to you introduced in Chapter 1. Give a one-sentence definition for each. Consult the glossary if you need help.

1.2 The following skills are introduced in Chapter 1. Write down a "plan of attack" for accomplishing each: (a) classifying samples of matter; (b) interpreting and drawing simple molecular pictures; (c) working with the density equation; (d) using scientific notation; (e) determining significant figures; (f) converting between different units; and (g) solving chemistry problems.

1.3 Go through the end-of-chapter problems and identify those that require the skills listed in Learning Exercise 1.2. Some problems may require more than one skill, and some require skills that are not listed.

1.4 "Memory bank" equations are those important enough for you to memorize. Start a list of these equations and learn them. Chapter 1 has only two memory bank equations.

Problems   <u>ilw</u> = interactive learning ware problem. Visit the website at www.wiley.com/college/olmsted

What Is Chemistry?

1.1 Describe three political problems for which knowledge of chemistry would be helpful.

1.2 Describe three environmental problems for which knowledge of chemistry would be helpful.

1.3 List three reasons why it is important for a pharmacy student to learn about chemistry.

1.4 List three reasons why it is important for an engineering student to learn about chemistry.

1.5 A chemist says, "My results must be wrong because they don't agree with the theory." Criticize this statement, and make recommendations for what the chemist should do next.

1.6 A leading manufacturer announces a new detergent ingredient that is "completely biodegradable and nontoxic." If you were a chemist working for a consumer products testing company, what recommendations would you make for testing this claim?

Atoms, Molecules, and Compounds

1.7 What is the elemental symbol for each of the following elements? (a) hydrogen; (b) helium; (c) hafnium; (d) nitrogen; (e) neon; and (f) niobium.

1.8 What is the elemental symbol for each of the following elements? (a) potassium; (b) platinum; (c) plutonium; (d) lead; (e) palladium; and (f) phosphorus.

1.9 What is the name of each of the following elements? (a) As; (b) Ar; (c) Al; (d) Am; (e) Ag; (f) Au; (g) At; and (h) Ac.

1.10 What is the name of each of the following elements? (a) Br; (b) Be; (c) B; (d) Bk; (e) Ba; and (f) Bi.

1.11 Examine the following molecular pictures and determine the corresponding molecular formulas.

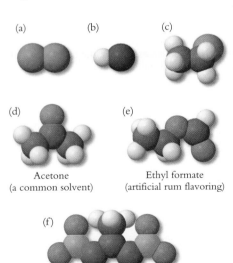

(a)   (b)   (c)

(d)   (e)

Acetone
(a common solvent)

Ethyl formate
(artificial rum flavoring)

(f)

2,4,6-Trinitrotoluene
(the explosive TNT)

1.12 Examine the following molecular pictures and determine the corresponding molecular formulas.

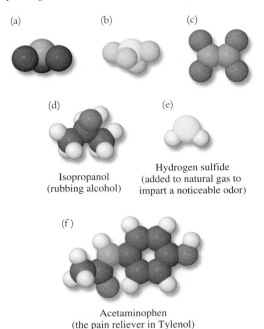

(a)   (b)   (c)

(d)   (e)

Isopropanol
(rubbing alcohol)

Hydrogen sulfide
(added to natural gas to
impart a noticeable odor)

(f)

Acetaminophen
(the pain reliever in Tylenol)

1.13 Write the chemical formula of the compound whose molecules contain: (a) four atoms of chlorine and one atom of carbon; (b) two atoms of oxygen and two atoms of hydrogen; (c) four phosphorus atoms and ten oxygen atoms; (d) two atoms of iron and three atoms of sulfur.

1.14 Write the chemical formula of the compound whose molecules contain: (a) five carbon atoms and twelve hydrogen atoms; (b) four atoms of fluorine and one atom of silicon; (c) two atoms of nitrogen and five atoms of oxygen; (d) one iron atom and three chlorine atoms.

The Periodic Table of the Elements

1.15 What element is just after xenon in the periodic table?

1.16 What element is just before sodium in the periodic table?

1.17 What are the names and chemical symbols of the elements that are vertical and horizontal neighbors of sulfur in the periodic table? Which of these have chemical properties similar to those of sulfur?

1.18 What are the names and chemical symbols of the elements that are vertical and horizontal neighbors of tin in the periodic table? Which of these have chemical properties similar to those of tin?

1.19 Name two metals that react with bromine to give compounds with the chemical formula MBr.

1.20 Name two elements that react with oxygen to give compounds with the chemical formula MO.

1.21 Write the names and symbols of all elements that occupy the same row of the periodic table as nitrogen.

1.22 Write the names and symbols of all elements that occupy the same column of the periodic table as nitrogen.

Characteristics of Matter

1.23 Classify each of the following as a pure substance, solution, or heterogeneous mixture: (a) block of iron; (b) cup of coffee; (c) glass of milk; (d) atmosphere, when free of dust; (e) atmosphere, when dusty; and (f) block of wood.

1.24 Classify each of the following as a pure substance, solution, or heterogeneous mixture: (a) blood; (b) dry ice; (c) krypton gas; (d) a rusty nail; (e) table salt; and (f) glass of lemonade.

1.25 Classify each of the following as a solid, liquid, or gas: (a) gasoline; (b) Teflon tape; (c) snow; and (d) water vapor.

1.26 Classify each of the following as a solid, liquid, or gas: (a) tree sap; (b) ozone; (c) dry ice; and (d) motor oil.

1.27 Classify each of the following as a chemical or a physical transformation: (a) formation of frost; (b) drying of clothes; and (c) burning of leaves.

1.28 Classify each of the following as a chemical or a physical transformation: (a) water boiling; (b) coffee brewing; and (c) photographic film being developed.

1.29 Classify each of the following as an element, a compound, or a mixture: (a) lake water; (b) distilled water; (c) mud; (d) helium inside a balloon; (e) rubbing alcohol; and (f) paint.

1.30 Classify each of the following as an element, a compound, or a mixture: (a) Earth's atmosphere; (b) beer; (c) iron magnet; (d) ice; (e) liquid bromine; and (f) mercury in a barometer.

Measurements in Chemistry

1.31 Convert to scientific notation: (a) 100,000, precise to ± 1; (b) ten thousand, precise to ± 1000; (c) 0.000400; (d) 0.0003; and (e) 275.3.

1.32 Convert to scientific notation: (a) 175,906; (b) 0.0000605; (c) two and a half million, precise to ± 100; and (d) two and a half billion, precise to $\pm$ one million.

1.33 Express each of the following in SI base units using scientific notation (example: 1.45 mm = 1.45×10^{-3} m): (a) 432 kg; (b) 624 ps; (c) 1024 ng; (d) 93,000 km, precise to ± 10; (e) 1 day; and (f) 0.0426 in.

1.34 Express each of the following in SI base units using scientific notation: (a) 1 week; (b) 1.35 mm; (c) 15 miles; (d) 4.567 μs; (e) 6.45 mL; and (f) 47 kg.

1.35 The mass unit most commonly used for precious stones is the carat: 1 carat = 3.168 grains, and 1 gram = 15.4 grains. Find the total mass in kilograms (kg) of a ring that contains a 5.0×10^{-1} carat diamond and 7.00 grams of gold.

1.36 What is the total mass in grams, expressed in scientific notation with the correct number of significant figures, of a solution containing 2.000 kg of water, 6.5 g of sodium chloride, and 47.546 g of sugar?

Calculations in Chemistry

1.37 What is the mass of 1 quart of water (1 L = 1.057 quarts)?

1.38 What is the mass of 1 quart of mercury (1 L = 1.057 quarts)?

1.39 A plastic block measures 15.5 cm by 4.6 cm by 1.75 cm, and its mass is 98.456 g. Compute the density of the plastic.

1.40 A penny has a diameter of 1.8 cm and a thickness of 0.15 cm, and its mass is 2.50 g. Compute the density of the penny (cylinder volume, $V = \pi r^2 h$).

1.41 Calculate the volume of an aluminum spoon whose mass is 15.4 g.

1.42 Calculate the volume of a quartz crystal of mass 0.246 g.

Chemical Problem Solving

1.43 A chemist who wished to verify the density of water constructed a cylindrical container of aluminum 4.500 inches high, whose inside radius measured 0.875 inch. The empty cylinder had a mass of 93.054 g. When filled with water, its mass was 270.064 g. Find the density of water from these data, expressing your result in SI units with the correct precision (cylinder volume, $V = \pi r^2 h$).

1.44 A chemist who prepared a new organic liquid wanted to determine its density. Having only a small sample to work with, the chemist had to use a small container. A tube whose volume was 8.00×10^{-3} cm³ weighed 0.4763 g when empty and 0.4827 g when filled with the liquid. Compute the density of the liquid in g/cm³.

ilw **1.45** Which possesses more mass, a sphere of gold with a diameter of 2.00 cm or a cube of lead measuring 2.00 cm on each side (sphere volume, $V = 4\pi r^3/3$)?

1.46 Which possesses more mass, a sphere of aluminum with a diameter of 2.00 cm or a cube of iron measuring 2.00 cm on each side (sphere volume, $V = 4\pi r^3/3$)?

1.47 Bromine is one of the two elements that is a liquid at room temperature (mercury is the other). The density of bromine at room temperature is 3.12 g/mL. What volume of bromine is required if a chemist needs 36.5 g for an experiment?

1.48 The density of gasoline at room temperature is 0.70 g/mL. If the gas tank of a car holds 12.0 gallons (45.4 L), what is the mass of a tankful of gasoline?

Additional Paired Problems

1.49 How many elements have symbols beginning with T? Give the name and symbol of each.

1.50 How many elements have symbols beginning with P? Give the name and symbol of each.

1.51 Draw a molecular picture that represents a homogeneous mixture of molecular fluorine and molecular chlorine in a 3:1 ratio of fluorine to chlorine. Both are gases, and both are diatomic molecules. The drawing should contain at least 20 molecules. Use Figure 1-4 as a guide.

1.52 Gasoline does not dissolve in water. Instead, molecules of gasoline float on the surface of water. Draw a molecular picture that shows a heterogeneous mixture of gasoline and water. Draw enough molecules of each to show the structure of the mixture clearly. Use circles for water molecules and ovals for gasoline molecules.

1.53 What are the name and symbol of the element of highest mass whose symbol and English name do not match?

1.54 What are the name and symbol of the element of lowest mass whose symbol and English name do not match?

1.55 An athlete runs the 100-yard dash in 10.17 s. At the same speed, how many seconds does it take the same athlete to run 100 meters?

1.56 An athlete runs the mile in 3 min 57 s. At the same speed, how many seconds does it take the same athlete to run 1500 meters?

1.57 Which elements would you expect to have chemical behavior similar to that of gold?

1.58 Which elements would you expect to have chemical behavior similar to that of sodium?

1.59 Draw a molecular picture of SO_2, in which each S atom is flanked by two O atoms in a bent arrangement. (SO_2 is a major contributor to acid rain.)

1.60 Draw a molecular picture of O_3, in which a central O atom is connected to the other two O atoms in a bent arrangement. (O_3 is a major contributor to smog.)

1.61 Match each photograph with the appropriate molecular picture. Write a short justification of your choices.

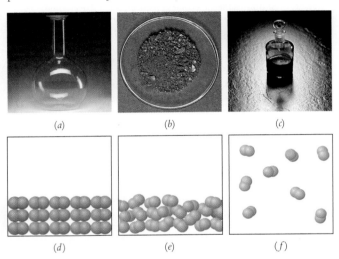

1.62 Decide whether each of the following molecular pictures represents a homogeneous solution, a single substance, or a heterogeneous mixture. If the picture represents a single substance, tell whether it is an element or a chemical compound. Explain your choices.

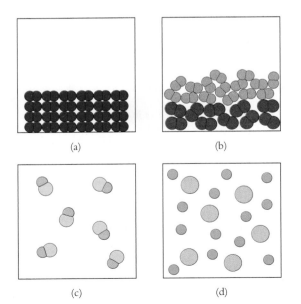

1.63 Ethylene glycol is used as antifreeze in car radiators. The freezing temperature of ethylene glycol is $-11.5\ °C$. Convert this freezing temperature to kelvins (K).

1.64 The melting point of silver metal is 1235 K. Convert this melting point to degrees Celsius ($°C$).

1.65 Perform the following calculations and report your answers with the correct number of significant figures.

(a) $$\frac{(6.531 \times 10^{13})(6.02 \times 10^{23})}{(435)(2.000)}$$

(b) $$\frac{4.476 + (3.44)(5.6223) + 5.666}{(4.3)(7 \times 10^4)}$$

1.66 Perform the following calculations, and report your answers with the correct number of significant figures.

(a) $$\frac{(3.14159)(4.599 \times 10^6) - (1.12 \times 10^7)}{(4.756 \times 10^8) + (3.67 \times 10^4)}$$

(b) $$\frac{(6.577 \times 10^{-6}) + 0.00369 + (8.234 \times 10^{-4})}{(0.0002567) + (6.9377 \times 10^{-5})}$$

1.67 For each of the following statements, determine whether the property in italics is extensive or intensive. (a) The *boiling point* of ammonia (NH_3) is 239.8 K. (b) The *mass* of a diamond is 2.34 carats. (c) The *density* of nickel is 8.90 g/cm³. (d) A copper wire is 3.2 cm in *length*. (e) Bromine is a *dark red* liquid.

1.68 For each of the following statements, determine whether the property in italics is extensive or intensive. (a) The *density* of iron is 7.86 g/cm³. (b) Liquid oxygen has a *pale blue color*. (c) The crankcase of an automobile holds 5 *quarts* of oil. (d) The *melting point* of gallium metal is 30 °C. (e) A recipe calls for 100 *grams* of sugar.

1.69 Examine the following molecular pictures and determine the chemical formulas of the compounds they represent.

(a)

Freon 21
(a refrigerant suspected of damaging the ozone layer in the stratosphere)

(b)
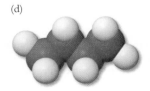

Formic acid
(the stinging compound in an ant bite)

(c)

Bromine trifluoride
(used in the production of fuel for nuclear reactors)

(d)
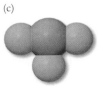

Butane
(the fluid in disposable lighters)

1.70 Examine the following molecular pictures and determine the chemical formulas of the compounds they represent.

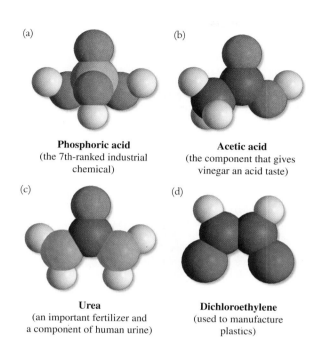

(a)
Phosphoric acid
(the 7th-ranked industrial chemical)

(b)
Acetic acid
(the component that gives vinegar an acid taste)

(c)
Urea
(an important fertilizer and a component of human urine)

(d)
Dichloroethylene
(used to manufacture plastics)

1.71 Convert each of the following into SI units: (a) engine displacement of 454 cubic inches; (b) car speed of 35 mph; (c) height of 6 feet 9 inches; and (d) boulder mass of 227 pounds.

1.72 Convert each of the following into SI units: (a) gold nugget mass of 1.5 ounces; (b) light speed of 6.71×10^8 mph; (c) hike length of 11 miles; and (d) car mileage of 32 miles/gallon.

1.73 From the elements Ne, Cs, Sr, Br, Co, Pu, In, and O, choose one that fits each of the following descriptions. (a) alkaline earth metal; (b) element whose properties are similar to those of aluminum; (c) element that reacts with potassium; (d) transition metal; (e) noble gas; (f) actinide.

1.74 From the elements Ar, K, Ca, Cl, Cu, U, P, and S, choose one that fits each of the following descriptions. (a) alkaline earth metal; (b) element whose properties are similar to those of nitrogen; (c) element that reacts with potassium; (d) transition metal; (e) noble gas; (f) actinide.

1.75 Draw appropriately scaled and colored molecular pictures of each of the following molecules: HCN (linear); H_2O (bent, O in the middle); CO; and NNO (linear).

1.76 Draw appropriately scaled and colored molecular pictures of each of the following molecules: CO_2 (linear, C in the middle); ClO_2 (bent, Cl in the middle); ClBr; and HOCl (linear). (Use Figure 1-4 as a guide.)

1.77 The distance from New York to Los Angeles is 2786 miles. How many minutes would it take an airplane flying at a constant speed of 685 km/hr to fly between the two cities?

1.78 The distance between Seattle, Washington and Portland, Oregon is 279 km. How many minutes would it take a traveler moving at a constant speed of 65 miles per hour to drive between the two cities?

1.79 Potassium metal has a lustrous silvery appearance. It melts at 336.8 K. It reacts with chlorine gas to give a compound whose formula is KCl. The density is 0.862 g/mL. Potassium is soft enough to cut with a knife. When added to water, it often bursts into flame. Which of these properties are physical properties, and which are chemical properties?

1.80 Elemental oxygen is a colorless gas. It exists as diatomic molecules. It reacts readily with metals to form oxides, such as FeO. When cooled below 90 K, it condenses to form a pale blue liquid. Categorize each of these properties as physical or chemical.

1.81 Write the names and chemical symbols for three examples of each of the following: (a) halogens; (b) alkaline earth metals; (c) actinides; and (d) noble gases.

1.82 Write the names and chemical symbols for three examples of each of the following: (a) transition metals; (b) lanthanides; (c) main group elements; and (d) alkali metals.

1.83 Determine how many kilometers through outer space a beam of light travels in one year (365.24 days).

1.84 The distance from the Earth to the sun is 9.3×10^7 miles. How long does it take for light from the sun to reach our planet?

More Challenging Problems

1.85 The diameter of a chlorine atom is 2.00×10^2 pm. How many chlorine atoms lined up end to end would form a line 1.0 inch long?

1.86 Some chemists refer to chemistry as the "central" science because of its importance to other sciences. Do you agree? Give your reasons.

1.87 One of Jules Verne's most famous novels is *20,000 Leagues Under the Sea*. The league is a nautical unit of distance. Here are some nautical distance conversion factors:

1 league = 3 nautical miles	1 nautical mile = 10 cable lengths
1 cable length = 100 fathoms	1 fathom = 6 ft

How deep is 20,000 leagues in feet and in kilometers?

1.88 The photo below shows blue copper sulfate between samples of elemental copper metal and elemental sulfur. The third element present in copper sulfate is oxygen, which exists as a colorless gas under terrestrial conditions. (a) What are the elemental symbols of the three elements present in copper sulfate? (b) In copper sulfate, copper, and sulfur combine in a 1:1 atomic ratio, and there are four oxygen atoms for every copper atom in the compound. What is the formula of copper sulfate?

1.89 The day is defined to contain exactly 24 hours of 60 minutes, and each minute contains exactly 60 seconds. The year is 365.24 days long. How many seconds are there in a century, which is 100 years? Express your answer in scientific notation with the correct number of significant figures.

1.90 A square of aluminum foil measuring 3.00 inches on each side weighs 255 mg. Find the thickness in micrometers (μm) of the aluminum foil.

1.91 We began this chapter by listing several questions that chemists are interested in answering. List five more.

1.92 Almost all the elements whose symbols come from Latin names are metals. Suggest a chemical reason why these elements were named many years ago.

1.93 Using everyday observations, decide which one in the following pairs of substances has the greater density. Explain your reasoning. (a) oil or vinegar; (b) table salt or water; and (c) iron or aluminum.

1.94 Carbon monoxide is a common pollutant in urban environments. On one particular day, the air contains 5.5 mg of carbon monoxide per 1.000 cubic meter of air. How many grams of carbon monoxide are present in a room whose dimensions are 12 feet $\times$ 9.5 feet $\times$ 10.5 feet?

Group Study Problems

1.95 On average, a sample of human blood whose dimensions are 0.1 mm $\times$ 0.1 mm $\times$ 0.1 mm contains 6.0×10^3 red blood cells. The volume of blood in a typical adult is about 5 L. How many red blood cells are there in an adult?

1.96 The diameter of metal wire is given by its wire gauge number. For example, 16 gauge wire has a diameter of 0.0508 in. Calculate the length in meters of a 5.00-pound spool of 16-gauge copper wire. The density of copper is 8.92 g/cm^3.

1.97 Water for irrigation is usually expressed in units of acre-feet. One acre-foot is enough water to cover one acre of land to a depth of 1 foot (1 acre = $43,560 \text{ ft}^2$). A water reservoir has a maximum capacity of 8.97×10^5 acre-feet. What is the lake's volume in (a) liters; (b) cubic feet; and (c) cubic meters?

1.98 The Greek scientist Archimedes was given the task of determining whether his king's crown was pure gold or an alloy of gold and silver. The king specified that no part of the crown could be destroyed. To do this, Archimedes invented volume determination by displacement. First, he measured the crown's mass to be 2.65 kg (they did not use SI units then, but the method does not depend on the units used). Next, he put the crown in a full basin of water and found the amount of overflow to be 145 cm^3. Was the crown pure gold or a gold-silver alloy? Do a calculation that supports your answer.

Answers to Section Exercises

1.1.1 Chemistry is involved when sugar dissolves in a glass of iced tea. A chemical process in yeast creates the gas that causes bread to rise. In making candy, "caramelizing" sugar causes a chemical change in the sugar molecules. Marinade tenderizes meats through chemical changes.

1.1.2 Chemistry applies to the automobile industry in many ways. To name just a few, chemistry is involved in the search for new additives that make gasoline more energy efficient and less polluting. Chemistry plays a role in the creation of new plastic materials for automobile interiors and bodies. Ceramics created by chemists and engineers are used to make engine parts. New paint formulations give automobiles coatings that inhibit corrosion of metal parts.

1.1.3 Develop a set of chemical reactions that can be performed on Earth and on Venus. Observe the results carefully on both planets. If the results obtained on Venus differ from the results obtained on Earth, the theory may be valid. However, if the results are the same, the theory cannot be valid.

1.2.1 Ce, Cs, Cu, Ca, and C

1.2.2 Zirconium, nickel, tin, tungsten, selenium, beryllium, and gold

1.2.3 (a) H_2O_2; (b) SO_2; (c) N_2O; (d) $C_2H_4O_2$; and (e) C_2H_4.

1.3.1 (a) $CaCl_2$; (b) CsI; (c) BaO; and (d) MgF_2

1.3.2 $AlBr_3$ and $GaCl_3$

1.3.3 Aluminum, main group; fluorine, halogen and main group; cobalt, transition metal; phosphorus, main group; krypton, noble gas; europium, lanthanide; thorium, actinide; barium, alkaline earth; and sodium, alkali metal

1.4.1 (a) Single substance, an element; (b) single substance, a compound; and (c) heterogeneous mixture

1.4.2 Solution, pure substance, heterogeneous mixture, pure substance, and solution

1.4.3 Physical, chemical, physical, and chemical

1.5.1 $4.63 \times 10^{-7} \text{ m}^3$, 1.7935×10^7 m, and 9.36×10^8 s

1.5.2 3.0×10^8 m/s and 9.5×10^{15} m

1.5.3 70.3 kg, 110. m, 311.6 K, and 3.155690×10^7 s

1.6.1 3.0 g

1.6.2 0.703 g/cm^3

1.6.3 (a) uncertain; (b) exact; (c) uncertain; (d) uncertain; (e) exact; and (f) exact

1.7.1 $4.8 \times 10^{-3} \text{ m}^3$

1.7.2 1.3 g/cm^3

2

The Atomic Nature of Matter

INTRODUCTION: THE EVOLUTION OF ATOMIC THEORY

Pure substances appear uniform in composition. A microscopic flake of silicon has the same uniform gray color and smooth appearance as a large bar of this element. We might expect this uniformity to continue down to the smallest imaginable scale. This view of matter was championed centuries ago by the Greek philosopher Aristotle (384–322 BC). In contrast, two other early philosophers, Leucippus (born c. 490 BC) and Democritus (c. 470–c. 380 BC), speculated that at sufficiently small scale, every substance is made up of tiny building blocks that have distinct properties of their own. Democritus used the word *atomos,* meaning "indivisible," for the smallest bit of matter that cannot be further divided. In the absence of experimental evidence for atoms, Aristotle's view of matter came to be widely accepted and remained dominant for the next 2000 years.

The modern atomic theory of matter was developed and accepted by scientists 200 years ago. The pioneer in this work was the Englishman John Dalton (1766–1844). Experiments had revealed that when two elements combine together, they do so in fixed proportions. For example, no matter what starting amounts of carbon and oxygen an experimenter uses, two specific reaction products are obtained. One product contains three grams of carbon for every four grams of oxygen, and the other contains three grams of carbon for every eight grams of oxygen. In other words, carbon and oxygen combine only in fixed proportions.

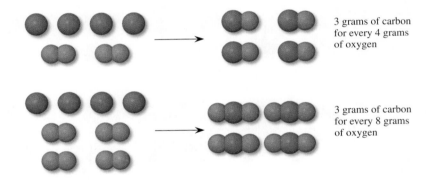

3 grams of carbon for every 4 grams of oxygen

3 grams of carbon for every 8 grams of oxygen

Dalton reasoned that if matter were continuous, elements would combine in all proportions, much as a cup of coffee can be made as sweet as we like by adding various amounts of sugar. He realized that fixed proportions result if matter is made up of indivisible atoms, with atoms of different elements (carbon and oxygen, for instance) having different characteristic masses. Dalton's theory stimulated further experiments to test its predictions. All experiments supported the theory, and by the middle of the nineteenth century, atomic theory was fully accepted as correct. Still, no one had actually observed atoms! More than 100 years would pass before images of atoms could be produced experimentally. Atomic theory was accepted because the macroscopic observations of chemistry could only be explained by postulating that matter is made of atoms.

Scientists eventually invented instruments that let us "see" individual atoms. Our inset image was obtained using scanning tunneling microscopy (STM). In essence, STM "takes a picture" of atomic layers that can be magnified billions of times, making it possible to observe individual atoms. The STM image of silicon in our inset shows how individual silicon atoms are arranged on the surface of the sample.

The theme of this chapter is the atomic perspective of matter, as expressed by atomic theory and the principles of atomic structure. Two important physical ideas, the force exerted between electrically charged objects and the role played by energy, are indispensable to this perspective. A comprehensive understanding of chemistry requires that these ideas become integral parts of your thinking.

2.1 ATOMIC THEORY

Recent advances allow chemists to view how atoms combine and rearrange to make new materials, but chemists have always observed changes in macroscopic appearance and properties as a chemical reaction occurs. For example, Figure 2-1 shows that when solid magnesium metal reacts with a liquid solution of an acid, the metal disappears and gas bubbles out of the solution. These macroscopic changes result from interactions among atoms and molecules, as shown by the inset views. An understanding of chemistry requires knowledge of how matter looks at the atomic level and how atomic-sized objects interact.

The modern view of atomic theory can be summarized in four general statements:

1. All matter is composed of tiny particles called atoms.
2. All atoms of a given element have identical chemical properties that are characteristic of that element.
3. Atoms form chemical compounds by combining in whole-number ratios.
4. Atoms can change how they are combined, but they are neither created nor destroyed in chemical reactions.

Figure 2-1
Magnesium metal reacts with aqueous acid solution to generate hydrogen gas. The expanded views show molecular views of the gas and metal.

The spectacular reaction shown in Figure 2-2 involved immense amounts of fire and heat, all resulting from the chemistry of changes at the atomic level. Hydrogen gas and oxygen gas, when combined and ignited, burn vigorously. The products are water molecules that contain hydrogen atoms and oxygen atoms bound to one another. If the reaction is performed with varying amounts of hydrogen and oxygen, the *amount* of water formed changes from one experiment to the next, but its *composition* does not. Water always contains 1.0 g of hydrogen for every 8.0 g of oxygen, and two atoms of hydrogen for every atom of oxygen.

It is possible to reverse this reaction between hydrogen and oxygen. An electrical current causes water molecules to decompose into hydrogen gas and oxygen gas. These gases can be captured, and their amounts measured as they bubble out of the solution. Repeated observations show that the two substances are always produced in the same mass ratio: Every 9.0 g of decomposed water produces 1.0 g of hydrogen and 8.0 g of oxygen. Chemists think of these processes in terms of atoms, the building blocks of all matter. Figure 2-3 shows how chemists visualize hydrogen gas, oxygen gas, and water. Notice that hydrogen gas contains hydrogen atoms (feature 1 of the atomic theory). Experiments on hydrogen reveal that all its atoms behave identically by combining into diatomic molecules (feature 2). Oxygen gas is made up of diatomic molecules containing oxygen atoms, but these molecules act differently than molecules made from hydrogen atoms (feature 2). When molecules of hydrogen and oxygen react to give water, atoms of hydrogen combine with atoms of oxygen in a 2 : 1 ratio (feature 3). Atoms are rearranged in this chemical process, but the total number of each type of atom remains the same (feature 4).

Figure 2-2
In May 1937, the Hindenburg, a lighter-than-air ship whose balloon was filled with hydrogen gas, exploded and burned. The spectacular fire was the result of hydrogen reacting with atmospheric oxygen.

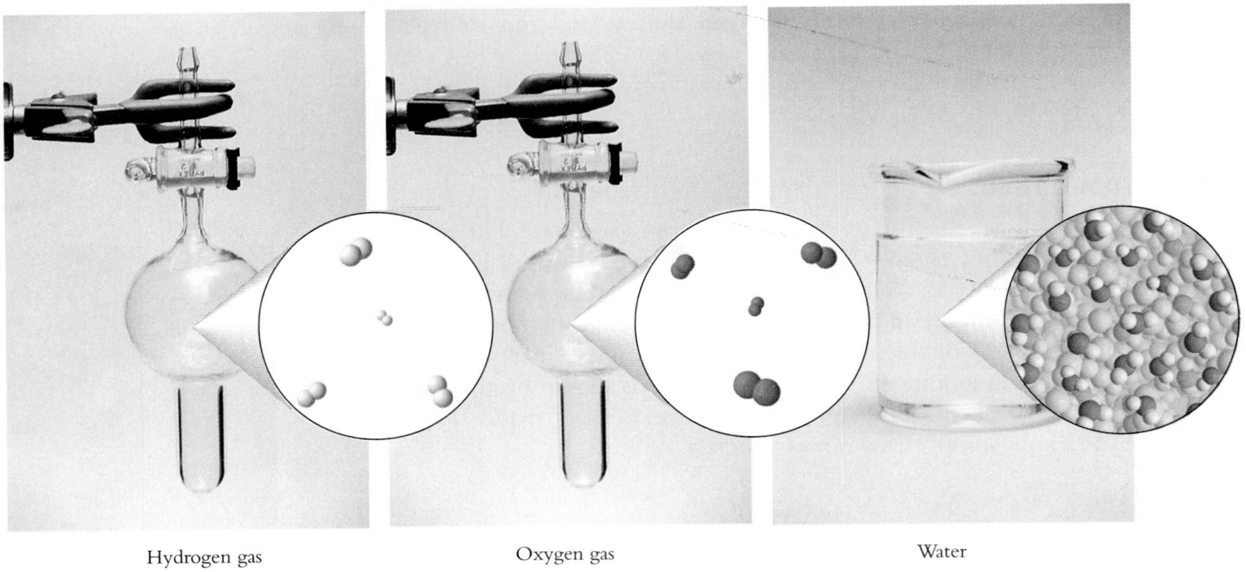

Hydrogen gas	Oxygen gas	Water

Figure 2-3
Hydrogen gas, oxygen gas, and liquid water are all composed of atoms, shown schematically in the "magnified" windows.

Conservation of Atoms and Mass

One of the cornerstones of the atomic theory is that atoms are neither created nor destroyed. In other words, the number of atoms of each type is constant and unchanging. When a quantity does not change, that quantity is said to be *conserved*. A statement that some quantity is conserved is a **conservation law.** Atoms are conserved in chemical and physical processes.

/// *Atoms are neither created nor destroyed during physical or chemical processes.*

The numbers of atoms of each element are conserved in any chemical reaction. When methane is burned to produce heat, for example, the carbon atom from each molecule of methane ends up in a molecule of carbon dioxide. The four hydrogen atoms of methane end up in two molecules of water. The atoms rearrange and recombine during the course of the reaction, but the numbers of each type of atom do not change. The chemical reaction of methane with molecular oxygen can be represented by the molecular picture shown in Figure 2-4. Counting the atoms, we see that there are four oxygen atoms (red), four hydrogen atoms (white), and one carbon atom (black) before and after the reaction.

Figure 2-4
Molecular picture of the reaction of methane (CH_4) with oxygen (O_2) to produce carbon dioxide (CO_2) and water (H_2O). Atoms of each element are conserved.

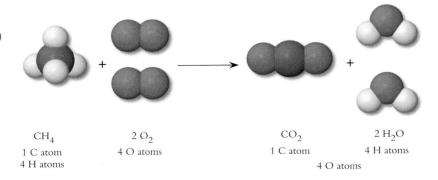

CH_4	$2\,O_2$	CO_2	$2\,H_2O$
1 C atom	4 O atoms	1 C atom	4 H atoms
4 H atoms			4 O atoms

Figure 2-5
It appears that mass is lost when wood burns, but the mass lost by the wood is transferred to the atmosphere as molecules of carbon dioxide and water.

Chemists keep track of individual atoms and electrons at the atomic level, but in the laboratory, chemists measure mass. Neither the numbers nor the masses of atoms and electrons change during chemical processes, so mass is also conserved. For example, 3 g of carbon dioxide and water are produced by the burning of 1 g of methane and 2 g of oxygen.

As Albert Einstein theorized in 1905, mass can be converted into energy and energy into mass. In chemical transformations, however, the amount of mass lost or gained is always too small to measure.

/// *Mass is neither created nor destroyed during physical or chemical processes.*

In some chemical transformations, it appears that mass is not conserved. When wood burns, for example (Figure 2-5), the mass of the ash is much less than the mass of the original wood. This is because the carbon, hydrogen, and oxygen atoms that make up most of the wood form carbon dioxide and water vapor, both of which are gases that escape into the atmosphere. Careful experiments in which all the products are captured and weighed show that the mass of ash plus gases equals the mass of wood plus oxygen. The total mass remains unchanged.

Atoms Combine to Make Molecules

Another essential feature of the atomic theory is that atoms combine in whole-number ratios to make molecules. In the spectacular reaction of hydrogen with oxygen that destroyed the Hindenburg, every two molecules of hydrogen combined with one molecule of oxygen to form two molecules of water (Figure 2-6). Even though this reaction occurred in the atmosphere in the presence of a large amount of oxygen, the product molecules contained hydrogen atoms and oxygen atoms in fixed, 2 : 1 ratios.

Many elements can combine with one another in more than one way. Hydrogen and oxygen provide a simple example. When these elements react directly, they form water molecules containing two hydrogen atoms and one atom of oxygen. Under the right conditions, the same two elements form a different compound,

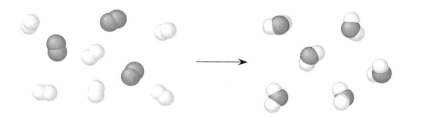

Figure 2-6
A molecular view of several oxygen molecules reacting with hydrogen molecules to form water. Note that atoms are conserved.

hydrogen peroxide. Each molecule of this substance contains two atoms of hydrogen and two atoms of oxygen. Although water and hydrogen peroxide are made from the same elements, their chemical properties are very different. Water is stable under most conditions. It is easy to form and difficult to destroy. Hydrogen peroxide, on the other hand, is an extremely reactive molecule. It is not difficult to make, but it can be destroyed in many ways. Water is life-sustaining; in fact, life depends completely on water. Hydrogen peroxide can be life-destroying. Because it kills microorganisms, hydrogen peroxide is used as a disinfectant.

From an understanding of how atoms join together to make molecules, chemists can explain why two compounds that seem so similar have profoundly different reactivity patterns. We describe how atoms link together in Chapters 8 and 9. Meanwhile, remember that chemists try to visualize chemical reactions at the molecular level. This molecular view is so ingrained that chemists often take it for granted. As you study chemistry, strive to attain this molecular point of view, for which Example 2-1 provides some practice.

| Example 2-1 | **Molecular Pictures** |

When charcoal burns in air, carbon atoms combine with oxygen atoms from molecular oxygen to form carbon dioxide. One molecule of carbon dioxide contains one carbon atom and two oxygen atoms. Experiments on carbon dioxide show that each molecule is linear, with a carbon atom in the middle. Draw a molecular picture that illustrates this reaction.

Strategy: This is a qualitative problem requiring you to visualize and represent molecules. In molecular pictures, atoms are represented with circles, and different colors or shadings are used to distinguish between elements. The problem does not state how many atoms and molecules to draw, so we can start with any convenient amounts. However, the number of atoms of each element must not change during the reaction.

Solution: Every carbon atom requires two oxygen atoms (one molecule of O_2) and generates one molecule of carbon dioxide. Here is how the molecular picture looks if we start with six atoms of carbon and six molecules of oxygen.

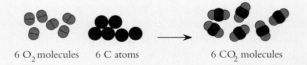

6 O_2 molecules   6 C atoms   6 CO_2 molecules

In our representation, six carbon atoms have combined with six oxygen molecules to form six molecules of carbon dioxide. For further practice, draw a molecular picture starting with five oxygen molecules and enough carbon atoms to react with these molecules.

Atoms and Molecules Are Continually in Motion

In all matter, even substances as solid as steel, individual atoms and molecules are continually in motion. In solids, atoms "rattle around"—vibrate—in the "cages" formed by the atoms that surround them. In liquids, atoms or molecules move past

Box 2-1	Chemistry and Technology: Molecular Machines?

A student using a molecular graphics computer program can take individual atoms and connect them together to make any desired molecular structure. Imagine this program interfaced to a molecular assembly line that could build actual molecules to match those structures. Science fiction? Perhaps not. Theoretical and computational models indicate that molecular manufacturing doesn't violate any known physical or chemical laws. Nanotechnology—the construction of molecular structures, one at a time—appears to be within the realm of possibility.

What uses might "molecular machines" have? Imaginative researchers have come up with numerous possibilities. Current medical technology cannot get inside the brain to remove the blood clots responsible for strokes. What if we could manufacture "nanosubmarines" that could propel themselves through the arteries, navigate to the proper site, and either delicately excise the clot or deliver a nanodose of a clot-dissolving drug? The arena of biotechnology poses many similar challenges that might be met by nanotechnology.

Alternatively, consider the field of computer technology. The computer "chip," the heart of modern computing, has been miniaturized nearly to the limits of current technology, resulting in amazingly fast machines with prodigious memories. Computer designers dream of further miniaturization, down to the "molecular wire" level. This would increase the speed and capacity of computers by several orders of magnitude. To construct computer chips with molecular dimensions will require nanotechnology.

Building things at the molecular level will require tools that are not much bigger than atoms and molecules. Designing and building such tools is a major challenge, but research is already underway. The figure shows a molecular gear based on tubes of carbon atoms studded with atomic appendages that mesh as the gear turns.

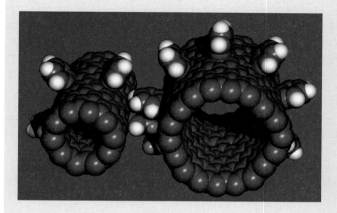

How will such fantastic molecular machines be built? One approach uses the scanning tunneling microscope

(STM) as a molecular "tweezers." This instrument, which we describe in Chapter 1, has been used to place single atoms onto a germanium surface and to place organic molecules precisely onto graphite surfaces. More spectacularly, as the photo shows, iron atoms have been arranged on a copper surface to form the Japanese word *atom*.

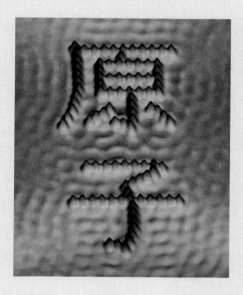

We are still years away from building molecular machines. In fact, using STM as a tool for molecular engineering has been compared to trying to build a wristwatch with a sharpened stick.

Supposing that molecular tools can be constructed, yet another obstacle must be overcome for nanotechnology to be effective. A medical nanosubmarine is likely to contain about a billion (10^9) atoms. At an assembly speed of 1 atom per second, it would take 10^9 seconds to construct one such device. That's almost 32 years! If the assembly rate can be increased to 1 atom per microsecond, the construction time for a 1-billion-atom machine drops to 1000 seconds, or just under 17 minutes. That's not bad if only a few machines are needed, but molecular machines are tiny, so large numbers of machines will be required for any practical application. Consequently, scientists will have to discover ways to mass-produce nanodevices.

To be practical, then, nanotechnology must be precise, extremely fast, and amenable to mass production. Perhaps this strikes you as definitely in the realm of science fiction rather than science fact; and perhaps it is. Nevertheless, scientists at universities such as Cornell and MIT are vigorously tackling the challenges of this field, and companies like Xerox and IBM have active research groups as well.

Figure 2-7
Atomic pictures of a monatomic solid *(left)*, liquid *(center)*, and gas *(right)*, showing how atoms move about in each phase.

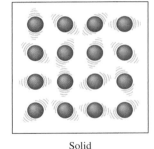 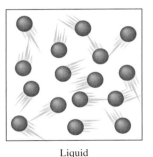 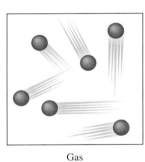

Solid   Liquid   Gas

Animation

Monatomic substances contain individual atoms that are not bound to any other atoms. An example is argon (Ar). Other substances, such as O_2, H_2, and H_2O, contain groups of atoms bound into molecules. At the atomic-molecular level, the species that move freely about are atoms for monatomic species and molecules for species composed of atomic groups.

one another continually, like minnows in a stream endlessly changing positions. In gases, atoms or molecules are free to move over large distances. Figure 2-7 is a schematic illustration of motion at the atomic-molecular level.

The pressures exerted by gases demonstrate molecular motion. Gases are collections of molecules, so the pressure exerted by a gas must come from these molecules. Just as the basketball in Figure 2-8 exerts a force when it hits a backboard, moving gas molecules exert forces when they hit the walls of their container. The collective effect of many molecular collisions generates a pressure.

The **diffusion** of one liquid into another also demonstrates molecular motion. Figure 2-9 shows that if a drop of ink is added to a beaker of still water, the color slowly but surely spreads throughout the water. The water molecules and the molecules that give ink its color move continuously. As they slide by one another, the ink molecules eventually become distributed uniformly throughout the volume of liquid.

Dynamic Molecular Equilibrium

Atoms and molecules are always moving, even when no visible changes take place. In our ink example, ink molecules move randomly in all directions. As the molecular views in Figure 2-9 indicate, however, once the molecules are evenly distributed, the total number of ink molecules and water molecules in any region of the liquid does not change. As a result, there is no further change in color.

This condition of balanced motion is called **dynamic equilibrium.** Although a dynamic system contains objects that move continuously, a system at equilibrium shows no change in its observable properties. Our example of ink in water is dynamic because the water and ink molecules continually move about. The mixture is at equilibrium when the color is uniform and unchanging. In any part of the solution, ink molecules continue to move, but the number of ink molecules in each region does not change. Dynamic equilibria occur frequently in chemical systems. Chemical processes reach a state of equilibrium if allowed to continue for a sufficient time. Nevertheless, molecular activity always goes on after equilibrium has been reached. The following example, illustrated schematically in Figure 2-10, should help you grasp this important idea.

Wet towels hung on a clothesline eventually dry because the continual motion of molecules in liquid water allows some molecules to escape from the liquid phase (Figure 2-10*a*). A wet towel left in a closed washing machine, however, stays wet for a long time. This is because water molecules that escape from the surface of the towel are trapped within the washing chamber (Figure 2-10*b*). The number of water molecules in the gas phase increases, and some of these molecules are recaptured by the towel when they collide with its surface. The system soon

Figure 2-8
When a basketball strikes a backboard, it exerts a force on the backboard. The gas molecules inside the basketball *(inset)* also exert forces when they strike the walls of the basketball. The net result of many collisions is gas pressure. We describe pressure in Chapter 5.

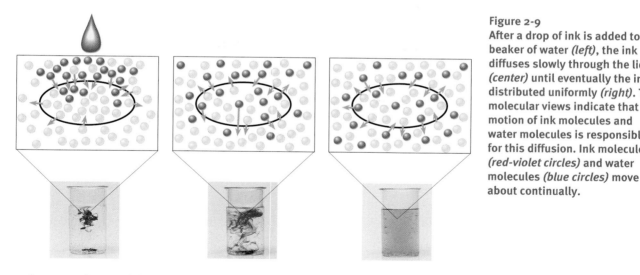

Figure 2-9
After a drop of ink is added to a beaker of water *(left)*, the ink diffuses slowly through the liquid *(center)* until eventually the ink is distributed uniformly *(right)*. The molecular views indicate that the motion of ink molecules and water molecules is responsible for this diffusion. Ink molecules *(red-violet circles)* and water molecules *(blue circles)* move about continually.

reaches a condition of dynamic equilibrium in which, for every water molecule that leaves the surface of the towel, one water molecule returns from the gas phase to the towel (Figure 2-10c). Under these conditions the towel remains wet indefinitely.

Summarizing, once this system has reached dynamic equilibrium, molecules continue to leave the liquid phase for the gas phase, but equal numbers of molecules are captured from the gas by the liquid. The amount of water in each phase remains the same (equilibrium) even though molecules continue to move back and forth between the gas and the liquid (dynamic). As with dye dispersed in water, no net change occurs after equilibrium is established.

Animation

Figure 2-10
(*a*), When water evaporates from an open container, molecules escape but are not recaptured. (*b*), If the container is closed, the number of water molecules in the gas above the liquid increases, and some are recaptured when they collide with the liquid surface. (*c*), The container reaches dynamic equilibrium when enough water molecules are present in the gas to make the rate of recapture equal to the rate of escape.

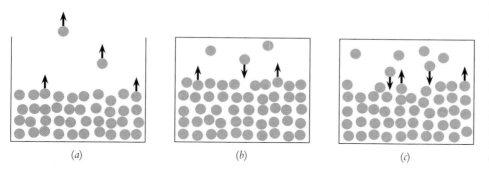

(*a*)   (*b*)   (*c*)

Section Exercises

2.1.1 Elemental bromine, chlorine, and iodine exist as diatomic molecules. Chlorine is a gas at room temperature, bromine is a liquid, and iodine is a solid. Draw molecular pictures that show the molecular distributions in samples of chlorine, bromine, and iodine.

2.1.2 Chlorine and hydrogen molecules can react with each other to form molecules of hydrogen chloride, which contain one atom of each element. Draw a molecular picture showing three molecules of chlorine reacting with enough molecules of hydrogen to convert all the chlorine into hydrogen chloride.

2.1.3 After a summer shower, rain puddles on the road quickly disappear. Describe what happens in molecular terms.

2.2 ATOMIC ARCHITECTURE: ELECTRONS AND NUCLEI

Atoms are the fundamental building blocks of chemistry, but are atoms made of other, still smaller particles? It turns out that atoms do have internal structures. Furthermore, atoms of a particular element have an internal structure that is different from that of every other element. These differences in structure are what make the chemistry of one element different from that of any other. The rich diversity of chemical behavior can be traced to the internal structure of atoms.

A series of elegant experiments, designed and performed between 1895 and 1915, revealed the essential structure of the atom. In this section we outline the main features of the most important experiments to give you an idea of how scientists probe matter at the atomic level.

Forces

Our intuition tells us that forces either hold things together or push them apart. To understand the atom, we have to know something about the nature of forces. Physicists have identified four types of forces: gravitational, electromagnetic, strong nuclear, and weak nuclear. Of these, the most familiar one for us is **gravitational force.**

All objects are pulled toward the center of the Earth by the force of gravity. Rain falls from the clouds, skydivers plunge toward the Earth, and balls thrown into the air return to the ground. The sun's gravitational force holds the planets in their regular orbits. In fact, every mass exerts a gravitational attraction on all other masses. The existence of gravitational force between any two bodies is a fundamental law of the universe. Gravitation is obviously an important force for large objects such as airplanes, baseballs, and humans.

For tiny objects such as atoms and molecules, electrical forces are most important. **Electrical force** can be either positive or negative. Studies show that when two electrically charged objects have opposite signs, electrical force pulls the objects together (Figure 2-11). When both charges have the same sign, however, the objects repel one another. Electrical force between two objects increases with the amount of charge on each, and it decreases as the objects move farther apart.

A charged object in motion is also subject to **magnetic force.** Objects that generate magnetic force are called magnets. Magnets usually have two ends, a north (N) pole and a south (S) pole. As with electrical force, magnetic force may attract or repel; opposite poles (N and S) attract, but like poles (N-N or S-S) repel.

To understand some of the early experiments that probed atomic structure, you need to know that a moving charged object is deflected along a curved path as it passes between the poles of a magnet (Figure 2-12). The amount of curvature reveals information about the properties of moving charges, so magnetic fields were very useful in studies of atomic structure.

Electrons

Important clues about the structure of atoms came from experiments that used electrical force. One of these experiments, shown schematically in Figure 2-13, used two perforated metal plates sealed inside a glass tube along with a sample of a gas. One plate was given a large positive electrical charge, and the other was given a large negative charge. When the charges became large enough, electrical forces caused an electrical discharge (similar to a lightning bolt) to leap across the space between the plates.

Gravitational force pulls objects toward the Earth.

Opposite charges pull objects together.

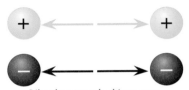

Like charges push objects apart.

Figure 2-11
Charged objects are attracted to one another if the charges have opposite signs. However, charged objects of the same sign, either both positive or both negative, repel one another.

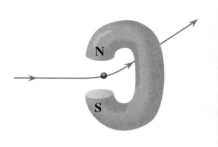

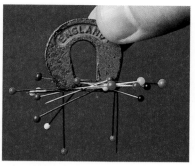

Figure 2-12
Magnetic force causes a moving charge to change direction when it passes between the poles of a magnet. The amount of bending can be related quantitatively to the charge, mass, and speed of the particle.

This high-energy discharge caused some of the atoms of the gas to break apart. The pieces of the broken atoms turned out to be charged particles. Particles with positive charges moved toward the negative plate, and particles with negative charges moved toward the positive plate. Because the plates had holes in them, some of these charged particles passed through the plates and were captured by collectors at the ends of the tube. This experiment showed that atoms are made up of smaller fragments that possess positive and negative charges.

Changing the gas in the tube changed the behavior of the positively charged particles, but the negatively charged particles always acted the same. These negatively charged fragments, which are common to all atoms, are called **electrons.** The atomic or molecular charged particles created by removing electrons are called positive *ions*. The gas discharge experiment showed that atoms can be decomposed to negatively charged electrons and positively charged ions.

The discovery of the electron prompted a series of more sophisticated experiments. J.J. Thomson experimented with a device called a cathode ray tube, which is illustrated in Figure 2-14. A cathode ray is a beam of electrons. Because an electron beam is a collection of moving electrical charges, the beam is affected by electrical and magnetic forces. When either type of force is applied at a right angle to the direction of the electrons' motion, the force causes the beam to bend. The amount of bending depends on the speed, charge, and mass of the electrons. In Thomson's experiment a cathode ray was subjected simultaneously to electrical and magnetic forces. By measuring the amount of magnetic force required to exactly counterbalance the deflection of the beam by a known electrical force, Thomson was able to calculate the ratio of the electron's charge to its mass:

$$\frac{\text{Charge}}{\text{Mass}} = \frac{e}{m} = -1.76 \times 10^{11} \text{ C/kg}$$

The fundamental unit of electrical charge is the coulomb (C), defined as the quantity of electricity transferred by a current of 1 ampere in 1 second.

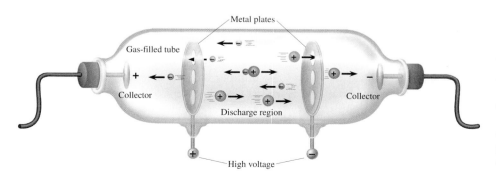

Metal plates

Gas-filled tube

Collector

Discharge region

Collector

High voltage

Figure 2-13
Schematic drawing of a gas discharge tube in operation. When a very high voltage is applied to the two perforated plates, an electrical discharge occurs between them. The positively charged and negatively charged particles that form in the gas then move in opposite directions.

Figure 2-14
Schematic drawing of a cathode ray tube. An electrical discharge generates electrons. These negative electrons form a beam. A pair of charged plates deflects the beam *(bent line)*, but if a magnetic force is also imposed, it is possible to counterbalance the effect of the electric force *(straight line)*.

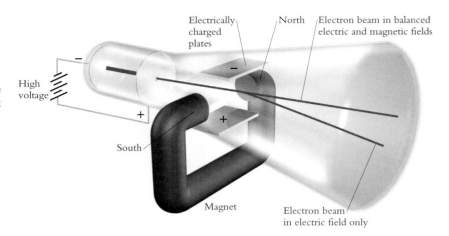

This experiment showed that one kilogram of electrons has a total charge of -1.76×10^{11} C, but Thomson was unable to find out how much charge resides on a single electron.

An American physicist, Robert A. Millikan, designed an experiment to determine the charge on the electron. As shown in Figure 2-15, Millikan set up electric plates to oppose the Earth's gravitational force. He sprayed a mist of oil droplets into the chamber between the two plates. Using a telescope, Millikan could watch the motion of the droplets as they drifted slowly downward under the force of gravity. Millikan then changed the behavior of the droplets in two ways. First he irradiated the droplets with X rays. This irradiation generated extra electrical charges on some of the droplets. By themselves, these excess electrical charges had no effect on the motion of the droplets. However, adding charge to the metal plates generated an upward force on any droplets that carried extra electrons. Millikan adjusted the amount of electrical charge on the plates until the upward-acting electrical force exactly counterbalanced the downward-acting gravitational force on a droplet. This stopped the downward drift of these negative particles.

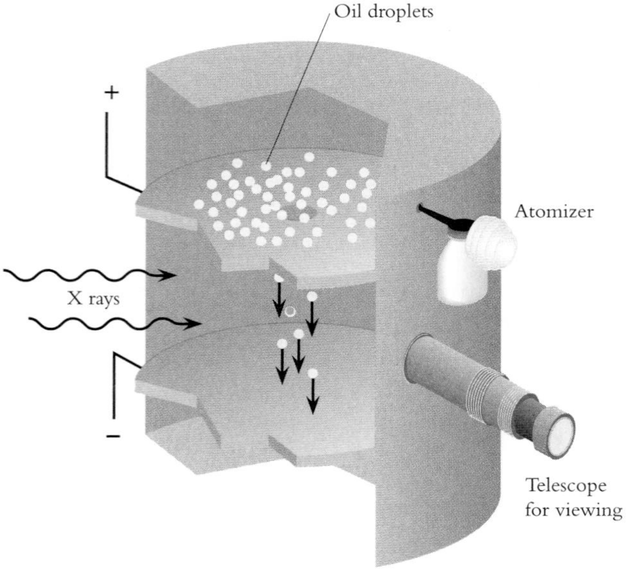

Gravitational force acted on the masses of the oil droplets. Millikan determined the mass of a droplet by observing its behavior when it moved under the force of gravity only. Electrical force acted on the negative charges on the oil droplets. Millikan measured the amounts of electrical force required to suspend the motion of many different droplets and calculated the charges on the droplets. He found several different values, but the charge was always equal to $n(-1.6 \times 10^{-19}$ C), where n was an integer (1, 2, 3 . . .).

Millikan concluded that n was the number of extra electrons carried by an oil droplet. Thus, the charge of an individual electron is -1.6×10^{-19} C. Combining this value with Thomson's measurement of charge/mass ratio, Millikan computed the mass of a single electron:

$$m_{electron} = \frac{e}{(e/m)} = \frac{-1.6 \times 10^{-19} C}{-1.76 \times 10^{11} C/kg} = 9.1 \times 10^{-31} \text{ kg}$$

Figure 2-15
Schematic view of Millikan's oil drop experiment. An atomizer generated a fine mist of oil droplets *(yellow circles)*. X rays gave some droplets extra negative charge *(orange circle)*. In the presence of sufficient electrical force, these negatively charged droplets could be suspended in space.

The Nucleus

By the early twentieth century, scientists had discovered that atoms contain electrons and positively charged particles. The nature of electrons had been elucidated by the experiments of Thomson and Millikan, but the nature of the positive particles was entirely unknown. Also, it was not known how the particles fit together to make an atom.

The experiment that showed how charges and masses are distributed in an atom was carried out in 1911 by Ernest Rutherford. J.J. Thomson had hypothesized that the atom was similar to a chocolate chip cookie, with negative electrons (the "chips") embedded in the positive "dough" of the atom in a way that balanced repulsion between like charges and attraction between unlike charges. Rutherford tested this model of the atom by using subatomic projectiles to bombard a target of atoms. The projectiles, called *alpha particles,* had been discovered during research on radioactivity. Alpha particles are high-energy, positively charged fragments of helium atoms emitted during radioactive decay of unstable elements such as uranium.

Rutherford's experiment and its results are represented in Figure 2-16. Alpha particles were directed at a thin film of gold metal. According to the "cookie model," the mass of each gold atom in the foil should have been spread evenly over the entire atom. Rutherford knew that alpha particles had enough energy to pass directly through such a uniform distribution of mass. He expected the particles to slow down and change direction only by a small amount as they passed through the foil. The results were quite unexpected. Most alpha particles passed directly through the gold film, and some were deflected slightly. To Rutherford's

Thomson, who was from England, used plum pudding with raisins as his analogy.

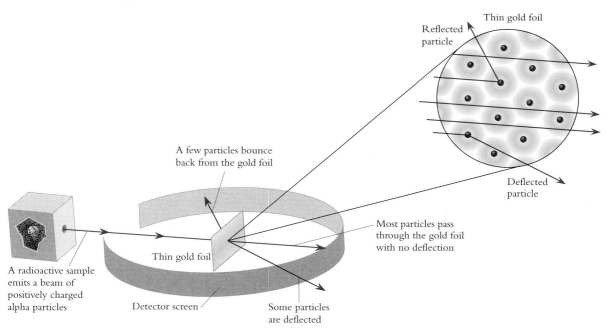

Figure 2-16
Schematic view of Rutherford's scattering experiment. When a beam of positively charged alpha particles was "shot" at a thin gold foil, most of them passed through without much effect. Some, however, were reflected backward.

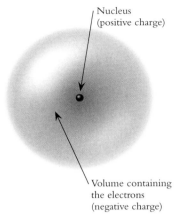

Nucleus
(positive charge)

Volume containing
the electrons
(negative charge)

Figure 2-17
Schematic drawing of an atom,
showing a central, positive
nucleus surrounded by a cloud of
electrons. This model of the atom
is consistent with the results of
Rutherford's scattering
experiments.

In the absence of other forces, a
nucleus that contains more than
one proton should fly apart
because the protons strongly repel
one another. The third type of
fundamental force, called the
strong nuclear force, acts within
nuclei and generates enough
attraction among nuclear particles
to hold nuclei together.

astonishment, however, a few particles bounced directly back. Rutherford said it was similar to shooting an artillery shell at a piece of tissue paper and having it bounce back at him. Somewhere within the atom there had to be a positively charged mass capable of blocking the path of high-energy, positively charged alpha particles. The cookie model of the atom crumbled.

To explain his obervations, Rutherford proposed a new hypothesis for atomic structure. He suggested that every atom has a tiny central core within which all the positive charge and most of the mass is concentrated. This central core, called the **nucleus,** is surrounded by the electrons, as shown schematically in Figure 2-17. Electrons occupy a volume that is huge compared with the size of the nucleus. However, each electron has such a small mass that alpha particles are not deflected by the electrons. Consequently, an alpha particle is deflected only when it passes very near a nucleus, and it bounces back only when it collides head-on with a nucleus. Because most of an atom's volume contains only electrons, most projectiles pass through the foil without being affected.

From the number of particles deflected and the pattern of deflection, Rutherford calculated the fraction of the atomic volume occupied by the positive nucleus. That fraction is 1 part in 10^{14}. To give you an idea of what that means, an atom the size of a baseball stadium would have a nucleus the size of a pea. The density of the nucleus is so great that a nucleus the size of a pea would have a mass of more than 250 million tons, as much as 33 million elephants!

Experiments on nuclei showed that the nucleus itself contains two types of subatomic particles called **protons** and **neutrons.** Protons account for the positive charges of nuclei, whereas neutrons contribute mass but are electrically neutral. A proton's positive charge is equal in magnitude to the negative charge of an electron. The mass of a proton, on the other hand, is almost 2000 times greater than the mass of an electron. The mass of a neutron is almost the same as the mass of a proton. The experiments leading to the discovery of protons and neutrons were as important to our understanding of matter as were the experiments leading to the discovery of the electron. Our interest is in chemistry, however, and we show in later chapters that the chemistry of atoms and molecules depends mainly on their electrons.

Our picture of atomic architecture is now complete. Three kinds of particles—electrons, protons, and neutrons—combine in various numbers to make the different atoms of all the elements of the periodic table. Table 2-1 summarizes the characteristics of these three atomic building blocks.

Table 2-1
Atomic Building Blocks

Name	Symbol	Charge	Mass
Electron	e	-1.6022×10^{-19} C	9.1094×10^{-31} kg
Proton	p	$+1.6022 \times 10^{-19}$ C	1.6726×10^{-27} kg
Neutron	n	0	1.6749×10^{-27} kg

■ **2.2.1** Draw a sketch (including appropriate signs for the electric plates) for a tube in which positive ions are accelerated into a deflection region. Use Figure 2-14 as a guide.

■ **2.2.2** In an experiment such as Millikan's (Refer to Figure 2-15), an oil droplet with a mass of 1.2 μg and carrying three extra electrons is held motionless by an electrical force. (a) In which direction is a droplet of mass 1.2 μg and carrying four electrons moving? (b) In which direction is a droplet of mass 2.0 μg and carrying three extra electrons moving? Explain your answers.

■ **2.2.3** (a) How many electrons are required to give a total mass of 1.0 μg (1 kg = 10^3 g; 1 μg = 10^{-6} g)? (b) What is the mass of the same number of neutrons?

2.3 ATOMIC DIVERSITY: THE ELEMENTS

According to the atomic theory, each element has unique properties. The differences among elements are caused by differences in their atoms. Each element is unique because its atoms contain characteristic numbers of protons, neutrons, and electrons.

Early experiments showed that atoms can be stripped of electrons by strong electrical forces. Atoms can also gain electrons under the influence of electrical force. In fact, much of the chemistry that takes place in the world around us involves electrons shifting from one chemical substance to another. Chemical reactions have no effect, however, on the structures of nuclei. All atoms of a particular element have the same number of protons in the nucleus, and these do not change during chemical processes. The defining feature of an element, therefore, is the charge carried by the protons in its nucleus.

/// An element is identified by the charge of its nucleus.

Every element has a unique nuclear charge and a specific and unchanging number of protons. The number of protons in the nucleus is called the **atomic number** and is symbolized Z. All atoms with the same value of Z belong to the same element. For example, all hydrogen atoms have Z of 1, all helium atoms have Z of 2, and all uranium atoms have Z of 92. The periodic table lists the elements in order of increasing atomic number. Each element has a unique name, symbol, and atomic number. The symbol H represents hydrogen ($Z = 1$), He represents helium ($Z = 2$), and U represents uranium ($Z = 92$).

As stated in Chapter 1, the periodic table lists elements in order of increasing masses, with a few exceptions. In fact, nuclear charge is the organizing feature of the periodic table. As nuclear charge increases, so also does nuclear mass (with few exceptions), so these two characteristics of elements are closely related.

Isotopes

The *identity* of an atom is determined by its nuclear charge, but its *mass* is the sum of the contributions from all its atomic building blocks. Recall from Table 2-1 that the mass of an electron is almost 2000 times smaller than the mass of a proton or a neutron. Consequently the mass of an atom is determined almost entirely by the mass of its nucleus. Thus the mass of an atom depends on the number of protons and neutrons in its nucleus.

Two atoms with the same number of protons but different numbers of neutrons are called **isotopes.** For example, every uranium atom has 92 protons in its nucleus, but whereas most uranium nuclei contain 146 neutrons, others contain only 143 neutrons. Naturally occurring uranium contains both isotopes, so any sample of pure uranium is a homogeneous mixture of uranium atoms with two slightly different masses.

An isotope is usually specified by its mass number. The **mass number** is the total number of protons and neutrons contained in a nucleus. Every isotope of a chemical element can be represented completely by writing its chemical symbol (X) preceded by a superscript giving its mass number (A) and a subscript giving its atomic number (Z):

> The subscript Z is redundant because each chemical symbol already defines a unique atomic number. The subscripts are nevertheless useful for isotopic bookkeeping.

$$\text{Mass number} \longrightarrow {}^A_Z X \longleftarrow \text{Elemental symbol}$$
$$\text{Atomic number} \longrightarrow$$

For example, the two isotopes of uranium are ${}^{235}_{92}U$ and ${}^{238}_{92}U$.

Another way to describe an isotope is to cite its elemental name and mass number. The isotopes of uranium are ${}^{238}U$, or uranium-238, and ${}^{235}U$, or uranium-235. Example 2-2 shows how to determine the composition of atoms.

Example 2-2	Determining the Composition of Atoms

TiO$_2$

Determine the number of protons, neutrons, and electrons in the following species: (a) ${}^{19}_9F$; (b) ${}^{54}Cr$; and (c) lead-207.

Strategy: The number of subatomic particles is determined from the atomic number (Z) and the mass number (A).

Solution:

a. F is the elemental symbol for fluorine. The subscript 9 is Z, which is the number of protons in the nucleus. The superscript 19 is A. The number of neutrons is found by subtracting Z from A: $A - Z = 19 - 9 = 10$ neutrons. Because this is a neutral atom, the number of electrons must equal the number of protons. Fluorine has 9 protons, 10 neutrons, and 9 electrons.

b. Cr is the symbol for chromium. A is 54, but Z is not given. The value of Z can be found in the periodic table. Chromium has Z of 24, which tells us the nucleus contains 24 protons. Subtracting Z from A, we find that there are 30 neutrons in this isotope. Finally, the atom is neutral, so 24 electrons are present.

c. Lead has Z of 82. A neutral atom of lead-207 has 82 protons, 82 electrons, and $207 - 82 = 125$ neutrons.

A few elements, such as fluorine and phosphorus, occur naturally with just one isotope, but most elements contain a mixture of isotopes. For example, element number 22 is titanium (Ti), a light and strong metal used in jet engines and in artificial human joints. There are five naturally occurring isotopes of Ti. Each one has 22 protons in its nucleus, but the number of neutrons varies from 24 to 28. In a chemical reaction, all isotopes of an element behave nearly identically. This means that the isotopic composition of an element remains essentially constant. The isotopic composition of Ti is:

$$\begin{array}{lll} {}^{46}Ti, \ 8.2\% & {}^{47}Ti, \ 7.4\% & {}^{48}Ti, \ 73.8\% \\ {}^{49}Ti, \ 5.4\% & {}^{50}Ti, \ 5.2\% \end{array}$$

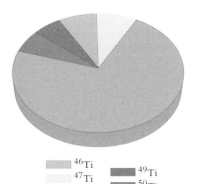

${}^{46}Ti$	${}^{49}Ti$
${}^{47}Ti$	${}^{50}Ti$
${}^{48}Ti$	

The titanium in TiO$_2$, the white solid used to make white paint opaque, has the same isotopic composition as the titanium metal in artificial knee joints.

Wherever titanium is found, for example in TiO$_2$, a compound used as a white pigment in paint, or pure Ti metal in an artificial knee joint, it has this isotopic composition.

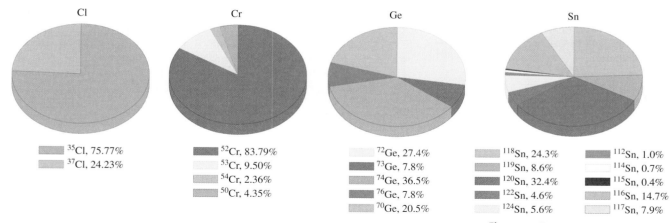

^{35}Cl, 75.77%		
^{37}Cl, 24.23%		

^{52}Cr, 83.79%		
^{53}Cr, 9.50%		
^{54}Cr, 2.36%		
^{50}Cr, 4.35%		

^{72}Ge, 27.4%		
^{73}Ge, 7.8%		
^{74}Ge, 36.5%		
^{76}Ge, 7.8%		
^{70}Ge, 20.5%		

^{118}Sn, 24.3%		^{112}Sn, 1.0%
^{119}Sn, 8.6%		^{114}Sn, 0.7%
^{120}Sn, 32.4%		^{115}Sn, 0.4%
^{122}Sn, 4.6%		^{116}Sn, 14.7%
^{124}Sn, 5.6%		^{117}Sn, 7.9%

Figure 2-18
The natural abundances of the isotopes of four elements (Cl, Cr, Ge, and Sn) illustrate the diversity of isotopic distributions. The mass number and percent abundance of each isotope are indicated.

Tin has the largest number of stable isotopes, ten. Figure 2-18 illustrates the range of isotopic compositions.

Hydrogen, the simplest of all chemical elements, has two naturally occurring isotopes. Its most common isotope, ^{1}H, contains a single electron and a single proton, and 99.98% of the hydrogen atoms on the Earth have this composition. The second isotope of hydrogen, ^{2}H, has a nucleus made up of both a proton and a neutron. The addition of this neutron doubles the mass of the atom. Doubling the mass is enough to make the chemical behavior of ^{2}H differ somewhat from that of ^{1}H. Thus the element hydrogen is an exception to our generalization about the common chemical properties of isotopes. Chemists give the "heavy" isotope its own name and symbol: **deuterium** (D). In naturally occurring substances the abundance of deuterium is so low that the subtle differences in its chemistry are rarely observed. It is possible to prepare molecules that are enriched in deuterium, however, and this allows chemists to take advantage of its special properties. The most common compound of deuterium is "heavy" water, D_2O, which is available from chemical supply houses at a price of about $50 for 100 g.

Mass Spectrometry

The existence of isotopes is demonstrated dramatically by research done with a **mass spectrometer,** one of which is represented in Figure 2-19. In this instrument, a sample of matter passes through an electrical discharge. The discharge knocks electrons off the atoms or molecules, changing them into positively charged

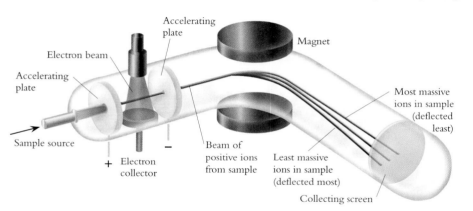

Figure 2-19
Schematic representation of one type of mass spectrometer. An electron beam fragments gas atoms or molecules into positively charged ions. The ions are accelerated and then deflected by a magnet. More massive particles are deflected by smaller amounts than less massive particles.

Figure 2-20
Mass spectrum of neon. (*a*), actual appearance, (*b*), bar graph representation.

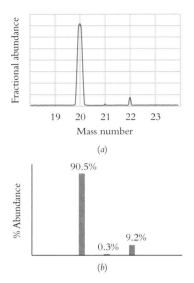

(a)

(b)

ions. Electrical force shapes these ions into a beam that passes between the poles of a magnet. Recall from Section 2.2 that charged particles moving through a magnetic field are deflected (see Figure 2-12). The curvature of the path depends on the mass of the ion: Ions with large masses are deflected less than ions with small masses. The mass spectrometer separates positively charged ions according to their masses and produces a graph of their abundance as a function of mass.

Figure 2-20 shows the mass spectrum of the element neon. The three peaks in the mass spectrum come from three different isotopes of neon, and the peak heights are proportional to the natural abundances of these isotopes. The most common form of neon has a mass number of 20, with 10 protons and 10 neutrons in its nucleus, whereas its two minor isotopes have 11 and 12 neutrons. Examples 2-3 and 2-4 illustrate how to read and interpret mass spectra, and the Tools for Discovery Box outlines various applications of the mass spectrometer.

Example 2-3	Identifying Isotopes

A sample of lead atoms is analyzed by mass spectrometry. The results are shown as a bar graph on the left. Use information from the graph to write the elemental symbol that represents each Pb isotope. List the number of protons and neutrons in each. Determine the isotopes' natural abundances.

Strategy: Each peak in a mass spectrum corresponds to an ion with a different mass. From the mass numbers of these peaks, we can determine the isotopic symbols and obtain a count of the protons and the neutrons. The height of each peak is proportional to the abundance of fragments of that particular mass.

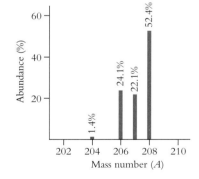

Solution: The four peaks in the mass spectrum represent four Pb isotopes. Their *A* values are 204, 206, 207, and 208. Consulting the periodic table, we find that *Z* of lead is 82. Thus the elemental symbols are as follows:

$$^{204}_{82}\text{Pb} \qquad ^{206}_{82}\text{Pb} \qquad ^{207}_{82}\text{Pb} \qquad ^{208}_{82}\text{Pb}$$

The atomic number (*Z*) gives the number of protons in the nucleus, so all four isotopes have 82 protons. On the other hand, *A* gives the sum of the number of protons and neutrons. Subtracting 82 from each *A* yields the number of neutrons in the nucleus of each isotope. Finally, the natural abundances of the isotopes are given by the heights of the peaks in the mass spectrum. These values are given on the spectrum. To summarize:

^{204}Pb	82 protons	122 neutrons	1.4% abundant
^{206}Pb	82 protons	124 neutrons	24.1% abundant
^{207}Pb	82 protons	125 neutrons	22.1% abundant
^{208}Pb	82 protons	126 neutrons	52.4% abundant

Molecular Mass Spectrum

Example 2-4

Elemental chlorine has two isotopes: ^{35}Cl, 75.77%, and ^{37}Cl, 24.23%. The mass spectrum of HCl (hydrogen chloride) gas is shown as a bar graph in the figure at right. Identify each of the peaks in the mass spectrum.

Strategy: Each peak in a mass spectrum corresponds to an ion with a different mass. From the mass numbers of these peaks, we can determine the isotopic compositions of different isotopic forms of HCl. The height of each peak is proportional to the abundance of fragments of that particular mass.

Solution: There are two large peaks in the mass spectrum of HCl. The larger of the two, seen at mass number 36, is due to the ion that contains a hydrogen atom paired with the most abundant isotope of chlorine, chlorine-35. The relative abundance of that ion (~76%) matches the natural abundance of ^{35}Cl. Likewise, the smaller peak at mass number 38 corresponds to an ion that pairs a hydrogen atom with chlorine-37 (~24% abundant).

In addition to the two large peaks, the spectrum shows two very small peaks at mass numbers 37 and 39. These low-abundance peaks are due to ions that contain deuterium, the minor isotope of hydrogen: ^{35}Cl-D and ^{37}Cl-D, respectively. Recall that deuterium, which has a mass number of 2, is only 0.02% abundant, so only a tiny fraction of all the hydrogen chloride molecules in the sample contain this isotope.

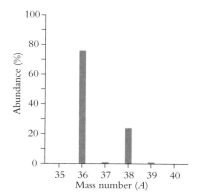

2.3.1 Use a periodic table (inside front cover) to fill in the missing information:

Name	Symbol	Z
Carbon	C	6
_____	Au	____
_____	____	33
Iron	____	____

2.3.2 Write isotopic symbols for (a) a cobalt atom with 30 neutrons; (b) the element with $Z = 3$ and $A = 7$; and (c) potassium with one more neutron than protons.

2.3.3 The isotopic composition of naturally occurring chromium (Cr) is given in Figure 2-18. Draw a bar graph that shows the mass spectrum of Cr.

2.4 CHARGED ATOMS: IONS

Experiments using mass spectrometers and electrical discharge tubes show that electrons can be stripped away from neutral atoms or molecules. An atom is neutral when it has equal numbers of protons and electrons. In a neutral atom the positive charge of the nucleus is balanced exactly by the negative charge of the electrons. Likewise, a molecule is neutral when it has the same number of electrons as the sum of the numbers of protons of all its atoms. Removing electrons from neutral

Box 2-2   Tools for Discovery: Applications of Mass Spectrometry

In addition to measuring the masses of isotopes, mass spectrometers can measure the masses of molecules with high accuracy. Scientists use this capability to solve a variety of problems. Mass spectrometers are essential instruments in laboratories that study topics as varied as drug identification and climate change.

Chemical analysis. The most widespread modern use of mass spectrometers is to identify chemical substances. When a molecule is placed in the mass spectrometer, the electrical discharge strips away one of its electrons. This so-called "parent ion" has virtually the same mass as the neutral molecule. If the mass of the parent ion is measured with high enough accuracy, the data can provide the molecular formula of the substance. The parent ion usually breaks apart into a collection of smaller pieces, many of which are also positively charged. The masses of these fragments provide a chemical "fingerprint" that indicates how the atoms in a molecule are connected together. A relatively simple example is the mass spectrum of methane, shown in the figure. There are peaks at mass numbers 16, 15, 14, 13, 12, and 1. Knowing that hydrogen atoms have a mass number of 1 and carbon atoms have a mass number of 12, these peaks can be identified as CH_4^+, CH_3^+, CH_2^+, CH^+, C^+, and H^+. From this fingerprint we can infer the formula and structure of methane.

The number of peaks in a mass spectrum grows rapidly with the complexity of the molecule, so each substance has a unique mass spectral pattern. These patterns help verify the presence of a particular compound in a mixture. In crime investigation laboratories, forensic chemists use the mass spectrometer to identify illegal drugs by comparing the mass spectrum of a sample with the known fragmentation pattern of a substance. Mass spectrometry is also used in analyzing urine samples for evidence of substance abuse. The figure at top right shows the fragmentation patterns for cocaine and heroin.

Isotopic ratios. Different isotopes differ in their atomic masses. The intensities of the signals from different isotopic ions allow isotopic abundances to be determined with high accuracy. Mass spectrometry reveals that the isotopic

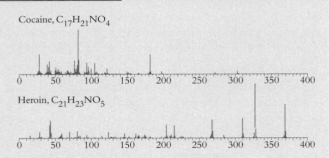

Cocaine, $C_{17}H_{21}NO_4$

Heroin, $C_{21}H_{23}NO_5$

abundances in samples from different sources have slightly different values. Isotopic ratios vary because different isotopes have slightly different properties; for example, they move at slightly different speeds. These differences have tiny effects because they are at the level of parts per ten thousand (0.0001). Nevertheless, the differences are large enough to cause slight variations in isotopic ratios.

For example, water molecules that contain ^{18}O evaporate from the oceans slightly more slowly than water molecules that contain ^{16}O. This rate difference is larger at low temperature than at high temperature. The difference is enough to monitor global warming and cooling. Marine carbonate sediments from different depths have different ages, and they also have slightly different values for $^{18}O/^{16}O$. The figure below shows global warming and cooling over the past 500,000 years.

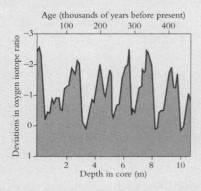

Sulfur also shows variations in the abundance ratio of ^{34}S and ^{32}S. These variations are caused by tiny differences in the rates of chemical reactions of the two isotopes. This measure can identify the source of sulfur contaminants in the atmosphere. For example, mining and smelting operations release harmful SO_2 into the atmosphere. Iron ores such as pyrites ($FeS2$) have different values of $^{34}S/^{32}S$ than ores of lead (PbS) and zinc (ZnS). Thus, the value of $^{34}S/^{32}S$ in atmospheric SO_2 can identify whether the source of pollutant was an iron or zinc smelting operation.

atoms and molecules leaves behind fragments that have positive electrical charges. Here are two examples:

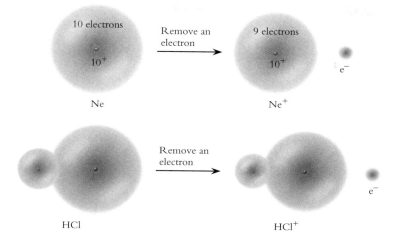

Electrically charged atomic or molecular particles are called **ions.** When the charge is positive, the particles are **cations.** Electrons are not classified as ions, but extra electrons can become attached to neutral atoms or molecules. The resulting negative particles are **anions.** Here is an example:

Ion is pronounced "eye′-un," cation "cat′-eye-un," and anion "an′-eye-un."

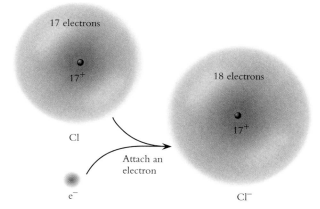

Ions can be generated in many ways, and in each of them electrons move from one location to another. Yet no matter how ions are formed, electrons are neither created nor destroyed. In other words, electrons are conserved. The loss of an electron by a neutral neon atom can be written in a way that makes it appear that an electron has been created: $Ne \rightarrow Ne^+ + e^-$. However, a neutral neon atom has 10 electrons, but a neon cation has only 9 electrons. Thus 10 electrons are present both before and after the formation of the cation.

The key to keeping track of electrons is net charge. Moving electrons around may generate positive charges in one location and negative charges in another location, but the net charge of the entire system remains unchanged. The ionization of neon atoms provides an illustration. The net charge of a neon cation plus an electron is zero, the same as the net charge of a neon atom.

Figure 2-21
Solid sodium metal reacts vigorously with chlorine gas to produce solid sodium chloride.

/// *Net electrical charge is always conserved.*

Electrical charges of the same type repel each other, so a collection of just cations or just anions is highly unstable. Even the cations in a mass spectrometer capture electrons and are neutralized as soon as they strike the collector. In contrast, opposite charges attract each other. This makes it possible for a collection containing both cations and anions to be stable, even though collections of either type alone are not. To be stable, a collection of cations and anions must be electrically neutral overall. That is, the amount of positive charge carried by the cations must be balanced exactly by the amount of negative charge carried by the anions.

Ionic Compounds

A solid that contains cations and anions in a balanced whole-number ratio is called an **ionic compound.** Sodium chloride, commonly known as *table salt,* is a simple example. Sodium chloride can form through the chemical reaction of elemental sodium and elemental chlorine. Figure 2-21 shows this vigorous reaction. The appearance and composition of these substances are very different (Figure 2-22). Sodium is a soft, silver-colored metal that is an array of Na atoms packed closely together. Chlorine is a pale-yellow toxic gas made up of diatomic, neutral Cl_2 molecules. These two elements react to form colorless crystals of NaCl that contain Na^+ and Cl^- ions in a $1:1$ ratio. This process can be represented by a chemical equation:

$$2\,Na(s) + Cl_2(g) \longrightarrow 2\,NaCl(s)$$

The (s) and (g) in the equation indicate phases; (s) designates solid, (g) designates gas.

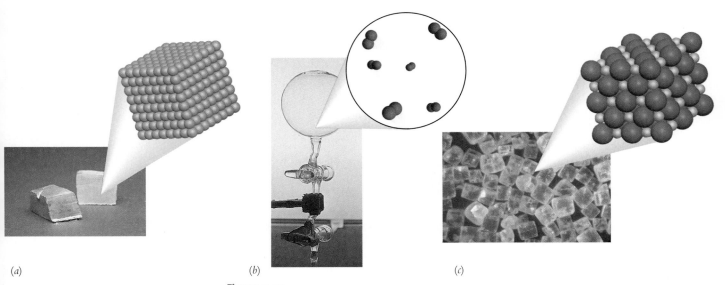

(a)   (b)   (c)

Figure 2-22
Photographs and molecular pictures of sodium metal (*a*), chlorine gas (*b*), and crystalline sodium chloride (*c*).

In the course of this reaction, ions are created as electrons are transferred from sodium atoms to chlorine atoms. Each sodium atom loses one electron and ends up with one less electron than the number of protons in the nucleus. This gives the sodium cation a charge of +1 unit, designated by a superscript plus ($^+$):

$$Na \longrightarrow Na^+ + e^-$$

The electron lost by a sodium atom becomes attached to a chlorine atom to produce an anion. We designate the negative charge by showing chlorine's elemental symbol with a superscript minus ($^-$):

$$Cl + e^- \longrightarrow Cl^-$$

A sodium chloride crystal contains equal numbers of Na^+ cations and Cl^- anions packed together in an alternating cubic array. A portion of the sodium chloride array is illustrated in Figure 2-22. The cations and anions are held in position by electrical forces. Each Na^+ cation is attracted to all the nearby Cl^- anions. Likewise, each Cl^- anion is attracted to all its Na^+ neighbors. Positive cations and negative anions group together in equal numbers to make the entire collection neutral.

Ionic compounds have chemical properties very different from those of the neutral atoms from which they form. Sodium metal reacts very violently with water, and chlorine gas is poisonous and highly corrosive. In contrast, sodium chloride simply dissolves in water and is a substance that most people use to season their food.

Several elements have a strong tendency to form ionic compounds, also called **salts.** The formulas of salts are dictated by the fact that the numbers of cations and anions must lead to overall charge neutrality. Sodium and the other Group 1 elements (alkali metals) form salts containing atomic cations with +1 charges. Chlorine and the other Group 17 elements (halogens) form salts containing atomic anions with −1 charges. The metals in Group 2 (alkaline earth metals) form salts containing atomic cations that have lost two electrons. Magnesium chloride, for example, contains Mg^{2+} cations and Cl^- anions. To maintain electrical neutrality, this compound has a $1:2$ ratio of cations to anions, giving it the chemical formula $MgCl_2$. Oxygen and sulfur form ionic compounds with metals. These compounds contain metal cations and atomic anions that have gained two electrons. Examples are calcium oxide (CaO), which contains Ca^{2+} cations and O^{2-} anions in a $1:1$ ratio, and potassium sulfide (K_2S), which contains K^+ cations and S^{2-} anions in a $2:1$ ratio.

Ionic Solutions

Solid sodium chloride dissolves in water to give a clear liquid solution that looks the same as it did before the salt was added. Despite their appearances, pure water and a solution of sodium chloride behave very differently. One striking difference is the ability of these liquids to conduct electricity. Figure 2-23 shows that although pure water does not conduct electricity, a solution of NaCl is a good conductor.

When NaCl dissolves in water, its ions enter the solution and move freely among the water molecules. Mixing leads to a uniform distribution of Na^+ and

Electrons are symbolized as an e with a superscript minus sign: e^-.

Figure 2-23
Pure water (*left*) and a solution of sugar (*right*) do not conduct electricity because they contain virtually no ions. A solution of salt (*center*) conducts electricity well because it contains mobile cations and anions.

Figure 2-24
Molecular picture of a solution of sodium chloride in water. All the molecules and ions move freely about, but overall electrical neutrality is maintained because the total amount of anionic charge equals the total amount of cationic charge.

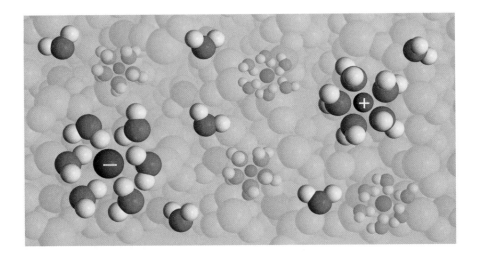

"Molecular" pictures may contain ions as well as atoms and molecules. A molecular picture is a schematic view of how matter appears at the molecular-atomic-ionic level.

Ionic compounds and ionic solutions may contain molecular as well as atomic ions. We introduce the most common molecular ions in Chapter 3 after describing the basic features of molecular composition.

Cl^- ions through the entire solution, with each ion surrounded by a sheath of water molecules as shown in Figure 2-24.

The presence of ions in solution is what gives a sodium chloride solution the ability to conduct electricity. If positively and negatively charged wires are dipped into the solution, the ions in the solution respond to the charges on the wires. Chloride anions move toward the positive wire, and sodium cations move toward the negative wire. This directed movement of ions in solution is a flow of electric current. Pure water, which has virtually no dissolved ions, does not conduct electricity. Any solution formed by dissolving an ionic solid in water conducts electricity. Ordinary tap water, for example, contains ionic impurities that make it an electrical conductor.

Section Exercises

2.4.1 Draw a molecular picture that shows a portion of a solution of calcium chloride ($CaCl_2$) in water that is electrically neutral overall. (Omit the water molecules.)

2.4.2 When magnesium reacts with oxygen, each magnesium atom loses two electrons, and each oxygen atom gains two electrons.
(a) What are the symbols that describe the resulting ions?
(b) What is the chemical formula of the ionic compound formed from magnesium and oxygen?
(c) Draw a molecular picture that shows the reaction between Mg metal and O_2 gas. Include 6 Mg atoms and enough O_2 to react completely.

2.4.3 What is the total number of protons in a Mg^{2+} ion? What is the total number of electrons?

Figure 2-25
Moving atoms, like speeding jet planes, possess kinetic energy.

2.5 ENERGY OF ATOMS AND MOLECULES

As noted in Section 2.1, atoms are continually moving. Atoms in motion possess kinetic energy, just as a moving jet plane has kinetic energy (Figure 2-25). The kinetic energy of an airplane changes with conditions; similarly, the kinetic energy of any particular type of atom or molecule changes with its conditions. For example, when cold water is heated to boiling, its molecules move faster and have substantially more kinetic energy. Atoms and molecules also possess stored energy. The combustion of gasoline and oxygen in an automobile engine is a chemical reaction that releases energy stored in the molecules. An understanding of chemistry requires knowledge about forms of energy and how energy exchanges take place.

Forms of Energy

Energy takes on several different forms. Kinetic, potential, thermal, and radiant energy all play important roles in chemistry. Figure 2-26 shows examples of energies that are important in chemistry. Working with energy at the molecular level requires a molecular perspective and an understanding of energy flow. Here we introduce types of energy with examples that do not require molecular thinking.

Kinetic energy is energy of motion. Any moving body has kinetic energy, which varies with the mass of the body and the speed at which it is moving. The

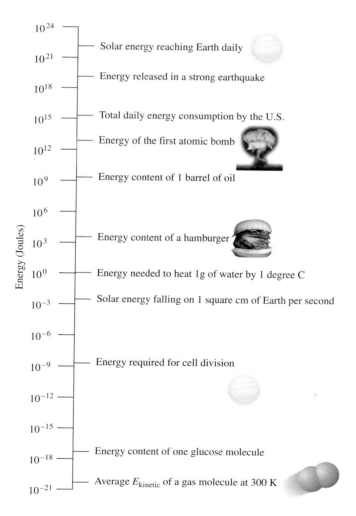

Figure 2-26
Energies span over 40 orders of magnitude, from 10^{-21} J (the kinetic energy of a typical gas molecule at room temperature) to 10^{22} J (the energy reaching the Earth from the sun daily).

exact relationship that links kinetic energy ($E_{kinetic}$) to mass (m) and speed (u) is given by Equation 2-1:

$$E_{kinetic} = \frac{1}{2} mu^2 \tag{2-1}$$

Because speed is measured in units of distance/time, the dimensions associated with kinetic energy are as follows:

$$(\text{Mass})(\text{Distance})^2(\text{Time})^{-2}$$

A negative power indicates that the quantity is divided:

$$(kg)(m)^2(s)^{-2} = \frac{(kg)(m)^2}{(s)^2}$$

so the SI unit for kinetic energy is $(kg)(m)^2(s)^{-2}$. No matter what form energy takes, it always can be described using these dimensions and units. Because energy is such a fundamental property, this unit is given a special name. It is called the *joule* (J), in honor of James Joule, a nineteenth century English physicist who did pioneering experiments on the nature of different forms of energy:

$$1 \text{ kg m}^2 \text{ s}^{-2} = 1 \text{ J}$$

Thermal energy is the total energy of motion of the atoms and molecules that make up an object. Although an object might be at rest, its atoms are continually moving. Consequently, each individual atom has kinetic energy given by Equation 2-1. When all these kinetic energies are added together, the result is the thermal energy of the object. When an object is heated, all its atoms and molecules move faster, leading to an increase in thermal energy. A higher temperature signals greater thermal energy; a lower temperature signals lesser thermal energy. For example, 100 mL of hot tea has more thermal energy than 100 mL of iced tea.

Potential energy is energy that is stored. An object at rest has no kinetic energy but usually possesses potential energy. For example, a rock teetering high on a ledge is about to release stored gravitational energy, and a cloud on the verge of "hurling" a thunderbolt earthward is about to release stored electrical energy. Although gravitational energy is most familiar to us, other types of potential energy are particularly important in the study of chemistry.

CHAPTERS 9, 12 →
We explore chemical energy in Chapters 9 and 12.

Chemical energy is one particular form of potential energy. Chemical energy is the result of the attractive forces between atoms that are bound in molecules. This energy can be released during chemical reactions. For example, gasoline has chemical energy that is released when the gasoline burns in an automobile engine.

Electrical energy is the result of electrical forces between charged objects. Recall that opposite electrical charges attract each other. This attraction leads to electrical potential energy. The electrical potential energy of two charges, q_1 and q_2, separated by distance r can be calculated using Equation 2-2:

$$E_{electrical} = k\frac{q_1 q_2}{r} \tag{2-2}$$

We use Equation 2-2 in later chapters to analyze the energetics of electrons, ions, and nuclei.

When distance is expressed in picometers (1 pm = 10^{-12} m) and charges are in electronic units, the constant in the equation is $k = 2.31 \times 10^{-16}$ J pm. To remove electrons from atoms or molecules, energy must be supplied to overcome this electrical energy.

Radiant energy is energy contained in electromagnetic radiation. The temperature of an object increases when it is placed in direct sunlight because sunlight possesses radiant energy. The radiant energy absorbed by an object is converted to thermal energy, thus the increase in temperature. Radiant energy is not restricted to light from the sun. Heat lamps, for example, give off infrared radiation, and microwave ovens cook food using the radiant energy of microwave radiation.

◀ **CHAPTER 6**
In Chapter 6 we examine radiation and radiant energy more closely.

Conservation of Energy

Energy can be transferred from one object to another, and energy can also be transformed from one type to another. One of the most fundamental laws of science is that the total amount of energy remains the same during transfers and transformations. That is, *energy is conserved,* as summarized in the law of conservation of energy:

> /// *Energy is neither created nor destroyed in any process, although it may be transferred from one body to another or changed from one form into another.*

This law has been confirmed by a multitude of experiments. Whenever the energy of one body increases, a compensating decrease must occur in the energy of some other body.

Conservation of energy can be shown using a baseball. A baseball that has been popped up has kinetic energy. As it rises in the air, the ball slows until it reaches its highest point. At that point, it is not moving, so it has no kinetic energy. What has happened to its initial kinetic energy? During its climb, the baseball slows down because gravitational force acts to convert its kinetic energy into potential energy. When the ball reaches its highest point, all the kinetic energy of its initial upward motion has been stored as gravitational potential energy. If the ball were to be trapped at this point (for example, if it landed on top of the screen behind home plate), the energy would remain stored until the ball fell back to the field. Then the stored energy would be released as kinetic energy.

When a catcher catches a fastball, the ball loses the kinetic energy it had while it was in flight because the ball is now at rest in the catcher's glove. This energy has not been transformed into gravitational potential energy because the ball is no higher above the Earth than before the pitch. It seems that energy has not been conserved, but careful temperature measurements would reveal that as the ball came to rest, the temperature of both the ball and the glove increased slightly. When a catcher catches a fastball, the kinetic energy lost by the ball shows up as thermal energy of the glove and the ball. This increase in thermal energy results in a higher temperature.

Extending the baseball analogy, a baseball pitcher is very tired after throwing fastballs for nine innings. That tired feeling is the body's signal that much stored chemical energy was consumed in throwing all those pitches. Each time your body moves, it does so by transforming chemical energy. The body releases this stored energy by breaking down complex molecules such as carbohydrates into carbon dioxide and water. To replenish our supply of stored chemical energy, we must eat.

The energy transfers that accompany chemical and physical transformations also conform to the conservation of energy. Energy is conserved when a green plant absorbs sunlight and manufactures carbohydrates and when polymeric plastics are manufactured in a chemical plant. The law of conservation of energy governs even such commonplace events as using ice cubes to cool soft drinks.

At the molecular level, conservation of energy helps interpret not only molecular behavior but also the very nature of atoms and molecules. In later chapters, we use energy diagrams to illustrate energy relationships among atoms and molecules. Such diagrams are constructed by combining measurements with the knowledge that energy must be conserved. The law of conservation of energy is applied to chemical processes and interactions throughout this book.

Section Exercises

2.5.1 Calculate the kinetic energy (in J) of an electron whose speed is 75 km/hr (consult Table 2-1 for data about the electron).

2.5.2 What transformations of energy take place during the following processes? (a) Water is heated in a microwave oven; (b) Snow breaks loose in an avalanche; (c) A speeding automobile brakes to a stop.

2.5.3 Humans replenish their supply of stored chemical energy by eating foods, which come from plants or other animals. Where do green plants obtain their supply of energy?

CHAPTER REVIEW

Summary and Key Terms

1. All matter is composed of atoms that possess mass, and atoms combine to form molecules. Atoms and mass are conserved in physical and chemical transformations. A statement that a quantity is conserved is a **conservation law.** Atoms and molecules are in constant motion, as indicated by gas pressure and **diffusion** of liquids. A chemical system that undergoes no net change is at **dynamic equilibrium.**

2. Forces, including **gravitational force, electrical force,** and **magnetic force,** hold things together or push them apart. Atoms contain **electrons** and **nuclei** held together by electrical forces. The relatively massive nuclei are composed of positive **protons** and neutral **neutrons.** Most of an atom's volume is occupied by its electrons. Neutral atoms have as many electrons around their nuclei as protons in their nuclei.

3. Every atom of a particular element has the same number of protons, its **atomic number.** Different **isotopes** of an element have different numbers of neutrons, giving them different **mass numbers.** Isotopic composition can be determined using a **mass spectrometer.** All isotopes of an element show virtually the same chemical behavior, except for **deuterium,** the heavy isotope of hydrogen, whose behavior differs measurably from that of the lighter isotope.

4. Neutral atoms can lose or gain electrons to become electrically charged **ions.** Positive **cations** and negative **anions** form **ionic compounds,** also called **salts.** Salts can dissolve in water to produce ionic solutions.

5. Energy, which plays an important role in chemistry, appears in many forms, among them **kinetic energy, potential energy, thermal energy, chemical energy, electrical energy,** and **radiant energy.** Energy can be transferred or transformed between forms but is always conserved.

Skills to Master

▶ Drawing molecular pictures

▶ Writing and interpreting atomic symbols

▶ Interpreting mass spectra

▶ Analyzing energy changes

Learning Exercises

2.1 List the conservation laws that appear in this chapter. Describe each one in your own words.

2.2 Describe what atoms "look like," what they are composed of, and how they behave.

2.3 Continue your list of "memory bank" equations, adding those from Chapter 2 that you need to memorize.

2.4 List all terms new to you that appear in this chapter. Use your own words to write a one-sentence definition of each. Consult the glossary if you need help.

2.5 List the new skills that were presented in this chapter. Go through the end-of-chapter problems and identify those that require skills from your list.

Problems

Atomic Theory

2.1 Draw molecular pictures that show part of a sample of each of the following: (a) helium, a monatomic gas; (b) tungsten, an atomic solid; and (c) gallium, an atomic liquid at body temperature.

2.2 Draw molecular pictures that show part of a sample of each of the following: (a) mercury, an atomic liquid; (b) iron, an atomic solid; and (c) neon, a monatomic gas.

2.3 The following is a molecular picture of carbon reacting with oxygen to form carbon monoxide, a deadly poisonous gas. Describe how this picture illustrates the features of atomic theory.

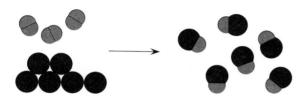

2.4 When lightning strikes, some N_2 reacts with O_2 to generate NO, as shown in the following molecular picture. Describe how this picture illustrates the features of atomic theory.

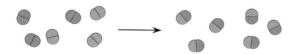

2.5 When a strip of magnesium metal burns in air, the mass of the resulting residue is greater than the mass of the original strip of metal. Explain this observation in terms of conservation of mass.

2.6 Liquid mercury metal can be obtained from a mercury ore called cinnabar simply by heating the ore in air. There is a significant mass loss when this process occurs. How can this mass loss be reconciled with conservation of mass?

2.7 Bromine, a diatomic molecule, is a liquid at room temperature but gaseous above 332 K and solid below 266 K. Draw molecular pictures that represent bromine in each of its three phases.

2.8 Carbon dioxide can exist as a liquid, solid (dry ice), or gas. Carbon dioxide is a linear molecule with a carbon atom in the middle and an oxygen atom on either side. Draw molecular pictures that represent carbon dioxide in each phase.

2.9 When we smell the odor of a rose, our olfactory nerves are sensing molecules of the scent. Explain how smelling a rose demonstrates that molecules are always moving.

2.10 Fog is a fine mist of water molecules suspended in air. Explain how a fog bank rolling in off the sea demonstrates that molecules are always moving.

2.11 The process in which a solid is converted directly into a gas is called sublimation. Iodine is an element that sublimes. A sample of solid iodine in a stoppered flask was allowed to stand undisturbed for several years. As the following photo shows, crystals of solid iodine grew on the sides of the flask. Use the principle of dynamic

equilibrium to explain at the molecular level what happened. Include an observation about the color of the atmosphere inside the flask.

2.12 The photo below shows a covered flask containing a highly concentrated salt solution with many salt crystals on the bottom. If the flask is allowed to stand for a long time, some crystals become smaller while others grow in size, even though the total mass of crystals is constant. Explain what is happening at the molecular level in terms of a dynamic equilibrium.

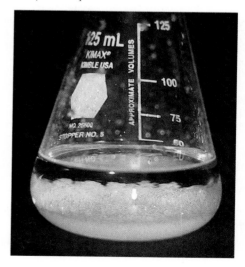

Atomic Architecture: Electrons and Nuclei

2.13 The collection screen of a cathode ray tube registers the total charge of a beam of electrons to be -1.00×10^{-6} coulombs. (a) How many electrons is this? (b) What mass does this number of electrons have?

2.14 In a gas discharge tube containing hydrogen gas, a total charge of 2.44×10^{-12} coulombs is measured at the collector. Assume that this charge is entirely due to protons. (a) How many protons is this? (b) What mass does this number of protons have?

2.15 In Millikan's oil drop experiment, some droplets have negative charges, so others must have positive charges. Suppose the upper electric plate in Figure 2-15 is made negative and the lower plate positive. Describe the results obtained under these conditions.

2.16 Describe the scattering pattern that would have been observed in Rutherford's experiment if atoms were like chocolate chip cookies.

2.17 How many protons does it take to give a mass of 1.5 g? What charge does this sample carry?

2.18 How many electrons does it take to give a mass of 1.5 g? What charge does this sample carry?

2.19 A helium atom is electrically neutral, with two protons and two neutrons in its nucleus. (a) How many electrons does it have? (b) What fraction of the mass of a He atom is due to its electrons?

2.20 A carbon atom is electrically neutral with six electrons and six neutrons. (a) How many protons does it have? (b) What fraction of a carbon atom's mass comes from its nucleus?

Atomic Diversity: The Elements

2.21 How many protons, neutrons, and electrons are contained in each of the following atoms or ions?

(a) $^{16}_{8}O^{2-}$; (b) $^{11}_{5}B$; (c) $^{55}_{25}Mn^{3+}$; (d) $^{35}_{17}Cl^{-}$; (e) $^{37}_{17}Cl^{+}$

2.22 How many protons, neutrons, and electrons are contained in each of the following atoms or ions?

(a) $^{66}_{30}Zn^{2+}$; (b) $^{15}_{7}N$; (c) $^{81}_{35}Br^{+}$; (d) $^{79}_{35}Br^{-}$; (e) $^{238}_{92}U$

2.23 Write the symbols of the following isotopes: (a) $Z = 26$ with 30 neutrons; (b) U-236; (c) argon with two more neutrons than protons; and (d) an atom with 9 protons, 10 neutrons, and 9 electrons.

2.24 Write the symbols of the following isotopes: (a) helium with 1 neutron; (b) $U - 236$; (c) zinc with $A = 66$; (c) element number 54 with 78 neutrons; and (d) nitrogen with the same number of protons and neutrons.

2.25 Elemental boron is 20.0% boron-10 and 80.0% boron-11. Sketch the mass spectrum of this element.

2.26 Elemental chlorine is a mixture of two isotopes: ^{35}Cl, 75.77% and ^{37}Cl, 24.23%. Sketch the mass spectrum of this element.

2.27 Platinum has four stable isotopes whose mass numbers and percentages are 194, 32.9%; 195, 33.8%; 196, 25.3%; and 198, 7.2%. Construct a pie chart illustrating these isotopic abundances.

2.28 Zinc consists of five isotopes whose mass numbers and percentages are 64, 48.6%; 66, 27.9%; 67, 4.1%; 68, 18.8%; and 70, 0.6%. Construct a pie chart illustrating these isotopic abundances.

Charged Atoms: Ions

2.29 Group the following species into cations, anions, and neutral species: C, Cl^-, CCl_4, Cl_2^+, CO_2, CO^+, $Cr_2O_7^{2-}$, and Cr^{3+}.

2.30 Group the following species into cations, anions, and neutral species: O_2, OH_3^+, O^{2-}, O^+, O_3, OH^-, O, O_2^-, and OH_2.

2.31 Write the chemical formula for the species resulting from each of the following processes: (a) A hydroxide ion (OH^-) gains H^+. (b) A sodium atom loses an electron. (c) An HCl molecule loses H^+. (d) An oxygen atom gains two electrons.

2.32 Write the chemical formula for the species resulting from each of the following processes: (a) An argon atom loses an electron. (b) A water molecule loses H^+. (c) An oxygen molecule gains an electron. (d) A water molecule gains H^+.

2.33 Draw molecular pictures illustrating the processes in Problem 2.31 (a) and (c).

2.34 Draw molecular pictures illustrating the processes in Problem 2.32 (b) and (d).

2.35 Based on their positions in the periodic table, decide what ion is likely to form from each of the following elements: (a) rubidium; (b) fluorine; and (c) barium.

2.36 Based on their positions in the periodic table, decide what ion is likely to form from each of the following elements: (a) cesium; (b) strontium; and (c) iodine.

2.37 Write the chemical formulas of all ionic compounds that can form between the elements listed in Problem 2.35.

2.38 Write the chemical formulas of all ionic compounds that can form between the elements listed in Problem 2.36.

2.39 Draw a molecular picture that illustrates a solution of NaI in water. Make sure your picture is electrically neutral.

2.40 Draw a molecular picture that illustrates a solution of $MgCl_2$ in water. Make sure your picture is electrically neutral.

2.41 Aluminum is one of the few elements that forms cations with +3 charge. What are the chemical formulas of aluminum oxide and aluminum fluoride?

2.42 Nitrogen forms a few compounds in which it exists as an anion with −3 charge, called "nitride." What are the chemical formulas of sodium nitride and magnesium nitride?

Energy of Atoms and Molecules

2.43 Explain how each of the following observations is consistent with conservation of energy: (a) An apple gains kinetic energy as it falls from a tree. (b) When that apple hits the ground, it loses its kinetic energy.

2.44 Explain how each of the following observations is consistent with conservation of energy: (a) Water running into a pond loses kinetic energy. (b) A rock lying in the sun increases in temperature.

2.45 Calculate the kinetic energy of an electron moving at a speed of 4.55×10^5 m/s.

2.46 Calculate the kinetic energy of a proton moving at a speed of 2.32×10^3 m/s.

2.47 What is the kinetic energy of a neutron moving with the same speed as the electron in Problem 2.45?

2.48 What is the kinetic energy of an electron moving with the same speed as the proton in Problem 2.46?

2.49 In each of the following processes, energy is transformed from one type to another. Identify what type of energy is consumed and what type of energy is produced. (a) Sunlight heats the roof of a house. (b) Packed snow breaks loose in an avalanche. (c) Wax burns in a candle flame.

2.50 In each of the following processes, energy is transformed from one type to another. Identify what type of energy is consumed and what type is produced. (a) Methane burns on the element of a stove. (b) An elevator carries passengers from the ground floor to the fourth floor. (c) A firefly produces light on a summer evening.

2.51 Compute the speed of a neutron whose kinetic energy is 3.75×10^{-23} J (see Table 2-1 for the mass of a neutron).

2.52 Compute the speed of a proton whose kinetic energy is 2.75×10^{-25} J (see Table 2-1 for the mass of a proton).

Additional Paired Problems

2.53 Which of the following processes obeys the requirement that charge be conserved? For those that do not, tell how many electrons must be added or removed: (a) One N_2 molecule decomposes into two N^{3-} ions. (b) An oxalate ion, $C_2O_4^{2-}$, decomposes to two CO_2 molecules. (c) A molecule of phosphoric acid, H_3PO_4, fragments into three H^+ ions and a PO_4^{3-} ion.

2.54 Which of the following processes obeys the requirement that charge be conserved? For those that do not, tell how many electrons must be added or removed: (a) An H_2O molecule decomposes into an H^+ ion and an OH^- ion. (b) A P_4 molecule decomposes into four P^{3+} ions. (c) An O_2 molecule fragments into two O^{2-} ions.

2.55 Air is mostly diatomic molecules of nitrogen and oxygen, in a molecular ratio of 4 to 1. Draw a molecular picture of a sample of air containing a total of 10 molecules.

2.56 One breathing mixture for deep-sea divers contains 25% molecular oxygen and 75% helium. Draw a molecular picture of a sample of this mixture that contains three molecules of oxygen.

2.57 The pie chart in Figure 2-18 shows the isotopic abundances for tin. Sketch the mass spectrum of this element.

2.58 The pie chart in Figure 2-18 shows the isotopic abundances for chromium. Sketch the mass spectrum of this element.

2.59 The following isotopes are useful for medical imaging. Determine the number of protons, neutrons, and electrons of each of them: (a) ^{99}Tc, used to study head function (shown in the figure); (b) ^{52}Fe, used in bone marrow scans; (c) ^{133}Xe, for studying lung function; (d) ^{131}I, for examination of the thyroid gland.

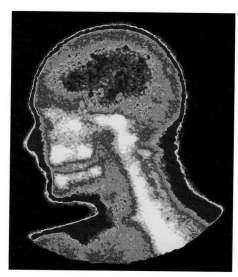

Radioactive image of a human head.

2.60 Determine the number of protons, neutrons, and electrons of each of the following isotopes: (a) cobalt-60, used in radiation treatments of cancer; (b) C-14, used in radiocarbon dating; (c) U-235, used in atomic bombs; and (d) uranium-238, used in "breeder" nuclear reactors.

2.61 What is the kinetic energy (in J) of a 2250-pound automobile traveling at 57.5 miles per hour?

2.62 What is the speed (in miles per hour) of a 60-kg runner whose kinetic energy is 345 J?

2.63 Draw a molecular picture that shows what happens when the salt KBr dissolves in water. Make sure your picture is electrically neutral.

2.64 Draw a molecular picture that shows what happens when the salt CaI_2 dissolves in water. Make sure your picture is electrically neutral.

2.65 Nuclei with the same A value but different Z values are called isobars. Each of the elements from chlorine to scandium has a nucleus with $A = 40$. Write correct atomic symbols for all these isobars.

2.66 Except for beryllium, each of the elements with Z values from 1 to 8 has a stable isotope with the same number of protons as neutrons. Write the correct atomic symbols for each of these isotopes.

2.67 Naturally occurring magnesium has three isotopes whose mass numbers and percent abundances are: 24, 78.99%; 25, 10.00%; and 26, 11.01%. Sketch the mass spectrum of Mg^+.

2.68 Naturally occurring nickel has five stable isotopes whose mass numbers and percent abundances are: 58, 68.27%; 60, 26.10%; 61, 1.13%; 62, 3.59%; and 64, 0.91%. Sketch the mass spectrum of Ni^+.

2.69 How many protons and electrons are present in the following ions? (a) Na^+; (b) N^{3-}; (c) Ti^{4+}; and (d) I^-.

2.70 How many protons and electrons are present in the following ions? (a) Al^{3+}; (b) Se^{2-}; (c) K^+; and (d) Ca^{2+}.

2.71 When calcium reacts with chlorine, each calcium atom loses two electrons, and each chlorine atom gains one electron. (a) What are the symbols of the resulting ions? (b) What is the chemical formula of the ionic compound formed from calcium and chlorine? (c) Draw a molecular picture that shows the reaction between five atoms of Ca metal and enough molecules of Cl_2 gas to react completely.

2.72 Approximately 20% of the iron and steel produced in the United States each year is used to replace rusted metal. Rust forms when iron reacts with oxygen. Each iron atom loses three electrons, and each oxygen atom gains two electrons. (a) What are the symbols of the resulting ions? (b) What is the chemical formula of the compound formed from iron and oxygen? (c) Draw a molecular picture that shows the reaction between four atoms of Fe metal and enough O_2 gas to react completely.

More Challenging Problems

2.73 Table salt, which has the formula NaCl, is made up of Na^+ and Cl^- ions arranged in a cubic crystal. When heated to a very high temperature, salt crystals melt to give a liquid which still contains these ions. Draw molecular pictures of table salt crystals and of liquid NaCl.

2.74 In the mass spectrum of chloroform ($CHCl_3$), one peak is due to CCl_2^+. (a) Write the reaction that converts chloroform to its molecular ion, $CHCl_3^+$. (b) Assuming that the molecular ion loses atoms to generate CCl_2^+, write the reaction for this process. (c) Draw a molecular picture illustrating the reaction of part (b).

2.75 The element bromine exists as diatomic molecules and is a liquid under normal conditions. Bromine evaporates easily, however, giving a red-brown color to the gas phase above liquid bromine, as shown in the photo. Draw molecular pictures showing liquid bromine and the gas above it.

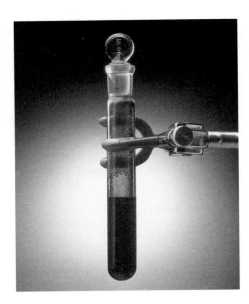

2.76 The ratio of neutrons to protons in stable nuclei varies from 1 : 1 to about 1.5 : 1. Write the isotopic symbols for all isotopes that contain 14 nuclear particles and have ratios in this range.

2.77 Give the atomic symbols for all the isotopes of elements between $Z = 20$ and $Z = 40$ that have exactly 1.25 times as many neutrons as protons.

2.78 A scientist claims to have prepared a sample of chlorine gas that is isotopically pure. Which scientific apparatus described in this chapter could be used to test this claim? Explain.

2.79 Is the following molecular picture of a reaction correct? If so, explain why. If not, redraw the right-hand portion so the picture is correct.

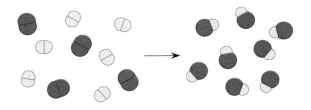

2.80 A student drew the following molecular pictures showing partial vaporization of solid I_2 to give gaseous I_2 molecules. Are these molecular pictures correct? If so, explain why. If not, redraw the right-hand portion of the figure so that it correctly represents this process.

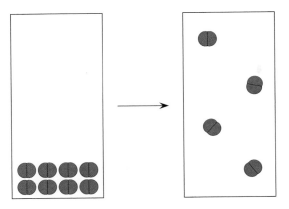

2.81 Elemental chlorine is a mixture of two isotopes: ^{35}Cl, 75.77% and ^{37}Cl, 24.23%. The mass spectrum of chlorine gas contains several peaks due to Cl_2^+ ions. How many different peaks are there? Which one is most intense? Which is least intense?

Group Study Problems

2.82 People with little knowledge of science often are cautious about "synthetic" versions of molecules produced in nature and used by humans. Vitamins, for example, can be either isolated from plants or they can be manufactured in the laboratory. Is pure vitamin C isolated from oranges "better for you" than pure vitamin C made in the laboratory? Use the atomic theory to support your answer.

2.83 Air is mostly diatomic molecules of nitrogen and oxygen. What positive and negative ions form when a lightning bolt, which is an immense electric discharge, passes through the atmosphere?

2.84 In discovering the electron, Thomson was able to measure the ratio of the electron's mass to its charge, but he was unable to determine the mass of a single electron. How did Millikan's experiment allow the mass of the electron to be determined?

2.85 The process in which a solid is converted directly to a gas is called sublimation. Solid carbon dioxide (CO_2), which is known as dry ice, is one common substance that sublimes. Draw a series of molecular pictures that show how a piece of dry ice in a closed container at low temperature illustrates the principle of dynamic equilibrium.

Answers To Section Exercises

2.1.1

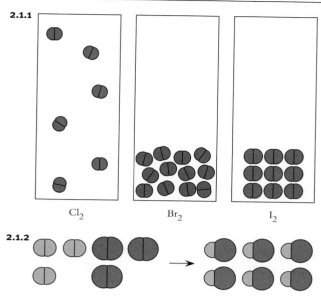

Cl$_2$ Br$_2$ I$_2$

2.1.2

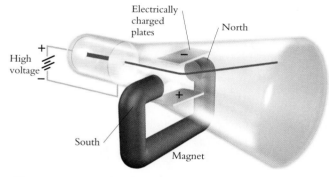

2.1.3 Water molecules in the warm rain puddle escape from the liquid phase to the gas phase, where they diffuse away or are swept away by the wind.

2.2.1

Electrically charged plates

North

High voltage

+

−

South

+

Magnet

(You are not expected to draw pictures as elaborate as this one, but your picture should include the essential features: an enclosed tube, two sets of charged plates, a magnet, and a beam that deflects away from the positively charged plate.)

2.2.2 (a) The larger charge increases the electric force, so this droplet moves upward. (b) The larger mass increases the gravitational force, so this droplet moves downward.

2.2.3 (a) 1.1×10^{24} electrons; (b) 1.8 g

2.3.1

Element	Symbol	Z
Carbon	C	6
Gold	Au	79
Arsenic	As	33
Iron	Fe	26

2.3.2 (a) ^{57}Co; (b) ^{7}Li; and (c) ^{39}K

2.3.3

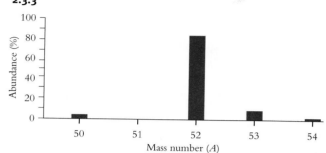

2.4.1

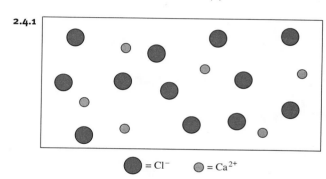

$\bullet$ = Cl$^-$ $\bullet$ = Ca^{2+}

2.4.2 (a) Mg^{2+} and O^{2-}; (b) MgO; and (c)

2.4.3 12 protons and 10 electrons
2.5.1 2.0×10^{-28} J
2.5.2 (a) Radiant energy is tranformed into thermal energy; (b) gravitational (potential) energy is transformed into kinetic energy; (c) kinetic energy is transformed into thermal energy.
2.5.3 Plants absorb radiant energy from sunlight and convert it into chemical energy through photosynthesis.

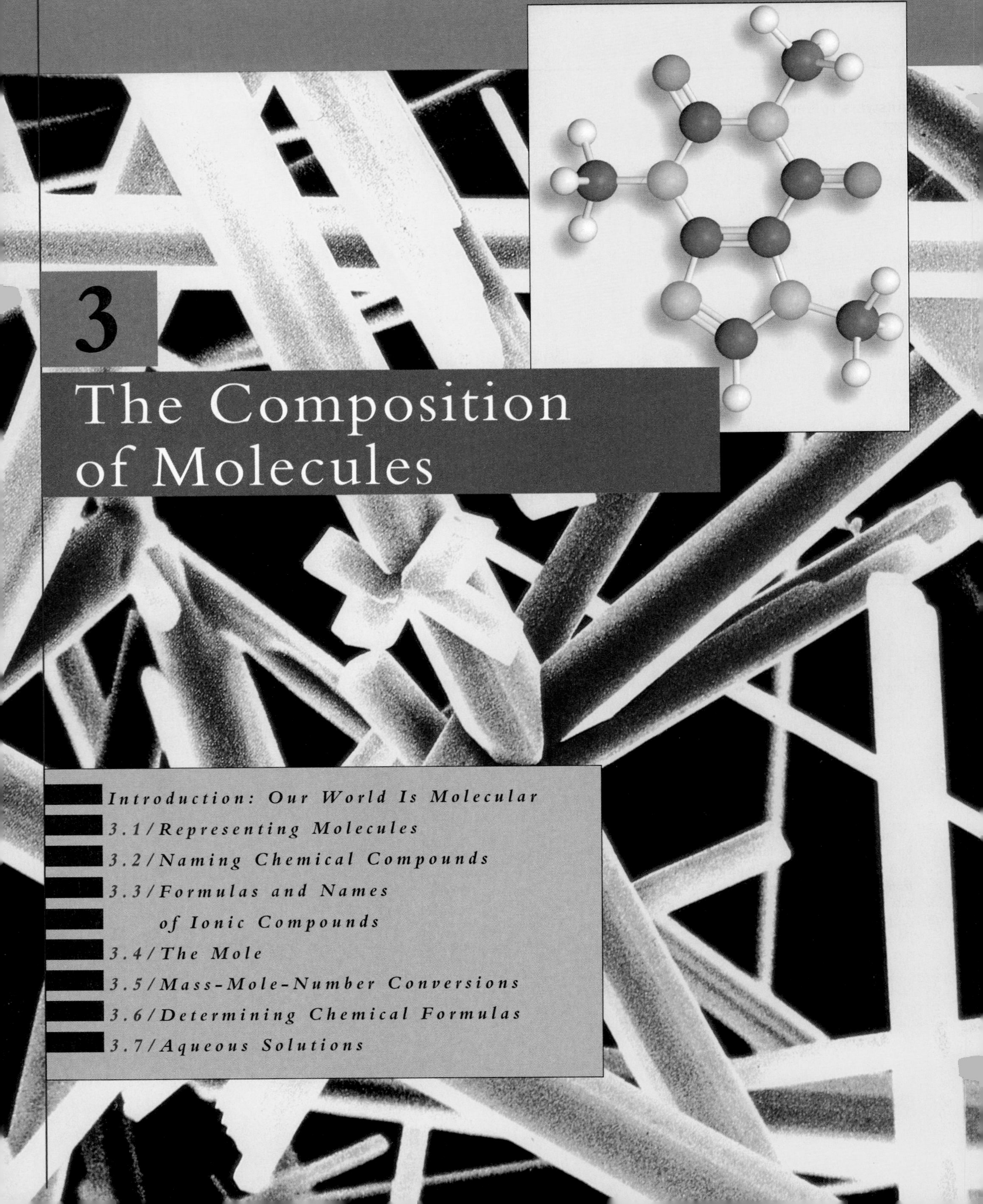

3

The Composition of Molecules

INTRODUCTION: OUR WORLD IS MOLECULAR

How would you describe the differences between a cup of coffee and a cup of hot water? What probably comes to mind are the rich aroma, the dark color, and the satisfying flavor of a good cup of coffee. Coffee's action as a stimulant is another obvious difference. These properties come from the chemical compounds that hot water dissolves from ground coffee beans. These compounds are molecules constructed from different atoms bound together in very specific arrangements. The molecule that makes coffee a stimulant is caffeine. Our background photo is a magnification of crystals of pure caffeine, and the inset is a ball-and-stick model of this molecule.

The properties of a particular molecule are due to the types and number of atoms it contains and upon how those atoms are arranged in space. Caffeine is a stimulant because it mimics the shape of one part of cyclic adenosine monophosphate (cyclic AMP), a molecule that regulates the supply of energy in the brain. When caffeine is absorbed into the blood and carried to the brain, it binds to an enzyme that normally controls the supply of cyclic AMP. As a result, the enzyme can no longer bind cyclic AMP, the brain's supply of this energy-regulating molecule is increased, and we feel stimulated.

Molecular sizes and shapes play key roles in determining chemical and physical properties. The immense variety of chemical and physical properties displayed by substances in the natural world mirrors an equally immense variety of different types of molecules. However, variety need not come from a large number of different elements. The molecules that make up a cup of coffee are made up almost entirely out of atoms of just five elements: hydrogen, carbon, oxygen, nitrogen, and sulfur. Carbon in particular is capable of combining in many different ways, generating molecules with elaborate structures.

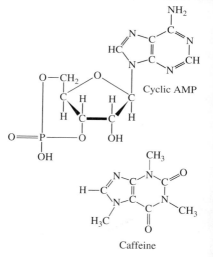

Cyclic AMP

Caffeine

Even simple molecules often have strikingly different properties. For example, carbon and oxygen form two different simple compounds. Whereas a molecule of carbon monoxide contains one oxygen atom and one carbon atom, carbon dioxide contains two atoms of oxygen and one atom of carbon. Although these molecules have some common properties (both are colorless, odorless gases), the difference in chemistry caused by a change of one atom is profound. We produce and exhale carbon dioxide as a natural byproduct of metabolism. This compound is relatively harmless to humans. In contrast, carbon monoxide is a deadly poison, even at very low concentrations. The dramatic difference in effect of these two substances arises because carbon monoxide is nearly identical in size and shape to molecular oxygen. In the blood, carbon monoxide binds very strongly to hemoglobin, preventing it from binding life-giving oxygen.

Chapter 3 examines molecules from two perspectives: "What is it?" and "How much is there?" As we describe in this chapter, chemists answer these questions in terms of *names* and *numbers* of molecules.

Molecular oxygen (O_2)

Carbon monoxide (CO)

3.1 REPRESENTING MOLECULES

The question, "What is it?" can be answered by a name or a picture (Figure 3-1). Chemists use both chemical names and molecular pictures to describe molecules. Molecular pictures take several forms, including structural formulas, ball-and-stick models, space-filling models, and line structures. These molecular representations can help you improve your ability to "think molecules."

Carbon dioxide (CO_2)

Figure 3-1
"It's a bird! It's a plane! It's Superman!" Answers to the question "What is it?" may take the form of words, pictures, or drawings.

Chemical Formulas

A **chemical formula** describes the composition of a substance by giving the relative numbers of atoms of each element. When a substance contains discrete molecules, a chemical formula is also a molecular formula. A chemical formula contains elemental symbols to represent atoms and subscripted numbers to indicate the number of atoms of each type. The simplest chemical formulas describe pure elements. The chemical formulas of most elements are their elemental symbols: helium is He, silicon is Si, copper is Cu. However, seven elements occur naturally as diatomic molecules (Figure 3-2), so their chemical formulas take the form X_2. A few other elements occur as atomic clusters, notably P_4 and S_8.

A chemical compound is a substance that contains a combination of atoms of different elements. Because a compound contains more than one element, there is more than one way to write its formula. For example, hydrogen chloride is a diatomic molecule with one atom each of hydrogen and chlorine. Its chemical formula might be written as HCl or ClH. To avoid possible confusion, chemists have standardized the writing of chemical formulas.

For **binary compounds**—those containing only two elements—the following guidelines apply:

Guidelines for Formulas of Binary Compounds

1. Except for hydrogen, the element farther to the left in the periodic table appears first: KCl, PCl_3, Al_2S_3, and Fe_3O_4.

2. If hydrogen is present, it appears last except when the other element is from Group 16 or 17 of the periodic table: LiH, NH_3, B_2H_6, and CH_4 but H_2O_2, H_2S, HCl, and HI.

3. If both elements are from the same group of the periodic table, the lower one appears first: SiC and BrF_3.

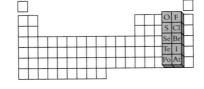

Figure 3-2
The seven chemical elements that normally exist as diatomic molecules are hydrogen, nitrogen, oxygen, fluorine, chlorine, bromine, and iodine. A mnemonic for these seven diatomic molecules is
I Bring Clay For Our New House.

Example 3-1 shows how to use these guidelines.

H_2 N_2 O_2 F_2
Cl_2 Br_2 I_2

| Writing Chemical Formulas | Example 3-1 |

Write the correct chemical formulas for the molecules containing sulfur whose molecular pictures follow:

(a) (b) (c) (d)

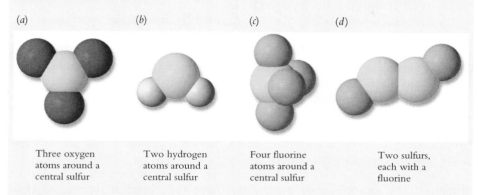

Three oxygen atoms around a central sulfur

Two hydrogen atoms around a central sulfur

Four fluorine atoms around a central sulfur

Two sulfurs, each with a fluorine

Strategy: To determine a chemical formula from a molecular picture, count atoms of each type and arrange the symbols in the correct order using the guidelines.

Solution:

a. Oxygen and sulfur are in the same group of the periodic table. According to the third guideline, the lower element appears first. Thus the correct formula is SO_3.

b. This compound contains hydrogen and sulfur, an element from Group 16 of the periodic table. According to the second guideline, hydrogen is placed before elements from this group. Thus the correct formula is H_2S.

c. The elements in this compound fall in two different groups of the periodic table. According to the first guideline, the element farther to the left appears first. Thus the correct formula is SF_4.

d. Once again, S is listed before F. The formula is S_2F_2. It is important to list all the atoms in the compound. The formula SF would be ambiguous because it could mean any combination of sulfur and fluorine in a 1:1 atomic ratio: SF, S_2F_2, S_3F_3, etc.

When three or more different elements occur in a compound, the order depends on whether or not the compound contains ions. We describe ionic compounds in Section 3.3. Many multiple-element compounds that do not contain ions contain carbon. The formulas of carbon-containing compounds start with carbon, followed by hydrogen. After that, any other elements appear in alphabetical order, as illustrated by the following examples: C_2H_6O, C_4H_9BrO, CH_3Cl, and $C_8H_{10}N_4O_2$.

Structural Formulas

The chemical formula of a substance gives only the number of atoms of each element present in one of its molecules. A **structural formula,** on the other hand, not only gives the number of atoms but also shows how the atoms are connected to one another. The atoms in molecules have specific arrangements because they are held together by attractive forces called **bonds.** In brief, a bond is the result of the electrical force of attraction between positively charged nuclei and negatively charged electrons. For now, it is enough to know that a pair of electrons shared between two atoms generates a chemical bond.

In a structural formula, bonds are represented by lines that connect the atoms. For example, the major component of bottled cooking gas is propane. Each molecule of propane contains three carbon atoms and eight hydrogen atoms, so its chemical formula is C_3H_8. The three carbon atoms of a propane molecule link to form a chain. Each outer carbon is bonded to three hydrogens, and the inner carbon is bonded to two hydrogens. This arrangement results in the structural formula of propane that is shown in Figure 3-3. Notice that the structural formula of propane contains the same information about the number of atoms as the chemical formula: three C atoms and eight H atoms. However, the structural formula also shows how the atoms are connected.

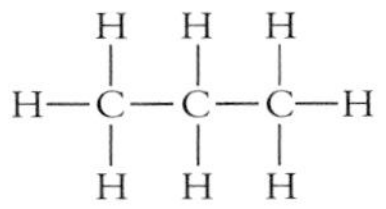

Quite often, molecules that have the same chemical formula have distinct structures. For example, two compounds have the chemical formula C_2H_6O (Figure 3-4). Dimethyl ether has a C—O—C linkage, while ethanol contains C—O—H. To distinguish between these two different substances, chemists write the formula of dimethyl ether as CH_3OCH_3 and the formula of ethanol as C_2H_5OH. The first formula indicates that the molecule has two CH_3 units linked to an oxygen atom. The second formula tells us that there is a C_2H_5 unit attached to an OH group. Molecules that have the same molecular formula but different arrangements of atoms are called *isomers.*

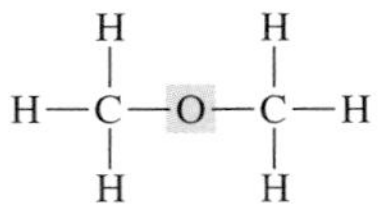

Each line in a structural formula represents one pair of shared electrons, but atoms can share more than one pair of electrons. When two atoms share one pair of electrons, the bond is called a "single bond," and the structural formula shows a single line. When two atoms share four electrons, the bond is called a "double bond," and the structural formula shows two lines between the atoms. Similarly, when two atoms share six electrons, the bond is called a "triple bond," and the structural formula shows three lines between the atoms. Two carbon atoms can bond to each other through any of these three kinds of bonds, as shown in Figure 3-5.

CHAPTERS 8 AND 9 →
We discuss chemical bonding in Chapters 8 and 9.

Figure 3-3
The structural formula of propane, C_3H_8.

Dimethyl ether
$(CH_3)_2O$

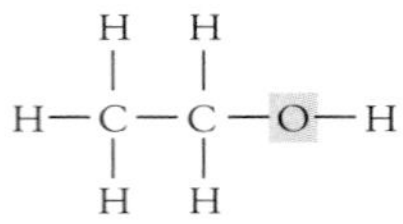

Ethanol
CH_3CH_2OH

Figure 3-4
Dimethyl ether and ethanol are isomers. Both have the formula C_2H_6O.

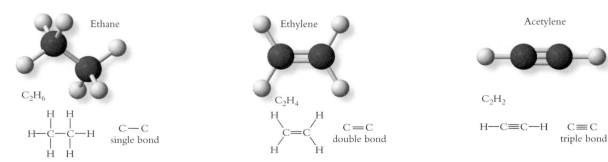

Figure 3-5
Ethane, a component of natural gas, contains a C—C single bond. Ethylene, widely used to make plastics, contains a C=C double bond. Acetylene, used as fuel for welding torches, contains a C≡C triple bond.

Three-Dimensional Models

A molecule is a three-dimensional array of atoms. In fact, many of a molecule's most important properties, such as its odor and chemical reactivity, depend on its three-dimensional shape. Although molecular and structural formulas describe the composition of a molecule, they do not represent the molecule's shape. To provide information about shapes, chemists frequently use ball-and-stick models or space-filling models.

In a **ball-and-stick model**, balls represent atoms, and sticks represent chemical bonds. The balls are labeled with elemental symbols or with different colors to distinguish among different elements. Figure 3-6 shows a ball-and-stick model for propane.

A **space-filling model** recognizes that a molecule is defined by the space occupied by its electrons. Recall from Rutherford's experiment that electrons make up nearly the entire volume of any atom. Therefore each atom in a space-filling model is shown as a distorted sphere representing its electrons. These spheres merge into one another to build up the entire molecule. Examples of space-filling models appear in Figure 3-2 and Figure 3-6. Figure 3-7 shows ball-and-stick and space-filling models for several chemical compounds common in everyday life.

Line Structures

The chemistry of living processes is complex, and many carbon-based molecules found in living organisms have extremely complicated structures. Because of this complexity, chemists have developed **line structures,** which are compact

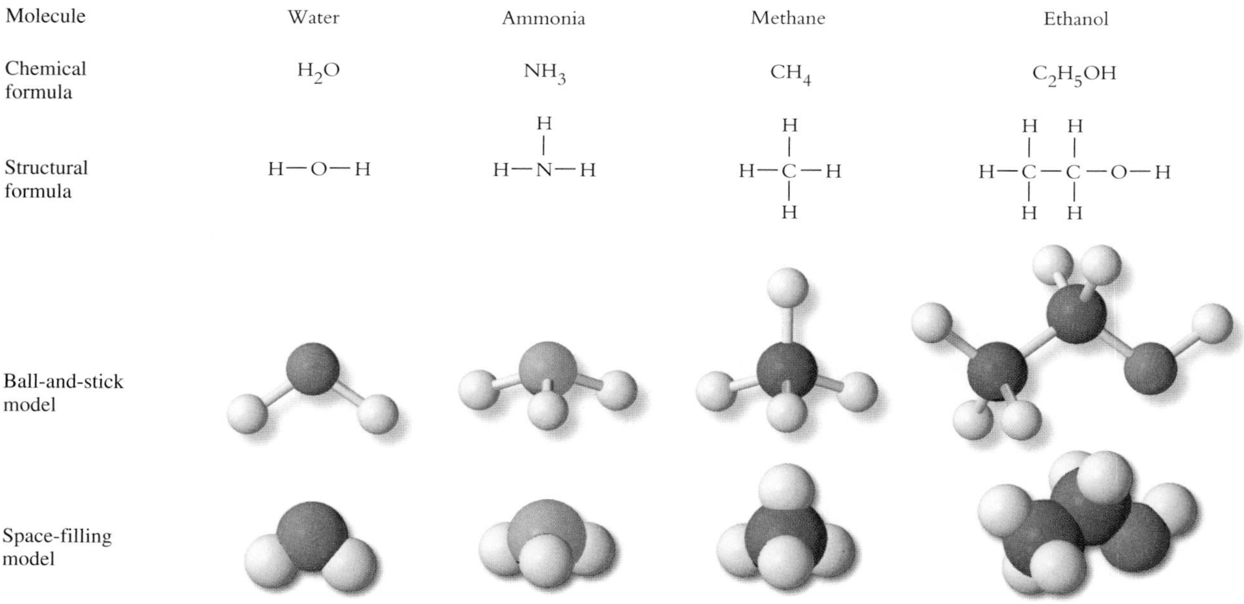

Ball-and-stick model

Space-filling model

**Figure 3-6
Three-dimensional models of propane, C$_3$H$_8$.**

Animation

◄ FIGURE 1-4
The color code used for atoms in this book appears in Figure 1-4.

Molecule	Water	Ammonia	Methane	Ethanol
Chemical formula	H$_2$O	NH$_3$	CH$_4$	C$_2$H$_5$OH
Structural formula	H—O—H	$\begin{array}{c} \text{H} \\ | \\ \text{H—N—H} \end{array}$	$\begin{array}{c} \text{H} \\ | \\ \text{H—C—H} \\ | \\ \text{H} \end{array}$	$\begin{array}{c} \text{H H} \\ |\ \ | \\ \text{H—C—C—O—H} \\ |\ \ | \\ \text{H H} \end{array}$
Ball-and-stick model				
Space-filling model				

**Figure 3-7
Different ways of representing some common chemical substances.**

representations of the structural formulas of carbon compounds. Line structures are constructed according to the following guidelines:

Guidelines for Line Structures

1. All bonds except C—H bonds are shown as lines.
2. C—H bonds are not shown in the line structure.
3. Single bonds are shown as single lines; double bonds are shown as two lines; triple bonds are shown as three lines.
4. Carbon atoms are not labeled.
5. All atoms except carbon and hydrogen are labeled with their elemental symbols.
6. Hydrogen atoms are labeled when they are attached to any atom other than carbon.

Example 3-2 illustrates these guidelines for three relatively simple molecules.

| Example 3-2 | Drawing Line Structures |

Construct line structures for the compounds with the following structural formulas.

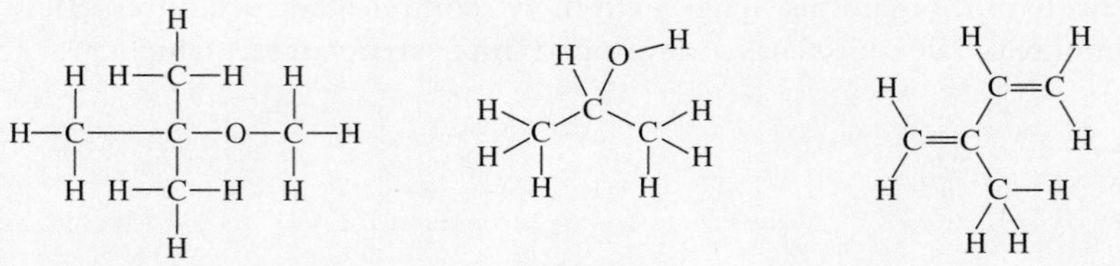

$C_5H_{12}O$	2-propanol	C_5H_8
Methyl tert-butyl ether	(rubbing alcohol)	Isoprene
(a key antiknock ingredient		(the building block of
in gasoline)		natural rubber)

Strategy: A line structure is built by using information about how atoms are connected in the molecule. The structural formula is the basis for the line structure. Apply the guidelines to convert the structural formula representation into a line-structure representation.

Solution: The first two guidelines state that bonds are represented by lines and that all C—H bonds are ignored. Thus we remove the C—H bonds from the structural formula, leaving a bond framework. According to the third guideline, the double bonds remain as two lines:

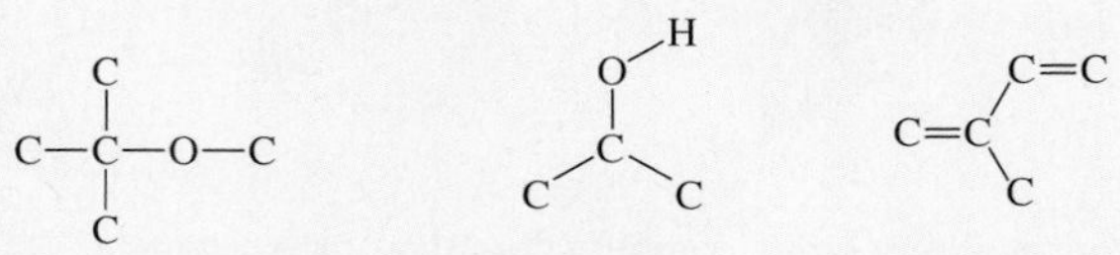

<table>
<tr><td>Drawing Line Structures (continued)</td><td>Example 3-2</td></tr>
</table>

According to the fourth guideline, carbon atoms are not labeled, so we remove the C's:

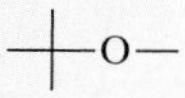

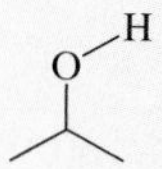

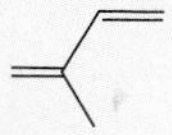

These are the line structures for the three molecules.

Although line structures are convenient, a chemist must be able to convert them back into chemical formulas. The reconstruction of a complete formula from a line structure relies on the most important general feature of carbon chemistry: In all neutral molecules containing carbon, each carbon atom has *four* chemical bonds. Look again at Figure 3-5 and count the bonds around each carbon atom. The total is always four. Each carbon atom in acetylene has three bonds to the other carbon atom and one C—H bond. Each carbon atom of ethylene has two bonds to its neighboring carbon atom and two bonds to hydrogen atoms. Finally, each carbon atom of ethane has one bond to the other carbon atom and three bonds to hydrogen atoms.

Keep in mind that carbon atoms are not shown in a line structure, so the first step in constructing a structural formula from a line structure is to place a C at every line intersection and at the end of every line. Then add singly bonded hydrogen atoms (—H) until every carbon atom has four bonds. Example 3-3 illustrates this procedure.

> Carbon monoxide (CO), which has three bonds to carbon, is the only common exception to the generalization that neutral carbon compounds have four bonds to carbon.

<table>
<tr><td>Converting Line Structures</td><td>Example 3-3</td></tr>
</table>

Construct the structural formulas and determine the chemical formulas from the following line drawings:

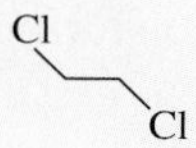

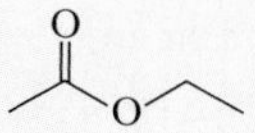

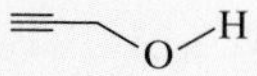

Strategy: Line drawings show all structural features except carbon atoms and C—H bonds. Convert a line drawing into a structural formula in two steps. First, place a C at any unlabeled line end and at each line intersection. Second, add C—H bonds until each carbon atom has a total of four bonds. The chemical formula is then obtained by counting the atoms of each element.

Solution: Place a C at each intersection and line end:

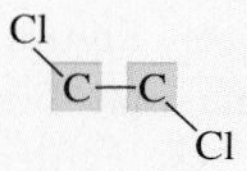

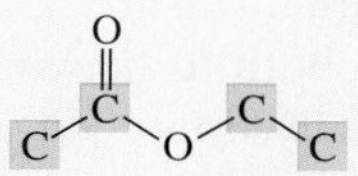

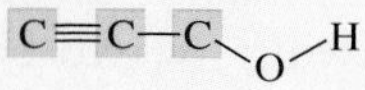

| Example 3-3 | Converting Line Structures (continued) |

Now add C—H bonds until each carbon atom has a total of four bonds.

Each carbon atom in the first structure has two bonds, so each carbon needs two C—H bonds to complete the structural formula:

The second structure has four carbon atoms. The two end carbons have just one bond, so each needs three C—H bonds. The carbon with the double bond to oxygen already has its complete set of four bonds. The fourth carbon atom has a bond to C and a bond to O, so it needs two C—H bonds:

The third structure contains a triple bond. The end carbon needs one C—H bond, but the other carbon atom of the triple bond already has four bonds. The next carbon has two bonds, one to its carbon neighbor and one to the oxygen. Two C—H bonds are needed to give this atom its usual set of four bonds:

These are the correct structural formulas. Now it is a simple matter to count the number of atoms of each element and write the chemical formulas of the compounds: $C_2H_4Cl_2$, $C_4H_8O_2$, and C_3H_4O.

The structural formulas of the compounds in these examples are not very complicated, yet their line structures are clearer than other representations. In ever-larger carbon-containing compounds, the simplification provided by line drawings is a great help to chemists. Line structures are not cluttered by the C—H bonds in the molecule, making them easier both to draw and to interpret. Figure 3-8 shows the line drawings of two biologically important molecules that may be familiar to you, and the Chemical Milestones Box describes unusual molecules made up entirely of carbon.

C an you imagine atoms connected together to form a molecule shaped like a minuscule soccer ball? How about connections that result in molecular tubes? Remarkably, the element carbon can form these molecular shapes. Perhaps even more remarkably, chemists did not discover this until late in the twentieth century.

Two elemental forms of carbon with different appearances and properties are well known. These are graphite and diamond. The most common form of carbon is graphite, a black, soft, brittle solid. When a candle burns in a limited supply of air, it generates black powdery soot that is mostly graphite. In graphite, carbon atoms are connected in the planar honeycomb arrangement shown below. Each carbon atom is connected to *three* neighbors in the same plane. The atoms of any plane are connected in a network, but there are no connections between adjacent planes. This allows layers to slide past each other, giving graphite useful lubricating properties. Graphite is brittle because, lacking connections between atoms in separate layers, it fractures easily along its layer planes.

The carbon atoms in a diamond are connected in a three-dimensional network, each atom connected to *four* others. Each atom is at the center of a regular tetrahedron, as shown below. We describe this geometry, which occurs in many compounds of carbon, in Chapter 8. The three-dimensional

connections result in a solid that is transparent, hard, and durable. The diamond structure forms naturally only at extremely high temperature and pressure, deep within the Earth. That's why diamonds are rare and precious.

A third form of carbon was observed in mass spectrometry experiments in 1985 and was isolated in 1990. Whereas transparent diamond and black graphite are extended lattices of interconnected carbon atoms, this red-violet form contains discrete molecules with the formula C_{60}. Sixty carbon atoms form the regular structure, shown below, that resembles a soccer ball. Five carbon atoms bond together in a pentagonal ring. Each side of the pentagonal ring is also a side of a hexagonal ring. These clusters of rings join to form a ball of 60 carbon atoms in which each carbon atom is at a corner of one pentagon and two hexagons. This type of carbon was named buckminsterfullerene, in honor of R. Buckminster Fuller. Fuller invented the geodesic dome, a similar structure of pentagons and hexagons that creates a strong and spacious building design.

Buckminsterfullerene is one of a group of molecules having the general formula C_n, with n having several possible integral values. This group is collectively called "fullerenes." Fullerenes can be made by heating graphite intensely in a helium atmosphere. Under these conditions, graphite vaporizes as molecular fragments, some of which recondense as fullerenes. Chemists propose a variety of applications for fullerenes, including molecular ball bearings, lubricants, and plastics. It has also been suggested that fullerenes could lead to new cancer drugs by encapsulating a radioactive atom inside the fullerene cage.

The newest addition to the forms of elemental carbon is the nanotube. A carbon nanotube is a long cylinder of carbon atoms, connected together in much the same way as in fullerene. Both the diameter and the length of carbon nanotubes can vary. Properties of nanotubes, such as their ability to conduct electrical charge, change dramatically with the dimensions of the tube. Carbon nanotubes are under intensive study. For example, a carbon nanotube laid down on a silicon chip forms a "molecular transistor." Such devices may eventually lead to further miniaturization of the "chips" that are at the heart of modern computers.

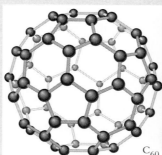

C_{60}

Figure 3-8
Structural formula and line drawing of caffeine, the stimulant found in coffee, cola, and tea, and line drawing of cholesterol, a substance that is one cause of heart disease. If you still are not convinced that line structures are an efficient way to represent molecules, try drawing the complete structural formula of cholesterol.

Caffeine
$C_8H_{10}N_4O_2$

Cholesterol
$C_{27}H_{46}O$

Section Exercises

3.1.1 Write the chemical formula for each of the following substances: (a) stearic acid, whose molecules contain 36 hydrogen atoms, 18 carbon atoms, and 2 oxygen atoms; (b) silicon tetrachloride, whose molecules contain one silicon atom and four chlorine atoms; and (c) freon–113, whose molecules contain three atoms each of fluorine and chlorine and two atoms of carbon.

3.1.2 Determine the molecular formula, structural formula, and line structure for each compound whose ball-and-stick model follows: (See Figure 1-4 for atom colors.)

3.1.3 The line structures that follow represent starting materials for making plastics. For each of them, draw the structural formula and give the chemical formula.

Isoprene
(used in nature to produce rubber)

Styrene
(used to make styrofoam)

Methyl methacrylate
(used to make plexiglass)

3.2 NAMING CHEMICAL COMPOUNDS

In the early days of chemistry, the list of known compounds was short, so chemists could memorize the names of all of them. New compounds were often named for their place of origin, physical appearance, or properties. As the science of chemistry grew, the number of known compounds quickly became too large for anyone to keep track of all of the common names. Today, more than 20 million compounds are known, and thousands of new ones are discovered or created each year. Consequently, chemists need systematic procedures for naming chemical compounds. The International Union of Pure and Applied Chemistry (IUPAC) has established uniform guidelines for naming various types of chemical substances, and chemists increasingly use IUPAC-approved names rather than their common counterparts. Table 3-1 gives some examples.

Systematic names are less colorful than common names, but they make chemistry less hectic because it is much easier to learn a few systematic guidelines than to memorize the names of thousands of individual compounds.

We focus on naming the substances that are most commonly encountered in general chemistry: binary compounds of metals and nonmetals and compounds containing ions. Because many useful and interesting chemical compounds contain carbon, we also describe a few principles for naming this rich array of substances. This section presents guidelines for compounds that do not contain metallic elements. The naming of compounds containing metals is described in Section 3.3.

A system for naming is called *nomenclature*, from two Latin words: *nomen* means "name" and *calare* means "call."

Naming Nonmetallic Binary Compounds

The guidelines presented in Section 3.1 for writing the chemical formulas of binary compounds describe the order in which the elements appear. This order also determines how the compound is named. Here are some guidelines for naming binary compounds:

Guidelines for Naming Binary Compounds

1. The element that appears first retains its elemental name.

2. The second element begins with a root derived from its elemental name and ends with the suffix-*ide*. Some common roots are listed in Table 3-2.

3. When there is more than one atom of a given element in the formula, the name of the element usually contains a prefix that specifies the number of atoms present. Common prefixes are given in Table 3-3.

Table 3-2
Common Roots for Naming Compounds

Element	Root
As	Arsen-
Br	Brom-
C	Carb-
Cl	Chlor-
Cr	Chrom-
F	Fluor-
H	Hydr-
I	Iod-
Mn	Mangan-
N	Nitr-
O	Ox-
P	Phosph-
S	Sulf-

Table 3-1
Common and IUPAC Names of Some Common Substances

Common name	Wood alcohol	Grain alcohol	Rubbing alcohol	Milk of magnesia
IUPAC name	Methanol	Ethanol	2-Propanol	Magnesium hydroxide
Common name	Caustic soda	Limestone	Baking soda	Laughing gas
IUPAC name	Sodium hydroxide	Calcium carbonate	Sodium hydrogen carbonate	Dinitrogen oxide

Table 3-3
Number Prefixes For Chemical Names

Number	Prefix	Example	Name
1	Mon(o)-*	CO	Carbon monoxide
2	Di-	SiO_2	Silicon dioxide
3	Tri-	NI_3	Nitrogen triiodide
4	Tetr(a)-	CCl_4	Carbon tetrachloride
5	Pent(a)-	PCl_5	Phosphorus pentachloride
6	Hex(a)-	SF_6	Sulfur hexafluoride
7	Hept(a)-	IF_7	Iodine heptafluoride

* If the numerical prefix ends with the letter "o" or "a" and the name of the element begins with a vowel, drop the last letter of the prefix.

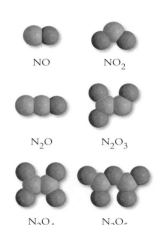

NO NO_2

N_2O N_2O_3

N_2O_4 N_2O_5

Numerical prefixes are essential in naming one of a series of similar binary compounds. For example, nitrogen and oxygen form six different molecules: NO, nitrogen oxide; NO_2, nitrogen dioxide; N_2O, dinitrogen oxide; N_2O_3, dinitrogen trioxide; N_2O_4, dinitrogen tetroxide; and N_2O_5, dinitrogen pentoxide. Example 3-4 shows some additional examples.

Example 3-4	**Naming Binary Compounds**

Name the following binary compounds: SO_2, CS_2, BCl_3, and BrF_5.

Strategy: None of these compounds contains a metallic element, so we apply the guidelines for binary compound nomenclature.

Solution: Name the first element, use a root plus –*ide* for the second element, and indicate the number of atoms with prefixes.
    Here are the correct names: sulfur dioxide, carbon disulfide, boron trichloride, and bromine pentafluoride.

Binary Compounds of Hydrogen

Hydrogen requires special consideration because it may appear first or second in the formula and name of a compound. With elements from Groups 1 and 17, hydrogen forms diatomic molecules named according to our guidelines. For example, LiH is lithium hydride, and HF is hydrogen fluoride. With elements from Groups 2 and 16, hydrogen forms compounds containing two atoms of hydrogen. Except for oxygen, there is only one commonly occurring binary compound for each element, so the prefix *di-* is omitted. Examples are H_2S, hydrogen sulfide, and CaH_2, calcium hydride. Oxygen forms two binary compounds with hydrogen. One is water, H_2O, and the other is hydrogen peroxide, H_2O_2. The

Table 3-4
Names of Some Hydrogen Compounds

Group 13	Group 14	Group 15
B_2H_6 Diborane	CH_4 Methane	NH_3 Ammonia
	SiH_4 Silane	PH_3 Phosphine

binary compounds of hydrogen with elements from Groups 13 through 15 have unsystematic names, some of which are listed in Table 3-4. Carbon, boron, and silicon form many different binary compounds with hydrogen; only the simplest is listed in the table.

Carbon-Based Compounds

The chemistry of carbon and its compounds is called *organic chemistry*. Biology and biochemistry build on the foundations of organic chemistry because the world of living matter is composed largely of carbon–based compounds. Hence, we present examples of organic molecules throughout this book, and here we list a few of the guidelines for naming organic compounds.

Carbon forms a huge number of binary compounds with hydrogen. Three major categories of these compounds are *alkanes, alkenes,* and *alkynes.* An alkane has only single bonds between carbon atoms. The four simplest alkanes, which are shown in Figure 3-9, are methane, ethane, propane, and butane. An alkene, on the other hand, contains one or more double bonds between carbons, and an alkyne has one or more triple bonds between carbon atoms. Figure 3-5 shows the structures of ethylene, the simplest alkene, and acetylene, the simplest alkyne.

One of the major organizing principles in organic chemistry is the presence of special arrangements of atoms. These so-called *functional groups* convey particular chemical properties. For example, a substance that contains an —OH group is called an *alcohol.* Methanol, ethanol, and 2-propanol, mentioned at the beginning of this section, are all alcohols, because each contains an —OH group. The systematic name of an alcohol is obtained by adding the suffix *-ol* to the name of the alkane with the same carbon framework. Thus CH_3OH has the carbon framework of methane and is called *methanol,* whereas C_2H_5OH has the carbon framework of ethane and is called *ethanol.*

The two different alcohols shown in Figure 3-10 are derivatives of propane. If one hydrogen atom on either of the *end* carbons of propane is replaced with OH, the alcohol is called *1-propanol.* If OH replaces a hydrogen atom on the *central* carbon, the alcohol is *2-propanol.* The numerical prefix specifies the carbon that bears the functional group. Notice that there is no such molecule as 3-propanol. The two end carbons of propane are identical, so replacing a hydrogen atom on either of them gives the same molecule. This situation is analogous to the train shown in Figure 3-10. We identify the engine as the first car of the train whether the train travels from left to right or from right to left.

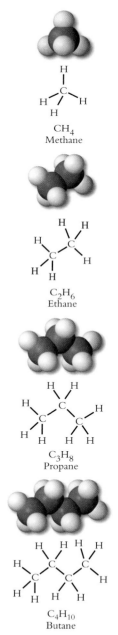

CH_4
Methane

C_2H_6
Ethane

C_3H_8
Propane

C_4H_{10}
Butane

Figure 3-9
The four simplest alkanes are methane, ethane, propane, and butane.

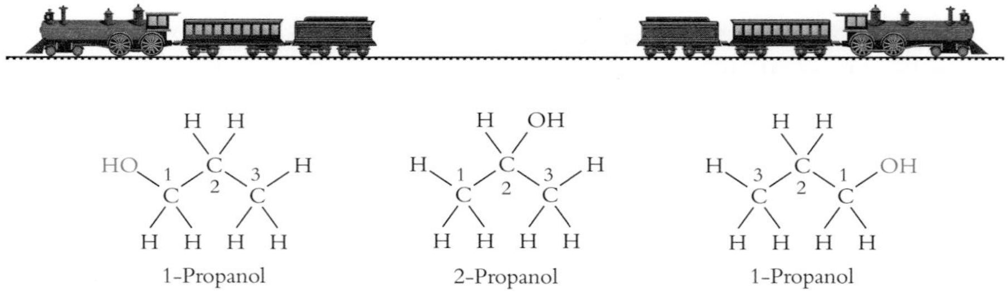

Figure 3-10
Structural formulas of 1-propanol and 2-propanol. When the —OH group is attached to an end carbon atom, the compound is 1-propanol. The molecule is the same whether viewed from left to right or from right to left, just as a train has its engine as its first car whether the train is moving from left to right or from right to left.

Many families of organic molecules are introduced as we proceed through the chapters of this book. Names are introduced as we need them.

The halogens make up another important organic functional group. The functional group is named by adding the *-o* suffix to the end of the halogen's root name: fluoro-, chloro-, bromo-, and iodo-. For these compounds the halogen is identified at the beginning of the name rather than at the end: 1-bromopropane or 2-chloropropane, for example.

Section Exercises

3.2.1 Write chemical formulas for the following compounds: chlorine monofluoride, xenon trioxide, hydrogen bromide, silicon tetrachloride, sulfur dioxide, and hydrogen peroxide.

3.2.2 Name the following compounds: ClF_3, H_2Se, ClO_2, $SbCl_3$, PCl_5, N_2O_5, N_2Cl_4, and NH_3.

3.2.3 Draw structural formulas of the following molecules: butane, 2-butanol, 1-butanol, and 1-bromobutane.

3.3 FORMULAS AND NAMES OF IONIC COMPOUNDS

Recall from Chapter 2 that ionic compounds contain positively charged cations and negatively charged anions. Some ionic compounds contain atomic cations and anions such as Na^+ and Cl^-. In many other cases, ionic compounds contain groups of atoms that have net electrical charges. Any stable sample of matter must be electrically neutral, and this requires that ionic compounds have the same amounts of positive and negative charges. This requirement for electrical neutrality, together with knowledge of the charges on various ions, allows us to determine the chemical formulas and names of ionic compounds.

Atomic Cations and Anions

The elements classified as metals have a strong tendency to lose electrons and form atomic cations. Almost every compound whose formula contains a metallic element from Group 1 or Group 2 is ionic. Other metals not only form ionic compounds in which they exist as cations but also commonly form compounds in which they share electrons.

An atomic ion forms when a neutral atom gains or loses one or more electrons: See Chapter 2.

Group

	1	2	3	4	5	6	7	8	9	10	11	12	13	14	15	16	17	18
1																		
2	Li^+															O^{2-}	F^-	
3	Na^+	Mg^{2+}											Al^{3+}			S^{2-}	Cl^-	
4	K^+	Ca^{2+}				Cr^{2+} Cr^{3+}	Mn^{2+} Mn^{3+}	Fe^{2+} Fe^{3+}	Co^{2+} Co^{3+}		Cu^+ Cu^{2+}	Zn^{2+}					Br^-	
5	Rb^+	Sr^{2+}									Ag^+	Cd^{2+}		Sn^{2+} Sn^{4+}			I^-	
6	Cs^+	Ba^{2+}										Hg^{2+}		Pb^{2+} Pb^{4+}				
7																		

Figure 3-11
The periodic table, showing the monatomic ions formed by the more common elements.

Cations formed from metals in Group 1 always have $+1$ charges, and cations formed from Group 2 metals always have $+2$ charges. Four other common metals have only one stable cation: Ag^+, Zn^{2+} Cd^{2+}, and Al^{3+}. Most transition metals, on the other hand, can exist in more than one cationic form. Figure 3-11 shows the important cations formed by the common elements.

Figure 3-11 also indicates that, while many elements form stable atomic cations, only six form stable atomic anions. Four elements in Group 17 form anions with -1 charge, and oxygen and sulfur form anions with -2 charge.

Polyatomic Ions

Sodium nitrate, $NaNO_3$, is an example of an ionic substance that contains a group of atoms with a net charge. Sodium is present as Na^+ atomic cations. The other atoms of sodium nitrate are grouped together in one structure, NO_3^-, which carries a -1 charge (Figure 3-12). This anion is a molecular ion, the nitrate ion. The nitrate ion displays some properties of ions and some properties of molecules: It has a negative electrical charge, but it also contains chemical bonds. Molecular ions are called **polyatomic ions** to distinguish them from neutral molecules and atomic ions.

Two important polyatomic *cations* appear in introductory chemistry. These are the ammonium ion, NH_4^+, and the hydronium ion, H_3O^+, both of which are shown in Figure 3-13. These cations always have $+1$ charges. The diatomic cation Hg_2^{2+} is the only relatively common polyatomic metal cation.

There are many different polyatomic *anions*, including several that are abundant in nature. Each is a stable chemical species that maintains its structure in the solid state and in aqueous solution. Polyatomic anions are treated as distinct units when writing chemical formulas, naming compounds, or drawing molecular pictures. The names, formulas, and charges of the more common polyatomic

Nitrate anion
NO_3^-

Figure 3-12
The nitrate anion is a group of four atoms held together by chemical bonds. The entire unit bears a -1 electrical charge.

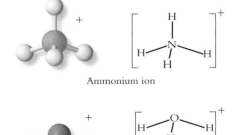

Ammonium ion

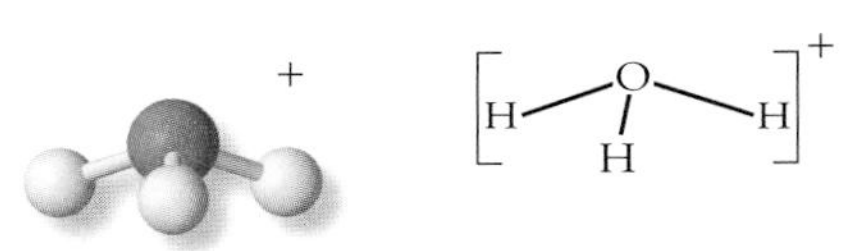

Hydronium ion

Figure 3-13
Structures of the ammonium and hydronium cations. Each always bears a $+1$ charge.

Table 3-5
Common Polyatomic Ions

Formula	Name	Formula	Name
Cations		**Oxyanions**	
NH_4^+	Ammonium	SO_4^{2-}	Sulfate
H_3O^+	Hydronium	SO_3^{2-}	Sulfite
Hg_2^{2+}	Mercury(I)	NO_3^-	Nitrate
		NO_2^-	Nitrite
Diatomic Anions		PO_4^{3-}	Phosphate
OH^-	Hydroxide	MnO_4^-	Permanganate
CN^-	Cyanide	CrO_4^{2-}	Chromate
Anions with Carbon		$Cr_2O_7^{2-}$	Dichromate
CO_3^{2-}	Carbonate	ClO_4^-	Perchlorate
$CH_3CO_2^-$	Acetate	ClO_3^-	Chlorate
$C_2O_4^{2-}$	Oxalate	ClO_2^-	Chlorite
		ClO^-	Hypochlorite

Tutorial

Species means "a distinct kind." *Chemical species* refers to a distinct kind of structure at the molecular level. Chemical species, which may be atoms, molecules, or ions, are represented by chemical formulas.

anions are listed in Table 3-5. You should memorize the common polyatomic ions because they appear regularly throughout this textbook.

Most polyatomic anions contain a central atom surrounded by one to four oxygen atoms. These species are called **oxyanions,** and they are named according to the following guidelines:

Guidelines for Naming Oxyanions

1. The name has a root taken from the name of the central atom (for example, carbo<u>nate</u>, CO_3^{2-}, and ni<u>trite</u>, NO_2^-).

2. When an element forms two different oxyanions, the one with fewer oxygen atoms ends in *-ite*, and the other ends in *-ate* (for example, SO_3^{2-}, sulf<u>ite</u>, and SO_4^{2-}, sulf<u>ate</u>).

3. Chlorine, bromine, and iodine each form four different oxyanions that are distinguished by prefixes and suffixes. The nomenclature of these ions is illustrated for bromine, but it applies to chlorine and iodine as well: BrO^-, <u>hypo</u>brom<u>ite</u>; BrO_2^-, brom<u>ite</u>; BrO_3^-, brom<u>ate</u>; and BrO_4^-, <u>per</u>brom<u>ate</u>.

4. A polyatomic anion with a charge more negative than -1 may add a hydrogen cation (H^+) to give another anion. These anions are named from the parent anion by adding the word *hydrogen*. For example, HCO_3^- is hydrogen carbonate, HPO_4^{2-} is hydrogen phosphate, and $H_2PO_4^-$ is dihydrogen phosphate.

Carbonate
CO_3^{2-}

Hydrogen carbonate
HCO_3^-

Recognizing Ionic Compounds

Ionic compounds show distinctive chemical behavior, so it is important to recognize which substances are classified as ionic. There is no single method that correctly identifies ionic compounds under all circumstances, but the following guideline is true for most ionic compounds encountered in general chemistry.

/// A compound is ionic if it contains a metal from Group 1 or Group 2 or one of the polyatomic ions. Binary metal oxides and sulfides also have ionic character.

We present further guidelines for identifying ionic compounds in Section 8-2, after we develop ideas about chemical bonding.

Example 3-5 provides practice in identifying ionic compounds.

Recognizing Ionic Compounds   **Example 3-5**

Determine which of the following substances are ionic: CCl_4, $SrCl_2$, Li, $Co(NO_3)_3$, KCN, $TiCl_4$, and KH_2PO_4.

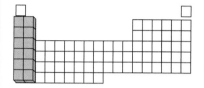

Strategy: To identify an ionic compound, look for a binary metal oxide or sulfide, a Group 1 or Group 2 metal, or one of the polyatomic ions.

Solution:

CCl_4: There is neither a Group 1 or Group 2 metal nor a polyatomic ion, so carbon tetrachloride is not ionic.

$SrCl_2$: The presence of strontium, a Group 2 metal, indicates that strontium chloride is an ionic compound.

Li: Lithium is a Group 1 metal, but there is no other element present to neutralize the charge that would accompany Li^+ ions in an ionic compound. This is lithium in its elemental, nonionic state.

$Co(NO_3)_3$: There is a polyatomic ion, nitrate, so this is an ionic compound. This is an example of an ionic compound containing a transition metal cation.

KCN: This compound contains a Group 1 cation and a polyatomic anion, so it is ionic.

$TiCl_4$: Titanium tetrachloride does not meet either guideline for ionic compounds. This is an example of a transition metal compound that is not ionic.

KH_2PO_4: This compound contains a Group 1 cation and a polyatomic anion, so it is ionic.

Ionic Formulas

Every ionic compound contains discrete ionic units with specific charges. In addition, ionic compounds must always contain equal amounts of positive and negative charge. These requirements dictate the ratio of cations to anions in an ionic substance. The following guidelines ensure uniformity in writing ionic formulas:

Guidelines for Ionic Formulas

1. The cation-anion ratio must give a net charge of zero.

2. The cation is always listed before the anion.

3. The formula of any polyatomic ion is written as a unit.

4. Polyatomic ions are placed in parentheses with a following subscript to indicate ratios different from 1 : 1.

Here are some specific examples illustrating chemical formulas of ionic compounds:

Ammonium nitrate: Ammonium cations are always charged $+1$, and nitrate anions are always -1, so these ions combine in $1:1$ ratio. The cation is listed before the anion: NH_4NO_3. Notice that we do not lump together the two nitrogen atoms because the ammonium and nitrate ions are distinct entities.

Sodium carbonate: Sodium, a Group 1 metal, always forms $+1$ atomic cations, and carbonate anions are always -2. Thus the compound formed from sodium and carbonate must contain two Na^+ ions for every CO_3^{2-}, and its formula is Na_2CO_3.

Calcium phosphate: Calcium ions, from Group 2 of the periodic table, always carry a $+2$ charge, whereas phosphate carries a charge of -3. There must be three cations (total charge $+6$) for every two anions (total charge -6) for a chemical formula of $Ca_3(PO_4)_2$.

Ionic compounds are named by using the same guidelines used for naming binary molecules; however, cations are named before anions. Thus NH_4NO_3 is ammonium nitrate, Na_2CO_3 is sodium carbonate, and $Ca_3(PO_4)_2$ is calcium phosphate. The subscripts are not specified in these names because the fixed ionic charges determine the cation-anion ratios unambiguously. Example 3-6 reinforces these guidelines by showing how to construct chemical formulas from chemical names.

Example 3-6	Formulas of Ionic Compounds

Determine the chemical formulas of calcium chloride, magnesium nitrate, and potassium dihydrogen phosphate.

Strategy: The names of the cations and anions identify the chemical formulas, including the charges of each ion. The charges, in turn, dictate the ion ratio that must appear in the chemical formula.

Solution:

Calcium chloride contains Ca^{2+} (Group 2) and Cl^- (Group 17), so there must be two anions for every cation: $CaCl_2$.

Magnesium nitrate contains Mg^{2+} (Group 2) and the polyatomic anion NO_3^-. Again, there must be two anions for every cation: $Mg(NO_3)_2$.

Potassium dihydrogen phosphate requires an extra step in reasoning. Potassium (Group 1) is K^+, and phosphate is PO_4^{3-}. However, dihydrogen means that two hydrogen $+1$ cations are attached to phosphate, reducing its charge to -1. The dihydrogen phosphate ion, $H_2PO_4^-$, is a single structural unit, so the two ions are present in a $1:1$ ratio: KH_2PO_4.

Cations of Variable Charge

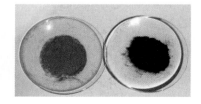

By convention, the chemical formulas of many ionic compounds do not explicitly state the charges of the ions. It is not necessary to do so because the charges can be deduced from knowledge of the various ions. However, many metals form more than one stable cation. For example, copper forms two different oxides, black CuO and red Cu_2O. The oxide ion has a -2 charge, so for the first compound to be neutral the copper cation must bear a $+2$ charge. In CuO, each copper ion must have $+1$ charge.

Without an additional guideline for nomenclature, each of these ionic compounds would be called *copper oxide*. These are different compounds, however, so we must introduce a way to distinguish between them. For any metal that forms *more than one* stable cation, the charge is specified by placing a Roman numeral in parentheses after the name of the metal. According to this guideline, CuO is copper(II) oxide, and Cu_2O is copper(I) oxide.

Example 3-7 shows how to apply the guidelines to a set of binary compounds.

Binary Nomenclature	**Example 3-7**

Name the chlorine compounds that have the following chemical formulas: $MgCl_2$, $CoCl_3$, PCl_3, $SnCl_2$, $SnCl_4$, and $GeCl_4$.

Strategy: Name compounds by applying the guidelines. Binary compounds that contain metals obey different guidelines than those with no metal. Unless a metal forms only one stable atomic cation, its charge must be specified with a Roman numeral in parentheses. Metals that form just one stable cation include M^+ from Group 1, M^{2+} from Group 2, Al^{3+}, Cd^{2+}, Zn^{2+}, and Ag^+.

Solution:

$MgCl_2$: Magnesium is a metal that always forms a +2 ion. This ionic compound is named without Roman numerals or prefixes: magnesium chloride.

$CoCl_3$: Cobalt, a transition metal, forms more than one stable cation: cobalt(III) chloride.

PCl_3: Phosphorus is not a metal, so use the guidelines for binary compounds: phosphorus trichloride.

$SnCl_2$: Tin is not one of the metals that forms a single cation. With two -1 chloride ions, this compound is tin(II) chloride.

$SnCl_4$: By the same reasoning used for tin(II) chloride: tin(IV) chloride.

$GeCl_4$: Germanium is not a metal, so the standard binary guidelines apply: germanium tetrachloride.

Hydrates

Many ionic compounds can have water molecules incorporated into their solid structures. Such compounds are called **hydrates.** To emphasize the presence of discrete water molecules in the chemical structure, the formula of any hydrate shows the waters of hydration separated from the rest of the chemical formula by a dot. A coefficient before H_2O indicates the number of water molecules in the formula. Copper(II) sulfate pentahydrate is a good example. The formula of this beautiful deep blue solid is $CuSO_4 \cdot 5H_2O$, indicating that five water molecules are associated with each $CuSO_4$ unit. Upon prolonged heating $CuSO_4 \cdot 5H_2O$ loses its waters of hydration along with its color. Other examples of hydrates include aluminum nitrate nonahydrate, $Al(NO_3)_3 \cdot 9H_2O$, with nine water molecules for every one Al^{3+} cation and three NO_3^- anions, and nickel(II) sulfate hexahydrate, $NiSO_4 \cdot 6H_2O$.

Unfortunately, the nature of the cations and anions in an ionic compound cannot be used to determine how many waters of hydration will be included when the ions form a solid. The number of water molecules, which can range from 0 to as high as 18, must be determined by doing experiments. In fact, some ionic substances exist in several different forms with different numbers of water molecules.

Examples 3-8 and 3-9 integrate the procedures used to convert between names and chemical formulas.

Example 3-8	Naming Chemical Compounds

Name the following compounds: CrO_3, ClF_3, Ag_2SO_4, and NH_4HSO_4.

Strategy: Apply the guidelines for naming compounds.

Solution:

CrO_3: As a transition metal, chromium forms more than one stable cation. Name the metal first, using a Roman numeral to designate chromium's charge. Each of the three oxide anions has a -2 charge. To maintain net charge neutrality, Cr must be $+6$, so the name of the compound is chromium(VI) oxide.

ClF_3: This compound contains two elements from Group 17 of the periodic table. Chlorine is named first because it is lower in the group, and we add a prefix that specifies the number of fluorine atoms: chlorine trifluoride.

Ag_2SO_4: The polyatomic sulfate ion indicates that this is an ionic compound. Silver is always $+1$, so no Roman numeral is needed: silver sulfate.

NH_4HSO_4: The polyatomic ammonium cation is combined with a hydrogen-containing anion: ammonium hydrogen sulfate.

Example 3-9	Chemical Formulas

Determine the correct chemical formulas of potassium permanganate, dinitrogen tetroxide, nickel(II) chloride hexahydrate, sodium hydrogen phosphate, and iron(III) oxide.

Strategy: Work from name to formula, using information about polyatomic ions and being careful to build a formula that is electrically neutral.

Solution:

Potassium permanganate: Permanganate is MnO_4^- (Table 3-5), and potassium always has a $+1$ charge (Group 1). The formula is $KMnO_4$.

Dinitrogen tetroxide: The prefixes identify the formula: N_2O_4.

Nickel(II) chloride hexahydrate: Nickel(II) is Ni^{2+}, and chloride is Cl^-. To achieve electrical neutrality, there must be two Cl^- and one Ni^{2+}. "Hexa" indicates six water molecules; $NiCl_2 \cdot 6H_2O$.

Sodium hydrogen phosphate: One hydrogen cation (H^+) attached to phosphate (PO_4^{3-}) leaves two negative charges, which requires two sodium ions for neutrality: Na_2HPO_4.

Iron(III) oxide: Iron(III) is Fe^{3+}, and oxide is O^{2-}. To be a neutral compound, there must be two Fe^{3+} ions (total charge $= +6$) for every three oxide ions (total charge $= -6$): Fe_2O_3.

■ **3.3.1** Name the following compounds: SCl_2, $CaCl_2$, $PbCl_2$, $NaNO_3$, MnO_2, $ZrCl_4$, NaH, and $NaIO_3$.

■ **3.3.2** Write correct molecular formulas for aluminum oxide, potassium dichromate, lead(II) nitrate, nitrogen dioxide, sodium sulfate, iodine pentafluoride, manganese(II) acetate, and sodium hypochlorite.

■ **3.3.3** Name the following compounds (Refer to Figure 1-4 for atom color codes):

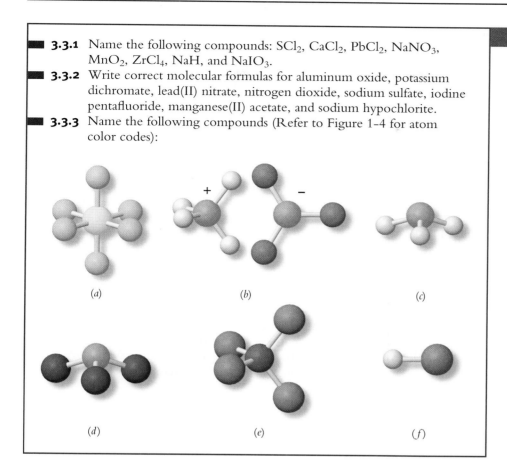

(a) (b) (c)

(d) (e) (f)

3.4 THE MOLE

Sections 3.1 through 3.3 address how chemists answer the question "What is it?" It is just as important to ask: "How much is there?" To answer this sort of question, chemists must be able to keep track of numbers of atoms and molecules. It is very difficult to count individual atoms and molecules, but we can easily measure the mass of a macroscopic sample of matter. Then, if we know the mass of an individual molecule, we can calculate the number of molecules in the sample. There is an enormous difference between the mass of one molecule and the masses of samples measured in the laboratory. For example, a good laboratory balance measures mass values from about 10^{-6} to 10^3 g. An atom, on the other hand, has a mass between 10^{-24} and 10^{-21} g. Even a tiny mass of 10^{-6} g contains approximately 10^{16} molecules.

The Mole and the Avogadro Constant

To avoid having to work with unimaginably large numbers, chemists use a convenient unit called the **mole (mol):**

/// *One mole is the number of atoms in exactly 12 g of the pure isotope carbon-12.*

Using mass spectrometers, scientists have determined that the mass of a ^{12}C atom is 1.992646×10^{-23} g. This experimental mass combined with the definition of the mole gives the number of atoms in one mole:

$$\left(\frac{12 \ \cancel{g \ ^{12}C}}{1 \ mol}\right)\left(\frac{1 \ atom}{1.992646 \times 10^{-23} \ \cancel{g \ ^{12}C}}\right) = 6.022142 \times 10^{23} \ atoms/mol$$

The number of items in one mole is important enough to have its own name and symbol. It is known as **the Avogadro constant,** commonly symbolized N_A. Although the Avogadro constant is known with eight-figure accuracy, four figures are enough for most calculations: $N_A = 6.022 \times 10^{23}$ items/mol.

The mole is a convenient unit for chemical amounts because the mass of most typical laboratory samples contains between 0.01 and 10 mol of atoms or molecules. For example, 1.20 g of ^{12}C contains 6.02×10^{22} atoms of carbon, which is 0.100 mol. The convenience of using a larger unit to describe many small items is not limited to atoms and molecules: Eggs are sold by the dozen, and paper is sold by the ream; chemists "sell" atoms and molecules by the mole. There are 12 eggs in a dozen eggs, 500 sheets of paper in a ream of paper and 6.022×10^{23} atoms in a mole of copper (Figure 3-14).

How large is a mole? If a computer were to count items at one million per second, it would take about 19 billion years to reach the Avogadro constant. If every person on the planet (population = 6 billion) were to spend 1 million dollars per second, it would take about 300 years to spend one mole of dollars. Why the mole was chosen to be this size is described in the Chemical Milestones Box.

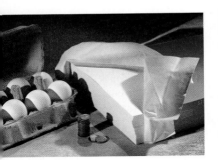

Figure 3-14
Different units are useful for measuring different-sized items. Eggs are sold by the dozen, paper is sold by the ream, and atoms are measured by the mole. Twenty-four pre-1982 pennies contain 1 mol of copper atoms.

Molar Mass

As described in Chapter 2, each isotope of a particular element has a different mass. Therefore, the mass of one mole of any isotope must also have its own unique value. The characteristic mass of one mole of any other isotope can be found by multiplying the mass of one atom of that isotope by the Avogadro constant. For example, mass spectrometry experiments reveal that one atom of carbon-13 has a mass of 2.15928×10^{-23} g, from which we can calculate the mass of one mole of ^{13}C atoms:

$$\left(\frac{2.15928 \times 10^{-23} \ g}{\cancel{atom \ of \ ^{13}C}}\right)\left(\frac{6.022142 \times 10^{23} \ \cancel{atoms \ of \ ^{13}C}}{1 \ mol}\right) = 13.0035 \ g/mol$$

A sample of a chemical element is usually a mixture of all of its stable isotopes. The hydrogen in a sample of water is 99.985% ^1H and 0.015% ^2H; elemental titanium is a mixture of five stable isotopes, and tin has ten isotopes, each with its own particular mass. Each of the isotopes contributes proportionally to the total mass of one mol of any naturally-occurring element. Table 3-6 summarizes the contribution of each isotope to the mass of one mole of titanium.

/// *The mass of one mole of any naturally occurring element is the sum of the contributions from its isotopes.*

The symbol Σ is the Greek letter sigma. It means "Take the sum of the following quantities."

Elemental molar mass = Σ (Fractional abundance)(Isotopic molar mass) **(3-1)**

For purposes of chemical bookkeeping, it is unnecessary to know the isotopic molar masses and isotopic compositions of the elements. Instead, we need to know

| Box 3-2 | Chemical Milestones: Story of the Mole |

The size of the mole—6.022142×10^{23} items—and the choice of the molar mass standard—1 mole of carbon-12 has a mass of exactly 12 g—may seem peculiar and arbitrary. After all, why not define a mole so that it contains exactly 10^{20} items, or base molar mass on the lightest element, hydrogen? Although the choices are indeed arbitrary, they were chosen for a combination of historical and logical reasons.

The *units* of mass were established well before the concept of *atomic* mass was developed. The scientific community had already agreed to use the gram as the fundamental mass unit when atomic theory became accepted. When nineteenth-century chemists determined the mass ratios of the elements in common chemical substances, their measurements were made in grams. The mole was introduced to relate mass ratios to numbers of atoms at a time when the mass of an individual atom could not be measured directly.

Oxygen was a natural choice on which to base mass ratios of the elements because oxygen forms compounds with most of the other elements. As chemists carried out measurements on oxygen-containing materials, they noticed that if 16 g of oxygen were used as a common amount, the masses of other elements were often very close to whole numbers. For example, 2 g of hydrogen combines with 16 g of oxygen, 14 g of nitrogen combines with 16 g of oxygen, and 137 g of barium combines with 16 g of oxygen. Furthermore, defining the mole to be exactly 16 g of oxygen made the mass of one mole of the lightest element, hydrogen, very close to 1 g.

Amedeo Avogadro

After the size of the molar mass unit had been assigned, it fixed the number of atoms in the mole. If 16 g of oxygen make 1 mol, then 1 mol contains however many oxygen atoms it takes to give 16 g; that number is the Avogadro constant, 6.022142×10^{23} atoms/mol. The name honors Amedeo Avogadro, an Italian physicist who first recognized that some of the common gases are not monatomic but instead composed of diatomic molecules. Avogadro's studies of gases led him to propose that equal volumes of different gases contain equal numbers of molecules, an important step in the development of the mole.

Whereas the measurements made by chemists compared masses of elements that reacted with one another, physicists developed methods to determine the mass of a single atom. Mass spectroscopy, for example, measures the mass of an individual isotope. Consequently, physicists developed a separate molar mass scale that set the molar mass of one isotope of oxygen, ^{16}O, equal to 16 g/mol. Because naturally-occurring oxygen contains small amounts of two heavier isotopes, the two scales are slightly different. The molar mass of fluorine, for example, was 18.9993 g/mol on the scale used by chemists but 19.0045 g/mol on the physicists' scale. These values are the same to four significant figures (19.00 g/mol), but as mass measurements became more and more accurate, the discrepancy between the scales became unacceptable.

As chemists and physicists engaged increasingly in collaborative research, such as in NASA's studies of planetary features, the need for a common standard became evident. Rather than using either oxygen-based scale, scientists agreed to base the molar mass scale on one isotope of carbon, defining the molar mass of ^{12}C to be exactly 12 g/mol.

This choice represented a compromise that offered advantages to both chemists and physicists. It was necessary to base the molar mass scale on a single isotope so that the high accuracy of physicists' measurements could be reflected in tabulated values. The choice of carbon resulted in values that are very nearly identical to the old chemists' scale. Using this standard, naturally occurring oxygen has an atomic molar mass of 15.999 rather than exactly 16, a difference of less than 0.01%. Thus a scale based on ^{12}C is not only very precise but also very close to one of the old standards.

As science has advanced, our knowledge of the value of the Avogadro constant has become more accurate. Early estimates, based on the properties of gases, were only good to one significant figure: $N_A \sim 1 \times 10^{24}$ items/mol. By the beginning of the twentieth century, indirect measurements had improved the accuracy to two significant figures: $N_A = 6.0 \times 10^{23}$ items/mol. Millikan's experiment in 1916, measuring the charge on the electron, extended the accuracy to three significant figures. In the 1930s, X-ray analysis allowed highly accurate determinations of the atomic spacing in crystals. From this, the number of atoms in a crystal of known mass could be calculated, leading to an accuracy of four-significant figures for N_A. Currently, mass spectrometric measurements have extended the accuracy of the mass of a ^{12}C atom, and thus our knowledge of the Avogadro constant, to seven significant figures:

$$N_A = 6.022142 \times 10^{23} \text{ items/mol.}$$

Table 3-6
Composition of I Mol of Titanium

Isotope	Isotopic Molar Mass (g/mol)	Fractional Abundance	Total Mass (g)
46	45.95263	0.082	3.77
47	46.9518	0.074	3.47
48	47.948	0.738	35.39
49	48.94787	0.054	2.64
50	49.9448	0.052	2.60
Naturally occurring Ti	—	1.000	47.87

the mass of one mole of an element containing its natural composition of isotopes (for example, the mass of one mole of naturally occurring elemental titanium is 47.87 g). These molar masses usually are included in the periodic table, and they appear on the endpapers of this textbook.

We will refer to the mass of one mole of any substance as its **molar mass (MM).** Elsewhere, however, you may find this same quantity called by other names such as *atomic mass* and *atomic weight*. These names can be misleading because atomic implies a single atom, but individual atoms do not have average masses. Instead, each atom has a mass characteristic of one isotope. In this textbook, we use *molar mass* to denote the mass (in grams) of one mole of substance containing isotopes in their naturally occurring abundances. We use *atomic mass* to denote the mass (in grams) of a single atom and *isotopic molar mass* for the mass (in grams) of one mole of an individual isotope.

Example 3-10 illustrates the use of isotopic molar masses and natural abundances to calculate the molar mass of elemental iron.

Example 3-10	**Calculating a Molar Mass**

The mass spectral analysis of an iron sample gives the following data. Calculate the molar mass of iron (Fe).

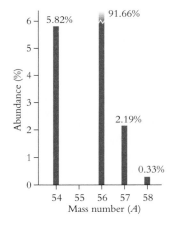

Isotope	Isotopic Molar Mass (g/mol)	Abundance
^{54}Fe	53.940	5.82%
^{56}Fe	55.935	91.66%
^{57}Fe	56.935	2.19%
^{58}Fe	57.933	0.33%

Strategy: The molar mass of a naturally occurring mixture of isotopes is the weighted average of the isotopic molar masses. Each isotope contributes to the total in proportion to its percentage (fractional) abundance, and the average is calculated using Equation 3-1.

Solution: First, the percentages must be converted to fractions by dividing by 100%. Then we multiply these fractional abundances by the isotopic masses and add the results:

> **Calculating a Molar Mass** *(continued)*   **Example 3-10**
>
> $MM = (0.0582)(53.940 \text{ g/mol } {}^{54}\text{Fe}) + (0.9166)(55.935 \text{ g/mol } {}^{56}\text{Fe})$
> $\quad + (0.0219)(56.935 \text{ g/mol } {}^{57}\text{Fe}) + (0.0033)(57.933 \text{ g/mol } {}^{58}\text{Fe})$
>
> $MM = (3.14 \text{ g/mol } {}^{54}\text{Fe}) + (51.27 \text{ g/mol } {}^{56}\text{Fe})$
> $\quad + (1.25 \text{ g/mol } {}^{57}\text{Fe}) + (0.19 \text{ g/mol } {}^{58}\text{Fe})$
>
> $MM = 55.85$ g/mol of natural Fe

Molar Masses of Chemical Compounds

Just as each element has a characteristic molar mass, so does every chemical compound. Chemical compounds are composed of atoms bound together into molecules or ions clustered together in electrically neutral aggregates. In either case a chemical formula describes the atomic composition of a compound.

One mole of any chemical compound is one mole of its chemical formula unit. Here are some examples:

- One mole of O_2 is one mole of O_2 molecules. Each molecule contains two oxygen atoms, so one mole of O_2 molecules contains two moles of oxygen atoms.
- One mole of methane (CH_4) is one mole of CH_4 molecules, so it contains one mole of carbon atoms and four moles of hydrogen atoms.
- One mole of sodium chloride ($NaCl$) contains one mole each of Na^+ cations and Cl^- anions.
- One mole of magnesium chloride ($MgCl_2$) contains one mole of Mg^{2+} and two moles of Cl^-.

When atoms combine to form molecules, they retain their atomic identities and characteristic molar masses.

/// The molar mass of a compound is found by adding together the molar masses of all of its elements, taking into account the number of moles of each element present.

The mass of an ion differs slightly from the mass of its parent atom because the ion has more or fewer electrons. However, the mass of an electron is so much smaller than the mass of a nucleus that the differences can be neglected. Consequently, molar masses of ions are calculated in the same way as molar masses of neutral compounds.

One mole of O_2 molecules contains two moles of O atoms, so the molar mass of molecular O_2 is calculated as follows:

$$\left(\frac{2 \text{ mol O}}{1 \text{ mol O}_2} \right)\left(\frac{15.999 \text{ g}}{1 \text{ mol O}} \right) = 31.998 \text{ g/mol O}_2$$

$$MM \text{ of } O_2 = 31.998 \text{ g/mol}$$

Verify for yourself that the molar masses of CH_4, $NaNO_3$, and $MgCl_2$ are 16.043 g/mol, 85.00 g/mol, and 95.211 g/mol, respectively.

As chemical formulas become more complicated, the calculation of molar mass requires more steps, as shown in Examples 3-11 and 3-12.

Example 3-11 | **Molar Masses of Ionic Compounds**

Sodium carbonate, ammonium nitrate, and ammonium sulfate are ionic substances that rank among the 30 most important industrial chemicals. Determine the molar mass of each compound.

Strategy: To calculate the molar mass of a compound, we must first know its chemical formula. The formula describes how many moles of each element are present in one mole of the compound. We use this information and the elemental molar masses to compute the molar mass of the compound.

Solution:

The formula of sodium carbonate is Na_2CO_3. Every mole of the salt contains two moles of Na^+ cations and one mole of CO_3^{2-} anions. The carbonate polyatomic ion is a single structural unit made up of one carbon atom and three oxygen atoms. Here is the molar mass calculation:

$$MM \; Na_2CO_3 = \left(\frac{2 \; \text{mol Na}}{1 \; \text{mol } Na_2CO_3} \right) \left(\frac{22.990 \text{ g}}{1 \; \text{mol Na}} \right) + \left(\frac{1 \; \text{mol C}}{1 \; \text{mol } Na_2CO_3} \right) \left(\frac{12.011 \text{ g}}{1 \; \text{mol C}} \right)$$

$$+ \left(\frac{3 \; \text{mol O}}{1 \; \text{mol } Na_2CO_3} \right) \left(\frac{15.999 \text{ g}}{1 \; \text{mol O}} \right) = 105.988 \text{ g/mol } Na_2CO_3$$

Ammonium nitrate is NH_4NO_3. This salt contains one NH_4^+ cation for every NO_3^- anion. Thus in every mole of NH_4NO_3 there are two moles of N, four moles of H, and three moles of O:

$$MM \; NH_4NO_3 = \left[\frac{2 \; \text{mol N}}{1 \; \text{mol } NH_4NO_3} (14.007 \text{ g/mol N}) \right]$$

$$+ \left[\frac{4 \; \text{mol H}}{1 \; \text{mol } NH_4NO_3} (1.008 \text{ g/mol H}) \right]$$

$$+ \left[\frac{3 \; \text{mol O}}{1 \; \text{mol } NH_4NO_3} (15.999 \text{ g/mol O}) \right]$$

$$= 80.043 \text{ g/mol } NH_4NO_3$$

Ammonium sulfate has the formula $(NH_4)_2SO_4$. To calculate the molar mass of this compound, we must count the numbers of atoms correctly. The subscript 2 after the parentheses around NH_4^+ indicates that this cation occurs twice in the formula, so there are two N atoms and eight H atoms in the chemical formula:

$$MM \; (NH_4)_2SO_4 = \left[\frac{2 \; \text{mol N}}{1 \; \text{mol } (NH_4)_2SO_4} (14.007 \text{ g/mol N}) \right]$$

$$+ \left[\frac{8 \; \text{mol H}}{1 \; \text{mol } (NH_4)_2SO_4} (1.008 \text{ g/mol H}) \right]$$

$$+ \left[\frac{4 \; \text{mol O}}{1 \; \text{mol } (NH_4)_2SO_4} (15.999 \text{ g/mol O}) \right]$$

$$+ \left[\frac{1 \; \text{mol S}}{1 \; \text{mol } (NH_4)_2SO_4} (32.066 \text{ g/mol S}) \right]$$

$$= 132.140 \text{ g/mol } (NH_4)_2SO_4$$

Molecular Molar Mass

Example 3-12

The line drawings and chemical formulas of two common insecticides are at right. Determine the molar mass of each insecticide.

Strategy: To determine the molar mass of a substance, use the chemical formula to find the number of moles of each element contained in one mole of the substance. These molar numbers are multiplied by the molar masses of the elements and added. The line structures demonstrate how the atoms are connected in these molecules, but the structures are not needed for these calculations because the chemical formulas are provided.

Sevin
$C_{12}H_{11}NO_2$

Malathion
$C_{10}H_{19}O_6PS_2$

Solution:

For Sevin:

$$\left(\frac{12 \text{ mol C}}{1 \text{ mol Sevin}}\right)\left(\frac{12.011 \text{ g C}}{1 \text{ mol C}}\right) = \frac{144.13 \text{ g C}}{1 \text{ mol Sevin}}$$

$$\left(\frac{11 \text{ mol H}}{1 \text{ mol Sevin}}\right)\left(\frac{1.008 \text{ g H}}{1 \text{ mol H}}\right) = \frac{11.09 \text{ g H}}{1 \text{ mol Sevin}}$$

$$\left(\frac{1 \text{ mol N}}{1 \text{ mol Sevin}}\right)\left(\frac{14.007 \text{ g N}}{1 \text{ mol N}}\right) = \frac{14.007 \text{ g N}}{1 \text{ mol Sevin}}$$

$$\left(\frac{2 \text{ mol O}}{1 \text{ mol Sevin}}\right)\left(\frac{15.999 \text{ g O}}{1 \text{ mol O}}\right) = \frac{31.998 \text{ g O}}{1 \text{ mol Sevin}}$$

$$MM\ C_{12}H_{11}NO_2 = 201.23 \text{ g/mol}$$

In this calculation, we determine molar mass by a complete analysis of units. Here is the calculation for Malathion, using abbreviated designations for the units:

One mole of Malathion contains:

$$(10 \text{ mol C})(12.011 \text{ g/mol}) = 120.11 \text{ g C}$$
$$(19 \text{ mol H})(1.008 \text{ g/mol}) = 19.15 \text{ g H}$$
$$(6 \text{ mol O})(15.999 \text{ g/mol}) = 95.994 \text{ g O}$$
$$(1 \text{ mol P})(30.974 \text{ g/mol}) = 30.974 \text{ g P}$$
$$(2 \text{ mol S})(32.066 \text{ g/mol}) = 64.132 \text{ g S}$$

$$MM\ C_{10}H_{19}O_6PS_2 = 330.36 \text{ g/mol}$$

The mass of 10 mol C is known to two decimal places, so the sum of the elemental molar masses is also known to two decimal places.

Chemists have traditionally called the mass of one mole of a chemical compound the "molecular weight" (*MW*). We prefer the term *molar mass* because *molecular weight* is misleading. First, the quantity being expressed is a *mass*, not a *weight*. Second, the mass expressed is the mass of one mole of substance, not one molecule. Moreover, ionic substances such as those in Example 3-12 do not contain molecules, yet they have well-defined molar masses. We use the term *molar mass* throughout this book, but you should be aware that *molecular weight* appears in many other books. Remember that both terms refer to the mass of one mole of a chemical substance.

Section Exercises

3.4.1 Use the data in the following table to calculate the molar mass of naturally-occurring sulfur.

Isotope	Isotopic Molar Mass	Abundance
^{32}S	31.97207 g/mol	95.02%
^{33}S	32.97146 g/mol	0.75%
^{34}S	33.96786 g/mol	4.21%
^{36}S	35.96709 g/mol	0.02%

3.4.2 Calculate the molar masses of the following compounds:

Compound	Name	Uses
(a) $NaAsO_2$	Sodium arsenite	Insecticide, herbicide, fungicide
(b) Na_4SiO_4	Sodium silicate	Soaps, adhesives, flame retardant
(c) $Ca(ClO)_2$	Calcium hypochlorite	Bactericide, bleach, pool cleaner
(d) $Ba(NO_3)_2$	Barium nitrate	Green fireworks
(e) Tl_2SO_4	Thallium sulfate	Rat and ant poison

3.4.3 Twenty different amino acids are the essential building blocks of proteins. Calculate the molar masses of these three.

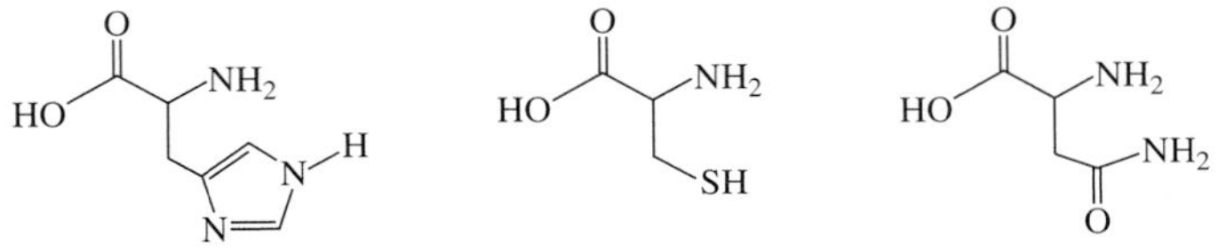

Histidine
$C_6H_9N_3O_2$

Cysteine
$C_3H_7NO_2S$

Asparagine
$C_4H_8N_2O_3$

3.5 MASS-MOLE-NUMBER CONVERSIONS

Many of the calculations in chemistry involve converting back and forth among the mass of a substance, the number of moles, and the number of atoms and/or molecules. These calculations are all centered around the mole.

Mass-Mole-Atom Conversions

Molar mass can be thought of as a conversion factor between mass in grams and number of moles. These conversions are essential in chemistry, because chemists count amounts of substances in moles but routinely measure masses in grams. To determine the number of moles in a sample, we measure the mass of the sample and then divide by the substance's molar mass. Similarly, to find the mass of a particular number of moles, we multiply that number of moles by the molar mass.

Suppose, for example, that a bracelet contains 168 g of silver (Ag). To determine the number of moles of silver in the bracelet, divide the mass by the molar mass of Ag, which is 107.87 g/mol:

$$n = \frac{m}{MM} \quad so \quad \frac{168 \text{ g Ag}}{107.87 \text{ g/mol}} = 1.56 \text{ mol Ag}$$

If a chemical reaction calls for 0.250 mol of silver, multiply moles by molar mass to determine the mass of Ag that should be used:

$$m = n\, MM \quad so \quad (0.250 \text{ mol Ag})(107.87 \text{ g/mol}) = 27.0 \text{ g Ag}$$

The Avogadro constant is the conversion factor that links number of moles with the number of individual particles. To determine the number of atoms in a sample of an element, we multiply the number of moles by the Avogadro constant. Likewise, if a sample contains a certain number of atoms, the number of moles in the sample can be calculated by dividing by the Avogadro constant.

The number of silver atoms in the bracelet that contains 1.56 mol of Ag is found by multiplying by the Avogadro constant.

$$\# = n\, N_A \quad so \quad (1.56 \text{ mol Ag})(6.022 \times 10^{23} \text{ atoms/mol}) = 9.39 \times 10^{23} \text{ atoms Ag}$$

Radioactivity detectors, which "count" the number of nuclei that decay, provide an example of a conversion from number of particles to moles. If a radioactivity measurement indicates that 175 ^{14}C nuclei decay in one hour, dividing by the Avogadro constant converts to moles:

$$n = \frac{\#}{N_A} \quad so \quad \frac{175 \text{ nuclei } ^{14}\text{C}}{6.022 \times 10^{23} \text{ nuclei/mol}} = 2.91 \times 10^{-22} \text{ mol } ^{14}\text{C}$$

The Avogadro constant and molar mass make it possible to convert readily among the mass of a pure element, the number of moles, and the number of atoms in the sample. These conversions are represented schematically in the flowchart shown in Figure 3–15.

Moles occupy the central position of the flowchart in Figure 3–15 because the mole is the unit that chemists use to move through almost all chemical calculations. When you set out to solve a chemical problem, first interpret the question on the molecular level. The second part of chemical problem solving often involves quantitative calculations, which usually require working with *moles*.

/// Chemical calculations are built around the mole.

Example 3–13 provides a practical application of these ideas.

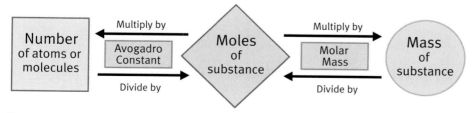

Figure 3-15
This flowchart shows how the number of moles is connected to the mass of a sample and the number of atoms or molecules.

Example 3-13	Mass-Mole-Atom Conversions

The eruption of Mount St. Helens on May 18, 1980, provided geologists with a unique opportunity to study the action of volcanos. Gas samples from the plume were collected and analyzed for toxic heavy metals. To collect mercury (Hg), gas samples were passed over a piece of gold metal, which binds Hg atoms very tightly. The mass of the metal increased as it absorbed Hg from the plume. From a plume-gas sample containing 200 g of ash, 3.60 μg of Hg was deposited on the gold. How many moles of mercury were present in the gas sample? How many atoms is this?

Strategy: This problem contains several pieces of information, not all of which are needed to answer the questions. The seven-step approach described in Chapter 1 is particularly useful for extracting the essential features from such a problem.

Solving Quantitative Problems

Solution:

1. *Determine what is asked for.* The problem asks for the number of moles and the number of atoms of Hg present in the plume sample.

2. *Visualize the problem.* Mercury atoms from the gas plume "stick" to the gold metal, causing an increase in mass. Thus the increase in mass of the piece of gold equals the mass of Hg in the plume sample.

3. *Organize the data.* The data available include what is given in the problem and what can be looked up in tables:

$$\text{Given in the problem: } 3.60 \ \mu g \text{ of Hg deposited}$$

$$\text{From tables: } 1 \ \mu g = 10^{-6} \text{ g}$$

$$MM \text{ of Hg} = 200.6 \text{ g/mol}$$

$$N_A = 6.022 \times 10^{23} \text{ atoms/mol}$$

Notice that the mass of the ash is not included in the organized data. The problem asks about the amount of mercury in the gas sample. The mass of ash in that sample is an extra piece of data that is not needed to solve the problem.

4. *Identify a process to solve the problem.* The mass of mercury is given. Conversion from mass to number of moles and number of atoms requires equations developed in this section:

$$n = m/MM \qquad \text{and} \qquad \text{Atoms} = n \, N_A$$

5. *If necessary, manipulate the equations.* Moles and atoms of mercury are the goal of the calculation. The equations listed in Step 3 do not require manipulation.

6. *Substitute and do calculations.* Pay careful attention to units.

Carry one extra digit and round to the correct number of significant figures after the final calculation.

$$\text{Moles Hg} = \frac{3.60 \ \mu g \text{ Hg}}{(200.6 \text{ g mol}^{-1})(10^6 \ \mu g \text{ g}^{-1})} = 1.795 \times 10^{-8} \text{ mol Hg}$$

$$\text{Atoms Hg} = (1.795 \times 10^{-8} \text{ mol Hg})(6.022 \times 10^{23} \text{ atoms/mol})$$
$$= 1.08 \times 10^{16} \text{ atoms}$$

Mass-Mole-Atom Conversions *(continued)*

Example 3-13

7. *Check your results for reasonableness.* First of all, the units cancel to give moles and atoms, which are the units for which the problem asked. Further, the mass of Hg is quite small, so we expect the number of moles to be small also. The number of atoms, 1.08×10^{16}, is large but much smaller than the Avogadro constant.

It may seem that 10^{-8} mol is not very much mercury, but consider that the initial eruption blanketed the Yakima Valley with 35 metric tons of volcanic ash per acre. Given that a metric ton is 1000 kg, determine for yourself how many grams, moles, and atoms of Hg the volcano deposited on every acre of the Yakima Valley.

Mass-Mole-Number Conversions for Compounds

The connections shown in Figure 3-15 apply to chemical compounds as well as to atoms of pure elements. For example, the molar mass of the insecticide Sevin was calculated in Exercise 3-12. Molar mass and the Avogadro constant provide links between mass of a sample, the number of moles, and the number of molecules. The following calculation shows how many molecules of Sevin are in a 5.0-g sample:

$$\frac{5.0 \text{ g Sevin}}{132.14 \text{ g/mol}} = 0.0378 \text{ mol Sevin}$$

$$(0.0378 \text{ mol Sevin})(6.022 \times 10^{23} \text{ molecules/mol}) = 2.3 \times 10^{22} \text{ molecules of Sevin}$$

Notice that once again moles are at the center of the scheme. To repeat, *calculations of chemical amounts center on the mole.*

Mass-mole-number calculations often involve atoms *within* a compound as well as the compound itself. As shown schematically in Figure 3-16, the chemical formula provides the link between moles of a compound and the number of moles of the compound's individual elements. Examples 3-14 and 3-15 illustrate mass-mole-number conversions.

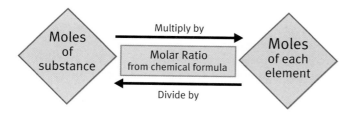

Figure 3-16
The chemical formula is the link between the moles of a compound and the moles of each of the compound's elements.

Elemental Content

Example 3-14

Ammonium nitrate (NH_4NO_3) is used as fertilizer because it is a good source of nitrogen atoms. It is such a good source, in fact, that it ranks among the top 15 industrial chemicals produced yearly in the U.S. How many moles of nitrogen atoms are present in a 1.00-pound bag of NH_4NO_3 fertilizer? How many atoms is this? (1 pound = 453.592 g)

| Example 3-14 | Elemental Content *(continued)* |

Strategy: First think about the chemistry of the problem, and then construct a mathematical solution. Use the seven-step procedure.

Solution:

1. The problem asks for the number of moles and atoms of nitrogen in a 1.00-pound sample of NH_4NO_3.

2. Visualize NH_4NO_3. (Always think atoms and molecules.) Ammonium and nitrate are common polyatomic ions whose chemical formulas you should remember.

3. The problem gives the mass of the sample, and we can anticipate that we will need the molar mass of ammonium nitrate:

$$\text{Mass of sample} = (1.00 \text{ pound})(453.592 \text{ g/pound}) = 453.6 \text{ g } NH_4NO_3$$

and

$$\text{Molar mass of } NH_4NO_3 = 80.043 \text{ g/mol (calculated in Example 3-11)}$$

4. The equations linking mass, moles, and atoms follow:

$$\text{mol} = m/MM \qquad and \qquad \text{Atoms} = (\text{mol})(N_A)$$

5. Often, it helps to draw a flowchart that organizes the steps necessary to analyze and solve a problem. A flowchart for this problem appears in the margin.

6. Now we are ready to work through the calculations. Begin by dividing the sample mass by the molar mass to determine moles of NH_4NO_3:

$$(453.6 \text{ g } NH_4NO_3)\left(\frac{1 \text{ mol}}{80.043 \text{ g}}\right) = 5.667 \text{ mol } NH_4NO_3$$

The chemical formula reveals that every mole of NH_4NO_3 contains 2 mol of nitrogen atoms:

$$(5.667 \text{ mol } NH_4NO_3)\left(\frac{2 \text{ mol N atoms}}{1 \text{ mol } NH_4NO_3}\right) = 11.33 \text{ mol N atoms}$$

As a final answer, this value should be rounded to three significant figures to match the data (1.00 pound): 11.3 mol N atoms. For use in the next step in the overall calculation, however, the value should retain a fourth digit to avoid possible rounding errors. Finish the problem by using the Avogadro constant to convert from the number of moles to the number of atoms:

$$(11.33 \text{ mol N})(6.022 \times 10^{23} \text{ atoms/mol}) = 6.82 \times 10^{24} \text{ N atoms}$$

The final result is rounded to three significant figures to match the three significant figures given for the mass of the bag of fertilizer.

7. Is this a reasonable result? Although 10^{24} is an immense number of atoms, we know that one mole contains 6×10^{23} atoms. Numbers of the order of 10^{24} are reasonable when calculating the number of atoms in a sample of everyday size.

Solving Quantitative Problems

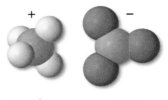

NH_4^+
Ammonium

NO_3^-
Nitrate

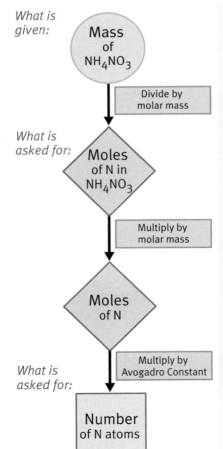

What is given:

Mass of NH_4NO_3

Divide by molar mass

What is asked for:

Moles of N in NH_4NO_3

Multiply by molar mass

Moles of N

Multiply by Avogadro Constant

What is asked for:

Number of N atoms

Mass–Mole–Number Conversions

Example 3-15

The thyroid gland produces hormones that help regulate body temperature, metabolic rate, reproduction, growth, the synthesis of blood cells, and more. Iodine must be present in our diet for these thyroid hormones to be produced. Iodine deficiency leads to sluggishness and weight gain and can cause severe problems in the development of fetuses.

One thyroid hormone is thyroxine, whose chemical formula is $C_{15}H_{11}I_4NO_4$. How many milligrams of $C_{15}H_{11}I_4NO_4$ can be produced from 2.1×10^2 mg of iodine atoms, the amount that a typical 180-pound adult consumes per day? How many molecules of $C_{15}H_{11}I_4NO_4$ is this?

Strategy: At first glance this problem might seem more challenging than Example 3-14. The formula of thyroxine, $C_{15}H_{11}I_4NO_4$, is much more complicated than NH_4NO_3. Also, the problem asks about one substance (thyroxine), but the data concern another (iodine). However, the standard approach shows that these two problems are similar. Again, follow the 7-step procedure.

Solution: What does the problem ask for? It asks for the number of milligrams and the number of molecules of $C_{15}H_{11}I_4NO_4$ that can be made from 210 mg of iodine atoms.

What is going on at the molecular level? The critical piece of information is the chemical formula of thyroxine: $C_{15}H_{11}I_4NO_4$. This tells us that four iodine atoms are needed to make one molecule of thyroxine. The line structure of $C_{15}H_{11}I_4NO_4$ is included to help you get used to visualizing molecules.

The relevant equations are the same ones used in previous problems:

$$\text{mol} = m/MM \qquad \text{Atoms} = (\text{mol})(N_A) \qquad \text{Molecules} = (\text{mol})(N_A)$$

The problem provides the mass of iodine, which we can convert to moles using the molar mass from the periodic table. The adult's mass is not needed.

We are asked for two pieces of information, namely the mass of thyroxine that can be prepared from a sample of iodine and the number of $C_{15}H_{11}I_4NO_4$ molecules contained in that mass. To find these values, we must first calculate moles of $C_{15}H_{11}I_4NO_4$.

How can we find moles of $C_{15}H_{11}I_4NO_4$ starting from milligrams of iodine? First, use the molar mass of iodine to convert milligrams to moles. Then use the information provided by the molecular formula to determine that each molecule of $C_{15}H_{11}I_4NO_4$ contains four atoms of iodine, so each mole of $C_{15}H_{11}I_4NO_4$ contains four moles of iodine atoms. The color screens on the line structure of thyroxine emphasize this 4:1 ratio.

In the second part of the problem, calculate the number of $C_{15}H_{11}I_4NO_4$ molecules from the moles of $C_{15}H_{11}I_4NO_4$ molecules and the Avogadro constant.

The flowchart in the margin summarizes the problem.

Once again, *moles* are the central feature of both calculations. The mathematical solution to this problem is left to you. The numerical answers are:

$MM = 776.9$ g/mol, $m = 320$ mg, $\# = 2.5 \times 10^{20}$ molecules

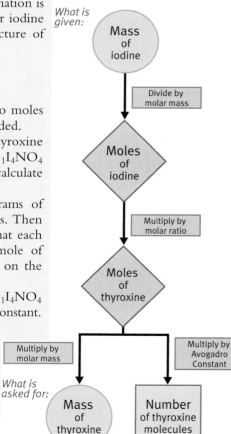

Thyroxine
$C_{15}H_{11}I_4NO_4$

What is given:

Mass of iodine

↓ Divide by molar mass

Moles of iodine

↓ Multiply by molar ratio

Moles of thyroxine

Multiply by molar mass

Multiply by Avogadro Constant

What is asked for:

Mass of thyroxine

Number of thyroxine molecules

Section Exercises

3.5.1 Calculate the number of atoms present in 2.55 g of each of these elements: (a) lithium; (b) titanium; and (c) gold.

3.5.2 Adenosine triphosphate (ATP) is the body's principal energy storage molecule. The formula of ATP is $C_{10}H_{16}N_5O_{13}P_3$. (a) How many moles of ATP are in a 50.0-mg sample? (b) How many molecules of ATP is this? (c) How many atoms of phosphorus are in the sample? (d) How many atoms of nitrogen? (e) What is the mass of oxygen in 50.0 mg of ATP? (f) How many total atoms are there in 50.0 mg of ATP?

3.5.3 Recently, there has been concern about pollution in the home from radon, a radioactive gas whose elemental molar mass is 222 g/mol. The Environmental Protection Agency says that a level of radon of 3.6×10^{-17} g/L of air is unhealthy. At this level, how many moles of radon would there be in a living room whose volume is 2455 L? How many atoms is this?

3.6 DETERMINING CHEMICAL FORMULAS

The chemical formula of a compound contains essential information about its composition. The formula identifies which elements are present, and it states the number of atoms of each kind present in one unit of the compound. We need the chemical formula of a substance to calculate its molar mass. In fact, almost all chemical calculations require the correct chemical formula. How are chemical formulas determined in the first place?

Every time a chemist prepares a "new compound" (one that has never been reported in the scientific literature), its chemical formula must be established beyond reasonable doubt. Usually, the molecule is analyzed by several methods, and each technique reveals information about the compound's identity.

To determine a chemical formula, we would like to count the atoms of each element in one molecule of the compound. Atoms are too small to count, but we might hope to measure the number of moles of each element present in one mole of the compound. Unfortunately, there is no direct experimental method for measuring moles. Instead, laboratory experiments give the *masses* of the various elements contained in some *total mass* of the compound.

Mass Percent Composition

The **mass percent composition** is a listing of the mass of each element present in 100 g of a compound. This percent by mass listing is also called the compound's **elemental analysis.**

We illustrate how the mass percent composition of a compound is related to its chemical formula using ammonium nitrate (NH_4NO_3), a compound whose formula is already known. The molar masses of NH_4NO_3 and its constituent elements can be used to convert the chemical formula into mass percentages.

The formula describes how many moles of each element are present in one mole of the substance. One mole of NH_4NO_3 contains two moles of nitrogen, four moles of hydrogen, and three moles of oxygen. Multiplying these numbers of moles by elemental molar masses gives the mass of each element contained in one mole of ammonium nitrate:

$$(2 \text{ mol N})(14.007 \text{ g/mol}) = 28.014 \text{ g N}$$

$$(4 \text{ mol H})(1.008 \text{ g/mol}) = 4.032 \text{ g H}$$

$$(3 \text{ mol O})(15.999 \text{ g/mol}) = 47.997 \text{ g O}$$

Summing, we see that one mole of NH_4NO_3 has a total mass of 80.043 g. Of this total, 28.014 g is nitrogen. The ratio of these masses is the mass fraction of nitrogen, and the percent nitrogen by mass is the mass fraction multiplied by 100%:

$$\frac{28.014 \text{ g}}{80.043 \text{ g}} = 0.35000 \quad \text{(mass fraction of nitrogen in } NH_4NO_3\text{)}$$

$$(0.35000)(100\%) = 35.000\% \quad \text{(mass percent of nitrogen in } NH_4NO_3\text{)}$$

In other words, 35.000% of the mass of NH_4NO_3 comes from its nitrogen atoms. This means that every 100 g of NH_4NO_3 contains 35.000 g of nitrogen. The mass percents of hydrogen and oxygen can be found in the same way:

$$\left(\frac{4.032 \text{ g}}{80.043 \text{ g}}\right)(100\%) = 5.037\% \text{ H} \qquad \left(\frac{47.997 \text{ g}}{80.043 \text{ g}}\right)(100\%) = 59.964\% \text{ O}$$

If the calculations have been done properly, the sum of the percent compositions of the individual elements will be 100%:

$$35.000\% \text{ N}$$

$$5.037\% \text{ H}$$

$$\underline{59.964\% \text{ O}}$$

$$100.001\%, \text{ rounds to } 100.00\%$$

This example shows how to compute mass percentages from a chemical formula. When the formula of a compound is unknown, chemists must work in the opposite direction. First, they do experiments to find the mass percentage of each element, and then they deduce what chemical formula matches those percentages.

The elemental analysis of a compound is usually determined by a laboratory that specializes in this technique. A chemist who has prepared a new compound sends a sample to the laboratory for analysis. The laboratory charges a fee that depends on the type and number of elements analyzed. The results are returned to the chemist as a listing of mass percent composition. The chemist must then figure out which chemical formula matches this composition. If a chemist has reason to expect a particular chemical formula, the observed percentages can be matched against the calculated percentages for the expected formula. This process is illustrated in Example 3-16.

Formulas and Mass Percentages	**Example 3-16**

A sample thought to be caffeine, the stimulant found in coffee, tea, and cola beverages, gave the following elemental analysis:

49.5% C   5.2% H   28.8% N   16.5% O

Does this elemental analysis agree with the chemical formula of caffeine, which is $C_8H_{10}N_4O_2$?

| Example 3-16 | Formulas and Mass Percentages *(continued)* |

Strategy: We are asked to compare an elemental analysis with a chemical formula. To do so, we can either convert mass percentages to mole amounts or convert the formula to mass percentages. It is easier to compute mass percentages and compare them with the measured values.

Solution: Begin by calculating the mass of one mole of caffeine (because the percentage analysis is reported only to three significant figures, we need the molar mass to no more than two decimals):

$$(8 \text{ mol C})(12.01 \text{ g/mol}) = 96.08 \text{ g C}$$

$$(10 \text{ mol H})(1.008 \text{ g/mol}) = 10.08 \text{ g H}$$

$$(4 \text{ mol N})(14.01 \text{ g/mol}) = 56.04 \text{ g N}$$

$$(2 \text{ mol O})(16.00 \text{ g/mol}) = 32.00 \text{ g O}$$

$$\text{TOTAL} = 194.20 \text{ g}$$

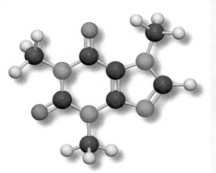

Caffeine
$C_8H_{10}N_4O_2$
$MM = 194.2$ g/mol

Divide the mass of each element by 194.20 g to obtain its mass fraction. Next multiply by 100% to convert mass fractions to mass percentages:

Element	Expected % for $C_8H_{10}N_4O_2$	% Found
C	49.47	49.5
H	5.19	5.2
N	28.86	28.8
O	16.48	16.5

Within the accuracy of the measured elemental analysis, the experimental percentages are the same as those expected for $C_8H_{10}N_4O_2$. Thus the data are consistent with caffeine.

Empirical Formula

Elemental analysis is a powerful tool for confirming molecular formulas, but it has its limitations. Consider caffeine again. Its chemical formula is $C_8H_{10}N_4O_2$, which means that the four elements are present in molar ratios $8:10:4:2$. Dividing this set of numbers by 2 does not change the relative molar amounts, so a compound whose formula is $C_4H_5N_2O$ has exactly the same elemental analysis as caffeine. A compound whose formula is $C_{12}H_{15}N_6O_3$ has the same elemental analysis, too. Elemental analysis cannot distinguish among these three possibilities. For that, we need to know the molar mass of the substance. Mass spectrometry, described in Chapter 2, is one way of measuring molar mass. If the mass spectrum of the substance shows that the molar mass is 194.2 g/mol, the formula is $C_8H_{10}N_4O_2$. On the other hand, a molar mass of 291.3 g/mol would correspond to $C_{12}H_{15}N_6O_3$. We describe other techniques for determining molar mass in later chapters.

If the molar mass of the compound is not known, the best we can do is to find the simplest formula that agrees with the elemental analysis. This simplest formula, or **empirical formula,** contains the smallest set of whole-number subscripts that match the elemental analysis. The empirical formula of caffeine is $C_4H_5N_2O$.

Sometimes, chemists have to analyze substances about which they know very little. A chemist may isolate an interesting molecule from a natural source, such as a plant or an insect. Under these conditions the chemical formula must be deduced from mass percentage data, without the help of an "expected" formula. A four-step procedure accomplishes this by using mass-mole conversions, the molar masses of the elements, and the fact that a chemical formula must contain integral numbers of atoms of each element.

1. Divide each mass percentage by the molar mass of the element. This gives the number of moles of each element in a 100-g sample.
2. Divide the results of Step 1 by whichever number of moles is the smallest. This maintains the mole ratios from Step 1 but bases them on one mole of the least abundant element.
3. If some results from Step 2 are far from integers, multiply through by a common factor that converts all molar amounts to integers or near-integers.
4. Round off each molar number to the nearest integer.

These guidelines for solving elemental analysis problems are applied in Example 3–17.

Chemical Formula from Composition	**Example 3-17**

Analysis of ibuprofen, the active ingredient in several over-the-counter pain relievers, shows that it contains 75.7% carbon, 8.8% hydrogen, and 15.5% oxygen. The mass spectrum of ibuprofen shows that its molar mass is less than 210 g/mol. Determine the chemical formula of this compound.

Strategy: First, follow the four-step process for finding the empirical formula. Then compare the empirical formula with the molar mass information to find the true formula.

Solution:

1. A 100-g sample of this compound would contain the following:

$$\left(\frac{75.7 \, g\,C}{100 \text{ g ibuprofen}} \right)\left(\frac{1 \text{ mol C}}{12.01 \, g\,C} \right) = 6.30 \text{ mol C/100 g ibuprofen}$$

The calculation of carbon is shown in detail to highlight the unit cancellation. Here are the condensed calculations for hydrogen and oxygen:

$$\left(\frac{8.8 \, g\,H/100 \text{ g}}{1.008 \, g\,H/\text{mol}} \right) = 8.7 \text{ mol H/100 g}$$

$$\left(\frac{15.5 \, g\,O/100 \text{ g}}{16.00 \, g\,O/\text{mol}} \right) = 0.969 \text{ mol O/100 g}$$

2. Dividing each molar number by the smallest gives the following:

$$\frac{6.30 \text{ mol C}}{0.969 \text{ mol O}} = 6.50 \text{ mol C/mol O} \qquad \frac{8.7 \text{ mol H}}{0.969 \text{ mol O}} = 9.0 \text{ mol H/mol O}$$

$$\frac{0.969 \text{ mol O}}{0.969 \text{ mol O}} = 1 \text{ mol O/mol O}$$

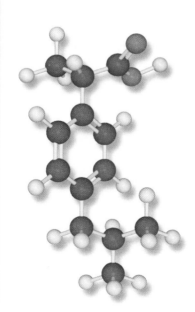

Ibuprofen

Example 3-17	Chemical Formula from Composition *(continued)*

3. The value for carbon is not an integer, but if we multiply all values by 2, we obtain integers: 13.0 mol C, 18.0 mol H, and 2.00 mol O.

4. The empirical formula of ibuprofen is $C_{13}H_{18}O_2$.

The mass spectrum of the compound indicates that its molar mass is less than 210 g/mol. The molar mass calculated from the empirical formula is 206.27 g/mol. This tells us that the chemical formula of ibuprofen is the same as its empirical formula.

We have shown how to convert mass percentages to formulas, but we have not yet shown how mass percentages are determined. Somehow a compound of unknown formula must be analyzed for the masses of each of its elements.

Analysis by Decomposition

Elemental analysis is relatively straightforward when compounds can be decomposed into pure elements. Figure 3-17, for example, shows that an orange-red solid compound of mercury decomposes on heating to yield elemental mercury, a silver-colored liquid. A colorless gas is also produced. To exploit this decomposition reaction for elemental analysis, the masses of these products must be measured carefully. Suppose we start with 5.00 g of the compound and collect the gas formed during the decomposition. Weighing the gas gives its mass as 0.37 g, and mass spectroscopy identifies the gas as molecular oxygen. The liquid mercury has a mass of 4.63 g. The sum of the masses of oxygen and mercury equals the original mass of the compound: 4.63 g + 0.37 g = 5.00 g. Because mass is always conserved and the entire 5.00 g has been accounted for, this tells us that the compound contains only mercury and oxygen.

Figure 3-17
When mercury(II) oxide is heated, it decomposes into liquid mercury, driving off oxygen gas in the process.

Because each chemical element is conserved, the masses of the products are equal to the masses of the elements contained in the original compound. This lets us complete the elemental analysis of the compound. The 5.00 g sample contained 4.63 g mercury (Hg) and 0.37 g O. The percent composition is found by dividing each elemental mass by the total mass and multiplying by 100:

$$\left(\frac{4.63 \text{ g Hg}}{5.00 \text{ g sample}}\right)(100\%) = 92.6\% \text{ Hg} \qquad \left(\frac{0.37 \text{ g O}}{5.00 \text{ g sample}}\right)(100\%) = 7.4\% \text{ O}$$

Therefore a 100.0-g sample would contain 92.6 g Hg and 7.4 g O. To get from elemental analysis to a chemical formula, we begin by dividing each mass by the appropriate molar mass:

$$\frac{92.6 \text{ g Hg}}{200.6 \text{ g/mol}} = 0.462 \text{ mol Hg} \qquad \frac{7.4 \text{ g O}}{16.0 \text{ g/mol}} = 0.46 \text{ mol O}$$

These amounts are the same to within the accuracy of the measurements, so we conclude that the compound contains mercury and oxygen in 1:1 molar ratio. The empirical formula is HgO.

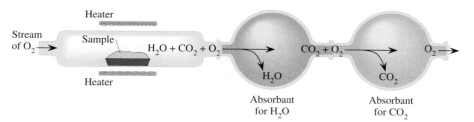

Figure 3-18
In combustion analysis a hot stream of oxygen gas reacts with a compound to form CO_2 and H_2O. The product gases are trapped, and their masses are determined by the gain in mass of each trap.

Combustion Analysis

Compounds that do not decompose cleanly into their elements must be analyzed by other means. **Combustion analysis** is often used to determine the empirical formula of an unknown substance. This method is particularly useful for carbon-containing compounds. In combustion analysis, an accurately known mass of a compound is burned in a stream of oxygen gas. Under carefully controlled conditions, the products of this burning process are water and carbon dioxide.

In a combustion reaction, all of the carbon in the sample is converted to carbon dioxide, and all of the hydrogen is converted to water. Certain other elements present in the sample are also converted to their oxides. Figure 3-18 shows a schematic view of an apparatus for combustion analysis. The stream of oxygen used to burn the substance carries the combustion products out of the reaction chamber and through a series of traps. Each trap is designed to collect just one combustion product. The mass of each trap is measured before and after combustion, and the difference is the mass of that particular product.

The flowchart in Figure 3-19 shows one way to analyze the data from a combustion analysis. The most important feature in this analysis is that atoms of each element are conserved. Every *carbon* atom in the original sample ends up in a CO_2 molecule, and every *hydrogen* atom in the original sample ends up in a molecule of H_2O. Mass is also conserved in a combustion reaction, so the mass of carbon contained in the carbon dioxide is the same as the mass of carbon in the original sample, and the mass of hydrogen in water is the same as the mass of hydrogen in the original sample. After the masses of CO_2 and H_2O produced in the combustion reaction have been determined, the mass percentages of carbon and hydrogen in these products can be used to calculate the masses of carbon and hydrogen present in the original sample. This leads to an elemental analysis of the unknown compound from which we can deduce the empirical formula.

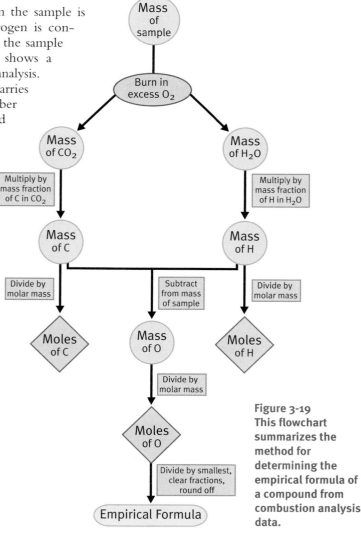

Figure 3-19
This flowchart summarizes the method for determining the empirical formula of a compound from combustion analysis data.

The products of a combustion reaction contain oxygen, carbon, and hydrogen. Did the oxygen in these products come from the original sample or from the oxygen used in the combustion process? To answer this question, we note that *all the mass of the original sample must be accounted for* because mass is never destroyed in a chemical reaction. The masses of CO_2 and H_2O tell us how much carbon and hydrogen were present in the original sample. Subtracting the combined masses of carbon and hydrogen from the mass of the original sample gives the total mass of any other elements that were present in the original sample. If the unknown compound contained only carbon and hydrogen, this difference in mass will be zero. If the mass difference is not zero, then other elements must have been present. If a compound is known to contain only carbon, hydrogen, and oxygen, then the mass difference must result from the mass of oxygen in the original sample.

Example 3-18 demonstrates how to use combustion analysis to determine the formula of a compound containing only C and H, and Example 3-19 shows how combustion analysis is conducted when the compound contains O.

Example 3-18	Combustion Analysis

A petroleum chemist isolated a component of gasoline and found its molar mass to be 114 g/mol. When a 1.55-g sample of this compound was burned in excess oxygen, 2.21 g of H_2O and 4.80 g CO_2 were produced. Find the empirical and molecular formulas of the compound.

Strategy: The flowchart in Figure 3-19 outlines the process. From masses of products, determine masses of elements. Then convert masses of elements to moles of elements. From moles of the elements, find the empirical formula. Finally, use information about the molar mass to obtain the molecular formula.

Solution: Begin by computing masses of carbon and hydrogen in the combustion products. This requires the mass fractions of C in CO_2 and H in H_2O:

$$C: (4.80 \text{ g } CO_2)\left[\frac{(1 \text{ mol C})(12.01 \text{ g/mol C})}{44.01 \text{ g}}\right] = 1.31 \text{ g C}$$

$$H: (2.21 \text{ g } H_2O)\left[\frac{(2 \text{ mol H})(1.008 \text{ g/mol H})}{18.02 \text{ g}}\right] = 0.247 \text{ g H}$$

(Because the final calculations will give integers, it is not necessary to carry an extra significant figure in this problem.)

Now we convert mass to moles:

$$\frac{1.31 \text{ g C}}{12.01 \text{ g/mol}} = 0.109 \text{ mol C} \qquad \frac{0.247 \text{ g H}}{1.008 \text{ g/mol}} = 0.245 \text{ mol H}$$

Next, compare the masses of C and H with the mass of the original sample to see whether the sample contained oxygen:

$$1.31 \text{ g C} + 0.247 \text{ g H} = 1.56 \text{ g} \qquad \text{Sample mass} = 1.55 \text{ g}$$

These are virtually the same, so C and H account for all of the mass in the original sample. Thus there are no other elements in the unknown compound.

Combustion Analysis *(continued)*

Example 3-18

We now have a molar relationship between hydrogen and carbon in the original compound: 0.245 mol H to 0.109 mol C. Divide each by the smaller value:

$$\frac{0.245 \text{ mol H}}{0.109 \text{ mol C}} = 2.25 \text{ mol H/mol C} \qquad \frac{0.109 \text{ mol C}}{0.109 \text{ mol C}} = 1.00 \text{ mol C/mol C}$$

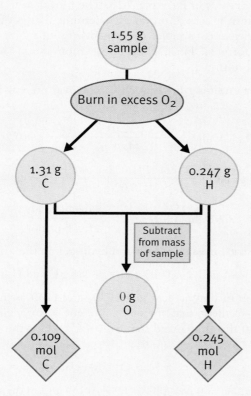

The mole ratio for hydrogen is not close to a whole number, so we must multiply each ratio by a common integer that will make the value for hydrogen a whole or near-whole number. In this case, 4 is the smallest integer that accomplishes this goal, with $(2.25)(4) = 9$ and $(1.00)(4) = 4$. This means that the empirical formula is C_4H_9.

Now, what is the molecular formula of the unknown? We must compare the empirical molar mass with the known molar mass, which is 114.2 g. Compute the empirical mass from the empirical formula and elemental molar masses:

$$\text{Empirical mass} = 4 \text{ (mol C)}(12.01 \text{ g/mol}) + 9 \text{ (mol H)}(1.008 \text{ g/mol})$$

$$= 48.04 \text{ g/mol} + 9.072 \text{ g/mol} = 57.11 \text{ g/mol}$$

$$\frac{MM}{\text{Empirical mass}} = \frac{114.2 \text{ g/mol}}{57.11 \text{ g/mol}} = 2$$

Because the molar mass is twice the empirical mass, we know that the molecular formula of this hydrocarbon is twice its empirical formula: C_8H_{18}. (Its common name is *octane*, from the eight carbon atoms in the molecule.)

Several decimals are readily converted to integers by multiplying by a small whole number. When you see any of these, think about multiplying rather than rounding:
$(0.20)(5) = 1$, $(0.25)(4) = 1$,
$(0.33)(3) = 1$, $(0.40)(5) = 2$,
$(0.50)(2) = 1$, $(0.60)(5) = 3$,
$(0.67)(3) = 2$, $(0.75)(4) = 3$,
and $(0.80)(5) = 4$.

| Example 3-19 | Empirical Formula |

Butyric acid, a component of rancid butter, has a vile stench. Burning 0.440 g of butyric acid in excess oxygen yields 0.882 g of CO_2 and 0.360 g of H_2O as the only products. The molar mass of butyric acid is 88 g/mol. What are its empirical formula and molecular formula?

Strategy: Proceed exactly as in Example 3-18. The flowchart in Figure 3-19 outlines the process. From masses of products, determine masses of elements. Then convert masses of elements to moles of elements. From moles of the elements, find the empirical formula. Finally, use information about the molar mass to obtain the molecular formula.

Solution: Start by determining how much carbon was present in the sample of butyric acid:

$$C: (0.882 \text{ g } CO_2) \left[\frac{(1 \text{ mol C})(12.01 \text{ g/mol C})}{44.01 \text{ g}} \right] = 0.241 \text{ g C}$$

$$H: (0.360 \text{ g } H_2O) \left[\frac{(2 \text{ mol H})(1.008 \text{ g/mol H})}{18.02 \text{ g}} \right] = 0.0403 \text{ g H}$$

$$\frac{0.241 \text{ g C}}{12.01 \text{ g/mol}} = 0.0201 \text{ mol C} \quad and \quad \frac{0.0403 \text{ g H}}{1.008 \text{ g/mol}} = 0.0400 \text{ mol H}$$

Next, compute the total mass of carbon and hydrogen in the original sample:

$$0.241 \text{ g C} + 0.0403 \text{ g H} = 0.281 \text{ g C and H}$$

The combustion of butyric acid gives CO_2 and H_2O as the only products, so the only other element that might be present is oxygen. The difference between the mass of the sample and the mass of C + H is the mass of O in the original sample:

$$\text{Mass C} + \text{Mass H} + \text{Mass O} = \text{Mass of sample} = 0.440 \text{ g}$$

$$0.440 \text{ g sample} - 0.281 \text{ g (C + H)} = 0.159 \text{ g O in the sample}$$

Divide this mass by the elemental molar mass of oxygen:

$$\frac{0.159 \text{ g O}}{16.00 \text{ g/mol}} = 0.00994 \text{ mol O}$$

Divide each number of moles by the smallest among them, 0.00994 mol O, to obtain moles of each element per mole of oxygen:

$$\frac{0.0201 \text{ mol C}}{0.00994 \text{ mol O}} = 2.02 \qquad \frac{0.0400 \text{ mol H}}{0.00994 \text{ mol O}} = 4.02 \qquad \frac{0.00994 \text{ mol O}}{0.00994 \text{ mol O}} = 1.00$$

Round each value to the nearest integer; 4.02 becomes 4, and 2.02 becomes 2. The empirical formula for butyric acid is C_2H_4O, giving an empirical mass of 44 g/mol. We are told that the molar mass of butyric acid is 88 g/mol. This is twice as large as the empirical mass. You should be able to use this information to show that the molecular formula of butyric acid is $C_4H_8O_2$.

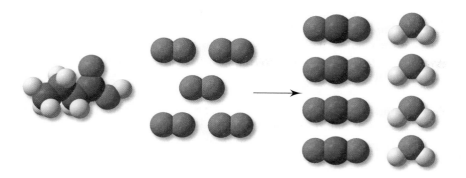

Figure 3-20
A molecular picture of the combustion of butyric acid to give carbon dioxide and water. Each molecule of butyric acid ($C_4H_8O_2$) produces four molecules of CO_2 and four molecules of H_2O. Notice that atoms are conserved in the reaction. There are 4 carbon atoms, 8 hydrogen atoms, and 12 oxygen atoms before and after the reaction.

The central feature of any combustion analysis is to keep track of all the atoms involved in the decomposition. Figure 3-20 illustrates this point with a molecular picture for Example 3-19.

If an unknown sample contains elements in addition to C, H, and O, then more than one experiment is required to determine the empirical formula. We examine analyses that require more than one experiment in Section 3.7.

Section Exercises

- **3.6.1** Determine the percentage composition of acetic acid, $C_2H_4O_2$, the ingredient that gives vinegar its tart taste.
- **3.6.2** One of the major iron ores is an oxide with the following percentage composition: Fe, 72.36% and O, 27.64%. What is the chemical formula of this ore?
- **3.6.3** Our bodies can neither make nor store much ascorbic acid, also known as vitamin C, so this essential compound must be supplied in our diets. Combustion of 7.75 mg of vitamin C gives 11.62 mg CO_2 and 3.17 mg H_2O. Vitamin C contains only C, H, and O, and its molar mass is between 150 and 200 g/mol. Determine its molecular formula.

3.7 AQUEOUS SOLUTIONS

We developed the concept of the mole in terms of pure chemical substances. Pure substances account for many interesting and important reactions, but many chemical reactions take place in solution. A pure substance dissolved in solution is a **solute.** A substance used to dissolve solutes is a **solvent.** Most of the time, the solvent is a liquid and is present in much larger quantities than any solutes.

Chemists use many different liquid solvents, but we focus most of our attention on water. When water is the solvent, the solution is said to be **aqueous.** A rich array of chemistry occurs in aqueous solution, including many geological and biochemical processes. Aqueous solutions dominate the chemistry of the Earth and the biosphere. The oceans, for instance, are rich broths of various cations and anions, sodium and chloride being the most abundant. The oceans can be thought of as huge aqueous solvent vessels for the remarkably complex chemistry of our world. Blood is an aqueous system that contains many ionic species, most notably carbonate, sodium, and potassium. In fact, the human body is mostly water, and much of the biochemistry of life takes place in aqueous solution.

Because aqueous solutions play a central role in the world around us, the chemistry of aqueous solutions is discussed in depth in several chapters of this book. To understand the chemistry of aqueous solutions, *it is essential that you learn to recognize the common ions at a glance*. This is especially true of the polyatomic ions.

Molarity

Any solution contains at least two chemical species, the solvent and one or more solutes. The mass of a solution is the sum of the masses of the solvent and all dissolved solutes. To answer questions such as "How much is there?"about solutions, we need to know the amount of each solute present in a specified volume of solution. The amount of a solute in a solution is given by the **concentration,** which is the ratio of the amount of solute to the amount of solution. In chemistry the most common measure of concentration is **molarity (*M*).** Molarity is the number of *moles of solute (n)* divided by the total *volume of the solution (V)* in liters:

$$\text{Molarity } (M) = \frac{\text{Moles of solute}}{\text{Total volume of solution}} = \text{mol/L} \quad or \quad M = n/V \quad \textbf{(3-2)}$$

Equation 3-2 defines molarity. When we need to calculate the number of moles of a substance in a solution of known molarity, we use a rearranged form of this equation: $n = MV$.

An aqueous solution is prepared by dissolving a measured quantity of solid or liquid in enough water to give some desired final volume. For example, to make a solution of common table sugar, or sucrose, we might dissolve 2.05 g of sugar in just enough water to make the final volume 25.0 mL. What is the molarity of this solution? The units of molarity are moles per liter (mol/L). To determine the molarity of the sugar solution, we must convert the mass of sucrose into moles and convert the solution volume into liters. These conversions involve no new ideas. To begin with, use the formula of sucrose, which is $C_{12}H_{22}O_{11}$, to calculate the molar mass, 342 g/mol. Next, convert mass to moles in the usual manner:

$$\frac{2.05 \text{ g sugar}}{342 \text{ g/mol}} = 5.994 \times 10^{-3} \text{ mol sugar}$$

Now convert the volume of the solution from milliliters to liters:

$$(25.0 \text{ mL})(10^{-3} \text{ L/mL}) = 2.50 \times 10^{-2} \text{ L}$$

Finally, divide moles of sugar by volume in liters to obtain molarity:

$$\frac{5.994 \times 10^{-3} \text{ mol}}{2.50 \times 10^{-2} \text{ L}} = 0.240 \text{ M}$$

Concentrations are used so frequently in chemistry that a shorthand notation for concentration is almost essential. Chemists represent the molar concentration of a species by enclosing its formula in brackets: $[C_{12}H_{22}O_{11}] = 0.240$ M for the sugar solution.

In the laboratory, solutions of known concentration are often prepared in a glass vessel called a *volumetric flask*. Volumetric flasks typically allow volumes to be measured with an accuracy of four significant figures. Figure 3-21 summarizes the general procedure for making a solution in a volumetric flask. This procedure is further described in Example 3-20.

Remember that molarity has units of mol/L, so volumes must be expressed in L.

We use an italicized M when molarity appears as a quantity in an equation, as in M = n/V. We use a Roman M when molarity is the unit associated with a numerical value, as in 2.40 × 10⁻¹ M.

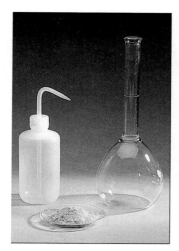

Figure 3-21
Solutions of known concentration are prepared using a volumetric flask. When the container is filled with liquid to the etched line, the volume of solution in the container is the amount specified for that volumetric flask.

| Preparing a Solution | Example 3-20 |

How many grams of nickel(II) chloride hexahydrate are required to prepare 250. mL of aqueous solution whose concentration is 0.255 M?

Strategy: As with all calculations of chemical amounts, we must work with moles. Because grams are asked for, we must do a mole–mass conversion; this requires the molar mass of the substance, which in turn requires that we know the chemical formula.

Solution: Begin by finding the chemical formula from the name, nickel(II) chloride hexahydrate. Chloride ion carries -1 charge, and the (II) indicates that nickel is a $+2$ cation. Electrical neutrality requires two chlorides for every nickel. The name also tells us that the salt contains six water molecules for each unit of nickel(II) chloride. Thus each formula unit of the salt contains one Ni^{2+} cation, two Cl^- anions, and six water molecules: $NiCl_2 \cdot 6H_2O$.

To determine the mass of $NiCl_2 \cdot 6H_2O$ required to prepare the solution, first calculate the number of moles of the salt required, and then use the molar mass to determine the number of grams:

$$\left(\frac{0.255 \text{ mol}}{\text{L}}\right)(250. \text{ mL})\left(\frac{10^{-3} \text{ L}}{1 \text{ mL}}\right) = 6.375 \times 10^{-2} \text{ mol } NiCl_2 \cdot 6H_2O$$

We need the molar mass of $NiCl_2 \cdot 6H_2O$ to convert moles to mass. By now, molar mass calculations should be routine. The molar mass of $NiCl_2 \cdot 6H_2O$ is 237.69 g/mol:

$$(6.375 \times 10^{-2} \text{ mol})(237.69 \text{ g/mol}) = 15.2 \text{ g}$$

To prepare the desired solution, we would first weigh 15.2 g of solid using a balance. We would transfer this solid into a 250.0 mL volumetric flask, add enough water to dissolve the solid, then continue adding water and mixing until the solution level matched the mark on the flask.

Ionic Solutions

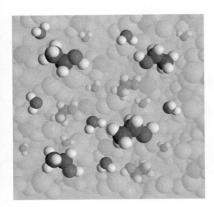

Figure 3-22
Molecular picture of an aqueous solution of ethanol, which contains ethanol molecules and water molecules.

To understand the chemical behavior of solutions, we must "think molecules." Before working any problem involving aqueous solutions, begin with the question, "What chemical species are present in the solution?" There will always be lots of molecules of the solvent, water. In addition, there will be solute species, which may be molecules or ions.

Aqueous solutions of molecular substances such as sugar ($C_{12}H_{22}O_{11}$) or ethanol (C_2H_5OH) contain individual molecules in a "sea" of water molecules (Figure 3-22). How do we know that these solutes dissolve as neutral molecules? Recall from Chapter 2 (Figure 2-23) that when ions are present in a solution, the solution conducts electricity. Solutions of sugar or ethanol do not conduct electricity, showing that these solutions contain no mobile charged particles. Sugar and ethanol dissolve as neutral molecules.

Aqueous solutions of salts do conduct electricity, showing that salts break apart to give cations and anions in solution. For example, a solution of NaCl contains Na^+ ions, Cl^- ions, and H_2O molecules. Furthermore, the ratio of cations and anions produced by a salt is given by its chemical formula. A 1.0 M solution of sodium chloride is 1.0 M in each of Na^+ and Cl^-. In contrast, $NiCl_2$ contains two moles of Cl^- anions for every one mole of Ni^{2+} cations. This ratio is maintained when nickel(II) chloride dissolves in water; a 1.0 M solution of $NiCl_2$ is 1.0 M in Ni^{2+} and 2.0 M in Cl^-. Notice that preserving the cation–anion ratio ensures that the solution is electrically neutral overall. The total positive charge carried by one mole of Ni^{2+} equals the total negative charge carried by two moles of Cl^-:

> The (s) and (aq) in the equation indicate phases; (s) designates solid, (aq) designates aqueous solution.

$$NiCl_2(s) \xrightarrow{H_2O} Ni^{2+}(aq) + \underbrace{Cl^-(aq) + Cl^-(aq)}$$

$$1 \text{ mol} \Uparrow \qquad 1 \text{ mol} \Uparrow \qquad 2 \text{ mol} \Uparrow$$

When a salt containing polyatomic ions dissolves in water, the cations separate from the anions, but each polyatomic ion remains intact. An example is ammonium nitrate, composed of NH_4^+ polyatomic cations and NO_3^- is polyatomic anions. Ammonium nitrate dissolves in water to give a solution containing NH_4^+ cations and NO_3^- anions.

> In later chapters, we describe situations in which species in solution transfer protons among one another. For example, NH_4^+ can lose a proton to neutral H_2O, creating neutral NH_3 and cationic H_3O^+.

/// Polyatomic ions remain intact in solution.

Figure 3-23 shows a molecular picture of an ammonium nitrate solution, and Example 3-21 treats an aqueous solution that contains polyatomic ions.

Example 3-21	**Molarity of Ions in Solution**

Find the molarities of the ionic species present in 250. mL of an aqueous solution containing 1.75 g of ammonium sulfate, $(NH_4)_2SO_4$.

Strategy: The chemical formula identifies the ions that are present in the final solution. The formula also tells us how many moles of each ion are present in one mole of the salt. Use mass, molar mass, and volume to calculate molarity.

Solution: Looking at the formula, we recognize the ammonium cation, NH_4^+, and the sulfate anion, SO_4^{2-}. When ammonium sulfate dissolves in water, it

| Molarity of Ions in Solution *(continued)* | Example 3-21 |

dissociates into its component polyatomic ions. Each mole of salt will produce a total of 3 mol of ions: 2 mol of NH_4^+ and 1 mol of SO_4^{2-}.

After identifying what species are present in solution, we can calculate molarities. Begin with the molar mass of the salt:

$$(2 \text{ mol N})(14.01 \text{ g/mol}) = 28.02 \text{ g N}$$

$$(8 \text{ mol H})(1.008 \text{ g/mol}) = 8.064 \text{ g H}$$

$$(1 \text{ mol S})(32.06 \text{ g/mol}) = 32.06 \text{ g S}$$

$$\underline{(4 \text{ mol O})(16.00 \text{ g/mol}) = 64.00 \text{ g O}}$$

$$1 \text{ mol } (NH_4)_2SO_4 = 132.14 \text{ g}$$

The molar mass is used to determine the number of moles of salt and, in turn, the number of moles of each ion. Notice that even though we are aiming for molarities of ionic species, we determine moles of the salt. This is because the mass that is provided is the *mass of the salt*. To find moles of salt (*n*), this mass must be divided by the *molar mass of the salt*:

$$n = \frac{m}{MM} = \frac{1.75 \text{ g}}{132.14 \text{ g/mol}} = 1.324 \times 10^{-2} \text{ mol } (NH_4)_2SO_4$$

> Remember to carry an extra significant figure until the final calculation.

From the number of moles of salt, calculate the number of moles of each ion:

$$(1.324 \times 10^{-2} \text{ mol } (NH_4)_2SO_4) \left[\frac{2 \text{ mol } NH_4^+}{1 \text{ mol } (NH_4)_2SO_4} \right] = 2.648 \times 10^{-2} \text{ mol } NH_4^+$$

$$(1.324 \times 10^{-2} \text{ mol } (NH_4)_2SO_4) \left[\frac{1 \text{ mol } SO_4^{2-}}{1 \text{ mol } (NH_4)_2SO_4} \right] = 1.324 \times 10^{-2} \text{ mol } SO_4^{2-}$$

Now find the molarities of the individual ions:

$$(250. \text{ mL})(10^{-3} \text{ L/mL}) = 0.250 \text{ L}$$

$$[NH_4^+] = \frac{2.648 \times 10^{-2} \text{ mol}}{0.250 \text{ L}} = 1.06 \times 10^{-1} \text{ M}$$

$$[SO_4^{2-}] = \frac{1.324 \times 10^{-2} \text{ mol}}{0.250 \text{ L}} = 5.30 \times 10^{-2} \text{ M}$$

> Use dimensional analysis to obtain correct conversion factors. Rearrange 1 mL = 10^{-3} L to give 1 = (10^{-3} L/mL), with mL in the denominator so it cancels mL in the numerator.

This solution is described as "0.0530 M $(NH_4)_2SO_4$", stating moles of salt per liter of solution. However, it is critical to understand that in saying this, we are describing a solution that actually contains *a mixture of ions, each with its own molarity.*

Dilutions

One of the common methods for preparing a solution of known concentration is to dilute an existing solution of higher concentration by adding more solvent. In a **dilution** the *amount of solute* remains the same, but the *volume of the solution* increases. Thus, a dilution results in a solution of lower molarity.

Figure 3-23
Molecular view of an aqueous
solution of ammonium nitrate.
Ammonium cations dissociate
from nitrate anions, but neither of
these polyatomic species breaks
down into smaller pieces.
Instead, they retain their
identities as polyatomic clusters.

Tutorial

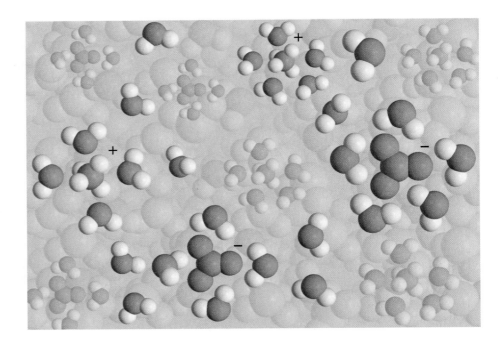

The most important feature of a dilution is that *the number of moles of solute does not change during the dilution*. Consider the dilution of aqueous solution of nickel(II) chloride, as shown in Figure 3-24. The fading of the green color, which is due to the Ni^{2+} ions, gives macroscopic evidence for the change in concentration. At the molecular level, the *number* of Ni^{2+} and Cl^- ions present before the dilution is the same as the number present after the dilution. However, because the volume of the solution increases, the *concentration* of the new solution is lower than the concentration of the original solution.

The quantitative aspects of dilutions can be found from the fact that the number of moles of solute does not change during a dilution:

$$Moles_{solute,initial} = Moles_{solute,final}$$

Figure 3-24
When a concentrated solution of
$NiCl_2$ (*left*) is diluted by adding
more solvent, the resulting
solution (*right*) contains the same
number of Ni^{2+} and Cl^- ions. The
solution is more dilute, however,
because the ions are spread
around in a larger volume of
solvent.

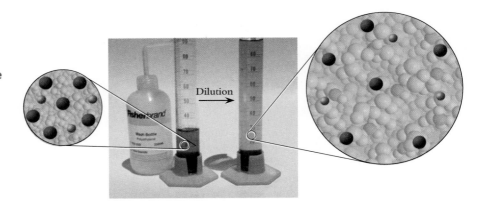

Dilution

Because Moles = (Molarity)(Volume), this leads to a simple equation:

$$M_i V_i = M_f V_f \qquad (3\text{-}3)$$

Equation 3-3 is very convenient for dilution calculations. If any three of the quantities are known, we can calculate the fourth, as Example 3-22 shows.

Dilution of a Solution	Example 3-22

Aqueous hydrochloric acid, HCl, is usually sold as a 12.0-M solution, commonly referred to as *concentrated HCl*. A chemist needs to prepare 2.00 L of 1.50 M HCl for a number of different applications. What volume of concentrated HCl solution should the chemist use in the dilution?

Solving Quantitative Problems

Strategy: The seven-step problem-solving approach is appropriate.

Solution:

1. The question asks for the volume of concentrated HCl required to prepare a dilute solution.
2. Visualization: A concentrated solution is diluted to a larger volume. The chemist will remove a sample from the bottle of concentrated HCl solution and mix it with water.
3. The information available is as follows:

   Concentrated HCl = 12.0 M   Diluted HCl = 1.50 M

   Final volume = 2.00 L

4. This is a dilution process, so we use Equation 3-3.
5. Rearrange Equation 3-3 to solve for V_i, the volume required: $V_i = M_f V_f / M_i$.
6. Do the calculation. In Step 4 we identified the appropriate values:

   $M_f = 1.50$ M $V_f = 2.00$ L $M_i = 12.0$ M $V_i = ?$

   $$V_i = \frac{(1.50\ \text{M})(2.00\ \text{L})}{(12.0\ \text{M})} = 0.250\ \text{L}$$

7. Check for reasonableness: The volume has the correct units, and the amount required is smaller than the total volume. This is reasonable.

The solution is prepared by using a volume-measuring device, such as a graduated cylinder, to measure 0.250 L (250. mL) of concentrated HCl. About 0.5 L of water is added to a 1.00-L volumetric flask, and then the concentrated acid from the graduated cylinder is poured slowly into the water. Filling the volumetric flask to the mark with additional water completes the dilution.

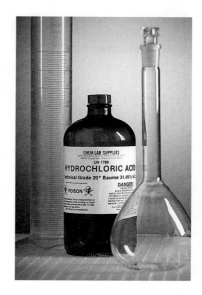

Water should never be added directly to concentrated acid; this causes a vigorous reaction that may lead to a dangerous acid splash. Instead, concentrated acid is always added to water. Additional water can then be added safely to the less concentrated solution that results.

Chemists often need to work with solutions with a variety of concentrations. Often, a relatively concentrated "stock" solution is prepared by weighing. Then, a portion of this solution is diluted volumetrically. In procedures of this kind, it is important to keep track of the volumes, as illustrated in Example 3-23 and shown in the flowchart in Figure 3-25.

| Example 3-23 | Preparing Solutions |

An agricultural chemist wished to study the effect of varying fertilizer applications on the growth of tomato plants. The chemist prepared a "stock" aqueous solution of urea, $(NH_2)_2CO$, by dissolving 1.75 g of this compound in water to make 1.00 L of solution. Then she prepared a series of more dilute solutions to apply to her tomato plants. One of these solutions contained 5.00 mL of "stock" solution diluted to give a final volume of 25.00 mL. What was the concentration of urea in this diluted solution?

Strategy: The problem describes solutions that are prepared and diluted, so the equations for solution concentrations apply. We are asked to find the final concentration of a dilute solution that is prepared in a two-step process. (a) Solid urea is dissolved to make a "stock" solution; (b) A portion of the "stock" solution is diluted. In solving multi-part problems, it is best to work with one part at a time.

Solution: The flowchart in Figure 3-25 helps visualize the process: After the "stock" solution is prepared, a 5.00-mL sample is removed and transferred to an empty 25.00 mL volumetric flask. Water is added to the mark to give the final solution. In order to calculate the concentration of the dilute solution, we need the molarity of the 5.00-mL sample of stock solution. This information comes from the first part of the problem.

Given the mass of urea and the total volume of the stock solution, we can calculate the stock solution's concentration:

$$m_{urea} = 1.75 \text{ g in } V = 1.00 \text{ L}; \quad MM_{urea} = 60.06 \text{ g/mol}$$

$$M = \frac{n}{V} \quad and \quad n = \frac{m}{MM} \quad so \quad M_{stock} = \frac{m}{V(MM)}$$

$$M_{stock} = \frac{1.75 \text{ g}}{(1.00 \text{ L})(60.06 \text{ g/mol})} = 2.914 \times 10^{-2} \text{ M}$$

We are now ready for the second part of the problem, the dilution step. The dilution equation, Equation 3.3, applies:

$$M_i V_i = M_f V_f \quad so \quad M_f = \frac{M_i V_i}{V_f}$$

The chemist removes a 5.00-mL sample of the stock solution; this is V_i. The molarity of the sample is the same as that of the stock solution stock, namely $M_i = 2.914 \times 10^{-2}$ M. Knowing that the final volume is 25.00 mL, we are ready to apply Equation 3-3.

$$M_f = \frac{(2.914 \times 10^{-2} \text{ M})(5.00 \text{ mL})}{(25.00 \text{ mL})} = 5.82 \times 10^{-3} \text{ M}$$

To check for reasonableness, make sure you have used the correct volumes. The *stock solution* was prepared in a volume of 1.00 L. The *dilution* was from 5.00 mL to 25.00 mL, a factor of 5. We can see by inspection that the final concentration is lower than the stock concentration by a factor of about 5.

Precipitation Analysis

Silver nitrate dissolves readily in water to give a solution that contains Ag^+ cations and NO_3^- anions. Sodium chloride also dissolves readily in water to give a solution that contains Na^+ cations and Cl^- anions. In both cases the dissolved cations and anions move freely through their respective solutions. Combining these two solutions mixes the four ions together. Silver chloride is not very soluble in water, so silver cations and chloride anions combine to give solid silver chloride. This white solid separates from the solution in a process called **precipitation.** Figure 3-26 illustrates the precipitation of AgCl.

Precipitation makes it possible to do elemental analysis of elements, such as chlorine, that cannot easily be determined by combustion. If a compound that contains C, H, O, and Cl is burned, the chlorine does not give a single product whose mass can be measured. To complete the elemental analysis, all the chlorine atoms in a separate sample of the compound are converted into chloride ions. These ions are dissolved in water, and an excess amount of silver nitrate solution is added. This causes all the Cl^- to precipitate as solid AgCl. The chlorine content in the original sample can be calculated from the mass of the AgCl precipitate. The key feature of the analysis is that the number of moles of chloride in the AgCl equals the number of moles of Cl in the original sample. Example 3-24 shows how precipitation analysis, mass spectroscopy, and combustion analysis can be used to determine the molecular formula of a compound.

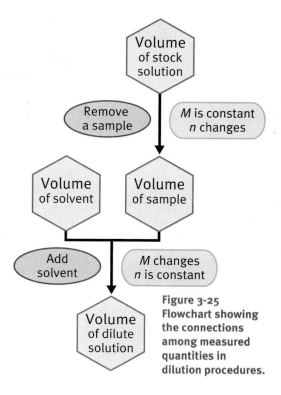

Figure 3-25
Flowchart showing the connections among measured quantities in dilution procedures.

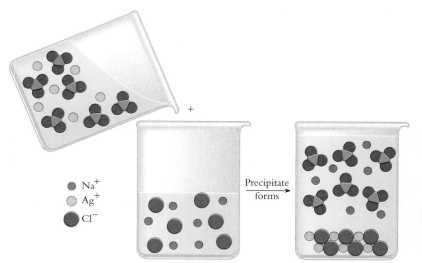

Na$^+$
Ag$^+$
Cl$^-$

Precipitate forms

Figure 3-26
A molecular view of the precipitation of silver chloride, which occurs when solutions of silver nitrate and sodium chloride are mixed (water molecules have been omitted for clarity). The photo shows white solid silver chloride precipitating.

Analysis Involving Precipitation	Example 3-24

An environmental waste-disposal company received a drum of a liquid from a cleaning firm. The label had been destroyed, making its identity uncertain. The disposal company analyzed the liquid using mass spectroscopy, combustion analysis, and precipitation analysis. The mass spectrum showed that

| Example 3-24 | Analysis Involving Precipitation *(continued)* |

the compound had a molar mass of 131.4 g/mol and that it contained chlorine. Combustion of a 1.75-g sample gave 0.121 g of H_2O and 1.17 g of CO_2. The chlorine in a separate 0.655-g sample was converted into chloride ions and precipitated by treatment with aqueous silver nitrate, giving 2.16 g of AgCl. Determine the molecular formula of the compound.

Strategy: We are asked to determine a molecular formula. The analysis has two parts, which can be visualized using a two-stage flowchart.

The results of the combustion analysis provide the carbon and hydrogen content of the unknown, and the precipitation results give the chlorine content. These two experiments lead to the empirical formula, which can be combined with the molar mass to determine the molecular formula of the unknown. It is best to analyze each experiment individually.

Solution: First, use the data from the combustion analysis to determine the mass percent of C and H:

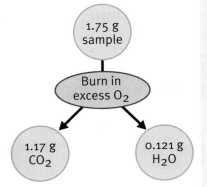

$$(0.121 \text{ g } H_2O)\left[\frac{(2 \text{ mol})(1.008 \text{ g/mol})}{18.02 \text{ g } H_2O}\right] = 1.35 \times 10^{-2} \text{ g H}$$

$$\left(\frac{1.35 \times 10^{-2} \text{ g H}}{1.75 \text{ g sample}}\right)(100\%) = 0.771\% \text{ H}$$

$$(1.17 \text{ g } CO_2)\left[\frac{(1 \text{ mol})(12.01 \text{ g/mol})}{44.01 \text{ g } CO_2}\right] = 3.19 \times 10^{-1} \text{ g C}$$

$$\left(\frac{3.19 \times 10^{-1} \text{ g C}}{1.75 \text{ g sample}}\right)(100\%) = 18.2\% \text{ C}$$

Use the same procedure to determine the mass percent of chlorine, but remember that the precipitation is a separate experiment. The mass of sample used for the chloride precipitation is different from the mass of sample used for the combustion analysis:

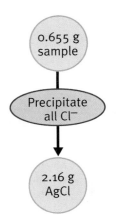

$$(2.16 \text{ g AgCl})\left[\frac{(1 \text{ mol Cl})(35.45 \text{ g/mol})}{143.3 \text{ g AgCl}}\right] = 5.34 \times 10^{-2} \text{ g Cl}$$

$$\left(\frac{5.34 \times 10^{-1} \text{ g Cl}}{0.655 \text{ g sample}}\right)(100\%) = 81.5\% \text{ Cl}$$

Next, add the percentages to see whether all the elements in the compound are accounted for:

$$81.5\% + 18.2\% + 0.77\% = 100.5\%.$$

The result is 100% within the accuracy of the measurements. This tells us that the compound contains no additional elements.

Now convert elemental percentages into moles per 100-g sample:

$$\frac{0.771 \text{ g H}}{1.008 \text{ g/mol}} = 0.765 \text{ mol H} \qquad \frac{18.2 \text{ g C}}{1.01 \text{ g/mol}} = 1.515 \text{ mol C}$$

$$\frac{81.5 \text{ g Cl}}{35.45 \text{ g/mol}} = 2.299 \text{ mol Cl}$$

Analysis Involving Precipitation *(continued)*

Example 3-24

Divide each by the smallest value, which is 0.765 mol H:

$$\frac{0.765 \text{ mol H}}{0.765 \text{ mol H}} = 1.00 \qquad \frac{1.515 \text{ mol C}}{0.765 \text{ mol H}} = 1.98, \text{ round to 2}$$

$$\frac{2.299 \text{ mol Cl}}{0.765 \text{ mol H}} = 3.01, \text{ round to 3}$$

The empirical formula of the liquid is C_2HCl_3. The mass that corresponds to this formula is 131.4 g/mol. This matches the molar mass determined by mass spectroscopy. The compound is trichloroethylene, used widely as a cleaning solvent.

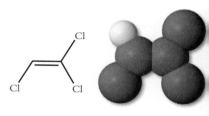

Trichloroethylene

Section Exercises

3.7.1   An urban gardener buys a 1.0-kg bag of ammonium nitrate to fertilize house plants. Compute the concentrations of the ionic species present in a stock solution prepared by dissolving 7.5 g of the salt in 0.300 L of water.

3.7.2   Plants can be badly damaged by solutions that are highly concentrated. A friend of the urban gardener recommends against using fertilizer solutions that are greater than 1.0 mM (1.0×10^{-3} M) in total nitrogen. How many milliliters of the stock solution from Section Exercise 3.7.1 will be needed to prepare 750 mL of 1.0 mM fertilizer solution?

3.7.3   Draw a molecular picture of a portion of the solution described in Section Exercise 3.7.2. Make sure the solution is electrically neutral (omit water molecules to simplify your drawing).

▪▪▪ CHAPTER REVIEW

Summary and Key Terms

1.   A **chemical formula** describes the atomic composition of a substance. Molecules are held together by **bonds.** The formula of a **binary compound** follows specific guidelines. A chemical substance can be represented by its **structural formula**, a **ball-and-stick model**, a **space-filling model**, or a **line structure**.

2.   Each class of compounds, such as binary, carbon-containing, or ionic, is named according to systematic guidelines.

3.   Atomic cations form when atoms lose electrons, and anions form when atoms gain electrons. **Polyatomic ions** are groups of atoms that carry electric charge. Polyatomic anions containing oxygen are called **oxyanions.** Ionic compounds usually contain Group 1 or Group 2 metals or polyatomic ions. **Hydrates** are solids that include water molecules.

4.   Amounts of atoms and molecules are measured using the **mole (mol)**, which is the number of atoms in exactly 12 g of ^{12}C. A mole contains the **Avogadro constant**, 6.022×10^{23}, of objects. The **molar mass (MM)** of a substance is the mass in grams of one mole of its atoms or molecules. Amount calculations in chemistry center around the mole. Chemical formulas are determined from mass percentages, which in turn are determined by

careful measurements of the masses of known compounds produced when a substance undergoes decomposition, combustion, or precipitation reactions.

5.   Mole-mass conversions are made using molar masses, and mole-number conversions are made using the Avogadro constant.

6.   **Elemental analysis** or **mass percent composition** describes the mass composition of a substance. The **empirical formula** of a compound can be determined from its mass composition, and vice versa. Decomposition analysis and **combustion analysis** are methods for finding the empirical formula of a compound.

7.   A solution contains a **solvent** and varying amounts of **solutes. Aqueous** (water as solvent) solutions dominate much of chemistry. Amounts of solutes in solution are expressed using the **concentration,** expressed as moles of solute per liter of solution, or **molarity (M)**. Ionic solutions contain cations and anions as solute species but are neutral overall. Solution concentrations can be varied by **dilution,** which is the addition of solvent to a solution. **Precipitation** of an insoluble salt can result when two solutions are mixed.

Skills to Master

▶ Drawing various types of chemical structures

▶ Naming binary and ionic compounds

▶ Drawing molecular pictures

▶ Recognizing ionic compounds

▶ Converting among mass, moles, and number of atoms or molecules

▶ Calculating molar masses

▶ Converting between mass percentages and chemical formulas

▶ Determining empirical and molecular formulas

▶ Calculating solution molarities

Learning Exercises

3.1 List the various ways of representing molecules. Describe how they differ from one another.

3.2 Explain how to determine whether a chemical compound is ionic or whether it shares electrons.

3.3 Write a paragraph that defines molar mass, mole, and molarity and explains the differences among them.

3.4 Diagram the process for converting from the mass of a compound of a known chemical formula to the number of atoms of one of its constituent elements. Include all necessary equations and conversion factors.

3.5 Explain in words the reasoning used to deduce an empirical formula from combustion analysis of a compound containing C, H, and O.

3.6 Describe what a solution of magnesium nitrate looks like to a molecular-sized observer.

3.7 Update your list of "memory bank" equations.

3.8 List all the terms in Chapter 3 that are new to you. Using your own words, write a one-sentence definition of each. Consult the glossary if you need help.

Problems   ilw = interactive learning ware problem. Visit the website at www.wiley.com/college/olmsted

Representing Molecules

3.1 Write chemical formulas for the molecules whose ball-and-stick models follow:

3.2 Write chemical formulas for the molecules whose ball-and-stick models follow:

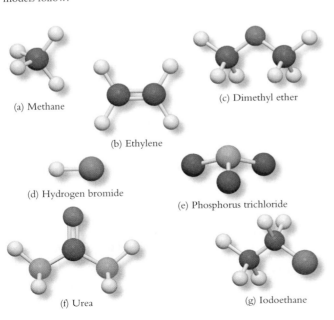

(a) Methane

(b) Ethylene

(c) Dimethyl ether

(d) Hydrogen bromide

(e) Phosphorus trichloride

(f) Urea

(g) Iodoethane

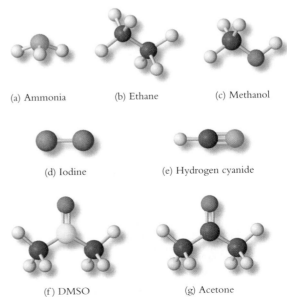

(a) Ammonia

(b) Ethane

(c) Methanol

(d) Iodine

(e) Hydrogen cyanide

(f) DMSO

(g) Acetone

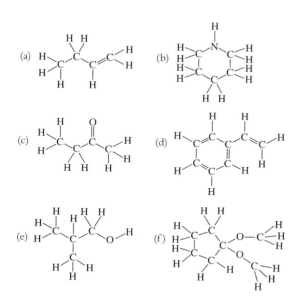

3.3 Write structural formulas for the molecules in Problem 3.1.

3.4 Write structural formulas for the molecules in Problem 3.2.

3.5 Convert the following structural formulas into line structures:

3.6 Convert the following structural formulas into line structures:

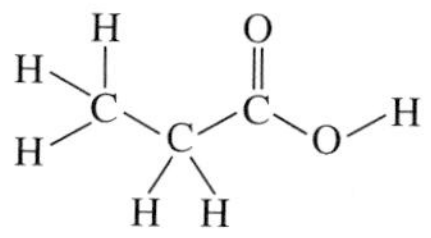

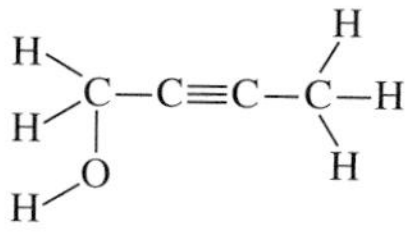

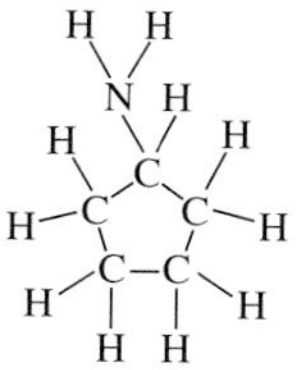

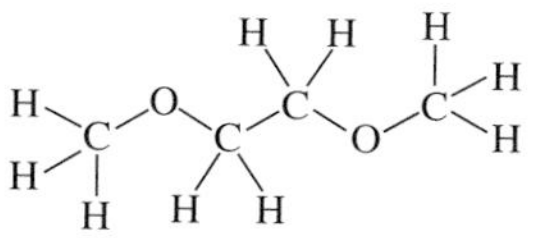

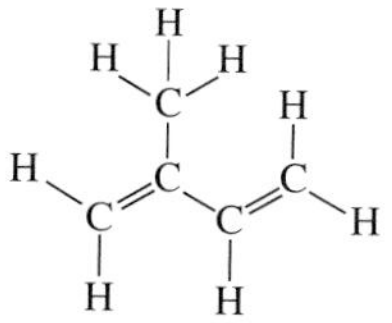

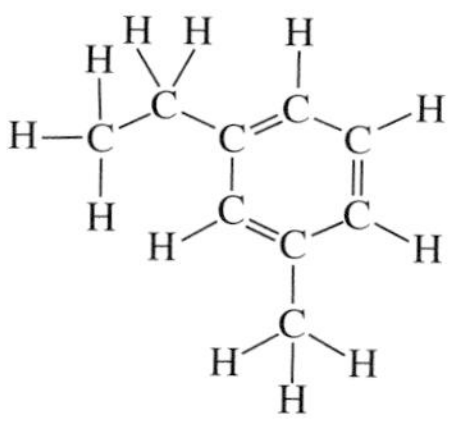

3.7 Convert the following line structures into structural formulas:

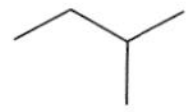

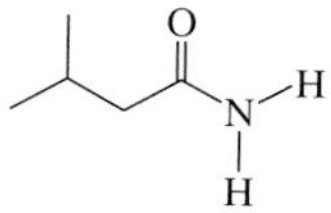

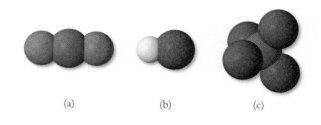

3.8 Convert the following line structures into structural formulas:

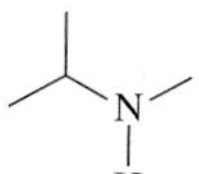

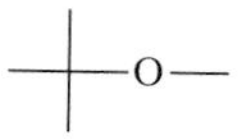

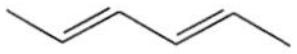

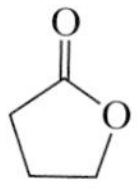

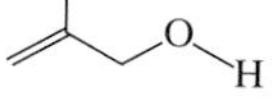

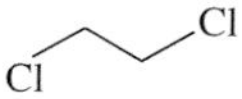

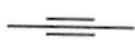

Naming Chemical Compounds

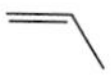

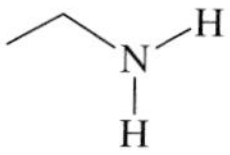

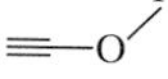

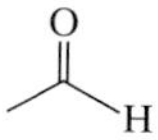

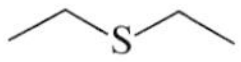

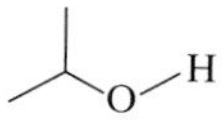

3.9 Write chemical formulas and names for the compounds whose space-filling models follow:

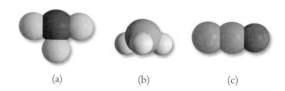

(a)   (b)   (c)

3.10 Write chemical formulas and names for the compounds whose space-filling models follow:

(a)   (b)   (c)

3.11 Write chemical formulas and names for the compounds whose ball-and-stick models follow:

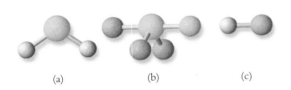

(a)   (b)   (c)

3.12 Write chemical formulas and names for the compounds whose ball-and-stick models follow:

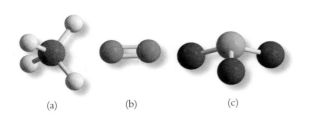

(a)   (b)   (c)

3.13 Write chemical formulas for these compounds: (a) methane; (b) hydrogen fluoride; (c) calcium hydride; (d) phosphorus trichloride; (e) dinitrogen pentoxide; (f) sulfur hexafluoride; and (g) boron trifluoride.

3.14 Write chemical formulas for these compounds: (a) ammonia; (b) hydrogen sulfide; (c) 2-chloropropane; (d) silicon dioxide; (e) molecular nitrogen; (f) xenon tetrafluoride; and (g) bromine pentafluoride.

3.15 Name the following compounds: (a) S_2Cl_2; (b) IF_7 ; (c) HBr; (d) N_2O_3; (e) SiC; and (f) CH_3OH.

3.16 Name the following compounds: (a) XeF_2; (b) $GeCl_4$; (c) N_2F_4; (d) LiH; (e) SeO_2; and (f) CH_3CH_2OH.

Formulas and Names of Ionic Compounds

3.17 Which of the following compounds are ionic? Write the formula of each compound. (a) hydrogen fluoride; (b) calcium fluoride; (c) aluminum sulfate; (d) ammonium sulfide; (e) sulfur dioxide; and (f) carbon tetrachloride.

3.18 Which of the following compounds are ionic? Write the formula of each compound. (a) manganese(II) acetate; (b) sodium hypochlorite; (c) silicon tetrachloride; (d) lithium periodate; (e) magnesium bromide; and (f) hydrogen selenide.

3.19 Which of the following compounds are ionic? Name each compound. (a) CH_2Cl_2; (b) CO_2; (c) CaO; (d) K_2CO_3; (e) PBr_3; (f) HF; and (g) Na_2HPO_4.

3.20 Which of the following compounds are ionic? Name each compound. (a) $(NH_4)_2SO_4$; (b) KBr; (c) SF_6; (d) H_2S; (e) Na_2S; (f) NH_3; and (g) C_2H_6.

3.21 Write chemical formulas for these compounds: (a) sodium sulfate; (b) potassium sulfide; (c) potassium dihydrogen phosphate; (d) cobalt(II) fluoride tetrahydrate; (e) lead(IV) oxide; (f) sodium hydrogen carbonate; and (g) lithium perbromate.

3.22 Write chemical formulas for these compounds: (a) potassium chlorate; (b) ammonium hydrogen carbonate; (c) iron(II) phosphate; (d) copper(II) nitrate hexahydrate; (e) aluminum chloride; (f) cadmium(II) chloride; and (g) potassium oxide.

3.23 Name the following compounds: (a) $CaCl_2 \cdot 6H_2O$; (b) $Fe(NH_4)_2(SO_4)_2$; (c) K_2CO_3; (d) $SnCl_2 \cdot 2H_2O$; (e) NaClO; (f) Ag_2SO_4; (g) $CuSO_4$; (h) KH_2PO_4; (i) $NaNO_3$; (j) $CaSO_3$; and (k) $KMnO_4$.

3.24 Name the following compounds: (a) $K_2Cr_2O_7$; (b) $NaNO_2$; (c) $Mg_3(PO_4)_2$; (d) $CrCl_3$; (e) V_2O_3; (f) $KHSO_4$; (g) CsBr; (h) $In(NO_3)_3 \cdot 5H_2O$; (i) $Al(ClO_4)_3$; (j) $SnCl_4$; and (k) $TaCl_5$.

The Mole

3.25 Calculate the number of moles in the following masses: (a) 7.85 g of Fe; (b) 65.5 µg of carbon; (c) 4.68 mg of Si; and (d) 1.46 metric tons of Al (1 metric ton = 10^3 kg).

3.26 Calculate the number of moles in the following masses: (a) 3.67 kg of titanium; (b) 7.9 mg of calcium; (c) 1.56 g of ruthenium; and (d) 9.63 pg of technetium.

3.27 Use the data in the following table to calculate the molar mass of naturally-occurring argon:

Isotope	Isotopic Molar Mass	Abundance
^{36}Ar	35.96755 g/mol	0.337%
^{38}Ar	37.96272 g/mol	0.063%
^{40}Ar	39.9624 g/mol	99.600%

3.28 Use the data in the following table to calculate the molar mass of naturally occurring silicon:

Isotope	Isotopic Molar Mass	Abundance
^{28}Si	27.97693 g/mol	92.23%
^{29}Si	28.97649 g/mol	4.67%
^{30}Si	29.97376 g/mol	3.10%

3.29 Calculate the molar mass of each of the following substances: (a) carbon tetrachloride; (b) potassium sulfide; (c) O_3 (ozone); (d) lithium bromide; (e) GaAs (a semiconductor); and (f) silver nitrate.

3.30 Calculate the molar mass of each of the following compounds: (a) ammonium carbonate; (b) N_2O (laughing gas); (c) calcium carbonate; (d) NH_3 (ammonia); (e) sodium sulfate; and (f) C_4H_{10} (butane).

3.31 Determine the molecular formula and calculate the molar mass of each of the following essential amino acids.

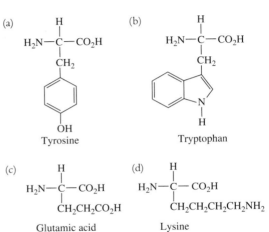

(a) Tyrosine   (b) Tryptophan
(c) Glutamic acid   (d) Lysine

3.32 Determine the molecular formula and calculate the molar mass of each of the following B vitamins.

(a) Biotin

(b) Nicotinamide

(c) Pyridoxamine

(d) Pantothenic acid

Mass-Mole-Number Conversions

3.33 Calculate the number of atoms present in 5.86 mg of each of the following elements: (a) beryllium; (b) phosphorus; (c) zirconium; and (d) uranium.

3.34 The U.S. Recommended Daily Allowances (RDAs) of several vitamins follow. In each case, calculate how many molecules are in the U.S. RDA. (a) 60. mg vitamin C, $C_6H_8O_6$; (b) 400. mg folic acid, $C_{16}H_{22}N_7O_5$; (c) 1.5 mg vitamin A, $C_{20}H_{30}O$; and (d) 1.70 mg vitamin B_2 (riboflavin), $C_{17}H_{20}N_4O_6$.

3.35 Calculate the mass of each of the following: (a) 3.75×10^5 molecules of methane; (b) 2.5×10^9 molecules of adrenaline, $C_9H_{13}NO_3$; and (c) one molecule of chlorophyll, $C_{55}H_{72}MgN_4O_5$.

3.36 Calculate the mass of each of the following: (a) 1.0×10^{15} molecules of ozone; (b) 5.000×10^3 molecules of cholesterol, $C_{27}H_{46}O$; and (c) one molecule of Vitamin B_{12}, $C_{63}H_{88}CoN_{14}O_{14}P$.

ilw 3.37 One carrot may contain 0.75 mg of vitamin A ($C_{20}H_{30}O$). How many moles of vitamin A is this? How many molecules? How many hydrogen atoms are in 0.75 mg of vitamin A? What mass of hydrogen is this?

3.38 A particular oral contraceptive contains 0.035 mg ethynyl estradiol in each pill. The formula of this compound is $C_{20}H_{24}O_2$. How many moles of ethynyl estradiol are there in one pill? How many molecules is this? How many carbon atoms are in a 0.035-mg sample of ethynyl estradiol? What mass of carbon is this?

Determining Chemical Formulas

3.39 Portland cement contains CaO, SiO_2, Al_2O_3, and Fe_2O_3. Calculate the mass percent composition of each compound.

3.40 Small amounts of the following compounds are added to glass to give it color: CaF_2 (milky white), MnO_2 (violet),

CoO (blue), Cu_2O (green). Calculate the mass percent composition of each additive.

3.41 Nicotine is an addictive compound found in tobacco leaves. Elemental analysis of nicotine gives the following data: C: 74.0%, H: 8.65%, N: 17.35%. What is the empirical formula of nicotine? The molar mass of nicotine is 162 g/mol. What is the molecular formula of nicotine?

3.42 Tooth enamel is composed largely of hydroxyapatite, which has the following mass percent composition: O: 41.41%, P: 18.50%, H: 0.20%, Ca: 39.89%. Calculate the empirical formula of hydroxyapatite. The molar mass of hydroxyapatite is 1004 g/mol. What is the molecular formula of hydroxyapatite?

3.43 A chemical compound was found to contain only Fe, C, and H. When 5.00 g of this compound was completely burned in O_2, 11.8 g of CO_2 and 2.42 g of H_2O were produced. Find the percent by mass of each element in the original compound and its empirical formula.

3.44 A 1.45-gram sample of a compound containing C, H, and O was burned completely in excess oxygen. It yielded 1.56 g of H_2O and 3.83 g of CO_2. Find the percent by mass of each element in the original compound and its empirical formula.

ilw 3.45 The medicinal properties of garlic have been known for many centuries. These properties, as well as garlic's obvious odor, can be attributed to a variety of sulfur-containing molecules. One such molecule is allicin, a potent antibacterial agent that is effective in the treatment of typhus. Allicin's clinical use has been abandoned, however, because of its potent odor. From 4.0 kg of garlic, a chemist isolates 6.0 g of allicin. Burning 5.00 mg of allicin produces 8.13 mg CO_2, 3.95 mg SO_2, and 2.76 mg H_2O. Mass spectroscopy shows that the molar mass of allicin is about 160 g/mol. Determine the molecular formula of allicin.

3.46 Hemlock is a poisonous herb of the carrot family. The ancient Greeks used hemlock extracts for state executions. Socrates was killed in this way. In a combustion experiment, 50.50 mg of the poisonous component of hemlock is burned in excess oxygen. The products are 139.9 mg CO_2 and 60.91 mg H_2O. In a second experiment, 75.62 mg of the compound was found to contain 8.35 mg of nitrogen. Mass spectroscopy indicates that the molar mass is less than 200 g/mol. Determine the molecular formula of the poisonous compound found in hemlock.

Aqueous Solutions

3.47 (a) Calculate the molarities of the major ionic species in 1.50×10^2 mL of aqueous solution that contains 4.68 g of magnesium chloride. (b) Draw a molecular picture that shows a portion of this solution, showing the relative proportions of the ions present.

3.48 A chemist places 3.25 g of sodium carbonate in a 250.-mL volumetric flask and fills it to the mark with water. (a) Calculate the molarities of the major ionic species. (b) Draw a molecular picture that shows a portion of this solution, making sure the portion is electrically neutral.

ilw 3.49 A student prepares a solution by dissolving 4.75 g of solid KOH in enough water to make 275 mL of solution. (a) Calculate the molarities of the major ionic species present. (b) Calculate the molarities of the major ionic species present if 25.00 mL of this solution is added to a 100.-mL volumetric flask, and water is added to the mark. (c) Draw molecular pictures of portions of the solutions in (a) and (b), showing how they differ.

3.50 A mildly antiseptic mouthwash can be prepared by dissolving sodium chloride in water. (a) Calculate the molarities of the ionic species present in 0.150 L of solution containing 27.0 g of sodium chloride. (b) Calculate the new molarities if 50.0 mL of this solution is diluted with water to give 450. mL of a new solution. (c) Draw molecular pictures of portions of the solutions described in (a) and (b), showing how they differ.

3.51 Concentrated hydrochloric acid is 12.1 M. What volume of this solution should be used to prepare 0.500 L of 0.125 M HCl?

3.52 Concentrated ammonia is 14.8 M. What volume of this solution should be used to prepare 1.25 L of 0.500 M NH_3?

3.53 Calculate the concentrations of the major ionic species present in each of the following solutions: (a) 4.55 g Na_2CO_3 in 245 mL of solution; (b) 27.45 mg of NH_4Cl in 1.55×10^{-2} L of solution; (c) 1.85 kg potassium sulfate in 5.75×10^3 L of solution.

3.54 Calculate the concentrations of the major ionic species present in each of the following solutions: (a) 1.54 g sodium hydrogen carbonate in 75.0 mL of solution; (b) 1.44 mg $FeCl_3$ in 2.75 mL of solution; (c) 8.75 kg KNO_3 in 235 L of solution.

3.55 What mass of silver chloride will be recovered if excess sodium chloride solution is added to 595 mL of an aqueous solution containing 1.75×10^{-2} M Ag^+?

3.56 What mass of barium sulfate will be recovered if excess barium chloride solution is added to 775 mL of an aqueous solution containing 4.66×10^{-1} M sulfate ions?

3.57 A pure substance was known to contain only C, H, and Cl. When a 4.00-g sample was burned, 4.34 g of CO_2 was produced. In a separate experiment, the chlorine contained in 0.125 g of the compound was converted to 0.334 g of AgCl. Determine the empirical formula of the compound.

3.58 A student discovered an old bottle labeled "copper sulfate, hydrated" on a laboratory shelf. When the student heated 0.500 g of this compound in an oven for 1 hour, it lost its waters of hydration. The resulting residue had a mass of 0.320 g. When the student dissolved 0.300 g of the compound in water and added excess $BaCl_2$ solution, 0.280 g of solid $BaSO_4$ was produced. Find the empirical formula of the hydrated copper sulfate.

Additional Paired Problems

3.59 Cinnabar is an ore of mercury known to contain only Hg and S. When a 0.350-g sample of cinnabar is heated in oxygen, the ore decomposes completely, giving 0.302 g of pure Hg metal. Find the empirical formula of cinnabar.

3.60 When galena, an ore of lead that contains only Pb and S, is heated in oxygen, the ore decomposes to produce pure lead. In one such process, 7.85 g of galena gave 6.80 g of lead. Find the empirical formula of galena.

3.61 The line structures of three different plant growth hormones are given below. For each one, write the chemical formula for the compound and calculate its molar mass.

(a)

Abscisic acid
(inhibits germination)

(b)

Indole acetic acid
(promotes growing shoots)

(c)

Zeatin
(promotes root growth)

3.62 The following molecules are known for their characteristic fragrances. For each one, convert the line structure into a complete structural formula, and calculate its molar mass.

(a)

Benzaldehyde
(cherry)

(b)

Methylbutyl acetate
(banana)

(c)

Jasmone
(jasmine)

(d)

Limonene
(lemon)

(e)

Vanillin
(vanilla)

3.63 The following pairs of substances are quite different despite having similar names. Write correct formulas for each. (a) sodium nitrite and sodium nitrate; (b) potassium carbonate and potassium hydrogen carbonate; (c) iron(II) oxide and iron(III) oxide; (d) iodine and iodide ion.

3.64 The following pairs of substances are quite different despite having similar names. Write correct formulas for each. (a) Sodium chloride and sodium hypochlorite; (b) nitrogen oxide and nitrogen dioxide; (c) potassium chlorate and potassium perchlorate; and (d) ammonia and ammonium ion.

3.65 Aluminum sulfate is $Al_2(SO_4)_3$. (a) Compute its molar mass. (b) Compute the number of moles contained in 25.0 g of this compound. (c) Determine its percent composition. (d) Determine the mass of this compound that contains 1.00 mol of O.

3.66 Nickel sulfate hexahydrate is $NiSO_4 \cdot 6H_2O$. (a) Compute its molar mass. (b) Compute the number of moles contained in 25.0 g of this compound. (c) Determine its percent composition. d) Determine the mass of this compound that contains 1.00 mol of O.

3.67 Answer the following questions about Sevin, whose chemical formula is $C_{12}H_{11}NO_2$: (a) How many moles of carbon atoms are there in 8.3 g of Sevin? (b) How many grams of oxygen are there in 4.5 g of Sevin? (c) The label on a 75-mL bottle of garden insecticide states that the solution contains 0.010 % Sevin and 99.99% inert ingredients. How many moles and how many molecules of Sevin are in the bottle? (Assume the density of the solution is 1.00 g/mL.) (d) The instructions for the bottle of insecticide from part (c) say to dilute the insecticide by adding 1.0 mL of the solution to 1.0 gallon of water. If you spray 15 gallons of the diluted insecticide mixture on your rose garden, how many moles of Sevin are dispersed?

3.68 Answer the following questions about Malathion, whose chemical formula is $C_{10}H_{19}O_6PS_2$: (a) How many sulfur atoms are there in 6.5 g of Malathion? (b) How many grams of oxygen are there in 17.8 g of Malathion? (c) A 200.0-mL container of Malathion is 50% by mass Malathion and 50% inert ingredients (mostly water). What is the molarity of Malathion in this solution (assume a density of 1.00 g/mL)? (d) When dispensed using a hose-end sprayer, this solution of Malathion becomes diluted by a factor of 10^4. If the sprayer delivers 25 mL of this diluted solution onto the leaves of a fruit tree, how many moles of Malathion have been applied?

3.69 A copper atom has a diameter of 127.8 pm. What is the length in meters of a line containing 1.000 mol of copper atoms?

3.70 A dollar bill is 0.10 mm thick. How high would a stack of 1.00 mol of dollar bills be?

3.71 In everyday life, we encounter chemicals with unsystematic names. What is the name a chemist would use for each of the following substances? (a) dry ice (CO_2); (b) saltpeter (KNO_3); (c) salt (NaCl); (d) baking soda ($NaHCO_3$); (e) soda ash (Na_2CO_3); (f) lye (NaOH); (g) lime (CaO); and (h) milk of magnesia ($Mg(OH)_2$).

3.72 A mineral is a chemical compound found in the Earth's crust. What are the chemical names of the following minerals? (a) TiO_2 (rutile); (b) PbS (galena); (c) Al_2O_3 (bauxite); (d) $CaCO_3$ (limestone); (e) $BaSO_4$ (barite); (f) $Mg(OH)_2$ (brucite); (g) HgS (cinnabar); and (h) Sb_2S_3 (stibnite).

3.73 Calculate the mass percentages of all of the elements in each of the following minerals that are found in meteorites: (a) fayalite: Fe_2SiO_4; (b) albite: $NaAlSi_3O_8$; (c) kaolinite: $Al_2Si_2O_5(OH)_4$; and serpentine silicate: $MgSi_4O_{10}(OH)_8$.

3.74 Calculate the mass percentages of all of the elements in the following semiprecious minerals: (a) lapis lazuli, $Na_4Al_3Si_3O_{12}Cl$;

(b) garnet, $Mg_3Al_2(SiO_4)_3$; (c) turquoise, $CuAl_6(PO_4)_4(OH)_8 \cdot 4H_2O$; and (d) jade, $Ca_2Mg_5Si_8O_{22}F_2$.

3.75 What species are present in solution when the following compounds are dissolved in water? (a) Ammonium sulfate; (b) carbon dioxide; (c) sodium fluoride; (d) potassium carbonate; (e) sodium hydrogen sulfate; and (f) chlorine.

3.76 What species are present in solution when the following compounds are dissolved in water? (a) Sodium dichromate; (b) copper(II) chloride; (c) barium hydroxide; (d) methanol; (e) sodium hydrogen carbonate; and (f) iron(III) nitrate.

3.77 Four commonly used fertilizers are ammonium nitrate (NH_4NO_3), ammonium sulfate [$(NH_4)_2SO_4$], urea [$(NH_2)_2CO$], and ammonium hydrogen phosphate [$(NH_4)_2HPO_4$]. How many kilograms of each of these would be required to provide 1.00 kg of nitrogen?

3.78 What mass percentage of carbon is contained in each of the following fuels: (a) propane (C_3H_8); (b) octane (C_8H_{18}); (c) ethanol (C_2H_5OH); and (d) methane (CH_4)?

3.79 Name the following compounds: (a) NH_4Cl; (b) XeF_4; (c) Fe_2O_3; (d) SO_2; and (e) $KClO_4$.

3.80 Name the following compounds: (a) $KClO_3$; (b) $KClO_2$; (c) KClO; (d) KCl; and (e) Na_2HPO_4.

3.81 Heart disease causes 37% of the deaths in the United States. However, the death rate from heart disease has dropped significantly in recent years, partly because of the development of new drugs for heart therapy by chemists working in the pharmaceutical industry. One of these new drugs is verapamil, used for the treatment of arrhythmia, angina, and hypertension. A tablet contains 120.0 mg of verapamil. Determine the following quantities: (a) the molar mass of verapamil; (b) the number of moles of verapamil in one tablet; and (c) the number of nitrogen atoms in one tablet.

Verapamil
$C_{27}H_{38}O_4N_2$

3.82 Penicillin was discovered in 1928 by Alexander Fleming, who was a bacteriologist at the University of London. This molecule was originally isolated from a mold that contaminated some of Fleming's experiments. Penicillin destroys bacterial cells without harming animal cells, so it has been used as an antibiotic and has saved countless lives. The molecular formula of penicillin G is $C_{16}H_{18}N_2O_4S$. (a) Calculate the molar mass of penicillin G. (b) How many moles of penicillin are in a 50.0-mg sample? (c) How many carbon atoms are in a 50.0-mg sample? (d) What is the mass of sulfur in a 75.0-mg sample?

Penicillin G
$C_{16}H_{18}N_2O_4S$

More Challenging Problems

3.83 Quinine is the bitter-tasting compound in tonic water, and it is also used to prevent malaria. Burning a 0.137-g sample of quinine in excess oxygen gives 0.372 g CO_2 and 0.0910 g H_2O. In a second experiment, a 0.183-g sample of quinine is found to contain 0.0158 g of nitrogen. Determine the chemical formula of quinine, given that its molar mass is between 300 and 350 g/mol.

3.84 Sea water at 25 °C contains 8.3 mg/L of oxygen. What molarity is this?

3.85 Vitamin B_{12} is a large molecule called cobalamin. There is one atom of cobalt in each molecule of vitamin B_{12}, and the mass percent of cobalt is 4.34%. Calculate the molar mass of cobalamin.

3.86 Concentrated acetic acid, CH_3CO_2H, is 17.4 M. A laboratory worker measured out 100.0 mL of concentrated acetic acid and added enough water to make 500.0 mL of solution. A 75.0-mL portion of the acetic acid solution was then mixed with enough water to make 1.50 L of dilute solution. What was the final molarity of acetic acid in the dilute solution?

3.87 Calculate the mass of lithium that has the same number of atoms as 5.75 g of platinum.

3.88 The waters of the oceans contain many elements in trace amounts. Rubidium, for example, is present at the level of 2.2 nM (1 nM = 10^{-9} M). How many ions of rubidium are present in 1.00 liter of sea water? How many liters would have to be processed to recover 1.00 kg of rubidium, assuming the recovery process was 100% efficient?

3.89 The seventh-ranked industrial chemical in U.S. production is phosphoric acid, whose chemical formula is H_3PO_4. One method of manufacture starts with elemental phosphorus, which is burned in air; the product of this reaction then reacts with water to give the final product. In 1995, the U.S. manufactured 26.19×10^7 pounds of phosphoric acid. How many moles is this? If 15% of this material was made by burning elemental phosphorus, how many moles and how many kilograms of phosphorus were consumed?

3.90 Police officers confiscate a packet of white powder that they believe contains heroin. Purification by a forensic chemist yields a 38.70-mg sample for combustion analysis. This sample gives 97.46 mg CO_2 and 20.81 mg H_2O. A second sample is analyzed for its nitrogen content, which is 3.8%. Show by calculations whether these data are consistent with the formula for heroin, $C_{21}H_{22}NO_5$.

3.91 When an unknown compound is burned completely in O_2, 1.23 g of CO_2 and 97.02 g of H_2O are recovered. What additional information is needed before the molecular formula of the unknown compound can be determined?

3.92 Aqueous solutions of potassium permanganate are often used in general chemistry laboratories. A laboratory instructor weighed out 474.1 g of $KMnO_4$, added the solid to a 2.00-L volumetric flask, and filled to the mark with water. A 50.00-mL sample was removed from the volumetric flask and mixed with enough water to make 1.500 L of solution. What was the final molarity of the $KMnO_4$ solution?

3.93 The density of water is 1.0 g/mL. Calculate the molarity of water molecules in pure water.

3.94 A worker in a biochemistry laboratory needed a solution that was 0.30 M in sodium acetate ($NaCH_3CO_2$) and 0.15 M in acetic acid (CH_3CO_2H). On hand were stock solutions of 5.0 M sodium acetate and 5.0 M acetic acid. Describe how the worker prepared 1.5 L of the desired solution.

3.95 Adenosine triphosphate (ATP) is used to generate chemical energy in plant and animal cells. The molecular formula of ATP is $C_{10}H_{16}N_5O_{13}P_3$. Determine the mass percent composition of ATP.

3.96 A sulfur-containing ore of copper releases sulfur dioxide when heated in air. A 5.26-g sample of the ore releases 2.12 g of SO_2 on heating. Assuming that the ore contains only copper and sulfur, what is the empirical formula?

3.97 Turquoise has the formula $CuAl_6(PO_4)_4(OH)_8 \cdot 4H_2O$. Answer the following questions about turquoise. (a) How many grams of aluminum are in a 7.25-g sample of turquoise? (b) How many phosphate ions are in a sample of turquoise that contains 5.50×10^{-3} g of oxygen? (c) What is the charge of the copper ion in turquoise?

3.98 A chemist needs a solution that contains aluminum ions, sodium ions, and sulfate ions. Around the lab she finds a large volume of 0.355 M sodium sulfate solution and a bottle of solid $Al_2(SO_4)_3 \cdot 18H_2O$. The chemist places 250. mL of the sodium sulfate solution and 5.13 g of aluminum sulfate in a 500.-mL volumetric flask. The flask is filled to the mark with water. Determine the molarity of aluminum ion, sodium ion, and sulfate ion in the solution.

3.99 Aluminum sulfate is used in the manufacture of paper and in the water purification industry. In the solid state, aluminum sulfate is a hydrate. The formula is $Al_2(SO_4)_3 \cdot 18H_2O$. (a) How many grams of sulfur are there in 0.570 moles of solid aluminum sulfate? (b) How many water molecules are there in a 5.1-g sample of solid aluminum sulfate? (c) How many moles of sulfate ions are there in a sample of solid aluminum sulfate that contains 12.5 moles of oxygen atoms? (d) An aqueous solution of aluminum sulfate contains 1.25% by mass aluminum and has a density of 1.05 g/mL. What is the molarity of aluminum ions in the solution?

3.100 Solution A is prepared by dissolving 90.0 g of Na_3PO_4 in enough water to make 1.5 L of solution. Solution B is 2.5 L of 0.705 M Na_2SO_4. (a) What is the molar concentration of Na_3PO_4 in Solution A? (b) How many mL of Solution A will give 2.50 g of Na_3PO_4? (c) A 50.0-mL sample of Solution B is mixed with a 75.00 mL sample of Solution A. Calculate the concentration of Na^+ ions in the final solution.

3.101 Concentrated aqueous sulfuric acid is 80.% by weight H_2SO_4 and has a density of 1.75 g/mL. (a) Calculate the molarity of the concentrated solution. (b) How many mL of the concentrated solution are needed to make 2.50 L of 0.65 M sulfuric acid?

3.102 Anhydrous copper(II) sulfate is $CuSO_4$, a pale blue solid. Hydrated copper(II) sulfate is a deep blue solid containing five water molecules for every $CuSO_4$ unit: $CuSO_4 \cdot 5H_2O$. How many moles of $CuSO_4$ are there in 125 g of anhydrous copper(II) sulfate? What mass of hydrated copper(II) sulfate contains the same number of moles?

Group Study Problems

3.103 Disodium aurothiomalate ($Na_2C_4H_3O_4SAu$) has the trade name Myochrysine and is used to treat arthritis. (a) What is its molar mass? (b) A solution is prepared by dissolving 0.25 g in 10.0 mL of water. The drug is diluted by adding 2.00 mL of the stock solution to a 10.00-mL volumetric flask and filling to the mark with water. Patients receive 0.40 mL of the dilute Myochrysine solution by intramuscular injection. Calculate the number of grams of Myochrysine administered in each injection. (c) How many moles of Myochrysine is this? (d) During treatment, the concentration of gold can be as high as 30.0 mg per 10.0 mL of serum. How many atoms of gold are in a 5.0-mL sample of serum?

3.104 In the United States, the federal government regulates the amounts of pollutants that industry emits. Emissions in excess of allotted levels are taxed. As an incentive to clean up emissions, companies that emit less than their allotted amounts are allowed to sell their unused portions (so-called emission credits) to other companies. Several years ago, the Tennessee Valley Authority (TVA), which operates 11 coal-fired electricity plants, purchased emission credits from Wisconsin Power and Light, one of the cleanest energy-producing companies in the nation. The TVA bought "pollution rights" for the emission of 1.00×10^4 tons of sulfur dioxide per year at a price of $275 per ton. How much will it cost to emit 1 mol of SO_2? How many molecules can be emitted for $1.00?

3.105 A sample of a component of petroleum was subjected to combustion analysis. An empty vial of mass 2.7534 g was filled with the sample, after which vial plus sample had a mass of 2.8954 g. The sample was burned in a combustion train whose CO_2 trap had a mass

of 54.4375 g and whose H_2O trap had a mass of 47.8845 g. At the end of the analysis, the CO_2 trap had a new mass of 54.9140 g and the H_2O trap had a new mass of 47.9961 g. Determine the empirical formula of this component of petroleum.

3.106 Describe how the results of the analysis in Problem 3.105 would be affected by the following errors in operation of the apparatus. Which measurements would change, and in what direction? How would that affect the calculation of the empirical formula? (a) A limited supply of oxygen was provided, rather than excess oxygen; (b) some of the H_2O leaked through the H_2O trap but was collected in the CO_2 trap; and (c) the H_2O trap functioned properly, but some of the CO_2 leaked through the CO_2 trap.

3.107 A compound was analyzed and found to contain C, H, N, O, and Cl. When a 0.150-g sample of this compound was burned, it produced 0.160 g of CO_2 and 0.0656 g of H_2O. All of the nitrogen in another sample with a mass of 0.200 g was converted to 0.0276 g of NH_3. Finally, the chlorine in a 0.500-g sample of the compound was converted to Cl^- and precipitated as AgCl. The AgCl precipitate had a mass of 0.580 g. Find the empirical formula of the compound.

3.108 A biologist needed to prepare a solution for growing a cell culture. For the cell culture, he needed to make 1.5 L of solution at the following final concentrations: $[KH_2PO_4] = 0.55$ M and $[K_2HPO_4] = 0.85$ M. He made two stock solutions. Solution A: 545 g of KH_2PO_4 was dissolved in enough water to give a volume of 2.0 L. Solution B: 1045 g of K_2HPO_4 was dissolved in enough water to give a total volume of 3.0 L. Describe how to make the solution needed for the cell culture using stock solutions A and B.

Answers to Section Exercises

3.1.1 (a) $C_{18}H_{36}O_2$; (b) $SiCl_4$; and (c) $C_2Cl_3F_3$

3.1.2

3.1.3

3.2.1 ClF, XeO₃, HBr, SiCl₄, SO₂, and H₂O₂

3.2.2 Chlorine trifluoride, hydrogen selenide, chlorine dioxide, antimony trichloride, phosphorus pentachloride, dinitrogen pentoxide, dinitrogen tetrachloride, and ammonia

3.2.3

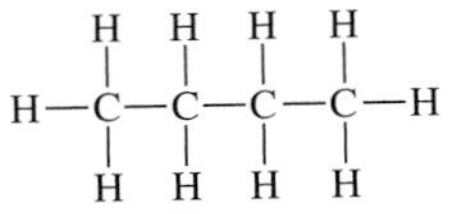

H H H H
| | | |
H—C—C—C—C—H
| | | |
H H H H
Butane

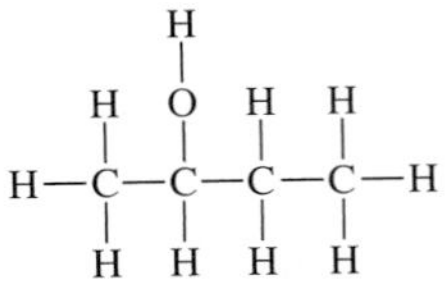

H O H H
| | | |
H—C—C—C—C—H
| | | |
H H H H
2-Butanol

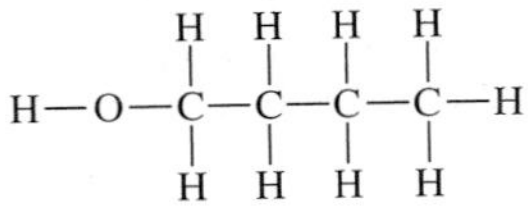

H H H H
| | | |
H—O—C—C—C—C—H
| | | |
H H H H
1-Butanol

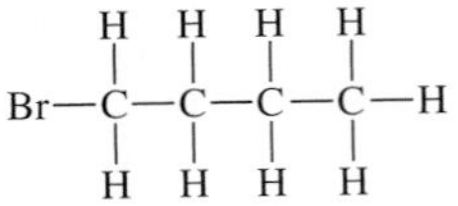

H H H H
| | | |
Br—C—C—C—C—H
| | | |
H H H H
1-Bromobutane

3.3.1 Sulfur dichloride, calcium chloride, lead(II) chloride, sodium nitrate, manganese(IV) oxide, zirconium(IV) chloride, sodium hydride, and sodium iodate

3.3.2 Al₂O₃, K₂Cr₂O₇, Pb(NO₃)₂, NO₂, Na₂SO₄, IF₅, Mn(CH₃CO₂)₂, and NaClO

3.3.3 (a) Sulfur hexafluoride, (b) ammonium nitrate, (c) ammonia, (d) nitrogen trichloride, (e) carbon tetrabromide, and (f) hydrogen iodide

3.4.1 32.06 g/mol

3.4.2 (a) 129.91 g/mol; (b) 184.04 g/mol; (c) 142.98 g/mol; (d) 261.34 g/mol; and (e) 504.83 g/mol

3.4.3 (a) 155.16 g/mol; (b) 121.16 g/mol; (c) 132.12 g/mol

3.5.1 (a) 2.21×10^{23} atoms; (b) 3.21×10^{22} atoms; and (c) 7.80×10^{21} atoms

3.5.2 (a) 9.86×10^{-5} mol; (b) 5.94×10^{19} molecules; (c) 1.78×10^{20} atoms; (d) 2.97×10^{20} atoms; (e) 20.5 mg; and (f) 2.79×10^{21} atoms

3.5.3 4.0×10^{-16} mol and 2.4×10^{8} atoms

3.6.1 C, 40.00%; H, 6.71%; O, 53.29%

3.6.2 Fe₃O₄

3.6.3 C₆H₈O₆

3.7.1 [NH₄⁺] = 0.31 M, [NO₃⁻] = 0.31 M

3.7.2 1.2 mL

3.7.3

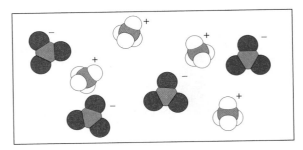

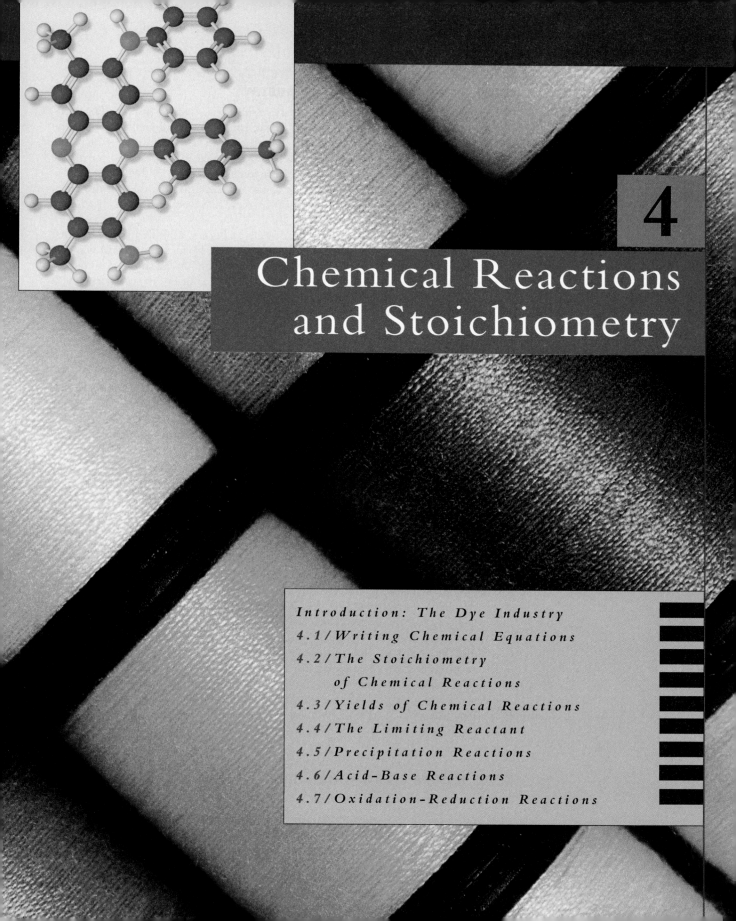

Chemical Reactions and Stoichiometry

4

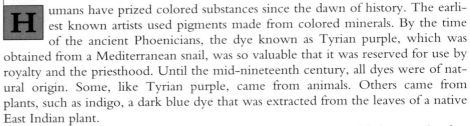

INTRODUCTION: THE DYE INDUSTRY

Humans have prized colored substances since the dawn of history. The earliest known artists used pigments made from colored minerals. By the time of the ancient Phoenicians, the dye known as Tyrian purple, which was obtained from a Mediterranean snail, was so valuable that it was reserved for use by royalty and the priesthood. Until the mid–nineteenth century, all dyes were of natural origin. Some, like Tyrian purple, came from animals. Others came from plants, such as indigo, a dark blue dye that was extracted from the leaves of a native East Indian plant.

In 1856, the young English chemist William Perkin stumbled upon the first synthetic dye. Following a speculative suggestion made by his boss, Perkin was trying to synthesize quinine, a valuable anti–malaria drug. In his home laboratory, Perkin treated a chemical obtained from coal tar with a variety of substances, with no success. As he was about to discard the residue from yet another failed reaction, Perkin noticed that it was colored with a purple tinge. He washed the residue with hot alcohol and obtained a purple solution from which strikingly beautiful purple crystals precipitated. Perkin had no idea what the substance was nor what reactions had created it, but he immediately saw its potential as a new dye.

Perkin turned his full attention to producing this new dye, using family money to establish a factory. Within six months he was producing the dye commercially. The dye was named "mauveine," and the color is known as "mauve." Today we know the structure of mauveine, which appears in the inset. We also know that Perkin was very lucky, because mauveine was produced from an impurity in the chemicals that Perkin was studying.

Mauveine was an immediate hit with French textile dyers, allowing Perkin to retire wealthy at the age of 35, having founded the first industry based on a synthetic chemical. Within 20 years of Perkin's discovery, the German chemist Johann Baeyer developed a method to synthesize indigo in the laboratory. Soon thereafter the German dye industry blossomed. Thus, both the English and the German chemical industries began with syntheses of dyes.

These early efforts to synthesize useful chemicals proceeded by a combination of experience, skill, and luck. Contemporary synthetic chemists know detailed information about molecular structures and can use sophisticated computer programs to simulate a synthesis before trying it in the laboratory. Still, designing a chemical synthesis is a challenging task that requires creativity and a thorough understanding of molecular structure and reactivity. No matter how complex, every chemical synthesis is built on the principles and concepts of general chemistry. One key principle is that quantitative relationships always connect the amounts of materials consumed and the amounts of products formed in a chemical reaction. We can use these relationships to calculate the amounts of materials needed to make a desired amount of product and to analyze the efficiency of a chemical synthesis. The quantitative description of chemical reactions is the focus of Chapter 4.

4.1 WRITING CHEMICAL EQUATIONS

Balanced Equations

In chemical reactions, *the amount of each element is always conserved.* This is one of the postulates of Dalton's atomic theory. In addition, *the total amount of electrical charge is always conserved.* This is the law of conservation of charge. A **balanced chemical**

equation describes a chemical reaction in which the amounts of all elements and of electrical charge are conserved. In addition to providing a list of ingredients for the reaction and identifying the products of the reaction, a balanced chemical equation tells us the relative amounts of all species involved in the reaction.

As an example, consider the industrial synthesis of ammonia. Ammonia (NH_3) is made by the Haber process, a single chemical reaction between molecules of hydrogen (H_2) and nitrogen (N_2). Although it is simple, this synthesis has immense industrial importance. The United States produces more than 16 billion kilograms of ammonia annually.

In this synthesis, the reactants are N_2 and H_2, which react to produce NH_3. Here is the balanced chemical equation:

Reactants   **Products**

N_2 + $3 H_2$ ⟶ $2 NH_3$

Nitrogen   plus   hydrogen   yields   ammonia.

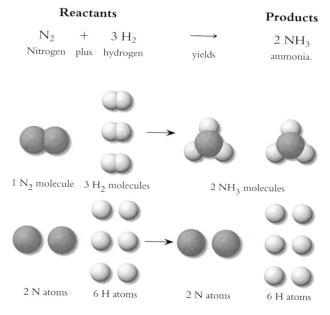

1 N_2 molecule   3 H_2 molecules   2 NH_3 molecules

2 N atoms   6 H atoms   2 N atoms   6 H atoms

All balanced chemical equations have the following features:

1. The **reactants** appear on the left, and the **products** appear on the right. The arrow joining them indicates the direction of reaction.

2. An integer precedes the formula of each substance. These numbers are the **stoichiometric coefficients**. When no number appears (as for N_2 in this equation), the stoichiometric coefficient is 1.

3. The stoichiometric coefficients in a chemical equation are the smallest integers that give a balanced equation.

4. Charge is conserved. In this equation, all participants are neutral species, so charge is conserved regardless of the stoichiometric coefficients.

In Section 4.5 we show how to apply charge conservation to reactions that include ions.

Stoichiometric coefficients describe the relative numbers of molecules involved in the reaction. In any actual reaction, immense numbers of molecules are involved, but the relative numbers are always related through the stoichiometric coefficients. Further, these coefficients describe both the relative numbers of *molecules* and the relative numbers of *moles* involved in the reaction. For example, the

Haber reaction always involves immense numbers of molecules, but the equation describing the synthesis of ammonia tells us the following:

a. Each molecule of nitrogen reacts with three molecules of hydrogen to give two molecules of ammonia.

b. Each mole of nitrogen reacts with three moles of hydrogen to give two moles of ammonia.

Keep in mind that the key feature of balanced chemical equations is the conservation law:

*/// **The number of atoms of each element is conserved in any chemical reaction.***

Balancing Equations

The stoichiometric coefficients in a balanced chemical equation must be chosen so that the atoms of each element are conserved. For now, we will balance chemical equations by inspection. "Balancing by inspection" means changing stoichiometric coefficients until the number of atoms of each element is the same on each side of the arrow. Usually, we can tell what changes need to be made by looking closely at the reaction and matching the numbers of atoms of each element on both sides of the equation. Consider the following example.

Propane, which is used as a fuel for gas barbecues, reacts with molecular oxygen to form carbon dioxide and water:

$$C_3H_8 + O_2 \longrightarrow CO_2 + H_2O$$

To determine if this equation is balanced, make a list of the elements and numbers of atoms on each side:

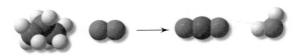

$$C_3H_8 + O_2 \longrightarrow CO_2 + H_2O$$

$$3\,C + 8\,H + 2\,O \longrightarrow 1\,C + 2\,H + 3\,O$$

The equation isn't balanced, because there are too many carbon and hydrogen atoms on the left and too many oxygen atoms on the right. We need to change the numbers of molecules by changing stoichiometric coefficients until the numbers of atoms of each element are equal. It is easiest to balance a chemical equation one element at a time, starting with the elements that appear in only one substance on each side. Notice that all of the carbon atoms in propane end up in carbon dioxide molecules and all of propane's hydrogen atoms appear in water molecules. This feature allows us to balance carbon and hydrogen easily.

To take care of the three carbon atoms per propane molecule, we need three molecules of CO_2. Thus the carbon atoms are balanced by changing the stoichiometric coefficient of CO_2 from 1 to 3. In this reaction the ratio of CO_2 to propane is 3:1. Similarly, we need four molecules of water for the eight hydrogen atoms in one molecule of propane. Using this information, we modify the equation as follows:

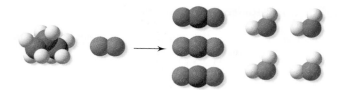

$$C_3H_8 + O_2 \longrightarrow 3\ CO_2 + 4\ H_2O$$
$$3\ C + 8\ H + 2\ O \longrightarrow 3\ C + 8\ H + 10\ O$$

The situation is looking better because atoms of carbon and hydrogen are now conserved. However, the equation still is not balanced because there are too few oxygen atoms on the left side. To balance oxygen, we must give O_2 a stoichiometric coefficient of 5:

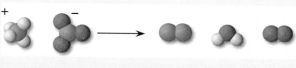

$$C_3H_8 + 5\ O_2 \longrightarrow 3\ CO_2 + 4\ H_2O$$
$$3\ C + 8\ H + 10\ O \longrightarrow 3\ C + 8\ H + 10\ O$$

Now the equation is balanced. Combustion of one molecule of propane gas produces three molecules of CO_2 and four molecules of H_2O and consumes five molecules of O_2. Also, combustion of one mole of propane produces three moles of CO_2 and four moles of H_2O and consumes five moles of O_2. Example 4-1 shows how to balance another reaction.

Balancing Chemical Reactions   Example 4-1

Ammonium nitrate, a colorless ionic solid used as a fertilizer, explodes when it is heated above 300 °C. The products are three gases: molecular nitrogen, molecular oxygen, and steam (water vapor). Write a balanced equation for the explosion of ammonium nitrate.

Strategy: The description in the problem tells us what happens to the starting material: NH_4NO_3 breaks apart into molecules of N_2, O_2, and H_2O. An unbalanced form of the equation can be written from this description. Then we must balance each element in turn by inspection.

$$NH_4NO_3 \longrightarrow N_2 + H_2O + O_2$$

Example 4-1	**Balancing Chemical Reactions** *(continued)*

Solution: Count atoms of each element to see if the elements are in balance:

$$2\,N \quad 4\,H \quad 3\,O \longrightarrow 2\,N \quad 2\,H \quad 3\,O$$

Focus first on the elements that appear in only one reactant and one product, nitrogen and hydrogen in this case. Nitrogen is already balanced. To balance the H atoms, change the stoichiometric coefficient of water from 1 to 2:

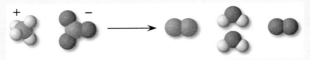

$$NH_4NO_3 \longrightarrow N_2 + 2\,H_2O + O_2$$

$$2\,N \quad 4\,H \quad 3\,O \longrightarrow 2\,N \quad 4\,H \quad 4\,O$$

Now atoms of nitrogen and hydrogen are conserved, but the equation is not balanced with respect to oxygen. We cannot change the $1:1$ ratio of NH_4NO_3 to N_2 and the $1:2$ ratio of NH_4NO_3 to H_2O, because these ratios must be retained to keep hydrogen and nitrogen atoms conserved. Thus we must adjust the coefficient of O_2 to balance with respect to oxygen. Numerically, we can take care of oxygen by changing the coefficient of O_2 to $\frac{1}{2}$:

$$NH_4NO_3 \longrightarrow N_2 + 2\,H_2O + \tfrac{1}{2}\,O_2$$

$$2\,N \quad 4\,H \quad 3\,O \longrightarrow 2\,N \quad 4\,H \quad 3\,O$$

Later, we show that it can be convenient to use fractional coefficients such as $\frac{1}{2}$. When this is the case, the coefficient refers to $\frac{1}{2}$ mol, not $\frac{1}{2}$ molecule.

This balances the equation, but it is unrealistic from a molecular perspective because there is no such thing as half a molecule. To get rid of the $\frac{1}{2}$ without unbalancing the reaction, we multiply *all* the coefficients by 2:

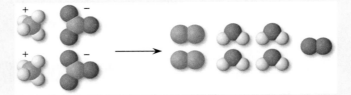

$$2\,NH_4NO_3 \longrightarrow 2\,N_2 + 4\,H_2O + O_2$$

$$4\,N \quad 8\,H \quad 6\,O \longrightarrow 4\,N \quad 8\,H \quad 6\,O$$

Now the reaction is balanced and reasonable from a molecular perspective.

Section Exercises

4.1.1 Although gasoline is a complex mixture of molecules, the chemical reaction that takes place in an automobile engine can be represented by using one of its components, octane (C_8H_{18}). Octane reacts with O_2 to produce CO_2 and H_2O. Such burning of fossil fuels releases millions of tons of carbon dioxide into the Earth's atmosphere each year. Write a balanced equation for the combustion of octane.

■ **4.1.2** Acrylonitrile is used to make synthetic fibers such as Orlon. About 1.5 billion kg of acrylonitrile are produced each year. Balance the following chemical equation, which shows how acrylonitrile is made from propene, ammonia, and oxygen:

$$C_3H_6 + NH_3 + O_2 \longrightarrow C_3H_3N + H_2O$$

Propene

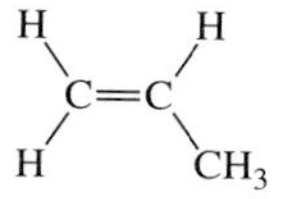

Acrylonitrile

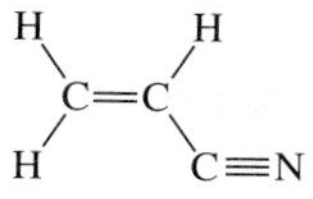

■ **4.1.3** Additional reactions accompany the combustion of octane. For example, molecular nitrogen reacts with molecular oxygen in an automobile cylinder to give nitrogen oxide. After leaving the engine, nitrogen oxide reacts with atmospheric oxygen to give nitrogen dioxide, the red-brown gas seen in the air over many urban areas. Write balanced equations for these two reactions.

4.2 THE STOICHIOMETRY OF CHEMICAL REACTIONS

Chapter 3 describes how the relationships among atoms, moles, and masses are used to answer "how much" questions about individual substances. Combining these ideas with the concept of a balanced chemical equation lets us answer "how much" questions about chemical reactions. The study of the amounts of materials consumed and produced in chemical reactions is called **stoichiometry**.

A chemical synthesis requires the proper amounts of starting materials for a successful outcome. Just as a cake recipe provides the amounts of ingredients needed for successful baking, a balanced chemical equation is a chemical recipe for successful synthesis. For example, how many grams of hydrogen do we need to produce 68 g of ammonia? To find out, we begin with the balanced chemical equation:

$$N_2 + 3\,H_2 \longrightarrow 2\,NH_3$$

According to the balanced equation, the synthesis of 2.0 moles of ammonia requires 3.0 moles of hydrogen and 1.0 mole of nitrogen. However, the balanced equation does not give us any direct information about the *masses* involved in the synthesis. One gram of N_2 plus three grams of H_2 *does not* make two grams of NH_3. Remember the most important lesson from Chapter 3:

/// *Calculations in chemistry are centered around the mole.*

In other words, a chemical recipe for making ammonia requires amounts in moles, not masses. Thus to prepare 68 g of ammonia, we need to know the number of moles of ammonia in 68 g. To do that, we must divide the mass of ammonia by the molar mass of ammonia, 17.0 g/mol, to convert grams to moles:

$$n = \frac{m}{MM} = \frac{68 \text{ g NH}_3}{17.0 \text{ g/mol}} = 4.0 \text{ mol NH}_3$$

Stoichiometry is pronounced "stoy-key-om'-etry." It combines two Greek words, *stoicheion* ("element") and *metron* ("measure").

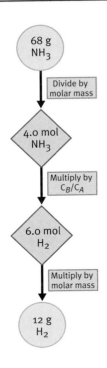

How many moles of H_2 are needed to make 4.0 moles of NH_3? According to the balanced equation, 2 mol of NH_3 requires 3 mol of H_2. The ratio of stoichiometric coefficients from a balanced equation is called the **stoichiometric ratio**. We now apply this ratio to the desired amount of NH_3:

$$(4.0 \text{ mol NH}_3)\left(\frac{3 \text{ mol H}_2}{2 \text{ mol NH}_3}\right) = 6.0 \text{ mol H}_2$$

↑ Stoichiometric ratio

All that remains is to convert from moles of H_2 to mass:

$$m = n(MM) = (6.0 \text{ mol H}_2)(2.02 \text{ g/mol}) = 12 \text{ g H}_2$$

The same procedure can be used to show that the synthesis of 68 g of NH_3 requires 56 g of N_2.

To summarize, the amounts of different reagents that participate in a chemical reaction are related through the stoichiometric coefficients in the balanced chemical equation. To convert from moles of one reagent to moles of any other reagent, multiply by the stoichiometric ratio that leads to proper cancellation of units:

$$\text{Moles}_B = \left(\frac{\text{Coefficient}_B}{\text{Coefficient}_A}\right)\text{Moles}_A \qquad \textbf{(4-1)}$$

Examples 4–2 and 4–3 illustrate this procedure.

Example 4-2	**How Much Product Can Be Made?**

Geranyl formate is used as a synthetic rose essence in cosmetics. The compound is prepared from formic acid and geraniol:

$$HCO_2H + C_{10}H_{18}O \longrightarrow C_{11}H_{18}O_2 + H_2O$$

Formic acid　　Geraniol　　Geranyl formate　Water

A chemist needs to make some geranyl formate for a batch of perfume. How many grams of geranyl formate can a chemist make from 375 g of geraniol?

Strategy: There are some exotic chemical names here, but they should not distract you from the basic principles of reaction stoichiometry. The stoichiometric coefficients state that one mole of each reactant will produce one mole of each product. The flowchart summarizes the steps used to convert the mass of geraniol into the mass of geranyl formate.

Solution: First, convert the mass of geraniol into moles by dividing by the molar mass:

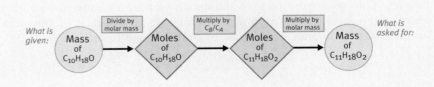

How Much Product Can Be Made? *(continued)*

Example 4-2

$$\frac{375 \text{ g geraniol}}{154.2 \text{ g/mol}} = 2.432 \text{ mol geraniol}$$

A 1:1 stoichiometric ratio links moles of geraniol to moles of geranyl formate:

$$\text{Moles geranyl formate} = \left(\frac{1 \text{ mol geranyl formate}}{1 \text{ mol geraniol}}\right) 2.432 \text{ mol geraniol}$$

If the chemist starts the reaction with 2.432 mol of geraniol, the synthesis can produce 2.432 mol of geranyl formate. To finish the problem, multiply by the molar mass of geranyl formate to determine the mass of geranyl formate produced in the reaction:

$$(2.432 \text{ mol geranyl formate})\left(\frac{182.3 \text{ g geranyl formate}}{1 \text{ mol geranyl formate}}\right) = 443 \text{ g geranyl formate}$$

The result has three significant figures, just like the initial mass of geraniol. The answer makes sense. Geranyl formate has a slightly larger molar mass than geraniol, and we can see that the mass of the product is slightly more than the mass of the starting material.

Amounts of Reactants and Products

Example 4-3

Poisonous hydrogen cyanide (HCN) is an important industrial chemical. It is produced from methane (CH_4), ammonia, and molecular oxygen. The reaction also produces water. An industrial manufacturer wants to convert 175 kg of methane into HCN. How much hydrogen cyanide can be produced in the reaction? What masses of ammonia and oxygen will be required?

Strategy: This problem looks complicated, so it is a good idea to apply the seven-step problem-solving method.

Solution:

1. This is a stoichiometry problem (how much?), in which we are asked to find masses of reactants and products.

2. To visualize this kind of problem, we need a balanced chemical equation:
   Begin by writing the unbalanced form of the equation from the information given, namely methane, ammonia, and molecular oxygen react to yield hydrogen cyanide and water:

$$CH_4 + NH_3 + O_2 \longrightarrow HCN + H_2O$$

Now balance the equation by inspection, one element at a time. Carbon and nitrogen are in balance when there are equal numbers of molecules of CH_4, NH_3, and HCN. Hydrogen and oxygen, however, are not in balance. There are seven hydrogen atoms on the left but only three on the right. The coefficient for HCN cannot be changed without unbalancing C and N. Thus, to balance H, we must multiply the coefficient for H_2O by 3:

$$CH_4 + NH_3 + O_2 \longrightarrow HCN + 3 H_2O$$

Solving
Quantitative
Problems

Example 4-3 | **Amounts of Reactants and Products** *(Continued)*

This gives three oxygen atoms on the right, which corresponds to $\frac{3}{2}$ O_2 molecules among the starting materials:

$$CH_4 + NH_3 + \tfrac{3}{2}\, O_2 \longrightarrow HCN + 3\, H_2O$$

To avoid the impossible $\frac{1}{2}$ molecule, multiply all of the stoichiometric co-efficients by 2. This gives the balanced equation:

$$2\, CH_4 + 2\, NH_3 + 3\, O_2 \longrightarrow 2\, HCN + 6\, H_2O$$

3. The problem gives only one piece of data: 175 kg of CH_4 will be converted into HCN. The problem asks about all three starting materials, so we will need their molar masses:

$$MM_{(CH_4)} = 16.04 \text{ g/mol}, \quad MM_{(NH_3)} = 17.03 \text{ g/mol}, \quad MM_{(O_2)} = 32.00 \text{ g/mol}.$$

4. Knowing the mass of one reagent, we are asked to find the masses of others. We need to determine molar amounts and then convert to kilograms. The calculations require two equations for interconversion of moles and mass:

$$n = \frac{m}{MM} \quad and \quad n_B = \left(\frac{\text{Coefficient}_B}{\text{Coefficient}_A} \right) n_A$$

5. After converting mass to moles and n_A to n_B, we need to convert moles back to mass: $m = n\, MM$

6. Begin by converting mass of methane to moles:

$$(175 \text{ kg } CH_4)\left(\frac{10^3 \text{ g}}{1 \text{ kg}} \right) = 1.75 \times 10^5 \text{ g } CH_4$$

$$\frac{1.75 \times 10^5 \text{ g } CH_4}{16.04 \text{ g/mol}} = 1.091 \times 10^4 \text{ mol } CH_4$$

Next use the ratio of coefficients to find the number of moles of O_2 that are required for the synthesis:

$$(1.091 \times 10^4 \text{ mol } CH_4)\left(\frac{3 \text{ mol } O_2}{2 \text{ mol } CH_4} \right) = 1.636 \times 10^4 \text{ mol } O_2$$

Finally, use the molar mass of O_2 to convert from moles back to mass:

$$(1.636 \times 10^4 \text{ mol } O_2)\left(\frac{32.00 \text{ g } O_2}{1 \text{ mol } O_2} \right)\left(\frac{1 \text{ kg}}{10^3 \text{ g}} \right) = 524 \text{ kg } O_2$$

7. Now check for reasonableness. The units of the result, kilograms, indicate that the calculations are in order, because the question asked for the mass of oxygen. The mass of O_2 required for the synthesis, 524 kg, is a large quantity, but it is reasonable because the reaction also involves a large quantity of methane (175 kg).

Similar calculations for ammonia and hydrogen cyanide show that 186 kg of NH_3 are required as a starting material and that the reaction produces 295 kg of HCN. You should carry out the appropriate calculations to confirm these values.

Box 4-1   Chemistry and Life: Feeding the World

Fertilizers are immensely important to humanity. Fertilization using animal products has been practiced since ancient times. Animal manure returns nutrients to the soil, replenishing elements that are depleted as crops are grown and harvested. It is likely that the use of animal fertilizers followed quickly on the heels of the domestication of goats, sheep, and cattle. Lacking means of fertilization, primitive people farmed a plot of land until the nutrients in the soil were exhausted. Then they moved to new ground, where they burned the natural vegetation and began farming again. This slash-and-burn method of growing food is still used extensively in South America, where farmers are destroying vast tracts of rain forest.

Agriculture requires fertilizers because growing plants remove various chemical elements from the soil. In a fully contained ecosystem, decaying organic matter replenishes these elements, but the elements contained in crops that are harvested and shipped elsewhere are not replenished. Thus intensive agriculture inevitably depletes the soil of essential elements, which must be replaced by fertilization.

Growing plants require a variety of chemical elements. Nitrogen, phosphorus, and potassium are required in the greatest amounts, but plants also need trace amounts of calcium, copper, iron, zinc, and other elements. By far the most substantial need is for nitrogen. The Earth's atmosphere is 80% molecular nitrogen, but plants cannot use N_2. Instead, most plants absorb nitrogen from the soil in the form of nitrate ions.

The production of nitrogen fertilizers is a major activity of the chemical industry. Every year, the top 15 chemicals in industrial production in the United States include several nitrogen-containing compounds whose major use is in fertilizers. Molecular nitrogen serves as the primary source of nitrogen for chemical production. Ammonia, which is synthesized from N_2 and H_2, can be injected directly into the ground, where it dissolves in moisture in the soil and serves as a fertilizer. Ammonia is more widely used in reactions with acids to produce other fertilizers: Ammonia and nitric acid produce ammonium nitrate, while ammonia and sulfuric acid produce ammonium sulfate. These chemicals and urea, $(H_2N)_2CO$, are among the most important fertilizers.

Crop yields can rise dramatically with the use of commercial fertilizers. For example, in 1800 an acre of land in the United States produced about 25 bushels of corn. In the 1980s the same acre of land produced 110 bushels. Worldwide, approximately 4 billion acres of land are used to grow food crops. This would probably be enough land to feed the world's population if the entire acreage could be fertilized commercially. It has been estimated that world crop production would increase by about 50% if about $40 per acre were spent to apply modern chemical fertilizers. However, it would cost about $160 trillion to produce this additional food. Furthermore, the use of chemical fertilizers can lead to the contamination of streams, lakes, and bays with phosphates and nitrates. In addition, the production of fertilizers from molecular nitrogen is very energy intensive. In the United States alone, hundreds of millions of barrels of oil are used every year to produce fertilizers. Finding ways to produce noncontaminating and inexpensive fertilizers without huge energy investment is a major challenge to today's chemical industry.

As world supplies of petroleum are depleted, and as the Earth's population steadily increases, society will be forced to develop more efficient ways to make fertilizer. Genetic engineering offers a promising solution. There is a remarkable bacterium that lives in the roots of leguminous plants such as soybeans, peas, and peanuts. This organism can convert molecular nitrogen into ammonia. The plant and the bacterium have a symbiotic relationship. Ammonia produced by the bacterium nourishes the plant, and the plant provides other nutrients to the bacterium. Exactly how the bacterium converts nitrogen to ammonia is the subject of vigorous research. Scientists hope eventually to transfer the bacterial gene responsible for the conversion of nitrogen to ammonia into the cells of nonleguminous plants.

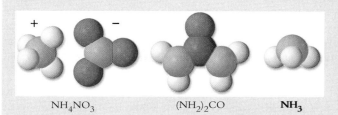

NH_4NO_3 $(NH_2)_2CO$ **NH_3**

Section Exercises

4.2.1 Phosphorus trichloride is produced from the reaction of solid phosphorus and chlorine gas:

$$P_4 + 6\ Cl_2 \longrightarrow 4\ PCl_3$$

  (a) What mass of phosphorus trichloride can be prepared from 75.0 g of phosphorus?

  (b) What mass of chlorine will be consumed in the reaction?

4.2.2 Tooth enamel consists, in part, of $Ca_5(PO_4)_3(OH)$. Tin(II) fluoride (toothpaste labels call it *stannous fluoride*) is added to some toothpastes because it exchanges F^- for OH^- to give a more decay-resistant compound, $Ca_5(PO_4)_3F$. In addition to fluoroapatite, this reaction produces tin(II) oxide and water. What mass of $Ca_5(PO_4)_3(OH)$ can be converted to $Ca_5(PO_4)_3F$ by reaction with 0.115 g of tin(II) fluoride?

4.2.3 Combustion reactions require molecular oxygen. In an automobile the fuel-injection system must be adjusted to provide the right "mix" of gasoline and air. Compute the number of grams of oxygen required to react completely with 1.00 L of octane (C_8H_{18}, $\rho = 0.80$ g/mL). What masses of water and carbon dioxide are produced in this reaction? (Hint: Recall the equation for density from Chapter 1: $\rho = m/V$.)

4.3 YIELDS OF CHEMICAL REACTIONS

When chemical reactions are performed under practical conditions, the amounts of products obtained are almost always less than the amounts predicted by stoichiometric analysis. There are three major reasons for this.

1. Many reactions stop before reaching completion. For example, solid sodium metal reacts with oxygen gas to give sodium peroxide (Na_2O_2). As the reaction proceeds, a solid crust of the product builds up on the surface of the metal, as shown in Figure 4-1. The crust prevents oxygen from reaching the sodium metal, and the reaction stops, even though both starting materials are still present.

   Other reactions do not go to completion because they reach dynamic equilibrium. While reactant molecules continue to form product molecules, product molecules also interact to re-form reactant molecules. The Haber reaction and many precipitation reactions, described later in this chapter, are examples of reactions that reach dynamic equilibrium rather than going to completion.

CHAPTERS 15–17 →
Quantitative calculations for reactions at equilibrium are treated in Chapters 15–17.

2. Competing reagents often consume some of the starting materials. Sodium metal also reacts with water to produce sodium hydroxide. If a sample of oxygen is contaminated with water vapor, both O_2 and H_2O will compete for the sodium metal. The more water present in the gas mixture, the less Na_2O_2 will be formed.

3. When the product of a reaction is purified and isolated, some of it is inevitably lost during the collection process. Gases may escape while being pumped out of a reactor. Liquids adhere to glass surfaces, making it impossible to transfer every drop of a liquid product. Likewise, it is impossible to scrape every trace of a solid material from a reaction vessel.

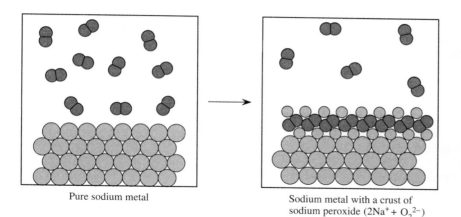

Pure sodium metal

Sodium metal with a crust of
sodium peroxide ($2Na^+ + O_2^{2-}$)

Figure 4-1
The reaction of sodium metal with molecular oxygen produces a crust of sodium peroxide. This crust prevents more sodium peroxide from forming, even though both starting materials are still present.

The amount of a product obtained from a reaction is often reported as a **yield**. The amount of product predicted by stoichiometry is the **theoretical yield,** whereas the amount actually obtained is the **actual yield.** The **percent yield** is the percentage of the theoretical amount that is actually obtained:

$$\text{Percent yield} = 100\% \left(\frac{\text{Actual amount}}{\text{Theoretical amount}} \right) \qquad (4\text{-}2)$$

When we calculate a percent yield, the amounts can be expressed in either moles or mass, provided both the actual and theoretical amounts are in the same units. Example 4–4 shows how to use Equation 4–2.

Calculating Percent Yield

Example 4-4

According to Example 4–2, it is possible to make 443 g of geranyl formate from 375 g of geraniol. A chemist making geranyl formate for the preparation of a perfume uses 375 g of starting material and collects 417 g of purified product. What is the percent yield of this synthesis?

Strategy: To calculate a percent yield, we need to compare the actual amount obtained in the synthesis with the theoretical amount that could be produced, using Equation 4–2.

Solution: The theoretical yield of geranyl formate is 443 g. This is the amount of product that would result from complete conversion of geraniol into geranyl formate. The actual yield, 417 g, is the quantity of the desired product that the chemist collects. The percent yield is their ratio multiplied by 100%:

$$\text{Percent yield} = (100\%) \left(\frac{417 \text{ g}}{443 \text{ g}} \right) = 94.1\%$$

If the percent yield of a reaction is already known, we can calculate how much of a product to expect from a synthesis that uses a known amount of starting material. For example, the Haber synthesis of ammonia stops when 13% of the starting materials have formed products. Knowing this, how much ammonia could an industrial producer expect to make from 2.0 metric tons of molecular hydrogen? First, calculate the theoretical yield:

$$\text{Theoretical yield} = \frac{(2.0 \text{ ton } H_2)(10^6 \text{ g/ton})}{(2.016 \text{ g/mol } H_2)}\left(\frac{2 \text{ mol } NH_3}{3 \text{ mol } H_2}\right) = 6.61 \times 10^5 \text{ mol } NH_3$$

Rearrange Equation 4–2 and substitute known values:

$$\text{Actual yield} = (\text{Theoretical yield})\left(\frac{\% \text{ yield}}{100\%}\right)$$

$$\text{Actual yield} = (6.61 \times 10^5 \text{ mol})\left(\frac{13\%}{100\%}\right) = 8.60 \times 10^4 \text{ mol } NH_3$$

Finally, convert back to mass:

$$\text{Actual mass} = (8.60 \times 10^4 \text{ mol})(17.03 \text{ g/mol})(10^{-6} \text{ ton/g}) = 1.5 \text{ metric ton}$$

Chemists often do yield calculations in reverse to determine the masses of reactants to use to obtain a desired amount of product. If the percent yield is known from previous experiments, we can calculate the amount of starting material needed to make a specific amount of product. Example 4-5 shows how to do this.

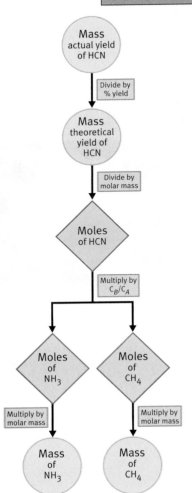

| Example 4-5 | Calculating Reactant Mass from Yield |

The industrial production of hydrogen cyanide is described in Example 4-3. If the yield of this synthesis is 97.5%, how many kilograms of methane and how many of ammonia should be used to produce 1.50×10^5 kg of HCN?

Strategy: This is a two-stage problem that requires a yield calculation and a conversion among molar amounts of products and reactants. This is typical of yield calculations. Use the percent yield to determine the theoretical yield, and then use the stoichiometric ratios to calculate masses of starting materials.

Solution: First, rearrange the equation for percent yield to find the theoretical yield that will give an actual yield of 1.50×10^5 kg:

$$\text{Theoretical yield} = (100\%)\left(\frac{\text{Actual yield}}{\text{Percent yield}}\right)$$

$$\text{Theoretical yield} = (100\%)\left(\frac{1.50 \times 10^5 \text{ kg}}{97.5\%}\right) = 1.538 \times 10^5 \text{ kg HCN}$$

$$(1.538 \times 10^5 \text{ kg HCN})(10^3 \text{ g/kg}) = 1.538 \times 10^8 \text{ g HCN}$$

$$\frac{1.538 \times 10^8 \text{ g HCN}}{27.03 \text{ g/mol}} = 5.690 \times 10^6 \text{ mol HCN}$$

Next, to convert this mass to moles apply the stoichiometric ratio, and multiply by the molar mass of methane:

$$(5.690 \times 10^6 \text{ mol HCN})\left(\frac{2 \text{ mol } CH_4}{2 \text{ mol HCN}}\right) = 5.690 \times 10^6 \text{ mol } CH_4$$

$$(5.690 \times 10^6 \text{ mol } CH_4)(16.04 \text{ g/mol})(10^{-3} \text{ kg/g}) = 9.13 \times 10^4 \text{ kg } CH_4$$

The flowchart summarizes the calculations. A similar calculation shows that the synthesis also requires 9.69×10^4 kg of NH_3.

Suppose a chemical reaction gives a 50% yield. Is that a good yield or a poor yield? The answer depends on the circumstances. Many valuable chemicals are manufactured in processes that require several steps. The overall yield of a multi-step synthesis is the product of the yields of the individual steps. For example, suppose that a reaction requires two steps. If the first reaction has a yield of 85% and the second a yield of 65%, the overall yield is $(100\%)(0.85)(0.65) = 55\%$. Notice that the overall yield is lower than the yield of the least efficient step. For this reason, the more steps there are in a synthesis, the more important the yields of the individual reactions become. For example, the synthesis of a growth hormone might require a sequence of 16 different chemical reactions. At an average yield per step of 95%, the overall yield of this synthesis is 44%, but at an average yield of 75% the overall yield of the synthesis falls to 1.0%!

Many-step reactions that have only moderate yields at each step are wasteful and expensive. For this reason, chemists devote much time, effort, and ingenuity to devise reaction sequences and conditions that improve the yields of chemical syntheses. Our most important industrial chemicals are produced in billion-pound quantities on an annual basis. Here, improving the synthesis yield by even a few tenths of a percent can save a company millions of dollars each year.

Section Exercises

4.3.1 When heated, potassium chlorate decomposes to potassium chloride and gaseous molecular oxygen:

$$2\ KClO_3 \longrightarrow 2\ KCl + 3\ O_2$$

What is the theoretical yield of oxygen when 5.00 g of potassium chlorate decomposes? Calculate the percent yield if a 5.00-g sample gives 1.84 g O_2 on decomposition. Give possible reasons why the actual yield is less than the theoretical yield.

4.3.2 Sulfuric acid is made from elemental sulfur in a three–step industrial process. How many tons of sulfur are required to produce 15 tons of H_2SO_4 if the three steps have yields of 94%, 92.5%, and 97%?

4.3.3 Aspirin is produced by treating salicylic acid with acetic anhydride:

$$2\ C_7H_6O_3 + C_4H_6O_3 \longrightarrow 2\ C_9H_8O_4 + H_2O$$

salicylic acid   acetic   aspirin
anhydride

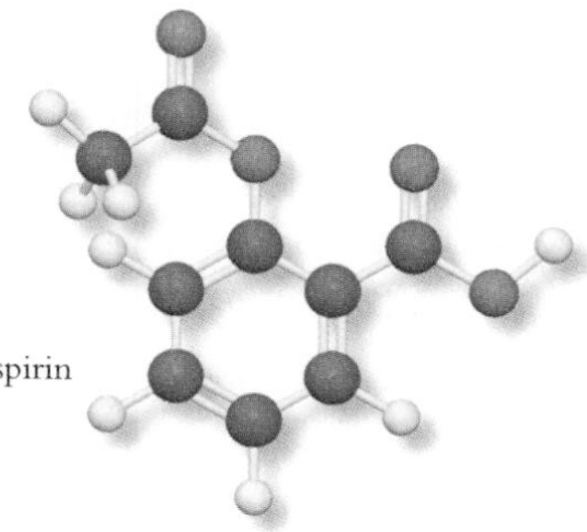

Aspirin

If the synthesis has an 87% yield, what mass of salicylic acid should be used to produce 75 g of aspirin?

4.4 THE LIMITING REACTANT

The reactions encountered so far in this chapter were set up so that all the starting materials were used up at the same time. In other words, there were no "leftovers" at the end of the reaction. Often, chemical reactions are run with an excess of one or more starting materials. This means that one reactant will "run out" before the others. The reactant that runs out is called the **limiting reactant** because it limits the amount of product that can be made. The other starting materials are said to be *in excess*.

We illustrate the limiting reactant concept with the bicycle analogy shown in Figure 4-2. It takes two wheels, a frame, and a chain to build a bicycle. Imagine that a bicycle shop has the following parts on hand: 8 wheels, 5 frames, and 5 chains. How many bicycles can be built with these parts? There are enough frames and chains to make five bicycles. However, because each bicycle needs two wheels, eight wheels are enough to make only four bicycles. Even though there are more wheels around the shop than any of the other parts, the shop will run out of wheels first. After four bicycles are built, the shop inventory will be 4 bicycles, 0 wheels, 1 frame, and 1 chain. There are not enough parts left to make a fifth bicycle because there is a shortage of two wheels. In this bicycle shop, wheels are the limiting reactant.

Chemical reactions must be analyzed in this same way. Instead of comparing numbers of bicycle parts, we compare the number of moles of each starting material on hand with the number of moles required to make the desired product. Consider the Haber synthesis once again. Figure 4-3 shows a limiting reactant situation for this synthesis. If we start with six molecules of H_2 and four molecules of N_2, the six molecules of H_2 will combine with two molecules of N_2 to make four molecules of NH_3. When all the H_2 molecules are consumed, two molecules of N_2

Figure 4-2
The number of bicycles that can be assembled is limited by whichever part runs out first. In the inventory shown in this figure, wheels are that part.

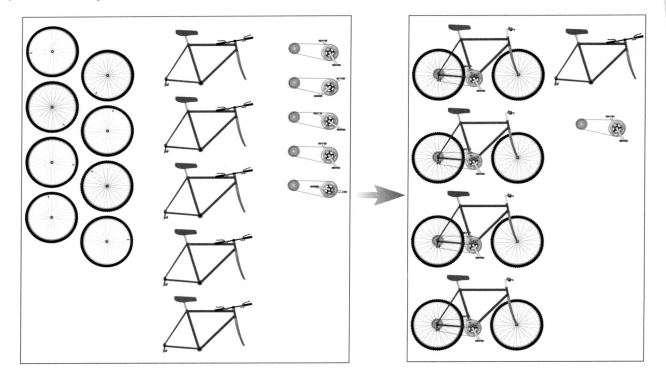

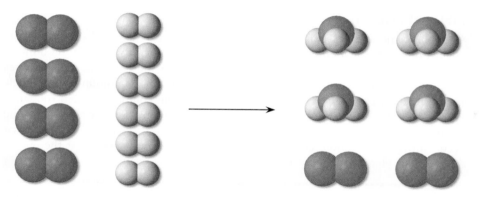

Figure 4-3
A molecular view of a limiting reactant situation for the ammonia synthesis. To make 4 molecules of NH_3 requires 2 molecules of N_2 and 6 molecules of H_2. If we start with 4 molecules of N_2 and 6 molecules of H_2, H_2 is the limiting reactant.

will be left over. Here, H_2 is the limiting reactant. This molecular analysis applies equally well to numbers of moles. In molar terms, if six moles of H_2 and four moles of N_2 react completely, all the H_2 is consumed, four moles of NH_3 are produced, and two moles of N_2 are left over.

To solve a quantitative limiting reactant problem, we identify the limiting reactant by working with moles of amounts and the stoichiometric coefficients from the balanced equation. For the ammonia synthesis, if we start with 84.0 g of molecular nitrogen and 24.2 g of molecular hydrogen, what mass of ammonia can be prepared? First, convert from masses to moles:

$$\frac{84.0 \text{ g } N_2}{28.01 \text{ g/mol}} = 3.00 \text{ mol } N_2 \qquad \frac{24.2 \text{ g } H_2}{2.016 \text{ g/mol}} = 12.0 \text{ mol } H_2$$

According to the balanced equation, the stoichiometric ratio of nitrogen to ammonia is $1:3$.

$$N_2 + 3 H_2 \longrightarrow 2 NH_3$$

There are 3.00 moles of N_2, which is enough to make 6.00 moles of NH_3. The stoichiometric ratio of hydrogen to ammonia is $3:2$. To make the same 6.00 moles of NH_3 would require:

$$6.00 \text{ mol } NH_3\left(\frac{3 \text{ mol } H_2}{2 \text{ mol } NH_3}\right) = 9.00 \text{ mol } H_2 \text{ required}$$

Because 12.0 moles of H_2 are provided, we have more than enough H_2 to use up all of the N_2. Nitrogen is the limiting reactant, and hydrogen is present in excess.

The limiting reactant can be identified by a simple procedure that takes into account the stoichiometry of a reaction. In the ammonia example, if we divide the number of moles of each reactant by its stoichiometric coefficient we find that the limiting reactant, N_2, has the smaller ratio of moles over coefficient:

$$\frac{3.00 \text{ mol } N_2}{1 \text{ mol } N_2} = 3.00 \qquad \frac{12.0 \text{ mol } H_2}{3 \text{ mol } H_2} = 4.00$$

This leads to the following generalization.

/// *The limiting reactant is the one whose number of moles divided by its stoichiometric coefficient has the smallest value.*

Tables of Amounts

A table of amounts is a convenient way to organize the data and summarize the calculations of a stoichiometry problem. Such a table helps to identify the limiting reactant, shows how much product will form during the reaction, and indicates how much of the excess reactant will be left over. A table of amounts has the balanced chemical equation at the top. The table has one column for each substance involved in the reaction and three rows listing amounts. The first row lists the starting amounts for all the substances. The second row shows the changes that occur during the reaction, and the last row lists the amounts present at the end of the reaction. Here is a table of amounts for the ammonia example:

Reaction	N_2	+	$3 H_2$	$\longrightarrow$	$2 NH_3$
Starting amount (mol)	3.00		12.0		0.00
Change in amount (mol)	−3.00		−9.00		+6.00
Final amount (mol)	0.00		3.0		6.00

A table of amounts has three key features, all of which involve the **changes in amounts**:

1. *Changes are negative for reactants and positive for products.* This is because the amounts of reactants decrease during the reaction and the amounts of products increase during the reaction.

2. *All changes are related by stoichiometry*. In other words, each ratio of changes in amount must equal the ratio of stoichiometric coefficients in the balanced equation. In the example above, the changes in amount for H_2 and N_2 are in the ratio $3:1$, the same as the stoichiometric ratio for the coefficients of H_2 and N_2 in the balanced equation.

3. In each column, *the starting amount plus the change in amount gives the final amount.* Each entry in the final amount row of the table is the sum of the previous entries in the same column.

The table of amounts for the ammonia example illustrates these features:

1. The changes are negative for N_2 and H_2, which are consumed, and positive for NH_3, which is produced.

2. The changes in amount for H_2 and N_2 are in the ratio of $3:1$, the same as the stoichiometric ratio for the coefficients of H_2 and N_2 in the balanced equation.

3. The change of −9.00 mol for H_2 leaves $(12.00 - 9.00) = 3.00$ mol as the final amount.

You can use these three key features to construct a table of amounts for any reaction.

When a reaction goes to completion, the limiting reactant is consumed completely, so its final amount must be zero. Two facts help identify the limiting reactant. First, final amounts can never be negative. If a negative amount appears in the bottom row of the table, a mistake has been made in identifying the limiting reactant. Second, the limiting reactant is always the one whose ratio of moles to stoichiometric coefficient is smallest. The smallest ratio indicates the limiting reactant, in this case N_2. Thus dividing starting amounts by coefficients is a quick way to identify the limiting reactant. Examples 4–6 and 4–7 illustrate the features of limiting reactants and amounts tables.

| Simple Limiting Reactant Calculation | Example 4-6 |

The synthesis of aspirin was described in Section Exercise 4.3.3:

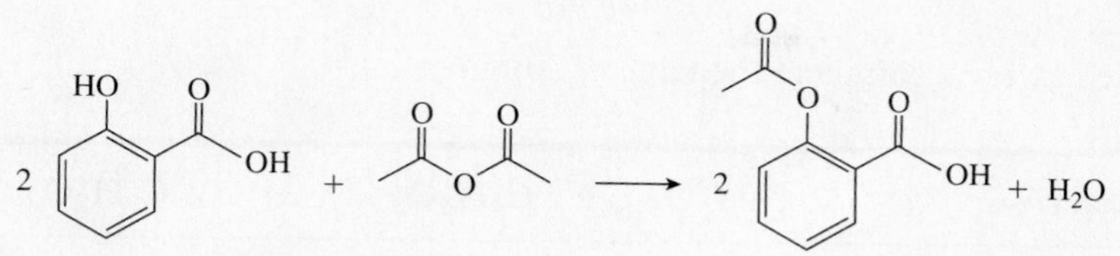

| Name: | Salicylic acid | Acetic anhydride | Aspirin |
| MM: | 138.1 g/mol | 102.1 g/mol | 180.2 g/mol |

Suppose a chemist started with 152 g of salicylic acid and 86.8 g of acetic anhydride. How many grams of aspirin could be prepared?

Strategy: The problem asks for the mass of a product of a chemical reaction. We recognize this as a limiting reactant situation because we are given the masses of both starting materials. First, identify the limiting reactant by working with moles and stoichiometric coefficients; then carry out standard stoichiometry calculations to determine the mass of product. A table of amounts helps organize the calculations.

Solution: Begin by converting mass to moles. The problem gives the molar masses.

$$\frac{152 \text{ g } C_7H_6O_3}{138.1 \text{ g/mol}} = 1.10 \text{ mol } C_7H_6O_3 \qquad \frac{86.8 \text{ g } C_4H_6O_3}{102.1 \text{ g/mol}} = 0.850 \text{ mol } C_4H_6O_3$$

Now divide moles by stoichiometric coefficients to identify the limiting reactant.

$$\frac{1.10 \text{ mol } C_7H_6O_3}{2 \text{ mol } C_7H_6O_3} = 0.555 \qquad \frac{0.850 \text{ mol } C_4H_6O_3}{1 \text{ mol } C_4H_6O_3} = 0.850$$

Because it has the smaller ratio, $C_7H_6O_3$ is the limiting reactant.
Next we organize the data by using a table of amounts.

Reaction	**2 $C_7H_6O_3$**	**+ $C_4H_6O_3$**	$\longrightarrow$	**2 $C_9H_8O_4$**	**+ H_2O**
Starting amount (mol)	1.10	0.850		0	0
Change in amount (mol)					
Final amount (mol)					

The entries in the change row are determined using the stoichiometric ratios.

$$1.10 \text{ mol } C_7H_6O_3 \left(\frac{1 \text{ mol } C_4H_6O_3}{2 \text{ mol } C_7H_6O_3} \right) = 0.550 \text{ mol } C_4H_6O_3 \text{ consumed}$$

$$1.10 \text{ mol } C_7H_6O_3 \left(\frac{2 \text{ mol } C_9H_8O_4}{2 \text{ mol } C_7H_6O_3} \right) = 1.10 \text{ mol } C_9H_8O_4 \text{ produced}$$

| Example 4-6 | Simple Limiting Reactant Calculation (Continued) |

$$1.10 \text{ mol } C_7H_6O_3\left(\frac{1 \text{ mol } H_2O}{2 \text{ mol } C_7H_6O_3}\right) = 0.550 \text{ mol } H_2O \text{ produced}$$

We can now complete the table of amounts.

Reaction	2 $C_7H_6O_3$ +	$C_4H_6O_3$	$\longrightarrow$	2 $C_9H_8O_4$ +	H_2O
Starting amount (mol)	1.10	0.850		0	0
Change in amount (mol)	-1.10	-0.550		$+1.10$	$+0.550$
Final amount (mol)	0.00	0.300		1.10	0.550

To finish the problem, convert moles of product into mass.

$$(1.10 \text{ mol } C_9H_8O_4)(180.2 \text{ g/mol}) = 198 \text{ g } C_9H_8O_4$$

Notice that neither the masses of the reactants nor the number of moles indicate directly which reactant is limiting. In this example, salicylic acid ($C_7H_6O_3$) is the limiting reactant even though the mass of salicylic acid and the number of moles of salicylic acid are both larger than the corresponding amounts of acetic anhydride ($C_4H_6O_3$).

Masses and number ratios cannot be used to determine the "limiting reactant" in bicycle manufacture either. Five chains are fewer in number than eight wheels and have less mass than eight wheels. Nevertheless, the wheels are used up before the chains.

| Example 4-7 | Limiting Reactant Calculation |

Nitric acid, a leading industrial chemical, is used in the production of fertilizers and explosives. One step in the industrial production of nitric acid is the reaction of ammonia with molecular oxygen to form nitrogen oxide:

$$4 NH_3 + 5 O_2 \longrightarrow 4 NO + 6 H_2O$$

In a study of this reaction, a chemist mixed 125 g of ammonia with 256 g of oxygen and allowed them to react to completion. What masses of NO and H_2O were produced, and what mass of which reactant was left over?

Solving Quantitative Problems

Strategy: We apply the seven-step problem-solving method to this problem.

Solution:

1. We are asked about masses of reagents in a chemical reaction. This is a stoichiometry problem.

2. Visualization involves the balanced chemical equation:

$$4 NH_3 + 5 O_2 \longrightarrow 4 NO + 6 H_2O$$

3. The data available are starting masses of the reactants, 125 g of NH_3 and 256 g of O_2. We also need the molar masses, which are easily calculated:

Limiting Reactant Calculation *(Continued)*

Example 4-7

$$MM_{NH_3} = 17.03 \text{ g/mol}; \quad MM_{O_2} = 32.00 \text{ g/mol};$$
$$MM_{NO} = 30.01 \text{ g/mol}; \quad MM_{H_2O} = 18.02 \text{ g/mol}.$$

4. From the fact that starting amounts of both reactants are given, we identify this as a limiting reactant problem (one reactant will be consumed before the other is consumed). A table of amounts will help organize the information.

5. As always in stoichiometric calculations, we need to work with moles

$$n_{NH_3} = \frac{125 \text{ g}}{17.03 \text{ g/mol}} = 7.34 \text{ mol} \qquad n_{O_2} = \frac{256 \text{ g}}{32.00 \text{ g/mol}} = 8.00 \text{ mol}$$

Determine the limiting reactant by dividing each starting amount by the stoichiometric coefficient for that reactant:

$$\frac{7.34 \text{ mol NH}_3}{4 \text{ mol NH}_3} = 1.84 \qquad \frac{8.00 \text{ mol O}_2}{5 \text{ mol O}_2} = 1.60$$

The ratio for O_2 is smaller, so oxygen is the limiting reactant.

6. Now set up and complete a table of amounts. Because O_2 is limiting, it will be completely consumed, so its change is -8.00 mol. Knowing this, we calculate all other changes in amounts using ratios of stoichiometric coefficients:

$$\text{Change in NH}_3 = (8.00 \text{ mol O}_2)\left(\frac{4 \text{ mol NH}_3}{5 \text{ mol O}_2}\right) = 6.40 \text{ mol NH}_3 \text{ used}$$

$$\text{Change in NO} = (8.00 \text{ mol O}_2)\left(\frac{4 \text{ mol NO}}{5 \text{ mol O}_2}\right) = 6.40 \text{ mol NO produced}$$

$$\text{Change in H}_2O = (8.00 \text{ mol O}_2)\left(\frac{6 \text{ mol H}_2O}{5 \text{ mol O}_2}\right) = 9.60 \text{ mol H}_2O \text{ produced}$$

Remember that changes in amounts are negative for reactants and positive for products. Here is the complete amounts table:

You should verify that if ammonia is selected as the limiting reactant, the amounts table will have a negative amount of oxygen at the end of the reaction. A negative final amount is impossible.

Reaction:	4 NH₃ +	5 O₂	⟶	4 NO +	6 H₂O
Starting amount (mol)	7.34	8.00		0	0
Change in amount (mol)	−6.40	−8.00		+6.40	+9.60
Final amount (mol)	0.94	0		6.40	9.60

All that remains is to convert back to masses using $m = n(MM)$. We leave it to you to verify that the masses present at the end of the reaction are 16 g NH_3, 192 g NO, and 173 g H_2O.

7. Are these masses reasonable? First, it is reasonable that none of the final amounts is negative. Second, we can check to see if total mass is conserved. We started with 125 g ammonia and 256 g oxygen, a total of 381 g. The masses present at the end total 381 g, too, indicating that the calculations are consistent, and the results are reasonable.

Section Exercises

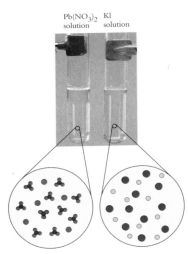

Pb(NO₃)₂ solution   KI solution

4.4.1 Suppose that the industrial synthesis of HCN described in Example 4-3 is carried out using 5.00×10^2 kg each of ammonia and methane in excess oxygen. What is the maximum mass of HCN that could be produced, and what mass of which reactant would be left over?

4.4.2 Lithium metal is one of the few substances that reacts directly with molecular nitrogen:

$$6 \, Li + N_2 \longrightarrow 2 \, Li_3N$$

What mass of the product, lithium nitride, can be prepared from 4.5 g of lithium metal and 9.5 g of molecular nitrogen?

4.4.3 Solid carbon reacts with chlorine gas to give carbon tetrachloride. If 150 g of carbon reacts with 250 g of Cl_2 with an 85% yield, what mass of CCl_4 is produced?

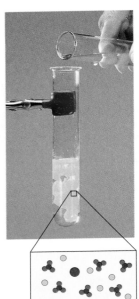

Figure 4-4
Mixing colorless aqueous solutions of potassium iodide and lead(II) nitrate results in the formation of a yellow precipitate of lead(II) iodide.

Tutorial

4.5 PRECIPITATION REACTIONS

The diversity of chemical reactions is immense. To make sense of this vast expanse of reactions, we need a system for grouping them into categories. The reactions within each category should share some characteristics or follow a common theme. One relatively simple category is **precipitation** reactions, in which cations and anions in aqueous solution combine to form neutral insoluble solids. We introduced precipitation reactions in Section 3.7; here we consider their stoichiometry.

Species in Solution

To understand precipitation reactions, it is essential to work with the ionic species that exist in aqueous solution. For instance, mixing colorless solutions of lead(II) nitrate and potassium iodide causes a brilliant yellow solid to precipitate from the mixture (Figure 4-4). To identify this yellow solid, we must examine the chemical species present in the solutions.

When a salt dissolves in water, it produces cations and anions. Lead(II) nitrate and potassium iodide are **soluble** salts. Lead(II) nitrate dissolves in water to generate Pb^{2+} cations and NO_3^- anions. Potassium iodide dissolves in water to generate K^+ and I^- ions. Mixing the solutions combines all four types of ions. A precipitate forms if any of the possible combinations of the ions forms a salt that is **insoluble** in water.

The balanced chemical equation for a precipitation reaction includes information about the physical states of the reagents. We use parentheses: (s) for solid, (l) for liquid, (g) for gas, and (aq) for aqueous, meaning dissolved in water. The ions in our example all are in the aqueous phase. They might combine in four different ways to produce a solid product: KI, KNO_3, PbI_2, and $Pb(NO_3)_2$.

$$K^+(aq) + I^-(aq) \longrightarrow KI(s)$$

$$K^+(aq) + NO_3^-(aq) \longrightarrow KNO_3(s)$$

$$Pb^{2+}(aq) + 2 \, I^-(aq) \longrightarrow PbI_2(s)$$

$$Pb^{2+}(aq) + 2 \, NO_3^-(aq) \longrightarrow Pb(NO_3)_2(s)$$

Figure 4-5
One liter of water can dissolve 354 g of KNO_3, a soluble salt. In contrast, PbI_2 is considered to be an insoluble salt because only 2 mg dissolves in 1 L of water.

Two of the possibilities, KI and $Pb(NO_3)_2$, can be eliminated immediately. We know these two salts are *soluble* because they were contained in the starting solutions. Thus the precipitate is either lead(II) iodide or potassium nitrate. If we were to conduct a second solubility experiment in which solutions of KI and $NaNO_3$ were mixed, we would find that no precipitate forms. This demonstrates that K^+ and NO_3^- ions do not form a solid precipitate, so the bright yellow precipitate must be lead(II) iodide, PbI_2.

As the two salt solutions mix, Pb^{2+} cations and I^- anions combine to produce lead(II) iodide, which precipitates from the solution. On standing, the yellow precipitate settles, leaving a colorless solution that contains potassium cations and nitrate anions. The molecular blow-ups in Figure 4-4 depict these solutions at the molecular level.

Actually, PbI_2 is not completely insoluble. Tiny amounts of Pb^{2+} and I^- ions remain in the aqueous solution after precipitation, but these amounts are so small that we consider PbI_2 to be an insoluble salt. Only 4.1×10^{-6} mol of PbI_2 dissolve in one liter of water at 25 °C, whereas 3.5 mol of KNO_3 dissolve in one liter of water at 25 °C. Figure 4-5 highlights this difference in solubility.

These ions in solution exist in dynamic equilibrium with the precipitate. The system is dynamic because lead ions and iodide ions continually move in and out of solution; it is at equilibrium because the net amount of dissolved ions remains constant.

← CHAPTER 2

Chapter 2 contains a definition of dynamic equilibrium.

Net Ionic Equations

In a precipitation reaction, reactant *ions* combine to form a *neutral ionic solid*. One reactant carries positive charge and the other carries negative charge, but the product is electrically neutral. Because electrical charge always is conserved, the positive charge of the reacting cations must be equal to the negative charge of the reacting anions. For example, if we mix aqueous solutions of potassium hydroxide (KOH) and iron(III) chloride ($FeCl_3$), a precipitate forms. Analysis shows the solid to be electrically neutral iron(III) hydroxide, $Fe(OH)_3$. This precipitation reaction can be described by a balanced equation that contains only those species involved in the reaction. A list of species present helps us to determine that balanced equation:

Species present: K^+, Fe^{3+}, Cl^-, OH^-, H_2O.

Substance formed: $Fe(OH)_3$.

The substance formed contains three OH^- anions combined with every Fe^{3+} cation, resulting in the neutral product. Thus the reactant species are Fe^{3+} cations and OH^- anions with stoichiometric coefficients of 1 and 3. The balanced chemical equation is called the **net ionic equation:**

$$Fe^{3+}(aq) + 3\ OH^-(aq) \longrightarrow Fe(OH)_3(s)$$

Even "insoluble" salts dissolve slightly, leading to dynamic equilibrium with some ions in solution. We explore solubility equilibria in Chapter 15.

A net ionic equation contains only those species that participate in a chemical reaction. Notice that neither K^+ nor Cl^- appears in the equation for the precipitation of $Fe(OH)_3$. Even though these two ions are present in the solution, they undergo no change during the precipitation reaction. Ions that are not involved in the chemical change are referred to as **spectator ions.** Spectator ions are omitted from the net ionic equation.

Solubility Guidelines

Through many years of experience and research, chemists have discovered patterns in the solubilities of ionic substances. Most salts are insoluble. The soluble salts are summarized in Table 4-1, and the flowchart in Figure 4-6 shows how to determine if a salt is soluble or insoluble.

The flowchart lets us predict whether a salt is soluble or insoluble. For example, will $Cu(OH)_2$ dissolve in water? The compound does not contain NH_4^+ or a Group 1 cation, nor does it contain any anion that confers solubility. Because $Cu(OH)_2$ does not appear among the exceptions (Table 4-1), it is insoluble. Example 4-8 further illustrates the solubility flowchart and net ionic equations.

Example 4-8	Salt Solubility

Will a precipitate form when solutions of magnesium sulfate ($MgSO_4$) and barium chloride ($BaCl_2$) are mixed? If so, write the net ionic equation for the reaction.

Strategy: The solution that results from mixing contains all the ions of the original solutions. If any cation-anion combination results in an insoluble salt, that salt will precipitate from solution. List the ions and then apply the flowchart to find out whether any new combination of cations and anions gives an insoluble salt. If there is an insoluble salt, write the net ionic equation that produces it.

Solution: The ions present are Mg^{2+} and SO_4^{2-} from one solution, Ba^{2+} and Cl^- from the other, so the possible new combinations are $MgCl_2$ and $BaSO_4$.

According to the flowchart, salts of chloride and sulfate are soluble. Therefore no precipitate will form unless one of these compounds is included among the exceptions listed in Table 4-1. Barium sulfate is one of these exceptions, so it is an insoluble salt.

Solid barium sulfate will form when the two solutions are combined. Each ion carries two units of charge, so they combine in a $1:1$ stoichiometric ratio:

$$Ba^{2+}(aq) + SO_4^{2-}(aq) \longrightarrow BaSO_4(s)$$

Precipitation Stoichiometry

Tables of amounts are useful in stoichiometry calculations involving precipitation. For example, a precipitate of $Fe(OH)_3$ forms when 50.0 mL of 1.50 M NaOH is mixed with 35.0 mL of 1.00 M $FeCl_3$ solution. We need a balanced chemical equation and amounts in moles to calculate how much precipitate forms. The balanced chemical equation is the net reaction for formation of $Fe(OH)_3$:

$$Fe^{3+}(aq) + 3\ OH^-(aq) \longrightarrow Fe(OH)_3(s)$$

Table 4-1
Soluble Salts

Cations whose salts are soluble: NH_4^+ and Group 1 (Li^+, Na^+, K^+, etc.)
Anions whose salts are soluble:

Anion:	NO_3^-	ClO_4^-	HSO_4^-	$CH_3CO_2^-$	Cl^-	Br^-	I^-	SO_4^{2-}
Exceptions:	None	None	None	None	$AgCl$	$AgBr$	AgI	
					Hg_2Cl_2	Hg_2Br_2	Hg_2I_2	Hg_2SO_4
					$PbCl_2$	$PbBr_2$	PbI_2	$PbSO_4$
								$CaSO_4$
								$SrSO_4$
								$BaSO_4$

Other soluble compounds: $Ca(OH)_2$, $Sr(OH)_2$, $Ba(OH)_2$.

Notice that the net ionic reaction focuses our attention on the ions that actually participate in the reaction. Now calculate the number of moles of Fe^{3+} and OH^-, using $n = MV$:

$$(1.00 \text{ mol } \cancel{FeCl_3}/\cancel{L})\left(\frac{1 \text{ Fe}^{3+}}{1 \cancel{FeCl_3}}\right)(35.0 \cancel{\text{ mL}})\left(\frac{10^{-3} \cancel{L}}{1 \cancel{\text{ mL}}}\right) = 3.50 \times 10^{-2} \text{ mol } Fe^{3+}$$

$$(1.50 \text{ mol NaOH/L})\left(\frac{1 \text{ OH}^-}{1 \text{ NaOH}}\right)(50.0 \text{ mL})\left(\frac{10^{-3} \text{ L}}{1 \text{ mL}}\right) = 7.50 \times 10^{-2} \text{ mol } OH^-$$

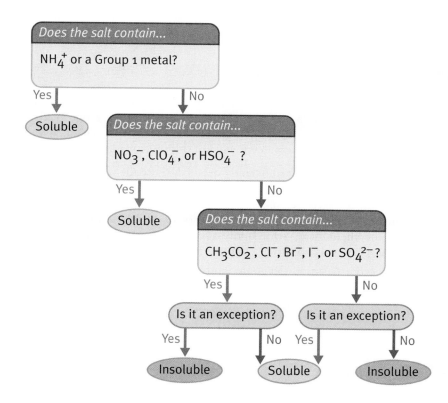

Figure 4-6
A flow chart can be used to predict whether a salt is soluble or insoluble in water.

The amounts of both Fe^{3+} and OH^- are known, so this is a limiting reactant situation. The ion that is consumed first determines how much $Fe(OH)_3$ precipitates. Dividing the numbers of moles by the stoichiometric coefficients identifies the limiting reactant:

$$\frac{3.50 \times 10^{-2} \text{ mol Fe}^{3+}}{1 \text{ mol Fe}^{3+}} = 3.50 \times 10^{-2}$$

$$\frac{7.50 \times 10^{-2} \text{ mol OH}^-}{3 \text{ mol OH}^-} = 2.50 \times 10^{-2}$$

Hydroxide has the smaller ratio, so it is the limiting reactant. A table of amounts helps determine the amount of $Fe(OH)_3$ that precipitates from solution:

Reaction	Fe^{3+}	+	$3\ OH^-$	$\longrightarrow$	$Fe(OH)_3$
Starting amount (mol)	3.50×10^{-2}		7.50×10^{-2}		0
Change in amount (mol)	-2.50×10^{-2}		-7.50×10^{-2}		$+2.50 \times 10^{-2}$
Final amount (mol)	1.00×10^{-2}		0		2.50×10^{-2}

Mixing the two solutions will produce 2.50×10^{-2} mol of $Fe(OH)_3$ precipitate, which is 2.67 g. The mixed solution contains Na^+ cations and Cl^- anions, too, but we can ignore these spectator ions in our calculations. Notice that this precipitation reaction is treated just like other limiting reactant problems. Examples 4-9 and 4-10 further illustrate the application of general stoichiometric principles to precipitation reactions.

Example 4-9	Percent Yield of Precipitation

Silver bromide is a major component of photographic films and paper. A film manufacturer mixed 75.0 L of a 1.25 M solution of silver nitrate with 90.0 L of a 1.50 M potassium bromide solution and obtained 17.0 kg of silver bromide precipitate. What was the percent yield of this precipitation reaction?

Strategy: This problem is sufficiently detailed to use the seven-step strategy.

Solving
Quantitative
Problems

Solution:

1. The problem asks for a percent yield, so we need to compare the actual yield and the theoretical yield. Information is provided about amounts of both starting materials, so this is a limiting reactant situation.

2. We need a balanced net ionic equation. The precipitate, AgBr, forms from two ions, Ag^+ and Br^-.

$$Ag^+(aq) + Br^-(aq) \longrightarrow AgBr(s)$$

3. The data available are 75.0 L of 1.25 M silver nitrate, 90.0 L of 1.50 M potassium bromide, and 17.0 kg of silver bromide product. The molar mass of AgBr is 187.8 g/mol.

Percent Yield of Precipitation (continued)

Example 4-9

4. Use $n = MV$ to calculate amounts of each ion present in the two solutions before mixing, and determine which is limiting. Then solve for the theoretical yield, and apply Equation 4-2 to calculate percent yield.

5. and 6.

$$\text{mol Ag}^+ = MV = (1.25 \text{ mol/L})(75.0 \text{ L}) = 93.8 \text{ mol}$$

$$\text{mol Br}^- = MV = (1.50 \text{ mol/L})(90.0 \text{ L}) = 135 \text{ mol}$$

The 1 : 1 stoichiometric ratio between Ag^+ and Br^- makes it easy to identify the limiting reactant: Ag^+ is limiting, so the precipitation reaction can produce no more than 93.8 mol of AgBr. This is the theoretical yield.

Now convert the mass of precipitate into moles:

$$n = \frac{m}{MM} = \frac{(17.0 \text{ kg})(10^3 \text{ g/kg})}{187.8 \text{ g/mol)}} = 90.5 \text{ mol (actual yield)]}$$

Use the theoretical and actual yields in Equation 4-2 to determine the percent yield:

$$\% \text{ Yield} = (100\%)\left(\frac{\text{Actual yield}}{\text{Theoretical yield}}\right) = (100\%)\left(\frac{90.5 \text{ mol}}{93.8 \text{ mol}}\right) = 96.5\%$$

7. A yield slightly less than 100% seems reasonable. Silver is an expensive substance, so manufacturers seek to maximize the yield of AgBr based on silver. Consequently, silver ion is always made the limiting reactant in this preparation. In other words, the precipitation is always carried out with an excess of bromide ions to ensure that as much Ag^+ precipitates as possible.

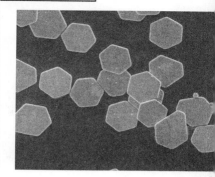

Precipitation Stoichiometry

Example 4-10

A white precipitate forms when 2.00×10^2 mL of 0.200 M potassium phosphate solution is mixed with 3.00×10^2 mL of 0.250 M calcium chloride solution. Write the net ionic equation that describes this process. Calculate the mass of the precipitate that forms, and identify the ions remaining in solution.

Strategy: We could again apply the seven-step process in detail. Instead, we take a more compact approach. Begin by determining what species are present in the reaction mixture. Next, use the solubility guidelines to identify the precipitate. After writing the balanced net ionic reaction, use solution stoichiometry and a table of amounts to find the required quantities.

Solution: The combined solutions contain four different ions: Ca^{2+}, K^+, Cl^-, and PO_4^{3-}. According to the solubility guidelines, potassium salts and chloride salts are soluble. The white precipitate must be calcium phosphate, the only combination that does not involve these ions. To balance the charges, three Ca^{2+} ions must combine with two PO_4^{3-} ions. Potassium and chloride are spectator ions that do not appear in the net ionic reaction:

If you had trouble writing the balanced equation, review Section 4.1.

$$3 \text{ Ca}^{2+}(aq) + 2 \text{ PO}_4^{3-}(aq) \longrightarrow \text{Ca}_3(\text{PO}_4)_2(s)$$

| Example 4-10 | Precipitation Stoichiometry *(continued)* |

To calculate the mass of the precipitate, we need to know the numbers of moles of Ca^{2+} and PO_4^{3-}.

$$(0.200 \text{ mol } K_3PO_4/L)\left(\frac{1 \text{ PO}_4^{3-}}{1 \text{ K}_3PO_4}\right)(2.00 \times 10^2 \text{ mL})\left(\frac{10^{-3} \text{ L}}{1 \text{ mL}}\right)$$

$$= 4.00 \times 10^{-2} \text{ mol PO}_4^{3-}$$

$$(0.250 \text{ mol } CaCl_2/L)\left(\frac{1 \text{ Ca}^{2+}}{1 \text{ CaCl}_2}\right)(3.00 \times 10^2 \text{ mL})\left(\frac{10^{-3} \text{ L}}{1 \text{ mL}}\right)$$

$$= 7.50 \times 10^{-2} \text{ mol Ca}^{2+}$$

Identify the limiting reactant by dividing the numbers of moles by the stoichiometric coefficients:

$$\frac{7.50 \times 10^{-2} \text{ mol Ca}^{2+}}{3 \text{ mol Ca}^{2+}} = 2.50 \times 10^{-2}$$

$$\frac{4.00 \times 10^{-2} \text{ mol PO}_4^{3-}}{2 \text{ mol PO}_4^{3-}} = 2.00 \times 10^{-2}$$

The smaller value identifies phosphate as the limiting reactant, so its final amount will be zero. Using this information, construct a table of amounts such that PO_4^{3-} is completely consumed:

Reaction	3 Ca^{2+}	$+ \; 2 \text{ PO}_4^{3-}$	$\longrightarrow \quad Ca_3(PO_4)_2$
Starting amount (10^{-2} mol)	7.50	4.00	0
Change in amount (10^{-2} mol)	−6.00	−4.00	+2.00
Final amount (10^{-2} mol)	1.50	0	2.00

The table of amounts shows that 2.00×10^{-2} mol of calcium phosphate will precipitate. Use this amount to compute the mass of product:

$$[2.00 \times 10^{-2} \text{ mol Ca}_3(PO_4)_2] \, (310.2 \text{ g/mol}) = 6.20 \text{ g Ca}_3(PO_4)_2$$

The second part of the problem asks us to identify the ions remaining in the solution. Two spectator ions are present: K^+ and Cl^-. In addition, the table of amounts reveals that some of the Ca^{2+} remains in solution after the precipitation reaction.

The ions remaining in solution are K^+, Cl^-, and excess Ca^{2+}.

Synthesis via Precipitation

The solubility guidelines can be used to design ways of making salts. Suppose that we want to prepare barium sulfate, $BaSO_4$. This substance is opaque to X rays, so it is often used to visualize the intestinal tract. Patients are given a "barium cocktail," and then the areas of interest are irradiated. Barium sulfate absorbs the X rays to give a "picture" of the intestines. Soluble barium salts are poisonous, but $BaSO_4$ is insoluble in water, so it can be administered safely. (Only 1.0×10^{-5} mol dissolves in 1 L of water at 25 °C.)

A source of soluble Ba^{2+} cations is required for the synthesis of barium sulfate. We can use any barium salt that is soluble, such as barium nitrate, barium acetate, or barium chloride. A soluble source of sulfate ions is also needed, such as Na_2SO_4 or $(NH_4)_2SO_4$. Because Ba^{2+} ions are toxic, it is important to precipitate all of the barium, so the two salts are measured out in quantities that will ensure that barium is the limiting reactant. Each salt is dissolved separately in water, and then the two solutions are mixed. After precipitation, the $BaSO_4$ is collected by filtration, the mass of the product is measured, and the yield of the reaction is calculated. It is important to know the yield of the reaction because any toxic barium ions left in the solution will have to be disposed of properly. Example 4-11 shows how to apply these principles to the synthesis of CaF_2.

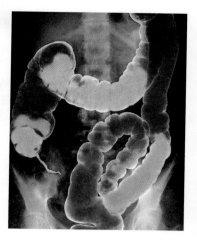

Synthesis via Precipitation — Example 4-11

Calcium fluoride is very insoluble in water and is transparent to light. Because of these properties, calcium fluoride is sometimes used to make windows for optical devices. Design a synthesis of 1.0 kg of CaF_2.

Strategy: The synthesis of calcium fluoride requires soluble sources of Ca^{2+} ions and F^- ions. We can choose any pair of appropriate salts that happen to be available in the laboratory. Once we have selected suitable salts, we use stoichiometric principles to calculate the masses of each that would be appropriate.

Solution: Common cations that confer solubility are Na^+ and K^+, so the fluoride of either would be an appropriate choice. As a source of Ca^{2+}, choose a salt containing an anion that confers solubility, such as Cl^- or NO_3^-. Two compounds that are inexpensive and found in many laboratories are NaF and $CaCl_2$.

To calculate the masses that should be used, begin by writing the net ionic equation for the synthesis:

$$Ca^{2+}(aq) + 2\ F^-(aq) \longrightarrow CaF_2(s)$$

By now, the conversion of 1.0 kg of CaF_2 to 13 mol should be straightforward. This means that we need 13 mol of Ca^{2+} and 26 mol of F^-. These ions can be provided by 13 mol of $CaCl_2$ and 26 mol of NaF. Use molar masses to determine the masses of the two starting materials:

$$(13\ mol\ CaCl_2)(111.0\ g/mol) = 1.4 \times 10^3\ g\ CaCl_2$$

$$(26\ mol\ NaF)(41.90\ g/mol) = 1.1 \times 10^3\ g\ NaF$$

To carry out the synthesis, weigh the proper amounts of calcium chloride and sodium fluoride and dissolve each in water. Then mix the two solutions to generate a precipitate, and filter the solution to isolate the product. One of the attractive features of this synthesis is that the ions left in the solution, Na^+ and Cl^-, are nontoxic.

Section Exercises

4.5.1 Determine whether the following salts are soluble or insoluble: (a) sodium acetate; (b) $AgNO_3$; (c) barium hydroxide; (d) CaO; (e) lead(II) sulfate; (f) $ZnCl_2$; and (g) manganese(II) sulfide.

4.5.2 Design a synthesis of 1.5 kg of calcium phosphate that starts with soluble salts.

4.5.3 Cadmium ions are environmental pollutants found in mining waste, metal plating, water pipes, and industrial discharge. Cadmium ions replace zinc ions in biochemistry and cause kidney damage, high blood pressure, and brittle bones. Dissolved Cd^{2+} impurities can be removed from a water sample by precipitation with sulfide ions. What is the minimum mass of ammonium sulfide required to precipitate all the Cd^{2+} from 5.0×10^3 L of water contaminated with 0.077 M cadmium ions?

4.6 ACID-BASE REACTIONS

One of the most fundamental chemical reactions is the combination of a hydroxide ion (OH^-) and a hydronium ion (H_3O^+) to produce two molecules of water:

$$OH^-(aq) + H_3O^+(aq) \longrightarrow 2H_2O(l)$$

Many books abbreviate the hydronium ion as H⁺(aq) or just H⁺. We prefer H_3O^+ because it serves as a reminder of the molecular structure of the hydronium ion and of the proton-transfer nature of acid-base reactions.

A molecular view of this reaction (Figure 4-7) shows that the hydroxide anion accepts one hydrogen atom from the hydronium cation. Taking account of charges, it is a hydrogen *cation* (H^+) that is transferred. The reaction occurs rapidly when H_3O^+ and OH^- ions collide. The hydroxide anion accepts a hydrogen cation from the hydronium cation, forming two neutral water molecules.

Proton Transfer

A hydrogen cation is a hydrogen atom that has lost its single electron, leaving a bare hydrogen nucleus. A bare hydrogen nucleus is a proton. Thus any reaction in which H^+ moves from one species to another is called a *proton transfer reaction*. Protons readily form chemical bonds. In aqueous solution, they associate with water molecules to form hydronium ions.

In Chapter 20, we introduce a second definition of acids and bases, the Lewis definition, which focuses attention on *electron* movement rather than *proton* movement. Until then, acid-base always means "proton transfer."

Any reaction in which a proton is transferred from one substance to another is an **acid-base reaction.** In an acid-base reaction, an acid donates a proton, and a base accepts that proton. The production of a water molecule from a hydroxide anion (a base) and a hydronium ion (an acid) is an example of an acid-base reaction. Defining acids and bases in terms of proton transfer is known as the Brønsted-Lowry definition of acids and bases. Acids and bases are abundant in chemistry. Any species that can give up a proton to another substance is an **acid,** and any substance that can accept a proton from another substance is a **base.**

Figure 4-7
Proton transfer between H_3O^+ and OH^-.

Animation

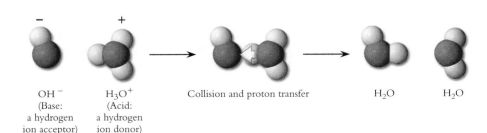

OH⁻
(Base: a hydrogen ion acceptor)

H_3O^+
(Acid: a hydrogen ion donor)

Collision and proton transfer

H_2O

H_2O

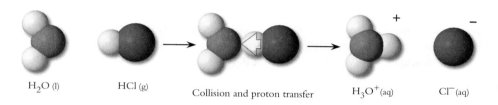

H₂O (l)   HCl (g)   Collision and proton transfer   H₃O⁺(aq)   Cl⁻(aq)

Figure 4-8
Proton transfer between HCl and
H₂O.

Animation

/// **Acid: A substance that donates protons.**
/// **Base: A substance that accepts protons.**

In Figure 4-7, the hydronium ion acts as an acid because it donates a proton to a base. The hydroxide anion acts as a base because it accepts a proton from an acid. When a hydronium ion with charge $+1$ transfers a proton to a hydroxide ion with charge -1, the two resulting water molecules have zero charges. The pair of charges becomes a neutral pair. A proton transfer reaction such as this one, in which water is one product and a pair of charges has been neutralized, is called a **neutralization reaction.**

Strong and Weak Acids

Hydrogen chloride, a colorless gas with a pungent, irritating odor, dissolves in water to give a solution of hydrochloric acid. In water, hydrogen chloride acts as an acid because it donates a proton to a water molecule, giving a hydronium ion and a chloride ion. As shown in Figure 4-8, water acts as the base, accepting a proton from the acid, hydrogen chloride. Proton transfer always involves an acid and a base.

Hydrogen chloride produces hydronium ions and chloride ions *quantitatively* when it dissolves in water, which means that virtually every molecule of HCl transfers its proton to a water molecule. As a result, the concentration of hydronium ions equals the concentration of the acid. The species present in an aqueous solution of HCl are Cl^-, H_3O^+, and of course, H_2O. Any acid that undergoes quantitative reaction with water to produce hydronium ions and the appropriate anion is called a **strong acid.** Table 4-2 gives the formulas and structures of six

**Table 4-2
Common Strong Acids**

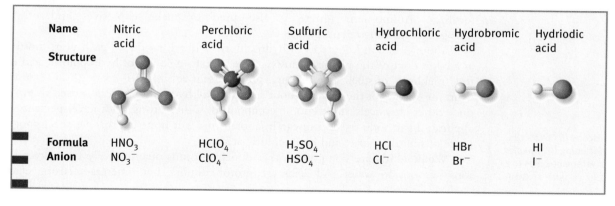

Name	Nitric acid	Perchloric acid	Sulfuric acid	Hydrochloric acid	Hydrobromic acid	Hydriodic acid
Structure						
Formula	HNO_3	$HClO_4$	H_2SO_4	HCl	HBr	HI
Anion	NO_3^-	ClO_4^-	HSO_4^-	Cl^-	Br^-	I^-

common strong acids, all of which are supplied commercially as concentrated aqueous solutions. These solutions are corrosive and normally are diluted for routine use in acid–base chemistry.

At the concentrations normally used in the laboratory, a solution of any strong acid in water contains H_3O^+ and anions that result from the loss of a proton. Example 4-12 shows a molecular view of the proton transfer reaction of a strong acid.

Example 4-12	Proton Transfer

In nitric acid, three oxygen atoms are bonded directly to a nitrogen atom, and a hydrogen is bonded to one of the oxygens. Write the net ionic equation and draw a molecular picture that illustrates the reaction between nitric acid and water.

Strategy: Nitric acid is one of the six common strong acids. This means that when nitric acid dissolves in water, each acid molecule transfers a proton to a water molecule, generating a hydronium ion and the appropriate anion. Both the reaction and its molecular representation must show this proton transfer.

Solution: In this example the anion is nitrate:

$$H_2O + HNO_3 \longrightarrow H_3O^+ + NO_3^-$$

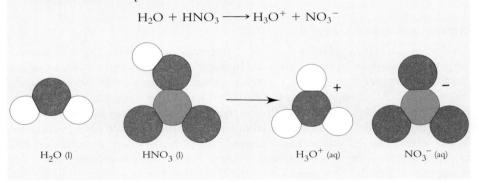

Among the strong acids, three are significant industrial chemicals. More than 40 billion kilograms of sulfuric acid are produced in the United States each year at a cost of about 8¢ per kg. About 60% is used in the production of fertilizers. Another 30% is used to manufacture detergents, drugs, explosives, dyes, paint, paper, and other chemicals. Most of the remaining 10% is used in petroleum refining and metallurgy. Nitric acid is also a valuable industrial chemical. About 90% of the nearly 8 billion kilograms produced each year is used to make NH_4NO_3, which is used as a fertilizer. Ammonium nitrate is also used to make explosives and other nitrogen-containing chemicals.

About 3 billion kilograms of hydrochloric acid are produced each year, mostly as a byproduct of the plastics industry. The largest single use of hydrochloric acid is the "pickling" of steel. The pickling process removes iron(III) oxide (Fe_2O_3, rust) from the surface of the metal. About a third of all hydrochloric acid is used to produce other chemicals, mostly ionic compounds. Other strong acids have specialized applications in industry and research laboratories, but none approaches the importance of sulfuric, nitric, and hydrochloric acids.

Weak acids are compounds that donate protons quantitatively to hydroxide ions but not to water. All acids are proton donors, but whereas a strong acid

As we describe in Chapter 16, weak acids, including HF, transfer protons to water molecules to a small extent.

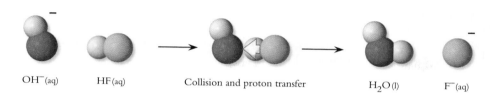

OH⁻(aq)   HF(aq)   Collision and proton transfer   H₂O(l)   F⁻(aq)

Figure 4-9
Proton transfer between HF and H₂O.

Animation

Formic acid
HCO₂H

Benzoic acid
C₆H₅CO₂H

quantitatively donates protons to water, a weak acid does not. Aqueous solutions of weak acids contain small concentrations of hydronium ions, making the solutions acidic, but nearly all the weak acid molecules remain intact. Representative of strong acids, HCl generates H_3O^+ and Cl^- quantitatively when dissolved in water, Representative of weak acids, HF remains predominantly as HF molecules when dissolved in water. However, HF donates protons quantitatively to OH^- ions to give H_2O molecules and F^- ions, as shown in Figure 4-9.

In contrast to the small number of strong acids, there are many weak acids. Some are biochemical substances produced by living organisms. For example, the sour tang of vinegar comes from acetic acid, CH_3CO_2H. Acetic acid has one acidic hydrogen atom, the one attached to the oxygen atom, as shown in Figure 4-10. Hydroxide ion will remove this proton to give the acetate ion. The hydrogen atoms bound to the carbon atom of acetic acid are not acidic. Neither water nor hydroxide will remove these hydrogen atoms. Acetic acid is a simple example of a very large family of organic compounds that contain the CO_2H grouping of atoms. Acids that contain the CO_2H group are called **carboxylic acids.** Carboxylic acids typically are weak acids. Two other common carboxylic acids are formic acid, HCO_2H and benzoic acid, $C_6H_5CO_2H$.

Sulfuric acid, H_2SO_4, is both a strong acid and a weak acid. The compound has two acidic hydrogen atoms. In an aqueous solution, one hydrogen atom undergoes quantitative proton transfer to water.

$$H_2SO_4(aq) + H_2O(l) \longrightarrow HSO_4^-(aq) + H_3O^+(aq)$$

However, the second hydrogen atom remains attached to HSO_4^-. Thus an aqueous solution of sulfuric acid contains H_3O^+ and HSO_4^- ions. Like other weak acids, the hydrogen sulfate ion reacts quantitatively with hydroxide to give sulfate ion and water.

$$HSO_4^-(aq) + OH^-(aq) \longrightarrow SO_4^{2-}(aq) + H_2O(l)$$

Another example of a weak acid is phosphoric acid, H_3PO_4, an important industrial compound used primarily to make fertilizer. All three of the hydrogen atoms of phosphoric acid will react with hydroxide ions. Example 4-13 illustrates the acid-base properties of phosphoric acid.

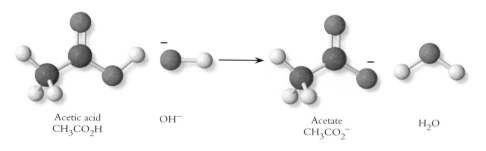

Acetic acid   OH⁻   Acetate   H₂O
CH₃CO₂H   CH₃CO₂⁻

Figure 4-10
Proton transfer between acetic acid and hydroxide.

| Example 4-13 | **Weak Acid Reaction with OH⁻ Ions** |

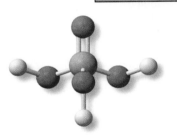

Phosphoric acid
H_3PO_4

Write the balanced net ionic equation for the reaction of phosphoric acid with an excess of aqueous potassium hydroxide.

Strategy: When an acid and a base react, the acid transfers a proton to the base. We must identify the species in solution and then evaluate the proton transfer chemistry.

Solution: An aqueous solution of potassium hydroxide contains K^+, OH^-, and H_2O. Phosphoric acid is a weak acid, so most of its molecules remain as H_3PO_4 in aqueous solution. The species present at the beginning of the reaction are K^+ and OH^- ions and molecules of H_3PO_4 and H_2O. The hydroxide ion is a powerful base that removes all of the acidic hydrogen atoms from both strong and weak acids. An excess of potassium hydroxide means that there are enough OH^- ions to remove all three acidic hydrogens from phosphoric acid. This reaction produces three molecules of water and one phosphate anion:

$$H_3PO_4(aq) + 3\,OH^-(aq) \longrightarrow PO_4^{3-}(aq) + 3\,H_2O(l)$$

Phosphoric acid, sulfuric acid, and the hydrogen sulfate ion represent a very large group of acids known as **oxyacids.** As the structures in Table 4-2 show, an oxyacid has a central atom bonded to a variable number of oxygen atoms and OH groups. Except for sulfuric acid, nitric acid, and perchloric acid, all of the oxyacids described in this textbook are weak acids. Chapter 16 describes in detail the chemistry of strong and weak acids, including carboxylic acids and oxyacids.

Strong and Weak Bases

The hydroxide ion is a powerful proton acceptor. Thus, any compound that generates stoichiometric quantities of hydroxide ions when it dissolves in water is called a **strong base.** All Group 1 hydroxides, AOH (where A = any of the Group 1 cations) are strong bases. Sodium hydroxide is one example:

$$NaOH(s) \longrightarrow Na^+(aq) + OH^-(aq)$$

Three of the Group 2 hydroxides, $Ca(OH)_2$, $Sr(OH)_2$, and $Ba(OH)_2$, are soluble strong bases.

Just as there are weak acids, there are also weak bases. A **weak base** does not readily accept protons from water molecules but does quantitatively accept protons from hydronium ions. Ammonia is the most common weak base. Ammonia exists predominantly as NH_3 molecules in aqueous solution, but it undergoes quantitative proton transfer with hydronium ions to generate ammonium ions.

$$NH_3(aq) + H_3O^+(aq) \longrightarrow NH_4^+(aq) + H_2O(l)$$

We describe other examples of weak bases in Chapter 16.

Acid-Base Stoichiometry

The quantitative aspects of acid-base chemistry obey the principles introduced earlier in this chapter. The common acid-base reactions that are important in general chemistry take place in aqueous solution, so acid-base stoichiometry uses molarities and volumes extensively. Example 4-14 illustrates the essential features of aqueous acid-base stoichiometry.

Acid–Base Stoichiometry	Example 4-14

What volume of 0.050 M nitric acid solution is needed to neutralize 0.075 L of 0.065 M $Ba(OH)_2$?

Strategy: Apply the seven-step method: Determine what's asked for, visualize the chemistry, organize the data, set up the appropriate molar equalities, and work toward a quantitative solution.

Solving
Quantitative
Problems

Solution:

1. The problem asks for the volume of acid that reacts stoichiometrically with a given amount of base.

2. To visualize and identify the reaction, start with species in solution. Nitric acid is one of the six strong acids, so an aqueous solution of nitric acid contains H_3O^+ cations and NO_3^- anions. A solution of $Ba(OH)_2$ contains Ba^{2+} and OH^- ions. (Of course, both solutions contain large quantities of H_2O molecules.) When the solutions are mixed, there might be precipitate formation and/or proton transfer. According to the solubility guidelines, all nitrate salts are soluble, so no precipitate forms. Hydronium ion is a strong acid, and hydroxide ion is a strong base. When a strong acid is mixed with a strong base, acid-base neutralization occurs rapidly. By the time the solutions are mixed thoroughly, the proton transfer reaction is complete:

$$H_3O^+(aq) + OH^-(aq) \longrightarrow 2\,H_2O(l)$$

3. The data provided are the volume and molarity of the base:

Base	Acid
$[Ba(OH)_2] = 0.065$ M	$[HNO_3] = 0.050$ M
$V = 0.075$ L	$V =$ asked for

4. The process to reach a quantitative solution to the problem requires working with moles. Thus we need the relationship linking moles to molarity and volume:

$$n = MV.$$

5. The equation has to be used in two ways.

   a. Calculate moles of OH^- using $n = MV$ and stoichiometric reasoning.
   b. Calculate volume of nitric acid solution using $V = n/M$.

6. First, determine how many moles of hydroxide must be neutralized:

$$[0.065 \text{ mol/L } Ba(OH)_2]\left(\frac{2 \text{ mol } OH^-}{1 \text{ mol } Ba(OH)_2}\right)(0.075 \text{ L}) = 9.75 \times 10^{-3} \text{ mol } OH^-$$

| Example 4-14 | Acid–Base Stoichiometry *(continued)* |

From the balanced net ionic equation, we see that neutralization requires one hydronium ion for every hydroxide ion, so we need 9.75×10^{-3} mol of hydronium ions. Knowing the concentration of the nitric acid, we can calculate the volume of solution that contains this number of moles:

$$V_{acid} = \frac{9.75 \times 10^{-3} \text{ mol H}_3\text{O}^+}{0.050 \text{ mol H}_3\text{O}^+/\text{L}} = 0.20 \text{ L}$$

7. The hydroxide ions are neutralized by 0.20 L of the acid solution. This is a bit more than twice the volume of the base solution. The answer is reasonable because each mole of $Ba(OH)_2$ contains two moles of hydroxide ions, but each mole of HNO_3 supplies just one mole of H_3O^+. Furthermore, the molarity of the base solution is larger than the molarity of the acid solution.

Titration

The amount of an acid in a sample can be determined by adding a solution of base of known concentration until the amount of base added exactly matches the amount of acid in the sample. Similarly, the amount of a base in a sample can be determined by adding a solution of acid of known concentration until the amount of acid added exactly matches the amount of base in the sample. This procedure is called **titration.** The solution to be analyzed is placed in a beaker or flask. A second solution of known concentration, called the **titrant,** is added slowly by means of a calibrated measuring vessel called a buret. Figure 4–11 shows a buret set up to carry out a titration.

In one type of titration, a solution of a strong base such as sodium hydroxide is added slowly to a solution that contains an unknown amount of an acid. Each hydroxide ion added to the acid solution accepts one proton from a molecule of acid. As the titration proceeds, fewer and fewer acid molecules remain in the beaker, but the solution is still acidic. At the **stoichiometric point,** just enough hydroxide ions have been added to react with every acidic proton present in the beaker before the titration was started. The hydroxide ions in the next drop of

Figure 4-11
Titration of HCl by NaOH.
Spectator ions have been omitted for clarity. (*a*) Initially, the HCl solution contains many H_3O^+ ions. (*b*) At the stoichiometric point, all the hydronium ions have been neutralized, and the solution contains only water molecules and spectator ions. (*c*) Past the stoichiometric point, the solution contains many OH^- ions.

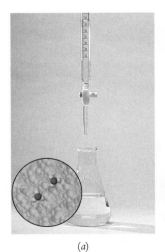

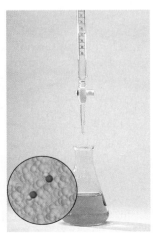

(*a*) (*b*) (*c*)

titrant do not react because acid molecules are no longer present in the solution. Before the stoichiometric point, the solution contains excess acid. After the stoichiometric point, the solution contains excess OH^-.

At the stoichiometric point, the amount of base added exactly matches the amount of acid originally present.

Stoichiometric point: mol OH^- added = mol acidic hydrogen present (4-3)

If we know the molarity of the titrant and measure the volume of titrant required to reach the stoichiometric point, we can calculate the number of moles of hydroxide required to react with all the acid. This allows us to determine the concentration of the unknown acid solution.

How do we know when the stoichiometric point of a titration has been reached? At the stoichiometric point, the nature of the acid-base species in the solution changes. We cannot "see" ions directly, but we can obtain a visual measure by placing a tiny amount of a substance called an **indicator** in the solution to be titrated. An indicator is a molecule whose color depends on the concentration of hydronium ions. A properly chosen indicator shows a distinct color change when the titration reaches the stoichiometric point and the acid in the solution has all reacted. In Figure 4-11, the indicator is phenolphthalein, a molecule that is colorless when hydronium ions are present but pink when hydroxide ions are present. This color change signals an end to the titration. As soon as we see the color change, we stop adding base and measure the volume that has been delivered from the buret. Example 4-15 illustrates how to calculate concentration from a titration volume.

Acid-Base Titration	**Example 4-15**

Industrial wastewater often is contaminated with strong acids. Environmental regulations require that such wastewater be neutralized before it is returned to the environment. A 1.50×10^2 mL sample of wastewater was titrated with 0.1250 M sodium hydroxide, and 38.65 mL of the base was required to reach the stoichiometric point. What was the molarity of hydronium ions in this sample of wastewater?

Strategy: We are asked for the molarity of an acid. The analysis is a titration. Knowing that the wastewater contains strong acid, we can write the general acid-base neutralization reaction:

$$H_3O^+ + OH^- \longrightarrow 2\,H_2O$$

The key feature of a titration is that at the stoichiometric point, the number of moles of OH^- added equals the number of moles of acid originally present. This lets us use the data for the base to calculate the number of moles of H_3O^+ in the sample.

Solution: Use the titration to calculate the number of moles of OH^- required to reach the stoichiometric point:

$$\text{mol } OH^- \text{ added} = MV$$

$$\text{mol } OH^- \text{ added} = (0.1250 \text{ mol/L})(38.65 \text{ mL})(10^{-3} \text{ L/mL}) = 4.831 \times 10^{-3} \text{ mol}$$

According to Equation 4-3, this is also the number of moles of acid in the original sample. Because the original sample had a volume of 1.50×10^2 mL, its molarity is as follows:

Example 4-15 | **Acid–Base Titration** *(continued)*

$$M = \frac{n}{V} = \frac{4.831 \times 10^{-3} \text{ mol}}{(1.50 \times 10^2 \text{ mL})(10^{-3} \text{ L/mL})} = 3.22 \times 10^{-2} \text{ mol/L}$$

Is this reasonable? An acid concentration around 10^{-2} M may not appear high, but it is more than 1000 times the concentration of hydronium ion in unpolluted streams.

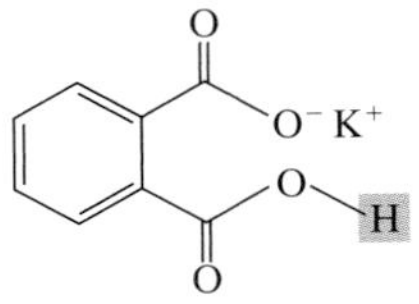

A titration requires a solution whose concentration is known. In Example 4-15 the NaOH solution used as the titrant was known to be 0.1250 M. A titrant of known concentration is known as a standard solution, and the concentration of such a solution is determined by a **standardization** titration. In a standardization titration, the solution being titrated contains a known amount of acid or base. An excellent acid for standardization is potassium hydrogen phthalate, $KHC_8H_4O_4$. This substance, a carboxylic acid that contains one weakly acidic hydrogen atom per molecule, is easily obtained as a highly pure solid. A known number of moles can be weighed on an analytical balance, dissolved in pure water, and then titrated with the base solution to be standardized. Example 4-16 illustrates the standardization procedure.

$KHC_8H_4O_4$

Example 4-16 | **Standardization**

A biochemist needed to standardize a solution of KOH. A sample of potassium hydrogen phthalate weighing 0.6745 g was dissolved in 100.0 mL of water and a drop of indicator was added. The solution was then titrated with the KOH solution. The titration required 41.75 mL of base to reach the stoichiometric point. Find the molarity of the KOH solution.

Strategy: First, we must identify the chemistry. This is an acid–base titration in which hydrogen phthalate anions (the acid) react with OH^- (the base). We use the molar equality of acid and base at the stoichiometric point together with the equations that link moles with mass and volume.

Solution: Hydrogen phthalate has one acidic hydrogen atom. As the net reaction shows, there is a 1:1 molar ratio between hydrogen phthalate and hydroxide:

$$HC_8H_4O_4^- + OH^- \longrightarrow C_8H_4O_4^{2-} + H_2O$$

The number of moles of hydrogen phthalate anions is the same as the number of moles of $KHC_8H_4O_4$, which is calculated from its mass and molar mass:

$$n_{HC_8H_4O_4^-} = n_{KHC_8H_4O_4} = \frac{0.6745 \text{ g}}{204.2 \text{ g/mol}} = 3.303 \times 10^{-3} \text{ mol}$$

At the stoichiometric point of the titration, the number of moles of hydroxide added from the buret equals the number of moles of hydrogen phthalate:

$$n_{OH^- \text{ added}} = 3.303 \times 10^{-3} \text{ mol}$$

$$\text{Volume added} = 41.75 \text{ mL} = 41.75 \times 10^{-3} \text{ L}$$

$$M = \frac{n}{V} = \frac{3.303 \times 10^{-3} \text{ mol}}{41.75 \times 10^{-3} \text{ L}} = 7.911 \times 10^{-2} \text{ M}$$

The technique of titration is equally useful for the titration of an unknown base by a solution of strong acid. The calculations proceed exactly as described previously. For the titration of a base, the stoichiometric point is reached when the number of moles of added acid in the titrant equals the number of moles of base in the unknown solution.

Stoichiometric point: Moles H_3O^+ added = Moles base present (4-4)

Section Exercises

4.6.1 Carbonic acid, H_2CO_3 (molecular model shown in the margin) is a weak oxyacid that forms when carbon dioxide dissolves in water. Carbonic acid contains two acidic hydrogen atoms. Write the net ionic reaction that occurs when carbonic acid reacts with an excess of hydroxide ions. Draw a molecular picture of the process.

4.6.2 The hydronium ion concentration of the wastewater in Example 4-15 is 3.22×10^{-2} M. What mass of NaOH would have to be added to 1.00×10^3 L of this wastewater to make it neutral?

4.6.3 While cleaning a laboratory, a technician discovers a large bottle containing a colorless solution. The bottle is labeled "$Ba(OH)_2$," but the molarity of the solution is not given. Concerned because of the toxicity of Ba^{2+} ions, the technician titrates with a solution of hydrochloric acid standardized at 0.1374 M. A 25.00-mL sample of the barium hydroxide solution requires 36.72 mL of the HCl solution to reach the stoichiometric point. What is the concentration of Ba^{2+} in the solution?

Carbonic acid

4·7 OXIDATION-REDUCTION REACTIONS

As described in Section 4.6, one important class of chemical reactions involves transfers of protons between chemical species. An equally important class of chemical reactions involves transfers of electrons between chemical species. These are **oxidation–reduction reactions.** Commonplace examples of oxidation-reduction reactions include the rusting of iron, the digestion of food, and the burning of gasoline. Paper manufacture, the subject of the Chemistry and the Environment Box on page 173, employs oxidation-reduction chemistry to bleach wood pulp. All metals used in the chemical industry and manufacturing are extracted and purified through oxidation-reduction chemistry, and many biochemical pathways involve the transfer of electrons from one substance to another.

The reaction of magnesium metal with aqueous strong acid, which is shown in Figure 4-12, illustrates the fundamental principles of oxidation-reduction. When a piece of magnesium is dropped into a solution of hydrochloric acid, a reaction starts almost immediately.

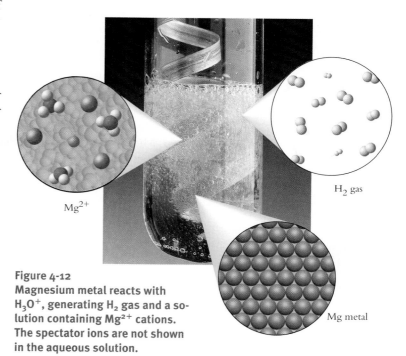

Figure 4-12
Magnesium metal reacts with H_3O^+, generating H_2 gas and a solution containing Mg^{2+} cations. The spectator ions are not shown in the aqueous solution.

The metal dissolves, and gas bubbles from the solution. The gas is H_2, and analysis of the solution reveals the presence of Mg^{2+} ions. A list of chemical species before and after the reaction indicates what has taken place:

Before Reaction		After Reaction	
$Mg(s)$	(reactant)	$Mg^{2+}(aq)$	(product)
$H_3O^+(aq)$	(reactant)	$H_2(g)$	(product)
$Cl^-(aq)$	(spectator)	$Cl^-(aq)$	(spectator)
$H_2O(l)$	(solvent)	$H_2O(l)$	(solvent)
		$H_3O^+(aq)$	(excess reactant)

Solid magnesium has been transformed into Mg^{2+} ions, and hydronium ions have decomposed to give H_2 gas and water molecules. Quantitative measurements reveal that for every mole of Mg consumed, the reaction also consumes two moles of H_3O^+, and it produces one mole of H_2 and two moles of water. The reaction can be summed up in the following balanced chemical equation:

$$Mg(s) + 2\ H_3O^+(aq) \longrightarrow Mg^{2+}(aq) + H_2(g) + 2\ H_2O(l)$$

By examining the changes occurring for magnesium and hydronium ions, we discover that Mg atoms lose electrons during this reaction and hydronium ions gain electrons:

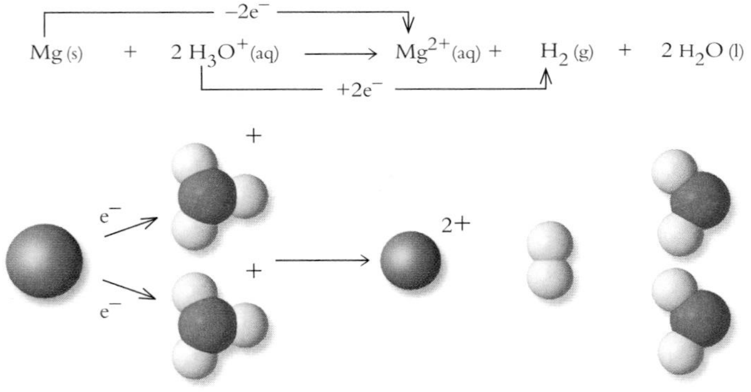

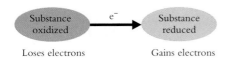

The loss of electrons by magnesium atoms to form Mg^{2+} cations indicates that this reaction between magnesium metal and hydronium ions involves oxidation and reduction. An atom of magnesium is oxidized, losing two electrons to form a Mg^{2+} cation. Because electrons must be conserved in every chemical process, electrons lost by Mg must be gained by some other species. In this example the electrons lost by Mg are gained by H_3O^+, which is reduced to form H_2 and H_2O.

Conservation of electrons is the basis of oxidation and reduction. Gains and losses of electrons always occur together.

/// Oxidation is the loss of electrons from a substance.
/// Reduction is the gain of electrons by a substance.

Because of this necessary connection of reduction with oxidation, the process is often referred to simply as a **redox reaction** (*re*duction–*ox*idation). Figure 4-13 summarizes the redox process.

Each side of this balanced equation has a net charge of +2. Remember, however, that the net ionic equation does not show all of the species present in solution. In this example, Cl^- ions balance out the positive charges of Mg^{2+} and H_3O^+. All solutions have a net charge of zero, so whenever a net ionic equation has an overall charge, there are spectator ions present that neutralize the charge.

Figure 4-13
Oxidation (loss of electrons) and reduction (gain of electrons) must always occur together.

Box 4-2	Chemistry and the Environment: Paper Without Pollution

Examine the page that is open before you: Not the words, but the material itself, paper. We often take this product for granted, but paper-making is one of the most important developments in the advance of civilization. According to legend, the first sheets of paper were made from mulberry leaves in China in AD 105. For many centuries paper was made in individual sheets, so it was a rare and expensive commodity. Paper-making machines were first developed in the early years of the nineteenth century. The development of machinery that allowed high-speed paper production was partially responsible for the increase in literacy and education of people around the world.

Modern paper-making is chemically intensive. Strong bases are used to convert wood fibers into pliable pulp and oxidizing agents are used to bleach the pulp. Inorganic materials are added as fillers that increase opacity, and polymer binders are added to increase strength. Finally, paper is treated with organic agents called "sizers" that improve its printing qualities.

The high need for chemicals in the various paper processing steps, however, has made the elimination of polluting wastes difficult to achieve.

Making paper without pollution requires that each part of the process be nonpolluting. The chemicals most commonly used in the production of pulp are $NaOH$ and Na_2S. In modern paper mills, sulfur-containing byproducts are scrubbed from the plant exhaust, and the aqueous sodium hydroxide is reclaimed and recycled. The fillers used to make paper opaque—titanium dioxide, calcium carbonate, and kaolin (a clay)—are natural, nonpolluting minerals. The polymer binders and sizers are relatively easily recaptured from the aqueous waste stream.

The bleaching process, in contrast, poses major difficulties. Traditional paper bleaching uses chlorine gas, which is reduced to chloride anions, Cl^-, as it oxidizes the colored pigments in wood pulp. The chloride anion is not a pollutant, as it is a major species in the oceans. Unfortunately, chlorine processing also generates small quantities of chlorine-containing dioxins such as 2,3,7,8-tetrachloro-dibenzo-p-dioxin, whose structure (left) appears less formidable than its name.

$C_{12}H_4Cl_4O_2$

Dioxins are of concern because they accumulate in the biosphere, where they have highly deleterious effects. Tests have shown that when the concentration of dioxins in the blood of laboratory animals reaches a critical level, reproductive and immune system defects result. Moreover, recent data indicate that the concentration of dioxins in the blood of the average U.S. resident has nearly reached that level. A major reason is that dioxins are not very water-soluble, so they accumulate in the body rather than being readily processed and excreted. Consequently, several groups, including the American Public Health Association, have issued calls for phase-out of the use of industrial chlorine.

The paper industry has responded by replacing chlorine gas with other oxidizing agents. Among the alternatives are sodium chlorate ($NaClO_3$) and hydrogen peroxide (H_2O_2). By 1997 it was estimated that half the U.S. paper industry had converted its bleaching process to chlorine dioxide generated from sodium chlorate. Although advocates of sodium chlorate claim that no detectable dioxins are produced by this bleaching agent, others contend that only completely chlorine-free bleaches such as hydrogen peroxide will be completely safe. It may be only a matter of time before all chlorine-based bleaches in paper-making are discontinued. If that occurs, the paper industry may be the first major chemical-intensive industry to succeed at sustainable development.

Redox reactions are more complicated than precipitation or proton transfer reactions because the electrons transferred in redox chemistry do not appear in the balanced chemical equation. Instead, they are "hidden" among the starting materials and products. However, we can keep track of electrons by writing two **half-reactions** that describe the oxidation and the reduction separately. A half-reaction is a balanced chemical equation that includes electrons and describes either the oxidation or reduction but not both. Thus a half-reaction describes half of a redox reaction. Here are the half-reactions for the redox reaction of magnesium and hydronium ions:

$$
\begin{array}{rcll}
Mg & \longrightarrow & Mg^{2+} + 2e^- & \text{Oxidation} \\
2\,H_3O^+ + 2e^- & \longrightarrow & H_2 + 2\,H_2O & \text{Reduction} \\
\hline
\text{Net:}\quad 2\,H_3O^+ + Mg & \longrightarrow & Mg^{2+} + H_2 + 2\,H_2O &
\end{array}
$$

Separating oxidation from reduction makes it possible to verify that electrons are conserved in a redox reaction. Note that the electrons produced in the oxidation of magnesium are consumed in the reduction of hydronium ions. The electrons required for a reduction must come from an oxidation.

A detailed discussion of redox reactions must wait until Chapter 18, after we explore the nature of the atom, periodic properties of the elements, and thermodynamics. For now, we focus on only a few types of redox reactions that are common and relatively simple.

Figure 4-14
Zinc metal displaces copper ions from aqueous solution. The blue color signals the presence of copper ions. The color fades and copper metal appears as the reaction proceeds.

Metal Displacement

When a strip of zinc metal is added to a solution of copper(II) sulfate, the blue color slowly fades, and the zinc metal is replaced by copper metal (Figure 4-14). Copper ions in the solution are reduced to copper metal, whereas zinc atoms are oxidized to Zn^{2+} cations. This is an example of a **metal displacement reaction,** in which a

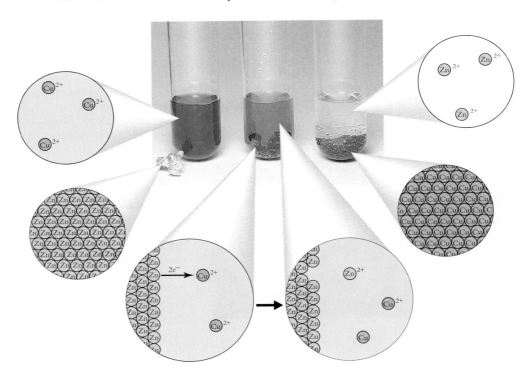

metal ion in solution (Cu^{2+}) is displaced by another metal (Zn) by means of a redox reaction. Figure 4–14 also shows molecular views of this displacement reaction.

$$\text{Oxidation: } Zn(s) \longrightarrow Zn^{2+}(aq) + 2\ e^- \qquad \text{Reduction: } Cu^{2+}(aq) + 2\ e^- \longrightarrow Cu(s)$$

$$\text{Redox: } Zn(s) + Cu^{2+}(aq) \longrightarrow Zn^{2+}(aq) + Cu(s)$$

Many other metal displacement reactions can be visualized, but not all of them occur. Some metals are oxidized readily, but others are highly resistant to oxidation. Likewise, some metal cations are highly susceptible to reduction, but others resist reduction. Zinc displaces copper ions from aqueous solutions, but copper will not replace zinc ions, because Cu^{2+} is easier to reduce than Zn^{2+}. Zinc will not displace Mg^{2+} ions, because magnesium cations are harder to reduce than zinc cations. Copper ions are displaced by aluminum but not by silver. After many pairs of elements are compared, it becomes possible to arrange the metals in order of their reactivity. Such a list, called an **activity series,** is shown in Table 4–3. Metals at the top of the list are easier to oxidize than metals at the bottom. Conversely, metal cations at the bottom of the list are easier to reduce than those at the top. A metal will transfer electrons to any cation that is lower on the list. Furthermore, the list is in order of reactivity, so the greater the separation between the species, the more vigorous the reaction. Example 4–17 shows how to use the activity series to predict the outcome of a metal displacement reaction.

The Activity Series	Example 4-17

Predict whether a displacement reaction will occur in the following instances. Explain your conclusions.

(a) A small piece of calcium is added to a solution of nickel(II) chloride.

(b) Iron pellets are added to a solution of magnesium chloride.

(c) A strip of aluminum foil is dipped in aqueous silver nitrate.

Strategy: A displacement reaction occurs when a metal high in the activity series is added to a solution containing a cation lower on the list. For each case, we need to identify the species present in the mixture and make an evaluation based on the activity series.

Solution:

(a) Solid Ca metal is added to a solution that contains Ni^{2+} and Cl^- ions. Ca, near the top of the activity series, is one of the most reactive metals. Ca will displace most other metal ions from solution, including Ni^{2+}:

$$\text{Oxidation: } Ca(s) \longrightarrow Ca^{2+}(aq) + 2\ e^- \qquad \text{Reduction: } Ni^{2+}(aq) + 2\ e^- \longrightarrow Ni(s)$$

$$\text{Redox: } Ca(s) + Ni^{2+}(aq) \longrightarrow Ca^{2+}(aq) + Ni(s)$$

(b) Solid iron metal is added to a solution that contains Mg^{2+} and Cl^-. Iron is below Mg^{2+} in the activity series, so no reaction will occur.

(c) Solid aluminum metal is dipped in a solution that contains Ag^+ cations and NO_3^- anions. Aluminum is in the middle of the metals, and Ag^+ is near the bottom of the ions. Thus aluminum displaces Ag^+:

$$\text{Oxidation: } Al(s) \longrightarrow Al^{3+}(aq) + 3\ e^- \qquad \text{Reduction: } 3\ [Ag^+(aq) + e^- \longrightarrow Ag(s)]$$

$$\text{Redox: } 3\ Ag^+(aq) + Al(s) \longrightarrow 3\ Ag(s) + Al^{3+}(aq)$$

Table 4-3
Activity Series of the Metals

		Ion	Metal	
	Cations difficult to displace	$K^+ + e^- \rightleftharpoons K$		Metals that react with both H_2O and H_3O^+
Increasing ease of reduction		$Ca^{2+} + 2e^- \rightleftharpoons Ca$		
		$Na^+ + e^- \rightleftharpoons Na$		
		$Mg^{2+} + 2e^- \rightleftharpoons Mg$		
		$Al^{3+} + 3e^- \rightleftharpoons Al$		
		$Zn^{2+} + 2e^- \rightleftharpoons Zn$		Metals that react with H_3O^+
		$Fe^{2+} + 2e^- \rightleftharpoons Fe$		
		$Ni^{2+} + 2e^- \rightleftharpoons Ni$		
		$Pb^{2+} + 2e^- \rightleftharpoons Pb$		
	Hydronium ion reduction	$H_3O^+ + e^- \rightleftharpoons \frac{1}{2}H_2 + H_2O$		
		$Cu^{2+} + 2e^- \rightleftharpoons Cu$		
	Easily displaced cations	$Ag^+ + e^- \rightleftharpoons Ag$		Unreactive metals
		$Au^{3+} + 3e^- \rightleftharpoons Au$		

(Left arrow: Increasing ease of reduction; Right arrow: Increasing ease of oxidation)

Oxidation of Metals by H_3O^+ and H_2O

The redox reaction opening this section is the oxidation of magnesium metal by hydronium ions. It turns out that most metals are oxidized by H_3O^+, as can be seen by the location of the hydronium ion reduction reaction in the activity series of Table 4–3. All the metals above hydrogen will displace hydronium ions from solution. In each case the reduction half-reaction is the same one involved in the magnesium reaction: Hydronium ions are reduced to water and hydrogen gas. The electrons required to reduce H_3O^+ are provided by the metal, which is oxidized to its most stable cation. Remember that Group 1 metals form cations with $+1$ charge, and Group 2 metals form cations with $+2$ charge. Aluminum forms Al^{3+}, zinc forms Zn^{2+}, cadmium forms Cd^{2+} and silver forms Ag^+. Other metals, such as iron and copper, form more than one stable cation. In these cases the cationic charge must be specified before a balanced redox reaction can be written. Table 4–3 lists reduction half-reactions of several metals. Here are some examples of metals reacting with H_3O^+:

$$2\,Li(s) + 2\,H_3O^+(aq) \longrightarrow 2\,Li^+(aq) + H_2(g) + 2\,H_2O(l)$$

$$Fe(s) + 2\,H_3O^+(aq) \longrightarrow Fe^{2+}(aq) + H_2(g) + 2\,H_2O(l)$$

$$2\,Al(s) + 6\,H_3O^+(aq) \longrightarrow 2\,Al^{3+}(aq) + 3\,H_2(g) + 6\,H_2O(l)$$

A few metals are so easy to oxidize that they react vigorously with water. Sodium, for example, reduces water to H_2 and OH^-. Here are the half-reactions:

$$2\,Na(s) \longrightarrow 2\,Na^+(aq) + 2\,e^-$$

$$2\,H_2O(l) + 2\,e^- \longrightarrow H_2(g) + 2\,OH^-(aq)$$

Net: $2\,Na(s) + 2\,H_2O(l) \longrightarrow 2\,Na^+(aq) + H_2(g) + 2\,OH^-(aq)$

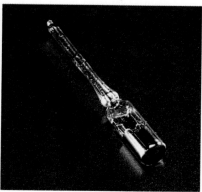

Figure 4-15
Sodium and other Group 1 metals react vigorously with water. The energy released in this redox reaction can ignite the hydrogen gas with spectacular and potentially dangerous results. Consequently, Group 1 metals must be protected from exposure to the environment. Shown here is a sample of cesium metal sealed in a glass ampule.

As Figure 4–15 shows, this redox reaction releases enough energy to ignite the highly flammable hydrogen gas. All the Group 1 metals react in a manner similar to sodium, so these elements must be protected from exposure to water and oxygen. The Group 2 metals also react with water, but much more slowly.

Oxidation by Molecular Oxygen

Almost all elements combine with molecular oxygen to form binary oxides. In fact, loss of electrons is called *oxidation* because elements lose electrons when they combine with oxygen. The metals of Groups 1 and 2 form oxides that are ionic solids composed of metal cations and oxide anions. The oxygen atoms of O_2 capture two electrons each to form a pair of oxide anions. Every O^{2-} anion requires a counterbalancing cationic charge of $+2$. Group 1 metals form A^+ cations, so their oxides have the chemical formula A_2O, whereas Group 2 metals form A^{2+} cations and oxides with the chemical formula AO.

The stoichiometry of transition metal oxides is more variable. Iron, for example, forms three binary oxides. In FeO the iron atoms have lost two electrons each (Fe^{2+}, O^{2-}), and in Fe_2O_3 they have lost three electrons each (Fe^{3+}, O^{2-}). In Fe_3O_4 one third of the cations are Fe^{2+} and the other two thirds are Fe^{3+}. Other transition metals lose anywhere from one to eight electrons in forming oxides. There is no simple pattern that determines which oxide is most stable, but we can tell how many electrons the metal has lost by assigning a gain of two electrons to each oxygen atom in the chemical formula. Thus the metal atoms in Cu_2O have lost one electron each, those in CrO_3 have lost six electrons each, and those in TiO_2 have lost four electrons each.

As we might expect, the more easily a metal is oxidized, the more likely it is to react with molecular oxygen. The least reactive metals—those at the bottom of the activity series—are among the few elements found in their elemental forms in

← CHAPTERS 18 & 19
Metallurgy and redox reactions that do not involve metal ions are discussed in Chapters 18 and 19.

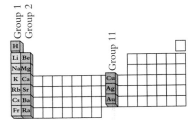

the Earth's crust. These metals fall in Group 11 of the periodic table. They are used for durable items such as coins and jewelry. In contrast, the metals at the top of the activity series (those in Groups 1 and 2 of the periodic table) react vigorously with oxygen. Magnesium, for example, can be set ablaze with a match, and cesium bursts into flame spontaneously when exposed to air. The resulting oxides react readily with water to generate aqueous solutions of the metal cations. Eventually, these cations precipitate as metal salts, so Group 1 and 2 metals are found in the Earth's crust as salts rather than oxides. Examples are rock salt (sodium chloride, $NaCl$), limestone (calcium carbonate, $CaCO_3$), and epsomite (magnesium sulfate heptahydrate, $MgSO_4 \cdot 7H_2O$).

Almost all other metals react relatively rapidly with oxygen to form insoluble oxides. These metals are found in nature as minerals. Metallurgical processes combine chemistry and engineering to reduce metals in minerals to their elemental forms. In the presence of air, these metals tend to return to their oxidized states: Objects made from iron must be protected from oxygen and water, or they react to form hydrated iron(III) oxide, better known as rust. Objects made of aluminum also react with O_2, but the product of this redox reaction, Al_2O_3, forms an impervious film on the metal surface that prevents O_2 from reaching the underlying metal. Moving farther up the activity series, the reactions with oxygen become more vigorous.

Compounds and nonmetallic elements also undergo redox reactions with oxygen. The nonmetals form binary oxides of varying stoichiometry. Once again, we can determine electron losses for oxygen's partner by viewing each oxygen atom as gaining two electrons during the redox reaction. Examples include CO_2 (C loses 4 e^-), SO_2 (S loses 4 e^-), SO_3 (S loses 6 e^-), and N_2O_5 (each N loses 5 e^-). These products do not contain ions, but they are redox reactions nonetheless. The combustion reactions described in Chapter 3 are examples of redox reactions of chemical compounds. In a combustion reaction, oxygen combines with carbon and hydrogen to give CO_2 and H_2O. In the process carbon atoms lose electrons, and oxygen atoms gain electrons.

Example 4-18 shows how to analyze the various reaction types discussed in Sections 4.5 to 4.7.

Example 4-18	**Types of Chemical Reactions**

Predict what will happen when the following pairs of substances are allowed to react. Write a balanced chemical equation for each reaction. When the reaction involves ions, write a net ionic equation. Identify each reaction as precipitation, as acid-base, or as redox. (a) $AgNO_3(aq)$ and $NaCl(aq)$; (b) $HCl(aq)$ and $Zn(s)$; (c) $NaOH(s)$ and $CH_3CO_2H(aq)$; (d) $Ca(s)$ and $H_2O(l)$; and (e) $K(s)$ and $O_2(g)$.

Strategy: Begin by identifying what chemical species are present in the mixture. Then classify the species and identify which category of reaction is involved. Recognizing a few key species can give important clues about chemical reactivity. For example, the presence of H_3O^+ signals either an acid-base reaction or the oxidation of a metal. Hydroxide ions suggest either an acid-base reaction or a precipitation. A solid metal indicates a redox reaction with either hydronium ion or molecular oxygen.

| **Types of Chemical Reactions** (*continued*) | **Example 4-18** |

Solution:

(a) Silver nitrate and sodium chloride are both salts. They dissolve in water to generate the ionic species Ag^+, NO_3^-, Na^+, and Cl^-. Neither solution contains H_3O^+ or OH^-, so this is not an acid–base reaction. There is no solid metal present, so a redox reaction is unlikely. A precipitation reaction is possible, so we apply the solubility guidelines. These indicate that $NaNO_3$ is soluble in water, but $AgCl$ is not. We conclude that this is a precipitation reaction:

$$Ag^+(aq) + Cl^-(aq) \longrightarrow AgCl(s)$$

(b) Hydrochloric acid contains H_3O^+ and Cl^-. The presence of hydronium ions and a solid metal suggests a redox reaction. According to the activity series, zinc is one of the metals that reacts with H_3O^+. We can write the appropriate half-reactions and then combine them to give the overall redox reaction:

Reduction: $2 H_3O^+(aq) + 2 e^- \longrightarrow H_2(g) + 2 H_2O(l)$

Oxidation: $Zn(s) \longrightarrow Zn^{2+}(aq) + 2 e^-$

Redox: $2 H_3O^+(aq) + Zn(s) \longrightarrow H_2(g) + 2 H_2O(l) + Zn^{2+}(aq)$

(c) Acetic acid is a weak acid, so its solution contains mostly CH_3CO_2H and H_2O molecules. Sodium hydroxide dissolves to produce Na^+ ions and OH^- ions. Hydroxide ion, a strong base, removes the acidic hydrogen from acetic acid in an acid–base reaction:

$$CH_3CO_2H(aq) + OH^-(aq) \longrightarrow CH_3CO_2^-(aq) + H_2O(l)$$

(d) The presence of solid calcium metal suggests a redox reaction. This Group 2 metal is oxidized by water according to the following half-reactions:

Oxidation: $Ca(s) \longrightarrow Ca^{2+}(aq) + 2 e^-$

Reduction: $2 H_2O(l) + 2 e^- \longrightarrow H_2(g) + 2 OH^-(aq)$

Redox: $Ca(s) + 2 H_2O(l) \longrightarrow Ca^{2+}(aq) + 2 OH^-(aq) + H_2(g)$

(e) Potassium, a Group 1 metal, is readily oxidized by molecular oxygen. The redox chemistry is best seen through the individual half-reactions:

Oxidation: $K(s) \longrightarrow K^+ + e^-$ Reduction: $O_2(g) + 4 e^- \longrightarrow 2 O^{2-}$

Redox: $4 K(s) + O_2(g) \longrightarrow 2 K_2O(s)$

Section Exercises

■ **4.7.1** Although aluminum cans are not attacked by water, strong acid oxidizes Al to Al^{3+} cations, liberating hydrogen gas in the process. How many moles of H_2 gas will be liberated if a 2.43-g sample of pure Al metal reacts completely with an excess of 3.00 M sulfuric acid solution?

■ **4.7.2** Metallic titanium is produced from rutile ore (TiO_2) by a direct replacement reaction with magnesium that frees titanium metal and produces MgO. (a) Write the balanced redox reaction for this process and identify the species that are oxidized and reduced. (b) How many kilograms of magnesium are required to manufacture 1.00×10^2 kg of titanium metal, assuming that the reaction has a 100% yield?

■ **4.7.3** Predict whether or not a reaction will occur, and if a reaction does take place, write the half-reactions and the balanced redox reaction: (a) A strip of nickel wire is dipped in 6.0 M HCl. (b) Aluminum foil is dipped in aqueous $CaCl_2$. (c) A lead rod is dipped in a beaker of water. (d) An iron wire is immersed in a solution of silver nitrate.

■■■ **CHAPTER REVIEW**

Summary and Key Terms

1 Chemical reactions, such as those used to synthesize useful chemical compounds, are described by **balanced chemical equations** that identify the **reactants**, the **products**, and the **stoichiometric coefficients**. In a properly balanced chemical equation, the number of atoms of each element is conserved, and so is electrical charge. Many chemical equations can be balanced by inspection.

2 **Stoichiometry** is the determination of the amounts of reagents that participate in a reaction. Stoichiometric calculations are centered around the mole and use the **stoichiometric ratios** expressed in the balanced chemical equation.

3 The amount of product predicted by stoichiometry is the **theoretical yield**, but products of a reaction are seldom formed quantitatively. Instead, every reaction has an **actual yield**, the amount actually obtained. The **percent yield** is 100% times the amount actually obtained divided by the amount predicted by stoichiometric ratios.

4 When supplies of two or more reactants are limited, a **limiting reactant** is consumed first. The limiting reactant determines the stoichiometry of a reaction. Limiting reactant problems are solved conveniently using tables of amounts, which summarize the amounts consumed and produced.

5 Salts are **soluble** or **insoluble** in water, and there are regular patterns of solubility. Mixing solutions of salts may result in a **precipitation reaction**. A **net ionic equation** describes precipitate formation and is used for stoichiometric calculations on precipitation reactions. **Spectator ions** do not participate in the net ionic reaction.

6 An **acid-base reaction** involves transfer of a proton from one species to another. **Acids** are proton donors; **bases** are proton acceptors. Acid-base reactions between cations and anions are **neutralization reactions**. A **strong acid** quantitatively transfers hydronium ions in water, a **weak acid** reacts quantitatively with hydroxide ions, a **strong base** quantitatively transfers hydroxide ions in water, and a **weak base** reacts quantitatively with hydronium ions. **Carboxylic acids** and **oxyacids** are common types of weak acid. Acid-base stoichiometry involves molarity-volume-mole conversions. Acid-base analysis can be carried out by **titration**, in which a **titrant** is added to a solution being studied. The **stoichiometric point,** at which the amount of titrant exactly matches the amount of substance being titrated, is often found using an **indicator** whose color depends on the hydronium ion concentration. The concentrations of solutions used in titrations can be determined by **standardization** titrations.

7 **Oxidation-reduction reactions**, also called **redox reactions,** involve the transfer of electrons between species. Oxidation, which is the loss of electrons, and reduction, which is the gain of electrons, always occur together. **Half-reactions** can be used to identify the oxidation and reduction portions of a redox reaction. **Metal displacement reactions** involve one metal being oxidized and another reduced. The **activity series** describes the relative ease of oxidation of the metals. Hydronium ions and molecular oxygen are particularly effective at causing oxidation of metals.

Skills to Master

▶ Balancing chemical equations by inspection

▶ Using stoichiometric ratios

▶ Determining and using reaction yields

▶ Identifying the limiting reactant

▶ Constructing tables of amounts

▶ Identifying species in aqueous solutions

▶ Writing net ionic equations

▶ Identifying soluble salts, acids, and bases

▶ Calculating amounts in titrations

▶ Recognizing simple redox reactions

Learning Exercises

4.1 Several examples of chemical reasoning are introduced in this chapter. Write out the reasoning steps that you will follow in (a) balancing a chemical equation; (b) identifying the limiting reactant; (c) determining whether a precipitate forms; and (d) computing a reaction yield.

4.2 List features that can be used to distinguish each of the three major types of reactions introduced in this chapter.

4.3 Draw one specific molecular picture that illustrates each of the reaction types introduced in this chapter.

4.4 Construct a flowchart that summarizes the problem solving strategy used for Example 4-18.

4.5 Define each of these terms: (a) percent yield; (b) limiting reactant; (c) spectator ion; (d) precipitate; (e) titration; and (f) oxidation.

4.6 Update your list of memory bank equations. For each equation, add a phrase that describes the kinds of chemical problems for which the equation is used.

4.6 List all terms new to you that appear in Chapter 4. In your own words, give a one-sentence definition of each. Consult the glossary if you need help.

Problems ilw = interactive learning ware problem. Visit the website at www.wiley.com/college/olmsted

Writing Chemical Equations

4.1 Balance the following chemical equations:
(a) $NH_4NO_3 \rightarrow N_2O + H_2O$
(b) $P_4O_{10} + H_2O \rightarrow H_3PO_4$
(c) $HIO_3 \rightarrow I_2O_5 + H_2O$
(d) $As + Cl_2 \rightarrow AsCl_5$

4.2 Balance the following chemical equations:
(a) $N_2O_5 + H_2O \rightarrow HNO_3$
(b) $KClO_3 \rightarrow KCl + O_2$
(c) $Fe + O_2 + H_2O \rightarrow Fe(OH)_2$
(d) $P_4 + Cl_2 \rightarrow PCl_3$

4.3 Draw a molecular picture that illustrates the reaction in Problem 4.1(d).

4.4 Draw a molecular picture that illustrates the reaction in Problem 4.2(d).

4.5 Balance the chemical equations for the following important industrial processes:
(a) Molecular hydrogen and carbon monoxide react to form methanol (CH_3OH).
(b) $CaO + C \rightarrow CO + CaC_2$
(c) $C_2H_4 + O_2 + HCl \rightarrow C_2H_4Cl_2 + H_2O$

4.6 The following reactions play roles in the manufacture of nitric acid. Write a balanced equation for each of them.
(a) $NH_3 + O_2 \rightarrow NO + H_2O$
(b) Nitrogen oxide and molecular oxygen combine to give nitrogen dioxide.
(c) Nitrogen dioxide and water react to form nitric acid and nitrogen oxide.
(d) $NH_3 + O_2 \rightarrow N_2 + H_2O$
(e) $NH_3 + NO \rightarrow N_2 + H_2O$

4.7 Draw a molecular picture that illustrates the reaction in Problem 4.5(a).

4.8 Draw a molecular picture that illustrates the reaction in Problem 4.6(b).

4.9 Here are some reactions involving water. Balance the equations:
(a) $Ca(OH)_2 + H_3PO_4 \rightarrow H_2O + Ca_3(PO_4)_2$
(b) $Na_2O_2 + H_2O \rightarrow NaOH + H_2O_2$
(c) $BF_3 + H_2O \rightarrow HF + H_3BO_3$
(d) $NH_3 + CuO \rightarrow Cu + N_2 + H_2O$

4.10 Here are some reactions of phosphorus and its compounds. Balance the equations:
(a) $P_4 + Na \rightarrow Na_3P$
(b) $Na_3P + H_2O \rightarrow PH_3 + NaOH$
(c) $PH_3 + O_2 \rightarrow P_4O_{10} + H_2O$
(d) $P_4O_{10} + H_2O \rightarrow H_3PO_4$

The Stoichiometry of Chemical Reactions

4.11 For each reaction in Problem 4.5, calculate the mass of the second reactant that is required to react completely with 5.00 g of the first reactant.

4.12 For each reaction in Problem 4.6, calculate the mass (in kg) of the first reactant that is required to react completely with 875 kg of the second reactant.

4.13 Iodine can be prepared by bubbling chlorine gas through an aqueous solution of sodium iodide:

$$2\,NaI(aq) + Cl_2(g) \longrightarrow I_2(s) + 2\,NaCl(aq)$$

What mass (in g) of sodium iodide is required to produce 1.50 kg of iodine?

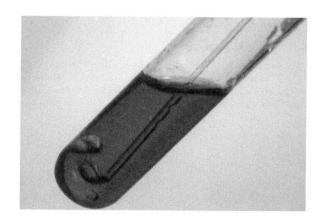

4.14 The fertilizer ammonium sulfate is prepared by the reaction between ammonia and sulfuric acid.

$$2\,NH_3(g) + H_2SO_4(aq) \longrightarrow (NH_4)_2SO_4(aq)$$

What mass (in kg) of NH_3 is needed to make 3.50 metric tons (1 metric ton = 1000 kg) of ammonium sulfate?

4.15 The fermentation of sugar to produce ethyl alcohol occurs by the following reaction:

$$C_6H_{12}O_6(s) \xrightarrow{\text{yeast}} 2\,C_2H_5OH(l) + 2\,CO_2(g)$$

What mass of ethyl alcohol can be made from 1.00 kg of sugar?

4.16 One starting material for the preparation of nylon is adipic acid. Adipic acid is produced from the oxidation of cyclohexane:

$$2\ \text{(cyclohexane)} + 5\,O_2 \longrightarrow 2\ \text{HO–(adipic acid)–OH} + 2\,H_2O$$

Cyclohexane
C_6H_{12}

Adipic acid
$C_6H_{10}O_4$

If 375 kg of cyclohexane reacts with an unlimited supply of oxygen, how much adipic acid (in kg) can be formed?

ilw 4.17 The use of freons is being phased out because of the damage these compounds do to the stratospheric ozone layer. One of the freons, CCl_2F_2, is manufactured from carbon tetrachloride and hydrogen fluoride. The other reaction product is hydrogen chloride. If a manufacturer wishes to convert 175 kg of carbon tetrachloride into CCl_2F_2, what is the minimum mass of hydrogen fluoride required? What masses of the two products will be obtained if this reaction is 100% efficient?

CCl_2F_2

4.18 One way to clean up emissions of sulfur dioxide formed during the burning of coal is to pass the exhaust gases through a water slurry of powdered limestone. In the reaction, sulfur dioxide reacts with calcium carbonate and oxygen gas to produce calcium sulfate and carbon dioxide. A coal-burning plant that uses coal with a typical sulfur content produces 2.00×10^4 metric tons (1 metric ton = 1000 kg) of sulfur dioxide in a month. How many metric tons of calcium carbonate will be consumed during the scrubbing process, assuming 100% efficiency? How many metric tons of calcium sulfate waste will the plant have to dispose of? (The process is actually about 90% efficient.)

Yields of Chemical Reactions

ilw 4.19 Approximately 12 billion kilograms of phosphoric acid are produced annually for fertilizers, detergents, and agents for water treatment. Phosphoric acid can be prepared by heating the mineral fluoroapatite with sulfuric acid in the presence of water.

$$Ca_5(PO_4)_3F + 5\,H_2SO_4 + 10\,H_2O \longrightarrow$$
Fluoroapatite

$$3\,H_3PO_4 + 5\,CaSO_4 \cdot 2H_2O + HF$$
Phosphoric acid

If every kilogram of fluoroapatite yields 4.00×10^2 g of phosphoric acid, what is the percent yield?

4.20 When HgO is heated, it decomposes into elemental mercury and molecular oxygen gas. If 60.0 g of Hg is obtained from 80.0 g of the oxide, what is the percent yield of the reaction?

4.21 Phenobarbital is a sleep-inducing drug whose chemical formula is $C_{12}H_{12}N_2O_3$. It is manufactured in an eight-step process starting from toluene, C_7H_8. Theoretically, each molecule of toluene yields one molecule of phenobarbital. If each of the steps has a yield of 88%, what mass of toluene is needed to manufacture 25 kg of phenobarbital?

4.22 Most of the ammonia produced by the Haber process is used as fertilizer. A second important use of NH_3 is in the production of nitric acid, a top-15 industrial chemical. Nitric acid is produced by a three-step synthesis called the *Ostwald process*:

$$4\,NH_3 + 5\,O_2 \longrightarrow 4\,NO + 6\,H_2O$$

$$2\,NO + O_2 \longrightarrow 2\,NO_2$$

$$3\,NO_2 + H_2O \longrightarrow 2\,HNO_3 + NO$$

The NO is recycled so that every mole of ammonia theoretically yields one mole of nitric acid. Starting with 7.50×10^2 kg of ammonia, what mass of nitric acid can be produced if each step is 94.5% efficient?

4.23 What is the yield of the reaction described in Problem 4.17 if the manufacturer obtains 105 kg of CCl_2F_2 from 175 kg of CCl_4? Given this reaction yield, what masses of CCl_4 and hydrogen fluoride should be used in order to make 155 kg of CCl_2F_2?

4.24 If the oxidation reaction described in Problem 4.16 is 76.5% efficient, what mass of cyclohexane is required to produce 3.50 kg of adipic acid?

The Limiting Reactant

4.25 For each of the reactions given in Problem 4.5, consider an industrial process that starts with 1.00 metric ton (1000 kg) of each reactant. Construct a table of amounts, identify the limiting reactant, and determine the maximum mass of each product that could be produced.

4.26 For each of the reactions given in Problem 4.6, consider an industrial process that starts with 7.50×10^3 kg of each reactant. Construct a table of amounts, identify the limiting reactant, and determine the maximum mass of each product that could be produced.

ilw 4.27 Ammonia is produced industrially using the Haber process:

$$N_2 + 3\,H_2 \longrightarrow 2\,NH_3$$

Suppose that an industrial reactor is charged with 75.0 kg each of N_2 and H_2. Use a table of amounts to determine what mass of ammonia could be produced if the reaction went to completion.

4.28 Acrylonitrile is an important building block for synthetic fibers and plastics. Over 1.4 billion kilograms of acrylonitrile are produced in the United States each year. The compound is synthesized from propene in the following reaction:

$$2\,C_3H_6(g) + 2\,NH_3(g) + 3\,O_2(g) \longrightarrow 2\,C_3H_3N(l) + 6\,H_2O(g)$$

Propene

Acrylonitrile

Use a table of amounts to determine how many kilograms of acrylonitrile can be prepared from 1.50×10^3 kg of propene, 6.80×10^2 kg of ammonia, and 1.92×10^3 kg of oxygen.

4.29 A fast-food restaurant makes double cheeseburgers using one hamburger roll, two quarter-pound patties of beef, one slice of cheese, a fourth of a tomato, and 15 g of shredded lettuce. At the start of the day, the restaurant manager has 12 dozen rolls, 40 lb of beef patties, 2 packages of sliced cheese containing 65 slices each, 40 tomatoes, and 1 kg of lettuce. Which ingredient will run out first, and how many cheeseburgers can be made?

4.30 A hardware store sells a "handyman assortment" of wood screws containing the following:

Type	Number of screws	Mass per screw (g)
#8, $\frac{1}{2}''$	10	12.0
#8, $1''$	8	21.5
#10, $\frac{1}{2}''$	8	14.5
#10, $1\frac{1}{2}''$	5	31.0
#12, $2''$	4	45.0

The store buys these screws in 1-pound boxes (1 lb = 454 g). From two boxes of each type of screws, how many "handyman assortments" can the store prepare?

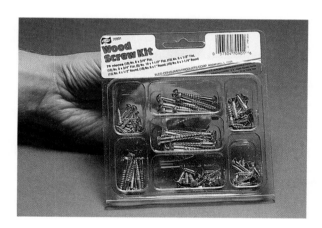

4.31 Elemental phosphorus, P_4, reacts vigorously with oxygen to give P_4O_{10}. Use a table of amounts to determine how much P_4O_{10} can be prepared from 3.75 g of P_4 and 6.55 g O_2, and how much of the excess reactant will remain at the end of the reaction.

4.32 The following unbalanced reaction is called the *thermite reaction*. It releases tremendous amounts of energy and is sometimes used to generate heat for welding.

$$Al + Fe_3O_4 \longrightarrow Fe + Al_2O_3$$

Use a table of amounts to determine the masses of all substances present after the reaction if 2.00×10^2 g of Al and 7.00×10^2 g of Fe_3O_4 react to completion.

Precipitation Reactions

4.33 What are the major species present in aqueous solutions of each of the following: (a) NH_4Cl; (b) $Fe(ClO_4)_2$; (c) Na_2SO_4; (d) Br_2; and (e) KBr?

4.34 What are the major species present in aqueous solutions of each of the following: (a) $(NH_4)_2CO_3$; (b) $NaHSO_4$; (c) $CoCl_2$; (d) $Mg(NO_3)_2$; and (e) CH_3OH (methanol)?

4.35 What are the major species present in aqueous solutions of each of the following: (a) potassium hydrogen phosphate; (b) acetic acid; (c) sodium acetate; (d) ammonia; and (e) ammonium chloride?

4.36 What are the major species present in aqueous solutions of each of the following: (a) carbonated water (water saturated with carbon dioxide); (b) lithium carbonate; (c) potassium sulfite; (d) hydrogen sulfide; (e) sodium hydrogen sulfate?

4.37 Consider the addition of each of the following aqueous solutions to each solution described in Problem 4.33. In each case, decide whether a precipitate will form. If so, identify the precipitate. (a) $AgNO_3$; (b) Na_2CO_3; (c) $Ba(OH)_2$.

4.38 Consider the addition of each of the following aqueous solutions to each solution described in Problem 4.34. In each case, decide whether a precipitate will form. If so, identify the precipitate. (a) $Pb(CH_3CO_2)_2$; (b) $Ca(OH)_2$; (c) KOH.

4.39 Write the balanced net ionic equation for each of the these precipitation reactions. Also, identify the spectator ions.
- (a) $AgNO_3(aq) + KOH(aq) \longrightarrow ?$
- (b) $Fe(ClO_4)_3(aq) + (NH_4)_2C_2O_4(aq) \longrightarrow ?$
- (c) $Pb(NO_3)_2(aq) + NaBr(aq) \longrightarrow ?$
- (d) $KOH(aq) + NiSO_4(aq) \longrightarrow ?$

4.40 Write the balanced net ionic equation for each of these precipitation reactions. Also, identify the spectator ions.
- (a) $LiCl(aq) + AgNO_3(aq) \longrightarrow ?$
- (b) $MgSO_4(aq) + Na_3PO_4(aq) \longrightarrow ?$
- (c) $Ba(OH)_2(aq) + Na_2SO_4(aq) \longrightarrow ?$
- (d) $AlCl_3(aq) + KOH(aq) \longrightarrow ?$

ilw 4.41 If 55.0 mL of a 5.00×10^{-2} M solution of silver nitrate is mixed with 95.0 mL of 3.50×10^{-2} M potassium carbonate, what mass of solid forms, and what ions remain in solution?

4.42 If 75.0 mL of a 0.750 M solution of lead(II) nitrate is mixed with 125 mL of 0.855 M ammonium chloride, what mass of solid forms, and what ions remain in solution?

Acid-Base Reactions

4.43 Write the balanced net ionic equation for each of the these acid-base reactions. Also, identify the spectator ions.
- (a) $HCl(aq) + Ca(OH)_2(aq) \longrightarrow ?$
- (b) $H_3PO_4(aq) + $ excess $LiOH(aq) \longrightarrow ?$
- (c) $NH_3(aq) + HNO_3(aq) \longrightarrow ?$
- (d) $CH_3CO_2H(aq) + KOH(aq) \longrightarrow ?$

4.44 Write the balanced net ionic equation for each of these acid-base reactions. In each case, write the formulas of the spectator ions.
- (a) $NH_3(aq) + HBr(aq) \longrightarrow ?$
- (b) $HClO(aq)$ (a weak acid) $+ NaOH(aq) \longrightarrow ?$
- (c) $HClO_4(aq) + Ca(OH)_2(aq) \longrightarrow ?$
- (d) $H_2SO_4(aq) + $ excess $KOH(aq) \longrightarrow ?$

4.45 One common component of antacids is $Al(OH)_3$. If an upset stomach contains 155 mL of 0.175 M HCl, what mass of $Al(OH)_3$ is required to completely neutralize the acid?

4.46 One common component of antacids is $Mg(OH)_2$. If an upset stomach contains 125 mL of 0.115 M HCl, what mass of $Mg(OH)_2$ is required to completely neutralize the acid?

4.47 Calculate the molarities of all ions present in a solution made by mixing 150.0 mL of 2.00×10^{-2} M $Ba(OH)_2$ solution with 100.0 mL of 5.00×10^{-2} M HCl solution.

4.48 Calculate the molarities of all ions present in a solution made by adding 1.53 g of solid NaOH to 215 mL of 0.150 M $HClO_4$.

ilw **4.49** A student wishes to determine the concentration of a solution of KOH. The student adds 5.00 mL of the KOH solution to 150 mL of water, adds an indicator, and titrates with 0.1206 M HCl. If the titration requires 27.35 mL of acid, what is the concentration of the KOH solution?

4.50 A student wishes to determine the concentration of a solution of HCl. The student adds 10.00 mL of the HCl solution to 150 mL of water, adds an indicator, and titrates with 0.0965 M NaOH. If the titration requires 32.45 mL of base, what is the concentration of the HCl solution?

4.51 A technician standardizes a solution of NaOH by titrating 0.6634 g of potassium hydrogen phthalate dissolved in 200 mL of water. It takes 36.55 mL of base solution to reach the stoichiometric point. Determine the concentration of the NaOH solution.

4.52 A technician standardizes a solution of KOH by titrating 0.7455 g of potassium hydrogenphthalate dissolved in 150 mL of water. It takes 39.20 mL of base solution to reach the stoichiometric point. Determine the concentration of the KOH solution.

Oxidation-Reduction Reactions

4.53 Some of the following react when added together, but others do not. For those that do, write a balanced net ionic redox equation. (a) Cu + HCl(aq); (b) Cu + $MgCl_2$(aq); (c) Cu + $AgNO_3$(aq); and (d) K + H_2O.

4.54 Some of the following react when added together, but others do not. For those that do, write a balanced net ionic redox equation. (a) Al + HCl(aq); (b) Zn + Au^{3+}(aq); (c) Ni + $MgCl_2$(aq); and (d) Na + H_2O.

4.55 Write the balanced redox reactions for the formation of each of the following oxides from the reaction of molecular oxygen with pure metal: (a) strontium oxide; (b) chromium(III) oxide; (c) tin(IV) oxide.

4.56 Write the balanced redox reactions for the formation of each of the following oxides from the reaction of molecular oxygen with pure metal: (a) chromium(VI) oxide; (b) zinc oxide; (c) copper(I) oxide.

4.57 Aluminum metal generates H_2 when dropped into 6 M HCl. Calculate the mass of H_2 that will form from the complete reaction of 0.355 g Al with 8.00 mL of 6.00 M HCl.

4.58 Iron metal reacts with hydrochloric acid to give H_2 gas and Fe^{2+} ions. Suppose that 5.8 g of iron is to be dissolved in 1.5 M HCl. What is the minimum volume of the acid solution required to react with all of the iron?

Additional Paired Problems

4.59 The Solvay process is a commercial method for producing sodium carbonate. In one step of this process, sodium bicarbonate is precipitated by mixing highly concentrated aqueous solutions of sodium chloride and ammonium bicarbonate:

$$NH_4HCO_3(aq) + NaCl(aq) \longrightarrow NaHCO_3(s) + NH_4Cl(aq)$$

(a) Write the net ionic reaction for this step of the Solvay process. (b) What are the spectator ions? (c) If 5.00×10^2 L of 1.50 M NH_4HCO_3 are treated with 5.00×10^2 L of 6.00 M NaCl, and 35.0 kg of $NaHCO_3$ are produced, what is the percent yield of the process? (d) Calculate the concentrations of all dissolved ions at the end of the reaction.

4.60 A white precipitate forms when aqueous calcium nitrate is mixed with aqueous ammonium sulfate. (a) Identify the precipitate and write the net ionic equation for the reaction. (b) What are the spectator ions? (c) Calculate the mass of the precipitate that forms when 100.0 mL of 1.50 M aqueous calcium nitrate is mixed with 75.0 mL of 3.00 M ammonium sulfate. (d) Calculate the concentrations of all dissolved ions at the end of the reaction.

4.61 Ruthenium forms four different binary oxides: Ru_2O_3, RuO_2, RuO_3, and RuO_4. Write a balanced chemical equation for molecular oxygen reacting with metallic ruthenium to form each of these oxides, and determine how many electrons a ruthenium atom loses in each case.

4.62 Nitrogen forms the binary oxides NO, N_2O, NO_2, N_2O_4, and N_2O_5. Write a balanced chemical equation for O_2 reacting with N_2 to form each of these oxides, and determine how many electrons a nitrogen atom loses in each case.

4.63 Pyrite (FeS_2) reacts with excess molecular oxygen to give iron(III) oxide and SO_2. If the extraction process is 85% efficient, what mass of iron(III) oxide can be formed in the processing of 175 metric tons of ore that contains 55% iron pyrite by mass?

4.64 The element titanium is commonly found as the ore ilmenite, $FeTiO_3$. Much of the world reserves of titanium is found in Canada. At a particular mine, a sample of earth was found to contain 15% ilmenite by mass. What mass of pure titanium metal can be isolated from 1.00×10^3 metric tons of earth, if the extraction process is 95% efficient?

4.65 Predict the product(s) of the following reactions by writing balanced equations. When the reaction involves ions, write a net ionic equation. Identify each reaction as precipitation, acid-base, or redox.

(a) Al(s) + O_2(g) $\longrightarrow$
(b) C_3H_8(g) + O_2(g) $\longrightarrow$
(c) Mg(s) + HBr(aq) $\longrightarrow$
(d) NaOH(aq) + HCl(aq) $\longrightarrow$
(e) $Pb(NO_3)_2$(aq) + $(NH_4)_2CO_3$(aq) $\longrightarrow$
(f) $Ca(OH)_2$(aq) + H_2SO_4(aq) $\longrightarrow$

4.66 Predict the product(s) of the following reactions by writing balanced equations. When the reaction involves ions, write a net ionic equation. Identify each reaction as precipitation, acid-base, or redox.

(a) $Ca(OH)_2$(s) + HCl(aq) $\longrightarrow$
(b) Ni(s) + HCl(aq) $\longrightarrow$
(c) $AgNO_3$(aq) + KOH(aq) $\longrightarrow$
(d) $Ba(OH)_2$(aq) + $HClO_4$(aq) $\longrightarrow$
(e) Ca(s) + H_2O(l) $\longrightarrow$
(f) HNO_3(aq) + NH_3(aq) $\longrightarrow$

4.67 Even though xenon is a "noble gas," it reacts with fluorine and oxygen. When xenon gas reacts with F_2 gas, XeF_4 is one of the products. If 5.00 g of xenon reacts with excess fluorine and generates 4.00 g of XeF_4, what is the percent yield and how much xenon remains unreacted, assuming no other product forms?

4.68 Hydrogen fluoride is produced industrially by the action of sulfuric acid on CaF_2. If 365 kg of CaF_2 are treated with excess sulfuric acid and 155 kg of HF are produced, what is the percent yield and how much CaF_2 remains unreacted, assuming no other fluorine-containing product forms?

4.69 The reaction of atom X (●) with atom Y (●) is represented in the following diagram. Write the balanced equation for this reaction.

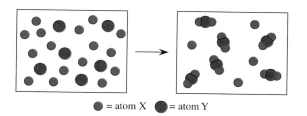

● = atom X   ● = atom Y

4.70 The reaction of element X (●) with element Y (●) is represented in the diagram shown below. Write the balanced equation for this reaction.

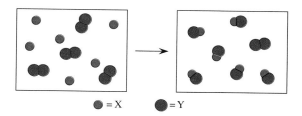

● = X   ● = Y

4.71 The gases leaving an automobile cylinder after combustion include CO_2, CO (from incomplete combustion), NO, H_2O, H_2, N_2, O_2, and unburned hydrocarbons. After leaving the engine, these exhaust gases are passed through a catalytic converter whose purpose is to change pollutants into less harmful substances. Many reactions occur in a catalytic converter, including those shown below. Write a balanced chemical equation for each.

    (a) $H_2 + NO \rightarrow NH_3 + H_2O$
    (b) $CO + NO \rightarrow N_2 + CO_2$
    (c) $NH_3 + O_2 \rightarrow N_2O + H_2O$
    (d) Nitrogen oxide and ammonia react to give nitrogen and water.
    (e) Water and nitrogen oxide react to give molecular oxygen and ammonia.
    (f) Molecular hydrogen and oxygen react to give water.

4.72 Recall that combustion is a reaction with O_2. Write balanced chemical equations for the combustion reactions of the following substances. Assume that the only products are H_2O and CO_2.

    (a) C_4H_{10} (butane, the substance used in lighter fluid)
    (b) C_6H_6 (benzene, a solvent that is carcinogenic)
    (c) C_2H_5OH (ethanol, the intoxicant in alcoholic drinks)
    (d) C_5H_{12} (pentane, a highly volatile liquid):
    (e) $C_6H_{11}OH$ (cyclohexanol, a useful solvent).

4.73 Devise a synthesis, write the net ionic reaction, and compute masses of each starting material needed to make 2.50 kg of each of the following solid ionic compounds: (a) $FePO_4$; (b) $Zn(OH)_2$; (c) $NiCO_3$.

4.74 Lithopone, a brilliant white pigment used in paints, paper, and white rubber products, is a mixture of two insoluble ionic solids, ZnS and $BaSO_4$. Suggest how 1.0 kg of lithopone could be prepared by a precipitation reaction.

4.75 A 4.6-g sample of ethanol (C_2H_5OH) is burned completely in air. (a) How many moles of H_2O are formed? (b) How many molecules? (c) How many grams?

4.76 An automobile engine completely burns 3.5 g of octane (C_8H_{18}). (a) How many moles of CO_2 are formed? (b) How many molecules? (c) How many grams?

4.77 The following diagram represents a small portion of a reaction vessel that contains the starting materials for the Haber synthesis of ammonia.

$$N_2 + H_2 \longrightarrow NH_3 \qquad \text{(unbalanced)}$$

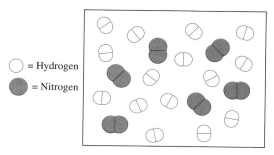

○ = Hydrogen

● = Nitrogen

(a) What is the limiting reactant in this reaction? (b) Draw a picture that shows how this portion of the vessel will look when the reaction is complete. (c) If each molecule in the picture represents one mole of compound, what mass of ammonia will the system produce, and what mass of the excess reactant will be left behind?

4.78 Carbon dioxide, which is used to carbonate beverages and as a coolant (dry ice), is produced from methane and water vapor:

$$CH_4(g) + H_2O(g) \longrightarrow CO_2(g) + H_2(g) \qquad \text{(unbalanced)}$$

The diagram shown below represents a small portion of a vessel that contains starting materials for this reaction.

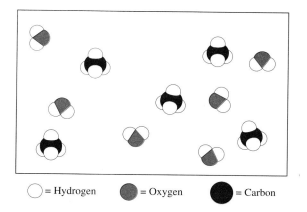

○ = Hydrogen   ● = Oxygen   ● = Carbon

(a) What is the limiting reactant in the reaction? (b) Draw a picture that shows what the vessel will look like when the reaction is complete. (c) If each molecule in the picture represents one mole of compound, what mass of each product will the system produce and what mass of the excess reactant will be left behind?

4.79 Identify the species present in the following aqueous solutions. In each case, write a net ionic equation that describes what reaction occurs upon mixing equimolar portions of the two solutions. (a) NH_3 and HCl; (b) $CaCl_2$ and Na_2SO_4; (c) KOH and HBr; and (d) HNO_2 (a weak acid) and KOH.

4.80 Identify the species present in the following aqueous solutions. In each case, write a net ionic equation that describes what reaction occurs upon mixing equimolar portions of the two solutions. (a) $HgCl_2$ and MgS; (b) HNO_3 and $Ba(OH)_2$; (c) NaOH and CH_3CO_2H; and (d) KOH and $FeCl_3$.

4.81 Write the net ionic reaction and draw a molecular picture that shows what happens when 50.0 mL of 0.010 M HCl solution is mixed with 50.0 mL of 0.0050 M $Ba(OH)_2$ solution.

4.82 Write the net ionic reaction and draw a molecular picture that shows what happens when 20.0 mL of 0.200 M NaOH solution are mixed with 40.0 mL of 0.100 M HCl solution.

4.83 Propylene oxide is used primarily in the synthesis of poly(propylene glycol), a polymer used in the manufacture of automobile seats, bedding and carpets. Around 2 billion kilograms of this compound are produced annually in the U.S. Propylene oxide is produced by the following reaction:

| *t*-Butyl hydroperoxide $C_4H_{10}O_2$ | Propene C_3H_6 | | Propylene oxide C_3H_6O | *t*-Butanol $C_4H_{10}O$ |

(a) How many kilograms of propylene oxide can be prepared from 75 kg of *t*-butyl hydroperoxide? (b) What mass of propene will be required for the synthesis?

4.84 Vinyl chloride, one of the top 20 industrial compounds, is used primarily to make the polymer poly(vinyl chloride), better known as "PVC." This versatile polymer is used to make piping, siding, gutters, floor tile, clothing, and toys. Vinyl chloride is made by oxychlorination of ethylene (C_2H_4). The overall balanced equation follows:

| Ethylene | Vinyl chloride |

Annual production of vinyl chloride is about 6.8 billion kilograms. What minimum masses (in billions of kilograms) of C_2H_4 and HCl are required to produce this much vinyl chloride?

More Challenging Problems

4.93 In the Haber synthesis of ammonia, N_2 and H_2 react at high temperature, but they never react completely. In a typical reaction, 24.0 kg of H_2 and 84.0 kg of N_2 react to produce 68 kg of NH_3. Using a table of amounts, find the theoretical yield, the percent yield, and the masses of H_2 and N_2 that remain unreacted, assuming that no other products form.

4.85 Magnesium metal burns with a bright flame in oxygen gas, and the product is solid white magnesium oxide. Draw a molecular picture showing six magnesium atoms and four oxygen molecules. Then draw another molecular picture of this same system after reaction occurs.

4.86 Sodium metal reacts vigorously with chlorine gas to form solid white sodium chloride. Draw a molecular picture showing ten sodium atoms and three chlorine molecules. Then draw another molecular picture of this same system after reaction occurs.

4.87 Calcium carbonate (limestone) reacts with hydrochloric acid to generate water and carbon dioxide gas. In a certain experiment, 5.0 g of $CaCO_3$ is added to 0.50 L of 0.10 M HCl. (a) Write a balanced equation for this reaction. (b) What mass of carbon dioxide is formed? (c) Calculate the concentrations of all species present in solution at the end of the reaction.

4.88 Magnesium metal reacts with HCl solution, liberating H_2 gas and generating Mg^{2+} cations in solution. A 1.215-g sample of Mg metal is added to 50.0 mL of a 4.0 M HCl solution, and the reaction goes to completion. (a) Write a balanced equation for this reaction. (b) What mass of H_2 is formed? (c) Calculate the concentrations of all ions present in the solution at the end of the reaction.

4.89 Predict the products of the following reactions by writing balanced equations. When the reaction involves ions, write a net ionic equation. Identify each as precipitation, acid–base, or redox.
(a) $HCl(aq) + Ca(s) \longrightarrow$
(b) $Li(s) + O_2(g) \longrightarrow$
(c) $HBr(aq) + NH_3(aq) \longrightarrow$
(d) $C_3H_8(g) + O_2(g) \longrightarrow$

4.90 Predict the products of the following reactions by writing balanced equations. When the reaction involves ions, write a net ionic equation. Identify each as precipitation, acid–base, or redox.
(a) $KOH(aq) + Co(NO_3)_2(aq) \longrightarrow$
(b) $HCl(aq) + Fe(s) \longrightarrow Fe^{3+}(aq)$
(c) $KI(aq) + Pb(CH_3CO_2)_2(aq) \longrightarrow$
(d) $HClO_4(aq) + NH_3(aq) \longrightarrow$
(e) $C_6H_5CO_2H(aq) + Ba(OH)_2(aq) \longrightarrow$

4.91 Compounds that undergo explosions typically produce large quantities of hot gases from much smaller volumes of highly reactive solids or liquids. Balance the reaction for the decomposition of nitroglycerine, a violent explosive:

$$C_3H_5N_3O_9(l) \longrightarrow N_2(g) + CO_2(g) + H_2O(g) + O_2(g)$$

4.92 Compounds that undergo explosions typically produce large quantities of hot gases from much smaller volumes of highly reactive solids or liquids. Balance the reaction for the decomposition of trinitrotoluene (TNT), a violent explosive:

$$C_7H_5N_3O_6(l) \longrightarrow N_2(g) + CO_2(g) + H_2O(g) + C(s)$$

4.94 A student prepared 1.00 L of a solution of NaOH for use in titrations. The solution was standardized by titrating a sample of potassium hydrogen phthalate whose mass was 0.7996 g. Before titration, the buret reading was 0.15 mL. When the indicator changed color, the buret reading was 43.75 mL. Calculate the molarity of the NaOH solution.

4.95 One set of reactants for rocket fuel is hydrazine and hydrogen peroxide, which react vigorously when mixed:

$$N_2H_4(l) + H_2O_2(l) \longrightarrow N_2(g) + H_2O(g) \qquad \text{(unbalanced)}$$

The density of liquid hydrazine is 1.44 g/mL, and that of hydrogen peroxide is 1.01 g/mL. What volume ratio of these two liquids should be used if both fuels are to be used up at the same time?

4.96 Sulfur dioxide is an atmospheric pollutant that is converted to sulfuric acid when it reacts with water vapor. This is one source of acid rain, a serious environmental problem. The sulfur dioxide content of an air sample can be determined as follows. A sample of air is bubbled through an aqueous solution of hydrogen peroxide to convert all of the SO_2 to H_2SO_4.

$$H_2O_2 + SO_2 \longrightarrow H_2SO_4$$

Titration of the resulting solution completes the analysis (both H atoms of H_2SO_4 are titrated). In one such case, the analysis of 1.55×10^3 L of Los Angeles air gave a solution that required 5.70 mL of 5.96×10^{-3} M NaOH to complete the titration. Determine the number of grams of SO_2 present in the air sample.

4.97 Vitamin C (also called *ascorbic acid*) is an acid whose formula is $HC_6H_7O_6$. When treated with strong base, it undergoes the following reaction:

$$HC_6H_7O_6 + OH^- \longrightarrow C_6H_7O_6^- + H_2O$$

A pharmacist suspects that the vitamin C tablets received in a recent shipment are not pure. When a single 500.0-mg tablet is dissolved in 200.0 mL of water and titrated with a standard base that is 0.1045 M, it takes 24.45 mL to reach the stoichiometric point. Are the tablets pure? If not, what is the mass percentage of impurities? (Assume any impurities in the tablets are neutral.)

4.98 Silicon tetrachloride is used in the electronics industry to make elemental silicon for computer chips. Silicon tetrachloride is prepared from silicon dioxide, carbon graphite, and chlorine gas.

$$SiO_2(s) + 2\ C(s) + 2\ Cl_2(g) \longrightarrow SiCl_4(l) + 2\ CO(g)$$

If the reaction goes in 95.7% yield, how much silicon tetrachloride can be prepared from 75.0 g of each starting material, and how much of each reactant remains unreacted?

4.99 The largest single use of sulfuric acid is for the production of phosphate fertilizers. The acid reacts with calcium phosphate in a 2:1 mole ratio to give calcium sulfate and calcium dihydrogen phosphate. The mixture is crushed and spread on fields, where the salts dissolve in rain water. (Calcium phosphate, commonly found in phosphate rock, is too insoluble to be a direct source of phosphate for plants.) (a) Write a balanced equation for the reaction of sulfuric acid with calcium phosphate. (b) How many kilograms each of sulfuric acid and calcium phosphate are required to produce 50.0 kg of the calcium sulfate–dihydrogen phosphate mixture? (c) How many moles of phosphate ion will this mixture provide?

4.100 Write the balanced equation and determine the number of moles of water produced when 2.95 mL of pyridine (C_5H_5N, $\rho = 0.982$ g/mL) reacts with excess O_2 to give water, carbon dioxide, and molecular nitrogen.

4.101 Silver jewelry is usually made from silver and copper alloys. The amount of copper in an alloy can vary considerably. The finest-quality alloy is sterling silver, which is 92.5% by mass silver. To determine the composition of a silver-copper alloy, a jeweler dissolved 0.135 g of metal shavings in 50 mL of concentrated nitric acid and then added 1.00 M KCl solution until no more precipitate formed. Filtration and drying yielded 0.156 g of AgCl precipitate. What was the mass composition of the silver alloy?

4.102 Surface deposits of elemental sulfur around hot springs and volcanoes are believed to come from a two-step redox process. Combustion of hydrogen sulfide (H_2S) produces sulfur dioxide and water. The sulfur dioxide reacts with more hydrogen sulfide to give elemental sulfur and water. Write balanced chemical equations for these two reactions, and determine the minimum mass of hydrogen sulfide that a volcano must emit in order to deposit 1.25 kg of sulfur.

4.103 A former antiknock ingredient in gasoline is a colorless liquid whose formula is $C_5H_{12}O$. Write the balanced equation, and determine the number of moles of carbon dioxide produced when 3.15 mL of the ingredient reacts with excess oxygen to give carbon dioxide and water (density of $C_5H_{12}O = 0.740$ g/mL).

4.104 The CO_2 exhaled by astronauts must be "scrubbed" (removed) from the spacecraft atmosphere. One way to do this is with solid LiOH:

$$CO_2(g) + 2\ LiOH(s) \longrightarrow Li_2CO_3(s) + H_2O(l)$$

The CO_2 output of an astronaut is about 1.0 kg/day. What is the minimum mass of LiOH required for a six-day space shuttle flight involving five astronauts?

4.105 Potassium hydroxide is considerably cheaper than lithium hydroxide and undergoes the analogous reaction with CO_2. Repeat the calculation of Problem 4.104 for KOH. Why is the more expensive substance used on the space shuttle?

4.106 Phosphorus is essential for plant growth, and it is often the limiting nutrient in aqueous ecosystems. However, too much phosphorus can cause algae to grow at an explosive rate. This process, known as *eutrophication*, robs the rest of the ecosystem of essential oxygen, often destroying all other aquatic life. One source of aquatic phosphorus pollution is the HPO_4^{2-} used in detergents in sewage plants. The simplest way to remove HPO_4^{2-} is to treat the contaminated water with lime, CaO, which generates Ca^{2+} and OH^- ions in water. The phosphorus precipitates as $Ca_5(PO_4)_3OH$. (a) Write the balanced equation for CaO dissolving in water. (b) Write the balanced equation for the precipitation reaction. (Hint: proton transfer occurs as well as solid formation.) (c) How many kilograms of lime are required to remove all the phosphorus from a 1.00×10^4 L holding tank filled with contaminated water that is 0.0156 M in HPO_4^{2-}?

Group Study Problems

4.107 Analysis of a sample of natural gas gives the following results: 74% (by mass) methane (CH_4), 18% ethane (C_2H_6), and 8% propane (C_3H_8). How many moles of CO_2 would be produced in the combustion of 750 g of this gas?

4.108 Bronze is an alloy of copper and zinc. When a 5.73-g sample of bronze was treated with excess aqueous HCl, 21.3 mg of H_2 was produced. What was the percentage by mass composition of the bronze? (Hint: Consult the activity series shown in Table 4–3.)

4.109 CF_3CH_2F (HFC-134a) has replaced chlorofluorocarbon compounds for use as refrigerants. HFC-134a is produced from trichloroethylene by the following reactions:

$$CCl_2=CHCl + HF \xrightarrow{\text{catalyst}} CF_3CH_2Cl + HCl$$
$$CF_3CH_2Cl + HF \xrightarrow{\text{catalyst}} CF_3CH_2F + HCl$$

In 1999, over 1.3 million metric tons of HFC-134a were produced, and production is increasing at a rate of nearly 20% annually. Manufacturers sell HFC-134a for \$5.50/kg. Assume that you are a chemist working for a company that sold 65 million kilograms of HFC-134a last year. (a) If the synthesis had a reaction yield of 47%, how much trichloroethylene was used to produce the HFC-134a? (b) You have developed a new catalyst that improves the yield of the synthesis of HFC-134a from 47% to 53%. If the company uses the same mass of trichloroethylene for next year's production, how many kilograms of HFC-134a can the company produce if they use your new process? (c) Assuming the selling price of HFC-134a does not change, how much more money will the company make next year because of your work?

4.110 As a final examination in the general chemistry laboratory, a student was asked to determine the mass of $Ca(OH)_2$ that dissolves in 1.000 L water. Using a published procedure, the student did the following: (1) About 1.5 mL of concentrated HCl (12 M) was added to 750 mL of distilled water. (2) A solution of KOH was prepared by adding approximately 1.37 g KOH to 1.0 L distilled water. (3) A sample of potassium hydrogen phthalate (185.9 mg) was dissolved in 100 mL of distilled water. Titration with the KOH solution required 25.67 mL to reach the stoichiometric point. (4) A 50.00-mL sample of the HCl solution prepared in step 1 was titrated with the KOH solution. The titration required 34.02 mL of titrant to reach the stoichiometric point. (5) The student was given a 25.00-mL sample of a saturated solution of $Ca(OH)_2$ for analysis. Titration with the HCl solution required 29.28 mL to reach the stoichiometric point. How many grams of calcium hydroxide dissolve in 1.00 L of water?

4.111 Decaborane, $B_{10}H_{14}$, was used as fuel for the Redstone rockets of the 1950s. Decaborane reacts violently with oxygen according to the following equation.

$$B_{10}H_{14} + O_2 \longrightarrow B_2O_3 + H_2O \quad \text{(unbalanced)}$$

The two starting materials are stored in separate containers. When mixed, they ignite spontaneously, releasing large amounts of energy. Both fuel materials should run out at the same time because this minimizes the excess mass that the rocket must carry. If the total mass of both components is to be 12.0×10^4 kg, what mass of liquid oxygen and what mass of decaborane should be used?

4.112 Although H_2SO_4 has two acidic hydrogen atoms, only one of them transfers readily to water. Thus an aqueous solution of sulfuric acid contains water molecules, hydronium cations, and hydrogen sulfate anions. The following drawing represents one part of an aqueous solution of sulfuric acid (the solvent water molecules have been omitted for clarity).

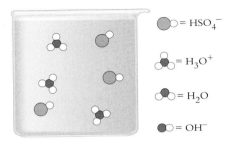

For each of the following situations, draw a new picture that shows what this portion of the solution looks like after the reaction is complete. Include any water molecules produced in the reaction. (Hint: Hydroxide reacts with hydronium ions in preference to HSO_4^-.) (a) Three OH^- ions are added to the solution. (b) Three additional OH^- ions are added to solution (a). (c) Two additional OH^- ions are added to solution (b).

Answers to Section Exercises

4.1.1 $2\,C_8H_{18} + 25\,O_2 \rightarrow 16\,CO_2 + 18\,H_2O$
4.1.2 $2\,C_3H_6 + 2\,NH_3 + 3\,O_2 \rightarrow 2\,C_3H_3N + 6\,H_2O$
4.1.3 $N_2 + O_2 \rightarrow 2\,NO \qquad 2\,NO + O_2 \rightarrow 2\,NO_2$
4.2.1 (a) 333 g PCl_3 can be prepared; (b) 258 g Cl_2 is consumed.
4.2.2 0.737 g
4.2.3 2.80×10^3 g O_2 is required; 1.14×10^3 g H_2O and 2.47×10^3 g CO_2 are produced.
4.3.1 Theoretical yield is 1.96 g, and percent yield is 94.0%. Possible reasons are that not all the $KClO_3$ decomposes or that some O_2 escapes and is not collected.
4.3.2 Start with 5.8 tons of sulfur.
4.3.3 66 g salicylic acid are required.
4.4.1 The maximum mass is 793 kg HCN, and 28.7 kg CH_4 are left.
4.4.2 7.5 g
4.4.3 2.3×10^2 g
4.5.1 d, e, and g are insoluble; a, b, c, and f are soluble.
4.5.2 Mix solutions of a soluble calcium salt and a soluble phosphate salt, for example, $Ca(NO_3)_2$ and Na_3PO_4. To make 1.5 kg of product requires 2.4 kg $Ca(NO_3)_2$ and 1.6 kg Na_3PO_4.

4.5.3 26 kg
4.6.1 Net ionic reaction:
$H_2CO_3(aq) + 2\,OH^-(aq) \rightarrow CO_3^{2-}(aq) + 2\,H_2O(l)$
Molecular picture:

4.6.2 1.29×10^3 g
4.6.3 0.1009 M
4.7.1 0.135 mol
4.7.2 (a) $2\,Mg + TiO_2 \rightarrow Ti + 2\,MgO$. Mg is oxidized and TiO_2 is reduced. (b) 1.02×10^2 kg
4.7.3 (a) Half-reactions are $Ni(s) \rightarrow Ni^{2+}(aq) + 2e^-$ and $2\,H_3O^+(aq) + 2e^- \rightarrow H_2(g) + 2\,H_2O(l)$; balanced redox equation is $Ni(s) + 2\,H_3O^+(aq) \rightarrow Ni^{2+}(aq) + H_2(g) + 2\,H_2O(l)$. (b) and (c) No reaction. (d) Half-reactions are $Fe(s) \rightarrow Fe^{2+}(aq) + 2e^-$ and $Ag^+(aq) + e^- \rightarrow Ag(s)$; balanced redox equation is $Fe(s) + 2\,Ag^+(aq) \rightarrow Fe^{2+}(aq) + 2\,Ag(s)$

5

The Behavior of Gases

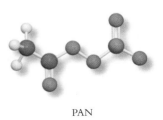

The photochemical smog of Mexico City causes not only reduced visibility but a host of human health problems.

PAN

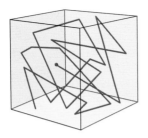

Figure 5-1
Molecules in a gas move freely throughout the entire volume of a container, changing direction whenever they collide with other molecules or with a wall. The line traces a possible path of a single molecule.

INTRODUCTION: EARTH'S ATMOSPHERE

Earth's atmosphere is a shroud of gas that surrounds our planet. Most of the atmosphere lies within 100 km of the surface. The atmosphere contains two major components: a bit less than 80% N_2 and a bit more than 20% O_2. In addition to these two species, the atmosphere contains small amounts of water and traces of other species. Our inset picture is a representation of the atmosphere.

In recent years scientists have discovered that some of the trace components of the atmosphere are changing in ways that may cause dramatic changes in the environment. All of us have heard the alarms: Acid rain damages ecosystems. Burning fossil fuels may lead to global warming. Human activities have caused a "hole" in the ozone layer. More immediately visible are the spectacular sunsets—such as the one in our background photo—that result when the atmosphere contains a significant concentration of fine particles.

How could molecules present in mere trace amounts have such extensive effects on the environment? Acid rain provides a good example. Burning gasoline and coal produces some gaseous NO_2 and SO_2. These soluble oxides dissolve in water vapor and eventually produce nitric acid and sulfuric acid. Rain that forms from such acid-containing water vapor is highly acidic and can damage forests and aquatic life. For example, rain falling in Europe now regularly registers 10 times "normal" acidity. In 1994 it was reported that 25% of the trees in Europe suffer from leaf loss or leaf damage, most likely as a result of acid rain.

Smog associated with Los Angeles and other major urban centers is caused by the action of sunlight on automobile exhaust gases. Los Angeles has reduced its smog levels through strict pollution controls, but large cities in the tropics, most notably Mexico City and Jakarta, Indonesia, are experiencing increasing levels of smog.

Smog contains nitrogen oxides, ozone, and more complex molecules such as peroxy acetyl nitrate (PAN). These and other trace contaminants of the atmosphere interact with sunlight in complicated ways that scientists do not yet fully understand. Consequently, chemists, meteorologists, computer modelers, and other researchers are studying how smog forms and how it can be prevented.

Acid rain and smog show that changes in the trace composition of the atmosphere can have dramatic consequences on the environment. The impact of ozone depletion in the upper atmosphere (treated in Chapter 14) and long-term global warming (treated in Chapter 6) are remote, but potentially even more damaging than smog or acid rain. Research may determine the effects of these changes in our atmosphere and may indicate what actions society must take to reverse their consequences.

We begin this chapter with a molecular description that explains features common to all gases. Next, we show how to do stoichiometric calculations for reactions involving gas-phase species. Finally, we return to the Earth's atmosphere and describe some aspects of its composition and chemical reactions.

5.1 MOLECULES IN MOTION

The atoms and molecules that make up a gas move freely about to fill the volume of its container, as shown schematically in Figure 5-1. Substances that exist as gases under normal conditions tend to be small molecules with molar masses less than 50 g/mol (Table 5-1).

Freedom of motion allows gas molecules to move easily from one place to another. Wind is an example of a gas flowing from one place to another. As another example, when a vase of freshly cut roses is placed on a table, the fragrance of the flowers soon fills the room. The molecules that give roses their odor escape from the

Table 5-1
Some Gaseous Substances*

Elemental Gases			Binary Gases		
Substance	Formula	MM (g/mol)	Substance	Formula	MM (g/mol)
Hydrogen	H_2	2.0	Methane	CH_4	16.0
Helium	He	4.0	Ammonia	NH_3	17.0
Neon	Ne	20.2	Carbon monoxide	CO	28.0
Nitrogen	N_2	28.0	Ethane	C_2H_6	30.0
Oxygen	O_2	32.0	Nitrogen oxide	NO	30.0
Fluorine	F_2	38.0	Hydrogen sulfide	H_2S	34.1
Argon	Ar	39.9	Hydrogen chloride	HCl	36.5
Ozone	O_3	48.0	Carbon dioxide	CO_2	44.0
Chlorine	Cl_2	71.0	Nitrogen dioxide	NO_2	46.0

* (Under "normal" conditions: 298 K, 1 atm pressure)

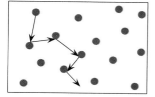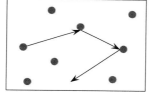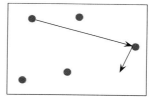

Figure 5-2
As molecular density decreases, the average distance traveled between molecular collisions increases.

flower into the gas phase. These fragrant gas molecules then move throughout the room's entire volume, moving among the nitrogen and oxygen molecules of the air in much the same way as a flower seller moves through a street crowded with people.

Any sample of a gas contains huge numbers of molecules that are in continuous motion and undergo frequent collisions with one another. Figure 5-2 shows that the frequency of collisions depends on the number of molecules per unit volume, in other words on the **molecular density** of the gas. At low molecular density, a molecule may move all the way across a container before it encounters another molecule. At high molecular density, a molecule travels only a very short distance before it collides with another molecule.

Molecular Speeds

How fast do gas molecules move? Molecular speeds can be measured using a molecular beam apparatus, which is shown schematically in Figure 5-3a. Gas

Figure 5-3
(a) Diagram of a molecular beam apparatus designed to measure the speeds of gas molecules.
(b) Distribution of molecules observed by the detector as a function of time after opening the shutter. Slow molecules take longer to reach the detector than fast molecules.

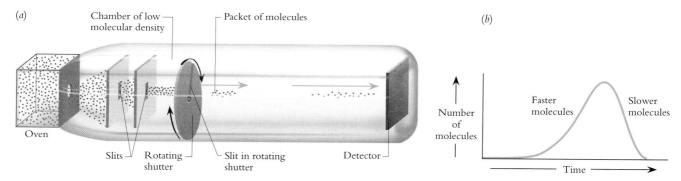

(a) Chamber of low molecular density — Packet of molecules
Oven
Slits — Rotating shutter — Slit in rotating shutter — Detector

(b) Number of molecules — Faster molecules — Slower molecules — Time

molecules escape from an oven through a small hole into a chamber in which the molecular density is very low. A set of slits blocks the passage of all molecules except those moving in the forward direction. The result is a beam of molecules, all moving in the same direction. A rotating shutter blocks the beam path except for a small slit that allows a packet—"pulse"—of molecules to pass through. Each molecule moves down the beam axis at its own speed, and the faster a molecule moves, the less time it takes to travel the length of the chamber. A detector at the end of the chamber measures the number of molecules arriving as a function of time, giving a profile of speeds.

When the speed profile of a gas is measured in this way, it always shows the distribution shown in Figure 5-3b. If all the molecules traveled at the same speed, they would reach the detector at the same time, in a single clump. Instead, faster molecules move ahead of the main packet, and slower molecules fall behind. This experiment shows that molecules in a gas have a distribution of speeds.

A pattern emerges when this molecular beam experiment is repeated for various gases at a common temperature: Molecules with small masses move faster than those with large masses. Figure 5-4 shows this for H_2, CH_4, and CO_2. Of these molecules, H_2 has the smallest mass and CO_2 the largest. The vertical line drawn for each gas shows the speed at which the distribution reaches its maximum height. More molecules have this speed than any other, so this is the most probable speed for molecules of that gas. The most probable speed for a molecule of hydrogen at 300 K is 1.57×10^3 m/s, which is 3.51×10^3 mi/hr. Example 5-1 describes an experiment with a molecular beam apparatus.

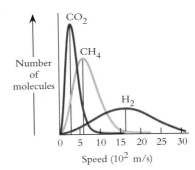

Figure 5-4
Molecular speed distributions for H_2, CH_4, and CO_2 at a temperature of 300 K.

Example 5-1	A Molecular Beam Experiment

The figures below represent mixtures of neon atoms and hydrogen molecules.

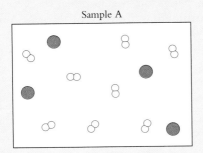

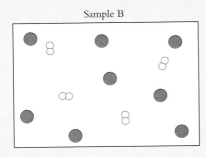

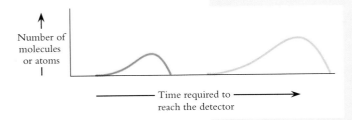

Time required to reach the detector

One of the gas mixtures was used in a pulsed molecular beam experiment. The result of the experiment is shown at left. Which of the two gas samples, A or B, was used for this experiment?

Strategy: Begin by taking an inventory of the two samples, then decide how each of them would behave in a beam experiment.

A Molecular Beam Experiment (continued)

Example 5-1

Solution: Sample A contains eight H_2 molecules and four Ne atoms, and Sample B contains four H_2 molecules and eight Ne atoms. In a beam experiment both samples would give two peaks in relative areas of $2:1$.

Which sample was used for this experiment? Particles with small mass move faster than particles with large mass, so we expect H_2 ($MM = 2.02$ g/mol) to reach the detector before Ne ($MM = 20.2$ g/mol). The data show that the first substance to reach the detector is present in the smaller amount. Consequently, the sample used in the beam experiment is the one with the smaller amount of the hydrogen molecules, Sample B.

Speed and Energy

The energy of a molecule is related to its speed. Recall from Chapter 2 that any moving object has kinetic energy ($E_{kinetic}$), as defined by Equation 2-1:

$$E_{kinetic} = \tfrac{1}{2}mu^2$$

where m is the object's mass and u its speed. The most probable speed of hydrogen molecules at 300 K is 1.57×10^3 m/s, but we need the mass of one H_2 molecule to calculate its most probable kinetic energy. The molar mass gives the mass of one mole of molecules, so dividing molar mass by the Avogadro constant (N_A) gives the mass per molecule. The fundamental unit of energy is the joule (J). One joule equals 1 kg m^2 s^{-2}, so mass (m) should be expressed in kilograms:

$$m(kg/molecule) = \frac{MM(g/mol)(1\ kg/10^3\ g)}{N_A(molecules/mol)}$$

$$m = \left(\frac{2.016\ g}{1\ mol}\right)\left(\frac{1\ mol}{6.022 \times 10^{23}\ molecules}\right)\left(\frac{1\ kg}{10^3\ g}\right)$$

$$= 3.348 \times 10^{-27}\ kg/H_2\ molecule$$

Now apply Equation 2-1:

$$E_{kinetic} = \tfrac{1}{2}(3.348 \times 10^{-27}\ kg\ molecule^{-1})(1.57 \times 10^3\ m\ s^{-1})^2$$
$$= 4.13 \times 10^{-21}\ kg\ m^2\ s^{-2}\ molecule^{-1}$$

$$E_{kinetic}\ (\text{most probable}) = 4.13 \times 10^{-21}\ J/molecule$$

The most probable speeds of methane and carbon dioxide are slower than the most probable speed of hydrogen, but CH_4 and CO_2 are more massive than H_2. When kinetic energy calculations are repeated for these gases, they show that the most probable kinetic energy is the same for all three gases.

For CH_4:

$$\frac{(16.04\ g/mol)(10^{-3}\ kg/g)(5.57 \times 10^2\ m/s)^2}{2(6.022 \times 10^{23}\ molecules/mol)} = 4.13 \times 10^{-21}\ J/CH_4\ molecule$$

For CO_2:

$$\frac{(44.01\ g/mol)(10^{-3}\ kg/g)(3.37 \times 10^2\ m/s)^2}{2(6.022 \times 10^{23}\ molecules/mol)} = 4.13 \times 10^{-21}\ J/CO_2\ molecule$$

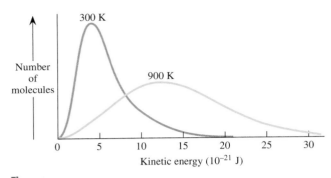

Figure 5-5
Distribution in molecular energies for a gas at 300 and 900 K.

Even though the speed distributions for these three gases peak at different values, the most probable kinetic energies are identical.

/// At a given temperature, all gases have the same molecular kinetic energy distribution.

Molecular beam experiments show that molecules move faster as temperature increases. Molecules escaping from the oven at 900 K take less time to reach the detector than molecules escaping at 300 K. Both speed and kinetic energy increase with temperature. Figure 5-5 shows kinetic energy distributions at 300 and 900 K. As the figure shows, molecules have a wide distribution of kinetic energies. Comparing this figure to Figure 5-4, we see that the distribution of molecular energies is similar to the distribution in molecular speeds. Unlike speeds, however, molecular energy distributions are the same for all gases at any particular temperature. The distributions shown in Figure 5-5 apply to any gas.

Average Kinetic Energy

More molecules have the most probable kinetic energy than any other value. By proportion, however, very few molecules actually have the most probable kinetic energy. This is because any sample of gas contains many molecules that move with many different kinetic energies. The average kinetic energy ($\overline{E}_{\text{kinetic}}$) per molecule can be found by adding all the individual molecular energies and dividing by the total number of molecules. Equation 5-1 shows that the result depends on the temperature of the gas:

$$\overline{E}_{\text{kinetic}} = \frac{3RT}{2N_A} \tag{5-1}$$

T is in kelvins, and N_A has units of molecules per mole. Therefore to give kinetic energy in J/molecule, R must have units of J/mol K.

where T is the temperature in kelvins, N_A is the Avogadro constant, and R is a quantity called the **gas constant**. The value of R is determined experimentally, and its units are those required to give kinetic energy in joules per molecule:

$$R = 8.314 \, \text{J mol}^{-1} \, \text{K}^{-1}$$

The average kinetic energy expressed by Equation 5-1 is kinetic energy *per molecule*. The total kinetic energy ($E_{\text{kinetic, molar}}$) of one mole of gas molecules is obtained by multiplying Equation 5-1 by the Avogadro constant:

$$E_{\text{kinetic, molar}} = (\text{No. of molecules})(\overline{E}_{\text{kinetic}}) = (N_A) \frac{3RT}{2N_A} = \tfrac{3}{2} RT$$

Thus 1 mole of any gas has a total molecular kinetic energy of $\tfrac{3}{2}RT$. Example 5-2 applies these equations to sulfur hexafluoride.

Molecular Kinetic Energies

Example 5-2

Compute the average molecular kinetic energy and molar kinetic energy of gaseous sulfur hexafluoride, SF_6, at 150 °C.

Strategy: The problem asks for a calculation of kinetic energies and provides a chemical formula and a temperature. We have two equations for the kinetic energy of a gas:

$$\overline{E}_{kinetic} = \frac{3RT}{2N_A} \quad and \quad E_{kinetic, molar} = \tfrac{3}{2}RT$$

To carry out the calculations, we need the constants R and N_A as well as T in kelvins. The constants are found in standard reference tables, but both are used so frequently in chemical calculations that it is wise to memorize them:

$$R = 8.314 \text{ J/mol K} \qquad N_A = 6.022 \times 10^{23} \text{ molecules/mol}$$

Solution: Convert the temperature from °C to K by adding 273.15:

$$T = 150 \text{ °C} + 273.15 = 423 \text{ K}$$

Now substitute and perform the calculations:

$$\overline{E}_{kinetic} = \frac{3(8.314 \text{ J/mol K})(423 \text{ K})}{2(6.022 \times 10^{23} \text{ molecules/mol})} = 8.76 \times 10^{-21} \text{ J/molecule}$$

$$E_{kinetic, molar} = \tfrac{3}{2}(8.314 \text{ J/mol K})(423 \text{ K}) = 5.28 \times 10^3 \text{ J/mol}$$

Does it bother you to find that neither the chemical formula nor the molar mass is needed for these calculations? Remember that not all data are necessarily required for any particular calculation. Because average kinetic energy depends on temperature but not on molar mass, we do not need mass information to do this problem.

Rates of Gas Movement

Figures 5–4 and 5–5 show that molecular speed depends on molecular mass and that molecular energy depends on temperature. We can use Equation 5–1 to link average molecular speed (u_{avg}), molar mass, and temperature. Because Equations 2–1 and 5–1 both describe kinetic energy, they can be set equal to each other:

$$\tfrac{1}{2}\, mu_{avg}^{2} = \frac{3RT}{2N_A}$$

Now solve this equality for u:

$$u_{avg}^{2} = \frac{3RT}{mN_A} = \frac{3RT}{MM}$$

$$u_{avg} = \left(\frac{3RT}{MM}\right)^{1/2} \tag{5-2}$$

This speed is called the **root-mean-square** speed, because it is found by taking the square root of the average (mean) value of u^2. According to Equation 5–2, the

Effusion

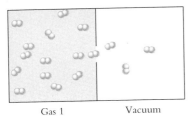

Gas 1   Vacuum

Diffusion

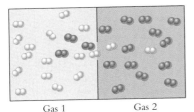

Gas 1   Gas 2

Figure 5-6
Effusion is the movement of a gas into a vacuum. Diffusion is the gradual mixing of two or more gases.

Animation

The movement of gases through the atmosphere, such as the fragrance of a rose moving from the flower to our noses, generally involves air currents as well as diffusion. Convection, the flow of gas in a current, moves molecules much more rapidly than diffusion.

average speed of gas molecules is directly proportional to the square root of the temperature and is inversely proportional to the square root of the molar mass.

Equation 5-2 can be directly applied to the movement of molecules escaping from a container into a vacuum. This process is **effusion.** Effusion is exemplified by the escape of molecules from the oven of Figure 5-3.

Usually, molecules do not escape into a vacuum. Instead, they move around by **diffusion,** the movement of one type of molecule through molecules of another type. Diffusing molecules undergo frequent collisions, so their paths are similar to that shown in Figure 5-1. Nevertheless, their average rate of movement depends on temperature and molar mass according to Equation 5-2. Figure 5-6 shows a molecular-level comparison of effusion and diffusion.

An example of diffusion is shown in Figure 5-7, in which aqueous solutions of hydrochloric acid (HCl) and ammonia (NH_3) are placed at opposite ends of a glass tube. Molecules of HCl gas and NH_3 gas escape from the solutions and diffuse through the air in the tube. When the two gases meet, they undergo an acid–base reaction to make ammonium chloride, a white solid salt:

$$HCl(g) + NH_3(g) \longrightarrow NH_4Cl(s)$$

$$\text{Strong acid} \quad \text{Weak base}$$

The lighter NH_3 molecules diffuse more rapidly than the heavier HCl molecules, so the white band of salt forms closer to the HCl end of the tube, as can be seen in Figure 5-7.

Rates of molecular motion are directly proportional to molecular speeds, so Equation 5-2 predicts that for any gas, rates of effusion and diffusion increase with the square root of the temperature in kelvins. Also, at any particular temperature effusion and diffusion are faster for molecules with small molar masses, as Figure 5-7 indicates.

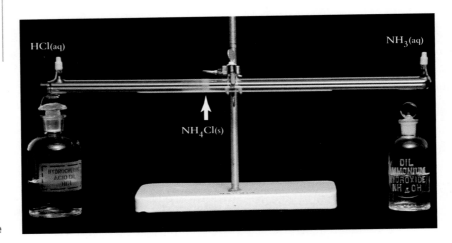

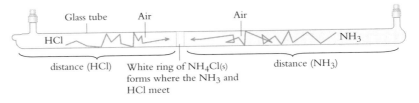

Figure 5-7
NH_3 diffuses through a glass tube faster than HCl. When the two gases meet, they form ammonium chloride (NH_4Cl), which appears as a white band closer to the end of the tube that contains HCl. (An aqueous solution of ammonia is also known as ammonium hydroxide.)

5.1.1 A molecular beam experiment of the type illustrated in Figure 5-3 is performed with an equimolar mixture of He and CO_2. Sketch the appearance of a graph of the number of molecules reaching the detector as a function of time.

5.1.2 Calculate each of the following energies: (a) total kinetic energy of 1.00 mol of He atoms at -100 °C; (b) total kinetic energy of 1.00 g of N_2 molecules at 0 °C; and (c) kinetic energy of a single molecule of SF_6 that has twice as much energy as the average molecular energy at 200 °C.

5.1.3 Calculate the average speeds of H_2O and D_2O molecules when $T = 500$ °C. (D represents deuterium, the isotope of hydrogen that contains one neutron.) Use these speeds to calculate the ratio of effusion rates for these two gases from an oven at 500 °C.

5.2 THE IDEAL GAS EQUATION

A gas can be described by its macroscopic properties as well as by the molecular description of Section 5-1. Any gas contains some number of moles of atoms or molecules, has a mass, occupies a volume, and is at some temperature. This section develops the link between the molecular description of a gas and its macroscopic properties.

When an object strikes a surface, a force is exerted against that surface. As an example, a basketball exerts force when it strikes a backboard. Gas molecules also exert forces. Wind moves a sailboat through the water as a result of force exerted on the sail. The gas inside a balloon prevents the balloon from collapsing by exerting force on the inner surface. At the molecular level, atoms and molecules exert forces through their never-ending collisions with the walls of their container, as shown for helium in the balloon in Figure 5-8.

The collective effect of molecular collisions is measured by **pressure (P)**, which is the force exerted per unit area. Pressure is caused by molecular collisions, but it is a macroscopic property. To move from our molecular view of collisions to a description of the pressure exerted by a gas, we must examine more closely the nature of collisions.

Ideal Gases

The behavior of a gas suggests that its molecules are not affected by one another. Gases are easy to compress, showing that there is lots of empty space in a gas. Gases also escape easily through any opening, indicating that their molecules are not strongly attracted to one another. We define an **ideal gas** to be one for which both the volume of molecules and the forces among the molecules are so small that they have no effect on the behavior of the gas. Thus an ideal gas can be defined by the following characteristics:

/// *The volume occupied by the molecules of an ideal gas is negligible compared with the volume of its container.*

/// *The energies generated by forces among ideal gas molecules are negligible compared with molecular kinetic energies.*

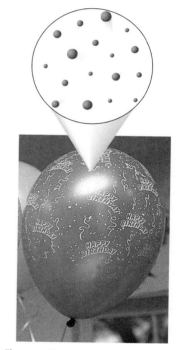

Figure 5-8
Most of the volume inside a helium-filled balloon is empty space. Helium atoms are in continual motion, colliding with the walls and with one another.

The ideal gas is a model for gas behavior. The behavior of any real gas departs somewhat from ideality because of molecular volumes and intermolecular forces. We consider how these factors affect real gas behavior in Chapter 10. In that chapter, we show that departures from ideality are small enough to neglect under many circumstances.

In an ideal gas each molecule is independent of all others. As a result, a molecule of an ideal gas travels freely through space until it strikes a wall of the container. Because of this independence, the macroscopic properties of an ideal gas can be calculated by summing the behavior of each individual molecule.

The pressure, P, exerted by a gas is the result of all the molecular collisions with the walls of the container. When a molecule strikes a wall, it exerts a force on the wall. During each second, many collisions exert many such forces. Pressure is the sum of all these forces per unit area. To obtain an equation for pressure, we can analyze the force exerted by each molecule-wall collision and then add them together.

The contribution of a single molecule to the total gas pressure depends on the size of the forces exerted in its collisions and on how often the molecule strikes the walls of its container. The size of the force exerted depends on the *mass* of the molecule and on its *speed*. A molecule with a small mass exerts a smaller force than a massive molecule traveling at the same speed. Similarly, a slow-moving molecule exerts a smaller force than the same molecule moving faster. At the same time, the number of collisions with the wall depends on the molecule's *speed*. Faster-moving molecules hit the walls more often than slower-moving molecules. Taken together, these generalizations indicate that total effect of a single molecule on the pressure of a gas is proportional to the mass of the molecule multiplied by the square of its speed. Recall that kinetic energy is also proportional to mass multiplied by the square of speed, so the effect of each molecule is proportional to its kinetic energy:

$$P_{\text{per molecule}} \propto E_{\text{kinetic}}$$

The symbol $\propto$ should be read as "is proportional to." It means that when the property on the right is multiplied by some factor, the property on the left is multiplied by the same factor.

Recall also that at a given temperature, the average molecular kinetic energy is the same for all gases, so this proportionality is independent of molar mass. However, the average molecular kinetic energy does depend on the absolute temperature (T, in kelvins), as expressed by Equation 5-1:

$$\overline{E}_{\text{kinetic}} = \frac{3RT}{2N_A}$$

Thus the average contribution of each molecule to the pressure of a gas is proportional to the temperature:

$$P_{\text{per molecule}} \propto \frac{3RT}{2N_A} \propto T$$

At this point, we might expect that simply multiplying the average effect per molecule by number of molecules would give the total pressure, but we must consider another factor. In any unit of time, only the molecules in the space near a wall are able to reach the wall. Thus the number of collisions per second depends on the molecular density of the gas, not the total number of molecules in the container. To relate pressure to other macroscopic properties, we must determine how macroscopic properties depend on molecular density.

First, consider what happens when a gas is compressed to half its original volume. The volume decreases, but the total number of molecules remains the same. Cutting the volume in half doubles the molecular density because there are twice as many molecules in any given space. This results in twice as many wall collisions, as shown in Figure 5-9.

Second, doubling the amount of gas in a fixed volume also doubles the molecular density, and doubling the molecular density doubles the number of collisions. This is shown in Figure 5-10.

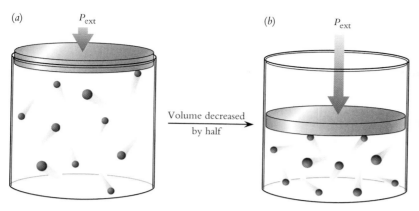

Figure 5-9
Schematic view of the effect of compressing a fixed quantity of gas into a smaller volume. Container (*b*) has the same number of molecules as (*a*) but only half its volume. Consequently, the molecular density is twice as great in (*b*), with twice as many collisions per second with the walls.

An increase in the molecular density of a gas gives a proportionate increase in number of collisions per second, which in turn increases the pressure of the gas. This proportionality can be expressed in terms of the molar density n/V (where n is the number of moles and V is the volume):

$$P \propto \frac{n}{V}$$

We now have two proportional relationships that describe gas behavior.

$$P \propto T \quad and \quad P \propto \frac{n}{V}$$

For a closed container, changing the temperature of the gas changes neither n nor V, so the molecular density of a gas is independent of its temperature. This independence lets us combine these two expressions into a single proportionality that gives the dependence of pressure on amount, volume, and temperature:

$$P \propto \frac{nT}{V}$$

A proportional relationship is converted to an equation by introducing a constant:

$$P = (\text{constant}) \frac{nT}{V}$$

Scientists have evaluated the constant in this equation experimentally by measuring the pressure exerted by gases when n, T, and V are known. The constant turns out to be R, the gas constant defined in Section 5.1:

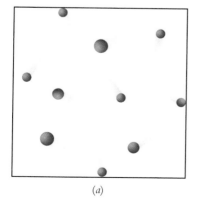

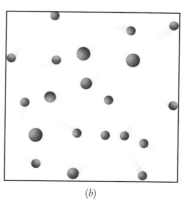

(*a*)   (*b*)

Figure 5-10
Schematic view of the effect of doubling the number of gas molecules in a fixed volume. Container (*b*) has twice as many molecules as (*a*). Consequently, the molecular density is twice as large in (*b*), with twice as many collisions per second with the walls.

$$P = R \frac{nT}{V}$$

This equality describes the macroscopic behavior of an ideal gas. Multiplying both sides by V gives the **ideal gas equation:**

$$PV = nRT \qquad\qquad\qquad \textbf{(5-3)}$$

The ideal gas equation predicts that every gas behaves identically. The constant R has the same value for all gases, so the equation makes no reference to any particular molecular property, such as size or mass. The only restrictions on the ideal gas equation are those needed for molecules to behave ideally. In other words, molecular sizes must be negligible compared to container volume, and forces between molecules must be negligible. These restrictions may appear to be drastic, but many gases show near-ideal behavior under normal conditions.

Section Exercises

5.2.1 According to the ideal gas equation, what will happen to the pressure of a gas if we make each of the following changes, holding all other conditions constant? (a) Double T (in kelvins); (b) reduce the container's volume by a factor of two; (c) triple the amount of gas in the container; and (d) replace the gas with an equal number of moles of another gas whose molar mass is twice as great.

5.2.2 The molecular picture represents a small portion of an ideal gas exerting a pressure P.
Redraw the picture in two ways, each of which gives a new pressure that is half as great.

5.2.3 Is a gas more likely to behave ideally at low molecular density or at high molecular density? Explain your answer in terms of the two assumptions made for ideal gases.

5-3 PRESSURE

The air around us is a huge reservoir of gas that exerts pressure on the Earth's surface. This pressure of the atmosphere can be measured with an instrument called a **barometer.** A simple mercury barometer is shown schematically in Figure 5-11. A long glass tube, closed at one end, is filled with liquid mercury. The filled tube is inverted carefully into a dish that is partially filled with more mercury. The force of gravity pulls downward on the mercury in the tube. With no opposing force, the mercury would all run out of the tube and mix with the mercury in the dish. The mercury does fall, but the flow stops at a fixed height. The column of mercury stops falling because the atmosphere exerts a pressure on the mercury in the dish, exerting a force that pushes the column up the tube. The column is in balance where the force of the atmosphere on the mercury in the dish exactly balances the downward force of gravity on the mercury column.

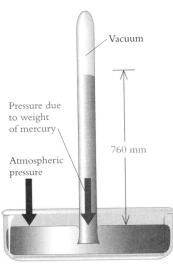

Figure 5-11
Diagram of a mercury barometer.

At sea level, atmospheric pressure supports a mercury column approximately 760 mm in height. As we describe later in this chapter, the pressure falls as altitude increases. Changes in weather cause fluctuations in atmospheric pressure. Nevertheless, at sea level the height of the column seldom varies by more than 10 mm, except under extreme conditions, such as in the eye of a hurricane, when the mercury in a barometer may fall below 740 mm.

A **manometer** is similar to a barometer, but in a manometer, gases exert pressure on *both* liquid surfaces. Consequently a manometer measures the *difference* in pressures exerted by two gases. A simple manometer, shown in Figure 5-12, is a U-shaped glass tube containing mercury. One side of the tube is exposed to the atmosphere and the other to a gas whose pressure we want to measure. In Figure 5-12, the pressure exerted by the atmosphere is less than the pressure exerted by the gas in the bulb. The difference in heights of mercury (Δh, in mm) between the two sides of the manometer depends on the difference in the pressures.

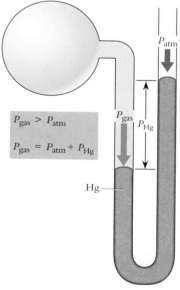

$$P_{gas} > P_{atm}$$

$$P_{gas} = P_{atm} + P_{Hg}$$

Units of Pressure

Traditionally, chemists define the units of pressure in terms of the Earth's atmosphere and the mercury barometer. The standard **atmosphere (atm)** is the pressure that will support a column of mercury 760 mm in height.

A second common pressure unit, the **torr,** is also based on the mercury barometer. One torr is the pressure exerted by a column of mercury 1 mm in height. Because the standard atmosphere supports a 760-mm column of mercury, the relationship between the atmosphere and the torr is 1 atm = 760 torr = 760 mm Hg.

The accepted SI unit for pressure is the **pascal (Pa)**. Recall that pressure is force per unit area, so the pascal can be expressed by combining the SI units for force and area. The SI unit of force is the newton (N), and area is measured in square meters (m^2). Thus the pascal is 1 N/m^2. Expressed in pascals, the numerical value of atmospheric pressure is quite large. By international agreement, 1 atm is defined exactly in terms of pascals: 1 atm $= 1.01325 \times 10^5$ Pa. Example 5-3 illustrates pressure measurement and unit conversions, and the Tools for Discovery Box explores ways to achieve very low pressures.

Figure 5-12
The difference in heights of liquid on the two sides of a manometer, Δh, is a measure of the difference in gas pressures (ΔP) applied to the two sides.

Δ (Delta) stands for "difference in."

Measuring Pressure	Example 5-3

A scientist collected an atmospheric gas sample. A manometer attached to the gas sample gave the reading shown in the figure (see also Figure 5-12), and the barometric pressure in the laboratory was 752 mm Hg. Calculate the pressure of the sample in atmospheres and in Pascals.

Strategy: We must combine the reading from the manometer with the barometric pressure to find the pressure of the gas sample. The manometer displays the pressure difference in millimeters of mercury, so conversion factors are needed to express the pressure in atmospheres and pascals.

Solution: The difference in mercury levels in the manometer is 20 mm. This is the pressure difference in torr between the gas sample and atmospheric pressure. The latter, as measured with a barometer, is 752 torr. Shall we add or subtract the 20 torr pressure difference? Notice that the mercury level in the manometer is lower on the side exposed to the atmosphere. Thus the atmosphere pushes on

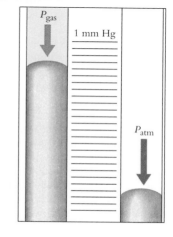

Example 5-3	Measuring Pressure *(continued)*

the mercury harder than does the gas sample, meaning that the pressure of the gas sample is lower than the pressure of the atmosphere. Subtract the pressure difference:

$$P_{gas} = P_{atm} - \Delta P = 752 \text{ torr} - 20 \text{ torr} = 732 \text{ torr}$$

Use the appropriate conversion factors to give the pressure in atm and pascals.

$$1 \text{ atm} = 760 \text{ torr} \quad and \quad 1 \text{ atm} = 1.01325 \times 10^5 \text{ Pa}$$

$$P \text{ (atm)} = 732 \text{ torr } (1 \text{ atm}/760 \text{ torr}) = 0.963 \text{ atm}$$

$$P \text{ (Pa)} = 0.963 \text{ atm } (1.01325 \times 10^5 \text{ Pa/atm}) = 9.76 \times 10^4 \text{ Pa}$$

Box 5-1	Tools for Discovery: High Vacuum

"Nature abhors a vacuum." True? On Earth, it is certainly difficult to remove the gas from a container to generate and maintain a vacuum—the absence of gas. In outer space, on the other hand, a vacuum is the rule rather than the exception. Most of the volume of the universe is nearly empty space, close to a perfect vacuum. If a spacecraft were to spring a leak, its gaseous atmosphere would quickly escape into that vacuum by effusion. Perhaps it would be better to say that conditions on Earth are unfavorable for vacuums.

Over the years, scientists have spent immense efforts to achieve high vacuums. Powerful and sophisticated pumps are required to remove the gas from a chamber. Once a high vacuum is established, the chamber must be leak-free at the molecular level in order to keep atmospheric gas from seeping back into an evacuated system. Moreover, the materials used to make a vacuum chamber must be carefully selected to avoid substances that give off gases. A modern high-vacuum instrument is an elaborate array of pumps, valves, seals, and windows, as the photograph shows.

Why bother to go to all this trouble, just to remove the gases from a small volume of space? High vacuum is used in chemistry and physics research to achieve any of these conditions: the absence of molecular collisions, the maintenance of an ultra-clean environment, or a simulation of conditions in outer space.

Mass spectrometers, workhorse instruments described in Chapter 2, require a vacuum to function. A mass spectrometer generates a beam of ions that is sorted according to specifications of the particular instrument. Usually, the sorting depends on differences in speed, trajectory, and mass. For instance, one type of mass spectrometer measures how long it takes ions to travel from one end of a tube to another. Residual gas must be removed from the tube to eliminate collisions between gas molecules and the ions that are being analyzed. As the diagram shows, collisions with unwanted gas molecules deflect the ions from their paths and change the expected mass spectral pattern.

The beam in a mass spectrometer is one example of a *molecular beam*, in this case a beam of molecular ions. Molecular beams are used in many studies of fundamental chemical interactions. In a high vacuum, a molecular beam allows chemists to study the reactions that take place through specifically designed types of collisions. For example, a "crossed-beam" experiment involves the intersection of two molecular beams of two different substances. The

As this example shows, the pressures encountered most frequently by humans are in the range of hundreds of kilopascals (kPa). For this reason, chemists find it more convenient to use torr, millimeters of mercury, or atmospheres to measure pressure. Because few chemists use the pascal, we use torr, mm Hg, and atm in this text, even though they are not part of the SI system.

$1 \text{ Pa} = 1 \text{ N m}^{-2}$, and
$1 \text{ N} = 1 \text{ kg m s}^{-2}$. Thus, the pascal is a collection of fundamental SI units: $1 \text{ Pa} = 1 \text{ kg m}^{-1} \text{ s}^{-2}$.

Pressure and the Ideal Gas Equation

With a set of units for pressure, we can begin quantitative applications of the ideal gas equation. However, using atmospheres rather than pascals as the unit of pressure means that we need a different value for the gas constant. The value of R introduced earlier, 8.314 J/mol K, is appropriate only when pressure is expressed in pascals and volume is expressed in m^3. When the pressure of a gas is expressed

Box 5-1	Tools for Discovery: High Vacuum (continued)

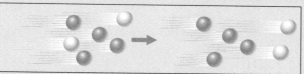

In a high-vacuum, collision-free tube, a beam of ions is sorted according to ion speed (mass).

If the vacuum is not high enough, residual gas molecules (black circle) collide with ions, deflecting them from their paths and destroying the sorting process.

pure surfaces of metals or semiconductors requires ultra-high vacuum, pressures on the order of 10^{-12} atm.

High vacuum on a large scale has long played a role in studies of nuclear reactions. To study fundamental nuclear processes, physicists have constructed very long tubes through which they can accelerate beams containing nuclei of various elements. The photo shows one of the longest of these, the high-energy linear accelerator at Stanford University. The interior of the accelerator is a one-mile-long pipe that is an obvious feature in this photo. The pipe must be maintained at high vacuum. Otherwise, the particles in the beam would be deflected by molecular collisions, ruining the experiments. This long pipe has massive vacuum pumps located at regular intervals along its entire length, pumping continually to keep the pressure inside the pipe at an acceptably low level.

types of substances, molecular speeds, and orientations of the beams can be changed systematically to give detailed information about how chemical reactions occur at the molecular level. Chemists also have learned how to create molecular beams in which the molecules have very little energy of motion. These isolated, low-energy molecules are ideal for studies of fundamental molecular properties.

An ultra-clean environment is another major reason for generating high vacuum. At atmospheric pressure, every atom on a solid surface is bombarded with gas molecules at a rate of trillions per second. Even under a reasonably high vacuum, 10^{-9} atm, every atom on a surface is struck by a gas molecule about once per second. If the surface is reactive, these collisions result in chemical reactions whose products are likely to contaminate the surface. The study of

in atmospheres and volume is expressed in liters, the gas constant becomes $R = 0.08206$ L atm mol^{-1} K^{-1}. Example 5-4 shows how to calculate pressure using the ideal gas equation.

| Example 5-4 | **Pressure and the Ideal Gas Equation** |

A 265-gallon steel storage tank contains 88.5 kg of methane (CH_4). If the temperature is 25°C, what is the pressure inside the tank?

Strategy: The problem asks for the pressure exerted by a gas, which can be calculated using the ideal gas equation. We need to make sure all the variables are expressed in their proper units. Temperature must be in kelvins, amount of methane in moles, and volume in liters.

Solution: Each conversion is straightforward:

$$V = (265 \text{ gallons})(3.7854 \text{ L/gallon}) = 1.003 \times 10^3 \text{ L} \quad T = 25 + 273.15 = 298 \text{ K}$$

$$n = \frac{(88.5 \text{ kg})(10^3 \text{ g/kg})}{(16.04 \text{ g/mol})} = 5.517 \times 10^3 \text{ mol CH}_4$$

Next, rearrange the ideal gas equation so that pressure is isolated on the left:

$$PV = nRT \quad \text{so} \quad P = \frac{nRT}{V}$$

Before we substitute the known values and solve for pressure, it is essential to choose the proper value of the gas constant. In common usage, the pressure is expressed in atmospheres, so we must use $R = 0.08206$ L atm mol^{-1} K^{-1}.

$$P = \frac{(5.517 \times 10^3 \text{ mol})(0.08206 \text{ L atm/mol K})(298 \text{ K})}{(1.003 \times 10^3 \text{ L})} = 135 \text{ atm}$$

This high value indicates why gases such as methane must be stored in tanks made of materials such as steel that can withstand high pressures.

Section Exercises

5.3.1 In the eye of a severe hurricane, the height of a mercury barometer may fall to 710 mm. Express this pressure in torr and atmospheres. What percentage change from standard atmospheric pressure is this?

5.3.2 Oxygen gas is sold in pressurized tanks. Calculate the pressure in atmospheres inside a tank that has a volume of 7.45 L, contains 0.500 kg of O_2, and has a temperature of 20 °C.

5.3.3 What pressure is exerted on the left side of the manometer shown in Figure 5-12 when the difference in column heights is 4.75 cm, the right side is open to the atmosphere, and barometric pressure is 752.8 torr? Express your result in torr, in atmospheres, and in kilopascals.

5.4 APPLYING THE IDEAL GAS EQUATION

The ideal gas model describes any gas for which molecular volumes and intermolecular forces are small enough to neglect. If these assumptions are valid, we should be able to apply the ideal gas equation to any change of conditions on any system of gases. To begin this section, we consider pressure–volume variations and temperature–volume variations, both of which were important in the historical development of the ideal gas model.

Pressure-Volume Variations

Quantitative measurements on gases were made as early as 1662 by the Englishman Robert Boyle, who carried out many experiments on gases confined in J-shaped glass tubes. As shown in Figure 5-13, these tubes were closed at one end. Boyle added liquid mercury to a tube, trapping a fixed quantity of air on the closed side. As the figure shows, when he added additional mercury to the open end of the tube, the mass of the additional mercury compressed the trapped gas to a smaller volume. Boyle found that the volume occupied by the trapped air was inversely proportional to the total pressure applied by the mercury plus the atmosphere. In other words, doubling the total pressure on the trapped air reduced its volume by a factor of two. The pressure–volume product, PV, always remained the same. Boyle concluded that the product of pressure and volume is a constant for a fixed quantity of gas at constant temperature.

This behavior is just what the ideal gas equation predicts. Boyle worked with a fixed amount of air trapped in the tube, so the number of moles of gas remained the same during his experiments. In other words, n was held constant. Boyle also worked at only one temperature ("room" temperature), so T was held constant. As a result, in Boyle's experiments the product nRT was constant. Thus the ideal gas equation agrees with Boyle's observations: $PV = nRT = $ constant when T and n are fixed. The relationship, $PV = $ constant, is known as *Boyle's Law*. Example 5-5 applies the ideal gas equation to a pressure–volume variation.

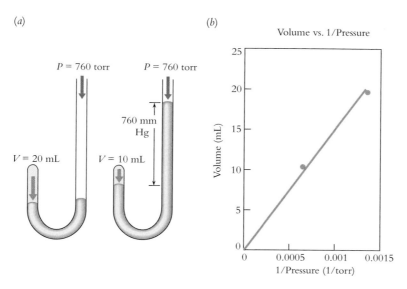

Figure 5-13
(*a*) Schematic illustration of Boyle's experiments. When the external pressure on a J-tube is increased by adding mercury, the gas on the closed side is compressed. If the pressure is doubled, the gas volume is halved. (*b*) When *V* is plotted against 1/*P*, the result is a straight line.

Example 5-5	Pressure–Volume Variations

A sample of helium gas is held at constant temperature inside a cylinder whose volume is 0.80 L when a piston exerts a pressure of 1.5 atm. If the external pressure on the piston is increased to 2.1 atm, what will be the new volume?

Solving
Quantitative
Problems

Strategy: We follow the seven-step strategy for problem solving.

Solution:

1. The question asks for the new volume.

2. Visualize the conditions by drawing a schematic diagram of the initial and final conditions:

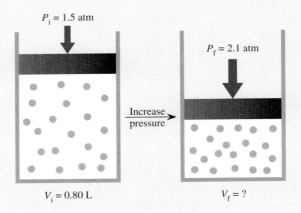

3. The data supplied are initial volume and pressure and final pressure:

$$P_i = 1.5 \text{ atm}, \quad P_f = 2.1 \text{ atm}, \quad V_i = 0.80 \text{ L}.$$

4. Gas behavior is involved, so the equation that applies is the ideal gas equation, $PV = nRT$.

5. Rearranging the gas equation to solve for V will not help because we do not know n, the number of moles of He present in the system. Nor do we know T, the temperature of the gas. We do know that n and T remain unchanged as the pressure increases.

To determine the final volume of the helium gas, apply the ideal gas equation to the initial (i) and final (f) conditions:

$$P_i V_i = n_i R T_i \quad \text{and} \quad P_f V_f = n_f R T_f$$

In this problem, the quantity of He inside the cylinder and the temperature of the gas are constant:

$$n_i = n_f \quad \text{and} \quad T_i = T_f \quad \text{so} \quad n_i R T_i = n_f R T_f$$

Therefore:

$$P_i V_i = P_f V_f \text{ (constant } n \text{ and } T)$$

Notice that this equality can be solved for V_f without knowing the values for n and T.

Pressure–Volume Variations *(continued)*

Example 5-5

6. Now substitute and calculate the final volume:

$$V_f = \frac{P_i V_i}{P_f} = \frac{(1.5 \text{ atm})(0.80 \text{ L})}{(2.1 \text{ atm})} = 0.57 \text{ L}$$

7. This answer is reasonable because a pressure increase has caused a volume decrease.

Temperature–Volume Variations

Boyle also observed that heating a gas causes it to expand in volume, but more than a century passed before Jacques-Alexandre-César Charles reported the first *quantitative* studies of gas volume as a function of temperature. Charles found that gas volume increases linearly with temperature as long as the pressure and number of moles are held constant. In other words, a graph of volume vs. temperature gives a straight line, as shown in Figure 5-14.

These data are consistent with the ideal gas model. To see this, rearrange the ideal gas equation to isolate volume:

$$PV = nRT \qquad so \qquad V = \frac{nRT}{P} = \left(\frac{nR}{P}\right)T$$

Grouping n, R, and P shows the direct relationship between temperature (T) and volume (V) when moles and pressure are constant. The linear relationship between V and T is known as *Charles' Law*.

Notice in Figure 5-14 that the volume of the gas always extrapolates to zero at 0 K. Experiments performed on a series of different gases all give this result. Does this make sense? The ideal gas model assumes that the volume occupied by the atoms and molecules is negligible when compared with the total volume of the gas. As a gas is cooled, the volume of the gas decreases, but the volume of its molecules remains fixed. As long as the fractional volume is negligible, the compression data fall on a straight line. For an ideal gas, volume would indeed shrink to zero at 0 K, as the extended straight lines predict. For any real gas, the fractional volume occupied by the molecules becomes significant at very low temperature when the gas volume becomes very small. Nonetheless, the gases listed in Table 5-1 behave ideally at "normal" temperature and pressure (around 298 K and 1 atm), because the fractional volume occupied by the molecules is insignificant under these conditions.

The direct proportionality between T and V applies only when T is expressed in kelvins. Recall Equation 5-1, relating temperature in kelvins to kinetic energy. Because kinetic energy is always positive, temperature in kelvins is also always positive. The minimum value for T is 0 K. This is absolute zero temperature, and the Kelvin scale is an absolute temperature scale. Other temperature scales (°C and °F, for example) have their zero points at positive absolute temperatures, so these scales have negative as well as positive values. Any calculation that uses the gas constant must involve temperatures in kelvins. Example 5-6 illustrates this point.

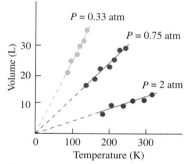

Figure 5-14
Plots of volume vs. temperature for 1 mol of air at three constant pressures.

← **CHAPTER 10**
In Chapter 10, we show that every gas condenses to a liquid when temperature is lowered sufficiently. Near its condensation temperature, a gas no longer obeys the ideal gas equation unless its pressure is substantially less than 1 atm.

Example 5-6	Temperature–Pressure Variations

A 1.54-L gas bulb in a chemistry laboratory contains oxygen gas at 21 °C and 758 torr (see the figure). The air conditioning in the laboratory breaks down, and the temperature rises to 31 °C. What pressure does the gauge show now?

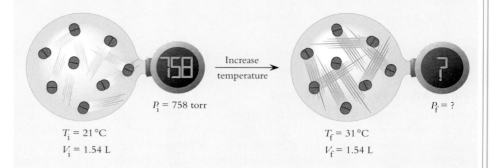

$T_i = 21\,°C$
$V_i = 1.54\ L$

$T_f = 31\,°C$
$V_f = 1.54\ L$

Strategy: We are asked to calculate a final pressure. What are the molecules doing? Neither the number of gas molecules nor the volume of the flask changes as the temperature rises. Instead, the molecules inside the flask move faster, resulting in more collisions at higher average energy. Thus the pressure of the gas rises as the temperature increases. The ideal gas can be used to calculate the new pressure, but the temperature must be expressed in kelvins.

Solution: As in Example 5-5, it is useful to organize the data into two sets, one for the initial conditions and one for the final conditions:

$P_i = 758$ torr $\qquad\qquad P_f = ?$
$n_i = ?$ $\qquad\qquad\qquad n_f = n_i$
$T_i = 21\ °C = 294\ K \qquad T_f = 31\ °C = 304\ K$
$V_i = 1.54\ L \qquad\qquad V_f = 1.54\ L$

The ideal gas equation applies to both sets of conditions; n and V are constant. Rearrange the equation so that all the constants are grouped together:

$$PV = nRT \qquad so \qquad \frac{P}{T} = \frac{nR}{V} = constant$$

This relationship applies to both sets of conditions, giving the following ratio:

$$\frac{P_i}{T_i} = \frac{P_f}{T_f} \qquad (constant\ n\ and\ V)$$

Now rearrange the expression and solve for the final pressure:

$$P_f = T_f\left(\frac{P_i}{T_i}\right) = (304\ K)\left(\frac{758\ torr}{294\ K}\right) = 784\ torr$$

Notice that temperature must be in kelvins (always true for gas calculations) and that the final pressure is in the same units as the initial pressure. Is the result reasonable? Temperature increased by a small fraction (294 to 304 K), and our calculated pressure is larger than the original pressure by a small fraction, so the answer seems reasonable.

Variations on the Gas Equation

Pressure-volume studies and temperature-volume studies verify that gases obey the ideal gas equation, $PV = nRT$. We can use this equation with confidence to calculate gas properties under various circumstances. During chemical and physical transformations, any of the four variables in the ideal gas equation (P, V, n, T) may change, and any of them may remain constant. A good strategy for organizing gas calculations is to determine which variables do not change and rearrange the gas equation to group them all on the right. Here are some examples:

For a fixed amount at constant temperature, $\qquad PV = nRT = \text{constant}$

For a fixed amount at constant volume, $\qquad \dfrac{P}{T} = \dfrac{nR}{V} = \text{constant}$

For a fixed amount at constant pressure, $\qquad \dfrac{V}{T} = \dfrac{nR}{P} = \text{constant}$

For a fixed volume at constant temperature, $\qquad \dfrac{P}{n} = \dfrac{RT}{V} = \text{constant}$

For a fixed pressure at constant temperature, $\qquad \dfrac{V}{n} = \dfrac{RT}{P} = \text{constant}$

Each calculation of gas properties proceeds from the ideal gas equation. Rather than memorize several versions of the ideal gas equation, we strongly recommend that you learn how to analyze gas problems, as we do in Examples 5-7 and 5-8.

Changing Gas Conditions	Example 5-7

A sample of carbon dioxide in a 10.0-L gas cylinder at 25 °C and 1.00 atm pressure is compressed and heated. The final temperature and volume are 50 °C and 5.00 L. Compute the final pressure.

Strategy: As in Example 5-6, we are asked to calculate the pressure of a gas after a change in conditions. Begin by analyzing the change at the molecular level. As the sample is heated, CO_2 molecules absorb energy and move faster. This leads to more frequent collisions at higher energy, hence an increase in pressure. At the same time, the gas is compressed into a smaller volume, so the molecular density of CO_2 rises accordingly. Higher molecular density also means more frequent collisions and higher pressure. Thus we expect the pressure to increase. The amount of increase is computed using the ideal gas equation.

Solution: Start the quantitative part of the problem by collecting and organizing the numerical data:

Initial	**Final**
$P_i = 1.00$ atm	$P_f = ?$
$V_i = 10.0$ L	$V_f = 5.00$ L
$T_i = 298$ K	$T_f = 323$ K

| Example 5-7 | Changing Gas Conditions *(continued)* |

The data tell us that P, V, and T all change. The number of moles of CO_2 is not mentioned directly, but the physical transformation performed on the gas does not affect the number of molecules, so we conclude that $n_i = n_f$. We rearrange $PV = nRT$, dividing both sides by T to group the constants n and R on one side:

$$\frac{P_i V_i}{T_i} = nR = \frac{P_f V_f}{T_f}$$

Rearrange this expression to solve for the final pressure, insert the known information, and perform the calculations:

$$P_f = \left(\frac{P_i V_i}{T_i}\right)\left(\frac{T_f}{V_f}\right) = \frac{(1.00 \text{ atm})(10.0 \text{ L})(323 \text{ K})}{(298 \text{ K})(5.00 \text{ L})} = 2.17 \text{ atm}$$

The final pressure is 2.17 atm. The numerical answer agrees with our qualitative analysis at the molecular level.

| Example 5-8 | Gas Law Calculations |

Two natural gas storage tanks, with volumes of 1.5×10^4 and 2.2×10^4 L, are at the same temperature. The tanks are connected by pipes that equalize their pressures. What fraction of the stored natural gas is in the larger tank?

Strategy: The problem asks us to determine how the gas is distributed between the two tanks. This is a gas problem, so we use the ideal gas equation.

Solution: There are not many data to summarize:

$$V_1 = 1.5 \times 10^4 \text{ L} \quad and \quad V_2 = 2.2 \times 10^4 \text{ L}$$

Examine the conditions to determine which variables are the same for both tanks. The problem states that the tanks have the same pressure and temperature. Rearrange the gas equation to group the constants and find an equation that applies:

$$PV = nRT \quad so \quad \frac{n}{V} = \frac{P}{RT} = \text{constant}$$

The problem asks about the fraction of gas in the larger tank. This fraction is the ratio of the number of moles in the larger tank to the total number of moles in both tanks. Because n/V is constant, we can write:

$$\frac{n_{\text{large tank}}}{V_{\text{large tank}}} = \frac{n_{\text{total}}}{V_{\text{total}}}$$

Now solve for the fraction of gas in the larger container:

$$\frac{n_{\text{large tank}}}{n_{\text{total}}} = \frac{V_{\text{large tank}}}{V_{\text{total}}} = \frac{2.2 \times 10^4 \text{ L}}{(2.2 \times 10^4 \text{ L}) + (1.5 \times 10^4 \text{ L})} = 0.59$$

A fraction of 0.59 means that 59% of the gas is in the larger tank. This seems reasonable because more than half the gas has to be in the larger tank.

Determination of Molar Mass

The ideal gas equation can be combined with the mole-mass relation to find the molar mass of an unknown gas:

$$PV = nRT \quad \text{(Ideal gas equation)} \qquad and \qquad n = \frac{m}{MM} \quad \text{(Mole-mass relation)}$$

First, solve the ideal gas equation for n: $\quad n = \dfrac{PV}{RT}$

Now, set the two expressions for n equal to each other:

$$\frac{m}{MM} = \frac{PV}{RT}$$

Finally, rearrange this equation to obtain an equation for molar mass:

$$MM = \frac{mRT}{PV}$$

Example 5-9 demonstrates how to use the properties of a gas to determine its molar mass.

Molar Mass Determination	Example 5-9

Calcium carbide (CaC_2) is a hard, gray-black solid that has a melting point of 2000 °C. This compound reacts strongly with water to produce a gas and a solution containing OH^- ions. A 12.8-g sample of CaC_2 was treated with excess water. The resulting gas was collected in an evacuated 5.00-L glass bulb with a mass of 254.49 g. The filled bulb had a mass of 259.70 g and a pressure of 0.988 atm when its temperature was 26.8 °C. Calculate the molar mass and determine the formula of the gas. Write the balanced chemical equation for the reaction.

Strategy: We will use the ideal gas equation to calculate the molar mass. The molar mass will help identify the correct molecular formula among a group of candidates selected using conservation of atoms principle. The problem involves a chemical reaction, so we must make a connection between the gas measurements and the chemistry that takes place.

Solution: Begin by analyzing the chemistry, then use the gas data to determine the molar mass. Because the reactants are known, we can write a partial equation that describes the chemical reaction:

$$CaC_2 \text{ (s)} + H_2O \text{ (l)} \longrightarrow Gas + ?$$

In any chemical reaction, atoms must be conserved, so the gas molecules can contain only H, O, C, and/or Ca atoms. The problem asks us to identify the gas from its properties.

The problem gives the following data about the unknown gas:

$$V_{bulb} = V_{gas} = 5.00 \text{ L} \qquad T = 26.8 \text{ °C} = 300.0 \text{ K} \qquad P = 0.988 \text{ atm}$$

$$m_{bulb + gas} = 259.70 \text{ g} \qquad m_{bulb} = 254.49 \text{ g}$$

$$m_{gas} = m_{bulb + gas} - m_{bulb} = 5.21 \text{ g}$$

Example 5-9	Molar Mass Determination *(continued)*

We use V, T, P, and the ideal gas equation to find the number of moles of gas. Then, with the mass of the gas sample, we can determine the molar mass:

$$\frac{PV}{RT} = n_{gas} = \frac{m_{gas}}{MM}$$

We could determine n and then solve for MM, but we can also solve the equality for MM and then substitute the data directly:

$$MM = \frac{RTm_{gas}}{PV}$$

$$MM = \frac{(0.08206 \text{ L atm mol}^{-1}\text{K}^{-1})(300.0 \text{ K})(5.21 \text{ g})}{(0.988 \text{ atm})(5.00 \text{ L})} = 26.0 \text{ g/mol}$$

To identify the gas, we examine the formulas and molar masses of known compounds that contain H, O, C, and Ca:

Formula	*MM*	Comment
Ca	40	A gas with $MM = 26.0$ g/mol cannot contain Ca
CO	28	This is close but too high
O_2	32	This is also too high, and H_2O_2 is even higher
H_2O	18	$H_{10}O$, $MM = 26.0$ g/mol, does not exist
CH_4	16	CH_{14}, $MM = 26.0$ g/mol, does not exist
C_2H_2	26	This substance has the observed molar mass

A little trial and error leads to the molecular formula: C_2H_2, commonly known as acetylene. Knowing the formula of the product, we can balance the equation for the gas-forming reaction:

$$CaC_2(s) + 2\ H_2O(l) \longrightarrow Ca(OH)_2(s) + C_2H_2(g)$$

Acetylene is a high-energy molecule used as a fuel for oxyacetylene welding.

Section Exercises

5.4.1 On a summer day the temperature inside a house rises to 37 °C (99 °F). In the evening the house may cool to 21 °C (68 °F). Does air enter or leave the house as it cools? If the volume of the house is 9.50×10^5 L and the barometric pressure has held steady at 768 mm Hg, what mass of air is transferred (average molar mass of air is 28.8 g/mol)?

5.4.2 Draw molecular pictures that show the behavior of a gas whose temperature is doubled at constant pressure.

5.4.3 A 2.96-g sample of a compound of mercury and chlorine is vaporized in a 1.000-L bulb at 307 °C, and the final pressure is found to be 394 torr. What are the molar mass and chemical formula of the compound?

5.5 GAS MIXTURES

Examples 5-5 through 5-9 involve gases that contain only one chemical substance, but many gases are mixtures of two or more species. The atmosphere, with its mixture of nitrogen, oxygen, and various trace gases, is an obvious example. Another example is the gas used by deep-sea divers, which contains a mixture of helium and oxygen. The ideal gas model provides guidance as to how we describe mixtures of gases.

In an ideal gas, all molecules act independently, so in an ideal gas mixture the molecules of each gas act independently. Gas behavior depends on the *number* of gas molecules but not on the *identity* of the gas molecules. The ideal gas equation applies to each gas in the mixture, as well as to the entire collection of molecules.

Suppose we pump 4.0 mol of helium into a deep-sea diver's tank. If we pump in another 4.0 mol of He, the container now contains 8.0 mol of gas. The pressure can be calculated using the ideal gas equation, with $n = 4.0 + 4.0 = 8.0$ mol. Now suppose that we pump in 4.0 mol of molecular oxygen. Now the container contains a total of 12.0 mol of gas. The total pressure still can be calculated using the ideal gas equation, with $n = 8.0 + 4.0 = 12.0$ mol. According to the ideal gas model, it does not matter whether we add the same gas or a different gas. Because all molecules in a sample of an ideal gas behave exactly the same way, the pressure increases in proportion to the increase in the total number of moles of gas.

How does a 1:2 mixture of O_2 and He appear on the molecular level? As O_2 is added to the container, its molecules move throughout the volume and become distributed uniformly. Diffusion quickly causes gas mixtures to become homogeneous. This is shown in Figure 5-15.

Dalton's Law of Partial Pressures

The pressure exerted by an ideal gas mixture is determined by the total number of moles:

$$P = \frac{n_{total}RT}{V}$$

We can express the total number of moles of gas as the sum of the amounts of the individual gases. For the diver's He and O_2 mixture:

$$n_{total} = n_{He} + n_{O_2}$$

Substitution gives a two-term equation for the total pressure:

$$P = \frac{(n_{He} + n_{O_2})RT}{V} = \frac{n_{He}RT}{V} + \frac{n_{O_2}RT}{V}$$

Notice that each term on the right resembles the ideal gas equation rearranged to express pressure. Each term therefore represents the **partial pressure (p)** of one of the gases. As Figure 5-16 illustrates, partial pressure is the pressure that would be present in a gas container if one gas were present by itself:

$$p_{He} = \frac{n_{He}RT}{V} \qquad p_{O_2} = \frac{n_{O_2}RT}{V}$$

The total pressure in the container is the sum of the partial pressures:

$$P_{total} = p_{He} + p_{O_2}$$

We have used He and O_2 to illustrate the behavior of a mixture of ideal gases, but the same result is obtained regardless of the number and identity of the gases.

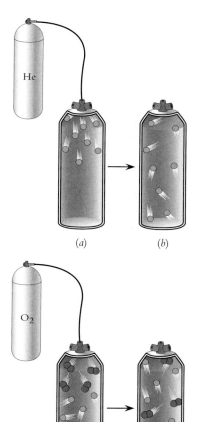

(a)   (b)

(c)   (d)

Figure 5-15
When 8.0 mol of He is added to a diver's tank (*a*) the atoms quickly become distributed uniformly throughout the tank (*b*) When 4.0 mol of O_2 is added to this tank (*c*) the molecules move about independently of the He atoms, causing the gases to mix uniformly (*d*).

Animation

The partial pressure of each gas is designated with a lower-case *p* to distinguish it from the total pressure of the mixture, *P*.

Figure 5-16
Molecular pictures of a sample of O_2, a sample of He, and a mixture of the two gases. Both components are distributed uniformly throughout the gas volume. Each gas behaves the same, whether it is pure or part of a mixture.

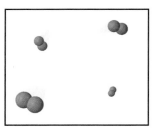 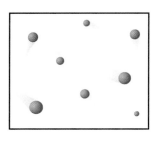 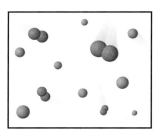

3.0 L at 273 K

4.0 mol O_2

P_{O_2} = 30 atm

3.0 L at 273 K

8.0 mol He

P_{He} = 60 atm

3.0 L at 273 K

12.0 mol gas

p_{O_2} = 30 atm

p_{He} = 60 atm

P_{total} = 90 atm

> /// *In a mixture of gases, each gas contributes to the total pressure the amount that it would exert if the gas were present in the container by itself.*

This is **Dalton's law of partial pressures.** To obtain a total pressure, simply add the contributions from all gases present:

John Dalton was the first to describe gas mixtures in this way as an application of his atomic theory.

$$P_{total} = p_1 + p_2 + p_3 + \cdots p_i$$

When doing calculations on a mixture of gases, we can apply the ideal gas equation to each component to find its partial pressure (p_i). Alternatively, we can treat the entire gas as a unit, using the total number of moles to determine the total pressure of the mixture (P).

Describing Gas Mixtures

There are several ways to describe the chemical composition of a mixture of gases. The simplest method is merely to list each component with its partial pressure or number of moles. Two other descriptions, mole fractions and parts per million, are also used frequently.

Chemists often express chemical composition in fractional terms, stating the number of moles of a substance as a fraction of the number of moles of all substances in the mixture. This way of stating composition is the **mole fraction (X):**

$$\text{Mole fraction of } A = X_A = \frac{n_A}{n_{total}}$$

Mole fractions provide a simple way to relate the partial pressure of one component to the total pressure of the gas mixture:

$$p_A = \frac{n_A RT}{V} \qquad and \qquad P_{total} = \frac{n_{total} RT}{V}$$

Dividing p_A by P_{total} gives:

$$\frac{p_A}{P_{total}} = \frac{\left(\dfrac{n_A RT}{V}\right)}{\left(\dfrac{n_{total} RT}{V}\right)} = \frac{n_A}{n_{total}} = X_A$$

Rearranging gives Equation 5-4:

$$p_A = X_A P_{total} \tag{5-4}$$

The partial pressure of a component in a gas mixture is its mole fraction times the total pressure. Example 5-10 illustrates calculations with gas mixtures.

Gas Mixtures **Example 5-10**

The amount of gas introduced into a diving tank can be determined by weighing the tank before and after charging the tank with gas. A diving shop placed 80.0 g of O_2 and 20.0 g of He in a 5.00-L tank at 298 K. Determine the total pressure of the mixture, and find the partial pressures and mole fractions of the two gases.

Strategy: We have a mixture of two gases in a container whose volume and temperature are known. The problem asks for pressures and mole fractions. Because molecular interactions are negligible, each gas can be described independently by the ideal gas equation. As usual, we need molar amounts for the calculations.

Solution: Begin with the data provided:

$$V = 5.00 \text{ L} \qquad T = 298 \text{ K} \qquad m_{He} = 20.0 \text{ g} \qquad m_{O_2} = 80.0 \text{ g}$$

Convert the mass of each gas into an amount in moles:

$$n = m/MM \qquad n_{He} = 5.00 \text{ mol} \qquad n_{O_2} = 2.50 \text{ mol}$$

Next, use the ideal gas equation to compute the partial pressure of each gas:

$$p_{He} = \frac{(5.00 \text{ mol})(8.206 \times 10^{-2} \text{ L atm/mol K})(298 \text{ K})}{(5.00 \text{ L})} = 24.5 \text{ atm}$$

The same calculation for O_2 gives $p_{O_2} = 12.2$ atm.

The total pressure is the sum of the partial pressures:

$$P_{total} = p_{He} + p_{O_2} = 24.5 \text{ atm} + 12.2 \text{ atm} = 36.7 \text{ atm}$$

Mole fractions can be calculated from moles or partial pressures:

$$X_{He} = \frac{n_{He}}{n_{total}} = \frac{5.00 \text{ mol}}{7.50 \text{ mol}} = 0.667 \qquad X_{O_2} = \frac{p_{O_2}}{P_{total}} = \frac{12.2 \text{ atm}}{36.7 \text{ atm}} = 0.333$$

To check for reasonableness, add the mole fractions, which must sum to 1.00.

Deep sea divers use mixtures of helium and oxygen in their breathing tanks.

When referring to very sparse components of a gas mixture, scientists typically use **parts per million (ppm)** to designate the relative number of molecules of a substance present in a sample. Parts per million measures how many molecules of substance are present in one million molecules of sample. Concentrations of atmospheric pollutants often are given in ppm. If a pollutant is present at a concentration of 1 ppm, there is one molecule of the pollutant in every one million molecules. In molar terms, 1 ppm means that there is one mole of pollutant for every one million moles of gas. Or, to put it another way, 10^{-6} mol of pollutant in every mole of air.

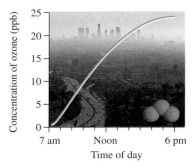

Concentration of ozone (ppb)

7 am   Noon   6 pm

Time of day

Air quality in Los Angeles, California on Sept. 11, 2000.

The city of Los Angeles issues a smog alert when the concentration of ozone in its atmosphere reaches 0.5 ppm. This means that in every mole of air, there is 0.5×10^{-6} mol of ozone. This may not seem like much, but ozone is a very toxic substance that is particularly damaging to soft tissue such as the lungs.

Even lower concentrations in a gas mixture are expressed in **parts per billion (ppb).** Parts per billion measures how many molecules of substance are present in one billion molecules of sample.

1 ppm = 1 molecule out of every 10^6 molecules

1 ppb = 1 molecule out of every 10^9 molecules

Mole fractions, parts per million, and parts per billion all are ratios of moles of a particular substance to total moles of sample. Mole fraction is moles per mole, ppm is moles per million moles, and ppb is moles per billion moles. These measures are related by scale factors: ppm $= 10^6 X$, and ppb $= 10^9 X$. In other words, a concentration of 1 ppm is a mole fraction of 10^{-6}, and a concentration of 1 ppb is a mole fraction of 10^{-9}. When the ozone concentration in the atmosphere reaches 0.5 ppm, the mole fraction of ozone is 0.5×10^{-6}, or 5×10^{-7}. Example 5-11 shows how to work with parts per million.

Example 5-11	**Working with ppm**

The exhaust gas from an average automobile contains 206 ppm of the pollutant nitrogen oxide, NO. If an automobile emits 125 L of exhaust gas at 1.00 atm and 350 K, what mass of NO has been added to the atmosphere?

Strategy: The question asks for mass of NO. Information about ppm tells us how many moles of NO are present in one mole of exhaust gas. We can use the ideal gas equation to determine the total number of moles of gas emitted, use the ppm information to find moles of NO, and do a mole–mass conversion to get the required mass.

Solution: Data: Concentration of NO = 206 ppm, V = 125 L, P = 1.00 atm, T = 350 K

$$n_{gas} = \frac{PV}{RT} = \frac{(1.00 \text{ atm})(125 \text{ L})}{(8.206 \times 10^{-2} \text{ L atm/mol K})(350 \text{ K})} = 4.35 \text{ mol gas}$$

$$X_{NO} = 10^{-6}(206 \text{ ppm}) = 2.06 \times 10^{-4}$$

$$n_{NO} = X_{NO}\, n_{gas} = (2.06 \times 10^{-4})(4.35 \text{ mol}) = 9.00 \times 10^{-4} \text{ mol}$$

$$m = n\, MM = (9.00 \times 10^{-4} \text{ mol})(30.0 \text{ g/mol}) = 2.7 \times 10^{-2} \text{ g}$$

The description presented in this section applies to a gas mixture that is not undergoing chemical reactions. As long as reactions do not occur, the number of moles of each gas is determined by the amount of that substance initially present. When reactions occur, the numbers of moles of reactants and products change as predicted by the principles of stoichiometry. Changes in composition must be taken into account before the properties of the gas mixture can be computed. Gas stoichiometry is described in the next section.

5.5.1 Find the partial pressures and mole fractions of a gas mixture that contains 1.00 g each of H_2 and N_2, if the total pressure of this mixture is 2.30 atm.

5.5.2 Draw a molecular picture that shows how a sample of the gas mixture in Section Exercise 5.5.1 appears at the molecular level.

5.5.3 An atmospheric chemist reported that the air in an urban area contained CO at a concentration of 35 ppm. The temperature of the air was 29 °C, and the atmospheric pressure was 745 torr. (a) What was the mole fraction of CO? (b) What was the partial pressure of CO? (c) How many molecules of CO were in 1.0×10^3 L of this air?

5.6 GAS STOICHIOMETRY

The principles of stoichiometry apply equally to solids, liquids, and gases. That is, no matter what phase substances are in, their chemical behavior can be described in molecular terms, and their transformations must be visualized and balanced using molecules and moles.

The ideal gas equation relates the number of moles of gas to the physical properties of that gas. When a chemical reaction involves a gas, the ideal gas equation provides the link between *P-V-T* data and molar amounts.

$$n_i = \frac{p_i V}{RT}$$

Stoichiometric calculations always require amounts in moles. For gases, amounts in moles are usually calculated from the ideal gas equation. Example 5-12 shows how to do this.

Gas Stoichiometry

Example 5-12

Example 5-9 describes the synthesis of acetylene (C_2H_2) from calcium carbide (CaC_2). Modern industrial production of acetylene is based on the decomposition of methane (CH_4) under carefully controlled conditions. At temperatures greater than 1600 K, two methane molecules rearrange to give three molecules of hydrogen and one molecule of acetylene:

$$2\ CH_4(g) \xrightarrow{1600K} C_2H_2(g) + 3\ H_2(g)$$

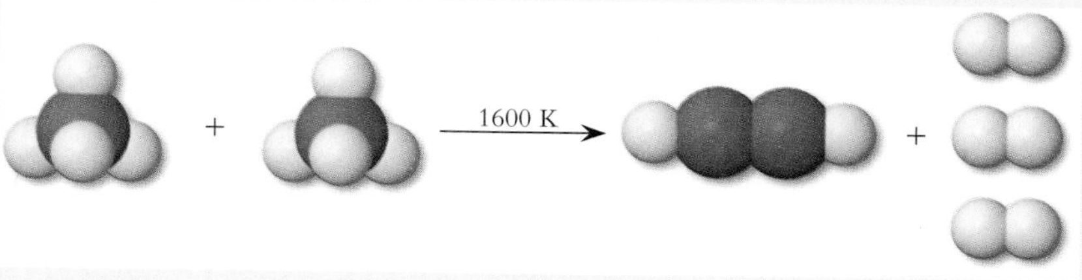

A 50.0-L steel vessel, filled with CH_4 to a pressure of 10.0 atm at 298 K, is heated to 1600 K to crack CH_4 and produce C_2H_2. What mass of C_2H_2 can be produced? What pressure does the reactor reach at 1600 K?

Example 5-12	Gas Stoichiometry *(continued)*

Solving
Quantitative
Problems

Strategy: We apply the seven-step approach to this problem.

Solution:

1. This is a stoichiometry problem involving gases. The mass of a product and the final pressure must be calculated.

2. Draw a diagram showing the process:

CH$_4$ (g)		C$_2$H$_2$ (g) + H$_2$ (g)
$V = 50.0$ L $T = 298$ K $P = 10.0$ atm	Heat $\longrightarrow$	$V = 50.0$ L $T = 1600$ K **$P = ?$ atm**

3. In any stoichiometry problem, work with moles. This problem involves gases, so use the ideal gas equation to convert P-V-T information into moles.

4. The diagram showing the process contains the data.

5. Use the ideal gas equation to calculate the amount of CH$_4$ present initially:

$$n = \frac{P_i V}{R T_i} = \frac{(10.0 \text{ atm})(50.0 \text{ L})}{(0.08206 \text{ L atm mol}^{-1}\text{K}^{-1})(298 \text{ K})} = 20.45 \text{ mol CH}_4$$

(Carry one additional significant figure to avoid errors caused by premature rounding.)

Now construct an amounts table to determine how much of the products form in complete reaction:

Reaction	2 CH$_4$ $\longrightarrow$	C$_2$H$_2$	+	3 H$_2$
Initial amount (mol)	20.45	0		0
Change in amount (mol)	-20.45	$+10.22$		$+30.67$
Final amount (mol)	0	10.22		30.67

6. Do a mole-mass conversion to determine the mass of acetylene formed:

$$m = n \text{ } MM = (10.22 \text{ mol})(26.04 \text{ g/mol}) = 266 \text{ g acetylene}$$

Use the ideal gas equation and the total number of moles present at the end of the reaction to calculate the final pressure:

$$n_{\text{total}} = 10.22 + 30.67 = 40.89 \text{ mol}$$

$$\text{Rearrange} \quad PV = nRT \quad to \quad P = \frac{nRT}{V}$$

$$P = \frac{(40.89 \text{ mol gas})(0.08206 \text{ L atm mol}^{-1} \text{ K}^{-1})(1600 \text{ K})}{(50.0 \text{ L})} = 107 \text{ atm}$$

7. The mass of acetylene, 266 g, is reasonable. The final pressure, 107 atm, seems high, but the temperature of the reactor has increased substantially and the amount of products is twice the amount of reactants, so a large pressure increase is to be expected in this reaction.

Any of the types of problems discussed in Chapters 3 and 4 can involve gases. The strategy for doing stoichiometric calculations is the same whether the species involved are solids, liquids, or gases. In this chapter, we add the ideal gas equation to our equations for converting measured quantities into moles. Example 5–13 is a limiting reactant problem that involves a gas.

Limiting Reactants in a Gas Mixture

Example 5-13

Margarine can be made from natural oils such as coconut oil by hydrogenation:

$$C_{57}H_{104}O_6(l) + 3\ H_2(g) \xrightarrow{200°\ C,\ 7\ atm,\ Ni\ catalyst} C_{57}H_{110}O_6(s)$$

Oil   Margarine

An industrial hydrogenator with a volume of 2.50×10^2 L is charged with 12.0 kg of oil and 7.00 atm of hydrogen (H_2) at 473 K (200 °C), and the reaction goes to completion. What is the final pressure of H_2 and how many kilograms of margarine will be produced?

Strategy: We have data for the amounts of both starting materials, so this is a limiting reactant problem. Given the chemical equation, the first step in a limiting reactant problem is to determine the number of moles of each starting material present at the beginning of the reaction. Next compute ratios of moles to coefficients to identify the limiting reactant. After that, a table of amounts summarizes the stoichiometry.

Solution: The ideal gas equation is used for the gaseous starting material:

$$\text{Moles } H_2 = \frac{PV}{RT} = \frac{(7.00\ \text{atm})(2.50 \times 10^2\ \text{L})}{(0.08206\ \text{L atm mol}^{-1}\ \text{K}^{-1})(473\ \text{K})} = 45.09\ \text{mol}$$

For the other starting material, do a mole-mass conversion:

$$\text{Moles oil} = \frac{m}{MM} = \frac{1.20 \times 10^4\ \text{g}}{885.4\ \text{g/mol}} = 13.55\ \text{mol}$$

A commercial hydrogenator.

Now divide each number of moles by the stoichiometric coefficient to identify the limiting reactant.

$$\frac{45.09\ \text{mol } H_2}{3\ \text{mol } H_2} = 15.03$$

$$\frac{13.55\ \text{mol oil}}{1\ \text{mol oil}} = 13.55$$

The reactant with the smaller ratio, oil, is limiting.

The table of amounts follows:

Reaction	3 H_2	+	$C_{57}H_{104}O_6$	$\longrightarrow$	$C_{57}H_{110}O_6$
Initial amount (mol)	45.09		13.55		0
Change in amount (mol)	−40.65		−13.55		+13.55
Final amount (mol)	4.4		0		13.6

Example 5-13	Limiting Reactants in a Gas Mixture *(continued)*

(The final amounts have been rounded off correctly.)

Use the ideal gas equation to calculate the pressure of hydrogen at the end of the reaction:

$$p = \frac{nRT}{V} = \frac{(4.4 \text{ mol})(0.08206 \text{ L atm/mol K})(473 \text{ K})}{(2.50 \times 10^2 \text{ L})} = 0.68 \text{ atm}$$

Now calculate the mass of margarine:

$$m = n \, MM = (13.6 \text{ mol})(891.5 \text{ g/mol}) = 1.21 \times 10^4 \text{ g} = 12.1 \text{ kg}$$

In this example, oil is the limiting reactant. Excess hydrogen is easily recovered from a gas-phase reactor, so margarine manufacturers make the oil the limiting reactant to ensure complete conversion of oil into margarine. The excess H_2 gas is recovered and used again in a subsequent reaction.

Summary of Mole Conversions

Because moles are the currency of chemistry, all stoichiometric computations require amounts in moles. In the real world, we measure mass, volume, temperature, and pressure. With the ideal gas equation, our catalog of relationships for mole conversion is complete. Table 5-2 lists the three equations; each applies to a particular category of chemical substances.

All three of these equations should be firmly embedded in your memory, *along with the substances to which they apply*. Using $PV = nRT$ on an aqueous solution gives impossible results. Example 5-14 uses all three relationships. Viewed as a whole, the example may seem quite complicated. As the solution illustrates, however, breaking the problem into separate parts allows each part to be solved using simple chemical and stoichiometric principles. Complicated problems are often simplified considerably by looking at them one piece at a time.

Table 5-2
Summary of Mole Relationships

Substance	Relationship	Equation
Pure liquid or solid	Moles = $\dfrac{\text{Mass}}{\text{Molar mass}}$	$n = \dfrac{m}{MM}$
Liquid solution	Moles = (Molarity)(Volume)	$n = MV$
Gas	Moles = $\dfrac{\text{(Pressure)(Volume)}}{R\text{(Temperature)}}$	$n = \dfrac{PV}{RT}$

R, Gas constant.

General Stoichiometry

Example 5-14

Redox reactions of metals with acids are described in Chapter 4. Oxidation of the metal generates hydrogen gas and an aqueous solution of ions. Suppose that 3.50 g of magnesium metal is dropped into 0.150 L of 6.00 M HCl in a 5.00-L cylinder at 25.0 °C whose initial gas pressure is 1.00 atm, and the cylinder is immediately sealed. Find the final partial pressure of hydrogen, the total pressure in the container, and the concentrations of all ions in solution.

Strategy: Data are given for all reactants, so this is a limiting reactant problem. We must balance the chemical equation and then work with a table of molar amounts.

Solution: Begin by analyzing the chemistry. First, list all major species present in the system before reaction: $Mg(s)$, $H_3O^+(aq)$, $Cl^-(aq)$, and $H_2O(l)$.

Magnesium, a Group 2 metal, reacts with acids to generate +2 cations:

$$Mg(s) + H_3O^+(aq) \longrightarrow Mg^{2+}(aq) + H_2(g) \qquad \text{(unbalanced)}$$

This reaction can be balanced by inspection. Charge balance requires that the +2 charge on the Mg^{2+} be matched by two hydronium ions among the starting materials. Two oxygen atoms on the left must then be balanced by two water molecules on the right:

$$Mg(s) + 2\,H_3O^+(aq) \longrightarrow Mg^{2+}(aq) + H_2(g) + 2\,H_2O(l)$$

The problem asks for pressures and ion concentrations. The final pressure can be determined from P-V-T data and n_{H_2}. Moles of hydrogen can be found from the mass of magnesium, the stoichiometric ratio, and a table of amounts. Here is a summary of the data:

$$V_{\text{container}} = 5.00 \text{ L} \qquad T = 298 \text{ K} \qquad m_{Mg} = 3.50 \text{ g}$$
$$V_{\text{solution}} = 0.150 \text{ L} \qquad P_{\text{air}} = 1.00 \text{ atm} \qquad [H_3O^+] = [Cl^-] = 6.00 \text{ M}$$

$Mg(s) + HCl(aq)$

Now we analyze the stoichiometry of the reaction. The starting amounts of H_3O^+ and Mg are given, but the data must be converted to moles before we can construct a table of amounts:

$$\text{Moles Mg} = n_{Mg} = \frac{m}{MM} = \frac{3.50 \text{ g Mg}}{24.31 \text{ g/mol}} = 0.1439 \text{ mol Mg}$$

$$\text{Moles } H_3O^+ = n_{H_3O^+} = MV = (6.00 \text{ mol/L})(0.150 \text{ L}) = 0.900 \text{ mol } H_3O^+$$

You should be able to use the numbers of moles and the stoichiometric ratio to show that magnesium is the limiting reactant. The table of amounts follows:

Reaction	Mg	+ 2 H₃O⁺	→ Mg²⁺	+ H₂ + 2 H₂O
Initial amount (mol)	0.1439	0.900	0	0
Change in amount (mol)	−0.1439	−0.2878	+0.1439	+0.1439
Final amount (mol)	0	0.6122	0.1439	0.1439

| Example 5-14 | General Stoichiometry *(continued)* |

Now that we know the final numbers of moles, we can calculate the final concentrations:

$$[Mg^{2+}] = \frac{0.1439 \text{ mol}}{0.150 \text{ L}} = 0.959 \text{ M}$$

$$[H_3O^+] = \frac{0.6122 \text{ mol}}{0.150 \text{ L}} = 4.08 \text{ M} \qquad [Cl^-] = 6.00 \text{ M}$$

Remember, chloride is a spectator ion so its concentration does not change.

Before calculating the pressures, we must visualize the reaction vessel. The container's *total* volume is 5.00 L, but 0.150 L is occupied by the aqueous solution. This leaves 4.85 L for the gas mixture. The partial pressure of hydrogen is calculated using the ideal gas equation and assuming that no H_2 remains in solution; this is a good assumption because hydrogen gas is not very soluble in water:

$$p_{H_2} = \frac{nRT}{V} = \frac{(0.1439 \text{ mol})(0.08206 \text{ L atm/mol K})(298 \text{ K})}{(4.85 \text{ L})} = 0.726 \text{ atm}$$

The amount of gas originally present does not change in the reaction, so this gas pressure remains constant at 1.00 atm. The final total pressure is the sum of the partial pressures:

$$P_{\text{total}} = p_{H_2} + p_{\text{initial}} = 0.726 \text{ atm } H_2 + 1.00 \text{ atm air} = 1.73 \text{ atm}$$

Section Exercises

5.6.1 In an oxyacetylene torch, acetylene (C_2H_2) burns in a stream of O_2. This combustion reaction produces a flame with a temperature exceeding 3000 K. A robotic welding machine welds the inside of a tank whose air volume is 1.50×10^2 L, at a temperature of 30 °C. What mass of C_2H_2 must be burned to raise the CO_2 partial pressure to 0.10 atm? Assume that a negligible amount of the CO_2 escapes from the tank before all the C_2H_2 has been burned. What mass of water will be produced during the burn?

5.6.2 A reaction vessel contains H_2 gas at 2.40 atm. If just enough O_2 is added to react completely with the H_2 to form H_2O, what will the total pressure in the vessel be before any reaction occurs? (H_2 and O_2 can also be used as a torch. The flame produced in an oxyhydrogen torch is about 4000 K.)

5.6.3 When ammonium nitrate (NH_4NO_3) explodes, all the products are gases:

$$2 \text{ NH}_4\text{NO}_{3(s)} \longrightarrow 4 \text{ H}_2\text{O(g)} + \text{O}_2\text{(g)} + 2 \text{ N}_2\text{(g)}$$

If 5.00 g of NH_4NO_3 explodes in a closed 2.00-L container that originally contains air at $P = 1.00$ atm and $T = 25$ °C, and the temperature rises to 205 °C, what total pressure develops?

5.7 CHEMISTRY OF THE ATMOSPHERE

The blanket of air that cloaks our planet behaves as an ideal gas, but the atmosphere is bound to the Earth by gravitational attraction, not by confining walls. The pressure exerted by the atmosphere can be thought of as the pressure of a column of air, just as the pressure exerted by mercury in a barometer is the pressure of the column of mercury. As altitude increases, the amount of air decreases, and so does the pressure. Lower pressure means lower molecular density, as indicated by the ideal gas equation:

$$\frac{n}{V} = \frac{P}{RT}$$

Figure 5-17 shows a molecular profile of a column of atmospheric air.

Pressure also varies with atmospheric conditions. For example, molecules in the atmosphere move faster when they are heated by the sun. This in turn causes an increase in pressure that is described by the ideal gas equation:

$$P = \frac{nRT}{V}$$

Molecules in high-pressure areas flow into regions of lower pressure, generating wind. This continual interplay of temperature and pressure plays a major role in determining our planet's weather.

Composition of the Lower Atmosphere

The atmosphere is a complex, dynamic mixture of gases. The composition and chemical reactivity patterns of the atmosphere change significantly as altitude above the surface of the Earth increases, as we explore in Chapter 6. Here, we focus on the layer closest to the surface, called the **troposphere.** The troposphere is a mixture of many chemical substances, but nitrogen and oxygen make up more than 99% of its composition. Only three others, H_2O, Ar, and CO_2, are present in amounts greater than 0.01%. Of the five principal constituents, all but water are present in nearly constant proportions. Because water can condense and evaporate readily, the water content of the atmosphere changes from day to day, depending on temperature and geography. The composition of dry air at sea level, expressed in mole fractions, is shown in Table 5-3.

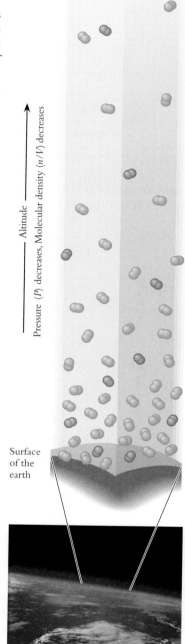

Surface of the earth

Figure 5-17
Molecular profile of the Earth's atmosphere, showing a column above some point on the Earth's surface. As altitude increases, both pressure (P) and molecular density (n/V) decrease.

Altitude —→
Pressure (P) decreases, Molecular density (n/V) decreases

Table 5-3
Composition of Dry Air at Sea Level*

Major Components	Mole Fraction	Trace Components	Abundance (ppm)
N_2	0.7808	Ne	18.2
O_2	0.2095	He	5.24
Minor Components	**Mole Fraction**	CH_4	1.4
Ar	9.34×10^{-3}	Kr	1.14
CO_2	3.25×10^{-4}	H_2	0.50
		NO	0.50
		N_2O	0.25

* Note: Figures are for water-free air. The H_2O content of air varies between $X = 0$ and $X = 0.04$

Vapor Pressure

After a rainfall, puddles of water slowly disappear. The higher the temperature, the faster the puddles vanish. Puddles disappear because water molecules move from the liquid phase to the gas phase in a process called **evaporation.** Evaporation is common to all substances in the condensed phases, not just water. We use the term **vapor** to describe a gaseous substance that forms from evaporation. The Earth's atmosphere always contains some water vapor.

CHAPTER 10 →

Intermolecular forces are discussed in detail in Chapter 10.

Evaporation can be understood by examining the kinetic energies of molecules. All molecules—solid, liquid, and gas—have some kinetic energy of motion, and any sample of molecules has a distribution of kinetic energies determined by its temperature (see Figure 5-4). Attractive forces between molecules confine liquids and solids to a fixed volume, but some molecules at the surface of that volume have enough kinetic energy to overcome the forces holding them in the condensed phase. These molecules evaporate from a liquid or solid and move into the vapor phase. The rate of evaporation increases with temperature because the fraction of molecules with sufficient energy to escape into the vapor phase increases as the sample warms up.

The water vapor that evaporates from a puddle moves away into the atmosphere. This process continues until the puddle has vaporized completely. The situation is quite different when a substance is confined to an enclosed space. If the vaporized molecules cannot escape, their numbers increase, and they exert pressure on the walls of the closed container. As evaporation proceeds, more and more molecules enter the vapor phase, and the partial pressure of the vapor rises accordingly. Figure 5-18 shows that as this happens, some gas molecules are recaptured when they collide with the surface of the liquid. Eventually, the partial pressure of the substance in the vapor phase reaches a level at which the number of molecules being recaptured (condensing) during any time interval equals the number of molecules escaping (evaporating) from the surface during that same time interval.

When the rate of evaporation equals the rate of condensation, the system is in a state of **dynamic equilibrium.** The system is dynamic because molecular transfers continue, and it has reached equilibrium because no further net change occurs. The pressure of the vapor at dynamic equilibrium is called the **vapor pressure (*vp*)** of the substance. The vapor pressure of any substance increases rapidly with temperature because the kinetic energies of the molecules increase as the temperature rises. Table 5-4 lists the vapor pressures for water at various temperatures.

Table 5-4
Vapor Pressures (*vp*) of Water at Various Temperatures (*T*)

T (°C)	vp (torr)
0	4.579
5	6.543
10	9.209
15	12.788
20	17.535
25	23.756
30	31.824
35	42.175
40	55.324
45	71.88
50	92.51
55	118.04
60	149.38
65	187.54
70	233.7
75	289.1
80	355.1
85	433.6
90	525.76
95	633.90
100	760.00

Figure 5-18
Schematic view of evaporation and condensation in a closed system. (*a*) Initial conditions. (*b*) Intermediate conditions. (*c*) Equilibrium conditions: partial pressure=vapor pressure.

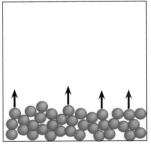
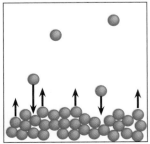
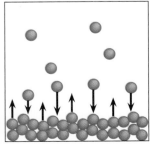

(*a*) No vapor present: escapes >> captures

(*b*) Small amount of vapor present: escapes > captures

(*c*) Equilibrium: escapes = captures

A vapor pressure is the pressure exerted by a gas in equilibrium with its condensed phase. When this equilibrium has been reached, the gas is saturated with that particular vapor. Notice in Table 5-4 that at 25 °C the atmosphere is saturated with water vapor when the partial pressure of H_2O is 23.756 torr. At this pressure, the molecular density of H_2O in the gas phase is sufficient to make the rate of condensation equal to the rate of evaporation. Any attempt to add more water molecules to the gas phase results in condensation to hold the partial pressure of H_2O fixed at 23.756 torr.

In most cases the atmosphere contains less water vapor than the maximum amount it can hold; that is, $p_{H_2O} < vp_{H_2O}$. The amount of water vapor actually present in the atmosphere is described by the **relative humidity,** which is the partial pressure of water present in the atmosphere, divided by the vapor pressure of water at that temperature and multiplied by 100% to convert to percentage:

$$\text{Relative humidity} = (100\%)\left(\frac{p_{H_2O}}{vp_{H_2O}}\right) \qquad (5\text{-}5)$$

Relative humidity describes how close the atmosphere is to being saturated with water vapor.

Because the vapor pressure of water varies with temperature, a given amount of water in the atmosphere represents a higher relative humidity as the temperature falls. For example, a partial pressure of H_2O of 6.54 torr at 15 °C corresponds to a relative humidity of about 50% ($vp = 12.788$ torr from Table 5-4):

$$\text{Relative humidity} = (100\%)\left(\frac{6.54 \text{ torr}}{12.788 \text{ torr}}\right) = 51.1\%$$

At 5 °C, however, this same partial pressure is 100% relative humidity.

The formation of dew and fog are consequences of this variation in relative humidity. Warm air at high relative humidity may cool below the temperature at which its partial pressure of H_2O equals the vapor pressure. When air temperature falls below this temperature, called the *dew point*, some H_2O must condense from the atmosphere. Example 5-15 shows how to work with vapor pressure variations with temperature, and the Chemistry and the Environment Box explores how variations in other trace gases affect climate.

Dew and fog form when water vapor condenses from the atmosphere.

Water Vapor Pressure

Example 5-15

Fog forms when humid warm air from above a body of water moves inland and cools. What is the highest temperature at which fog could form from air that is at 65% relative humidity when its temperature is 27.5 °C?

Strategy: This problem asks about the partial pressure of water vapor in the atmosphere. Fog forms when that partial pressure exceeds the vapor pressure. Partial pressures are not given among the data, but relative humidity describes how close the partial pressure of water vapor is to its vapor pressure at the given temperature. Because vapor pressure varies strongly with temperature, we must use the information given in Table 5-4.

| Example 5-15 | Water Vapor Pressure *(continued)* |

Solution: To convert relative humidity at 27.5 °C into a partial pressure, we need the vapor pressure of water at that temperature. Table 5-4 lists vapor pressures in 5 °C increments, but we can interpolate to find the correct value. Because 27.5 °C is halfway between 25 °C and 30 °C, the vapor pressure is about halfway between its values at 25 °C and 30 °C:

T (°C)	25	30	27.5
vp (torr)	23.756	31.824	½(23.756 + 31.824) = 27.790

Next, we find the partial pressure from a rearranged version of Equation 5-5:

$$p_{H_2O} = \frac{(\text{Relative humidity})(vp_{H_2O})}{100\%} = \frac{(65\%)(27.790 \text{ torr})}{100\%} = 18 \text{ torr}$$

This partial pressure remains constant as the temperature drops, meaning that fog can form below the temperature at which the vapor pressure of water is 18 torr. Examining Table 5-4, we find that the first listed temperature at which $vp_{H_2O} < 18$ torr is $T = 20$ °C, where $vp_{H_2O} = 17.535$ torr. Thus under these conditions, fog can form if the temperature drops below 20 °C.

Chemistry in the Troposphere

Every year humanity releases millions of tons of gaseous and particulate pollutants into the Earth's atmosphere. The main source of atmospheric pollution is the burning of fossil fuels. For example, diesel fuel and gasoline provide energy for automobiles, trucks, and buses, but burning these mixtures also produces vast quantities of CO_2 and lesser amounts of CO. Carbon monoxide is highly toxic, and carbon dioxide is a principal agent in global warming, as we discuss in Chapter 6. In addition, internal combustion engines are rather inefficient, so significant amounts of unburned hydrocarbons are emitted in exhaust gases. Combustion also generates nitrogen oxides and sulfur oxides, both of which undergo reactions in the atmosphere that contribute to pollution, including acid rain.

Oxides of Nitrogen

Normally, N_2 is a very stable, unreactive molecule, as noted in Chapter 4. However, under the extreme conditions found in an automobile cylinder, nitrogen molecules may react with oxygen molecules to produce nitrogen oxide:

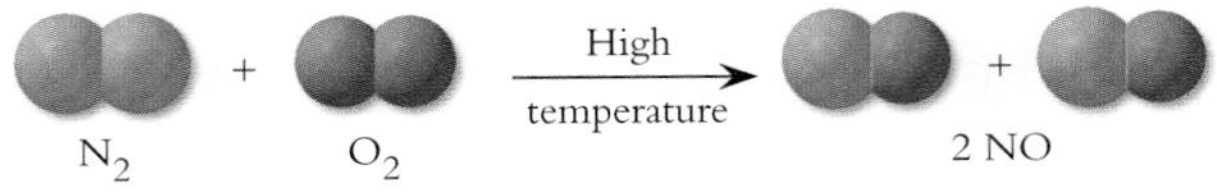

Nitrogen oxide reacts in the atmosphere with O_2 to form nitrogen dioxide:

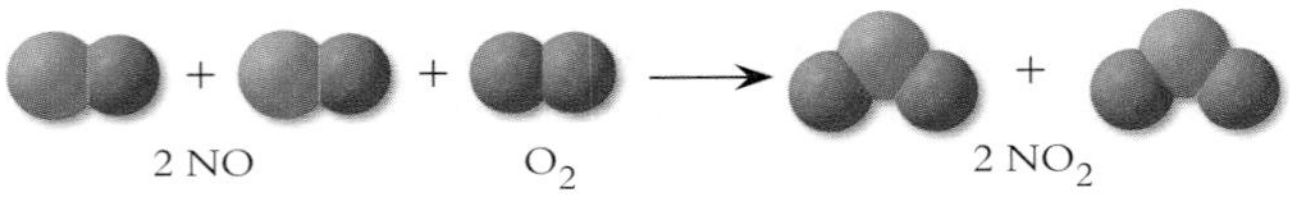

Box 5-2	Chemistry and the Environment: Does Human Activity Change the Weather?

I n 1997–98, meteorologists predicted successfully that an observed increase in temperature of Pacific Ocean waters, known as El Niño, would lead to unusual and devastating global weather patterns. As predicted, rain and snowfall in California were more than double the normal values, and spring tornados in the South and Midwest were more devastating than usual. Weather elsewhere was also significantly affected. A relatively small change in the temperature of one portion of the Pacific Ocean resulted in substantial changes in global weather patterns. The El Niño phenomenon highlights the fact that the Earth's weather patterns can be changed by variations in the environment.

Can human activities modify the environment in ways that cause changes in weather patterns? Growing evidence suggests that they can. The activities of modern society may be changing the average temperature of the atmosphere. The temperature of the atmosphere is determined by the balance between the amount of sunlight reaching the Earth's surface and the amount of heat radiated back into space. Water, CO_2, and CH_4 in the atmosphere act as "greenhouse" gases that reduce the planet's heat loss. Increased amounts of these trace components thus are expected to increase the average temperature of our world.

Atmospheric concentrations of both CO_2 and CH_4 are increasing. The graph shows data for CO_2, whose amount has increased by 30% as humans consume more and more fossil fuel. The graph also shows that the global mean temperature has risen by nearly 1 °C.

The three main variables that most influence the weather are air temperature, pressure, and water content. Their intricate interplay, energized by sunlight, creates weather patterns. Given this complexity, how can we determine if the increase in global temperature is due to the higher levels of greenhouse gases, and how can we predict the effects of further increases?

One approach is computer modeling to simulate the atmosphere and project how the weather might change as a result of changes in conditions. These models predict that if the current output of greenhouse gases continues,

average global temperature will rise by at least 1 °C over the next 50 to 100 years. The models also predict the consequences of global warming, including increases in extreme weather events such as heat waves and tornados.

A rise in average temperature of 1 °C may not seem enough to cause dramatic shifts in the weather, but the historical record shows otherwise. The period from 1500 to 1850 is called the Little Ice Age, because there were extensive increases in the sizes of the glaciers in all alpine regions. During that period, the average global temperature was just 0.5 °C lower than in 1900.

While current trends in climate lead experts to predict further global warming, natural events could change the balance. The geological record shows a series of Ice Ages alternating with eons of global warming. The cataclysmic eruption of the Indonesian volcano Krakatau in 1883 spewed huge quantities of volcanic dust and ash into the atmosphere. This caused enough global cooling that the following year was known in Europe, half a globe away, as "the year without a summer."

In the light of such uncertainties, calls to curb the pace of greenhouse emissions are controversial. Proponents point to the potentially devastating consequences of predicted global warming. Opponents cite the high current costs of reducing the use of fossil fuels, coupled with the uncertainties of how natural events affect the weather. Meanwhile, climate experts continue to collect data on current weather trends and refine their computer models, hoping to make more definitive projections about what lies ahead for our planet's weather.

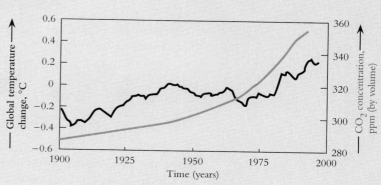

This compound is a red-brown gas that can be seen in the atmosphere over many large cities, where levels of NO_2 often reach 0.9 ppm. Tolerable limits for this toxic, irritating gas are around 3 to 5 ppm.

Nitrogen dioxide absorbs energy from sunlight and decomposes into NO molecules and oxygen atoms. Remember that oxygen is one of the elements that exists as a diatomic molecule in its natural state, so oxygen atoms are very reactive. Molecular nitrogen is too stable to react with oxygen atoms, but molecules of oxygen react with oxygen atoms to produce ozone (O_3), a toxic and highly reactive molecule:

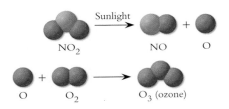

CHAPTER 6 and 14 →

Later in the text, we discuss ozone depletion (Chapters 6 and 14) as well as the greenhouse effect (Chapter 6).

Both ozone and oxygen atoms react with unburned hydrocarbons to produce many compounds that are harmful to the respiratory system. The mixture of all these pollutants is sometimes called **photochemical smog.**

To control photochemical smog, governments have set limits on the levels of NO and unburned hydrocarbons allowed in automobile exhaust. Today, chemists and engineers are researching cleaner-burning fuels and more efficient engines. Catalytic converters convert much of the NO and NO_2 from exhaust gases into N_2 and O_2 before they are released into the atmosphere. These have helped alleviate pollution from nitrogen oxides at only a small additional cost. We could reduce pollution emissions even further, but consumers and manufacturers are reluctant to pay the higher costs required to develop and produce cleaner fuels and engines.

CHAPTER 14 →

Catalysis is treated in Chapter 14.

Oxides of Sulfur

About half the electricity produced in the United States is generated by burning coal. Like petroleum, coal was formed millions of years ago from decaying plants under conditions of high temperature and pressure. Coal is a network of carbon atoms that also contains small amounts of other elements, including hydrogen, oxygen, nitrogen, and sulfur.

When coal burns, its sulfur combines with O_2 to produce sulfur dioxide:

$$S(s, \text{from coal}) + O_2(g) \longrightarrow SO_2(g)$$

In the presence of dust particles or ultraviolet (UV) light, atmospheric SO_2 reacts further with O_2 to form SO_3:

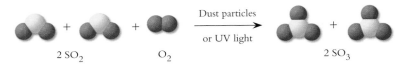

The combustion products from sulfur impurities in coal are particularly damaging to the environment. In humans, prolonged exposure to sulfur dioxide diminishes lung capacity and aggravates respiratory problems such as asthma, bronchitis, and emphysema. Concentrations as low as 0.15 ppm can incapacitate persons with these diseases, and at about 5 ppm everyone experiences breathing difficulties. In 1952 a particularly serious episode of SO_2 pollution in London caused approximately 4000 deaths over several days.

One particularly troubling aspect of SO_2 and SO_3 pollution is **acid rain,** which occurs when these gases combine with water to produce acid mists:

$$SO_2(g) + H_2O(g) \longrightarrow H_2SO_3(mist)$$

$$SO_3(g) + H_2O(g) \longrightarrow H_2SO_4(mist)$$

Raindrops passing through such mists become acidic, increasing the acidity of rainwater as much as 1000-fold. Acid rain is common in the heavily industrialized areas of the United States, Canada, and Europe. In areas where acid rain is particularly severe, lakes are becoming so acidic that fish can no longer survive. In addition to the devastating effect on aquatic life, acid rain accelerates the leaching of nutrients from the soil, alters the metabolism of organisms in the soil, accelerates the corrosion of metals, and damages important building materials such as limestone and marble (Figure 5-19).

Sulfur dioxide can be removed from power plant exhaust gas by a scrubber system. One common method involves the reaction of SO_2 with calcium oxide (lime) to form calcium sulfite:

$$SO_2(g) + CaO(s) \longrightarrow CaSO_3(s)$$

Unfortunately, scrubber systems are expensive to operate, and the solid $CaSO_3$ is generated in large enough quantities to create significant disposal problems.

Sulfur pollution may become increasingly significant in the United States, if increasing amounts of energy production come from coal-fired plants. Because different deposits of coal have different sulfur concentrations, the amount of pollution will depend on the type of coal used. The oldest coal, anthracite, is 90% carbon and has a sulfur content lower than 1%, whereas bituminous coal can contain as much as 5% sulfur by mass. Unfortunately, much of the vast reserve of coal in North America is high-sulfur bituminous coal. Thus increased dependence on coal as a source of energy will come with a high cost to society. The choices are to find other energy sources; to spend more money to develop efficient scrubber systems; to reduce consumption through conservation; or to face potentially disastrous consequences for our environment.

Figure 5-19
Statues are slowly decomposed by acid rain.

Section Exercises

5.7.1 What are the concentrations in parts per million of the three most abundant trace atmospheric constituents listed in Table 5–3?

5.7.2 Argon gas can be recovered from the atmosphere by appropriate cooling and liquefaction processes. How many liters of dry air at 1.00 atm and 27 °C must be processed to collect 10.0 mol of Ar?

5.7.3 A weather report gives the current temperature at 18 °C and sets the dew point at 10 °C. Using data from Table 5-4, determine the partial pressure of water vapor in the atmosphere and calculate the relative humidity.

▐▐▐▐ CHAPTER REVIEW

Summary and Key Terms

1. Gases contain molecules moving independently with a distribution of molecular energies. Collision frequency depends on **molecular density** and temperature. The average kinetic energy is the same for all gases but increases linearly with temperature, with the proportionality given by the **gas constant.** Gas transport from one place to another occurs by **effusion, diffusion,** and convection, at rates that depend on the **root-mean-square** speeds of the molecules.

2. Collisions of gas molecules with the walls of the container generate a **pressure.** For an **ideal gas,** one whose molecules move completely independently of one another, pressure is related to other variables through the **ideal gas equation,** $PV = nRT$.

3. Pressure is measured using a **barometer** or a **manometer.** Pressure is measured in **atmospheres, torr,** or **pascals.**

4. The ideal gas equation can be rearranged to give equations for pressure-volume and temperature-volume variations. It also can be used to calculate molar masses.

5. In a gas mixture, each component exerts a **partial pressure** that is given by the ideal gas equation. **Dalton's law of partial pressures** states that the total pressure of a gas mixture is the sum of its components' partial pressures. The composition of a gaseous mixture can be specified by partial pressures, **mole fractions, parts per million,** or **parts per billion.** Stoichiometric calculations use the ideal gas equation to find numbers of moles of gases.

6. Apart from the use of the ideal gas equation to determine number of moles, stoichiometric calculations for gases are handled in the same way as stoichiometric calculations for liquids, solids, and solutions.

7. The Earth's lower atmosphere is the **troposphere.** The amount of water **vapor** in the troposphere varies and depends on **evaporation** of liquid water. **Dynamic equilibrium** occurs when water evaporates and condenses at the same rate. The partial pressure of water vapor at equilibrium is its **vapor pressure.** **Relative humidity** is a common way to describe the amount of water in the atmosphere. Chemical reactions involving trace gases added to the atmosphere by human activities can lead to **photochemical smog** or **acid rain.**

Skills to Master

▶ Visualizing molecular motion in gases

▶ Calculating gas properties using the ideal gas equation

▶ Relating final conditions to initial conditions

▶ Determining partial pressures

▶ Solving stoichiometry problems that include gases

▶ Working with vapor pressures

Learning Exercises

5.1 List the macroscopic properties and the microscopic properties of gases.

5.2 Update and reorganize your list of memory bank equations so that all equations that involve moles are grouped together.

5.3 Summarize the strategy for calculating final pressures of gases in a limiting reactant problem.

5.4 Define partial pressure, vapor pressure, and relative humidity. Explain how they are related.

5.5 List all terms new to you that appear in Chapter 5. Write a one-sentence definition of each in your own words. Consult the glossary if you need help.

Problems   ilw = interactive learning ware problem. Visit the website at www.wiley.com/college/olmsted

Molecules in Motion

5.1 Redraw Figure 5-3 to show the distribution of molecules if the temperature of the oven is doubled.

5.2 Redraw Figure 5-4 to show a temperature of 200 K.

5.3 Draw a single graph that shows the *speed* distributions of: N_2 at 200 K, N_2 at 300 K, and He at 300 K.

5.4 Draw a single graph that shows the *speed* distributions of: Cl_2 at 500 K, Br_2 at 500 K, and Cl_2 at 300 K.

5.5 Redraw the graph in Problem 5.3 as a graph of *energy* distributions for the three gases.

5.6 Redraw the graph in Problem 5.4 as a graph of *energy* distributions for the three gases.

5.7 Calculate the most probable speed and average kinetic energy per mole for each of the following: (a) He at 627 °C; (b) O_2 at 27 °C; and (c) SF_6 at 627 °C.

5.8 Calculate the most probable speed and average kinetic energy per mole for each of the following: (a) Ar at 127 °C; (b) Xe at 127 °C; and (c) C_3H_8 at 327 °C.

The Ideal Gas Equation

5.9 Which of the following exerts a greater force when it collides with a wall, and why? (a) Neon atom with average speed at 2000 K or xenon atom with average speed at 2000 K; (b) methane molecule with average speed at 390 K or methane molecule with average speed at 700 K; and (c) H_2 molecule with a speed of 10^5 m/s or F_2 molecule with the same speed.

5.10 Which of the following exerts a greater force when it collides with a wall, and why? (a) O_2 molecule traveling 250 m/s or Cl_2 molecule traveling at the same speed; (b) argon atom with average speed at 300 K or helium atom with average speed at 300 K; and (c) N_2 molecule with average speed at 298 K or N_2 molecule with average speed at 200 K.

5.11 Explain in molecular terms why each of these statements is true: (a) At very high pressure, no gas behaves ideally. (b) At very low temperature, no gas behaves ideally.

5.12 A cylinder with a movable piston contains a sample of gas. Describe in molecular terms the effect on pressure exerted by the gas for each of the following changes: (a) The piston is pushed in. (b) Some gas is removed while the piston is held in place. (c) The gas is heated while the piston is held in place.

5.13 The figure represents an ideal gas in a container with a movable friction-free piston.

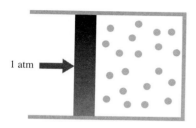

(a) The external pressure on the piston exerted by the atmosphere is 1 atm. If the piston is not moving, what is the pressure inside the container? Explain in terms of molecular collisions.
(b) Redraw the sketch to show what would happen if the temperature of the gas in the container is doubled. Explain in terms of molecular collisions.

5.14 Consider the figure appearing in Problem 5.13. (a) The monatomic gas shown in the figure is replaced by an equal number of molecules of a diatomic gas, all other conditions remaining the same. What is the pressure inside the container? Explain in terms of molecular collisions. (b) The external pressure is reduced to 0.75 atm, and the piston moves as a result. Redraw the sketch to show the new situation.

Pressure

5.15 Describe what would happen to the barometer in Figure 5–11 if the tube holding the mercury had a pinhole at its top.

5.16 Redraw the U-tube shown in Figure 5–12 to show how the manometer appears if the pressure on the side exposed to the atmosphere is greater than the pressure on the side exposed to the gas by an amount $2\Delta h$.

5.17 Express the following in SI units: (a) 455 torr; (b) 2.45 atm; (c) 0.46 torr; and (d) 1.33×10^{-3} atm.

5.18 Convert the following to torr: (a) 1.00 Pa; (b) 125.6 kPa; (c) 75.0 atm; and (d) 4.55×10^{-10} atm.

5.19 A sample of air was compressed to a volume of 20.0 L. The temperature was 298 K, and the pressure was 5.00 atm. How many moles of gas were in the sample? If the sample had been collected from air at $P = 1.00$ atm, what was the original volume of the gas?

5.20 A bicycle pump inflates a tire whose volume is 565 mL until the internal pressure is 647 kPa at a temperature of 21.7 °C. How many moles of air does the tire contain? What volume of air at 101 kPa and 21.7 °C did the pump transfer?

Applying the Ideal Gas Equation

5.21 Rearrange the ideal gas equation to give the following expressions: (a) An equation that relates P_i, T_i, P_f and T_f when n and V are constant; (b) $V = ?$; and (c) an equation that relates P_i, V_i, P_f and V_f when n and T are constant.

5.22 Rearrange the ideal gas equation to give the following expressions: (a) $n = ?$; (b) an equation that relates V_i, n_i, V_f, and n_f when P and T are constant; and (c) $n/V = ?$

ilw 5.23 It requires 0.255 L of air to fill a metal foil balloon to 1.000 atm pressure at 25 °C. The balloon is tied off and placed in a freezer at −15 °C. What is the new volume of air in the balloon?

5.24 A pressurized can of whipping cream has an internal pressure of 1.075 atm at 25 °C. If it is placed in a freezer at −15 °C, what is the new value for its internal pressure?

5.25 Describe the molecular changes that account for the result in Problem 5.23.

5.26 Describe the molecular changes that account for the result in Problem 5.24.

5.27 Under which of the following conditions could you use the equation $P_1 V_1 = P_2 V_2$? (a) A gas is compressed at constant T. (b) A gas-phase chemical reaction occurs. (c) A container of gas is heated. (d) A container of liquid is compressed at constant T.

5.28 Under which of the following conditions could you *not* use the equation, $P_1 V_1/T_1 = P_2 V_2/T_2$? (a) P is expressed in torr. (b) T is expressed in °C. (c) V is changing. (d) n is changing.

5.29 Freons (CFCs) are compounds that contain carbon, chlorine, and fluorine in various proportions. They are used as foaming agents, propellants, and refrigeration fluids. Freons are controversial because of the damage they do to the ozone layer in the stratosphere. A 2.55-g sample of a particular freon in a 1.50-L bulb at 25.0 °C has a pressure of 262 torr. What is the molar mass and formula of the compound?

5.30 Gaseous hydrocarbons, which contain only carbon and hydrogen, are good fuels because they burn in air to generate large amounts of heat. A sample of hydrocarbon with $m = 1.65$ g exerts a pressure of 1.50 atm in a 945-mL bulb at 21.5 °C. Determine the molar mass and chemical formula of this hydrocarbon.

Gas Mixtures

ilw 5.31 The amount of NO_2 in a smoggy atmosphere was measured to be 0.78 ppm. The barometric pressure was 758.4 torr. Compute the partial pressure of NO_2 in atmospheres.

5.32 California's automobile emission standards require that exhaust gases contain less than 220 parts per million hydrocarbons and less than 1.2% CO (both these values are in moles per mole of air). At standard atmospheric pressure, what are the partial pressures, in torr and in atmospheres, that correspond to these values?

5.33 In dry atmospheric air, the four most abundant components are N_2, $X = 0.7808$; O_2, $X = 0.2095$; Ar, $X = 9.34 \times 10^{-3}$; and CO_2, $X = 3.25 \times 10^{-4}$. Calculate the partial pressures of these four gases, in torr, under standard atmospheric conditions.

5.34 Find the partial pressures, total pressure, and mole fractions of a gas mixture in a 4.00-L container at 375 °C if it contains 1.25 g each of Ar, CO, and CH_4.

5.35 The following questions refer to Figure 5.15d: (a) Which gas exerts a higher pressure, O_2 or He? (b) Which has a higher mole fraction? (c) Compute the mole fraction of He.

5.36 The figures shown below represent very small portions of three gas mixtures, all at the same volume and temperature.

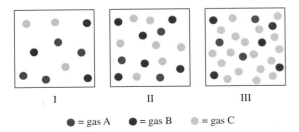

● = gas A   ● = gas B   ● = gas C

(a) Which sample has the highest partial pressure of gas A ?
(b) Which sample has the highest mole fraction of gas B? (c) In sample III, what is the concentration of gas A in ppm?

5.37 The natural gas in a storage tank is analyzed. A 2.00-g sample of the gas contains 1.57 g CH_4, 0.41 g C_2H_6, and 0.020 g C_3H_8. Calculate the partial pressures of each gas in the storage tank if the total pressure in the tank is 2.35 atm.

5.38 A sample of automobile exhaust gas is analyzed at a smog station. The gas contains 487.4 ppm CO_2, 10.3 ppb NO, and 4.2 ppb CO. Calculate the partial pressures of these gases if the exhaust is emitted at 113.1 kPa pressure.

Gas Stoichiometry

ilw 5.39 Humans consume glucose to produce energy. The products of glucose consumption are CO_2 and H_2O.

$$C_6H_{12}O_6(s) + O_2(g) \longrightarrow CO_2(g) + H_2O(l) \qquad \text{(unbalanced)}$$

What volume of CO_2 is produced under body conditions (37 °C, 1.00 atm) during the consumption of 4.65 g of glucose?

5.40 Oxygen gas can be generated by heating $KClO_3$ in the presence of a catalyst.

$$KClO_3 \longrightarrow KCl + O_2 \qquad \text{(unbalanced)}$$

What volume of O_2 gas will be generated at $T = 25$ °C and $P = 765.1$ torr from 1.57 g of $KClO_3$?

5.41 Sodium metal reacts with molecular chlorine gas to form sodium chloride. A closed container of volume 3.00×10^3 mL contains chlorine gas at 27 °C and 1.25×10^3 torr. Then 6.90 g of solid sodium is introduced, and the reaction goes to completion. What is the final pressure if the temperature rises to 47 °C?

5.42 Carbon monoxide and molecular oxygen react to form carbon dioxide. A 50.0-L reactor at 25.0 °C is charged with 1.00 atm of CO. The gas is then pressurized with O_2 to give a total pressure of 3.56 atm. The reactor is sealed, heated to 350 °C to drive the reaction to completion, and cooled back to 25.0 °C. Compute the final partial pressure of each gas.

5.43 Ammonia is produced industrially by reacting N_2 with H_2 at elevated pressure and temperature in the presence of a catalyst.

$$N_2 + H_2 \longrightarrow NH_3 \qquad \text{(unbalanced)}$$

Assuming 100% yield, what mass of ammonia would be produced from a 1:1 mole ratio mixture in a reactor which has a volume of 8.75×10^3 L, under a total pressure of 275 atm at $T = 455$ °C?

5.44 Ethylene oxide is produced industrially from the reaction of ethylene with oxygen at atmospheric pressure and 280 °C, in the presence of a silver catalyst:

$$C_2H_4(g) + O_2(g) \longrightarrow C_2H_4O(g) \qquad \text{(unbalanced)}$$

Assuming a 100% yield, how many kg of ethylene oxide would be produced from 5.00×10^4 L of a mixture containing ethylene and oxygen in 1:1 mole ratio?

5.45 In actual practice, the reaction of Problem 5.43 gives a yield of only 13%. Repeat the calculation of Problem 5.43 taking this into account.

5.46 In actual practice, the ethylene oxide reaction of Problem 5.44 gives only a 65% yield under optimal conditions. Repeat the calculation of Problem 5.44 assuming a 65% yield.

Chemistry of the Atmosphere

5.47 How much does the mole fraction of O_2 in the atmosphere change at 20 °C as the relative humidity varies from 100% to 0%?

5.48 How much does the partial pressure of N_2 gas in the atmosphere change at 30 °C and 1.00 atm as the relative humidity varies from zero to 100%?

5.49 Find the dew point for relative humidity 78% at 25 °C.

5.50 In the tropics, water will condense in human lungs when the temperature and relative humidity are too high. Using Table 5-4, estimate the vapor pressure of water at body temperature of 37 °C. If atmospheric temperature is 40 °C, at what relative humidity does this life-threatening process occur?

5.51 Neon was discovered by cooling dry air until it is liquefied and then boiling off the components one at a time. What volume of dry air at 298 K and 1.00 atm must be treated in this way to obtain 10 mg of neon (see Table 5-3)?

5.52 Compute the total mass of krypton contained in 1.00 cubic kilometer of dry air, assuming constant atmospheric pressure of 1.00 atm. (See Table 5-3.)

Additional Paired Problems

5.53 At an altitude of 150 km above the Earth's surface, atmospheric pressure is 10^{-10} atm and the temperature is 310 K. What is the molecular density at this altitude?

5.54 Molecular clouds composed mostly of hydrogen molecules have been detected in interstellar space. The molecular density in these clouds is about 10^{10} molecules m^{-3}, and their temperature is around 25 K. What is the pressure in such a cloud?

5.55 A 3.00-g sample of an ideal gas occupies 0.963 L at 22 °C and 0.969 atm. What will be its volume at 15 °C and 1.00 atm?

5.56 A 96.0-g sample of O_2 gas at 0.0 °C and 380 torr is compressed and heated until the volume is 3.00 L and the temperature is 27 °C. What is the final pressure in torr?

5.57 The gas SF_6 is used to trace air flows because it is nontoxic and can be detected selectively in air at a concentration of 1 ppb. What partial pressure is this? At this concentration, how many molecules of SF_6 are contained in 1.0 cm^3 of air at $T = 21$ °C?

5.58 In 1990, carbon dioxide levels at the South Pole reached 351.5 parts per million by volume (The 1958 reading was 314.6 ppm by volume). Convert this reading to a partial pressure in atmospheres. At this level, how many CO_2 molecules are there in 1.0 L of dry air at −45 °C?

5.59 What mass of krypton is present in a 725-mL container at 925 °C in which the pressure of krypton is 10.0 atm? How many atoms is this?

5.60 Nitrogen gas is available commercially in pressurized 9.50-L steel cylinders. If a tank has a pressure of 145 atm at 298 K, how many moles of N_2 are in the tank? What is the mass of N_2?

5.61 The figure shows three chambers with equal volumes, connected by closed valves. Each chamber contains helium gas, with amounts proportional to the number of atoms shown. Answer each of the following questions, briefly stating your reasoning: (a) Which of the three has the highest pressure? (b) If the pressure of B is 1.0 atm, what is the pressure of A? (c) If the pressure of A starts at 1.0 atm, and then all of the atoms from B and C are added to A's container, what will be the new pressure? (d) If the pressure in B is 0.50 atm, what will the pressure be after the valves are opened?

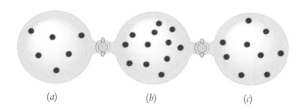

(a)   (b)   (c)

5.62 The following figure shows two tanks connected by a valve. Each tank contains a different gas, both at 0.0 °C.

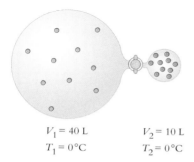

$V_1 = 40$ L $V_2 = 10$ L
$T_1 = 0$ °C $T_2 = 0$ °C

(a) Redraw the system to show how it appears after the valve has been opened. (b) If each molecule in the diagram represents 1.0 mol of gas, what is the system's total pressure after the valve is opened? (Assume that the valve has negligible volume.) (c) If all the gas is

pumped into the smaller tank, what are the partial pressures of the two gases?

5.63 At an altitude of 40 km above the Earth's surface the temperature is about −25 °C, and the pressure is about 3.0 torr. Calculate the average molecular speed of ozone (O_3) at this altitude.

5.64 The average kinetic energy of a 1.55-g sample of argon gas in a 5.00-L bulb is 1.02×10^{-20} J/atom. What is the pressure of the gas? What is the average speed of the argon atoms under these conditions?

5.65 Compute the density (g/L) of SF_6 gas at 755 torr and 27 °C.

5.66 Compute the density (g/L) of H_2 gas at 380 torr and −23 °C.

5.67 Use data from Table 5-4 to estimate the dew point for air that has: (a) relative humidity of 80% at 35 °C; (b) relative humidity of 50% at 15 °C; and (c) relative humidity of 30% at 25 °C.

5.68 A sample of warm moist air (100% relative humidity) is collected in a container at $P = 756$ torr and $T = 30.0$ °C. Find (a) the new pressure after a drying agent is added to remove the water vapor and (b) the mole fraction and concentration in grams per liter of water vapor in the original air sample.

5.69 Describe a gas experiment that would show that the element oxygen exists naturally as diatomic molecules.

5.70 Describe how gas measurements can be used to obtain the molar mass of an unknown gas.

5.71 A mixture of cyclopropane gas (C_3H_6) and oxygen (O_2) in 1.00 : 4.00 mole ratio is used as an anesthetic. What mass of each of these gases is present in a 2.00-L bulb at 23.5 °C if the total pressure is 1.00 atm?

5.72 A gas cylinder of volume 5.00 L contains 1.00 g of Ar and 0.0500 g of Ne. The temperature is 275 K. Find the partial pressures, total pressure, and mole fractions.

5.73 Consider two gas bulbs of equal volume, one filled with H_2 gas at 0 °C and 2 atm, the other containing O_2 gas at 25 °C and 1 atm. Which bulb has (a) more molecules; (b) more mass; (c) higher average kinetic energy of molecules; (d) higher average molecular speed?

5.74 Consider two gas bulbs of equal volume, one filled with Cl_2 gas at 100 °C and 2 atm, the other containing N_2 gas at 25 °C and 1 atm. Which bulb has (a) more molecules; (b) more mass; (c) higher average kinetic energy of molecules; (d) higher average molecular speed?

5.75 Refer to Figure 5-5 to determine the following for O_2 gas at 300 K: (a) What is the most probable kinetic energy? (b) What is the most probable speed?

5.76 Molecular beam experiments on ammonia at 425 K give the speed distribution shown in the figure: (a) What is the most probable speed? (b) What is the most probable kinetic energy?

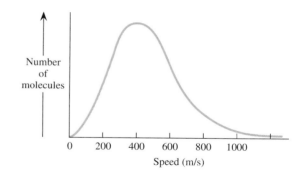

5.77 Explain in your own words how sulfur impurities in coal can lead to acid rain. Use balanced equations when appropriate.

5.78 Explain in your own words how the reactions occurring in the cylinders of automobile engines contribute to the production of photochemical smog. Use balanced equations when appropriate.

5.79 A sample of gas is found to exert a pressure of 525 torr when it is in a 3.00-L flask at 0.00 °C. Compute (a) the new volume if P becomes 755 torr and T is unchanged; (b) the new pressure if V becomes 2.00 L and T is unchanged; and (c) the new pressure if the temperature is raised to 50.0 °C and V is unchanged.

5.80 A sample of gas is found to exert a pressure of 322 torr when it is in a 2.00-L flask at 100.00 °C. Compute (a) the new volume if P becomes 525 torr and T is unchanged; (b) the new pressure if the volume is reduced to 1.50 L and T becomes 50.0 °C; and (c) the new pressure if half the gas is removed but V and T remain the same.

5.81 Redraw Figure 5-8 to show each of the following changes in conditions (you may show only half the container to illustrate the changes): (a) The temperature increases by 50%. (b) One-third of the helium atoms are removed. (c) Hydrogen molecules replace half the helium atoms.

5.82 Redraw Figure 5-8 to show each of the following changes in conditions (you may show only half the container to illustrate the changes): (a) Half of the helium atoms are removed. (b) The helium atoms are replaced with an equal number of nitrogen molecules. (c) The temperature is lowered by 50%.

5.83 Determine whether each of the following statements is true or false. If false, rewrite the statement so that it is true. (a) At constant T and V, P is inversely proportional to the number of moles (n) of gas. (b) At constant V, the pressure of a fixed amount of gas is directly proportional to T. (c) At fixed n and V, the product of P and T is constant.

5.84 Determine whether each of the following statements is true or false. If false, rewrite the statement so that it is true. (a) At fixed n and P, V is independent of T. (b) At fixed n and T, the product of P and V is constant. (c) When gas is added to a chamber at fixed V and T, the pressure increases as n^2.

More Challenging Problems

5.85 Find the partial pressures in atmospheres of the eight most abundant atmospheric components listed in Table 5-3 at 25 °C, 50% relative humidity and $P = 765$ torr.

5.86 A 0.1054-g mixture of $KClO_3$ and a catalyst was placed in a quartz tube and heated vigorously to drive off all the oxygen as O_2. The O_2 was collected at 25.17 °C and a pressure of 759.2 torr. The volume of gas collected was 22.96 mL. (a) How many moles of O_2 were produced? (b) How many moles of $KClO_3$ were in the original mixture? (c) What was the mass percent of $KClO_3$ in the original mixture?

5.87 Elemental analysis of an organic liquid with a fishy odor gives the following elemental mass percentages: H, 14.94%; C, 71.22%; and N, 13.84%. Vaporization of 250 mg of the compound in a 150-mL bulb at 150 °C gives a pressure of 435 torr. What is the molecular formula of the compound?

5.88 When heated to 150 °C, $CuSO_4 \cdot 5H_2O$ loses its water of hydration as gaseous H_2O. A 2.50-g sample of the compound is placed in a sealed 4.00-L steel vessel containing dry air at 1.00 atm and 27 °C and the vessel is then heated to 227 °C. What are the final partial pressure of H_2O and the final total pressure?

5.89 A 15.00-g piece of dry ice (solid CO_2) is dropped into a bottle with a volume of 0.750 L, and the air is pumped out. The bottle is sealed and the CO_2 is allowed to evaporate. What will be the final pressure in the bottle if the final temperature is 0.0 °C?

5.90 Two chambers are connected by a valve. One chamber has a volume of 15 L and contains N_2 gas at a pressure of 2.0 atm. The other has a volume of 1.5 L and contains O_2 gas at 3.0 atm. The valve is opened, and the two gases are allowed to mix thoroughly. The temperature is constant at 300 K throughout this process. (a) How many moles each of N_2 and O_2 are present? (b) What are the final pressures of N_2 and O_2, and what is the total pressure? (c) What fraction of the O_2 is in the smaller chamber after mixing?

5.91 Calculate the density of (a) dry air at 1 atm and 298 K and (b) air of 100% relative humidity at 1 atm and 298 K.

5.92 On a smoggy day, the ozone content of air over Los Angeles reaches 0.50 ppm. Compute the partial pressure of ozone and the number of molecules of ozone per cubic centimeter if atmospheric pressure is 762 torr and the temperature is 28 °C.

5.93 Liquid oxygen, used in some large rockets, is produced by cooling dry air to −183 °C. How many liters of dry air at 25 °C and 750 torr would have to be processed to produce 150 L of liquid oxygen (density = 1.14 g/mL)?

5.94 Benzaldehyde is a fragrant molecule used in artificial cherry flavoring. Combustion of 125 mg of benzaldehyde gives 363 mg of carbon dioxide and 63.7 mg of water. In another experiment, a 110-mg sample is vaporized at 150 °C in a 0.100-L bulb. The vapor gives a pressure of 274 torr. Determine the molecular formula of benzaldehyde.

5.95 The methanation reaction, $3 H_2 + CO \rightarrow CH_4 + H_2O$, is used commercially to prepare methane. A gas reactor with a volume of 1.00×10^2 L is pressurized at 575 K with 20.0 atm of H_2 gas and 10.0 atm of CO gas; 145 g of CH_4 is produced. What is the percent yield of the synthesis?

5.96 In an explosion, a compound that is a solid or a liquid decomposes very rapidly, producing large volumes of gas. The force of the explosion results from the rapid expansion of the hot gases. For example, TNT (trinitrotoluene) explodes according to the following balanced equation:

TNT
$C_7H_5(NO_2)_3$

$2 C_7H_5(NO_2)_{3\,(s)} \longrightarrow 12 CO_{(g)} + 2 C_{(s)} + 5 H_{2\,(g)} + 3 N_{2\,(g)}$

(a) How many moles of gas are produced in the explosion of 1.0 kg of TNT? (b) What volume will these gases occupy if they expand to a total pressure of 1.0 atm at 25 °C? (c) At 1.0 atm total pressure, what would be the partial pressure of each gas?

5.97 Many of the transition metals form complexes with CO; these complexes are called *metal carbonyls* and have the general formula $M(CO)_x$. A 0.500-g sample of gaseous nickel carbonyl in a 0.100-L bulb generates a pressure of 552 torr at 30 °C. What is the formula of nickel carbonyl?

5.98 When a sealed bulb containing a gas is immersed in an ice bath, it has a gas pressure of 345 torr. When the same bulb is placed in an oven, the pressure of the gas rises to 745 torr. What is the temperature of the oven in Celsius?

5.99 The Haber synthesis of ammonia is a gas phase reaction that takes place at high temperature (400 to 500 °C) and pressure (100 to 300 atm). The starting materials for the Haber synthesis are placed inside a container, in proportions shown in the figure. Assuming 100% yield, draw a sketch that illustrates the system at the end of the reaction.

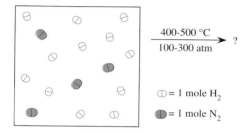

5.100 At low temperature nitrogen dioxide molecules join together to form dinitrogen tetroxide.

$$2\,NO_2 \longrightarrow N_2O_4 \qquad \text{(low temperature)}$$

A sample of NO_2 sealed inside a glass bulb at 23 °C gave a pressure of 691 torr. Lowering the temperature to −5 °C converted the NO_2 to N_2O_4. What was the final pressure inside the bulb? (Hint: Pay particular attention to the stoichiometry of the reaction.)

5.101 People often remark that the "air is thin" at higher elevation. Explain this comment in molecular terms using the fact that the atmospheric pressure atop Mount Everest is only about 250 mm.

5.102 The figures shown below represent mixtures of argon atoms and hydrogen molecules. The volume of container B is twice the volume of container A.

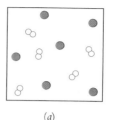 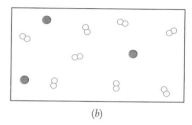

(a)   (b)

(a) Which container has a higher total gas pressure? Explain. (b) Which container has a higher partial pressure of molecular hydrogen? Explain. (c) One of the two gas mixtures shown above was used in a pulsed molecular beam experiment. The result of the experiment is shown below. Which of the two gas samples, A or B, was used for this experiment? Explain.

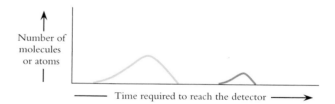

Group Study Problems

5.103 A balloon filled with helium gas at 1.0 atm and 25 °C is to lift a 350-kg payload. What is the minimum volume required for the balloon? Assume dry air at the same pressure and temperature make up the atmosphere. (Hint: The total mass of payload plus He must be less than the mass of air displaced by the balloon.)

5.104 What temperature must the air be if the balloon described in Problem 5.103 is going to use hot air instead of helium?

5.105 A mouse is placed in a sealed chamber filled with air at 765 torr and equipped with enough solid KOH to absorb any CO_2 and H_2O produced. The gas volume in the chamber is 2.05 L, and its temperature is held at 298 K. After 2 hours, the pressure inside the chamber has fallen to 725 torr. What mass of oxygen has the mouse consumed?

5.106 Liquid helium at 4.2 K has a density of 0.147 g/mL. A 2.00-L metal bottle at 95 K contains air at 1.0 atm pressure. We introduce 0.100 L of liquid helium, seal the bottle, and allow the entire system to warm to room temperature (25 °C). What is the pressure inside the bottle?

5.107 You are on vacation in Hawaii, where the temperature is 27 °C. You wish to go for a scuba dive. Your tank, with a volume of 12.5 L, is filled with dry air to a pressure of 2425 psi (165 atm). If your body requires you to consume 14.0 g O_2 per minute, how long can you dive at a depth of 70 feet if you must allow 6.0 minutes to return to the surface?

5.108 If 1.00 metric ton of soft coal containing 4.55% (by mass) sulfur is burned in an electric power plant, what volume of SO_2 is produced at 55 °C and 1.00 atm? What mass of CaO is required to "scrub" this SO_2? What mass of $CaSO_3$ must be disposed of?

5.109 Construct the following graphs for an ideal gas when $T = 298$ K: (a) V vs. P; (b) PV vs. P; and (c) $1/V$ vs P. Describe in words the shape of each of these graphs.

Answers to Section Exercises

5.1.1

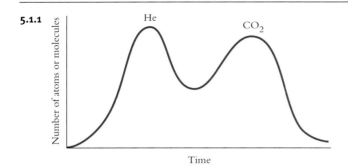

5.1.2 (a) 2.16×10^3 J; (b) 122 J; and (c) 1.96×10^{-20} J
5.1.3 Speeds are 1.04×10^3 m/s for H_2O and 9.82×10^2 m/s for D_2O. Ratio of rates is 1.05.
5.2.1 (a) and (b) Pressure doubles; (c) pressure triples; (d) there is no change.
5.2.2

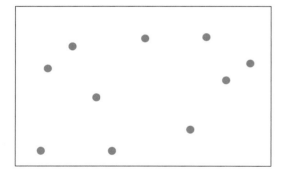

5.2.3 Low molecular density because there are fewer molecular interactions

5.3.1 710 torr, 0.93 atm, and 6.6% change
5.3.2 50.4 atm
5.3.3 800.3 torr, 1.053 atm, and 1.067×10^2 kPa
5.4.1 59.1 kg of air enters as the house cools.
5.4.2

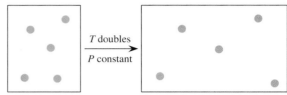

5.4.3 272 g/mol, $HgCl_2$
5.5.1 H_2, $X = 0.933$ and $p = 2.15$ atm; N_2, $X = 0.067$ and $p = 0.15$ atm
5.5.2

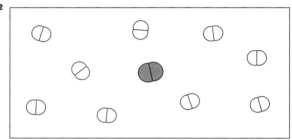

5.5.3 (a) 3.5×10^{-5}; (b) 2.6×10^{-2} torr; and (c) 8.3×10^{20} molecules
5.6.1 7.8 g C_2H_2 and 5.4 g H_2O
5.6.2 3.60 atm
5.6.3 5.89 atm (remember to calculate the new partial pressure of air at the higher temperature)
5.7.1 Ne, 1.82×10^{-5}; He, 5.24×10^{-6}; and CH_4, 1.4×10^{-6}
5.7.2 2.64×10^4 L
5.7.3 $vp = 9.209$ torr, 59% relative humidity

6

Atoms and Light

INTRODUCTION: LASERS

In the right combination, atoms and light combine to create one of the most remarkable tools of modern technology, the laser. Lasers have many commercial applications, ranging from laser light shows to CD scanners, eye surgery, and fiber optics communications. Lasers have also become versatile tools for advanced scientific research. To give just one example, lasers can be finely tuned to deposit vapor-phase atoms in regular patterns. Our inset shows chromium atoms deposited in a dot array using laser focusing. Each peak is 13 nm high.

The word *laser* stands for *light amplification by stimulated emission of radiation*. In a laser, atoms and light interact to generate a beam of light that has precisely defined properties. In simple terms, when a sample of atoms (or molecules) is "pumped" uphill in energy, light of a specific wavelength (color) may interact with the atoms or molecules to produce more light of that wavelength. Mirrors are used to reflect the light back into the sample so the process can repeat itself many times, amplifying the intensity of the light immensely.

Light from a laser differs substantially from conventional light. Laser light is monochromatic—has a single color—whereas conventional light sources typically produce light of many colors. A laser is highly directional, whereas conventional sources send light in all directions. As a result, laser light is more highly organized than light from normal sources. Many of the applications of lasers take advantage of this high degree of organization.

Laser beams are monochromatic and highly directional.

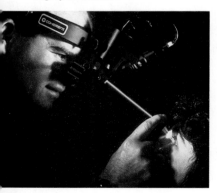

Bar code scanners, for example, exploit the directionality of a laser beam. The laser light reflects more strongly from white stripes than from black. Because of the directionality of the laser beam, the pattern of the bar code is mirrored faithfully in the reflected laser beam. These variations are read by a sensor that converts the light pattern into an electronic representation of the bar code that the sensor transmits to a computer.

The combination of high intensity and directionality is the basis for laser surgery. Essentially, a laser acts as a "light knife." In eye surgery, a laser beam can be focused to a tiny spot at the back of the retina. This allows the surgeon to carry out delicate operations such as repairing a detached retina without physically invading the eyeball.

Lasers come in a variety of sizes, from pencil-sized optical pointers to industrial-scale optical "saws." Regardless of its size, every laser is based on the principles of interactions between light and atoms or molecules. We describe these principles in this chapter. First, we summarize the properties of atoms and light. Next we discuss the energy changes that accompany the interactions between electrons and light. Finally, we describe the properties of electrons bound to atoms and construct a picture of atomic structure. Differences in atomic structure distinguish the elements from one another and generate the rich array of chemical behavior exhibited by different types of atoms.

6.1 CHARACTERISTICS OF ATOMS

We begin with a review of the fundamental characteristics of atoms, many of which were introduced in Chapter 2:

Atoms possess mass. Matter possesses mass, and matter is made up of atoms, so atoms possess mass. This property was already recognized in the time of John Dalton, who made it one of the postulates of his atomic theory.

Atoms contain positive nuclei. Recall from Chapter 2 what Rutherford's scattering experiment demonstrated. Every atom contains a tiny central core where all the positive charge and most of the mass are concentrated. Subsequent experiments have shown that, while the masses of nuclei can have various values, the mass of every nucleus is at least 2000 times larger than the mass of the electron. Thus more than 99.9% of the mass of an atom is contained in its nucleus.

Atoms contain electrons. For an atom to be neutral, the number of electrons that it contains must equal the total positive charge on its nucleus. Because each element has a characteristic positive charge associated with its nucleus, ranging from +1 for hydrogen to greater than +100 for the heaviest elements, atoms of different elements have different numbers of electrons.

Atoms occupy volume. Matter occupies space, and matter is made up of atoms, so atoms occupy space. It is extremely difficult to compress a solid such as copper or a liquid such as mercury, because the electron cloud of each atom occupies some volume that no other atom is able to penetrate because of electron-electron repulsion. Example 6–1 shows how to estimate the volume of an atom from the density of a sample, the molar mass of the substance, and the Avogadro constant.

Atomic Volumes	**Example 6-1**

The density of lithium, the lightest metal, is 0.534 g/cm^3. Estimate the volume occupied by a single atom in solid lithium.

Strategy: Our task is to estimate the volume of one atom of lithium. As usual, the mole is a convenient place to begin the calculations. Visualize a piece of lithium containing one mole of atoms. The molar mass, taken from the periodic table, tells us the number of grams of Li in one mole. The density equation can be used to convert from mass to volume. Once we have the volume of one mole of lithium, we divide by the number of atoms per mole to find the volume of a single atom.

Solution: First, assemble the data:

$$\rho = 0.534 \text{ g/cm}^3 \text{ (stated in the problem)}$$

$$MM = 6.941 \text{ g/mol (from periodic table)}$$

$$N_A = 6.022 \times 10^{23} \text{ atom/mol}$$

Now rearrange the density equation and solve for the molar volume of lithium.

$$\rho = \frac{m}{V} \quad so \quad V_{molar} = \frac{m}{\rho} = \frac{6.941 \text{ g/mol}}{0.534 \text{ g/cm}^3} = 13.00 \text{ cm}^3/\text{mol}$$

Finally, divide by the Avogadro constant to find volume per atom:

$$V_{atom} = \frac{V_{molar}}{N_A} = \frac{13.00 \text{ cm}^3/\text{mol}}{6.022 \times 10^{23} \text{ atom/mol}} = 2.16 \times 10^{-23} \text{ cm}^3/\text{atom}$$

Finally, check for reasonableness. The units are correct. This is a very tiny volume, as we would expect, knowing that atoms are very tiny. A cube with this volume has sides measuring 2.78×10^{-10} m in length (0.278 nm).

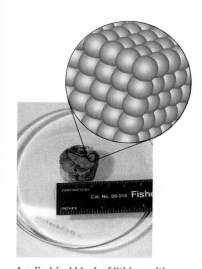

A cylindrical block of lithium with diameter and height of 2.4 cm contains approximately 1 mol of Li atoms.

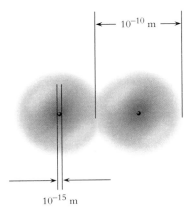

10^{-10} m

10^{-15} m

Figure 6-1
Depiction of two atoms, with their electron clouds in contact.

CHAPTER 10 →
We describe the forces that hold atoms and molecules together in the liquid and solid phases in Chapter 10.

CHAPTERS 8 & 9 →
In Chapters 8 and 9, we describe how electron clouds interact as atoms combine to make molecules.

The volume of an atom is determined by its electron cloud. Example 6-1 demonstrates that atomic dimensions are a little over 10^{-10} m, whereas Rutherford's experiments showed that nuclear dimensions are only about 10^{-15} m. This is 100,000 times smaller than atomic dimensions, so the nucleus is buried deep within the electron cloud. If an atom were the size of a sports stadium, its nucleus would be the size of a pea. Figure 6-1 shows a schematic view of two atoms with their electron clouds in contact with each other.

Atoms have various properties. The periodic table is a catalog of the elements, each with its own unique set of physical and chemical properties. Each element has a unique value for Z, the positive charge on its nucleus, and the number of electrons possessed by a neutral atom of that element is also equal to Z. The different properties of elements arise from these variations in nuclear charges and numbers of electrons.

Atoms attract one another. Every gas changes into a liquid if the pressure is high enough and the temperature is low enough. The atoms or molecules of a liquid or solid stick together in a finite volume rather than expanding, as a gas does, to fill all available space. This cohesiveness comes from electrical forces of attraction between the negative electron cloud of each atom and the positive nuclei of other atoms.

Atoms can combine with one another. Atoms can combine with one another to form molecules. As we pointed out in Chapter 2, this is one of the fundamental points of the atomic theory.

What does an atom experience in an encounter with another atom? The nucleus, which contains most of the atom's mass, is confined to a tiny volume. Electrons, on the other hand, are spread out through space. Therefore a collision between two atoms is a collision of their electron clouds. The electron clouds repel each other but are attracted by the nuclei. Chemists describe molecular structure, properties of materials, and chemical reactions in terms of how electrons respond to these electrical forces.

Our catalog of atomic characteristics emphasizes electrons, because electrons determine the chemical properties of atoms. For the same reason, the next several chapters examine electrons and the way they influence chemical properties. First, however, we describe light and its interaction with atoms, because light is an essential tool for probing properties of electrons.

Section Exercises

6.1.1 The density of gold metal is 18.9 g/mL. (a) What is the volume occupied by one gold atom? (b) If a gold nucleus is 1/100,000 times as large as a gold atom, what is the volume of one gold nucleus?

6.1.2 (a) Use the volume occupied by a gold atom to estimate the thickness of one atomic layer of gold. (b) Estimate how many layers of gold atoms there are in a strip of gold foil that is 1.0 μm thick (1 μm = 10^{-6} m).

Like electromagnetic radiation, water waves vary in time (frequency) and space (wavelength).

6.2 CHARACTERISTICS OF LIGHT

By far the most useful tool for studying the structure of atoms is **electromagnetic radiation.** What we call **light** is one form of this radiation. It is important to know the properties of light if we are to understand what electromagnetic radiation reveals about atomic structure.

Light Has Wave Aspects

Light has properties that are wave-like. A wave is a regular oscillation in some particular property, such as the up-and-down variation in position of water waves. Water waves vary with time. A surfer waiting for a "big one" bobs up and down as "small ones" pass by. Light waves vary with time, too. This variation is characterized by the wave's **frequency (ν),** which is the number of wave crests passing a point in space in one second. The frequency unit is s^{-1}, also designated Hertz (Hz). Water waves also vary in space; that is, wave height differs from one place to another. Light waves vary in space in a manner illustrated in Figure 6-2. This variation in space is characterized by the **wavelength (λ),** which is the distance between successive wave crests. Wavelengths are measured in units of distance, such as meters or nanometers. The height of a wave is called its **amplitude.** The amplitude of a light wave measures the **intensity** of the light. As Figure 6-3 shows, a bright light is more intense than a dim one.

Light waves always move through a vacuum—empty space—at the same speed. The speed of light is a fundamental constant, denoted by the symbol c: $c = 2.99792458 \times 10^8$ m/s. For any wave, its wavelength (in units of m) multiplied by its frequency (in units of s^{-1}) equals its speed (m/s). Thus light obeys Equation 6–1:

$$\lambda \nu = c \qquad (6\text{--}1)$$

Example 6-2 provides an example of how Equation 6-1 is used.

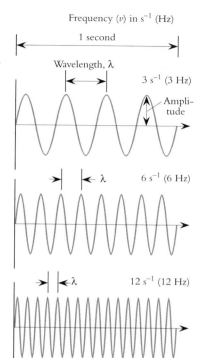

Frequency (ν) in s^{-1} (Hz)

1 second

Wavelength, λ

$3\ s^{-1}$ (3 Hz)

Amplitude

$6\ s^{-1}$ (6 Hz)

$12\ s^{-1}$ (12 Hz)

Figure 6-2
A light wave can be described by its wavelength and frequency. Notice that as the wavelength increases, the frequency of the light decreases.

The value of c can be rounded to 2.998×10^8 m/s for most calculations.

λ is the Greek letter *lambda*, and ν is the Greek letter *nu*.

Wavelength–Frequency Conversion	Example 6-2

An FM radio station transmits its signal at 88.1 MHz. What is the wavelength of the radio signal?

Strategy: This is a simple conversion problem. The link between wavelength (λ) and frequency (ν) is given by Equation 6-1.

Solution: First, summarize the data:

$$c = 2.998 \times 10^8 \text{ m/s} \qquad \nu = 88.1 \text{ MHz}$$

Next, rearrange Equation 6-1 to solve for wavelength.

$$\nu\lambda = c \qquad so \qquad \lambda = \frac{c}{\nu}$$

To obtain equivalent units, convert the frequency units from MHz to Hz. The prefix "M" stands for "mega," which is a factor of 10^6. Remember that Hz is equivalent to s^{-1}.

$$\lambda = \frac{2.998 \times 10^8 \text{ m/s}}{88.1 \times 10^6/\text{s}} = 3.40 \text{ m}$$

Figure 6-3
A bright light is more intense and has greater amplitude than a dim light. The two waves shown here have the same wavelength, but different amplitudes. Amplitude is represented by the height of the wave at its crest.

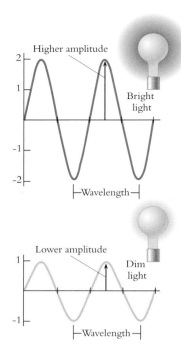

The wavelengths and frequencies of electromagnetic radiation cover an immense range. What we call light is the tiny part of the spectrum that can be detected by the human eye. Figure 6-4 shows that the visible spectrum of light covers the wavelength range from about 400 nm (violet) to 700 nm (red). The center of this range is yellow light, with a wavelength around 580 nm and a frequency of around $5.2 \times 10^{14} \text{ s}^{-1}$. Although visible light is extremely important to living creatures for seeing, the gamma ray, X-ray, ultraviolet and infrared, microwave, and radio frequency portions of the electromagnetic spectrum have diverse effects and applications in our lives.

Radiation with short wavelengths, in the X-ray and gamma-ray regions, can generate ions by removing electrons from atoms and molecules. These ions are highly reactive and can cause serious damage to the material that absorbs the light. However, under closely controlled conditions, X rays are used in medical imaging, and gamma rays are used to treat cancer.

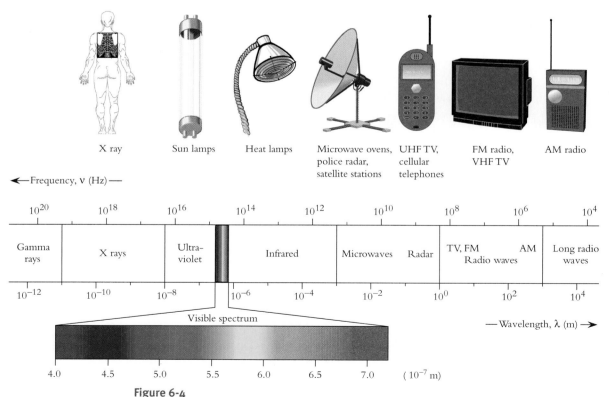

Figure 6-4
The electromagnetic spectrum, showing its various regions and the wavelengths and frequencies associated with each.

Radiations with long wavelengths fall in the infrared, microwave, and radio frequency regions. Heat lamps make use of infrared radiation, microwave ovens cook with microwave radiation, and radio and television signals are transmitted by radio waves.

What we perceive as white light actually contains a range of wavelengths. Figure 6-5 shows that these wavelengths are revealed when the light passes through a prism or when sunlight passes through raindrops. The prism or rain-drops bend different wavelengths of light through different angles, so the light that passes through spreads out in space, with each wavelength appearing at its own characteristic angle.

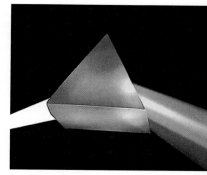

Figure 6-5
When white light passes through a prism or raindrops, each wavelength is bent through a different angle, spreading out the different wavelengths and generating the color pattern of a rainbow.

The Photoelectric Effect

Light carries energy. When our bodies absorb sunlight, for example, we feel warm because the energy of the sunlight has been transferred to our skin. More-over, the total energy of a beam of light depends on its intensity (how bright the light is). Sunlight at midday is more intense than the rays of the setting sun, which are reduced in intensity as they pass through a larger thickness of the atmosphere.

The photoelectric effect is the basis for many light-sensing devices, such as automatic door openers and camera exposure meters.

A phenomenon known as the **photoelectric effect** shows how the energy of light depends on its frequency and intensity. An apparatus for studying the photoelectric effect is illustrated schematically in Figure 6-6. In a photoelectric experiment, a beam of light strikes the surface of a metal. Under the right conditions, the light causes electrons to be ejected from the metal's surface. These electrons strike a detector that measures the number of electrons and their ki-netic energy.

A detailed study of the photoelectric effect reveals how the behavior of the electrons is related to the characteristics of the light:

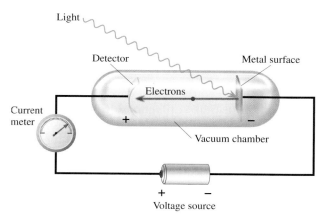

Figure 6-6
An apparatus for studying the photoelectric effect.

1. Below a characteristic threshold frequency, ν_0, no electrons are observed, regardless of the light's intensity.

2. Above the threshold frequency, the maximum kinetic energy of ejected electrons increases linearly with the frequency of the light, as shown in Figure 6-7.

3. Above the threshold frequency, the *number* of emitted electrons increases with the light's intensity, but the *kinetic energy* per electron does not depend on the light's intensity.

4. All metals exhibit the same pattern, but as Figure 6-7 shows, each metal has a different threshold frequency.

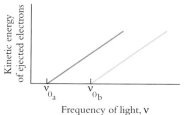

Figure 6-7
Variation in the maximum kinetic energy of electrons ejected from two different metal surfaces (a and b) by light of various frequencies.

In 1905, Albert Einstein provided an elegant explanation of the photoelectric effect. Einstein postulated that light comes in packets or bundles, called **photons.** Each photon has an energy that is directly proportional to the frequency:

$$E_{photon} = h\nu_{photon} \tag{6-2}$$

In this equation, E is the energy of light and ν is its frequency. The proportionality constant between energy and frequency is known as **Planck's constant (h)** and has a value of $6.62606876 \times 10^{-34}$ J s. Example 6-3 illustrates the use of Equation 6-2.

Planck's constant (h) is named in honor of Max Planck, who first introduced this constant in 1900 to explain another phenomenon of light.

Example 6-3	The Energy of Light

What is the energy of a photon of red light of wavelength 655 nm?

Strategy: This conversion problem requires two steps. Equations 6-1 and 6-2 relate the energy of a photon to its frequency and wavelength.

Solution: Summarize the data:

$$h = 6.626 \times 10^{-34}\,\text{J s} \qquad \lambda = 655\,\text{nm} \qquad c = 2.998 \times 10^{8}\,\text{m/s}$$

Combine the equations into an equation that relates energy to wavelength:

$$E = h\nu \quad and \quad \nu\lambda = c \quad so \quad E = \frac{hc}{\lambda}$$

Substitute and evaluate:

$$E_{photon} = \frac{(6.626 \times 10^{-34}\,\text{J s})(2.998 \times 10^{8}\,\text{m/s})}{(655\,\text{nm})(10^{-9}\,\text{m/nm})} = 3.03 \times 10^{-19}\,\text{J}$$

Albert Einstein is considered the greatest modern physicist because his theories influenced topics in physics ranging from the basis for lasers to the expanding universe. *Albert Einstein™ Licensed by Hebrew University of Jerusalem. Represented by The Roger Richman Agency, Inc., www.albert-einstein.net. Photo provided courtesy of the Archives of the California Institute of Technology.*

Einstein applied the law of conservation of energy to the photoelectric effect, as shown schematically in Figure 6-8. When a metal surface absorbs a photon, the energy of the photon is transferred to an electron:

$$\Delta E_{electron} = E_{photon}$$

Some of this energy is used to overcome the forces that bind the electron to the metal, and the remainder shows up as kinetic energy of the ejected electron.

The energy of a photon that has the threshold frequency (ν_0) corresponds to the binding energy of the electron. In other words, the energy of a photon at the threshold frequency equals the minimum energy needed to overcome the forces that bind the electron to the metal. Putting these ideas together, Einstein obtained a simple linear equation that matches the graph in Figure 6-7:

Electron kinetic energy = Photon energy − Binding energy

$$E_{kinetic}\,(\text{electron}) = h\nu - h\nu_0 \tag{6-3}$$

Figure 6-8
Diagram of energy balance for the photoelectric effect.

Einstein's explanation accounts for the observed properties of the photoelectric effect. First, when the energy of the photon is less than $h\nu_0$ (low-frequency light), there is not enough energy per photon to overcome the electron's binding energy. Under these conditions, no electrons can escape from the metal surface, no matter how intense the light. Second, after the energy of the photon exceeds the threshold value ($h\nu > h\nu_0$), electrons are ejected. The "extra" energy of the photon is transferred to the ejected electron as kinetic energy; this extra kinetic energy increases linearly with ν. Third, the intensity of a light beam is a measure of the number of photons; light with higher amplitude carries more photons than light of lower amplitude. The intensity of the light *does not* determine the amount of energy per photon. "Higher intensity" means more photons but not more energy per photon. More photons striking the metal result in more electrons being ejected, but the energy of each photon and each electron is unchanged. Finally, each metal has its own characteristic threshold frequency because electrons are bound more tightly to some metals than to others.

Example 6-4 shows how to apply Einstein's analysis of the photoelectric effect.

The Photoelectric Effect	**Example 6-4**

The minimum energy needed to remove an electron from potassium metal is 3.7×10^{-19} J. Will photons of frequencies 4.3×10^{14} s^{-1} (red light) and of 7.5×10^{14} s^{-1} (blue light) trigger the photoelectric effect? If so, what is the maximum kinetic energy of the ejected electrons?

Strategy: This problem asks if red and blue photons can cause potassium metal to lose electrons. We must analyze the energy requirements for ejection of an electron. No electrons will be ejected unless the energy of the photons exceeds some threshold value characteristic of the metal. If the photon energy exceeds this threshold value, electrons will be ejected with kinetic energy given by Equation 6-3. An important part of this problem is the conversion of photon frequency to photon energy.

Solution: The threshold energy of potassium metal, 3.7×10^{-19} J, is given in the problem. To determine whether the red and blue photons will eject electrons, we convert the frequencies of these photons to their corresponding energies:

$$E_{\text{red photon}} = h\nu = (6.626 \times 10^{-34} \text{ J s})(4.3 \times 10^{14} \text{ s}^{-1}) = 2.8 \times 10^{-19} \text{ J}$$

$$E_{\text{blue photon}} = h\nu = (6.626 \times 10^{-34} \text{ J s})(7.5 \times 10^{14} \text{ s}^{-1}) = 5.0 \times 10^{-19} \text{ J}$$

According to these calculations, a photon of red light does not have enough energy to overcome the forces that bind electrons to the metal. A blue photon, however, will eject an electron from the surface of potassium metal because its energy exceeds the threshold value.

What happens to the "extra" energy of a blue photon? Equation 6-3 indicates that it is transferred to the electron as kinetic energy:

$$E_{\text{kinetic}} \text{ (electron)} = h\nu - h\nu_0$$

$$E_{\text{kinetic}} \text{ (electron)} = (5.0 \times 10^{-19} \text{ J}) - (3.7 \times 10^{-19} \text{ J}) = 1.3 \times 10^{-19} \text{ J}$$

Light Has Particle Aspects

Before 1905, the properties of light had been explained with a wave picture. Einstein's explanation of the photoelectric effect was simple, but revolutionary, because it showed that light also has properties of particles. Light consists of photons, each of which is a bullet with the discrete energy $E = h\nu$. Thus light can be described as discrete particles of energy, and a complete description of light includes wave-like and particle-like properties.

Neither the particle nor the wave view of light is wrong or right. Light has some properties of waves and some properties of particles. When light interacts with a relatively large body such as a raindrop or a prism, its wave properties dominate the interaction. On the other hand, when light interacts with a small body such as an atom or an electron, particle properties dominate the interaction. Each view provides different information about the properties of light, and when we think about light, we must think of *wave-particles* that combine both types of features.

Light and Atoms

When a photon is absorbed, its energy is transferred to whatever absorbs the photon. In the photoelectric effect, this energy is transferred to an electron at the metal surface. In addition to revealing the particle nature of light, photoelectric experiments can be used to determine the binding energies of electrons to metal surfaces. On the other hand, when light interacts with free atoms rather than atoms bound to a metal, the interaction reveals information about electrons bound to individual atoms.

When an atom absorbs a photon, the atom is transformed to a higher energy state called an **excited state.** Atoms in excited states subsequently give up their excess energy to return to lower energy states. The lowest energy state of an atom, which is its most stable state, is called the **ground state.** Atomic energy transformations can be represented by diagrams, as shown in Figure 6-9. In such an **energy level diagram,** energy increases along the vertical axis. Each energy state of the atom is represented by a horizontal line. Absorption of a photon, shown by an upward arrow, causes a transition from the ground state to one of the excited states. Excited states may lose some or all of their excess energy by emitting photons, shown by downward arrows. We develop the idea of energy level diagrams in more detail in Section 6.3.

The key feature in the exchange of energy between atoms and light is that the change in energy of the atom exactly equals the energy of the photon, as required by the principle of conservation of energy:

$$\Delta E_{atom} = \pm\, h\nu_{photon} \qquad (6\text{-}4)$$

When an atom *absorbs* a photon, the atom gains the photon's energy. When an atom *emits* a photon, the atom loses the photon's energy. As an atom returns from an excited state to the ground state, it must lose exactly the amount of energy that it originally gained. However, excited atoms usually lose excess energy in several steps involving small energy changes, so the frequencies of emitted photons often are lower than those of absorbed photons. Example 6-5 provides an example of energy changes associated with the emission of light.

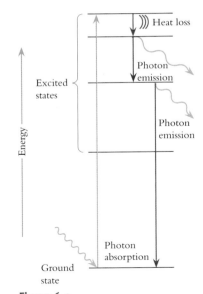

Figure 6-9
When an atom in its ground state absorbs a photon, the atom's energy increases, converting the atom to an excited state. Later, when the atom loses this extra energy, it may do so in more than one step, emitting one or more lower-energy photons. An excited-state atom also can lose some or all of its excess energy in collisions with other atoms.

Emission Energies

Example 6-5

A sodium-vapor street lamp emits yellow light at wavelength $\lambda = 589$ nm. What is the energy change for a sodium atom involved in this emission? How much energy is released per mole of sodium atoms?

Strategy: This problem relates energies of photons to energy changes of atoms. The solution requires a conversion involving wavelength and energy.

Solution: Sodium atoms emit photons of light ($\lambda = 589$ nm) as they are transformed from an excited state to a lower energy level. We can use Equations 6-1 and 6-2 to relate the wavelength of one of these photons to its energy:

$$E_{\text{photon}} = h\nu = \frac{hc}{\lambda}$$

Energy is conserved, so the energy of the emitted photon must exactly equal the energy lost by the atom:

$$\Delta E_{\text{atom}} = -E_{\text{photon}} = -\frac{hc}{\lambda}$$

The negative sign in the equation is included because the atom *loses* energy.

To calculate the energy of a photon, we need the following data:

$$h = 6.626 \times 10^{-34} \text{ J s} \qquad c = 2.998 \times 10^8 \text{ m/s} \qquad \lambda = 589 \text{ nm} = 589 \times 10^{-9} \text{ m}$$

$$\Delta E_{\text{atom}} = -E_{\text{photon}} = \frac{(6.626 \times 10^{-34} \text{ J s})(2.998 \times 10^8 \text{ m/s})}{(589 \times 10^{-9} \text{ m})}$$

$$= -3.37 \times 10^{-19} \text{ J}$$

This calculation gives the energy change for one sodium atom emitting one photon. We use the Avogadro constant to convert from energy per atom to energy per mole of atoms:

$$\Delta E_{\text{mol}} = (\Delta E_{\text{atom}})(N_A) = (-3.37 \times 10^{-19} \text{ J})(6.022 \times 10^{23} \text{ mol}^{-1})$$

$$= -2.03 \times 10^5 \text{ J/mol} = -203 \text{ kJ/mol}$$

When light is absorbed, as in the photoelectric effect, photons are destroyed, and the decrease in energy of the light beam is matched by an increase in energy of the system that absorbed the light. Einstein's analysis of the photoelectric effect is an example of this energy balance. Conversely, when light is emitted, as in a light bulb, the energy of the photons must be matched by a decrease in energy of the system that generated the light. We describe in subsequent sections of this chapter how these transitions between energy levels provide valuable information about the structures of atoms.

Section Exercises

6.2.1 A compact disc player uses light of frequency 3.85×10^{14} s^{-1} to read the information on the disc. (a) What is this light's wavelength? (b) In what portion of the electromagnetic spectrum does this wavelength fall (visible, ultraviolet, and so on)? (c) What is the energy of one mole of photons at this frequency?

6.2.2 The light reaching us from distant stars is extremely dim, so astronomers use instruments capable of detecting small numbers of photons. An infrared photon detector registered a signal at 1250 nm from Alpha Centauri with a total energy of 1.20×10^{-16} J. How many photons were detected?

6.2.3 In a photoelectric effect experiment, a metal absorbs photons with $E = 6.00 \times 10^{-19}$ J. The maximum kinetic energy of the ejected electrons is 2.00×10^{-19} J. Calculate the frequency of the light and the binding energy in joules per electron, and convert this energy into kilojoules per mole.

6.3 ABSORPTION AND EMISSION SPECTRA

When a light beam passes through a tube containing a gas, the atoms absorb specific and characteristic frequencies of the light. As a result, the beam emerging from the sample tube contains fewer photons at these specific frequencies. We can see where the frequencies are missing in the visible portion of the spectrum by passing the emerging light through a prism. The prism deflects the light, with different frequencies deflecting through different angles. After leaving the prism the beam strikes a screen, where it shows the missing frequencies as gaps or dark bands. These are the frequencies of light absorbed by the atoms in the sample tube. The resulting pattern, shown schematically in Figure 6-10, is called an **absorption spectrum.**

An absorption spectrum measures the frequencies of the photons that an atom *absorbs.* A similar experiment can be performed to measure the energies of the

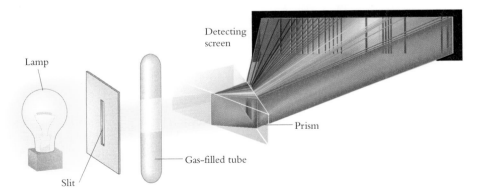

Figure 6-10
Schematic representation of an apparatus that measures the absorption spectrum of a gaseous element. The gas in the tube absorbs light at specific wavelengths, called "lines," so the intensity of transmitted light is low at these particular wavelengths.

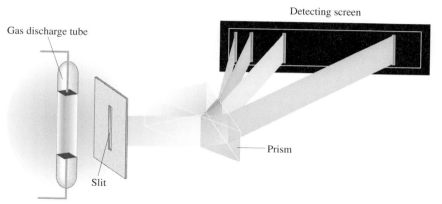

Figure 6-11
Schematic representation of an apparatus that measures the emission spectrum of a gaseous element. Emission lines appear bright against a dark background. The spectrum shown is the emission spectrum for hydrogen atoms.

photons *emitted* by atoms in excited states. An apparatus that measures these emitted photons is shown schematically in Figure 6-11. An electrical discharge excites a collection of atoms from their ground state into higher-energy states. These excited atoms lose all or part of their excess energy by emitting photons. This emitted light can be analyzed by passing it through a prism to give an **emission spectrum.** This is a plot of the intensity of light emitted as a function of frequency. The emission spectrum for hydrogen, shown in Figure 6-11, shows several sharp emission lines of high intensity. The frequencies of these lines correspond to photons emitted by the hydrogen atoms as they return to their ground state.

As Figure 6-12 shows, each element has unique absorption and emission spectra. That is, each element has its own set of characteristic frequencies of light that it can absorb or emit. Although Figures 6-10 and 6-11 show absorption and emission spectra in the visible region, electron transitions also take place in several regions of the electromagnetic spectrum that the human eye cannot detect. Instruments allow scientists to "see" into these regions. Each frequency absorbed or emitted by an atom can be converted into an energy value. These characteristic patterns of energy gains and losses provide information about atomic structure.

Quantization of Energy

When an atom absorbs light of frequency ν, the light beam loses energy $h\nu$, and the atom gains that amount of energy. What happens to the energy that the atom gains? A clue is that when the frequency of the bombarding light is high enough, it

Figure 6-12
The emission spectra from gaseous samples of Na, Hg, and Ne. These unique emission patterns provide valuable clues about atom structure.

produces cations and free electrons. In other words, a photon with high enough energy can cause an atom to lose one of its electrons. This implies that absorption of a photon results in a gain in energy for an electron in the atom. Consequently, the energy change for the atom equals the energy change for an atomic electron:

$$\Delta E_{atom} = \Delta E_{electron} = h\nu$$

The atomic spectra of most elements are complex and show little regularity. However, the emission spectrum of the hydrogen atom is sufficiently simple to be described by a single formula:

$$\nu_{emission} = (3.29 \times 10^{15} \text{ s}^{-1})\left(\frac{1}{n_1^2} - \frac{1}{n_2^2}\right)$$

The Swiss mathematician and physicist Johann Balmer proposed a form of this equation in 1885. At that time, the link between frequency and energy was not known. Balmer did not understand the significance of n_1 and n_2, both of which are integers (1, 2, 3, and so on). Then, in 1913, Niels Bohr used the discovery that $E = h\nu$ to interpret Balmer's observations. Bohr realized that the emission frequencies have specific values because the electron in a hydrogen atom is restricted to specific energies described by an equation containing an integer n:

$$E_n = -\frac{2.18 \times 10^{-18} \text{ J}}{n^2} \qquad (6\text{-}5)$$

Niels Bohr (1885–1962) was a Danish physicist whose discovery of the quantization of atomic energy levels won him a Nobel Prize in 1922. Bohr headed a world-renowned institute for atomic studies in Copenhagen in the 1920s and 1930s.

The constants in Bohr's equation and Balmer's equation are related through $E = h\nu$.

Equation 6-5 contains a negative sign because a free electron, removed from an atom, is defined to have zero energy. When an electron binds to a nucleus, energy is released, so any bound electron has a lower (more negative) energy than this defined zero point.

Bohr's idea of restricted energy levels was revolutionary, because scientists at that time thought that the electron in a hydrogen atom could have any energy, not just the ones described by Equation 6-5. In contrast, Bohr interpreted the hydrogen emission spectrum to mean that electrons bound to atoms can have only certain specific energy values. When a property is restricted to specific values, that property is said to be **quantized.** The atomic energy levels of hydrogen (and other elements) are quantized. In Equation 6-5, each integral value of n describes one of the allowed energy levels of the hydrogen atom. For example, the energy of an electron in hydrogen's fourth level is:

$$E_4 = -\frac{2.18 \times 10^{-18} \text{ J}}{4^2} = -1.36 \times 10^{-19} \text{ J}$$

When an electron changes energy levels, the change is an electronic transition between quantum levels. When a hydrogen atom absorbs or emits a photon, its electron changes from one energy level to another. Thus the change in energy of the atom is the difference between the two levels:

$$\Delta E_{atom} = E_{final} - E_{intial}$$

Photons always have positive energies, but energy changes (ΔE) can be positive or negative. When absorption occurs, an atom gains energy, ΔE for the atom is positive, and a photon disappears:

$$E_{absorbed \ photon} = \Delta E_{atom}$$

When emission occurs, an atom loses energy, ΔE for the atom is negative, and a photon appears:

$$E_{\text{emitted photon}} = -\Delta E_{\text{atom}}$$

We can combine these two equations by using absolute values:

$$E_{\text{photon}} = |\Delta E_{\text{atom}}| \tag{6-6}$$

Example 6-6 shows how Equations 6-5 and 6-6 are applied to the hydrogen atom.

Hydrogen Energy Levels

Example 6-6

How much energy does a hydrogen atom lose when its electron changes from the fourth energy state to the second energy state? What is the wavelength of the photon emitted?

Strategy: The problem asks about energy and the wavelength of a photon emitted by a hydrogen atom. Wavelength is related to energy, and photon energy is determined by the difference in energy between the two levels involved in the transition. In this case the electron moves from the fourth to the second energy level. The energy difference between these two states is given by:

$$\Delta E_{\text{atom}} = E_{\text{final}} - E_{\text{initial}} = (E_2 - E_4)$$

Solution: As described in the text, $E_4 = -1.36 \times 10^{-19}$ J. A similar calculation with Equation 6-5 gives the second energy level:

$$E_2 = -\frac{2.18 \times 10^{-18} \text{ J}}{2^2} = -5.45 \times 10^{-19} \text{ J}$$

This lower energy level has a more negative energy. The energy difference is:

$$\Delta E_{\text{atom}} = E_{\text{final}} - E_{\text{initial}} = (-5.45 \times 10^{-19} \text{ J}) - (-1.36 \times 10^{-19} \text{ J})$$
$$= -4.09 \times 10^{-19} \text{ J}$$

This energy change is *negative* because the atom *loses* energy. This lost energy appears as a photon whose energy is given by Equation 6-6:

$$E_{\text{photon}} = |\Delta E_{\text{atom}}| = |-4.09 \times 10^{-19} \text{ J}| = 4.09 \times 10^{-19} \text{ J}$$

To determine the wavelength of the photon, we use Equations 6-1 and 6-2:

$$E_{\text{photon}} = h\nu = \frac{hc}{\lambda}$$

Solving for λ gives

$$\lambda_{\text{photon}} = \frac{hc}{E_{\text{photon}}}$$
$$= \frac{(6.626 \times 10^{-34} \text{ J s})(2.998 \times 10^8 \text{ m/s})(10^9 \text{ nm/m})}{(4.09 \times 10^{-19} \text{ J})} = 486 \text{ nm}$$

According to Figure 6-4, a photon with a wavelength of 486 nm is in the blue-green region of the spectrum.

Figure 6-13
A ball on a staircase shows some properties of quantized energy states.

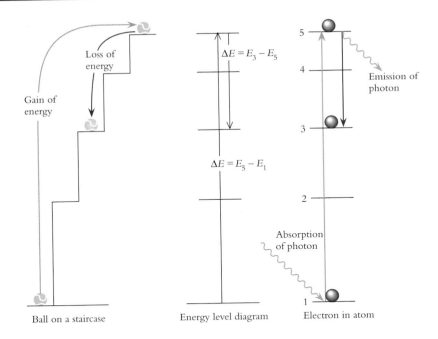

Loss of energy

Gain of energy

$\Delta E = E_3 - E_5$

$\Delta E = E_5 - E_1$

5

4

Emission of photon

3

2

Absorption of photon

1

Ball on a staircase

Energy level diagram

Electron in atom

Energy Level Diagrams

The quantum levels of an electron bound to an atom are crudely analogous to the potential energies available to a ball on a staircase. As illustrated in Figure 6-13, a ball may sit on any of the steps, with potential energy higher than it would have at the bottom of the stairs. To move a ball from the bottom of the staircase to step 5 requires the addition of a specific amount of energy, $\Delta E = E_5 - E_1$. If too little energy is supplied, the ball cannot reach this step. Conversely, if a ball moves down the staircase, it releases specific amounts of energy. If a ball moves from step 5 to step 3, it loses energy, $\Delta E = E_3 - E_5$. Although a ball may rest squarely on any step, it cannot be suspended at some position between the stair-steps. Electrons in atoms, like balls on steps, cannot exist "between steps" but must occupy one of the specific, quantized energy levels. (Remember that this is an analogy; atomic energy levels are not at all similar to staircases except in being quantized.)

According to Equation 6-5, a hydrogen atom has a regular progression of quantized energy levels. Figure 6-14 shows the energy level diagram for hydrogen atoms; some of the absorption and emission transitions are represented by arrows. Notice that the energies of absorption from the lowest energy level are identical to the energies of emission to the lowest energy level, meaning that the wavelengths of light absorbed in these upward transitions are identical to the wavelengths of light emitted in the corresponding downward transitions.

Elements other than hydrogen also have quantized energy levels, but they lack the regular spacing described by Equation 6-5. Scientists use experimental values for observed absorption and emission lines to calculate the allowed energy levels for each different element. As an example, Figure 6-15 shows the energy level diagram for mercury. The absorption spectrum of mercury shows two dominant lines (upward arrows). Its emission spectrum, generated by an electrical discharge, contains light of many different wavelengths (downward arrows). Energy level diagrams such as Figures 6-14 and 6-15 summarize the quantized energy levels of an atom as

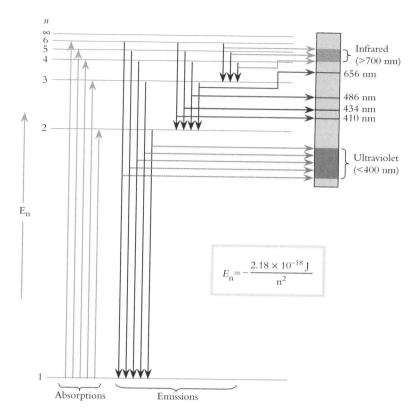

Figure 6-14
Energy levels for the hydrogen atom and some of the transitions that occur between levels. Upward arrows represent absorption transitions, and downward arrows represent emissions.

$$E_n = -\frac{2.18 \times 10^{-18}\ J}{n^2}$$

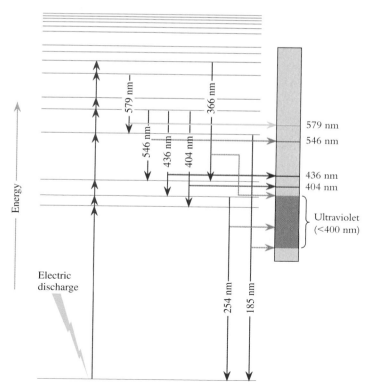

Figure 6-15
Energy level diagram for mercury (Hg) atoms showing the most prominent absorption and emission lines. The numbers accompanying the arrows are the wavelengths in nanometers (nm) of the photons associated with these transitions.

deduced from light measurements and energy conservation. Example 6-7 shows another example of this type of reasoning.

| Example 6-7 | **Energy Level Diagrams** |

Ruby lasers use crystals of Al_2O_3. The crystals contain small amounts of Cr^{3+} ions, which absorb light between 400 and 560 nm. These excited-state ions lose some energy as heat. After losing heat, the Cr^{3+} ions return to the ground state by emitting red light of wavelength 694 nm. Calculate (a) the molar energy of the 500-nm radiation used to excite the Cr^{3+} ions, (b) the molar energy of the emitted light, and (c) the fraction of the excitation energy emitted as red photons and the fraction lost as heat. (d) Draw an energy level diagram, in kilojoules per mole, that summarizes these processes.

Strategy: This problem asks about energies, light, and atoms. With multi-part problems, the best strategy is to work through the parts one at a time. Parts (a) and (b) concern the link between light and energy, which was discussed in Section 6.1. Once the transition energies have been determined, we can calculate the fraction of the excited-state energy lost as heat, part (c), and we can draw an energy level diagram that shows how the levels are related, part (d).

Solution: We solve part (a) using Equations 6-1 and 6-2 and the Avogadro constant:

$$E_{photon} = h\nu \quad and \quad \lambda\nu = c \quad so \quad E_{photon} = \frac{hc}{\lambda}$$

$$E_{photon\ absorbed} = \frac{(6.626 \times 10^{-34} \text{ J s})(2.998 \times 10^8 \text{ m/s})}{(500 \text{ nm})(10^{-9} \text{ m/nm})} = 3.97 \times 10^{-19} \text{ J}$$

This is the energy change per atom. To calculate the changes per mole, we multiply by the Avogadro constant. We also need to convert from J to kJ:

$$E_{photon\ absorbed} = (3.97 \times 10^{-19} \text{ J})(6.022 \times 10^{23} \text{ mol}^{-1})(10^{-3} \text{ kJ/J}) = 239 \text{ kJ/mol}$$

Apply the same reasoning to part (b). The result gives the energy of the red photon.

$$E_{photon\ emitted} = 172 \text{ kJ/mol}$$

To solve part (c), remember that energy is conserved. The sum of the emitted heat and the emitted photon must equal the energy absorbed by the ion:

$$E_{photon\ absorbed} = E_{photon\ emitted} + E_{heat\ emitted}$$

Because 239 kJ/mol is absorbed and 172 kJ/mol is emitted, the fraction of the excitation energy re-emitted is 172/239 = 0.721. The fraction converted to heat is the difference between this value and 1.000, or 0.279. In other words, 72.1% of the energy absorbed by the chromium ion is emitted as red light, and the other 27.9% is lost as heat.

Part (d) asks for an energy level diagram for this process. The electron starts in the ground state. On absorption of a photon, the electron moves to an energy level that is higher by 239 kJ/mol. The chromium ion loses 27.9% of its excited-state energy as heat as the electron moves to a different level that is 172 kJ/mol above

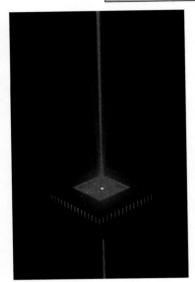

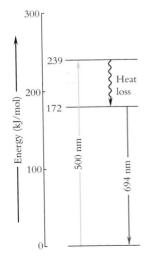

Light from a ruby laser is generated from Cr^{3+} ions.

| Energy Level Diagrams *(continued)* | Example 6-7 |

the ground state. Finally, emission of the red photon returns the Cr^{3+} ion to the ground state. The numerical values allow us to construct an accurate diagram.

Are these results reasonable? Yes. When Cr^{3+} ions absorb light, they are pumped to an energy level higher than the energy that they later emit. If calculations had shown that the emitted light had a larger energy than the absorbed light, the result would have been unreasonable because the system would violate the law of conservation of energy.

The Tools for Discovery Box discusses how spectroscopy—the observation of spectra—can be used to study otherwise inaccessible objects.

Section Exercises

■ **6.3.1** When minerals absorb invisible ultraviolet light from a mercury lamp, they emit visible light of a longer wavelength, converting the remaining energy into heat. How much energy per mole is converted to heat by a mineral that absorbs ultraviolet light at 366 nm and emits green light at 545 nm?

■ **6.3.2** Gaseous helium atoms absorb X rays of wavelength 53.7 nm. After absorbing an X ray of this wavelength, a helium atom may emit light of wavelength 501.6 nm. What is the net energy change for a helium atom that has gone through this absorption-emission sequence? Draw an energy level diagram that shows the sequence.

■ **6.3.3** Calculate the wavelengths that hydrogen atoms in the fifth energy level can emit.

6.4 PROPERTIES OF ELECTRONS

The energy of the electron plays a central role in determining chemical behavior. Several other properties of electrons also influence the physical and chemical characteristics of atoms and molecules.

Properties Shared by All Electrons

Some properties are characteristic of all electrons, but others arise only when electrons are bound to atoms or molecules. First, we describe the properties possessed by all electrons.

Each electron has the same mass and charge. Every electron has a mass measuring 9.109×10^{-31} kg and a charge of 1.602×10^{-19} C, as the experiments described in Chapter 2 demonstrate.

Electrons behave like magnets. Some types of atoms behave like tiny magnets. The best-known example is iron, the material used to make many permanent magnets. Experiments have shown that the magnetic behavior of atoms is caused by magnetic properties of component parts, especially electrons.

The magnetic properties of electrons arise from a property called **spin,** which we describe in more detail in Chapter 7. All electrons have spin of the same magnitude, but electron spin can respond to a magnet in two different ways. Most magnetic effects associated with atoms are caused by the spins of their

Box 6-1   Tools for Discovery: Spectroscopy, Observing from Afar

How do we know the composition of the sun and other stars? How can we measure the temperature inside a flame so hot that any thermometer would melt? How can we explore chemical reactions among molecules that are much too tiny to see directly? Light allows us to do all these things. The study of matter with electromagnetic radiation is called *spectroscopy*.

Astronomers use spectroscopy to identify the composition of the sun and other stars. A striking example is the discovery of the element helium. In 1868, astronomers viewing a solar eclipse observed emission lines that did not match any known element. The English astronomer Joseph Lockyer attributed these lines to a new element that he named *helium,* from *helios,* the Greek word for the sun. For 25 years the only evidence for the existence of helium was these solar spectral lines.

In 1894, the Scottish chemist William Ramsay removed nitrogen and oxygen from air through chemical reactions. From the residue, Ramsay isolated argon, the first noble gas to be discovered. A year after discovering argon, Ramsay obtained an unreactive gas from uranium-containing mineral samples. The gas exhibited the same spectral lines that had been observed in the solar eclipse of 1868. After helium was shown to exist on Earth, this new element was studied and characterized.

The tip of the flame of a gas stove is much too hot to measure using physical probes, but its temperature can be measured spectroscopically. The core of a gas flame glows bright blue. Other hot objects emit different colors: A candle flame is yellow, and the heating element of a hairdryer glows red. All hot objects lose energy by giving off light,

and the hotter the object, the higher the average energy of the emitted light. An object at around 1000 K gives off red light. At around 2000 K, an object emits predominantly yellow light, and blue light indicates a temperature of around 3000 K. The colors of stars range from distinctly red, through yellow, to quite blue, indicating that different stars have different temperatures at their surfaces.

The previous examples illustrate the use of light given off—emission spectra—to explore inaccessible objects. Absorption spectroscopy is a powerful complementary technique. In this application, light shines on the system to be studied. The researcher then examines how that light changes during the interaction with the system. Absorption spectroscopy has been enhanced greatly by the development of powerful lasers. For example, by using a combination of a pulsed laser (laser #1 in the illustration below) and a continuous laser (laser #2), chemists can observe chemical bonds in the process of breaking. The light from laser #1 is absorbed by a specific chemical bond, which stretches and breaks. The molecule absorbs light from laser #2 only while it is in the process of breaking. The diagram shows a graph of the observed intensity of the second laser beam as a function of time.

If the pulse from laser #1 is ultrafast, the bond-energizing step occurs in a very short time, 150×10^{-15} s. As the bond stretches through the specific length at which it absorbs photons from laser #2, the molecule can absorb a photon from the second laser beam. This absorption causes the transmitted intensity of laser #2 to fall rapidly as the bond stretches. When the bond breaks, photons from laser #2 are no longer absorbed and the transmitted intensity returns to its original value. By measuring the time it takes for this to occur, chemists have determined how fast a chemical bond breaks.

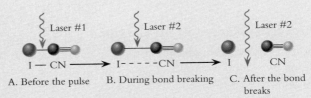

A. Before the pulse   B. During bond breaking   C. After the bond breaks

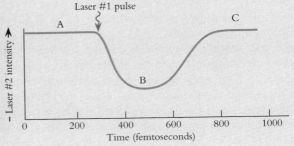

electrons. Iron and nickel are permanent magnets because of the cooperative effect of many electrons.

Electrons have wave properties. We are used to thinking of electrons as particles. As it turns out, electrons display both particle properties and **wave properties.** The French physicist Louis de Broglie first suggested that electrons display wave-particle duality like that exhibited by photons. De Broglie reasoned from nature's tendency toward symmetry: If things that behave like waves (light) have particle characteristics, then things that behave like particles (electrons) should also have wave characteristics.

Experiments had shown that a beam of light shining on an object exerts a pressure, and this, in turn, implies that a photon has momentum. Quantitative measurements of the pressure exerted by light showed that the momentum of light (p) is related to its energy through a simple equation:

$$E = pc$$

As we have already described, light energy also is related to its wavelength: $E = h\nu = hc/\lambda$. Setting these two energy expressions equal to each other gives an expression relating p to λ:

$$pc = \frac{hc}{\lambda}$$

The speed of light cancels, leaving an equation that de Broglie suggested should apply to electrons and other particles as well as to photons:

$$p = \frac{h}{\lambda}$$

The momentum of a particle is the product of its mass and speed, $p = mu$. Making this substitution and solving for λ gives a form of the de Broglie equation that links the wavelength of a particle with its mass and speed:

$$\lambda_{\text{particle}} = \frac{h}{mu} \tag{6-7}$$

De Broglie's theory predicted that electrons are wave-like. How might this be confirmed? Waves display characteristic intensity patterns, examples of which are shown in Figure 6-16. In Figure 6-16a, water waves radiate away from two bobbing floats and form a standing pattern. In Figure 6-16b, a similar wave pattern is formed by X rays. Here, high-energy photons have passed through the regular array of atoms in a crystal, whose nuclei scatter the photon waves. If electrons have wave properties, they should display regular wave patterns like these.

In 1927, American physicists Clinton Davisson and Lester Germer and British physicist George Thomson carried out experiments in which they exposed metal films to electron beams with well-defined kinetic energies. Both experiments generated patterns like those shown in Figure 6-16b, confirming the validity of the de Broglie equation for electron wavelengths. This established the wave nature of electrons.

In recent years, scanning tunneling electron microscopes have produced pictures of electron waves, an example of which is shown in Figure 6-16c. Here, two atoms on an otherwise smooth metal surface act like the floats in Figure 6-16a, and cause the electrons in the metal to set up the standing wave pattern shown in the figure.

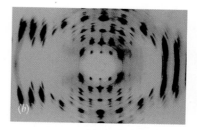

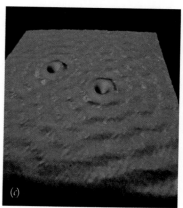

Figure 6-16
Examples of wave patterns.
(*a*) Floats produce standing water waves. (*b*) X rays generate wave interference patterns.
(*c*) Protruding atoms on a metal surface generate standing electron waves.

De Broglie received the Nobel Prize in physics in 1929, only two years after experiments confirmed his theory. Davisson, a student of Nobel laureate Robert Millikan, and Thomson, the son and student of J.J. Thomson (who won the Nobel prize for discovering the electron), shared the Nobel Prize in physics in 1937.

Table 6-1
Equations for Photons and Free Electrons

Property	Photon Equation	Electron Equation
Energy	$E = h\nu$	$E_{kinetic} = \dfrac{mu^2}{2}$
Wavelength	$\lambda = \dfrac{hc}{E}$	$\lambda = \dfrac{h}{mu}$
Speed	$c = 3 \times 10^8$ m/s	$u = \sqrt{\dfrac{2E_{kinetic}}{m}}$

h, Planck's constant; ν, frequency; m, mass; u, speed.

Both photons and electrons are particle-waves, but different equations describe their properties. Table 6-1 summarizes the properties of photons and free electrons, and Example 6-8 shows how to use these equations.

Example 6-8	**Wavelengths**

The structure of a crystal can be studied by observing the wave interference patterns that result from passing particle-waves through the crystal lattice. To generate well-defined patterns, the wavelength of the particle-wave must be about the same as the distance between atomic nuclei in the crystal. In a typical crystal, this distance is 0.25 nm. Determine the energy of (a) a photon particle-wave beam with this wavelength and (b) an electron particle-wave beam with this wavelength.

Strategy: This problem has two parts, one dealing with photons and the other with electrons. We are asked to relate the wavelengths of the particle-waves to their corresponding energies. Table 6-1 emphasizes that photons and electrons have different relationships between energy and wavelength. Thus, we use different equations for parts (a) and (b).

Solution:
(a) Photon energy is $E = h\nu = hc/\lambda$. We substitute and evaluate, being careful with units:

$$E_{photon} = \frac{(6.626 \times 10^{-34} \text{ J s})(2.998 \times 10^8 \text{ m/s})}{(0.25 \text{ nm})(10^{-9} \text{ m/nm})} = 7.9 \times 10^{-16} \text{ J}$$

(b) For an electron, we need to work with two equations. The de Broglie equation links the speed of an electron with its wavelength:

$$\lambda_{particle} = \frac{h}{mu}$$

The kinetic energy equation links the speed of an electron with its kinetic energy:

$$E_{kinetic} = \tfrac{1}{2} mu^2$$

Wavelengths (*continued*)

Example 6-8

Begin by determining the speed of the electron:

$$u_{electron} = \frac{h}{m\lambda} = \frac{(6.626 \times 10^{-34} \text{ kg m}^2/\text{s})}{(9.109 \times 10^{-31} \text{ kg})(0.25 \times 10^{-9} \text{ m})} = 2.91 \times 10^6 \text{ m/s}$$

Next use the speed to find the kinetic energy of the electron:

$$E_{kinetic} = \frac{1}{2}\, mu^2 = \frac{(9.109 \times 10^{-31} \text{ kg})(2.91 \times 10^6 \text{ m/s})^2}{2}$$

$$E_{kinetic} = 3.9 \times 10^{-18} \text{ kg m}^2/\text{s}^2 = 3.9 \times 10^{-18} \text{ J}$$

The joule, J, is the SI unit for energy. The joule is a derived unit: $1\text{ J} = 1 \text{ kg m}^2 \text{ s}^{-2}$

The de Broglie equation predicts that every particle has wave characteristics. The wave properties of subatomic particles such as electrons and neutrons play important roles in their behavior, but larger particles such as Ping-Pong balls or automobiles do not behave like waves. The reason is the scale of the waves. For all except subatomic particles, the wavelengths involved are so short that we are unable to detect the wave properties. This is illustrated in Example 6-9.

Matter Waves

Example 6-9

Compare the wavelengths of an electron traveling at 1.00×10^5 m/s and a Ping-Pong ball of mass 11 g traveling at 2.5 m/s.

Strategy: This problem deals with particle-waves that have mass. Equation 6-7, the de Broglie equation, relates the mass and speed of an object to its wavelength.

Solution: For the electron: $m_e = 9.109 \times 10^{-31}$ kg, $u = 1.00 \times 10^5$ m/s

$$\lambda_{electron} = \frac{h}{mu} = \frac{(6.626 \times 10^{-34} \text{ kg m}^2/\text{s})}{(9.109 \times 10^{-31} \text{ kg})(1.00 \times 10^5 \text{ m/s})} = 7.27 \times 10^{-9} \text{ m}$$

For the Ping-Pong ball: $m_{ball} = 11$ g, $u = 2.5$ m/s

$$\lambda_{ball} = \frac{h}{mu} = \frac{6.626 \times 10^{-34} \text{ kg m}^2/\text{s})(10^3\text{g}/1\text{kg})}{(11 \text{ g})(2.5 \text{ m/s})} = 2.4 \times 10^{-32} \text{ m}$$

The wavelength of the electron is about the same size as the radius of an atom, but the wavelength of the Ping-Pong ball is inconsequential compared to its size.

Heisenberg's Uncertainty Principle

A particle occupies a particular location, but a wave has no exact position. A wave extends over some region of space. Because of their wave properties, electrons are always spread out rather than located in one particular place. As a result, *the position of an electron cannot be precisely defined*. We describe electrons as *delocalized* because their waves are spread out rather than pinpointed.

Instead of electrons having exact locations and motions, they are distributed over some volume. Werner Heisenberg, a German physicist, found in the 1920s that the motion and position of a particle-wave cannot be simultaneously pinned down. If a particular particle-wave can be pinpointed in a specific location, its motion cannot be known. Conversely, if the motion of a particle-wave is known precisely, its location cannot be known. Heisenberg summarized this uncertainty in what has become known as the **uncertainty principle:** The more accurately we know position, the more uncertain we are about motion, and vice versa. Uncertainty is a feature of all objects, but it becomes noticeable only for very tiny objects like electrons.

Heisenberg's uncertainty principle forced a change in thinking about how to describe the universe. In a universe subject to uncertainty, many things cannot be measured exactly, and it is never possible to predict with certainty exactly what will occur next. This uncertainty has become accepted as a fundamental feature of the universe at the scale of electrons, protons, and neutrons.

Section Exercises

6.4.1 Calculate the wavelength associated with a photon whose energy is 1.00×10^{-19} J and the wavelength associated with an electron having a kinetic energy of 1.00×10^{-19} J.

6.4.2 The smallest distance that can be resolved by a microscope is typically 1.25λ, where λ is the wavelength of light or electrons used. The microscope can be greatly improved if a beam of electrons is used instead of a beam of light. If an electron microscope is to successfully resolve objects 0.600 nm apart, what kinetic energy must the electrons have?

6.4.3 Describe the differences between the properties of a free electron and those of a Ping-Pong ball.

6.5 QUANTIZATION AND QUANTUM NUMBERS

The properties of electrons described so far (mass, charge, spin, and wave nature) apply to all electrons. Electrons traveling freely in space, electrons moving in a copper wire, and electrons bound to atoms all have these characteristics. Bound electrons, those held in a specific region in space by electrical force, have additional important properties relating to their energies and the shapes of their waves. These additional properties can take on only certain specific values, so they are said to be **quantized.**

Energies of bound electrons are quantized. As described in Section 6.3, atoms of each element have unique, quantized electronic energy levels (see Figures 6-13 and 6-14). This quantization of energy is a property of *bound* electrons. The absorption and emission spectra of atoms consist of discrete energies because electrons undergo transitions from one bound state to another. In contrast, if an atom absorbs enough energy to remove an electron completely, the electron is no longer bound and can take on any amount of kinetic energy. *Bound* electrons have quantized energy states; *free* electrons have continuous energy states.

Absorption and emission spectroscopy provides experimental values for the quantized energies of atomic electrons. The theory of quantum mechanics provides a mathematical explanation that links quantized energies to the wave characteristics of electrons. These wave properties of atomic electrons are described by the

Schrödinger equation, a complicated mathematical equation with numerous terms describing the kinetic and potential energies of the atom.

The Schrödinger equation has solutions only for specific energy values. In other words, the energy of an atom is quantized, restricted to certain values. For each quantized energy, the Schrödinger equation generates a wave function that describes how the electrons are distributed in space.

To picture the spatial distribution of an electron around a nucleus, we must try to visualize a three-dimensional wave. Scientists have coined a name for these three-dimensional waves that characterize electrons: they are called **orbitals.** The word comes from *orbit*, which describes the path that a planet follows when it moves about the sun. An orbit, however, is two-dimensional (circle or ellipse), whereas an orbital is a three-dimensional volume (sphere or hourglass). The shape of a particular orbital shows how an atomic or a molecular electron fills three-dimensional space. Just as energy is quantized, orbitals have specific shapes and orientations. We describe the details of orbitals in Section 6.6.

Each quantized property can be identified, or indexed, using a number called a **quantum number.** These are integers that specify the values of the electron's quantized properties. Each electron in an atom has three quantum numbers that specify its three variable properties. A set of three quantum numbers is a shorthand notation that describes a particular energy, orbital shape, and orbital orientation. A fourth quantum number, with a value of $+\frac{1}{2}$ or $-\frac{1}{2}$, describes spin orientation. To describe an atomic electron completely, chemists specify a value for each of its four quantum numbers.

Principal Quantum Number

The most important quantized property of an atomic electron is its energy. The quantum number that indexes energy is the **principal quantum number (n).** For the simplest atom, hydrogen, we can calculate the energy of the electron if we know n, as described by Equation 6-5. However, that equation applies only to the hydrogen atom.

No known equation provides the exact energies of an atom that has more than one electron, but each electron in an atom can be assigned a value of n that is a positive integer and that correlates with the energy of the electron. The most stable energy for an atomic electron corresponds to $n = 1$, and each successively higher value of n describes a less stable energy state.

/// *The principal quantum number must be a positive integer:*
$n = 1, 2, 3, \ldots 8, 9$, etc.

Values such as $n = 0$, $n = -3$, and $n = \frac{5}{2}$ are unacceptable because they do not represent solutions to the Schrödinger equation, meaning that they do not correspond to reality.

The principal quantum number also tells us something about the size of an atomic orbital, because the energy of an electron is correlated with its distribution in space. That is, the more energy the electron has, the more it is spread out in space. As illustrated by Figure 6–17, this situation is roughly analogous to a bouncing tennis ball: The more kinetic energy the ball has, the higher it bounces and the greater its average distance above the court. Similarly, the higher the principal quantum number, the more energy the electron has and the greater its average distance from the nucleus.

Free stationary electrons are defined to have zero energy. Zero energy, in this case, is a reference value rather than an absolute value. A bound electron is held within an atom by attractive electrical force. Bound electrons are lower in energy than free stationary electrons, so they have negative energy values. Therefore the term "most stable" in this context corresponds to "most negative."

$n = 1$ $n = 2$

Sizes of atomic orbitals

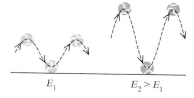

E_1 $E_2 > E_1$

Average heights of bouncing balls

Figure 6-17
The orbital size of an atomic electron is correlated with its orbital energy (*top*), just as the bounce height of a tennis ball is correlated with its kinetic energy (*bottom*).

Figure 6-18
Some everyday objects have their masses concentrated along preferred axes. A basketball has no such axes, but a football has one, and a tire iron has two.

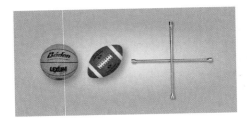

Summarizing, the principal quantum number (n) can have any positive integral value. It indexes the energy of the electron and is correlated with orbital size. As n increases, the energy of the electron increases, its orbital gets bigger, and the electron is less tightly bound to the atom.

When the Schrödinger equation for a one-electron atom is solved mathematically, the restrictions on n and l emerge as quantization conditions that correlate with energy and the shape of the wave function.

Azimuthal Quantum Number

In addition to size, an atomic orbital also has a specific shape. The solutions for the Schrödinger equation and experimental evidence show that orbitals have a variety of shapes. A second quantum number indexes the shapes of atomic orbitals. This quantum number is the **azimuthal quantum number (l)**.

Shapes of objects—such as the basketball, football, and tire iron shown in Figure 6-18—can be categorized according to their preferred axes. A basketball has no preferred axis because its mass is distributed equally in all directions about its center. A football has one preferred axis, with more mass along this axis than in any other direction. A tire iron has two preferred axes, at right angles to each other. In analogous fashion, electron density in an orbital can be concentrated along preferred axes.

The value of l correlates with the number of preferred axes in a particular orbital and thereby identifies the orbital shape. According to quantum theory, orbital shapes are highly restricted. These restrictions are linked to energy, so the possible values of l are limited by the value of the principal quantum number, n. The smaller n is, the more compact the orbital and the more restricted its possible shapes:

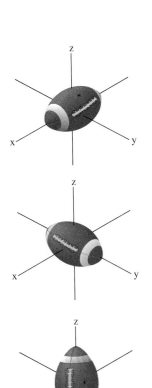

/// *The azimuthal quantum number (l) can be zero or any positive integer smaller than n: $l = 0, 1, 2, \ldots (n - 1)$.*

Historically, orbital shapes have been identified with letters rather than numbers. These letter designations correspond to the values of l as follows:

Value of l	0	1	2	3	4
Orbital letter	s	p	d	f	g

An orbital is named by listing the numerical value for n, followed by the letter that corresponds to the numerical value for l. Thus an electron with quantum numbers $n = 3$, $l = 0$ is a $3s$ orbital. A $5f$ orbital has $n = 5$, $l = 3$. Notice that the restrictions on l mean that many $n–l$ combinations do not correspond to orbitals that exist. For example, when $n = 1$, l can only be zero. In other words, $1s$ orbitals exist, but there are no $1p$, $1d$, $1f$ or $1g$ orbitals. Similarly, there are $2s$ and $2p$ orbitals but no $2d$, $2f$, or $2g$ orbitals. Remember that n restricts l, but l does not restrict n. Thus a $10d$ orbital ($n = 10$, $l = 2$) is rather high in energy but perfectly legitimate, but there is no orbital with $n = 2$, $l = 10$.

Magnetic Quantum Number

A sphere has no preferred axis, so it has no directionality in space. When there is a preferred axis, as for a football, Figure 6-19 shows that the axis can point in many

Figure 6-19
A football has directionality and shape. The figure shows three of the many ways in which a football can be oriented relative to a set of x-, y-, and z-axes.

different directions relative to an *xyz* coordinate system. Thus objects with preferred axes have directionality as well as shape.

Among atomic orbitals, *s* orbitals are spherical and have no directionality. Other orbitals are nonspherical, so in addition to having shape, every orbital points in some direction. Like energy and orbital shape, orbital direction is quantized. That is, *p, d,* and *f* orbitals have restricted orientations, whereas footballs are free to point in any direction. These restrictions are indexed by the **magnetic quantum number (m_l).**

Just as orbital size (*n*) limits the number of preferred axes (*l*), the number of preferred axes (*l*) limits the orientations of the preferred axes (m_l). When *l* = 0, there is no preferred axis and there is no orientation, so $m_l = 0$. One preferred axis (*l* = 1) can orient in any of three directions, giving three possible values for m_l: +1, 0, and −1. Each time *l* increases in value by one unit, two additional values of m_l become possible:

> /// *The magnetic quantum number (m_l) can have any positive or negative integral value between 0 and l:* $m_l = 0, \pm1, \pm2, \ldots \pm l$

The magnetic quantum number derives its name from the fact that different orbital orientations generate different behaviors in the presence of magnetic fields.

Spin Orientation Quantum Number

As described in Section 6-4, an electron has magnetism associated with a property called spin. Magnetism is directional, so the spin of an electron is directional, too. Spin orientation is quantized: Electron spin must be oriented in one of two ways, labeled "up" or "down." The **spin orientation quantum number (m_s)** indexes this behavior. The two possible values of m_s are $+\frac{1}{2}$ (up) and $-\frac{1}{2}$ (down).

A complete description of an atomic electron requires a set of four quantum numbers, *n, l,* m_l, and m_s, which must meet all the restrictions summarized in Table 6-2. Any set of quantum numbers that does not obey these restrictions does not correspond to an orbital and cannot describe an electron.

An atomic orbital is designated by its *n* and *l* values, such as 1*s*, 4*p*, 3*d*, and so on. For each type of orbital the two orientation quantum numbers, m_l and m_s, can have several different values. Thus there is more than one set of quantum numbers for each orbital designation. The 3*s* orbital, for example, has two valid sets of quantum numbers:

$$n = 3, l = 0, m_l = 0, m_s = +\tfrac{1}{2} \quad and \quad n = 3, l = 0, m_l = 0, m_s = -\tfrac{1}{2}$$

Spin "up" Spin "down"

As *n* increases, so does the number of valid sets of quantum numbers. Example 6-10 shows this.

Table 6-2
Restrictions on Quantum Numbers for Atoms

Quantum Number	Restrictions	Range
n	Positive integers	1, 2 … ∞
l	Positive integers less than *n*	0, 1 … (*n* − 1)
m_l	Integers between *l* and −*l*	−*l* … −1, 0, +1 … +*l*
m_s	Half-integers, $+\frac{1}{2}$ or $-\frac{1}{2}$	$-\frac{1}{2}, +\frac{1}{2}$

Example 6-10	Valid Quantum Numbers

Determine how many valid sets of quantum numbers exist for $4d$ orbitals, and give two examples.

Strategy: The question asks for the sets of quantum numbers that have $n = 4$ and $l = 2$. Each set must meet all the restrictions listed in Table 6-2. The easiest way to see how many valid sets there are is to list all the valid quantum numbers.

Solution: Because this is a $4d$ orbital, n and l are specified and cannot vary. The other two quantum numbers, however, have several acceptable values. For each value of m_l, either value of m_s is acceptable, so the total number of possibilities is the product of the number of possible values for each quantum number:

Quantum number	n	l	m_l	m_s
Possible values	4	2	2, 1, 0, -1, -2	$+\frac{1}{2}$, $-\frac{1}{2}$
Number of possible values	1	1	5	2

Possible sets of values for a $4d$ electron: $(1)(1)(5)(2) = 10$

There are ten sets of quantum numbers that describe a $4d$ orbital. Here are two of them, chosen randomly:

$$n = 4, l = 2, m_l = 1, \quad m_s = +\tfrac{1}{2}$$

$$n = 4, l = 2, m_l = -2, m_s = -\tfrac{1}{2}$$

You should be able to list the other eight sets.

Section Exercises

6.5.1 List all the valid sets of quantum numbers for $n = 3$. Give each its proper orbital name (for example, the orbital with $n = 1$, $l = 0$ is $1s$).

6.5.2 Which of the following sets of quantum numbers describe actual orbitals, and which are nonexistent? For each one that is nonexistent, list the restriction that makes it forbidden.

	n	l	m_l	m_s
(a)	4	1	1	0
(b)	4	4	1	$+\frac{1}{2}$
(c)	4	0	1	$+\frac{1}{2}$
(d)	4	2	2	$-\frac{1}{2}$

6.5.3 Determine the number of different allowable sets of quantum numbers that have $n = 4$.

6.6 SHAPES OF ATOMIC ORBITALS

Bound electrons are described by waves with characteristic shapes. Although most of the mass of an atom is concentrated in its nucleus, most of its volume comes from its electrons. The volume of an atom is "filled" with wave-particle electrons. This distribution can be described using the notion of electron density: Where electrons are most likely to be found, there is high electron density. The wave characteristics of electrons cause them to be smeared out rather than located at an exact position in space. Orbitals describe this delocalization. Each electron, rather than being a point charge, is a three-dimensional particle-wave that is distributed throughout space in an orbital. Moreover, when the energy of an electron changes, the size and shape of its distribution in space change as well.

/// Each atomic energy level can be associated with a specific three-dimensional atomic orbital.

An atom that contains many electrons can be described by superimposing (adding together) the orbitals for all of its electrons to obtain the overall size and shape of the atom.

The chemical properties of atoms are determined by the behavior of their electrons. Because atomic electrons are described by orbitals, the interactions of electrons can be described in terms of orbital interactions. The two characteristics of orbitals that determine how electrons interact are their shapes and their energies. Orbital shapes, the subject of this section, describe the distribution of electrons in three-dimensional space. Orbital energies, which we describe in Chapter 7, determine how easily electrons can be moved.

The quantum numbers n and l determine the size and shape of an orbital. As n increases, the size of the orbital increases, and as l increases, the shape of the orbital becomes more elaborate.

Orbital Depictions

Recall that an electron is a particle-wave delocalized in three-dimensional space. An orbital picture provides a map of how the electron wave is distributed in space.

There are several ways to represent these three-dimensional maps. Each one shows some important orbital features, but none shows all of them. We use three different representations: plots of electron density, pictures of electron density, and pictures of electron contour surfaces.

An **electron density plot** is useful because it represents the electron distribution in an orbital as a two-dimensional plot. These graphs show electron density along the y-axis and distance from the nucleus, r, along the x-axis. Figure 6-20a shows an electron density plot for the 2s orbital.

Electron density plots are useful because those for several orbitals can be superimposed to indicate the relative sizes of various orbitals. The simplicity of such a graph is also a drawback, however, because the three-dimensionality of an orbital has been reduced to a two-dimensional representation.

Orbital pictures have an advantage over electron density plots in that they can indicate the three-dimensional nature of orbitals. One type of orbital picture is a two-dimensional dot pattern in which the density of dots represents electron density. Such an **orbital density picture** of the 2s orbital is shown in Figure 6-20b. This two-dimensional pattern of dots shows a cross-sectional slice through the middle of the orbital.

Orbital density pictures are probably the most comprehensive views we can draw, but they require much time and care. An **electron contour drawing** provides a simplified orbital picture. In this representation, we draw a contour surface that encloses almost all the electron density. Commonly, "almost all" means 90%. Thus the electron density is high inside the contour surface but very low outside the surface. Figure 6-20c shows a contour drawing of the 2s orbital.

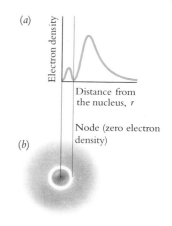

(a)

(b)

(c)

Figure 6-20
Different depictions of the 2s orbital. (*a*) A plot of electron density vs. distance from the nucleus. (*b*) An orbital density picture. (*c*) An electron contour drawing.

A useful analogy for understanding the value of contour surfaces is a swarm of bees around a hive. At any one time, some bees will be off foraging for nectar, so a contour surface drawn around *all* the bees might cover several acres. This would not be a very useful map of bee density. A contour surface containing 90% of the bees, on the other hand, would be just a bit bigger than the hive itself. This would be a very useful map of bee density, because anyone inside that contour surface would surely interact with bees.

The drawback of contour drawings is that all details of electron density *inside* the surface are lost. Thus if we want to convey the maximum information about orbitals, we must use combinations of the various types of depictions.

The advantages and disadvantages of the three types of plots are highlighted by how they show one characteristic feature of orbitals. Figure 6-20*a* shows clearly that there is a value for *r* where the electron density falls to zero. A place where electron density is zero is called a **node.** Figure 6-20*b* shows the node for the 2*s* orbital as a white ring. In three dimensions, this node is a spherical surface. Figure 6-20*c* does not show the node, because the spherical nodal surface is hidden inside the 90% contour surface. The graph shows the location of the node most clearly, the orbital density picture gives the best sense of the shape of the node, while the contour drawing fails to show it at all.

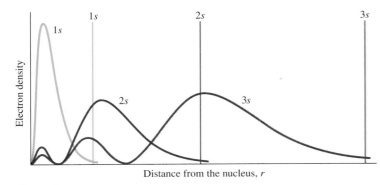

Figure 6-21
Electron density plots for the 1s, 2s, and 3s atomic orbitals of the hydrogen atom. The vertical lines indicate the values of r where the 90% contour surface would be located.

Figure 6-22
Electron density plots for the 3s (*maroon line*), 3p (*lavender line*), and 3d (*teal line*) orbitals for the copper atom. All three orbitals are nearly the same size.

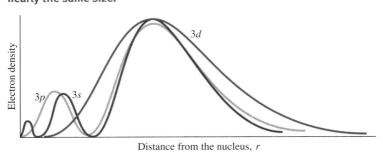

Orbital Size

How large are orbitals? Experiments that measure atomic radii provide information about the size of an orbital. Also, theoretical models of the atom predict how the electron density of a particular orbital changes with distance from the nucleus, *r*. When these sources of information are combined, they reveal several regular features about orbital size.

*In any particular atom, orbitals get larger as the value of **n** increases.* For any particular atom, the **n** = 2 orbitals are larger than the 1*s* orbital, the **n** = 3 orbitals are larger than the **n** = 2 orbitals, and so on. The electron density plots in Figure 6-21 show this trend for the first three *s* orbitals of the hydrogen atom. This plot also shows that the number of nodes increases as **n** increases.

In any particular atom, all orbitals with the same principal quantum number are similar in size. For example, Figure 6-22 shows that the **n** = 3 orbitals of the copper atom have their maximum electron densities at similar distances from the nucleus. The same regularity holds for all other atoms. The quantum numbers other than **n** affect orbital size only slightly. We describe these small effects in the context of orbital energies in Chapter 7.

Each orbital becomes smaller as nuclear charge increases. As the positive charge of the nucleus increases, the electrical force exerted by the nucleus on the negatively charged electrons increases, too, and electrons become more tightly bound. This in turn reduces the radius of the orbital. As a result, each orbital shrinks in size as atomic number increases. For example, the 2*s* orbital steadily decreases in size across the second row of the periodic table from Li ($Z = 3$) to Ne ($Z = 10$). The atomic number, Z, is equal to the number of protons in the nucleus, so increasing Z means increasing nuclear charge.

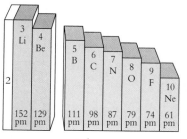

pm = picometers

Details of Orbital Shapes

The shapes of orbitals strongly influence chemical interactions. Hence, we need to have detailed pictures of orbital shapes to understand the chemistry of the elements.

The quantum number $l = 0$ corresponds to an s orbital. According to the restrictions on quantum numbers, there is only one s orbital for each value of the principal quantum number. All s orbitals are spherical, with radii and number of nodes that increase as n increases.

The quantum number $l = 1$ corresponds to a p orbital. A p electron can have any of three values for m_l, so for each value of n there are three different p orbitals. The p orbitals, which are not spherical, can be shown in various ways. The most convenient representation shows the three orbitals with identical shapes but pointing in three different directions. Figure 6-23 shows electron contour drawings of the $2p$ orbitals. Each p orbital has high electron density in one particular direction, perpendicular to the other two orbitals, with the nucleus at the center of the system. The three different orbitals can be represented so that each has its electron density concentrated on both sides of the nucleus along a preferred axis. We can write subscripts on the orbitals to distinguish the three distinct orientations: p_x, p_y, and p_z. Each p orbital also has a nodal plane that passes through the nucleus: The nodal plane for the p_x orbital is the yz plane, for the p_y orbital the nodal plane is the xz plane, and for the p_z orbital it is the xy plane.

As n increases, the detailed shapes of the p orbitals become more complicated (the number of nodes increases, just as for s orbitals). Nevertheless, the *directionality* of the orbitals does not change. Each p orbital is perpendicular to the other two in its set, and each p orbital has its lobes along its preferred axis, where electron density is high. To an approaching atom, therefore, an electron in a $3p$ orbital presents the same characteristics as one in a $2p$ orbital, except that the $3p$ orbital is bigger. Consequently, the shapes and relative orientations of the $2p$ orbitals in Figure 6-23 represent the prominent spatial features of all p orbitals.

The quantum number $l = 2$ corresponds to a d orbital. A d electron can have any of five values for m_l (-2, -1, 0, $+1$, and $+2$), so there are five different orbitals in each set. Each d orbital has two nodal planes. Consequently, the shapes

Tutorial

Figure 6-23
Contour drawings of the three $2p$ orbitals. The three orbitals have the same shape, but each is oriented perpendicular to the other two.

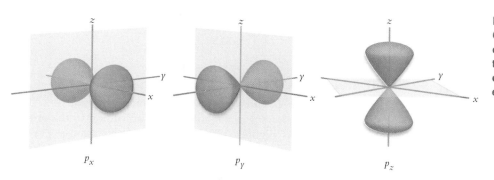

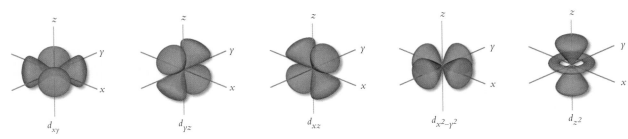

d_{xy} d_{yz} d_{xz} $d_{x^2-y^2}$ d_{z^2}

Figure 6-24
Contour drawings of the *d* orbitals. Each has two preferred axes.

Tutorial

The fifth *d* orbital has only one preferred axis, but it also has a "preferred ring" of significant electron density in the *xy* plane. This orbital has two nodes, each of which is a cone, one above and the other below the *xy* plane.

of the *d* orbitals are more complicated than their *s* and *p* counterparts. The most convenient way to show them is depicted by the contour drawings in Figure 6-24. In these drawings, three orbitals look like three-dimensional cloverleafs, each lying in a plane with the lobes pointed between the axes. A subscript is used to identify the plane in which each lies: d_{xy}, d_{xz}, and d_{yz}. A fourth orbital is also a cloverleaf in the *xy* plane, but its lobes point along the *x* and *y* axes. This orbital is designated $d_{x^2-y^2}$. The fifth orbital looks quite different. Its major lobes point along the *z* axis, but there is also a "doughnut" of electron density in the *xy* plane. This orbital is designated d_{z^2}.

The chemistry of all the common elements can be described completely using *s, p,* and *d* orbitals, so we need not extend our catalog of orbital shapes to the *f* orbitals and beyond.

Section Exercises

6.6.1 Construct contour drawings of *s, p,* and *d* orbitals. Label the coordinate axes.

6.6.2 Draw orbital pictures that show a 2*p* orbital on one atom interacting with a 2*p* orbital on a different atom: (a) end-on (preferred axes pointing toward each other) and (b) side by side (preferred axes parallel to each other).

6.6.3 Construct an accurately scaled composite contour drawing on which you superimpose a 2*s* orbital, a $2p_x$ orbital, and the outermost portion of a $3p_x$ orbital of the same atom. (Use different colors to distinguish the different orbitals.) What does your contour drawing tell you about the importance of the *n* = 2 orbitals for chemical interactions when the $3p_x$ orbital contains electrons?

6.7   SUNLIGHT AND THE EARTH

The sun bathes the Earth in a never-ending flux of electromagnetic radiation. This light interacts with atoms and molecules in the atmosphere according to the principles developed in this chapter. As Figure 6-25 shows, the Earth's atmosphere can be divided into regions that blend into one another without distinct boundaries.

Each region has its own distinguishing chemistry that arises from the interaction of light with the atoms and molecules of the atmosphere.

Reactions in the Thermosphere

The **thermosphere** is the outermost part of the atmosphere, above an altitude of about 85 km. Here, molecules of nitrogen and oxygen absorb X-ray radiation coming from the sun. These photons have enough energy to ionize molecules and to break chemical bonds. Here are two examples:

$$N_2 + h\nu \longrightarrow N_2^+ + e^- \qquad O_2 + h\nu \longrightarrow O + O$$

The products of these reactions are unstable, so they eventually recombine, releasing energy in the form of heat:

$$N_2^+ + e^- \longrightarrow N_2 + \text{Heat} \qquad O + O \longrightarrow O_2 + \text{Heat}$$

The thermosphere, which gets its name from the heat released in these reactions, is a complex mixture of atoms, ions, and molecules, including a high mole fraction of oxygen atoms. At the same time, however, the total atmospheric pressure in the thermosphere is less than 10^{-7} atm, which means that the density of atoms and molecules in the thermosphere is quite low.

The aurora borealis, a spectacular atmospheric light show shown in Figure 6-26, originates in the thermosphere. In addition to electromagnetic radiation, the sun emits a steady stream of protons and electrons. The Earth's magnetic field deflects most of these particles, but some reach the thermosphere above the north and south poles of the planet, particularly during times of solar storms when their emission from the sun is at its highest. As they enter the atmosphere, these high-energy particles collide with the atoms and molecules in the thermosphere, transferring energy and generating highly excited states. As these excited atoms and molecules relax back to their ground states, they emit visible light. The color of the emission depends on the species involved. For example, nitrogen molecules emit red light during an aurora event. Oxygen atoms can emit either red or green light, and nitrogen atoms emit blue light. In other words, the aurora is based on the same principles as emission spectra: Atoms and molecules that reach excited states by gaining energy from some external source return to their ground states by emitting photons.

As solar radiation passes through the thermosphere, its highest-energy photons are removed progressively through absorption by atoms and molecules. The result is that the intensity of high-energy light decreases as sunlight moves down through the atmosphere toward the Earth's surface. At the bottom of the thermosphere, about 85 km above the Earth's surface, almost all the ionizing radiation has been removed. This marks the boundary between the thermosphere and the mesosphere. Sunlight not absorbed in the thermosphere passes through the mesosphere, which is about 35 km thick, with little absorption.

Reactions in the Ozone Layer

Near the top of the **stratosphere,** solar radiation generates an abundance of ozone (O_3) molecules that has important consequences for life on Earth. These molecules

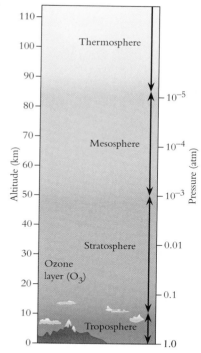

Figure 6-25
The Earth's atmosphere is divided into several regions. Note that the scale for pressure is logarithmic.

Figure 6-26
The aurora borealis is due to the emission of photons by excited state atoms and molecules in the thermosphere. In the northern hemisphere the aurora is called the Northern Lights.

form in two steps. First, a photon with a wavelength between 180 and 240 nm breaks an O_2 molecule into two atoms of oxygen:

$$O_2 + h\nu_{(\lambda\,=\,180-240\ \text{nm})} \longrightarrow O + O$$

An oxygen molecule captures one of these oxygen atoms to form an ozone molecule:

$$O_2 + O \longrightarrow O_3 + \text{Heat}$$

The second step occurs twice for each O_2 fragmentation, giving the overall balanced process for ozone formation:

$$3\ O_2 + h\nu_{(\lambda\,=\,180-240\ \text{nm})} \longrightarrow 2\ O_3 + \text{Heat}$$

The production of ozone requires both a source of oxygen atoms and frequent collisions between the atoms and the molecules that make up the atmosphere. A high fraction of the atmosphere is oxygen atoms at high altitudes, but not enough collisions occur to form ozone in significant amounts. This is because at this high altitude the pressure, the molecular density and the rate of molecular collisions are very low. Below 20 km, on the other hand, all the light energetic enough to split oxygen molecules into oxygen atoms has already been absorbed. Consequently, below this altitude, insufficient oxygen atoms are present to generate ozone. Therefore the **ozone layer** is between 20 and 35 km above the Earth's surface.

Ozone strongly absorbs UV light in the 200- to 340-nm region. The energies of these photons are high enough to break ozone apart into O_2 molecules and oxygen atoms:

$$O_3 + h\nu_{(\lambda\,=\,200-340\ \text{nm})} \longrightarrow O + O_2$$

The interactions of molecules and light in the ozone layer result in a delicate balance that holds ozone concentration at a relatively constant value. This balance is maintained by the three reactions already mentioned:

1. Photons with wavelengths of 180 to 240 nm break apart O_2 molecules.
2. Photons with wavelengths of 200 to 340 nm break apart O_3 molecules.
3. Oxygen atoms combine with O_2 molecules to produce O_3 molecules and heat.

In Chapter 14, we examine the ozone layer in more detail and discuss how this delicate balance is being endangered by human activities.

The absorption of UV light by ozone and oxygen molecules is critical for life on Earth. If this radiation reached the Earth's surface, it would cause severe biological damage because light of wavelengths around 300 nm has enough energy to break biological molecules apart. If we were continually bathed in such light, death rates would increase dramatically for almost all living species. Our Chemistry and the Environment Box explores the effects of ultraviolet light.

The Greenhouse Effect

The energy balance between light absorbed and emitted determines the Earth's temperature. The sunlight that reaches the Earth's surface is in the ultraviolet, visible, and infrared (IR) regions. The Earth's surface absorbs almost all of this sunlight, which heats the Earth's surface to comfortable temperatures, heats and evaporates water from the oceans that later falls as rain or snow, and drives photosynthesis, the chemical engine of life. Warm bodies emit radiation, and the Earth is

U ltraviolet photons have enough energy to break chemical bonds. When UV light breaks bonds in biochemical molecules, the products can undergo chemical reactions that lead to cell damage.

The sunlight that normally reaches the Earth's surface has very little UV light with wavelengths shorter than about 360 nm. As the diagram shows, photons with shorter wavelengths are absorbed in the thermosphere and stratosphere.

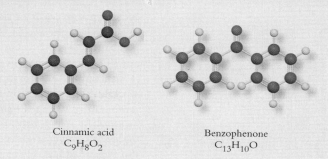

Cinnamic acid
$C_9H_8O_2$

Benzophenone
$C_{13}H_{10}O$

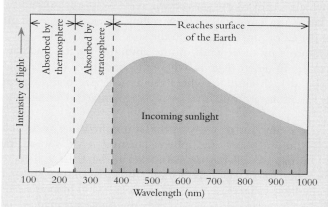

Ultraviolet light in the 360- to 400-nm range reaches the Earth's surface. This long-wavelength UV light, while not as damaging as shorter-wavelength radiation, nevertheless can cause damage to skin cells. Our bodies have developed a mechanism to protect us from such damage. When exposed to sunlight, skin produces melanin molecules, which absorb UV light and convert its energy into heat. Every photon that is absorbed by a melanin molecule is prevented from being absorbed by some other component of skin.

While melanin provides some protection from the damaging effects of ultraviolet light, studies have shown that prolonged exposure to sunlight increases the risk that skin cells will become cancerous. Skin cancer is much more prevalent among people who have a long history of exposure to the sun. Fair-skinned people are more prone to skin cancer, because their bodies produce less melanin and therefore less protection from the damage caused by UV light.

Sun worshipers and those who work outside can protect themselves from ultraviolet exposure with skin creams that contain UV-absorbing molecules. The UV-absorbing molecule in the first sunscreens was PABA, para-aminobenzoic acid. However, this compound may have toxic effects of its own. Current sunscreens contain a range of chemical ingredients, the UV-absorbing components being derivatives of cinnamic acid and benzophenone.

Ultraviolet light is also harmful to the eyes. Although our eyes do not detect UV photons, the lens of the eye is particularly susceptible to damage by UV light. The lens focuses visible photons, but it does not absorb them. On the other hand, the lens does absorb ultraviolet photons, causing cell damage that eventually can result in the formation of cataracts. Optometrists and ophthalmologists recommend that eyeglasses contain UV-absorbers to protect the eyes from exposure to ultraviolet light.

In recent years, the amount of ozone in the stratosphere has diminished through reactions with human-generated pollutants, as we describe in Chapter 14. As the concentration of ozone in the stratosphere falls, the amount of ultraviolet light that reaches the Earth's surface increases. There is great concern about the biological effects of this higher-energy UV light. In the lower southern hemisphere, where ozone depletion has been greatest, the incidence of eye damage among sheep has shown a sharp increase in recent years, perhaps because of increased UV exposure.

Sunlight is not the only source of UV light. Ultraviolet light is emitted by mercury vapor black lights. These high-energy UV photons can be absorbed by nearby atoms and molecules. The electrons excited by this UV absorption often lose energy by emitting visible light. For example, metal ions in certain minerals absorb UV photons and emit visible light, as shown in the photo.

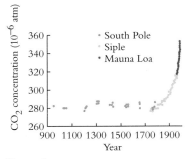

Figure 6-27
The concentration of CO_2 in the atmosphere has increased dramatically in the course of the last century, as humans have increased their consumption of fossil fuels.

no exception. Most of the radiation emitted by the Earth lies in the infrared region of the spectrum.

Nitrogen and oxygen, the major components of the atmosphere, are transparent to infrared radiation. However, some trace gases in the **troposphere,** the layer of atmosphere closest to the surface of the Earth, are strong IR absorbers. Foremost among these are carbon dioxide, water vapor, and methane. These gases moderate temperature changes from day to night by absorbing some of the infrared photons emitted by the Earth. Following absorption, the gases re-emit still longer wavelength photons, some of which return to the surface where they are reabsorbed by the planet. This process keeps the temperature from falling dramatically at night when the ground is no longer absorbing energy from the sun.

The degree of temperature control depends on the amount of water and carbon dioxide in the atmosphere. In desert regions, where the amount of H_2O in the atmosphere is very low, and at high altitude, where the atmosphere is less dense, day-to-night temperature variations are much more severe than at lower altitude and higher humidity.

Over the past century, the amount of CO_2 in the atmosphere has increased dramatically, as Figure 6-27 shows. Many atmospheric chemists have warned for years that this increase is causing a small but dramatic rise in the Earth's average temperature. In 1995, the report of the United Nations Intergovernmental Panel on Climate Change documented an increase in average global air temperature of between 0.3 and 0.6 °C over the past century. More dramatically, an increase of between 1.0 and 3.5 °C is predicted for the current century.

A temperature change of 3 °C may seem trivial, but it has been estimated that the average global temperature during the last major ice age was only a few degrees lower than it is today. Moreover, computer models predict that increased average temperatures lead to increased heat waves and more severe weather patterns such as hurricanes. Additional predicted consequences include a rise in sea levels that could inundate coastal urban centers and disturbances of the ecological balance in many parts of the globe.

Among the components of our atmosphere, the concentration of carbon dioxide is a mere 325 parts per million (ppm). In other words, 999,675 of every million molecules in the air are *not* CO_2. (Almost all the molecules are N_2 or O_2.) At such a low concentration, how could CO_2 possibly cause a measurable change in the Earth's surface temperature? The answer lies in the role that minor atmospheric species play in the global energy balance.

The Earth's temperature is determined by the balance between solar energy absorbed by the planet and the energy lost through infrared radiation. Even a small change in the concentration of CO_2 in the atmosphere can alter this energy balance. Carbon dioxide does not affect the energy *input* to the planet because CO_2 is transparent to most of the incoming solar radiation. In contrast, CO_2 is extremely effective at absorbing infrared radiation, so the energy *output* from the planet decreases when the amount of carbon dioxide in the atmosphere rises by even a small amount. In this respect, CO_2 acts like a glass windowpane in a greenhouse, letting in almost all the solar energy but recapturing a significant amount of the infrared energy emitted by the Earth. A little CO_2 goes a long way. Figure 6-28 shows this process schematically.

At present, the atmospheric CO_2 content is increasing by about 1.5 ppm/year. Moreover, concentrations of other so-called **greenhouse gases** such as methane and N_2O have also increased during the industrial age. Methane has more than doubled, CO_2 has increased by 30%, and N_2O has increased by 15%. The presence of each of these gases in the atmosphere means that less energy escapes from the

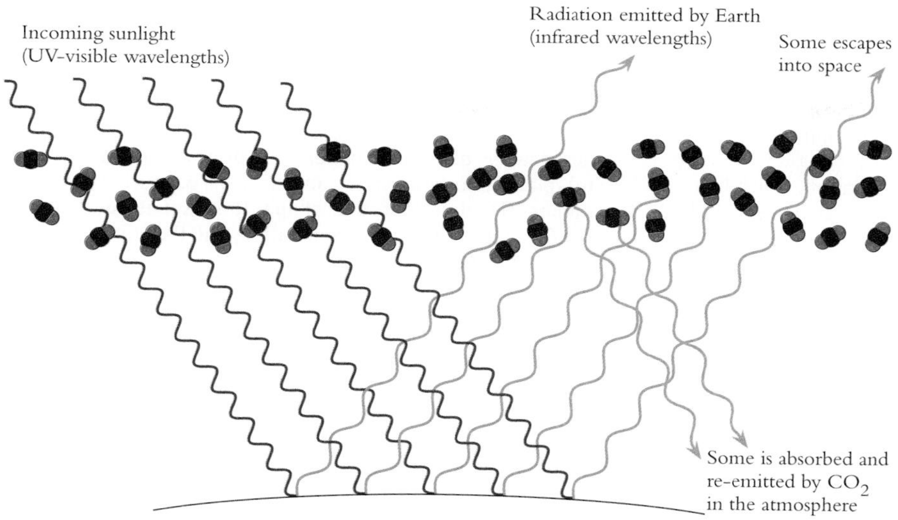

Figure 6-28
The role of atmospheric CO_2 in the greenhouse effect. Carbon dioxide is transparent to incoming sunlight, but it absorbs and re-emits a significant amount of the infrared radiation emitted by the Earth. This alters Earth's energy balance, raising its average temperature.

Earth's surface, causing the surface to heat up until the amount of radiant energy that escapes once again balances the incoming amount of solar energy.

The increase in greenhouse gases over the past century is a documented reality, but several uncertainties make it difficult to say how much the greenhouse effect will change Earth's climate. For example, the CO_2 buildup may not continue at the accelerating pace of recent times, because the biosphere and the oceans may be able to absorb much of the additional CO_2 that results from human activities. Whether or not increased greenhouse gases will inevitably result in global warming is also controversial because the Earth's temperature slowly changes as a result of other factors. The ice ages demonstrate, for example, that the Earth undergoes significant warming and cooling cycles without human intervention. Nevertheless, the human species is running a huge meteorological experiment on our planet, the long-term outcome of which might be very damaging. By the time we know the results of this experiment, it may be too late to do anything about it.

Section Exercises

■ **6.7.1** Volcanic eruptions change the Earth's energy balance by adding large amounts of smoke particles to the troposphere and stratosphere. Spectacular red sunsets result, because these particles scatter shorter-wavelength light so that less of it reaches the Earth's surface. What effect does this have on the Earth's average temperature? Draw a graph similar to the one in the Box to support your answer. On this graph, show the effect on incoming sunlight and the resulting shift, if any, in outgoing radiation.

■ **6.7.2** The ozone cycle is a delicate balance described by the 3 equations mentioned in the text. Combine these 3 equations in appropriate numbers with cancellation of common species to show that the overall ozone cycle converts UV light into heat:

$$h\nu_{(\lambda = 180 - 340 \text{ nm})} \longrightarrow \text{Heat}$$

(Hint: You need to show one of the three reactions running backward.)

CHAPTER REVIEW

Summary and Key Terms

1. Atoms possess mass, occupy volume, and contain electrons and nuclei. They attract one another and can combine to form molecules.

2. **Light** is one form of **electromagnetic radiation,** which has wave properties characterized by **frequency, wavelength,** and **amplitude** or **intensity.** Light moves in vacuum with a constant speed, *c*. The **photoelectric effect** shows that light also behaves like particles of individual **photons,** each with energy given by $E = h\nu$, where *h* is **Planck's constant.** Atoms can absorb light and increase in energy from their **ground states** to **excited states,** and excited states can emit light and decrease in energy. An energy level diagram summarizes these changes in atomic energies.

3. Each element has a characteristic **absorption spectrum** and **emission spectrum** that are plots of the intensities of light absorbed or emitted as a function of wavelength. Because energy levels are **quantized,** these spectra show sharp lines at specific energies. The energy levels of the hydrogen atom can be described by a simple mathematical formula.

4. All electrons have the same mass and charge, possess **spin** which gives them magnetism, and have **wave properties** that are described by the de Broglie equation. The **uncertainty principle,** that position and motion cannot be simultaneously determined, is a consequence of the wave properties.

5. Bound electrons are delocalized in **orbitals,** and their energies are **quantized.** Wave functions for atomic electrons are characterized by **quantum numbers.** The **principal quantum number** (*n*) relates to orbital energy and size. The **azimuthal quantum number** (*l*) relates to orbital shape. The **magnetic quantum number** (*m_l*) relates to spatial orientation, and the **spin orientation quantum number** (*m_s*) relates to spin orientation. Each quantum number can take on a restricted set of values.

6. Orbital shapes can be depicted by **electron density plots, orbital density pictures,** and **electron contour drawings.** Orbitals have locations of zero electron density, called **nodes.** Orbital size depends on *n* and **Z,** and shape depends on *l*.

7. Light from the sun provides the energy that warms the Earth. The Earth also gives off energy in the form of light. High in the **thermosphere,** ultraviolet light breaks molecules apart. At the top of the **stratosphere,** interactions of photons with oxygen molecules creates the **ozone layer,** which absorbs ultraviolet light that would have lethal consequences if it reached the Earth's surface. Water vapor and CO_2 in the **troposphere** act as greenhouse gases, keeping the Earth's surface warm by absorbing and re-emitting infrared light. Human activities have increased the concentration of carbon dioxide in the atmosphere, which may lead to global warming.

Skills to Master

▶ Calculating photon energies, wavelengths, and frequencies

▶ Sketching and interpreting energy level diagrams

▶ Calculating energies of hydrogen atoms

▶ Analyzing the photoelectric effect

▶ Using the de Broglie equation

▶ Determining sets of quantum numbers

▶ Drawing shapes of *s, p,* and *d* orbitals

▶ Understanding how sunlight interacts with the Earth

Learning Exercises

6.1 List the properties of electrons and of photons, including the equations used to describe each.

6.2 Write a short description of (a) the photoelectric effect, (b) wave-particle duality, (c) electron spin, and (d) the uncertainty principle.

6.3 Describe an atomic energy level diagram and the information it incorporates.

6.4 Update your list of memory bank equations. For each new equation, specify the conditions under which it can be used.

6.5 Make a list of all terms in this chapter that are new to you. Write a one-sentence definition of each in your own words. Consult the glossary if you need help.

Problems

Characteristics of Atoms

6.1 The density of silver is 1.050×10^4 kg/m³, and the density of lead is 1.134×10^4 kg/m³. For each metal, (a) calculate the volume per atom; (b) estimate the atomic diameter; and (c) using this estimate, calculate the thickness of a metal foil containing 6.5×10^6 atomic layers of the metal.

6.2 The density of aluminum metal is 2.700×10^3 kg/m³, and the density of copper is 8.960×10^3 kg/m³. For each metal, (a) calculate the volume per atom; (b) estimate the atomic diameter; and (c) using this estimate, calculate the thickness of a metal foil containing 6.5×10^6 atomic layers of the metal.

6.3 Describe evidence that indicates that atoms have mass.

6.4 Describe evidence that indicates that atoms have volume.
6.5 Draw an atomic picture of a layer of aluminum metal atoms.
6.6 Stainless steel is iron containing a small amount of carbon, with the relatively small carbon atoms occupying "holes" between larger iron atoms. Draw a picture that shows what a layer of stainless steel looks like on the atomic level.

Characteristics of Light

6.7 Convert the following wavelengths into frequencies (Hz, using power-of-ten notation): (a) 4.33 nm; (b) 2.35×10^{-10} m; (c) 735 mm; (d) 4.57 μm.
6.8 Convert the following wavelengths into frequencies (Hz, using power-of-ten notation): (a) 2.76 km; (b) 1.44 cm; (c) 3.77×10^{-7} m; (d) 348 nm.
6.9 Convert the following frequencies into wavelengths, expressing the result in the indicated units: (a) 4.77 GHz (m); (b) 28.9 kHz (cm); (c) 60 Hz (mm); (d) 2.88 MHz (μm).
6.10 Convert the following frequencies into wavelengths, expressing the result in the indicated units: (a) 2.77 MHz (mm); (b) 90.1 kHz (m); (c) 50 Hz (km); (d) 8.88 GHz (μm).
6.11 Calculate the energy in joules per photon and in kilojoules per mole of the following: (a) blue-green light with a wavelength of 490.6 nm; (b) X rays with a wavelength of 25.5 nm; and (c) microwaves with a frequency of 2.5437×10^{10} Hz.
6.12 Calculate the energy in joules per photon and in kilojoules per mole of the following: (a) red light with a wavelength of 665.7 nm; (b) infrared radiation whose wavelength is 1255 nm; and (c) ultraviolet light with a frequency of 4.5528×10^{15} Hz.
6.13 A nitrogen laser puts out a pulse containing 10 mJ of energy at a wavelength of 337.1 nm. How many photons is this?
6.14 A dye laser emits a pulse at 450 nm that contains 2.75×10^{15} photons. What is the energy content of this pulse?
6.15 What are the wavelength and frequency of photons with the following energies: (a) 745 kJ/mol; (b) 3.55×10^{-19} J/photon?
6.16 What are the wavelength and frequency of photons with the following energies: (a) 355 J/mol; (b) 2.50×10^{-18} J/photon?
6.17 When light of frequency 1.30×10^{15} s^{-1} shines on the surface of cesium metal, electrons are ejected with a maximum kinetic energy of 5.2×10^{-19} J. Calculate (a) the wavelength of this light; (b) the binding energy of electrons to cesium metal; and (c) the longest wavelength of light that will eject electrons.
6.18 The binding energy of electrons to chromium metal is 7.21×10^{-19} J. Calculate (a) the longest wavelength of light that will eject electrons from chromium metal; (b) the frequency required to give electrons with kinetic energy of 2.5×10^{-19} J; and (c) the wavelength of the light in part (b).
6.19 Draw energy level diagrams that illustrate the difference in electron binding energy between cesium metal and chromium metal. Refer to Problems 6.17 and 6.18.
6.20 A phototube delivers an electrical current when a beam of light strikes a metal surface inside the tube. Phototubes do not respond to infrared photons. Draw an energy level diagram for electrons in the metal of a phototube and use it to explain why phototubes do not respond to infrared light.
6.21 Refer to Figure 6-4 to answer the following questions: (a) What is the wavelength range for radio waves? (b) What color is light whose wavelength is 5.8×10^{-7} m? (c) In what region does radiation with frequency of 4.5×10^8 Hz lie?

6.22 Refer to Figure 6-4 to answer the following questions: (a) What is the wavelength range for infrared radiation? (b) What color is light whose wavelength is 4.85×10^{-7} m? (c) In what region does radiation with frequency of 4.5×10^{18} Hz lie?

Absorption and Emission Spectra

6.23 From Figure 6-15, calculate the energy difference in kilojoules per mole between the excited state of mercury that emits 404-nm light and the ground state.
6.24 From Figure 6-15, determine the wavelength of light needed to excite an electron from the ground state to the lowest excited state of the mercury atom.
6.25 Determine the wavelengths that hydrogen atoms absorb to reach the $n = 8$ and $n = 9$ states from the ground state. In what region of the electromagnetic spectrum do these photons lie?
6.26 Determine the frequencies that hydrogen atoms emit in transitions from the $n = 6$ and $n = 5$ levels to the $n = 3$ level. In what region of the electromagnetic spectrum do these photons lie?
6.27 Using Figure 6-14, explain why more lines appear in emission spectra than in absorption spectra.
6.28 Using Figure 6-15, explain why mercury emits photons at longer wavelengths than the wavelengths that mercury absorbs.

Properties of Electrons

6.29 What is the mass of one mole of electrons?
6.30 What is the charge of one mole of electrons?
6.31 Determine the wavelengths of electrons with the following kinetic energies: (a) 1.15×10^{-19} J; (b) 3.55 kJ/mol; (c) 7.45×10^{-3} J/mol.
6.32 Determine the wavelengths of electrons with the following kinetic energies: (a) 76.5 J/mol; (b) 4.77×10^{-18} J; (c) 3.21×10^{-11} J.
6.33 Determine the kinetic energies of electrons with the following wavelengths: (a) 3.75 nm; (b) 4.66 m; (c) 8.85 mm.
6.34 Determine the kinetic energies of electrons with the following wavelengths: (a) 3.75 m; (b) 4.66 μm; (c) 2.85 mm.

Quantization and Quantum Numbers

6.35 List all the valid sets of quantum numbers for a $6p$ electron.
6.36 List all the valid sets of quantum numbers for a $4f$ electron.
6.37 If an electron has $n = 3$, list the restrictions on its other quantum numbers.
6.38 If an electron has $m_l = -2$, list the restrictions on its other quantum numbers.
6.39 For the following sets of quantum numbers, determine which describe actual orbitals and which are nonexistent. For each one that is nonexistent, list the restriction that forbids it:

	n	l	m_l	m_s
(a)	5	3	-2	-1
(b)	5	3	-3	$1/2$
(c)	3	3	-3	$1/2$
(d)	3	0	0	$-1/2$

6.40 For the following sets of quantum numbers, determine which describe actual orbitals and which are nonexistent. For each one that is nonexistent, list the restriction that forbids it:

	n	l	m_l	m_s
(a)	3	-1	-1	$\frac{1}{2}$
(b)	3	1	-1	$-\frac{1}{2}$
(c)	3	1	2	$\frac{1}{2}$
(d)	3	2	2	$\frac{1}{2}$

6.41 List the values for the quantum numbers for a $3d$ electron that has spin up and the largest possible value for its magnetic quantum number.

6.42 List the values for the quantum numbers for a $4p$ electron that has spin down and the most negative possible value for its magnetic quantum number.

Shapes of Atomic Orbitals

6.43 Refer to Figures 6-20 and 6-23. Draw the analogous set of three depictions for an orbital that has $n = 2, l = 1$.

6.44 Refer to Figures 6-20 and 6-22. Draw the analogous set of three depictions for an orbital that has $n = 3, l = 0$.

6.45 Draw pictures showing how the p_y orbital looks when viewed along the z-axis, the y-axis, and the x-axis (some views may look the same).

6.46 Draw pictures showing how the d_{xz} orbital looks when viewed along the z-axis, the y-axis, and the x-axis (some views may look the same).

6.47 Identify each of the following orbitals, and state the values for n and l for each. (Hint: Use the size of the $3s$ orbital to identify the other three.)

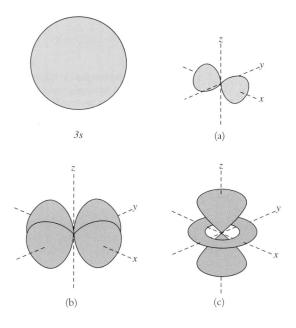

6.48 Shown below are electron density pictures and electron density plots for the $1s$, $2s$, $2p$, and $3p$ orbitals. Assign the various depictions to their respective orbitals.

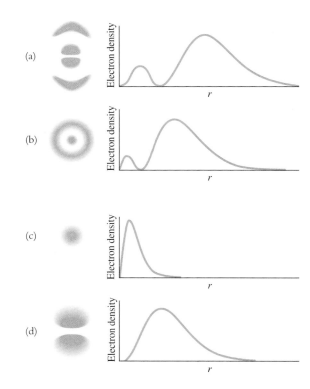

6.49 What are the limitations of plots of electron density vs. r?

6.50 The conventional method of showing the three-dimensional shape of an orbital is an electron contour surface. What are the limitations of this representation?

6.51 Construct contour drawings for the orbitals graphed in Figure 6-21, appropriately scaled to illustrate the size differences among these orbitals.

6.52 Draw contour drawings for the orbitals graphed in Figure 6-22, illustrating the shape differences among these orbitals.

Sunlight and the Earth

6.53 When molecules of nitrogen and oxygen in the thermosphere absorb short-wavelength light, N_2 molecules ionize and O_2 molecules break into atoms. Draw molecular pictures that illustrate these processes.

6.54 When ozone molecules in the mesosphere absorb UV light, they fragment into oxygen atoms and oxygen molecules. Draw a molecular picture that illustrates this process.

6.55 Light of wavelength 340 nm or shorter is required to fragment ozone molecules. What is the minimum energy in kJ/mol for this process? If an ozone molecule absorbs a 250-nm photon, how much excess kinetic energy will the fragments possess?

6.56 It requires 496 kJ/mol to break O_2 molecules into atoms and 945 kJ/mol to break N_2 molecules into atoms. Calculate the maximum wavelengths of light that can break these molecules apart. What part of the electromagnetic spectrum contains these photons?

6.57 List the region of the atmosphere and the atmospheric gases that absorb light in each of the following spectral regions: (a) less than 200 nm; (b) 240 to 310 nm.

6.58 List the region of the atmosphere and the atmospheric gases that absorb light in each of the following spectral regions: (a) 200 to 240 nm; (b) 700 to 2000 nm.

6.59 A high-altitude balloon equipped with a transmitter and pressure sensor reports a pressure of 10^{-3} atm (see Fig. 6-25). (a) What altitude has the balloon reached? (b) In what region of the atmosphere is the balloon? (c) What chemical processes take place in this region?

6.60 At one stage of the return to Earth of a space shuttle flight, its instruments report that the atmospheric pressure is 10^{-6} atm (see Fig. 6-25). (a) What is the altitude? (b) In what region of the atmosphere is the shuttle? (c) What chemical processes take place in this region?

Additional Paired Problems

6.61 Calculate each of the following for a photon of frequency 4.5×10^{13} Hz which is reflected off the moon: (a) its wavelength; (b) its energy; and (c) how long it takes to reach the Earth, which is 2.86×10^5 miles from the moon.

6.62 Calculate the following for a photon ($\lambda = 525$ nm) emitted by the sun: (a) its frequency; (b) its energy; and (c) the time it takes to reach the Earth, which is 93 million miles from the sun.

6.63 It requires 243 kJ/mol to fragment Cl_2 molecules into Cl atoms. What is the longest wavelength (in nm) of sunlight that could accomplish this? Will Cl_2 molecules in the troposphere fragment?

6.64 It requires 364 kJ/mol to break the chemical bond in HBr molecules. What is the longest wavelength (in nm) of light that has enough energy to cause this bond to break? Will HBr molecules in the troposphere be fragmented into atoms?

6.65 How many sets of quantum number values are there for a $4p$ electron? List them.

6.66 How many sets of quantum number values are there for a $3d$ electron? List them.

6.67 Redraw the first light wave in Figure 6-3 to show a wave whose frequency is twice as large as that shown in the figure, but whose amplitude is the same.

6.68 Redraw the first light wave in Figure 6-3 to show a wave whose wavelength and amplitude are each twice as large as that shown in the figure.

6.69 It requires a minimum of 216.4 kJ/mol to remove an electron from a potassium metal surface. What is the longest wavelength of light that can do this?

6.70 It requires a minimum of 216.4 kJ/mol to remove an electron from a potassium metal surface. If UV light at 255 nm strikes this surface, what is the maximum speed of the ejected electrons?

6.71 In a photoelectric effect experiment, light of energy 6.00×10^{-19} J is absorbed by a metal, and the ejected electrons have $kE_{max} = 2.70 \times 10^{-19}$ J. Calculate (a) the binding energy of electrons in the metal; (b) the wavelength of the light; and (c) the wavelength of the electrons.

6.72 In a photoelectric effect experiment, the minimum frequency needed to eject electrons from a metal is 7.5×10^{14} s^{-1}. Suppose that a 366-nm photon from a mercury discharge lamp strikes the metal. Calculate (a) the binding energy of the electrons in the metal; (b) the maximum kinetic energy of the ejected electrons; and (c) the wavelength associated with those electrons.

6.73 One frequency of a CB radio is 27.3 MHz. Calculate the wavelength and energy of photons at this frequency.

6.74 Microwave ovens use radiation whose wavelength is 12.5 cm. What is the frequency and energy in kJ/mol of this radiation?

6.75 Refer to Figure 6-25 to answer the following questions: (a) What is the pressure at an altitude of 60 km? (b) What atomic and molecular species are present at that altitude? (c) At what altitude is the pressure 8 torr? (d) What region of the atmosphere is this?

6.76 Refer to Figure 6-25 to answer the following questions: (a) What is the pressure at an altitude of 100 km? (b) Describe the chemistry that takes place at that altitude. (c) What are the atomic and molecular species present at that altitude?

6.77 Barium salts in fireworks generate a yellow-green color. Ba^{2+} ions emit light with $\lambda = 487$, 514, 543, 553, and 578 nm. Convert these wavelengths into frequencies and into energies in kJ/mol.

6.78 The bright-red color of highway safety flares comes from strontium ions in salts such as $Sr(NO_3)_2$ and $SrCO_3$. Burning a flare produces strontium ions in excited states, which emit red photons at 606 nm and several wavelengths between 636 and 688 nm. Calculate the frequency and energy (kJ/mol) of 606-, 636-, and 688-nm emissions.

6.79 A hydrogen atom emits a photon as its electron changes from $n = 5$ to $n = 1$. What is the wavelength of the photon? In what region of the electromagnetic spectrum is this photon found?

6.80 A hydrogen atom emits a photon as its electron changes from $n = 7$ to $n = 3$. What is the wavelength of the photon? In what region of the electromagnetic spectrum is this photon found?

More Challenging Problems

6.81 The photoelectric effect for magnesium metal has a threshold frequency of 8.95×10^{14} s^{-1}. Can Mg be used in photoelectric devices that sense visible light? Do a calculation in support of your answer.

6.82 Energetic free electrons can transfer their energy to bound electrons in atoms. In 1913, James Franck and Gustav Hertz passed electrons through mercury vapor at low pressure to determine the minimum kinetic energy required to produce the excited state that emits ultraviolet light at 253.7 nm. What is that minimum kinetic energy? What wavelength is associated with electrons of this energy?

6.83 The radius of a typical atom is 10^{-10} m, and the radius of a typical nucleus is 10^{-15} m. Compute typical atomic and nuclear volumes and determine what fraction of the volume of a typical atom is occupied by its nucleus.

6.84 Neutrons, like electrons and photons, are particle-waves whose diffraction patterns can be used to determine the structures of molecules. Calculate the kinetic energy of a neutron with a wavelength of 75 pm.

6.85 The human eye can detect as little as 2.35×10^{-18} J of green light of wavelength 510 nm. Calculate the minimum number of photons that can be detected by the human eye.

6.86 Gaseous lithium atoms absorb light of wavelength 323 nm. The resulting excited lithium atoms lose some energy through collisions with other atoms. The atoms then return to their ground state by emitting two photons with $\lambda = 812.7$ and 670.8 nm. Draw an energy level diagram that shows this process. What fraction of the energy of the absorbed photon is lost in collisions?

6.87 Calculate the wavelengths associated with an electron and a proton, each traveling at 5.000% of the speed of light.

6.88 One hydrogen emission line has a wavelength of 486 nm. Identify the values for n_{final} and $n_{initial}$ for the transition giving rise to this line.

6.89 An atomic energy level diagram, shown to scale, follows:

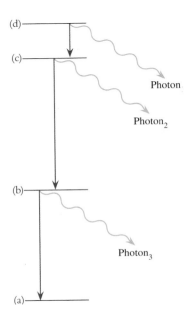

An excited state atom emits photons when the electron moves in succession from level d to level c, from level c to level b, and from level b to the ground state (level a). The wavelengths of the emitted photons are 565 nm, 152 nm, and 121 nm (not necessarily in the proper sequence). Match each emission with the appropriate wavelength and calculate the energies of levels b, c, and d relative to level a.

6.90 Small helium-neon lasers emit 1.0 mJ/s of light at 634 nm. How many photons does such a laser emit in one minute?

6.91 The argon-ion laser has two major emission lines, at 488 and 514 nm. Each of these emissions leaves the Ar^+ ion in an energy level that is 2.76×10^{-18} J above the ground state. (a) Calculate the energies of the two emission wavelengths in joules. (b) Draw an energy level diagram (in joules per atom) that illustrates these facts. (c) What frequency and wavelength radiation is emitted when the Ar^+ ion returns to its lowest energy level?

6.92 The series of emission lines that results from excited hydrogen atoms undergoing transitions to the $n = 3$ level is called the "Paschen series." Calculate the energies of the first five lines in this series of transitions, and draw an energy level diagram that shows them to scale.

6.93 It takes 486 kJ/mol to remove electrons completely from sodium atoms. Sodium atoms absorb and emit light of wavelengths 589.6 and 590.0 nm. (a) Calculate the energies of these two wavelengths in kJ/mol. (b) Draw an energy level diagram for sodium atoms that shows the levels involved in these transitions and the ionization energy. (c) If a sodium atom has already absorbed a 590.0-nm photon, what is the wavelength of the second photon a sodium atom must absorb in order to remove an electron completely?

6.94 As a general rule, the temperature drops more on clear nights than it does on cloudy nights. What feature of clouds accounts for this fact?

6.95 The graph below shows the results of photoelectron experiments on two metals, using light of the same energy:

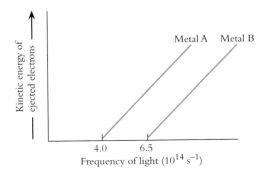

(a) Calculate the binding energy of each metal. Which has the higher binding energy? Explain. (b) Calculate the kinetic energies of electrons ejected from each metal by photons with wavelength of 125 nm. (c) Calculate the wavelength range over which photons can eject electrons from one metal but not from the other.

Group Study Problems

6.96 For the $3d_{yz}$ orbital, draw graphs of electron density vs. r.
(a) for r lying along the z-axis; (b) for r lying along the x-axis;
(c) for r pointing halfway between the y- and z-axes in the yz plane. (Hint: Consult Figure 6-24.)

6.97 A hydrogen atom undergoes an electronic transition from the $n = 4$ to the $n = 2$ state. In the process the H atom emits a photon, which then strikes a cesium metal surface and ejects an electron. It takes 3.23×10^{-19} J to remove an electron from Cs metal. Calculate (a) the energy of the $n = 4$ state of the H atom; (b) the wavelength of the emitted photon; (c) the energy of the ejected electron; and (d) the wavelength of the ejected electron.

6.98 The sun's atmosphere contains vast quantities of He$^+$ cations. These ions absorb some of the sun's thermal energy, promoting electrons from the He$^+$ ground state to various excited states. A He$^+$ ion in the fifth energy level may return to the ground state by emitting three successive photons: an infrared photon ($\lambda = 1014$ nm), a green photon ($\lambda = 469$ nm), and an X ray ($\lambda = 26$ nm), (a) Calculate the excitation energies of each of the levels occupied by the He$^+$ ion as it returns to the ground state. (b) Draw an energy level diagram for He$^+$ cations that illustrates these processes.

6.99 Gaseous Ca atoms absorb light at 422.7, 272.2, and 239.9 nm. After the 272.2-nm absorption, an emission at 671.8 nm is observed. Absorption of light at 239.9 nm is followed by emission at 504.2 nm. Construct an energy level diagram for Ca atoms and answer the following questions: (a) After the 504.2-nm emission, what wavelength would have to be emitted to return to the lowest energy state? (b) What wavelength of light corresponds to the energy difference between the state reached using 422.7-nm light and that reached using 272.2-nm light? (c) Do any sequences described by the data lead to a common energy level? If so, which ones?

6.100 Design a figure that summarizes the chemistry described in Section 6.7. Begin your figure with the regions of the atmosphere. Add the key atomic, molecular, and ionic species found in each region. Include three downward arrows that represent the penetration of sunlight into the atmosphere. Label two of these arrows "Ionizing radiation" and "High-energy UV light". Label the third arrow with the components of sunlight that reach the Earth's surface. Show on your figure the ozone layer, the region associated with the greenhouse effect, and the location of the aurora borealis. All of the information you need can be found in the text and figures of Section 6.7.

Answers to Section Exercises

6.1.1 (a) 1.73×10^{-29} m^3; and (b) 1.73×10^{-34} m^3
6.1.2 (a) 2.59×10^{-10} m; and (b) 3.9×10^3 layers
6.2.1 (a) 779 nm; (b) infrared; and (c) 154 kJ
6.2.2 754 photons
6.2.3 $\nu = 9.04 \times 10^{14}$ s^{-1};
$E_{binding} = 4.00 \times 10^{-19}$ J $= 241$ kJ/mol
6.3.1 108 kJ/mol
6.3.2 3.30×10^{-18} J

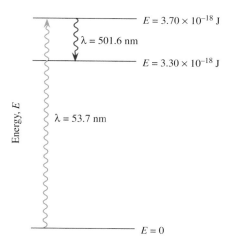

6.3.3 4070 nm, 1280 nm, 434 nm, and 94.9 nm
6.4.1 Photon: 1990 nm; electron: 1.56 nm
6.4.2 1.05×10^{-18} J
6.4.3 Their masses are very different. A free electron carries electrical charge, is a small magnet, and sometimes displays wave behavior; a Ping-Pong ball is electrically neutral and nonmagnetic and displays particle behavior.

6.5.1

$n = 3$	$l = 0$	$m_l = 0$	$m_s = +\frac{1}{2}$	$3s$
$n = 3$	$l = 0$	$m_l = 0$	$m_s = -\frac{1}{2}$	$3s$
$n = 3$	$l = 1$	$m_l = 0$	$m_s = +\frac{1}{2}$	$3p$
$n = 3$	$l = 1$	$m_l = 0$	$m_s = -\frac{1}{2}$	$3p$
$n = 3$	$l = 1$	$m_l = +1$	$m_s = +\frac{1}{2}$	$3p$
$n = 3$	$l = 1$	$m_l = +1$	$m_s = -\frac{1}{2}$	$3p$
$n = 3$	$l = 1$	$m_l = -1$	$m_s = +\frac{1}{2}$	$3p$
$n = 3$	$l = 1$	$m_l = -1$	$m_s = -\frac{1}{2}$	$3p$
$n = 3$	$l = 2$	$m_l = 0$	$m_s = +\frac{1}{2}$	$3d$
$n = 3$	$l = 2$	$m_l = 0$	$m_s = -\frac{1}{2}$	$3d$
$n = 3$	$l = 2$	$m_l = +2$	$m_s = +\frac{1}{2}$	$3d$
$n = 3$	$l = 2$	$m_l = +2$	$m_s = -\frac{1}{2}$	$3d$
$n = 3$	$l = 2$	$m_l = -2$	$m_s = +\frac{1}{2}$	$3d$
$n = 3$	$l = 2$	$m_l = -2$	$m_s = -\frac{1}{2}$	$3d$
$n = 3$	$l = 2$	$m_l = +1$	$m_s = +\frac{1}{2}$	$3d$
$n = 3$	$l = 2$	$m_l = +1$	$m_s = -\frac{1}{2}$	$3d$
$n = 3$	$l = 2$	$m_l = -1$	$m_s = +\frac{1}{2}$	$3d$
$n = 3$	$l = 2$	$m_l = -1$	$m_s = -\frac{1}{2}$	$3d$

6.5.2 (a) Nonexistent, m_s must be $+\frac{1}{2}$ or $-\frac{1}{2}$; (b) nonexistent, l must be less than n; (c) nonexistent, m_l cannot exceed l; and (d) actual
6.5.3 32
6.6.1 Refer to Figures 6-20, 6-23, and 6-24.

6.6.2

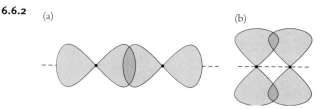

(a) (b)

6.6.3

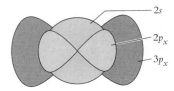

The $n = 2$ orbitals are not important when $3p_x$ is occupied.

6.7.1 Decreases the temperature, because amount of incoming blue light is reduced.

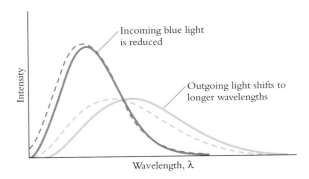

Incoming blue light is reduced

Outgoing light shifts to longer wavelengths

Intensity

Wavelength, λ

6.7.2

$$h\nu + O_2 \longrightarrow 2O$$

$$O + O_2 \longrightarrow O_3 + \text{Heat}$$

$$O + O_2 \longrightarrow O_3 + \text{Heat}$$

$$h\nu + O_3 \longrightarrow O_2 + O$$

$$h\nu + O_3 \longrightarrow O_2 + O$$

$$2O \longrightarrow O_2 + \text{Heat}$$

$$\overline{h\nu \longrightarrow \text{Heat}}$$

7

Atomic Energies
and Periodicity

INTRODUCTION: THE CHEMISTRY OF FIREWORKS

People love the spectacle of fireworks. From Bastille Day in France to Guy Fawkes Day in Britain, from Chinese New Year to Canada Day, fireworks bring joy to celebrations all around the world. In the United States, about $100 million worth of fireworks are discharged every year in honor of Independence Day.

Fireworks date back more than 1000 years to the discovery of black powder in China. Black powder is a mixture of potassium nitrate (KNO_3), charcoal, and sulfur. This first gunpowder was brought to Europe during the Middle Ages and was used widely in weapons, in construction, and for fireworks.

Typically, fireworks contain oxygen sources (oxidizers) and fuel. Ignition produces a redox reaction that releases large amounts of energy. This energy causes the components of the fireworks to emit light. To produce white light in fireworks, a reactive metal such as magnesium is combined with potassium perchlorate ($KClO_4$) as the oxidizer. The redox reaction produces particles of solid magnesium oxide that are heated to temperatures in excess of 3000 °C during the explosion of the fireworks. These extremely hot particles appear white by emitting light across much of the visible spectrum. Larger particles stay hot longer than smaller ones and can continue to burn using O_2 in the atmosphere as the oxidizer, so large particles emit long-lived sparks rather than short flashes. Iron particles, which burn at a lower temperature ($\sim$1500 °C) than magnesium, generate golden sparks.

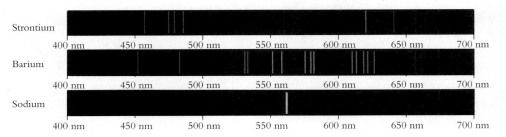

The colors of fireworks displays are produced by emission from atomic ions as described in Chapter 6. The explosions of fireworks promote electrons to excited states. The resulting excited ions emit light of colors that depend on the energies of the atomic orbitals. The energy level scheme of every element is different, so fireworks manufacturers can change colors by incorporating different elements. Sodium ions emit yellow light, strontium ions produce red light, and green comes from barium ions. Blue light is emitted by excited copper ions, but good blue fireworks require precisely optimized conditions, so a fireworks show can be judged by the quality of its blue explosions.

Aerial fireworks shells are launched using fast-burning fuses that ignite charges of black powder. After a shell is high above

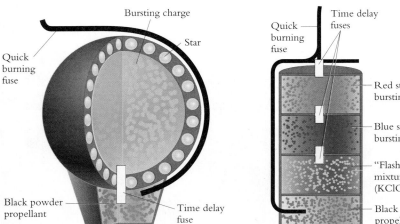

the ground, a time–delay fuse sets off a "bursting charge." Embedded in this charge are pellets that contain the oxidizer and the fuel. Placing the pellets randomly throughout the bursting charge results in an irregular spray of light, while arranging the pellets in a sphere gives a highly symmetrical explosion. Multiple explosions are achieved by placing pellets and bursting charges in separate containers linked by a series of time delay fuses.

The colors of fireworks depend on the energies of the atomic orbitals of the various atomic ions, but orbital energy levels have much more far-reaching consequences. Orbital energies determine the stabilities of atoms and how atoms react. The structure of the periodic table is based on orbital energy levels. Chapter 7 explores the details of orbital energies, to provide the foundation for interpreting chemical behavior patterns.

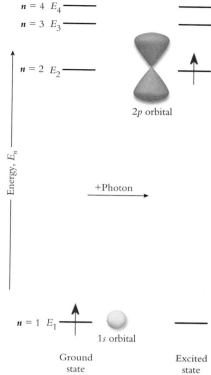

7.1 ORBITAL ENERGIES

A hydrogen atom can absorb a photon and change from its most stable state (ground state) to a less stable state (excited state), as described in Section 6.3. We can account for this process in terms of atomic orbitals. When a hydrogen atom absorbs a photon, its electron transfers to a larger, higher–energy orbital with a larger principal quantum number. This process is illustrated in Figure 7-1.

The atomic orbital model explains perfectly the spectra and the energy levels of the hydrogen atom. Does this model apply to other atoms? Experiments show that although the details are different for each kind of atom, the underlying principles are the same. Variations in nuclear charge and in the number of electrons change the electrical forces that hold electrons in their orbitals. These changes in orbital energies can be understood qualitatively using forces of electrical attraction and repulsion, as we describe later in this chapter.

Figure 7-1
When a hydrogen atom absorbs light, the energy of the photon converts it from the ground state to an excited state. In the process, its electron transfers to an orbital that is larger and less stable.

The Effect of Nuclear Charge

A helium +1 cation, like a hydrogen atom, has just one electron. Absorption and emission spectra show that He^+ has energy levels that depend on n, just like the hydrogen atom. Nevertheless, Figure 7-2 shows that the emission spectra of He^+

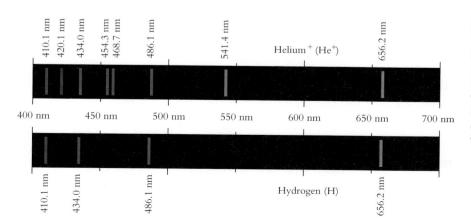

Figure 7-2
The emission spectra of He^+ and H reveal transitions at characteristic energies. The emitted photons have different wavelengths and energies because He^+ has quantized energy levels that are different from those of H.

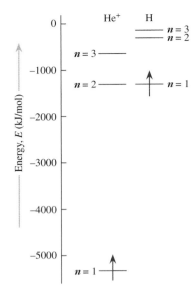

Figure 7-3
An energy level diagram for He⁺ and H. Each species has just one electron, but each He⁺ orbital is four times more stable than the corresponding H orbital.

and H are very different, which means that these two species must have different energy levels. We conclude that something besides **n** influences orbital energy. The important difference between He⁺ and H is their nuclear charge, Z. Whereas a hydrogen nucleus is a single proton with a $+1$ charge, a helium nucleus contains two protons and two neutrons and has a charge of $+2$. The larger charge of He⁺ attracts the single electron more strongly, binding it more tightly. Thus any given energy level in He⁺ is more stable (lower in energy) than the corresponding level in H.

The stability of an orbital can be determined by measuring the amount of energy required to remove an electron completely. This is the **ionization energy (IE)**:

$$H \longrightarrow H^+ + e^- \qquad IE_H = 2.18 \times 10^{-18}\,J$$
$$He^+ \longrightarrow He^{2+} + e^- \qquad IE_{He^+} = 8.72 \times 10^{-18}\,J$$

The ionization energy of He⁺ is four times as large as the *IE* of H. Thus the ground state orbital for He⁺ must be four times as stable as the ground state orbital for H. Spectral analysis shows that each orbital of a helium cation is four times more stable than its counterpart orbital in a hydrogen atom, showing that orbital stability increases with Z^2. The relationship among the energy levels of He⁺ and H is shown in Figure 7-3. The diagram is in exact agreement with calculations based on the Schrödinger equation.

Effect of Other Electrons

A hydrogen atom or a helium cation contains just one electron, but nearly all other atoms and ions contain *collections* of electrons. In a multielectron atom, each electron affects the properties of all the other electrons. These electron-electron interactions make the orbital energies of every element unique.

A given orbital is less stable in a multielectron atom than it is in the single-electron ion with the same nuclear charge. For instance, Table 7-1 shows that it takes more than twice as much energy to remove the electron from He⁺ (one electron) as it does to remove one of the electrons from a neutral He atom (two electrons). This demonstrates that the 1s orbital in He⁺ is more than twice as stable as the 1s orbital in neutral He. The nuclear charge of both species is $+2$, so the smaller ionization energy for He must result from the presence of the second electron. A negatively charged electron in a multielectron atom is attracted to the positively charged nucleus, but it is repelled by the other negatively charged electrons. This electron-electron repulsion accounts for the lower ionization energy of the helium atom.

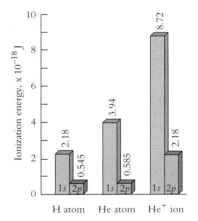

Table 7−1
Comparative Ionization Energies

Orbital	H Atom	He Atom	He⁺ Ion
1s	2.18×10^{-18} J	3.94×10^{-18} J	8.72×10^{-18} J
2p	0.545×10^{-18} J	0.585×10^{-18} J	2.18×10^{-18} J

Screening

Figure 7-4 shows a free electron approaching a helium cation. The incoming electron is attracted to the +2 charge of the nucleus, but the electron is also repelled by the negative charge on the He$^+$ 1s electron. This electron–electron repulsion cancels a portion of the attraction between the nucleus and the incoming electron. Chemists call this partial cancellation **screening.**

When an approaching electron is far enough away from a He$^+$ ion, it feels an attraction due to the net charge on the ion, +1. As the electron approaches more closely, however, the delocalized charge of the 1s electron no longer offsets completely the +2 nuclear charge. In other words, the 1s electron screens only part of the total nuclear charge. Consequently, an approaching electron feels a net attraction resulting from some **effective nuclear charge (Z_{eff})** less than +2 but greater than +1.

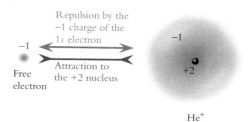

Figure 7-4
When a free electron approaches a He$^+$ cation, it is attracted to the +2 charge on the nucleus but repelled by the −1 charge on the 1s electron. When it is far from the cation, the electron experiences a net charge of +1.

With its −1 charge, each bound electron could reduce Z_{eff} by a maximum of one charge unit. However, screening is never complete, because electrons are delocalized over the space around a nucleus. Furthermore, electron density close to the nucleus is very effective at screening the nuclear charge, but electron density relatively far from the nucleus is not effective. Incomplete screening can be seen in the ionization energies of hydrogen atoms, helium atoms, and helium ions (Table 7-1). Without *any* screening, the ionization energy of a helium atom would be the same as that of a helium ion; both would be 8.72×10^{-18} J. With *complete* screening, one helium electron would compensate for one of the protons in the nucleus, making $Z_{eff} = +1$. The energy required to remove an electron from a helium atom would then be the same as the energy required to remove an electron from a hydrogen atom, 2.18×10^{-18} J. The actual ionization energy of a helium atom is 3.94×10^{-18} J, which is approximately halfway between the fully screened and totally unscreened values. Screening is incomplete because both helium electrons occupy an extended region of space, so neither is completely effective at shielding the other from the +2 charge of the nucleus.

Electrons in compact orbitals are packed around the nucleus more tightly than electrons in large, diffuse orbitals. As a result, the effectiveness in screening nuclear charge decreases with orbital size. Because the size of an orbital increases with *n*, an electron's ability to screen decreases as *n* increases. In a multielectron atom, lower-*n* electrons are concentrated between the nucleus and higher-*n* electrons. The negative charges of these inner electrons counteract most of the positive charge of the nucleus.

The efficient screening by electrons with small values of *n* can be appreciated by comparing the ionization energies of the 2p orbitals listed in Table 7-1. Consider an excited-state helium atom that has one of its electrons in the 1s orbital and its other electron in a 2p orbital. It takes 0.585×10^{-18} J to remove the 2p electron from this excited-state helium atom. This value is almost the same as that of an excited hydrogen atom with its lone electron in a 2p orbital, 0.545×10^{-18} J. These data show that Z_{eff} is quite close to +1 for the 2p orbital of an excited atom. In the excited He atom, the electron in the 1s orbital is very effective at screening the electron in the 2p orbital from the full +2 charge of the nucleus.

The screening effect of a 1s electron also can be seen by comparing an excited He atom with an excited He$^+$ ion. For the ion, no screening occurs, and the 2p orbital ionization energy is 2.18×10^{-18} J. For the excited atom, the 2p electron is

Figure 7-5
Inner electrons screen outer electrons very effectively. Screening makes the ionization energy of a He 2p electron nearly the same as that of a 2p electron in an excited-state H atom. (The 1s energy level for He is more stable than shown in this figure: See Figure 7-3.)

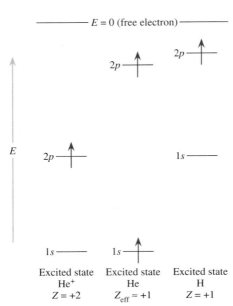

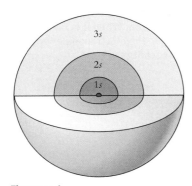

Figure 7-6
Cutaway view of the first three s orbitals. The 1s orbital screens the 2s and 3s orbitals. The 2s orbital screens the 3s orbital but not the 1s orbital. The 3s orbital is ineffective in screening the 1s and 2s orbitals.

screened very effectively by the 1s electron, resulting in a much smaller ionization energy, 0.585×10^{-18} J. Figure 7-5 compares the stabilities of the 2p orbital in these three species.

In multielectron atoms, electrons with any given value of *n* provide effective screening for any orbital with a larger value of *n*. That is, $n = 1$ electrons screen the $n = 2$, $n = 3$, and larger orbitals, whereas $n = 2$ electrons provide effective screening for the $n = 3$, $n = 4$, and larger orbitals but provide little screening for the $n = 1$ orbital. Figure 7-6 illustrates this for the 1s, 2s, and 3s orbitals.

The amount of screening also depends on the shape of the orbital. The shaded area of Figure 7-7 emphasizes that the 2s orbital has a region of significant electron density near the nucleus. The 2p orbital lacks this inner layer, so virtually all of its electron density lies outside the region occupied by the 1s orbital. As a result, the 2p orbital is more effectively screened by the 1s orbital than is the 2s orbital, even though both $n = 2$ orbitals are about the same size. Consequently, a 2s electron feels a larger effective nuclear charge than a 2p electron. This results in stronger electrical attraction, which makes the 2s orbital more stable than the 2p orbitals. The 2p orbitals of any multielectron atom are always less stable (higher in energy) than the 2s orbital.

The screening differences experienced by the 2s and 2p orbitals also extend to larger values of *n*. The 3s orbital is more stable than the 3p orbital, the 4s is more stable than the 4p, and so on. Similar effects are observed for orbitals with higher *l* values. The 3d orbitals are always less stable than the 3p orbitals, and the 4d orbitals are less stable than the 4p orbitals. These effects can be summarized in a single general statement:

/// ***The higher the value of the l quantum number, the more that orbital is screened by electrons in smaller, more stable orbitals.***

In a one-electron system (H, He$^+$, Li^{2+}, and so on) the stability of the orbitals depends only on Z and *n*. In multielectron systems, orbital stability depends primarily on Z and *n*, but it also depends significantly on *l*. In a sense, *l* fine tunes orbital energies.

Figure 7-7
Electron density plots for the 1s, 2s, and 2p orbitals. Unlike the 2p orbital, the 2s orbital has significant electron density very near the nucleus (shaded region).

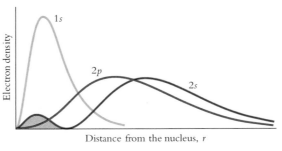

Electrons with the same *l* value but different values of m_l do not screen one another effectively. For example, when electrons are placed in different p orbitals, the amount of mutual screening is slight. This is because screening is effective only when much of the electron density of one orbital lies between the nucleus and the electron density of another. As Figure 6-23 shows, however, the p orbitals are perpendicular to one another, with high electron densities in different regions of space. The electron density of the $2p_x$ orbital does not lie between the $2p_y$ orbital and the nucleus, so there is little screening. The d orbitals also occupy different regions of space from one another, so mutual screening among electrons in these orbitals is also small. Example 7-1 provides another look at screening.

Screening	Example 7-1

Make an electron density plot showing the $1s$, $2p$, and $3d$ orbitals to scale. Label the plot in a way that summarizes the screening properties of these orbitals.

Strategy: This is a qualitative problem that asks us to combine information about three different orbitals on a single plot. We need to extract the necessary information from appropriate sources and combine them to scale on a single figure.

Solution: The electron density plots of the $n = 1$, $n = 2$, and $n = 3$ orbitals are shown in Figures 6-21, 6-22, and 7-7. Combining them gives this result:

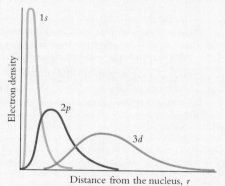

Our next task is to add labels that summarize the screening properties of these orbitals. Screening is provided by small orbitals whose electron density is concentrated inside larger orbitals. In this case, $1s$ screens both $2p$ and $3d$; $2p$ screens $3d$, but not $1s$; and $3d$ screens neither $1s$ nor $2p$. The screening patterns can be labeled as shown.

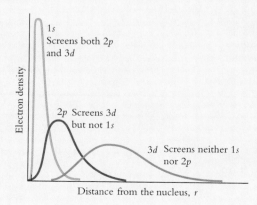

Box 7-1 | **Tools for Discovery: Photoelectron Spectroscopy**

Absorption and emission spectra of atoms and ions yield information about energy differences *between* orbitals, but they do not give an orbital's *absolute* energy. The most direct measurements of orbital energies come from a technique called *photoelectron spectroscopy*.

Photoelectron spectroscopy works like the photoelectric effect described in Chapter 6, except that the sample is the gas phase. Light shines on the sample, a photon is absorbed, and the photon's energy is transferred to an electron. If the photon energy is high enough, the electron is ejected from the sample. The kinetic energy of the ejected electron is equal to the difference between the photon energy and the binding energy (ionization energy) of the electron, as the diagram shows.

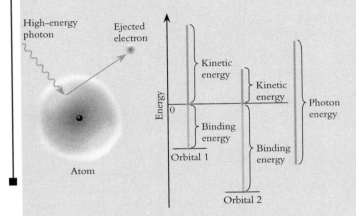

A photoelectron spectrometer uses high-energy photons, typically 11,900 kJ/mol (emitted by excited Mg atoms). These photons have more than enough energy to knock an electron out of an atom or molecule of the sample. The kinetic energy of the departing electron can be determined by measuring its speed. Knowing the photon energy and the electron's kinetic energy, we can find the binding energy of the orbital from which the electron came:

$$\text{Orbital binding energy} = E_{\text{photon}} - E_{\text{kinetic}} \text{ (electron)}$$

The photoelectron spectrum of a monatomic gas is a set of peaks representing the energies of orbitals. The figure shows the spectrum of neon. Two peaks correspond to orbital energies of 2080 and 4680 kJ/mol. The smaller value is the ionization energy of the 2p orbital and the larger

value is the ionization energy of the 2s orbital. The ionization energy of neon's 1s electrons is so great that they cannot be ejected by the photons used in photoelectron experiments. The ionization energy of an electron measures its stability when the electron is bound to the atom. In other words, a 2p electron of a neon atom is 2080 kJ/mol more stable than a free electron in a vacuum.

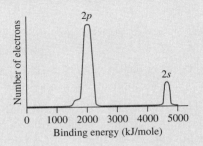

Photoelectron spectra of the first five elements, shown schematically in the graph, illustrate how atomic energy levels change with atomic number and with quantum numbers. The peaks in black are due to 1s electrons. As Z increases, the stability of the 1s orbital increases dramatically (for Be and B, these orbitals have ionization energies greater than 10^4 kJ/mol). Peaks in blue are due to 2s electrons, and the peak in magenta is attributed to a 2p electron. The table below gives values for orbital ionization energies (all in kJ/mol) for elements 11–21 obtained from photoelectron spectra.

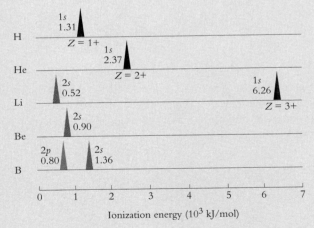

Element	Na	Mg	Al	Si	P	S	Cl	Ar	K	Ca	Sc
2s	6840	9070									
2p	3670	5310	7790								
3s	500	740	1090	1460	1950	2050	2440	2820	3930	4650	5440
3p			580	790	1010	1000	1250	1520	2380	2900	3240
3d											770
4s									420	590	630

Quantitative information about energies of atomic orbitals is obtained using **photoelectron spectroscopy,** which applies the principles of the photoelectric effect to gaseous atoms. Our Tools for Discovery Box explores this powerful spectroscopic technique.

There are many different atomic orbitals, and each has a characteristic energy and shape. How the electrons of an atom distribute themselves among the atomic orbitals is the subject of the next two sections.

Section Exercises

7.1.1 Electron density plots of $n = 1$, $n = 2$, and $n = 3$ orbitals are shown in Figures 7–7, 6–20, and 6–21. Draw a plot that shows the $n = 1$ and $n = 3$ orbitals to scale. Use different colors to keep the figure as clear as possible. Shade the regions of the 3s and 3p plots where screening is ineffective.

7.1.2 The outer layers of the sun contain He atoms in various excited states, one of which contains one 1s electron and one 3p electron. Based on the effectiveness of screening, estimate the ionization energy of the 3p electron in this excited atom.

7.1.3 Redraw Figure 7–3 so that it includes the energy levels for Li^{2+}.

7.2 STRUCTURE OF THE PERIODIC TABLE

The periodicity of chemical properties, which is summarized in the periodic table, is one of the most useful organizing principles in chemistry. Periodic patterns also provide information about electron arrangements in atoms.

The periodic table lists the elements in order of increasing atomic number. Because every neutral atom has the same number of electrons as its atomic number, this list is also in order of increasing number of atomic electrons. Hydrogen, with $Z = 1$ and one electron, appears first, followed sequentially by helium (two electrons), lithium (three electrons), and so on.

A list of the elements in one long row is not a periodic table, because it does not reveal periodic patterns. To convert this single, long row into a periodic table, the "ribbon" of elements is cut at appropriate points to generate strips that are placed in rows. The rows are positioned so that each column contains elements with similar chemical properties. Figure 7–8 illustrates the positioning for the first 20 elements. The first cut comes between helium (two electrons) and lithium (three electrons); the next comes between neon (10 electrons) and sodium; and the third is between argon (18 electrons) and potassium. When the resulting rows are placed one under the other (see Figure 7–8), elements with similar chemical properties appear in the same column. Two examples are He, Ne, and Ar—all unreactive gases—and Li, Na, and K—all highly reactive metals.

To determine the connection between atomic orbitals and the periodic table, we must first describe two additional features of atomic structure: the Pauli exclusion principle and the aufbau principle.

The Pauli Exclusion Principle

It is a general principle of nature that any system tends to find its most stable arrangement. Atoms obey this principle, so hydrogen atoms are normally in

Figure 7-8
The first 20 elements organized in a linear ribbon of increasing atomic number. To generate the periodic table, the ribbon is cut, and the segments are placed in rows so that elements with similar chemical properties fall in the same column. For example, Li, Na, and K all are highly reactive metals.

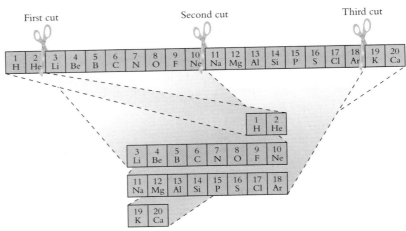

Wolfgang Pauli won the Nobel Prize for Physics in 1945 for discovering electron spin and the exclusion principle.

Aufbau is a German word meaning "construction."

their most stable state, the **ground state.** A hydrogen atom in its ground state has its one electron in the 1s orbital. We might expect a multielectron atom in its ground state to have all its electrons in the most stable, 1s orbital. However, studies show that this is not the case. Instead, bound electrons obey a fundamental law of quantum mechanics called the **Pauli exclusion principle:** *Each electron in an atom has a unique set of quantum numbers.* Named for the Austrian physicist Wolfgang Pauli, this principle can be derived from the mathematics of quantum mechanics, but it cannot be rationalized in a simple way. Nevertheless, all experimental evidence upholds the idea. When one electron in an atom has a particular set of quantum numbers, no other electron in the atom is described by that same set. There are no exceptions to the Pauli exclusion principle.

The Aufbau Principle

The ground state of an atom is, by definition, the most stable arrangement of its electrons. *Most stable* means that the electrons occupy the lowest-energy orbitals available. We construct the **ground-state configuration** of an atom by placing electrons in the orbitals starting with the most stable in energy and moving progressively upward. In accordance with the Pauli principle, each successive electron is placed in the *most stable orbital* whose quantum numbers *are not already assigned* to another electron. This is the **aufbau principle.**

In applying the aufbau principle, remember that a full description of an electron requires four quantum numbers: n, l, m_l, and m_s. Each combination of the quantum numbers n and l describes one quantized energy level. Moreover, each level is made up of a set of orbitals, each with a different value of m_l. Within a set, all orbitals with the same value of l have the same energy. Orbitals with identical energy are called **degenerate orbitals.** For example, the 2p energy level ($n = 2$, $l = 1$) is "triply degenerate": There are three distinct p orbitals ($m_l = -1$, 0, and +1), all with the same energy. In addition, different values of the spin orientation quantum number, m_s, describe the different spin orientations of an electron. When the two possible values of m_s are taken into account, we find that six different sets of quantum numbers can be used to describe an electron in a 2p energy level:

$n = 2$	$l = 1$	$m_l = +1$	$m_s = +\frac{1}{2}$
$n = 2$	$l = 1$	$m_l = +1$	$m_s = -\frac{1}{2}$
$n = 2$	$l = 1$	$m_l = 0$	$m_s = +\frac{1}{2}$
$n = 2$	$l = 1$	$m_l = 0$	$m_s = -\frac{1}{2}$
$n = 2$	$l = 1$	$m_l = -1$	$m_s = +\frac{1}{2}$
$n = 2$	$l = 1$	$m_l = -1$	$m_s = -\frac{1}{2}$

In other words, the $2p$ energy level can hold as many as six electrons without violating the Pauli exclusion principle. The same is true of every set of p orbitals ($3p$, $4p$, etc). Each set has six equivalent descriptions and can hold six electrons. A similar analysis for other values of l shows that each s orbital can hold up to two electrons, a set of d orbitals up to 10 electrons, and a set of f orbitals up to 14 electrons.

The Pauli and aufbau principles dictate where the cuts occur in the ribbon of elements. After two electrons have been placed in the $1s$ orbital (He), the next electron must go in a less stable, $n = 2$ orbital (Li). After eight additional electrons have been placed in the $2s$ and $2p$ orbitals (Ne), the next electron must go in a less stable, $n = 3$ orbital (Na). The ends of the rows in the periodic table are the points at which the next electron must be placed in an orbital of next higher principal quantum number.

Which $n = 2$ orbital is occupied by the third electron in a lithium atom? Screening causes the orbitals with the same principal quantum number to decrease in stability as l increases. Consequently, the $2s$ orbital, being more stable than the $2p$ orbital, fills first. Similarly, $3s$ fills before $3p$, which fills before $3d$, and so on.

Here is a summary of the conditions for atomic ground states:

1. Each electron in an atom occupies the most stable available orbital.
2. No two electrons can have identical descriptions.
3. Orbital capacities are as follows: s, 2 electrons; p set, 6 electrons; d set, 10 electrons; f set, 14 electrons.
4. The higher the value of n, the less stable the orbital.
5. For equal n, the higher the value of l, the less stable the orbital.

Armed with these conditions, we can correlate the rows and columns of the periodic table with values of the quantum numbers n and l. This correlation appears in the periodic table shown in Figure 7-9. Remember that the elements are arranged so that Z increases one unit at a time from left to right across a row. At the end of each row, we move down one row, to the next higher value of n, and return to the left side to the next higher Z value. Inspection of Figure 7-9 reveals that the ribbon of elements is cut after elements 2, 10, 18, 36, 54, and 86. As the atomic number increases, the length of ribbon between cuts increases, too. The first segment contains only hydrogen and helium. Then there are two 8-element segments, followed by two 18-element segments, and finally two 32-element pieces. The last segment stops before reaching its full length, however, because these elements have not been discovered in nature, nor have they been prepared in the laboratory.

Elements with very high Z are unstable, as we show in Chapter 21.

In Figure 7-9, each *row* is labeled with the highest principal quantum number of its occupied orbitals. For example, elements of the third row (Na $\longrightarrow$ Ar) have electrons in orbitals with $n = 3$ (in addition to electrons with $n = 1$ and 2). Each

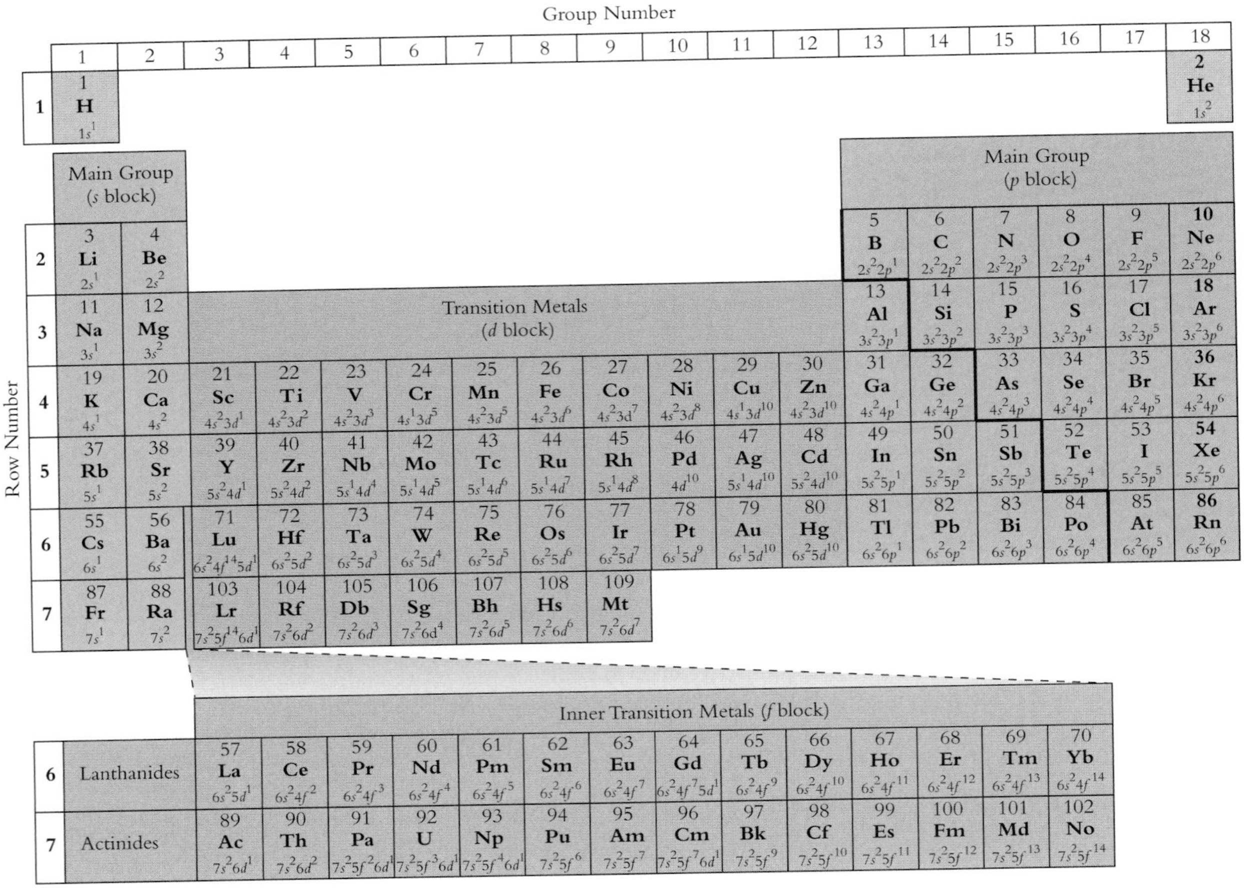

Group Number

Row Number

Group	1	2	3	4	5	6	7	8	9	10	11	12	13	14	15	16	17	18
1	1 H $1s^1$																	2 He $1s^2$
2	3 Li $2s^1$	4 Be $2s^2$		Main Group (s block)									5 B $2s^2 2p^1$	6 C $2s^2 2p^2$	7 N $2s^2 2p^3$	8 O $2s^2 2p^4$	9 F $2s^2 2p^5$	10 Ne $2s^2 2p^6$
3	11 Na $3s^1$	12 Mg $3s^2$				Transition Metals (d block)							13 Al $3s^2 3p^1$	14 Si $3s^2 3p^2$	15 P $3s^2 3p^3$	16 S $3s^2 3p^4$	17 Cl $3s^2 3p^5$	18 Ar $3s^2 3p^6$
4	19 K $4s^1$	20 Ca $4s^2$	21 Sc $4s^2 3d^1$	22 Ti $4s^2 3d^2$	23 V $4s^2 3d^3$	24 Cr $4s^1 3d^5$	25 Mn $4s^2 3d^5$	26 Fe $4s^2 3d^6$	27 Co $4s^2 3d^7$	28 Ni $4s^2 3d^8$	29 Cu $4s^1 3d^{10}$	30 Zn $4s^2 3d^{10}$	31 Ga $4s^2 4p^1$	32 Ge $4s^2 4p^2$	33 As $4s^2 4p^3$	34 Se $4s^2 4p^4$	35 Br $4s^2 4p^5$	36 Kr $4s^2 4p^6$
5	37 Rb $5s^1$	38 Sr $5s^2$	39 Y $5s^2 4d^1$	40 Zr $5s^2 4d^2$	41 Nb $5s^1 4d^4$	42 Mo $5s^1 4d^5$	43 Tc $5s^1 4d^6$	44 Ru $5s^1 4d^7$	45 Rh $5s^1 4d^8$	46 Pd $4d^{10}$	47 Ag $5s^1 4d^{10}$	48 Cd $5s^2 4d^{10}$	49 In $5s^2 5p^1$	50 Sn $5s^2 5p^2$	51 Sb $5s^2 5p^3$	52 Te $5s^2 5p^4$	53 I $5s^2 5p^5$	54 Xe $5s^2 5p^6$
6	55 Cs $6s^1$	56 Ba $6s^2$	71 Lu $6s^2 4f^{14} 5d^1$	72 Hf $6s^2 5d^2$	73 Ta $6s^2 5d^3$	74 W $6s^2 5d^4$	75 Re $6s^2 5d^5$	76 Os $6s^2 5d^6$	77 Ir $6s^2 5d^7$	78 Pt $6s^1 5d^9$	79 Au $6s^1 5d^{10}$	80 Hg $6s^2 5d^{10}$	81 Tl $6s^2 6p^1$	82 Pb $6s^2 6p^2$	83 Bi $6s^2 6p^3$	84 Po $6s^2 6p^4$	85 At $6s^2 6p^5$	86 Rn $6s^2 6p^6$
7	87 Fr $7s^1$	88 Ra $7s^2$	103 Lr $7s^2 5f^{14} 6d^1$	104 Rf $7s^2 6d^2$	105 Db $7s^2 6d^3$	106 Sg $7s^2 6d^4$	107 Bh $7s^2 6d^5$	108 Hs $7s^2 6d^6$	109 Mt $7s^2 6d^7$									

Main Group (p block)

Inner Transition Metals (f block)

6 Lanthanides		57 La $6s^2 5d^1$	58 Ce $6s^2 4f^2$	59 Pr $6s^2 4f^3$	60 Nd $6s^2 4f^4$	61 Pm $6s^2 4f^5$	62 Sm $6s^2 4f^6$	63 Eu $6s^2 4f^7$	64 Gd $6s^2 4f^7 5d^1$	65 Tb $6s^2 4f^9$	66 Dy $6s^2 4f^{10}$	67 Ho $6s^2 4f^{11}$	68 Er $6s^2 4f^{12}$	69 Tm $6s^2 4f^{13}$	70 Yb $6s^2 4f^{14}$
7 Actinides		89 Ac $7s^2 6d^1$	90 Th $7s^2 6d^2$	91 Pa $7s^2 5f^2 6d^1$	92 U $7s^2 5f^3 6d^1$	93 Np $7s^2 5f^4 6d^1$	94 Pu $7s^2 5f^6$	95 Am $7s^2 5f^7$	96 Cm $7s^2 5f^7 6d^1$	97 Bk $7s^2 5f^9$	98 Cf $7s^2 5f^{10}$	99 Es $7s^2 5f^{11}$	100 Fm $7s^2 5f^{12}$	101 Md $7s^2 5f^{13}$	102 No $7s^2 5f^{14}$

Figure 7-9
The periodic table with its rows and blocks labeled to show the relationship between sectors of the table and ground-state configurations. Rows are labeled with the highest principal quantum number of the occupied orbitals, and blocks are labeled with the letter (*s, p, d, f*) indicating the last orbital filled.

column is labeled with its group number, starting with Group 1 on the left and proceeding to Group 18 on the right (the *f* block does not have group numbers). In general, elements in the same group of the periodic table have the same arrangement of electrons in their least stable occupied orbitals.

Order of Orbital Stability

One feature of the periodic table is not explained by the conditions listed so far. Each cut in the ribbon of the elements falls at the end of the *p* block. This indicates that when the **n**p orbitals are full, the next orbital to accept electrons is the (**n** + 1)s orbital. For example, after filling the 3p orbitals from Al (*Z* = 13) to Ar (*Z* = 18), the next element, potassium, has its final electron in the 4s orbital rather than one of the 3d orbitals. According to the aufbau principle, this shows that the 4s orbital is more stable than the 3d orbitals in the potassium atom. A similar situation exists at the end of the next row. When the 4p orbital is full (Kr, $Z = 36$), the next ele-ment (Rb, $Z = 37$) has an electron in the 5s orbital rather than either a 4d orbital or a 4f orbital. In fact, electrons are not added to the 4f orbitals until element 58, after the 5s, 5p, and 6s orbitals have filled.

This aspect of the periodic table can be understood by comparing the two main features that affect the order of orbital stability in a multielectron atom: the energy difference between orbitals with different values of n, and the energy difference between orbitals with the same value of n but different values of l. These features can act in opposing directions. Consider, for example, the 4s and 3d orbitals. In a one-electron atom, there is no screening, so 4s, with a larger value of n, is less stable than the 3d orbitals. In a multielectron atom, however, the $n = 1$ and $n = 2$ electrons screen the 3d orbitals more effectively than they screen 4s. As Z increases, this effect is magnified: The stability of the s orbitals increases more rapidly than the stability of the d orbitals. When $Z = 19$, the increase in sta-bility of the 4s relative to the 3d is enough to make the 4s orbital more stable than 3d for an isolated neutral atom. Figure 7-10 shows the order of orbital stability for the $n = 3$ and $n = 4$ orbitals for isolated atoms of the elements from $Z = 19$ to $Z = 30$. Notice that the most stable $n = 4$ orbital (4s) is more stable than the least stable $n = 3$ orbitals (3d). This places potassium ($Z = 19$) and calcium ($Z = 20$) in the s block. The 3d orbitals fill starting with scandium ($Z = 21$), after the 4s orbital is full.

This pattern repeats for the 5s and 4d pair of orbitals for neutral atoms between $Z = 37$ and $Z = 48$. Here, the 5s orbital is more stable than the 4d orbital. Furthermore, the 4f orbitals are screened very effectively by $n = 3$ electrons, so the 4f orbitals do not fill until after the 5s, 4d, 5p, and 6s orbitals. The ns orbital is always more stable than its $(n − 1)d$ counterpart for elements in Groups 1 and 2 of the periodic table. Radium, for example, has its last two electrons in the 7s orbital.

The arrangement of the periodic table provides a simple way to determine the filling order of the elements, as shown in Figure 7-11 and carried out in Example 7-2.

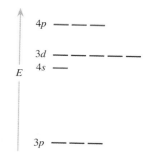

Figure 7-10
Qualitative energy level diagram for the 3p, 3d, 4s, and 4p atomic orbitals for $Z = 19$ to 30. All other orbitals have been omitted for clarity.

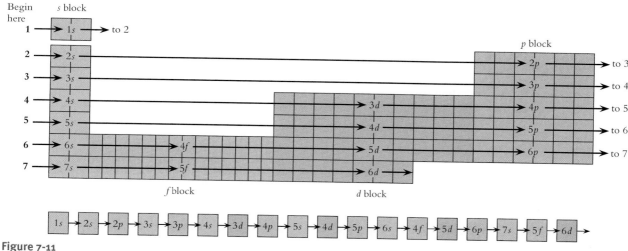

Figure 7-11
The periodic table in block form, showing the filling sequence of the atomic orbitals. Filling proceeds from left to right across each row and from the right end of each row to the left end of the succeeding row.

| Example 7-2 | Orbital Filling Sequence |

Which orbitals are filled, and which set of orbitals is partially filled, in a germanium atom?

Strategy: For this qualitative problem, use the periodic table to determine the order of orbital filling. Locate the element in a block and identify its row and column. Move along the ribbon of elements to establish the sequence of filled orbitals.

Solution: Germanium is element 32. Consult Figure 7-9 to determine that Ge is in Group 14, row 4 of the *p* block:

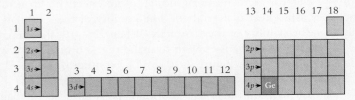

Starting from the top left of the periodic table and working left to right across the rows until we reach Ge, we identify the filled orbitals: 1*s*, 2*s*, 2*p*, 3*s*, 3*p*, 4*s*, 3*d*. Germanium is in row 4 of the *p* block, so the 4*p* set of orbitals is partially filled.

Valence Electrons

The chemical behavior of an atom is determined by the electrons that are accessible to an approaching chemical reagent. Accessibility, in turn, has a spatial component and an energetic component. An electron is accessible *spatially* when it occupies one of the largest orbitals of the atom. Electrons on the perimeter of the atom, farthest from the nucleus, are the first ones encountered by an incoming chemical reagent. An electron is accessible *energetically* when it occupies one of the least stable occupied orbitals of the atom. Electrons in less stable orbitals are more chemically active than electrons in more stable orbitals.

Similar electron accessibility generates similar chemical behavior. For example, iodine has many more electrons than chlorine, but these two elements display very similar chemical behavior, as reflected by their placement in the same group of the periodic table. This is because the chemistry of chlorine and iodine is determined by the number of electrons in their largest and least stable occupied orbitals: 3*s* and 3*p* for chlorine and 5*s* and 5*p* for iodine. Each of these elements has seven accessible electrons, and this accounts for the chemical similarities.

Accessible electrons are called **valence electrons,** and inaccessible electrons are called **core electrons.** Valence electrons participate in chemical reactions, but core electrons do not. Orbital size increases and orbital stability decreases as the principal quantum number *n* gets larger. Therefore the valence electrons for most atoms are the ones in orbitals with the largest value of *n*. Electrons in orbitals with lower *n* values are core electrons. In chlorine, valence electrons have *n* = 3, and core electrons have *n* = 1 and *n* = 2. In iodine, valence electrons have *n* = 5, and all others are core.

Box 7-2   Chemical Milestones: History of the Periodic Table

Today we work confidently with the rows and columns of the periodic table. Yet less than 150 years ago, only about half of all elements known today had been discovered, and these presented a bewildering collection of chemical and physical properties. The discovery of the patterns that underlie this apparent randomness is a tale of inspired chemical detective work.

One early attempt to organize the elements clustered them into groups of three, called *triads*, whose members display similar chemical properties. Lithium, sodium, and potassium, for example, have many common properties and were considered to be a triad. This model was severely limited, for many elements could not be grouped into triads. The triad model is just one of nearly 150 different periodic arrangements of the elements that have been proposed.

Our modern periodic table was developed independently in the late 1860s by Dimitri Mendeleev (Russian) and Julius Lothar Meyer (German). At that time, about 60 elements had been discovered, but nothing was known about atomic structure. Lothar Meyer and Mendeleev had to work with elemental molar masses and other known elemental properties.

Lothar Meyer, a physicist, examined atomic volumes. He plotted atomic volume against molar mass and observed the pattern shown in the figure. There is a clear pattern of "waves," cresting successively at Li, Na, K, Rb, and Cs.

Mendeleev was bolder in his interpretation than Lothar Meyer, and for this reason we honor him as the primary discoverer of the modern periodic table. A few elements did not fit the pattern of variation in combining numbers with molar mass. Mendeleev proposed that these irregularities meant that the element's molar mass had been measured incorrectly. For example, Mendeleev predicted that the correct molar mass of indium is 113 g/mol, not 75 g/mol, the value assigned at that time on the assumption that the formula for indium oxide is InO. Later experiments showed that the correct formula is In_2O_3, and indium's true molar mass is 114.8 g/mol.

Mendeleev also predicted the existence of elements that had not yet been discovered. His arrangement of the then-known elements left some obvious holes in the periodic table. For instance, between zinc (combines with 2 Cl) and arsenic (combines with 5 Cl) were holes for one element that would combine with three chlorine atoms and another that would combine with four. Mendeleev assigned these holes to two new elements. He predicted that one element would have a molar mass of 68 and chemical properties like those of aluminum, while the other would have a molar mass of 72 and chemical properties similar to silicon. These elements, gallium ($Z = 31$, $MM = 69.7$ g/mol) and germanium ($Z = 32$, $MM = 72.6$ g/mol), were discovered within 15 years. Chemists soon verified that gallium resembles aluminum in its chemistry, while germanium resembles silicon, just as Mendeleev had predicted.

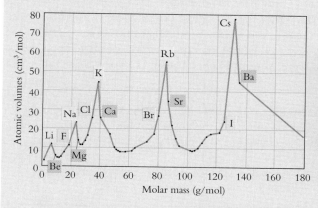

Ge

Si

Mendeleev, a chemist, examined the relative numbers of the atoms of different elements that combine in chemical compounds. In $MgCl_2$, for example, each magnesium atom combines with two chlorine atoms. When he matched combining ability against molar mass, Mendeleev found the same sort of pattern as Lothar Meyer, with Li, Na, K, Rb, and Cs all combining 1 : 1 with Cl. Thus each scientist was led to propose a table in which elements are arranged in rows of increasing mass, with breaks so that these five elements fall in the same column.

Mendeleev's table and the predictions it allowed him to make provide an excellent example of how a scientific theory allows far-reaching predictions of as-yet-undiscovered phenomena. Today's chemists still use the periodic table as a predictive tool. For example, modern semiconductor materials such as gallium arsenide were developed in part by predicting that elements in the appropriate rows and columns of the periodic table should have the desired properties. At present, scientists seeking to develop new superconducting materials rely on the periodic table to identify elements that are most likely to confer superconductivity.

The near-degeneracy of ns and $(n - 1)d$ orbitals creates some ambiguity about valence and core electrons for elements in the d and f blocks. For example, titanium forms a chloride and an oxide whose chemical formulas are consistent with four valence electrons: $TiCl_4$ and TiO_2. This shows that the two $3d$ electrons of titanium participate in its chemistry, even though they have a lower principal quantum number than its two $4s$ electrons. On the other hand, zinc forms compounds such as $ZnCl_2$ and ZnO, indicating that only two electrons are involved in its chemistry. This chemical behavior indicates that zinc's ten $3d$ electrons, which completely fill the $3d$ orbitals, are inaccessible for chemical reactions. When the d orbitals are partially filled, the d electrons participate in chemical reactions. However, when the d orbitals are completely filled, the d electrons do not participate in reactions. Analogous behavior is observed for the f orbitals, leading to a general rule for identifying valence electrons:

/// *Valence electrons are all those of highest principal quantum number plus those in partially filled d and f orbitals.*

Applying this general rule, we find that the number of valence electrons can be determined easily from group numbers. For Groups 1-8, the number of valence electrons equals the group number. As examples, potassium and rubidium, members of Group 1, have just one valence electron each. Tungsten, in Group 6, has six valence electrons: two $6s$ electrons and four $5d$ electrons. For Groups 12–18, the number of valence electrons equals the group number minus 10 (the number of electrons it takes to fill the d orbitals). Thus antimony and nitrogen, in Group 15, have $15 - 10 = 5$ valence electrons each (two s and three p). For Groups 9–11, the number of valence electrons cannot be stated unambiguously, because the d electrons may or may not participate in bonding.

We have described the layout of the periodic table in terms of the orbital descriptions of the various elements. As our Chemical Milestones Box describes, the periodic table was first proposed well before quantum theory was developed, when the only guidelines available were patterns of chemical and physical behavior.

Section Exercises

7.2.1 What is the atomic number of the element that would occupy the position in Row 7, Column 17 of the periodic table?

7.2.2 How many valence electrons are in each of the following elements: C, Br, Cr, and Ta?

7.2.3 List all elements that have two valence electrons and specify the orbitals to which the valence electrons belong.

7.3 ELECTRON CONFIGURATIONS

The location of an element in the periodic table tells us how many valence electrons it has. Although this is sufficient to describe much of the chemical behavior of the element, it does not provide a complete description of the way in which electrons are distributed among the atomic orbitals. A complete specification of how electrons are distributed is called an **electron configuration.** An electron configuration provides enough information to assign each of the electrons in the atom to a specific orbital.

There are three common ways to represent electron configurations. One is a complete specification of quantum numbers. The second is a shorthand notation from which the quantum numbers can be inferred. The third is a diagrammatic representation of orbital energy levels and their occupancy.

A list of the values of all quantum numbers is easy for the single electron in a hydrogen atom: $n = 1$, $l = 0$, $m_l = 0$, and $m_s = +\frac{1}{2}$ or $n = 1$, $l = 0$, $m_l = 0$, and $m_s = -\frac{1}{2}$. Either designation is equally valid, because under normal conditions these two states are equal in energy. In a large collection of hydrogen atoms, the electrons in half the atoms have one designation and the electrons in the other half have the other designation.

For an iron atom, which has 26 electrons, a listing of all quantum numbers is tedious. To save time and space, chemists have devised a shorthand notation to write electron configurations. The orbital symbols ($1s$, $2p$, $4d$, etc.) are followed by superscripts designating how many electrons are in each set of orbitals. The compact configuration for a hydrogen atom is $1s^1$, indicating one electron in the $1s$ orbital.

The third way to represent an atomic configuration uses an energy level diagram similar to the one shown in Figure 7-10 to designate orbitals. Each electron is represented by an arrow and is placed in the appropriate orbital. The direction of the arrow indicates the value of m_s. For $m_s = +\frac{1}{2}$ the arrow points upward, and for $m_s = -\frac{1}{2}$ it points downward. The configuration of hydrogen can be represented by a single arrow in a $1s$ orbital.

$1s$

A neutral helium atom has two electrons. To write the ground-state electron configuration of He, we apply the aufbau principle. One unique set of quantum numbers is assigned to each electron, moving from the most stable orbital upward until all electrons have been assigned. The most stable orbital is always $1s$ ($n = 1$, $l = 0$, $m_l = 0$). Both helium electrons can occupy the $1s$ orbital, provided one of them has $m_s = +\frac{1}{2}$ and the other has $m_s = -\frac{1}{2}$. Here are the three representations of helium's ground-state electron configuration:

$$n = 1, l = 0, m_l = 0, m_s = +\frac{1}{2}$$
$$n = 1, l = 0, m_l = 0, m_s = -\frac{1}{2} \qquad 1s^2$$

The two electrons in this configuration are said to be *paired* electrons, meaning that they are in the same energy level, with opposing spins. Opposing spins cancel, so paired electrons have zero net spin.

A lithium atom has three electrons. The first two electrons fill lithium's lowest possible energy level, the $1s$ orbital, and the third electron occupies the $2s$ orbital. The three representations for the ground-state electron configuration of a lithium atom are as follows:

$$n = 1, l = 0, m_l = 0, m_s = +\frac{1}{2}$$
$$n = 1, l = 0, m_l = 0, m_s = -\frac{1}{2} \qquad 1s^2 \, 2s^1$$
$$n = 2, l = 0, m_l = 0, m_s = +\frac{1}{2}$$

The set $n = 2$, $l = 0$, $m_l = 0$, $m_s = -\frac{1}{2}$ is equally valid for the third electron.

The next atoms of the periodic table are beryllium and boron. You should be able to write the three different representations for the ground-state configurations of these elements. The filling principles are the same as we move to higher atomic numbers. Example 7-3 shows how these principles are applied to aluminum.

When a strong magnetic field is imposed on a collection of atoms, the two spin orientations of the electron take on different energies, and it is possible for electrons in the state of lower energy to undergo transitions to the state of higher energy by absorbing radiation. This is the basis for electron spin resonance spectroscopy.

Example 7-3	An Electron Configuration

Construct an energy level diagram and the shorthand representation of the ground-state configuration of aluminum. Provide one set of valid quantum numbers for the highest-energy electron.

Strategy: First consult the periodic table to locate aluminum and determine how many electrons are present in a neutral atom. Then construct the electron configuration using the patterns of the periodic table.

Solution: Aluminum has $Z = 13$, so a neutral atom of Al has 13 electrons. The 13 electrons are placed sequentially, using arrows, into the most stable orbitals available. Two electrons fill the $n = 1$ orbital, eight electrons fill the $n = 2$ orbitals, two electrons fill the $3s$ orbital, and one electron goes in a $3p$ orbital.

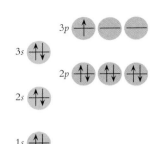

The last electron could be placed in any of the $3p$ orbitals, because these three orbitals are equal in energy. The final electron also could be given either spin orientation. By convention, we place electrons in unfilled orbitals starting with the left-hand side, with spins pointed up.

The shorthand configuration is $1s^2\, 2s^2\, 2p^6\, 3s^2\, 3p^1$.

The least stable electron is in a $3p$ orbital, meaning $n = 3$ and $l = 1$. The value of m_l can be any of three values: $+1$, -1, or 0. The spin quantum number, m_s, can be $+\frac{1}{2}$ or $-\frac{1}{2}$. One valid set of quantum numbers is:

$$n = 3,\ l = 1,\ m_l = 1,\ \text{and}\ m_s = +\tfrac{1}{2}$$

You should be able to write the other five possible sets.

Electron configurations become longer as the number of electrons increases. To make the writing of a configuration even more compact, chemists make use of the regular pattern for the electrons with lower principal quantum numbers. Compare the configurations of neon and aluminum:

Ne (10 electrons) $1s^2\, 2s^2\, 2p^6$

Al (13 electrons) $1s^2\, 2s^2\, 2p^6\, 3s^2\, 3p^1$

The description of the first 10 electrons in the configuration of aluminum is identical to that of neon, so we can represent that portion as [Ne]. With this notation, the configuration of Al becomes [Ne] $3s^2\, 3p^1$. The element at the end of each row of the periodic table has a **noble gas configuration.** These configurations can be written in the following shorthand notation:

Notation		Configuration	Element
[He]	=	$1s^2$	He (2 electrons)
[Ne]	=	[He] $2s^2\, 2p^6$	Ne (10 electrons)
[Ar]	=	[Ne] $3s^2\, 3p^6$	Ar (18 electrons)
[Kr]	=	[Ar] $4s^2\, 3d^{10}\, 4p^6$	Kr (36 electrons)
[Xe]	=	[Kr] $5s^2\, 4d^{10}\, 5p^6$	Xe (54 electrons)
[Rn]	=	[Xe] $6s^2\, 5d^{10}\, 4f^{14}\, 6p^6$	Rn (86 electrons)

To write the configuration of any other element, we first consult the periodic table to find its location relative to the noble gases. Then we specify the noble gas configuration and build the remaining portion of the configuration according to the aufbau principle. Example 7-4 applies this procedure to strontium.

A Shorthand Electron Configuration	Example 7-4

Determine the configuration of strontium, first in shorthand form and then in full form.

Strategy: Locate the element in the periodic table, and find the nearest noble gas with smaller atomic number. Start with the configuration of that noble gas, and add enough additional electrons to the next filling orbitals to give the neutral atom.

Solution: Strontium (Sr, $Z = 38$) is in Row 5, Group 2. The nearest noble gas of smaller Z is Kr ($Z = 36$). Thus the configuration of Sr has 36 electrons in the Kr configuration and two additional electrons. The last orbital to fill in Kr is $4p$, and the periodic table shows that the next orbital to fill is the $5s$ orbital:

Sr configuration: $[Kr] 5s^2$

To write the full configuration, decompose the krypton configuration:

$$[Kr] = [Ar] 4s^2 3d^{10} 4p^6 = 1s^2 2s^2 2p^6 3s^2 3p^6 4s^2 3d^{10} 4p^6$$

Full configuration of Sr (38 electrons):

$$1s^2 2s^2 2p^6 3s^2 3p^6 4s^2 3d^{10} 4p^6 5s^2.$$

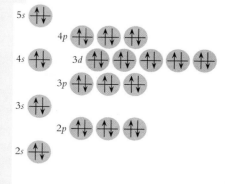

Electron–Electron Repulsion

The aufbau principle allows us to assign quantum numbers to aluminum's 13 electrons without ambiguity. The first 12 electrons fill the $1s$, $2s$, $2p$, and $3s$ energy levels, and the last electron can occupy any $3p$ orbital with either spin orientation. But what happens when more than one electron must be placed in a p energy level? Carbon atoms, for example, have six electrons, two of which occupy $2p$ orbitals. How should these two electrons be arranged in the $2p$ orbitals? As Figure 7-12 shows, three different arrangements of these electrons obey the Pauli principle and appear to be consistent with the aufbau principle:

1. The electrons could be paired in the same $2p$ orbital (same m_l value but different m_s values).

2. The electrons could occupy different $2p$ orbitals with the same spin orientation (different m_l values but the same m_s value).

3. The electrons could occupy different $2p$ orbitals with opposite spin orientations (different m_l values and different m_s values).

These three arrangements have different energies, because electrons that are close together repel each other more than electrons that are far apart. As a result, for two or more otherwise equal-energy orbitals, greatest stability results when electrons occupy the orbitals that keep them farthest apart. Placing two electrons in different p orbitals keeps them relatively far apart, so an atom is more stable with the two electrons in different p orbitals. Thus arrangements 2 and 3 are more stable than arrangement 1.

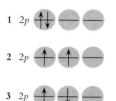

Figure 7-12
Three different arrangements of two $2p$ electrons obey the Pauli and the aufbau principles.

Arrangements 2 and 3 look spatially equivalent, but experiments show that a configuration that gives unpaired electrons the same spin orientation is always more stable than one that gives them opposite orientations. The way in which electrons occupy orbitals of equal energies is summarized by **Hund's rule:**

/// *The most stable configuration is the one with the maximum number of unpaired electron spins.*

According to Hund's rule, the ground-state configuration for carbon atoms is arrangement 2. Example 7-5 provides practice in the application of Hund's rule.

Example 7-5	Applying Hund's Rule

Write the shorthand electron configuration and draw the ground-state orbital energy level diagram for the valence electrons in a sulfur atom.

Strategy: From the periodic table, we see that sulfur has 16 electrons and is in the p-block, Group 16. To build the ground-state configuration, apply the normal filling rules and then apply Hund's rule if needed.

Solution: The first 12 electrons fill the four lowest-energy orbitals:

$$1s^2\ 2s^2\ 2p^6\ 3s^2$$

Sulfur's remaining 4 electrons occupy the three $3p$ orbitals. The complete configuration is $1s^2\ 2s^2\ 2p^6\ 3s^2\ 3p^4$, or $[Ne]\ 3s^2\ 3p^4$.

To minimize electron–electron repulsion, put three of the $3p$ electrons in different orbitals, all with the same spin, and then place the fourth electron, with opposite spin, in the first orbital. In accord with Hund's rule, this gives the same value of m_s to all electrons that are not paired. Here is the energy level diagram for the valence electrons:

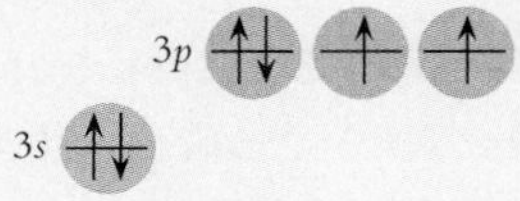

Near-Degenerate Orbitals

The filling order embodied in the periodic table predicts a regular progression of ground-state configurations. Experimentally, however, some elements are found to have ground-state configurations that do not match the regular progression. Among the first 40 elements there are only two exceptions: copper and chromium. Chromium ($Z = 24$) is in Group 6, four elements into the d block. We would predict that chromium's valence configuration should be $4s^2\ 3d^4$. Instead, experiments show that the ground-state configuration of this element is $4s^1\ 3d^5$. Likewise, the configuration of copper ($Z = 29$) is $4s^1\ 3d^{10}$ rather than the predicted $4s^2\ 3d^9$. Look again at Figure 7-10, which shows the effect of screening on the $3d$ and $4s$ orbitals. Notice that screening makes these two sets of orbitals nearly the same in energy. Orbitals with nearly the same energy are **near-degenerate orbitals.**

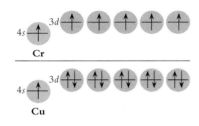

Among elements whose valence electrons are filling near-degenerate orbitals, the ground-state configuration is determined by a balance of several factors. The details of this balance are beyond the scope of General Chemistry, except to note that even a subtle change can cause variations in the filling pattern predicted by the periodic table.

Table 7–2
Near-Degenerate Atomic Orbitals

Orbitals	Atomic Numbers Affected	Example
4s, 3d	24, 29	Cr: [Ar] $4s^1 3d^5$
5s, 4d	41–47	Ru: [Kr] $5s^1 4d^7$
6s, 5d, 4f	57, 58, 64, 78, 79	Au: [Xe] $6s^1 4f^{14} 5d^{10}$
6d, 5f	89, 91–93, 96	U: [Rn] $7s^2 5f^3 6d^1$

Above element 40, as the valence orbitals become closer together in energy, the occurrence of near-degenerate orbitals becomes more common. Table 7-2 lists the near-degenerate orbitals and the atomic numbers for which the filling sequence differs from the expected pattern. These configurations are also indicated in Figure 7-9. Often, an s orbital contains only one electron rather than two. Five of the exceptional ground-state configurations have a common pattern and are easy to remember: Cr and Mo are $s^1 d^5$, and Cu, Ag, and Au are $s^1 d^{10}$. The other exceptional cases follow no recognizable pattern, because they are generated by subtle interactions among all the electrons.

Configurations of Ions

The electron configurations of atomic ions are written using the same procedure as for neutral atoms, taking into account the proper number of electrons. An anion has one *additional* electron for each unit of negative charge. A cation has one *less* electron for each unit of positive charge.

For most atomic ions, the filling order of orbitals is the same as that of neutral atoms. For example, Na^+, Ne, and F^- all contain 10 electrons, and each has the configuration $1s^2 2s^2 2p^6$. Atoms and ions that have the same number of electrons are said to be **isoelectronic.**

The near-degeneracy of ns and $(n - 1)d$ orbitals causes the configurations of some cations to differ from the configurations predicted by the filling pattern of the periodic table. This feature is particularly important for the transition metals. Experiments show that in transition metal *cations* the $(n - 1)d$ orbitals are *always* more stable than the $(n)s$ orbitals. For example, an Fe^{3+} cation contains 23 electrons. The first 18 electrons fill the 1s, 2s, 2p, 3s, 3p, and 3d orbitals, as predicted by the periodic table. However, the five remaining electrons populate the 3d set. Thus the configuration of the Fe^{3+} cation is [Ar] $3d^5$.

Vanadium atoms ([Ar] $4s^2 3d^3$) and Fe^{3+} cations ([Ar] $3d^5$) have different configurations, even though each has 23 electrons. Remember that the energy ranking of near-degenerate orbitals such as 4s and 3d depends on a balance of several factors, and that even a small variation in that balance can change the filling order of the orbitals. That is the case for transition metal atoms and cations, as shown in Figure 7-13.

In neutral transition metal atoms, the ns orbital is slightly more stable than the $(n - 1)d$ orbital. In transition metal cations, however, the $(n - 1)d$ orbitals are always more stable than the ns orbital.

Example 7-6 shows how to write the configuration of a transition metal cation.

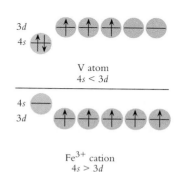

V atom
$4s < 3d$

Fe^{3+} cation
$4s > 3d$

Figure 7-13
Transition metal atoms and cations have different energy rankings for the near-degenerate 4s and 3d orbitals, even when the species may contain the same number of electrons.

| Example 7-6 | Configuration of a Cation |

What is the ground-state electron configuration of a Cr^{3+} cation?

Strategy: Use the aufbau approach, remembering that because Cr^{3+} is a transition metal cation its $3d$ orbital is more stable than the near-degenerate $4s$.

Solution: A neutral chromium atom has 24 electrons, so the corresponding Cr^{3+} cation has 21 electrons. The first 18 electrons follow the usual filling order to give the argon core configuration: $1s^2\ 2s^2\ 2p^6\ 3s^2\ 3p^6$, or [Ar]. The remaining three electrons are placed in the $3d$ set of orbitals, following Hund's rule: [Ar] $3d^3$

For any cation, the empty $4s$ orbital is slightly higher in energy than the partially filled $3d$ orbital. Thus, the isoelectronic V^{2+} and Cr^{3+} cations both have the [Ar] $3d^3$ configuration. On the other hand, the isoelectronic neutral atom scandium has the configuration [Ar] $4s^2\ 3d^1$.

Here are the guidelines for building atomic or ionic electron configurations:

1. Count the total number of electrons.
   1a. Add electrons for anions
   1b. Subtract electrons for cations
2. Fill orbitals to match the nearest noble gas of smaller atomic number.
3. Add remaining electrons to the next filling orbitals according to Hund's rule.
   3a. For neutral atoms and anions, $(n-1)d < ns$
   3b. For cations, $ns < (n-1)d$
4. Look for near-degenerate exceptions and correct the configuration, if necessary.

Magnetic Properties of Atoms

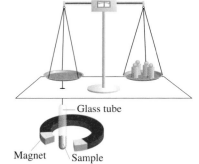

Figure 7-14
The number of unpaired electrons in a paramagnetic substance can be measured with a Gouy balance. The paramagnetic sample is attracted into the magnetic field, creating a downward force on the left pan. Weights are added to the pan on the right until the force is balanced. The magnetic strength of the sample is proportional to the mass required to balance the pans.

How do we know that an Fe^{3+} ion in its ground state has the configuration [Ar] $3d^5$ rather than the [Ar] $4s^2\ 3d^3$ configuration predicted by the periodic table? Remember from Chapter 6 that electron spin gives rise to magnetic properties. Consequently, any atom or ion with unpaired electrons has nonzero net spin and is attracted by a strong magnet. We can divide the electrons of an atom or ion into two categories with different spin characteristics. In the filled orbitals, all the electrons are paired. Each electron with spin orientation $+\frac{1}{2}$ has a partner with spin orientation $-\frac{1}{2}$. The spins of these electrons cancel each other, giving a net spin of zero. An atom or ion with all electrons paired is not attracted by strong magnets and is termed **diamagnetic.** In contrast, spins do not cancel when unpaired electrons are present. An atom or ion with unpaired electrons is attracted to strong magnets and is termed **paramagnetic.** The spins of all the unpaired electrons are additive, so the amount of paramagnetism shown by an atom or ion is proportional to the number of unpaired spins. The magnetic properties of chemical species can be measured with an instrument known as a Gouy balance, as shown schematically in Figure 7-14.

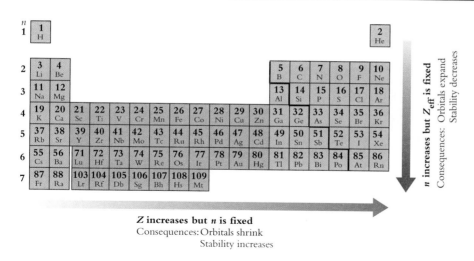

Figure 7-15
Underlying features of the
periodic table.

Moving from left to right across a row of the periodic table, the *n* value of the least stable occupied orbital remains the same while Z increases. A larger nuclear charge exerts a stronger coulombic attraction on the electron cloud, and this stronger attraction results in smaller orbitals. Furthermore, electrons closer to the nucleus are energetically more stable than those farther from the nucleus. Moving from *left to right* across a row, orbitals become *smaller and more stable*.

Proceeding down a column of the periodic table, *n* and Z both increase. As *n* increases, orbitals become larger and less stable, but as Z increases, orbitals become smaller and more stable. Which trend dominates here? Recall that the number of core electrons increases as we move downward from one row to the next. For example, sodium ($Z = 11$) has 10 core electrons and 1 valence electron. In the next lower row, potassium ($Z = 19$) has 18 core electrons in addition to its 1 valence electron. The screening provided by potassium's additional eight core electrons largely cancels the effect of the additional eight protons in its nucleus. Consequently, the increase in Z value from one row to the next is offset by increased screening. As a result, *n* is the most important factor in determining orbital size and stability within a column. From *top to bottom* of a column, orbitals become *larger and less stable*.

Atomic Radii

Because most of the volume of an atom is occupied by its electron cloud, the size of an atom is determined by the sizes of its orbitals. Atomic size follows these periodic trends:

///*Atomic size decreases from left to right and increases from top to bottom of the periodic table.*

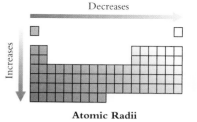

Atomic Radii

A convenient measure of atomic size is the radius of the atom. Figure 7–16 shows the trends in atomic radii. For example, the atomic radius decreases smoothly across row 4, from 227 pm for potassium to 114 pm for bromine. The atomic radius increases smoothly down Group 1, from 152 pm for lithium to

Figure 7-16
The radius of gaseous atoms varies with atomic number. Atomic radius decreases from left to right within any row (blue lines) and increases from top to bottom within any group (red lines).

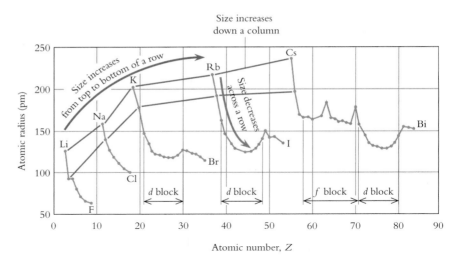

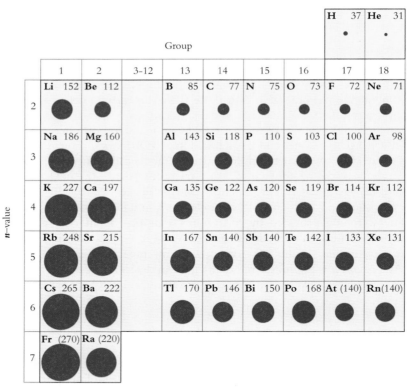

265 pm for cesium. Notice, however, that the atomic radius changes very little across the d and f blocks of the table. This is due to screening. For these elements, the largest orbital is the filled ns orbital. Moving left to right across a row, Z increases, but electrons add to the smaller $(n-1)d$ or $(n-2)f$ orbitals. An increase in Z by one unit is matched by the addition of one screening electron. From the perspective of the outlying s orbital, the increases in Z are balanced by increased screening from the added d or f electrons. Thus the electron in the outermost

occupied orbital, *ns*, feels an effective nuclear charge that changes very little across these blocks. As a consequence, atomic size remains nearly constant across each row of the *d* and *f* blocks.

It is important to be familiar with periodic trends in physical and chemical properties, but it is just as important to understand the principles that give rise to these trends. Example 7-8 shows how to analyze trends in terms of the underlying principles.

Trends in Atomic Radii	**Example 7-8**

For each of the following pairs, predict which atom is larger and why: Si or Cl, S or Se, and Mo or Ag.

Strategy: Qualitative predictions about atomic size can be made on the basis of electron configurations and the effects of Z and *n* on size.

Solution: Silicon and chlorine are in the third row of the periodic table:

	p^1	p^2	p^3	p^4	p^5	p^6
n = 3		Si			Cl	
Z:	13	14	15	16	17	18

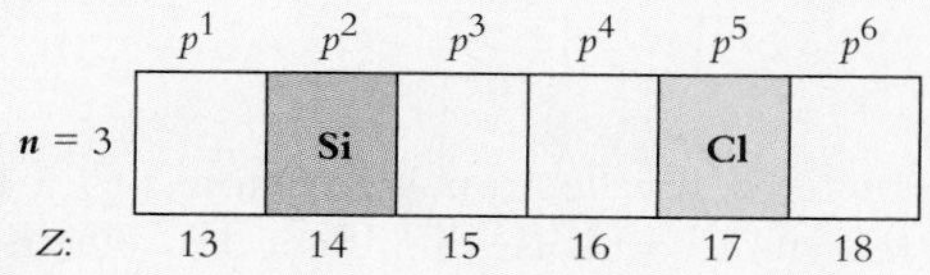

Chlorine's nuclear charge (+17) is larger than silicon's (+14), so chlorine's nucleus exerts a stronger pull on its electron cloud. Chlorine also has three more electrons than silicon, which raises the possibility that screening effects could counter the extra nuclear charge. Remember, however, that electrons in the same type of orbital do a poor job of screening one another from the nuclear charge. For Si and Cl, screening comes mainly from the core electrons, not from the electrons in the 3*p* orbitals. Because screening effects are similar for these elements, nuclear charge determines which of the two atoms is larger. Therefore we conclude that chlorine, with its greater nuclear attraction for the electron cloud, is the smaller atom. In terms of trends, chlorine lies to the right of silicon in the same row of the periodic table. Size decreases from left to right in any row; thus chlorine is smaller than silicon.

Sulfur and selenium are in Group 16 of the periodic table.

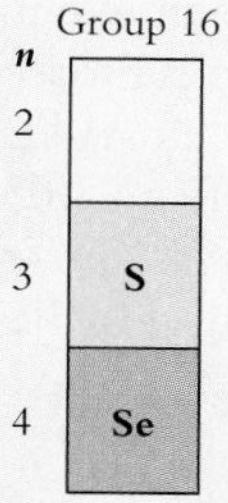

Although they both have the s^2p^2 valence configurations, selenium's least stable electrons are in orbitals with a larger *n* value. Orbital size increases with *n*. Selenium also has a greater nuclear charge than sulfur, which raises the possibility that nuclear attraction could offset increased *n*. Remember, however, that much of this extra nuclear charge is offset by the screening influence of the core electrons. Selenium has 18 core electrons, and sulfur has 10.

	Group 16
n	
2	
3	S
4	Se

Thus we conclude that selenium, with its larger *n* value, is larger than sulfur. In terms of trends, selenium is immediately below sulfur in the same column of the periodic table. Size increases down a column; thus selenium is larger than sulfur.

| Example 7-8 | Trends in Atomic Radii *(continued)* |

Molybdenum and silver are in the same row of the *d* block:

		Mo			Ag

They have the following configurations:

$$Mo = [Kr]\ 5s^1\ 4d^5 \qquad Ag = [Kr]\ 5s^1\ 4d^{10}$$

In each case, $5s$ is the largest occupied orbital. The $4d$ orbitals are smaller, with their electron density located mostly inside the $5s$ orbital. Consequently, $4d$ is effective at screening $5s$. The nuclear charge of silver is five units larger than that of molybdenum, but silver also has five extra screening electrons. These offset the extra nuclear charge, making Mo and Ag nearly the same size. In terms of trends, molybdenum and silver occupy the same row of the d block of the periodic table, across which size changes very little; thus molybdenum and silver are nearly the same size.

Ionization Energy

When an atom absorbs light, an electron is promoted to a less stable orbital. As electrons move into less stable orbitals, they have less electrical attraction for the nucleus. If the absorbed photon has enough energy, an electron can be ejected from the atom, as occurs in photoelectron spectroscopy.

The minimum amount of energy needed to remove an electron from a neutral atom is the first ionization energy (IE_1). Variations in ionization energy mirror variations in orbital stability, because an electron in a less stable orbital is easier to remove than one in a more stable orbital:

/// *First ionization energy increases from left to right across each row and decreases from top to bottom of each column of the periodic table.*

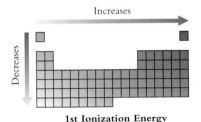

1st Ionization Energy

Figure 7–17 shows how the first ionization energy of gaseous atoms varies with atomic number. Notice the trends in ionization energy: A regular increase from left to right across each row (Row 3: 496 kJ/mol for Na to 1520 kJ/mol for Ar), and a regular decrease from top to bottom of each column (Group 18: 2372 kJ/mol for He to 1037 kJ/mol for Rn). As with atomic radius, ionization energy is fairly constant for elements in the d and f blocks, because increases in Z are offset by increased screening from the d and f orbitals.

Appendix C gives the first three
ionization energies for the first 36
elements.

Higher Ionizations

A multielectron atom can lose more than one electron, but ionization becomes more difficult as cationic charge increases. The first three ionization energies for a magnesium atom in the gas phase provide an illustration. (Ionization energies are measured on gaseous elements to ensure that the atoms are isolated from one another.)

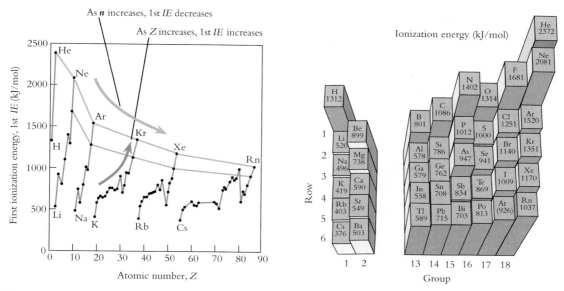

Figure 7-17
The first ionization energy increases from left to right and decreases from top to bottom of the periodic table.

Process	Configurations	IE
$Mg(g) \longrightarrow Mg^+(g) + e^-$	$[Ne]\,3s^2 \longrightarrow [Ne]\,3s^1$	738 kJ/mol
$Mg^+(g) \longrightarrow Mg^{2+}(g) + e^-$	$[Ne]\,3s^1 \longrightarrow [Ne]$	1450 kJ/mol
$Mg^{2+}(g) \longrightarrow Mg^{3+}(g) + e^-$	$[Ne] \longrightarrow [He]\,2s^2\,2p^5$	7730 kJ/mol

Notice that the second ionization energy of magnesium is almost twice as large as the first, even though each electron is removed from a $3s$ orbital. This is because Z_{eff} increases as the number of electrons decreases. That is, the positive charge on the magnesium nucleus remains the same throughout the ionization process, but the net charge of the electron cloud decreases with each successive ionization. As the number of electrons decreases, each electron feels a greater electrical attraction to the nucleus, resulting in a larger ionization energy.

The third ionization energy of magnesium is much larger than the first two. In addition to the positive charge of the cation being larger, the electron that is removed is a core electron having a lower principal quantum number ($2p$ rather than $3s$). Removing core electrons from any atom requires much more energy than removing valence electrons.

Electron Affinity

A neutral atom can add an electron to form an anion. The energy change when an electron is added to an atom is called the **electron affinity (EA).** Both ionization energy (*IE*) and electron affinity measure the stability of a bound electron, but for different species. Here, for example, are the values for fluorine:

$$F \longrightarrow F^+ + e^- \qquad IE_1 = 168 \text{ kJ/mol}$$

$$F + e^- \longrightarrow F^- \qquad EA = -322 \text{ kJ/mol}$$

Tables in reference sources often give electron affinities as positive values when the negative ion is more stable than the neutral atom. This convention is contrary to the sign convention for other energetic processes, which uses negative values when energy is released.

Energy is released when an electron is added to a fluorine atom to form a fluoride anion. In other words, a fluoride anion is more stable than a fluorine atom plus a free electron. Another way of saying this is that fluorine atoms have an affinity for electrons.

The energy associated with removing an electron to convert an anion to a neutral atom (that is, the reverse of electron attachment) has the same magnitude as the electron affinity, but the opposite sign. Removing an electron from F^-, for example, requires energy, giving a positive energy change:

$$F^- \longrightarrow F + e^- \qquad \Delta E = 322 \text{ kJ/mol}$$

The aufbau principle must be obeyed when an electron is added to a neutral atom, so the electron goes into the most stable orbital available. Hence, we expect trends in electron affinity to parallel trends in orbital stability. However, electron-electron repulsion and screening are more important for negative ions than for neutral atoms, so there is no clear trend in electron affinities as n increases. Thus there is only one general pattern:

/// *Electron affinity tends to become more negative from left to right across a row of the periodic table.*

The plot in Figure 7-18 shows how electron affinity changes with atomic number. The blue line highlights the increasing trend in the magnitude of electron affinity across each row of the periodic table. This trend is due to increasing effective nuclear charge, which binds the added electron more tightly to the nucleus.

Figure 7-18
The electron affinity of atoms varies with atomic number. In moving across any main group (blue lines) the electron affinity becomes more negative, but this is the only clear trend.

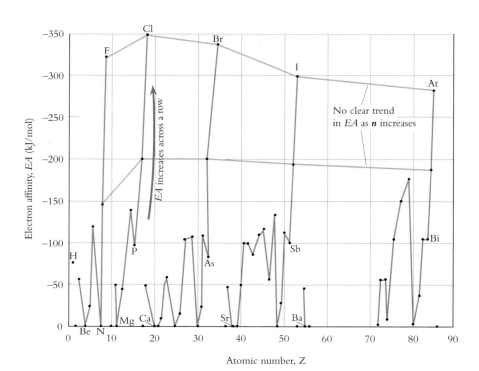

Notice that in contrast to the pattern for ionization energies (Figure 7.17), values for electron affinities remain nearly constant among elements occupying the same column of the periodic table.

The electron affinity values for many of the elements shown in Figure 7-18 appear to lie on the x-axis. Actually, these elements have positive electron affinities, meaning the resulting anion is less stable than the neutral atom. Moreover, the second electron affinity of every element is large and positive. Positive electron affinities cannot be measured directly. Instead, these values are estimated by other methods, as we show in Section 7.5.

Irregularities in Orbital Stability

Ionization energies and electron affinities deviate somewhat from smooth periodic behavior. These deviations can be attributed to screening effects and electron-electron repulsion. Aluminum, for example, has a smaller ionization energy than either of its neighbors in Row 3:

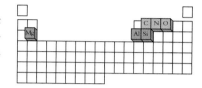

Element	Z	Atom Configuration	IE_1	Cation Configuration
Mg	12	[Ne] $3s^2$	735 kJ/mol	[Ne] $3s^1$
Al	13	[Ne] $3s^2 3p^1$	577 kJ/mol	[Ne] $3s^2$
Si	14	[Ne] $3s^2 3p^2$	787 kJ/mol	[Ne] $3s^2 3p^1$

The configurations of these elements show that a 3s electron is removed to ionize magnesium, whereas a 3p electron is removed to ionize aluminum or silicon. Screening makes the 3s orbital significantly more stable than a 3p orbital, and this difference in stability more than offsets the increase in nuclear charge in going from magnesium to aluminum.

As another example, nitrogen has a positive electron affinity, whereas its neighbors in Row 2 have significantly negative values:

Element	Z	Atom Configuration	EA	Anion Configuration
C	6	$1s^2 2s^2 2p^2$	−122 kJ/mol	$1s^2 2s^2 2p^3$
N	7	$1s^2 2s^2 2p^3$	>0 kJ/mol	$1s^2 2s^2 2p^4$
O	8	$1s^2 2s^2 2p^4$	−141 kJ/mol	$1s^2 2s^2 2p^5$

Electron affinity generally becomes more negative across a row because progressively higher nuclear charge makes the orbitals more stable. Making an anion out of a nitrogen atom, however, requires pairing two electrons in one of the p orbitals. The increased nuclear charge on moving from carbon to nitrogen is not enough to overcome the repulsion generated by confining two negatively charged electrons in one p orbital. Consequently, neutral nitrogen atoms are more stable than N⁻ anions.

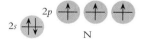

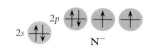

Forming an anion from an oxygen atom also requires pairing electrons in a p orbital, but for oxygen the destabilization caused by electron-electron repulsion is offset by the progressive increase in nuclear charge. In other words, increasing nuclear charge from 6 to 7 (C ⟶ N) is not enough to overcome electron-electron repulsion, but increasing Z from 6 to 8 (C ⟶ O) does provide enough orbital stabilization to offset the destabilization involved in electron pairing. Keep in mind, as well, that screening of one 2p orbital does not increase much as a result of adding electrons to other 2p orbitals.

Table 7–3
Trends in an Isoelectronic Series

Species	O^{2-}	F^-	Ne	Na^+	Mg^{2+}
Z	8	9	10	11	12
Radius (pm)	126	119	—*	116	86
Ionization energy (kJ/mol)	<0 ($-EA_2$)	322 ($-EA$)	2100 (IE_1)	4560 (IE_2)	7730 (IE_3)

* Radii of neutral atoms cannot be measured by the methods used for ionic species.

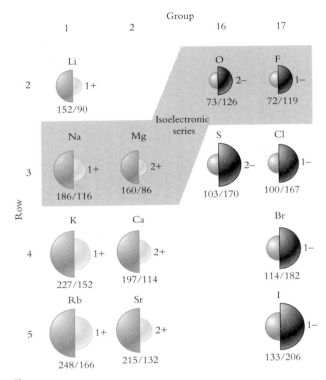

Figure 7-19
Comparison of the sizes of neutral atoms and their ions for some representative elements.

Sizes of Ions

An atomic cation is always smaller than the corresponding neutral atom. Conversely, an atomic anion is always larger than the neutral atom. These trends, which are illustrated in Figure 7-19, can be explained by electron–electron repulsion. A cation has fewer electrons than its parent neutral atom. This reduction in the number of electrons means that the cation's remaining electrons experience less electron–electron repulsion. An anion has more electrons than its parent neutral atom. This increase in the number of electrons means that there is greater electron–electron repulsion in the anion than in the parent neutral atom.

Figure 7-19 also highlights the relationships among isoelectronic species, those possessing equal numbers of electrons. As noted in Section 7.3, the F^- anion and the Na^+ cation are isoelectronic, each having 10 electrons and the configuration [He] $2s^2 2p^6$. For members of an isoelectronic series, properties change regularly with Z. For example, Table 7-3 shows two properties of the 10-electron isoelectronic series. A progressive increase in nuclear charge results in a corresponding decrease in ionic radius, a result of stronger electrical force between the nucleus and the electron cloud. For the same reason, as Z increases, it becomes progressively more difficult to remove an electron.

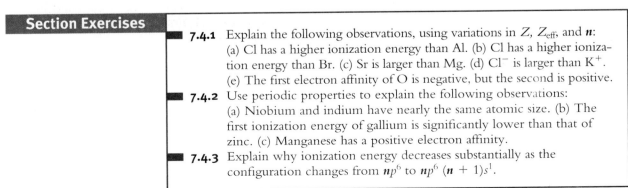

Section Exercises

7.4.1 Explain the following observations, using variations in Z, Z_{eff}, and n: (a) Cl has a higher ionization energy than Al. (b) Cl has a higher ionization energy than Br. (c) Sr is larger than Mg. (d) Cl^- is larger than K^+. (e) The first electron affinity of O is negative, but the second is positive.

7.4.2 Use periodic properties to explain the following observations: (a) Niobium and indium have nearly the same atomic size. (b) The first ionization energy of gallium is significantly lower than that of zinc. (c) Manganese has a positive electron affinity.

7.4.3 Explain why ionization energy decreases substantially as the configuration changes from np^6 to np^6 $(n + 1)s^1$.

7.5 ENERGETICS OF IONIC COMPOUNDS

The trends in ionization energies and electron affinities indicate that some elements form ions more easily than others. Moreover, we know that ions with opposite charges attract each other. Ionic compounds form when the stabilization gained through ionic attraction exceeds the energy required to create ions from neutral atoms. In this section, we perform energy analyses to show which elements tend to form ionic compounds.

One example is illustrated in Figure 7-20. Soft, lustrous sodium metal reacts vigorously with yellow-green chlorine gas to form sodium chloride, a white crystalline solid that we know as common table salt. Sodium chloride contains sodium and chlorine in a 1:1 elemental ratio. As shown in Figure 7-21, a crystal of sodium chloride is made up of equal numbers of Na^+ and Cl^- ions organized in a regular, alternating array called a **lattice.** Using features of ionization energy, electron affinity, and electrical forces, we can analyze the energetics of this reaction.

To make use of ionization energies and electron affinities, we must describe a path for the reaction in which the ionization steps take place in the gas phase. Here is one such path:

$$Na(s) \longrightarrow Na(g)$$
$$Na(g) \longrightarrow Na^+(g) + e^-$$
$$\tfrac{1}{2}Cl_2(g) \longrightarrow Cl(g)$$
$$Cl(g) + e^- \longrightarrow Cl^-(g)$$
$$Na^+(g) + Cl^-(g) \longrightarrow NaCl(s)$$

$$\overline{Na(s) + \tfrac{1}{2}Cl_2(g) \longrightarrow NaCl(s)}$$

Notice that adding all the individual steps of this path gives a net reaction that matches the overall stoichiometry of the reaction. Energy changes for reactions do not depend on the path by which starting materials are converted to products. Applying the principle of conservation of energy, we see that the net energy change for the reaction is the sum of the energies of the individual steps.

Step 1: Vaporization. Sodium metal must be vaporized to form sodium gas. Experimental measurements have determined that 108 kJ/mol must be supplied to overcome the forces that hold sodium atoms in the solid state.

$$Na(s) \longrightarrow Na(g) \qquad \Delta E_{vaporization} = 108 \text{ kJ/mol}$$

Step 2: Ionization of Na. Ionizing sodium atoms requires that energy be supplied, the amount being the first ionization energy for sodium.

$$Na(g) \longrightarrow Na^+(g) + e^- \qquad \Delta E = IE = 495.5 \text{ kJ/mol}$$

Step 3: Bond Breakage. Chlorine molecules must be broken apart into chlorine atoms. As we describe in Chapter 9, the energy required to break one mole of chlorine molecules into individual atoms is the *bond energy*. The bond energy of molecular chlorine is 240 kJ/mol. We need ½ mole of Cl_2 to form 1 mole of NaCl.

$$\tfrac{1}{2}Cl_2(g) \longrightarrow Cl(g) \qquad \Delta E = \tfrac{1}{2}(\text{bond energy}) = 120 \text{ kJ/mol}$$

Step 4: Ionization of Cl. The electron affinity of chlorine is negative, which means that energy is released when a chlorine atom gains an electron.

$$Cl(g) + e^- \longrightarrow Cl^-(g) \qquad \Delta E = EA = -348.5 \text{ kJ/mol}$$

Instead of ions, most substances contain molecules, in which electrons are shared between atoms to form chemical bonds. We treat bonding in Chapters 8 and 9.

Figure 7-20
Sodium and chlorine react vigorously and spontaneously to form sodium chloride. The salt can be seen as the cloud rising from the flask.

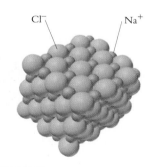

Figure 7-21
Sodium chloride is made up of Na^+ cations and Cl^- anions organized in a three-dimensional cubic array.

← **SECTIONS 9.2 & 12.5**
Bond energies are tabulated in Section 9.2, and energies of vaporization are described in Section 12.5.

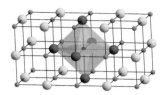

Figure 7-22
In a sodium chloride crystal each sodium ion has six closest chloride neighbors and many other chlorides further away. The figure highlights one set of closest neighbors.

Step 5: Condensation. Individual Na^+ and Cl^- ions must condense into a three-dimensional lattice of ions. The energy released in the condensation represents all the ion-ion attractive and repulsive interactions in the lattice and is called the **lattice energy.**

To calculate a lattice energy, we need an equation for the energy resulting from the interaction of two electrically charged objects. Such an equation can be derived from Coulomb's law stating the magnitude of electrical force between charged particles:

$$E_{coulomb} = k \frac{(q_1)(q_2)}{r} \tag{7-1}$$

In Equation 7-1, q_1 and q_2 are the electrical charges of the objects and r is the distance between them. The value of k is 1.389×10^5 kJ pm/mol when distances are in picometers, charges are expressed in units of electron charge ($q_1 = -1$ for an electron and $q_2 = +Z$ for a nucleus), and the calculation is done for one mole of objects. If the two objects have opposite charges, one positive and one negative, the sign of $E_{coulomb}$ is negative. A negative sign indicates that energy is released, which means opposite charges become more stable as they get closer together. If the objects have the same charge, both positive or both negative, Equation 7-1 gives a positive energy, meaning that energy must be added to move the objects toward one another. Thus Coulomb's law is a restatement of the principle of electrical force, which we describe in Chapter 2.

For the sodium chloride pair of ions, $q_1 = +1$ and $q_2 = -1$. To complete the calculation we need to know how closely the ions approach each other before their mutual attraction is balanced by electron cloud repulsion. In the sodium chloride crystal this distance is 313 pm. Using this value for r, we can calculate the energy released in forming one mole of ion pairs:

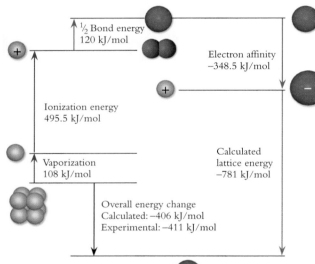

Figure 7-23
The reaction of sodium metal with chlorine gas to produce solid sodium chloride can be analyzed by breaking the overall process into a series of steps involving ions in the gas phase.

Tutorial

$$Na^+(g) + Cl^-(g) \longrightarrow NaCl(g, \text{ion pair})$$

$$E_{coulomb} = \frac{(1.389 \times 10^5 \text{ kJ pm/mol})(+1)(-1)}{(313 \text{ pm})} = -444 \text{ kJ/mol}$$

This is only part of the lattice energy, because coulombic interactions do not stop at individual ion pairs. In solid sodium chloride, each sodium cation is attracted to all the surrounding chloride anions. As Figure 7-22 shows, one Na^+ ion has six nearby Cl^- ions. Moreover, there are many other chloride ions further away, and all of them contribute to the lattice energy in proportion to their distance from a particular sodium ion. At the same time, the lattice energy includes repulsive interactions from ions of the same charge that are close to one another. Coulomb's law can be expanded to include all these ion-ion interactions in all three dimensions. When this is done for sodium chloride, the calculated value of the lattice energy is -781 kJ/mol.

$$Na^+(g) + Cl^-(g) \longrightarrow NaCl(s) \qquad \Delta E_{\text{lattice energy, calculated}} = -781 \text{ kJ/mol}$$

The overall energy change for the sodium chloride reaction is obtained by summing the energies of the five steps. Figure 7-23 summarizes the process.

$$\Delta E_{\text{calculated}} = \Delta E_{\text{vaporization}} + IE + \tfrac{1}{2} \text{ (bond energy)} + EA + \Delta E_{\text{lattice energy, calculated}}$$

$$\Delta E_{\text{calculated}} = 108 + 495.5 + 120 + (-348.5) + (-781) = -406 \text{ kJ/mol}$$

This calculated energy is very close to the experimental value for the energy released in the actual reaction:

$$Na(s) + \tfrac{1}{2} Cl_2(g) \longrightarrow NaCl(s) \qquad \Delta E_{\text{experimental}} = -411 \text{ kJ/mol}$$

There is close agreement between the actual energy released in the reaction and the energy calculated by assuming that the final product is composed of ions. This is strong evidence that solid sodium chloride is best described as made up of sodium cations and chloride anions.

The energy cycle illustrated in Figure 7-23 is called a *Born-Haber cycle*. Example 7-9 shows that a Born-Haber cycle can be used to estimate an electron affinity.

Born-Haber Cycle　　　　　　　　　　　　　　　　　　　**Example 7-9**

Magnesium metal burns in air to produce magnesium oxide, a white solid that contains Mg^{2+} and O^{2-}. The oxide anion, O^{2-}, is not stable except in a crystalline solid such as MgO. This makes it impossible to measure directly the second electron affinity of oxygen. Use a Born-Haber cycle and the following data to calculate oxygen's second electron affinity.

$$\text{Vaporization energy of Mg} = \Delta E_{\text{vap}} = 148 \text{ kJ/mol}$$
$$1^{\text{st}} \text{ ionization energy of Mg} = IE_1 = 738 \text{ kJ/mol}$$
$$2^{\text{nd}} \text{ ionization energy of Mg} = IE_2 = 1451 \text{ kJ/mol}$$
$$\text{Bond energy of } O_2 = \Delta E_{\text{BE}} = 495 \text{ kJ/mol}$$
$$1^{\text{st}} \text{ electron affinity of O} = EA_1 = -141 \text{ kJ/mol}$$
$$\text{Calculated lattice energy MgO} = \Delta E_{\text{lattice}} = -3963 \text{ kJ/mol}$$
$$\text{Experimental reaction energy} = \Delta E_{\text{reaction}} = -602 \text{ kJ/mol}$$

Strategy: We are asked to find the second electron affinity of oxygen:

$$O^-(g) + e^- \longrightarrow O^{2-}(g) \qquad EA_2 = ?$$

As noted in the previous section, second electron affinities are all large and positive.

The different steps of a Born-Haber cycle can be connected together as shown in Figure 7-23. The overall energy change for the reaction described by the cycle is equal to the sum of the energy changes for the individual steps. A diagram of the steps, similar to Figure 7-23, helps sort out the calculation. Remember that the overall reaction consumes ½ mole of O_2 for each mole of Mg. Therefore, our calculation requires ½ the bond energy of molecular oxygen.

Solution: All the energy changes for the cycle are found among the data, except for the electron affinity of O^-, which is EA_2. Set up the sum of the individual energy changes and then solve for EA_2.

$$\Delta E_{\text{reaction}} = \Delta E_{\text{vap}} + IE_1 + IE_2 + \tfrac{1}{2}\Delta E_{\text{BE}} + EA_1 + EA_2 + \Delta E_{\text{lattice}}$$
$$EA_2 = \Delta E_{\text{reaction}} - \Delta E_{\text{vap}} - IE_1 - IE_2 - \tfrac{1}{2}\Delta E_{\text{BE}} - EA_1 - \Delta E_{\text{lattice}}$$
$$EA_2 = [-602 - 148 - 738 - 1451 - 248 - (-141) - (-3963)] \text{ kJ/mol}$$
$$EA_2 = 917 \text{ kJ/mol}$$

The result is a large positive value, reflecting the fact that an isolated O^{2-} anion is very unstable.

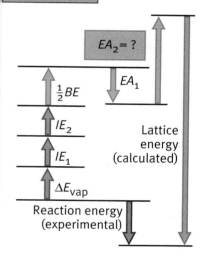

Tutorial

Why Not $Na^{2+}\ Cl^{2-}$?

The coulombic attraction between doubly charged ions is significantly greater than the attraction between singly charged ions, because q_1 and q_2 are twice as large. This suggests that the transfer of a second electron might give an even more stable crystal composed of Na^{2+} and Cl^{2-} ions. We can extend our calculations to see if this is true.

The energy of attraction between Na^{2+} and Cl^{2-} ions in one $Na^{2+}Cl^{2-}$ ion pair would be about four times larger than that of one Na^+Cl^- pair. This works out to a calculated lattice energy for $Na^{2+}Cl^{2-}$ of around -2200 kJ/mol. However, to collect this energy the ion pair must pay the price of forming the doubly charged ions. Appendix C provides the first two ionization energies for sodium:

$$Na \longrightarrow Na^+ + e^- \qquad 1s^2\ 2s^2\ 2p^6\ 3s^1 \longrightarrow 1s^2\ 2s^2\ 2p^6 \qquad IE_1 = 495.5\ \text{kJ/mol}$$

$$Na^+ \longrightarrow Na^{2+} + e^- 1s^2\ 2s^2\ 2p^6 \longrightarrow 1s^2\ 2s^2\ 2p^5 \qquad IE_2 = 4562\ \text{kJ/mol}$$

We do not know the second electron affinity, but the result of Example 7-9 suggests that the value will be large and positive.

We can see why sodium and chlorine do not form $Na^{2+}Cl^{2-}$ even without taking chlorine's second electron affinity into account. Forming Na^{2+} requires more than 4500 kJ/mol, much more energy than the 2200 kJ/mol released in the formation of the $Na^{2+}Cl^{2-}$ lattice.

The second ionization energy of sodium is much larger than its first ionization energy because a *core* $2p$ electron must be removed to create Na^{2+} from Na^+. Removal of a core electron always requires a great deal of energy, so it is a general feature of ionic systems that ions formed by removing core electrons are not found in stable ionic compounds.

Cation Stability

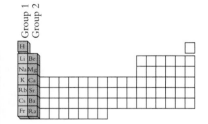

Knowing that the energy cost of removing core electrons is always excessive, we can predict that the ionization process will stop when all valence electrons have been removed. Thus a knowledge of ground-state configurations is all that we need to make qualitative predictions about cation stability.

Each element in Group 1 of the periodic table has one valence electron. These elements form ionic compounds containing A^+ cations. Examples are KCl and Na_2CO_3. All elements in Group 2 of the periodic table have two valence electrons and form ionic compounds containing A^{2+} cations. Examples are $CaCO_3$ and $MgCl_2$.

Beyond these two columns, the removal of all valence electrons is usually not energetically possible. For example, iron has eight valence electrons but forms only two stable cations, Fe^{2+} and Fe^{3+}. Compounds of iron containing these ions are abundant in the Earth's crust. Pyrite (FeS_2) and iron(II) carbonate ($FeCO_3$, or siderite) are examples of Fe^{2+} salts. Iron(III) oxide (Fe_2O_3, or hematite) can be viewed as a network of Fe^{3+} cations and O^{2-} anions. One of the most abundant iron ores, magnetite, has the chemical formula Fe_3O_4 and contains a $2:1$ ratio of Fe^{3+} and Fe^{2+} cations. The formula of magnetite can also be written as $FeO \cdot Fe_2O_3$ to emphasize the presence of two different cations.

Other metallic elements form ionic compounds with cation charges ranging from $+1$ to $+3$. Aluminum nitrate nonahydrate, $Al(NO_3)_3 \cdot 9H_2O$, is composed of Al^{3+} cations, NO_3^- anions, and water molecules. Silver nitrate ($AgNO_3$), which contains Ag^+ cations, is a soluble silver salt that is used in silver plating.

Compounds of Iron
(clockwise from top) $FeCO_3$, Fe_2O_3, Fe_3O_4, FeS_2.

Anion Stability

Halogens, the elements in Group 17 of the periodic table, have the largest electron affinities of all the elements, so halogen atoms (ns^2np^5) readily accept electrons to produce halide anions (ns^2np^6). As a result, many metals form binary compounds, called *halides,* that contain metal cations and halide anions. Examples include NaCl (chloride anion), CaF_2 (fluoride anion), AgBr (bromide anion), and KI (iodide anion).

Isolated atomic anions with charges more negative than -1 are always unstable, but oxide (O^{2-}, $1s^2\ 2s^2\ 2p^6$) and sulfide (S^{2-}, [Ne] $3s^2\ 3p^6$) are found in many ionic solids, such as CaO and PbS. The -2 anions in these solids are stabilized by a combination of large lattice energies and a three-dimensional array of surrounding cations.

The ionic model describes a number of metal halides, oxides, and sulfides, but it does not describe adequately most other chemical substances. Whereas substances such as CaO, NaCl, and MgF_2 behave like simple cations and anions held together by coulombic attraction, substances such as CO, Cl_2, and HF do not. In a crystal of MgF_2, electrons have been *transferred* from magnesium atoms to fluorine atoms, but the stability of HF molecules arises from the *sharing* of electrons between hydrogen atoms and fluorine atoms. We describe electron sharing, which is central to molecular stability, in Chapters 8 and 9.

Section Exercises

- **7.5.1** Magnesium fluoride forms from the elements as follows:

  $$Mg(s) + F_2(g) \longrightarrow MgF_2(s) \quad \Delta E_{reaction} = -1123 \text{ kJ/mol}$$

  The energy of vaporization of Mg is 148 kJ/mol, and the bond energy of F_2 is 155 kJ/mol . Use this information and data from Appendix C to calculate the lattice energy of MgF_2.

- **7.5.2** Iron and cobalt form compounds that can be viewed as containing A^{3+} cations, but nickel does not. Use the ionization energies in Appendix C to predict which transition metal elements are unlikely to form stable cations with charges greater than $+2$.

- **7.5.3** From the location of each element in the periodic table, predict which ion of each of the following elements will be found in ionic compounds: Ca, Cs, Al, and Br.

7.6 IONS AND CHEMICAL PERIODICITY

The elements that form ionic compounds are found in specific places in the periodic table. Atomic anions are restricted to the halogens and their close neighbors on the right side of the table, oxygen and sulfur. Atomic cations in compounds, on the other hand, can form from all elements in the *s, d,* and *f* blocks.

Ion formation is only one pattern of chemical behavior. Many other chemical trends can be traced ultimately to valence electron configurations, but the description of chemical bonding that appears in Chapters 8 and 9 is needed to explain such periodic properties. Nevertheless, important patterns in chemical behavior can be related to the ability of some elements to form ions. One example is the subdivision of the periodic table into metals, nonmetals, and metalloids, first introduced in Chapter 1. The elements that can form cations relatively easily are metals.

All metals have similar properties, in part because their outermost *s* electrons are relatively easy to remove. All elements in the *s* block have ns^1 or ns^2 valence configurations. The *d*-block elements have one or two *s* electrons and various numbers of *d* electrons. Examples are titanium ($4s^2\ 3d^2$) and silver ($5s^1\ 4d^{10}$). Elements in the *f* block have two *s* electrons and a number of *f* electrons of lower principal quantum number. Samarium, for example, has the valence configuration

$6s^2 4f^6$. As we describe in Chapter 9, the metallic behavior of these elements occurs partly because the *s* electrons are shared readily among all atoms. Metals form ionic salts because *s* electrons (and some *d, p,* and *f* electrons) can be readily removed from the metal atoms to form cations.

Elemental properties vary widely within the *p* block. Elements toward the left or bottom of this block can lose *p* electrons easily and therefore have metallic properties. Examples are aluminum ($3s^2 3p^1$), tin ($5s^2 5p^2$), and bismuth ($6s^2 6p^3$). In contrast, the halogens and noble gases on the right of this block are distinctly nonmetallic. The noble gases, Group 18 of the periodic table, are monatomic gases that resist chemical attack, because their electron configurations contain completely filled *s* and *p* orbitals.

Elements in the intermediate columns of the *p* block have very different chemical properties even though they have the same valence configurations. Carbon, silicon, germanium, and tin all have $ns^2 np^2$ valence configurations; yet carbon is a nonmetal, silicon and germanium are metalloids, and tin is a metal.

Qualitatively, we can understand this variation by recalling that as the principal quantum number increases, the valence orbitals become less stable. In tin, the four $n = 5$ valence electrons are bound relatively loosely to the atom, resulting in the metallic properties associated with electrons that are easily removed. In carbon, the four $n = 2$ valence electrons are bound relatively tightly to the atom, resulting in nonmetallic behavior. Silicon ($n = 3$) and germanium ($n = 4$) fall in between these two extremes. Example 7-10 describes the elements with five valence electrons.

Example 7-10	**Classifying Elements**

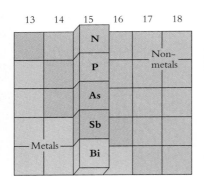

Nitrogen is a colorless diatomic gas. Phosphorus has several elemental forms, but the most common is a red solid that is used for matches. Arsenic and antimony are gray solids, and bismuth is a lustrous solid. Classify these elements of Group 15 as metals, nonmetals, or metalloids.

Strategy: All elements except those in the *p* block are metals. Group 15, however, is part of the *p* block, within which elements display all forms of elemental behavior. To decide the classifications of these elements, we must examine this group relative to the diagonal arrangement of the metalloids:

Solution: We see that Group 15 passes through all three classes of elements. The elements with the lowest *Z* values, nitrogen and phosphorus, are nonmetals. The element with highest *Z* value, bismuth, is a metal, and the two elements with intermediate *Z* values, arsenic and antimony, are metalloids.

s-Block Elements

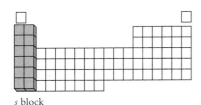

s block

The electronic configuration of any element in Groups 1 and 2 of the periodic table contains a core of tightly bound electrons and one or two *s* electrons that are loosely bound. The **alkali metals** (Group 1, ns^1 configuration) and the **alkaline earth metals** (Group 2, ns^2 configuration) form stable ionic salts because their valence electrons are easily removed. Nearly all salts of alkali metals and many salts of alkaline earth metals dissolve readily in water, so naturally occurring sources of water frequently contain these ions.

The four most abundant *s*-block elements in the Earth's crust are sodium, potassium, magnesium, and calcium; their occurrence is summarized in Table 7-4. These elements are found in nature in salts such as NaCl, KNO_3, $MgCl_2$, $MgCO_3$,

Table 7 – 4
Abundance of *s*-Block Elements

Element	Abundance in Crust (% by Mass)	Abundance in Seawater (mol/L)	Abundance in Plasma (mol/L)
Na	2.27	0.462	0.142
K	1.84	0.097	0.005
Mg	2.76	0.053	0.003
Ca	4.66	0.100	0.005

and $CaCO_3$. Portions of these solid salts dissolve in rainwater as it percolates through the Earth's crust. The resulting solution of anions and cations eventually finds its way to the oceans. When water evaporates from the oceans, the ions are left behind. Over many eons the continual influx of river water containing these ions has built up the substantial salt concentrations found in the Earth's oceans.

Table 7-4 shows that each of the four common *s*-block ions is abundant not only in sea water but also in body fluids, where these ions play essential biochemical roles. Sodium is the most abundant cation in fluids that are outside of cells, and proper functioning of body cells requires that sodium concentrations be maintained within a very narrow range. One of the main functions of the kidneys is to control the excretion of sodium. Whereas sodium cations are abundant in the fluids outside of cells, potassium cations are the most abundant ions in the fluids inside cells. The difference in ion concentration across cell walls is responsible for the generation of nerve impulses that drive muscle contraction. If the difference in potassium ion concentration across cell walls deteriorates, muscular activity, including the regular muscle contractions of the heart, can be seriously disrupted.

The cations Mg^{2+} and Ca^{2+} are major components of bones. Calcium occurs as hydroxyapatite, a complicated substance whose chemical formula is $Ca_5(PO_4)_3(OH)$. The structural form of magnesium in bones is not fully understood. In addition to being essential ingredients of bone, these two cations also play key roles in various biochemical reactions, including photosynthesis, the transmission of nerve impulses, and the formation of blood clots.

Some compounds of the *s*-block elements are important industrial chemicals, too. For example, more than 1.4 billion kilograms of potassium carbonate, commonly known as "potash" (K_2CO_3), is produced in the United States each year. This compound, which is obtained from mineral deposits, is the most common source of potassium for fertilizers. Fertilization with potassium is necessary because this element is essential for healthy plant growth. Moreover, potassium salts are highly soluble in water, so potassium quickly becomes depleted from the soil. Consequently, agricultural land requires frequent addition of potassium fertilizers.

Three other compounds of *s*-block elements — calcium oxide (CaO, known as "lime"), sodium hydroxide (NaOH), and sodium carbonate (Na_2CO_3) — are among the top 15 industrial chemicals in annual production. Lime is perennially in the top 10 because it is the key ingredient in construction materials such as concrete, cement, mortar, and plaster. Two other compounds, calcium chloride ($CaCl_2$) and sodium sulfate (Na_2SO_4), rank just below the top 50 in industrial importance.

Many industrial processes make use of chemically useful anions such as hydroxide (OH^-), carbonate (CO_3^{2-}), and chlorate (ClO_3^-). These anions must be sup-

Beryllium behaves differently from the other *s*-block elements because the $n = 2$ orbitals are more compact than orbitals with higher principal quantum number. The first ionization energy of beryllium, 899 kJ/mol, is comparable with those of nonmetals, so beryllium does not form compounds that are clearly ionic.

The calcium compound that makes up bones is very durable.

The first US patent was for a method of making potash, issued in 1790 to Samuel Hopkins of Pittsford, Vermont. The patent examiner was Thomas Jefferson and the signator was George Washington.

plied by chemical compounds that include cations. Sodium is most frequently used as this spectator cation because it is abundant, inexpensive, and nontoxic. Hydroxide ion is industrially important because it is a strong base. Sodium hydroxide is used to manufacture other chemicals, textiles, paper, soaps, and detergents. Sodium carbonate and sand are the major starting materials in the manufacture of glass. Glass contains sodium and other cations embedded in a matrix of silicate (SiO_3^{2-}) anions. About half the sodium carbonate produced in the world is used in glass making.

p-Block Elements

The properties of elements in the p block vary across the entire spectrum of chemical possibilities. The elements in Group 13, with a single electron in a p orbital as well as two valence s electrons, display chemical reactivity characteristic of three valence electrons. Except for boron, the elements of this group are metals that form stable cations with +3 charge. Examples are $Al(OH)_3$ and GaF_3. Metallic character diminishes rapidly as additional p electrons are added. This change culminates in the elements in Group 18. With filled p orbitals, these elements are so unreactive that for many years they were thought to be completely inert. Xenon is now known to form compounds with the most reactive nonmetals, oxygen, fluorine, and chlorine; krypton forms a few highly unstable compounds with these elements.

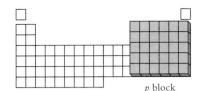

p block

Although the nonmetals do not readily form cations, many of them combine with oxygen to form polyatomic oxyanions. These anions have various stoichiometries, but there are some common patterns. Two second-row elements form oxyanions with three oxygen atoms: carbon (four valence electrons) forms carbonate, CO_3^{2-}, and nitrogen (five valence electrons) forms nitrate, NO_3^-. In the third row, the most stable oxyanions contain four oxygen atoms: SiO_4^{4-}, PO_4^{3-}, SO_4^{2-}, and ClO_4^-.

Many of the minerals that form the Earth's crust contain oxyanions. Limestone, $CaCO_3$, and dolomite, $MgCa(CO_3)_2$, are carbonates; barite, $BaSO_4$, is a sulfate; fluoroapatite, $Ca_5(PO_4)_3F$, contains phosphate; and olivine, a mixture of $MgSiO_4$ and $FeSiO_4$, and zircon, $ZrSiO_4$, are silicates.

Several leading industrial chemicals contain these anions or are acids resulting from addition of H^+ to the anions. Sulfuric acid, H_2SO_4, is the #1 industrial chemical in the United States. Two other industrially important acids are nitric acid, HNO_3, and phosphoric acid, H_3PO_4. A key ingredient in concrete is lime, CaO, made by heating $CaCO_3$ to drive off CO_2. Industrially important salts include ammonium sulfate, $(NH_4)_2SO_4$; aluminum sulfate, $Al_2(SO_4)_3$; sodium carbonate, Na_2CO_3; and ammonium nitrate, NH_4NO_3.

Section Exercises

7.6.1 Consult the table of first ionization energies in Appendix C and calculate the average values for the nonmetals, metalloids, and s-block elements. How does the trend in these averages relate to the ionic chemistry of these elements?

7.6.2 Classify each of the following elements as a metal, nonmetal, or metalloid: S, Si, Sr, Se, Sc, Sg, and Sn.

7.6.3 Some metals form oxyanions with the same relationship among charge, valence electrons, and number of oxygen atoms as described in this section. Consult Table 3-5 and identify the oxyanions of metals that satisfy this relationship.

CHAPTER REVIEW

Summary and Key Terms

1. Orbital energies can be measured using **photoelectron spectroscopy** and by determining **ionization energy (*IE*)**. Orbital energies depend on *n, Z*, and, through screening generated by the presence of other electrons, *l*. **Screening** results in electrons experiencing an **effective nuclear charge** (Z_{eff}) less than *Z*.

2. According to the **aufbau principle**, electrons fill the most stable orbitals available. The **Pauli exclusion principle** states that no two electrons have the same quantum numbers. A **ground-state configuration** of an atom describes the most stable arrangement of its electrons. Orbitals with identical energy are **degenerate orbitals**. The form of the periodic table is determined by common features of the accessible electrons in ground-state configurations, which in turn are dictated by the relative stability of the various orbitals. Accessible electrons are called **valence electrons**, and inaccessible electrons are called **core electrons**.

3. An **electron configuration** specifies completely the distribution of electrons in an atom. A configuration includes the orbital designation (1s, 3d, etc) and the number of electrons in each orbital. By **Hund's rule**, electrons are always oriented so that spin is maximized for partially filled orbitals. This results in **paramagnetic** species with unpaired electrons. Species with all electrons paired are **diamagnetic**. Some atomic configurations depart from the normal filling sequence because of **near-degenerate orbitals**. This also results in some differences in configurations among **isoelectronic** species.

4. Atomic properties show regular periodic variations: As *Z* increases across a row, orbitals shrink and become more stable, and as *n* increases down a column, orbitals become larger and less stable. Atomic size, ionization energy, and **electron affinity** all show systematic periodic variations with *n* and *Z*.

5. Ionic compounds form between metals and halogens, oxygen, sulfur or polyatomic anions. In the solid state, ions are arranged in regular patterns in a three-dimensional **lattice**. The electronic forces that hold ions in the solid state are summarized in the **lattice energy**. The stability of binary ionic compounds can be analyzed in terms of a Born-Haber cycle of energy changes accompanying fundamental processes.

6. Metals easily form cations. Among nonmetals, the Group 18 noble gases are particularly unreactive. Metals in the *s* block are **alkali metals** (Group 1) and **alkaline earth metals** (Group 2). Elements in the *p* block show a wide range of different chemical properties.

Skills to Master

▶ Arranging orbitals in order of energy

▶ Writing electron configurations for atoms and ions

▶ Recognizing the effects of near-degenerate orbitals

▶ Correlating configurations with periodicity

▶ Counting valence electrons

▶ Predicting periodic variations in properties

▶ Calculating energies for ion formation

Learning Exercises

7.1 Draw appropriately scaled pictures of all the occupied orbitals in a krypton atom.

7.2 Write brief explanations of (a) screening; (b) the Pauli exclusion principle; (c) the aufbau principle; (d) Hund's rule; (e) degeneracy; and (f) valence electrons.

7.3 Construct an orbital energy level diagram for all orbitals with *n* < 8 and *l* < 4. Use the periodic table to help determine the correct order of the energy levels.

7.4 Describe patterns in periodic properties; explain how they affect ionization energy and electron affinity.

7.5 List the 18 elements classified as nonmetals. Give the name, symbol, and atomic number for each.

7.6 Update your list of memory bank equations by adding principles for atomic configurations.

7.7 List all terms new to you that appear in Chapter 7. Write a one-sentence definition of each, using your own words. Consult the glossary if you need help.

Orbital Energies

Problems   <u>ilw</u> = interactive learning ware problem. Visit the website at www.wiley.com/college/olmsted

7.1 For each pair of orbitals, determine which is more stable and explain why: (a) He 1s and He 2s; (b) Kr 5p and Kr 5s; and (c) He 2s and He$^+$ 2s.

7.2 For each pair of orbitals, determine which is more stable and explain why: (a) C 2s and C 2p; (b) Ar 5p and Ar$^+$ 5p; and (c) Ar 4s and Ar 5s.

7.3 In a hydrogen atom the 3s, 3p, and 3d orbitals all have the same energy. In a helium atom, however, the 3s orbital is lower in energy than the 3p orbital, which in turn is lower in energy than the 3d orbital. Explain why the energy rankings of hydrogen and helium are different.

7.4 The energy of the *n* = 2 orbital of the He$^+$ ion is the same as the energy of the *n* = 1 orbital of the H atom. Explain this fact.

7.5 Refer to Table 7-1 to answer the following questions. In each case, provide a brief explanation of your choice. (a) Which ionization energies show that an electron in a $1s$ orbitals provides nearly complete screening of an electron in a $2p$ orbital? (b) Which ionization energies show that the stability of $n = 2$ orbitals increases with Z^2?

7.6 Refer to Table 7-1 to answer the following questions. In each case, provide a brief explanation of your choice. (a) Which ionization energies show that stability of $n = 1$ orbitals increase with Z? (b) Which ionization energies show that an electron in a $1s$ orbital incompletely screens another electron in the same orbital?

Structure of the Periodic Table

7.7 Draw the periodic table in block form, and outline and label each of the following sets: (a) Elements that are one electron short of filled p orbitals; (b) elements for which $n = 3$ orbitals are filling; (c) elements with half-filled d orbitals; and (d) the first element that contains a $5s$ electron.

7.8 Draw the periodic table in block form, and outline and label each of the following sets: (a) Elements for which $n = 1$ orbitals are filling; (b) elements for which the $5f$ orbitals are filling; (c) elements with $s^2 p^4$ configurations; and (d) elements with filled valence s orbitals but empty valence p orbitals.

7.9 Determine the atomic number and valence orbital that is filling for the element below lead in the periodic table.

7.10 Determine the atomic number and valence orbital that is filling for the element below francium in the periodic table.

7.11 Predict the location in the periodic table (row and column) of element 111.

7.12 Predict the location in the periodic table (row and column) of element 113.

7.13 How many valence electrons does each of the following atoms have? O, V, Rb, Sn, and Cd.

7.14 How many valence electrons does each of the following elements have? P, Cr, Kr, I, and Ba.

Electron Configurations

7.15 List a correct set of values of the quantum numbers for each of the valence electrons in the ground-state configurations of Be, O, Ne, and P.

7.16 List a correct set of values of the quantum numbers for each of the valence electrons in the ground-state configurations of Li, C, F, and Mg.

7.17 Which of the atoms of Problem 7.15 are paramagnetic? Draw orbital energy level diagrams to support your answer.

7.18 Which of the atoms of Problem 7.16 are paramagnetic? Draw orbital energy level diagrams to support your answer.

7.19 The following are hypothetical configurations for a beryllium atom. Which use nonexistent orbitals, which are forbidden by the Pauli principle, which are excited states, and which is the ground-state configuration? (a) $1s^3 2s^1$; (b) $1s^1 2s^3$; (c) $1s^1 2p^3$; (d) $1s^2 2s^1 2p^1$; (e) $1s^2 2s^2$; (f) $1s^2 1p^2$; and (g) $1s^2 2s^1 2d^1$.

7.20 None of the following hypothetical configurations describes the ground state of a fluorine atom. For each, state the reason why it is not correct: (a) $1s^2 2s^2 2p^4$; (b) $1s^2 2s^1 2p^6$; (c) $1s^3 2s^2 2p^4$; and (d) $1s^2 2s^2 1p^5$.

7.21 The ground state of Mo has higher spin than that of Tc. Construct energy level diagrams for the valence electrons that show how electron configurations account for this difference.

7.22 The ground state of V has lower spin than that of Cr. Construct energy level diagrams for the valence electrons that show how electron configurations account for this difference.

7.23 Use the periodic table to find and list (a) all elements whose ground-state configurations indicate near-degeneracy of the $4s$ and $3d$ orbitals; (b) the elements in the column that has two elements with one valence configuration and two with another valence configuration; and (c) all adjacent pairs of elements whose valence configurations indicate near-degeneracy of the $6d$ and $5f$ orbitals.

7.24 Use the periodic table to find and list (a) all elements whose ground-state configurations indicate near-degeneracy of the $5s$ and $4d$ orbitals; (b) two elements whose configurations indicate near-degeneracy of the $5d$ and $4f$ orbitals; and (c) three elements in the same group that have different valence configurations.

7.25 For nitrogen, how many different excited-state configurations are there in which no electron has $n > 2$? Write all of them.

7.26 For fluorine, how many different excited-state configurations are there in which no electron has $n > 2$? Write all of them.

Periodicity of Atomic Properties

7.27 Arrange the following atoms in order of decreasing first ionization energy (smallest last): Ar, Cl, Cs, and K.

7.28 Arrange the following atoms in order of increasing size (largest last): Cl, F, P, and S.

7.29 One of the elements has these ionization energies and electron affinity (all in kJ/mol): $IE_1 = 376$, $IE_2 = 2420$, $IE_3 = 3400$, $EA = -45.5$. In what column of the periodic table is this element found? Give your reasoning. Refer to Appendix C if necessary.

7.30 One of the elements has these ionization energies and electron affinity (all in kJ/mol): $IE_1 = 503$, $IE_2 = 965$, $IE_3 = 3600$, $EA = 46$. In what group of the periodic table is this element found? Give your reasoning. Refer to Appendix C if necessary.

7.31 According to Appendix C, each of the following elements has a positive electron affinity. For each one, construct its valence orbital energy level diagram and use it to explain why the anion is unstable: N, Mg, and Zn.

7.32 According to Appendix C, each of the following elements has a positive electron affinity. For each one, construct its valence orbital energy level diagram and use it to explain why the anion is unstable: Be, Ar, and Mn.

7.33 List the atomic ions that are isoelectronic with Br^- and have net charges (absolute values) that are less than 4 units. Arrange these in order of increasing size.

7.34 List the ionic species that are isoelectronic with Ar and have net charges (absolute values) that are less than 4 units. Arrange these in order of increasing size.

Energetics of Ionic Compounds

7.35 Use periodic trends and the electron affinities in Appendix C to list the elements you would expect to find as atomic anions with -1 charge in ionic compounds.

7.36 Use periodic trends and the electron affinities in Appendix C to list the elements you would expect to find as atomic anions with -2 charge in ionic compounds.

7.37 Given the following data, estimate the energy released when gaseous K and I atoms form $[K^+I^-](g)$.

Element	EA (kJ/mol)	IE₁ (kJ/mol)	IE₂ (kJ/mol)	Ion Radius (pm)
K	−48.4	418.8	3051	133 (cation)
I	−295.3	1008.4	1845.9	220 (anion)

7.38 Repeat the calculation in Problem 7.37 for K^{2+} and I^{2-}, using 500 kJ/mol as the estimated second electron affinity of iodine and assuming no change in distance of closest approach.

7.39 From the following list, select the elements that form ionic compounds: Ca, C, Cu, Cs, Cl, and Cr. Indicate whether each forms a stable cation or a stable anion.

7.40 From the following list, select the elements that form ionic compounds: B, Ba, Be, Bi, and Br. Indicate whether each forms a stable cation or a stable anion.

ilw 7.41 Calculate the overall energy change for the formation of lithium fluoride from lithium metal and fluorine gas. In addition to data found in Appendix C, the following information is needed: The lattice energy of LiF is −1036 kJ/mol; the bond energy of F_2 is 155 kJ/mol; lithium's energy of vaporization is 159 kJ/mol.

7.42 Calculate the overall energy change for the formation of calcium bromide from calcium metal and liquid bromine. In addition to data found in Appendix C, the following information is needed: The lattice energy of $CaBr_2$ is −2176 kJ/mol; the bond energy of Br_2 is 224 kJ/mol; bromine's energy of vaporization is 30.9 kJ/mol; calcium's energy of vaporization is 178 kJ/mol.

7.43 Draw a Born-Haber diagram similar to Figure 7-23 for the reaction described in Problem 7.41.

7.44 Draw a Born-Haber diagram similar to Figure 7-23 for the reaction described in Problem 7.42.

Ions and Chemical Periodicity

7.45 Classify each of the elements from Group 16 of the periodic table as a metal, a nonmetal, or a metalloid.

7.46 Classify each of the elements from Group 14 of the periodic table as a metal, a nonmetal, or a metalloid.

7.47 Classify each of the elements listed in Problem 7.39 as a metal, a nonmetal, or a metalloid.

7.48 Classify each of the elements listed in Problem 7.40 as a metal, a nonmetal, or a metalloid.

7.49 We list polonium as a metal, but some chemists classify it as a metalloid. List other metals that might be expected to show properties in between those of metals and metalloids.

7.50 What is the maximum number of valence p electrons possessed by a metallic element? Which metal(s) have this configuration?

Additional Paired Problems

7.51 Write the correct ground-state electronic configurations for C, Cr, Sb, and Br.

7.52 Write the correct ground-state electronic configurations for N, Ti, As, and Xe.

7.53 Write the correct electronic configuration for the Mn^{2+} ground state, and give a correct set of quantum numbers for all electrons in the least stable occupied orbital.

7.54 Write the correct electronic configuration for the Co^{3+} ground state, and give a correct set of quantum numbers for all electrons in the least stable occupied orbital.

7.55 Predict the total electron spin for P, Br^-, and Cu^+.

7.56 Predict the total electron spin for Gd, Sr, and Ag^+.

7.57 Write correct ground-state electron configurations for the neutral atoms with atomic numbers 9, 20, and 33.

7.58 Write correct ground state electron configurations for the neutral atoms and doubly positive ions with atomic number 14, 34, and 92.

For Problems 59–62, explain your rankings in terms of quantum numbers and Coulombic interactions.

7.59 Arrange the following in order of increasing ionization energy: N, O, Ne, Na, and Na^+.

7.60 Arrange the following in order of decreasing ionization energy: Br, Ar, Ar^+, and Cl.

7.61 Arrange the following in order of decreasing size (radius): Cl^-, K^+, Cl, and Br^-.

7.62 Arrange the following in order of increasing size: K, K^+, Ar, and Ca.

7.63 Write a brief explanation for each of the following:
(a) In a hydrogen atom the $2s$ and $2p$ orbitals have identical energy. (b) In a helium atom the $2s$ and $2p$ orbitals have different energies.

7.64 Write a brief explanation for each of the following:
(a) All three $2p$ orbitals of a helium atom have identical energy. (b) The $2p$ orbital in He^+ has nearly the same energy as the $2p$ orbital in H.

7.65 Refer to Figure 7-17 to answer the following questions about first ionization energies: (a) Which element shows the greatest decrease from its neighbor of next lower Z? (b) What is the atomic number of the element with the lowest value? (c) Identify three ranges of Z across which the value changes the least. (d) List the atomic numbers of all elements whose values are between 925 and 1050 kJ/mol.

7.66 Refer to Figure 7-18 to answer the following questions about electron affinities: (a) Which column of the periodic table has the second-highest set of values? (b) Which element has the highest value? (c) Apart from the end of a row, which element shows the greatest decrease compared with its neighbor of immediately lower Z? (d) What do elements 6, 14, and 32 have in common? (e) List the atomic numbers of all elements whose values are within 10% of 50 kJ/mol.

7.67 Draw energy level diagrams that show the ground-state valence electron configurations for Cu^+, Mn^{2+}, and Au^{3+}.

7.68 Draw energy level diagrams that show the ground-state valence electron configurations for Ir^+, Cd^{2+}, and V^{2+}.

7.69 Which has the most unpaired electrons, S^+, S, or S^-? Use electron configurations to support your answer.

7.70 Which has the most unpaired electrons, Si, P, or S? Use electron configurations to support your answer.

More Challenging Problems

7.71 Consider the three atomic orbitals that follow. They are drawn to scale, and orbital c has $n = 3$.

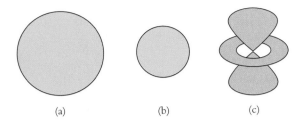

(a)   (b)   (c)

(a) Rank the orbitals in order of stability for a multielectron atom. (b) Provide two sets of quantum numbers for an electron in orbital a. (c) Give the atomic number of an element that has two electrons in orbital a but no electrons in orbital c. (d) Give the name of a cation that has one electron in orbital c. (e) How many other orbitals have the same principal quantum number as orbital c? (f) An element has its two least stable electrons in orbital b. If an atom of that element loses one electron, will orbital b become larger or smaller or remain the same size?

7.72 From its location in the periodic table, predict some of the physical and chemical properties of francium. What element does it most closely resemble?

7.73 What would be the next two orbitals to fill after the $7p$ orbital?

7.74 Which has the more stable $2s$ orbital, a lithium atom or a Li^{2+} cation? Explain your reasoning.

7.75 Use the data in Appendix C to explain why the noble gases seldom take part in chemical reactions.

7.76 Use electron-electron repulsion and orbital energies to explain the following irregularities in first ionization energies: (a) Boron has a lower ionization energy than beryllium. (b) Sulfur has a lower ionization energy than phosphorus.

7.77 Write the correct electronic configuration for the lowest-energy *excited* state of each of the following: Be, O^{2-}, Br^-, Ca^{2+}, and Sb^{3+}.

7.78 Are the ground-state configurations the same for the isoelectronic species Ce^{2+}, La^+, and Ba ? What features of orbital energies account for this?

7.79 The ionization energy of lithium atoms in the gas phase is about half as large as the ionization energy of beryllium atoms in the gas phase. In contrast, the ionization energy of Li^+ is about four times larger than the ionization energy of Be^+. Explain the difference between the atoms and the ions.

7.80 The figures show Cl^-, Ar, and K^+ drawn to scale. Decide which figure corresponds to which species and explain your reasoning.

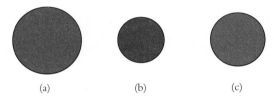

(a)   (b)   (c)

7.81 Make an electron density plot that shows how the $3s$ and $3p$ orbitals are screened effectively by the $2p$ orbitals. Provide a brief explanation of your plot.

Group Study Problems

7.82 Consider the following three atomic orbitals. They are drawn to scale, and orbital a has $n = 2$.

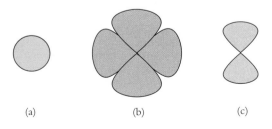

(a)   (b)   (c)

(a) How many electrons can be placed in orbital b? (b) Provide three sets of quantum numbers that describe an electron in orbital c. (c) Arrange these orbitals in order of increasing stability for a multielectron atom. (d) In a sodium atom, how many other orbitals have exactly the same energy as orbital b? (e) Give the name of an element that has orbital a filled but orbitals b and c empty. (f) Give the name of an element that has both an empty and a partially filled orbital c. (g) Give the name of a common anion that has one electron in orbital c.

7.83 No elements have ground-state configurations with electrons in g ($l = 4$) orbitals, but excited states can have such electrons.

(a) How many different g orbitals are there? (b) What are the possible values of m_l? (c) What is the lowest principal quantum number for which there are g orbitals? (d) Which orbitals may be near-degenerate with the lowest-energy g orbitals? (e) What is the atomic number of the first element that has a g electron in its ground-state configuration, assuming the g orbital begins to fill after an s orbital?

7.84 Use the data in Appendix C and the following information to calculate the overall energy changes for the formation of CaCl and $CaCl_2$. Use your results to determine which compound is more likely to form in the reaction of solid calcium with chlorine gas.

Vaporization energy of Ca = 178 kJ/mol;
Cl_2 bond energy = 240 kJ/mol;
lattice energy for $CaCl_2$ = −2258 kJ/mol;
lattice energy for CaCl (estimated) = −720 kJ/mol.

7.85 In the galaxy Topsumturvum, all of the rules for atomic configurations are the same as ours, except that orbital stability increases slightly as l increases. As a result, the $3s$ orbital is more stable than the $4d$ orbitals, but the $4f$ orbitals are slightly more stable than the $3s$ orbital. (a) Draw an orbital energy ladder for Topsumturvum. (b) Determine the ground-state configuration for the Topsumturvum atom containing 27 electrons. (c) Draw a block diagram that shows the first 50 elements in the periodic table for Topsumturvum.

Questions 86–89 refer to a hypothetical universe named Morspin, where electrons have three spins ($m_s = +1/2, -1/2, 0$) rather than the two spins ($m_s = +1/2, -1/2$) in our own universe. All other physical laws in the two universes are the same. The periodic table of Morspin follows:

1	2														3
4	5	6			7	8	9	10	11	12	13	14	15		
16	17	18			19	20	21	22	23	24	25	26	27		
28	29	30	31–45	46	47	48	49	50	51	52	53	54			
55	56	57	58–72	73	74	75	76	77	78	79	80	81			

7.86 (a) Which would have the larger first ionization energy, Morspin element 18 or Morspin element 30? Explain. (b) Which would have the larger radius, Morspin element 15 or cation 17^{2+}? Explain.

(c) Which would have the larger electron affinity, Morspin element 47 or 48? Explain.

7.87 (a) What would be the electron configuration of Morspin cation 38^{3+}? (b) Sketch the d_{xy} orbital as it would appear in this alternative universe. Be sure to include the coordinate axes. (c) What would be the correct ground-state configurations for Morspin atoms of atomic numbers 3 and 14?

7.88 (a) How many electrons would the first three noble gases in Morspin have? (b) Write acceptable sets of quantum numbers for the valence electrons of Morspin elements 4, 7, and 32. (c) Which would screen a $4s$ electron more effectively, the $n = 3$ electrons of Morspin or the $n = 3$ electrons of our own universe? Explain.

7.89 (a) What would be the electron configuration of Morspin element 19? (b) How many unpaired electrons would there be in element 28? Explain. (c) What would be the highest positive charge of cations found in ionic salts that could form from Morspin element 30? Explain.

Answers to Section Exercises

7.1.1

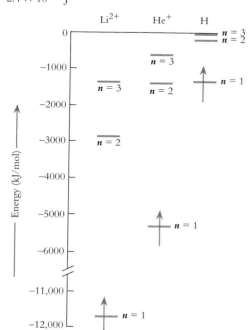

7.1.2 2.4×10^{-19} J

7.1.3

7.2.1 117
7.2.2 C, 4; Br, 7; Cr, 6; and Ta, 5
7.2.3 Be, Mg, Ca, Sr, Ba, Ra, Zn, Cd, and Hg; all have s^2.
7.3.1 (a) [Xe] $4f^{14} 5d^{10}$, zero spin; (b) [Rn] $7s^2 5f^{14} 6d^{10} 7p^6$, zero spin; and (c) [Ar], zero spin
7.3.2

7.3.3 Mo, [Kr] $5s^1 4d^5$; I, [Kr] $5s^2 4d^{10} 5p^5$; and Hg^+, [Xe] $6s^1 5d^{10} 4f^{14}$
7.4.1 (a) Cl has larger Z; (b) Cl has lower n; (c) Sr has higher n; (d) Cl^- has smaller Z; and (e) O^- has smaller Z_{eff}.
7.4.2 (a) Between Nb and In, a d block fills, so Z_{eff} remains nearly constant; (b) Ga loses a $4p$ electron, which is easier to remove than the $4s$ electrons of Zn; and (c) Mn has d^5 valence configuration, so the next electron must pair with an existing electron.
7.4.3 Stability decreases markedly with n.
7.5.1 Energy is required: 178.3 kJ/mol.
7.5.2 Cu and Zn
7.5.3 Ca^{2+}, Cs^+, Al^{3+}, and Br^-
7.6.1 Nonmetals, 1390; metalloids, 824; s block, 610 (all in kJ/mol); low ionization energy leads to formation of ions
7.6.2 S, nonmetal; Si, metalloid; Sr, metal; Se, nonmetal; Sc, metal; Sg, metal; and Sn, metal
7.6.3 MnO_4^- (7 valence electrons, charge -1, total 8); CrO_4^{2-} (6 valence electrons, charge -2, total 8); and $Cr_2O_7^{2-}$ (12 valence electrons, charge -2, total 14)

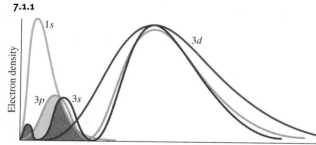

8

Fundamentals of Chemical Bonding

INTRODUCTION: NANOTECHNOLOGY

A gear-shaped molecule spins freely on a copper metal surface. A molecular "soccer ball" nestles snugly in a cavity on a protein. It sounds like science fiction, but these creations are real developments in the blossoming field of nanotechnology. The term *nanotechnology* refers to science carried out at the nanometer scale (1 nm = 10^{-9} m), by manipulating individual atoms and molecules.

Our background image, taken by a scanning tunneling microscope, shows a single layer of gear-shaped organic molecules lying on a copper metal surface. Most of the molecules are locked in place by nesting close to their neighbors. One of them, however, appears distinctly fuzzy, which suggests that the molecule may be spinning rapidly. This particular molecule was "unlocked" by moving it a fraction of a nanometer with the tip of a scanning tunneling microscope. (Our Tools for Discovery Box in Chapter 1 gives a brief discussion of scanning tunneling microscopy.)

The inset image is a computer simulation of this system. On the left, the molecule in the middle is immobilized by its four surrounding neighbors. On the right, a vacant space has been created by a slight change in the position of the middle molecule. Computer-based calculations verify that this small change in position allows the molecule to rotate freely, just as is observed experimentally.

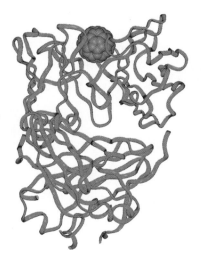

A spinning molecule on a copper surface may seem no more useful than a spinning ice-skater: elegant and beautiful, but hardly practical. Nevertheless, many leading scientists believe that nanotechnology is the next great frontier of science and technology. Nobel laureate physicist Horst Stormer, for example, has said, "Nanotechnology has given us the tools . . . to play with the ultimate toy box of nature—atoms and molecules. The possibilities to create new things appears limitless." The National Aeronautics and Space Administration and the National Science Foundation agree and are sponsoring extensive research in nanotechnology.

Another example of nanotechnology research is an attempt to develop biological molecules that can interact with buckminsterfullerene, C_{60}. By themselves, C_{60} molecules are difficult to manipulate because they are "greasy" and inert. Scientists envision using proteins bound to C_{60}, like the one illustrated here, as molecular machines that can deliver C_{60} units to build larger carbon structures.

Advocates of nanotechnology cite a wealth of potential applications for this new field, including tailored synthetic membranes that can collect specific toxins from industrial waste and computers that process data much faster than today's best models. The list of possible benefits from nanotechnology is limited only by our imaginations.

Nanotechnology involves designing and constructing specific molecular structures, one atom at a time. This will require a thorough understanding of chemical bonding, because every molecule is held together by chemical bonds in which electrons are shared between atoms. The basic principles of bond formation are always the same, but different elements have different bond-forming abilities, and chemical bonds show a variety of forms and strengths.

This chapter and the next describe chemical bonding. First we explore the interactions among electrons and nuclei that account for bond formation. Then we show how atoms are connected together in simple molecules such as water (H_2O). We show how these connections lead to a number of characteristic molecular geometries. In Chapter 9 we discuss more elaborate aspects of bonding that account for the properties of materials as diverse as deoxyribonucleic acid (DNA) and transistors.

8.1 OVERVIEW OF BONDING

A body of water is subject to the force of gravity. Without a barrier to block its path, water runs downhill to its lowest possible point, where it has lowest gravitational energy. In a similar way, a collection of electrons and nuclei are subject to electrical force. Electrons and nuclei move continually, but in this motion they arrange themselves in ways that optimize the net attractive forces among electrons and nuclei. This process is driven by Coulomb's law, as we describe in Chapter 7, and the electrical energy is given by Equation 7-1:

$$E_{coulomb} = k \frac{q_1 q_2}{r} \qquad (7\text{-}1)$$

If, instead of dealing with molar quantities as was the case in Chapter 7, we consider two individual charges expressed in units of electron charge, the value of k is 2.31×10^{-16} J pm.

1 picometer = 1 pm = 10^{-12} m.

Equation 7-1 describes the energy of one pair of charges, such as one electron attracted to one nucleus. Molecules, however, contain two or more nuclei and two or more electrons. To obtain the total electrical energy of a molecule, Equation 7-1 must be applied to every pair of charged species. These pair-wise interactions are of three types. First, oppositely charged electrons and nuclei attract one another. Attractive interactions release energy, so an electron attracted to a nucleus is at lower energy and therefore more stable than a free electron. Second, electrons repel each other, raising the energy and reducing the stability of a molecule. Third, nuclei repel each other, so nucleus–nucleus interactions reduce the stability of a molecule.

The electrons and nuclei in a molecule balance these three interactions in a way that gives the molecule its greatest possible stability. This balance is achieved when the electrons are concentrated between the nuclei. We view the electrons as shared between the nuclei and call this sharing a **covalent bond.** In any covalent bond, the attractive energy between the nuclei and electrons exceeds the repulsive energy arising from nuclear–nuclear and electron–electron interactions.

Electrical interactions also account for the shapes of molecules. According to Coulomb's law, the bonds of a molecule are arranged in space in a **molecular geometry** that minimizes electron-electron repulsion. We describe these two features in greater detail in the remainder of this chapter and in Chapter 9.

The Hydrogen Molecule

Molecular hydrogen, H_2, contains just two nuclei and two electrons. This simple molecule illustrates how coulombic interactions generate covalent bonds. Begin by thinking about what happens as two hydrogen atoms approach each other. As the atoms come together, each electron is attracted by the opposite nucleus, pulling the two atoms closer together. At the same time, the two nuclei repel each other, and so do the two electrons. These repulsive interactions drive the atoms apart.

For H_2 to be a stable molecule, the sum of the attractive energies must exceed the sum of the repulsive energies. Figure 8-1 displays a static arrangement of electrons and nuclei in which the electron-nucleus distances are shorter than the electron-electron and nucleus–nucleus distances. In this arrangement, attractive interactions are stronger than repulsive interactions, leading to a stable molecule. Notice that the two electrons occupy the region between the two nuclei, where they can interact with both nuclei at once. In other words, the electrons are shared between the atoms in a covalent bond.

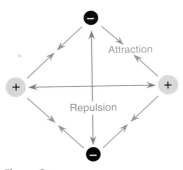

Figure 8-1
Schematic illustration of two electrons and two nuclei arranged so that attractive coulombic interactions (blue lines) are greater than repulsive coulombic interactions (red lines).

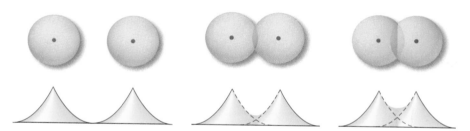

Figure 8-2
As two hydrogen atoms approach each other, the overlap of their 1s atomic orbitals increases. Constructive interference generates a new orbital with high electron density between the nuclei.

Orbital Overlap

As described in Chapter 6, electrons are not point charges as shown in Figure 8-1. Instead, electrons are "smeared out" in space in ways that can be represented by orbitals. Thus, while the figure shows that electrons positioned between nuclei maximize coulombic attraction, an accurate view of a hydrogen molecule requires orbitals rather than point charges. In a hydrogen molecule, electrons are smeared throughout the molecule in a manner that puts high electron density between the two nuclei. To understand chemical bonding, we must develop new orbital pictures that show this distribution of electrons between nuclei. In other words, we need to develop a set of **bonding orbitals.**

Although determining the distribution of electrons in any molecule is a very difficult mathematical task, we can build a good picture of the bonding orbitals of H_2 using the atomic orbitals of its two hydrogen atoms. The atomic orbital that describes the electron in a hydrogen atom is the spherical 1s orbital. When two hydrogen atoms come together to form a molecule, their 1s orbitals change in response to the coulombic forces between the atoms. These modifications to the orbitals can be described by superimposing the wave functions, as shown in a simplified way in Figure 8-2.

When two waves are superimposed, they interact to generate a new wave. When two orbitals are superimposed, their interaction is called **orbital overlap.** Orbital overlap can be described simply: The amplitude of the new wave function at every point is the sum of the amplitudes of the overlapping wave functions. As shown in Figure 8-2, the orbital overlap of the 1s wave functions of two hydrogen atoms gives a new wave function with high electron density between the nuclei. This bonding orbital replaces the atomic orbitals as the description of the electrons in a hydrogen molecule.

/// A bonding orbital, constructed by combining atomic orbitals from adjacent atoms, represents the electron density of a chemical bond.

Bond Length and Bond Energy

As two hydrogen atoms come together, attractive forces between the nuclei and the electrons make the molecule more stable. The amount of energy lowering depends on the separation between the nuclei, as shown in Figure 8-3. At distances greater than 300 pm, there is almost no interaction between the atoms, orbital overlap is nearly zero, and the energy of the two atoms is just the sum of their atomic energies. At closer distances the attraction between the electrons and the nuclei increases, orbital overlap becomes substantial, and energy is released. Moving the nuclei closer together generates greater stability until the two are 74 pm

Figure 8-3
The interaction energy of a pair of hydrogen atoms varies with internuclear separation.

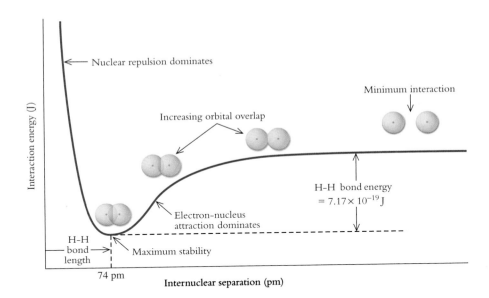

Nuclear repulsion dominates

Minimum interaction

Increasing orbital overlap

Interaction energy (J)

H-H bond energy
$= 7.17 \times 10^{-19}$ J

Electron-nucleus
attraction dominates

H-H
bond
length

Maximum stability

74 pm

Internuclear separation (pm)

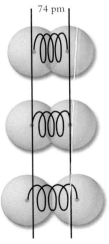

74 pm

Figure 8-4
Molecules vibrate continually about their bond length, like two balls attached to a spring.

apart. At distances smaller than 74 pm, however, the nucleus–nucleus repulsion increases more rapidly than the electron–nucleus attraction. Thus at a separation distance of 74 pm, the hydrogen molecule is at the bottom of an "energy well" in its lowest energy configuration, hence most stable. The atoms are combined in a molecule, sharing two electrons in a covalent bond.

Experimental studies of molecular motion reveal that nuclei vibrate continuously, oscillating about their optimum separation distance like two balls attached to opposite ends of a spring. Figure 8-4 shows this in schematic fashion for a hydrogen molecule vibrating about its optimum separation distance of 74 pm.

Figure 8-3 shows two characteristic features of chemical bonds. The separation distance where the molecule is most stable (74 pm for H_2) is known as the **bond length,** and the amount of stability at this separation distance is known as the **bond energy,** or strength of the bond. Bond lengths and strengths are important properties of bonds that are used frequently to describe the characteristics of chemical bonding.

Experiments show that it takes 7.17×10^{-19} J of energy to pull apart a hydrogen molecule into individual H atoms. In other words, the bond energy of H_2 is 7.17×10^{-19} J. To convert to energy/mol we multiply this energy by the Avogadro constant (6.022×10^{23} atoms/mol), giving 435 kJ/mol. This means that one mole of H_2 is more stable than two moles of H atoms by 435 kJ.

Other Diatomic Molecules: HF and F_2

Bond formation in H_2 is relatively easy to describe, because we need to account for the orbital interactions of just two orbitals and the placement of just two electrons. Bond formation in other molecules, however, involves interactions among many orbitals and configurations that contain many electrons. Nevertheless, we can describe bond formation in various diatomic molecules using simple orbital overlap as we illustrate using HF and F_2.

As hydrogen and fluorine atoms approach each other, their electron clouds interact. Figure 8-5 indicates that of the five occupied atomic orbitals in a fluorine

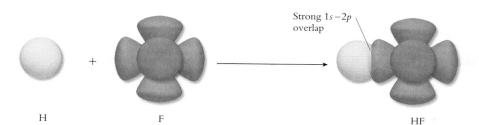

Figure 8-5
Bond formation in HF can be represented by orbital overlap between a hydrogen 1s atomic orbital and the fluorine 2p atomic orbital that points toward the approaching hydrogen atom.

atom, one $2p$ orbital preferentially interacts with the $1s$ orbital of the approaching hydrogen atom. The fluorine $2s$ orbital is less effectively screened from the nuclear charge than the $2p$ orbitals, so it is more compact than the $2p$ orbitals. The fluorine $1s$ orbital has a lower principal quantum number and is very compact. Two of the $2p$ orbitals point in the wrong directions to overlap effectively. Only the fluorine $2p$ orbital that points directly at the approaching hydrogen atom shows significant orbital overlap. Consequently, bonding in HF can be described by one covalent bond that results from overlap between the hydrogen $1s$ orbital and the fluorine $2p$ orbital that points along the bond axis.

A similar situation arises when two fluorine atoms approach each other. The first valence atomic orbitals to overlap are the $2p$ orbitals pointing along the axis that joins the atoms. Bond formation in fluorine is the result of the strong overlap of these two $2p$ atomic orbitals.

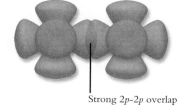

Strong $2p$-$2p$ overlap

Approaches to Bonding

Any orbital can hold two electrons with opposite spins without violating the Pauli exclusion principle, so a single bonding orbital can describe both electrons of the covalent bond in a diatomic molecule such as H_2, HF, or F_2. Other molecules, however, require entire sets of bonding orbitals that describe how all the valence electrons are packaged amid the framework of nuclei.

In this chapter and in Chapter 9 we describe two different ways to think about bonding orbitals: localized orbitals and delocalized orbitals. These two approaches to bonding share the following features:

1. Each electron is described by a wave function. In other words, each electron in a molecule is assigned to a specific orbital.

2. No two electrons in a molecule have identical wave functions because the Pauli exclusion principle applies to electrons in molecules as well as in atoms.

3. In the description of a molecule in its ground state, electrons fill the orbitals that are most stable because the aufbau principle applies to molecules as well as atoms.

4. Even though every atom has an unlimited number of atomic orbitals, the valence orbitals are all that are needed to describe bonding.

Bonding involves the valence orbitals almost exclusively. These orbitals have the appropriate size and energy to interact strongly. Consider, for example, the orbital interactions between fluorine atoms combined to form an F_2 molecule.

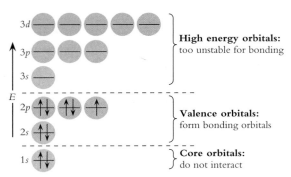

Figure 8-6
The energy level diagram for a fluorine atom ($1s^2 2s^2 2p^5$).

The localized orbital model using valence atomic orbitals is often called the *valence bond* theory.

Delocalized orbitals are also called *molecular orbitals,* and this approach is often called *molecular orbital theory.*

Each atom has the energy level diagram shown in Figure 8-6. Recall that the sizes of the orbitals increase substantially as the principal quantum number (n) increases. As a result, fluorine's core $1s$ orbitals are much smaller than any of the $n = 2$ orbitals and do not participate effectively in orbital overlap. On the other hand, orbitals with $n > 2$ lie at considerably higher energy than the $n = 2$ orbitals and are too unstable to form strong covalent bonds. The only orbitals of fluorine that form chemical bonds are the valence orbitals, those with $n = 2$.

The orbital features illustrated by molecular fluorine apply to all molecules. That is, core orbitals are too compact to overlap significantly, so they do not participate in bond formation. Moreover, all orbitals beyond the valence shell are too high in energy to give bonding orbitals that will be occupied by electrons. Only the valence orbitals are large enough to overlap strongly and stable enough to generate bonding orbitals that are more stable than the occupied atomic orbitals.

In this chapter we discuss the **localized orbital model** of covalent bonding. In this approach, we make the approximation that bonding electrons are localized between two atoms. Valence electrons not involved in chemical bonds are said to be **nonbonding electrons** and remain in atomic orbitals on individual atoms.

The localized orbital model is easy to apply, even to very complex molecules, and it does an excellent job of explaining much chemical behavior. In many instances, however, localized bonds are insufficient to explain molecular properties and chemical reactivity. In Chapter 9, we show how to construct delocalized orbitals, which spread over several atoms. Delocalized orbitals explain chemical properties that localized orbitals cannot, but delocalized orbitals require more complicated analysis than localized orbitals.

These two approaches to bonding are complementary because each describes some of the features of how a molecule is put together. Remember that the underlying molecule is the reality and that the molecule does not change when we modify the way we describe its bonding. Bonding has many features, and each approach is well suited to describe some molecules but less well suited for others. Thus a comprehensive understanding of bonding requires more than one approach.

Section Exercises

8.1.1 Draw a figure similar to Figure 8-3 that shows changes in the wave functions as a hydrogen atom approaches a fluorine atom and forms an H—F bond.

8.1.2 Describe bond formation between hydrogen atoms and chlorine atoms to form HCl molecules.

8.1.3 Consult a periodic table and determine which orbitals of selenium atoms and of phosphorus atoms are used to form covalent bonds.

8.2 UNEQUAL ELECTRON SHARING

In a hydrogen molecule each nucleus has a unit charge of $+1$. Consequently, electrons are attracted equally to both nuclei. The result is a symmetrical orbital in which the electron density near one nucleus is the same as the density near the other nucleus. The coulombic description for a fluorine molecule is more complicated but still symmetrical. Each nucleus has a charge of $+9$ units and is screened from the bonding orbital by the core electrons. The bonding electrons experience the same net attraction toward both nuclei. In a chemical bond between two identical atoms, such as H_2 and F_2, the bonding electrons are shared equally between the two nuclei.

In contrast to the symmetrical forces in H_2 and F_2, the bonding electrons in HF molecules experience *unsymmetrical* attractive forces. The bonding electrons are attracted in one direction by the $+1$ charge on the hydrogen nucleus. In the other direction, the same electrons are attracted by the $+9$ charge of the fluorine nucleus, screened by electrons in $1s$, $2s$, and $2p$ orbitals. Although the $1s$ and $2s$ electrons provide highly effective shielding, the $2p$ orbitals point in the wrong directions to screen the fluorine nucleus effectively from the bonding electrons. As a result, the effective nuclear charge of the fluorine atom is greater than $+1$, but less than $+9$.

Unsymmetrical attractive forces lead to an unsymmetrical distribution of electrons. The electron density in an unsymmetrical bond is concentrated closer to the nucleus with the larger effective nuclear charge. The HF molecule reaches lowest energy when the electron density in the bonding orbital is concentrated closer to the fluorine atom than to the hydrogen atom. This is an example of unequal sharing of electrons. Although two electrons are shared between the nuclei, bond formation has withdrawn electron density from the vicinity of the hydrogen nucleus and added electron density in the vicinity of the fluorine nucleus. As a result, the hydrogen end of the molecule bears a partial positive charge, and the fluorine atom bears a partial negative charge of equal magnitude. We say that the electron density in the H—F bond is polarized toward the fluorine atom and that HF has a **polar covalent bond.**

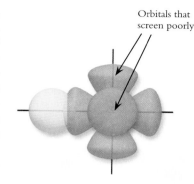

Orbitals that screen poorly

Electronegativity

Any covalent bond between two different atoms is polar to some extent, because no two elements have identical effective nuclear charges. The extent to which an element attracts bonding electrons is called its **electronegativity** and is symbolized by the Greek letter chi (χ). When two atoms have different electronegativities, the bond between them is polar; the greater the difference, the more polar the bond.

Electronegativity is related to electron affinity and ionization energy. Electronegativity measures how strongly an atom attracts the *bound* electrons in a bonding orbital to make a polar covalent bond. In contrast, electron affinity measures how strongly an atom attracts a *free* electron to make an atomic anion, and ionization energy measures how strongly an atom attracts one of its own bound electrons.

Electronegativities are estimated by using combinations of properties that depend on effective charge, including ionization energy, electron affinity, and bond energy. One commonly used set of electronegativities was developed by the American chemist Linus Pauling. These values, which are unitless, are presented in the periodic table shown in Figure 8-7. Modern X-ray techniques make it possible to make experimental measurements of the electron density distributions of chemical bonds. The distributions obtained in this way agree with those predicted from estimated electronegativities, indicating that the estimated values are reliable.

Notice in Figure 8-7 that electronegativities increase from the lower left to the upper right of the periodic table. Cesium ($\chi = 0.7$) has the lowest value, and fluorine ($\chi = 4.0$) has the highest value. Notice also that electronegativities decrease down most columns and increase from left to right across the s and p blocks. Like the periodic trends in ionization energies and electron affinities that we described in Chapter 7, these trends can be explained in terms of variations in nuclear charge and principal quantum number.

Metals generally have low electronegativities. Metals in the s block have $\chi < 1$, except for magnesium ($\chi = 1.2$) and beryllium ($\chi = 1.5$). Those in the f block have $\chi = 1.3 \pm 0.2$. Most other metals have $\chi < 2$.

Chi is pronounced *kai* (rhymes with *eye*).

Ionization energy

Electronegativity

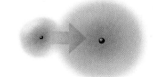

Electron affinity

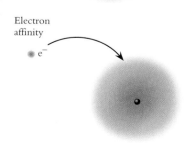

Figure 8-7
Pauling values for
electronegativities. Values for
Group 18 have not been
determined.

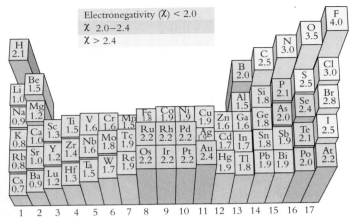

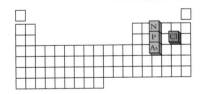

Linus Pauling (1901–1994) won the
Nobel Prize in chemistry in 1954 for
his ideas about chemical bonds.
Pauling was a leader in the movement
to limit nuclear weapons, for which he
was awarded the Nobel Prize for peace
in 1963.

Nonmetals have high electronegativities, ranging from $\chi = 2.1$ for phosphorus and hydrogen to $\chi = 4.0$ for fluorine. According to Figure 8-7, all nonmetals except phosphorus and hydrogen have $\chi > 2.5$. The electronegativities of metalloids vary between 1.8 for silicon and 2.0 for arsenic, larger than those of most metals and smaller than those of most nonmetals. Electronegativity differences ($\Delta\chi$) between bonded atoms indicate where any particular bond lies on the continuum of bond polarities. The two extremes in polarities are represented by F_2 and CsF. The chemical bond in F_2 ($\Delta\chi = 0$) is nonpolar because electrons are shared equally between the two fluorine atoms. In contrast, CsF ($\Delta\chi = 3.3$) is an ionic compound in which electrons have been fully transferred to give Cs^+ cations and F^- anions. Most bonds, for example the bonds in HF ($\Delta\chi = 1.9$) and ClF ($\Delta\chi = 0.5$), fall between these extremes. These are polar covalent bonds, in which electrons are shared unequally but are not fully transferred. Example 8-1 illustrates the periodicity of electronegativity and bond polarity variations.

Example 8-1	Electronegativities

Use the periodic table, without looking up electronegativity values, to rank the following bonds from least polar to most polar: (a) P—Cl, As—Cl, N—Cl and (b) C—S, C—O, C—F.

Strategy: The larger the difference in electronegativity, the more polar the bond. Therefore we can use the trends in electronegativity to arrange these bonds in order of polarity. First locate the elements in the periodic table, then use periodic trends to determine relative electronegativity differences and rank the bond polarities.

Solution: (a) Because each of these three bonds contains chlorine, the trend in bond polarity matches the trend in the electronegativities of P, As, and N. These three elements are in the same column of the periodic table (Group 15), and electronegativity decreases from top to bottom of a column. Chlorine, in Group 17, has a higher electronegativity than the elements in Column 15, so the electronegativity difference increases from N to P to As. Thus the least polar bond is N—Cl, and the most polar is As—Cl:

$$N—Cl < P—Cl < As—Cl$$

Electronegativities *(continued)*

Example 8-1

(b) Because each of these bonds contains carbon, the trend in bond polarities depends on the trend in electronegativities for S, O, and F. Sulfur is beneath oxygen in Group 16 of the periodic table, so sulfur is less electronegative than oxygen. Fluorine is to the right of oxygen in the same row, so fluorine is more electronegative than oxygen because electronegativity increases across a row. All three have larger electronegativities than carbon, so the electronegativity difference and bond polarity increases as electronegativity increases:

$$C—S < C—O < C—F$$

You can use tabulated electronegativity values (Figure 8-7) to verify that these rankings are correct.

Although bond polarities vary continuously from zero to fully ionic, it is convenient to classify substances as ionic or polar. Table 8-1 indicates that a compound is classified as polar, when $\Delta\chi < 1.6$, and a compound is classified as ionic when $\Delta\chi > 2.0$. When electronegativity differences are between 1.6 and 2.0, the classification depends on whether or not the compound is a metal halide. Metal halides in this range, such as $MgCl_2$ and KI, are classified as ionic. Metal oxides such as Fe_2O_3 and nonmetallic compounds such as HF and SiO_2 are classified as polar despite the large differences in their electronegativities.

Section Exercises

8.2.1 For each of the following pairs, identify which element tends to attract electron density from the other in a covalent bond: (a) Si and O; (b) C and H; (c) As and Cl; and (d) Cl and Sn.

8.2.2 List the bonds of Exercise 8.2.1 from least polar to most polar.

8.2.3 Using electronegativity differences, determine which of the following compounds are clearly ionic and which should be considered to have some covalent bond character: LiCl, $FeCl_3$, AgCl, Al_2O_3, Na_2S, and ZnS.

Table 8-1
Representative Electronegativity Differences

Compound	$\Delta\chi$	Bond Type
Cl_2	$3.0 - 3.0 = 0$	Nonpolar
IBr	$2.8 - 2.5 = 0.3$	Slightly polar
NO	$3.5 - 3.0 = 0.5$	Slightly polar
CO	$3.5 - 2.5 = 1.0$	Polar
HCl	$3.0 - 2.1 = 0.9$	Polar
HF	$4.0 - 2.1 = 1.9$	Highly polar
KI	$2.5 - 0.8 = 1.7$	Ionic
CaO	$3.5 - 1.0 = 2.5$	Ionic
LiF	$4.0 - 1.0 = 3.0$	Ionic

Increasing polarity

8.3 LEWIS STRUCTURES

According to the localized orbital model of bonding, there are two types of valence electrons: (1) bonding electrons, which are shared in bonding orbitals between two nuclei, and (2) nonbonding electrons, which are localized in atomic orbitals on individual atoms. In this section, we develop a process for constructing schematic drawings of molecules called **Lewis structures.** A Lewis structure shows how the atoms in a molecule are bonded together. A Lewis structure also reveals a molecule's distribution of bonding and nonbonding valence electrons. In a sense, a Lewis structure is a molecular blueprint that shows how a molecule is laid out. From this perspective, writing a Lewis structure is the first step in developing an orbital overlap description of a molecule. Lewis structures are named after their inventor, G. N. Lewis, who is the subject of the Chemical Milestones Box at the end of this section.

The Conventions

Lewis structures are constructed according to the following conventions, which we illustrate in Figure 8-8 for the HF molecule:

1. *Each atom is represented by its elemental symbol.* In this respect, a Lewis structure resembles a chemical formula.

2. *Only the valence electrons appear in a Lewis structure.* The core electrons occupy compact atomic orbitals that are not involved in bonding.

3. *A line joining two elemental symbols represents one pair of electrons shared between two atoms.* Two atoms may share up to three pairs of electrons in **single bonds** (two shared electrons, one line), **double bonds** (four shared electrons, two lines), or **triple bonds** (six shared electrons, three lines).

4. *Dots placed next to an elemental symbol represent nonbonding electrons on that atom.* Nonbonding electrons are usually found in pairs because orbitals usually contain two electrons with opposing spins.

Figure 8-8
The Lewis structure conventions for hydrogen fluoride.

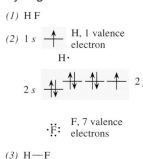

These conventions divide molecular electrons into three groups. Core electrons are associated with individual atoms and are not shown in a Lewis structure. Bonding valence electrons are shared between atoms and are shown as lines. Nonbonding valence electrons are localized on atoms and are shown as dots.

Writing Lewis Structures

Lewis structures can be written by following a logical seven-step procedure. We develop this seven-step procedure by working through specific examples, each one highlighting one or more of the steps.

Step 1. *If the substance is ionic, treat each ion separately.* The reason is that ions do not share electrons with other ions. Thus each ion is an individual chemical species with its own Lewis structure. An ionic compound has two Lewis structures, one for its cation and one for its anion. A compound is ionic if it contains a polyatomic anion and a metal or an ammonium

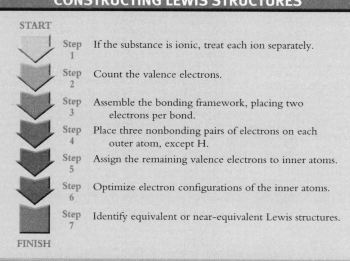

CONSTRUCTING LEWIS STRUCTURES

START

Step 1 — If the substance is ionic, treat each ion separately.

Step 2 — Count the valence electrons.

Step 3 — Assemble the bonding framework, placing two electrons per bond.

Step 4 — Place three nonbonding pairs of electrons on each outer atom, except H.

Step 5 — Assign the remaining valence electrons to inner atoms.

Step 6 — Optimize electron configurations of the inner atoms.

Step 7 — Identify equivalent or near-equivalent Lewis structures.

FINISH

cation. (See Table 3-5 for the common polyatomic ions.) If there is a metal atom but no polyatomic cation, apply electronegativity differences to identify whether or not the compound is ionic.

Step 2. Count the valence electrons. A Lewis structure shows *all* valence electrons and *only* valence electrons, so a count of valence electrons is essential. Recall from Section 7.5 that the number of valence electrons of an atom can be found from its position in the periodic table. Add the contributions from all atoms to obtain a total count of valence electrons. Finally, if ions are present, *add* one electron for each negative charge on the ion and *subtract* one electron for each positive charge on the ion.

Examples 8-2 and 8-3 illustrate steps 1 and 2.

> Anions have excess negative charge because they have gained negatively charged electrons. Cations have excess positive charge because they have lost electrons.

Lewis Structure Involving Ions	**Example 8-2**

Determine the Lewis structure of sodium hydroxide, NaOH.

Strategy: The strategy for determining a Lewis structure is always the same: Follow the seven-step procedure.

Constructing
Lewis
Structures

Solution:

1. Is the compound ionic? We recognize the presence of the hydroxide anion and a metal, indicating an ionic substance that contains Na^+ cations and OH^- anions, each of which must be treated separately.

2. Use the periodic table to count valence electrons: Sodium, in Group 1, has one valence electron (s^1 configuration). In forming the Na^+ cation, sodium loses its only valence electron. Thus Na^+ has zero valence electrons.

   For the hydroxide anion, oxygen, in Group 16, has six valence electrons (s^2p^4), and hydrogen has a single electron. Because the species is an anion, we add one extra electron to account for the negative charge.

   For OH^-

   $$1\ O = (1)(6\ e^-) = 6\ e^-$$
   $$1\ H = (1)(1\ e^-) = 1\ e^-$$
   $$-1\ ion = 1\ e^-$$
   $$\overline{\text{Total valence } e^- = 8\ e^-}$$

3. The sodium cation contains a single atom, so there is no bonding framework. The Lewis structure of Na^+ is simply that: Na^+.

   To build the bonding framework for the hydroxide anion, draw a line indicating one bond between the atoms: $O{-}H$

4. The hydroxide anion has eight valence electrons, two of which form the $O{-}H$ bond. Place the remaining six electrons around the oxygen atom:

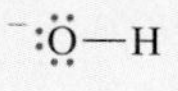

   $$^-:\ddot{O}{-}H$$

All eight valence electrons have now been placed. At this point we have built a provisional Lewis structure of the hydroxide ion. In this case the provisional Lewis structure is also the final Lewis structure.

| Example 8-3 | Lewis Structure of Carbon Tetrachloride |

Determine the Lewis structure of carbon tetrachloride, which has a central C atom bonded to four Cl atoms.

Constructing
Lewis
Structures

Strategy: Follow the seven-step procedure.

Solution:

1. Is the compound ionic? There are neither polyatomic ions nor metals in the formula, so the answer is "no." Treat the entire compound as a unit.

2. Use the periodic table to count valence electrons. Carbon is in Group 14, so it contributes four valence electrons (s^2p^2). Chlorine is in Group 17, so each atom contributes seven valence electrons (s^2p^5).

$$1\ C = (1)(4\ e^-) = 4\ e^-$$
$$4\ Cl = (4)(7\ e^-) = 28\ e^-$$
$$\overline{\text{Total Valence } e^- = 32\ e^-}$$

3. Build the bonding framework. The problem gives the required information: The carbon atom is in the center, bonded to each of the four chlorine atoms:

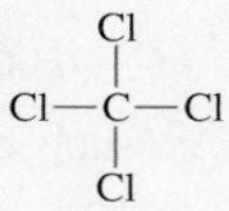

$$\begin{array}{c} Cl \\ | \\ Cl-C-Cl \\ | \\ Cl \end{array}$$

4. With four bonds, the bonding framework uses eight valence electrons, leaving 24 to be distributed. Placing six electrons around each chlorine atom uses up all the valence electrons, giving a provisional Lewis structure. As in Example 8-2, this is also the Lewis structure:

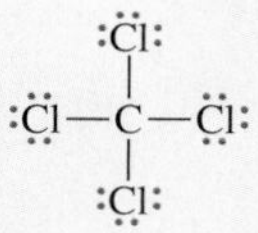

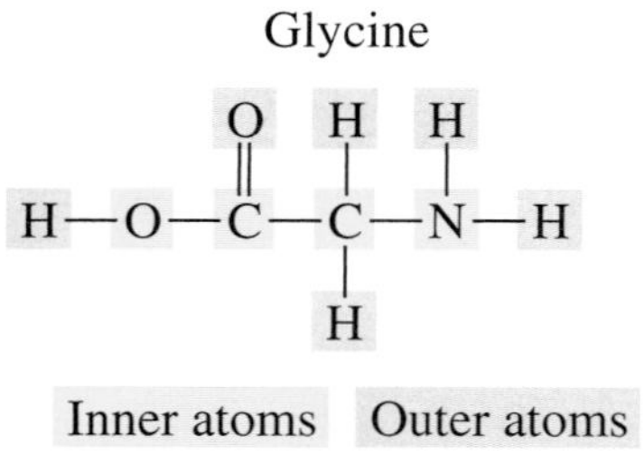

$$\begin{array}{c} :\ddot{Cl}: \\ | \\ :\ddot{Cl}-C-\ddot{Cl}: \\ | \\ :\ddot{Cl}: \end{array}$$

Glycine

$$\begin{array}{c} \quad O \quad H \quad H \\ \quad \| \quad | \quad | \\ H-O-C-C-N-H \\ \quad \quad \quad | \\ \quad \quad \quad H \end{array}$$

Inner atoms   Outer atoms

Figure 8-9
Outer atoms bond to only one other atom. Inner atoms bond to two or more atoms.

Step 3. *Assemble the bonding framework.* The bonding frameworks for our first two examples are straightforward. In Example 8-2, the two-atom hydroxide anion can have only one bonding framework. In Example 8-3, the statement of the problem describes the bonding framework of carbon tetrachloride. For more complicated molecules, there is no foolproof method for putting the atoms together in the correct arrangement. The following guidelines lead from the chemical formula to the correct arrangement of atoms *in many cases*. Other molecules are so complex, however, that the framework of atoms must be provided before the Lewis structure can be assembled.

a. An *outer atom* bonds to only one other atom. An *inner atom* bonds to more than one other atom. Figure 8-9 uses the amino acid glycine to show the difference between inner atoms and outer atoms.

b. Hydrogen atoms are always outer atoms in the compounds discussed in this text, because hydrogen can form just one localized bond. Notice that all five of glycine's hydrogen atoms are outer atoms.

c. Outer atoms other than hydrogen usually are the ones with the highest electronegativities. Here are some examples. In sulfur dioxide (SO_2), two outer oxygen atoms ($\chi = 3.5$) are bonded to a central sulfur atom ($\chi = 2.5$). Silicon tetrachloride ($SiCl_4$) contains an inner silicon atom ($\chi = 1.8$) bonded to four outer chlorine atoms ($\chi = 3.0$). In chlorine trifluoride (ClF_3), chlorine bonds to three outer fluorine atoms ($\chi = 4.0$):

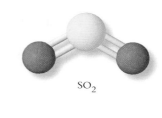

SO_2

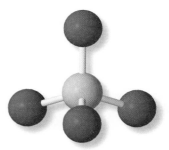

$SiCl_4$

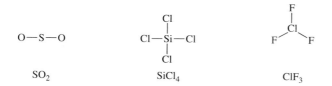

These three frameworks raise an important point about Lewis structures. Although Lewis structures show how atoms are connected to one another, a Lewis structure generally does *not* show the three-dimensional shape of a molecule. Ball-and-stick models show, for instance, that silicon tetrachloride is not a flat square molecule. We describe how to determine the shapes of molecules later in this chapter.

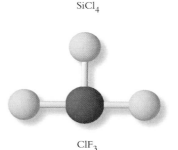

ClF_3

d. The order in which atoms are listed in the formula often indicates the bonding pattern. For example, in HCN the framework is H—C—N; the atoms in the OCN^- anion are in the order O—C—N; and CH_3NH_2 has the framework shown here:

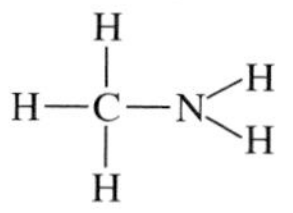

$$
\begin{array}{c}
\quad\;\; H \\
\quad\;\; | \\
H - C - N \overset{\displaystyle H}{\underset{\displaystyle H}{\diagup}} \\
\quad\;\; | \\
\quad\;\; H
\end{array}
$$

Notice in this case that the hydrogen atoms are grouped in the formula with their bonding partners, three hydrogen atoms with carbon and two more with nitrogen.

e. Atoms enclosed in parentheses are bonded together, and the entire group can be bound to the preceding or the following atom. Both conventions are widespread, and it takes experience and practice to sort it all out. Example 8-4 illustrates this guideline.

Parentheses and Lewis Structures   **Example 8-4**

Determine the Lewis structure of $(CH_3)_4C$ (2,2-dimethylpropane, also known as neopentane).

Strategy: Again we follow the seven-step procedure, using the chemical formula to guide our determination of the bonding framework.

| Example 8-4 | Parentheses and Lewis Structures *(continued)* |

Constructing
Lewis
Structures

Solution:

1. There are no polyatomic ions or metals present, so we treat the entire molecule as a unit.

2. You should be able to verify that the valence electron count is 32 e^-.

3. The chemical formula provides important clues about the bonding framework. There are four CH_3 units, each one bonded to another carbon atom. Because hydrogen must be an outer atom, we conclude that each of the four carbon atoms from the CH_3 units bonds to the fifth carbon atom:

4. The framework contains 16 bonds, which use up all 32 of the valence electrons. This provisional structure is also the final Lewis structure.

The framework guidelines do not always give an unambiguous structure, but additional information may come from the way the chemical formula is written or from chemical intuition. Consider phosphoric acid, H_3PO_4. The electronegativity guideline indicates that the less electronegative phosphorus atom ($\chi = 2.1$) is surrounded by four more electronegative oxygen atoms ($\chi = 3.5$). As usual, hydrogen atoms are outer atoms, but are they bonded to oxygen, phosphorus, or both? Recall from Chapter 4 that phosphoric acid is an example of an oxyacid, in which an inner atom is bonded to a variable number of oxygen atoms and OH groups. Having learned something about oxyacids, we can write the framework of H_3PO_4.

Once we know the bonding framework of a molecule or ion, we use steps 4 and 5 of the procedure to place any remaining valence electrons, first on outer atoms and then on inner atoms.

Step 4. Place three nonbonding pairs of electrons on each outer atom, except H. Hydrogen uses only the $1s$ orbital for bonding, so a hydrogen atom can accommodate no more than two electrons. Any other atom uses all four of its s and p valence orbitals to accept four pairs of electrons. Three nonbonding electron pairs, also called **lone pairs,** plus one bonding pair, give each outer atom four electron pairs. Such a set of four pairs of electrons associated with an atom is often referred to as an **octet.**

/// An outer atom other than hydrogen is most stable when it is associated with an octet of electrons.

In diatomic molecules there may not be enough leftover valence electrons to place three nonbonding pairs on each atom. Assign three pairs to the more electronegative atom, then place the remaining electrons on the other atom.

Step 5. Assign the remaining valence electrons to inner atoms. This step is straightforward if there is only one inner atom. If the molecule has more than one

H—O—P—O—H
(with O above and O—H below the P)

Bonding framework of H_3PO_4

inner atom, place nonbonding pairs around the most electronegative atom until it has an octet of electrons. If there are still unassigned electrons, do the same for the next most electronegative atom. Continue in this manner until all the electrons have been assigned.

In some cases, electrons are left over after each inner atom has a total of four nonbonding pairs and bonds. This occurs only when an inner atom has a valence shell with a principal quantum number greater than 2. In such cases, we place any remaining electrons on any inner atom that has $n > 2$. This is because atoms with $n > 2$ have valence d orbitals, which allow them to accommodate more than eight electrons. For example, phosphorus, sulfur, and chlorine have nine valence orbitals (one $3s$, three $3p$, and five $3d$), so they can accommodate as many as 18 valence electrons. Lewis structures of compounds with inner atoms such as P, S, and Cl may have between 8 and 12 electrons associated with these atoms. Example 8-5 illustrates the application of Steps 4 and 5.

Lewis Structure of ClF_3	**Example 8-5**

Chlorine trifluoride is used to recover uranium from nuclear fuel rods in a high-temperature reaction that produces gaseous uranium hexafluoride:

$$2\ ClF_3(g) + U(s) \longrightarrow UF_6(g) + Cl_2(g)$$

Determine the Lewis structure of ClF_3.

Strategy: "Lewis structure" means "seven-step procedure."

Solution:

1. Since ClF_3 contains no metals or common polyatomic ions, we treat the molecule as a single unit.

2. Seven valence electrons from each of four atoms gives the molecule a total of 28 valence electrons.

3. Build the framework. Chlorine, with lower electronegativity, is the inner atom. Make a single bond to each of the three fluorine atoms:

$$\begin{array}{c} F \\ | \\ F-Cl-F \end{array} \qquad \text{Remaining electrons: } 28 - 6 = 22$$

4. Add three nonbonding pairs of electrons to each outer atom:

$$\begin{array}{c} :\ddot{F}: \\ | \\ :\ddot{F}-Cl-\ddot{F}: \end{array} \qquad \text{Remaining electrons: } 28 - 6 - 3(6) = 4$$

5. Four electrons are yet to be assigned. They must be placed on the inner chlorine atom, giving it three bonds and two nonbonding pairs.

$$\begin{array}{c} :\ddot{F}: \\ | \\ :\ddot{F}-\ddot{Cl}-\ddot{F}: \end{array}$$

This puts 10 electrons around chlorine, but chlorine has d orbitals available to accept valence electrons, so this is a legitimate Lewis structure.

Constructing Lewis Structures

The first five steps in our procedure lead to a Lewis structure that contains the correct bonding framework and the correct number of valence electrons. As stated above, this represents a provisional Lewis structure for the molecule. Although the provisional structure is the correct structure in some cases, in other cases it does not represent the best Lewis structure for the molecule. This is because the *distribution* of electrons in the provisional structure may not be the one that gives the molecule its lowest possible energy. Step 4 of the procedure places electrons preferentially on outer atoms. This ensures that *outer* atoms all have full complements of electrons but may not give the optimal configuration for the *inner* atoms. Step 6 addresses this need.

Step 6. *Optimize electron configurations of the inner atoms.* The most straight-forward indication that an inner atom has less than an optimal complement of electrons is when it lacks an octet. In this situation, the atom has not made full use of its valence *s* and *p* orbitals. A more stable distribution of electrons can be reached by moving some of the electrons from adjacent outer atoms to make double or triple bonds to the inner atoms. Example 8-6 illustrates this procedure.

Example 8-6	Lewis Structure of Formaldehyde

Aqueous solutions of formaldehyde, H_2CO, are used to preserve biological specimens. Determine the Lewis structure of formaldehyde.

Strategy: We apply the seven-step procedure.

Constructing
Lewis
Structures

Solution:

1–5. You should be able to follow the first five steps of the procedure to obtain the following structure:

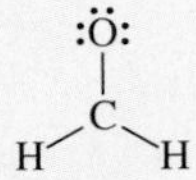

6. In this structure only six electrons (three single bonds) are associated with the carbon atom. A more stable distribution of electrons is obtained by completing carbon's octet. No additional electrons are available, so it is necessary to move one pair of electrons from the outer oxygen to make a second bond between carbon and oxygen. In the completed Lewis struc-ture, the double bond signifies four electrons shared between carbon and oxygen:

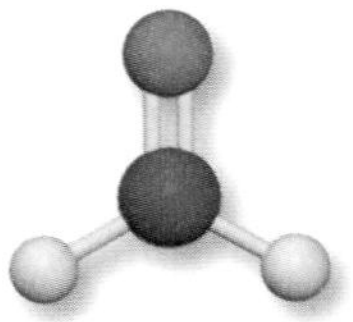

Sharing four electrons allows carbon and oxygen each to have an octet of electrons associated with it, thus making full use of their valence orbitals. The right-hand Lewis structure above is the optimal Lewis struc-ture for formaldehyde.

Valence d orbitals allow elements beyond the second row of the periodic table to have more than eight electrons associated with them. Thus an octet of electrons does not guarantee that an atom beyond the second row is in its most stable configuration. If Step 5 of the procedure leaves a positive charge on an inner atom with $n > 2$, we shift electrons to form double bonds, even if this gives the inner atom more than eight electrons.

To determine if a provisional Lewis structure contains an inner atom that appears to have a positive charge, we need a way to assess the charge distribution associated with a Lewis structure. We do this using **formal charge (FC)** of an atom, the difference between the number of valence electrons in the free atom and the number of electrons located on that atom in the Lewis structure:

$$\text{Formal charge} = (\text{Valence electrons of free atom}) - (\text{Valence electrons assigned in Lewis structure}) \qquad (8\text{-}1)$$

Here, the number of valence electrons in the *free atom* is the number assigned to that atom in Step 2 of the Lewis structure procedure. The number of valence electrons *assigned in the Lewis structure* is found from the Lewis structure by counting the electrons around each atom.

Formal charge assignments assume that all nonbonding electrons are localized on the atoms to which they are assigned and that one electron of each bonding pair is associated with each bonded atom. Thus we can determine electrons assigned to an atom by counting dots and lines around that atom. Each dot is a nonbonding electron, and each line is a bonding pair, one of whose electrons is assigned to the atom.

Example 8-7 shows how to find the optimal configuration for a compound containing an element with $n > 2$.

Lewis Structure of Sulfur Dioxide	**Example 8-7**

As described in Chapter 5, sulfur dioxide, a byproduct of burning fossil fuels, is the primary contributor to acid rain. Determine the Lewis structure of SO_2.

Strategy: Use the seven-step procedure.

Solution:

1. The formula contains neither a metal nor a polyatomic ion, so the compound is molecular.

2. All three atoms are from Group 16 of the periodic table, so each has six valence electrons, giving a total of 18 for the molecule.

3. The more electronegative oxygen atoms are outer atoms: O—S—O

4. The two bonds use 4 valence electrons, leaving 14 to place. Six go on each of the outer oxygen atoms, using 12: $:\ddot{O}$—S—$\ddot{O}:$

5. The last two valence electrons are placed on the sulfur atom: $:\ddot{O}$—$\ddot{S}$—$\ddot{O}:$

6. Sulfur is from Row 3, so we find its formal charge. Sulfur has six valence electrons (Group 16) and four assigned electrons (two bonds + one lone pair of electrons):

$$FC_S = 6 - 4 = +2$$

Constructing
Lewis
Structures

| Example 8-7 | Lewis Structure of Sulfur Dioxide *(continued)* |

With an empty set of $3d$ orbitals, sulfur can accommodate additional electrons to eliminate the $+2$ formal charge. We move two electrons from each of the outer oxygen atoms. Notice that moving electrons to make $S=O$ double bonds also eliminates the -1 formal charge on the outer oxygen atoms:

$$\ddot{O} - \ddot{S} - \ddot{O} \Longrightarrow \ddot{O} = \ddot{S} = \ddot{O}$$

FC = +2
FC = −1   FC = −1   All FC = 0

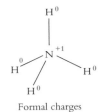

Formal charges

Actual charges

Figure 8-10
The nitrogen atom in the ammonium ion has a formal charge of +1, but the actual positive charge of the ion is distributed around the hydrogen atoms and the nitrogen atom has a partial negative charge.

To summarize, the provisional Lewis structure reached after Step 5 may not allocate an optimum number of electrons to one or more of the inner atoms. The electron distribution must be optimized when any inner atom does not have at least eight electrons or when an inner atom from beyond the second row has a positive formal charge. In either of these situations, a more stable structure can be achieved by transferring nonbonding electrons from outer atoms to inner atoms to create double bonds (four shared electrons) or triple bonds (six shared electrons).

Formal charge calculations do not indicate how charge is *actually* distributed in a molecule. Remember that most bonds are polar, meaning that the electrons in a bond are skewed toward the more electronegative atom. In contrast, we determine formal charges by *assuming* that all bonding electrons are shared equally. As a result, formal charge calculations are extremely useful for assessing whether a valence electron distribution is reasonable, but they do not reliably predict bond polarity or the distribution of actual charge.

The ammonium cation illustrates this distinction between formal charge and actual charge. The nitrogen atom of NH_4^+ has a formal charge of $+1$, but electronegativity values indicate that nitrogen attracts electrons more strongly than hydrogen ($\chi_N = 3.0$, $\chi_H = 2.1$). Thus the actual electron distribution in the N—H bond is skewed toward the N atom, leaving each H atom with a partial positive charge and the N atom with a partial negative charge, as shown in Figure 8-10.

Step 7. *Identify equivalent or near-equivalent Lewis structures.* The final step in the procedure for constructing Lewis structures is needed when there are different ways to shift electron pairs in Step 6. Example 8-8 shows how this step is applied.

| Example 8-8 | Equivalent Lewis Structures |

Determine the Lewis structure of potassium nitrate, KNO_3.

Strategy: Follow the seven-step procedure.

Solution:

1. The formula contains potassium, a metal, and the polyatomic nitrate ion. This is an ionic compound, so we must treat each species independently.

Constructing
Lewis
Structures

Equivalent Lewis Structures *(continued)*   Example 8-8

2. The potassium ion, K^+, has no valence electrons.

   The valence electron count for the nitrate ion follows:

   $1\ N = (1)(5\ e^-) = 5\ e^-$
   $3\ O = (3)(6\ e^-) = 18\ e^-$
   $\underline{\quad -1\ \text{anion} = 1\ e^- \quad}$
   $\text{Total} = 24\ \text{valence}\ e^-$

3. With no bonds and no valence electrons, the Lewis structure of the potassium ion is simply K^+.

   In the nitrate anion, the more electronegative oxygen atoms are outer atoms.

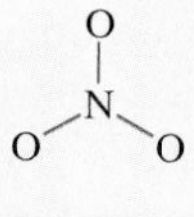

4. The three bonds use six valence electrons, leaving 18 to place. Six go on each of the outer oxygen atoms, which distributes all of the remaining valence electrons.

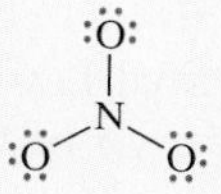

5. All of the valence electrons have been placed, so Step 5 does not apply to this problem.

6. Nitrogen is a second-row element, so its valence shell is limited to eight electrons. We optimize the configuration by completing nitrogen's octet. This can be accomplished by making one double bond between the nitrogen and any of the three oxygen atoms. The choice is arbitrary.

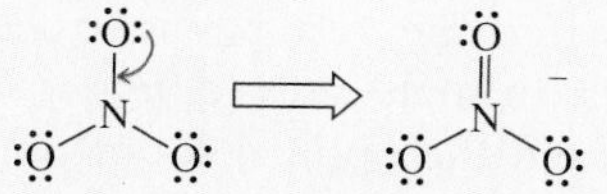

   $6e^-$ around N $8e^-$ around N

7. There are two other options for completing the nitrogen octet in the nitrate ion:

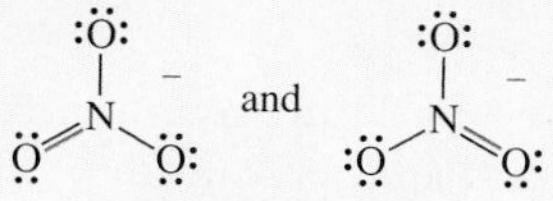

   Actually, no single Lewis structure for the nitrate anion describes its bonding adequately. Any single Lewis structure of NO_3^- shows one $N{=}O$ double bond and two $N{-}O$ single bonds. In Chapter 9, we show that single bonds and double bonds between the same types of atoms have different lengths and different energies. In contrast, experiments show that all three bonds in the nitrate anion are identical. To show that all the $N{-}O$ bonds in

| Example 8-8 | Equivalent Lewis Structures *(continued)* |

nitrate are alike, we must use a composite of the three equivalent Lewis structures. Equivalent structures are connected by double-headed arrows to emphasize that a complete depiction includes all of them:

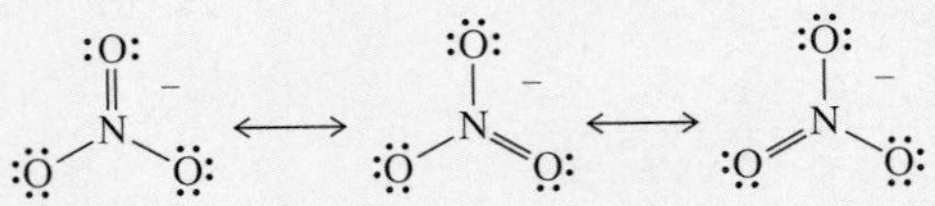

It is essential to realize that electrons in the nitrate anion *do not* flip back and forth between the three bonds as implied by equivalent structures. The true character of the anion is a blend of the three, in which all three nitrogen–oxygen bonds are equivalent. The need to draw several structures for such species reflects the fact that Lewis structures are approximate representations. They reveal much about how electrons are distributed in a molecule or ion, but they are imperfect instruments that cannot describe the entire story of chemical bonding. In Chapter 9, we show how these structures, which traditionally are called **resonance structures,** can be interpreted from an orbital perspective.

The nitrate anion illustrates that formal charge calculations are not particularly useful for second-row atoms. In the final Lewis structure, the inner nitrogen atom retains a formal charge of $+1$, while two of the outer oxygen atoms have formal charges of -1. Although shifting an additional pair of electrons would neutralize two of these formal charges, this would give the nitrogen atom more than an octet. Optimal structures of species containing inner second-row atoms are reached when the inner atom has an octet, regardless of formal charges.

Note also that the sum of formal charges for the nitrate anion totals -1. *The sum of the formal charges on all atoms equals the charge of the species.* For a neutral molecule, the sum of the formal charges is zero. For a cation or anion, the sum of the formal charges equals the charge on the ion.

Formal charge helps us find the best Lewis structures when resonance structures involve different atoms. In such cases, different structures may have different formal charge distributions, and the optimal set of nonequivalent resonance structures includes those forms with the least amount of formal charge. Example 8-9 treats a molecule with near-equivalent resonance structures.

| Example 8-9 | Lewis Structure of NNO |

Determine the Lewis structure of dinitrogen monoxide, NNO, a gas used as an anesthetic, a foaming agent, and a propellant for whipped cream.

Constructing
Lewis
Structures

Strategy: Once again, follow the seven-step procedure.

Solution:

1. Neither a metal nor a polyatomic ion is present, so treat NNO as a unit.

2. Each nitrogen atom contributes five valence electrons, and the oxygen atom contributes six valence electrons, for a total of 16.

Lewis Structure of NNO *(continued)* Example 8-9

3. Guidelines for structures indicate that the atoms are linked together in the order listed in the formula, giving the framework N—N—O. The two bonds require four valence electrons, leaving 12 to be placed.

4. Each outer atom is assigned three pairs of nonbonding electrons. This uses the 12 remaining valence electrons: :N̈—N—Ö:

5. All the valence electrons have been assigned, so there are no lone pairs to place on the inner nitrogen atom.

6. The inner nitrogen atom has only four electrons, so we complete its octet by transferring two pairs of electrons from the outer atoms. We can choose any two pairs, for instance one pair from each outer atom: :N̈=N=Ö:

7. There are two other options for completing the octet of the inner nitrogen:

$$:N\equiv N—\ddot{O}: \quad and \quad :\ddot{N}—N\equiv O:$$

The three potential resonance forms can be evaluated on the basis of their formal charges:

+1 FC	+1 FC	+1 FC
:N̈=N=Ö:	:N≡N—Ö:	:N̈—N≡O:
−1 FC 0 FC	0 FC −1 FC	−2 FC +1 FC

The third structure shows more accumulation of formal charge than the first two. Thus the optimal Lewis structure of the NNO molecule is a composite of the first two structures but not the third:

$$:N̈=N=Ö: \quad \longleftrightarrow \quad :N\equiv N—\ddot{O}:$$

Experimental studies support our conclusion about the resonance structures of NNO: The length of the nitrogen-oxygen bond is between those of an N—O single bond and an N=O double bond.

Examples 8-2 through 8-9 illustrate several of the features of Lewis structures. Examples 8-10 and 8-11 reinforce the ideas developed in this section.

Lewis Structure of a Diatomic Molecule Example 8-10

Molecular nitrogen is the most abundant component of the Earth's atmosphere and an important industrial chemical used in the Haber synthesis of ammonia. Determine the Lewis structure of N_2.

Strategy: The Lewis structure is determined by following the stepwise procedure.

Solution:

1. Molecular nitrogen is a covalent compound.

2. There are ten valence electrons, five from each nitrogen atom.

3. A diatomic molecule has a simple bonding framework: N—N

Constructing Lewis Structures

Example 8-10 | **Lewis Structure of a Diatomic Molecule** *(continued)*

4. Two electrons are used to construct the bonding framework, leaving eight to be placed around the outer atoms. Three electron pairs are placed around one nitrogen atom, leaving one pair to be placed on the other nitrogen atom to give this structure: $:N—\ddot{N}:$

5. Since there are no inner atoms, this step does not apply.

6. One nitrogen atom has an octet, but the other has only four electrons. Complete the second octet by making a triple bond between the two atoms.

$$:N≡N:$$

7. There is only one way to complete the octet of the second nitrogen atom, so there are no resonance structures for the N_2 molecule.

    The N≡N triple bond is one of the strongest known chemical bonds.

Example 8-11 | **Lewis Structure of an Organic Molecule**

Acetic acid (CH_3CO_2H) is an important industrial chemical and is the sour ingredient in vinegar. Determine its Lewis structure.

Constructing Lewis Structures

Strategy: Although this is a more elaborate molecule than earlier examples, the same stepwise procedure can be followed.

Solution:

1. Acetic acid is not ionic, so treat the entire molecule as a unit.

2. The count of valence electrons follows:

$$(2 \text{ C}) (4 \text{ e}^-) = 8 \text{ e}^-$$
$$(2 \text{ O}) (6 \text{ e}^-) = 12 \text{ e}^-$$
$$\underline{(4 \text{ H}) (1 \text{ e}^-) = 4 \text{ e}^-}$$
$$\text{Total} = 24 \text{ e}^-$$

3. The way the chemical formula is written helps us construct the correct framework. Three of the hydrogen atoms are connected to one of the carbon atoms, the two oxygen atoms are connected to the other carbon atom, and the fourth hydrogen atom is connected to one of the oxygen atoms:

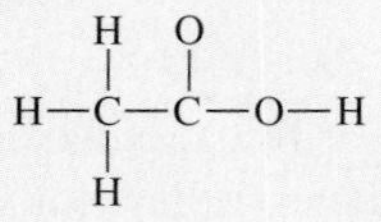

Remaining electrons: $24 – 2(7) = 10$

4. Only one outer atom is not a hydrogen atom. This outer oxygen atom requires three pairs of nonbonding electrons:

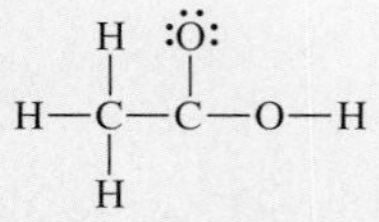

Remaining electrons: $24 – 2(7) – 6 = 4$

Lewis Structure of an Organic Molecule *(continued)*

Example 8-11

5. Four valence electrons must be placed around the three inner atoms. The most electronegative inner atom is oxygen, so two pairs of nonbonding electrons are placed on oxygen, completing its octet. All the valence electrons have now been placed:

$$\begin{array}{c}
\quad\quad H \quad\quad :\ddot{O}_a: \\
\quad\quad | \quad\quad\quad \| \\
H-\underset{\underset{H}{|}}{C_a}-C_b-\ddot{\underset{..}{O}}_b-H
\end{array}$$

(The different atoms of each element are labeled with subscripts so that we can identify them in Steps 6 and 7.)

6. Among the inner atoms, C_a and O_b have eight electrons each, but C_b has only six. Both oxygen atoms have lone pairs that could be transferred, but O_a has a formal charge of -1, so using one of its pairs of electrons to form a double bond gives C_b its octet and reduces all formal charges in the molecule to zero:

$$\begin{array}{c}
\quad H \quad :\ddot{O}: \\
\quad | \quad\quad \| \\
H-\underset{\underset{H}{|}}{C}-C-\ddot{\underset{..}{O}}-H
\end{array}$$

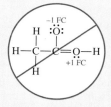

7. An alternative way to fill the valence shell of C_b would be to make a double bond to the inner oxygen, O_b. Notice, however, that doing so creates formal charge on both oxygen atoms:

$$\begin{array}{c}
\quad\quad\quad -1\ FC \\
\quad H \quad :\ddot{O}: \\
\quad | \quad\quad | \\
H-\underset{\underset{H}{|}}{C}-C=\underset{..}{O}-H \\
\quad\quad\quad\quad +1\ FC
\end{array}$$

This is not a realistic depiction, particularly when compared with a structure with no formal charges. There are no resonance structures for acetic acid.

The Lewis structure of a molecule shows how its valence electrons are distributed. These structures present simple, yet information-filled views of the bonding in chemical species. In the remaining sections of this chapter and in Chapter 9, we build on Lewis structures to develop our orbital overlap models of chemical bonding.

Box 8-1   Chemical Milestones: G. N. Lewis

Every chemist and every student of chemistry learns how to construct and interpret Lewis structures. Nearly all descriptions of molecules rely heavily on Lewis structures. We use Lewis structures so widely because they not only are easy to construct but also are accurate.

The idea of describing chemical bonding in terms of pairs of valence electrons was first proposed by G. N. Lewis in a paper published in 1916. The starting point in Lewis' reasoning was the two rows of eight elements at the beginning of the periodic table, from Li to Ne and Na to Ar. While teaching general chemistry in 1902, long before the underlying theories of chemical bonding had been worked out, Lewis drew diagrams, shown below, showing the electrons of these elements arranged in cubes. Group 1 elements had one electron at one corner of such a cube. Group 8 elements had all corners of the cube filled. Lewis supposed that completely filled or completely empty cubes were particularly stable, and this accounted for the existence of the stable ions with zero or eight valence electrons: Na^+, Mg^{2+}, O^{2-}, Cl^-.

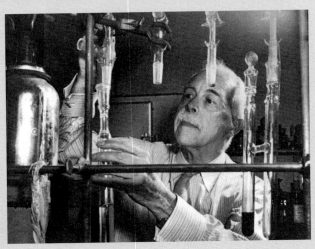

The basic idea of the Lewis structure was quickly adopted and championed by Irving Langmuir, a physical chemist who later won a Nobel Prize for his research on the chemistry of surfaces. As the mathematical theories and techniques of quantum mechanics were developed in the 1930s, chemists became able to compute details of bonding in ways that were unimaginable when Lewis put forth his ideas. Still, today we describe the bonding of molecules such as methane, water, and acetic acid in a way that mirrors the original Lewis description.

While remembered today primarily for his insights into chemical bonding, G. N. Lewis made equally insightful contributions to chemical thermodynamics (we describe thermodynamics in Chapters 12 and 13). He introduced yet a third general concept in an entirely different area, acid-base chemistry. We describe the Lewis acid-base concept in Chapter 20.

You would think that a chemist who introduced three powerful generalizations, two of which are identified with his name, would be certain to win a Nobel Prize in chemistry; yet that never happened. Perhaps this was because Lewis' ideas were so simple that they seemed naive. A structure that shows bonding electrons as sticks and dots can, after all, be drawn by any student. Whatever the reasons, it is unfortunate that the best-known—and perhaps the most original—American chemist of the early twentieth century was not awarded this prize.

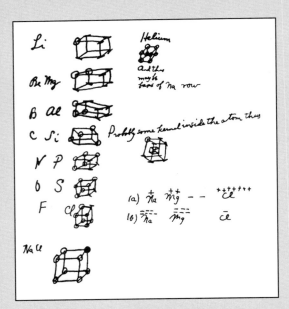

Working from this idea of the particular stability of a set of eight electrons, Lewis realized that if a chemical bond is shown as a pair of electrons, halogen molecules such as Cl_2 can be shown with eight electrons around each chlorine atom, as in the diagram on the right. This insight, that chemical bonds can be described as pairs of electrons shared between bonded atoms, was a stroke of genius. Using this simple yet elegant and powerful approach, a whole host of bonding patterns could be explained.

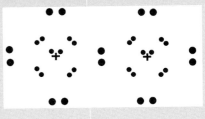

8.3.1 Determine Lewis structures for the following molecules: (a) Br_2; (b) SO_3; (c) SF_4; and (d) $CaCO_3$.

8.3.2 Determine Lewis structures for the following molecules: (a) CF_2Cl_2 (Freon™ 12, being discontinued as a coolant in air conditioners; both Cl and both F atoms are bonded to C); (b) HCCH (acetylene, used in welding torches); and (c) $(H_3C)_2CO$ (acetone, an organic solvent).

8.3.3 One of the products of the reaction described in Example 8-5 is gaseous UF_6. Determine the Lewis structure of this substance.

8.4 TETRAHEDRAL SYSTEMS: CARBON

The Lewis structure of a molecule shows how the valence electrons are distributed among the atoms. This gives a useful qualitative picture, but a more thorough understanding of chemistry requires more detailed descriptions of molecular bonding and molecular shapes. In particular, the *three-dimensional* structure of a molecule, which plays a key role in determining chemical reactivity, is not directly shown by a Lewis structure. In the next few sections, we develop an orbital overlap picture of localized bonding that includes molecular shapes.

The Shape of Methane

The Lewis structure of methane (CH_4) shows that the molecule contains four C—H single covalent bonds. How are these bonds arranged in three-dimensional space? On the basis of methane's chemical behavior, chemists concluded more than a century ago that methane's molecules are highly symmetrical. Modern experiments show that the carbon atom is at the center of the molecule, with the four hydrogen atoms located at the corners of a regular **tetrahedron.** Figure 8-11 shows four representations of the tetrahedral methane molecule. Each H—C—H set makes a bond angle of 109.5° (Figure 8-11*b*), and all C—H bond

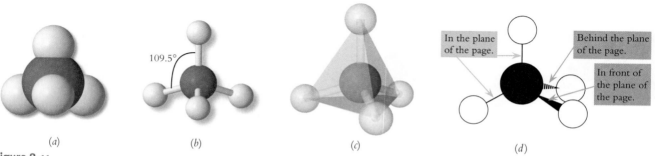

Figure 8-11
Tetrahedral methane can be drawn several ways. (*a*) Space-filling model; (*b*) Ball-and-stick model, showing the 109.5° bond angles; (*c*) Ball-and-stick model, showing the tetrahedral faces; (*d*) Ball-and-stick model using wedge representations for out-of-plane bonds.

lengths are 109 pm. A regular tetrahedron has the shape of a pyramid with four identical faces and four identical corners (Figure 8-11*c*). Each face is an equilateral triangle.

Chemists use a variation on the ball-and-stick model to depict more clearly the three-dimensional character of molecules, as shown for methane in Figure 8-11*d*. The central carbon atom is placed in the plane of the paper. In these models, solid lines represent bonds lying in the plane of the paper, solid wedges represent bonds that protrude outward from the plane of the paper, and dashed lines represent bonds extending backward, behind the plane.

Methane is the simplest molecule with a tetrahedral shape, but many molecules contain atoms with tetrahedral geometry. Because tetrahedral geometry is so prevalent in chemistry, it is important to be able to visualize the shape of a tetrahedron.

Why a Tetrahedron?

The most stable shape for any molecule maximizes electron-nucleus attractions while minimizing nuclear-nuclear and electron-electron repulsions. The distribution of electron density in each chemical bond is the result of attractions between the electrons and the nuclei. The distribution of chemical bonds relative to one another, on the other hand, is dictated by coulombic repulsion between electrons in different bonds. The spatial arrangement of bonds must minimize electron-electron repulsion. This is accomplished by keeping chemical bonds as far apart as possible. This principle of minimizing electron-electron repulsion is called **valence shell electron pair repulsion,** often abbreviated **VSEPR.**

Methane contains four pairs of valence electrons, each shared in a chemical bond between the carbon atom and one of the four hydrogen atoms. Most of the electron density in any one bond must lie between the two nuclei. At the same time, these four pairs of electrons all repel one another. Electron-electron repulsion is minimized by keeping the four C—H bonds as far apart as possible. Consider building methane by sequential addition of H^- anions to a C^{4+} ion, as shown schematically in Figure 8-12. The first anion can approach from any direction. To stay as far away from the first as possible, the second anion approaches from the opposite side of the carbon atom, generating a linear array. The third anion approaches this structure from one side and repels the two existing C—H bonds to make a triangular shape. The fourth anion approaches from above or below the plane of the existing bonds and repels the three existing C—H bonds. This converts the triangular shape into the tetrahedral geometry of methane.

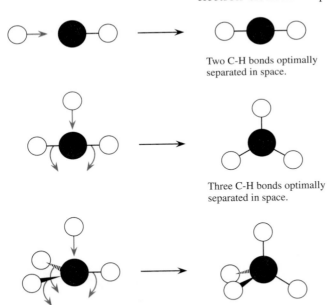

Two C-H bonds optimally separated in space.

Three C-H bonds optimally separated in space.

Four C-H bonds optimally separated in space.

Figure 8-12
The methane tetrahedron can be visualized as built up by sequential additions of C—H bonds, always keeping the bonds separated as far as possible.

Hybridization: *sp*³ Orbitals

We cannot generate a tetrahedron by simple overlap of atomic orbitals, because atomic orbitals do not point toward the corners of a tetrahedron. The valence 2*s* orbital is shaped like a sphere, so it has no directionality. Moreover, as Figure 8–13 shows, the $2p$ orbitals point at right angles to one another, so simple overlap of carbon $2p$ orbitals with hydrogen $1s$ orbitals would give 90° H—C—H bond angles, not the 109.5° angles found in tetrahedral methane.

One convenient way to visualize tetrahedral geometry for the valence orbitals of a carbon atom is to imagine the four orbitals mixing together to generate a set of four new orbitals. This can be done in such a way that each new orbital points toward a different corner of a tetrahedron. Combining an atom's atomic orbitals to form a special set of directional orbitals is referred to as **hybridization.** Bond formation in methane can then be viewed as the overlap of a hydrogen $1s$ orbital with one of these hybrid orbitals of the carbon atom to form a bonding orbital.

Any hybrid orbital is named from the atomic valence orbitals from which it is constructed. In this case the hybrids are called **sp^3 hybrid orbitals** because the set of four is constructed from one s orbital and three p orbitals.

Each hybrid orbital is directional, with a lobe of high electron density pointing in one specific direction. The detailed shape of a hybrid orbital is rather elaborate and difficult to draw in two dimensions. A cross-sectional view is shown in Figure 8–14a. We use a stylized hybrid orbital, shown in Figure 8–14b, to depict orbital overlap to form a bonding orbital. In this representation, the small backside lobe is omitted, and the orbital is slimmed down so that several orbitals can be shown clearly around an atom.

A complete orbital overlap picture of methane appears in Figure 8–14c. This depiction shows that all four sp^3 hybrids have the same shape, but each points to one corner of a regular tetrahedron. This gives each orbital a strongly favored direction for overlap with an orbital from an approaching atom. Overlap of each hybrid orbital with the atomic $1s$ orbital of a hydrogen atom generates a new bonding orbital that can hold two electrons. Four such interactions generate four localized bonds that use all the valence electrons of the five atoms involved.

Outer Atoms

All the outer atoms in methane are hydrogen atoms, which use $1s$ orbitals to form bonds to the carbon atom. The carbon-chlorine bonds in CCl_4 illustrate how other

Spherical $2s$ orbital

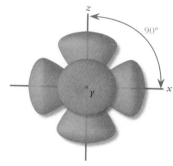

The three $2p$ orbitals

Figure 8-13
The atomic $2s$ and $2p$ orbitals point in the wrong directions to form 109.5° bond angles. The $2p$ orbitals are at right angles to one another, so $1s$ – $2p$ overlap would lead to 90° bond angles.

Tutorial

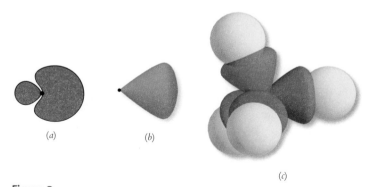

(a) (b)

(c)

Figure 8-14
(a) Cross-sectional drawing of the shape of an sp^3 hybrid orbital. (b) Depiction of hybrid orbitals that we use in this text. The "tails" of the sp^3 hybrids are omitted, and the orbital is slimmed down for clarity. (c) Drawing showing the four hybrid orbitals interacting with hydrogen $1s$ orbitals to generate the four bonds of methane.

Tutorial

outer atoms form bonds. Chlorine has four occupied valence orbitals that might participate in bonding. Which one is used to form a C—Cl bond? Remember that overlap is strongest when orbitals point in the appropriate direction in space. For an outer atom, all we need is one orbital pointing toward its bonding partner. Any one of the $3p$ orbitals of chlorine meets this requirement, so each C—Cl bond can be visualized as resulting from overlap between a carbon sp^3 hybrid and a chlorine $3p$ atomic orbital to form a bonding orbital.

Hybridization is a useful way to describe orbital interactions for *inner* atoms because inner atoms must form bonds in directions that minimize electron–electron repulsion. *Outer* atoms bond in only one direction, however, so bond formation can always be represented adequately using one particular atomic valence orbital. Consequently, it is *never* necessary to invoke hybridization for *outer* atoms. Instead, hydrogen atoms use atomic $1s$ orbitals to form bonds, and bonding for all other outer atoms can be described using atomic p orbitals.

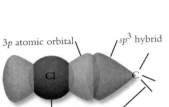

3p atomic orbital sp^3 hybrid

Cl

Core orbitals   Other sp^3 hybrids

Alkanes

Methane is the smallest member of a huge class of compounds called *hydrocarbons*, whose molecules contain only carbon and hydrogen. Hydrocarbons in which each carbon atom is bonded to four other atoms are called *alkanes*. Alkanes contain chains of carbon atoms held together by covalent single bonds, all of which can be visualized as forming from sp^3 hybrid orbitals.

The bonding of alkanes is illustrated by ethane (C_2H_6), which is shown in Figure 8-15. We can think of ethane as a methane molecule with one hydrogen atom replaced by a CH_3 (methyl) group. The Lewis structure of ethane shows each carbon atom surrounded by four pairs of bonding electrons. The bonds around each carbon are arranged in a tetrahedron to keep the four bonding pairs as far apart as possible. Each of the six C—H bonds can be viewed as resulting from the overlap of a $1s$ hydrogen orbital with one carbon sp^3 hybrid, and the carbon–carbon bond forms from sp^3-sp^3 overlap.

Alkanes have the general formula C_nH_{2n+2}, where n is an integer. The structures of alkanes that contain two and three carbon atoms are unambiguous because the carbon atoms can be placed only in a row. Figure 8-16 demonstrates, however, that there are two possible carbon backbones when $n = 4$. Starting from propane, we can replace a hydrogen atom on a terminal carbon with a methyl group to form

In *n*-decane, each C atom has tetrahedral geometry

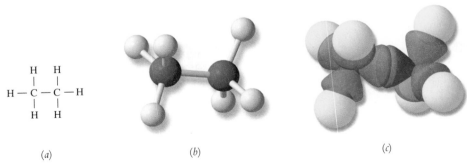

(a)   (b)   (c)

Figure 8-15
The Lewis structure (*a*), ball-and-stick model (*b*), and orbital overlap picture (*c*) for ethane, C_2H_6.

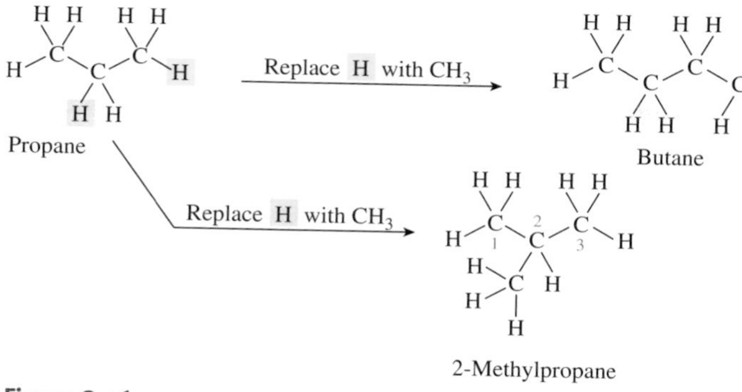

Figure 8-16
Structural representations of alkanes having formula C_4H_{10}. These are examples of structural isomers. They have the same chemical formula but different bonding arrangements.

butane, an alkane with four carbon atoms in a row. The 109.5° bond angles about each carbon give the molecular backbone a zigzag shape. Alternatively, we can replace either hydrogen atom on the inner carbon of propane to give a different compound, 2-methylpropane. In this compound, three carbon atoms are in a row, but the fourth carbon atom is off to one side.

When two or more compounds have the same molecular formula but different arrangements of atoms, they are called *structural isomers*. Butane and 2-methylpropane are structural isomers with the formula C_4H_{10}. As the number of carbon atoms in the alkane increases, so does the number of possible structural isomers. Thousands of different alkanes exist, because there are no limits on the length of the carbon chain.

Petroleum, a complex mixture of many different hydrocarbons, is the main source of alkanes. Petroleum can be processed into various fractions by boiling a mixture in huge distilling towers, as shown in Figure 8-17. Each fraction contains alkanes with a relatively narrow range of molar masses. Table 8-2 lists these major fractions and some of their uses. As the table shows, these compounds are the primary source of energy in our society.

Figure 8-17
Oil refineries use immense distillation towers to separate crude petroleum into its useful fractions.

Table 8-2
Major Fractions of Petroleum and Their Uses

Fraction	Formulas	Boiling Point Range (°C)	Uses
Natural gas	CH_4 to C_4H_{10}	−160 to +20	Fuel, cooking gas
Petroleum ether	C_5H_{12} to C_6H_{14}	30 to 60	Solvent for organic compounds
Gasoline	C_6H_{14} to $C_{12}H_{26}$	60 to 180	Fuel, solvent
Kerosene	$C_{12}H_{26}$ to $C_{16}H_{34}$	170 to 275	Rocket and jet engine fuel, domestic heating
Heating oil	$C_{15}H_{32}$ to $C_{18}H_{38}$	250 to 350	Industrial heating, fuel for electricity production
Lubricating oil	$C_{16}H_{34}$ to $C_{24}H_{50}$	300 to 370	Lubricants for automobiles and machines
Residue	$C_{20}H_{42}$ and up	over 350	Asphalt, paraffin

8.4.1 The Freon compounds are implicated in the reduction of the ozone layer. One Freon has the formula CF_2Cl_2, the Lewis structure of which is the subject of Section Exercise 8.3.2. Describe the bonding of this molecule in terms of hybrids and atomic valence orbitals. Draw a ball-and-stick model that shows the shape of this molecule.

8.4.2 The alkane that results when each of the hydrogen atoms of a CH_4 molecule is replaced by a methyl group (CH_3) is the subject of Example 8-4. Draw a ball-and-stick figure of the molecule, and describe the bonding and geometry around the carbon atoms.

8.4.3 Draw ball-and-stick models of the three structural isomers with chemical formula C_5H_{12}. Then determine the number of structural isomers for molecules whose chemical formula is $C_5H_{11}Cl$.

8.5 OTHER TETRAHEDRAL SYSTEMS

In addition to carbon atoms, tetrahedral geometry is common among molecules containing nitrogen, oxygen, and silicon atoms. For all these atoms, bonding can often be described using sp^3 hybrid orbitals.

Tetrahedral Nitrogen and Oxygen

The Lewis structure of NH_4^+ shows the cation with four N—H single covalent bonds, so the ammonium ion, with its four pairs of bonding electrons spaced as far apart as possible, has the same electronic structure as methane. (When one element is substituted for another but the number and distribution of valence electrons remain the same, the species are called *structural analogs*.) The cation has tetrahedral molecular geometry and can be viewed as an sp^3 hybridized nitrogen atom in which each hybrid orbital forms a bond with a hydrogen $1s$ orbital.

SECTION 4.6 →
Proton-transfer reactions are defined and described in Section 4.6.

The ammonium ion can be formed by proton transfer from a hydronium ion (H_3O^+) to ammonia (NH_3). Here are the Lewis structures of the substances involved in this reaction:

$$H-\underset{\underset{\displaystyle H}{|}}{\overset{\overset{\displaystyle H}{|}}{N}}-H \;+\; H-\overset{\overset{\displaystyle H}{|}}{\underset{..}{O}}-H \quad\longrightarrow\quad H-\underset{\underset{\displaystyle H}{|}}{\overset{\overset{\displaystyle H}{|}}{N}}-H^{+} \;+\; \overset{\overset{\displaystyle H}{}}{\underset{\displaystyle H}{\ddot{O}}}_{\,H}$$

Notice that the inner atoms in these four structures are bonded to different numbers of outer atoms. These differences are described by the **coordination number** of the atom, the number of other atoms to which an atom is bonded. In the ammonium ion, nitrogen has a coordination number of 4 because it is bonded to four hydrogen atoms, whereas the coordination number of nitrogen in ammonia is 3. The oxygen atom in water has a coordination number of 2, but the oxygen atom in a hydronium ion has a coordination number of 3. These species also have different numbers of lone pairs of electrons: H_2O has two, NH_3 and H_3O^+ each have one, and NH_4^+ has none. We define the **steric number** of an inner atom to be the sum of its coordination number and the number of its lone pairs. Each species involved in this proton-transfer reaction has a steric number of 4.

The steric number identifies how many groups of electrons must be widely separated in space, so it also determines orbital geometry. In ammonia, for example, the nitrogen atom bonds to three hydrogen atoms, and it has one lone pair of electrons.

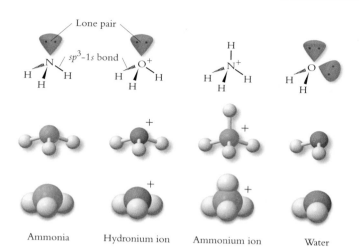

Figure 8-18
Representations of an ammonia molecule, a hydronium ion, an ammonium ion, and a water molecule.

Ammonia   Hydronium ion   Ammonium ion   Water

How are the three hydrogen atoms and the lone pair oriented in space? Just as in methane and the ammonium ion, the four groups of electrons are positioned as far apart as possible, thus minimizing electron–electron repulsion. Electrons repel one another regardless of whether they are bonding pairs or lone pairs, so the orientation of orbitals is determined by the sum of bonding electron pairs plus lone pairs.

/// *An inner atom with a steric number of 4 has tetrahedral orbital geometry, and its bonding can be described using sp^3 hybrid orbitals.*

Three-dimensional drawings of the four species involved in the protonation of ammonia are shown in Figure 8-18. Notice that, although each has tetrahedral orbital orientation, there are three different shapes. The steric number dictates orbital geometry because it describes the number of electron pairs that must be arranged to minimize electron-electron repulsion. The *shape* of a molecule, on the other hand, describes how *atomic nuclei*, not orbitals, are arranged in space. Lone pairs do not appear in a molecular shape. Nonetheless, the shape of any molecule can be derived from its orbital geometry by ignoring the orbitals that contain lone pairs.

For example, the shape of ammonia is found by ignoring the tetrahedral arm occupied by the lone pair. What remains is a nitrogen atom atop three N—H legs. This shape is called a **trigonal pyramid.** Nitrogen is at the apex of the pyramid, and the three hydrogen atoms make up the triangular base of the pyramid. To find the shape of a water molecule, we ignore the two tetrahedral arms that contain lone pairs. This leaves a planar H—O—H atomic system with a **bent shape.**

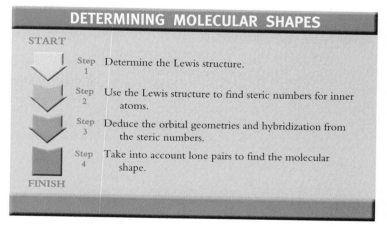

DETERMINING MOLECULAR SHAPES

START

Step 1 Determine the Lewis structure.

Step 2 Use the Lewis structure to find steric numbers for inner atoms.

Step 3 Deduce the orbital geometries and hybridization from the steric numbers.

Step 4 Take into account lone pairs to find the molecular shape.

FINISH

Our approach to these molecules illustrates the general strategy for determining the shape and bonding pattern for a molecule. The process has four steps, beginning with the Lewis structure and ending with molecular shape. Example 8-12 illustrates this strategy using the hydronium ion, and Table 8-3 summarizes the characteristics of steric number 4.

Example 8-12	Shape of the Hydronium Ion

Describe the bonding and shape of the hydronium ion (H_3O^+).

Determining
Molecular
Shapes

Strategy: Follow the four-step process for molecular shapes. Begin with the Lewis structure. Use this structure to determine the steric number, which indicates the orbital geometry and hybridization. From the orbital geometry and presence of lone pairs, identify the molecular shape.

Solution:

1. Determine the Lewis structure. A hydronium ion has eight valence electrons. Six are used to make three O—H single bonds, and two are placed as a lone pair on the oxygen atom.
   Oxygen, a Row 2 element, is surrounded by eight electrons, so this is the correct Lewis structure.

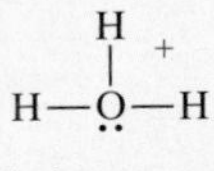

2. From the Lewis structure, determine the steric number of the inner atom. The sum of the lone pairs (1) and the coordination number (3) yields a steric number of 4.

3. A steric number of 4 indicates that the orbital geometry is tetrahedral and that the bonding can be described with sp^3 hybrids for the oxygen atom. Three hybrid orbitals overlap with hydrogen $1s$ atomic orbitals to form bonds. The fourth sp^3 hybrid contains a lone pair of electrons.

4. The tetrahedral arm containing the lone pair is ignored in visualizing the shape of the hydronium ion, which, like ammonia, is a trigonal pyramid. The ball-and-stick model and an orbital overlap sketch complete the description:

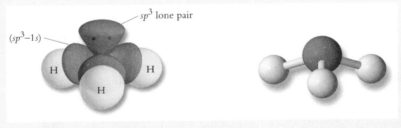

Table 8-3
Characteristics of Steric Number 4

Hybridization: sp^3
Tetrahedral orbital geometry with 109.5° angles

Coordination Number	Lone Pairs	Molecular Shape	Example	Picture
4	0	Tetrahedron	CH_4	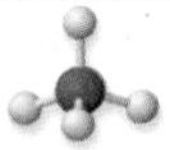
3	1	Trigonal pyramid	NH_3	
2	2	Bent	H_2O	

Many other chemical substances also have tetrahedral orbital arrangements. Figure 8-19 shows three examples. Silicon reacts with chlorine gas to form silicon tetrachloride ($SiCl_4$), a regular tetrahedron like methane. Two important industrial solvents are dichloromethane (CH_2Cl_2) and methanol (CH_3OH), both of which have tetrahedral inner carbon atoms. These molecules are asymmetrical because there are atoms of different elements at different corners. The oxygen atom in methanol also has a tetrahedral orbital arrangement, but two lone pairs give it a bent shape. Consequently, methanol can be visualized either as methane with one H atom replaced by an OH group or as water with one H atom replaced by a CH_3 group.

Silicon

One class of silicon compounds is the *silanes*, structural analogs of alkanes that contain only silicon and hydrogen. Less structural diversity exists among silanes than among alkanes, because chain lengths seldom exceed eight silicon atoms. Unlike alkanes, silanes are extremely reactive and ignite or explode spontaneously in air. The differences between the properties of silanes and alkanes result from the different sizes of the valence orbitals of silicon and carbon. The $3s$ and $3p$ orbitals of a silicon atom are much more diffuse (that is, spread out in space) than the $2s$ and $2p$ orbitals of a carbon atom. Consequently, Si—Si bonds are significantly longer and weaker than C—C bonds. Weak bonds are more reactive than strong bonds.

Compounds of silicon with oxygen are prevalent in the Earth's crust. About 95% of crustal rock and its various decomposition products (sand, clay, soil) are composed of silicon oxides. In fact, oxygen is the most abundant element in the Earth's crust (45% by mass) and silicon is second (27%). Near Earth's surface, four of every five atoms are silicon or oxygen.

Silicon forms an oxide, *silica*, whose empirical formula is SiO_2. Silica consists of a continuous network of Si—O bonds rather than individual SiO_2 molecules. Figure 8-20 shows part of this network. Each silicon atom is at the center of a regular tetrahedron, bonded to four oxygen atoms. As in water molecules, each

Silicon tetrachloride

Dichloromethane

Methanol

Figure 8-19
Three-dimensional ball-and-stick models of $SiCl_4$, CH_2Cl_2, and CH_3OH. The inner atoms of each molecule have tetrahedral geometries.

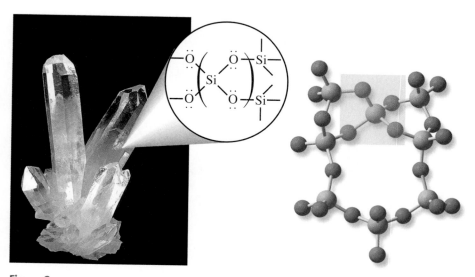

Figure 8-20
A quartz crystal consists of silica with a network structure. The ball-and-stick representation of part of the silica network shows its tetrahedral arrangement. The colored screen highlights the tetrahedral arrangement around one Si atom.

Lewis structure

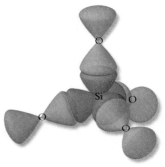

Ball-and-stick model

Orbital overlap model

Figure 8-21
Three views of the orthosilicate ion: a Lewis structure, a ball-and-stick model, and an orbital overlap model.

oxygen atom has two bonding and two lone pairs. The valence orbitals of both elements can be described by sp^3 hybridization.

Closely related to silica are the silicate minerals, all of which contain polyatomic anions made of silicon and oxygen. The simplest silicates, called *orthosilicates*, contain SiO_4^{4-} anions. The SiO_4^{4-} anion, which is illustrated in Figure 8-21, has a central silicon atom bonded to four outer oxygen atoms. The silicon atom has tetrahedral geometry, and its bonding can be described with sp^3 hybridization, like the carbon atom in methane. The Si—O bonds are described by the overlap of an atomic $2p$ orbital from oxygen with one of silicon's sp^3 hybrids.

Different minerals contain different metal cations to balance the -4 charge on the orthosilicate ion. Examples include calcium silicate (Ca_2SiO_4), an important ingredient in cement, and zircon ($ZrSiO_4$), whose crystals are often sold as artificial diamonds.

Orthosilicates are ionic, containing metal cations and discrete SiO_4^{4-} anions. Orthosilicate minerals have 4:1 ratios of oxygen atoms to silicon atoms, and every oxygen atom is a negatively charged outer atom. In silica, on the other hand, all oxygen atoms are inner atoms bonded to two silicon atoms. Silica has no ionic units and has a 2:1 ratio of oxygen atoms to silicon atoms. Nevertheless, the silicon atoms in both structures have tetrahedral shapes.

Other silicate minerals contain inner *and* outer oxygen atoms. In *metasilicates*, for example, two oxygen atoms bonded to each silicon atom are inner and two are outer, with negative formal charges. Metasilicate networks can be linear chains or rings, examples of which are shown in Figure 8-22. The examples show that the ratio of silicon to oxygen in metasilicates is 3:1. The models also show that metasilicates consist of large networks of Si—O—Si linkages bearing negatively charged outer O atoms. These charges must be counterbalanced by metal cations. In jade ($NaAlSi_2O_6$), which has a linear chain structure, the metal cations are Na^+ and Al^{3+}. Beryl ($Be_3Al_2Si_6O_{18}$), the main commercial source of beryllium metal, has six silicon tetrahedra linked in a ring and contains Be^{2+} and Al^{3+} cations. Pure beryl is colorless, but if 2% of the Al^{3+} ions in beryl are replaced by Cr^{3+} ions, the mineral becomes a brilliant green emerald. Figure 8-23 shows these colorful silicate minerals.

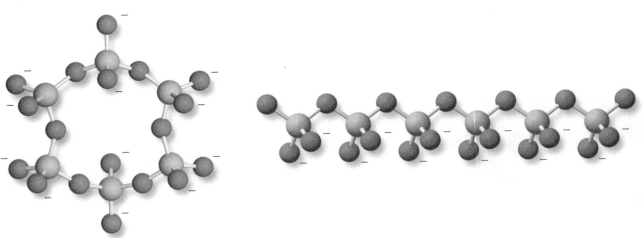

Figure 8-22
Ball-and-stick representations of two forms of metasilicates, the six-Si ring of beryl (left) and a portion of the linear chain of jade (right).

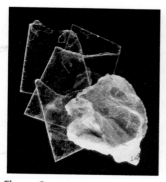

Figure 8-23
Photos of four silicates, mica, asbestos, beryl, and emerald. Silicates take on many forms, depending on the detailed structure of the Si—O bonding network.

Silicates also exist in which each silicon atom bonds to one outer oxygen and to three oxygens that are inner. The result is a linked network in which every silicon atom forms three Si—O—Si links, giving a planar, sheetlike structure. The empirical formula of this silicate is $Si_2O_5^{2-}$. In many minerals, some of the silicon atoms are replaced by aluminum atoms to give aluminosilicates. The *micas*—one has the chemical formula $KMg_3(AlSi_3O_{10})(OH)_2$—contain sheets in which every fourth silicon atom is replaced by an aluminum atom. Because of the planar arrangement of its aluminosilicate network, mica is easily broken into flakes. Figure 8-23 includes a photograph of mica.

Even more complicated chemical formulas result when some silicon atoms have one outer oxygen atom, whereas others in the same mineral have two outer oxygen atoms. The *asbestos* minerals—one is crocidolite, $Na_2Fe_5(OH)_2(Si_4O_{11})_2$—have this type of structure. The photograph in Figure 8-23 shows that this silicate contains long fibrous chains. At the molecular level, the chains can be visualized as linked $Si_4O_{11}^{6-}$ units.

The silicon atoms in silicates always have tetrahedral geometry and bonding that can be represented by sp^3 hybridization. Moreover, those oxygen atoms that link two silicon atoms can be viewed as sp^3 hybridized, whereas atomic $2p$ orbitals are sufficient to describe the bonding for oxygen atoms that occupy outer positions.

When asbestos is handled, microscopic fibers become suspended in the atmosphere and are breathed into the lungs. There, they lodge in lung tissue, where they remain for many years, causing irritation that eventually leads to loss of lung function. Asbestos, which was once used extensively as insulation, is now recognized as a significant health hazard.

Section Exercises

8.5.1 Tin compounds are used to stabilize certain plastics against thermal breakdown. They are also important agricultural pesticides because their toxic action is selective for microorganisms, allowing for control of pests at minimum risk for higher life forms. Tin tetrachloride ($SnCl_4$) is an important starting material for the preparation of a variety of tin compounds. Write a Lewis structure for $SnCl_4$, and give a complete description of its bonding, including a sketch of the orbital overlap model.

8.5.2 Hydrazine, N_2H_4, is occasionally used as a rocket fuel. Write the Lewis structure of hydrazine, describe its bonding, and draw a picture that shows the shape of the molecule.

8.5.3 Draw a ball-and-stick model of the $Si_2O_5^{2-}$ unit found in mica. Your sketch should have one Si—O—Si linkage and should show that it connects to four other $Si_2O_5^{2-}$ units to form an interlocking sheet.

8.6 OTHER MOLECULAR SHAPES

Tetrahedral geometry may be the most common shape in chemistry, but several other shapes also occur frequently. This section describes four additional orbital geometries and their associated molecular shapes.

The Shape of Triethylaluminum: *sp*² Hybridization

CHAPTER 9 →

The structure and bonding of alkenes appear in Chapter 9.

The CH_2CH_3 group is called *ethyl* because it is an *eth*ane molecule with one H atom replaced.

Triethylaluminum, $Al(C_2H_5)_3$, has long been used in the chemical industry in the production of alkenes—hydrocarbons that have $C\!=\!\!C$ double bonds. In the presence of triethylaluminum, two or more ethylene molecules link together to form straight-chain hydrocarbons that contain an even number of carbon atoms and one double bond. For example, four ethylene molecules form octene:

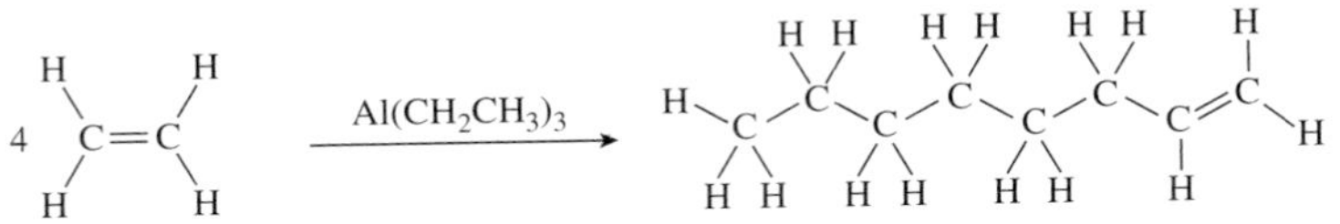

A description of the bonding in triethylaluminum begins with the Lewis structure. The chemical formula, $Al(C_2H_5)_3$, indicates that aluminum bonds to three carbon atoms, one from each ethyl group. There are 42 valence electrons, all of which are used to complete the bonding framework:

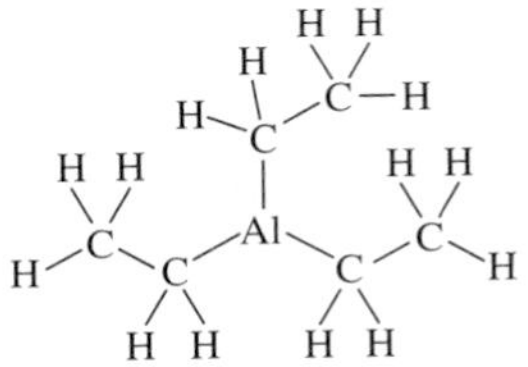

Each of the six carbon atoms in triethylaluminum has an octet of electrons and a steric number of 4. Thus each ethyl group of $Al(C_2H_5)_3$ can be described exactly as in ethane (see Figure 8-15), except that one C—H bond is replaced by a C—Al bond.

Notice that the aluminum atom is associated with only six valence electrons. Although aluminum has less than an octet of electrons, the adjacent carbon atoms have no lone pairs available to form multiple bonds. In addition, the formal charge on the Al atom is zero. Thus the Al atom has only six electrons in the optimal Lewis structure of triethylaluminum.

With three bonds and no lone pairs, the aluminum atom has a steric number of 3. The three pairs of electrons must be as far apart as possible to minimize electron-electron repulsion. Recall Figure 8-12, which shows that three pairs are as far apart as possible when they are arranged in a triangular array or a **trigonal plane,** with bond angles of 120°. The three Al—C bonds of triethylaluminum lie in such a triangular plane. Figure 8-24*a* shows the molecule in ball-and-stick fashion, and Figure 8-24*b* is another view with the ethyl groups shown schematically in order to emphasize the trigonal planar geometry around the aluminum atom.

The valence orbitals of the Al atom have the wrong geometry to form a trigonal plane. Furthermore, a set of sp^3 hybrids is not appropriate, because these would give C—Al—C bond angles of 109.5° rather than 120°. We need a different set of hybrid orbitals to represent an atom with trigonal planar orbital geometry. Proper mixing of the *s* orbital and two *p* orbitals gives a set of three new orbitals, called ***sp*² hybrid orbitals.** A set of sp^2 hybrid orbitals uses two of the valence *p* orbitals but not the

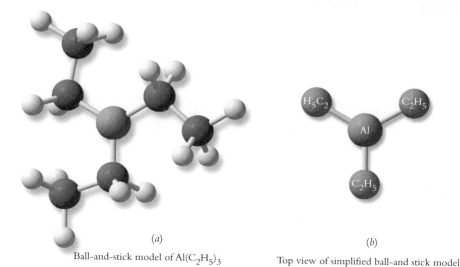

Figure 8-24
Ball-and-stick model (*a*) and stylized model for Al(C₂H₅)₃ (*b*).
(*b*) shows the trigonal planar geometry around the Al atom.

(*a*)
Ball-and-stick model of Al(C₂H₅)₃

(*b*)
Top view of simplified ball-and-stick model

third. An individual sp^2 hybrid orbital looks very much like its sp^3 counterpart, with high electron density along one direction of one axis. However, these hybrid orbitals differ from the sp^3 set in their orientations, as Figure 8-25 shows. The three sp^2 hybrid orbitals point to the three corners of an equilateral triangle. Figure 8-25 also shows that the unused p orbital is perpendicular to the plane of the three hybrids.

sp^2 sp^2

Cross-sectional views
of an sp^2 hybrid orbital

> /// *An inner atom with a steric number of 3 has trigonal planar orbital geometry, and its bonding can be described using sp² hybrid orbitals.*

In triethylaluminum, each Al—C bond can be visualized as an sp^2 hybrid on aluminum overlapping with an sp^3 hybrid on a carbon atom. Figure 8-26 shows this bonding representation, with three equivalent Al—C bonds and the unused $3p$ orbital on the aluminum atom.

As we describe in Chapter 9, energy is released when a bond forms. Consequently, atoms that form covalent bonds tend to use all their valence s and p orbitals to make as many bonds as possible. We might expect the sp^2-hybridized aluminum atom to form a fourth bond with its unused $3p$ orbital. A fourth bond does not form in Al(C₂H₅)₃ because the carbon atoms bonded to aluminum have neither orbitals nor electrons available for additional bond formation. The potential to form a fourth bond makes triethylaluminum a very reactive molecule, and this explains the ability of Al(C₂H₅)₃ to promote the joining of ethylene into longer alkenes.

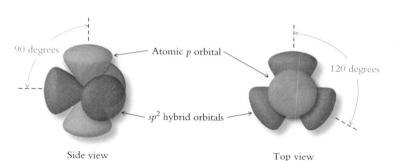

90 degrees

Atomic p orbital

120 degrees

sp^2 hybrid orbitals

Side view

Top view

Figure 8-25
An *sp²*-hybridized atom has three coplanar hybrid orbitals separated by 120° angles. One *p* orbital, oriented perpendicular to the plane of the hybrids, is left unchanged.

Tutorial

Figure 8-26
An orbital overlap picture of triethylaluminum, showing schematically its three equivalent bonds and the unused 3*p* valence orbital. For clarity, only one hybrid orbital from each carbon atom is shown.

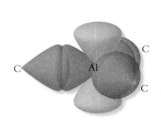

Side view   Top view

The Shape of Dimethylzinc: *sp* Hybridization

Zinc forms both ionic and covalent compounds. Dimethylzinc, $Zn(CH_3)_2$, is a covalent compound containing two Zn—C bonds. A gas that has been used in synthesis reactions since the mid-1800s, dimethylzinc finds current use as a reagent to modify surfaces of catalysts and semiconductors. We can readily determine the Lewis structure of dimethylzinc. Zinc is in Group 12 of the periodic table (configuration [Ar] $4s^2\ 3d^{10}$), so it has only two valence electrons. Each CH_3 group contributes seven electrons, giving the molecule a total of 16 valence electrons. All are used in the bonding framework:

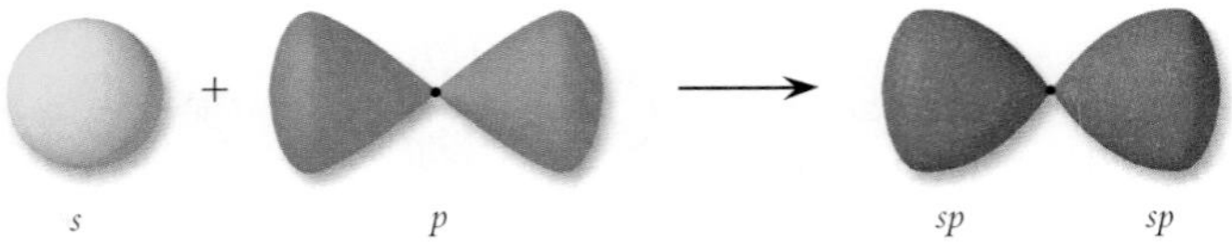

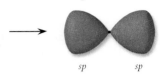

s   *p*   *sp*   *sp*

Figure 8-27
A pair of *sp* hybrid orbitals is formed by mixing together an *s* atomic orbital and one *p* atomic orbital. The two hybrid orbitals point in opposite directions.

$$H{\Large{\gt}}C-Zn-C{\Large{\lt}}H$$

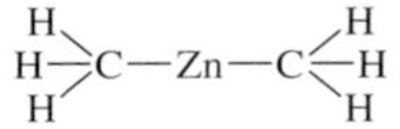

This Lewis structure shows two pairs of bonding electrons and no lone pairs on the zinc atom, giving a steric number of 2. Two pairs of electrons are kept farthest apart when they are arranged along a line. Thus the C—Zn—C bond angle is 180°, and **linear** geometry exists around the zinc atom.

To describe linear orbital geometry, we need a hybridization scheme that generates two orbitals pointing in opposite directions. This new scheme is a pair of *sp* **hybrid orbitals,** formed from the zinc 4*s* orbital and one of its 4*p* orbitals. Figure 8-27 shows the shape and orientation of the two *sp* hybrid orbitals.

Each carbon-zinc bond in dimethylzinc can be described as a localized orbital resulting from the overlap of an sp^3 hybrid on the carbon atom and an *sp* hybrid on the zinc atom. As with Al in triethylaluminum, the bonding description of Zn in dimethylzinc is unusual because it includes vacant *p* orbitals as well as *sp* hybrid orbitals. As shown in Figure 8-28, a set of *sp* hybrid orbitals makes use of only one *p* orbital. The remaining two *p* orbitals are perpendicular to each other and perpendicular to the pair of hybrids.

Compounds like triethylaluminum and dimethylzinc that have metal-carbon bonds are rather uncommon. Nevertheless, trigonal planar geometry (sp^2) and linear geometry (*sp*) occur frequently in nature. As we show in Chapter 9, these geometries occur in molecules with double bonds and triple bonds.

The other elements in Group 12, Cd and Hg, also form dimethyl compounds. Dimethylmercury, which can be synthesized by bacteria from industrial wastes, is fatal in very small quantities.

Figure 8-28
An *sp*-hybridized atom has two hybrid orbitals pointing at 180° to each other. The two remaining *p* orbitals are perpendicular to the hybrids and perpendicular to each other.

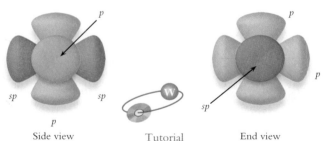

sp   *p*
p
sp   *sp*
p   *p*
Side view   Tutorial   End view

Participation of *d* Orbitals: Trigonal Bipyramidal Geometry

The elements beyond Row 2 of the periodic table have valence *d* orbitals that can participate in bonding. The participation of *d* orbitals results in steric numbers greater than 4.

Phosphorus pentachloride exemplifies the orbital geometry associated with a steric number of 5. The Lewis structure of PCl_5 shows that this molecule has five P—Cl bonds. The five pairs of bonding electrons must be localized in areas that keep the chlorine atoms as far apart as possible. The arrangement that accomplishes this is a triangular plane of three bonds combined with a linear arrangement of two bonds. The resulting geometry, shown in Figure 8-29, is called **trigonal bipyramidal** because it can be viewed as two pyramids with a common triangular base.

Four valence 3*s* and 3*p* orbitals can form no more than four covalent bonds, but phosphorus can form additional bonds using the 3*d* orbitals in its valence shell. One way to involve a 3*d* orbital in bond formation is by combining the 3*s* orbital, the set of 3*p* orbitals, and one 3*d* orbital to form a set of five *sp³d* **hybrid orbitals.** In PCl_5, each P—Cl bond can be visualized as resulting from overlap between a chlorine 3*p* orbital and an *sp³d* hybrid on the phosphorus atom.

> /// *An inner atom with a steric number of 5 has trigonal bipyramidal orbital geometry, and its bonding can be described using sp³d hybrid orbitals.*

Unlike the geometries for other steric numbers, the five positions in a trigonal bipyramid are not all equivalent, as shown in Figure 8-29*a*. Three positions lie at the corners of an equilateral triangle around the phosphorus atom, separated by 120° bond angles. Atoms in the trigonal plane are in *equatorial positions*. The other two positions lie along an axis above and below the trigonal plane, separated from equatorial positions by 90° bond angles. Atoms in these sites are in *axial positions*.

The difference between equatorial and axial positions determines the arrangement of bonding pairs and lone pairs around an atom with a steric number of 5. An

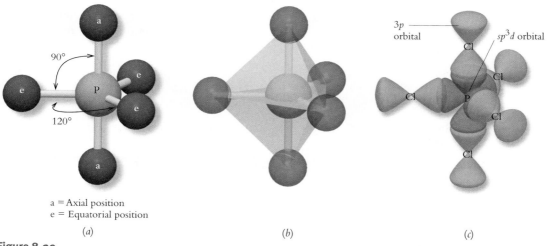

a = Axial position
e = Equatorial position

(*a*)   (*b*)   (*c*)

Figure 8-29
Views of the trigonal bipyramidal geometry of phosphorus pentachloride. (*a*) Ball-and-stick model with axial (a) and equatorial (e) positions labeled; (*b*) Ball-and-stick model showing the equatorial plane and the two pyramids; (*c*) an orbital representation showing overlap of the *sp³d* hybrids on P and 3*p* atomic orbitals on Cl.

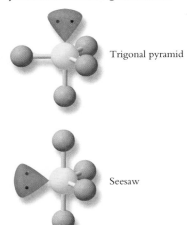

$$:\ddot{F}:$$
$$:\ddot{F}-\overset{\cdot\cdot}{S}-\ddot{F}:$$
$$:\ddot{F}:$$

Figure 8-30
Sulfur tetrafluoride has two possible molecular geometries:

Trigonal pyramid

Seesaw

example is provided by sulfur tetrafluoride, a colorless gas that has industrial uses as a potent fluorinating agent. The Lewis structure of SF_4 shows four S—F bonds and one lone pair of electrons on the sulfur atom. These five pairs of electrons are distributed in a trigonal bipyramid around the sulfur atom, whose bonding can be represented with sp^3d hybrid orbitals.

Because equatorial and axial positions differ, two molecular geometries are possible for SF_4. As Figure 8-30 shows, placing the lone pair in an axial position gives a trigonal pyramid, whereas placing the lone pair in an equatorial position gives a **seesaw** shape.

Experiments show that SF_4 has the seesaw geometry, which means that this shape is more stable than the trigonal pyramid. This is explained by the fact that lone pairs are attracted to just one nucleus while bonding pairs are attracted to two nuclei. As a result, lone pairs are more spread out in space than electrons in bonds. Consequently, energy is minimized when lone pairs are placed as far as possible from other lone pairs and from bonding pairs. The greater stability resulting from placing the lone pair in an equatorial position indicates that less net electron-electron repulsion is associated with this arrangement. Studies of the geometries about other atoms with steric number 5 show that lone pairs *always* occupy equatorial positions.

The trigonal bipyramid (PCl_5) and the seesaw (SF_4) are two of the four geometries for an atom with steric number 5. Example 8-13 introduces a third.

Example 8-13 | **Geometry of ClF_3**

The Lewis structure of chlorine trifluoride is treated in Example 8-5. Determine the molecular geometry, draw a three-dimensional picture of the molecule, and describe the bonding with an appropriate set of hybrid orbitals.

Strategy: Use the Lewis structure of ClF_3 to determine the steric number of the chlorine atom. The molecular shape is obtained from the orbital geometry after placing lone pairs in appropriate positions. Use the steric number to select the appropriate hybrid orbitals:

$$:\ddot{F}-\overset{\cdot\cdot}{\underset{}{Cl}}\overset{\displaystyle\ddot{F}:}{\underset{\displaystyle\ddot{F}:}{\Big<}}$$

Solution: The steric number for chlorine is 5, leading to a trigonal bipyramidal orbital geometry.

When the steric number is 5, there are two distinct positions, equatorial and axial. Placing lone pairs in equatorial positions always leads to the greatest stability. Thus ClF_3 is **T-shaped** with two equatorial lone pairs:

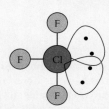

Table 8-4
Characteristics of Steric Number 5

Hybridization: *sp³d*
Trigonal bipyramidal orbital geometry with 120° and 90° angles

Coordination Number	Lone Pairs	Molecular Shape	Example	Picture
5	0	Trigonal bipyramid	PCl_5	
4	1	Seesaw	SF_4	
3	2	T-shaped	ClF_3	
2	3	Linear	I_3^-	

The fourth molecular shape arising from a steric number of 5 is represented by the polyatomic anion I_3^-. You should be able to show that this anion has a linear shape, with three lone pairs in the equatorial positions.

Table 8-4 summarizes the characteristics of atoms with steric number 5.

Octahedral Geometry

Sulfur hexafluoride is a colorless, odorless, tasteless, nontoxic, unreactive, nonflammable gas. It is prepared commercially by burning sulfur in the presence of excess fluorine. Another way to make SF_6 is by treating SF_4 with F_2 at high temperature. Because of its unusual stability, sulfur hexafluoride is used as an insulating gas for high-voltage electrical devices.

The Lewis structure of SF_6, shown in Figure 8-31a, indicates that sulfur has six S—F bonds and no lone pairs. The molecular geometry that keeps the six fluorine atoms as far apart as possible is **octahedral** shape, shown in Figure 8-31b. Figure 8-31c shows that an octahedron has eight triangular faces.

To form six bonds, the sulfur atom must use six valence orbitals. Six equivalent orbitals can be constructed by combining the 3s orbital, the three 3p orbitals, and two of the five 3d orbitals to form a set of six *sp³d²* **hybrid orbitals.** One hybrid orbital points toward each of the octahedron's six corners. We can visualize each S—F bond forming from the overlap of one of these sulfur *sp³d²* hybrid orbitals with a fluorine 2p atomic orbital.

/// An inner atom with a steric number of 6 has octahedral orbital geometry, and its bonding can be described using sp³d² hybrid orbitals.

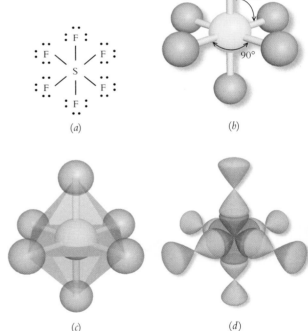

(a)   (b)

(c)   (d)

Figure 8-31
Views of sulfur hexafluoride.
(*a*) Lewis structure; (*b*) Ball-and-stick model showing 90° bond angles; (*c*) Ball-and-stick model with triangular faces shown shaded; (*d*) an orbital representation showing overlap of *sp³d²* hybrids on S and 2p atomic orbitals on F.

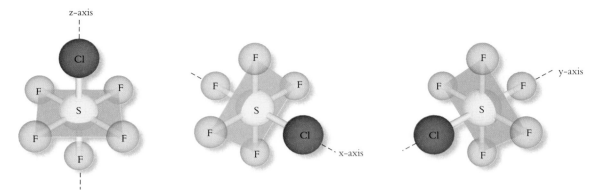

Figure 8-32
Replacing any of the six fluorine atoms of sulfur hexafluoride with a chlorine atom gives a molecule with a square of fluorines capped by one fluorine and one chlorine. This shows that all six positions in an octahedron are equivalent.

The six positions around an octahedron are equivalent, as demonstrated in Figure 8-32. Replacing one fluorine atom in SF_6 with a chlorine atom gives SF_5Cl. No matter which fluorine is replaced, the SF_5Cl molecule has four fluorine atoms in a square, with the fifth fluorine and the chlorine atom on opposite sides, at right angles to the plane of the square.

Three common molecular geometries are associated with an octahedral set of sp^3d^2 hybrids. Most often an inner atom with a steric number of 6 has octahedral molecular geometry with no lone pairs. Example 8-14 uses a compound of xenon, whose chemical behavior is described in the Chemical Milestones Box on page 370, to show a second common molecular shape, **square planar.**

| **Example 8-14** | **Structure and Bonding of XeF₄** |

Describe the geometry and bonding, and draw a ball-and-stick sketch, of xenon tetrafluoride.

Determining Molecular Shapes

Strategy: We follow the usual procedure. Determine the Lewis structure, use it to find the steric number for xenon, deduce the orbital geometry and hybridization, and look for lone pairs to identify the molecular shape.

Solution:

1. The xenon atom contributes eight valence electrons to the molecule ($5s^2\ 5p^6$). Four fluorine atoms add 28 more for a total of 36 valence electrons. Eight electrons are used to make the four Xe—F bonds, and 24 more fill the valence shells of the fluorine atoms. We are left with four electrons that must be placed as lone pairs on the xenon atom:

   8 e⁻ used to make Xe—F bonds
   24 e⁻ used to fill the F valence shells
   4 e⁻ used in two Xe lone pairs
   ─────────────────────────────────
   36 total valence electrons

   $$:\ddot{F}:$$
   $$\overset{|}{:}\ddot{F}\!-\!\ddot{Xe}\!-\!\ddot{F}:$$
   $$:\ddot{F}:$$

2. The steric number equals the sum of bonded atoms (four) and lone pairs (two): six.

Structure and Bonding of XeF₄ *(continued)*   **Example 8-14**

3. The electron pairs must be located at the corners of an octahedron to minimize electron–electron repulsion. To give the greatest stability, the two lone pairs must be as far apart as possible. Placing them at opposite ends of one axis, 180° apart, minimizes their mutual repulsion. This leaves the four fluorine atoms in a square plane around xenon. Each bond can be represented by the overlap between an sp^3d^2 hybrid orbital on the xenon atom and a $2p$ atomic orbital on the fluorine atom:

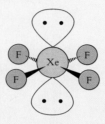

The third molecular shape arising from an octahedral set of orbitals is **square pyramidal.** A square pyramidal molecule appears in Section Exercise 8.6.1. Table 8–5 summarizes the characteristics of atoms with steric number 6.

Table 8-5
Characteristics of Steric Number 6

Hybridization: sp^3d^2
Octahedral orbital geometry with 90° angles

Coordination Number	Lone Pairs	Molecular Shape	Example	Picture
6	0	Octahedron	SF_6	
5	1	Square pyramid	ClF_5	
4	2	Square plane	XeF_4	

8.6.1 Chlorine pentafluoride is a gas that must be stored in stainless steel vessels because it reacts with glass. The molecule has square pyramidal molecular geometry. Determine the Lewis structure of ClF_5 and draw a ball-and-stick model that shows the shape of the molecule.

8.6.2 Describe the bonding in gallium triiodide.

8.6.3 Describe the bonding in the $AsCl_4^-$ anion.

Box 8-2	Chemical Milestones: Do Noble Gases React?

The noble gases are the only elements that exist naturally as individual atoms. Electron configurations make it clear why noble gas atoms prefer to remain as single atoms. Each noble gas has an $s^2 p^6$ filled shell configuration. All electrons are paired, and there are no vacant valence orbitals. Moreover, the ionization energies of these elements are extremely high—over 2000 kJ/mol for He and Ne, and over 1000 kJ/mol for the other noble gases.

Nevertheless, as early as 1933 it was suggested that xenon might form stable compounds with the most electronegative elements, fluorine and oxygen. Early attempts to react xenon directly with fluorine were unsuccessful, and chemists ignored this possibility for the next thirty years.

In 1962, the English chemist Neil Bartlett overturned the conventional wisdom. Bartlett was exploring the reactions of platinum hexafluoride, an extremely reactive molecule. He found that PtF_6 reacted cleanly and rapidly with molecular oxygen.

$$O_2 + PtF_6 \longrightarrow O_2^+ \, PtF_6^-$$

Bartlett knew that the ionization energies of O_2 and Xe are nearly identical (1180 and 1170 kJ/mol, respectively). He reasoned that if PtF_6 reacted with molecular oxygen, it might also react with xenon. Sure enough, mixing Xe gas with PtF_6 resulted in an immediate reaction that formed a yellow solid. With this simple experiment, the field of noble gas chemistry was inaugurated. Within a year of Bartlett's first experiment, eight different compounds of xenon had been made and studied. For example, when heated or illuminated, xenon reacts with fluorine to form a mixture of three different xenon fluorides: XeF_2, XeF_4, and XeF_6. The hexafluoride reacts with water to give an oxyfluoride.

$$XeF_6 + H_2O \longrightarrow XeOF_4 + 2\,HF$$

Reaction with excess water produces xenon trioxide.

$$XeF_6 + 3\,H_2O \longrightarrow XeO_3 + 6\,HF$$

Noble gas chemistry is almost entirely restricted to xenon's reactions with the most electronegative elements, fluorine and oxygen. Chlorine, the next most electronegative element, reacts with xenon to form transient species that decompose at room temperature. Krypton forms KrF_2 and a few other compounds, but the chemistry of krypton is much more restricted than that of xenon.

The ability of xenon to react is easily explained. Although it has a closed-shell configuration, [Kr] $5s^2\,4d^{10}\,5p^6$, xenon's outermost electrons can be promoted relatively easily into $5d$ orbitals, [Kr] $5s^2\,4d^{10}\,5p^{6-n}\,5d^n$, making it possible to form bonds that can be described in terms of sp^3d and sp^3d^2 hybrid orbitals.

Until recently, noble gas chemistry had almost no practical applications. In the past few years, however, lasers have been developed that are based on the chemical reactions of noble gases.

An excimer laser contains a mixture of about 2% noble gas and 0.2% halogen in an inert gas such as neon. The

mixtures include Ar-F_2, Kr-F_2, and Xe-Cl_2. In a strong electric discharge, the noble gas atoms become excited and ionized and the halogen molecules fragment into atoms. Excited states of the halogen atoms and rare gas atoms are created. These species collide and form excited diatomic molecules: ArF*, KrF*, XeCl*. These excited molecules emit ultraviolet light and break up into atoms in the process:

$$KrF^* \longrightarrow Kr + F + h\nu$$

Excimer lasers emit ultraviolet light that can vaporize solid substances like semiconductor surfaces and polymers, making it possible for such a laser to serve as a cutting device. Laser light can be focused to very small areas, so an excimer laser can perform very delicate cutting tasks.

A recent application of the excimer laser is eye surgery to correct for nearsightedness or astigmatism. In this technique, known as PRK or photorefractive keratectomy, a surgeon wields a computer-controlled excimer laser to selectively vaporize a portion of the corneal lens. Thinning the lens in just the right way can improve the ability of the eye to focus light correctly without the need for glasses or contact lenses. This procedure can be performed in just a few minutes with only a local anesthetic.

The capability to control an excimer laser beam also is exploited in the semiconductor industry, where these lasers are used to etch elaborate features during the fabrication of semiconductor chips. Neil Bartlett probably never dreamed that his explorations of the chemistry of xenon would lead to such exotic applications.

8.7 CONFIRMATION OF MOLECULAR SHAPES

The bonding model developed in Sections 8.4 to 8.6 predicts molecular shapes on the principle that electron-electron repulsion is minimized when electron pairs are placed as far apart as possible. Are the predictions of this model accurate? One way to find out is by examining bond angles and dipole moments. These properties provide revealing evidence about molecular shapes.

Bond Angles

Each of the steric numbers described in Sections 8.4 to 8.6 results in electron pairs separated by well-defined angles. If our bonding model is accurate, the actual bond angles found by experimental measurements on real molecules should match the optimal angles predicted by orbital geometry.

Experimental results agree with the predictions of the model. For instance, measurements show bond angles of 109.5° in CH_4, 120° in $Al(C_2H_5)_3$, and 90° in SF_6. Moreover, when the steric number of an atom changes, bond angles change exactly as the model predicts. Phosphorus pentachloride provides a good example. Experiments show that gaseous PCl_5 is a trigonal bipyramid with two sets of bond angles, 120° and 90°. When this compound solidifies, however, it forms ionic crystals that contain PCl_4^+ cations and PCl_6^- anions. Molecular structure determination reveals that the bond angles in the cation are 109.5°, whereas those in the anion are 90°. These are exactly what the bonding model predicts for tetrahedral and octahedral geometries, respectively.

Symmetrical molecules without lone pairs on their inner atoms, such as CH_4, PCl_5, and SF_6, have exactly the bond angles of regular geometric shapes, but lone pairs alter bond angles somewhat. In ammonia, for example, the experimental H—N—H angles are 107°, and in water the H—O—H bond angle is only 104.5°. These distortions from ideal 109.5° angles of tetrahedral geometry are caused by electron-electron repulsion. The electrons in a bonding hybrid orbital are attracted to both nuclei involved in the bond, but the lone pair electrons in a nonbonding hybrid orbital are attracted to just one nucleus. As a result, lone pair electron density is spread more widely over space than electron density in a bonding orbital. This means that the electron-electron repulsion generated by nonbonding pairs is always greater than that generated by bonding pairs.

Distorting the tetrahedral geometry to increase the angles around the lone pair minimizes the overall electron-electron repulsion in an ammonia molecule. Thus the three hydrogen atoms move slightly closer together to minimize the repulsion between their bonding pairs and the larger lone pair. Similarly, the angle between the two lone pairs in a water molecule increases by a few degrees, pushing the two hydrogen atoms closer together. The magnitude of distortion depends on the details of electron distribution in the molecule, but the qualitative effect is always the same.

/// *Lone pairs in a molecule cause bond angles to be a few degrees smaller than predicted for symmetrical geometry.*

Example 8-15 examines bond angles in another molecule.

Example 8-15	Bond Angles

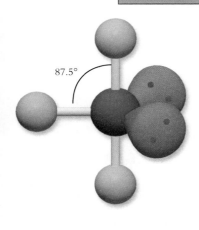

Experiments show that chlorine trifluoride has bond angles of 87.5°. Give an interpretation of these bond angles.

Strategy: To interpret bond angles, we must construct a model of the bonding using Lewis structures and steric numbers.

Solution: Refer to Example 8-13, where we describe chlorine trifluoride. The chlorine atom has a steric number of 5. There are two lone pairs that lie in the equatorial plane of the trigonal bipyramid.

The model predicts that ClF_3 is a T-shaped molecule, with F—Cl—F bond angles of 90°. However, two of the three equatorial positions contain lone pairs, which extend over more space than bonding pairs do. To reduce overall electron-electron repulsion, the two axial fluorine atoms move slightly away from the lone pairs and toward the equatorial fluorine. This explains why the bond angle is slightly less than 90°.

The triatomic molecules, water, ozone, and carbon dioxide, provide particularly strong experimental evidence that our bonding model is valid.

As discussed earlier, the H—O—H bond angle in water is distorted slightly from its tetrahedral value to minimize repulsion between the lone pairs on the oxygen atom. The inner oxygen atom of ozone has just one lone pair and a steric number of 3, giving a predicted bond angle of 120°. The bond angle in ozone is reduced by 3° to minimize electron-electron repulsion between the lone pair and the two bonding pairs. In contrast, the carbon atom in CO_2 has no lone pairs and a steric number of 2. In agreement with the observed bond angle of 180°, the model predicts that the two C=O bonds point in opposite directions along a line.

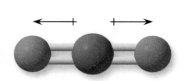

Dipole Moments

As described in Section 8.2, most chemical bonds are polar, meaning that one end is slightly negative and the other is slightly positive. Bond polarities, in turn, tend to give a molecule a negative end and a positive end. A molecule with this type of asymmetrical distribution of electron density is said to have a **dipole moment.**

Some molecules contain polar bonds but have no dipole moment. In these molecules, symmetrical shape causes polar bonds to cancel one another. A linear triatomic molecule such as CO_2 is a simple example. Both bonds in CO_2 are polar, but they point in opposite directions along a line. This is shown in Figure 8-33, in which each polar bond is shown by an arrow with a crossbar at one end. The head of the arrow points to the negative end of the polar bond, and the "+" end of the arrow points toward the positive end. For CO_2, the two arrows point in opposite directions, and the effect of one polar bond exactly cancels the effect of the other. In H_2O, in contrast, the effects of the two polar bonds do not cancel. Water has

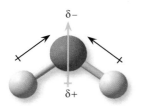

Figure 8-33
When identical polar bonds point in opposite directions, the effects of their polarities cancel, giving no net dipole moment. When they do not point in opposite directions, there is a net effect and a net molecular dipole moment, designated δ.

a partial negative charge on its oxygen atom and partial positive charges on its hydrogen atoms. The net dipole moment of the molecule is shown by the colored arrow.

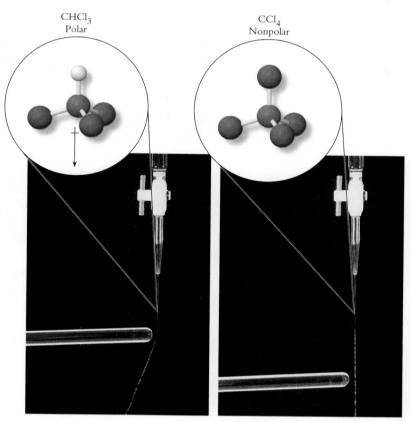

No fully symmetrical molecule has a dipole moment. Phosphorus pentachloride, for example, has no dipole moment. The two axial P—Cl bonds point in opposite directions, and although three P—Cl bonds arranged in a trigonal plane have no counterparts pointing in opposite directions, trigonometric analysis shows that the polar effects of three identical bonds in a trigonal plane cancel exactly. Likewise, the bonds in a tetrahedron are arranged so that their polarities cancel exactly, so neither CH_4 nor CCl_4 has a dipole moment.

The perfect symmetry of these geometric forms is disrupted when a lone pair replaces a bond, giving a molecule with a dipole moment. Examples are SF_4 (seesaw), ClF_3 (T shape), NH_3 (trigonal pyramid), and H_2O (bent), all of which have dipole moments. Replacing one or more bonds with a bond to a different kind of atom also introduces a dipole moment. Thus, our model predicts that chloroform ($CHCl_3$) has a dipole moment but CCl_4 does not. The carbon atom of chloroform forms four bonds in a near-regular tetrahedron, but the four bonds are not identical. The C—Cl bonds are more polar than the C—H bond, so the polarities of the four bonds do not cancel.

Figure 8-34
An electrically charged rod attracts a stream of chloroform but has no effect on a stream of carbon tetrachloride.

How can we test this prediction? Molecular dipole moments can be measured experimentally, because molecules that have asymmetrical charge distributions are attracted to an electrically charged object. Figure 8-34 shows this phenomenon for chloroform and carbon tetrachloride. A charged rod deflects a stream of liquid chloroform, showing that its molecules have dipole moments. There is no deflection for liquid carbon tetrachloride, the molecules of which have no dipole moments.

In a symmetrical octahedral system such as SF_6, each polar S—F bond has a counterpart pointing in the opposite direction. The bond polarities cancel in pairs, leaving this molecule without a dipole moment. Example 8-16 examines molecular variations on octahedral geometry.

Predicting Dipole Moments	Example 8-16

Do ClF_5 and XeF_4 have dipole moments?

Strategy: Molecules have dipole moments unless their symmetries are sufficient to cancel their bond polarities. Therefore we must examine the structure and layout of bonds in each molecule.

| Example 8-16 | Predicting Dipole Moments *(continued)* |

Solution: Chlorine pentafluoride and xenon tetrafluoride are listed in Table 8-5. Each has an inner atom with a steric number of 6, but their octahedral molecular arrangements include lone pairs. As a result, ClF_5 has a square pyramidal shape, whereas XeF_4 has a square planar shape. Pictures can help us determine whether or not the bond polarities cancel:

Each molecule has four fluorine atoms at the corners of a square. The Xe—F bond polarities cancel in pairs, leaving XeF_4 with no dipole moment. Four bond polarities also cancel in ClF_5, but the fifth Cl—F bond has no counterpart in the opposing direction, so ClF_5 has a dipole moment that points along the axis containing the lone pair.

Summary of Geometries and Hybridization

The relationships among steric number, hybridization, orbital orientation, and shapes are summarized in Table 8-6. If you remember the orbital orientation and hybridization scheme associated with each steric number, you can deduce molecular shapes, bond angles, and existence of dipole moments.

Table 8-6
Features of Molecular Geometries

Steric Number	Orbital Geometry	Hybrid-ization	Lone Pairs	Molecular Shape	Bond Angles	Picture
2	Linear	sp	0	Linear	180°	
3	Trigonal plane	sp^2	0	Trigonal plane	120°	
			1	Bent	<120°	
4	Tetrahedron	sp^3	0	Tetrahedron	109.5°	
			1	Trigonal pyramid	<109.5°	
			2	Bent	<109.5°	
5	Trigonal bipyramid	sp^3d	0	Trigonal bipyramid	120°, 90°	
			1	Seesaw	<120°, <90°	
			2	T-shape	<90°	
			3	Linear	180°	
6	Octahedron	sp^3d^2	0	Octahedron	90°	
			1	Square pyramid	<90°	
			2	Square plane	90°	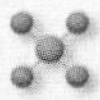

This relatively small catalog of molecular shapes accounts for a remarkable number of molecules. Even complicated molecules such as proteins and other polymers have shapes that can be traced back to these relatively simple templates. The overall shape of a large molecule results from the orbital geometry around each of its inner atoms.

Section Exercises

8.7.1 Predict the bond angles for BCl_3, SF_4, and $SnCl_4$.

8.7.2 Determine whether or not BCl_3, SF_4, and $SnCl_4$ have dipole moments.

8.7.3 Experimental evidence shows that PF_3Cl_2 has a dipole moment, whereas PCl_3F_2 does not. Determine the structures of these two compounds and explain how the structures minimize total repulsion.

CHAPTER REVIEW

Summary and Key Terms

1. Chemical bonds form when electrons are shared between atoms to form **covalent bonds.** Electron density is distributed between the nuclei to maximize electron-nucleus attractions. Bonds can be represented by **orbital overlap** of atomic orbitals. Bonds are arranged in a **molecular geometry** that maximizes stability. Each bond reaches its maximal stability, the **bond energy,** at an internuclear separation called the **bond length.** In the **localized orbital model** of covalent bonding, bonding electrons are localized between atoms, and other valence electrons are **nonbonding electrons.**

2. In bonds between atoms of different elements, differences in effective nuclear charge lead to unequal sharing of electrons and **polar covalent bonds. Electronegativity** indicates the ability of an atom to attract electrons in a bond.

3. The **Lewis structure** of a molecule is a compact representation of the distribution of its valence electrons. Lewis structures show **single bonds, double bonds, triple bonds,** and **lone pairs.** A seven-step process leads to the construction of correct Lewis structures for molecules of many types. Many atoms are most stable when they are surrounded by an **octet** of electrons. The most stable structure minimizes **formal charge.** Some molecules have more than one equivalent Lewis structure, called **resonance structures.**

4. In methane and many other compounds of carbon, the geometry about the carbon atoms is that of a **tetrahedron.** This shape minimizes **valence shell electron pair repulsion (VSEPR). Hybridization** accounts for tetrahedral geometry using localized orbitals: *sp*³ **hybrid orbitals** point toward the corners of a tetrahedron.

5. The combination of the **coordination number** and the **steric number,** both of which can be deduced from a Lewis structure, determine molecular geometries. A steric number of 4 means tetrahedral orbital shape, from which the **trigonal pyramid** and **bent shape** arise when nonbonding electrons are present. Compounds containing nitrogen, oxygen, and silicon frequently display these shapes.

6. A steric number of 3 gives rise to a **trigonal planar** geometry that can be described using *sp*² **hybrid orbitals.** A steric number of 2 gives rise to **linear** geometry and can be described using *sp* **hybrid orbitals.** The participation of one *d* orbital results in steric number 5 (*sp*³*d* **hybrid orbitals**), characterized by **trigonal bipyramidal** geometry. Nonbonding electrons change the shape to a **seesaw** or a **T-shape.** The participation of two *d* orbitals results in steric number 6 (*sp*³*d*² **hybrid orbitals**), characterized by **octahedral** geometry. Nonbonding electrons modify this geometry to a **square pyramid** or a **square planar** molecule.

7. Experimental values for bond angles and molecular **dipole moments** provide confirmation of the geometries predicted by the valence bonding model.

Skills to Master

▶ Drawing sketches that show orbital overlap

▶ Predicting and depicting molecular shapes

▶ Assessing bond polarity using electronegativities

▶ Predicting and explaining bond angles

▶ Writing Lewis structures

▶ Identifying molecules that have dipole moments

▶ Determining steric number, orbital arrangements, and hybridization

Learning Exercises

8.1 Design a flow chart that shows how to determine the Lewis structure of a molecule.

8.2 Design a flow chart that shows how to determine the shape of a molecule.

8.3 Describe the role that Coulomb's law plays in determining each of the following properties: (a) bond length; (b) bond polarity; (c) bond angle; and (d) molecular geometry.

8.4 Write a paragraph that explains why formal charges do not match actual charges.

8.5 Draw as many different pictures as you can that illustrate geometric shapes of molecules that have no dipole moments.

8.6 Explain in your own words the meaning of each of these terms: (a) orbital overlap; (b) electronegativity; (c) bonding framework; (d) inner atom; (e) multiple bond; (f) hybridization; and (g) alkane.

8.7 In your own words, write a one-sentence definition of each term in Chapter 8 that is new to you. Consult the glossary if you need help.

Problems   ilw = interactive learning ware problem. Visit the website at www.wiley.com/college/olmsted

Overview of Bonding

8.1 What is the electron configuration of a beryllium atom? Which of its electrons will be involved in bond formation?

8.2 What is the electron configuration of a silicon atom? Which of its electrons will be involved in bond formation?

8.3 It is possible to synthesize Na_2 molecules in the gas phase at low pressure. Describe the bonding in Na_2, and include a picture of the overlapping orbitals.

8.4 Describe bond formation between two bromine atoms to form a Br_2 molecule.

8.5 Which orbitals participate in bond formation for aluminum, arsenic, fluorine, and tin?

8.6 Which orbitals participate in bond formation for carbon, sulfur, mercury, and xenon?

8.7 Hydrogen forms diatomic molecules with elements from Group 1 of the periodic table. Describe the bonding in LiH and include a picture of the overlapping orbitals.

8.8 Describe bond formation between a hydrogen atom and an iodine atom to form a molecule of HI.

Unequal Electron Sharing

8.9 On the basis of the Pauling electronegativity values in Figure 8-7, compile a list of pairs of elements from the *d* block whose electronegativities do not obey the normal periodic trends.

8.10 On the basis of the Pauling electronegativity values in Figure 8-7, compile a list of pairs of elements from the *p* block whose electronegativities do not obey the normal periodic trends.

8.11 Classify the following substances as nonpolar, polar, or ionic: F_2, NaF, HF, NaH, and CaO.

8.12 Classify the following substances as nonpolar, polar, or ionic: KCl, HCl, Cl_2, KH, and CO.

8.13 For each of the following pairs, identify which element tends to attract electron density from the other in a covalent bond: (a) C and N; (b) S and H; (c) Zn and I; and (d) S and As.

8.14 For each of the following pairs, identify which element tends to attract electron density from the other in a covalent bond: (a) C and O; (b) O and H; (c) Hg and C; and (d) Si and Cl.

8.15 Arrange the following molecules in order of increasing bond polarity: H_2O, NH_3, PH_3, and H_2S.

8.16 Arrange the following bonds in order of increasing polarity: Al—Br, Al—Cl, C—Br, Al—F, and Cl—Cl.

Lewis Structures

8.17 Use the standard procedures to determine the Lewis structures of HBr, KBr, and CBr_4.

8.18 Use the standard procedures to determine the Lewis structures of NH_4Br, PCl_5, and Br_2.

8.19 Use the standard procedures to determine the Lewis structures of Na_2SO_3, H_2Se, and $AlCl_3$.

8.20 Use the standard procedures to determine the Lewis structures of PCl_3, SO_3, and NH_4PCl_6.

8.21 Determine the Lewis structures of NH_3, NH_4NO_3, and HNO_3.

8.22 Detemine the Lewis structures of KNO_2, NCl_3, and NO_2.

ilw **8.23** Determine the Lewis structures of the polyatomic ions in LiOH and KH_2PO_4.

8.24 Determine the Lewis structures of the polyatomic ions in $NaBF_4$ and $LiIO_4$.

8.25 Determine the Lewis structures of the polyatomic ions in NaCN and $(NH_4)_2CrO_4$.

8.26 Determine the Lewis structures of the polyatomic ions in $NaHSO_4$ and $NaClO_3$.

8.27 Determine the Lewis structures of the polyatomic ions in $LiNO_3$ and $(NH_4)_2CO_3$.

8.28 Determine the Lewis structures of the polyatomic ions in KPF_6 and $KMnO_4$.

Tetrahedral Systems: Carbon

8.29 Describe the bonding and geometry of dichloromethane, CH_2Cl_2.

8.30 Describe the bonding and the geometry about each inner atom of ethyl bromide, C_2H_5Br.

8.31 Draw a ball-and-stick model that shows the geometry of 1,2-dichloroethane, ClH_2CCH_2Cl.

8.32 Draw a ball-and-stick model that shows the geometry of propane, C_3H_8.

8.33 Draw all possible structural isomers for hexane, the alkane with the formula C_6H_{14}.

8.34 Draw all possible structural isomers for molecules with the formula C_4H_9F.

Other Tetrahedral Systems

8.35 Write the Lewis structure of ethanol, C_2H_5OH. Determine its geometry and hybridization. Draw a ball-and-stick model of C_2H_5OH, showing it as a methane molecule with one hydrogen atom replaced by a CH_3 group and another hydrogen atom replaced by an OH group.

8.36 Write the Lewis structure of dimethyl ether, $(CH_3)_2O$. Determine its geometry and hybridization. Draw a ball-and-stick model of $(CH_3)_2O$, showing it as a water molecule with each hydrogen atom replaced by a CH_3 group.

8.37 Silicon forms a tetramethyl compound with the formula $(CH_3)_4Si$. Determine the Lewis structure, geometry, and hybridization about each inner atom of this substance, and then draw a ball-and-stick model of the compound.

8.38 The second simplest silicate ion is the disilicate anion, $Si_2O_7^{6-}$, in which one oxygen bridges between the two silicon atoms. Determine the Lewis structure of this anion and draw a ball-and-stick model that shows its geometry.

Other Molecular Shapes

ilw **8.39** Determine the Lewis structure and molecular geometry, of $XeOF_4$. What is the hybridization of the xenon atom (Xe is the only inner atom)?

8.40 When PCl_5 solidifies, it ionizes as follows:

$$2 \; PCl_5 \longrightarrow [PCl_4]^+ + [PCl_6]^-$$

Determine the Lewis structure, molecular geometry, and hybridization of the phosphorus atom in each of these ions.

8.41 Determine the Lewis structures, geometries, and hybridizations of the inner atoms of GeF_4, SeF_4, and XeF_4.

8.42 Iodine forms three compounds with chlorine: ICl, ICl_3, and ICl_5. Determine the Lewis structures and molecular shapes of these three compounds. What is the hybridization of the iodine atom in each?

Confirmation of Molecular Shapes

8.43 Determine the Lewis structures of the following compounds, and determine which ones have dipole moments: (a) SiF_4; (b) H_2S; (c) XeF_2; (d) $GaCl_3$; and (e) NF_3.

8.44 Determine the Lewis structures of the following compounds, and determine which ones have dipole moments: (a) CH_4; (b) $CHCl_3$; (c) CH_2Cl_2; (d) CH_3Cl; and (e) CCl_4.

8.45 Predict the bond angles for the compounds in Problem 8.43.

8.46 Predict the bond angles for the compounds in Problem 8.44.

Additional Paired Problems

8.47 Draw Lewis structures and ball-and-stick structures showing the correct geometries for molecules of the following substances: (a) Cl_2O, dichlorine oxide (used for bleaching wood pulp and water treatment, about 10^5 tons produced each year); (b) C_6H_6, benzene, which contains a ring of six carbon atoms, each bonded to one hydrogen atom (one of the top 20 industrial chemicals, used in production of polymers); and (c) C_2H_4O, ethylene oxide, which contains a C—C—O triangular ring (one of the top 50 industrial chemicals, used in polymer production).

8.48 Draw Lewis structures and ball-and-stick structures showing the geometries of molecules of the following substances: (a) CH_3NCO, methyl isocyanate (toxic compound responsible for thousands of deaths in Bhopal, India, in 1984); and (b) F_2NNF_2, tetrafluorohydrazine (colorless liquid used as rocket fuel).

8.49 Determine the molecular geometries of the following molecules: $SiCl_4$, SeF_4, and CI_4.

8.50 Determine the molecular geometries of the following ions: ClF_2^-, BF_4^-, and PF_4^+.

8.51 How many different structural isomers are there for octahedral molecules with the general formula AX_3Y_3? Draw three-dimensional structures of each.

8.52 Identify which of the four octahedral molecules shown here are equivalent:

(a)   (b)   (c)   (d)

8.53 In the gas phase, aluminum trichloride exists as individual $AlCl_3$ molecules. Describe the bonding in gaseous $AlCl_3$.

8.54 $SbCl_3$, a soft, colorless solid, is used as a flame retardant. Describe the bonding and geometry in this compound.

8.55 Describe the bonding and determine the empirical chemical formula of the silicon-oxygen network of zircon.

8.56 Describe the bonding and determine the empirical chemical formula of the silicon-oxygen network of asbestos.

8.57 Write Lewis structures and calculate formal charges for the following polyatomic ions: (a) bromate; (b) nitrite; (c) phosphate; and (d) hydrogen carbonate.

8.58 Carbon, nitrogen, and oxygen form two different polyatomic ions: cyanate ion, NCO^-, and isocyanate ion, CNO^-. Write Lewis structures for each anion, including near-equivalent resonance structures and indicating formal charges.

8.59 Amines are organic substances derived from ammonia. Determine the Lewis structure, predict the bond angles, and draw a ball-and-stick model of methylamine, H_3CNH_2.

8.60 Determine the Lewis structure and predict all the bond angles in the glycine molecule, $H_2NCH_2CO_2H$, the simplest amino acid.

8.61 Sulfur and fluorine form three different molecules containing one S atom: SF_2, SF_4, and SF_6. Draw the Lewis structure of each.

8.62 Sulfur and fluorine form four different molecules containing an S—S bond: SSF_2, $FSSF$, F_3SSF, and F_5SSF_5. Draw the Lewis structure of each.

8.63 For each molecule in Problem 8.61, determine the molecular geometry and hybridization associated with each inner atom.

8.64 For each molecule in Problem 8.62, determine the molecular geometry and hybridization associated with each inner atom.

8.65 Carbocations are unstable high-energy ions that contain positively charged carbon atoms. Carbocations form during the course of a reaction but are usually consumed rapidly. Describe the bonding and geometry of the carbocation $(CH_3)_3C^+$.

8.66 The methylene fragment, CH_2, has been identified as a reactive species in some gas phase chemical reactions. Determine its Lewis structure, describe its bonding, and predict the H—C—H bond angle.

8.67 The H—O—H bond angle in a water molecule is $104.5°$. However, in the isoelectronic analog hydrogen sulfide, the H—S—H bond angle is only $92.2°$. Explain these variations in bond angles, using orbital sizes and electron-electron repulsion arguments. Draw space-filling models to illustrate your explanation.

8.68 The bond angles are 107.3° in NH_3, 100.3° in PCl_3, and 93.3° in PH_3. Explain these variations in bond angles, using orbital sizes and electron-electron repulsion arguments. Draw space-filling models to illustrate your explanation.

8.69 Determine the Lewis structures of HNO_3, $HClO_2$, and H_2SO_4. In each of these acids, all hydrogen atoms are bonded to oxygen atoms.

8.70 Determine the Lewis structures of H_3PO_4, $HClO_3$, and H_2SO_3. In each of these acids, all hydrogen atoms are bonded to oxygen atoms.

8.71 List the following X—H bonds from smallest bond polarity to largest bond polarity: C—H, F—H, N—H, O—H, and Si—H.

8.72 List the following X—O bonds from smallest bond polarity to largest bond polarity: C—O, N—O, S—O, O—H, and Br—O.

8.73 Do these structures represent the same compound or are they different compounds? Explain.

8.74 Do these structures represent the same compound or are they different compounds? Explain.

More Challenging Problems

8.75 Both PF_3 and PF_5 are known compounds. NF_3 also exists, but NF_5 does not. Show that these facts are consistent with this chapter's orbital model.

8.76 Write the Lewis structure of borazine, $B_3N_3H_6$. The molecule contains a planar ring of alternating boron and nitrogen atoms, with a hydrogen atom attached to each ring atom.

8.77 Imagine a square planar molecule, XY_2Z_2, in which X is the central atom and Z is more electronegative than Y. Two structural isomers are possible. (a) Draw a ball-and-stick model of each isomer. (b) Suppose you have made one isomer, but do not know which one. Explain how measuring the dipole moment of the substance would identify the isomer.

8.78 Recall from Section 4.6 that H_3O^+ transfers a hydrogen ion to the base in an acid-base reaction. A lone pair of electrons on the base is used to form the new bond to the hydrogen atom. Draw orbital overlap pictures that illustrate the bonding changes that occur when a hydronium ion reacts with a hydroxide ion to produce water.

8.79 Tellurium compounds, which are toxic and malodorous, must be handled with extreme care. Predict the formulas of the tellurium-fluorine molecules (or ions) with the following molecular geometries: (a) bent; (b) T-shaped; (c) square pyramid; (d) trigonal bipyramid; (e) octahedron; and (f) seesaw.

8.80 Cyclopropane, C_3H_6, which has three carbon atoms in a ring, is far more reactive than other alkanes. Determine the Lewis structure of this molecule and the hybridization of the C atoms. Suggest a reason for the reactivity of C_3H_6. (Hint: What is the bond angle between the carbon atoms?)

8.81 Among the halogens, only one known molecule has the formula XY_7. It has pentagonal bipyramidal geometry, with five Y atoms in a pentagon around the central atom X. The other two Y atoms are in axial positions. Based on our discussion of electron-electron repulsion and atomic size, determine the identities of atoms X and Y. Explain your reasoning. (Astatine is not involved. This element is radioactive and highly unstable.)

8.82 Indium triiodide exists in the gas phase as individual InI_3 molecules. In the liquid phase, on the other hand, two InI_3 molecules combine to give In_2I_6, in which two iodine atoms bridge between indium atoms, and there are four outer iodine atoms. Describe the bonding in InI_3 and In_2I_6. Draw a ball-and-stick model of In_2I_6.

8.83 Write Lewis structures for the following molecules: (a) $OPCl_3$; (b) HN_3 (contains four atoms in a row); (c) $SeCl_6$; and (d) $H_2S_2O_6$ (has an S—S bond).

8.84 Determine the Lewis structures for the two possible arrangements of the N_2O molecule, N—N—O and N—O—N. Experiments show that the molecule is linear and has a dipole moment. What is the arrangement of atoms? Justify your choice.

8.85 Spiroalkanes are hydrocarbons in which two carbon-containing rings share one carbon atom. One of the simplest spiroalkanes is called "spiroheptane." Describe the bonding in spiroheptane. Draw a ball-and-stick model that shows the geometry of the molecule.

Spiroheptane
C_7H_{12}

Group Study Problems

8.86 With one exception, sulfur forms compounds with all steric numbers. Find an example of each, describe the geometry of each, and identify which steric number is not found.

8.87 Convert each of the following line structures into a complete Lewis structure. For each atom, identify the atomic or hybrid orbitals used for bonding. Draw ball-and-stick pictures that show the molecular geometries accurately.

8.88 Describe the bonding differences among silica, silicate minerals and metasilicate minerals.

8.89 Many transition metal cations form metal complexes with geometries that match those described in this chapter. Determine the Lewis structure and describe the geometry around each inner atom of $[Sc(H_2O)_6]^{3+}$.

8.90 In the following reactions, phosphorus forms a bond to a Row 2 element. In one reaction, phosphorus donates two electrons to make the fourth bond, but in the other reaction, phosphorus accepts two electrons to make the fourth bond. Use Lewis structures of starting materials and products to determine in which reaction phosphorus is a donor and in which it acts as an acceptor.

$$PCl_3 + N(CH_3)_3 \longrightarrow Cl_3PN(CH_3)_3$$

$$PCl_3 + BBr_3 \longrightarrow Cl_3PBBr_3$$

8.91 The molecule dichlorobenzene, $C_6H_4Cl_2$, which contains a ring of six carbon atoms, has three structural isomers. These differ in the relative positions of the Cl atoms around the ring. Determine the Lewis structures and geometries of these three isomers. Which of the isomers are polar and which are nonpolar? Which isomer has the largest dipole moment?

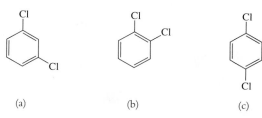

(a)   (b)   (c)

8.92 Determine the Lewis structures, orbital geometries, and molecular shapes of the following compounds, which contain odd numbers of valence electrons (Hint: treat the odd electron like a lone pair). (a) NO_2, nitrogen dioxide (red-brown gas that pollutes the air over many cities); (b) ClO_2, chlorine dioxide (highly explosive gas used as an industrial bleach).

Answers to Section Exercises

8.1.1

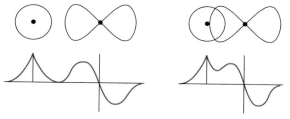

8.1.2 The $1s$ orbital of H overlaps with the $3p$ orbital of Cl pointing along the bond axis. Two electrons occupy this orbital.

8.1.3 Se, $4s$, $4p$; and P, $3s$, $3p$

8.2.1 (a) O; (b) C; (c) Cl; and (d) Cl

8.2.2 $C-H < As-Cl < Sn-Cl < Si-O$

8.2.3 Ionic: LiCl, Na_2S; and some covalent character: $FeCl_3$, AgCl, ZnS, Al_2O_3

8.3.1

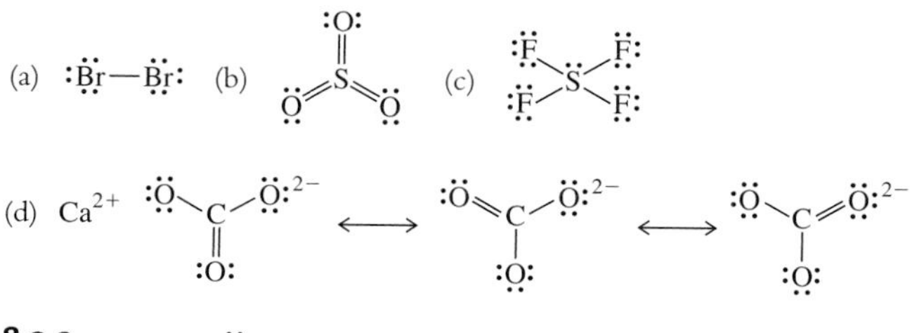

(a) $:\ddot{B}r-\ddot{B}r:$ (b) (c)

(d) Ca^{2+}

8.3.2

(a)   (b) $H-C\equiv C-H$

(c)

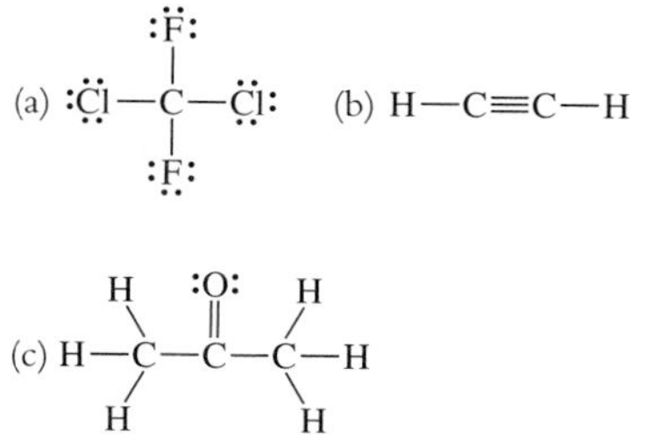

8.3.3

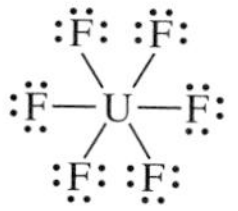

8.4.1 The carbon atom uses sp^3 hybrids and each halogen atom uses a p orbital.

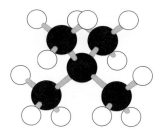

8.4.2 Each carbon atom uses sp^3 hybrids and has tetrahedral geometry. The hydrogen atoms use $1s$ orbitals.

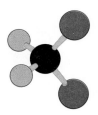

8.4.3 The arrows indicate different types of H atoms. There are eight structural isomers with the chemical formula $C_5H_{11}Cl$.

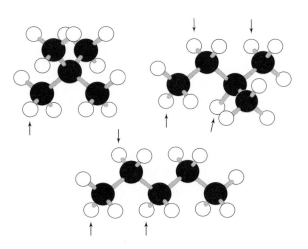

8.5.1 Each bond forms from an sp^3 hybrid from Sn and a $3p$ atomic orbital from Cl.

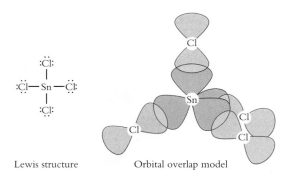

Lewis structure   Orbital overlap model

8.5.2 Each N atom uses sp^3 hybrids to form an N—N bond and two N—H bonds. There is a lone pair in the remaining hybrid, and the shape around each N atom is a trigonal pyramid.

8.5.3 The atoms inside the circle make up one $Si_2O_5^{2-}$ unit:

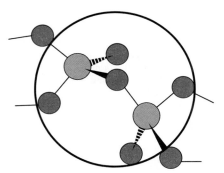

8.6.1

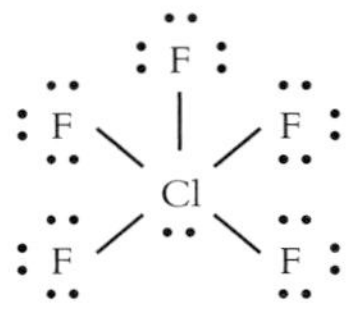

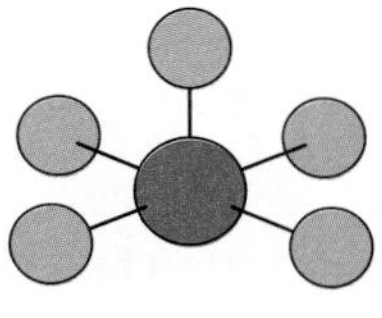

Lewis structure   Square pyramid

8.6.2 The bonding about Ga is similar to that around Al in $Al(C_2H_5)_3$. Gallium uses sp^2 hybrid orbitals to overlap with p atomic orbitals of iodine. The molecule is trigonal planar in shape.

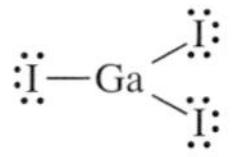

8.6.3 The bonding is similar to that in SF_4. Arsenic uses sp^3d hybrids to overlap with p atomic orbitals of chlorine. The fifth hybrid holds a lone pair. The ion has a seesaw shape.

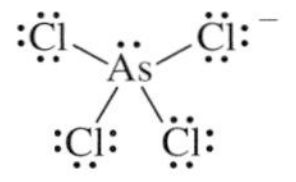

8.7.1 BCl_3, 120°; SF_4, < 90° and < 120°; and $SnCl_4$, 109.5°

8.7.2 Only SF_4 has a dipole moment.

8.7.3 The chlorine atoms, being larger than fluorine atoms, occupy the equatorial positions because this minimizes atom-atom repulsion.

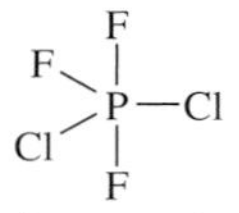

Asymmetric 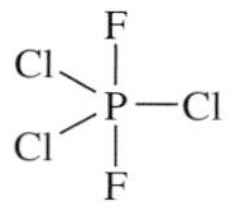   Symmetric

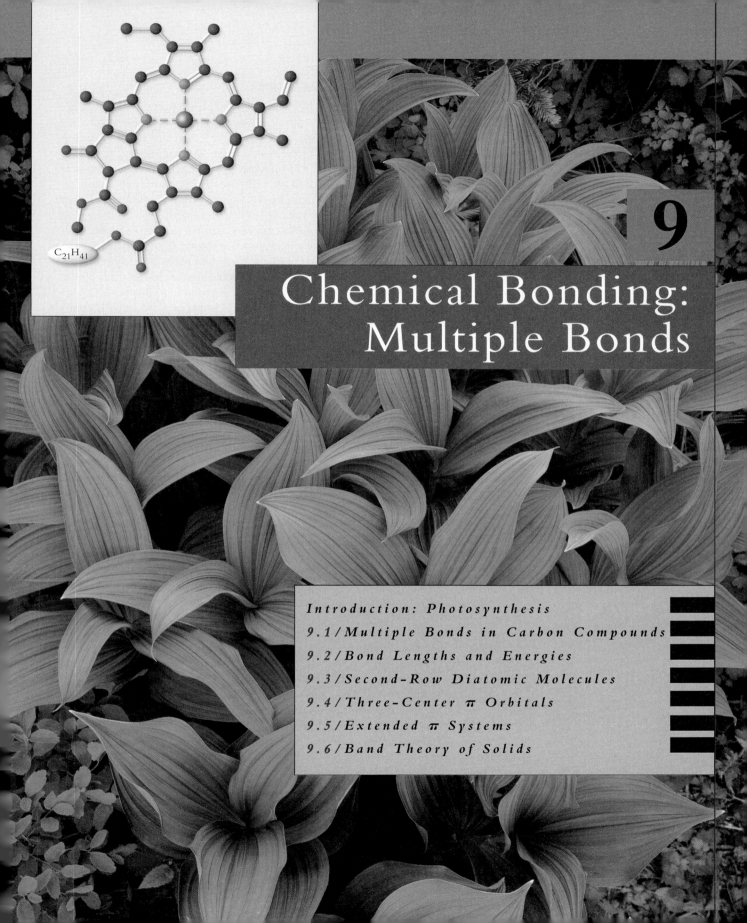

$C_{21}H_{41}$

9

Chemical Bonding: Multiple Bonds

INTRODUCTION: PHOTOSYNTHESIS

Photosynthesis is the process by which green plants harvest light energy from the sun and store it as chemical energy. Through a complex sequence of chemical reactions, plants use solar energy to convert water and carbon dioxide into energy-rich glucose and molecular oxygen. Depending on the type of plant and its particular biochemistry, the glucose produced by photosynthesis is converted to more complex sugars, starch, or cellulose. These compounds, in turn, are available to plant-eating animals as sources of energy and building blocks for biochemical synthesis.

Photosynthesis begins with chlorophyll, a green pigment that gives plants their luscious green color. Chlorophyll molecules absorb photons of visible light, converting solar energy into electronic energy in the form of a molecular excited state. The light-absorbing portion of the chlorophyll molecule appears in the inset figure. The process is analogous to the atomic excited states described in Chapter 6.

Chlorophyll molecules are packed together in chloroplasts, subcellular components present in the cells of higher plants and green algae. Within the membranes of a chloroplast, the excited state energy of chlorophyll is used to move an electron from the chlorophyll molecule to a plastoquinone molecule, which in turn donates an electron to a cytochrome molecule. This cascade of electrons from chlorophyll to plastoquinone to cytochrome is an essential portion of the energy-harvesting segment of photosynthesis. The line structure of plastoquinone is shown in the margin. The electron-shuttling portion of the cytochrome looks much the same as chlorophyll, with iron rather than magnesium at the center of the molecule. Chlorophyll, plastoquinone, and cytochrome are complicated molecules, but all three rely upon extended networks of single bonds alternating with double bonds for their roles in photosynthesis. Molecules that contain such networks are particularly good at absorbing light and at undergoing reversible oxidation-reduction reactions. Both these properties are essential for photosynthesis.

In this chapter we develop a model of bonding that can be used to describe the characteristics of double and triple chemical bonds. In molecules with alternating single and double bonds, some of the electrons are spread over the entire network. In many cases these delocalized electrons absorb photons in the visible region of the spectrum, and such absorption is what gives some substances their colors. This helps explain why molecules like chlorophyll and plastoquinone are key participants in photosynthesis.

We begin with an orbital description of double and triple bonds in carbon-containing molecules. Next, we examine how bond lengths and bond strengths correlate with the bonding model developed in Chapter 8 and in Section 9.1. Then we present additional orbital characteristics that apply to bonding in diatomic molecules and to larger molecules that contain extended systems of multiple bonds. Finally, we show how the bonding model can be generalized to describe the electronic structures of metals and semiconductors. This model successfully explains the conducting properties of the materials that form the basis for all of modern-day solid state electronics.

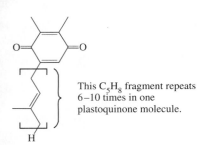

This C_5H_8 fragment repeats 6–10 times in one plastoquinone molecule.

Plastoquinone

9.1 MULTIPLE BONDS IN CARBON COMPOUNDS

Many of the Lewis structures shown in Chapter 8 and elsewhere in this book represent molecules that contain double bonds and triple bonds. From simple molecules such as ethylene and acetylene to complex biochemical compounds such as chlorophyll and plastoquinone, multiple bonds are abundant in chemistry. Double bonds and triple bonds can be described by extending the orbital overlap model introduced in Chapter 8. We begin with ethylene, a simple hydrocarbon with the formula C_2H_4.

Ethylene is a colorless, flammable gas with a boiling point of -104 °C. More than 22 billion kilograms of ethylene are produced annually in the United States, making it one of the top five industrial chemicals. The manufacture of plastics (polyethylene is the most common example) consumes 75% of this output, and much of the rest is used to make antifreeze. Because ethylene stimulates the breakdown of cell walls, it is used commercially to hasten the ripening of fruit, particularly bananas.

Bonding in Ethylene

Every description of bonding starts with a Lewis structure. Ethylene has 12 valence electrons. The bond framework of the molecule has one C—C bond and four C—H bonds, requiring ten of these electrons. After placing the final two electrons as a lone pair on one of the carbon atoms, the second carbon atom is left with only six electrons. Making a double bond between the carbon atoms gives both carbon atoms octets and completes the Lewis structure:

The C–H bonds form from the overlap of hydrogen's $1s$ atomic orbital with one of carbon's sp^2 hybrid orbitals.

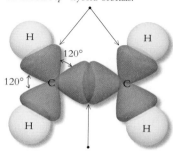

What orbital overlap picture best describes ethylene? Both carbon atoms are bonded to three other atoms, and neither has a lone pair. Thus each carbon atom has a steric number of 3, which predicts trigonal planar orbital geometry. Experiments show that C_2H_4 is a planar molecule with 120° bond angles. These features are consistent with sp^2 hybridization, so a bonding framework can be constructed using carbon sp^2 hybrid orbitals and hydrogen $1s$ orbitals. According to this model, each C—H bond involves overlap of an sp^2 hybrid on carbon with a $1s$ atomic orbital on hydrogen, and the C—C bond forms from the overlap of two sp^2 hybrid orbitals. Figure 9-1 shows the bonding framework for ethylene.

The C–C bond forms from the overlap of two sp^2 hybrid orbitals.

Figure 9-1
Schematic drawing of the bonding framework of ethylene. The orbitals of each carbon atom can be represented as sp^2 hybrids.

Figure 9-1 shows one bond joining the carbon atoms in ethylene, but the Lewis structure indicates that a *double* bond exists between the carbon atoms. By examining carbon's full set of valence orbitals, we can see how to construct a second bond between the carbon atoms of ethylene. Three of the valence orbitals are required for the sp^2 hybrid orbitals, leaving one valence p orbital that is not part of the hybrid set. This p orbital has high electron density above and below the sp^2 plane. As shown in Figure 9-2, these "leftover" p orbitals on adjacent carbon atoms overlap in a side-by-side fashion. This produces *one* bond with two regions of high electron density, one on either side of the plane defined by the bonding framework.

Tutorial

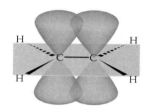

σ Bonds and π Bonds

In the bond framework shown in Figure 9-1, all the bonds form as a result of end-on overlap of orbitals directed toward each other. As illustrated by the three examples in Figure 9-3, this type of overlap gives high electron density distributed

Figure 9-2
Formation of an sp^2 hybrid set leaves one unused valence p orbital. In ethylene, these orbitals overlap to form a second bond between the carbon atoms.

Figure 9-3
Schematic views of the σ bonds in HF and C₂H₄. All have high electron density concentrated along the internuclear axis and axial symmetry, so their end-on profiles are circles.

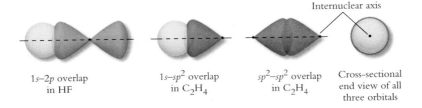

Internuclear axis

$1s$–$2p$ overlap in HF | $1s$–sp^2 overlap in C₂H₄ | sp^2–sp^2 overlap in C₂H₄ | Cross-sectional end view of all three orbitals

The symbol σ is the Greek letter corresponding to the English letter s.

symmetrically *along* the internuclear axis. A bond of this type is called a **sigma (σ) bond,** and a bonding orbital that describes a σ bond is a σ orbital.

/// *A σ bond has high electron density distributed symmetrically along the bond axis.*

Bonds that form from the side-by-side overlap of atomic p orbitals have different electron density profiles than σ bonds. A p orbital has zero amplitude in a plane passing through the nucleus, so bonds that form from side-by-side overlap have zero electron density directly on the bond axis. High electron density exists between the bonded atoms, but it is concentrated *above and below* the bond axis. A bond of this type is called a **pi (π) bond,** and a bonding orbital that describes a π bond is a π orbital. Figure 9-4 shows a π bond from four different perspectives.

The symbol π is the Greek letter corresponding to the English letter p.

/// *A π bond has high electron density between the nuclei, concentrated above and below the bond axis.*

The double bond in ethylene contains one σ bond and one π bond. The σ bond forms from the end-on overlap of two hybrid orbitals, and the π bond forms from the side-by-side overlap of two atomic p orbitals. Figure 9-5 illustrates the complete orbital picture of the bonding in ethylene. Ethylene is the simplest of a class of molecules, the alkenes, all of which contain C═C double bonds. The alkenes are the subject of our Chemistry and Life Box on page 385.

The availability of a leftover valence p orbital to form π bonds is characteristic of sp^2 hybridization and is not restricted to carbon atoms. Example 9-1 shows how

Figure 9-4
Four views of a π bond.

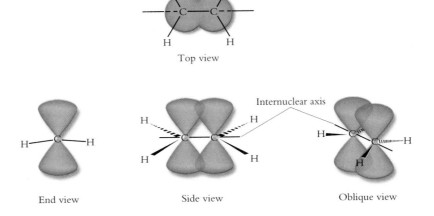

Top view

Internuclear axis

End view | Side view | Oblique view

Box 9-1 — Chemistry and Life: The Alkenes

Alkenes are hydrocarbons that contain one or more carbon-carbon double bonds. The simplest alkene, ethylene, is produced by plants to stimulate the ripening of fruit. Larger alkenes are abundant in the biological world and play a wide variety of roles.

Animals communicate by excreting minute amounts of chemical signaling agents known as *pheromones*. Some pheromones repel predators, others mark trails, still others signal alarm. Pheromones have a wide variety of chemical structures, but many of them are alkenes. For example, the trail marker used by termites is neocembrene, an alkene with four carbon-carbon double bonds. Animals produce pheromones to attract members of the opposite sex. One such pheromone is released by female moths to signal that they are ready to mate. A male moth that gets a sniff of this pheromone immediately seeks out the female.

Remarkably, the same molecule has been isolated from the urine of female elephants that are about to ovulate. The presence of this pheromone in her urine sends a signal to bull elephants that the mating season has arrived. The similar function of this same compound in two different species is one of the surprising coincidences of evolution.

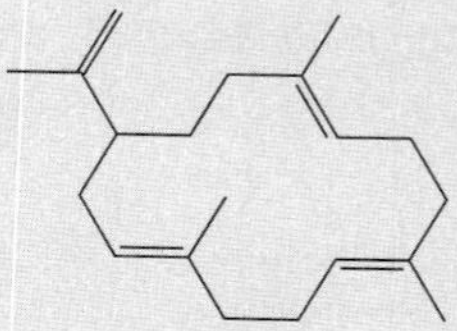

Neocembrene ($C_{20}H_{32}$)
(a termite trail marker)

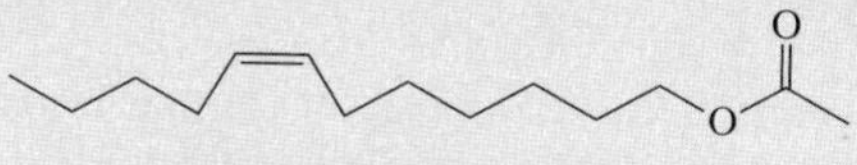

Moth/elephant sex attractant

Plants produce a vast array of *terpenes,* alkenes built in multiples of five carbon atoms. Many terpenes have characteristic fragrances. For example, the fresh odor of a pine forest is due to pinene, a ten-carbon molecule with a ring structure and one double bond. The fragrances of terpenes make them important in the flavor and fragrance industry. Limonene, another ten-carbon molecule with a ring and two double bonds, is the principal component of lemon oil. Geraniol, a chainlike molecule with two double bonds, is one of the molecules that is responsible for the fragrance of roses and is used in many perfumes. Many other terpenes have important medicinal properties.

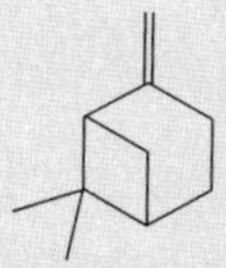

Pinene ($C_{10}H_{16}$) 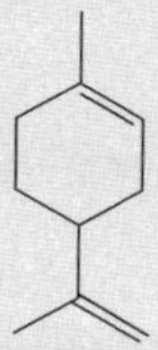   Limonene ($C_{10}H_{16}$) 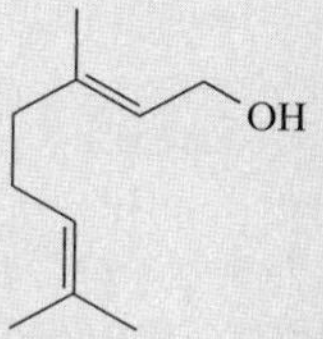   Geraniol ($C_{10}H_{18}O$)

Alkenes with many double bonds in a row are colored. Some plant pigments are alkenes of this kind. One example is carotene, which gives carrots their distinctive orange color. Animals break down carotene into a smaller alkene, Vitamin A, which is essential for vision. Xanthin molecules, relatives of carotene that contain oxygen atoms, occur in corn, orange juice, and shellfish. The xanthin shown here gives the flamingo its brilliant pink plumage.

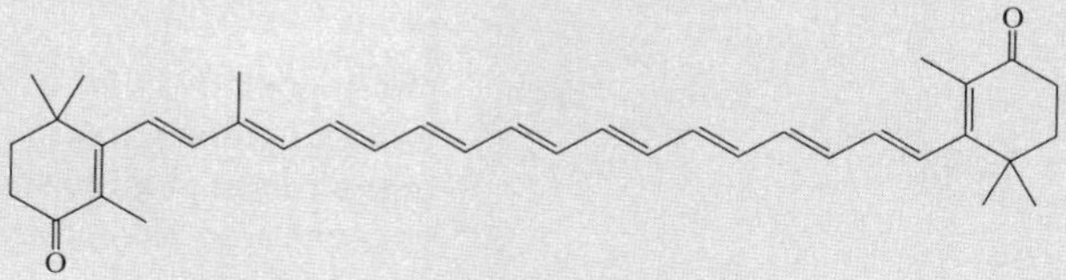

Xanthin ($C_{38}H_{48}O_2$)

Isoprene may be the naturally-occurring alkene with the greatest economic impact. This compound, a major component of the sap of the rubber tree, is used to make the long-chain molecules of natural rubber (polyisoprene). As we describe in Chapter 11, the synthetic rubbers that make up most of today's tires are made from other alkenes.

Used tires create serious environmental problems, because they do not degrade easily and release noxious contaminants when they burn. Recently, a process has been developed that breaks tires down into a polyisoprene oil that can then be further decomposed into limonene. The yield of limonene is only a few percent, but if this process can be made more efficient it may become possible to turn ugly, smelly old tires into fragrant oil of lemon.

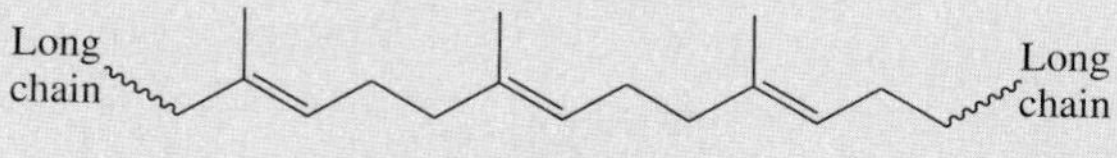

Long chain   Long chain

Polyisoprene

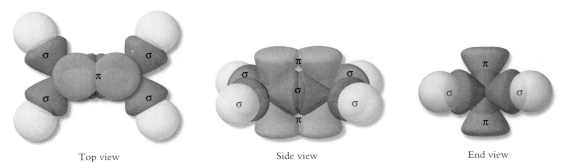

Top view Side view End view

Figure 9-5
Orbital pictures of the bonding in ethylene, viewed from the top, from the side, and from one end. Notice that the electron density in the π bond is distributed above and below the plane containing the six nuclei, whereas the electron densities of the σ bonds are concentrated along the internuclear axes.

Tutorial

the same bonding picture applies to a compound that has a carbon-nitrogen double bond. This bonding picture is constructed using the following procedure:

1. Determine the Lewis structure.
2. Use the Lewis structure to determine steric numbers and hybridizations.
3. Construct a σ bond framework.
4. Add in the π bonds.

| **Example 9-1** | **Orbital Overlap in Double Bonds** |

An imine is a molecule that contains a carbon-nitrogen double bond. Describe the bonding of the simplest possible imine, H_2CNH, by sketching the σ and π bonding systems.

Strategy: Follow the four-step procedure. Use the Lewis structure to identify the appropriate hybridization. Then draw sketches of the various orbitals involved in the bonding.

Solution: Each molecule of H_2CNH has twelve valence electrons. Eight electrons are used to construct the bonding framework, leaving four electrons. Placing these on nitrogen, the more electronegative inner atom, leaves the carbon atom with only three pairs of electrons. Give the carbon atom an octet by transferring one of nitrogen's lone pairs to make a C=N bond:

$$\text{Incomplete octet} \longrightarrow \quad \begin{array}{c} H \\ \diagdown \\ C-\ddot{N}: \\ \diagup \quad \diagdown \\ H \qquad H \end{array} \qquad\qquad \begin{array}{c} H \\ \diagdown \\ C=\ddot{N} \\ \diagup \quad \diagdown \\ H \qquad H \end{array}$$

Complete Lewis structure

The carbon and nitrogen atoms have steric numbers of 3, so their bonding frameworks can be represented using sp^2 hybrid orbitals. Now we can sketch the σ bonding system of imine:

Orbital Overlap in Double Bonds *(continued)*

Example 9-1

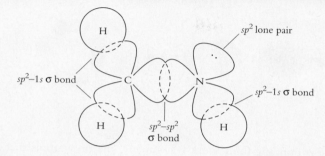

The carbon and nitrogen atoms each have a *p* orbital left over after construction of the *sp²* hybrids. These two *p* orbitals are perpendicular to the plane that contains the five nuclei. Side-by-side overlap of the *p* orbitals gives a *π* bond that completes the bonding description of this molecule.

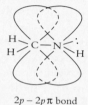

π Bonds Involving Oxygen Atoms

In addition to carbon and nitrogen, oxygen is a common participant in the formation of *π* bonds. The bonding patterns for oxygen are illustrated by acetic acid (CH_3CO_2H), whose Lewis structure was determined in Example 8-11. Figure 9-6 shows that acetic acid has a carbon atom and an inner oxygen whose steric numbers are 4, indicating tetrahedral orbital geometry and bonding that can be described using *sp³* hybrids. The second carbon atom has a steric number of 3, indicating trigonal planar orbital geometry and bonding that can be described using *sp²* hybrids. Remember that outer atoms do not require hybrid orbitals, so acetic acid's outer oxygen atom uses a *2p* orbital to form a *σ* bond to carbon. The *σ* bond framework appears in Figure 9-6*b*.

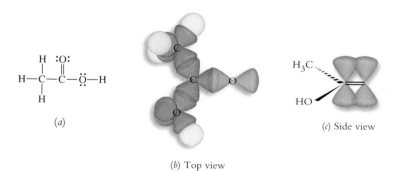

(b) Top view

Figure 9-6
(*a*) The Lewis structure of acetic acid. (*b*) The bonding framework, showing the orbitals used to build the *σ* bonds. (*c*) The outer O atom has a *2p* orbital that overlaps with the remaining *2p* orbital on the adjacent C atom to form a *π* bond.

Tutorial

Figure 9-7
The bond framework for acetylene. The side view (*a*) shows the σ-bonding system, whereas the end view (*b*) calls attention to the unused valence *p* orbitals that lie at right angles to the molecular skeleton.

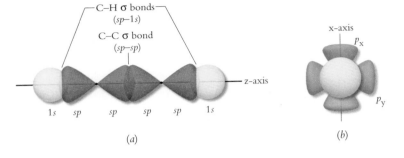

Tutorial

Ozone, discussed in Section 9.4, is the only exception to this generalization about inner oxygen atoms.

Once the σ bonding framework is complete, the only atoms that have unused valence orbitals are the sp^2-hybridized carbon atom and the outer oxygen atom. As Figure 9-6*c* shows, each of these has a $2p$ atomic orbital positioned perfectly for side-by-side overlap, just as in ethylene and imine.

We complete the description of acetic acid by identifying the two orbitals that contain the two lone pairs on the outer oxygen atom. The σ bond and the π bond account for two of oxygen's valence $2p$ orbitals. This leaves the third $2p$ orbital and the $2s$ orbital for the lone pairs.

The bonding pattern of the oxygen atoms in acetic acid repeats in most other oxygen-containing compounds. Outer oxygen atoms frequently have some π character in their bonding. Inner oxygen atoms, in contrast, do not participate in π bonding. Any inner oxygen atom has a steric number of 4 and can be described appropriately using sp^3 hybrid orbitals.

Acetylene: Formation of a Triple Bond

The Lewis structure of acetylene (C_2H_2) shows a triple bond between the carbon atoms:

$$H-C\equiv C-H$$

Each carbon atom has a steric number of 2, indicating that acetylene is a linear molecule and that sp hybrid orbitals can be used to construct the bonding orbital framework. Figure 9-7*a* shows the σ bonding system of acetylene.

The sp hybrid set requires just one of the valence *p* orbitals. Figure 9-7*b* shows that the remaining two *p* orbitals are oriented at right angles to the axis of the

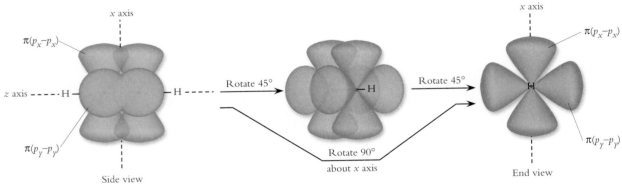

Figure 9-8
Three views of the π bonding in acetylene. *Left*, molecule viewed from the side. *Middle*, molecule viewed at a 45° angle. *Right*, molecule viewed from one end. Notice that the π bonds are perpendicular to each other and have electron density off the internuclear axis.

acetylene molecule and at right angles to each other. Each *p* orbital can overlap with its counterpart on the other carbon atom, just as in ethylene. Thus two π bonds of acetylene form from the side-by-side overlap of the two *p* orbitals on each carbon atom. This is illustrated in Figure 9-8. According to this description, a triple bond is made up of one σ bond and two π bonds. Figure 9-9, which is a superimposition of the views in Figures 9-7 and 9-8, depicts all the bonds in the acetylene molecule. In order to make all the bonding orbitals visible, the π orbitals have been moved outward, displaced slightly from their true locations

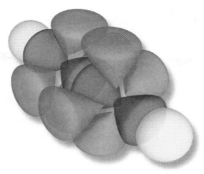

Figure 9-9
Composite orbital overlap view of the σ and π bonds of acetylene. The π bonds are shown in an "exploded" view in order to make the C—C σ bond visible.

Tutorial

Example 9-2 treats another molecule with bonding that can be described by *sp* hybrid orbitals.

Orbital Overlap in Triple Bonds	**Example 9-2**

Hydrogen cyanide (HCN) is an extremely poisonous gas with an odor resembling almonds. Approximately one billion pounds of HCN are produced each year, most of which are used to prepare starting materials for polymers. Construct a complete bonding picture for HCN.

Strategy: Use the four-step procedure described earlier. Begin with the Lewis structure for the molecule and then identify the appropriate hybrid orbitals. Construct a σ bond framework, and complete the bonding picture by assembling the π bonds from the unhybridized *p* orbitals.

Solution:

1. The first five steps of our procedure for writing Lewis structures leave the carbon atom with only two pairs of electrons. Complete the carbon's octet by shifting two electron pairs to make a triple bond:

$$H-C-\ddot{N}: \qquad\qquad H-C\equiv N:$$

Incomplete octet $\qquad\qquad$ Lewis structure

2. Hydrogen and nitrogen are outer atoms, so they use atomic orbitals to form covalent bonds. The carbon atom has a steric number of 2, so it can be described in terms of *sp* hybrids. With this information, we can construct the σ bonding network for HCN:

$$\sigma(1s{-}sp) \qquad \sigma(sp{-}2p)$$

$$H \quad C \quad N$$

The triple bond includes two π bonds formed by the pair-wise overlap of the *p* orbitals that remain on the carbon and nitrogen atoms. Two figures from different perspectives help illustrate the triple bond:

| Example 9-2 | Orbital Overlap in Triple Bonds (continued) |

Side view   End view

The nonbonding pair on the nitrogen atom occupies the 2s orbital, which is the only valence orbital of this atom not used for bonding.

The σ and π bonding networks are shown separately for clarity. Keep in mind, however, that the molecule is a *composite* of both networks, so visualize the HCN molecule with the two bonding networks superimposed.

Section Exercises

9.1.1 Determine the Lewis structure and describe the bonding completely (including an orbital sketch) for the formaldehyde molecule (H_2CO). Formaldehyde is a pungent, colorless gas that is highly toxic and is a suspected carcinogen.

9.1.2 Propene is the three-carbon alkene, and propyne is the three-carbon hydrocarbon that contains a triple bond:

Propene   Propyne

Determine the chemical formulas and Lewis structures of these two substances. Describe their bonding completely, including the geometry and hybridization for each carbon atom.

9.1.3 Convert the line structure of limonene (see Chemistry and Life Box) into a Lewis structure. Determine how many carbon atoms participate in π bonding and draw orbital sketches for a π bond, a C—C σ bond, and a C—H σ bond in this molecule.

9.2 BOND LENGTHS AND ENERGIES

The electron density distribution in a chemical bond can be represented by orbital overlap drawings of the type developed in Chapter 8 and Section 9.1. Electron density distributions, in turn, lead to observable properties of chemical bonds. Three of these properties are particularly important. First, the stability of a chemical bond is measured by its **bond energy.** Second, the atoms are separated by a characteristic distance called the **bond length.** Third, the electrons are distributed between the atoms in a characteristic pattern that can generate **bond polarity.** In this section, we describe trends in bond lengths and examine bond energies in detail.

SECTION 8.2 →
Section 8.2 presents the relationship between bond polarity and electronegativity.

Bond Length

The H—H bond length in molecular hydrogen is 74 picometers (pm). At this distance, attractive interactions are maximized relative to repulsive interactions (see Figure 8-3). This is the shortest possible chemical bond because it results from overlap between the smallest atomic orbitals, two $1s$ orbitals. Bond length increases as the atomic valence orbitals become larger. For example, the H—F bond of hydrogen fluoride forms by overlap of a hydrogen $1s$ orbital with a fluorine $2p$ orbital (see Figure 8-5). A $2p$ orbital is larger than a $1s$ orbital, giving this bond a length of 92 pm. The F—F bond of molecular fluorine, which forms by overlap of two $2p$ orbitals, has a length of 142 pm. These three bond lengths demonstrate a general trend: Bond length is influenced strongly by the principal quantum number (n) of the valence orbitals. Because size increases with n, bond length varies predictably with the principal quantum number of the valence orbitals.

Table 9-1 lists average bond lengths for the most common chemical bonds. Bonds that involve valence orbitals with the same value of n are grouped to emphasize the connection between bond length and the principal quantum number of the valence electrons.

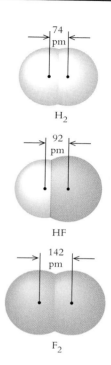

Table 9–1
Average Bond Lengths*

H—X Bonds									
n_a	n_b								
1	1	H—H	74						
1	2	H—C	109	H—N	101	H—O	96	H—F	92
1	3	H—Si	148	H—P	144	H—S	134	H—Cl	127
1	4							H—Br	141

Larger Elements									
n_a	n_b								
2	2	C—C	154	C—N	147	C—O	143	C—F	135
2	2			N—N	145	O—O	148	F—F	142
2	3	C—Si	185	C—P	184	C—S	182	C—Cl	177
2	3			O—P	163	O—Si	166	F—Si	157
2	3					F—S	156	N—Cl	175
2	4, 5			F—Xe	190	C—Br	194	C—I	214

Second-Row Elements									
n_a	n_b								
3	3	Si—Si	235	P—P	221	S—S	205	Cl—Cl	199
3	3	Si—Cl	202	P—Cl	203				
4	4							Br—Br	228
5	5							I—I	267

Multiple Bonds							
C=C	133	C=N	138	C=O	120	O=O	121
C≡C	120	C≡N	116	C≡O	113	P=O	150

* All values are in picometers; 1 pm = 10^{-12} m.

Within any valence shell, nuclear charge also affects bond length. Recall from Section 7.4 that atomic size decreases from left to right across a row of the periodic table. Larger orbitals form longer bonds, so P—P bonds are longer than S—S bonds, and S—S bonds are longer than Cl—Cl bonds. Bond polarity contributes to bond length, too, because partial charges generate coulombic attraction that pulls the atoms closer together. For example, notice in Table 9-1 that C—O bonds are slightly shorter than O—O bonds. This is a result of the polarity of the C—O bond.

The bond lengths in Table 9-1 show one final feature: A multiple bond is shorter than the corresponding single bond between the same two atoms. This is because placing additional electrons between atoms increases coulombic attraction and shrinks the distance between the atoms. Thus triple bonds are the shortest of all bonds among second-row elements.

To summarize, the following factors influence bond lengths:

1. The smaller the principal quantum numbers of the valence orbitals, the shorter the bond.
2. The higher the bond multiplicity, the shorter the bond.
3. The higher the effective nuclear charges of the bonded atoms, the shorter the bond.
4. The larger the electronegativity difference of the bonded atoms, the shorter the bond.

Example 9-3 provides practice in the use of these factors.

| Example 9-3 | Bond Lengths |

What factor accounts for each of the following differences in bond length?
   (a) I_2 has a longer bond than Br_2.
   (b) C—O bonds are shorter than C—C bonds.
   (c) H—C bonds are shorter than the C≡O bond.
   (d) The carbon–oxygen bond in formaldehyde, H_2C=O, is longer than the bond in carbon monoxide, C≡O.

Strategy: Bond lengths are controlled by four factors, some of which are more influential than others. To explain a difference in bond length, we need to determine the way that the factors are balanced.

Solution:
   (a) I—I > Br—Br. These elements are in different rows of the periodic table, with $n = 5$ for the valence orbitals of I and $n = 4$ for the valence orbitals of Br. Thus the I_2 bond is longer than the Br_2 bond because iodine has larger valence orbitals.
   (b) C—O < C—C. Carbon and oxygen are second-row elements, so $n = 2$ for both. Oxygen has a higher nuclear charge than carbon, however, so its valence orbitals are somewhat smaller than carbon's valence orbitals. This makes C—O bonds shorter than C—C bonds. In addition, a C—O bond is polar, which contributes to the shortening of the C—O bond.
   (c) H—C < C≡O. Here, we are comparing bonds in which the principal quantum number and the amount of multiple bonding influence bond length. The experimental fact that H—C bonds are shorter than the triple C≡O bond

Bond Lengths *(continued)*

Example 9-3

indicates that the compactness of the $n = 1$ valence orbital of hydrogen is a more important factor than the presence of multiple bonding.

(d) $C=O > C\equiv O$. Both bonds are between carbon and oxygen, so n, atomic number (Z), and electronegativity difference ($\Delta\chi$) are the same. However, carbon monoxide contains a triple bond, whereas formaldehyde has a double bond. The triple bond in CO is shorter than the double bond in H_2CO because more shared electrons means a shorter bond.

To π Bond or Not to π Bond: Carbon vs. Silicon

The single bond between silicon and oxygen in the silicate minerals can be described using sp^3 hybrid orbitals. In fact, silicon forms single bonds with tetrahedral geometry in almost all its compounds. This is because the $3p$ valence orbitals of silicon atoms form much more stable σ bonds than π bonds. In contrast, the $2p$ valence orbitals of carbon form stable π and σ bonds, and carbon atoms can have linear, trigonal planar, and tetrahedral geometries. Carbon's versatility in bonding gives rise to the vast, intricate worlds of organic chemistry and biochemistry.

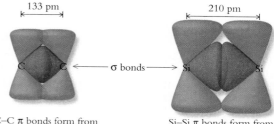

C–C π bonds form from the overlap of $2p$ orbitals

Si–Si π bonds form from the overlap of $3p$ orbitals

Figure 9-10
Comparison of π bonding between two carbon and two silicon atoms.

Why does silicon not form π bonds as readily as its Group 14 neighbor? The answer can be found by examining the sizes of the valence orbitals of silicon and carbon. Figure 9-10 compares side-by-side p orbital overlap for carbon atoms and silicon atoms. Notice that for the larger silicon atom, the $3p$ orbitals are too far apart for strong side-by-side overlap. The result is a very weak π bond. Consequently, silicon makes more effective use of its valence orbitals to form four σ bonds, which can be described using sp^3 hybrid orbitals.

In general, π bonding requires the presence of a second-row element. Carbon, nitrogen, and oxygen readily form π bonds with one another. Oxygen also readily forms π bonds with third- and fourth-row elements. Phosphate (PO_4^{3-}), sulfate (SO_4^{2-}), perchlorate (ClO_4^{-}), and other similar anions contain O—X π bonds, which we describe in Section 9.4. Many other examples of π bonding exist, but we confine our descriptions to bonds between second-row elements and in oxyanions.

Bond Energy

A chemical bond is a stable arrangement of electrons shared between the bonded atoms. Therefore, a bond cannot be broken without an input of energy. On the other hand, energy is released when a bond forms. As described in Section 8.1, one mole of hydrogen molecules has a total bond energy (*BE*) of 435 kJ. In other words, the formation of one mole of H_2 from two moles of H atoms releases 435 kJ of energy:

$$H(g) + H(g) \longrightarrow H_2(g) \qquad \Delta E = -435 \text{ kJ/mol}$$

Molar bond energies are tabulated as positive quantities because a bond energy is the amount of energy required to break one mole of chemical bonds:

$$H_2(g) \longrightarrow H(g) + H(g) \qquad \Delta E = BE = +435 \text{ kJ/mol}$$

To understand the idea of bond energy, think about grasping a molecule and pulling it apart at a junction between two particular atoms. The energy that must be supplied to accomplish this separation is the bond energy. Usually we are interested in molar quantities, so bond energies normally are tabulated in energy per mole. Bond energies differ greatly from substance to substance, as illustrated by the bond energies of the following diatomic molecules:

$$Br_2(g) \longrightarrow Br(g) + Br(g) \qquad BE = 190 \text{ kJ/mol}$$

$$HCl(g) \longrightarrow H(g) + Cl(g) \qquad BE = 430 \text{ kJ/mol}$$

$$N_2(g) \longrightarrow N(g) + N(g) \qquad BE = 945 \text{ kJ/mol}$$

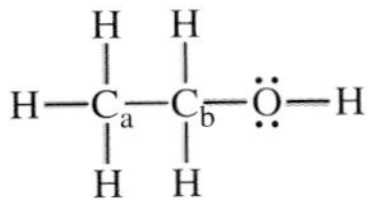

Bond energy depends both on the types of atoms in the bond and on the rest of the molecular structure. As a consequence, molecules often have several different types of bonds, each with its own energy. For example, we can imagine five different bond-breaking processes for an ethanol molecule, each leading to a different set of neutral species. A different amount of energy is required to break each of these bonds. To break the C—C bond requires 345 kJ/mol, to break the C—O bond takes 360 kJ/mol, and 460 kJ/mol is needed to break the O—H bond. It takes 410 kJ/mol to break a C—H bond of the CH_3 group in C_2H_5OH but only 393 kJ/mol to break a C—H bond of the CH_2 group.

Although bond energy depends on the entire molecule, many bonds between specific atoms have energies that cluster around an average value. For example,

H—C$_a$—C$_b$—Ö—H (structure with H H / H H substituents)

$CH_2CH_2OH + H$	C$_a$—H bond energy
$CH_3 + CH_2OH$	C—C bond energy
$CH_3CHOH + H$	C$_b$—H bond energy
$C_2H_5 + OH$	C—O bond energy
$C_2H_5O + H$	O—H bond energy

Table 9–2
Average Bond Energies*

Diatomic Molecules (Dissociation Energy of Gaseous Molecules)

H—H	435	F—F	155	O=O	495
H—F	565	Cl—Cl	240	N=O	605
H—Cl	430	Br—Br	190	N≡N	945
H—Br	360	I—I	150	C≡O	1070
H—I	295				

Single Covalent Bonds (Average Values)

H—C	415	C—C	345	Si—Si	225	N—N	160	O—O	145
H—N	390	C—N	305	Si—F	565	N—O	200	O—Si	450
H—O	460	C—O	360	Si—Cl	390	N—F	285	O—P	335
H—Si	320	C—Si	300	Si—Br	310	N—Cl	200	O—F	190
H—P	320	C—P	265	Si—N	320	P—P	210	O—Cl	220
H—S	365	C—S	270	Sn—Sn	145	P—F	490	O—Br	200
H—Te	240	C—F	485	Sn—Cl	315	P—Cl	320	S—O	265
		C—Cl	330			P—Br	270	S—F	285
		C—Br	275			P—I	185	S—S	240
		C—I	215			As—Cl	295	S—Cl	255

Multiple Covalent Bonds (Average Values)

C=C	615		N=N	420	C≡C	835	
C=N	615		N=O	605	C≡N	890	
C=O	750	800†	O=P	545			
C=S	575		O=S	515			

* All values are in kilojoules per mole (kJ/mol) and are rounded to the nearest 5 kJ/mol.
† C=O bond energy in CO_2

C—H bond energies have an average value of 415 kJ/mol. Notice that both C—H bond energies for ethanol are close to the average value, but neither matches exactly.

Table 9-2 gives average bond energies for a variety of chemical bonds. Note the following features of this table:

1. Each value is the energy per mole required to break a gaseous molecule into a pair of neutral gaseous fragments. In addition to bond energies, molecules in liquids and solids experience attractive intermolecular forces that modify the energies of individual bonds. Average bond energies are less reliable when they are applied to substances in the liquid and solid phases.

2. All the values (other than those for diatomic molecules) are averages over many different compounds. A bond in any given compound may have a value that differs from the average by 5 kJ/mol or more.

3. The values are given to the nearest 5 kJ/mol.

Bond energies are difficult to measure accurately. Most molecules are like ethanol in that they contain many different bonds. Consequently, measuring one specific bond energy in a molecule is never easy. Furthermore, the fragments that result when molecules are torn apart are unstable. They recombine quickly, releasing energy as the new chemical bonds form. In short, it is difficult to keep track of every energy change that occurs in experiments with molecules. Because direct measurements of bond energies are difficult to make, most of the values in Table 9-2 come from experiments that measure bond energies *indirectly*.

← **CHAPTER 10**
In Chapter 10 we examine forces that exist between molecules in liquids, solids, and solutions.

Factors Affecting Bond Energy

Bond energies, like bond lengths, vary in ways that can be traced to atomic properties. There are three consistent trends in bond strengths:

1. *Bond strength increases as more electrons are shared between the atoms.* Shared electrons are the glue of chemical bonding, so sharing more electrons strengthens the bond. Carbon-carbon bonds demonstrate this trend:

C—C	Single bond	$BE = 345$ kJ/mol
C=C	Double bond	$BE = 615$ kJ/mol
C≡C	Triple bond	$BE = 835$ kJ/mol

2. *Bond strength increases as the electronegativity difference ($\Delta\chi$) between the bonded atoms increases.* Polar bonds gain stability from the coulombic attraction between the negative and positive fractional charges around the bonded atoms. This trend is exemplified by bonds between oxygen and other second-row elements:

O—O	$\Delta\chi = 0.0$	$BE = 145$ kJ/mol
O—N	$\Delta\chi = 0.5$	$BE = 200$ kJ/mol
O—C	$\Delta\chi = 1.0$	$BE = 360$ kJ/mol

3. *Bond strength decreases as orbital overlap decreases.* As the valence orbitals become larger and more diffuse, the electron density of a bond is spread out more. This reduces the amount of overlap along the bond axis, which in turn decreases the net coulombic attraction between the electrons and the nuclei. The following bond energies illustrate this effect:

H—F	$n_F = 2$	$BE = 565$ kJ/mol
H—Cl	$n_{Cl} = 3$	$BE = 430$ kJ/mol
H—Br	$n_{Br} = 4$	$BE = 360$ kJ/mol
H—I	$n_I = 5$	$BE = 295$ kJ/mol

The bonds in these molecules are formed by overlap of the hydrogen $1s$ orbital with a valence p orbital from the halogen. As the p orbitals become larger, their electron densities become more spread out in space, the amount of overlap decreases, and the bonds become weaker.

Bond energies result from the interplay of several factors, including nuclear charge, principal quantum number, coulombic forces, and electronegativity. Thus it should not be surprising that there are numerous exceptions to the three trends in bond energies. Although it is possible to *explain* many differences in bond energy, it frequently is not possible to *predict* differences with confidence.

Reaction Energy

Chemical reactions involve breaking bonds in the starting materials and making bonds in the products. Breaking a bond *always* requires an input of energy, whereas forming a bond *always* results in a release of energy. Thus bond energies play major roles in determining the outcomes of chemical reactions. The balance between the energy absorbed in bond breakage and the energy released in bond formation determines whether an overall chemical reaction releases energy or absorbs energy.

The reaction of molecular hydrogen with molecular oxygen to produce gaseous water illustrates how energy changes accompany a chemical reaction. The balanced equation indicates that two molecules of hydrogen react with one molecule of oxygen.

$$2\,H_2(g) + O_2(g) \longrightarrow 2\,H_2O(g)$$

Imagine this reaction occurring in steps. First, two H_2 molecules break their single bonds to give four free atoms of hydrogen. Next, one oxygen molecule breaks its double bond to form two free atoms of oxygen. Then these six atoms recombine, forming four O–H single bonds to produce two molecules of water:

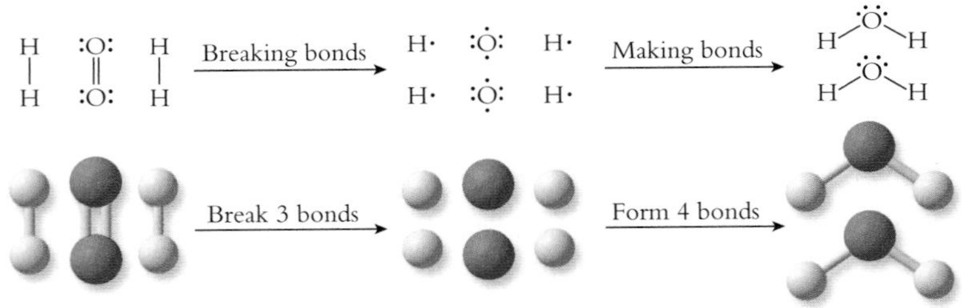

The energy change in the first step, $\Delta E_{\text{bond breaking}}$, is the sum of all the bond energies of reactant molecules:

$$\Delta E_{\text{bond breaking}} = (2\text{ mol }H_2)(BE\ H_2) + (1\text{ mol }O_2)(BE\ O_2)$$

$$\Delta E_{\text{bond breaking}} = (2\text{ mol})(435\text{ kJ/mol}) + (1\text{ mol})(495\text{ kJ/mol}) = 1365\text{ kJ}$$

Energy required to break all
bonds of starting materials

This energy change is *positive* because the chemical system *gains* the energy required to break the chemical bonds.

The energy change in the second step, $\Delta E_{\text{bond formation}}$, is the sum of all the energies released during bond formation of the products. The reaction produces two moles of water containing four moles of O—H bonds. Bond formation always releases energy, so the chemical system loses energy to the surroundings. Consequently, the energy change is negative:

$$\Delta E_{\text{bond formation}} = -(4 \text{ mol O—H bonds})(460 \text{ kJ/mol}) = -1840 \text{ kJ}$$

Energy released during formation
of bonds of products

The overall energy change for the reaction is the sum of these two terms:

$$\Delta E_{\text{reaction}} = \Delta E_{\text{bond breaking}} + \Delta E_{\text{bond formation}} = (1365 \text{ kJ}) + (-1840 \text{ kJ})$$

$$= -475 \text{ kJ}$$

The energy released in the reaction between H_2 and O_2 is used to drive the main engines of the space shuttle.

The negative sign of $\Delta E_{\text{reaction}}$ indicates that the reaction of hydrogen with oxygen to form water releases energy to the surroundings. To understand why, consider the nature of the bonds involved in the reaction. Eight bonding electrons are distributed in three bonds among the reactants, two H—H bonds and one O=O bond. These eight electrons form four O—H bonds in the products. The H—H bond (435 kJ/mol) and the O—H bond (460 kJ/mol) have similar energies, but the bond strength of O=O *per electron pair* is only 495/2 = 248 kJ/mol. Thus replacing one O=O bond with two O—H bonds yields a large amount of energy. Many reactions of molecular oxygen release energy because the O=O bond is relatively weak compared with two O—X bonds formed with most other elements.

The analysis of this reaction is an example of reasoning that can be applied to any chemical reaction. The energy change for a chemical reaction can be estimated as the sum of all bond energies of the reactants minus the sum of all bond energies of the products:

$$\Delta E_{\text{reaction}} = \Sigma \, BE_{\text{reactants}} - \Sigma \, BE_{\text{products}} \qquad (9\text{-}1)$$

The Greek letter Σ means "the sum of." Thus $\Sigma \, BE$ means "the sum of all the bond energies."

As a second example, consider the energy change that takes place during the combustion reaction of propane with oxygen. Propane (C_3H_8) contains only C—H and C—C bonds. Note from Table 9-2 that C—H and C—C are among the strongest single bonds. Strong bonds are difficult to break, so they are liabilities in starting materials. As a result, propane is not particularly reactive at low temperature. At high temperature, however, propane reacts with oxygen in a combustion reaction that produces carbon dioxide and water. This reaction releases considerable energy for two reasons. As already noted, reactions of O_2 often release energy because of the relative weakness of the O=O bond. In addition, many reactions that generate CO_2 release energy because each electron pair in a C=O bond contributes more stability than the same pair in a C—X or O—Y bond. Each C=O bond in CO_2 has a bond energy of 800 kJ/mol, for a contribution per electron pair of 400 kJ/mol, whereas a C—C bond, for example, has an average bond energy of 345 kJ/mol. Although they are liabilities in a starting material, strong bonds are assets when they are present among the products.

← CHAPTER 3 & 4
The stoichiometry of combustion reactions appears in Chapters 3 and 4.

Lewis structures show the changes that occur during combustion of propane:

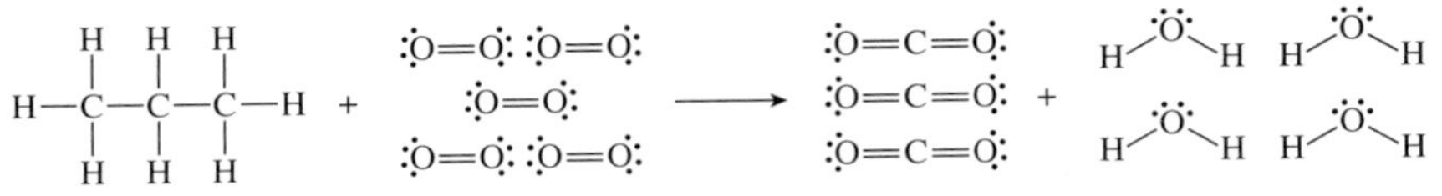

$$C_3H_8 \qquad 5\,O_2 \qquad\qquad 3\,CO_2 \qquad\qquad 4\,H_2O$$

To estimate the amount of energy absorbed or released in this reaction, we must compile an inventory of all the bonds that break and all the bonds that form. Eight C—H bonds and two C—C bonds break in each propane molecule, and one O═O bond breaks in each oxygen molecule. Two C═O bonds form in each CO_2 molecule, and two O—H bonds form in each H_2O molecule. In summary:

$$C_3H_8(g) + 5\,O_2(g) \longrightarrow 3\,CO_2(g) + 4\,H_2O(g)$$

Bonds broken: Bonds formed:

2 C—C bonds, 8 C—H bonds   6 C═O bonds, 8 H—O bonds

5 O═O bonds

Now we apply Equation 9-1, using bond energy values from Table 9-2:

$$\Delta E_{reaction} = \Sigma\, BE_{reactants} - \Sigma\, BE_{products}$$

$$\Delta E_{reaction} = [(2 \text{ mol C—C})(345 \text{ kJ/mol}) + (8 \text{ mol C—H})(415 \text{ kJ/mol})$$

$$+ (5 \text{ mol O═O})(495 \text{ kJ/mol})]$$

$$- [(6 \text{ mol C═O})(800 \text{ kJ/mol}) + (8 \text{ mol O—H})(460 \text{ kJ/mol})]$$

$$\Delta E_{reaction} = 6485 \text{ kJ} - 8480 \text{ kJ} = -1995 \text{ kJ}$$

A calculation using average bond energies indicates that burning one mole of propane releases 1995 kJ of energy. The energy change accompanying this reaction has been measured accurately to be −2044 kJ/mol of propane. Average bond energies predict a value that is off by about 50 kJ/mol, an error of about 2.5%. This error illustrates that calculations using tabulated bond energies provide only an *estimate* of the energy change of a chemical reaction. That is because the values listed in Table 9-2 are *averages* taken from measurements on many different substances containing each particular type of bond. As the description of bond energies in ethanol shows, the bond energies in any given molecule usually differ somewhat from the average values. These variations are typically only a few percent, so calculations using average bond energies reliably indicate the magnitude of energies released or absorbed in the course of chemical reactions.

As the analysis for propane shows, burning a compound that contains carbon and hydrogen releases a great amount of energy that can be used to heat homes, cook food, power vehicles, or produce electricity. Most of the energy requirements of modern society are provided by the energy released during the combustion of hydrocarbons derived from petroleum.

Energy analyses can be extended to reactions other than combustion, as shown in Example 9-4.

| Example 9-4 | Reaction Energies |

The United States produces more than 7 billion kilograms of vinyl chloride per year. Most is converted to the polymer poly(vinyl chloride) (PVC) that is used to make piping, siding, gutters, floor tiles, clothing, and toys. Vinyl

Reaction Energies *(Continued)*   **Example 9-4**

chloride is made in a two-step process. The balanced overall equation is:

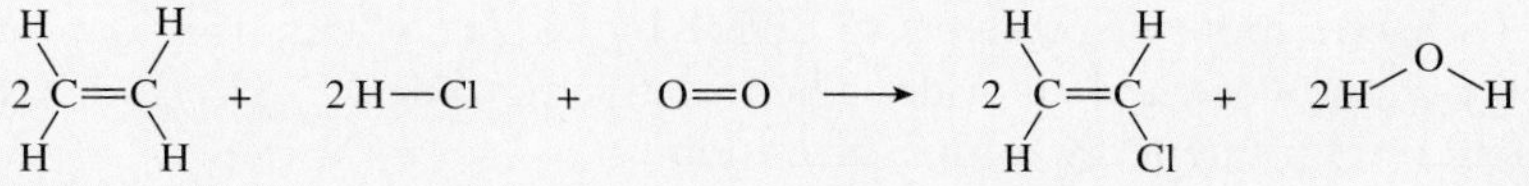

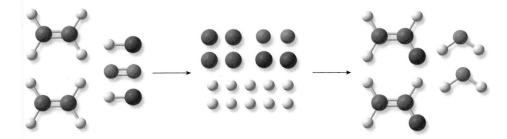

Ethylene   Vinyl chloride

Based on average bond energies, what energy change accompanies the formation of one mole of vinyl chloride? Does the synthesis require an input of energy?

Strategy: First, use the balanced equation and Lewis structures to construct a "bond inventory" for the synthesis. Then consult Table 9-2 for average bond energies and substitute in Equation 9-1.

Solution: In this case the structures are given in the equation. Here is the inventory:

Starting Materials	**Atom Inventory**	**Products**
8 C—H bonds	10 H atoms	6 C—H bonds
2 C=C bonds	4 C atoms	2 C=C bonds
2 H—Cl bonds	2 Cl atoms	2 C—Cl bonds
1 O=O bond	2 O atoms	4 O—H bonds

Look up bond energies from Table 9-2:

C—H = 415 kJ/mol   C=C = 615 kJ/mol   H—Cl = 430 kJ/mol
O=O = 495 kJ/mol   C—Cl = 330 kJ/mol   O—H = 460 kJ/mol

Now use Equation 9-1:

$$\Delta E_{reaction} = \Sigma\ BE_{reactants} - \Sigma\ BE_{products}$$

Sum the bond energies for reactants and products and subtract the latter from the former:

Reactants	**Products**
(8 mol C—H)(415 kJ/mol)	(6 mol C—H)(415 kJ/mol)
(2 mol C=C)(615 kJ/mol)	(2 mol C=C)(615 kJ/mol)
(2 mol H—Cl)(430 kJ/mol)	(2 mol C—Cl)(330 kJ/mol)
(1 mol O=O)(495 kJ/mol)	(4 mol O—H)(460 kJ/mol)

$$\Delta E_{reaction} = (5905\ kJ) - (6220\ kJ) = -315\ kJ$$

Example 9-4	Reaction Energies *(Continued)*

Our calculation predicts that 315 kJ of energy is released for every two moles of vinyl chloride. However, the problem asks for the energy change associated with one mole of product, so divide by two to obtain 158 kJ of energy released per mole of vinyl chloride. The negative sign for $\Delta E_{reaction}$ indicates that the reaction releases energy to the surroundings.

As another example of this type of analysis, consider the differences in energy and reactivity among carbon-carbon single, double, and triple bonds. Double bonds are stronger but more reactive than single bonds. Triple bonds, in turn, are stronger but more reactive than double or single bonds. This trend can be illustrated with hydrogenation, a reaction of vital importance in the chemical industry. During hydrogenation, molecular hydrogen adds to a molecule with multiple bonds. Ethylene and acetylene, for example, react with hydrogen under appropriate conditions to produce ethane:

$$H-C\equiv C-H + 2\,H_2 \longrightarrow H-\underset{\underset{H}{|}}{\overset{\overset{H}{|}}{C}}-\underset{\underset{H}{|}}{\overset{\overset{H}{|}}{C}}-H \qquad \underset{H}{\overset{H}{}}C=C\underset{H}{\overset{H}{}} + H_2 \longrightarrow H-\underset{\underset{H}{|}}{\overset{\overset{H}{|}}{C}}-\underset{\underset{H}{|}}{\overset{\overset{H}{|}}{C}}-H$$

Acetylene   Ethylene

The energy changes that accompany hydrogenation can be estimated from average bond energies. According to the balanced equation, hydrogenation of ethylene creates two new C—H bonds at the expense of one H—H bond and converts a C=C bond into a C—C bond. The remaining C—H bonds do not change and can be left out of the calculation:

$$\Delta E_{reaction} = [(1 \text{ mol } H_2)(435 \text{ kJ/mol}) + (1 \text{ mol C=C})(615 \text{ kJ/mol})]$$
$$- [(2 \text{ mol C—H})(415 \text{ kJ/mol}) + (1 \text{ mol C—C})(345 \text{ kJ/mol}) = -125 \text{ kJ}$$

A similar calculation for acetylene gives −300 kJ/mol for the hydrogenation energy of C_2H_2. The estimated difference, 175 kJ/mol, is in excellent agreement with the experimental difference, 174.5 kJ/mol.

These two reactions give the same product, ethane. Thus the difference in hydrogenation energy must be caused by the difference in energy between the starting materials. Because more energy was released from the hydrogenation of acetylene, the $C_2H_2 + 2\,H_2$ system is higher in energy than the $C_2H_4 + H_2$ system. The energy level diagram in Figure 9-11 summarizes these relationships.

It may seem odd that hydrogenation of acetylene releases more energy than hydrogenation of ethylene, even though the former has the stronger bond between carbon atoms. To see why this is, compare the bond energies *per electron pair*. The bond energy of acetylene is 835/3 = 278 kJ/mol for each pair of electrons. For ethylene, the bond energy for each pair of electrons is 615/2 = 308 kJ/mol. Acetylene has a stronger *overall* bond energy, but it is a relatively weak bond *per pair of electrons*.

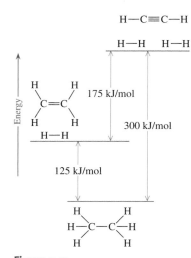

Figure 9-11
An energy level diagram for the hydrogenation of acetylene and ethylene, showing the difference in energy of the two sets of starting materials.

9.2.1 Arrange the following bonds in order of increasing length (shortest first). State the factors responsible for the position of each bond in your sequence: C—C, C=O, C=C, C—H, and C—Cl.

9.2.2 Methanol (CH_3OH) can be mixed with gasoline to give gasohol, an alternative fuel for automobiles. Estimate how much energy is released when one mole of methanol is burned in air.

9.2.3 Methyl chloride (CH_3Cl) can be produced from methane in either of the following reactions:

$$CH_4(g) + HCl(g) \longrightarrow CH_3Cl(g) + H_2(g)$$

$$CH_4(g) + Cl_2(g) \longrightarrow CH_3Cl(g) + HCl(g)$$

(a) Draw the Lewis structures of all substances in these reactions and create a bond inventory. (b) Estimate the energy changes that accompany each reaction. On the basis of these values, which reaction is more suitable for industrial production of methyl chloride? Explain.

9.3 SECOND-ROW DIATOMIC MOLECULES

Several common gases are diatomic molecules formed by elements from the second row of the periodic table: C, N, and O. Two such molecules, N_2 and O_2, dominate the Earth's atmosphere. Other examples include CO, a poisonous by-product of combustion reactions, and NO, one of the components of photochemical smog. These second-row diatomic molecules all have double or triple bonds. We use N_2 to develop an orbital overlap picture that we can then apply to the bonding of all second-row diatomic molecules.

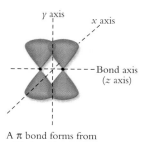

A σ bond forms
from end-on overlap

A π bond forms from
side-by-side overlap

Figure 9-12
Orbital sketches show how the overlap of 2p valence orbitals forms bonds. End-on overlap gives a σ bond with electron density along the bond axis, whereas side-by-side overlap gives a π bond with electron density above and below the bond axis.

Bonding in N_2

The Lewis structure of molecular nitrogen shows the two atoms connected by a triple bond, with a nonbonding electron pair on each nitrogen atom: :N≡N:. Both atoms in N_2 are outer atoms, so there is no need for hybridization. Atomic valence 2p orbitals can be used to construct the three bonds.

← **EXAMPLE 8-10**
See Example 8-10 for the construction of the Lewis structure for N_2.

Figure 9-12 shows how the three bonds of N_2 can be constructed from the overlap of the 2p atomic orbitals to form bonding orbitals. One of the bonds is formed from the $2p_z$ orbitals, which point toward each other along the bond axis. This end-on overlap gives a σ orbital that concentrates electron density between the two nitrogen nuclei. The second and third bonds of N_2 are a pair of π bonds formed from side-by-side overlap of the remaining 2p orbitals. One of these π bonds comes from the $2p_y$ orbitals, as shown in Figure 9-12. The $2p_x$ orbitals, which we do not show in Figure 9-12, overlap in the same side-by-side manner,

The convention is to identify the z-axis of a diatomic molecule to be its bond axis.

Figure 9-13
Three views of the two side-by-side π bonds of N_2.

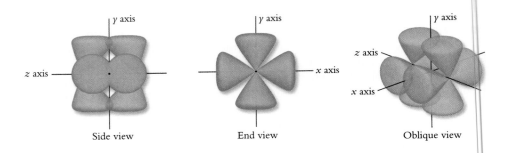

Side view   End view   Oblique view

but the π orbital formed by the $2p_x$–$2p_x$ overlap is perpendicular to the one formed by the $2p_y$ orbitals. Figure 9-13 shows the two π bonding orbitals formed by the $2p_x$ and $2p_y$ orbitals from three different perspectives. The π orbitals of N_2 closely resemble those of acetylene.

Molecular Orbitals

A complete bonding picture for second-row diatomic molecules must account for all the valence orbitals, including $2s$. To understand how the $2s$ orbitals interact, we must return to the wavelike behavior of electrons.

Waves interact by addition of their amplitudes. Two waves that occupy the same region of space become superimposed, generating a new wave that is a composite of the original waves. Wave amplitudes add when both waves have the same sign, as Figure 9-14a shows. The new wave has a larger amplitude than either original wave. In contrast, wave amplitudes subtract when the two waves have opposite signs, as illustrated in Figure 9-14b. This new wave has a smaller amplitude than either original wave. Moreover, there is a point where the amplitudes exactly cancel to give a *node*—a point where the wave amplitude is zero.

Because electrons have wavelike properties, orbital interactions involve similar addition or subtraction of amplitudes. So far, we have described only the additive orbital interactions. The wave amplitudes add in the overlap region, generating a new bonding wave function with larger amplitude between the nuclei. However, a complete mathematical treatment of orbital overlap requires that the number of wave functions be conserved. In other words, whenever several wave functions interact, they must generate an equal number of new wave functions. The hybrid orbitals introduced in Chapter 8 illustrate this. One s and one p orbital form two sp hybrids; one s and two p orbitals form three sp^2 hybrids, and so on.

Figure 9-14
(a) When two waves (*dashed lines*) are added, the resulting new wave (*solid line*) has a large amplitude in the overlap region. (b) When one wave is subtracted from another (*dashed lines*), the resulting new wave (*solid line*) has a smaller amplitude in the overlap region.

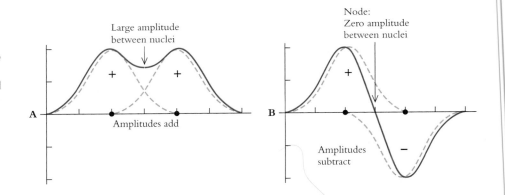

When two atomic orbitals from *different* atoms interact, they generate two new molecular orbitals. One interaction is additive, leading to an orbital with high electron density between the nuclei. High electron density between the nuclei results in more attraction than repulsion (recall Figure 8–1), so this combination is a **bonding molecular orbital.** The second interaction is subtractive, leading to an orbital with low electron density between the nuclei. A low electron density between nuclei results in more nucleus–nucleus repulsion than electron–nucleus attraction. This combination is destabilizing and is referred to as an **antibonding molecular orbital.** An antibonding orbital is distinguished from a bonding orbital by a superscript asterisk (*).

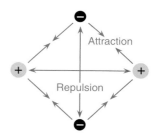

Electron contour drawings can be used to show the three-dimensionality of molecular orbitals, just as they depict the three-dimensionality of atomic orbitals. As their names suggest, however, molecular orbitals span an entire molecule while atomic orbitals are distributed about a single nucleus. Figure 9-15 shows the bonding and antibonding molecular orbitals that result from the interaction of the 2s orbitals in second-row diatomic molecules. Notice that the molecular orbitals encompass both the nuclei.

The pattern of additive and subtractive overlap also applies to the overlap of p orbitals. Whereas Figure 9–12 showed the additive overlap of atomic 2p orbitals to form bonding molecular orbitals, Figure 9–16 shows the subtractive interactions that generate σ* and π* molecular orbitals.

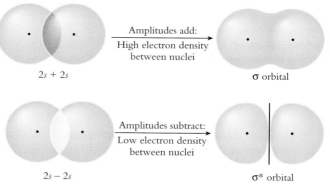

When additive and subtractive interactions are included, eight molecular orbitals can be constructed from the eight atomic valence orbitals of a diatomic molecule. The two 2s orbitals generate one bonding and one antibonding molecular orbital, and the six 2p orbitals generate three bonding and three antibonding molecular orbitals.

Figure 9-15
When two 2s orbitals interact, they generate two new orbitals, one bonding molecular orbital and one antibonding (*) molecular orbital.

The relative energies of these molecular orbitals depend on several related features. The details are challenging, and ranking molecular orbitals usually involves intuition coupled with computer-based calculations. Nevertheless, the orbital overlap model and the properties of electrons allow us to draw some conclusions:

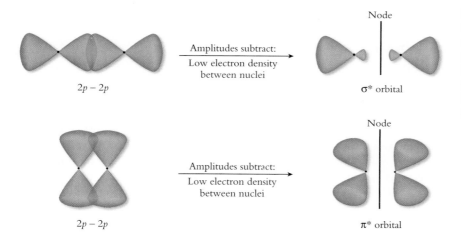

Figure 9-16
Overlap between 2p orbitals results in antibonding (*) orbitals when the amplitudes of the two interacting wave functions subtract in the overlap region, leading to low electron density and a node between the nuclei.

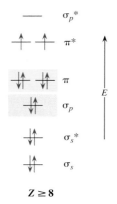

Figure 9-17
The molecular orbital energy level diagram for O_2. This diagram applies to diatomic molecules with $Z \geq 8$.

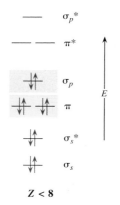

Figure 9-18
The molecular orbital energy level diagram for N_2. This diagram applies to diatomic molecules with $Z < 8$.

1. The bonding and antibonding σ_s orbitals are more stable than any of the six molecular orbitals derived from the $2p$ orbitals. This is because the $2s$ orbitals that give rise to σ_s and σ_s^* are more stable than the $2p$ atomic orbitals.

2. The two π bonding orbitals have identical energies, because the atomic orbitals from which they are constructed have identical energies. Likewise, the two π^* orbitals have identical energies.

3. The antibonding orbitals formed from the atomic $2p$ orbitals are the least stable molecular orbitals, with the σ_p^* orbital less stable than the π^* orbitals.

After applying these features we are left with the bonding σ_p and π molecular orbitals. These must be placed between σ_s^* and π^*, but which one is more stable? Orbital overlap arguments suggest that σ_p should be more stable than π, because end-on orbital overlap is more effective than side-by-side overlap. This ranking is observed for molecular oxygen. Figure 9-17 shows how the valence electrons of O_2 are placed in accordance with the Pauli and aufbau principles.

We can write valence shell configurations for diatomic molecules by naming each orbital (σ_s, π_x etc.), using superscripts to indicate how many electrons each orbital contains. Here is the configuration for the ground state of O_2:

$$(\sigma_s)^2 \, (\sigma_s^*)^2 \, (\sigma_p)^2 \, (\pi_x)^2 \, (\pi_y)^2 \, (\pi_x^*)^2$$

The molecular orbital diagram shown in Figure 9-17 applies only for $Z \geq 8$. Experiments show that for second-row diatomic molecules with $Z < 8$, the π set of molecular orbitals is lower energy than σ_p. As a result, the molecular orbital diagram for molecular nitrogen appears as shown in Figure 9-18. Here is the ground state configuration of N_2:

$$(\sigma_s)^2 \, (\sigma_s^*)^2 \, (\pi_x)^2 \, (\pi_y)^2 \, (\sigma_p)^2$$

This change in the order of energy levels is analogous to the changes in energy ranking of atomic orbitals described in Chapter 7. Just as the $3d$ and $4s$ atomic orbitals are near-degenerate, so too are σ_p and π molecular orbitals. Subtle features such as electron–electron repulsion between electrons in different orbitals destabilize the σ_p orbital relative to the π orbitals. The details are beyond the scope of General Chemistry, but the end result is that the destabilization is greater for small Z than for large Z. Consequently, for $Z < 8$, σ_p is destabilized enough to make it higher in energy than π.

To summarize, the features of molecular orbital theory lead to two energy level diagrams. Figure 9-17 applies to molecules with $Z \geq 8$, as illustrated for O_2. Figure 9-18 applies to second-row diatomic molecules with $Z < 8$, as shown for N_2.

Bond Order

The orbital configuration of N_2 shows four pairs of bonding electrons, while the Lewis structure indicates that the molecule has a triple bond, not the quadruple bond that four bonding pairs suggests. The reason is that antibonding electrons cancel the stabilizing effect of bonding electrons. Thus the two antibonding electrons of N_2 cancel the contributions of two bonding electrons. This leaves *net* bonding resulting from three pairs of electrons; in other words, a triple bond. Equation 9-2 defines the **bond order;** which is the net amount of bonding between two atoms:

$$\text{Bond order} = \tfrac{1}{2} \, (\text{Number of bonding electrons} - \text{Number of antibonding electrons}) \tag{9-2}$$

Lewis structures also provide information about bond order, from the number of lines joining two atoms. Note that the Lewis structure of N_2, which contains three lines joining the two atoms, predicts the bond order of nitrogen accurately, even though a more detailed bonding description reveals that N_2 contains anti-bonding electrons.

Example 9-5 describes molecular fluorine as another illustration of bond order.

Bonding in Fluorine Molecules

Example 9-5

In Chapter 8 the bonding in molecular fluorine is described simply in terms of two overlapping $2p$ orbitals. Determine the configuration of F_2 using bonding and antibonding molecular orbitals. Do the two treatments predict different bond properties?

Strategy: To build a configuration, we need an orbital energy level diagram. Then we add electrons according to the Pauli and aufbau principles.

Solution: Each F atom has $Z = 9$, so we use the molecular orbital diagram shown in Figure 9-17. Fluorine has 14 valence electrons that fill the seven most stable orbitals. Every orbital except the highest-energy σ_p^* orbital is filled, giving the following configuration:

$$(\sigma_s)^2 \ (\sigma_s^*)^2 \ (\sigma_p)^2 \ (\pi_x)^2 \ (\pi_y)^2 \ (\pi_x^*)^2 \ (\pi_y^*)^2$$

Bond order = ½ [8 bonding electrons − 6 antibonding electrons] = 1

According to this treatment, F_2 has a bond order of one because its six anti-bonding electrons nullify the bonding effect of six bonding electrons, leaving net bonding from two electrons. This is a single bond, which is the same as predicted by the Lewis structure and the simple overlap picture of Chapter 8.

Evidence for Antibonding Orbitals

The molecular orbital model developed in this section is more elaborate than the simple overlap pictures used to describe bond frameworks in Chapter 8. Is this more complicated model necessary to give a thorough picture of chemical bonding? Experimental evidence suggests that the answer is "Yes."

Figure 9-19 shows that liquid oxygen adheres to the poles of a magnet. Attraction to a magnet field shows that molecular oxygen is paramagnetic. Recall from Chapter 7 that paramagnetism arises when a species has unpaired electrons. Neither the Lewis structure of O_2 nor the simple orbital overlap model reveals the presence of unpaired electrons. The molecular orbital description, however, shows that the least stable occupied orbitals for O_2 are the pair of equivalent π^* antibonding orbitals, each of which contains one electron (Figure 9-17).

Bond length and bond energy measurements on O_2 and its cation also indicate that the highest-energy occupied orbital is antibonding. When a bonding electron is removed from a molecule, the resulting cation has a weaker, longer bond than the neutral molecule. In contrast, removing an electron from O_2 to form O_2^+ increases the bond energy and decreases the bond length:

Species	Bond length	Bond energy	Configuration	Bond Order
O_2	121 pm	495 kJ/mol	$\ldots (\pi)^4(\pi^*)^2$	2
O_2^+	112 pm	643 kJ/mol	$\ldots (\pi)^4(\pi^*)^1$	2.5

$$:\ddot{F}-\ddot{F}:$$

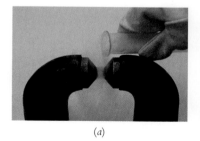

(a)

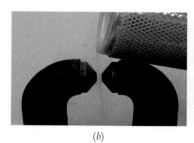

(b)

Figure 9-19
Molecular oxygen is paramagnetic, so it clings to the poles of a magnet (a). Molecular nitrogen is diamagnetic and passes directly through the magnet (b).

These data show that the least stable occupied orbital of O_2 is antibonding in character: removing an electron reduces the amount of antibonding and strengthens the bond.

Heteronuclear Diatomic Molecules

There are two stable second-row heteronuclear diatomic molecules, CO and NO. Because the qualitative features of orbital overlap do not depend on the identity of the atoms, the bonding in CO and NO can be described by the same sets of orbitals that describe the bonding in N_2 and O_2.

The CO molecule ($Z_{average} = 7$) has orbital energies arranged as for N_2 and has ten valence electrons, so its orbital configuration is identical with that of N_2, giving CO a bond order of 3. The NO molecule ($Z_{average} = 7.5$) also has orbital energies arranged as for N_2, but this is one of the few stable molecules that contains an odd number of valence electrons. It has eleven valence electrons, one less than O_2. Thus NO has one electron in an antibonding π^* orbital and this valence configuration:

$$(\sigma_s)^2 \, (\sigma_s{}^*)^2 \, (\sigma_p)^2 \, (\pi_x)^2 \, (\pi_y)^2 \, (\pi_x{}^*)^1$$

There are eight bonding electrons and three antibonding electrons, giving a bond order of 2.5 for NO. Table 9-2 verifies that this bond is stronger than the double bond of O_2 but weaker than the triple bond of N_2.

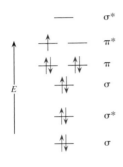

:N̈=Ö:

Lewis structure

Section Exercises

9.3.1 The cyanide ion (CN^-) is a diatomic anion found in ionic compounds such as potassium cyanide. Determine the Lewis structure and describe the bonding in this ion, including its electron configuration and bond order.

9.3.2 Draw an orbital sketch of each of the occupied valence orbitals for the CN^- ion.

9.3.3 The first ionization energy of NO is 891 kJ/mol, that of N_2 is 1500 kJ/mol, and that of CO is 1350 kJ/mol. Use electron configurations to explain why NO ionizes so much more easily than either N_2 or CO.

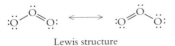

Lewis structure

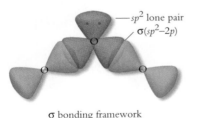

—sp^2 lone pair
$\sigma(sp^2$–$2p)$

σ bonding framework

Figure 9-20
The Lewis structure and the σ bonding framework of the ozone molecule. Two resonance structures exist.

9.4 THREE-CENTER π ORBITALS

The π molecular orbitals described so far are localized between two atoms. In many molecules, including the organic molecules described in the introduction to this chapter, π orbitals are spread over three or more atoms. Such **delocalized π orbitals** can form when *more than two p* orbitals overlap in the appropriate geometry. In this section we develop this idea for three-atom systems; in the following sections we apply it to larger molecules.

Ozone

We begin our exploration of delocalized π bonds with the ozone molecule, O_3. As described in Chapter 6, ozone in the upper stratosphere protects plants and animals from hazardous ultraviolet radiation. Ozone has 18 valence electrons and a Lewis structure as shown in Figure 9-20. Experimental measurements show that ozone is a bent molecule with a bond angle of 118°.

The bond framework of ozone can be represented by the localized σ bonds shown in Figure 9-20. The inner oxygen atom, with a steric number of 3, is trigonal planar and can be described using sp^2 hybrid orbitals. Two of these sp^2 orbitals form σ bonds with $2p$ orbitals from the outer oxygen atoms. The third sp^2 hybrid contains a lone pair of electrons. The Lewis structure also indicates the presence of a π bond and five nonbonding electron pairs distributed among the two outer oxygen atoms.

The inner oxygen atom has one unused $2p$ orbital, which is oriented perpendicular to the plane defined by the sp^2 hybrids. This orbital is oriented perfectly to form π orbitals through side-by-side overlap with similarly oriented p orbitals on the outer oxygen atoms. The simplest picture would show a localized π bond, as seen in either of the two resonance structures. As Figure 9-21 shows, however, the p orbitals on *both* outer oxygen atoms overlap with the inner atom's p orbital. Thus all three p orbitals interact, producing delocalized π orbitals spread over all three atoms.

It is a general feature of orbital interactions that when p orbitals overlap side by side, the resulting number of bonding π molecular orbitals equals the number of antibonding π^* molecular orbitals. Thus ozone, with three overlapping atomic p orbitals, has one π orbital and one π^* orbital. As mentioned earlier, the number of molecular orbitals must equal the number of interacting atomic orbitals, so three interacting atomic orbitals generate three molecular orbitals. What is the third orbital?

Clues to the nature of the third orbital can be found in the placement of electrons in ozone's two resonance structures (Figure 9-22a). Notice that in one resonance structure, the left outer atom has three lone pairs and a single bond, while the right outer atom has two lone pairs and a double bond. In the other resonance structure, the third lone pair is on the *right* outer atom, with the double bond to the *left* outer atom. The double bond appears in different positions in the two structures, and one of the lone pairs also appears in different positions. These variations signal delocalized π orbitals.

The double bond shown in different locations in the two different resonance structures represents a pair of electrons in a delocalized π bonding molecular orbital that spans all three of the oxygen atoms, as shown in Figure 9-21. One of ozone's lone pairs also appears in different positions in the two resonance structures, again signaling a delocalized orbital. This lone pair is spread over both *outer* atoms but not across the *inner* atom, as shown in Figure 9-22b. This new type of π orbital has no electron density along the bond axes and is neither bonding nor anti-bonding. We designate it as a nonbonding molecular orbital, π_n. The lone pair shown in different positions in the resonance structures occupies the delocalized π_n orbital. To summarize, the interactions among the three p orbitals of ozone generate three molecular orbitals, of which the π and π_n are filled. The third orbital, a π^* antibonding orbital, is empty. Except in diatomic molecules, ground-state configurations seldom include occupied antibonding π orbitals.

The bond lengths in ozone provide evidence that the π electrons extend over the entire molecule rather than being isolated between two atoms. Both bond lengths are 128 pm, intermediate between the length of the double bond in O_2 (121 pm) and the length of the single O—O bond in H_2O_2 (148 pm).

Ozone contains three adjacent atoms, each with a p orbital positioned for side-by-side overlap. This example highlights the requirement for delocalized π systems.

Figure 9-21
The *p* orbital on the inner oxygen atom of ozone overlaps equally well with *p* orbitals on its two outer neighbors, generating a delocalized π bond.

(a)

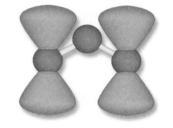

π_n orbital
(b)

Figure 9-22
(*a*) The resonance structures of ozone show a delocalized π bond (*shaded in gold*) in addition to a delocalized lone pair (*shaded in grey*). (*b*) The delocalized lone pair occupies a π_n orbital.

Molecules do not flip back and forth between resonance structures. Resonance structures are used to represent molecules that contain delocalized π electrons.

///A delocalized π system is present whenever p orbitals on three or more adjacent atoms are in position for side-by-side overlap.

Delocalized π systems often can be recognized in the Lewis structure of a molecule by the presence of resonance forms, as Example 9-6 illustrates.

Example 9-6 | Bonding in Acetate

The acetate anion ($CH_3CO_2^-$) forms when acetic acid, the acid present in vinegar, reacts with hydroxide ion:

$$CH_3CO_2H + OH^- \longrightarrow CH_3CO_2^- + H_2O$$

Describe the bonding of this anion. Sketch the σ bonding system and the occupied π orbitals.

Strategy: As with any description of bonding, the procedure begins with construction of the Lewis structure, from which we can determine the steric numbers and hybridizations. Then we can describe the σ and π bonds in the species.

Solution:

1. Count valence electrons and determine the Lewis structure.

   2 C (4 valence electrons) + 2 O (6 valence electrons)
   + 3 H (1 valence electron) + 1 (negative charge) = 24 valence electrons.

2. From the Lewis structure, we see that the CH_3 carbon has a steric number of 4, so its bonding can be described by sp^3 hybrid orbitals. The other carbon has a steric number of 3, indicating trigonal planar geometry and sp^2 hybrid orbitals. Three of the sp^3 hybrid orbitals form σ bonds by overlapping with $1s$ orbitals on the hydrogen atoms. The fourth sp^3 hybrid forms a σ bond with an sp^2 hybrid from the neighboring carbon atom. The two remaining sp^2 hybrids bond to $2p$ orbitals from the outer oxygen atoms. This gives a total of six σ bonds and uses 12 of the 24 valence electrons:

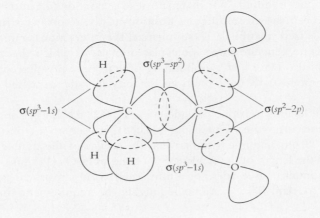

Bonding in Acetate *(continued)*

Example 9-6

The resonance structures indicate that acetate has a delocalized π system. The sp^2 hybridized carbon atom and the two outer oxygen atoms all possess $2p$ orbitals available for side-by-side overlap. As in the ozone molecule, these three orbitals form a set of three delocalized π orbitals. The π system for acetate has the same orbital assignments as the π system for ozone. Here are sketches of the occupied orbitals, π and π_n:

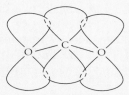

Bonding orbital   Nonbonding orbital

Carbon Dioxide

Carbon dioxide is a crucial molecule in photosynthesis because plants use the carbon atoms of CO_2 for the synthesis of glucose. Furthermore, animals and plants use fats and carbohydrates for energy production and eliminate the carbon atoms from these molecules as CO_2.

Carbon dioxide has 16 valence electrons. Its Lewis structure shows that the molecule has two double bonds. The molecule is linear, a geometry predicted by the steric number of 2 for the carbon atom. Figure 9-23 shows the two σ bonds formed by end-on overlap between sp hybrids on the carbon atom and $2p_z$ atomic orbitals of oxygen.

The $2p_x$ and $2p_y$ orbitals of the carbon atom are not used in forming the hybrid set. Each of these orbitals is oriented perfectly to overlap side by side with p orbitals of each oxygen neighbor. As in ozone, the three $2p_y$ orbitals interact to generate three delocalized π orbitals. One of these orbitals is bonding, one is nonbonding, and the third is antibonding. The bonding and nonbonding orbitals, shown in Figure 9-24, have the same forms as their counterparts in ozone but follow the linear shape of the bond framework. The $2p_x$ atomic orbitals combine in exactly the same way, except that the resulting three orbitals point at right angles to the p_y set. These two sets of delocalized π orbitals have exactly the same energies, but they do not interact with one another.

The 16 valence electrons in the CO_2 molecule occupy the eight most stable valence orbitals. The Lewis structure indicates that there are four bonding pairs. Two pairs form the σ bonds shown in Figure 9-23. The other two bonding pairs are in the two delocalized π orbitals. Two nonbonding pairs occupy the two π_n nonbonding orbitals. The remaining two are in the $2s$ atomic orbitals on the outer oxygen atoms. This completes the valence configuration, leaving the π^* antibonding orbitals vacant.

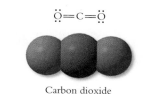

Carbon dioxide

σ bonding system

Figure 9-23
The σ bonding system of CO_2. Two σ bonds form from the overlap of carbon sp hybrids with oxygen $2p_z$ atomic orbitals.

Figure 9-24
A delocalized π and π_n orbital of CO_2. Notice the similarity to the π and π_n orbitals of ozone.

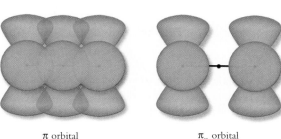

π orbital π_n orbital

Other Second-Row Triatomics

Triatomic species can be linear, like CO_2, or bent, like O_3. The principles of orbital overlap do not depend on the identity of the atoms involved, so all second-row triatomic species with 16 valence electrons have the same bonding scheme as CO_2 and are linear. For example, dinitrogen oxide (N_2O) has 16 valence electrons. As a consequence, it has an orbital configuration identical to that of CO_2. Each molecule is linear with an inner atom whose steric number is 2. As in CO_2, the bonding framework of N_2O can be represented with sp hybrid orbitals. Eight electrons occupy four delocalized π orbitals, two that are bonding and two that are nonbonding. The resonance structures of N_2O, described in Example 8-9, reflect these delocalized orbitals. Second-row triatomic ions with 16 valence electrons (N_3^-, NCO^-, and CNO^-) also are linear, with inner atoms whose bonding can be represented using sp hybrid orbitals and two sets of delocalized p orbitals.

Ozone, which has 18 valence electrons, exemplifies bent molecules. Another example is the nitrite anion. The bonding of NO_2^- can be represented using sp^2 hybrid orbitals for the inner nitrogen atom and one set of delocalized π orbitals.

Section Exercises

9.4.1 Construct the bonding description (Lewis structure, hybridization, geometry, and orbital assignments and shapes) for the azide anion (N_3^-). Explain why this anion has a different bonding description from the nitrite anion.

9.4.2 Explain the difference between an isolated π orbital and a delocalized π orbital.

9.4.3 Construct the bonding description for the formate anion (HCO_2^-) and compare its π system with that of the acetate anion.

9.5 EXTENDED π SYSTEMS

Ozone and carbon dioxide demonstrate that p orbitals can overlap side by side with more than one neighbor. This feature can lead to an extended system of π bonds, with side-by-side overlap extending over many atoms. To see how this comes about, we first look at four-atom systems.

Four-Atom Delocalized Orbitals

The Lewis structure of the carbonate anion has three resonance structures:

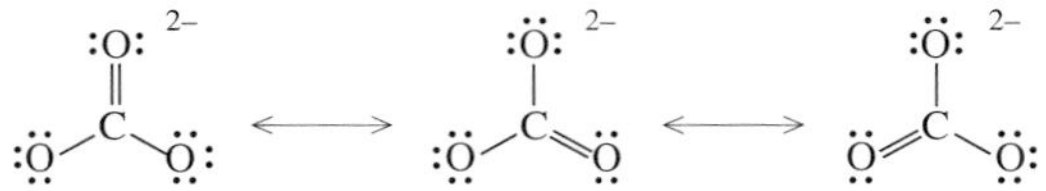

Carbonate anion (CO_3^{2-})

The existence of resonance structures indicates that a complete description of the carbonate ion requires delocalized orbitals. The carbon atom has a steric number of 3, meaning that the σ bonding framework for CO_3^{2-} can be described using sp^2 hybrid orbitals on the carbon atom. These hybrid orbitals overlap end-on with $2p$ orbitals of the three oxygen atoms, as shown in Figure 9-25a.

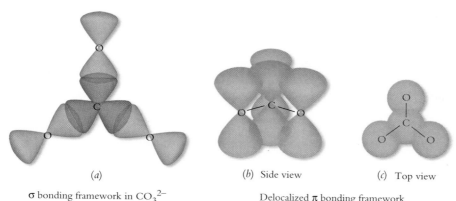

(a)

σ bonding framework in CO_3^{2-}

(b) Side view

(c) Top view

Delocalized π bonding framework

Figure 9-25
(a) The σ bonding framework of carbonate forms from carbon sp^2 hybrids and oxygen $2p$ orbitals. (b) The leftover p orbital on the inner carbon atom overlaps with p orbitals from all three outer oxygen atoms, giving delocalized π orbitals. (c) Viewed from above, the bonding π orbital has a propeller shape.

The carbon atom has one unused $2p$ orbital perpendicular to the molecular plane. As Figure 9-25b shows, all three oxygen atoms also have perpendicular p orbitals, oriented properly for side-by-side overlap. These four p orbitals give four π molecular orbitals delocalized over the entire anion. Figure 9-25c shows that the most stable π molecular orbital has a propeller shape when viewed from above.

Four atoms with appropriately oriented p orbitals can also line up in a row, as Example 9-7 illustrates.

Delocalized π Orbitals

Example 9-7

Describe the geometry and π bonding of methyl methacrylate. This compound is an important industrial chemical, ranking in the second 50 in annual production. It is used mainly to make plastics such as poly(methyl methacrylate) (PMMA).

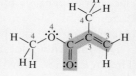

Methyl methacrylate
$C_5H_8O_2$

Strategy: First, translate the line drawing into a Lewis structure. Use the Lewis structure to identify the steric number, hybridization, and orbital geometry of each inner atom. Then determine which atoms have p orbitals in the proper positions to form delocalized π orbitals.

Solution: The Lewis structure is shown here with the steric number for each inner atom. The atoms with p orbitals that can overlap side by side are highlighted in purple.

Each atom with a steric number of 4 has tetrahedral geometry and can be described by sp^3 hybridization. Carbon atoms with steric numbers of 4 form four σ bonds each. The inner oxygen atom also has a steric number of 4, forms two σ bonds, has two sp^3 lone pairs, and has a bent shape. The three adjacent carbon atoms with steric numbers of 3 have trigonal planar orbital geometry and can be described by sp^2 hybrid orbitals that form σ bonds. Each of these adjacent atoms has a leftover p orbital that can form π bonds. These three p orbitals plus one p orbital on the outer oxygen atom are oriented in a row. Through side-by-side overlap they form a set of four delocalized π molecular orbitals, two bonding and two antibonding. Each of the π bonding orbitals contains a pair of electrons; the antibonding orbitals are empty.

← CHAPTER 11
Plastics such as PMMA are described in Chapter 11.

| Example 9-7 | Delocalized π Orbitals *(continued)* |

A ball-and-stick model of methyl methacrylate shows the planar part of the molecule covered by the delocalized π orbitals. The atoms that contribute p orbitals to the delocalized π system are marked with asterisks, and the other atoms that lie in the same plane are marked with dots.

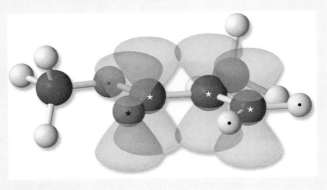

Rings and Chains with Delocalized π Systems

Planar rings with delocalized π bonding are common in biological systems. Look again at the structures of plastoquinone and chlorophyll in the introduction to this chapter. Plastoquinone contains a planar ring of six carbon atoms, with two carbon atoms double-bonded to outer oxygen atoms. Chlorophyll contains four five-atom rings (four C and one N) connected in a larger ring that forms a plane. Both these molecules contain extended delocalized π systems.

Benzene (C_6H_6) is an example of a stable ring molecule with a delocalized π system. This compound has a ring of six carbon atoms with alternating double and single bonds, as the Lewis structure in Figure 9-26a shows. The figure also shows that benzene has two resonance structures. The molecule has six adjacent sp^2 hybridized atoms, each with a p orbital perpendicular to the plane of the ring (Figure 9-26b). These six p orbitals are in perfect position for side-by-side overlap. Together they form a set of six molecular orbitals: three bonding π orbitals and three antibonding π^* orbitals. There are six π electrons in benzene, just enough to fill the bonding molecular orbitals. The antibonding orbitals remain empty. The π electrons of benzene form circular orbital rings above and below the plane of the carbon atoms. This circular orbital is highly resistant to chemical attack, making the benzene ring remarkably stable.

Chains with delocalized π systems also occur commonly in biological systems. Three examples, shown in Figure 9-27, are Vitamin A, retinal, and β-carotene. β-carotene is an orange compound that gives carrots their color. Our bodies convert β-carotene into Vitamin A. This compound, in turn, is used to make retinal, a molecule that plays an essential role in the chemistry of vision. Notice that each of these molecules contains a sequence of carbon atoms with alternating single and double bonds. Vitamin A has a row of 10 carbon atoms with sp^2 hybridization, and carotene has 22. The leftover $2p$ valence orbital on each of these atoms interacts with those on its neighbors to form extended delocalized π bonding orbitals. Retinal, in addition to a row of 11 such carbon atoms, has a terminal oxygen atom that contributes to the delocalized π system.

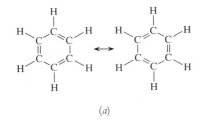

(a)

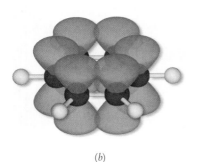

(b)

Figure 9-26
The Lewis structure (a) and orbital overlap picture (b) of benzene.

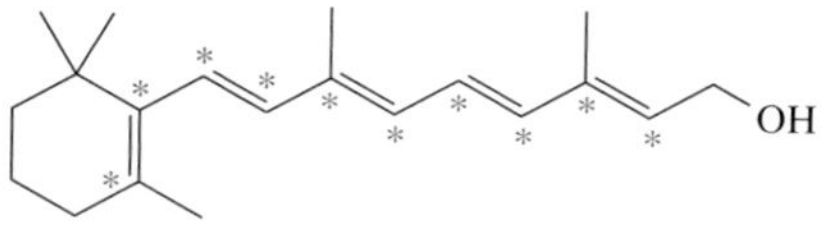

Figure 9-27
The line structures of three biochemically important molecules that contain extended π bonding systems. The atoms that contribute to the π systems are indicated with asterisks.

Vitamin A ($C_{20}H_{30}O$)

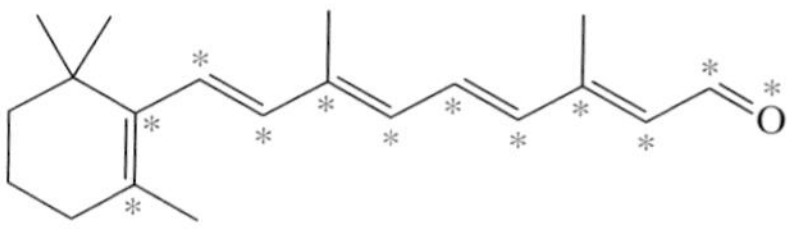

Retinal ($C_{20}H_{28}O$)

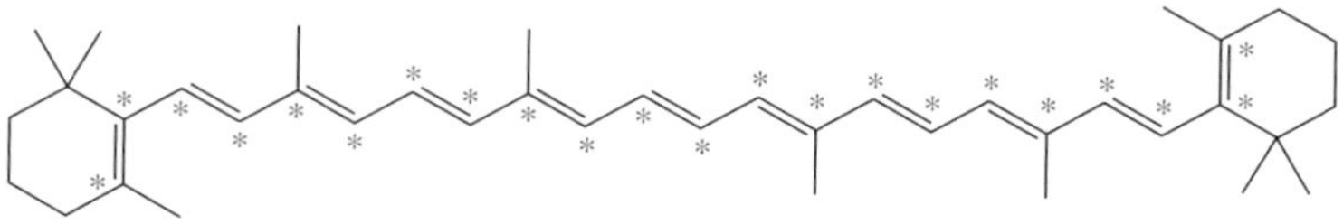

β-Carotene ($C_{40}H_{56}$)
(an orange plant pigment, and the
precursor to vitamin A)

Consequences of Delocalized Orbitals

In a localized π bond, a p orbital on one atom overlaps with a p orbital on a neighboring atom in side-by-side fashion. In delocalized systems, on the other hand, one or more of the p orbitals overlaps with two, sometimes three, adjacent p orbitals. With more overlap, a molecule containing a delocalized π system is more stable than a molecule with similar bonding but with isolated double bonds. This has consequences for bond stability, for absorption spectra and color, and for redox behavior.

The most obvious effect of delocalization is on the absorption spectrum of a molecule. Many organic molecules that have delocalized π orbitals are colored. A molecule is colored when it absorbs photons in the visible portion of the electromagnetic spectrum. Recall from Chapter 6 that substances can absorb only photons whose energies match differences among energy levels. Visible photons have wavelengths between 700 and 400 nm and energies between 180 kJ/mol and 330 kJ/mol. Thus, in a colored substance, the energy needed to promote an electron from the highest filled orbital to the lowest unoccupied orbital is between 180 kJ/mol and 330 kJ/mol.

It is a general feature of π orbital energies that, the larger the number of atoms over which the orbital extends, the smaller the energy difference between orbitals. Molecules with localized π bonds (such as ethylene and acetone) have energy gaps much greater than 330 kJ/mol and are colorless. Figure 9-28 shows the absorption spectra of three molecules that contain highly delocalized π systems. Benzene (6 ring atoms) absorbs light in the ultraviolet spectral region and is colorless. Anthracene (14 ring atoms) absorbs light at the 400-nm edge of the visible region. Anthracene itself is colorless, but many derivatives of anthracene are pale yellow. Chlorophyll (24 ring atoms) absorbs light at both ends of the visible spectrum and is intensely green.

The effect of delocalization on bond stability is illustrated by comparing 1,3-pentadiene and 1,4-pentadiene. Both have eight C—H bonds, two C—C bonds, and two C=C bonds. The only significant difference between these molecules is that the π system of 1,3-pentadiene is delocalized because there are four adjacent sp^2-hybridized carbon atoms. In contrast, 1,4-pentadiene has an sp^3-hybridized CH$_2$ group acting as a spacer to separate two localized π bonds.

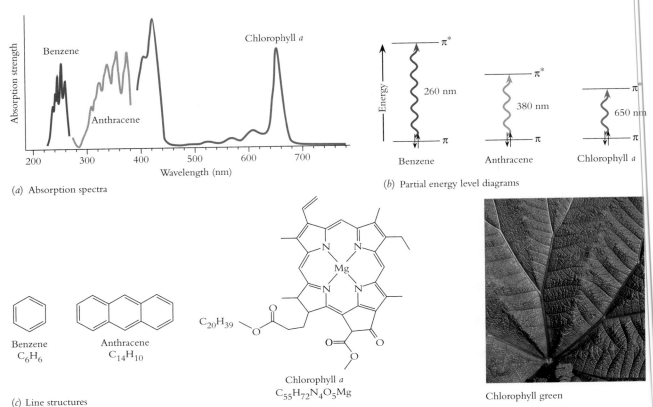

(a) Absorption spectra

(b) Partial energy level diagrams

(c) Line structures

Benzene
C_6H_6

Anthracene
$C_{14}H_{10}$

Chlorophyll a
$C_{55}H_{72}N_4O_5Mg$

Chlorophyll green

Figure 9-28
The absorption spectra (a), energy levels of the highest occupied and lowest unoccupied orbitals (b), and line structures (c) of benzene, anthracene, and chlorophyll, three organic molecules with delocalized π systems. The energy gap between highest occupied and lowest unoccupied orbital becomes smaller as the size of the delocalized π system grows.

Does the delocalized π system of 1,3-pentadiene stabilize the molecule relative to 1,4-pentadiene? Each molecule reacts with two molecules of H_2 to give a common product, pentane. Thus, comparing their energies of hydrogenation allows us to determine the stabilities of two delocalized π bonds relative to two isolated π bonds. The hydrogenation of 1,4-pentadiene releases 252 kJ/mol, but the hydrogenation of 1,3-pentadiene releases only 224 kJ/mol. As Figure 9-29 demonstrates, these hydrogenation energies show that 1,3-pentadiene is 28 kJ/mol *more stable* than 1,4-pentadiene. Because the bonding in these two molecules is otherwise the same, we can conclude that the extra stability comes from the delocalized π system.

Delocalized π systems also affect the oxidation-reduction characteristics of organic substances. As we pointed out in the Introduction to this chapter, plastoquinone and cytochrome play important redox roles in photosynthesis. As another example, β-carotene, with its delocalized π system, is thought to act as an antioxidant, protecting the body from cancer-causing oxidation processes. Oxidation and reduction reactions involve the transfer of electrons from occupied orbitals in one substance to empty orbitals in another substance. Whether or not a redox reaction takes place depends on orbital energies, so the stabilization of π systems by delocalization has a direct effect on redox behavior.

Remember that *lower* on an energy level diagram means *more stable*.

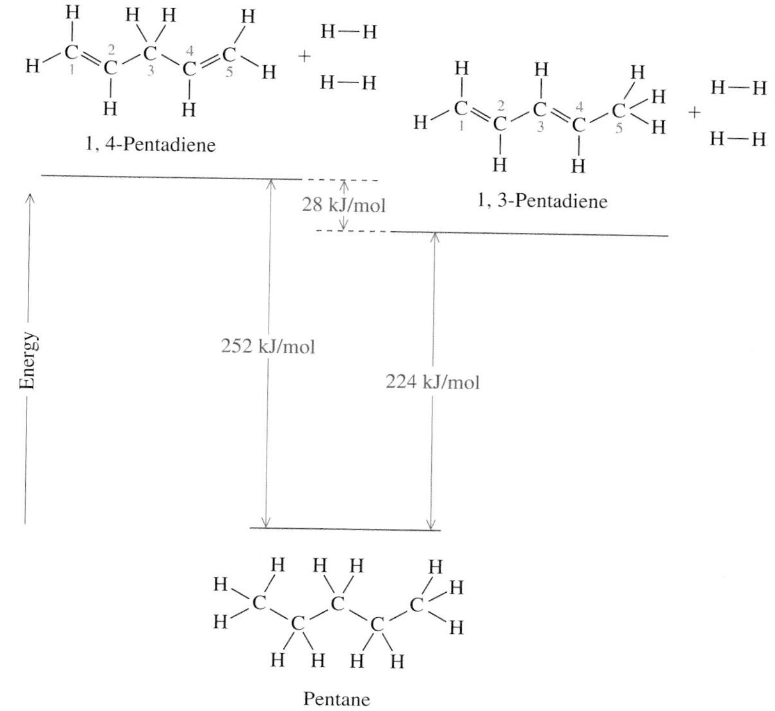

1, 4-Pentadiene

1, 3-Pentadiene

28 kJ/mol

Energy

252 kJ/mol

224 kJ/mol

Pentane

Figure 9-29
Energy level diagram for the hydrogenation of 1,4-pentadiene and 1,3-pentadiene. The difference in these energies, 28 kJ/mol, represents the additional stability that 1,3-pentadiene gains from delocalization of its π system.

π Bonding Beyond the Second Row

Several common polyatomic anions, including sulfate, perchlorate, and phosphate, have inner atoms from the third row of the periodic table. In these anions, valence d orbitals are available to participate in bonding. As shown in Chapter 8, d orbitals can hybridize, and the resulting hybrids participate in σ bond formation. The d orbitals can also participate in π bonding. Figure 9-30 shows how a π orbital can form through side-by-side orbital overlap of a $3d$ orbital on one atom and a $2p$ orbital on another atom.

Several elements in row 3 of the periodic table can form π bonds to oxygen through side-by-side overlap of $3d$ and $2p$ orbitals. An example is the sulfate anion, whose Lewis structure is shown in Figure 9-31. The steric number of S in SO_4^{2-} is 4, indicating that the species has tetrahedral geometry. We can describe the bonding framework using sp^3 hybrid orbitals. However, the Lewis structure shows that the sulfur atom is associated with *six* electron pairs, signalling that two valence d orbitals are involved in bonding. In sulfate and other tetrahedral oxyanions, d orbitals participate in delocalized π bonding, the details of which are beyond the scope of this text. You need only recognize that they include π orbitals formed through the side-by-side overlap of d orbitals and p orbitals.

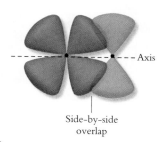

Figure 9-30
A localized π bond can form through side-by-side overlap between a d and a p orbital. As with other π bonds, electron density is concentrated between the nuclei, above and below the internuclear axis.

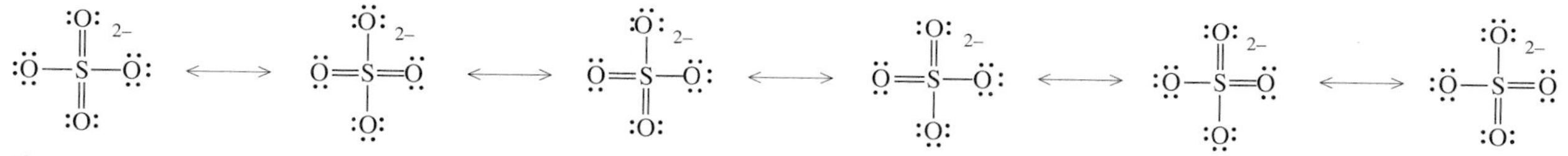

Figure 9-31
All four oxygen atoms are equivalent, so six resonance structures are needed to represent the π bonding of the sulfate anion. These structures indicate that delocalized orbitals span the entire anion.

Section Exercises

9.5.1 Draw sketches (side and top views) showing the π bonding system in 1,3-butadiene, industrially important as the primary starting material in the manufacture of synthetic rubber. This hydrocarbon has the chemical formula C_4H_6, and its Lewis structure shows two double bonds separated by a single bond.

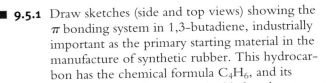

9.5.2 Shown at right is the line structure of azulene, a deep purple, almost black solid. How many delocalized orbitals are in an azulene molecule? How many of these orbitals are occupied?

Azulene
$C_{10}H_8$

9.5.3 The chromate ion ($CrO_4{}^{2-}$) has the same number of valence electrons as the sulfate anion. Describe the bonding in $CrO_4{}^{2-}$.

A solution of azulene has a blue color.

9.6 BAND THEORY OF SOLIDS

Copper, iron, and aluminum are three very common metals in modern society. Copper wires carry the electricity that powers most appliances, including the lamp by which you may be reading. The chair in which you are sitting may have an iron frame, and you may be sipping a soft drink from an aluminum can. The properties that allow metals to be used for such a wide range of products can be traced to the principles of bonding and electronic structure.

Iron and other metals have tremendous mechanical strength, which suggests that the bonds between their atoms must be strong. At the same time, most metals are malleable, which means they can be shaped into thin sheets to make objects such as aluminum cans. Metals are also ductile, which means they can be drawn into wires. The properties of malleability and ductility suggest that atoms in metals can be moved about without weakening the bonding. Finally, metals conduct electricity, which suggests that the electrons in a metal are free to move throughout the solid.

All these properties of metals are consistent with a bonding description that places the valence electrons in delocalized orbitals. This section describes the bonding in metals and many other solids in terms of the **band theory of solids,** an extension of the delocalized orbital ideas that are the theme of this chapter. Band theory accounts for the properties of metals, and it also explains the properties of metalloids such as silicon.

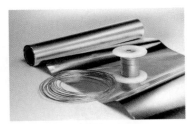

Metals are malleable and ductile.

Delocalized Orbitals in Lithium Metal

Lithium, the lightest metal, can be used to demonstrate the principles of band theory. Solid lithium contains atoms held together in a three-dimensional crystal lattice. Bonding interactions among these atoms can be described by orbital overlap. To see how this occurs, consider building an array of lithium atoms one at a time.

The bonding in Li_2 can be described with overlapping valence orbitals. The valence $2s$ orbitals of two lithium atoms interact to form a σ bonding orbital and a σ^* antibonding orbital. The bonding orbital is filled, and the antibonding orbital is empty. The single bond in Li_2 is relatively weak, 105 kJ/mol, because compact $2s$ orbitals do not overlap strongly.

Although Li_2 molecules are unstable relative to metallic lithium, the Li—Li bond is strong enough that these diatomic molecules can exist as long as they are isolated in the gas phase.

Clusters of lithium atoms such as Li_4, Li_6, and Li_{18} can be generated in the gas phase by bombarding a piece of lithium metal with gaseous metal atoms in a chamber at very low pressure. The Li_4 molecule is bonded together by four $2s$ orbitals that interact to give two filled delocalized bonding orbitals and two empty delocalized antibonding orbitals. Likewise, Li_6 has 6 delocalized orbitals, Li_{18} has 18 delocalized orbitals, and so on. As with all molecular orbitals, these interactions generate equal numbers of bonding and antibonding orbitals. In 1 cm^3 of Li metal there are approximately 4×10^{22} atoms whose $2s$ atomic orbitals generate approximately 4×10^{22} delocalized orbitals. This is a huge number of bonding orbitals and antibonding orbitals; yet they all obey the principles developed in this chapter. Most important, each orbital extends over the entire lattice of atoms.

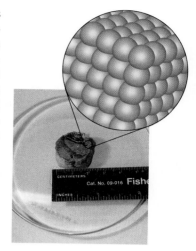

Figure 9-32 shows that the energy spacing between orbitals decreases as the number of delocalized orbitals increases. For the huge number of atoms in a piece of metal, the orbitals are spaced so closely that they behave as if they were merged into an energy band. This bonding model is called *band theory* because energy bands are one of its main features.

To keep our picture simple, we include only the $2s$ orbitals in Figure 9-32. For most metals, however, a comprehensive picture of the bonding includes interactions among all valence s, p, d, and even f orbitals. Each group of orbitals interacts to generate an energy band. Interactions among bands is complex, and often the bands overlap. Despite this complexity, however, the fundamental features illustrated in Figure 9-32 remain valid.

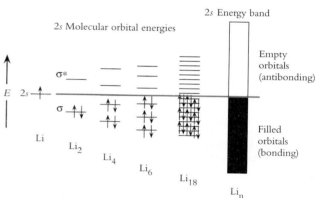

Figure 9-32
Schematic view of the energies of delocalized orbitals made from the $2s$ orbitals of lithium atoms.

Electrical Conductivity

The most important feature of metallic bonding is the energy bands that arise because of the close spacing between orbital energies. The countless orbitals within the band are separated by infinitesimally small energy gaps. The valence electrons of the metal atoms occupy these orbitals according to the aufbau and Pauli principles. In a metal, the occupied orbitals of highest energy are so close in energy to the unoccupied orbitals of lowest energy that it takes little energy to transfer an electron from an occupied to an unoccupied orbital.

When an electrical potential is applied to a metal, the negative pole repels electrons and the positive pole attracts them. In energy terms, the occupied orbitals near the negative pole are pushed higher in energy than unoccupied orbitals near the positive pole. This "tilts" the energy levels, as shown in Figure 9-33, and electrons can become more stable by moving downhill out of filled orbitals into empty orbitals at lower energy. As a result, electrons flow through the metal from the negative end to the positive end, generating an electrical current.

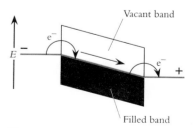

Figure 9-33
An electrical potential shifts the energy levels in a metal. The arrows show the direction of electron flow.

Insulators and Conductors: Carbon vs. Lead

Metallic lead is dark in color and is an electrical conductor. Diamond, the most valuable form of carbon, is transparent and is an electrical insulator. These properties are very different; yet both lead and carbon are in Group 14 of the periodic table and have the same valence configuration, s^2p^2. Why, then, are diamonds transparent insulators, whereas lead is a dark-colored conductor?

Both diamond and lead can be described using delocalized orbitals and energy bands, but as Figure 9-34 shows, the energy distributions of the bands are quite different. Lead is metallic because the energy separation between its filled and vacant orbitals is infinitesimally small. One continuous band of orbitals results in highly mobile electrons and electrical conductivity. The valence orbitals of carbon, on the other hand, form two distinct bands. The filled bonding orbitals are well separated in energy from the empty, antibonding orbitals. The energy difference between these two bands is called the **band gap** (E_g). For diamond, $E_g = 580$ kJ/mol.

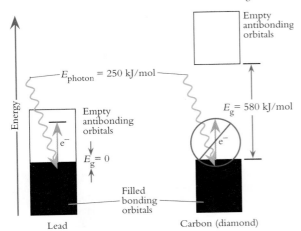

C = [He] $2s^2\ 2p^2$
Pb = [Xe] $4f^{14}\ 5d^{10}\ 6s^2\ 6p^2$

$E_{photon} = 250$ kJ/mol

Empty antibonding orbitals

Empty antibonding orbitals

$E_g = 580$ kJ/mol

$E_g = 0$

e^-

e^-

Filled bonding orbitals

Lead

Carbon (diamond)

Energy

Figure 9-34
Carbon and lead have different properties because the energy differences between their filled bonding orbitals and their empty antibonding orbitals are different.

Lead and diamond have distinctly different band structures for two reasons. First, the smaller size of carbon's valence orbitals results in stronger overlap. This results in a large stabilization energy and a large band gap. Second, lead has valence d and f orbitals that are involved in band formation. These d and f bands overlap the s and p bands, contributing to the continuous distribution of orbital energies that gives lead its metallic properties.

Carbon in the form of diamond is an electrical insulator because of its huge band gap. In fact, its gap of 580 kJ/mol exceeds the C—C bond energy of 345 kJ/mol. In other words, it requires more energy to promote an electron from band to band in diamond than to break a covalent bond. Lead, in contrast, is a metallic conductor because it has $E_g = 0$.

The differences in their band structures also explain why diamond is transparent and lead is dark. Substances can absorb only photons whose energies match differences between energy levels, and a colored substance absorbs visible photons, whose energies are between 180 kJ/mol and 330 kJ/mol. For diamond, the energy difference between the top of the filled band and the bottom of the empty band is 580 kJ/mol. Diamond cannot absorb visible light because there are no energy levels between 180 and 330 kJ/mol above the filled band. Visible light passes straight through a diamond. Lead, on the other hand, has many vacant orbitals into which visible photons can promote electrons. Thus lead can absorb all visible wavelengths, making it gray.

Metalloids

Orbital energies are determined largely by the amount of spatial overlap. Because atomic orbitals become increasingly diffuse (spread out in space) as their principal quantum number (n) increases, the spatial overlap of the valence orbitals decreases as the valence shell n value increases (see Figure 9-10). Less overlap means less difference between the energy of the bonding and antibonding levels, so the band gap shrinks from carbon to lead. Carbon ($n = 2$) has a large band gap and is a nonmetal, but tin ($n = 5$) and lead ($n = 6$) have $E_g = 0$ and are metals. Silicon and germanium ($n = 3$ and $n = 4$, respectively) have intermediate band gap values, which cause them to behave as **metalloids.**

Like carbon, silicon and germanium have low-energy bands of bonding orbitals that are completely filled with electrons and higher-energy bands that are empty. For silicon, $E_g = 105$ kJ/mol, whereas $E_g = 64$ kJ/mol for germanium. Electrons cannot make the transition between these bands unless sufficient energy is supplied. Thus silicon and germanium are nonconductors in the absence of an energy source. However, electrons can move between the bands at higher temperatures or in the presence of photons with energy that matches E_g. When electrons are transferred into the high-energy band, they can move freely among the many vacant near-degenerate orbitals. In addition, electron transfer creates vacancies (holes) among the bonding orbitals in the low-energy band. The holes and high-energy electrons allow electricity to flow. Thus silicon and germanium become conductors in the presence of an energy source. They are called **semiconductors** to indicate their ability to conduct electricity under special conditions. Figure 9-35 uses band gap diagrams to summarize a semiconductor's behavior.

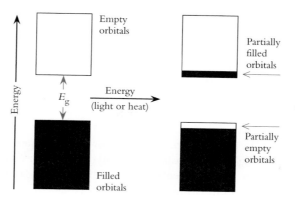

Figure 9-35
When a semiconductor absorbs heat or light, electrons are excited from the filled band to the empty band, and the semiconductor conducts electricity.

Transfer of electrons from one band to another requires energy equal to the band gap. This energy can be provided by heating (exploited in thermistors) or by absorption of light (used in photoconductors). Example 9-8 deals with photoconductors.

Properties of Photoconductors　　　　　　　　　　　　　　　　**Example 9-8**

"Electric eye" door openers use photoconductors that respond to infrared light with a wavelength (λ) of 1500 nm. Which semiconductor material is suitable for photoconductors operating at this wavelength, germanium ($E_g = 64$ kJ/mol) or silicon ($E_g = 105$ kJ/mol)?

Strategy: To conduct electricity, a semiconductor must be provided with energy that is *at least* equal to its band gap. In a photoconductor, this energy must come from photons. Thus we need to compare the band gaps with the energy of the infrared photons.

Solution: First, calculate the energy of 1500-nm light, recalling from Chapter 6 that photon energy is related to wavelength:

$$E_{photon} = h\nu = \frac{hc}{\lambda}$$

$$E_{photon} = \frac{(6.626 \times 10^{-34}\,\text{J s})(2.998 \times 10^8\,\text{m/s})}{(1500\,\text{nm})(10^{-9}\,\text{m/nm})} = 1.32 \times 10^{-19}\,\text{J}$$

Multiply by the Avogadro constant to convert the energy of one photon into the energy of one mole of photons in kilojoules per mole:

$$(1.32 \times 10^{-19}\,\text{J/photon})(6.022 \times 10^{23}\,\text{photons/mol})(10^{-3}\,\text{kJ/J}) = 79\,\text{kJ/mol}$$

This energy is insufficient to overcome the band gap of silicon, so silicon could not be used for an infrared photoconductor. Germanium, however, with its smaller band gap ($E_g = 64$ kJ/mol), becomes an electrical conductor when illuminated by infrared light with $\lambda = 1500$ nm.

Doped Semiconductors

The relatively large band gaps of silicon and germanium limit their usefulness in electrical devices. Fortunately, the conductive properties of these solid elements can be altered by adding tiny amounts of other elements that have different valence electron configurations. When a specific impurity is deliberately added to a pure substance, the resulting material is said to be *doped*. A **doped semiconductor** has almost the same band structure as the pure material, but it has different electron populations in its bands.

In pure silicon, all orbitals of the low-energy band are filled, and all orbitals of the high-energy band are empty. When pure silicon is doped with arsenic atoms, the As atoms replace some of the Si atoms in the crystal lattice. Each arsenic atom contributes five valence electrons instead of the four contributed by a silicon atom. The fifth valence electron cannot enter the low-energy band of the solid because that band is already filled. The fifth valence electron from each As atom must therefore occupy the higher-lying energy band. Silicon or germanium doped with atoms from Group 15 is called an ***n*-type semiconductor** because extra *negative* charges exist in its high-energy band. These few electrons in a partially occupied band make the doped material an electrical conductor.

Silicon can also be doped with gallium, a Group 13 element. Gallium has three valence electrons, so each Ga atom in the Si lattice has one less electron than the atom that the Ga replaces. In gallium-doped silicon, there are not quite enough electrons to fill all the bonding orbitals. Silicon and germanium doped with atoms from Group 13 are known as ***p*-type semiconductors** because their low-energy bands have *positive* vacancies. Electrons move through the crystal by flowing from filled orbitals into these vacant ones.

Semiconductors such as silicon are extremely sensitive to impurities. Replacing just 0.00001% of the Si atoms with a dopant can cause as much as a 100,000-fold increase in electrical conductivity.

Figure 9-36 shows band gap diagrams of *n*-type and *p*-type semiconductors. Electrical current flows in a doped semiconductor in the same way as current flows in a metal (see Figure 9-33). Only a small energy difference exists between the top of the filled band and the next available orbital, so the slightest applied potential tilts the bands enough to allow electrons to move and current to flow.

Not all semiconductors are made from silicon or germanium. Compounds made from equimolar amounts of Group 13 and Group 15 elements are also semiconductors. Gallium arsenide (GaAs) is a typical example. One Ga atom and one As atom have a total of eight valence electrons, so gallium arsenide is isoelectronic with germanium. Gallium arsenide can be doped with zinc atoms to make a *p*-type semiconductor or with tellurium atoms to make an *n*-type semiconductor. Other semiconductors contain equimolar compositions of an element such as zinc that has two valence electrons and a Group 16 element (S, Se, or Te) with six valence electrons. Zinc sulfide is an example. Our Chemistry and Technology Box describes the light-emitting characteristics of this type of semiconductor.

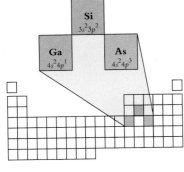

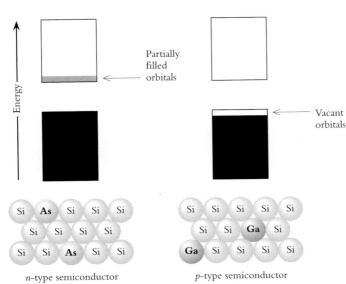

Figure 9-36
Schematic views of an *n*-type and a *p*-type semiconductor.

Box 9-2 | Chemistry and Technology: Light-Emitting Diodes

One application of modern solid-state electronic devices is semiconductor materials that convert electrical energy into light. These *light-emitting diodes* (LEDs) are used for visual displays and solid-state lasers. Many indicator lights are LEDs, and diode lasers read compact discs in a CD player. The field of diode lasers is expanding particularly rapidly, driven by such applications as fiber optic telephone transmission.

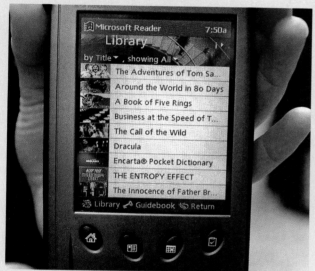

The heart of a light-emitting diode is a junction between a *p*-type semiconductor and an *n*-type semiconductor. The different semiconductor types have different electron populations in their bands. The lower-energy band of a *p* semiconductor is deficient in electrons, while the upper-energy band of an *n* semiconductor has a small population of electrons. The band structure in the junction region is shown schematically in the figure below.

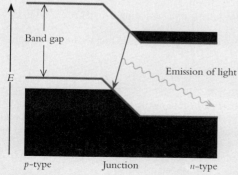

When an electrical potential is applied to an LED device, the *p* region is made positive relative to the *n*

region. The positive potential attracts electrons, which can jump from the higher-energy band of the *n* region to the lower-energy band of the *p* region. In doing so, these electrons emit their excess energy as photons of light.

The color of the light depends on the energy loss for the electrons and increases with the size of the band gap of the semiconductor. Thus red LEDs use semiconductors with relatively small band gaps, whereas green LEDs have larger band gaps.

Semiconductors containing arsenic are particularly versatile. A GaAs semiconductor can be made a *p* type by doping with a small excess of Ga or an *n* type by doping with a small excess of As. The band gap of these semiconductors can be varied by replacing some Ga with Al or some As with P. Since GaAs forms solid solutions with both AlAs and GaP, the composition can be varied continuously, from pure AlAs to pure GaAs and from pure GaAs to pure GaP. These materials have nonstoichiometric chemical formulas, $Al_xGa_{1-x}As$ and GaP_xAs_{1-x}.

The band gap of the LED varies with composition for both these solid solutions, as shown in the figure below. The cause of the variation is different for the two substances. Semiconductor band gaps increase when orbital overlap decreases. A decrease in orbital overlap can arise from increased spacing between atoms or increased ionic character of the bonds.

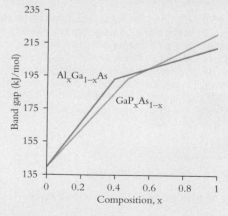

Phosphorus and arsenic have nearly identical electronegativities, so in GaP_xAs_{1-x}, the dominant effect is the smaller atomic radius of P relative to As. Substituting P atoms for As atoms shrinks the dimensions of the semiconductor lattice. This leads to greater overlap of the valence orbitals, increased stability of the bonding orbitals (valence band), and an increased band gap.

Substituting Al atoms for Ga atoms does not change the lattice dimensions, which are determined by the size of As, the largest atom of this set. Here, the lower electronegativity of Al relative to Ga leads to more ionic character and a smaller band gap.

Section Exercises

9.6.1 CdS, a semiconductor studied for use in solar cells, absorbs blue light at 470 nm. Calculate its band gap in kJ/mol.

9.6.2 Draw a band gap diagram that illustrates a semiconductor made of gallium arsenide doped with zinc atoms.

9.6.3 You have been asked to develop a *p*-type semiconductor using Al, Si, and/or P. Which two elements would you choose, and what role would each element play in the semiconductor?

CHAPTER REVIEW

Summary and Key Terms

1. A multiple bond contains one **sigma (σ) bond** formed from end-on overlap of *s*, *p*, or hybrid orbitals and **pi (π) bonds** formed from side-by-side overlap of atomic *p* or *d* orbitals on adjacent atoms. All bonds have high electron density along the bond axis, but π bonds have high electron density between the atoms but off the axis.

2. The properties of **bond length** and **bond energy** depend on the *n* values of the valence orbitals, **bond polarity,** bond multiplicity, and effective nuclear charge. The greater length of bonds between atoms with *n* > 2 restricts π bond formation to second-row elements and oxyanions. Average bond energy values can be used to estimate the energy changes accompanying chemical reactions.

3. A detailed description of bonding in diatomic molecules requires **bonding molecular orbitals** generated by additive orbital interactions and **antibonding molecular orbitals** generated by subtractive orbital interactions. Antibonding orbitals have low electron density between atoms, including nodes where electron density is zero. Bond stability is given by the **bond order,** which is the number of bonding pairs minus the number of antibonding pairs.

4. Electrons can occupy **delocalized π orbitals** extending over three or more atoms, when *p* orbitals on three or more adjacent atoms are in position for side-by-side overlap. Such delocalized π systems gain stability as a result of multiple side-by-side interactions. A Lewis structure with resonance structures signals delocalized π orbitals.

5. Extended delocalized π orbitals can span many atoms. Rings with delocalized π orbitals play important roles in biological systems. Extended delocalized π orbitals stabilize a molecule, lead to colored substances that absorb visible light, and modify the molecule's redox properties.

6. According to the **band theory of solids,** delocalized orbitals give rise to energy bands that explain the properties of metals and **metalloids (semiconductors).** Metals have partially filled orbital bands, among which electrons can move freely. Semiconductors have small **band gaps** between filled and empty energy bands. **Doped semiconductors** can be ***n*-type semiconductors** or ***p*-type semiconductors.** These contain deliberately added elements as impurities that add electrons to otherwise empty bands or remove electrons from otherwise filled bands, generating partially filled bands and leading to charge mobility.

Skills to Master

▶ Describing side-by-side orbital overlap

▶ Explaining trends in bond lengths and energies

▶ Estimating reaction energies from average bond energies

▶ Describing bonding in diatomic molecules

▶ Recognizing and sketching delocalized π orbitals

▶ Describing energy bands in metals and semiconductors

Learning Exercises

9.1 Design a detailed flow chart that shows how to describe the bonding of a second-row diatomic molecule.

9.2 Design a detailed flow chart that shows how to describe the bonding of a polyatomic species that has multiple bonds.

9.3 Prepare a table of the factors that affect bond lengths and strengths.

9.4 Prepare a list of molecules and ions mentioned in this chapter that contain electrons in delocalized orbitals. Organize your list by the number of atoms over which delocalization occurs.

9.5 Describe in your own words your understanding of each of the following terms: (a) σ bond; (b) π bond; (c) antibonding molecular orbital; (d) nonbonding electrons; (e) bond order; and (f) delocalized orbital.

9.6 Summarize the types of experimental evidence that support the idea of delocalized orbitals.

9.7 Prepare a list of all terms new to you that appear in Chapter 9. Write a one-sentence definition of each. Consult the glossary if you need help.

Problems   ilw = interactive learning ware problem. Visit the website at www.wiley.com/college/olmsted

Multiple Bonds in Carbon Compounds

9.1 Describe the bonding in the common solvent acetone, $(CH_3)_2CO$. Include sketches of all the bonding orbitals.

9.2 Describe the bonding in the solvent used in nail polish, ethyl acetate, $CH_3CO_2C_2H_5$. Include sketches of all the bonding orbitals.

ilw **9.3** Describe the steric number, geometry, and hybridization of each of the atoms of neocembrene, whose line structure appears in the Chemistry and Life Box.

9.4 Describe the steric number, geometry, and hybridization of each of the atoms of geraniol, whose line structure appears in the Chemistry and Life Box.

9.5 The following compounds all have molecular formula C_5H_8. Use steric numbers and hybridization to develop bonding pictures of them.

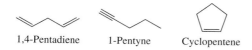

1,4-Pentadiene   1-Pentyne   Cyclopentene

9.6 The following compounds differ in the bonding between the second and third carbon atoms in the chain.

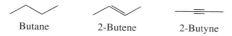

Butane   2-Butene   2-Butyne

Use steric numbers and hybridization to develop bonding pictures of these three molecules.

Bond Lengths and Energies

9.7 Arrange the following bonds in order of increasing length (shortest first). List the factors responsible for each placement: H—N, N—N, Cl—N, N≡N, and C=O.

9.8 Arrange the following bonds in order of increasing bond length (shortest first). List the factors responsible for each placement: Cl—Cl, Br—Br, O—Cl, P=O, and H—F.

9.9 Use Table 9-2 to arrange the following bonds in order of increasing bond strength (weakest first). List the single most important factor for each successive increase in strength: C=C, H—N, C=O, N≡N, and C—C.

9.10 Use Table 9-2 to arrange the following bonds in order of increasing bond strength (weakest first). List the single most important factor for each successive increase in strength: Si—Si, C—C, C—Si, H—C, and Sn—Sn.

9.11 Use average bond energies to estimate the energy change for the reaction of N_2 and O_2 to give N_2O_4.

9.12 Acrylonitrile, an important starting material for the manufacture of plastics and synthetic rubber, is made from propene:

$$2\ C_3H_6 + 2\ NH_3 + 3\ O_2 \longrightarrow 2\ H_2C{=}CH{-}C{\equiv}N + 6\ H_2O$$

Use average bond energies to estimate the energy change for the acrylonitrile synthesis.

9.13 Estimate the reaction energy for these reactions:

$$2\ H_2 + O_2 \longrightarrow 2\ H_2O \qquad 3\ H_2 + N_2 \longrightarrow 2\ NH_3$$

$$3\ H_2 + CO \longrightarrow CH_4 + H_2O$$

9.14 Estimate the energy change for the following reaction when X is (a) F; (b) Cl; (c) Br; and (d) I.

$$2\ HX(g) \longrightarrow H_2(g) + X_2(g)$$

Second-Row Diatomic Molecules

9.15 Use a molecular orbital diagram to show that the Na_2 molecule is stable.

9.16 Use a molecular orbital diagram to show that the Ne_2 molecule is unstable.

9.17 In each of the following pairs, which has the stronger bond? Use orbital configurations to justify your selections: (a) CO or CO^+; (b) N_2 or N_2^+; and (c) CN^- or CN.

9.18 In each of the following pairs, which has the stronger bond? Use orbital configurations to justify your selections: (a) O_2 or O_2^+; (b) CO or CO^-; and (c) F_2 or F_2^+.

9.19 The active ingredient in commercial laundry bleach is sodium hypochlorite (NaClO). Describe the bonding in the hypochlorite anion, ClO^-. (Hint: In this ion, chlorine's valence orbitals behave like $n = 2$ orbitals.)

9.20 The superoxide ion, O_2^-, is a reactive species that may play a role in the chemistry of aging. Describe the bonding in this ion.

9.21 Draw orbital sketches of all of the occupied bonding and antibonding orbitals of CO.

9.22 Draw orbital sketches that show the shapes of all of the occupied bonding and antibonding orbitals of O_2.

Three-Center π Orbitals

9.23 Describe the bonding of carbon disulfide, CS_2, in terms of orbital overlap and delocalized electrons. How many valence electrons occupy delocalized π orbitals in CS_2?

9.24 Describe the bonding of sulfur dioxide, SO_2, in terms of orbital overlap and delocalized electrons. How many valence electrons occupy delocalized π orbitals in SO_2?

9.25 Describe the bonding in NO_2^+ and NO_2^-. Include sketches of the σ bonding systems and include any lone pairs on the nitrogen atoms. Explain why these two ions have different geometries.

9.26 Detonators for explosive charges containing trinitrotoluene (TNT) contain mercury fulminate, $Hg(CNO)_2$, which explodes when struck. Write a Lewis structure for the CNO^- anion and describe its bonding, including any delocalized orbitals.

9.27 Hydrazoic acid has the formula HN_3, with the three nitrogen atoms in a row. Determine the Lewis structure, hybridization, and bond angles of this compound. Describe its π bonding network.

9.28 Two molecules that have been detected in interstellar space are HNCO and OCS. Describe the bonding in each molecule, identify the hybrid orbitals used to make σ bonds, and identify their delocalized π systems.

Extended π Systems

9.29 Which molecules whose line structures appear in the Chemistry and Life Box are stabilized by delocalized π electrons? For those that are, identify the atoms that contribute to the delocalized orbitals.

9.30 Which of the following molecules are stabilized by delocalized π orbitals? For those that are, identify the atoms over which delocalization extends.

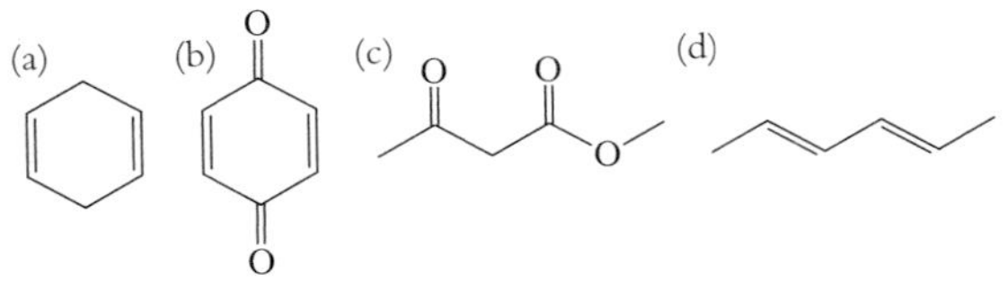

(a) (b) (c) (d)

9.31 Vitamin C has the structure on the right.
(a) Identify the hybridization of each carbon and oxygen atom. (b) How many π bonds are there? (c) How many electrons are in delocalized π orbitals? (d) Redraw the structure and circle the largest continuous plane of atoms.

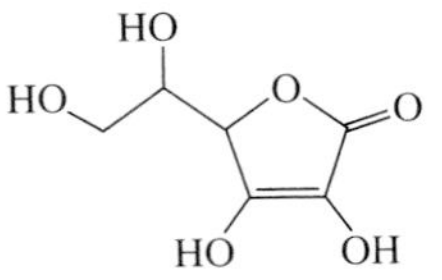

Vitamin C
($C_6H_8O_6$)

9.32 Carvone, whose line structure appears at right, is the principal flavor and fragrance ingredient in spearmint.
(a) Identify the hybridization of each carbon atom. (b) How many π bonds are there? (c) How many electrons are in delocalized π orbitals? (d) Redraw the structure and circle the largest continuous plane of atoms.

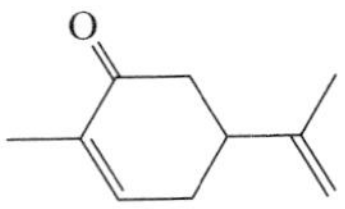

Carvone
($C_{10}H_{14}O$)

9.33 The two common oxides of carbon are CO and CO_2, which are linear. A third oxide of carbon, which is rare, is carbon suboxide, OCCCO. Draw a Lewis structure for carbon suboxide. Determine its geometry and identify all orbitals involved in delocalized π orbitals. Make separate sketches of the σ and π bonding systems.

9.34 The oxalate anion ($C_2O_4^{2-}$) has a C—C bond and two outer oxygen atoms bonded to each carbon atom. Determine the Lewis structure of this anion, including all resonance structures. What is the geometry about the carbon atoms, and how many atoms contribute p orbitals to the delocalized π system?

9.35 Determine the Lewis structure and describe the bonding of the perchlorate anion (ClO_4^-).

9.36 The permanganate ion (MnO_4^-) gives potassium permanganate solutions a rich burgundy color. Determine the Lewis structure, including all resonance structures, and describe the bonding in this ion.

Band Theory of Solids

9.37 Identify the following as p-type, n-type, or undoped semiconductors: (a) GaP; (b) InSb containing some Te; and (c) CdSe.

9.38 Identify the following as p-type, n-type, or undoped semiconductors: (a) Ge; (b) Ge doped with P; and (c) AlAs doped with Zn.

9.39 Use band theory to explain why iron is harder and melts at a higher temperature than potassium. (Hint: Consider the number of valence electrons in each element.)

9.40 The following energies measure the strength of bonding in sodium, magnesium, and aluminum metals:

$$Na(s) \longrightarrow Na(g) \qquad \Delta E = 99 \text{ kJ/mol}$$
$$Mg(s) \longrightarrow Mg(g) \qquad \Delta E = 127 \text{ kJ/mol}$$
$$Al(s) \longrightarrow Al(g) \qquad \Delta E = 291 \text{ kJ/mol}$$

Use band theory to explain these data. (Hint: Consider the number of valence electrons in each element.)

Additional Paired Problems

9.41 Shorter bonds are usually stronger bonds, but this is not always the case. Using Tables 9-1 and 9-2, find and list any X—Y bonds, for X and Y both $n = 2$ elements, that are (a) shorter but weaker and (b) longer but stronger than the corresponding X—X bond.

9.42 Using Tables 9-1 and 9-2, prepare a list of bonds to fluorine that are both longer and stronger than F—F bonds.

9.43 Use average bond energies to compare the stabilities of ethanol, C_2H_5OH, and dimethyl ether, $(CH_3)_2O$, which have the same empirical formula, C_2H_6O.

9.44 Use average bond energies (Tables 9-2) to compare the stabilities of allyl alcohol, CH_2CHCH_2OH, and acetone, $(CH_3)_2CO$, which have the same empirical formula, C_3H_6O.

9.45 What kind of semiconductor is each of the following: (a) Ge doped with Sb; (b) As doped with Si; and (c) Si doped with In?

9.46 On the basis of periodic trends, arrange the following in order of increasing band gap and classify them as insulators, semiconductors, or conductors: Ga, As, and S.

9.47 Four different bonds are found in H_2O, CO_2, and HCN. Which is the shortest? Which is the longest? What factors are responsible for each?

9.48 Four different bonds are found in NH_3, CO, and HOCl. Which is the shortest? Which is the longest? What factors are responsible for each?

9.49 Arrange the four different bonds in problem 9.47 in order of increasing bond strength. Give the reasons for your placements.

9.50 Arrange the four different bonds in problem 9.48 in order of increasing bond strength. Give the reasons for your placements.

9.51 Sulfur forms two stable oxides, SO_2 and SO_3. Describe the bonding and geometry of these compounds.

9.52 Nitrogen forms two stable oxyanions, NO_2^- and NO_3^-. Describe the bonding and geometry of these compounds.

9.53 The following partial structure is that of azodicarbonamide: When heated, azodicarbonamide breaks apart into gaseous carbon monoxide, nitrogen, and ammonia. Azodicarbonamide is used as a foaming agent in the polymer industry.

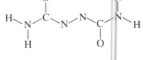

(a) Add nonbonding electron pairs and multiple bonds as required to complete the Lewis structure of this molecule. (b) Determine the geometry and hybridization of each inner atom. (c) Is there a delocalized π system? If so, what atoms are involved and how many electrons participate?

9.54 In the lower atmosphere, NO_2 participates in a series of reactions in air that is also contaminated with unburned hydrocarbons. One product of these reactions is peroxyacetyl nitrate (PAN). Here is the skeletal arrangement of PAN: (a) Complete the Lewis structure of this compound. (b) Determine the hybridization for the atoms marked with asterisks. (c) Give the approximate values of the bond angles indicated with arrows. (d) If delocalized π orbitals exist, describe them.

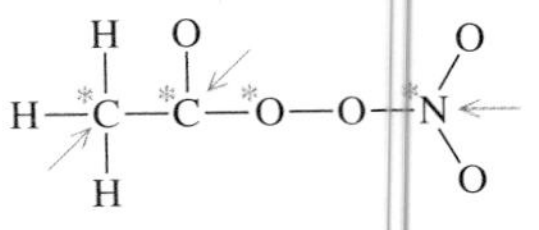

9.55 Use average bond energies (Table 9-2) to compare the combustion energies of ethane, ethylene, and acetylene. Calculate which of these hydrocarbons releases the most energy per gram.

9.56 Isooctane, C_8H_{18}, is the basis of the octane rating system for gasoline because it burns smoothly, with minimal engine knocking. Pure isooctane is assigned a value of 100. Octane ratings are assigned to gasoline mixtures based on their ability to prevent knocking relative to isooctane. Estimate the energy released during the combustion of 1 mol of pure isooctane.

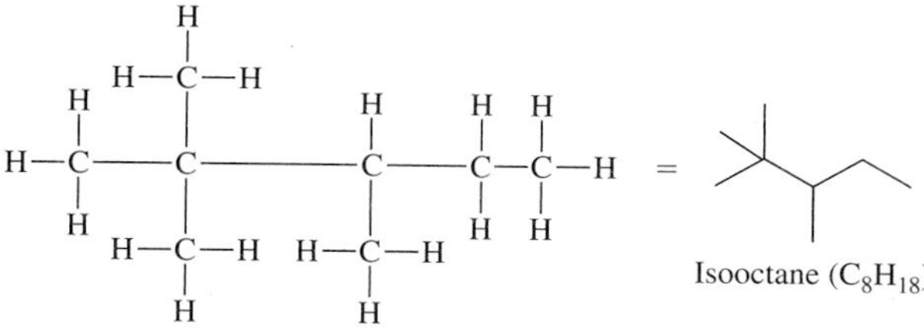

Isooctane (C_8H_{18})

9.57 Draw a band gap diagram of silicon doped with antimony.
9.58 Draw a band gap diagram of silicon doped with indium.
9.59 Use tabulated bond energies to estimate the energy change when HCl adds to ethylene (C_2H_4) to produce CH_3CH_2Cl and when Cl_2 adds to ethylene to produce $ClCH_2CH_2Cl$.
9.60 Phosgene (Cl_2CO) is a highly toxic gas that was used for chemical warfare during World War I. Use the bond energies in Table 9-2 to estimate the energy change that occurs when carbon monoxide and chlorine combine to make phosgene.

$$CO(g) + Cl_2(g) \longrightarrow Cl_2CO(g)$$

9.61 Determine the type of orbitals (atomic, sp^3, or sp^2) used by each atom in the molecules shown at the right:
9.62 Determine the type of orbitals (atomic, sp^3, or sp^2) used by each atom in the following molecules:

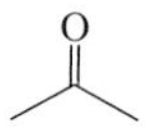

H_3CNH_2

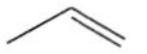

N_2H_4

9.63 Acrolein has the line structure at right. Determine its Lewis structure and describe its bonding, including geometry, hybridization, and delocalized π orbitals. 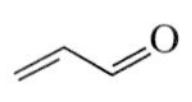
9.64 1,3-Butadiene has the line structure at right. Determine its Lewis structure and describe its bonding, including geometry, hybridization, and delocalized π orbitals. 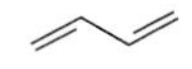
9.65 Saxitoxin is a fatally toxic natural product produced by certain marine dinoflagellates. The molecule occasionally accumulates in clams and mussels during conditions of red tide. (a) How many π bonds does saxitoxin have? (b) What orbitals are used for bonding by atoms A, B, C, and D? (c) What are the bond angles about atoms D, E, and F? (d) Redraw the structure of saxitoxin adding the lone pairs.

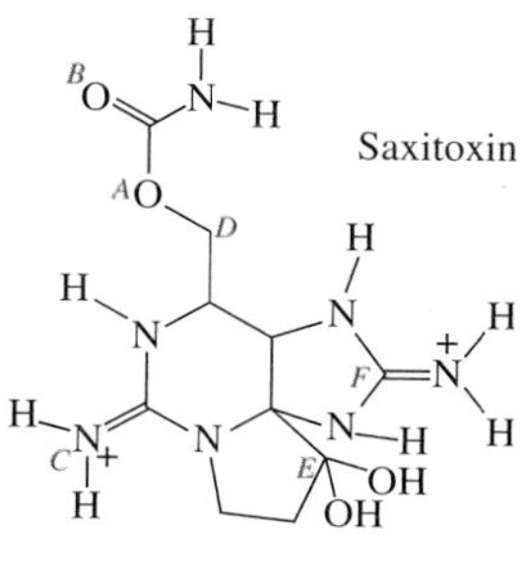

Saxitoxin

9.66 Capsaicin is the molecule responsible for the hot spiciness of chili peppers.

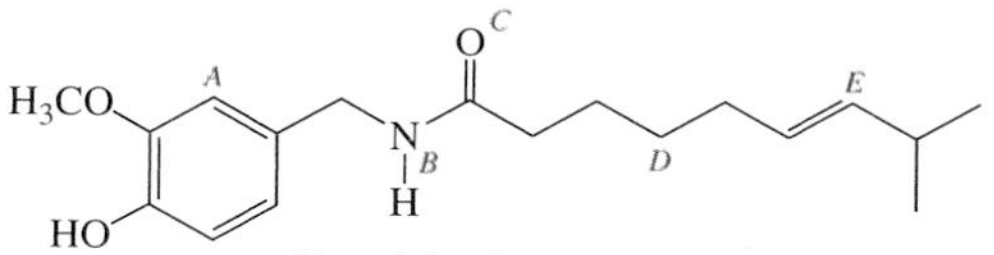

H_3CO
HO
Capsaicin ($C_{18}H_{27}O_3N$)

(a) How many π bonds does capsaicin have? (b) What orbitals are used for bonding by each of the labeled atoms? (c) What are the bond angles about each of the labeled atoms? (d) Redraw the structure of capsaicin adding the lone pairs.

More Challenging Problems

9.67 Describe the bonding in ketene (H_2CCO). Make separate sketches of the σ and π bonding systems.
9.68 Carbon monoxide has the strongest known chemical bond, 1070 kJ/mol. Explain why this bond is stronger than any other.
9.69 Consider the bond lengths of the following molecules: N_2, 110 pm; O_2, 121 pm; and F_2, 143 pm. Explain the variation in length in terms of the orbital descriptions of these molecules.
9.70 Nitrogen molecules can absorb photons to generate excited-state molecules. Construct an energy level diagram and place the valence electrons so that it describes the most stable *excited* state of an N_2 molecule. Is the N—N bond in this excited-state N_2 molecule stronger or weaker than the N—N bond in ground-state nitrogen? Explain your answer.
9.71 Combine the features of Figures 9-33 and 9-35 in a single drawing that shows how electricity is conducted in silicon and germanium.
9.72 The bond strength of diatomic Cl—F is 250 kJ/mol, and the bond strength of Br—Cl is 215 kJ/mol. List these, along with F—F, Cl—Cl, and Br—Br bonds, in order of increasing bond strength. State the factors that account for the ordering of bonds in this list.
9.73 Use average bond energies (see Table 9-2) to estimate the net energy change per mole of silicon for the conversion of a silicon chain into an Si—O—Si chain. Repeat this calculation to estimate the net energy change per mole of carbon for the conversion of a carbon chain into a C—O—C chain.

9.74 Write the electron configuration and determine the bond order of NO, NO^+, and NO^-. Which display magnetism?
9.75 Oxygen forms three different ionic compounds with potassium: potassium oxide (K_2O), potassium superoxide (KO_2), and potassium peroxide (K_2O_2). Write electron configurations for the superoxide and peroxide anions. Compare the configurations with the configuration of the oxygen molecule. Rank the three species in order of increasing bond order, bond energy, and bond length. Which of the three are magnetic? Which has the largest magnetism?
9.76 In the stratosphere, chlorofluorocarbons (Freons) absorb high-energy ultraviolet light, breaking C—Cl bonds. The chlorine atom then enters a reaction cycle that converts ozone to molecular oxygen. What is the minimum frequency of the photons that can break a C—Cl bond? (Refer to Table 9-2 and Chapter 6 for help.)
9.77 Explain how a C≡C triple bond can be stronger but more reactive than a C=C bond.
9.78 The ionization energy of molecular oxygen is smaller than that of atomic oxygen (1314 kJ/mol vs. 1503 kJ/mol). In contrast, the ionization energy for molecular nitrogen is larger than that of atomic nitrogen (1503 kJ/mol vs. 1402 kJ/mol). Explain these data, using the electron configurations of the diatomic molecules.

9.79 Chlorine forms several oxyanions, including ClO_3^- and ClO_2^-. Determine Lewis structures and describe the bonding for each of these anions, including delocalized π bonds.

9.80 When an oxalate anion, $C_2O_4^{2-}$, adds two protons to form oxalic acid, two C—O bonds get longer and two get shorter than the bonds in oxalate anions. Which bonds get longer and which shorter? Use bonding principles to explain these changes.

9.81 What is the relationship among bond length, bond energy, and bond order? Which of them can be measured?

9.82 Buckminsterfullerene contains 60 carbon atoms, each of which uses sp^2 hybrid orbitals and contributes one $2p$ orbital to make delocalized molecular orbitals. How many π orbitals does Buckminsterfullerene have and how many π bonding electrons does this molecule have?

9.83 The bond strengths in the H—X molecules (X = F, Cl, Br, I) show that bonds become weaker as orbitals become more diffuse. Make a series of orbital overlap sketches for the H—X molecules to illustrate this. Shade your figures in a way that illustrates the changes in electron density.

9.84 The concept of threshold energy appears in Chapter 6 as part of the discussion of the photoelectric effect. What feature of band theory would you associate with the threshold energy? Consider electrons ejected by photons with a frequency significantly higher than the threshold frequency. Would these electrons all have the same kinetic energy? Explain.

9.85 Elements in the same group of the periodic table usually have similar chemical properties. Nitrogen and oxygen are unique among elements in their groups, however, because they are stable as diatomic molecules. Phosphorus and sulfur, the next two members of these groups, are stable as long-chain molecules. Use the properties of orbitals to account for these differences in chemical behavior.

9.86 Use orbital sketches to illustrate the bonding of allene (H_2CCCH_2). Identify all orbitals in the bonding scheme. Make sure your sketches show the three-dimensional arrangement of the hydrogen atoms. Make separate sketches of the σ and π bonding systems.

9.87 For each of the three trends in bond energies, find an example in Table 9-2 involving nitrogen that illustrates the trend. (Do not use N—O, which is used in the text.)

9.88 Silicon carbide (carborundum) is a hard crystalline material similar to diamond. It is widely used for cutting tools. Describe the bonding in silicon carbide, whose empirical formula is SiC. Draw an energy band picture showing the band gap for SiC compared with those of diamond and silicon.

9.89 Acrylonitrile is used to manufacture polymers for synthetic fibers. Describe the bonding of acrylonitrile. (a) Identify the hybrid and/or atomic orbitals involved in bonding for each atom. (b) Draw sketches that show the σ and π bonding systems. (c) If there are delocalized π orbitals, identify and describe them.

9.90 Use orbital energy diagrams to decide whether NF would be stabilized or destabilized by adding one electron to make the corresponding anion and by removing an electron to form the corresponding cation. Sketch the orbital involved in these changes.

9.91 The molecule CN is not stable in our atmosphere, but it has been detected in the interstellar regions. Describe its bonding, and sketch its least stable occupied molecular orbital.

9.92 Explain how the following bond lengths (in picometers) support the existence of delocalized π orbitals in the carbonate ion: C—O (avg.), 143; C=O (avg.), 122; and CO_3^{2-}, 129.

9.93 Chlorine forms one neutral oxide, ClO_2. Describe the bonding in this unusual compound. Explain why it is considered unusual.

Group Study Problems

9.94 Imagine making an O_4 molecule by attaching an oxygen atom to the nonbonding pair of electrons on the central oxygen of ozone (O_3). Describe the bonding in O_4, which is isoelectronic with NO_3^- and CO_3^{2-}. Compare the expected stability of the O_4 molecule with that of two O_2 molecules. Does the simple picture of bonding explain why O_4 does not exist?

9.95 The pentadienyl cation ($C_5H_7^+$) is a high-energy species that can be generated in the laboratory under carefully controlled conditions. (a) Describe the bonding orbitals used by each of the carbon atoms. (b) How many p orbitals contribute to the π bonding system? How many molecular orbitals do they form? (c) How many delocalized π electrons are there in $C_5H_7^+$? (d) Would you expect this ion to have a more or less stable π bonding system than 1,3-pentadiene or 1,4-pentadiene (give reasons)?

9.96 One of the first compounds discovered for the treatment of AIDS was azidothymidine (AZT). Here is the line structure of AZT: (a) Identify the type of orbitals used for σ bonding by each of the nitrogen atoms. (b) What is the bond angle around the oxygen atom in the five-

membered ring? (c) How many carbon atoms can be described as using sp^3 hybrid orbitals? (d) How many π bonds are there? What orbitals are used to form these bonds?

9.97 Metal carbonyls are compounds in which molecules of carbon monoxide form covalent bonds to transition metals. Examples are $Cr(CO)_6$, $Fe(CO)_5$, and $Ni(CO)_4$. The principal bonding interaction in a metal carbonyl is a σ bond formed between the carbon atom of CO and one of the valence d orbitals of the metal. (a) What is the hybridization of the C atoms in metal carbonyls? (b) Use the d_{z^2} orbital to make an orbital overlap sketch of the M—CO σ bond. (c) Draw structures showing the geometries of these carbonyls. (d) Metal carbonyls also have π bonds that result from overlap of the π^* orbital of CO with one of the metal d orbitals. Use the d_{xz} orbital to make an orbital overlap sketch of the π bond in a metal carbonyl.

9.98 Use the principles of orbital overlap to explain why two different isomers of 1,2-dichloroethylene exist but 1,2-dichloroethane exists in only one form.

trans-1,2-Dichloro-ethylene   *cis*-1,2-Dichloro-ethylene   1,2-Dichloro-ethane

Use bond strength arguments to estimate the amount of energy required to interconvert the two isomers.

Answers to Section Exercises

9.1.1

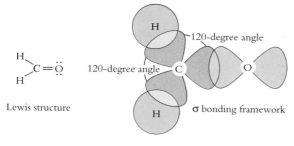

Lewis structure

120-degree angle

σ bonding framework

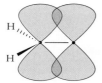

π bonding system

The carbon atom uses sp^2 hybrids to form three σ bonds. Its unused p orbital overlaps with an oxygen p orbital to form a π bond.

9.1.2

Propene (C_3H_6)   Propyne (C_3H_4)

The CH_3 carbon in each compound uses sp^3 hybrids and has tetrahedral geometry. In propene, the other two carbon atoms have trigonal planar geometry, use sp^2 hybrids for three σ bonds, and have a π bond between them. In propyne, the other two carbon atoms have linear geometry, use sp hybrids for two σ bonds, and have two π bonds between them.

9.1.3 Four carbon atoms participate in π bonding. For appropriate orbital sketches, see Figures 9-2 and 9-3.

9.2.1 C—H ($1s$) < C=O (double, polar) < C=C (double) < C—C (single) < C—Cl ($3p$)

9.2.2 -630 kJ

9.2.3 (a)

$$H-\underset{\underset{H}{|}}{\overset{\overset{H}{|}}{C}}-H \quad H-\ddot{\underset{}{C}}\ddot{l}: \quad H-\underset{\underset{H}{|}}{\overset{\overset{H}{|}}{C}}-\ddot{\underset{}{C}}\ddot{l}: \quad H-H \quad :\ddot{\underset{}{C}}\ddot{l}-\ddot{\underset{}{C}}\ddot{l}:$$

CH_4, four C—H bonds; CH_3Cl, three C—H bonds, one C—Cl bond; others have one bond each. (b) First reaction releases 80 kJ; second reaction absorbs 105 kJ. The first reaction is used in industry because it does not require an energy input.

9.3.1 CN^- has 10 valence electrons, isoelectronic with N_2. Its bonding is identical to that of N_2.

9.3.2 See Figures 9-12 and 9-13.

9.3.3 The electron that is removed to ionize NO comes from a π^* orbital, which is less stable than the π orbitals from which electrons must be removed to ionize N_2 or CO.

9.4.1 With 16 valence electrons, N_3^- is isoelectronic with CO_2. Its bonding is identical with that of CO_2. It is linear, with the inner atom having a steric number of 2.

9.4.2 In an isolated π orbital, electron density is spread between a pair of atoms. In a delocalized π orbital, electron density is spread over more than two atoms.

9.4.3 The formate anion is just like acetate, except it has a hydrogen atom in place of the methyl (CH_3) group of acetate. In the bonding framework, replace the sp^3 hybrid–sp^2 hybrid σ bond with a $1s$–sp^2 hybrid σ bond. The π bonding system is identical to that of acetate.

9.5.1

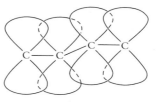

Side view   Top view

9.5.2 Azulene has 10 delocalized π orbitals; 5 of the orbitals are occupied.

9.5.3 The Lewis structure and bonding are the same as in the sulfate anion. The ion is tetrahedral. The Cr atom uses sp^3 hybrids to form four σ bonds and uses d orbitals to form delocalized π bonds with the O atoms.

9.6.1 250 kJ/mol

9.6.2 This is a p-type semiconductor because Zn has one less valence electron than Ga.

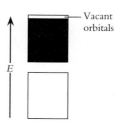

Vacant orbitals

E

9.6.3 Si would be the semiconductor and Al would be the dopant. Aluminum has one less valence electron than silicon.

10

Effects of Intermolecular Forces

Na$^+$

INTRODUCTION: THE CHEMISTRY OF SHAMPOO

Personal care is big business: Americans spend close to $30 billion annually on shampoos and hair conditioners, skin care, toothpaste, and other products to make us look, feel, and smell better. Of this total, 40% goes for shampoos, conditioners, and styling gels.

The ingredients in shampoo can appear like hieroglyphics, even to a chemist. Here are some examples: sodium lauryl sulfate (shown in our molecular inset), glycol distearate, panthenol, methyl paraben, and dimethicone. It's enough to make your scalp itch, but despite their exotic-sounding names, the active ingredients in a shampoo play three fundamental roles. Some ingredients wash away the substances that make hair "dirty"; others adhere to hair to impart a desirable feel and texture, and still others are emulsifiers that keep the mixture from separating into its components.

To accomplish these roles, the components of a hair product must have the right combinations of properties. Some need to have strong attractions to water (hydrophilic), while others must have weak attractions to water (hydrophobic).

Water can wash away hydrophilic substances but cannot remove hydrophobic dirt and grease. Shampoos contain molecules that are simultaneously hydrophilic and hydrophobic, such as our inset molecule, sodium lauryl sulfate. This molecule contains a hydrophilic sulfate group attached to a hydrophobic hydrocarbon chain. The chain is attracted to grease and dirt, whereas the anionic portion is attracted to water. Molecules like these associate with hydrophobic dirt particles to form hydrophilic clumps that water can wash away.

Washing removes the natural coating of protective oil that gives hair its body and shine. To counter this, shampoos and conditioners contain hydrophobic oils. These oils cling to the surface of hair and remain in place upon rinsing. Dimethicone is an artificial oil that contains silicon-oxygen chains and methyl groups.

Hydrophobic and hydrophilic substances do not mix well with one another. Oil and vinegar are a common example. To keep their various ingredients mixed together uniformly, shampoos contain emulsifiers, which combine hydrophobic and hydrophilic components. Panthenol is one such emulsifier; it has hydrophobic regions interspersed with hydrophilic polar groups.

No single substance can provide all the properties desired in a shampoo. Besides, the ingredients must be nontoxic and biodegradable. As a result, chemists who formulate hair-care products face challenging chemical design problems. That is why a typical list of ingredients contains a multitude of exotic-sounding chemicals.

This chapter begins with descriptions of intermolecular forces. Then we show how these forces, even though considerably weaker than the chemical bonds described in Chapters 8 and 9, are important in determining the behavior of solids, liquids, and solutions. Our focus in the final sections is on the effects of intermolecular interactions on aqueous solutions.

Dimethicone
($n = 200–300$)

Panthenol
$C_9H_{19}NO_4$

Hydrophilic

Hydrophobic

10.1 THE NATURE OF INTERMOLECULAR FORCES

We begin exploring intermolecular forces by considering the properties of the elements. At room temperature and pressure, all but 13 of the elements are solids (Figure 10-1). Two others, mercury and bromine, are liquids, leaving 11 elements that are gases. Only for these 11 gaseous elements are intermolecular forces small enough to neglect at room temperature.

1 2 3 4 5 6 7 8 9 10 11 12 13 14 15 16 17 18

Figure 10-1
At room temperature and pressure, He, Ne, Ar, Kr, Xe, and Rn are monatomic gases; H$_2$, N$_2$, O$_2$, F$_2$, Cl$_2$ are diatomic gases; and Hg and Br$_2$ are liquids. All other elements are solids.

Figure 10-2
Under normal conditions, chlorine is a pale yellow-green gas, bromine is a dark red liquid, and iodine is a purple crystalline solid.

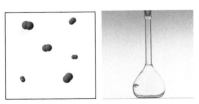
Gaseous Cl$_2$

Liquid Br$_2$

Solid I$_2$

The Halogens

The halogens, the elements from Group 17 of the periodic table, provide an introduction to intermolecular forces. The halogens are most stable as diatomic molecules: F$_2$, Cl$_2$, Br$_2$, and I$_2$. The bonding patterns of the four halogens are identical. Each molecule contains two atoms held together by a single covalent bond that can be described by end-on overlap of valence p orbitals.

Although their bonding patterns are the same, bromine and iodine differ from chlorine and fluorine in their macroscopic physical appearance and in their molecular behavior, as Figure 10-2 shows. At room temperature and pressure, F$_2$ and Cl$_2$ are gases, Br$_2$ is a liquid, and I$_2$ is a solid.

Molecules of F$_2$ or Cl$_2$ move freely throughout their gaseous volume, traveling many molecular diameters before colliding with one another or with the walls of their container. Because much of the volume of a gas is empty space, samples of gaseous F$_2$ and Cl$_2$ readily expand or contract in response to changes in pressure. This freedom of motion exists because the intermolecular forces between these molecules are small.

Molecules of liquid bromine also move about relatively freely, but there is not much empty space between molecules. Because its molecules are already in close contact with one another, a liquid cannot be compressed significantly by increasing the pressure. Neither does a liquid expand significantly if the pressure is reduced, because the intermolecular forces in a liquid are large enough to prevent the molecules from breaking away from one another.

Solid iodine has even less empty space between molecules than does liquid bromine. Furthermore, molecules in a solid do not move freely past one another. A sample of solid iodine contains highly regular crystals in which I$_2$ molecules are arranged in ordered arrays. Each molecule vibrates back and forth about a single lowest-energy position, but it cannot slide easily past its neighbors. Like liquids, solids have sufficiently strong intermolecular forces that they do not expand or contract significantly when pressure changes.

Intermolecular attractive forces pull molecules toward one another, while kinetic energy of motion makes molecules move away from one another. Whether a substance is a gas, liquid, or solid depends on the balance between the energy of motion of its molecules and the stabilization energy generated by its intermolecular forces. When the average kinetic energy of motion exceeds the energy of intermolecular attraction, molecules remain separated from one another and the substance is a gas. When the average kinetic energy of motion is smaller than the energy of intermolecular attraction, molecules remain close to one another and the substance is a liquid or solid.

The bar graph in Figure 10-3 shows that the intermolecular attractive forces are greater for Br$_2$ than for F$_2$. At room temperature, the molecular kinetic energies of fluorine molecules exceed the attractive forces of F$_2$—F$_2$ interactions. In contrast, bromine molecules have enough kinetic energy to move freely about, but their energy of motion is insufficient to overcome the intermolecular forces that hold them together in the liquid phase. At room temperature, iodine molecules are locked in position in the solid state because the attractive forces between I$_2$ molecules is even larger than that between Br$_2$ molecules.

Real Gases

As we describe in Chapter 5, the ideal gas model is based on two assumptions: A gas has negligible forces between its molecules, and gas molecules have negligible volumes. The behavior of bromine shows that neither of these assumptions is completely correct for a real gas. Although bromine is a liquid at room temperature, the red color above the liquid shown in Figure 10-4 suggests that the gas above a sample of liquid Br_2 contains a substantial number of gaseous Br_2 molecules. Under these conditions, intermolecular forces are strong enough to condense most, but not all of the molecules to the liquid phase. Moreover, exerting pressure on liquid bromine hardly changes its volume. The molecules in liquid bromine are packed close together, and intermolecular repulsive forces strongly resist further reduction of the volume. The volume of a sample of liquid bromine is determined by the volume of its molecules. The molecules in the colored gas above the liquid in Figure 10-4 have the same intermolecular forces and molecular volumes as do the molecules in the liquid; in other words, both these properties have finite values and the gas is not ideal.

How close do real gases come to ideal behavior? To answer this question, we rearrange the ideal gas equation to examine the ratio PV/nRT. Figure 10-5 shows how PV/nRT varies with pressure for chlorine gas at room temperature. If chlorine were ideal, the PV/nRT ratio would always be 1, as shown by the red line on the graph. Instead, chlorine shows deviations from 1 as the pressure increases.

Notice in the inset of Figure 10-5 that chlorine is nearly ideal at pressures around one atmosphere. In fact, PV/nRT deviates from 1.0 by less than 4% at pressures below 4 atm. As the pressure increases, however, the deviations become increasingly significant. At first, the PV/nRT ratio for chlorine drops below 1. This is because the chlorine molecules are close enough together for attractive forces to play a significant role. These intermolecular attractions hold molecules together and reduce the forces exerted when the molecules strike the walls of the container. As a result, the pressure of the real gas is slightly lower than the ideal value.

Figure 10-5 also shows that at pressures greater than 375 atm, PV/nRT becomes *larger* than 1. This is the effect of molecular size. At high enough pressure, the molecules are packed so close together that the total volume of the molecules is no longer negligible compared to the overall volume of the gas. The added volume of the molecule means that the volume of the real gas is slightly greater than the volume of the ideal gas.

Every gas shows deviations from ideal behavior at high pressure. Figure 10-6 shows PV/nRT for He, F_2, CH_4, and N_2, all of which are gases at room temperature. Notice that PV/nRT for helium increases steadily as pressure increases. Interatomic forces for helium are too small to reduce the ratio below 1, but the finite size of the helium atom generates deviations from ideality that become significant at pressures above 100 atm.

Given that every gas deviates from ideal behavior, can we use the ideal gas model to discuss the properties of real gases? The answer is "yes," as long as conditions do not become too extreme. The gases with which chemists usually work, such as He, Cl_2, and N_2, are nearly ideal at room temperature at pressures below about 10 atm.

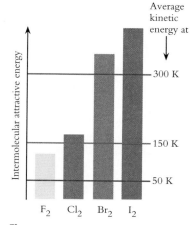

Figure 10-3
The intermolecular forces acting on the four halogens are different.

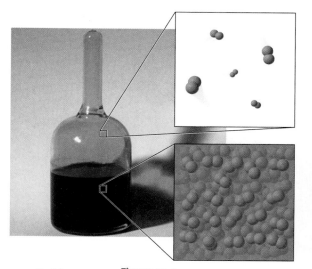

Figure 10-4
Even though bromine is a liquid at room temperature and pressure, enough molecules escape into the gas phase to give the gas above a liquid sample a distinct red color.

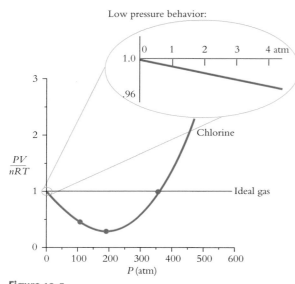

Figure 10-5
Variation in PV/nRT with pressure for chlorine gas at room temperature. The inset shows the low-pressure region on an expanded scale.

The van der Waals Equation

Graphs like those in Figures 10-5 and 10-6 show how real gases deviate from ideal behavior. It would be useful to have an equation that describes the relationship between pressure and volume for a real gas, just as $PV = nRT$ describes an ideal gas. One way to approach real gas behavior is to modify the ideal gas equation to take into account attractive forces and molecular volumes. The result is the **van der Waals equation,** named for the scientist who first proposed it in 1873, Johannes van der Waals:

$$\left(P + \frac{n^2 a}{V^2}\right)(V - nb) = nRT \qquad \textbf{(10-1)}$$

The van der Waals equation adds two correction terms to the ideal gas equation. Each correction term includes a constant that has a specific value for each different substance. The first correction term, $n^2 a / V^2$, adjusts for attractive intermolecular forces. The van der Waals constant a measures the strength of intermolecular forces for the gas; the stronger the forces, the larger the value of a. The second correction term, nb, adjusts for molecular sizes. The van der Waals constant b measures the size of molecules of the gas; the larger the molecules, the larger the value of b.

The van der Waals constants for a number of gases appear in Table 10-1, and the magnitude of van der Waals corrections is explored in Example 10-1.

Figure 10-6
Variation in PV/nRT for He, F_2, CH_4, and N_2 at 300 K.

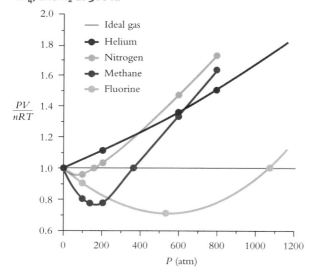

Table 10-1
van der Waals Constants

Substance	a (L²atm/mol²)	b (L/mol)
He	0.0341	0.0237
Ne	0.211	0.0171
Ar	1.345	0.0322
H_2	0.244	0.0266
N_2	1.390	0.0391
O_2	1.360	0.0318
CO	1.453	0.0395
F_2	1.156	0.0290
Cl_2	6.493	0.0562
CO_2	3.592	0.0427
H_2O	5.464	0.0305
NH_3	4.170	0.0371
CH_4	2.253	0.0428
C_2H_6	5.489	0.0638
C_6H_6	18.00	0.0115

Magnitudes of van der Waals Corrections

Example 10-1

Gases such as methane are sold and shipped in compressed gas cylinders. A typical cylinder has a volume of 15.0 L and, when full, contains 62.0 mol of CH_4. After prolonged use, 0.620 mol of CH_4 remains in the cylinder. Use the van der Waals equation to calculate the pressures in the cylinder when full and after use, and compare the values to those obtained from the ideal gas equation. Assume a temperature of 27 °C.

Strategy: The van der Waals equation is Equation 10-1, and van der Waals constants must be looked up in tables. To calculate the pressures we rearrange the van der Waals equation.

$$P = \left(\frac{nRT}{V - nb} \right) - \left(\frac{n^2 a}{V^2} \right)$$

Solution: The van der Waals constants for CH_4 appear in Table 10-1. For CH_4, $a = 2.253 \ L^2 atm/mol^2$ and $b = 0.0428 \ L/mol$; $V = 15.0$ L and $T = 300$ K.

When the tank is full, $n = 62.0$ mol:

$$P = \left(\frac{(62.0 \ \text{mol})(0.08206 \ \text{L atm/mol K})(300 \ \text{K})}{[15.0 \ \text{L} - (62.0 \ \text{mol})(0.0428 \ \text{L/mol})]} \right)$$

$$- \left(\frac{(62.0 \ \text{mol})^2 (2.253 \ L^2 \ \text{atm/mol})^2}{(15.0 \ \text{L})^2} \right)$$

$$P = \left(\frac{1526 \ \text{L atm}}{(15.0 \ \text{L}) - (2.654 \ \text{L})} \right) - (38.49 \ \text{atm}) = 85.1 \ \text{atm}$$

$$P_{\text{ideal}} = \frac{nRT}{V} = \frac{(62.0 \ \text{mol})(0.08206 \ \text{L atm/mol K})(300 \ \text{K})}{(15.0 \ \text{L})} = 102 \ \text{atm}$$

After use, $n = 0.620$ mol:

$$P = \left(\frac{(0.620 \ \text{mol})(0.08206 \ \text{L atm/mol K})(300 \ \text{K})}{[15.0 \ \text{L} - (0.620 \ \text{mol})(0.0428 \ \text{L/mol})]} \right)$$

$$- \left(\frac{(0.620 \ \text{mol})^2 (2.253 \ L^2 \ \text{atm/mol})^2}{(15.0 \ \text{L})^2} \right)$$

$$P = \left(\frac{15.26 \ \text{L atm}}{(15.0 \ \text{L}) - (0.0265 \ \text{L})} \right) - (0.00385 \ \text{atm}) = 1.02 \ \text{atm}$$

$$P_{\text{ideal}} = \frac{nRT}{V} = \frac{(0.620 \ \text{mol})(0.08206 \ \text{L atm/mol K})(300 \ \text{K})}{(15.0 \ \text{L})} = 1.02 \ \text{atm}$$

Notice that the van der Waals correction is appreciable (16.6 %) at high pressure but is negligible for a pressure near 1 atm.

Vaporization and Condensation

Because the energy of motion depends on temperature, changing the temperature of a substance changes the balance between intermolecular forces and energy of motion. At some point, temperature changes convert a substance from one phase to another. For example, bromine is a liquid below 59 °C at atmospheric pressure.

Figure 10-7
Molecular views of vaporization, condensation, and dynamic equilibrium within a closed container.

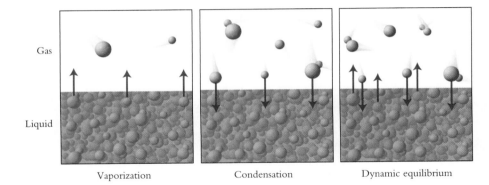

Gas

Liquid

Vaporization   Condensation   Dynamic equilibrium

Table 10-2
Melting and Boiling Points

Substance	mp (K)	bp (K)
He	0.95*	4.2
H_2	14.0	20.3
N_2	63.3	77.4
F_2	53.5	85.0
Ar	83.8	87.3
O_2	54.8	90.2
Cl_2	172	239
Br_2	266	332
I_2	387	458
P_4	317	553
Na	371	1156
Mg	922	1363
Si	1683	2628
Fe	1808	3023

* Under high pressure.
He remains liquid at 0 K under normal pressure.

CHAPTER 5 →
We describe the link between kinetic energy and temperature.

Above this temperature, bromine exists as a gas. Similarly, gaseous chlorine condenses to a liquid when it is cooled below $-34\ °C$ at atmospheric pressure. Liquid chlorine becomes a solid at $-101\ °C$. Fluorine liquefies at $-188\ °C$ and solidifies at $-220\ °C$.

Vaporization is the conversion of a liquid into a gas, and condensation is the reverse process. A liquid vaporizes when molecules leave the liquid phase faster than they are captured from the gas. A gas condenses when molecules leave the gas phase more rapidly than they escape from the liquid. As Figure 10-7 shows, these two processes are in dynamic equilibrium at the boiling point, at which the rate of escape from the liquid is matched exactly by the rate of capture from the gas.

Similarly, melting is the conversion of a solid into a liquid, and freezing is the conversion of a liquid into a solid. A solid melts when molecules leave the solid phase more rapidly than they are captured from the liquid; a liquid freezes when molecules are captured from the liquid phase faster than they escape from the solid. The melting point represents a dynamic equilibrium between the two opposing processes of escape and capture.

Boiling points and melting points provide good qualitative indications of the strengths of intermolecular forces. This is because the rates of escape and capture depend on the balance between molecular kinetic energies and intermolecular forces of attraction. A substance with large intermolecular forces must be raised to a high temperature before its molecules have sufficient kinetic energies to overcome those forces. A substance with small intermolecular forces must be cooled to a low temperature before its molecules have small enough kinetic energies to coalesce into a condensed phase. Table 10-2 lists the boiling and melting points of some representative elemental substances.

Section Exercises

10.1.1 On the basis of the behavior of the other elements of Group 17, predict whether At_2 will be a gas, liquid, or solid at room temperature. Redraw the bar graph of Figure 10-3 to include At_2.

10.1.2 From Figure 10-6, determine which of the four gases has the largest intermolecular forces and which has the smallest. State your reasoning.

> **10.1.3** Use the data in Table 10-2 to answer the following:
> (a) Predict whether Xe boils at a higher or lower temperature than Ar.
> (b) List F_2, N_2, H_2, and Cl_2 in order of increasing intermolecular forces.
> (c) Which group in the periodic table has the smallest intermolecular forces?

10.2 TYPES OF INTERMOLECULAR FORCES

There are three general types of intermolecular forces. **Dispersion forces** describe the attraction between the negatively charged electron cloud of one molecule and the positively charged nuclei of neighbor molecules. All substances have dispersion forces. **Dipolar forces** describe the attraction between the negatively charged end of a polar molecule and the positively charged ends of neighboring polar molecules. Dipolar forces exist only for compounds that possess permanent dipole moments (see Section 8.7). **Hydrogen bonding** involves lone pairs of electrons on an electronegative atom of one molecule and a polar bond to hydrogen in another molecule. Hydrogen bonds are confined to molecules that contain O, N, and F atoms.

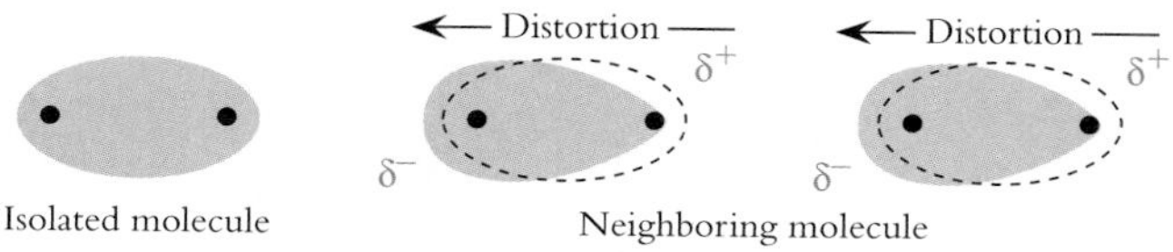

Isolated molecule   Neighboring molecule

Figure 10-8
Exaggerated view of how dispersion forces arise.

Dispersion Forces

We can explore the nature of dispersion forces by examining how the boiling points of the halogens vary. Fluorine boils at 85 K, chlorine at 239 K, bromine at 332 K, and iodine at 458 K. The data indicate that as the sizes of halogen molecules increase, the magnitudes of their intermolecular forces become larger. To understand this trend, consider what happens when two halogen molecules approach each other. Each molecule contains positive nuclei surrounded by a cloud of negative electrons. As two molecules approach, the nuclei of one molecule attract the electron cloud of the other. Electrons are highly mobile, so their orbitals change shape in response to this attraction. At the same time, the two electron clouds repel each other, which leads to further distortion to minimize electron-electron repulsion. As Figure 10-8 indicates, this distortion of the electron cloud creates a charge imbalance, giving the molecule a slight positive charge at one end and a slight negative charge at the other. Dispersion forces arise from the net attractive forces among molecules generated by all these induced charge imbalances.

The magnitude of dispersion forces depends on how easy it is to distort the electron cloud of a molecule. This ease of distortion is called the **polarizability** because distortion of an electron cloud generates a temporary polarity within the molecule. As Figure 10-9 illustrates, the large electron cloud of I_2 distorts more readily than the small electron cloud of F_2. Both molecules contain 14 valence electrons, but those of F_2 are in relatively compact $n = 2$ orbitals, whereas those of I_2 occupy highly diffuse $n = 5$ orbitals. The larger valence orbitals of I_2 distort much more readily than do those of F_2, generating larger dispersion forces. Consequently, iodine is a solid at room temperature while fluorine is a gas.

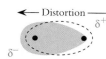

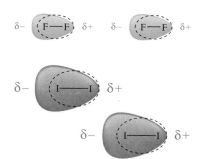

Figure 10-9
As a result of its larger _n_ value, iodine's electron cloud is much larger and more polarizable than fluorine's.

The boiling point trend for the halogens reflects this behavior: The boiling point increases progressively as the molecules become larger and more polarizable. This reasoning is extended to include the elemental noble gases in Example 10-2.

Example 10-2 | **Boiling Point Trends**

Neon and xenon are gases at room temperature, but both become liquids if the temperature is low enough. Draw a molecular picture showing the relative sizes and polarizabilities of atoms of neon and xenon, and use the picture to determine which substance has a lower boiling point.

Strategy: The boiling point of a substance depends on the magnitude of its intermolecular forces, which in turn depends on the polarizability of its electron cloud. Monatomic gases contain atoms rather than molecules, so we must assess interatomic forces for these substances.

Solution: The dispersion forces that act between atoms of the noble gases depend on the polarizabilities of their electron clouds. The valence electrons of neon occupy small, $n = 2$ atomic orbitals that have low polarizabilities, whereas those of xenon occupy relatively large, highly polarizable $n = 5$ orbitals. When two atoms approach each other, the smaller electron cloud of neon distorts less than the larger electron cloud of xenon, as a molecular picture illustrates:

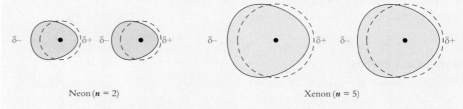

Neon ($n = 2$)   Xenon ($n = 5$)

Less polarizability means smaller partial charges and weaker intermolecular forces. Thus neon has the lower boiling point. The experimental values are 166.1 K for Xe and 27.1 K for Ne.

It is easier to distort the electron cloud of a large molecule than of a small molecule. Figure 10-10 shows how the boiling points of alkanes change as the carbon chain gets longer. As alkanes get longer, their electron clouds become larger and more polarizable, making dispersion forces larger and raising the boiling point. As examples, methane (CH_4) is a gas, pentane (C_5H_{12}) is a low-boiling liquid, decane ($C_{10}H_{22}$) is a high-boiling liquid, and eicosane ($C_{20}H_{42}$) is a waxy solid.

Dipolar Forces

Dispersion forces exist among all molecules, but some substances remain liquid at much higher temperatures than can be accounted for by dispersion forces alone. Consider 2-methylpropane and acetone, whose structures are shown in Figure 10-11. These two molecules have the same molar mass, similar shape, and nearly the same number of valence electrons (26 vs. 24). They are so similar that we might expect them to have similar boiling points, but acetone is a liquid at room temperature, whereas 2-methylpropane is a gas.

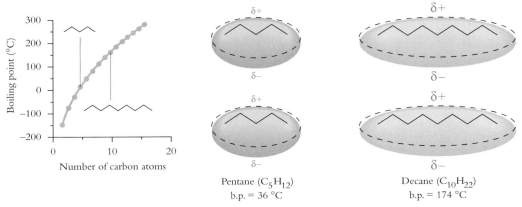

Pentane (C_5H_{12})
b.p. = 36 °C

Decane ($C_{10}H_{22}$)
b.p. = 174 °C

Figure 10-10
The boiling points of alkanes increase with the length of the carbon chain. Long-chain alkanes have larger dispersion forces because of the increased polarizability of their larger electron clouds.

Why does acetone remain a liquid at temperatures well above the boiling point of 2-methylpropane? The reason is that acetone has a dipole moment. Remember from Chapter 8 that chemical bonds are polarized toward the more electronegative atom. Thus the C=O bond of acetone is highly polarized, with a partial negative charge on the O atom ($\chi = 3.5$) and a partial positive charge on the C atom ($\chi = 2.5$).

When two polar acetone molecules approach each other, they align with the positive end of one molecule close to the negative end of the other (Figure 10-11). In a liquid array, this repeating pattern of head-to-tail alignment gives rise to significant net attractive dipolar forces among the molecules.

The dispersion forces in acetone are about the same as those in 2-methylpropane, but the addition of dipolar forces makes the total amount of intermolecular attraction between acetone molecules substantially greater than that between molecules of 2-methylpropane. Consequently, acetone boils at a considerably higher temperature than 2-methylpropane. Example 10-3 provides some additional comparisons of dispersion forces and dipole forces.

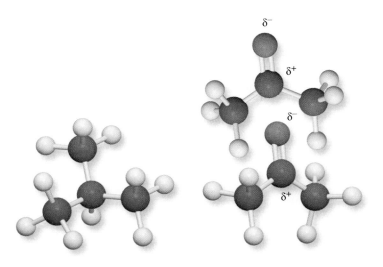

Figure 10-11
Ball-and-stick models of acetone and 2-methylpropane show that they have similar molecular shapes. The important difference between them is the dipole moment resulting from the polar bond in acetone. In liquid acetone the permanent dipoles tend to align with positive ends nearer negative ones and negative ends nearer positive ones.

Example 10-3 | **Boiling Points and Structure**

The line structures of butane, methyl ethyl ether, and acetone follow. Explain the trend in boiling points: butane (0 °C), methyl ethyl ether (8 °C), and acetone (56 °C).

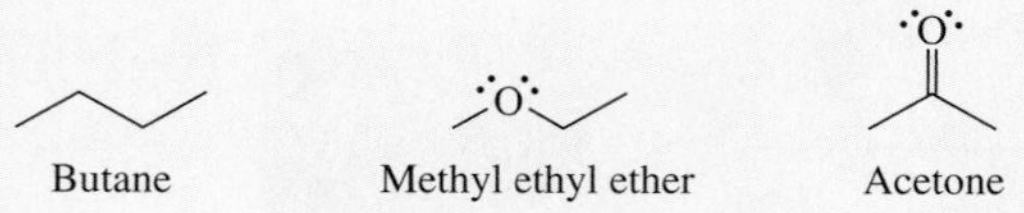

| Butane | Methyl ethyl ether | Acetone |

Strategy: These boiling points can be explained in terms of dispersion forces and dipolar forces. First, assess the magnitudes of dispersion forces, which are present in all substances, and then look for molecular polarity.

Solution: A table helps organize the available information:

Substance	Boiling point	Valence electrons	Dipole moment
Butane	0 °C	26	No
Methyl ethyl ether	8 °C	26	Yes
Acetone	56 °C	24	Yes

The table shows that dispersion forces alone cannot account for the range in boiling temperatures. Methyl ethyl ether and butane have the same number of electrons and similar shapes; yet their boiling points are different. Acetone, which has fewer electrons than the other compounds, has smaller dispersion forces; yet it boils at a higher temperature. The order of boiling points indicates that acetone is a more polar molecule than methyl ethyl ether, which in turn is more polar than butane.

We expect butane to be nonpolar because of the small electronegativity difference between carbon and hydrogen. Acetone and methyl ethyl ether, on the other hand, contain polar carbon-oxygen bonds. Molecular geometry reveals why acetone is more polar than methyl ethyl ether. The full Lewis structures of these molecules show that the oxygen atom in the ether has a steric number of 4 and bent geometry. The two C—O bond dipoles in methyl ethyl ether partially cancel each other, leaving a relatively small molecular dipole moment. On the other hand, the polar C═O bond in acetone is unopposed, so acetone has a larger dipole moment and is more polar than methyl ethyl ether.

The arrows show the charge displacement for each polar bond.

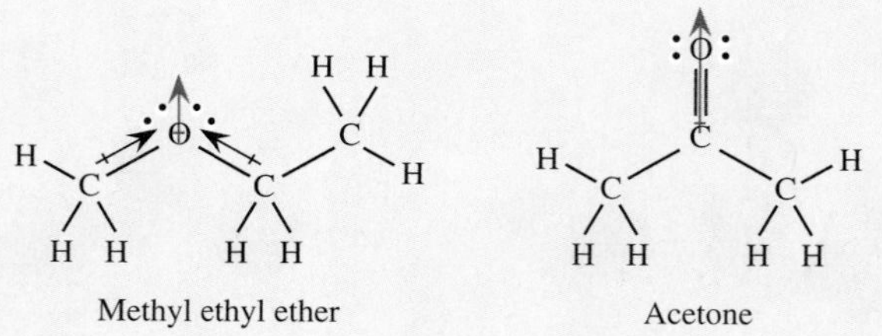

| Methyl ethyl ether | Acetone |

Hydrogen Bonding

Methyl ethyl ether is a gas at room temperature (boiling point = 8 °C), but 1-propanol, shown in Figure 10-12, is a liquid (boiling point = 97 °C). Both compounds have the same molecular formula, C_3H_8O, and both have chains of four atoms, C—O—C—C and O—C—C—C. Consequently, the electron clouds of these two molecules are about the same size, and their dispersion forces are comparable. Each molecule has an sp^3-hybridized oxygen atom with two polar single bonds, so their dipolar forces should be similar. The very different boiling points of 1-propanol and methyl ethyl ether make it clear that dispersion and dipolar forces do not reveal the entire story of intermolecular attractions.

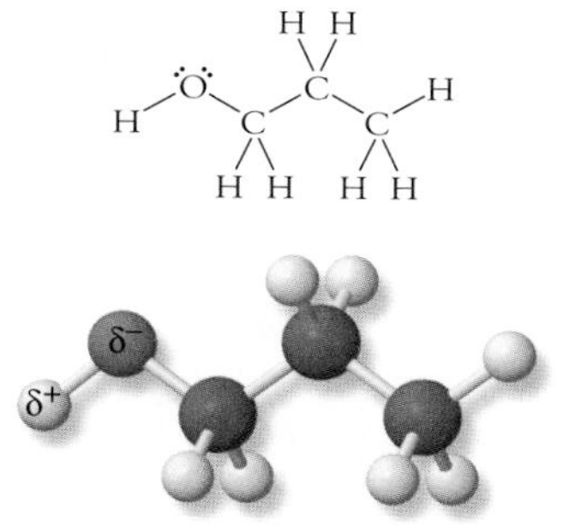

The forces of attraction between 1-propanol molecules are stronger than those between methyl ethyl ether molecules because of an intermolecular interaction called a **hydrogen bond.** A hydrogen bond occurs when a highly electronegative atom with a lone pair of electrons shares its nonbonding electrons with a positively polarized hydrogen atom. Hydrogen bonds are only 5% to 10% as strong as covalent bonds, but they are comparable to and sometimes stronger than dipolar and dispersion interactions.

There are two requirements for hydrogen bond formation. First, there must be an electron-deficient hydrogen atom to act as electron-pair acceptor. Hydrogen atoms in O—H, F—H, and N—H bonds meet this requirement. Second, there must be a highly electronegative atom to act as an electron-pair donor. Three second-row elements, O, N, and F, meet this requirement. Figure 10-13 shows representative examples of hydrogen bonding. The hydrogen bonds, which are highlighted, are shown as dashed lines to indicate the weakly bonding nature of these interactions.

Notice from the examples shown in Figure 10-13 that hydrogen bonds can form between different molecules (for example, H_3N----H_2O) or between identical molecules (for example, HF----HF). Also notice that molecules can form more than one hydrogen bond (glycine, for example) and that hydrogen bonds can form within a molecule (salicylic acid, for example) as well as between molecules. Example 10-4 explores the possibilities for hydrogen bond formation.

Figure 10-12
The Lewis structure and ball-and-stick model of 1-propanol.

Chlorine and sulfur atoms are also sufficiently electronegative to participate in hydrogen bonding, and there is some evidence for such bonding in HCl. However, the nonbonding electrons on these atoms are in diffuse $3p$ orbitals that do not interact as strongly with a hydrogen atom as electrons in more compact $2p$ orbitals.

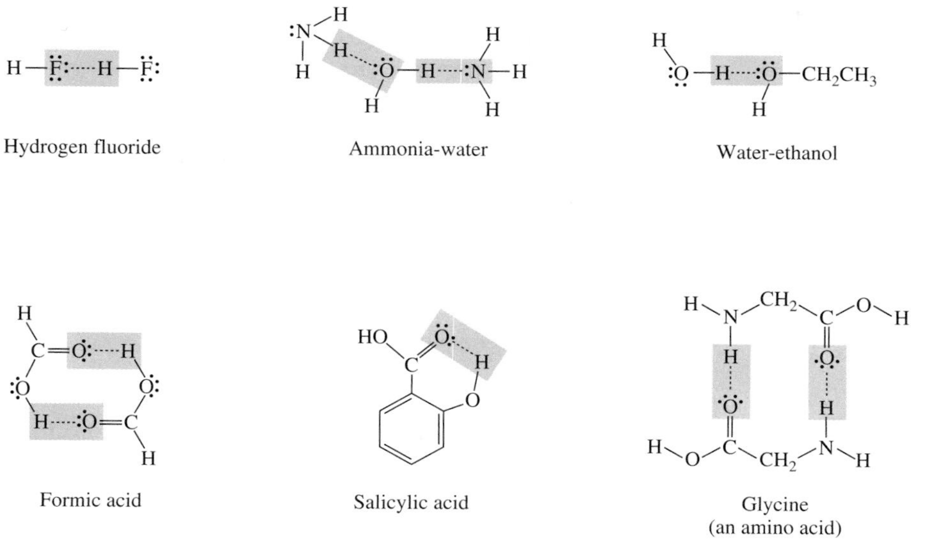

Figure 10-13
Examples of hydrogen bonding.

Hydrogen fluoride

Ammonia-water

Water-ethanol

Formic acid

Salicylic acid

Glycine
(an amino acid)

| **Example 10-4** | **Formation of Hydrogen Bonds** |

In which of the following systems will hydrogen bonding play an important role: CH_3F, $(CH_3)_2CO$ (acetone), CH_3OH, and NH_3 dissolved in $(CH_3)_2CO$?

Strategy: Hydrogen bonds occur when both electron-deficient hydrogen atoms in polar H—X bonds and highly electronegative atoms with nonbonding pairs of electrons are present. Use Lewis structures to determine whether or not these requirements are met.

Solution: Here are the Lewis structures of the four molecules:

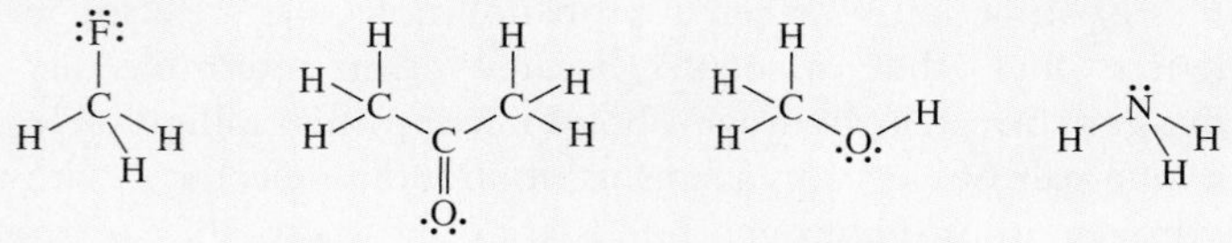

Acetone and CH_3F contain electronegative atoms with nonbonding pairs, but neither has any highly polar H—X bonds. Thus there is no hydrogen bonding between molecules of these substances.

The O—H bond in CH_3OH meets the require-ments for hydrogen bonding. The O—H hydrogen atom on one molecule interacts with the oxygen atom of a neighboring molecule:

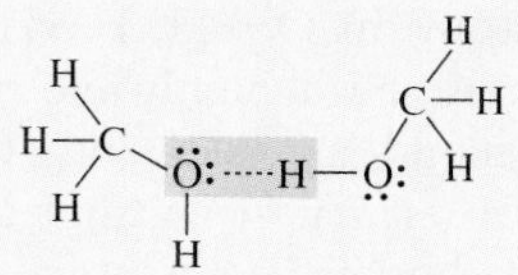

For a solution of ammonia in acetone, we must ex-amine both components. Acetone has an electronega-tive oxygen atom with nonbonding pairs, whereas NH_3 has a polar N—H bond. Consequently, a mixture of these two compounds displays hydrogen bonding be-tween ammonia's hydrogen atoms and acetone's oxy-gen atoms:

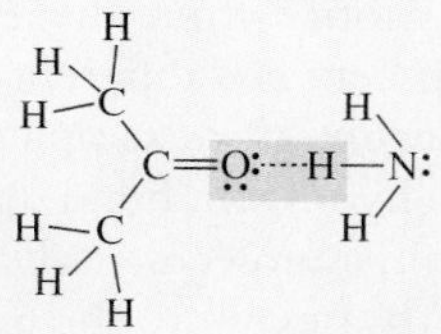

For more practice, draw a similar picture that shows the hydrogen bonding in a solution of acetone in water.

Hydrogen bonding is particularly important in biochemical systems, because bio-molecules contain many oxygen and nitrogen atoms that participate in hydrogen bonding. For example, the amino acids from which proteins are made contain NH_2 (amino) and CO_2H (carboxylic acid) groups. Four different types of hydrogen bonds exist in these systems: O—H----N, N—H----O, O—H----O, and N—H----N. Hydrogen bonding between glycine molecules is shown in Figure 10-13, and we examine more details of hydrogen bonding in biomolecules in Chapter 11.

Binary Hydrogen Compounds

Boiling points of the binary hydrogen compounds illustrate the interplay among different types of intermolecular forces. The graph in Figure 10-14 shows that there are regular periodic trends in the boiling points of these compounds. In gen-eral, the boiling points of the binary hydrogen compounds increase from top to bottom of each column of the periodic table. This trend is due to increasing disper-sion forces: The larger the molecular valence orbitals, the stronger the dispersion forces and the higher the boiling point. In Group 16, for example, H_2S ($n = 3$)

boils at −60 °C, H$_2$Se (n = 4) at −41 °C, and H$_2$Te (n = 5) at −4 °C.

As Figure 10-14 shows, ammonia, water, and hydrogen fluoride depart dramatically from this periodic behavior. This is because their molecules experience particularly large intermolecular forces resulting from hydrogen bonding. In hydrogen fluoride, for instance, partial donation of an electron pair from the highly electronegative fluorine atom of one HF molecule to the electron-deficient hydrogen atom of another HF molecule creates a hydrogen bond. Similar interactions among many HF molecules result in a network of hydrogen bonds that gives HF a boiling point much higher than those of HCl, HBr, and HI.

Fluorine has the highest electronegativity, so the strongest *individual* hydrogen bonds are those in HF. Every hydrogen atom in liquid HF is involved in a hydrogen bond, but there is only one polar hydrogen atom per molecule. Thus each HF molecule participates in two hydrogen bonds with two other HF partners. There is one hydrogen bond involving the partially positive hydrogen atom and a second involving the partially negative fluorine atom.

Water has a substantially higher boiling point than hydrogen fluoride, which indicates that the total amount of hydrogen bonding in H$_2$O is greater than that in HF. The higher boiling point of water reflects the fact that it forms more hydrogen bonds *per molecule* than hydrogen fluoride. A water molecule has two hydrogen atoms that can form hydrogen bonds and two nonbonding electron pairs on each oxygen atom. This permits every water molecule to form four hydrogen bonds.

Hydrogen bonding in solid ice creates a three-dimensional network that puts each oxygen atom at the center of a distorted tetrahedron. Figure 10-15 shows

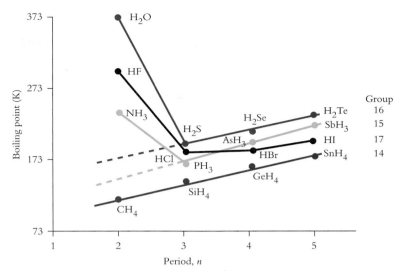

Figure 10-14
Periodic trends in the boiling points of binary hydrogen compounds. Notice that H$_2$O, HF, and NH$_3$ are exceptions to the trends.

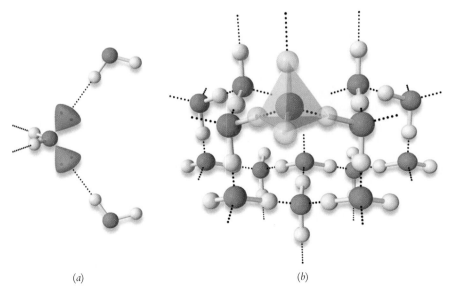

(a) (b)

Figure 10-15
The structure of ice. (*a*) Each oxygen atom is at the center of a distorted tetrahedron of hydrogen atoms. The tetrahedron is composed of two short covalent O—H bonds and two long H—O hydrogen bonds. (*b*) Water molecules are held in a network of these tetrahedra.

that two arms of the tetrahedron are regular covalent O—H bonds, whereas the other two arms of the tetrahedron are hydrogen bonds to two different water molecules.

Dispersion forces, dipole interactions, and hydrogen bonds all are much weaker than covalent intramolecular bonds. For example, the average C—C bond energy is 345 kJ/mol, whereas dispersion forces are just 0.1 to 5 kJ/mol for small alkanes such as propane. Dipolar interactions between polar molecules such as acetone range between 5 and 20 kJ/mol, and hydrogen bonds range between 5 and 50 kJ/mol.

> **Recall from Chapter 9 that bond energy is the amount of energy required to break 1 mol of a particular bond. Table 9-2 lists bond energies.**

Section Exercises

10.2.1 Explain the following differences in boiling points:
(a) Kr boils at −152 °C, and propane boils at −42 °C.
(b) C(CH$_3$)$_4$ boils at 10 °C, and CCl$_4$ boils at 77 °C.
(c) N$_2$ boils at −196 °C, and CO boils at −191.5 °C.

10.2.2 Acetone and methanol have nearly equal boiling points. What types of intermolecular forces does each exhibit? What does the similarity in boiling points tell you about the relative magnitudes of each type of force in these two compounds?

10.2.3 There are nine important hydrogen bonding interactions. One of them is O—H----O. Draw the other eight. For each of the nine, draw a Lewis structure of a specific example using real molecules.

10.3 FORCES IN SOLIDS

One of the most active areas of research in modern chemistry, physics, and engineering is the development of new solid materials. Solids play an ever-larger role in society, from high-temperature superconductors, to heat-resistant tiles for the outer "skin" of the space shuttle, to new tissue-compatible solids for surgical implants. In this section, we describe the various forces that exist in solids.

Magnitudes of Forces

The melting points of the elements presented in Table 10-2 span an immense range, from <15 K (H$_2$) to >1500 K (Si and Fe). The table shows that the forces holding solids together range from very small to extremely large. This is because in addition to the intermolecular forces described in Section 10.2, solids can be bound together by covalent bonds, metallic bonding (with delocalized electrons), and ionic interactions. There are four distinct types of solids, molecular, metallic, network, and ionic, each characterized by a different type of force. Table 10-3 contrasts the forces and energies in these four types of solids.

The molecules (or atoms, for noble gases) of a **molecular solid** are held in place by the types of forces already discussed in this chapter: dispersion forces, dipolar interactions, and/or hydrogen bonds. The atoms of a **metallic solid** are held in place by the delocalized bonding described in Chapter 9. A **network solid** contains an array of covalent bonds linking every atom to its neighbors. An **ionic solid** contains cations and anions, attracted to one another by coulombic interactions.

Table 10-3
Forces in Different Types of Solids

Solid type	Molecular	Molecular	Molecular	Metallic	Network	Ionic
Attractive forces	Dispersion	Dispersion + dipolar	Dispersion + dipolar + H bonding	Delocalized bonding	Covalent	Coulombic
Energy (kJ/mol)	0.05–40	5–25	10–40	75–1000	150–500	400–4000
Example	Ar	HCl	H_2O	Cu	SiO_2	NaCl
Melting point (K)	84	158	273	1357	1983	1074
Molecular picture		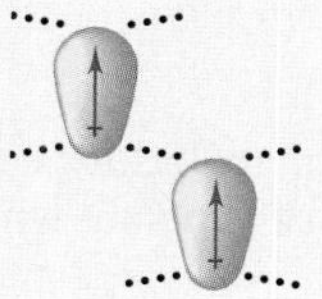	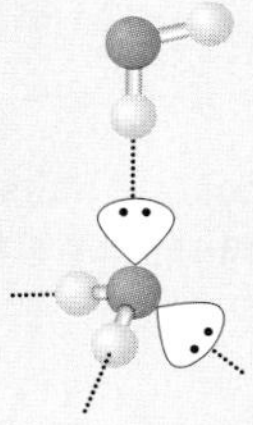	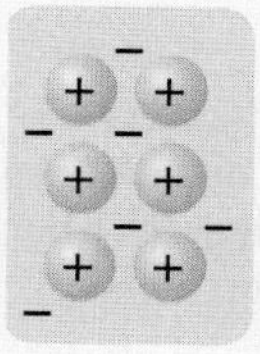	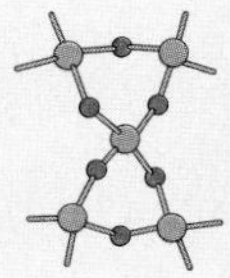	

Molecular Solids

Molecular solids are aggregates of molecules bound together by intermolecular forces. Substances that are gases under normal conditions form molecular solids when they condense at low temperature. Many larger molecules have sufficient dispersion forces to exist as solids at room temperature. One example is naphthalene ($C_{10}H_8$), a white solid that melts at 80 °C. Naphthalene has a planar structure like that of benzene (see Chapter 9), with a cloud of ten delocalized π electrons that lie above and below the molecular plane. Naphthalene molecules are held in the solid state by strong dispersion forces among these highly polarizable π electrons.

In addition to dispersion forces, molecular solids often involve dipolar interactions and hydrogen bonding. Benzoic acid ($C_6H_5CO_2H$) is a particularly clear-cut example, as illustrated in Figure 10-16. Molecules of benzoic acid are held in place by a combination of dispersion forces among the π electrons and hydrogen bonding among the CO_2H groups. With fewer π electrons, benzoic acid has weaker dispersion forces than naphthalene, but its hydrogen bonding gives benzoic acid a higher melting point: 122 °C.

The effect of extensive hydrogen bonding is revealed by the relatively high melting point of glucose ($C_6H_{12}O_6$), the sugar found in blood and human tissue. Glucose melts at 155 °C, because each of its molecules has five —OH groups that form hydrogen bonds to neighboring molecules. Although glucose lacks the highly polarizable π electrons found in naphthalene and benzoic acid, the sugar has the highest melting point of the three compounds.

The molecules of a molecular solid retain their individual properties. Solid I_2, for example, is purple because individual I_2 molecules absorb visible light, as shown by the fact that I_2 vapor also is violet. Naphthalene protects wool against moths because naphthalene molecules that escape into the gas phase are toxic to

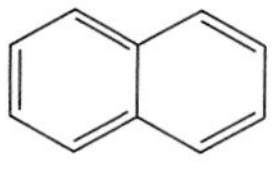

Naphthalene
$C_{10}H_8$

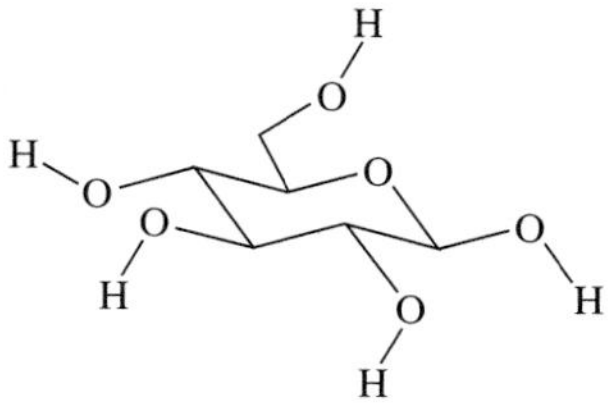

Glucose
$C_6H_{12}O_6$

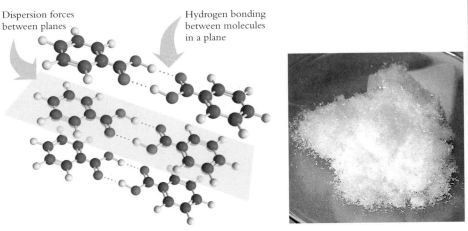

Dispersion forces between planes

Hydrogen bonding between molecules in a plane

Figure 10-16
Crystals of benzoic acid contain pairs of molecules held together head to head by hydrogen bonds. These pairs then stack in planes which are held together by dispersion forces.

moths, and glucose tastes sweet because the taste buds recognize the shape of the glucose molecule.

Network Solids

In sharp contrast to molecular solids, network solids have very high melting points. Compare, for example, the behavior of phosphorus and silicon, which are third-row neighbors in the periodic table. As listed in Table 10-2, phosphorus melts at 317 K, but silicon melts at 1683 K. Phosphorus is a molecular solid that contains individual P_4 molecules, but silicon is a network solid in which covalent bonds among Si atoms connect all the atoms. The vast array of covalent bonds in a network solid compels us to think of the entire structure as one giant "molecule."

Phosphorus is a molecular solid.

The cause of the great difference in the melting points of these two elements is evident from Table 10-3: Energies of covalent bonds are much greater than the attractive energies generated by intermolecular forces. Solid silicon cannot melt until a significant fraction of the Si—Si covalent bonds have broken. The average Si—Si bond energy is 225 kJ/mol, whereas intermolecular forces are less than 40 kJ/mol.

Bonding patterns determine the properties of network solids. Diamond and graphite, the two forms of elemental carbon that occur naturally on Earth, are both network solids, but they have very different physical and chemical properties. Diamond contains a three-dimensional array of σ bonds, with each tetrahedral (sp^3) carbon atom linked to all the others through a network of covalent bonds. This three-dimensional network of strong covalent bonds makes diamonds extremely strong and excellent abrasives. Graphite, in contrast, has trigonal planar (sp^2) carbon atoms in a two-dimensional array of σ bonds. The structure is supplemented by delocalized π bonding above and below the plane of the σ bonds. Each two-dimensional layer is attracted to its neighboring layers only by dispersion forces among the π electrons. As a result, planes of carbon atoms easily slide past one another, making graphite brittle and a lubricant.

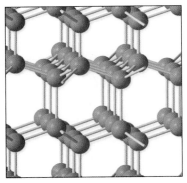

Silicon is a network solid.

Several oxides and sulfides display the characteristics of network solids. The bond network of silica appears in Section 8.5. Other examples are titania (TiO_2) and alumina (Al_2O_3). These materials have extremely high melting points because of the networks of strong X—O bonds holding their atoms together. Like graphite, MoS_2 is a two-dimensional network solid that serves as a solid lubricant.

Covalent bonds make network solids extremely durable. Geological examples include the "everlasting sands" and granite formations such as the Rock of Gibraltar. Like diamonds, other valuable gemstones are network solids, too. Rubies and sapphires are covalent crystals of aluminum oxide with small amounts of colored transition metal ion impurities. Carborundum is a 1:1 network solid of silicon and carbon that has the same lattice structure as diamond. Carborundum is much less expensive than diamond but almost as strong and wear resistant, so it is used for the edges of cutting tools. Example 10-5 compares the structures and properties of a network solid and a molecular solid.

Diamond has three-dimensional bonding.

Graphite has a planar bond network.

Network and Molecular Solids

Example 10-5

Whereas SiO_2 melts at 1710 °C, other nonmetal oxides melt at much lower temperatures. For example, P_4O_6 melts at 25 °C. Referring to the accompanying bonding pictures, describe the forces that hold these solids together.

Strategy: Identify the type of solid. Solids may be network, ionic, metallic, or molecular. Different forces account for the stability of each type.

Solution: Because these are nonmetal oxides, they cannot be described as metallic. Neither oxide contains ions, so they must be network or molecular. The bonding patterns provide the information needed to categorize the oxides and explain their melting temperatures.

SiO_2: The bonding pattern in silica is a three-dimensional array of strong covalent bonds. Many of these bonds must be broken before silica melts. Silica melts at 1710 °C because its three-dimensional covalent bonding network is highly stable.

P_4O_6: The molecular structure shows that P_4O_6 is composed of individual molecules rather than an array of covalent bonds. Strong covalent bonding holds the atoms in each molecule together, but each molecule is attracted to the others only by dispersion forces. In this molecular solid, little energy is required to overcome dispersion forces and allow P_4O_6 solid to melt.

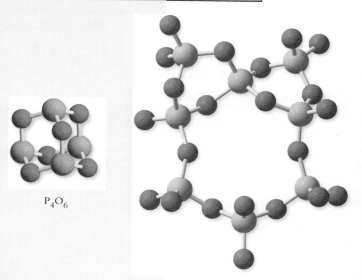

P_4O_6

Metallic Solids

As described in Section 9.6, the bonding in solid metals comes from electrons in highly delocalized valence orbitals. There are so many such orbitals that they form energy bands, giving the valence electrons high mobility. Consequently, each metal atom can be viewed as a cation embedded in a "sea" of mobile valence electrons. The properties of metals can be explained on the basis of this picture. The most obvious of these properties, electrical conductivity, is discussed in Section 9.6.

Metals display a range of melting points, indicating that the strength of metallic bonding is variable. The alkali metals of Group 1 are quite soft and melt near room temperature. Sodium, for example, melts at 98 °C and cesium at 28.5 °C. Bonding is weak in these metals because each atom of a Group 1 metal contributes only one valence electron to the bond-forming energy band. Metals near the middle of the d block, on the other hand, are very hard and have some of the highest known melting points. Tungsten melts at 3407 °C, rhenium at 3180 °C, and chromium at 1857 °C. Atoms of these metals contribute two s electrons and several d electrons to the bond-forming energy band, leading to extremely strong metallic bonding.

Metals are ductile and malleable, meaning they can be drawn into wires or hammered into thin sheets. When a piece of metal is formed into a new shape, its atoms change position. Because the bonding electrons are fully delocalized, however, changing the positions of the atoms does not cause corresponding changes in the energy levels of the electrons. The "sea" of electrons is largely unaffected by the pattern of metal cations, as Figure 10-17 illustrates. Thus metals can be forced into many shapes, including sheets and wires, without destroying the bonding nature of their filled bands of delocalized orbitals.

Sodium is so soft that it can be cut with a knife.

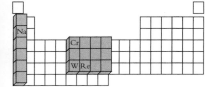

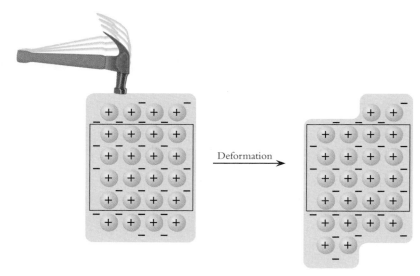

Figure 10-17
When a metal changes shape, its atoms shift position. However, because the valence electrons are fully delocalized, the energy of these electrons is unaffected.

The transition metals display a range of properties. Copper and silver are much better electrical conductors than chromium. Tungsten has very low ductility. Mercury is a liquid at room temperature. All these differences result from variations in numbers of valence electrons. A transition metal such as vanadium or chromium has five or six valence electrons per atom; all occupy bonding orbitals in the metal lattice. As a result, there are strong attractive forces among the metal atoms, and vanadium and chromium are strong and hard. Beyond the middle of the transition metal series, the additional valence electrons occupy antibonding orbitals, which reduces the net bonding. This effect is most pronounced at the end of the d block, where the number of antibonding electrons nearly matches the number of bonding electrons. Zinc, cadmium, and mercury, with d^{10} configurations, have melting temperatures that are more than 600 °C lower than those of their immediate neighbors.

Ionic Solids

As described in Chapter 7, ionic solids contain cations and anions strongly attracted to each other through coulombic interactions. Such forces are *interionic* rather than *intermolecular*. Ionic solids must be electrically neutral, so their stoichiometries are determined by the charges carried by the positive and negative ions. These in turn are determined by the positions of the elements in the periodic table.

Many ionic solids contain metal cations and polyatomic anions. Here again, the stoichiometry of the solid is dictated by the charges on the ions and the need for the solid to maintain electrical neutrality. Some examples of 1:1 ionic solids containing polyatomic anions are $NaOH$, KNO_3, $CuSO_4$, $BaCO_3$, and $NaClO_3$. Some metallic ores have 1:1 stoichiometry but contain more exotic anions, such as scheelite, $CaWO_4$ (contains WO_4^{2-}), zircon, $ZrSiO_4$ (contains SiO_4^{4-}), and ilmenite, $FeTiO_3$ (contains TiO_3^{2-}).

Ionic solids can contain more than one cation or anion. For example, the ore apatite, $Ca_5(PO_4)_3F$, contains both phosphate and fluoride anions. The mineral beryl, $Be_3Al_2Si_6O_{18}$, contains beryllium and aluminum cations as well as the $Si_6O_{18}^{12-}$ polyatomic anion. An even more complicated example is garnierite, $(Ni,Mg)_6Si_4O_{10}(OH)_2$, which has a variable composition of Ni^{2+} and Mg^{2+}. Although the relative proportions of Mg^{2+} and Ni^{2+} vary, garnierite always has six cations for every anion of $Si_4O_{10}^{10-}$ and two anions of OH^-.

One of the most exciting developments in materials science in recent years involves mixed oxides containing rare earth metals. Some of these compounds have been found to be *superconductors*, as described in our Chemistry and Technology Box. Below a certain temperature, a supercondutor can carry an immense electrical current without losses resulting from resistance. Before 1986, it was thought that this property was limited to a few metals at temperatures below 25 K. Then it was found that a mixed oxide of lanthanum, barium, and copper showed superconductivity at around 30 K, and since then the temperature threshold for superconductivity has been advanced to 135 K.

Ceramic superconductor materials display a characteristic that is unexpected in traditional chemistry but relatively common in solids: nonstoichiometric composition. When $YBa_2Cu_3O_{7-x}$ contains a stoichiometric amount of oxygen ($x = 0$ or 1), it has inferior superconducting properties. A slight deviation from stoichiometric composition is required for optimal superconducting behavior. In addition, the copper cations display deviations from integral stoichiometry. Assigning characteristic charges to the other elements (+3 for Y, +2 for Ba, and −2 for O), we find

Box 10-1   Chemistry and Technology: Superconductors

Trains that run on frictionless tracks and computer chips smaller yet faster and with much larger capacities than the present generation—these are potential applications of room-temperature superconductors. Research groups around the world are developing new materials in hopes of reaching this spectacular goal.

Metals conduct electricity because their valence electrons are highly mobile, allowing easy flow of charge. Normal electrical conductors always have significant resistivity. Qualitatively, electrons are slowed by bumping into atoms along their paths. In contrast to normal conductivity, superconductivity is characterized by zero electrical resistance. Electrons move freely through a superconductor without any friction. A superconductor can carry immense amounts of current without losses resulting from heating.

A material that displays superconductivity does so only below a critical temperature, T_c. Above T_c, the material has normal resistivity, but as the temperature drops below T_c, its resistance abruptly disappears as the graph shows.

The transition between normal and superconductivity occurs at different temperatures for different materials. In general, T_c is near 4.2 K, the boiling point of liquid helium. For

this reason, any device that makes use of superconductivity must be immersed in a bath of liquid helium. Little wonder that superconductivity was not discovered until early in this century and remained a laboratory curiosity until the mid-1980s.

Early work on superconductors concentrated on metals or metal mixtures (alloys). Niobium alloys are particularly good superconductors, and in 1973 a niobium alloy, Nb_3Ge, was found to have $T_c = 23$ K, the highest known value for a metal superconductor. In 1986, a ceramic oxide with formula

$La_{2-x}Ba_xCuO_4$ was found to show superconductivity at 30 K. Through intense research efforts on ceramic oxides, $YBa_2Cu_3O_{7-x}$, with $T_c = 93$ K, was discovered in 1987.

A temperature above 77 K brought superconductivity to a temperature that can be achieved using inexpensive liquid nitrogen as a coolant. Moreover, the compound is easy to make: Three oxides, Y_2O_3, BaO, and CuO, are ground together in the correct stoichiometric ratio and heated to 950 °C. The mixture is cooled and pressed into pellets, after which the pellets are heated to just below their melting point to bind the grains in the pellets tightly together. Finally, the pellets are heated in oxygen at around 550 °C.

Ceramic oxide superconductors have distinct atomic layers. The Cu-containing superconductors contain planes of copper and oxygen atoms, as the molecular view shows. These planes alternate with layers containing oxygen and the other metals that make up the superconductor. Superconductivity takes place in the Cu—O planes.

The record for the highest superconducting temperature in the year 2001 is held by $HgBa_2Ca_2Cu_3O_9$, $T_c = 135$ K. This is still far below room temperature, but research continues on ceramic oxides and other materials.

For many applications, a superconductor must first be drawn into a wire. This has recently been accomplished. The photo shows a superconductor ribbon wrapped around the copper wires that it could replace.

Flexible superconducting tapes provide promise of uses for superconductors in motors, generators, and even electric transmission lines. Meanwhile, superconducting magnets cooled to the temperature of liquid helium already are in use. High-field nuclear magnetic resonance (NMR) spectrometers have become standard instruments in chemical research laboratories, and the same type of machine (called an MRI spectrometer) is used for medical diagnosis in hospitals worldwide.

Graph: Resistivity (10^{-6} Ω-cm) vs Temperature (K). Curves labeled $Bi_2Sr_2CuO_{6+x}$ and $YBa_2Cu_3O_{7-x}$, with T_c markers.

O   Cu
La, Sr, or Ba

that Cu must have an average charge of $+6.8/3 = 2.27$ when $x = 0.1$. Because fractional charges are impossible, this means that most of the Cu atoms in the substance bear $+2$ charges, but a few have $+3$ charge. We return to nonstoichiometric compositions of solid materials in Section 10.4.

Section Exercises

■ **10.3.1** Describe the forces that exist in (a) solid CO_2 (dry ice); (b) crystalline yellow elemental sulfur, S_8; (c) tin (soft, malleable, mp = 232 °C); and (d) Li_2O.
■ **10.3.2** Arrange the following in order of increasing melting point and explain the reasons for your rankings: C (diamond), F_2, K, Co.
■ **10.3.3** Determine the chemical formulas of the following ionic substances: (a) magnesium phosphate; (b) spodumene, a mineral containing lithium and aluminum cations and $Si_2O_6^{4-}$; and (c) the superconducting oxide that contains Bi^{3+}, Si^{2+}, and Cu^{2+} in $2:2:1$ atomic ratio.

10.4 ORDER IN SOLIDS

In a solid, atoms, molecules, or ions are in fixed positions, their motions restricted to vibrations about preferred locations. This contrasts with gases and liquids, whose molecules change position continuously. Many solids display highly regular repeating patterns of organization; these solids are classified as crystalline. Diamonds, sugar crystals, quartz, and table salt are examples of **crystalline solids.** Some of the most valued materials of society are precious gemstones, which are crystals of rare and richly colored minerals. In other solids, the arrangement of particles is irregular, with no observable pattern. These **amorphous solids** include cotton candy, glass, and wax.

Whether a solid sample of a particular substance is crystalline or amorphous can depend on how the solid is prepared. Silicon dioxide, for example, exists in both forms. Quartz is a crystalline form of silicon dioxide found in minerals all over the world. Each tetrahedral silicon atom is bonded to a total of four oxygens in a highly symmetrical three-dimensional network. This strong bonding network gives silicon dioxide its high melting point of 1710 °C. Slow cooling of molten SiO_2 gives crystalline quartz, but rapid cooling gives an amorphous glass. Figure 10-18 shows molecular representations of quartz and glass. Both structures are three-dimensional networks, but the two-dimensional representations shown in the figure are sufficient to show the differences between crystalline and amorphous solids.

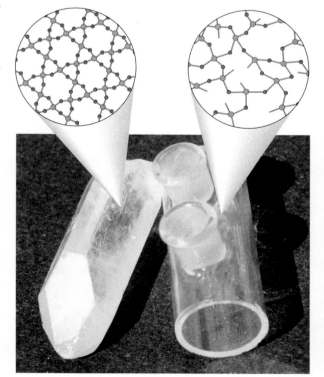

Figure 10-18
Quartz (left) is a crystalline form of silicon dioxide containing a three-dimensional array of SiO_2 units linked by highly regular covalent bonding. Silica glass (right) also has a three-dimensional array of SiO_2 units, but in this case the bond arrangement is irregular, and the solid is amorphous.

Figure 10-19
Shown here are vanadinite ($Pb_5(VO_4)_3Cl$) (*left*) quartz (SiO_2) (*center*), pyromorphite ($Pb_5(PO_4)_3Cl$) (*top right*), and pyrites (FeS_2) (*bottom right*).

The following characteristics of crystalline solids are illustrated by the photos in Figure 10-19:

1. The shape of a crystal is characterized by parallel faces and edges. The edges of a crystal usually intersect at characteristic angles, 90° in the case of vanadinite.

2. When a crystal breaks into smaller pieces, fragmentation occurs along crystal edges. Small crystals have the same characteristic angles as larger crystals.

3. A crystal has a high degree of symmetry, cubic in the case of vanadinite.

Close-Packed Crystals

A solid is most stable when each atom, molecule, or ion has as many close neighbors as possible, thus maximizing intermolecular attractions. An arrangement that accomplishes this is described as a **close-packed structure.** Close-packed structures have atoms or molecules arranged so that the empty space around them is minimized.

To visualize a close-packed atomic solid, think of the atoms as spheres placed as compactly as possible. Begin by assembling a single planar layer as shown in Figure 10-20a. Notice that the most compact planar arrangement places each sphere within a regular hexagon formed by six others. Each sphere has six nearest neighbors in the same plane. Now add a second layer of spheres. To achieve a most compact arrangement, each sphere sits in one of the "dimples" between a trio of spheres in the first layer, as shown in Figure 10-20b. This adds three more nearest neighbors for each sphere. As additional spheres are added, the second layer eventually looks identical to the lower layer, except that it is offset slightly to allow the spheres to nestle in the dimples formed by the layer below.

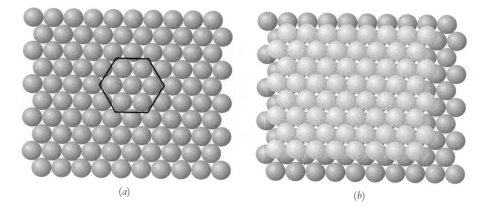

Figure 10-20
(***a***) Spheres close-packed in a layer generate a hexagonal pattern. (***b***) When a second layer is packed on top of the first, each sphere in the second layer nestles in the dimple created by three adjacent spheres in the lower layer. The second layer has two different sets of dimples, one directly above the spheres in the first layer (***blue***) and the other offset from the spheres in the layer below it (***unshaded***).

Now consider adding a third layer of close-packed spheres. This new layer can be placed in two different ways because there are two sets of dimples in the second layer. One set is located directly above the spheres in the first layer (shown in blue in Figure 10-20*b*). If spheres in the third layer lie in these dimples, the third layer is directly above the first, and the resulting three-dimensional structure is a **hexagonal close-packed** structure. If spheres in the third layer lie in the other set of dimples (unhighlighted in Figure 10-20*b*), the third layer is offset from both of the lower layers. This arrangement is a **cubic close-packed** structure. Either way, addition of a third layer adds three more nearest neighbors for each sphere. In these close-packed structures, each sphere has twelve nearest neighbors: six in the same plane, three above, and three below.

The expanded views in Figure 10-21 show the different arrangements of the hexagonal and cubic close-packed crystalline types. In the hexagonal close-packed structure, notice that the third layer lies directly above the first, the fourth above the second, and so on. The layers can be labeled ABAB . . . and so on. In the cubic close-packed structure, the third layer is offset from the other two, but the fourth layer is directly above the first. This arrangement can be labeled ABCABC . . . and so on.

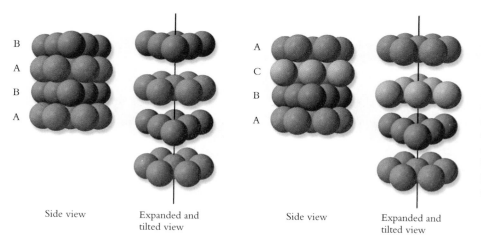

Figure 10-21
Side and expanded views of hexagonal and cubic close-packed crystal types. In the hexagonal close-packed structure, spheres on both sides of any plane are in the same positions, and the *third* layer is directly above the first. In the cubic close-packed structure, layers take up three different positions, and the *fourth* layer is directly above the first.

Animation

Figure 10-22
Three views of the hexagonal layers contained within a face-centered cubic array. (**a**) A view from above of three hexagonal layers, with all but one atom removed from the top layer. (**b**) A view from above showing an outline of the cube and atoms from three successive layers that make up the cubic array. (**c**) A view of the cube from the side, with the atoms of the middle hexagonal layer screened for emphasis.

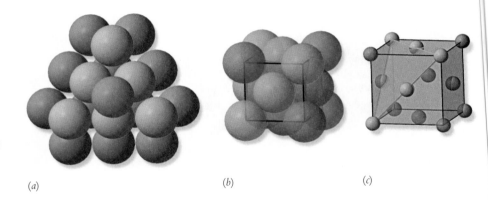

(a)　　　　　　(b)　　　　　　(c)

Animation

To see the cubes in a cubic close-packed structure, we need to remove all but one of the atoms from the top layer, as shown in Figure 10-22a. The remaining atom from this top layer (shown in mauve in the figure) defines one corner of a cube. Each face of the cube contains an atom at each corner and one atom in the center, as shown by Figure 10-22b. Three atoms from the second layer (shown in gold in the figure) lie at corners of the cube, and three other atoms from the second layer fall in the centers of three faces of the cube. Three additional corners of the cube are occupied by atoms from the third layer (shown in teal in the figure). Figure 10-22c shows that the atoms from one hexagonal layer fall along a diagonal plane through the cube. Because each face of this cube contains an atom at its center, this array is also known as the **face-centered cubic** structure.

Atoms and molecules with spherical symmetry often form crystals with hexagonal or cubic close-packed geometry. For instance, magnesium and zinc crystallize with their atoms in a hexagonal close-packed array, while silver, aluminum, and gold crystallize in the cubic close-packed arrangement. Argon solidifies at low temperature as a cubic close-packed crystal, and neon can solidify in either form.

The packing in ionic crystals requires that ions of opposite charges alternate with one another to maximize interionic attraction. For many 1:1 ionic crystals such as NaCl, the most stable arrangement is two interlocking face-centered cubic arrays, as is illustrated in Figure 10-23.

Figure 10-23
Ionic crystals such as NaCl contain face-centered cubic arrangements of each ion. In this view a cube is drawn with the cations, shown as yellow spheres, at its corners and in the centers of the faces. The anions, shown as green spheres, occupy positions at the center of each edge of the cube. A view from an oblique angle reveals that this structure contains alternating hexagonal planar arrays of cations and anions.

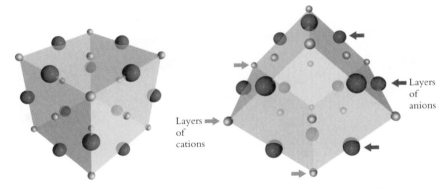

Layers of cations

Layers of anions

Face-centered cube　　　　**Oblique view of face-centered cube**

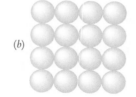
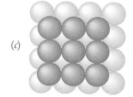

(a)   (b)   (c)

Figure 10-24
Views of the body-centered cubic array. (a) the basic cube, with one sphere removed to show the central sphere. (b) One planar layer. (c) Placement of a second layer.

Another type of arrangement, which is shown in Figure 10-24a, is a **body-centered cubic** structure. A body-centered cube can be constructed by placing a set of spheres in a square array, as shown in Figure 10-24b. A second set of spheres is nestled in the dimples of the first set, as shown in Figure 10-24c. This arrangement gives a second square array, on which yet another set can be nested. In a body-centered cubic structure, each sphere has eight nearest neighbors rather than twelve. The packing is not quite as compact as in close-packed arrays, but some metals are more stable when bonded to eight neighbors rather than to twelve. All the alkali metals, as well as iron and the transition metals from Groups 5 and 6, adopt this atomic arrangement.

Unit Cells

Any crystal is a vast array of atoms, molecules, or ions arranged in some regular repeating pattern. Because the pattern repeats exactly, every crystal has one smallest unit from which the entire pattern can be assembled. This minimum unit is called a **unit cell.** The idea of the unit cell is illustrated in two dimensions by the art of M. C. Escher, as shown in Figure 10-25. Escher often used symmetrical patterns aligned together to create an overall design. The repeating units can be visualized as tiles placed edge to edge. In Figure 10-25, each tile is a parallelogram. A unit cell in a crystal is a three-dimensional fragment stacked together like a set of blocks. The faces of unit cells can be squares, rectangles, or parallelograms.

The body-centered cubic crystal provides one of the simplest illustrations of three-dimensional unit cells. The unit cell for such a crystal is a cube (Figure 10-26a).

Figure 10-25
A drawing by M. C. Escher that contains a repeating pattern. One two-dimensional unit cell is highlighted. The entire drawing can be generated by repeated use of this unit cell.

(a)   (b)

Figure 10-26
(a) The unit cell of iron, which forms body-centered cubic crystals. (b) Unit cells stacked together.

Figure 10-27
Nonspherical objects such as bananas require more elaborate packing schemes than spherical objects such as oranges.

Notice that the unit cell has one complete atom in the exact center of the cube. The corners of the cube are defined by the centers of eight other atoms. Thus this cube contains a central atom and portions of eight additional atoms. The body-centered cubic crystal is built by stacking together many unit cells, as shown in Figure 10-26*b*. It takes eight stacked unit cells to complete one of the corner atoms, so each unit cell contains one eighth of an atom at each of its corners. The cell has eight such corners, giving eight one-eighth parts of atoms in addition to the complete atom in the center. Adding all these together gives a total of two atoms in each unit cell.

Up to now, we have described the crystalline arrays favored by spherical objects such as atoms, but most molecules are far from spherical. The photos of stacks of produce in Figure 10-27 illustrate that nonspherical objects require more elaborate arrays to achieve maximal stability. Compare the stack of bananas with the stack of oranges. Just as the stacking pattern for bananas is less symmetrical than that for oranges, the structural patterns for most molecular crystals are less symmetrical than those for crystals of spherical atoms, reflecting the lower symmetry of the molecules that make up the crystal.

Amorphous Solids

Solid materials are most stable in crystalline form. When a liquid is cooled slowly, it generally solidifies as crystals. When solids form rapidly, on the other hand, their atoms or molecules may become locked into positions other than those of a regular crystal, giving amorphous materials. Ordinary cane sugar is crystalline, and crystalline rock candy forms when melted sugar is cooled slowly. In contrast, an amorphous solid forms when melted sugar is cooled rapidly: Cotton candy contains long threads of amorphous sugar.

What we call **glass** is an entire family of amorphous solids based on silica (SiO_2). Pure silica is usually found as crystals containing the regular array of covalent bonds shown in Figure 10-18*a*. Quartz contains crystals of pure silica. When quartz is melted and then quickly cooled, however, it forms fused silica, an amorphous solid glass. Silica glass has many desirable properties. It resists corrosion, transmits light well, and withstands wide variations in temperature. Unfortunately, pure silica is very difficult to work with because of its high melting point (1710 °C). As a result, silica glass is used only for special applications.

Sodium oxide (Na_2O) is mixed with silica to make glass that can be shaped at a lower tem-

Cotton candy is amorphous.

Rock candy is crystalline.

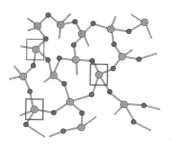

Quartz glass   Sodium-containing glass

Figure 10-28
Adding sodium oxide to quartz glass breaks some of the Si—O covalent bonds (*left view,* red outlines) to create terminal O^- ions associated with Na^+ cations (*right view*). This softens the glass, reducing its melting point.

perature. Sodium oxide is ionic, and it breaks the Si—O—Si chain of covalent bonds, as shown in Figure 10-28. This weakens the lattice strength of the glass, lowers its melting point, and reduces the viscosity of the resulting liquid. However, the weakened lattice also means that glass made from mixed sodium and silicon oxides is vulnerable to chemical attack.

A desirable glass melts at a reasonable temperature, is easy to work with, and yet is chemically inert. Such a glass can be prepared by adding a third component that has bonding characteristics intermediate between those of purely ionic sodium oxide and those of purely covalent silicon dioxide. Several different components are used, depending on the properties desired in the glass.

The glass used for windowpanes and bottles is soda-lime-silica glass, a mixture of sodium oxide, calcium oxide, and silicon dioxide. The addition of CaO strengthens the lattice enough to make the glass chemically inert to most common substances. (Strong bases and HF, however, attack this glass.) Pyrex, the glass used in coffeepots and laboratory glassware, can withstand rapid temperature changes that would crack soda-lime-silica glass. Pyrex is a composite of B_2O_3, CaO, and SiO_2. Lenses and other optical components are made from glass that contains PbO. Light rays are strongly bent as they pass through lenses made of this glass. Colored glasses contain small amounts of colored metal oxides such as Cr_2O_3 (amber), NiO (green), or CoO (brown).

Many contemporary materials are amorphous solids composed of extremely large molecules called *polymers*. Polymeric solids are intermediate between molecular and covalent solids. They have discrete but extremely large molecules held together by dispersion forces and/or hydrogen bonds. Because polymer molecules are so large, their covalent bonding plays an important role in determining the properties of the solid. Plastic polymers can be shaped and molded because the intermolecular forces between polymer molecules are weaker than chemical bonds, but a plastic polymer also has relatively high strength because its long-chain molecules are held together by strong covalent bonds.

Crystal Imperfections

The two extremes of ordering in solids are perfect crystals with complete regularity and amorphous solids that have little order. Most solid materials are crystalline but contain defects. **Crystalline defects** can profoundly alter the properties of a solid material, often in ways that have useful applications. Doped semiconductors, described in Section 9.6, are solids into which impurity "defects" are introduced deliberately in order to modify electrical conductivity. Gemstones are crystals containing impurities that give them their color. Sapphires and rubies are imperfect crystals of colorless Al_2O_3; Ti^{3+} or Fe^{3+} makes sapphires blue; and Cr^{3+} makes rubies red.

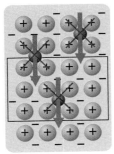

Figure 10-29
The presence of even a few carbon atoms (*gray spheres*) in the interstitial holes in an iron lattice prevents adjacent layers of iron atoms from sliding past one another and hardens the iron into steel.

As noted earlier, yttrium-barium-copper superconductors have optimal properties when they have a slight deficiency of oxygen. This departure from stoichiometric composition is created by defects in the crystal structure. The superconductor crystal has oxygen anions missing from some positions in the crystal lattice, and the number of missing anions can vary sufficiently to give the material a nonstoichiometric overall composition. The solid remains electrically neutral when its anion content varies, because some of the cations take on different charges. The copper cations in the superconductor can have a +2 or +3 charge. The relative number of Cu^{3+} ions decreases as oxygen anions are removed from the structure.

Substitutional impurities replace one metal atom with another, while interstitial impurities occupy the spaces between metal atoms. Interstitial impurities represent imperfections that play important roles in the properties of metals. For example, small amounts of impurities are deliberately added to iron to improve its mechanical properties. Pure iron is relatively soft and easily deformed, but the addition of a small amount of carbon creates steel, a much harder material. Carbon atoms fill some of the open spaces between iron atoms in the crystalline structure of steel. Although they fit easily into these spaces, their presence reduces the ability of adjacent layers of iron atoms to slide past each other, as Figure 10-29 illustrates.

This section describes only a few of the features of order and disorder in solid materials. The rapidly growing field of materials science addresses these and many other atomic and molecular aspects of solids that determine the technologically useful properties of materials.

Section Exercises

10.4.1 Predict the angles found in the crystalline fragments broken from a hexagonal close-packed crystal. What additional angles would be found in fragments from cubic close-packed crystals? (Hint: Think about the angles found in hexagons and cubes.)

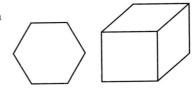

10.4.2 The unit cell of cubic close-packed crystals is a cube defined by the centers of spheres at its eight corners; there is an additional sphere embedded in the center of each face (see Figure 10-22*c*). Draw this unit cell. Determine what fraction of each type of atom (corner and center of face) is within the unit cell. (Hint: How many unit cells must be stacked together to give one complete face atom?)

10.4.3 Carbon atoms in steel occupy interstitial positions, as shown in Figure 10-29. Manganese atoms, which are also added to steel to increase its hardness, occupy substitutional positions, replacing iron atoms in the crystalline structure. Draw an atomic picture of a portion of manganese-containing steel.

10.5 LIQUIDS AND SOLUTIONS

Our discussion so far has focused primarily on gases and pure solids. However, much of the chemistry that occurs in the world takes place in solutions. Recall from Chapter 1 that a solution is a homogeneous mixture of chemical substances.

Most people think of solutions as liquid mixtures. The oceans are solutions in which ions, gases, and molecules are dissolved in vast amounts of water. Vinegar is a liquid solution of acetic acid in water, and gasoline is a liquid solution that contains many different hydrocarbons. Gases can be solutions too, and so can solids. The most familiar gaseous solution is the Earth's atmosphere, a mixture of nitrogen, oxygen, and small amounts of several other gases. Brass is a solid solution of copper and zinc, and steel is a solid solution of small amounts of carbon and manganese dissolved in iron. The following sections describe the properties of solutions. In order to understand solution chemistry we must first explore the properties of pure liquids.

Liquid Properties

In a liquid, intermolecular forces are strong enough to confine the molecules to a specific volume, but they are not strong enough to keep molecules from moving from place to place within the liquid. The relative freedom of motion of liquid molecules leads to three liquid properties arising from intermolecular forces: surface tension, capillary action, and viscosity.

A small amount of a liquid tends to take a spherical shape: For example, mercury drops are nearly spherical and water drips from a faucet in nearly spherical liquid droplets. **Surface tension**, which measures the resistance of a liquid to an increase in its surface area, is the physical property responsible for this behavior.

Figure 10-30 illustrates at the molecular level why liquids exhibit surface tension. A molecule in the interior of a liquid is completely surrounded by other molecules, each of which exerts attractive forces on its neighbors. A molecule at a liquid surface, on the other hand, has other molecules beside it and beneath it but very few above it in the gas phase. As a result, the net intermolecular force on molecules at the surface pulls them toward the interior of the liquid.

The net attraction into the liquid means that molecules at the surface are less stable than molecules fully within the liquid. As a result, a liquid is most stable when the fewest molecules are at its surface. This occurs when the liquid has minimal surface area. Spheres have less area per unit volume than any other shape, so small drops of a liquid tend to be spheres. Large drops are distorted from ideal spheres by the force of gravity.

Molecules in contact with the surface of their container experience two sets of intermolecular forces. Molecules in the liquid are attracted to one another by *cohesive forces*. In addition, molecules in the liquid are attracted to the molecules of the container walls by *adhesive forces*.

One result of adhesive forces is the curved surface, called a **meniscus,** of a liquid contained in a narrow tube. As Figure 10-31 shows, water in a glass tube forms a concave meniscus that increases the number of water molecules contacting the walls. This is because adhesive forces of water to glass are stronger than the cohesive forces among water molecules.

Figure 10-31 shows another result of imbalances between adhesive and cohesive forces. For tubes of small enough diameter, water actually climbs the walls, pulled upward by the strong adhesive forces. This upward movement of water against the force of gravity is **capillary action.** Capillary action between sap and the cellulose walls of wood fiber plays a role in how trees transport sap from their roots to their highest branches.

Water can be poured very quickly from one container to another, salad oil pours more slowly, and honey sometimes seems to take forever. A liquid's resis-

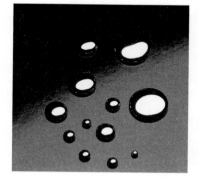

Molecule is pulled away from surface

No net "pull" in any direction

Figure 10-30
In the interior of a liquid (*bottom*), each molecule experiences equal forces in all directions.
A molecule at the surface of a liquid (*top*) is pulled back into the liquid by intermolecular forces.

tance to flow is called its **viscosity;** the greater its viscosity, the more slowly the liquid pours. Viscosity measures how easily molecules slide by one another, and this depends strongly on molecular shapes. Liquids such as water, acetone, and benzene, whose molecules are small and compact, have low viscosity. In contrast, large molecules such as the sugars in honey and the hydrocarbons found in oils tend to get tangled up with each other. Tangling inhibits the flow of molecules and leads to high viscosity. In addition, strong intermolecular cohesive forces make it harder for molecules to move about.

Molecules move faster as temperature increases, and this allows them to slide by one another more easily. Thus, viscosity decreases as temperature increases. This dependence is quite noticeable for highly viscous substances such as honey and syrup, which are much easier to pour when hot than when cold.

The Nature of Solutions

A solution is characterized by its components. The substance that determines the state of the solution is the **solvent.** Normally, the solvent is the component present in the greatest quantity. Water is the most common solvent for liquid solutions. Nitrogen is the solvent for our planet's atmosphere. All other substances in a solution are called the **solutes.**

Except for gaseous solutions, there is usually an upper limit to the amount of solute that will dissolve in a given amount of solvent. When that limit has been reached, the solution can hold no more solute and is saturated. The concentration of a saturated solution is the solubility of the substance in that particular solvent. Concentration is usually expressed as a molarity or as the mole fraction of the solute.

Whether or not a particular substance dissolves in a liquid depends largely on intermolecular forces of attraction. These interactions are of three types: those between ions or molecules of the pure solute, those between solvent molecules, and those between solvent and solute in the solution. The balance among these forces of attraction results in wide variations in solubility. For example, the solubility of NaCl in water is about 6 M, that of AgCl is only 10^{-5} M, and that of NaCl in gasoline is virtually zero. Some substances form solutions in all proportions and are said to be completely **miscible.** Acetone and water are liquids that can be mixed in any proportion from pure water to pure acetone.

Miscibility of Liquids

Substances that dissolve in each other usually have similar types of intermolecular interactions, a generalization that can be summarized by the expression *like dissolves like.* As a result, one liquid dissolves in another if the forces of attraction between molecules of the two different substances are comparable to the interactions between molecules of the same type.

When two liquids are mixed, all three sets of intermolecular interactions must be assessed. Consider water, methanol, octane, and cyclohexane, all of which are

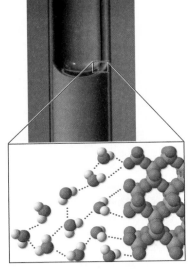

Figure 10-31
Water in a small-diameter glass tube has a concave meniscus and rises in the tube by capillary action.

SECTION 3.7 & 5.5
Molarity (M) is defined in Section 3.7, and mole fraction (**X**) is defined in Section 5.5.

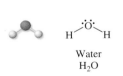

Water
H_2O

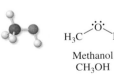

Methanol
CH_3OH

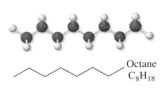

Octane
C_8H_{18}

Cyclohexane
C_6H_{12}

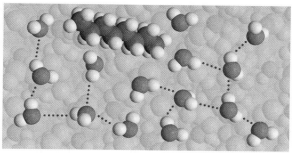

Figure 10-32
Left, A collection of water molecules has an intricate hydrogen-bonding network (dotted lines). *Right*, The presence of an octane molecule disrupts some of the hydrogen bonds.

liquids at room temperature. Water and methanol are miscible because the molecules in both the pure liquids and their mixtures form many hydrogen bonds. Octane and cyclohexane are miscible because the molecules in both the pure liquids and their mixtures interact through the dispersion forces caused by their polarizable electron clouds. In contrast, water and octane are nearly insoluble in each other. When mixed, octane and water partition into distinct layers, one of nearly pure water and the other of nearly pure octane. Water and cyclohexane are also insoluble in each other and form two layers when mixed.

The solubilities of water, methanol, octane, and cyclohexane in one another are examples of like dissolving like. Water and methanol form hydrogen bonds. When these liquids are mixed, the degree of hydrogen bonding in the solution is about the same as in either of the pure liquids. Likewise, dispersion forces in solutions of octane and cyclohexane are about the same as in the pure liquids. As Figure 10-32 shows, however, octane molecules cannot dissolve in water unless they disrupt water's hydrogen bonding network. Octane does not form hydrogen bonds, so the only forces of attraction between water molecules and octane molecules are dispersion forces. Because hydrogen bonds are stronger than dispersion forces, the cost of disrupting water's hydrogen bonding network is far greater than the stability gained from octane-water dispersion forces.

Some liquids can interact with other substances in multiple ways. Acetone, for instance, has a polar $C{=}O$ bond and a three-carbon bonding framework (Figure 10-33). The bonding framework is similar to that of a hydrocarbon, so acetone mixes with cyclohexane. The polar $C{=}O$ group makes acetone miscible with other polar molecules such as acetonitrile (CH_3CN). Finally, the polar oxygen atom in acetone has lone pairs of electrons that can form hydrogen bonds with H atoms of ammonia or water. Because of its versatility, acetone is an important industrial solvent. It is also used in laboratories to clean and rinse glassware. Example 10-6 treats several alcohols that also display multiple types of interactions.

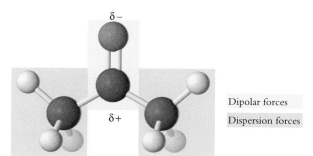

δ−

δ+

Dipolar forces

Dispersion forces

Figure 10-33
Acetone is a versatile solvent because it will dissolve many substances. The three-carbon chain is compatible with nonpolar molecules such as octane and cyclohexane, whereas the polar $C{=}O$ bond is compatible with polar molecules such as methanol and water.

Example 10-6	Solubility Trends

Give a molecular explanation for the following trend in alcohol solubilities in water:

Propanol	$CH_3CH_2CH_2OH$	Completely miscible
Butanol	$CH_3CH_2CH_2CH_2OH$	1.1 M
Pentanol	$CH_3CH_2CH_2CH_2CH_2OH$	0.30 M
Hexanol	$CH_3CH_2CH_2CH_2CH_2CH_2OH$	0.056 M

Strategy: Solubility limits depend on the stabilization generated by solute-solvent interactions balanced against the destabilization that occurs when solvent bonding networks are disrupted by solute. Thus we must examine the intermolecular interactions involving water and alcohol molecules.

Solution: When an alcohol dissolves in water, the hydrogen-bonding network of water is disrupted by the nonpolar hydrocarbon part of the alcohol. Counterbalancing this disruption of solvent, hydrogen-bonding interactions are generated between the —OH groups of the alcohol and water molecules.

As the nonpolar region of an alcohol grows longer, each solute molecule disrupts more and more of the solvent's hydrogen bonds. At the same time, each alcohol listed has only one —OH group, so the amount of compensating solute-solvent hydrogen bonding is the same for all the alcohols.

Longer-chain alcohols are progressively less soluble in water because as the hydrocarbon chain gets longer, more destabilization is involved in inserting the alcohol into the water matrix.

Zinc metal dissolves in aqueous acids by chemical reaction.

SECTION 4.6 & 4.7 →
Recall from Section 4.6 that strong acids generate hydronium ions in aqueous solution. The reaction of metals with acids is discussed in Section 4.7.

Solubility of Solids

Like dissolves like also describes the solubility properties of solids. As we have already described, there are four different kinds of solids: network, ionic, molecular, and metallic. Each is held together by a different kind of interaction, so each has its own solubility characteristics.

Network solids such as diamond, graphite, or silica cannot dissolve without breaking covalent chemical bonds. Because intermolecular forces of attraction are always much weaker than covalent bonds, solvent-solute interactions are never strong enough to offset the energy cost of breaking bonds. Covalent solids are insoluble in all solvents, but they may be attacked chemically by some liquids or vapors.

Metals also are difficult solids to dissolve because they contain extensive delocalized bonding networks that must be disrupted before the metal can dissolve.

When an alkali metal contacts water or when metals such as Ca, Zn, or Fe are treated with aqueous acid, the metal reacts with the solution, producing hydrogen gas and a solution of the metal cation (for example, Na^+ and Ca^{2+}). This process differs from other solution processes in that the solute undergoes a chemical transformation. The aqueous medium dissolves the metal by a chemical reaction that converts the insoluble metal into soluble cations. Zinc metal, for example, reacts with hydrochloric acid to generate H_2 gas and Zn^{2+} cations in solution:

$$Zn(s) + 2\,H_3O^+(aq) \longrightarrow Zn^{2+}(aq) + H_2(g) + 2\,H_2O(l)$$

The solution produced when zinc reacts with aqueous HCl is an aqueous solution of Zn^{2+} ions, not a solution of Zn metal in water. If this solution is boiled to dryness, the remaining solid is $ZnCl_2$, not zinc metal.

A few metals *react* with water, and several *react* with aqueous acids, but no metal will simply *dissolve* in water. Likewise, metals do not dissolve in nonpolar liquid solvents.

Metals can dissolve in each other (like dissolves like). A mixture of substances with metallic properties is called an *alloy*. Some alloys are solutions, but others are heterogeneous mixtures. Brass, for instance, is a homogeneous solution of copper (20% to 97%) and zinc (80% to 3%), but common plumber's solder is a heterogeneous alloy of lead (67%) and tin (33%). When solder is examined under a microscope, separate regions of solid lead and solid tin can be seen. When brass is examined, no such regions can be detected.

Mercury, the only metal that is a liquid at room temperature, dissolves a number of metals to give liquid solutions. Any solution of another metal dissolved in mercury is called an *amalgam*. Metals close to mercury in the periodic table, such as silver, gold, zinc, and tin, are particularly soluble in mercury. An amalgam of silver, tin, and mercury was used until recently to make dental fillings. When the intermetallic compound Ag_3Sn is ground with mercury, it forms a semisolid amalgam that can be shaped to fill a cavity. On standing, mercury binds with the other metals to form a hard solid mixture of Ag_5Hg_8 and Sn_7Hg_8. The mixture expands slightly during reaction, forming a tight fit within the cavity. The mercury atoms in dental fillings are chemically bound and do not dissolve, so they are safe for the wearer, despite the fact that mercury is highly toxic.

Ionic solids contain cations and anions held in three-dimensional lattices by strong coulombic attractions, as described in Section 7-6. Thus ionic solids do not dissolve unless considerable solvent-ion interactions exist to counterbalance the energy cost of breaking the ions free from the lattice. There are no simple ionic liquids at room temperature, so at first we might think there are no solvents suitable for ionic solids. Some ionic solids dissolve in water, however, because water is a highly polar liquid in which strong ion-dipole interactions exist between water molecules and ions in aqueous solution. Figure 10-34 illustrates the solvation of Na^+ and Cl^- ions as NaCl dissolves in water.

Recall that in assessing solubility, three sets of forces must be considered. Strong attractive cation-anion interactions in the ionic solid have to be overcome; these impede solubility and make many ionic salts insoluble. Counterbalancing this are attractive forces between water dipoles and solute ions; these enhance solubility

New resin materials are replacing mercury amalgams as the material of choice for dental fillings.

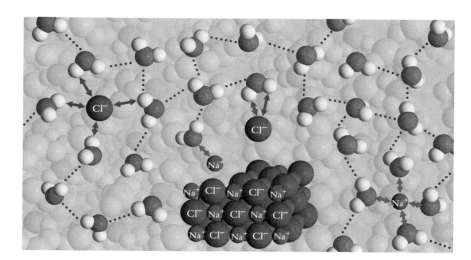

Figure 10-34
A molecular picture showing the ion-dipole interactions that help a solid ionic crystal dissolve in water. The arrows indicate ion-dipole interactions.

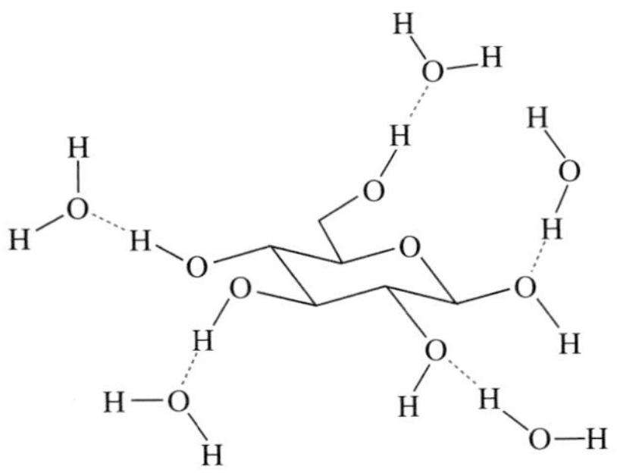

Figure 10-35
All five polar O—H groups can form hydrogen bonds of glucose with water molecules (red dotted lines).

SECTION 11.5 →
Glucose and other sugars are discussed in more detail in Section 11.5.

and help account for the existence of soluble salts. Hydrogen-bonding interactions in the solvent also have to be taken into account, but ions often are compact enough in size that their presence does not significantly alter the hydrogen-bonding network in water. Besides energetics, order-disorder considerations, which are discussed in Chapter 13, also play a major role in the solubility behavior of ionic solids.

The interplay among these several effects makes it difficult to predict the solubilities of ionic salts. As a simplifying generalization, the solubility guidelines presented in Chapter 4 categorize ionic solids as soluble or insoluble. For soluble salts the stabilization resulting from ion-dipole interactions compares favorably with the coulombic forces of the ionic solid. For insoluble salts, ion-dipole interactions provide too little stabilization to overcome the forces that hold the ions in the solid lattice.

Molecular solids are held together by dispersion forces, dipole forces, and, sometimes, hydrogen bonds. Such solids dissolve readily in solvents with similar types of intermolecular forces. Nonpolar I_2, for instance, is soluble in nonpolar liquids such as carbon tetrachloride (CCl_4). Many organic compounds are molecular solids that dissolve in organic liquids such as cyclohexane and acetone.

Hydrogen bonding in the aqueous environment allows water to dissolve materials that form hydrogen bonds. Hydrogen bonding makes sugars such as sucrose and glucose very soluble in water. Glucose, whose structure is shown in Figure 10-35, is an organic molecule with five polar O—H groups, each of which forms hydrogen bonds with water molecules. Thus glucose ($C_6H_{12}O_6$) is quite soluble in water. On the other hand, naphthalene, a solid hydrocarbon limited to dispersion forces, is nearly insoluble in water.

The best solvent for a molecular solid is one whose intermolecular forces match the forces holding the molecules in the crystal. For a solid held together by dispersion forces, good solvents are nonpolar liquids such as CCl_4 and C_6H_{12}. For polar solids, a polar solvent such as acetone works well. The best solvent for ionic salts is water. This does not mean, however, that *every* polar solid dissolves in acetone or that *every* ionic salt dissolves in water. A substance dissolves only when the balance of coulombic forces favors solubility. We can predict that the balance will be unfavorable for a salt in a nonpolar solvent or a nonpolar organic molecule in water, but there is no foolproof method for predicting a favorable balance for potentially favorable cases. Example 10-7 provides some practice in recognizing solubility types. Our Tools for Discovery Box describes chromatography, a separation technique that exploits solubility differences.

| Example 10-7 | **Solubilities of Vitamins** |

Vitamins, organic molecules required by the body for proper function but not synthesized by the body, must be present in the foods people eat. Vitamins can be grouped into two categories: fat-soluble, which dissolve in fatty hydrocarbon-like tissues, and water-soluble. The structures of several vitamins follow. Assign each one to the appropriate category.

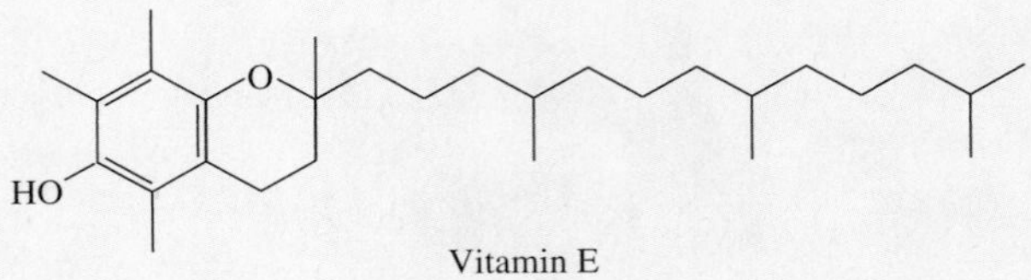

Vitamin E

Solubilities of Vitamins *(continued)*   Example 10-7

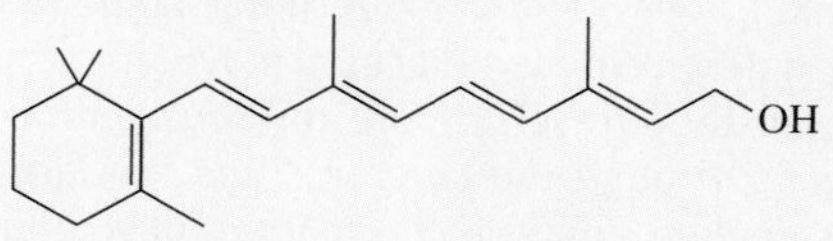

Vitamin A

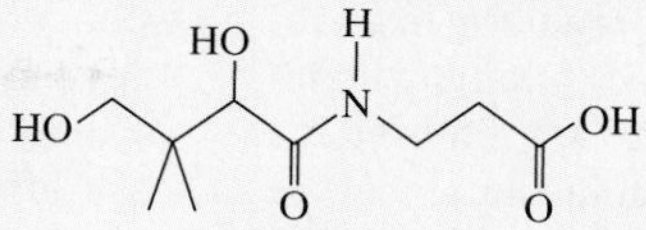

Pantothenic acid
(Vitamin B$_5$)

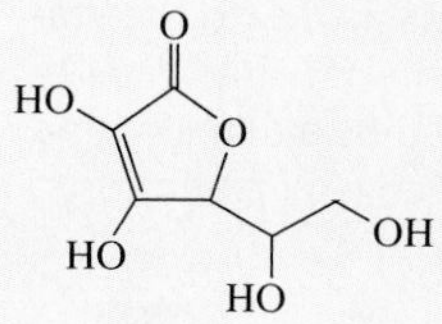

Vitamin C

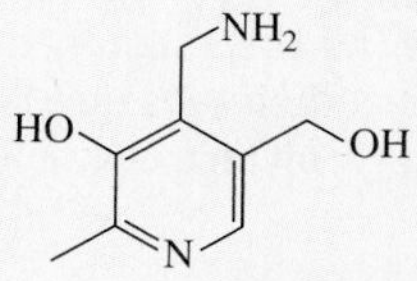

Pyridoxamine
(Vitamin B$_6$)

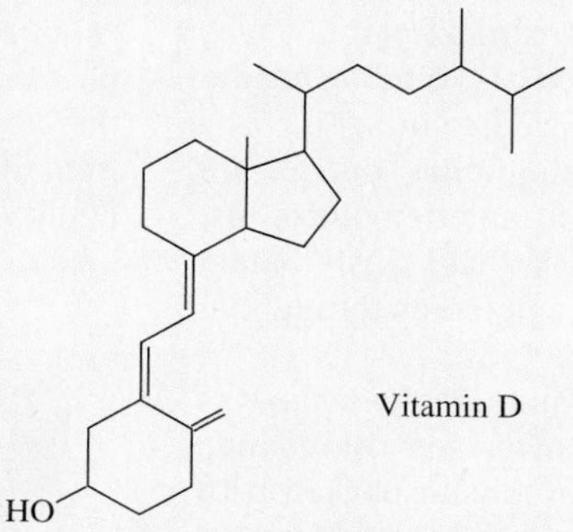

Vitamin D

Strategy: At first glance it may seem that the like-dissolves-like guideline does not apply here. Certainly, none of these complex molecules look like water, and the resemblance to simple hydrocarbons such as cyclohexane is also remote. Keep in mind, however, that the basis for the like-dissolves-like principle is that similar compounds dissolve in each other because they have common patterns of intermolecular interactions. Example 10-7 indicates that alcohols with large nonpolar segments do not dissolve well in water. We can categorize vitamins similarly by the amount of the structure that can be stabilized by hydrogen bonding to water molecules.

Solution: A hydrogen-bond donor must have a hydrogen atom bonded to F, O, or N, and a hydrogen-bond acceptor is an electronegative atom with a lone pair of electrons. By these criteria, all the vitamins shown are capable of some hydrogen bonding. However, vitamins A, D, and E have large regions containing only nonpolar C—C and C—H bonds. Like the longer alcohols in Example 10-7, these molecules have too few hydrogen-bonding sites for them to be soluble in water. As a result, these are fat-soluble vitamins.

The remaining molecules, vitamin C, pantothenic acid, and pyridoxamine, have a comparatively large number of O—H and N—H groups. These groups allow each of these vitamins to form many hydrogen bonds, so they are all water-soluble. (In fact, all the B vitamins are soluble in water.)

The different solubilities of these two kinds of vitamins have important metabolic consequences. Fat-soluble vitamins can be stored in fatty body tissue for a long time because they do not dissolve in aqueous body fluids. As a result, too much of a fat-soluble vitamin can overload the storage capabilities and lead to a toxic reaction. Water-soluble vitamins, on the other hand, cannot be stored, and the body excretes anything more than the amount it can use immediately. People must therefore have a regular supply of water-soluble vitamins in their diets to remain healthy.

Box 10-2	Tools for Discovery: Chromatography

One type of procedure that occurs regularly in a laboratory is the separation and purification of chemical substances. Synthesis reactions often produce mixtures of two or more products; biomolecules must be extracted from their natural sources; confiscated substances must be purified and identified by criminologists. Among the various techniques for the purification of chemical compounds, one of the most versatile and powerful is *chromatography*. The earliest work in chromatography was the separation of colored plant materials, leading to the name of the technique (which means "color writing").

There are many types of chromatography, but all are based on the same principles. A mobile solution phase carries the compounds to be separated, over a stationary insoluble phase, which binds these compounds through intermolecular forces.

The figure below shows how chromatography separates compounds. The mobile phase dissolves the compounds of interest and percolates through a column packed with the stationary phase. The rates of movement of the compounds depend on their strengths of interaction with the stationary phase. Because solutes move only when in the mobile phase, molecules that have low affinities for the stationary phase (shown as blue circles) move quickly, whereas those that bind more tightly to the stationary phase (shown as red circles) lag behind.

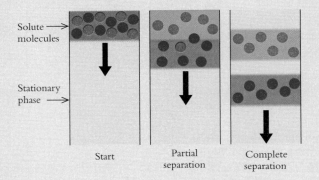

Solute molecules

Stationary phase

Start — Partial separation — Complete separation

After the materials have traveled a sufficient distance, they become separated into distinct bands; each band may contain one pure material. As the mobile phase leaves the column, it can be collected in small fractions. When the separations are complete, the various components of the original mixture are found in different fractions.

Chromatography is versatile because the phases can be varied to match the compounds that need to be separated. Some stationary phases separate solutes according to solute polarity. Polar groups on the stationary phase attract polar solutes, slowing their movement through the column. Other stationary phases contain particles called molecular sieves. These contain molecule-sized channels, much like a sponge. Large molecules cannot enter the channels, so they move along quickly. Small molecules move in and out of the channels, so they travel a more tortuous path and take longer to pass through the column. Thus the largest solute molecules emerge from the chromatography column first, and the smallest molecules emerge last.

Gas chromatography (GC) is used to separate mixtures of gases or volatile liquids. The mixture to be separated is vaporized in an oven, and the gaseous mobile phase passes through a long, narrow column packed with the stationary phase. As the components of the mixture emerge from the column, their presence is sensed by a detector and displayed as a graph on a computer screen (see graph).

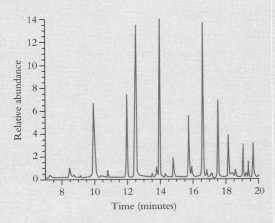

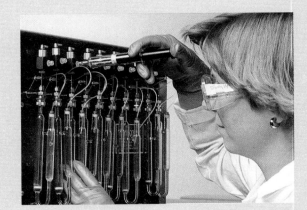

The most widespread use of GC is in identifying trace components of a mixture. Among other things, GC is used to test urine for the presence of illegal drugs, identify pollutants and measure their concentrations in groundwater, assay the purity of a volatile compound isolated in the laboratory, and follow the progress of a chemical reaction by monitoring the disappearance of starting materials or the appearance of products.

10.5.1 Explain in terms of attractive forces why salts containing Zn^{2+} are soluble in water but Zn metal is insoluble.

10.5.2 List the types of intermolecular interactions that stabilize a solution of acetone in methanol, and draw molecular pictures that illustrate any dipole–dipole and hydrogen-bonding interactions that exist between molecules of these substances.

10.5.3 On the basis of their molecular structures, predict which of the following silicon-containing materials are water-soluble: elemental Si, SiO_2, Na_4SiO_4, and $Si(CH_3)_4$.

10.6 EFFECTS OF SOLUTES ON AQUEOUS SOLUTIONS

Solute molecules alter many properties of a liquid. For instance, adding table salt to water gives a solution that boils at a slightly higher temperature than pure water, and adding ethylene glycol to the water in an automobile radiator gives a solution that protects against freezing. Changes such as these can be understood from a molecular perspective if we first describe phase changes from a molecular viewpoint and then examine the effect of added solute molecules.

Phase Equilibria

The freezing point of pure water is 0 °C under a pressure of 1 atm. Below this temperature, liquid water freezes to solid ice; above 0 °C, ice melts to give liquid water. At exactly 0 °C, solid ice and liquid water can coexist indefinitely in a condition of dynamic equilibrium.

The molecular view shown in Figure 10-36a reveals that two processes occur in a mixture of ice and water at 0 °C. Water molecules in the liquid sometimes are captured and added to the solid phase when they collide with the ice crystals. At the same time, molecules on the surface of the ice crystals sometimes become detached and enter the surrounding liquid. A state of dynamic equilibrium is reached when equal numbers of molecules move in each direction in any given time. Then there is no net change, even though individual molecules continue to move back

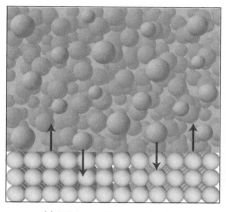
(a) Solid-pure liquid equilibrium

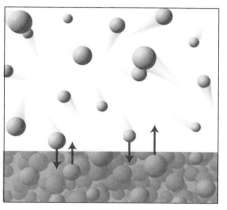
(b) Pure liquid-gas equilibrium

Figure 10-36
Molecular views of the dynamic equilibria between pure phases. (*a*) **The equilibrium between liquid and solid.** (*b*) **The equilibrium between liquid and gas.**

Radioactive ice is made from water in which a small fraction of the hydrogen atoms have been replaced by the radioactive isotope tritium. Tritium and radioactivity are discussed in Chapter 20.

and forth between the phases. One proof of this molecular movement comes from radioactivity studies. If radioactive ice is placed in nonradioactive water at 0 °C, the water slowly becomes radioactive because of the transfer of radioactive water molecules between phases.

The ice–water equilibrium exists only at 0 °C (when the pressure is 1 atm). Lowering the temperature decreases the rate at which molecules escape from the surface of the ice, and the liquid freezes. Raising the temperature increases the rate of escape, and the solid melts.

A similar dynamic equilibrium exists between liquid water and water vapor. When the pressure exerted on the liquid is 1 atm, the temperature at which liquid water is in equilibrium with gaseous water is 100 °C (Figure 10–36b). Some molecules at the liquid surface have sufficient energy to escape into the gas phase, and some molecules in the gas phase are captured when they strike the liquid surface. Under conditions of dynamic equilibrium, equal numbers of molecules move in each direction at any given time. When the pressure is 1 atm, this equilibrium exists only at 100 °C. Lowering the temperature reduces the rate at which molecules escape from the liquid phase, and condensation occurs. Raising the temperature increases the rate of escape from the liquid phase, and the liquid boils.

When water boils in an open container, the steam diffuses into the surrounding atmosphere, leading to a continual escape of molecules. Consequently, the liquid-vapor equilibrium can be observed only when the gas is confined to a closed space.

The **normal freezing point (fp)** of a substance is the temperature at which solid and liquid coexist at equilibrium under a pressure of one atmosphere. The **normal boiling point (bp)** of a liquid is the temperature at which liquid and vapor coexist at equilibrium under a pressure of one atmosphere.

Effect of Solutes

Four common properties of solutions, known as **colligative properties,** are modified by the presence of solute molecules. They are freezing point, boiling point, vapor pressure, and osmotic pressure.

The molecular view of freezing and boiling can be used to show how dissolved substances influence melting points and boiling points. In a solution, solute molecules replace some of the solvent molecules, so a given volume of a solution contains a smaller number of solvent molecules than the same volume of pure solvent. Consequently, the presence of solute molecules reduces the rate at which solvent molecules leave the liquid phase. Figure 10-37 shows how this affects a

Figure 10-37
Molecular views of the rates of solid-liquid phase transfer of a pure liquid and a solution at the normal freezing point. The addition of solute disrupts the dynamic equilibrium between escape and capture.

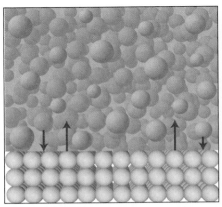

Dynamic equilibrium:
Two solid molecules escape,
two liquid molecules are captured

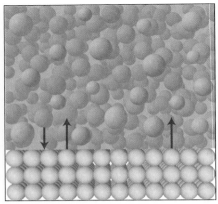

Solute disrupts equilibrium:
Two solid molecules escape,
one liquid molecule is captured

liquid–solid equilibrium: Changing one rate without changing the other rate throws the dynamic equilibrium out of balance.

The addition of solutes decreases the freezing point of a solution. In the solution, solvent molecules collide with crystals of solid solvent less frequently than they do in the pure solvent. Consequently, fewer molecules are captured by the solid phase than escape from the solid to the liquid. Cooling the solution restores dynamic equilibrium because it simultaneously reduces the number of molecules that have sufficient energy to break away from the surface of the solid and increases the number of molecules in the liquid with low enough kinetic energy to be captured by the solid.

Experiments show that at low solute concentration, the change in freezing point of a solution, ΔT_f, depends on the concentration of the solution:

$$\Delta T_f = K_f \, c_m \qquad \qquad (10\text{--}2)$$

Here, c_m is a new concentration measure, the **molality,** which is defined to be the number of moles of solute divided by the mass of solvent in kilograms. This way of expressing concentration is preferred to molarity (moles of solute divided by volume of solution in liters) because molality, unlike molarity, is independent of temperature. The constant K_f is called the **freezing point depression constant.** The constant is different for different solvents but does not depend on the identity of the solutes. For water, K_f is 1.858 °C kg/mol. Example 10-8 illustrates the use of Equation 10-2.

Equation 10-2 can be derived from our simple molecular picture and kinetic molecular theory. The derivation is independent of the nature of solute and solvent, so Equation 10-2 is valid for other solvents besides water, except that K_f has a different value for each solvent.

Freezing Point Depression	**Example 10-8**

Ethylene glycol (1,2-ethanediol) is added to automobile radiators to prevent cooling water from freezing. Estimate the freezing point of radiator coolant that contains 2.00 kg of ethylene glycol and 5.00 L of water.

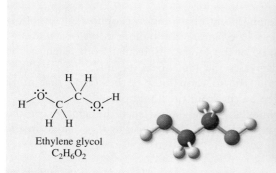

Ethylene glycol
$C_2H_6O_2$

Strategy: The question asks for the freezing point of a solution. The phrase *to prevent the water from freezing* reveals that we are dealing with the depression of the freezing point of water. Equation 10-2 describes this process for a dilute solution: $\Delta T_f = K_f \, c_m$. A coolant solution is quite concentrated, so this equation will not apply exactly, but we can use it to obtain an estimate of the freezing point.

Example 10-8	Freezing Point Depression *(continued)*

Solution: The freezing point depression constant for water is known from experiments and can be found in tables: $K_f = 1.858 \,°C \, kg/mol$. To calculate the freezing point, we must first determine the molality of the solute in this solution.

Molality is defined to be moles of solute divided by mass of solvent in kilograms. The number of moles of solute is found from the mole–mass relationship, and the number of moles of solvent can be found from the density of water. The molar mass of ethylene glycol is obtained from its chemical formula, $C_2H_6O_2$: $MM = 62.07 \, g/mol$:

$$n_{C_2H_6O_2} = \frac{(2.00 \, kg)(10^3 \, g/kg)}{62.07 \, g/mol} = 32.22 \, mol$$

We find the mass of water from its density, $\rho = 1.00 \, g/mL$, and then convert to kg.

$$m_{water} = \rho V = (1.00 \, g/mL)(5.00 \, L)(10^3 \, mL/L)(10^{-3} \, kg/g) = 5.00 \, kg$$

$$c_m = \frac{32.22 \, mol}{5.00 \, kg} = 6.444 \, mol/kg$$

Substitute into Equation 10-2 to estimate the difference between the freezing point of the solution and that of pure water:

$$\Delta T_f = K_f \, c_m = (1.858 \,°C \, kg/mol)(6.444 \, mol/kg) = 12 \,°C$$

The result is rounded to two significant figures because the equation is not expected to be accurate at high molality.

This is the amount by which the freezing point of the solution differs from that of pure water. Because the freezing point of water is 0 °C and freezing points are depressed by adding solutes, the new freezing point is below 0 °C: $T_f = -12 \,°C$.

The effect of a solute on the boiling point of a solution is opposite to its effect on the freezing point. A nonvolatile solute *increases* the boiling point of a solution. This is because the solute blocks some of the solvent molecules from reaching the surface of the solution and thus decreases the rate of escape into the gas phase. To get back to dynamic equilibrium, the solution must be heated so that more molecules acquire sufficient energy to escape from the liquid phase.

A nonvolatile solute is one that has a negligible vapor pressure at the boiling point of the solution.

Molecular analysis and experimental studies show that the change in the boiling point of a solution obeys the same type of equation as the change in the freezing point:

$$\Delta T_b = K_b \, c_m \tag{10-3}$$

In Equation 10-3, ΔT_b is the elevation of the boiling point, c_m is the molality of the solute, and K_b is a constant called the **boiling point elevation constant.** The constant depends on the identity of the solvent but not on the identities of the nonvolatile solutes. Thus there is a different boiling point elevation constant for every solvent; for water, $K_b = 0.512 \,°C \, kg/mol$. Equation 10-3 is used in the same way as Equation 10-2.

The temperature changes described by Equations 10-2 and 10-3 are caused by the number of solute particles present in a given amount of solvent. When an ionic salt dissolves in water, each mole of salt produces two or more moles of ions. A dilute solution of sodium chloride (NaCl), for example, contains two moles of ions for every mole of NaCl—one mole of Na^+ cations and one mole of Cl^- anions. The proper application of Equations 10-2 and 10-3 takes this into account by viewing c_m as the *total* number of moles of particles per kilogram of solvent. When this is done, however, the predicted values of ΔT_f or ΔT_b are higher than experimental values. This is because cations and anions in aqueous solutions form ion pairs, reducing the total number of solute particles.

Osmosis

Water molecules can pass through cell membranes, but most solutes cannot. These are semipermeable membranes, and the movement of water through them is called **osmosis.**

If a semipermeable membrane separates two identical solutions, solvent molecules move in both directions at the same rate, and there is no net osmosis. The solutions on the two sides of the membrane are at dynamic equilibrium. The situation changes when the solutions on the two sides of the membrane are different. Consider the membrane in Figure 10-38a, which has pure water on one side and a solution of sugar in water on the other. The sugar molecules reduce the concentration of solvent molecules in the solution. Consequently, more solvent molecules pass through the membrane from the pure solvent to the solution than from the solution to the pure solvent. Now water flows from the pure-solvent side to the solution side, and there is a net rate of osmosis.

In the absence of other forces, osmosis continues until the concentration of solvent is the same on both sides of the membrane. However, pressure can be used

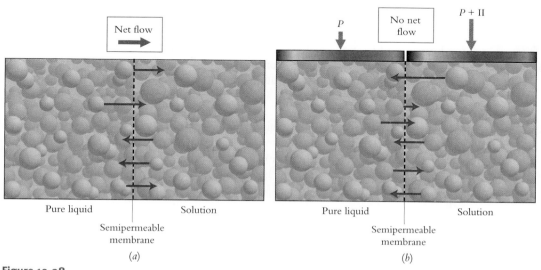

Figure 10-38
(*a*) The solvent concentration in a solution is lower than the solvent concentration in pure liquid, leading to an imbalance in flow through a semipermeable membrane. (*b*) By increasing the pressure on the solution, the rate of solvent flow out of the solution can be increased until it equals the rate of flow out of the pure solvent.

to stop this process. An increase in pressure on the solution side pushes solvent molecules against the membrane and thereby increases the rate of transfer of water molecules from the solution side to the solvent side.

Figure 10-38*b* shows that dynamic equilibrium can be established by increasing the pressure on the solution until the rate of solvent transfer is equal in both directions. The pressure increase needed to equalize the transfer rates is called the **osmotic pressure** (Π). Osmotic pressure is a pressure *difference*. Pressure is exerted on both sides of a semipermeable membrane, but Π is the extra pressure that must be exerted on the solution to maintain dynamic equilibrium.

Like freezing point depression and boiling point elevation, osmotic pressure is proportional to the concentration of solute molecules. Experiments show that the osmotic pressure is proportional to both concentration (expressed as molarity, M) and temperature:

$$\Pi = MRT \qquad\qquad (10\text{-}4)$$

In Equation 10-4, M is the total molarity of all solutes, T is the temperature in kelvins, and R is the gas constant. We must express R in appropriate units. If osmotic pressure is expressed in atmospheres, the fact that molarity is in moles per liter requires us to use $R = 0.08206$ L atm/mol K.

Osmotic pressure effects can be substantial. For example, ocean waters contain dissolved salts at a total ionic molarity of about 1.13 M. The osmotic pressure of ocean water can be calculated:

$$\Pi = MRT = (1.13\ \text{mol/L})(0.08206\ \text{L atm/mol K})(298\ \text{K}) = 27.6\ \text{atm}$$

This means that if pure water and ocean water were placed on opposite sides of a semipermeable membrane, an external pressure of 27.6 atm would have to be applied to the ocean water side to prevent osmosis. The osmotic pressure of ocean water is more than 25 times atmospheric pressure. By comparison, the freezing point of ocean water is depressed by only about 1% from the freezing point of pure water, from 273 K to about 271 K (-2 °C).

Osmotic pressure plays a key role in biological chemistry because the cells of the human body are encased in semipermeable membranes and bathed in body fluids. Under normal physiological conditions, the body fluid outside the cells has the same total solute molarity as the fluid inside the cells, and there is no net osmosis across cell membranes. Solutions with the same solute molarity are called *isotonic* solutions.

The situation changes if a molarity imbalance is created. Figure 10-39 shows red blood cells immersed in solutions of different molarities. When the fluid outside the cell is at higher solute molarity, transport of water across the membrane into the cell slows. The net result is that water leaves the cell, causing it to shrink. When the fluid outside the cell is at lower molarity, movement of water into the cell increases. The extra water in the cell causes an increase in internal pressure. Eventually, this internal pressure of the cell would match the osmotic pressure, and water transport would reach dynamic equilibrium. Unfortunately, osmotic pressures are so large that cells can burst under the increased pressure before they reach equilibrium.

Red blood cells are particularly susceptible to these potentially damaging concentration changes because they are suspended in the aqueous medium of the blood. Consequently, solutions used for intravenous feeding must be isotonic. Example 10-9 deals with isotonic solutions.

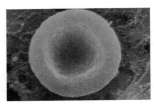
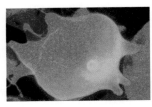

Figure 10-39
When bathed in isotonic solution (*top*), red blood cells retain their normal shape because there is no net osmosis across their membranes. In a solution at higher concentration (*center*), the osmotic flow removes water from the cell interior, causing cells to shrink and wrinkle. In a solution at lower concentration (*bottom*), the net osmotic flow pumps water into cells, expanding them until they may rupture.

Isotonic Solutions	Example 10-9

Isotonic intravenous solutions contain 49 g/L of glucose ($C_6H_{12}O_6$). What is the osmotic pressure of blood?

Strategy: Isotonic solutions, by definition, exert equal osmotic pressure. Therefore Π for blood is the same as Π for the glucose solution. We can calculate Π from Equation 10-4 after converting the concentration into moles per liter:

$$\Pi = MRT \quad \text{and} \quad M = \frac{n}{V} = \frac{m/MM}{V}$$

Solution: From the formula of glucose, $MM = 180$ g/mol. Substitute to find the molarity of the glucose solution:

$$M = \frac{(49 \text{ g})}{(180 \text{ g/mol})(1 \text{ L})} = 0.272 \text{ mol/L}$$

Because we are working with blood in the human body, T is human body temperature, which is 37 °C.

$$T = 37 °C = 37 + 273 = 310 \text{ K} \qquad R = 0.08206 \text{ L atm/mol K}$$

$$\Pi = (0.272 \text{ mol/L})(0.08206 \text{ L atm/mol K})(310 \text{ K}) = 6.9 \text{ atm}$$

The result is rounded to two significant figures to match the initial data (49 g/L).

If a sample of red blood cells is added to pure water, osmosis carries water into the cells. This process would continue until the internal pressure of the cell was 6.9 atm higher than the pressure on the outside of the cell. However, 6.9 atm is much more than the cell membrane can tolerate. Consequently, red blood cells burst when immersed in pure water.

When the additional pressure applied to a solution is less than the osmotic pressure, solvent molecules flow from pure solvent into the solution. When the additional pressure equals the osmotic pressure, equilibrium is established, and there is no net flow of solvent molecules. What if we apply an additional pressure that is greater than the osmotic pressure? Now the pressure on the solution is sufficient to drive solvent molecules from the solution into the pure solvent. Osmosis now transfers solvent in the opposite direction; **reverse osmosis** occurs.

Reverse osmosis can be used to purify water, because the liquid passing through the semipermeable membrane is pure solvent. A reverse osmosis water purifier requires semipermeable membranes that do not rupture under the high pressures required for reverse osmosis. This presents formidable engineering difficulties. Recall that seawater has an osmotic pressure of nearly 28 atm and that red blood cells rupture by 7 atm. Nevertheless, in recent years, membranes have been developed that make it feasible to purify water using this technique.

Determination of Molar Mass

The magnitude of osmotic pressure is large enough that measurements of Π provide a convenient way to determine the molar mass of a compound. The osmotic pressure equation (Equation 10-4) can be solved for molar mass after molarity is expressed in terms of mass and molar mass:

$$\Pi = MRT \qquad \Pi = \frac{mRT}{V(MM)}$$

A simple rearrangement gives an equation for calculating molar mass:

$$MM = \frac{mRT}{\Pi V} \tag{10-5}$$

For determination of the molar mass of an unknown compound, a measured mass of material is dissolved to give a measured volume of solution. The system is held at constant temperature, and the osmotic pressure is determined. Osmotic pressure measurements are particularly useful for determining the molar masses of large molecules such as polymers and biological materials, as Example 10-10 illustrates.

Example 10-10 | **Determining Molar Mass**

A 25.00-mL aqueous solution containing 0.420 g of hemoglobin has an osmotic pressure of 4.6 torr at 27 °C. What is the molar mass of hemoglobin?

Strategy: Equation 10-5 is used to calculate the molar mass by osmometry.

Solution: All of the necessary data are given in the problem:

$$m_{solute} = 0.420 \text{ g} \quad R = 0.08206 \text{ L atm/mol K} \quad T = 27 \text{ °C} + 273 = 300 \text{ K}$$

$$\Pi = (4.6 \text{ torr})(1 \text{ atm}/760 \text{ torr}) = 6.05 \times 10^{-3} \text{ atm}$$

$$V_{solution} = (25.00 \text{ mL})(1 \text{ L}/1000 \text{ mL}) = 2.500 \times 10^{-2} \text{ L}$$

$$MM = \frac{mRT}{\Pi V} = \frac{(0.420 \text{ g})(8.206 \times 10^{-2} \text{ L atm/mol K})(300 \text{ K})}{(6.05 \times 10^{-3} \text{ atm})(2.500 \times 10^{-2} \text{ L})}$$

$$MM = 6.8 \times 10^4 \text{ g/mol}$$

This is a large value for a molar mass, but it is reasonable because biological molecules often are quite large. As in any calculation, be careful to express all data in appropriate units. The osmotic pressure was measured to two significant figures, so the result has two significant figures.

Section Exercises

10.6.1 A water-soluble protein molecule has a molar mass of 985 g/mol. Calculate the freezing point depression, boiling point elevation, and osmotic pressure at 27 °C of an aqueous solution containing 0.750 g/L of this protein. (Assume that the solution has a density of 1.000 g/mL.)

10.6.2 Redraw Figure 10-38*b* using arrows to represent water movement across the membrane during reverse osmosis.

10.6.3 In your own words, write a detailed, molecular-level description of how reverse osmosis can be used to desalinate seawater.

10.7 DUAL-NATURE MOLECULES: SURFACTANTS AND BIOLOGICAL MEMBRANES

As described in our chapter introduction, substances that do not dissolve in water, such as organic fats and oils, are called **hydrophobic.** Substances that are miscible with water, including hydrogen-bonding molecules such as methanol and acetone, are called **hydrophilic.** Some molecules contain both hydrophilic and hydrophobic regions. Such a dual-nature molecule may have a polar or ionic head that is compatible with water and a long hydrocarbon tail that is incompatible with water. Sodium stearate, whose structure is shown in Figure 10-40, is a dual-nature molecule. The head of the stearate anion resembles the water-soluble acetate anion, and the tail is a hydrocarbon chain containing 17 carbon atoms.

Dual-nature molecules such as sodium stearate can form three different structures when they are placed in water (Figure 10-41). They may form a molecular *monolayer* on the surface, in which the polar head groups are immersed in the water while the nonpolar tails are aggregated together on the surface. Agitating the solution may cause the molecules to arrange into spherical aggregates called *micelles,* in which the hydrophobic tails point inward and the polar heads lie on the outside of the structure, where they interact with the aqueous solvent. Dual-nature molecules may also form enclosed bilayers, called *vesicles,* which have two parallel rows of molecules oriented so that their hydrocarbon tails are clustered.

All these arrangements obey the principle of like dissolving like. The hydrocarbon tails aggregate through dispersion forces because they are incompatible with the aqueous medium. The hydrogen-bonding network of the solvent would be disrupted by incorporating these tails into the solution. The polar heads, on the other hand, interact strongly with water to maximize hydrogen bonding and ion-dipole interactions.

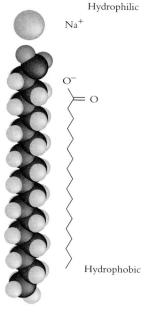

Sodium stearate
$NaC_{17}H_{35}CO_2$

Figure 10-40
Sodium stearate is a typical dual-nature molecule. It has an ionic, hydrophilic polar head and a hydrophobic nonpolar tail.

Surfactants

Dual-nature molecules are widely used in industry to modify the behavior of aqueous solutions. In this context they are called **surfactants.** Common surfactant head groups include carboxylate ($-CO_2^-$), sulfonate ($-SO_3^-$), sulfate ($-OSO_3^-$), and ammonium ($-NH_3^+$). The negative charge of anionic head groups usually is neutralized by Na^+, and the positive charge of ammonium usually is neutralized by

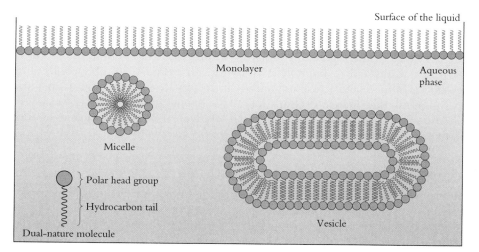

Figure 10-41
Cross-sectional molecular views of the structures that can form when dual-nature molecules are placed in water. The molecules may form a monolayer at the surface, spherical clusters called micelles, or bilayer structures called vesicles.

Cl$^-$. These counterions are used because they are nontoxic and their salts are highly soluble.

The most familiar surfactants are soaps and detergents. Clothing becomes soiled by a wide variety of substances; some are water-soluble, and others are not. A soap or detergent removes water-insoluble grease (for example, butter, fat, and oil) from the clothing surface. Dispersion forces stabilize the attachment of grease particles to the hydrocarbon tails of surfactant molecules, which form aggregates. Agitation removes these aggregates from the fabric, suspending them in solution as tiny micelles with grease particles trapped inside. The micelles do not redeposit on the fabric because the hydrophilic heads of the surfactants hold them in solution. When water is drained from the washing machine, the grease-containing micelles are swept away, leaving clean clothes behind.

Soaps are derived from natural sources such as animal fats that contain stearic acid and other long-chain organic acids. These carboxylate surfactants form insoluble salts with Ca^{2+} and Mg^{2+}. In regions where water is hard, these soaps precipitate calcium and magnesium stearate scum that inhibits cleansing action and is responsible for bathtub rings. Detergents such as sodium lauryl sulfate, on the other hand, are "synthetic" compounds, originally prepared in the laboratory, that contain sulfonate and sulfate head groups. These groups do not form precipitates with 2+ cations. The cleaning action of soaps and detergents is similar, but detergents have largely replaced soaps because of their superior behavior in hard water.

Surfactants are used in such a wide variety of ways that billions of dollars are spent on them every year. They appear in many household products, including cleansing agents and shampoos (recall the chapter introduction). Some surfactants are used as emulsifiers in processed foods such as bottled salad dressing. An emulsifier causes normally incompatible liquids such as the oil and water in salad dressing to disperse in each other, by forming molecular connections between the liquids. The emulsifier's hydrophobic tail interacts with oil molecules, and its hydrophilic head interacts with water molecules.

Approximately half of the surfactants produced in the United States are used in household and industrial cleaning products, but the remaining half are used in a wide range of industries. In agriculture, surfactants are used as wetting agents that assist in the uniform application of sprayed pesticides. They also are used to prevent caking of fertilizers. Additives to agricultural products must not interfere with the active agents and must be biodegradable and environmentally benign. In the food industry, various surfactants are used as emulsifiers, cleaners, foaming agents, and antifoaming agents. Paints are dispersions of dyes, binding agents, and fillers. Most paints contain surfactants that improve their flow and mixing properties. Surfactants are used widely in the plastics industry as foaming agents to assist in the production of plastic foams and to improve moldability and extrudability of specially shaped products. In the manufacture of textiles, surfactants are used to clean natural fibers, as lubricants that reduce friction during the spinning and weaving processes, as emulsifiers that improve the application of dyes and finishes, and as antistatic agents.

Cell Membranes

It may seem like a huge conceptual leap from industrial surfactants to biological cell membranes, but the same principles apply to both sets of substances.

Every biological cell is surrounded by a membrane only a few molecules thick. Among the major components of membranes are molecules called *phospholipids*, which are dual-nature molecules. Although their chemical structures are more

Soap made by boiling animal fat in an alkaline solution obtained from ashes has been known since the time of the ancient Sumerians, 2500 BC.

complex than simple surfactants such as sodium stearate, phospholipids nevertheless have hydrophilic heads and hydrophobic tails. Figure 10-42 shows the structure of one membrane phospholipid, lecithin. The hydrophilic end of lecithin has a cationic $N(CH_3)_3^+$ group and eight oxygen atoms with nonbonding pairs of electrons, all of which form hydrogen bonds with water molecules. The hydrophobic portion of lecithin consists of two hydrocarbon tails.

Phospholipids form bilayers in aqueous media. The molecules align in two approximately parallel rows with tails on the inside and heads toward the outside, in contact with the solution. This arrangement, shown in Figure 10-43, is analogous to the vesicles in Figure 10-41. The bilayer forms a closed sac that contains the aqueous cytoplasm and all the cellular components. Thus a cell can be viewed as a large and complex vesicle.

One purpose of a cellular lipid bilayer is to control which molecules pass into and out of the cell. Uncharged small molecules such as water, ammonia, and oxygen can diffuse through the membrane. Hydrophobic molecules such as hydrocarbons can also pass through, because they are soluble in the overlapping tails that make up the interior of the bilayer. Ions and water-soluble polar molecules such as glucose and urea, on the other hand, cannot get through the membrane.

For cells to carry out their functions, glucose and other nutrients must be brought in, and urea and other waste products must be expelled. This would be an

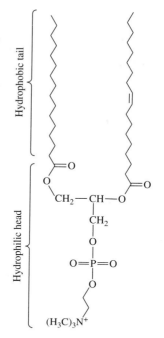

Figure 10-42
The chemical structure of lecithin. Lecithin is one of the most common phospholipids present in cell membranes. It is also used as a "natural" emulsifier in beauty products.

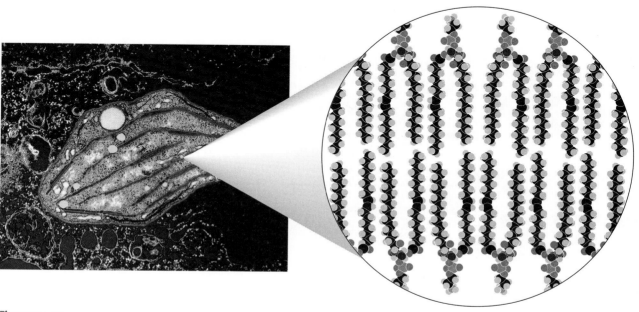

Figure 10-43
A lipid bilayer contains two layers of dual-nature molecules arranged tail to tail. The polar head groups face outward and are stabilized by dipolar and hydrogen-bonding interactions with water molecules.

Figure 10-44
A schematic representation of a cell membrane. Proteins embedded in the lipid bilayer play a variety of roles. Among other tasks, protein membranes act as gates for transporting ions and molecules into and out of the cell.

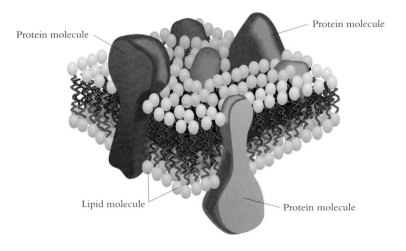

Protein molecule

Protein molecule

Protein molecule

Lipid molecule

impossible task if cell membranes were composed only of phospholipids. Specific large biomolecules act as molecular gates through the membranes. These proteins are embedded in the bilayers but protrude into the surrounding water or into the cell interiors, as shown schematically in Figure 10-44.

CHAPTER 11 →
The structures of proteins are described in Chapter 11.

Section Exercises

■ **10.7.1** Line drawings of some molecules follow. Identify the hydrophilic and hydrophobic regions of each and determine which are surfactants.

Tetramethylammonium chloride

Benzalkonium chloride
(used as a disinfectant)

Dipentyl ether

■ **10.7.2** Explain why glucose and other large, water–soluble molecules cannot pass through a lipid bilayer. (The structure of glucose is shown in Figure 10-35.)

■ CHAPTER REVIEW

Summary and Key Words

1. Attractive forces between molecules cause most substances to be liquids or solids under normal conditions, and they lead to nonideal behavior of gases at high pressure and low temperature. Real gases deviate from ideality because of intermolecular attractions and finite molecular sizes. These deviations can be described quantitatively using the **van der Waals equation.**

2. Intermolecular forces determine the boiling point of a substance. These interactions include **dispersion forces, dipolar forces,** and **hydrogen bonding.** Disperson forces arise from the **polarizability** of the molecule's electron cloud. Hydrogen bonding involves partial sharing of electrons between a lone pair of electrons on a fluorine, oxygen, or nitrogen atom and a hydrogen atom in a highly polar bond.

3. Solids are categorized according to the type of forces holding them together. **Molecular solids** have dispersion, dipole, and hydrogen bonding forces, **metallic solids** have delocalized bonding, **network solids** have covalent bonding, and **ionic solids** have ion-ion interactions. Solid properties vary widely because of the variety of these forces.

4. Any **crystalline solid** is composed of a repeating pattern whose smallest complete part is a **unit cell.** The simplest of these repeat patterns are **close-packed structures**, adopted by many atomic and metallic solids: **hexagonal close-packed, face-centered cubic,** and **body-centered cubic** structures. **Amorphous solids,** such as **glasses,** lack regular structures. **Crystalline defects** can impart useful properties to solids.

5. The molecules in liquids cohere but move freely. Liquid droplets tend to form spheres because of **surface tension,** and in a confined space they form a curved **meniscus** because they adhere to the surface of their container. Adhesive forces also generate **capillary action,** by which liquids rise in narrow tubes. Liquids resist flow to varying degrees, measured by **viscosity.** A solution is a homogeneous mixture of varying amounts of solutes contained in a **solvent.** Substances that are subject to similar intermolecular forces tend to dissolve in each other and may be completely **miscible.** When intermolecular forces are dissimilar, substances are likely to be insoluble. In general, "like dissolves like."

6. The **normal freezing point** and **normal boiling point** are the temperatures at which a pure liquid freezes or boils under 1 atm pressure. Solutes generate **colligative properties:** They depress the freezing point and vapor pressure, raise the boiling point, and generate an osmotic pressure of a solution. The **freezing point depression constant** and **boiling point elevation constant** relate changes in these temperatures to solution **molality.** The **osmotic pressure** is a pressure difference between a pure solvent and a solution. **Reverse osmosis** can be used to purify solutions.

7. Substances that do not dissolve in water are **hydrophobic,** while those that dissolve in water are **hydrophilic. Surfactants** are molecules that contain water-compatible and water-incompatible structures. Surfactants form monolayers, micelles, and vesicles in aqueous media. Similar dual-nature molecules make up cell membranes.

Skills to Master

▶ Explaining variations in boiling points

▶ Identifying hydrogen bonds

▶ Describing surface tension and viscosity

▶ Recognizing types of solids

▶ Depicting simple crystal types

▶ Drawing molecular pictures of solutions

▶ Predicting solubility patterns

▶ Calculating colligative properties

▶ Describing surfactant properties

Learning Exercises

10.1 List all the types of interactions that can act to hold a solid together. Organize the list from strongest to weakest.

10.2 Draw molecular pictures that show every type of hydrogen bond that exists in a solution containing methanol, water, and ammonia.

10.3 Write a paragraph that describes the phenomenon of superconductivity.

10.4 Define and give an example of each of the following: (a) close-packed structure; (b) unit cell; (c) molecular solid; (d) covalent solid; (e) amorphous solid; and (f) surfactant.

10.5 Update your list of memory-bank equations. Be sure to mention how the equations in this chapter are used.

10.6 Write a paragraph that describes the types of substances that form monolayers, micelles, and vesicles in water. Explain the differences among these structures.

10.7 Prepare a list of the terms in Chapter 10 that are new to you. Write a one-sentence definition for each, using your own words. If you need help, consult the glossary.

Problems ilw = interactive learning ware problem. Visit the website at www.wiley.com/college/olmsted

The Nature of Intermolecular Forces

10.1 Xenon condenses at 166 K, krypton condenses at 121 K, and argon condenses at 87 K. Draw a bar graph similar to Figure 10-3 showing the relative magnitudes of interatomic attractive energies for these three noble gases. Include two horizontal lines representing the average kinetic energy at 140 K and 100 K.

10.2 Methane condenses at 121 K, but carbon tetrachloride boils at 350 K. Draw a bar graph similar to that in Figure 10-3 showing the relative attractive energies between molecules for these two substances. Include a horizontal line representing the average kinetic energy at room temperature (298 K).

10.3 Predict whether the effects of molecular volume and of intermolecular attractions become more or less significant when the following changes are imposed: (a) A gas is expanded to a larger volume at constant temperature. (b) More gas is introduced into a container of constant volume at constant temperature.

10.4 Predict whether the effects of molecular volume and of intermolecular attractions become more or less significant when the following changes are imposed: (a) The temperature of a gas is lowered at constant volume. (b) A gas is compressed into a smaller volume at constant temperature.

10.5 Draw pictures showing the atomic arrangements of samples of $Ag(s)$, $Ar(g)$ and $Hg(l)$.

10.6 Draw pictures showing the atomic arrangements of samples of Zn(s), Kr(g) and Ga(l).

10.7 From the following experimental data, calculate the percent deviation from ideal behavior: 1.00 mol CO_2 in a 1.20-L container at 40.0 °C exerts 19.7 atm pressure.

10.8 From the following experimental data, calculate the percent deviation from ideal behavior: 3.000 g H_2 at 0.00 °C and 193.5 atm occupies a volume of 189.18 cm^3.

10.9 Chlorine gas is commercially produced electrochemically from sea water and then stored under pressure in metal tanks. A typical tank has a volume of 15.0 L and contains 1.25 kg of Cl_2. Use the van der Waals equation to calculate the pressure in this tank if the temperature is 295 K, and compare the result with the ideal gas value.

10.10 Chlorine gas used in chemical syntheses is transferred from tanks under high pressure to reactor flasks at considerably lower pressure. If 10.5 g of Cl_2 is introduced into a 5.00-L reaction flask at 145 °C, calculate the pressure of Cl_2 gas using the van der Waals equation and using the ideal gas equation.

Types of Intermolecular Forces

10.11 Arrange the following in order of ease of liquefaction: CCl_4, CH_4, and CF_4. Explain your ranking.

10.12 Arrange the following in order of increasing boiling point: Ar, He, Ne, and Xe. Explain your ranking.

10.13 Draw molecular pictures of CCl_4 and CH_4 showing the relative amount of polarizability of each.

10.14 Draw molecular pictures of Ne and Xe showing the relative amount of polarizability of each.

10.15 List ethanol (CH_3CH_2OH), propane ($CH_3CH_2CH_3$), and *n*-pentane ($CH_3CH_2CH_2CH_2CH_3$) in order of increasing boiling point, and explain what features determine this order.

10.16 List *n*-propanol ($CH_3CH_2CH_2OH$), dimethyl ether (CH_3OCH_3) and diethyl ether ($CH_3CH_2OCH_2CH_3$) in order of increasing boiling point and explain what features determine this order.

10.17 Which of the following will form hydrogen bonds with another molecule of the same substance? Draw molecular pictures illustrating such hydrogen-bond formation. (a) CH_2Cl_2; (b) H_2SO_4; (c) H_3COCH_3; (d) $H_2NCH_2CO_2H$.

10.18 Which of the following form hydrogen bonds with water? Draw molecular pictures illustrating such hydrogen-bond formation. (a) CH_4; (b) I_2; (c) HF; (d) H_3COCH_3; and (e) $(CH_3)_3COH$.

10.19 Draw Lewis structures showing all possible hydrogen-bonding interactions for (a) two NH_3 molecules; (b) one NH_3 molecule and one H_2O molecule.

10.20 Draw Lewis structures that show the hydrogen bonding interactions for (a) two CH_3OH molecules; and (b) One HF molecule and one acetone molecule [$(CH_3)_2C=O$].

Forces in Solids

10.21 Classify each of the following as an ionic, network, molecular, or metallic solid: Sn, S_8, Se, SiO_2, and Na_2SO_4.

10.22 Classify each of the following as an ionic, network, molecular, or metallic solid: Pt, P_4, Ge, As_2O_3, and $(NH_4)_3PO_4$.

10.23 Describe the differences in (a) bonding characteristics and (b) macroscopic properties between metals and network solids.

10.24 Describe the differences in (a) interparticle forces and (b) macroscopic properties between molecular and ionic solids.

10.25 Indicate what type of solid (ionic, network, metallic, or molecular/atomic) each of the following forms on solidification: (a) Br_2; (b) KBr; (c) Ba; (d) SiO_2; and (e) CO_2.

10.26 Indicate what type of solid (ionic, covalent, metallic, or molecular/atomic) each of the following forms on solidification: (a) HCl; (b) KCl; (c) NH_4NO_3; (d) Mn; and (e) Si.

Order in Solids

10.27 The density of graphite is 2260 kg m^{-3}, whereas that of diamond is 3513 kg m^{-3}. Describe the bonding features that cause these two forms of carbon to have different densities.

10.28 Amorphous silica has a density of around 2.3 g/cm^3, and crystalline quartz has a density of 2.65 g/cm^3. Describe the bonding features that cause these two forms of the same substance to have different densities.

10.29 The unit cell of the mineral perscovite follows. What is the formula of perscovite?

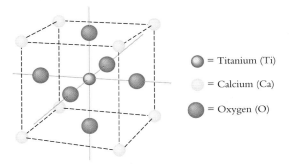

= Titanium (Ti)
= Calcium (Ca)
= Oxygen (O)

10.30 The unit cell of a compound of xenon and fluorine follows. Determine the chemical formula of the compound.

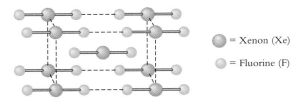

= Xenon (Xe)
= Fluorine (F)

10.31 Construct part of the Lewis structure of carborundum, the diamond–like compound of empirical formula SiC.

10.32 Construct part of the Lewis structure of quartz, the network solid of empirical formula SiO_2.

Liquids and Solutions

10.33 Pentane is a C_5 hydrocarbon; gasoline contains mostly C_8 hydrocarbons, and fuel oil contains hydrocarbons in the C_{12} range. List these three hydrocarbons in order of increasing viscosity, and explain what molecular feature accounts for the variation.

10.34 Given that a lubricant must flow easily in order to perform its function, which grade of motor oil is preferred for winter use, high or low viscosity? Why?

10.35 Salad oil (whose major ingredients are long–chain fatty acids such as $C_{17}H_{35}CO_2H$) and vinegar (a dilute aqueous solution of acetic acid, CH_3CO_2H) are not miscible. Explain.

$C_{17}H_{35}CO_2H$

10.36 Acetonitrile, CH_3CN, is miscible with water and miscible with cyclohexane (C_6H_{12}), but water and cyclohexane are nearly insoluble in each other. Explain.

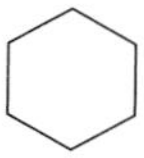

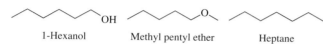

Acetonitrile   Cyclohexane

10.37 Ammonia can be condensed to a liquid at low temperature. What kinds of solids would you expect to be soluble in liquid ammonia?

10.38 Mercury is the only metallic element that is a liquid at room temperature. What kinds of solids would you expect to be soluble in liquid mercury?

10.39 How many hydrogen bonds to water molecules can one glycerol molecule ($HOCH_2CHOHCH_2OH$) form? Draw Lewis structures that show the hydrogen bonding of a glycerol molecule dissolved in water.

10.40 How many hydrogen bonds to water molecules can be formed by one molecule of glycine in its ionic form ($H_3NCH_2CO_2$)? Draw Lewis structures that show the hydrogen bonding of a glycine molecule dissolved in water.

10.41 A sample for gas chromatography contains the following compounds:

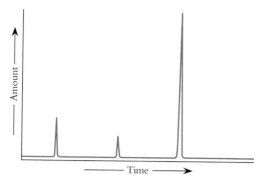

1-Hexanol   Methyl pentyl ether   Heptane

The sample is passed through a gas chromatography column whose condensed phase attracts polar compounds. The chromatograph looks like this:

(a) Match each peak with the appropriate compound. (b) Determine which compound is present in largest amount and which in smallest amount.

10.42 You have prepared a sample of polymer and have performed liquid chromatography using a molecular sieve column to determine its molecular size. The chromatogram follows:

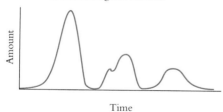

(a) How many components does your sample contain? (b) Is there a larger amount of long-, medium-, or short-chain polymer molecules in the sample? Explain.

Effects of Solutes on Aqueous Solutions

ilw 10.43 Estimate the freezing point of a wine that is 12% by mass ethanol. (Ignore all other solutes.)

10.44 Calculate the freezing point of a solution that contains 12.50 g sucrose ($C_{12}H_{22}O_{11}$) in 155 mL of water.

10.45 Do you have enough information to calculate the boiling point of the wine in Problem 10.43? If so, calculate it. If not, explain what feature of wine prevents you from doing this calculation.

10.46 Do you have enough information to calculate the boiling point of the solution in Problem 10.44? If so, calculate it. If not, explain what feature of the solution prevents you from doing this calculation.

10.47 An aqueous solution contains 1.00 g/L of a derivative of the detergent lauryl alcohol. The osmotic pressure of this solution at 25.0 °C is measured to be 64.8 torr. (a) What is the molar mass of the detergent? (b) The hydrocarbon portion of the molecule is an 11-carbon chain. What is the molar mass of the polar portion?

10.48 An aqueous solution containing 1.00 g of a sugar in 1.00×10^2 mL of solution has an osmotic pressure of 1.36 atm at 25 °C. What is the molar mass of this sugar?

Dual-Nature Molecules

10.49 Of the following compounds, which will be the best and which will be the worst surfactant? Support your choices with molecular pictures. (a) Propanoic acid, $H_3CCH_2CO_2H$; (b) sodium lauryl sulfate, $H_3C(CH_2)_{11}OSO_3^- Na^+$; and (c) lauryl alcohol, $H_3C(CH_2)_{11}OH$.

10.50 Of the following compounds, which will be the best and which will be the worst surfactant? Support your choices with molecular pictures. (a) Sodium alkylbenzenesulfonate, $C_{18}H_{29}SO_3Na$; (b) octadecane, $C_{18}H_{38}$; and (c) decanoic acid, $H_3C(CH_2)_8CO_2H$.

10.51 The figure below represents a monolayer of the surfactant sodium stearate on a liquid surface. Decide if the liquid is polar or nonpolar. Explain your reasoning.

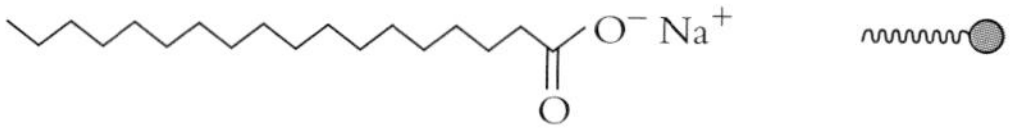

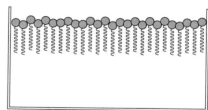

Sodium stearate   Simplified depiction of the stearate anion

10.52 Stearic acid forms a monolayer on the surface of gasoline. Draw a molecular picture that shows how stearic acid molecules are arranged in this monolayer.

Additional Paired Problems

10.53 You have been hired by the local utility company as a quality control chemist. They ask you to analyze samples of their natural gas, which they buy from several sources. What technique would you recommend? Describe how this technique works.

10.54 You have prepared a new polymer and want to determine if it has components with different molar masses or is uniform in molar mass. What technique would provide this information most conveniently? Describe how the technique works.

10.55 What molality of ethylene glycol is required to protect the water in an automobile cooling system from freezing at $-20\ °C$?

10.56 Compute the molar mass of vitamin C, if a solution containing 22.0 g in 1.00×10^2 g of water freezes at $-2.33\ °C$.

10.57 Rank the following substances in order of increasing solubility in water, and state the reasons for your rankings: C_6H_6 (benzene), $HOCH_2CH(OH)CH(OH)CH_2OH$ (erythritol), and $C_5H_{11}OH$ (pentanol).

10.58 Rank the following substances in order of increasing solubility in cyclohexane (C_6H_{12}) and state the reasons for your rankings: KCl, C_2H_5OH, and C_3H_8.

10.59 Dichlorodiphenyltrichloroethane (DDT) has the following structure:

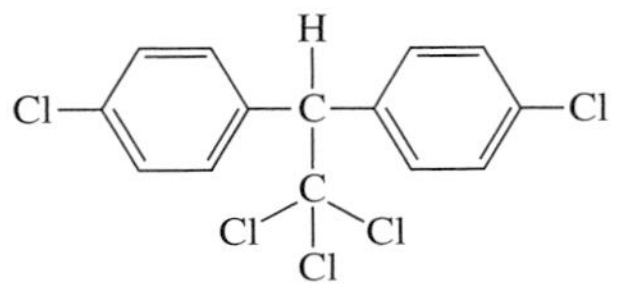

Is this compound hydrophilic or hydrophobic? Is it readily excreted by animals, or will it concentrate in fatty tissues? Does your answer explain why DDT has been banned as a pesticide?

10.60 Butylated hydroxytoluene (BHT) is used as a food preservative. It has the following molecular structure:

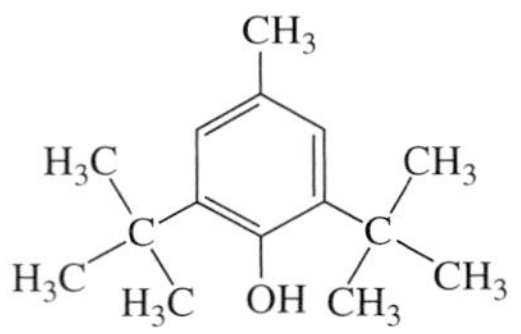

Would you expect to find this compound to be excreted in urine or stored in body fat? Explain your reasoning. BHT is nontoxic to humans.

10.61 Water and carbon tetrachloride are not miscible. When mixed together, they form two layers, like water and oil. If an aqueous solution of I_2 is shaken with CCl_4, the iodine is "extracted" into the CCl_4 layer. Explain this behavior based on your knowledge of intermolecular forces.

10.62 If some benzene is shaken with a mixture of water and carbon tetrachloride, the resulting mixture contains two layers (See Problem 10.61). Which layer contains the benzene? Explain this behavior based on your knowledge of intermolecular forces.

10.63 List all of the different kinds of forces that must be overcome to convert each of the following from a liquid to a gas: (a) NH_3; (b) $CHCl_3$; (c) CCl_4; and (d) CO_2.

10.64 List all the intermolecular forces that stabilize the liquid phase of each of the following: (a) Xe; (b) SF_4; (c) CF_4; and (d) CH_3CO_2H (acetic acid).

10.65 For each of the following pairs, identify which has the higher boiling point, and identify the type of force that is responsible: (a) H_3COCH_3 and CH_3OH; (b) SO_2 and SiO_2; (c) HF and HCl; and (d) Br_2 and I_2.

10.66 The boiling points of the Group 16 binary hydrides are as follows: H_2O, 100 °C; H_2S, $-60\ °C$; H_2Se, $-41\ °C$. Explain in terms of intermolecular forces why H_2S has a lower boiling point than either H_2O or H_2Se.

10.67 The compound 1,2-dichloroethylene (ClCHCHCl) exists as *cis* and *trans* isomers. One isomer boils at 47 °C, the other at 60 °C. Draw Lewis structures of the two isomers. Use dipole moments and symmetry arguments to assign the boiling points to the isomers.

10.68 Why do the two isomers shown have very different melting points?

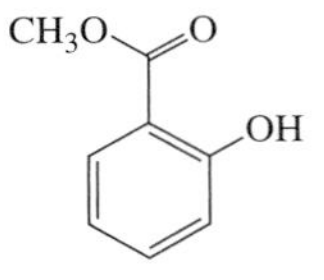

$C_8H_8O_3$
Methyl 2-hydroxybenzoate
(Oil of wintergreen)
mp = $-8\ °C$

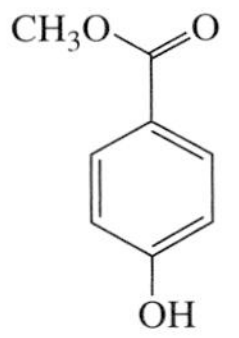

$C_8H_8O_3$
Methyl 4-hydroxybenzoate
mp = 127 °C

10.69 To make a good solder joint, the liquid metal solder must adhere well to the metal surfaces being joined. Flux is used to clean the metal surfaces. What types of substances must flux remove?

10.70 A pipet is considered to be dirty when water forms beads on its walls rather than forming a thin film that drains well. Which of the following on the surface of a pipet wall will make it dirty? In each case, explain the intermolecular forces underlying your classification. (a) Grease, (b) Mg^{2+} ions, (c) acetone, (d) SiO_2.

10.71 Which gas deviates more from ideal PV/nRT behavior, F_2 or Cl_2? Explain your choice.

10.72 Which gas deviates more from ideal PV/nRT behavior, CH_4 or SnH_4? Explain your choice.

More Challenging Problems

10.73 Does the boiling point of HCl (see the graph in Figure 10-14) suggest that it may form hydrogen bonds? Explain your answer and draw a molecular picture that shows the possible hydrogen bonds between HCl molecules.

10.74 One of the earliest methods of preserving fish was by salting. Explain what happens when fish is placed in concentrated salt solution.

10.75 Would water dissolve salts as well as it does if it had a linear structure (such as CO_2) instead of a bent one? Explain.

10.76 When an aqueous solution is cooled to low temperature, part of the water freezes as pure ice. What happens to the freezing point of the remaining solution when this occurs? A glass of wine placed in a freezer at $-10\ °C$ for a very long time forms some ice crystals but does not completely freeze. Compute the molality of ethanol in the remaining liquid phase.

10.77 Molecular hydrogen and atomic helium have two electrons, but He boils at 4.2 K, whereas H_2 boils at 20 K. Neon boils at 27.1 K, whereas methane, which has the same number of electrons, boils at 114 K. Explain why molecular substances boil at a higher temperature than atomic substances with the same number of electrons.

10.78 Quartz and glass are both forms of silicon dioxide. A piece of quartz breaks into a collection of smaller regular crystals with smooth faces. A piece of glass breaks into irregular shards. Use molecular structures to explain why the two solids break so differently.

10.79 The structures and boiling points of *n*-pentane and 2,2-dimethylpropane follow:

Pentane	2,2-Dimethylpropane
C_5H_{12}	C_5H_{12}
bp = 36 °C	bp = −10 °C

Use the boiling point data along with molecular drawings to explain how shape affects the magnitude of dispersion forces. (Hint: See Figure 10–10.)

10.80 A 0.570-g sample of dry ice (solid carbon dioxide) was sealed inside a 25.00-mL evacuated glass bulb. The CO_2 was allowed to evaporate at a temperature of 27 °C. Calculate the pressure inside the bulb using the ideal gas equation and using the van der Waals equation. Compare the two results in terms of intermolecular forces.

10.81 Fish have blood that is isotonic with seawater, which freezes at −2.3 °C. What is the osmotic pressure of fish blood at 15 °C?

10.82 Explain using intermolecular forces whether or not water can dissolve gasoline. Knowing that gasoline is less dense than water, would you use water to fight a gasoline fire? Explain.

10.83 Some surfactants form membranes that span small holes between two aqueous solutions. These membranes are liquid bilayers two molecules thick. Draw a molecular picture of one of these membranes.

10.84 Draw the unit cell of the NaCl crystal and determine the number of nearest neighbors of opposite charge for each ion in this unit cell.

10.85 In the two figures shown, the green molecule is about to strike the wall of its container. Assuming all other conditions are identical, which collision will exert greater pressure on the wall? Explain in terms of intermolecular interactions.

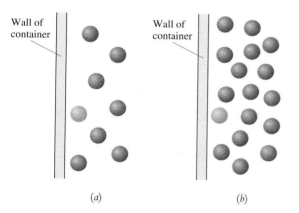

(a)   (b)

10.86 Brackish water, with a salt content around 0.5% by mass, is found in semiarid regions such as the American southwest. Assuming that brackish water contains only sodium chloride, estimate the osmotic pressure of brackish water at 298 K. (Hint: The total molarity is the sum of the molarities of Na^+ and Cl^-).

10.87 Arrange the following liquids in order of increasing viscosity, and state the factors that determine the ranking: butanol, $CH_3CH_2CH_2CH_2OH$; *n*-pentane, $CH_3CH_2CH_2CH_2CH_3$; propane-1,3-diol, $HOCH_2CH_2CH_2OH$; and 2,2-dimethylpropane, $(CH_3)_4C$.

Group Study Problems

10.88 Classify each of the following solids as network, metallic, ionic, or molecular: (a) a solid that conducts electricity; (b) a solid that does not conduct electricity but dissolves in water to give a conducting solution; and (c) a solid that does not conduct electricity and melts below 100 °C to give a non-conducting liquid.

10.89 Describe the similarities and differences between hexagonal close-packed and body-centered cubic structures.

10.90 Drinking seawater can be dangerous because it is more concentrated than a person's body fluids. Discuss the consequences, at both the cellular and the molecular levels, of drinking too much seawater.

10.91 Recently, a new group of solids was prepared that can act as superconductors at temperatures near the boiling point of liquid nitrogen. The unit cell of one of these new superconductors follows. Identify the formula of the compound.

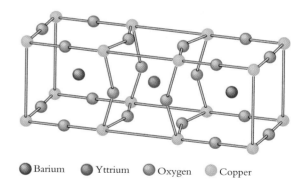

⬤ Barium   ⬤ Yttrium   ⬤ Oxygen   ⬤ Copper

10.92 Polystyrene is a plastic that dissolves in benzene. When a benzene solution (ρ = 0.88 g mL^{-1}) containing 8.5 g/L of polystyrene is placed in an osmometer at 25 °C, the solution reaches equilibrium when the liquid inside the osmometer is 10.9 cm higher than outside. Calculate the molar mass of the polystyrene.

Answers to Section Exercises

10.1.1 At_2 is a solid, with stronger intermolecular forces than I_2. Continuing the trends shown in Figure 10-3, the bar for At_2 should be higher than that for I_2.

10.1.2 Intermolecular forces cause PV/nRT to be smaller than 1.0. Fluorine reaches the smallest value, so it has the largest intermolecular forces. Helium never dips below 1.0, indicating very small intermolecular forces.

10.1.3 (a) Ar boils at a higher temperature than He, indicating that boiling point increases moving down Group 18. Thus Xe boils at a higher temperature than Ar. (b) The lower the boiling point of the element, the smaller its intermolecular forces. The order is H_2, N_2, F_2, Cl_2. (c) Group 18, which has the lowest boiling points, has the smallest intermolecular forces.

10.2.1 (a) Propane has larger polarizability because its electrons are spread over a larger volume than those in krypton. (b) CCl_4 has a larger polarizability because it has more valence electrons and because of the larger size of chlorine's $n = 3$ valence orbitals. (c) CO has a dipole moment, generating dipole-dipole attractions not present in N_2.

10.2.2 Acetone has dispersion forces and dipole-dipole forces. Methanol has dispersion forces, dipole-dipole forces, and hydrogen bonding forces. Their boiling points are similar because the dipolar and dispersion forces in acetone are larger than the analogous forces in methanol, counterbalancing the hydrogen bonding in methanol.

10.2.3

O—H---N   H—Ö—H---:N—H (with H atoms)   O—H---F   :Ö—H---:F—H (with H)

F—H---F   H—F:---H—F: F—H---N   :F—H---:N—H (with H atoms)

F—H---O   :F—H---:Ö—H (with H)   N—H---F   :N—H---:F—H (with H atoms)

N—H---N   :N—H---:N—H (with H atoms)   N—H---O   :N—H---:Ö—H (with H atoms)

10.3.1 (a) CO_2 is a molecular solid, with dispersion forces between the molecules. (b) S_8 is another molecular solid, with dispersion forces between the molecules. (c) Tin is a metal, with delocalized metallic bonding holding the atoms together. (d) Li_2O is ionic, with coulombic attractions between Li^+ cations and O^{2-} anions.

10.3.2 The order of melting points is F_2, K, Co, C (diamond). F_2 is molecular, with dispersion forces only. K and Co are metals, but Co has more valence electrons, resulting in stronger bonds. Diamond is a network solid with strong covalent bonds.

10.3.3 (a) $Mg_3(PO_4)_2$; (b) $LiAlSi_2O_6$; (c) $Bi_2Sr_2CuO_6$.

10.4.1 All angles in a hexagonal crystal are multiples of 60°, so we expect angles of 60°, 120°, and so on. A cubic close-packed crystal also has right angles, so we expect 90° angles as well.

10.4.2 An eighth of each corner atom and half of each face atom is in the unit cell.

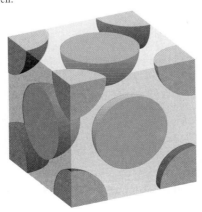

10.4.3

10.5.1 Zn^{2+} is soluble because the solution is stabilized by ion-dipole interactions between partially negative O atoms in water molecules and the Zn^{2+} cations, which replace ion-ion interactions in a zinc salt. Zn metal is insoluble because the cations in the metal are strongly bound by the "sea" of electrons in delocalized orbitals.

10.5.2 Intermolecular interactions are dispersion, dipole-dipole, and hydrogen bonding.

Dipole-dipole   Hydrogen bonding

10.5.3 Only Na_4SiO_4, which is an ionic solid, is water-soluble.

10.6.1 Freezing point depression = 1.44×10^{-3} °C; boiling point elevation = 3.96×10^{-4} °C; and osmotic pressure = 1.87×10^{-2} atm = 14.2 torr.

10.6.2 The figure should look the same, except there should be more arrows moving from right to left than from left to right.

10.6.3 Your description should include the features shown in your figure for 10.6.2 and the prevention of ion passage through the pores of the semipermeable membrane.

10.7.1

H_3C—N^+(—CH_3)(—CH_3)—CH_3 Cl^- All hydrophilic (ions)

(benzene ring)—N^+(—CH_3)(—CH_3)—(long chain) Cl^-

Hydrophilic   Hydrophobic
(This molecule is a surfactant.)

All hydrophobic

10.7.2 Because they are large, they cannot slip easily between the molecules of a lipid bilayer membrane. Because they are hydrophilic, they do not dissolve in the bilayer.

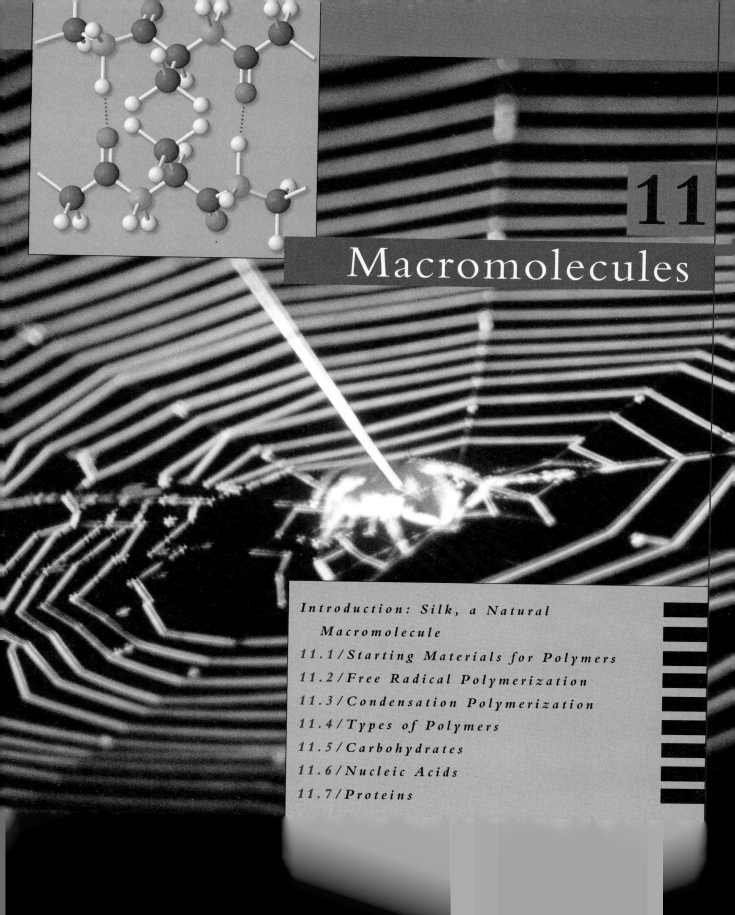

11

Macromolecules

INTRODUCTION: SILK, A NATURAL MACROMOLECULE

Stumble into a spider web and you will experience a natural material that is one of the strongest substances known. Spiders spin their webs from silk, a huge molecule made up of a vast chain of carbon and nitrogen atoms, as shown in our inset picture. The spider's web is both amazingly strong and highly flexible: Scientists have studied the properties of spider silk and have determined that it is at least five times stronger than steel. What's more, spider silk is waterproof and flexible. These properties come from the chemical bonding described in Chapters 8 and 9, as well as the intermolecular forces described in Chapter 10.

The sensation of a spider web across your face may be unsettling, but a similar natural material has been used for centuries to make silk, a fabric that is prized for its smooth texture. Silkworms produce the silk fibers used to make clothing. They feast on mulberry leaves and convert the molecules from these leaves into silk, from which they spin cocoons.

According to Chinese legend, a princess strolling among the mulberry trees in a palace garden plucked a cocoon from a leaf. She accidentally dropped the cocoon into her cup of steaming-hot tea, from which she fished out a long silken thread. Whether or not the legend is true, the ancient Chinese discovered how to harvest silkworm cocoons, boil them to loosen the tangle, and unravel the silk into a fiber from which elegant clothing could be produced. A single silkworm cocoon can yield nearly a mile-long filament of silk, but the filament is so fine that nine of them are required to make the thread from which silk fabric is woven. It takes around 30 mulberry trees to yield enough cocoons to make one kilogram of silk.

The domestication of the silkworm in China occurred many centuries BC, and by the time of the Roman empire, silk fabric was a prized trade commodity. The caravan routes across Asia, through such exotic-sounding cities as Tashkent and Samarkand, became known as the Silk Roads. It is estimated that nearly 90% of the imports into the Roman Empire consisted of silk goods.

As we describe in this chapter, silk is a macromolecule, an extremely long molecule built from relatively simple molecular building blocks. Like many other naturally occurring macromolecules, silk is made up of proteins—long sequences of amino acids linked together into near-infinite chains. Spiders and silkworms secrete liquid proteins that solidify into strands of silk.

Chemists learned the basic composition of silk many years ago, but the reasons why this macromolecule is so strong, yet flexible, are still not fully understood. Recent studies indicate that the secret lies in the way the chains of this protein nestle together, with a pattern that is regular but not fully crystalline. Current research efforts focus on using techniques of genetic engineering to replicate natural spider silk on a useful scale.

Silk is just one example of a macromolecule, also known as a polymer. Macromolecules are the subject of this chapter. The principles introduced in Chapters 8–10 help to explain the properties of these super-large molecules, many of which are carbon-based. In this chapter, we outline the principles of the structure and synthesis of the major classes of macromolecules and describe the properties that give these chemical substances central roles in industrial chemistry and biochemistry. We describe the components from which macromolecules are constructed, some important industrial polymers, and the intricate macromolecules found in living systems.

11.1 STARTING MATERIALS FOR POLYMERS

A **macromolecule** is any very large (macro) molecule. A diamond, for example, is a macromolecule made entirely of carbon atoms. A grain of sand is a macromolecule made of silicon and oxygen atoms, in 1:2 ratio, bonded together in a three-dimensional array. A **polymer** is a macromolecule constructed by linking together many copies of much smaller molecules called **monomers.** All the examples of macromolecules that we describe in this chapter are polymers. We begin with a description of the various monomers that are the most important starting materials for industrial and natural polymers.

Polymers form from monomers. Monomers are organic molecules characterized by their **functional groups,** specialized groups of atoms that impart a specific chemical function. In this section, we introduce functional groups that influence the chemistry of macromolecules.

Any functional group is only part of an organic molecule. When chemists wish to emphasize a functional group, they commonly use the symbol R to indicate a portion of the molecule that plays an unimportant role in the molecule's chemical behavior. An R group in these structures can be H or any organic fragment containing a carbon atom that bonds directly to the functional group. An organic fragment can be as simple as a methyl group (CH_3) or a phenyl group (C_6H_5), or it can represent an elaborate structure containing dozens of atoms.

The most important functional groups that participate in polymerization reactions are listed in Table 11-1. All these groups have electron pairs that can be incorporated into new chemical bonds relatively easily. The π electrons in a double bond are more reactive than electrons in σ bonds, and the lone pairs of electrons on O, N, and S atoms are available for bond formation.

C=C Double Bonds

A carbon-carbon double bond is a reactive functional group because of the π electrons. Remember from Chapter 9 that ethylene has a C=C bond made up of one σ bond plus one π bond. As shown in Figure 11-1, the electrons in the π bond are located off the bond axis, making them more readily available for chemical reactions. Moreover, π electrons are less tightly bound than σ electrons. Consequently, the reactivity patterns of ethylene are dominated by the chemistry of its π electrons. We describe the polymerization reaction of ethylene and other monomers containing C=C bonds in Section 11.2.

—OH and —SH

An **alcohol** is a molecule that contains a hydroxyl group (—OH) covalently bonded to carbon. Methanol and ethanol are the simplest alcohols. Sugar molecules, which

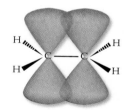

Figure 11-1
The π electrons are readily available for chemical reactions that form bonds to other molecules.

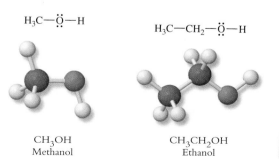

CH₃OH
Methanol

CH₃CH₂OH
Ethanol

The hydroxyl group of an alcohol, which is covalently linked to a carbon atom, has completely different chemical properties than the hydroxide ion, OH⁻, which is present in inorganic hydroxides such as NaOH and Mg(OH)₂.

Table 11-1
Polymerizable Functional Groups

Formula	Name	Class	Example	Reaction Partners
$C{=}C$	Double bond	Alkene	Ethylene	Alkenes
$-\overset{..}{\underset{..}{O}}-H$	Hydroxyl	Alcohol	Ethanol	Acids, Amines, Aldehydes, Alcohols
aldehyde structure	Aldehyde	Aldehyde	Formaldehyde	Alcohols, Amines
carboxyl structure	Carboxyl	Acid	Acetic acid	Alcohols, Amines
amine structure	Amine	Amine	Ethylenediamine	Acids, Alcohols
phosphate structure	Phosphate	Phosphate	Phosphoric acid	Phosphates, Alcohols
$-\overset{..}{\underset{..}{S}}-H$	Sulfhydryl	Thiol	Cysteine	Thiols

3—methylbutanethiol

we describe in Section 11.5, contain several hydroxyl groups bonded to carbon. Some chemical reactions break the O—H bond, while others break the C—O bond of an alcohol or sugar. The O—H bond is highly polar, so alcohols and sugars readily form hydrogen bonds. We describe later how hydrogen bonds play important roles in the properties of polymers, especially those that have biochemical importance.

A **thiol** contains an —SH group covalently bonded to carbon. Sulfur is just below oxygen in the periodic table, so a thiol is somewhat similar to an alcohol. Still, the chemical and physical properties of thiols differ significantly from those of alcohols. For example, alcohols have inoffensive odors, but thiols smell bad. The stench of skunk scent is due to thiols, including 3-methylbutanethiol. Thiols are important in proteins because of their abilities to form S—S linkages, which we describe in Section 11-7.

—NH2

An **amine** is a derivative of ammonia in which hydrogen atoms are replaced by carbon atoms:

Methylamine Ammonia Methylamine Dimethylamine Trimethylamine

The NH_2 group is polar and forms hydrogen bonds. Moreover, the lone pair of electrons on the nitrogen atom can accept a proton from a hydronium ion:

$$CH_3\text{-amine} + H_3O^+ \longrightarrow \text{protonated amine} + H_2O$$

When they are protonated, amines are compatible with water, a property critical in biochemical processes because it helps biological macromolecules dissolve in water. The N—H bond in an amine is fairly easy to break, and this property leads to several important polymerization reactions.

The Carbonyl and Carboxyl Groups

An oxygen atom doubly bonded to carbon, known as the *carbonyl group,* is very important in organic chemistry. The C=O linkage occurs in **aldehydes, ketones,** and **carboxylic acids.**

The chemistry of aldehydes and ketones is dominated by the polarity of the carbonyl group. The difference in electronegativity between carbon and oxygen polarizes the double bond, creating a partial positive charge on the carbon and a partial negative charge on the oxygen.

Aldehydes are more important than ketones in polymerization reactions, but both groups are important in all aspects of chemistry. The simplest aldehyde, formaldehyde, is a top-50 industrial chemical that is a crucial monomer for polymer synthesis.

Carboxylic acids, which can be represented as $R—CO_2H$, appear again and again in organic chemistry and biochemistry. A carboxylic acid is highly polar and can give up H^+ to form a carboxylate anion, $R—CO_2^-$. The carboxyl group also readily forms hydrogen bonds. These properties enhance the solubility of carboxylic acids in water, a particularly important property for biochemical macromolecules. Like aldehydes and ketones, carboxylic acids react in many different ways, but breaking the C—OH bond is the only reaction that is important in polymer formation.

Aldehyde Ketone

Carboxylic acid

The O atom in the carbonyl group has two nonbonding pairs of valence electrons, but chemists frequently omit them from drawings of molecular structures.

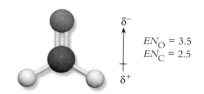

$EN_O = 3.5$
$EN_C = 2.5$

Formaldehyde

Phosphates

The **phosphate** group is a dianion obtained from phosphoric acid (H_3PO_4) by removing two protons and replacing the third hydrogen with a bond to C or P. Phosphate is an important functional group in biochemistry, being involved in cellular energy production as well as acting as an important monomer in biopolymers, particularly in DNA. Bonds to phosphate groups form or break in the course of a number of important biochemical reactions.

To summarize from the perspective of polymer formation, the most important role of functional groups in polymerization is to provide bonds that are relatively easy to break. Because C—H and C—C σ bonds are relatively strong and do not break easily, polymerization requires monomers that contain reactive functional groups. To form polymers, bonds in these groups must break, and new bonds that link monomers into macromolecules must form.

Linkage Groups

When two monomers react by combining their functional groups, the result is a new functional group called a *linkage group*. The three linkage groups listed in Table 11-2 are particularly important in polymerization reactions. As a linkage forms, a small molecule is produced, often but not always water. A reaction that produces a linkage group and eliminates a small molecule is called a **condensation reaction.**

 Tutorial

An **ester** is an organic compound formed from the condensation reaction between a carboxylic acid and an alcohol. A water molecule is eliminated as the oxygen atom of the alcohol links to the carbon atom of the carbonyl group. For example, ethanol reacts with acetic acid to give ethyl acetate:

$$CH_3CH_2\text{—}O\text{—}H \quad HO\underset{\overset{\displaystyle O}{\|}}{C}CH_3 \longrightarrow CH_3CH_2\text{—}O\underset{\overset{\displaystyle O}{\|}}{C}CH_3 + H_2O$$

Ethanol	Acetic acid	Ethyl acetate
(an alcohol)	(a carboxylic acid)	(an ester)

Carboxylic acids undergo condensation reactions with amines to form **amides.** Water is eliminated as the new N—C bond forms:

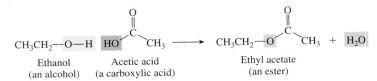

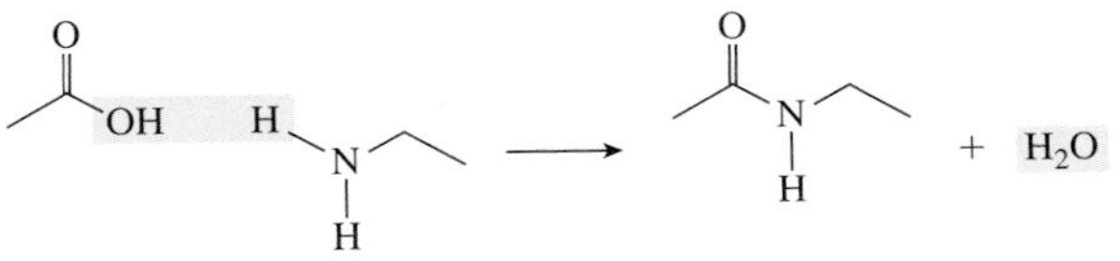

CH$_3$CO$_2$H	H$_2$NC$_2$H$_5$	CH$_3$C(O)NHCH$_2$CH$_3$
(a carboxylic acid)	(an amine)	(an amide)

Table 11-2
Important Polymer Linkage Groups

Linkage	Name	Precursors	Polymer Type
(ester structure)	Ester	Acid + alcohol	Polyesters
(amide structure)	Amide	Acid + amine	Polyamides, proteins
C—O—C	Ether	Alcohol + alcohol	Cellulose, starch

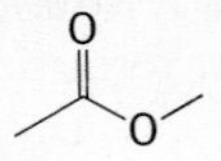

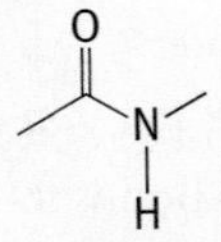

(Side margin figures and caption:)

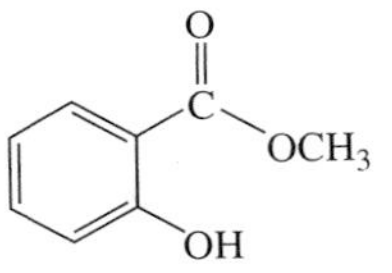

Methyl salicylate

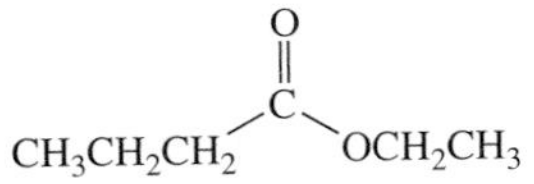

Ethyl butyrate

Many esters have pleasant smells and tastes and are important components of flavorings and fragrances. Methyl salicylate has the fragrance of wintergreen, and ethyl butyrate occurs in pineapples.

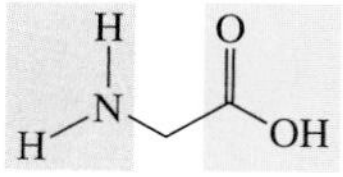

The amide linkage occurs in nature in proteins, biological macromolecules with amazing structural and functional diversity. Silk, described in the introduction to this chapter, is one such molecule. The monomers from which proteins form are amino acids, each of which contains both an amine group and a carboxyl group. The amino acids are strung together through a series of condensation reactions that give amide linkages.

Amine   Carboxylic acid

The condensation reaction of two alcohol molecules to eliminate water and form a C—O—C bond sequence is yet another linkage reaction. A molecule that contains a C—O—C linkage is called an **ether**:

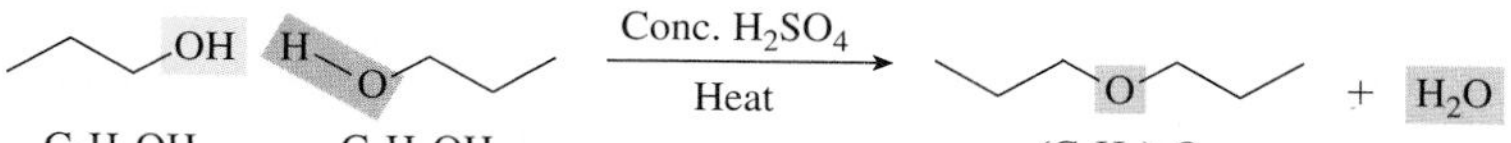

C₃H₇OH (an alcohol)   C₃H₇OH (an alcohol)   (C₃H₇)₂O (an ether)   + H₂O

This direct condensation of alcohols has limited use as a route to polymers because many alcohols undergo other reactions under the conditions required for condensation. The most important polymers containing ether linkages are starch and cellulose, which are biochemical macromolecules. Nature forges ether linkages in a more roundabout way than by eliminating water directly.

Two molecules of phosphoric acid can undergo a condensation reaction, eliminating a water molecule and forming a P—O—P linkage:

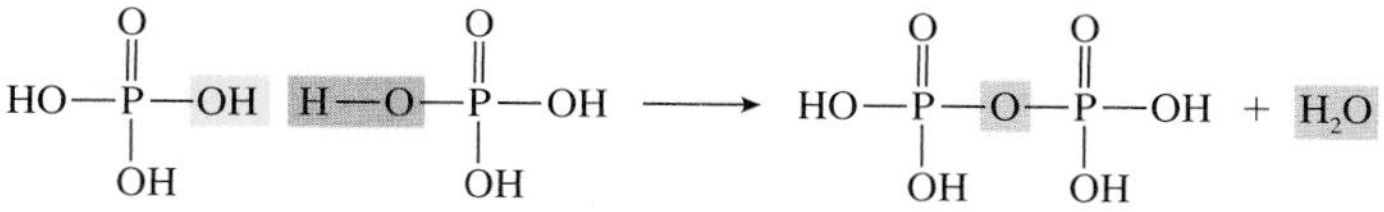

Phosphoric acid or phosphate can also condense with an alcohol to generate a P—O—C linkage. As described later, such linkages are found in the structure of DNA.

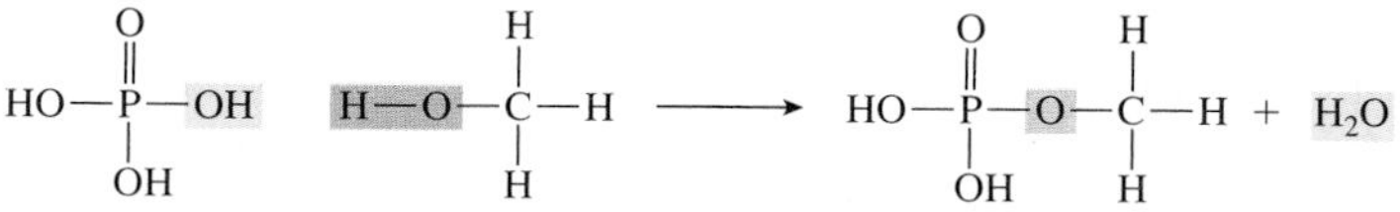

Section Exercises

11.1.1 Molecular pictures of several molecules appear below. Identify the functional groups present in each. (Some contain more than one functional group.)

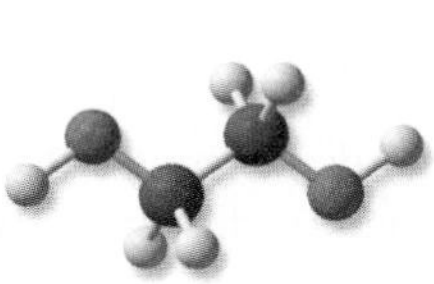

Ethylene glycol 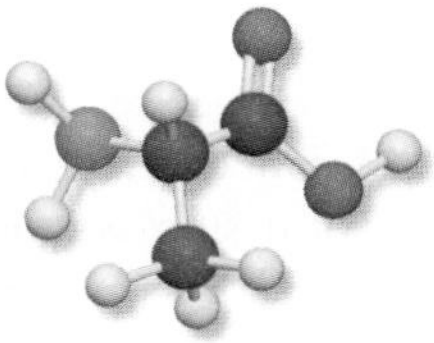   Alanine 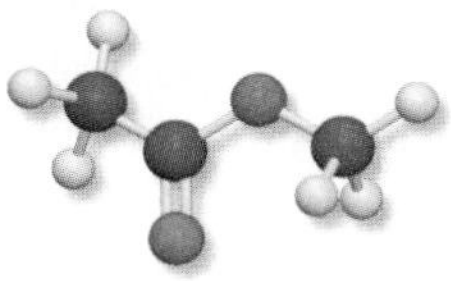   Methyl acetate

11.1.2 Amino acids are the molecular building blocks of proteins. Any two amino acids can condense in two ways; each creates an amide linkage. Draw the structures of the two condensation products that link these two amino acids:

Alanine
$C_3H_7NO_2$

Phenylalanine
$C_9H_{11}NO_2$

11.1.3 Draw one complete specific line structure for each of the following: (a) an amine with the formula $C_6H_{15}N$; (b) an ester with at least seven carbon atoms; (c) an alcohol with a molar mass of 74 g/mol; and (d) a molecule with at least ten carbons that contains an aldehyde group and a phenyl ring (C_6H_5).

11.2 FREE RADICAL POLYMERIZATION

The simplest of all polymers forms when many ethylene molecules bond together to form long chains containing 500 or more repeating CH_2 groups. This polymer is called **polyethylene** because it is made by joining together many ethylene molecules.

An electron from the π bond of one ethylene molecule pairs with an electron from the π bond of another ethylene molecule to form a new σ bond between the two molecules. The second electron from each π bond pairs with a π electron from another ethylene molecule to continue the chain.

Writing the structural formula of a macromolecule such as polyethylene with thousands of atoms would be very time-consuming and tedious. Fortunately, the entire structure of a polyethylene molecule can be represented by simply specifying its repeat unit, as shown in Figure 11-2.

Synthesis of Polyethylene

Ethylene forms a polymer through a three-step sequence. First is an **initiation** step, in which a reactive chemical substance attacks the π bond of a single ethylene molecule. The product of the initiation step is able to react readily with the π bond of another ethylene molecule in a **propagation** step. Many propagation steps occur, each one lengthening the growing polymer chain by adding one monomer unit. Finally, chain growth comes to an end through a **termination**

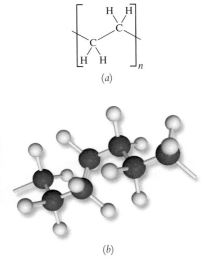

Figure 11-2
(a) Line structure of the repeat unit of polyethylene, which is made up of repeating CH_2 units.
(b) Ball-and-stick model showing a portion of polyethylene.

step. Initiation, propagation, and termination are important steps in the synthesis of many polymers.

The polymerization of ethylene starts with the thermal decomposition of an initiator molecule, whose general formula is R—O—O—R. Heating breaks the weak O—O single bond to form a pair of R—O· **free radicals.** Free radicals are highly reactive molecules that contain unpaired electrons. A free radical will attack any bond that has exposed electron density. In this case, a free radical attacks the π bond of an ethylene molecule:

> Molecules with O—O bonds are called *peroxides*. Because the average O—O bond energy is only 145 kJ/mol, these are among the easiest covalent bonds to break.

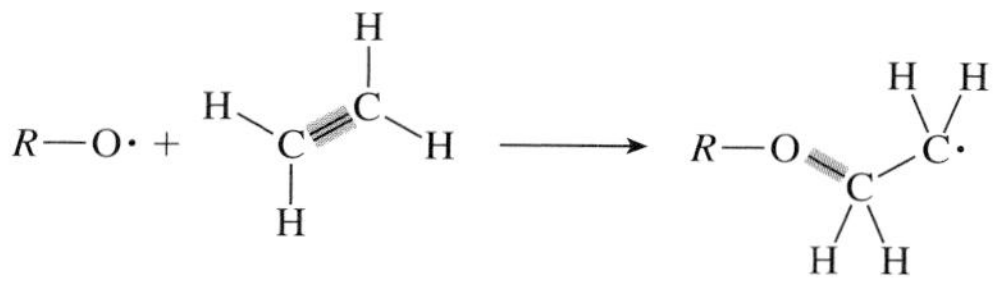

In the initiation reaction, one π electron of the ethylene molecule pairs with the single electron on the oxygen atom of the free radical. Together, these two electrons create a new C—O σ bond (shown highlighted in purple). The second electron from the ethylene π bond remains as an unpaired electron on the outermost carbon atom of the product.

The carbon atom with the unpaired electron is another free radical, so the stage is set for propagation. The unpaired electron on the carbon atom attacks the π bond of another molecule of ethylene, making a new C—C σ bond and leaving yet another carbon free radical:

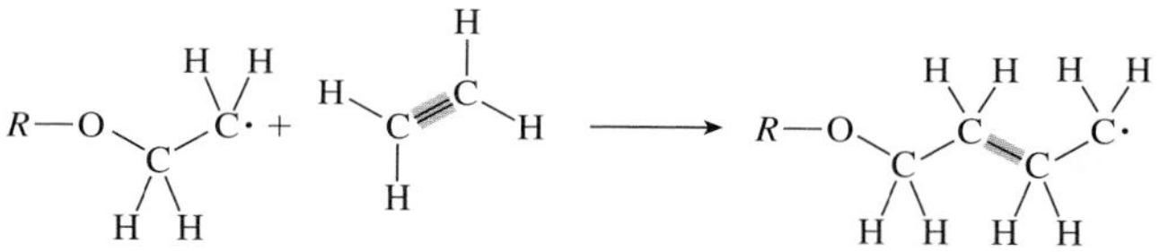

This addition reaction can be repeated over and over in the propagation process, and each step adds two CH_2 groups to the growing polymer chain. The propagation can be written in the following general form:

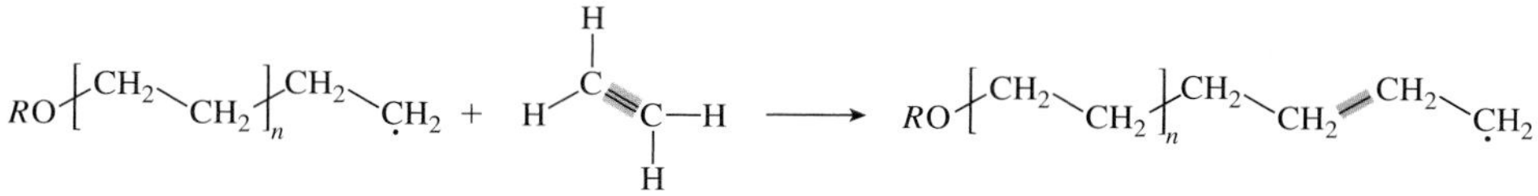

Because chain growth involves free radicals, this type of polymerization is called **free radical polymerization.**

Free radical polymerization does not go on indefinitely, because sometimes two free radicals collide and react. This destroys two free radicals and terminates the chain. For example, another initiator fragment can react with the carbon atom at the end of the chain:

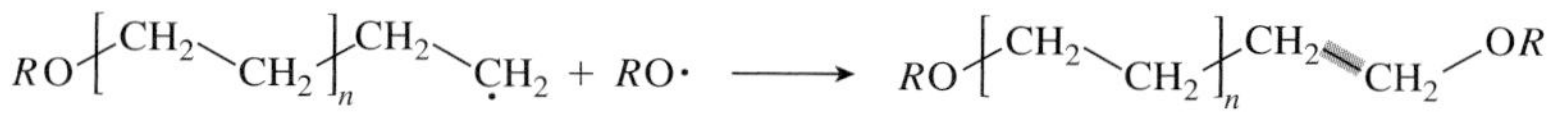

Table 11-3
Important Polymers Made from Alkenes

Monomer	Polymer	Uses	Production (10⁹ kg)*
Ethylene	Polyethylene	Piping, bottles, toys	8.1
Vinyl chloride	Poly (vinyl chloride) (PVC)	Piping, floor tile, clothing	3.9
Propene	Polypropylene	Carpets, labware, toys	3.3
Styrene	Polystyrene (Styrofoam)	Containers, heat insulators	2.3
Acrylonitrile	Polyacrylonitrile (PAN)	Carpets, knitware	0.14
Tetrafluoroethylene	Teflon (PTFE)	Cookware, bearings	<0.05

* 1999 figures.

The product of the last reaction may not look like polyethylene, but remember that n is a large number, around 250 or more. Thus the OR ends of the macromolecule are insignificant compared with the many CH_2—CH_2 repeat units that make up the bulk of the polymer.

Other Polyalkenes

A variety of polymers can be made using derivatives of ethylene in which one or more of the hydrogen atoms is replaced with other groups of atoms. Changing the structure of the monomer makes it possible to adjust the properties of the polymer. Table 11-3 lists six important alkene monomers and their corresponding polymers. You may recognize some of their common names. Figure 11-3 shows the molecular structures of five of these monomers. Notice the variety of groups that can replace hydrogen atoms: CH_3, Cl, F, CN, and C_6H_5. Example 11-1 shows how to write the structure of a polyethylene derivative.

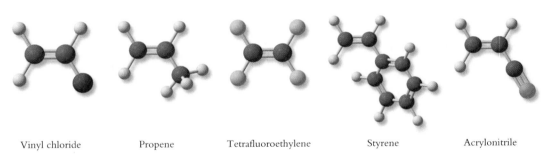

| Vinyl chloride | Propene | Tetrafluoroethylene | Styrene | Acrylonitrile |

Figure 11-3
Structures of alkene monomers used in the manufacture of important polymers.

| Drawing the Structure of a Polymer | Example 11-1 |

Polyacrylonitrile, known commercially as *Orlon,* is made by polymerizing acrylonitrile (see Figure 11-3). Orlon is used to make fibers for carpeting and clothing. Draw the Lewis structure of polyacrylonitrile, showing at least three repeat units.

Strategy: Polyalkenes form by linking carbon atoms in a free radical polymerization. The polymer structure is constructed by connecting monomer units. The structure of the acrylonitrile monomer is shown in Figure 11-3.

Solution: Acrylonitrile polymerizes in the same way as ethylene. A free radical initiator attacks the π electrons of the C=C bond. The chain grows through propagation steps until a termination reaction occurs. As with polyethylene, propagation creates a long chain of C—C σ bonds. In polyacrylonitrile the CN groups dangle from the chain. Here is the structure, showing three repeat units:

Notice that this polymer has the same structure as polyethylene, except that a CN group is attached to every second carbon atom. A line structure of polyacrylonitrile eliminates the clutter caused by the hydrogen atoms. A stylized ball-and-stick model of the same polymer segment is included for comparison.

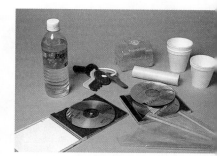

Polyalkenes are used to make many of the items used in everyday life.

Rubber

The first alkene polymer to be used in society was polyisoprene, a natural product extracted from the sap of rubber trees. See our Chemical Milestones Box for a description of the history of rubber. The monomer from which this polymer is constructed is isoprene, an alkene with two double bonds. The line structure of isoprene and the repeat unit of polyisoprene are shown in Figure 11-4. When isoprene polymerizes, attack on one π bond causes the second π bond to shift, giving a macromolecule containing a four-carbon repeat unit with a double bond in its center: Butadiene, whose structure is also shown in Figure 11-4, polymerizes in the same fashion to give a synthetic rubber.

Natural rubber and polybutadiene account for under 40% of the 2.1×10^9 kg of rubber produced in the United States each year. The rest, more than 60%,

One of the first uses of natural rubber was for erasers, called *rubbers* by the British. Eventually, this name came to mean any substance made of polymerized dienes.

← **SECTION 9.5**
The bonding of dienes is discussed in Section 9.5.

| Box 11-1 | Chemical Milestones: The Natural Rubber Industry |

T he rubber industry has a long and colorful history. Natural rubber is produced from latex, a milky fluid found in cells that lie between the bark and the wood of many plants. You may have seen latex flow from the broken stalks of milkweed plants, but the source of commercial rubber is the Hevea tree, a native of Brazil. When the bark of this tree is slashed, its milky white sap oozes out and can be collected in cups mounted on the tree's trunk. The people of the Amazon jungle made bouncing balls, shoes, and water jars out of rubber, and Portuguese explorers sent waterproof boots and a rubber-coated coat back to their king. The first commercial exports included some rubber shoes shipped to Boston in 1823.

Entrepreneurs in Europe and the United States experimented with rubber hats, coats, erasers, life preservers, and a number of other products. Through the first half of the nineteenth century, however, rubber items were impractical because natural rubber becomes sticky in hot weather and brittle in cold weather. In 1830 the world consumed just 156 tons of rubber, all from Brazil.

Nine years later, Charles Goodyear discovered how to stabilize rubber through vulcanization, and rubber quickly became an essential part of the world economy. Goodyear's discovery set off a rubber boom that rivaled the California gold rush. Adventurers swarmed to the jungles to amass fortunes as the price of rubber soared. The Amazon people were captured and forced to work at harvesting the latex fluid. Many of them died under the harsh conditions. New towns sprang to life along the river as paddle-wheel steamers carried luxury goods to rubber barons, who built grand mansions up the tributaries of the Amazon. The pinnacle of the rubber boom was the town of Manaus, which is located a thousand miles up the Amazon system on the Negro river. At its height, Manaus teemed with mansions, cafes, and fine hotels. It had a streetcar system and even an enormous ornate opera house.

The Amazon rubber industry collapsed almost overnight. In 1876 the English botanist Henry Wickham shipped 70,000 Hevea seeds to the Royal Botanic Gardens in London. New strains of Hevea were developed that produced three to four times as much rubber and were more disease-resistant than their wild Amazonian cousins. Soon, seedlings were sent to Malaya, Java, and other islands of the East Indies. Thirty-five years later, rubber plantations on these islands took control of the industry.

In 1910 the price of rubber was $2.88 a pound, world production was 94,000 tons, and wild rubber from Brazil accounted for 83,000 tons. In 1912 the price collapsed, and by 1932 a pound of rubber cost 2.5¢. In 1937 the world bought more than 1.1 million tons of rubber, less than 2% of which came from Brazil. The rubber cities of the Amazon became near ghost towns as the market for wild rubber vanished. The opulent opera house in Manaus had barely opened its doors when the curtain came down on the Brazilian rubber economy.

The rubber industry changed again when the Japanese captured the East Indian rubber plantations during World War II. The resulting shortage of rubber prompted an intensive research program to produce synthetic rubber. Today, more than 2 million tons of synthetic rubber is produced each year in the United States. Natural rubber is still produced in the tropics, but its importance pales compared to the glory days of the Brazilian rubber plantations.

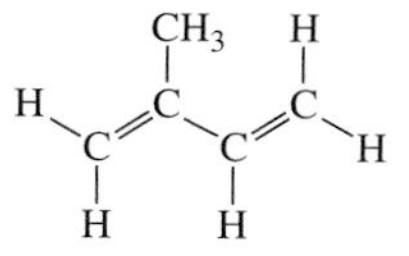

Isoprene 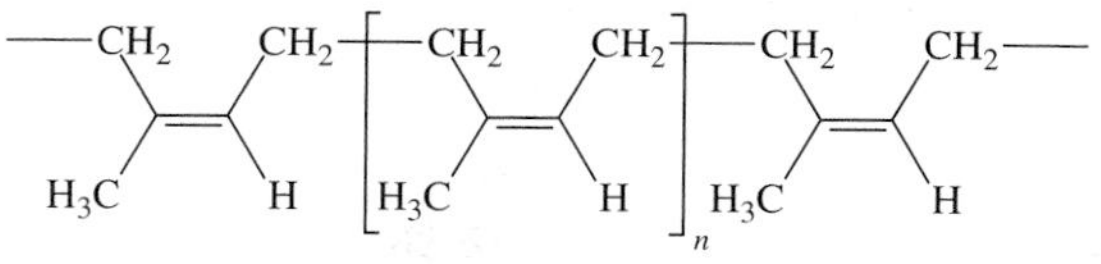   Polyisoprene (natural rubber) 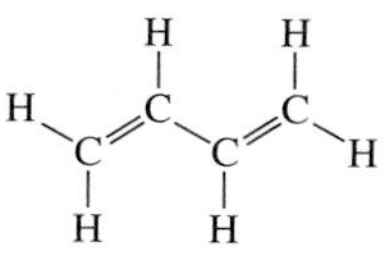   Butadiene

Figure 11-4
The structures of isoprene and butadiene, two dienes that form rubber-like polymers. The repeat unit of natural rubber is shown also.

are **copolymers,** made by polymerizing mixtures of two different monomers. The largest production, accounting for 8×10^8 kg of total rubber production, is a polymer of three parts butadiene to one part styrene. During polymerization of butadiene and styrene, the chain grows by adding whichever monomer happens to collide with the chain end. The propagation process is random, so the exact sequence varies, but the resulting polymer contains three butadiene units for every styrene unit. Here is the line structure of a representative portion of such a polymer, with styrene units highlighted in light brown and butadiene units in gold:

Butadiene-styrene copolymer

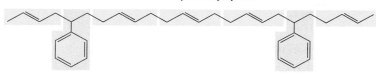

At any point in the chain, there is a 75% probability of finding a butadiene unit and a 25% probability of finding a styrene unit.

The long-chain molecules described so far are used for flexible items such as rubber bands, but they do not have the durability and strength associated with rubber products such as automobile tires. To achieve these properties, rubber must be treated chemically to create chemical bonds between long-chain molecules. This process is called **cross-linking** because links are formed *across* the chains in addition to bonds *along* the chains. Vulcanization, the formation of cross-links in rubber, was discovered in 1839 by Charles Goodyear, founder of the first U.S. rubber company.

Vulcanization is a complicated chemical process whose details are still not fully understood. Nevertheless, the starting materials and end product are well characterized. Sulfur is added to the polymer, and the mixture is heated under controlled conditions. Some of the polymer C—H bonds break and are replaced by C—S bonds. The cross-links consist of C—(S_n)—C chains, with n being 2 or more. Figure 11-5 represents the overall scheme as it applies to polybutadiene.

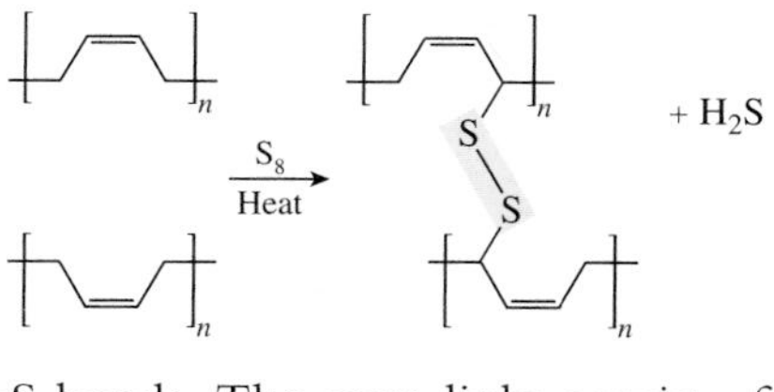

While cross-links increase the tensile strength of rubber, they also increase its rigidity. When the sulfur content increases beyond 10% by mass, vulcanized rubber becomes hard and brittle. Thus, the amount of cross-linking must be controlled to optimize tensile strength without generating a brittle product.

$\xrightarrow[\text{Heat}]{S_8}$ + H$_2$S

Figure 11-5
Schematic view of the vulcanization of polybutadiene. When heated in the presence of sulfur, rubber forms cross-links between polymer chains.

Sulfur cross-links also play an important role in determining the structures of proteins, as discussed in Section 11.7.

Section Exercises

11.2.1 Draw the structure of the polymer formed from each of these monomers:

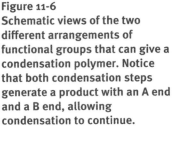

(a) Vinyl fluoride C_2H_3F

(b) Butadiene C_4H_6

(c) Methyl methacrylate $C_5H_8O_2$

11.2.2 In a $3:1$ copolymer of butadiene and styrene, the placement of butadiene and styrene fragments along the chain is random. Draw a line structure for a portion of the copolymer that has this sequence: —butadiene—styrene—styrene—butadiene—.

11.2.3 Neoprene is a synthetic rubber used to make gaskets. A section of neoprene follows. Draw the structure of the monomer used to make neoprene.

Figure 11-6
Schematic views of the two different arrangements of functional groups that can give a condensation polymer. Notice that both condensation steps generate a product with an A end and a B end, allowing condensation to continue.

11.3 CONDENSATION POLYMERIZATION

In a condensation reaction, two molecules form bonds between their functional groups by eliminating a small molecule, usually water. If no additional functional groups are present, that is as far as the linkage process goes. Polymer formation via condensation requires that each monomer contain *two* linkage-forming functional groups. After the two molecules link together, there are still two functional groups that can link to two more monomers, allowing the chain to grow into a polymer.

Figure 11-6 illustrates schematically that there are two arrangements of functional groups that make it possible for condensation polymers to form. A single monomer may contain two *different* functional groups capable of linking, or one monomer may contain two of one particular functional group while a second monomer contains two of another functional group that can link with the first.

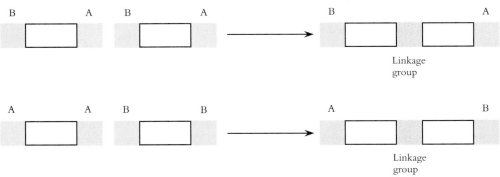

Amino acids belong to the first category. As its name implies, an amino acid contains an amine and a carboxylic acid group. Amino acids join through amide linkages:

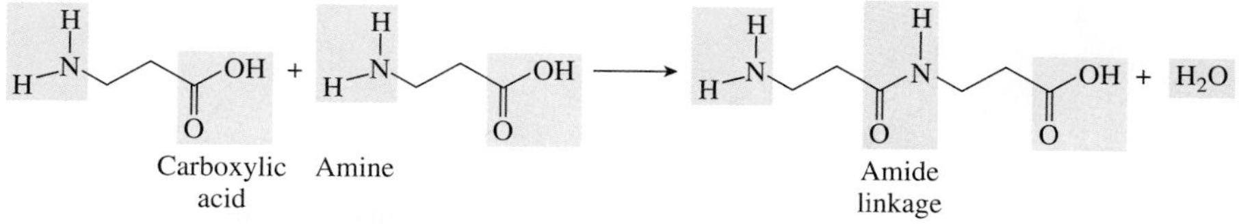

The product still has an amine group at one end and a carboxylic acid group at the other. As a result, each end can form another amide linkage, extending the chain.

The condensation of a diacid with a diol (di–alcohol) belongs to the second category. Here, a monomer with two carboxylic acid groups links to a monomer with two hydroxyl groups to give an ester:

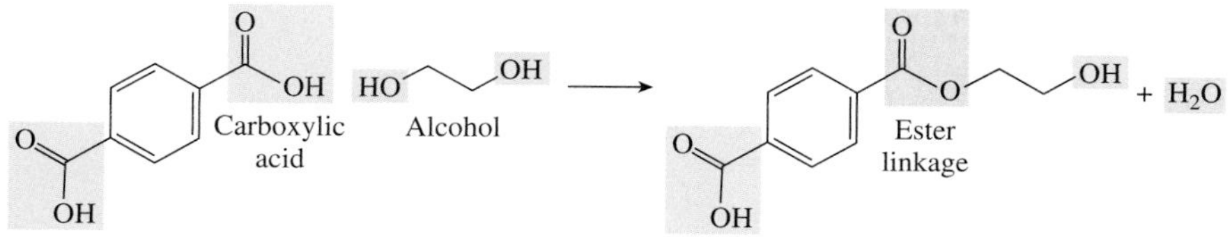

The product is a bifunctional molecule with a carboxylic acid group at one end and a hydroxyl group at the other. Chain growth can continue through repeated condensation reactions at both ends of the molecule.

Most polymers in production today are made from two different monomers because experience has shown that this synthetic route offers advantages over the use of bifunctional monomers. First, monomers with two identical functional groups are easier and less expensive to produce than monomers with two different groups. Second, the properties of the polymer can be varied simply by changing the structure of one of the monomers.

Polyamides

Any polymer that contains amide linkage groups is called a **polyamide.** Proteins, biological polyamides, are described in Section 11.7. In this section and the next, we focus on three commercially important polyamides: Nylon 66, Nylon 6, and Kevlar.

Nylon 66 was the first polyamide to be produced commercially. Developed by Wallace Carothers at the DuPont Chemical Company in 1935, it still leads the polymer industry in annual production. Nylon 66, made from adipic acid and hexamethylenediamine, is so easy to make that it is often used for a classroom demonstration, as shown in Figure 11-7.

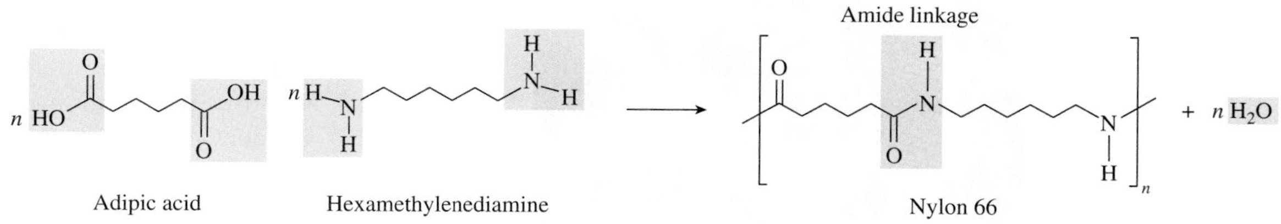

Figure 11-7
Nylon 66 can be drawn from a beaker containing the two starting materials. The bottom layer is an aqueous solution of hexamethylenediamine, and the top layer is adipoyl chloride (a more reactive derivative of adipic acid) dissolved in hexane. Nylon 66 forms at the interface between the two liquids.

Terephthalic acid   Phenylenediamine   + H_2O

Figure 11-8
A portion of the structure of Kevlar, a nylon made from terephthalic acid and phenylenediamine. Pound for pound, this polymer is stronger than steel.

Many other useful polymers can be made by combining different diamines and diacids. One example is Kevlar, a polyamide made by condensing terephthalic acid with phenylenediamine.

Figure 11-8 shows a portion of the structure of Kevlar, which is so strong that it is used to make bulletproof vests.

The tremendous strength of Kevlar comes from a combination of intramolecular and intermolecular forces. The stiff phenyl rings make each polymer chain strong and rigid, while the chains are held together by hydrogen bonds. Interchain hydrogen bonds enhance the strength and durability of all polyamides, but Kevlar has a particularly large number of hydrogen bonds per unit volume because of its compact, highly regular molecular structure.

We can identify the monomers from which a condensation polymer is made by visualizing the condensation reaction operating in reverse. The linkage bond breaks, and a small molecule is inserted. Example 11-2 illustrates this type of reasoning for another important polyamide.

The Structure of a Polyamide

Example 11-2

Qiana, a polyamide that feels much like silk, has the following structure:

Identify the monomers used to make Qiana.

Strategy: Visualize the condensation reaction that forms a polymer. Two monomers with appropriate functional groups combine, forming a new bond and eliminating water. Reverse this process to see what monomers make Qiana.

Solution: A polyamide is made from the condensation reaction of a diamine and a dicarboxylic acid. To identify the monomers, separate the amide linkage group and add water across the C—N bond:

Now construct the monomers:

Qiana cloth looks and feels much like silk.

Polyesters

Nylon was the first commercial polymer to make a substantial impact on the textile industry, but **polyesters** now comprise the largest segment of the market for synthetic fibers. In fact, polyesters account for 40% of the more than 4 billion kilograms of synthetic fibers produced in the United States each year. The leading polyester, by far, is poly(ethylene terephthalate), or PET. This polymer is made from terephthalic acid and ethylene glycol in an acid–alcohol condensation reaction:

| Terephthalic acid | Ethylene glycol | PET |

A typical polyester molecule has 50 to 100 repeat units and a molar mass between 10,000 and 20,000 g/mol. PET can be formed into fibers (such as Dacron) or films (such as Mylar). Mylar films, which can be rolled into sheets 30 times thinner

Dacron is used for synthetic blood vessels.

than a human hair, are used to make magnetic recording tape and packaging for frozen food. Dacron is best known for its use in clothing, but it has many other applications. For example, tubes of Dacron are used as synthetic blood vessels in heart bypass operations because Dacron is inert, nonallergenic, and noninflammatory.

More than 2 billion kilograms of PET are produced in the United States each year. This polymer is used to make tire cord, beverage bottles, home furnishings, small appliances, and many other common items.

Urea-Formaldehyde Polymers

Urea, $(NH_2)_2C{=}O$, and formaldehyde, $H_2C{=}O$, form several important condensation polymers. Polymerizations that involve formaldehyde proceed by way of alcohol intermediates, which then condense with other functional groups. For example, polymers useful as glues and coatings are made from formaldehyde and urea. The first step is an addition reaction to give an alcohol:

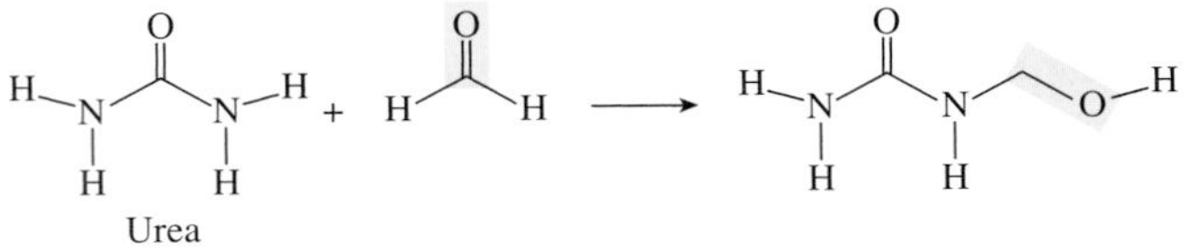

The concentration of formaldehyde is kept low so that only one NH_2 group of urea reacts. Polymerization follows this first step because every N—H hydrogen can undergo a condensation reaction with a hydroxyl group to give a chain of alternating carbon and nitrogen atoms:

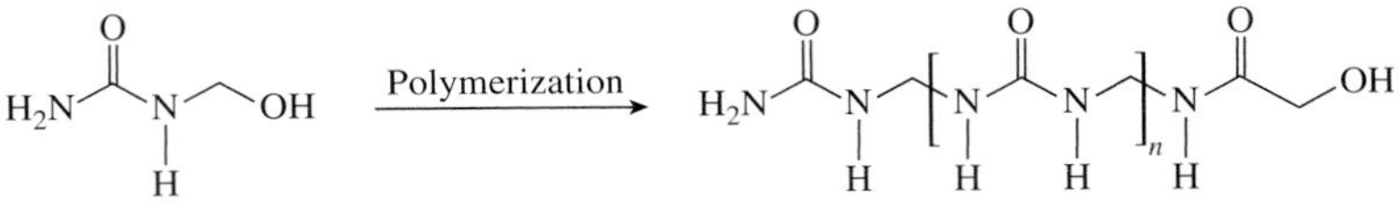

Urea and formaldehyde react in another fashion to form a strong, lightweight, and heat-resistant polymer. To prepare the polymer, chemists first condense urea to give melamine, a cyclic compound:

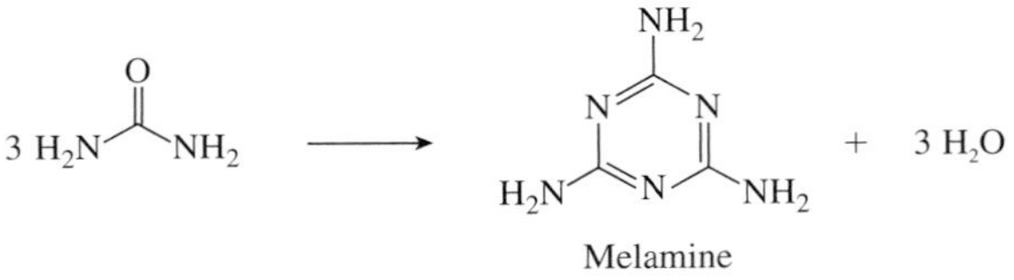

Each molecule of melamine then reacts with three molecules of formaldehyde:

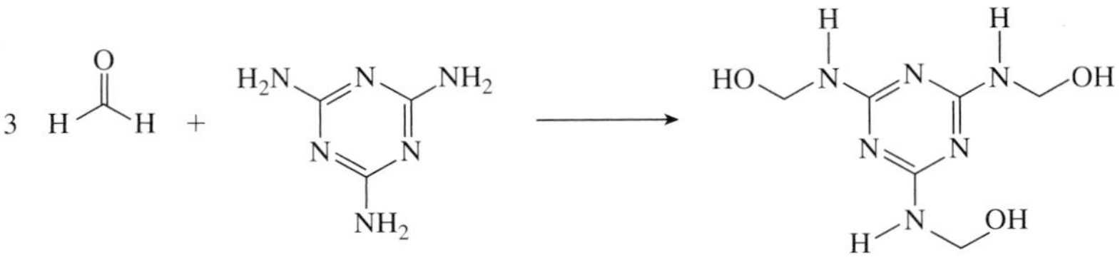

Finally, alcohol-amine condensation gives a polymer with a regular array of cross-links, as shown in Figure 11-9. Melmac dinnerware and Formica countertops contain this polymer.

Figure 11-9
A portion of the structure of the polymer made from melamine and formaldehyde, with one of the monomer units highlighted. This material is used to make Formica countertops because it is strong, lightweight, and heat-resistant.

Section Exercises

11.3.1 The following monomers can be used to synthesize polymers:

Draw structures that show the repeat units of the polyamide and the polyester arising from these monomers.

11.3.2 Polycarbonates are colorless polymers nearly as tough as steel. One of the most common polycarbonates, Lexan, is used in bulletproof windows and as face plates in the helmets worn by astronauts. Lexan is made by condensing phosgene and a compound that contains two —OH functional groups, with the elimination of HCl:

Phosgene

Write the structure of Lexan. Show at least two complete repeat units of the polymer.

11.3.3 The following polymer is used to make carpets because its fibers have a single, most stable, three-dimensional structure. Footprints quickly disappear from a carpet made of these fibers because compressed fibers quickly return to their most stable orientation. Identify the monomer used to make the polymer and write a balanced equation that shows the condensation of two of these monomers.

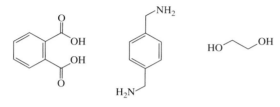

11.4 TYPES OF POLYMERS

To a chemist, a polymer is described by its chemical structure. To a manufacturer or a consumer, the important features of a polymer are its macroscopic properties. A tire manufacturer needs a flexible, tough, and durable polymer. A shirtmaker wants to spin a polymer into fibers that wear well, hold dyes, and are free of wrinkles. Makers of lenses look for transparent polymers that can be molded into precise shapes. Each of these polymer applications has somewhat different requirements.

Although they have an endless variety of properties, polymers can be divided into three categories based on their form and resistance to stretching. These are **plastics, fibers,** and **elastomers.** Plastics differ in form from fibers: whereas plastics exist as blocks or sheets, fibers have been drawn into long threads. Unlike plastics or fibers, elastomers can be stretched without breaking. Formica tabletops and polyethylene packaging films are examples of plastics. Orlon carpets are made from polymer fibers, and rubber bands are elastomers. Some polymers, such as nylon, can be formed into both plastics and fibers.

Plastics

To the general public the term *plastic* has become synonymous with *polymer*. More precisely, a plastic is the type of polymer that hardens on cooling or on evaporation of solvent, allowing it to be molded or extruded into specific shapes or spread into thin films.

Plastics fall into two groups based on their response to heating. Those that melt or deform on heating are classed as **thermoplastic** materials, whereas plastics that retain their structural integrity are said to be **thermosetting.** Cross-linking makes it harder for a polymer to deform, so highly cross-linked polymers are thermosetting. Thermoplastic polymers have small amounts of cross-linking. For example, polyethylene consists of huge alkane molecules without cross-linking, held together only by dispersion forces between the chains. Polyethylene is a thermoplastic polymer that melts when heated because individual molecules acquire enough kinetic energy to overcome dispersion forces and slide past one another. In contrast, Formica (see Figure 11-9) is so extensively cross-linked that it can be viewed as an immense single molecule. To melt Formica would require the breakage of covalent bonds. This is a thermosetting polymer, which may decompose irreversibly if it is heated to a high enough temperature but will not reversibly melt or deform.

Polyethylene, the best known thermoplastic material, exists in two general forms with different properties. High-density polyethylene is a rigid, strong polymer used to make bottlecaps, toys, pipes, and cabinets for electronic devices such as computers and televisions. Low-density polyethylene is a soft, semirigid polymer used to make plastic bags, squeeze bottles, food packaging films, and other common items. Both kinds of polyethylene have the same repeat unit, CH_2, but different structures of individual polyethylene molecules, as shown in Figure 11-10.

High-density polyethylene forms under conditions that produce polymers made of straight chains of CH_2 units. These linear molecules maximize attraction resulting from dispersion forces by "lining up" in rows that create crystalline regions within the polymer. Maximizing attraction between the chains imparts strength and rigidity to the polymer. In contrast, low-density polyethylene has chains of CH_2 groups that branch off the main backbone of the polymer. These branches prevent the polymer molecules from packing closely together, thus decreasing dispersion forces and weakening the attraction between the chains. The result is an amorphous polymer that is flexible and melts at a relatively low temperature.

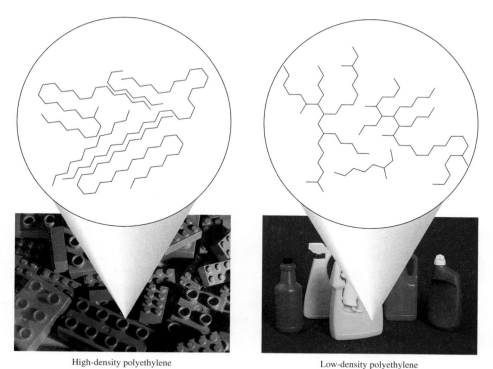

High-density polyethylene

Low-density polyethylene

Figure 11-10
High-density polyethylene has aligned chains of CH$_2$ units, giving a tough, rigid polymer with a high melting point. Low-density polyethylene has branching chains that give a soft polymer with a low melting point.

The flexibility of some plastics can be improved by the addition of small molecules called plasticizers. For example, pure poly (vinyl chloride) (PVC) turns brittle and cracks too easily to make useful flexible plastic products. With an added plasticizer, however, PVC can be used to make seat covers for automobiles, raincoats, garden hoses, and other flexible plastic objects. Plasticizers must be liquids that mix readily with the polymer. In addition, they must have low volatility so that they do not escape rapidly from the plastic. Dioctylphthalate is a liquid plasticizer that is formed by condensing two alcohol molecules with one molecule of phthalic acid, as illustrated in Figure 11-11.

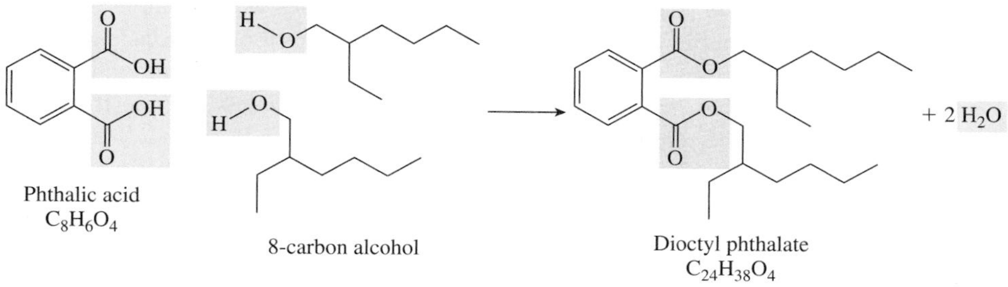

Phthalic acid
$C_8H_6O_4$

8-carbon alcohol

$+ 2\ H_2O$

Dioctyl phthalate
$C_{24}H_{38}O_4$

Figure 11-11
Dioctylphthalate, the leading plasticizer, is synthesized by condensing two molecules of an eight-carbon alcohol with phthalic acid.

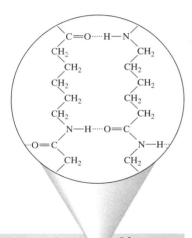

Fibers

For centuries, humans have woven and spun cotton, wool, and silk into fabrics for clothing and other purposes. These natural fibers are described later in this chapter. Today, synthetic fibers are equally important components of the fabrics industry. Nylon 6, a polyamide with six carbon atoms in its repeat unit, makes resilient fibers that are used in carpeting. Part of the reason for the strength of these fibers is hydrogen bonding between neighboring strands of the nylon polymer, as Figure 11-12 illustrates.

Fibers are thin threads of polymer made by forcing a fluid thermoplastic material through a set of tiny pores, as shown in Figure 11-13. This process requires that the polymer be in the liquid phase, so a fiber-forming polymer must melt at low temperature or dissolve in a convenient volatile solvent. Most synthetic fibers are polyesters, polyamides, or polyacrylonitrile. The polar functional groups in these polymers produce strong intermolecular forces that add significant tensile strength to the material.

Figure 11-12
One of the polymers used for carpet fibers is Nylon 6, in which neighboring polymer strands are held together by hydrogen bonds.

Elastomers

An **elastomer** is a polymer that is flexible, allowing it to be distorted from one shape to another. Polyisoprene (natural rubber), polybutadiene, and butadiene-styrene copolymer are the most important commercial elastomers. All contain some C=C bonds, and their bulk properties are affected by the varying geometries about the carbon atoms that make up the polymer backbone.

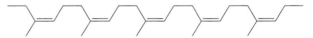

The geometry about the carbon atoms of the CH_2 groups is tetrahedral, whereas that about the carbon atoms involved in double bonds is trigonal. Because of these varied shapes, the molecules lack the structural regularity required to form a polymer with crystalline properties. Consequently, elastomers such as polyisoprene are amorphous solids with individual polymer strands tangled together. This lack of structural regularity keeps the molecules from approaching too closely, thus minimizing attraction resulting from dispersion forces. This in turn allows individual polymer molecules to slide past one another and makes the polymer flexible.

The π bonds in polyisoprene introduce further structural variations that influence the properties of the elastomer. A molecule cannot readily undergo rotation about a double bond because of the directional character of π bonding. Consequently, there are two distinct geometries about the double bonds in polyisoprene, as shown in Figure 11-14. In the *cis* configuration, the carbon atoms of the polymer chain are both on the same side of the double bond. In the *trans* configuration, the carbon atoms of the polymer chain are on opposite sides of the double bond. In natural rubber, the double bonds are all in the *cis* configuration and adjacent chains cannot get close together, giving a flexible polymer. A second form of polyisoprene, gutta-percha, has *trans* double bonds. Gutta-percha is less flexible than natural rubber because the regular arrangement of carbon atoms along its chain allows adjacent chains to nest closely together. This leads to large dispersion forces and a hard but brittle material that is used, for example, for golf ball covers.

Table 11-4 shows that the flexibility and strength of an elastomer are highly dependent on its amount of cross-linking. Linking chains increases the strength of the elastomer but reduces its flexibility. Cross-linking also increases the ability of the elastomer to return to its original shape after distortion, as Figure 11-15 illustrates. For uses such as automobile tires that require both strength and flexibility, the amount of cross-linking must be carefully controlled.

Figure 11-13
Fibers form when a fluid thermoplastic polymer is forced through tiny pores into long, thin threads, which solidify as they cool or as solvent evaporates.

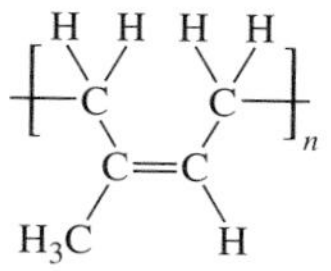

Polyisoprene
(*cis* configuration)

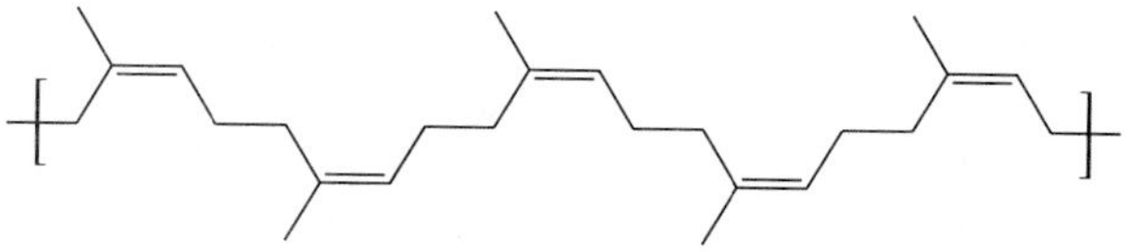

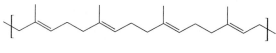

Natural rubber

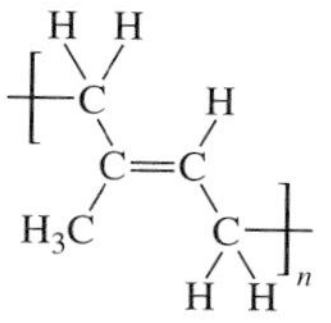

Polyisoprene
(*trans* configuration)

Gutta-percha

Figure 11-14
Rotation about double bonds is energetically hindered, so polyisoprene exists with two distinct internal geometries, termed *cis* and *trans* configurations. Natural rubber has an all-*cis* chain configuration, and gutta-percha has an all-*trans* chain configuration.

Table 11-4
Effects of Cross-Linking on Rubber

Mononer Units Between Cross-Links	Degree of Flexibility	Strength	Product
5–10	Small	Very high	Casing for calculators
10–20	Restricted	High	Tires
20–30	Moderate	Moderate	Tire tubes
30–40	Moderate	Moderate	Artificial heart membrane
50–80	Moderate	Low	Kitchen gloves
100–150	High	Low	Surgical gloves

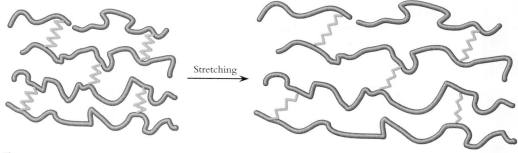

Stretching

Figure 11-15
The amount of cross-linking in an elastomer determines its strength and flexibility. A modest number of cross-links (shown schematically in blue) allows a material to stretch but then pulls it back to its original shape.

Polymer Stability

Any polymer that has commercial value must be stable under a variety of conditions. This means that it must not degrade when exposed to light, heat, or a variety of chemicals, including acids, bases, and oxidizing agents. Over time, polymers with superior stability have replaced less stable materials. For example, the first synthetic plastic, celluloid, is so highly flammable that it is no longer an important commercial polymer.

Among elastomers, artificial rubbers have replaced natural rubber for many uses because of their high resistance to chemical attack by ozone, an atmospheric pollutant. When ozone reacts with polymer chains, it breaks $C=C$ π bonds and introduces additional cross-linking. Breaking π bonds causes the rubber to soften, and cross-linking makes it more brittle. Both changes eventually lead to rupture of the polymer structure.

Among fibers, nylon and polyester have captured large segments of the clothing and floor-covering markets, in part because they last longer than natural fibers such as silk and wool. Moth holes in a wool sweater demonstrate that living creatures degrade natural fibers. Artificial fibers, by contrast, have no such natural enemies.

The stability of polymers is advantageous while they are being used, but it becomes a liability when these materials need to be discarded. Up to 15% of the volume of municipal waste dumps is polymeric material. Unlike natural materials, polymeric trash does not readily decompose after it is discarded. Thus polymer disposal has become a cumulative problem.

Recycling polymers is one way to minimize the disposal problem, but not much recycling occurs at present. In 1996, only 23% of the plastic made in the United States was recycled, compared with 52% of the aluminum and 41% of the paper. A major obstacle to recycling plastics is the great variation in the composition of polymeric material. Polyethylene and polystyrene have different properties, and a mixture of the two is inferior to either. Recyclers must either separate different types of plastics or process the recycled material for less specialized uses. Recently, manufacturers have begun labeling plastic containers with numbers that indicate their polymer type and make it easier to recycle these materials. Table 11-5 shows the recycling number scheme.

Thermoplastics are more suitable for recycling than elastomers or thermosetting polymers. Thermoplastics can be heated above their melting temperatures and then recast into new shapes. Elastomers and thermosets, on the other hand, have extensive cross-linking networks that must be destroyed and then reformed in the

Many early motion pictures were recorded on film made from celluloid. Ping-Pong balls are among the few products still made from celluloid.

Table 11-5
Recycling Categories for Plastics

Number	Polymer	Abbreviation
1	Poly(ethylene terephthalate)	PET
2	High-density polyethylene	HDPE
3	Poly (vinyl chloride) (PVC)	P
4	Low-density polyethylene	LDPE
5	Polypropylene	PP
6	Polystyrene	PS
7	Other	–

process of recycling. Processes that destroy cross-linking, however, generally break down the polymer beyond the point at which it can be easily reconstituted.

One way to reduce disposal problems is to make polymers degradable. This poses a substantial challenge to polymer chemists: to fine-tune polymer properties so that the materials are stable while in use but degrade readily to innocuous materials when their useful life is over. One promising approach is to use polymers structurally similar to those found in nature. The goal is to develop polymers that can be degraded by biological organisms. Starch, which is described in the next section, is a natural polymer whose properties can be tailored to various commercial requirements.

Section Exercises

- **11.4.1** Both polystyrene (see Figure 11-3) and Kevlar (Figure 11-8) contain phenyl rings; yet polystyrene is a thermoplastic material like polyethylene, whereas Kevlar is tough and rigid. Explain why the presence of phenyl rings in polystyrene does not lead to the rigidity characteristic of Kevlar.
- **11.4.2** Polyethylene (Figure 11-10) and polyisoprene (Figure 11-14) each has two distinct structures with differing properties. Compare and contrast the structures of the two forms of each polymer.
- **11.4.3** Make a drawing similar to Figure 11-10 that illustrates how cross-linking reduces the stretchability of a polymer.

11.5 CARBOHYDRATES

Cellulose and starch are macromolecules with empirical formulas that resemble "hydrated carbon," $C_x(H_2O)_y$, where x and y are integers. The monomers from which these macromolecules are constructed are sugars such as glucose and fructose. These monomers and macromolecules are the **carbohydrates.** Structurally, carbohydrates are very different from simple combinations of carbon and water. Even the smallest carbohydrates contain carbon chains with hydrogen atoms, OH groups, and occasional ether linkages.

Carbohydrates are an important food source for most organisms. Glucose, fructose, and sucrose are small carbohydrate molecules that can be broken down rapidly to provide quick energy for cells. Large amounts of energy are stored in carbohydrate macromolecules called **polysaccharides.** For example, glycogen is a polysaccharide used by humans for long-term energy storage. Other polysaccharides, such as cellulose and chitin, are building materials for plants and animals.

Large carbohydrates are organized into chains composed of smaller carbohydrate units, called **monosaccharides.** A monosaccharide contains three to six carbon atoms. Oligosaccharides are small chains of two to ten monosaccharide units, and polysaccharides are long-chain polymers of monosaccharides.

Monosaccharides

The simplest carbohydrates have the formula $(CH_2O)_n$, where n is between 3 and 6. Of the 70 or so monosaccharides that are known, 20 occur in nature. The most important naturally occurring monosaccharides contain five carbons

The word *saccharide* comes from the Latin term *saccharum*, meaning "sugar."

Figure 11-16
Ribose and α-glucose are two monosaccharides that occur in nature. As is characteristic of carbohydrates, each has a ring structure containing several HCOH groups. We number the carbon atoms for identification purposes.

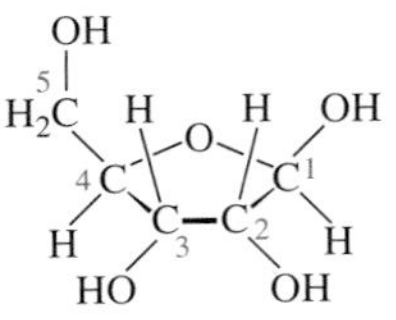

Ribose
$C_5H_{10}O_5$

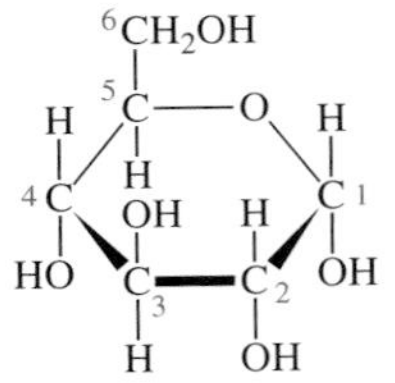

α-Glucose
$C_6H_{12}O_6$

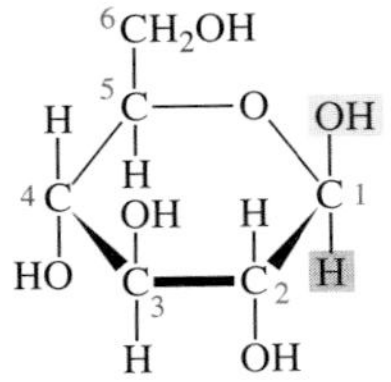

β-Glucose

Figure 11-17
The structure of β-glucose, which differs from α-glucose in the placement of the H atom and the OH group on carbon atom number 1 (see shading).

(pentoses) or six carbons (hexoses). Structures of ribose, an important pentose, and α-glucose, a hexose that is the most common monosaccharide, are shown in Figure 11-16. As shown in the figure, it is customary to number the carbon atoms in a monosaccharide, beginning with the HCOH group adjacent to the ether linkage.

The structures of ribose and α-glucose exemplify the characteristics of most monosaccharides. Each is a cyclic compound with an oxygen atom forming an ether linkage in one of the ring positions. Monosaccharides exist in several structural forms, but their ring forms are the basic building blocks of polysaccharides. Ribose has four hydroxyl groups bonded to the ring, and α-glucose has five. Monosaccharides are soluble in water because all have several hydroxyl groups that can form hydrogen bonds with water.

Because the six-membered ring of α-glucose is an important structural component in many polysaccharides, we examine its structure in more detail. As Figure 11-16 shows, each carbon atom in the glucose ring is bonded to one hydrogen atom. Four of the ring carbons are bonded to hydroxyl groups (—OH), and the fifth ring carbon atom is bonded to a —CH₂OH fragment.

Glucose exists as two different isomers, α- and β-glucose. The β-glucose isomer, which is slightly more stable, is shown in Figure 11-17. The two isomers differ only in the placement of —H and —OH on carbon atom number 1. Whereas the —OH on carbon 1 of α-glucose lies below the ring in the view shown in Figure 11-16, that —OH lies above the ring in β-glucose. This difference may not appear significant, but it has profound importance, especially for glucose polymers. When α- and β-glucose form polysaccharides, the shape of one polymer is quite different from the shape of the other.

Every atom in the glucose ring has a steric number of 4 and approximately tetrahedral geometry, so the glucose ring is puckered rather than planar. This gives glucose a "lawn chair" shape. Although the ring is not planar, we can define a molecular plane that passes through the midpoints of the ring bonds, as shown in Figure 11-18. Each carbon atom of the ring is bonded to two non-ring

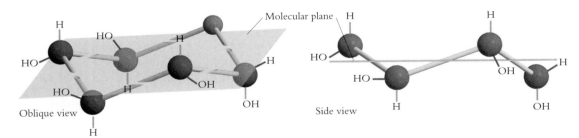

Figure 11-18
The six-membered rings of α-glucose and other hexoses consist of atoms with tetrahedral geometries, giving a puckered ring structure. A molecular plane can be defined that passes through the midpoints of all the bonds of the ring. Bonds to non-ring atoms are either perpendicular to this plane (shown in green) or roughly parallel to the plane (shown in brown).

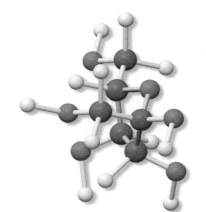

Figure 11-19
Fructose is a hexose monosaccharide that forms a five-membered ring structure. In the ball-and-stick model, the ring is highlighted in blue.

The structures here and on subsequent pages of this chapter are line drawings. Recall that C labels and all C—H bonds are omitted from line drawings to show underlying structural features more clearly.

β–Fructose:
$C_6H_{12}O_6$

Fructose and glucose are present in fruit juice and honey. The suffix *-ose* signifies that the molecule is a carbohydrate.

atoms that lie in different positions relative to this molecular plane. The bond to one of the two atoms is perpendicular to the plane, whereas the other lies roughly parallel to the plane.

Hexose monosaccharides can form both five- and six-membered rings. In most cases, the six-membered ring structure is more stable, but fructose is an important example of a hexose that is more stable as a five-membered ring. The structure of β-fructose is shown in Figure 11-19. Notice that there are —CH_2OH fragments bonded to two positions of this five-membered ring. Examples 11-3 and 11-4 explore the structures of monosaccharides in more detail.

Monosaccharide Structures **Example 11-3**

Describe the differences in the structures of ribose and fructose.

Strategy: Monosaccharides can differ in their formulas, their ring sizes, and the spatial orientations of their hydroxyl groups. To analyze the differences between two monosaccharides, begin with structural drawings of the molecules, oriented so the ether linkages are in comparable positions. Then examine the structures to locate differences in constituents and bond orientations.

Solution: Here are the structures of ribose (Figure 11-16) and β-fructose (Figure 11-19):

Ribose
$C_5H_{10}O_5$

β-Fructose
$C_6H_{12}O_6$

Although the chemical formulas indicate that ribose is a pentose and fructose is a hexose, the ring portions of the structures are identical. Proceeding clockwise around the rings from the oxygen atom, we see that the structures differ at the first two positions. In the first position, ribose has a C—H bond, and β-fructose has a C—CH_2OH linkage. In the second position the molecules have the same two bonds but in different orientations. The OH group points "up" in β-fructose and "down" in ribose.

| Example 11-4 | Drawing Monosaccharides |

The six-carbon sugar α-galactose is identical to α-glucose except at carbon atom number 4, where the orientations are different. Draw the molecular structure of α-galactose. Simplify the structure by using flat rings rather than the true three-dimensional forms.

Strategy: When a carbohydrate with a six-membered ring is drawn, it is best to start with the ring itself. Next, use α-glucose (see Figure 11-16) as a convenient template to obtain the proper orientations for the groups that galactose and glucose have in common. Finally, switch positions of groups as needed to obtain the correct final structure for galactose.

Solution: Here is the flat-ring structure of α-glucose:

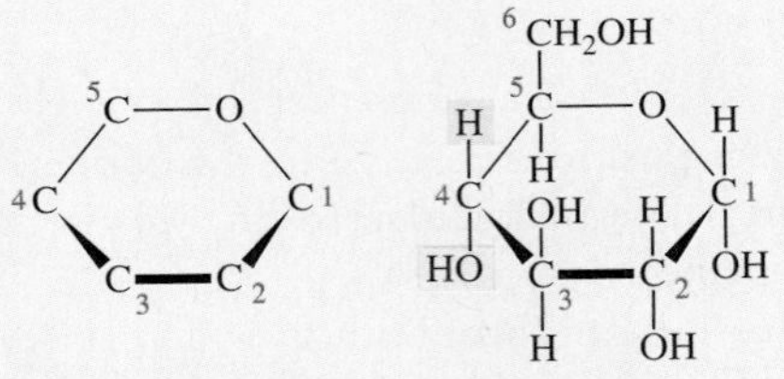

6-membered ring α-Glucose

The problem states that α-galactose differs from α-glucose at carbon atom number 4. This position is highlighted in the structure. Starting from α-glucose, exchange the hydroxyl group with the hydrogen on this carbon atom:

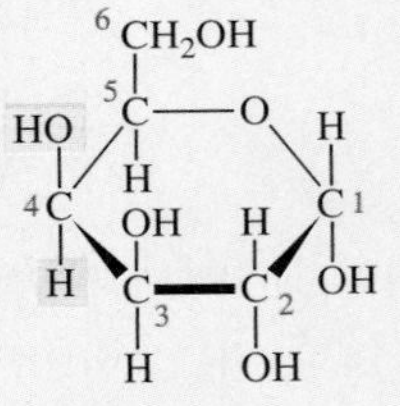

α-Galactose

Disaccharides

Two monosaccharides can combine by a condensation reaction between two hydroxyl groups. This reaction forms a linkage in which an oxygen atom connects two saccharide rings:

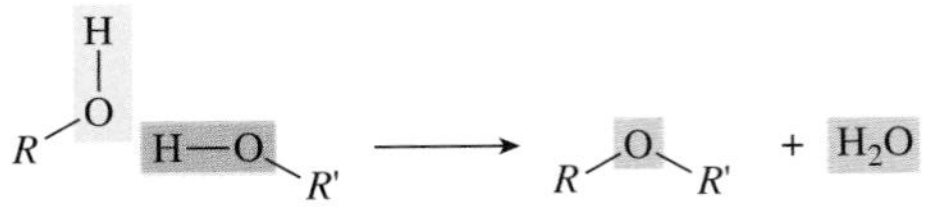

A C—O—C linkage between two sugar molecules is termed a *glycosidic bond*. Monosaccharides contain several hydroxyl groups, so many glycosidic bonds are possible. Nevertheless, in all natural glycosidic bonds the linkage uses a hydroxyl group on a carbon atom next to a ring oxygen atom of one sugar. The second sugar can link to the first through any of its hydroxyl groups. For example, α-maltose forms from two molecules of α-glucose. The glycosidic bond occurs at the position adjacent to the ring oxygen for one glucose molecule and a different position on the second glucose, as shown in Figure 11-20. Maltose is formed from starch and decomposes in the presence of yeast, first to give glucose and then ethanol and water.

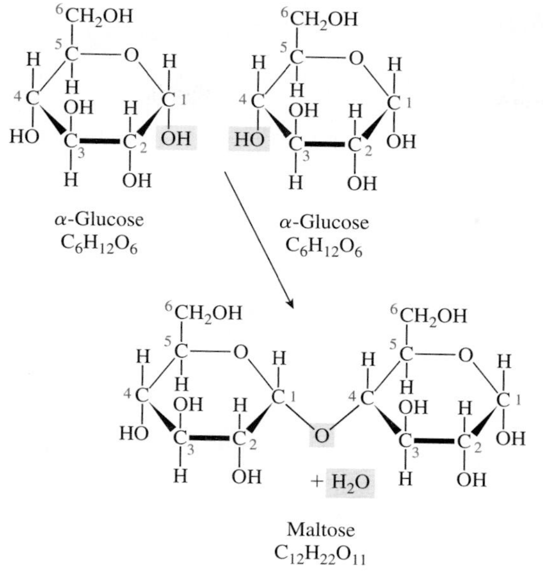

Some disaccharides serve as soluble energy sources for animals and plants, whereas others are important because they are intermediates in the decomposition of polysaccharides. A major energy source for humans is sucrose, which is common table sugar. About 80 million tons of sucrose are produced each year. Of that, 60% comes from sugar cane and 40% comes from sugar beets. Sucrose contains α-glucose linked to β-fructose. To visualize the glycosidic linkage, rotate the fructose molecule 180° from the view shown in Figure 11-19, as shown in Figure 11-21. This places the linking hydroxyl groups in the appropriate positions for making a glycosidic bond. Example 11-5 shows how to draw another glycosidic bond.

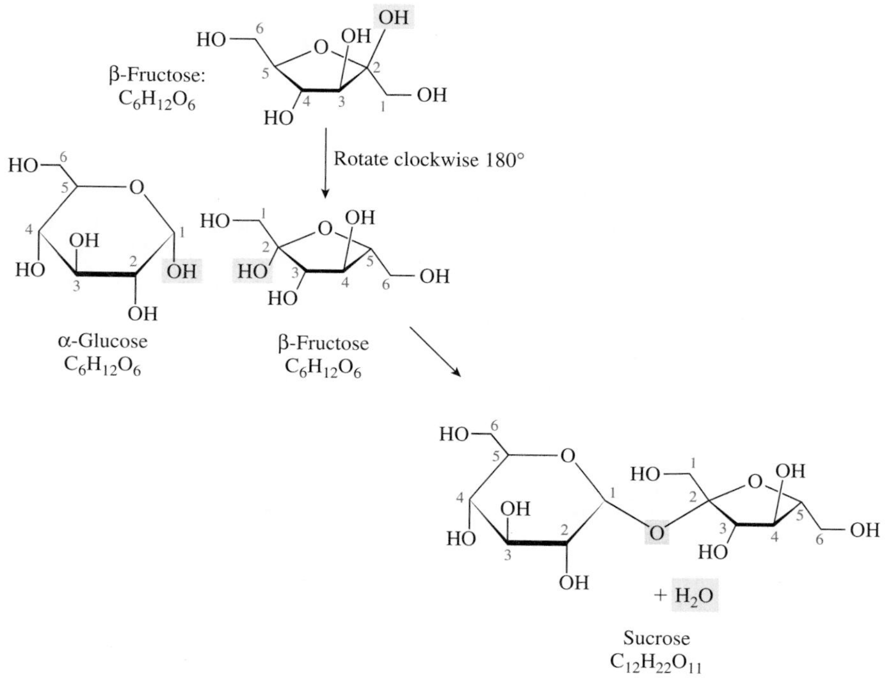

Figure 11-20
The structure of maltose. A glycosidic bond links different ring carbon atoms of the two glucose molecules.

Figure 11-21
Sucrose is a sugar in which an α-glucose and a β-fructose molecule are linked through a condensation reaction.

Example 11-5	Decomposing a Sugar

Glucose is the principal sugar in the circulatory systems of mammals. This "blood sugar" is burned by the body to produce energy. Insects obtain energy from trehalose, whose line structure follows. Identify the monosaccharides from which trehalose is constructed.

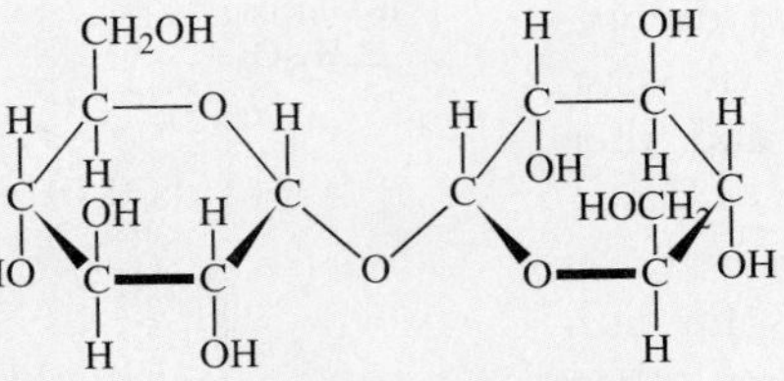

Insects store energy in the form of trehalose.

Trehalose
$C_{12}H_{22}O_{11}$

Strategy: Identifying monomer building blocks requires us to visualize a condensation reaction operating in reverse. The linkage bond breaks, and water is added.

Solution: Break the glycosidic bond of trehalose and add water to generate these two monosaccharides:

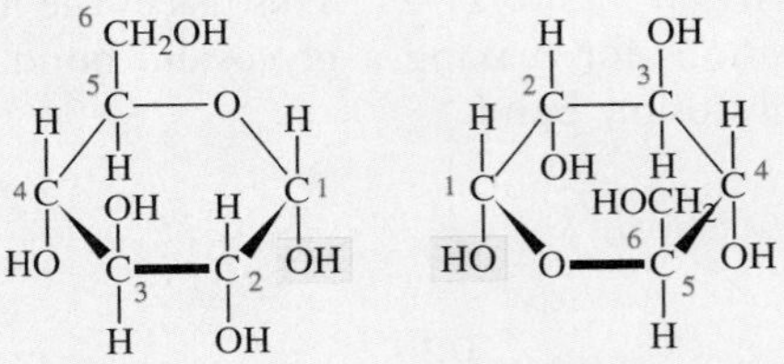

α-Glucose
$C_6H_{12}O_6$

α-Glucose
$C_6H_{12}O_6$

The monosaccharide on the left is oriented with its ring oxygen atom in the back right position. Compare the locations of its —OH groups with those in the structures in Figure 11–18 to see that this sugar is α-glucose.

The monosaccharide on the right must be rotated clockwise about the plane of the ring to bring its ring oxygen atom to the back right position. When that is done, you should recognize that this sugar is identical to the other: Both the monosaccharides in trehalose are α-glucose.

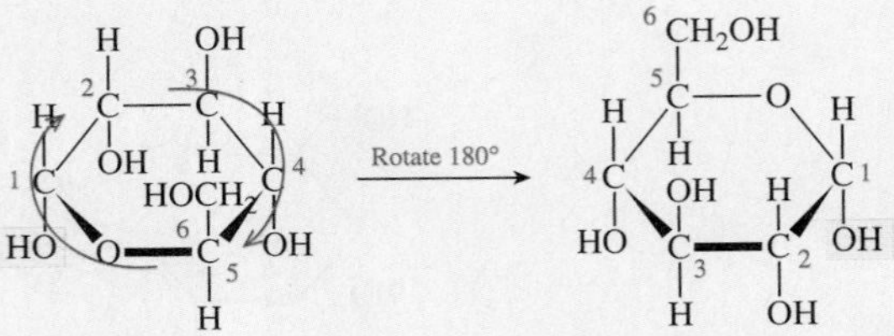

α-Glucose

Rotate 180°

α-Glucose

Polysaccharides

Polysaccharides, macromolecules made up of linked monosaccharides, play two major roles in biological organisms. Some, such as cellulose, are structural materials. Others, including glycogen, act as reservoirs for energy storage. All carbohydrates are good sources of chemical energy, because they release energy upon reacting with oxygen to produce water and carbon dioxide.

Polysaccharides formed from α-glucose are called **starches.** A starch stores sugar until it is needed for energy production. Three important starches are glycogen, which animals produce in their livers, and amylose and amylopectin, produced by plants through photosynthesis. On average, plant starch is about 20% amylose and 80% amylopectin. Each of these polysaccharides contains glucose as its monomer, but they differ in how the monosaccharide units are linked.

The simplest starch is amylose, which consists of long chains of α-glucose linked end to end by glycosidic bonds. One molecule of amylose contains about 200 glucose units. Amylopectin and glycogen are predominantly linear chains, too, but in these polysaccharides, some of the rings form a second glycosidic linkage, generating branches along the main chain. Figure 11-22 shows a portion of the structure of amylopectin. Amylopectin contains about 1000 glucose monomers. Branches that are about 30 units long occur at intervals of 20 to 25 glucose units.

Starch is the principal carbohydrate reservoir in plants and is the major component of rice, grains, corn, and potatoes. The long chains are sparingly soluble in water, so they collect in cells as granules. Starch is digested by a series of enzymes. One enzyme breaks the polymer into smaller, more soluble pieces. Then another enzyme cleaves these fragments into maltose. Finally, maltose is cleaved into glucose molecules, which are metabolized for energy production. Humans, other animals, and plants have the enzymes necessary to digest plant starch.

Animals store glucose in the form of glycogen. This form of starch is structured like amylopectin but with more frequent, shorter branches. Glycogen is synthesized and stored as granules in the liver and in muscle tissue, where the sugar is readily available for rapid energy production. The liver has a limited capacity to store glycogen. When that capacity is exceeded, the body activates alternative biochemical pathways that convert sugar into fat.

The most abundant organic molecule in the biosphere is **cellulose,** a polysaccharide that is the principal building material for plants. Like amylose, cellulose is a

An enzyme (see Section 11.7) speeds up a biochemical reaction.

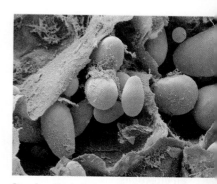

Starch collects in granules in plant cells and in animal liver cells.

Figure 11-22
Stylized view showing how glucose rings are linked in the structure of amylopectin. Long chains of glucose molecules are linked via glycosidic bonds. Branching occurs through additional linkages at some glucose rings. The hydroxyl groups that are not involved in linkages have been omitted for clarity.

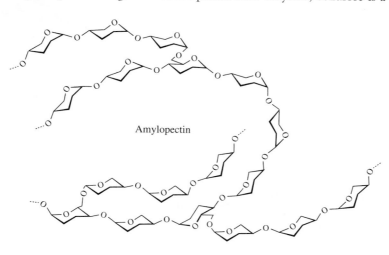

Amylopectin

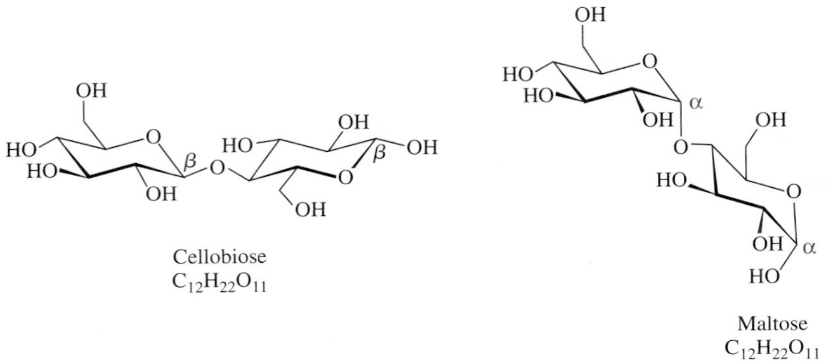

Figure 11-23
Maltose (two α-glucose units) and cellobiose (two β-glucose units) have different shapes, even though the two disaccharides have identical chemical formulas.

Cellobiose
$C_{12}H_{22}O_{11}$

Maltose
$C_{12}H_{22}O_{11}$

linear polymer of glucose. Unlike amylose, however, the glucose monomers in cellulose are in the β configuration.

Figure 11-23 shows the structure of cellobiose, the disaccharide formed from two β-glucose molecules linked head to tail. Cellobiose and maltose are made from very similar monomers, but the figure shows that they have distinctly different shapes. In maltose, whose repeat unit is two molecules of α-glucose, the glycosidic linkage imparts a distinct kink to the structure. As a result, a chain of maltose units can coil on itself, resulting in the granular shape of starch deposits. In contrast, cellobiose, which is made entirely from β-glucose, has a nearly flat structure.

Linking a series of cellobiose units to make cellulose gives a long, ribbon-like chain of sugar units. As shown in Figure 11-24a, the planar arrangement of cellulose makes it possible for hydrogen bonds to form between polysaccharide chains, generating extended packages of cellulose ribbons.

A typical cellulose molecule contains 2000 to 3000 glucose units in an unbranched linear chain. Individual chains group together through hydrogen bonds in bundles. The cell walls of plants are made up of these bundles laid down in a cross-hatched pattern that gives cellulose strength in all directions. This cross-hatched structure is shown in Figure 11-24b. The figure also shows the chain-like microscopic structure of cellulose.

Figure 11-24
(a) Line structure showing the planar arrangement of cellulose. Hydrogen bonding between the chains gives a sheetlike structure. (b) Plant cell walls are made of bundles of cellulose chains laid down in a cross-hatched pattern that gives cellulose strength in all directions.

(a)

(b)

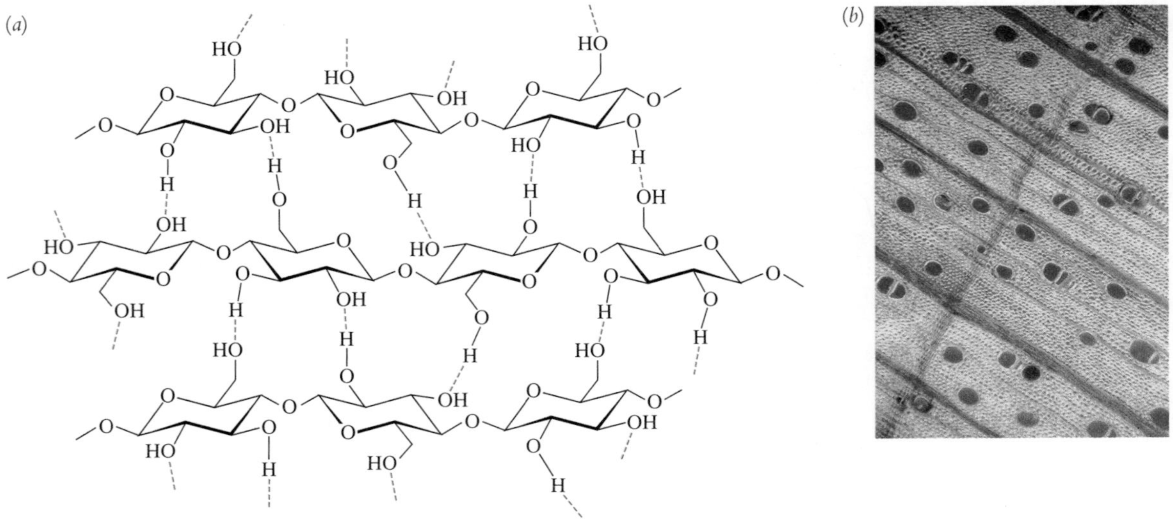

Amide

OH

O

N—H

HO

O

O

HO

N—H

O

OH

O

Chitin

Figure 11-25
Chitin is a structural polysaccharide found in the exoskeletons of arthropods. The macromolecule is a derivative of cellulose in which one of the hydroxyl groups has been replaced by an amide.

The enzymes that cleave α-glucose units from starches cannot attack the β-glucose linkages of cellulose because the geometry of the glycosidic linkage is different. As a result, cows and other ruminants must rely on bacteria in their digestive tracts to break down cellulose. These microorganisms, which are also present in termites, use a different group of enzymes to cleave β-glucose units.

Cellulose is the most abundant structural polysaccharide, but it is by no means the only one. Plants and animals use a variety of polysaccharides for a wide range of structural applications. Some polysaccharides contain sugars other than glucose, and others are derivatives of cellulose in which some of the hydroxyl groups on the ring have been converted into other functional groups. One important example is chitin, a derivative of cellulose present in many animals. Chitin makes up the exoskeletons of crustaceans, spiders, insects, and other arthropods. The structure of chitin, shown in Figure 11-25, is the same as that of cellulose, except that one hydroxyl group is replaced by an amide group.

Section Exercises

11.5.1 Gulose is a six-carbon sugar that differs from glucose in two positions. In gulose the hydroxyl orientations at positions 2 and 3 are reversed from their orientations in glucose. Draw the structures of α- and β-gulose.

^{6} CH$_2$OH

^{5}C — O

^{4}C   C^1

C — C
$_3$ $_2$

11.5.2 Approximately 5% of milk is lactose, which is made from β-galactose and β-glucose. The hydroxyl group adjacent to the ring oxygen atom in galactose links to the hydroxyl group in glucose that is in position 4. Draw the structure of lactose using the flattened views of the monosaccharide rings. (The structure of galactose appears in Example 11-4.)

11.5.3 Many humans suffer from lactose intolerance because, although they can digest sucrose, they lack the enzyme that decomposes lactose. An enzyme that breaks down a sugar molecule recognizes the shape of the molecule; thus the enzyme that breaks down sucrose cannot break down lactose. Examine the structures of lactose and sucrose (see Figure 11-21) and identify the differences in their shapes that might account for this.

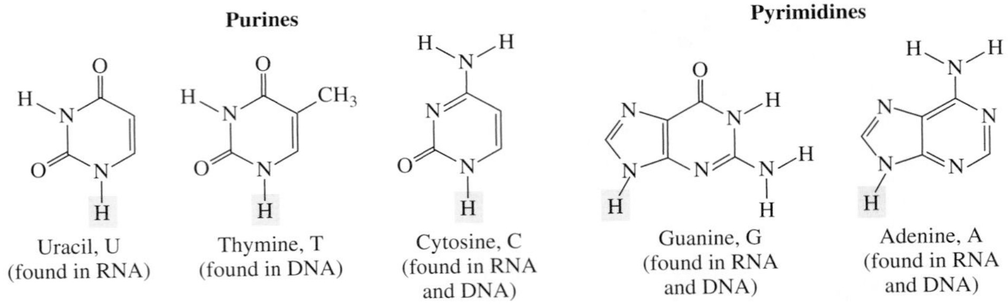

Purines

Pyrimidines

Uracil, U
(found in RNA)

Thymine, T
(found in DNA)

Cytosine, C
(found in RNA
and DNA)

Guanine, G
(found in RNA
and DNA)

Adenine, A
(found in RNA
and DNA)

Figure 11-26
The organic bases found in DNA and RNA are of two types. Uracil, thymine, and cytosine are purines with single rings. Guanine and adenine are pyrimidines with two rings. The hydrogen atoms that are eliminated during condensation are highlighted.

11.6 NUCLEIC ACIDS

All biological organisms have the ability to reproduce themselves. The instructions for self-replication are stored and transmitted by macromolecules called **nucleic acids.** There are two types of nucleic acids, one that stores genetic information and one that transmits the information. Genetic information is *stored* in molecules of **deoxyribonucleic acid (DNA),** which are located in cell nuclei. These huge molecules have molar masses as large as a trillion g/mol. The information stored in DNA is *transmitted* by **ribonucleic acid (RNA).** There are several kinds of RNA; each has its own role in the operation of a cell. Molecules of RNA have molar masses of 20,000 to 40,000 g/mol, so they are much smaller than their DNA counterparts. In this section, we examine the structures of nucleic acids and survey their biochemical functions.

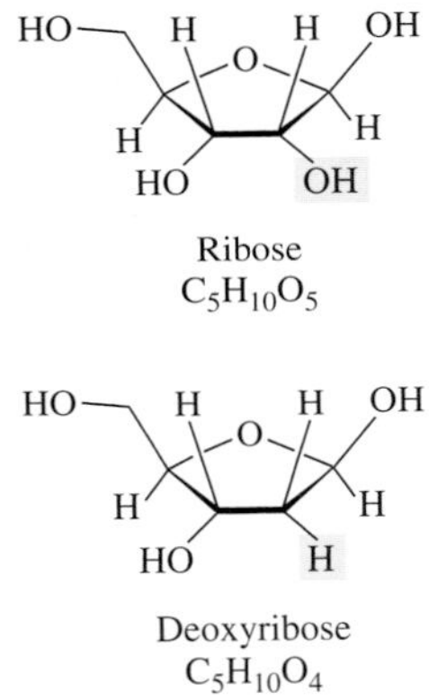

The Building Blocks

Nucleic acids are macromolecules made of three component parts:

1. A nitrogen-containing organic base. There are five such bases. **Adenine** and **guanine** are *pyrimidines,* two-ring structures. **Thymine, cytosine,** and **uracil** are *purines,* one-ring structures. Thymine occurs only in DNA, and uracil occurs only in RNA. Structures are shown in Figure 11-26.

2. A pentose sugar. In RNA the sugar is ribose, and in DNA the sugar is deoxyribose, a ribose in which one OH group has been replaced with one H atom (Figure 11-27).

3. A phosphate linkage derived from phosphoric acid.

These three components are linked through condensation reactions. As Figure 11-28 shows, the sugar is connected to the base when the —OH group on the

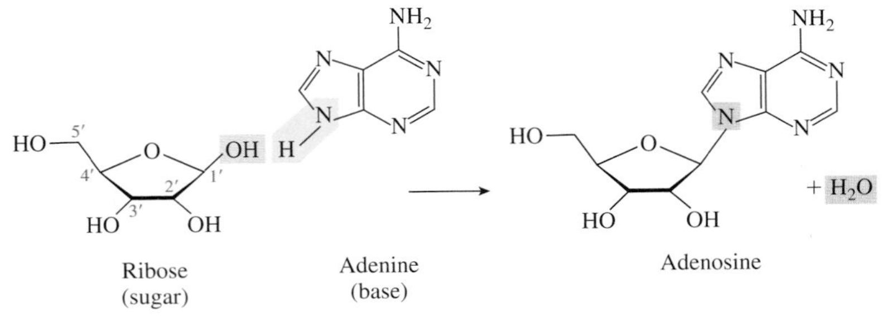

Ribose
$C_5H_{10}O_5$

Deoxyribose
$C_5H_{10}O_4$

Figure 11-27
Structures of ribose and deoxyribose.

Figure 11-28
Line structures showing the formation of adenosine by condensation of adenine and ribose.

Ribose
(sugar)

Adenine
(base)

Adenosine

$+ H_2O$

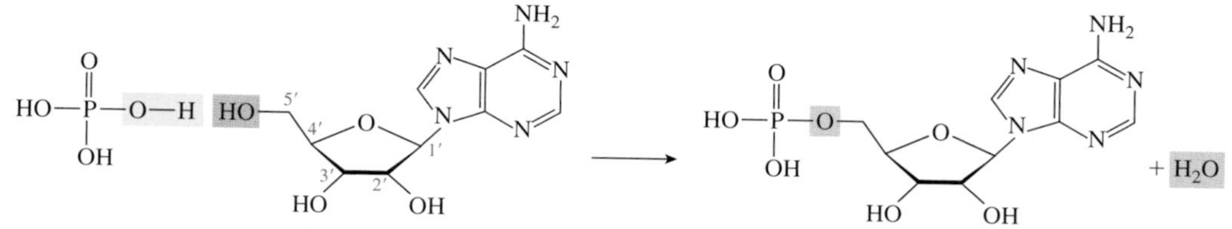

Phosphoric acid (phosphate group) Adenosine Adenosine monophosphate (nucleotide)

Figure 11-29
The formation of adenosine monophosphate (AMP) by condensation of adenosine and phosphoric acid. The three linked units form the nucleotide building block required for nucleic acid synthesis.

The carbon atoms in the sugars of DNA and RNA are numbered with primes (1′, 2′, 3′, 4′, 5′) to distinguish these numbers from numbers used for the positions around the rings of the organic bases.

1-position in the ribose ring condenses with an N—H from the base, thus connecting the rings with a C—N bond. The combination of a base and a sugar is named for its base: cytidine, uridine, thymidine, guanosine, and adenosine. Phosphoric acid condenses with the hydroxyl group on the 5-position (the CH_2OH hydroxyl group) of the sugar to complete the linkage and form a nucleotide. Adenosine monophosphate (AMP), the nucleotide that contains adenine, is shown in Figure 11-29. Example 11-6 shows the formation of another nucleotide.

Drawing Nucleotides

Example 11-6

Draw the structure of uridine monophosphate (UMP).

Strategy: When molecules made from component parts are constructed, a good approach is to draw each piece separately and then combine them in the proper order. The pieces are linked by condensation reactions.

Solution: UMP contains uracil, ribose, and one phosphate group. The structures of uracil and the hydrogen eliminated during the condensation appear in Figure 11-26.

Here are the component parts, drawn in position to eliminate water molecules and link:

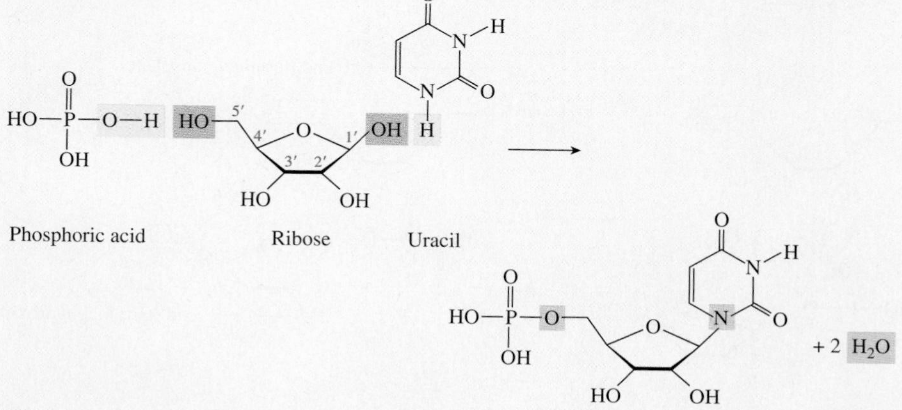

Uridine monophosphate

| Example 11-6 | Drawing Nucleotides *(continued)* |

After the components have been placed next to one another, drawing the final structure requires removal of the H_2O units. UMP has three components. From right to left in the drawing, they are uracil, a single-ring base; ribose, the sugar; and one phosphate group.

The Primary Structure of Nucleic Acids

A nucleic acid polymer contains nucleotide chains in which the phosphate group of one nucleotide links to the sugar ring of a second. The resulting backbone is an alternating sequence of sugars and phosphates, as shown in Figure 11-30. As the figure indicates, the backbone has directionality. Each phosphate unit forms a bridge from the 5′ position of one ribose to the 3′ position of another ribose.

Each position along a nucleic acid sequence is identical except for the identity of its base. The sequence of bases is called the **primary structure** of a polynucleotide chain. By convention the listing always begins with the nucleotide that has the terminal phosphate group (5′ end), and it continues to the opposite end of the chain, where the sugar has an unreacted hydroxyl group in the linkage position (3′ end). The nucleotides are listed as their one-letter abbreviations, A, C, G, and T or U. For example, ACGT stands for the following sequence:

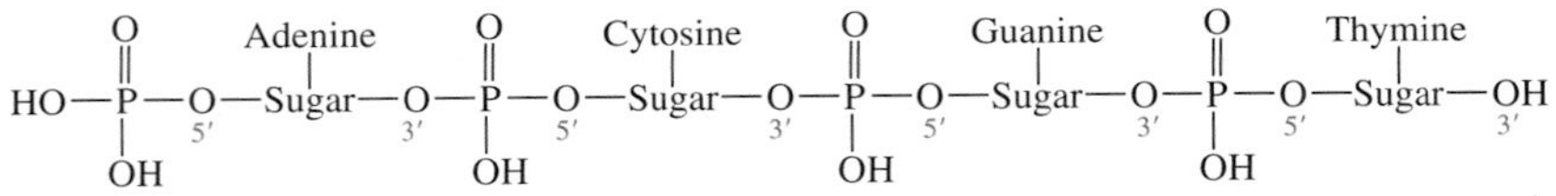

Secondary Structure of DNA: The Double Helix

Although DNA was first isolated in 1868, the nature of nucleic acids remained a mystery for more than 50 years. Not until the 1920s were the structures of nucleotides determined. By then, scientists suspected that DNA was the genetic material, but no one could fathom how sequences of just four different nucleotides could store immense quantities of genetic information. During the 1940s, the British chemist Alexander Todd performed research on nucleic acids that eventually won him a Nobel Prize. Todd discovered the basic composition

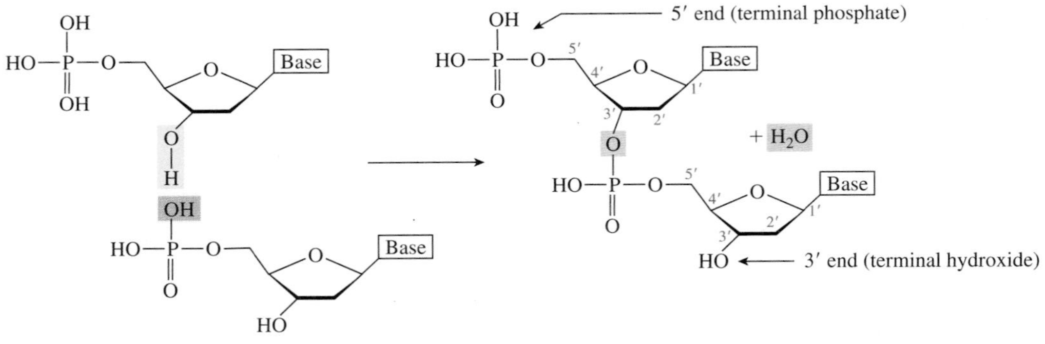

Figure 11-30
The backbone of a nucleic acid is formed by condensation reactions between nucleotides.

of the polynucleotide chain, with sugars, phosphates, and bases linked as shown in Figure 11-30. Another crucial step was made early in the 1950s by American chemist Edwin Chargaff, who studied the composition of DNA from a variety of plants and animals. Chargaff found that the relative amounts of different bases changed from one species to another. In every species he examined, however, the molar ratios of guanine to cytosine and of adenine to thymine were always very close to 1.0. Chargaff concluded that these constant ratios could not be coincidence. Somehow, adenine and thymine are paired in DNA, and so are guanine and cytosine.

By now the stage was set for the discovery of the three-dimensional structure of DNA, the so-called **secondary structure** of the molecule. Some of the best minds in science were working on the problem, driven in part by the expectation of winning a Nobel Prize. In 1953, James D. Watson and Francis Crick of Cambridge University announced that they had discovered the structure of DNA. Watson and Crick relied on X-ray diffraction patterns of crystalline DNA. The diffraction photographs, among the finest available at the time, were taken by Rosalind Franklin, a researcher in the laboratory of Maurice Wilkins at King's College in London. Combining Franklin's data with the earlier insights of Chargaff, Todd, and many others, Watson and Crick concluded that DNA must consist of two helices wound around one another in a **double helix.** In the double helix, shown schematically in Figure 11-31, the hydrophilic sugars and the phosphate groups lie on the outside of the molecule, with the hydrophobic bases tucked inside the structure. For their brilliant insight, Watson, Crick, and Wilkins shared the 1963 Nobel Prize in medicine and physiology.

According to the Watson-Crick model, hydrogen bonding holds the double helix together. The bases of one strand of DNA form hydrogen bonds to the bases of the second strand. Because of their matching structures, adenine pairs with thymine through two hydrogen bonds, and guanine pairs with cytosine through three hydrogen bonds. These sets are said to be **complementary base pairs.** As shown in Figure 11-32, complementary base pairs fit together like matching gears. Matching bases in this way keeps the distance between the two strands constant through the entire length of the DNA molecule.

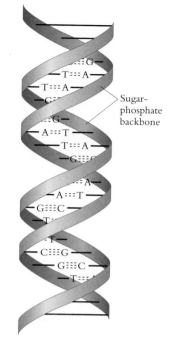

Figure 11-31
DNA consists of two strands of sugar-phosphate backbones wound one around the other in a double helix. The two helices are connected by hydrogen bonds between bases that pair within the molecule.

The structure of DNA was discovered by (shown left to right) Francis Crick, James Watson, Maurice Wilkins, and Rosalind Franklin.

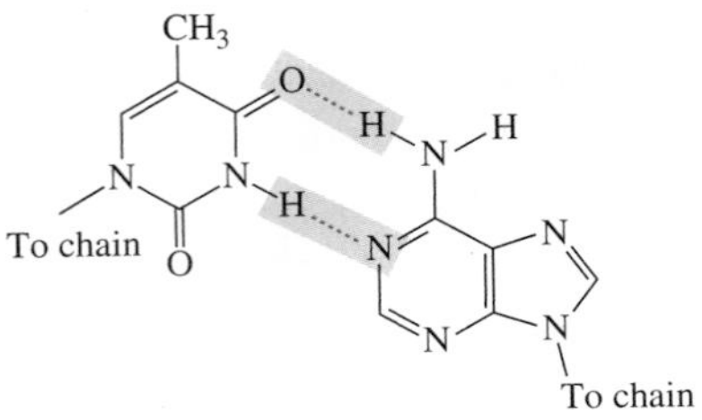

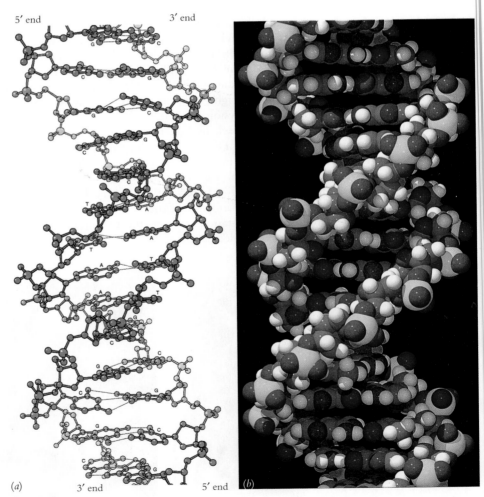

Thymine · · · · · Adenine

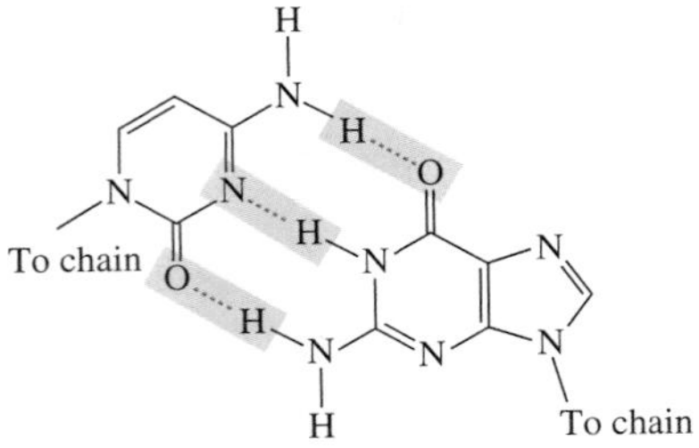

Cytosine · · · · · Guanine

Figure 11-32
The complementary base pairs of DNA. Thymine fits with adenine, and cytosine fits with guanine.

The nucleotide bases are flat molecules. Each base pair is parallel to the one below it, with 340 picometers separating the two. There is a rotation of 36° between pairs, giving ten base pairs per complete turn of the helix. The two sugar-phosphate backbone strands wind around these stacked pairs, as shown in Figure 11–33. The two strands of DNA run in opposite directions, with the terminal phosphate end of one polynucleotide matched with the free hydroxyl end of the other.

In recent years, our detailed understanding of the structure of DNA and how it functions has led to many new methods of genetic manipulation. Our Chemistry and Technology Box on page 522 explores one aspect of this subject.

The Structure of RNA

DNA and RNA have similar structures in that each has a sugar-phosphate backbone with one organic base bound to each sugar. However, there are four distinct differences between RNA and DNA:

Figure 11-33
The structure of DNA. (a) A ball-and-stick model, with the sugar-phosphate backbone colored blue and the bases colored red. (b) A space-filling model, showing C atoms in blue, N atoms in dark blue, H atoms in white, O atoms in red, and P atoms in yellow.

1. The sugar in RNA is ribose, not deoxyribose.

2. RNA uses uracil instead of thymine. The common bases in RNA are adenine, uracil, guanine, and cytosine.

3. RNA is much smaller than DNA. The molecules of RNA range in molar mass between 20,000 and 40,000 g/mol instead of as much as 10^9 g/mol.

4. RNA is usually single-stranded, not double-stranded.

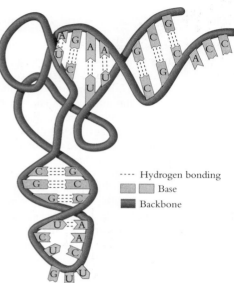

Although RNA is usually single-stranded, an RNA molecule often has distinct double-stranded regions. Intramolecular base pairing between guanine and cytosine and between adenine and uracil creates loops and kinks in the RNA molecule. The structure of one kind of RNA molecule is shown in Figure 11-34.

Whereas DNA has a single role as the storehouse of genetic information, RNA plays many roles in the operation of a cell. There are several different types of RNA, each having its own function. The principal job of RNA is to provide the information needed to synthesize proteins. Protein synthesis requires several steps, each assisted by RNA. One type of RNA copies the genetic information from DNA and carries this blueprint out of the nucleus and into the cytoplasm, where construction of the protein takes place. The protein is assembled on the surface of a ribosome, a cell component that contains a protein and a second type of RNA. The protein is constructed by sequential addition of amino acids in the order specified by the DNA. The individual amino acids are carried to the growing protein chain by yet a third type of RNA. The details of protein synthesis are well understood, but the process is much too complex to be described in an introductory course in chemistry.

Figure 11-34
The structure of an RNA molecule. Notice the folding caused by the intrastrand base pairing.

Section Exercises

■ **11.6.1** Draw the structure of the RNA nucleotide that contains guanine.
■ **11.6.2** Part of a DNA sequence is G-C-C-A-T-A-G-G-T. What is its complementary sequence?
■ **11.6.3** Nucleotides can contain more than one phosphate group. For example, energy-storing adenosine triphosphate (ATP) is discussed in Chapter 13. An ATP molecule is formed in two sequential condensation reactions between phosphoric acid and the phosphate group of AMP. Draw the structure of ATP.

$$AMP + 2\ H_3PO_4 \longrightarrow ATP + 2\ H_2O$$

11.7 PROTEINS

Life is organized around the cell, the smallest functioning unit of an organism. The most important biochemicals in cells, including enzymes, antibodies, hormones, transport molecules, and the structural materials that make up the cell itself, are all proteins. Proteins protect organisms from disease; extract energy from food; move essential cellular components from place to place; and are responsible for vision, taste, and smell. Proteins even synthesize the genetic material contained in all cells. In other words, proteins are the molecular machinery of the cell. The major structural materials in most animals—hair, skin, muscle, tendons, cartilage,

| Box 11-2 | Chemistry and Technology: DNA "Chips" |

Every gene contains DNA with a unique sequence of bases forming a genetic code containing the information an organism uses to live and replicate itself. Many years of research have resulted in an understanding of how the information content of DNA is translated into particular biochemical substances and how DNA replicates. The processes include unwinding of the DNA double helix so its code can be read or duplicated.

In an unwound helix, the bases are exposed and can form new hydrogen-bonding interactions. In protein synthesis, the unwound DNA sequence serves as a template to build a molecule of RNA whose base sequence is complementary to that of the DNA sequence. The RNA molecule, in turn, serves as a blueprint for protein manufacture. In replication, as the illustration shows, unwinding and duplication generates two identical DNA helices from a single helix.

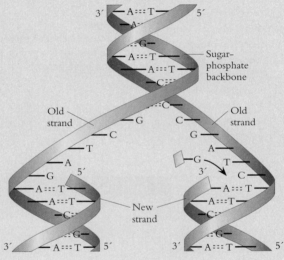

DNA replication

In both RNA synthesis and DNA replication, the bases of the original DNA strand are matched to a set of complementary bases through strong hydrogen-bonding interactions. This process is highly specific because each base interacts only with its complementary base. In DNA replication, the growing chain incorporates the correct complementary base by binding that base, then forming the backbone linkages.

The promise of being able to predict and modify the genetic characteristics of an organism has fuelled massive efforts to determine the base sequences of human genes. The human genome project has now reached the goal of sequencing all the important DNA carried by humans.

A developing application of DNA technology uses a DNA "chip" that contains many small segments of bases of known sequence. Such DNA chips are being used to attack cancers. Cancers occur when defective genes cause cells to divide uncontrollably, but the process of protein synthesis also typically is modified. Different types of cancer result in different modifications in protein synthesis, depending on how the DNA sequence of the cancerous cell differs from that of a normal cell.

In the new technology, single strands of DNA extracted from a cancerous cell are attached to a DNA chip and the chip bathed in a solution of the DNA that is involved in protein synthesis from normal cells. As the schematic illustration shows, a particular base sequence on the DNA molecule will bind to an area of the chip that contains its complementary sequence, but it will not bind to any area that has a mismatched sequence.

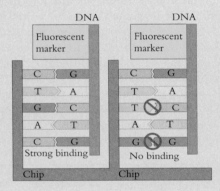

When the chip is rinsed, fluorescent DNA remains bound to all those areas of the chip that have complementary sequences. Areas of the chip whose sequences are not complementary do not bind to the DNA. When the chip is exposed to ultraviolet light, the areas with bound DNA fluoresce, generating light from each area to which the DNA has bound. The fluorescence pattern reveals which protein synthesis processes have been modified in the cancerous cell, and this in turn helps physicians determine what treatment is most likely to be effective against the cancer.

The power of this new technique comes from the large number of DNA sequences that can be placed on a single chip, several tens of thousands on a single 1.3 cm × 1.3 cm chip. This gives a richly detailed fluorescence pattern that can be interpreted using a high resolution optical scanner and computer analysis. Such detailed information makes it possible to distinguish among a large number of different types of cancerous tissues.

claws, nails, horns, hooves, and feathers—are also made of proteins. How can this single group of molecules play so many different roles in the chemistry of life? To answer that question, we need to look at the structure of proteins. Remarkably, all proteins have amino acids as their common structural components.

Amino Acids

As the name implies, an **amino acid** is a bifunctional molecule with a carboxylic acid group at one end and an amino group at the other. All proteins are polyamides made from condensation reactions of amino acids. The amino acids in proteins all have a central carbon atom bonded to one hydrogen atom and to a side chain group, symbolized in Figure 11–35 as R.

A total of 20 different amino acids are used to build proteins in living cells. The simplest is glycine, where R = H. In each of the 19 amino acids other than glycine, the side chain begins with a carbon atom. The side chains in naturally occurring amino acids are shown in Figure 11–36.

Among the common amino acids, some have side chains that contain polar functional groups such as hydroxyl, amino, and carboxylic acid. These hydrophilic amino acids are commonly found on the outside of a protein, where their interactions with water molecules increase the solubility of the protein. Other amino acids

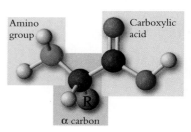

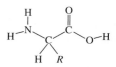

Figure 11-35
Ball-and-stick representation and structural formula showing the H_2N—CH—CO_2H chain that forms the common structure of all naturally occurring amino acids.

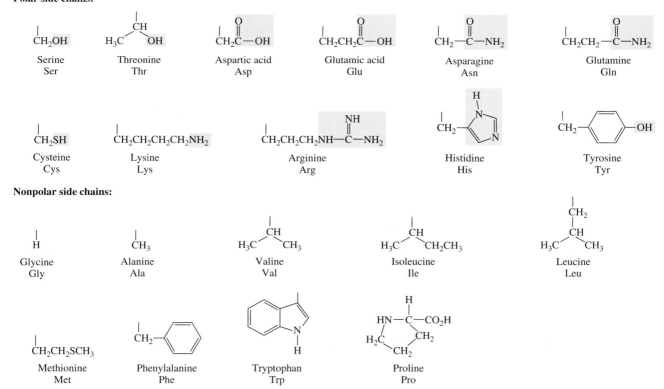

Figure 11-36
Of the 20 amino acids, 11 have side chains containing polar groups (color screened), and 9 have nonpolar side chains. One, proline, has a unique ring structure. Under the name of each amino acid is its three-letter abbreviation.

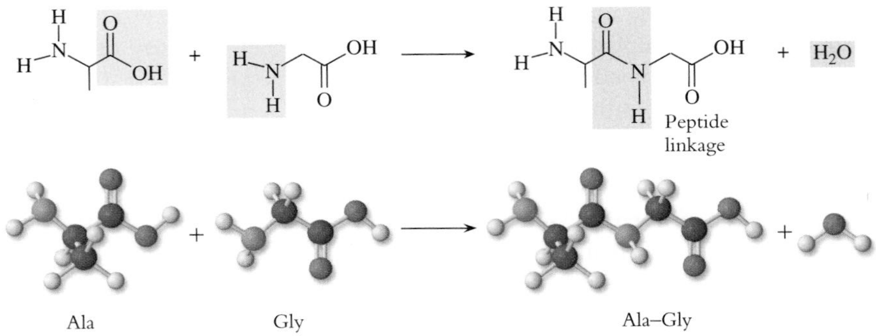

Figure 11-37
Ball-and-stick models of alanine (Ala) and glycine (Gly) and of the condensation reaction between them to form a dipeptide.

Ala   Gly   Ala–Gly

have nonpolar hydrophobic side chains containing mostly carbon and hydrogen atoms. These amino acids are often tucked into the inside of a protein, away from the aqueous environment of the cell.

Biochemists represent each amino acid with a three-letter abbreviation. These abbreviations appear under the names of the amino acids in Figure 11-36. For example, the abbreviation for glycine is Gly.

CHAPTER 10 →
The hydrophilic and hydrophobic properties of molecules are discussed in Chapter 10.

Polypeptides

Proteins form in a sequence of condensation reactions in which the amino end of one amino acid combines with the carboxylic acid end of another, eliminating a water molecule to create an amide linkage. The amide group that connects two amino acids is called a **peptide linkage.** When two amino acids are linked, the product is a dipeptide. A dipeptide formed from alanine and glycine is shown in Figure 11-37.

Notice that this dipeptide is also an amino acid because the molecule retains an amino group at one end and a carboxylic acid group at the other end. Consequently, an additional amino acid can add to either end of a dipeptide to form a new peptide that also has an amino terminal group at one end and a carboxylic acid terminal group at its other end. Figure 11-38 shows the peptide that results from addition of another alanine molecule at the amino end and a cysteine molecule at the carboxyl end of the Ala-Gly dipeptide.

Protein synthesis in cells occurs by sequential condensations, always at the carboxylic acid end of the growing chain, eventually leading to a macromolecule

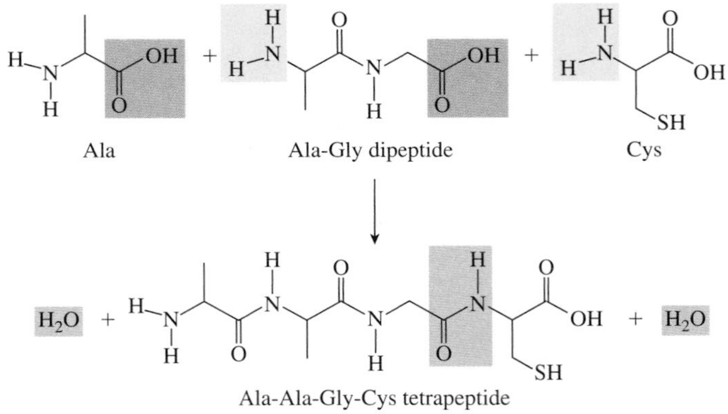

Figure 11-38
Condensation reactions can take place at both ends of a dipeptide. The product can continue to grow into a polypeptide or a protein. The figure illustrates alanine and cysteine adding to opposite ends of the Ala-Gly dipeptide.

Ala   Ala-Gly dipeptide   Cys

Ala-Ala-Gly-Cys tetrapeptide

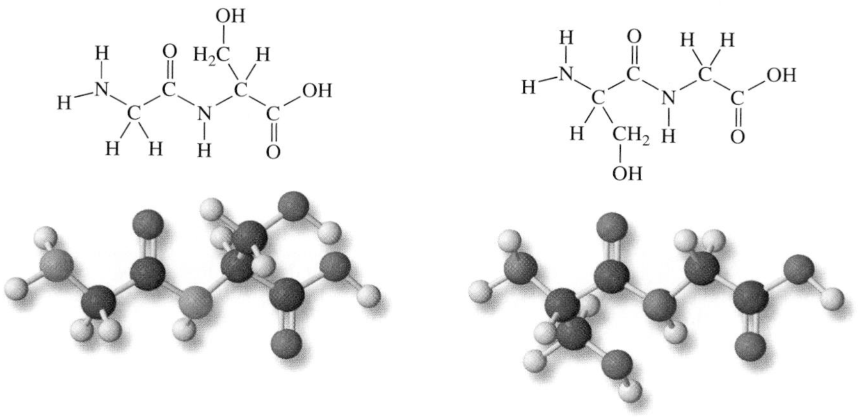

Figure 11-39
**Glycine and serine form two
dipeptides, Gly-Ser and Ser-Gly.
As the ball-and-stick models
show, even dipeptides have
distinctive shapes that depend on
their primary structures.**

Glycine–Serine
Gly–Ser

Serine–Glycine
Ser–Gly

called a **polypeptide.** All proteins are macromolecular polypeptides, but the lengths of proteins vary tremendously. Some small hormone proteins contain as few as eight or nine amino acids. Myosin, a very large muscle protein, has approximately 1750 amino acid units. Proteins exist with amino acid chain lengths of all values in between these extremes.

The sequence of amino acids in a polypeptide is its **primary structure.** It is customary to write primary structures of polypeptides using the three-letter abbreviation for each amino acid. By convention the structure is written so that the amino acid on the left bears the terminal amino group of the polypeptide and the amino acid on the right bears the terminal carboxylic acid group. Figure 11-39 shows the two dipeptides that can be made from glycine and serine. Although they contain the same amino acids, they are different molecules whose chemical and physical properties differ. Example 11-7 shows how to draw the primary structure of a peptide.

The Primary Structure of a Peptide	**Example 11-7**

Write the line structure of the peptide Asp-Met-Val-Tyr.

Strategy: We are asked to translate a shorthand designation into a line drawing. First, construct a backbone containing three amide linkages, putting the terminal NH_2 group on the left end and the terminal CO_2H group on the right end of the peptide. Then attach the appropriate side groups, as determined from the molecular structures of the amino acids in Figure 11-36.

Solution: The four amino acids are joined with peptide linkages in the order given. Put the backbone in place as a line structure. Remember, however, that carbon atoms are not shown in line structures and hydrogens are included only for atoms other than carbon.

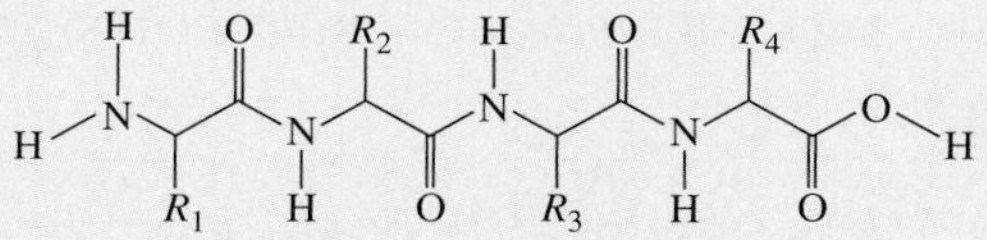

Example 11-7	The Primary Structure of a Peptide *(continued)*

Next, add the four side chains in the positions marked R_1, R_2, R_3, and R_4 to give the final structure of the peptide. The leftmost amino acid, aspartic acid, is at the amino end. Aspartic acid is followed by methionine, valine, and tyrosine. Tyrosine is at the carboxylic acid end. Here are the individual amino acids, with their R groups highlighted:

To finish the structure, we replace the R groups on the peptide backbone with their appropriate line structures:

Proteins with similar primary structures can serve very different functions. For example, Figure 11-40 shows the primary structures of two pituitary hormones, vasopressin and oxytocin. These structures differ by just two amino acids (those highlighted in color), but changing these two amino acids has a profound effect on the biochemistry of these molecules. Vasopressin regulates the rate at which water is reabsorbed by the kidneys and intestine. Alcohol suppresses the release of vasopressin, which is why consumption of alcoholic beverages leads to excessive urine production and dehydration. Oxytocin, on the other hand, induces contractions of the uterus during childbirth, and it triggers the release of milk by contracting muscles around the ducts that come from the mammary glands.

The enormous diversity of protein structure and function comes from the many ways in which 20 amino acids can combine into polypeptide chains. Consider the tetra-peptide chains that can be made from the two amino acids, cysteine and aspartic acid:

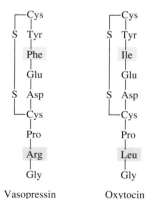

Figure 11-40
The primary structures of the pituitary hormones vasopressin and oxytocin differ by just two amino acids. (The sulfur cross-links between the cysteine residues are described later in this section.)

Cys-Cys-Cys-Cys	Cys-Cys-Cys-Asp	Cys-Cys-Asp-Cys	Cys-Cys-Asp-Asp
Cys-Asp-Cys-Cys	Cys-Asp-Cys-Asp	Cys-Asp-Asp-Cys	Cys-Asp-Asp-Asp
Asp-Cys-Cys-Cys	Asp-Cys-Cys-Asp	Asp-Cys-Asp-Cys	Asp-Cys-Asp-Asp
Asp-Asp-Cys-Cys	Asp-Asp-Cys-Asp	Asp-Asp-Asp-Cys	Asp-Asp-Asp-Asp

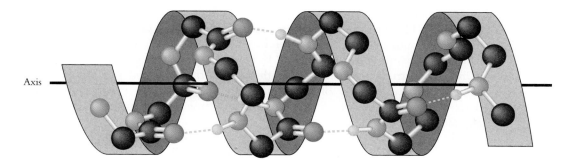

Figure 11-41
A helical peptide backbone. The side chains are omitted to emphasize the shape of the helix. Notice the hydrogen bonding between N—H and C═O groups.

In all, there are $2^4 = 16$ different ways to combine two amino acids in chains of four. The 20 common amino acids combine to give 20^4 tetrapeptides, which is 1.6×10^5 different molecules. Each time the chain adds an amino acid, the number of possibilities is multiplied by 20, so a set of n amino acids can form 20^n different polypeptides. Because proteins contain hundreds of amino acids, each of which can be any of the 20 possibilities the number of possible structures becomes immense. In fact, 20^{100}, the number of ways to construct a protein of 100 units, is greater than the estimated number of atoms in the universe.

Secondary Protein Structure

A primary structure represents a polypeptide as a simple linear string of amino acids. Actually, within long polypeptides, certain sections fold into sheets or twist into coils. These regions with specific structural characteristics constitute the **secondary structure** of the protein. Figures 11-41 and 11-42 show the two most common secondary structures.

The secondary structure of a protein is determined by hydrogen bonding between C═O and N—H groups of the peptide linkages that make up the backbone of the protein. Hydrogen bonds can exist within the same protein chain or between different chains. Hydrogen bonding within a single protein chain gives a **helix** (Figure 11-41). Notice that the helix is held together by hydrogen bonds between the O atom of the carbonyl group in one amino acid and the H atom from an amide nitrogen three amino acids away. One complete turn of the helix contains 3.6 amino acids.

Figure 11-42
Two views of a pleated sheet. (*a*) The three-dimensional ball-and-stick view emphasizes the hydrogen bonding (dotted lines) in the pleated sheet. (*b*) The structural formula view emphasizes the sheets.

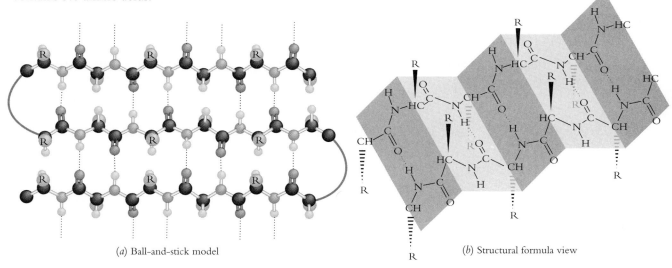

(*a*) Ball-and-stick model

(*b*) Structural formula view

Another common form of secondary protein structure is a **pleated sheet.** (Figure 11-42). The chains in a sheet are fully extended rather than coiled, and hydrogen bonds exist between different portions of the protein chains, with one part of the chain bonded to another. In pleated regions the oxygen and the amide hydrogen protrude at 90° angles to the axis of the extended protein backbone. This allows row on row of polypeptides to form hydrogen bonds and make a sheet of protein. The pleats in the sheet are caused by the bond angles of the peptide linkages.

Several other secondary protein structures have roles in determining the shapes of proteins, but they are far less common than the helix and pleated sheet. A discussion of these less common secondary structures is beyond the scope of this text.

Tertiary Protein Structure

Each protein has a unique three-dimensional shape called its **tertiary structure.** The tertiary structure is the result of the bends and folds that a polypeptide chain adopts to achieve the lowest possible energy. As an analogy, consider the cord in Figure 11-43 that connects the computer to its keyboard. The cord can be pulled out so that it is long and straight; this corresponds to its primary structure. Normally, the cord has a helical region in its center; this is its secondary structure. In addition, the helix may be twisted and folded on top of itself. This three-dimensional character of the cord is its tertiary structure.

The tertiary structure of a protein is determined primarily by the way in which water interacts with the side chains on the amino acids in the polypeptide chain. When the polypeptide folds into a three-dimensional shape that arranges its hydrophobic regions inside the overall structure, polar interactions with water molecules are maximized, making the system most stable. The folding of the protein is further directed and strengthened by a large number of hydrogen bonds. Hydrophilic side chains form hydrogen bonds among themselves and with the protein backbone. In addition, water forms hydrogen bonds with the protein backbone and with hydrophilic side chains such as those in arginine, aspartic acid, and tyrosine as Example 11-8 shows.

Figure 11-43
The cord that connects this computer to its keyboard illustrates the primary, secondary, and tertiary levels of protein structure.

| **Example 11-8** | **Hydrogen Bonding in Proteins** |

Draw a line structure that shows the various ways in which water molecules form hydrogen bonds with a protein backbone.

Strategy: Remember from Chapter 10 that partial positive charges on water's hydrogen atoms lead to hydrogen bonding with electronegative O and N atoms, whereas the partial negative charge on the oxygen atom of a water molecule forms hydrogen bonds with highly polar N—H and O—H bonds.

Solution: Begin by drawing a section of protein backbone. Because the problem asks only about hydrogen bonds to the backbone, the side chains are not involved in this problem and may be designated simply as R.

Hydrogen Bonding in Proteins *(continued)*   Example 11-8

Now add water molecules to illustrate the hydrogen bonding interactions. Each hydrogen bond is shown as a dotted line. Because the problem does not ask that all hydrogen bonds be shown, it is sufficient to show one or two examples of each type.

The amino acid cysteine plays a unique role in tertiary protein structure. The —SH groups of two cysteine side chains can cross-link through an S—S bond called a **disulfide bridge,** as shown below.

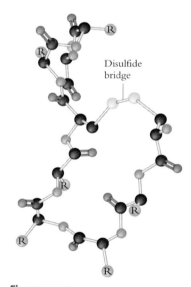

Figure 11-44
The three-dimensional structure of oxytocin. The sulfur atoms of the disulfide bridge are in yellow. The side chains have been simplified as green spheres to emphasize the shape of the polypeptide chain.

Notice in Figure 11-40 that the primary structures of oxytocin and vasopressin have two cysteine residues connected by a disulfide bridge. The three-dimensional structure of the oxytocin backbone is shown in Figure 11-44. Notice that the disulfide bridge locks the peptide chain into a compact cyclic structure.

Globular Proteins

A huge and diverse group of molecules called **globular proteins** carries out most of the work done by cells, including synthesis, transport, and energy production. Globular proteins have compact, roughly spherical tertiary structures containing folds and grooves. **Enzymes** are globular proteins that speed up biochemical reactions. The reactions of metabolism proceed too slowly to be of use to living organisms unless enzymes make them go faster. Antibodies, the agents that protect humans from disease, are also globular proteins. Other globular proteins transport smaller molecules through the blood. Hemoglobin, the macromolecule that carries oxygen in the blood, is a transport protein. Globular proteins, including oxytocin and vasopressin, also act as hormones. Others are bound in cell membranes, facilitating passage of nutrients and ions into and out

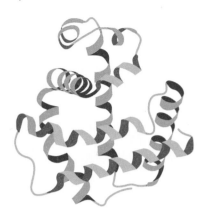

Immunoglobulin   Myoglobin   G-Actin

Figure 11-45
Ribbon views of proteins with varying amounts of helices and pleated sheets.
Immunoglobulin, an antibody, is made up almost entirely of pleated sheets (magenta).
Myoglobin, which stores oxygen in muscle tissue, is composed of about 70% helical
(blue). G-Actin, a component of muscle protein fibers, is a complex mixture of helices and
sheets. Regions with no specific secondary structure are shown in orange.

of the cell. Many globular proteins have hydrophilic side chains over the outer
surface of their tertiary structures, making them soluble in the aqueous environment of the cell.

The secondary structures of globular proteins include helices and pleated sheets
in varying proportions. The unique primary structure of each globular protein leads
to a unique distribution of secondary structures and to a specific tertiary structure.
The stylized ribbon views of globular proteins in Figure 11-45 illustrate the diversity and intricate structures of these molecules. Regions that are helical are shown
in blue, regions that are pleated are magenta, and sections that have no specific secondary structure are orange.

Fibrous Proteins

Structural components of cells and tissue are made of proteins that form fibers.
These **fibrous proteins** are the cables, girders, bricks, and mortar of organisms.

The helix is the prevalent secondary structure in keratins, which include wool,
hair, skin, fingernails, and fur. These molecules are long strands of helical protein that
lie with their axes parallel to the axis of the fiber. In hair, individual keratin molecules
are wound like the strands of a rope. Because of its compact helical chains, keratin
has a certain amount of elasticity. Stretching the fibers stretches and breaks the relatively weak hydrogen bonds of the helix, but as long as none of the stronger covalent
bonds is broken, the protein relaxes to its original length when released. Hydrogen
bonds can be broken and re-formed easily, but covalent bonds cannot.

A "permanent wave" changes the shape of a person's hair by changing the tertiary structure of hair protein. A solution that breaks the disulfide bridges between
the protein chains is applied to the hair. The hair is then set in curlers to give the
desired shape. A second solution that recreates S—S bonds is applied to the hair.
When the curlers are removed, the new disulfide bridges hold the hair in its new
configuration.

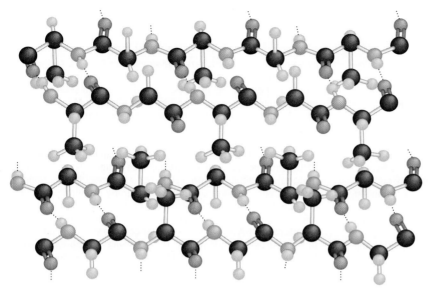

Figure 11-46
Ball-and-stick model of fibroin.

Other fibrous proteins contain extensive regions of pleated sheets. The fibers spun by a silkworm, for example, are made almost entirely of fibroin, a protein composed primarily of just three amino acids: glycine (45%), alanine (30%), and serine (12%). Each chain of fibroin contains extensive regions where a sequence of six amino acids occurs repeatedly: . . . Gly–Ser–Gly–Ala–Gly–Ala . . . . Notice that every other amino acid is glycine, which is the smallest amino acid. This alternating arrangement allows the sheets to pack together as shown in Figure 11-46.

Section Exercises

■ **11.7.1** Write the three-letter shorthand form of this peptide sequence:

■ **11.7.2** Draw a line structure of the peptide sequence Ser–Gly–Lys–Asp and show how the side chains form hydrogen bonds to water molecules.

■ **11.7.3** Write a few sentences that explain the difference among primary, secondary, and tertiary protein structure. The description should identify which types of intramolecular and intermolecular forces are responsible for each.

▮▮▮▮ CHAPTER REVIEW

Summary and Key Terms

1. A **polymer** is a **macromolecule** constructed from **monomers** that contain reactive **functional groups.** Functional groups include C==C double bonds, **alcohols, thiols, amines, aldehydes, ketones, carboxylic acids,** and **phosphates.** When monomers combine via a **condensation reaction,** they form a new linkage group. Linkage groups include **esters, amides,** and **ethers.**

2. **Polyethylene** is a long-chain hydrocarbon polymer made from ethylene monomers. Polyethylene forms in a **free radical polymerization reaction** that begins with **initiation,** proceeds by **propagation,** and **ends** when two polymer fragments combine in a **termination** step. Substituted derivatives of ethylene also polymerize. Rubbers are polymers formed from dienes or **copolymers** of dienes and alkenes, such as butadiene and styrene. Rubbers are strengthened by introducing **cross-linking** through sulfur-sulfur bonds.

3. Many polymers form from condensation reactions between monomers containing functional groups. Alcohols, carboxylic acids, and amines are the functional groups that participate most frequently in condensation polymerization. The most important commercial condensation polymers are **polyamides** and **polyesters.**

4. Polymers are classed as **plastics, fibers,** or **elastomers. Thermoplastic** polymers are flexible and melt on heating, whereas **thermosetting** polymers are rigid and heat-resistant. Fibers can be drawn from a melt or solution but harden into durable filaments. Elastomers are flexible yet resilient.

5. **Carbohydrates** are compounds of carbon, hydrogen, and oxygen in 1 : 2 : 1 atomic ratios. **Monosaccharides** are carbohydrates constructed around five- and six-membered rings. Monosaccharides link into disaccharides such as sucrose and **polysaccharides** such as **starches** and **cellulose.** Many carbohydrates differ from one another only in the orientations of OH groups about their rings, but these differences lead to different shapes and biological functions.

6. **Nucleic acids** exist in two main types, **deoxyribonucleic acid (DNA)** and **ribonucleic acid (RNA).** These molecules, which store and transmit genetic information, contain three types of building blocks: pentose sugars, nitrogen-containing organic bases, and phosphate linkages. There are five nitrogen-containing bases: **adenine, guanine, thymine, cytosine,** and **uracil.** The **primary structure** of DNA or RNA is its sequence of bases. Each base forms hydrogen bonds to a partner, forming a **complementary base pair,** and DNA is organized into a **secondary structure** of these pairs that forms a **double helix.** RNA has a less regular secondary structure.

7. Proteins are **polypeptides** built from combinations of 20 different **amino acids;** some are hydrophilic, and others are hydrophobic. Bonding in proteins is through **peptide linkages.** The **primary structure** of a protein is its amino acid sequence, its **secondary structure** is the geometric organization of neighboring peptide units, and its **tertiary structure** is the bends and folds in the polypeptide chain. Protein secondary structure includes coiled **helices** and planar **pleated sheets.** There are compact **globular proteins,** including **enzymes,** and extended **fibrous proteins.**

Skills to Master

▶ Recognizing functional groups

▶ Recognizing linkage groups

▶ Describing free radical polymerization

▶ Drawing structures of polyalkenes and rubbers

▶ Identifying alkene monomers

▶ Describing condensation polymerization

▶ Drawing structures of condensation polymers

▶ Drawing structures of saccharides

▶ Visualizing geometries of polysaccharides

▶ Drawing DNA and RNA structures

▶ Determining complementary DNA sequences

▶ Correlating polymer structures and properties

▶ Describing primary, secondary, and tertiary protein structure

▶ Recognizing hydrophilic and hydrophobic amino acids

▶ Describing hydrogen bonding in biological macromolecules

Learning Exercises

11.1 Prepare a list of the types of polymer linkage reactions described in this chapter. List at least one important polymer that forms from each type of reaction.

11.2 Define primary, secondary, and tertiary structure. Give examples of each type of structure for a protein and for DNA.

11.3 Draw molecular pictures that illustrate the linkages in each of the following polymer types: polyethylene, polyester, polyamide, polyether, and silicone.

11.4 Draw pictures that illustrate the essential features of the helix, the pleated structure, and the double helix.

11.5 Write a paragraph that describes the role of hydrogen bonding in the structures of biological macromolecules.

11.6 List all terms new to you that appear in Chapter 11. Write a one-sentence definition of each, using your own words. Consult the index and glossary if you need help.

Problems

Starting Materials for Polymers

11.1 Draw structural formulas and circle and identify the functional groups in the following molecules:

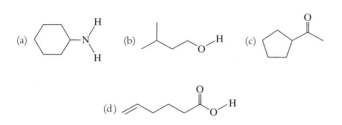

11.2 Draw structural formulas and circle and identify the functional groups in the following molecules:

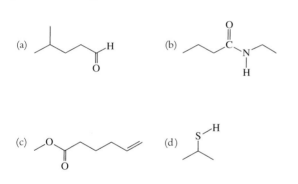

11.3 Convert the ball-and-stick structures to line drawings and identify the functional groups in the following molecules:

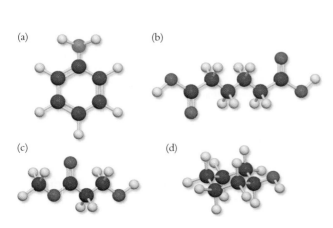

11.4 Convert the ball-and-stick structures to line drawings and identify the functional groups in the following molecules:

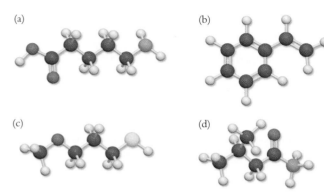

11.5 Draw structural formulas and circle and identify the linkage groups in the following molecules:

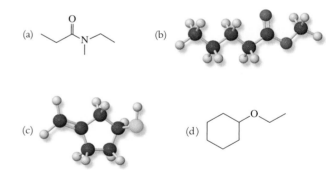

11.6 Draw structural formulas and circle and identify the linkage groups in the following molecules:

11.7 Draw one example of (a) an amine with the formula $C_5H_{13}N$; (b) an ester with eight carbon atoms; (c) an aldehyde whose molar mass is at least 80 g/mol; and (d) an ether that contains a phenyl ring (C_6H_5).

11.8 Draw one example of (a) an alcohol with the formula $C_4H_{10}O$; (b) a carboxylic acid with a molar mass of at least 60 g/mol; (c) a thiol with 6 C atoms; and (d) an amide that contains at least 10 carbons, including a phenyl ring (C_6H_5).

11.9 Write balanced equations that show how the following molecules are produced in condensation reactions:

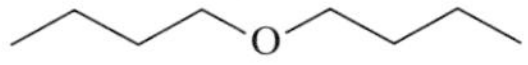

11.10 Write balanced equations that show how the following molecules are produced in condensation reactions:

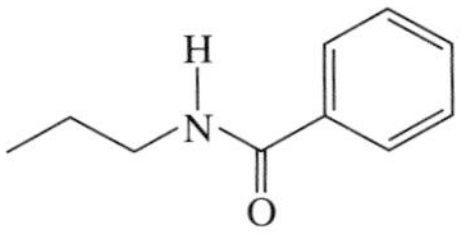

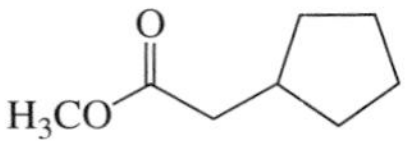

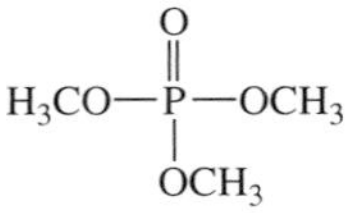

11.11 Draw the structures of all possible products resulting from condensation reactions between aspartic acid and isoleucine:

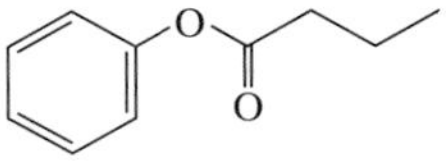

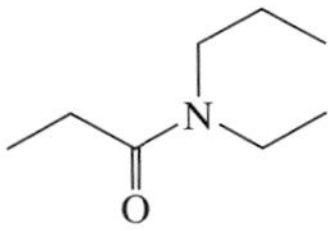

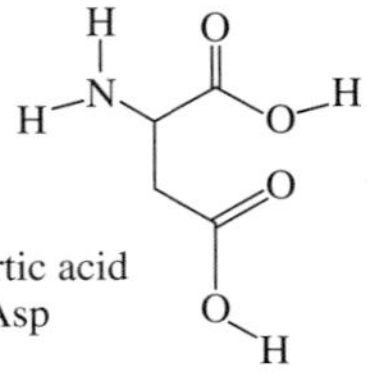

Aspartic acid
Asp

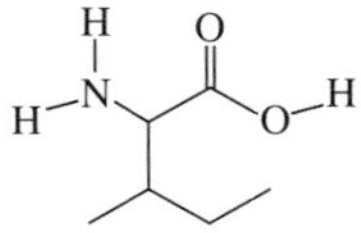

Isoleucine
Ile

11.12 Draw the structures of the condensation products of the following reactions:

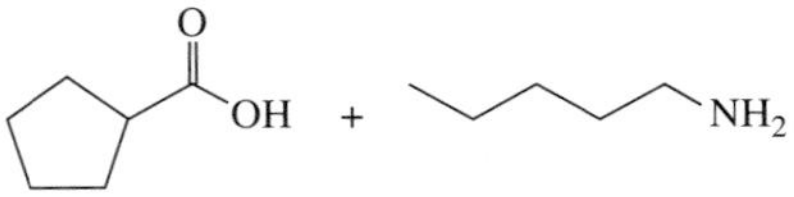

Free Radical Polymerization

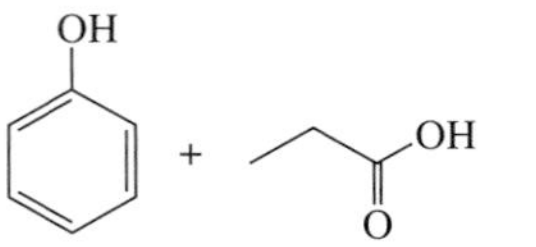

11.13 Saran is a copolymer made from vinyl chloride and vinylidene chloride ($H_2C=CCl_2$). Draw the structure of this polymer, showing at least four repeat units in the polymer.

11.14 Draw the structure of polypropylene, showing at least five repeat units.

11.15 Identify the monomers used to make the following polymers:

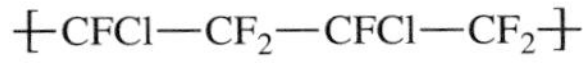

11.16 Nitrile rubber, whose structure follows, is a copolymer made from two monomers. The polymer is used to make automotive hoses and gaskets. Draw the structures of the monomers and name them.

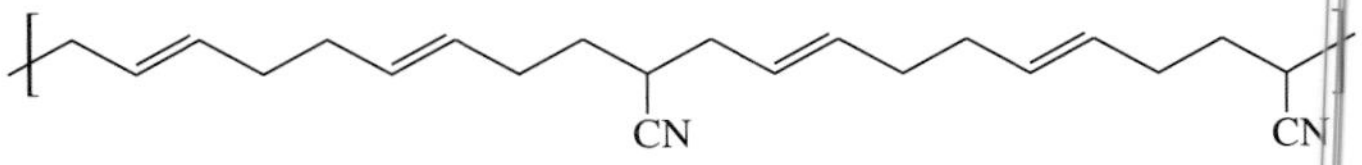

11.17 Draw a section of the polymer chain for polybutadiene and describe how it differs from polyethylene.

11.18 Draw a section of the polymer chain showing at least four repeat units for each of the following polymers: (a) Teflon; (b) PVC; and (c) Styrofoam.

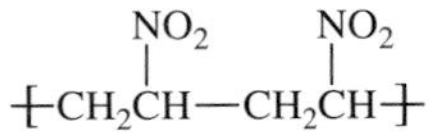

Condensation Polymerization

11.19 The structure of Nylon 11 follows. Draw a line structure of each monomer used to make this polymer.

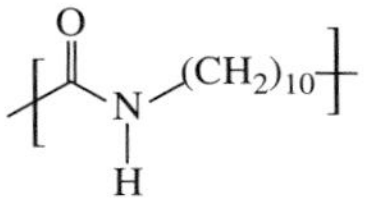

11.20 Polybutylene terephthalate, used to make countertops and sinks, has the following structure. Draw the structural formulas of the monomers from which this polymer is made.

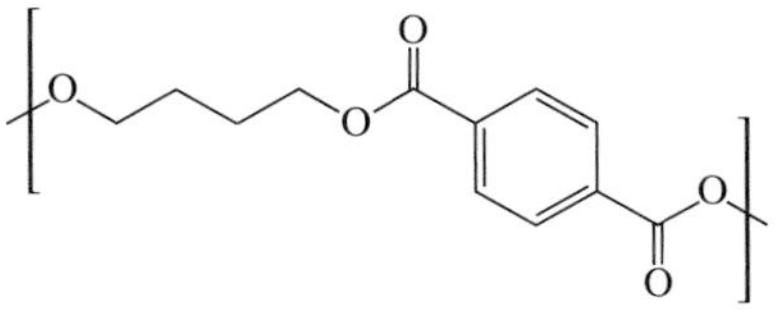

11.21 Ethylene oxide forms a polyether by ring opening followed by chain formation. Draw the structure of this polymer.

Ethylene oxide

11.22 Kodel is a polyester fiber made from terephthalic acid and cyclohexanedimethanol. Draw a segment of Kodel that contains at least four repeat units.

Cyclohexanedimethanol $HOCH_2-$$-CH_2OH$

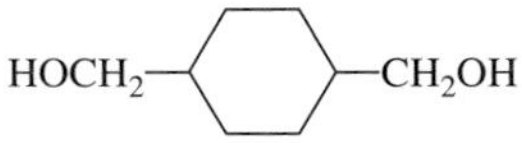

Types of Polymers

11.23 Which is more extensively cross-linked, the rubber in an automobile tire or the rubber in a pair of surgical gloves? Explain.

11.24 Plastic wrap can be stretched slightly to fit snugly over a food container. Is plastic wrap a thermoplastic or a thermosetting polymer? Explain.

11.25 Describe what category of polymer to use to make each of the following items: (a) a balloon; (b) rope; and (c) a camera case.

11.26 Describe what category of polymer to use to make each of the following items: (a) a countertop; (b) artificial turf; and (c) a bungee cord.

11.27 Describe the changes in polymer properties that occur on adding dioctylphthalate to the reaction mixture.

11.28 Describe the changes in polymer properties that occur on cross-linking.

Carbohydrates

11.29 The six-carbon sugar talose differs from glucose in the orientations of the hydroxyl groups at the 2 and 4 positions. Draw the structure of α-talose.

α-Glucose

11.30 The five-carbon sugar ribose can form a six-membered ring, ribopyranose, that differs from glucose in the orientation of the —OH at position 3 and in having —H instead of —CH$_2$OH in position 6. Draw the ring structure of α-ribopyranose.

11.31 Draw the structure of β-talose (refer to Problem 11.29).

11.32 Draw the structure of β-ribopyranose (refer to Problem 11.30).

11.33 Glycogen is a glucose polymer that collects in granules in the liver. Cellulose is also a polymer of glucose, but cellulose forms sheetlike arrangements that are used in the cell walls of plants. What are the distinguishing structural features that allow glycogen to form granules whereas cellulose forms sheets? Illustrate your answer with structural drawings.

11.34 Glycogen and cellulose are both polymers of glucose. Explain why humans can use glycogen but not cellulose as an energy source. Why can cows digest cellulose, but humans cannot?

Nucleic Acids

11.35 Part of a DNA sequence is A-A-T-G-C-A-C-T-G. What is its complementary sequence?

11.36 Part of a DNA sequence is C-G-T-A-G-G-A-A. What is its complementary sequence?

11.37 Draw the complete structure of the following segment of DNA: A-T-C-G.

11.38 Draw the complete structure of the following segment of RNA: G-U-A-C.

11.39 One strand of DNA contains the base sequence A-T-C. What is the base sequence of the complementary strand? Draw a structure of this section of DNA that shows the hydrogen bonding between the base pairs.

11.40 Although RNA is a single-stranded molecule, it can have extensive regions of double-stranded structure resulting from intramolecular hydrogen bonding between guanine and cytosine and between adenine and uracil. Draw the hydrogen-bonding interaction between adenine and uracil.

Proteins

11.41 Draw the structures of the following amino acids: (a) Tyr; (b) Phe; (c) Glu; and (d) Met.

11.42 Draw the structures of the following amino acids: (a) Cys; (b) His; (c) Leu; and (d) Pro.

11.43 Assign each amino acid in Problem 11.41 as possessing hydrophobic or hydrophilic side chains. Explain each assignment.

11.44 Assign each amino acid in Problem 11.42 as possessing hydrophobic or hydrophilic side chains. Explain each assignment.

11.45 Identify the following amino acids and characterize them as possessing hydrophilic or hydrophobic side chains:

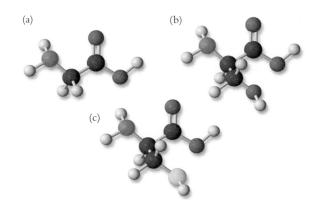

(a)   (b)

(c)

11.46 Identify the following amino acids and characterize them as possessing hydrophilic or hydrophobic side chains:

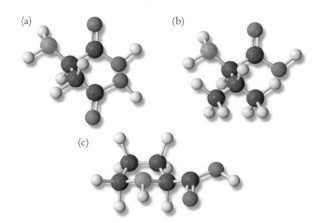

(a)   (b)

(c)

11.47 Draw the line structures of all possible dipeptides that can form in condensation reactions among alanine, glutamic acid, and methionine.

11.48 Draw the line structures of all possible dipeptides that can form from Cys, Gly, and Asn.

Additional Paired Problems

11.49 Compute the molar mass of a polyethylene molecule that has 744 monomer units.

11.50 Compute the molar mass of a polystyrene molecule that has 452 monomer units.

11.51 There are four different bases in DNA strands. How many different 12-base combinations are there?

11.52 The smallest proteins contain about 50 amino acids. How many different 50-amino acid proteins can be formed from the 20 common amino acids? Express the answer in power of 10 notation.

11.53 Hair spray is a solution of a polymer in a volatile solvent. When the solvent evaporates, a thin film of polymer that holds the hair in place is left behind. Many hair sprays contain a copolymer made of the following monomers:

(a) Assume the polymer is made from a 1:1 mixture of the two monomers. Will the two monomers alternate in the polymer or will the arrangement be random? Explain. (b) Draw a section of the copolymer containing at least six repeat units. (c) What structural features of the copolymer allow it to hold hair in place?

11.54 Chewing gum is mostly polyvinylacetate. The monomer used to make chewing gum is vinyl acetate:
(a) Draw a portion of the polyvinylacetate poly-mer. Show at least three repeat units.
(b) Is polyvinylacetate a plastic, a fiber, or an elastomer? Use the behavior of chewing gum to justify your choice. (c) Chewing gum can be removed from clothing by cooling the polymer with a piece of ice. At low temperature, the gum crumbles easily and can be removed from the fabric. Explain this procedure at the molecular level.

Vinyl acetate
$C_4H_6O_2$

11.55 One nucleotide and its hydrogen-bonded partner in double-stranded DNA is called a *duplex*. Identify the duplex formed by guanine and draw its structure.

11.56 One nucleotide and its hydrogen-bonded partner in double-stranded DNA is called a *duplex*. Identify the duplex formed by thymine and draw its structure.

11.57 Use the polymerization of acrylonitrile to describe each of the three steps of free radical polymerization. Write structures that illustrate the steps.

11.58 Condensation polymers are usually made from two different monomers; each has two of the same functional group. Give three reasons that this strategy is used so often in polymer synthesis.

11.59 Almost 1000 g of α-glucose will dissolve in 1 L of water. Draw the structure of α-glucose and include enough hydrogen-bonded water molecules to account for its tremendous solubility.

11.60 Nylon 6 has the following repeat structure:

Draw two strands of Nylon 6, each with four repeat units. Show the two strands connected by hydrogen bonds.

11.61 What features do nylon and proteins have in common? In what ways are they distinctly different?

11.62 Complete the following table:

	RNA	DNA
Sugar		
One-ring bases		
Two-ring bases		

11.63 Identify the monomers used to make the following polymers: (a) Kevlar; (b) PET; and (c) Styrofoam.

11.64 Identify the monomers used to make the following polymers: (a) Dacron; (b) PVC; and (c) gutta-percha.

11.65 High-density polyethylene has more CH_2 groups per unit volume than low-density polyethylene. Explain why this is so in terms of the structures of the two forms of the polymer.

11.66 Explain the major differences between fibrous proteins and globular proteins. What role does each type of protein play in biological organisms?

11.67 For a cell to synthesize a particular protein, the information about that protein, which is stored in the nucleus in DNA, must be transmitted to the cytoplasm, where protein synthesis takes place. This shuttling of genetic information is accomplished by a type of RNA, messenger RNA (mRNA). The DNA is used as a template to make mRNA, so the bases of mRNA must be complementary to DNA. Assuming that mRNA is made from strand A of the DNA that follows, identify bases 1, 2, 3, and 4 in the mRNA.

11.68 Transfer-RNA molecules (tRNA) have a set of three bases at their tip that are exposed and can bind their complementary bases. What sequence of bases will be recognized by the following tRNA sequences: (a) GAU, (b) AGG, and (c) CCU?

11.69 In the 1950s, Edwin Chargaff of Columbia University studied the composition of DNA from a variety of plants and animals. He found that the relative amounts of different bases changed from one species to another. However, in every species studied, the molar ratios of guanine to cytosine and of adenine to thymine were found to be very close to 1.0. Explain Chargaff's observations in terms of the Watson-Crick model of DNA structure.

11.70 The melting point of DNA, which is the temperature at which the double helix unwinds, increases as the amount of guanine and cytosine increases and the amount of adenine and thymine decreases. Explain this observation.

11.71 Describe how interactions with solvents affect the tertiary structure of a protein. Include explanations of the roles of regions containing mostly hydrophobic side groups and of regions containing mostly hydrophilic side groups. State where hydrophobic and hydrophilic side groups would be located on a water-soluble globular protein.

11.72 Globular proteins adopt three-dimensional structures that place some of the amino acids on the inside of the molecule, out of contact with the aqueous environment, and others on the outer surface of the molecule. Which of the following amino acids would be most likely to be found on the inside of a globular protein: (a) Arg; (b) Val; (c) Met; (d) Thr; and (e) Asp. Explain your choices.

More Challenging Problems

11.73 The first step of glucose metabolism is an enzyme-catalyzed condensation reaction between phosphoric acid and the CH_2OH hydroxyl on glucose. Draw the structure of this glucose phosphate.

11.74 Proteins are synthesized in the cell by adding one amino acid at a time to the growing polypeptide chain. Each amino acid is carried to the protein in a form in which the amino acid is linked to adenosine monophosphate. The amino acid is joined to AMP by a condensation of its carboxylic acid with the phosphate group. Draw the structure of Ala-AMP.

11.75 Copolymerization of styrene with a small amount of divinylbenzene gives a cross-linked polymer that is hard and insoluble. Draw a picture of this polymer that shows at least two cross-links. (Hint: The cross-linking starts when divinylbenzene is incorporated into the growing polystyrene chain.)

Divinylbenzene

11.76 According to Table 11-1, amines can form polymers by reacting with alcohols. The polymerization reaction is condensation with elimination of water. Draw the structure of the repeat unit of the polymer that forms from ethylene glycol and *p*-phenylene diamine.

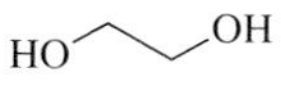

Ethylene glycol    *p*-Phenylene diamine

11.77 Fungal laccase is an enzyme found in fungi that live on rotting wood. The enzyme is blue and contains 0.40% by mass copper. The molar mass of the enzyme is approximately 64,000 g/mol. How many copper atoms are there in one molecule of fungal laccase?

11.78 One of the problems encountered in the polymerization of monomers that contain two different functional groups is that the molecules tend to cyclize rather than polymerize. Draw the structure of the cyclized product that would be produced from the following monomer:

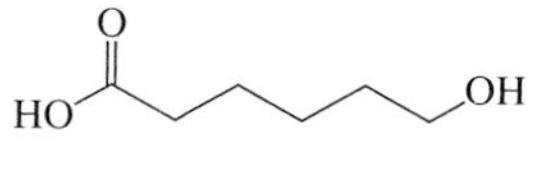

11.79 In the synthesis of glycogen, an enzyme catalyzes the transfer of a glucose molecule from glucose-UDP to the growing end of a glycogen polysaccharide. Glucose-UDP is a uridine diphosphate molecule linked to the hydroxyl group on the carbon adjacent to the ring oxygen atom of α-glucose. The glucose is at the end of UDP's phosphate chain. Draw the structure of glucose-UDP.

11.80 The artificial sweetener aspartame (NutraSweet) is the methyl ester of the following dipeptide:

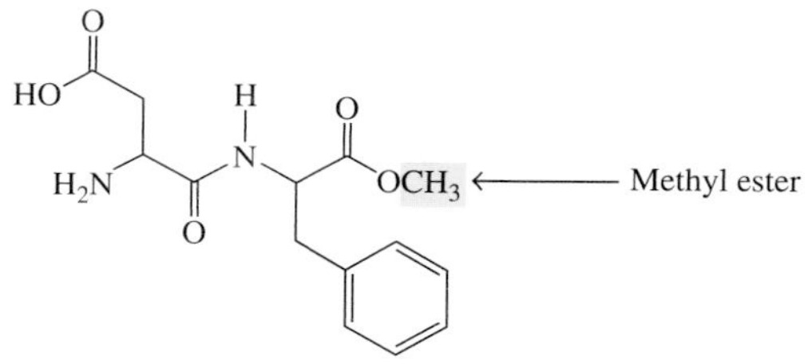

What two amino acids are used to make aspartame?

11.81 The nitrogen atom of a peptide linkage has trigonal planar geometry. What is the hybridization of the nitrogen atom in a peptide linkage? Explain why nitrogen adopts this form of hybridization.

11.82 The development of artificial substances compatible with human tissue is an important area of research. One example is a polymer of lactic acid, which is used to make body implants needed for only a short time. Eventually the polymer is converted back to lactic acid, which is metabolized to CO_2 and water in the same manner as natural lactic acid. Thus the body absorbs the polymer without leaving any permanent residue. Draw at least four repeat units of the structure of the polymer made from lactic acid.

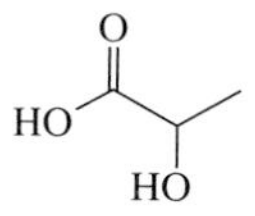

Lactic acid

11.83 Gentobiose is a disaccharide found bonded to a number of biological molecules. Gentobiose contains two linked β-glucose molecules. One molecule links through the hydroxyl group in position 1, and the other links through the CH_2OH hydroxyl group. Draw the structure of gentobiose.

11.84 Suppose a polypeptide is constructed with alanine as the only monomer. What is the empirical formula of this polypeptide? If the polypeptide has a molar mass of 1.20×10^3 g/mol, how many repeat units of alanine does it contain?

Group Study Problems

11.85 Automobiles and major appliances such as refrigerators and washing machines require very tough, long-lasting paints that are baked onto the surface of the object. One group of such paints is known as *alkyds,* which stands for *alc*ohol and ac*id*. These polyesters have extensive cross-linking that characterizes a tough coating material. One of the simplest alkyds is formed from glycerol and phthalic acid. Heating at 130 °C for about an hour maximizes the amount of cross-linking. Draw the structure of the condensation polymer that forms from these two monomers:

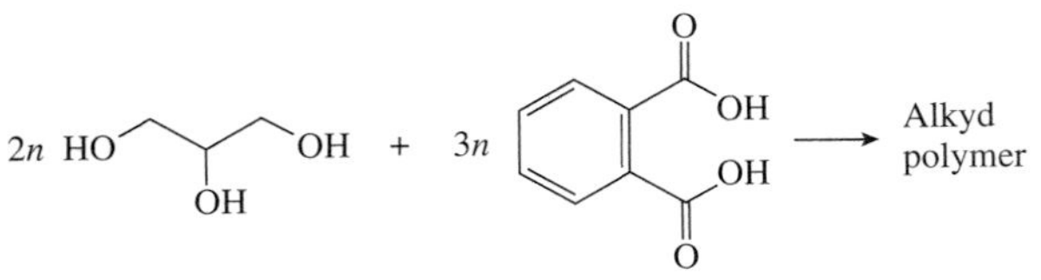

$2n$ HO—⟨⟩—OH + $3n$ ⟨⟩ → Alkyd polymer

Glycerol   Phthalic acid

11.86 Draw the structure of the nucleotide formed from cytosine, ribose, and a phosphate.

11.87 Glyptal is a highly cross-linked polymer made by heating glycerol and phthalic anhydride. Show the structure of glyptal. (Hint: The highlighted section reacts with water in a reverse condensation reaction to give the monomer that polymerizes with glycerol.)

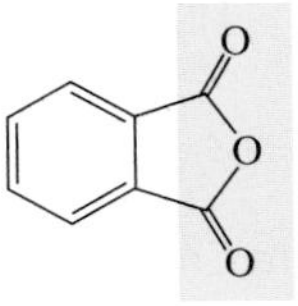

Phthalic anhydride

11.88 Draw the structures of polyethylene and the copolymer of butadiene and styrene, showing at least six repeat units for each polymer. On the basis of their molecular structures, explain why polyethylene is more rigid than butadienestyrene copolymer.

Answers to Section Exercises

11.1.1 (a) Hydroxyl; (b) amine and carboxyl; and (c) ester

11.1.2

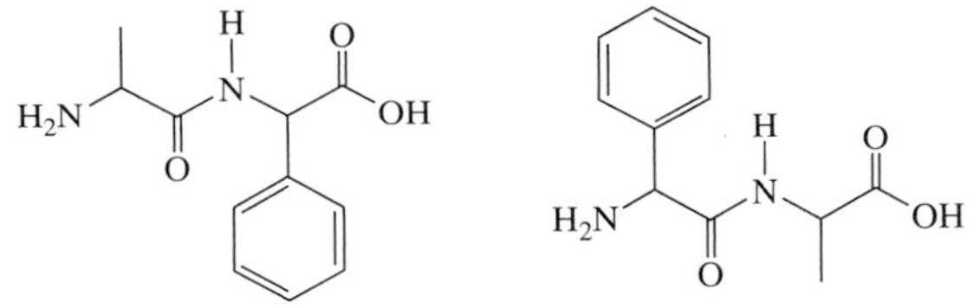

11.2.2

11.2.3

11.1.3 (In each case there are several correct answers.)

(a)   (b)

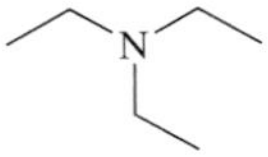

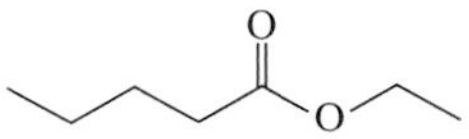

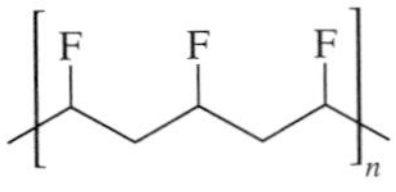

(c)    (d)

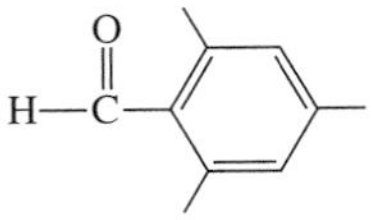

11.3.1

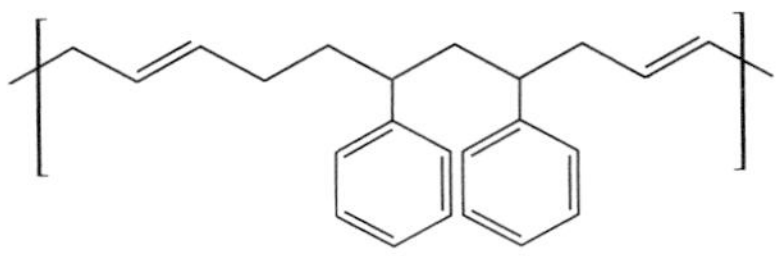

11.2.1

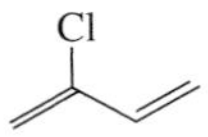

(a)   (b)

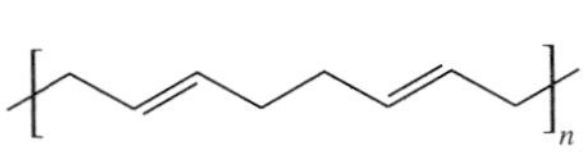

(c)

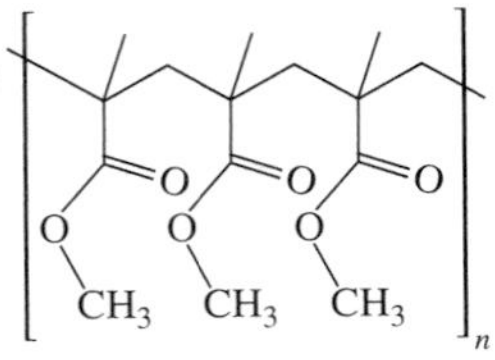

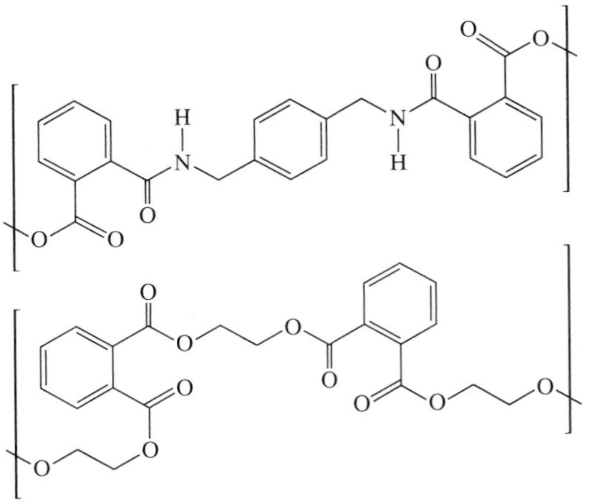

11.3.2

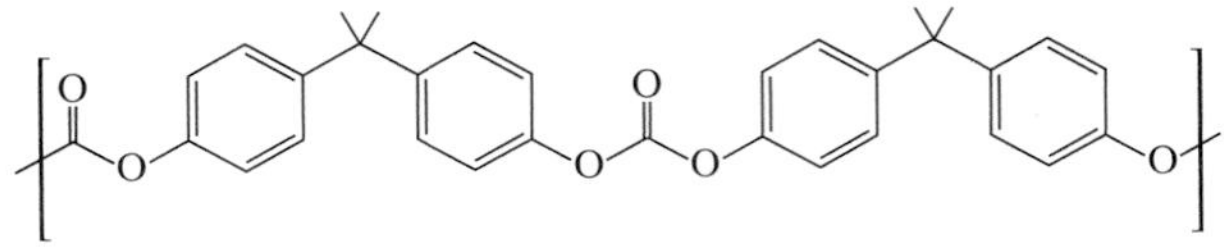

11.3.3

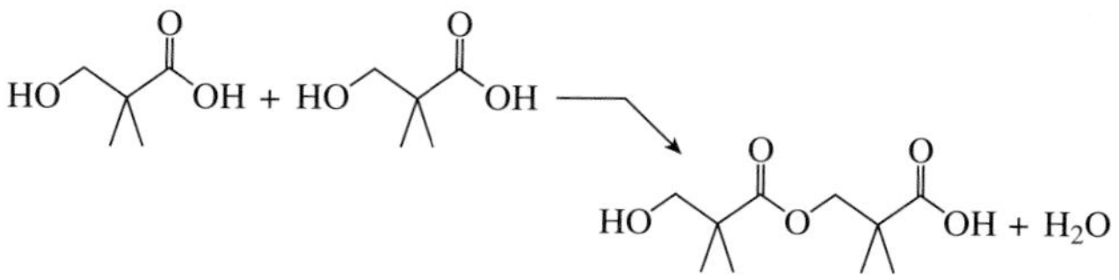

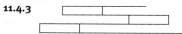

11.4.1 In Kevlar, hydrogen bonding between the amide linkage groups holds the polymer chains in a sheetlike arrangement. This gives the phenyl rings a specific ordered orientation and makes the polymer highly crystalline. On the other hand, polystyrene chains are held together by weaker dispersion forces. The polymer is less ordered and more flexible than Kevlar.

11.4.2 Polyethylene and isoprene are both made from alkenes by free radical polymerization. Polyethylene has long chains of CH_2 groups; in some applications, there are branches off the main chain of the polymer. The chains of CH_2 groups can "line up" in rows that create regions of crystallinity within the polymer. The degree of crystallinity, which depends on the amount of branching, determines whether polyethylene is a thermosetting or thermoplastic polymer. In contrast, polyisoprene is made by polymerizing a diene, so the repeat unit of the polymer retains one $C=C$ bond. Because of the specific geometry of the double bonds, polyisoprene lacks the structural regularity required to form a crystalline polymer. Instead, polyisoprene chains are tangled with no specific arrangement. As a consequence, the polymer is an elastomer.

11.4.3

Crystallinity comes from dispersion forces that are easily overcome by stretching the polymer.

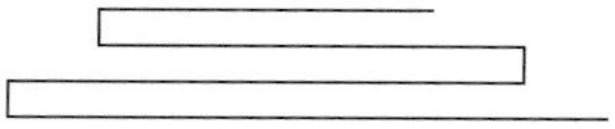

The ordered structure of the polymer comes from strong covalent cross-links that prohibit stretching of the chain.

11.5.1

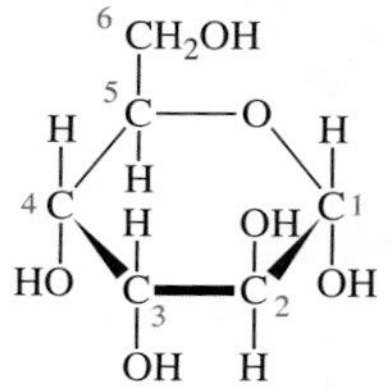

α-Gulose 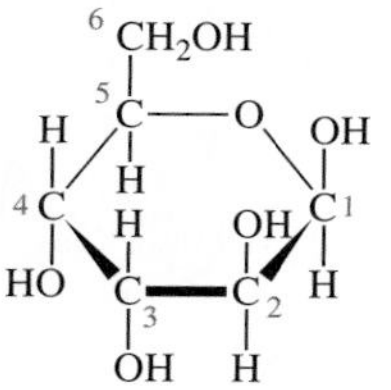 β-Gulose

11.5.2

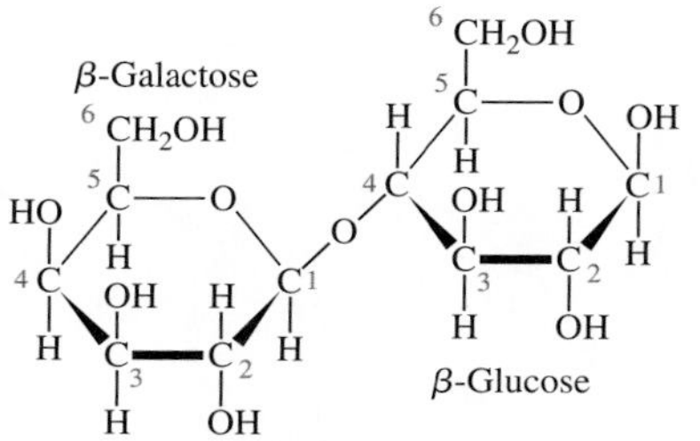

β-Galactose

β-Glucose

11.5.3 Sucrose is fructose linked to α-glucose. Lactose is galactose linked to β-glucose. The different glycosidic linkages give different shapes; lactose is nearly flat, whereas sucrose is to some extent bent around the glycosidic linkage. Different-shaped sugars require different-shaped enzymes for the cleavage reaction.

11.6.1

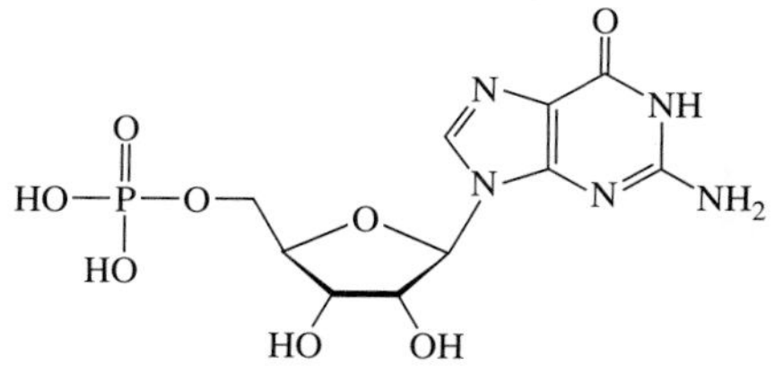

11.6.2 C-G-G-T-A-T-C-C-A

11.6.3

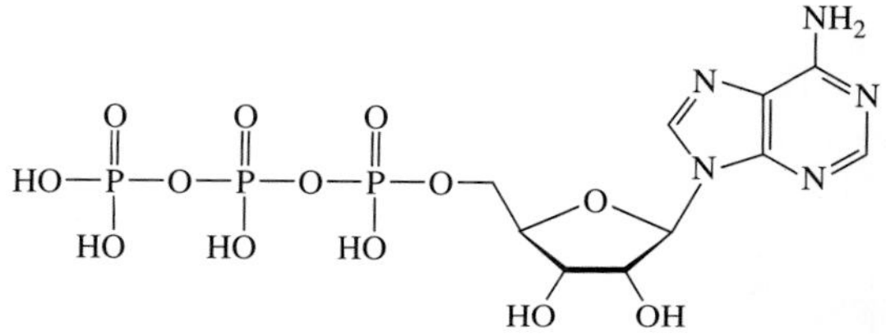

11.7.1 Ala-Thr-Val-Gly-His

11.7.2 (Lone pairs are shown only on hydrogen-bonded atoms.)

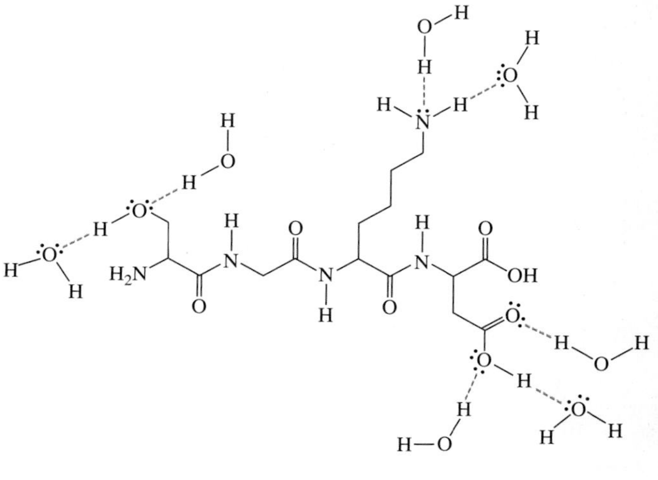

11.7.3 The primary structure of a protein is the linear sequence of amino acids, which are held together by covalent bonds. Secondary structure involves regions of the protein chain that twist into coils or fold into sheets. Hydrogen bonding between the amide residues of the protein backbone accounts for secondary structure. The overall shape of a protein is its tertiary structure. Dispersion forces, hydrogen bonding, and covalent cross-links between cysteine side chains contribute to the tertiary structure.

12

Chemical Energetics

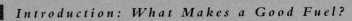

INTRODUCTION: WHAT MAKES A GOOD FUEL?

T
he intense heat of a raging fire comes from the energy released when wood combines with oxygen from the atmosphere. Even though a fire out of control can be devastating, a wood fire can supply energy for a host of useful applications. Wood has served as a fuel ever since our prehistoric ancestors discovered fire. The main component of wood is cellulose, long chains of glucose molecules. Our molecular inset shows the repeat unit of cellulose.

Smaller carbohydrates such as glucose serve as the fuels of the biosphere. Green plants use solar energy to drive the formation of carbohydrates from CO_2 and H_2O:

$$6 \, CO_2 + 6 \, H_2O \xrightarrow{\text{Sunlight}} C_6H_{12}O_6 + 6 \, O_2$$

When carbohydrates react with molecular oxygen, 2820 kJ/mol of energy is released:

$$C_6H_{12}O_6 + 6 \, O_2 \longrightarrow 6 \, CO_2 + 6 \, H_2O \qquad \text{Energy released: 2820 kJ/mol}$$

Both plants and animals use this reaction as their chemical source of energy.

Carbohydrates are fuels for the biosphere, but they are insufficient to meet the energy requirements of modern society. For this purpose, humans turn to other chemical fuels. A chemical fuel must react with some readily available substance to release a large amount of energy. In other words, a fuel is a chemical substance in a high-energy state for which there is a "downhill" reaction path.

The most common fuels react with oxygen in combustion reactions that generate carbon dioxide and water:

Natural gas: $CH_4 + 2 \, O_2 \longrightarrow CO_2 + 2 \, H_2O$ Energy released: 810 kJ/mol

Gasoline: $2 \, C_8H_{18} + 25 \, O_2 \longrightarrow 16 \, CO_2 + 18 \, H_2O$ Energy released: 11,000 kJ/mol

Space shuttle fuel: $2 \, H_2 + O_2 \longrightarrow 2 \, H_2O$ Energy released: 484 kJ/mol

Combustion releases energy because $C = O$ and $O - H$ bonds are more stable than $C - H$ bonds and $O = O$ bonds.

Another requirement of a fuel is that it must be possible to control the reaction. This involves a delicate balance. A fuel that is too reactive cannot be stored safely, but a substance that is not reactive enough may release too little energy to be useful. Even with the best of fuels, combustion can get out of control, with disastrous results—for example, oil well fires can be extremely difficult to extinguish. A particularly tragic incident was the hydrogen-oxygen explosion that destroyed the space shuttle Challenger.

Our major chemical fuels—natural gas, petroleum products, and coal—are "fossil" fuels, formed from the decomposition of plants over many millions of years. We are consuming these fuels much more rapidly than they are being formed. Developing fuel sources to replace nonrenewable fossil fuels is a major challenge for our technological society.

Chemical fuels are examples of the importance of energy transformations. Chapters 12 and 13 introduce chemical thermodynamics, a powerful tool for analyzing such transformations. The focus of Chapter 12 is the balance and flow of energy. After developing some important definitions, we take a close look at energy flows and transformations. To end this chapter, we introduce enthalpy, an energy-related concept, and describe its usefulness. Chapter 13 examines the conditions needed for processes to be spontaneous.

12.1 THERMODYNAMIC DEFINITIONS

Thermodynamics describes the flow of energy. The quantitative nature of thermodynamics requires careful definitions. Energy flows between a *system* and its *surroundings*. When a system gains or loses energy, the properties of that system undergo a change. Scientists describe such an occurrence as a change in the *state* of the system. System, surroundings, and state have precise thermodynamic definitions.

System

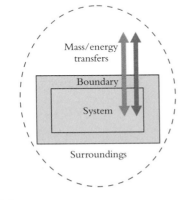

A thermodynamic **system** is anything that we want to describe and study by itself. Once we select something as a system, everything else is its **surroundings.** A system is separated from its surroundings by a **boundary,** across which matter and/or energy can be transferred. An aquarium is a good example of a system. The aquarium and its contents are separated from the surroundings by glass walls and by the interface between air and water. Matter and energy move between the aquarium system and the surroundings as water evaporates, as food is added, and as the aquarium warms or cools during the course of a day.

A system must be self-contained, so that we can specify its properties independently of the surroundings. The boundaries of a system must also be well defined so that we can measure any transfers that may take place. Figure 12-1 shows several examples involving automobiles that can be defined as thermodynamic systems: an automobile engine, an entire automobile, or even an automobile assembly line.

If we choose to define an automobile engine as a system, its boundary is the outer surface of the engine. Matter (gasoline and air) is transferred into the engine from the surroundings, and different matter (exhaust gases) is transferred from the engine into the surroundings. Mechanical energy is transferred from the engine to the crankshaft, and heat energy is transferred from the engine to the cooling system.

The choice of what to define as the system depends on our particular interests. An efficiency expert would think in terms of the assembly line in Figure 12-1, a design engineer would regard an entire automobile as the system, a power engineer might pick the engine as the system, and a fuel efficiency specialist could define an individual cylinder and piston as the system. A chemist would be inclined to define the chemicals in the cylinder as the system. Each of these would be an acceptable and useful choice for a system.

A system may be open, closed, or isolated, depending on what kinds of transfers take place. An *open system* exchanges matter and energy with its surroundings. A *closed system* exchanges energy but not matter, and an *isolated system* exchanges neither matter nor energy with the surroundings. Figure 12-2 shows the same system, a carbonated soft drink, in each of these three situations.

Figure 12-1
Some examples of thermodynamic systems. Systems can be large or small, depending on how they are defined.

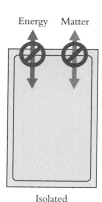

Open   Closed   Isolated

Figure 12-2
Different types of systems: A glass of a soft drink is an open system, an unopened can of a soft drink is a closed system, and a stoppered Thermos flask of a soft drink approximates an isolated system.

Almost all real systems are open. Automobiles, for example, take in gasoline and air and expel exhaust gases and heat. Open systems are very difficult to describe quantitatively, however, so much of our discussion of thermodynamics concerns closed and isolated systems. Isolated systems represent ideal situations that can be approached but never actually attained in the real world. For example, no matter how good the manufacturer makes the insulation for the thermos flask shown in Figure 12-2, it will slowly exchange energy (heat) with its surroundings.

State and Changes of State

The conditions that describe a system are collectively called its **state.** As an example, the state of the soft drink in Figure 12-2 could be given by its mass, volume, and temperature. Conditions that must be specified to establish the state of a system are called **state variables.** The state variables for chemical systems are familiar quantities: pressure (P), volume (V), temperature (T), and amounts of substances (n).

A change in conditions is a **change of state.** Changes of state occur during most thermodynamic processes. For instance, a gas initially described by conditions P_i, V_i, T_i, and n_i may undergo a transformation to a different set of final conditions P_f, V_f, T_f, and n_f. It is not necessary for all the conditions to change during a change of state. If one mole of helium expands at constant temperature, as shown in Figure 12-3, the volume of the gas increases and its pressure decreases, but the temperature and number of moles remain constant. Any process in which at least one state variable changes its value is a change of state.

A subscript i denotes initial conditions, and a subscript f denotes final conditions.

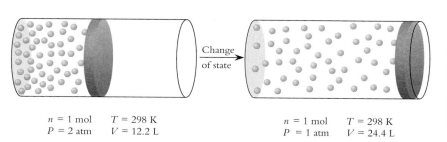

$n = 1$ mol $T = 298$ K
$P = 2$ atm $V = 12.2$ L

Change of state

$n = 1$ mol $T = 298$ K
$P = 1$ atm $V = 24.4$ L

Figure 12-3
A schematic example of a change of state. When one mole of helium gas expands at constant temperature, the volume and the pressure change, but the number of moles and the temperature stay the same.

Physical and Chemical Changes of State

In a physical change of state some of the state variables change, but the chemical composition of the system stays the same. The expansion of helium gas illustrated in Figure 12-3 is a physical change of state. The atoms are distributed over a larger volume, but the atoms themselves do not change during the expansion. No chemical change occurs in the system as the gas expands.

Most applications of thermodynamics that interest chemists are chemical changes of state. In a chemical change of state, the amounts of reactants and products change. Physical variables usually change as well. To describe a chemical change of state, we must specify not only the changes in the physical variables (P, V, and T), but also changes in the identities and the amounts of reactants and products. This requires a chemical equation and information about the amount (n) of each species.

A simple example of a chemical change of state is the decomposition of calcium carbonate (limestone) at high temperature to form calcium oxide (quicklime) and carbon dioxide:

$$CaCO_3(s) \xrightarrow{\text{Heat}} CaO(s) + CO_2(g)$$

To specify this system and its change of state completely, we must describe the number of moles of $CaCO_3$ that decompose (Δn) as well as any temperature and pressure changes that accompany the reaction.

In a chemical reaction, some bonds break, and new bonds form. Bond breakage and formation are always accompanied by a change in energy. In the limestone example, energy in the form of heat must be added to drive the reaction. In other cases, chemical reactions release energy to the surroundings. Thermodynamics helps us understand the flows of energy in chemical processes.

When treating a system thermodynamically, we focus our attention on the initial and final conditions, ignoring the details of what happens in between. An example is provided by the burning of gasoline in an automobile cylinder. This is a complicated process involving a chemical change of state. A spray of air and gasoline is injected into the cylinder. After the gases are compressed into a small volume, a spark ignites the mixture, and the fuel burns. The energy released in the combustion reaction heats the product gases, which expand against a piston. Energy is transferred to the surroundings, with several results: The engine becomes hot, the piston turns a crankshaft, and the crankshaft causes the automobile to move.

To analyze this process using thermodynamics, we select an appropriate system and identify its initial and final states. We can define the system to be the chemicals in the cylinder (air, gasoline, and the products of combustion) and consider everything else (cylinder, piston, crankshaft, and driver) to be part of the surroundings. This lets us focus on the chemical reaction and its energy change. The chemistry can be simplified by assuming that the cylinder contains only octane (a principal component of gasoline) and oxygen gas. Thus the system is octane and oxygen molecules reacting to produce water and carbon dioxide in proportions described by the balanced combustion equation:

Water produced in the combustion reaction is a gas under the high-temperature conditions of the automobile's cylinder.

$$2\ C_8H_{18}(l) + 25\ O_2(g) \longrightarrow 16\ CO_2(g) + 18\ H_2O(g)$$

To describe the chemical change of state that occurs when gasoline is burned, we must specify the initial state and the final state in terms of how much material reacts. As an example, we consider 2 mmol of octane allowed to burn completely in 35 mmol (an excess) of oxygen, as depicted in Figure 12-4.

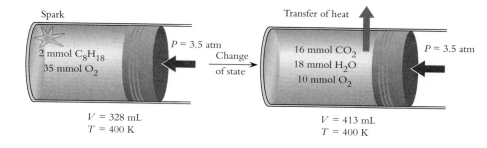

Figure 12-4
An idealized schematic diagram of the change of state that occurs when gasoline is burned in an automobile cylinder.

Figure 12-4 also specifies the initial conditions and final conditions for this process. The gas mixture expands during the combustion, so there is a change in volume. Pressure inside the cylinder also might change, but to keep our example as simple as possible, we assume that the gases expand at a rate that keeps the pressure inside the piston equal to the constant external pressure of 3.5 atm. We also assume that the automobile's coolant (part of the surroundings) holds the system at a constant temperature of 400 K.

As described in the chapter introduction, chemical energy is released when octane burns. Because energy is conserved, all the chemical energy released in the reaction must appear elsewhere. Some of this energy is transferred to the gas molecules as kinetic energy. Collisions between the energetic molecules and the piston push back the piston, expanding the volume of the gases. The rest of the chemical energy heats the automobile's coolant. The quantitative accounting for how much energy goes where is a major concern of thermodynamics.

Example 12-1 gives another example of a system and its surroundings.

The primary objective of an automobile engine is to transform chemical energy into energy of motion of the automobile, but much energy is transferred to the surroundings as heat. We show in Chapter 13 that some of this "waste" is unavoidable.

Changes in a System

Example 12-1

A manufacturer has developed a new thermal coffee mug. As part of product testing, the company needs to determine how long a cup of coffee stays hot in the mug. Define an appropriate system and list the variables needed to specify the initial and final conditions.

Strategy: We need to identify a system and surroundings, then define the initial and final conditions of the system.

Solution: Choose the system to contain the components of interest. In this case the system is the mug, plus a sample of hot coffee. The system also must contain a thermometer to measure the temperature of the coffee, and we need a clock to measure time. For this experiment the initial conditions include the volume and temperature of the coffee, V_i and T_i. The final conditions are V_f, T_f, and the elapsed time of the experiment. A figure summarizes the conditions.

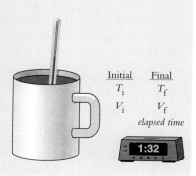

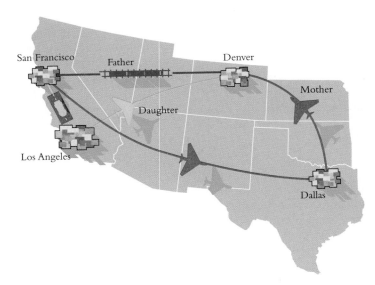

Figure 12-5
Distance between two cities is a state function, but distance traveled is not a state function.

State and Path Functions

The properties that describe a system and its transformations can be grouped in two broad categories. Properties such as energy depend only on the state of the system. A change in energy depends on the difference between the initial conditions and the final conditions (changes in temperature, pressure, etc.), but it does not depend on how the change of state is accomplished. A property that depends only on the state of the system is called a **state function.**

Other properties depend on *how* the change occurs. A property that depends on how a change takes place is called a **path function.** We show in the next section that heat transferred and work done are path functions.

An everyday example illustrates the distinction between state functions and path functions. The Dalton family, who live in San Francisco, decide to visit relatives in Denver. They travel different routes, as shown in Figure 12-5. Mr. Dalton takes a train directly from San Francisco to Denver, but Ms. Dalton goes to Dallas for a business meeting and then flies from Dallas to Denver. The Daltons' daughter drives to Los Angeles, where she catches a flight to Denver. On arrival, each of the Daltons is asked two questions by their relatives: "How far is Denver from San Francisco?" and "How far did you travel to get here?" Each answers "950 miles" to the first question because the distance between the two cities is a difference in values of a *state function.* Each answers differently to the second question, however, because each Dalton traveled a different distance to reach Denver. Distance traveled depends on the path and is a change in a *path function.*

The travels of the Dalton family illustrate two features of state functions:

1. The change in a state function can be determined without knowing the details of a process, because that change is independent of path. Denver is 950 miles from San Francisco, no matter how the Daltons travel between the two cities. This means that state function values can be tabulated and looked up when needed. The distance from San Francisco to Denver can be found in an atlas of the United States. Changes in energy accompanying standard chemical reactions can be found in Appendix D of this textbook.

2. The change in a state function can be measured using the most convenient method available. To determine the distance between two cities, modern geographers make use of satellite mapping. In Section 12.3, we describe convenient methods for determining energy changes of chemical reactions.

These two features of state functions are more important for scientists than for cross-country travelers like the Daltons. Chemists determine values of state functions by doing careful experiments using some convenient path. These values are collected in tables, just as distances between cities are collected in an atlas. Energy is one such state function, and chemists use tabulated energy values to analyze chemical processes from a thermodynamic point of view. For instance, recall from Chapter 9 how tabulated bond strengths are used to determine the energy changes that accompany chemical reactions.

12.1.1 Define an appropriate thermodynamic system for studying each of the following transformations:
(a) 100 g of water is heated in an open container.
(b) A mixture of H_2 and O_2 is detonated in a closed container.
(c) A balloon filled with air is placed in a freezer.

12.1.2 Which of the variables (P, V, T, and n) change during the transformations described in Exercise 12.1.1?

12.1.3 Which of the following is a change in a state function: (a) the change in energy when one gallon of gasoline is burned completely or (b) the miles traveled in an automobile when one gallon of gasoline is burned completely?

12.2 ENERGY CHANGES

Energy (E) is the foundation of thermodynamics. Energy comes in various forms, including chemical energy, molecular energy of motion, and radiant energy. The energy associated with motion is termed *kinetic energy*. Stored energy is also described as *potential energy:* Water stored behind a dam possesses gravitational potential energy. Although energy can be transferred among objects and converted from one form to another, it is always conserved. In this chapter, we consider energy changes that accompany chemical reactions. In particular, we examine how chemical energy can be converted into other forms.

← **SECTION 2.5**
See Section 2.5 for a description of different forms of energy.

Energy Absorption and Release

Recall from Chapter 9 that breaking a bond always consumes energy and that making a bond always releases energy. The balance between these two processes results in two classes of reactions, which are illustrated in Figure 12-6. When the energy released by bond forming is greater than the energy consumed by bond breaking, there is a net release of chemical energy by the system (Figure 12-6a). In

(a) Energy-releasing process

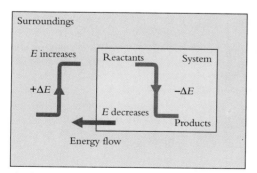

(b) Energy-absorbing process

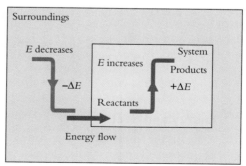

Figure 12-6
Schematic diagrams showing energy flows that accompany reactions. (a) When a reaction releases energy, the surroundings gain energy. For these reactions, change in energy (ΔE) is negative for the system and positive for the surroundings. (b) When the reaction absorbs energy, the surroundings lose energy. For these reactions, ΔE is positive for the system and negative for the surroundings.

an energy–releasing reaction, the chemical system transfers energy to the surroundings as reactants are converted to products. An example is the reactions of fuels with oxygen gas, which release large amounts of energy that can be used to heat a home or to drive an engine.

When the energy needed for bond breaking is greater than the energy released by bond forming, the chemical reaction must be driven by energy absorbed from the surroundings (Figure 12-6b). In an energy-absorbing reaction, the surroundings transfer energy to the chemical system as reactants are converted to products.

When a system gains energy, the surroundings lose the same amount of energy, and vice versa. Consequently, the sign of ΔE for a system is always different from the sign of ΔE for its surroundings. If a system *gains* energy, ΔE is *positive* for the system and *negative* for the surroundings. Conversely, if a system *loses* energy, ΔE is *negative* for the system and *positive* for the surroundings. These sign relationships apply to all quantities that can be transferred between a system and its surroundings. For example, if water is added to a glass (system) from a faucet (surroundings), Δm is positive for the system and negative for the surroundings.

If a reaction releases energy when going in one direction, it must absorb an equal amount of energy to go in the opposite direction. Reversing the *direction* of a reaction changes the *sign* of its energy change. This is true not only for changes in energy, but also for all other state functions. If a change is "uphill" going in one direction, it must be "downhill" by an equal amount in the other direction.

The reaction of two molecules of NO_2 to form one molecule of N_2O_4 provides a simple chemical example. When two NO_2 molecules collide, they may form a chemical bond in a reaction that releases energy:

Uphill: energy absorbed

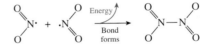

Energy-releasing reaction

The energy flow for NO_2 reacting to form N_2O_4 is like the one in Figure 12-6a. When N_2O_4 is heated, on the other hand, it can decompose into NO_2 molecules:

Downhill: energy released

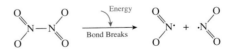

Energy-absorbing reaction

The decomposition reaction absorbs energy, with an energy profile similar to the one in Figure 12-6b.

Energy and Heat

One of the most important forms of energy for chemistry is molecular energy of motion. Temperature is a measure of this form of energy: The molecules in a system have larger molecular energy of motion when the temperature of the system is high than when the temperature is low. Furthermore, a change in the amount of molecular energy of motion of a system results in a change of the system's temperature.

When a system's energy changes, there must be an exchange of energy with the surroundings. One way to exchange energy is by a flow of **heat (q).** Heat flows are changes in energy, so they are measured in joules (J). Temperature changes often accompany heat flows. Although temperature does not have energy units, we can use temperature changes to calculate energy changes. Experiments show that a change in an object's temperature (ΔT) depends on four factors:

1. ΔT depends on q, the *amount of heat transferred*. That is, the transfer of 50 J of heat to an object causes an increase in temperature that is twice as large as the increase caused by 25 J of heat.

2. ΔT depends on the *direction in which heat is transferred*. If a system absorbs heat, ΔT will be positive; but if a system releases heat, ΔT will be negative.

3. ΔT depends inversely on the *amount of material*. That is, the transfer of 50 J of heat to 1 mol of a substance causes a temperature increase that is twice as large as the increase caused by the transfer of 50 J of heat to 2 mol of the same substance.

4. ΔT depends on the *identity of the material*. For instance, 50 J of heat increases the temperature of 1 mol of gold more than it increases the temperature of 1 mol of water.

> Although temperatures in Celsius and Kelvin differ, temperature changes have the same magnitude in the two scales because the size of the temperature unit is the same. Thus heat capacities have the same values whether expressed in J/mol K or J/mol °C.

The dependence on identity is expressed by the **molar heat capacity** (C; units, J mol^{-1} K^{-1}), which states the amount of heat needed to raise the temperature of 1 mol of substance by 1 kelvin (1 K). Every substance has a different value for C.

Using C, all four factors can be expressed in one equation that describes the temperature change resulting from a heat transfer:

$$\Delta T = T_{\text{final}} \overset{\downarrow}{-} T_{\text{initial}} = \frac{\overset{\downarrow}{q}}{nC}$$

Specifies direction · Amount of heat transferred · Amount of material · Depends on identity of material

In this equation, q is the amount of heat transferred, n is the number of moles of material involved, and C is the molar heat capacity of the substance. The molar heat capacities of several chemical substances are listed in Table 12-1.

Table 12-1
Molar Heat Capacities (C) of Representative Substances at 25 °C

Gases		Liquids		Solids	
Substance	C (J mol^{-1} K^{-1})	Substance	C (J mol^{-1} K^{-1})	Substance	C (J mol^{-1} K^{-1})
Ar	20.786	Hg	27.983	C(s, graphite)	8.527
He	20.786	H_2O	75.291	Si	20.00
H_2	28.824	Br_2	75.689	S	22.64
HCl	29.12	CH_3OH	81.6	Al	24.35
N_2	29.125	CH_3CN	91.5	Cu	24.435
CO	29.142	C_2H_5OH	111.46	Fe	25.10
O_2	29.355	CH_3CO_2H	124.3	Ag	25.351
NO	29.844	CCl_4	131.75	Na	28.24
Cl_2	33.907			KCl	51.30
H_2S	34.23			NaOH	59.54
NH_3	35.06			$BaCl_2$	75.14
CO_2	37.11			Al_2O_3	79.04
				$CaCO_3$	81.88
				NH_4Cl	84.1
				$MgSO_4$	96.48
				$K_2Cr_2O_7$	219.24

Note that neither temperature nor a change in temperature equates directly with the heat flow, q. We can relate q to ΔT only if we know the identity and amount of the material that undergoes a change of state. Example 12-2 provides some practice.

| Example 12-2 | Heat Capacity and Temperature Change |

Calculate the temperature change that results from adding 25 J of heat energy to each of the following: (a) 0.75 mol of Hg; (b) 0.35 mol of Hg; (c) 0.35 mol of H_2O.

Strategy: This calculation is straightforward, provided we specify the appropriate values for q, C, and n:

Solution:
(a) $q = 25$ J, $n = 0.75$ mol, C (from Table 12-1) = 27.983 J/mol K

$$\Delta T = \frac{q}{nC} = \frac{(25 \text{ J})}{(0.75 \text{ mol})(27.983 \text{ J mol}^{-1} \text{ K}^{-1})} = 1.2 \text{ K}$$

(b) $q = 25$ J, $n = 0.35$ mol, C (from Table 12-1) = 27.983 J/mol K

$$\Delta T = \frac{q}{nC} = \frac{(25 \text{ J})}{(0.35 \text{ mol})(27.983 \text{ J mol}^{-1} \text{ K}^{-1})} = 2.6 \text{ K}$$

(c) $q = 25$ J, $n = 0.35$ mol, C (from Table 12-1) = 75.291 J/mol K

$$\Delta T = \frac{q}{nC} = \frac{(25 \text{ J})}{(0.35 \text{ mol})(75.291 \text{ J mol}^{-1} \text{ K}^{-1})} = 0.95 \text{ K}$$

In each case the sign of ΔT is positive because the system (chemical substance) absorbs heat from the surroundings.

Heat capacities can also be expressed in terms of *mass* rather than *moles*. When expressed in terms of mass, heat capacity is called the *specific heat* of a substance, and it has the units $J g^{-1} K^{-1}$. Specific heats are not used in this book, but you may encounter them in other sources.

Rearranging the temperature expression used in Example 12-2 gives:

$$q = nC\Delta T \tag{12-1}$$

Relating q and ΔT in this manner demonstrates that any change of temperature is accompanied by a heat transfer. We can use Equation 12-1 to compute the amount of heat transferred in a change of state that involves a temperature change. Example 12-3 does this for an everyday process.

| Example 12-3 | Heat Transfer and Temperature Change |

An aluminum frying pan that weighs 745 g is heated on a stove from 25 °C to 205 °C. What is q for the frying pan?

Strategy: A temperature change signals a heat flow between system and surroundings. The system is the aluminum pan, and the stove is the part of the surroundings that supplies the energy. Equation 12-1 lets us calculate how much heat flows, provided we know n, C, and ΔT.

Heat Transfer and Temperature Change *(continued)*

Example 12-3

Solution: The mass of the frying pan and the molar mass of aluminum are used to calculate n:

$$n = \frac{m}{MM} = \frac{745 \text{ g Al}}{26.98 \text{ g/mol}} = 27.61 \text{ mol Al}$$

Molar heat capacities are tabulated in Table 12-1:

$$C_{\text{aluminum}} = 24.35 \text{ J mol}^{-1} \text{ K}^{-1}$$

Temperatures are given in the problem:

$$T_i = 25 \text{ °C} = 298 \text{ K} \qquad T_f = 205 \text{ °C} = 478 \text{ K}$$

$$\Delta T = T_f - T_i = 478 \text{ K} - 298 \text{ K} = 180. \text{ K}$$

Now we substitute and compute, converting to kilojoules to avoid large numbers:

$$q = nC\Delta T = (27.61 \text{ mol})(24.35 \text{ J mol}^{-1} \text{ K}^{-1})(180. \text{ K})(10^{-3} \text{ kJ/J}) = 121 \text{ kJ}$$

When a frying pan is heated, it loses some energy as heat to the surrounding air at the same time as it gains energy as heat from the burner. The calculation in Example 12-3 gives the *net* heat gain for the pan.

Work and Energy

Some of the energy released when gasoline burns in an automobile engine is converted to heat that must be dissipated by the engine's cooling system. The rest of the energy released by burning gasoline is what moves the automobile from one place to another. Energy used to move an object against an opposing force is called **work (*w*).** For an automobile to move, work must be done to overcome the forces of friction and gravity. The amount of work depends on the magnitude of the force that must be overcome and the amount of movement or displacement. A simple equation connects work, force (F), and displacement (d):

$$w = Fd \tag{12-2}$$

The SI unit of force is the newton (N), which can be broken down into fundamental units by examining Newton's first law, $F = ma$. Mass (m) has units of kilograms, and acceleration (a) is in meters per second squared (m s^{-2}):

$$1 \text{ N} = 1 \text{ kg m s}^{-2}$$

Displacements are measured in meters, so the product of force multiplied by displacement has units of newton meters (N m). Work is energy transferred between objects, so it is measured in joules. This means that 1 J and 1 N m are equivalent:

$$1 \text{ J} = 1 \text{ N m} = 1 \text{ kg m}^2 \text{ s}^{-2}$$

This equivalence is useful as a conversion factor in some calculations of work.

Humans convert chemical energy into work for many purposes. Automobile engines accomplish the work of movement by burning gasoline. Many other machines also rely on chemical energy to accomplish work. In addition, humans themselves require chemical energy to fuel their activities, many of which entail

Figure 12-7
Whether at work or at play, human activities involve doing thermodynamic work, which the body supplies from the chemical energy stored in foods.

thermodynamic work (Figure 12-7). Whether at work or at play, a person expends energy to accomplish movement, and this energy comes from the chemical energy stored in food. Our Chemistry and Life Box explores human energy requirements.

Expansion Work

Many chemical changes of state are accompanied by changes in volume, particularly when gases are involved. A gas in a container with movable walls can expand, and this expansion results in a movement against the force exerted by the walls. Thus, work is done in moving the walls of the container.

Expansion work is determined conveniently from the system's pressure and volume, as shown by the piston in Figure 12-8. The expanding gas inside the cylinder pushes the piston through a displacement (d). The opposing force can be

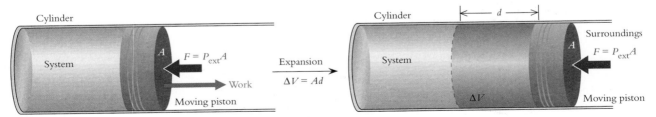

Figure 12-8
Expanding gas in an automobile cylinder displaces the piston. In this displacement, work is done against the force exerted on the area of the piston by the external pressure.

BOX 12-1 | Chemistry and Life: Human Energy Requirements

A person's weight depends on how much energy is taken in and how much energy is expended. Food brings stored chemical energy into the body, and digestion releases this chemical energy. Some energy is transferred out of the body as work and heat, but when energy intake exceeds the immediate needs of the body, the extra energy is stored as fat.

To lose weight, people must take in less energy than their bodies require. When this happens, the body "burns" fat to meet its energy requirements. Individuals can modify their bodies' energy balance by eating less and by exercising more.

"Eat less and exercise more" is overly simplistic for three reasons. First, foods of various types have different energy contents. Second, exercise of various types requires different amounts of energy. Third, individuals of various metabolic types process foods with different efficiencies.

Foods can be grouped into types, each with a characteristic energy content. The first table lists average energy contents in kilojoules per gram (1 kJ/g = 6.78 Cal/oz) for some foods. Fats (energy content = 39 kJ/g) are most energy-rich. Carbohydrates (16 kJ/g) and proteins (17 kJ/g) are the other main sources of energy. Vegetables and fruits contain much water but very little fat, so their energy con-

tent per gram is low. Margarine is mostly fat, so its energy content per gram is extremely high.

Type of Food	Energy (kJ/g)	Type of Food	Energy (kJ/g)
Green vegetables	1.2	Regular yogurt	10
Beer	2.0	Bread, cheese	12
Fruits	2.5	Ground beef	16
Low-fat yogurt	4.5	Sugar	16
Broiled chicken	6.0	Margarine	30

What we eat is as important as *how much* we eat. For example, 10 g of margarine provides the same energy content as 250 g of green vegetables. A 100-g serving of low-fat yogurt, moreover, contains less than half the energy of a 100-g serving of regular yogurt. The easiest way to reduce energy intake is by eliminating fats and greasy foods such as hamburgers and pizza.

Energy intake is only part of the equation. We also can adjust our energy balance by exercising. Various forms of exercise require different average energy outputs. Exercise involves doing thermodynamic work, and as the following table indicates, the amount of work depends on the type of exercise and the amount of mass being displaced.

Energy Consumed, kJ/hr			
	Body Weight		
Activity	55 kg (120 lb)	70 kg (155 lb)	85 kg (185 lb)
Resting	290	335	380
Doing housework	650	750	850
Walking at 4 km/hr	770	880	990
Walking at 6 km/hr	1090	1250	1410
Bicycling at 9 km/hr	775	880	985
Bicycling at 20 km/hr	2400	2760	3120
Playing volleyball	1380	1590	1800
Skiing at 16 km/hr	2175	2510	2825
Running at 16 km/hr	3285	3770	4245

A single example illustrates the relative importance of diet and exercise in weight control. A typical can of a soft drink contains about 31 g of sugar. Sugar provides 16 kJ/g, so the total energy content of a can of soft drink is 490 kJ. To expend 490 kJ of energy, a 70-kg person must walk for 23 minutes at 6 km/hr or play volleyball for 18 minutes. A 100-g (about a ¼-pound) hamburger on a bun contains 1600 kJ of energy. Thus a 70-kg person has to play volleyball continuously for over an hour to burn off the energy contained in a hamburger and soft drink.

This comparison demonstrates that it is much easier to control energy balance through diet than by exercise. The human body stores excess energy in fatty tissues. When half a kilogram of this fatty tissue is consumed, it releases 20,000 kJ of energy, which is enough to fuel about 9 hours of tennis, 6 hours of skiing, or 4 hours of running.

related to the *external* pressure (P_{ext}). As described in Chapter 5, pressure is force per unit area:

$$\text{Pressure} = \frac{\text{Force}}{\text{Area}} = \frac{F}{A} \qquad or \qquad F = PA$$

Substituting this expression into Equation 12-2, we see how the work done on the surroundings (w_{surr}) is related to external pressure (P_{ext}):

$$w_{surr} = Fd = P_{ext}Ad$$

During expansion, the gas changes its volume. As Figure 12-8 illustrates, this volume change is the product of area multiplied by displacement:

$$\Delta V = (\text{Area})(\text{Displacement}) = Ad$$

This lets us write the expression for expansion work transferred to the surroundings in terms of the opposing pressure and the change of volume of the system:

$$w_{surr} = P_{ext}\Delta V_{sys}$$

Every energy transfer has a *direction*. When a system does work on the surroundings, the system transfers energy to the surroundings. Thus w is positive for the surroundings and negative for the system in an expansion:

$$w_{sys} = -w_{surr}$$

$$w_{sys} = -P_{ext}\Delta V_{sys} \tag{12-3}$$

Equation 12-3 is used to calculate the amount of work done during the expansion of any system against a constant pressure.

In a contraction, on the other hand, the pressure is still positive, but ΔV_{sys} is negative; thus Equation 12-3 gives a positive w_{sys} for a contraction. Example 12-4 shows how to apply Equation 12-3.

> The units of pressure multiplied by volume are the same as the units of force multiplied by displacement:
> Units of PV: $(N\ m^{-2})(m^3) = N\ m = J$
> Units of Fd: $N\ m = J$

Example 12-4 | **Calculating Work**

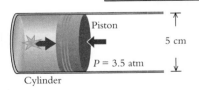

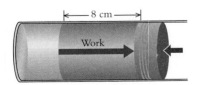

An automobile piston of diameter 5.00 cm is displaced 8.00 cm when the cylinder fires. If the gas expands against an external pressure of 3.50 atm, how much work is done in each stroke of the piston?

Strategy: When the cylinder fires, the molecules in the hot gas push the piston back against the external pressure. As a result, the volume of the system increases, and the system does work on the surroundings. We expect the sign of w to be negative. Equation 12-3 is used to calculate expansion work:

$$w_{sys} = -P_{ext}\Delta V_{sys}$$

Solution: The external pressure, 3.50 atm, is given in the problem, but the volume change must be calculated from the displacement of the piston (8.00 cm) and the dimensions of the cylinder (diameter = 5.00 cm, radius = 2.50 cm):

$$\Delta V = Ad$$

$$A = \pi r^2 = (3.1416)(2.50\ \text{cm})^2 = 19.64\ \text{cm}^2$$

| Calculating Work (continued) | Example 12-4 |

$$\Delta V_{sys} = Ad = (19.64 \text{ cm}^2)(8.00 \text{ cm}) = 157.1 \text{ cm}^3$$

$$w_{sys} = -P_{ext}\Delta V_{sys} = -(3.50 \text{ atm})(157.1 \text{ cm}^3) = -5.50 \times 10^2 \text{ cm}^3 \text{ atm}$$

These units look wrong. How do the units of "cm^3 atm" relate to energy? Recall from Chapter 5 that the SI measure of pressure is the pascal (Pa). To find the proper energy units, make the appropriate substitutions:

$$1 \text{ atm} = 1.013 \times 10^3 \text{ Pa} = 1.013 \times 10^5 \text{ kg m}^{-1} \text{ s}^{-2}$$

As noted earlier, $1 \text{ J} = 1 \text{ kg m}^2 \text{ s}^{-2}$, so $1 \text{ atm} = 1.013 \times 10^5 \text{ J m}^{-3}$:

$$\text{Work} = (-5.50 \times 10^2 \text{ cm}^3 \text{ atm})(1 \text{ m}/10^2 \text{ cm})^3(1.013 \times 10^5 \text{ J m}^{-3}/\text{atm}) = -55.7 \text{ J}$$

The negative sign is a reasonable result, because we know that an expanding piston must do work on the surroundings, so the surroundings gain energy and the system loses energy.

> Chemists often measure volumes in liters: $1 \text{ L} = 10^{-3} \text{ m}^3$. Work calculations then give results in L atm, which can be converted to J using $1 \text{ L atm} = 101.325 \text{ J}$.

First Law of Thermodynamics

A closed system can exchange energy with its surroundings in two ways: Energy can be transferred as heat or as work. Furthermore, a system can either gain or lose energy. That is, either heat or work can be transferred into or out of a system. During a chemical reaction, changes in chemical energy may cause heat transfer (for example, a gas flame heats an aluminum frying pan) and/or work transfer (for example, burning gasoline drives an automobile piston). Because energy must be conserved, the energy change of the system is linked to the flow of heat and work:

$$\Delta E_{sys} = q_{sys} + w_{sys} \qquad (12\text{-}4)$$

This equation summarizes the observation that the change of energy of a system equals the heat transferred into or out of the system plus the work transferred into or out of the system. Scientists have found that heat and work transfers are sufficient to account for the energy changes that accompany any process. Equation 12-4 applies to any process and is called the **first law of thermodynamics.**

Energy transfer is directional, and for this reason it is essential to keep track of the signs associated with heat and work. When heat flows into a system, q_{sys} is positive and the energy of the system increases. When heat flows out of a system, q_{sys} is negative and the energy of the system decreases. Similarly, when work is done on a system, w_{sys} is positive and the energy of the system increases. When a system does work on the surroundings, w_{sys} is negative and the energy of the system decreases.

Furthermore, whatever the system loses, the surroundings gain, and vice versa. When heat is transferred from a system to its surroundings, the system loses energy and the surroundings gain energy. Conversely, when heat is transferred from the surroundings to a system, the system gains energy and the surroundings lose energy. Similar sign relationships exist for work. These relationships can be summarized in two equalities:

$$q_{surr} = -q_{sys} \qquad and \qquad w_{surr} = -w_{sys}$$

These equalities allow us to write an equation for the energy change of the surroundings as follows:

$$\Delta E_{surr} = q_{surr} + w_{surr} = (-q_{sys}) + (-w_{sys}) = -(q_{sys} + w_{sys}) = -\Delta E_{sys}$$

This expression is a restatement of the law of conservation of energy: Any change in the energy of a system must be counterbalanced by an opposite change in the energy of the surroundings.

For convenience, the subscript denoting the *system* usually is omitted in thermodynamic equations. Thus an energy term without a subscript always refers to the system. To avoid ambiguity, we *never* omit the subscript indicating the surroundings.

The first law of thermodynamics states that ΔE, the energy change of any system, is equal to the heat gained by the system plus the work done on the system. Any heat absorbed by the system increases the system's energy, and any work done on the system likewise increases the system's energy. The surroundings must provide this energy so that the total energy of the universe is conserved. Thus as the system absorbs heat, the surroundings lose energy. Moreover, as work is done on the system, the surroundings lose the energy needed to do that work. Whenever a system undergoes any change in energy, the surroundings undergo an equal and opposite change in energy, so the total energy of the universe remains unchanged.

We can now return to the example of an automobile cylinder and express energy flows in terms of the first law of thermodynamics. Burning gasoline inside the cylinder releases chemical energy. The energy change for the system is balanced by a corresponding energy change for the surroundings. Some energy is transferred to the engine block as heat, and some energy is transferred to the piston as work. For the combustion of gasoline in a cylinder, q and w are negative.

Energy is a state function, as we can realize by looking at a chemical process from a molecular perspective. For example, the change in Figure 12-4 is the conversion of 2 mmol of octane and 25 mmol of oxygen gas into 16 mmol of carbon dioxide and 18 mmol of water. The value of ΔE for this process is the difference between the energy required to break all the bonds in the reactant molecules and the energy released in forming all the bonds in the product molecules. This energy difference depends only on the strengths of the various chemical bonds, not on how the system interacts with its surroundings. Consequently, ΔE depends on the net change in the conditions of the system but not on *how* that change occurs. Whether this combustion process occurs in an automobile cylinder or in a gasoline stove, the net energy change is the same.

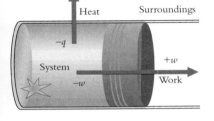

State Functions vs. Path Functions

Energy is a state function, but heat and work are path functions. To illustrate this, Figure 12-9 describes two different paths for the combustion of 2 mmol of octane. The first path represents what happens in an automobile engine: As octane burns, the system does work on its surroundings by driving back the piston. At the same time, some heat is transferred from the chemicals to the engine block:

$$\Delta E_{engine} = q_{engine} + w_{engine}$$

What happens in a gasoline stove is represented by the second path: The heat released from burning octane is transferred completely to the surroundings:

$$\Delta E_{stove} = q_{stove}$$

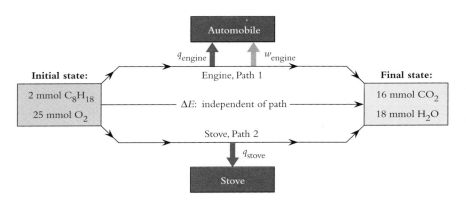

Figure 12-9
A block diagram of two different paths for the combustion of octane. Occurring inside an engine (*path 1*), the process transfers heat and work. Occurring in a stove burner (*path 2*), the process does no work and transfers only heat.

Notice that although $w_{stove} = 0$, $w_{engine} > 0$, showing that the work done is different for these two paths. In other words, work is a path function, not a state function. Energy is a state function, however, and the same quantity of gasoline burns in each case, so the change in energy is the same for both paths:

$$\Delta E_{engine} = \Delta E_{stove}$$

In the engine, some of this energy accomplishes work, and the rest is transferred as heat. In the stove, all of this energy is transferred as heat. In other words:

$$q_{engine} < q_{stove}$$

The heat transferred is different for the two paths, so q is a path function, not a state function. As with the distance each Dalton traveled in going from San Francisco to Denver (see Figure 12-5), q and w are path functions. As with the distance between San Francisco and Denver, ΔE is a state function.

Section Exercises

12.2.1 How much heat is required to raise the temperature of 25.0 g of water from 25 °C to 65 °C on an octane-burning stove (see Table 12-1)?

12.2.2 In a steam engine, water vapor drives a piston to accomplish work. What amount of work, in joules, is done when 25.0 g of water at 100.0 °C is vaporized to steam, pushing a piston against 1.00 atm of external pressure? (Hint: Use the ideal gas equation to find the volume change when liquid water becomes water vapor.)

12.2.3 Bicyclists in the Tour de France average well over 20 km/hr and the riders are on the road an average of 4.0 hr/day. How much ground beef must a 55-kg bicyclist consume daily to maintain his weight? If the cyclist ate fruit rather than ground beef, how much fruit would be required? (Consult the tables in the Chemistry and Life Box.)

12.3 HEAT MEASUREMENTS: CALORIMETRY

In a chemical reaction such as the combustion of octane, it is not possible to make direct measurements of the changes in chemical bond energies. Instead, chemists measure transfers of heat (q) and work (w) to determine the energy change that

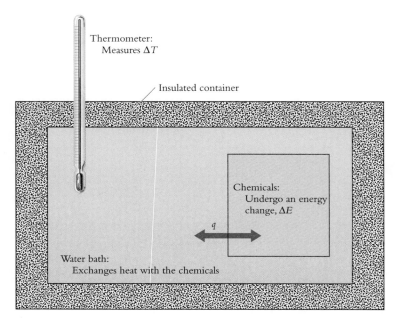

Figure 12-10
Block diagram of a typical calorimeter.

Exo is Greek for "out," *endo* is Greek for "within," and *therm* is Greek for "heat."

accompanies a chemical change of state. In this section, we focus on the measurement of heat transfers accompanying such changes.

Calorimeters

A device that measures heat flow is called a **calorimeter.** The basic features of a calorimeter are shown in Figure 12-10. They include an insulated container and a thermometer that monitors the temperature of the calorimeter. In a calorimetry experiment, chemicals are placed in the container and made to react. If the reaction involves an energy change, heat flows between the chemicals and the calorimeter. As this heat flow takes place, the insulation prevents any heat exchange between the calorimeter and its surroundings. Consequently, the calorimeter and the chemicals act as an isolated system during the short time required for the chemical process to occur.

The heat flow between the chemicals and the calorimeter causes the temperature of the calorimeter to rise or fall. If the chemicals release heat, this heat gain raises the temperature of the calorimeter. Such a process is termed **exothermic.** Conversely, if the chemicals absorb heat, this heat loss lowers the temperature of the calorimeter. Such a process is termed **endothermic.**

The law of conservation of energy allows us to account for all the heat flow within the isolated system:

$$q_{sys} = q_{calorimeter} + q_{chemicals} = 0$$

Thus the amount of heat released or absorbed in a chemical process can be determined by measuring $q_{calorimeter}$:

$$q_{chemicals} = -q_{calorimeter}$$

We can determine $q_{calorimeter}$ from its temperature change using an equation similar to Equation 12-1:

$$q_{calorimeter} = C_{cal}\Delta T \qquad (12\text{-}5)$$

Here, C_{cal} is the total heat capacity of the calorimeter. That is, C_{cal} is the amount of heat required to raise the temperature of the entire calorimeter (water bath, container, and thermometer) by 1 K.

A *total* heat capacity has units of J/K, whereas a *molar* heat capacity has units of J/mol K. If we have *n* moles of a pure substance, its total heat capacity is
$C_{total} = nC_{molar}.$

To know C_{cal}, we must calibrate the calorimeter by measuring the temperature change resulting from the transfer of a known amount of heat. For example, an electrical heater can supply heat ($q_{electric}$) that can be measured very accurately by measuring current, voltage, and time. Then, $q_{electric}$ and the measured temperature increase can be used in Equation 12-5 to calculate C_{cal}:

$$C_{cal} = \frac{q_{electric}}{\Delta T_{electric}} \quad \substack{\longrightarrow \text{ Calculated from electrical measurements} \\ \longrightarrow \text{ Measured directly}}$$

Example 12-5 illustrates this technique.

| Determining Total Heat Capacity | Example 12-5 |

A calorimeter is calibrated with an electrical heater. Before the heater is turned on, the calorimeter temperature is 23.55 °C. The addition of 10.00 kJ of electrical energy raises the temperature to 24.67 °C. Determine the total heat capacity of this calorimeter.

Strategy: Follow the standard seven-step problem-solving procedure.

Solving Quantitative Problems

Solution:

1. We are asked to determine the total heat capacity of a calorimeter.

2. A sketch of the process helps to identify what takes place. The electrical heater produces energy that flows into the calorimeter and raises the temperature of the water bath.

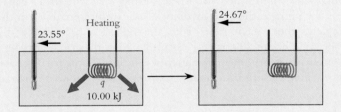

3. We are provided with heat and temperature data:

$$q_{electric} = 10.00 \text{ kJ} = q_{calorimeter}$$

$$\Delta T_{electric} = T_f - T_i = (24.67 - 23.55 \text{ °C})$$
$$= 1.12 \text{ °C}$$

4. Equation 12-5 relates the quantities: $q_{calorimeter} = C_{cal}\Delta T$

5. Rearrange this equation to solve for C_{cal}:

$$C_{cal} = \frac{q_{electric}}{\Delta T_{electric}}$$

6. Now, substitute and evaluate:

$$C_{cal} = \frac{10.00 \text{ kJ}}{1.12 \text{ °C}} = 8.93 \text{ kJ/°C} = 8.93 \text{ kJ/K}$$

Remember that temperature difference (ΔT) has the same magnitude in °C and K.

7. The result has the correct units, is positive, and is rather large; this is what we expect for a calorimeter, so this is a reasonable result.

Types of Calorimeters

Volume, pressure, temperature, and amounts of substances may change during a chemical reaction. When scientists make experimental measurements, however, they prefer to control volume, pressure, or both, to simplify the interpretation of their results. In general, it is possible to hold volume or pressure constant, but not both. In *constant-volume calorimetry,* the volume of the system is fixed, whereas in *constant-pressure calorimetry,* the pressure of the system is fixed. Whichever type of calorimetry is used, temperature changes are used to calculate q.

Figure 12-11
A constant-pressure calorimeter can be constructed from two Styrofoam cups, a cover, a stirrer, and a thermometer.

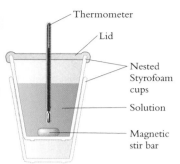

Constant-pressure calorimetry requires only a thermally insulated container and a thermometer. A simple, inexpensive constant-pressure calorimeter can be made using two nested Styrofoam cups. An example is shown in Figure 12-11. The inner cup holds the water bath, a magnetic stir bar, and the reactants. The thermometer is inserted through the cover. The outer cup provides extra thermal insulation.

In one use of this calorimeter, aqueous solutions containing the reactants are mixed in the cup, and the thermometer registers the resulting temperature change. The heat capacity of the calorimeter must be determined independently. One way to do this would be through electrical heating, as described in Example 12-5. However, satisfactory accuracy is often obtained by assuming that the heat capacity of the calorimeter is the same as the heat capacity of the water contained in the calorimeter. This neglects the contribution from the nested cups and the thermometer, which may be only about 1% of C_{cal}. Example 12-6 illustrates another application of constant-pressure calorimetry. As described in our Tools for Discovery Box on page 562, constant-pressure calorimetry also has been extended to studies of biological systems.

| Example 12-6 | **Constant-Pressure Calorimetry** |

Ammonium nitrate (NH_4NO_3, $MM = 80.05$ g/mol) is a salt used in cold packs to "ice" injuries. When 20.0 g of this salt dissolves in 125 g of water in a coffee-cup calorimeter, the temperature falls from 23.5 to 13.4 °C. Determine q for the calorimeter. Is the chemical process exothermic or endothermic?

Strategy: Follow a procedure similar to the one in Example 12-5.

Solving
Quantitative
Problems

Solution:

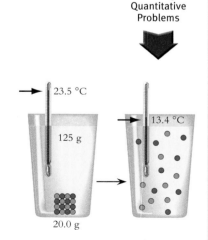

1. We are asked to find q and determine its sign for the chemical process.

2. A sketch helps to identify the process. The temperature of the calorimeter changes in response to a heat flow that accompanies the dissolving of NH_4NO_3.

3. We are given temperature data and information about the calorimeter's contents:

$$\Delta T = T_f - T_i = 13.4\ °C - 23.5\ °C = -10.1\ °C = -10.1\ K$$

$$NH_4NO_3: 20.0\ g,\ MM = 80.05\ g/mol;\ H_2O: 125\ g$$

4. Equation 12-5 lets us determine q after heat capacity and temperature change are known:

$$q_{calorimeter} = C_{cal}\Delta T$$

5. Given no additional information, we make the approximation that C_{cal} is the heat capacity of its water content:

$$C_{cal} \cong n_{water}\ C_{water}$$

Constant-Pressure Calorimetry *(continued)*	**Example 12-6**

$$n = \frac{m}{MM} = \frac{125 \text{ g}}{18.016 \text{ g/mol}} = 6.938 \text{ mol}$$

$$C_{\text{water}} = (6.938 \text{ mol})(75.291 \text{ J/mol K}) = 522.4 \text{ J/K}$$

6. Now we are ready to solve for the quantities asked for:

$$q = C_{\text{cal}}\Delta T$$

$$q = (522.4 \text{ J/K})(-10.1 \text{ K}) = -5276 \text{ J} = -5.28 \text{ kJ}$$

Is the solution process exothermic or endothermic? The temperature of the calorimeter falls during the process, which means that heat is transferred from the calorimeter to the reactants. In other words, the water in the calorimeter loses heat as the ammonium nitrate dissolves. The ions gain energy, so the dissolving process is endothermic.

7. Are these results reasonable? The units are kJ, which is appropriate for an energy calculation. The drop in temperature of the calorimeter indicates a heat-absorbing process, so the negative q is reasonable.

Figure 12-12 illustrates a constant-volume calorimeter, which is often used to measure q for combustion reactions. A sample of the substance to be burned is placed inside the sealed calorimeter in the presence of excess oxygen gas. When the sample burns, energy flows from the chemicals to the calorimeter. The temperature change of the calorimeter, with the calorimeter's heat capacity, gives the amount of heat released in the reaction. Example 12-7 illustrates this technique.

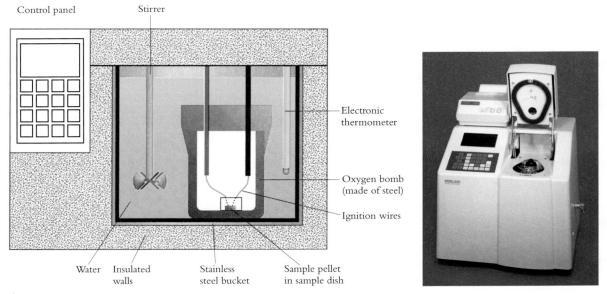

Figure 12-12
A commercially produced device for constant-volume calorimetry. This is called a *bomb calorimeter* because the container in which the reaction occurs resembles a bomb.

Box 12-2 | Tools for Discovery: Calorimetry in Biology

Calorimetry has played a central role in chemistry from its emergence as an experimental science in the eighteenth century. It is also an important technique in biology. The inventor of the calorimeter was Antoine Lavoisier, shown in the illustration. Lavoisier was a founder of modern chemistry, but he also carried out calorimetric measurements on biological materials. Lavoisier and Pierre Laplace reported in 1783 that respiration is a very slow form of combustion. Thus calorimetry has been applied to biology virtually from its invention.

Calorimetry is valuable in biological studies because every living thing is an energy-processing system. An analysis of the energy flows in a biological system can provide extensive information about how organisms use energy for growth, reproduction, and many other processes.

Despite Lavoisier's early studies on the essential link between energy and life, calorimetric measurements played a rather minor role in biology until recent years, primarily because of practical obstacles. Every organism is an open system that must take in and give off matter as part of its normal function, and it is very difficult to make accurate heat-flow measurements on open systems. Moreover, the sizes of many organisms are poorly matched to the sizes of calorimeters. Although a chemist can adjust the amount of a substance on which to carry out calorimetry, a biologist often cannot.

Improvements in calorimeters have led to a blossoming of biology-related studies within the last two decades. One is the differential scanning calorimeter, in which the temperature difference is measured between two matched chambers, only one of which contains the material being studied. A second is improved sensitivity of temperature detection using differential amplifiers and microscopic-sized sensors. These and other advances in instrumentation make possible the accurate detection of temperature

changes of less than 0.001 K and heat flows as small as 1 mJ. The ability to measure small amounts of heat transfers has been exploited imaginatively by biologists, as the following examples illustrate.

Calorimetry shows that the rates of metabolism of plant tissues vary widely with species, with cell types, and with environmental conditions. This provides a means of exploring the mechanisms by which various agents influence the health of a plant community. Studies are being done on beneficial agents such as growth promoters and detrimental ones such as atmospheric pollutants. For example, a correlation has been found between the metabolic heat rates and the extent of damage to pine needles by ozone.

Muscle activity is accompanied by cellular pumping of sodium ions. The energy requirements of the sodium pump have been studied on an individual cardiac muscle mounted inside a tiny differential calorimeter and stimulated by electrical impulses. The heat evolved was different in the presence and absence of a known inhibitor of the sodium pump.

The metabolism of microbes is relatively easy to study by calorimetry because microbial growth is accompanied by energy release in the form of heat. Pharmaceutical calorimetry uses the heat profiles of microbial cultures to assess how various drugs and drug combinations affect the growth of the culture. A graph of heat profiles shows that a drug that acts as a growth inhibitor reduces the heat output of a culture, whereas a lethal drug kills the culture and eliminates its heat production. A flow calorimeter, in which a bathing solution is slowly passed through the cell culture, is particularly well suited for these studies, because it mimics a living system and provides information about how quickly a given drug acts.

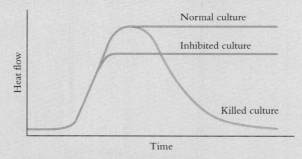

These are just some of the ways in which calorimetry is used in contemporary biological research. Our examples highlight studies at the cellular level, but ecologists also use calorimetry to explore the energy balances in ecosystems, and whole-organism biologists have found ways to carry out calorimetric measurements on fish, birds, reptiles, and mammals, including humans.

Constant-Volume Calorimetry   **Example 12-7**

A 1.250-g sample of octane (C_8H_{18}, MM = 114.2 g/mol) is burned in excess O_2 in the constant-volume calorimeter described in Example 12-5. The temperature of the calorimeter rises from 21.05 to 27.78 °C. Find the heat transferred to the calorimeter during the combustion of the octane.

Strategy: Once again, the seven-step process serves us well.

Solution:

1. We are asked to find q for the calorimeter.

2. A sketch helps to identify the process. The heat given off in the combustion reaction is absorbed by the calorimeter.

3. The value of C_{cal} is calculated in Example 12-5, and ΔT can be found from the data:

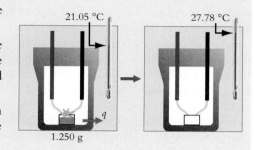

Solving Quantitative Problems

$$C_{cal} = 8.93 \text{ kJ/°C} \quad and \quad \Delta T_{reaction} = (27.78 - 21.05) \text{ °C} = 6.73 \text{ °C}$$

The mass of octane is 1.250 g and its molar mass is 114.2 g/mol.

4 and 5. Equation 12-5 can be used directly to find q: $q = C_{cal}\Delta T$

6. Now substitute and do the calculation:

$$q = C_{cal}\Delta T = (8.93 \text{ kJ/°C})(6.73°C) = 60.1 \text{ kJ}$$

7. The temperature of the calorimeter rises during the process, so a positive q is a reasonable result. The reaction releases heat and is exothermic, also a reasonable result, as we know that octane is a good fuel.

Energy Changes and Calorimetry

In a calorimetry experiment, the heat flow resulting from a process is determined by measuring the temperature change of the calorimeter. Then q can be related to energy change through the first law of thermodynamics (Equation 12-4):

$$\Delta E = q + w$$

Because heat and work are path functions, however, the conditions of the experiment must be specified. When the experiment is performed at constant volume, a subscript v is used. When the experiment is performed at constant pressure, a subscript p is used. This gives different expressions for the two types of calorimeters:

$$\Delta E = q_v + w_v \quad or \quad \Delta E = q_p + w_p$$

To calculate a state function change ΔE using measured values of q, we must evaluate w. Calorimeters are designed so that only expansion work occurs. Recall that Equation 12-3 describes work done by expansion:

$$w_{sys} = -P_{ext}\Delta V_{sys} \quad\quad\quad (12\text{-}3)$$

For a constant-volume calorimeter, $\Delta V_{sys} = 0$, so there is no expansion work: $w_v = 0$. Thus:

$$\Delta E = q_v \qquad \text{(Constant-volume calorimeter)}$$

For a constant-pressure calorimeter, the volume of the system may change, so $w_p \neq 0$ and Equation 12-3 must be evaluated. We do this in Section 12-4.

Molar Energy Change

Energy change is an extensive quantity, which means that the amount of energy released or absorbed depends on the amount of substances that react. For example, as more octane burns, more energy is released. Thus when we report an energy change, we must also report the amounts of the chemical substances that generate the energy change. For tabulation purposes, all changes in thermodynamic functions such as ΔE are given *per mole,* just as heat capacities are conveniently tabulated as molar values. We use ΔE_{molar} for the energy change that accompanies the reaction of one mole of a particular substance. Experimental measurements by calorimetry usually involve amounts different from one mole. The molar energy change can be found from an experimental energy change by dividing by the number of moles that reacted:

$$\Delta E_{molar} = \frac{\Delta E}{n} \qquad (12\text{-}6)$$

Example 12-8 shows how to evaluate ΔE_{molar} for the combustion of octane.

Example 12-8	Molar Energy Change

What is ΔE_{molar} for the combustion of octane described in Example 12-7?

Strategy: Although the problem does not provide any data, it refers to an example where appropriate data are provided. As with other calorimetry problems, the seven-step approach works well. In this example, however, we will not enumerate the individual steps.

Solution: We are asked for ΔE_{molar}, the amount of energy released in the combustion of 1 mol of octane.

To visualize the process, refer to the illustration provided with Example 12-7. Note that the reaction consumes 1.250 g of octane, information that was not required in the previous calculations.

We could start with the original data, but $q_{calorimeter}$ is known from Example 12-7: $q_{calorimeter} = 60.1$ kJ when 1.250 g of octane burns in a constant-volume calorimeter. Because the calorimeter and its contents are an isolated system, none of the energy released in the combustion reaction escapes to the surroundings during the experiment. The calorimeter absorbs all of the energy released in the reaction:

$$q_{system} = q_{reaction} + q_{calorimeter} = 0 \qquad so \qquad q_{reaction} = -q_{calorimeter}$$

Thus $q_{reaction} = -60.1$ kJ. This heat transfer occurs at constant volume, so:

$$q_{reaction} = q_v = \Delta E = -60.1 \text{ kJ}$$

This is the energy released during the combustion of 1.250 g of octane. To obtain the molar energy change, we need to use Equation 12-6:

Molar Energy Change *(continued)* | **Example 12-8**

$$\Delta E_{molar} = \frac{\Delta E}{n}$$

Determine how many moles reacted, converting from mass to moles using molar mass:

$$n = \frac{m}{MM} = \frac{1.250 \text{ g}}{114.2 \text{ g/mol}} = 0.010946 \text{ mol}$$

Now use Equation 12-6 to determine the molar energy change:

$$\Delta E_{molar} = \frac{\Delta E}{n} = \frac{-60.1 \text{ kJ}}{0.010946 \text{ mol}} = -5.49 \times 10^3 \text{ kJ/mol octane}$$

The units are correct, and we know that octane is a good fuel, so the large negative value also is reasonable.

The examples presented in this section illustrate the fundamental features of calorimetry.

Section Exercises

■ **12.3.1** In a constant-pressure calorimeter, 1.530×10^3 J of electrical energy changes the water temperature from 20.50 °C to 21.85 °C. When 1.75 g of a solid salt is dissolved in the water, the temperature falls from 21.85 °C to 21.44 °C. Find the value of q_p for the solution process.

■ **12.3.2** In a constant-volume calorimeter, 3.56 g of solid sulfur is burned in excess oxygen gas:

$$S(s) + O_2(g) \longrightarrow SO_2(g)$$

The calorimeter has a total heat capacity of 4.32 kJ/°C. The combustion reaction causes the temperature of the calorimeter to increase from 25.93 °C to 33.56 °C. Calculate the heat released per mole of sulfur burned.

■ **12.3.3** Imagine a calorimeter with a sliding piston that makes it possible to perform constant-pressure calorimetry experiments on a mixture of liquids and gases.

> Consider burning 1.250 g of octane in this calorimeter, which initially is at 25 °C.
> (a) Will the calorimeter temperature rise or fall?
> (b) Use the ideal gas equation to determine whether the volume of the system will increase or decrease. (Hint: The temperature change is almost the same as in Example 12-7.)
> (c) Is the work positive, negative, or zero?
> (d) Will ΔE be the same, more negative, or more positive than ΔE calculated in Example 12-8?

12.4 ENTHALPY

In our world, most chemical processes occur in contact with the Earth's atmosphere at a virtually constant pressure of one atmosphere. For example, plants convert carbon dioxide and water into carbohydrates; animals metabolize food; water heaters and stoves burn fuel; and running water dissolves minerals from the soil. All these processes involve energy changes at constant pressure. Virtually all aqueous-solution chemistry also occurs at constant pressure. Thus the heat flow measured using constant-pressure calorimetry, q_p, closely approximates heat flows in many real-world processes.

How is q_p related to energy changes? Unlike the constant-volume situation, a process at constant pressure involves both heat and work, as already described:

$$\Delta E = q_p + w_p \qquad so \qquad q_p = \Delta E - w_p$$

In a typical constant-pressure calorimeter, the work done is caused by expansion against the constant external pressure, and the pressure exerted by the system equals this constant external pressure. Thus we can use Equation 12-3 to relate work to pressure and volume:

$$w_p = -P_{ext}\Delta V_{sys} = -P\Delta V$$

We omit the subscripts for P and V because both quantities refer to the chemical system. When we substitute $-P\Delta V$ in place of w_p, we have

$$q_p = \Delta E + P\Delta V$$

Now we define a new quantity, called **enthalpy (H)**, and see how this quantity relates to q_p:

$$H = E + PV \tag{12-7}$$

Because it is composed entirely of state functions, enthalpy is also a state function. From the equation defining enthalpy, we can relate enthalpy changes to changes in energy, pressure, and volume:

$$\Delta H = \Delta(E + PV) = \Delta E + \Delta(PV) = \Delta E + P\Delta V + V\Delta P$$

For a process occurring at constant pressure, $\Delta P = 0$, making the last term zero:

$$\Delta H = \Delta E + P\Delta V \qquad \text{(Constant pressure)}$$

As shown above, the quantity on the right is q_p, so a simple equality links enthalpy change and the heat flow in a constant-pressure process:

$$\Delta H = q_p \qquad \text{(12-8)}$$

Equation 12-8 lets us describe enthalpy in words: Enthalpy is a thermodynamic quantity whose change equals the heat flow at constant pressure. Just as the heat flow in a constant-volume process gives the energy change, the heat flow in a constant-pressure process gives the enthalpy change. Consequently, enthalpy changes can be measured directly using a constant-pressure calorimeter.

The units of enthalpy are the same as the units of energy: joules or kilojoules. Tables give enthalpy changes per mole of substance, just as they give molar energy changes. Molar enthalpy changes are usually expressed in kilojoules per mole.

Energy is a *fundamental* thermodynamic property. Enthalpy is a *defined* thermodynamic property that is convenient when working at constant pressure. Enthalpy changes and energy changes are related through the PV product:

$$\Delta H = \Delta E + \Delta(PV)$$

For solids and liquids, volume changes during chemical reactions are very small:

$$\Delta(PV)_{\text{condensed phases}} \cong 0$$

Thus enthalpy changes and energy changes are essentially equal for processes that involve only liquids and solids.

An example of such a process is the dissolving of ammonium nitrate in water to produce aqueous ions:

$$NH_4NO_3(s) \longrightarrow NH_4^+(aq) + NO_3^-(aq)$$

As Example 12-6 shows, 20.0 g of this salt absorbs 5.28 kJ of energy when it dissolves in a constant-pressure calorimeter. This information lets us calculate the molar enthalpy change when NH_4NO_3 dissolves:

$$\Delta H = q_p = 5.28 \text{ kJ} \qquad n = \frac{20.0 \text{ g}}{80.05 \text{ g/mol}} = 0.2498 \text{ mol}$$

$$\Delta H_{\text{molar}} = \frac{\Delta H}{n} = \frac{5.28 \text{ kJ}}{0.2498 \text{ mol}} = 21.1 \text{ kJ/mol}$$

For this process, none of the reactants or products is a gas, so the energy and enthalpy changes are essentially equal:

$$\Delta E_{\text{molar}} \cong 21.1 \text{ kJ/mol}$$

Pressure and/or volume change significantly when gases are produced or consumed during a reaction. Consequently, the enthalpy change is likely to be different than the energy change for any chemical reaction involving gaseous reagents. To determine the difference, we can use the ideal gas equation to relate the change in the pressure-volume product to the change in number of moles:

$$PV_{\text{gas}} = nRT \qquad \text{so} \qquad \Delta(PV)_{\text{gas}} = \Delta(nRT)$$

Combining the equations for $\Delta(PV)$, we find that the enthalpy and energy changes for any chemical reaction are related according to Equation 12-9:

$$\Delta H_{\text{reaction}} = \Delta E_{\text{reaction}} + \Delta(PV)_{\text{reaction}} \cong \Delta E_{\text{reaction}} + \Delta(PV)_{\text{gases}}$$

$$\Delta H_{\text{reaction}} \cong \Delta E_{\text{reaction}} + \Delta(nRT)_{\text{gases}} \qquad \text{(12-9)}$$

Example 12-9 demonstrates a typical magnitude for this difference.

| Example 12-9 | ΔE and ΔH for Combustion |

Find the difference between the values for the molar ΔH and ΔE for the combustion of octane at 298 K.

Strategy: Enthalpy changes and energy changes for reactions are related by Equation 12-9:

$$\Delta H_{reaction} \cong \Delta E_{reaction} + \Delta(nRT)_{gases}$$

To apply this equation to molar quantities, we must write the combustion reaction in which one mole of octane burns. The balanced chemical equation for combustion applies to two moles of octane:

$$2\,C_8H_{18}(l) + 25\,O_2(g) \longrightarrow 16\,CO_2(g) + 18\,H_2O(l)$$

Divide each coefficient by 2 to obtain a reaction in which one mole of octane burns at 298 K:

$$C_8H_{18}(l) + {}^{25}\!/_2\,O_2(g) \longrightarrow 8\,CO_2(g) + 9\,H_2O(l)$$

To calculate the difference between $\Delta H_{reaction}$ and $\Delta E_{reaction}$, we need information about the $\Delta(nRT)$ term as it applies to this reaction. Since R and T are constant, this term can be expanded:

$$\Delta(nRT)_{gases} = n_f RT - n_i RT = RT\,(n_f - n_i) = RT\,\Delta n_{gases}$$

Here, n_f and n_i are the numbers of moles of gaseous products and gaseous reactants in the balanced chemical equation and Δn is the difference between them.

Solution: Here are the values needed to complete the calculation:

$$n_f = \text{mol gas (products)} = \text{mol } CO_2 = 8$$

$$n_i = \text{mol gas (reactants)} = \text{mol } O_2 = {}^{25}\!/_2$$

$$\Delta n_{gases} = 8 \text{ mol } CO_2 - {}^{25}\!/_2 \text{ mol } O_2 = -{}^{9}\!/_2 \text{ mol gas}$$

$$T = 298 \text{ K}$$

To find the difference between ΔH and ΔE, we rearrange Equation 12-9:

$$\Delta H - \Delta E \cong \Delta(nRT) = RT\,\Delta n$$

$$\Delta H - \Delta E = (8.314 \text{ J/mol K})(298\text{K})(-{}^{9}\!/_2 \text{ mol}) = -1.11 \times 10^4 \text{ J} = -11.1 \text{ kJ}$$

In Example 12-8, we used calorimetry data to determine that the energy change when one mole of octane burns is -5.49×10^3 kJ/mol. The present calculation shows that the enthalpy change for 1 mol of octane differs from this by -11.1 kJ; ΔH is -5.50×10^3 kJ/mol. Although the value of Δn_{gases} is significant, the fractional difference between ΔH_{molar} and ΔE_{molar} is only ${}^{11.1}\!/_{5490}$, or about 0.2%.

As Example 12-9 illustrates, reaction energies and reaction enthalpies are usually about the same, even when reactions involve gases. For this reason, chemists often use $\Delta E_{reaction}$ and $\Delta H_{reaction}$ interchangeably. Because many everyday processes occur at constant pressure, thermodynamic tables usually give values for enthalpy changes.

Heats of Formation

How do we determine the energy and enthalpy change for a chemical reaction? We could perform calorimetry experiments and analyze the results, but to do this for every chemical reaction would require a tremendous effort. Furthermore, it turns out to be unnecessary. Using the first law of thermodynamics and the idea of a state function, we can calculate enthalpy changes for almost any reaction using experimental values for one set of reactions, the **formation reactions.**

> /// *A formation reaction produces 1 mol of a chemical substance from the elements in their most stable forms at 298 K and 1 atm.*

Formation reactions exhibit the following features:

1. There is a *single product* with a stoichiometric coefficient of *1*.
2. All the reactants are *elements,* and each is in the form that is most stable when $T = 298$ K and $P = 1$ atm.

Here are the formation reactions for mercury(II) chloride and for methanol:

$$Hg(l) + Cl_2(g) \longrightarrow HgCl_2(s)$$

$$C(s, graphite) + 2\,H_2(g) + \tfrac{1}{2}\,O_2(g) \longrightarrow CH_3OH(l)$$

Neither of these formation reactions occurs readily under normal conditions, but Figure 12-13 shows examples that do. Notice from these examples that fractional stoichiometric coefficients are common in formation reactions because the reaction must generate exactly one mole of product.

The **standard enthalpy of formation (ΔH_f°)** is the enthalpy change accompanying the formation of one mole of a chemical substance from pure elements in their most stable forms under standard conditions, which are defined to be

The *most stable state* refers to both *phase* and *chemical form.* For example, chlorine is most stable as Cl_2, not Cl, and as a gas, not a liquid or a solid. The most stable state of carbon is C(s, graphite), not C(s,diamond), C(g), or C(l).

(a) (b) (c)

Figure 12-13
Some formation reactions proceed readily when the elements are brought together. The formation reactions shown are Fe $+ \frac{1}{2}$ O$_2 \rightarrow$ FeO (left), P$_4 +$ 5 O$_2 \rightarrow$ P$_4$O$_{10}$ (center), and K $+ \frac{1}{2}$ Cl$_2 \rightarrow$ KCl (right).

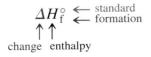

ΔH_f° $\leftarrow$ standard
$\uparrow\uparrow$ $\leftarrow$ formation
change enthalpy

In 1982 the International Union of Pure and Applied Chemistry recommended that the standard state pressure be changed from 1 atm to 1 bar, which is 10^5 pascals. This SI unit is slightly smaller than 1 atm. (1 atm = 0.9869 bar) However, many tables of thermodynamic data pre-date 1982 and are based on 1 atm as the standard state pressure.

298 K and 1 atm pressure. The superscript ° means "under standard conditions of 298 K and 1 atm." The subscript $_f$ means "formation of one mole." Every chemical substance has a characteristic ΔH_f° that can be measured and tabulated. Standard enthalpies of formation, expressed in kilojoules per mole, can be negative, positive, or zero.

For an element in its most stable form, the formation "reaction" involves no change at all, so ΔH_f° for any element in its most stable form is zero. The element oxygen, for example, exists as diatomic molecules under standard conditions, so the "formation reaction" of diatomic oxygen involves no change:

$$O_2(g) \longrightarrow O_2(g) \qquad \text{No change: } \Delta H_f^\circ = 0$$

An element that is *not* in its standard state has a nonzero ΔH_f°. For example, *atomic* oxygen has a positive enthalpy of formation because energy must be added to break diatomic molecules apart into oxygen atoms:

$$\tfrac{1}{2} O_2(g) \longrightarrow O(g) \qquad \Delta H_f^\circ = 249 \text{ kJ/mol}$$

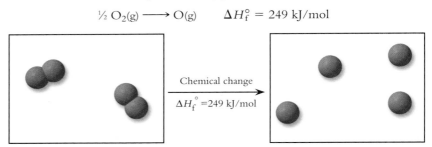

Over the years, scientists have measured standard enthalpies of formation of many chemical substances. The values are tabulated in chemical reference books such as the *CRC Handbook of Chemistry and Physics*. Values for many common substances are listed in Appendix D of this text.

Enthalpy Changes for Chemical Reactions

Standard enthalpies of formation are particularly useful because they can be used to find the enthalpy change for any reaction that occurs under standard conditions. As an example, consider nitrogen dioxide reacting to form N_2O_4:

$$2 NO_2(g) \longrightarrow N_2O_4(g)$$

Imagine the reaction following the two-step pathway that appears in Figure 12-14. In the first step, two moles of NO_2 molecules decompose into N_2 molecules and O_2 molecules:

$$2 NO_2(g) \longrightarrow N_2(g) + 2 O_2(g)$$

In the second step, these molecules react to produce N_2O_4:

$$N_2(g) + 2 O_2(g) \longrightarrow N_2O_4(g)$$

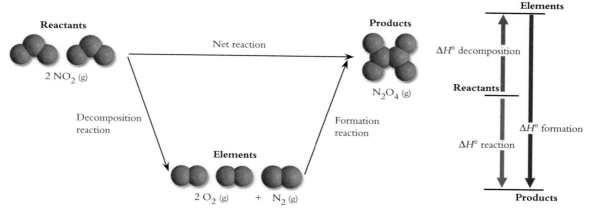

Figure 12-14
The enthalpy change for the reaction of NO_2 to produce N_2O_4 can be determined by using a two-step path. In the first step, NO_2 decomposes to N_2 and O_2. In the second step, the elements react to form N_2O_4.

The pathway shown in Figure 12-14 is not how the reaction actually occurs, but it is a perfectly valid path for calculating the enthalpy change. Enthalpy is a state function, and the change of any state function is independent of the path of the reaction. This feature can be summarized as a general statement that is known as **Hess' law:**

← **SECTION 7.5**
Hess' law is the basis for the analysis in Section 7.5 of the energies involved in the formation of ionic salts.

> /// *The enthalpy change for any overall process is equal to the sum of enthalpy changes for any set of steps that leads from the reactants to the products.*

In this example, the enthalpy change for the overall reaction can be determined by adding the enthalpy changes of the two steps of Figure 12-14:

$$\Delta H^{\circ}_{\text{reaction}} = \Delta H^{\circ}_{\text{decomposition}} + \Delta H^{\circ}_{\text{formation}}$$

$$2\,NO_2(g) \longrightarrow \cancel{N_2}(g) + 2\,\cancel{O_2}(g) \qquad \Delta H^{\circ}_{\text{decomposition}}$$

$$\cancel{N_2}(g) + 2\,\cancel{O_2}(g) \longrightarrow N_2O_4(g) \qquad \Delta H^{\circ}_{\text{formation}}$$

$$\overline{2\,NO_2(g) \longrightarrow N_2O_4(g) \qquad \Delta H^{\circ}_{\text{reaction}}}$$

The decomposition step is related to the formation reaction of NO_2. First, decomposition is the opposite of formation. Second, two moles of NO_2 break apart in this decomposition reaction. Using these features, we can express the enthalpy change of the decomposition reaction for NO_2 in terms of the standard enthalpy of formation of NO_2. The enthalpy change of the decomposition reaction has the opposite sign as the corresponding formation reaction, and the total enthalpy of decomposition requires that we multiply by the stoichiometric coefficient:

$$\Delta H^{\circ}_{\text{decomposition}} = -2\,\Delta H^{\circ}_{\text{f}}(NO_2)$$

The enthalpy change of the formation reaction is just $\Delta H^{\circ}_{\text{f}}$ of N_2O_4. Thus the molar enthalpy change of the overall reaction can be expressed entirely in terms of standard formation reactions:

$$\Delta H^{\circ}_{\text{reaction}} = \Delta H^{\circ}_{\text{f}}(N_2O_4) - 2\,\Delta H^{\circ}_{\text{f}}(NO_2)$$

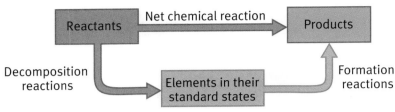

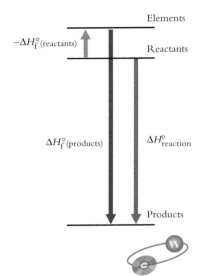

Figure 12-15
Any chemical reaction can be imagined to proceed in two stages. First, reactants decompose into their constituent elements. Second, these elements recombine to form products.

The enthalpy change of the overall reaction is the sum of the formation enthalpies of the products minus the sum of the formation enthalpies of the reactants.

Using standard enthalpies of formation from Appendix D, we can calculate the enthalpy change for the overall reaction:

$$\Delta H^\circ_{\text{reaction}} = (1 \text{ mol } N_2O_4)(11.1 \text{ kJ/mol } N_2O_4) - (2 \text{ mol } NO_2)(33.2 \text{ kJ/mol } NO_2)$$
$$= -55.3 \text{ kJ}$$

The negative value for the enthalpy change indicates that this reaction is exothermic. Qualitatively, this is logical because during the reaction a new bond forms between the nitrogen atoms, but none of the bonds of the reactants breaks.

Our analysis of the reaction of nitrogen dioxide molecules is not unique. The same type of path can be visualized for any chemical reaction, as is shown diagrammatically in Figure 12-15. The reaction enthalpy for any chemical reaction can be found from the standard enthalpies of formation for all the reactants and products. Multiply each standard enthalpy of formation by the appropriate stoichiometric coefficient and then add the values for the products and subtract the values for the reactants. This procedure is summarized in Equation 12-10:

$$\Delta H^\circ_{\text{reaction}} = \Sigma \text{ coeff}_p \, \Delta H^\circ_f \text{ (products)} - \Sigma \text{ coeff}_r \, \Delta H^\circ_f \text{ (reactants)} \qquad \textbf{(12-10)}$$

As can be seen from Equation 12-10, the value of $\Delta H^\circ_{\text{reaction}}$ depends on the coefficients in the balanced equation. This means that the enthalpy change for one mole of any particular reagent may differ from $\Delta H^\circ_{\text{reaction}}$. Returning to our earlier example, the enthalpy change of -55.3 kJ is for two moles of NO_2 reacting to form one mole of N_2O_4. To find the enthalpy change per mole of NO_2, we must divide by 2: $(-55.3 \text{ kJ/2 mol}) = -27.6 \text{ kJ/mol}$. We can generalize this calculation: To find the molar enthalpy change for any component of the reaction, divide $\Delta H^\circ_{\text{reaction}}$ by the appropriate stoichiometric coefficient.

To illustrate the use of Equation 12-10, we calculate the enthalpy of combustion of methane, the principal component of natural gas. We must begin with the balanced chemical equation:

$$CH_4(g) + 2 \, O_2(g) \longrightarrow CO_2(g) + 2 \, H_2O(l)$$

The enthalpy change that accompanies this reaction could be evaluated step by step by first decomposing methane into its elemental constituents and then recombining the elements into carbon dioxide and liquid water, as shown schematically in Figure 12-16. There is no need to do this in detail, however, because the outcome of this set of processes is summarized by Equation 12-10. The products are one mole of CO_2 and two moles of H_2O; the reactants are one mole of CH_4 and two moles of O_2:

$$\Delta H^\circ_{\text{reaction}} = [\Delta H^\circ_f(CO_2) + 2 \, \Delta H^\circ_f(H_2O)] - [\Delta H^\circ_f(CH_4) + 2 \, \Delta H^\circ_f(O_2)]$$

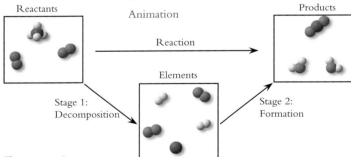

Figure 12-16
The enthalpy change for the combustion of methane can be evaluated for the path that decomposes the reactants into elements and then recombines those elements to form the products.

We make use of data from Appendix D. The value for O_2 is zero because this is an element in its standard state:

$$\Delta H^{\circ}_{\text{reaction}} = \left[(1 \text{ mol } CO_2)\left(\frac{-393.5 \text{ kJ}}{\text{mol } CO_2}\right) + (2 \text{ mol } H_2O)\left(\frac{-285.8 \text{ kJ}}{\text{mol } H_2O}\right)\right]$$

$$- \left[(1 \text{ mol } CH_4)\left(\frac{-74.6 \text{ kJ}}{\text{mol } CH_4}\right) + (2 \text{ mol } O_2)\left(\frac{0 \text{ kJ}}{\text{mol } O_2}\right)\right]$$

$$\Delta H^{\circ}_{\text{reaction}} = [-965.1 \text{ kJ}] - [-74.6 \text{ kJ}] = -890.5 \text{ kJ}$$

Equation 12-10 can also be used to calculate the standard enthalpy of formation of a substance whose formation reaction does not proceed cleanly and rapidly. The enthalpy change for some other chemical reaction involving the substance can be determined by calorimetric measurements. Then Equation 12-10 can be used to calculate the unknown standard enthalpy of formation. Example 12-10 shows how to do this.

Standard Enthalpy of Formation	Example 12-10

The calorimetry experiment described in Example 12-9 shows that the enthalpy of combustion of octane (C_8H_{18}) is -5.50×10^3 kJ/mol. Using tabulated standard enthalpies of formation in Appendix D, compute the standard enthalpy of formation of octane.

Strategy: Once again, it is convenient to follow the seven-step procedure to solve this problem.

Solving
Quantitative
Problems

Solution:

1. We are asked to find an enthalpy of formation.

2. Because enthalpy is a state function, we can use the fact that ΔH for a reaction is independent of the reaction path and visualize the reaction as occurring through decomposition and formation reactions.

3. The data that are available are standard enthalpies of formation (Appendix D) and the experimental heat of combustion from Example 12-9.

4. We need the balanced chemical equation for octane combustion and Equation 12-10, which links enthalpies of formation with the enthalpy of a reaction.

5. The balanced combustion equation must be written for one mole of octane because the enthalpy of combustion refers to one mole. Under standard conditions, octane and water are liquids:

$$C_8H_{18}(l) + \tfrac{25}{2} O_2(g) \longrightarrow 8 CO_2(g) + 9 H_2O(l)$$

Use Equation 12-10 to set the heat of combustion, ΔH_{molar}, equal to the sum of the enthalpies of formation and solve for the value for octane:

$$\Delta H_{\text{molar}} = \Sigma \text{ coeff}_p \Delta H^{\circ}_f(\text{products}) - \Sigma \text{ coeff}_r \Delta H^{\circ}_f(\text{reactants})$$

$$\Delta H_{\text{molar}} = 8 \Delta H^{\circ}_f(CO_2) + 9 \Delta H^{\circ}_f(H_2O) - \Delta H^{\circ}_f(\text{octane}) - \tfrac{25}{2} H^{\circ}_f(O_2)$$

$$\Delta H^{\circ}_f(\text{octane}) = 8 \Delta H^{\circ}_f(CO_2) + 9 \Delta H^{\circ}_f(H_2O) - \tfrac{25}{2} \Delta H^{\circ}_f(O_2) - \Delta H_{\text{molar}}$$

| Example 12-10 | Standard Enthalpy of Formation *(continued)* |

6. Substitute heats of formation (Appendix D) and the heat of combustion:

$$\Delta H_f^\circ(\text{octane}) = (8 \text{ mol } CO_2)(-393.5 \text{ kJ/mol}) + (9 \text{ mol } H_2O)(-285.8 \text{ kJ/mol})$$

$$-(\tfrac{25}{2} \text{ mol } O_2)(0 \text{ kJ/mol}) - (5.50 \times 10^3 \text{ kJ/mol})$$

$$\Delta H_f^\circ(\text{octane}) = -2.2 \times 10^2 \text{ kJ/mol}$$

7. This is a reasonable value. The units are correct and the negative value indicates that octane is more stable than the elements from which it forms.

The formation enthalpies found in tables are values under standard conditions, $T = 298$ K and $P = 1$ atm. What if a reaction occurs under nonstandard conditions? Energies and enthalpies of substances change as temperature and pressure change, so we must expect that ΔH for a reaction depends on temperature and pressure as well. These dependencies can be calculated, but such calculations are beyond the scope of our present coverage. Fortunately, for most chemical reactions, ΔH changes slowly with temperature or pressure. Therefore calculations of reaction enthalpies using standard enthalpies of formation are reliable even when the temperature and pressure depart somewhat from standard conditions. Example 12-11 applies tabulated enthalpies to a reaction that proceeds under nonstandard conditions.

| Example 12-11 | Molar Energy Change |

Nitric acid (HNO_3), which is produced from ammonia in the gas phase at elevated temperature and pressure, is an important chemical in the fertilizer industry because it can be converted into ammonium nitrate. The industrial reaction takes two steps, which can be summarized by the following chemical equation:

$$12 \, NH_3(g) + 21 \, O_2(g) \longrightarrow 8 \, HNO_3(g) + 4 \, NO(g) + 14 \, H_2O(g)$$

Determine whether this overall process is exothermic or endothermic and estimate the energy change per mole of HNO_3 formed, ΔE_{molar}.

Strategy: Again, we follow the seven-step procedure, this time without numbering the steps.

Solution: The first question asks if this reaction is exothermic (releases energy) or endothermic (absorbs energy). The second question asks us to estimate the energy change for the reaction per mole of one reagent. Even though the reaction conditions differ from standard conditions, ΔH varies slowly with temperature and pressure, so we can use Equation 12-10 and tabulated standard enthalpies of formation to estimate the reaction enthalpy. Reaction enthalpy is almost the same as reaction energy, so this calculation will give a satisfactory estimate of the reaction energy.

Appendix D lists the following standard heats of formation, all in kJ/mol:

$NH_3(g)$	-45.9	$HNO_3(g)$	-133.9
$H_2O(g)$	-241.8	$NO(g)$	$+91.3$

| Molar Energy Change (continued) | Example 12-11 |

Substitute these values into Equation 12-10 and use the stoichiometric coefficients from the balanced chemical equation:

$$\Delta H_{reaction} \cong [(8 \text{ mol})(-133.9 \text{ kJ/mol}) + (4 \text{ mol})(91.3 \text{ kJ/mol})$$

$$+ (14 \text{ mol})(-241.8 \text{ kJ/mol})]$$

$$- [(12 \text{ mol})(-45.9 \text{ kJ/mol}) + (21 \text{ mol})(0 \text{ kJ/mol})]$$

$$= [-1071.2 \text{ kJ} + 365.2 \text{ kJ} - 3385.2 \text{ kJ}] - [-550.8 \text{ kJ} + 0 \text{ kJ}]$$

$$\Delta H_{reaction} \cong -[4091.2 \text{ kJ}] - [-555.2 \text{ kJ}] = -353.6 \text{ kJ} = -3.536 \times 10^3 \text{ kJ}$$

It is important to pay careful attention to the signs, which we have done by carrying out the calculations in sequence. First, multiply each standard enthalpy, including its sign, by the appropriate coefficient. Then combine values for reactants and values for products, adding or subtracting according to the sign of each enthalpy. Finally, subtract the sum for reactants from the sum for products to obtain the final result.

The negative sign for $\Delta H_{reaction}$ indicates that this reaction is exothermic. This is the value for a reaction that generates eight moles of $HNO_3(g)$. Therefore the value *per mole* of nitric acid produced is:

$$\Delta H_{molar} = \frac{-3.536 \times 10^3 \text{ kJ}}{8 \text{ mol } HNO_3} = -442 \text{ kJ/mol } HNO_3$$

In Example 12-10, we use -285.8 kJ/mol, ΔH_f° for liquid water. However, $H_2O(g)$ is a different species, so it has a different heat of formation than $H_2O(l)$. This demonstrates how important it is to keep track of the phase of each species.

In Chapter 9 the energies of chemical reactions are estimated from bond energies rather than calculated from formation reactions. Equation 9-1, like Equation 12-10, contains sums of energies for reactants and products:

$$\Delta E_{reaction} = \Sigma \, BE_{reactants} - \Sigma \, BE_{products} \qquad (9\text{-}1)$$

The similarity can be traced to the fact that both equations describe similar processes and are consequences of Hess' law. For Equation 12-10, we imagine a process in which reactants decompose into elements in their standard states, which then recombine to give products. For Equation 9-1, we imagine a process in which reactants break up entirely into gaseous atoms, which then recombine to give products.

It is preferable to use Equation 12-10 when standard enthalpies of formation are available for the reagents taking part in a chemical reaction, because this equation uses actual experimental values for standard enthalpies of formation. Consequently, Equation 12-10 is exact under standard conditions and provides very good estimates even when conditions are not standard. Equation 9-1, on the other hand, uses tabulated bond energies, which are average values for each chemical bond rather than exact values for the bonds in any specific substance. Consequently, Equation 9-1 provides estimates of reaction energies rather than exact values. The advantage of the use of average values is that, when confronted with a chemical reaction involving a substance for which good thermodynamic measurements have not yet been made, we can nevertheless make a reasonable estimate of the sign and magnitude of ΔE and ΔH for that reaction.

Section Exercises

12.4.1 Which of the following substances have zero values for their standard enthalpy of formation: (a) ozone, $O_3(g)$; (b) solid mercury, $Hg(s)$; (c) liquid bromine, $Br_2(l)$; (d) graphite, $C(s)$; (e) atomic fluorine, $F(g)$; and (f) solid sulfur, $S_8(s)$?

12.4.2 Hydrogen gas is prepared industrially from methane and steam:

$$CH_4(g) + H_2O(g) \longrightarrow CO(g) + 3\ H_2(g)$$

Use standard enthalpies of formation to determine $\Delta H_{reaction}$ and $\Delta E_{reaction}$ for this process.

12.4.3 In the smelting of iron ore, iron oxides are decomposed in the presence of carbon. The process involves several steps, but the overall stoichiometry can be summarized in a single chemical equation:

$$Fe_2O_3(s) + C(s,\ graphite) \longrightarrow Fe(s) + CO_2(g) \qquad (unbalanced)$$

Compute the standard enthalpy change for the balanced equation.

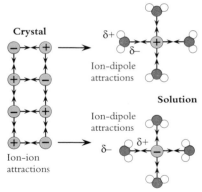

Instant hot and cold packs rely on heats of solution to generate instant heating or cooling.

12.5 ENTHALPY AND INTERMOLECULAR FORCES

The energy changes that accompany chemical reactions result largely from making and breaking chemical bonds. When substances move from one phase to another, energy is released or absorbed because of changes in intermolecular forces. In this section we discuss the energy and enthalpy changes that accompany the formation of solutions and phase changes.

Heat of Solution

Recall from Chapter 10 what happens when an ionic salt dissolves in water. As shown schematically in Figure 12-17, coulombic forces must be overcome to break the ions loose from the solid lattice, and ion–dipole forces bind the ions to water molecules in the solution. The **molar heat of solution (ΔH_{soln})** measures the net energy flow that occurs as a substance dissolves.

Figure 12-17
Whether a salt dissolves with absorption of energy (endothermic) or with release of energy (exothermic) depends on the balance between the ion-ion forces that must be overcome in the solid crystal and the ion-dipole forces that stabilize the ions in solution.

Heats of solution can be exothermic or endothermic. If the crystal binds ions more tightly than the solution, the salt must absorb energy as it dissolves, giving a process with a positive enthalpy change. Many salts with positive ΔH_{soln} values are insoluble, but one soluble salt with a relatively large positive ΔH_{soln} is NH_4NO_3, $\Delta H_{soln} = 21.1$ kJ/mol. On the other hand, if the solution binds ions more tightly than the crystal, energy is released as the salt dissolves, giving an exothermic process with negative ΔH_{soln}. Many salts release large amounts of energy when they dissolve in water. Examples are $CaCl_2$, $\Delta H_{soln} = -83$ kJ/mol and $MgSO_4$, $\Delta H_{soln} = -91.2$ kJ/mol.

Heats of solution are the basis for instant cold packs and instant hot packs used for the first-aid treatment of minor sprains and pulled muscles. These packs contain two separate compartments. One contains water, and the other contains a salt: NH_4NO_3 for cold packs and $MgSO_4$ or $CaCl_2$ for hot packs. Kneading the pack breaks the wall between the compartments, allowing the salt to mix with water. As the salt dissolves to form an aqueous solution, the temperature of the pack changes. Heat is absorbed or released only as the salt dissolves, however, so after all the salt has dissolved, the pack gradually returns to room temperature. Further manipulation of the pack has no effect. Example 12-12 addresses the enthalpy changes in a cold pack.

Enthalpies of Solution

Example 12-12

If an instant cold pack contains 75 g of ammonium nitrate and 250 g of water, how much will the temperature of the bag decrease when the chemicals are mixed? Assume that all the energy absorbed during the solution process is provided by the water.

Strategy: This problem requires more detailed analysis than earlier examples in this chapter. The seven-step process allows us to solve the problem systematically.

Solving
Quantitative
Problems

Solution:

1. The problem asks for the temperature change when ammonium nitrate dissolves in water.

2. This process involves a heat transfer. The energy required to break up the NH_4NO_3 lattice is supplied by the water in the form of heat. The energy lost by the water causes a drop in temperature. Because energy is conserved, the sum of the energy needed to dissolve the salt and the energy lost by water is zero. A block diagram of the process helps clarify the relationships:

The solution in a cold pack contains NH_4^+ and NO_3^- ions as well as water, but the molar heat capacities of these ions contribute less than 1% percent to the heat capacity of the solution.

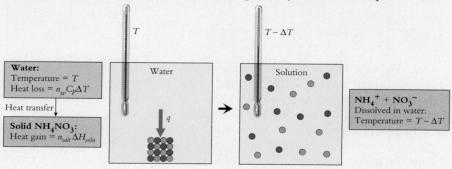

3. The data include the masses of water and ammonium nitrate (given in the problem) and the heat of solution of ammonium nitrate:

$$m\,(NH_4NO_3) = 75\text{ g},\ m\,(H_2O) = 250\text{ g},\ \Delta H_{soln}\,(NH_4NO_3) = 21.1\text{ kJ/mol}$$

4. The total amount of heat absorbed by the salt is the number of moles of salt multiplied by the molar heat of solution:

$$q_{salt} = n_{salt}\Delta H_{soln}$$

This heat comes from the water, so the water cools as it gives up heat:

$$q_{water} = n_{water}C_{water}\Delta T$$

Energy is conserved, so the sum of these two heats is zero:

$$q_{water} + q_{salt} = 0$$

5. Combining the three equations for q gives a single equation:

$$n_{salt}\Delta H_{soln} + n_{water}C_{water}\Delta T = 0$$

This equation can be solved to give an expression for ΔT:

$$\Delta T = \frac{-n_{salt}\Delta H_{soln}}{n_{water}C_{water}}$$

| **Example 12-12** | **Enthalpies of Solution** (*continued*) |

6. In addition to the data listed in step 3 we need the heat capacity of water, which is listed in Table 12-1:

$$C_{water} = 75.3 \text{ J/mol K}$$

We also need to convert masses to moles.

$$n_{salt} = \frac{m}{MM} = \frac{75 \text{ g}}{80.05 \text{ g/mol}} = 0.937 \text{ mol NH}_4\text{NO}_3$$

$$n_{water} = \frac{m}{MM} = \frac{250 \text{ g}}{18.02 \text{ g/mol}} = 13.87 \text{ mol H}_2\text{O}$$

$$\Delta T_{water} = \frac{-n_{salt}\Delta H_{soln}}{n_{water}C_{water}} = \frac{-(0.937 \text{ mol})(21.1 \text{ kJ/mol})(10^3 \text{ J/kJ})}{(13.87 \text{ mol})(75.3 \text{ J/mol °C})}$$

$$\Delta T_{water} = -19 \text{ °C}$$

7. The result has only 2 significant figures, matching the data (75 g). If the cold pack is initially at room temperature (around 25 °C), its temperature will fall to about 6 °C. This value seems reasonable.

Complete solvation of ions involves from two to eight water molecules per ion. In highly concentrated aqueous solutions, there are not enough water molecules to fully solvate all the ions. Because of this, the addition of water to a concentrated solution may increase ion–solvent interactions. As a result, when a concentrated solution is diluted by adding water, energy is released and the temperature increases. The molar energy change resulting from the dilution is called the **heat of dilution.**

Aqueous strong acids often have large heats of dilution, because proton transfer from a strong acid to water is highly exothermic. This can lead to problems when a concentrated acid is diluted with water. When water contacts the acid solution, the dilution process releases a large amount of energy. This energy heats the solution, and the heating can occur rapidly enough to form a hot spot whose temperature is greater than the boiling temperature of water. If this occurs, droplets of acid will splatter from the solution. Hot acid is a caustic material that rapidly burns the skin and attacks many other materials, including clothing. Sulfuric acid is particularly likely to splatter because H_2SO_4 has a high viscosity and does not mix readily with added water. Therefore acids should always be diluted by slowly adding the concentrated acid to a larger volume of water. In this way the transformation of concentrated acid into a more dilute solution occurs relatively slowly, and no local hot spots of high temperature can develop.

Phase Changes

Phase changes are accompanied by enthalpy changes, even though a phase change involves neither a chemical reaction nor a temperature change. Consider a teakettle filled with water and placed on a hot stove. The burner supplies energy that heats the water until the temperature reaches the boiling point of 100 °C. As Figure 12-18 shows, the temperature then stops rising even though heat is still being transferred to the water. The added energy now causes molecules of water to move from the liquid phase into the gas phase, and the temperature remains constant as the water boils away. Steam escapes into the atmosphere as the stove (surroundings) transfers heat to the water molecules (system).

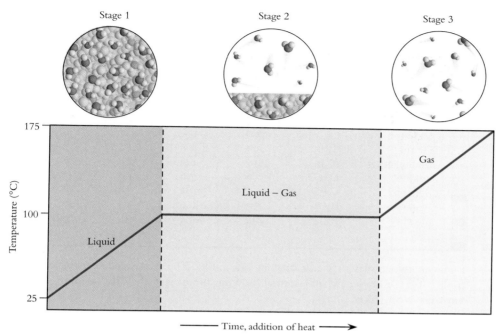

Stage 1   Stage 2   Stage 3

Figure 12-18
When heat is supplied at a constant rate to a sample of liquid water, the temperature rises until it reaches 100 °C (*stage 1*). The system then remains at that temperature as water molecules move from the liquid to the gas (*stage 2*). Once all the liquid has evaporated, the temperature can again rise (*stage 3*).

The molecular perspective reveals why energy must be supplied to boil water. Liquid water is held together by hydrogen bonds between water molecules (see Chapter 10). These forces stabilize liquid water by about 40 kJ/mol. A molecule cannot escape the liquid phase unless it has enough energy of motion to overcome these attractive forces. Heat must be supplied to increase the molecular energy of motion.

Normally, phase changes occur at constant pressure, so the heat needed to cause a phase change is equal to an enthalpy change: $q_p = \Delta H$. Thus a change of phase is always accompanied by a change in enthalpy. The magnitude of this enthalpy change depends on the strength of intermolecular forces in the substance undergoing the phase change. For example, the many intermolecular hydrogen bonds among water molecules, shown in Figure 12-19, generate large intermolecular forces that hold water molecules in the liquid phase. For an organic liquid such as acetone, the intermolecular dipole–dipole forces are significantly smaller. Thus it requires less heat to vaporize one mole of acetone than it does to vaporize one mole of water.

The amount of heat required to vaporize a substance also depends on the size of the sample. Twice as much energy is required to overcome all the intermolecular

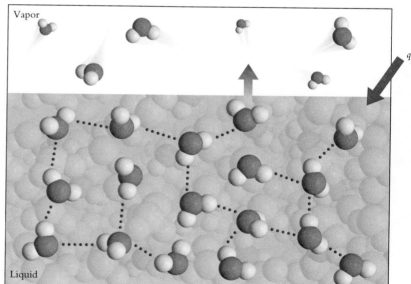

Figure 12-19
In order for a molecule of water to leave the liquid phase and enter the gas phase (*blue arrow*), it must absorb energy (*red arrow*) to overcome intermolecular hydrogen bonding in the liquid phase (*dotted lines*).

Table 12-2
Molar Heats of Phase Change

Substance	Formula	T_{fus} (K)	ΔH_{fus} (kJ/mol)	T_{vap} (K)	ΔH_{vap} (kJ/mol)
Argon	Ar	83	1.3	87	6.3
Oxygen	O_2	54	0.45	90	9.8
Methane	CH_4	90	0.84	112	9.2
Ethane	C_2H_6	90	2.85	184	15.5
Diethyl ether	$(C_2H_5)_2O$	157	6.90	308	26.0
Ethanol	C_2H_5OH	156	7.61	351	39.3
Benzene	C_6H_6	278.5	10.9	353	31.0
Water	H_2O	273	6.01	373	40.79
Mercury	Hg	234	23.4	630	59.0

forces in two moles of water than in one mole. The heat needed to vaporize one mole of a substance at its normal boiling point is called the **molar heat of vaporization, (ΔH_{vap}).**

Energy must also be provided to melt a solid substance. This energy is used to overcome the intermolecular forces that hold molecules or ions in fixed positions in the solid phase. Thus the melting of a solid also has a characteristic enthalpy change. The heat needed to melt one mole of a substance at its normal melting point is called the **molar heat of fusion, (ΔH_{fus}).**

Table 12-2 lists values of ΔH_{fus} and ΔH_{vap} for different chemical substances, and Example 12-13 provides practice in using these quantities.

> Vaporization is designated by the subscript *vap*, fusion is designated by the subscript *fus*, and sublimation is designated by the subscript *subl*.

Example 12-13	**Heat of Phase Change**

A swimmer emerging from a pool is covered with a film containing about 75 g of water. How much heat must be supplied to evaporate this water?

Strategy: Energy in the form of heat is required to evaporate the water from the swimmer's skin. The energy needed to vaporize the water can be found using the molar heat of vaporization and the number of moles of water.

Solution: The process can be shown with a simple block diagram:

$$75 \text{ g } H_2O \text{ liquid} \xrightarrow{n_{H_2O}\Delta H_{vap}} 75 \text{ g } H_2O \text{ vapor}$$

The ΔH_{vap} of water is 40.79 kJ/mol (see Table 12-2). The molar mass of water is 18.02 g/mol, so 75 g of water is 4.16 mol. Therefore the heat that must be supplied is:

$$q_p = n\,\Delta H_{vap} = (4.16 \text{ mol})(40.79 \text{ kJ/mol}) = 1.7 \times 10^2 \text{ kJ}$$

If the swimmer's body must supply all this heat, a substantial chilling effect occurs. Thus swimmers usually towel off (to reduce the amount of water that must be evaporated) or lie in the sun (to let the sun provide most of the heat required).

Fusion and vaporization are the most familiar phase changes, but **sublimation** is also common. Sublimation is a phase change in which a solid converts directly to a vapor without passing through the liquid phase. Dry ice (solid CO_2) sublimes at 195 K with $\Delta H_{subl} = 25.2$ kJ/mol. Mothballs contain naphthalene ($C_{10}H_8$, $\Delta H_{subl} = 73$ kJ/mol), a crystalline white solid that slowly sublimes to produce vapor whose odor repels moths. The purple color of the gas above iodine crystals in a closed container provides visible evidence that this solid also sublimes at room temperature.

Phase changes can go in either direction. When steam is cooled, it condenses; when liquid water is cooled, it freezes. Each of these is *exothermic* because each is the reverse of an endothermic phase change. That is, heat is released as a gas condenses to a liquid and as a liquid freezes to a solid. To make ice cubes, for instance, water is placed in the freezer compartment of a refrigerator, where the refrigerator absorbs the heat that is released as ice forms. A phase change that is exothermic has a negative enthalpy change:

The purple vapor above the solid iodine in this flask is due to I_2 molecules in the gas phase.

Much of the awesome energy released in a thunderstorm comes from the exothermic phase change of water vapor condensing to liquid.

$$\Delta H_{solidification} = -\Delta H_{fus} \quad and \quad \Delta H_{condensation} = -\Delta H_{vap}$$

By convention, tabulated values of heats of phase changes are always specified in the endothermic direction:

$$
\begin{array}{lll}
\text{Solid} & \longrightarrow \text{Liquid} & \Delta H_{fus} \\
\text{Liquid} & \longrightarrow \text{Vapor} & \Delta H_{vap} \\
\text{Solid} & \longrightarrow \text{Vapor} & \Delta H_{subl}
\end{array}
$$

A solid must absorb heat to melt. This makes ice a good substance to use for cooling. As ice melts, it absorbs heat from its surroundings, lowering the temperature. Because energy is conserved, the amount of heat given up by the surroundings is equal to the amount of heat absorbed by melting ice. This concept is illustrated by Example 12-14.

Phase and Temperature Changes	Example 12-14

A thirsty marathon runner pours 225 mL of Gatorade into a cup at 27 °C. What is the minimum mass of ice at 0 °C that must be added to cool the drink to 0 °C?

Strategy: The seven-step strategy can be used, with emphasis on steps 2 and 4. We need to visualize this problem correctly in order to identify and apply the equations of thermodynamics.

Solving Quantitative Problems

Solution:

1. We are asked to find the mass of ice needed to cool a drink.

2. This problem involves an energy balance. The Gatorade must be cooled by removing energy, so ice must melt to absorb that energy. The heat released

Example 12-14 | **Phase and Temperature Changes** *(continued)*

in cooling the Gatorade must equal the heat absorbed to melt the amount of ice. A block diagram helps visualize the computational path:

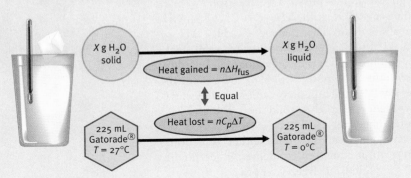

3. Cooling involves heat capacities, and melting involves heats of fusion. Thus we list these quantities along with the other data given in the problem. The volume of Gatorade is given: 225 mL. Assuming that Gatorade is essentially water (a reasonable thermodynamic assumption), we use the heat capacity of water from Table 12-1: $C = 75.29$ J/mol K

The temperature change can be determined from the data given:

$$\Delta T = (0\ °C - 27\ °C) = -27\ °C = -27\ K$$

The molar heat of fusion of ice (see Table 12-2) is 6.01 kJ/mol.

4. Two equations apply to this process, one for cooling and the other for melting:

$$\text{Cooling: } \Delta H_{\text{Gatorade}} = q_p = nC\Delta T$$

$$\text{Melting: } q_{\text{ice}} = \Delta H_{\text{ice}} = n_{\text{ice}}\,\Delta H_{\text{fus}}$$

5. We need to know the number of moles of Gatorade. Using the density and molar mass of water, the volume of Gatorade can be converted to moles:

$$n_{\text{Gatorade}} = \frac{(225\ \text{mL})(1.00\ \text{g/mL})}{(18.02\ \text{g/mol})} = 12.5\ \text{mol}$$

6. Now calculate the heat lost by the Gatorade:

$$q_{\text{Gatorade}} = (12.5\ \text{mol})(75.29\ \text{J/mol K})(-27\ \text{K})(10^{-3}\ \text{kJ/J}) = -25.4\ \text{kJ}$$

Energy is conserved, so 25.4 kJ of heat must be absorbed by the ice:

$$q_{\text{Gatorade}} + q_{\text{ice}} = 0 \quad so \quad q_{\text{ice}} = -q_{\text{Gatorade}} = 25.4\ \text{kJ}$$

The heat absorbed causes some of the ice to melt:

$$q_{\text{ice}} = \Delta H_{\text{ice}} = n_{\text{ice}}\,\Delta H_{\text{fus}} = 25.4\ \text{kJ} \quad so \quad n_{\text{ice}} = \frac{25.4\ \text{kJ}}{6.01\ \text{kJ/mol}} = 4.22\ \text{mol}$$

The problem asked for the mass of ice, so we finish with a mole-mass conversion:

$$m_{\text{ice}} = (n)(MM) = (4.22\ \text{mol})(18.02\ \text{g/mol}) = 76\ \text{g ice required}$$

7. Does that seem like a reasonable result to you?

The graph in Figure 12-18 shows that adding heat to boiling water does not cause the temperature of the water to increase. Instead, the added energy is used to overcome intermolecular attractions as molecules leave the liquid phase and enter the gas phase. Other two-phase systems, such as an ice-water mixture, show similar behavior. Heat added to an ice-water mixture melts some of the ice, but the mixture remains at 0 °C. Similarly, when an ice-water mixture in a freezer loses heat to the surroundings, the energy comes from some liquid water freezing, but the mixture remains at 0 °C until all the water has frozen. This behavior can be used to hold a chemical system at a fixed temperature. A temperature of 100 °C can be maintained by a boiling water bath, and an ice bath holds a system at 0 °C. Lower temperatures can be achieved with other substances. Dry ice maintains a temperature of −78 °C; a bath of liquid nitrogen has a constant temperature of −196 °C (77 K); and liquid helium, which boils at 4.2 K, is used for research requiring ultracold temperatures.

Energies of Phase Transfer

Phase changes normally occur at constant pressure, so the heat absorbed or released is equal to the enthalpy change, $q_p = \Delta H$. The volume change that accompanies conversion between solid and liquid is very small, so the amount of work done is negligible. Thus for a phase transition between the solid and liquid state, the energy change is almost identical to the enthalpy change, $\Delta E \cong \Delta H$. However, if the phase change involves a gas, the process includes a change in volume, so work cannot be neglected. This results in a significant difference between ΔH and ΔE.

The change in volume for a transformation from a condensed phase to the gas phase is almost equal to the volume of the resulting gas. Boiling water provides an example. According to the ideal gas equation, one mole of water vapor at 373 K has a volume of 30.6 L:

$$\frac{V}{n} = \frac{RT}{P} = \frac{(0.08206 \text{ L atm mol}^{-1}\text{K}^{-1})(373 \text{ K})}{(1.00 \text{ atm})} = 30.6 \text{ L/mol}$$

In contrast, the same amount of liquid water at 373 K has a volume of only 18 mL, or 0.018 L. This small volume can be neglected compared with the larger volume of the gas.

Remember that at constant pressure, ΔH and ΔE are related by $P\Delta V$. Thus for any vaporization process:

$$\Delta H = \Delta E + P\Delta V \cong \Delta E + PV_{gas} = \Delta E + nRT_{vap}$$

This leads to a general equation that relates the energy of vaporization or sublimation to the corresponding molar enthalpy:

$$\Delta E_{vap} = \Delta H_{vap} - RT_{vap}$$

> For thermodynamic calculations, we express R in units that include an energy:
> $R = 8.314 \text{ J mol}^{-1}\text{K}^{-1}$

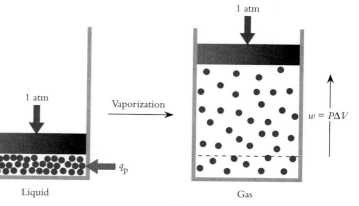

Figure 12-20
This schematic view of vaporization illustrates that work accompanies vaporization at constant pressure. As liquid molecules escape to the gas phase, they exert pressure on the piston, moving it back against a constant external pressure exerted by the atmosphere.

Vaporization and sublimation energies are typically about 10% less than their corresponding enthalpies. Using water as an example, $\Delta H_{vap} = 40.79$ kJ/mol and $RT_{vap} = 3.10$ kJ/mol, giving $\Delta E_{vap} = 37.69$ kJ/mol. This difference arises because vaporization at constant pressure accomplishes work against the surroundings in addition to overcoming intermolecular forces in the condensed phase. The escaping vapor must do some work by expanding against the constant external pressure of the atmosphere. Figure 12-20 shows this concept in schematic fashion.

Table 12-3
Summary of Enthalpy Changes

	Symbol	Example of Process
Heat of formation	ΔH_f°	$H_2(g) + \frac{1}{2} O_2(g) \longrightarrow H_2O(l)$
Heat of reaction	$\Delta H_{reaction}$	$CO(g) + H_2O(l) \longrightarrow CO_2(g) + H_2(g)$
Heat of solution	ΔH_{soln}	$NaCl(s) \xrightarrow{H_2O} Na^+(aq) + Cl^-(aq)$
Heat of vaporization	ΔH_{vap}	$H_2O(l) \longrightarrow H_2O(g)$
Heat of fusion	ΔH_{fus}	$H_2O(s) \longrightarrow H_2O(l)$
Heat of sublimation	ΔH_{subl}	$CO_2(s) \longrightarrow CO_2(g)$

Sections 12.4 and 12.5 describe enthalpy changes associated with several types of processes. To conclude this chapter, Table 12-3 provides a summary list of these processes.

Section Exercises

12.5.1 The heat of fusion of sodium chloride is 27.2 kJ/mol, but its heat of solution in water is only 3 kJ/mol. Use intermolecular forces to explain the difference between these two molar enthalpies.

12.5.2 To make iced tea, 100.0 g of ice at 273 K is added to 500.0 mL of tea at 298 K in an insulated glass. What is the temperature after all the ice melts? Assume that the heat capacity of tea is the same as that of water.

12.5.3 If dry ice is heated under a pressure greater than 10 atm, it melts instead of subliming. If the pressure is then reduced to 1 atm, the liquid boils to give gaseous CO_2. Is the enthalpy change for this two–step process larger than, smaller than, or the same as the enthalpy of sublimation for CO_2? Give reasoning to support your answer.

CHAPTER REVIEW

Summary and Key Terms

1. Thermodynamics examines the energy transfers across a **boundary** between a **system** and its **surroundings**. A system is characterized by its **state**, and described by **state variables**. When one or more variables changes, the system has undergone a **change of state**. State changes can be physical, chemical, or both. A **state function** is one whose value is independent of the path by which a change of state occurs, but a **path function** depends on the path. State functions are particularly important in thermodynamics.

2. **Energy (E)** plays a central role in chemistry. Chemical energy is stored in chemical bonds and is released or absorbed during chemical reactions. Chemical processes release stored energy as **heat (q)**, **work (w)**, or both. These quantities are linked through the **first law of thermodynamics**, $\Delta E = q + w$. When a substance gains or loses heat, its temperature changes by an amount that depends on its **molar heat capacity (C)**. Expansion work accompanies many chemical processes.

3. A heat-releasing process is **exothermic**, and a heat-absorbing process is **endothermic**. Heat flows are conveniently measured using **calorimeters**, which generally operate at either constant volume or constant pressure. The heat transferred in a constant-volume calorimeter equals the change of energy: $\Delta E = q_v$.

4. In many everyday chemical processes, pressure is constant, but volume is not. For these situations, **enthalpy** is a useful function. Enthalpy is defined as $H = E + PV$. At constant pressure, the change in enthalpy can be found directly from the heat flow: $\Delta H = q_p$. For most chemical reactions, $\Delta E \cong \Delta H$. In a **formation reaction**, a pure substance forms from elements in their

standard states. The enthalpy change for such a reaction is the **standard enthalpy of formation (ΔH_f°)**. According to **Hess' law,** the enthalpy change for any reaction can be calculated from ΔH_f° values.

5. The energy flow accompanying the dissolving of a substance is its **molar heat of solution (ΔH_{soln})**, and the energy flow accompanying a change in concentration of a solution is the **heat of dilution.** Changes of phase such as melting, vaporization, and solution formation usually occur at constant pressure and can be characterized by their enthalpy changes. The **molar heat of fusion, molar heat of vaporization,** and **molar heat of sublimation** are always positive.

Skills to Master

▶ Defining system and surroundings

▶ Calculating q and w

▶ Working with standard enthalpies of formation

▶ Relating energy and enthalpy changes

▶ Making calorimetric calculations

▶ Calculating enthalpies of phase changes and solution

Learning Exercises

12.1 List all the properties that may change as a chemical reaction occurs in a closed system. Draw a "before-and-after" block diagram that schematically represents such a process.

12.2 Write a paragraph that explains the relationships among energy, heat, work, and temperature.

12.3 Describe in your own words what enthalpy is and why it is preferred to energy in describing the thermodynamics of many chemical processes.

12.4 Energy accounting on open systems is more difficult than on closed systems. Explain the feature of an open system that creates difficulties, and describe how you would set up an energy balance for an open system.

12.5 Update your list of memory bank equations. Write a sentence that describes the restrictions on each equation. (For example, $H = E + PV$ is a definition that has no restrictions.)

12.6 Prepare a list of the terms in Chapter 12 that are new to you. Write a one-sentence definition for each, using your own words. If you need help, consult the glossary.

Problems   ilw = interactive learning ware problem. Visit the website at www.wiley.com/college/olmsted

Thermodynamic Definitions

12.1 Is each of the following an open system, a closed system, or an (almost) isolated system? (a) human being; (b) coffee in a thermos; (c) ice-cube tray filled with water; and (d) helium-filled balloon.

12.2 Categorize each of the following as open, closed, or (almost) isolated systems from the thermodynamic perspective. Explain your reasoning in each case: (a) can of tomato soup; (b) freezer chest full of ice; (c) the Earth; and (d) a satellite in orbit.

12.3 Which of the following properties of a sodium chloride solution are state variables? (a) temperature at which it was prepared; (b) its current temperature; (c) mass of NaCl used in its preparation; and (d) time when it was prepared.

12.4 Which of the following are state functions? (a) height of a mountain; (b) distance traveled in climbing that mountain; (c) energy consumed in climbing the mountain; and (d) gravitational potential energy of a climber on top of the mountain.

12.5 Consider driving an automobile. Define as many systems as you can that might be used to discuss thermodynamic properties involving your car and the automobile.

12.6 Consider the human body. Define as many systems as you can that might be used to discuss the thermodynamics of human life.

Energy Changes

12.7 An electric heater is used to supply 25.0 joules of energy to each of the following samples. Compute the final temperature in each case: (a) 10.0-g block of Al originally at 15.0 °C; (b) 25.0-g block of Al originally at 295 K; (c) 25.0-g block of Ag originally at 295 K; and (d) 25.0-g sample of H_2O originally at 22.0 °C.

12.8 Each of the following is placed in an ice bath until it has lost 65.0 J of energy. Compute the final temperature in each case: (a) 35.0-g block of Al originally at 65.0 °C; (b) 50.0-g block of Al originally at 65.0 °C; (c) 50.0-g block of Ag originally at 65.0 °C; and (d) 50.0-g sample of H_2O originally at 325.0 K.

12.9 An iron kettle weighing 1.35 kg contains 2.75 kg of water at 23.0 °C. The kettle and water are heated to 95.0 °C. How many joules of energy are absorbed by the water and by the kettle?

12.10 A piece of silver whose mass is 15.0 g is immersed in 25.0 g of water. This system is heated electrically from 24.0 °C to 37.6 °C. How many joules of energy are absorbed by the silver and how many by the water?

ilw **12.11** A silver coin weighing 27.4 g is heated to 100.0 °C in boiling water. It is then dropped into 37.5 g of water initially at 20.5 °C. Find the final temperature of coin + water.

12.12 A stainless steel spoon weighs 24.7 g and is at a temperature of 18.5 °C. It is immersed in 85.0 mL of hot coffee ($T = 84.0$ °C) in a thermos flask. What is the final temperature of spoon + coffee? (Assume that the heat capacity of the spoon is the same as that of pure Fe and that the heat capacity of the coffee is the same as that of pure water.)

12.13 A pot containing 475 mL of water at 21.5 °C is heated on a stove until its temperature is 87.6 °C. What is q for the water?

12.14 A dish containing 145 g of water at 54.0 °C is put in a refrigerator to cool. It is removed when its temperature is 5.50 °C. What is q for the water?

12.15 How much work is done in blowing up a balloon from zero volume to a volume of 2.5 L, assuming that $P = 1.00$ atm and no work is required to stretch the rubber? (In reality, the work that goes into stretching the rubber is substantial.)

12.16 A typical hot-air balloon has a volume of 19.5 m^3 when inflated. How much work must be done to inflate such a balloon when atmospheric pressure is 755 torr?

12.17 Refer to the data in the Chemistry and Life Box and calculate how far a person weighing 55 kg must walk at a rate of 6.0 km/hr to consume the additional energy contained in 250 g of ground beef relative to 250 g of broiled chicken.

12.18 Refer to the data in the Chemistry and Life Box and calculate how far a person weighing 85 kg must run at 16 km/hr to consume the energy contained in 1 lb (0.455 kg) of sugar.

Heat Measurements: Calorimetry

12.19 Constant-volume calorimeters are sometimes calibrated by running a combustion reaction of known ΔE and measuring the change in temperature. For example, the combustion energy of glucose is 15.57 kJ/g. When a 1.7500-g sample of glucose burns in a constant-volume calorimeter, the calorimeter temperature increases from 21.45 °C to 23.34 °C. Find the total heat capacity of the calorimeter.

12.20 Constant-pressure calorimeters can be calibrated by electrical heating. When a calorimeter containing 125 mL of water is supplied with 1150. J of electrical energy, its temperature rises from 23.45 °C to 25.25 °C. What is the total heat capacity of the calorimeter, and what percentage of this is due to the water?

12.21 When 1.350 g of benzoic acid ($C_7H_6O_2$) burns completely in excess O_2 gas at constant volume and 298 K, it releases 35.61 kJ of energy. (a) What is the balanced chemical equation for this reaction? (b) What is the molar energy of combustion of benzoic acid? (c) How much energy is released per mole of O_2 consumed?

12.22 Acetylene (C_2H_2) is used in welding torches because it has a high heat of combustion. When 1.00 g of acetylene burns completely in excess O_2 gas at constant volume, it releases 48.2 kJ of energy. (a) What is the balanced chemical equation for this reaction? (b) What is the molar energy of combustion of acetylene? (c) How much energy is released per mole of O_2 consumed?

ilw **12.23** A 1.35-g sample of caffeine ($C_8H_{10}N_4O_2$) is burned in a constant-volume calorimeter that has a heat capacity of 7.85 kJ/K. The temperature increases from 297.65 K to 302.04 K. Determine the amount of heat released and the molar energy of combustion of caffeine.

12.24 An electric heater is used to add 19.75 kJ of heat to a constant-volume calorimeter. The temperature of the calorimeter increases by 4.22 °C. When 1.75 g of methanol is burned in the same calorimeter, the temperature increases by 8.47 °C. Calculate the molar energy of combustion of methanol.

Enthalpy

12.25 Determine the standard enthalpy change for each of the following reactions:

    (a) $C_2H_4(g) + 3\ O_2(g) \longrightarrow 2\ CO_2(g) + 2\ H_2O(l)$
    (b) $2\ NH_3(g) \longrightarrow N_2(g) + 3\ H_2(g)$
    (c) $5\ PbO_2(s) + 4\ P(s,\ white) \longrightarrow P_4O_{10}(s) + 5\ Pb(s)$
    (d) $SiCl_4(l) + 2\ H_2O(l) \longrightarrow SiO_2(s) + 4\ HCl(g)$

12.26 Determine the standard enthalpy change for each of the following reactions:

    (a) $2\ Al(s) + 3\ Cl_2(g) \longrightarrow 2\ AlCl_3(s)$
    (b) $3\ NO_2(g) + H_2O(l) \longrightarrow 2\ HNO_3(g) + NO(g)$
    (c) $2\ C_2H_2(g) + 5\ O_2(g) \longrightarrow 4\ CO_2(g) + 2\ H_2O(l)$

12.27 Find $\Delta E^\circ_{reaction}$ for each of the reactions in Problem 12.25.

12.28 Find $\Delta E^\circ_{reaction}$ for each of the reactions in Problem 12.26.

12.29 Write a balanced equation for the formation reaction of each of the following substances: (a) $K_3PO_4(s)$; (b) acetic acid, $CH_3CO_2H(l)$; (c) trimethylamine, $(CH_3)_3N(g)$; and (d) bauxite, $Al_2O_3(s)$.

12.30 Write the balanced equation for the formation reaction of each of the following substances: (a) butanol, $C_4H_9OH(l)$; (b) sodium carbonate, $Na_2CO_3(s)$; (c) ozone, $O_3(g)$; and (d) rust, $Fe_3O_4(s)$.

ilw **12.31** Using standard heats of formation, determine the heats of the following reactions:

    (a) $4\ NH_3(g) + 5\ O_2(g) \longrightarrow 4\ NO(g) + 6\ H_2O(l)$
    (b) $4\ NH_3(g) + 3\ O_2(g) \longrightarrow 2\ N_2(g) + 6\ H_2O(l)$

12.32 Using standard heats of formation, determine the heats of the following reactions:

    (a) $Fe_2O_3(s) + 3\ H_2O(l) \longrightarrow 2\ Fe(OH)_3(s)$
    (b) $B_2O_3(s) + 3\ H_2O(l) \longrightarrow 2\ H_3BO_3(s)$

Enthalpy and Intermolecular Forces

ilw **12.33** When a 4.75-g sample of $CaCl_2$ dissolves in 110.0 g of water in a coffee-cup calorimeter, the temperature rises from 22.0 °C to 29.7 °C. Calculate the molar heat of solution of $CaCl_2$.

12.34 When a 1.00-g sample of $KClO_3$ dissolves in 50.0 g of water in a coffee-cup calorimeter, the temperature drops from 298.00 K to 296.36 K. Calculate the molar heat of solution of $KClO_3$.

12.35 For the following reaction:

$$Mg^{2+}(aq) + SO_4^{2-}(aq) \longrightarrow MgSO_4(s) \quad \Delta H = 91.3\ \text{kJ/mol}$$

(a) When $MgSO_4$ dissolves in water, is heat absorbed or released by the water? (b) Calculate q_{water} for 2.55 g of solid $MgSO_4$ dissolving in 5.00×10^2 mL of water. (c) Calculate ΔT for the process in (b), assuming that the container is completely insulated.

12.36 For the following reaction:

$$NH_4^+(aq) + NO_3^-(aq) \longrightarrow NH_4NO_3(s) \quad \Delta H = -21.1\ \text{kJ/mol}$$

(a) When NH_4NO_3 dissolves in water, does the water absorb or release heat? (b) Calculate q_{water} for 25.0 g of solid NH_4NO_3 dissolving in 2.50×10^2 mL of water. (c) Calculate ΔT for the process in (b), assuming that the container is completely insulated and the ions contribute negligibly.

12.37 Explain, based on interionic and intermolecular forces, why the reaction in Problem 12.35 has a positive enthalpy change.

12.38 Explain, based on interionic and intermolecular forces, why the reaction in Problem 12.36 has a negative enthalpy change.

12.39 Referring to Table 12-2, explain the following in terms of intermolecular forces: (a) Methane has a lower heat of vaporization than ethane. (b) Ethanol has a significantly higher heat of vaporization than diethyl ether. (c) Argon has a higher heat of fusion than methane.

12.40 Referring to Table 12-2, explain the following in terms of intermolecular forces: (a) Water has a higher heat of vaporization than methane. (b) Benzene has a higher heat of fusion than ethane. (c) Oxygen has a higher heat of vaporization than argon.

Additional Paired Problems

12.41 Does more heat have to be removed from an automobile engine when it burns one gram of gasoline while idling in a traffic jam or when it burns one gram of gasoline while accelerating? Explain in terms of ΔE, q, and w.

12.42 One way to vaporize a liquid is to inject a droplet into a high vacuum. If this is done, is the heat absorbed by the droplet equal to ΔE or ΔH for the phase change? Explain in terms of q and w.

12.43 The amount of heat produced in an "ice calorimeter" is determined from the quantity of ice that melts. Suppose that a 12.7-g copper block at 200.0 °C is dropped into an ice calorimeter. How many grams of ice will melt?

12.44 A gold coin ($C(Au) = 25.4$ J/mol K) whose mass is 7.65 g is heated to 100.0 °C in a boiling water bath and then quickly dropped into an ice calorimeter. What mass of ice melts?

12.45 The human body "burns" glucose ($C_6H_{12}O_6$) for energy. Burning 1.00 g of glucose produces 15.7 kJ of heat. (a) Write the balanced equation for the combustion (burning) of glucose. (b) Determine the molar heat of combustion of glucose. (c) Using appropriate thermodynamic data, determine the heat of formation of glucose.

12.46 Solid urea, $(NH_2)_2CO$, burns to give CO_2, N_2, and liquid H_2O. Its heat of combustion is -632.2 kJ/mol. (a) Write the balanced combustion equation. (b) Calculate the heat generated per mole of H_2O formed. (c) Using this heat of combustion and the appropriate thermodynamic data, determine the heat of formation of urea.

12.47 Explain what will happen to any living organism if we try to make it an isolated thermodynamic system.

12.48 What energy transformations take place when a moving automobile brakes and skids to a stop?

12.49 It takes 100.0 J of heat to raise the temperature of 52.5 g of Pb from 280.0 K to 299.6 K. What is the molar heat capacity of Pb?

12.50 A piece of rhodium metal whose mass is 4.35 g is heated to 100 °C and then dropped into an ice calorimeter. When the Rh metal has cooled to 0 °C, 0.316 g of ice has melted. What is the molar heat capacity of rhodium?

12.51 An ideal gas is initially at $T = 300$. K, $P = 20.0$ atm pressure, and $V = 30.0$ L. The gas is compressed at constant temperature by a constant external pressure of 4.00 atm until its volume is 20.0 L. Compute q, w, and ΔE for the gas.

12.52 A helium-filled balloon is at 325 K, contains 0.197 g He, and has a volume of 1.31 L. It is placed in a freezer ($T = 255$ K), and its volume decreases to 1.01 L. Find ΔE for the gas. (C of He = 20.8 J/mol K.)

12.53 Use standard enthalpies of formation to determine $\Delta H_{reaction}$ for the following reactions:

(a) $2\ SO_2(g) + O_2(g) \longrightarrow 2\ SO_3(g)$
(b) $2\ NO_2(g) \longrightarrow N_2O_4(g)$
(c) $Fe_2O_3(s) + 2\ Al(s) \longrightarrow Al_2O_3(s) + 2\ Fe(s)$

12.54 Using standard enthalpies of formation, calculate $\Delta H_{reaction}$ for the following reactions (all reagents are gases):

(a) $2\ NH_3 + 3\ O_2 + 2\ CH_4 \longrightarrow 2\ HCN + 6\ H_2O$
(b) $2\ C_2H_2 + 5\ O_2 \longrightarrow 4\ CO_2 + 2\ H_2O$
(c) $C_2H_4 + O_3 \longrightarrow CH_3CHO + O_2$

12.55 "Strike anywhere" matches contain P_4S_3, a compound that ignites when heated by friction. It reacts vigorously with oxygen, as follows:

$$P_4S_3(s) + 8\ O_2(g) \longrightarrow P_4O_{10}(s) + 3\ SO_2(g)\ \Delta H^\circ_{reaction} = -3677\ kJ$$

Use data from Appendix D to determine ΔH°_f for $P_4S_3(s)$.

12.56 When a corpse decomposes, much of the phosphorus in the body is converted to phosphine, PH_3, a colorless gas with the odor of rotting fish. Phosphine is a highly reactive molecule that ignites spontaneously in air. In the graveyard, phosphine that escapes from the ground ignites in air, giving small flashes of flame. These flashes are sometimes attributed to supernatural causes, such as a will-o'-the-wisp. Determine ΔH° for the combustion of phosphine.

$$PH_3(g) + O_2(g) \longrightarrow P_4O_{10}(s) + H_2O(g) \quad \text{(unbalanced)}$$

12.57 A 44.0-g sample of an unknown metal at 100.0 °C is placed in a constant-pressure calorimeter containing 80.0 g of water at 24.8 °C. Assume the heat capacity of the calorimeter is negligible. The final temperature is 28.4 °C. Calculate the heat capacity of the metal and use the result to identify the metal: Al = 0.903 J/g K; Cr = 0.616 J/g K; Co = 0.421 J/g K; or Cu = 0.385 J/g K.

12.58 A coin dealer, offered a rare silver coin, suspected that it might be a counterfeit nickel copy. The dealer heated the coin, which weighed 15.5 g, to 100.0 °C in boiling water and then dropped the hot coin into 21.5 g of water at $T = 15.5$ °C in a coffee-cup calorimeter. The temperature of the water rose to 21.5 °C. Was the coin made of silver or nickel (C (Ni) = 26.1 J/mol K)?

12.59 For the constant-temperature process that follows, give the sign ($+$, $-$, or 0) for each of the specified thermodynamic functions. In each case give a brief account of your reasoning: (a) ΔH_{sys}; (b) ΔE_{surr}; and (c) ΔE_{univ}.

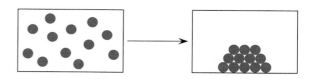

12.60 For the constant-temperature process that follows, give the sign ($+$, $-$, or 0) for each of the specified thermodynamic functions. In each case give a brief account of your reasoning: (a) w_{sys}; (b) q_{sys}; and (c) ΔE_{surr}.

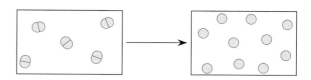

12.61 Using standard heats of formation, determine $\Delta H_{reaction}$ for the "reaction,"

$$H_2O(l, 298\ K) \longrightarrow H_2O(g, 298\ K)$$

Compare your result with ΔH_{vap} for H_2O. What differences in conditions account for the difference in values?

12.62 Chlorine trifluoride reacts readily with ammonia:

$$2\ ClF_3(g) + 2\ NH_3(g) \longrightarrow N_2(g) + 6\ HF(g) + Cl_2(g)$$

$$\Delta H_{reaction} = -1196\ kJ$$

Use standard heats of formation to determine ΔH°_f for $ClF_3(g)$.

12.63 In some liquid-fuel rockets, such as the lunar lander module of the Apollo moon missions, the fuels are liquid hydrazine (N_2H_4) and dinitrogen tetroxide gas (N_2O_4). The two chemicals ignite on contact to release very large amounts of energy:

$$2\ N_2H_4(l) + N_2O_4(g) \longrightarrow 3\ N_2(g) + 4\ H_2O(g)$$

(a) Calculate the enthalpy change that takes place when one mole of hydrazine is burned in a lunar lander. (b) If O_2 was used in the lander instead of N_2O_4, would the reaction give off more heat per mole of hydrazine or less heat per mole of hydrazine? Explain.

12.64 For spacecraft fuels, the energy content *per gram* of fuel should be as large as possible. Which of the following has the largest energy content per gram and which has the smallest?
(a) dimethylhydrazine, $(CH_3)_2NNH_2$, $\Delta H_{combustion} = -1694$ kJ/mol;
(b) methanol, CH_3OH, $\Delta H_{combustion} = -726$ kJ/mol; or
(c) octane, C_8H_{18}, $\Delta H_{combustion} = -5590$ kJ/mol.

12.65 For each of the reactions that follows, estimate the difference between $\Delta H_{reaction}$ and $\Delta E_{reaction}$:
(a) $Ni^{2+}(aq) + Cu(s) \longrightarrow Ni(s) + Cu^{2+}(aq)$
(b) $C(s, graphite) + H_2O(l) \longrightarrow CO(g) + H_2(g)$
(c) Combustion of liquid butanol (C_4H_9OH) in excess $O_2(g)$.

12.66 Determine ΔE_{vap} for argon, ethane, and mercury. Which has the largest percentage difference between ΔE_{vap} and ΔH_{vap}?

12.67 Write the balanced chemical equation associated with each of the following enthalpy changes: (a) heat of sublimation of I_2 (b) heat of formation of gaseous atomic iodine; (c) heat of formation of $C_2H_3Cl(g)$; and (d) heat of solution of sodium sulfate.

12.68 Write the balanced chemical equation associated with each of the following enthalpy changes: (a) heat of vaporization of Br_2; (b) heat of formation of gaseous atomic bromine; (c) heat of formation of $CH_3OH(g)$; and (d) heat of solution of magnesium chloride.

More Challenging Problems

12.69 Steam causes more severe burns than boiling water. To show why, calculate the following: (a) energy released when 2.50 g of boiling water is cooled to body temperature (37.5 °C) and (b) energy released when 2.50 g of steam at 100.0 °C is condensed and cooled to body temperature.

12.70 The five stable oxides of nitrogen are NO, NO_2, N_2O, N_2O_4, and N_2O_5. Balance each of the following oxidation reactions, and then use standard formation enthalpies to calculate the heat of reaction per mole of *atomic* nitrogen for each reaction:
(a) $N_2 + O_2 \longrightarrow 2\ NO$
(b) $N_2O + O_2 \longrightarrow NO$
(c) $NO + O_2 \longrightarrow NO_2$
(d) $NO_2 + O_2 \longrightarrow N_2O_5$

12.71 Gases are sold and shipped in metal tanks under high pressure. A typical tank of compressed air has a volume of 30.0 L and is pressurized to 15.0 atm at $T = 298$ K. What work had to be done in filling this tank? (Hint: What volume did the air occupy before it was compressed?)

12.72 A 70-kg person uses 220 kJ of energy to walk 1.0 km. This energy comes from "burning" glucose (see Problem 12.45), but only about 30.% of the heat of combustion of glucose can be used for propulsion. The rest is used for other bodily functions or is "wasted" as heat. Assuming that a sugar-coated breakfast cereal contains 35% sugar (which can be considered glucose) and no other energy source, calculate how many grams of cereal provide enough energy to walk 1.0 km.

12.73 A home swimming pool contains 155 m^3 of water. At the beginning of swimming season, the water must be heated from 20 °C to 30 °C. (a) How much heat energy must be supplied? (b) If this energy is supplied by a natural gas heater with an 80% heat transfer efficiency, how many grams of methane must be burned? The heat of combustion of methane is -803 kJ/mol.

12.74 Light is radiant energy (see Chapter 6). Solar water heaters use the radiant energy of sunlight. How many moles of photons from the middle of the solar spectrum (515 nm) would be required to heat 40.0 L of water from 25.0 °C to 50.0 °C, assuming 80% efficiency of heating?

12.75 One way to cool a hot beverage is with a cold spoon. A silver spoon weighing 99 g is placed in a Styrofoam cup containing 205 mL of hot coffee at 350 K. Find the final temperature of the coffee, assuming that the initial temperature of the spoon is 280 K and that coffee has the same heat capacity as water. Would an aluminum spoon of the same mass cool the coffee more or less effectively?

12.76 Calculate how many grams of methane must be burned to convert 2.50 kg of water at 25.0 °C into steam at 100.0 °C, assuming 100% efficiency of heat transfer.

12.77 A room in a home measures 3.0 m by 5.0 m by 4.0 m. Assuming no heat or material losses, how many grams of natural gas (methane, CH_4) must be burned to heat the air in this room from 15 °C to 25 °C? Assume that air is 78% N_2 and 22% O_2, and use data from Table 12-1 and Appendix D.

12.78 Only one of the following expressions describes the heat of a chemical reaction under all possible conditions. Which is it? For each of the others, give an example for which the expression gives the wrong value for the heat. (a) ΔE; (b) ΔH; (c) q_v; (d) q_p; and (e) $\Delta E - w$.

12.79 For the following constant-temperature process, give the sign $(+, -, \text{or } 0)$ for each of the specified thermodynamic functions. In each case, give a brief account of your reasoning: (a) w_{sys}; (b) ΔE_{surr}; and (c) q_{sys}.

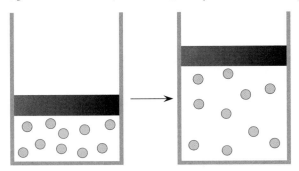

12.80 The following figure represents a piston and cylinder containing a collection of gas molecules. The piston can move in either direction.

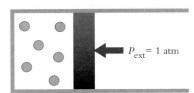

Assume the gas molecules are the system. (a) Redraw the figure to show what happens when some work is done on the system. (b) Redraw the figure to show what happens when the gas undergoes an exothermic chemical reaction.

12.81 The heat required to sustain animals that hibernate comes from the biochemical combustion of fatty acids, one of which is arachidonic acid ($\Delta H_f^\circ = -636$ kJ/mol):

Arachidonic acid ($C_{20}H_{32}O_2$)

$$C_{20}H_{32}O_2(s) + 27\ O_2(g) \longrightarrow 20\ CO_2(g) + 16\ H_2O(l)$$

Calculate the mass of arachidonic acid needed to warm a 500-kg bear from 5 to 25 °C. Assume that the average heat capacity of bear flesh is 4.18 J/g K.

12.82 In metric terms, a typical automobile averages 6.0 km/L of gasoline burned. Gasoline has a heat of combustion of 48 kJ/g and a density of 0.68 g/mL. How much energy is consumed in driving an automobile 1.0 km?

12.83 A 9.50-g copper block, initially at 200.0 °C, is dropped into a thermos flask containing 200 mL of water initially at 5.00 °C. What is the final temperature of the system?

12.84 Construct a graph similar to the one in Figure 12-18 that summarizes the energy changes that accompany the following process. A 75.0-mL sample of water at 35 °C is cooled until it freezes. Cooling continues until the temperature reaches −25 °C. Plot temperature along the y-axis and kJ of heat along the x-axis. Table 12-3 will be helpful, and C (ice) = 37.7 J/mol K.

12.85 Liquefied ammonia is used sometimes as a solvent for chemical reactions. At the end of a reaction, the liquid ammonia is allowed to evaporate in an exhaust hood. Calculate q, w, ΔH, and ΔE for the evaporation of 275 mL $NH_3(l)$ at a pressure of 1.00 atm using the following information about liquid ammonia: density = 0.81 g/mL, boiling point = −33 °C, $\Delta H_{vap} = 23.2$ kJ/mol.

Group Study Problems

12.86 For each of the following, make an appropriate choice for the system and define it completely. State whether the system absorbs or releases heat. (a) Water on your skin evaporates after you shower. (b) Methane burns in a Bunsen burner, heating a beaker of water. (c) Strong acid and strong base solutions are mixed in a thermos flask. The temperature of the resulting solution increases.

12.87 A student studying the properties of gaseous $C_2Cl_2F_2$, a chlorofluorocarbon refrigerant, cooled a 1.25-g sample at a constant atmospheric pressure of 1.00 atm from 50.0 to 20.0 °C. During the cooling, the sample volume decreased from 274 to 248 mL. Calculate ΔH and ΔE for the chlorofluorocarbon for this process. For $C_2Cl_2F_2$, $C = 80.7$ J/mol K.

12.88 According to Table 12-1, molar heat capacities of monatomic gases (He, Ar) are significantly smaller than those of diatomic gases (N_2, O_2, H_2). Explain in molecular terms why more heat must be supplied to raise the temperature of 1 mol of diatomic gas by 1 K than to raise the temperature of 1 mol of monatomic gas by 1 K.

12.89 Suppose 100.0 mL of 1.00 M HCl and 100.0 mL of 1.00 M NaOH, both initially at 25.0 °C, are mixed in a Thermos flask. When the reaction is complete, the temperature is 31.8 °C. Assuming that the solutions have the same heat capacity as pure water, compute the heat released. Use this value to evaluate the molar heat of the neutralization reaction:

$$H_3O^+(aq) + OH^-(aq) \longrightarrow 2\ H_2O(l)$$

12.90 Ethanol, CH_3CH_2OH, is used as a gasoline additive because it boosts octane ratings. Gasoline that contains ethanol is known as gasohol. Calculate the amount of energy released by burning one gallon of ethanol. The density of ethanol is 0.787 g/mL. Use the data in Appendix D and average bond energies from Table 9-2.

Answers to Section Exercises

12.1.1 (a) The 100 g of water; (b) the H_2-O_2 mixture; and (c) the air in the balloon

12.1.2 (a) V, T, and n; (b) P, V, T, and n; and (c) V and T

12.1.3 (a) State function change and (b) not a state function change

12.2.1 $q_{sys} = 4.2 \times 10^3$ J

12.2.2 $w_{sys} = -4.30 \times 10^3$ J

12.2.3 0.60 kg of ground beef or 3.8 kg of fruit

12.3.1 $q_p = +4.6 \times 10^2$ J

12.3.2 297 kJ/mol

12.3.3 (a) Temperature will rise. (b) Volume will decrease because although the temperature rises slightly, the number of moles of gas decreases. (c) Work is positive and (d) E is a state function, so ΔE is the same.

12.4.1 c, d, and f

12.4.2 $\Delta H_{reaction} = +205.9$ kJ and $\Delta E_{reaction} = +200.9$ kJ

12.4.3 $2\ Fe_2O_3(s) + 3\ C(s,\ graphite) \longrightarrow 4\ Fe(s) + 3\ CO_2(g)$

$$\Delta H_{reaction} = +467.9\ kJ$$

12.5.1 In both cases, interionic forces must be overcome to break apart the crystals, but when NaCl dissolves in water, energy is released through ion-dipole interactions.

12.5.2 Melting all the ice yields 100 mL of water at 0 °C and cools the tea to 9.07 °C. When these mix and equilibrate, the resulting temperature is 7.55 °C.

12.5.3 The same because enthalpy is a state function, so its change is independent of path.

13

Spontaneity
of Chemical Processes

INTRODUCTION: ORDER AND DISORDER

Perhaps you have noticed that there is a natural tendency for ordered structures to become disordered. Death is followed by decay. The highly ordered biological structures of the once-living creature become degraded. Human monuments crumble and decompose. The inscription on a gravestone erodes and becomes illegible. The gravestone tilts and eventually falls over. Left untended, any structure built by humans disintegrates and "returns to nature," to a state of disorder.

Despite this natural tendency, living organisms are highly ordered. At the molecular level, a DNA double helix is a regular spiral of repeating units carrying specific genetic messages. The order of a DNA sequence is very precise, and the slightest deviation can have disastrous consequences. Macroscopically, plants grow in regular, ordered ways. Animals create ordered structures, too, from the simple hexagonal shape of a honeycomb to the elaborate network of a highway bridge.

Chemists have been challenged to duplicate the extremely regular synthesis patterns found in living matter. It has been a long-standing goal to invent molecules that assemble themselves into specific, ordered arrays. Recently there have been some exciting successes. Our inset is a molecular view of a self-assembled "molecular wreath." This highly ordered structure forms when four molecular chains weave themselves together in the presence of 12 copper cations. Chemists are exploring self-assembling molecular systems with the goal of creating new materials with diverse and useful properties. They also hope to learn the secrets of the complicated self-organizing molecules that form the basis for life.

We see all around us that ordered structures tend to become disordered, and we also see that life creates order from disorder. In the absence of living organisms, the preferred direction of processes is from order to disorder. Why is that? From the perspective of energy conservation, any process in which energy is conserved could proceed equally well in either direction. The first law of thermodynamics says nothing about the preferred direction of a process.

The explanation for why processes have preferred directions lies in the second law of thermodynamics, which we introduce in this chapter. This second law is as important to chemists as the first, because chemists need to know in advance whether a particular product can form when a set of reactants is mixed. Furthermore, if a chemical reaction cannot proceed under one set of conditions, chemists need to know whether it can proceed under some other set of conditions.

The second law of thermodynamics deals with order and disorder. To treat these concepts quantitatively, a new property is needed. This property, which measures the amount of disorder in a system, is called *entropy*. For analysis of how the disorder of a system and its surroundings changes during a process, a combination of energy and entropy—free energy—proves useful. Order and disorder, entropy and free energy, and the second law of thermodynamics are the subjects of this chapter.

13.1 SPONTANEITY

Every process has a preferred direction, which is referred to in thermodynamics as the **spontaneous** direction. A process follows its spontaneous direction unless acted on by some external agent. For example, the spontaneous direction for water movement is downhill, from higher altitude to lower altitude. Water runs uphill only if it forced in that direction by a pump.

The word *spontaneous* has a different meaning in thermodynamics than it does in everyday speech. Ordinarily, spontaneous refers to an event that takes place without any effort or premeditation. For example, a crowd cheers spontaneously for an outstanding performance. In contrast, thermodynamic spontaneity refers to the direction that a process will take if left alone and given sufficient time. A thermodynamic process may be spontaneous, and yet the process may not occur. The reaction of hydrogen and oxygen to form water is highly spontaneous. In fact, this reaction releases so much energy that it is used to drive the main engines of the space shuttle:

$$\text{Spontaneous:} \qquad 2\,H_2(g) + O_2(g) \longrightarrow 2\,H_2O(l)$$

Mixtures of hydrogen and oxygen must be handled very carefully, because the slightest spark triggers a violent reaction. Nevertheless, hydrogen–oxygen mixtures can be stored indefinitely as long as there are no sparks to initiate the reaction. From the thermodynamic perspective, the reaction of hydrogen and oxygen is spontaneous whether or not the spark is present.

A process that does not occur may be spontaneous but very slow, or it may be nonspontaneous. An example of a nonspontaneous process is the decomposition of water into molecules of hydrogen and oxygen. Water will not decompose under normal conditions, even in the presence of a spark:

$$\text{Nonspontaneous:} \qquad 2\,H_2O(l) \longrightarrow 2\,H_2(g) + O_2(g)$$

When a process is spontaneous, its reverse process is nonspontaneous under the same conditions. The spontaneous direction is for H_2 and O_2 to combine to form water molecules; the reverse decomposition process is nonspontaneous.

Chemical manufacturers design elaborate, expensive plants to produce valuable chemical products spontaneously from starting materials. The furnaces in a steel mill, for example, mix chemicals under controlled conditions to convert iron ore into steel. Because of the expense involved in building a chemical plant, a manufacturer must be able to determine in advance whether or not a reaction will be spontaneous. Fortunately, this can be done using thermodynamics.

CHAPTER 14 →
We describe the factors that determine how fast a chemical reaction goes in Chapter 14.

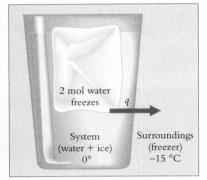

Figure 13-1
Schematic view of the spontaneous process for a water-and-ice mixture in a freezer. The energy-releasing process, freezing, is spontaneous under these conditions.

Energy and Spontaneity

One characteristic of a chemical process is its energy change (see Chapter 12). Some processes release energy to the surroundings ("downhill" for the system), others absorb energy from the surroundings ("uphill" for the system).

Knowing that a reaction is energetically downhill or uphill is not enough to determine whether it is spontaneous. We can show this using an everyday example: a mixture of liquid water and ice at 0 °C. As diagrammed in Figure 13-1, ice forms spontaneously if an ice and water mixture is placed in a freezer at −15 °C:

$$H_2O(l,\ 0\ ^\circ C) \xrightarrow{\text{Freezer, } -15\ ^\circ C} H_2O(s,\ 0\ ^\circ C)$$

Suppose that we leave this mixture in the freezer until 2.00 mol of water has frozen. We can calculate the enthalpy change from the heat of fusion of ice:

$$\Delta H = -n\Delta H_{fus} = -(2.00\ \text{mol})(6.01\ \text{kJ/mol}) = -12.0\ \text{kJ}$$

The energy change is also −12.0 kJ because $\Delta E \cong \Delta H$ for processes that occur in condensed phases. The negative values for ΔE and ΔH reveal that in this spontaneous process, the ice–water system releases energy to its colder surroundings.

Now consider what happens to the same mixture when it is in contact with the atmosphere at 25 °C: ice melts spontaneously, as shown schematically in Figure 13-2:

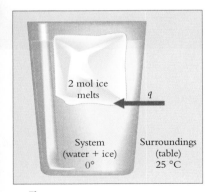

Figure 13-2
Schematic view of the spontaneous process for a water-and-ice mixture on a table top. The energy-absorbing process, melting, is spontaneous under these conditions.

$$H_2O(s, 0\ ^\circ C) \xrightarrow{\text{Table, 25\ }^\circ C} H_2O(l, 0\ ^\circ C)$$

We calculate the enthalpy and energy changes for the melting of 2.00 mol of ice using its heat of fusion:

$$\Delta E \cong \Delta H = n\Delta H_{fus} = (2.00\ \text{mol})(6.01\ \text{kJ/mol}) = +12.0\ \text{kJ}$$

The positive signs for ΔH and ΔE indicate that in this spontaneous process the ice-water system absorbs energy from its warmer surroundings.

Both these everyday processes are spontaneous, but one has a negative ΔE and the other a positive ΔE. These processes demonstrate that knowing ΔE or ΔH is not enough to predict whether a process will occur spontaneously. If we hope to use thermodynamics to determine when a process will be spontaneous, we need another state function besides energy and enthalpy.

Molecular Disorder

To understand spontaneity, we need to examine what happens at the molecular level during spontaneous chemical processes. Before looking at molecular processes, however, consider some large-scale events that go in one particular direction:

1. A box of marbles dropped on the floor will scatter, but the marbles will not spontaneously roll back into the box.
2. An untended wooden fence eventually falls apart, but piles of wood will not spontaneously assemble into fences.
3. A completed jigsaw puzzle can be disassembled by a sweep of the hand, but un-worked jigsaw puzzles never assemble themselves by random events.

Each process has a spontaneous direction, from ordered (organized) to disordered (disorganized). This can be summarized in a common-sense law:

/// *Things tend to become disorganized.*

Disorder and *randomness* are synonyms for *disorganization*. We use these terms interchangeably.

Does this common-sense law also apply to events at the molecular level? Consider the two glass bulbs shown in Figure 13-3. One of the bulbs contains

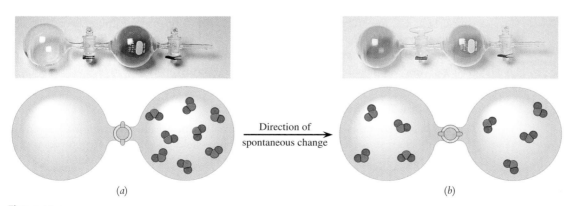

| (a) | | (b) |

Figure 13-3
When the valve is opened between two bulbs, one of which is empty and the other filled with nitrogen dioxide gas, molecules of NO_2 move spontaneously from the filled bulb until both bulbs have equal concentrations of NO_2. The reverse process is never observed.

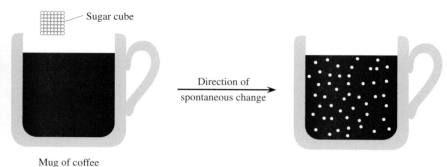

Figure 13-4
Sugar dissolving in coffee is a disordering process. Sugar molecules are distributed more randomly in solution than they are in the solid crystal.

nitrogen dioxide, a red-brown gas, but the second bulb is empty. When the valve that connects the bulbs is opened, the red-brown gas expands to fill both bulbs. The opposite process never occurs. That is, if both bulbs contain NO_2 at the same pressure, opening the valve never causes the pressure to rise in one bulb and fall in the other.

The molecular view of what happens in the two bulbs is shown in the second part of Figure 13-3. The NO_2 molecules in the filled bulb are always moving, continually colliding with one another and with the walls of their container. When the valve between the two bulbs is opened, some molecules move into the empty bulb, and eventually the concentration of molecules in each bulb is the same. At this point, the gas molecules are in a state of dynamic equilibrium. Molecules still move back and forth between the two bulbs, but the concentration of molecules in each bulb remains the same.

The gas molecules in a bulb behave similarly to a handful of marbles. Molecules, like marbles, are more ordered when they occupy a small volume than when spread over a large volume. Thus our common-sense law applies to this molecular example.

For a second example, consider Figure 13-4, which shows sugar dissolving in coffee from a molecular perspective. Before the solid sugar dissolves, the sugar molecules are organized in a crystal. As the molecules dissolve, they become distributed randomly and uniformly throughout the liquid coffee. The opposite process never occurs. That is, sugar cubes do not form from sweet coffee. Once again, the direction of spontaneous change at the molecular level conforms to the common-sense law, because sugar molecules are distributed more randomly when they are dissolved in coffee than when they are part of a solid crystal. A sugar cube is the molecular equivalent of an assembled jigsaw puzzle. The molecular organization of a sugar cube is replaced with disorganization as it dissolves in coffee.

Energy and Disorder

Liquid water contains molecules that are free to move about through their entire volume, whereas ice contains water molecules held in a highly structured three-dimensional lattice. Thus when water freezes, the spontaneous conversion leads to *less* disorder for the system. This seems like a violation of the common-sense law, but notice that there is a flow of heat in addition to the order-disorder change. To understand how this transfer of energy relates to order-disorder, consider the energy conversion shown in Figure 13-5.

When a baseball pitcher throws a fastball, the kinetic energy of the ball is highly organized in that all the molecules of the ball move together through space. As described in Chapter 2, when the ball comes to rest in the catcher's glove, the

CHAPTER 5 →
See Chapter 5 for a description of the molecular motion of gas molecules and see Section 2.1 for an introduction to dynamic equilibrium.

Figure 13-5
As a thrown baseball is caught, its directional kinetic energy is transformed from the organized kinetic energy of the ball to the disorganized kinetic energy of many individual molecules of the catcher's glove.

conversion of kinetic energy to thermal energy causes an increase in the temperature of the glove. As explained in Chapter 5, a higher temperature means that the molecules in the sample (the glove in this case) move with larger average speeds. Moreover, these increased molecular speeds point randomly in all directions because the molecules in the glove have no particular organization in their movement. In other words, the organized, directional character of the kinetic energy of the ball is replaced with random kinetic energy of molecules in the glove. In this process, heat flow from the ball to the glove causes energy to become disorganized.

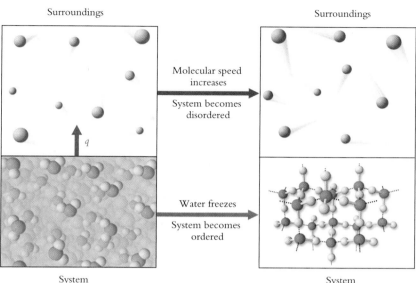

Figure 13-6
When water freezes, its molecules become more ordered, but the accompanying energy flow increases the temperature and disorder of the surroundings.

Now consider what happens to energy in the water-ice example. As Figure 13-6 illustrates, the molecules become more ordered when water freezes. At the same time, heat flows from the system to the surroundings, which is an energy disordering process. If the common-sense law of spontaneity is correct, then the energy disorder caused by the heat flow must be larger than the molecular order created by the fixed positions of the ice molecules.

The spontaneity of any process must be evaluated in terms of the organization of energy as well as the organization of matter. Chemists and physicists have measured these changes for many spontaneous processes. They always obtain the same result, which is the **second law of thermodynamics:**

/// Any spontaneous process increases the disorder of the universe.

This result may not seem to be a law of thermodynamics, because it does not refer directly to energy. However, our observation that heat flow causes changes in the organization of energy establishes a qualitative link between energy and disorder. Section 13.2 examines the quantitative relationship between heat flow and disorder.

Section Exercises

13.1.1 Explain the following observations in terms of organization and disorganization:
(a) Untended fences eventually fall down.
(b) A glass is easy to break but very difficult to mend.
(c) A wine cooler does not spontaneously separate into alcohol, water, and fruit juice.

13.1.2 Solid sugar can be recovered from coffee by boiling off the water. (Coffee candy can be made in this way.) Sugar molecules become more organized in this process. What else must occur for total disorder to increase?

13.1.3 Draw molecular pictures of liquid water and water vapor that show what happens to the amount of order among H_2O molecules when water evaporates.

13.2 ENTROPY: THE MEASURE OF DISORDER

Increasing disorder is a qualitative criterion for the preferred direction of a chemical process. Measuring disorder quantitatively allows accurate predictions about the spontaneity of reactions. Scientists measure disorder in two ways, by counting the amount of disorder or by linking the change in amount of disorder with the flow of heat between a system and its surroundings.

Entropy

The amount of disorder in a system depends on the state of that system. As an example, one mole of liquid water at 0 °C has more disorder than one mole of ice at 0 °C. However, the amount of disorder is independent of how the system got to that state. In other words, one mole of water at 0 °C has the same amount of disorder whether it was obtained by melting an ice cube or by cooling water from a higher temperature. The state function that provides a quantitative measure of disorder is called **entropy** and is symbolized **S**. The second law of thermodynamics can be restated in terms of entropy:

/// *The total entropy of the universe increases in spontaneous processes.*

The entropy of a substance can be defined by an equation that was first proposed by Ludwig Boltzmann.

$$S = k \ln W$$

According to this equation, entropy is equal to a constant (the Boltzmann constant, $k = 1.3806 \times 10^{-23}$ J/K) multiplied by the natural logarithm of the number of ways (W) the system can be described. To get a feel for W, consider two marbles of different colors placed in a box containing nine compartments of equal size. Figure 13-7 shows two different ways to distribute the two marbles. We could place both marbles into any of the nine compartments, so there are nine different ways to place both marbles in the *same* compartment. That is, the distribution that puts both marbles in the same compartment has $W = 9$. Another type of distribution places each marble in a *different* compartment. After placing the first marble in one of nine compartments, the second can go into any of the other eight. The nine possible starting points for the first marble multiplied by the eight possible compartments for the second gives a total of 72 possible arrangements, so $W = 72$. The fact that W is larger for the second distribution than for the first means that the second distribution has a larger value of S.

Although the equation appears simple, applying it to a molecular system always is challenging, because there are so many molecules in any realistic molecular system that we must count huge numbers of possibilities to determine the value of W. Fortunately, there are easier ways to measure *changes* in entropy (ΔS). The change of entropy for a process is important because the second law of thermodynamics states that the entropy of the universe always increases in a spontaneous process; that is, $\Delta S_{universe} > 0$. Thus scientists are more concerned with changes in S than with the actual value of S.

Changes in entropy can be linked to heat flows. We restrict our treatment of entropy changes to situations in which a flow of heat occurs without a temperature change. Heat flow at constant temperature may seem paradoxical, because we usually associate a flow of heat with a change in temperature. However, three

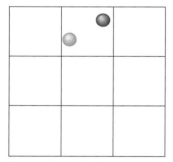

Two marbles confined to one box

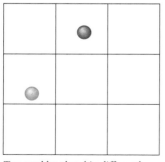

Two marbles placed in different boxes

Figure 13-7
A collection of marbles distributed among compartments provides a simple example illustrating the sense of W.

important types of chemical processes occur in which heat flows take place without a temperature change:

1. Phase changes take place at constant temperature, absorbing or releasing heat in the process.

2. A chemical reaction may occur under conditions in which temperature is held constant, such as in the human body or a thermostated automobile engine. The exothermicity or endothermicity of the reaction generates a heat flow, even though the temperature is constant.

3. The surroundings may be so large that they can absorb or release significant amounts of heat before the temperature changes by a measurable amount. An experiment performed in a constant-temperature water bath is a common example of this category.

Whereas $S = k \ln W$ defines entropy in terms of molecular disorder, entropy can also be defined in terms of heat flows. This definition of entropy involves an equation linking the change in entropy of a system with heat flow and temperature:

$$\Delta S = \frac{q_T}{T} \tag{13-1}$$

> Computing the entropy change for a process that is not at constant temperature requires the use of calculus and is beyond the scope of this text.

To put this in words, when heat flows at constant temperature, the entropy change is equal to the heat transferred (q_T) divided by the temperature (T) in kelvins. The subscript T in Equation 13-1 is a reminder that this equation is restricted to processes that occur at constant temperature. Because q is an energy term, the units of ΔS are energy/temperature, or J/K.

The presence of q in this equation is consistent with the earlier observation that heat flows change the organization of energy. Heat is divided by temperature in the equation because temperature is a measure of how much thermal disorder is already present in a substance. At low temperature, there is little disorder, so the addition of a given amount of heat increases the disorder significantly. At high temperature, in contrast, there is considerable disorder, so adding the same amount of heat increases total disorder by a smaller amount.

The example in Section 13.1 involving ice and water can provide a good test of whether Equation 13-1 is consistent with the second law of thermodynamics. Example 13-1 applies the entropy equation to ice forming in a freezer.

Entropy Change During Freezing

Example 13-1

What is the total entropy change when 2.00 mol of liquid water freezes at 0.0 °C in a freezer compartment whose temperature is held at -15 °C? (For H_2O, $\Delta H_{fus} = 6.01$ kJ/mol.)

Strategy: The seven-step problem-solving approach served us well in doing the thermodynamics examples of Chapter 12 and is equally valuable for this type of problem.

Solving Quantitative Problems

Solution:

1. The problem asks for the *total* entropy change, which includes ΔS for the water and ΔS for the surroundings.

| Example 13-1 | Entropy Change During Freezing *(continued)* |

2. When water freezes, heat flows from the system into the surroundings (See Figure 13-1). Thus the entropy of the water decreases, and the entropy of the surroundings increases.

3. The problem states that $n = 2.00$ mol and $\Delta H_{fus} = 6.01 \times 10^3$ J/mol. The system remains at a constant temperature of 0.0 °C (273.15 K) while the phase change occurs. The ice might later cool to -15 °C, but the question asks only about the freezing process. The freezer holds the immediate surroundings at constant temperature of -15 °C $= 258$ K.

4 and 5. For these constant-temperature processes, entropy changes can be computed using Equation 13-1.

> **Remember that in thermodynamic calculations, the temperature must always be in kelvins.**

$$\Delta S = \frac{q_T}{T}$$

The amount of heat lost by the water can be calculated from the heat of fusion:

$$q = -n\,\Delta H_{fus}$$

6. Remember that ΔH_{fus}, which is positive, refers to ice melting. Heat flows out of the system as the water freezes, so q has a negative value. Thus:

$$q_{H_2O} = -(2.00 \text{ mol})(6.01 \times 10^3 \text{ J/mol}) = -1.202 \times 10^4 \text{ J}$$

$$\Delta S_{H_2O} = \frac{q_{H_2O}}{T_{H_2O}} = \frac{-1.202 \times 10^4 \text{ J}}{273.15 \text{ K}} = -44.0 \text{ J/K}$$

The freezer absorbs the heat released by the water, but it does so at *its* temperature, 258 K. Because the freezer absorbs heat, $q_{freezer}$ has a positive sign. Thus for the freezer:

$$q_{freezer} = -q_{H_2O} = +1.202 \times 10^4 \text{ J}$$

$$\Delta S_{freezer} = \frac{q_{freezer}}{T_{freezer}} = \frac{1.202 \times 10^4 \text{ J}}{258 \text{ K}} = 46.6 \text{ J/K}$$

The total entropy change is the sum of these changes:

$$\Delta S_{total} = \Delta S_{H_2O} + \Delta S_{freezer} = (-44.0 \text{ J/K}) + (46.6 \text{ J/K}) = +2.6 \text{ J/K}$$

7. The negative sign for the entropy change of the system is consistent with our qualitative picture of greater disorder in a liquid than in a solid. A positive value for the net entropy change is a reasonable result because the formation of ice in a freezer is spontaneous and the second law of thermodynamics states that total entropy must increase in a spontaneous process.

Equation 13–1 gives a positive value for the entropy change for the spontaneous freezing of water in a freezer. The total entropy change for 2.00 mol of ice melting in a room at 25 °C (See Figure 13-2) can be analyzed in exactly the same way. The heat required to melt the ice has the same magnitude but the opposite sign as the heat used to freeze the water in Example 13-1: 1.202×10^4 J/mol. Ice melts at the same temperature as water freezes, 273.15 K, so the entropy change for the water is also equal in magnitude but is opposite in sign: $\Delta S = 44.0$ J/K. The entropy change of the surroundings, on the other hand, must be calculated using the temperature of the surroundings, 25 °C = 298 K. The heat required to melt the ice comes from the surroundings, but the amount of heat is so small and the surroundings are so large that the temperature remains constant:

$$\Delta S_{\text{surroundings}} = \frac{-1.202 \times 10^4\,\text{J}}{298\,\text{K}} = -40.3\,\text{J/K}$$

$$\Delta S_{\text{total}} = 44.0 - 40.3 = +3.7\,\text{J/K}$$

The net change in entropy is a positive quantity for both these spontaneous processes, even though one process is exothermic and the other endothermic. This is in agreement with the second law of thermodynamics.

When this type of calculation is carried out for other processes, the same result is always obtained. For any spontaneous process, the total change of entropy is a positive quantity. Thus this new state function of entropy provides a thermodynamic criterion for spontaneity.

/// **In any spontaneous process, $\Delta S_{total} > 0$.**

The ice and water example, summarized in diagrammatic form in Figure 13-8, illustrates potential pitfalls in evaluating entropy changes. First, be careful about the sign of q. The sign is different for the system and the surroundings and depends on whether the process is endothermic or exothermic. Table 13-1 summarizes the signs. Second, always use the temperature of the component whose entropy change is being evaluated. In our examples, the freezer, the room, and the water-ice mixture all have different temperatures. Third, *total* entropy change must be found by summing the entropy changes of the system and the surroundings. This total entropy change is more generally known as the *entropy change of the universe*:

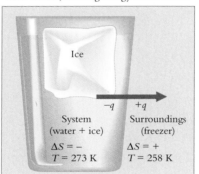
Water freezes, releasing energy to the freezer

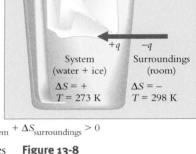
Ice melts, absorbing energy from the room

In both cases, $\Delta S_{\text{total}} = \Delta S_{\text{system}} + \Delta S_{\text{surroundings}} > 0$

Figure 13-8
In calculating entropy changes: Be careful about the sign of q, use the appropriate temperatures, and sum the changes for system and surroundings.

$$\Delta S_{\text{universe}} = \Delta S_{\text{system}} + \Delta S_{\text{surroundings}} \qquad (13\text{-}2)$$

Example 13-2 does another entropy calculation.

Table 13–1
Signs Associated with Heat Transferred (q)

Process	System	Surroundings	Example
Exothermic	−	+	Water freezing
Endothermic	+	−	Ice melting

Example 13-2	**Entropy Change of a Refrigerant**

In a refrigerator, a liquid refrigerant absorbs heat from the contents of the refrigerator. This heat vaporizes the refrigerant, which is later recondensed to a liquid by pressure supplied by a mechanical pump. One common refrigerant is HFC-134a, CH_2FCF_3 (MM = 102.0 g/mol). HFC-134a boils at −27 °C (246 K) with a heat of vaporization of 22.0 kJ/mol. Calculate the entropy change of the universe when 1.50×10^2 g of HFC-134a vaporizes at 246 K, exchanging heat with contents of the refrigerator at 4 °C (277 K).

Solving
Quantitative
Problems

Strategy: The problem describes a relatively complicated process, so again we apply the seven-step problem-solving process.

Solution:

1. The problem asks for the entropy change of the universe ($\Delta S_{universe}$) caused by the refrigeration process. We must determine the entropy change of the liquid refrigerant, HFC-134a, as well as the entropy change of the contents of the refrigerator.

2. A block diagram helps to visualize the thermodynamic processes. The liquid refrigerant evaporates at constant temperature as it absorbs heat from the contents of the refrigerator, which are at a constant temperature.

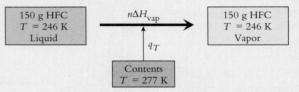

3. Here are the data: For CH_2FCF_3, $m = 1.50 \times 10^2$ g, MM = 102.0 g/mol, T_b = 246 K, ΔH_{vap} = 22.0 kJ/mol. T_{refrig} = 277 K.

4. and 5. Equation 13-2 gives the entropy change of the universe:

$$\Delta S_{universe} = \Delta S_{system} + \Delta S_{surroundings}$$

It is convenient to define the HFC as the system and to assume that the refrigerator isolates its contents from the rest of the universe on a short time scale, so we can approximate $\Delta S_{surroundings}$ as $\Delta S_{contents}$.

$$\Delta S_{universe} = \Delta S_{HFC} + \Delta S_{contents}$$

Use Equation 13-1 and the data to find ΔS for the refrigerator and for the HFC-134a. For HFC-134a boiling at 246 K and 1 atm pressure,

$$q = n\,\Delta H_{vap}$$

6. Begin by calculating the number of moles of HFC-134a:

$$n_{HFC} = \frac{1.50 \times 10^2 \text{ g}}{102.0 \text{ g/mol}} = 1.471 \text{ mol}$$

$$q_{HFC} = (1.471 \text{ mol})(22.0 \text{ kJ/mol})(10^3 \text{ J/kJ}) = 3.235 \times 10^4 \text{ J}$$

$$\Delta S_{HFC} = \frac{q_{HFC}}{T_{HFC}} = \frac{3.235 \times 10^4 \text{ J}}{246 \text{ K}} = 132 \text{ J/K}$$

| **Entropy Change of a Refrigerant** (*continued*) | **Example 13-2** |

This is the entropy change of the refrigerant fluid. The contents of the refrigerator give up heat equal to the heat absorbed by the fluid:

$$q_{contents} = -q_{HFC}$$

$$\Delta S_{contents} = \frac{-q_{HFC}}{T_{contents}} = \frac{-3.235 \times 10^4 \, J}{277 \, K} = -117 \, J/K$$

The overall entropy change is calculated using Equation 13-2:

$$\Delta S_{universe} = \Delta S_{HFC} + \Delta S_{contents} = 132 - 117 = 15 \, J/K$$

$\uparrow$ System $\quad\uparrow$ Surroundings

7. A positive entropy change for the universe is a reasonable outcome.

We can state both the first and the second laws of thermodynamics in terms of the universe:

First law: $\quad \Delta E_{universe} = 0$ (always)

Second law: $\Delta S_{universe} > 0$ (always)

Our Chemistry and Life Box on page 602 describes how entropy is linked with time and life.

Direction of Heat Flow

Equation 13-1 states that an entropy change is heat flow divided by temperature, and according to the second law of thermodynamics, the total entropy change is always positive. Taken together, these two requirements dictate that spontaneous heat flow between two bodies at different temperatures always goes from the warmer body to the colder body. For example, when a kettle of boiling water sits on the heating element of a stove, heat always flows from the hot burner to the relatively cooler boiling water.

To understand why this must occur, consider the entropy changes that would accompany heat transfer in the *opposite* direction. Suppose the burner is at 455 K and the water is at 373 K. We can calculate the entropy change that would occur if 1.00 J of heat flowed from the water to the burner. In this scenario, q for the burner is positive, so it gains entropy. For the water, q is negative, so it loses entropy:

$$\Delta S_{burner} = 1.00 \, J/455 \, K = 2.20 \times 10^{-3} \, J/K$$

$$\Delta S_{water} = -1.00 \, J/373 \, K = -2.68 \times 10^{-3} \, J/K$$

$$\Delta S_{universe} = \Delta S_{burner} + \Delta S_{water} = (2.20 - 2.68) \times 10^{-3} \, J/K = -4.8 \times 10^{-4} \, J/K$$

A spontaneous "uphill" flow of heat, from the cooler water to the hotter burner, would result in a *decrease* in the entropy of the universe, which is forbidden by the second law of thermodynamics. According to all observations, which are summarized in the second law, heat never flows spontaneously from a cold to a hot body.

Box 13-1   Chemistry and Life: Entropy, Time, and Life

A ccording to the second law of thermodynamics, events go only in the direction that increases the entropy of the universe. Time, like entropy, is unidirectional. Despite the time machines invented by novelists, time always marches on. You are older as you read this sentence than you were when you started reading the chapter.

Entropy and time are two unidirectional parameters that have been connected by philosophers. Some have characterized entropy as "time's arrow." The idea is that entropy changes can be used to determine which of two states is "before" and which is "after," even if no clock is present. For example, the following figure shows two views of a chamber at two different times. The chamber contains two different gases. In the upper view the two gases are separated, and in the lower view they are mixed uniformly. If the chamber is isolated so that the surroundings cannot contribute to the entropy changes, the spontaneous change is from upper to lower because the lower arrangement is less ordered than the upper arrangement. Using entropy as time's arrow, we conclude that the lower arrangement must be at a later time than the upper one. If this were not the case, we would have a violation of the second law of thermodynamics.

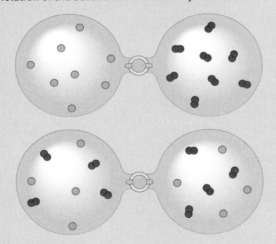

Although living things cannot escape the unidirectional aspect of time, it may seem at times that they successfully reduce entropy. As an example, ask yourself which of the two views of the same house is "before" and which is "after." Left to itself, any structure degenerates into disorder, but with human intervention, structures can be made "as good as new."

To analyze the spontaneous direction of events that are influenced by living beings, we must consider the surroundings as well as the system. The house on the left is obviously more disordered than the one on the right. If this house has been left alone, the right picture is "before" and the left is "after." On the other hand, if the

house on the right is the same house after repainting, calculating the total entropy change of the universe shows that the painter generated disorder in the surroundings to make this house more ordered. Then the picture of the less-ordered house may predate the picture of the more-ordered house.

If humans ever travel to distant galaxies, how will they recognize the presence of new life forms? One likely way is to search for ordered structures whose construction appears to violate the second law of thermodynamics. If a visitor from outer space landed on Earth with no one in sight and found a fence running in a straight line across the landscape, that visitor would surely infer the existence of intelligent life from the improbable order of the fence.

Even without leaving our planet, we search for signs of extraterrestrial life by searching for improbable patterns of order. Our planet is bombarded continually by electromagnetic radiation of many wavelengths, coming to us across the far reaches of space from elsewhere in the universe. If intelligent beings are trying to make their presence known, they may be transmitting some highly ordered pattern of electromagnetic signals that other intelligent life could recognize as too ordered to occur except by design.

■ **13.2.1** Benzene (C_6H_6) has $\Delta H_{fus} = 10.9$ kJ/mol and a freezing point of 5.5 °C, whereas water has $\Delta H_{fus} = 6.01$ kJ/mol and freezes at 0.0 °C. Suppose that a sealed jar containing a mixture of solid and liquid benzene at 5.5 °C is immersed in a mixture of ice and water at 0.0 °C. What will happen?

■ **13.2.2** Compute $\Delta S_{universe}$ for 10.0 g of benzene changing phase as described in Section Exercise 13.2.1.

■ **13.2.3** Suppose that a mixture of ice and water is placed in a refrigerator that is held at the freezing point of water, exactly 0.0 °C. Calculate ΔS for the ice and water mixture, for the refrigerator and for the universe when 5.00 g of ice forms. Is this process spontaneous? Is the reverse process spontaneous?

13.3 ABSOLUTE ENTROPIES

Chapter 12 describes how to determine changes in enthalpy or energy, but as noted there, no practical method exists to determine the *absolute* enthalpy or energy of a substance. Unlike energy and enthalpy, entropy has a well-defined zero point, where disorder is at a minimum. Having a zero point for molecular disorder makes it possible to determine the **absolute entropy** of any substance.

Disorder and Temperature

As described in Section 13.2, removing heat from any substance decreases its disorder. The relationship between heat and disorder can be explored in more detail by considering one mole of argon initially at $T = 300$ K and $P = 1$ atm pressure. Under these conditions, argon is a near-ideal gas that occupies a volume of 24.4 L. The argon atoms exhibit substantial disorder because they are distributed randomly throughout the entire volume, moving in all directions with a wide range of molecular speeds.

Now consider what happens to this sample of argon gas when the temperature is lowered to 90 K. Heat flows from the system to the surroundings, reducing the entropy of the gas. Because the temperature of the argon changes from 300 K to 90 K, we cannot use Equation 13-1 to calculate ΔS. However, the entropy change can be described qualitatively. As shown in Figure 13-9, the gas volume decreases to 7.3 L. Compare this molecular view with the views of marbles in Figure 13-7: When two marbles are confined to one compartment, the system has a lower value of S than when the marbles are in two compartments. The situation with the argon sample is similar. The reduction in gas volume reduces disorder and entropy by confining the atoms to a smaller space. The atoms also move more slowly at the lower temperature, which makes an additional contribution to the decrease in disorder.

At a temperature of 87.3 K, argon gas condenses to the liquid phase, as illustrated in Figure 13-10. This change occurs at constant temperature (87.3 K) and is accompanied by a large reduction in entropy, which can be calculated using Equation 13-1 and the molar enthalpy of vaporization of argon:

$$q_{Ar} = -n\,\Delta H_{vap} = -(1.00 \text{ mol})(6.53 \text{ kJ/mol})(10^3 \text{ J/kJ}) = -6.53 \times 10^3 \text{ J}$$

$$\Delta S_{condensation} = \frac{q_{Ar}}{T_{condensation}} = \frac{-6.53 \times 10^3 \text{ J}}{87.3 \text{ K}} = -74.8 \text{ J/K}$$

The entropy change for this cooling process can be determined using calculus.

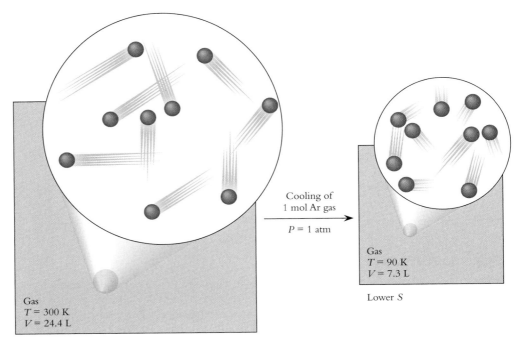

Cooling of
1 mol Ar gas

$P = 1$ atm

Gas
$T = 90$ K
$V = 7.3$ L

Lower S

Gas
$T = 300$ K
$V = 24.4$ L

Higher S

Animation

Figure 13-9
A molecular view of the changes that occur when a sample of argon gas at 1 atm is cooled from 300 K to 90 K. The disorder of the gas decreases because of the volume change and the smaller average speed of the atoms.

This decrease in entropy is caused primarily by the decrease in volume that accompanies the condensation. At 87.3 K the molar volume of argon gas is 7.17 L, but as a liquid the same amount of argon occupies just 0.29 L.

When the temperature reaches 83.8 K, argon freezes to form a regular cubic crystal. Again, Equation 13-1 can be used to calculate ΔS for this phase change at constant temperature (83.8 K), this time using the molar enthalpy of fusion:

Figure 13-10
A molecular view of the changes that accompany the condensation of argon from a gas to a liquid at 87.3 K. The atoms are confined to a much smaller space, causing a large reduction in entropy.

Animation

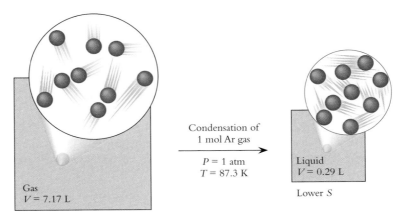

Condensation of
1 mol Ar gas

$P = 1$ atm
$T = 87.3$ K

Liquid
$V = 0.29$ L

Lower S

Gas
$V = 7.17$ L

Higher S

$$q_{Ar} = -n \, \Delta H_{fus} = -(1.00 \text{ mol})(1.21 \text{ kJ/mol})(10^3 \text{ J/kJ}) = -1.21 \times 10^3 \text{ J}$$

$$\Delta S_{solidification} = \frac{q_{Ar}}{T_{solidification}} = \frac{-1.21 \times 10^3 \text{ J}}{83.8 \text{ K}} = -14.4 \text{ J/K}$$

As shown in Figure 13-11, atoms in a solid are organized in a highly ordered, regular crystalline array, whereas atoms in a liquid move freely about, forming less-ordered patterns.

Even as a solid, argon is still significantly disordered. As shown in Figure 13-12, argon atoms vibrate back and forth within the crystal, so there is still some randomness in their individual positions. However, as the solid is cooled to even lower temperature, the vibrational motion of the atoms continues to diminish. If we could cool the solid to the lowest possible temperature, the atoms would have minimum energy of motion, and each atom would be locked in place in the crystal. That temperature is 0 K. The entropy of the system is now at a minimum, provided that there are no impurities in the sample or imperfections in the solid crystal. Impurities are a source of disorder because molecules of the impurity are distributed randomly throughout the crystal. Imperfections in a crystal are a source of disorder because atoms are out of place. A crystal that contains impurities or imperfections contains some residual entropy.

When the disorder in the sample is at a minimum, the entropy of the sample is defined to be zero. The **third law of thermodynamics** states that a pure, perfect crystal at 0 K has zero entropy. In equation form the third law is:

$$S_{(\text{pure, perfect crystal}; \; T \,=\, 0 \text{ K})} = 0 \qquad\qquad \textbf{(13-3)}$$

Our development of the third law is based on an analysis of how disorder decreases as the temperature drops. The law has also been tested experimentally by comparing measured entropy changes with calculated values based on $S = 0$ J/mol at $T = 0$ K. The results, some of which are described in our Chemistry and Technology Box, confirm our qualitative analysis.

The third law of thermodynamics establishes a "starting point" for entropies. At 0 K, any pure perfect crystal is completely ordered and has $S = 0$ J/mol. At any higher temperature, that substance is disordered, so it has a positive entropy. The entropy of a sample depends on the conditions. The molar entropies of many pure substances have been measured under standard thermodynamic conditions of 1 atm and 298 K. These absolute molar entropies, designated $S°$, are found in the same thermodynamic tables that list heats of formation. A few values of $S°$ are listed in Table 13-2 to give you an idea of the magnitudes of absolute entropies. A more extensive list can be found in Appendix D.

The representative values of $S°$ listed in Table 13-2 reveal the following general trends in absolute entropies:

1. Unlike enthalpies of formation, $S°$ values are *never* zero. Absolute entropies are zero only at 0 K, and $S°$ values refer to substances at 298 K.

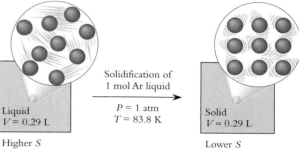

Figure 13-11
A molecular view of the changes that occur when argon freezes. From liquid to solid, disorder decreases because of the high degree of regularity in the solid phase.

Solidification of
1 mol Ar liquid

$P = 1$ atm
$T = 83.8$ K

Liquid
$V = 0.29$ L

Higher S

Solid
$V \approx 0.29$ L

Lower S

Animation

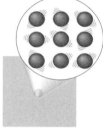

Higher temperature   Lower temperature

Figure 13-12
The vibrational motion in a solid crystal creates some randomness in the positions of the atoms. As the temperature is reduced, the disorder of the sample is reduced because the magnitude of vibrational motions decreases.

Animation

Box 13-2 Chemistry and Technology: Seeking Absolute Zero

At a temperature of absolute zero, atoms and molecules truly would be frozen in place. Because all excess energy would be removed, the inherent properties of atoms and molecules would be sharply defined. For this reason, chemists and physicists have worked extensively to develop techniques to reduce the temperature of a sample as close as possible to absolute zero. Along the way, they have made a number of astonishing discoveries, including superconductivity and superfluidity.

Reducing the temperature becomes ever more difficult, the nearer we approach absolute zero. If we wish to reduce the temperature of a sample below the temperature of its surroundings, we must find some way other than a heat flow to remove energy of motion, because "uphill" heat flows are never spontaneous. To cool a refrigerator below the temperature of its surroundings, we use a refrigerating fluid that condenses when compressed and absorbs energy from its surroundings as it vaporizes. Expanding nitrogen gas cools and will eventually condense at 77 K. Still lower temperatures require even more elaborate procedures.

Low-temperature research is very difficult, but the scientific rewards are quite large. In fact, five Nobel prizes in physics and chemistry have been awarded for low-temperature research. The first, in 1913, went to the Dutch physicist Heike Kamerlingh-Onnes, who discovered how to cool He gas to 4.2 K and convert it into a liquid. The American William Giauque received the 1949 prize in chemistry and the Russian Pjotr Kapitza was awarded the 1978 prize in physics. Each was honored for a variety of discoveries resulting from low-temperature research, and each developed a new technique for achieving low temperature.

Giauque exploited order-disorder phenomena to cool solids to ultra-low temperatures. His technique, improved versions of which still are used in contemporary low-temperature research, begins with a sample containing paramagnetic ions (Fe^{3+}, for example). The sample is bathed in liquid helium under reduced pressure to chill it to a temperature below 4.2 K. The sample is held in a strong magnetic field, which causes all the paramagnetic ions to align in a highly ordered arrangement. Then the liquid helium is removed and the magnet turned off. The paramagnetic ions become disordered, and as they do, they absorb heat, leading to a drop in temperature. Temperatures

in the milliKelvin to microKelvin range can be achieved using this technique.

The 1996 Nobel prize in physics went to a trio of Americans who, working at temperatures of 2 milliKelvin, discovered and characterized superfluid helium. A superfluid behaves completely unlike conventional liquids. Liquids normally are viscous because their molecules interact with one another to reduce fluid motion. Superfluid helium has zero viscosity, meaning that its atoms move nearly independently of one another. Superfluid liquid helium can conduct heat perfectly, so heating a sample at one particular spot results in the same increase in temperature throughout the entire volume. A superfluid also flows extremely easily, so it can form a fountain, shown in the photo, in apparent defiance of gravity. Superfluidity is just one of the surprising new properties discovered through low temperature research. Another example is superconductivity, described in our Chemistry and Technology Box in Chapter 10.

The most recent Nobel prize for low-temperature research, awarded in 1997, went to three American physicists working in different laboratories. Their work focused on gases rather than liquids and solids. When an atomic gas expands through a small opening, it forms a beam of atoms travelling in one direction (recall the molecular beams described in Section 5.1). Normally, these atoms have speeds around 1 km/s. When they are bombarded with photons from an intense laser beam with just the right frequency, however, the interaction between photons and atoms slows down the atoms, much as a stiff headwind slows down a runner. A sophisticated combination of six laser beams can slow a beam of sodium atoms to a speed of 30 cm/s, which corresponds to a temperature of 2.4×10^{-4} K. At this temperature, the sodium atoms would instantaneously condense to form a solid if they collided with one another, but the use of atomic beams ensures that the atoms do not collide, so they remain in the gas phase. Refinements of this technique have made it possible to cool helium atoms to temperatures around 2×10^{-7} K. At this temperature, the atoms move at a leisurely 2 cm/s. Just as studies of condensed phases at low temperature revealed new types of behavior, gaseous arrays at very low temperature have unique properties that are just beginning to be explored.

Table 13 – 2
Absolute Molar Entropies ($S°$) of Selected Substances*

Substance	Phase	$S°$ (J/mol K)	Substance	Phase	$S°$ (J/mol K)
C	Solid (Diamond)	2.4	He	Gas	126.153
C	Solid (Graphite)	5.74	Ar	Gas	154.846
Si	Solid	18.8	Xe	Gas	169.685
Al	Solid	28.3	H_2	Gas	130.680
Cu	Solid	33.2	CO	Gas	197.660
Ag	Solid	42.6	F_2	Gas	202.79
SiO_2	Solid	41.5	Cl_2	Gas	223.08
NaCl	Solid	72.1	CO_2	Gas	213.78
I_2	Solid	116.1	C_2H_2	Gas	200.9
H_2O	Liquid	69.95	C_2H_4	Gas	219.3
Hg	Liquid	75.9	C_2H_6	Gas	229.2
Br_2	Liquid	152.2	SiF_4	Gas	282.7

★ $T = 298$ K, and $P = 1$ atm.

2. The absolute entropy of a substance depends on its phase. For example, compare three substances that are quite similar except for their phase under standard conditions (Figure 13-13):

> The superscript ° for absolute entropies always designates standard thermodynamic conditions, meaning 1 atm pressure and 298 K.

$$I_2(s), \ S° = 116 \text{ J/mol K} \quad Br_2(l), \ S° = 152 \text{ J/mol K} \quad Cl_2(g), \ S° = 223 \text{ J/mol K}$$

The substance in the more restricted phase has the lower entropy. For substances that are otherwise similar, $S_{solid} < S_{liquid} < S_{gas}$.

3. Among substances that share a phase and similar structure, $S°$ values increase with molar mass. For example:

Substance	MM (g/mol)	$S°$ (J/mol K)	Substance	MM (g/mol)	$S°$ (J/mol K)
He(g)	4.00	126	Cu(s)	63.6	33
Ar(g)	39.9	155	Ag(s)	108	43

Iodine (solid)

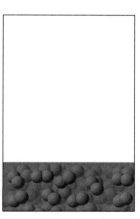

Bromine (liquid)

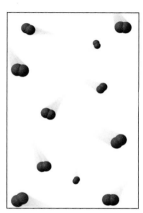

Chlorine (gas)

Figure 13-13
The absolute entropies of otherwise similar substances depend on their phases:
$S°\ I_2(s) < S°\ Br_2(l) < S°\ Cl_2(g)$

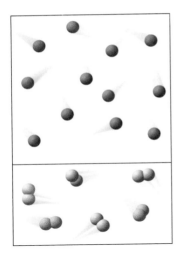

Figure 13-14
At standard temperature and pressure, a 2-mol sample of gaseous He (*top*) is more disordered than a 1-mol sample of H$_2$ (*bottom*).

4. Molar entropies increase as the size of the molecule increases. A molecule with many atoms has more ways of arranging its atoms in space than a molecule with only a few atoms. Thus larger molecules have a greater amount of intramolecular disorder. Compare, for example, the standard molar entropies of acetylene, ethylene, and ethane:

Acetylene	C$_2$H$_2$	S° = 200.9 J/mol K
Ethylene	C$_2$H$_4$	S° = 219.2 J/mol K
Ethane	C$_2$H$_6$	S° = 229.2 J/mol K

One important feature not revealed by Table 13-2 is that binding atoms into molecules always reduces the amount of disorder. As illustrated in Figure 13-14, a sample containing 2 mol He has considerably more entropy (252 J/mol K) than 1 mol H$_2$ (131 J/mol K), even though both samples contain the same total number of atoms.

Example 13-3 provides practice in identifying how absolute entropies vary with molecular properties.

Example 13-3 | **Absolute Entropies**

For each of the following pairs of substances under standard conditions, predict which has the larger standard entropy and give a reason why: (a) 1 mol each of Hg and Au; (b) 1 mol each of NO and NO$_2$; (c) 2 mol of NO$_2$ and 1 mol of N$_2$O$_4$; and (d) 1 mol each of Xe and Kr.

Strategy: This is a qualitative example, so the seven–step procedure is not appropriate. Analyze each pair, keeping in mind that several features affect molecular disorder and entropy, including phase, number of atoms or molecules, amount of bonding, and molar mass.

Solution:
(a) At 298 K, mercury is a liquid metal, but gold is a solid metal. In general, a liquid has more disorder than a solid. The two elements have almost the same molar mass, so Hg should have a larger S° value than Au. Values in Appendix D confirm the prediction: Hg, 76 J/mol K; Au, 47 J/mol K.

(b) Both NO and NO$_2$ are gases under standard conditions, but each molecule of NO$_2$ has three atoms, and each molecule of NO has two atoms. Thus NO$_2$ should have a higher standard molar entropy than NO, and experimental values confirm this: NO$_2$, 240 J/mol K; NO, 211 J/mol K.

(c) Two moles of NO$_2$ contain the same number of atoms as 1 mol of N$_2$O$_4$. In N$_2$O$_4$, however, pairs of NO$_2$ units are tied together by a bond that reduces their randomness. Thus 2 mol of NO$_2$ should have more entropy than 1 mol of N$_2$O$_4$, as confirmed by experimental values: NO$_2$, (240 J/mol K) (2 mol) = 480 J/K; N$_2$O$_4$, (304 J/mol K)(1 mol) = 304 J/K.

(d) Both Xe and Kr are monatomic gases from Group 18 of the periodic table. Because Xe has a higher molar mass than Kr, Xe is expected to have greater entropy: Xe, 170 J/mol K; Kr, 164 J/mol K.

Entropy and Concentration

The molar entropy of a gas is larger than the molar entropy of the same substance as a liquid or a solid. This is because molecules are more disordered in a dispersed sample than in a compact sample. This link between dispersion and disorder also means that entropy increases when a gas expands into a larger volume. Consider, for example, a gas expanding as shown in Figure 13-15a. As the gas expands, its molecules become more randomly distributed in space, increasing the amount of disorder and entropy.

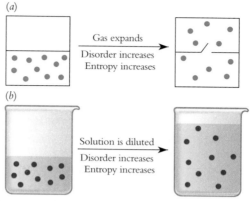

Figure 13-15
(a) When a gas expands, its disorder and entropy increase. (b) When a solution is diluted by a factor of two, the increase in disorder of the solute is the same as when the pressure of a gas is reduced by a factor of two.

When a solution is diluted by adding more solvent, there is a similar increase in dispersion. Figure 13-15b illustrates that the dilution of a *solution* has the same disordering effect as the expansion of a *gas* into a larger volume. Whenever the same number of molecules is distributed throughout a larger volume, the amount of disorder increases. This means that the less concentrated the sample, the more random its molecular distribution and the greater its disorder. Entropy increases as concentration decreases.

The mathematical machinery of thermodynamics allows this qualitative statement to be expressed quantitatively. Experiments and theory show that the molar entropy of a gas or solute varies logarithmically with concentration:

$$S = S° - R \ln (c/c°)$$

The ideal gas equation (Equation 5-3) can be rearranged to show that the concentration of a gas, which is moles of gas per unit volume, is conveniently measured by the pressure (P):

$$c = \frac{n}{V} = \frac{P}{RT}$$

Applying this equality to both c and $c°$, we find that for a gas, the concentration ratio equals the pressure ratio:

$$\frac{c}{c°} = \frac{P}{P°}$$

Noting that $c° = 1$ M and $P° = 1$ atm, we can omit them from the equations for entropies of gases and solutes when their concentrations differ from standard conditions:

$$S_{(P \neq 1 \text{ atm})} = S° - R \ln P \qquad S_{(c \neq 1 \text{ M})} = S° - R \ln c \qquad \textbf{(13-4)}$$

The concentration always must be expressed using the same units as the standard state of the substance: atmospheres for gases and molarity for solutes.

Equation 13-4 refers to molar quantities. To obtain the total entropy of a sample, its molar entropy must be multiplied by n, the number of moles. Example 13-4 illustrates the calculation of ΔS for a change in concentration.

| Example 13-4 | Entropy Change on Expansion |

Oxygen gas is used in many applications, from welders' torches to respirators. The gas is sold commercially in pressurized steel tanks. Suppose a tank contains O_2 at $P = 6.50$ atm and $T = 298$ K. Using standard thermodynamic data, compute the molar entropy of the gas in the tank at 6.50 atm and the change in entropy of a 0.155-mol sample of gas withdrawn from the tank at 1.10 atm and constant temperature.

Strategy: This is a quantitative thermodynamics problem that asks us to determine entropy under nonstandard conditions using standard thermodynamic data, so we turn to the seven-step approach.

Solving Quantitative Problems

Solution:

1. The problem asks for two quantities: the molar entropy of O_2 gas in the pressurized tank and the entropy change when a sample of that gas undergoes a pressure change.

2. and 3. Begin with a diagram of the process that includes a summary of all the information given in the problem:

Tank of O_2 gas

$P = 6.50$ atm
$T = 298$ K

$S_{molar} = ?$

Remove gas

$\Delta S = ?$

Sample of O_2 gas

$P = 1.10$ atm
$T = 298$ K
$n = 0.155$ mol

From Appendix D: $S° (O_2(g)) = 205$ J/mol K

4. and 5. The gas stored in the tank is not at standard pressure, so apply Equation 13-4 to calculate its molar entropy. As the gas expands, its entropy increases. The final pressure is not standard pressure, so again use Equation 13-4 to calculate S at the final pressure. Then calculate the entropy change for the expansion, taking the difference in molar entropies between initial and final conditions and multiplying by the number of moles undergoing the expansion.

6. $S_{(P = 6.50 \text{ atm})} = S° - R \ln P = 205$ J/mol K $- (8.314$ J/mol K$) \ln 6.50$

$= 205$ J/mol K $- 15.562$ J/mol K $= 189$ J/mol K

$S_{(P = 1.10 \text{ atm})} = S° - R \ln P = 205$ J/mol K $- (8.314$ J/mol K$) \ln 1.10$

$= 205$ J/mol K $- 0.7924$ J/mol K $= 204$ J/mol K

$\Delta S = n(S_{final} - S_{initial}) = (0.155 \text{ mol})(204$ J/mol K $- 189$ J/mol K$) = 2.3$ J/K

7. The positive value for this entropy change reflects the fact that the expansion is a spontaneous process.

Standard Reaction Entropies

Entropy changes are important in every process. However, chemists are interested primarily in the effects of entropy on chemical reactions. If a reaction occurs under standard conditions, its entropy change can be calculated from absolute entropies in exactly the same way as reaction enthalpies can be calculated from standard heats of formation.

The products of the reaction have molar entropies, and so do the reactants. The total entropy of the products is the sum of the molar entropies of the products multiplied by their stoichiometric coefficients in the balanced chemical equation. The total entropy of the reactants is a similar sum for the reactants. The entropy change is the difference between these two quantities:

$$\Delta S^{\circ}_{\text{reaction}} = \Sigma \text{ coeff}_p \, S^{\circ} \text{ (products)} - \Sigma \text{ coeff}_r \, S^{\circ} \text{ (reactants)} \qquad (13\text{-}5)$$

Notice that Equation 13-5 is similar to Equation 12-10, which is used to calculate reaction *enthalpies*.

$$\Delta H^{\circ}_{\text{reaction}} = \Sigma \text{ coeff}_p \, \Delta H^{\circ}_f \text{ (products)} - \Sigma \text{ coeff}_r \, \Delta H^{\circ}_f \text{ (reactants)} \qquad (12\text{-}10)$$

As Example 13-5 shows, the applications of these equations follow parallel paths.

Entropy and Enthalpy Changes

Example 13-5

Acrylonitrile is an essential monomer in the polymer industry because it is used to make polyacrylonitrile for synthetic fibers:

Polyacrylonitrile   Many repeat units

Acrylonitrile is made from propene:

2 ⌇ + 2 NH₃ + 3 O₂ ⟶ 2 ⌇≡N + 6 H₂O

Propene
(C₃H₆)

Acrylonitrile
(CH₂CHCN)

Calculate $\Delta H^{\circ}_{\text{reaction}}$ and $\Delta S^{\circ}_{\text{reaction}}$ for the synthesis of acrylonitrile from standard thermodynamic data. (All three starting materials are gases, and both products are liquids.)

Strategy: We use a short version of the seven-step method.

Solution: Standard heats of formation and standard entropies are tabulated in Appendix D. Here are the values for the substances involved in this reaction:

Substance	$C_3H_6(g)$	$NH_3(g)$	$O_2(g)$	$C_3H_3N(l)$	$H_2O(l)$
ΔH°_f (kJ/mol)	20.0	−45.9	0	172.9	−285.83
S° (J/mol K)	226.9	192.8	205.15	188	69.95
Coefficient	2	2	3	2	6

Standard thermodynamic changes for a reaction can be calculated using Equations 12-10 and 13-5:

$$\Delta H^{\circ}_{\text{reaction}} = \Sigma \text{ coeff}_p \, \Delta H^{\circ}_f \text{ (products)} - \Sigma \text{ coeff}_r \, \Delta H^{\circ}_f \text{ (reactants)}$$

$$\Delta S^{\circ}_{\text{reaction}} = \Sigma \text{ coeff}_p \, S^{\circ} \text{ (products)} - \Sigma \text{ coeff}_r \, S^{\circ} \text{ (reactants)}$$

| Example 13-5 | Entropy and Enthalpy Changes *(continued)* |

We leave it to you to plug in the numbers and calculate the answers. The results are as follows:

$$\Delta H^\circ_{reaction} = -1.317 \times 10^3 \text{ kJ} \qquad \Delta S^\circ_{reaction} = -659 \text{ J/K}$$

According to these values, the reaction is exothermic, but the order in the system increases. Because liquids are formed from gases, this increase in order is a reasonable result.

Section Exercises

13.3.1 Of the following pairs, which has the greater entropy? Explain your choice. (a) 1 g of dew or 1 g of frost; (b) 1 mol of gaseous hydrogen atoms or 0.5 mol of gaseous hydrogen molecules; (c) "perfect" diamond or flawed diamond, each ¼ carat; and (d) 5 mL of liquid ethanol at 0 °C or 5 mL of liquid ethanol at 50 °C.

13.3.2 Explain the following differences in entropies in molecular terms (substances are at standard conditions unless otherwise noted):
(a) 1 mol of O_2 has less entropy than 1 mol of O_3; (b) 3 mol of O_2 has more entropy than 2 mol of O_3; (c) 1 mol of I_2 has less entropy than 1 mol of O_2; and (d) 1 mol of HCl(aq) in concentrated solution (12 M) has less entropy than 1 mol of HCl(aq) in dilute solution (0.100 M).

13.3.3 Draw molecular pictures to illustrate your answers to part b of Section Exercise 13.3.1 and part b of Section Exercise 13.3.2.

13.3.4 Compute the standard entropy change for the following reaction:

$$12 \text{ NH}_3(g) + 21 \text{ O}_2(g) \longrightarrow 8 \text{ HNO}_3(g) + 4 \text{ NO}(g) + 12 \text{ H}_2O(l)$$

13.4 SPONTANEITY AND FREE ENERGY

The second law of thermodynamics states that the entropy of the universe always increases during a spontaneous process. That is, the sign of $\Delta S_{universe}$ is positive for every spontaneous chemical process. A calculation of $\Delta S_{universe}$ determines whether a particular chemical process is spontaneous. Unfortunately, this is not practical for most processes. It is usually possible to calculate the entropy change for a *system*, but the *surroundings* may undergo complicated changes of state for which ΔS is very difficult to determine. The surroundings include virtually all the universe, and keeping track of changes in the universe is a tricky matter. It would be much more convenient to have a way to determine the direction of spontaneous change using *just the system*, not the surroundings.

There is no single criterion that can be applied to *all* processes. However, with sufficient restrictions placed on the conditions, there is a state function whose change for the *system* predicts spontaneity. This new state function is called **free energy (G)** and is defined by Equation 13-6:

$$\text{Free energy} = G = H - TS \tag{13-6}$$

where H is enthalpy, T is temperature, and S is entropy.

The definition of free energy gives an equation that relates the change in free energy for any system to other thermodynamic changes:

$$\Delta G_{sys} = \Delta H_{sys} - \Delta(TS)_{sys}$$

To use ΔG_{sys} to predict spontaneity, this equation must be linked to the total entropy change of the universe. This cannot be done unless some restrictions are placed on the conditions. First, the process must occur at *constant temperature*. This lets us relate $\Delta(TS)_{sys}$ to ΔS_{sys}:

$$\Delta(TS) = (TS)_{final} - (TS)_{initial} = T(S_f - S_i) = T\,\Delta S_{sys}$$

$$\Delta G_{sys} = \Delta H_{sys} - T\,\Delta S_{sys} \qquad \text{(constant } T\text{)} \qquad\qquad (13\text{-}7)$$

Next, the *enthalpy* change for the system can be related to the *entropy* change for the surroundings by restricting the conditions to *constant pressure*. Recall that ΔH equals q when P is constant:

$$\Delta H_{sys} = q_{sys} \qquad \text{(constant } P\text{)}$$

Remember also that the heat flow for a system is always equal in magnitude but opposite in sign to the heat flow of the surroundings:

$$q_{sys} = -q_{surr} \qquad so \qquad \Delta H_{sys} = -q_{surr}$$

Furthermore, the restriction of constant temperature has already been imposed, so the heat flow of the surroundings measures the entropy change of the surroundings. That is:

$$\frac{q_{surr}}{T} = \Delta S_{surr} \qquad and \qquad q_{surr} = T\,\Delta S_{surr}$$

Combining these equalities gives an equation that relates ΔH_{sys} and ΔS_{surr}:

$$\Delta H_{sys} = -T\,\Delta S_{surr} \qquad \text{(constant } P \text{ and } T\text{)}$$

Now substitute this result into Equation 13-7:

$$\Delta G_{sys} = -T\,\Delta S_{surr} - T\,\Delta S_{sys} = -T(\Delta S_{surr} + \Delta S_{sys})$$

Finally, because $\Delta S_{surr} + \Delta S_{sys} = \Delta S_{universe}$:

$$\Delta G_{sys} = -T\,\Delta S_{universe} \qquad \text{(constant } P \text{ and } T\text{)}$$

This powerful equality states that the free energy of the *system* changes in a way that mirrors the entropy change of the *universe* in any process that occurs at constant T and P. Defining a new function and imposing some restrictions provides a way to use properties of a system to determine whether a process is spontaneous. Because T is always positive and $\Delta S_{universe}$ is positive for any spontaneous process:

/// ΔG_{sys} *is negative for all spontaneous processes under conditions of constant temperature and pressure.*

Although the restrictions of constant T and P are stringent, they apply to many important chemical processes. Some examples are shown in Figure 13-16. A particularly important example is the human body, which has a nearly constant temperature of 37 °C and nearly constant pressure of 1 atm. Any biochemical reaction that occurs in the body occurs under conditions in which the immediate surroundings are at constant T and P.

Figure 13-16
Many real-life situations operate under nearly-constant temperature and pressure and thus are subject to $\Delta G_{sys} < 0$ as the condition for spontaneity. Examples shown here are a laboratory synthesis, a biological system, and an industrial process.

Because free energy is a state function, its values can be tabulated for use in chemical calculations. As with standard heats of formation, the **standard molar free energy of formation (ΔG_f°)** for any substance is defined to be the change of free energy when one mole of that substance is formed from elements in their standard states. The same reasoning that is used for enthalpy changes leads to an equation for calculating the free energy change for any chemical reaction:

$$\Delta G_{reaction}^\circ = \Sigma \, \text{coeff}_p \, \Delta G_f^\circ \, (\text{products}) - \Sigma \, \text{coeff}_r \, \Delta G_f^\circ \, (\text{reactants}) \qquad \textbf{(13-8)}$$

The form of Equation 13-8 should look familiar to you, because it is analogous to Equation 12-10 for reaction enthalpies and Equation 13-5 for reaction entropies.

The standard free energy change for a reaction can also be calculated from ΔH° and ΔS° for the reaction by making use of Equation 13-7 under standard conditions:

$$\Delta G^\circ = \Delta H^\circ - T \, \Delta S^\circ \qquad \textbf{(13-9)}$$

Either of these equations can be used to find standard free energy changes. Which equation to use depends on the available data. Example 13-6 illustrates both types of calculations.

Example 13-6	Free Energy of Reaction

Find the standard free energy change for the acrylonitrile synthesis discussed in Example 13-5.

Strategy: There are two ways to calculate $\Delta G_{reaction}^\circ$. The first method uses standard free energies of formation and Equation 13-8. The second method uses Equation 13-9 and the values of ΔH° and ΔS° calculated earlier. Either method requires the balanced chemical equation:

| Free Energy of Reaction *(continued)* | Example 13-6 |

$$2\ C_3H_6(g) + 2\ NH_3(g) + 3\ O_2(g) \longrightarrow 2\ C_3H_3N(l) + 6\ H_2O(l)$$

We perform both calculations to show that they give the same result.

Solution:

(a) Using Equation 13-8 and ΔG_f° values from Appendix D:

Equation 13-8: $\Delta G_{reaction}^\circ = \Sigma$ coeff$_p\ \Delta G_f^\circ$ (products) $-\ \Sigma$ coeff$_r\ \Delta G_f^\circ$ (reactants)

Substitute values from Appendix D:

$$\Delta G_{reaction}^\circ = [(6\ mol)(-237.1\ kJ/mol) + (2\ mol)(208.6\ kJ/mol)]$$
$$- [(3\ mol)(0\ kJ/mol) + (2\ mol)(-16.4\ kJ/mol) + (2\ mol)(74.62\ kJ/mol)]$$

$$\Delta G_{reaction}^\circ = -1122\ kJ$$

(b) Using Equation 13-9 and the enthalpy and entropy results of Example 13-5:

$$\text{Equation 13-9: } G_{reaction}^\circ = \Delta H^\circ - T\,\Delta S^\circ$$

Results of Example 13-5:

$$\Delta H_{reaction}^\circ = -1.317 \times 10^3\ kJ \quad and \quad \Delta S_{reaction}^\circ = -659\ J/K$$

$$G_{reaction}^\circ = [-1317\ kJ] - [(298\ K)(-659\ J/K)(10^{-3}\ kJ/J)] = -1121\ kJ$$

The large negative $\Delta G_{reaction}^\circ$ indicates that the production of acrylonitrile is highly spontaneous under standard conditions.

Pay close attention to units when using Equation 13-9. The values of ΔH° and ΔG° are usually given in kJ or kJ/mol, but S° is usually expressed in J/K or J/mol K. Thus entropies must be multiplied by 10^{-3} kJ/J before adding the two terms.

Change in Free Energy under Nonstandard Conditions

Standard conditions refers to unit concentrations and 298 K, but chemical reactions occur at many different concentrations and temperatures. To use ΔG as a measure of spontaneity under nonstandard conditions, we must know how G depends on temperature and concentration.

First, consider temperature. Equation 13-9 is valid at *any* temperature as long as the temperature is *constant*. However, application of the equation at a temperature different from 298 K requires the appropriate values of $\Delta H_{reaction}$ and $\Delta S_{reaction}$ at the nonstandard temperature.

Chemical substances become more disordered as temperature increases. These changes in entropy as a function of temperature can be calculated, but the techniques require calculus. Fortunately, temperature affects the entropies of reactants and products in the same way. Although the absolute entropy of every substance increases with temperature, the amount of disorder for the reactants often changes with temperature by almost the same amount as the amount of disorder for the products. This means that the temperature effect on the entropy *change* for a reaction is usually small enough that we can consider $\Delta S_{reaction}$ to be independent of temperature.

As pointed out in Section 12.4, $\Delta H_{reaction}$ also does not change rapidly with temperature. As a result, free energy changes at temperatures other than 298 K can be estimated by assuming that standard enthalpies and entropies at 298 K also apply at any other temperature:

$$\Delta G_{reaction, T}^\circ \cong \Delta H_{reaction, 298\ K}^\circ - T\,\Delta S_{reaction, 298\ K}^\circ \quad \textbf{(13-10)}$$

In Equation 13-10 the superscript ° denotes standard concentrations (1 atm and 1 M) of all reagents, even though temperature is nonstandard ($T \neq 298$ K). When the temperature is not specified, ° means standard concentrations *and* 298 K. Therefore ΔG° means "free energy change with all reagents at unit concentration and $T = 298$ K," whereas ΔG_{500}° means "free energy change with all reagents at unit concentration and $T = 500$ K."

An immediate consequence of Equation 13-10 is that the spontaneity of any reaction with a large $\Delta S°$ is very sensitive to temperature. Example 13-7 shows this.

| Example 13-7 | Temperature and Spontaneity |

Dinitrogen tetroxide can decompose into two molecules of nitrogen dioxide:

$$N_2O_4(g) \longrightarrow 2\ NO_2(g)$$

(a) Show that this reaction is not spontaneous under standard conditions.
(b) Find the temperature at which the reaction becomes spontaneous at standard pressure.

Strategy: What's asked for? The key word here is *spontaneous*, which suggests that free energies are involved in this problem. The criterion for spontaneity at constant temperature and pressure is $\Delta G_{reaction} < 0$, so the problem asks us to show that $\Delta G_{reaction} > 0$ under standard conditions. Then we can use Equation 13-10 to calculate the temperature that makes $\Delta G_{reaction} < 0$.

Solution:

(a) Because standard free energies of formation are available in Appendix D, we can use Equation 13-8 to determine whether or not the decomposition reaction is spontaneous under standard conditions:

$$\Delta G° = (2\ mol)(51.3\ kJ/mol) - (1\ mol)(99.8\ kJ/mol) = 2.8\ kJ$$

The positive value for $\Delta G°$ indicates that this reaction is *not* spontaneous under standard conditions. In fact, the calculation tells us that the reaction will be spontaneous in the opposite direction. Under standard conditions, NO_2 reacts to form N_2O_4:

$$2\ NO_2(g) \longrightarrow N_2O_4(g) \qquad \Delta G° = -2.8\ kJ$$

(b) For temperatures different from 298 K, we use Equation 13-10 to estimate $\Delta G°_{reaction}$. This equation lets us determine the temperature at which N_2O_4 decomposition becomes spontaneous when the partial pressures of both gases are 1 atm:

$$\Delta G°_{reaction, T} \cong \Delta H°_{reaction, 298\ K} - T\Delta S°_{reaction, 298\ K}$$

As temperature changes, ΔG must become zero before it becomes negative. The reaction will be spontaneous at any temperature farther from 298 K than the temperature at which $\Delta G = 0$. To find the temperature at which $\Delta G° = 0$, begin by calculating $\Delta H°$ and $\Delta S°$ using tabulated data:

$$\Delta H° = (2\ mol)(33.2\ kJ/mol) - (1\ mol)(11.1\ kJ/mol) = 55.3\ kJ$$

$$\Delta S° = (2\ mol)(240.1\ J/mol\ K) - (1\ mol)(304.4\ J/mol\ K) = 175.8\ J/K$$

Now set ΔG equal to zero and rearrange the equation to solve for temperature:

$$\Delta G = \Delta H° - T\Delta S° = 0 \qquad or \qquad T\Delta S° = \Delta H°$$

$$T = \frac{\Delta H°}{\Delta S°} = \frac{55.3\ kJ}{(175.8\ J/K)(10^{-3}\ kJ/J)} = 315\ K$$

At 315 K, $\Delta G° = 0$. Therefore at all temperatures greater than 315 K, $\Delta G°$ is negative, and decomposition of N_2O_4 is spontaneous at 1 atm partial pressures.

Changing Concentration

Substances that participate in chemical reactions typically are at concentrations different from one molar and may be at pressures different from one atmosphere. For example, a biochemist who wants to know what processes are spontaneous under physiological conditions will find that the substances dissolved in biological fluids are rarely at one molar concentration. How does ΔG vary with changes in molarity and pressure? Recall that enthalpy is virtually independent of concentration but that entropy obeys Equation 13-4.

$$S_{(P \neq 1\ atm)} = S° - R \ln P \qquad S_{(c \neq 1\ M)} = S° - R \ln c \qquad \textbf{(13-4)}$$

To see how concentration affects ΔS, consider the decomposition of limestone ($CaCO_3$) to produce lime (CaO), carried out industrially in kilns in which the partial pressure of CO_2 (p) is different from 1 atm:

$$CaCO_3(s) \longrightarrow CaO(s) + CO_2(g)$$

The entropy change for the reaction is the difference in entropy between products and reactants:

$$\Delta S_{reaction} = S(CaO) + S(CO_2) - S(CaCO_3)$$

The entropy of each solid is its standard entropy, $S°$, but the entropy of the gas must be corrected for the deviation of pressure from standard conditions. Equation 13-4 gives the molar entropy of carbon dioxide as a function of its partial pressure:

$$S_{(p \neq 1\ atm)} = S° - R \ln p$$

$$S(CO_2) = S°(CO_2) - R \ln(p_{CO_2})$$

Now substitute the corrected entropy of CO_2 into the equation for $\Delta S_{reaction}$:

$$\Delta S_{reaction} = S°(CaO) + [S°(CO_2) - R \ln(p_{CO_2})] - S°(CaCO_3)$$

Next, rearrange the equation so that all the standard entropies are together:

$$\Delta S_{reaction} = S°(CaO) + S°(CO_2) - S°(CaCO_3) - R \ln(p_{CO_2})$$

The first three terms are the standard entropy change for the reaction, allowing simplification:

$$\Delta S_{reaction} = \Delta S°_{reaction} - R \ln(p_{CO_2})$$

In this example, only one of the reagents has a concentration that can vary, and each stoichiometric coefficient is one. What happens for a more complicated reaction? Consider the synthesis of ammonia carried out in a pressurized reactor containing N_2, H_2, and NH_3 at partial pressures different from 1 atm:

$$N_2(g) + 3\ H_2(g) \longrightarrow 2\ NH_3(g)$$

The entropy change for the reaction is the difference in entropy between products and reactants, obtained by multiplying each corrected entropy by the appropriate stoichiometric coefficient:

$$\Delta S_{reaction} = 2\ S(NH_3) - 3\ S(H_2) - S(N_2)$$

Applying Equation 13-4 gives the molar entropy of each gas:

$$S_{(p \neq 1\ atm)} = S° - R \ln p$$

$$S(N_2) = S°(N_2) - R \ln(p_{N_2})$$

$$S(H_2) = S°(H_2) - R \ln(p_{H_2})$$

$$S(NH_3) = S°(NH_3) - R \ln(p_{NH_3})$$

Now substitute the corrected entropies:

$$\Delta S_{reaction} = 2 \left[S°(NH_3) - R \ln(p_{NH_3}) \right] - 3 \left[S°(H_2) - R \ln(p_{H_2}) \right]$$
$$- \left[S°(N_2) - R \ln(p_{N_2}) \right]$$

Next, rearrange the equation so that all the logarithmic terms are together:

$$\Delta S_{reaction} = 2\, S°(NH_3) - 3S°(H_2) - S°(N_2)$$
$$-2\,R \ln(p_{NH_3}) + 3\,R \ln(p_{H_2}) + R \ln(p_{N_2})$$

As before, the first three terms are the standard entropy change for the reaction:

$$\Delta S_{reaction} = \Delta S°_{reaction} - 2\,R \ln(p_{NH_3}) + 3\,R \ln(p_{H_2}) + R \ln(p_{N_2})$$

The properties of logarithms can be used to combine the logarithmic terms. First, $a \ln x = \ln x^a$, giving:

$$2\,R \ln(p_{NH_3}) = R \ln(p_{NH_3})^2 \qquad and \qquad 3\,R \ln(p_{H_2}) = R \ln(p_{H_2})^3$$

With these changes, the equation becomes:

$$\Delta S_{reaction} = \Delta S°_{reaction} - R \left[\ln(p_{NH_3})^2 - \ln(p_{H_2})^3 - \ln(p_{N_2}) \right]$$

A second logarithmic property, $\ln u - \ln v = \ln(u/v)$, lets us put all the logarithmic terms into a single ratio:

$$\Delta S_{reaction} = \Delta S°_{reaction} - R \ln \left[\frac{(p_{NH_3})^2}{(p_{N_2})(p_{H_2})^3} \right]$$

Note that the pressure ratio has product concentrations raised to their stoichiometric coefficients in the numerator and reactant concentrations raised to their stoichiometric coefficients in the denominator. The form of this equation applies to all reactions, not just the synthesis of ammonia. A general reaction can be written as

$$aA + bB \longrightarrow dD + eE$$

Here, A and B represent any reactants, D and E represent any products, and the lower case letters represent the stoichiometric coefficients. The expression for the entropy change accompanying the reaction is

$$\Delta S_{reaction} = \Delta S°_{reaction} - R \ln \left[\frac{(c_D)^d (c_E)^e}{(c_A)^a (c_B)^b} \right]$$

The ratio in the logarithmic term is called the **reaction quotient (Q):**

$$\left[\frac{(c_D)^d (c_E)^e}{(c_A)^a (c_B)^b} \right] = Q \tag{13-11}$$

Each concentration in the reaction quotient is measured relative to the standard concentrations (1 atm for gases, 1 M for solutes) so gas concentrations must be in atmospheres and solute concentrations in moles per liter.

The entropy change for a reaction under nonstandard concentrations can be expressed in terms of the standard entropy change and Q:

$$\Delta S_{reaction} = \Delta S°_{reaction} - R \ln Q$$

This general expression describing how $\Delta S_{\text{reaction}}$ varies with concentrations leads to an equation for the free energy change when concentrations are nonstandard:

$$\Delta G_{\text{reaction}} = \Delta H^\circ_{\text{reaction}} - T(\Delta S^\circ_{\text{reaction}} - R \ln Q)$$

Because $\Delta H^\circ - T\,\Delta S^\circ = \Delta G^\circ$, this equation reduces to Equation 13-12:

$$\Delta G_{\text{reaction}} = \Delta G^\circ_{\text{reaction}} + RT \ln Q \qquad \text{(13-12)}$$

An immediate consequence of Equation 13-12 is that the direction of spontaneity of a reaction depends heavily on the concentrations of reactants and products. Product concentrations appear in the numerator of Q, so when the concentration of a product increases, $\ln Q$ increases as well. Increasing the $\ln Q$ term makes ΔG less negative, so a reaction becomes less spontaneous as product concentrations increase. Conversely, reactant concentrations appear in the denominator of Q, so increasing the concentration of a reactant decreases $\ln Q$, which in turn makes ΔG more negative. Thus a reaction becomes more spontaneous as reactant concentrations increase.

If a reaction is not spontaneous at unit concentrations, it can be forced to go forward by increasing the concentrations of reactants or by removing products as they form. If a reaction has a very positive $\Delta G^\circ_{\text{reaction}}$, however, it may not be possible to change the value of Q sufficiently to make $\Delta G_{\text{reaction}} < 0$. Remember that $\ln Q$ changes more slowly than Q itself. A tenfold change in Q, for instance, changes $\ln Q$ only by a factor of 2.3. Nonetheless, reactions that are almost spontaneous under standard conditions can be driven forward by making appropriate changes in concentration. Example 13-8 shows how this is done.

← **CHAPTER 15**
The reaction quotient plays an important role in equilibrium calculations as well as in thermodynamics, as we show in Chapter 15.

| **Effect of Concentration on Spontaneity** | **Example 13-8** |

The decomposition of dinitrogen tetroxide produces nitrogen dioxide:

$$N_2O_4(g) \longrightarrow 2\ NO_2(g)$$

(a) Find the minimum partial pressure of N_2O_4 at which the reaction is spontaneous if $p_{NO_2} = 1$ atm and $T = 298$ K.

(b) Find the maximum partial pressure of NO_2 at which the reaction is spontaneous if $p_{N_2O_4} = 1$ atm and $T = 298$ K.

Strategy: This is a two-part problem, so each part should be solved independently. Both parts relate change in free energy to concentrations, so Equation 13-12 must be used:

$$\Delta G = \Delta G^\circ + RT \ln Q = \Delta G^\circ + RT \ln \left(\frac{p^2_{NO_2}}{p_{N_2O_4}} \right)$$

The standard free energy change was determined in Example 13-7:

$$\Delta G^\circ = 2.8 \text{ kJ/mol}$$

Example 13-8	**Effect of Concentration on Spontaneity** (*continued*)

Solution:

(a) The problem asks for the partial pressure of N_2O_4 that will make the decomposition spontaneous when $T = 298$ K and p of $NO_2 = 1$ atm. The value of ΔG must be zero before it can become negative. Therefore to find the threshold pressure of N_2O_4 that makes the decomposition spontaneous, set $\Delta G = 0$ and $p_{NO_2} = 1$ atm and then rearrange to solve for the partial pressure of N_2O_4:

> The mol unit in the value of $\Delta G°$ refers to "per mole of reaction" and is included in the calculation to cancel the mole unit in R.

$$0 = \Delta G° + RT \ln \left[\frac{(1 \text{ atm})^2}{(p_{N_2O_4})} \right]$$

$$\ln \left[\frac{1}{(p_{N_2O_4})} \right] = -\frac{\Delta G°}{RT} = \frac{-(2.8 \text{ kJ/mol})(10^3 \text{ J/kJ})}{(8.314 \text{ J/mol K})(298 \text{ K})} = -1.13$$

> Recall that $y = \ln x$ implies that $x = e^y$, where e is the base for natural logarithms.

$$\frac{1}{(p_{N_2O_4})} = e^{-1.13} = 0.323 \quad and \quad p_{N_2O_4} = 1/0.323 = 3.1 \text{ atm}$$

This decomposition is spontaneous as long as the pressure of N_2O_4 is greater than 3.1 atm. As always, we have to be careful about units. $\Delta G°$ is converted to joules to use the appropriate value of the gas constant (R) in J/mol K.

(b) The second part of the problem asks for the maximum partial pressure of NO_2 below which the decomposition is spontaneous when $T = 298$ K and p of $N_2O_4 = 1$ atm. The procedure is analogous to the one developed for part (a). You should be able to show that the maximum pressure is 0.57 atm.

This example shows that a reaction with a small positive $\Delta G°$ can be made spontaneous by relatively small changes in concentrations.

If neither temperature nor concentrations are at standard values, free energy calculations must be done in two steps. First, correct for temperature to obtain $\Delta G_T°$ using Equation 13-10.

$$\Delta G°_{\text{reaction}, T} \cong \Delta H°_{\text{reaction}, 298 \text{ K}} - T \Delta S°_{\text{reaction}, 298 \text{ K}}$$

Second, use the value of $\Delta G_T°$ in Equation 13-12 to complete the calculation of ΔG.

$$\Delta G_{\text{reaction}} = \Delta G°_{\text{reaction}, T} + RT \ln Q$$

Influencing Spontaneity

Suppose a chemist wants to carry out a particular chemical synthesis, but the reaction has a positive value for $\Delta G°$. The thermodynamic calculation indicates that the reaction is spontaneous in the wrong direction under *standard* conditions, but this does not prevent it from occurring under *all* conditions. What can be done to make the reaction go in the desired direction?

Example 13-8 illustrates that changing the concentration quotient changes ΔG for a reaction by altering the entropy change of the system. In particular, reducing the pressure of NO_2 below 0.57 atm or increasing the pressure of N_2O_4 above 3.1 atm would cause spontaneous decomposition of N_2O_4, even though this reaction is not spontaneous under standard conditions. Reactions in liquid solutions likewise may be induced to proceed spontaneously by increasing the concentrations of reactants or by reducing the concentrations of products.

Changing the temperature of the system is another way to influence the spontaneity of a reaction. The equation for ΔG has two parts, ΔH and $T \Delta S$, which can work together or in opposition:

$$\Delta G_T^\circ = \Delta H^\circ - T \Delta S^\circ$$

A *positive* ΔS° promotes spontaneity because it makes ΔG° more negative. This reflects the fact that a positive ΔS° means the system becomes more disordered during the reaction. A *negative* ΔH° promotes spontaneity as well because it also makes ΔG° more negative. This reflects the fact that the surroundings become more disordered when a reaction releases energy. Thus a reaction that has a positive ΔS° and a negative ΔH° is spontaneous at any temperature.

The combustion of propane is an example of a reaction that is spontaneous at all temperatures:

$$C_3H_8(g) + 5\ O_2(g) \longrightarrow 3\ CO_2(g) + 4\ H_2O(g)$$

$$\Delta H^\circ = -897\ kJ \qquad \Delta S^\circ = +145\ J/K$$

The products of this reaction are more disordered than the reactants, and the reaction releases energy. Consequently, ΔG° is negative at all temperatures, and the reverse reaction cannot be made spontaneous by changing the temperature.

By the same reasoning, a *negative* ΔS° and a *positive* ΔH° oppose spontaneity, so a reaction in which the order of the system increases and energy is absorbed is nonspontaneous regardless of temperature. The system and its surroundings both would experience decreases in entropy if such a process were to occur, and this would violate the second law of thermodynamics.

A reaction that has the same sign for ΔS° and ΔH° will be spontaneous at some temperatures but nonspontaneous at others. At low temperature, ΔS° is multiplied by a small value for T, so at sufficiently low temperature, ΔH° contributes more to ΔG° than $T \Delta S^\circ$ does. At high temperature, ΔS° is multiplied by a large value for T, so at sufficiently high temperature, ΔS° contributes more to ΔG° than does ΔH°.

Reactions that have positive ΔH° and positive ΔS° are favored by entropy but disfavored by enthalpy. Such reactions are spontaneous at high temperature, where the $T \Delta S^\circ$ term dominates ΔG°. The reactions are nonspontaneous at low temperature, where the ΔH° term dominates ΔG°. These reactions are spontaneous at high temperature by virtue of the increased disorder in the system.

The opposite situation holds for reactions that have negative values for ΔH° and ΔS°. These reactions are spontaneous at low temperature because their release of heat increases the disorder in the surroundings. The favorable ΔH° dominates ΔG° as long as T does not become too large. At high temperature, however, the unfavorable ΔS° dominates ΔG°, and the reaction is no longer spontaneous. The effects of temperature on spontaneity are summarized in Table 13-3.

Calcium sulfate, the substance used to absorb water in desiccators, provides an example of this temperature sensitivity. Anhydrous calcium sulfate absorbs water vapor from the atmosphere to give the hydrated salt. The reaction has a negative ΔS° because the system becomes more ordered when gaseous water molecules move into the solid state. The reaction also has a negative ΔH° because of the coulombic forces of attraction between the ions of the salt and the polar water molecules.

Desiccators are used to store chemicals that react slowly with water. Calcium sulfate chips in the bottom of the desiccator absorb water vapor from the atmosphere. The blue chips contain an indicator that turns pink when the calcium sulfate is saturated with water.

Table 13–3
Influence of Temperature on Spontaneity

$\Delta H°$	$\Delta S°$	$\Delta G°$ (high T)	$\Delta G°$ (low T)	Spontaneous
−	+	−	−	All T
+	−	+	+	No T
+	+	−	+	High T
−	−	+	−	Low T

$$CaSO_4(s) + 2\,H_2O(g) \longrightarrow CaSO_4 \cdot 2H_2O(s)$$

$$\Delta H° = -104.9 \text{ kJ} \qquad \Delta S° = -290.2 \text{ J/K}$$

At 298 K, the favorable $\Delta H°$ contributes more to $\Delta G°$ than the unfavorable $\Delta S°$:

$$\Delta G°_{298\text{ K}} = (-104.9 \text{ kJ}) - (298 \text{ K})(-290.2 \text{ J/K})(10^{-3} \text{ kJ/J}) = -18.4 \text{ kJ}$$

Thus at room temperature, anhydrous calcium sulfate acts as a "chemical sponge," trapping water vapor spontaneously to form calcium sulfate dihydrate.

The calcium sulfate in a desiccator is effective at removing water vapor only as long as some anhydrous salt remains. When all the anhydrous salt has been converted to the dihydrate, the desiccator can no longer maintain a dry atmosphere. Fortunately, the thermodynamics of this reaction makes it possible to regenerate the drying agent. At 100 °C (373 K), $\Delta S°$ contributes more to $\Delta G°$ than does $\Delta H°$:

$$\Delta G°_{373\text{ K}} = (-104.9 \text{ kJ}) - (373 \text{ K})(-290.2 \text{ J/K})(10^{-3} \text{ kJ/J}) = +3.3 \text{ kJ}$$

At this temperature the reverse reaction is spontaneous. Calcium sulfate dihydrate can be converted to anhydrous calcium sulfate in a drying oven at 100 °C. Then it can be cooled and returned to a desiccator, ready once more to act as a chemical sponge for water.

Section Exercises

13.4.1 Estimate $\Delta G°$ for the formation of gaseous water at $T = 373$ K.

13.4.2 Using Examples 13-7 and 13-8, find the minimum partial pressure of N_2O_4 at which decomposition of N_2O_4 occurs, if T is 127 °C and p of $NO_2 = 0.50$ atm.

13.4.3 Does a temperature exist at which the water formation reaction is nonspontaneous under standard pressure? If so, compute this temperature. If not, explain why in molecular terms.

13.5 SOME APPLICATIONS OF THERMODYNAMICS

Nitrogen Fixation

The distribution of nitrogen between the Earth's crust and the atmosphere is very uneven. In the crust, nitrogen is present at the level of 19 parts per million (ppm) by mass, four orders of magnitude less than oxygen (4.55×10^5 ppm) and silicon (2.72×10^5 ppm). In contrast, 80% of the atmosphere is molecular nitrogen. Paradoxically, nitrogen is absolutely essential for all life, but the sea of atmospheric nitrogen is virtually inaccessible to higher life forms. Most biochemical systems lack the ability to break the strong triple bond between the nitrogen atoms in N_2. Molecular nitrogen must be converted to some other form, usually ammonia (NH_3) or nitrate (NO_3^-), before most life forms can incorporate nitrogen atoms into their biochemical molecules. This process is known as **nitrogen fixation.**

In nature, nitrogen fixation is accomplished by nitrogenase, an enzyme that binds N_2 and weakens its bonding sufficiently to break the triple bond. Only a few algae and bacteria contain nitrogenase, and despite concerted studies by biochemists, the structure of the enzyme is only partially known.

The thermodynamics of nitrogen chemistry helps explain why N_2 is so abundant in the atmosphere, and yet the element remains inaccessible to most life forms. Table 13-4 shows that most of the abundant elements react with O_2 spontaneously under standard conditions. This is why many of the elements are encountered in the Earth's crust as their oxides. However, N_2 is resistant to oxidation, as shown by the positive ΔG_f° for NO_2.

Because of their resistance to chemical attack, nitrogen atoms are not "locked up" in solid or liquid substances as are other elements such as Si, Al, Fe, and H. The most stable form of the element nitrogen is a gaseous diatomic molecule with a very strong triple bond. Therefore the element nitrogen is concentrated in the Earth's gaseous atmosphere even though it is only a trace element in overall abundance.

Every breath of air we take is 80% nitrogen, but our bodies must rely on the nitrogen found in the proteins we eat to supply the elemental nitrogen required for biosynthesis. In the plant kingdom, the most important sources of nitrogen are NH_3 and the ammonium cation (NH_4^+).

Why is the nitrogen atom in NH_3 accessible to living organisms? Consider an organism synthesizing the amino acid glycine from its nitrogen-deficient precursor, acetic acid. In elemental terms, the net reaction is:

$$CH_3CO_2H + N + H \longrightarrow H_2NCH_2CO_2H$$

← **CHAPTER 14**
We describe what is known about the structure of nitrogenase, the enzyme contained in nitrogen-fixing bacteria, in Chapter 14.

Table 13 – 4
Surface-Abundant Elements and Their Oxides

Element	% by Mass	Oxide	ΔG_f° (kJ/mol)
O	49.1	O_2	0
Si	26.1	SiO_2	−856
Al	7.5	Al_2O_3	−1582
Fe	4.7	Fe_3O_4	−1015
Ca	3.4	CaO	−603
Na	2.6	Na_2O	−376
K	2.4	KO_2	−239
Mg	1.9	MgO	−569
H	0.88	H_2O	−237
Ti	0.58	TiO_2	−885
Cl	0.19	Cl_2O	+98
C	0.09	CO_2	−394
N	<0.1	NO_2^*	+51

* Several other oxides of nitrogen exist. All have even more positive free energies of formation than NO_2.

According to Table 13-4, chlorine is also resistant to oxidation. Unlike nitrogen, however, chlorine reacts spontaneously with metals to generate salts such as NaCl and $MgCl_2$. Thus among abundant elements on Earth, nitrogen is uniquely stable in its elemental form.

$$:N\equiv N:$$
Bond energy = 945 kJ/mol

Acetic acid

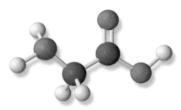

Glycine

Neither nitrogen nor hydrogen exists as free atoms, so the synthesis of glycine from acetic acid requires molecular sources of nitrogen and hydrogen. Atmospheric nitrogen and water are the most abundant sources of these two elements, but the production of glycine from N_2 and H_2O is significantly nonspontaneous under standard conditions:

$$4\ CH_3CO_2H + 2\ N_2 + 2\ H_2O \longrightarrow 4\ H_2NCH_2CO_2H + O_2$$

$$\Delta G^\circ_{reaction} = (1\ mol)(0\ kJ/mol) + (4\ mol)(-367\ kJ/mol) - (4\ mol)(-389\ kJ/mol)$$
$$-(2\ mol)(-237\ kJ/mol) - (2\ mol)(0\ kJ/mol) = +564\ kJ$$

Per mole of glycine, this reaction is "uphill" by 141 kJ.

On the other hand, NH_3 can provide both nitrogen and hydrogen atoms. Each NH_3 molecule also contains two extra hydrogen atoms that can be "burned" to produce water:

$$CH_3CO_2H + NH_3 + O_2 \longrightarrow H_2NCH_2CO_2H + H_2O$$

$$\Delta G^\circ_{reaction} = [(1\ mol)(-237\ kJ/mol) + (1\ mol)(-367\ kJ/mol)]$$
$$- [(1\ mol)(-389\ kJ/mol) + (1\ mol)(-17\ kJ/mol)]$$
$$+ (0.5\ mol)(0\ kJ/mol)] = -198\ kJ$$

This reaction is significantly spontaneous under standard conditions because it couples the production of water with the formation of the N—C bond in glycine.

The intensive agriculture characteristic of industrialized countries requires much more fixed nitrogen than is readily available from natural sources. Consequently, one of the major products of the chemical industry is nitrogen-containing fertilizers. Among the top 50 industrial chemicals, nitrogen gas (separated from air by cooling and liquefaction) perennially ranks number 2, with ammonia, nitric acid (HNO_3), urea [$(NH_2)_2CO$], and ammonium nitrate (NH_4NO_3) all in the top 15. All these chemicals are produced in huge amounts because of their roles in the fertilizer industry as feedstocks (nitrogen gas and nitric acid), as fertilizers (urea and ammonium nitrate), or as both (ammonia).

Free energies of formation suggest that ammonia and ammonium nitrate could be produced spontaneously in nature:

$$\tfrac{1}{2}N_2(g) + \tfrac{3}{2}H_2(g) \longrightarrow NH_3(g) \qquad \Delta G^\circ_f = -16.4\ kJ/mol$$

$$N_2(g) + \tfrac{3}{2}O_2(g) + 2\ H_2(g) \longrightarrow NH_4NO_3(s) \qquad \Delta G^\circ_f = -183.9\ kJ/mol$$

These reactions might occur in an atmosphere containing significant amounts of molecular hydrogen, but on Earth the only readily available source of hydrogen is water. The free energies of the reactions that yield NH_3 and NH_4NO_3 from gaseous nitrogen and water are highly unfavorable:

$$N_2 + 3\ H_2O \longrightarrow 2\ NH_3 + \tfrac{3}{2}O_2 \qquad \Delta G^\circ_{reaction} = +679\ kJ$$

$$N_2 + 2\ H_2O + \tfrac{1}{2}O_2 \longrightarrow NH_4NO_3 \qquad \Delta G^\circ_{reaction} = +290\ kJ$$

Thus large energy costs are involved in generating fixed nitrogen.

The strategy used in fertilizer production is to synthesize molecular hydrogen from methane:

$$CH_4 + H_2O \longrightarrow CO + 3\ H_2$$

$$\Delta G^\circ = +142\ kJ \qquad \Delta H^\circ = +205.9\ kJ \qquad \Delta S^\circ = +214.6\ J/K$$

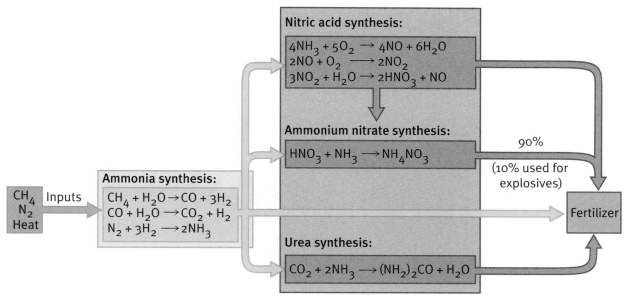

Figure 13-17

Block diagram of the industrial routes from nitrogen gas to fertilizers. Methane and energy in the form of heat are key ingredients in the first step.

Although this reaction is not spontaneous at room temperature, it becomes thermodynamically favorable at a temperature of 1000 K:

$$\Delta G^\circ_{1000\ K} = \Delta H^\circ - T\,\Delta S^\circ = +205.9\ kJ - (1000\ K)(214.6\ J/K)(10^{-3}\ kJ/J)$$
$$= -8.7\ kJ$$

To keep the temperature from falling while the endothermic reaction proceeds, heat energy must be supplied continuously. Otherwise, the temperature would quickly fall below the minimum value at which H_2 synthesis is spontaneous. For this reason, the production of H_2 (and ultimately of fertilizer) requires considerable amounts of energy. The energy cost is even greater if we consider that methane is obtained from nonrenewable reservoirs of natural gas, created over countless eons through the photosynthetic storage of solar energy.

With an ample supply of hydrogen, production of ammonia from N_2 becomes feasible:

$$N_2 + 3\ H_2 \longrightarrow 2\ NH_3 \qquad \Delta G^\circ_{reaction} = -32.8\ kJ$$

Figure 13-17 shows other spontaneous reactions that lead from ammonia to nitric acid, urea, and ammonium nitrate.

Phase Diagrams

A phase change converts a substance from one phase to another. Phase changes have characteristic thermodynamic properties: Any change from a more ordered phase to a less ordered phase increases both the enthalpy and the entropy of the substance. Enthalpy increases because the molecules of the substance gain energy to overcome the intermolecular forces that hold them in the more ordered phase.

Figure 13-18
Schematic view of the three phase changes leading from more ordered to less ordered phases, illustrated by the phase changes for water. Each is accompanied by positive enthalpy and entropy changes for the substance.

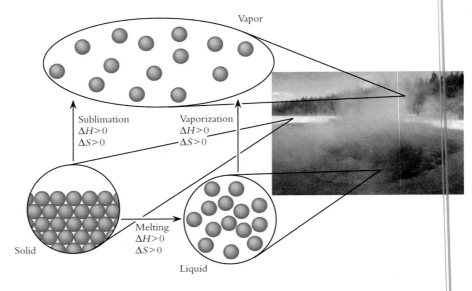

Entropy increases because the molecules are more randomly distributed in the less ordered phase. Figure 13-18 shows that when a solid melts or sublimes or a liquid vaporizes, both ΔH and ΔS are positive.

As described in Section 13.4, when ΔH and ΔS have the same sign, the spontaneous direction of a process depends on T. For a phase change, enthalpy dominates ΔG at low temperature, and the formation of the more ordered phase is spontaneous. In contrast, entropy dominates ΔG at high temperature, and the formation of the less ordered phase is spontaneous. At one characteristic temperature, $\Delta G = 0$ and the system is in a state of dynamic equilibrium. At equilibrium, the phase change proceeds in both directions at the same rate and the two phases coexist.

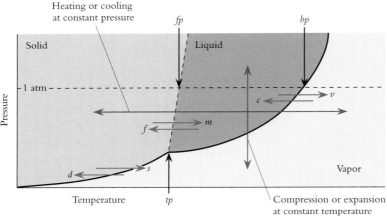

Figure 13-19
The general form of a phase diagram. Any point within the diagram corresponds to a specific value of pressure and temperature. The blue arrows show the six different phase transitions: sublimation (*s*), deposition (*d*), melting (*m*), freezing (*f*), vaporization (*v*), and condensation (*c*).

CHAPTER 5 →
See Chapter 5 for information about vapor pressure. The vapor pressure of water at various temperatures appears in Table 5-4.

The spontaneous direction of a phase change also depends on pressure, primarily because the molar entropy of a gas exhibits strong pressure-dependence:

$$S_{gas} = S° - R \ln P_{gas}$$

$$\Delta G_{vap} = \Delta H_{vap} - T \Delta S_{vap} = \Delta H_{vap} - T \Delta S°_{vap} + RT \ln P_{gas}$$

The result of this pressure dependence is that at any temperature, there is some pressure at which $\Delta G_{vap} = 0$. This is the vapor pressure (*vp*) of the substance at that temperature. For example, the vapor pressure of water at 25 °C is 23.76 torr.

Phase diagrams show how the stability of a phase depends on temperature and pressure. That is, a phase diagram is a map of the pressure-temperature world showing the phase behavior of a substance. As Figure 13-19 shows, a phase diagram is a *P-T* graph that shows the ranges of temperature and pressure over which each phase is stable. Pressure is plotted along the *y*-axis, and temperature is plotted along the x-axis. In the upper left-hand region (low T, high P), the substance is stable as a solid. In the lower right-hand region (high T, low P), the substance is stable as a gas. In some intermediate range of T and P, the substance is stable as a liquid.

As illustrated by Figure 13-19, phase diagrams have several important features:

1. Boundary lines between phases establish the regions where each phase is most stable.

2. Movement across a boundary line corresponds to a phase change. The arrows on the figure show six different phase changes.

3. At any point along a boundary line, the two phases on either side of the line coexist in a state of dynamic equilibrium. The normal freezing point (fp) and normal boiling point (bp) of a substance are the points where the phase boundary lines intersect the horizontal line that represents $P = 1$ atm.

4. The three boundary lines meet in a single point, called the **triple point (tp).** All three phases are simultaneously stable at this unique combination of temperature and pressure. Notice that, although *two* phases are stable under any of the conditions specified by the boundary lines, *three* phases can be simultaneously stable only at the triple point.

5. What happens to a substance as temperature changes at constant pressure can be determined by drawing a horizontal line at the appropriate pressure on the phase diagram.

6. What happens to a substance as pressure changes at constant temperature can be determined by drawing a vertical line at the appropriate temperature on the phase diagram.

7. The temperature at which vapor condenses is highly dependent on pressure. Increasing the pressure on a gas decreases its entropy, which in turn results in a decrease in ΔS_{phase}. Along the boundary lines, $\Delta G_{phase} = 0$. Thus a decrease in ΔS_{phase} must be accompanied by an increase in temperature to maintain the condition $\Delta G_{phase} = 0 = \Delta H_{phase} - T \Delta S_{phase}$.

8. The temperature at which melting occurs is almost independent of pressure because neither ΔH_{phase} nor ΔS_{phase} shows a significant P dependence for transformations between condensed phases. Thus ΔG_{phase} is almost independent of P, and the boundary line between solid and liquid is nearly vertical.

9. The solid-vapor boundary line extrapolates to $P = 0$ and $T = 0$. This is a consequence of the direct link between temperature and molecular energy. At $T = 0$ K, molecules have minimum energy, so they cannot escape from the solid lattice. At 0 K, the vapor pressure of every substance would be 0 atm.

The phase diagram for water, shown in Figure 13-20, illustrates these features for a familiar substance. The figure shows that liquid water and solid ice coexist at the normal freezing point, $T = 273.15$ K and $P = 1.00$ atm. Liquid water and water vapor coexist at the normal boiling point, $T = 373.15$ K and $P = 1.00$ atm. The triple point of water occurs at $T = 273.16$ K and $P = 0.0060$ atm. The figure shows that when P is lower than 0.0060 atm, there is no temperature at which water is stable as a liquid. At low pressure, ice sublimes but it does not melt.

Figure 13-20
The phase diagram for water. Arrows indicate the triple point (*tp*), normal freezing point (*fp*), and normal boiling point (*bp*).

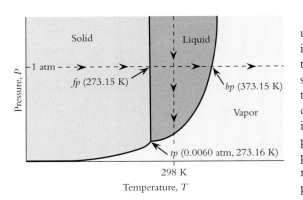

The dashed lines on Figure 13-20 show two paths that involve phase changes for water. The *horizontal dashed line* shows what happens as the temperature is changed at a constant pressure of 1 atm. As ice is warmed from a low temperature, it remains in the solid phase until the temperature reaches 273.15 K. At that temperature solid ice melts to liquid water, and water remains liquid as the temperature increases until the temperature reaches 373.15 K. At 373.15 K, liquid water changes to water vapor. Water is most stable in the gas phase at all higher temperatures. The *dashed vertical line* shows what happens as the pressure on water is reduced at a constant temperature of 298 K (approximately room temperature). Water remains in the liquid phase until the pressure drops to 0.03 atm. At 298 K water is most stable in the gas phase at any pressure lower than 0.03 atm. Notice that at 298 K there is no pressure at which water is most stable in the solid state.

All phase diagrams share the common features listed above, but the detailed appearance changes from substance to substance according to the strength of intermolecular interactions. Figure 13-21 shows two examples, the phase diagrams for molecular nitrogen and carbon dioxide. Both these substances are gases under normal conditions. Unlike H_2O, whose triple point lies close to 298 K, N_2 and CO_2 have triple points well below room temperature. Although both are gases at room temperature and pressure, they behave differently when cooled at $P = 1$ atm. Molecular nitrogen liquefies at 77.4 K and then solidifies at 63.3 K, whereas carbon dioxide condenses directly to the solid phase at 195 K. This difference in behavior arises because the triple point of CO_2, unlike the triple points of H_2O and N_2, occurs at a pressure greater than one atmosphere. The phase diagram of CO_2 shows that at a pressure of one atmosphere, no temperature exists at which the liquid phase is stable.

Figure 13-21
Phase diagrams for nitrogen and carbon dioxide, two substances that are gases at room temperature and pressure. Arrows indicate the triple point (*tp*), normal freezing point (*fp*), normal boiling point (*bp*), and normal sublimation point (*sp*).

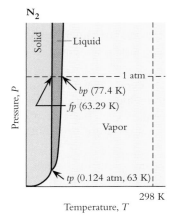

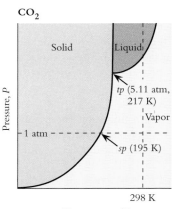

Phase diagrams are constructed by measuring the temperatures and pressures at which phase changes occur, reinforced by calculations of temperatures and pressures at which $\Delta G_{phase} = 0$. Approximate phase diagrams such as those shown in Figures 13-20 and 13-21 can be constructed from the triple point, normal melting point, and normal boiling point of a substance. Example 13-9 illustrates this procedure.

| Constructing a Phase Diagram | Example 13-9 |

Ammonia is a gas at room temperature and pressure. Its normal boiling point is 239.8 K, and it freezes at 195.5 K. The triple point for NH_3 is $P = 0.060$ atm and $T = 195.4$ K. Use this information to construct an approximate phase diagram for NH_3.

Strategy: The normal melting, boiling, and triple points give three points on the phase boundary curves. To construct the curves from knowledge of these three points, use the common features of phase diagrams: the vapor–liquid and vapor–solid boundaries of phase diagrams slope upward, the liquid–solid line is nearly vertical, and the vapor–solid line begins at $T = 0$ K and $P = 0$.

Solution: Begin by choosing appropriate scales, drawing the $P = 1$ atm line and locating the given data points. An upper temperature limit of 300 K encompasses all the data:

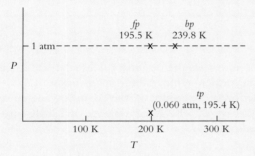

Next, connect the points and label the domains:

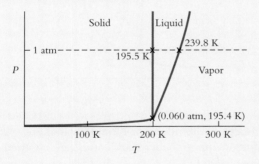

Phase diagrams can be used to determine what phase of a substance is stable at any particular pressure and temperature. They also summarize how phase changes occur as either condition is varied. Example 13–10 provides an illustration.

| Interpreting a Phase Diagram | Example 13-10 |

A chemist wants to perform a synthesis in a vessel at $P = 0.50$ atm using liquid NH_3 as the solvent. What temperature range would be suitable? When the synthesis is complete, the chemist wants to boil off the solvent without raising T above 220 K. Is this possible?

Example 13-10	Interpreting a Phase Diagram *(continued)*

Strategy: The phase diagram for NH_3 shows the boundary lines for the liquid domain. These boundary lines can be used to determine the conditions under which phase changes occur.

Solution: Because the chemist wants to work at $P = 0.50$ atm, draw a horizontal line across the phase diagram at $P = 0.50$ atm. Here is an expanded view of the phase diagram between 150 K and 300 K:

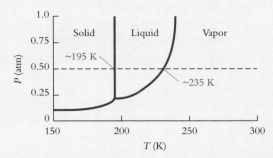

The horizontal line intersects the boundary lines at about 235 and 195 K. Liquid NH_3 is stable between these temperatures at this pressure.

At the completion of the synthesis, the chemist wants to remove the solvent without raising T above 220 K. A vertical line at 220 K on the phase diagram represents this condition:

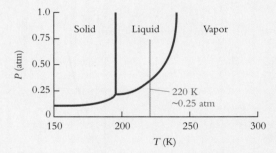

The line intersects the liquid–vapor boundary at about 0.25 atm. A vacuum pump capable of reducing P below 0.25 atm can be used to vaporize and remove the NH_3 while keeping the temperature below 220 K.

Thermal Pollution

One of the most important practical consequences of the second law of thermodynamics is that heat must be transferred from the system to its surroundings whenever a process is driven in the direction that increases order in the system. This heat raises the temperature and increases the entropy of the surroundings, thus offsetting the entropy decrease of the system. Refrigerators, automobile engines, nuclear power plants, and humans are subject to this requirement. All these objects increase the temperature of their surroundings as they operate. This increase in temperature is easily handled for small systems. For industrial-scale operations such as nuclear power plants, however, the amount of heat generated can have serious consequences. This type of heating is called **thermal pollution.**

Any thermally powered electrical power plant uses heat flows to convert stored energy into work. The stored energy may be chemical (coal, oil, or natural gas) or nuclear (uranium). When this potential energy is released, it creates a heat flow that is used to generate steam, which drives turbines to generate electricity. Electricity is the organized motion of electrons, all going in the same direction in a wire. Generating this order among the electrons requires a decrease in entropy that must be paid for by creating greater disorder somewhere else. Consequently, much of the energy used for thermal generation of electrical power has to be "wasted" by dumping it into the surroundings as heat.

The increase in temperature that results from thermal power generation creates biological stresses on the local ecosystem. For example, fish are particularly susceptible to changes in water temperature. As water temperature increases, the metabolic rate of the fish increases. At the same time, however, the amount of oxygen available to supply the metabolic demands of the fish decreases because oxygen is less soluble in warm water than in cold water. Species that thrive in cold water, such as trout, die if the water temperature rises by more than a few degrees. The ecology of simpler forms of aquatic life is also highly sensitive to temperature. Plankton that provide the food base for many aquatic ecosystems thrive at temperatures between 14 °C and 24 °C. When water temperatures exceed 24 °C, blue-green algae crowd out these essential plankton, leading to serious disruption of the ecological balance.

Hydroelectric generation of electricity need not lead to thermal pollution, because gravitational energy can drive turbines with minimal generation of heat. Unfortunately, there are other detrimental ecological consequences of hydroelectric power plants, such as the flooding of vast natural habitats, disruption of rivers and streams, and changes in local weather patterns caused by the creation of a huge body of water. Although these effects are perhaps less destructive than thermal pollution, no truly "clean" power source seems to exist. The second law of thermodynamics states that a price must always be paid for generating order. Inevitably, human existence generates order, so we pay this price continually. Conservation measures minimize the thermodynamic costs, but these costs can never be eliminated completely.

Section Exercises

■ **13.5.1** Atmospheric nitrogen can be transformed to ammonia by various bacteria that fix nitrogen. The overall process is complicated, but the net reaction can be written as:

$$N_2 + 3 H_2O \longrightarrow 2 NH_3 + \tfrac{3}{2} O_2$$

Assuming standard conditions (not quite true but close enough for approximations), how much free energy must the bacteria consume to fix one mole of N atoms?

■ **13.5.2** Refer to the phase diagrams in this chapter to answer the following questions:
(a) What is the maximum pressure at which solid N_2 can sublime?
(b) What happens if the pressure above ice at -10 °C is reduced to 0.5 atm, to 0.05 atm, and finally to 0.005 atm?

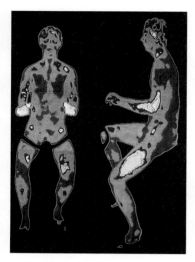

As this infrared photo shows, a person exercising emits heat to his/her surroundings

(c) What occurs if dry ice (solid CO_2) is heated from 180 K to room temperature in a container held at constant pressure of 7 atm?

■ **13.5.3** All automobile engines operate at relatively high temperature and require continual cooling. Using our description of thermal pollution, explain how the second law of thermodynamics requires automobile engines to "dump" heat to their surroundings.

13.6 BIOENERGETICS

Life creates order out of chaos, but thermodynamics shows that any spontaneous process must increase the total amount of disorder in the universe. To generate order in a system, it is necessary to release stored energy as heat, which increases the disorder of the surroundings. Consequently, living things must use large amounts of energy to survive. The plant kingdom extracts energy from sunlight to fuel its organizational processes. The animal kingdom, on the other hand, uses the chemical energy stored in plants or other animals to drive its organizational efforts. In this section, we describe representative energetic processes that operate in plants and animals.

Biochemical Energy Production

SECTION 11.5 →
Glucose and other important carbohydrates are discussed in Section 11.5.

Palmitic acid:

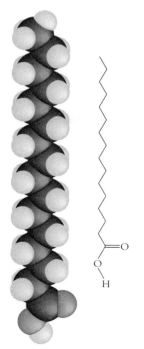

Living organisms use carbohydrates as their source of energy. Plants make their own carbohydrates through photosynthesis. Animals, on the other hand, obtain carbohydrates by eating plants or other animals. Plants and animals transform carbohydrates into fats, which also can be used as sources of energy. The extraction of chemical energy from these compounds is an important component of metabolism, the chemical processes of an organism. This component of metabolism involves highly spontaneous oxidation reactions, as illustrated by glucose (a carbohydrate) and palmitic acid (one component of a fat):

Glucose: $\qquad C_6H_{12}O_6 + 6\,O_2 \longrightarrow 6\,CO_2 + 6\,H_2O \qquad \Delta G° = -2870 \text{ kJ}$

Palmitic acid: $\quad C_{15}H_{31}CO_2H + 23\,O_2 \longrightarrow 16\,CO_2 + 16\,H_2O \quad \Delta G° = -9790 \text{ kJ}$

The negative standard free energy changes of these reactions arise because the relatively weak O=O bond in molecular oxygen is converted into stronger O—H and C=O bonds in H_2O and CO_2. Entropy also favors these reactions because gaseous oxygen converts a solid into gaseous CO_2 and liquid H_2O. Not only are the products in more disordered phases, but there are also more molecules on the product side of the equation than on the side of the starting materials.

The large amount of energy stored in molecules such as glucose and palmitic acid means that a little fat or carbohydrate goes a long way as a fuel for life processes. However, a living cell would be destroyed quickly if all the energy stored in these molecules were released in a single reaction. To use energy-rich molecules without being destroyed, cells use elaborate chains of sequential reactions that allow this stored energy to be harvested a little at a time. Part of this energy is released as heat that maintains body temperature as it is dissipated to the surroundings. Another portion of the energy is stored in other high-energy molecules that the body uses as power sources for the many reactions that occur within cells. In addition to storing the energy produced in the oxidation reactions, these

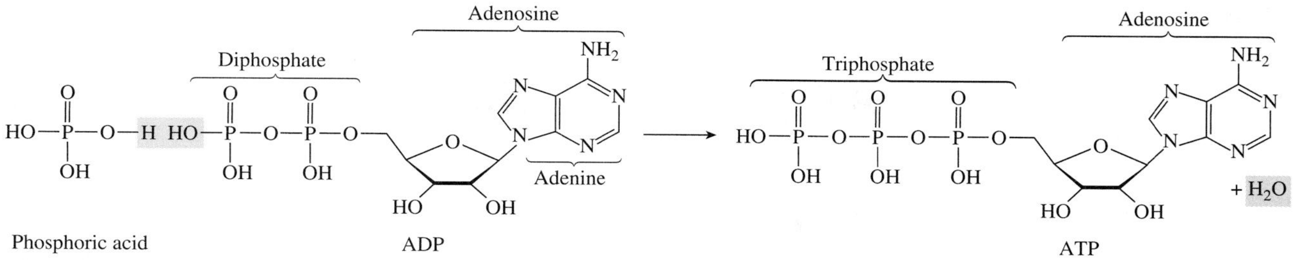

Figure 13-22
Phosphoric acid reacts with ADP to produce water and ATP. Because a P-O-P linkage is weaker than a P-O-H linkage, this reaction is endothermic by about 30 kJ/mol. Notice that the adenosine portion of the molecule remains intact during this reaction.

high-energy species serve as energy transport molecules, moving to different regions of the cell where energy is required for cell functions.

Adenosine triphosphate (ATP) is the most important of these energy transport molecules. Some of the energy released during the oxidation of glucose is used to drive a condensation reaction in which adenosine diphosphate (ADP) and phosphoric acid link together and eliminate water. This reaction stores chemical energy, as indicated by its positive standard free energy change:

$$ADP + H_3PO_4 \longrightarrow ATP + H_2O \qquad \Delta G° = +30.6 \text{ kJ}$$

The molecular details of this reaction are illustrated in Figure 13-22. Although ATP is a complex molecule, the adenosine portion does not change during the condensation reaction. The reaction merely adds a third phosphate group to the end of an existing chain.

The exact processes by which carbohydrates and fats are converted to CO_2 and H_2O depend on the conditions and the particular needs of the cell. Each possible route involves a complex series of chemical reactions, many of which are accompanied by the conversion of ADP to ATP. Glucose, for example, can convert as many as 36 ADP molecules into ATP molecules as it is oxidized to CO_2 and H_2O:

$$C_6H_{12}O_6 + 6\,O_2 + 36\,ADP + 36\,H_3PO_4 \longrightarrow 6\,CO_2 + 36\,ATP + 42\,H_2O$$

Coupled Reactions

Cells use the energy stored in ATP molecules to drive reactions that would otherwise be nonspontaneous under physiological conditions. This is accomplished by coupling the nonspontaneous reaction with the highly spontaneous conversion of ATP back to ADP and phosphoric acid:

$$ATP + H_2O \longrightarrow ADP + H_3PO_4 \qquad \Delta G° = -30.6 \text{ kJ}$$

Coupled reactions share a common intermediate that transfers energy from one reaction to the other. For example, the amino acid glutamine is synthesized in cells by the reaction of ammonia with another amino acid, glutamic acid:

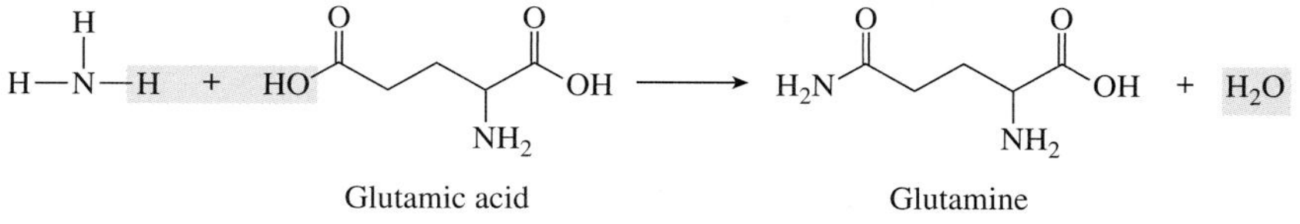

Glutamic acid   Glutamine

This reaction is thermodynamically unfavorable, $\Delta G° = +14$ kJ, but it can be driven by coupling it with the conversion of ATP into ADP:

$$\text{Glutamic acid} + \text{NH}_3 \longrightarrow \text{Glutamine} + \text{H}_2\text{O} \qquad \Delta G°_{\text{glutamine}} = +14 \text{ kJ}$$

$$\underline{\text{ATP} + \text{H}_2\text{O} \longrightarrow \text{ADP} + \text{H}_3\text{PO}_4 \qquad \Delta G°_{\text{ATP}} = -30.6 \text{ kJ}}$$

Net: Glutamic acid + ATP + NH$_3$ $\longrightarrow$
$$\text{Glutamine} + \text{ADP} + \text{H}_3\text{PO}_4 \qquad \Delta G°_{\text{reaction}} = -17 \text{ kJ}$$

The net energy change for the coupled process is the sum of the $\Delta G°$ values for the individual reactions:

$$\Delta G°_{\text{reaction}} = \Delta G°_{\text{glutamine}} + \Delta G°_{\text{ATP}} = 14 \text{ kJ} + (-30.6 \text{ kJ}) = -17 \text{ kJ}$$

The negative value of $\Delta G°_{\text{reaction}}$ indicates that the free energy released in the ATP reaction is more than enough to drive the conversion of glutamic acid into glutamine.

Although the coupled reactions can be represented by the net reaction, this process actually occurs in steps. In the first step of the coupled reaction, a phosphate group is transferred from ATP to glutamic acid:

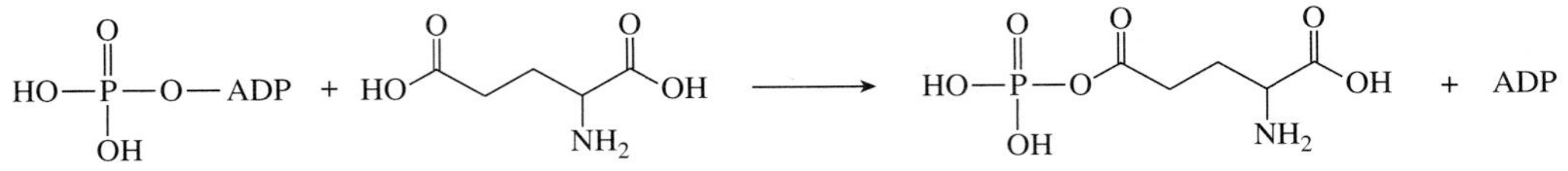

Glutamic acid

Next, an ammonia molecule reacts with the phosphorylated form of glutamic acid, producing phosphoric acid and glutamine:

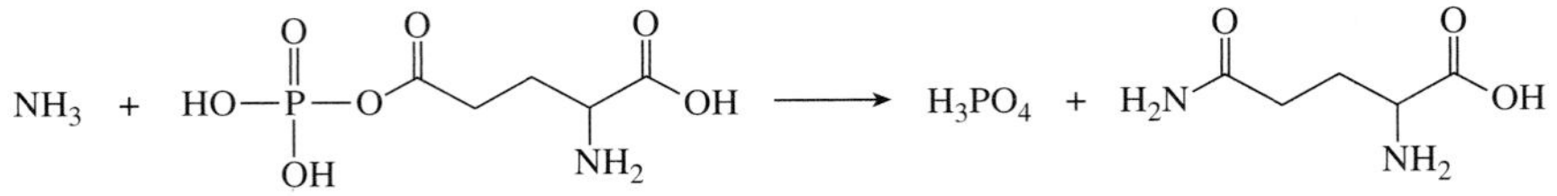

Phosphorylated glutamic acid

Overall, one molecule of ATP is converted to ADP and phosphoric acid for each molecule of glutamine produced from glutamic acid.

Coupled biochemical reactions occur on the surfaces of enzymes. As described in Chapter 11, enzymes are proteins. The surface area of an enzyme has a particular shape that accommodates some particular molecule that participates in a coupled reaction. Each different enzyme acts as an enabler, or catalyst, for a particular biochemical reaction.

Coupled reactions are also involved in the synthesis of ATP. Example 13-11 illustrates one of these energy-storing reactions.

CHAPTER 14 →
A catalyst is a chemical species that makes a reaction go faster than it would in the absence of a catalyst. Enzymes and other catalysts are discussed in Chapter 14.

Example 13-11	**ATP-Forming Reactions**

One of the biochemical reactions that produces ATP involves the conversion of acetyl phosphate to acetic acid and phosphoric acid:

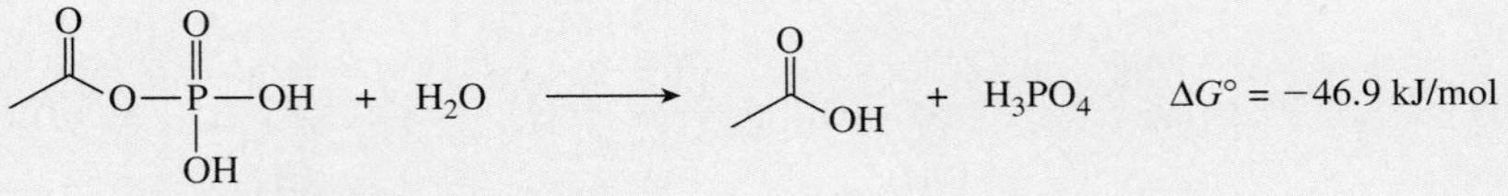

Acetyl phosphate   Acetic acid

| **ATP-Forming Reactions** (continued) | **Example 13-11** |

Write the overall balanced equation and show that the coupled reaction is spontaneous.

Strategy: A coupled process links a spontaneous reaction with a nonspontaneous one. In this case the energy released in the acetyl phosphate reaction provides the energy needed to drive the conversion of ADP to ATP.

Solution: Combining the two reactions gives the overall balanced equation:

$$CH_3CO_2PO_3H_2 + \cancel{H_2O} \longrightarrow CH_3CO_2H + \cancel{H_3PO_4}$$

$$ADP + \cancel{H_3PO_4} \longrightarrow ATP + \cancel{H_2O}$$

$$\overline{Net: CH_3CO_2PO_3H_2 + ADP \longrightarrow CH_3CO_2H + ATP}$$

The net energy change for the coupled process is the sum of the $\Delta G°$ values for the individual reactions:

$$\Delta G°_{net} = \Delta G°_{acetyl\ phosphate} + \Delta G°_{ATP} = -46.9\ kJ + 30.6\ kJ = -16.3\ kJ$$

The negative value of $\Delta G°_{net}$ shows that the free energy released in the acetyl phosphate reaction is more than enough to drive the conversion of ADP to ATP.

Energy Efficiency

Cells store the energy released during the oxidation of glucose by converting ADP into ATP. The storage process cannot be perfectly efficient, however, because each step in the reaction sequence must have a negative free energy change. In practical terms, this requires that some energy be released to the surroundings as heat.

Here is the complete balanced equation for glucose oxidation coupled with ATP production under normal physiological conditions:

$$C_6H_{12}O_6 + 6\ O_2 + 36\ ADP + 36\ H_3PO_4 \longrightarrow 6\ CO_2 + 42\ H_2O + 36\ ATP$$

According to this equation, the oxidation of one mole of glucose yields 36 moles of ATP. The overall free energy change for this process can be determined from the values for its uncoupled parts:

$$C_6H_{12}O_6 + 6\ O_2 \longrightarrow 6\ CO_2 + 6\ H_2O \qquad \Delta G° = -2870\ kJ$$

$$36\ (ADP + H_3PO_4 \longrightarrow ATP + H_2O)\ \Delta G° = (36\ mol)(+30.6\ kJ/mol) = +1100\ kJ$$

$$\Delta G_{overall} = -2870\ kJ + 1100\ kJ = -1770\ kJ$$

Although 1100 kJ of energy is stored in this coupled process, 1770 kJ of energy is "wasted." Thus cells harness 38% of the chemical energy stored in glucose to drive the biochemical machinery of metabolism. The remaining 62% is dissipated as heat, increasing the entropy of the surroundings as the living cell organizes itself and its immediate environment.

Fats such as palmitic acid are metabolized through pathways similar to the ones used for the oxidation of glucose. The complete oxidation of one mole of palmitic

Although the standard free energy change of the ADP-ATP reaction is 30.6 kJ/mol under *standard* conditions, the concentrations of the phosphate species in cells are far from 1 M. This causes the free energy to vary from its standard value. Typically, the conversion of ATP to ADP in a living cell releases about 50 kJ/mol of energy.

acid molecules liberates 9790 kJ of free energy and produces 130 ATP molecules. You should be able to verify that this metabolic process has about the same efficiency as the oxidation of glucose.

Oxidizing one mole of glucose releases 2870 kJ of free energy, whereas oxidizing one mole of palmitic acid releases more free energy, 9790 kJ. Although some of this extra energy results from its larger molecular size, palmitic acid also releases more energy per atom of carbon than glucose. Glucose releases about 480 kJ/mol of carbon atoms, and palmitic acid releases about 610 kJ/mol of carbon atoms. Organisms convert carbohydrates into fats because fats store more energy per unit mass.

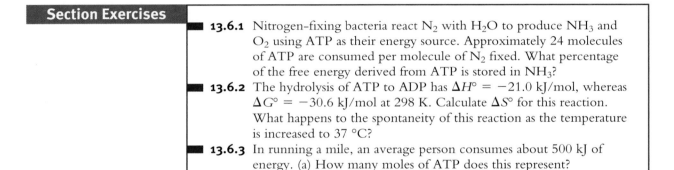

Section Exercises

13.6.1 Nitrogen-fixing bacteria react N_2 with H_2O to produce NH_3 and O_2 using ATP as their energy source. Approximately 24 molecules of ATP are consumed per molecule of N_2 fixed. What percentage of the free energy derived from ATP is stored in NH_3?

13.6.2 The hydrolysis of ATP to ADP has $\Delta H° = -21.0$ kJ/mol, whereas $\Delta G° = -30.6$ kJ/mol at 298 K. Calculate $\Delta S°$ for this reaction. What happens to the spontaneity of this reaction as the temperature is increased to 37 °C?

13.6.3 In running a mile, an average person consumes about 500 kJ of energy. (a) How many moles of ATP does this represent? (b) Assuming 38% conversion efficiency, how many grams of glucose must be "burned"?

▬▬▬ CHAPTER REVIEW

Summary and Key Terms

1. In thermodynamics, **spontaneous** refers to the preferred direction of a process. Spontaneous processes may be exothermic or endothermic, but the **second law of thermodynamics** states that the amount of disorder in the universe increases in every spontaneous process.

2. **Entropy (S)** is a measure of the amount of disorder, and the second law of thermodynamics states that the total entropy of the universe always increases. The entropy change for a constant-temperature process equals the heat flow divided by the temperature. Heat always flows from warmer objects to cooler objects.

3. The **third law of thermodynamics** states that at 0 K, pure, perfect crystals are perfectly ordered and have $S = 0$. Based on this zero point, it is possible to determine the **absolute entropy** of a pure substance at any temperature. The entropy of any substance decreases as the temperature decreases, but absolute entropies are always positive. Entropies also vary with the logarithm of concentration. Standard molar entropies can be used to calculate the entropy change that accompanies a chemical reaction.

4. **Free energy (G)** is defined by $G = H - TS$. The free energy change for a system at constant temperature and pressure is always negative. The **standard molar free energy of formation ($\Delta G_f°$)** can be used to calculate the free energy change that accompanies a chemical reaction. The free energy change for a chemical reaction depends strongly on the **reaction quotient (Q)**, which is the ratio of product concentrations to reactant concentrations, all raised to powers equal to their stoichiometric coefficients. The spontaneity of a process can be influenced by changing the temperature or concentrations.

5. There are many applications of thermodynamics. **Nitrogen fixation,** the formation of nitrogen compounds from molecular nitrogen, is nonspontaneous unless it is coupled with other highly spontaneous processes. **Phase diagrams** summarize the thermodynamics of phase behavior, including the normal boiling point, normal freezing point, and **triple point**. Large-scale conversion of fuel into electricity inevitably results in **thermal pollution** around the power plant.

6. Many biological processes lead to increased order and are nonspontaneous unless coupled to reactions that release energy. Living organisms use energy-rich substances, including fats, carbohydrates, and **adenosine triphosphate (ATP)**.

Skills to Master

▶ Assessing molecular disorder and entropy

▶ Calculating entropy changes at constant temperature

▶ Calculating entropy changes with concentration

▶ Calculating reaction entropies

▶ Determining spontaneity from change in free energy (ΔG)

▶ Estimating changes in ΔG with temperature

▶ Calculating changes in ΔG with concentration

▶ Drawing and interpreting phase diagrams

▶ Estimating energy efficiencies of biological reactions

Learning Exercises

13.1 Describe the thermodynamic criteria for spontaneity. Describe in your own words what entropy is and why it is not a conserved quantity.
13.2 Make a list of everyday processes that illustrate that disorder always increases.
13.3 Write a paragraph that explains why living creatures must convert some stored energy into heat in the course of their activities.

13.4 Update your list of memory bank equations. Describe the restrictions on each equation.
13.5 Make a list of all terms new to you that appear in Chapter 13. Using your own words, write a one-sentence definition for each. Consult the Glossary if you need help.

Problems   ilw = interactive learning ware problem. Visit the website at www.wiley.com/college/olmsted

Spontaneity

13.1 Describe the order-disorder qualities of the following processes: (a) Ocean waves wash away a sand castle. (b) Water and acetone, two liquids, mix to form a homogeneous liquid solution. (c) In the child's game "pick up sticks," a bundle of sticks is dropped to the floor. (d) Water evaporates from a puddle after a summer shower.
13.2 Describe the order-disorder aspects of the following processes: (a) A secretary "straightens up" the boss' desk. (b) Wood burns, producing CO_2 and H_2O vapors. (c) I_2 crystals form as a hot solution of I_2 in CCl_4 cools. (d) A skilled mechanic reassembles a torn-down engine.
13.3 The following figure shows what happens when a drop of ink is added to a beaker of water. Using ideas of organization and disorganization, explain what is occurring.

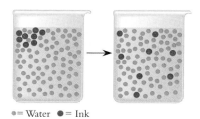

● = Water   ● = Ink

13.4 "All the king's horses and all the king's men couldn't put Humpty together again." Describe Humpty-Dumpty's fate in terms of organization and disorganization.
13.5 Explain each of the following observations from the order-disorder perspective: (a) A puncture causes a tire to deflate. (b) An open bottle of perfume on a table eventually fills the room with the fragrance of the perfume.

13.6 Explain each of the following observations from the order-disorder perspective: (a) A glass dropped on the floor shatters into many pieces. (b) The wind scatters raked leaves.

Entropy: The Measure of Disorder

13.7 A tic-tac-toe game contains nine compartments. What is W for all possible first moves by each player (one X and one O placed in two different compartments)?
13.8 Consider a box of 16 equal-sized compartments into which two different colored marbles are to be placed. Calculate W for placing each in a different compartment and for placing both in the same compartment.
13.9 Ice melts at 273.15 K with $\Delta H_{fus} = 6.01$ kJ/mol. An ice cube whose mass is 13.8 g is dropped into a swimming pool whose temperature is held at 27.5 °C. (a) What is the entropy change (ΔS) for melting the ice? (b) What is ΔS of the pool? (c) What is the overall ΔS?
13.10 Solid CO_2 (dry ice) sublimes with $\Delta H_{subl} = 25.2$ kJ/mol at 195 K. A block of dry ice whose mass is 27.5 g sublimes in a room whose temperature is 26.5 °C. (a) What is the entropy change (ΔS) of the CO_2? (b) What is ΔS of the room? (c) What is the overall ΔS?
13.11 What are the signs of ΔS for the system and for the surroundings in each of the following processes? (a) Water boils in a teakettle on a hot stove. (b) Ice in an ice cube tray, left on a counter top, melts. (c) A cup of coffee is reheated in a microwave oven.
13.12 What are the signs of ΔS for the system and for the surroundings in each of the following processes? (a) Ice forms on the surface of a bird bath in winter. (b) A hot cup of coffee cools when left to stand. (c) A Popsicle melts when left on a table.
ilw **13.13** Calculate the entropy change of 15.5 g of steam that condenses to liquid water at 373.15 K. Without doing additional calculations, what can you say about the entropy change of the surroundings?

13.14 Calculate the entropy change of 75.4 g of water that freezes to ice at 273.15 K. Without doing additional calculations, what can you say about the entropy change of the surroundings?

13.15 Molar heats of fusion of several substances can be found in Table 12-2. Calculate the molar entropy of fusion at its normal melting point for each of the following: (a) argon; (b) methane; (c) ethanol; and (d) mercury.

13.16 Molar heats of vaporization of several substances can be found in Table 12-2. Calculate the molar entropy of vaporization at its normal boiling point for each of the following: (a) molecular oxygen; (b) ethane; (c) benzene; and (d) mercury.

Absolute Entropies

13.17 For each of the following pairs of substances, determine which has the larger molar entropy at 298 K and state the main reason for the difference: (a) $NaCl(aq)$ and $MgCl_2(aq)$; (b) $HgO(s)$ and $HgS(s)$; and (c) $Br_2(l)$ and $I_2(s)$.

13.18 For each of the following pairs of substances, determine which has the larger molar entropy at 298 K and state the main reason for the difference: (a) Br_2 and Cl_2; (b) Ni and Pt; (c) C_5H_{12} (pentane, liquid) and C_8H_{18} (octane, liquid); and (d) SiF_4 and CH_4 (both gases).

13.19 Oxygen, ozone, and methane are all gases at standard temperature. Their molar entropies are in the sequence $CH_4 < O_2 < O_3$. Using molecular properties, explain why ozone is more disordered than oxygen but methane is more ordered than either.

13.20 Mercury, water, and bromine are liquids at standard temperature. Their molar entropies are in the sequence $H_2O < Hg < Br_2$. Using molecular properties, explain why bromine is more disordered than mercury but water is the most highly ordered of these three.

13.21 Using tabulated values of $S°$, calculate the standard entropy per mole of atoms for He, H_2, CH_4, and C_3H_6. Explain the trend that you find in terms of what bond formation does to atomic order.

13.22 Using tabulated values of $S°$, calculate the standard entropy per mole of atoms for Ar, O_2, O_3, and C_2H_6. Explain the trend that you find in terms of what bond formation does to atomic order.

13.23 Compute the standard entropy change for the following reactions:

(a) $N_2(g) + 3\ H_2(g) \longrightarrow 2\ NH_3(g)$
(b) $3\ O_2(g) \longrightarrow 2\ O_3(g)$
(c) $PbO_2(s) + 2\ Ni(s) \longrightarrow Pb(s) + 2\ NiO(s)$
(d) $C_2H_4(g) + 3\ O_2(g) \longrightarrow 2\ CO_2(g) + 2\ H_2O(l)$

13.24 Compute the standard entropy change for the following reactions:

(a) $2\ H_2(g) + O_2(g) \longrightarrow 2\ H_2O(l)$
(b) $C(s) + 2\ H_2(g) \longrightarrow CH_4(g)$
(c) $C_2H_5OH(l) + 3\ O_2(g) \longrightarrow 2\ CO_2(g) + 3\ H_2O(l)$
(d) $Fe_2O_3(s) + 2\ Al(s) \longrightarrow Al_2O_3(s) + 2\ Fe(s)$

13.25 For each reaction in Problem 13.23, explain what features of the disorder of reactants and products account for the magnitude and sign of $\Delta S°$.

13.26 For each reaction in Problem 13.24, explain what features of the disorder of reactants and products account for the magnitude and sign of $\Delta S°$.

13.27 Compute the absolute entropy of the following: (a) 2.50 mol of Ar gas at $p = 0.25$ atm; (b) 0.75 mol of O_3 gas at $p = 2.75$ atm; and (c) 0.45 mol of a mixture of N_2 ($X = 0.78$) and O_2 ($X = 0.22$) at $P_{tot} = 1.00$ atm.

13.28 Compute the absolute entropy of the following: (a) 1.00 mol of molecular hydrogen gas at $p = 5.0$ atm; (b) 0.25 mol of ethane gas

at $p = 0.10$ atm; and (c) 1.00 mol of a mixture of N_2 ($p = 125$ atm) and H_2 ($p = 375$ atm).

Spontaneity and Free Energy

13.29 Each of the following statements is false. Rewrite each so that it makes a correct statement about free energy: (a) $\Delta G_{universe} > 0$ for any spontaneous process. (b) ΔG_{system} increases in any process at constant T and P. (c) $\Delta H = \Delta G - T\Delta S$.

13.30 Each of the following statements is false. Rewrite each so that it makes a correct statement about free energy: (a) In any process at constant T and P, the free energy of the universe decreases. (b) $\Delta G_{sys} = 0$ for any spontaneous process. (c) $\Delta G = \Delta H + \Delta S$.

13.31 Compute the standard free energy change for each reaction in Problem 13.23.

13.32 Compute the standard free energy change for each reaction in Problem 13.24.

13.33 Estimate $\Delta G_{425}°$ for each of the reactions in Problem 13.23. (Assume that there are no phase changes.)

13.34 Calculate $\Delta G_{reaction}$ at 298 K for each reaction in Problem 13.24 if $p = 0.25$ atm for each gaseous substance.

Some Applications of Thermodynamics

13.35 Compute $\Delta H°$, $\Delta S°$, and $\Delta G°$ for the production of NH_4NO_3 from ammonia and oxygen:

$$2\ NH_3(g) + O_2(g) \longrightarrow NH_4NO_3(s) + H_2O(l)$$

(This reaction is not feasible industrially because NH_3 combustion cannot be controlled to give NH_4NO_3 as a product.)

13.36 Compute $\Delta H°$, $\Delta S°$, and $\Delta G°$ for the production of urea from ammonia (see Figure 13-17).

13.37 Sketch the approximate phase diagram for Br_2 from the following information: normal melting point is 265.9 K, normal boiling point is 331.9 K, triple point at $P = 5.79 \times 10^{-2}$ atm and $T = 265.7$ K. Label the axes and the area where each phase is stable.

13.38 Oxygen has a normal melting point of 55 K and a normal boiling point of 90.2 K, and its triple point occurs at $P = 0.0015$ atm and $T = 54$ K. Sketch the approximate phase diagram for oxygen, and label the axes and the area where each phase is stable.

13.39 Using the phase diagram of Problem 13.37, describe what happens to a sample of Br_2 as the following processes take place. Draw lines on the phase diagram showing each process. (a) A sample at $T = 400$ K is cooled to 250 K at constant $P = 1.00$ atm. (b) A sample is compressed at constant $T = 265.8$ K from $P = 1.00 \times 10^{-3}$ atm to $P = 1.00 \times 10^3$ atm. (c) A sample is heated at constant $P = 2.00 \times 10^{-2}$ atm from 250 K to 400 K.

13.40 Using the phase diagram of Problem 13.38, describe what happens to a sample of O_2 as the following processes take place. Draw lines on the phase diagram showing each process. (a) A sample at $T = 125$ K is cooled to 25 K at constant $P = 1.00$ atm. (b) A sample is compressed at constant $T = 54.5$ K, starting at $P = 1.00 \times 10^{-2}$ atm, to $P = 1.00 \times 10^2$ atm. (c) A sample is heated at constant $P = 1.00 \times 10^{-3}$ atm from 25 K to 100 K.

Bioenergetics

13.41 Two children on opposite ends of a see-saw can be used as an analogy for a coupled reaction. Describe the coupling of spontaneous and nonspontaneous processes during the actions of a see-saw.

13.42 A pair of weights and a pulley can be used as an analogy for a coupled reaction. Use the figure to describe a coupled process. Identify spontaneous and nonspontaneous portions of the process.

13.43 Glucose and fructose combine to make sucrose in a condensation reaction for which $\Delta G° = 23.0$ kJ/mol:

$$C_6H_{12}O_6 + C_6H_{12}O_6 \longrightarrow C_{12}H_{22}O_{11} + H_2O$$

Glucose   Fructose   Sucrose

Write the balanced equation for this reaction coupled with the ADP-ATP reaction and operating in the direction of overall spontaneity and calculate $\Delta G°$ for the overall process.

13.44 Although the ATP-ADP reaction is the principal energy shuttle in metabolic pathways, many other examples of coupled reactions exist. For example, the glutamic acid–glutamine reaction discussed in the text can couple with the acetyl phosphate hydrolysis shown in Example 13-11. Write the balanced equation for the coupled reaction operating in the direction of overall spontaneity and calculate $\Delta G°$ for the overall process.

Additional Paired Problems

13.45 A teaspoon from a freezer is placed in a glass of water, and the two equilibrate to the same temperature. In this process, what can you deduce about each of the following: (a) $\Delta E_{universe}$; (b) $\Delta E_{teaspoon}$; (c) $\Delta S_{universe}$; (d) ΔS_{water}; and (e) $q_{teaspoon}$.

13.46 A pie is removed from a hot oven and allowed to cool to room temperature. In this process, what can you deduce about the changes in each of the following: (a) $\Delta E_{universe}$; (b) ΔE_{pie}; (c) $\Delta S_{universe}$; (d) ΔS_{pie}; and (e) q_{pie}.

13.47 Arrange the following in order of increasing entropy, from smallest to largest value: 1.0 mol H_2O (liquid, 373 K), 0.50 mol H_2O (liquid, 298 K), 1.0 mol H_2O (liquid, 298 K), 1.0 mol H_2O (gas, 373 K, 1.0 atm), and 1.0 mol H_2O (gas, 373 K, 0.1 atm).

13.48 Arrange the following in order of increasing entropy, from smallest to largest value: 1.00 g Br_2 (gas, 331.9 K, 0.10 atm), 1.00 g Br_2 (gas, 331.9 K, 1.00 atm), 1.00 g Br_2 (liquid, 331.9 K, 1.00 atm), and 1.00 g Br atoms (gas, 331.9 K, 0.10 atm).

13.49 For the chemical reaction:

$$Al_2O_3(s) + 3 H_2(g, p = 1 \text{ atm}) \longrightarrow 2 Al(s) + 3 H_2O(l)$$

at $T = 298$ K, answer each of the following by doing a quantitative calculation of the appropriate thermodynamic function: (a) Is this a spontaneous reaction? (b) Does the reaction absorb or release heat? (c) Are the products more or less ordered than the reactants?

13.50 For the chemical reaction:

$$3 Fe(s) + 4 H_2O(l) \longrightarrow Fe_3O_4(s) + 4H_2(g, p = 1 \text{ atm})$$

at $T = 298$ K, answer each of the following by doing a quantitative calculation of the appropriate thermodynamic function: (a) Is this a spontaneous reaction? (b) Does the reaction absorb or release heat? (c) Are the products more or less ordered than the reactants?

13.51 Use data from Appendix D to compute $\Delta S°$ for each of the following reactions:

(a) $2 ClO_2^-(aq) + O_2(g) \longrightarrow 2 ClO_3^-(aq)$
(b) $4 FeCl_3(s) + 3 O_2(g) \longrightarrow 2 Fe_2O_3(s) + 6 Cl_2(g)$
(c) $3 N_2H_4(l) + 4 O_3(g) \longrightarrow 6 NO(g) + 6 H_2O(l)$

13.52 Calculate the standard entropy change at 298 K of each of the following reactions, which are important in the chemistry of coal. Assume that coal has the same thermodynamic properties as graphite.

(a) $C(s, coal) + H_2O(g) \longrightarrow CO(g) + H_2(g)$
(b) $C(s, coal) + O_2(g) \longrightarrow CO_2(g)$
(c) $C(s, coal) + ½ O_2(g) \longrightarrow CO(g)$
(d) $CO(g) + H_2O(g) \longrightarrow CO_2(g) + H_2(g)$

13.53 Use data from Appendix D to compute $\Delta G°$ for each reaction in Problem 13.51.

13.54 Use data from Appendix D to compute $\Delta G°$ for each reaction in Problem 13.52.

13.55 Use data from Appendix D to compute $\Delta G°$ at 350.0 °C for each reaction in Problem 13.51.

13.56 Use data from Appendix D to compute $\Delta G°$ at 200.0 °C for each reaction in Problem 13.52.

13.57 For the following constant-temperature process, give the sign (+, −, or 0) for each of the specified thermodynamic functions. In each case give a brief account of your reasoning: (a) w_{sys}; (b) q_{sys}; and (c) ΔS_{surr}.

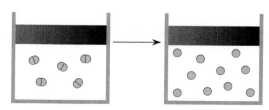

13.58 For the following constant-temperature process, give the sign (+, −, or 0) for each of the specified thermodynamic functions. In each case give a brief account of your reasoning: (a) q_{sys}; (b) ΔS_{sys}; and (c) $\Delta E_{universe}$.

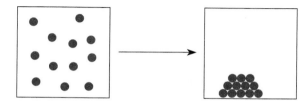

13.59 One possible source of acid rain is the reaction between NO_2, a pollutant from automobile exhausts, and water:

$$3 NO_2(g) + H_2O(l) \longrightarrow 2 HNO_3(g) + NO(g)$$

Determine whether this is thermodynamically feasible (a) under standard conditions and (b) at 298 K, with each product gas present at $p = 1.00 \times 10^{-6}$ atm.

13.60 Plants and a few bacteria use sunlight as an energy source to produce glucose from CO_2 and water. Other bacteria that live very deep in the sea, where there is no sunlight, use hydrogen sulfide to generate the free energy required to drive glucose synthesis:

$$8 H_2S(g) + 4 O_2(g) \longrightarrow 8 H_2O(l) + S_8(s)$$

Write a balanced equation for the coupled process and do a calculation that shows the process is spontaneous under standard conditions.

13.61 An ice cube tray containing 155 g of liquid H_2O at 0.0 °C is placed in a freezer whose temperature is −20.0 °C. As soon as all the H_2O has frozen, the tray is removed from the freezer. (a) Find ΔS for the H_2O. (b) Find ΔS for the universe. (c) If the tray were left in the freezer until its temperature reached −20 °C, would there be an additional entropy change for the universe? Explain.

13.62 Find ΔS for the system, surroundings, and universe when 25.0 g of liquid H_2O is evaporated at 100. °C, if the heat required is provided by a hot plate whose temperature is 315 °C.

13.63 Use data from Appendix D to answer quantitatively the following questions about the dissolving of table salt in water under standard conditions:

$$NaCl(s) \longrightarrow Na^+(aq) + Cl^-(aq)$$

(a) Is this a spontaneous reaction? (b) Does it release energy? (c) Does the amount of order in the chemical system increase?

13.64 Repeat the calculations of Problem 13.63 for silver chloride, AgCl, dissolving in water. What do the results reveal about the difference in solubility of these two salts and the reasons for the difference?

13.65 ATP and ADP have been referred to as the "energy currency" of the cell. Explain this analogy.

13.66 Explain why a nuclear power plant must contribute thermal pollution to the environment.

13.67 Refer to Figure 13-21. Describe in detail what occurs when each of the following is carried out: (a) The pressure on a sample of liquid N_2 is reduced from 1.00 to 0.010 atm at a constant temperature of 70 K. (b) A sample of N_2 gas is compressed from 1.00 to 50.0 atm at a constant temperature of 298 K. (c) A sample of N_2 gas is cooled at a constant pressure of 1.00 atm from 298 K to 50 K.

13.68 Refer to Figure 13-21. Describe in detail what occurs when each of the following is carried out: (a) A sample of CO_2 gas is compressed from 1.00 atm to 50.0 atm at $T = 298$ K. (b) Dry ice at 195 K is heated to 350 K at $P = 6.00$ atm. (c) A sample of CO_2 gas at $P = 1.00$ atm is cooled from 298 K to 50 K.

13.69 Even though ammonia can be used directly as a fertilizer, much of it is converted to urea or ammonium nitrate before being applied in the field. Considering the physical properties of these substances, suggest why ammonia is not preferred as a fertilizer.

13.70 What are the four most important nitrogen-containing chemicals, and for what are they used?

13.71 Determine $\Delta G_{combustion}$ per gram of palmitic acid and of glucose. On a per-gram basis, which is the better energy source?

13.72 The most important commercial process for generating hydrogen gas is the water-gas shift reaction:

$$CH_4(g) + H_2O(g) \longrightarrow CO(g) + 3 H_2(g)$$

Use tabulated thermodynamic data to find the following: (a) $\Delta G°$ and (b) $\Delta G°_{1300}$.

More Challenging Problems

13.73 A sample of liquid Br_2 (0.080 mol) at 332 K is placed in a cylinder equipped with a piston at $P_{ext} = 1.0$ atm. The cylinder is immersed in a constant temperature bath at 332 K, the boiling point of Br_2 ($\Delta H_{vap} = 30.9$ kJ/mol). The system undergoes a constant-temperature, constant-pressure expansion to a final volume of 2.2 L. The sketch shows the apparatus immediately after the cylinder is placed in the bath. (a) Draw a sketch that shows how the system looks when it reaches its final state. (b) Determine ΔS for the universe.

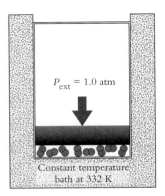

$P_{ext} = 1.0$ atm

Constant temperature bath at 332 K

13.74 The notion of thermodynamic coupling of a nonspontaneous process with a spontaneous process is not restricted to chemical reactions. Identify the spontaneous and nonspontaneous portions of the following coupled processes: (a) Water behind a dam passes through

a turbine and generates electricity. (b) A gasoline engine pumps water from a valley to the top of a hill.

13.75 Crystalline KCl has $S° = 83$ J/mol K, and crystalline CaO has $S° = 55$ J/mol K. What accounts for the larger disorder of KCl crystals?

13.76 Both CCl_4 (carbon tetrachloride) and CS_2 (carbon disulfide) are liquids used as solvents in special industrial applications. Using data from Appendix D, calculate $\Delta H°$ and $\Delta G°$ for combustion of these liquids:

$$CCl_4(l) + 5 O_2(g) \longrightarrow CO_2(g) + 4 ClO_2(g)$$

$$CS_2(l) + 3 O_2(g) \longrightarrow CO_2(g) + 2 SO_2(g)$$

Based on your results, would you recommend special precautions against fires for industrial plants using either solvent? Explain your recommendations.

13.77 In a system operating without any restrictions, heat is not a state function. However, heat flow describes a state function change under certain restricted conditions. For each of the following conditions, identify the state function that corresponds to heat flow: (a) q_V; (b) q_p; and (c) q_T.

13.78 A piece of dry ice at $T = 195$ K is dropped into a beaker containing water at $T = 273.15$ K. You may assume that all of the resulting CO_2 vapor escapes into the atmosphere. (a) Describe the process that takes place at the molecular level. (b) Calculate the overall entropy change if 12.5 g of dry ice undergoes this process.

13.79 Determine the sign of ΔS for the system for each of the following changes: (a) A soft drink is chilled in an ice chest. (b) The air in a bicycle pump is compressed. (c) A carton of juice concentrate is mixed with water.

13.80 The following diagram represents two flasks connected by a valve. Each flask contains a different gas.

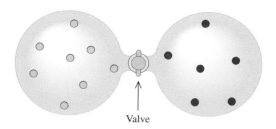

Valve

(a) If the valve is opened, the two gases can move back and forth between the two flasks. Redraw the figure to show the system at maximum entropy. (b) Redraw the figure in a way that shows the state of the system at its lowest possible entropy. (The substances need not remain gaseous.)

13.81 Without doing any calculations, predict the signs (+, −, or 0) of ΔH and ΔS for the following processes, all occurring at constant T and P. Explain your predictions.
(a) $2 NO_2(g) \longrightarrow N_2O_4(g)$
(b) A block of gold is melted in a jeweler's crucible
(c) $CH_4(g) + 2 O_2(g) \longrightarrow CO_2(g) + 2 H_2O(g)$

13.82 Heat pumps extract heat from a cold reservoir (usually the ground) and use it to maintain the temperature of a warmer reservoir (for example, a home). Explain why heat pumps cannot operate without converting some additional energy into heat.

13.83 For the ATP $\longrightarrow$ ADP reaction, $\Delta G° = -30.6$ kJ/mol. Under cellular conditions, this reaction releases more than 50 kJ/mol. List the ways in which cellular conditions differ from standard conditions and describe the effect that each difference has on the free energy change.

13.84 On a hot day, one of your friends suggests opening the door of your refrigerator to cool your kitchen. Will this strategy work? Explain your answer using principles of thermodynamics.

13.85 What is the efficiency of the metabolic conversion of palmitic acid to ATP? Compute the number of grams of palmitic acid that would have to be metabolized to provide the heat to warm a swimmer from whose skin 75 g of water evaporates.

13.86 Few of the elements are found in the Earth in their pure form. Most elements exist as oxides or sulfides. Explain why this is so, and explain why nitrogen exists as a pure element.

13.87 Lightning in the atmosphere causes the following reaction:

$$N_2 + O_2 + lightning \longrightarrow 2 NO$$

A firefly produces light by converting the compound luciferin to dehydroluciferin:

$$luciferin \longrightarrow dehydroluciferin + light$$

For each of these processes, decide whether ΔG is negative or positive. Explain your reasoning.

13.88 Humans perspire as a way of keeping their bodies from overheating during strenuous exercise. The evaporation of perspiration transfers heat from the body to the surrounding atmosphere. Calculate ΔS of the universe for evaporation of 1.0 g of water if the skin is at 37.5 °C and air temperature is 23.5 °C. Use ΔH_{vap} of water to find q. (Hint: Be careful about your choice for the system.)

13.89 At its triple point, a dynamic equilibrium can exist among all three phases of matter. Draw a molecular picture of argon that shows what happens at the triple point.

13.90 Two moles of HCl gas have a larger standard entropy, 374 J/mol K, than 1 mol each of H_2 gas and Cl_2 gas, 354 J/mol K. What feature of the molecular structures of these substances accounts for this difference?

Group Study Problems

13.91 In the upper atmosphere, ozone is produced from oxygen:

$$3 O_2(g) \longrightarrow 2 O_3(g)$$

(a) Compute $\Delta H°$, $\Delta S°$, and $\Delta G°$ for this reaction. (b) Is there a temperature at which this reaction becomes spontaneous at 1 atm pressures? If so, find it. If not, explain why one does not exist. (c) Assume an atmosphere with $p_{O_2} = 0.20$ atm and $T = 298$ K. Below what pressure of O_3 is O_3 production spontaneous? (d) In view of your answers to parts (a) through (c), how can the ozone layer form? (You may need to review Section 6.4.)

13.92 $S°$ of graphite is 3 times larger than $S°$ of diamond. Explain why this is so. (You may need to review the structures and properties of graphite and diamond in Chapter 9.) A recently discovered form of elemental carbon, buckminsterfullerene, is a solid that consists of individual molecules with formula C_{60}. Is the molar entropy of buckminsterfullerene larger or smaller than that of graphite? Explain.

13.93 Here are thermodynamic data for fusion of NH_3:

$$NH_3(s) \longrightarrow NH_3(l) \qquad \Delta H° = 5.65 \text{ kJ/mol}, \Delta S° = 28.9 \text{ J/mol K}$$

(a) Calculate $\Delta G°$ for the melting of 1.00 mol of NH_3. (b) Calculate the freezing point of NH_3.

13.94 Methane gas has a lower $S°$ (186.3 J/mol K) than gaseous ammonia ($S° = 192.8$ J/mol K), whereas $S°$ of gaseous H_2O is 188.8 J/mol K. Does this surprise you? Why or why not? Suggest a reason why methane has a lower absolute entropy.

13.95 Phosphorus forms white crystals made up of P_4 molecules. There are two forms of white crystalline phosphorus, called α and β. The difference between $P_{4\alpha}$ and $P_{4\beta}$ is determined by the way the P_4 molecules pack together in the crystal lattice. The α form is always obtained when liquid phosphorus freezes. However, at temperatures below −77 °C, the $P_{4\alpha}$ crystals change spontaneously to $P_{4\beta}$:

$$P_{4\alpha} \xrightarrow{-77 \text{ °C}} P_{4\beta}$$

(a) Which form of phosphorus has a more ordered crystalline structure? (b) Determine the signs of ΔH and ΔS for this process. Explain your reasoning.

Answers to Section Exercises

13.1.1 (a) A fence represents order; when it falls down, it becomes disordered; (b) a glass has a relatively ordered structure, whereas a broken glass is highly disordered; and (c) any solution of several substances is disordered compared with the separate components.

13.1.2 Increased organization in one place (formation of sugar crystals) must be accompanied by even greater disorganization elsewhere. In this example, the water molecules that boil away are much more disordered than water molecules in the liquid.

13.1.3

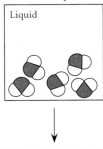

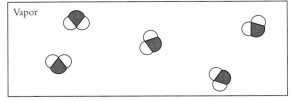

13.2.1 Heat will flow from the warmer mixture to the cooler mixture, so benzene will freeze and ice will melt.

13.2.2 0.10 J/K

13.2.3 $\Delta S_{\text{mixture}} = -6.11$ J/K, $\Delta S_{\text{refrigerator}} = 6.11$ J/K, $\Delta S_{\text{universe}} = 0$; neither process is spontaneous.

13.3.1 (a) 1 g of dew because liquids are more disordered than solids of the same material; (b) 1 mol of gaseous hydrogen atoms because although both samples contain the same number of atoms, binding atoms into molecules reduces the amount of disorder; (c) the flawed diamond because flaws represent disorder; and (d) the sample at 50 °C because increasing the temperature of a sample increases its molecular motion and disorder.

13.3.2 (a) There are more atoms in the sample of O_3, so there are more opportunities for randomness in the molecular arrangement;

(b) both samples have the same number of atoms, but there are more molecules in 3 mol of O_2, so there is more randomness in its molecular arrangement; (c) I_2 is a solid under standard conditions, so it is more ordered than gaseous O_2; and (d) diluting a solution spreads its solute particles (ions in this case) over a larger region, giving it greater disorder.

13.3.3

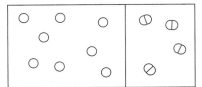

Hydrogen atoms   Hydrogen molecules

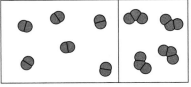

Oxygen molecules (O_2)   Ozone (O_3)

13.3.4 -2803.99 J/K

13.4.1 -224.2 kJ/mol

13.4.2 4.4×10^{-2} atm

13.4.3 Yes, $T = 5.43 \times 10^3$ K

13.5.1 340 kJ

13.5.2 (a) 0.124 atm; (b) ice remains solid until the pressure drops to about 0.005 atm, when it sublimes; and (c) at 7 atm, dry ice melts at around 217 K and is liquid at room temperature.

13.5.3 When an automobile engine moves an automobile, it does work and generates order (all parts of the automobile move together in the same direction). To do this without violating the second law of thermodynamics, enough heat must be transferred to the surroundings to generate an amount of disorder greater than the amount of order accompanying the work.

13.6.1 92%

13.6.2 $+32.2$ J/mol K; the spontaneity increases as temperature increases.

13.6.3 (a) 16.3 mol of ATP; and (b) 82.5 g of glucose

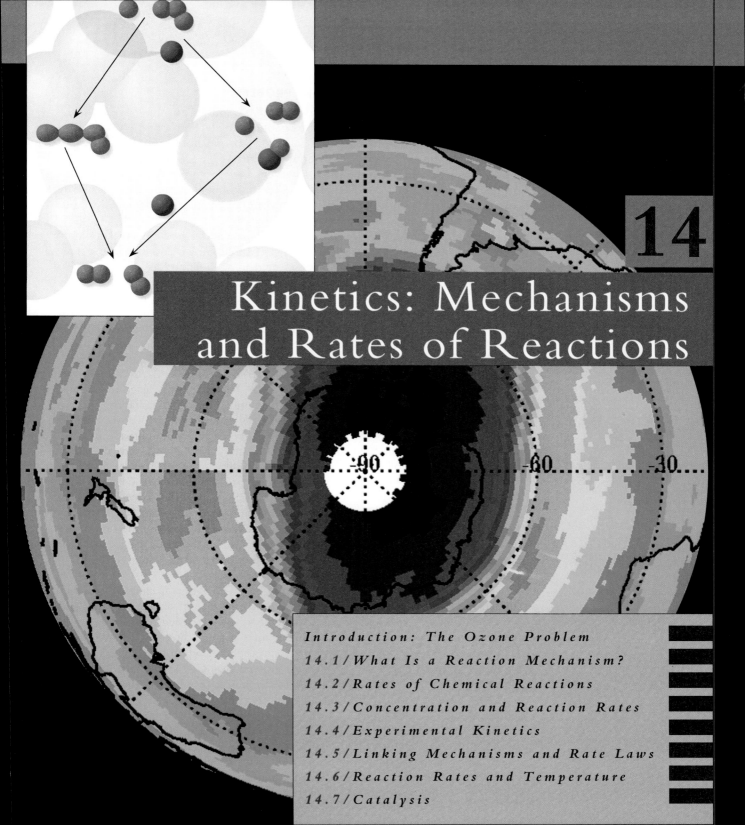

14

Kinetics: Mechanisms and Rates of Reactions

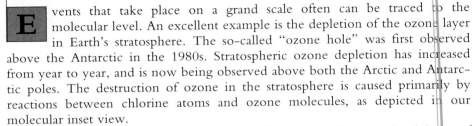

INTRODUCTION: THE OZONE PROBLEM

Events that take place on a grand scale often can be traced to the molecular level. An excellent example is the depletion of the ozone layer in Earth's stratosphere. The so-called "ozone hole" was first observed above the Antarctic in the 1980s. Stratospheric ozone depletion has increased from year to year, and is now being observed above both the Arctic and Antarctic poles. The destruction of ozone in the stratosphere is caused primarily by reactions between chlorine atoms and ozone molecules, as depicted in our molecular inset view.

The chlorine atoms in the upper atmosphere come from the breakdown of CF_2Cl_2 and other similar chlorofluorocarbons (CFCs), known commercially as Freons. Production of these compounds was more than one million tons in 1988, largely for use in refrigerators and air conditioners. Once released from these or other sources, CFCs diffuse slowly upward in the atmosphere until they reach the ozone layer. As shown by laboratory experiments, ultraviolet light from the sun, which is abundant in the ozone layer, splits off chlorine atoms that then react with ozone. The result has been dramatic: Experimental data reveal annual ozone decreases of more than 50% above Antarctica and smaller but similar decreases in the Arctic. The background photo shows the Antarctic hole (red-violet) on September 20, 2000.

This depletion raised international concerns, because, as discussed in Chapter 6, ozone in the stratosphere removes high-energy ultraviolet radiation coming from the sun. Because this radiation is lethal, loss of ozone in the stratosphere imperils life on Earth. In 1987, world nations agreed to a 50% cutback in CFC production by the year 2000; three years later, they agreed on essentially complete abandonment of CFC production by 2000; and in 1992 the ban was moved forward to 1996. Despite dramatic reductions in the use of CFCs, recovery of the ozone layer is not assured, because CFCs and other ozone-depleting pollutants will remain in the atmosphere for many years to come. Some simulations indicate that Arctic ozone depletion may exceed 50% before the end of the next decade, and the Antarctic ozone hole was so large in the year 2000 that populated regions of Chile and Australia were exposed to increased amounts of ultraviolet radiation.

The ozone hole would almost certainly be much worse if chemists had not studied the reactions of CFCs with atmospheric gases before ozone depletion was discovered. The 1995 Nobel Prize in Chemistry was awarded to the three pioneers in this effort. A German chemist, Paul Crutzen, discovered how ozone concentration is regulated in a "normal" stratosphere, while two Americans, F. Sherwood Rowland and Mario Molina, showed that CFCs can destroy ozone. These studies of molecular reactions allowed quick determination that CFCs are a likely cause of ozone depletion and led to the international restrictions described above.

The story of the ozone hole is just one example of the importance of studies of the molecular details of chemical reactions. Some chemists use information about how reactions occur to design and synthesize useful new compounds. Others explore how to modify reaction conditions so as to minimize the cost of producing industrial chemicals. This chapter provides an introduction to how chemical reactions occur at the molecular level. We show how to describe a reaction from the molecular perspective, introduce the basic principles that govern these processes, and describe some experimental methods used to study chemical reactions.

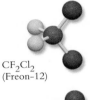

CF_2Cl_2
(Freon-12)

$CFCl_3$
(Freon-11)

$C_2F_3Cl_3$
(Freon-113)

14.1 WHAT IS A REACTION MECHANISM?

Atoms, ions, and molecules rearrange and recombine during chemical reactions. These processes usually do not occur all at once. Instead, each reaction consists of a sequence of molecular events called a **reaction mechanism.**

> /// *A reaction mechanism is the* **exact molecular pathway** *that starting materials follow on their way to becoming products.*

We introduce the principles of mechanisms using simple chemical reactions. After presenting the principles, we will be able to return to the more complicated mechanisms involved in the ozone problem.

Example of a Mechanism: Formation of N_2O_4

Nitrogen dioxide, a red-brown gas, is stable at room temperature. However, Figure 14-1 shows that reducing the temperature causes a sample of NO_2 to become colorless. The color change that happens at low temperature occurs because two NO_2 molecules combine to form one N_2O_4 molecule, and N_2O_4 is a colorless gas.

The mechanism of this reaction describes what happens at the molecular level. An N_2O_4 molecule forms when two NO_2 molecules collide and stick together by forming a bond between the two nitrogen atoms.

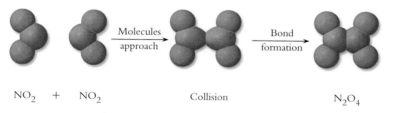

$$NO_2 \quad + \quad NO_2 \qquad\qquad \text{Collision} \qquad\qquad N_2O_4$$

The formation of N_2O_4 requires more than a simple collision between two NO_2 molecules. The product contains a bond between the nitrogen atoms, so the collision must bring the two nitrogen atoms into contact. A collision between the nitrogen atom of one NO_2 molecule and an oxygen atom of a second will not lead to N_2O_4.

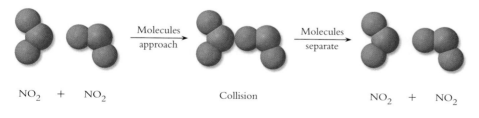

$$NO_2 \quad + \quad NO_2 \qquad\qquad \text{Collision} \qquad\qquad NO_2 \quad + \quad NO_2$$

Elementary Reactions

A mechanism is a description of the actual molecular events that occur during a chemical reaction. Each such event is an **elementary reaction.** Elementary reactions involve one, two, or occasionally three reactant molecules or atoms. In other words, elementary reactions can be unimolecular, bimolecular, or termolecular. A typical mechanism consists of a sequence of elementary reactions. Although an *overall* reaction describes the starting materials and final products, it usually is not

Figure 14-1
When a flask containing orange NO_2 gas at room temperature (*top*) is cooled in a low-temperature bath (*bottom*), NO_2 molecules combine to form colorless N_2O_4 molecules.

The Lewis structure of NO_2 indicates that the nitrogen atom has an unpaired electron. Two NO_2 molecules combine by using their unpaired electrons to form an N—N bond.

elementary because it does not represent the individual steps by which the reaction occurs.

The most common type of elementary process is a **bimolecular reaction** that results from the collision of two molecules, atoms, or ions. The collision of two NO_2 molecules to give N_2O_4 is a bimolecular reaction. Here is another example:

Animation

$$OH^- + H_3O^+ \qquad \text{Collision} \qquad H_2O + H_2O$$

Notice that the characteristic feature of a bimolecular elementary reaction is a collision between two species that results in a rearrangement of chemical bonds. The two reaction partners stick together by forming a new bond, or they form two new species by transferring one or more atoms from one partner to another.

In a **unimolecular reaction,** a single molecule fragments into two pieces or rearranges to a new isomer. A simple example of a unimolecular reaction is the decomposition of N_2O_4 into two molecules of NO_2. The N—N bond cleaves in this reaction, as shown in Figure 14-2. A vibration of the molecule stretches the N—N bond. If the molecule has sufficient energy, the bond breaks, much like a spring that has been stretched too far. A stable molecule such as N_2O_4 does not decompose unless it first acquires the energy needed to break one of its bonds. This energy can come from molecular collisions that transfer energy from one molecule to another.

In a **termolecular reaction,** three chemical species collide simultaneously. Termolecular reactions are rare because they require a collision of three species at the *same time* and in exactly the right orientation to form products. The odds against such a three-body collision are high. Most chemical reactions, including all those introduced in this book, can be described at the molecular level as sequences of bimolecular and unimolecular elementary reactions.

To summarize, the mechanism of a reaction converts starting materials to products through a specific sequence of bimolecular collisions and unimolecular rearrangements.

Any collection of gas molecules has a distribution of kinetic energies. For a review, see Section 5.1 and Figure 5-5.

Alternative Mechanisms

Very few chemical reactions are as simple as the low-temperature reaction of NO_2 to form N_2O_4. For example, at high temperature, NO_2 undergoes a different reaction, decomposing into nitrogen oxide and molecular oxygen:

$$2\,NO_2 \longrightarrow 2\,NO + O_2$$

Later, we explore *why* NO_2 molecules can react in two different ways. We are now interested only in *how* these reactions occur.

Animation

Figure 14-2
A unimolecular elementary reaction may involve bond breakage. If an N_2O_4 molecule possesses enough energy, the vibration can break the N—N bond to produce two NO_2 molecules.

N_2O_4 $\quad$ Bond stretches $\quad$ Bond breaks $\quad$ NO_2 $\quad$ NO_2

One possible sequence that explains the decomposition of NO_2 molecules starts with a unimolecular reaction. At high temperature, collisions may transfer enough energy to allow some of the NO_2 molecules to break apart. That is, an N—O bond in NO_2 may break to produce a molecule of NO and an oxygen atom. Because oxygen atoms are highly reactive, a subsequent bimolecular collision between an oxygen atom and an NO_2 molecule would then generate NO and O_2. This reaction sequence can be summarized as Mechanism I for NO_2 decomposition:

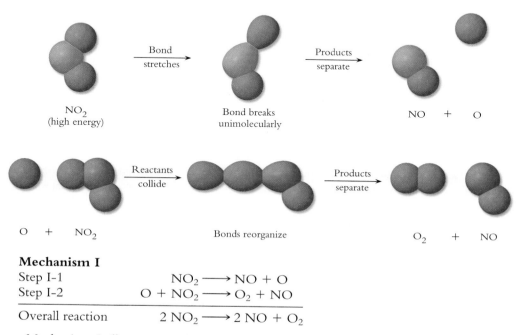

NO₂
(high energy)

Bond
stretches

Bond breaks
unimolecularly

Products
separate

NO   +   O

O   +   NO₂

Reactants
collide

Bonds reorganize

Products
separate

O₂   +   NO

Mechanism I

Step I-1	$NO_2 \longrightarrow NO + O$
Step I-2	$O + NO_2 \longrightarrow O_2 + NO$

Overall reaction	$2\,NO_2 \longrightarrow 2\,NO + O_2$

Mechanism I illustrates an important requirement for reaction mechanisms. Because a mechanism is a summary of events at the molecular level, a mechanism *must* lead to the correct stoichiometry to be an accurate description of the chemical reaction. The sum of the steps of a mechanism must give the balanced stoichiometric equation for the overall chemical reaction; if it does not, the proposed mechanism must be discarded. In Mechanism I the net result of two sequential elementary reactions is the observed reaction stoichiometry.

A second possible mechanism for NO_2 decomposition starts with a bimolecular reaction. When two fast-moving NO_2 molecules collide, an oxygen atom may be transferred between them to form molecules of NO_3 and NO. Molecules of NO_3 are unstable and readily break apart into NO and O_2. This reaction sequence can be summarized as Mechanism II for NO_2 decomposition:

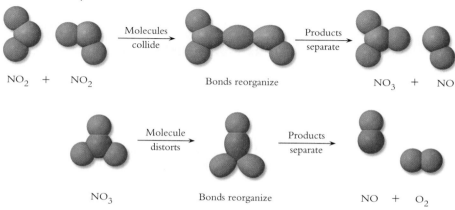

NO₂   +   NO₂

Molecules
collide

Bonds reorganize

Products
separate

NO₃   +   NO

NO₃

Molecule
distorts

Bonds reorganize

Products
separate

NO   +   O₂

Mechanism II

Step II-1	$2\,NO_2 \longrightarrow NO_3 + NO$
Step II-2	$NO_3 \longrightarrow NO + O_2$

Overall reaction	$2\,NO_2 \longrightarrow 2\,NO + O_2$

Mechanism II gives the observed overall stoichiometry because the NO_3 molecule produced in the first step is consumed in the second step.

Either Mechanism I or Mechanism II can account for the decomposition of NO_2. Each consists of elementary steps, each generates a highly reactive species that undergoes further chemistry, and each accounts for the observed reaction stoichiometry. Does either mechanism represent what really happens at the molecular level? Before describing how to test whether or not a mechanism is realistic, we must introduce some additional features of mechanisms.

Intermediates

Both proposed mechanisms for NO_2 decomposition contain chemical species produced in the first step and consumed in the second step. This is the defining characteristic of an **intermediate.** An intermediate is a chemical species *produced* in an early step of a mechanism and *consumed* in a later step. Intermediates never appear in the overall chemical equation. Notice that neither the O atoms of Mechanism I nor the NO_3 molecules of Mechanism II appear in the balanced chemical equation for NO_2 decomposition.

Intermediates are reactive chemical species that usually exist only briefly. They are consumed rapidly by bimolecular collisions with other chemical species or by unimolecular decomposition. The intermediate in Mechanism I is an oxygen atom that reacts rapidly with NO_2 molecules. The intermediate in Mechanism II is an unstable NO_3 molecule that rapidly decomposes.

The most direct way to test the validity of a mechanism is to determine what intermediates are present during the reaction. If oxygen atoms were detected, we would know that Mechanism I is a reasonable description of NO_2 decomposition. Likewise, the observation of NO_3 molecules would suggest that Mechanism II is reasonable. In practice, the detection of intermediates is quite difficult because they are usually reactive enough to be consumed as rapidly as they are produced. As a result, the concentration of an intermediate in a reaction mixture is very low. Highly sensitive measuring techniques are required for the direct detection of chemical intermediates.

The Rate-Determining Step

How fast do chemical reactions occur? The speed of a reaction is described by its **rate.** Rate is the number of events per unit time, such as the number of molecules reacting per second. Every elementary reaction has a characteristic rate. Some reactions are so fast that they are complete in the smallest measurable fraction of a second, whereas others are so slow that they require almost an eternity to reach completion. The observed rate of an *overall* chemical reaction is determined by the rates of the elementary reactions that make up the mechanism.

The various elementary reactions in a mechanism rarely occur at exactly the same rate. Instead, each step has a unique rate, and every mechanism has one elementary reaction that is slower than all the others. The slowest elementary step in a mechanism is called the **rate-determining step.** The rate-determining step governs the rate of the overall chemical reaction because no net chemical reaction can go faster than its slowest step. The idea of the rate-determining step is central to the study of reaction mechanisms.

Anyone who has flown on a commercial airline is familiar with the consequences of rate-determining steps, as illustrated by Figure 14-3. At a busy airport, for example, airplanes queue up while waiting for a runway to become available. After the pilot receives permission to take off, the plane becomes airborne very quickly. No matter how quickly an airplane is loaded and ready to leave the gate, the rate of takeoffs is determined by availability of runways. That is, the rate-determining step for airplane takeoff is the rate of takeoffs from the runways. Baggage claim is the rate-determining step at the end of a flight. No matter how quickly the passengers leap from their seats, push through the aisle, and race through the terminal, passengers who checked their baggage cannot leave the airport any faster than the baggage is delivered.

Chemical knowledge often can be used to assess which step of a mechanism is likely to be rate-determining. For example, in both proposed mechanisms for the decomposition of NO_2, chemical knowledge suggests that the first step is much slower than the second step. In Mechanism I an oxygen atom is produced in the first step, $NO_2 \rightarrow NO + O$. Oxygen atoms are known to be highly reactive, so it is reasonable to predict that this intermediate reacts rapidly with NO_2 molecules. Compared with the fast second step of this mechanism, the step that forms the oxygen atoms is expected to be slow and rate-determining. Similarly, the first step of Mechanism II, $2 NO_2 \rightarrow NO_3 + NO$, produces NO_3. Its odd number of electrons indicates that this intermediate is unstable and will decompose in the second step of Mechanism II almost as soon as it forms. Again, the second step of the mechanism is expected to be fast, so the step that forms the reactive intermediate is slow and rate-determining. Later in this chapter we discuss experiments that make it possible to distinguish between Mechanisms I and II.

Figure 14-3
The slowest step in any sequential process is rate-determining. Airplanes take off at the rate that runways become available, and arriving passengers leave the terminal at the rate their baggage is delivered.

Section Exercises

14.1.1 Nitrogen dioxide (NO_2) in polluted air forms from NO and CO_2 by a single bimolecular reaction. Identify the second molecular product of this reaction, and draw a molecular picture similar to the ones in this section that shows how this chemical reaction occurs.

14.1.2 The following mechanism has been proposed for the reaction of NO with H_2:

$$2 NO \longrightarrow N_2O_2$$

$$N_2O_2 + H_2 \longrightarrow H_2O + N_2O$$

$$N_2O + H_2 \longrightarrow N_2 + H_2O$$

Determine the balanced chemical equation for the reaction that occurs by this mechanism and identify any chemical intermediates.

14.1.3 One common analogy for the rate-determining step involves automobiles on a highway passing through a tollbooth. In your own words, explain rate-determining steps using the tollbooth analogy.

14.2 RATES OF CHEMICAL REACTIONS

Some of the most important information about a mechanism comes from experiments that determine how fast a chemical reaction occurs under various conditions. In chemical reactions, amounts of reactants and products change, so reaction rates are given in units of amount per unit time, for example molecules per second. Amounts also can be expressed as concentrations, so rates can be measured in units of concentration per unit time; for example, molar per minute.

Recall that M is the symbol for molarity, expressed as mol/L.

The study of the rates of chemical reactions is called **kinetics.** Chemists study reaction rates for many reasons. To give just one example, Rowland and Molina used kinetic studies to show the destructive potential of CFCs. Kinetic studies are essential to the explorations of reaction mechanisms, because a mechanism can never be determined by calculations alone. Kinetic studies are important in many areas of science, including biochemistry, synthetic chemistry, biology, environmental science, engineering, and geology. The usefulness of chemical kinetics in elucidating mechanisms can be understood by examining the differences in rate behavior of unimolecular and bimolecular elementary reactions.

A Molecular View

To obtain a molecular perspective of reaction rates, consider the unimolecular reaction shown in Figure 14-4. At elevated temperature, the compound *cis*-2-butene can rearrange to form its isomer, *trans*-2-butene. The reaction occurs after collisions transfer enough energy to a *cis*-2-butene molecule to break the C—C π bond. Once the bond breaks, rotation around the C—C σ bond takes place very rapidly until the π bond reforms.

For a molecular view of rates, consider a sample containing 12 molecules of *cis*-2-butene, which represents a tiny portion of an immense reaction mixture. Figure 14-5 illustrates schematically how this reaction progresses with time. During the first minute, six *cis*-2-butene molecules isomerize to *trans*-2-butene. The average rate for this time period is 6 molecules/min. As the reaction proceeds, however, the rate of reaction decreases. During the second minute, only three molecules isomerize, giving an average rate for that period of 3 molecules/min.

Why does the rate become slower as this reaction proceeds? Dividing the number of *cis*-2-butene molecules that react by the total number of *cis*-2-butene molecules present at the beginning of the interval reveals the answer:

$$\text{First minute: Fraction reacting} = \frac{6 \; \textit{cis}\text{-2-butene molecules react}}{12 \; \textit{cis}\text{-2-butene molecules present}} = 0.5$$

$$\text{Second minute: Fraction reacting} = \frac{3 \; \textit{cis}\text{-2-butene molecules react}}{6 \; \textit{cis}\text{-2-butene molecules present}} = 0.5$$

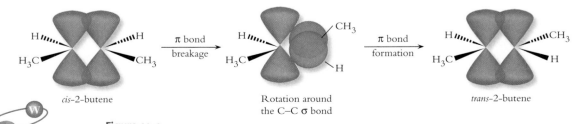

cis-2-butene π bond breakage   Rotation around the C–C σ bond π bond formation   trans-2-butene

Animation

Figure 14-4
cis-2-**Butene can rearrange in a unimolecular reaction if collisions give the molecule enough extra energy to break the C—C π bond.**

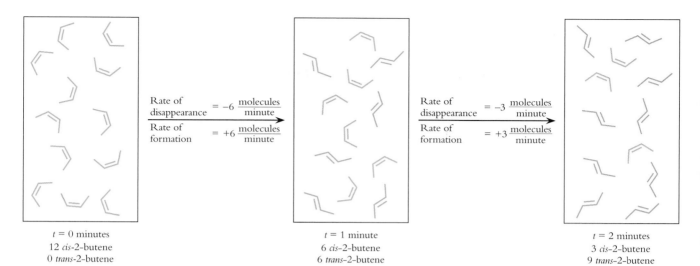

Figure 14-5
A molecular view of the isomerization of *cis*-2-butene illustrates that the rate of isomerization decreases as the number of *cis*-2-butene molecules decreases.

The fraction of the *cis*-2-butene molecules present that react during each one-minute interval is the same. That is, the rate of reaction is constant on a *per molecule* basis. As the reaction proceeds, however, fewer *cis*-2-butene molecules remain, causing the overall rate of reaction to decrease. This is characteristic of unimolecular elementary reactions. The rate per molecule is constant, but if the number of reactant molecules is cut in half, the rate of reaction is cut in half, as well.

Bimolecular elementary reactions also slow down as the number of reactant molecules decreases, as can be seen for the formation of N_2O_4 from NO_2. Figure 14-6 shows 12 molecules of starting material, again representing a tiny fraction of a much larger sample. Four NO_2 molecules combine during the first minute, so the

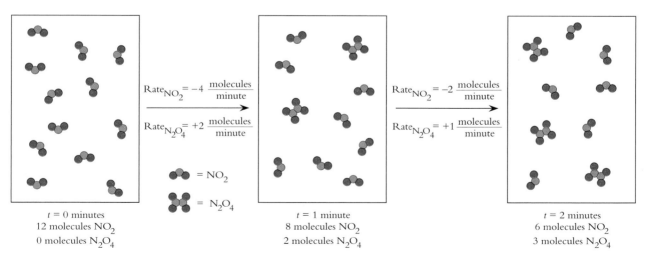

Figure 14-6
A molecular view of the formation of N_2O_4 from NO_2. The rate of reaction decreases as the number of molecules decreases because fewer NO_2 molecules are present to undergo collisions.

average rate during this period is 4 molecules/min. As the reaction proceeds, however, the rate decreases. During the second minute, only two molecules react, giving an average rate for that interval of 2 molecules/min.

Dividing the number of molecules reacting during each time interval by the number of reactant molecules present shows that this reaction behaves differently from a unimolecular reaction:

$$\text{First minute: Fraction reacting} = \frac{4 \ NO_2 \text{ molecules react}}{12 \ NO_2 \text{ molecules present}} = 0.33$$

$$\text{Second minute: Fraction reacting} = \frac{2 \ NO_2 \text{ molecules react}}{8 \ NO_2 \text{ molecules present}} = 0.25$$

In this case, not only do fewer molecules react, but also the rate of reaction per molecule is slower. If it were possible to count the number of collisions, we would find that the reaction slows down not only because fewer reactant molecules are present, but also because fewer collisions occur. For all bimolecular reactions, the rate *per collision* is constant, but as the reaction proceeds, fewer reactant molecules remain to collide with one another, so the rate per molecule decreases.

Figure 14-6 also shows that each species has its own rate, but the individual rates are linked by the stoichiometric coefficients. The reaction generates one molecule of N_2O_4 for every two molecules of NO_2 that are destroyed. That is, the $1:2$ stoichiometry of this reaction results in a $1:2$ relationship between the rate of disappearance of NO_2 and the rate of appearance of N_2O_4. The ratio of rates for different species is always equal to the ratio of their stoichiometric coefficients.

A Macroscopic View: Concentration Changes

Molecules are too small and much too numerous to follow on an individual basis. Therefore a chemist interested in measuring the rate of a reaction monitors the *concentration* of a particular compound as a function of time. The concentrations of reactants, products, or both may be monitored. For example, Figure 14-7 shows some experimental data obtained from a series of concentration measurements on the decomposition of NO_2:

$$2 \ NO_2 \longrightarrow 2 \ NO + O_2$$

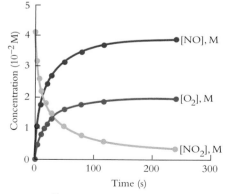

Time	(s)	0	5.0	10	15	20	30	50	80	120	240
[NO$_2$]	(10^{-2} M)	4.1	3.1	2.5	2.1	1.8	1.4	1.0	0.70	0.50	0.30
[NO]	(10^{-2} M)	0	1.0	1.6	2.0	2.3	2.7	3.1	3.4	3.6	3.8
[O$_2$]	(10^{-2} M)	0	0.50	0.80	1.0	1.1	1.3	1.6	1.7	1.8	1.9

Figure 14-7
A plot of [NO$_2$], [O$_2$], and [NO] as a function of time (seconds) for the decomposition reaction of NO$_2$. The concentration data are shown in the table.

The rate of the reaction at any specific time is given by how fast the concentration changes. The plots in Figure 14-7 show the same features that the molecular pictures of Figures 14-5 and 14-6 show: Rates of reaction decrease as starting materials are consumed, and rates for different species are linked by stoichiometry.

The average rate of a reaction can be expressed as the change in concentration (Δc) over some time interval (Δt). For oxygen formation during the decomposition of NO_2:

$$\text{Rate of production of } O_2 = \frac{\Delta[O_2]}{\Delta t}$$

Because rates change continuously with time, accurate rate determinations must use small time intervals. At the outset of the NO_2 decomposition experiment shown in Figure 14-7, the time interval between measurements is 5 seconds. For the first 5 seconds, the average rate of O_2 production is as follows:

$$\text{Rate } O_2 = \frac{\Delta[O_2]}{\Delta t} = \frac{(0.50 \times 10^{-2} \text{ M} - 0 \text{ M})}{(5.0 \text{ s} - 0 \text{ s})} = 1.0 \times 10^{-3} \text{ M/s}$$

At later times the rate is different. Verify this for yourself by calculating the average rates for the second and third 5-second intervals.

We can also write a rate expression for the consumption of NO_2:

$$\text{Rate of consumption of } NO_2 = -\frac{\Delta[NO_2]}{\Delta t}$$

The negative sign appears in this equation to make the rate positive, even though NO_2 concentration decreases with time. By convention, rate expressions are always written in a way that gives positive rates. Over the first 5 seconds of this reaction, the average rate of NO_2 consumption is as follows:

$$\text{Rate } NO_2 = -\frac{\Delta[NO_2]}{\Delta t} = -\frac{(3.1 \times 10^{-2} \text{ M} - 4.1 \times 10^{-2} \text{ M})}{(5.0 \text{ s} - 0 \text{ s})} = 2.0 \times 10^{-3} \text{ M/s}$$

Notice that over the same period the rate of O_2 formation is only half the rate of NO_2 consumption. This follows from the molecular view of the mechanism and from the stoichiometry of the reaction. The rate relationship among the three species involved in NO_2 decomposition is given by the following expression:

$$\text{Reaction rate} = \frac{\Delta[O_2]}{\Delta t} = \left(\frac{1}{2}\right)\left(\frac{\Delta[NO]}{\Delta t}\right) = \left(-\frac{1}{2}\right)\left(\frac{\Delta[NO_2]}{\Delta t}\right)$$

The relationship among reaction rates and stoichiometric coefficients can be applied to any reaction. For a reaction of the form, $a \text{ A} + b \text{ B} \rightarrow d \text{ D} + e \text{ E}$, the rate expressions are related as follows:

$$\text{Reaction rate} = \left(-\frac{1}{a}\right)\left(\frac{\Delta[A]}{\Delta t}\right) = \left(-\frac{1}{b}\right)\left(\frac{\Delta[B]}{\Delta t}\right)$$
$$= \left(\frac{1}{d}\right)\left(\frac{\Delta[D]}{\Delta t}\right) = \left(\frac{1}{e}\right)\left(\frac{\Delta[E]}{\Delta t}\right)$$

(14-1)

Here, lower-case letters represent stoichiometric coefficients and upper-case letters represent chemical substances.

Equation 14-1 is one of the few rate statements that follow from knowledge of the reaction stoichiometry. Almost all other rate information must be determined by carrying out experiments on how concentrations change with time. Example 14-1 illustrates the application of Equation 14-1.

Example 14-1 | **Relative Rates of Reaction**

Acrylonitrile is produced from propene, ammonia, and oxygen by the following balanced equation (see Example 13-5):

$$2\ C_3H_6 + 2\ NH_3 + 3\ O_2 \longrightarrow 2\ CH_2CHCN + 6\ H_2O$$

Relate the rates of reaction of starting materials and products.

Strategy: The rates for different species participating in a chemical reaction are related by the stoichiometric coefficients of the balanced chemical equation. Equation 14-1 provides the exact relationship.

Solution: The balanced equation shows that three molecules of oxygen are consumed for two molecules of propene and two molecules of ammonia. Thus, the rate of C_3H_6 and NH_3 consumption is only two thirds the rate of O_2 consumption. Those seven molecules of starting materials produce two molecules of CH_2CHCN and six molecules of H_2O. Thus CH_2CHCN is produced at the same rate as C_3H_6 is consumed, whereas H_2O is produced three times as fast as CH_2CHCN. The link between relative reaction rates and reaction stoichiometry is Equation 14-1. Therefore:

$$\text{Reaction rate} = \left(-\frac{1}{2}\right)\left(\frac{\Delta[C_3H_6]}{\Delta t}\right) = \left(-\frac{1}{3}\right)\left(\frac{\Delta[O_2]}{\Delta t}\right)$$

$$= \left(-\frac{1}{2}\right)\left(\frac{\Delta[NH_3]}{\Delta t}\right)\left(\frac{1}{2}\right)\left(\frac{\Delta[CH_2CHCN]}{\Delta t}\right) = \left(\frac{1}{6}\right)\left(\frac{\Delta[H_2O]}{\Delta t}\right)$$

Section Exercises

14.2.1 Ammonia is produced by the following reaction:

$$N_2 + 3\ H_2 \longrightarrow 2\ NH_3$$

Consider a portion of a flask that contains five molecules of N_2 and nine molecules of H_2. Draw a molecular picture showing the contents of this portion of the flask after two molecules of NH_3 have formed. Represent nitrogen atoms as dark circles and hydrogen atoms as light circles.

14.2.2 In the ammonia synthesis from the previous problem, which of the starting materials is consumed at a faster rate?

14.2.3 State the relative rates for the consumption of starting materials and the formation of products for the ammonia synthesis.

14.3 CONCENTRATION AND REACTION RATES

Both concentration and temperature affect the rate of a chemical reaction. This section examines how changes in the concentrations of starting materials and products affect the rate of a chemical reaction. Temperature effects are described in Section 14.6.

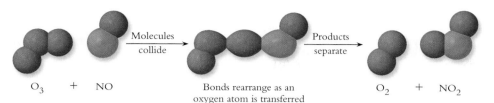

Figure 14-8
The reaction between O_3 and NO is believed to occur by a mechanism that consists of the single bimolecular step shown here in a molecular view.

Concentration Effects

A reaction of ozone provides an example of concentration effects. Ozone in the atmosphere near the Earth's surface is a serious pollutant that damages soft tissues such as the lungs. In major urban areas, smog alerts are issued whenever there are elevated concentrations of ozone in the lower atmosphere. Nitrogen oxide, another component of photochemical smog, is a colorless gas produced in a side reaction in automobile engines. One of the many reactions that links these species is the reaction of NO and O_3 to produce O_2 and NO_2:

$$NO + O_3 \longrightarrow NO_2 + O_2$$

Experiments indicate that this reaction occurs through a single bimolecular collision, as shown in Figure 14-8.

For a better understanding of the effect of changing concentrations on the rate of a chemical reaction, it helps to visualize the reaction at the molecular level. Figure 14-9 is a molecular view of portions of two containers of equal volume, each containing NO and O_3. The container in Figure 14-9b has the same number of O_3 molecules but twice as many NO molecules as the container in Figure 14-9a. As with all bimolecular reactions, the collision rates in the two containers determine the relative rates of reaction.

In this one-step bimolecular reaction, a single collision in the proper orientation leads to the transfer of an oxygen atom from O_3 to NO. As with the

◄ SECTION 5.7
The reactions that generate photochemical smog are introduced in Section 5.7.

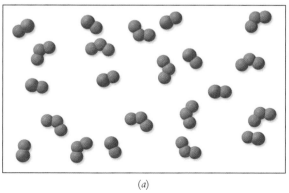

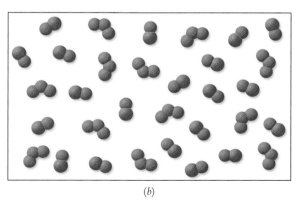

(a) (b)

Figure 14-9
A schematic illustration of two containers of equal volume containing different amounts of O_3 and NO. Container b has the same amount of O_3 but twice as much NO as container a. With twice as many NO molecules, twice as many NO- O_3 collisions occur, so the reaction in container b proceeds twice as fast as the reaction in container a.

formation of N_2O_4, the rate of a bimolecular reaction is proportional to the number of collisions: The more collisions there are, the more often the reaction occurs. The container in Figure 14-9b holds twice as many NO molecules as the container in Figure 14-9a, so an O_3 molecule in Figure 14-9b has twice as many opportunities to collide with an NO molecule. The rate of reaction in the flask in Figure 14-9b is twice the rate in the flask in Figure 14-9a. Example 14-2 extends this reasoning to O_3 molecules.

Example 14-2	Rates and Number of Molecules

The container that follows contains O_3 and NO. Compared with the container in Figure 14-9a, how fast will the reaction proceed?

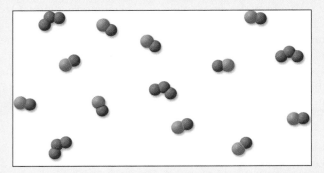

Strategy: The mechanism of this reaction requires a collision between an NO molecule and an O_3 molecule, as shown in Figure 14-8. The rate of the reaction therefore depends on the number of collisions that occur, and the collision frequency is proportional to the number of molecules.

Solution: In Figure 14-9a there are 12 molecules of O_3 and 10 of NO. The container shown in this example contains 4 molecules of O_3 and 10 molecules of NO. The concentration of O_3 is only one third as great, so any one molecule of NO will encounter an O_3 molecule three times less often than it would in the container in Figure 14-9a. This new mixture will react one third as fast as the one in Figure 14-9a.

Rate Laws

The effect of concentration on the rate of a particular chemical reaction can be summarized in an algebraic expression known as a rate law. A **rate law** links the rate of a reaction with the concentrations of the reactants through a **rate constant (k).** Also, as we show later in this chapter, the rate law may contain concentrations of chemical species that are not part of the balanced overall reaction.

Although every reaction has its own unique rate law, many rate laws have a general form:

$$\text{Rate} = k[A]^y[B]^z \qquad (14\text{-}2)$$

In general, each concentration has some exponent (y and z). These exponents are called the *orders* of the reaction. In Equation 14-2, y is the order of reaction with respect to species A, and z is the order with respect to species B. When the value of y is 1, the reaction is called **first order** in A; when the value of z is 2, the reaction is called **second order** in B, and so on. Orders of reaction are small integers or simple fractions. The most common orders are 1 and 2. The sum of the exponents is known as the *overall order* of the reaction.

The exponents in a rate law depend on the reaction mechanism rather than on the stoichiometry of the overall reaction. Consequently, a rate law must always be determined by conducting experiments; it can never be derived from the stoichiometry of the overall chemical reaction.

/// The order of a reaction must be determined by experiments.

Experiments show that the rate law for O_3 reacting with NO is first order in each of the starting materials and second order overall:

$$NO + O_3 \longrightarrow NO_2 + O_2 \qquad Rate = k[NO][O_3]$$

This rate law is fully consistent with the molecular view of the mechanism. If the concentration of either O_3 or NO is doubled, the number of collisions between starting material molecules doubles too, and so does the rate of reaction. If the concentrations of *both* starting materials are doubled, the collision rate and the reaction rate increase by a factor of four.

The order of reaction often differs from the stoichiometric coefficient. For example, experimental studies show that the rate law for the reaction of O_3 with NO_2 to give N_2O_5 and O_2 is first order in each reactant:

$$2 NO_2 + O_3 \longrightarrow N_2O_5 + O_2 \qquad Rate = k[NO_2][O_3]$$

Notice that for this reaction, the order of reaction with respect to NO_2 is 1, whereas the stoichiometric coefficient is 2. Thus the order of a reaction for a particular species cannot be predicted by looking at the overall balanced equation. We describe additional examples in Section 14.5.

Predicting a Rate Law

The relationship between a mechanism and its rate law can be illustrated for the decomposition of NO_2. Experiments that we describe in Section 14.4 reveal that the rate of this reaction is proportional to the square of NO_2 concentration:

$$2 NO_2 \longrightarrow 2 NO + O_2 \qquad Rate = k[NO_2]^2$$

As described earlier, this reaction might be described by either Mechanism I or Mechanism II:

Mechanism I

$NO_2 \longrightarrow NO + O$	Step I-1 (rate-determining)
$O + NO_2 \longrightarrow O_2 + NO$	Step I-2 (fast)

Mechanism II

$2 NO_2 \longrightarrow NO_3 + NO$	Step II-1 (rate-determining)
$NO_3 \longrightarrow NO + O_2$	Step II-2 (fast)

Each mechanism *predicts* a rate behavior that can be compared with the experimental rate law. If the prediction differs from the experimental observation, the mechanism is incorrect.

Remember that the overall rate of a reaction is determined by the rate of the slowest step. In other words, no reaction can proceed faster than the rate-determining step. Any step that comes *after* the rate-determining step cannot influence the overall rate of reaction. In this example, the first step of each mechanism is rate-determining. That is, each proposed mechanism predicts an overall rate of NO_2 decomposition that is the same as the rate of the first step in the mechanism.

The first step in Mechanism I is the unimolecular decomposition of NO_2. Our molecular analysis shows that the rate of a unimolecular reaction is constant on a *per molecule* basis. Thus if the concentration of NO_2 is doubled, twice as many molecules decompose in any given time. In quantitative terms, if NO_2 decomposes by Mechanism I, the rate law will be:

$$\text{Rate (Mechanism I)} = k[NO_2]$$

Once an NO_2 molecule decomposes, the O atom that results from decomposition very quickly reacts with another NO_2 molecule.

The rate-determining step of Mechanism II is a bimolecular collision between two identical molecules. A bimolecular reaction has a constant rate on a *per collision* basis. Thus if the number of collisions between NO_2 molecules increases, the rate of decomposition increases accordingly. Doubling the concentration of NO_2 doubles the number of molecules present, and it also doubles the number of collisions for *each* molecule. Each of these factors doubles the rate of reaction, so doubling the concentration of NO_2 increases the rate for this mechanism by a factor of *four*. Consequently, if NO_2 decomposes by Mechanism II, the rate law will be:

$$\text{Rate (Mechanism II)} = k[NO_2][NO_2] = k[NO_2]^2$$

The two proposed mechanisms for this reaction predict different rate laws. Whereas Mechanism I predicts that the rate is proportional to NO_2 concentration, Mechanism II predicts that the rate is proportional to the *square* of NO_2 concentration. Experiments agree with the prediction of Mechanism II, so Mechanism II is consistent with the experimental behavior of the NO_2 decomposition reaction. Mechanism I predicts rate behavior contrary to what is observed experimentally, so Mechanism I cannot be correct.

We return to the relationship between rate laws and mechanisms in Section 14.5, after discussing experimental methods for determining the rate law of a reaction.

Rate Constants

The value of a rate constant is independent of concentration, but it does depend on temperature. The details of temperature effects are discussed in Section 14.6.

Every reaction has its own characteristic rate constant that depends on the intrinsic speed of that particular reaction. For example, the value of k in the rate law for NO_2 decomposition is different from the value of k for the reaction of O_3 with NO. Rate constants are independent of concentration and time.

Rates of reaction have units of $(\text{concentration})(\text{time})^{-1}$. Because time does not appear in any other term on the right-hand side of the rate law, the units of k must always include time in the denominator. The concentration units of the rate constant depend on the overall order of the rate law, however, because the units of the rate constant must cancel concentration units to give the proper units for the rate of reaction. For example, the reaction of NO with O_3 is described by the following rate law:

$$\text{Rate} = k[NO][O_3]$$

To agree with the units of $(concentration)(time)^{-1}$ for the rate of reaction, the rate constant must have units of $(concentration)^{-1}(time)^{-1}$ (such as $M^{-1} s^{-1}$). These units fit into the rate law as follows:

Rate law:	Rate	=	k	[NO]	[O$_3$]
Units:	$(M\ s^{-1})$	=	$(M^{-1}\ s^{-1})$	(M)	(M)

Example 14-3 provides practice in deducing the units of a rate constant.

Units of the Rate Constant · **Example 14-3**

Reactions in aqueous solution can have complicated kinetics. An example is the reaction between arsenic acid and iodide ions:

$$H_3AsO_4(aq) + 3\ I^-(aq) + 2\ H_3O^+(aq) \longrightarrow H_3AsO_3(aq) + I_3^-(aq) + 3\ H_2O(l)$$

The rate law for this reaction has been found experimentally to be as follows:

$$Rate = k[H_3AsO_4][I^-][H_3O^+]$$

What are the units of the rate constant when the time is expressed in minutes?

Strategy: Units on the left must be the same as units on the right. The rate constant must have units that achieve this. Analyze the units for each component of the rate equation to find the units for k.

Solution: On the left side of the rate law, the rate of reaction is expressed in terms of changes in concentration and time:

$$Rate\ of\ reaction = \frac{\Delta c}{\Delta t}$$

For species in solution, concentration (c) is in molarity (M). According to the conditions of the problem, time (t) is in minutes. Therefore the units of the rate must be:

$$Rate\ of\ reaction = M\ min^{-1}$$

The right-hand side of the rate law contains a product of three concentrations, so it shows third-order behavior overall. This gives concentration units of M^3. This requires units of $M^{-2}\ min^{-1}$ for the rate constant:

$$Rate = (M^{-2}\ min^{-1})(M)(M)(M) = M\ min^{-1}$$

units of k units of $[I^-]$ units of $[H_3O^+]$

units of $[H_3AsO_4]$

Once again, we emphasize that the order of reaction and the value of the rate constant must be determined by doing experiments. Knowing the order of reaction then makes it possible to write the specific rate law for the chemical process. In the next three sections, we discuss how chemists determine orders of reactions and we continue to explore how rate laws are related to chemical mechanisms.

Section Exercises

14.3.1 Draw a container with numbers of NO and O_3 molecules that would react at half the rate of the one in Figure 14-9a.

14.3.2 The isomerization reaction of *cis*-2-butene, shown in Figure 14-4, is found experimentally to follow first-order kinetics. What is the rate law, and what are the units of the rate constant?

14.3.3 The reaction between hydrogen and bromine is as follows:

$$H_2(g) + Br_2(g) \longrightarrow 2\,HBr(g)$$

Experiments show that this reaction is first order in H_2 and one-half order in Br_2. What is the rate law, what is the overall order of the reaction, and what are the units of the rate constant?

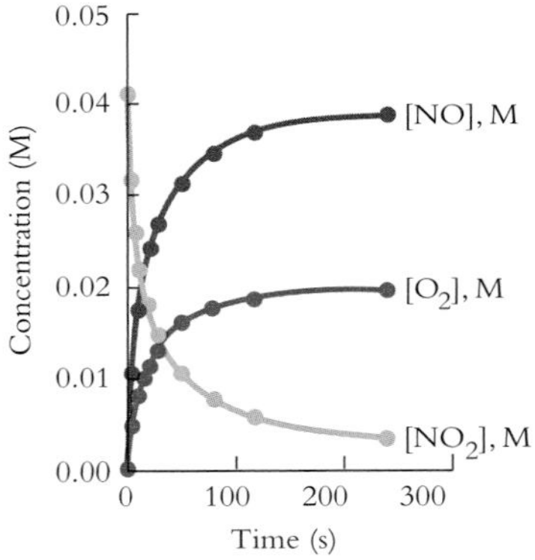

14.4 EXPERIMENTAL KINETICS

Every rate law must be determined experimentally. A chemist may imagine a reasonable mechanism for a reaction, but that mechanism must be tested by comparing the *actual* rate law for the reaction with the rate law *predicted* by the mechanism. To determine a rate law, chemists observe how the rate of a reaction changes with concentration. The graph of the data for the NO_2 decomposition reaction shown in Figure 14-7 is an example of such observations.

Regardless of the rate law, the rate of a reaction generally decreases with time because the concentrations of reactants decrease. The form of a rate law is determined by exploring the details of how the rate decreases with time. Because rate laws describe how rates vary with *concentration*, it is necessary to do mathematical analysis to convert rate laws into equations that describe how concentrations vary with *time*. These equations are different for different reaction orders.

First-Order Reactions

Recall that Mechanism I for the decomposition of NO_2 and the mechanism for the isomerization reaction of *cis*-2-butene both predict a simple first-order rate law:

$$\text{Rate} = k[NO_2] \qquad and \qquad \text{Rate} = k[\textit{cis}\text{-2-butene}]$$

The general form of a first-order rate law is rate $= k[A]$, where A is a reactant in the overall reaction. Mathematical treatment converts this general form into an equation relating concentration and time. For a first-order reaction, a logarithmic relationship links concentration and time:

> Students who have taken calculus will recognize that Equation 14-3 results from integration of the first-order rate law.

$$\ln\left(\frac{[A]_0}{[A]}\right) = kt \tag{14-3}$$

In this expression, [A] is the concentration of reactant A at time t, and $[A]_0$ is the concentration of reactant A at the beginning of the reaction ($t = 0$).

The most important feature of Equation 14-3 is that a graph of $\ln\left(\dfrac{[A]_0}{[A]}\right)$ vs. t gives a straight line whose slope is k. Consequently, experimental data can be tested for first-order behavior by preparing such a graph and observing whether or not it

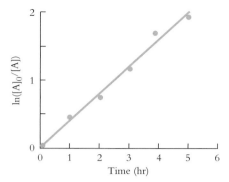

Time (hr)	$[C_5H_{11}Br]$ (M)	$[C_5H_{11}Br]_0/[C_5H_{11}Br]$	$\ln([A]_0/[A])$
0	0.150	1.00	0
1.00	0.099	1.52	0.42
2.00	0.073	2.05	0.72
3.00	0.047	3.19	1.16
4.00	0.028	5.36	1.68
5.00	0.022	6.82	1.92

Figure 14-10
A plot of $\ln([A_0]/[A])$ vs. time in hours for the conversion of an alkyl bromide to an alkene. The experimental data are shown in the table. The linearity of this plot verifies that this reaction obeys a first-order rate law.

is linear. Figure 14–10 illustrates this procedure for measurements done on the conversion of an alkyl bromide to an alkene:

$$\text{Alkyl bromide} \ (C_5H_{11}Br) + H_2O \longrightarrow \text{Alkene} \ (C_5H_{10}) + H_3O^+ + Br^-$$

> An alkyl bromide is a molecule in which one of the hydrogen atoms of an alkane has been replaced with a bromine atom. Alkenes, which contain C=C bonds, are discussed in Chapter 9.

The linear appearance of the plot shows that this reaction obeys a first-order rate law. Additional mechanistic studies suggest that alkene formation proceeds in a two-step sequence. In the first step, which is rate-determining, the C—Br bond breaks to generate a bromide anion and an unstable cationic intermediate. In the second step, the intermediate transfers a proton (H^+) to a water molecule, forming the alkene and H_3O^+:

$$C_5H_{11}Br \xrightarrow{\text{Slow}} \text{Intermediate} \ (C_5H_{11}^+) + Br^-$$

$$\text{Intermediate} \ (C_5H_{11}^+) + H_2O \xrightarrow{\text{Fast}} C_5H_{10} + H_3O^+$$

This example shows that first-order kinetics are not restricted to reactions with just one starting material, such as the decomposition of nitrogen dioxide or the isomerization of *cis*-2-butene.

The rate constant for the alkyl bromide reaction is equal to the slope of the line. The slope is determined by using any two points on the line:

$$\text{Rate constant} = k = \text{Slope} = \frac{\Delta y}{\Delta x} = \frac{1.16 - 0}{3.00 \ \text{hr} - 0 \ \text{hr}} = \frac{1.16}{3.00 \ \text{hr}} = 0.387 \ \text{hr}^{-1}$$

Equation 14-3 is used to test whether or not a reaction is first-order overall. The concentration of reactant A is monitored as a function of time. If a plot of $\ln\left(\dfrac{[A]_0}{[A]}\right)$ vs. t is linear, the reaction is first-order in A. If the plot is not linear, the reaction is not first-order. In Example 14-4, we apply this test to the data for the decomposition reaction of NO_2 presented in Figure 14-7.

Example 14-4	First-Order Kinetic Analysis

According to proposed Mechanism I, the decomposition of NO_2 should follow first-order kinetics. Do the experimental data of Figure 14-7 support this mechanism?

Strategy: According to the concentration-time form of first-order rate laws, a plot of $\ln\left(\dfrac{[A]_0}{[A]}\right)$ vs. t must be a straight line if Mechanism I is correct.

Solution: Convert the concentration data into ratios, take the logarithms, and then prepare a logarithmic plot to determine whether the reaction is first-order:

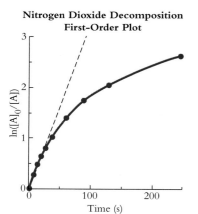

Nitrogen Dioxide Decomposition First-Order Plot

Time (s)	0	5	10	15	20	30	50	80	120	240
$[NO_2]$ (10^{-2} M)	4.1	3.1	2.5	2.1	1.8	1.4	1.0	0.070	0.050	0.030
$\dfrac{[NO_2]_0}{[NO_2]}$	1.00	1.32	1.64	1.95	2.28	2.93	4.10	5.86	8.20	13.7
$\ln\left(\dfrac{[NO_2]_0}{[NO_2]}\right)$	0.00	0.28	0.50	0.67	0.82	1.08	1.41	1.77	2.10	2.62

The graph is not linear, so we conclude that the decomposition of NO_2 does not follow first-order kinetics. Consequently, Mechanism I, which predicts first-order behavior, cannot be correct.

Data must be collected over a relatively long period in order to determine whether a first-order plot is linear. Notice that the first five points on the plot in Example 14-4 fall reasonably close to the dashed straight line. Only after more than 50% of the reactant has been consumed does this plot deviate substantially from linearity.

Another characteristic of first-order reactions is that the time it takes for half the reactant to disappear is the same, no matter what the concentration. This time is called the **half-life** ($t_{1/2}$). An equation for $t_{1/2}$ is obtained by applying Equation 14-3 to a time interval equal to the half-life. When half the original concentration has been consumed, $[A] = 0.5[A]_0$:

$$kt_{1/2} = \ln\left(\frac{[A]_0}{0.5[A]_0}\right)$$

The $[A]_0$ terms cancel:

$$\ln\left(\frac{[A]_0}{0.5[A]_0}\right) = \ln 2$$

Rearranging gives an equation for $t_{1/2}$:

$$t_{1/2} = \frac{\ln 2}{k} \qquad (14\text{-}4)$$

Equation 14-4 does not contain the concentration of A, so the half-life of a first-order reaction is a constant that is independent of how much A is present. The decomposition reactions of radioactive isotopes provide excellent examples of first-order processes, as Example 14-5 illustrates.

Half-Lives	**Example 14-5**

Carbon-14 (^{14}C) is a radioactive isotope with a half-life of 5.73×10^3 years. The amount of ^{14}C present in an object can be used to determine its age. Calculate the rate constant for decay of ^{14}C and determine how long is required for 90% of the ^{14}C in a sample to decompose.

Strategy: We are asked to calculate two quantities, so this is a quantitative problem to which we can apply the seven-step approach.

Solving Quantitative Problems

Solution:

1. We are asked to determine a rate constant and the time required for material to be consumed.

2. Visualize the process: atoms of ^{14}C decompose as time passes.

3. The only piece of data is the half-life: $t_{1/2} = 5.73 \times 10^3$ years.

4. Equation 14-3 links time and concentration through the rate constant, and Equation 14-4 links the rate constant to the half-life. Knowing a half-life, we can calculate the rate constant using Equation 14-4. Then the value of the rate constant and Equation 14-3 can be used to determine the time required to reach a certain concentration.

5. We need to break the problem into its two parts. To find the value of the rate constant, rearrange Equation 14-4 to isolate k:

$$t_{1/2} = \frac{\ln 2}{k} \qquad so \qquad k = \frac{\ln 2}{t_{1/2}}$$

To find the value for time, rearrange Equation 14-3 to isolate t:

$$\ln\left(\frac{[A]_0}{[A]}\right) = kt \qquad so \qquad t = \frac{\ln\left(\dfrac{[A]_0}{[A]}\right)}{k}$$

6. Now substitute to calculate the two quantities:

$$k = \frac{0.6931}{5.73 \times 10^3 \text{ yr}} = 1.21 \times 10^{-4} \text{ yr}^{-1}$$

When 90% has decayed, 10% remains, or $[A] = 0.10[A]_0$:

$$\ln\left(\frac{[A]_0}{[A]}\right) = \ln\left(\frac{[A]_0}{0.10[A]_0}\right) = \ln(10)$$

The ages of ancient pots can be determined by ^{14}C dating of charcoal found with the pots.

← **CHAPTER 21**
The use of radioactive isotopes and their half-lives to determine the age of an object is discussed in detail in Chapter 21.

Example 14-5	Half-Lives *(continued)*

$$t = \frac{\ln(10)}{k} = \frac{2.303}{1.21 \times 10^4 \text{ yr}^{-1}} = 1.90 \times 10^4 \text{ yr}$$

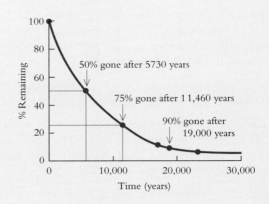

7. It requires 19,000 years for 90% of a sample of ^{14}C to decompose. This is "reasonable" given that the half-life of this isotope is 5730 years.

Second-Order Reactions

Students who have taken calculus will recognize that Equation 14-5 results from integration of the second-order rate law.

Mechanism II for the decomposition of NO_2 predicts that the reaction should have a second-order rate law:

$$\text{Rate} = \frac{\Delta[NO_2]}{\Delta t} = k[NO_2]^2$$

The general form of the second-order rate expression is: Rate $= k[A]^2$. Mathematical treatment converts this general form into a time-dependent rate law in which the reciprocal of concentration varies linearly with time:

$$\frac{1}{[A]} - \frac{1}{[A]_0} = kt \tag{14-5}$$

As before, $[A]$ is the concentration of reactant A at time t, and $[A]_0$ is the concentration of reactant A at the beginning of the reaction ($t = 0$).

The most important feature of Equation 14-5 is that a graph of $\left(\dfrac{1}{[A]} - \dfrac{1}{[A]_0}\right)$ vs. t gives a straight line whose slope is k. Consequently, experimental data can be tested for second-order behavior by preparing such a graph and observing whether or not it is linear. Figure 14–11 shows the decomposition data of Figure 14-7 plotted in this way. The graph is linear, showing that the reaction is second-order in NO_2. Thus the kinetic behavior of the decomposition of NO_2 matches the predicted rate law for Mechanism II (see earlier discussion). Chemists have accepted Mechanism II as the correct mechanism for this decomposition because it predicts the correct rate law and consists of plausible molecular processes.

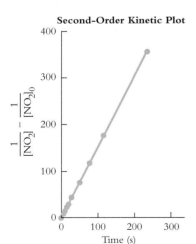

Second-Order Kinetic Plot

Figure 14-11
The decomposition data for NO_2 plotted as the reciprocal of concentration vs. time. This graph is linear, with a slope equal to the second-order rate constant.

Example 14-6 provides another example of the analysis of rate data.

Analysis of Rate Data	Example 14-6

The Diels–Alder reaction, in which two alkenes combine to give a new product, is one of the most frequently used reactions for the synthesis of organic compounds. Thousands of examples are found in the chemical literature. The reaction of butadiene is a simple example:

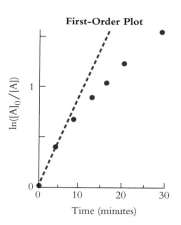

Otto Diels and Kurt Alder were awarded the Nobel Prize in Chemistry in 1950 for showing the synthetic usefulness of this reaction.

$$2\ C_4H_6$$
(Butadiene)

$$C_8H_{12}$$

Use the data in the table to determine the rate law and the rate constant for the Diels–Alder reaction of butadiene:

Time (min)	0	4.0	8.0	12.0	16.0	20.0	30.0
$[C_4H_6]$ (M)	0.130	0.0872	0.0650	0.0535	0.0453	0.0370	0.0281

Strategy: When asked to determine a rate law and rate constant from concentration-time data, we must determine the order of the reaction. The rate law for this reaction has some order x with respect to butadiene ($C_4H_6 = A$), which we must determine:

$$\text{Rate} = k[A]^x$$

We can use the data provided in the problem to test graphically whether the reaction is first- or second-order in butadiene. If the reaction is first-order, a plot of $\ln\left(\dfrac{[A]_0}{[A]}\right)$ vs. t is linear. If the reaction is second-order, a plot of

$\dfrac{1}{[A]} - \dfrac{1}{[A]_0}$ vs. t is linear. If neither plot is a straight line, the reaction is something other than first- or second-order.

Solution: Begin by calculating the values for the logarithms and the reciprocals of the concentrations:

Time (min)	0	4.0	8.0	12.0	16.0	20.0	30.0
$\ln\left(\dfrac{[A]_0}{[A]}\right)$	0	0.40	0.69	0.89	1.05	1.26	1.53
$\dfrac{1}{[A]} - \dfrac{1}{[A]_0}$	0	3.8	7.7	11.0	14.4	19.3	27.9

Next, prepare first- and second-order plots to see whether either gives a straight line. The graphs are shown in the margin.

The first-order plot is not linear, so the reaction cannot be first-order. The straight line in the $\dfrac{1}{[A]} - \dfrac{1}{[A]_0}$ vs. t plot shows that the reaction is second-order in butadiene.

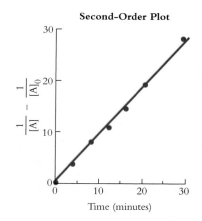

| Example 14-6 | Analysis of Rate Data *(continued)* |

This is the rate law for this Diels–Alder reaction:

$$\text{Rate} = k[C_4H_6]^2$$

Two data points are required to evaluate the rate constant. Here is the calculation using the first and last points:

$$k = \text{Slope} = \frac{\Delta y}{\Delta x} = \frac{27.9\ M^{-1} - 0\ M^{-1}}{30.0\ \text{min} - 0\ \text{min}} = \frac{27.9\ M^{-1}}{30.0\ \text{min}} = 0.93\ M^{-1}\ \text{min}^{-1}$$

"Isolation" Experiments

The first- and second-order rate laws in terms of concentration and time describe changes in concentration of a *single* reactant. However, most reactions involve concentration changes for more than one species. Although it is possible to develop equations relating concentration and time for such reactions, such equations are more complicated and more difficult to interpret than the equations that involve just one reactant. Fortunately, it is often possible to simplify the experimental behavior of a reaction.

One way to simplify the behavior of a reaction is to adjust the conditions so that the initial concentration of one starting material is much smaller than the initial concentrations of the others. This establishes experimental conditions under which the concentration of only one of the starting materials changes significantly during the reaction. This concentration is then said to be *isolated*. We use another reaction that is important in atmospheric chemistry to illustrate the **isolation method** in detail.

Trees and shrubs contain a group of fragrant compounds called *terpenes*. The simplest terpene is isoprene. All other terpenes are built around carbon skeletons constructed from one or more isoprene units. Plants emit terpenes into the atmosphere, as anyone who has walked in a pine or eucalyptus forest will have noticed. The possible effect of terpenes on the concentration of ozone in the troposphere has been the subject of much debate and has led to careful measurements of rates of reaction with ozone.

The reactions of terpenes with ozone lead to a complicated array of products, but the rate behavior for ozone reacting with a terpene can be studied by measuring the concentration of the terpene as a function of time:

$$O_3 + \text{terpene} \longrightarrow \text{products} \qquad \text{Rate} = k[O_3]^y[\text{terpene}]^z$$

Because the reaction rate depends on two concentrations rather than one, neither y nor z can be determined easily unless one concentration is isolated. Isolation is achieved by making the initial concentration of one reactant much smaller than the initial concentration of the other. Data collected under these conditions can then be analyzed using Equations 14-3 and 14-5, which relate concentration to time.

For example, an experiment could be performed on the reaction of ozone with isoprene with the following initial concentrations:

$$[O_3]_0 = 5.40 \times 10^{-4}\ M \qquad [\text{isoprene}]_0 = 3.0 \times 10^{-6}\ M$$

Isoprene
C_5H_8

The familiar odors of a pine forest are caused by terpenes.

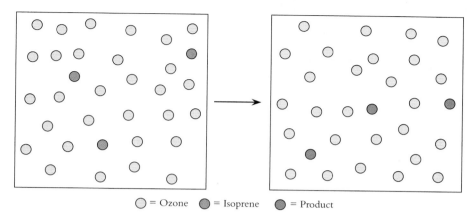

Figure 14-12
When the initial concentration of ozone (*green circles*) is much greater than that of isoprene (*lavender circles*), isoprene reacts completely to give products (*orange circles*) without changing the concentration of ozone appreciably.

○ = Ozone   ● = Isoprene   ● = Product

Under these initial conditions, the isoprene is consumed entirely before there is an appreciable change in the concentration of ozone, as Figure 14-12 illustrates. For 1:1 reaction stoichiometry and an initial ozone concentration 200 times larger than the initial isoprene concentration, the ozone concentration changes by less than 1%:

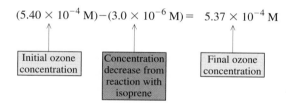

$$(5.40 \times 10^{-4}\ \text{M}) - (3.0 \times 10^{-6}\ \text{M}) = 5.37 \times 10^{-4}\ \text{M}$$

| Initial ozone concentration | Concentration decrease from reaction with isoprene | Final ozone concentration |

In other words, the reaction is flooded with a large excess of ozone. Because the concentration of ozone does not change appreciably, the near–constant concentration of ozone can be grouped with the rate constant:

$$\text{Rate} = \underbrace{k[O_3]^y}[\text{isoprene}]^z$$

This quantity remains constant over the course of the reaction because $[O_3] \gg [\text{isoprene}]$.

Therefore let:

$$k[O_3]^y = k_{obs}$$

The term k_{obs} is short for "observed rate constant."

Now the rate law simplifies to rate $= k_{obs}[\text{isoprene}]^z$. It is important to remember that k_{obs} is an *experimental* rate constant, but it is not the *true* rate constant (k) that appears in the general rate law.

Isolating isoprene as the only reactant whose concentration changes simplifies the general rate law to a form that can be tested against the single concentration rate laws:

First-order case:
$$\ln\left(\frac{[\text{isoprene}]_0}{[\text{isoprene}]}\right) = k_{obs}t$$

Second-order case:
$$\frac{1}{[\text{isoprene}]} - \frac{1}{[\text{isoprene}]_0} = k_{obs}t$$

First-Order Plot

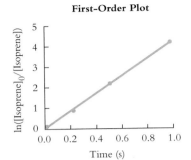

Second-Order Plot

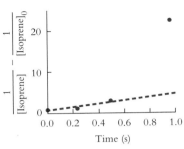

Time (s)	$[\text{Isoprene}]$ $(10^{-6}\,\text{M})$	$\ln\left(\dfrac{[\text{Isoprene}]_0}{[\text{Isoprene}]}\right)$	$\dfrac{1}{[\text{Isoprene}]} - \dfrac{1}{[\text{Isoprene}]_0}$
0.000	3.00	0.00	0.00
0.215	1.33	0.817	0.419
0.495	0.364	2.109	2.41
0.955	0.0443	4.216	22.2

Figure 14-13
Data for the reaction of isoprene with ozone under conditions when the concentration of isoprene is isolated. The experimental behavior is first-order under these conditions.

Experimental data collected under these conditions are plotted in Figure 14−13. Notice that the first-order plot is linear but the second-order plot is not, demonstrating that the reaction is first-order in isoprene. The slope of the first-order plot gives the observed rate constant:

$$\text{Slope} = \frac{4.216 - 0.00}{0.955\,\text{s} - 0.00\,\text{s}} = 4.4\,\text{s}^{-1} = k_{\text{obs}} = k[\text{O}_3]^\gamma$$

Before we can calculate the true rate constant (k), we must determine γ, the order with respect to ozone. One way to determine γ is to repeat the isolation experiment with a different initial concentration of ozone. For example, the experiment could be repeated with the same initial concentration of isoprene but an initial ozone concentration of 2.70×10^{-4} M, half as large as the concentration in the first experiment. When all the isoprene has been consumed, the ozone concentration is 2.67×10^{-4} M, a decrease of about 1%. When this experiment is done, the result is $k_{\text{obs}} = 2.2\,\text{s}^{-1}$, half as large as k_{obs} in the first experiment:

$$k_{\text{obs, I}} = k[\text{O}_3]_{\text{I}}^\gamma = 4.4\,\text{s}^{-1} \qquad k_{\text{obs, II}} = k[\text{O}_3]_{\text{II}}^\gamma = 2.2\,\text{s}^{-1}$$

There are two unknowns to be determined, k and γ, but if we take the ratio of the observed rate constants, k cancels:

$$\frac{k_{\text{obs, I}}}{k_{\text{obs, II}}} = \frac{\cancel{k}[\text{O}_3]_{\text{I}}^\gamma}{\cancel{k}[\text{O}_3]_{\text{II}}^\gamma} = \frac{4.4\,s^{-1}}{2.2\,s^{-1}} = 2.0$$

Now substitute the initial concentrations to obtain a numerical expression for γ:

$$2.0 = \frac{[\text{O}_3]_{\text{I}}^\gamma}{[\text{O}_3]_{\text{II}}^\gamma} = \left(\frac{5.4 \times 10^{-4}\,\text{M}}{2.7 \times 10^{-4}\,\text{M}}\right)^\gamma = (2.0)^\gamma$$

The ratio of experimental rate constants is 2.0, and the ratio of the ozone concentration terms is $(2.0)^\gamma$. We can see by inspection that $\gamma = 1$. One way to see this is by taking the logarithm of each side of the equality:

$$\ln (2.0) = \gamma \ln (2.0)$$

Solving for γ gives $\gamma = \ln (2.0)/\ln (2.0) = 1$, so the reaction is first-order in ozone. Thus the complete experimental rate law for the reaction is

$$\text{Rate} = k[\text{O}_3][\text{isoprene}]$$

Finally, the true rate constant can be calculated from k_{obs} using the data from either experiment:

$$k_{obs} = k[O_3] \qquad so \qquad k = \frac{k_{obs}}{[O_3]} = \frac{4.4 \text{ s}^{-1}}{5.4 \times 10^{-4} \text{ M}} = 8.1 \times 10^3 \text{ M}^{-1} \text{ s}^{-1}$$

Example 14-7 provides another illustration of the isolation technique.

Rate Law from Isolation Experiments

Example 14-7

The reaction of hydrogen and bromine produces hydrogen bromide, a highly corrosive gas:

$$H_2(g) + Br_2(g) \longrightarrow 2 HBr(g)$$

To determine the rate law for this reaction, a chemist performed two isolation experiments using different initial concentrations. Both experiments gave linear graphs of $\ln\left(\dfrac{[H_2]_0}{[H_2]}\right)$ vs. t, but with different slopes. Here are the details:

$[Br_2]_0$	$[H_2]_0$	Slope of graph
3.50×10^{-5} M	2.50×10^{-7} M	8.87×10^{-4} s^{-1}
2.00×10^{-5} M	2.50×10^{-7} M	6.71×10^{-4} s^{-1}

Determine the rate law and the rate constant for the reaction.

Strategy: When asked to determine a rate law and rate constant, we must determine the order of the reaction. The rate law for this reaction can be expected to contain the concentrations of H_2 and Br_2 raised to powers y and z that must be determined:

$$\text{Rate} = k[H_2]^y[Br_2]^z$$

Because the rate law contains more than one species, isolation experiments are required to determine the orders of reaction. In the experiments whose data are shown, the initial concentrations of Br_2 are about 100 times the initial concentrations of H_2, so the Br_2 concentration remains essentially constant over the course of the reaction. Thus, the rate law can be rewritten with $[Br_2]$ included as part of the observed rate constant:

$$\text{Rate} = k_{obs}[H_2]^y \qquad where \qquad k_{obs} = k[Br_2]^z$$

Solution: In each experiment a first-order graph gives a straight line, which means that the reaction is first-order in hydrogen ($y = 1$). For a first-order reaction, the slope gives k_{obs}:

$$k_{obs, I} = \text{Slope}_I = 8.87 \times 10^{-4} \text{ s}^{-1} \qquad and \qquad k_{obs, II} = \text{Slope}_{II} = 6.71 \times 10^{-4} \text{ s}^{-1}$$

To find the order of reaction with respect to bromine, we must compare the ratio of the observed rate constants to the ratio of the initial concentrations of Br_2:

$$\frac{k_{obs, I}}{k_{obs, II}} = \frac{k[Br_2]_I^z}{k[Br_2]_{II}^z}$$

| Example 14-7 | Rate Law from Isolation Experiments *(continued)* |

One ratio is obtained from the slopes of the graphs:

$$\frac{k_{obs,\,I}}{k_{obs,\,II}} = \frac{8.87 \times 10^{-4}\ s^{-1}}{6.71 \times 10^{-4}\ s^{-1}} = 1.32$$

A second ratio comes from the two different initial concentrations of Br_2:

$$\frac{[Br_2]_I}{[Br_2]_{II}} = \frac{3.50 \times 10^{-5}\ M}{2.00 \times 10^{-5}\ M} = 1.75$$

Putting these two together yields an equality where z is the only unknown:

$$1.32 = \frac{k_{obs,\,I}}{k_{obs,\,II}} = \frac{[Br_2]_I^z}{[Br_2]_{II}^z} = (1.75)^y$$

This cannot be solved by inspection, but the exponent z can be isolated by taking the logarithm of both sides:

$$\ln(1.32) = z \ln(1.75)$$

$$0.278 = z(0.560) \qquad \text{from which} \qquad z = 0.496$$

Reaction orders must be integers or simple fractions, so y is rounded up to 0.5. The reaction is half-order in bromine, and the rate law is as follows:

$$\text{Rate} = k[H_2][Br_2]^{1/2}$$

To calculate the true rate constant (k), use the relationship between the observed rate constant and k, substituting data from either of the two experiments:

$$k_{obs} = k[Br_2]^{1/2} \qquad so \qquad k = \frac{k_{obs}}{[Br_2]^{1/2}}$$

$$k = \frac{8.87 \times 10^{-4}\ s^{-1}}{(3.50 \times 10^{-5}\ M)^{1/2}} = \frac{8.87 \times 10^{-4}\ s^{-1}}{5.92 \times 10^{-3}\ M^{1/2}} = 0.150\ M^{-1/2}\ s^{-1}$$

We explain the half-order dependence on bromine concentration in Section 14.5.

Section Exercises

14.4.1 The reaction of N_2O_5 with gaseous H_2O is a potentially important reaction in the chemistry of acid rain:

$$N_2O_5(g) + H_2O(g) \longrightarrow 2\ HNO_3(g)$$

It is possible to monitor the concentration of N_2O_5 with time. Describe a set of isolation experiments from which the rate law and rate constant for this reaction could be determined.

14.4.2 In addition to reacting with water vapor, N_2O_5 can decompose to nitrogen dioxide and oxygen:

$$2\ N_2O_5(g) \longrightarrow 4\ NO_2(g) + O_2(g)$$

Here are some data for this decomposition reaction:

Time (min)	0	20.0	40.0	60.0	80.0
$[N_2O_5]$ (10^{-2} M)	0.92	0.50	0.28	0.15	0.08

Determine the order and the rate constant by constructing appropriate graphs using these data.

14.4.3 NO_2 reacts with CO at an elevated temperature:

$$NO_2(g) + CO(g) \longrightarrow NO(g) + CO_2(g)$$

Use the following information to determine the rate law:

(a) When an experiment is performed with initial conditions, $[NO_2]_0 = 1.75 \times 10^{-3}$ M and $[CO]_0 = 0.15$ M,

a plot of $\dfrac{1}{[NO_2]} - \dfrac{1}{[NO_2]_0}$ vs. t gives a straight line.

(b) When this experiment is repeated with initial conditions, $[NO_2]_0 = 1.75 \times 10^{-3}$ M and $[CO]_0 = 0.25$ M,

the slope of the $\dfrac{1}{[NO_2]} - \dfrac{1}{[NO_2]_0}$ vs. t plot is identical to the slope of the plot for experiment (a).

14.5 LINKING MECHANISMS AND RATE LAWS

The mechanism of any chemical reaction has the following characteristics:

1. The mechanism is one or more elementary reactions that describes how the chemical reaction occurs. These elementary reactions may be unimolecular, bimolecular, or (very rarely) termolecular.

2. The sum of the individual steps in the mechanism must give the overall balanced chemical equation. Sometimes a step may occur more than once in the mechanism.

3. The reaction mechanism must be consistent with the experimental rate law.

In this section we show that every mechanism predicts a rate law. If the rate law predicted by a proposed mechanism matches the experimental rate law, the mechanism is a possible description of how the reaction proceeds. On the other hand, if the rate law predicted by the proposed mechanism differs from experimental rate law, the proposed mechanism must be wrong.

Remember that a bimolecular elementary reaction is a collision between two molecules; its rate law contains the concentrations of both reactants:

/// **Bimolecular elementary reaction:** $A + B \longrightarrow products$
$Elementary\ rate = k[A][B]$

A unimolecular elementary reaction is a fragmentation or rearrangement of one chemical species, so its rate law contains the concentration of only that species:

/// **Unimolecular elementary reaction:** $C \longrightarrow products$
$Elementary\ rate = k[C]$

Table 14–1
Some Mechanisms and Rate Laws*

Stoichiometry	Mechanism	Rate Law
$2\,NO_2 \rightarrow N_2O_4$	$NO_2 + NO_2 \rightarrow N_2O_4$	Rate $= k[NO_2]^2$
$2\,NO_2 \rightarrow 2\,NO + O_2$	$2\,NO_2 \rightarrow NO_3 + NO$ (slow)	Rate $= k[NO_2]^2$
	$NO_3 \rightarrow NO + O_2$	
$NO + O_3 \rightarrow NO_2 + O_2$	$NO + O_3 \rightarrow NO_2 + O_2$	Rate $= k[NO][O_3]$
cis-2-butene $\rightarrow$	cis-2-butene $\rightarrow$	Rate $=$
trans-2-butene	trans-2-butene	k[cis-2-butene]
$C_5H_{11}Br + H_2O \rightarrow$	$C_5H_{11}Br \rightarrow C_5H_{11}^{+} + Br^{-}$ (slow)	Rate $= k[C_5H_{11}Br]$
$C_5H_{10} + H_3O^{+} + Br^{-}$	$C_5H_{11}^{+} + H_2O \rightarrow C_5H_{10} + H_3O^{+}$	

*Examples from Sections 14.1 through 14.4.

These rate expressions for elementary steps are related to the experimental rate law for the overall reaction in a way that depends on which step in the mechanism is rate-determining.

Rate-Determining First Step

Each example we have used to introduce the concepts of chemical mechanisms has a first step that is rate-determining. These mechanisms and their rate laws are summarized in Table 14-1.

In each of these mechanisms, the first elementary reaction is the rate-determining step. Because the overall reaction can go no faster than its rate-determining step, no elementary reaction that occurs after the rate-determining step affects the overall rate of reaction. Therefore:

/// *When the first step of a mechanism is rate-determining, the predicted rate law for the overall reaction is the rate expression for that first step.*

The predicted rate law is first-order for a reaction whose first step is unimolecular and rate-determining. The predicted rate law is second-order overall for a reaction whose first step is bimolecular and rate-determining. For example, the first step of the mechanism for the $C_5H_{11}Br$ reaction is unimolecular and slow, so the rate law predicted by this mechanism is first-order: Rate $= k[C_5H_{11}Br]$. Example 14-8 treats the rate law of another mechanism whose first step is rate-determining.

Example 14-8	**Predicted Rate Laws**

At elevated temperature, NO_2 reacts with CO to produce CO_2 and NO:

$$CO + NO_2 \longrightarrow CO_2 + NO$$

In one possible mechanism for this reaction, products form directly in a one-step bimolecular collision that transfers an oxygen atom from NO_2 to CO:

Predicted Rate Laws *(continued)*

Example 14-8

In another possible mechanism, a pair of NO_2 molecules collide in the rate-determining step to form NO and NO_3. In a second and faster step, the highly reactive NO_3 intermediate transfers an oxygen atom to CO in a bimolecular collision:

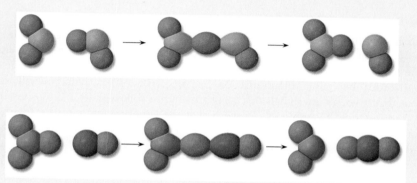

What is the predicted rate law for each of these mechanisms?

Strategy: It is best to work with each mechanism separately. When the first step is rate-determining, the predicted rate law matches the rate expression for that first elementary step.

Solution: First, analyze the one-step mechanism:

$$CO + NO_2 \longrightarrow CO_2 + NO$$

This process is analogous to the reaction of NO and O_3 discussed in Section 14.3. In a simple one-step atom transfer, the reaction is first-order in each of the starting materials and second-order overall:

One-step mechanism: $\quad$ Rate $= k[CO][NO_2]$

The second mechanism occurs in two steps:

$$2\,NO_2 \longrightarrow NO_3 + NO \quad \text{(slow, rate-determining)}$$
$$NO_3 + CO \longrightarrow NO_2 + CO_2 \quad \text{(fast)}$$

In this mechanism the first step is slow and rate-determining. Because the collision partners are identical, this reaction is second-order in NO_2 and second-order overall:

Two-step mechanism: $\quad$ Rate $= k[NO_2]^2$

Carbon monoxide does not appear in this rate law because it participates in the mechanism *after* the rate-determining step. Remember, any reaction after the rate-determining step does not affect the overall rate of reaction.

Example 14-8	Predicted Rate Laws *(continued)*

Experiments show that this reaction is second-order in NO_2, as predicted by the second proposed mechanism. The one-step mechanism can be ruled out because it is not consistent with the experimental rate law. Agreement with the rate law does not prove that the second mechanism is the correct one, however, because other mechanisms may predict the same rate law. It is one strong piece of evidence that supports this particular two-step process.

Rate-Determining Later Step

For many reaction mechanisms, the rate-determining step occurs after one or more faster steps. In such cases the reactants in the early steps may or may not appear in the rate law. Furthermore, the rate law is likely to depart from simple first- or second-order behavior. Fractional orders, negative orders, and overall orders greater than two, all are signals that a fast first step is followed by a slow subsequent step. For instance, Example 14-7 reveals that the experimental rate law for the reaction of H_2 gas with Br_2 gas depends on the square root of the Br_2 concentration, and the reaction also is first-order in H_2:

$$H_2 + Br_2 \longrightarrow 2\ HBr \qquad \text{Rate} = k[H_2][Br_2]^{1/2}$$

Although the reaction follows simple 1:1 stoichiometry, its experimental rate law cannot be explained by a simple mechanism. In order for the first step of the mechanism for this reaction to be rate-determining, it would have to include a half-molecule of Br_2. There is no such species, so the first step cannot be rate-determining. Instead, some later step in the mechanism must be rate-determining.

Figure 14-14 shows the accepted mechanism for this reaction using molecular pictures. The reaction begins with the dissociation of Br_2 molecules into Br atoms:

Figure 14-14
A molecular view of the accepted mechanism for the reaction of H_2 and Br_2 to produce 2 HBr.

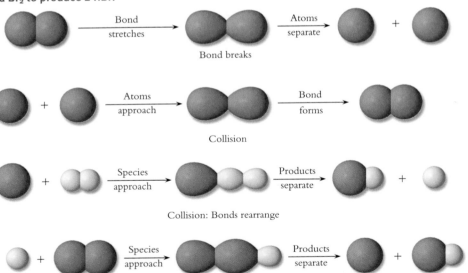

Bond stretches — Bond breaks — Atoms separate — +

+ — Atoms approach — Collision — Bond forms

+ — Species approach — Collision: Bonds rearrange — Products separate — +

+ — Species approach — Collision: Bonds rearrange — Products separate — +

Typically, elementary reactions in a mechanism are written with their rate constants above or below the reaction arrow. A reaction that is the reverse of another is labeled accordingly.

$$Br_2 \xrightarrow{k_1} 2\ Br$$

Almost all the Br atoms produced in the first step simply recombine to regenerate Br_2 molecules:

$$2\ Br \xrightarrow{k_{-1}} Br_2$$

This is an example of a **reversible reaction,** one that occurs rapidly in both directions. For simplicity, we combine these two elementary reactions in a single expression:

$$Br_2 \underset{k_{-1}}{\overset{k_1}{\rightleftharpoons}} 2\ Br \quad \text{(fast, reversible)}$$

← **CHAPTERS 15–16**
We explore chemical reversibility in detail in Chapters 15 and 16.

Although most bromine atoms produced in the first step recombine, occasionally a Br atom reacts with a molecule of H_2 to form one molecule of HBr and a hydrogen atom. This step is slow and rate-determining:

$$Br + H_2 \xrightarrow{k_2} HBr + H \quad \text{(slow, rate-determining)}$$

The hydrogen atom produced in the rate-determining step undergoes a third elementary reaction, a bimolecular collision with a Br_2 molecule. The H atom rapidly forms a bond with one Br atom to form one molecule of HBr and a Br atom:

$$H + Br_2 \xrightarrow{k_3} HBr + Br \quad \text{(fast)}$$

The second step in this mechanism is rate-determining, so the overall rate of the reaction is governed by the rate of this step:

$$\text{Rate} = k_2[Br][H_2]$$

This rate law describes the rate behavior predicted by the proposed mechanism accurately, but the law cannot be tested against experiments because it contains the concentration of Br atoms, which are intermediates in the reaction. As mentioned earlier, an intermediate has a short lifetime and is hard to detect, so it is difficult to make accurate measurements of its concentration. Furthermore, it is not possible to adjust the experimental conditions in a way that changes the concentration of an intermediate by a known amount. Therefore if this proposed rate law is to be tested against experimental behavior, the concentration of the intermediate must be expressed in terms of reactants and products.

Equality of Rates

The proposed mechanism begins with the decomposition of Br_2 molecules into Br atoms, most of which recombine rapidly to give Br_2 molecules. At first, Br_2 molecules decompose faster than Br atoms recombine, but the Br atom concentration quickly becomes large enough for recombination to occur at the same rate as decomposition. When the two rates are equal, so are their rate expressions:

$$\text{Rate of decomposition} = \text{Rate of recombination}$$

$$k_1[Br_2] = k_{-1}[Br]^2$$

This equality can be used to express the concentration of the intermediate in terms of concentrations of the reactants. First solve for [Br] by rearranging the rate expressions and taking the square root of both sides:

$$[Br]^2 = \frac{k_1}{k_{-1}}[Br_2] \quad \textit{from which} \quad [Br] = \left(\frac{k_1}{k_{-1}}\right)^{1/2}[Br_2]^{1/2}$$

Now substitute this equation for [Br] into the expression for the rate-determining step:

$$\text{Rate} = k_2[Br][H_2] = k_2\left(\frac{k_1}{k_{-1}}\right)^{1/2}[H_2][Br_2]^{1/2} = k[H_2][Br_2]^{1/2}$$

In this expression, k, the rate constant for the overall reaction, is related to the elementary rate constants:

$$k = k_2 \left(\frac{k_1}{k_{-1}} \right)^{1/2}$$

The predicted rate law is first-order in hydrogen and one-half-order in bromine, in agreement with the experimental rate law determined in Example 14-7. In Example 14-9, we show that this mechanism also meets the other criteria for a satisfactory mechanism.

Example 14-9 | **The H_2—Br_2 Mechanism**

The proposed mechanism for the reaction between H_2 and Br_2 predicts the experimental rate law. Does it meet the other criteria for a satisfactory mechanism?

Strategy: A satisfactory mechanism must be made up entirely of elementary steps, must give the correct stoichiometry of the reaction, and must predict the experimental rate law.

Solution: As already shown, the proposed mechanism predicts the rate law.

This mechanism has four steps. The first step is the unimolecular decomposition of Br_2. The remaining steps all are simple bimolecular collisions. Because all the reactions are unimolecular or bimolecular, the mechanism meets the first criterion.

The easiest way to determine whether the steps lead to the correct stoichiometry is to write all four reactions in the same direction. Then cancel those species that appear both on the left and on the right:

$$Br_2 \longrightarrow 2\,Br$$
$$2\,Br \longrightarrow Br_2$$
$$Br + H_2 \longrightarrow HBr + H$$
$$H + Br_2 \longrightarrow HBr + Br$$

Net reaction: $H_2 + Br_2 \longrightarrow 2\,HBr$

Combining the steps of the mechanism leads to the balanced equation for the overall reaction, so the second criterion also is satisfied. The proposed mechanism satisfies the three requirements.

As the mechanism for the reaction between H_2 and Br_2 shows, the observation of an order other than first or second indicates that some step beyond the first one in the mechanism is rate-determining. The one-half-order rate dependence for Br_2 comes about because of the initial rapid cleavage reaction into a *pair* of Br atoms.

Our Chemistry and the Environment Box discusses reactions in the stratosphere, and the connection between mechanisms and rate laws is illustrated further by Example 14-10.

Box 14-1 Chemistry and the Environment: Reactions of Ozone

T he ozone layer is found $30-45$ km above the surface of the Earth. Nowhere else in the atmosphere is ozone found in significant amounts. Ozone is limited to this narrow band because only in the ozone layer can O_3 form through sunlight interacting with O_2.

Ozone is formed in the stratosphere by the action of ultraviolet light on oxygen molecules:

$$3\,O_2 + h\nu_{(\lambda < 240\,nm)} \longrightarrow 2\,O_3$$

An oxygen molecule absorbs a photon ($h\nu$) of high-energy ultraviolet light ($\lambda < 240$ nm). The energy of the photon breaks O_2 into two O atoms.

$$O_2 + h\nu_{(\lambda < 240\,nm)} \longrightarrow O + O \qquad Step\ 1$$

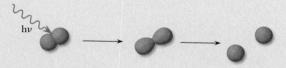

An O atom is reactive and adds to O_2 to form O_3.

$$O + O_2 \longrightarrow O_3 \qquad Step\ 2$$

This ozone molecule contains excess energy and can decompose again. The stratosphere contains enough N_2, however, that an ozone molecule usually collides with a nitrogen molecule and gives up its excess energy before it can break apart.

Step one of the mechanism generates two oxygen atoms, each of which reacts with an O_2 molecule. Thus, each time the first step occurs, the second step occurs twice:

$$O_2 + h\nu_{(\lambda < 240\,nm)} \longrightarrow \cancel{O} + \cancel{O}$$
$$\cancel{O} + O_2 \longrightarrow O_3$$
$$\cancel{O} + O_2 \longrightarrow O_3$$

Net reaction: $\quad 3\,O_2 + h\nu_{(\lambda < 240\,nm)} \longrightarrow 2\,O_3$

The first step is the rate-determining step because as soon as an oxygen atom forms, it is snapped up by the nearest available O_2 molecule.

The formation of O_3 goes on continuously, but O_3 is also decomposed by ultraviolet light between 240 and 340 nm:

$$O_3 + h\nu_{(\lambda = 240 - 340\,nm)} \longrightarrow O_2 + O$$

The O atom can react with a second O_3 molecule:

$$O + O_3 \longrightarrow 2\,O_2$$

The result is a reduction in the amount of ozone:

Net: $\quad 2\,O_3 + h\nu_{(\lambda = 240 - 340\,nm)} \longrightarrow 3\,O_2$

Short-wavelength UV light forms O_3, whereas long-wavelength UV light decomposes O_3. The reactions form a delicate balance in which the rate of O_3 decomposition matches the rate of O_3 production. The resulting ozone layer absorbs nearly all the solar photons in the 240- to 340-nm range. Photons in this wavelength range have enough energy to damage and even destroy living cells, so the ozone layer protects life on Earth.

Unfortunately, chemical species produced by humans can react in ways that reduce the O_3 concentration.

For example, NO changes the oxygen balance through the following reactions:

$$NO + O_3 \longrightarrow NO_2 + O_2$$

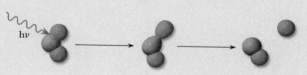

$$NO_2 + h\nu_{(\lambda < 400\,nm)} \longrightarrow NO + O$$

$$NO_2 + O \longrightarrow NO + O_2$$

The first step in this process occurs twice to generate the two molecules of NO_2 needed for the second and third steps. The result of this sequence is that NO in the stratosphere increases the rate of O_3 decomposition:

Net: $\quad 2\,O_3 + h\nu_{(\lambda < 400\,nm)} \longrightarrow 3\,O_2$

As mentioned in the introduction to this chapter, CFCs also have potentially devastating effects on the ozone layer. In Section 14.7, we describe how CFCs destroy stratospheric ozone.

Example 14-10	Reaction Between NO and O_3

Nitrogen oxide converts ozone into molecular oxygen, as follows:

$$O_3 + NO \longrightarrow O_2 + NO_2$$

The experimental rate law is rate $= k[O_3][NO]$. Which of the following mechanisms are consistent with the experimental rate law?

Mechanism I

$$O_3 + NO \xrightarrow{k_1} O + NO_3 \qquad \text{(slow)}$$

$$O + O_3 \xrightarrow{k_2} 2\,O_2 \qquad \text{(fast)}$$

$$NO_3 + NO \xrightarrow{k_3} 2\,NO_2 \qquad \text{(fast)}$$

Mechanism II

$$O_3 \underset{k_{-1}}{\overset{k_1}{\rightleftharpoons}} O_2 \qquad \text{(fast, reversible)}$$

$$NO + O \xrightarrow{k_2} NO_2 \qquad \text{(slow)}$$

Mechanism III

$$O_3 + NO \xrightarrow{k_1} O_2 + NO_2$$

Strategy: We need to determine whether the rate law predicted by each mechanism matches the experimental rate law by calculating the rate law predicted by each.

Solution: Mechanism I is a three-step process in which the first step is rate-determining. When the first step of a mechanism is rate-determining, the predicted rate law is the same as the rate expression for that first step. Here, the rate-determining step is a bimolecular collision. The rate expression for a bimolecular collision is first-order in each collision partner:

$$\text{Rate} = k_1[O_3][NO]$$

Mechanism I is consistent with the experimental rate law.

Mechanism II begins with fast reversible ozone decomposition followed by a rate-determining bimolecular collision of an oxygen atom with a molecule of NO. The rate of the slow step is as follows:

$$\text{Rate} = k_2[NO][O]$$

This rate expression contains the concentration of an intermediate, atomic oxygen. To convert the rate expression into a form that can be compared with the experimental rate law, assume that the rate of the first step is equal to the rate of its reverse process. Then solve the equality for the concentration of the intermediate:

$$k_1[O_3] = k_{-1}[O][O_2] \qquad \textit{from which} \qquad [O] = \frac{k_1[O_3]}{k_{-1}[O_2]}$$

$$\text{Rate} = \frac{k_1 k_2 [O_3][NO]}{k_{-1}[O_2]}$$

| Reaction Between NO and O_3 *(continued)* | Example 14-10 |

Mechanism II is inconsistent with the experimental rate law. Notice that this mechanism predicts a rate that is inversely proportional to the concentration of molecular oxygen. That means that oxygen would slow down the reaction: As the concentration of oxygen *increases*, the rate of the reaction *decreases*.

Mechanism III is a simple one-step bimolecular collision. Its predicted rate law is as follows:

$$Rate = k_1[O_3][NO]$$

Mechanism III is consistent with the experimental rate law.

Therefore the O_3 + NO reaction might go by Mechanism I or Mechanism III, but not by Mechanism II.

Section Exercises

14.5.1 A student proposes the following mechanism for the gas–phase decomposition of N_2O_5, (all substances are gases):

Overall reaction:	$2\ N_2O_5 \longrightarrow 4\ NO_2 + O_2$
Step 1:	$N_2O_5 \longrightarrow N_2O_4 + O$
Step 2:	$O + N_2O_5 \longrightarrow N_2O_4 + O_2$

(a) Propose a third step to complete this mechanism.

(b) Experiments show that this decomposition reaction is first-order. Which step in the mechanism must be rate-determining?

14.5.2 What order for N_2O_5 does the mechanism in Section Exercise 14.5.1 predict if the first step is fast and reversible and the second step is rate-determining?

14.5.3 The oxidation of NO by O_2 is an example of a third-order reaction:

$$2\ NO(g) + O_2(g) \longrightarrow 2\ NO_2(g) \qquad Rate = k[NO]^2[O_2]$$

Although a single elementary termolecular mechanism is consistent with the rate law, three-body collisions are rare. It is more likely that the reaction begins with a bimolecular collision:

$$NO + NO \underset{k_{-1}}{\overset{k_1}{\rightleftarrows}} N_2O_2 \qquad (fast)$$

(a) Propose a slow second step to complete this mechanism.

(b) Show that this mechanism is consistent with the observed rate law.

14.6 REACTION RATES AND TEMPERATURE

Common experience tells us that chemical reactions proceed faster at higher temperature. Food is stored in refrigerators because food spoils more quickly at high temperature than at low temperature. Wood does not burn at room temperature, but it burns vigorously at high temperature. At the macroscopic level, higher temperature means faster reactions, but how does temperature affect the rate of a reaction at the molecular level? Because temperature is a measure of the energy of

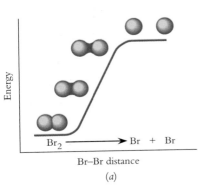

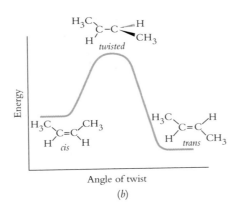

Figure 14-15
Energy profiles for two unimolecular processes. (*a*) The unimolecular decomposition of a bromine molecule. (*b*) The unimolecular isomerization of *cis*-2-butene.

motion of molecules, we need to explore the relationship between reaction rates and molecular energy to answer this question.

Energy Changes in a Unimolecular Reaction

In a unimolecular reaction, a molecule fragments into two pieces or rearranges to a different isomer. In either case, a chemical bond breaks. For example, in the fragmentation of bromine molecules, breaking a σ bond gives a pair of bromine atoms:

$$Br_2 \longrightarrow 2\ Br$$

Recall that this unimolecular process is the first step of the reaction between molecular hydrogen and molecular bromine to give HBr.

Figure 14-15*a* shows the energy changes that occur during the fragmentation of a bromine molecule. The reaction begins with a stretching of the Br—Br bond. As the bond becomes longer, the system moves to higher energy. This decreased stability results from the loss of orbital overlap as the bromine atoms move apart. If the bond is stretched far enough, the molecule fragments into two bromine atoms. When a bond breaks, the fragment products are always at higher energy than the starting materials.

Like fragmentations, unimolecular rearrangements are always "uphill" at the beginning of the process, because a bond breaks. Unlike fragmentations, rearrangements are "downhill" at the end as a new bond forms. An example of this kind of energy profile for the isomerization reaction of *cis*-2-butene appears in Figure 14-15*b*.

Energy Changes During Bimolecular Reactions

In a bimolecular reaction, some bonds break and other bonds form. For example, in the reaction between NO and O_3, an O atom is transferred from O_3 to NO:

$$O_3 + NO \longrightarrow O_2 + NO_2$$

In the collision process the ozone molecule distorts as the O—O bond stretches and weakens, and NO must also undergo distortion as it bonds with the incoming oxygen atom from ozone. Distorting a molecule from its most favored configuration always requires energy. Thus at the outset of the reaction, the system is destabilized by the molecular distortions required for atom transfer. At the end of the process, on the other hand, the system is stabilized as the products adopt their lowest-energy configuration. Figure 14-16 shows a schematic representation of the energy profile for this reaction.

Figure 14-16
A schematic representation of the energy profile for the bimolecular reaction between O_3 and NO molecules.

Activation Energy

The energy profiles for these elementary reactions share several common features, one of which is particularly important for a discussion of reaction rates: In each example the system must move "uphill" from the energy of the reactants to an energy maximum. In almost all chemical reactions, the molecules must overcome an energy barrier before starting materials can become products. This energy barrier, the minimum energy that must be supplied before reaction can take place, is the **activation energy (E_a)** of the chemical reaction. Activation energies arise because chemical bonds in reactant molecules must distort or break before new bonds can form in product molecules. The activation energy for any reaction mechanism is independent of both *reactant concentrations* and *temperature*.

Figures 14-15 and 14-16 are **activation energy diagrams** for elementary reactions. They are plots of the change in energy as reactants are transformed into products. Figure 14-17 shows a generalized activation energy diagram labeled with its characteristics. The *x*-axis represents the course of the reaction as starting materials are transformed into products. Because these transformations may involve complicated combinations of rotations, vibrations, and atom transfers, accepted practice is to label this axis as the **reaction coordinate** without specifying further details about what changes occur.

The *y*-axis of an activation energy diagram is the energy of the molecular system. The graph shows how chemical energy changes during the course of the reaction. Low values of energy represent high molecular stability, and high values represent low molecular stability. The reaction shown in the activation energy diagram in Figure 14-17 releases energy, so its products are lower in energy than its reactants. For a reaction that absorbs energy, in contrast, the activation energy diagram shows the energy of the products to be higher than the energy of the reactants. Whether the net reaction releases or absorbs energy, however, there always is a positive activation energy because of the requirement to break or distort one chemical bond before another can form.

The molecular arrangement at the point of highest energy along the diagram is known as the **activated complex.** The difference in energy between the reactants and this activated complex is the activation energy and is generally labeled E_a. An activated complex is *not* considered to be an intermediate. An intermediate is stable for a short time, perhaps milliseconds, and may be observed with highly sensitive instruments. An activated complex forms and disappears in the time required for a molecular collision, which is about 10^{-13} second. In other words, the activated complex represents the configuration of atoms at the moment of collision.

The magnitude of an activation energy depends on the details of the bonding changes that occur during the formation of the activated complex. As described in Chapter 9, the energy change for a reaction (ΔE) can be estimated from average bond energies. In contrast, the activation energy (E_a) is not so easily estimated. Activation energies must be determined by measuring how rate constants vary with temperature.

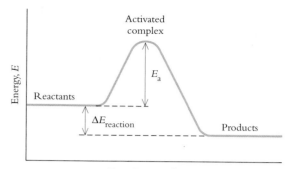

Figure 14-17
A generalized activation energy diagram for an elementary reaction. The figure shows a reaction that releases energy, resulting in a negative overall energy change (ΔE).

The activated complex is also known as the *transition state*.

Activation Energy and the Rate Constant

A chemical reaction cannot occur unless the starting materials have enough energy to overcome the activation energy barrier. Where do molecules obtain this energy? In addition to its chemical potential energy, every molecule has energy associated with translation, rotation, and vibration. During a chemical reaction, some of this energy is transformed into chemical energy as a molecule distorts or as two molecules collide.

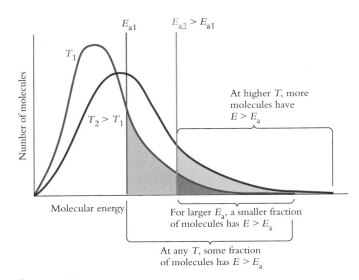

Figure 14-18
The distribution of molecular energies, shown at two different temperatures. Notice that the energy needed for reaction to occur is a fixed value that is independent of T.

CHAPTER 16 →
We describe the accomplishments of Svante Arrhenius, who first described activation energies, in our Chemical Milestones Box in Chapter 16.

Any collection of molecules spans a wide range of molecular energies. Some molecules have much energy, but others have very little. This distribution of molecular energies, described in Section 5.1, depends on the temperature of the sample, as shown in Figure 14-18. Heating the system increases the translational, rotational, and vibrational energy of the molecules. Thus as the temperature increases, the distribution broadens and shifts toward higher energy.

The energy distributions in Figure 14-18 show that only some of the molecules in any reaction system have energies greater than that required for reaction. How large this fraction is depends on E_a and temperature:

1. At any T, a larger fraction of molecules can react if E_a is small than if E_a is large.
2. For any reaction, a larger fraction of molecules can react if T is high than if T is low.

The larger the fraction of molecules that can react, the faster a reaction will proceed. Therefore:

/// **The rate of a reaction increases with temperature.**

The Arrhenius Equation

The value of the rate constant for a particular reaction depends on the activation energy of the reaction, the temperature of the system, and how often a collision occurs in which the atoms are in the required orientation. All these factors can be summarized in a single equation, called the **Arrhenius equation:**

$$k = A e^{-E_a/RT} \qquad (14\text{-}6)$$

In this equation, A is the value that the rate constant would have if all molecules had enough energy to react. The exponential term contains the dependence on activation energy (E_a in J/mol) and temperature (T in K). The gas constant (R in J/mol K) serves as a unit conversion factor. The negative sign of the exponential means that the larger the value of the exponent, the smaller the rate constant. Qualitatively, this means that as E_a gets larger, the rate constant gets smaller, and that as T gets larger, the rate constant gets larger.

Equation 14-6 can be converted from exponential form to a form that is easier to treat graphically by taking the natural logarithm of both sides and making use of the fact that the natural logarithm of an exponential is the exponent:

$$\ln k = \ln (A\, e^{-E_a/RT}) = \ln A + \ln (e^{-E_a/RT})$$

$$\ln k = \ln A - \frac{E_a}{RT} \qquad (14\text{-}7)$$

Equation 14-7 leads to a linear graph: plotting $\ln k$ along the y-axis vs. $1/T$ along the x-axis gives a straight line with a slope of $-E_a/R$ and an intercept of $\ln A$. Thus A and E_a can be evaluated after k has been measured at several different temperatures. Example 14-11 shows how this is done.

Graphing to Determine E_a

Example 14-11

At high temperature cyclopropane isomerizes to propene:

Cyclopropane   Propene
(C_3H_6)   (C_3H_6)

When this reaction is studied at different temperatures, the following rate constants are obtained.

$T(°C)$	477	523	577	623
$k(s^{-1})$	1.8×10^{-4}	2.7×10^{-3}	3.0×10^{-2}	2.6×10^{-1}

What is the activation energy for the isomerization of cyclopropane?

Strategy: The variation of a rate constant with temperature is described by the Arrhenius equation. According to its logarithmic form (Equation 14-7), a plot of $\ln k$ vs. $1/T$, with temperature expressed in Kelvins, should be a straight line.

Solution: We need to convert the data into the appropriate form and then prepare a graph. Here are the values that should be graphed:

$1/T$ (10^{-3} K)	1.33	1.25	1.18	1.11
$\ln k$	-8.62	-5.92	-3.51	-1.35

A graph of these data has the expected linear form, with the following slope:

$$\text{Slope} = -3.27 \times 10^4 \text{ K.}$$

According to Equation 14-7,

$$\text{Slope} = -\frac{E_a}{R}$$

$$E_a = -R \text{ (Slope)}$$

$$E_a = -(8.314 \text{ J/mol K})(-3.27 \times 10^4 \text{ K})$$
$$\times (10^{-3} \text{ kJ/J}) = 273 \text{ kJ/mol}$$

A C—C bond must break for isomerization to occur, and typical C—C bond energies are around 350 kJ/mol, so 273 kJ/mol is a reasonable value for the activation energy.

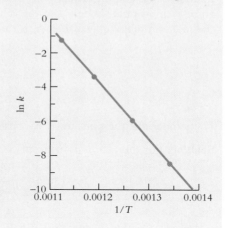

Graphing, using rate constants measured at several temperatures, is the most accurate method to find the value of E_a. However, a good estimate of E_a can be calculated from rate constants measured at just two temperatures:

$$\ln k_1 = \ln A - \frac{E_a}{RT_1} \qquad and \qquad \ln k_2 = \ln A - \frac{E_a}{RT_2}$$

The value of A changes slowly enough with T that A usually can be treated as a constant for temperature changes of 50 K or less. Thus the A term can be eliminated by subtracting one equation from the other:

$$\ln k_2 - \ln k_1 = \left(\ln A - \frac{E_a}{RT_2} \right) - \left(\ln A - \frac{E_a}{RT_1} \right) = \frac{E_a}{R} \left(\frac{1}{T_1} - \frac{1}{T_2} \right)$$

The logarithmic terms can be combined and the equation solved for E_a:

$$E_a = \frac{R \ln\left(\dfrac{k_2}{k_1}\right)}{\left(\dfrac{1}{T_1} - \dfrac{1}{T_2}\right)} \tag{14-8}$$

Example 14-12 shows how to use Equation 14-8.

Example 14-12	Calculating an Activation Energy

The reactions of NO_2 have been studied as a function of temperature. For the following decomposition reaction, the rate constant is 2.7×10^{-2} M^{-1} s^{-1} at 227 °C and 2.4×10^{-1} M^{-1} s^{-1} at 277 °C:

$$2\,NO_2 \longrightarrow 2\,NO + O_2$$

Studies of the conversion of NO_2 to N_2O_4 give $k = 5.2 \times 10^8$ M^{-1} s^{-1} at both 298 and 350 K:

$$2\,NO_2 \longrightarrow N_2O_4$$

Calculate the activation energies of these two reactions.

Strategy: The graphical treatment of the variation of rate constants with temperature is appropriate when a set of values is available. When only two values of the rate constant are available, E_a is calculated using Equation 14-8.

$$E_a = \frac{R \ln\left(\dfrac{k_2}{k_1}\right)}{\left(\dfrac{1}{T_1} - \dfrac{1}{T_2}\right)}$$

Remember that the units must be consistent, so T must be in K and R in J/mol K.

Solution: For the decomposition of NO_2:

$$E_a = \frac{(8.314\,\text{J mol}^{-1}\,\text{K}^{-1}) \ln\left(\dfrac{2.4 \times 10^{-1}\,\text{M}^{-1}\,\text{s}^{-1}}{2.7 \times 10^{-2}\,\text{M}^{-1}\,\text{s}^{-1}}\right)}{\left(\dfrac{1}{500.\,\text{K}} - \dfrac{1}{550.\,\text{K}}\right)}$$

$$E_a = \frac{(8.314\,\text{J mol}^{-1}\,\text{K}^{-1}) \ln(8.89)}{(2.000 \times 10^{-3}\,\text{K}^{-1} - 1.818 \times 10^{-3}\,\text{K}^{-1})} = \frac{(8.314\,\text{J mol}^{-1}\,\text{K}^{-1})(2.18)}{1.82 \times 10^{-4}\,\text{K}^{-1}}$$

As usual, we carry one additional significant figure until we round the final answer to two significant figures:

$$E_a = (9.96 \times 10^4\,\text{J/mol})(10^{-3}\,\text{kJ/J}) = 1.0 \times 10^2\,\text{kJ/mol}$$

For the formation of N_2O_4:

$$k_2 = k_1 \quad so \quad \ln(k_2/k_1) = \ln 1 = 0$$

Thus $\quad E_a = 0$ kJ/mol

An activation energy of 0 kJ/mol indicates that no bonds need to be distorted or broken in this reaction. We discuss this in more detail shortly.

After the activation energy of a reaction has been determined, we can use the Arrhenius equation to estimate values of the rate constant for the reaction at temperatures where experiments have not been carried out. This is particularly useful for temperatures at which a reaction is too slow or too fast to be studied conveniently. Example 14-13 illustrates this application.

Calculating k from E_a	**Example 14-13**

When two NO_2 molecules collide and react, they can form a bond or exchange an oxygen atom.
For the oxygen exchange, $E_a = 1.0 \times 10^2$ kJ/mol, and the rate constant at 250 °C = 8.6×10^{-2} M^{-1} s^{-1}. Estimate the rate constant for oxygen exchange at room temperature (25 °C).

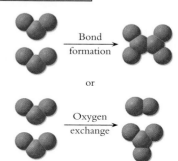

Strategy: We can calculate a rate constant at any temperature using the Arrhenius equation provided that we know E_a and the rate constant for the reaction at some other temperature.

Solution: Rearranging Equation 14-8 gives an equation that can be solved for the rate constant at a given temperature:

$$\ln\left(\frac{k_2}{k_1}\right) = \frac{E_a}{R}\left(\frac{1}{T_1} - \frac{1}{T_2}\right)$$

The values that we need to substitute into this equation are given in the problem:

$$k_1 = 8.6 \times 10^{-2}\ M^{-1}\ s^{-1} \qquad T_1 = 250\ °C = 523\ K \qquad T_2 = 25\ °C = 298\ K$$

After substituting, we solve for k_2, the rate constant of the decomposition reaction at 298 K:

$$\ln\left(\frac{k_2}{8.2 \times 10^{-2}\ M^{-1}\ s^{-1}}\right) = \frac{(1.0 \times 10^2\ \text{kJ/mol})(10^3\ \text{J/kJ})}{(8.314\ \text{J/mol K})}\left(\frac{1}{523\ K} - \frac{1}{298\ K}\right)$$

$$= -17.4$$

$$\frac{k_2}{8.6 \times 10^{-2}\ M^{-1}\ s^{-1}} = e^{-17.4} = 2.8 \times 10^{-8}$$

$$k_2 = 2 \times 10^{-9}\ M^{-1}\ s^{-1}$$

Notice that at room temperature, the rate constant for exchange of oxygen atoms, 2×10^{-9} M^{-1} s^{-1}, is 17 powers of ten smaller than the rate constant for bond formation between two N atoms, 5.2×10^8 M^{-1} s^{-1}. This immense difference reflects the fact that exchange of atoms involves a large energy input to break an N—O bond, whereas bond formation between two NO_2 molecules requires no energy input.

The final result has only one significant figure, because the ln term has only one decimal place.

Values of Activation Energy

Most reactions between stable molecules have activation energies of 100 kJ/mol or greater, even when the overall reaction is exothermic. As an example, consider the reaction of hydrogen and oxygen:

$$2\ H_2(g) + O_2(g) \longrightarrow 2\ H_2O(g)$$

$$\Delta H° = -484\ \text{kJ mol}^{-1} \qquad \Delta G° = -457\ \text{kJ mol}^{-1}$$

Figure 14-19
A balloon filled with a mixture of hydrogen and oxygen is stable until a source of energy, such as a spark or flame, initiates the reaction. After the reaction begins, the energy released as water forms is enough to cause an explosion.

This combustion reaction is so spontaneous and exothermic that it is used to drive the main engines of the space shuttle. Nevertheless, mixtures of hydrogen and oxygen are stable indefinitely at room temperature. This reaction does not occur at room temperature because strong bonds must be broken to transform H_2 and O_2 into water molecules. In other words, the bonds in the reactant molecules generate a large activation energy for the reaction. As Figure 14-19 shows, a violent explosion occurs when a spark is applied to a mixture of H_2 and O_2. The spark gives some of the molecules enough energy to overcome the activation barrier, and after the reaction starts, the formation of water molecules releases enough energy to drive the reaction to completion very quickly.

The combustion reactions of hydrocarbons such as natural gas and gasoline, which also release large amounts of energy, have negligible rates at room temperature because they have large activation energy barriers, typically $140-200$ kJ/mol. In order for a hydrocarbon molecule to combine with oxygen to form CO_2 and H_2O, one of its C—H bonds first must be weakened substantially. This accounts for the large activation energy barrier for combustion.

Even elementary reactions can have large activation energies. For example, the isomerization of *cis*-2-butene to *trans*-2-butene is a unimolecular rotation whose activation energy is 273 kJ/mol. This high value arises because a C—C π bond must be broken during the course of the isomerization process. (See Figure 14-4 on page 650.)

Example 14-12 shows that no energy barrier exists for the combination of two NO_2 molecules to form N_2O_4. The activation energy for this reaction is zero because NO_2 is an odd-electron molecule with a lone electron readily available for bond formation:

$$
\underset{O}{\overset{O}{\|}}{N}\cdot \;+\; \cdot{N}\underset{O}{\overset{O}{\|}} \longrightarrow \underset{O}{\overset{O}{\|}}{N}-{N}\underset{O}{\overset{O}{\|}}
$$

The recombination of two bromine atoms to form a Br_2 molecule has zero activation energy for the same reason.

■ **14.6.1** Use bonding arguments to predict whether the ammonia synthesis reaction has a high or a low activation energy:

$$N_2(g) + 3 H_2(g) \longrightarrow 2 NH_3(g) \qquad \Delta H° = -91.8 \text{ kJ}$$

■ **14.6.2** The reaction of H_2 with Br_2 discussed in Section 14.5 has been studied extensively as a function of temperature. Here are some experimental results:

$T(10^2 \text{ K})$	5.00	5.25	5.50	5.75
$k(\text{M}^{-1/2} \text{ s}^{-1})$	8.1×10^{-7}	6.5×10^{-6}	4.0×10^{-5}	2.1×10^{-4}

(a) Determine E_a by graphical analysis, and (b) estimate k at 425 K.

■ **14.6.3** Use standard thermodynamic data to determine ΔE and construct the activation energy diagram for the following reaction. (Hint: $\Delta(PV) = 0$ for this reaction, so enthalpies and energies can be used interchangeably.)

$$CO + NO_2 \longrightarrow CO_2 + NO \qquad E_a = 133 \text{ kJ}$$

14.7 CATALYSIS

According to the discussion in Section 14.6, one way to make a reaction go faster is to run it at higher temperature. Many industrial reactions are carried out at high temperature to maximize the amount of product that can be synthesized in a given time. Greater rates of production mean more income for the company, but the extra energy required to run a reaction at higher temperature adds to production costs. Furthermore, high-temperature reactions introduce safety concerns, and many chemical species are not stable at high temperature.

Another way to make a reaction go faster is to add a substance called a catalyst. A **catalyst** functions by changing the mechanism of a reaction in a manner that lowers activation energy barriers. Although the catalyst changes the mechanism of a reaction, it is not part of the overall stoichiometry of the reaction. A catalyst is consumed in an early step of a reaction mechanism but regenerated in a later step.

Catalysts are vital in the chemical industry. The market for catalysts in the United States exceeds $2.0 billion, including more than $600 million for petroleum refining and more than $750 million for chemical production. Although these are large sums of money, the products made available by catalysts are far more valuable than the catalysts themselves. The total value of fuels and chemicals produced by catalysts exceeds $900 billion.

Catalysis and the Ozone Problem

Catalysts are immensely beneficial in industry, but accidental catalysis in the atmosphere can be disastrous. Recall from Box 14-1 that the chemistry of ozone in the stratosphere involves a delicate balance of reactions that maintain a stable concentration of O_3. Chlorofluorocarbons (CFCs) shift that balance by acting as catalysts for the destruction of O_3 molecules.

Because CFCs are highly resistant to chemical attack, they are very stable in the lower atmosphere, where they can exist for up to 100 years. This stability gives CFCs time to diffuse up through the troposphere and into the stratosphere. There,

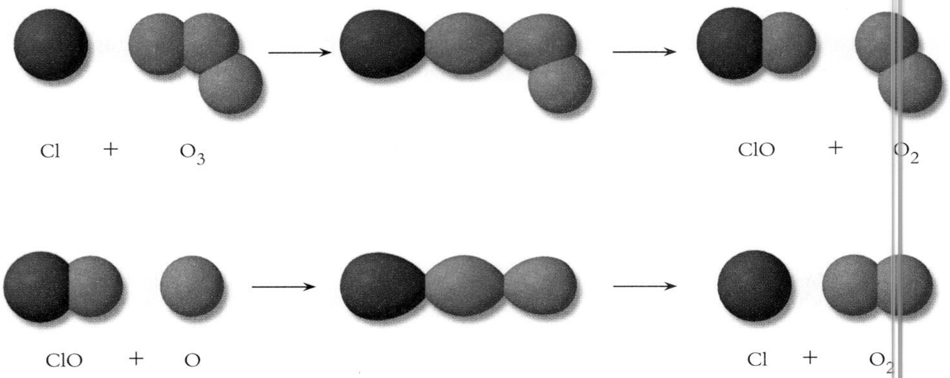

Cl + O$_3$

ClO + O$_2$

ClO + O

Cl + O$_2$

Figure 14-20
Chlorine atoms catalyze the reaction of ozone with oxygen atoms by forming an unstable ClO intermediate that readily reacts with an O atom, forming another O$_2$ molecule and regenerating the Cl atom.

CFCs absorb short-wavelength ultraviolet light from the sun that breaks carbon-chlorine bonds and produces chlorine atoms:

$$CF_2Cl_2 \xrightarrow{h\nu} CF_2Cl + Cl$$

Chlorine atoms react with O$_3$ molecules to produce O$_2$ and ClO, as shown by the molecular pictures in Figure 14-20. This is a catalytic process because chlorine monoxide reacts with an oxygen atom to produce a second O$_2$ molecule and *regenerate the chlorine atom*:

$$Cl + O_3 \longrightarrow ClO + O_2$$
$$ClO + O \longrightarrow Cl + O_2$$

Net: $\quad O_3 + O \xrightarrow{Cl} 2\,O_2$

Because a catalyst does not appear in the overall stoichiometry of a reaction, its presence is noted by writing the formula of the catalyst above or below the reaction arrow.

The net reaction for this two-step mechanism is the conversion of an O$_3$ molecule and an oxygen atom into two O$_2$ molecules. In this mechanism, chlorine atoms catalyze ozone decomposition. They participate in the mechanism, but they do not appear in the overall stoichiometry. Although chlorine atoms are consumed in the first step, they are regenerated in the second. The cyclical nature of this process means that each chlorine atom can catalyze the destruction of many O$_3$ molecules. It has been estimated that each chlorine atom produced by a CFC molecule in the upper stratosphere destroys about 100,000 molecules of ozone before it is removed by other reactions such as recombination:

$$CF_2Cl + Cl \longrightarrow CF_2Cl_2$$

It is important to recognize the difference between a *catalyst* and an *intermediate*. A catalyst is *consumed* in an early step of a mechanism and regenerated in a later step. The opposite is true of an intermediate. Intermediates are transient species that are *formed* early in a mechanism and consumed again in a later step:

$$Catalyst + Reactant_1 \longrightarrow Intermediate$$

$$Intermediate + Reactant_2 \longrightarrow Products + Catalyst$$

Chlorine monoxide is an intermediate in chlorine-catalyzed decomposition of ozone. Unlike chlorine atoms, each molecule of ClO participates in just one ozone decomposition reaction.

Figure 14-21, an activation energy diagram for the ozone–oxygen atom reaction, shows that the direct reaction between O_3 and O has a substantially higher activation energy than the chlorine-catalyzed reaction sequence. In other words, the activated complex between Cl atoms and O_3 molecules lies at a lower energy than the activated complex between O atoms and O_3 molecules, making chlorine atoms much more effective than oxygen atoms at destroying O_3 molecules. At the temperature of the ozone layer, 220 K, the rate data for the two pathways are as follows:

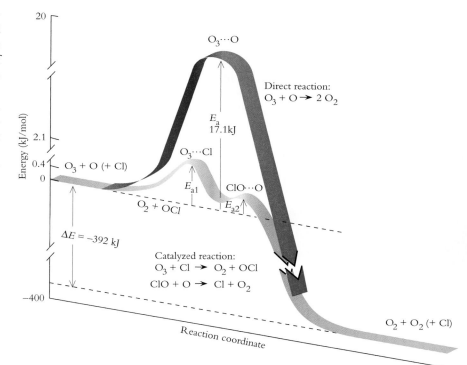

Figure 14-21
The activation energy diagram for the reaction between O_3 and O. Notice the break in the vertical scale: the overall reaction is exothermic by −392 kJ.

$$O_3 + O \longrightarrow 2\,O_2 \qquad E_a = 17.1 \text{ kJ/mol} \qquad k = 4.1 \times 10^5 \text{ M}^{-1}\text{ s}^{-1}$$
$$O_3 + Cl \longrightarrow O_2 + OCl \qquad E_{a1} = 2.1 \text{ kJ/mol} \qquad k_1 = 5.2 \times 10^9 \text{ M}^{-1}\text{ s}^{-1}$$
$$ClO + O \longrightarrow Cl + O_2 \qquad E_{a2} = 0.4 \text{ kJ/mol} \qquad k_2 = 2.2 \times 10^{10} \text{ M}^{-1}\text{ s}^{-1}$$

Notice that both steps of the chlorine-catalyzed reaction appear on the activation energy diagram. Each step of a mechanism has its own activation energy, so the diagram has two activation barriers. Experimental data indicate that the first barrier is higher than the second. Because the uncatalyzed reaction involves just one step, it has only one activation barrier.

The activation energy diagram in Figure 14-21 provides a clear picture of the role of a catalyst. Chlorine atoms provide an alternative mechanism for the reaction of ozone with oxygen atoms. This lower-energy pathway breaks down ozone in the stratosphere at a significantly faster rate than in the absence of the catalyst. As a result, the delicate balance among ozone, oxygen atoms, and oxygen molecules is changed in a way that poses a serious threat to the life-protecting ozone layer.

Homogeneous and Heterogeneous Catalysts

The catalysis of ozone decomposition by chlorine atoms occurs entirely in the gas phase. Chlorine is classified as a **homogeneous catalyst** because the catalyst and the reactants are present in the *same phase*, in this case the gas phase. A **heterogeneous catalyst,** on the other hand, is in a *different phase* than the one where the reaction occurs. A heterogeneous catalyst is usually a solid, and the reactants are gases or are dissolved in a liquid solvent.

The development of the ozone hole over Antarctica is accelerated by heterogeneous catalysis on microcrystals of ice. These microcrystals form in abundance in the Antarctic spring, which is when the ozone hole appears. Ice microcrystals are less common in the Arctic atmosphere, so ozone depletion has not been as extensive in the northern hemisphere.

Heterogeneous catalysts are the active ingredients in automobile catalytic converters. When combustion occurs in an automobile engine, side reactions generate small amounts of undesired products. Some carbon atoms end up as poisonous CO rather than CO_2. Another reaction that takes place at the high temperatures and pressures in automobile engines is the conversion of N_2 to NO. Furthermore, the combustion process fails to burn all the hydrocarbons. Hydrocarbons, CO, and NO all are undesirable pollutants that can be removed from exhaust gases by the action of heterogeneous catalysts:

$$2\ CO + O_2 \xrightarrow{\text{Catalyst}} 2\ CO_2$$

$$\text{Hydrocarbons} + O_2 \xrightarrow{\text{Catalyst}} CO_2 + H_2O$$

$$2\ NO \xrightarrow{\text{Catalyst}} N_2 + O_2$$

A mixture of catalysts is needed to catalyze the variety of reactions that must be carried out in the converter. The most important catalyst is platinum metal, but palladium and rhodium are used as well, as are transition metal oxides such as CuO and Cr_2O_3.

Heterogeneous reactions are complicated because the reacting species must be transferred from one phase to another before the reaction can occur. Despite much research, chemists still have limited knowledge about the mechanisms of reactions that involve heterogeneous catalysts. However, it is known that heterogeneous catalysis generally proceeds in four steps, as illustrated in Figure 14-22 for the conversion of NO into N_2 and O_2.

1. The starting materials bind to the surface of the catalyst. This process is known as **adsorption.** When a substance is adsorbed, its internal bonds are weakened or broken in favor of bonds to the catalyst.

Figure 14-22
Schematic molecular pictures of the four steps for the conversion of NO to N_2 and O_2 on a platinum metal surface.

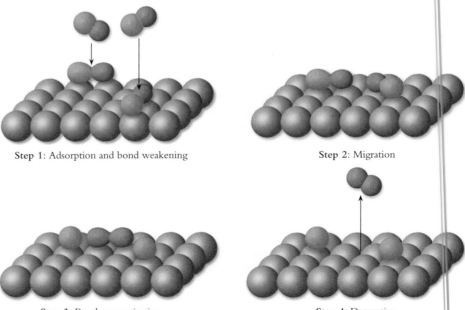

Step 1: Adsorption and bond weakening

Step 2: Migration

Step 3: Bond reorganization

Step 4: Desorption

2. Bound materials migrate over the surface of the catalyst.

3. Bound substances react to form products.

4. Products escape from the surface of the catalyst. This step is **desorption.**

Catalysis in Industry

The business of the chemical industry is to transform inexpensive substances into more valuable ones. In many cases, catalysts play important roles in these processes. Here, we describe the roles of catalysts in some important industrial reactions. Other catalyzed industrial reactions are considered in Chapter 15, where we describe properties of chemical equilibria.

Petroleum refining. Petroleum is the source not only of fuels but also of carbon compounds used for synthesis in the chemical industry. Crude petroleum is a complex mixture containing thousands of compounds, most of which are alkanes, alkenes, and derivatives of benzene. Petroleum refining converts this black goo into a variety of useful products, from gasoline to lubricants (see Figure 14-23). Refining begins with distillation, which separates the various components into fractions with different boiling point ranges. Large, high-boiling hydrocarbons are less versatile than smaller ones, so a further step in refining is the catalytic cracking of large hydrocarbons into smaller ones. Cracking is performed at high temperature in the presence of a heterogeneous catalyst composed of silicon dioxide and aluminum oxide:

$$C_{12} \text{ and higher} \xrightarrow[500\,°C]{SiO_2/Al_2O_3} C_5 \text{ to } C_{10}$$
$$\text{Kerosene} \qquad\qquad \text{Alkanes, alkenes}$$

Chemicals from coal. Heterogeneous catalysts are used to convert solid coal into gasoline and other chemicals. Solid coal is not easily transformed into hydrocarbon chains, so the conversion requires two general steps: gasification followed by catalytic hydrocarbon-forming reactions. Coal is first gasified by reaction with steam:

$$C(coal) + H_2O(g) \longrightarrow CO(g) + H_2(g) \qquad \Delta H = 1131 \text{ kJ/mol}$$

This reaction is significantly endothermic, so it must be driven by added energy. The energy required to drive the gasification, as well as energy needed to produce and maintain the flow of steam, is supplied by burning coal:

$$C(coal) + O_2(g) \longrightarrow CO_2(g) \qquad \Delta H = -407 \text{ kJ/mol}$$

The reaction of steam with coal generates a 1:1 gas mixture of H_2 and CO. Hydrocarbon formation requires a 2:1 mixture, so additional H_2 is produced by the reaction of steam with some of the CO:

$$CO(g) + H_2O(g) \longrightarrow CO_2(g) + H_2(g) \qquad \Delta H = -42 \text{ kJ/mol}$$

A variety of solid metallic catalysts containing oxides of iron, chromium, or copper can be used for hydrocarbon formation. High gas pressure (20 atm) and elevated temperature (250 to 350 °C) are also required. Several hydrocarbon-forming reactions occur; the most important is the stepwise production of alkanes.

Most catalysts used in the chemical industry are solid metal halides, metal oxides, or pure metals.

← **SECTION 8.4**
For more information about petroleum, see Section 8.4.

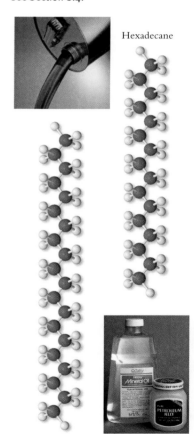

Hexadecane

Dodecane

Figure 14-23
Crude petroleum is refined to produce a variety of useful products. Shown are motor oil and lubricant jelly.

Here are two simple examples:

$$CO + 3 H_2 \xrightarrow[\text{High } P,T]{\text{Catalyst}} CH_4 + H_2O$$

$$CO + 2 H_2 + CH_4 \xrightarrow[\text{High } P,T]{\text{Catalyst}} C_2H_6 + H_2O$$

Another important chemical produced from CO and H_2 is methanol, CH_3OH. The synthesis is carried out at high temperature (200 to 300 °C) and pressure (50 to 100 atm) in the presence of copper (II) oxide or some other metal oxide catalyst:

$$CO + 2 H_2 \xrightarrow[\text{High } P,T]{\text{Catalyst}} CH_3OH$$

Approximately 5 billion kilograms of methanol are produced annually in the United States.

Mixtures of CO and H_2 have immense potential as a feedstock for the chemical industry. The variety of products available from CO and H_2 is virtually unlimited, and the Earth has huge coal reserves. Unfortunately, the direct conversion of coal is energy-intensive and highly polluting. For all of these reasons, coal conversion has been studied intensively. Despite considerable research, however, coal conversion is currently not economical compared with petroleum refining. The catalytic process is not the main stumbling block. Instead, the problem is the high cost of gasification, both in energy and in capital investment. The development of a catalyst that is effective at low temperature would solve this problem.

Acetic acid production. Although most industrial catalysts are heterogeneous, a growing number of industrial reactions use homogeneous catalysts. One example is the production of acetic acid. Most of the 2.1 billion kilograms of acetic acid produced annually is used in the polymer industry. The reaction of methanol and carbon monoxide to form acetic acid is catalyzed by a rhodium compound that dissolves in methanol:

$$CH_3OH + CO \xrightarrow[\text{175 °C, 700 atm}]{[Rh(CO)_2I_3]} CH_3CO_2H$$

These are just a few examples of catalytic reactions used industrially. Catalyst research is one of the most active areas of chemistry and chemical engineering.

Biocatalysis: Enzymes

Living organisms carry out an astonishing variety of chemical processes. Organisms organize small molecules into complex biopolymers such as deoxyribonucleic acid (DNA) and proteins. Conversely, organisms break down large, energy-rich molecules in many steps to extract chemical energy in small portions to drive their many activities. Further, organisms produce antibodies, medium-sized molecules that combat bacterial invaders. In many of these processes, organisms break chemical bonds selectively without resorting to high temperature.

Most of these reactions are regulated by biochemical catalysts called enzymes. An **enzyme** is a specialized protein that catalyzes a specific biochemical reaction. Some enzymes are found in extracellular fluids such as saliva and gastric juices, but most are found inside cells. Each type of cell has a different array of enzymes that act together to determine what role the cell plays in the overall biochemistry of the organism.

CHAPTER 11 →
Protein structure is described in Chapter 11.

Enzymes are complicated molecules. Biochemists have determined the molecular structures of some enzymes, but the structures of many enzymes are not yet known. Our Chemistry and Life Box explores one of the most important of these, nitrogenase.

Even though there is a wide diversity of structures, most enzyme activity follows a general mechanism that has several reversible steps. In the first step, a

Box 14-2	Chemistry and Life: How Does Nitrogenase Work?

I n the epic poem, *Rime of the Ancient Mariner*, a becalmed sailor laments "Water, water, everywhere, nor any drop to drink." With regard to the element nitrogen, we might say "N_2, N_2, everywhere, nor any N to eat." Even though the atmosphere of our planet is 80% molecular nitrogen, plants are unable to break the extremely strong $N \equiv N$ triple bond. Instead, plants must rely on other sources for the nitrogen atoms that are essential for the synthesis of biomolecules such as amino acids and DNA.

In order for nitrogen to be available to life forms, it must be either oxidized to nitrate (NO_3^-) or reduced to ammonia (NH_3). Either process is known as *nitrogen fixation*. Atmospheric nitrogen is oxidized to nitrate by lightning, but lightning accounts for only about 1% of the fixed nitrogen required by the biosphere. The fertilizer industry reduces N_2 to NH_3 by reaction with H_2, but fertilizers only supply about 30% of fixed nitrogen. The bulk—about 70%—comes from the activity of one group of bacteria that can convert N_2 to NH_3. These bacteria contain an enzyme, nitrogenase, that catalyzes this difficult chemical transformation.

Researchers are working to understand how this enzyme works. Research on nitrogenase takes two main forms. One is an examination of the structure and operation of the enzyme to determine the mechanism by which N_2 is converted to ammonia. The other form is the synthesis of artificial catalysts that mimic the operation of nitrogenase.

Nitrogenase is a complex protein with two metal-containing parts that "dock" together during the fixation of nitrogen. Part 1 contains an interconnected group of seven iron atoms, nine sulfur atoms, and a single molybdenum atom. It is believed that nitrogen binds to the molybdenum atom, where it gains the six electrons and six protons needed to become two ammonia molecules. However, exactly how N_2 binds to the [7Fe-9S-Mo] group and how it is reduced to NH_3 remain a mystery.

Part 1 of the nitrogenase protein contains another interconnected group of Fe-S atoms, this one with eight iron atoms and seven sulfur atoms. This [8Fe-7S] group collects electrons and transmits them to the binding center.

Part 2 of nitrogenase contains a third Fe-S group, this one made up of four iron atoms and four sulfur atoms. This part of the enzyme also binds two molecules of ATP.

When the two parts of the enzyme dock with one another, the [4Fe-4S] group of Part 2 ends up close to the [8Fe-7S] group of Part 1. It is thought that when the parts of the enzyme join together, two electrons are pumped from the ATP molecules to the 4Fe-4S site, then to the 8Fe-7S site, and finally to the [7Fe-9S-Mo] binding site. Next, the parts separate, and Part 2 is recharged with two new ATP molecules. This process must occur four times to provide the eight electrons that are consumed during the fixation reaction. The entire fixation process consumes eight electrons and protons, six of each to form the NH_3 molecule and two more that are "wasted" by forming an H_2 molecule.

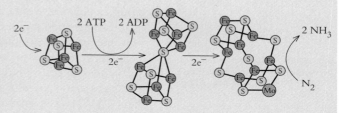

Although the mechanistic details of N_2 reduction have not been determined, chemists have been able to make a molybdenum-containing complex that can bind and fragment molecular nitrogen. Two molecules of a trigonal molybdenum compound form a "sandwich" complex with N_2 in the center. The complex slowly fragments into two pieces, each with a $Mo \equiv N$ triple bond. Although this reaction is slow and does not produce ammonia, it nevertheless breaks the $N \equiv N$ triple bond, suggesting that further exploration of Mo-N_2 chemistry could lead to success in mimicking nitrogenase, or at least understanding better how this remarkable enzyme works.

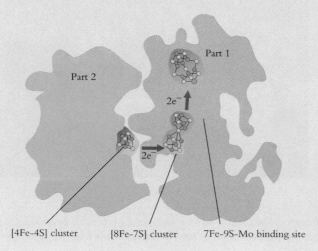

[4Fe-4S] cluster   [8Fe-7S] cluster   7Fe-9S-Mo binding site

reactant molecule known as a *substrate* (S) binds to a specific location on the enzyme (E), usually a groove or a pocket on the surface of the protein:

$$E + S \rightleftharpoons [E\!-\!S]$$

The substrate binds to the active site through intermolecular interactions that usually include significant amounts of hydrogen bonding.

Binding causes subtle changes in the shape of the enzyme, which in turn distorts the structure of the substrate:

$$[E\!-\!S] \rightleftharpoons [E\!-\!S]_{Distorted}$$

Distortion allows the substrate to react more easily with another reactant (R) to form the desired product (P). Once the product forms, it no longer binds strongly to the enzyme:

$$[E\!-\!S]_{Distorted} + R \longrightarrow [E\!-\!P] \longrightarrow E + P$$

The enzyme is regenerated at the end of this sequence, so it is once again available to bind another substrate molecule.

Note that the steps in this enzyme-catalyzed biochemical mechanism are similar to the steps in chemical heterogeneous catalysis: binding with bond weakening, reaction at the bound site, and release of products.

A specific illustration of enzymatic catalysis is the addition of a phosphate group to glucose, which is the first step in glycolysis. Virtually all cells use glycolysis to produce adenosine triphosphate (ATP), the energy-rich compound that supplies energy for many biological processes. This addition of phosphate is catalyzed by an enzyme, hexokinase. The first step is binding of glucose by the enzyme.

When glucose binds to hexokinase, chemical interactions between the two molecules cause the enzyme-glucose complex to change its shape, as is shown in Figure 14-24. The enzyme folds in on the substrate, isolating the bound glucose

The names of enzymes usually end in *-ase*, for example hexokinase and nitrogenase.

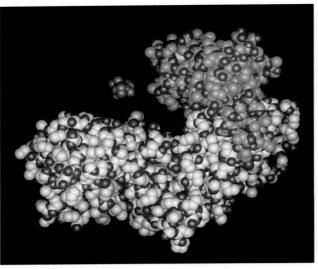
(a)

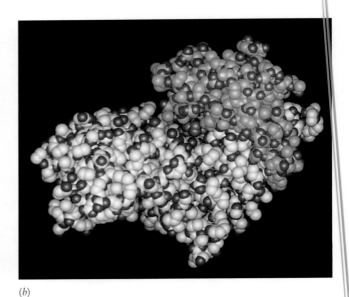
(b)

Figure 14-24
Computer models showing the shape of hexokinase (*a*) without and (*b*) with bound glucose. The enzyme folds around the substrate to bind it and isolate it from its aqueous environment.

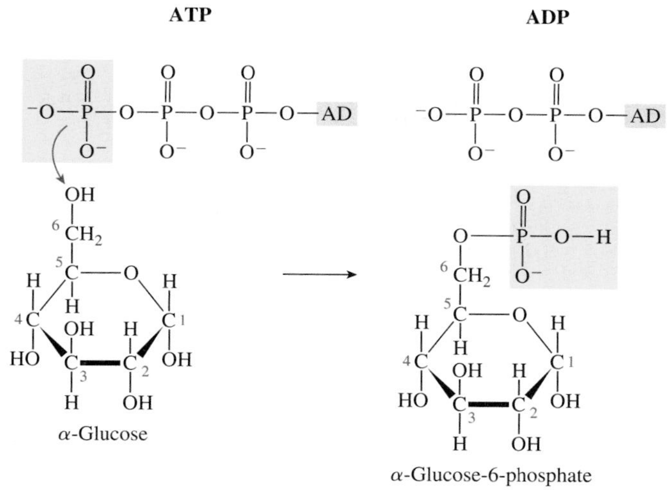

Figure 14-25
Hexokinase catalyzes the phosphorylation of α-glucose by adenosine triphosphate (ATP). "AD" represents the adenosine portion of ATP and ADP. The complete structure of ATP appears in Figure 13-22.

molecule from its aqueous environment. This allows an ATP molecule to transfer a phosphate group to the hydroxyl group in position 6 of glucose, as shown in Figure 14-25. The product, glucose–6–phosphate, is too big to bind tightly to hexokinase. Consequently, the enzyme opens up, releases the product molecule, and is ready to bind another glucose molecule.

Although the molecular details of enzyme mechanisms are complex, the kinetic behavior of many enzymatic processes is first-order in both the substrate and the enzyme. Example 14-14 shows that the mechanism just outlined is consistent with this kinetic behavior.

Enzyme Kinetics   **Example 14-14**

Derive the predicted rate law for the general mechanism for enzyme catalysis, assuming that the distortion step is rate-determining:

$$E + S \underset{k_{-1}}{\overset{k_1}{\rightleftarrows}} [E-S] \xrightarrow{k_2, \text{ slow}} [E-S]_D$$

$$[E-S]_D + R \xrightarrow{k_3} E + P$$

Strategy: The rate law predicted by a mechanism can be derived by setting the overall rate of reaction equal to the rate of the slowest step. As described in Section 14.5, any earlier steps are assumed to have equal forward and reverse rates.

Solution: The mechanism includes a rate-determining distortion step, for which the rate expression is:

$$\text{Rate} = k_2[E-S]$$

This rate equation contains the concentration of an intermediate, so use the forward and reverse rates of the first step to derive an expression for $[E-S]$ in terms of $[E]$ and $[S]$:

Example 14-14 | **Enzyme Kinetics** (*continued*)

$$k_1[E][S] = k_{-1}[E\text{—}S] \quad so \quad [E\text{—}S] = \frac{k_1}{k_{-1}}[E][S]$$

Now substitute into the original rate expression:

$$\text{Rate} = \frac{k_1 k_2}{k_{-1}}[E][S] = k_{obs}[E][S] \quad where \quad k_{obs} = \frac{k_1 k_2}{k_{-1}}$$

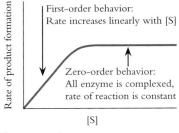

First-order behavior:
Rate increases linearly with [S]

Zero-order behavior:
All enzyme is complexed,
rate of reaction is constant

Rate of product formation

[S]

Figure 14-26
The typical profile of how the rate of an enzyme-catalyzed reaction varies with substrate concentration.

Enzyme reactions are almost always first-order in enzyme, as one would expect for a reaction in which one molecule of enzyme participates in the rate-determining step. As Figure 14-26 illustrates, the rate dependence on substrate concentration generally shows two distinct regions of kinetic behavior. At low substrate concentration, the binding or distortion step is rate-determining, and the reaction is first-order in substrate. However, when substrate concentration is high, all the enzyme molecules are bound either to substrate or to product. Under these conditions, the enzyme is catalyzing the reaction as fast as it can, so adding more substrate cannot make the enzyme work any faster. At high substrate concentrations, the reaction is zero-order in substrate.

This has been a brief overview of a rich field. Details of enzyme structure and catalytic activity are studied in laboratories worldwide. Moreover, genetic engineering makes it possible to "manufacture" key enzymes in large quantities, so enzymes may become industrial catalysts that accomplish reactions rapidly and selectively.

Section Exercises

14.7.1 When metals such as platinum, palladium, and rhodium are used as catalysts, they are usually deposited as thin layers over a highly porous material such as charcoal. Explain why these materials are less expensive and more effective catalysts than pieces of pure metal.

14.7.2 Which of the following statements are true? Correct the untrue statements so that they are true: (a) The concentration of a homogeneous catalyst appears in the rate law. (b) A catalyst changes an endothermic reaction into an exothermic reaction. (c) A catalyst lowers the activation energy of the rate-determining step. (d) A catalyst is consumed in an early step of a mechanism and regenerated in a later step.

14.7.3 Why would it be advantageous to develop a catalyst for the following reaction?

$$4\,CO + 2\,NO_2 \longrightarrow N_2 + 4\,CO_2$$

CHAPTER REVIEW

Summary and Key Terms

1. A chemical reaction occurs in a sequence of **elementary reactions** that make up the **reaction mechanism**. These include **bimolecular reactions, unimolecular reactions**, and (rarely) **termolecular reactions**. Mechanisms can involve **intermediates**, which are produced in early steps and consumed in later steps. The rate of a reaction is set by the slowest, **rate-determining step** in its mechanism.

2. **Kinetics** is the study of the **rates** of reactions—how fast they occur. Rates, which can be expressed as changes of concentration with time, depend on how quickly molecules interact and rearrange.

3. A **rate law** expresses how the rate of a reaction depends on concentrations. Many rate laws have the form rate = $k[A]^y[B]^z$, where k is the **rate constant**. Often, the exponents have values of 1 (**first-order**) or 2 (**second-order**). Every mechanism has a predicted rate law given by the rate expression for its rate-determining step.

4. The specific form of a rate law must be determined by experimental measurements of how concentrations vary with time. For a first-order reaction, $\ln\left(\dfrac{[A]_0}{[A]}\right)$ varies linearly with time, and first-order reactions have constant **half-lives**. For a second-order reaction, $\dfrac{1}{[A]} - \dfrac{1}{[A]_0}$ varies linearly with time. A reaction displaying elaborate rate behavior can often be made to display first-order or second-order behavior by the **isolation method**, in which one reactant concentration is made much smaller than all other reactant concentrations.

5. A satisfactory mechanism must be a set of elementary steps that add up to the correct overall stoichiometry and predict the experimentally observed rate law. When the first step in a mechanism is rate-determining, the rate law for that first step is the rate law for the reaction. When a step beyond the first is rate-determining, the concentrations of intermediates may appear in the rate expression. These must be eliminated from the rate law before it can be compared with experimental data. A technique for doing this assumes that early steps are **reversible reactions** and equates their forward and reverse rates.

6. An elementary reaction usually requires an energy input to drive the rearrangements of chemical bonds. This **activation energy** (E_a) of the reaction can be shown schematically on **activation energy diagrams**, which plot energy as a function of the **reaction coordinate**. The point of highest energy on such a diagram is the **activated complex**. Because the fraction of molecules with sufficient energy to react increases with temperature, rate constants increase with increasing temperature. The **Arrhenius equation** expresses the dependence of k on E_a and T and allows E_a to be calculated from measurements of rate constants at different temperatures.

7. Reaction rates are altered by **catalysts**, which change the mechanisms of reactions, thereby reducing their activation energies. **Homogeneous catalysts** act in the same phase where the reaction normally occurs, whereas **heterogeneous catalysts**, which are usually solids, provide sites where reactants can undergo **adsorption**, react, and then undergo **desorption**. Catalysts play substantial roles in industrial syntheses. Biological catalysts, **enzymes**, function by binding a substrate in such a way that reaction becomes easier.

Skills to Master

▶ Visualizing elementary reactions

▶ Completing reaction mechanisms

▶ Predicting rate laws from mechanisms

▶ Determining reaction orders from amount/time data

▶ Evaluating rate constants from experimental data

▶ Testing the consistency of a mechanism

▶ Applying the Arrhenius equation

▶ Interpreting activation energy diagrams

▶ Explaining the action of catalysts

Learning Exercises

14.1 List the requirements of a satisfactory mechanism, and describe the characteristics of the steps of a mechanism.

14.2 Write one-sentence explanations for the following kinetic terms: (a) first order; (b) second order; (c) isolation method; (d) elementary step; (e) catalyst; (f) intermediate; and (g) enzyme.

14.3 Describe the differences among the rate, the rate law, and the rate constant for a chemical reaction.

14.4 Write a paragraph that describes what the activation energy is and how it affects the kinetic behavior of a reaction.

14.5 Update your list of memory bank equations. Include information about how to apply each new equation.

14.6 Make a list of terms new to you that appear in Chapter 14. In your own words, write a one-sentence definition of each. Consult the Glossary if you need help.

Problems   **ilw** = interactive learning ware problem. Visit the website at www.wiley.com/college/olmsted

What Is a Reaction Mechanism?

14.1 What is the rate-determining step in each of the following processes? (a) A line of people get coffee from a large coffee urn. (b) You go through the express line (10 items, no checks) at a supermarket. (c) A squad of parachutists makes a "jump" from an airplane cargo door.

14.2 What is the rate-determining step in each of the following processes? (a) People buy their lunches at a cafeteria. (b) Music enthusiasts enter an amphitheater for a concert. (c) You leave a pay-as-you-exit parking lot.

14.3 The reaction of H_2 with I_2 is:

$$H_2 + I_2 \longrightarrow 2\ HI$$

Use reactant molecules to write appropriate elementary reactions that satisfy the following criteria: (a) a unimolecular decomposition that generates I; (b) a bimolecular collision that forms a square H_2I_2 complex; and (c) a bimolecular collision in which a hydrogen atom is transferred between reactants.

14.4 The reaction of NO with Cl_2 is:

$$2\ NO + Cl_2 \longrightarrow 2\ NOCl$$

Use reactant molecules to write appropriate elementary reactions that satisfy the following criteria: (a) a unimolecular decomposition that generates Cl; (b) a bimolecular collision in which a Cl atom is transferred between reactants; and (c) a termolecular collision leading to the observed products.

14.5 Draw molecular pictures illustrating each part of Problem 14.3.

14.6 Draw molecular pictures illustrating each part of Problem 14.4.

14.7 Write three different satisfactory mechanisms for the reaction in Problem 14.3, one having your elementary reaction (a) as its first step, one having your elementary reaction (b) as its first step, and one having elementary reaction (c) as its first step.

14.8 Write three different satisfactory mechanisms for the reaction in Problem 14.4, one having your elementary reaction (a) as its first step, one having your elementary reaction (b) as its first step, and one having elementary reaction (c) as its first step.

Rates of Chemical Reactions

14.9 For the reaction $2\ NO + Cl_2 \longrightarrow 2\ NOCl$, do the following: (a) Express the rate in terms of the disappearance of Cl_2. (b) Relate the rate of NOCl formation to the rate of Cl_2 disappearance. (c) If Cl_2 reacts at a rate of 47 M s^{-1}, state how fast NOCl will form.

14.10 Do the following for the ozone decomposition mechanism: (a) Express the rate in terms of O_2 formation. (b) Relate the rate of O_3 consumption to the rate of O_2 production. (c) If O_2 forms at a rate of 2.7×10^{-6} M s^{-1}, state how fast ozone disappears.

14.11 For the reaction, $2\ NO + Cl_2 \longrightarrow 2\ NOCl$, do the following: (a) Draw a molecular picture showing a sample that contains 12 NO molecules and 5 Cl_2 molecules. (b) Redraw the picture to show the result after four molecules of NO have reacted. (c) Redraw the picture to show the result when one reactant has been consumed completely.

14.12 For the reaction, $O_3 + NO \longrightarrow O_2 + NO_2$, do the following: (a) Draw a molecular picture showing a sample that contains 6 NO molecules and 8 O_3 molecules. (b) Redraw the picture to show the result after three molecules of NO have reacted. (c) Redraw the picture to show the result when one reactant has been consumed completely.

14.13 Calcium oxide, an important ingredient in cement, is produced by decomposing calcium carbonate at high temperature:

$$CaCO_3(s) \longrightarrow CaO(s) + CO_2(g)$$

In one reaction, 3.5 kg of calcium carbonate is heated at 550 °C in a 5.0-L vessel. The pressure of CO_2 is 0.15 atm after 5.0 minutes. (a) What is the average rate of CO_2 production in mol/min during the 5-minute interval? (b) How many moles of $CaCO_3$ decompose in the 5-minute interval?

14.14 A chemist is studying the rate of the Haber synthesis: $N_2 + 3\ H_2 \longrightarrow 2\ NH_3$. Starting with a closed reactor containing 1.25 mol/L of N_2 and 0.50 mol/L of H_2, the chemist finds that the H_2 concentration has fallen to 0.25 mol/L after 30 seconds. (a) What is the average rate of reaction over this time? (b) What is the average rate of NH_3 production? (c) What is the N_2 concentration after 30 seconds?

14.15 At high temperature, cyclopropane isomerizes to propene:

Cyclopropane —Heat→ Propene
(C_3H_6)   (C_3H_6)

In a small sample containing 10 molecules of cyclopropane, the reaction proceeds at an average rate of 0.25 molecules/min for 20 minutes. (a) Use line drawings to illustrate the small sample before the reaction begins. (b) Redraw the picture after 20 minutes of reaction.

14.16 Under the appropriate conditions, *cis*-dichloroethylene isomerizes to *trans*-dichloroethylene:

cis-Dichloroethylene   *trans*-Dichloroethylene
($C_2H_2Cl_2$)   ($C_2H_2Cl_2$)

In a small sample of gas containing 12 molecules of *cis*-dichloroethylene, the reaction proceeds at an average rate of 1.5 molecules/s for 2.0 seconds. (a) Use line drawings to illustrate the small sample before the reaction begins. (b) Redraw the picture after 2.0 seconds of reaction.

Concentration and Reaction Rates

14.17 Radioactive isotopes decay according to first-order kinetics. For one particular isotope, 1.00 mol registers 12 decays in 1.00 min. (a) How many decays will occur in 1.00 min if 5.00 mol of this isotope are present? (b) What fraction of the isotope decays per minute in each case? (c) Explain the relationship between your answers to (a) and (b).

14.18 Popcorn kernels pop independently (that is, "unimolecularly"). At constant temperature, 6 kernels pop in 5 seconds when 150 kernels are present. (a) After 50 kernels have popped, how many kernels pop in 5 seconds? (b) Is there a change in the fraction of kernels popping per second? If so, by how much? (c) Explain the relationship between your answers to (a) and (b).

14.19 For the net reaction, 2 AB + 2 C ⟶ A₂ + 2 BC, the following slow first step has been proposed:

$$C + AB \longrightarrow BC + A$$

(a) What rate law is predicted by this step? (b) What units are associated with the rate constant for this rate law? (c) Write additional steps that complete the mechanism.

14.20 For the net reaction in Problem 14.19, another possible slow first step follows:

$$2 AB \longrightarrow A_2 + 2 B$$

(a) What rate law is predicted by this step? (b) What units are associated with the rate constant for this rate law? (c) Write additional steps that complete the mechanism.

14.21 For the net reaction in Problem 14.19, another possible slow first step follows:

$$2 C + AB \longrightarrow AC + BC$$

(a) What rate law is predicted by this step? (b) What units are associated with the rate constant for this rate law? (c) Write additional steps that complete the mechanism.

14.22 For the net reaction in Problem 14.19, another possible slow first step follows:

$$AB \longrightarrow A + B$$

(a) What rate law is predicted by this step? (b) What units are associated with the rate constant for this rate law? (c) Write additional steps that complete the mechanism.

14.23 The two pictures shown below represent starting conditions for the following reaction:

$$O_3 + NO \longrightarrow O_2 + NO_2 \qquad Rate = k[O_3][NO]$$

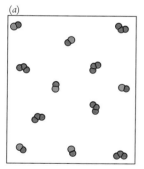

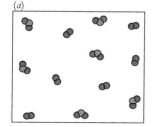

Will Flask *b* react faster or slower or at the same rate as Flask *a*? By how much? Explain your reasoning in terms of molecular collisions.

14.24 The two pictures shown below represent starting conditions for the following reaction:

$$O_2 + NO_2 \longrightarrow O_3 + NO \qquad Rate = k[O_2][NO_2]$$

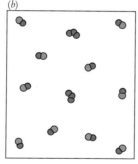

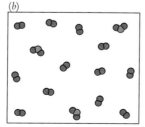

Will Flask *b* react faster or slower or at the same rate as Flask *a*? By how much? Explain your reasoning in terms of molecular collisions.

Experimental Kinetics

14.25 The following data are obtained for the decomposition of N₂O₅ at 45 °C:

$$2 N_2O_5(g) \longrightarrow 4 NO_2(g) + O_2(g)$$

t (10^2 s)	0	2.00	4.00	6.00	8.00
[N₂O₅] (atm)	2.50	2.22	1.96	1.73	1.53

(a) Use these data to determine the order of the decomposition process. Show your reasoning. (b) Determine the rate constant, including appropriate units. (c) What is the pressure of N₂O₅ after 1600 s? (d) How long will it take for the pressure to decrease to 0.500 atm?

14.26 Radioactive iodine (¹³¹I) is used frequently in biological studies. A radiation biologist studies the rate of decomposition of this substance and obtains the following data:

Time (days)	0	4.0	8.0	12.0	16.0
Mass ¹³¹I (μg)	12.0	8.48	6.0	4.24	3.0

(a) Use these data to determine the order of the decomposition process. Show your reasoning. (b) Determine the rate constant, including appropriate units. (c) How many micrograms will be left after 32 days? (d) How many days will it take for the sample size to decrease to 1.2 μg?

14.27 The decomposition reaction of NOBr is second-order in NOBr, with a rate constant at 20 °C of 25 L/mol-min. If the initial concentration of NOBr is 0.025 M, find (a) the time at which the concentration will be 0.010 M and (b) the concentration after 125 min of reaction.

14.28 The condensation reaction of butadiene, C₄H₆, is second-order in C₄H₆, with a rate constant of 0.93 L/mol-min (See Example 14-6). If the initial concentration of C₄H₆ is 0.240 M, find (a) the time at which the concentration will be 0.100 M and (b) the concentration after 25 min of reaction.

14.29 The conversion of C₅H₁₁Br into C₅H₁₀ follows first-order kinetics, with a rate constant of 0.385 hr⁻¹. If the initial concentration of C₅H₁₁Br is 0.125 M, find (a) the time at which the concentration will be 1.25×10^{-3} M and (b) the concentration after 3.5 hr of reaction.

14.30 The light-emitting decay of excited mercury atoms is first-order, with a rate constant of 1.65×10^6 s⁻¹. A sample contains 4.5×10^{-6} M of excited mercury atoms. Find (a) the time at which the concentration will be 4.5×10^{-7} M and (b) the concentration after 2.5×10^{-6} s.

14.31 Azomethane decomposes into nitrogen and ethane at high temperature:

$$H_3C-N=N-CH_3 \longrightarrow N_2 + C_2H_6$$

A chemist studying this reaction at 300 °C obtains the following data:

Time (10^2 s)	0	1.00	1.50	2.00	2.50	3.00
[Azomethane] (mM)	7.94	6.15	5.40	4.75	4.20	3.69

Prepare first- and second-order graphic plots of these data, determine the order of the reaction, and calculate its rate constant.

14.32 Gas A decomposes by the reaction $3\ A \rightarrow B + C$ at 45 °C, and its concentration changes as follows:

Time (10^2 s)	0	2.00	4.00	6.00	8.00
[A] (M)	2.50	2.22	1.96	1.73	1.53

Find the rate law and determine the rate constant.

Linking Mechanisms and Rate Laws

14.33 Problem 14.21 asks about a mechanism for the hypothetical reaction of AB with C. Assume that the second step is rate-determining and that the first step is reversible. Derive the rate law under these assumptions.

14.34 Problem 14.8 asks for mechanisms for the reaction of NO with Cl_2. For parts (a) and (b) of that problem, assume that the first step in your mechanism is fast and reversible and the second step is rate-determining. Derive the rate laws under these assumptions.

ilw **14.35** A student proposes the following mechanism for the atmospheric decomposition of ozone to molecular oxygen:

$$O_3 + O_2 \rightleftharpoons O_5 \qquad \text{(fast, reversible)}$$
$$O_5 \longrightarrow 2\ O_2 + O \qquad \text{(slow, rate-determining)}$$

(a) Propose a third step that completes this mechanism. (b) Determine the rate law predicted by this mechanism. (c) Atmospheric chemists would consider this mechanism to be molecularly unreasonable. Explain why.

14.36 The reaction of CO with Cl_2 gives phosgene ($COCl_2$), a nerve gas used in World War I. Even though the stoichiometry is simple, the mechanism has several steps:

$$Cl_2 \rightleftharpoons 2\ Cl \qquad \text{(fast, reversible)}$$
$$Cl + CO \longrightarrow COCl \qquad \text{(slow, rate-determining)}$$
$$COCl + Cl_2 \longrightarrow COCl_2 + Cl \qquad \text{(fast)}$$

(a) Show that this mechanism gives the correct overall stoichiometry. (b) What rate law does this mechanism predict? (c) Identify any reactive intermediates in the mechanism.

14.37 The reaction of NO with O_2 to give NO_2 is an important process in the formation of smog in Los Angeles:

$$2\ NO + O_2 \longrightarrow 2\ NO_2$$

Experiments show that this reaction is third-order overall. The following mechanism has been proposed:

$$NO + NO \rightleftharpoons N_2O_2 \qquad k_1,\ k_{-1}$$
$$N_2O_2 + O_2 \longrightarrow NO_2 + NO_2 \qquad k_2$$

(a) If the second step is rate-determining, what is the rate law? (b) Is this rate law consistent with the overall third-order behavior? Explain. (c) Draw molecular pictures that show different ways the intermediate species might bind together, and identify the one that is most reasonable with respect to the second step of the mechanism.

14.38 Gaseous N_2O_5 decomposes according to the following equation:

$$2\ N_2O_5 \longrightarrow 4\ NO_2 + O_2$$

Much evidence suggests that the mechanism is as follows:

$$N_2O_5 \rightleftharpoons NO_2 + NO_3 \qquad \text{(fast decomposition)}$$
$$NO_2 + NO_3 \longrightarrow NO + NO_2 + O_2 \qquad \text{(slow)}$$
$$NO + NO_3 \longrightarrow 2\ NO_2 \qquad \text{(fast)}$$

(a) Show that this mechanism gives the correct overall stoichiometry. (b) Determine the rate law predicted by this mechanism. (c) This decomposition is first-order experimentally. Does this information prove that step 2, rather than step 1, is rate-determining? Explain.

Reaction Rates and Temperature

14.39 If a reaction has an activation energy of zero, how will its rate constant change with temperature? Explain in molecular terms what $E_a = 0$ means.

14.40 If a reaction has an activation energy of zero, how is ΔE for the forward reaction related to E_a for the reverse reaction? Draw an activation energy diagram illustrating your answer.

14.41 Consider the exothermic reaction $AC + B \rightarrow AB + C$. (a) Draw an activation energy diagram for this reaction. (b) Label the energies of reactants and products. (c) Show $\Delta E_{\text{reaction}}$ by a double-headed arrow. (d) Show E_a for the forward reaction by a single-headed arrow. (e) Label and draw a molecular picture of the activated complex.

14.42 Consider the endothermic reaction $AB + C \rightarrow AC + B$. (a) Draw an activation energy diagram for this reaction. (b) Label the energies of reactants and products. (c) Show $\Delta E_{\text{reaction}}$ by a double-headed arrow. (d) Show E_a for the forward reaction by a single-headed arrow. (e) Label and draw a molecular picture of the activated complex.

14.43 For the isomerization of cyclopropane to propene, $\Delta H_{\text{reaction}} = -33$ kJ and $E_a = 273$ kJ/mol. (a) Draw a molecular picture of cyclopropane converting into propene. (b) Draw the activation energy diagram for the reaction, and label it completely.

14.44 Nitrogen dioxide in smog can combine in an elementary reaction to form N_2O_4 molecules. The combination reaction is exothermic by 57 kJ/mol. For the reverse reaction, the dissociation of N_2O_4, $E_a = 70$ kJ/mol. (a) Draw a molecular picture of NO_2 combining to form N_2O_4. (b) Draw an activation energy diagram that shows the energy relationships as quantitatively as possible. Label the diagram completely.

ilw **14.45** Fireflies "flash" at a rate that depends on the temperature. At 29 °C, the average rate is 3.3 flashes every 10 seconds, whereas at 23 °C, the average rate falls to 2.7 flashes every 10 seconds. Calculate the "energy of activation" for the flashing process.

14.46 The rate at which tree crickets chirp is 190/min at 28 °C but only 39.6/min at 5 °C. From these data, calculate the "energy of activation" for the chirping process.

Catalysis

14.47 The industrial process for forming methanol involves a catalyst:

$$CO + 2\ H_2 \xrightarrow[\text{High } P, T]{\text{Catalyst}} CH_3OH$$

Draw molecular pictures showing the bond breakage and formation that must occur in the course of this reaction. Suggest how the catalyst might make it easier for these reactions to occur.

14.48 The industrial process for forming acetic acid involves a catalyst:

$$CH_3OH + CO \xrightarrow[\text{175 °C, 700 atm}]{[\text{Rh(CO)}_2\text{I}_3]} CH_3CO_2H$$

Draw molecular pictures showing the bond breakage and formation that must occur in the course of this reaction. Suggest how the catalyst might make it easier for these reactions to occur.

14.49 The addition of molecular hydrogen to ethylene is extremely slow unless Pd metal is present. In the presence of the metal, however, H_2 is adsorbed on the metal surface as H atoms, which then add to C_2H_4 when it strikes the surface:

$$H—H \;+\; \overset{\displaystyle H \quad\; H}{\underset{\displaystyle H \quad\; H}{C=C}} \xrightarrow{\;Pd\;} \overset{\displaystyle H \quad H}{\underset{\displaystyle H \quad H}{H—C—C—H}}$$

(a) Draw an activation energy diagram for this mechanism that illustrates the effect of the catalyst. (b) Identify any catalysts or intermediates in the mechanism. Explain. (c) Draw a molecular picture that illustrates how this addition takes place.

Additional Paired Problems

14.51 What happens to the rate of a reaction involving H_2 if the concentration of H_2 is tripled and the reaction is (a) second-order in H_2; (b) zero-order in H_2; and (c) 3/2-order in H_2?

14.52 What happens to the rate of a reaction involving CO if the concentration is doubled and the reaction is (a) first-order in CO; (b) half-order in CO; and (c) inverse first-order in CO?

14.53 Oxygen reacts with NO to form NO_2 as the only product. Here is a proposed first step of the mechanism:

$$O_2 + NO \longrightarrow NO_2 + O$$

(a) What second step is required to give a satisfactory mechanism? (b) Draw molecular pictures to illustrate each step.

14.54 Oxygen reacts with CO to form CO_2 as the only product. Here is a proposed first step of the mechanism:

$$O_2 + CO \longrightarrow CO_2 + O$$

(a) What second step is required to give a satisfactory mechanism? (b) Draw molecular pictures to illustrate each step.

14.55 With appropriate catalysts, it is possible to make benzene from acetylene. The reaction has simple overall stoichiometry and can be studied in kinetics experiments:

$$3\,C_2H_2(g) \longrightarrow C_6H_6(g)$$

(a) Write the rate *expression* (not the rate law) for this reaction in terms of the disappearance of C_2H_2 and in terms of the appearance of C_6H_6. (b) If enough information is given to find the rate law, find it. If not, describe the data that you would need and how you would analyze the data to determine the rate law.

14.56 The first step in the Ostwald process for the synthesis of nitric acid is the combustion of ammonia:

$$4\,NH_3 + 5\,O_2 \xrightarrow{\;Pt\ gauze\;} 4\,NO + 6\,H_2O$$

In a catalytic experiment, the rate of reaction is found to be:

$$Rate = \frac{\Delta[NO]}{\Delta t} = 1.5 \times 10^{-3}\ M\ s^{-1}$$

What is the rate expressed in terms of NH_3, O_2, and H_2O?

14.57 Use the following information to construct an activation energy diagram for the following reaction:

$$S(s) + O_2(g) \longrightarrow SO_2(g)$$

$$\Delta H_f^\circ(SO_2) = -296.1\ kJ/mol,\ E_a = 150\ kJ/mol$$

14.50 NO is an atmospheric pollutant that destroys ozone in the stratosphere. Here is the accepted mechanism:

$$
\begin{array}{ll}
O_3 + NO \longrightarrow NO_2 + O_2 & \text{(slow)} \\
NO_2 + O \longrightarrow NO + O_2 & \text{(fast)} \\
\hline
O_3 + O \longrightarrow 2\,O_2 &
\end{array}
$$

(a) Draw an activation energy diagram for this mechanism. Include as many details as possible. (b) Identify any catalysts or intermediates in the mechanism. Explain. (c) The activation energy for decomposition of ozone promoted by NO is 11.9 kJ/mol. The activation energy for decomposition of ozone promoted by Cl is 2.1 kJ/mol. Which pollutant is a more serious threat to the ozone layer, Cl or NO? Explain.

14.58 The reaction of ethylene (C_2H_4) with H_2 to form ethane (C_2H_6) (see Problem 14.49) is exothermic by 137 kJ/mol; has $E_a = 200$ kJ/mol; and is catalyzed by Pd metal, which reduces E_a by 60 kJ/mol. Draw an activation energy diagram for this reaction, and label it clearly. Be sure all the given facts are illustrated in the diagram.

14.59 The hypothetical reaction $X_2 + 2\,Y \rightarrow 2\,XY$ goes by the following mechanism:

$$
\begin{array}{ll}
X_2 \longrightarrow 2\,X & \text{(slow)} \\
X + Y \longrightarrow XY & \text{(fast)}
\end{array}
$$

The pictures that follow represent starting conditions for the reaction. What would the rate of reaction be for Flask *b* compared with that for Flask *a*? Explain.

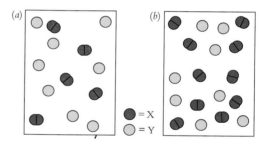

14.60 Here is another set of starting conditions for the reaction described in Problem 14.59. What would the rate of reaction be for Flask *b* compared with that for Flask *a*? Explain.

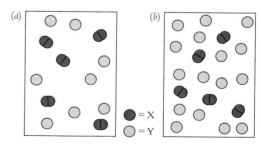

14.61 Cyclopropane converts to propene by first-order kinetics with a rate constant at 500 °C of $5.5 \times 10^{-4}\ s^{-1}$. Calculate how long it takes for 10.0% of a sample to decompose, for 50.0% to decompose, and for 99.9% to decompose.

14.62 The reaction of Cl_2 with H_2S in aqueous solution is first-order in each reactant, with $k = 3.5 \times 10^{-2}$ $M^{-1}s^{-1}$ at 28 °C. A solution has $[Cl_2] = 0.035$ M and $[H_2S] = 5.0 \times 10^{-5}$ M. (a) Find the H_2S concentration after 225 s of reaction. (b) Find the time at which the H_2S concentration has fallen to 1.0×10^{-5} M.

14.63 Phosphine (PH_3) decomposes into phosphorus and molecular hydrogen:

$$4\ PH_3(g) \longrightarrow P_4(g) + 6\ H_2(g)$$

Experiments show that this is a first-order reaction, with a rate constant of $1.73 \times 10^{-2}\ s^{-1}$ at 650 °C. Which of the following statements are true? (a) The overall reaction is elementary. (b) The data allow E_a to be evaluated. (c) The rate constant will be smaller than $1.73 \times 10^{-2}\ s^{-1}$ at 500 °C. (d) The reaction $PH_3 \rightarrow PH_2 + H$ might be the first, rate-determining step in the mechanism of this reaction.

14.64 The Haber reaction for the manufacture of ammonia is:

$$N_2 + 3\ H_2 \rightarrow 2\ NH_3$$

Without doing any experiments, which of the following can you say *must* be true? (a) Reaction rate $= -\Delta[N_2]/\Delta t$. (b) The reaction is first-order in N_2. (c) The rate of disappearance of H_2 is three times the rate of disappearance of N_2. (d) The rate of disappearance of N_2 is three times the rate of disappearance of H_2. (e) $\Delta[H_2]/\Delta t$ has a positive value. (f) The reaction is not an elementary reaction. (g) The activation energy is positive.

14.65 For each of the following reactions, what is the order with respect to each starting material and what is the overall order of the reaction:

(a) $2\ N_2O_5 \rightarrow 4\ NO_2 + O_2$ Rate $= k[N_2O_5]$

(b) $2\ NO + 2\ H_2 \rightarrow N_2 + 2\ H_2O$ Rate $= k[NO]^2[H_2]$

(c) Glucose + ATP $\xrightarrow{\text{Enzyme}}$ Rate $= k[enzyme]$
       glucose, phosphate + ADP

14.66 For each of the following reactions, what is the order with respect to each starting material, and what is the overall order of the reaction?

(a) $N_2 + 3\ H_2 \xrightarrow{\text{Catalyst}} 2\ NH_3$ Rate $= k$

(b) Sucrose + $H_2O \rightarrow$ glucose + fructose
              Rate $= k[\text{sucrose}][H_2O][H_3O^+]$

(c) $CHCl_3 + Cl_2 \rightarrow CCl_4 + HCl$ Rate $= k[CHCl_3][Cl]^{1/2}$

14.67 One possible mechanism for the reaction between H_2 and NO follows:

$$2\ NO \rightleftharpoons N_2O_2 \qquad \text{(fast)}$$

$$N_2O_2 + H_2 \longrightarrow N_2O + H_2O \qquad \text{(slow)}$$

$$N_2O + H_2 \longrightarrow N_2 + H_2O \qquad \text{(fast)}$$

(a) What is the overall stoichiometry of this reaction? (b) What rate law is predicted by this mechanism?

14.68 A chemical reaction is thought to proceed by the following mechanism:

$$A + B \underset{k_{-1}}{\overset{k_1}{\rightleftharpoons}} C$$

$$C + D \xrightarrow{k_2} A + E$$

$$E + B \xrightarrow{k_3} 2\ F$$

(a) What is the net stoichiometry of the reaction? (b) If the second step is rate-determining, what is the rate law? (c) Identify any catalysts and intermediates.

14.69 Write chemical equations that show how CF_2Cl_2 contributes to the destruction of ozone in the stratosphere.

14.70 Write the chemical equations for the reactions that maintain the balance among O_3, O_2, and O in the unpolluted stratosphere.

14.71 What is the difference between speed and spontaneity of a chemical reaction?

14.72 Can the assumption that the forward and reverse rates are equal ever be exact for a reacting system? Explain your answer.

14.73 Use molecular arguments to explain why each of the following factors speeds up a chemical reaction: (a) a catalyst; (b) an increase in temperature; and (c) an increase in concentration.

14.74 Explain in molecular terms why the Haber synthesis cannot proceed in a single-step elementary reaction.

14.75 Molecule A decomposes to give B and C. The following figures represent two experiments conducted to study this decomposition reaction. Flask 2 is found to react four times faster than Flask 1. What is the rate law for the decomposition of A? Explain your reasoning in terms of molecular collisions.

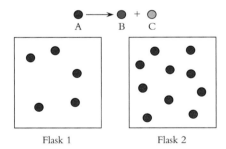

Flask 1   Flask 2

14.76 Phosphorus-32 is a radioactive isotope that decomposes in a unimolecular first-order process. The following figures represent portions of two flasks, each containing six atoms of ^{32}P. Will the time required for the six atoms in Flask 2 to decompose to three atoms be faster, slower, or the same as that in Flask 1? Explain.

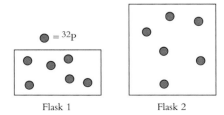

Flask 1   Flask 2

More Challenging Problems

14.77 The enzyme urease catalyzes the hydrolysis of urea to ammonia and carbon dioxide:

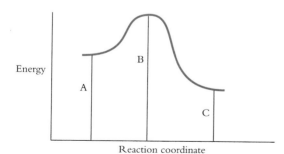

The uncatalyzed reaction has an activation energy of 125 kJ/mol. The enzyme catalyzes a mechanism that has an activation energy of 46 kJ/mol. By what factor does urease increase the rate of urea hydrolysis at 21°C? (Hint: Use the Arrhenius equation.)

14.78 An atmospheric scientist interested in how NO is converted to NO_2 in urban atmospheres carries out two experiments to measure the rate of this reaction. The data are tabulated below. Find the rate law and rate constant for this reaction.

A: $[NO]_0 = 9.63 \times 10^{-3}$ M, $[O_2]_0 = 4.1 \times 10^{-4}$ M:

t (s)	0	3.0	6.0	9.0	12.0
$[O_2]$ (10^{-4} M)	4.1	2.05	1.02	0.51	0.25

B: $[NO]_0 = 4.1 \times 10^{-4}$ M, $[O_2]_0 = 9.75 \times 10^{-3}$ M:

t (10^2 s)	0	1.00	2.00	3.00	4.00
$[NO]$ (10^{-4} M)	4.1	2.05	1.43	1.02	0.82

14.79 Photographers use a rule of thumb, that development time is cut in half for a 10 °C temperature rise, to determine how to modify film development time as the temperature varies. (a) Calculate the activation energy for the chemistry of film developing, assuming that "normal" temperature is 20 °C. (b) If a certain film takes 10 minutes to develop at 20 °C, how long will it take at 25 °C? (Hint: The answer is not 7.5 minutes.)

14.80 If chlorofluorocarbons catalyze O_3 decomposition, which is exothermic by 392 kJ/mol, they must also catalyze O_3 production from O_2. Using the energies shown in Figure 14-21, explain why catalysis results in net O_3 destruction even though reactions in both directions are accelerated.

14.81 For the activation energy diagram shown, which of the following statements are true:

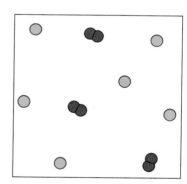

Energy

Reaction coordinate

(a) $\Delta E_{reaction} = A - C$
(b) $\Delta E_{reaction} = B - C$
(c) $E_a(\text{forward}) = E_a(\text{reverse})$
(d) A represents the energy of the starting materials.
(e) $E_a(\text{forward}) = B - C$
(f) E_a (forward) $< E_a$ (reverse)

14.82 Here are some data for the decomposition of molecule A. Determine the rate law and rate constant:

t (s)	0	10.0	20.0	30.0
$[A]$ (M)	0.64	0.52	0.40	0.28

14.83 At least three possible reaction mechanisms exist for the reaction of hydrogen with halogens, $H_2 + X_2 \rightarrow 2$ HX. Determine the rate law predicted by each of them:

(a) $H_2 + X_2 \longrightarrow 2$ HX   (bimolecular, direct)
(b) $X_2 \longrightarrow X + X$ (slow)
   $X + H_2 \longrightarrow HX + H$ (fast)
   $H + X \longrightarrow HX$ (fast)
(c) $X_2 \rightleftharpoons X + X$ (fast, reversible)
   $X + H_2 \longrightarrow HX + H$ (slow)
   $H + X \longrightarrow HX$ (fast)

14.84 Consider the following hypothetical reaction, which is 3/2-order overall and half-order in  :

$$2 \bigcirc + \bullet\bullet \longrightarrow 2 \bullet\bigcirc$$

The following figure represents one set of initial conditions for the reaction. Draw a similar figure that represents a set of initial conditions for which the rate would be twice as fast as the first case:

14.85 The following mechanism has been proposed for the gas-phase reaction of $CHCl_3$ and Cl_2:

$$Cl_2 \underset{k_{-1}}{\overset{k_1}{\rightleftharpoons}} 2\ Cl \qquad \text{(fast, reversible)}$$

$$Cl + CHCl_3 \xrightarrow{k_2} HCl + CCl_3 \qquad \text{(slow)}$$

$$CCl_3 + Cl \xrightarrow{k_3} CCl_4 \qquad \text{(fast)}$$

(a) Write the net balanced equation. (b) What are the intermediates in the reaction? (c) What rate law is predicted by this mechanism?

14.86 Ammonium cyanate (NH_4NCO) rearranges in water to urea (NH_2CONH_2). The following data are obtained at 50°C:

Time (hr)	0	1.0	2.0	3.0	5.0	7.0	9.0
Conc. (M)	0.500	0.375	0.300	0.250	0.188	0.150	0.125

At 25 °C, the concentration falls from 0.500 M to 0.300 M in 6.0 hours. (a) Determine the rate law. (b) Determine the rate constant at 50 °C. (c) Determine the activation energy.

14.87 How can you identify intermediates and catalysts in a proposed mechanism for a reaction?

14.88 For the reaction $NO_2 + CO \rightarrow NO + CO_2$, the experimental rate law follows:

$$Rate = k[CO][NO_2]$$

Which of the following sets of conditions will give the fastest rate? Explain your choice. (a) 0.5 mol of NO_2 and 0.5 mol of CO in a 2.0-L vessel; (b) 0.5 mol of NO_2 and 0.5 mol of CO in a 1.0-L vessel; and (c) 2.0 mol of NO_2 and 0.1 mol of CO in a 1.0-L vessel.

14.89 Acetaldehyde decomposes to methane and carbon monoxide:

Kinetic data for the decomposition of acetaldehyde follow:

Time (10^3 s)	0	1.000	2.000	3.000	4.000
[CH_3CHO] (mol L^{-1})	0.250	0.118	0.0770	0.0572	0.0455

(a) Determine the rate law for the decomposition of acetaldehyde. (b) What is the value of the rate constant? (c) How long does it take for 75% of the acetaldehyde to decompose?

14.90 A chemist interested in a reaction having overall stoichiometry

$$3A + 2B \longrightarrow Products$$

carries out two sets of experiments, both at 25 °C. In an experiment in which [B] is 1.00 M and [A] is 0.050 M at the start of the reaction, the chemist obtains these data:

Time (10^2 s)	0	1.0	2.0	3.0	4.0	5.0
[A] (M)	0.050	0.040	0.032	0.025	0.020	0.016

Then the chemist performs another experiment with [B] = 1.50 M and [A] = 0.050 M. Again, it takes 3.0×10^2 seconds for the concentration of A to fall to 0.025 M. (a) What is the rate law of the reaction? Show your reasoning. (b) Calculate the rate constant.

Group Study Problems

14.91 According to the induced-fit model of enzyme activity, binding a reactant to the enzyme causes a distortion in the conformation of the reactant itself. This distortion decreases the activation energy of the catalyzed reaction. For the hypothetical unimolecular reaction A → B + C, draw a reaction coordinate diagram for the uncatalyzed reaction and for the same reaction catalyzed by an enzyme. Use the enzyme model to explain the differences in the two diagrams.

14.92 In the preparation of cobalt complexes, one reaction involves displacement of H_2O by Cl^-:

$$[Co(NH_3)_5H_2O]^{3+} + Cl^- \longrightarrow [Co(NH_3)_5Cl]^{2+} + H_2O$$

A proposed mechanism for this displacement starts with rapid reversible dissociation of water from the complex:

$$[Co(NH_3)_5H_2O]^{3+} \rightleftharpoons [Co(NH_3)_5]^{3+} + H_2O$$

(a) Propose a slow second step that completes the mechanism and gives the correct overall stoichiometry. (b) Derive the rate law that this mechanism predicts. (c) When the rate is studied in 1 M aqueous HCl solution that is 1 mM in $[Co(NH_3)_5H_2O]^{3+}$, first-order experimental kinetics are observed. Is this observation consistent with the proposed mechanism? State your reasoning clearly and in detail.

14.93 In your own words, describe the induced-fit model of enzyme specificity. Illustrate with diagrams, using a hypothetical enzyme that catalyzes the decomposition of a square but cannot catalyze the decomposition of a triangle:

14.94 Red and white Ping-Pong balls with small Velcro patches are placed in an air-blowing machine like the ones used to scramble numbered balls in lottery drawings on TV. If red and white balls collide at the Velcro points, they stick together. When 20 red and 10 white balls are put in the machine, 4 pairs form after 2 minutes of blowing. (a) If 10 balls of each color had been placed in the machine, how many pairs would form under the same conditions? (b) What if there were 20 red and 15 white balls? (c) What if there were 10 red and 20 white balls? (d) What if the machine had been crammed with 40 red and 20 white balls? (e) Explain why the rates change with the numbers of balls.

14.95 Bromide ions catalyze the decomposition of aqueous H_2O_2:

$$2 H_2O_2(aq) \longrightarrow 2 H_2O(l) + O_2(g)$$

by abstracting an oxygen atom to form BrO^-:

$$H_2O_2(aq) + Br^-(aq) \longrightarrow H_2O(l) + BrO^-(aq)$$

The BrO^- ion then quickly reacts with another H_2O_2 molecule:

$$BrO^-(aq) + H_2O_2(aq) \longrightarrow H_2O(l) + A + B$$

(a) Identify A and B. (b) Construct an activation energy diagram for this reaction. Consult thermodynamic tables to get an estimate of ΔE. Clearly show the effect of bromide ion catalysis and locate on your diagram the energy of the BrO^- intermediate.

14.96 The reaction between ozone and nitrogen dioxide follows:

$$2 NO_2(g) + O_3(g) \longrightarrow N_2O_5(g) + O_2(g) \qquad \Delta H° = -200 \text{ kJ}$$

The reaction proceeds according to the experimental rate law, rate = $k[NO_2][O_3]$, and NO_3 has been identified as an intermediate in the reaction. The activation energy is 50 kJ/mol. (a) Devise a two-step mechanism for the reaction that is consistent with the experimental observations. Identify the rate-determining step. (b) Draw a molecular picture that illustrates your rate-determining step. (c) Sketch the activation energy diagram for this reaction and label it as completely as you can.

Answers to Section Exercises

14.1.1 Second product is CO: $NO + CO_2 \rightarrow NO_2 + CO$.

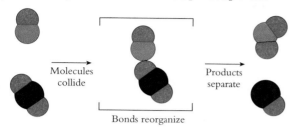

Molecules collide — Bonds reorganize — Products separate

14.1.2 Chemical equation is $2\ NO + 2\ H_2 \rightarrow 2\ H_2O + N_2$, and the intermediates are N_2O_2 and N_2O.

14.1.3 Your description should include highway speeds and the rate at which toll-takers take tolls.

14.2.1

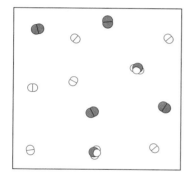

14.2.2 Hydrogen

14.2.3

$$\text{Relative rate} = -\frac{\Delta[N_2]}{\Delta t} = -\frac{1}{3}\frac{\Delta[H_2]}{\Delta t} = -\frac{1}{2}\frac{\Delta[NH_3]}{\Delta t}$$

14.3.1

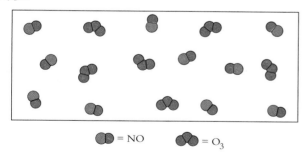

$\bigcirc\!\bigcirc$ = NO $\qquad$ $\bigcirc\!\!\bigcirc\!\!\bigcirc$ = O_3

14.3.2 The rate law is rate = k[butene], and the rate constant has units of time^{-1}.

14.3.3 The rate law is rate = $k[H_2][Br_2]^{1/2}$, the overall order is $3/2$, and the rate constant has units of $M^{1/2}$ time^{-1}.

14.4.1 Do two experiments; in each $[H_2O]_0 > 100\ [N_2O_5]_0$ but with two different values for $[H_2O]_0$. Plot

$$\ln\left(\frac{[N_2O_5]_0}{[N_2O_5]}\right) \text{ and } \frac{1}{[N_2O_5]} - \frac{1}{[N_2O_5]_0} \text{ vs. } t \text{ to determine order}$$

with respect to N_2O_5, and use the ratio of slope values and H_2O concentrations to determine order with respect to H_2O.

14.4.2 A plot of $\ln\left(\frac{[N_2O_5]_0}{[N_2O_5]}\right)$ vs. t is linear, so the reaction is first-order. $k = 3.05 \times 10^{-2}$ min^{-1}.

14.4.3 The rate law is rate = $k[NO_2]^2$. The fact that the rate does not change when the initial concentration of CO changes shows that the rate law does not depend on [CO].

14.5.1 (a) $N_2O_4 \rightarrow 2\ NO_2$ (occurs twice); and (b) the first step must be rate-determining.

14.5.2 If the second step is rate-determining, the rate law is second-order in N_2O_5.

14.5.3 (a) $N_2O_2 + O_2 \rightarrow 2\ NO_2$; and (b) rate = $k_2[N_2O_2][O_2]$, but using equality of rates, $[N_2O_2] = k_1[NO]^2/k_{-1}$, resulting in rate = $k[NO]^2[O_2]$.

14.6.1 The N_2 triple bond is extremely strong. The H_2 bond, 435 kJ/mol, is a strong single bond. From these bond strengths, we predict a high activation energy for the reaction.

14.6.2 (a) Graph $\ln k$ vs. $1/T$:

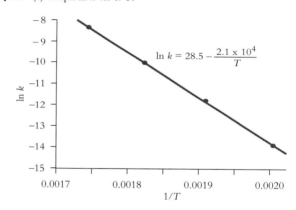

$$\ln k = 28.5 - \frac{2.1 \times 10^4}{T}$$

$E_a = 1.8 \times 10^2$ kJ/mol; and (b) 8.3×10^{-10} s^{-1}.

14.6.3

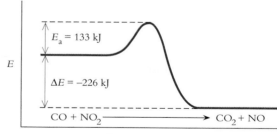

$E_a = 133$ kJ

$\Delta E = -226$ kJ

$CO + NO_2 \longrightarrow CO_2 + NO$

Reaction coordinate

14.7.1 They are less expensive because only a thin layer of the expensive metal, rather than solid pieces, is required. They are more effective because a layer on porous charcoal has a larger surface area (where the catalyzed reaction occurs) than the relatively smooth surface of a pure piece of metal.

14.7.2 (a) True; (b) a catalyst does not change the overall energy change of a reaction; (c) a catalyst lowers the activation energy of an overall reaction by changing its mechanism; and (d) true.

14.7.3 The reaction converts pollutants (CO and NO_2) into benign constituents of the atmosphere (CO_2 and N_2), so catalyzing this reaction could reduce atmospheric pollution levels.

15

Principles of Chemical Equilibrium

INTRODUCTION: AGRICULTURE AND NITROGEN FIXATION

Growing plants require a constant supply of carbon, hydrogen, oxygen, and nitrogen to synthesize living tissue. Water, atmospheric carbon dioxide, and molecular oxygen supply carbon, hydrogen, and oxygen, but nitrogen is not readily available. Even though molecular nitrogen is the main component of the atmosphere, the nitrogen atoms in N_2 are held together by an exceptionally strong triple bond. As noted in our Chemistry and Life Box in Chapter 14, very few organisms can make direct use of N_2. Instead, plants extract nitrogen from nitrate and ammonium ions in the soil. The availability of usable nitrogen, known as "fixed" nitrogen, often limits the growth rate of vegetation.

In a naturally regulated ecosystem, most of the supply of fixed nitrogen comes from the recycling of decaying animal and plant matter. In addition, lightning strikes convert atmospheric nitrogen into compounds containing fixed nitrogen, and a few types of bacteria are capable of using N_2 directly. These natural sources fall well short of providing enough nitrogen for intensive agriculture. Farmers must supply nitrogen-containing fertilizers, as our opening photo illustrates. Before World War I, the main source of fertilizer was natural deposits of sodium nitrate, but these supplies were quite limited. As early as 1898, the English chemist Sir William Ramsay predicted that the limited supply of fixed nitrogen would lead to world famine by the middle of the twentieth century.

In the absence of a new source of fixed nitrogen, Ramsay's prediction might have come true, but during the years leading up to World War I, German chemists developed the Haber process for producing ammonia from molecular nitrogen and hydrogen:

$$N_2 + 3\,H_2 \longrightarrow 2\,NH_3$$

Even today, more than eighty years after its development, the Haber process thoroughly dominates the modern fertilizer industry, because it provides a plentiful and relatively inexpensive industrial source of fixed nitrogen.

As an indispensable source of fertilizer, the Haber process is one of the most important reactions in industrial chemistry. Nevertheless, even under optimal conditions the yield of the ammonia synthesis in industrial reactors is only about 13%. This is because the Haber process does not go to completion; the net rate of producing ammonia reaches zero when substantial amounts of N_2 and H_2 are still present.

The Haber process tells us that chemical reactions can stop long before they reach completion. In fact, even a reaction that goes nearly to completion may reach a balance point before the limiting reactant is consumed. At balance, the concentrations no longer change even though some of each starting material is still present. This balance point represents dynamic chemical equilibrium.

In this chapter, we present basic features of chemical equilibrium. We explain why reactions such as the Haber process cannot go to completion. We also show why using catalysts and elevated temperatures can accelerate the rate of this reaction but cannot shift its equilibrium position in favor of ammonia and why elevated temperature shifts the equilibrium in the wrong direction. In Chapters 16 and 17, we turn our attention specifically to applications of equilibria, including acid-base chemistry.

Lightning strikes can convert N_2 into fixed nitrogen.

Ammonia for fertilizer is produced in reaction towers.

15.1 DESCRIBING CHEMICAL EQUILIBRIA

The detailed chemistry of the ammonia synthesis is complex, so we introduce the principles of equilibrium using the chemistry of nitrogen dioxide. As described in Chapter 14, molecules in a sample of nitrogen dioxide are always colliding with

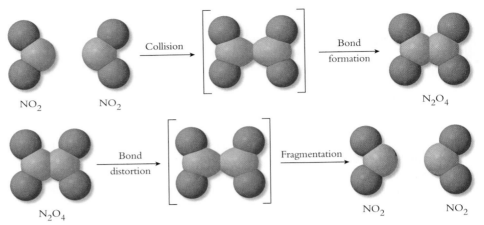

Figure 15-1
When two NO_2 molecules collide in the proper orientation, an N—N bond forms to produce a molecule of N_2O_4. When a molecule of N_2O_4 acquires sufficient energy through molecular collisions, its N—N bond distorts and eventually fragments to produce two molecules of NO_2.

one another. A collision in the correct orientation can result in formation of an N_2O_4 molecule:

$$2\ NO_2 \longrightarrow N_2O_4$$

In a vessel that contains only NO_2 molecules, the formation of N_2O_4 is the only reaction that takes place. However, once N_2O_4 molecules are present, the reverse reaction begins to occur. An N_2O_4 molecule can fragment after collisions give it sufficient energy to break the N—N bond. These fragmentations regenerate NO_2:

$$N_2O_4 \longrightarrow 2\ NO_2$$

Figure 15-1 depicts these two processes from the molecular perspective.

Dynamic Equilibrium

Collisions between NO_2 molecules produce N_2O_4 and consume NO_2. At the same time, fragmentation of N_2O_4 produces NO_2 and consumes N_2O_4. When the concentration of N_2O_4 is very low, the first reaction occurs more often than the second. As the N_2O_4 concentration increases, however, the rate of fragmentation increases. Eventually, the rate of N_2O_4 production equals the rate of its decomposition. Even though individual molecules continue to combine and decompose, the rate of one reaction is exactly balanced by the rate of the other. This is a **dynamic equilibrium.** At dynamic equilibrium, the rates of the forward and reverse reactions are equal. The system is *dynamic* because individual molecules react continuously. It is *at equilibrium* because there is no *net* change in the system.

Experimental Confirmation

The reactions that occur in a mixture of NO_2 and N_2O_4 lead to changes in concentrations. The concentration of NO_2 can be determined experimentally, because NO_2 is orange and N_2O_4 is colorless. The color intensity of the gas mixture is proportional to the concentration of NO_2. Figure 15-2 summarizes the results of quantitative experiments on the NO_2/N_2O_4 system. The data show that the concentrations of both gases level off to constant values as the reaction reaches equilibrium. If the reaction to form N_2O_4 went to completion, the concentration of NO_2 would drop to zero. If the reaction to decompose N_2O_4 went to completion, the concentration of N_2O_4 would drop to zero. Instead, these reactions reach equilibrium when substantial amounts of both gases are present.

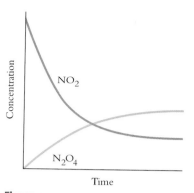

Figure 15-2
Changes in the concentrations of NO_2 and N_2O_4 with time. The system reaches equilibrium before either substance is fully consumed.

The Equilibrium Constant

When a mixture contains both NO_2 and N_2O_4, both the formation reaction and the decomposition reaction occur:

Formation: $$NO_2 + NO_2 \xrightarrow{k_f} N_2O_4$$

Decomposition: $$N_2O_4 \xrightarrow{k_d} NO_2 + NO_2$$

Each of these reactions is the reverse of the other, so they can be written as a single equation with double arrows to show that the reaction proceeds in both directions:

$$NO_2 + NO_2 \underset{k_d}{\overset{k_f}{\rightleftharpoons}} N_2O_4$$

> Recall from Chapter 14 that k designates a rate constant. The subscripts on these rate constants refer to formation and decomposition.

At equilibrium the reactions continue, even though the concentrations of NO_2 and N_2O_4 no longer change. That is, the rate of formation of N_2O_4 is exactly counterbalanced by the rate of its decomposition, so no *net* change occurs. The rates of these elementary reactions can be expressed as shown in Chapter 14:

$$\text{Rate of formation of } N_2O_4 = k_f[NO_2]^2$$

$$\text{Rate of decomposition of } N_2O_4 = k_d[N_2O_4]$$

At equilibrium, the rate at which N_2O_4 decomposes equals the rate at which N_2O_4 forms:

$$k_f[NO_2]_{eq}^2 = k_d[N_2O_4]_{eq}$$

> We designate a concentration at equilibrium with a subscript eq.

This equality can be rearranged to group the rate constants on one side and the concentrations on the other, with product concentrations in the numerator:

$$\frac{k_f}{k_d} = \frac{[N_2O_4]_{eq}}{[NO_2]_{eq}^2}$$

The ratio of rate constants is called the **equilibrium constant (K_{eq}):**

$$K_{eq} = \frac{[N_2O_4]_{eq}}{[NO_2]_{eq}^2}$$

> Notice that this analysis of the NO_2/N_2O_4 equilibrium follows the same logic as the assumption for equality of rates introduced in Chapter 14.

The equilibrium constant expresses the relationship between the concentration of products and reactants at equilibrium.

Reactions in aqueous solution behave in a similar way to those in the gas phase. As the concentration of products increases, the products react to regenerate reactants, and eventually the reaction reaches equilibrium. Example 15-1 treats an equilibrium in aqueous solution.

| Example 15-1 | Aqueous Equilibrium |

Molecular iodine dissolves in a solution containing iodide anions to form triiodide:

$$I_2(aq) + I^-(aq) \longrightarrow I_3^-(aq)$$

Describe the additional reaction that occurs as this system approaches equilibrium, draw molecular pictures illustrating the reactions that occur at equilibrium, and write the equilibrium constant expression for this reaction.

Strategy: The problem asks for a qualitative analysis of a chemical equilibrium. We must visualize what takes place at the molecular level, describe the system in words, draw pictures that summarize the reactions, and then use the ideas developed for the NO_2/N_2O_4 reaction to write an expression for the equilibrium constant.

Solution: Initially, the forward reaction to form triiodide anions is the only process that occurs. As the concentration of the product increases, however, the reverse reaction also takes place:

$$I_3^-(aq) \longrightarrow I_2(aq) + I^-(aq)$$

We need two molecular pictures to illustrate the chemistry. One shows an I_2 molecule colliding with an I^- anion to form I_3^-. A second picture illustrates the reverse process:

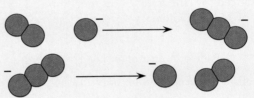

The equilibrium constant expression is determined from the rates for the forward and reverse reactions:

$$\text{Rate of formation of } I_3^- = k_f[I_2]\,[I^-]$$

$$\text{Rate of decomposition of } I_3^- = k_d[I_3^-]$$

Setting these rates equal to each other gives the equilibrium condition, which can be solved for the equilibrium constant expression:

$$I_2(aq) + I^-(aq) \rightleftharpoons I_3^-(aq)$$

$$k_f[I_2]_{eq}\,[I^-]_{eq} = k_d[I_3^-]_{eq} \qquad so \qquad K_{eq} = \frac{[I_3^-]_{eq}}{[I_2]_{eq}\,[I^-]_{eq}}$$

Reversibility

In the NO_2/N_2O_4 system, molecules of NO_2 combine to give N_2O_4 molecules, and N_2O_4 molecules decompose to give NO_2 molecules. This is an example of the **reversibility** of molecular reactions. Look again at Figure 15-1: If two NO_2 molecules can form a bond when they collide, then that bond also can break apart when

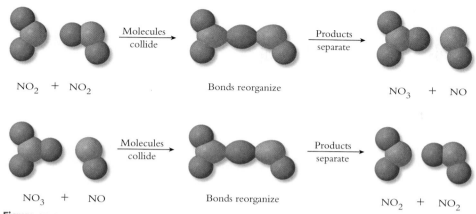

Figure 15-3
If two molecules of NO_2 can exchange an O atom when they collide, forming NO_3 and NO, then a collision of NO_3 and NO can also result in exchange of an O atom, forming two molecules of NO_2.

an N_2O_4 molecule distorts. The concept of reversibility is a general principle that applies to all molecular processes. Every *elementary* reaction that goes in the forward direction also can go in the reverse direction. As a consequence of reversibility, we can write each step in a chemical mechanism using a double arrow to describe what happens at chemical equilibrium.

To illustrate the generality of reversibility and the equilibrium expression, we extend our kinetic analysis to a chemical reaction that has a two-step mechanism. At elevated temperature NO_2 decomposes into NO and O_2 instead of forming N_2O_4. The mechanism for the decomposition reaction, which appears in Chapter 14, is illustrated in Figure 15-3:

$$NO_2 + NO_2 \underset{k_{-1}}{\overset{k_1}{\rightleftharpoons}} NO + NO_3$$

$$NO_3 \overset{k_2}{\longrightarrow} NO + O_2$$

Just as the decomposition of N_2O_4 is reversible, the decomposition of NO_3 is also reversible. That is, if an O_2 molecule and an NO molecule collide in the proper orientation, they can "stick together" to form NO_3:

$$NO + O_2 \overset{k_{-2}}{\longrightarrow} NO_3$$

This reaction plays a negligible role early in the decomposition of NO_2, when the concentrations of O_2 and NO are too small for collisions between these products to occur frequently. As the products accumulate, however, such collisions become more and more likely. Eventually, product concentrations become large enough that the rate of the reverse reaction matches the rate of the forward reaction. At this time the system has reached equilibrium, and each step in the mechanism has a forward rate that equals its reverse rate:

$$NO_2 + NO_2 \underset{k_{-1}}{\overset{k_1}{\rightleftharpoons}} NO + NO_3$$

$$NO_3 \underset{k_{-2}}{\overset{k_2}{\rightleftharpoons}} NO + O_2$$

$$k_1[NO_2]_{eq}^2 = k_{-1}[NO]_{eq}[NO_3]_{eq} \qquad and \qquad k_2[NO_3]_{eq} = k_{-2}[NO]_{eq}[O_2]_{eq}$$

Rearranging each of these rate equalities gives equilibrium expressions for both steps of the mechanism:

$$\frac{k_1}{k_{-1}} = \frac{[NO]_{eq}\,[NO_3]_{eq}}{[NO_2]_{eq}^2} \quad \textit{and} \quad \frac{k_2}{k_{-2}} = \frac{[NO]_{eq}\,[O_2]_{eq}}{[NO_3]_{eq}}$$

If we multiply one of these equations by the other, the concentration of the intermediate, $[NO_3]_{eq}$, cancels to produce an equilibrium expression entirely in terms of concentrations of reactants and products:

$$\frac{k_1}{k_{-1}}\frac{k_2}{k_{-2}} = \frac{[NO]_{eq}\,\cancel{[NO_3]_{eq}}}{[NO_2]_{eq}^2} \frac{[NO]_{eq}\,[O_2]_{eq}}{\cancel{[NO_3]_{eq}}}$$

The two concentration terms containing NO can be combined, and the ratio of rate constants can be replaced by a single constant, the equilibrium constant for the reaction:

$$K_{eq} = \frac{[NO]_{eq}^2\,[O_2]_{eq}}{[NO_2]_{eq}^2}$$

Each equilibrium expression described so far contains a ratio of concentrations of products and reactants. Moreover, each concentration is raised to a power equal to its stoichiometric coefficient in the balanced equation for the overall reaction. Concentration ratios always have products in the numerator and reactants in the denominator:

$$2\,NO_2 \rightleftharpoons 2\,NO + O_2 \qquad\qquad 2\,NO_2 \rightleftharpoons N_2O_4$$

$$K_{eq} = \frac{[NO]_{eq}^2\,[O_2]_{eq}}{[NO_2]_{eq}^2} \qquad\qquad K_{eq} = \frac{[N_2O_4]_{eq}}{[NO_2]_{eq}^2}$$

It is possible to carry out this type of kinetic analysis whether a mechanism is simple or elaborate. That is, an equilibrium expression always can be derived by applying reversibility and setting forward and reverse rates equal to one another at equilibrium. It is unnecessary to go through this procedure for every chemical equilibrium. As our two examples suggest, inspection of the overall stoichiometry *always* gives the correct expression for the equilibrium constant. That is, a reaction of the form:

$$aA + bB \rightleftharpoons dD + eE$$

has an equilibrium constant expression given by Equation 15–1:

$$K_{eq} = \frac{[D]_{eq}^d\,[E]_{eq}^e}{[A]_{eq}^a\,[B]_{eq}^b} \qquad\qquad (15\text{–}1)$$

Equation 15–1 expresses the fact that the equilibrium expression for any reaction can be written from its overall stoichiometry. Equation 15–1 is applied to the Haber process in Example 15–2.

Example 15-2	Equilibrium Constant Expression

What is the equilibrium expression for the Haber synthesis of ammonia?

$$N_2(g) + 3\,H_2(g) \rightleftharpoons 2\,NH_3(g)$$

Strategy: If the mechanism for this equilibrium were given, we could derive an equilibrium expression using the principle of reversibility, but Equation 15–1

| Equilibrium Constant Expression *(continued)* | Example 15-2 |

shows that we can write the equilibrium expression directly without using mechanistic information.

Solution: The product concentration, raised to a power equal to its stoichiometric coefficient, appears in the numerator. The concentrations of the two starting materials, each raised to a power equal to its stoichiometric coefficient, appear in the denominator:

$$K_{eq} = \frac{[NH_3]_{eq}^2}{[N_2]_{eq}[H_2]_{eq}^3}$$

Equilibrium constant expressions appear similar to the rate laws described in Chapter 14. For example, compare the equilibrium constant expression and the rate law for the NO_2/NO reaction:

$$2\,NO_2 \longrightarrow 2\,NO + O_2 \qquad K_{eq} = \frac{[NO]_{eq}^2[O_2]_{eq}}{[NO_2]_{eq}^2} \qquad Rate = k[NO_2]^2$$

Both relationships include a constant and both involve concentrations raised to exponential powers. However, a rate law and an equilibrium expression describe fundamentally different aspects of a chemical reaction. A *rate law* describes how the *rate of a reaction changes* with concentration. As we describe in this chapter, an *equilibrium expression* describes the concentrations of reactants and products when *the net rate of the reaction is zero.*

Section Exercises

15.1.1 One possible mechanism for the O_2/O_3 equilibrium in the stratosphere is relatively simple:

$$O_3 \underset{k_{-1}}{\overset{k_1}{\rightleftharpoons}} O_2 + O$$

$$O + O_3 \underset{k_{-2}}{\overset{k_2}{\rightleftharpoons}} 2\,O_2$$

Use this mechanism to derive an expression for the equilibrium constant.

15.1.2 Draw molecular pictures like the ones shown in Figure 15-1 to illustrate all the reactions that occur when the O_3/O_2 system is at equilibrium.

15.1.3 Write the equilibrium expression for the combustion of NH_3 during the synthesis of nitric acid:

$$4\,NH_3(g) + 5\,O_2(g) \rightleftharpoons 4\,NO(g) + 6\,H_2O(g)$$

15.2 PROPERTIES OF EQUILIBRIUM CONSTANTS

The equilibrium constant expression given by Equation 15-1 is completely general and can be applied to any chemical reaction. Three features of equilibrium constants are especially important:

/// K_{eq} is related to the stoichiometry of the balanced net reaction.

The numerator contains only concentrations of products, and the denominator contains only concentrations of reactants. Each concentration is raised to a power equal to its stoichiometric coefficient.

/// K_{eq} applies only at equilibrium.

We use subscripts (eq) to emphasize that the concentrations of reactants and products used in the ratio must be concentrations at *equilibrium*.

/// K_{eq} is independent of initial conditions.

The equilibrium constant is a constant for any particular reaction at a given temperature. Whether initial conditions include pure reactants, pure products, or any composition in between, the system reaches a state of *equilibrium* that is determined by the value of K_{eq}.

Concentration Units and K_{eq}

In Chapter 13, we describe how to deal with concentration units when working with a reaction quotient, Q. In Section 15.3 we explore in detail the link between Q and K_{eq}. Here we reintroduce Q in order to address the issue of concentration units and the equilibrium constant.

In Chapter 13 concentrations are represented as c_X. In Chapter 14 and so far in this chapter, concentrations are expressed as [X]. By either notation, concentrations refer to standard thermodynamic units, molarity for species in aqueous solution and atmospheres for gases. Henceforth, to remind you of this convention, we represent gas concentrations as p_X.

Recall that every concentration is measured *relative to a defined standard concentration*. For gases, the defined standard is 1 atm pressure; for solutes in aqueous solution, the defined standard is 1 M. We treat equilibrium constant expressions in this same way, as demonstrated for the N_2O_4/NO_2 equilibrium:

$$K_{eq} = \frac{(p_{N_2O_4})_{eq}}{(p_{NO_2})_{eq}^2}$$

1. Equilibrium constants are dimensionless numbers. At 25 °C, experiments give $K_{eq} = 3.10$ for the N_2O_4/NO_2 equilibrium.

2. An equilibrium constant expression contains concentrations (pressures of gases, molarities of solutes). In the expressions for Q and for K_{eq}, the concentration of each gas in the N_2O_4/NO_2 system is expressed as its partial pressure in atmospheres.

3. Although not stated explicitly, each concentration in a reaction quotient and in an equilibrium constant expression has been divided by standard concentration (1 atm for gases, 1 M for solutes) to make the equilibrium constant dimensionless. For example,

$$(p_{N_2O_4})_{eq} = \frac{(p_{N_2O_4})_{eq}}{1 \text{ atm}}$$

Direction of a Reaction at Equilibrium

For a reaction at equilibrium, the rate of the forward reaction is balanced exactly by the rate of the reverse reaction. For this reason, any equilibrium reaction can be written in either direction. The equilibrium constant for the Haber synthesis of ammonia, for example, can be expressed in two ways:

$$N_2(g) + 3\,H_2(g) \rightleftharpoons 2\,NH_3(g) \qquad K_{eq,\,f} = \frac{(p_{NH_3})^2_{eq}}{(p_{N_2})_{eq}\,(p_{H_2})^3_{eq}}$$

$$2\,NH_3(g) \rightleftharpoons N_2(g) + 3\,H_2(g) \qquad K_{eq,\,r} = \frac{(p_{N_2})_{eq}\,(p_{H_2})^3_{eq}}{(p_{NH_3})^2_{eq}}$$

The two equilibrium constants are reciprocals of each other:

$$K_{eq,\,f} = \frac{1}{K_{eq,\,r}} \qquad\qquad (15\text{-}2)$$

$\uparrow$ $\uparrow$

Equilibrium constant for forward reaction   Equilibrium constant for reverse reaction

The direction chosen for the equilibrium reaction is determined by convenience. A scientist interested in producing ammonia from N_2 and H_2 would use $K_{eq,\,f}$. On the other hand, someone studying the decomposition of ammonia on a metal surface would use $K_{eq,\,r}$. Either choice works as long as the *products* of the net reaction appear in the *numerator* of the equilibrium constant expression and the *reactants* appear in the *denominator*. Example 15-3 applies this reasoning to the iodine-triiodide reaction.

Formation and Decomposition	Example 15-3

Write the equilibrium constant expression for the decomposition equilibrium for the triiodide anion, $I_3^-(aq) \rightleftharpoons I_2(aq) + I^-(aq)$, and relate this expression to the equilibrium constant expression for formation of triiodide anions.

Strategy: Equilibrium constant expressions are always written with the concentrations of products in the numerator and the concentrations of reactants in the denominator, and when a reaction is reversed, its equilibrium constant expression is inverted.

Solution: We can write the equilibrium constant expression by inspection of the decomposition equilibrium:

$$K_{eq}\,(\text{decomposition}) = \frac{[I_2]_{eq}\,[I^-]_{eq}}{[I_3^-]_{eq}}$$

Inverting this expression gives the equilibrium constant expression for the formation reaction (Example 15-1):

$$I_2(aq) + I^-(aq) \rightleftharpoons I_3^-(aq)$$

$$K_{eq}\,(\text{formation}) = \frac{1}{K_{eq}\,(\text{decomposition})} = \frac{[I_3^-]_{eq}}{[I_2]_{eq}\,[I^-]_{eq}}$$

Pure Liquids, Pure Solids, and Solvents

Chemical equilibria often involve pure liquids and solids in addition to gases and solutes. As pointed out in Chapter 13, the concentration of a pure liquid or solid does not vary significantly. That is, the concentrations of pure liquids or solids are always equal to their standard concentrations. Thus division by standard concentration results in a value of 1 for any pure liquid or solid. This allows us to omit pure liquids and solids from equilibrium constant expressions. For a general reaction:

$$aA + bB \rightleftharpoons dD + sS$$

where S is a pure solid or liquid:

$$K_{eq} = \frac{[D]_{eq}^d}{[A]_{eq}^a [B]_{eq}^b}$$

Water often is a reagent in an aqueous equilibrium. For example, when carbon dioxide dissolves in water, it reacts with a water molecule to form carbonic acid:

$$CO_2(g) + H_2O(l) \rightleftharpoons H_2CO_3(aq) \qquad K_{eq} = \frac{[H_2CO_3]_{eq}}{(p_{CO_2})_{eq} [H_2O]_{eq}}$$

Recall that mole fraction is defined in Section 5.5 as the ratio of the number of moles of the substance divided by the total number of moles of all substances present.

Here, carbonic acid concentration is expressed in molarity and carbon dioxide concentration in atmospheres. What units are appropriate for the concentration of water, which is neither a solute nor a pure liquid? The thermodynamic convention is to express the concentration of solvent as its mole fraction, X. The mole fraction of water varies slightly as solutes are added, but in most aqueous solutions the concentration of water is much greater than that of any solute. For instance, when CO_2 is bubbled through water at room temperature, the equilibrium concentration of H_2CO_3 is only 3.4×10^{-2} M. One liter of water contains 1000 g of H_2O, which is 55.5 mol. Compared to 55.5 mol, the 3.4×10^{-2} mol of carbonic acid in this solution is negligible, so the mole fraction of water is negligibly different from 1.00.

Even in relatively concentrated solutions, the mole fraction of water remains close to 1.00. At a solute concentration of 0.50 M, for example, $X_{H_2O} = 0.99$, only 1% different from 1.00. Equilibrium calculations are seldom accurate to better than 5%, so this small deviation from 1.00 can be neglected. Consequently, we treat solvent water just like a pure substance: Its concentration is essentially invariant, so it is omitted from the equilibrium constant expression.

Example 15-4 provides practice in manipulating equilibrium constant expressions.

Example 15-4	Writing an Equilibrium Constant Expression

Write the equilibrium constant expression for the reaction of iron metal with strong aqueous acid, and indicate the concentration units for each reagent:

$$2\,Fe(s) + 6\,H_3O^+(aq) \rightleftharpoons 2\,Fe^{3+}(aq) + 6\,H_2O(l) + 3\,H_2(g)$$

Strategy: The stoichiometry of the reaction determines the form of the equilibrium constant expression. Pure solids, liquids, or solvents do not appear in the expression, since their concentrations are constant.

Writing an Equilibrium Constant Expression *(continued)*

Example 15-4

Solution: In this reaction, Fe is a pure solid and H_2O is the solvent, so they do not appear in the equilibrium expression. The other reagents have exponents equal to their stoichiometric coefficients:

$$K_{eq} = \frac{[Fe^{3+}]_{eq}^2 (p_{H_2})_{eq}^3}{[H_3O^+]_{eq}^6}$$

The concentration of the two ionic solutes are expressed as molarities, and the concentration of H_2 is expressed as a partial pressure in atmospheres.

Magnitudes of Equilibrium Constants

The magnitudes of equilibrium constants vary over a tremendous range and depend on the nature of the reaction as well as on the temperature of the system. Many reactions have very large equilibrium constants. For example, the reaction between H_2 and Br_2 to form HBr has a huge equilibrium constant:

$$H_2(g) + Br_2(g) \rightleftharpoons 2\, HBr(g) \qquad K_{eq} = \frac{(p_{HBr})_{eq}^2}{(p_{H_2})_{eq}\,(p_{Br_2})_{eq}} = 5.4 \times 10^{18}$$

The large value for this equilibrium constant indicates that the reaction goes virtually to completion. If the initial pressures are 1 atm each for H_2 and Br_2, then the pressure of HBr will be 2 atm when the system reaches equilibrium. The partial pressures of H_2 and Br_2 at equilibrium will be about 10^{-9} atm, which is negligible compared with the initial pressures.

Other reactions have extremely small equilibrium constants. For example, elemental fluorine, a diatomic molecule under standard conditions, is nevertheless at equilibrium with fluorine atoms:

$$F_2(g) \rightleftharpoons 2\, F(g) \qquad K_{eq} = \frac{(p_F)_{eq}^2}{(p_{F_2})_{eq}} = 2.1 \times 10^{-22}$$

The tiny value of this equilibrium constant indicates that a sample of fluorine at 25 °C consists almost entirely of F_2 molecules. If the partial pressure of F_2 is 1.0 atm at equilibrium, the partial pressure of fluorine atoms is 2.5×10^{-11} atm, which is negligible compared with 1.0 atm. Nevertheless, the equilibrium constant is not zero, indicating that some fluorine atoms are present in the gas.

Some equilibrium constants are neither large nor small. As already mentioned, the equilibrium constant for the formation of N_2O_4 from NO_2 at 25 °C is 3.10. This moderate value indicates that an equilibrium mixture of NO_2 and N_2O_4 at 25 °C contains measurable amounts of each molecule.

Physical Equilibria

The examples presented so far are *chemical* reactions that reach dynamic equilibrium. *Physical* transformations can also lead to dynamic equilibria. An important example is the phase change of a pure substance between a condensed phase and the gas phase. Figure 15-4 shows that the gas phase above liquid bromine has a red-brown hue. This color signals the presence of Br_2 molecules in the vapor phase. Another example is the vaporization equilibrium of water, which has consequences that appear in everyday life. The fogging of bathroom mirrors, also shown in

Figure 15-4
The color of the gas above liquid bromine and the "fogging" of bathroom mirrors are visual evidence for the liquid-vapor equilibria.

Figure 15-4, occurs when the water vapor in warm air comes into contact with the cooler surface of the mirror, causing vapor to condense. Solids also have vaporization equilibria. The sublimation of solid carbon dioxide (dry ice) is a common example.

The condensed phase is a pure liquid or a pure solid, so its concentration does not appear in equilibrium constant expressions:

$$Br_2(l) \rightleftharpoons Br_2(g) \qquad K_{eq} = (p_{Br_2})_{eq} = \text{Vapor pressure}$$
$$H_2O(l) \rightleftharpoons H_2O(g) \qquad K_{eq} = (p_{H_2O})_{eq} = \text{Vapor pressure}$$
$$CO_2(s) \rightleftharpoons CO_2(g) \qquad K_{eq} = (p_{CO_2})_{eq} = \text{Vapor pressure}$$

For any vaporization equilibrium, K_{eq} at any particular temperature is equal to the equilibrium pressure of the substance in the gas phase. This equilibrium pressure is the **vapor pressure.**

CHAPTER 10 →
Intermolecular forces are treated in Chapter 10.

Another important example of dynamic physical equilibria is the dissolving of gases in liquids. Many gases are only slightly soluble in liquids. This is because substances that are gases under standard conditions have very small intermolecular forces. Molecular oxygen is a typical example. Oxygen molecules are nonpolar, so there are no dipole-dipole or hydrogen bonding interactions between these molecules. Also, the valence orbitals in O_2 molecules have $n = 2$, which means the valence electrons occupy compact orbitals that generate small dispersion interactions. Intermolecular interactions between oxygen molecules and molecules of a solvent such as water are minimal, so oxygen is not very soluble in water. Water in contact with the Earth's atmosphere contains O_2 at a concentration of only about 3×10^{-4} M. Nevertheless, this small concentration is essential for aquatic life. Fish and other aquatic life would die if O_2 gas in the atmosphere were not at equilibrium with O_2 dissolved in water:

$$O_2(g) \rightleftharpoons O_2(aq) \qquad K_{eq} = \frac{[O_2(aq)]_{eq}}{(p_{O_2})_{eq}} = K_H$$

Plants and fish depend on the equilibrium between oxygen in the atmosphere and oxygen dissolved in water. Home aquariums must have air bubblers to replenish the oxygen consumed by fish and plants.

This expression is often solved for concentration, a form that is known as **Henry's law:**

$$[O_2(aq)]_{eq} \text{ (units of mol/L)} = K_H(pO_2)_{eq} \text{ (units of atm)}$$

Thus the equilibrium constant for a gas-solution equilibrium is designated K_H and is called the **Henry's law constant.** This equilibrium constant has a different value for each combination of gas and solvent, and it also varies with temperature. Values for K_H for representative gases dissolving in water appear in Table 15-1.

Table 15-1
Henry's Law Constants* (K_H)

Gas	0.0 °C	25 °C	30 °C
N_2	1.1×10^{-3}	6.7×10^{-4}	4.0×10^{-4}
O_2	2.5×10^{-3}	1.3×10^{-3}	8.9×10^{-4}
CO	1.6×10^{-3}	9.6×10^{-4}	4.4×10^{-4}
Ar	2.5×10^{-3}	1.5×10^{-3}	1.0×10^{-3}
He	4.1×10^{-4}	3.8×10^{-4}	3.8×10^{-4}
CO_2	7.8×10^{-2}	3.4×10^{-2}	1.6×10^{-2}

*Aqueous solutions.

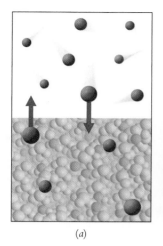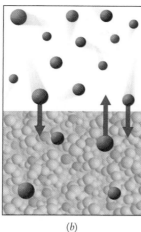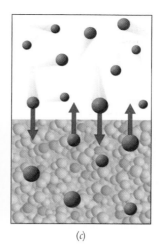

(a)   (b)   (c)

Figure 15-5
Schematic molecular view of a gas-solution equilibrium. (*a*) At equilibrium, the rate of escape of gas molecules from the solution equals the rate of capture of gas molecules by the solution. (*b*) An increase in pressure throws the system out of equilibrium. (*c*) More gas molecules dissolve until the rates of escape and capture once again balance.

The equilibrium expression for K_H indicates that gas solubility increases with the partial pressure of the gas in contact with the solution. The molecular view of a solution in Figure 15-5 shows why this is so. Gas molecules that collide with a liquid surface can be captured into solution; as the partial pressure increases, the number of collisions between gas molecules and the solution surface also increases. This causes more gas molecules to be captured by the solution and increases the concentration of dissolved gas.

Example 15-5 uses Henry's law to determine the concentrations of atmospheric gases that dissolve in water, and our Chemistry and Life Box addresses an important consequence of Henry's law.

Solubilities of Atmospheric Gases	**Example 15-5**

The Earth's atmosphere is 78% N_2, 21% O_2, and minor amounts of other gases, including CO_2 (0.0325%). Determine the concentrations of N_2, O_2, and CO_2 in water at equilibrium with the Earth's atmosphere at 25 °C.

Strategy: Each gas establishes its own dynamic equilibrium with water. The concentration depends on the partial pressure of the gas in the atmosphere and on the value of its Henry's law constant at 25 °C. Recall from Chapter 5 that the partial pressure of any gas in a mixture is given by the mole fraction (X_i) multiplied by total pressure.

Solution: Using 1.00 atm for the total pressure:

$$p_{O_2} = X_{O_2}P = \left(\frac{21\% \; O_2}{100\%}\right)(1.00 \text{ atm}) = 0.21 \text{ atm } O_2$$

The partial pressure of N_2 is 0.78 atm and that of CO_2 is 3.25×10^{-4} atm.

Now use the equilibrium constant expression and values of K_H from Table 15-1 to calculate the concentration of each dissolved gas D.

$$K_H = \frac{[D(aq)]_{eq}}{(p_D)_{eq}} \qquad so \qquad [D(aq)]_{eq} = K_H(p_D)_{eq}$$

$$[N_2(aq)]_{eq} = (6.7 \times 10^{-4} \text{ M/atm})(0.78 \text{ atm}) = 5.2 \times 10^{-4} \text{ M } N_2$$

$$[O_2(aq)]_{eq} = (1.3 \times 10^{-3} \text{ M/atm})(0.21 \text{ atm}) = 2.7 \times 10^{-4} \text{ M } O_2$$

$$[CO_2(aq)]_{eq} = (3.4 \times 10^{-2} \text{ M/atm})(3.25 \times 10^{-4} \text{ atm}) = 1.1 \times 10^{-5} \text{ M } CO_2$$

Box 15-1 Chemistry and Life: Scuba Diving and Henry's Law

According to Henry's law, gases become more soluble as pressure increases. This solubility property has minimum effects on everyday life, because changes in altitude or weather cause only modest variations in atmospheric pressure. Scuba divers, however, must pay careful attention to the solubility equilibria of gases. The pressure exerted on a diver increases by one atmosphere for every 30 feet of descent. Increasing pressure causes more and more gas to dissolve in a diver's blood. When the diver returns to the surface, the amount of dissolved gas far exceeds its solubility. Consequently, ascending too rapidly can be disastrous.

Carbonated beverages illustrate what happens when a dissolved gas undergoes a rapid drop in pressure. Soft drinks, soda water, and champagne are bottled under several atmospheres pressure of carbon dioxide. When a bottle is opened, the total pressure quickly falls to 1 atm. At this lower pressure, the concentration of CO_2 in the solution is much higher than its solubility, so the excess CO_2 forms gas bubbles and escapes from the liquid. As the photo shows, this process can be dramatic.

Scuba divers experience similar pressure changes. The amount of air dissolved in the blood increases significantly as the diver descends. If a diver returns to the surface too quickly, nitrogen gas dissolved in the blood forms bubbles in the same way as the CO_2 in a freshly-opened carbonated drink. These bubbles interfere with the transmission of nerve impulses and restrict the flow of blood. The effect is extremely painful and can cause paralysis or death. The bubbles tend to collect in the joints where they cause severe contractions, giving rise to the name of this dangerous condition—the bends.

Divers avoid the bends by returning to the surface slowly, taking short "decompression stops" at intermediate depths to allow excess gas to escape from their blood without forming bubbles. Another way divers reduce the risk of the bends is by using helium-oxygen gas mixtures instead of compressed air. Helium is only half as soluble as nitrogen, so less gas dissolves in blood.

Scuba divers face other hazards from the effects of Henry's law. At depths between 80 and 130 feet divers experience an intoxicating feeling from the high concentration of N_2 in the blood. This dissolved N_2 is thought to interfere with nerve transmission, giving rise to feelings of euphoria known as "rapture of the deep," which leads to a loss of judgment that can be deadly. The famous oceanographer Jacques-Yves Cousteau wrote: "I am personally quite receptive to nitrogen rapture. I like it and fear its doom. It destroys the instinct of life." Still another hazard associated with Henry's law is oxygen toxicity. In high enough concentration, molecular oxygen is poisonous. Breathing pure oxygen at one atmosphere pressure for just a few hours causes pain in the chest and coughing. Lung congestion and permanent tissue damage result after 24 hours of breathing pure O_2, and longer exposure causes death. Breathing compressed air (21% O_2) at a dive depth of 125 feet is equivalent to breathing pure oxygen at sea level. At 200 feet ($p_{O_2} = 1.3 - 1.5$ atm) oxygen toxicity causes muscle twitching, vomiting, and dizziness, and can result in deadly seizures. Although recreational divers seldom reach depths of 200 feet, those who do venture this far into the sea must breathe mixtures of gas that contain less than 10% oxygen.

Scuba divers experience an undersea world filled with mystery and beauty, but diving can be a perilous hobby. Little wonder that novice divers must undergo rigorous training courses before they are free to explore the depths.

A few gases form concentrated aqueous solutions through chemical interactions. For example, a 12 M solution can be prepared by bubbling HCl gas through water, because proton transfer occurs to generate H_3O^+ ions:

$$HCl(g) + H_2O(l) \rightleftharpoons Cl^-(aq) + H_3O^+(aq)$$

Ammonia is another gas that is very soluble in water, giving solutions as concentrated as 14.8 M. Ammonia dissolves because it forms hydrogen bonds with water molecules. When a molecule of ammonia displaces a water molecule, one hydrogen-bonding interaction is exchanged for another.

Section Exercises

■ **15.2.1** Write expressions for the equilibrium constants for the following reactions:
(a) $PCl_5(s) \rightleftharpoons PCl_3(l) + Cl_2(g)$
(b) $CaCO_3(s) \rightleftharpoons CaO(s) + CO_2(g)$
(c) $Ca_3(PO_4)_2(s) \rightleftharpoons 3\ Ca^{2+}(aq) + 2\ PO_4^{3-}(aq)$
(d) $HCN(g) + H_2O(l) \rightleftharpoons H_3O^+(aq) + CN^-(aq)$
(e) $PCl_3(l) + Cl_2(g) \rightleftharpoons PCl_5(s)$
(f) $3\ Ca^{2+}(aq) + 2\ PO_4^{3-}(aq) \rightleftharpoons Ca_3(PO_4)_2(s)$

■ **15.2.2** Draw a schematic molecular picture similar to the ones in Figure 15-5 that illustrates the following equilibrium:

$$Br_2(g) \rightleftharpoons Br_2(l)$$

■ **15.2.3** Some gases can be collected by liquid displacement. The gas is bubbled through water into an inverted container. If 0.18 mol of CO_2 at $P = 0.98$ atm is bubbled through 450 mL of water at 298 K, what fraction of the gas dissolves in the water? (See Table 15-1 for K_H.)

15.3 THERMODYNAMICS AND EQUILIBRIUM

Why do some reactions go virtually to completion, whereas others reach equilibrium when hardly any of the starting materials have been consumed? At the molecular level, bond energies and molecular organization are the determining factors. These features correlate with the thermodynamic state functions of enthalpy and entropy. As discussed in Chapter 13, free energy (G) is the state function that combines these properties. This section establishes the connection between thermodynamics and equilibrium.

Free Energy and the Equilibrium Constant

Recall from Chapter 13 that the free energy change for a chemical process, ΔG, is a signpost for spontaneity. Equations 13-12 and 13-11 relate ΔG to concentrations through the reaction quotient Q. For the general reaction:

$$aA + bB \longrightarrow dD + eE$$

$$\Delta G_{reaction} = \Delta G°_{reaction} + RT \ln Q \qquad where \qquad Q = \frac{(c_D)^d(c_E)^e}{(c_A)^a(c_B)^b}$$

A negative value for ΔG indicates that a process is spontaneous in the direction written, provided the system remains at constant temperature and pressure. For the reaction of NO_2 to form N_2O_4:

$$2\,NO_2 \longrightarrow N_2O_4 \qquad Q_{formation} = \frac{(p_{N_2O_4})}{(p_{NO_2})^2}$$

In a flask that contains only NO_2, the spontaneous reaction is the formation of N_2O_4. In other words, the formation reaction of N_2O_4 has a negative value for ΔG:

$$2\,NO_2 \longrightarrow N_2O_4 \qquad \Delta G_{formation} < 0 \qquad when \qquad p_{N_2O_4} = 0$$

Now consider what happens to ΔG as the reaction proceeds. An increase in the amount of N_2O_4 causes an increase in the numerator of Q while a decrease in amount of NO_2 causes a decrease in the denominator of Q. Both these changes increase the overall value of Q, and as a result, ΔG becomes progressively less negative. Eventually, Q becomes large enough to make $\Delta G_{formation} = 0$. At this point the reaction is at equilibrium. Now $Q = K_{eq}$, because each concentration in the expression for Q is then an equilibrium concentration.

Starting from the other end of the reaction gives an analogous result. In a flask containing only N_2O_4, the formation reaction cannot proceed because there is no NO_2 present. The decomposition reaction of N_2O_4 now has a negative value for ΔG, so the decomposition reaction proceeds to form NO_2.

$$N_2O_4 \longrightarrow 2\,NO_2 \qquad Q_{decomposition} = \frac{(p_{NO_2})^2}{p_{N_2O_4}}$$

$$\Delta G_{decomposition} < 0 \qquad when \qquad p_{NO_2} = 0$$

Decomposition of N_2O_4 causes Q to increase; ΔG becomes less negative, and eventually the reaction quotient reaches a value that makes $\Delta G = 0$.

Figure 15-6 shows schematically how ΔG for the NO_2/N_2O_4 system varies with Q. The formation of N_2O_4 is spontaneous and the reaction proceeds to the right if the relative amount of NO_2 is large. This causes $\Delta G_{formation}$ to increase until $\Delta G_{formation} = 0$ and the equilibrium position is reached. Decomposition of N_2O_4 is spontaneous and reaction proceeds to the left if the relative amount of N_2O_4 is large. This causes $\Delta G_{decomposition}$ to increase until $\Delta G_{decomposition} = 0$ and the equilibrium position is reached. To summarize, regardless of the initial concentrations in a reaction system, the system reacts in the direction for which ΔG is negative. This reduces the magnitude of ΔG, and eventually $\Delta G = 0$ when the system has reached chemical equilibrium.

As this example indicates, the direction in which a reaction proceeds depends on the relationship between Q and K_{eq}:

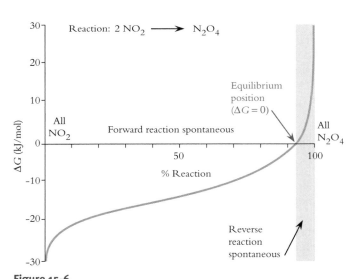

Figure 15-6
Variation in ΔG with Q. At high NO_2 concentration (*left*), the spontaneous reaction is $2\,NO_2 \rightarrow N_2O_4$; at high N_2O_4 concentration (*right*), the spontaneous reaction is $N_2O_4 \rightarrow 2\,NO_2$. As the spontaneous reaction proceeds, Q and ΔG change until the equilibrium condition is reached and $\Delta G = 0$.

/// **When $Q < K_{eq}$, the reaction goes to the right to make products.**

/// **When $Q = K_{eq}$, the reaction is at equilibrium and there is no net change.**

/// **When $Q > K_{eq}$, the reaction goes to the left to make reactants.**

Notice that if the concentration of any *product* is zero, $Q = 0$, and the direction of the reaction must be to the *right*, toward products. Conversely, if any *reactant* concentration is zero, $Q = \infty$, and the direction of the reaction must be to the *left*, toward reactants.

The relationship between Q and K_{eq} signals the direction of a chemical reaction. The free energy change, ΔG, also signals the direction of a chemical reaction. These two criteria can be compared:

/// Reaction Goes Right when	Equilibrium when	Reaction Goes Left when
$Q < K_{eq}$	$Q = K_{eq}$	$Q > K_{eq}$
$\Delta G_{reaction} < 0$	$\Delta G_{reaction} = 0$	$\Delta G_{reaction} > 0$

Calculating K_{eq} from $\Delta G°$

The relationship between the equilibrium constant and free energy provides a straightforward method for calculating equilibrium constants. We begin with Equation 13-12, which links the standard free energy change for a reaction with its free energy change under nonstandard conditions:

$$\Delta G = \Delta G° + RT \ln Q \qquad (13\text{-}12)$$

At equilibrium, $\Delta G = 0$ and $Q = K_{eq}$. We can substitute these equalities into Equation 13-12 and then rearrange to relate K_{eq} to the standard free energy change:

$$0 = \Delta G° + RT \ln K_{eq}$$

$$\Delta G° = -RT \ln K_{eq} \qquad (15\text{-}3)$$

Equation 15-3 is extremely important because it links thermodynamic data with equilibrium constants. Equation 13-8 makes it possible to use tabulated values for ΔG_f° to calculate the value for $\Delta G°$ for many reactions:

$$\Delta G_{reaction}° = \Sigma \text{ coeff}_p \, \Delta G_f^{\circ} \text{ (products)} - \Sigma \text{ coeff}_r \, \Delta G_f^{\circ} \text{ (reactants)} \qquad (13\text{-}8)$$

Then K_{eq} can be calculated from Equation 15-3. Such calculations are applied to the Haber process in Example 15-6.

K_{eq} from Thermodynamics	Example 15-6

Use standard thermodynamic data to calculate the value of K_{eq} at 298 K for the Haber reaction:

$$N_2(g) + 3 H_2(g) \rightleftharpoons 2 NH_3(g)$$

Solving Quantitative Problems

Strategy: To show the logic of this kind of calculation, we apply the seven-step approach to problem-solving.

Example 15-6	K_{eq} from Thermodynamics (continued)

Solution:

1. We are asked to calculate K_{eq} at 298 K for the Haber reaction.

2. We visualize the process by remembering that there is a connection between free energy, $\Delta G^{\circ}_{reaction}$, and equilibrium, K_{eq}.

3. The problem provides only the balanced equation and the temperature. Any other necessary data will be found in Tables and Appendices. Specifically, Appendix D contains the appropriate values for ΔG°_f, which we will need to calculate $\Delta G^{\circ}_{reaction}$.

$$\Delta G^{\circ}_f (kJ/mol): \quad N_2(g): 0 \quad H_2(g): 0 \quad NH_3(g): -16.4 \ kJ/mol$$

4. We need two equations. We calculate the free energy change for the reaction using Equation 13-8, and Equation 15-3 links free energy with the equilibrium constant.

$$\Delta G^{\circ}_{reaction} = \Sigma \ coeff_P \ \Delta G^{\circ}_f \ (products) - \Sigma \ coeff_r \ \Delta G^{\circ}_f \ (reactants) \quad \textbf{(13-8)}$$

$$\Delta G^{\circ} = -RT \ln K_{eq} \quad \textbf{(15-3)}$$

5. Rearrange Equation 15-3 to isolate $\ln K_{eq}$:

$$\ln K_{eq} = \frac{-\Delta G^{\circ}_{reaction}}{RT}$$

6. Now substitute and calculate $\Delta G^{\circ}_{reaction}$ and K_{eq}:

$$\Delta G^{\circ}_{reaction} = (2 \ mol \ NH_3)(-16.4 \ kJ/mol \ NH_3) - 3(0) - 1(0) = -32.8 \ kJ$$

$$\ln K_{eq} = \frac{-(-32.8 \ kJ)(10^3 \ J/kJ)}{(8.314 \ J/K)(298 \ K)} = 13.24$$

$$K_{eq} = antiln \ (13.24) = e^{13.24} = 5.6 \times 10^5$$

7. This is a reasonable but relatively large value, which indicates that at room temperature, the equilibrium position for the Haber reaction favors the product, NH_3.

Equilibrium Constants and Temperature

Experimental studies of the effect of temperature on equilibria reveal a consistent pattern. The equilibrium constant of an *exothermic* reaction *decreases* with increasing temperature, whereas the equilibrium constant of an *endothermic* reaction *increases* with increasing temperature. We can use two equations for ΔG°, Equations 13-9 and 15-3, to provide a thermodynamic explanation for this behavior:

$$\Delta G^{\circ} = \Delta H^{\circ} - T \Delta S^{\circ} \quad \textbf{(13-9)}$$

$$\Delta G^{\circ} = -RT \ln K_{eq} \quad \textbf{(15-3)}$$

We set these two expressions for $\Delta G°$ equal to each other:

$$\Delta H° - T \Delta S° = -RT \ln K_{eq}$$

Solving for $\ln K_{eq}$ gives an equation relating K_{eq} to standard enthalpy and entropy changes:

$$\ln K_{eq} = -\frac{\Delta H°}{RT} + \frac{\Delta S°}{R} \qquad (15\text{-}4)$$

An exothermic reaction has a negative $\Delta H°$, making the first term on the right of Equation 15-4 positive. As T increases, this term decreases, causing K_{eq} to decrease. An endothermic reaction, in contrast, has a positive $\Delta H°$, making the first term on the right of Equation 15-4 negative. As T increases, this term becomes less negative, causing K_{eq} to increase. These variations in K_{eq} with temperature, which often are substantial, can be estimated using Equation 15-4. Example 15-7 applies Equation 15-4 to the Haber synthesis.

K_{eq} and Temperature	Example 15-7

Use tabulated thermodynamic data (see Appendix D) to estimate K_{eq} for the Haber reaction at 500 °C.

Strategy: Again, we follow the seven-step procedure.

Solving
Quantitative
Problems

Solution:

1. This example is similar to Example 15-6, but note we are asked to find K_{eq} at a temperature that is different from 298 K.

2. Because the temperature is not 298 K, we cannot use Equation 15-3 directly. Instead, the process is as follows:

   a. Calculate values for $\Delta H°_{reaction}$ and $\Delta S°_{reaction}$ using tabulated thermodynamic values.

   b. Use Equation 15-4 to determine the value of the equilibrium constant at 500 °C.

3. Appendix D contains the appropriate values:

Substance	$N_2(g)$	$H_2(g)$	$NH_3(g)$
$\Delta H°_f$ (kJ/mol)	0	0	−45.9
$S°$ (J/mol K)	191.61	130.680	192.8

4. and 5. We need three equations:

$$\Delta H°_{reaction} = \Sigma \ coeff_p \ \Delta H°_f(products) - \Sigma \ coeff_r \ \Delta H°_f(reactants) \qquad (12\text{-}10)$$

$$\Delta S°_{reaction} = \Sigma \ coeff_p \ S°(products) - \Sigma \ coeff_r \ S°(reactants) \qquad (13\text{-}5)$$

$$\ln K_{eq} = -\frac{\Delta H°}{RT} + \frac{\Delta S°}{R} \qquad (15\text{-}4)$$

| Example 15-7 | K_{eq} and Temperature (continued) |

6. First, determine $\Delta H^\circ_{reaction}$ and $\Delta S^\circ_{reaction}$:

$$\Delta H^\circ_{reaction} = (2)(-45.9 \text{ kJ/mol}) - 3(0) - 0 = -91.8 \text{ kJ/mol}$$

$$\Delta S^\circ_{reaction} = (2)(192.8 \text{ J/mol K}) - (3)(130.680 \text{ J/mol K}) - 191.61 \text{ J/mol K}$$

$$\Delta S^\circ_{reaction} = -198.1 \text{ J/mol K}$$

Now calculate K_{eq} at 500 °C:

$$\ln K_{eq} = \frac{(-91.8 \text{ kJ/mol})(10^3 \text{ J/kJ})}{(8.314 \text{ J/mol K})(500 + 273 \text{ K})} + \frac{(-198.1 \text{ J/mol K})}{(8.314 \text{ J/mol K})}$$

$$\ln K_{eq} = 14.28 - 23.83 = -9.55$$

Taking the antilogarithm, or e^x, of -9.55 gives the estimated equilibrium constant at 500 °C:

$$K_{eq} = 7.1 \times 10^{-5}$$

7. At this temperature, the equilibrium constant has a reasonable but small value, indicating that the reactants are favored. This is consistent with the observation that the Haber reaction has only a 13% yield at elevated temperature.

Examples 15-6 and 15-7 underscore a dilemma faced by industrial chemists and engineers. Example 15-6 shows that at 298 K the equilibrium position of the Haber reaction strongly favors the formation of ammonia. Why then is the Haber synthesis not carried out at 298 K? The reason is that even with a catalyst, the reaction is much too slow to be useful at this temperature. For the reaction to proceed at a practical rate, the temperature must be increased. Unfortunately, the equilibrium constant falls dramatically as temperature increases. Example 15-7 shows that at 773 K, a realistic temperature for the Haber reaction, the equilibrium position does not favor NH_3:

$$N_2 + 3 H_2 \rightleftharpoons 2 NH_3$$

$$K_{eq, 298 \text{ K}} = 5.6 \times 10^5 \qquad K_{eq, 773 \text{ K}} = 7.1 \times 10^{-5}$$

As we described in the Chemistry and Life Box in Chapter 14, chemists are searching for a catalyst that will allow this essential reaction to be carried out close to room temperature.

The Haber reaction is a practical example of the effect of temperature on an *exothermic* reaction. A practical example of an *endothermic* reaction is the use of methane and steam to produce the molecular hydrogen needed for the Haber reaction. The production of hydrogen in this way is highly endothermic, so the reaction is carried out at temperatures much greater than 1000 K to force the equilibrium toward the products:

$$CH_4(g) + H_2O(g) \rightleftharpoons CO(g) + 3 H_2(g)$$

$$\Delta H^\circ = +206 \text{ kJ/mol} \qquad K_{eq, 298 \text{ K}} = 1.3 \times 10^{-25} \qquad K_{eq, 1500 \text{ K}} = 1.1 \times 10^4$$

You should be able to verify the value of the equilibrium constant at 1500 K.

15.3.1 Use thermodynamic data (see Appendix D) to determine K_{eq} for the oxidation of ClO at 298 K:

$$2 \, ClO(g) + O_2(g) \rightleftharpoons 2 \, ClO_2(g)$$

15.3.2 Does K_{eq} for the reaction described in Section Exercise 15.3.1 increase or decrease when the temperature is raised above room temperature? State your reasoning.

15.3.3 Use thermodynamic data to estimate K_{eq} at 1000 °C for the reaction in Section Exercise 15.3.1.

15.4 SHIFTS IN EQUILIBRIUM

What happens when the conditions change for a system that is already at equilibrium? Suppose additional amounts of one or more substances are introduced or the temperature of the system changes. How does a system that had been at equilibrium respond to these changes?

Le Châtelier's Principle

A simple principle gives a quick *qualitative* indication of how a system at equilibrium responds to a change in conditions. **Le Châtelier's principle**—first formulated in 1884 by Henri-Louis Le Châtelier, a French industrial chemist—states:

/// *When a change is imposed on a system at equilibrium, the system will react in the direction that reduces the amount of change.*

Le Châtelier's principle is a compact summary of how several factors influence equilibrium. Introducing a reagent causes a reaction to proceed in the direction that consumes the reagent. Reducing the temperature removes heat from the system and causes the reaction to produce some heat by proceeding in the exothermic direction.

Changes in Amounts of Reagents

A change in the amount of any substance in the reaction quotient displaces the system from its equilibrium position. Consider an industrial reactor containing a mixture of methane, hydrogen, steam, and carbon monoxide at equilibrium:

$$CH_4(g) + H_2O(g) \rightleftharpoons CO(g) + 3 \, H_2(g) \qquad Q = \frac{(p_{CO})(p_{H_2})^3}{(p_{CH_4})(p_{H_2O})}$$

How does this system respond if more steam is injected into the reactor? Adding one of the starting materials decreases the value of Q, making $Q < K_{eq}$. Because the system responds in a way that restores equilibrium, adding more steam causes the reaction to proceed to the right, consuming CH_4 and H_2O and producing CO and H_2 until equilibrium is restored. Figure 15-7 provides a graphical illustration of how concentrations change when steam is added.

Removing a product from a system at equilibrium also makes $Q < K_{eq}$ and leads to the formation of additional products. This behavior is commonly used to advantage in chemical synthesis. For example, calcium oxide (lime), an important

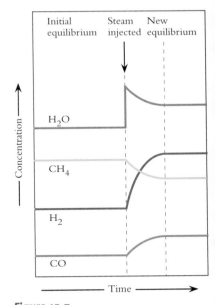

Figure 15-7
Adding some steam to a reactor that is at chemical equilibrium changes the value of Q, so the reaction is no longer at equilibrium. The reaction proceeds in the direction that consumes some of the added reagent in order to re-establish equilibrium.

Figure 15-8
(*a*) If CO_2 cannot escape, the decomposition reaction of $CaCO_3$ quickly establishes equilibrium. (*b*) Removing the CO_2 gas from a limestone kiln maintains Q at a value below the equilibrium value, allowing the reaction to continue. (*c*) As long as CO_2 gas can escape, the reaction will proceed to completion.

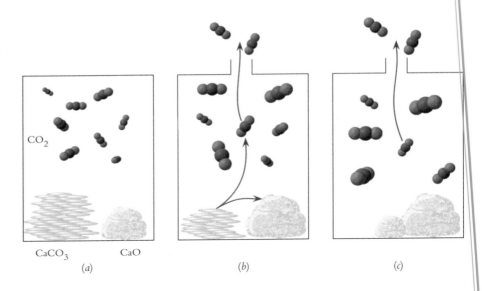

$CaCO_3$ CaO

(*a*)   (*b*)   (*c*)

material in the construction industry, is made by heating calcium carbonate in a furnace to about 1100 K:

$$CaCO_3(s) \rightleftharpoons CaO(s) + CO_2(g) \qquad K_{eq} = (p_{CO_2})_{eq} = 1.00 \text{ atm at } 1100 \text{ K}$$

If the pressure of CO_2 in the furnace were to reach 1.00 atm, the system would attain equilibrium, and no additional products would form. The CO_2 is allowed to escape from the reactor as it forms, and this drives the reaction to completion. Figure 15–8 shows this from a molecular perspective. Continuous removal of a product maintains the pressure of CO_2 below 1.00 atm, so Q has a smaller value than K_{eq}, and the reaction continues until all the $CaCO_3$ has been converted to CaO.

A change in the amount of a chemical species that has no impact on the reaction quotient will not disturb a chemical equilibrium. Thus adding or removing air has no effect on the equilibrium conditions of the $CaO/CaCO_3/CO_2$ system, because the equilibrium constant depends only on the partial pressure of CO_2, not on the total pressure of gas in the system. Likewise, adding more $CaCO_3$ to a limestone reactor does not result in a higher equilibrium pressure of CO_2, because the concentration of pure solid $CaCO_3$ has a fixed value that is independent of how much solid is present.

Vapor pressure provides a convenient illustration of why adding a pure liquid or solid does not change equilibrium concentrations. The vapor pressure of water above a small puddle is the same as the vapor pressure above a large pond at the same temperature. Figure 15-9 shows why: More molecules escape from the larger surface of the pond, but more molecules are captured, too. The balance between captures and escapes is the same for both puddle and pond.

The effects of changes in amounts on a system at equilibrium can be summarized in accordance with Le Châtelier's principle:

1. Any change in conditions that *increases* the value of Q causes the reaction to consume products and produce reactants until equilibrium is reestablished.

2. Any change in conditions that *decreases* the value of Q causes the reaction to form products and consume reactants until equilibrium is reestablished.

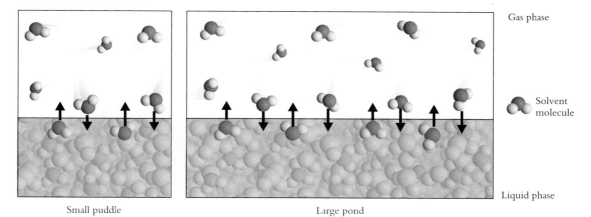

Figure 15-9
The equilibrium pressure (vapor pressure) above a liquid is independent of the amount of liquid, because the ratio of evaporating to condensing molecules is independent of amount.

3. Any change in amounts that has no effect on the value of Q has no effect on the equilibrium position.

Example 15-8 provides practice in applying Le Châtelier's principle at the molecular level.

Effects of Concentration Changes   Example 15-8

The left-hand view in the accompanying figure is a molecular picture showing a very small portion of an iodine-triiodide solution at equilibrium. The right-hand view shows this same solution, no longer at equilibrium because additional I_2 has been dissolved in the solution. Redraw this molecular picture to show qualitatively how the system responds to this change.

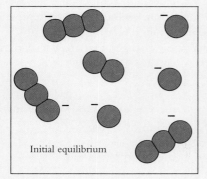

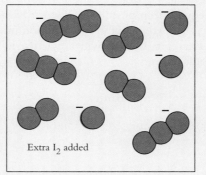

Initial equilibrium   Extra I_2 added

Strategy: Because this problem asks for a qualitative answer, we do not need to do calculations. It is sufficient to apply Le Châtelier's principle to determine the

| Example 15-8 | Effects of Concentration Changes *(continued)* |

direction of change and draw the new picture that shows the result of the change.

Solution: The change that has been imposed is addition of a reactant, I_2. The system responds by moving in the direction that reduces the concentration of this added reactant. We show this by combining one I_2 molecule with an I^- anion to make one more I_3^- anion. The new view contains one less I_2 molecule, one less I^- anion, and one more molecule of the I_3^- product:

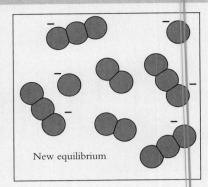

New equilibrium

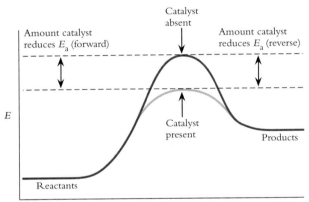

Catalyst absent

Amount catalyst reduces E_a (forward)

Amount catalyst reduces E_a (reverse)

E

Catalyst present

Products

Reactants

Reaction coordinate

Figure 15-10
A catalyst changes the mechanism of a reaction and lowers the net activation energy barrier in both directions. It has no effect on the overall $\Delta G°$ and K_{eq} for the reaction.

Effect of Catalysts

Catalysts *do not* affect the equilibrium constant. Figure 15-10 provides a reminder that a catalyst changes the mechanism of the reaction in a way that reduces the net activation energy barrier, but it does not alter the thermodynamic changes that accompany the reaction. In other words, a catalyst reduces the forward activation energy and the reverse activation energy, so the rate of reaction is increased in *both* directions. However, the catalyst does not affect the nature of the reactants or the products, so the standard free energy change, $\Delta G°$, does not change, and neither does K_{eq}. A catalyst allows a reaction to reach equilibrium *more rapidly,* but it does not alter the equilibrium *position.*

Effect of Temperature

As shown in Section 15.3, K_{eq} varies with temperature in a way that can be understood using the principles of thermodynamics. Temperature is the *only* variable that causes a change in the value of K_{eq}. The effect of temperature on K_{eq} depends on the enthalpy change of the reaction, ΔH. An *increase* in temperature always shifts the equilibrium position in the *endothermic* direction, and a *decrease* in temperature always shifts the equilibrium position in the *exothermic* direction.

Example 15-9 provides practice in analyzing changes that may lead to shifts in equilibrium.

| Example 15-9 | Shifts in Equilibrium |

Consider a saturated solution of $CaCl_2$ at equilibrium with excess $CaCl_2(s)$:

$$CaCl_2(s) \rightleftharpoons Ca^{2+}(aq) + 2\ Cl^-(aq) \qquad \Delta H°_{reaction} = +585\ kJ$$

How do the following changes affect the amount of dissolved $CaCl_2$?

(a) More $CaCl_2(s)$ is added.

(b) Some NaCl is dissolved in the solution.

Shifts in Equilibrium *(continued)*

Example 15-9

(c) Some $NaNO_3$ is dissolved in the solution.

(d) Some pure water is added.

(e) The solution is heated.

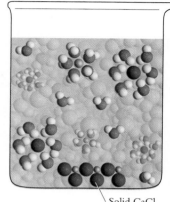

Solid $CaCl_2$

Strategy: According to Le Châtelier's principle, the system will respond in the direction that reduces the amount of change. It will only do so, however, if an appropriate response exists. Changes in quantities of substances must be analyzed for their effect, if any, on the value of Q. If Q becomes smaller, the reaction will proceed to the right, and more $CaCl_2$ will dissolve. If Q becomes larger, the reaction will proceed to the left, and $CaCl_2$ will precipitate.

Solution:

(a) More $CaCl_2(s)$ is added. Because $Q = [Ca^{2+}][Cl^-]^2$, adding $CaCl_2(s)$ does *not* change Q. The system remains at equilibrium. There is no effect on the amount of dissolved $CaCl_2$.

(b) Some NaCl is dissolved in the solution. This change increases $[Cl^-]$, making $Q > K_{eq}$. The reaction proceeds to the left, and some $CaCl_2$ precipitates.

(c) Some $NaNO_3$ is dissolved in the solution. There is no effect on Q, so there is no effect on the amount of dissolved $CaCl_2$.

(d) Some pure water is added. Adding water dilutes the solution and lowers $[Ca^{2+}]$ and $[Cl^-]$. Because Q is now less than K_{eq}, the reaction proceeds to the right. More $CaCl_2$ dissolves.

(e) The solution is heated. Because the solubility reaction is endothermic ($\Delta H°$ is positive), K_{eq} increases as T increases. More $CaCl_2$ dissolves.

15.4.1 The following solubility reaction is exothermic by about 70 kJ/mol:

$$Ba(OH)_2(s) \rightleftharpoons Ba^{2+}(aq) + 2\,OH^-(aq)$$

Will more $Ba(OH)_2$ dissolve, or will some precipitate, or will no change occur, when each of the following changes are made on a saturated solution of $Ba(OH)_2$?

(a) More $Ba(OH)_2(s)$ is added to the solution.

(b) More water is added to the solution.

(c) Some HCl is added to the solution.

(d) The solution is cooled on an ice bath.

15.4.2 Refer to Examples 15-6 and 15-7. List four changes in conditions that might be used to increase the yield of ammonia in the Haber process.

15.4.3 Chemists are optimistic that a catalyst will be found for the production of ammonia from hydrogen and nitrogen under standard conditions. In contrast, no hope exists of developing a catalyst for the production of hydrogen from methane and steam under standard conditions. Explain.

15.5 WORKING WITH EQUILIBRIA

The quantitative treatment of a reaction equilibrium usually involves one of two things. Either the equilibrium constant must be computed from a knowledge of concentration, or equilibrium concentrations must be determined from a knowledge of initial conditions and K_{eq}. In this section, we describe the basic reasoning and techniques needed to solve equilibrium problems. Stoichiometry plays a major role in equilibrium calculations, so you may want to review the techniques described in Chapter 4, particularly Section 4.4 on limiting reactants.

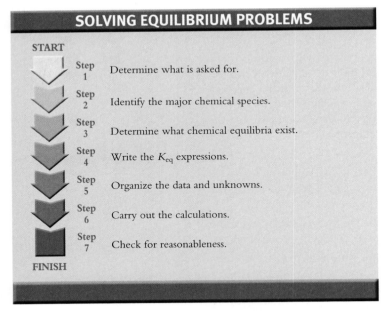

SOLVING EQUILIBRIUM PROBLEMS

START

Step 1 Determine what is asked for.

Step 2 Identify the major chemical species.

Step 3 Determine what chemical equilibria exist.

Step 4 Write the K_{eq} expressions.

Step 5 Organize the data and unknowns.

Step 6 Carry out the calculations.

Step 7 Check for reasonableness.

FINISH

Chemistry of Equilibria

Equilibrium conditions are determined by the chemical reactions that occur in a system. Consequently, it is necessary to analyze the chemistry of the system before doing *any* calculations. After the chemistry is known, a mathematical solution to the problem can be developed. We can modify the seven-step approach to problem-solving so that it applies specifically to equilibrium problems, proceeding from the chemistry to the equilibrium constant expression to the mathematical solution.

We illustrate this approach using the equilibrium shown in Figure 15-11. When solid LiF is added to water, a small amount of the salt dissolves, leading to an equilibrium between the solid and a solution. Chemical analysis reveals that the equilibrium concentration of F^- ions in the solution is 6.16×10^{-2} M. We want to determine the equilibrium constant for this process.

1. Determine what is asked for.

This is the usual first step. We need to know our destination before we can map out a route. We are asked for the value of the equilibrium constant for a solid dissolving to form an aqueous solution.

2. Identify the major chemical species.

This step is important because chemical equilibria are dynamic interactions among *molecules and ions.* The key to success in working with equilibria is to "think molecules (and ions)." When solid LiF is added to water (Figure

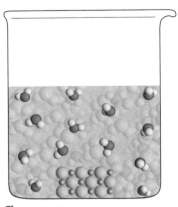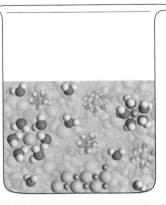

Figure 15-11
(*a*) When solid LiF is added to water, initially there are no ions in solution. (*b*) At equilibrium, Li^+ cations and F^- anions are present.

15-11*a*), the species present before any reaction occurs are LiF(s) and H_2O(l).

3. Determine what chemical equilibria exist.

Sometimes the net chemical reaction is provided, but in other cases you have to examine the species present and determine what reactions can occur among them. The statement of the problem indicates that fluoride anions are present in solution when LiF dissolves in water. To maintain electrical neutrality, Li^+ ions must also be present in equal number. Here is the net reaction:

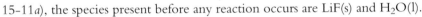

$$LiF(s) \rightleftharpoons Li^+(aq) + F^-(aq)$$

4. Write the K_{eq} expressions.

This is the equation (or equations) that we must use to complete the calculations. The equilibrium constant expression for LiF dissolving in water is:

$$K_{eq} = [Li^+]_{eq} [F^-]_{eq}$$

5. Organize the data and unknowns.

In some problems, concentrations at equilibrium are provided. In other problems concentrations at equilibrium must be calculated, usually by using amounts tables. According to chemical analysis, a solution of LiF at chemical equilibrium has $[F^-]_{eq} = 6.16 \times 10^{-2}$ M. The stoichiometric ratio of LiF is 1:1, so an equal amount of Li^+ dissolves: $[Li^+]_{eq} = 6.16 \times 10^{-2}$ M.

← CHAPTER 4

Amounts tables are described in Chapter 4.

6. Carry out the calculations.

Refer to Step 1 to find out what must be calculated, the equilibrium constant or the equilibrium concentrations. In this example, we are after an equilibrium constant, which we calculate using the equilibrium concentrations:

$$K_{eq} = [Li^+][F^-] = (6.16 \times 10^{-2})(6.16 \times 10^{-2}) = 3.79 \times 10^{-3}$$

7. Check for reasonableness.

As always, we ask if the result makes sense. The concentration of ions in the saturated solution is in the range of 10^{-2} M, so an equilibrium constant in the 10^{-3} range is reasonable.

As the LiF example illustrates, the most direct way to determine the value of an equilibrium constant is to mix substances that can undergo a chemical reaction, wait until the system reaches equilibrium, and measure the concentrations of the species present once equilibrium is established. Although the calculation of an equilibrium constant requires knowledge of the equilibrium concentrations of *all* species whose concentrations appear in the equilibrium constant expression, stoichiometric analysis often can be used to deduce the concentration of one species from the known concentration of another species. Example 15-10 shows how to approach an equilibrium constant problem from a molecular perspective.

K_{eq} From a Molecular View	Example 15-10

The figure represents a molecular view of a gas-phase reaction that has reached equilibrium. Assuming that each molecule in the molecular view represents a partial pressure of 1 atm, determine K_{eq} for this reaction.

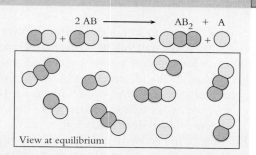

View at equilibrium

Strategy: Our seven-step approach to equilibrium problems will lead to the correct result.

Solving Equilibrium Problems

| Example 15-10 | K_{eq} From a Molecular View *(continued)* |

Solution:

1. **Determine what is asked for.** The problem asks us to calculate an equilibrium constant, K_{eq}.

2. **Identify the major chemical species.** The chemical species present are the hypothetical gas–phase molecules, AB_2, A, and AB.

3. **Determine what chemical equilibria exist.** The forward reaction is provided. We know that at equilibrium the reverse reaction proceeds at an equal rate. We represent the equilibrium reaction with a double–headed arrow:

$$2\,AB \rightleftharpoons AB_2 + A$$

4. **Write the equilibrium constant expression.** We use the chemical reaction to determine the equilibrium constant expression:

$$K_{eq} = \frac{[AB_2]_{eq}\,[A]_{eq}}{[AB]_{eq}^2}$$

5. **Organize the data and unknowns.** The problem states that each molecule represents a partial pressure of 1.0 atm, so we can determine the equilibrium concentrations of each reagent by counting molecules in the molecular picture:

$$[AB_2]_{eq} = 4.0 \text{ atm}, \qquad [A]_{eq} = 1.0 \text{ atm}, \qquad [AB]_{eq} = 4.0 \text{ atm}$$

6. **Carry out the calculations.** We substitute each concentration into the equilibrium constant expression and evaluate.

$$K_{eq} = \frac{[AB_2]_{eq}\,[A]_{eq}}{[AB]_{eq}^2} = \frac{(4.0 \text{ atm})(1.0 \text{ atm})}{(4.0 \text{ atm})^2} = 0.25$$

7. **Check for reasonableness.** There are comparable numbers of molecules of reactants and products present in this system at equilibrium, so an equilibrium constant close to 1 is reasonable.

Initial Conditions and Concentration Tables

When we do equilibrium calculations, we are usually interested in the concentrations of species present *at equilibrium*. In many cases, however, we have information about what we call **initial concentrations,** *before* any net change has occurred. Initial concentrations are the concentrations that would be present if it were possible to mix all the reactants but block the reactions that lead to equilibrium. These concentrations are easy to calculate from the initial conditions, but they seldom exist in reality because substances begin to react as soon as they are mixed.

A chemical system reacts, often rapidly, from initial conditions to equilibrium. As this occurs, concentrations of starting materials decrease, and concentrations of products increase. These concentration changes are related in two ways. First, the concentration of each reagent at equilibrium is its initial concentration plus the change that has occurred:

$$\begin{bmatrix} \text{Equilibrium} \\ \text{concentration} \end{bmatrix} = \begin{bmatrix} \text{Initial} \\ \text{concentration} \end{bmatrix} + \begin{bmatrix} \text{Change in concentration} \\ \text{during reaction} \end{bmatrix}$$

$$[A]_{eq} = [A]_i + \Delta[A] \qquad (15\text{-}5)$$

Second, the changes in concentration of the various reagents are related by their stoichiometric ratios:

$$\begin{bmatrix} \text{Change in} \\ \text{concentration of B} \end{bmatrix} = \begin{bmatrix} \text{Stoichiometric} \\ \text{ratio of B to A} \end{bmatrix}\begin{bmatrix} \text{Change in} \\ \text{concentration of A} \end{bmatrix}$$

$$\Delta[B] = \frac{\text{Coeff. B}}{\text{Coeff. A}} \Delta[A] \qquad (15\text{-}6)$$

As we show in upcoming examples, the changes in concentration may be positive or negative, depending on whether a reagent is being produced or consumed.

These relationships provide complete stoichiometric information about the equilibrium. Just as amounts tables are useful in doing stoichiometric calculations, a **concentration table** that provides initial concentrations, changes in concentrations, and equilibrium concentrations is an excellent way to organize Step 5 of the problem-solving procedure for equilibrium problems. Figure 15-12 shows how to complete a concentration table: Equation 15-5 applies to every column, but Equation 15-6 applies only across the "change" row.

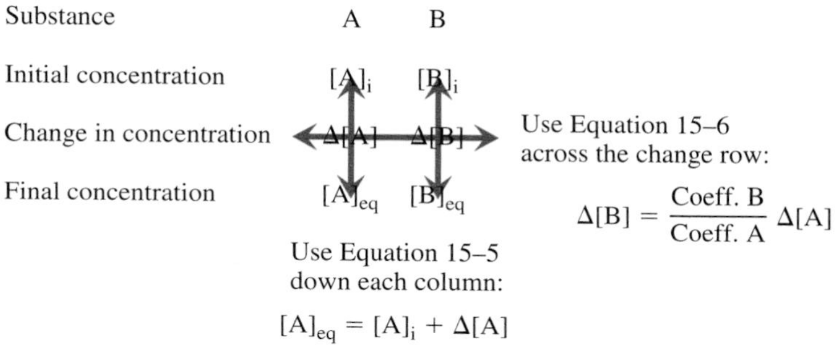

Example 15-11 shows how to use stoichiometric reasoning and a concentration table to calculate an equilibrium constant. This is the first main type of equilibrium problem.

Figure 15-12
Schematic view of a concentration table. The arrows indicate how the relationships expressed in Equations 15-5 and 15-6 are used to complete the table.

Substance	A	B
Initial concentration	$[A]_i$	$[B]_i$
Change in concentration	$\Delta[A]$	$\Delta[B]$
Final concentration	$[A]_{eq}$	$[B]_{eq}$

Use Equation 15–6 across the change row:
$$\Delta[B] = \frac{\text{Coeff. B}}{\text{Coeff. A}} \Delta[A]$$

Use Equation 15–5 down each column:
$$[A]_{eq} = [A]_i + \Delta[A]$$

← SECTION 4-4
Review Section 4.4 for information on amounts tables.

Calculating an Equilibrium Constant　　　**Example 15-11**

Benzoic acid is a weak acid that undergoes proton transfer with water. When a 0.125 M aqueous solution of benzoic acid ($C_6H_5CO_2H$) reaches equilibrium, $[H_3O^+] = 0.0028$ M. What is K_{eq} for the proton transfer reaction?

Strategy: Again, the seven-step approach to equilibrium problems will lead to the correct result. This is a more complicated example than Example 15-10, so a concentration table as part of Step 5 helps keep track of the stoichiometric relationships.

Solution:

1. The problem asks us to calculate an equilibrium constant, K_{eq}.
2. The chemical species present initially are benzoic acid and water, $C_6H_5CO_2H$ and H_2O. The H_3O^+ mentioned in the problem is formed as the solution comes to equilibrium.

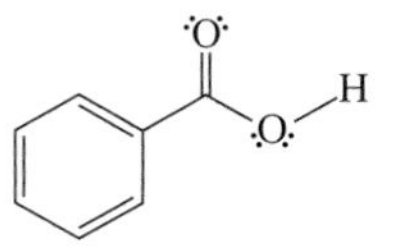

Solving
Equilibrium
Problems

Benzoic acid
($C_6H_5CO_2H$)

| Example 15-11 | Calculating an Equilibrium Constant *(continued)* |

3. Benzoic acid is a proton donor, and the equilibrium concentration of hydronium ions is provided, which suggests that one of the products is H_3O^+. This leads to the correct chemical reaction, proton transfer from benzoic acid to water:

$$C_6H_5CO_2H(aq) + H_2O(l) \rightleftharpoons C_6H_5CO_2^-(aq) + H_3O^+(aq)$$

4. Use the chemical reaction to determine the equilibrium constant expression. Water is the solvent, so it is omitted from the equilibrium expression:

$$K_{eq} = \frac{[H_3O^+]_{eq}\,[C_6H_5CO_2^-]_{eq}}{[C_6H_5CO_2H]_{eq}}$$

5. Construct a table of initial concentrations, changes in concentration, and equilibrium concentrations for each species that appears in the equilibrium constant expression. The equilibrium concentrations from the last row of the table are needed to find K_{eq}. Start by entering the data given in the problem. The initial concentration of benzoic acid is 0.125 M. Pure water contains no benzoate ions and a negligible concentration of hydronium ions. The problem also states the equilibrium concentration of hydronium ions, 0.0028 M.

Reaction $H_2O(l)$ +	$C_6H_5CO_2H(aq)$ $\rightleftharpoons$	$C_6H_5CO_2^-(aq)$ +	$H_3O^+(aq)$
Initial concentration (M)	0.125	0	0
Change in concentration (M)			
Equilibrium concentration (M)	*Need to find*	*Need to find*	0.0028

We see in the last column of the table that two of the concentrations for H_3O^+ are known, so we can use Equation 15-5 to determine the change in concentration for this species:

Equilibrium concentration = Initial concentration + Change in concentration

$$0.0028\text{ M} = 0\text{ M} + \Delta[H_3O^+]$$

$$\Delta[H_3O^+] = 0.0028\text{ M}$$

Once we know one change, Equation 15-6 allows us to complete the change row by calculating the changes in the other concentrations.

$$\Delta[B] = \frac{\text{Coeff. B}}{\text{Coeff. A}}\,\Delta[A]$$

All of the stoichiometric coefficients are equal to 1, so all three changes have the same magnitude, but they have different *signs*. As the reaction comes to equilibrium, concentrations of products *increase*, so the sign for changes in products is *positive*. The concentrations of starting materials *decrease* in the reaction, so the sign for the change in benzoic acid is *negative*.

| Calculating an Equilibrium Constant (*continued*) | | | **Example 15-11** |

Reaction $H_2O(l)$	+	$C_6H_5CO_2H(aq)$ $\rightleftharpoons$ $C_6H_5CO_2^-(aq)$	+	$H_3O^+(aq)$
Initial concentration (M)		0.125	0	0
Change in concentration (M)		−0.0028	+0.0028	+0.0028
Equilibrium concentration (M)		*Need to find*	*Need to find*	0.0028

Two of the three entries in each of the other columns are known. To complete the table, apply Equation 15-5 for these columns to determine the concentrations at equilibrium for these species:

$$[C_6H_5CO_2^-]_{eq} = 0 + 0.0028 \text{ M} = 0.0028 \text{ M}$$

$$[C_6H_5CO_2H]_{eq} = 0.125 \text{ M} - 0.0028 \text{ M} = 0.122 \text{ M}$$

Reaction $H_2O(l)$	+	$C_6H_5CO_2H(aq)$ $\rightleftharpoons$ $C_6H_5CO_2^-(aq)$	+	$H_3O^+(aq)$
Initial concentration (M)		0.125	0	0
Change in concentration (M)		−0.0028	+0.0028	+0.0028
Equilibrium concentration (M)		0.122	0.0028	0.0028

6. Substitute the values in the last row of the concentration table into the equilibrium constant expression and evaluate the result:

$$K_{eq} = \frac{(2.8 \times 10^{-3})(2.8 \times 10^{-3})}{(0.122)} = 6.4 \times 10^{-5}$$

7. The concentration of hydronium ion at equilibrium is in the 10^{-3} range, so a value of K_{eq} in the 10^{-5} range is reasonable.

Calculating Equilibrium Concentrations

The second main type of equilibrium problem asks for values of equilibrium concentrations. We also use concentration tables for this type of problem, with one additional feature. In such problems, a variable x is assigned to one unknown concentration, and then the equilibrium constant is used to find the value of x by standard algebraic techniques. Examples 15-12 and 15-13 illustrate this use and manipulation of unknowns.

| Gas Equilibrium Concentrations | **Example 15-12** |

The molecular view represents a set of initial conditions for the reaction described in Example 15-10. Each molecule represents a partial pressure of 1.0 atm. Determine the equilibrium conditions and redraw the picture to illustrate those conditions.

Example 15-12	Gas Equilibrium Concentrations *(continued)*

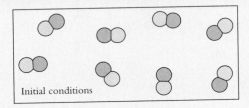

Initial conditions

Solving
Equilibrium
Problems

Strategy: We can apply the seven-step strategy.

Solution:

1. The problem asks for the equilibrium pressures and a molecular view illustrating the equilibrium conditions.

2. The only species present initially is the diatomic gas AB. From Example 15–10, we know that this gas reacts to form A and AB_2.

3. The chemical equilibrium is the same one that we identified in Example 15–10:

$$2\,AB \rightleftharpoons AB_2 + A$$

4. The equilibrium constant expression is the same as in Example 15–10:

$$K_{eq} = \frac{[AB_2]_{eq}\,[A]_{eq}}{[AB]_{eq}^2} = 0.25$$

5. From the molecular picture, we can determine the initial pressure of AB: $p(\text{initial, AB}) = 8.0$ atm. To relate this to equilibrium pressures, we need a concentration table. The only information available to us is the initial pressures:

Reaction (substance)	2 AB $\rightleftharpoons$ AB$_2$ + A		
Initial pressure (atm)	8.0	0	0
Change in pressure (atm)			
Equilibrium pressure (atm)			

We assign the unknown x to represent the *change* in pressure of A during the reaction. Then we can find the changes for the other reagents by applying Equation 15–6 across the change row. The stoichiometric ratio indicates that two molecules of AB must be consumed for each molecule of A produced. Because AB molecules are *consumed* as the reaction comes to equilibrium, the change in pressure of AB has a *negative* value. Thus if the change in pressure of A is $+x$ atm, the change in pressure of AB is $-2x$ atm. By stoichiometry, the change for AB_2 is also $+x$:

Reaction (substance)	2 AB $\rightleftharpoons$ AB$_2$ + A		
Initial pressure (atm)	8.0	0	0
Change in pressure (atm)	$-2x$	$+x$	$+x$
Equilibrium pressure (atm)	$8.0 - 2x$	x	x

Gas Equilibrium Concentrations *(continued)*   **Example 15-12**

6. To evaluate x, we substitute the equilibrium pressures into the equilibrium constant expression:

$$0.25 = \frac{[AB_2]_{eq}\,[A]_{eq}}{[AB]^2_{eq}} = \frac{x^2}{(8.0 - 2x)^2}$$

In general, we would have to solve for x using the quadratic expression, but this particular expression can be simplified by taking the square root of each side:

$$0.50 = \frac{x}{(8.0 - 2x)} \quad so \quad 0.50(8.0 - 2x) = x$$

$$4.0 - x = x, \quad 4.0 = 2x, \quad and \quad x = 2.0$$

This lets us calculate the equilibrium pressures:

$$p(AB) = 8.0 - 2(2.0) = 4.0 \text{ atm}$$

$$p(AB_2) = p(A) = 2.0 \text{ atm}$$

To complete the problem, we draw a molecular view that shows four molecules of AB, two molecules of AB_2, and two atoms of A:

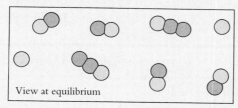

View at equilibrium

7. We conclude that this result is reasonable: there is less starting material and more products than were present at the beginning, and atoms of A and B have been conserved.

Equilibrium Concentrations   **Example 15-13**

The equilibrium constant for the proton transfer reaction of benzoic acid, determined in Example 15-11, is 6.4×10^{-5}. Calculate the equilibrium concentration of benzoic acid and benzoate anions in a 5.0×10^{-2} M solution of the acid.

Strategy: We can apply the seven-step procedure.

Solving Equilibrium Problems

Solution:

1. This problem asks about equilibrium concentrations and provides the value of an equilibrium constant. Under these circumstances, we expect to have to define an appropriate unknown to represent the change in concentrations of one reagent.

2. Initially, benzoic acid and water are the only species present.

3. The proton transfer reaction is the same as in Example 15-11:

$$C_6H_5CO_2H(aq) + H_2O(l) \rightleftharpoons C_6H_5CO_2^-(aq) + H_3O^+(aq)$$

Example 15-13 | **Equilibrium Concentrations** *(continued)*

4. The equilibrium constant expression is the same as in Example 15-11:

$$K_{eq} = \frac{[H_3O^+]_{eq}\,[C_6H_5CO_2^-]_{eq}}{[C_6H_5CO_2H]_{eq}} = 6.4 \times 10^{-5}$$

5. Construct a concentration table. The only quantitative information provided in the problem is the value for the initial concentration of benzoic acid (5.0×10^{-2} mol/L):

Reaction $H_2O(l)$	$+$ $C_6H_5CO_2H(aq)$ $\rightleftharpoons$	$C_6H_5CO_2^-(aq)$ $+$	$H_3O^+(aq)$
Initial concentration (M)	0.050	0	0
Change in concentration (M)			
Equilibrium concentration (M)	*Need to find*	*Need to find*	*Need to find*

Completing the table requires an appropriate unknown. If we let x represent the *change* in concentration of H_3O^+ during the reaction, the changes in the other concentrations are $-x$ for benzoic acid and $+x$ for benzoate anion.

Reaction $H_2O(l)$	$+$ $C_6H_5CO_2H(aq)$ $\rightleftharpoons$	$C_6H_5CO_2^-(aq)$ $+$	$H_3O^+(aq)$
Initial concentration (M)	0.050	0	0
Change in concentration (M)	$-x$	$+x$	$+x$
Equilibrium concentration (M)	*Need to find*	*Need to find*	*Need to find*

Applying Equation 15-5 to each column completes the table.

Reaction $H_2O(l)$	$+$ $C_6H_5CO_2H(aq)$ $\rightleftharpoons$	$C_6H_5CO_2^-(aq)$ $+$	$H_3O^+(aq)$
Initial concentration (M)	0.050	0	0
Change in concentration (M)	$-x$	$+x$	$+x$
Equilibrium concentration (M)	$0.50 - x$	x	x

6. Next we substitute the values in the last row of the concentration table into the equilibrium constant expression:

$$K_{eq} = \frac{[H_3O^+]_{eq}\,[C_6H_5CO_2^-]_{eq}}{[C_6H_5CO_2H]_{eq}} = 6.4 \times 10^{-5} = \frac{x^2}{(0.050 - x)}$$

Equilibrium Concentrations *(continued)* Example 15-13

Multiplying through to clear the fraction gives:

$$x^2 = (3.2 \times 10^{-6}) - (6.4 \times 10^{-5})x \quad or$$

$$x^2 + (6.4 \times 10^{-5})x - (3.2 \times 10^{-6}) = 0$$

This is a quadratic equation in the form $ax^2 + bx + c = 0$. In this case,

$$a = 1, b = 6.4 \times 10^{-5}, \text{ and } c = 3.2 \times 10^{-6}$$

Use the quadratic formula to find the value of x:

$$x = \frac{-b \pm \sqrt{b^2 - 4ac}}{2a} = \frac{-(6.4 \times 10^{-5}) \pm \sqrt{(6.4 \times 10^{-5})^2 - 4(1)(-3.2 \times 10^{-6})}}{2(1)}$$

$$x = \frac{-(6.4 \times 10^{-5}) \pm \sqrt{(4.096 \times 10^{-9}) - (1.28 \times 10^{-5})}}{2(1)}$$

There are two solutions to this equation, but one gives a negative value for x. We know that x must be positive, because it represents the increase in hydronium ion concentration. Hence we solve for the positive solution:

$$x = \frac{-(6.4 \times 10^{-5}) + (3.58 \times 10^{-3})}{2} = 1.76 \times 10^{-3}$$

Now, we can solve for the concentrations of the ions at equilibrium:

$$[C_6H_5CO_2^-]_{eq} = x \text{ M} = 1.76 \times 10^{-3} \text{ M}$$

$$[C_6H_5CO_2H]_{eq} = (5.0 \times 10^{-2} \text{ M} - 1.76 \times 10^{-3} \text{ M}) = 4.8 \times 10^{-2} \text{ M}$$

7. Both concentrations are reasonable values: they are positive and somewhat less than the initial concentration of benzoic acid.

Examples 15-12 and 15-13 involve equilibrium constants with moderate values. However, many chemical reactions have equilibrium constants that are either very small ($K_{eq} < 10^{-8}$) or very large ($K_{eq} > 10^8$). For these equilibria we can usually use approximations to simplify the calculations.

A *very small* value of K_{eq} means that concentrations of *products* at equilibrium are *very low* compared with reactant concentrations. As Figure 15-13a shows, when K_{eq} is small reactions barely get started before they reach equilibrium. One example is the equilibrium that results when solid AgBr is placed in water:

$$AgBr(s) \rightleftharpoons Ag^+(aq) + Br^-(aq) \quad K_{eq} = 5.35 \times 10^{-13}$$

The equilibrium constant for this reaction is small, indicating that only a small amount of AgBr will dissolve in water.

At the other extreme, a *very large* value of K_{eq} means that concentrations of *products* at equilibrium are *very high* compared with reactant concentrations. As Figure 15-13b shows, when K_{eq} is large, the reaction proceeds almost to completion before equilibrium is reached. One example is the equilibrium that results from the mixing of aqueous solutions of $AgNO_3$ and KBr:

$$Ag^+(aq) + Br^-(aq) \rightleftharpoons AgBr(s) \quad K_{eq} = 1/(5.35 \times 10^{-13}) = 1.87 \times 10^{12}$$

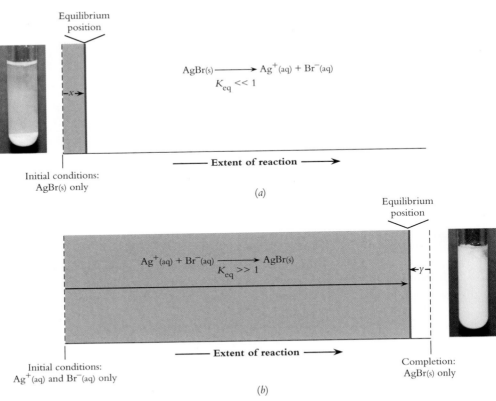

Figure 15-13
Schematic representation of the relationship between K_{eq} and the extent of reaction.
(*a*) When $K_{eq} \ll 1$, the difference from initial conditions, *x*, is small. (*b*) When $K_{eq} \gg 1$, the difference from completion, *y*, is small.

The equilibrium constant for this reaction is large, indicating that nearly all of the aqueous ions end up as solid precipitate.

Working with Small Equilibrium Constants

When K_{eq} is small, the changes in concentration are likely to be small compared with initial concentrations. Example 15-14 illustrates the use of approximations to simplify the mathematical calculations in such cases.

Example 15-14	Small Equilibrium Constant

Sodium benzoate is a common food preservative. This salt dissolves in water to produce benzoate anions, which accept protons from water:

Benzoate anion
$(C_6H_5CO_2^-)$

Small Equilibrium Constant *(continued)* | **Example 15-14**

$$H_2O(l) + C_6H_5CO_2^-(aq) \rightleftharpoons OH^-(aq) + C_6H_5CO_2H(aq)$$

$$K_{eq} = \frac{[OH^-]_{eq}[C_6H_5CO_2H]_{eq}}{[C_6H_5CO_2^-]_{eq}} = 1.5 \times 10^{-10}$$

What is the equilibrium concentration of hydroxide ion in a 0.135 M solution of sodium benzoate?

Strategy: We follow the step-by-step procedure for working with equilibria.

Solution:

1. The problem asks for the equilibrium concentration of OH^-.

2. Under initial conditions, three chemical species are present: benzoate anions, Na^+, and H_2O.

3. The net reaction is given. A proton is transferred from a water molecule to a benzoate anion. A small amount of benzoic acid forms as the reaction reaches equilibrium. Notice that Na^+ ions do not participate in the net reaction.

4. The equilibrium constant expression is given in the problem.

5. Set up a concentration table. The concentration of water is not required because it does not appear in the equilibrium constant expression. A convenient choice for the unknown x is the amount of benzoic acid formed in the reaction. This choice makes several other entries in the table equal to x.

Solving
Equilibrium
Problems

Reaction	H_2O +	$C_6H_5CO_2^-$	$\rightleftharpoons$	$C_6H_5CO_2H$ +	OH^-
Initial concentration, (M)		0.135		0	0
Change to equilibrium, (M)		$-x$		$+x$	$+x$
Equilibrium concentration, (M)		$0.135 - x$		x	x

6. Substitute the equilibrium concentrations in the expression for K_{eq} and solve for the unknown:

$$K_{eq} = \frac{[OH^-]_{eq}[C_6H_5CO_2H]_{eq}}{[C_6H_5CO_2^-]_{eq}} = 1.5 \times 10^{-10} = \frac{(x)(x)}{(0.135 - x)}$$

The equilibrium constant has a small value (10^{-10}). This suggests that we can make an approximation by assuming that the change required to reach equilibrium (x) is very small. This situation is like that shown in Figure 15-13a. As shown in the concentration table, the concentration of benzoate ion is $(0.135 - x)$ M at equilibrium. Knowing that x is small, we make the approximation that x can be neglected relative to 0.135:

$$x \ll 0.135 \quad so \quad 0.135 - x \cong 0.135$$

This simplifies the equilibrium expression:

$$1.5 \times 10^{-10} = \frac{x^2}{0.135 - x} \cong \frac{x^2}{0.135}$$

| Example 15-14 | Small Equilibrium Constant *(continued)* |

$$x^2 \cong (0.135)(1.5 \times 10^{-10}) = 0.203 \times 10^{-10}$$

$$x \cong 0.45 \times 10^{-5} = 4.5 \times 10^{-6}$$

We use the value of x to calculate the concentrations at equilibrium:

$$[C_6H_5CO_2^-] = (0.135 - x) \text{ M} = (0.135 - 4.5 \times 10^{-6}) \text{ M} = 0.135 \text{ M}$$

$$[C_6H_5CO_2H] = [OH^-] = x \text{ M} = 4.5 \times 10^{-6} \text{ M}$$

7. When we make an approximation, we must check that it is reasonable. The approximation in this example is $x \ll 0.135$ M. We find $x = 4.5 \times 10^{-6}$ M, which is more than four orders of magnitude less than 0.135, so the approximation is a good one.

Whenever we make an approximation, we must verify that it is valid by comparing the value calculated using the approximation with the approximation itself. Most equilibrium constants are uncertain by about 5%, so x can be neglected whenever its value is two or more orders of magnitude smaller than the value from which it is subtracted or added.

Notice, however, that we do not neglect the lone x in determining the equilibrium concentrations of hydroxide ions and benzoic acid. We can neglect x *only* when it appears in a sum or difference and *never when it stands alone*.

Usually, x in a sum or difference is small enough to neglect if the equilibrium constant is smaller than 10^{-4} and the initial concentrations of starting materials are equal to or greater than 0.1 M. Nevertheless, the validity of an approximation must always be verified by comparing the result with the approximation. In other words, after calculating the value of x using an approximation such as $(A - x) \cong A$, check to see whether the calculated value for x indeed is less than 5% of A.

Working with Large Equilibrium Constants

When K_{eq} is very large, the situation is that represented by Figure 15-13b. In such a situation, the reaction proceeds *nearly to completion,* so the difference between completion and equilibrium, represented by y in the figure, is very small. For this reason, it is easier mathematically to work from completion *backward* to equilibrium rather than from initial concentrations *forward* to equilibrium. Thus when an equilibrium constant is very large, we use stoichiometry to determine what the concentrations *would be at completion.* Then we define y to be the amount of back-reaction leading from completion to equilibrium. Solving for y leads to the true equilibrium concentrations. The equilibrium position of the system lies very near the completion point of the reaction, so y will be very small, allowing us to make approximations.

This is a two-step process, so for a reaction with a large K_{eq} we can construct two concentration tables: the first to find the concentrations *at completion* and the second to find concentrations *at equilibrium.* Example 15-15 illustrates an equilibrium position close to completion.

Large Equilibrium Constant

Example 15-15

One of the steps in the industrial production of nitric acid is the combustion of NO at room temperature:

$$2\,NO(g) + O_2(g) \rightleftharpoons 2\,NO_2(g) \qquad K_{eq} = 4.2 \times 10^{12}$$

Suppose a reactor is charged with 10.0 atm each of NO and O_2. Find the partial pressure of each gas at equilibrium.

Strategy: Use the step-by-step approach for an equilibrium problem. Because the magnitude of the equilibrium constant is very large, *almost all* of the starting materials will be converted to products. Because of this, the problem will be easier to solve by taking the reaction to completion and then returning to equilibrium.

Solving
Equilibrium
Problems

Solution:

1. The problem asks us to calculate equilibrium pressures of all reagents.

2. The species present before reaction are NO gas and O_2 gas.

3. The reaction is given in the problem:

$$2\,NO(g) + O_2(g) \rightleftharpoons 2\,NO_2(g)$$

4. We obtain the equilibrium constant expression from the reaction:

$$K_{eq} = \frac{(p_{NO_2})_{eq}^2}{(p_{NO})_{eq}^2\,(p_{O_2})_{eq}} = 4.2 \times 10^{12}$$

5. Next, we set up a concentration table. Because K_{eq} is very large, we first take the reaction to completion. Finding the concentrations at completion is a limiting reactant problem. Because we have equal amounts of the two reactants and the coefficient for NO is 2, NO would be used up first, so it is the limiting reactant, and its partial pressure would be zero at completion. This requires a change in pressure of -10.0 atm for NO. By stoichiometry, the change of pressure for NO_2 is $+10.0$ atm and the change for O_2 is half the change for NO, -5.0 atm.

Reaction (species)	2 NO(g) +	O_2(g) $\rightleftharpoons$	2 NO_2(g)
Initial pressure (atm)	10.0	10.0	0
Change to completion (atm)	-10.0	-5.0	$+10.0$
Pressure at completion (atm)	0	5.0	10.0

Now we work from completion to equilibrium. The equilibrium constant for this reaction is very large, but the partial pressure of NO cannot be zero at equilibrium. We define y to represent the change in NO pressure on going from completion to equilibrium. Then the stoichiometric coefficients and Equation 15-6 give $+0.5y$ for the change in pressure of O_2 and $-y$ for the change in NO_2.

We choose y as the variable rather than x to highlight the fact that the system is coming back to equilibrium from completion as opposed to working forward from initial conditions.

Example 15-15	Large Equilibrium Constant *(continued)*

Reaction (species)	2 NO(g)	+	O_2(g)	$\rightleftharpoons$	2 NO_2(g)
Pressure at completion (atm)	0		5.0		10.0
Change to equilibrium (atm)	$+ y$		$+ 0.5y$		$- y$
Equilibrium pressure (atm)	y		$5.0 + 0.5y$		$10.0 - y$

6. Having obtained expressions for the equilibrium pressures, we substitute these into K_{eq}:

$$K_{eq} = 4.2 \times 10^{12} = \frac{(10.0 - y)^2}{(y)^2 \, (5.0 + 0.5y)}$$

Because K_{eq} is very large, the amount of the limiting reactant present at equilibrium, y, will be very small compared with the amounts of materials present at completion. That is the point of taking the reaction to completion and then working back toward equilibrium. This leads to two approximations:

$$10.0 - y \cong 10.0 \quad and \quad 5.0 + 0.5y \cong 5.0$$

Now solve for y:

$$4.2 \times 10^{12} \cong \frac{(10.0)^2}{(y)^2 \, (5.0)}$$

$$y^2 \cong \frac{(10.0)^2}{(5.0)(4.2 \times 10^{12})}$$

$$y^2 \cong 4.76 \times 10^{-12} \quad and \quad y \cong 2.2 \times 10^{-6}$$

Use the value of y to determine the equilibrium pressures:

$$(p_{NO})_{eq} = y \text{ atm} = 2.2 \times 10^{-6} \text{ atm}$$

$$(p_{O_2})_{eq} = (5.0 + 0.5y) \text{ atm} = 5.0 \text{ atm} \quad (p_{NO_2})_{eq} = (10.0 - y) \text{ atm} = 10.0 \text{ atm}$$

7. The results show that y is more than five orders of magnitude smaller than 5.0 and 10.0 atm, so the approximations are valid.

The examples of this section illustrate the general approach to equilibrium problems. Notice that these examples include gas phase, precipitation, and acid–base chemistry. We chose a variety of equilibrium examples to emphasize that the general strategy for working with equilibria is always the same, no matter what type of equilibrium is involved. In Chapters 16 and 17 we apply these ideas in more detail to important types of equilibria. Our Chemistry and Technology Box describes the role of equilibrium in the chemical industry.

Box 15-2   Chemistry and Technology: Industrial Equilibria

Reactions that reach equilibrium well before completion are not very attractive to the industrial chemist, because unreacted starting materials must be separated from the products, an expensive process. Nevertheless, gas-phase equilibria are involved in the industrial syntheses of H_2SO_4, NH_3, and HNO_3.

Sulfuric acid is produced from elemental sulfur, which is burned in air to give SO_2. This gas reacts with additional O_2 to produce SO_3. The reaction requires a catalyst such as V_2O_5 and a temperature around 700 K. Even so, this reaction reaches equilibrium well before completion.

$$2\ SO_2(g) + O_2(g) \xrightleftharpoons{V_2O_5,\ 700\ K} 2\ SO_3(g)$$

To form sulfuric acid, SO_3 is added to H_2SO_4, and the resulting solution is treated with water:

$$SO_3(g) + H_2SO_4(l) \longrightarrow H_2S_2O_7(l)$$

$$H_2S_2O_7(l) + H_2O(l) \longrightarrow 2\ H_2SO_4(l)$$

The Haber synthesis for the conversion of N_2 into NH_3 is outlined below. The process uses N_2 from the atmosphere, but H_2 must be generated from natural gas (methane) and steam:

$$CH_4(g) + H_2O(g) \xrightleftharpoons{Ni,\ 800\ ^\circ C} CO(g) + 3\ H_2(g)$$

The equilibrium constant for this reaction is small at 800 °C, but the CO is used to make more hydrogen:

$$CO(g) + H_2O(g) \xrightleftharpoons{Fe_2O_3,\ Cr_2O_3,\ 250\ ^\circ C} CO_2(g) + H_2(g)$$

This reaction is run at 250 °C, a temperature at which it has a large equilibrium constant.

The Haber reaction must be catalyzed, and even then it does not approach completion:

$$N_2(g) + 3\ H_2(g) \xrightleftharpoons{Fe,\ K_2O,\ Al_2O_3,\ 450\ ^\circ C} 2\ NH_3(g)$$

Recycling increases the yield. The product gas is chilled below −10 °C, causing ammonia to liquefy. The remaining gases are recycled back into the reactor to make more ammonia.

Much of the ammonia synthesized by the Haber process is used to make HNO_3. Ammonia and air are heated and passed over Pt gauze:

$$4\ NH_3(g) + 5\ O_2(g) \xrightarrow{Pt\ gauze,\ 1200\ K} 4\ NO(g) + 6\ H_2O(g)$$

The combustion reaction goes nearly to completion, but the further reaction of NO and O_2 does not. As it leaves the reactor, NO is at equilibrium with NO_2:

$$2\ NO(g) + O_2(g) \rightleftharpoons 2\ NO_2(g)$$

Because this reaction is exothermic, lowering the temperature favors nitrogen dioxide, so as the products are cooled, more NO is converted into NO_2.

Finally, NO_2 is bubbled through water to give a concentrated aqueous solution of nitric acid, HNO_3:

$$3\ NO_2(g) + H_2O(l) \longrightarrow 2\ HNO_3(aq) + NO(g)$$

Because of the expense that accompanies inefficient syntheses, most industrial processes have been designed using reactions that go virtually to completion.

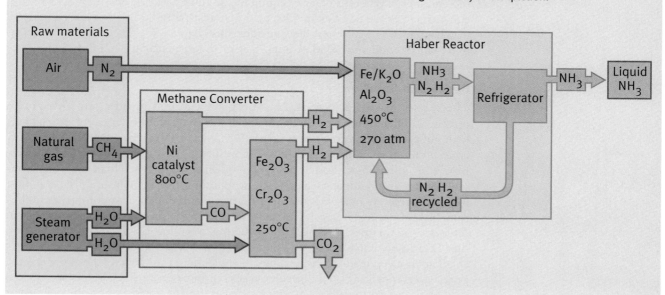

Section Exercises

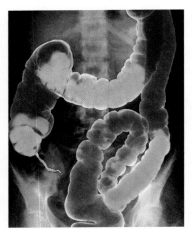

Barium sulfate is opaque to X rays.

15.5.1 Barium sulfate is a relatively insoluble salt used for medical radiographs of the gastrointestinal tract:

$$BaSO_4(s) \rightleftharpoons Ba^{2+}(aq) + SO_4^{2-}(aq)$$

The concentration of Ba^{2+} ions in a saturated aqueous solution is 1.05×10^{-5} M. Determine K_{eq} for barium sulfate dissolving in water.

15.5.2 A solution of acetic acid in water is described by the following equilibrium:

$$CH_3CO_2H(aq) + H_2O(l) \rightleftharpoons CH_3CO_2^-(aq) + H_3O^+(aq)$$

Acetic acid   Acetate

The equilibrium constant for this acid-base reaction is $K_{eq} = 1.8 \times 10^{-5}$. Calculate the equilibrium concentrations of acetic acid, acetate ion, and hydronium ion in a 2.5 M solution of acetic acid.

15.5.3 Silver bromide, a solid used in photographic film, can be prepared by mixing solutions of silver nitrate and potassium bromide.

$$Ag^+(aq) + Br^-(aq) \rightleftharpoons AgBr(s) \qquad K_{eq} = 1.87 \times 10^{12}$$

Calculate the concentrations of Ag^+ and Br^- ions remaining in solution after mixing 2.50 L of 0.100 M $AgNO_3$ solution with 2.50 L of 0.500 M KBr solution.

15.6 EQUILIBRIA IN AQUEOUS SOLUTIONS

Equilibria that occur in aqueous solution are of particular interest, because water is the medium of life and a major influence on the geography of our planet. Many substances dissolve in water, and the solutes in an aqueous solution may participate in a number of different types of equilibria. Solubility itself is one important type of equilibrium, as we describe in Chapter 17. Acid-base reactions, considered in detail in Chapter 16, are another. To conclude this chapter, we describe how to determine which equilibria are most important in any particular aqueous solution.

Species in Solution

Tutorial

The species in an aqueous solution can be categorized broadly into two groups present at different relative concentrations. We designate those present in relatively high concentrations as **major species.** We refer to those present in relatively low concentrations as **minor species.** Minor species in aqueous solutions generally have concentrations at least three orders of magnitude lower than the concentrations of the major solute species. In most solutions, one equilibrium plays the most important role. We call this the **dominant equilibrium.** The dominant equilibrium always has major species as its *reactants*, and the products of that equilibrium often are minor species. Thus, we focus our attention on major species because they determine which equilibria are most important in the solution.

The first step in analyzing an aqueous equilibrium is to identify the major species. Pure water contains H_2O molecules at a concentration of 55.5 M, so an aqueous solution always contains H_2O as a major species. In *pure* water, H_2O is the only major species, but an aqueous *solution* contains two or more major species: H_2O and the solute species present at highest concentration.

As discussed in Section 3.7, whenever an ionic solid dissolves in water, the salt breaks apart to give a solution of cations and anions. Thus in any aqueous salt solution, the major species are *water molecules* and the *cations* and *anions* generated by the salt. For example, a solution of potassium chloride contains K^+ and Cl^- ions and H_2O molecules as major species. Likewise, the major species in a solution of ammonium nitrate are NH_4^+, NO_3^-, and H_2O.

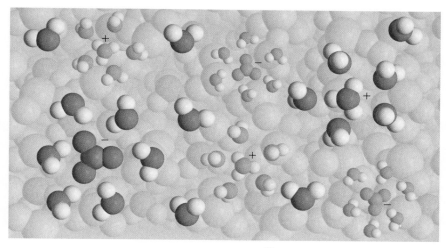

Figure 15-14
The major species present in an aqueous solution of nitric acid are water molecules, hydronium ions, and nitrate anions. The concentration of HNO_3 molecules is negligible.

A strong acid is a substance that reacts virtually completely with water to produce hydronium ions. For example, when nitric acid (HNO_3) dissolves in water, its molecules react with water to produce hydronium ions:

$$H_2O + HNO_3 \longrightarrow H_3O^+(aq) + NO_3^-(aq)$$

This reaction is written with a single arrow to designate that it goes virtually to completion: the equilibrium constant for this proton transfer reaction is so large that the concentration of unreacted acid is negligible. The major species in an aqueous solution of any *strong acid* are H_2O, H_3O^+, and the anion produced by removing a proton from the acid. This situation is illustrated in Figure 15-14, which shows a molecular picture of aqueous nitric acid, and Table 15-2 lists the most important strong acids.

A strong base generates hydroxide ions when it dissolves in water. The most common examples of strong bases (see Table 15-2) are soluble metal hydroxides such as NaOH and KOH. These ionic substances separate into ions when they dissolve in water:

$$KOH(s) \longrightarrow K^+(aq) + OH^-(aq)$$

The major species in any aqueous solution of a *strong base* are H_2O, OH^-, and the *cation* generated by the base.

In addition to salts, strong acids, and strong bases, many molecular substances dissolve in water. These solutes retain their structures, dissolving in aqueous solution to give intact molecules. Vinegar, for example, is a dilute aqueous solution of acetic acid (CH_3CO_2H). The major species present in vinegar are H_2O and CH_3CO_2H. Figure 15-15 shows a molecular picture of the major species in a vinegar solution.

To identify the major species in any aqueous solution, first categorize the solutes. A soluble salt, strong acid, or strong base generates the appropriate cations and anions as major species. Every other solute generates its molecular species in solution. In addition, H_2O is *always* a major species in aqueous solutions. Example 15-16 provides practice in identifying the major species in solution.

← **CHAPTER 4**
Strong acids and bases are first described in Chapter 4.

Table 15-2
Common Strong Acids and Bases

Acids of Industrial Importance	
H_2SO_4	Sulfuric acid
HNO_3	Nitric acid
HCl	Hydrochloric acid
Acids with Laboratory Applications	
$HClO_4$	Perchloric acid
HBr	Hydrobromic acid
HI	Hydriodic acid
Bases of Industrial Importance	
NaOH	Sodium hydroxide
$Ca(OH)_2$	Calcium hydroxide
KOH	Potassium hydroxide
Bases with Laboratory Applications	
LiOH	Lithium hydroxide
$Ba(OH)_2$	Barium hydroxide

Figure 15-15
A molecular view of the major species present in a solution of acetic acid (vinegar).

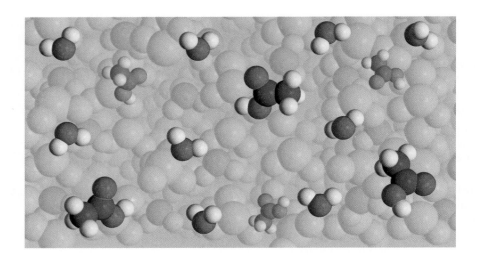

Example 15-16	Major Species in Solution

Identify the major species in each of the following aqueous solutions: (a) $NaCH_3CO_2$ (sodium acetate); (b) $HClO_4$ (perchloric acid); (c) $C_6H_{12}O_6$ (glucose, used for intravenous feeding); and (d) NH_3 (ammonia, used for household cleaning).

Strategy: Identify the nature of each solute. If it is a salt, strong acid, or strong base, it generates ions in aqueous solution. All other solutes give aqueous solutions that contain molecules as the major species.

Solution:

(a) From its formula, you should recognize that $NaCH_3CO_2$ is a salt. It contains the alkali metal cation Na^+ and the polyatomic acetate anion, $CH_3CO_2^-$. According to the solubility guidelines from Chapter 4, all sodium salts are soluble. Thus this salt dissolves, generating an aqueous solution in which the major species are H_2O, Na^+, and $CH_3CO_2^-$:

(b) $HClO_4$ is a strong acid, so it transfers a proton to water quantitatively. (See the list of strong acids in Table 15-2.) The major species in aqueous perchloric acid are H_2O, H_3O^+, and ClO_4^-:

Major Species in Solution *(continued)*   **Example 15-16**

(c) Glucose is not a strong acid, a strong base, or a salt. It dissolves in water without reacting, so the major species in intravenous feeding solutions are molecules of $C_6H_{12}O_6$ and H_2O:

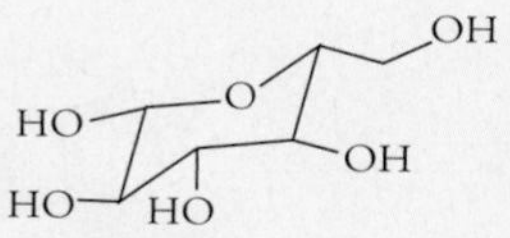

Glucose ($C_6H_{12}O_6$)

(d) NH_3 is not a strong acid, a strong base, or a salt. The major species in household ammonia are molecules of NH_3 and H_2O:

Types of Aqueous Equilibria

Most aqueous equilibria fall into three broad categories: proton transfer, solubility, or complexation. The nature of the major species in the solution determines which category of equilibrium we need to consider.

One of the most important types of aqueous equilibrium involves proton transfer from an acid to a base. In aqueous solutions, water can act as an acid or a base. In the presence of an acid, symbolized HA, water acts as a base by accepting a proton. The equilibrium constant for transfer of a proton from an acid to a water molecule is called the **acid ionization constant (K_a):**

$$HA(aq) + H_2O(l) \rightleftarrows A^-(aq) + H_3O^+(aq) \qquad K_{eq} = \frac{[A^-]_{eq}\,[H_3O^+]_{eq}}{[HA]_{eq}} = K_a$$

For instance, when acetic acid, an example of a weak acid, dissolves in water, some acetic acid molecules transfer protons to water molecules to produce acetate anions and hydronium ions:

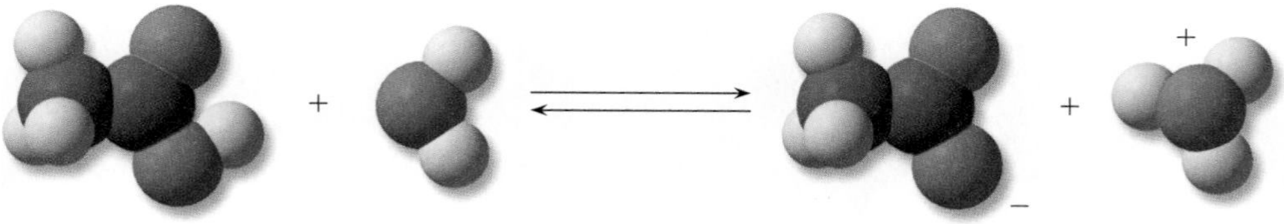

$$K_a = \frac{[CH_3CO_2^-]_{eq}[H_3O^+]_{eq}}{[CH_3CO_2H]_{eq}}$$

In the presence of a base B, water acts as an acid by donating a proton, and the equilibrium constant for the transfer of a proton from water to a base is called the **base ionization constant (K_b):**

$$B(aq) + H_2O(l) \rightleftharpoons BH^+(aq) + OH^-(aq) \qquad K_{eq} = \frac{[BH^+]_{eq}[OH^-]_{eq}}{[B]_{eq}} = K_b$$

When ammonia, an example of a weak base, dissolves in water, some ammonia molecules accept protons from water to produce ammonium cations and hydroxide ions:

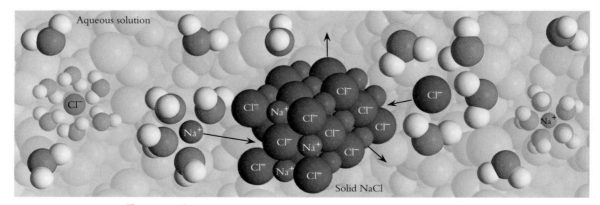

$$NH_3(aq) + H_2O(l) \rightleftharpoons NH_4^+(aq) + OH^-(aq) \qquad K_b = \frac{[NH_4^+]_{eq}[OH^-]_{eq}}{[NH_3]_{eq}}$$

Notice that the expressions for K_a and K_b do not include the water molecules that act as starting materials for the proton transfer reactions. As described in Section 15.2, water, as the solvent, is always present in huge excess, so its concentration is virtually constant. Thus the concentration of water does not change significantly during an acid–base reaction and is omitted from K_a and K_b.

Many substances that participate in aqueous reactions are soluble salts. These ionic solids dissolve in water to give solutions of cations and anions. For almost all salts, there is an upper limit to the amount that will dissolve in water. A salt solution is *saturated* when the amount dissolved has reached this upper limit of solubility. Any additional salt added to a saturated solution remains undissolved at the bottom of the vessel. When excess solid salt and a saturated solution are present, a solubility equilibrium exists. At equilibrium, ions dissolve continually, but other ions precipitate out at exactly the same rate. This equilibrium is illustrated for sodium chloride by the molecular picture in Figure 15-16.

By convention, a solubility equilibrium is written in the direction of a solid dissolving to give aqueous ions, and the equilibrium constant for this reaction is called the **solubility product (K_{sp}).**

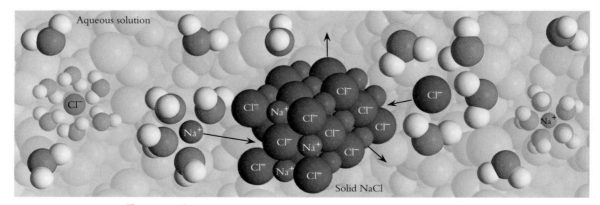

Figure 15-16
A molecular view of the solubility equilibrium for a solution of sodium chloride in water. At equilibrium, ions dissolve from the crystal surface at the same rate they are captured, so the concentration of ions in the solution remains constant.

Here, for example, is the reaction describing the solubility equilibrium of copper(II) chloride:

$$CuCl_2(s) \rightleftharpoons Cu^{2+}(aq) + 2\,Cl^-(aq) \qquad K_{eq} = [Cu^{2+}]_{eq}\,[Cl^-]^2_{eq} = K_{sp}$$

The evaporation of water from a saturated solution leaves a solution in which the ion concentrations exceed the solubility limit. To return to equilibrium, the salt must precipitate from the solution. Evaporation is used to "mine" sodium chloride and other salts from the highly salty waters of inland seas such as Great Salt Lake in Utah and Israel's Dead Sea.

Metal cations in aqueous solution often form chemical bonds to anions or neutral molecules that have lone pairs of electrons. A silver cation, for example, can associate with two ammonia molecules to form a silver–ammonia complex:

$$H\!-\!\overset{\displaystyle H}{\underset{\displaystyle H}{N}}\!:\; +\; Ag^+ \;+\; :\!\overset{\displaystyle H}{\underset{\displaystyle H}{N}}\!-\!H \;\rightleftharpoons\; H\!-\!\overset{\displaystyle H}{\underset{\displaystyle H}{N}}\!-\!Ag^+\!-\!\overset{\displaystyle H}{\underset{\displaystyle H}{N}}\!-\!H$$

The resulting species is called a **complex ion.** The equilibrium constant for the formation of a complex ion is called its **formation constant (K_f).** Tabulated values of K_f always refer to the equilibrium constant for the complex forming from the metal cation. Here, for example, is the reaction describing the complexation equilibrium between Ag^+ and NH_3:

$$Ag^+(aq) + 2\,NH_3(aq) \rightleftharpoons [Ag(NH_3)_2]^+(aq) \qquad K_f = \frac{[Ag(NH_3)_2{}^+]_{eq}}{[Ag^+]_{eq}\,[NH_3]^2_{eq}}$$

Notice that square brackets are used in a new way for the silver–ammonia complex in the chemical reaction. Chemists use square brackets to identify a complex ion such as $[Ag(NH_3)_2]^+$, because the species involved in complexation, NH_3 in this instance, are set off by parentheses. When square brackets appear in an equilibrium constant expression, they always designate concentration; when they appear in a chemical reaction, they designate a complex.

Identifying Types of Equilibria

The major species in an aqueous solution determine the categories of equilibria that are important for that solution. Each major species present in the solution must be examined in light of these general categories. Are any of the major species weak acids or weak bases? Are there ions present that combine to form an insoluble salt? Do any of the major species participate in more than one equilibrium? Any equilibrium may be approached from either direction, so there are six different types of aqueous equilibria in which major species are reactants:

1. An acid may donate a proton to water or some other base.
2. A base may accept a proton from water or some other acid.
3. A salt may dissolve in water.
4. Cations and anions in a solution may react to form a solid.
5. A cation may associate with electron-donating species to form a complex.
6. A complex may dissociate.

We consider each of these in more detail in subsequent chapters, but being able to identify types of equilibria helps greatly in solving equilibrium problems. Example 15-17 provides practice in identifying equilibria.

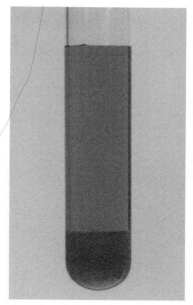

The color of an aqueous solution of a copper salt shows that copper cations are present in the aqueous phase at equilibrium with the solid salt.

← **CHAPTER 19**
We consider the formation of metal complexes in more depth in Chapter 19.

| Example 15-17 | Types of Aqueous Equilibria |

Write expressions for the equilibrium constants and decide which general category applies to each of these equilibria:

(a) $H_2CO_3(aq) + H_2O(l) \rightleftharpoons HCO_3^-(aq) + H_3O^+(aq)$
(b) $Fe^{3+}(aq) + 3\ OH^-(aq) \rightleftharpoons Fe(OH)_3(s)$
(c) $NH_4^+(aq) + OH^-(aq) \rightleftharpoons NH_3(aq) + H_2O(l)$

Strategy: Examine each reaction closely, and look for the characteristic features of the general equilibria. The equilibrium constant expression can be written by inspecting the overall stoichiometry.

Solution:

(a) In this reaction, water accepts a proton from H_2CO_3. The equilibrium constant for a proton transfer reaction that consumes H_2O and produces H_3O^+ is called K_a:

$$\frac{[H_3O^+]_{eq}\ [HCO_3^-]_{eq}}{[H_2CO_3]_{eq}} = K_a$$

(b) This is a precipitation reaction. Solid $Fe(OH)_3$ forms in the forward direction. In the reverse direction, solid $Fe(OH)_3$ dissolves. This reverse reaction is a solubility reaction for which $K_{eq} = K_{sp}$. According to Equation 15-2, K_{eq} must be $1/K_{sp}$ for the reaction in the direction written:

$$\frac{1}{[Fe^{3+}]_{eq}\ [OH^-]_{eq}^3} = \frac{1}{K_{sp}}$$

(c) This reaction involves proton transfer. In the *forward* direction, a hydroxide ion removes a proton from an ammonium ion. In the *reverse* direction, a proton is transferred from water to ammonia. This reverse reaction is included among the general categories; it is the transfer of a proton from a water molecule to a base. The equilibrium constant for this reaction is K_b. In the direction the reaction is written, the equilibrium constant is $1/K_b$:

$$\frac{[NH_3]_{eq}}{[NH_4^+]_{eq}\ [OH^-]_{eq}} = \frac{1}{K_b}$$

Each chemical equilibrium has a characteristic equilibrium constant. Many of these have been measured and tabulated. Representative K_a, K_b, and K_{sp} values appear in Appendix E, and more extensive tables can be found in *The Handbook of Chemistry and Physics*.

Figure 15-17
Mixing aqueous solutions of potassium iodide and lead(II) nitrate results in the formation of a yellow precipitate. The precipitate is lead(II) iodide.

Spectator Ions

With all the possible equilibria in aqueous systems, most species might be expected to participate in at least one equilibrium. Nonetheless, in many solutions some of the ionic species undergo no significant reactions. These species are classified as **spectator ions.**

Recall from Section 4.5 what happens when a solution of potassium iodide is mixed with a solution of lead(II) nitrate. The major species present are $K^+(aq)$, $I^-(aq)$, $Pb^{2+}(aq)$, $NO_3^-(aq)$, and H_2O. As shown in Figure 15-17, a bright yellow solid precipitates from the mixture. This solid is insoluble PbI_2. The Pb^{2+} and I^- ions that remain in solution are at equilibrium with the precipitate:

$$PbI_2(s) \rightleftharpoons Pb^{2+}(aq) + 2\ I^-(aq) \qquad K_{sp} = 9.8 \times 10^{-9}$$

Notice that two of the major species are not involved in the solubility equilibrium. Potassium ions and nitrate ions are neither acids nor bases, and their salts are soluble. Thus they are spectator ions in this system. Example 15-18 provides practice in identifying spectator ions.

Spectator Ions

Example 15-18

When a 0.100 M solution of sodium bromide is mixed with an equal volume of a 0.100 M solution of silver nitrate, a white solid precipitates from the solution. Identify the precipitate, write the net ionic reaction for the solubility equilibrium, and identify any spectator ions.

Strategy: Analyze the problem from the molecular perspective. Begin by identifying the major species present. Then identify the equilibria in which these ions participate.

Solution: Sodium bromide and silver nitrate are both ionic compounds, so the major species are Na^+, Ag^+, Br^-, NO_3^-, and H_2O molecules.
The solubility guidelines presented in Section 4.5 identify the precipitate as silver bromide:

$$AgBr(s) \rightleftharpoons Ag^+(aq) + Br^-(aq) \qquad K_{sp} = 5.35 \times 10^{-13}$$

Major ionic species that do not participate in aqueous equilibria are classified as spectator ions. In this case the spectator ions are Na^+ and NO_3^-.

The types of aqueous equilibria described in this section have been given special names, and it is essential that you be able to recognize them. Keep in mind, however, that the principles described in the previous sections apply to all chemical equilibria. Chemists categorize equilibria for convenience, but they treat all equilibria the same way.

Section Exercises

■ **15.6.1** What are the major species present in each of the following solutions: (a) 1.00 M perchloric acid; (b) 0.25 M ammonia; (c) 0.50 M potassium hydrogen carbonate; and (d) 0.010 M hypochlorous acid, HClO?

■ **15.6.2** Relate each of the following equilibrium reactions to equilibrium constants of standard types: K_{sp}, K_a, and K_b:
(a) $Al^{3+}(aq) + 3\ OH^-(aq) \rightleftharpoons Al(OH)_3(s)$
(b) $HCN(aq) + H_2O(l) \rightleftharpoons CN^-(aq) + H_3O^+(aq)$
(c) $NH_3(aq) + H_2O(l) \rightleftharpoons NH_4^+(aq) + OH^-(aq)$

■ **15.6.3** A green precipitate forms when a solution of sodium hydroxide is added to a solution of nickel(II) sulfate. Identify the precipitate, write the net ionic equation for the solubility equilibrium, and identify the spectator species.

■ CHAPTER REVIEW

Summary and Key Terms

1. A chemical reaction establishes a **dynamic equilibrium** in which all steps in the mechanism of the reaction proceed with equal rates in both directions. The **equilibrium constant, K_{eq},** describes the relationship among concentrations of reactants and products at equilibrium. It contains a constant on the left-hand side and the ratio of equilibrium concentrations, raised to powers equal to their stoichiometric coefficients, on the right-hand side. Because of the **reversibility** of molecular reactions, dynamic equilibria can be reached from either direction. The reaction quotient, Q, is related to the equilibrium constant: $Q = K_{eq}$ when $\Delta G = 0$.

2. The equilibrium constant expression can be derived from the balanced chemical reaction. Equilibria can be written in either direction. Equilibrium constants are dimensionless numbers, because each concentration in the reaction quotient is divided by the standard concentration for that substance. By convention, the concentrations of solutes are expressed as molarities, and the concentrations of gases are expressed as partial pressures in atmospheres. The concentrations of pure liquids, pure solids, and solvents are not included in K_{eq}. Vaporization equilibrium is characterized by the equilibrium **vapor pressure.** Gas-solution equilibria are described by the **Henry's law constant, K_H.**

3. Equilibrium constants are directly related to standard free energies of reaction, so K_{eq} can be calculated from standard thermodynamic properties. The magnitude of Q relative to K_{eq} determines the direction in which a reaction proceeds. Thermodynamics also explains the variations in K_{eq} with temperature.

4. When a system is at equilibrium, a change in conditions may disturb the equilibrium. **Le Châtelier's principle** asserts that the system reacts in the direction that reduces the effect of the change.

5. Calculations of equilibrium concentrations are best done using the step-by-step procedure of identifying the chemistry, determining the equilibrium that exists, writing the equilibrium constant expression, completing a **concentration table,** and then substituting equilibrium concentrations into the equilibrium constant expression. The entries in concentration tables are **initial concentrations,** changes to equilibrium, and equilibrium concentrations. Equilibrium calculations can often be simplified by making appropriate approximations about equilibrium concentrations. When a reaction has a very large equilibrium constant, equilibrium concentrations are most conveniently found from the difference between concentrations at completion and concentrations at equilibrium.

6. The **dominant equilibrium** in an aqueous solution has **major species** as reactants and often generates **minor species** as products. Solutions of salts, strong acids and strong bases contain ions as major species. Acid-base equilibria are described by the **acid ionization constant (K_a)** and the **base ionization constant (K_b).** Salt solubilities are described by the **solubility product (K_{sp}),** and **complex ion** formation is described by a **formation constant (K_f).** Ions that do not participate in an aqueous equilibrium are classified as **spectator ions.**

Skills to Master

▶ Visualizing molecular processes at dynamic equilibrium

▶ Writing equilibrium constant expressions

▶ Calculating K_{eq} from thermodynamic properties

▶ Predicting changes in K_{eq} with temperature

▶ Predicting the effects of changes in conditions on equilibrium

▶ Calculating K_{eq} from equilibrium concentrations

▶ Calculating equilibrium concentrations

▶ Completing concentration tables

▶ Using approximations in equilibrium calculations

▶ Recognizing major species in aqueous solution

▶ Recognizing types of equilibria

▶ Relating equilibrium constant expressions to tabulated values

Learning Exercises

15.1 List the types of chemical equilibrium introduced in this chapter and give a specific example of each.

15.2 Draw a molecular picture that illustrates each of the examples that you gave in Learning Exercise 15-1.

15.3 Describe in your own words how to set up and solve a problem that asks for concentrations at equilibrium.

15.4 Describe how to calculate K_{eq} from equilibrium concentrations and from thermodynamic functions.

15.5 Write a paragraph that describes the logic, process, and justification for using approximations in solving equilibrium problems.

15.6 Use your own words to define (a) initial concentrations, (b) change to equilibrium, (c) change to completion, and (d) equilibrium concentrations.

15.7 Summarize in writing the connections between equilibrium constants and thermodynamic functions.

15.8 Update your list of memory bank equations and their uses.

15.9 List all terms new to you that appear in Chapter 15. Write a one-sentence definition for each in your own words. Consult the Glossary if you need help.

Problems   ilw = interactive learning ware problem. Visit the website at www.wiley.com/college/olmsted

Describing Chemical Equilibria

15.1 The alkene 2-butene has two isomers, *cis*-butene and *trans*-butene, which interconvert at high temperature or in the presence of a catalyst:

cis-Butene (g)   *trans*-Butene (g)

Suppose that the equilibrium constant for this reaction is 3.0. Draw a qualitative graph that shows how the pressure of each gas changes with time if the system initially contains pure *cis*-butene.

15.2 Redraw the graph for the reaction in Problem 15.1, showing what happens if the system initially contains pure *trans*-butene.

15.3 Aqueous solutions of sodium hypochlorite undergo decomposition according to the following two-step mechanism:

$$ClO^-(aq) + ClO^-(aq) \xrightarrow{k_1} Cl^-(aq) + ClO_2^-(aq)$$

$$ClO^-(aq) + ClO_2^-(aq) \xrightarrow{k_2} Cl^-(aq) + ClO_3^-(aq)$$

(a) What additional reactions will be important as the reaction approaches equilibrium? (b) What is the equilibrium constant expression for the overall reaction? (c) Express the equilibrium constant in terms of the rate constants of the elementary reactions.

15.4 The reaction of H_2 gas with CO gas to give formaldehyde has been described by this mechanism (all species in the gas phase):

$$H_2 \underset{k_{-1}}{\overset{k_1}{\rightleftharpoons}} 2\,H$$

$$H + CO \xrightarrow{k_2} HCO$$

$$HCO + H \xrightarrow{k_3} H_2CO$$

(a) What reactions describe the decomposition of H_2CO under these same conditions? (b) What is the equilibrium constant expression for this reaction? (c) Express the equilibrium constant in terms of rate constants for the elementary reactions.

15.5 Draw molecular pictures that illustrate the reversibility of the reactions involved in the hypochlorite decomposition given in Problem 15.3. (See Figure 15-1 for examples.)

15.6 Draw molecular pictures that illustrate the reversibility of the reactions involved in the reaction of H_2 with CO to form H_2CO given in Problem 15.4. (See Figure 15-1 for examples.)

15.7 Describe experiments that could be done to demonstrate that the decomposition reaction of ClO^- anions (Problem 15.3) is reversible.

15.8 Describe experiments that could be done to demonstrate that the reaction of H_2 with CO (Problem 15.4) is reversible.

Properties of Equilibrium Constants

15.9 Write the equilibrium constant expression for each reaction:
(a) $I_2(s) + 5\,F_2(g) \rightleftharpoons 2\,IF_5(g)$
(b) $P_4(s) + 5\,O_2(g) \rightleftharpoons P_4O_{10}(s)$
(c) $BaCO_3(s) + C(s) \rightleftharpoons BaO(s) + 2\,CO(g)$
(d) $CO(g) + 2\,H_2(g) \rightleftharpoons CH_3OH(l)$
(e) $H_3PO_4(aq) + 3\,H_2O(l) \rightleftharpoons PO_4^{3-}(aq) + 3\,H_3O^+(aq)$

15.10 Write the equilibrium constant expression for each reaction:
(a) $2\,H_2S(g) + 3\,O_2(g) \rightleftharpoons 2\,H_2O(l) + 2\,SO_2(g)$
(b) $Fe_2O_3(s) + 3\,CO(g) \rightleftharpoons 2\,Fe(s) + 3\,CO_2(g)$
(c) $Cl_2(g) + 3\,F_2(g) \rightleftharpoons 2\,ClF_3(g)$
(d) $NH_3(g) + H_3O^+(aq) \rightleftharpoons NH_4^+(aq) + H_2O(l)$
(e) $SnO_2(s) + 2\,H_2(g) \rightleftharpoons Sn(s) + 2\,H_2O(l)$

15.11 Write each of the chemical reactions of Problem 15.9 in the opposite direction and determine the equilibrium constant expressions for these reactions.

15.12 Write each of the chemical reactions of Problem 15.10 in the opposite direction and determine the equilibrium constant expressions for these reactions.

15.13 State the standard (reference) concentration for each substance appearing in the equilibria of Problem 15.9.

15.14 State the standard (reference) concentration for each substance appearing in the equilibria of Problem 15.10.

15.15 Draw a molecular picture similar to the one in Figure 15-5 that shows how reducing the pressure of O_2 above a body of water leads to a reduced concentration of dissolved O_2.

15.16 Using mercury as an example, draw a molecular picture that shows how an increase in temperature leads to an increase in the vapor pressure above a liquid. Mercury is monatomic in both liquid and vapor phase.

15.17 One of the reasons that different aquatic life forms thrive in water of different temperatures is the variation with temperature in the concentration of dissolved oxygen. Using data in Table 15-1, calculate the percentage change in oxygen concentration when water is cooled from 25.0 to 0.0 °C.

15.18 One detrimental effect of "thermal pollution" of water supplies is that a rise in temperature reduces the amount of dissolved O_2 available for fish. Using information in Table 15-1, calculate how many liters of water a fish requires at 30.0 °C to obtain the same amount of O_2 that it could obtain from 1.00 L of water at 25.0 °C.

Thermodynamics and Equilibrium

15.19 Using standard thermodynamic data from Appendix D, calculate the equilibrium constant at 298 K for each of the following chemical equilibria:
(a) $CO(g) + H_2O(l) \rightleftharpoons CO_2(g) + H_2(g)$
(b) $2\,CO(g) + O_2(g) \rightleftharpoons 2\,CO_2(g)$
(c) $BaCO_3(s) + C(s, graphite) \rightleftharpoons BaO(s) + 2\,CO(g)$
(d) $3\,CO(g) + 6\,H_2(g) \rightleftharpoons C_3H_6(g) + 3\,H_2O(l)$

15.20 Using standard thermodynamic data from Appendix D, calculate the equilibrium constant at 298 K for each of the following chemical equilibria:
(a) $CH_4(g) + H_2O(l) \rightleftharpoons CO(g) + 3\,H_2(g)$
(b) $4\,NH_3(g) + 5\,O_2(g) \rightleftharpoons 4\,NO(g) + 6\,H_2O(l)$
(c) $SnO_2(s) + 2\,H_2(g) \rightleftharpoons Sn(s, white) + 2\,H_2O(l)$
(d) $3\,Fe(s) + 4\,H_2O(l) \rightleftharpoons Fe_3O_4(s) + 4\,H_2(g)$

15.21 Estimate K_{eq} for the equilibria in Problem 15.19 b and c at 250 K.

15.22 Estimate K_{eq} for the equilibria in Problem 15.20 c and d at 350 K.

15.23 Determine K_{eq} for the equilibria in Problem 15.19 a and d at 395 K if H_2O is present as a gas rather than as a liquid.

15.24 Estimate K_{eq} for the equilibria in Problem 15.20 c and d at 425 K if H_2O is present as a gas rather than as a liquid.

Shifts in Equilibrium

15.25 For each equilibrium in Problem 15.19, predict the effect of injecting additional CO(g) into the system.

15.26 For each equilibrium in Problem 15.20, predict the effect of injecting additional $H_2O(l)$ into the system.

15.27 Consider the following reaction at equilibrium in water:

$$PbCl_2(s) \rightleftharpoons Pb^{2+}(aq) + 2\ Cl^-(aq)$$

Predict whether $PbCl_2$ will dissolve, precipitate, or do neither after each of the following changes: (a) More $PbCl_2(s)$ is added. (b) More H_2O is added. (c) Solid NaCl is added. (d) Solid KNO_3 is added.

15.28 The following exothermic gas–phase reaction is at equilibrium:

$$2\ SO_2(g) + O_2(g) \rightleftharpoons 2\ SO_3(g)$$

Predict what happens to the amount of SO_3 in the system when each of the following changes is made: (a) The temperature is raised. (b) More O_2 is added. (c) Some Ar gas is introduced.

15.29 Consider the following gas-phase reaction:

$$SO_2(g) + Cl_2(g) \rightleftharpoons SO_2Cl_2(g) \qquad \text{(exothermic)}$$

Describe four changes that would drive the equilibrium to the left.

15.30 Consider the following gas-phase reaction:

$$PCl_5(g) \rightleftharpoons PCl_3(g) + Cl_2(g) \qquad \text{(endothermic)}$$

Describe four changes that would drive the equilibrium to the left.

Working with Equilibria

ilw **15.31** An industrial chemist puts 1.00 mol each of $H_2(g)$ and $CO_2(g)$ in a 1.00-L container at constant temperature of 800 °C. This reaction occurs:

$$H_2(g) + CO_2(g) \rightleftharpoons H_2O(g) + CO(g)$$

When equilibrium is reached, 0.49 mol of CO(g) is in the container. Find the value of K_{eq} for the reaction.

15.32 At high temperature, HCl and O_2 react to give Cl_2 gas:

$$4\ HCl(g) + O_2(g) \rightleftharpoons 2\ Cl_2(g) + 2\ H_2O(g)$$

If HCl at 2.30 atm and O_2 at 1.00 atm react at 750 K, the equilibrium pressure of Cl_2 is measured to be 0.93 atm. Determine the value of K_{eq} at 750 K.

15.33 When 0.0500 mol of propanoic acid ($C_2H_5CO_2H$) is dissolved in 0.500 L of water, proton transfer occurs:

$$C_2H_5CO_2H + H_2O \rightleftharpoons C_2H_5CO_2^- + H_3O^+$$

The equilibrium concentration of H_3O^+ ions is measured to be 1.15×10^{-3} M. Evaluate K_{eq}.

15.34 Cyanic acid, HCNO, is a weak acid:

$$HCNO + H_2O \rightleftharpoons CNO^- + H_3O^+$$

In a 0.20 M aqueous solution of HCNO, the concentration of H_3O^+ cations is 6.5×10^{-3} M. Evaluate K_{eq}.

ilw **15.35** The equilibrium constant for the following reaction is 1.6×10^5 at 1024 K:

$$H_2(g) + Br_2(g) \rightleftharpoons 2\ HBr(g)$$

Find the equilibrium pressures of all gases if 10.0 atm of HBr is introduced into a sealed container at 1024 K.

15.36 At 100 °C, $K_{eq} = 1.5 \times 10^8$ for the following reaction:

$$CO(g) + Cl_2(g) \rightleftharpoons COCl_2(g)$$

Using appropriate approximations, calculate the partial pressure of CO at 100 °C at equilibrium in a chamber that initially contains $COCl_2$ at a pressure of 0.250 atm.

15.37 At 1000 °C, $K_{eq} = 0.403$ for the following reaction:

$$FeO(s) + CO(g) \rightleftharpoons Fe(s) + CO_2(g)$$

If CO gas at 5.0 atm is injected into a container at 1000 °C that contains excess FeO, what are the partial pressures of all gases present at equilibrium?

15.38 The equilibrium constant for the dissociation of Cl_2 into atomic chlorine at 1200 K is 2.5×10^{-5}. If 5.0 g of Cl_2 gas is placed in a 3.0-L container and heated to 1200 K, what is the equilibrium pressure of Cl atoms? Use appropriate approximations.

Equilibria in Aqueous Solutions

15.39 Identify the major species present in an aqueous solution of each of the following substances: (a) CH_3CO_2H; (b) NH_4Cl; (c) KCl; (d) $NaCH_3CO_2$; and (e) NaOH.

15.40 Identify the major species present in an aqueous solution of each of the following substances: (a) HClO; (b) $CaBr_2$; (c) KClO; (d) HNO_3; and (e) HCN.

15.41 Identify the acid-base equilibria, if any, for the major species in Problem 15.39.

15.42 Identify the acid-base equilibria, if any, for the major species in Problem 15.40.

15.43 When the following substances dissolve in water, what major species are present: (a) acetone; (b) potassium bromide; (c) lithium hydroxide; and (d) sulfuric acid?

15.44 When the following substances dissolve in water, what major species are present: (a) sodium hydrogen carbonate; (b) methanol; (c) hydrogen bromide; and (d) benzoic acid?

15.45 Write the equilibrium constant expression for each of the following reactions and relate it to equilibrium constants of standard types: K_{sp}, K_a, and K_b:
(a) $HClO_2(aq) + H_2O(l) \rightleftharpoons ClO_2^-(aq) + H_3O^+(aq)$
(b) $Fe^{3+}(aq) + 3\ OH^-(aq) \rightleftharpoons Fe(OH)_3(s)$
(c) $CN^-(aq) + H_3O^+(aq) \rightleftharpoons HCN(aq) + H_2O(l)$

15.46 Write the equilibrium constant expression for each of the following reactions and relate it to equilibrium constants of standard types: K_{sp}, K_a, and K_b:
(a) $Ag_2SO_4(s) \rightleftharpoons 2Ag^+(aq) + SO_4^{2-}(aq)$
(b) $H_3PO_4(aq) + H_2O(l) \rightleftharpoons H_2PO_4^-(aq) + H_3O^+(aq)$
(c) $HCO_2^-(aq) + H_3O^+(aq) \rightleftharpoons HCO_2H(aq) + H_2O(l)$

15.47 Identify the spectator ions present when the following are mixed:
(a) a solution of CH_3CO_2H and a solution of NaOH
(b) a solution of $CaCl_2$ and a solution of K_3PO_4
(c) a solution of HNO_3 and a solution of KOH

15.48 Identify the spectator ions present when the following are mixed:
(a) a solution of $FeCl_3$ and a solution of NaOH
(b) a solution of NH_3 and a solution of HNO_3
(c) a solution of KCH_3CO_2 and a solution of $HClO_4$

Additional Paired Problems

15.49 Write the equilibrium constant expressions for the following:
(a) $2\,NaHCO_3(s) \rightleftharpoons Na_2CO_3(s) + CO_2(g) + H_2O(g)$
(b) $2\,N_2(g) + 6\,H_2O(l) \rightleftharpoons 4\,NH_3(g) + 3\,O_2(g)$
(c) $2\,C_2H_4(g) + O_2(g) \rightleftharpoons 2\,CH_3CHO(g)$
(d) $Ag_2SO_4(s) \rightleftharpoons 2\,Ag^+(aq) + SO_4^{2-}(aq)$
(e) $NH_4HS(s) \rightleftharpoons H_2S(g) + NH_3(g)$

15.50 Write the equilibrium constant expressions for the following:
(a) $4\,NH_3(g) + 5\,O_2(g) \rightleftharpoons 4\,NO(g) + 6\,H_2O(l)$
(b) $HCN(aq) + H_2O(l) \rightleftharpoons CN^-(aq) + H_3O^+(aq)$
(c) $2\,NH_4^+(aq) + SO_4^{2-}(aq) \rightleftharpoons (NH_4)_2SO_4(s)$
(d) $SO_4^{2-}(aq) + H_2O(l) \rightleftharpoons OH^-(aq) + HSO_4^-(aq)$
(e) $2\,O_3(g) \rightleftharpoons 3\,O_2(g)$

15.51 Phosgene is a deadly gas whose use in warfare has been outlawed. The gas decomposes at elevated temperature:

$$COCl_2(g) \rightleftharpoons CO(g) + Cl_2(g) \qquad K_{eq} = 8.3 \times 10^{-4} \text{ at } 360\,°C$$

What are the pressures at equilibrium if a 2.55-L metal container containing 0.31 mol of phosgene is heated to 360 °C?

15.52 K_{eq} for the Haber reaction is 2.81×10^{-5} at 472 °C. If a reaction starts with 3.0 atm of H_2 and 5.0 atm of N_2 at 472 °C, what is the equilibrium pressure of NH_3?

15.53 A test tube and a petri dish, each containing water, are placed in a bell jar. Water evaporates from both samples until gas–liquid equilibrium is reached. More water evaporates from the petri dish than from the test tube, but at equilibrium, the vapor pressures above the two vessels are equal. Explain in molecular terms.

15.54 A student mixes sodium chloride and water until the solid and the aqueous solution are at equilibrium. If more solid sodium chloride is added to this mixture, does the concentration of sodium chloride in the aqueous solution change? Explain your answer in molecular terms.

15.55 The figure represents two chambers containing gases that react as follows:

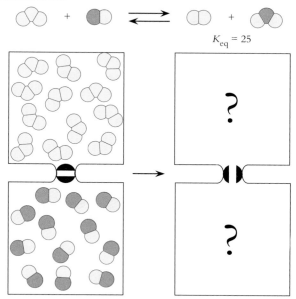

Draw a molecular picture that shows what the system looks like if the stopcock is opened and the reaction proceeds to equilibrium.

15.56 The figure represents two chambers containing gases that react as follows:

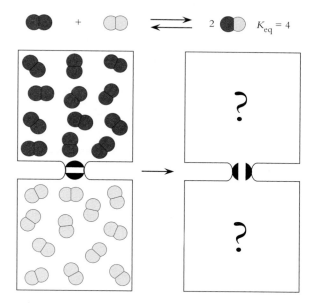

Draw a molecular picture that shows what the system looks like if the stopcock is opened and the reaction proceeds to equilibrium.

15.57 At 350 K, K_{eq} for the following reaction is 322:

$$Br_2(g) + I_2(g) \rightleftharpoons 2\,IBr(g)$$

Suppose the equilibrium partial pressure of bromine is 0.512 atm and that of iodine is 0.327 atm. What is the pressure of IBr?

15.58 Coal (solid carbon plus a collection of impurities) can be converted to a mixture of carbon dioxide and hydrogen gas by the following process:

$$C(s) + 2\,H_2O(g) \rightleftharpoons CO_2(g) + 2\,H_2(g)$$

$$K_{eq} = 0.38 \text{ at } 1300\,K$$

At an industrial plant, a mixture of coal and water is placed inside a steel reaction vessel. When the mixture is heated to 1300 K, the equilibrium partial pressure of water is 2.80×10^2 atm. Calculate the partial pressures of H_2 and CO_2.

15.59 Using tabulated standard thermodynamic data from Appendix D, calculate K_{eq} for the reaction of NO_2 to form N_2O_4 at 298 K and at 525 K:

$$2\,NO_2(g) \rightleftharpoons N_2O_4(g)$$

15.60 Using tabulated standard thermodynamic data from Appendix D, calculate K_{eq} for the following reaction at 298 K and at 825 K:

$$2\,N_2O(g) + O_2(g) \rightleftharpoons 4\,NO(g)$$

15.61 Write the equilibrium reaction and equilibrium constant expression for each of the following processes: (a) Trimethylamine, $(CH_3)_3N$, a weak base, is added to water. (b) Hydrofluoric acid, HF, a weak acid, is added to water. (c) Solid calcium sulfate, $CaSO_4$, a sparingly soluble salt, is added to water.

15.62 Write the equilibrium reaction and equilibrium constant expression for each of the following processes: (a) Some carbon monoxide gas is placed above water in a sealed vessel. (b) Aniline, $(C_6H_5)NH_2$, a weak base, is dissolved in water. (c) Solid calcium hydroxide, $Ca(OH)_2$, a sparingly soluble salt, is added to water.

15.63 Predict the effect of each of the following changes on the equilibrium position of the reaction in Example 15-15: (a) The partial pressure of NO_2 is cut in half. (b) The volume of the reactor is doubled. (c) A total of 10.0 atm of Ar gas is added to the reactor.

15.64 Predict the effect each of the following changes will have on the equilibrium position of the reaction in Example 15-13: (a) Some sodium benzoate is dissolved in the solution. (b) An additional 1.0 L of water is added to the solution. (c) Some NaCl is dissolved in the solution.

More Challenging Problems

15.65 $K_{eq} = 1.07 \times 10^{-33}$ for the following reaction at 298 K:

$$Sn(s) + 2\ H_2(g) \rightleftharpoons SnH_4(g)$$

Find the equilibrium pressure of $SnH_4(g)$ in a container at 298 K containing 10.0 g of Sn(s) and 2.00×10^2 atm of H_2. What volume of container is expected to contain a single molecule of SnH_4?

15.66 Hydrogen fluoride is a highly reactive gas. It has many industrial uses, but the most familiar property of HF is its ability to react with glass. As a result, HF is used to etch glass and frost the inner surfaces of light bulbs. Hydrogen fluoride gas must be stored in stainless steel containers, and aqueous solutions must be stored in plastic bottles. Hydrogen fluoride can be produced from H_2 and F_2:

$$H_2(g) + F_2(g) \rightleftharpoons 2\ HF(g) \qquad K_{eq} = 115$$

In a particular experiment, 3.00 atm each of H_2 and F_2 are added to a 1.50-L flask. Calculate the equilibrium partial pressures of all species.

15.67 At elevated temperature, carbon tetrachloride decomposes to its elements:

$$CCl_4(g) \rightleftharpoons C(s) + 2\ Cl_2(g)$$

At 700 K, if the initial pressure of CCl_4 is 1.00 atm, the total pressure at equilibrium is 1.35 atm. Use these pressures to calculate K_{eq} at 700 K.

15.68 Suppose that the equilibrium system described in Problem 15.67 is expanded to twice its initial volume. Find the new equilibrium pressure.

15.69 Consider the following reaction:

$$CO_2(g) + 2\ OH^-(aq) \rightleftharpoons CO_3^{2-}(aq) + H_2O(l)$$

(a) Write the equilibrium constant expression for this reaction. (b) What will happen to the pressure of CO_2 in this equilibrium system if some Na_2CO_3 solid is dissolved in the solution? (c) What will happen to the pressure of CO_2 in this equilibrium system if some HCl gas is bubbled through the solution?

15.70 For the following reaction, K_{eq} is 1.83×10^{-3} at 395 K:

$$PCl_5(g) \rightleftharpoons PCl_3(g) + Cl_2(g)$$

If 2.00 g of PCl_5 is placed in a 3.00-L bulb at 395 K, what is the equilibrium pressure of Cl_2?

15.71 Using tabulated thermodynamic data from Appendix D, compute (a) K_{eq} at 298 K; (b) the temperature at which the equilibrium pressure is 1.00 atm; and (c) K_{eq} at 1050 K for the following reaction:

$$Hg(g) + HgCl_2(s) \rightleftharpoons Hg_2Cl_2(s)$$

15.72 Benzene can be sulfonated by concentrated sulfuric acid:

$$C_6H_6 + H_2SO_4 \rightleftharpoons C_6H_5SO_3^- + H_3O^+$$

This reaction occurs by the following three-step mechanism:

$$H_2SO_4 \rightleftharpoons SO_3 + H_2O$$

$$SO_3 + C_6H_6 \rightleftharpoons C_6H_6SO_3$$

$$C_6H_6SO_3 + H_2O \rightleftharpoons C_6H_5SO_3^- + H_3O^+$$

Determine the relationship between the equilibrium constant for the net reaction and the rate constants for the various elementary steps.

15.73 At 1020 °C, K_{eq} for the conversion of $CO_2(g)$ to $CO(g)$ by solid graphite, C(s), is 167.5. A 1.00-L, high-pressure chamber containing excess graphite powder is charged with 0.500 mol each of CO_2 and CO and then is heated to 1020 °C. What is the equilibrium total pressure?

15.74 A chemist claims to have discovered a new gaseous element, effluvium (Ef), which reacts with atmospheric nitrogen to form effluvium nitride:

$$2\ Ef(g) + 3\ N_2(g) \rightleftharpoons 2\ EfN_3(g)$$

In a container initially containing N_2 at 1.00 atm pressure and Ef gas at 0.75 atm, the total gas pressure at equilibrium, according to the chemist's measurements, is 0.85 atm. Compute K_{eq} for this reaction.

15.75 The following figure represents a system coming to equilibrium:

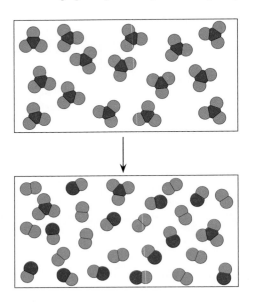

(a) Use molecular pictures similar to those in Problem 15.55 to write a balanced equation for the equilibrium reaction. (b) What is K_{eq} for the reaction?

15.76 In the gas phase, acetic acid is in equilibrium with a dimer held together by a pair of hydrogen bonds.
(a) If the total pressure of acetic acid gas in a glass bulb is 0.75 atm, what is the partial pressure of the dimer? (b) Is the equilibrium constant for this reaction higher or lower at 200 °C? (Hint: Hydrogen bonds must be broken for the dimer to decompose.)

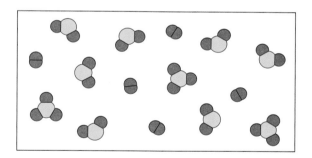

Dimer

Group Study Problems

15.77 (a) A container contains CO_2 at $P = 0.464$ atm. When graphite is added to the container, some CO_2 is converted to CO, and at equilibrium, the total pressure is 0.746 atm. Compute K_{eq} for the reaction:

$$CO_2(g) + C(s) \rightleftharpoons 2\ CO(g)$$

(b) Suppose that the equilibrium system is compressed to one third its initial volume. Find the new equilibrium pressure.

15.78 When 1.00 mol of gaseous HI is sealed in a 1.00-L flask at 225 °C, it decomposes until the equilibrium amount of I_2 present is 0.182 moles:

$$2\ HI(g) \rightleftharpoons H_2(g) + I_2(g)$$

(a) Use these data to calculate K_{eq} for this reaction at 225 °C.
(b) Using standard thermodynamic data from Appendix D, estimate K_{eq} at 625 °C.
(c) Using the result of (b), calculate the equilibrium partial pressures of all reagents under these conditions.

15.79 Using the information in Table 15-1, calculate the following:
(a) The number of grams of CO_2 that can be dissolved in 225 mL of a carbonated beverage at 1.10 atm pressure and 25 °C.
(b) The partial pressure of CO_2 in the gas space above the liquid if a bottle of this carbonated beverage is stored in an ice chest at 0.0 °C.

15.80 One reaction that plays a role in photochemical smog is the following:

$$O_3(g) + NO(g) \rightleftharpoons O_2(g) + NO_2(g) \qquad K_{eq} = 6.0 \times 10^{34}$$

Automobile engines produce both O_3 and NO. If the morning rush hour results in an atmosphere containing 1.5 ppm NO and 6.5 ppb O_3, calculate the equilibrium partial pressures of the three pollutants (O_2 is 21% of the atmosphere).

15.81 The combustion of sulfur dioxide is one of the steps in the synthesis of sulfuric acid:

$$2\ SO_2 + O_2 \rightleftharpoons 2\ SO_3$$

The following diagram represents an equilibrium mixture of SO_2, O_2, and SO_3 at 1100 K:

= SO$_2$ = SO$_3$ = O$_2$

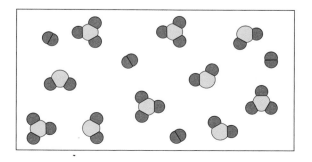

(a) The equilibrium mixture is changed to the following configuration. What happens to the position of the equilibrium? Explain.

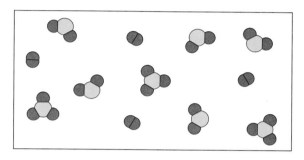

(b) When the temperature of the system is increased to 1300 K, the equilibrium shifts to the following configuration:

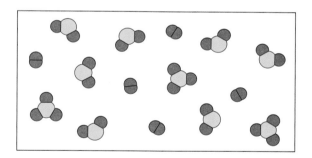

Is the reaction exothermic or endothermic? Use equilibrium arguments to explain your answer.

(c) The original equilibrium mixture is changed to the following configuration:

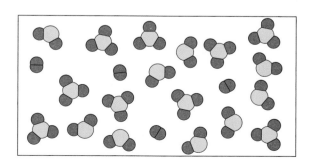

What happens to the position of the equilibrium? Explain.

15.82 The key step in the synthesis of sulfuric acid is the combustion of SO_2 to give sulfur trioxide:

$$2 SO_2(g) + O_2(g) \rightleftharpoons 2 SO_3(g) \qquad K_{eq} = 5.60 \times 10^4 \text{ at } 350 \text{ °C}$$

(a) Sulfur dioxide and oxygen are mixed initially at 0.350 and 0.762 atm, respectively, at 350 °C. What are the partial pressures of the three gases when the mixture reaches equilibrium?

(b) Use thermodynamic data (see Appendix D) to calculate K_{eq} for the sulfur trioxide reaction at 298 K. Based on this result, should this industrial synthesis be run at a higher or lower temperature than 350 °C to improve the yield?

(c) Suggest reasons why this reaction is run industrially at 700 K.

Answers to Section Exercises

15.1.1 $K_{eq} = \dfrac{[O_2]^3}{[O_3]^2}$

15.1.2

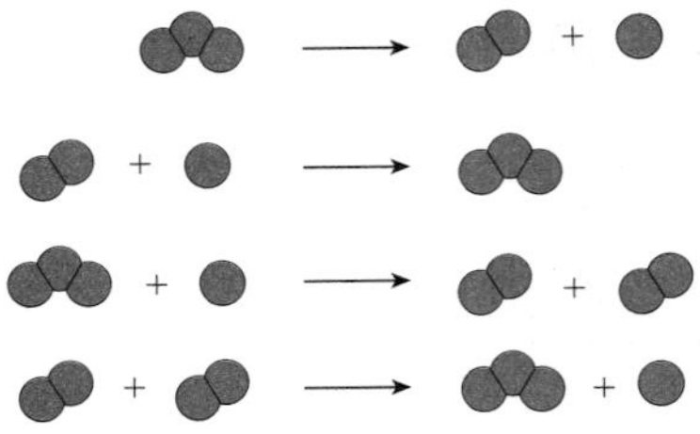

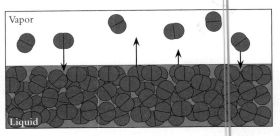

15.1.3 $K_{eq} = \dfrac{[H_2O]^6 \, [NO]^4}{[NH_3]^4 \, [O_2]^5}$

15.2.1 (a) $K_{eq} = (p_{Cl_2})_{eq}$

(b) $K_{eq} = (p_{CO_2})_{eq}$

(c) $K_{eq} = [PO_4{}^{3-}]_{eq}^2 \, [Ca^{2+}]_{eq}^3$

(d) $K_{eq} = \dfrac{[H_3O^+][CN^-]}{(p_{HCN})_{eq}}$

(e) $K_{eq} = \dfrac{1}{(p_{Cl_2})_{eq}}$

(f) $K_{eq} = \dfrac{1}{[PO_4{}^{3-}]_{eq}^2 \, [Ca^{2+}]_{eq}^3}$

15.2.2

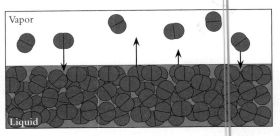

15.2.3 The fraction that dissolves is 8.3×10^{-2} (8.3%). This is not an efficient way to collect CO_2 because its solubility in water is significant.

15.3.1 1.4×10^{-8}

15.3.2 Increases because the reaction is endothermic.

15.3.3 2.4×10^{-8}

15.4.1 (a) No change in Q, thus no change; (b), (c), and (d) more $Ba(OH)_2$ dissolves.

15.4.2 Lower T, remove NH_3, and increase pressures of H_2 and/or N_2.

15.4.3 The H_2-N_2 reaction has a large K_{eq} at room temperature, whereas the CH_4-H_2O reaction has a small K_{eq} at room temperature.

15.5.1 $K_{eq} = 1.10 \times 10^{-10}$

15.5.2 $[CH_3CO_2{}^-]_{eq} = [H_3O^+]_{eq} = 6.7 \times 10^{-3}$ M and $[CH_3CO_2H]_{eq} = 2.5$ M

15.5.3 $[Ag^+]_{eq} = 2.7 \times 10^{-12}$ M; $[Br^-]_{eq} = 0.200$ M

15.6.1 (a) H_3O^+, $ClO_4{}^-$ and H_2O; (b) NH_3 and H_2O; (c) K^+, $HCO_3{}^-$ and H_2O ; and (d) $HClO$ and H_2O

15.6.2 (a) $1/K_{sp}$; (b) K_a; and (c) K_b

15.6.3 Precipitate is $Ni(OH)_2$; net reaction is

$$Ni^{2+}(aq) + 2 \, OH^-(aq) \longrightarrow Ni(OH)_2(s);$$

spectators are $SO_4{}^{2-}$ and Na^+.

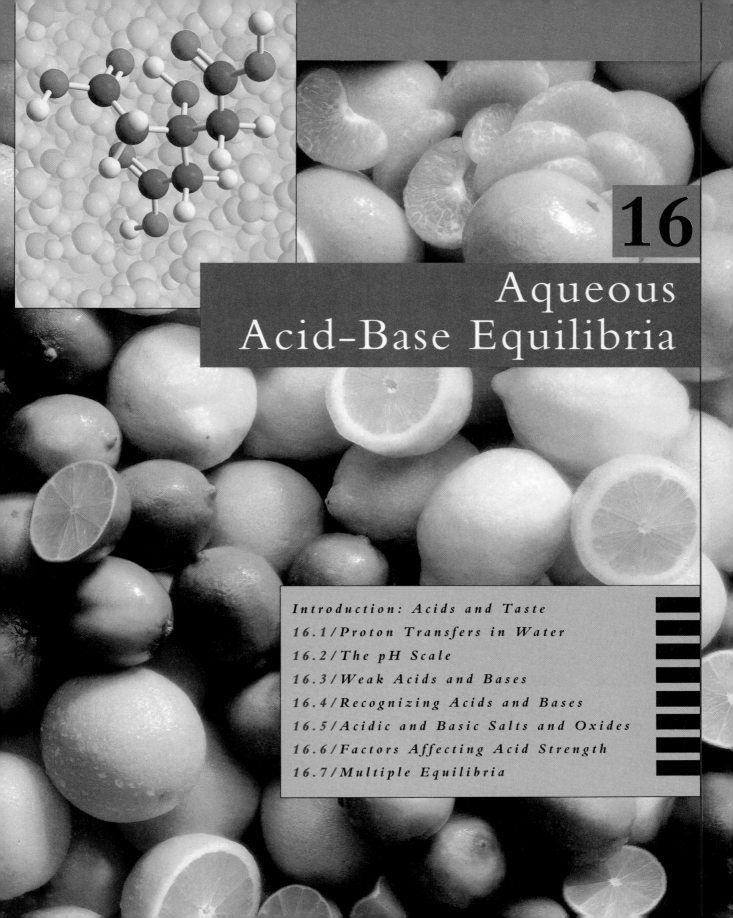

16

Aqueous Acid–Base Equilibria

INTRODUCTION: ACIDS AND TASTE

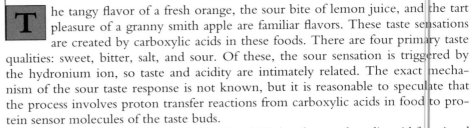

The tangy flavor of a fresh orange, the sour bite of lemon juice, and the tart pleasure of a granny smith apple are familiar flavors. These taste sensations are created by carboxylic acids in these foods. There are four primary taste qualities: sweet, bitter, salt, and sour. Of these, the sour sensation is triggered by the hydronium ion, so taste and acidity are intimately related. The exact mechanism of the sour taste response is not known, but it is reasonable to speculate that the process involves proton transfer reactions from carboxylic acids in food to protein sensor molecules of the taste buds.

Our molecular view shows citric acid, which has three carboxylic acid functional groups, —CO_2H. The acidity of lemons is caused by citric acid. Lemons can contain as much as 3% citric acid by mass. Citric acid is present in all citrus fruits as well as many other tart-tasting foods, including berries, pineapples, pears, and tomatoes.

Another fruit whose carboxylic acids contribute to its taste is the grape, which contains several carboxylic acids as well as sugar molecules. When grapes are fermented to make wine, most of their sugar is converted into ethanol, but the carboxylic acids remain unchanged. A wine is "sweet" if its residual sugar content masks the sour taste of these residual acids and "dry" if its residual sugar content is less than about 1%.

Vegetables also contain carboxylic acids that contribute to their flavors. One example is oxalic acid, prevalent in spinach and rhubarb. Raw rhubarb leaves are mildly poisonous, and folklore holds that the toxic substance is oxalic acid. However, raw spinach can be eaten safely despite its equally high content of this acid.

Whereas fermentation converts sugars into alcohol, bacteria feed on the lactose (milk sugar) in milk to produce lactic acid, a carboxylic acid. Under controlled conditions, this process gives yogurt. Similarly, the proper combination of yeast and bacteria acts on flour to form just enough lactic acid and acetic acid (CH_3CO_2H) to give sourdough bread its distinctive tang.

In addition to conveying acidity, the carboxylic acid group is highly polar, so smaller carboxylic acids such as acetic acid are water-soluble. Vinegar is a solution of acetic acid in water, typically about 5% by mass of the acid. Acetic acid can be produced from sugar-containing foods by a combination of fermentation and oxidation, which breaks down glucose molecules into ethanol and then converts ethanol to acetic acid. The word *vinegar* comes from the French term for "sour wine"; indeed, poor wine tastes sour because some of its ethanol has become acetic acid.

This survey highlights a few simple carboxylic acids that contribute to the tastes of foods. In addition, long-chain carboxylic acids, the fatty acids, are responsible for much of the energy content of nuts, seeds, vegetable oils, and animal fats, as described in Section 13.6. Furthermore, carboxylic acids are produced and consumed during the metabolic processes that the body uses to extract chemical energy from carbohydrates.

This chapter describes the details of acid-base chemistry. We begin with a molecular view and a way of measuring acidity. Then we look at acid-base equilibrium calculations, making use of the general approaches to equilibrium developed in Chapter 15. We describe the important applications of buffers and titrations in Chapter 17.

Oxalic acid

Lactic acid

16.1 PROTON TRANSFERS IN WATER

In an acid-base reaction, a proton (H^+) is transferred from one chemical species to another. A species that donates a proton is an acid, and a species that accepts a proton

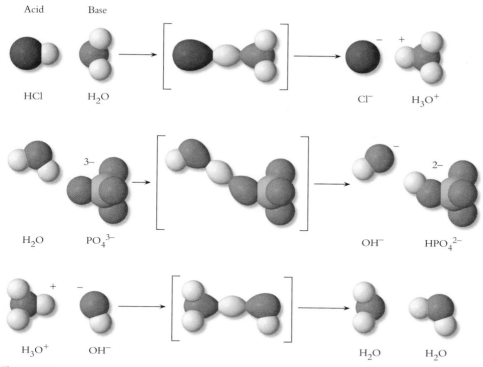

Figure 16-1
Every acid-base reaction involves transfer of a proton from an acid to a base.

is a base. This identification of acids and bases is the **Brønsted–Lowry** definition of acid-base reactions. From this perspective, every acid-base reaction has two reactants, an acid and a base. Every acid-base reaction also forms two products:

Acid		**Base**				
$HCl(g)$	$+$	$H_2O(l)$	$\longrightarrow$	$Cl^-(aq)$	$+$	$H_3O^+(aq)$
$H_2O(l)$	$+$	$PO_4^{3-}(aq)$	$\longrightarrow$	$OH^-(aq)$	$+$	$HPO_4^{2-}(aq)$
$H_3O^+(aq)$	$+$	$OH^-(aq)$	$\longrightarrow$	$H_2O(l)$	$+$	$H_2O(l)$
$CH_3CO_2H(aq)$	$+$	$OH^-(aq)$	$\longrightarrow$	$CH_3CO_2^-(aq)$	$+$	$H_2O(l)$

Although these reactions look different, they all involve the same chemical process: an acid collides with a base and transfers H^+. Figure 16-1 illustrates this process at the molecular level.

Notice that water molecules appear in all of these examples. Aqueous acid-base chemistry usually includes water as either a starting material or a product.

Dissociation of Water

These examples of acid-base reactions show that water can act as either an acid or a base: Water accepts a proton from an HCl molecule, but it donates a proton to a PO_4^{3-} anion. As an acid, water donates a proton to a base and becomes a hydroxide anion. As a base, water accepts a proton from an acid and becomes a hydronium cation. A chemical species that can both donate and accept protons is said to be **amphiprotic.** Water is an amphiprotic molecule.

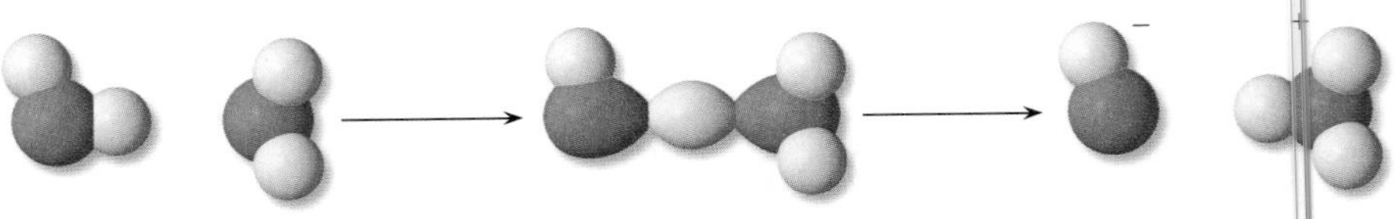

H$_2$O
(Acid: A hydrogen ion donor)

H$_2$O
(Base: A hydrogen ion acceptor)

Collision and transfer

OH$^-$ H$_3$O$^+$

Figure 16-2
The water equilibrium illustrates the amphiprotic nature of H$_2$O. In this reaction, one water molecule acts as a proton donor (acid), and another acts as a proton acceptor (base).

This ability to serve as both a proton donor and a proton acceptor suggests that a proton transfer equilibrium exists for pure water, as Figure 16-2 illustrates.

$$2 \ H_2O(l) \rightleftharpoons OH^-(aq) + H_3O^+(aq)$$

$$K_{eq} = K_w = [H_3O^+][OH^-] = 1.0 \times 10^{-14} \quad \text{(at 298 K)}$$

The equilibrium constant for this reaction is the **water equilibrium constant (K_w).** Its value is 1.0×10^{-14} at 25 °C.

In pure water, the hydronium and hydroxide ion concentrations are equal: According to K_w, the concentrations of H$_3$O$^+$ ions and OH$^-$ ions in pure water at 298 K (25 °C) are 1.0×10^{-7} M:

$$[H_3O^+]_{eq} = [OH^-]_{eq} = 1.0 \times 10^{-7} \text{ M} \quad \text{(pure water at 298 K)}$$

Equal concentrations of these two ions means that pure water is neither acidic nor basic.

The water equilibrium describes a balance between $[H_3O^+]_{eq}$ and $[OH^-]_{eq}$. When an acid dissolves in water, the hydronium ion concentration increases, so the hydroxide ion concentration must decrease to maintain the product of the concentrations at 1.0×10^{-14}. Similarly, when a base dissolves in water the hydroxide ion concentration increases, and the hydronium ion concentration decreases.

Strong Acids

As noted in Chapter 4, acids that donate protons to water molecules quantitatively are called **strong acids.** Table 4-2 lists the six most common strong acids: HNO$_3$, HClO$_4$, H$_2$SO$_4$, HCl, HBr, and HI. In an aqueous solution of a strong acid, the hydronium ion concentration is equal to the concentration of the acid solution. The concentration of hydroxide ions in a solution of a strong acid can be calculated from the concentration of H$_3$O$^+$ and K_w. Example 16-1 provides an illustration.

Example 16-1	Ion Concentrations in a Solution of Strong Acid

Determine the ion concentrations in a 1.00×10^{-2} M aqueous solution of HClO$_4$, a strong acid.

Solving Equilibrium Problems

Strategy: The seven-step method described in Chapter 15 provides the strategy for working any quantitative problem that involves equilibria. As a reminder, we detail each step in this first example of acid-base equilibria.

| Ion Concentrations in a Solution of Strong Acid *(continued)* | Example 16-1 |

Solution:

1. **Determine what is asked for.** This problem asks for concentrations of all ions in an acid solution.

2. **Identify the major chemical species.** Because $HClO_4$ is a strong acid, it transfers protons to water molecules quantitatively. Thus the major species in solution are H_3O^+, ClO_4^-, and H_2O:

$$HClO_4 + H_2O \longrightarrow H_3O^+(aq) + ClO_4^-(aq)$$

3. **Determine what chemical equilibria exist.** The water equilibrium always exists in aqueous solution:

$$2\,H_2O(l) \rightleftharpoons OH^-(aq) + H_3O^+(aq)$$

4. **Write the K_{eq} expressions.** The strong acid dissociates completely, so the water equilibrium is the only one that applies to this problem:

$$K_{eq} = K_w = [H_3O^+][OH^-] = 1.0 \times 10^{-14}$$

5. **Organize the data and unknowns.** The solution is 1.00×10^{-2} M $HClO_4$, so the concentrations of ions generated by the quantitative reaction are $[H_3O^+] = [ClO_4^-] = 1.00 \times 10^{-2}$ M. The final concentrations are found using an equilibrium analysis. Set up a concentration table for the water dissociation equilibrium, and define the change in hydronium ion concentration as x:

Species (reaction)	$2\,H_2O(l)$	$\rightleftharpoons$	$H_3O^+(aq)$	$+$	$OH^-(aq)$
Initial concentration (M)			1.00×10^{-2}		0
Change in concentration (M)			$+x$		$+x$
Equilibrium concentration (M)			$1.00 \times 10^{-2} + x$		x

In any solution of an acid, the total hydronium and hydroxide ion concentrations include the 10^{-7} M contribution from the water reaction. This contribution is negligible except in highly dilute solutions, which we do not discuss in this text.

6. **Carry out the calculations.** The equilibrium constant is much smaller than the initial concentration, 1.00×10^{-2}, so it is reasonable to assume that $x \ll 1.00 \times 10^{-2}$. This leads to an approximation:

$$(1.00 \times 10^{-2} + x) \cong 1.00 \times 10^{-2}$$

Solving for x gives the equilibrium concentration of hydroxide ions:

$$K_w = [H_3O^+]_{eq}[OH^-]_{eq} = (1.00 \times 10^{-2})\,x = 1.0 \times 10^{-14}$$

$$x = \frac{K_w}{[H_3O^+]} = \frac{1.0 \times 10^{-14}}{1.00 \times 10^{-2}} = 1.0 \times 10^{-12} \quad so$$

$$[OH^-]_{eq} = 1.0 \times 10^{-12}\ M$$

7. **Check for reasonableness.** Each of the ion concentrations is equal to or smaller than the starting concentration of the acid, which is reasonable. Also, the approximation $x \ll 1.00 \times 10^{-2}$ is valid, because $x = 1.0 \times 10^{-12}$.

This example illustrates that the change in hydronium ion concentration due specifically to the water equilibrium is negligibly small in an aqueous solution of a strong acid at acid concentrations greater than 10^{-5} M. Consequently, the hydronium ion concentration equals the concentration of the strong acid, and

$$[OH^-] = \frac{(1.0 \times 10^{-14})}{[H_3O^+]} \text{ M.}$$

Strong Bases

A substance that generates hydroxide ions quantitatively in aqueous solution is a **strong base.** The most common strong bases are the soluble metal hydroxides, among which NaOH perennially ranks among the top ten industrial chemicals. When a soluble metal hydroxide dissolves in water, it breaks apart into the metal cation and hydroxide anion:

$$NaOH(s) \xrightarrow{H_2O} Na^+(aq) + OH^-(aq)$$

For any aqueous strong base, the hydroxide ion concentration can be calculated directly from the overall solution molarity. As for aqueous strong acids, the hydronium and hydroxide ion concentrations are linked through the water equilibrium, as shown by Example 16-2.

Example 16-2	Ion Concentrations in a Strong Base Solution

What are the ion concentrations in 0.500 L of an aqueous solution that contains 5.00 g of NaOH?

Strategy: For this example, we summarize the first four steps of the method: The problem asks for the concentration of ions. Sodium hydroxide is a strong base that dissolves in water to generate Na^+ cations and OH^- anions quantitatively. The concentration of hydroxide ion equals the concentration of the base. The water equilibrium links the concentrations of OH^- and H_3O^+, so an equilibrium calculation is required to determine the concentration of hydronium ion. What remains is to organize the data, carry out the calculations, and check for reasonableness.

Solution: The overall molarity of NaOH can be calculated from mass, molar mass, and volume.

$$MM_{NaOH} = 40.00 \text{ g/mol}$$

$$n_{NaOH} = \frac{m}{MM} = \frac{5.00 \text{ g}}{40.00 \text{ g/mol}} = 0.125 \text{ mol}$$

$$M_{NaOH} = \frac{n}{V} = \frac{0.125 \text{ mol}}{0.500 \text{ L}} = 0.250 \text{ mol/L}$$

The solution is 0.250 M in NaOH, so the initial concentrations of the two ions are $[Na^+] = [OH^-] = 0.250$ M.

Use the water equilibrium to determine the concentration of H_3O^+.

$$2 H_2O(l) \rightleftharpoons OH^-(aq) + H_3O^+(aq) \qquad K_w = [H_3O^+][OH^-] = 1.0 \times 10^{-14}$$

Set up a concentration table, defining the change in hydronium ion concentration as x:

| **Ion Concentrations in a Strong Base Solution** *(continued)* | | **Example 16-2** |

Species (reaction)	$2\ H_2O(l) \rightleftharpoons$	$H_3O^+(aq)$ +	$OH^-(aq)$
Initial concentration (M)		0	0.250
Change in concentration (M)		$+x$	$+x$
Equilibrium concentration (M)		x	$0.250 + x$

The equilibrium constant is much smaller than 0.250, so it is reasonable to assume that $x \ll 0.250$ and make the approximation $0.250 + x \cong 0.250$. Solving for x gives the equilibrium concentration of hydronium ions:

$$K_w = 1.0 \times 10^{-14} = x\,(0.250 + x) \cong 0.250\,x$$

$$x \cong \frac{1.0 \times 10^{-14}}{0.250} = 4.0 \times 10^{-14} \qquad so$$

$$[H_3O^+]_{eq} = 4.0 \times 10^{-14}\ M \qquad and \qquad [OH^-]_{eq} = 0.250\ M$$

Note that x is indeed much smaller than 0.250.

This example illustrates that the change in hydroxide ion concentration due specifically to the water equilibrium is always negligibly small in an aqueous solution of a strong base. Consequently, the hydroxide ion concentration equals the concentration of the strong base, and $[H_3O^+] = \dfrac{(1.0 \times 10^{-14})}{[OH^-]}$ M.

| | **Section Exercises** |

16.1.1 Write the net proton transfer reaction and draw a molecular picture to illustrate the reaction between H_3O^+ and SO_4^{2-}.

16.1.2 If a sample of acid rain has $[H_3O^+] = 2.5 \times 10^{-4}$ M, what is $[OH^-]$?

16.1.3 A student prepares 500.0 mL of aqueous solution containing 5.61 g of KOH. Determine the concentrations of all ions in this solution.

16.2 THE pH SCALE

The hydronium ion concentration in aqueous solution ranges from extremely high to extremely low. As examples, in 6.0 M aqueous HCl, the concentration of H_3O^+ is 6.0 M. Pure water has a hydronium ion concentration of 10^{-7} M. As we calculated in Example 16–2, in 0.25 M aqueous NaOH the concentration of H_3O^+ is 4.0×10^{-14} M. Chemists use logarithms to express this immense range of concentrations. The **pH** scale of acid concentration is defined as:

$$pH = -\log[H_3O^+] \qquad\qquad (16\text{-}1)$$

The pH of a solution is obtained by taking the logarithm of the hydronium ion concentration and then changing the sign. For example, the pH of pure water is:

$$pH = -\log[H_3O^+] = -\log\,(1.0 \times 10^{-7}) = -(-7.00) = 7.00$$

| Recall that $\log(10^x) = x$.

Litmus paper turns red when exposed to acid.

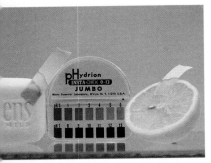

pH paper gives a rapid qualitative measurement of pH.

The conversion from pH to $[H_3O^+]$ uses powers of 10. For example, lemon juice has a pH of about 2.0:

$$[H_3O^+] = 10^{-pH} = 10^{-2.0} = 1 \times 10^{-2}\text{ M}$$

A logarithm carries no units, but the result of this calculation is a concentration in mol/L (M). The pH scale has these features:

■ pH = 7.00 defines a "neutral" solution, $[H_3O^+] = [OH^-] = 10^{-7}$ M, neither acidic nor basic. Acid solutions have pH < 7, basic solutions have pH > 7.

■ The *more acidic* the solution, the *lower* its pH: Lemon juice, at pH ~ 2, is considerably more acidic than acid rain, pH ~ 4.

■ A change in pH of one unit reflects a ten-fold change in hydronium ion concentration: Normal rainfall has pH ~ 5, but acid rain has ten times larger hydronium ion concentration, pH ~ 4.

■ The immense range of concentrations from > 1 M to < 10^{-14} M is compressed into a more convenient range, from ~ −1 to ~ +15. Figure 16-3 shows the range of pH and hydronium ion concentrations.

The measurement of pH is a routine operation in most laboratories. Acidity measurements are done using indicators, pH paper, or pH meters. Indicators and pH paper change colors in response to changes in pH. These methods give quick but qualitative visual measurement of the acid-base properties of a solution. Litmus paper is a form of pH paper that turns red when dipped in acidic solution and blue when dipped in basic solution. Universal pH paper displays a range of colors in response to different solution pH values.

Figure 16-3
The pH values of commonly encountered aqueous solutions range from 0 to 14.

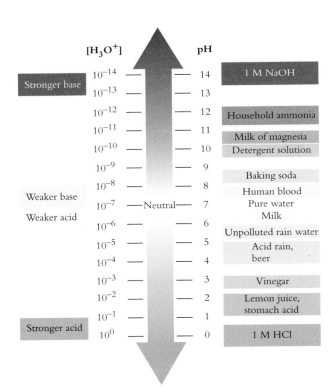

For quantitative pH determinations, scientists use pH meters (Figure 16-4). These instruments use probes to make electrical measurements between a reference solution containing acid at known concentration and the sample being tested. The pH meter is calibrated by dipping the probe in a solution of known pH and adjusting the meter to read the correct value. Then the probe is dipped into the sample solution, and the pH appears on a digital display.

Other 'p' Scales

A logarithmic scale is useful not only for expressing hydronium ion concentrations, but also for expressing hydroxide ion concentrations and equilibrium constants. That is, the pH definition can be generalized to other quantities:

$$pOH = -\log[OH^-] \qquad pK_a = -\log K_a \qquad pK_b = -\log K_b$$

Figure 16-4
Quantitative pH measurements are made with digital pH meters. These meters are calibrated with standard solutions of known pH.

Use of these definitions and the properties of logarithms leads to a statement of the water equilibrium constant in terms of pH:

$$K_w = [H_3O^+][OH^-] = 1.0 \times 10^{-14}$$

$$\log\{[H_3O^+][OH^-]\} = \log(1.0 \times 10^{-14}) = -14.00$$

$$\log[H_3O^+] + \log[OH^-] = -14.00$$

$$-\log[H_3O^+] - \log[OH^-] = 14.00$$

$$pH + pOH = 14.00 \qquad\qquad \textbf{(16–2)}$$

> A logarithm has two parts. The numbers preceding the decimal point determine the power of 10. The numbers after the decimal point determine the numerical value. Therefore when two digits follow the decimal place of the logarithm, the corresponding numerical value has two significant figures.

Equation 16-2 connects pH and pOH for any aqueous solution. Example 16-3 demonstrates the usefulness of this equation.

---

pH and pOH   **Example 16-3**

What is the pH of a 0.25 M solution of NaOH?

Strategy: To find pH of a solution, first compute $[H_3O^+]$ or $[OH^-]$ and then apply Equation 16-1 or 16-2. Because NaOH is a strong base, this example does not require explicit use of an equilibrium expression.

Solution: In this solution of strong base, $[OH^-]$ is equal to the concentration of sodium hydroxide:

$$[OH^-] = 0.25 \text{ M}$$

$$pOH = -\log(0.25) = -(-0.60) = 0.60$$

$$pH = 14.00 - pOH = 14.00 - 0.60 = 13.40$$

Notice how much easier this calculation is than the analysis in Example 16-2. The detailed reasoning of the concentration table, K_w, and an approximation have been compressed into two steps.

Verify that this result is the same as the one obtained in Example 16-2 by converting from pH to $[H_3O^+]$:

$$13.40 = pH = -\log[H_3O^+]$$

$$\log[H_3O^+] = -13.40 \quad so \quad [H_3O^+] = 10^{-13.40} = 4.0 \times 10^{-14} \text{ M}$$

Section Exercises

- **16.2.1** Determine the pH values of 1.5×10^{-3} M HCl and 2.5×10^{-2} M NaOH.
- **16.2.2** The most acidic rainfall ever recorded in the U.S. had a pH of 1.80. What were the hydronium and hydroxide ion concentrations in this rainfall?

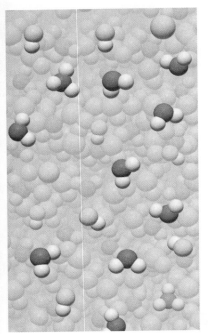

Figure 16-5
In a solution of the weak acid HF, the position of proton-transfer equilibrium favors the reactants. H$_2$O and HF are present at highest concentrations, and H$_3$O$^+$ and F$^-$ are present in much lower concentrations.

16.3 WEAK ACIDS AND BASES

The examples in Sections 16.1 and 16.2 describe solutions of strong acids and bases. Although strong acids and bases have important applications, the vast majority of acids and bases are weak. This means that their proton transfer reactions in aqueous solution do not go to completion, so we need to deal with equilibrium expressions and calculations when analyzing the contents of these solutions.

Weak Acids: Proton Transfer to Water

Measurements of the pH of acids show that, in general, proton transfer reactions reach equilibrium well before the reaction goes to completion. For example, the pH of a 0.25 M solution of HF is 1.92, giving $[H_3O^+] = 1.2 \times 10^{-2}$ M. This is substantially lower than 0.25 M, the expected result from quantitative proton transfer. In other words, when HF dissolves in water, equilibrium is established when only a fraction of the HF molecules have transferred protons to water molecules:

$$H_2O(l) + HF(aq) \rightleftharpoons H_3O^+(aq) + F^-(aq)$$

An acid that reaches equilibrium when only a small fraction of its molecules has transferred protons to water is called a **weak acid.** In a solution of a weak acid, the *major* species are water molecules and the acid, HA. The products of the proton transfer reaction, H$_3$O$^+$ and A^-, are present in smaller concentrations as *minor* species. Figure 16-5 provides a molecular view.

The strength of a weak acid is measured by its acid ionization constant, K_a. This equilibrium constant can be calculated from the measured pH of the solution, as illustrated in Example 16-4.

Example 16-4	Calculating K_a

The pH of a 0.25 M aqueous HF solution is 1.92. Calculate K_a for this weak acid.

Strategy: Use the seven-step method. We are asked to evaluate an equilibrium constant. Visualize the molecular situation: What are the species involved in this equilibrium? After identifying the chemistry, set up and complete the appropriate concentration table and solve for the desired quantity, in this case K_a.

Solution: The problem states that hydrogen fluoride is a weak acid, so the major species in solution are H$_2$O and HF molecules. In aqueous solution, HF transfers protons to H$_2$O:

$$H_2O(l) + HF(aq) \rightleftharpoons H_3O^+(aq) + F^-(aq)$$

Calculating K_a (continued)

Example 16-4

From the pH, calculate the concentration of hydronium ions at equilibrium:

$$[H_3O^+] = 10^{-pH} = 10^{-1.92} = 1.2 \times 10^{-2} \text{ M}$$

Now construct the concentration table. The key is to use the initial and equilibrium concentrations of H_3O^+ to complete the "change" row. Shading indicates the concentrations that provide the starting points:

Species (reaction)	H_2O + HF	$\rightleftharpoons$	H_3O^+	+	F^-
Initial concentration (M)	0.25		0		0
Change in concentration (M)	-1.2×10^{-2}		$+1.2 \times 10^{-2}$		$+1.2 \times 10^{-2}$
Equilibrium concentration (M)	0.24		1.2×10^{-2}		1.2×10^{-2}

Now substitute numerical values of the concentrations and solve for K_a:

$$K_a = \frac{(1.2 \times 10^{-2})^2}{0.24} = 6.0 \times 10^{-4}$$

The equilibrium concentration of hydronium ions is in the 10^{-2} M range, so this value for K_a appears reasonable.

Solutions of hydrogen fluoride are used to etch glass.

The problem identified this solution as 0.25 M HF, but at equilibrium, [HF] = 0.24 M. Conventionally, a solution's concentration is stated as its initial concentration, even though the equilibrium concentrations may differ slightly from this initial concentration.

Although it is not strongly reactive as an acid, HF has other unique reactive properties. It is one of the few common substances that reacts with glass. For this reason, artists use aqueous solutions of HF to etch glass, and it is used industrially to "frost" light bulbs. Because it destroys glass bottles, aqueous HF must be stored in plastic containers. Besides attacking glass, HF has a devastating effect on human nerve tissues. It is absorbed through the skin readily, so people who work with HF must always wear strong plastic gloves.

Even though proton transfer is never complete for a weak acid dissolved in water, some proton transfer always occurs, so the hydronium ion concentration in any solution of a weak acid is greater than that in pure water. Consequently, the pH of an aqueous solution of a weak acid is always less than 7.00.

Aqueous solutions of strong acids and weak acids both have pH < 7, but their molecular compositions are quite different, as Figure 16-6 displays. Both solutions are acidic, because both contain hydronium ion concentrations greater than 10^{-7} M. The strong acid transfers protons quantitatively, generating a high concentration of hydronium ions and leaving a negligible number of undissociated acid molecules. In contrast, the weak acid undergoes only a small amount of proton transfer, generating a much lower concentration of hydronium ions and remaining almost entirely undissociated.

The behavior of weak acids played an important role in the development of the ionic view of aqueous solutions, as described in our Chemical Milestones Box.

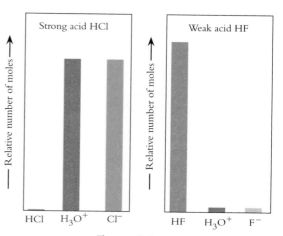

Figure 16-6
The concentrations of molecular and ionic species are dramatically different in a solution of a strong acid (HCl) and a solution of a weak acid (HF).

Box 16-1 Chemical Milestones: Arrhenius and the Ionic Theory

The existence of ions in aqueous solutions was first proposed by Svante Arrhenius, a young Swedish chemist, during the 1880s, well before the electronic structure of atoms had been discovered. Arrhenius' insight came while he was pursuing his PhD in chemistry, exploring why aqueous solutions conduct electricity.

The types of conductivity measurements carried out by Arrhenius were much like the demonstrations of electrical conductivity used today. Using a voltage source and a current meter, one can determine how much current a solution carries, and this is proportional to its conductivity. A light bulb can serve as a qualitative current meter. As the photos show, solutes fall into three categories. A solution of sugar (left) does not conduct, a solution of NaCl (right) conducts extremely well, and a solution of acetic acid (center) conducts only a little. In Arrhenius' time, ions were called "electrolytic molecules," so the categories were termed nonelectrolytes, strong electrolytes, and weak electrolytes.

acetic acid solution and pure water, separated by a thin layer of concentrated sugar solution, show low conductivity. When the layers are mixed, the conductivity *increases*. To Arrhenius, this demonstrated that diluting acetic acid leads to more dissociation into ions. Layers of solutions of silver sulfate and barium chloride conduct well. When these layers are mixed, precipitates of silver chloride and barium sulfate form, and the light goes out. To Arrhenius, this showed that ions were present in solutions of salts, and that these ions were removed by precipitation.

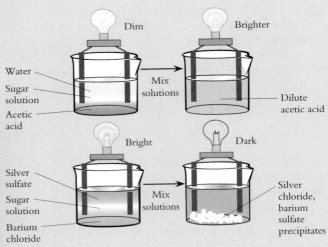

While Arrhenius was studying conductivity, others were characterizing colligative properties of solutions. The Dutch chemist J. T. van't Hoff, studied osmotic pressure and derived the law of osmotic pressure, $\pi V = nRT$. Van't Hoff noted the parallel between this law and the ideal gas equation, and he proposed that solute molecules in solution act independently of one another. Van't Hoff's law worked for solutions of nonelectrolytes and weak electrolytes, but for strong electrolytes, van't Hoff had to multiply n by a coefficient, i. For HCl and NaCl the value of i was close to 2, and for $CaCl_2$, i was close to 3. For this reason, strong electrolytes were considered to be exceptions to van't Hoff's law.

From his conductivity measurements on solutions, Arrhenius concluded that strong electrolytes are not exceptions. Instead they dissociate into ions. When $i = 2$, it meant that each solute species dissociated to give two ions. A compound with $i = 3$ dissociated to give three ions. Moreover, by doing experiments at varying levels of concentration, Arrhenius concluded that at sufficiently high dilution, every electrolyte becomes fully dissociated.

Arrhenius devised demonstrations to support his ideas. Two of these are illustrated by the diagrams. Layers of

Arrhenius knew that these ideas were revolutionary. He avoided mentioning "ions" in his thesis, for fear his examiners would reject the thesis. Even so, his committee gave him the lowest possible passing grade. In desperation, Arrhenius sent copies of his work to famous professors in several countries. Wilhelm Ostwald, a brilliant chemist, recognized the genius of Arrhenius' ideas and traveled to Sweden to meet him.

Still, the chemical establishment remained opposed to the notion of ions in solution. In an attempt to convince "the wild army of Ionians" of how wrong their ideas were, the British Association scheduled a discussion titled "Theories of Solution," and invited van't Hoff, Arrhenius, and Ostwald to present their views. The rest of the discussion was packed with conservative older chemists, the idea being that reason would prevail and the Ionians would give up their views. Instead, the younger chemists sought out the Ionians for spirited exchanges, while the old chemists delivered their lectures to nearly empty rooms.

By 1895, Arrhenius' views were accepted by nearly all chemists, and Arrhenius finally was awarded a professorship in Stockholm. In 1903, the triumph of Arrhenius became complete, when he was awarded the Nobel Prize for Chemistry.

Weak Bases: Proton Transfer from Water

When ammonia dissolves in water, the resulting solution has a pH greater than 7.00, indicating that the solution is basic (Figure 16-3). According to Equation 16-1, the concentration of hydroxide ions in aqueous ammonia must be greater than that in pure water. Ammonia molecules accept protons from water molecules, generating ammonium cations and hydroxide anions:

$$NH_3(aq) + H_2O(l) \rightleftharpoons NH_4^+(aq) + OH^-(aq)$$

Ammonia is an example of a **weak base.** A weak base generates hydroxide ions by accepting protons from water but reaches equilibrium when only a fraction of its molecules have done so. The equilibrium constant for this type of equilibrium is designated K_b:

$$K_b = \frac{[NH_4^+]_{eq} \, [OH^-]_{eq}}{[NH_3]_{eq}}$$

Example 16–5 explores the ammonia equilibrium in more detail.

---

pH of a Weak Base

Example 16-5

What is the pH of 0.25 M aqueous ammonia ($K_b = 1.8 \times 10^{-5}$)?

Strategy: To calculate a pH, we first need to know the equilibrium concentration of either H_3O^+ or OH^-. We use the seven-step procedure.

Solving Equilibrium Problems

Solution: The major species present in aqueous ammonia are molecules of NH_3 and H_2O. Both of these compounds produce hydroxide ions as minor species in solution:

$$NH_3(aq) + H_2O(aq) \rightleftharpoons NH_4^+(aq) + OH^-(aq) \qquad K_b = 1.8 \times 10^{-5}$$

$$2\,H_2O(aq) \rightleftharpoons H_3O^+(aq) + OH^-(aq) \qquad K_w = 1.0 \times 10^{-14}$$

These two equilibria are linked by the fact that the solution can have only one hydroxide ion concentration. Both equilibria must be satisfied, but which reaction should we use to calculate $[OH^-]_{eq}$? Notice that both equilibrium constants are much less than 1, but K_b for ammonia is more than a billion times larger than K_w. This indicates that almost all the hydroxide ions in solution come from the ammonia equilibrium. Thus the ammonia reaction is the appropriate choice for the calculation.

Determine $[OH^-]_{eq}$ by setting up a concentration table, solving the equilibrium expression for the unknown, and finding $[OH^-]_{eq}$. We know the initial concentrations but must identify x as the change in concentration needed to reach equilibrium.

| Example 16-5 | pH of a Weak Base (continued) |

Species (reaction)	H_2O	+	NH_3	$\rightleftharpoons$	$NH_4^+(aq)$	+	$OH^-(aq)$
Initial concentration (M)			0.25		0		0
Change in concentration (M)			$-x$		$+x$		$+x$
Equilibrium concentration (M)			$0.25 - x$		x		x

$$K_b = 1.8 \times 10^{-5} = \frac{x^2}{(0.25 - x)} \cong \frac{x^2}{0.25}$$

$$x^2 = 4.5 \times 10^{-6} \quad \text{so} \quad x = 2.1 \times 10^{-3}$$

Since the equilibrium concentration is x, $[OH^-]_{eq} = 2.1 \times 10^{-3}$ M.

Complete the problem using Equation 16-2 to calculate the pH of the solution:

$$pOH = -\log[OH^-]_{eq} = -\log(2.1 \times 10^{-3}) = 2.68$$

$$pH + pOH = 14.00 \quad \text{so} \quad pH = 14.00 - 2.68 = 11.32$$

Note that $x \ll 0.25$, validating the approximation, $(0.25 - x) \cong 0.25$. A pH greater than 7 but less than 14 indicates a moderately basic solution, so this is a reasonable result.

The difference between strong and weak bases is highlighted by the results of Examples 16-2 and 16-5:

Strong base
0.25 M NaOH(aq)
$[OH^-]_{eq} = 0.25$ M
pH = 13.40

Weak base
0.25 M NH_3(aq)
$[OH^-]_{eq} = 2.1 \times 10^{-3}$ M
pH = 11.30

The strong base is a soluble hydroxide that ionizes completely in water, so the concentration of OH^- matches the 0.25 M concentration of the base. For the weak base, in contrast, the equilibrium concentration of OH^- is substantially smaller than the 0.25 M concentration of the base. At any instant, only 0.8% of the ammonia molecules have accepted protons from water molecules, producing a much less basic solution in which OH^- is a minor species. The equilibrium concentration of unprotonated ammonia is nearly equal to the initial concentration.

Examples 16-4 and 16-5 illustrate the two main types of equilibrium calculations as they apply to solutions of acids and bases. Notice that the techniques are exactly the same as those introduced in Chapter 15. We can calculate values of equilibrium constants from a knowledge of concentrations at equilibrium (Example 16-4), and we can calculate equilibrium concentrations from a knowledge of equilibrium constants and initial concentrations (Example 16-5).

- **16.3.1** A 4.8×10^{-2} M solution of hypochlorous acid (HClO) has a pH of 4.36. Compute K_a for this acid.
- **16.3.2** Draw a molecular picture showing the proton transfer reaction that takes place in an aqueous solution of HClO.
- **16.3.3** Construct a table comparing the concentrations of species in solutions of HCl and HClO, each of which is 4.8×10^{-2} M.

16.4 RECOGNIZING ACIDS AND BASES

How do the structural features of a molecule correlate with its acid–base properties? A Brønsted acid transfers a hydrogen atom as H^+, so we expect an acid to have a hydrogen atom bonded to an electronegative element, X, such as N, O, or Cl. The electronegative atom polarizes the H—X bond, leaving the hydrogen atom with a partial positive charge that facilitates proton transfer. This charge distribution is shown in Figure 16-7 for the HCl molecule, a strong acid.

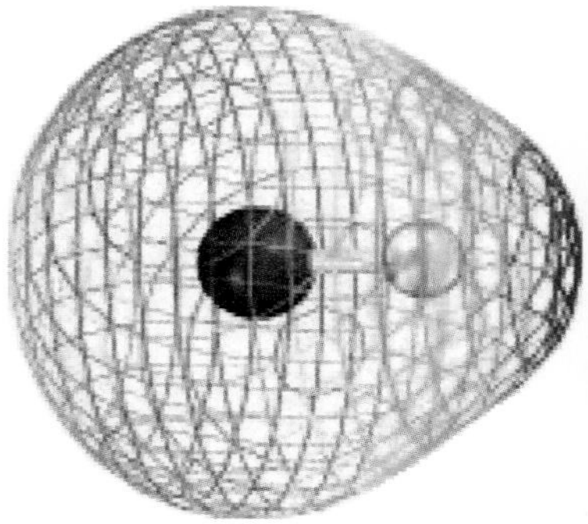

/// Acids contain polar H—X bonds.

Brønsted bases are H^+ acceptors, so a base must supply a pair of electrons to form a new bond to hydrogen. Thus, a base must contain an atom with a lone pair of electrons.

/// Bases contain lone pairs of electrons.

Electron configurations and bond polarities determine whether a chemical substance is a proton donor, a proton acceptor, both, or neither. Recognition is made easier by the fact that the common acids and bases fall into a small number of structural categories.

Figure 16-7
The colored mesh around this HCl molecule shows how electrical charge is distributed in this molecule. Regions of net positive charge are represented in *blue*; regions of net negative charge are represented in *red*. Notice that the region of net positive charge encompasses the acidic hydrogen atom.

Oxyacids

One general type of acid contains an inner atom bonded to a variable number of oxygen atoms and acidic OH groups. These molecules are called **oxyacids**. The chemical formulas of oxyacids usually are written with the hydrogen atoms listed first, in order to emphasize that these compounds are acids. The general formula is $H_x EO_y$, where x and $y = 1-4$ and E, the inner atom, can be a number of different elements including B, C, N, P, S, and Cl. Three of the strong acids, HNO_3, H_2SO_4, and $HClO_4$, are oxyacids. Most other oxyacids are weak, as shown by the examples in Table 16-1. Notice from the structures shown in this table that the acidic hydrogen atom always bonds to an oxygen atom, *not* to the inner atom E.

The names of the oxyacids are based on the common polyatomic ions listed in Table 3-5.

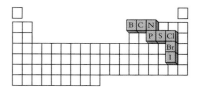

Table 16-1
Some Weak Oxyacids

Name	Formula	K_a	Structure	Model
Chlorous acid	$HClO_2$	1.1×10^{-2}	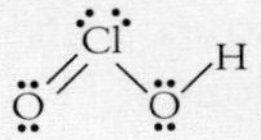	
Nitrous acid	HNO_2	5.6×10^{-4}	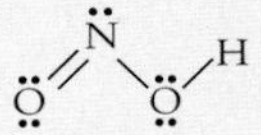	
Hypochlorous acid	$HClO$	4.0×10^{-8}	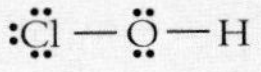	
Hypoiodous acid	HIO	3.2×10^{-11}	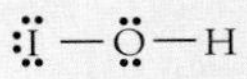	

Oxyacids are named by the following rules:

1. Each oxyacid has an identifying name followed by the word "acid."

2. An oxyacid that forms from a polyatomic anion whose name ends in *-ate* has a name ending in *-ic*. For example, HNO_3 forms by adding a proton to the nitrate polyatomic anion, so HNO_3 is nitric acid. Likewise $HClO_4$ is perchloric acid from the perchlorate anion.

3. An acid that forms from a polyatomic anion whose name ends in *-ite* has a name ending in *-ous*. For example, HNO_2 forms by adding a proton to the nitrite polyatomic anion, so HNO_2 is nitrous acid. Likewise $HClO$ is hypochlorous acid from the hypochlorite anion.

Carboxylic Acids

A carboxylic acid contains the carboxyl group with a hydrogen atom bonded to oxygen. All carboxylic acids are weak. Their chemical formulas usually are written with the $-CO_2H$ group at the end of the formula, and their names always end in $-ic$. Figure 16-8 shows three examples: formic acid, $K_a = 1.8 \times 10^{-4}$, acetic acid, $K_a = 1.8 \times 10^{-5}$ and benzoic acid, $K_a = 6.3 \times 10^{-5}$.

In an aqueous solution, a small fraction of carboxylic acid molecules transfer their acidic hydrogen atoms to water, generating hydronium ions and a molecule that contains the $-CO_2^-$ anionic group. This group is *carboxylate,* and the name of the anion ends in *-ate:* HCO_2^- is formate, $CH_3CO_2^-$ is acetate, and $C_6H_5CO_2^-$ is benzoate.

$$C_6H_5CO_2H(aq) \quad + \quad H_2O(l) \quad \rightleftharpoons \quad C_6H_5CO_2^-(aq) \quad + \quad H_3O^+(aq)$$

Benzoic acid   Benzoate anion

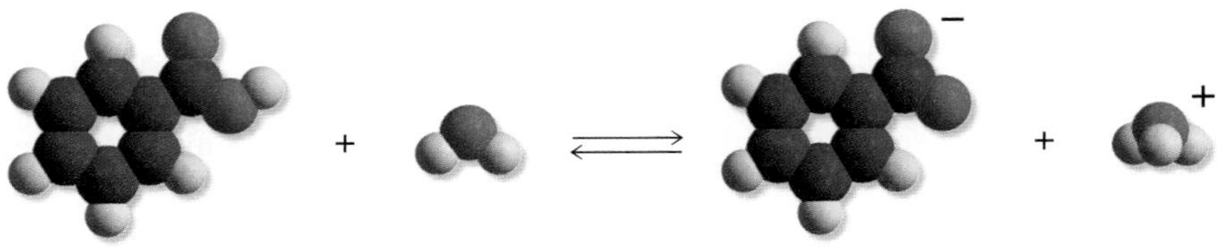

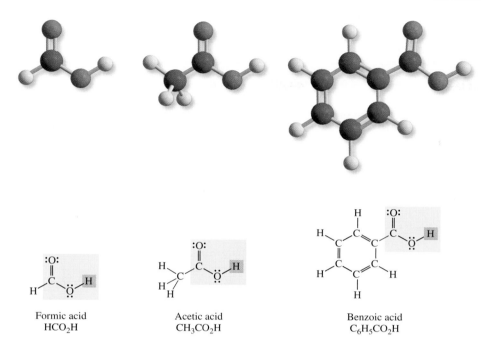

Figure 16-8
Ball-and-stick models and Lewis structures of three carboxylic acids, with acidic hydrogens and the carboxyl groups highlighted.

Formic acid
HCO₂H

Acetic acid
CH₃CO₂H

Benzoic acid
C₆H₅CO₂H

Carboxylic acids almost always contain hydrogen atoms bonded to carbon in addition to hydrogen atoms in their carboxyl groups. Notice in Figure 16–8 that these hydrogen atoms are not highlighted. That is because hydrogen atoms bonded to carbon are *not* acidic. The C—H bond is only slightly polar, so these hydrogen atoms do not participate in aqueous proton transfer reactions.

Polyprotic Acids

An acid that contains more than one acidic hydrogen atom is called a **polyprotic acid.** Sulfuric acid (H_2SO_4) and phosphoric acid (H_3PO_4), shown in Figure 16-9,

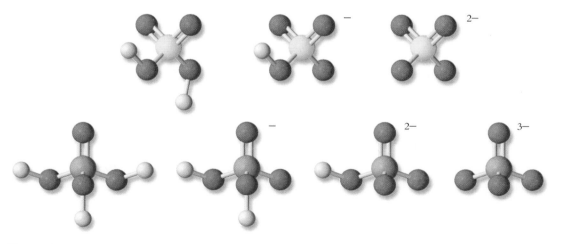

Figure 16-9
Ball-and-stick models of two polyprotic oxyacids, sulfuric acid and phosphoric acid, and their anions.

Table 16-2
Representative Polyprotic Acids

Name	Formula	K_{a1}	Monoanion	K_{a2}
Sulfuric acid	H_2SO_4	Strong	HSO_4^-	1.0×10^{-2}
Oxalic acid	HO_2CCO_2H	5.6×10^{-2}	$HO_2CCO_2^-$	1.5×10^{-4}
Sulfurous acid	H_2SO_3	1.4×10^{-2}	HSO_3^-	6.3×10^{-8}
Phosphoric acid (triprotic: $K_{a3} = 4.8 \times 10^{-13}$)	H_3PO_4	6.9×10^{-3}	$H_2PO_4^-$	6.3×10^{-8}
Phthalic acid	$HO_2CC_6H_4CO_2H$	1.1×10^{-3}	$HO_2CC_6H_4CO_2^-$	3.7×10^{-6}
Carbonic acid	H_2CO_3	4.5×10^{-7}	HCO_3^-	4.7×10^{-11}

are two common polyprotic oxyacids. Many carboxylic acids are polyprotic, too. Table 16-2 lists several examples of polyprotic acids.

> A molecule with two acidic hydrogen atoms is *diprotic*, and one with three acidic hydrogen atoms is *triprotic*.

As Table 16-2 shows, each successive proton transfer reaction of a polyprotic acid has its own value for K_a, and each successive value is smaller than its predecessor. We explain the reasons for this trend in Section 16.6 and show how to treat calculations involving polyprotic acids in Section 16.7.

Other Acids

A number of small molecules that are acids do not fall into any of the categories mentioned above. These acids have no clear patterns in their structure, so it is best simply to learn their names and structures. Three hydrogen halides, HCl, HBr and HI, are strong acids, but the fourth, HF, is a weak acid. Hydrogen sulfide, H_2S, and hydrocyanic acid, HCN, are weak acids that also happen to be quite poisonous. Table 16-3 summarizes the acid-base properties of these and other relatively small weak acids.

Table 16-3
Acid-Base Properties of Some Weak Acids

Name	K_a	Structure	Model
Hydrofluoric acid	6.3×10^{-4}	H—F̈:	
Hydrazoic acid	2.5×10^{-5}	H—N̈=N=N̈	
Hydrogen sulfide (diprotic: $K_{a2} = 1 \times 10^{-19}$)	8.9×10^{-8}	H—S̈—H	
Hydrocyanic acid	6.2×10^{-10}	H—C≡N:	
Phenol	1.0×10^{-10}	⬡—Ö—H	
Hydrogen peroxide	2.4×10^{-12}	H—Ö—Ö—H	

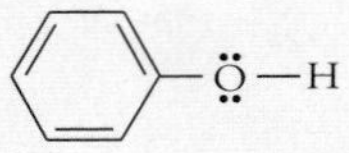

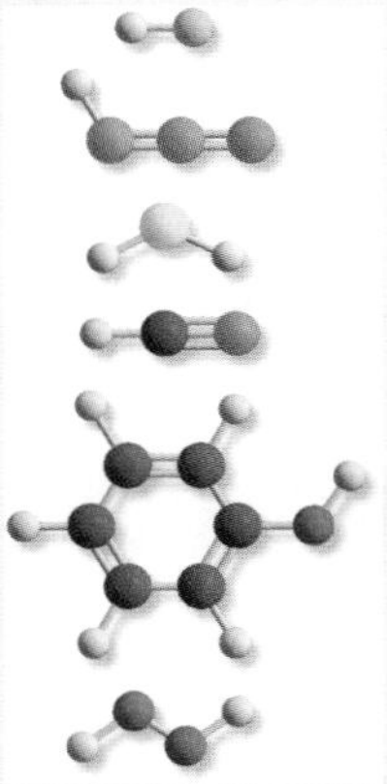

Table 16-4
Representative Organic Bases

Name	Formula	K_b	Line Structure
Dimethylamine	$(CH_3)_2NH$	5.4×10^{-4}	
Methylamine	CH_3NH_2	4.6×10^{-4}	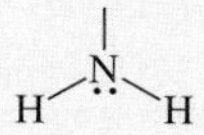
Trimethylamine	$(CH_3)_3N$	6.5×10^{-5}	
Pyridine	C_5H_5N	1.7×10^{-9}	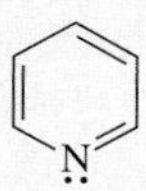
Aniline	$C_6H_5NH_2$	7.4×10^{-10}	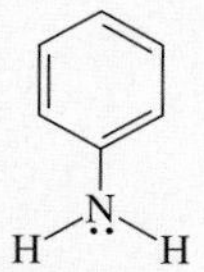

Weak Bases

Besides water, the most common weak base is ammonia, NH_3. Ammonia acts as a base because the lone electron pair on the nitrogen atom can form a bond to a proton. **Amines** are organic compounds in which one or more of the N—H bonds in ammonia have been replaced with N—C bonds. The nitrogen atom in an amine, like its counterpart in ammonia, has a lone pair of electrons that can bond a proton, so all amines are weak bases. Table 16-4 lists examples of bases derived from ammonia.

Example 16-6 shows how the acid-base properties of a molecule can be determined from its structure and the knowledge that acids are proton donors and bases are proton acceptors.

Recognizing Acids and Bases	Example 16-6

Glycine, the simplest amino acid found in proteins, is $H_2NCH_2CO_2H$. Determine the acid-base characteristics of this compound.

Strategy: The Lewis structure of glycine reveals any functional groups that result in acidic or basic properties. To determine the acid-base properties of this compound, look for polar H—X bonds and atoms with lone pairs of electrons. In particular, O—H hydrogen atoms are likely to be acidic, and nitrogen atoms with lone pairs convey base character.

Box 16-2	Chemistry and Life: Drugs and the Brain

For thousands of years, humans have used plants to make medicines and drugs. Many of the biologically active ingredients in plant extracts contain at least one nitrogen atom with a lone pair of electrons, making them weak bases. Such naturally occurring weak bases are classified as alkaloids. Modern pharmaceutical chemists have built upon nature's chemistry by synthesizing new compounds that are important medicines.

The profound physiological effects of alkaloids have been known for centuries. For example, Socrates was put to death with an extract of hemlock, which contains a poisonous alkaloid, coniine. Other alkaloids have long been valued for their beneficial medical effects. Examples include morphine (a pain-killer), quinine (used to treat malaria) and atropine (used to treat Parkinson's disease and in eye drops that dilate the pupils).

Coniine
$C_8H_{17}N$

Morphine
$C_{17}H_{19}NO_3$

Quinine
$C_{20}H_{24}N_2O_2$

Atropine
$C_{17}H_{23}NO_3$

Because of their basic properties, alkaloids were among the first natural substances that early chemists extracted and purified. Morphine was isolated from poppies in 1805 and was the first alkaloid to be characterized. When treated with aqueous strong acid, alkaloids accept protons to produce water-soluble cations. The protonated alkaloids dissolve, leaving the rest of the plant materials behind. Adding strong base to the aqueous extract reverses the proton transfer reaction, converts the alkaloid back to its neutral base form, and causes pure alkaloid to precipitate from the solution:

$$\text{Alkaloid(s)} + H_3O^+(aq) \rightleftharpoons \text{AlkaloidH}^+(aq) + H_2O(l)$$

$$\text{AlkaloidH}^+(aq) + OH^-(aq) \rightleftharpoons \text{Alkaloid(s)} + H_2O(l)$$

Most of us have had to endure the discomfort of dental work, be it repairing cavities or more elaborate oral surgery. Dentists reduce the discomfort immensely by the use of a local pain-killer, Novocain, to block the nerves in the

mouth. Novocain is a weak base containing two nitrogen atoms. A related analgesic, lidocaine (also known as Xylocaine), is sufficiently strong that applying it to the skin causes the nerves in the immediate area to shut down temporarily.

A variety of drugs have been developed that act as sedatives, antidepressants, or stimulants; some of these are effective in treating psychiatric disorders. Many of these drugs are weak bases. Examples are barbiturates such as phenobarbital, tranquilizers like diazepam (Valium), and amphetamines derived from phenylethylamine.

Although nitrogen-containing drugs can be highly beneficial, others—both natural and synthetic—are notorious narcotics. Among alkaloids, morphine is responsible for the narcotic effects of opium and is also the starting material for the synthesis of codeine and heroin. Three other well-known narcotic alkaloids are nicotine, mescaline, and cocaine. Synthetic drugs as well as alkaloids can be abused. Overuse of barbiturates, tranquilizers, or amphetamines can lead to addiction or even death. The darkest side of these substances may be the so-called "designer drugs" such as Ecstasy, which alter brain chemistry in unpredictable and sometimes devastating ways.

Procaine
(Novocain)
$C_{13}H_{20}N_2O_2$

Pentobarbital
(Nembutal)
$NaC_{11}H_{17}N_2O_3$

Diazepam
(Valium)
$C_{16}H_{13}ClN_2O$

Mescaline
$C_{11}H_{17}NO_3$

Nicotine
$C_{10}H_{14}N_2$

Cocaine
$C_{17}H_{21}NO_4$

Ecstasy or MDMA
$C_{11}H_{15}NO_2$

Example 16-6

Recognizing Acids and Bases (*continued*)

Solution: Use the principles developed in Chapter 8 to construct the Lewis structure from the molecular formula:

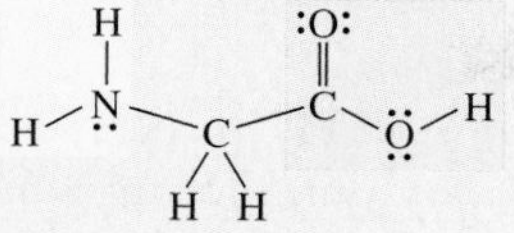

The Lewis structure reveals that glycine contains a carboxyl group, CO_2H, at one end and an amine group, NH_2, at the other. Consequently, the molecule has the properties of both acids and bases.

Amine group *Basic* — Carboxyl group *Acidic*

The amino acids, basic building blocks of proteins, all share this dual acid-base character. See Chapter 11 for a description of the amino acids and their biological chemistry. Organic bases also have a long and varied history as pain killers and narcotics, as our Chemistry and Life Box describes.

Conjugate Acid-Base Pairs

In a solution of hydrofluoric acid, HF molecules donate protons to water molecules, generating H_3O^+ and F^- ions. As discussed in Chapter 15, the forward reaction and the reverse reaction must occur at equal rates in a solution that is at equilibrium. In an aqueous solution of HF that is at equilibrium, hydronium ions donate protons to fluoride ions to generate molecules of H_2O and HF:

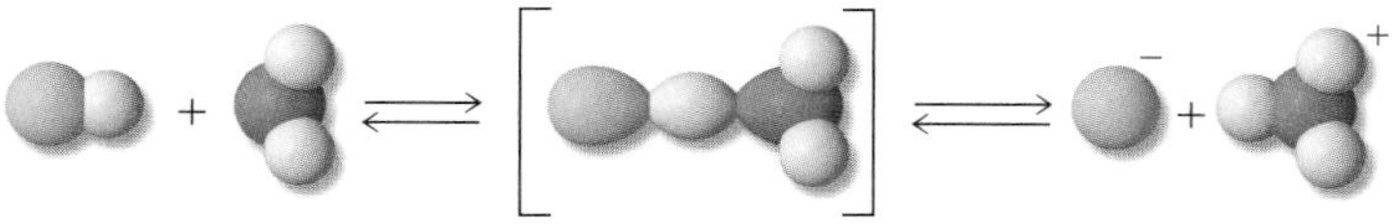

HF H_2O Collision and proton transfer F^- H_3O^+

The reaction involves proton transfer in both directions, so there are two proton donors, HF and H_3O^+. There are also two proton acceptors, H_2O and F^-. In other words, HF is an acid, and F^- is a base. When an H_2O molecule accepts a proton, the resulting cation is a proton donor. In other words, H_3O^+ is an acid, and H_2O is a base. This is a general feature of proton transfer equilibria. Any proton donor and the species generated by removing one of its protons are called a **conjugate acid-base pair.** Hydronium ion is the conjugate acid of H_2O, and fluoride ion is the conjugate base of HF. Thus, HF and F^- are a conjugate acid-base pair; H_3O^+ and H_2O are another conjugate acid-base pair.

Water can act as an acid or a base, so there are two conjugate acid-base pairs for water: H_3O^+ and H_2O are a conjugate acid-base pair, and H_2O and OH^- are another conjugate acid-base pair. Example 16-7 reinforces the structural relationships of conjugate acid-base pairs.

Conjugate Acid–Base Pairs

Example 16-7

Write the chemical formula and the Lewis structure and draw a molecular picture of each of the following: (a) the conjugate acid of NH_3; (b) the conjugate base of HCO_2H; and (c) the conjugate acid of HSO_4^-.

Strategy: Conjugate acid-base pairs are related through proton transfer. Removing a proton from an acid generates its conjugate base. Adding a proton to a base

| Example 16-7 | Conjugate Acid–Base Pairs (continued) |

generates its conjugate acid. We use the procedures of Chapter 8 to construct the Lewis structures.

Solution:

(a) Add H^+ to NH_3 to obtain its conjugate acid, NH_4^+. With a steric number of four, the ammonium ion has tetrahedral geometry.

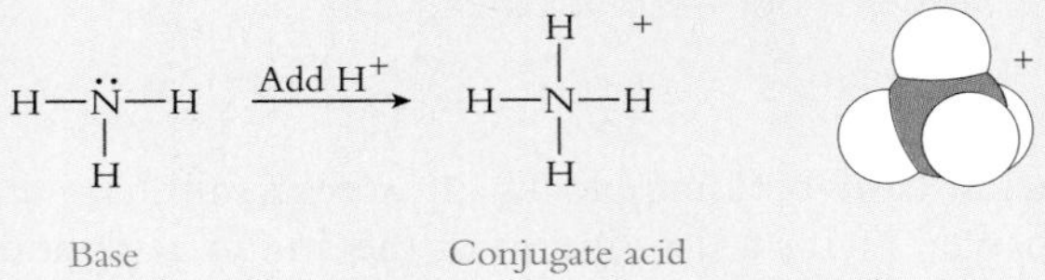

(b) Remove H^+ from HCO_2H to obtain its conjugate base, HCO_2^-:

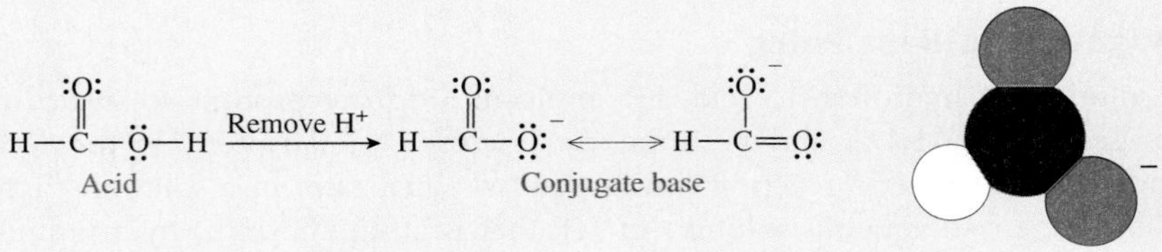

(c) Add H^+ to HSO_4^- to obtain its conjugate acid, H_2SO_4:

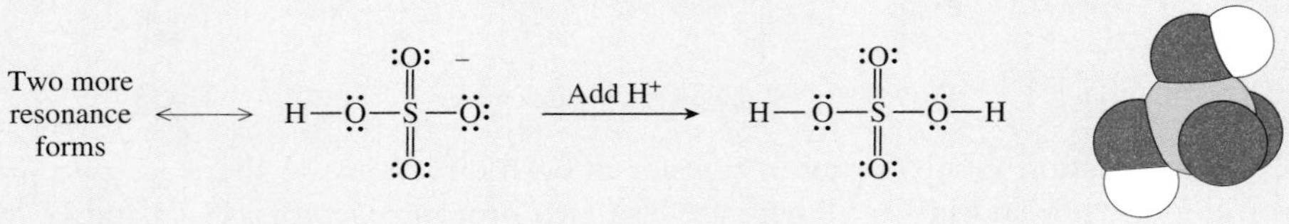

HSO_4^-, like H_2O, has a conjugate base (SO_4^{2-}) and a conjugate acid (H_2SO_4).

The introduction of conjugate acid-base pairs completes our inventory of acids and bases. In addition to strong bases, ammonia, and amines, the anions of weak acids act as bases.

Section Exercises

16.4.1 Classify each of the following substances as a weak acid, strong acid, weak base, strong base, or neither an acid nor a base: (a) $HClO_4$; (b) $NaOH$; (c) CH_3OH; (d) $C_2H_5CO_2H$; and (e) $C_2H_5NH_2$.

16.4.2 Draw ball-and-stick models for the conjugate acid-base pair of each weak acid and weak base in Section Exercise 16.4.1.

16.4.3 Identify the acid-base properties of nicotine and atropine, whose structures are on the next page.

■ **16.4.3** Continued

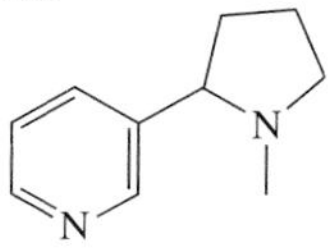

Nicotine
$C_{10}H_{14}N_2$

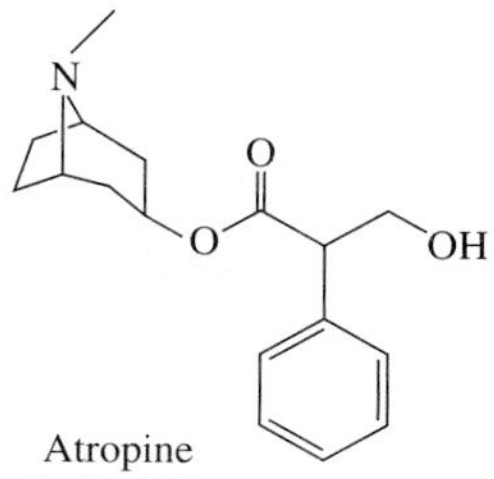

Atropine
$C_{17}H_{23}NO_3$

16.5 ACIDIC AND BASIC SALTS AND OXIDES

Some compounds that are not immediately recognizable as acids and bases nevertheless display acid–base properties. In this section we describe the acid–base chemistry of salts and oxides.

Basic and Acidic Salts

An aqueous solution of a soluble salt contains cations and anions. These ions often have acid–base properties. Anions that are conjugate bases of weak acids make the solution basic, and cations that are conjugate acids of weak bases make the solution acidic. For example, sodium fluoride dissolves in water to give Na^+, F^-, and H_2O as major species. The fluoride anion is the conjugate base of the weak acid HF. This anion establishes a proton transfer equilibrium with water:

$$H_2O(l) + F^-(aq) \rightleftharpoons HF(aq) + OH^-(aq) \qquad K_{eq} = K_b = \frac{[OH^-]_{eq}[HF]_{eq}}{[F^-]_{eq}}$$

The reaction generates hydroxide anions, so the solution is basic. Fluoride acts as a base, so the equilibrium constant is a base dissociation constant, K_b.

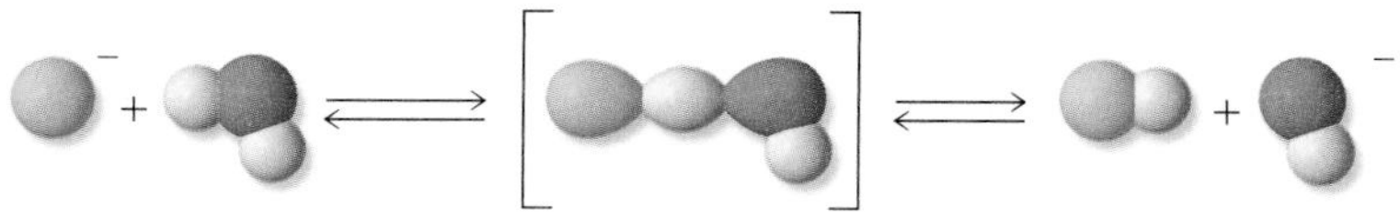

Collision and proton transfer

The fluoride ion equilibrium is linked to two other proton transfer equilibria. The relationship is revealed by combining the proton transfer equilibrium for HF with the proton transfer equilibrium for F^-:

$$\begin{array}{ll} \cancel{HF(aq)} + H_2O(l) \rightleftharpoons \cancel{F^-(aq)} + H_3O^+(aq) & K_a \\ H_2O(l) + \cancel{F^-(aq)} \rightleftharpoons OH^-(aq) + \cancel{HF(aq)} & K_b \\ \hline 2\,H_2O(l) \rightleftharpoons OH^-(aq) + H_3O^+(aq) & K_w \end{array}$$

Combining these two equilibria leads to cancellation of HF and F^-, so the sum of the two is the water equilibrium. How are the equilibrium constants for these three equilibria related?

$$K_a = \frac{[H_3O^+]_{eq}\,[F^-]_{eq}}{[HF]_{eq}} \qquad K_b = \frac{[OH^-]_{eq}\,[HF]_{eq}}{[F^-]_{eq}} \qquad K_w = [H_3O^+]_{eq}\,[OH^-]_{eq}$$

Multiplying the concentration quotients for K_a and K_b leads to cancellation of the concentrations of F^- and HF, just as they cancel when the individual reactions are added. Whenever two or more equilibria are added, the equilibrium constant for the net reaction is the product of the individual equilibrium constants of the summed reactions:

$$\underbrace{\frac{[H_3O^+]_{eq}\,\cancel{[F^-]_{eq}}}{\cancel{[HF]_{eq}}}}_{K_a}\;\underbrace{\frac{[OH^-]_{eq}\,\cancel{[HF]_{eq}}}{\cancel{[F^-]_{eq}}}}_{K_b} = \underbrace{[H_3O^+]_{eq}[OH^-]_{eq}}_{K_w}$$

$$K_a K_b = K_w \qquad\qquad (16\text{-}3)$$

Equation 16-3 applies to any acid and its conjugate base. The equation can also be expressed in logarithmic form using pK notation and pK_w = 14.00:

$$pK_a + pK_b = 14.00 \qquad\qquad (16\text{-}4)$$

Because K_a for HF and K_w are known, Equation 16-3 can be used to determine K_b for fluoride anion. From Example 16-4, $K_a = 6.0 \times 10^{-4}$ for HF, and as always, $K_w = 1.0 \times 10^{-14}$. Thus:

$$K_b = \frac{K_w}{K_a} = \frac{1.0 \times 10^{-14}}{6.0 \times 10^{-4}} = 1.7 \times 10^{-11}$$

Notice that both K_b and K_a are much smaller than 1. This reflects the fact that the conjugate base of a weak acid is itself weak: HF is a weak acid, and F^- is a weak base.

Example 16-8 shows how to deal quantitatively with the basic nature of a salt solution.

| Example 16-8 | Salt of a Weak Acid |

Sodium hypochlorite (NaOCl) is the active ingredient in laundry bleach. Typically, bleach contains 5.0% of this salt by mass, which is a 0.67 M solution. Determine the concentrations of all species and compute the pH of laundry bleach.

Strategy: This is a quantitative acid-base equilibrium problem, so we use the seven-step method introduced in Chapter 15.

Solving
Equilibrium
Problems

Solution:

1. We are asked to determine pH, which requires that we find the hydronium ion concentration.

2. Sodium hypochlorite is a salt, so its aqueous solution contains the ionic species Na^+ and OCl^-. Thus the major species in solution are Na^+, OCl^-, and H_2O.

3. Water is a proton donor and a proton acceptor. The sodium ion is neither an acid nor a base, so it is a spectator ion in this solution. Because HOCl is a

Salt of a Weak Acid *(continued)*

Example 16-8

weak oxyacid, OCl^- is its conjugate base. The hypochlorite ion accepts a proton from a water molecule:

$$H_2O + OCl^- \rightleftharpoons HOCl + OH^- \qquad K_{eq} = K_b$$

4. Determine K_b for OCl^- from the information given in Table 16-1:

$$K_a \text{ (HOCl)} = 4.0 \times 10^{-8}$$

$$K_b \text{ (OCl}^-) = \frac{K_w}{K_a} = \frac{1.0 \times 10^{-14}}{4.0 \times 10^{-8}} = 2.5 \times 10^{-7}$$

5. Set up a concentration table for this equilibrium. The initial concentration of OCl^- is given, and x can represent the change in concentration of OCl^-:

Species (reaction)	H_2O +	OCl^-	$\rightleftharpoons$	$HOCl$ +	OH^-
Initial concentration (M)		0.67		0	0
Change in concentration (M)		$-x$		$+x$	$+x$
Equilibrium concentration (M)		$0.67 - x$		x	x

6. Solve the equilibrium constant expression to find the desired concentrations.

$$K_b = \frac{[HOCl]_{eq} [OH^-]_{eq}}{[OCl^-]_{eq}} = \frac{(x)(x)}{(0.67 - x)} = 2.5 \times 10^{-7}$$

Now make the approximation that $x \ll 0.67$:

$$0.67 - x \cong 0.67 \qquad 2.5 \times 10^{-7} \cong \frac{x^2}{0.67}$$

$$x^2 \cong 1.68 \times 10^{-7} \qquad \textit{from which} \qquad x \cong 4.1 \times 10^{-4}$$

The value of x provides three of the required concentrations:

$$[OCl^-] = (0.67 - x) \text{ M} = (0.67 - 4.1 \times 10^{-4}) \text{ M} = 0.67 \text{ M}$$

$$[OH^-] = [HOCl] = x \text{ M} = 4.1 \times 10^{-4} \text{ M}$$

The sodium ion concentration is simply equal to the initial concentration of the salt: $[Na^+] = 0.67$ M

To find $[H_3O^+]$, rearrange the expression for K_w:

$$[H_3O^+] = \frac{K_w}{[OH^-]} = \frac{1.0 \times 10^{-14}}{4.1 \times 10^{-4}} = 2.4 \times 10^{-11} \text{ M}$$

Finally, use $[H_3O^+]$ to determine the pH of the bleach solution:

$$pH = -\log[H_3O^+] = -\log(2.4 \times 10^{-11}) = 10.62$$

7. The value of x is three orders of magnitude smaller than 0.67, so the approximation is valid. We know that the hypochlorite anion is the conjugate base of hypochlorous acid, so a pH > 7 is a reasonable result.

Salts that contain cations of weak bases are acidic. For example, the ammonium cation is the conjugate *acid* of ammonia. When ammonium salts dissolve in water, NH_4^+ ions transfer protons to H_2O molecules, generating H_3O^+ and making the solution slightly acidic:

$$NH_4^+(aq) + H_2O(l) \rightleftharpoons NH_3(aq) + H_3O^+(aq)$$

The equilibrium constant for this reaction can be calculated from Equation 16-3 and K_b for ammonia:

$$K_a = \frac{K_w}{K_b} = \frac{1.0 \times 10^{-14}}{1.8 \times 10^{-5}} = 5.6 \times 10^{-10}$$

In contrast to the acid-base character of ions arising from weak acids or bases, cations and anions resulting from strong acids or bases do not affect solution pH. The sodium ions in a solution of NaF do not participate in proton transfer reactions, nor do the chloride ions in a solution of ammonium chloride. Consequently, the water equilibrium determines the pH of a solution of a salt containing neither weak acid nor weak base components. Consider as an example an aqueous solution of sodium chloride. The major ionic species present are Na^+ and Cl^-. Neither of these ions contains a hydrogen atom, so they are not proton donors. The sodium cation has no lone pairs of electrons, so it is not a base. The chloride anion does have lone pairs, but Cl^- is the conjugate base of a *strong* acid, HCl. Remember that a strong acid donates its protons to water quantitatively. This means that the chloride anion has a negligible tendency to accept protons from hydronium ions in aqueous solution. Because water is a much weaker acid than the hydronium ion, Cl^- will not accept protons from water, either. Both Na^+ and Cl^- are spectator ions, and the only important proton transfer equilibrium in an aqueous solution of NaCl is the water equilibrium.

The salt of a weak base, pyridinium chloride, is the subject of Example 16-9.

| Example 16-9 | Salt of a Weak Base |

What are the important acid-base equilibria in an aqueous solution of pyridinium chloride (C_5H_5NHCl)? What are the values of their equilibrium constants?

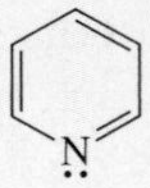

Pyridine
C_5H_5N

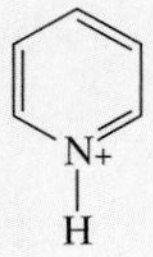

Pyridinium ion
$C_5H_5NH^+$

Strategy: This is a qualitative problem. First identify the major species, then list the equilibria in which they participate. Values for equilibrium constants can be found in tables in this chapter and in Appendix E.

Solution: Pyridinium chloride is a salt that generates ions in solution. The major species are the pyridinium cation ($C_5H_5NH^+$), Cl^-, and H_2O. Its formula identifies the pyridinium cation as the conjugate acid of the weak base, pyridine.

The chloride anion is the anion of a strong acid, so Cl^- is a spectator ion. There are two acid-base equilibria with major species as reactants:

$$H_2O + C_5H_5NH^+ \rightleftharpoons H_3O^+ + C_5H_5N \qquad K_a$$

$$H_2O + H_2O \rightleftharpoons H_3O^+ + OH^- \qquad K_w = 1.0 \times 10^{-14}$$

Salt of a Weak Base *(continued)*   **Example 16-9**

K_b for pyridine, listed in Table 16-4, is used to calculate K_a:

$$K_b \ (C_5H_5NH_2) = 1.7 \times 10^{-9}$$

$$K_a = \frac{K_w}{K_b} = \frac{1.0 \times 10^{-14}}{1.7 \times 10^{-9}} = 5.9 \times 10^{-6}$$

This K_a value is much greater than the value of K_w, so the pH of a pyridinium chloride solution is determined by the weak acid.

Basic and Acidic Oxides

Most elements form binary compounds with oxygen, and many binary oxides display acid-base chemistry. In general, metal-containing oxides are bases, and oxides of the non–metals are acids.

Soluble metal oxides are ionic compounds that act as strong bases, because the oxide anion has four lone pairs of electrons and readily accepts a proton from a water molecule.

$$O^{2-} + H_2O(l) \longrightarrow 2 \ OH^-(aq)$$

Collision and proton transfer

Sodium oxide is representative of soluble metal oxides:

$$Na_2O(s) + H_2O(l) \longrightarrow 2 \ Na^+(aq) + 2 \ OH^-(aq)$$

Insoluble metal oxides do not dissolve in pure water but react readily with aqueous strong acids. Two hydronium ions donate protons to the oxide anion, generating three water molecules and freeing the metal cation to dissolve:

$$MgO(s) + 2 \ H_3O^+(aq) \longrightarrow Mg^{2+}(aq) + 3 \ H_2O(l)$$

$$CaO(s) + 2 \ H_3O^+(aq) \longrightarrow Ca^{2+}(aq) + 3 \ H_2O(l)$$

Nonmetal oxides react with water to produce oxyacids. If the resulting acid is strong, the reaction goes to completion and is accompanied by proton transfer to water to produce quantitative amounts of hydronium ions:

$$SO_3(g) + 2 \ H_2O(l) \longrightarrow H_3O^+(aq) + HSO_4^-(aq)$$

$$N_2O_5(g) + 3 \ H_2O(l) \longrightarrow 2 \ H_3O^+(aq) + 2 \ NO_3^-(aq)$$

$$Cl_2O_7(g) + 3 \ H_2O(l) \longrightarrow 2 \ H_3O^+(aq) + 2 \ ClO_4^-(aq)$$

Gaseous oxides that produce weak acids establish equilibrium between the aqueous acid and the gaseous oxide:

$$SO_2(g) + H_2O(l) \rightleftharpoons H_2SO_3(aq)$$

$$CO_2(g) + H_2O(l) \rightleftharpoons H_2CO_3(aq)$$

Section Exercises

16.5.1 Determine the concentrations of all species present in a 0.35 M solution of KCN. (Potassium cyanide is a deadly poison.)

16.5.2 Write the acid-base equilibrium that determines the pH of aqueous solutions of each of the following salts, and state whether the resulting solution is acidic, basic, or neither: (a) NH_4I, (b) $NaClO_4$, and (c) $NaCH_3CO_2$.

16.5.3 Determine what reactions occur when the following oxides dissolve in water, and predict whether the resulting solutions are acidic or basic: (a) Li_2O; (b) Cl_2O.

16.6 FACTORS AFFECTING ACID STRENGTH

Why is NaOH a strong base, CH_3OH neither an acid nor a base, and HNO_3 a strong acid? Each of these compounds contains O—H bonds, yet their proton transfer properties are strikingly different. In this section we examine the effect of molecular structure on acid strength.

Effect of Charge

The charge on a species has a major effect on its ability to donate or accept protons. Remember that opposite electric charges attract, and like charges repel. An anion is both a better proton acceptor and a poorer proton donor than is a neutral molecule. Likewise, a cation is a poorer proton acceptor and a better proton donor.

The effect of charge is shown clearly by the acid-base properties of water and its ions (See Figure 16-10). The water molecule is both a weak acid and a weak base. Hydroxide anion is a strong base because its negative charge strongly attracts protons. Despite having an O—H bond, hydroxide is not acidic because the negative charge of OH^- makes it exceedingly difficult to remove the remaining proton. H_3O^+ is a strong acid because its positive charge enhances the removal of one of its protons. Hydronium shows negligible basicity even though it has a lone pair of electrons on its oxygen atom. The positive charge of H_3O^+ makes it exceedingly difficult to bind another proton.

Polyprotic acids also show clearly the effect of charge on acidity. As mentioned earlier, the successive K_a values for a polyprotic oxyacid decrease by from three to five

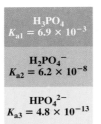

$$H_3PO_4$$
$$K_{a1} = 6.9 \times 10^{-3}$$

$$H_2PO_4^-$$
$$K_{a2} = 6.2 \times 10^{-8}$$

$$HPO_4^{2-}$$
$$K_{a3} = 4.8 \times 10^{-13}$$

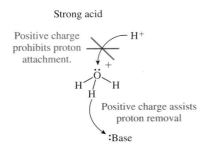

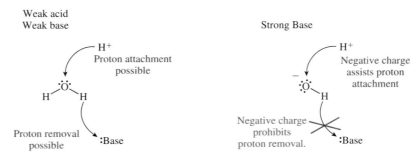

Figure 16-10
Hydronium cations, neutral water molecules, and hydroxide anions illustrate the effect of charge on acid-base behavior.

orders of magnitude. The neutral parent acid always is a stronger acid than is the anion produced by removing one proton.

Structural Factors

A proton transfer reaction involves breaking a covalent bond. For an acid, H—X bond breaking transfers a proton to the base as the bonding electrons are converted to a lone pair on X. Breaking the H—X bond becomes easier to accomplish as the bond energy becomes weaker and as the bonding electrons become more polarized toward X. Bond strengths and bond polarities help explain trends in acidity among neutral molecules.

For example, the acidity of the binary hydrides increases moving across a row of the periodic table. This trend follows bond polarity: As the electronegativity of the inner atom, X, increases, the H—X bond polarity increases and so does acidity. The C—H bonds of methane are essentially nonpolar, so the hydrogens are not at all acidic. On the other hand, with an electronegativity difference of 1.9, the electrons of the H—F bond are highly polarized toward the fluorine atom. The resulting partial positive charge on the hydrogen atom assists proton transfer.

← **CHAPTER 9**
Trends in bond energies are discussed in Chapter 9.

Acid strength ⇒			
CH_4	NH_3	H_2O	HF
2.5	3.0	3.5	4.0
Electronegativity ⇒			

Acid strength ⇒			
HF	HCl	HBr	HI
565	430	360	295
⇐ Bond energy (kJ/mol)			

Acidity increases moving down a column of the periodic table. The difference between HF (weak acid) and the other hydrogen halides (strong acids) exemplifies the vertical trend. Moving down a column of the periodic table, the principal quantum number of the valence orbitals of the halogen increases, orbital overlap in bonding decreases, and bond strength decreases. Thus the vertical trend follows bond strength. The H—F bond is substantially stronger than the H—Cl bond, so even though the H—F bond is more polar, HF is a weaker acid than HCl.

Among oxyacids, the strengths of O—H bonds depend on the amount of electron density in the bond. This, in turn, depends on how strongly the rest of the atoms in the molecule attract electrons. The halogen oxyacids show this effect in two ways, as shown in Figure 16-11. The acidity of the hypohalous acids increases as the electronegativity of the halogen increases. This is because the more electronegative the atom, the more it attracts electrons. The acidity of the chlorine oxyacids increases as the number of oxygen atoms increases. This is because each oxygen atom attracts electron density from the rest of the molecule, so the more oxygen atoms there are, the smaller is the electron density in the O—H bond.

⇐ Acid strength		
K_a 4.0×10^{-8}	2.8×10^{-9}	3.2×10^{-11}
HClO	HBrO	HIO
3.0	2.8	2.5
⇐ Electronegativity		

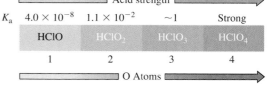

Acid strength ⇒			
K_a 4.0×10^{-8}	1.1×10^{-2}	~1	Strong
HClO	$HClO_2$	$HClO_3$	$HClO_4$
1	2	3	4
O Atoms ⇒			

Increasing acid strength

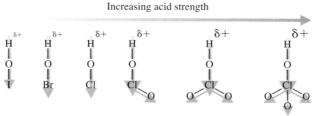

Figure 16-11
The strength of an oxyacid increases as more and more electron density is withdrawn from the O—H bond.

The presence of a second oxygen atom also explains why carboxylic acids like formic acid (HCO_2H) are acidic, but alcohols like methanol (CH_3OH) are not:

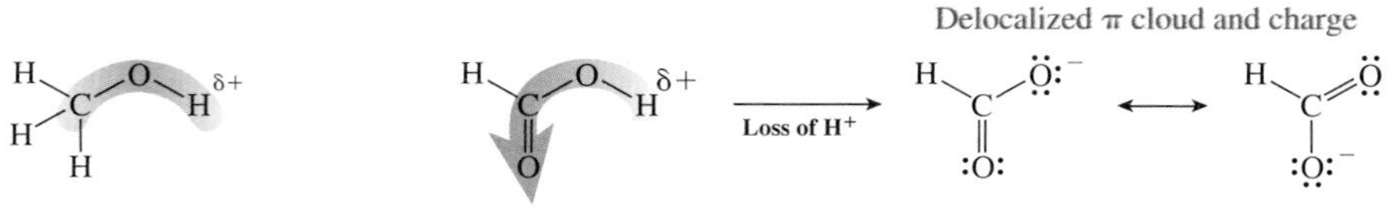

The second oxygen atom in a carboxyl group withdraws electron density from the O—H bond, weakening the bond and enhancing its acidity. The nonpolar C—H bonds in an alcohol do not have the same effect. At least equally important, the carboxylate conjugate base is stabilized by delocalization of two types: both the π electron cloud and the negative charge are spread over both oxygen atoms rather than being localized on one atom.

Example 16-10 explores another feature of relative acid strengths.

Example 16-10 **Acidities of Simple Carboxylic Acids**

Oxalic acid, $HO_2C—CO_2H$, has $K_{a1} = 1.3 \times 10^{-2}$ and $K_{a2} = 1.5 \times 10^{-4}$. Formic acid, HCO_2H, has $K_a = 1.8 \times 10^{-4}$. Explain why the first proton of oxalic acid is substantially more acidic than the proton on formic acid, but the second proton is less acidic.

Strategy: To explain features of acid strength, we need to look at bond strength and polarity effects.

Solution: Examine the structures of these three species to gain insight into their relative acid strengths:

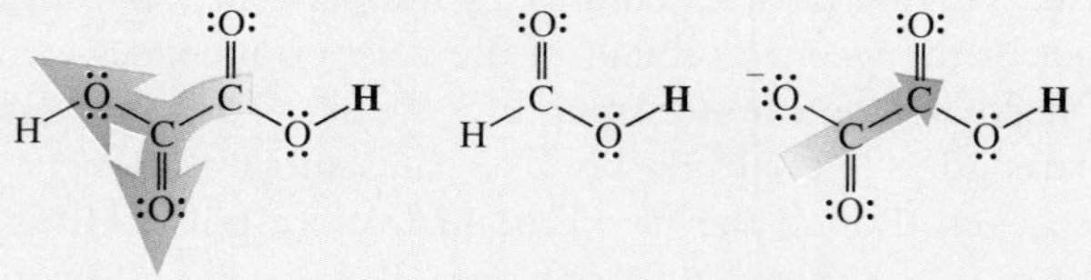

Oxalic acid $K_a = 1.3 \times 10^{-2}$ | Formic acid $K_a = 1.8 \times 10^{-4}$ | Hydrogen oxalate $K_a = 1.5 \times 10^{-4}$

As the left-hand structure indicates, the second CO_2H group in oxalic acid is better at withdrawing electron density than the H in formic acid. This makes the O—H bonds in oxalic acid more polar than the O—H bond in formic acid, so the first proton of oxalic acid is more easily transferred, making oxalic acid stronger than formic acid. The resulting anion has a negative charge that hinders removal of the second proton and makes hydrogen oxalate a weaker acid than formic acid.

Section Exercises

16.6.1 Sodium metal reacts with ammonia to form an ionic salt, $NaNH_2$, that contains NH_2^- anions. Is this salt acidic or basic, and why? What reaction occurs if this salt is placed in water?

16.6.2 Among the following pairs of acids, which is stronger and why? (a) H_3PO_4 and H_3PO_3; (b) H_2S and H_2Se; (c) H_2SO_3 and HSO_3^-.

16.6.3 Rank the following acids from weakest to strongest: HBr, HBrO, HClO, and $HClO_2$.

16.7 MULTIPLE EQUILIBRIA

Most of the examples in previous sections appear to involve a single acid-base equilibrium. A closer look reveals that nearly any solution that displays acid-base properties has at least two acid-base equilibria. Look again at Example 16-5, where we list two equilibria as potentially important:

$$NH_3(aq) + H_2O(l) \rightleftharpoons NH_4^+(aq) + OH^-(aq) \qquad K_b = 1.8 \times 10^{-5}$$

$$2\,H_2O(l) \rightleftharpoons H_3O^+(aq) + OH^-(aq) \qquad K_w = 1.0 \times 10^{-14}$$

The water equilibrium always exists in aqueous solution. In general, we can focus our initial attention on the equilibria involving other major species (NH_3 in this example). Nevertheless, the water equilibrium does exert its effect on the concentrations of OH^- and H_3O^+. In this example, the concentration of hydroxide anion is established by the ammonia equilibrium, but the concentration of hydronium cations must be found by applying the water equilibrium. In several of our earlier examples, we made implicit use of this feature.

Because many species exhibit acid-base properties, it is often possible to write several proton transfer equilibrium expressions for an aqueous solution. Each such expression is valid if the reactants are species that are actually present in the solution. We have already seen how to focus on the dominant equilibrium: consider only those expressions that have major species as reactants, and look for the one with the largest equilibrium constant.

After completing our analysis of the effects of the dominant equilibrium, we may need to consider the effects of other equilibria. The calculation of $[H_3O^+]$ in a solution of weak base illustrates circumstances where this secondary consideration is necessary. Here, the dominant equilibrium does not include the species, H_3O^+, whose concentration we wish to know. In such cases we must turn to an equilibrium expression that has the species of interest as a product. The reactants should be species that are involved in the dominant equilibrium, because that means that we know the initial concentrations of these new reactants. Look again at Example 16-8 for another application of this idea. In that example, the dominant equilibrium is the reaction between hypochlorite anions and water molecules:

$$H_2O(l) + OCl^-(aq) \rightleftharpoons HOCl(aq) + OH^-(aq)$$

Working with this equilibrium, we can determine the concentrations of OCl^-, HOCl, and OH^-. To find the concentration of hydronium ions, however, we must invoke a second equilibrium, the water equilibrium:

$$2\,H_2O(l) \rightleftharpoons H_3O^+(aq) + OH^-(aq)$$

Other than the water equilibrium, polyprotic acids provide the most common examples of secondary equilibria that play roles in determining concentrations of minor species in aqueous solution. As noted earlier, the K_a value for the second proton transfer is smaller than K_a for the first, so the first equilibrium dominates the chemistry of polyproticoxy acids. Consequently, when a base is added to a solution that contains both a polyproticoxy acid and its anion, the base accepts protons preferentially from the neutral acid. Only after the neutral acid has been consumed completely does the anion participate significantly in proton transfer. Example 16-11 provides molecular pictures of this feature.

Tutorial

| Example 16-11 | Molecular View of a Polyprotic Acid |

The drawing shows a molecular view of a very small region of an aqueous solution of oxalic acid. For clarity, water molecules are not shown. Redraw this molecular picture to show the solution (a) after two hydroxide ions react with these molecules and (b) after four hydroxide ions react with these molecules. Your drawings should include water molecules formed as products.

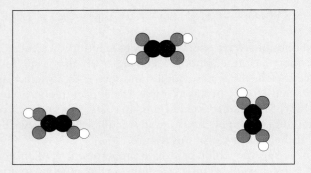

Tutorial

Strategy: This is a qualitative problem, so the seven-step method does not apply. To draw molecular pictures illustrating a proton transfer process, we must visualize the chemical reactions that occur, see what products result, then draw the resulting solution. When a strong base is added to a weak acid, hydroxide ions remove protons from the molecules of weak acid. When more than one acidic species is present, the stronger acid loses protons preferentially.

Solution: The oxalic acid solution contains water molecules and $H_2C_2O_4$ molecules as major species. Added hydroxide ions remove protons from the strongest acid, $H_2C_2O_4$:

$$H_2C_2O_4(aq) + OH^-(aq) \rightleftharpoons HC_2O_4^-(aq) + H_2O(l)$$

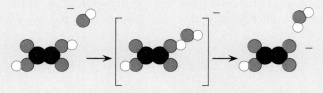

Collision and transfer

Each proton transfer reaction creates $HC_2O_4^-$, which is itself a weak acid. The K_a values show that $H_2C_2O_4$ is substantially more acidic than $HC_2O_4^-$:

$H_2C_2O_4$ $\qquad K_a = 5.6 \times 10^{-2}$ $\qquad\qquad HC_2O_4^-$ $\qquad K_a = 1.5 \times 10^{-4}$

Consequently, hydroxide ions react preferentially with $H_2C_2O_4$. Any added OH^- ions react with $H_2C_2O_4$ until every oxalic acid molecule has lost a proton.

(a) The first two hydroxide ions convert two $H_2C_2O_4$ molecules into $HC_2O_4^-$ ions. Thus the molecular picture should contain two $HC_2O_4^-$ anions, two water molecules and one unreacted $H_2C_2O_4$ molecule:

Molecular View of a Polyprotic Acid *(continued)*   **Example 16-11**

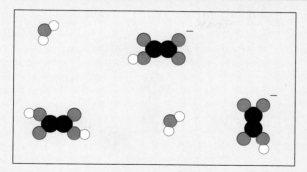

(b) When four OH^- ions are added, the first three consume all the $H_2C_2O_4$ molecules. The remaining OH^- ion reacts with the weaker acid, $HC_2O_4^-$:

$$HC_2O_4^-(aq) + OH^-(aq) \rightleftharpoons C_2O_4^{2-}(aq) + H_2O(l)$$

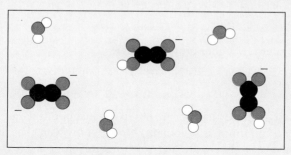

Collision and transfer

This reaction goes virtually to completion, so at equilibrium this system contains two $HC_2O_4^-$, one $C_2O_4^{2-}$, and four water molecules.

Keep in mind that the water molecules produced in these proton transfer reactions join the immense pool of water molecules already present.

Example 16-11 shows qualitatively how to deal with more than one acid-base equilibrium in a solution of a polyprotic acid. To obtain quantitative results, it may be necessary to carry out more than one calculation, using more than one concentration table, to find the concentrations of all the ions in these solutions. When multiple equilibria must be taken into account, it is particularly important to identify the dominant equilibrium, which will have major species as reactants. After solving for the concentrations of ions resulting from the dominant equilibrium, we then consider how other equilibria affect the concentrations of the minor species. Example 16-12 shows how this works.

| Example 16-12 | Ion Concentrations in a Polyprotic Acid Solution |

Carbonated water contains carbonic acid, a diprotic acid that forms when carbon dioxide dissolves in water.

$$CO_2(g) + H_2O(l) \rightleftharpoons H_2CO_3(aq)$$

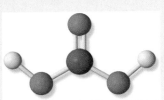

A typical carbonated beverage contains 0.050 M H_2CO_3. Determine the concentrations of the ions present in this solution.

Strategy: The problem describes a weak diprotic acid and asks for ion concentrations. In order to determine concentrations of all ions, we need to consider more than one equilibrium. This is done in stages, starting with the dominant equilibrium. We apply the seven-step strategy.

Solving Equilibrium Problems

Solution:

1. The problem asks us for the concentrations of the ions in an aqueous solution.

2. The major species in carbonated water are H_2CO_3 and H_2O.

3. There are two chemical equilibria involving these major species. The acid undergoes proton transfer with equilibrium constant K_{a1}:

$$H_2O(l) + H_2CO_3(aq) \rightleftharpoons H_3O^+(aq) + HCO_3^-(aq)$$

$$K_{a1} = 4.5 \times 10^{-7}$$

As always, water undergoes proton transfer:

$$H_2O(l) + H_2O(l) \rightleftharpoons H_3O^+(aq) + OH^-(aq)$$

$$K_w = 1.0 \times 10^{-14}$$

Since K_{a1} is more than seven orders of magnitude larger than K_w, we identify this equilibrium as dominant. Notice, however, that the water equilibrium generates some hydroxide ions in the solution, so this equilibrium must be used to find the concentration of hydroxide.

Because carbonic acid is diprotic, a second proton transfer equilibrium has an effect on the ion concentrations:

$$H_2O(l) + HCO_3^-(aq) \rightleftharpoons H_3O^+(aq) + CO_3^{2-}(aq)$$

$$K_{a2} = 4.7 \times 10^{-11}$$

This reaction involves a *minor* species, HCO_3^-, as a reactant, so it cannot be the dominant equilibrium. However, just as the water equilibrium generates some hydroxide ions, the hydrogen carbonate equilibrium generates some carbonate anions, whose concentration will have to be determined.

4. We can write an equilibrium constant expression for each equilibrium. Because the initial calculations involve the dominant equilibrium, we start with its expression:

Ion Concentrations in a Polyprotic Acid Solution *(continued)*   Example 16-12

$$K_{a1} = \frac{[H_3O^+]_{eq}\,[HCO_3^-]_{eq}}{[H_2CO_3]_{eq}}$$

5. and 6. Now we are ready to organize the data and the unknowns and do the calculations. There are multiple equilibria affecting ion concentrations, so we must work with more than one concentration table, starting with the dominant equilibrium. Set up a concentration table to determine concentrations of the ions generated by this reaction:

Species (reaction) $H_2O(l)$ +	$H_2CO_3(aq)$	$\rightleftharpoons$ $H_3O^+(aq)$	+ $HCO_3^-(aq)$
Initial concentration (M)	0.050	0	0
Change in concentration (M)	$-x$	$+x$	$+x$
Equilibrium concentration (M)	$0.050 - x$	x	x

Substitute the equilibrium concentrations into the equilibrium constant expression and solve for x, making the approximation that $x \ll 0.050$:

$$K_{a1} = 4.5 \times 10^{-7} = \frac{(x)(x)}{(0.050 - x)} = \frac{x^2}{0.050 - x} \cong \frac{x^2}{0.050}$$

$$x^2 = (0.050)(4.5 \times 10^{-7}) = 2.25 \times 10^{-8} \quad \textit{from which} \quad x = 1.5 \times 10^{-4}$$

We round to two significant figures because the K value has two significant figures. The concentrations are:

$$[H_3O^+] = [HCO_3^-] = 1.5 \times 10^{-4}\,M$$

$$[H_2CO_3] = 0.050 - 1.5 \times 10^{-4} = 0.050 - 0.00015 = 0.050\,M.$$

Note that x is about 0.3% of 0.050, so the approximation is valid.

Next, we take into account the proton transfer equilibrium involving hydrogen carbonate anion. To do this, we complete a second concentration table, using as "initial" concentrations those calculated for the first equilibrium:

Species (reaction) $H_2O(l)$ +	$HCO_3^-(aq)$	$\rightleftharpoons$ $H_3O^+(aq)$	+ $CO_3^{2-}(aq)$
Initial conc. (M)	1.5×10^{-4}	1.5×10^{-4}	0
Change in conc. (M)	$-x$	$+x$	$+x$
Equilibrium conc. (M)	$1.5 \times 10^{-4} - x$	$1.5 \times 10^{-4} + x$	x

Substitute equilibrium concentrations into the equilibrium constant expression and solve for x, making the approximation that $x \ll 1.5 \times 10^{-4}$:

| Example 16-12 | Ion Concentrations in a Polyprotic Acid Solution (*continued*) |

$$K_{a2} = 4.7 \times 10^{-11} = \frac{[H_3O^+]_{eq}\,[CO_3^{-2}]_{eq}}{[HCO_3^-]_{eq}}$$

$$4.7 \times 10^{-11} = \frac{(x)(1.5 \times 10^{-4} + x)}{1.5 \times 10^{-4} - x} \cong \frac{x(1.5 \times 10^{-4})}{1.5 \times 10^{-4}} \cong x$$

This value is too small to cause a measurable change in the concentrations already calculated, but it does tell us the concentration of carbonate anions in the solution: $[CO_3^{2-}] = 4.7 \times 10^{-11}$ M.

One more ion remains, OH^-, generated from the water equilibrium. It is possible to set up a concentration table, but it is easier to apply the water equilibrium expression directly:

$$K_w = 1.0 \times 10^{-14} = [H_3O^+][OH^-] \qquad so$$

$$[OH^-] = \frac{(1.0 \times 10^{-14})}{[H_3O^+]} = \frac{(1.0 \times 10^{-14})}{(1.5 \times 10^{-4})} = 6.7 \times 10^{-11} \text{ M}$$

Notice that the ions produced in the dominant equilibrium have substantially larger concentrations than those generated by other equilibria:

Products of dominant equilibria:

$$[H_3O^+] = [HCO_3^-] = 1.5 \times 10^{-4} \text{ M}$$

Products of other equilibria:

$$[CO_3^{2-}] = 4.7 \times 10^{-11} \text{ M} \qquad [OH^-] = 6.7 \times 10^{-11} \text{ M}$$

7. Are these results reasonable? The concentration of the acid that is responsible for the various ions in the solution is $[H_2CO_3] = 0.050$ M. Note that the ions generated when this acid undergoes proton transfer with water have concentrations that are two orders of magnitude smaller than the concentration of the parent acid. Note further that the products generated by other equilibria have concentrations that are more than six orders of magnitude smaller than the concentrations of ions produced in the reaction corresponding to the dominant equilibrium. The major species have highest concentrations; the species generated directly by reactions of major species have next highest concentrations; and species generated by secondary reactions have very low concentrations. This is a reasonable outcome.

When the value of K_{eq} for a secondary equilibrium is relatively close to the value of K_{eq} for the dominant equilibrium, equilibrium concentrations must be calculated using a more elaborate approach. Such calculations are beyond the scope of our coverage.

Section Exercises

16.7.1 Calculate the hydroxide ion concentration in a 0.333 M aqueous solution of NH_4Cl.

16.7.2 Write all the acid-base equilibrium reactions that have major species as reactants for a solution of sodium bicarbonate, $NaHCO_3$.

16.7.3 Determine which of the equilibria in Section Exercise 16.7.2 will be the most important and predict whether this solution is acidic or basic.

CHAPTER REVIEW

Summary and Key Words

1. Brønsted-Lowry acids and bases are proton donors and acceptors. Water is **amphiprotic** because it acts as both an acid and a base, as expressed by the **water equilibrium constant** (K_w). There are several **strong acids** that quantitatively donate protons to water and several **strong bases** that quantitatively generate hydroxide ions in water.

2. Aqueous hydronium ion concentrations span over 14 orders of magnitude and are conveniently expressed in logarithmic form using **pH**. Logarithmic scales can also be used to express hydroxide ion concentration (pOH) and equilibrium constants (pK). The pH of a solution can be measured using pH paper or a pH meter.

3. A solution of a **weak acid** contains the acid and water as major species. Its pH is determined by the acid equilibrium constant, K_a. A solution of a **weak base** contains the base and water as major species, and its pH is determined by the base equilibrium constant, K_b.

4. Acids contain polar H—X bonds, and bases contain lone pairs of electrons. **Oxyacids** contain polyatomic oxygen-containing groups, **carboxylic acids** contain the —CO_2H group, and **polyprotic acids** contain more than one acidic hydrogen atom. **Amines** are weak bases. Loss of a proton from a weak acid generates a weak base, and gain of a proton by a weak base generates a weak acid. A weak acid and its corresponding weak base are a **conjugate acid-base pair**.

5. A salt that contains an anion of a weak acid is basic, and a salt that contains the cation of a weak base is acidic. The anion of a strong acid is not acidic or basic. Metal oxides act as bases, and nonmetal oxides act as acids.

6. Positive charge increases acid strength and negative charge decreases acid strength. Strong H—X bonds are less acidic that weak H—X bonds, and molecular structural features that withdraw electron density from an O—H bond increase acid strength.

7. Most aqueous acid-base systems are subject to more than one equilibrium expression, that for the solute and the water equilibrium. Solutions of polyprotic acids have multiple acid-base equilibria.

Skills to Master

▶ Identifying weak and strong acids and bases

▶ Identifying the relevant equilibria

▶ Converting among pH, pOH, and concentrations

▶ Calculating K_a from equilibrium pH

▶ Calculating equilibrium concentrations of weak acids and bases

▶ Understanding factors that affect acid-base strength

▶ Dealing with multiple acid-base equilibria

Learning Exercises

16.1 Outline the procedure for working an equilibrium problem for a weak acid-base system.

16.2 Prepare a table listing the various types of acids and bases and the identifying features of each.

16.3 Write a paragraph describing the conjugate acid-base pair and explaining how each interacts with water.

16.4 Update your list of memory bank equations.

16.5 Prepare a list of all terms in Chapter 16 that are new to you. In your own words, write a one-sentence definition of each. Consult the Glossary if you need help.

Problems   ilw = interactive learning ware problem. Visit the website at www.wiley.com/college/olmsted

Proton Transfers in Water

16.1 Draw a set of molecular pictures that show the proton transfer reaction occurring when HBr dissolves in water.

16.2 Draw a set of molecular pictures that show the proton transfer reaction occurring when HNO_3 dissolves in water.

16.3 Determine the concentrations of hydronium and hydroxide ions in 1.25×10^{-3} M aqueous perchloric acid.

16.4 Determine the concentrations of hydronium and hydroxide ions in 3.65×10^{-2} M aqueous sodium hydroxide.

ilw 16.5 Concentrated aqueous HCl has a concentration of 12.1 M. Calculate the concentrations of all ions present in a solution prepared by pipetting 1.00 mL of concentrated HCl into a 100. mL volumetric flask and filling to the mark.

16.6 Concentrated aqueous $HClO_4$ has a concentration of 14.8 M. Calculate the concentrations of all ions present in a solution prepared by pipetting 5.00 mL of concentrated HCl into a 1000. mL volumetric flask and filling to the mark.

16.7 Calculate the concentrations of hydronium and hydroxide ions in a solution prepared by dissolving 0.488 g of HCl gas in enough water to make 325 mL of solution.

16.8 Calculate the concentrations of hydronium and hydroxide ions in a solution prepared by dissolving 0.345 g of solid NaOH in enough water to make 225 mL of solution.

The pH Scale

16.9 Convert each of the following H_3O^+ concentrations to pH: (a) 4.0 M; (b) 3.75×10^{-6} M; (c) 4.8 mM; and (d) 0.000255 M.

16.10 Convert each of the following H_3O^+ concentrations to pH: (a) 1.25 M; (b) 2.95 μM; (c) 0.0366 M; and (d) 7.45×10^{-4} M.

16.11 Convert each of the following OH^- concentrations to pH: (a) 4.0 M; (b) 3.75×10^{-6} M; (c) 4.8 mM; and (d) 0.000255 M.

16.12 Convert each of the following OH^- concentrations to pH: (a) 1.25 M; (b) 2.95 μM; (c) 0.0366 M; and (d) 7.45×10^{-4} M.

16.13 Convert each of the following pH values into a hydronium ion concentration: (a) 0.66; (b) 7.85; (c) 3.68; and (d) 14.33.

16.14 Convert each of the following pH values into a hydronium ion concentration: (a) 1.56; (b) 3.85; (c) 9.75; and (d) 11.22.

16.15 Convert each of the following pH values into a hydroxide ion concentration: (a) 0.66; (b) 7.85; (c) 3.68; and (d) 14.33.

16.16 Convert each of the following pH values into a hydroxide ion concentration: (a) 1.56; (b) 3.85; (c) 9.75; and (d) 11.22.

Weak Acids and Bases

16.17 Calculate the pH of a 1.5 M solution of each of the following compounds (See Appendix E for K values): (a) NaOH; (b) C_5H_5N; (c) $HONH_2$; and (d) HCO_2H.

16.18 Calculate the pH of a 2.5×10^{-2} M solution of the following compounds (See Appendix E for K values): (a) NH_3; (b) HClO; (c) HCN; and (d) $Ba(OH)_2$.

16.19 Draw molecular pictures illustrating the proton transfer reaction responsible for determining the pH of the solutions in Problem 16.17 (c) and (d).

16.20 Draw molecular pictures illustrating the proton transfer reaction responsible for determining the pH of the solutions in Problem 16.18 (a) and (c).

ilw **16.21** For a 1.50 M aqueous solution of hydrazoic acid, HN_3, do the following: (a) Identify the major and minor species. (b) Compute concentrations of all species present. (c) Find the pH. (d) Draw a molecular picture illustrating the equilibrium reaction that determines the pH.

16.22 For a 2.75×10^{-2} M aqueous solution of cyanic acid, HCNO, do the following: (a) Identify the major and minor species. (b) Compute concentrations of all species present. (c) Find the pH. (d) Draw a molecular picture illustrating the equilibrium reaction that determines the pH.

16.23 For a 0.350 M aqueous solution of trimethylamine, $N(CH_3)_3$, do the following: (a) Identify major and minor species. (b) Compute concentrations of all species. (c) Find the pH. (d) Draw a molecular picture illustrating the equilibrium reaction that determines the pH.

16.24 For a 1.85×10^{-3} M aqueous solution of aniline, $C_6H_5NH_2$, do the following: (a) Identify major and minor species. (b) Compute concentrations of all species. (c) Find the pH. (d) Draw a molecular picture illustrating the equilibrium reaction that determines the pH.

Recognizing Acids and Bases

16.25 Identify each of the following substances as a weak acid, strong acid, weak base, strong base, both weak acid and weak base, or neither an acid nor a base: (a) NH_3; (b) HCNO; (c) HClO; and (d) $Ba(OH)_2$.

16.26 Identify each of the following substances as a weak acid, strong acid, weak base, strong base, both weak acid and weak base, or neither an acid nor a base: (a) HNO_3; (b) CH_3CO_2H; (c) HOH; (d) HOCl; and (e) NH_2OH.

16.27 For each weak acid in Problem 16.17, identify the conjugate base. For each weak base, identify the conjugate acid.

16.28 For each weak acid in Problems 16.18 and 16.26, identify the conjugate base. For each weak base, identify the conjugate acid.

16.29 Write all the conjugate acid-base equilibrium expressions that apply to an aqueous solution of each of the substances in Problem 16.25, and identify each conjugate acid and base.

16.30 Write all the conjugate acid-base equilibrium expressions that apply to an aqueous solution of each of the substances in Problem 16.26, and identify each conjugate acid and base.

Acidic and Basic Salts and Oxides

ilw **16.31** For a 0.45 M solution of Na_2SO_3, do the following: (a) Identify the major species. (b) Identify the equilibrium that determines the pH. (c) Compute the pH.

16.32 For a 6.75×10^{-3} M solution of sodium benzoate, $NaC_6H_5CO_2$, do the following: (a) Identify the major species. (b) Identify the equilibrium that determines the pH. (c) Compute the pH.

16.33 For a solution that is 0.0100 M in NH_4NO_3, do the following: (a) Identify the major species. (b) Identify the equilibrium that determines the pH. (c) Compute the pH.

16.34 For a solution that is 4.75×10^{-2} M in NH_4Br, do the following: (a) Identify the major species. (b) Identify the equilibrium that determines the pH. (c) Compute the pH.

16.35 Write the reactions that occur when SO_2 dissolves in water, and draw a set of molecular pictures illustrating the equilibrium reaction that determines the pH of the resulting solution.

16.36 Write the reactions that occur when CO_2 dissolves in water, and draw a set of molecular pictures illustrating the equilibrium reaction that determines the pH of the resulting solution.

16.37 When insoluble FeO is treated with strong acid, the oxide dissolves. Write the chemical reaction that occurs.

16.38 When insoluble Fe_2O_3 is treated with strong acid, the oxide dissolves. Write the chemical reaction that occurs.

Factors Affecting Acid Strength

16.39 Among the following pairs of acids, which is stronger and why? (a) H_2SO_4 and HSO_4^-; (b) HClO and HIO; (c) HClO and $HClO_2$.

16.40 Among the following pairs of acids, which is stronger and why? (a) $HBrO_3$ and $HBrO_2$; (b) H_2S and H_2O; (c) H_2S and HS^-.

16.41 Draw Lewis structures of the acids in Problem 16.39 (b) and (c), and use arrows to show electron density shifts that account for their different acid strengths.

16.42 Draw Lewis structures of the acids in Problem 16.40 (a) and (b), and use arrows to show electron density shifts that account for their different acid strengths.

Multiple Equilibria

ilw **16.43** Use two concentration tables to calculate the concentrations of all species present in a 0.250 M solution of sodium acetate, $NaCH_3CO_2$. (See Appendix E for K values.)

16.44 Use two concentration tables to calculate the concentrations of all species present in a 3.45×10^{-2} M solution of KBrO. (See Appendix E for K values.)

16.45 Determine the concentrations of the ionic species present in a 1.55×10^{-2} M solution of the diprotic acid H_2CO_3 (see Table 16-2 for K values).

16.46 Determine the concentrations of the ionic species present in a 0.355 M solution of the diprotic acid H_2SO_3 (see Table 16-2 for K values).

Additional Paired Problems

16.47 Putrescine, a substance with a vile odor, has two basic amino functional groups as shown in the line structure that follows. Use Lewis structures to show the two proton transfer reactions of putrescine and water.

$$H_2N\diagup\diagdown\diagup NH_2 \qquad \text{Putrescine} \atop C_4H_{12}N_2$$

16.48 Draw Lewis structure sketches showing the aqueous-phase reaction between acetic acid and ammonia.

16.49 Determine the pH of a 0.250 M solution of aqueous NaF.

16.50 Calculate the pH of a 0.025 M aqueous solution of NH_4Cl.

16.51 Determine the concentrations of all species present in 2.00 M H_2SO_4.

16.52 Determine the concentrations of all species present in 0.200 M aqueous $NaNO_2$.

16.53 For each of the following compounds, identify the major species in solution and write a balanced equation for the equilibrium reaction that determines the pH: (a) H_2SO_4; (b) Na_2SO_4; (c) CO_2; and (d) NH_4Cl.

16.54 Identify major species and proton transfer equilibria for aqueous solutions of the following: (a) NH_4NO_3; (b) KH_2PO_4; (c) Na_2O; and (d) HCO_2H.

16.55 Use tabulated K_{eq} values to find the value of K_{eq} for the following reaction:

$$HPO_4{}^{2-}(aq) + OH^-(aq) \rightleftharpoons PO_4{}^{3-}(aq) + H_2O(l)$$

16.56 Use tabulated K_{eq} values to find the value of K_{eq} for the following reaction:

$$HPO_4{}^{2-}(aq) + H_3O^+(aq) \rightleftharpoons H_2PO_4{}^-(aq) + H_2O(l)$$

16.57 Boric acid, H_3BO_3, $K_a = 5.4 \times 10^{-10}$, is frequently used as an eyewash. (a) Use Lewis structures to illustrate the equilibrium reaction of K_a. (b) Calculate the pH of 0.050 M boric acid solution.

16.58 Hydrazine, N_2H_4, has $K_b = 1.3 \times 10^{-6}$. (a) Use Lewis structures to illustrate the equilibrium reaction of K_b. (b) Calculate the pH of a 2.00×10^{-1} M solution of N_2H_4.

16.59 In the mid-1930s a substance was isolated from a fungus that is a parasite of ryes and other grasses. This alkaloid, lysergic acid, has

been of great interest to chemists because of its strange, dramatic action on the human mind. Many derivatives of lysergic acid are known, some with medicinal applications. Perhaps the best known derivative of lysergic acid is the potent hallucinogen lysergic acid diethylamide (LSD):

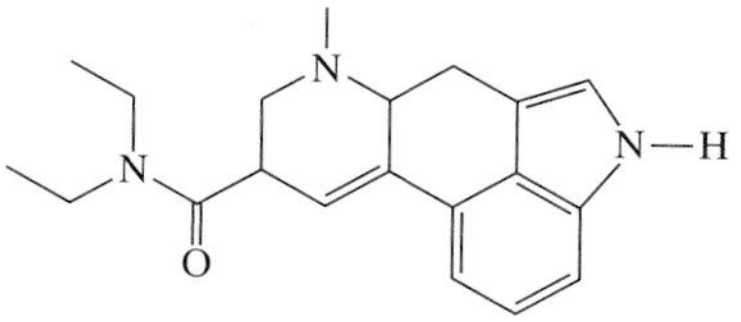

LSD $C_{20}H_{25}N_3O$

Like other alkaloids, LSD is a weak base, $K_b = 7.6 \times 10^{-7}$. What is the pH of a 0.55 M solution of LSD?

16.60 The addictive painkiller morphine, $C_{17}H_{19}NO_3$, is the principal molecule in the milky juice that exudes from unripe poppy seed capsules.

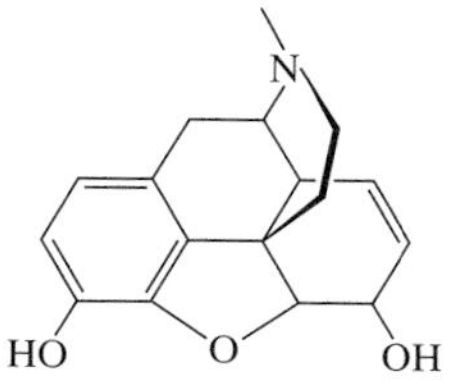

Morphine $C_{17}H_{19}NO_3$

Calculate the pH of a 0.015 M solution of morphine, given that $K_b = 7.9 \times 10^{-7}$.

16.61 The pH of a 0.060 M solution of a weak acid is 2.71. Calculate K_a and identify the acid from among those listed in Appendix E.

16.62 The pH of a 0.0100 M solution of the sodium salt of a weak acid is 11.00. What is the K_a of the weak acid?

More Challenging Problems

16.63 Oxalic acid, a diprotic carboxylic acid found in many plants, including rhubarb, is an effective stain remover. Consider the following diagram to be a small section of an aqueous solution of oxalic acid:

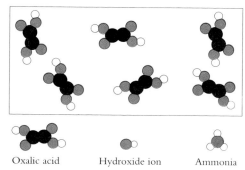

Oxalic acid   Hydroxide ion   Ammonia

(a) Draw a new picture that shows the appearance of the solution at equilibrium after four hydroxide ions enter the region.

(b) Draw a new picture that shows the appearance of the solution at equilibrium after eight hydroxide ions enter the region.

(c) Draw a new picture that shows the appearance of the solution at equilibrium after four ammonia molecules enter the region ($K_b = 1.8 \times 10^{-5}$).

16.64 By coincidence, the K_a of CH_3CO_2H and the K_b of NH_3 both are 1.8×10^{-5}. Write the appropriate acid–base equilibrium reactions and use them to determine whether each of the following 1.00 M solutions is acidic, basic, or neutral: (a) CH_3CO_2H; (b) NH_3; (c) NH_4Cl; (d) $NaCH_3CO_2$; and (e) $NH_4CH_3CO_2$.

16.65 A solution is prepared by dissolving 3.5 g of P_4O_{10} in 1.50 L of water. The oxide reacts quantitatively to form phosphoric acid. (a) Identify the major species in the solution. (b) Identify the minor species in the solution and rank them in order of concentration, highest first. (c) Calculate the pH of the solution.

16.66 Aqueous solutions of Na_2SO_3 and CH_3CO_2H are mixed. (a) List the major species in each solution. (b) Write the net ionic reaction that occurs on mixing. (c) Identify the acid, base, conjugate acid, and conjugate base.

16.67 In aqueous solution, amino acids exist as zwitterions (German for "double ions"), compounds in which internal proton transfer gives a molecule with two charged functional groups. Use Lewis structures to illustrate the proton transfer equilibrium between the uncharged form of glycine ($NH_2CH_2CO_2H$) and its zwitterion form.

16.68 Vinegar is a dilute aqueous solution of acetic acid. A sample of vinegar has a pH of 2.39 and a density of 1.07 g/mL. What is the mass percentage of acetic acid in the vinegar?

16.69 For each of the following reactions, write a balanced net ionic equation. Use different sizes of arrows to indicate whether the reaction goes nearly to completion or proceeds to only a small extent. (Hint: You may need to compare K_a values.)
  (a) $NaOH(aq) + C_6H_5CO_2H(s) \rightleftharpoons$?
  (b) $(CH_3)_3N(aq) + HNO_3(aq) \rightleftharpoons$?
  (c) $Na_2SO_4(aq) + CH_3CO_2H(aq) \rightleftharpoons$?
  (d) $NH_4Cl(aq) + Ca(OH)_2(aq) \rightleftharpoons$?
  (e) $K_2HPO_4(aq) + NH_3(aq) \rightleftharpoons$?

16.70 Identify each of the following as an acid, base, both, or neither and then write balanced reactions that illustrate the acidic or basic properties of the compound: (a) H_2CO_3; (b) $KHCO_3$; (c) NH_3; (d) $NaCl$; (e) Na_2SO_4; (f) SO_2; and (g) Li_2O.

Group Study Problems

16.71 According to its label, each tablet of Alka-Seltzer contains 1.916 g (0.0228 mol) of sodium hydrogen carbonate. When an Alka-Seltzer tablet dissolves in 150 mL of water: (a) What equilibrium determines the pH? (b) What is the pH of the solution?

16.72 When ammonium acetate dissolves in water, both the resulting ions undergo proton transfer reactions with water, but the net reaction can be written without using water:

$$NH_4{}^+(aq) + CH_3CO_2{}^-(aq) \rightleftharpoons NH_3(aq) + CH_3CO_2H(aq)$$

(a) Combine proton transfer equilibrium reactions involving water so that the net result is the above reaction, and use this sequence of reactions to derive an expression for K_{eq} for this reaction in terms of K_a, K_b, and K_w. (b) Make a list of all the proton-transfer reactions that occur in this solution. (c) Use K_a and K_b to calculate the concentrations of H_3O^+ and OH^- that result from the reactions of $NH_4{}^+$ and $CH_3CO_2{}^-$ with water in a 0.25 M solution of ammonium acetate. (d) Based on the results of part (c), what pH would you expect to find for this solution?

16.73 For each of the following reactions, write a balanced net ionic equation. Use different sizes of arrows to indicate whether the reaction goes nearly to completion or proceeds to only a small extent. (Hint: You may need to compare K_a values.)

  (a) $H_2S(aq) + NH_3(aq) \rightleftharpoons$?
  (b) $C_2H_5NH_2(aq) + KHSO_4(aq) \rightleftharpoons$?
  (c) $C_5H_5N(aq) + HCN(aq) \rightleftharpoons$?
  (d) $NH_4Br(aq) + Na_3PO_4(aq) \rightleftharpoons$?
  (e) $HClO(aq) + HONH_2(aq) \rightleftharpoons$?

16.74 In each of the following situations, one reaction goes essentially to completion. In each case, identify the major species in solution under initial conditions and write a balanced net ionic equation for the reaction that goes essentially to completion. (a) Gaseous HBr is bubbled through a 0.015 M solution of $Ca(OH)_2$. (b) 4.0 g of NaOH is added to 0.75 L of 0.055 M $NaHSO_4$. (c) 25.0 mL of saturated NH_4I is mixed with 50.0 mL of 0.95 M $Pb(NO_3)_2$.

16.75 Pure sulfuric acid (H_2SO_4) is a viscous liquid that causes severe burns when it contacts the skin. Like water, sulfuric acid is amphoteric, so a proton transfer equilibrium exists in pure sulfuric acid. (a) Write this proton transfer equilibrium reaction. (Hint: H_2O is NOT involved.) (b) Construct the Lewis structure of sulfuric acid and identify the features that allow this compound to function as a base. (c) Perchloric acid ($HClO_4$) is a stronger acid than sulfuric acid. Write the proton transfer reaction that takes place when perchloric acid dissolves in pure sulfuric acid.

Answers to Section Exercises

16.1.1 $H_3O^+(aq) + SO_4{}^{2-}(aq) \rightleftharpoons H_2O(aq) + HSO_4{}^-(aq)$

16.3.2

16.1.2 4.0×10^{-11} M
16.1.3 $[K^+] = [OH^-] = 0.200$ M; $[H_3O^+] = 5.0 \times 10^{-14}$ M
16.2.1 1.5×10^{-3} M HCl, pH = 2.82; 2.5×10^{-2} M NaOH, pH = 12.94
16.2.2 $[H_3O^+] = 1.6 \times 10^{-2}$ M; $[OH^-] = 6.3 \times 10^{-13}$ M
16.3.1 $K_a = 4.0 \times 10^{-8}$

16.3.3

Acid:	HCl	HClO
$[HA]$	0	4.8×10^{-2} M
$[H_3O^+]$	4.8×10^{-2} M	3.8×10^{-5} M
$[A^-]$	4.8×10^{-2} M	3.8×10^{-5} M

16.4.1 (a) Strong acid; (b) strong base; (c) neither; (d) weak acid; and (e) weak base

16.4.2

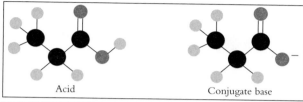

Acid   Conjugate base

Base   Conjugate acid

16.4.3 Nicotine is a weak base, having two N atoms with a lone pair of electrons. Atropine also is a weak base, having one N atom with a lone pair of electrons. Both compounds also contain O atoms, but they lack the CO_2H group, so they are not acidic.

16.5.1 $[HCN] = [OH^-] = 2.4 \times 10^{-3}$ M; $[CN^-] = [K^+] = 0.35$ M; $[H_3O^+] = 4.2 \times 10^{-12}$ M

16.5.2 (a) $NH_4^+(aq) + H_2O(l) \rightleftharpoons NH_3(aq) + H_3O^+(aq)$, acidic; (b) $2 H_2O(l) \rightleftharpoons OH^-(aq) + H_3O^+(aq)$, neither; and (c) $CH_3CO_2^-(aq) + H_2O(aq) \rightleftharpoons CH_3CO_2H(aq) + OH^-(aq)$, basic.

16.5.3 (a) basic: $Li_2O + H_2O \longrightarrow 2 Li^+(aq) + 2 OH^-(aq)$; and (b) acidic: $Cl_2O + H_2O \longrightarrow 2 HClO(aq)$

16.6.1 The salt is basic, because the anion of NH_3 is more basic than NH_3. The anion accepts a proton from water: $NH_2^- + H_2O \rightarrow NH_3 + OH^-$.

16.6.2 (a) H_3PO_4 is stronger, more O atoms to withdraw electrons; (b) H_2Se is stronger, higher principal quantum number means weaker $H\!-\!X$ bond; and (c) H_2SO_3 is stronger, neutral acid is stronger than anion.

16.6.3 HBrO is weakest, then HClO, $HClO_2$, and HBr, a strong acid, is strongest.

16.7.1 $[OH^-] = 7.3 \times 10^{-10}$ M

16.7.2 The major species are HCO_3^-, H_2O, and Na^+. There are three acid-base equilibria with these species as reactants:

$$HCO_3^-(aq) + H_2O(aq) \rightleftharpoons CO_3^{2-}(aq) + H_3O^+(aq)$$
$$HCO_3^-(aq) + H_2O(aq) \rightleftharpoons H_2CO_3(aq) + OH^-(aq)$$
$$2 H_2O(aq) \rightleftharpoons H_3O^+(aq) + OH^-(aq)$$

16.7.3 The most important equilibrium is the one with the largest K_{eq}. Here are the values:

$$HCO_3^-(aq) + H_2O(aq) \rightleftharpoons CO_3^{2-}(aq) + H_3O^+(aq)$$
$$K_{a2} = 4.7 \times 10^{-11}$$
$$HCO_3^-(aq) + H_2O(aq) \rightleftharpoons H_2CO_3(aq) + OH^-(aq)$$
$$K_b = K_w/K_{a1} = 2.2 \times 10^{-8}$$
$$2 H_2O(aq) \rightleftharpoons H_3O^+(aq) + OH^-(aq) \qquad K_w = 1.0 \times 10^{-14}$$

The value of K_b is larger than the others, so this reaction is the most important. The solution is basic, since this reaction produces OH^- anions.

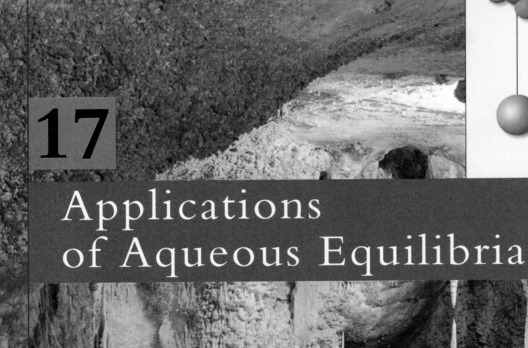

17

Applications of Aqueous Equilibria

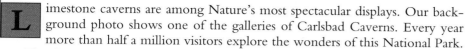

INTRODUCTION: LIMESTONE CAVERNS

L imestone caverns are among Nature's most spectacular displays. Our background photo shows one of the galleries of Carlsbad Caverns. Every year more than half a million visitors explore the wonders of this National Park.

The formation of Carlsbad Caverns began over 250 million years ago, during an age when an inland sea covered this region of the American southwest. In this sea, corals, mollusks, and other marine organisms flourished, incorporating calcium ions dissolved in seawater into solid calcium carbonate shells and skeletons. As these sea creatures died, their remains accumulated on the ocean floor. Over many eons, the shells and skeletons were compressed into limestone rock, eventually forming a deposit over 400 miles long.

When the climate became drier, the ancient sea evaporated, and this limestone deposit was buried beneath a bed of silt. In time, uplift and erosion exposed the limestone deposits to Nature's elements—primarily the action of running water.

Limestone is nearly insoluble in pure water, but rainwater generally is slightly acidic because of carbon dioxide that dissolves from the atmosphere:

$$CO_2(aq) + 2 H_2O(l) \rightleftharpoons HCO_3^-(aq) + H_3O^+(aq)$$

When exposed to acidic water, limestone slowly dissolves by forming hydrogen carbonate anions:

$$CaCO_3(s) + H_3O^+(aq) \rightleftharpoons Ca^{2+}(aq) + HCO_3^-(aq) + H_2O(l)$$

Over thousands of years, water percolating through limestone dissolved some of the rock, forming large underground chambers.

Inside newly formed caverns, calcium carbonate precipitates from water dripping from the ceiling of the chambers. This happens when water that is saturated with carbon dioxide and calcium hydrogen carbonate comes into contact with air. Some of the dissolved CO_2 escapes into the gas phase. This shifts the two equilibria to the left, and solid calcium carbonate precipitates:

$$CaCO_3(s) + CO_2(g) + H_2O(l) \xleftarrow{\text{Shifts to left}} Ca^{2+}(aq) + 2 HCO_3^-(aq)$$

(with an upward arrow labeled "Escapes" above $CO_2(g)$)

The amount of calcium carbonate precipitating from any particular drop is imperceptibly small. Nevertheless, over the years, these deposits grow into translucent hollow tubes of $CaCO_3$ called soda straws (see photo inset). Soda straws lengthen as water drops fall from their tips. These delicate structures can reach lengths of several feet. In time, water flowing over the outside of the tube adds width to the growing formation, and the soda straw matures into the familiar stalactite.

Over the eons, the flow and evaporation of water inside a cavern creates a stunning array of rock sculptures. Stalagmites grow upward from the floor, sometimes joining stalactites to form massive columns. Limestone dams create beautiful pools of water. Limestone draperies fall like curtains from water flowing around overhanging rock. Delicate mineral flowers sprout from the walls. All these features result from the aqueous solubility equilibrium of calcium carbonate.

This chapter describes several important applications of equilibria. We begin by describing buffers and titrations, two acid-base applications with significant laboratory applications. Then we describe solubility and complexation equilibria, two general categories that are important in nature.

17.1 BUFFER SOLUTIONS

Human blood contains a variety of acids and bases that maintain the pH very close to 7.4 at all times. Close control of blood pH is critical because death results if the pH of human blood drops below 7.0 or rises above 7.8. This narrow pH range corresponds to only a five-fold change in the concentration of hydronium ions. Chemical equilibria work in the blood to hold the pH within this narrow window. Close control of pH is achieved by a **buffer solution,** so called because it protects, or buffers, the solution against pH variations.

The Composition of Buffer Solutions

In order to protect a solution against pH variations, there must be a species present that reacts with added hydronium ions, and there must be another species present that reacts with added hydroxide ions. The conjugate base of a weak acid will react readily with hydronium ions, and the weak acid itself will react readily with hydroxide ions. This leads us to a definition of a buffer solution in terms of its composition:

> /// *A buffer solution contains both a weak acid and its conjugate base as major species in solution.*

As one example, dissolving sodium acetate in a solution of acetic acid produces a buffer solution in which both acetic acid and acetate anions are major species. Example 17-1 describes an acetic acid–acetate buffer solution.

Example 17-1	Concentrations in a Buffer Solution

A solution contains 0.125 mol of solid sodium acetate dissolved in 1.00 L of 0.250 M acetic acid. Determine the concentrations of hydronium ions, acetate ions, and acetic acid.

Strategy: The seven-step procedure described in Chapter 15 can be applied to most equilibrium problems, including this one.

Solving Equilibrium Problems

Solution:

1. **Determine what is asked for.** This problem asks about concentrations of species in an aqueous solution.

2. **Identify the major chemical species.** The original solution contains water and acetic acid molecules. Adding sodium acetate introduces two new major species, acetate anions and sodium cations. Thus the resulting buffer solution has four major species: H_2O, Na^+, $CH_3CO_2^-$, and CH_3CO_2H.

3. **Determine what chemical equilibria exist.** Acetic acid is a weak acid, acetate anion is a weak base, water can act as an acid or a base, and Na^+ is a spectator ion. These species are reactants in three acid-base equilibria:

$$CH_3CO_2H(aq) + H_2O(l) \rightleftharpoons CH_3CO_2^-(aq) + H_3O^+(aq) \quad K_a = 1.8 \times 10^{-5}$$

$$CH_3CO_2^-(aq) + H_2O(l) \rightleftharpoons CH_3CO_2H(aq) + OH^-(aq) \quad K_b = \frac{K_w}{K_a} = 5.6 \times 10^{-10}$$

$$H_2O(l) + H_2O(l) \rightleftharpoons H_3O^+(aq) + OH^-(aq) \quad K_w = 1.0 \times 10^{-14}$$

Concentrations in a Buffer Solution *(continued)* | **Example 17-1**

4. **Write the K_{eq} expressions.** Among these reactions, the first one has the largest equilibrium constant, so the acetic acid equilibrium will generate the largest changes from initial concentrations:

$$K_a = 1.8 \times 10^{-5} = \frac{[H_3O^+]_{eq}[CH_3CO_2^-]_{eq}}{[CH_3CO_2H]_{eq}}$$

5. **Organize the data and unknowns.** Set up a concentration table for this equilibrium:

Species (reaction)	CH_3CO_2H	$\rightleftharpoons$ $CH_3CO_2^-$	+ H_3O^+
Initial concentration (M)	0.250	0.125	0
Change in concentration (M)	$-x$	$+x$	$+x$
Concentration at equilibrium (M)	$0.250 - x$	$0.125 + x$	x

6. **Carry out the calculations.** Because the initial concentrations are much larger than K_{eq}, we make the approximation that x is negligible compared with the initial concentrations:

$$0.250 - x \cong 0.250 \qquad 0.125 + x \cong 0.125$$

To determine the concentration of hydronium ions, rearrange the expression for K_a and substitute these concentrations:

$$K_a = \frac{[H_3O^+]_{eq}[CH_3CO_2^-]_{eq}}{[CH_3CO_2H]_{eq}} \quad so \quad [H_3O^+]_{eq} = \frac{K_a[CH_3CO_2H]_{eq}}{[CH_3CO_2^-]_{eq}}$$

$$x = \frac{(1.8 \times 10^{-5})(0.250)}{(0.125)} = 3.6 \times 10^{-5}$$

$$[H_3O^+]_{eq} = 3.6 \times 10^{-5} \text{ M}$$

$$[CH_3CO_2H]_{eq} = 0.250 \text{ M} \qquad [CH_3CO_2^-]_{eq} = 0.125 \text{ M}$$

7. **Check for reasonableness.** Notice that the H_3O^+ concentration is much smaller than the concentrations of acetic acid and acetate, so the approximation that x is negligibly small is valid. The concentrations of acetic acid and acetate are reasonable, being comparable to the starting concentrations.

The analysis carried out in Example 17-1 reveals one of the key features of buffer solutions: the equilibrium concentrations of both the weak acid and its conjugate base are essentially the same as their initial concentrations.

In the laboratory, chemists prepare buffer solutions in three different ways. Each results in a solution containing a weak acid and its conjugate base as major species. The most straightforward way to produce a buffer solution is by dissolving a salt of a weak acid in a solution of the same weak acid, as described in Example 17-1.

A second way is by adding some strong base to a solution of a weak acid. This produces a buffer solution if the number of moles of strong base is about half the number of moles of weak acid. As a simple example, if 1 L of 0.5 M NaOH is mixed with 1 L of 2.0 M CH_3CO_2H, hydroxide anions react quantitatively with acetic acid molecules:

$$OH^- + CH_3CO_2H \rightleftharpoons H_2O + CH_3CO_2^- \qquad K_{eq} = \frac{1}{K_b} = 1.8 \times 10^9$$

The 0.5 mol of added hydroxide converts 0.5 mol of acetic acid molecules into acetate anions, producing a buffer solution containing 0.5 mol of acetate anions and 1.5 mol of acetic acid.

Just as addition of strong base to weak acid can generate a buffer solution, a third approach is to add some strong acid to a solution of a weak base. This produces a buffer solution if the amount of strong acid is less than the amount of weak base. Continuing with our examples of acetic acid–acetate buffers, if a solution of hydrochloric acid is added to a solution of sodium acetate, hydronium ions react quantitatively with acetate anions:

$$H_3O^+ + CH_3CO_2^- \rightleftharpoons H_2O + CH_3CO_2H \qquad K_{eq} = \frac{1}{K_a} = 5.6 \times 10^4$$

Example 17-2 describes the detailed treatment of a buffer made in this way.

Example 17-2	Strong Acid–Weak Base Buffer

Determine if a solution prepared by mixing 5.0 mL of concentrated HCl (12 M) with 75 mL of 1.0 M sodium acetate is a buffer solution.

Strategy: The question asks if this is a buffer solution. A buffer solution contains both a weak acid and its conjugate base as major species. Thus, to answer the question, we must calculate the concentrations of acetate anions and acetic acid in the solution. Obtaining these concentrations requires quantitative calculations, for which we use a compressed version of the seven-step method.

Solution: The species in the two solutions that are mixed together are H_2O and the following:

$$HCl: H_3O^+ \text{ and } Cl^- \qquad \text{Sodium acetate: } CH_3CO_2^- \text{ and } Na^+$$

Because the hydronium ion is a strong acid and acetate is a weak base, mixing the solutions results in near-quantitative reaction between these two ions:

$$H_3O^+ + CH_3CO_2^- \rightleftharpoons H_2O + CH_3CO_2H \qquad K_{eq} = \frac{1}{K_a} = 5.6 \times 10^4$$

We calculate the initial amounts of the species from the volumes and concentrations, using $n = MV$:

$$n\,(H_3O^+) = (12 \text{ M})(5.0 \text{ mL}) = 60. \text{ mmol}$$

$$n\,(CH_3CO_2^-) = (1.0 \text{ M})(75 \text{ mL}) = 75 \text{ mmol}$$

When the solutions are mixed, H_3O^+ reacts quantitatively with $CH_3CO_2^-$, consuming all the hydronium ions and generating acetic acid molecules, so after mixing we have the following amounts:

Strong Acid–Weak Base Buffer *(continued)* **Example 17-2**

$$n \, (CH_3CO_2^-) = 75 \text{ mmol} - 60 \text{ mmol} = 15 \text{ mmol}$$

$$n \, (CH_3CO_2H) = 60 \text{ mmol}$$

We see that both the weak acid and its conjugate base are major species in the mixture, so this is indeed a buffer solution.

Molecular View of a Buffer Solution

The purpose of a buffer solution is to maintain the pH within a very narrow range. The reactions that occur when H_3O^+ or OH^- is added to a buffer solution show how this is accomplished. Consider what happens when a small amount of hydroxide ion is added to the acetic acid–acetate buffer solution described in Example 17-1. The hydroxide ion is a strong base and acetic acid is a weak acid, so proton transfer from CH_3CO_2H to OH^- goes essentially to completion:

Tutorial

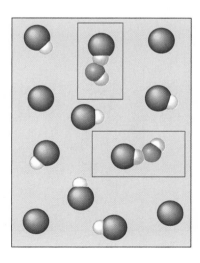

Collision and proton transfer

$$OH^- + CH_3CO_2H \rightleftharpoons H_2O + CH_3CO_2^- \qquad K_{eq} = \frac{1}{K_b} = 1.8 \times 10^9$$

As long as the buffer solution contains acetic acid as a major species, a small amount of hydroxide ion added to the solution will be neutralized completely. Figure 17-1 shows this schematically. The figure shows two hydroxide ions added

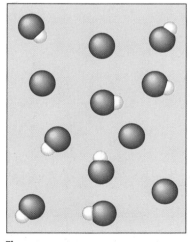
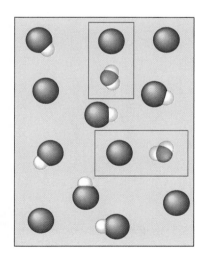

Figure 17-1
When a strong base is added to a buffer solution, the weak acid *HA* donates protons to hydroxide ions to form the conjugate base *A⁻*, preventing a large increase in hydroxide ion concentration. (All water molecules except those produced in the proton transfer process are omitted for clarity.)

to a portion of the buffer solution described in Example 17-1. When a hydroxide ion collides with a molecule of weak acid, proton transfer forms a water molecule and the conjugate base of the weak acid.

Provided that there are more weak acid molecules in the solution than the number of added hydroxide ions, the proton transfer reaction goes virtually to completion. Weak acid molecules change into acetate anions as they "mop up" added hydroxide.

The same molecular reasoning shows that a buffer solution can absorb added hydronium ions. Consider what happens when a small amount of hydronium ion is added to the acetic acid–acetate buffer solution described in Example 17-1. The hydronium ion is a strong acid and acetate anion is a weak base, so proton transfer from $CH_3CO_2^-$ to H_3O^+ goes essentially to completion:

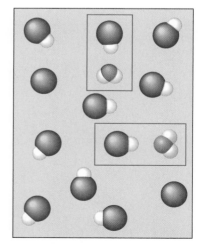

Collision and proton transfer

$$H_3O^+ + CH_3CO_2^- \rightleftharpoons H_2O + CH_3CO_2H \qquad K_{eq} = \frac{1}{K_a} = 5.6 \times 10^4$$

As long as the buffer solution contains acetate as a major species, a small amount of hydronium ion added to the solution will be neutralized completely. Figure 17-2 shows this schematically. The figure shows two hydronium ions added to a portion of the buffer solution described in Example 17-1. When a hydronium ion collides with a molecule of weak base, proton transfer occurs quantitatively, forming a water molecule and the conjugate acid of the weak base.

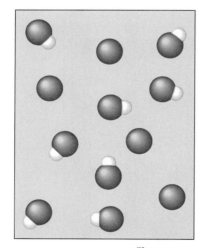
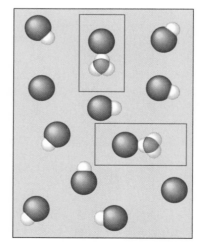
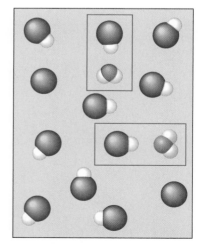

Figure 17-2
When a strong acid is added to a buffer solution, the conjugate base *A*⁻ accepts protons from hydronium ions to form the weak acid *HA*, preventing a large increase in hydronium ion concentration. (All water molecules except those produced in the proton transfer process are omitted for clarity.)

The Buffer Equation

The pH of a buffer solution depends on the weak acid equilibrium constant and the concentrations of the weak acid and its conjugate base. To show this, we begin by taking the logarithm of the acid equilibrium constant:

$$K_a = \frac{[H_3O^+]_{eq}[A^-]_{eq}}{[HA]_{eq}} \qquad so \qquad \log K_a = \log\left(\frac{[H_3O^+]_{eq}[A^-]_{eq}}{[HA]_{eq}}\right)$$

The logarithm of the concentration quotient can be separated by recalling that $\log xy = \log x + \log y$.

$$\log K_a = \log [H_3O^+]_{eq} + \log\left(\frac{[A^-]_{eq}}{[HA]_{eq}}\right)$$

Multiplying both sides of the equation by -1 allows the use of pK_a and pH instead of $\log K_a$ and $\log [H_3O^+]_{eq}$:

$$-\log K_a = -\log [H_3O^+]_{eq} - \log\left(\frac{[A^-]_{eq}}{[HA]_{eq}}\right) \qquad and \qquad pK_a = pH - \log\left(\frac{[A^-]_{eq}}{[HA]_{eq}}\right)$$

Now we rearrange this equation to solve for pH:

$$pH = pK_a + \log\left(\frac{[A^-]_{eq}}{[HA]_{eq}}\right)$$

This equation is exact, but a feature of buffer solutions allows us to simplify it. Any buffer solution contains both members of a conjugate acid-base pair as major species. In other words, both the weak acid and its conjugate base are present in relatively large amounts. As a result, the change to equilibrium, x, is small relative to the initial concentrations, and their equilibrium concentrations are virtually the same as their initial concentrations.

$$[A^-]_{eq} = [A^-]_{initial} - x \cong [A^-]_{initial}$$

$$[HA]_{eq} = [HA]_{initial} + x \cong [HA]_{initial}$$

Thus instead of using equilibrium concentrations in this equation, we can substitute the initial concentrations to give the **buffer equation:**

$$pH = pK_a + \log\left(\frac{[A^-]_{initial}}{[HA]_{initial}}\right) \qquad\qquad (17\text{-}1)$$

The buffer equation, which is often called the *Henderson-Hasselbalch equation,* is used to calculate the equilibrium pH of a buffer solution directly from initial concentrations. The approximation is valid as long as the difference between initial concentrations and equilibrium concentrations is negligibly small. As a rule of thumb, the buffer equation can be applied when initial concentrations of HA and A^- differ by less than a factor of 10. Example 17-3 provides an illustration of the use of the buffer equation.

Example 17-3 | The Buffer Equation

Buffer solutions with a pH of about 10 are prepared using sodium carbonate (Na_2CO_3) and sodium hydrogen carbonate ($NaHCO_3$). What is the pH of a solution prepared by dissolving 10.0 g each of these two salts in enough water to make 0.250 L of solution?

Solving Equilibrium Problems

Strategy: We use the seven-step strategy for equilibrium problems, except that we identify this as a buffer solution. This allows us to use the buffer equation in place of an equilibrium constant expression.

Solution:

1. We are asked to calculate the pH of a buffer solution.

2. Both compounds are salts that dissolve in water to give their constituent ions, so the major species in this buffer solution are H_2O, Na^+, HCO_3^-, and CO_3^{2-}.

3. A buffer solution must contain a weak acid and its conjugate base as major species. In this solution, HCO_3^- is the weak acid, and CO_3^{2-} is the conjugate base:

$$H_2O(l) + HCO_3^-(aq) \rightleftharpoons CO_3^{2-}(aq) + H_3O^+(aq)$$

4. This proton transfer reaction involves the second acidic hydrogen atom of carbonic acid, so the appropriate equilibrium constant is K_{a2}, whose pK is found in Appendix E: pK_{a2} = 10.33. Because this is a buffer solution, we apply the buffer equation:

$$pH = pK_a + \log\left(\frac{[A^-]_{initial}}{[HA]_{initial}}\right)$$

5. We find the initial concentrations from the masses of the salts and the volume of the solution:

$$[HCO_3^-]_{initial} = \frac{m}{(MM)V} = \frac{(10.0\ g)}{(84.01\ g/mol)(0.250\ L)} = 0.476\ M$$

$$[CO_3^{2-}]_{initial} = \frac{m}{(MM)V} = \frac{(10.0\ g)}{(106.0\ g/mol)(0.250\ L)} = 0.377\ M$$

6. Now we substitute the appropriate values into the buffer equation and evaluate:

$$pH = pK_a + \log\left(\frac{[A^-]_{initial}}{[HA]_{initial}}\right) = 10.33 + \log\left(\frac{0.377\ M}{0.476\ M}\right)$$

$$pH = 10.33 + (-0.101) = 10.23$$

7. The pH is close to the pK_a of the conjugate acid–base pair, so this is a reasonable result.

Buffer Action

When a strong acid is added to a buffer solution, hydronium ions react with the conjugate base A^-, lowering the concentration of A^- and increasing the concentration of HA. This reaction lowers the pH of the buffer solution, but the increase in H_3O^+ concentration is much smaller than would be generated by the same amount of strong acid in an unbuffered solution. Similarly, when a strong base is added to a buffer solution, hydroxide ions react with the acid HA, lowering the concentration of HA and increasing the concentration of A^-. In this case the pH of the buffer solution rises, but the increase in hydroxide ion concentration is much smaller than would be generated by the same amount of strong base added to an unbuffered solution. Example 17-4 shows a quantitative calculation.

Change in Buffer pH	**Example 17-4**

By how much does the pH of the buffer solution of Example 17-3 change on the addition of 3.50 mL of 6.0 M HCl?

Strategy: The first four steps of the seven-step strategy are identical to the ones in Example 17-3. In this example, addition of a strong acid or base modifies the concentrations that go into the buffer equation. We need to determine the new concentrations (step 5) and then apply the buffer equation (step 6).

Solution: In dealing with changes in amounts of acid and base, it is often convenient to work with moles rather than molarities. The units cancel in the concentration term of the buffer equation, so the ratio of concentrations can be expressed as a ratio of moles as well as a ratio of molarities:

$$\frac{[A^-]}{[HA]} = \frac{(\text{mol}_{A^-}/V)}{(\text{mol}_{HA}/V)} = \frac{\text{mol}_{A^-}}{\text{mol}_{HA}}$$

We do this problem using moles. First, determine the amounts present in the solution before addition of the HCl:

$$\text{mol}_{HA} = \frac{10.0 \text{ g}}{84.01 \text{ g/mol}} = 1.19 \times 10^{-1} \text{ mol}$$

$$\text{mol}_{A^-} = \frac{10.0 \text{ g}}{106.0 \text{ g/mol}} = 9.43 \times 10^{-2} \text{ mol}$$

Next, calculate the amount of hydronium ions added:

$$\text{mol}_{H_3O^+} = MV = (6.0 \text{ mol/L})(3.5 \text{ mL})(10^{-3} \text{ L/mL}) = 2.1 \times 10^{-2} \text{ mol}$$

The hydronium ions react completely with conjugate base anions, increasing the amount of HCO_3^- and reducing the amount of CO_3^{2-}:

$$\text{mol}_{HA} = 1.19 \times 10^{-1} \text{ mol} + 2.1 \times 10^{-2} = 1.40 \times 10^{-1} \text{ mol}$$

$$\text{mol}_{A^-} = 9.43 \times 10^{-2} - 2.1 \times 10^{-2} = 7.3 \times 10^{-2} \text{ mol}$$

Finally, substitute these new amounts into the buffer equation to compute the new pH:

$$\text{pH} = \text{p}K_a + \log \frac{\text{mol}_{A^-}}{\text{mol}_{HA}} = 10.33 + \log \left(\frac{0.073 \text{ mol}}{0.140 \text{ mol}} \right) = 10.33 - 0.28 = 10.05$$

Note that the pH changes by a small amount (0.18), a reasonable outcome for a buffer solution.

Figure 17-3
Addition of strong acid to a buffer solution changes the pH by much less than does addition of strong acid to water.

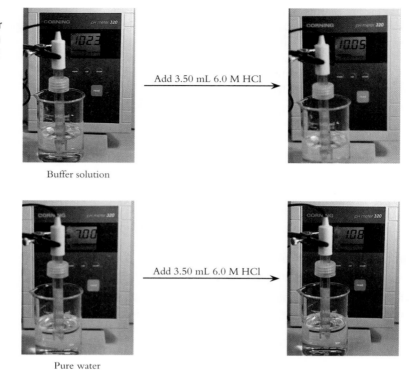

Buffer solution

Add 3.50 mL 6.0 M HCl

Pure water

Add 3.50 mL 6.0 M HCl

From Example 17-3, the pH of the buffer before adding the acid was 10.23, so the addition of 2.1×10^{-2} mol of acid reduces the pH by 0.18 pH units. This is tiny compared with the change upon adding the same amount of acid to 250 mL of water, which would change the pH from 7.00 to 1.08, a reduction of 5.92 pH units. Figure 17-3 illustrates and contrasts the two changes.

Section Exercises

17.1.1 Which of the following sets of chemicals can be used to prepare buffer solutions? For each one that can, specify the weak acid and its conjugate base that are major species in the buffer solution: (a) HCl + KCl; (b) HCl + KNO_2; (c) HCl + NH_4Cl; (d) NaOH + Na_2HPO_4; and (e) NaCl + $NaC_2H_3O_2$.

17.1.2 A buffer solution made from NH_4Cl and NH_3 is used to control pH in the range of pH = 8–10. Write balanced equations that show how this buffer system neutralizes H_3O^+ and OH^-.

17.1.3 A 1.50-L buffer solution is prepared from 0.200 mol NH_4Cl and 0.112 mol NH_3. a) What is the pH of the buffer solution? b) What would be the pH after adding 1.70 g of NaOH to the buffer described in part a? c) What would be the pH of a solution made by adding 1.70 g NaOH to 1.50 L of water?

17.2 CAPACITY AND PREPARATION OF BUFFER SOLUTIONS

Buffer solutions are practical and commonplace. In fact, many chemists and biologists use buffer solutions on a daily basis. Thus it is important to know how to make buffer solutions and to know the limit of a buffer solution's capacity to control pH.

Buffer Capacity

When small amounts of hydronium or hydroxide ions are added to a buffer solution, pH changes are very small. There is a limit, however, to the amount of protection that a buffer solution can provide. After either buffering agent is consumed, the solution loses its ability to maintain near-constant pH. The **buffer capacity** of a solution is the amount of added H_3O^+ or OH^- the buffer solution can tolerate without exceeding a specified pH range.

Buffer capacity is determined by the amounts of weak acid and conjugate base present in the solution. If enough H_3O^+ is added to react completely with the conjugate base, the buffer is destroyed. Likewise, the buffer is destroyed if enough OH^- is added to consume all of the weak acid. Consequently, buffer capacity is dependent on both the overall concentration and volume of the buffer solution. A buffer solution whose overall concentration is 0.50 M has five times the capacity as an equal volume of a buffer solution whose overall concentration is 0.10 M. Two liters of 0.10 M buffer solution has twice the capacity as one liter of the same buffer solution. Example 17-5 includes a calculation involving buffer capacity.

Buffer Capacity	Example 17-5

Biochemists and molecular biologists use phosphate buffers to match physiological conditions. A buffer solution that contains $H_2PO_4^-$ as the weak acid and HPO_4^{2-} as the weak base has a pH value very close to 7.0. A biochemist prepares 0.250 L of a buffer solution that contains 0.225 M HPO_4^{2-} and 0.330 M $H_2PO_4^-$. What is the pH of this buffer solution? Is the buffering action of this solution destroyed by addition of 0.40 g NaOH?

Strategy: Use the seven-step strategy to calculate the pH of the buffer solution using the buffer equation. Then compare the amount of acid in the solution with the amount of added base. Buffer action is destroyed if the amount of added base is sufficient to react with all the acid.

Solution: The buffering action of this solution is created by the weak acid $H_2PO_4^-$ and its conjugate base HPO_4^{2-}. The equilibrium constant for this pair is K_{a2} of phosphoric acid:

$$H_2O(l) + H_2PO_4^-(aq) \rightleftharpoons HPO_4^{2-}(aq) + H_3O^+(aq) \qquad K_{eq} = K_{a2}$$

Substitute the tabulated value of pK_{a2} and the given concentrations into the buffer equation to calculate pH:

$$pH = pK_a + \log\left(\frac{[A^-]_{initial}}{[HA]_{initial}}\right) = 7.21 + \log\left(\frac{0.225\ M}{0.330\ M}\right) = 7.04$$

The question asks whether addition of base destroys the buffering action, so compare the amount of acid present with the amount of base added. As solid NaOH is added, each hydroxide ion that enters the buffer solution consumes one $H_2PO_4^-$ and produces one ion of HPO_4^{2-} and one water molecule:

| Example 17-5 | Buffer Capacity *(continued)* |

$$OH^-(aq) + H_2PO_4^-(aq) \longrightarrow H_2O(l) + HPO_4^{2-}(aq)$$

$$\text{mol } OH^- \text{ added} = m/MM = 0.40 \text{ g}/40.00 \text{ g/mol} = 0.010 \text{ mol}$$

$$\text{mol } H_2PO_4^- \text{ present} = MV = (0.330 \text{ M})(0.250 \text{ L}) = 0.0825 \text{ mol}$$

The amount of base added is considerably less than the amount of acid initially present in the buffer solution, so the solution will still act as a buffer. The added base raises the pH, however, because reaction of OH^- with the buffer increases the amount of HPO_4^{2-} and decreases the amount of $H_2PO_4^-$. We leave it to you to use the procedure illustrated in Example 17-4 to show that the pH after addition of this solid is 7.17, an increase of 0.13 pH units.

Buffer Preparation

The buffer equation indicates that the pH of a buffer solution is close to the pK_a of the acid used to prepare the buffer:

$$pH = pK_a + \log \left(\frac{[A^-]_{\text{initial}}}{[HA]_{\text{initial}}} \right)$$

Every weak acid has a specific pK_a that determines the pH range over which it can serve as a buffering agent. Remember that a buffer solution must contain a weak acid and its conjugate weak base as *major* species. This condition is met when the ratio of weak base to weak acid is between 0.1 and 10. The buffer equation translates this restriction into a pH range over which the acid and its conjugate base can serve as an effective buffer:

$$pH_{\text{low}} = pK_a + \log 0.1 = pK_a - 1 \qquad pH_{\text{high}} = pK_a + \log 10 = pK_a + 1$$

$$pH \text{ range} = pK_a \pm 1$$

With a given weak acid, a buffer solution can be prepared at any pH within about one unit of its pK_a value. Suppose, for example, that a biochemist needs a buffer system to maintain the pH of a solution close to 5.0. What reagents should be used? According to the previous analysis, the weak acid can have a pK_a between 4.0 and 6.0. As the pK_a deviates from the desired pH, however, the solution has a reduced buffer capacity. Thus a buffer has maximum capacity when its acid has its pK_a as close as possible to the target pH. Table 17-1 lists some acid-base pairs often

Table 17-1
Common Buffer Systems

Acid	Conjugate Base	pK_a	pH Range
H_3PO_4	$H_2PO_4^-$	2.16	1–3
HCO_2H	HCO_2^-	3.75	3–5
CH_3CO_2H	$CH_3CO_2^-$	4.75	4–6
$H_2PO_4^-$	HPO_4^{2-}	7.21	6–8
NH_4^+	NH_3	9.25	8–10
HCO_3^-	CO_3^{2-}	10.33	9–11
HPO_4^{2-}	PO_4^{3-}	12.32	11–13

used as buffer solutions. For a pH = 5.0 buffer, acetic acid (pK_a = 4.75) and its conjugate base, acetate, would be a good choice.

A buffer solution must contain both the acid and its conjugate base, so at least two reagents must be added to water to prepare a buffer solution. An acetate buffer can be prepared, for example, from pure water, concentrated acetic acid, and an acetate salt. The cation contained in the salt should not have acid–base properties of its own, so sodium acetate would be an appropriate choice, but ammonium acetate would not.

The chemist must prepare a solution that contains acetic acid and acetate in amounts that generate a buffer with pH = 5.00. The buffer equation is used to calculate the desired molar ratio:

$$5.00 = 4.75 + \log \frac{[\text{Acetate}]}{[\text{Acetic acid}]} \qquad so \qquad \log \frac{[\text{Acetate}]}{[\text{Acetic acid}]} = 0.25$$

$$\frac{[\text{Acetate}]}{[\text{Acetic acid}]} = 10^{0.25} = 1.8$$

For the preparation of a buffer solution of pH = 5.00, sodium acetate and acetic acid should be added to pure water in a molar ratio of 1.8 : 1.0. The exact amounts of the reagents must be calculated using the desired volume and concentration of the solution, as Example 17-6 illustrates.

Buffer Preparation	Example 17-6

What mass of sodium acetate ($NaCH_3CO_2 \cdot 3H_2O$, MM = 136.08 g/mol) and what volume of concentrated acetic acid (17.45 M) should be used to prepare 1.5 L of a buffer solution at pH = 5.00 that is 0.150 M overall?

Strategy: Because we know we are dealing with a buffer solution made from a specific conjugate acid–base pair, we can skip the seven-step strategy and work with the buffer equation. We need to calculate the ratio of concentrations of conjugate base and acid that will produce a buffer solution of the desired pH. Then we use mole–mass–volume relationships to translate the ratio into actual quantities.

Solution: The problem specifies a solution volume of 1.5 L with a total molarity of 0.150 mol/L. The total molarity is the combined concentration of the two buffer components:

$$M_{\text{acetate}} + M_{\text{acetic acid}} = 0.150 \text{ M}$$

Use the total volume of the solution, 1.5 L, to determine the total number of moles in the system:

$$(0.150 \text{ mol/L})(1.5 \text{ L}) = 0.225 \text{ mol}$$

$$n_{\text{acetate}} + n_{\text{acetic acid}} = 0.225 \text{ mol}$$

The calculation involving the buffer equation appears in the text: a buffer solution with pH = 5.00 requires an acetate–acetic acid molar ratio of 1.8. This ratio can be rewritten as a molar equality:

$$\frac{n_{\text{acetate}}}{n_{\text{acetic acid}}} = 1.8 \qquad n_{\text{acetate}} = 1.8 \, n_{\text{acetic acid}}$$

| Example 17-6 | Buffer Preparation *(continued)* |

Now substitute and calculate the required moles of acetic acid:

$$1.8\, n_{\text{acetic acid}} + n_{\text{acetic acid}} = 0.225 \text{ mol}$$

$$n_{\text{acetic acid}} = \frac{0.225 \text{ mol}}{2.8} = 8.04 \times 10^{-2} \text{ mol}$$

The rest of the 0.225 mol must be acetate:

$$n_{\text{acetate}} + 8.04 \times 10^{-2} \text{ mol} = 0.225 \text{ mol}$$

$$n_{\text{acetate}} = 0.145 \text{ mol}$$

Finally, use molarity and molar mass to convert from moles to measurable amounts:

$$\text{Mass}_{\text{sodium acetate}} = (0.145 \text{ mol})(136.08 \text{ g/mol}) = 20. \text{ g}$$

$$V_{\text{acetic acid}} = \frac{(8.04 \times 10^{-2} \text{ mol})(10^3 \text{ mL/L})}{(17.45 \text{ mol/L})} = 4.6 \text{ mL}$$

The final values are rounded to two significant figures to match the precision of the mole ratio and the total volume of the solution.

Recall that the mole ratio is equal to the concentration ratio, because both species are present in the same volume of solution.

As described in Section 17.1, there are two other ways to prepare this buffer solution. Concentrated acetic acid could be added to pure water, followed by enough sodium hydroxide to generate the required 1.8:1.0 ratio of acetate to acetic acid:

$$CH_3CO_2H(aq) + OH^-(aq) \longrightarrow CH_3CO_2^-(aq) + H_2O(l)$$

Alternatively, a solution of sodium acetate could be prepared and a strong acid added to reach the proper acetate–acetic acid ratio:

$$CH_3CO_2^-(aq) + H_3O^+(aq) \longrightarrow CH_3CO_2H(aq) + H_2O(l)$$

Buffer preparation requires detailed, step-by-step calculations. Example 17-7 illustrates the complete procedure.

| Example 17-7 | Preparing a Buffer |

A biochemist asks a technician to prepare a buffer solution at pH = 9.00 with an overall concentration of 0.125 mol/L. The technician has solutions of 1.00 M HCl and NaOH and bottles of all common salts. What reagents should be used, and in what quantities, to prepare 1.00 L of a suitable buffer?

Strategy: A practical problem in solution preparation usually requires a different strategy than our standard seven-step procedure. The technician must first identify a suitable conjugate acid-base pair and decide what reagents to use. Then the concentrations needed must be determined, using pH and total concentration. Finally, the technician must determine the amounts of starting materials.

Solution: The technician needs a buffer at pH = 9.00. Of the buffer systems listed in Table 17-1, the combination of NH_3 and NH_4^+ has the proper pH range for the required buffer solution.

Preparing a Buffer *(continued)*

Example 17-7

Apparently, no bottles of aqueous ammonia are present in the laboratory, so the components of the buffer solution must come from the salts. The technician needs an ammonium salt with a counter anion that has no acid–base properties. Ammonium chloride (NH_4Cl) would be an appropriate choice. This salt contains the conjugate acid, NH_4^+, and the technician can generate NH_3 by adding strong base to the ammonium chloride solution:

$$NH_4^+(aq) + OH^-(aq) \longrightarrow NH_3(aq) + H_2O(l)$$

What concentrations of NH_4^+ and NH_3 are required? First use the buffer equation to find the proper ratio of base to acid, then use the total molarity of the solution to determine the concentrations of NH_3 and NH_4^+. The ratio of ammonia to ammonium ion must produce a pH of 9.00:

$$pH = pK_a + \log \frac{[Base]}{[Acid]} \qquad 9.00 = 9.25 + \log \frac{[NH_3]}{[NH_4^+]}$$

$$\log \frac{[NH_3]}{[NH_4^+]} = -0.25 \qquad \textit{from which} \qquad \frac{[NH_3]}{[NH_4^+]} = 0.56$$

$$\textit{and} \qquad [NH_3] = 0.56\,[NH_4^+]$$

Now the desired total molarity of the buffer solution is used to find the actual concentrations:

$$[NH_3] + [NH_4^+] = 0.125 \text{ M};$$

Substitute for $[NH_3]$: $0.56\,[NH_4^+] + [NH_4^+] = 0.125$ M

$$[NH_4^+] = \frac{0.125 \text{ M}}{1.56} = 0.0801 \text{ M}$$

$$[NH_3] = 0.125 \text{ M} - 0.0801 \text{ M} = 0.045 \text{ M}$$

Finally, use stoichiometry to calculate actual amounts. The buffer solution contains both ammonia and ammonium ions, but both species are derived from ammonium chloride. For the preparation of the solution, some of the NH_4^+ ions in an aqueous solution of ammonium chloride must be converted into NH_3 molecules. Thus the amount of salt required is found from the total molarity of the buffer solution:

$$n_{NH_4Cl} = MV = (0.125 \text{ mol/L})(1.00 \text{ L}) = 0.125 \text{ mol}$$
$$m_{NH_4Cl} = (n)(MM) = (0.125 \text{ mol})(53.49 \text{ g/mol}) = 6.69 \text{ g}$$

Ammonia is generated from NH_4^+ by adding sodium hydroxide solution:

$$n_{OH^-} = n_{NH_3} = (0.045 \text{ M})(1.00 \text{ L}) = 0.045 \text{ mol}$$

$$\text{Volume of } 1.00 \text{ mol NaOH(aq)} = \frac{0.045 \text{ mol}}{1.00 \text{ mol/L}} = 0.045 \text{ L}$$

To make the buffer, the technician should mix together 6.69 g NH_4Cl, 45 mL 1.00 M NaOH, and enough water to make 1.00 L of solution.

Biological systems use buffer solutions to maintain pH within close limits. One example, described in the Chemistry and Life Box, is the pH of blood, which is kept under close control by chemical equilibria.

Box 17-1   Chemistry and Life: The pH of Blood

The human body generates a steady flow of acidic by-products during its normal metabolic processes. Foremost among these is carbon dioxide, which is a major product of the reactions the body uses to produce energy (see Section 13.6). An average person produces from 10 to 20 mol (440 to 880 g) of CO_2 every day. Blood carries CO_2 from the cells to the lungs to be exhaled. In aqueous solution, dissolved CO_2 is in equilibrium with carbonic acid:

$$CO_2(aq) + H_2O(l) \rightleftharpoons H_2CO_3(aq)$$

In the absence of some regulatory mechanism, the continuous production of CO_2 would result in a pH of about 5.6 in blood. The proper functioning of blood chemistry, however, requires a pH of 7.40. Severe illness or death can result from changes of blood pH by only a few tenths of a unit. Blood pH is tightly regulated at 7.40 ± 0.05 by a buffer whose principal acid-base pair is H_2CO_3 and HCO_3^-.

The pK_{a1} for H_2CO_3 is 6.35, which is considerably less than the regulated pH of blood. The buffer equation indicates the acid-base ratio needed for a pH of 7.40:

$$7.40 = 6.35 + \log\left(\frac{[HCO_3^-]}{[H_2CO_3]}\right) \quad so \quad \frac{[HCO_3^-]}{[H_2CO_3]} = 10^{1.05} = 11$$

The H_2CO_3 concentration in blood is close to 13 mM, so to maintain a pH of 7.40, blood must contain 150 mM hydrogen carbonate. Because of the high ratio of HCO_3^- to H_2CO_3, the buffer capacity of blood is quite low. One more application of the buffer equation gives the carbonic acid concentration at which blood pH is 7.35:

$$7.35 = 6.35 + \log\left(\frac{150\ mM}{[H_2CO_3]}\right) \quad so \quad [H_2CO_3] = 15\ mM$$

Thus an 11% increase in CO_2 content would exceed the capacity of the carbonate buffer system. Blood contains other acid-base pairs, including acidic proteins, that help control pH. However, when the acid-base chemistry of the blood is disturbed by disease, the control system may fail. Then the blood pH is likely to fall, and a patient may require an intravenous infusion of hydrogen carbonate solution to raise the pH.

Despite buffering, blood pH would quickly become acidic were it not for rapid elimination of CO_2 from the body. In the lungs, CO_2 is transferred from blood cells to the gas phase (see photo above) and then exhaled from our bodies.

When blood pH falls below normal limits, the condition is termed *acidosis*. A number of body malfunctions can lead to acidosis, including diabetes, kidney failure, and persistent diarrhea. Temporary acidosis can result from prolonged vigorous exercise. These situations stimulate production of CO_2, increasing the acidity of the blood.

The opposite condition, in which blood pH rises above tolerable limits, is called *alkalosis*. Alkalosis can occur as a

result of hyperventilation or severe vomiting. Mountain climbers put themselves at risk of alkalosis by climbing too high without allowing proper time to become acclimated to high elevation. At high altitude there is less oxygen in the air, and climbers must compensate by breathing faster. Under these conditions the body can lose CO_2 too rapidly, causing a decrease in blood acidity and a corresponding rise in pH. This problem is compounded by decreased oxygen uptake caused by oxygen's lower partial pressure. The result can be severe altitude sickness, which results in death if the climber is not given supplemental oxygen or moved quickly to lower elevation.

17.2.1 A student adds 30. mL of 5.00 M HCl to the buffer solution described in Section Exercise 17.1.3. Is the buffering capacity of the solution destroyed? What is the final pH of the solution?

17.2.2 Determine the mass of solid sodium formate ($NaHCO_2$) and the volume of 0.500 M HCl solution required to generate 250 mL of buffer solution with pH = 3.50 and a total concentration (conjugate acid plus base) of 0.225 M.

17.2.3 Describe how you would prepare 2.5 L of a pH = 10.50 buffer solution with a total concentration of 0.15 M, using an appropriate salt and solid NaOH. Use Table 17-1 to choose the buffer system.

17.3 ACID-BASE TITRATIONS

As described in Chapter 4, acid–base reactions that go to completion can be exploited in chemical analysis using the method of titration. Titrations can be understood in greater detail from the perspective of acid–base equilibria. Protonation of a weak base by a strong acid is a reaction that goes virtually to completion by virtue of its large equilibrium constant:

$$B + H_3O^+ \longrightarrow BH^+ + H_2O \qquad K_{eq} \gg 1$$

↑ Any base, including OH^-

Likewise, when strong base is added to a solution containing an acid, the reaction goes essentially to completion, again because of a large equilibrium constant:

$$OH^- + HA \longrightarrow H_2O + A^- \qquad K_{eq} \gg 1$$

↑ Any acid, including H_3O^+

Even though these reactions go essentially to completion, equilibria determine the pH of the solution before, during, and at the stoichiometric point of the titration.

Titrations are treated like any other equilibrium analysis, but we must pay special attention to the major species present in the solution, because these change during the titration. The most common titrations are analysis of a weak acid using a solution of strong base and analysis of a weak base using a solution of strong acid.

Titration of a Weak Acid by OH⁻ Ions

The titration of a solution of a weak acid with a solution of a strong base is shown graphically in Figure 17-4. This titration curve has four distinct regions, each characterized by different major species:

1. Near the beginning of the titration, HA and H_2O are the only major species.

2. During most of the titration, both HA and its conjugate base, A^-, are major species. As a result, the solution is buffered, and the pH changes relatively slowly with added hydroxide.

3. When nearly all of the HA molecules have reacted with added OH^-, the only major species are A^- and H_2O. Further addition of OH^- ions causes a sharp increase in pH.

4. After all the HA molecules have reacted, the solution contains excess OH^- ions, which determine the pH.

Figure 17-4
Schematic profile of the titration curve for a weak acid HA titrated with hydroxide ions. The titration can be divided into four regions that differ in the major species present in solution. The pH values are those for titration of 0.500 M acetic acid.

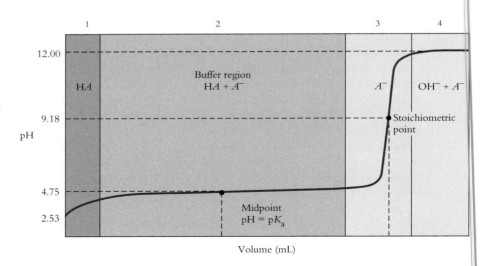

As an example, consider the titration of 0.150 L of 0.500 M acetic acid solution with 2.50 M potassium hydroxide. Very early in the titration, the major species are water and acetic acid, and the equilibrium that determines solution pH is proton transfer between these major species:

$$H_2O(l) + CH_3CO_2H(aq) \rightleftharpoons H_3O^+(aq) + CH_3CO_2^-(aq) \qquad K_a = 1.8 \times 10^{-5}$$

Using the standard method for solving equilibrium problems, you should be able to show that the initial solution contains 0.0750 mol of acetic acid, has a hydronium ion concentration of 2.97×10^{-3} M, and has a pH of 2.53.

As the titration proceeds, the hydroxide ions in each added volume of titrant convert acetic acid molecules to acetate ions:

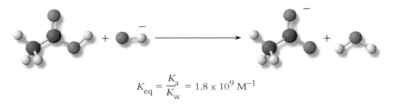

$$K_{eq} = \frac{K_a}{K_w} = 1.8 \times 10^9 \ M^{-1}$$

As a result, both acetate ions and acetic acid molecules are present as major species in solution. The presence of an acid and its conjugate base means that in this region of the titration, the solution is buffered, so the pH changes slowly as hydroxide ions are added to the solution.

This buffer region contains the **midpoint** of the titration, the point at which the amount of added OH⁻ is equal to exactly half the weak acid originally present. In the current example, the solution at the midpoint contains 0.0375 mol each of acetic acid and acetate. Applying the buffer equation reveals the key feature of the midpoint:

$$pH = pK_a + \log \frac{n_{acetate}}{n_{acetic\ acid}} = 4.75 + \log 1 = 4.75$$

/// *At the midpoint of a titration of a weak acid by a strong base, the pH of the solution equals the pK_a of the weak acid.*

Beyond the buffer region, when nearly all of the acetic acid has been consumed, the pH increases sharply with each added drop of hydroxide solution. The titration curve passes through an almost vertical region before leveling off again. Recall from Chapter 4 that the **stoichiometric point** of an acid titration (also called the *equivalence point*) is the point at which the number of moles of added base is exactly equal to the number of moles of acid present in the original solution. At the stoichiometric point of a weak acid titration, the conjugate base is a major species in solution, but the weak acid is not.

Example 17-8 illustrates that the pH at the stoichiometric point in a titration of a weak acid is *not* 7.0.

---

pH at the Stoichiometric Point	Example 17-8

What is the pH at the stoichiometric point of the titration of 0.150 L of 0.500 M acetic acid with 2.50 M KOH solution?

Strategy: This is another equilibrium calculation to which the standard seven-step procedure applies. Special attention must be given, however, to analyzing the initial conditions *at the stoichiometric point*, bearing in mind that the reaction between hydroxide ions and a weak acid goes essentially to completion.

Solving
Equilibrium
Problems

Solution:

1. We are asked to calculate the pH at the stoichiometric point.

2. At the stoichiometric point, the amount of added hydroxide ions equals the amount of acetic acid that was originally present:

$$\text{mol OH}^- = \text{mol CH}_3\text{CO}_2\text{H originally present}$$

This added hydroxide reacts essentially to completion with acetic acid

$$\text{CH}_3\text{CO}_2\text{H(aq)} + \text{OH}^-\text{(aq)} \longrightarrow \text{CH}_3\text{CO}_2^-\text{(aq)} + \text{H}_2\text{O(l)}$$

Thus, the major acid-base species present at the stoichiometric point are CH_3CO_2^- and H_2O.

3. The pH is determined by the proton transfer from water to acetate ions:

$$\text{CH}_3\text{CO}_2^-\text{(aq)} + \text{H}_2\text{O(l)} \longrightarrow \text{CH}_3\text{CO}_2\text{H(aq)} + \text{OH}^-\text{(aq)}$$

4. The equilibrium constant expression for this equilibrium is related to K_a for acetic acid:

$$K_{eq} = K_{b \text{ acetate}} = \frac{K_w}{K_{a \text{ acetic acid}}} = 5.6 \times 10^{-10}$$

5. We need to calculate "initial" conditions for the stoichiometric point. Recall that the added base has reacted to form acetate anions:

$$\text{mol OH}^- \text{ added} = M_{acid}V_{acid} = (0.500 \text{ mol/L})(0.150 \text{ L}) = 0.0750 \text{ mol}$$

$$\text{mol CH}_3\text{CO}_2\text{H} = 0, \quad \text{mol OH}^- = 0, \quad and \quad \text{mol CH}_3\text{CO}_2^- = 0.0750 \text{ mol}$$

Before constructing a concentration table, convert moles of acetate to molarity. The volume at the stoichiometric point is the original volume plus the volume of added titrant:

| Example 17-8 | pH at the Stoichiometric Point *(continued)* |

$$V_{\text{initial}} = 0.150 \text{ L}$$

$$V_{\text{titrant}} = \frac{n}{M} = \frac{0.0750 \text{ mol}}{2.50 \text{ mol/L}} = 0.0300 \text{ L}$$

$$V_{\text{total}} = 0.150 \text{ L} + 0.0300 \text{ L} = 0.180 \text{ L}$$

$$[CH_3CO_2^-]_{\text{initial}} = \frac{0.0750 \text{ mol}}{0.180 \text{ L}} = 0.417 \text{ M}$$

Species (reaction)	$H_2O(l) + CH_3CO_2^-(aq) \rightleftharpoons CH_3CO_2H(aq) + OH^-(aq)$		
Initial conc. (M)	0.417	0	0
Change in conc. (M)	$-x$	$+x$	$+x$
Equilibrium conc. (M)	$0.417 - x$	x	x

6. We have done this type of calculation many times, so you should be able to show that the pH of the solution at the stoichiometric point is 9.18.

7. It may surprise you that the pH at the stoichiometric point is not 7.00, but this is a reasonable outcome. We are adding *strong* base to a *weak* acid, so the resulting solution is basic in nature.

Although the exact pH at the stoichiometric point depends on what weak acid is being titrated, the qualitative result of Example 17.8 is reproduced for every titration of a weak acid with a strong base. At the stoichiometric point of a weak acid titration, the exact value of the pH is determined by K_b for the conjugate base, but it is *always greater than 7.0*.

Beyond the stoichiometric point, in the final region of the titration curve, the concentration of acetic acid is very close to zero. There are no acid molecules to react with any further hydroxide ions, so excess hydroxide ions are present in solution as a major species. Beyond the stoichiometric point, the pH of the solution is determined by the amount of excess hydroxide ion.

Titration of a Weak Base with H_3O^+ Ions

The principles that describe the titration of a weak acid also describe the titration of a weak base with hydronium ions. The titration curve for a weak base is shown in Figure 17-5.

Notice that the titration curve for a weak base has the same four regions seen in the titration curve of a weak acid:

1. At the beginning of the titration, the base B and water are the only major species.

2. In the long, flat buffer region, both B and its conjugate acid BH^+ are major species. The midpoint of the titration occurs in the buffer region. At this point the pH of the solution is equal to the pK_a of the conjugate acid of the base.

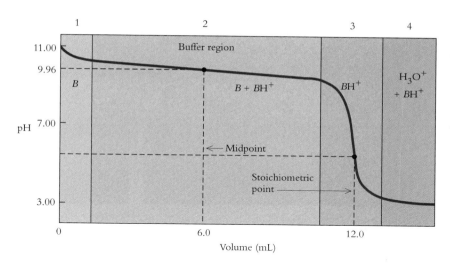

Figure 17-5
Schematic profile of the titration curve for a weak base *B* titrated with hydronium ions. The pH values are those for titration of ephedrine, a weak base that is the active ingredient in many decongestants.

3. When nearly all the *B* molecules have been protonated, the solution is no longer buffered. Further addition of hydronium ions causes a sharp drop in pH.
4. Beyond the stoichiometric point the solution contains excess H_3O^+ ions, which determine the pH.

Example 17-9 demonstrates that at the stoichiometric point in any weak base titration, the pH is *less than 7.0*.

Titration of a Weak Base	Example 17-9

Ephedrine, a weak base, is the active ingredient in many commercial decongestants. To analyze a sample of ephedrine dissolved in 0.200 L of water, a chemist carries out a titration with 0.900 M HCl, monitoring the pH continuously. The data obtained in this titration are shown in Figure 17-5. Calculate K_b for ephedrine and determine the pH of the solution at the stoichiometric point.

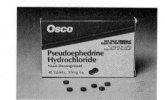

Strategy: This problem asks for two different quantities, requiring two separate analyses. Thus, the problem should be solved in two stages. The reaction that takes place during the titration is proton transfer from hydronium ions (added from the buret) to ephedrine molecules (in the solution). For simplicity, we designate ephedrine as Ep and its conjugate acid as EpH^+.

The problem provides a titration curve (Figure 17-5) and asks about two different equilibrium results, K_b and pH at the stoichiometric point. Ephedrine is not listed in Appendix E, so K_b must be determined by analyzing the titration curve. The pH calculation requires the standard procedure using K_b.

Solution:
Determination of K_b:

The titration curve does not give K_b directly. However, at the midpoint of the titration the concentration of Ep and EpH^+ are identical. Use this information in the buffer equation to show that at the midpoint of the titration, the pH of the solution equals the pK_a for EpH^+.

Ephedrine
$C_{10}H_{15}NO$

Example 17-9 | **Titration of a Weak Base** *(continued)*

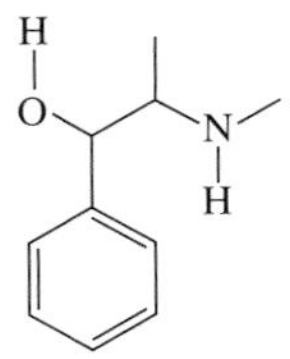

$$EpH^+(aq) + H_2O(l) \rightleftharpoons Ep(aq) + H_3O^+(aq)$$

$$pH = pK_a + \log \frac{[Base]}{[Acid]} = pK_a + \log(1) = pK_a$$

We are asked to determine K_b, which is related to pK_a through Equation 16-4:

$$pK_a + pK_b = 14.00$$

To locate the midpoint of the titration, use the volume needed to reach the stoichiometric point. At the stoichiometric point, all the ephedrine has been converted to EpH^+ ions, whereas at the midpoint, exactly half the ephedrine has been converted to EpH^+ ions. Thus the volume of the HCl solution at the midpoint of the titration is exactly half the volume needed to reach the stoichiometric point. According to the titration curve, the volume needed to reach the stoichiometric point is 12.0 mL. The midpoint, therefore, corresponds to the addition of 6.0 mL, and the pH at the midpoint is the pH reading when 6.0 mL of solution has been added. Reading from Figure 17-5:

$$pH_{midpoint} = pK_a = 9.96$$

Now apply Equation 16-4 to find K_b for ephedrine:

$$pK_a + pK_b = 14.00$$

$$pK_b = 14.00 - pK_a = 14.00 - 9.96 = 4.04$$

$$K_b = 10^{-4.04} = 9.1 \times 10^{-5}$$

pH at the Stoichiometric Point:

At the stoichiometric point, virtually all the ephedrine molecules have been converted to EpH^+ ions by proton transfer from H_3O^+.

$$Ep(aq) + H_3O^+(aq) \longrightarrow EpH^+(aq) + H_2O(l)$$

At this point, the solution contains EpH^+, H_2O, and Cl^- as major species. The chloride ion is the conjugate base of a *strong* acid, HCl, so the pH of the solution is determined by the acid–base equilibrium of water and the acid, EpH^+:

$$EpH^+(aq) + H_2O(aq) \rightleftharpoons Ep(aq) + H_3O^+(aq) \qquad pK_a = 9.96$$

$$K_a = \frac{[EP]_{eq}\,[H_3O^+]_{eq}}{[EpH^+]_{eq}} = 10^{-9.96} = 1.1 \times 10^{-10}$$

The amount of weak acid present at the stoichiometric point is equal to the amount of added hydronium ion:

$$mol\ EpH^+ = mol\ H_3O^+ \text{ needed to reach the stoichiometric point}$$

$$mol\ H_3O^+ \text{ needed} = (0.0120\ L)(0.900\ mol/L) = 0.0108\ mol$$

$$mol\ EpH^+ = 0.0108\ mol$$

The initial concentration of EpH^+ is found by dividing the number of moles by the total volume of the original solution plus the volume of added HCl solution:

Titration of a Weak Base *(continued)*

Example 17-9

$$[EpH^+]_{initial} = \frac{0.0108 \text{ mol}}{0.200 \text{ L} + 0.0120 \text{ L}} = 0.0509 \text{ M}$$

You should be able to set up a concentration table and use the K_a expression to solve for $[H_3O^+]_{eq}$. The pH of the solution at the stoichiometric point is 5.63.

Are these two results reasonable? We know that ephedrine is a weak base, so $K_b = 9.1 \times 10^{-5}$ is a reasonable value. In analogy to the strong base–weak acid titration, we expect the stronger member of the acid-base pair to influence the pH at the stoichiometric point, so the acidic pH of 5.63 is also reasonable.

You may wonder why we did not use the titration curve to determine the pH at the stoichiometric point as we did to find the pH at the midpoint. Notice from Figure 17-5 that near the stoichiometric point, the pH changes very rapidly with added H_3O^+. At this point, the curve is nearly vertical. Thus there is much uncertainty in reading a graph to determine the pH at the stoichiometric point. In contrast, a titration curve is nearly flat in the vicinity of the midpoint, minimizing uncertainty caused by errors in graph reading.

Indicators

An acid-base titration can be performed without measuring the pH of the solution, provided that an **indicator** is present that gives a color change when the titration reaches the stoichiometric point. An indicator is a weak organic acid that contains a highly delocalized π-bonding network. At pH $<$ pK_a of the weak organic acid, the indicator is in its protonated form, which we designate H*In*. At pH $>$ pK_a of the weak organic acid, the indicator gives up its proton and is converted to its conjugate base, designated *In*$^-$.

Deprotonation of H*In* changes the structure of the indicator, which alters the pattern of electron delocalization in the π system. Changing the indicator's delocalized π system also changes its color. Figure 17-6 shows these changes for one acid-base indicator, phenol red. The pK_a of phenol red is 7.9. When the pH is less than 7.9, the acid dominates, giving a yellow solution. When the pH is higher than 7.9, the conjugate base dominates, giving a red solution:

$$\text{H}In + \text{H}_2\text{O} \rightleftharpoons \text{H}_3\text{O}^+ + In^- \qquad \text{p}K_{In} = 7.9$$

Yellow   Red

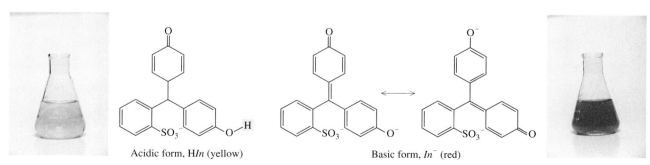

Acidic form, H*In* (yellow)   Basic form, *In*$^-$ (red)

Figure 17-6
The line structures of phenol red in its acidic and basic forms.

Table 17–2
Indicators and Their pH Ranges

Indicator	pK_{In}	pH Range	Colors Acid	Base
Thymol blue*	1.75	1.2–2.8	Red	Yellow
Methyl orange	3.40	3.1–4.4	Red	Yellow
Bromocresol green	4.68	4.0–5.6	Yellow	Blue
Methyl red	4.95	4.4–6.2	Red	Yellow
Bromocresol purple	6.3	5.2–6.8	Yellow	Purple
Phenol red	7.9	6.4–8.0	Yellow	Red
Thymol blue*	8.9	8.0–9.6	Yellow	Blue
Phenolphthalein	9.4	8.0–10.0	Colorless	Red
Thymolphthalein	10.0	9.4–10.6	Colorless	Blue

*Thymol blue has two acidic hydrogen atoms, so it can be used as an indicator for two different pH regions.

The acid dissociation constant of an indicator is designated K_{In}.

The pH of a solution changes during titration. As the pH passes the value of pK_{In}, the indicator changes color. A good indicator changes color very near the pH of the stoichiometric point. Because this pH depends on the pK_a of the substance being titrated, different titrations require different indicators. The best indicator for a titration has a pK_{In} that is the same as the pH of the solution at the stoichiometric point:

$$pK_{In} \cong pH_{\text{stoichiometric point}}$$

Because the pH changes rapidly near the stoichiometric point, an indicator is suitable for a titration if the pH at the stoichiometric point is within one unit of pK_{In}:

$$pH_{\text{stoichiometric point}} = pK_{In} \pm 1 \qquad (17\text{–}2)$$

Table 17-2 lists a selection of acid-base indicators, and Example 17-10 shows how to select an appropriate indicator.

Example 17-10 | **Selecting an Indicator**

A student wants to titrate a solution of ammonia whose approximate concentration is 10^{-2} M. What indicator would be appropriate?

Solving Equilibrium Problems

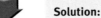

Strategy: This is a qualitative problem (selection of an indicator). Nevertheless, we follow the seven-step procedure, because a quantitative analysis is needed before the appropriate indicator can be identified.

Solution:

1. The student needs an indicator whose color changes as close to the stoichiometric point as possible:

$$pH_{\text{stoichiometric point}} = pK_{In} \pm 1$$

Thus the student must calculate the pH at the stoichiometric point and then consult Table 17-2.

Selecting an Indicator *(continued)* | **Example 17-10**

2. The titration reaction is:

$$NH_3 + H_3O^+ \longrightarrow NH_4^+ + H_2O$$

At the stoichiometric point, all the ammonia molecules have been converted to ammonium ions, so the major species present are NH_4^+ and H_2O.

3. The pH of the solution is thus determined by the acid–base equilibrium of NH_4^+ ions:

$$NH_4^+ + H_2O \rightleftharpoons NH_3 + H_3O^+$$

4. $K_{eq} = \dfrac{[NH_3]_{eq}\,[H_3O^+]_{eq}}{[NH_4^+]_{eq}} = \dfrac{K_w}{K_b} = 5.6 \times 10^{-10}$

5. Set up a concentration table at the stoichiometric point:

Species (reaction)	H_2O +	NH_4^+ $\rightleftharpoons$	NH_3 +	H_3O^+
Initial concentration (M)		10^{-2}	0	0
Change in concentration (M)		$-x$	$+x$	$+x$
Equilibrium concentration (M)		$10^{-2} - x$	x	x

6. Now substitute and solve:

$$5.6 \times 10^{-10} = \frac{(x)(x)}{(10^{-2} - x)} \cong \frac{x^2}{10^{-2}}$$

$$x^2 = 5.6 \times 10^{-12} \qquad x = 2.4 \times 10^{-6}$$

$$pH = 5.62$$

Table 17–2 shows that either bromocresol purple ($pK_{In} = 6.3$) or methyl red ($pK_{In} = 4.95$) would be suitable. Methyl red is typically used for titrations of weak bases by strong acids. The figure shows its color behavior.

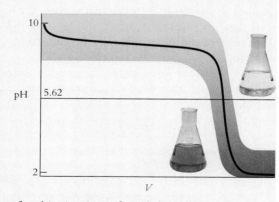

7. Checking for reasonableness, we see that x is small relative to 10^{-2}, and an acidic pH at the stoichiometric point is what we expect for this titration of weak base by strong acid.

The objective of an acid–base titration is to determine the amount of an acid or base in a solution. Because an indicator is itself a weak acid, it may appear that adding an indicator would alter equilibrium concentrations and influence the titration. However, a useful indicator gives a noticeable color to a solution at a concentration of 10^{-6} M. This is negligible compared with the concentration of the solution being titrated, which is usually in the range of 10^{-2} to 10^{-3} M.

The low concentration of an indicator also explains why the presence of this weak acid does not change the pH of the solution. Indicators are always present as minor species in solution, never major species. Thus the dominant equilibrium that determines solution pH never involves the indicator. The K_a of the substance being titrated establishes the equilibrium concentration of hydronium ions. This, in turn, establishes whether the indicator is in its acidic or basic form.

Section Exercises

17.3.1 Glycolic acid ($HOCH_2CO_2H$), a constituent of sugar cane juice, has a pK_a of 3.9. Sketch the titration curve for the titration of 60.0 mL of 0.010 M glycolic acid with 0.050 M KOH. Indicate the stoichiometric point, the buffer region, and the point of the titration where $pH = pK_a$. Sketch the curve qualitatively without doing any quantitative calculations.

17.3.2 What is the pH of the glycolic acid solution at the stoichiometric point? What is the pH of the solution when 3.0 mL of the KOH solution has been added?

17.3.3 Choose an indicator suitable for the titration of glycolic acid. What is the color of the solution at the stoichiometric point?

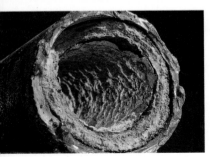

17.4 SOLUBILITY EQUILIBRIA

The limestone deposits that decorate Carlsbad and other caverns are the result of gradual precipitation of calcium carbonate from ground water. The solubility equilibrium is expressed by K_{sp}, as described in Chapter 15.

$$Ca^{2+}(aq) + CO_3^{2-}(aq) \rightleftharpoons CaCO_3(s) \qquad K_{sp} = [Ca^{2+}]_{eq}[CO_3^{2-}]_{eq}$$

Solubility equilibria are important in the natural environment and in our daily lives. Coral reefs are built up over many years through deposits of calcium carbonate from tiny sea organisms. Black smokers on the ocean's floor spew out insoluble metal sulfides and other minerals. Deposits from hard water can wreak havoc with plumbing.

In an aqueous solubility reaction, a salt dissolves to yield ions in solution. The amount of a salt that dissolves in water varies over a large range, as the following examples show:

$$AgI(s) \rightleftharpoons Ag^+(aq) + I^-(aq) \qquad K_{sp} = 8.5 \times 10^{-17}$$

$$Mg(OH)_2(s) \rightleftharpoons Mg^{2+}(aq) + 2\,OH^-(aq) \qquad K_{sp} = 5.6 \times 10^{-12}$$

$$CaCO_3(s) \rightleftharpoons Ca^{2+}(aq) + CO_3^{2-}(aq) \qquad K_{sp} = 9.8 \times 10^{-9}$$

$$Ag_2SO_4(s) \rightleftharpoons 2\,Ag^+(aq) + SO_4^{2-}(aq) \qquad K_{sp} = 1.2 \times 10^{-5}$$

$$NaCl(s) \rightleftharpoons Na^+(aq) + Cl^-(aq) \qquad K_{sp} = 6.2$$

Substances that have $K_{sp} \ll 1$, such as calcium carbonate, magnesium hydroxide, and iron(III) sulfide, are classified as **insoluble.** Salts that have K_{sp} between 10^{-2} and 10^{-5}, such as silver sulfate, are **slightly soluble,** and solids such as NaCl, which have $K_{sp} > 10^{-2}$, are **soluble.**

Solubility equilibria can be treated using our standard approach to equilibrium, as demonstrated in Examples 17-11, 12, and 13.

Chapter 4 →
This classification provides a more specific description of soluble and insoluble than the qualitative solubility guidelines presented in Chapter 4.

Solubility Products

Example 17-11

Gypsum is a relatively soft rock made of calcium sulfate. Rainwater percolates through gypsum, dissolves some of the rock, and eventually becomes saturated with Ca^{2+} ions and SO_4^{2-} ions. A geochemist takes a sample of ground water from a cave and finds that it contains 8.4×10^{-3} M SO_4^{2-} and 5.8×10^{-3} M Ca^{2+}. (The ratio is not $1:1$ because other sulfate rock contributes some of the SO_4^{2-} ions to the solution.) Use these data to determine the solubility product of calcium sulfate.

Strategy: This problem is straightforward, so a truncated version of the seven-step strategy is appropriate.

Solution: We are asked to evaluate K_{sp} of $CaSO_4$:

$$CaSO_4(s) \rightleftharpoons Ca^{2+}(aq) + SO_4^{2-}(aq) \qquad K_{sp} = [Ca^{2+}]_{eq}\,[SO_4^{2-}]_{eq}$$

The problem gives the required equilibrium concentrations, since it is stated that the ground water sample is saturated (that is, at equilibrium) with respect to these ions.

$$K_{sp} = (5.8 \times 10^{-3})(8.4 \times 10^{-3}) = 4.9 \times 10^{-5}$$

Remember that although equilibrium calculations require concentrations in molarities for solutes, the equilibrium constant expression is dimensionless.

Example 17–12 shows a calculation in which the stoichiometry differs from $1:1$.

Calculating K_{sp}

Example 17-12

When solid PbI_2 is added to pure water at 25 °C, the salt dissolves until the concentration of Pb^{2+} reaches 1.35×10^{-3} M. After this concentration is reached, excess solid remains undissolved. What is K_{sp} for this salt?

Strategy: The problem asks for an equilibrium constant, which means we need to find equilibrium concentrations of the species involved in the solubility reaction. Use the seven-step strategy, which we present here without step numbers.

Solving
Equilibrium
Problems

Solution: The starting materials are solid PbI_2 and H_2O. These are the only major species present before any solid dissolves.

The only equilibrium involving these reactants is the solubility reaction. Solid PbI_2 dissolves in water to produce its ions in solution, Pb^{2+} and I^-:

$$\text{Chemical reaction: } PbI_2(s) \rightleftharpoons Pb^{2+}(aq) + 2\,I^-(aq)$$

$$\text{Equilibrium constant expression: } K_{eq} = K_{sp} = [Pb^{2+}]_{eq}\,[I^-]_{eq}^2$$

The equilibrium concentration of Pb^{2+} ions is stated to be 1.35×10^{-3} M, but the equilibrium concentration of iodide ions is not stated:

$$[Pb^{2+}]_{eq} = 1.2 \times 10^{-3}\ M \qquad [I^-]_{eq} = ?$$

The equilibrium concentration of I^- is determined by stoichiometric analysis. Initially, the system contains only pure water and solid lead(II) iodide. Enough PbI_2 dissolves to make $[Pb^{2+}]_{eq} = 1.35 \times 10^{-3}$ M. One formula unit of PbI_2 contains one Pb^{2+} cation and *two* I^- anions. Thus twice as many iodide

| Example 17-12 | Calculating K_{sp} *(continued)* |

ions as lead ions enter the solution. The concentration of I^- at equilibrium is double that of Pb^{2+} cations:

$$[I^-]_{eq} = 2 [Pb^{2+}]_{eq} = 2.70 \times 10^{-3} \text{ M}$$

Substitute the values of the concentrations at equilibrium into the equilibrium expression and solve for K_{sp}:

$$K_{sp} = [Pb^{2+}]_{eq} [I^-]^2_{eq} = (1.35 \times 10^{-3})(2.70 \times 10^{-3})^2 = 9.84 \times 10^{-9}$$

This is a small value, which is reasonable given the small concentrations, 10^{-3} M, of the ions at equilibrium.

Example 17-13 deals with the second type of calculation, determining a concentration at equilibrium when the value of the solubility product is known.

| Example 17-13 | Solubility Calculations |

Cadmium is an extremely toxic metal that finds its way into the aqueous environment as a result of some human activities. A major cause of cadmium pollution is zinc mining and processing, because natural deposits of ZnS ores usually also contain CdS. During the processing of these ores, highly insoluble cadmium sulfide ($K_{sp} = 7.9 \times 10^{-27}$) may be converted into considerably less insoluble cadmium hydroxide ($K_{sp} = 7.2 \times 10^{-15}$). What mass of $Cd(OH)_2$ will dissolve in 1.00×10^2 L of an aqueous solution?

Strategy: *What mass will dissolve* requires that we find equilibrium concentrations, so we apply our standard approach.

Solving
Equilibrium
Problems

Solution:

Major species present: $Cd(OH)_2(s)$ and $H_2O(l)$

Equilibrium: $Cd(OH)_2(s) \rightleftharpoons Cd^{2+}(aq) + 2\,OH^-(aq)$

$$K_{sp} = [Cd^{2+}]_{eq} [OH^-]^2_{eq} = 7.2 \times 10^{-15}$$

Concentration table (change in $Cd^{2+} = x$):

Species (reaction)	$Cd(OH)_2(s)$	$\rightleftharpoons$	$Cd^{2+}(aq)$	+	$2\,OH^-(aq)$
Initial concentration (M)	(Solid)		0		0
Change in concentration (M)	(Solid)		$+x$		$+2x$
Equilibrium concentration (M)	(Solid)		x		$2x$

Evaluation:

$$7.2 \times 10^{-15} = (x)(2x)^2 = 4x^3 \quad so \quad x^3 = 1.8 \times 10^{-15} \quad and \quad x = 1.2 \times 10^{-5}$$

Solubility Calculations (continued)   **Example 17-13**

The concentration of Cd^{2+} ions in a saturated solution is 1.2×10^{-5} M. Therefore in 1.00×10^2 L of solution, there will be:

$$(1.2 \times 10^{-5} \text{ mol/L})(1.00 \times 10^2 \text{ L}) = 1.2 \times 10^{-3} \text{ mol}$$

Cadmium hydroxide generates Cd^{2+} in a $1:1$ stoichiometric ratio, so the number of moles of Cd^{2+} in solution equals the number of moles of $Cd(OH)_2$ that dissolves. Find the amount in grams using the molar mass of $Cd(OH)_2$:

$$m = n\,MM = (1.2 \times 10^{-3} \text{ mol})(146 \text{ g/mol}) = 0.18 \text{ g}$$

In contrast, only 1.3×10^{-9} g of CdS dissolves in the same volume of water.

Precipitation Equilibria

The solubility product (K_{sp}) describes a solubility equilibrium from the perspective of a salt dissolving in water. In the laboratory and in industry, solubility equilibria are often exploited in the opposite direction. Two solutions are mixed to form a new solution in which the solubility product of one substance is exceeded. This salt precipitates and can be collected by filtration. Example 17-14 illustrates how precipitation techniques can be used to remove toxic heavy metals from aqueous solutions.

Precipitation Reactions   **Example 17-14**

As illustrated in Example 17-13, wastewater resulting from metal processing often contains significant amounts of toxic heavy metal ions that must be removed before the water can be returned to the environment. One method uses sodium hydroxide solution to precipitate insoluble metal hydroxides. Suppose that 1.00×10^2 L of wastewater containing 1.2×10^{-5} M Cd^{2+} is treated with 1.0 L of 6.0 M NaOH solution. What is the residual concentration of Cd^{2+} after treatment, and what mass of $Cd(OH)_2$ precipitates?

Strategy: Again, we can use the standard approach to an equilibrium calculation. In this case, however, the reaction is precipitation, for which the equilibrium constant is quite large. Thus taking the reaction to completion is likely to be the appropriate approach to solving for equilibrium concentrations.

Solving
Equilibrium
Problems

Solution: The wastewater contains Cd^{2+}, so an anion must also be present in the solution to balance the charge of the cadmium ions. Other species may exist as well. The problem asks only about the cadmium in the wastewater, so assume that any other ions are spectators. The sodium hydroxide solution contains Na^+ and OH^-, so the major species in the treated wastewater include H_2O, Cd^{2+}, OH^-, and Na^+.

The equilibrium constant for the precipitation reaction is the inverse of K_{sp} for $Cd(OH)_2$:

$$Cd^{2+}(aq) + 2\,OH^-(aq) \rightleftharpoons Cd(OH)_2(s)$$

$$K_{eq} = \frac{1}{[Cd^{2+}]_{eq}\,[OH^-]_{eq}^2} = \frac{1}{K_{sp}} = \frac{1}{7.2 \times 10^{-15}} = 1.4 \times 10^{14}$$

Example 17-14	Precipitation Reactions *(continued)*

The problem asks for the residual concentration of cadmium ions. In other words, what is the Cd^{2+} ion concentration in the solution after the NaOH is added? Before the concentration table can be completed, initial ion concentrations must be determined. The volume of the mixed solutions is 1.01×10^2 L, which is less than 5% different from 1.00×10^2 L. This means the initial Cd^{2+} concentration is 1.2×10^{-5} M. Do a dilution calculation to find the "initial" concentration (after mixing but before reaction) of hydroxide ions:

$$M_2 = \frac{M_1 V_1}{V_2} = \frac{(6.0 \text{ M})(1.0 \text{ L})}{(1.00 \times 10^2 \text{ L})} = 6.0 \times 10^{-2} \text{ M} = [OH^-]_{initial}$$

Now we set up a concentration table. The equilibrium constant for precipitation is very large, so imagine the precipitation in two steps (see Example 15-15). First, take the reaction to completion. Then "switch on" the solubility equilibrium:

Species (reaction)	$Cd^{2+}(aq)$	$+\ 2\ OH^-(aq)$	$\rightleftharpoons$	$Cd(OH)_2(s)$
Initial concentration (M)	1.2×10^{-5}	6.0×10^{-2}		(solid)
Change to completion (M)	-1.2×10^{-5}	-2.4×10^{-5}		(solid)
Concentration at completion (M)	0	6.0×10^{-2}		(solid)
Change to equilibrium (M)	$+\gamma$	$+2\gamma$		(solid)
Equilibrium concentration (M)	γ	$6.0 \times 10^{-2} + 2\gamma$		(solid)

A tiny amount of $Cd(OH)_2$ dissolves in this approach to equilibrium, so we use the solubility equilibrium and K_{sp}:

$$K_{sp} = [Cd^{2+}]_{eq}[OH^-]^2_{eq} = 7.2 \times 10^{-15}$$

The variable γ represents the change to equilibrium in a reaction with a small equilibrium constant, so make the approximation that γ will be small compared with 6.0×10^{-2}:

$$K_{sp} = (\gamma)(6.0 \times 10^{-2} + 2\gamma)^2 \cong (\gamma)(6.0 \times 10^{-2})^2 = 7.2 \times 10^{-15}$$

$$\gamma = \frac{(7.2 \times 10^{-15})}{(6.0 \times 10^{-2})^2} = 2.0 \times 10^{-12}$$

$$[Cd^{2+}]_{eq} = 2.0 \times 10^{-12} \text{ M}$$

The problem also asks for the mass of $Cd(OH)_2$ that precipitates. The number of moles of precipitate equals the number of moles of Cd^{2+} ions removed by hydroxide treatment. The necessary information is contained in the concentration table in the row that gives the completion conditions. The precipitation equilibrium constant is so large ($> 10^{14}$) that equilibrium considerations can be neglected for this calculation:

$$\text{mol } Cd(OH)_2 = \text{mol } Cd^{2+} = MV$$
$$= (1.2 \times 10^{-5} \text{ M})(1.00 \times 10^2 \text{ L}) = 1.2 \times 10^{-3} \text{ mol}$$

$$\text{Mass } Cd(OH)_2 = (\text{mol})(\text{MM}) = (1.2 \times 10^{-3} \text{ mol})(146 \text{ g/mol}) = 0.18 \text{ g}$$

Notice that virtually all the mass that dissolved originally (Example 17-13) has been recovered.

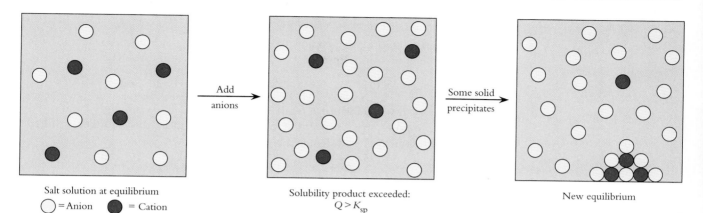

Figure 17-7

When additional anions (*yellow circles*) are added to a solution of a salt at equilibrium (*left*), the solubility product is exceeded (*center*), and some solid precipitates, reducing the concentration of cations (*blue circles*) in the solution. At the new equilibrium (*right*), there is a higher concentration of anions, but a lower concentration of cations. Water molecules have been omitted for clarity.

Common Ion Effect

The concentration of Cd^{2+} ions in the saturated solution of Example 17-13 is 1.2×10^{-5} M, whereas the concentration of this ion in the solution after the addition of hydroxide ion (Example 17-14) is only 2.0×10^{-12} M. This reduction in concentration is an example of what is called the **common ion effect.** The solubility product of a salt contains the concentrations of both cations and anions. By Le Châtelier's principle, adding more of the same anion to a saturated solution of a salt shifts the equilibrium position in the direction that reduces this added concentration: The concentration of cations present at equilibrium drops in response to the increase in concentration of anions. Figure 17-7 provides a schematic molecular illustration.

The common ion effect is quite general. Adding either the cation or the anion to a solution at equilibrium results in $Q > K_{sp}$, so the reaction proceeds to the right to return to equilibrium. At the final equilibrium position, the concentration of whichever ion was added is higher than before addition, but the concentration of the other ion is lower. As illustrated by Example 17-14, this reduction in the concentration of one ion at the expense of another has practical applications when one ion of the pair is undesirable.

Quantitative calculations of the common ion effect, such as Example 17–14, require no new principles, but it is necessary to pay close attention to "initial" concentrations. Also it often is convenient to take the reaction to completion and then allow it to return to equilibrium.

Effects of pH

The introduction to this chapter states that calcium carbonate is insoluble in pure water but dissolves in weakly acidic water. We can understand this effect by examining the acid-base properties of the species involved in the solubility equilibrium.

$$CaCO_3(s) \rightleftharpoons Ca^{2+}(aq) + CO_3^{2-}(aq)$$

The carbonate anion is the conjugate base of hydrogen carbonate, so we can write an acid–base equilibrium for a saturated solution of calcium carbonate:

$$CO_3^{2-}(aq) + H_2O(l) \rightleftharpoons HCO_3^-(aq) + OH^-(aq)$$

As a result of this equilibrium, a saturated solution of $CaCO_3$ is slightly basic.

Now consider what happens if we add some acid to this saturated solution. Hydronium ions react with hydroxide ions to form water:

$$OH^-(aq) + H_3O^+(aq) \rightleftharpoons 2 H_2O(l)$$

By Le Châtelier's principle, the system responds in the direction that will reduce the effect of this added acid, meaning that additional carbonate ions accept protons from water molecules.

$$CO_3^{2-}(aq) + H_2O(l) \longrightarrow HCO_3^-(aq) + OH^-(aq)$$

Responding to the reduction in carbonate ion concentration, additional calcium carbonate dissolves:

$$CaCO_3(s) \longrightarrow Ca^{2+}(aq) + CO_3^{2-}(aq)$$

Qualitatively, the amount of calcium carbonate that dissolves in water is increased by making the solution more acidic.

The analysis can be made quantitative by writing the various equilibria and their K_{eq} values. The reactions can be added to obtain the net reaction that occurs when calcium carbonate is exposed to acidic water, and the equilibrium constant for the net reaction is the product of the individual K_{eq}'s:

$$CaCO_3(s) \rightleftharpoons Ca^{2+}(aq) + CO_3^{2-}(aq) \qquad K_{eq} = K_{sp}$$

$$CO_3^{2-}(aq) + H_2O(l) \rightleftharpoons HCO_3^-(aq) + OH^-(aq) \qquad K_{eq} = K_b = \frac{K_w}{K_a}$$

$$OH^-(aq) + H_3O^+(aq) \rightleftharpoons 2 H_2O(l) \qquad K_{eq} = \frac{1}{K_w}$$

Net: $\quad CaCO_3(s) + H_3O^+(aq) \rightleftharpoons Ca^{2+}(aq) + HCO_3^-(aq) + H_2O(l)$

$$K_{acidic} = K_{sp}\left(\frac{K_w}{K_a}\right)\left(\frac{1}{K_w}\right) = \frac{K_{sp}}{K_a}$$

Substituting the numerical values gives the equilibrium constant in acidic solution:

$$K_{acidic} = \frac{K_{sp}}{K_a} = \frac{(3.36 \times 10^{-9})}{(4.7 \times 10^{-11})} = 71$$

Taking the ratio of the equilibrium constants shows that calcium carbonate is ten orders of magnitude more soluble in acidic solution than in water:

$$\frac{K_{acidic}}{K_{neutral}} = \frac{71}{K_{sp}} = \frac{71}{3.36 \times 10^{-9}} = 2.1 \times 10^{10}$$

Our Chemistry and the Environment Box explores the effects of equilibria on CO_2 in the atmosphere, and Example 17-15 shows that the acidity of the solvent places an upper limit on the amount of salt that will dissolve.

Box 17-2	Chemistry and the Environment: The Carbon Cycle

Carbon accounts for only 0.08% of our planet's mass. Nevertheless, life on Earth is based on carbon. Living organisms take in carbon, process it through biochemical reactions, and expel it as waste.

Carbon dioxide is especially important in the exchange of carbon between the biosphere and its environment. Green plants store energy from sunlight by converting CO_2 and H_2O into carbohydrates. Most carbohydrates have the general formula $(CH_2O)_n$, so the chemical equation for carbohydrate formation appears simple, even though carbohydrates are complex (as described in Chapter 11):

$$n\,CO_2 + n\,H_2O \xrightarrow{h\nu} (CH_2O)_n + n\,O_2$$

Plants and animals use enzymes to catalyze the conversion of carbohydrates back to CO_2 and H_2O, a process that releases chemical energy and powers life processes:

$$(CH_2O)_n + n\,O_2 \xrightarrow{Enzymes} n\,CO_2 + n\,H_2O + energy$$

These two processes cycle carbon between the atmosphere, where it is found primarily as CO_2, and the biosphere, where it is found primarily as carbohydrates.

The carbon cycle is complicated by several reactions that involve CO_2. These reactions transfer carbon between the atmosphere, the hydrosphere (Earth's surface waters), and the lithosphere (Earth's crustal solids). The processes that move carbon from one sphere to another are illustrated schematically in the figure below.

Carbon dioxide is divided between the atmosphere and the hydrosphere. This division strongly favors the hydrosphere, because the large volume of water in the oceans has an immense capacity to absorb CO_2.

Almost all the Earth's carbon is found in the lithosphere as carbonate sediments that have precipitated from the oceans. Shells of aquatic animals also contribute $CaCO_3$ to the lithosphere. Carbon returns to the hydrosphere as carbonate minerals dissolve in water percolating through the

Earth's crust. This process is limited by the solubility products for carbonate salts, so lithospheric carbonates represent a relatively inaccessible storehouse of carbon.

Until recently, the distribution of carbon among the different terrestrial spheres was stable. When humans began burning fossil fuels, however, such burning transferred carbon into the atmosphere as CO_2. This has become a rapidly changing feature of the overall carbon cycle. Over the last quarter century, the atmospheric concentration of CO_2 has grown by more than 10%.

Why should there be concern about an increase in atmospheric CO_2 whose concentration in air is only 350 parts per million? Even a doubling of CO_2 concentration still leaves it well below 0.1% of the atmosphere. Despite its low concentration, CO_2 exerts a major effect on the average temperature of Earth's surface. Recall from Chapter 6 that CO_2 is a "greenhouse" gas that traps outgoing radiation. Increasing CO_2 in the atmosphere may contribute to global warming that could have catastrophic consequences. In a warmer climate, for example, much of the polar ice caps would melt, and the oceans would expand as their temperatures increased. Sea level could rise by several meters and inundate many of the world's most populous regions, such as Bangladesh, the Netherlands, and the East and Gulf Coasts of the United States. In addition, even small temperature fluctuations may alter global weather patterns and perhaps lead to serious droughts.

Our planet is such a complicated, dynamic set of interconnected systems that scientists do not know the result of doubling the concentration of atmospheric CO_2. Scientists are studying the carbon cycle in hopes of learning what lies ahead for the planet. There is some urgency to such studies because recent global weather patterns show some of the characteristics predicted from computer models of global warming: average temperatures higher than those in previous years and an increased incidence of extreme weather such as hurricanes and heavy rainfall.

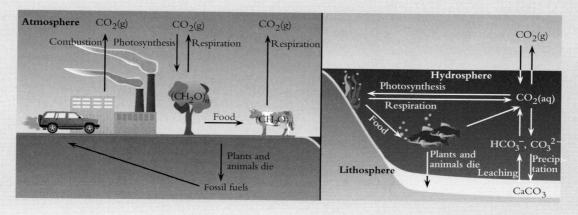

| Example 17-15 | Salt Dissolving in Acid |

The most acid rain on record had pH $= 2.00$. Calculate the concentration of Ca^{2+} cations in a solution formed when this rain becomes saturated with calcium carbonate.

Solving
Equilibrium
Problems

Strategy: We are asked to calculate a concentration at equilibrium, so we apply the seven-step method. The key feature is to identify the dominant equilibrium in this acidic solution. We do that by paying close attention to the major species present initially.

Solution:

1. The problem asks for $[Ca^{2+}]_{eq}$.

2. The major species present initially, other than water, are solid calcium carbonate and aqueous hydronium ions: $CaCO_3(s)$ and $H_3O^+(aq)$

3. These major species participate in the equilibrium described in the text:

$$CaCO_3(s) + H_3O^+(aq) \rightleftharpoons Ca^{2+}(aq) + HCO_3^-(aq) + H_2O(l)$$

4. The equilibrium constant expression is taken from the balanced equation. The value of the equilibrium constant was determined in the text:

$$K_{acidic} = \frac{[Ca^{2+}]_{eq}[HCO_3^-]_{eq}}{[H_3O^+]_{eq}} = 71$$

5. Organize the data using a concentration table. The large value for K_{acidic} indicates that the reaction should be taken to completion and then brought back to equilibrium. H_2O and $CaCO_3$ are not included in the table, because their concentrations do not appear in the equilibrium constant expression.

Species	H_3O^+ (aq)	Ca^{2+}(aq)	HCO_3^-(aq)
Initial concentration (M)	0.010	0	0
Change to completion (M)	-0.010	$+0.010$	$+0.010$
Concentration at completion (M)	0	0.010	0.010
Change to equilibrium (M)	$+y$	$-y$	$-y$
Equilibrium concentration (M)	y	$0.010 - y$	$0.010 - y$

6. We expect y to be small, so we make the following approximation:

$$(0.010 - y) \cong 0.010$$

Then we substitute and solve for y:

$$71 = \frac{[Ca^{2+}]_{eq}[HCO_3^-]_{eq}}{[H_3O^+]_{eq}} = \frac{(0.010)^2}{y}$$

$$y = \frac{(0.010)^2}{71} = 1.4 \times 10^{-6}$$

| **Salt Dissolving in Acid** (continued) | **Example 17-15** |

This is quite small relative to 0.010, so:

$$[Ca^{2+}]_{eq} = (0.010 \text{ M} - y) = 0.010 \text{ M}$$

7. The approximation is reasonable, as seen by the fact that the calculated value of y is much smaller than 0.010. To determine if the overall calculation is reasonable, notice that the hydronium concentration at equilibrium has fallen from 0.010 M to 1.4×10^{-6} M. The resulting solution is not far from neutral pH, a reasonable result for the salt of a weak base dissolving in acid solution.

The equilibrium constant expression derived for calcium carbonate dissolving in acidic solution is a general one that applies to any salt containing the conjugate base of a weak acid:

$$K_{acidic} = \frac{K_{sp}}{K_a}$$

Thus, many salts are more soluble in acidic solution than in pure water. In addition to the carbonates, important examples include the phosphates (PO_4^{3-}) and the sulfides (S^{2-}). Insoluble hydroxides also dissolve in acidic solution, as illustrated by magnesium hydroxide:

$$Mg(OH)_2(s) \rightleftharpoons Mg^{2+}(aq) + 2\,OH^-(aq) \qquad K_{sp} = 5.6 \times 19^{-12}$$

$$2[OH^-(aq) + H_3O^+(aq) \rightleftharpoons 2\,H_2O(l)] \qquad\qquad K = \left(\frac{1}{K_w}\right)^2$$

$$\text{Net: } Mg(OH)_2(s) + 2\,H_3O^+(aq) \rightleftharpoons Mg^{2+}(aq) + 2\,H_2O(l) \qquad K = \frac{K_{sp}}{K_w^2}$$

Knowing that $K_w = 1.0 \times 10^{-14}$, you should be able to show that the hydroxide salt is 28 orders of magnitude more soluble in acid solution than in pure water.

Section Exercises

■ **17.4.1** What is the concentration of Mg^{2+} ions in a saturated aqueous solution of $Mg(OH)_2$?

■ **17.4.2** Determine the concentration of Mg^{2+} ions remaining in solution after 0.150 L of 0.125 M $MgCl_2$ solution is treated with 0.100 L of 0.65 M NaOH solution. Does NaCl also precipitate from this solution?

■ **17.4.3** Determine the net reaction and the value of K_{eq} for $Al(OH)_3$ dissolving in acid solution, given that for this hydroxide, $K_{sp} = 4.6 \times 10^{-33}$.

17.5 COMPLEXATION EQUILIBRIA

In Section 15-6, we describe how metal cations in aqueous solution can form bonds to anions or neutral molecules that have lone pairs of electrons. This leads to formation of complex ions and to chemical equilibria involving complexation. The complexation equilibrium between Ag^+ and NH_3 is an example:

$$Ag^+(aq) + 2\ NH_3(aq) \rightleftharpoons [Ag(NH_3)_2]^+(aq) \qquad K_f = \frac{[Ag(NH_3)_2{}^+]_{eq}}{[Ag^+]_{eq}\,[NH_3]^2_{eq}}$$

The treatment of complexation equilibria is more complicated than treatments for acid–base or solubility equilibria, both because the stoichiometries of complexes are not obvious and because complex formation and dissociation frequently involve multiple steps.

Stoichiometry of Complexes

A species that bonds to a metal cation to form a complex is known as a **ligand.** Any species that has a lone pair of electrons has the potential to be a ligand, but in this section, we confine our description to a few of the most common ligands: ammonia, compounds derived from ammonia, cyanide, and halides. We describe additional examples in Chapter 19, which addresses the chemistry of the transition metals.

Any metal complex has a well-defined stoichiometry that is described by its chemical formula. For example, each cation of the silver-ammonia complex contains one Ag^+ cation bound to two neutral NH_3 ligands and carries a net charge of $+1$. The formula of a complex ion is enclosed in square brackets, as in $[Ag(NH_3)_2]^+$. The **coordination number,** which is the number of ligands bonded directly to the metal, provides a shorthand notation for the stoichiometry of a metal complex. Coordination numbers of 6 and 4 are most common, but 5 and 2 are not unusual, and there are no simple and reliable guidelines for predicting coordination numbers. Thus, coordination numbers must be determined by experiment. In this section, we provide the coordination number, chemical formula, or name for each complex that we describe.

Complexation Calculations

In principle, the calculation of concentrations of species that participate in a complexation equilibrium is no different from any other calculation involving equilibrium constant expressions. In practice, we have to consider multiple equilibria whenever a complex is present. This is because each ligand associates with the complex in a separate process with its own equilibrium expression. For instance, the silver-ammonia equilibrium is composed of two steps:

$$Ag^+(aq) + NH_3(aq) \rightleftharpoons [Ag(NH_3)]^+(aq) \qquad K_1 = \frac{[Ag(NH_3)^+]_{eq}}{[Ag^+]_{eq}\,[NH_3]_{eq}} = 2.1 \times 10^3$$

$$[Ag(NH_3)]^+(aq) + NH_3(aq) \rightleftharpoons [Ag(NH_3)_2]^+(aq)$$

$$K_2 = \frac{[Ag(NH_3)_2{}^+]_{eq}}{[Ag(NH_3)^+]_{eq}\,[NH_3]_{eq}} = 8.2 \times 10^3$$

Because the two equilibrium constant expressions have similar magnitudes, a solution of the silver-ammonia complex generally has a significant concentration of each of the species that participate in the equilibria. The details of such calculations are beyond our scope.

When the solution contains a large excess of ligand, the complexation equilibria can be handled using our standard seven-step approach. Under these conditions, each step in the complexation process proceeds nearly to completion, and we work with a single expression that describes the formation reaction for the complete complex. Example 17-16 treats a situation of this sort.

Gold Complex **Example 17-16**

The small amounts of gold contained in low-grade ores can be extracted using a combination of oxidation and complexation. Gold is oxidized to Au^+, which forms a very strong complex with cyanide anions:

$$Au^+(aq) + 2\, CN^-(aq) \rightleftharpoons [Au(CN)_2]^-(aq) \qquad K_f = 2 \times 10^{38}$$

Suppose that a sample of ore containing 2.5×10^{-3} mol of gold is extracted with 1.0 L of 4.0×10^{-2} M aqueous KCN solution. Calculate the concentrations of the three species involved in the complexation equilibrium.

Strategy: "Calculate the concentrations" indicates a quantitative equilibrium problem, so we apply the seven-step strategy. The complexation equilibrium is given in the problem:

$$Au^+(aq) + 2\, CN^-(aq) \rightleftharpoons [Au(CN)_2]^-(aq)$$

Solving
Equilibrium
Problems

Solution:

1. The problem asks for the equilibrium concentrations of the three ionic species: $[Au^+]_{eq}$, $[CN^-]_{eq}$, and $[Au(CN)_2^-]_{eq}$.

2. Potassium cyanide is a salt that dissociates completely into ions, and the problem states that gold is oxidized, so the major species present are Au^+, CN^-, K^+, and H_2O.

3. Both CN^- and H_2O have acid-base properties, but the problem asks only about the species involved in the complexation equilibrium, so the important equilibrium is the complexation reaction.

4. Now we write the equilibrium constant expression for formation of the complex ion:

$$K_f = \frac{[Au(CN)_2^-]_{eq}}{[Au^+]_{eq}\,[CN^-]_{eq}^2} = 2 \times 10^{38}$$

5. Organize the data using a concentration table. Initial concentrations can be found from the data stated in the problem, and because the equilibrium constant is quite large, we take the reaction to completion and then allow back-reaction to equilibrium:

Example 17-16	Gold Complex *(continued)*

Species (reaction)	$Au^+(aq)$	$+$	$2\,CN^-(aq)$	$\rightleftharpoons$	$[Au(CN)_2]^-(aq)$
Initial concentration (M)	2.5×10^{-3}		4.0×10^{-2}		0
Change to completion (M)	-2.5×10^{-3}		-5.0×10^{-3}		$+2.5 \times 10^{-3}$
Concentration at completion (M)	0		3.5×10^{-3}		2.5×10^{-3}
Change to equilibrium (M)	$+y$		$+2y$		$-y$
Equilibrium concentration (M)	y		$3.5 \times 10^{-3} + 2y$		$2.5 \times 10^{-3} - y$

6. We expect y to have a small value, so we make the two approximations,

$$(3.5 \times 10^{-3} + 2y) \cong 3.5 \times 10^{-3} \quad and \quad (2.5 \times 10^{-3} - y) \cong 2.5 \times 10^{-3}$$

$$K_f = 2 \times 10^{38} = \frac{(2.5 \times 10^{-3})}{(y)(3.5 \times 10^{-3})^2}$$

$$(3.5 \times 10^{-3})^2 \, (2 \times 10^{38}) \, y = (2.5 \times 10^{-3})$$

$$y = \frac{(2.5 \times 10^{-3})}{(2 \times 10^{38})(3.5 \times 10^{-3})^2} = 1 \times 10^{-36}$$

$$[Au^+]_{eq} = 1 \times 10^{-36} \text{ M}, \; [CN^-]_{eq} = 3.5 \times 10^{-3} \text{ M}, \quad and$$
$$[Au(CN)_2^-]_{eq} = 2.5 \times 10^{-3} \text{ M}$$

7. The calculated concentration for Au^+ is too small to measure, but this is reasonable in light of the very large value of K_f. The tiny value of y shows that the approximations are reasonable.

The result of Example 17-16 demonstrates one of the characteristics of complexation equilibria. In the presence of excess concentration of a complexing ligand, formation of a complex often reduces the concentration of a free metal cation essentially to zero.

The Chelate Effect

Ligands that have two or more donor atoms are **chelating** ligands. One of the most common chelating ligands is ethylenediamine (often abbreviated "en"), $H_2NCH_2CH_2NH_2$. Each nitrogen atom has a lone pair of electrons, so both can be donor atoms at the same time. Thus ethylenediamine is a bidentate ligand that can coordinate to a metal at two sites simultaneously.

Chelating ligands bind much more tightly to their metal cations than do monodentate ligands. A good example is the ethylenediamine complex with Ni^{2+}, as shown in Figure 17-8. The ethylenediamine complex is much more stable than the analogous ammonia complex:

$$Ni^{2+}(aq) + 6\,NH_3(aq) \rightleftharpoons [Ni(NH_3)_6]^{2+}(aq) \qquad K_f = 2.0 \times 10^8$$
$$Ni^{2+}(aq) + 3\,en(aq) \rightleftharpoons [Ni(en)_3]^{2+}(aq) \qquad K_f = 4.1 \times 10^{17}$$

The word *chelate* comes from the Greek *chele*, meaning "claw". The word *dentate* is derived from the Latin *dentis*, meaning "tooth." A monodentate ligand has "one tooth" for binding to a metal; a bidentate ligand has "two teeth," and so on.

Even though both ligands form Ni—N bonds of similar strength, ethylenediamine binds to Ni^{2+} many orders of magnitude more strongly than does NH_3. Why is this? Think about complexation at the molecular level. One of the Ni—N bonds in either complex can be broken fairly easily. When this happens to $[Ni(NH_3)_6]^{2+}$ the ammonia molecule drifts away and is replaced by another ligand, typically a water molecule:

$$[Ni(NH_3)_6]^{2+} \longrightarrow NH_3 + [Ni(NH_3)_5]^{2+}$$

$$[Ni(NH_3)_5]^{2+} + H_2O \longrightarrow [Ni(NH_3)_5H_2O]^{2+}$$

The result is the exchange of ligands.

This simple ligand exchange reaction is less likely for $[Ni(en)_3]^{2+}$. As shown in Figure 17-9, when one Ni—N bond breaks and one NH_2 group of an ethylenediamine ligand dissociates from the metal complex, the ligand is still bound to the metal cation by the second NH_2 group. Consequently, the en ligand is not free to float away in solution, so the first NH_2 group binds again to the metal rather than being replaced by a competing ligand. For an ethylenediamine to be replaced, both ends of the ligand must be released at almost the same time. This stabilization of a metal complex by a multidentate ligand is known as the *chelate effect*.

The chelate effect has important practical applications. As we discuss in Chapter 20, heavy metal ions such as Hg^{2+} and Pb^{2+} are extremely poisonous because of the way they interact with proteins. The effect is cumulative because the body has no mechanism for excreting heavy metal ions. Fortunately, chelating ligands can be used to treat poisoning by mercury and lead. One important example of a ligand used to treat metal ion poisoning is ethylenediaminetetraacetate ion, abbreviated $[EDTA]^{4-}$ or simply EDTA. With six donor atoms that can bond to a single metal ion, EDTA is a hexadentate ligand.

EDTA cannot be administered in uncomplexed form because it would form a strong complex with iron, removing this essential metal from the blood. Instead, the calcium disodium salt, $Na_2[Ca(EDTA)]$, is used. The presence of calcium prevents EDTA from complexing with iron, but heavy metal ions such as Hg^{2+} preferentially displace calcium cations from the complex ion:

$$[Ca(EDTA)]^{2-} + Hg^{2+} \longrightarrow [Hg(EDTA)]^{2-} + Ca^{2+}$$

The kidneys are able to pass chelated heavy metal ions into the urine, so treatment with $[Ca(EDTA)]^{2-}$ cleanses mercury from the blood, counteracting its poisoning effects.

Ethylenediamine
(en)

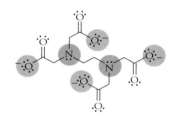

$[Ni(en)_3]^{2+}$ with the hydrogen atoms removed to emphasize the chelating en ligands

Figure 17-8
Ethylenediamine has two nitrogen atoms, each of which bonds to a central Ni^{2+} ion in $[Ni(en)_3]^{2+}$.

Ethylenediaminetetraacetate
$[EDTA]^{4-}$

$[Co(EDTA)]^-$ with the hydrogen atoms removed to emphasize the geometry around Co^{3+}

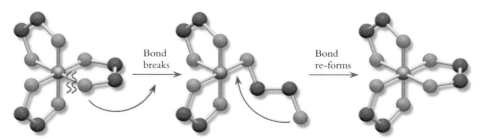

Figure 17-9
Breaking one bond in an ethylenediamine complex leaves the en ligand dangling but "tethered" to the metal by the second nitrogen atom.

Bond breaks

Bond re-forms

Complex Formation and Solubility

The effect of complex formation on the solubility of a solid can be observed in the kitchen. Copper cookware eventually builds up an unsightly black tarnish that is composed of a mixture of CuO and CuS, both of which are highly insoluble in water. If a tarnished copper pan is rubbed with a moist lemon, the black tarnish dissolves, returning the copper to its shiny metallic orange color. This is the result of a complexation reaction between citric acid, the molecule that gives lemons their sour taste, and the Cu^{2+} cation.

A complete description of the chemistry of copper–citrate complexation is complicated because it involves two insoluble copper salts and the proton transfer reaction of citric acid. Silver salts and their complexes provide a simpler illustration of how complexation enhances solubility. Figure 17-10 shows that silver will precipitate or dissolve in aqueous solution depending on the species that are present. Starting with a solution of silver nitrate, addition of aqueous NaCl causes AgCl to precipitate:

$$AgCl(s) \rightleftharpoons Ag^+(aq) + Cl^-(aq) \qquad K_{sp} = 1.8 \times 10^{-10}$$

If aqueous ammonia is added to the mixture, the precipitate dissolves by forming the silver–ammonia complex:

$$Ag^+(aq) + 2\,NH_3(aq) \rightleftharpoons [Ag(NH_3)_2]^+(aq) \qquad K_f = 1.6 \times 10^7$$

$$AgCl(s) + 2\,NH_3(aq) \rightleftharpoons [Ag(NH_3)_2]^+(aq) + Cl^-(aq) \qquad K_{eq} = K_{sp}K_f = 2.9 \times 10^{-3}$$

Although still less than 1, this equilibrium constant is large enough to allow solid AgCl to dissolve in strong aqueous ammonia.

Addition of a solution of NaBr results in a new precipitate, because AgBr is significantly less soluble than AgCl:

$$AgBr(s) \rightleftharpoons Ag^+(aq) + Br^-(aq) \qquad K_{sp} = 5.35 \times 10^{-13}$$

$$[Ag(NH_3)_2]^+(aq) + Br^-(aq) \rightleftharpoons AgBr(s) + 2\,NH_3(aq)$$

$$K_{eq} = \frac{K_f}{K_{sp}} = \frac{(1.6 \times 10^7)}{(5.35 \times 10^{-13})} = 3.0 \times 10^5$$

This large positive value indicates that the equilibrium position lies well to the right, favoring formation of solid AgBr.

Upon addition of sodium thiosulfate solution, this precipitate dissolves, because thiosulfate complexes more strongly with Ag^+ than does ammonia:

$$Ag^+(aq) + 2\,S_2O_3^{2-}(aq) \rightleftharpoons [Ag(S_2O_3)_2]^{3-}(aq) \qquad K_f = 2.0 \times 10^{13}$$

This stronger complexing agent is capable of dissolving AgBr:

$$AgBr(s) + 2\,S_2O_3^{2-}(aq) \rightleftharpoons [Ag(S_2O_3)_2]^{3-}(aq) + Br^-(aq) \qquad K_{eq} = K_{sp}K_f = 11$$

Far from being mere laboratory curiosities, these reactions find practical applications in photography. Most photographic films contain silver halides, primarily silver bromide. Exposure to light causes some Ag^+ cations to gain electrons, forming Ag granules that, after developing, generate the images captured on the film. But what about the unreacted silver bromide? If it were left in the film, further exposure to light would cause further reaction and eventual blackening of the image. To prevent this, film must be "fixed" by removing all unreacted AgBr. This is done using thiosulfate, which forms a complex with otherwise insoluble AgBr and allows it to be washed from the film. As Example 17-17 shows, fixer solution has a

(a)   AgCl(s),
$K_{sp} = 1.8 \times 10^{-10}$

(b) $[Ag(NH_3)_2]^+$(aq)

(c)   AgBr(s),
$K_{sp} = 5.35 \times 10^{-13}$

(d) $[Ag(S_2O_3)_2]^{3-}$(aq)

Figure 17-10
Silver precipitates from solution when a solution is added that contains an anion that forms an insoluble silver compound (*a* and *c*). The precipitates redissolve when a solution is added that contains a species that can form a strong complex with Ag^+ (*b* and *d*).

limited capacity to accept AgBr, so after repeated use, a solution must be discarded and a new solution prepared.

Capacity of Thiosulfate Fixer	**Example 17-17**

What mass of AgBr can be dissolved by 1.5 L of "fixer" solution that contains 0.50 M of thiosulfate anions?

Strategy: "What mass can be dissolved?" can be translated into "what is the concentration at equilibrium?" This is a problem in equilibrium concentrations, so we apply our seven-step method.

Solving Equilibrium Problems

Solution:

1. The problem asks for mass of AgBr that will dissolve in the solution. To find this, we must calculate the equilibrium concentration of silver-containing ions in the solution.

2. The species present initially are AgBr(s), $S_2O_3^{2-}$(aq), and an unspecified cation that is not part of the problem.

3. The equilibrium is the one described in the text, by which thiosulfate dissolves AgBr through complexation:

$$AgBr(s) + 2\ S_2O_3^{2-}(aq) \rightleftharpoons [Ag(S_2O_3)_2]^{3-}(aq) + Br^-(aq)$$

4. This reaction is the sum of the solubility reaction of AgBr and the complexation reaction between Ag and thiosulfate, so its K_{eq} is the product of the two equilibrium constants:

$$K_{eq} = \frac{[Ag(S_2O_3)_2^{3-}]_{eq}\ [Br^-]_{eq}}{[S_2O_3^{2-}]_{eq}^2} = K_{sp}K_f = 11$$

| Example 17-17 | Capacity of Thiosulfate Fixer *(continued)* |

5. Now we set up a concentration table for the aqueous species, using the initial concentration of thiosulfate, 0.50 M:

Species	2 $S_2O_3^{2-}$	$[Ag(S_2O_3)_2]^{3-}$	Br^-
Initial concentration (M)	0.50	0	0
Change to equilibrium (M)	$-2x$	$+x$	$+x$
Equilibrium concentration (M)	$0.50 - 2x$	x	x

6. Substitute the equilibrium concentrations into the equilibrium constant expression and solve for x:

$$\frac{[Ag(S_2O_3)_2{}^{3-}]_{eq}\,[Br^-]_{eq}}{[S_2O_3{}^{2-}]_{eq}^2} = \frac{x^2}{(0.50-2x)^2} = 11$$

We could solve by the quadratic equation, but notice that we can simplify this expression by taking the square root of each side:

$$\frac{x}{(0.50-2x)} = \sqrt{11} = 3.3 \qquad \textit{from which} \qquad x = 1.66 - 6.6\,x$$

$$7.6\,x = 1.66, \qquad \textit{and} \qquad x = 0.22$$

Thus, the equilibrium concentration of silver complex and bromide ion in the solution is 0.22 M. Multiplying by the volume of the solution gives the amounts in moles: mol = (0.22 M)(1.5 L) = 0.33 mol.

The stoichiometric ratio between AgBr and the ions in solution is 1:1, so this number is also the amount of AgBr that dissolves. Convert to mass by multiplying by the molar mass of AgBr:

$$m = (0.33 \text{ mol})(187.8 \text{ g/mol}) = 62 \text{ g}$$

7. Is this a reasonable mass? Notice that all the variables for the solution are somewhat less than 1 M, and the solution volume is somewhat greater than 1 L, so a result that is somewhat less than the molar mass of AgBr appears reasonable.

Section Exercises

17.5.1 Determine the concentration of the species present at equilibrium if 0.10 mol of ethylenediamine is dissolved in 2.00 L of a solution that contains 6.5×10^{-3} M Ni^{2+}. ($K_f = 4.1 \times 10^{17}$ for $[Ni(en)_3]^{2+}$).

17.5.2 When Zn^{2+} is added to aqueous ammonia, four ammonia ligands complex in stepwise fashion. Write all the equilibria involving Zn^{2+} and NH_3.

17.5.3 Calculate the mass of AgCl that will dissolve in 250 mL of 1.0 M aqueous ammonia.

■ CHAPTER REVIEW

Summary and Key Terms

1. A solution that contains comparable quantities of a weak acid and its conjugate base resists changes in pH and is called a **buffer solution.** The pH of a buffer solution can be calculated from the **buffer equation.**

2. A buffer solution has a limited ability to tolerate added acid or base, its **buffer capacity.** Buffer solutions are prepared by mixing two reagents that will produce a weak acid and its conjugate weak base as major species in solution.

3. Weak acids can be titrated with strong bases, and weak bases can be titrated with strong acids. The pH variations during such titrations can be determined using the standard approach to equilibrium calculations. At the **midpoint** of a titration, pH = pK_a. The **stoichiometric point** of a titration is the point where the amount of added titrant exactly matches the amount of substance being titrated. **Indicators** are weak acids or weak bases whose colors change as the pH varies.

4. Salts can be categorized as **insoluble, slightly soluble,** or **soluble,** depending on the magnitudes of their solubility products. The solubility product can be applied to salts dissolving and to ions combining to form precipitates. The **common ion effect** describes the reduction in solubility caused by addition of a common ion to a solution. The addition of acid increases the solubility of an insoluble salt by reacting with any anion that is the conjugate base of a weak acid.

5. Complex ions form when **ligands** bind to metal cations. The **coordination number** of the metal ion determines the stoichiometry of a complex. A **chelating** ligand binds to a metal ion through two or more interactions. Calculations involving complexation can be handled like any other equilibrium calculations if the ligand is present in excess. The solubility of a metal ion increases when a ligand is added that forms a complex. The equilibrium constant for the overall reaction is a composite of equilibrium constants for the individual reactions.

Skills to Master

▶ Using the buffer equation

▶ Preparing buffer solutions

▶ Analyzing titration curves

▶ Choosing an indicator

▶ Working solubility equilibrium problems

▶ Determining the effects of acids on solubility

▶ Working complexation equilibrium problems

▶ Determining the effects of complexing agents on solubility

Learning Exercises

17.1 Write a paragraph that explains the chemical principles of buffer action.

17.2 Explain why the pH is not necessarily 7.0 at the stoichiometric point in an acid–base titration.

17.3 List all the types of calculations described in Chapter 17 in which acid-base equilibrium expressions play a role.

17.4 Use your own words to define (a) buffer solution, (b) stoichiometric point, (c) common ion effect, and (d) chelate.

Problems   ilw = interactive learning ware problem. Visit the website at www.wiley.com/college/olmsted

Buffer Solutions

ilw **17.1** Which of the following solutions show buffer properties? Compute the pH of each solution that is buffered.
  (a) 0.100 L of 0.25 M NH_4Cl + 0.150 L of 0.25 M NaOH
  (b) 0.100 L of 0.25 M NH_4Cl + 0.050 L of 0.25 M NaOH
  (c) 0.100 L of 0.25 M NH_4Cl + 0.050 L of 0.25 M HCl
  (d) 0.100 L of 0.25 M NH_3 + 0.050 L of 0.25 M HCl

17.2 Which of the following solutions show buffer properties? Compute the pH of each solution that is buffered.
  (a) 0.100 L of 0.25 M $NaCH_3CO_2$ + 0.150 L of 0.25 M HCl
  (b) 0.100 L of 0.25 M $NaCH_3CO_2$ + 0.050 L of 0.25 M HCl
  (c) 0.100 L of 0.25 M $NaCH_3CO_2$ + 0.050 L of 0.25 M NaOH
  (d) 0.100 L of 0.25 M CH_3CO_2H + 0.050 L of 0.25 M NaOH

17.3 Compute the change in pH resulting from the addition of 5.0 mmol of strong acid to each solution in Problem 17.1 that is buffered.

17.4 Compute the change in pH resulting from the addition of 2.5 mmol of strong base to each solution in Problem 17.2 that is buffered.

17.5 Compute how many moles of base it takes to change the pH of each buffered solution in Problem 17.1 by 0.10 pH unit.

17.6 Compute how many moles of acid it takes to change the pH of each buffered solution in Problem 17.2 by 0.20 pH unit.

Capacity and Preparation of Buffer Solutions

17.7 From Table 17–1, select the best conjugate acid–base pairs for buffer solutions at pH 3.50 and 12.60. If you were going to add HCl solution as part of the buffer preparation, what other substance should you use in each case?

17.8 From Table 17-1, select the best conjugate acid–base pairs for buffer solutions at pH 6.85 and 10.80. If you were going to add NaOH solution as part of the buffer preparation, what other substance should you use in each case?

ilw 17.9 What mass of ammonium chloride must be added to 1.25 L of 0.25 M ammonia to make a buffer solution whose pH = 8.90?

17.10 What mass of sodium acetate must be added to 2.50 L of 0.55 M acetic acid to make a buffer solution whose pH = 5.75?

17.11 What is the maximum volume of 2.0 M HCl(aq) the buffer described in Problem 17.9 can tolerate without showing a pH change greater than 0.25 units?

17.12 What is the maximum volume of 2.5 M NaOH(aq) the buffer described in Problem 17.10 can tolerate without showing a pH change greater than 0.25 units?

Acid–Base Titrations

17.13 For each of the following, decide whether the pH at the stoichiometric point is greater than, less than, or equal to 7. In each case, identify the equilibrium that determines the pH. (a) NH_3(aq) titrated with $HClO_4$(aq); (b) $HClO_4$(aq) titrated with KOH(aq); and (c) $NaCH_3CO_2$(aq) titrated with HCl(aq).

17.14 For each of the following, decide whether the pH at the stoichiometric point is greater than, less than, or equal to 7. In each case, identify the equilibrium that determines the pH. (a) NaClO(aq) titrated with HCl(aq); (b) HNO_3(aq) titrated with KOH(aq); and (c) NH_4Cl(aq) titrated with NaOH(aq).

17.15 A laboratory technician wants to determine the aspirin content of a headache pill by acid–base titration. Aspirin has a K_a of 3.0×10^{-4}. The pill is dissolved in water to give a solution that is about 10^{-2} M and is then titrated with KOH solution. Find the pH at each of the following points, neglecting dilution effects: (a) before titration begins; (b) at the stoichiometric point; and (c) at the midpoint of the titration.

17.16 Sleeping pills often contain barbital, which is weakly acidic ($pK_a = 8.0$). For analysis of the barbital content of a sleeping pill, a titration is carried out with strong base. It takes 12.00 mL of 0.200 M base to reach the stoichiometric point. If the initial acid concentration is 0.0120 M and the solution volume is 200 mL, what is the pH of the solution after adding the following volumes: (a) 0.0 mL; (b) 5.0 mL; (c) 12.00 mL; and (d) 13.0 mL?

17.17 Sketch the titration curve for the titration described in Problem 17.15, and choose an appropriate indicator for this titration.

17.18 Sketch the titration curve for the titration described in Problem 17.16, and choose an appropriate indicator for this titration.

ilw 17.19 Calculate the pH at the second stoichiometric point when 150 mL of a 0.015 M solution of phthalic acid ($K_{a1} = 1.1 \times 10^{-3}$, $K_{a2} = 3.9 \times 10^{-6}$) is titrated with 1.00 M NaOH. What is a suitable indicator for this titration?

17.20 Calculate the pH at the second stoichiometric point when 250 mL of a 0.025 M solution of tartaric acid ($K_{a1} = 9.2 \times 10^{-4}$, $K_{a2} = 4.3 \times 10^{-5}$) is titrated with 1.00 M NaOH. What is a suitable indicator for this titration?

Solubility Equilibria

17.21 For the following salts, write a balanced equation showing the solubility equilibrium and write the solubility product expression for each: (a) silver chloride; (b) barium sulfate; (c) iron(II) hydroxide; and (d) calcium phosphate.

17.22 For the following salts, write a balanced equation showing the solubility equilibrium and write the solubility product expression for each: (a) lead(II) chloride; (b) magnesium carbonate; (c) nickel(II) hydroxide; and (c) silver acetate.

17.23 For each of the salts in Problem 17.21, determine the mass that dissolves in 475 mL of water at 25 °C.

17.24 For each of the salts in Problem 17.22, determine the mass that dissolves in 375 mL of water at 25 °C. (For $Ag(C_2H_3O_2)$, $K_{sp} = 1.94 \times 10^{-3}$)

17.25 Only 6.1 mg of calcium oxalate (CaC_2O_4) dissolves in 1.0 L of water at 25 °C. What is the solubility product of calcium oxalate?

17.26 The solubility of sodium sulfate (Na_2SO_4) in water is 9.5 g/100 mL. What is K_{sp} for sodium sulfate?

ilw 17.27 How many grams of $PbCl_2$ will dissolve in 0.750 L of 0.650 M $Pb(NO_3)_2$ solution? $K_{sp} = 1.7 \times 10^{-5}$.

17.28 How many grams of BaF_2 ($K_{sp} = 1.8 \times 10^{-7}$) will dissolve in 0.500 L of 1.00×10^{-1} M NaF solution?

17.29 Determine the value of K_{eq} for $CaSO_3$ ($K_{sp} = 1.0 \times 10^{-4}$) dissolving in acid solution, given that $K_{a2} = 6.3 \times 10^{-8}$ for H_2SO_3.

17.30 Determine the value of K_{eq} for $FeCO_3$ ($K_{sp} = 3.1 \times 10^{-11}$) dissolving in acid solution, given that $K_{a2} = 4.7 \times 10^{-11}$ for H_2CO_3.

17.31 Determine the concentration of Ca^{2+} in a solution obtained by treating solid $CaSO_3$ with 0.125 M HCl.

17.32 Determine the concentration of Fe^{2+} in a solution obtained by treating solid $FeCO_3$ with 0.255 M HNO_3.

Complexation Equilibria

17.33 Write the chemical formulas, including charge, of the complexes that form between the following metal cations and ligands: (a) Fe^{3+} and CN^-, coordination number = 6; (b) Zn^{2+} and NH_3, coordination number = 4; (c) V^{3+} and ethylenediamine (en), coordination number = 6.

17.34 Write the chemical formulas, including charge, of the complexes that form between the following metal cations and ligands: (a) Co^{2+} and SCN^-, coordination number = 4; (b) Hg^{2+} and NH_3, coordination number = 2; (c) Mg^{2+} and $EDTA^{4-}$, coordination number = 6.

ilw 17.35 Determine the concentration of the species present after 0.25 mol of NaCl is dissolved in 1.50 L of a solution that contains 7.5×10^{-3} M Pb^{2+}. ($K_f = 2.5 \times 10^{15}$ for $[PbCl_4]^{2-}$)

17.36 Determine the concentration of the species present after 0.15 mol of bipyridyl (bipy, a bidentate ligand) is dissolved in 2.50 L of a solution that contains 4.5×10^{-2} M Fe^{2+}. ($K_f = 1.6 \times 10^{17}$ for $[Fe(bipy)_3]^{2+}$)

17.37 When a solution of ethylenediamine is added to a solution containing Mn^{2+}, three en ligands bond to the metal in stepwise fashion. Write all the equilibria involving Mn^{2+} and en.

17.38 When a solution containing Co^{2+} is mixed with a solution containing oxalate anions ($C_2O_4^{2-}$), three bidentate oxalate ligands bond to the metal in stepwise fashion. Write all the equilibria involving Co^{2+} and $C_2O_4^{2-}$.

17.39 Calculate the mass of $CaSO_3$ that will dissolve in 0.50 L of 0.25 M solution of the tridentate ligand nitrilotriacetate (NTA^{3-}).

$$Ca^{2+} + 2\ NTA^{3-} \rightleftharpoons [Ca(NTA)_2]^{4-} \quad K_f = 3.2 \times 10^{11}$$

17.40 Calculate the mass of $FeCO_3$ that will dissolve in 0.33 L of a solution that is 0.20 M in the oxalate anion ($C_2O_4^{2-}$).

$$Fe^{2+} + 3\ C_2O_4^{2-} \rightleftharpoons [Fe(C_2O_4)_3]^{3-} \quad K_f = 3.3 \times 10^{20}$$

Additional Paired Problems

17.41 The pH of a formic acid/formate buffer solution is 4.04. Calculate the acid/conjugate base ratio for this solution. Draw a molecular picture that shows a small region of the buffer solution. (You may omit spectator ions and water molecules.) Use the following symbols:

17.42 The pH of an NH_4^+/NH_3 buffer solution is 8.77. Calculate the acid/conjugate base ratio for this solution. Draw a molecular picture that shows a small region of the buffer solution. (You may omit spectator ions and water molecules.) Use the following symbols:

17.43 The solubility product for $PbCl_2$ is 2×10^{-5}. If 0.50 g of solid $PbCl_2$ is added to 300 mL of water, will all the solid dissolve? (Hint: Calculate Q if all the solid dissolves.)

17.44 The solubility product for $MgCO_3$ is 7×10^{-6}. If 1.55 g of solid $MgCO_3$ is added to 255 mL of water, will all the solid dissolve? (Hint: Calculate Q if all the solid dissolves.)

17.45 Consider a buffer solution that contains 0.50 M NaH_2PO_4 and 0.20 M Na_2HPO_4. (a) Calculate its pH. (b) Calculate the change in pH if 0.120 g of solid NaOH is added to 150 mL of this solution. (c) If the acceptable buffer range of the solution is ± 0.10 pH units, calculate how many moles of H_3O^+ can be neutralized by 250 mL of the buffer.

17.46 Consider a buffer solution that contains 0.45 M H_3PO_4 and 0.55 M NaH_2PO_4. (a) Calculate its pH. (b) Calculate the change in pH if 0.260 g of solid KOH is added to 250 mL of this buffer solution. (c) If the acceptable buffer range of the solution is ± 0.15 pH units, calculate how many moles of OH^- can be neutralized by 150 mL of the buffer.

17.47 If 150 mL of 1.50×10^{-2} M $MgSO_4$ solution is mixed with 350 mL of 2.00×10^{-2} M $CaCl_2$ solution, will solid $CaSO_4$ form? ($K_{sp} = 4.9 \times 10^{-5}$)

17.48 If 200 mL of 2.50×10^{-2} M NaCl solution is mixed with 300 mL of 4.00×10^{-2} M $Pb(NO_3)_2$ solution, will solid $PbCl_2$ form? ($K_{sp} = 1.7 \times 10^{-5}$)

17.49 Bromocresol purple is an indicator that changes color from yellow to purple in the pH range from 5.2 to 6.8. Without doing a calculation, determine the color of the indicator in each of the following solutions. (a) 0.15 M HCl; (b) 0.25 M NaOH; (c) 1.0 M KCl; and (d) 0.55 M NH_3.

17.50 A solution has a pH of 8.5. What would be the color of the solution if the following indicators were present in the solution: (a) methyl orange; (b) phenol red; (c) bromocresol green; and (d) thymol blue (see Table 17.2)?

17.51 Leucine is an amino acid. A solution of leucine in water can be titrated with strong base. When this is done, the pH before titration is 1.85, the pH at the midpoint of the titration is 2.36, and the pH at the stoichiometric point is 6.00. Determine the value of K_a for leucine.

17.52 The following figure shows the data obtained in the pH titration of a biochemical substance that is a weak acid. From the information provided, determine the pK_a of this compound.

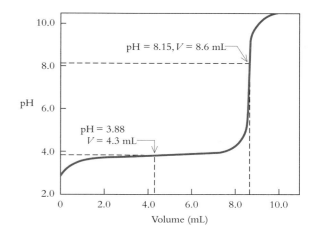

17.53 When lead(II) fluoride is dissolved in water at 25 °C, the equilibrium concentration of Pb^{2+} is 2.0×10^{-3} M. Use this information to calculate K_{sp} for PbF_2.

17.54 When silver sulfate dissolves in water at 25 °C, the equilibrium concentration of Ag^+ is 3.0×10^{-2} mol/L. Use this information to calculate K_{sp} for Ag_2SO_4.

17.55 Ammonia is a convenient buffer system in the slightly basic range. (a) What is the pH of a buffer solution containing 35.0 g of NH_4Cl dissolved in 1.00 L of 1.00 M NH_3? (b) How many moles of acid are required to change the pH of this solution by 0.05 pH units? (c) Suppose 5.0 mL of 12.0 M HCl solution is added to 250 mL of the solution of part (a). Calculate the new pH.

17.56 Acetic acid is a convenient buffer system in the slightly acidic range. (a) How many grams of sodium acetate must be added to 0.750 L of 0.100 M solution of acetic acid to make a buffer of pH = 5.00? (b) How many moles of acid are required to change the pH of this solution by 0.05 pH units? (c) Suppose 5.0 mL of 12.0 M HCl solution is added to 250 mL of the solution of part (a). Calculate the new pH.

17.57 The solubility of calcium phosphate, $Ca_3(PO_4)_2$, in water is 3.5×10^{-5} g/L. Use this information to calculate K_{sp} for this salt.

17.58 The solubility of calcium arsenate, $Ca_3(AsO_4)_2$, in water is 0.036 g/L. Use this information to calculate K_{sp} for this salt.

17.59 Suppose you titrate 0.300 L of a 0.200 M solution of sodium formate with 6.0 M HCl. (K_a for formic acid is 1.8×10^{-4}.) (a) What is the pH of the solution before beginning the titration? (b) What is the pH of the solution halfway through the titration? (c) What is the pH at the stoichiometric point? (d) What is a suitable indicator for this titration?

17.60 Suppose you titrate 0.200 L of a 0.150 M solution of $NaNO_2$ with 6.0 M HCl. (K_a for HNO_2 is 5.6×10^{-4}.) (a) What is the pH of the solution before beginning the titration? (b) What is the pH of the solution halfway through the titration? (c) What is the pH at the stoichiometric point? (d) What is a suitable indicator for this titration?

17.61 One way to remove heavy metal ions from water is by treatment with sodium carbonate. Calculate the mass of the precipitate that will form and the concentration of Mn^{2+} remaining in solution after 0.750 L of a solution that contains Mn^{2+} at 2.50×10^{-2} M is mixed with 0.150 L of 0.500 M Na_2CO_3 solution. (K_{sp} for $MnCO_3$ is 2.2×10^{-11}.)

17.62 Calculate the mass of the precipitate that will form and the concentration of Ag^+ remaining in solution after 0.250 L of 0.200 M $AgNO_3$ solution is mixed with 0.350 L of 0.300 M Na_2CO_3 solution. (K_{sp} for Ag_2CO_3 is 8.46×10^{-12}.)

17.63 If 0.10 g of $Zn(OH)_2$ is added to 1.00 L of water, what mass will dissolve? (See Appendix E for K_{sp} values.)

17.64 If 0.025 g of $Fe(OH)_3$ is added to 1.50 L of water, what mass will dissolve? (See Appendix E for K_{sp} values.)

17.65 In a biochemistry laboratory, you are asked to prepare a buffer solution to be used as a solvent for isolation of an enzyme. On the shelf you find the following solutions, all 1.00 M: formic acid ($K_a = 1.8 \times 10^{-4}$), acetic acid ($K_a = 1.8 \times 10^{-5}$), benzoic acid ($K_a = 6.3 \times 10^{-5}$), sodium benzoate ($NaC_6H_5CO_2$), sodium formate ($NaHCO_2$), and sodium acetate ($NaCH_3CO_2$). Describe how you would prepare 1.0 L of a pH = 4.80 buffer solution whose total molarity is 0.35 M.

17.66 You are doing undergraduate research for a biology professor. Your first assignment is to prepare a pH = 7.50 phosphate buffer solution to be used in the isolation of DNA from a cell culture. The buffer must have a total concentration of 0.500 M. On the shelf you find the following chemicals: NaOH; concentrated HCl (12.0 M); concentrated H_3PO_4 (14.7 M); KH_2PO_4 (MM = 136.1 g/mol); and K_2HPO_4. Write a quantitative detailed set of instructions that describe how you would prepare 1.5 L of the buffer solution.

More Challenging Problems

17.67 At 25 °C, $[Mg^{2+}] = 1.14 \times 10^{-3}$ M in a saturated aqueous solution of MgF_2. (a) Write the equilibrium reaction and the equilibrium constant expression. (b) Use the concentration of the saturated solution to compute K_{sp} for MgF_2 at 25 °C. (c) Using standard thermodynamic data from Appendix D, calculate K_{eq} at 100.0 °C.

17.68 Using the appropriate K_{sp} values from Appendix E, find the concentrations of all ions in the solution at equilibrium after 0.500 L of 0.300 M aqueous $Cu(NO_3)_2$ solution is mixed with 0.500 L of 0.400 M aqueous KOH solution.

17.69 Calculate the concentration of Hg^{2+} cations in a saturated solution of HgS, $K_{sp} = 4.0 \times 10^{-53}$. What volume of saturated solution can be expected to contain a single Hg^{2+} cation?

17.70 A biochemist wants to use x-ray diffraction to determine the structure of the protein that causes diphtheria. The biochemist must isolate crystals of the protein, and these crystals must be grown in carefully buffered solutions. Write instructions for preparing 1.0 L of a buffer that holds the pH of the protein solution at 5.20. Make the solution 0.50 M in an appropriate acid, and decide what else must be added to complete the preparation.

17.71 Phosphate ions are a major pollutant of water supplies. They can be removed by precipitation using solutions of Ca^{2+} ions because the K_{sp} of calcium phosphate is 2.0×10^{-33}. Suppose that 3.00×10^3 L of waste water containing PO_4^{3-} at 2.2×10^{-3} M is treated by adding 120 moles of solid $CaCl_2$ (which dissolves completely). (a) What is the concentration of phosphate ions after treatment? (b) What mass of calcium phosphate precipitates?

17.72 The pH of an acetic acid–acetate buffer solution is 4.27. (a) Calculate the $HA:A^-$ concentration ratio. (b) Draw a molecular picture that shows a small region of the buffer solution. (You may omit spectator ions and water molecules.) Use the following symbols:

 = Acetic acid   = Acetate

17.73 Strontium iodate, $Sr(IO_3)_2$, is considerably more soluble in hot than in cold water. At 25 °C, 0.030 g of this compound dissolves in 0.100 L of water, but 0.80 g dissolves at 100.0 °C. Calculate K_{sp} and $\Delta G°$ at each temperature.

17.74 The titration of 25.0 mL of a mixture of hydrochloric acid and formic acid with 0.578 M NaOH gave the titration curve below. What are molarities of HCl and HCO_2H in the solution?

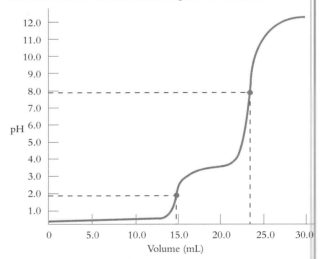

17.75 Limestone caverns are formed by the reaction of H_2O and CO_2 with natural deposits of calcium carbonate:

$$CaCO_3(s) + CO_2(g) + H_2O(l) \rightleftharpoons Ca^{2+}(aq) + 2\ HCO_3^-(aq)$$

$$K_{eq} = 1.56 \times 10^{-8}$$

The partial pressure of CO_2 in the atmosphere is 3.2×10^{-4} atm. What is the equilibrium concentration of calcium ions in ground water?

17.76 The solubility of Ni(II) hydroxide can be increased by adding either $NH_3(aq)$ or HCl(aq). Explain, including balanced equations. Will $NH_3(aq)$ or HCl(aq) increase, decrease or have no effect on the solubility of Ni(II) chloride? Explain your answer, including balanced equations.

17.77 Divide the silver salts listed in Appendix E into two sets: those that are more soluble in acidic solution than in pure water and those whose solubility is independent of pH.

17.78 One of the most common buffers used in protein chemistry is a weak base called TRIS: (The nitrogen atom of the amino group is the basic portion of the molecule.)

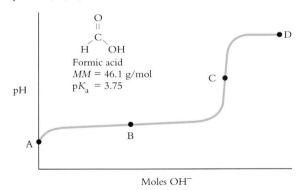

(a) Write a balanced equation that shows how TRIS acts as a weak base in water. (b) Buffer solutions are prepared from TRIS by adding enough 12 M HCl to produce concentrations appropriate for the desired pH. Write a balanced equation that shows what happens when 12 M HCl is added to an aqueous solution of TRIS. (c) Write a balanced equation that shows what happens when hydroxide ions are added to a TRIS buffer solution. (d) Write a balanced equation that shows what happens when hydronium ions are added to a TRIS buffer solution.

17.79 A biochemist prepares a buffer solution by adding enough TRIS (See Problem 17.78) and 12 M HCl to give 1.0 L of a buffer solution whose concentrations are [TRIS] = 0.30 M and [TRISH$^+$] = 0.60 M. (pK_b of TRIS is 5.91.) (a) Calculate the pH of the buffer solution. (b) Suppose 5.0 mL of 12 M HCl is added to 1.0 L of the buffer solution. Calculate the new pH of the solution.

17.80 A technician accidentally pours 35 mL of 12 M HCl into the 1.0 L of buffer solution freshly prepared as described in Problem 17.79. (a) Do a calculation to determine whether the buffer has been ruined. (b) Is it possible to bring the buffer solution back to the original pH calculated in Problem 17.79? If so, what reagent, and how much, must be added to restore the buffer?

17.81 The graph shows the titration curve for a solution of formic acid, which is a principal component in the venom of stinging ants. Identify the major species in the solution at equilibrium, identify the dominant equilibrium, and write the equilibrium constant expression for points A, B, C, and D on the curve.

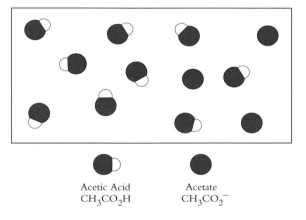

17.82 Quinine, an alkaloid derived from a tree that grows in tropical rain forests, is used in the treatment of malaria. Like all alkaloids, quinine is a sparingly soluble weak base: 1.00 g of quinine will dissolve in 1.90×10^2 L of water. (a) What is the pH of a saturated solution of quinine? (b) A 100.0-mL sample of saturated quinine is titrated with 0.0100 M HCl solution. What is the pH at the stoichiometric point of the titration?

Quinine
$C_{20}H_{24}N_2O_2$
$pK_b = 5.1$
$MM = 324$ g/mol

17.83 The following figure represents a small portion of an acetic acid—acetate buffer. Solvent molecules of water have been omitted for clarity. The pK_a of acetic acid is 4.75.

Acetic Acid
CH_3CO_2H

Acetate
$CH_3CO_2^-$

(a) What is the pH of the buffer solution? (b) Redraw the original figure to show the equilibrium condition that is established when one hydroxide ion enters the region. (c) Redraw the original figure to show the equilibrium condition that is established when one ion of HSO_4^- enters the region (pK_a of $HSO_4^- = 2.0$).

Group Study Problems

17.84 Virtually all investigations in cell biology and biochemistry must be carried out in buffered aqueous solutions. Imagine that you are studying an enzyme that is active only between pH = 7.1 and 7.4 and that you need to prepare 1.5 L of a phosphate buffer at pH = 7.25 whose total phosphate concentration is 0.085 M. On the laboratory shelves, you find the following reagents: concentrated

H_3PO_4 (14.75 M), solid KH_2PO_4, and solid K_2HPO_4. (a) Which of these will you use to prepare the buffer solution? (b) What quantities of each will you use? (c) The enzyme generates H_3O^+ as it functions. If you are running an experiment in 250 mL of the buffer, how many moles of H_3O^+ can the enzyme generate before it loses its activity?

17.85 The figure below shows the titration curves for 50.0-mL samples of weak bases A, B, and C. The titrant was 0.10 M HCl. (a) Which is the strongest base? (b) Which base has the largest pK_b? (c) What are the initial concentrations of the three bases? (d) What are the approximate pK_a values of the conjugate acids of the three bases? (e) Which of the bases can be titrated quantitatively using indicators? Explain your answers, and identify an appropriate indicator for each base that can be titrated successfully.

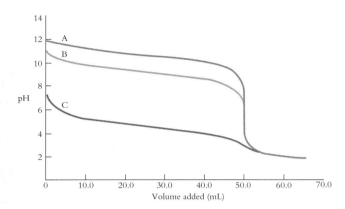

17.86 The K_{sp} of $Fe(OH)_3$ is 2.8×10^{-39}, and the K_{sp} of $Zn(OH)_2$ is 3×10^{-17}. (a) Can Fe^{3+} be separated from Zn^{2+} by the addition

of an NaOH solution to an acidic solution that contains 0.300 M Zn^{2+} and 0.100 M Fe^{3+}? (b) At what hydroxide ion concentration will the second cation precipitate? (c) What is the concentration of the other cation at that hydroxide ion concentration?

17.87 Consider the titration of a diprotic acid, H_2A. The graph shows the titration curve for this acid being titrated with strong base. Identify the major species present in the solution at points 1–6 on the titration curve.

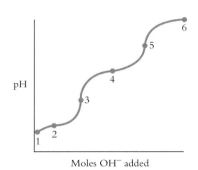

Moles OH$^-$ added

17.88 Seawater is approximately 0.5 M each in Na$^+$ and Cl$^-$ ions. By evaporation, NaCl can be precipitated from this solution. If 1.00×10^2 L of seawater is evaporated, at what volume will the first solid NaCl appear? ($K_{sp} = 6.2$)

Answers to Section Exercises

17.1.1 (a) Not suitable; (b) HNO_2 and NO_2^-; (c) not suitable; (d) HPO_4^{2-} and PO_4^{3-}; and (e) not suitable

17.1.2 $NH_4^+ + OH^- \rightarrow NH_3 + H_2O$ and $NH_3 + H_3O^+ \rightarrow NH_4^+ + H_2O$

17.1.3 (a) 9.00; (b) 9.24; and (c) 12.45

17.2.1 Buffering capacity is destroyed, final pH = 1.6

17.2.2 3.8 g sodium formate and 72 mL of 0.500 M HCl solution

17.2.3 Dissolve 9.0 g NaOH and 35 g NaHCO$_3$ in enough water to make 2.5 L of solution.

17.3.1

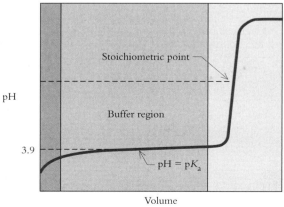

17.3.2 pH = 8.41 at stoichiometric point and 3.4 after 3.0 mL of KOH is added.

17.3.3 Phenol red would be suitable; it would be red at the stoichiometric point.

17.4.1 1.7×10^{-4} M

17.4.2 1.5×10^{-9} M. NaCl does not precipitate.

17.4.3 $Al(OH)_3 (s) + 3 H_3O^+(aq) \rightleftarrows 3 Al^{3+}(aq) + 6 H_2O (l)$, $K_{eq} = 4.6 \times 10^9$

17.5.1 $[Ni^{2+}] = 5.6 \times 10^{-16}$ M; $[en] = 3.0 \times 10^{-2}$ M; $[Ni(en)_3^{2+}] = 6.5 \times 10^{-3}$ M

17.5.2 $Zn^{2+}(aq) + NH_3(aq) \rightleftarrows [Zn(NH_3)]^{2+}(aq);$

$$[Zn(NH_3)]^{2+}(aq) + NH_3(aq) \rightleftarrows [Zn(NH_3)_2]^{2+}(aq);$$

$$[Zn(NH_3)_2]^{2+}(aq) + NH_3(aq) \rightleftarrows [Zn(NH_3)_3]^{2+}(aq);$$

$$[Zn(NH_3)_3]^{2+}(aq) + NH_3(aq) \rightleftarrows [Zn(NH_3)_4]^{2+}(aq)$$

17.5.3 1.8 g of AgCl dissolves.

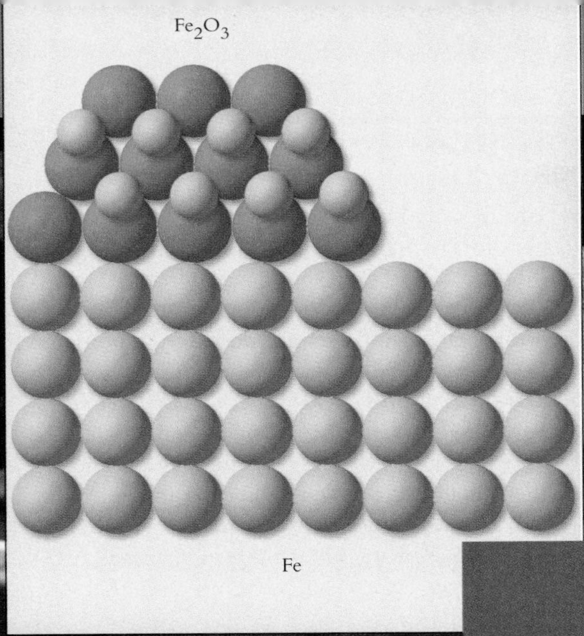

Fe₂O₃

Fe

18

Electron Transfer Reactions

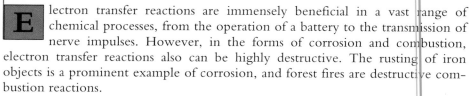

INTRODUCTION: CORROSION

E lectron transfer reactions are immensely beneficial in a vast range of chemical processes, from the operation of a battery to the transmission of nerve impulses. However, in the forms of corrosion and combustion, electron transfer reactions also can be highly destructive. The rusting of iron objects is a prominent example of corrosion, and forest fires are destructive combustion reactions.

Metallic iron is made up of neutral iron atoms held together by shared electrons (see Section 9.6). When iron rusts, electron transfer reactions take place. Iron atoms lose three electrons each, forming Fe^{3+} cations. At the same time, molecular oxygen gains electrons from the metal, each molecule adding four electrons to form a pair of oxide anions. As our inset figure shows, the Fe^{3+} cations combine with O^{2-} anions to form insoluble Fe_2O_3, rust. Over time, the surface of an iron object becomes pitted from loss of iron atoms and covered with flaky iron(III) oxide.

Corrosion of iron has a serious economic impact. Approximately 20% of the iron and steel produced in the United States is used to replace rusted metal, and the cost of replacing corroded materials is several billion dollars per year. Consequently, the prevention of corrosion is a major business.

Corrosion can be prevented by making the rate of reaction negligible. This can be accomplished by covering the metal surface with a protective coating. Paint is a liquid mixture that sets to form a solid film on a surface. Structures such as San Francisco's Golden Gate Bridge are always being painted, ensuring an unbroken layer to protect steel cables and towers from corrosion.

A forest fire is another example of a catastrophic electron transfer reaction. Wood is primarily cellulose, a carbohydrate polymer that reacts extremely slowly with molecular oxygen at room temperature. However, the electron transfer reaction between molecular oxygen and carbohydrates to generate CO_2 and H_2O is highly exothermic. Once ignited by a heat source such as a lightning bolt or a smoldering cigarette, the reaction generates enough heat to become self-sustaining and rapid.

Electron transfer reactions occur all around us. Objects made of iron become coated with rust when they are exposed to moist air. Animals obtain energy from the reaction of carbohydrates with oxygen to form carbon dioxide and water. Turning on a flashlight generates a current of electricity from a chemical reaction in the batteries. In an aluminum refinery, huge quantities of electricity drive the conversion of aluminum oxide into aluminum metal. These different chemical processes share one common feature: Each is an oxidation-reduction reaction, commonly called a redox reaction, in which electrons are transferred from one chemical species to another.

Redox reactions can proceed by direct transfer of electrons between chemical species. Examples include the rusting of iron and the metabolic breakdown of carbohydrates. Redox processes also can take place by indirect electron transfer from one chemical species to another via an electrical circuit. When a chemical reaction is coupled with electron flow through a circuit, the process is electrochemical. Flashlight batteries and aluminum smelters involve electrochemical processes.

We begin this chapter with a discussion of the principles of redox reactions, including redox stoichiometry. Then we introduce the principles of electrochemistry. Practical examples of redox chemistry, including corrosion, batteries, and metallurgy, appear throughout the chapter.

18.1 RECOGNIZING REDOX REACTIONS

Recall from Section 4.7 that in a redox reaction, one species loses electrons while another species gains electrons.

/// *Oxidation is the loss of electrons from a substance.*
/// *Reduction is the gain of electrons by a substance.*

A chemical species that loses electrons is oxidized, and a chemical species that gains electrons is reduced.

Remember also that electrons are conserved in all chemical processes. That is, electrons can be *transferred* from one species to another, but they are neither created nor destroyed. Thus it is impossible for oxidation to occur without reduction also occurring, because when one species gains electrons, another species must lose electrons.

/// *Oxidation and reduction always occur together.*

A simple example of a redox reaction is the violent reaction of magnesium metal with molecular oxygen to generate magnesium oxide, shown in Figure 18-1:

$$2\,Mg(s) + O_2(g) \longrightarrow 2\,MgO(s)$$

In magnesium metal, each atom is neutral; the electrons exactly counterbalance the nuclear charge. When a magnesium atom reacts with an oxygen molecule, the metal atom loses its valence electrons to produce a $+2$ cation:

$$Mg \longrightarrow Mg^{2+} + 2\,e^-$$

Magnesium atoms are oxidized in this reaction, so some other species must be reduced. Oxygen molecules accept the electrons lost by the magnesium atoms. Each oxygen *atom* gains two electrons from a magnesium atom, generating two oxide anions:

$$O_2 + 4\,e^- \longrightarrow 2\,O^{2-}$$

Magnesium cations and oxide anions attract each other strongly, forming the ionic solid, MgO. Notice that in the balanced redox reaction there is no net change in the number of electrons; two Mg atoms lose four electrons and one O_2 molecule gains four electrons.

The species that loses electrons in a redox process causes the reduction of some other species. Consequently, the species that loses electrons is called a **reducing agent.** Magnesium metal acts as a reducing agent in the presence of atmospheric oxygen. Similarly, the species that gains electrons causes the oxidation of some other chemical species and is called an **oxidizing agent.** Molecular oxygen acts as an oxidizing agent in the presence of magnesium metal. Every redox reaction has both an

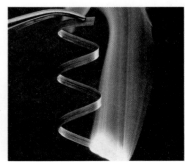

Figure 18-1
When set ablaze with a match, magnesium burns with an intense flame. The reaction involves electron transfer from the metal to oxygen.

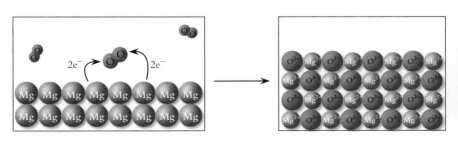

−2(2e⁻)
Reducing agent:
Loses electrons and is oxidized
2 Mg + O2 → 2 MgO
Oxidizing agent:
Gains electrons and is reduced
+2(2e⁻)

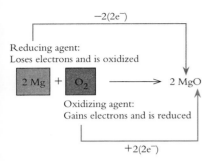

Figure 18-2
Schematic view of the processes involved in redox reactions.

oxidizing agent and a reducing agent. Figure 18-2 summarizes these features using the magnesium-oxygen reaction as an example.

Electron transfer is often difficult to recognize, because electrons normally do not appear in chemical formulas and equations. The formula of a molecule or ion specifies the atomic composition and the net charge but does not account for the electrons. Likewise, electrons never appear in a balanced chemical equation. The electrons are "hidden" in the chemical formulas of the species involved in the reaction. For example, the formulas Mg, O_2, and MgO do not suggest immediately that oxygen is reduced to oxide anions and magnesium oxidized to magnesium cations when magnesium burns. To recognize when electron transfer occurs, we need a method to keep track of electrons in chemical species.

Oxidation Numbers

The difficulty in recognizing redox reactions is illustrated by two of the reactions that occur during the extraction of iron from iron ores:

$$FeO(s) + CO(g) \xrightarrow{\text{Heat}} Fe(l) + CO_2(g)$$

$$FeCO_3(s) \xrightarrow{\text{Heat}} FeO(s) + CO_2(g)$$

CHAPTER 19 →
The recovery of iron and other metals from their ores is described in Chapter 19.

One of these is a redox reaction, but the other is not; how can we determine which one involves electron transfer?

To determine whether or not electrons are transferred in chemical reactions, chemists use a procedure that assigns an **oxidation number** (also known as *oxidation state*) to each atom in each chemical species. In a redox reaction, electron transfer causes some of the atoms to change their oxidation numbers. Thus redox reactions can be identified by noting changes in oxidation numbers.

In assigning oxidation numbers, electrons always are assigned to specific atoms in a compound. The electrons in a bond between atoms of two different elements, such as FeO or CO, are polarized toward the more electronegative atom (see Figure 18-3 for electronegativities). For oxidation number purposes, we imagine that these electrons are transferred *completely* to the more electronegative atom.

The oxidation number of an atom is the charge it would have if the compound were composed of ions. For example, the bonding in FeO is highly polar, as indicated by the electronegativities of Fe (1.8) and O (3.5). If the Fe—O bonds

Figure 18-3
Electronegativities of the elements.

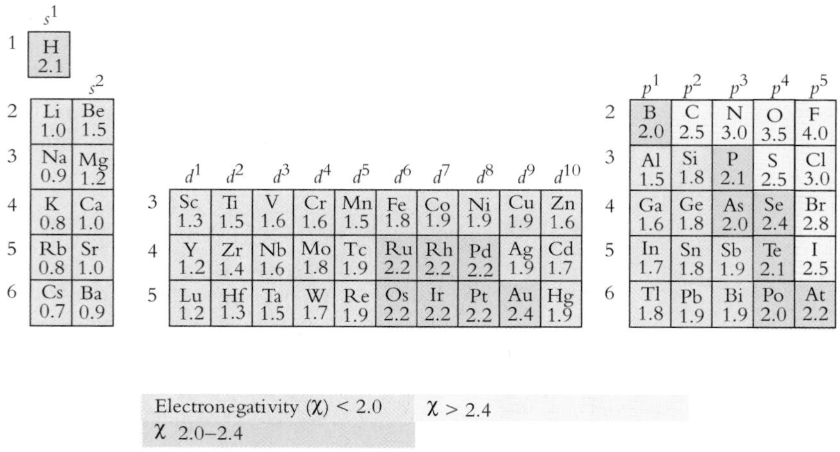

s1 s2
1 H 2.1
2 Li 1.0 Be 1.5
3 Na 0.9 Mg 1.2
4 K 0.8 Ca 1.0
5 Rb 0.8 Sr 1.0
6 Cs 0.7 Ba 0.9
d1 d2 d3 d4 d5 d6 d7 d8 d9 d10
3 Sc 1.3 Ti 1.5 V 1.6 Cr 1.6 Mn 1.5 Fe 1.8 Co 1.9 Ni 1.9 Cu 1.9 Zn 1.6
4 Y 1.2 Zr 1.4 Nb 1.6 Mo 1.8 Tc 1.9 Ru 2.2 Rh 2.2 Pd 2.2 Ag 1.9 Cd 1.7
5 Lu 1.2 Hf 1.3 Ta 1.5 W 1.7 Re 1.9 Os 2.2 Ir 2.2 Pt 2.2 Au 2.4 Hg 1.9
p1 p2 p3 p4 p5
2 B 2.0 C 2.5 N 3.0 O 3.5 F 4.0
3 Al 1.5 Si 1.8 P 2.1 S 2.5 Cl 3.0
4 Ga 1.6 Ge 1.8 As 2.0 Se 2.4 Br 2.8
5 In 1.7 Sn 1.8 Sb 1.9 Te 2.1 I 2.5
6 Tl 1.8 Pb 1.9 Bi 1.9 Po 2.0 At 2.2
Electronegativity (χ) < 2.0
χ 2.0–2.4
χ > 2.4

were fully ionic, FeO would consist of Fe^{2+} cations and O^{2-} anions. Thus we assign iron an oxidation number of $+2$ and oxygen an oxidation number of -2. The bonding in CO and CO_2 is much less polar, as shown by the electronegativities of C (2.5) and O (3.5). Nevertheless, if the C—O bonds were fully ionic, CO would consist of C^{2+} cations and O^{2-} anions, and CO_2 would contain C^{4+} cations and O^{2-} anions. Thus we assign O an oxidation number of -2 in both these compounds, whereas C has an oxidation number of $+2$ in CO and $+4$ in CO_2.

It is always possible to determine oxidation numbers starting from electronegativity differences. A more systematic method for determining oxidation numbers uses the following four guidelines:

1. Treat ions separately.
2. The sum of all the oxidation numbers must equal the charge of the species.
3. Hydrogen usually has an oxidation number of $+1$.
4. The most electronegative atom in a polyatomic species has a negative oxidation number equal to the number of electrons needed to complete its valence octet.

Some examples illustrate these guidelines.

1. In an ionic compound, each ion is an individual chemical species with its own set of oxidation numbers. For example, we treat ammonium nitrate, NH_4NO_3, as NH_4^+ cations and NO_3^- anions.
2. The sum of the oxidation numbers for NO_3^- must be -1, and the sum for NH_4^+ must be $+1$. For a neutral molecule such as NH_3, the sum of the oxidation numbers must be zero. By this guideline the oxidation numbers of atoms in pure neutral elements are zero. As examples, each atom in H_2, Fe, O_2, O_3, P_4, and C_{60} has an oxidation number of zero. Also, the oxidation number of any monatomic ion equals its charge: K^+ has a $+1$ oxidation number, Al^{3+} is $+3$, and Cl^- is -1.
3. Hydrogen has a $+1$ oxidation number in most of its compounds. In NH_3, for example, the hydrogen atoms are assigned an oxidation number of $+1$, leaving a -3 value for nitrogen. There is one common exception to this guideline. In the metal hydrides such as NaH and CaH_2, hydrogen is assigned an oxidation number of -1. These assignments are consistent with electronegativities: Na (0.9) and Ca (1.0) $<$ H (2.1) $<$ C (2.5) and Cl (3.0).
4. Nitrogen, which is in Group 15 of the periodic table, has five valence electrons and needs three more to complete its valence octet. Thus in the cyanide anion, CN^-, the more electronegative atom is assigned an oxidation number of -3. In the nitrate ion, NO_3^-, oxygen is the more electronegative atom, so each oxygen atom is assigned an oxidation number of -2 (Group 16, six valence electrons). Thus nitrogen must have a $+5$ oxidation number in order to satisfy Guideline 2. The only important exceptions to Guideline 4 are the peroxides, species that contain an O—O bond. Oxygen has a -1 oxidation number in the peroxides. Examples include hydrogen peroxide, H_2O_2 (H is $+1$ by Guideline 3, so O is -1 by Guideline 2), and sodium peroxide, Na_2O_2 (treated as Na^+ and O_2^{2-} by Guideline 1, making O -1 by Guideline 2).

Oxidation numbers do not indicate how charge is *actually* distributed in a molecule. Remember that most bonds are polar, meaning that the electrons in a bond are skewed toward the more electronegative atom. In contrast, we determine oxidation numbers by *assuming* that all bonds are ionic. As a result, oxidation numbers

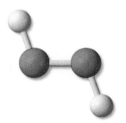

Hydrogen peroxide

are useful for assessing whether or not electron exchange occurs during a reaction, but they do not represent actual charge distributions. A good example is CN^-, where we assign oxidation numbers of -3 to N and $+2$ to C. In reality, the ion carries -1 net charge, and both atoms have slightly negative charges.

Examples 18-1 and 18-2 illustrate the application of these guidelines.

| Example 18-1 | Assigning Oxidation Numbers |

Assign oxidation numbers to all atoms in these two reactions involving iron:

$$FeO(s) + CO(g) \xrightarrow{Heat} Fe(l) + CO_2(g)$$

$$FeCO_3(s) \xrightarrow{Heat} FeO(s) + CO_2(g)$$

Strategy: Consider each substance separately, and follow the four guidelines.

Solution:

FeO: Oxygen has an oxidation number of -2 (Guideline 4). For the sum of the oxidation numbers to be zero (Guideline 2), iron must be $+2$.

CO: Oxygen has a value of -2 (Guideline 4). For the sum of the oxidation numbers to be zero (Guideline 2), carbon must have an oxidation number of $+2$.

Fe: This is a neutral element, so iron has an oxidation number of zero (Guideline 2).

CO_2: Each oxygen is -2, so carbon must be $+4$.

$FeCO_3$: This is an ionic compound containing the carbonate polyatomic anion, CO_3^{2-}, and the Fe^{2+} cation (Guideline 1).

In the Fe^{2+} cation, iron has an oxidation number equal to the charge, $+2$ (Guideline 2).

In the carbonate anion, each oxygen atom is -2, for a total of -6. The oxidation numbers must add up to the net charge on the anion (-2), so the carbon atom must have an oxidation number of $+4$.

Here are the reactions again, showing all oxidation numbers:

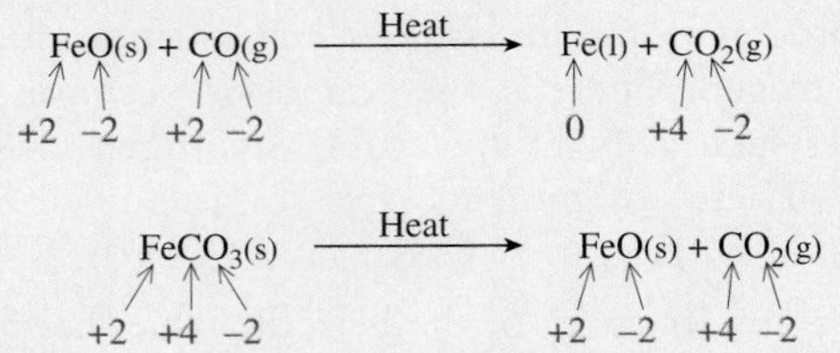

$$FeO(s) + CO(g) \xrightarrow{Heat} Fe(l) + CO_2(g)$$
$$\quad +2 \;\; -2 \;\;\; +2 \;\; -2 \qquad\qquad 0 \quad +4 \;\; -2$$

$$FeCO_3(s) \xrightarrow{Heat} FeO(s) + CO_2(g)$$
$$\quad +2 \;\; +4 \;\; -2 \qquad\qquad +2 \;\; -2 \;\; +4 \;\; -2$$

The first reaction is a redox reaction because carbon's oxidation number increases (oxidation), and iron's oxidation number decreases (reduction).

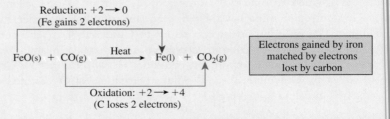

In the second reaction, none of the atoms changes its oxidation number, so this is not a redox process.

SECTIONS 8.2 and 8.3 →
Sections 8.2 and 8.3 describe how to recognize ionic compounds.

Examples of Oxidation Numbers

Example 18-2

Assign oxidation numbers to each element in the following substances: H_3PO_4, O_3, NaOCl, ClF_3, and KH.

Strategy: Apply the guidelines for oxidation numbers to each substance.

Solution:

H_3PO_4: Phosphoric acid is a covalent compound with a net charge of zero. Each hydrogen atom has an oxidation number of $+1$ (Guideline 3), and each oxygen has an oxidation number of -2 (Guideline 4). Now add the contributions from these atoms: $3(+1) + 4(-2) = -5$. For the oxidation numbers to sum to zero (Guideline 2), the phosphorus atom of phosphoric acid must have an oxidation number of $+5$.

O_3: Ozone is a neutral form of the element, so the oxidation number of its oxygen atoms is zero (Guideline 2).

NaOCl: Sodium hypochlorite is an ionic compound in which Na^+ is $+1$ and OCl^- is -1 (Guideline 1). Sodium cation has an oxidation number of $+1$ (Guideline 2). In the anion, oxygen is -2 (Guideline 4), so chlorine must be $+1$ for the sum of the oxidation numbers to match the -1 charge of the hypochlorite anion (Guideline 2).

ClF_3: Fluorine is the more electronegative atom, so each fluorine atom has an oxidation number of -1 (Guideline 4). For the sum of the oxidation numbers to be zero (Guideline 2), chlorine must be $+3$.

KH: In most cases hydrogen has an oxidation number of $+1$. In compounds with metals, however, the metal has a positive oxidation number and the oxidation number of hydrogen is -1. This requires potassium to be $+1$. The assignments are consistent with the electronegativites of the two elements: $\chi_K = 0.8$, $\chi_H = 2.1$.

Identifying a process as redox can have practical applications. Our Chemistry and the Environment Box describes a detoxification problem that was solved by applying redox chemistry.

18.1.1 Determine the oxidation numbers of all atoms in each of the following substances: (a) $KMnO_4$ (potassium permanganate, a powerful oxidizing agent found in many general chemistry laboratories); $SiCl_4$ (used as a precursor to ultrapure silicon for computer chips); (c) Na_2SO_3 (sodium sulfite, a food preservative); (d) S_8, the most stable form of sulfur; (e) $LiAlH_4$ (a reducing agent in organic chemistry. Hint: This is an ionic compound with a lithium cation and a polyatomic anion.).

18.1.2 Determine which of the following reactions are redox. For each redox reaction, state which species is oxidized and which is reduced.
(a) $CH_4(g) + 2\,O_2(g) \rightarrow CO_2(g) + 2\,H_2O(l)$
(b) $HCl(g) + NH_3(aq) \rightarrow NH_4Cl(aq)$
(c) $SiCl_4(l) + 2\,Mg(s) \rightarrow Si(s) + 2\,MgCl_2(s)$
(d) $PbS(s) + 4\,H_2O_2(aq) \rightarrow PbSO_4(s) + 4\,H_2O(l)$
(e) $3\,O_2(g) \rightarrow 2\,O_3(g)$

Box 18-1   Chemistry and the Environment: Purifying Groundwater

Industrial operations use large quantities of chemicals that pose contamination threats to groundwater. In the recent past, industrial wastes such as chlorinated hydrocarbons (used as solvents and for cleaning) and chromium-containing solutions (used in metal plating) were routinely dumped into storage tanks from which they eventually leaked, leading to groundwater contamination. Heightened public awareness of the health and environmental hazards of such practices resulted in laws restricting the dumping of chemical waste, but preventing groundwater contamination remains a major problem. Many former dump sites continue to leak hazardous wastes, and industrial processes continue to generate chemical wastes whose safe disposal is costly and difficult.

The ideal disposal method is a chemical treatment that can convert hazardous waste into environmentally benign materials. For example, trichloroethylene ($Cl_2C{=}CHCl$) is highly toxic to aquatic life, but this compound can be made nontoxic by chemical treatment that converts its chlorine atoms into chloride anions. Similarly, the chromium-containing waste from electroplating operations contains highly toxic CrO_4^{2-} anions, but a chemical treatment that converts CrO_4^{2-} into Cr^{3+} causes the chromium to precipitate from the solution as insoluble $Cr(OH)_3$. This removal of chromium detoxifies the water.

Both these conversion processes involve the addition of electrons to the toxic substances. The trichloroethylene molecule is electrically neutral and must gain electrons in reactions that generate negatively charged chloride anions.

The chromium atom in CrO_4^{2-} has an oxidation number of $+6$, so it must gain three electrons to be converted into Cr^{3+}. Because electrons are conserved, chemical detoxification of water containing either of these contaminants requires a substance that easily donates electrons. A suitable electron donor for groundwater cleanup must also be relatively inexpensive and must not itself be a source of toxicity.

These are stringent requirements; yet a very common material has recently been discovered to do this job spectacularly well: iron filings. Metallic iron contains neutral iron atoms that readily donate electrons to become Fe^{2+} or Fe^{3+} cations, neither of which is significantly toxic. Thus when iron reacts with trichloroethylene or chromate, all the products are relatively nontoxic.

How does this process work in practice? As the figure indicates, it turns out to be amazingly simple, taking advantage of the same natural water flow that leads to contamination. First, the source of contamination and direction of flow of the groundwater are determined. A pit is then dug downstream from the contamination source. This pit is filled with a mixture of sand and iron filings. The contaminant reacts with iron filings as the groundwater flows through the pit, and the water emerging downstream from the pit is contaminant-free. In December, 1994, the first full-scale operation of this type was installed in a trichloroethylene-contaminated site in Sunnyvale, California. At a cost of $1.5 million, a site was recovered that had been predicted to remain too contaminated for human use for up to 30 years.

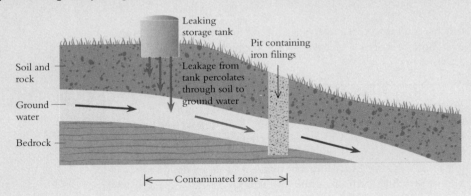

18.1.3 Nitrogen can have many different oxidation states, as illustrated by the following examples. Assign oxidation states to the nitrogen atoms in each species. (a) NH_3; (b) N_2; (c) N_2H_4; (d) HNO_3; (e) $NaNO_2$; (f) NO; (g) NO_2.

18.2 BALANCING REDOX REACTIONS

Some redox reactions have relatively simple stoichiometry and can be balanced by inspection. Others are much more complicated. Redox reactions involve the transfer of electrons from one species to another, so electrical charges must be considered explicitly when balancing redox equations.

The key to balancing complicated redox equations is to balance electrons as well as atoms. Because electrons do not appear in chemical formulas or balanced net reactions, the number of electrons transferred in a redox reaction often is not obvious. To balance complicated redox reactions, therefore, we need a procedure that shows the electrons involved in the oxidation and the reduction. The procedure described here separates redox reactions into two parts, an oxidation and a reduction. Each part is a **half-reaction** that describes half of the overall redox process.

Half-Reactions

In any redox reaction, some species are oxidized and others are reduced. Examining the oxidation and reduction parts of a redox reaction separately reveals the number of electrons transferred. This is most easily accomplished by dividing the redox reaction into two half-reactions, one describing the oxidation and the other describing the reduction. In half-reactions, electrons appear as reactants or products.

A starting material loses electrons in an oxidation, so electrons appear among the *products* of the oxidation half-reaction. A starting material gains electrons in a reduction, so electrons appear among the *reactants* of the reduction half-reaction. The reaction of magnesium metal with hydronium ions to produce hydrogen gas provides an example.

$$Mg(s) + 2\ H_3O^+(aq) \longrightarrow Mg^{2+}(aq) + H_2(g) + 2\ H_2O(l)$$

Here are the half-reactions for this redox reaction:

Oxidation: $\qquad\qquad Mg(s) \longrightarrow Mg^{2+}(aq) + 2\ e^-$

Reduction: $\quad 2\ H_3O^+(aq) + 2\ e^- \longrightarrow H_2(g) + 2\ H_2O(l)$

The half-reactions show that two electrons are required to produce each H_2 molecule in the reduction, and that the oxidation of one magnesium atom releases two electrons. In other words, electron transfer is balanced when one magnesium atom reacts for every H_2 molecule that forms. This example reveals an essential feature of redox reactions.

> /// *The number of electrons lost in oxidation must equal the number of electrons gained in reduction.*

The first step in balancing a redox reaction is to divide the unbalanced equation into half-reactions. Identify the participants in each half-reaction by remembering that the elements in each half-reaction eventually must be conserved. That is, any element that appears as a reactant in a half-reaction must also appear among the products. Hydrogen and oxygen frequently appear in both half-reactions, but other elements usually appear on both sides of a single half-reaction. Some reactions do not divide clearly into an oxidation and a reduction, but changes in oxidation numbers reveal how to divide the starting materials into half-reactions. Example 18-3 illustrates the separation process.

← **SECTION 4.7**

We describe the oxidation of metals by H_3O^+ in Section 4.7.

| Example 18-3 | Identifying Half-Reactions |

Separate the following unbalanced redox processes into half-reactions. All occur in aqueous solution:

(a) $Cr_2O_7^{2-}(aq) + Fe^{2+}(aq) \longrightarrow Cr^{3+}(aq) + Fe^{3+}(aq)$

(b) $Cu(s) + NO_3^-(aq) \longrightarrow Cu^{2+}(aq) + NO(g)$

(c) $H_2O_2(aq) + SO_3^{2-}(aq) \longrightarrow SO_4^{2-}(aq)$

(d) $Br_2(l) \longrightarrow BrO_3^-(aq) + Br^-(aq)$

Strategy: A redox equation usually can be separated by placing elements other than hydrogen and oxygen in separate half-reactions. If this does not work, oxidation numbers are used to identify the oxidized and reduced species. Occasionally, an element other than O or H may be involved in both half-reactions.

Solution:

(a) This reaction can be divided by inspection. Iron must appear in one half-reaction and chromium in the other. Oxygen, which is present only in $Cr_2O_7^{2-}$, will be balanced in a later step.

$$Fe^{2+} \longrightarrow Fe^{3+} \qquad Cr_2O_7^{2-} \longrightarrow Cr^{3+}$$

(b) Copper must be present in one half-reaction, and the nitrogen-containing species appear in the other.

$$Cu \longrightarrow Cu^{2+} \qquad NO_3^- \longrightarrow NO$$

(c) In this reaction, sulfur is the only element other than hydrogen and oxygen. Thus sulfur must appear on both sides of one half-reaction. Furthermore, since sulfur appears in only one of the starting materials and in only one of the products, the second half-reaction must contain only hydrogen and oxygen. No product containing hydrogen is given, but water is always present in aqueous solution. The second half-reaction is the conversion of hydrogen peroxide to water:

$$SO_3^{2-} \longrightarrow SO_4^{2-} \qquad H_2O_2 \longrightarrow H_2O$$

Oxidation numbers confirm this second half-reaction. Recall that whereas the oxygen atom in H_2O has an oxidation number of -2, the oxygen atoms in H_2O_2 have oxidation number -1.

(d) How can a reaction that contains just one reactant be divided into half-reactions? The only way is for this reactant to act as *both* the oxidizing agent and the reducing agent. Molecules of Br_2 react with one another; some are reduced and others are oxidized. Therefore Br_2 must appear in *both* half-reactions:

$$Br_2 \longrightarrow BrO_3^- \qquad Br_2 \longrightarrow Br^-$$

Oxidation numbers verify the half-reactions. The oxidation number of the bromine atoms in Br_2 is zero. Among the products, bromine is $+5$ in BrO_3^- and -1 in Br^-. Thus in one half-reaction molecular bromine is oxidized to bromate ions. In the reduction half-reaction molecular bromine is reduced to bromide ions.

Balancing Half-Reactions

After a redox reaction is divided into half-reactions, each half-reaction can be balanced independently. This is done in four steps, which *must* be done *in the order listed:*

a. Balance all elements except oxygen and hydrogen by adjusting stoichiometric coefficients.

b. Balance oxygen by adding H_2O to the side that is deficient in oxygen.

c. Balance hydrogen as follows: If the solution is acidic, add H_3O^+ cations to the side that is deficient in hydrogen and an equal number of H_2O molecules to the other side. If the solution is basic, add H_2O to the side that is deficient in hydrogen and an equal number of OH^- anions to the other side.

d. Balance net charge by adding electrons to the side that is deficient in negative charge.

These four steps result in a balanced half-reaction, in which elements, electrons, and total charge are all conserved. Example 18-4 illustrates the procedure.

Balancing Half-Reactions	**Example 18-4**

Balance the redox equation presented in Example 18-3(a):

$$Cr_2O_7^{2-} + Fe^{2+} \longrightarrow Cr^{3+} + Fe^{3+}$$

Strategy: The half-reactions were identified in Example 18-3. Balance each half-reaction using the four-step procedure.

Solution: Iron half-reaction: $Fe^{2+} \longrightarrow Fe^{3+}$

Steps a, b, c: Iron is already balanced, and neither oxygen nor hydrogen is present.

Step d: The net charge on the left is $+2$, and on the right it is $+3$. The charge on the right needs to be made less positive: To balance the charges, add one electron on the right:

$$Fe^{2+} \longrightarrow Fe^{3+} + 1e^-$$

Now for the chromium half-reaction: $Cr_2O_7^{2-} \longrightarrow Cr^{3+}$

Step a: To balance chromium, give Cr^{3+} a coefficient of 2:

$$Cr_2O_7^{2-} \longrightarrow 2\ Cr^{3+}$$

Step b: There are seven oxygen atoms on the left, so add seven water molecules on the right:

$$Cr_2O_7^{2-} \longrightarrow 2\ Cr^{3+} + 7\ H_2O$$

Step c: There are 14 hydrogen atoms on the right, requiring 14 H_3O^+ ions on the left and 14 additional H_2O molecules on the right:

$$14\ H_3O^+ + Cr_2O_7^{2-} \longrightarrow 2\ Cr^{3+} + 21\ H_2O$$

Step d: The right-hand side has $2(+3) = +6$ charge. The left-hand side has $14(+1) + 1(-2) = +12$ charge. Charge balance requires six electrons on the left:

$$14\ H_3O^+ + Cr_2O_7^{2-} + 6\ e^- \longrightarrow 2\ Cr^{3+} + 21\ H_2O$$

For convenience, we omit the phase designation during the intermediate steps in balancing redox reactions.

Balancing Redox Equations

After oxidation and reduction half-reactions are balanced, they can be combined to give the balanced chemical equation for the overall redox process. Although electrons are reactants in reduction half-reactions and products in oxidation half-reactions, they must cancel in the overall redox equation. To accomplish this, multiply each half-reaction by an appropriate integer that makes the number of electrons in the reduction half-reaction equal to the number of electrons in the oxidation half-reaction. The entire half-reaction must be multiplied by the integer to maintain charge balance.

Example 18-5 illustrates this procedure.

Example 18-5	Combining Half-Reactions

Complete the balancing of the reaction presented in Example 18-3(a).

$$Cr_2O_7^{2-} + Fe^{2+} \longrightarrow Cr^{3+} + Fe^{3+}$$

Strategy: The balanced half-reactions appear in Example 18-4. Balance the overall equation by combining the half-reactions in such a way that electrons cancel.

Solution: The half-reaction for iron contains one electron on the right, whereas the half-reaction for chromium contains six electrons on the left. To combine these half-reactions so that the electrons cancel, multiply the iron half-reaction by 6 and add it to the chromium half-reaction:

$$6\ Fe^{2+} \longrightarrow 6\ Fe^{3+} + 6\ e^-$$
$$\underline{14\ H_3O^+ + Cr_2O_7^{2-} + 6\ e^- \longrightarrow 2\ Cr^{3+} + 21\ H_2O}$$
$$14\ H_3O^+ + Cr_2O_7^{2-} + 6\ Fe^{2+} \longrightarrow 6\ Fe^{3+} + 2\ Cr^{3+} + 21\ H_2O$$

An "inventory" of the reaction verifies that atoms and charge are balanced:

Reactants	Products
42 H	42 H
2 Cr	2 Cr
21 O	21 O
6 Fe	6 Fe
$14(+1) + (-2) + 6(+2) = +24$ charge	$6(+3) + 2(+3) = +24$ charge

Both the elements and the total charge are balanced, so the overall equation is balanced.

When half-reactions are combined, there is often a duplication of some chemical species, particularly H_2O and H_3O^+ or OH^-. The overall equation is "cleaned up" by combining species that appear twice on the same side. Also, when a species appears on both sides of the balanced equation, equal numbers of the species are subtracted from each side.

To summarize, the balancing of a redox equation occurs in several steps that are carried out in sequence.

BALANCING REDOX REACTIONS

START

Step 1   Break the unbalanced equation into half-reactions by inspection.

Step 2   Balance each half-reaction, following the four-step procedure.

a   Balance all elements except oxygen and hydrogen by adjusting stoichiometric coefficients.

b   Balance oxygen by adding H_2O to the side that is deficient in oxygen.

c   Balance hydrogen:
If the solution is neutral or acidic, add H_3O^+ to the side that is deficient in H and H_2O to the other side. If the solution is basic, add H_2O to the side that is deficient in H and OH^- to the other side.

d   Balance net charge by adding electrons to the side that is deficient in negative charge.

Step 3   Multiply the half-reactions by integers that will lead to cancellation of electrons.

Step 4   Recombine the half-reactions, and simplify by combining and canceling duplicated species.

FINISH

Example 18-6 illustrates the entire balancing process, and Example 18-7 illustrates the balancing procedure in basic solution.

Balancing a Redox Equation

Example 18-6

Concentrated aqueous sulfuric acid, H_2SO_4, is a strong oxidizing agent that can react with elemental carbon:

$$H_2SO_4 + C \longrightarrow CO_2 + SO_2 \qquad \text{(unbalanced)}$$

What is the balanced equation for this process?

Strategy: Divide the reaction into half-reactions, balance each using the stepwise procedure, combine the half-reactions, and then clean up the result to eliminate duplicated species.

Balancing
Redox
Reactions

Solution:

1. The half-reactions can be identified by inspection:

$$C \longrightarrow CO_2 \qquad H_2SO_4 \longrightarrow SO_2$$

2. Each half-reaction is balanced separately using the four-step process. Concentrated sulfuric acid contains some water, so H_3O^+ and H_2O can be used as needed. The new feature in each step is shown with yellow background.

Example 18-6	Balancing a Redox Equation *(continued)*

Step a: C is already balanced   S is already balanced

Step b: $C + 2\,H_2O \longrightarrow CO_2$ $H_2SO_4 \longrightarrow SO_2 + 2\,H_2O$

Step c: $C + 6\,H_2O \longrightarrow CO_2 + 4\,H_3O^+$

$H_2SO_4 + 2\,H_3O^+ \longrightarrow SO_2 + 4\,H_2O$

Step d: $C + 6\,H_2O \longrightarrow CO_2 + 4\,H_3O^+ + 4\,e^-$

$2\,H_3O^+ + H_2SO_4 + 2\,e^- \longrightarrow SO_2 + 4\,H_2O$

3. Multiply the sulfur equation by 2 so the electrons will cancel when the half-reactions are added:

$$C + 6\,H_2O + 4\,H_3O^+ + 2\,H_2SO_4 + \cancel{4\,e^-} \longrightarrow$$
$$CO_2 + 4\,H_3O^+ + \cancel{4\,e^-} + 2\,SO_2 + 8\,H_2O$$

4. After adding, cancel common terms: $4\,H_3O^+$ on each side and $6\,H_2O$ on each side:

$$C + 2\,H_2SO_4 \longrightarrow CO_2 + 2\,SO_2 + 2\,H_2O$$

An inventory verifies that this equation is balanced:

Reactants	Products
1 C	1 C
4 H	4 H
2 S	2 S
8 O	8 O
0 charge	0 charge

Example 18-7	Redox in Basic Solution

Household ammonia should never be mixed with bleach, because a redox reaction occurs that generates toxic chlorine gas and hydrazine:

$$NH_3 + OCl^- \longrightarrow Cl_2 + N_2H_4 \qquad \text{(unbalanced)}$$

Balance this equation.

Strategy: Follow the step-by-step procedure.

Balancing
Redox
Reactions

Solution:

1. Break into half-reactions, one containing Cl and the other containing N.

$$NH_3 \longrightarrow N_2H_4 \qquad OCl^- \longrightarrow Cl_2$$

2. Balance the half-reactions individually.
   Step a. Balance N and Cl by inspection:

$$2\,NH_3 \longrightarrow N_2H_4 \qquad 2\,OCl^- \longrightarrow Cl_2$$

   Step b. Add H_2O to balance oxygen atoms:

$$2\,NH_3 \longrightarrow N_2H_4 \qquad 2\,OCl^- \longrightarrow Cl_2 + 2\,H_2O$$

Redox in Basic Solution (*continued*)

Example 18-7

Step c. Add H_2O and OH^- to balance hydrogen atoms:

$$2\ NH_3 + 2\ OH^- \longrightarrow N_2H_4 + 2\ H_2O$$

$$2\ OCl^- + 4\ H_2O \longrightarrow Cl_2 + 2\ H_2O + 4\ OH^-$$

In the second half-reaction two water molecules can be cancelled from each side:

$$2\ OCl^- + 2\ H_2O \longrightarrow Cl_2 + 4\ OH^-$$

Step d. Add electrons to balance the charges. The nitrogen half-reaction needs two electrons on the right, and the chlorine reaction needs two electrons on the left:

$$2\ NH_3 + 2\ OH^- \longrightarrow N_2H_4 + 2\ H_2O + 2e^-$$

$$2\ OCl^- + 2\ H_2O + 2e^- \longrightarrow Cl_2 + 4\ OH^-$$

3. Because there are two electrons in each half-reaction, adding the two half-reactions leads to cancellation of the electrons.

4. Now recombine the half-reactions and cancel common terms:

$$2\ NH_3 + 2\ OH^- + 2\ H_2O + 2\ OCl^- + 2e^- \longrightarrow$$
$$N_2H_4 + 2e^- + 2\ H_2O + Cl_2 + 4\ OH^-$$

Two OH^- ions also can be cancelled:

$$2\ NH_3 + 2\ OCl^- \longrightarrow N_2H_4 + Cl_2 + 2\ OH^-$$

Make sure the equation is balanced by taking an inventory:

Reactants	Products
2 N	2 N
6 H	6 H
2 Cl	2 Cl
2 O	2 O
2 − charge	2 − charge

The equation is balanced. Although we could have balanced this equation by inspection, the detailed procedure reveals its redox nature.

■ **18.2.1** Balance the following redox reactions from Example 18.3:
    (a) $Cu + NO_3^- \rightarrow Cu^{2+} + NO$ (acidic solution)
    (b) $H_2O_2 + SO_3^{2-} \rightarrow H_2O + SO_4^{2-}$ (basic solution)
    (c) $Br_2 \rightarrow BrO_3^- + Br^-$ (acidic solution)

■ **18.2.2** The tarnish that collects on objects made of silver is silver sulfide, a black solid. Tarnish can be removed by heating the tarnished object in an aluminum pan containing mildly basic water. Balance this equation:

$$Al_{(s)} + Ag_2S_{(s)} \longrightarrow Ag_{(s)} + Al(OH)_{3(s)} + H_2S_{(g)}$$

■ **18.2.3** Copper refining traditionally involves "roasting" sulfide ores with oxygen, which produces large quantities of polluting SO_2. An alternative process uses aqueous nitric acid to convert the CuS into soluble copper(II) hydrogen sulfate without generating SO_2. Balance this redox equation:

$$CuS(s) + NO_3^-(aq) \longrightarrow NO(g) + Cu^{2+}(aq) + HSO_4^-(aq)$$

18.3 SPONTANEOUS REDOX REACTIONS AND GALVANIC CELLS

The reaction of potassium metal with water to generate hydrogen gas, shown in Figure 18-4, is highly spontaneous:

$$2\,K(s) + 2\,H_2O(l) \xrightarrow{\text{Rapid}} 2\,K^+(aq) + 2\,OH^-(aq) + H_2(g)$$

In contrast, no reaction occurs when a silver spoon is dipped into water. The balanced redox equation for the possible oxidation of silver by water looks similar to that for potassium:

$$2\,Ag(s) + 2\,H_2O(l) \xrightarrow{\text{No reaction}} 2\,Ag^+(aq) + 2\,OH^-(aq) + H_2(g)$$

One metal reacts with water so vigorously that a fire or explosion may result, whereas the other does not react with water, even at high temperature. Despite their apparent similarities, the reaction of potassium is spontaneous, but that of silver is not. Thermodynamic calculations verify these observations. Remember from Chapter 13 that spontaneous reactions have negative values for ΔG. Tabulated free energies of formation can be used to calculate the standard free energy changes for these reactions. The $\Delta G°$ for the reaction of potassium has a large negative value, but the value for the reaction of silver is large and positive:

$$2\,K(s) + 2\,H_2O(l) \longrightarrow 2\,K^+(aq) + 2\,OH^-(aq) + H_2(g) \qquad \Delta G°_{reaction} = -407\ kJ$$

$$2\,Ag(s) + 2\,H_2O(l) \longrightarrow 2\,Ag^+(aq) + 2\,OH^-(aq) + H_2(g) \qquad \Delta G°_{reaction} = +314\ kJ$$

Experiments and calculations both indicate that electron transfer is easy for potassium but difficult for silver. In redox terms, potassium oxidizes easily, but silver resists oxidation. Because oxidation involves the loss of electrons, these differences in reactivity of silver and potassium can be traced to how easily each metal loses electrons to become an aqueous cation. The main factor is their first ionization

Only Rb, Cs, and Fr have lower first ionization energies than K. All react violently with water, as does Na, which has a first ionization energy of 495 kJ/mol.

Figure 18-4
Potassium reacts spontaneously and violently with water, but silver does not react with water at all.

energies, which show that it takes much more energy to remove an electron from silver than from potassium: 731 kJ/mol for Ag and 419 kJ/mol for K.

Direct and Indirect Electron Transfer

The reactivities of potassium and silver with water represent extremes in the spontaneity of electron transfer reactions. The redox reaction between two other metals illustrates less drastic differences in reactivity. Figure 18-5 shows the reaction that occurs between zinc metal and an aqueous solution of copper(II) sulfate: zinc slowly dissolves, and copper metal precipitates. As for the reaction of potassium with water, the standard free energy change is negative for this spontaneous reaction:

$$Cu^{2+}(aq) + Zn(s) \longrightarrow Cu(s) + Zn^{2+}(aq) \qquad \Delta G^{\circ}_{reaction} = -213 \text{ kJ}$$

Each zinc atom transfers two electrons to a copper cation. Balanced half-reactions show the electron gains and losses more clearly:

$$\text{Reduction: } Cu^{2+}(aq) + 2 \text{ e}^- \longrightarrow Cu(s)$$

$$\text{Oxidation: } Zn(s) \longrightarrow Zn^{2+}(aq) + 2 \text{ e}^-$$

The oxidation of Zn metal by Cu^{2+} ions is an example of *direct* electron transfer. A copper ion accepts two electrons when it collides with the surface of the zinc strip. Direct electron transfer occurs when electrons are transferred during a collision between the species being oxidized and the species being reduced, as shown by the molecular views in Figure 18-5.

Figure 18-5
When a strip of zinc metal is dipped in a solution of copper(II) sulfate, zinc is oxidized to $Zn^{2+}(aq)$, and $Cu^{2+}(aq)$ is reduced to copper metal. The insoluble metal precipitates from the solution. In the molecular views, water molecules and spectator anions have been omitted for clarity.

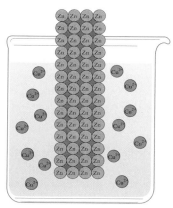
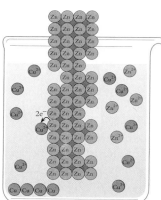
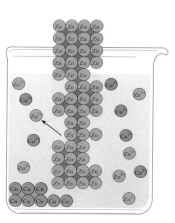

Figure 18-6
If the zinc- and copper-containing portions of this redox system are physically separated, electron transfer can occur only through an external wire.

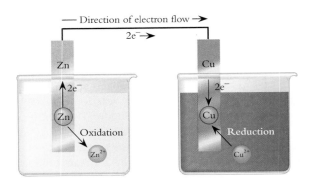

Spontaneous redox reactions can also occur by *indirect* electron transfer. In an indirect electron transfer, species involved in the redox chemistry are not allowed to come into direct contact with one another. Instead, the oxidation occurs at one end of a wire, generating free electrons. Reduction occurs at the other end of the wire, consuming free electrons. The wire conducts electrons between the oxidation site and the reduction site.

Figure 18-6 is a schematic view of two beakers set up for indirect electron transfer. The beaker on the left contains an aqueous solution of zinc sulfate and a strip of zinc metal. The beaker on the right contains an aqueous solution of copper(II) sulfate and a strip of copper metal. A wire connects the two metal strips to allow indirect electron transfer. The oxidation half-reaction generates free electrons and releases Zn^{2+} ions into the solution containing the zinc electrode:

$$Zn(s) \longrightarrow Zn^{2+}(aq) + 2\ e^-(wire)$$

Electrons released in the oxidation flow from the zinc strip through the wire into the copper strip. Copper cations can collect electrons when they collide with the surface of the copper strip:

$$Cu^{2+}(aq) + 2\ e^-(wire) \longrightarrow Cu(s)$$

The arrangement shown in Figure 18-6 does not generate a sustained flow of electrons, however, because as electrons are transferred, the solutions in both beakers quickly become unbalanced in electrical charge. Oxidation of the zinc strip releases Zn^{2+} ions into the solution, generating excess positive charge. On the other side, reduction of Cu^{2+} ions to copper metal produces a solution that is deficient in positive charge. This charge imbalance generates a coulombic force that stops the further flow of electrons.

Ion Transport

Redox by way of indirect electron transfer cannot continue unless there is a way to remove the charge imbalance created by electron flow. This charge imbalance can be removed by the movement of ions: Anions can move from the region of excess negative charge to the region of excess positive charge, and cations can move in the opposite direction. Such ion migration cannot allow the oxidation-reduction pair to come into direct contact, however, because then direct electron transfer would replace indirect electron transfer.

In our zinc-copper example, electrons flowing through the wire from left to right generate surplus positive charge on the left and surplus negative charge on the right. Sulfate anions moving from right to left can remove the surplus charges, but this flow of ions must occur in a way that prevents the two solutions from mixing freely. Figure 18-7 shows an arrangement that meets these needs. The two vessels are connected by a porous glass plate. This plate allows ions to migrate between the two compartments, but the mixing process is too slow to lead to direct electron transfer. Electrons flow through the wire, and ions diffuse through the plate. Adding the porous plate completes a circuit that can be used to generate a sustained flow of electrons through the external wire.

Salt bridges and membranes also can serve as separators that allow ion migration without free mixing of solutions. Schematic views of these devices appear in Figure 18-8. A porous plate contains many tiny channels that allow water molecules and ions to diffuse slowly from one side to the other. A salt bridge contains gelatin impregnated with cations and anions, usually K^+ and Cl^- or NH_4^+ and NO_3^-. The gelatin prevents solvent from moving freely but allows the ions to diffuse slowly. A membrane functions like a porous plate but is very thin, often only microns in width.

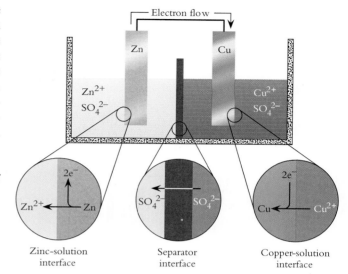

Figure 18-7
Schematic view of an arrangement that allows a sustained redox reaction accompanied by an external flow of electrons.

Electrodes

The spontaneous redox reaction shown in Figure 18-7 takes place at the surfaces of metal plates, where electrons are gained and lost by metal atoms and ions. These metal plates are examples of **electrodes.** At an electrode, redox reactions transfer electrons between the aqueous phase and the external circuit. An oxidation half-reaction releases electrons to the external circuit at one electrode. A reduction half-reaction withdraws electrons from the external circuit at the other electrode. The electrode where oxidation occurs is the **anode,** and the electrode where reduction occurs is the **cathode.**

Some electrodes are made of substances that participate in the redox reactions that transfer electrons. These are **active** electrodes. Other electrodes serve only to supply or accept electrons but are not part of the redox chemistry; these are **passive** electrodes. In Figure 18-7, both metal strips are active electrodes. As the redox reaction progresses, zinc metal dissolves off of the anode while copper metal precipitates at the cathode. The reactions that take place at these active electrodes are conversions between the metals contained in the electrodes and their aqueous cations.

At another type of active electrode, found in many batteries, the reaction is the conversion between a metal and an insoluble salt. At the surface of this type of electrode, metal cations combine with anions from the solution to form the salt. One example is the lead anode of an automobile battery, at whose surface lead metal loses electrons and

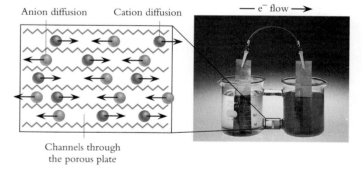

Channels through the porous plate

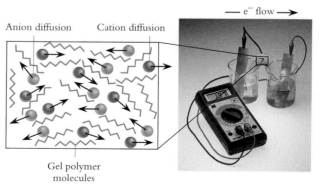

Gel polymer molecules

Figure 18-8
Porous plates and salt bridges allow slow diffusion of ions but prevent rapid mixing.

forms Pb^{2+} cations which combine immediately with sulfate ions in solution to form insoluble lead(II) sulfate:

$$Pb(s) + SO_4^{2-}(aq) \longrightarrow PbSO_4(s) + 2\ e^-$$

A redox half-reaction at an active electrode also may convert one metal salt into another. For example, in rechargeable nickel-cadmium batteries, a nickel(IV) oxide cathode is reduced to nickel(II) hydroxide. The half-reaction reduces Ni(IV) to Ni(II):

$$NiO_2(s) + 2\ H_2O(l) + 2\ e^- \longrightarrow Ni(OH)_2(s) + 2\ OH^-(aq)$$

A passive electrode conducts electrons to and from the external circuit but does not participate chemically in the half-reactions. Figure 18-9 shows a redox setup that contains passive electrodes. One compartment contains an aqueous solution of iron(III) chloride in contact with a platinum electrode. Electron transfer at this electrode reduces $Fe^{3+}(aq)$ to $Fe^{2+}(aq)$:

$$\text{Reduction:} \qquad Fe^{3+} + e^- \longrightarrow Fe^{2+}$$

In the second compartment, hydrogen gas bubbles over the surface of a platinum electrode. At this passive surface, hydrogen molecules are oxidized, giving up two electrons to form pairs of protons, which bond to water molecules to give H_3O^+ ions:

$$\text{Oxidation:} \qquad H_2 + 2\ H_2O \longrightarrow 2\ H_3O^+ + 2\ e^-$$

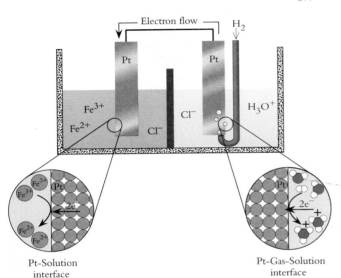

Figure 18-9
Diagram of a galvanic cell containing passive electrodes. The two platinum electrodes do not take part in the redox chemistry of this cell. They only conduct electrons to and from the interfaces.

Each half-reaction occurs at a platinum metal electrode, but neither electrode takes part in the redox chemistry. Platinum metal is often used as a passive electrode because platinum is one of the least reactive elements. Platinum has a large ionization energy, so it can act as an "electron shuttle" without participating in redox chemistry.

The combination of hydrogen gas, H_3O^+ ions, and a platinum electrode is referred to as a *hydrogen electrode*. This electrode appears in the right-hand portion of Figure 18-9. When a hydrogen electrode operates under standard conditions, $p_{H_2} = 1.00$ atm and $[H_3O^+] = 1.00$ M, it is a standard hydrogen electrode (SHE). The standard hydrogen electrode is particularly important in electrochemistry, as we describe in Section 18.4.

Galvanic Cells

Electrochemistry is the coupling of a chemical redox process with electron flow through a wire. The process represented in Figure 18-7 is electrochemical because the redox reaction releases chemical energy that is converted into an electrical current through an external wire. On the other hand, Figure 18-5 shows a redox process that is not electrochemical, because direct electron transfer cannot generate an electrical current through a wire.

Galvanic cells use redox reactions to generate electrical current. Galvanic cells find wide usage in the form commonly called *batteries*. To give just one example, flashlight batteries make use of electrochemical reactions to generate electrical current. When a flashlight is turned on, chemical redox reactions occur in the battery and generate an electron flow through the light bulb.

The cell sketched in Figure 18-10 uses the Cu^{2+}/Zn reaction to illustrate the essential components of a galvanic cell:

1. *A spontaneous redox reaction.* In Figure 18-10, the spontaneous reaction is electron transfer between zinc and copper:

$$Zn(s) + Cu^{2+}(aq) \longrightarrow Zn^{2+}(aq) + Cu(s)$$

2. *A physical barrier blocking direct electron transfer.* Here, the barrier is provided by placing the solutions in two separate compartments.

3. *Physical contact between chemical and electrical parts of the cell.* This is the role of the electrodes. The electrodes in Figure 18-10 are strips of copper and zinc.

4. *An external electrical circuit.* This circuit may simply transfer electrons, as in the Zn/Cu example. In useful applications, the flow of electrons in the external circuit provides power to operate a flashlight, radio, or other device.

5. *Some means of completing the circuit.* In Figure 18-10, the porous glass plate allows ions to diffuse slowly between the two solutions, completing the circuit.

6. *Ions present in the solutions.* All parts of the circuit must allow the passage of charge. When an aqueous phase is present, it must contain ions, which are supplied by a soluble salt or strong acid.

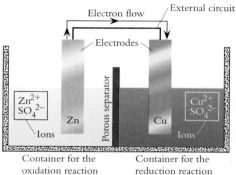

Figure 18-10
The essential components of a galvanic cell.

To summarize these features that characterize galvanic cells, Example 18-8 describes the lead storage battery.

Describing a Galvanic Cell	**Example 18-8**

The electrical current needed to start an automobile engine is provided by a lead storage battery. This battery contains aqueous sulfuric acid in contact with two electrodes. One electrode is metallic lead, and the other is solid PbO_2. Each electrode becomes coated with solid $PbSO_4$ as the battery operates. Determine the balanced half-reactions and the overall redox reaction and identify the anode and cathode in this galvanic cell.

Strategy: We are asked to identify the redox chemistry occurring in this battery. The problem provides a description of the chemical composition of a galvanic cell. To determine what redox reactions take place, examine the species present at each electrode. Then use the standard procedure to balance the half-reactions.

Solution: The aqueous solution contains H_2SO_4, a strong acid. Thus the major species in solution are H_2O, H_3O^+, and HSO_4^-. The description of the battery states that one electrode is Pb in contact with $PbSO_4$, and the other electrode is PbO_2 in contact with $PbSO_4$. This information identifies the two half-reactions:

$$Pb(s) \longrightarrow PbSO_4(s) \qquad PbO_2(s) \longrightarrow PbSO_4(s)$$

Lead is already balanced. Use the other sulfur-containing species, HSO_4^-, to balance sulfur:

$$Pb + HSO_4^- \longrightarrow PbSO_4 \qquad PbO_2 + HSO_4^- \longrightarrow PbSO_4$$

Balancing
Redox
Reactions

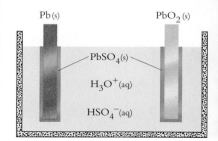

| Example 18-8 | Describing a Galvanic Cell *(continued)* |

Now all elements other than hydrogen and oxygen are balanced. Here are the additional steps for the left half-reaction:

Oxygen is already balanced. Add H_3O^+ and H_2O to balance hydrogen:

$$Pb + HSO_4^- + H_2O \longrightarrow PbSO_4 + H_3O^+$$

Add electrons to balance charge:

$$Pb + HSO_4^- + H_2O \longrightarrow PbSO_4 + H_3O^+ + 2\ e^-$$

Electrons appear as products, so this half-reaction is the oxidation, which takes place at the anode. The lead electrode is an active anode in a lead storage battery.

Now balance the second half-reaction:

Add water to balance oxygen:

$$PbO_2 + HSO_4^- \longrightarrow PbSO_4 + 2\ H_2O$$

Add H_3O^+ and H_2O to balance hydrogen:

$$PbO_2 + HSO_4^- + 3\ H_3O^+ \longrightarrow PbSO_4 + 5\ H_2O$$

Add e^- to balance charge:

$$PbO_2 + HSO_4^- + 3\ H_3O^+ + 2\ e^- \longrightarrow PbSO_4 + 5\ H_2O$$

Electrons appear as starting materials, so this half-reaction is the reduction, which takes place at the cathode. Lead(IV) oxide is an active cathode in a lead storage battery.

Because two electrons are transferred in each half-reaction, the overall redox reaction can be obtained by adding the half-reactions and cleaning up:

$$PbO_2(s) + Pb(s) + 2\ HSO_4^-(aq) + 2\ H_3O^+(aq) \longrightarrow 2\ PbSO_4(s) + 4\ H_2O(l)$$

> To remember how electrodes are named, notice that *oxidation* and *anode* both begin with vowels; *reduction* and *cathode* both begin with consonants.

Section Exercises

The thermite reaction.

18.3.1 The thermite reaction between aluminum metal and iron oxide is so rapid and exothermic that it generates a fountain of sparks (see photo) and can melt the container in which it takes place. The spontaneity of this reaction suggests the possibility of a galvanic cell involving aluminum and iron:

$$2\ Al + 6\ OH^- \longrightarrow Al_2O_3 + 3\ H_2O + 6\ e^-$$

$$Fe_2O_3 + 3\ H_2O + 6\ e^- \longrightarrow 2\ Fe + 6\ OH^-$$

Draw a schematic diagram, similar to Figure 18–7, showing the operation of this galvanic cell. Include the directions of all charge flows.

18.3.2 Use tabulated thermodynamic data to verify that the galvanic cell of Section Exercise 18.3.1 is spontaneous in the direction written.

18.3.3 Draw molecular pictures similar to the one shown in Figure 18–5 to illustrate the difference between an active electrode and a passive electrode. Use an iron anode in contact with a solution containing Fe^{2+} ions to illustrate an active electrode. Use a platinum cathode immersed in a solution containing Fe^{2+} and Fe^{3+} ions to illustrate a passive electrode.

18.4 CELL POTENTIALS

Batteries supply electrical current that consists of electrons flowing from one place to another. In some ways, flowing electrons are like flowing water. Because of gravitational force, water always flows downhill, from higher altitude to lower altitude. Similarly, electrons flow under the influence of coulombic force. Electrons flow from regions of more negative electrical potential energy to regions of more positive electrical potential energy.

Water at high altitude has gravitational potential energy that can be converted to other forms. As water flows, the gravitational potential energy that it loses is converted into kinetic energy in a waterfall, mechanical energy in a waterwheel, or electrical energy in a hydroelectric turbine. Similarly, as electrons flow, electrical potential energy can be converted to other forms: light energy in a flashlight, mechanical energy in the starter motor of an automobile, or thermal energy in an electrical heater.

A potential energy difference is referred to simply as a *potential*. Continuing our analogy with water, differences in gravitational potential are measured by height differences. The top and bottom of the spillway of a dam are at different heights and have different gravitational potentials: water falling down the spillway converts gravitational potential energy into kinetic energy of motion. Analogously, one electrode generally will be at higher electrical potential than another. Whereas gravitational potentials are measured in heights, electrical potential differences are measured in volts. For electrons, a point at a more negative electrical potential is "uphill" from a point at a more positive electrical potential. The parallel between gravitational potential and electrical potential is summarized in Figure 18-11.

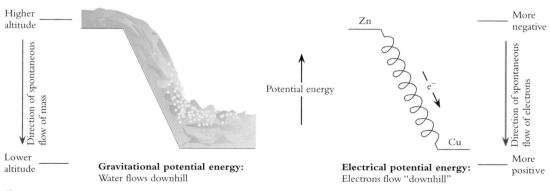

Gravitational potential energy:
Water flows downhill

Electrical potential energy:
Electrons flow "downhill"

Figure 18-11
Gravitational potential and electrical potential have common features. Water flows downhill from higher to lower gravitational potential when a suitable path, such as a spillway, is provided. Analogously, electrons flow "downhill" from higher to lower negative electrical potential when a suitable path, such as a conducting wire, is provided.

Electrode Equilibrium

To visualize how electrochemical cells generate electrical potential differences, consider a zinc electrode dipped into a solution of zinc sulfate. From the macroscopic perspective, nothing happens. At the molecular level, however, some of the zinc atoms of the electrode are oxidized to Zn^{2+} ions:

$$Zn(s) \longrightarrow Zn^{2+}(aq) + 2\ e^-(metal)$$

The ions move off into the solution, leaving an excess of electrons in the metal strip. The reverse process also occurs. A Zn^{2+} ion that collides with the metal surface may capture two electrons from the metal and be reduced to a neutral zinc atom:

$$Zn^{2+}(aq) + 2\ e^-(metal) \longrightarrow Zn(s)$$

When a zinc strip is dipped into the solution, the initial rates of these two processes are different. The different rates of reaction lead to a charge imbalance across the metal-solution interface. If the concentration of zinc ions in solution is low enough, the initial rate of oxidation is more rapid than the initial rate of reduction. Under these conditions, excess electrons accumulate in the metal, and excess cationic charges accumulate in the solution. As excess charge builds, however, the rates of reaction change until the rate of reduction is balanced by the rate of oxidation. When this balance is reached, the system is at dynamic equilibrium. Oxidation and reduction continue, but the net rate of exchange is zero:

$$Zn(s) \rightleftharpoons Zn^{2+}(aq) + 2\ e^-(metal)$$

Figure 18-12 illustrates the zinc equilibrium at the molecular level. At equilibrium, the charge imbalance in the zinc strip is about one excess electron for every 10^{14} zinc atoms, which is negligible from the macroscopic perspective but significant at the molecular level.

An analogous dynamic equilibrium is established when a strip of copper is dipped in a dilute solution of copper sulfate:

$$Cu(s) \rightleftharpoons Cu^{2+}(aq) + 2\ e^-(metal)$$

Once again, nothing appears to happen at the macroscopic level. At the molecular level, however, some of the copper atoms lose electrons and enter the solution as Cu^{2+} ions, and some of the Cu^{2+} ions capture electrons from the metal and deposit on the metal as Cu atoms. As with zinc, dipping a strip of copper metal in the solution generates a small charge imbalance.

Although both are extremely small, the charge imbalances for copper and zinc have different values. Zinc is easier to oxidize than copper. As a consequence, zinc creates a greater charge imbalance than copper. The concentration of excess electrons in the zinc electrode is greater than the concentration of excess electrons in the copper electrode, giving the zinc electrode more excess charge than the copper electrode.

When two electrodes contain different amounts of excess charge, there is a difference in electrical potential between them. Because it has more excess electrons, the zinc electrode is at a higher electrical potential than the copper electrode. In a galvanic cell, the difference in electrical potential causes electrons to flow from a region where the concentration of electrons is higher to a region where the concentration of electrons is lower. In this case, electrons flow from the zinc electrode toward the copper electrode, as shown at the molecular level in Figure 18-13.

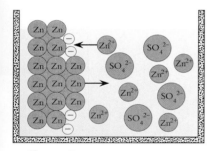

Figure 18-12
A strip of zinc metal immersed in a solution of zinc sulfate reaches a dynamic equilibrium when a few excess charges occur in each phase and the rates of oxidation and reduction are equal.

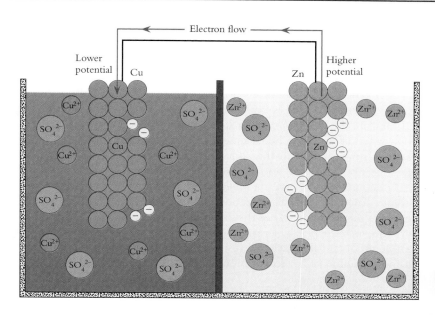

Electron flow

Lower potential   Cu

Higher potential   Zn

Animation

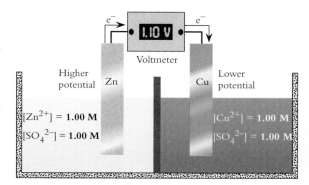

By convention, the positive terminal of a battery refers to the electrode at lower potential. Electrons flow from the negative terminal (anode) to the positive terminal (cathode).

The difference in electrical potential between two electrodes is designated E and is measured in volts (V). The magnitude of E increases as the amount of charge imbalance between the two electrodes increases. For any galvanic cell, the value of E and the direction of electron flow can be determined experimentally by inserting a voltmeter in the external circuit.

Electrons always flow spontaneously "downhill" from higher electrical potential to lower electrical potential. In a galvanic cell, the electrode with the higher potential is designated by convention as the "negative electrode." The electrode with the lower potential is designated as the "positive electrode." The difference in electrical potential between the two electrodes is the **cell potential.**

Standard Electrical Potential

Electrochemical cells can be constructed using an almost limitless combination of electrodes and solutions, and each combination generates a specific potential. Keeping track of the electrical potentials of all cells under all possible situations would be extremely tedious without a set of standard reference conditions. By definition, the standard electrical potential is the potential developed by a cell in which all chemical species are present under standard thermodynamic conditions. Recall that standard conditions for thermodynamic properties include concentrations of 1 M for solutes in solution and pressures of 1 atm for gases. Chemists use the same standard conditions for electrochemical properties. As in thermodynamics, standard conditions are designated with a superscript °. A standard electrical potential is designated $E°$.

The zinc-copper galvanic cell is under standard conditions when the concentrations of both ions are 1.00 M, as shown in Figure 18-14. The cell potential under these conditions can be determined by connecting the electrodes to a

Higher potential   Zn

Voltmeter

Lower potential   Cu

$[Zn^{2+}] = 1.00 \text{ M}$
$[SO_4^{2-}] = 1.00 \text{ M}$

$[Cu^{2+}] = 1.00 \text{ M}$
$[SO_4^{2-}] = 1.00 \text{ M}$

Figure 18-14
The copper/zinc electrochemical cell. The voltmeter measures the difference in electrical potentials between the two electrodes.

voltmeter. The measured potential is 1.10 V, with the Zn electrode at the higher (more negative) potential, so Zn gives up electrons and $E_{cell}^\circ = 1.10$ V:

$$Zn(s) + Cu^{2+}(aq, 1.00 \text{ M}) \longrightarrow Zn^{2+}(aq, 1.00 \text{ M}) + Cu(s) \qquad E_{cell}^\circ = 1.10 \text{ V}$$

Any redox reaction can be broken into two distinct half-reactions, an oxidation and a reduction. It is convenient to assign a potential to every half-reaction and tabulate E° values for all half-reactions. The standard cell potential for any redox reaction can then be obtained by combining the potentials for its two half-reactions. In order to do this, one particular half-reaction has to be selected as a reference reaction with zero potential. This is necessary because an experimental measurement always gives a *difference* between two potentials rather than an absolute potential. The standard potential of 1.10 V for the Zn/Cu cell, for example, is the difference between the E° values of its two half-reactions.

Once a reference half-reaction has been selected, all other half-reactions can then be assigned values relative to this reference value of 0 V. Chemists have chosen the reduction of hydronium ions to hydrogen gas as the reference half-reaction:

$$2 \text{ H}_3\text{O}^+(aq, 1.00 \text{ M}) + 2 \text{ e}^- \rightleftharpoons \text{H}_2(g, 1.00 \text{ atm}) + 2 \text{ H}_2\text{O}(l) \quad E^\circ = 0 \text{ V by definition}$$

This standard hydrogen electrode (SHE) is shown schematically in Figure 18-15.

In addition to defined standard conditions and a reference potential, tabulated half-reactions have a defined reference *direction*. As the double arrow in the previous equation indicates, E° values for half-reactions refer to electrode equilibria. Just as the value of an equilibrium constant depends on the direction in which the equilibrium reaction is written, the values of E° depend on whether electrons are reactants or products. For half-reactions, the conventional *reference* direction is *reduction*, with electrons always appearing as *reactants*. The conventional tabulated E° values for half-reactions are called **standard reduction potentials**. It is important to remember that every redox reaction includes both an oxidation and a reduction. Half-reactions are tabulated as reductions by convention, but in any real redox process, one half-reaction occurs as a reduction and the other as an oxidation.

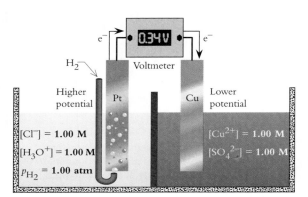

Figure 18-15
Schematic diagram of a cell for measuring E° of the Cu^{2+}/Cu half-reaction relative to the standard hydrogen electrode (SHE).

Defining a reference value for the SHE makes it possible to determine E° values of all other redox half-reactions. As an example, Figure 18-15 shows a cell in which a standard hydrogen electrode is connected to a copper electrode in contact with a 1.00 M solution of Cu^{2+}. In this cell the SHE is at higher electrical potential than the copper electrode, indicating that in this cell, reduction of Cu^{2+} and oxidation of H_2 occur spontaneously, and electrons flow from the SHE to the copper electrode:

$$Cu^{2+}(aq, 1.00 \text{ M}) + \text{H}_2(g, 1.00 \text{ atm}) + 2 \text{ H}_2\text{O} \longrightarrow Cu(s) + 2 \text{ H}_3\text{O}^+(aq, 1.00 \text{ M})$$

In other words, oxidation takes place at the SHE (anode), and reduction occurs at the copper electrode (cathode).

Measurements show that the Cu/H_2 cell has a voltage of 0.34 V. Because the potential of the SHE is defined to be 0 V, this means that the reduction half-reaction Cu^{2+}/Cu has $E^\circ = +0.34$ V *relative to the SHE*:

$$Cu^{2+}(aq, 1.00 \text{ M}) + 2 \text{ e}^- \rightleftharpoons Cu(s) \qquad E^\circ = 0.34 \text{ V (relative to SHE)}$$

Table 18–1
Representative Standard Reduction Potentials*

$$2\,H_3O^+ + 2\,e^- \rightleftharpoons H_2(g) + 2\,H_2O(l) \qquad E^\circ = 0\ V\ (\text{defined})$$

Reaction	E° (V)	Reaction	E° (V)
Positive Values		**Negative Values**	
$F_2(g) + 2\,e^- \rightleftharpoons 2\,F^-$	2.866	$Fe^{3+} + 3\,e^- \rightleftharpoons Fe$	−0.037
$PbO_2 + SO_4^{2-} + 4\,H_3O^+ + 2\,e^- \rightleftharpoons PbSO_4 + 6\,H_2O(l)$	1.6913	$Pb^{2+} + 2\,e^- \rightleftharpoons Pb$	−0.1262
$Au^{3+} + 3\,e^- \rightleftharpoons Au$	1.498	$Sn^{2+} + 2\,e^- \rightleftharpoons Sn$	−0.137
$Cl_2(g) + 2\,e^- \rightleftharpoons 2\,Cl^-$	1.35827	$PbSO_4 + 2\,e^- \rightleftharpoons Pb + SO_4^{2-}$	−0.3588
$Cr_2O_7^{2-} + 14\,H_3O^+ + 6\,e^- \rightleftharpoons 2\,Cr^{3+} + 21\,H_2O$	1.232	$Fe^{2+} + 2\,e^- \rightleftharpoons Fe$	−0.447
$O_2(g) + 4\,H_3O^+ + 4\,e^- \rightleftharpoons 6\,H_2O$	1.229	$Ni(OH)_2 + 2\,e^- \rightleftharpoons Ni + 2\,OH^-$	−0.72
$Pt^{2+} + 2\,e^- \rightleftharpoons Pt$	1.18	$Cr^{3+} + 3\,e^- \rightleftharpoons Cr$	−0.744
$Br_2(l) + 2\,e^- \rightleftharpoons 2\,Br^-$	1.066	$Zn^{2+} + 2\,e^- \rightleftharpoons Zn$	−0.7618
$Hg^{2+} + 2\,e^- \rightleftharpoons Hg(l)$	0.851	$Cd(OH)_2 + 2\,e^- \rightleftharpoons Cd + 2\,OH^-$	−0.860
$Ag^+ + e^- \rightleftharpoons Ag$	0.7996	$Al^{3+} + 3\,e^- \rightleftharpoons Al$	−1.662
$I_2 + 2\,e^- \rightleftharpoons 2\,I^-$	0.5355	$Mg^{2+} + 2\,e^- \rightleftharpoons Mg$	−2.37
$Cu^+ + e^- \rightleftharpoons Cu$	0.521	$Na^+ + e^- \rightleftharpoons Na$	−2.71
$O_2(g) + 2\,H_2O(l) + 4\,e^- \rightleftharpoons 4\,OH^-$	0.401	$Ca^{2+} + 2\,e^- \rightleftharpoons Ca$	−2.868
$Cu^{2+} + 2\,e^- \rightleftharpoons Cu$	0.3419	$Ca(OH)_2 + 2\,e^- \rightleftharpoons Ca + 2\,OH^-$	−3.02
$AgCl + e^- \rightleftharpoons Ag + Cl^-$	0.22233	$Li^+ + e^- \rightleftharpoons Li$	−3.0401

*Note: All ions are aqueous, all neutrals are solid unless noted.

Over the years, chemists have carried out many measurements similar to the one depicted in Figure 18-15. As a result, numerous standard reduction potentials are tabulated in reference sources such as the *Handbook of Chemistry and Physics*. Many values appear in Appendix F, and some representative values appear in Table 18-1.

Every standard reduction potential has a specific sign. When a substance is easier to reduce than hydronium ions under standard conditions, its E° is positive. When a substance is more difficult to reduce than hydronium ions under standard conditions, its E° is negative.

Standard Cell Voltages

In any galvanic cell that is under standard conditions, electrons are produced by the half-reaction with the more negative standard reduction potential and consumed by the half-reaction with the more positive standard reduction potential. In other words, the half-reaction with the more negative E° value occurs as the oxidation, and the half-reaction with the more positive E° value occurs as the reduction. Figure 18–16 summarizes the conventions used to describe galvanic cells.

The cell potential is positive for any spontaneous redox reaction. Tabulated standard reduction potentials allow us to determine the potential of any cell under standard conditions. This net standard cell potential is obtained by *subtracting* the more negative standard reduction potential from the more positive standard reduction potential, giving a positive overall potential. To summarize:

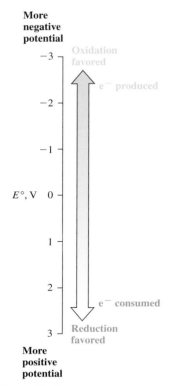

Figure 18-16
The half-reaction with the more negative $E°$ produces electrons at higher electrical potential through oxidation. The half-reaction with the more positive $E°$ consumes electrons at lower electrical potential through reduction.

/// *The half-reaction with the more negative reduction potential occurs at the anode as oxidation.*
/// *The half-reaction with the more positive reduction potential occurs at the cathode as reduction.*

Combining these features gives an equation that summarizes the calculation of the standard potential for a galvanic cell:

$$E°_{cell} = E°_{cathode} - E°_{anode} \qquad \textbf{(18-1)}$$

Here is this procedure for the copper-hydrogen cell illustrated in Figure 18-15. We write the two half-reactions as reductions:

$$Cu^{2+}(aq, 1.00\ atm) + 2\ e^- \rightleftharpoons Cu(s) \qquad\qquad E° = +0.34\ V$$

$$2\ H_3O^+(aq, 1.00\ M) + 2\ e^- \rightleftharpoons H_2(g, 1.00\ atm) + 2\ H_2O \qquad E° = 0\ V$$

The overall reaction is obtained by reversing the half-reaction that has the more negative potential, and $E°_{cell}$ is found by subtracting that $E°$ from the more positive $E°$.

$$H_2(g, 1.00\ atm) + 2\ H_2O + Cu^{2+}(aq, 1.00\ M) \longrightarrow 2\ H_3O^+(aq, 1.00\ M) + Cu(s)$$

$$E°_{cell} = (E°_{Cu^{2+}/Cu}) - (E°_{H_3O^+(aq)/H_2}) = 0.34\ V - 0\ V = 0.34\ V$$

In this spontaneous redox reaction, oxidation of H_2 releases electrons at the platinum anode. Excess electrons in the anode create a negative electrical potential. In the other compartment of the cell, Cu^{2+} ions are reduced to copper metal by capturing electrons from the copper cathode. In the wire connecting the electrodes, electrons flow "downhill" from the region of more negative electrical potential (the anode, $E° = 0\ V$) to the region of less negative electrical potential (the cathode, $E° = 0.34\ V$).

Consider another galvanic cell, one in which a standard Zn^{2+}/Zn half-reaction combines with the standard hydrogen electrode. According to Table 18-1, $E°$ for the standard Zn^{2+}/Zn half-reaction is $-0.76\ V$. The Zn^{2+}/Zn reduction potential is more negative than the $H_3O^+(aq)/H_2$ reduction potential ($-0.76\ V$ vs. $0\ V$), so zinc is the anode in this cell.

$$Zn^{2+}(aq, 1.00\ atm) + 2\ e^- \rightleftharpoons Zn(s) \qquad\qquad E° = -0.76\ V$$

$$2\ H_3O^+(aq, 1.00\ atm) + 2\ e^- \rightleftharpoons H_2(g, 1.00\ atm) + 2\ H_2O \qquad E° = 0\ V$$

Zinc is oxidized and hydronium ions are reduced, and electrons flow from the more negative zinc electrode to the less negative SHE. Once again, we reverse the direction of the half-reaction with the more negative potential and find $E°_{cell}$ by appropriate subtraction of half-cell potentials:

$$2\ H_3O^+(aq, 1.00\ atm) + Zn(s) \longrightarrow H_2(g, 1.00\ atm) + 2\ H_2O + Zn^{2+}(aq, 1.00\ atm)$$

$$E°_{cell} = (E°_{H_3O^+(aq)/H_2}) - (E°_{Zn^{2+}/Zn}) = 0\ V - (-0.76\ V) = 0.76\ V$$

The overall voltage generated by a standard galvanic cell is always obtained by *subtracting* one standard reduction potential from the other in the way that gives a positive value for $E°_{cell}$. Example 18-9 applies this reasoning to zinc and iron.

Standard Cell Potential

Example 18-9

A galvanic cell can be constructed from a zinc electrode immersed in a solution of zinc sulfate and an iron electrode immersed in a solution of iron(II) sulfate. What is the standard potential of this cell, and what is its spontaneous direction under standard conditions?

Strategy: First, the half-reactions must be identified. Then the standard reduction potentials can be looked up in a table, and the more negative value subtracted from the more positive value.

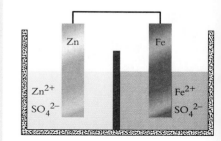

Solution: Write both half-reactions as reductions, and then look up the $E°$ values in Table 18-1:

$$Zn^{2+}(aq) + 2\ e^- \rightleftharpoons Zn(s) \qquad E°_{Zn^{2+}/Zn} = -0.7618\ V$$

$$Fe^{2+}(aq) + 2\ e^- \rightleftharpoons Fe(s) \qquad E°_{Fe^{2+}/Fe} = -0.447\ V$$

Now, subtract the more negative value (-0.7618 V) from the more positive value (-0.447 V) to obtain the standard cell potential:

$$E°_{cell} = (E°_{Fe^{2+}/Fe}) - (E°_{Zn^{2+}/Zn}) = -0.447\ V - (-0.7618\ V) = 0.315\ V$$

Under standard conditions, the iron–zinc cell operates spontaneously in the direction that reduces Fe^{2+} and oxidizes Zn.

$$Fe^{2+}(aq) + Zn(s) \longrightarrow Fe(s) + Zn^{2+}(aq) \qquad E°_{cell} = 0.315\ V$$

The calculation shows that zinc is oxidized preferentially over iron. Later in this chapter we describe the use of zinc as a sacrificial anode to prevent corrosion of iron.

Conventions for Standard Reduction Potentials

Here is a summary of the definitions and conventions for working with electrochemical potentials:

1. Standard conditions for electrochemical half-reactions are the same as those in thermodynamics: 25 °C (298.15 K), 1 M for solutes, and 1 atm for gases.
2. Reduction is the reference direction for electrochemical half-reactions: Electrons are reactants.
3. The standard potential for reducing $H_3O^+(aq)$ to H_2 is defined to be 0 V.
4. The standard potential for any galvanic cell is determined by subtracting the more negative standard reduction potential from the more positive standard reduction potential. A positive $E°$ indicates spontaneity under standard conditions.
5. In any galvanic cell, the half-reaction with the more negative reduction potential occurs as the oxidation at the anode, and the half-reaction with the more positive reduction potential occurs as the reduction at the cathode.

Example 18-10 provides another illustration of standard cell potentials.

| Example 18-10 | Half-Cell Potentials |

The following half-reactions occur in the rechargeable nickel-cadmium battery:

$$Cd(OH)_2(s) + 2\,e^- \rightleftharpoons Cd(s) + 2\,OH^-(aq)$$

$$NiO(OH)(s) + H_2O(l) + e^- \rightleftharpoons Ni(OH)_2(s) + OH^-(aq)$$

This battery has a potential of 1.35 V under standard conditions, with nickel as the cathode. Determine the net reaction, and use the tabulated standard reduction potential for the cadmium half-reaction to find $E°$ for the nickel half-reaction.

Strategy: We need to determine which reaction is the reduction and which is the oxidation when this battery is operating in its spontaneous direction. Then we can use Equation 18-1 to determine $E°$ for the nickel half-reaction.

Solution: The problem states that the nickel electrode is the cathode. Because reduction takes place at the cathode, we know that the nickel half-reaction is the reduction. In the oxidation half-reaction, cadmium is oxidized at the anode.

The problem states that that $E°_{cell} = 1.35$ V. The $E°$ value for the cadmium reaction appears in Table 18-1, but the value for the nickel half-reaction must be determined:

$$E°_{cell} = E°_{cathode} - E°_{anode}$$

$$E°_{cell} = (E°_{NiO(OH)/Ni(OH)_2}) - (E°_{Cd(OH)_2/Cd})$$

$$1.35\ V = (E°_{NiO(OH)/Ni(OH)_2}) - (-0.860\ V)$$

$$E°_{NiO(OH)/Ni(OH)_2} = 1.35\ V + (-0.860\ V) = +0.49\ V$$

To balance the net reaction, multiply the nickel reduction half-reaction by two to balance the electrons and combine the result with the cadmium oxidation half-reaction. Electrons and hydroxide ions cancel:

$$2\,NiO(OH)(s) + 2\,H_2O(l) + \cancel{2\,e^-} \longrightarrow 2\,Ni(OH)_2(s) + \cancel{2\,OH^-}(aq)$$

$$Cd(s) + \cancel{2\,OH^-}(aq) \longrightarrow Cd(OH)_2(s) + \cancel{2\,e^-}$$

$$\overline{2\,NiO(OH)(s) + 2\,H_2O(l) + Cd(s) \longrightarrow 2\,Ni(OH)_2(s) + Cd(OH)_2(s)}$$

The calculation of $E°$ for this cell illustrates an important feature of cell potentials. Cell potentials are simple differences between two standard reduction potentials. This is unchanged when one half-reaction is multiplied by 2 to cancel electrons in the overall redox reaction.

/// *When a reaction is multiplied by any integer, its cell potential remains unchanged.*

To understand why this is so, recall that cell potentials are analogous to altitude differences for water. Whether 1000 or 2000 L of water flows down a spillway, the altitude difference between the top and bottom of the spillway is the same. In the same way, multiplying a reaction by some integer changes the total number of moles of electrons transferred, but it does not change the potential difference through which the electrons are transferred. We return to this point in Section 18.5.

18.4.1 Use standard reduction potentials to determine the net reaction and standard cell potential for cells of two compartments, each containing a 1.00 M solution of the indicated cation in contact with an electrode of that neutral metal (that is, cells similar to the one shown in Figure 18-14): (a) Fe^{2+} and Cr^{3+}; (b) Cu^{2+} and Pb^{2+}; and (c) Au^{3+} and Ag^{+}.

18.4.2 Draw molecular pictures illustrating the charge transfer process taking place in the cell in Section Exercise 18.4.1(b).

18.4.3 Using the appropriate values from Table 18-1, calculate $E°$ for one cell of a lead storage battery. (Six of these cells are connected in series in an automobile storage battery.)

$$PbO_2(s) + Pb(s) + 2\,HSO_4^-(aq) + 2\,H_3O^+(aq) \longrightarrow$$
$$2\,PbSO_4(s) + 4\,H_2O(l)$$

18.5 FREE ENERGY AND ELECTROCHEMISTRY

A spontaneous electrochemical reaction generates a positive cell potential. Thermodynamically, the spontaneous direction of a redox reaction is determined by its free energy change: A spontaneous reaction has a negative ΔG. Thus a reaction that has a negative change in free energy generates a positive cell potential:

$$\Delta G_{\text{reaction}} < 0: \quad \text{Spontaneous redox reaction}$$

$$E_{\text{cell}} > 0: \quad \text{Spontaneous redox reaction}$$

Cell Potential and Free Energy

The linkage between free energy and cell potentials can be made quantitative. The more negative the value of $\Delta G°$ for a reaction, the more positive its standard cell potential, as the following two examples illustrate:

$$Cu^{2+}(aq) + Pb(s) \longrightarrow Cu(s) + Pb^{2+}(aq)$$
$$\Delta G° = -71.0 \text{ kJ/mol} \qquad E_{\text{cell}}° = +0.368 \text{ V}$$

$$Pb^{2+}(aq) + Sn(s) \longrightarrow Pb(s) + Sn^{2+}(aq)$$
$$\Delta G° = -2.1 \text{ kJ/mol} \qquad E_{\text{cell}}° = +0.011 \text{ V}$$

For further insight into how these two quantities are related, consider again the analogy between gravitational and electrical potentials. When water tumbles down a spillway, the total energy change depends on the height through which the water falls and how much water flows. Similarly, when electrons flow in an electrochemical cell, the change in free energy, ΔG, depends on the potential difference (E) and on the amount of charge that flows through the cell (n, the number of moles of electrons).

To complete the connection between electrical potential and free energy, electrical units must be converted to energy units. The charge on a single electron (e) has been measured to a very high degree of accuracy: $e = 1.60217646 \times 10^{-19}$ C.

The fundamental unit of electrical charge is the *coulomb* (C), defined as the quantity of electricity transferred by a current of 1 ampere in 1 second. Charge, potential, and energy are related: $1 V = 1 J/C$ and $1 J = 1 V C$

Michael Faraday, shown here delivering a popular lecture on the "Chemical History of a Candle," developed many of the fundamental principles of electricity.

The amount of charge provided by one mole of electrons is obtained by multiplying the charge on a single electron by the Avogadro constant (N_A). This quantity, the **Faraday constant (F),** is the link between chemical amounts (moles) and electrical amounts (coulombs):

$$F = eN_A = (1.60217646 \times 10^{-19} \text{ C})(6.022142 \times 10^{23} \text{ mol}^{-1})$$

$$F = 96{,}485.34 \text{ C mol}^{-1}$$

Putting these factors together, we find that the molar free energy change is the product of the electrical potential, the number of electrons transferred, and the Faraday constant:

$$\Delta G = -nFE \tag{18-2}$$

The Faraday constant (F) appears in Equation 18-2 to convert the amount of electrical charge (C) that "falls" through a potential difference (E) into a chemical amount (mol). The negative sign in Equation 18-2 is there because a spontaneous reaction has a negative value for ΔG, but a positive value for E. In this equation, n is dimensionless because it is a *ratio*, the number of electrons transferred per atom reacting. The product of potential and charge gives energy, and electrical units are defined in such a way that 1 J = 1 V C, so both sides of this equation have units of J/mol.

Remember that the number of electrons transferred is not explicitly stated in a net redox equation. This means that any overall redox reaction must be broken down into its balanced half-reactions to determine n, the ratio between the number of electrons transferred and the stoichiometric coefficients for the chemical reagents.

Equation 18-2 applies under all conditions, including standard conditions. At standard conditions, $\Delta G = \Delta G°$ and $E = E°$:

$$\Delta G° = -nFE° \tag{18-3}$$

Equation 18-3 expresses an important link between two standard quantities. The equation lets us calculate standard electrical potentials from tabulated values for standard free energies. Equally important, accurate potential measurements on galvanic cells yield experimental values for standard potentials that can be used to calculate standard free energy changes for reactions.

The cell shown in Figure 18-17 can serve as an example for calculations using Equation 18-3. One cell contains aqueous 1.00 M iron(III) chloride in contact with an iron metal electrode, and the other cell contains 1.00 M KCl in contact with a silver–silver chloride (AgCl/Ag) electrode. The half-reactions for these electrodes follow:

$$\text{AgCl(s)} + \text{e}^- \rightleftarrows \text{Ag(s)} + \text{Cl}^-\text{(aq)} \qquad E° = 0.2223 \text{ V}$$

$$\text{Fe}^{3+}\text{(aq)} + 3 \text{ e}^- \rightleftarrows \text{Fe(s)} \qquad E° = -0.037 \text{ V}$$

In the spontaneous redox reaction under standard conditions, AgCl is reduced and Fe is oxidized:

$$\text{Fe(s)} + 3 \text{ AgCl(s)} \longrightarrow \text{Fe}^{3+}\text{(aq)} + 3 \text{ Ag(s)} + 3 \text{ Cl}^- \text{(aq)}$$

$$E° = 0.2223 \text{ V} - (-0.037 \text{ V}) = 0.259 \text{ V}$$

Equation 18-3 can be used to calculate the standard free energy change for this net reaction, applying the appropriate conversion factors:

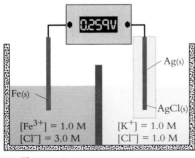

Figure 18-17
A galvanic cell can be constructed from a silver–silver chloride electrode in contact with a solution containing chloride anions and an iron electrode in contact with a solution containing iron(III) cations.

$$\Delta G^\circ = -nFE^\circ = -(3)(9.6485 \times 10^4 \text{ C mol}^{-1})(0.259 \text{ V})$$

$$\Delta G^\circ = (-7.50 \times 10^4 \text{ V C mol}^{-1})\left(\frac{1 \text{ kJ}}{10^3 \text{ J}}\right)\left(\frac{1 \text{ J}}{1 \text{ V C}}\right) = -75.0 \text{ kJ/mol}$$

Cell Potentials and Chemical Equilibrium

Equation 18-3 links ΔG° for a reaction with E°. From Chapter 15, Equation 15-3 links ΔG° for a reaction to K_{eq}:

$$\Delta G^\circ = -nFE^\circ \tag{18-3}$$

$$\Delta G^\circ = -RT \ln K_{eq} \tag{15-3}$$

Combining these two equalities and grouping the constants gives a new equation that links standard potentials directly to equilibrium constants:

$$-RT \ln K_{eq} = -nFE^\circ$$

$$E^\circ = \frac{RT}{nF} \ln K_{eq} \tag{18-4}$$

Many calculations using Equation 18-4 refer to standard temperature, 298.15 K. Furthermore, because K_{eq} often has a very large or very small value, calculations using log rather than ln are more convenient: $\log x = 2.302585 \ln x$. We can substitute these values and the value for F and then evaluate the multiplier for the log term at standard temperature:

$$\frac{2.302585 \, RT}{F} = \frac{(2.302585)(8.31451 \text{ J/mol K})(298.15 \text{ K})(1 \text{ V C/1 J})}{(96,485.34 \text{ C/mol})}$$

$$= 0.0592 \text{ V}$$

$$E^\circ = \left(\frac{0.0592 \text{ V}}{n}\right) \log K_{eq} \tag{18-5}$$

Equations 18-4 and 18-5 provide a method for calculating equilibrium constants from tables of standard reduction potentials. Example 18-11 illustrates the technique.

Cell Potentials and Equilibrium	Example 18-11

Use tabulated standard reduction potentials to determine K_{eq} for the following redox reaction:

$$Cu(s) + Br_2(l) \rightleftharpoons Cu^{2+}(aq) + 2 Br^-(aq)$$

Strategy: This is a quantitative calculation, so it is appropriate to use the seven-step problem-solving strategy.

Solving
Quantitative
Problems

Solution:

1. We are asked to determine an equilibrium constant from standard reduction potentials.

| Example 18-11 | Cell Potentials and Equilibrium *(continued)* |

2. The reaction is a redox process involving copper and bromine, which is easily broken into two half-reactions:

$$Br_2(l) + 2\ e^- \rightleftharpoons 2\ Br^-(aq) \quad \text{and} \quad Cu(s) \rightleftharpoons Cu^{2+}(aq) + 2\ e^-$$

3. Table 18-1 provides standard reduction potentials for the two reactions:

$$Br_2(l) + 2\ e^- \rightleftharpoons 2\ Br^-(aq) \qquad E° = 1.066\ V$$

$$Cu^{2+}(aq) + 2\ e^- \rightleftharpoons Cu(s) \qquad E° = 0.3419\ V$$

4. The link between standard reduction potentials and the equilibrium constant is Equation 18-5.

$$E° = \left(\frac{0.0592\ V}{n}\right) \log K_{eq}$$

5. and 6. To obtain the cell potential, subtract the less positive standard reduction potential from the more positive standard reduction potential. The spontaneous direction is reduction of Br_2 and oxidation of Cu.

$$E° = E°_{cathode} - E°_{anode} = 1.066\ V - (0.3419\ V) = 0.724\ V$$

Now use Equation 18-5 with the appropriate value of n, found from the half-reactions:

$$E° = \left(\frac{0.0592\ V}{n}\right) \log K_{eq}$$

$$\log K_{eq} = \frac{nE°}{0.0592\ V}$$

$$\log K_{eq} = \frac{(2)(0.724\ V)}{0.0592\ V} = 24.5$$

$$K_{eq} = 10^{24.5} = 3 \times 10^{24}$$

7. The magnitude of this equilibrium constant indicates that the redox reaction goes essentially to completion. This reflects the fact that bromine is a potent oxidizing agent and copper is relatively easy to oxidize.

The Nernst Equation

In most laboratories, electrochemistry is practiced under nonstandard conditions. That is, concentrations of dissolved solutes often are not 1 M, and gases are not necessarily at 1 atm. Recall from Chapter 13 that ΔG is very sensitive to concentration and pressure. The equation that links $\Delta G°$ with free energy changes under nonstandard conditions is Equation 13-12:

The term $RT \ln Q$ in Equation 13-12 describes an entropy effect. See Section 13.4 for a review.

$$\Delta G = \Delta G° + RT \ln Q \qquad (13\text{-}12)$$

Here, Q is the reaction quotient.

We can see how concentration and pressure affect cell potentials by substituting for ΔG and $\Delta G°$ using Equations 18-2 and 18-3.

$$\Delta G = -nFE \qquad and \qquad \Delta G^\circ = -nFE^\circ$$

$$-nFE = -nFE^\circ + RT \ln Q$$

Dividing both sides by $(-nF)$ gives the **Nernst equation:**

$$E = E^\circ - \frac{RT}{nF} \ln Q \qquad\qquad (18\text{-}6)$$

The Nernst equation is used to convert between standard cell potentials and potentials of electrochemical cells operating under nonstandard concentration conditions.

When measurements are made at standard temperature, 298 K, the numerical value calculated earlier can replace RT/F:

$$E = E^\circ - \left(\frac{0.0592 \text{ V}}{n}\right) \log Q \qquad when \qquad T = 298 \text{ K}$$

Recall that the numerator of Q contains concentrations of products, and the denominator contains concentrations of reactants, all raised to powers equal to their stoichiometric coefficients. Solvents and pure solids and liquids do not appear in the concentration quotient. Thus only solutes and gases appearing in the cell reaction affect its cell potential. Nevertheless, most cell reactions involve solutes, so a typical cell potential differs from E°. Example 18-12 provides an illustration.

Nonstandard Cell Conditions

Example 18-12

The permanganate ion is a powerful oxidizing agent that oxidizes water to oxygen under standard conditions:

$$MnO_4^-(aq) + 8 H_3O^+(aq) + 5 e^- \rightleftharpoons Mn^{2+}(aq) + 12 H_2O \qquad E^\circ = 1.507 \text{ V}$$

$$O_2(g) + 4 H_3O^+(aq) + 4 e^- \rightleftharpoons 6 H_2O \qquad E^\circ = 1.229 \text{ V}$$

What is the potential of a permanganate-oxygen cell operating at pH $= 7.00$, oxygen at $p = 0.200$ atm, 0.100 M MnO_4^-, and 0.100 M Mn^{2+}?

Strategy: This is a quantitative problem, so we follow the standard strategy.

Solution:

Solving
Quantitative
Problems

1. The problem asks about an actual potential under nonstandard conditions.

2. Before we determine the potential, we must visualize the electrochemical cell and determine the balanced chemical reaction. The half-reactions are given in the problem. To obtain the balanced equation, reverse the direction of the reduction half-reaction with the less positive standard potential, balance the electrons, and combine the two half-reactions:

$$4 MnO_4^- + 32 H_3O^+ + 20 e^- \longrightarrow 4 Mn^{2+} + 48 H_2O$$

$$30 H_2O \longrightarrow 5 O_2(g) + 20 H_3O^+ + 20 e^-$$

$$\overline{\rule{0pt}{1.2em}4 MnO_4^- + 12 H_3O^+(aq) \longrightarrow 4 Mn^{2+} + 5 O_2 + 18 H_2O}$$

| Example 18-12 | Nonstandard Cell Conditions (continued) |

3. The data are the $E°$ values for the half-reactions, which are given in the problem. Under standard conditions, permanganate is reduced to Mn^{2+} at the cathode, and water is oxidized at the anode:

$$E°_{cell} = E°_{cathode} - E°_{anode} = 1.507\ V - 1.229\ V = 0.278\ V$$

The balanced half-reactions show that $n = 20$.

4. The cell operates under nonstandard conditions, so use the Nernst equation.

$$E = E° - \left(\frac{0.0592\ V}{n}\right) \log Q$$

The form of the reaction quotient comes from the balanced overall redox equation.

$$Q = \frac{[p_{O_2}]^5\ [Mn^{2+}]^4}{[MnO_4{}^-]^4\ [H_3O^+]^{12}}$$

5. The Nernst equation is in the form needed to find what is asked for, E. Use the concentrations to evaluate Q:

$$Q = \frac{[p_{O_2}]^5\ [Mn^{2+}]^4}{[MnO_4{}^-]^4\ [H_3O^+]^{12}} = \frac{(0.200)^5\ (0.100)^4}{(0.100)^4\ (1.00 \times 10^{-7})^{12}} = 3.20 \times 10^{80}$$

$$\log Q = \log (3.20 \times 10^{80}) = 80.505$$

6. Substitute into the Nernst equation to find the cell potential:

$$E = E° - \left(\frac{0.0592\ V}{n}\right) \log Q = 0.278\ V - \left(\frac{0.0592\ V}{20}\right)(80.505) = 0.040\ V$$

This result is accurate to three decimal places because the $E°$ value is accurate to three decimal places and the correction is a subtraction.

7. The low concentration of H_3O^+ in pure water reduces this cell potential nearly to zero, but the reaction is still spontaneous, so permanganate can oxidize water. Solutions of potassium permanganate slowly deteriorate and cannot be stored for long times. The reaction is slow enough, however, that significant oxidation does not occur over days or weeks.

The pH Meter

As the Nernst equation describes, cell potentials are linked quantitatively to concentrations. One practical consequence of this relationship is that potential measurements can be used to determine concentrations of ions in solution. The most common example is the **pH meter,** which scientists use to measure the acidity of aqueous solutions. Figure 18-18 illustrates this device: A probe is dipped into the sample, and the pH appears on a digital display. A pH meter measures the electrical potential difference between the sample and a reference solution containing acid at known concentration.

Contemporary pH meters use probes made up of two reference electrodes, shown diagrammatically in Figure 18-18. One electrode contains a buffer solution of known pH. This buffer solution is separated from the solution whose pH is to be measured by a glass membrane, so this electrode is called a *glass electrode*. Because hydronium ions participate in the cell reaction of the glass electrode, the overall cell potential depends on the hydronium ion concentration in the solution whose pH is being measured.

Glass pH electrodes are simple to use and maintain. They respond selectively to hydronium ion concentration and provide accurate measurements of pH values between about 0 and 10. They can be small enough to be implanted into blood vessels or even inserted into individual living cells. In precision work, these electrodes are calibrated before each use, because their characteristics change somewhat with time and exposure to solutions. The electrode is dipped into a buffer solution of known pH, and the meter is electronically adjusted until it reads the correct value.

Electrochemical Stoichiometry

The coefficients of any balanced redox equation describe the stoichiometric ratios between chemical species, just as for other balanced chemical equations. In redox reactions, moles of chemical change can also be related to moles of electrons. Because electrons always cancel in a balanced redox equation, however, we need to look at half-reactions to determine the stoichiometric coefficients for the electrons. A balanced half-reaction provides the stoichiometric coefficients needed to compute the number of moles of electrons transferred for every mole of reagent.

Electricity is normally measured in units of charge, the coulomb (C), or as rate of electric current flow, the ampere (A: 1 A = 1 C/s). The total amount of charge is the product of the current flow, symbolized by I, and the time for which this current flows:

$$\text{Charge} = It$$

Just as molar mass provides the link between mass and moles, the Faraday constant provides the link between charge and moles. The number of moles of electrons transferred in a specific amount of time is the charge in coulombs divided by the charge per mole, F:

$$\text{Moles of electrons} = n = \frac{It}{F} \qquad (18\text{-}7)$$

$$I = \text{Current (C s}^{-1}), \qquad t = \text{Time (s)}, \qquad and \qquad F = 96{,}485 \text{ C mol}^{-1}$$

Equation 18-7 links the stoichiometry of a redox reaction with the characteristics of an electrochemical cell, and Example 18-13 shows how to apply this equation.

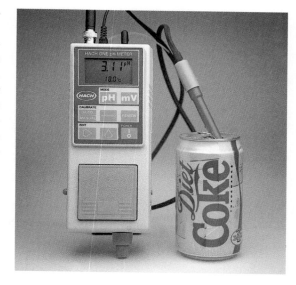

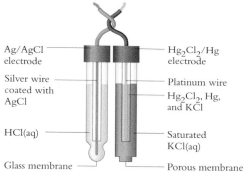

Figure 18-18
Modern pH meters are compact digital electronic devices. The probe consists of two electrodes whose compositions are shown in the diagram.

Here *n* refers to moles of electrons reacting, whereas in Equations 18-3 through 18-6, *n* is the moles of electrons participating in the balanced chemical reaction.

| Electron Stoichiometry | Example 18-13 |

An automobile's headlights typically draw 5.9 amperes of current. The galvanic cell of a lead storage battery, described in Example 18-8, consumes Pb and PbO_2 as it operates. A typical electrode contains about 250 g of PbO_2. Assuming that the battery can supply 5.9 A of current until all the PbO_2 has been consumed, how long will it take for a battery to run down if the lights are left on after the engine is turned off?

Example 18-13	Electron Stoichiometry *(continued)*

Strategy: This is an electrochemical stoichiometry problem, in which an amount of a chemical substance is consumed as electrical current flows. We use the seven–step strategy in summary form.

Solution: The question asks how long the battery can continue to supply current.
Current flows as long as there is lead(IV) oxide present to accept electrons, and the battery dies when all the lead(IV) oxide is consumed. We need to have a balanced half-reaction to provide the stoichiometric relationship between moles of electrons and moles of PbO_2.

There are 250 g of PbO_2, and the headlights draw 5.9 A of current.

Equation 18-7 links current with moles of electrons. Moles of electrons and moles of PbO_2 are related as described by the balanced half-reaction, determined in Example 18-8:

$$PbO_2(s) + HSO_4^-(aq) + 3\,H_3O^+(aq) + 2\,e^- \longrightarrow PbSO_4(s) + 5\,H_2O(l)$$

This gives us the stoichiometric ratio needed to relate moles of electrons to moles of chemical substance:

$$\text{mol e}^- = \text{mol } PbO_2 \left(\frac{2 \text{ mol e}^-}{1 \text{ mol } PbO_2} \right)$$

Use the molar mass of PbO_2 to convert mass of PbO_2 to moles:

$$\text{mol } PbO_2 = \left(\frac{m_{PbO_2}}{MM_{PbO_2}} \right) = \left(\frac{250 \text{ g}}{239.2 \text{ g/mol}} \right) = 1.045 \text{ mol}$$

$$\text{mol e}^- = \left(\frac{2 \text{ mol e}^-}{1 \text{ mol } PbO_2} \right) (1.045 \text{ mol } PbO_2) = 2.090 \text{ mol electrons}$$

We rearrange Equation 18-7, which provides the link between moles of electrons and the time it takes for the battery to become discharged:

$$\text{mol e}^- = \frac{It}{F} \qquad or \qquad t = \frac{(\text{mol e}^-)\,F}{I}$$

$$t = \frac{(96{,}485 \text{ C/mol})(2.090 \text{ mol})}{(5.9 \text{ C/s})(60 \text{ s/min})(60 \text{ min/hr})} = 9.5 \text{ hours}$$

This is the maximum time that the battery could supply current. In practice, the battery runs down before all the PbO_2 is consumed, because the voltage drops as the battery operates.

When an automobile engine is running, it drives an electrical generator or alternator that provides current for the headlights and other needs. The lead storage battery does not run down as long as the generator or alternator is functioning properly.

Redox reactions may involve solids, solutes, gases, or charge flows. Therefore you must be prepared for all the various conversions from molar amounts to measurable variables. As a reminder, Table 18-2 lists the four relationships used for mole calculations.

Galvanic cells use redox reactions to generate electrical current. Electrical current can also drive redox reactions, and the same stoichiometric relationships apply to such processes, as we describe in Section 18.7.

Table 18–2
Relationships Used to Calculate Moles

Type of Material	Equation	Section Reference
Pure Substance	$mol = m/MM$	Section 3.5
Solutes	$mol = MV$	Section 3.7
Gases	$mol = PV/RT$	Section 5.6
Electrons	$mol = It/F$	Section 18.5

Section Exercises

18.5.1 Using standard reduction potentials, determine $\Delta G°$ and K_{eq} at 25 °C for each of the following reactions:
(a) $2\ Cu(s) + Hg^{2+}(aq) \rightleftharpoons 2\ Cu^+(aq) + Hg(l)$
(b) $4\ Cr^{3+} + 21\ H_2O \rightleftharpoons 2\ Cr + Cr_2O_7^{2-} + 14\ H_3O^+$

18.5.2 Use the Nernst equation to determine the potential at 298 K for each of the cells described in Section Exercise 18.4.1, if all cation concentrations are 2.5×10^{-2} M rather than 1.00 M: (a) Fe^{2+} and Cr^{3+}; (b) Cu^{2+} and Pb^{2+}; and (c) Au^{3+} and Ag^+.

18.5.3 In mercury batteries, the spontaneous cell reaction is:

$$HgO + Zn \longrightarrow ZnO + Hg$$

Suppose that a mercury battery contains 250 mg each of HgO and Zn.
(a) Which reagent will be consumed first?
(b) How many coulombs of charge can the battery deliver before this reagent is consumed?
(c) What is the lifetime of the battery (in days) if it powers a digital watch requiring 1.50 microamperes of current?

18.6 REDOX IN ACTION

Electrochemical reactions have many practical applications. Some are spontaneous, and others are driven "uphill" by applying an external potential. In this section, we present practical examples of spontaneous redox processes and in Section 18.7, we describe the principles and applications of externally driven redox reactions.

Batteries

A battery is a galvanic cell that generates electrical current to power a practical device. Batteries range in size from the mercury "buttons" that power cameras and hearing aids to the charge storage banks of electric automobiles (Figure 18–19).

A battery must use cell reactions that generate and maintain a large electrical potential difference. This requires two half-reactions with substantially different standard reduction potentials. The ideal battery would be compact, inexpensive, rechargeable, and environmentally safe. This is a stringent set of requirements. No battery meets all of them, and only a few come close.

Figure 18-19
Batteries range in size from small "buttons" that power hearing aids to large battery packs for electric automobiles.

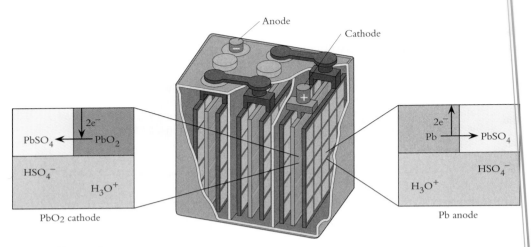

Figure 18-20
Schematic view of a lead storage cell. Lead is oxidized and lead(IV)oxide reduced during the operation of this battery.

The **lead storage battery** provides electrical power in automobiles. It is well suited for this use because it supplies the large current needed to drive starter motors and headlights and can be recharged easily. The lead storage cell is shown in a schematic view in Figure 18-20. The half-reactions were balanced in Example 18-8:

$$PbO_2(s) + HSO_4^-(aq) + 2\ e^- + 3\ H_3O^+(aq) \rightleftharpoons PbSO_4(s) + 5\ H_2O(l)$$
$$E° = +1.6913\ V$$

$$PbSO_4(s) + 2\ e^- + H_3O^+(aq) \rightleftharpoons Pb(s) + HSO_4^-(aq) + H_2O(l)$$
$$E° = -0.3588\ V$$

Here is the net reaction that occurs in the battery:

$$Pb(s) + 2\ HSO_4^-(aq) + PbO_2(s) + 2\ H_3O^+(aq) \longrightarrow 2\ PbSO_4(s) + 4\ H_2O(l)$$

$$E° = 1.6913 - (-0.3588) = 2.0501\ V$$

The anode is a lead plate that becomes coated with $PbSO_4$ as the battery discharges. The cathode is lead containing PbO_2, which also becomes coated with $PbSO_4$ as the battery discharges. Both electrodes are immersed in a bath of sulfuric acid that is approximately 1 M. Thus the working potential of the cell is close to its $E°$ value of about 2 V. Automobile batteries use six such cells, connected in series to generate a total electrical potential of 12 V. This battery does not require a porous separator between its electrodes because the starting materials and products for both half-reactions are insoluble in sulfuric acid. The lead-containing materials never come into contact with each other, so there is no possibility of direct electron transfer.

An automobile battery supplies electrical current when the motor is not running. While the auto is running, its motor drives an electrical generator that meets the vehicle's electrical needs. The generator also recharges the battery by supplying the energy that is required to drive its redox reactions in the reverse direction (see Section 18.7). When the electrical system is functioning properly, the battery is maintained at its optimum charge level, giving these batteries long lifetimes.

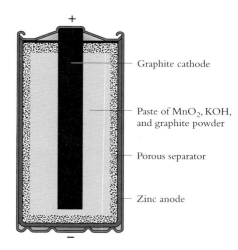

Graphite cathode

Paste of MnO$_2$, KOH, and graphite powder

Porous separator

Zinc anode

Figure 18-21
Schematic diagram (*left*) and examples (*right*) of alkaline dry cells. These batteries provide electrical current for many portable devices.

Flashlight batteries are usually **alkaline dry cells,** so called because they are basic (alkaline) but do not contain aqueous solutions (dry). In these batteries, zinc and manganese dioxide are the working materials (see schematic diagram in Figure 18-21). The dry cell has a zinc anode in contact with a moist paste of MnO$_2$, KOH, and graphite powder. The paste contacts a passive graphite cathode. Both half-reactions involve multiple steps, but the chemistry can be approximated by single reactions:

Cathode: $2\,MnO_2(s) + H_2O(l) + 2\,e^- \longrightarrow Mn_2O_3(s) + 2\,OH^-(aq)$

Anode: $Zn(s) + 2\,OH^-(aq) \longrightarrow Zn(OH)_2(s) + 2\,e^-$

Net: $2\,MnO_2(s) + H_2O(l) + Zn(s) \longrightarrow Mn_2O_3(s) + Zn(OH_2)(s)$

A dry cell generates a potential of about 1.5 V. These cells run irreversibly, and their cell potential slowly decreases with time. Nevertheless, dry cells have many uses because they are compact and made of inexpensive materials with low toxicities. Approximately a billion of these batteries are produced annually in the United States.

Mercury batteries are more expensive than dry cells and contain toxic mercury compounds. These disadvantages are offset by high capacity and a cell potential that does not vary with use. Zinc and mercury are the working chemicals in these batteries, whose cell reactions follow:

Cathode: $HgO(s) + H_2O(l) + 2\,e^- \longrightarrow Hg(l) + 2\,OH^-(aq)$

Anode: $Zn(s) + 2\,OH^-(aq) \longrightarrow ZnO(s) + H_2O(l) + 2\,e^-$

Net: $HgO(s) + Zn(s) \longrightarrow Hg(l) + ZnO(s)$

Figure 18-22 shows the molecular processes that occur in a mercury battery. In this cell, none of the components of the redox reaction changes concentration as the battery operates. The zinc- and mercury-containing components are pure liquids or pure solids, and hydroxide ions consumed at the cathode are regenerated at the anode. The aqueous medium allows free diffusion of hydroxide ions from the anode to the cathode. Because none of the concentrations change as the battery functions, a mercury battery delivers a constant voltage of approximately 1.35 V

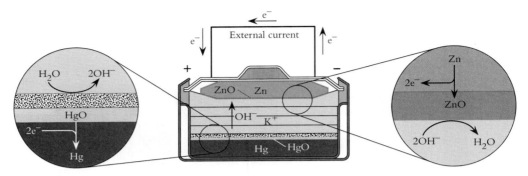

Figure 18-22
In a mercury battery, zinc is oxidized and mercury is reduced, but no net change occurs in the concentrations of any species in solution. The migration of OH⁻ from the Hg electrode to the Zn electrode carries current and maintains a uniform concentration.

throughout its lifetime. Mercury batteries are used widely in devices that require constant voltage, including watches, calculators, hearing aids, and cameras.

The major disadvantage of the standard dry cell, its irreversibility, has been overcome in the **nickel–cadmium battery.** As its name indicates, nickel and cadmium are the working substances in this battery:

Cathode: $NiO(OH)(s) + H_2O(l) + e^- \longrightarrow Ni(OH)_2(s) + OH^-(aq)$

Anode: $Cd(s) + 2\ OH^-(aq) \longrightarrow Cd(OH)_2(s) + 2\ e^-$

Net: $2\ NiO(OH)(s) + Cd(s) + 2\ H_2O(l) \longrightarrow 2\ Ni(OH)_2(s) + Cd(OH)_2(s)$

$$E = E° = 1.35\ V$$

The metal hydroxides generated when the cell discharges adhere tightly to the electrodes, much as lead sulfate adheres to the electrodes of a lead storage battery. As a result, the reactions are easily reversed. This cell, like the lead storage battery, can be recharged by supplying an external electrical potential.

Another recent development in batteries is the fuel cell, which is described in our Chemistry and Technology Box.

Corrosion

Batteries are a beneficial application of spontaneous redox chemistry. **Corrosion,** the destructive oxidation of metals, is harmful. Corrosion is a natural process that returns refined metals to their more stable metal oxides. The oxidizing agent in corrosion chemistry is atmospheric oxygen.

Oxygen is a potent oxidizing agent, particularly in the presence of aqueous acids:

$$O_2(g) + 2\ H_2O(l) + 4\ e^- \rightleftharpoons 4\ OH^-(aq) \qquad E° = +0.401\ V$$

$$O_2(g) + 4\ H_3O^+(aq) + 4\ e^- \rightleftharpoons 6\ H_2O(l) \qquad E° = +1.229\ V$$

Atmospheric O_2 has a partial pressure of 0.20 atm, and atmospheric water vapor is saturated with carbon dioxide. This dissolved CO_2 forms carbonic acid, which generates a hydronium ion concentration of about 2.0×10^{-6} M. The Nernst equation allows calculation of the half-cell potential for the reduction of $O_2(g)$ under these conditions:

Example 16-12 describes the chemistry of carbonic acid.

$$E = E° - \frac{0.0592 \text{ V}}{n} \log \left(\frac{1}{p_{O_2}[H_3O^+_{(aq)}]^4} \right)$$

$$E = 1.229 \text{ V} - \frac{0.0592 \text{ V}}{4} \log \left(\frac{1}{(0.20)(2.0 \times 10^{-6})^4} \right) = 0.88 \text{ V}$$

In damp air, materials with standard reduction potentials less than 0.88 V oxidize spontaneously. Iron and aluminum, the most important structural metals, are easily oxidized by atmospheric O_2:

$$4 \text{ Fe(s)} + 3 \text{ O}_2(g) + 12 \text{ H}_3O^+(aq) \longrightarrow 4 \text{ Fe}^{3+}(aq) + 18 \text{ H}_2O(l)$$

$$E = 0.88 \text{ V} - (-0.037 \text{ V}) = +0.92 \text{ V}$$

$$4 \text{ Al(s)} + 3 \text{ O}_2(g) + 12 \text{ H}_3O^+(aq) \longrightarrow 4 \text{ Al}^{3+}(aq) + 18 \text{ H}_2O(l)$$

$$E = 0.88 \text{ V} - (-1.662 \text{ V}) = +2.54 \text{ V}$$

Both Fe^{3+} and Al^{3+} form highly insoluble oxides that precipitate from solution. Overall, then, each metal is oxidized by O_2 to give its metal oxide:

$$4 \text{ Fe(s)} + 3 \text{ O}_2(g) \longrightarrow 2 \text{ Fe}_2O_3(s)$$

$$4 \text{ Al(s)} + 3 \text{ O}_2(g) \longrightarrow 2 \text{ Al}_2O_3(s)$$

The formation of rust on a steel surface, shown in our chapter opener illustrations, is an obvious manifestation of corrosion. Even more damaging effects of corrosion take place beneath the surface of the metal, where tiny cracks weaken the metal.

The corrosion of iron occurs particularly rapidly when an aqueous solution is present. This is because water that contains ions provides an oxidation pathway with a much lower activation energy barrier than the direct reaction of iron with oxygen gas. As illustrated schematically in Figure 18-23, oxidation and reduction occur at different locations on the metal's surface. In the absence of dissolved ions to act as charge carriers, a complete electrical circuit is missing, so the redox reaction is slow. In contrast, when dissolved ions are present, such as in salt water and acidic water, corrosion can be quite rapid.

Various strategies are employed to prevent corrosion. We described the use of paint as a protective coating in our chapter introduction. A metal surface can also be protected by coating it with a thin film of a second metal. When the second metal is easier to oxidize than the first, the process is *galvanization*. Objects made of iron, including automobile bodies and steel girders, are dipped in molten zinc to provide a sacrificial coating. If a scratch penetrates the zinc film, the iron is still protected because zinc oxidizes preferentially:

$$Fe^{2+} + 2 \text{ e}^- \rightleftharpoons Fe \qquad E° = -0.447 \text{ V}$$

$$Zn^{2+} + 2 \text{ e}^- \rightleftharpoons Zn \qquad E° = -0.7618 \text{ V}$$

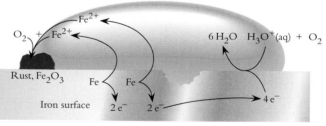

Oxidation: $Fe \longrightarrow Fe^{2+} + 2 \text{ e}^-$

Reduction: $O_2 + 4 \text{ H}_3O^+(aq) + 4 \text{ e}^- \longrightarrow 6 \text{ H}_2O$

Further Redox: $2 \text{ Fe}^{2+} + 4 \text{ OH}^- + \frac{1}{2} O_2 \longrightarrow Fe_2O_3 + 2 \text{ H}_2O$

Figure 18-23
A water droplet on an iron surface is a miniature electrochemical cell. The oxidation of iron occurs in an interior region of the droplet, whereas the reduction of oxygen preferentially occurs near the air-droplet interface. Ionic charge carriers are required to complete the circuit and allow the redox reactions to proceed.

Box 18-2	Chemistry and Technology: Fuel Cells

I magine an automobile that runs in silence and without polluting emissions. Such an automobile, long a dream of the environmentally conscious, has recently become a reality. The power source is a fuel cell, an electrochemical cell that uses a combustion reaction to produce electricity. Hydrocarbons such as natural gas and propane can be used in fuel cells, but the "cleanest" fuel is molecular hydrogen.

The advantage of a fuel cell over a conventional battery is that the fuel for electrical power can be replenished easily. Just as we pull into a service station to refill the gas tank, owners of automobiles powered by fuel cells will refill their fuel tanks with hydrogen or butane.

Although the principle was first proposed in 1839, making a practical fuel cell eluded scientists for a century and a half. The concept is simple, but the chemistry is difficult. A hydrogen fuel cell must cleanly convert H_2 into H_3O^+ at one electrode, cleanly convert O_2 into OH^- at the other electrode, and it must contain a medium that allows these ions to diffuse and combine stoichiometrically.

In a hydrogen fuel cell, oxidation of H_2 at the anode releases electrons into the circuit and produces aqueous H_3O^+. Reduction of O_2 at the cathode consumes electrons and generates OH^-, which combine with H_3O^+ to produce H_2O. The schematic diagram shows these processes.

Anode: $2 H_2(g) + 4 H_2O(l) \longrightarrow 4 H_3O^+(aq) + 4 e^-$

Cathode: $O_2(g) + 2 H_2O(l) + 4 e^- \longrightarrow 4 OH^-(aq)$

Net reaction: $2 H_2(g) + O_2(g) \longrightarrow 2 H_2O(l)$

A hydrogen fuel cell is environmentally friendly, but H_2 is much more difficult to store than liquid fuels. The production, distribution, and storage of hydrogen present major difficulties, so researchers are working on fuel cells that use liquid hydrocarbon fuels. One such fuel cell is composed of layers of yttria-stabilized zirconia (YSZ), solid ZrO_2

stabilized with around 5% Y_2O_3. This cell uses the combustion of a hydrocarbon as its energy source:

Anode: $2 [C_4H_{10} + 13 O^{2-} \longrightarrow 4 CO_2 + 5 H_2O + 26 e^-]$

Cathode: $13 [O_2 + 4 e^- \longrightarrow 2 O^{2-}]$

Net reaction: $2 C_4H_{10} + 13 O_2 \longrightarrow 8 CO_2 + 10 H_2O$

As the reaction shows, a hydrocarbon fuel cell produces carbon dioxide, a "greenhouse" gas. These fuel cells would nevertheless be less polluting than internal-combustion engines, because fuel cells are more efficient and do not generate polluting byproducts such as CO and NO_x.

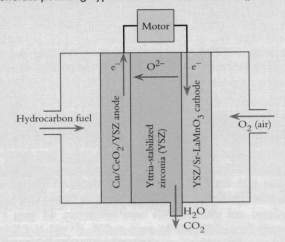

Alkali fuel cells containing KOH and platinum- and gold-coated electrodes were developed for the space program, but these are too expensive for down-to-earth vehicles. In addition, these cells require pure oxygen rather than CO_2-containing air.

Two new technologies have reduced the cost of alkali fuel cells to the point where a European company markets taxis that use them. One is the use of CO_2 "scrubbers" to purify the air supply, making it possible to use atmospheric O_2 rather than purified oxygen. The other is the development of ultrathin films of platinum so that a tiny mass of this expensive metal can provide the catalytic surface area needed for efficient fuel cell operation.

Automobile manufacturers have invested heavily in fuel cells. Buses powered by fuel cells are on the road, and prototype cars are being tested. The chances are excellent that there is a fuel-cell car in your future.

Many iron objects, such as the hulls of ships or oil drilling platforms, are too large to dip in a pot of molten zinc. Instead, these objects are protected from corrosion by connecting them to blocks of some more easily oxidized metal. Magnesium and zinc frequently are used for this purpose. As in galvanization, the more active metal is oxidized preferentially. The sacrificial metal must be replaced as it is consumed, but the cost of replacing a block of zinc or magnesium is minuscule compared with the cost of replacing a sophisticated iron structure.

Another way to protect a metal uses an impervious metal oxide layer. This process is known as *passivation*. In some cases, passivation is a natural process. Aluminum oxidizes readily in air, but the result of oxidation is a thin protective layer of Al_2O_3 through which O_2 cannot readily penetrate. Aluminum oxide adheres to the surface of unoxidized aluminum, protecting the metal from further reaction with O_2. Passivation is not effective for iron because iron oxide is porous and does not adhere well to the metal. Rust continually flakes off the surface of the metal, exposing fresh iron to the atmosphere. Alloying iron with nickel or chromium, whose oxides adhere well to metal surfaces, can be used to prevent corrosion. For example, stainless steel contains small amounts of nickel and chromium.

Section Exercises

- ■ **18.6.1** Draw molecular pictures to illustrate the oxidation and reduction processes that take place in a mercury battery.
- ■ **18.6.2** In Section Exercise 18.3.1, an aluminum–iron galvanic cell was proposed. Using information presented in this section, explain why this cell is unlikely to make a good battery.
- ■ **18.6.3** "Tin" cans are made of iron, coated with a thin film of tin. After a break occurs in the film, a tin can corrodes much more rapidly than zinc-coated iron. Using standard reduction potentials, explain this behavior.

18.7 ELECTROLYSIS

In a galvanic cell, a spontaneous chemical reaction generates an electrical current. It is also possible to use an electrical current to drive a chemical reaction that is nonspontaneous. A "dead" rechargeable battery can be recharged by connecting the battery to an external source of current. In this recharging process an electrical current drives a chemical reaction "uphill."

An electrochemical reaction can be driven in the nonspontaneous direction by an external electrical power source sufficient to offset the potential developed by the cell. Compare this relationship with the flow of water under the influence of gravity. The spontaneous direction for the flow of water is "downhill," from higher to lower gravitational potential. Nevertheless, we can pump water uphill—in the "nonspontaneous" direction—if we exert an opposing force that is greater than the force of gravity. Similarly, we can reverse the direction of a

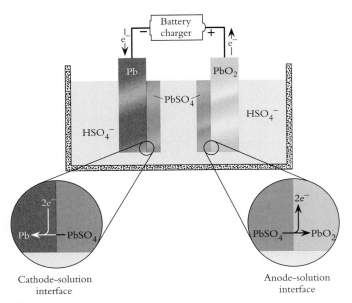

Figure 18-24
A lead storage battery is recharged by forcing electrons to flow in the opposite direction from galvanic operation.

galvanic cell by imposing an external potential to drive electrons "uphill." In the presence of an opposing voltage that is greater than its cell potential, the cell reaction is reversed. This is an **electrolytic cell** rather than a galvanic cell. The process of using electrical current to drive redox reactions is **electrolysis.**

Electrolysis is illustrated by the lead storage battery. The electrolytic (recharging) process is shown in Figure 18-24. As a galvanic cell, this battery oxidizes lead metal and reduces lead(IV) oxide spontaneously, generating an electrical potential of 2.05 V. Electrons flow from the Pb electrode to the PbO_2 electrode. The reaction can be driven in the opposite direction by an external potential of *opposite sign and greater than 2.05 V.* Under these conditions, the external potential forces electrons through the external wire from PbO_2 toward Pb. At the $PbSO_4$–PbO_2 interface, lead sulfate is oxidized to lead oxide:

$$PbSO_4(s) + 5\ H_2O(l) \longrightarrow PbO_2(s) + HSO_4^-(aq) + 3\ H_3O^+(aq) + 2\ e^-$$

The electrons released in this reaction are pushed into the lead electrode, and at the $PbSO_4$–Pb interface, lead sulfate captures electrons and is reduced to lead:

$$PbSO_4(s) + H_3O^+(aq) + 2\ e^- \longrightarrow Pb(s) + HSO_4^-(aq) + H_2O(l)$$

Whether the cell operates galvanically or electrolytically, the electrode where oxidation occurs is always called the anode. Thus under recharging conditions the PbO_2 electrode is the anode, and the Pb electrode is the cathode.

Electrolysis of Water

CHAPTER 15 →
The production of hydrogen gas from methane is described in Chapter 15.

As world deposits of petroleum and coal are exhausted, new sources of hydrogen will have to be developed for use as a fuel and in the production of ammonia for fertilizer. At present, most hydrogen gas is produced from hydrocarbons, but hydrogen gas can also be generated by the electrolysis of water. Figure 18-25 shows

Figure 18-25
A photograph and schematic diagram of a cell for the electrolysis of water. The redox reaction generates hydrogen and oxygen in a 2 : 1 ratio.

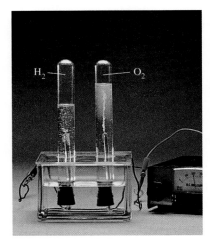

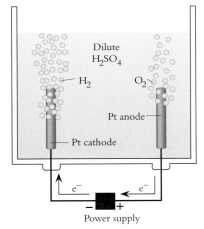

an electrolytic cell set up to decompose water. Two platinum electrodes are dipped in a dilute solution of sulfuric acid. The cell requires just one compartment because hydrogen and oxygen escape from the cell much more rapidly than they react with each other.

During the electrolysis of water, hydronium ions capture electrons from the cathode, producing hydrogen gas. Water molecules lose electrons to the anode, producing oxygen gas:

Oxidation at anode: $\qquad$ $6 H_2O(l) \longrightarrow O_2(g) + 4 H_3O^+(aq) + 4 e^-$

Reduction at cathode: $\quad$ $4 H_3O^+(aq) + 4 e^- \longrightarrow 2 H_2(g) + 4 H_2O(l)$

Overall: $\qquad$ $2 H_2O(l) \longrightarrow 2 H_2(g) + O_2(g)$

Whereas the potential supplied to an electrolytic cell determines whether or not electrolysis can occur, the amount of material electrolyzed is determined by the current flow and the time of electrolysis. Recall Equation 18-7 from Section 18.5:

$$\text{Moles of electrons} = n = \frac{It}{F} \qquad \textbf{(18-7)}$$

This equation provides the link between electrical measurements and amount of electrons, and the balanced half-reactions for the electrolytic process provide the link between the amount of electrons and amounts of chemical substances. Example 18-14 shows a calculation regarding electrolytic stoichiometry.

Electrolysis and Stoichiometry	**Example 18-14**

If the electrolytic cell shown in Figure 18-25 draws a current of 0.775 ampere for 45.0 minutes, calculate the volumes of H_2 and O_2 produced if these gases are collected at 25 °C and $P = 1.00$ atm.

Strategy: "Volumes of H_2 and O_2" identifies this as a stoichiometry problem, for which our seven-step approach, presented here in compact form, is appropriate.

Solution: The problem asks about amounts of chemicals produced in electrolysis and provides data about current and time. Visualize the electrolysis shown in Figure 18-25. Calculating amounts in electrochemistry requires the use of Equation 18-7 and knowledge of the numbers of electrons transferred in the half-reactions.

We calculate moles of electrons using Equation 18-7. Recall that 1 A = 1 C/s:

$$n_{electrons} = \frac{It}{F} = \frac{(0.775 \text{ C/s})(45.0 \text{ min})(60 \text{ s/min})}{(96,485 \text{ C/mol})} = 2.169 \times 10^{-2} \text{ mol}$$

To convert from moles of electrons to moles of chemical species, we need the molar ratios between electrons and chemical species, which are determined from the half-reactions:

$$6 H_2O(l) \longrightarrow O_2(g) + 4 H_3O^+(aq) + 4 e^-$$

$$4 H_3O^+(aq) + 4 e^- \longrightarrow 2 H_2(g) + 4 H_2O(l)$$

Example 18-14	Electrolysis and Stoichiometry *(continued)*

The transfer of four moles of e^- produces two moles of H_2 and one mole of O_2:

$$n_{H_2} = (2.169 \times 10^{-2} \text{ mol } e^-) \frac{(2 \text{ mol } H_2)}{(4 \text{ mol } e^-)} = 1.084 \times 10^{-2} \text{ mol } H_2$$

$$n_{O_2} = (2.169 \times 10^{-2} \text{ mol } e^-) \frac{(1 \text{ mol } O_2)}{(4 \text{ mol } e^-)} = 0.5423 \times 10^{-2} \text{ mol } O_2$$

To complete the calculation, moles of gas must be converted into volumes using the ideal gas equation:

$$V = \frac{nRT}{P}$$

We leave this stoichiometric calculation to you. The volume of H_2 is 0.265 L, and that of O_2 is 0.133 L.

Competitive Electrolysis

When an electrolytic cell is designed, care must be taken in the selection of the cell components. For example, consider what happens when an aqueous solution of sodium chloride is electrolyzed using platinum electrodes. Platinum is used for passive electrodes, because this metal is resistant to oxidation and does not participate in the redox chemistry of the cell. There are three major species in the solution: H_2O, Na^+, and Cl^-. Chloride ions cannot be reduced further, so there are just two candidates for reduction: water molecules and sodium ions:

$$Na^+ + e^- \rightleftharpoons Na \qquad E° = -2.71 \text{ V}$$

$$2\,H_2O + 2\,e^- \rightleftharpoons H_2 + 2\,OH^- \qquad E° = -0.828 \text{ V}$$

In an electrolytic cell, the most easily oxidized species is oxidized, and the most easily reduced species is reduced. The standard potentials show that water is much easier to reduce than sodium ions, so the electrolysis will produce hydrogen and hydroxide ions rather than sodium metal.

The possibilities for oxidation are more varied in this solution. Water and chloride ions can be oxidized, but sodium ions cannot. The various possibilities for oxidation can be found in Appendix F. Because oxidation is the reverse of reduction, the half-reactions chosen must have *products* that include Cl^- or H_2O. Although several possibilities exist, only the two with the lowest reduction potentials need to be considered:

$$O_2 + 4\,H_3O^+ + 4\,e^- \rightleftharpoons 6\,H_2O \qquad E° = 1.229 \text{ V}$$

$$Cl_2 + 2\,e^- \rightleftharpoons 2\,Cl^- \qquad E° = 1.358 \text{ V}$$

In general, we expect that an external potential will drive the pair of reactions whose spontaneous reaction has the least positive cell potential. In this case, combining the reduction of water to hydrogen and hydroxide with the oxidation of water to H_3O^+ and oxygen gas gives the least positive $E°$:

$$2\,H_2(g) + O_2(g) \longrightarrow 2\,H_2O(l) \qquad E°_{cell} = (1.229 \text{ V}) - (-0.828 \text{ V}) = +2.057 \text{ V}$$

However, the cell potential for the reaction involving Cl^- is only slightly larger:

$$H_2(g) + 2\,OH^-(aq) + Cl_2(g) \longrightarrow 2\,H_2O(l) + 2\,Cl^-(aq)$$

$$E^\circ_{cell} = (1.358\ V) - (-0.828\ V) = +2.186\ V$$

Experimentally, when an opposing potential sufficient to cause electrolysis is applied to a cell containing aqueous NaCl, the electrolytic reaction produces chlorine gas rather than oxygen gas at the anode:

Cathode: $2\,H_2O(l) + 2\,e^- \longrightarrow H_2(g) + 2\,OH^-(aq)$

Anode: $2\,Cl^-(aq) \longrightarrow Cl_2(g) + 2\,e^-$

The oxidation of water is favored by thermodynamics, but the oxidation of chloride is a much faster reaction. Notice that the oxidation of water requires four electrons per water molecule, but the oxidation of chloride anions requires only two electrons per chloride ion. When reactions have similar potentials, the one that requires smaller numbers of electrons often is favored.

Electroplating

Electrolysis can be used to deposit one metal on top of another, a process known as **electroplating.** Electroplating is done for cosmetic reasons as well as for protection against corrosion and wear. A thin layer of gold or silver is often plated on jewelry and tableware made from inexpensive metals such as iron. Figure 18-26 shows a schematic diagram of an electroplating apparatus. In this example, the spoon acts as a cathode where silver cations are reduced. The anode is a silver rod. An external power source removes electrons from the silver rod, causing oxidation to Ag^+ cations, which enter the solution. Electrons flow into the spoon, where silver ions in solution capture the electrons to become neutral silver atoms, which adhere to the iron surface as a layer of silver metal.

The quantity of metal deposited during a plating process is linked stoichiometrically to the current flow through Equation 18-7, as Example 18-15 illustrates.

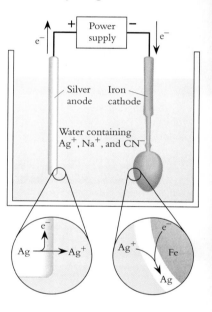

Figure 18-26
The electroplating of silver.

| **Electroplating** | **Example 18-15** |

An electroplating apparatus is used to coat jewelry with gold. What mass of gold can be deposited from a solution that contains $Au(CN)_4^-$ ions if a current of 5.0 amperes flows for 30.0 minutes?

Strategy: This, like Example 18-14, is an electrochemical stoichiometry problem. It asks about the amount of gold deposited in an electrolytic reaction. The method is the same as that of Example 18-14.

Solution:

$$n_{electrons} = \frac{It}{F} = \frac{(5.0\ C/s)(30.0\ min)(60\ s/min)}{(96{,}485\ C/mol)} = 0.0933\ mol\ e^-$$

To determine moles of gold, we need a balanced half-reaction. The reactant is $Au(CN)_4^-$, and the product is Au. Apply the usual procedure for balancing a half-reaction, recognizing the presence of cyanide ions:

$$Au(CN)_4^- + 3\,e^- \longrightarrow Au + 4\,CN^-$$

The stoichiometric coefficients, 1 and 3, indicate that one mole of gold will be plated for every three moles of electrons:

| Example 18-15 | Electroplating *(continued)* |

$$(0.0933 \text{ mol e}^-)\left(\frac{1 \text{ mol Au}}{3 \text{ mol e}^-}\right) = 0.0311 \text{ mol Au}$$

Finally, the molar mass of gold gives the mass of gold in grams:

$$(0.0311 \text{ mol Au})(197 \text{ g/mol}) = 6.1 \text{ g Au}$$

Chromium is plated on the surface of iron to prevent corrosion and to improve appearance.

Steel objects are often protected from corrosion by electroplating with chromium. The most straightforward process would be to electrolyze a solution of Cr^{3+} cations. This fails because aqueous Cr^{3+} ions are not reduced at a useful rate. Instead, solutions containing chromate anions are used:

$$8 \text{ H}_3\text{O}^+(aq) + \text{CrO}_4^{2-}(aq) + 6 \text{ e}^- \rightleftharpoons \text{Cr}(s) + 12 \text{ H}_2\text{O}(l)$$

This reaction proceeds much faster than the direct reduction of Cr^{3+}. It requires twice as much electrical current, however, since 6 mol of electrons must be supplied to deposit 1 mol of Cr rather than 3 mol for direct electrolysis of Cr^{3+}.

Section Exercises

18.7.1 Fluorine is manufactured by the electrolysis of hydrogen fluoride dissolved in molten KF: $2 \text{ HF}(\text{KF melt}) \longrightarrow \text{H}_2(g) + \text{F}_2(g)$
(a) Identify the half-reactions. (The only species present in the molten phase are HF, K^+, and F^-.)
(b) $E° = -0.187$ V for reducing HF to H_2. What is the minimum external potential required to drive the electrolysis of HF at standard concentrations?

18.7.2 If a commercial reactor for the electrolysis of HF operates continuously at a current of 1500 amperes, how many kilograms of F_2 are produced in 24 hours of operation? What is the volume of the tank required to store the F_2 at 298 K and 60 atm pressure?

18.7.3 Explain why neither aqueous KF nor pure liquid HF can be used for the electrolytic production of fluorine, even though both liquids are easier to handle than molten potassium fluoride.

CHAPTER REVIEW

Summary and Key Terms

1. Oxidation and reduction, the loss and gain of electrons, always occur together in proportions that conserve electrons. The **reducing agent** loses electrons and the **oxidizing agent** gains electrons. **Oxidation numbers** change during redox reactions, making them useful in recognizing these reactions.

2. Electrons do not appear explicitly in redox reactions. To balance a redox reaction, it is convenient to separate the overall reaction into reduction and oxidation **half-reactions,** which can be balanced independently using a step-wise procedure.

3. Spontaneous redox reactions can occur directly or indirectly. Indirect transfer involves electron flow through a wire and is electrochemical. Electrochemical processes require complete electrical circuits. The transfer of electrons occurs at **electrodes,** an **anode** where oxidation occurs and a **cathode** where reduction occurs. Whereas **active** electrodes participate in the redox chemistry, **passive** electrodes serve only as reservoirs of electrons. In a **galvanic cell,** or battery, a reduction reaction and an oxidation reaction take place at separate electrodes, generating an electrical voltage and a flow of current.

4. The spontaneity of a redox reaction is measured by the cell potential in volts. Each half-reaction has a **standard reduction potential ($E°$)** that reflects the ease with which the reactants accept electrons. The standard potential for any galvanic cell is the more positive $E°$ minus the more negative $E°$.

5. The negative free energy change of a redox reaction is directly proportional to its cell potential. Potentials and energies are related through the **Faraday constant.** Equilibrium constants for redox reactions also are linked with standard cell potentials. Cell potentials vary logarithmically with chemical concentrations in a manner described by the **Nernst equation.** The **pH meter** is one application of the Nernst equation.
6. Batteries are galvanic cells that generate useful voltages and currents. Examples include the **lead storage battery, alkaline dry**

cells, mercury batteries, and **nickel-cadmium batteries.** Metals, particularly aluminum and iron, are susceptible to **corrosion,** which is oxidation by atmospheric oxygen.
7. The application of a sufficiently large external voltage converts a galvanic cell to an **electrolytic cell** and causes **electrolysis,** which is a redox reaction driven by externally supplied electrons. Battery recharging, electrolysis of water, and **electroplating** are important applications of electrolysis.

Skills to Master

▶ Determining oxidation numbers

▶ Balancing redox reactions using half-reactions

▶ Describing galvanic cells

▶ Calculating standard cell potentials

▶ Relating cell potentials and free energies

▶ Relating cell potentials and equilibrium constants

▶ Doing Nernst equation calculations

▶ Doing charge-mole conversions

▶ Describing examples of batteries

▶ Analyzing electrolytic cells

Learning Exercises

18.1 Describe the steps that must be followed in (a) balancing a redox equation; (b) drawing a molecular diagram of a galvanic cell; (c) calculating the potential of a galvanic cell operating under nonstandard conditions; and (d) determining what reactions occur in an electrolytic cell.
18.2 Describe the mechanism for charge movement in each of the following components of an electrochemical cell: (a) the external wire; (b) the electrode-solution interface; (c) the solution; and (d) the porous barrier.

18.3 Write a paragraph explaining the linkages among cell potential, free energy, and the equilibrium constant.
18.4 List at least six practical examples of redox chemistry.
18.5 Update your list of memory bank equations. For each new equation, specify the type of calculations for which it is useful.
18.6 List all terms new to you that appear in Chapter 18, and write a one-sentence definition of each in your own words. Consult the Glossary if you need help.

Problems ilw = interactive learning ware problem. Visit the website at www.wiley.com/college/olmsted

Recognizing Redox Reactions

18.1 Determine the oxidation numbers of all atoms in the following: (a) $Fe(OH)_3$; (b) NF_3; (c) CH_3OH; (d) K_2CO_3; (e) NH_4NO_3; (f) $TiCl_4$; (g) $PbSO_4$; and (h) P_4.
18.2 Determine the oxidation numbers of all atoms in the following: (a) H_2CO; (b) $AlCl_3$; (c) XeF_4; (d) F_2O; (e) $K_2Cr_2O_7$; (f) $NaIO_3$; (g) P_2O_5; and (h) Na_2O_2.
18.3 Which of the following are redox reactions?
   (a) $HBr(g) + H_2O(l) \rightarrow H_3O^+(aq) + Br^-(aq)$
   (b) $2\,Fe^{2+}(aq) + H_2O_2(aq) \rightarrow 2\,Fe^{3+}(aq) + 2\,OH^-(aq)$
   (c) $Fe^{2+}(aq) + 2\,OH^-(aq) \rightarrow Fe(OH)_2(s)$
   (d) $2\,CH_3OH(l) + 3\,O_2(g) \rightarrow 2\,CO_2(g) + 4\,H_2O(l)$
   (e) $N_2(g) + 3\,H_2(g) \rightarrow 2\,NH_3(g)$
18.4 Which of the following are redox reactions?
   (a) $CO(g) + H_2O(g) \rightarrow CO_2(g) + H_2(g)$
   (b) $CO_2(g) + 2\,H_2O(l) \rightarrow H_3O^+(aq) + HCO_3^-(aq)$
   (c) $2\,CuS(s) + O_2(g) \rightarrow 2\,Cu(s) + SO_2(g)$
   (d) $2\,AgNO_3(aq) + Cu(s) \rightarrow Cu(NO_3)_2(aq) + 2\,Ag(s)$
   (e) $2\,AgNO_3(aq) + Na_2SO_4(aq) \rightarrow 2\,NaNO_3(aq) + Ag_2SO_4(s)$
18.5 Chlorine displays a wide range of oxidation numbers. Determine chlorine's oxidation number in each of the following species: (a) ClF_5; (b) $MgCl_2$; (c) $NaClO_4$; (d) Cl_2; (e) $KClO_2$; and (f) $NaClO$.

18.6 Sulfur displays a wide range of oxidation numbers. Determine the oxidation number of sulfur in each of the following species: (a) H_2S; (b) S_2Cl_2 (c) Li_2SO_4 (d) S_8; (e) Na_2SO_3 (f) SF_4.

Balancing Redox Reactions

18.7 Determine the half-reactions for the following redox processes:
   (a) Sodium metal reacts with water to give hydrogen gas.
   (b) Gold metal dissolves in "aqua regia"(a mixture of HCl and HNO_3) to give $[AuCl_4]^-$ and NO.
   (c) Acidic potassium permanganate reacts with aqueous $K_2C_2O_4$, producing carbon dioxide and Mn^{2+}.
   (d) Coal (solid carbon) reacts with steam to produce molecular hydrogen and carbon monoxide.
18.8 What are the half-reactions for these redox processes?
   (a) Aqueous hydrogen peroxide acts on Co^{2+}, and the products are hydroxide and Co^{3+}, in basic solution.
   (b) Methane reacts with oxygen gas and produces water and carbon dioxide.
   (c) To recharge a lead storage battery, lead(II) sulfate is converted to lead metal and to lead(IV) oxide.
   (d) Zinc metal dissolves in aqueous hydrochloric acid to give Zn^{2+} ions and hydrogen gas.

18.9 Balance the following half-reactions:

(a) $Cu^+ \rightarrow CuO$ (Acidic solution)
(b) $S \rightarrow H_2S$ (Acidic solution)
(c) $AgCl \rightarrow Ag$ (Basic solution)
(d) $I^- \rightarrow IO_3^-$ (Basic solution)
(e) $IO_3^- \rightarrow IO^-$ (Basic solution)
(f) $H_2CO \rightarrow CO_2$ (Acidic solution)

18.10 Balance the following half-reactions:

(a) $SbH_3 \rightarrow Sb$ (Acidic solution)
(b) $AsO_2^- \rightarrow As$ (Basic solution)
(c) $BrO_3^- \rightarrow Br_2$ (Acidic solution)
(d) $Cl^- \rightarrow ClO_2^-$ (Basic solution)
(e) $Sb_2O_5 \rightarrow Sb_2O_3$ (Acidic solution)
(f) $H_2O_2 \rightarrow O_2$ (Basic solution)

18.11 Balance the net redox reaction resulting from combining the following half-reactions in Problem 18.9 (the first listed in each case is the reduction): (a) 9a and 9b; (b) 9d and 9c; (c) 9d and 9e; and (d) 9f and 9b.

18.12 Balance the net redox reaction resulting from combining the following half-reactions in Problem 18.10 (the first listed in each case is the reduction): (a) 10a and 10c; (b) 10a and 10e; (c) 10b and 10d; and (d) 10b and 10f.

18.13 Consider the redox reaction:

$$2\ MnO_4^- + 10\ Cl^- + 16\ H_3O^+ \longrightarrow 2\ Mn^{2+} + 5\ Cl_2 + 24\ H_2O$$

(a) Which species is oxidized? (b) Which species is reduced? (c) Which species is the oxidizing agent? (d) Which species is the reducing agent? (e) Which species gains electrons? (f) Which species loses electrons?

18.14 Consider the redox reaction:

$$4\ NO + 3\ O_2 + 6\ H_2O \longrightarrow 4\ NO_3^- + 4\ H_3O^+$$

(a) Which species is oxidized? (b) Which species is reduced? (c) Which species is the oxidizing agent? (d) Which species is the reducing agent? (e) Which species gains electrons? (f) Which species loses electrons?

18.15 Balance the following redox equations:

(a) $CN^- + MnO_4^- \rightarrow CNO^- + MnO_2$ (Basic)
(b) $O_2 + As \rightarrow HAsO_2 + H_2O$ (Acidic)
(c) $Br^- + MnO_4^- \rightarrow MnO_2 + BrO_3^-$ (Basic)
(d) $NO_2 \rightarrow NO_3^- + NO$ (Acidic)
(e) $ClO_4^- + Cl^- \rightarrow ClO^- + Cl_2$ (Acidic)
(f) $AlH_4^- + H_2CO \rightarrow Al^{3+} + CH_3OH$ (Basic)

18.16 Balance the following redox equations:

(a) $H_5IO_6 + Cr \rightarrow IO_3^- + Cr^{3+}$ (Acidic)
(b) $Se + Cr(OH)_3 \rightarrow Cr + SeO_3^{2-}$ (Basic)
(c) $HClO_2 + Co \rightarrow Cl_2 + Co^{2+}$ (Acidic)
(d) $CH_3COH + Cu^{2+} \rightarrow CH_3CO_2^- + Cu_2O$ (Basic)
(e) $NO_3^- + H_2O_2 \rightarrow NO + O_2$ (Basic)
(f) $BrO_3^- + Fe^{2+} \rightarrow Br^- + Fe^{3+}$ (Acidic)

Spontaneous Redox Reactions and Galvanic Cells

18.17 Use standard thermodynamic values to determine whether or not each of the following redox reactions is spontaneous under standard conditions:

(a) $O_2 + 2\ Cu \rightarrow 2\ CuO$
(b) $O_2 + 2\ Hg \rightarrow 2\ HgO$
(c) $CuS + O_2 \rightarrow Cu + SO_2$
(d) $FeS + O_2 \rightarrow Fe + SO_2$

18.18 Use standard thermodynamic values to determine whether or not each of the following redox reactions is spontaneous under standard conditions:

(a) $H_2O + CO \rightarrow CO_2 + H_2$
(b) $2\ Al + 3\ MgO \rightarrow 3\ Mg + Al_2O_3$
(c) $PbS + Cu \rightarrow CuS + Pb$
(d) $N_2 + 2\ O_2 \rightarrow 2\ NO_2$

18.19 Draw a sketch that shows a molecular view of the charge transfer processes that take place at a silver–silver chloride electrode in contact with aqueous HCl, undergoing reduction:

$$AgCl(s) + e^- \longrightarrow Ag(s) + Cl^-(aq)$$

18.20 Draw a sketch that shows a molecular view of the charge transfer processes that take place at a nickel electrode in contact with aqueous nickel(II) chloride, undergoing oxidation:

$$Ni^{2+}(aq) + 2e^- \longrightarrow Ni(s)$$

18.21 Draw a sketch of a cell that could be used to study the following redox reaction:

$$H_2(g) + 2\ H_2O(l) + Cl_2(g) \rightleftharpoons 2\ H_3O^+(aq) + 2\ Cl^-(aq)$$

18.22 Draw a sketch of a cell that could be used to study the following redox reaction:

$$2\ Cu^+(aq) \rightleftharpoons Cu(s) + Cu^{2+}(aq)$$

18.23 Which of the electrodes in Problems 18.19 and 18.21 are active, and which are passive?

18.24 Which of the electrodes in Problems 18.20 and 18.22 are active, and which are passive?

Cell Potentials

18.25 Use standard reduction potentials in Appendix F to calculate $E°$ for the reactions in Problem 18.15, a, b, and e.

18.26 Use standard reduction potentials in Appendix F to calculate $E°$ for the reactions in Problem 18.16, a, b, and e.

ilw 18.27 Balance and calculate the standard potential for the following reaction in acidic solution:

$$NO \longrightarrow N_2O + NO_3^-$$

18.28 Balance and calculate the standard potential for the following reaction in acidic solution:

$$ClO_3^- \longrightarrow ClO_4^- + Cl^-$$

18.29 Describe a cell that could be used to measure the $E°$ of the F_2/F^- reduction reaction. Include a sketch similar to that shown in Figure 18-9. Which electrode would be the anode?

18.30 A cell is set up with two Cu wire electrodes, one immersed in a 1.0 M solution of $CuNO_3$, the other in a 1.0 M solution of $Cu(NO_3)_2$. Determine $E°$ of this cell, identify the anode, and draw a picture that shows the direction of electron flow at each electrode and in the external circuit.

18.31 Consult Appendix F and list the metals that can reduce Be^{2+} to Be under standard conditions. What characteristics do these metals have in common?

18.32 Consult Appendix F and list the elements that can oxidize H_2O to O_2 under standard conditions. What characteristics do these elements have in common?

Free Energy and Electrochemistry

18.33 Use standard reduction potentials in Appendix F to calculate $\Delta G°$ for the reactions in Problem 18.15, a, b, and e.

18.34 Use standard reduction potentials in Appendix F to calculate $\Delta G°$ for the reactions in Problem 18.16, a, b, and e.

ilw 18.35 Example 18-10 describes the nickel-cadmium battery. What potential does this battery produce if its hydroxide ion concentration is 1.50×10^{-2} M?

18.36 If the cell illustrated in Figure 18-15 contains 1.00×10^{-3} M concentrations of HCl and $CuSO_4$, what potential does it produce?

18.37 If it takes 15 seconds to start your car engine and the battery provides 5.9 amperes of current to the starter motor, what masses of Pb and PbO_2 are used in each battery cell?

18.38 Suppose that automobiles were equipped with "thermite" batteries as described in Section Exercise 18.3.1. What masses of Al and Fe_2O_3 would be consumed in the process described in Problem 18.37?

18.39 A car's alternator recharges the car battery when the engine is running. If the alternator produces 1.750 amperes of current and the operation of the engine requires 1.350 amperes of this current, how long will it take to convert 0.850 g of $PbSO_4$ back into Pb metal?

18.40 A digital watch draws 0.20 mA of current provided by a mercury battery, whose net reaction is:

$$HgO(s) + Zn(s) \longrightarrow ZnO(s) + Hg(l)$$

If a partially used battery contains 1.00 g of each of these four substances, for how many more hours will the watch run?

18.41 Dichromate ions, $Cr_2O_7{}^{2-}$, oxidize acetaldehyde, CH_3CHO, to acetic acid, CH_3CO_2H, and are reduced to Cr^{3+}. The reaction takes place in acidic solution. Balance the redox reaction and determine how many moles of electrons are required to oxidize 1.00 g of acetaldehyde. What mass of sodium dichromate would be required to deliver this many electrons?

18.42 Permanganate ions, $MnO_4{}^-$, oxidize acetaldehyde, CH_3CHO, to acetic acid, CH_3CO_2H, and are reduced to MnO_2. The reaction takes place in acidic solution. Balance the redox reaction and determine how many moles of electrons are required to oxidize 1.00 g of acetaldehyde. What mass of $KMnO_4$ would be required to deliver this many electrons?

Redox in Action

18.43 Set up the Nernst equation for the standard dry cell. Using this equation, show that the voltage of a dry cell must decrease with use.

18.44 Set up the Nernst equation for the lead storage cell, and use it to show that the voltage of this cell must decrease with use.

18.45 If a chromium-plated steel bicycle handlebar is scratched, breaking the protective film, will the chromium or the steel corrode? Use standard potentials to support your prediction.

18.46 Using standard potentials, explain why the steel propeller of an ocean-going yacht has a zinc collar.

18.47 Explain why mercury batteries find extensive use for cameras and pacemakers but are not used to start automobiles.

18.48 Explain why the lead storage battery, despite being the battery of choice for automobiles, is not suitable for space flights.

Electrolysis

18.49 In the electrolysis of aqueous NaCl, what mass of Cl_2 is generated by a current of 4.50 amperes flowing for 200.0 min?

18.50 A portable CD player that draws 150 milliamperes of current is powered by Ni-Cd rechargeable batteries. Compute the masses of Cd and NiO_2 consumed when a disk is played whose length is 65 minutes.

18.51 The Zn-Cu galvanic cell shown in Figure 18-7 is not rechargeable because of competitive electrolysis. Describe the reactions that take place if an opposing potential sufficient to reverse the reactions is applied to this galvanic cell.

18.52 The alkaline dry cell is not rechargeable. The solid products separate from the electrodes, so the reverse reactions cannot occur. What reactions may take place if an opposing potential sufficient to reverse the reactions is applied to a dry cell?

ilw 18.53 After use, a nickel-cadmium battery has 1.55 g of $Cd(OH)_2$ deposited on its anode. It is inserted in a recharger that supplies 125 mA of current at a voltage of 1.45 V. (a) To which electrode, Ni or Cd, should the anode from the charger be connected? Write the half-reaction occurring at this electrode during charging. (b) Compute the time in hours needed to convert all 1.55 g of $Cd(OH)_2$ back to Cd metal.

18.54 A "dead" 12-V lead storage battery has 4.80 g of $PbSO_4$ deposited on each of its anodes. It is connected to a "trickle charger" that supplies 0.120 ampere of current at a voltage of 13 V. (a) To which electrode, Pb or PbO_2, should the anode from the charger be connected? Write the half-reaction occurring at this electrode during charging. (b) Compute the time in hours needed to convert all 4.80 g of $PbSO_4$ back to $HSO_4{}^-$ ions in solution.

18.55 Electrochemistry can be used to measure electrical current in a silver coulometer, in which a silver cathode is immersed in a solution containing Ag^+ ions. The cathode is weighed before and after passage of current. A silver cathode initially has a mass of 10.77 g, and its mass increases to 12.89 g after current has flowed for 15.0 minutes. Compute the quantity of charge in coulombs and the current in amperes.

18.56 When Thomas Edison first sold electricity, he used zinc coulometers to measure charge consumption (See Problem 18.55). If the zinc plate in one of Edison's coulometers increased in mass by 7.55 g, how much charge had been consumed?

18.57 In a silver coulometer (see Problem 18.55), both electrodes are silver metal. Draw a molecular picture that illustrates the reactions that occur during operation of this coulometer.

18.58 In Edison's zinc coulometer (see Problem 18.56), both electrodes were zinc metal. Draw a molecular picture that illustrates the reactions that occur during operation of this coulometer.

Additional Paired Problems

18.59 Here are two standard reduction potentials:

$$O_2(g) + 2\,H_2O + 4\,e^- \rightleftharpoons 4\,OH^-(aq) \qquad E° = 0.401 \text{ V}$$

$$O_2(g) + 4\,H_3O^+(aq) + 4\,e^- \rightleftharpoons 6\,H_2O \qquad E° = 1.229 \text{ V}$$

Are these data sufficient to allow calculation of K_w at 298 K? If so, do the calculation. If not, explain in detail what additional data (thermodynamic or electrochemical) you would need to do the calculation.

18.60 Here are two standard reduction potentials:

$$SO_4{}^{2-}(aq) + 4 H_3O^+(aq) + 2 e^- \rightleftharpoons SO_2(g) + 6 H_2O(l)$$

$$E° = 0.20 \text{ V}$$

$$SO_4{}^{2-}(aq) + 4 H_3O^+(aq) + 2 e^- \rightleftharpoons H_2SO_3(aq) + 5 H_2O(l)$$

$$E° = 0.18 \text{ V}$$

Are these data sufficient to allow calculation of the equilibrium vapor pressure of $SO_2(g)$ over a 1.00 M solution of $H_2SO_3(aq)$ at 298 K? If so, do the calculation. If not, explain in detail what additional data (thermodynamic or electrochemical) you would need to do the calculation.

18.61 Use data in Appendix F to calculate the equilibrium constant of the reaction,

$$Zn^{2+}(aq) + 4 NH_3(aq) \rightleftharpoons Zn(NH_3)_4{}^{2+}(aq)$$

18.62 Use data in Appendix F to calculate K_{sp} for AgI.

18.63 For the reduction of $Cr(OH)_3$ by H_2 in basic solution to give Cr and H_2O, do the following: (a) Write the balanced net equation; (b) Compute $E°$; and (c) Compute $\Delta G°$.

18.64 Consider the following redox reaction:

$$MnO_2 \longrightarrow MnO_4{}^- + Mn^{2+} \quad \text{(unbalanced)}$$

(a) Balance this reaction in acidic solution; (b) write the expression for Q; and (c) determine n, $E°$, $\Delta G°$, and K_{eq} at 298 K.

18.65 Potassium chromate, K_2CrO_4, dissolves in acidic solution to generate strongly oxidizing $Cr_2O_7{}^{2-}$ ions:

$$K_2CrO_4(s) + H_3O^+(aq) \longrightarrow 2 K^+(aq) + HCrO_4{}^-(aq) + H_2O(l)$$

$$2 HCrO_4{}^-(aq) \longrightarrow Cr_2O_7{}^{2-}(aq) + H_2O(l)$$

This oxidizing agent is reduced to Cr^{3+} as it oxidizes other substances. (a) Balance the reduction half-reaction. (b) What mass of K_2CrO_4 produces 0.250 mol of electrons?

18.66 Breathalyzers determine the alcohol content in a person's breath by a redox reaction using dichromate ions:

$$C_2H_5OH + Cr_2O_7{}^{2-} \longrightarrow CH_3CO_2H + Cr^{3+} \quad \text{(unbalanced)}$$

(a) Balance this reaction. (b) If a breath sample generates a concentration of 4.5×10^{-4} M Cr^{3+} in 50.0 mL of solution, how many milligrams of alcohol were in the sample?

18.67 The figure shows a schematic sketch of a galvanic cell.

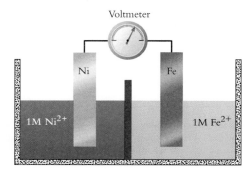

Voltmeter

Ni

Fe

1M Ni²⁺

1M Fe²⁺

(a) Identify the two half-reactions; (b) Determine the potential of this cell; (c) Identify the anode and cathode; (d) Redraw the sketch to show the direction of electron flow and the molecular processes occurring at each electrode.

18.68 The figure shows a schematic sketch of a galvanic cell.

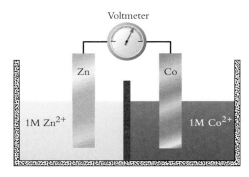

Voltmeter

Zn

Co

1M Zn²⁺

1M Co²⁺

(a) Identify the two half-reactions; (b) Determine the potential of this cell; (c) Identify the anode and cathode; (d) Redraw the sketch to show the direction of electron flow and the molecular processes occurring at each electrode.

18.69 For the galvanic cell in Problem 18.67, which solution concentration would have to be reduced, and to what concentration, to reduce the cell potential to 0.0 V?

18.70 For the galvanic cell in Problem 18.68, which solution concentration would have to be reduced, and to what concentration, to reduce the cell potential to 0.0 V?

18.71 Balance the redox reactions between $MnO_4{}^-$ and each of the following sulfur-containing species. The final products are Mn^{2+} and $HSO_4{}^-$ and the solution is acidic: (a) H_2SO_3; (b) SO_2; (c) H_2S; and (d) $H_2S_2O_3$.

18.72 Balance the redox reactions between $Cr_2O_7{}^{2-}$ and each of the following nitrogen-containing species. The final products are Cr^{3+} and $NO_3{}^-$ and the solution is acidic: (a) NO; (b) NO_2; (c) HNO_2; and (d) $NH_4{}^+$.

18.73 How long would it take to electroplate all the Cu^{2+} in 0.250 L of 0.245 M $CuSO_4$ solution with an applied potential difference of 0.225 V and a current of 2.45 amperes?

18.74 Calcium metal is obtained by the direct electrolysis of molten $CaCl_2$. If a metallurgical electrolysis apparatus operates at 27.6 A, what mass of calcium metal will it produce in 24 hours of operation?

18.75 Draw a sketch of the electroplating apparatus that illustrates the process occurring in Problem 18.73. Include arrows showing the direction of electron flow, label the anode and cathode, and draw molecular pictures showing the processes occurring at each electrode.

18.76 Draw a sketch of an electroplating apparatus that illustrates the process occurring in Problem 18.74. Include arrows showing the direction of electron flow, label the anode and cathode, and draw molecular pictures showing the processes occurring at each electrode.

18.77 For each of the following pairs of species, select the one that is the better oxidizing agent under standard conditions (see Appendix F): (a) In acidic solution, $Cr_2O_7{}^{2-}$ or $MnO_4{}^-$; (b) in basic solution, O_2 or $NO_3{}^-$; and (c) Fe^{2+} or Sn^{2+}.

18.78 For each of the following pairs of species, select the one that is the better reducing agent under standard conditions (see Appendix F): (a) Cu or Ag; (b) Fe^{2+} or Co^{2+}; and (c) H_2 or I^- (acid solution).

More Challenging Problems

18.79 Determine whether or not O_2 is capable of oxidizing each of the substances in Problem 18.78, under standard conditions (basic).

18.80 The following are spontaneous reactions. List the species in these reactions in order of oxidizing strength, from strongest oxidizing agent to weakest oxidizing agent:

$$2\ Cr^{2+} + Sn^{2+} \longrightarrow Sn + 2\ Cr^{3+}$$

$$Fe + Sn^{2+} \longrightarrow Sn + Fe^{2+}$$

$$Fe + U^{4+} \longrightarrow No\ reaction$$

$$U^{3+} + Cr^{3+} \longrightarrow U^{4+} + Cr^{2+}$$

$$Fe + 2\ Cr^{3+} \longrightarrow 2\ Cr^{2+} + Fe^{2+}$$

18.81 A cell is set up using two zinc wires and two solutions, one containing 0.250 M $ZnCl_2$ solution and the other containing 1.25 M $Zn(NO_3)_2$ solution. (a) What electrochemical reaction occurs at each electrode? (b) Draw a molecular picture showing spontaneous electron transfer processes at the two zinc electrodes. (c) Compute the potential of this cell.

18.82 Using standard reduction potentials, determine K_{eq} for the decomposition reaction of hydrogen peroxide:

$$2\ H_2O_2 \rightleftharpoons 2\ H_2O + O_2.$$

What does this value tell you about the stability of H_2O_2?

18.83 The first battery to find widespread commercial use was the Leclanche cell, in which the cathode reaction is:

$$2\ MnO_2(s) + Zn^{2+}(aq) + 2\ e^- \longrightarrow ZnMn_2O_4(s).$$

In a flashlight, one of these batteries provides 0.0048 A. If the battery contains 4.0 g of MnO_2 and fails after 90% of its MnO_2 is consumed, calculate the operating life of the flashlight.

18.84 Electrolytic reactions, like other chemical reactions, are not 100% efficient. In a copper purification apparatus depositing Cu from $CuSO_4$ solution, operation for 5.0 hours at constant current of 5.8 A deposits 32 g of Cu metal. What is the efficiency?

18.85 A galvanic cell consists of a Pt electrode immersed in a solution containing Fe^{2+} at 1.00 M and Fe^{3+} at unknown concentration, as well as a Cu electrode immersed in a 1.00 M solution of Cu^{2+}. The cell voltage is 0.00 V. What is the concentration of Fe^{3+} ions?

18.86 Use standard reduction potentials from Table 18-1 to determine K_{sp} for as many metal hydroxides as the table allows.

18.87 The sketch below shows a cell set up to electrolyze molten NaCl.

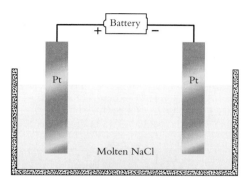

(a) What reactions occur? (b) Identify the anode and cathode. (c) Redraw the sketch showing the direction of electron flow, and include molecular pictures showing the processes at each electrode.

18.88 Given that $E° = -0.34$ V for the reduction of Tl^+ to Tl, find the voltage developed by a cell consisting of Tl metal dipping in an aqueous solution that is 0.050 M in Tl^+, connected by a porous bridge to a 0.50 M aqueous solution of HCl in contact with a Pt electrode over which H_2 gas is bubbling at 0.90-atm pressure.

18.89 Draw a sketch of the cell in Problem 18.88. Include molecular views of the processes taking place at the electrodes.

18.90 List all the metals that could be used as sacrificial anodes for iron. Which of these could also be sacrificial anodes for aluminum?

18.91 For the reaction between strontium and magnesium, $K_{eq} = 2.69 \times 10^{12}$:

$$Sr(s) + Mg^{2+}(aq) \rightleftharpoons Sr^{2+}(aq) + Mg(s)$$

Calculate $E°$ for a strontium-magnesium battery.

18.92 A galvanic cell is constructed with a silver–silver chloride and a nickel strip immersed in a beaker containing 1.50×10^{-2} M $NiCl_2$. (a) Determine the balanced cell reaction. (b) Calculate the potential of the cell. (c) Draw a sketch showing the electron transfer reaction occurring at each electrode.

Group Study Problems

18.93 Consider an electrochemical cell consisting of two vessels connected by a porous separator. One vessel contains 0.500 M HCl solution and an Ag wire electrode coated with AgCl solid. The other vessel contains 1.00 M $MgCl_2$ solution and an Mg wire electrode. (a) Determine the net reaction. (b) Calculate E for the cell (see Appendix F). (c) Draw a molecular picture showing the reactions at each electrode.

18.94 Use data from Appendix F for the reaction,

$$5\ I^-(aq) + IO_3^-(aq) + 6\ H_3O^+(aq) \longrightarrow 3\ I_2(s) + 9\ H_2O(l)$$

(a) Determine the spontaneous direction at pH = 2.00 and $[I^-] = [IO_3^-] = 0.100$ M; (b) Repeat the calculation at pH = 11.00; (c) At what pH is this redox reaction at equilibrium at these concentrations of I^- and IO_3^-?

18.95 From the standard reduction potentials appearing in Table 18-1, identify reaction pairs that are candidates for batteries that would produce more than 5 V of electrical potential under standard conditions. Suggest chemical reasons why no such battery has been commercially developed.

18.96 A chemist wanted to determine $E°$ for the Ru^{3+}/Ru reduction reaction. The chemist had all the equipment needed to make potential measurements, but the only chemicals available were $RuCl_3$, a piece of ruthenium wire, sulfuric acid, water, and a lead electrode from an old lead storage battery. Describe and sketch a cell that the chemist could set up in order to determine this $E°$. Show how the measured voltage would be related to $E°$ of the half-reaction. If the cell has a measured voltage of 0.745 V, with the lead electrode acting as the cathode, determine $E°$ for Ru^{3+}/Ru.

Answers to Section Exercises

18.1.1 (a) K = +1, O = −2, Mn = +7; (b) Cl = −1, Si = +4;
(c) Na = +1, O = −2, S = +4; (d) S = 0; and (e) Li = +1,
H = −1, Al = +3
18.1.2 (a) redox, C oxidized, O_2 reduced; (b) not redox; (c) redox,
Mg oxidized, Si reduced; (d) redox, S oxidized, O reduced; and
(e) not redox
18.1.3 (a) −3; (b) 0; (c) −2; (d) +5; (e) +3; (f) +2; and (g) +4
18.2.1 (a) $3 Cu + 2 NO_3^- + 6 H_3O^+ \rightarrow$
$$2 NO + 3 Cu^{2+} + 9 H_2O$$
(b) $SO_3^{2-} + H_2O_2 \rightarrow SO_4^{2-} + H_2O$
(c) $6 Br_2 + 18 H_2O \rightarrow 10 Br^- + 2 BrO_3^- + 12 H_3O$
18.2.2 $2 Al + 3 Ag_2S + 6 H_2O \rightarrow 6 Ag + 2 Al(OH)_3 + 3 H_2S$
18.2.3 $3 CuS + 11 H_3O^+(aq) + 8 NO_3^- \rightarrow$
$$8 NO + 15 H_2O + 3 Cu^{2+} + 3 HSO_4^-$$

18.3.1

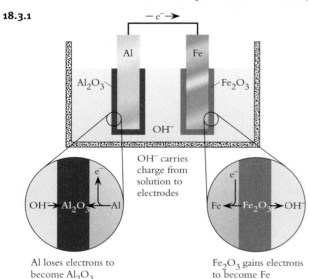

$-e^- \rightarrow$

Al   Fe

Al_2O_3 Fe_2O_3

OH^-

OH^- carries
charge from
solution to
electrodes

$OH \rightarrow Al_2O_3 \leftarrow Al$ $Fe \leftarrow Fe_2O_3 \rightarrow OH^-$

Al loses electrons to
become Al_2O_3

Fe_2O_3 gains electrons
to become Fe

18.3.2 $\Delta G° = -840.1$ kJ; negative value verifies spontaneity.
18.3.3

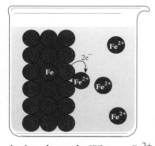

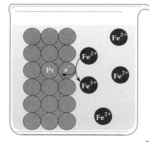

Active electrode: When an Fe^{2+}
ion collides with the surface of
the active electrode, Fe^{2+} gains
two electrons and sticks to the
electrode as an Fe atom.

Passive electrode: When an Fe^{2+}
ion collides with the surface of the
passive electrode, Fe^{2+} loses one
electron and becomes an Fe^{3+} ion,
which remains dissoved in
solution.

18.4.1 (a) $2 Cr + 3 Fe^{2+} \rightarrow 2 Cr^{3+} + 3 Fe$, $E° = 0.297$ V
(b) $Pb + Cu^{2+} \rightarrow Cu + Pb^{2+}$, $E° = 0.4681$ V
(c) $Au^{3+} + 3 Ag \rightarrow Au + 3 Ag^+$, $E° = 0.698$ V
18.4.2

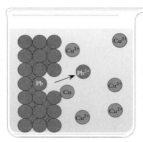

When a Cu^{2+} ion collides with the
Pb surface, Cu^{2+} is reduced to Cu
(elemental copper), and a Pb atom
is oxidized to Pb^{2+}. Cu is insoluble
in water, so it sticks to the Pb
surface; Pb^{2+} is soluble in water, so
it dissolves.

18.4.3 2.0501 V
18.5.1 (a) $\Delta G° = -63.7$ kJ, $K_{eq} = 1.45 \times 10^{11}$
(b) $\Delta G° = 1.14 \times 10^3$ kJ, $K_{eq} = 10^{-200}$
18.5.2 (a) 0.281 V; (b) 0.4681 V; and (c) 0.761 V
18.5.3 (a) HgO; (b) 2.23×10^2 C; and (c) 1720 days
18.6.1

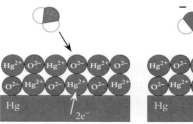

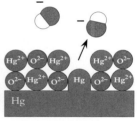

HgO captures 2 electrons
to form Hg and O^{2-}

O^{2-} reacts with H_2O
to form 2 OH^-

18.6.2 At the aluminum electrode, the deposit of aluminum oxide
is impervious to penetration by ions, so it blocks the passage of
current.
18.6.3 $E°$ for tin is less negative than that for iron, whereas $E°$
for zinc is more negative than that for iron. Thus iron
oxidizes preferentially to tin, but zinc oxidizes preferentially
to iron.
18.7.1 (a) $2 HF + 2 e^- \rightarrow H_2 + 2 F^-$ and $2 F^- \rightarrow F_2 + 2 e^-$; and
(b) 3.053 V
18.7.2 25.5 kg, requiring a tank whose volume is 274 L
18.7.3 In aqueous KF, water would be electrolyzed rather than KF,
generating O_2 instead of F_2; liquid HF does not conduct electricity
and support electrolysis.

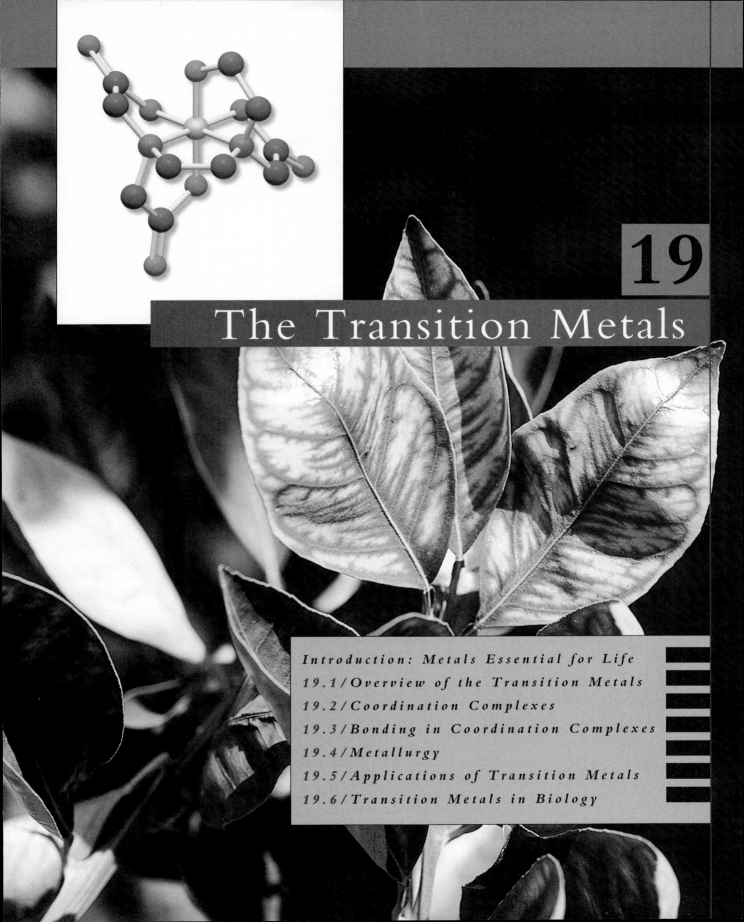

19

The Transition Metals

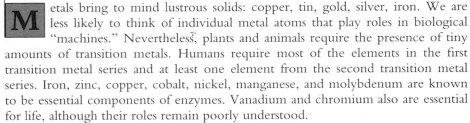

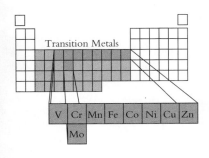

etals bring to mind lustrous solids: copper, tin, gold, silver, iron. We are less likely to think of individual metal atoms that play roles in biological "machines." Nevertheless, plants and animals require the presence of tiny amounts of transition metals. Humans require most of the elements in the first transition metal series and at least one element from the second transition metal series. Iron, zinc, copper, cobalt, nickel, manganese, and molybdenum are known to be essential components of enzymes. Vanadium and chromium also are essential for life, although their roles remain poorly understood.

Among the transition metals, the biochemistry of iron is known in the most detail. Iron is an essential component of chlorophyll, the green pigment that drives photosynthesis. Our background photo shows citrus trees suffering from the yellowing of leaves that is characteristic of iron deficiency. To combat iron deficiency, gardeners use fertilizers that contain iron chelates, octahedral complexes that slowly release their iron into the soil. A typical iron chelate appears in our molecular inset.

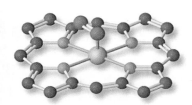

Iron is the transition metal present in largest quantities in the human body. There are about four grams of iron in the body of an average person. The best known biological iron-containing compound is the protein hemoglobin, the red component of blood that is responsible for the transport of oxygen. The central feature of hemoglobin is the porphyrin structure, a plane of carbon and nitrogen atoms that forms multiple rings with an extensive delocalized π system. In hemoglobin, an Fe^{2+} cation bonds to four nitrogen atoms that are part of the planar porphyrin ring; this structure is known as *heme*. An oxygen molecule can bind to the iron at a site perpendicular to the plane of the porphyrin ring. Hemoglobin exploits one of the special bonding properties of transition metal ions: their ability to form bonds by accepting electron pairs from atoms, such as nitrogen and oxygen, that contain lone pairs.

Variations of the heme structure are found in all life forms including the most primitive bacteria, indicating that iron-containing structures have been essential parts of life almost from its first appearance.

Another group of iron-containing proteins that is present in almost all forms of life is the iron-sulfur proteins, which contain clusters of iron and sulfur atoms. The iron atoms in these clusters have tetrahedral bonding geometry. Iron-sulfur proteins are essential for electron transfer processes, in which the clusters of iron and sulfur atoms gain and lose electrons. Here, another special property of transition metal ions is exploited: their ability to take on several different oxidation states.

Metals display a remarkable range of chemical properties. The individual atoms or cations in biological macromolecules readily form and break bonds to other species as well as gain or lose electrons. Geologically, metals typically occur in minerals, where metal cations are associated with oxygen, sulfur, or polyatomic anions such as carbonate and silicate. In pure metals and alloys, metals participate in delocalized bonding that generates high strength and electrical conductivity.

In this chapter, we survey the diversity of transition metals, beginning with an overview. Then we describe the structure and bonding in transition metal complexes. We describe metallurgy, the processes by which pure metals are extracted from mineral ores. The chapter ends with a presentation of some properties of transition metals and their biological roles.

◼ 19.1 ◼ OVERVIEW OF THE TRANSITION METALS

The *d* block of elements lies at the center of the periodic table, as shown in Figure 19-1. The ground state valence configuration of a gas-phase neutral transition metal atom from Group *x* generally is $s^2 d^{(x-2)}$. Titanium for example, is in Group 4 and has valence configuration $4s^2 3d^2$. Recall from Chapter 7, however, that the $(n + 1)s$ orbital of these elements has nearly the same energy as the nd orbitals. Because of this near-equality, the ground-state valence configuration of Group 11 elements is $s^1 d^{10}$, and a few other transition metals have $s^1 d^{(x-1)}$ configurations. Recall also that for transition metal *cations*, the nd orbital *always* is more stable than the $(n + 1)s$ orbital, so transition metal cations have vacant $(n + 1)s$ orbitals.

The elements of the *d* block, which lie between the reactive metals of the *s* block and the elements in the *p* block, are known as **transition metals.** A pure transition metal element is best described by the band theory of solids (recall Chapter 9). In this model, the valence *s* and *d* electrons form extended bands of orbitals that are delocalized over the entire network of metal atoms. These valence electrons are easily removed, so most elements in the *d* block react readily to form compounds rather than pure elements: oxides such as Fe_2O_3, sulfides such as ZnS, and mineral salts such as zircon, $ZrSiO_4$.

Between barium (Group 2, element 56) and lutecium (Group 3, element 71), the $4f$ orbitals fill, giving rise to a set of 14 metals known as the **lanthanides,** after its first member, lanthanum. The lanthanides are also called the *rare earths*, although except for promethium they are not particularly rare. Cerium, for example, is about five times as abundant as lead and half as abundant as chlorine. Between radon (Group 2, element 88) and lawrencium (Group 3, element 103), are the 14 elements known as the **actinides,** named for the first member of the set, actinium. The *f* block elements are also known as the **inner transition metals.**

(margin figure: $4s$ $3d$ Ti^{2+}, [Ar])

Physical Properties

As pure elements, almost all the transition metals are lustrous solids that conduct heat and electricity and are malleable and ductile. Although they share these general properties, transition metals display variations in other properties that can be traced to their different numbers of valence electrons.

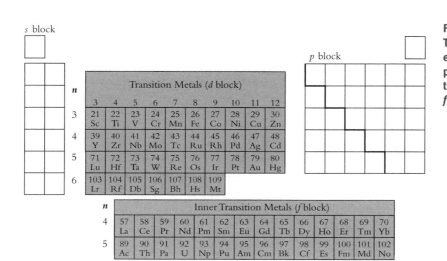

Figure 19-1
The transition metals are the elements in the *d* block of the periodic table. The inner transition metals are those in the *f* block.

Figure 19-2
Samples of metals from the first transition series. The metals are Cu (wire), Fe (block), Ni (balls), and Zn (sticks).

Figure 19-3
The melting points of the transition metals increase from Group 3 to 6 and then decrease to the end of the block.

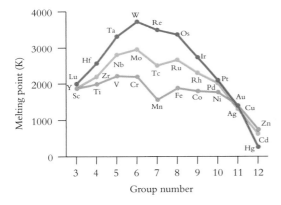

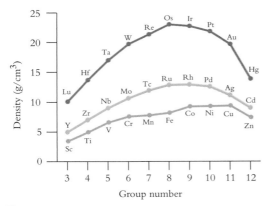

Figure 19-4
The densities of transition metals increase with atomic mass except at the end of the block.

As illustrated in Figure 19-2, most pure transition metals have the lustrous gray appearance that is termed "silvery" because of the appearance of silver metal. However, some transition metals have other colors, for example the orange color of copper and the golden hue of gold.

The melting points of the transition metals show a periodic variation that is displayed in Figure 19-3. Up to Group 6, the melting point increases with each added valence electron. This is because the first six valence electrons occupy orbital bands that are highly bonding in character, increasing the overall bond strength of the metal. Beyond Group 6, in contrast, the melting point decreases, falling precipitously for the elements of Group 12. This decrease is sufficient to make mercury unique among metals. This element has a melting point of -39 °C (234 K), making Hg a liquid at room temperature. Melting points decrease with added electrons beyond Group 6 because the added electrons occupy delocalized orbitals that are antibonding in character, weakening the overall bond strength of the metal.

The densities of transition metals also display regular periodic trends, as Figure 19-4 shows. Density increases moving down each column of the periodic table and increases smoothly across the first part of each row. Increasing atomic mass accounts for both these trends. The volume occupied by an individual atom in the metallic lattice varies relatively slowly within the d block, so the more massive the nucleus, the greater the density of the metal. Toward the end of each row, density decreases for the same reason that melting point decreases. The added electrons occupy antibonding orbitals, and this leads to a looser array of atoms, larger atomic volume, and decreased density.

Transition metals are good conductors of both electricity and heat. In general, these two properties are linked, because mobile electrons can transport both charge (electricity) and energy (heat) through bulk materials. The metals of Group 11 (copper, silver, and gold) have the highest electrical conductivity of any of the elements. Silver is the best conductor, but copper is more abundant and consequently less expensive, so it is the metal of choice for electrical wiring. The Group 11 elements also are the best conductors of heat among the metals.

Redox Behavior

The chemistry of the transition metals is determined in part by their atomic ionization energies, which measure how easily gaseous atoms of these elements lose electrons to form cations. The 3d and 4d series show a relatively gradual increase in ionization energy with atomic number (Z), whereas the trend for the 5d series is more pronounced (Figure 19-5). Values for transition metals in the 3d and 4d series are between 650 and 750 kJ/mol, somewhat higher than the first ionization energies of Group 2 alkaline earth metals but lower than the typical values for nonmetals in the p block.

Ionization energies indicate that transition metals can be relatively easily oxidized, but they do not reveal the variations that exist in the redox behavior of these metals. For example, except for the elements at the ends of the rows, each transition metal exists in several different oxidation states. The oxidation states displayed by the 3d transition metals are shown in Table 19-1.

Many of these oxidation states exist but are unstable relative to higher or lower oxidation states. Those oxidation states that are particularly stable are highlighted in the table.

Each element from vanadium to manganese has a particularly stable oxidation state corresponding to donation of all its valence electrons ($s^0 d^0$ configuration). All the elements except scandium display the $+2$ oxidation state, and these are particularly stable for the elements from chromium to zinc. Chromium, iron, and cobalt are particularly stable in the $+3$ oxidation state, and for vanadium and manganese, the $+4$ oxidation state is stable.

Cations with charges greater than $+3$ are generally unstable even in an ionic environment, because the energy required to remove an electron from an ion with $+3$ charge is very large. To convert Ti^{3+} to Ti^{4+}, for instance, requires an energy input of 4175 kJ/mol. Consequently, oxidation states greater than $+3$ are found in covalently bonded oxides, oxyanions, and halides but not in ionic compounds such as sulfates and phosphates. Table 19-2 lists representative examples of compounds displaying the important oxidation states of the $3d$ transition metals.

The relative stabilities of the various oxidation states of transition metals can be determined by examining standard reduction potentials. Here, for example, are the $E°$ values for half-reactions connecting the four most stable oxidation states of manganese:

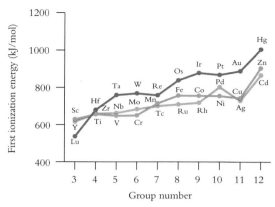

Figure 19-5
The first ionization energies of transition metals show gradual upward trends across each row of the periodic table.

$$Mn^{2+} + 2\ e^- \longrightarrow Mn \qquad E° = -1.185\ V$$

$$MnO_2 + 4\ H_3O^+ + 2\ e^- \longrightarrow Mn^{2+} + 6\ H_2O \qquad E° = 1.224\ V$$

$$MnO_4^- + 8\ H_3O^+ + 5\ e^- \longrightarrow Mn^{2+} + 12\ H_2O \qquad E° = 1.504\ V$$

$$MnO_4^- + 4\ H_3O^+ + 3\ e^- \longrightarrow MnO_2 + 6\ H_2O \qquad E° = 1.679\ V$$

Table 19-1
Oxidation States Displayed by the 3d Transition Metals*

Element	Sc	Ti	V	Cr	Mn	Fe	Co	Ni	Cu	Zn
Group	3	4	5	6	7	8	9	10	11	12
Oxidation State			**Valence Configuration**							
+1			d^4	d^5	d^6		d^8	d^9	d^{10}	
+2		d^2	d^3	d^4	d^5	d^6	d^7	d^8	d^9	d^{10}
+3	d^0	d^1	d^2	d^3	d^4	d^5	d^6	d^7	d^8	
+4		d^0	d^1	d^2	d^3	d^4	d^5	d^6		
+5			d^0	d^1	d^2		d^4			
+6				d^0	d^1	d^2				
+7					d^0					

■ $= +2$ ■ $= +3, +4$ ■ $= d^0$

*The table lists the configuration of the ion corresponding to each observed oxidation state. The most important oxidation states of each element are color screened.

Table 19-2
Representative Compounds of 3d Transition Metals in Their Commonly Occurring Oxidation States

Element	Ox. State	Compound	Ox. State	Compound	Ox. State	Compound
Sc	+3	$Sc(NO_3)_3 \cdot 6H_2O$				
Ti			+4	$TiBr_4, FeTiO_3$		
V			+4	VCl_4, VO_2	+5	$Pb_2(VO_4)Cl$
Cr	+2	$Cr(CH_3CO_2)_2$	+3	$Cr(OH)_3$	+6	$CrF_6, K_2Cr_2O_7$
Mn	+2	$MnCO_3, MnCl_2$	+4	MnO_2, MnF_4	+7	$KMnO_4$
Fe	+2	$FeSO_4 \cdot H_2O, FeS_2$	+3	$FeCl_3, FePO_4 \cdot 2H_2O$		
Co	+2	$CoS, Co(ClO_4)_2$	+3	$Co(OH)_3, CoF_3$		
Ni	+2	$NiSO_4 \cdot 7H_2O$				
Cu	+2	$Cu(NO_3)_2 \cdot 6H_2O$				
Zn	+2	$ZnS, ZnCO_3$				

Ox., Oxidation.

The negative value for the first half-reaction indicates that Mn metal is relatively easily oxidized to Mn^{2+}. The positive values for the remaining half-reactions show that the higher +4 oxidation state in MnO_2 is less stable than the +2 state, and the +7 state in MnO_4^- is less stable than either the +4 or the +2 state. The following values show that under standard conditions, O_2, which is considered to be a good oxidizing agent, can barely oxidize Mn^{2+} and is not capable of oxidizing MnO_2:

$$O_2 + 4 H_3O^+ + 4 e^- \longrightarrow 6 H_2O \qquad E° = 1.229 \text{ V}$$

$$O_2 + 2 Mn^{2+} + 6 H_2O \longrightarrow 2 MnO_2 + 4 H_3O^+ \qquad E° = 0.005 \text{ V}$$

$$3 O_2 + 4 MnO_2 + 6 H_2O \longrightarrow 4 MnO_4^- + 4 H_3O^+ \qquad E° = -0.450 \text{ V}$$

Similar patterns exist for other transition metals. To give just one example, O_2 can oxidize Fe to Fe^{2+} or Fe^{3+} under standard conditions, but it cannot oxidize iron to the +6 oxidation state in FeO_4^{2-}. Consequently, the later transition metals are found in nature in either the +2 or the +3 oxidation state. The early transition metals, in contrast, occur in their highest oxidation states. Titanium is found in the +4 state in TiO_2 (rutile) and $FeTiO_3$ (ilmenite). Vanadium has the +5 oxidation state in the vanadate anion, VO_4^{3-}, which occurs in minerals such as $Pb_5(VO_3)Cl$ (vanadinite) and $K(UO_2)(VO_4)$ (carnotite). The latter also contains UO_2^{2+} cations, in which uranium has the +6 oxidation state. Tungsten also occurs naturally in the +6 oxidation state in $FeWO_4$ (scheelite) and $CaWO_4$ (wolframite). In these mixed oxides, W has oxidation state +6 in the WO_4^{2-} anion, and Fe and Ca are in the +2 oxidation state.

Transition Metal Compounds

A few of the transition metals, including gold, platinum, and iridium, are found in nature as pure elements, but most of the others are found associated with either sulfur or oxygen. Iron, manganese, and the metals of Groups 3 to 6 (except for Mo) are most often found as oxides or occasionally as sulfates or carbonates.

Molybdenum and the metals of Groups 7 to 12 (except for Mn and Fe) are most often found as sulfides.

→ **Chapter 20**
We describe why transition metals tend to be combined with either oxygen or sulfur in Chapter 20.

Although naturally occurring compounds of transition metals are restricted in scope, a wide variety of compounds can be synthesized in the laboratory. As illustrated by the representative compounds listed in Table 19-2, these compounds fall into three general categories. There are many binary halides and oxides in a range of oxidation numbers. Ionic compounds containing transition metal cations and polyatomic oxyanions also are common; these include nitrates, carbonates, sulfates, phosphates, and perchlorates. Finally, there are numerous ionic compounds in which the transition metal is part of an oxyanion. The synthesis of transition metal compounds must be tailored to a particular product. The details are complex, and existing patterns of synthesis contain so many exceptions that an overall view is beyond the scope of an introductory course in chemistry.

Section Exercises

■ **19.1.1** Predict which element of the following pairs will have the higher melting point and support your prediction using periodic trends: (a) Cr or W; (b) W or Os; (c) Pd or Ag; and (d) Y or Nb.

■ **19.1.2** Predict which element of the following pairs will have the higher density and support your prediction using periodic trends: (a) Cr or W; (b) W or Os; (c) Pd or Ag; and (d) Y or Nb.

■ **19.1.3** The following compounds can be purchased from chemical companies. Determine the oxidation state of the transition metal in each of them. (a) $TaCl_5$; (b) $Fe(NO_3)_3 \cdot 9H_2O$; (c) Rh_2O_3; (d) CrO_2Cl_2; and (e) $Cu_2(OH)PO_4$.

19.2 COORDINATION COMPLEXES

Earlier chapters described how metal salts dissolve in water to give solutions of aqueous ions. Here are three examples:

$$Cu(ClO_4)_2(s) \xrightarrow{H_2O} Cu^{2+}(aq) + 2\,ClO_4^-(aq)$$

$$NiSO_4(s) \xrightarrow{H_2O} Ni^{2+}(aq) + SO_4^{2-}(aq)$$

$$FeCl_3(s) \xrightarrow{H_2O} Fe^{3+}(aq) + 3\,Cl^-(aq)$$

In Section 17-5, we described the formation of complexes between metal ions and ligands. For example, $Ni^{2+}(aq)$ forms a complex with ammonia:

$$Ni^{2+}(aq) + 6\,NH_3(aq) \rightleftharpoons [Ni(NH_3)_6]^{2+}(aq)$$

Although these chemical equations show the aqueous metal ions as simple particles in solution, the dissolved ions actually form chemical bonds to water molecules of the solvent. Recall that a **ligand** is a species that has lone pairs of electrons available to donate to a metal cation. Water molecules possess lone pairs of electrons, so water is a ligand that readily forms complex ions with metal cations. We write a solubility reaction as though the cations were simple particles in solution, but the dissolved ions actually form chemical bonds to water molecules of the solvent. For example, $Ni^{2+}(aq)$ bonds to six water molecules in octahedral geometry, as illustrated in Figure 19-6. The aqueous $[Ni(H_2O)_6]^{2+}$ cation contains Ni^{2+} bonded to six water ligands.

Figure 19-6
The Ni^{2+} cations in $[Ni(H_2O)_6]^{2+}$ and $[Ni(NH_3)_6]^{2+}$ are bonded to six ligands, one at each vertex of an octahedron.

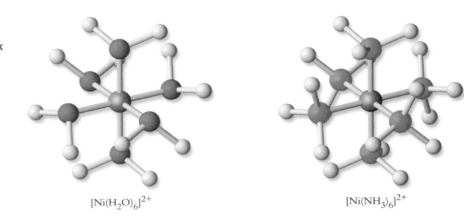

$[Ni(H_2O)_6]^{2+}$ $[Ni(NH_3)_6]^{2+}$

Color changes often provide evidence for the interaction of ligands and metal cations, particularly for the transition metals. The Ni^{2+} cation provides an example. Figure 19-7 shows that nickel(II) sulfate, a white crystalline solid, dissolves in water to give a green solution. The green color cannot be due to Ni^{2+} or SO_4^{2-}, which the white solid shows to be colorless. Rather, the color comes from the octahedral complex that forms when each nickel ion binds to six water molecules:

SECTION 19.3 ➡
Colors of transition metal complexes are described in Section 19.3.

$$NiSO_4(s) + 6\,H_2O(l) \longrightarrow [Ni(H_2O)_6]^{2+}(aq) + SO_4^{2-}(aq)$$

Replacing the water ligands with other ligands also can result in color changes. Figure 19-7 shows the color change that occurs when ammonia is added to a solution of hydrated Ni^{2+}. Ammonia molecules replace water ligands to give the blue species $[Ni(NH_3)_6]^{2+}$, whose ball-and-stick structure appears in Figure 19-6:

$$[Ni(H_2O)_6]^{2+} + 6\,NH_3 \longrightarrow [Ni(NH_3)_6]^{2+} + 6\,H_2O$$
green   blue

Figure 19-7
Solid nickel(II) sulfate, nearly colorless, dissolves in water to give a green solution containing $[Ni(H_2O)_6]^{2+}$ cations. The addition of ammonia produces a blue solution of $[Ni(NH_3)_6]^{2+}$ cations. Solvent evaporation gives a blue-violet precipitate of $[Ni(NH_3)_6]SO_4$.

Complex ions, also called **coordination complexes,** have well-defined stoichiometries and structural arrangements. Usually, the formula of a coordination complex is enclosed in brackets to show that the metal and all its ligands form a single structural entity. When an ionic coordination complex can be isolated from aqueous solution as a solid salt, the salt is composed of the complex cation associated with anions. In the chemical formula, the anions are shown outside the brackets. An example is the sulfate formed by the $[Ni(NH_3)_6]^{2+}$ complex, $[Ni(NH_3)_6]SO_4$.

Nature of Ligands

Metal cations are electron-deficient, so they generally do not have valence electrons that can be shared in chemical bonds to ligands. Consequently, a ligand must provide both electrons for the metal-ligand bond. This means that a ligand must have a lone pair of electrons that it donates to form a bond to the metal ion. The ammonia ligands in $[Ni(NH_3)_6]^{2+}$ coordinate to the metal by donation of lone electron pairs from nitrogen to form Ni—N bonds. Each bond forms by overlap of an empty valence orbital on the metal cation with the orbital on the nitrogen atom that contains a lone pair of electrons. The water ligands in $[Ni(H_2O)_6]^{2+}$ coordinate to the metal in a similar manner, with the oxygen atoms donating lone pairs to form Ni—O bonds.

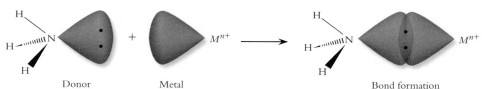

Donor   Metal   Bond formation

Hundreds of different ligands are known. Several of the most common are listed in Table 19-3. The simplest ligands coordinate to the metal through a single donor atom. The most common donor atoms in coordination chemistry are nitrogen, oxygen, and the halogens, but other important ligands have carbon, phosphorus, or sulfur donor atoms. A ligand with one donor atom is called **monodentate.** Ammonia is a monodentate ligand with a nitrogen donor atom, and water is a monodentate ligand with an oxygen donor atom. Other examples include halide ions, cyanide ion, hydroxide ion, and carbon monoxide.

As described in Chapter 17, **chelating** ligands contain two or more donor atoms in structures that allow the ligand to wrap around the metal. Examples featured in Chapter 17 are ethylenediamine ($H_2NCH_2CH_2NH_2$), a bidentate ligand, and ethylenediaminetetraacetate (EDTA), a hexadentate ligand.

Structures of Coordination Complexes

Coordination complexes are a remarkably diverse group of molecules that form from virtually all transition metals in a variety of oxidation states. These compounds involve an extensive array of ligands, and they adopt several molecular geometries.

The molecular geometry of a complex depends on the coordination number (see Section 17.5), which is the number of ligand atoms coordinated to the metal. The most common coordination number is six, and almost all metal complexes with coordination number 6 adopt octahedral geometry. This preferred geometry can be traced back to Coulomb's law: The donor atoms take up positions around the metal as far apart as possible to minimize electron–electron repulsion.

Although six is most prevalent, a coordination number of four is also common, and several important complexes have a coordination number of two. In addition, a few complexes display coordination numbers of three, five, and seven. Examples of coordination number two include the silver-ammonia complex and the gold-cyanide complex, both described in Chapter 17. To minimize ligand-ligand repulsions, a complex with a coordination number of two is invariably linear, as Figure 19-8 shows.

Four-coordinate complexes may be either square planar or tetrahedral, as shown in Figure 19-9. Tetrahedral geometry is most common among the first-row transition metals. Examples include $[Zn(NH_3)_4]^{2+}$, $[Cd(en)_2]^{2+}$, $[FeCl_4]^-$, and $Ni(CO)_4$. Square planar geometry is characteristic of transition metal ions such as palladium(II),

**Table 19-3
Common Monodentate Ligands**

Ligand	Name
Halides	
F^-	Fluoro
Cl^-	Chloro
Br^-	Bromo
I^-	Iodo
Donor C Atom	
CO	Carbonyl
CN^-	Cyano
Donor O Atom	
H_2O	Aqua
CO_3^{2-}	Carbonato
OH^-	Hydroxo
Donor N Atom	
NH_3	Ammine
NO_2^-	Nitrito

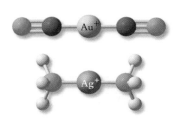

**Figure 19-8
Complexes with coordination number two always adopt linear geometry about the metal cation.**

**Figure 19-9
Four-coordinate complexes adopt either square planar geometry, as in $[AuCl_4]^-$, or tetrahedral geometry, as in $[NiCl_4]^-$.**

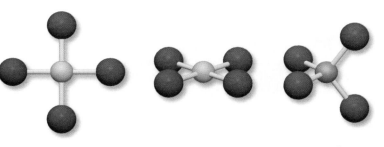

| $[AuCl_4]^-$ top view | $[AuCl_4]^-$ oblique view | $[NiCl_4]^{2-}$ |
| Square planar geometry | Square planar geometry | Tetrahedral geometry |

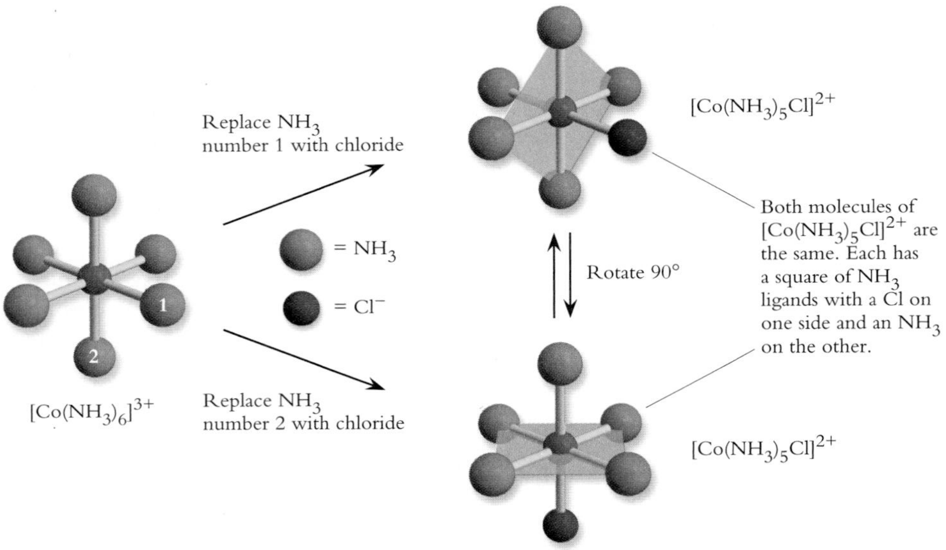

Replace NH$_3$
number 1 with chloride

$[Co(NH_3)_5Cl]^{2+}$

= NH$_3$

= Cl$^-$

Rotate 90°

Both molecules of
$[Co(NH_3)_5Cl]^{2+}$ are
the same. Each has
a square of NH$_3$
ligands with a Cl on
one side and an NH$_3$
on the other.

$[Co(NH_3)_6]^{3+}$

Replace NH$_3$
number 2 with chloride

$[Co(NH_3)_5Cl]^{2+}$

Figure 19-10
There is only one isomer of $[Co(NH_3)_5Cl]^{2+}$. The top structure is identical to the bottom structure because the two can be superimposed on each other after a 90-degree rotation.

Animation

Cis means "next to," and *trans*
means "across from."

gold(III), and others with eight d electrons in the valence shell. We describe the differences between square planar and tetrahedral geometry in more detail in Section 19.3.

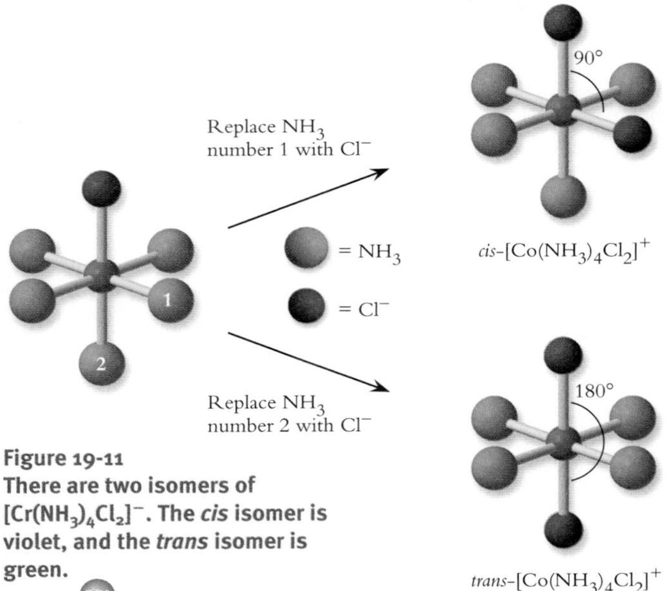

Isomers

In $[Co(NH_3)_6]^{3+}$, one ammonia ligand occupies each corner of an octahedron. Replacement of one of the six NH$_3$ ligands with a chloride ion generates $[Co(NH_3)_5Cl]^{2+}$. Regardless of which ligand is replaced, the geometry of the resulting complex is the same, because the six positions around an octahedron are equivalent by symmetry, as illustrated in Figure 19-10. Any octahedral coordination complex of the general formula ML_5X has just one possible structure.

Now suppose a second ammonia ligand is replaced with a chloride ion. If an ammonia ligand *adjacent to* the first chloride ion is replaced, the two chloride ions are separated by a bond angle of 90°. This is called the *cis* isomer (Figure 19-11). If the ammonia ligand *opposite* the chloride ion is replaced, the two chloride ions are separated by a bond angle of 180°. This is called the *trans* isomer. These two isomers of $[Co(NH_3)_4Cl_2]^-$ have the same chemical bonds but display different geometries, so they are **geometric isomers.** Geometric isomers have different properties. For example, whereas the *cis* isomer is violet, the *trans* isomer is green. Example 19-1 introduces two more geometric isomers of coordination complexes.

Replace NH$_3$
number 1 with Cl$^-$

= NH$_3$

= Cl$^-$

90°

cis-$[Co(NH_3)_4Cl_2]^+$

180°

trans-$[Co(NH_3)_4Cl_2]^+$

Replace NH$_3$
number 2 with Cl$^-$

Figure 19-11
There are two isomers of $[Cr(NH_3)_4Cl_2]^-$. The *cis* isomer is violet, and the *trans* isomer is green.

Animation

| Isomers of Coordination Complexes | Example 19-1 |

Draw ball-and-stick models of all possible isomers of the octahedral compound $[Cr(NH_3)_3Cl_3]$.

Strategy: It is best to approach this problem by starting with $[Cr(NH_3)_6]^{3+}$ and replacing one ligand at a time, considering all possible structures. The first substitution gives only one compound. Introducing a second chloride ion creates *cis* and *trans* isomers of $[Cr(NH_3)_4Cl_2]^+$. Choose each of these isomers in turn, and consider all possible structures that result from introducing a third chloride ligand. This is best accomplished by drawing pictures.

Solution: For the *trans* isomer, replacing any of the four remaining NH_3 ligands gives the same product, in which three of the same ligand are arranged in a T-shape about the metal ion:

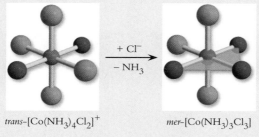

trans-$[Co(NH_3)_4Cl_2]^+$ *mer*-$[Co(NH_3)_3Cl_3]$

Starting with *cis*-$[Cr(NH_3)_4Cl_2]^+$, it is possible to make two isomers of the final product. This can be seen most clearly by labeling the four NH_3 ligands. Replacing an NH_3 ligand at site *a* or *b* with Cl^- generates the T-shaped arrangement. However, replacement at position *c* or *d* gives a new isomer in which the three chloride ligands occupy the corners of a triangle:

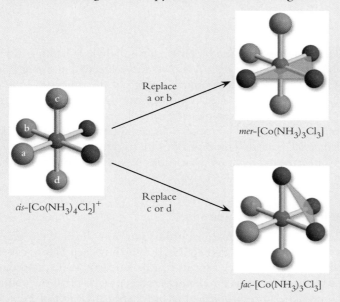

cis-$[Co(NH_3)_4Cl_2]^+$

Replace a or b

Replace c or d

mer-$[Co(NH_3)_3Cl_3]$

fac-$[Co(NH_3)_3Cl_3]$

In this isomer, three of the same ligands lie on an equatorial circle about the complex. Such a circle is called a *meridian*, and this isomer is the meridianal isomer, abbreviated *mer*.

Animation

The three like ligands lie in a plane that forms one face of the octahedron, shown in outline in the drawing. This is called the *facial* isomer, abbreviated *fac*.

This exhausts all possible structures of $[Cr(NH_3)_3Cl_3]$. Octahedral coordination complexes of general formula ML_3X_3 have two isomers.

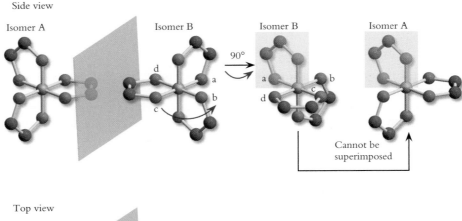

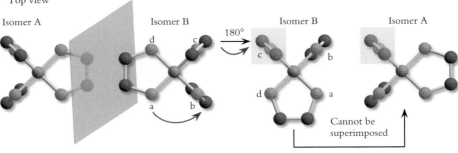

Figure 19-12
There are two possible bond arrangements for [Co(en)₃]³⁺ that are mirror images of each other. If one form is rotated so one part of the structure can be superimposed on its mirror image *(beige screens)*, the other parts cannot be superimposed, showing that the mirror images are two different isomers.

Octahedral complexes of the bidentate ligand ethylenediamine display a subtler type of isomerism. As Figure 19-12 shows, there are two possible bond arrangements for $[Co(en)_3]^{3+}$. Each is the mirror image of the other, and the two cannot be superimposed on one another. Nonsuperimposable mirror images are called *enantiomers*. A macroscopic example of enantiomers is your left and right hands, which are mirror images of each other that cannot be superimposed. Enantiomers are also called *optical isomers*. This type of isomerism plays important roles in molecular biology, because many biological molecules display optical isomerism, with only one enantiomer biologically active.

Naming Coordination Compounds

Originally, compounds containing coordination complexes were given common names such as Prussian blue ($KFe[Fe(CN)_6]$), which is deep blue, or Reinecke's salt ($NH_4[Cr(NH_3)_2(NCS)_4]$), named for its first maker. Eventually, coordination compounds became too numerous for chemists to keep track of all the common names. To solve the nomenclature problem, the International Union of Pure and Applied Chemistry (IUPAC) created a systematic procedure for naming coordination compounds. The following guidelines are used to determine the name of a coordination compound from its formula, or vice versa:

1. As with all salts, name the cation before the anion.

2. Within the complex, first name the ligands in alphabetical order, and then name the metal.

3. If the ligand is an anion, add the suffix -o to the stem name: *bromo* (Br^-), *cyano* (CN^-), and *hydroxo* (OH^-). The simplest neutral ligands have special names: *aqua* (H_2O), *ammine* (NH_3), and *carbonyl* (CO). Other neutral ligands retain their usual names. Some common ligands and their names appear in Table 19-3.

4. Use a Greek prefix (*di-*, *tri-*, *tetra-*, *penta-*, *hexa-*) to indicate the number of identical ligands. If the name of the ligand already incorporates one of these prefixes (as in ethylenediamine), enclose the ligand name in parentheses and use the alternative prefixes *bis-* (two), *tris-* (three), and *tetrakis-* (four).

5. If the coordination complex is an anion, add the suffix *-ate* to the stem name of the metal.

6. After the name of the metal, give the oxidation number of the metal in parentheses as a Roman numeral or as 0 if the oxidation number is zero.

Example 19-2 applies these guidelines.

Name of a Coordination Compound	Example 19-2

What is the IUPAC name for each of the following coordination compounds? (a) $[Co(NH_3)_6]Cl_3$; (b) $[Mn(CO)_5Br]$; and (c) $[Cr(NH_3)_3Cl_3]$

Strategy: Follow the guidelines for naming coordination compounds. Break the complex ion down one piece at a time.

Solution:

(a) The cation is the complex ion. Name the ligand and include the appropriate prefix: hexaammine. The formula lists three Cl^- anions outside the brackets. These ions are not part of the complex, so the net charge on the metal-containing cation is +3. Ammine is neutral, requiring that Co be in the +3 oxidation state: hexaamminecobalt(III) chloride. The anion requires no prefix, because the number of chloride anions can be deduced from the charge of the cation.

(b) There are five carbonyl ligands and one bromo ligand. The compound is neutral and contains one anionic ligand, Br^-, with a −1 charge, so manganese must be in the +1 oxidation state: bromopentacarbonylmanganese(I).

(c) The chromium-containing species is neutral. Three neutral ammine ligands, three negative chloro ligands, and no overall charge leaves chromium with an oxidation state of +3: triamminetrichlorochromium(III).

The following examples illustrate further applications of the IUPAC guidelines:

$Na_2[FeEDTA]$	Sodium ethylenediaminetetraacetatoferrate(II)
$[Co(en)_3]Cl_3$	Tris(ethylenediamine)cobalt(III) chloride
$[Zn(NH_3)_4](NO_3)_2$	Tetraamminezinc nitrate
$K[Ag(CN)_2]$	Potassium dicyanoargentate(II)
$[Rh(NH_3)_5Br]Br_2$	Pentaamminebromorhodium(III) bromide

These examples show that when information is not needed to identify the compound, it is omitted from the name. In the first name, for instance, it is not necessary to tell how many sodium ions are present, because we can deduce the number from the name of the complex anion. In the third name, the oxidation state of zinc is omitted because it is always $+2$. In the fifth name, the single bromo ligand is not preceded by the prefix mono. The last example shows that the numerical prefix is ignored in alphabetizing the ligands.

The first and fourth examples illustrate a nuance of the naming rules. Silver in an anionic complex is named by its Latin root argent-, from which the symbol Ag is derived. Other metals taking their Latin names in anionic coordination complexes are Fe (ferrate), Cu (cuprate), Sn (stannate), Au (aurate), and Pb (plumbate).

Example 19-3 provides more practice in working with the names of coordination compounds. Our Chemical Milestones Box describes the detective work that led to the birth of coordination chemistry.

Example 19-3	**Formula of a Coordination Compound**

Determine the formulas of the following coordination compounds:
- (a) *trans*-tetraamminedichlorocobalt(III) chloride
- (b) tris(ethylenediamine)manganese(II) sulfate
- (c) sodium hexacyanoferrate(II)

Strategy: To obtain the formula, break the name down, one piece at a time.

Solution:

(a) Determine the number of each type of ligand: tetraammine = 4 NH_3; dichloro = 2 Cl^-. Identify the metal (cobalt) and its oxidation state (III): Co^{3+}. Calculate the charge on the complex ion to determine the number of chloride ions present. Ammonia is neutral, but each chloride ligand contributes a -1 charge to the complex. Overall, then, the complex ion has charge of $+1$. Therefore one chloride anion is required to give a neutral salt: $[trans\text{-}Co(NH_3)_4Cl_2]Cl$. Notice that a particular isomer is indicated by an italicized prefix.

(b) Tri(ethylenediamine) = 3 $NH_2CH_2CH_2NH_2$ ligands. Recall that en is the conventional abbreviation for this ligand. The metal is Mn^{2+}. Because en ligands are neutral, one sulfate ion is required for overall electrical neutrality: $[Mn(en)_3]SO_4$.

(c) Sodium is the cation. The metal is Fe^{2+}, and there are six cyano anions. Overall, then, the complex anion has a net charge of -4. Four Na^+ cations are required to balance the charge: $Na_4[Fe(CN)_6]$.

Section Exercises	

- **19.2.1** For each of the following, determine the charge on the complex ion, the oxidation state of the metal, and the coordination number of the metal: (a) $K_3[Fe(CN)_6]$; (b) $[V(NH_3)_4Cl_2]$; and (c) $[Ni(en)_2]SO_4$.
- **19.2.2** Draw all of the isomers of $[Cr(en)(NH_3)_2I_2]^+$.
- **19.2.3** Name the following coordination compounds: (a) $[Cr(NH_3)_5I]SO_4$; (b) $K_4[PtCl_6]$; (c) $[cis\text{-}Fe(CO)_4Cl_2]$; and (d) $[Fe(H_2O)_6]Cl_2$.

Box 19-1   Chemical Milestones: The Birth of Coordination Chemistry

L ate in the nineteenth century, just as the principles of chemical bonding were being discovered, chemists carried out many studies of the interactions of ammonia with cations such as Cr^{3+}, Co^{3+}, Pt^{4+}, and Pd^{2+}. The most intriguing results were obtained for cobalt(III) chloride. By 1890, several ammonia compounds of $CoCl_3$ had been isolated. These coordination compounds differed in several of their properties, the most striking of which were their beautiful colors. At the time, the formulas of these cobalt complexes were written as follows:

$CoCl_3 \cdot 6NH_3$	Yellow-orange
$CoCl_3 \cdot 5NH_3$	Purple
$CoCl_3 \cdot 4NH_3$	Green
$CoCl_3 \cdot 3NH_3$	Green

Chemists were convinced that all the chlorine atoms had to be bonded to the cobalt in some way, and since ammonia is a gas at room temperature they could not understand why the ammonia molecules did not evaporate away. One of the first proposed explanations was the chain theory, in which the ammonia molecules were assumed to form chains between the metal and the chloride. For example, a prominent coordination chemist of the time proposed the following structure for $CoCl_3 \cdot 6NH_3$:

$$
\begin{array}{c}
NH_3-Cl \\
\diagup \\
Co-NH_3-NH_3-NH_3-NH_3-Cl \\
\diagdown \\
NH_3-Cl
\end{array}
$$

During the 1890s, these cobalt complexes attracted the attention of a Swiss chemist, Alfred Werner (1866–1919). Only in his early 20s, Werner had just earned his Ph.D. in organic chemistry. He studied the cobalt complexes in detail and developed the basis for our understanding of coordination chemistry.

Werner found that adding aqueous $AgNO_3$ to solutions of the various cobalt ammine complexes gave different amounts of silver chloride precipitate. For $CoCl_3 \cdot 6NH_3$, all three chloride ions precipitated as AgCl. Only two chloride ions precipitated for $CoCl_3 \cdot 5NH_3$, just one precipitated for $CoCl_3 \cdot 4NH_3$, and there was no precipitate for $CoCl_3 \cdot 3NH_3$. Werner explained these results by proposing that the cobalt ion in all four compounds had a coordination number of 6. He reformulated the compounds as follows:

$$CoCl_3 \cdot 6NH_3 = [Co(NH_3)_6]Cl_3$$
$$CoCl_3 \cdot 5NH_3 = [Co(NH_3)_5Cl]Cl_2$$
$$CoCl_3 \cdot 4NH_3 = [Co(NH_3)_4Cl_2]Cl$$
$$CoCl_3 \cdot 3NH_3 = [Co(NH_3)_3Cl_3]$$

The chloride ions that appear outside the brackets represent chloride anions that balance the positive charge on the coordination compound. When a coordination compound dissolves in water, the ligands (inside the brackets) remain bound to the metal cation, but the nonligands (outside the brackets) exist as individual ions. These chloride ions precipitate in the presence of silver ions. The chloride ions inside the brackets, which are ligands bonded to the cobalt center, do not precipitate as AgCl.

For this elegant insight into chemical structure, based on simple stoichiometric relations, Werner was awarded the Nobel Prize in Chemistry in 1913. Since then, a long path of discovery has uncovered many remarkable properties of coordination complexes. To give just one example, there are two isomers of the square planar platinum complex $[Pt(NH_3)_2Cl_2]$. The *cis* isomer, known as cisplatin, is an effective anticancer drug, but the *trans* isomer shows no anticancer activity at all. The mechanism by which the *cis* isomer destroys cancer cells is not fully understood, but research indicates that cisplatin disrupts cell duplication by inserting into the DNA double helix. It appears that DNA binds this platinum complex by replacing the two chloride ions. In a sense, the DNA molecule becomes a huge ligand for the platinum atom. Apparently, the coordination complex can form bonds to DNA only when the chloride ions are adjacent to each other in the *cis* configuration.

cis-$[Pt(NH_3)_2Cl_2]$

19.3 BONDING IN COORDINATION COMPLEXES

In the simplest view, a metal–ligand σ bond can be described as the overlap of a filled donor orbital on the ligand with an empty acceptor orbital on the metal. There is one such interaction for each ligand donor atom, so an octahedral complex of the general formula $[ML_6]^{n+}$ has six σ bonds. Although simple, this view of metal-ligand bonding does not explain the colors and other properties of coordination complexes. Molecular orbital theory provides the most complete description of the bonding in coordination complexes, but this approach is beyond the scope of introductory chemistry. Instead, we use a model called **crystal field theory** to explain the colors and magnetic properties of coordination compounds.

Crystal field theory focuses on coulombic interactions between the transition metal ion and the electron pairs of the donor ligands. In this view, the complex is held together by attractive forces between the negatively charged electrons of the lone pairs and the positive charge of the metal ion. Consequently, bonding can be described much like the bonding in an ionic crystal.

Although such attractive forces can account for metal-ligand bonding, there are also repulsive interactions between the electrons on the ligands and the valence d electrons of the metal ion. These repulsive interactions can account for many properties of coordination complexes. Thus crystal field theory can explain several important properties of transition metal complexes. Nevertheless, all metal-ligand bonds have covalent character, which limits the usefulness of an ionic model.

Orbital Stability in Octahedral Complexes

In a free metal ion without any ligands, all five d orbitals have identical energies, but what happens to the d orbitals when six ligands are placed around a metal in octahedral geometry? The complex is stabilized by coulombic attractions between the positive charge of the metal ion and negative electrons of the ligands. At the same time, the electrons in the metal d orbitals repel the electrons on the ligands. Electron–electron repulsion affects some of the d orbitals more than others. Figure 19-13 shows that two orbitals, $d_{x^2-y^2}$ and d_{z^2}, point directly toward the ligands. As a consequence, electrons in these orbitals experience greater electron–electron repulsion than do d_{xz}, d_{yz}, and d_{xy}, which point between the ligands. This means that metal valence electrons are more stable when they occupy the d orbitals that point away from the ligands.

Figure 19-14 summarizes the coulombic interactions of an octahedral complex ion. The three orbitals that are more stable are called t_{2g} orbitals, and the two less stable orbitals are called e_g orbitals. The difference in energy between the two sets is known as the **crystal field splitting energy,** symbolized by the Greek letter Δ.

The names t_{2g} and e_g are derived from symmetry properties of the orbitals that are not important for general chemistry.

Populating the d Orbitals

Electron configurations of transition metal complexes are governed by the principles described in Chapter 7. The Pauli exclusion principle states that no two electrons can have identical descriptions, and Hund's rule requires that all unpaired electrons have the same spin orientation. These concepts are used in Chapter 7 for atomic configurations and in Chapters 8 and 9 to describe the electron configurations of molecules. The same ideas also apply to transition metal complexes.

The number of electrons in the d orbitals depends on the electron configuration of the metal. That configuration can be found from the oxidation state of the metal and its atomic number. As an example, consider $[Cr(NH_3)_6]^{3+}$. The charge

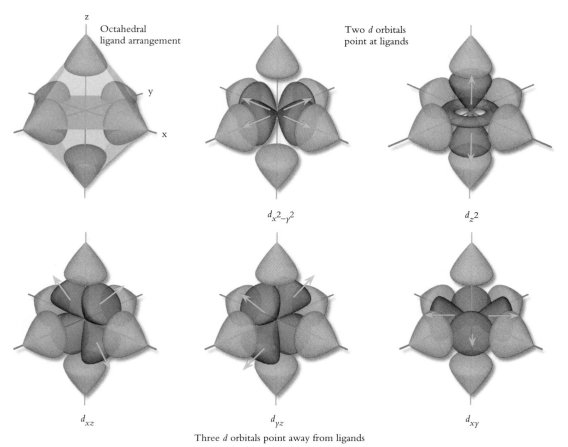

Figure 19-13
The five *d* orbitals and their relationship to an octahedral set of ligands. Whereas two orbitals point directly at the ligands, the other three orbitals point between the ligands.

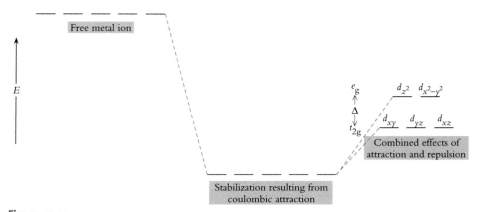

Figure 19-14
The crystal field energy level diagram for octahedral coordination complexes. The energies of the *d* orbitals differ because of differing amounts of electron-electron repulsion. The t_{2g} orbitals point between the ligands, have less electron repulsion, and are more stable than the e_g orbitals, which point directly at the ligands.

on the complex is $+3$, and because ammonia is a neutral ligand, all the charge must arise from the metal ion. In other words, chromium has an oxidation state of $+3$, which means that three of its valence electrons have been removed. Chromium is in Group 6 of the periodic table, and Group 6 atoms have six valence electrons. Thus Cr^{3+} has three valence electrons, all in $3d$ orbitals: [Ar] $3d^3$. One d electron occupies each of the three t_{2g} orbitals, and all three have the same spin. The crystal field electron configuration of $[Cr(NH_3)_6]^{3+}$ can be summarized as $(t_{2g})^3(e_g)^0$. Example 19-4 describes the electron configuration of another coordination complex.

Example 19-4	**Electron Configurations**

Draw an energy level diagram and write the electron configuration of $[Pt(en)_3]Cl_2$.

Strategy: Identify the ligands and the geometry of the coordination complex, construct the crystal field energy level diagram, count d electrons from the metal and place them according to the Pauli principle and Hund's rule.

Solution: Ethylenediamine is a bidentate ligand, so there are six donor atoms, giving the complex ion octahedral geometry. The two chlorides outside the bracket indicate that the complex ion is a $+2$ cation. Since en is a neutral ligand, Pt is in its $+2$ oxidation state. Platinum (Group 10) has ten valence electrons, two of which are removed to give the $+2$ oxidation state, leaving eight d electrons to be placed in the energy level diagram. The Pauli principle and Hund's rule dictate the result:

The electron configuration is $(t_{2g})^6(e_g)^2$.

The examples given so far lead to unambiguous electron configurations, but not all configurations are this straightforward. Consider any octahedral complex containing a metal atom with four d electrons. Following the standard filling procedure, the first three electrons are placed in the t_{2g} orbitals. The fourth electron might also be placed in a t_{2g} orbital, but there is a price to pay in energy: Two electrons in the same orbital are destabilized because two negative charges are confined to the same region of space (Coulomb's law again). This destabilization is called the **pairing energy (P)**. There is no pairing energy if the fourth electron is placed in an e_g orbital, but these orbitals are higher in energy than the t_{2g} orbitals. The energy difference between the two sets is the crystal field splitting energy, Δ.

Whether $(t_{2g})^4(e_g)^0$ or $(t_{2g})^3(e_g)^1$ is more stable depends on the relative magnitudes of P and Δ. If the energy required to pair electrons in a t_{2g} orbital is less than the energy required to populate an e_g orbital $(P < \Delta)$, the ground state configuration is $(t_{2g})^4(e_g)^0$. On the other hand, if $P > \Delta$, the most stable configuration is $(t_{2g})^3(e_g)^1$. Both possibilities are shown in Figure 19-15 for the four-electron, octahedral case.

In conformity with Hund's rule, a metal complex with the configuration $(t_{2g})^3(e_g)^1$ has four unpaired electrons. This configuration maximizes the number of electrons with unpaired spins, so it is described as **high spin**. The alternative configuration, $(t_{2g})^4(e_g)^0$, has just two unpaired electrons and is described as **low spin**.

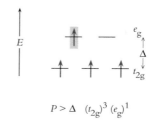

Figure 19-15
The two possible electron configurations for a d^4 metal complex. Notice that whereas Δ can be shown on this diagram, P cannot, because P measures electron-electron repulsion, not an orbital energy.

Magnetic Properties of Coordination Complexes

What experiments can be done to reveal whether a complex with four d electrons is $(t_{2g})^4(e_g)^0$ or $(t_{2g})^3(e_g)^1$? One of the most common ways to determine the electron configuration of a coordination complex is to measure its magnetic properties.

Recall from Chapters 6 and 7 that the spin of an electron generates magnetism. When electrons are paired, their spins point in opposite directions, causing their magnetism to cancel. A molecule with all electrons paired has no magnetism from its electron spins and is diamagnetic, but a molecule that has unpaired electrons is paramagnetic. The amount of magnetism in a paramagnetic molecule depends on the number of unpaired electrons: The more unpaired electrons in the d orbitals of a complex, the greater the magnetism. The magnetic properties of a complex ion can be measured with an instrument known as a *Gouy balance*, as shown schematically in Figure 19-16. The paramagnetic sample is attracted into the magnetic field, creating a downward force on the left pan. Weights are added to the pan on the right until the downward force is balanced. The magnetic strength of the sample is proportional to the mass required to balance the pans. This provides an experimental method for determining the ground-state electron configuration of a coordination complex.

Example 19-5 illustrates the magnetic properties of another configuration.

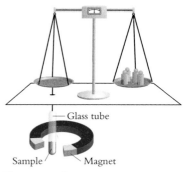

Figure 19-16
The number of unpaired electrons in a paramagnetic complex can be determined with a Gouy balance.

High- and Low-Spin Complexes	Example 19-5

$[Fe(NH_3)_6]^{2+}$ is paramagnetic, but $[Co(NH_3)_6]^{3+}$ is not. Write the electron configuration for each of these metal complexes and draw energy level diagrams showing which has the higher Δ.

Strategy: Use the formula of a complex to determine its geometry, the form of its energy level diagram, and the number of d electrons. The spin information indicates how many of these electrons are paired.

Solution: Both complexes have six monodentate ligands, so both are octahedral. Because ammine ligands are neutral molecules, the oxidation state of each metal is the same as the charge on the complex. Iron loses two of its eight valence electrons to reach the $+2$ oxidation state, leaving six electrons for the d orbitals. Likewise, cobalt in its $+3$ oxidation state has six d electrons.

The iron complex is paramagnetic, which means that it has unpaired electrons. This happens when the crystal field splitting energy is less than the pairing energy, so the energy level diagram should show a relatively small Δ. By Hund's rule, we place one electron in each of the five d orbitals before pairing any. Iron's sixth d electron pairs up with an electron in any one of the t_{2g} orbitals.

The cobalt complex is not paramagnetic, which means that it has all of its valence electrons paired. This happens when Δ is larger than P, so the energy level diagram should show a relatively large Δ. In a low-spin complex, the lower-energy t_{2g} orbitals are filled completely before putting electrons in e_g.

| Example 19-5 | High- and Low-Spin Complexes *(continued)* |

Here are the correct diagrams:

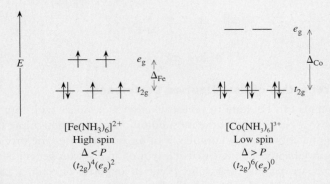

$[Fe(NH_3)_6]^{2+}$
High spin
$\Delta < P$
$(t_{2g})^4(e_g)^2$

$[Co(NH_3)_6]^{3+}$
Low spin
$\Delta > P$
$(t_{2g})^6(e_g)^0$

The paramagnetic configuration is $(t_{2g})^4(e_g)^2$, with four unpaired electrons. The other configuration is $(t_{2g})^6(e_g)^0$, with no unpaired electrons and zero magnetism.

Contributions to Crystal Field Splitting Energy

Whether a complex is low spin or high spin depends on the balance between pairing energy and crystal field splitting energy. Pairing energy changes very little from one coordination complex to the next. Consequently, electron configurations of coordination complexes are governed by the crystal field splitting energy, Δ.

The most important factor that influences the value of Δ is the identity of the ligands. Compare $[Fe(CN)_6]^{4-}$ and $[FeCl_6]^{4-}$, two coordination complexes similar in every respect except for their ligands. Both contain Fe^{2+} ions with six d electrons, both have six ligands arranged in an octahedral geometry, and both have the same ionic charge. Despite these similarities, the cyano complex is low spin and diamagnetic, whereas the chloro complex is high spin and paramagnetic. For $[Fe(CN)_6]^{4-}$, the crystal field splitting energy is larger than the pairing energy, so the molecule is low spin. In contrast, $P > \Delta$ for $[FeCl_6]^{4-}$. The chloro ligand generates a small energy gap between the t_{2g} and e_g orbitals, giving a high-spin coordination complex.

Studies of many coordination complexes reveal a common pattern in field strengths of the various ligands. This pattern is described by the **spectrochemical series,** in which ligands are listed in order of increasing energy level splitting:

$$I^- < Br^- < Cl^- < F^- < OH^- < H_2O < NH_3 < en < NO_2^- < CN^- < CO$$

Smaller splitting ⟶ Larger splitting

The oxidation state of the metal also contributes to the crystal field splitting energy. For a given ligand, Δ increases as the oxidation state of the metal increases. This trend is easy to interpret using Coulomb's law. As the charge on the metal increases, so does the attraction between the metal and its ligands. The ligands approach the metal more closely, resulting in stronger orbital interactions and a larger crystal field splitting (Figure 19-17). The complexes of Example 19-5, $[Fe(NH_3)_6]^{2+}$ (high spin) and $[Co(NH_3)_6]^{3+}$ (low spin) illustrate this feature. These

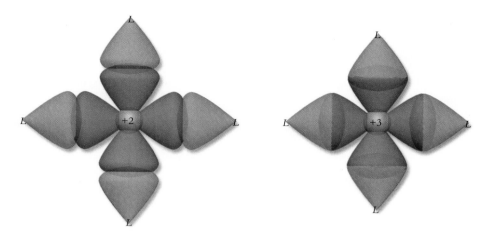

Figure 19-17
As the charge on the metal ion increases, coulombic attraction pulls the ligands closer to the cation. This leads to greater repulsive interactions between valence *d* electrons and ligand electrons.

two complexes have the same ligands and the same molecular geometry, and both are d^6 metal ions, but Co^{3+} generates a stronger crystal field than Fe^{2+}.

Another influence on the magnitude of the crystal field splitting is the position of the metal in the periodic table. Crystal field splitting energy increases substantially as valence orbitals change from $3d$ to $4d$ to $5d$. Again, orbital shapes explain this trend. Orbital size increases as n increases, and this means that the d orbital set becomes more exposed to orbital interactions with approaching ligands. For example, the value of Δ increases by about 50% for complexes of rhodium compared with analogous cobalt complexes, and there is another 25% increase on moving from rhodium to iridium. As a result of this trend, coordination complexes of $4d$ and $5d$ elements are almost always low spin. In contrast, the $3d$ transition metals have both low- and high-spin complexes.

Example 19-6 applies the principles of crystal field splitting to a series of coordination complexes.

Crystal Field Splitting Energy	**Example 19-6**

Arrange the following complexes in order of increasing crystal field splitting: $[Fe(H_2O)_6]^{2+}$, $[Fe(H_2O)_6]^{3+}$, $[FeCl_6]^{4-}$, and $[Ru(H_2O)_6]^{3+}$.

Strategy: The strength of splitting depends on the ligand, the charge on the metal, and the position of the metal in the periodic table. Examine these factors independently.

Solution:

Ligand: Cl^- is lower in the spectrochemical series than H_2O. Thus $[FeCl_6]^{4-}$ has low splitting relative to the others.

Cation charge: In $[FeCl_6]^{4-}$ and $[Fe(H_2O)_6]^{2-}$, iron is in its $+2$ oxidation state, but the metal is $+3$ in the other two complexes. Complexes with $+3$ oxidation states have larger splitting:

$$[FeCl_6]^{4-} < [Fe(H_2O)_6]^{2+} < [Fe(H_2O)_6]^{3+}, [Ru(H_2O)_6]^{3+}$$

Valence orbitals: Three complexes contain iron, a metal with $3d$ valence electrons, but the fourth contains ruthenium, with $4d$ valence electrons. All

| Example 19-6 | Crystal Field Splitting Energy *(continued)* |

other things being equal, a $4d$ metal has a crystal field approximately twice as large as a $3d$ metal.

Applying all three trends gives an unambiguous arrangement of these complexes:

$$[FeCl_6]^{4-} < [Fe(H_2O)_6]^{2+} < [Fe(H_2O)_6]^{3+} < [Ru(H_2O)_6]^{3+}$$

Lowest splitting   Highest splitting

Color in Coordination Complexes

Color is a spectacular property of coordination complexes. Figure 19-18 shows hexaaqua cations of several $3d$ transition metals. The origin of these colors lies in the d orbital energy differences and can be understood using crystal field theory.

Color is caused by absorption of some of the light from the visible spectrum (Figure 19-19). The wavelengths of light absorbed by a collection of molecules are lost from the rest of the spectrum. Light that is not absorbed is reflected by an opaque sample and is transmitted through a transparent sample. In either case, an observer sees all wavelengths except those absorbed by the molecules. If a sample absorbs all wavelengths of visible light except blue, that sample will appear blue to the observer. A substance that appears black absorbs all wavelengths of visible light. A substance that absorbs no visible wavelengths appears white if light is reflected by the surface and colorless if the light is transmitted through the sample.

Figure 19-18
Coordination complexes often have beautiful colors. Shown here are solutions of hexaaqua complexes of first row transition metals.

Molecules that absorb a small wavelength region have a color that is different from the color absorbed. For example, a molecule that absorbs only orange light appears blue. Likewise, a molecule that absorbs only blue appears orange. Orange and blue are said to be **complementary colors.** The complementary colors are listed in Table 19-4.

When a molecule absorbs light, it gains energy from the absorbed photons. Energy conservation requires that the energy change for the molecule equal the energy of the absorbed photon:

$$\Delta E_{molecule} = E_{photon} = h\nu$$

This requirement means that when a coordination complex absorbs light, the crystal field splitting energy, Δ, must match the energy of the absorbed light. Thus the crystal field splitting energy of a complex can be determined from the wavelength of visible light that the complex absorbs:

$$\Delta E_{molecule} = \Delta = h\nu = hc/\lambda$$

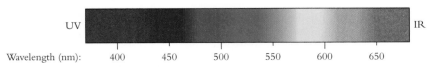

UV   IR

Wavelength (nm):   400   450   500   550   600   650

Figure 19-19
The visible spectrum, showing the colors as a function of wavelength.

Table 19-4
Relationships Among Wavelength, Color, and Crystal Field Splitting Energy (Δ)

Wavelength (nm)	Color Absorbed	Complementary Color	Δ (kJ/mol)
>720	Infrared	Colorless	<165
720	Red	Green	166
680	Red-orange	Blue-green	176
610	Orange	Blue	196
580	Yellow	Indigo	206
560	Yellow-green	Violet	214
530	Green	Purple	226
500	Blue-green	Red	239
480	Blue	Orange	249
430	Indigo	Yellow	279
410	Violet	Lemon-yellow	292
<400	Ultraviolet	Colorless	>299

Spectroscopic properties of coordination compounds are highlighted by a series of d^3 Cr^{3+} complexes, as shown in Figure 19-20. Each molecule has a ground state configuration with three unpaired electrons in the t_{2g} orbitals. Absorption of a photon excites one electron into an e_g orbital, giving an excited-state configuration, $(t_{2g})^2(e_g)^1$. These three complexes have different values of Δ, so they absorb light of different wavelengths. The cyano ligand is high in the spectrochemical series, generating a large crystal field splitting for $[Cr(CN)_6]^{3-}$. A photon of high-energy

SECTION 6.3 & 7.5
See Sections 6.3 and 7.5 for a review of spectroscopy, ground states, and excited states.

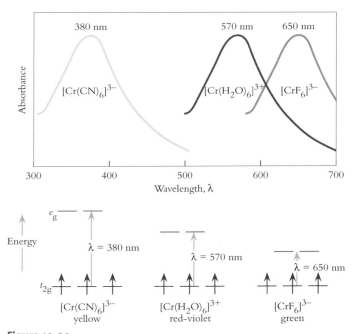

Figure 19-20
The colors of Cr^{3+} coordination complexes depend on the positions of the ligands in the spectrochemical series.

violet light is required to promote an electron from the t_{2g} set to the e_g set. Absorption of violet light gives $[Cr(CN)_6]^{3-}$ a yellow color. The fluoro ligand is near the bottom of the spectrochemical series, generating a weak crystal field. The fluoro complex absorbs red light, generating a green color for $[CrF_6]^{3-}$. When H_2O is the ligand, there is an intermediate energy gap. Yellow-green light is absorbed, so $[Cr(H_2O)_6]^{3+}$ is violet. Example 19-7 shows how to determine the value of Δ from an absorption spectrum.

Example 19-7	Determining the Value of Δ

Titanium(III) chloride dissolves in water to give $[Ti(H_2O)_6]^{3+}$. This complex ion has the absorption spectrum shown. From the wavelength at which maximum absorption occurs, predict the color of the solution and calculate Δ in kilojoules per mole.

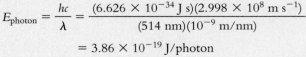

Strategy: The spectrum shows wavelengths of light absorbed by the metal complex. The wavelength absorbed most strongly corresponds to Δ, but appropriate conversions are needed to find the molar energy associated with this wavelength. The color of the solution is the complementary color of the most strongly absorbed wavelength.

Solution: Interpolate on the graph to find that the maximum intensity occurs at 514 nm. Use this wavelength to find the energy of one photon:

$$E_{photon} = \frac{hc}{\lambda} = \frac{(6.626 \times 10^{-34}\ J\ s)(2.998 \times 10^8\ m\ s^{-1})}{(514\ nm)(10^{-9}\ m/nm)}$$

$$= 3.86 \times 10^{-19}\ J/photon$$

The Avogadro constant is used to convert to kilojoules per mole:

$(3.86 \times 10^{-19}\ J/photon)(10^{-3}\ kJ/J)(6.022 \times 10^{23}\ photons/mol) = 232\ kJ/mol = \Delta$

The $[Ti(H_2O)_6]^{3+}$ ion absorbs at 514 nm, which is in the green to blue-green region of the visible spectrum. According to Table 19-4, the solution should be reddish purple. A photo of a solution of $[Ti(H_2O)_6]^{3+}$ is shown in the margin.

Square Planar and Tetrahedral Complexes

In a square planar complex, the four ligands lie along the x and y axes. As shown in Figure 19-21, $d_{x^2-y^2}$ is the only d orbital that points directly at the four ligands. Thus in square planar geometry $d_{x^2-y^2}$ is significantly higher in energy than the other four d orbitals. Although the d_{xy} orbital also lies in the metal-ligand plane, it points *between* the ligands and is subject to less electron–electron repulsion than $d_{x^2-y^2}$. The remaining three d orbitals point out of the metal-ligand plane, so are

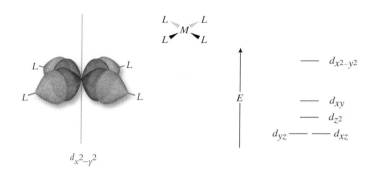

Figure 19-21
In square planar geometry, the $d_{x^2 - y^2}$ orbital is most strongly destabilized by interactions with the ligands. The d_{xy} orbital also lies in the ligand plane, but it points *between* the ligands.

repelled less by the ligands and are lower in energy than d_{xy}. The d_{z^2} orbital is slightly less stable than d_{xz} and d_{yz} because of the small band of electron density that circles the metal in the xy plane. The energy level diagram shown in Figure 19-21 shows the splitting pattern for a square planar complex. Square planar geometry is most common for d^8 metal ions with large crystal field splitting energies. The d^8 configuration fills the four lower-energy d orbitals, but it leaves the high-energy $d_{x^2-y^2}$ empty.

In tetrahedral complexes the ligands lie at the corners of a tetrahedron rather than at the corners of a square. The symmetry relationships between the d orbitals and these ligands are not easy to visualize, but the splitting pattern of the d orbitals can be determined using geometry. The result is the opposite of the pattern found in octahedral complexes: The d_{xz}, d_{xy}, and d_{yz} orbitals are higher in energy than $d_{x^2-y^2}$ and d_{z^2}. Figure 19-22 shows the splitting pattern for tetrahedral coordination. Because there are only four ligands rather than six, and because none of the d orbitals point directly at the ligands, the crystal field splitting energy for tetrahedral geometry is only about half that for octahedral geometry. As a result, the pairing energy is almost always smaller than the splitting energy ($P < \Delta$), and therefore almost all tetrahedral complexes are high spin.

Figure 19-22
The crystal field energy level diagram for tetrahedral complexes. The d orbitals are split into two sets, with three orbitals destabilized relative to the two others.

Section Exercises

19.3.1 Draw crystal field splitting diagrams that show the electron configurations for the following complex ions: (a) $[Cr(H_2O)_6]^{2+}$; (b) $[IrCl_6]^{3-}$; (c) $[V(en)_3]^{3+}$; and (d) $[NiCl_4]^{2-}$ (tetrahedral).

19.3.2 Explain why hexacyano complexes of metals in their +2 oxidation state are usually yellow, but the corresponding hexaaqua compounds are often blue or green.

19.3.3 The value of Δ for $[RhCl_6]^{3-}$ is 243 kJ/mol. What wavelength of light will promote an electron from the t_{2g} set to the e_g set? What color is the complex?

19.4 METALLURGY

Metallurgy is the production and purification of metals from naturally occurring deposits called *ores*. It has an ancient history and may represent the earliest useful application of chemistry. Metallurgical advances have had a profound influence on the course of human civilization, so much so that historians speak of the Bronze Age (ca. 3000 to 1000 BC) and the Iron Age (starting ca. 1000 BC). Except for

Figure 19-23
Metallurgy includes separation, conversion, reduction, and refining steps. The starting material is an impure ore, and the end product is pure metal.

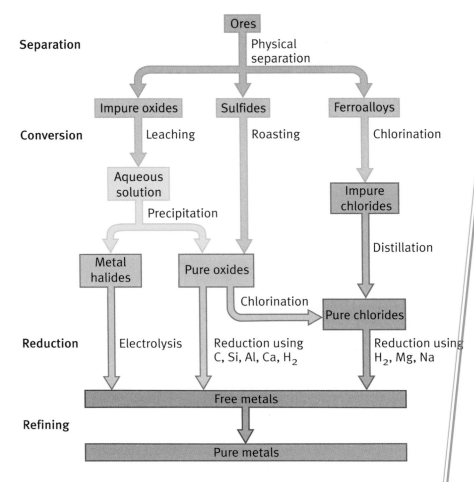

aluminum and tin, the story of metallurgy is primarily about the extraction and purification of transition metals from their ores.

Nearly all transition metals are oxidized readily, so most ores are compounds in which the metals have positive oxidation numbers. Examples include oxides (TiO_2, rutile; Fe_2O_3, hematite; Cu_2O, cuprite), sulfides (ZnS, sphalerite; MoS_2, molybdenite), phosphates ($CePO_4$, monazite; YPO_4, xenotime; both found mixed with other rare earth metal phosphates), and carbonates ($FeCO_3$, siderite). Other minerals contain oxyanions ($MnWO_4$, wolframite) and even more complex structures such as carnotite, $K_2(UO_2)_2(VO_4)_2 \cdot 3H_2O$. In this section, we discuss some of the techniques of metallurgy.

Figure 19-23 shows in schematic fashion some of the alternative paths leading from ores to pure metals. These paths include four general processes. First is *separation*. Generally, a metal ore obtained from a mine contains a particular compound of some desired metal mixed with various other materials. The mineral must be separated from these other contaminants. Separation often is followed by *conversion*, in which the mineral is treated chemically to convert it into a form that can be easily reduced. The third step is *reduction*. After a suitable compound has been obtained, it is reduced to free metal by chemical reaction with a reducing agent or by electrolysis. Free metal obtained by reduction often contains small amounts of impurities, so the final step is *refining* to purify the metal.

Overview of Metallurgical Processes

Ore obtained from a mining operation contains a desired mineral contaminated with other components, which may include sand, clay, and organic matter. This economically valueless portion of the ore, which is called gangue (pronounced "gang"), must be removed before the metal can be extracted and refined. Ores can be separated into components by physical or chemical methods.

Flotation is a common physical separation process in which the ore is crushed and mixed with water to form a thick slurry. As shown in Figure 19-24, the slurry is transferred to a flotation vessel and mixed with oil and a surfactant. The polar head groups of the surfactant coat the surface of the mineral particles, but the non-polar tails point outward, making the surfactant-coated mineral particles hydrophobic. Air is blown vigorously through the mixture, carrying the oil and the coated mineral to the surface, where they become trapped in the froth. The gangue has a much lower affinity for the surfactant; it becomes "wetted" by the water and sinks to the bottom of the flotation vessel. The froth is removed at the top, and the gangue is removed at the bottom.

A second separation technique is **leaching,** which uses solubility properties to separate the components of an ore. For example, modern gold production depends on the extraction of tiny particles of gold from gold-bearing rock deposits. After the rock is crushed, it is treated with an aerated aqueous basic solution of sodium cyanide, which leaches the gold into solution by oxidizing the metal and forming a soluble coordination complex:

$$4\,Au(s) + 8\,CN^-(aq) + O_2(g) + 2\,H_2O(l) \longrightarrow 4\,[Au(CN)_2]^-(aq) + 4\,OH^-(aq)$$

The aqueous gold-containing solution is then treated with zinc dust, which reduces gold back to the free metal:

$$Zn(s) + 2\,[Au(CN)_2]^-(aq) \longrightarrow 2\,Au(s) + 4\,[Zn(CN)_4]^-(aq)$$

Many ores are treated chemically before extraction to give the free metals. In particular, transition metals that occur naturally as sulfide ores often are converted into oxides. The process of oxidizing an ore by heating to a high temperature in the presence of air is known as **roasting.** In the roasting of a sulfide ore, sulfide is oxidized, and oxygen is reduced. The conversion of galena is a typical example:

$$2\,ZnS(s) + 3\,O_2(g) \longrightarrow 2\,ZnO(s) + 2\,SO_2(g)$$

Unfortunately, roasting produces copious amounts of highly polluting SO_2 gas that has seriously damaged the environment around smelters for sulfide ores.

Today, zinc and other metals can be extracted from sulfides by aqueous-phase conversion processes that avoid the generation of SO_2. Aqueous acid reacts with the sulfides to generate free sulfur or sulfate ions rather than SO_2. Here are two examples:

$$2\,ZnS(s) + 4\,H_3O^+(aq) + O_2(g) \longrightarrow 2\,Zn^{2+}(aq) + 2\,S(s) + 6\,H_2O(l)$$

$$3\,CuS(s) + 8\,NO_3^-(aq) + 8\,H_3O^+(aq) \longrightarrow$$
$$8\,NO(g) + 3\,Cu^{2+}(aq) + 3\,SO_4^{2-}(aq) + 12\,H_2O(l)$$

When the cost of preventing SO_2 escape into the atmosphere is taken into account, these more elaborate aqueous separation procedures are economically competitive with conversion by roasting.

Once an ore is in suitably pure form, it can be reduced to free metal. This is accomplished either chemically or electrolytically. Electrolysis is costly because it requires huge amounts of electrical energy. For this reason, chemical reduction is used unless the metal is too reactive for chemical reducing agents to be effective.

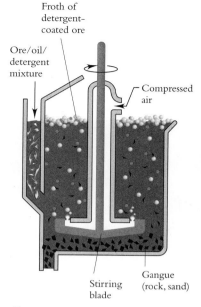

Figure 19-24
In the flotation process, surfactant-coated mineral particles float to the surface of the mixture, where they become trapped in the froth. The gangue settles to the bottom.

← **SECTION 10.7**
We describe surfactants in Section 10.7.

Vegetation around the nickel smelter in Sudbury, Ontario, Canada has been devastated by SO2 emitted during the roasting of NiS ores.

Mercury and lead are sufficiently easily reduced that roasting the sulfide ore frees the metal. Sulfide ion is the reducing agent, and both O_2 and the metal ion gain electrons:

$$HgS(s) + O_2(g) \longrightarrow Hg(l) + SO_2(g)$$

This reduction produces SO_2, which must be removed from the exhaust gases.

One of the most common chemical reducing agents for metallurgy is coke, a form of carbon made by heating coal at high temperature until all of the volatile impurities have been removed. Metals whose cations have moderately negative reduction potentials—Co, Ni, Fe, and Zn—are reduced by coke. For example, direct reaction with coke in a furnace frees nickel from its oxide:

$$NiO(s) + C(s) \longrightarrow Ni(l) + CO(g)$$

Chemical reduction of an ore usually gives metal that is not pure enough for its intended use. Further **refining** of the metal removes undesirable impurities. Several important metals, including Cu, Ni, Zn, and Cr, are refined by electrolysis, either from an aqueous solution of the metal salt or from anodes prepared from the impure metal. To give one example, Zn^{2+} ions, obtained by dissolving ZnS or ZnO in acidic solution, can be reduced while water is oxidized:

$$Zn^{2+}(aq) + 2\ e^- \longrightarrow Zn(s)$$

$$\underline{3\ H_2O(l) \longrightarrow \tfrac{1}{2}\ O_2(g) + 2\ H_3O^+(aq) + 2\ e^-}$$

$$Zn^{2+}(aq) + 3\ H_2O(l) \longrightarrow Zn(s) + \tfrac{1}{2}\ O_2(g) + 2\ H_3O^+(aq)$$

A unique chemical refining process is used to purify nickel. Nickel sulfide ores are concentrated by flotation and converted to nickel oxide by roasting. Reduction with coke gives nickel metal contaminated with various amounts of copper, iron, and cobalt. This impure mixture is refined by reaction with gaseous CO to form a coordination complex, $[Ni(CO)_4]$, which is a colorless volatile liquid:

$$Ni(s) + 4\ CO(g) \xrightarrow{40-90\ °C} [Ni(CO)_4](g) \qquad (bp = 43\ °C)$$

Iron and cobalt also form complexes with CO, but these byproducts are less volatile than $[Ni(CO)_4]$. Consequently, the nickel complex can be purified by distillation. Then the nickel complex is heated above 200 °C, causing decomposition to carbon monoxide and nickel metal that is 99.97% pure:

$$[Ni(CO)_4](g) \xrightarrow{>200\ °C} Ni(s) + 4\ CO(g)$$

One important drawback to this process is that tetracarbonylnickel(0) is extremely poisonous, so the concentration of the complex in the atmosphere of the refining plant must be kept below the part-per-billion level.

Table 19-5 provides a summary of the chemical species and processes involved in the metallurgy of many transition metals. A survey of several metals provides further examples of the four phases of metallurgy.

Iron and Steel

Iron has been the dominant structural material of modern times, and despite the growth in importance of aluminum and plastics, iron still ranks first in total use. Worldwide production of steel (iron strengthened by additives) is on the order of 700 million tons per year. The most important iron ores are two oxides, hematite (Fe_2O_3) and magnetite (Fe_3O_4). The production of iron from its ores involves several chemical processes that take place in a blast furnace. As shown in Figure 19-25,

Table 19-5
Metallurgy of Transition Metals

Metal	Ore	Separation Method	Intermediate*	Reducing Agent
Ti	TiO_2	Chlorination	$TiCl_4$	Mg
Cr	$FeCr_2O_4$		Cr_2O_3	Al
Mn	MnO_2		Mn_2O_3	Al
Fe	Fe_3O_4	Slag formation	$(CaSiO_3)$	C
Co	CoAsS	Roasting	CoO	C
Ni	Ni_9S_8	Complexation	$Ni(CO)_4$	H_2
Cu	$CuFeS_2$	Leaching	Cu^{2+}, (SO_4^{2-})	
Zn	ZnS	Roasting	ZnO	C
Mo	MoS_2			
Ag	Ag_2S			
W	$CaWO_4$	Leaching	WO_4^{2-}, WO_3	H_2
Au	Au	Leaching	$[Au(CN)_2]^-$	Zn
Hg	HgS	Roasting	(SO_2)	S^{2-}

*Intermediates that represent impurities are shown in parentheses.

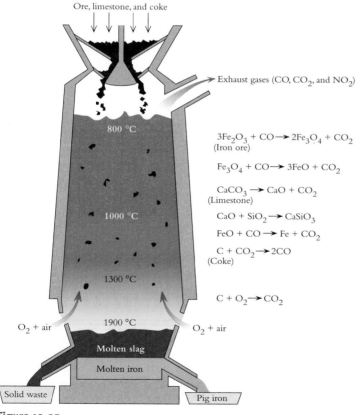

Ore, limestone, and coke

Exhaust gases (CO, CO_2, and NO_2)

800 °C

$3Fe_2O_3 + CO \longrightarrow 2Fe_3O_4 + CO_2$
(Iron ore)

$Fe_3O_4 + CO \longrightarrow 3FeO + CO_2$

$CaCO_3 \longrightarrow CaO + CO_2$
(Limestone)

$CaO + SiO_2 \longrightarrow CaSiO_3$

1000 °C

$FeO + CO \longrightarrow Fe + CO_2$

$C + CO_2 \longrightarrow 2CO$
(Coke)

1300 °C

$C + O_2 \longrightarrow CO_2$

$O_2 + air$ 1900 °C $O_2 + air$

Molten slag

Molten iron

Solid waste   Pig iron

Figure 19-25
A diagrammatic view showing the chemical reactions occurring within a blast furnace, which operates continuously at fiery temperatures (see photo).

this is an enormous chemical reactor where heating, reduction, and purification all occur together.

The raw materials placed in the blast furnace include the ore (usually hematite) and coke, which serves as the reducing agent. The ores always contain various amounts of silicon dioxide (SiO_2), which is removed chemically by reaction with limestone ($CaCO_3$). To begin the conversion process, pellets of ore, coke, and limestone are mixed and fed into the top of the furnace, and a blast of hot air is blown in at the bottom. As the starting materials fall through the furnace, the burning coke generates intense heat:

$$2C(s) + O_2(g) \longrightarrow 2\,CO(g) \qquad \Delta H^\circ = -110 \text{ kJ}$$

The result is a temperature gradient ranging from about 800 °C at the top of the furnace to 1900 °C at the bottom.

The reduction of iron oxide takes place in several stages in different temperature zones within the furnace. The reducing agent is CO produced from burning coke. Here are the key reactions:

$$3\,Fe_2O_3(s) + CO(g) \longrightarrow 2\,Fe_3O_4(s) + CO_2(g)$$

$$Fe_3O_4(s) + CO(g) \longrightarrow 3\,FeO(s) + CO_2(g)$$

$$FeO(s) + CO(g) \longrightarrow Fe(l) + CO_2(g)$$

Once liberated from its oxides, the iron melts when the temperature reaches 1500 °C. Molten iron collects in a pool at the bottom of the furnace.

At the same time that heating and reduction occur, limestone decomposes into calcium oxide and CO_2. The CaO then reacts with SiO_2 impurities in the ore to generate calcium silicate:

$$CaCO_3(s) \xrightarrow{\text{Heat}} CaO(s) + CO_2(g)$$

$$CaO(s) + SiO_2(s) \xrightarrow{\text{Heat}} CaSiO_3(l)$$

At blast furnace temperatures, calcium silicate is a liquid, called *slag*. Being less dense than iron, slag pools on the surface of the molten metal. Both products are drained periodically through openings in the bottom of the furnace.

Even though this chemistry is complex, the basic process is reduction of iron oxide by carbon in an atmosphere depleted of oxygen. Archaeologists have found ancient smelters in Africa (in what is now Tanzania) that exploited this chemistry to produce iron in prehistoric times. Early African peoples lined a hole with a fuel of termite residues and added iron ore. Charred reeds and charcoal provided the reducing substance. Finally, a chimney of mud was added. When this furnace was "fired," a pool of iron collected in the bottom.

The iron formed in a blast furnace, called *pig iron*, contains impurities that make the metal brittle. These include phosphorus and silicon from silicate and phosphate minerals that contaminated the original ore, as well as carbon and sulfur from the coke. This iron is refined in a converter furnace. Here, a stream of O_2 gas blows through molten impure iron. Oxygen reacts with the nonmetal impurities, converting them to oxides. As in the blast furnace, CaO is added to convert SiO_2 into liquid calcium silicate, in which the other oxides dissolve. The molten iron is analyzed at intervals until its impurities have been reduced to satisfactory levels. Then the liquid metal, now in the form called *steel,* is poured from the converter and allowed to solidify.

Most steels contain various amounts of other elements that are added deliberately to give the metal particular properties. These additives may be introduced during the converter process or when the molten metal is poured off. One of the most important additives is manganese, which is added to nearly every form of steel in amounts ranging from less than 1% to higher than 10%. More than 80% of manganese production ends up incorporated into steel.

Other Metals

Titanium. The metallurgy of titanium illustrates the purification of one metal by another. The major titanium ores are rutile (TiO_2) and ilmenite ($FeTiO_3$). Either is converted to titanium(IV) chloride by a redox reaction with chlorine gas and coke. For rutile:

$$TiO_2(s) + C(s) + 2\ Cl_2(g) \xrightarrow{500\ ^\circ C} TiCl_4(g) + CO_2(g)$$

In this reaction, carbon is oxidized and chlorine is reduced. When the hot gas cools, titanium tetrachloride (bp = 140 °C) condenses to a liquid that is purified by distillation.

Titanium metal is obtained by reduction of $TiCl_4$ with molten magnesium metal at high temperature. The reaction gives solid titanium metal (mp = 1660 °C) and liquid magnesium chloride (mp = 714 °C):

$$TiCl_4(g) + 2\ Mg(l) \xrightarrow{850\ ^\circ C} Ti(s) + MgCl_2(l)$$

Tungsten. Molecular hydrogen is an important reducing agent in metallurgy, as illustrated by the purification of tungsten. The two main ores of tungsten contain the tungstate anion, WO_4^{2-}, combined either with Ca^{2+} (scheelite) or with a combination of Fe^{2+} and Mn^{2+} in various proportions (wolframite). The treatments are somewhat different for these two ores, as the flowchart in Figure 19-26 illustrates. Leaching of scheelite ($CaWO_4$) with strong hydrochloric acid removes phosphorus, arsenic, and sulfur impurities, leaving insoluble tungstic acid, H_2WO_4.

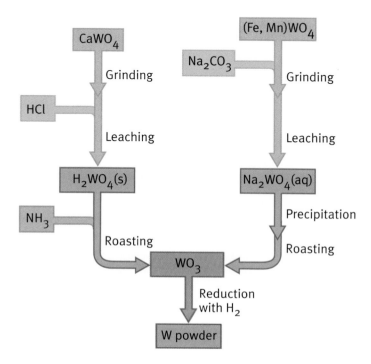

Figure 19-26
Flowchart showing the metallurgy of tungsten. Each major ore is treated to yield a tungstate, which is recovered as pure WO_3. Tungsten(VI) oxide is reduced with H_2 to yield metal powder.

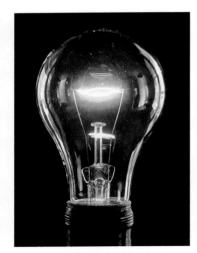

Pure tungsten metal is used as the filaments in electric light bulbs.

Iron and manganese in wolframite, $(Fe, Mn)WO_4$, are removed by roasting the ore with sodium carbonate followed by leaching with water to give an aqueous solution of sodium tungstate, Na_2WO_4. Both tungstic acid and sodium tungstate are converted to tungsten(VI) oxide on treatment with ammonia followed by roasting in air at 900 °C.

Once pure WO_3 is obtained, treatment with hydrogen gas at elevated temperature leads to the pure metal:

$$WO_3(s) + 3\ H_2(g) \xrightarrow{800\ °C} W(s) + 3\ H_2O(g)$$

Much of the tungsten produced in industry is converted to tungsten carbide, WC, an extremely hard and wear-resistant material used for tips of cutting and drilling tools. The most important use of pure tungsten metal is as the filaments in electric light bulbs.

Copper. Copper is found mainly in the sulfide ore chalcopyrite ($FeCuS_2$), but chalcocite (Cu_2S), cuprite (Cu_2O), and malachite ($Cu_2CO_3(OH)_2$) are also important. Copper ores often have concentrations of copper less than 1% by mass, so achieving economic viability requires mining operations on a huge scale. The extraction and purification of copper is complicated by the need to remove iron from chalcopyrite. The first step in the process is flotation, which concentrates the ore to around 15% Cu by mass. In the next step, the concentrated ore is roasted to convert $FeCuS_2$ to CuS and FeO. Copper(II) sulfide is unaffected if the temperature is kept below 800 °C:

$$2\ FeCuS_2(s) + 3\ O_2(g) \longrightarrow 2\ CuS(s) + 2\ FeO(s) + 2\ SO_2(g)$$

Copper mining is carried out on a huge scale.

Heating the mixture of CuS and FeO to 1400 °C in the presence of silica (SiO_2) causes the material to melt and separate into two layers. The top layer is molten $FeSiO_3$ formed from the reaction of SiO_2 with FeO. As this takes place, the copper in the bottom layer is reduced from CuS to Cu_2S. This bottom layer consists of molten Cu_2S contaminated with FeS. Reduction of the Cu_2S takes place in a converter furnace following the same principle that converts impure iron into steel. Silica is added, and oxygen gas is blown through the molten mixture. Iron impurities are converted first to FeO and then to $FeSiO_3$, which is a liquid that floats to the surface. At the same time, Cu_2S is converted to Cu_2O, which reacts with more Cu_2S to give copper metal and SO_2:

$$2\ Cu_2S(l) + 3\ O_2(g) \longrightarrow 2\ Cu_2O(l) + 2\ SO_2(g)$$

$$2\ Cu_2O(l) + Cu_2S(l) \longrightarrow 6\ Cu(l) + SO_2(g)$$

Copper metal obtained from the converter furnace must be refined to better than 99.95% purity before it can be used to make electrical wiring. This is accomplished by electrolysis, as illustrated in Figure 19-27. The impure copper is formed into slabs that serve as anodes in electrolysis cells. The cathodes are constructed from thin sheets of very pure copper. These electrodes are immersed in a solution of $CuSO_4$ dissolved in dilute sulfuric acid. Application of a controlled voltage causes oxidation in which copper, along with iron and nickel impurities, is oxidized to its cations. Less reactive metal contaminants, including silver, gold, and platinum, are not oxidized; as the electrolysis proceeds and the anode dissolves, these and other insoluble impurities fall to the bottom of the cell. This sludge is a valuable source of the precious metals, as described in Section 19.5.

The metal cations released from the anode migrate through the solution to the cathode. Because Cu^{2+} is easier to reduce than Fe^{2+} and Ni^{2+}, careful control of the applied voltage makes it possible to reduce Cu^{2+} to Cu metal, leaving Fe^{2+} and Ni^{2+} dissolved in solution.

Copper is used to make electrical wire.

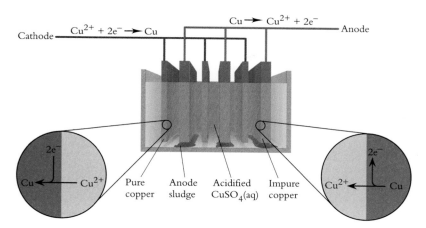

Cathode — $Cu^{2+} + 2e^- \longrightarrow Cu$

$Cu \longrightarrow Cu^{2+} + 2e^-$ — Anode

$2e^-$

$Cu \longleftarrow Cu^{2+}$

Pure copper | Anode sludge | Acidified $CuSO_4$(aq) | Impure copper

$Cu^{2+} \longleftarrow Cu$

$2e^-$

Figure 19-27
Copper is refined by electrolysis. Slabs of impure metal are used as the anodes. Oxidation releases Cu^{2+} ions into the electrolyte solution. The ions migrate to the cathode, where they are reduced to pure copper metal.

Section Exercises

■ **19.4.1** Construct a flowchart that summarizes the chemistry that takes place in a blast furnace.

■ **19.4.2** Examine Appendix F and explain why ZnO can be reduced with coke but Cr_2O_3 requires a more reactive metal such as aluminum.

■ **19.4.3** Which of the reactions listed in Figure 19-25 are redox reactions? Identify the reducing agent in each case.

An engine of a jet airplane contains approximately 5000 kg of titanium metal.

19.5 APPLICATIONS OF TRANSITION METALS

A complete discussion of all the transition metals is beyond the scope of an introductory course in chemistry. Instead, we provide a brief survey of several metals that highlights the diversity and utility of this group of elements.

Titanium

Titanium, the ninth most abundant element in the Earth's crust, is characterized by its high strength, its low density (~57% that of steel), and its stability at very high temperature. When alloyed with small amounts of aluminum or tin, titanium has the highest strength-to-weight ratio of all the engineering metals. Its major use is in the construction of aircraft frames and jet engines. Because titanium is also highly resistant to corrosion, it is used in the construction of pipes, pumps, and vessels for the chemical industry.

Because it is difficult to purify and fabricate, titanium is an expensive metal. For example, although titanium bicycles are highly prized by avid riders, they are quite expensive, averaging well over $1000 for just the frame.

The most important compound of titanium is titanium(IV) oxide, TiO_2. More than 2 million tons of TiO_2 are produced every year, much of it by the controlled combustion of $TiCl_4$:

$$TiCl_4(g) + O_2(g) \xrightarrow{1200\,°C} TiO_2(s) + 2\,Cl_2(g)$$

The Cl_2 produced in the combustion reaction is recycled to produce more $TiCl_4$ from rutile ore.

Titanium dioxide is brilliant white, highly opaque, chemically inert, and nontoxic. Consequently, it finds wide uses as a pigment in paints and other coatings, in

The "smoke" produced by aerial sky writers is titanium dioxide.

Chromium is used as a decorative and protective coating.

Figure 19-28
Chromium compounds display a striking range of beautiful colors. Shown here are Na_2CrO_4 (yellow), $K_2Cr_2O_7$ (orange), $CrCl_3$ (green), and CrO_3 (dark purple).

Chapter 11 →
Cross-linking of polymers is described in Chapter 11.

paper, sunscreens, cosmetics, and toothpaste. Almost all white-colored commercial products contain TiO_2.

Chromium

Chromium makes up just 0.012% of the Earth's crust, yet it is an important industrial metal. The main use of chromium is in metal alloys. Stainless steel, for example, contains as much as 20% chromium. Nichrome, a 60:40 alloy of nickel and chromium, is used to make heat-radiating wires in electrical devices such as toasters and hair dryers. Another important application of chromium metal is as a decorative and protective coating for the surface of metal objects, as described in Chapter 18.

The only important ore of chromium is chromite, $FeCr_2O_4$. Reduction of chromite with coke gives ferrochrome, an iron-chromium compound:

$$FeCr_2O_4(s) + 4\ C(s) \longrightarrow FeCr_2(s) + 4\ CO(g)$$

Ferrochrome is mixed directly with molten iron to form chromium-containing stainless steel.

Chromium compounds of high purity can be produced from chromite ore without reduction to the free metal. The first step is the roasting of chromite ore in the presence of sodium carbonate:

$$4\ FeCr_2O_4 + 8\ Na_2CO_3 + 7\ O_2 \xrightarrow{1100\ °C} 8\ Na_2CrO_4 + 2\ Fe_2O_3 + 6\ CO_2$$

The product is converted to sodium dichromate by reaction with sulfuric acid.

$$2\ Na_2CrO_4 + H_2SO_4 \longrightarrow Na_2Cr_2O_7 + Na_2SO_4 + H_2O$$

When the resulting solution is concentrated by evaporation, $Na_2Cr_2O_7 \cdot 2H_2O$ precipitates from the solution. This compound is the most important source of chromium compounds for the chemical industry. It is the starting material for most other chromium-containing compounds of commercial importance, including ammonium dichromate $((NH_4)_2Cr_2O_7)$, chromium(III) oxide (Cr_2O_3), and chromium(VI) oxide (CrO_3).

Pure chromium metal is made by a two-step reduction sequence. First, sodium dichromate is reduced to chromium(III) oxide by heating in the presence of charcoal:

$$Na_2Cr_2O_7 + C \xrightarrow{Heat} Cr_2O_3 + Na_2CO_3 + CO$$

Dissolving Cr_2O_3 in sulfuric acid gives an aqueous solution of Cr^{3+} cations:

$$Cr_2O_3(s) + 6\ H_3O^+(aq) \longrightarrow 2\ Cr^{3+}(aq) + 9\ H_2O(l)$$

Electrolysis of this solution reduces the cations to pure Cr metal, which forms a hard, durable film on the surface of the object serving as the cathode.

Chromium's name is derived from the Greek word *chroma*, for color. Indeed, this metal forms a wide variety of compounds with beautiful colors, as shown in Figure 19-28. Chromium compounds have been used for many years as pigments in paints and other coatings: $Na_2Cr_2O_7$ is bright orange, Cr_2O_3 is green, and the Zn and Pb salts of CrO_4^{2-} are bright yellow. However, in recent years the use of chromium pigments has diminished because chromium in the +6 oxidation state is highly toxic.

Chromium is also important in converting animal hides into leather. In the tanning process, hides are treated with basic solutions of Cr(III) salts, which causes cross-linking of collagen proteins. The hides toughen and become pliable and resistant to biological decay.

Copper, Silver, and Gold

The first three pure metals known to humanity probably were copper, silver, and gold, known as the coinage metals because they found early use as coins. All three are found in nature in their pure elemental form, and all have been highly valued throughout civilization. The oldest known gold coins were used in Egypt around 3400 BC. At about the same time, copper was obtained in the Middle East from charcoal reduction of its ores. The first metallurgy of silver was developed in Asia Minor (Turkey) about 500 years later.

All three of these metals are excellent electrical conductors and are highly resistant to corrosion. These properties, coupled with its relatively low cost, make copper one of the most useful metals in modern society. About half of all copper produced is for electrical wiring, and it is also widely used for plumbing pipes. Copper is used to make several important alloys, the most important of which are bronze and brass. Both alloys contain copper mixed with lesser amounts of tin and zinc in various proportions. In bronze the amount of tin exceeds that of zinc, and the opposite is true for brass. The discovery of bronze sometime around 3000 BC launched the advance of civilization known today as the *Bronze Age*. Because bronze is harder and stronger than other metals known in antiquity, it became a mainstay of the civilizations of India and the Mediterranean, used for tools, cookware, weapons, coins, and objects of art. Today the principal use of bronze is for bearings, fittings, and machine parts.

Copper is resistant to oxidation, but over the course of time the metal acquires a coating of green corrosion called *patina*. The green compound is a mixed salt of Cu^{2+}, hydroxide, sulfate, and carbonate that is formed by air oxidation in the presence of carbon dioxide and small amounts of sulfur dioxide:

$$3\,Cu + 2\,H_2O + SO_2 + 2\,O_2 \longrightarrow Cu_3(OH)_4(SO_4)$$

$$2\,Cu + H_2O + CO_2 + O_2 \longrightarrow Cu_2(OH)_2(CO_3)$$

Although trace amounts of copper are essential for all life forms, the metal is toxic in large amounts. Thus copper(II) salts, particularly $CuSO_4 \cdot 5H_2O$, are used as pesticides and wood preservatives. Wood soaked in solutions of Cu^{2+} or coated with paints containing Cu^{2+} resist degradation resulting from bacteria, algae, and fungi.

Silver is usually found as a minor component of ores of more abundant metals such as copper and zinc. Most commercial silver is produced as a byproduct of the production of these common metals. For example, electrolytic refining of copper generates a solid anodic residue that is rich in silver and other precious metals. The silver from this residue can be isolated by oxidizing the metal into nitrate-containing solutions, silver nitrate being one of the few soluble silver salts. The pure metal is then deposited electrolytically. Silver is used for tableware in the form of sterling silver, an alloy containing small amounts of copper to make the metal harder. Silver is also used in jewelry, mirrors, and batteries, but the single most important use of the metal, accounting for about a third of all production, is in photography, described in our Chemistry and Technology Box.

Silver does not form a simple oxide by direct oxidation in air, but the metal does form a black tarnish with oxygen and trace amounts of hydrogen sulfide in the atmosphere:

$$4\,Ag + 2\,H_2S + O_2 \longrightarrow 2\,Ag_2S + 2\,H_2O$$
$$\text{black}$$

The coffin of Tutankhamun contained over 100 kg of gold. Tutankhamun was only a minor pharaoh who died at age 18.

The name *copper* and the symbol Cu are derived from the Latin *cuprum*, after the island of Cyprus, where the Romans first obtained copper metal. The symbols Ag and Au for silver and gold come from the Latin names for these elements: *argentum* and *aurum*.

The green coating on copper objects such as the Statue of Liberty is a mixture of copper salts, $Cu_3(OH)_4(SO_4)$ and $Cu_2(OH)_2(CO_3)$.

Box 19-2 — Chemistry and Technology: How Are Images Captured Chemically?

Photography has become an almost routine part of life. The front page of every newspaper contains several photos, often in color. Tourists the world over capture memories of their trips on photographic film. All photographic films rely on the redox chemistry of silver to "capture" an image on film.

Photographic film is a transparent plastic coated with a gelatinous film containing a silver salt, usually AgBr. When the film is exposed to light, individual photons absorbed by the Ag^+ ions cause an electron-transfer reaction that produces neutral silver atoms:

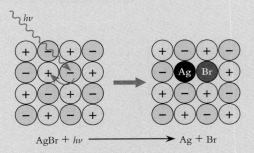

$$AgBr + h\nu \longrightarrow Ag + Br$$

Exposed film is developed with an aqueous reducing agent such as hydroquinone, which can reduce Ag^+ cations when catalyzed by Ag atoms. The reducing agent reacts selectively with those Ag^+ cations located next to already-reduced Ag atoms, generating clusters of Ag:

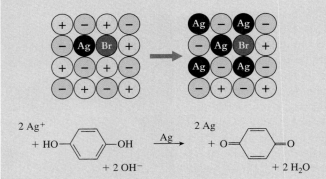

$$2\,Ag^+ + HO{-}\langle{-}\rangle{-}OH \xrightarrow{Ag} 2\,Ag + O{=}\langle{-}\rangle{=}O$$
$$+ 2\,OH^- \qquad\qquad + 2\,H_2O$$

Without intervention, this process would continue until all the Ag^+ had been reduced, giving a completely black film. To prevent this, developing is stopped after an appropriate time, and the film is treated with an aqueous solution containing thiosulfate anions, $S_2O_3^{2-}$. This anion forms soluble complex ions with silver cations, so rinsing with water removes all the remaining silver cations from the film, leaving islands of black clumps of Ag atoms wherever photons were absorbed.

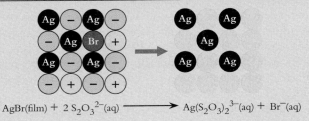

$$AgBr(film) + 2\,S_2O_3^{2-}(aq) \longrightarrow Ag(S_2O_3)_2^{3-}(aq) + Br^-(aq)$$

Color photography uses the photosensitivity of AgBr combined with dyes of different colors. A color image uses various proportions of red, blue, and green to create any other color of the visible spectrum.

A color film makes use of sensitizer molecules that absorb photons and then reduce silver ions. A color film contains three emulsions overlaid on one another, each emulsion containing a different sensitizer. One sensitizer selectively absorbs red light, one selectively absorbs blue light, and the third selectively absorbs green light. In each layer, absorption of photons of the selected color results in clumps of neutral silver atoms. These clumps occur in different places in each layer, as determined by where photons of each color were absorbed.

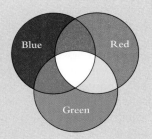

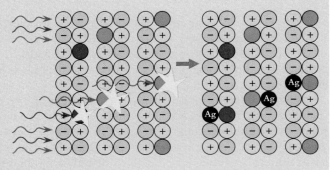

In addition to the sensitizer, each layer contains molecules that combine during the developing process to generate a colored dye. The combining reactions occur only where there are silver cations, so regions of the film that absorbed light and converted the silver cations into silver metal do not become colored. After development, the free silver atoms are removed from the film by treating with an oxidizing agent that converts Ag back to Ag^+ cations. What is left are islands of colored dye that match the colors of the scene that was photographed.

The extraction of gold by leaching is described in the previous section. Gold is used extensively in the manufacture of jewelry. Interestingly, Au(I) compounds are very effective in the treatment of rheumatoid arthritis, and there is recent evidence that certain gold-containing compounds have anticancer properties.

Zinc and Mercury

Zinc and mercury are found in the Earth's crust as sulfide ores, the most common of which are sphalerite (ZnS) and cinnabar (HgS). The extraction and purification of these metals are discussed in the previous section.

Most of the world's zinc output is used to prevent the corrosion of steel. Zinc is easier to oxidize than iron, as shown by the more negative reduction potential of Zn^{2+}.

$$Fe^{2+} + 2\ e^- \rightleftharpoons Fe \qquad E° = -0.447\ V$$

$$Zn^{2+} + 2\ e^- \rightleftharpoons Zn \qquad E° = -0.7618\ V$$

Consequently, a zinc coating oxidizes preferentially and protects steel from corrosion. Zinc coatings are applied in several ways: by immersion in molten zinc, by paint containing powdered zinc, or by electroplating.

Zinc is also combined with copper and tin to make brass and bronze. Finally, large amounts of zinc are used to make several types of batteries, as discussed in Chapter 18.

Zinc oxide is the most important zinc compound. The principal industrial use of zinc oxide is as a catalyst to shorten the time of vulcanization in the production of rubber. The compound also is used as a white pigment in paints, cosmetics, and photocopy paper. In everyday life, ZnO is also a common sunscreen.

The use of mercury for extracting silver and gold from their ores has been known for many centuries. Gold and silver form amalgams with liquid mercury, which is then distilled away to leave the pure precious metal. The Romans mined the mineral cinnabar (HgS) from deposits in Spain 2000 years ago, and in the sixteenth century the Spanish shipped mercury obtained from the same ore deposits to the Americas for the extraction of silver. Mercury is an important component of street lamps and fluorescent lights. It is used in thermometers and barometers and in gas-pressure regulators, electrical switches, and electrodes.

Zinc oxide is a very effective sunscreen.

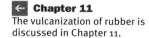 **Chapter 11**
The vulcanization of rubber is discussed in Chapter 11.

The Platinum Metals

Six of the transition metals—Ru, Os, Rh, Ir, Pd, and Pt—are known as the *platinum metals*. The group is named for the most familiar and most abundant of the six. These elements are usually found mingled together in ore deposits, and they share many common features. Although they are rare (total annual production is only about 200 tons), the platinum metals play important roles in modern society.

The platinum metals are valuable byproducts from the extraction of common metals such as copper and nickel. The anodic residue that results from copper refining is a particularly important source. The chemistry involved in their purification is complex and is not discussed here, except to note that the final reduction step involves reaction of molecular hydrogen with metal halide complexes.

By far the most important use of the platinum metals is for catalysis. The largest single use is in automobile catalytic converters. Platinum is the principal catalyst, but catalytic converters also contain rhodium and palladium. These elements

The combustion of ammonia is catalyzed by a gauze of platinum metal.

also catalyze a wide variety of reactions in the chemical and petroleum industry. For example, platinum metal is the catalyst for ammonia oxidation in the production of nitric acid, as described in Chapter 15:

$$4\,NH_3 + 5\,O_2 \xrightarrow{\text{Pt gauze, 1200 K}} 4\,NO + 6\,H_2O$$

Palladium is used as a catalyst for hydrogenation reactions in the food industry, and a rhodium catalyst is used in the production of acetic acid:

$$CH_3OH + CO \xrightarrow[\substack{175\ °C,\ 1\ atm}]{\substack{\text{Rh-containing}\\ \text{catalyst}}} CH_3CO_2H$$

Section Exercises

19.5.1 In recent years, copper has replaced galvanized iron as the material of choice for plumbing pipes. Explain why copper is a better material for pipes than zinc-coated iron. Justify your answer with balanced equations.

19.5.2 The chemistry of chromium and zinc can be used to remove traces of oxygen from bottled gases such as nitrogen and argon. The gas is bubbled through a solution of Cr^{2+} in the presence of zinc metal. Trace oxygen in the gas reacts quickly with Cr^{2+} to give Cr^{3+} and water. The zinc in turn reduces Cr^{3+} back to Cr^{2+}. Write balanced equations for these reactions.

19.5.3 Refer to the standard reduction potentials for various metal cations (see Appendix F) and use these to predict which metals will be found in the anodic residue that results from the electrolytic refining of copper.

19.6 TRANSITION METALS IN BIOLOGY

Almost 90% of the atoms that make up a human body are either hydrogen or oxygen. Most of these are in water, which constitutes around 70% of a human. The organic structures that make up the body, as well as the molecules involved in biosynthesis and energy production, are made almost entirely of C, H, N, and O. These four elements account for 99% of all the atoms present in a human. Another seven elements—Na, K, Ca, Mg, P, S, and Cl—are essential for all known life forms. These seven add another 0.9% to the total atom count of a human being. The remaining 0.1%, the so-called "trace elements," are required by most biological organisms. Although these elements are present in only minute amounts, they are essential for healthy function. Figure 19-29 shows a periodic table summarizing the elemental composition of living organisms.

Metalloproteins

The trace elements include nine transition metals: all members of the first row from vanadium to zinc and molybdenum from the second row. Most transition metals in the body are natural constituents of proteins, biological macromolecules made of long chains of amino acids that we described in Chapter 11. These **metalloproteins** play three essential roles in biochemistry. Some act as transport

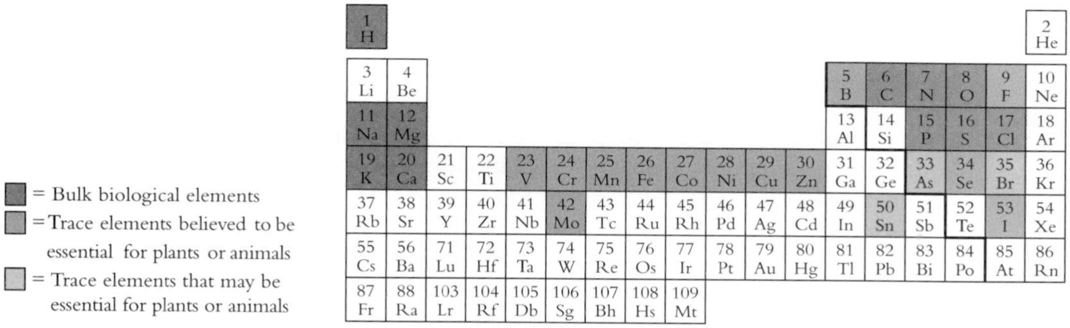

Figure 19-29
Elements essential for life.

■ = Bulk biological elements

■ = Trace elements believed to be essential for plants or animals

■ = Trace elements that may be essential for plants or animals

and storage agents, moving small molecules from place to place within an organism. Others are enzymes, catalysts for a diverse group of biochemical reactions. Both transport and catalysis depend on the ability of transition metals to bind and release ligands. The third role of metalloproteins is to serve as redox reagents, adding or removing electrons in countless different reactions. Transition metals are ideal for this purpose because of their capacity to shuttle between two or more oxidation states.

Metalloproteins typically are macromolecules containing many thousands of atoms. Determining the bonding, structure, and geometries of these macromolecules is a challenging task. Nevertheless, it is known that transition metals are bound to proteins through ligand-metal interactions of the sort described in this chapter. Among the 20 amino acids found in proteins, several have side chains that act as ligands: histidine, tyrosine, cysteine, glutamic acid, aspartic acid, and methionine bind to metals through lone pairs on nitrogen, oxygen, or sulfur.

← **FIGURE 11-38**
Structures of the amino acid side chains appear in Figure 11-38.

Transport and Storage

Organisms extract energy from food by using molecular oxygen to oxidize fats and carbohydrates. For most animals, the movement and storage of O_2 is accomplished by the iron-containing proteins *hemoglobin* and *myoglobin*. The iron atom in hemoglobin binds O_2 and transports this vital molecule from the lungs or the gills to various parts of the body where oxidation takes place. Hemoglobin makes O_2 about 70 times more soluble in blood than in water. Whereas hemoglobin transports O_2, myoglobin stores O_2 in tissues such as muscle that require large amounts of oxygen.

Hemoglobin and myoglobin have closely related structures. Both contain the heme structure shown in the introduction to this chapter. In myoglobin the heme is bound to a polypeptide chain of 153 amino acids arranged in helical arrays. The ribbon structure of myoglobin is shown in Figure 19-30. The polypeptide chain folds in a manner that creates a "pocket" in the protein for a heme group. Hemoglobin is made up of four polypeptide chains, each of which is similar in shape and structure to a myoglobin molecule.

Each heme unit in myoglobin and hemoglobin contains one Fe^{2+} ion bound to four nitrogen donor atoms in a square planar arrangement. This leaves the metal with two axial coordination sites to bind other ligands. One of these sites

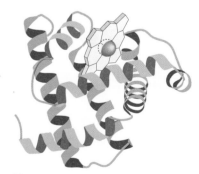

Figure 19-30
Myoglobin is a globular protein with a heme unit embedded in a pocket in the protein. See the chapter introduction for a more detailed view of the heme group.

is bound to a histidine side chain that holds the heme in the pocket of the protein. The other axial position is where reversible binding of molecular oxygen takes place.

The binding and release of oxygen by hemoglobin can be represented as a ligand-exchange equilibrium at the sixth coordination site on the Fe^{2+} ion. Each of the four polypeptide chains of hemoglobin contains one heme unit, so a molecule of hemoglobin can bind as many as four oxygen molecules:

$$[Heme]_4 + 4\ O_2 \rightleftharpoons [Heme(O_2)]_4$$

$$\text{deoxyhemoglobin} \qquad\qquad \text{oxyhemoglobin}$$

In the absence of oxygen, the iron center in each heme remains five-coordinate, with square pyramidal geometry, as shown in Figure 19-31a. In this form of hemoglobin, the d^6 metal ion is in a high-spin environment with four unpaired electrons in a $(t_{2g})^4(e_g)^2$ arrangement. Deoxyhemoglobin has a bluish color because the energy separation between the valence d orbitals of the iron cations is small, and the molecule absorbs red light at the low-energy end of the visible spectrum. When oxygen moves into the pocket of the protein, it binds to the sixth coordination site on the metal, as shown in Figure 19-31b. One of the effects of the O_2 ligand is to increase the crystal field splitting energy. As a result, in oxyhemoglobin the metal is in a low-spin $(t_{2g})^6(e_g)^0$ environment and is diamagnetic. In this form, hemoglobin absorbs light at the blue end of the visible spectrum and appears bright red.

In the lungs, hemoglobin "loads" its four oxygen molecules and then moves through the bloodstream. In the tissues, oxygen concentration is very low, but there is plenty of carbon dioxide, the end product of metabolism. The concentration of CO_2 has an important effect on hemoglobin-oxygen binding. Like oxygen, carbon dioxide can bind to hemoglobin. However, carbon dioxide binds to specific amino acid side chains of the protein, not to the heme group. Binding carbon dioxide to the protein causes the shape of the hemoglobin molecule to change in ways that reduce the equilibrium constant for O_2 binding. The reduced binding constant allows hemoglobin to unload its O_2 molecules in oxygen-deficient, CO_2-rich tissue. The bloodstream carries this deoxygenated hemoglobin back to the lungs, where it releases carbon dioxide and loads four more molecules of oxygen. This CO_2 effect does not operate in myoglobin, which binds and stores the oxygen released by hemoglobin.

Carbon monoxide seriously impedes transport of oxygen. The deadly effect of inhaled CO results from its reaction with hemoglobin. A CO molecule is almost the same size and shape as O_2, so it fits into the binding pocket of the hemoglobin molecule. In addition, the carbon atom of CO forms a stronger bond to Fe^{2+} than does O_2. Under typical conditions in the lungs, hemoglobin binds carbon monoxide 230 times more strongly than it binds O_2. Hemoglobin complexed to CO cannot transport oxygen, so when a significant fraction of hemoglobin contains CO, oxygen "starvation" occurs at the cellular level, leading to loss of consciousness and then to death.

Many deaths occur through accidental carbon monoxide poisoning. Burning fossil fuels generates some CO, particularly in an oxygen-depleted environment. Automobile engines, gas heaters, and charcoal braziers are all sources of CO. The presence of colorless, odorless CO goes undetected, and in a poorly ventilated room carbon monoxide may build up to lethal concentrations without the occupants being aware of its presence.

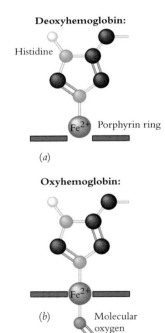

Deoxyhemoglobin:

Histidine

Fe²⁺ Porphyrin ring

(a)

Oxyhemoglobin:

Fe²⁺

(b) Molecular oxygen

Figure 19-31
(a) In deoxyhemoglobin, iron is five-coordinate in a square pyramidal shape. (b) On coordination of molecular oxygen, the metal adopts octahedral geometry.

An adult human contains about 4 g of iron, most of it in the form of heme-containing proteins. Yet the daily requirement of iron in the diet is only about 1 mg, indicating that the body recycles iron rather than excreting it. The recycling of iron requires a transport system and a storage mechanism. Transport is accomplished by a protein called *transferrin*. Transferrin collects iron in the spleen and liver, where hemoglobin is degraded, and carries it to the bone marrow where fresh red blood cells are synthesized. Transferrin is a protein with a molar mass of about 80,000 g/mol, folded into two distinct regions, each of which binds one Fe^{3+} ion in an octahedral environment.

The protein that stores iron in the body is called *ferritin*. A ferritin molecule consists of a protein coat and an iron-containing core. The outer coat is made up of 24 polypeptide chains, each with about 175 amino acids. As Figure 19-32 shows, the polypeptides pack together to form a sphere. The sphere is hollow, and channels through the protein coat allow movement of iron in and out of the molecule. The core of the protein contains hydrated iron(III) oxide, $Fe_2O_3 \cdot H_2O$. The protein retains its shape whether or not iron is stored on the inside. When filled to capacity, one ferritin molecule holds as many as 4500 iron atoms, but the core is only partially filled under normal conditions. In this way, the protein has the capacity to provide iron as needed for hemoglobin synthesis or to store iron if an excess is absorbed by the body.

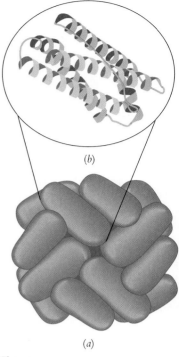

(b)

(a)

Figure 19-32
Schematic representation of ferritin, the iron storage protein. (a) The protein consists of 24 nearly identical polypeptides packed together to form a hollow sphere. (b) A ribbon structure of one of the polypeptide chains. As many as 4500 iron atoms can be stored inside the protein coat as $Fe_2O_3 \cdot H_2O$.

Enzymes

As described in Section 14-7, enzymes are the catalysts of biological reactions. Without enzymes, most of the reactions that occur in a cell would be imperceptibly slow. Transition metal ions play essential roles in the mechanisms of many enzyme-catalyzed reactions, as indicated by two representative examples, carboxypeptidase and superoxide dismutase.

Carboxypeptidase. During the digestion of foodstuffs, proteins must be broken down from long polypeptide chains into individual amino acids. Each peptide bond is cleaved, and a water molecule is broken apart and added to the ends of the cleaved peptide:

Many amino acids leading to the amino terminal group

Many amino acids leading to the carboxyl terminal group

Peptide linkage

This is an example of a *hydrolysis reaction*, in which a chemical bond is broken by reaction with water.

In the body, protein hydrolysis is catalyzed by several different enzymes. One such enzyme is carboxypeptidase, which cleaves amino acids one at a time from a protein, starting at the carboxyl end. The enzyme consists of a chain of 307 amino acids plus one Zn^{2+} ion. As shown in Figure 19-33, the metal ion is bound to

The term *hydrolysis* comes from the Greek words *hydro* ("water") and *lysis* ("to cut"). Hydrolysis reactions are the reverse of the condensation reactions discussed in Chapter 11.

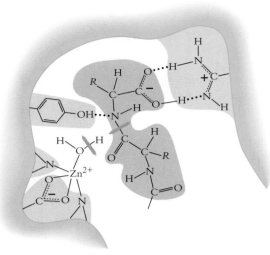

Figure 19-33
Schematic view of the catalytic site of carboxypeptidase, showing how the enzyme interacts with a protein substrate.

three amino acids: two histidines and a bidentate glutamic acid. The zinc center also binds the water molecule needed for the hydrolysis reaction.

Studies of carboxypeptidase have provided an overall picture of the hydrolysis mechanism, but several important details remain to be discovered. It is known that the water molecule required for the hydrolysis is coordinated to zinc before the cleavage of the C—N bond. The zinc ion is thought to position the water molecule properly for formation of the new bonds and to enhance the reactivity of the water molecule.

Superoxide dismutase. O_2 is essential to life, but reduced oxygen species such as superoxide, O_2^-, damage cells and are thought to play a role in the aging process. Cells contain enzymes that destroy these contaminants. One such enzyme, superoxide dismutase (SOD), is abundant in virtually every type of aerobic organism. Figure 19-34 shows the structure of the catalytic portion of the SOD enzyme. There is one copper atom and one zinc atom, each bound in a tetrahedral arrangement. As Figure 19-34 shows, the two metals are held close together by a histidine ligand that forms a bridge between Cu and Zn.

The role of SOD is to convert superoxide ion to hydrogen peroxide:

$$2\ O_2^- + 2\ H_3O^+ \xrightarrow{SOD} O_2 + H_2O_2 + 2\ H_2O$$

The mechanism is believed to be a two-step process involving reduction and oxidation of the copper center:

$$O_2^- + Cu^{2+}(SOD) \longrightarrow O_2 + Cu^+(SOD)$$

$$O_2^- + Cu^+(SOD) + 2\ H_3O^+ \longrightarrow Cu^{2+}(SOD) + H_2O_2 + 2\ H_2O$$

First, a superoxide ion transfers an electron to Cu^{2+}, giving molecular oxygen and Cu^+. In the second step, another superoxide reoxidizes the copper center back to Cu^{2+}. The resulting peroxide dianion is protonated rapidly to give hydrogen peroxide. The source of the two protons is still unclear. It appears that the role of the Zn^{2+} ion is to provide structural stability to the protein, because if zinc is removed, the protein degrades quite easily.

Interest in superoxide dismutase has increased in recent years with the discovery that a mutation in the gene coding for SOD is linked to certain types of the neurodegenerative disease amyotrophic lateral sclerosis (ALS), commonly known as Lou Gehrig's disease. Exactly how mutant forms of SOD are involved in ALS is a subject of intense research.

Figure 19-34
The structure of the enzymatic site of superoxide dismutase. The Zn^{2+} and Cu^{2+} cations lie in close proximity, with a histidine side chain (color screened) acting as a bridge between the metals.

Histidine bridge

Electron Transfer Proteins

From biochemical synthesis to bioenergetics, redox reactions are fundamental parts of the life process. Molecular oxygen is the oxidizing agent in most of these redox reactions. It has been estimated that the reduction of oxygen to water accounts for 90% of all the O_2 consumed in the biosphere:

$$O_2 + 4\ H_3O^+ + 4\ e^- \longrightarrow 6\ H_2O \qquad E° = 1.23\ V$$

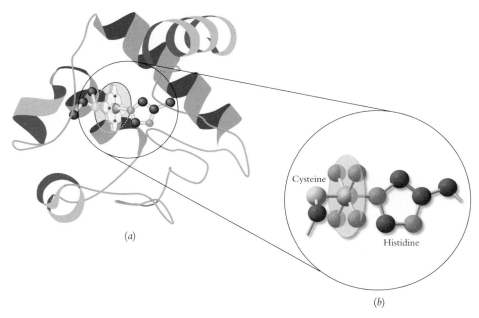

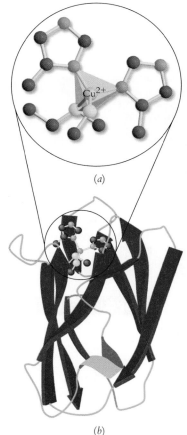

The many redox reactions that take place within a cell make use of metalloproteins with a wide range of electron transfer potentials. To name just a few of their functions, these proteins play key roles in respiration, photosynthesis, and nitrogen fixation. Some of them simply shuttle electrons to or from enzymes that require electron transfer as part of their catalytic activity. In many other cases, a complex enzyme may incorporate its own electron transfer centers. There are three general categories of transition metal redox centers: cytochromes, blue copper proteins, and iron–sulfur proteins.

Cytochromes. A cytochrome is a protein containing a heme with an iron cation bonded to four donor nitrogen atoms in a square planar array. Figure 19-35 shows the structure of cytochrome *c*, in which the fifth and sixth coordination sites of the octahedral iron center are occupied by a histidine nitrogen atom and a cysteine sulfur atom.

The electron transfer properties of the cytochromes involve cycling of the iron between the +2 and +3 oxidation states:

$$\text{(cytochrome)}Fe^{3+} + e^- \rightleftharpoons \text{(cytochrome)}Fe^{2+} \qquad E° = -0.3 \text{ V to } +0.4 \text{ V}$$

Different cytochromes have different side groups attached to the porphyrin ring. These side groups donate or withdraw electron density from the delocalized π system of the porphyrin, which in turn changes the redox potential of the iron cation in the heme.

Blue copper proteins. A typical blue copper redox protein contains a single copper atom in a distorted tetrahedral environment. Usually the metal binds to two N atoms and two S atoms through a pair of histidines, a methionine, and a cysteine. An example is plastocyanin, shown in Figure 19-36.

Figure 19-36
The structure of the redox protein plastocyanin. (*a*) Ribbon structure of the protein. The "flat arrows" represent regions of pleated sheet. (*b*) The copper center of plastocyanin is approximately tetrahedral.

Figure 19-37
There are three general types of iron-sulfur redox centers. (*a*) A single iron atom (brown) surrounded by four cysteine sulfur atoms (yellow). (*b*) Two iron atoms bound to two cysteines and a pair of bridging sulfide ligands. (*c*) A cubelike structure consisting of four irons and four sulfurs.

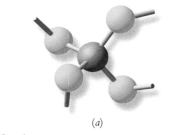

(*a*)

Cysteine

Bridging sulfides

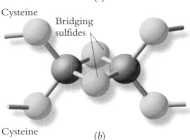

Cysteine   (*b*)

Bridging sulfide

Bridging sulfide

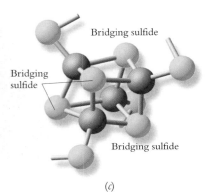

Bridging sulfide

(*c*)

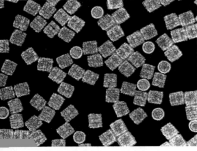

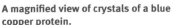

A magnified view of crystals of a blue copper protein.

Copper performs the redox function of the protein by cycling between Cu^+ and Cu^{2+}. As their name implies, these molecules have a beautiful deep blue color that is attributed to photon-induced charge transfer from the sulfur atom of cysteine to the copper cation center.

Iron-sulfur proteins. In an iron-sulfur protein, the metal center is surrounded by a group of sulfur donor atoms in a tetrahedral environment. Figure 19-37 shows the metal center structures of three different types of iron-sulfur redox centers. One type (Figure 19-37*a*) contains a single iron atom bound to four cysteine ligands. The electron transfer reactions at a one-iron center involve cycling of the metal between Fe^{2+} and Fe^{3+}. A second type (Figure 19-37*b*) contains two iron atoms, each bound to two cysteine ligands. The metals are connected through a pair of sulfide ligands. The most complicated redox center of the iron-sulfur proteins contains four irons and four sulfurs arranged in a distorted cube (Figure 19-37*c*). Each iron is bonded to one cysteine and three bridging sulfides. Individual iron cations in an iron-sulfur cluster do not undergo simple oxidation state changes. Instead, it appears that the electron transfer reactions involve orbitals that are delocalized over the entire cluster.

Section Exercises

19.6.1 Make a table that summarizes the biological chemistry of iron discussed in this section. Your table should include the names of the metalloproteins, their functions, and the coordination environment around the metal.

19.6.2 Most iron-containing proteins have a reddish color, and most copper proteins are blue. In contrast, zinc proteins such as carboxypeptidase are colorless. Why are zinc proteins colorless?

CHAPTER REVIEW

Summary and Key Terms

1. The metals of the *d* block, known as **transition metals**, share many common properties because of their similar orbital configurations. Those of the *f* block, the **inner transition metals**, are the **lanthanides** (4*f*) and **actinides** (5*f*).

2. Transition metals form many **coordination complexes** containing **ligands** that donate electrons to form metal-ligand bonds. Ligands can be **monodentate** or, if they have two or more donor atoms, can be **chelating** ligands. Complexes with two or more different ligands can exist as different **geometric isomers.** Optical isomers (enantiomers) are nonsuperimposable mirror image structures. Transition metal complexes are named using systematic rules.

3. Bonding in transition metal complexes can be described using **crystal field theory,** which predicts that electron pairs on ligands modify the stability of metal d orbitals. In octahedral geometry, the d orbitals split into two sets separated in energy by the **crystal field splitting energy,** Δ. The relative magnitudes of Δ and the **pairing energy** determine the details of d electron configurations. When Δ is relatively small, the complex has the maximum number of unpaired electrons and is **high spin,** whereas when Δ is relatively large, electrons pair up and the complex is **low spin.** The magnitude of Δ varies across the **spectrochemical series,** a listing of ligands in order of increasing Δ. The colors of complexes are also determined by Δ. When a complex absorbs light of a particular color, it appears to have the **complementary color** of the absorbed color.

4. Metallurgy is the purification of metals from their ores. It involves separation, conversion, reduction, and refining. Separation can be by **flotation** or **leaching.** Conversion often involves **roasting** in air. Metallurgy uses many redox reactions, particularly because the metal must be reduced from a positive oxidation number to zero. **Refining** removes unwanted impurities.

5. Transition metals have many uses in modern society. Applications include bulk structural metals, additive metals for alloys, catalysts, photography, pigments, preservatives, and medicines.

6. Metal-containing proteins, or **metalloproteins,** play diverse roles in biology. Transition metal cations in these proteins are responsible for the transport and storage of molecular oxygen, the shuttling of electrons from one location to another, and the catalysis by enzymes of a wide range of reactions.

Skills to Master

▶ Predicting properties of transition metals

▶ Determining configurations of transition metal ions

▶ Drawing structures of coordination compounds

▶ Naming coordination compounds

▶ Comparing orbital stabilities in complexes

▶ Correlating properties of complexes with their orbital configurations

▶ Describing metallurgical processes

▶ Describing the applications of transition metals to modern society

▶ Describing the roles that transition metals play in biology

Learning Exercises

19.1 Write a description of the features that transition metals have in common.

19.2 Describe the features of the transition metals that are exploited in biology. Use an example to illustrate each feature.

19.3 Draw ball-and-stick models of all possible isomers of linear, tetrahedral, square planar, and octahedral complexes containing two different ligands.

19.4 Make a list of the transition metals discussed in this chapter and summarize their applications.

19.5 Write a summary of the biological chemistry of iron and copper described in this chapter.

19.6 List all terms new to you that appear in Chapter 19, and write a one-sentence definition of each in your own words. Consult the Glossary if you need help.

Problems

Overview of the Transition Metals

19.1 The following compounds can be purchased from chemical supply companies. Determine the oxidation states of the transition metals in each: (a) $MnCO_3$; (b) $MoCl_5$; (c) Na_3VO_4; (d) Au_2O_3; and (e) $Fe_2(SO_4)_3 \cdot 5H_2O$.

19.2 The following compounds can be purchased from chemical supply companies. Determine the oxidation state of the transition metal in each: (a) $NiSO_4$; (b) $KMnO_4$; (c) $(NH_4)_2WO_4$; (d) $PbCrO_4$; and (e) $ZrOCl_2 \cdot 8H_2O$.

19.3 Give the names and symbols for the elements that have the following valence configurations: (a) $4s^1 3d^5$; (b) $5s^2 4d^{10}$; and (c) $4s^1 3d^{10}$.

19.4 Give the names and symbols for the elements that have the following valence configurations: (a) $5s^1 4d^8$; (b) $4s^2 3d^2$; and (c) $5s^1 4d^4$.

19.5 In each of the following pairs of transition metals, select the one with the higher value for the indicated property and give the reason: (a) melting points of Pd and Cd; (b) densities of Cu and Au; and (c) first ionization energies of Cr and Co.

19.6 In each of the following pairs of transition metals, select the one with the higher value for the indicated property and give the reason: (a) melting points of Zr and Mo; (b) densities of Ti and Cr; and (c) first ionization energies of Zr and Ag.

Coordination Complexes

19.7 What are the oxidation states and d electron counts for the metal ions in the following coordination complexes? (a) $[Ru(NH_3)_6]Cl_2$; (b) $trans$-$[Cr(en)_2I_2]I$; (c) cis-$[PdCl_2(P(CH_3)_3)_2]$, $(P(CH_3)_3 = \text{trimethylphosphine})$; (d) fac-$[Ir(NH_3)_3Cl_3]$; and (e) $[Ni(CO)_4]$.

19.8 What are the oxidation states and d electron counts for the metal ions in the following coordination complexes? (a) $[Rh(en)_3]Cl_3$; (b) cis-$[Mo(CO)_4Br_2]$; (c) $Na_3[IrCl_6]$; (d) mer-$[Ir(NH_3)_3Cl_3]$; and (e) $[Mn(CO)_5Cl]$.

19.9 Name the compounds in Problem 19.7.

19.10 Name the compounds in Problem 19.8.

19.11 Draw structures for the metal complexes in Problem 19.7.

19.12 Draw structures for the metal complexes in Problem 19.8.

19.13 Write the formulas of the following complex ions: (a) *cis*-tetraamminechloronitrocobalt(III); (b) amminetrichloroplatinate(II); (c) *trans*-diaquabis(ethylenediamine)copper(II); and (d) tetrachloroferrate(III).

19.14 Write the formulas of the following compounds: (a) potassium tetrachloroplatinate(II); (b) pentaammineaquachromium(III) iodide; (c) tris(ethylenediamine)manganese(II) chloride; and (d) pentaammineiodocobalt(III) nitrate.

19.15 Draw the structure for each complex ion in Problem 19.13.

19.16 Draw the structure for each complex ion in Problem 19.14.

Bonding in Coordination Complexes

19.17 For an octahedral complex of each of the following metal ions, draw a crystal field energy diagram that shows the electron populations of the various d orbitals. Where appropriate show both the high spin and low spin configurations: (a) Ti^{2+}; (b) Cr^{3+}; (c) Mn^{2+}; and (d) Fe^{3+}.

19.18 For an octahedral complex of each of the following metal ions, draw a crystal field energy diagram that shows the electron populations of the various d orbitals. Where appropriate show both the high spin and low spin configurations: (a) Zn^{2+}; (b) Cr^{2+}; (c) Co^{2+}; and (d) V^0.

19.19 Determine whether the following complexes are diamagnetic or paramagnetic. If a complex is paramagnetic, give the number of unpaired electrons. (a) $[Ir(NH_3)_6]^{3+}$; (b) $[Cr(H_2O)_6]^{2+}$; (c) $[PtCl_4]^{2-}$; and (d) $[Pd(P(CH_3)_3)_4]$.

19.20 Determine whether the following complexes are diamagnetic or paramagnetic. If a complex is paramagnetic, give the number of unpaired electrons. (a) $[Mo(CO)_6]$; (b) $[Co(NH_3)_6]^{3+}$; (c) $[CoBr_4]^{2-}$ (tetrahedral); and (d) $[Pt(en)Cl_2]$.

19.21 Compounds of Zr^{2+} are dark purple, but most Zr^{4+} compounds are colorless. Explain.

19.22 Of the coordination complexes $[Cr(H_2O)_6]^{3+}$ and $[Cr(NH_3)_6]^{3+}$, one is violet, and the other is orange. Decide which is which and explain your reasoning.

Metallurgy

19.23 Write balanced chemical equations for the following metallurgical processes: (a) roasting $CuFeS_2$; (b) removal of silicon from steel in a converter; and (c) reduction of titanium tetrachloride using sodium metal.

19.24 Write balanced chemical equations for these metallurgical processes: (a) NiS is heated in air; (b) Co_3O_4 is chemically reduced by Al metal; and (c) MnO_2 is reduced by coke.

19.25 A copper ore contains 2.37% Cu_2S by mass. Suppose that 5.60×10^4 kg of this ore is heated in air. Compute the mass of copper metal that is obtained and the volume of SO_2 gas produced at ambient conditions, 755 torr and 23.5 °C.

19.26 What mass of limestone, in kilograms, should be added for every kilogram of iron ore processed in a blast furnace if the limestone is 95.5% $CaCO_3$ and the iron ore contains 9.75% SiO_2?

19.27 Calculate the standard free energy change at 25 °C for reduction of ZnO to Zn using carbon and using carbon monoxide.

19.28 Determine $\Delta G°$ for each oxidation reaction that occurs in a steel-making converter. Compare your values with $\Delta G°$ for the reaction of iron with O_2 to give Fe_2O_3.

19.29 Write balanced chemical equations for all the processes used in converting scheelite ($CaWO_4$) to tungsten metal. Describe the metallurgical role of each reaction.

19.30 Write balanced chemical equations for all the processes used in converting the iron form of wolframite ($FeWO_4$) to tungsten metal. Describe the metallurgical role of each reaction.

Applications of Transition Metals

19.31 Identify the coinage metals and describe some of their applications.

19.32 Identify the platinum metals and describe some of their applications.

19.33 What features of titanium account for its use as an engineering metal?

19.34 Explain why titanium(IV) oxide is used extensively as a white pigment.

Transition Metals in Biology

19.35 Summarize the differences between hemoglobin and myoglobin.

19.36 What are the common features of myoglobin and the cytochromes?

19.37 Draw a crystal field splitting diagram that illustrates the electron transfer reaction of the simple iron redox protein shown in Figure 19-37a.

19.38 Draw a crystal field splitting diagram that illustrates the electron transfer reaction of a cytochrome.

Additional Paired Problems

19.39 Draw all possible isomers of the following compounds: (a) $[Ir(NH_3)_3Cl_3]$; (b) $[Pd(P(CH_3)_3)_2Cl_2]$; (c) $[Cr(CO)_4Br_2]$; and (d) $[Cr(en)(NH_3)_2I_2]$.

19.40 Draw the structures of all possible isomers of the following coordination compounds: (a) tetraamminedibromocobalt(III) bromide; (b) triamminetrichlorochromium(III); and (c) dicarbonylbis(trimethylphosphine)platinum(0) (trimethylphosphine = $P(CH_3)_3$).

19.41 Several commercial rust removers contain the bidentate ligand oxalate, $(O_2CCO_2)^{2-}$. Explain how these household products remove rust. Include a structural drawing of the species that forms.

19.42 The carbonate ion can be either a monodentate or a bidentate ligand. Make sketches that show the ligand binding to a metal cation in both modes.

19.43 Write electron configurations for the following: (a) Cr, Cr^{2+}, and Cr^{3+}; (b) V^-, V, V^+, V^{2+}, V^{3+}, V^{4+}, and V^{5+}; and (c) Ti, Ti^{2+}, and Ti^{4+}.

19.44 Write electron configurations for the following: (a) Au, Au^+, and Au^{3+}; (b) Ni, Ni^{2+}, and Ni^{3+}; and (c) Mn^-, Mn, and Mn^+.

19.45 Name the following coordination compounds:

(a) $\left[\begin{array}{c} \text{Cl} \cdots \overset{\displaystyle OH_2}{\underset{\displaystyle Cl}{\overset{\displaystyle |}{\underset{\displaystyle |}{Cr}}}} \cdots OH_2 \\ H_2O \end{array} \right]$ Cl

(b) $\left[\begin{array}{c} \text{OC} \cdots \overset{\displaystyle Br}{\underset{\displaystyle CO}{\overset{\displaystyle |}{\underset{\displaystyle |}{Mn}}}} \cdots CO \\ OC \end{array} \right]$

(c) $\begin{array}{c} \text{Cl} \cdots \underset{H_3N}{\overset{}{Pt}} \cdots Cl \\ NH_3 \end{array}$

19.46 Name the following coordination compounds:

(a) $\left[\begin{array}{c} H_3N \\ \underset{H_3N}{\overset{}{Zn}} \cdots NH_3 \\ NH_3 \end{array} \right]$ SO_4

(b) $K_3 \left[\begin{array}{c} \text{NC} \cdots \overset{\displaystyle CN}{\underset{\displaystyle CN}{\overset{\displaystyle |}{\underset{\displaystyle |}{Fe}}}} \cdots CN \\ NC \end{array} \right]$

(c) $\left[\begin{array}{c} \text{H}_3N \cdots \overset{\displaystyle NH_3}{\underset{\displaystyle NH_3}{\overset{\displaystyle |}{\underset{\displaystyle |}{Co}}}} \cdots Cl \\ Cl \end{array} \right]$ Cl

19.47 Write a balanced equation for the reaction catalyzed by the enzyme superoxide dismutase. Make a sketch of the coordination environment around the metal centers.

19.48 Write a balanced equation for the reaction catalyzed by the enzyme carboxypeptidase. Make a sketch of the coordination environment around the metal center.

19.49 Cu^{2+} forms tetrahedral complexes with some anionic ligands. When $CuSO_4 \cdot 5H_2O$ dissolves in water, a blue solution results. The addition of aqueous KF solution results in a green precipitate, but the addition of aqueous KCl results in a bright green solution. Identify each green species and write chemical reactions for these processes.

19.50 Cu^{2+} forms tetrahedral complexes with some anionic ligands. When $CuSO_4 \cdot 5H_2O$ dissolves in water, a blue solution results. The addition of aqueous KCN solution at first results in a white precipitate, but the addition of more KCN causes the precipitate to dissolve. Identify the precipitate and the dissolved species and write chemical reactions for these processes.

19.51 Both vanadium and silver are lustrous silvery metals. Suggest why silver is widely used for jewelry, but vanadium is not.

19.52 Titanium is nearly 100 times more abundant in the Earth's crust than copper; yet copper was exploited as a metal in antiquity, and titanium has found applications only in recent times. Explain.

19.53 Draw a ball-and-stick model of the *mer* isomer of $[NiCl_3F_3]^{4-}$, oriented so that the fluoride ions are in the top, bottom, and left forward positions.

19.54 Draw a ball-and-stick model of the *fac* isomer of $[NiCl_3F_3]^{4-}$, oriented so that the fluoride ions are in the top and two rear positions.

19.55 What is the name and the formula of the black tarnish that accumulates on objects made of silver? Write a balanced equation that shows how this black tarnish forms.

19.56 What is the name and the composition of the green tarnish that accumulates on objects made of copper? Write balanced equations that show how this green tarnish forms.

More Challenging Problems

19.57 Tetracarbonylnickel(0) is $[Ni(CO)_4]$, and tetracyanozinc(II) is $[Zn(CN)_4]^{2-}$. Predict the geometry and color of each complex.

19.58 Carbon monoxide and tungsten form an octahedral complex. The molecule is colorless even though $W(0)$ is d^6. Sketch the ligand field diagram for hexacarbonyltungsten(0), and suggest why the complex is colorless.

19.59 Some researchers have coined the term *the brass enzyme* to describe superoxide dismutase. Can you suggest a reason for this nickname?

19.60 Blue copper proteins are blue when they contain Cu^{2+} but colorless as Cu^+ compounds. The color comes from an interaction in which a photon causes an electron to transfer from a sulfur lone pair on a cysteine ligand to the copper center. Why does this charge transfer interaction occur for Cu^{2+} but not Cu^+?

19.61 The complex $[Ni(CN)_4]^{2-}$ is diamagnetic, but $[NiCl_4]^{2-}$ is paramagnetic. Propose structures for the two complexes and explain why they have different magnetic properties.

19.62 Predict whether the following complexes are high spin or low spin: (a) $[Fe(CN)_6]^{4-}$; (b) $[MnCl_4]^{2-}$; (c) $[Rh(NH_3)_6]^{3+}$; and (d) $[Co(H_2O)_6]^{2+}$.

19.63 The iron storage protein ferritin usually is neither empty of iron nor filled to capacity. Why is this situation advantageous for an organism?

19.64 Explain how liquid mercury is used to purify gold and silver.

19.65 One of the most common approaches to the investigation of metalloproteins is to replace the naturally occurring metal ion with a different one that has a property advantageous for chemical studies. For example, zinc proteins are often studied by UV-visible spectroscopy after Co^{2+} has been substituted for Zn^{2+}. Explain, using crystal field energy diagrams, why Co^{2+} is a better metal than Zn^{2+} for UV-visible spectroscopy.

19.66 Draw a molecular picture of a surfactant coating a mineral particle in a flotation process. (See Chapter 10 for a review of surfactants.)

19.67 In Zn purification by electrochemistry, ZnO is dissolved in sulfuric acid. Before deposition, zinc powder is added to displace less active metals, such as cadmium. Use $E°$ values and balanced equations to show how this is accomplished.

19.68 Determine the Lewis structure and draw a ball-and-stick model showing the geometry of the dichromate anion, which contains one bridging oxygen atom.

Group Study Problems

19.69 Oxyhemoglobin is bright red, but deoxyhemoglobin is blue. In both cases the iron is in the +2 oxidation state. Give a detailed explanation for the difference in color. How would you test your hypothesis? Based on your explanation, what color would you predict for a sample of blood that is saturated with carbon monoxide?

19.70 A portion of the absorption spectrum of a complex ion, $[Cr(H_2O)_4Cl_2]^+$, is represented by the following graph:

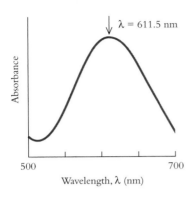

(a) Estimate the crystal field splitting energy Δ (in kilojoules per mole). (b) What color is the complex? (c) Name the complex cation. (d) Draw all possible isomers of the complex. (e) Draw the crystal field energy level diagram and show the electronic transition that gives the complex its color.

19.71 On Earth, two posttransition metals, Al and Pb, are used when low- and high-density metals are desired. Suppose you are transported to a planet elsewhere in our galaxy, where all transition metals are readily available but posttransition metals are rare. Where among the transition metals would you seek a replacement for Al for low-density uses, and where would you seek a replacement for Pb for high-density uses? Explain your reasoning.

19.72 Often, a detailed figure can give an informative summary of a complex concept. The figure in Box 15-2 summarizing the carbon cycle is one example. Design a figure that summarizes the transport and storage of oxygen by hemoglobin and myoglobin.

19.73 In the 1890s, Alfred Werner prepared several platinum complexes that contained both ammonia and chlorine. He determined the formulas of these species by precipitating the chloride ions with Ag^+. The empirical formulas and number of chloride ions that precipitate per formula unit follow:

Empirical formula	Number of Cl⁻ ions that precipitate per formula unit
$PtCl_4 \cdot 2NH_3$	0
$PtCl_4 \cdot 3NH_3$	1
$PtCl_4 \cdot 4NH_3$	2
$PtCl_4 \cdot 5NH_3$	3
$PtCl_4 \cdot 6NH_3$	4

Determine the molecular formulas, name these compounds, and draw the structures of the platinum complexes.

19.74 Design a flowchart similar to the one shown in Figure 19-23 that summarizes the metallurgy of copper from chalcopyrite.

Answers to Section Exercises

19.1.1 (a) W, mp increases down a column; (b) W, mp decreases after Column 6; (c) Pd, mp decreases after Column 6; and (d) Nb, mp increases up to Column 6.

19.1.2 (a) W, density increases with Z; (b) Os, density increases with Z; (c) Pd, density decreases at the end of a row; and (d) Nb, density increases with Z.

19.1.3 (a) Ta(V); (b) Fe(III); (c) Rh(III); (d) Cr(VI); and (e) Cu(II)

19.2.1 (a) +3, Fe(III), coordination number = 6; (b) 0, V(II), coordination number = 6; (c) +2, Ni(II), coordination number = 4

19.2.2

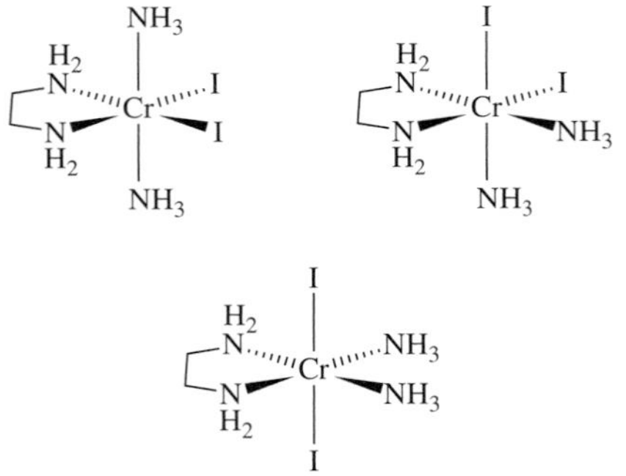

19.2.3 (a) Pentaammineiodochromium(III) sulfate; (b) potassium hexachloroplatinate(II); (c) tetracarbonyldichloroiron(II); and (d) hexaaquairon(II) chloride

19.3.1 (a)   (b)   (c)

$[Cr(H_2O)_6]^{2+}$
Cr(II), d^4
High spin

$[IrCl_6]^{3-}$
Ir(III), d^6
Low spin

$[V(en)_3]^{3+}$
V(III), d^2

(d)

$[NiCl_4]^{2-}$
Ni(II), d^8

19.3.2 Because cyanide is near the top of the spectrochemical series, it generates a relatively large energy gap between the two sets of d orbitals. The hexacyano complexes are yellow because they absorb high-energy indigo light. The corresponding aqua complexes have a much smaller crystal field splitting energy. They absorb orange or red light, thus appearing blue or green.

19.3.3 492 nm, orange

19.4.1

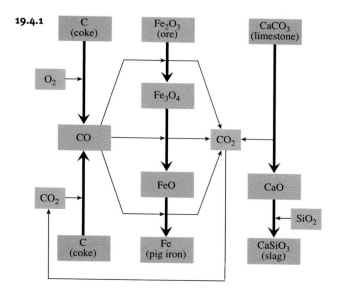

19.4.2 The standard reduction potentials are $Cr^{3+}, -0.744$ V and $Zn^{2+}, -0.7618$ V, suggesting that these oxides should be equally easily reduced. However, Cr^{2+} has a potential of -0.913 V, so reduction of chromium(III) oxide using coke stops at Cr^{2+}.

19.4.3

$$3 Fe_2O_3 + CO \longrightarrow 2 Fe_3O_4 + CO_2; \text{ reducing agent is CO.}$$

$$Fe_3O_4 + CO \longrightarrow 3 FeO + CO_2; \text{ reducing agent is CO.}$$

$$FeO + CO \longrightarrow Fe + CO_2; \text{ reducing agent is CO.}$$

$$CO_2 + C \longrightarrow 2 CO; \text{ reducing agent is C.}$$

$$C + O_2 \longrightarrow CO_2; \text{ reducing agent is C.}$$

19.5.1 Iron pipes are coated with zinc to prevent corrosion, zinc being easier to oxidize than iron. Copper is much more difficult to oxidize than either zinc or iron. Thus copper makes better

piping than galvanized iron because it is much less susceptible to corrosion.

19.5.2 Trace oxygen in the gas reacts with Cr^{2+}:

$$4 Cr^{2+}(aq) + O_2(g) + 4 H_3O^+(aq) \longrightarrow 4 Cr^{3+}(aq) + 6 H_2O(l)$$

Zinc reduces Cr^{3+} back to Cr^{2+}:

$$Zn(s) + 2 Cr^{3+}(aq) \longrightarrow Zn^{2+}(aq) + 2 Cr^{2+}(aq)$$

19.5.3 Metals with large positive standard reduction potentials will not electrolyze and will remain in the anodic residue: Au, Hg, Pt, Ag.

19.6.1

Table: Biological Roles of Iron

Protein	Function	Geometry	Coordination
Hemoglobin	O_2 transport	Octahedral and square pyramid	4N (porphyrin), 1N (His), O_2
Myoglobin	O_2 storage	Octahedral and square pyramid	4N (porphyrin), 1N (His), O_2
Transferrin	Fe transport	Octahedral	Not specified in text
Ferritin	Fe storage	Not specified in text	$Fe_2O_3 \cdot H_2O$
Cytochromes	Electron transfer	Octahedral	4N (porphyrin), 2N (His) or 1N (His), 1S (Met)
Fe-S Proteins	Electron transfer	Tetrahedral	4S (Cys) or 2S (Cys), 2S (sulfide)

19.6.2 Colors in coordination compounds come from electronic transitions, either from a lower energy d orbital to a higher energy d orbital or from a ligand to an empty d orbital (charge transfer). In zinc proteins, Zn^{2+} ion has a d^{10} configuration with all the d orbitals filled. Thus electronic transitions in the visible portion of the spectrum are impossible.

20

The Main Group Elements

Interior of
blood vessel

Platelet

N O

O
N   Enzyme

INTRODUCTION: NEW DISCOVERIES ABOUT AN OLD COMPOUND

One might think that everything had long since been learned about a molecule as simple as nitrogen oxide (NO). After all, Joseph Priestley made the compound for the first time more than 200 years ago. Yet, within the last decade there has been a resurgence of interest, triggered by discoveries that NO plays intriguing roles in a number of biochemical processes. Biological chemists have discovered that a multitude of animals synthesize NO, including humans, fruit flies, barnacles, chickens, and trout. This little molecule is now known to play roles in nerve function, regulation of blood pressure, blood clotting, and immune system responses.

In many of these processes, nitrogen oxide serves as a messenger. That is, NO is produced in one part of an organism and moves to another part, where its presence triggers a biochemical reaction. Other biochemical messengers generally function by virtue of specific molecular shapes, but NO has such a simple structure that it can trigger biochemical processes only through its chemical properties.

Nitrogen oxide helps to adjust the interactions between blood and blood vessels. One example is blood clotting, shown in our background photo. When a blood vessel's wall is cut, an enzyme in the cell wall sends chemical signals that interact with blood platelets, causing clotting. Recent studies have shown that NO inhibits this process, an effect exploited by mosquitoes, whose saliva contains NO-generating chemicals that prevent their victims' blood from clotting while the mosquito feeds. Another example is the dilation of blood vessels. Nitroglycerin has been prescribed for more than a century as a treatment of angina because it dilates blood vessels, thereby decreasing blood pressure and increasing blood flow. Only recently has it been discovered that NO is involved in the mechanism of this process. Biochemists speculate that NO binds to iron in the heme unit of the enzyme that causes dilation, as our molecular inset shows, but detailed knowledge is still lacking.

A recent research area targets the role of NO in transmitting nerve responses from one neural cell to another. There is evidence that the neural processes leading to memory include the production and diffusion of NO. An enzyme in the brain produces NO from an amino acid. A topic of intense current interest is how this NO interacts with more complex molecules as the brain functions. One hypothesis describes NO as a "retrograde messenger" that supplies positive feedback. In this hypothesis, once a neural trigger fires, an adjacent neuron produces NO, which diffuses back to the original neuron and stimulates further activity. This could be how long-term memory develops.

Chemical reactions display a near-infinite variety, and the story of nitrogen oxide illustrates that there are still exciting discoveries to be made in chemistry. Many of these are likely to involve the elements in the *p* block of the periodic table, because the *p*-block elements are remarkably diverse. Aluminum, one of the most reactive metals, is a member of this group. So too, are neon and argon, noble gases that are chemically inert. The elements of Groups 14 to 16 include nonmetals (C, N, O, P, S), metals (Sn, Pb, Bi), and metalloids (Si, Ge, Te). In this chapter, we describe selected features of this diverse chemistry. The chapter opens with sections describing a system for organizing chemical reactions, including those of *p*-block elements. Then we survey metals in the *p* block, metalloids, phosphorus, sulfur, and the halogens.

20.1 LEWIS ACIDS AND BASES

Because the breadth of chemical behavior can be bewildering in its complexity, chemists search for general ways to organize chemical reactivity patterns. Two familiar patterns are Brønsted acid–base (proton transfer) and oxidation–reduction (electron transfer) reactions. A related pattern of reactivity can be viewed as the donation of a pair of electrons to form a new bond. One example is the reaction between gaseous ammonia and trimethyl boron, in which the ammonia molecule uses its nonbonding pair of electrons to form a bond between nitrogen and boron:

$$NH_3 + B(CH_3)_3 \longrightarrow H_3NB(CH_3)_3$$

Viewed from the perspective of electrons, this reaction is similar to the transfer of a proton to an ammonia molecule:

In both cases the nitrogen atom uses its pair of nonbonding electrons to make a new covalent bond. This similarity led G.N. Lewis to classify ammonia as a base in its reaction with $B(CH_3)_3$ as well as in its reaction with H_3O^+. Whereas the Brønsted definition focuses on proton transfer, the Lewis definition of acids and bases focuses on electron pairs:

/// **Any chemical species that acts as an electron pair donor is a Lewis base. Any chemical species that acts as an electron pair acceptor is a Lewis acid.**

The Lewis description of acids and bases is the subject of this section.

Ammonia is a prime example of a **Lewis base.** In addition to its three N—H bonds, this molecule has a lone pair of electrons on its nitrogen atom, as shown in Figure 20-1. Although all of the valence orbitals of the nitrogen atom in NH_3 are occupied, the nonbonding pair can form a fourth covalent bond with a bonding partner that has a vacant valence orbital available.

Trimethylboron is an example of one type of **Lewis acid.** This molecule has trigonal planar geometry in which the boron atom is sp^2 hybridized with a vacant $2p$ orbital perpendicular to the plane of the molecule (Figure 20-1). Recall from Chapter 8 that atoms tend to use all their valence s and p orbitals to form covalent bonds. Second-row elements such as boron and nitrogen are most stable when surrounded by eight valence electrons divided among covalent bonds and lone pairs. The boron atom in $B(CH_3)_3$ can use its $2p$ orbital to form a fourth covalent bond to a new partner, provided that the new partner supplies both electrons. Trimethyl boron is a Lewis acid because it forms an additional bond by accepting a pair of electrons from some other chemical species.

Nonbonding sp^3 pair

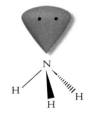

Vacant valence $2p$ orbit

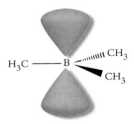

Figure 20-1
Ammonia, which has a pair of nonbonding valence electrons, is a typical Lewis base. Trimethylboron, which has a vacant valence orbital, represents one type of Lewis acid.

Formation of Lewis Acid-Base Adducts

The simplest type of Lewis acid-base reaction is the combination of a Lewis acid and a Lewis base to form a compound that is called an **adduct.** The reaction of ammonia and trimethyl boron is an example of adduct formation. A new bond forms between boron and nitrogen, with both electrons supplied by the lone pair of ammonia (see Figure 20-2). Forming an adduct with ammonia allows boron to use all its valence orbitals to form covalent bonds. As this occurs, the geometry about the boron atom changes from trigonal planar to tetrahedral, and the hybrid description of boron's valence orbitals changes from sp^2 to sp^3.

A general equation summarizes the formation of Lewis acid–base adducts:

$$A \ + \ :B \ \longrightarrow \ A{-}B$$

Lewis   Lewis   Adduct
acid   base

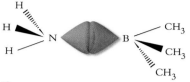

Figure 20-2
When an adduct forms between NH_3 and $B(CH_3)_3$, the N—B bond can be described by the overlap of sp^3 orbitals on each atom, with the two electrons supplied by the lone pair of the N atom.

Here are three more examples of adduct formation:

$$SiF_4 \ + \ :\ddot{\underset{..}{F}}:^- \ \longrightarrow \ SiF_5^-$$
Acid   Base   Adduct

$$AlCl_3 \ + \ :PCl_3 \ \longrightarrow \ Cl_3Al{-}PCl_3$$
Acid   Base   Adduct

$$PCl_5 \ + \ PCl_5 \ \longrightarrow \ [PCl_4]^+[PCl_6]^-$$
Acid   Base   Adduct

The third of these examples demonstrates two additional features of Lewis acid–base reactions. First, notice that some substances can act as both Lewis acids and Lewis bases. The chlorine atoms in PCl_5 have electron pairs to donate, imparting Lewis base properties to the molecule. The phosphorus atom in PCl_5 has valence d orbitals that accept an electron pair, imparting Lewis acid properties. Second, notice that in addition to formation of a new P—Cl bond, this reaction also involves P—Cl bond breaking. Here is a more detailed view of the process, using curved arrows to indicate shifts of electrons:

PCl₅   PCl₅   PCl₆⁻   PCl₄⁺

Recognizing Lewis Acids and Bases

A Lewis *base* must have valence electrons available for bond formation. Any molecule whose Lewis structure shows nonbonding electrons can act as a Lewis base. Ammonia, phosphorus trichloride, and dimethyl ether, each of which contains lone pairs, are Lewis bases. Anions can also act as Lewis bases. In the first example of adduct formation above, the fluoride ion, with eight valence electrons in its $2s$ and $2p$ orbitals, acts as a Lewis base.

A Lewis *acid* must be able to accept electrons to form a new bond. Because bond formation can occur in several ways, compounds with several different structural characteristics can act as Lewis acids. Nevertheless, most Lewis acids fall into the following categories:

1. *A molecule that has vacant valence orbitals.* A good example is $B(CH_3)_3$, which uses a vacant $2p$ orbital to form an adduct with ammonia. The elements in the p block beyond the second row of the periodic table have empty valence d orbitals that allow them to act as Lewis acids. The silicon atom in SiF_4 is an example.

2. *A molecule with delocalized π bonds involving oxygen.* Examples are CO_2, SO_2, and SO_3. Each of these molecules can form a σ bond between its central atom and a Lewis base, at the expense of a π bond. For example, the hydroxide anion, a good Lewis base, attacks the carbon atom of CO_2 to form hydrogen carbonate:

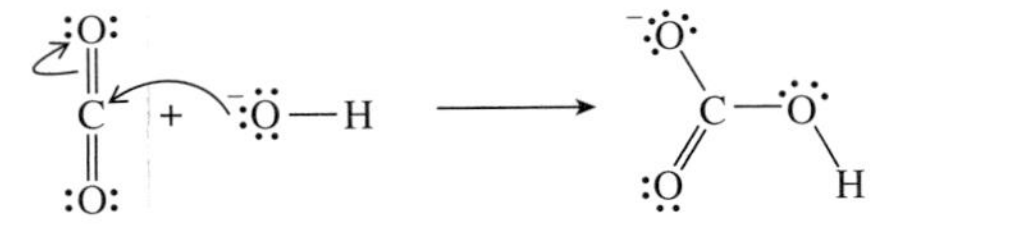

Lewis acid   Lewis base   Hydrogen carbonate adduct

In this reaction the oxygen atom of the hydroxide ion donates a pair of electrons to make a new C—O bond. Because all the valence orbitals of the carbon atom in CO_2 are involved in bonding to oxygen, one of the C—O π bonds must be broken to make an orbital available to overlap with the occupied orbital of the hydroxide anion.

3. *A metal cation.* Removing electrons from a metal atom always generates vacant valence orbitals. As described in Chapter 19, many transition metal cations form complexes with ligands in aqueous solution. In these complexes, the ligands act as Lewis bases, donating pairs of electrons to form metal-ligand bonds. The metal cation accepts these electrons, so it acts as a Lewis acid. Metal cations from the p block also act as Lewis acids. For example, Pb^{2+}(aq) forms a Lewis acid–base adduct with four CN^- anions, each of which donates a pair of electrons:

$$Pb^{2+}(aq) + 4\ CN^-(aq) \longrightarrow [Pb(CN)_4]^{2-}(aq)$$

Each of the four lead-carbon bonds forms by the overlap of an empty valence orbital on the metal ion with a lone pair on a carbon atom.

Example 20-1 provides practice in recognizing Lewis acids and bases.

Example 20-1	**Lewis Acids and Bases**

Identify the Lewis acids and bases in each of the following reactions and draw structures of the resulting adducts:

    (a) $AlCl_3 + Cl^- \rightarrow AlCl_4^-$

    (b) $Co^{3+} + 6\ NH_3 \rightarrow [Co(NH_3)_6]^{3+}$

    (c) $SO_2 + OH^- \rightarrow HSO_3^-$

Strategy: Every Lewis base has one or more lone pairs of valence electrons. A Lewis acid can have vacancies in its valence shell, or it can sacrifice a π bond to make a valence orbital available for adduct formation. To decide whether a molecule or ion acts as a Lewis acid or base, examine its Lewis structure for these features.

SECTION 8.3 →
Review Section 8.3 for procedures used to determine Lewis structures.

Solution: On the Lewis structures, we show possible transfers of electron pairs (blue) with curved red arrows.

| Lewis Acids and Bases (continued) | Example 20-1 |

(a) Both species of this pair contain chlorine atoms with lone pairs, so either might act as a Lewis base if a suitable Lewis acid is present. The aluminum atom of $AlCl_3$ has a vacant $3p$ orbital perpendicular to the molecular plane. The empty p orbital accepts a pair of electrons from the Cl^- anion to form the fourth Al—Cl bond. The Lewis acid is $AlCl_3$, and the Lewis base is Cl^-.

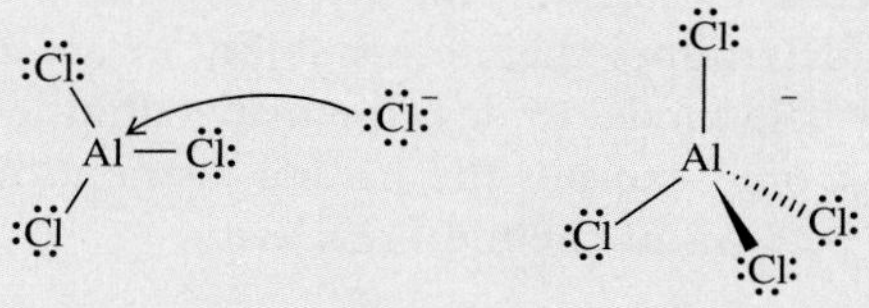

Acid   Base   Adduct

(b) As already noted, ammonia is a Lewis base because it has a donor pair of electrons on the nitrogen atom. Like other transition metal cations, Co^{3+} is a Lewis acid. The cation uses vacant $3d$ orbitals to form bonds to NH_3:

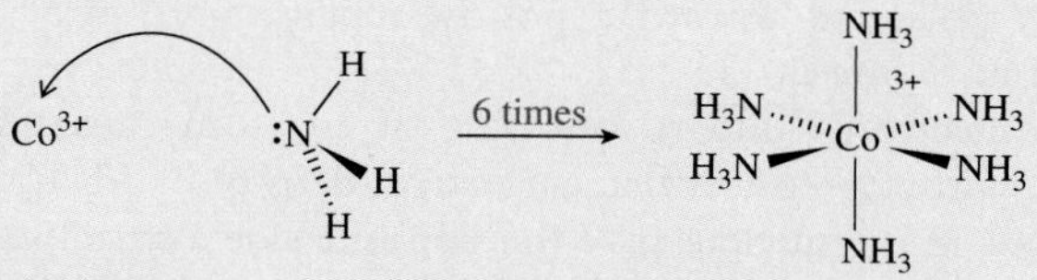

Acid   Base   Adduct

(c) Sulfur dioxide has delocalized π bonds, indicating Lewis acidity. The sulfur atom of SO_2 has a set of $3d$ orbitals that can be used to form an adduct. In this case, the hydroxide ion acts as a Lewis base. The anion uses one lone pair of electrons to form a new bond to sulfur:

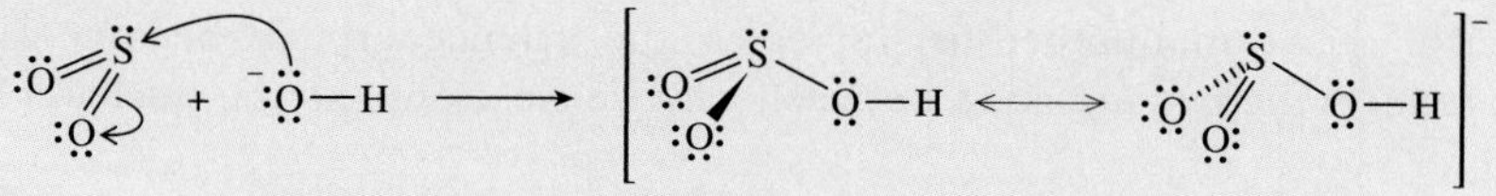

Acid   Base   Adduct

20.1.1 Draw the Lewis structures of each of the reactants in the following reactions:
(a) $SO_3 + OH^- \rightarrow HSO_4^-$
(b) $SnCl_2 + Cl^- \rightarrow [SnCl_3]^-$
(c) $AsF_3 + SbF_5 \rightarrow [AsF_2]^+ [SbF_6]^-$

20.1.2 Identify the Lewis acid and the Lewis base in each of the reactions that appears in Section Exercise 20.1.1.

20.1.3 Draw structures that show the donation of electrons that takes place in each reaction in Section Exercise 20.1.1 and draw the Lewis structures of the products.

20.2 HARD AND SOFT LEWIS ACIDS AND BASES

Many years ago, geochemists recognized that whereas some metallic elements are found as sulfides in the Earth's crust, others are usually encountered as oxides, chlorides, or carbonates. Copper, lead, and mercury are most often found as sulfide ores; Na and K are found as their chloride salts; Mg and Ca exist as carbonates; and Al, Ti, and Fe are all found as oxides. Today chemists understand why this differentiation among metal compounds exists. The underlying principle is how tightly an atom binds its valence electrons. The strength with which an atom holds its valence electrons also determines the atom's ability to act as a Lewis base, so we can use the Lewis acid-base model to describe many affinities that exist among elements. This notion not only explains the natural distribution of minerals, but also can be used to predict patterns of chemical reactivity.

Polarizability

Polarizability is discussed in Chapter 10 as a measure of how tightly electrons are bound to an atom or molecule. The polarizability of an atom, ion, or molecule is the ease with which its electron cloud can be distorted by an electrical field. An electron cloud is polarized toward a positive charge and away from a negative charge, as shown in Figure 20-3.

Pushing the electron cloud to one side of an atom causes a polarization of charge. The side with the concentrated electron density builds up a small negative charge; the protons in the nucleus give the opposite side a small positive charge.

Polarizability shows periodic variations that correlate with periodic trends in how tightly valence electrons are bound to the nucleus:

1. Polarizability decreases moving across a row of the periodic table. As the effective nuclear charge (Z_{eff}) increases, the nucleus holds the valence electrons more tightly.
2. Polarizability increases moving down a column of the periodic table. As the principal quantum number (***n***) increases, the valence orbitals become larger. This reduces the net attraction between valence electrons and the nucleus.

Polarizability increases (arrow)

Polarizability increases (vertical arrow)

Figure 20-3
The electron cloud of an atom is polarized toward a positive charge and away from a negative charge. A smaller atom (*top*) binds its valence electrons more tightly and is less polarizable than a larger atom (*bottom*), which binds its electrons loosely.

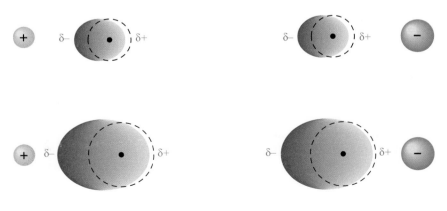

The electron cloud of an atom is attracted by a positive charge

The electron cloud of an atom is repelled by a negative charge

The Hard-Soft Concept

Lewis acids and bases can be organized according to their polarizability. If polarizability is low, the species is categorized as "hard." If polarizability is high, the species is "soft."

A **hard Lewis base** has electron pairs of low polarizability. This characteristic correlates with high electronegativity. Fluoride, the anion of the most electronegative element, is the hardest base because it contains a small, dense sphere of negative charge. Molecules and ions that contain oxygen or nitrogen atoms are also hard bases, although not as hard as fluorine. Examples include H_2O, CH_3OH, OH^-, NH_3, and H_2NCH_3.

A **soft Lewis base** has a large donor atom of high polarizability and low electronegativity. Iodide ion has its valence electrons in large $n = 5$ orbitals, making this anion highly polarizable and a very soft base. Other molecules and polyatomic anions with donor atoms from rows 3 to 6 are also soft bases. To summarize, the donor atom becomes softer from top to bottom of a column of the periodic table.

A **hard Lewis acid** has an acceptor atom with low polarizability. Most metal atoms and ions are hard acids. In general, the smaller the ionic radius and the larger the charge, the harder the acid. The Al^{3+} ion, with an ionic radius of only 67 pm, is a prime example of a hard Lewis acid. The nucleus exerts a strong pull on the compact electron cloud, giving the ion very low polarizability.

The designation of hard acids is not restricted to metal cations. For example, in BF_3 the small boron atom in its $+3$ oxidation state is bonded to three highly electronegative fluorine atoms. All the B—F bonds are polarized away from a boron center that is already electron deficient. Boron trifluoride is a hard Lewis acid.

A **soft Lewis acid** has a relatively high polarizability. Large atoms and low oxidation states often convey softness. Contrast Al^{3+} with Hg^{2+}, a typical soft acid. The ionic radius of Hg^{2+} is 116 pm, almost twice the size of Al^{3+}, because the valence orbitals of Hg^{2+} have a high principal quantum number, $n = 6$. Consequently, Hg^{2+} is a highly polarizable, very soft Lewis acid. The relatively few soft transition metal ions are located around gold in the periodic table.

The terms *hard* and *soft* are relative, so there is no sharp dividing line between the two, and many Lewis acids and bases are intermediate between hard and soft. Example 20-2 shows how to categorize Lewis acids and bases in order of their hard-soft properties.

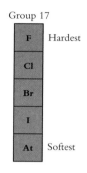

Group 17

F — Hardest

Cl

Br

I

At — Softest

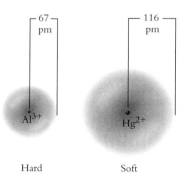

$\lvert$— 67 —$\rvert$ pm

$\lvert$— 116 —$\rvert$ pm

Al^{3+}

Hg^{2+}

Hard

Soft

Ranking Hardness and Softness	**Example 20-2**

Rank the following groups of Lewis acids and bases from softest to hardest: (a) H_2S, H_2O, and H_2Se; (b) H_2O, NH_3, and PH_3; (c) BCl_3, $GaCl_3$, and $AlCl_3$; and (d) Fe^0, Fe^{3+}, and Fe^{2+}.

Strategy: The first task is to decide whether the members of a given group are Lewis acids or bases. Then evaluate the relative softness and hardness based on polarizability, taking into account correlations with electronegativity, size, and charge.

Solution:

(a)   These three molecules have lone pairs, so they are Lewis bases. The central

| Example 20-2 | **Ranking Hardness and Softness** *(continued)* |

atoms are in the same column of the periodic table, so their polarizability and the softness of the molecules increases moving down the column. Thus H_2Se is softer than H_2S, which is softer than H_2O.

(b)  Again, these molecules have lone pairs, so they are Lewis bases. Ammonia and water are both hard bases, but H_2O is the harder of the two because oxygen is more electronegative than nitrogen. Because phosphorus is below nitrogen in the periodic table, phosphine, PH_3, is softer than ammonia. Thus PH_3 is softer than NH_3, which is softer than H_2O.

(c) These three molecules have trigonal planar geometry with sp^2 hybridized central atoms. Each has a vacant valence p orbital perpendicular to the molecular plane, making the molecules Lewis acids. The size, polarizability, and softness of the central acceptor atom increases going down the column. A gallium atom is larger and more polarizable than an aluminum atom. Thus $GaCl_3$ is softer than $AlCl_3$, which is softer than BCl_3. We have already noted that aluminum is hard. Thus gallium trichloride is a soft Lewis acid, whereas $AlCl_3$ and BCl_3 are both hard.

(d) Fe^0 Fe^{2+} Fe^{3+}

Metal atoms and cations are Lewis acids. As valence electrons are removed from a metal atom, the remaining electron cloud undergoes an ever-larger pull from the nuclear charge. This decreases the size of the ion as well as its polarizability. Thus Fe^0 is softer than Fe^{2+}, which is softer than Fe^{3+}.

The Hard-Soft Acid-Base (HSAB) Principle

The concept of hard and soft acids and bases can be used to interpret many trends in chemical reactivity. The hard-soft acid-base principle (HSAB principle) is an empirical summary of results collected from many chemical reactions studied through decades of research.

> /// *Hard Lewis acids tend to combine with hard Lewis bases.*
> *Soft Lewis acids tend to combine with soft Lewis bases.*

The geochemical distribution of metals conforms to the HSAB principle. Metals that form hard acid cations have strong affinities for hard bases such as oxide, fluoride, and chloride. Most metal ions that are hard acids are found bonded to the oxygen atoms of various silicate anions. These elements are concentrated in the Earth's mantle. Hard acid metals also occur in combinations with other hard bases, including oxides or, less often, halides, sulfates, and carbonates. Examples include rutile (TiO_2), limestone ($CaCO_3$), gypsum ($CaSO_4$), and sylvite (KCl).

Metals that are soft acids, such as gold and platinum, have a low affinity for hard oxygen atoms, so they are not affected by O_2 in the atmosphere. Consequently, these metals, including Ru, Rh, Pd, Os, Ir, Pt, and Au, are found in the crust of the Earth in their elemental form.

Other elements occur in nature as sulfides. The sulfide anion is a soft base, so this category includes some soft acids and many intermediate cases. Soft acids also may occur either in elemental form or as arsenide or telluride minerals. Examples include galena (PbS), cinnabar (HgS), chalcopyrite ($CuFeS_2$), argentite (Ag_2S), calaverite ($AuTe_2$), and sperrylite ($PtAs_2$).

The halides formed by Group 13 elements illustrate the trends in hardness and softness among Lewis acids and bases. Boron and aluminum are hard Lewis bases, because their valence orbitals are relatively compact and weakly polarizable. Gallium and indium, in contrast, are soft Lewis acids. All four elements have similar patterns of reactivity, but Ga and In show the expected preference for soft bases such as sulfur. The order of reactivity for the trihalides depends on the Lewis base. For a hard base such as ammonia, the reactivity trend for adduct formation is $BCl_3 > AlCl_3 > GaCl_3 > InCl_3$. The order is reversed when the trihalides form adducts with dimethylsulfide, $(CH_3)_2S$, a soft base.

Metathesis Reactions

A **metathesis reaction** is an exchange of bonding partners. Lewis acids and bases often undergo such exchanges, as exemplified by the following reaction:

$$BI_3 + GaF_3 \longrightarrow BF_3 + GaI_3$$

In this reaction, boron and gallium trade their halogen partners. According to the HSAB principle, metathesis reactions proceed in the direction that couples the harder acid with the harder base. Here, boron is harder than gallium, and fluoride is harder than iodide.

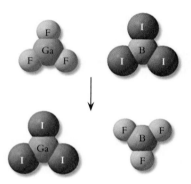

Metathesis is an important method for making molecules that have boron-carbon and aluminum-carbon bonds. These organoboron and organoaluminum species are valuable reagents in organic chemistry. Such compounds are often synthesized using alkyllithium reagents, powerful Lewis bases that are used widely in chemistry as a source of carbon atoms. Although highly reactive (they ignite spontaneously in air), alkyllithium reagents are available commercially, and they are easy to prepare in the laboratory by reacting alkyl halides with lithium metal:

$$ClCH_2CH_3 + 2\,Li \longrightarrow LiCH_2CH_3 + LiCl$$

Aluminum halides and boron halides undergo metathesis reactions with alkyllithium reagents according to the HSAB principle. Both boron and aluminum are softer acids than lithium, and carbon in these compounds is a very soft base. Metathesis generates lithium chloride and an organoaluminum or organoboron compound. To give a specific example, triethylaluminum, a colorless liquid that burns spontaneously in air, can be prepared by treating one mole of $AlCl_3$ with three moles of ethyllithium, $LiCH_2CH_3$:

$$3\,LiCH_2CH_3 + AlCl_3 \longrightarrow Al(CH_2CH_3)_3 + 3\,LiCl$$
Triethylaluminum

Many other alkyl groups can be used in place of CH_2CH_3. Organoaluminum compounds, including triethylaluminum, are used to make catalysts for the polymer industry.

Group 14 metal halides also undergo metathesis reactions. For instance, organotin compounds are prepared on an industrial scale using organoaluminum reagents. These reactions take place because tin is a softer Lewis acid than

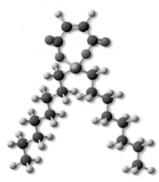

Dioctyltinmaleate

Figure 20-4
Organotin compounds such as dioctyltinmaleate are used to stabilize polymers such as poly(vinyl chloride), PVC, a starting material for many plastic products.

aluminum, and carbon (in an alkyl group, represented by R) is a softer Lewis base than chlorine:

$$3\ SnCl_4 + 4\ AlR_3 \longrightarrow 3\ SnR_4 + 4\ AlCl_3$$

Organotin compounds are important industrial chemicals. The greatest use is as stabilizers for poly(vinyl chloride) (PVC) plastics. These additives, such as dioctyltinmaleate (Figure 20-4), inhibit degradation of the polymer by heat, light, and oxygen. In the absence of these tin compounds, PVC yellows and becomes brittle. Organotin compounds are also used extensively in agriculture. More than a third of the world's food crops are lost to fungi, bacteria, insects, or weeds. Organotin compounds such as tributylhydroxytin, $(CH_3CH_2CH_2CH_2)_3SnOH$, inhibit growth of fungi among crops such as potatoes, peanuts, sugar beets, and rice. Marine paints for wooden boats also contain organotin compounds that inhibit the attachment of barnacles. Organotins are added to cellulose and wool to inhibit attack by moths. The tin compounds used in these applications are specific in their toxicity, so they present little danger to mammalian life.

Section Exercises

20.2.1 Rank the following from softest to hardest: (a) NCl_3, NH_3, and NF_3; (b) Pb^{2+}, Pb^{4+}, and Zn^{2+}; (c) ClO_4^-, ClO_2^-, and ClO_3^-; and (d) PCl_3, $SbCl_3$, and PF_3.

20.2.2 Iron is always found in nature in compounds, often with oxygen. The other members of Column 8 in the periodic table, ruthenium and osmium, occur in elemental form. Explain these observations using the HSAB principle.

20.2.3 Sulfur tetrafluoride fluorinates boron trichloride according to the following unbalanced equation: $BCl_3 + SF_4 \rightarrow BF_3 + SCl_2 + Cl_2$. This is both a redox reaction and a metathesis reaction. (a) Balance the equation. (b) Identify the elements that change oxidation state. (c) Explain the metathesis portion of the reaction using hard-soft acid-base arguments.

20.3 THE MAIN GROUP METALS

Figure 20-5 shows that the elements in the lower left portion of the p block of the periodic table are metals. These are the **main group metals.** Even though the most important metals of technological society are found in the d block, three main group metals, aluminum, lead, and tin, have considerable technological importance.

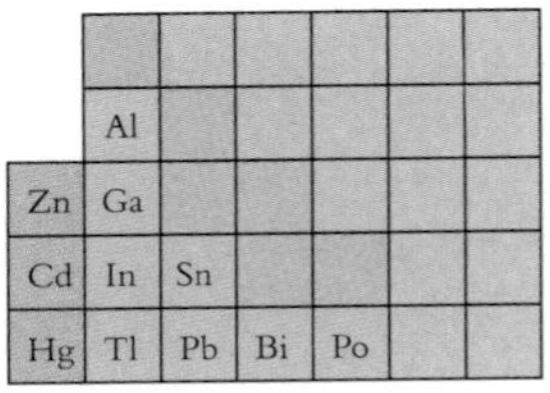

Production of Aluminum

Aluminum is unique among the main group metals. All other p block metals have filled valence d orbitals. As a consequence, these metals have much in common with their transition metal neighbors. They tend to be soft Lewis bases. Aluminum, on the other hand, lacks a filled d orbital set and is a hard Lewis acid that has more in common with its nearest neighbor, magnesium. Highly reactive, aluminum is

Al					
Zn	Ga				
Cd	In	Sn			
Hg	Tl	Pb	Bi	Po	

Figure 20-5
The main group metals are found in the corner of the p block closest to the transition metals.

found naturally in the $+3$ oxidation state and is reduced to the pure metal with difficulty. Thus although two main group metals, tin and lead, have been known since antiquity, aluminum was not discovered until 1825 and did not become a common commodity until more than 60 years later.

Bauxite, the main aluminum ore, is a mixed oxide-hydroxide, $Al(O)OH$, contaminated with SiO_2, Fe_2O_3, clay, and other hydroxide salts. To isolate the aluminum-containing material, the ore is treated with a strongly basic solution, whose high hydroxide concentration causes aluminum to form a soluble complex ion, $[Al(OH)_4]^-$:

$$Al(O)OH(s) + OH^-(aq) + H_2O \longrightarrow [Al(OH)_4]^-(aq)$$

The impurities in bauxite are not soluble in strong base, so the impurities remain behind when the solution is separated from the undissolved solids.

Although $[Al(OH)_4]^-$ is soluble, $Al(OH)_3$ is not, so diluting the strongly basic solution causes solid aluminum hydroxide to precipitate:

$$[Al(OH)_4]^-(aq) \longrightarrow Al(OH)_3(s) + OH^-(aq)$$

After this separation is complete, $Al(OH)_3$ is heated to drive off water, converting it into pure aluminum oxide:

$$2\ Al(OH)_3(s) \xrightarrow{1250\ °C} Al_2O_3(s) + 3\ H_2O(g)$$

Aluminum metal is produced from aluminum oxide by electrolysis using the **Hall–Héroult process,** whose story is detailed in our Chemical Milestones Box. The melting point of Al_2O_3 is too high (2015 °C) and its electrical conductivity too low to make direct electrolysis commercially viable. Instead, Al_2O_3 is mixed with cryolite (Na_3AlF_6) containing about 10% CaF_2. This combination has a melting point of 1000 °C, still a high temperature but not prohibitively so. Aluminum forms several complex ions with fluoride and oxide, so the molten mixture contains a variety of species, including AlF_4^-, AlF_6^{3-}, and $AlOF_3^{2-}$. These and other ions move freely through the molten mixture as electrolysis occurs.

Figure 20-6 shows a schematic representation of an electrolysis cell for aluminum production. An external electrical potential drives electrons into a graphite cathode, where Al^{3+} ions are reduced to Al metal:

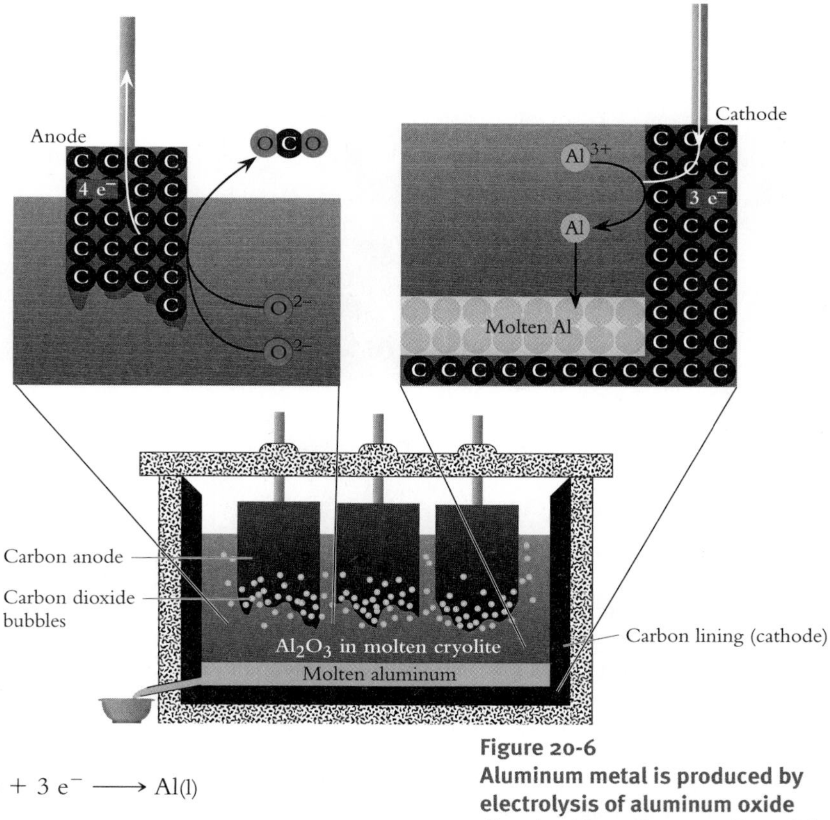

$$\text{Cathode:}\quad Al^{3+}(melt) + 3\ e^- \longrightarrow Al(l)$$

The anode, which is also made of graphite, is oxidized during electrolysis. Carbon from the anode combines with oxide ions to form CO_2 gas:

$$\text{Anode:}\quad 2\ O^{2-}(melt) + C(s) \longrightarrow CO_2(g) + 4\ e^-$$

Figure 20-6
Aluminum metal is produced by electrolysis of aluminum oxide dissolved in molten cryolite. Al^{3+} is reduced to Al at the cathode, and C is oxidized to CO_2 at the anode.

Box 20-1 Chemical Milestones: The Story of the Hall-Héroult Process

Aluminum is the third most abundant element in the Earth's crust and the most abundant metal. However, aluminum was not discovered until 1825 and was still a precious rarity 60 years later. The reason for aluminum's elusiveness is its high stability as Al^{3+}. The reduction of aluminum compounds to the free metal requires stronger reducing power than common chemical reducing agents can provide. Aluminum's discovery had to await the birth of electrochemistry and development of electrolysis.

Twenty-five years after its discovery, aluminum was a precious metal. Then a French chemist developed procedures for reducing aluminum compounds using sodium metal. The price of the metal dropped 100-fold. Still, in the 1880s aluminum was a semiprecious metal used for esoteric purposes such as a prince's baby rattle and the cap for the Washington Monument.

Aluminum's conversion from the stuff of princes' toys into recyclable kitchen foil required an inexpensive electrolytic reduction process. Two 22-year-old scientists, the American chemist Charles Hall and the French metallurgist Paul Héroult, discovered the same process independently in 1886. Both became famous as founders of the aluminum industry, Hall in the U.S. and Héroult in Europe.

Charles Hall was inspired by his chemistry professor at Oberlin College, who observed that whoever perfected an inexpensive way of producing aluminum would become rich and famous. After his graduation, Hall set to work in his home laboratory, trying to electrolyze various compounds of aluminum. He was aided by his sister Julia, who had stud-

ied chemistry and shared Charles' interests. Julia helped to prepare chemicals and witnessed many of the electrolysis experiments. After only 8 months of work, Hall had successfully produced globules of the metal. Meanwhile, Héroult was developing the identical process in France.

Hall capitalized on his discovery by founding a company for the manufacture of aluminum. That company became immensely successful, eventually growing into Alcoa. It made the Halls very rich.

Successful electrolysis of aluminum requires a liquid medium other than water that can conduct electricity. The key to the Hall-Héroult process is the use of molten cryolite, Na_3AlF_6, as a solvent. Cryolite melts at an accessible temperature, it dissolves Al_2O_3, and it is available in good purity. A second important feature is the choice of graphite to serve as the anode. Graphite provides an easy oxidation process, the oxidation of carbon to CO_2.

How did two persons working independently on two different continents come up with an identical process at the same time? The reasoning that led Hall and Héroult to the identical process was probably similar. Electrolysis was recognized as a powerful reducing method. All attempts to reduce aqueous aluminum cations failed, making it clear that some molten salt would have to be used. Experimenting with various salts, no doubt guided by the principle that "like dissolves like," the two young men eventually tried Na_3AlF_6, a mineral whose constituent elements should not interfere with the reduction of aluminum. Graphite electrodes were already in use, so experimenting with them would have been a natural choice.

The ingredients for this invention may have all been in place, but that does not detract from its brilliance. Hall and Héroult had the courage to explore new procedures in homemade laboratories without the support of research grants. They explored the possibilities systematically to find a process that was a spectacular success. In more than 100 years of growth in the aluminum industry, the only significant change to the Hall-Héroult process has been the addition of CaF_2 to the melt to lower the operating temperature.

Because of the variety of ionic species present in the melt, the reactions that take place at the electrodes are considerably more complex than the simple representations given here. Nevertheless, the net reaction is the one given by these simplified reactions:

$$4 \, Al^{3+}(melt) + 6 \, O^{2-}(melt) + 3 \, C(s) \longrightarrow 3 \, CO_2(g) + 4 \, Al(l)$$

The electrolysis apparatus operates well above aluminum's melting point of 660 °C, and liquid aluminum has a higher density than the molten salt mixture, so pure liquid metal settles to the bottom of the reactor. The pure metal is drained through a plug and cast into ingots.

Aluminum refining consumes huge amounts of electricity. Approximately 5% of all electricity consumed in the United States is used to produce aluminum.

Uses of Aluminum

Aluminum is one of society's most important structural metals. It is light yet very strong, and it resists corrosion by forming a thin layer of aluminum oxide on the surface of the metal. Alloys of aluminum are used for aircraft bodies, trailers, cooking utensils, highway signs, storage tanks, beverage cans, and many other objects.

Although the major use of aluminum by far is as a metal, some aluminum compounds also are economically important. Foremost among these is aluminum chloride, which is an important industrial catalyst.

The aluminum atom in aluminum chloride behaves as a Lewis acid, and the chlorine atoms are Lewis bases. As a consequence, two $AlCl_3$ molecules form a Lewis acid-base adduct, Al_2Cl_6. As shown in Figure 20-7, the compound has two tetrahedral aluminum atoms held together by a pair of chlorine atoms that bridge the two metal atoms. The 91° Al—Cl—Al bond angle suggests that the chlorine atom uses two valence p orbitals to form the bridge. There is an equilibrium between $AlCl_3$ and Al_2Cl_6 molecules, but except in the gas phase above 750 °C, the formation of Al_2Cl_6 goes essentially to completion.

In the solid state, aluminum chloride exists in a crystalline lattice. Each aluminum atom is surrounded by six chlorine atoms arranged around the metal atoms at the corners of an octahedron. Aluminum bromide and aluminum iodide form Al_2X_6 molecules in all three phases.

The aluminum trihalides are particularly important Lewis acids in the chemical industry. They promote or catalyze a large variety of reactions. One of the

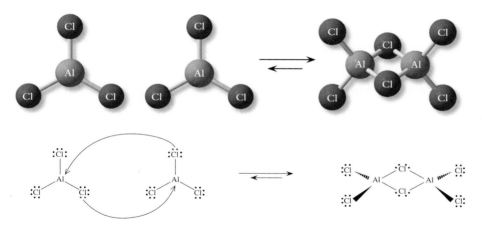

Figure 20-7
The Al atom in AlCl$_3$ has Lewis acid character, whereas the Cl atoms are Lewis bases. The compound forms a self-adduct in which two Cl atoms donate electron pairs to form additional Al—Cl bonds.

An acid chloride is a derivative of a carboxylic acid in which the hydroxyl group is replaced by a chlorine atom.

most important applications is a very general, widely used process called the Friedel-Crafts reaction. Aluminum chloride or some other Lewis acid is used to promote the reaction between an acid chloride and benzene or one of its derivatives. The simplest example of the Friedel-Crafts reaction is the formation of acetophenone:

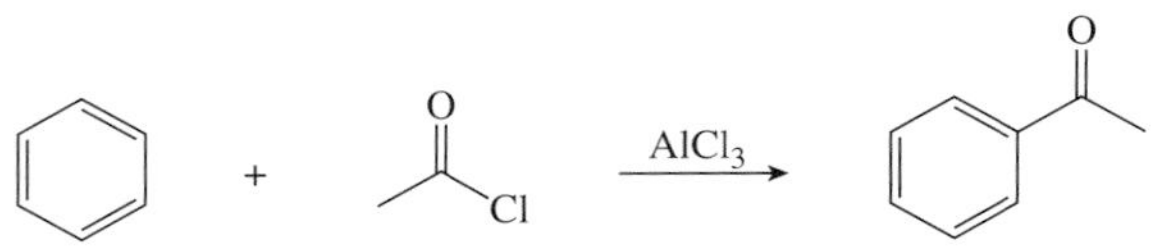

Benzene   Acetyl chloride   Acetophenone

The Lewis acid activates the acid chloride by forming an adduct with a chlorine atom bridge between carbon and aluminum. This chlorine bridge is similar to the one found in Al_2Cl_6:

Lewis base   Lewis acid   Adduct

The adduct fragments into an $[AlCl_4]^-$ anion and a very reactive cation.

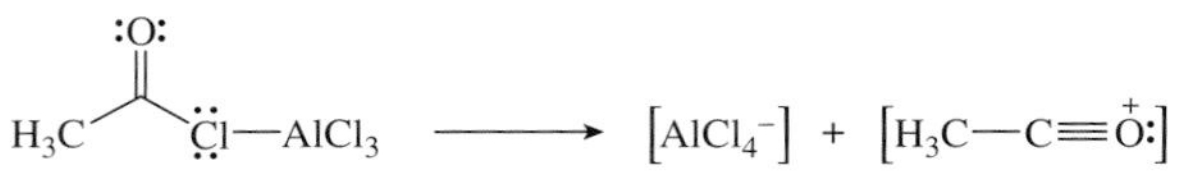

Adduct   Reactive cation

Reaction with a benzene molecule produces acetophenone and HCl and regenerates the $AlCl_3$ catalyst.

Friedel-Crafts chemistry is used in the synthesis of dyes, flavorings, fragrances, surfactants, pesticides, and many other types of organic compounds. Many different Lewis acids, including $AlCl_3$, BF_3, $TiCl_4$, $SnCl_4$, SbF_5, and $ZnCl_2$, are used.

Aluminum sulfate, $Al_2(SO_4)_3$, is widely used in water purification to remove finely divided particulate matter. When added to water, aluminum sulfate forms a precipitate of aluminum hydroxide that has a very open structure and large surface area. This precipitate, called a *gel,* traps dispersed particulate matter as it settles out of the liquid phase.

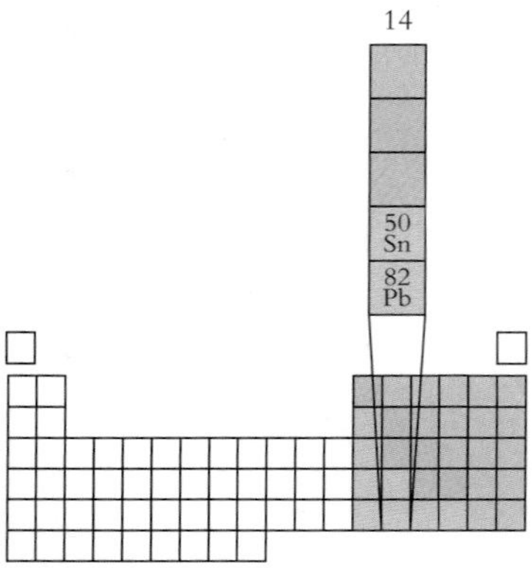

Tin and Lead

Tin and lead are neighbors in Group 14 of the periodic table. Each metal displays two common oxidation states, +2 and +4, and each can be reduced to the free metal with relative ease. The major ore of lead is galena, PbS, in which lead is in the +2 state. In contrast, the major ore of tin is cassiterite, SnO_2, where tin is in the +4 state. The Pb^{2+} state is substantially softer than the Sn^{4+} state, both because of the higher-valence orbitals and because of the lower charge. This accounts for lead being found associated with the softer base, S^{2-}, whereas tin combines with the hard base, O^{2-}.

Carbon in the form of charcoal is a sufficiently powerful reducing agent to convert lead and tin ores to the free metals. In the case of galena, the first step is roasting in air to form PbO:

$$2\,PbS + 3\,O_2 \xrightarrow{\text{Heat}} 2\,PbO + 2\,SO_2$$

$$PbO + C \xrightarrow{\text{Heat}} Pb + CO$$

$$SnO_2 + 2\,C \xrightarrow{\text{Heat}} Sn + 2\,CO$$

A bed of hot charcoal can supply both the heat and the reducing agent, and the simplicity of the process accounts for the early metallurgical development of these metals.

In former times, tin was used widely as a constituent of metal alloys, of which bronze, solder, and pewter are common examples. Bronze is an alloy of copper containing approximately 20% tin and smaller amounts of zinc. Pewter is also a Cu–Sn alloy, but in this case, tin is the major component ($\approx$85%), with roughly equal portions of copper, bismuth, and antimony. Solder consists of 67% lead and 33% tin.

Because of its relatively low melting point (232 °C) and good resistance to oxidation, tin was once used to provide a protective coating to metals such as iron that oxidize more readily. "Tin cans," now largely replaced by aluminum cans, are iron cans dipped in molten tin to provide a thin surface film of tin. Traditional metalsmiths still use a similar process, coating copperware with a thin film of tin.

Although tin is a heavy metal, it is not nearly as toxic as cadmium, mercury, and lead, all of which are extremely hazardous to living organisms. One reason is that tin oxide is highly insoluble, so tin seldom is found at measurable levels in aqueous solution. Perhaps more important, toxic heavy metals generally act by binding to sulfur in essential enzymes. Tin is a harder Lewis acid than the other heavy metals, so it has a lower affinity for sulfur, a relatively soft Lewis base.

The history of lead is almost as ancient as that of tin. In Roman times, lead was formed into pipes that were used for water supplies (hence our word *plumbing*, derived from the same Latin word, *plumbum*, that gives us the symbol Pb). Lead is a component of pewter and also was used as a glaze on drinking vessels. "White lead," $Pb_3(OH)_2(CO_3)_2$, is an enduring white substance that was used for many years as a paint pigment and even as a component of cosmetics.

More recently, the major use for lead has been in the automobile industry. The lead storage battery, described in Chapter 18, generates a high potential, can deliver large currents, and is easily recharged. Tetraethyllead, $(C_2H_5)_4Pb$, is excellent at preventing the pre-ignition of gasoline in automobile cylinders. Once used widely as an "antiknock" fuel additive, $(C_2H_5)_4Pb$ is no longer added to gasoline because of lead's high toxicity.

Unfortunately, lead easily enters the biosphere. Lead poisoning in humans causes learning impairment and behavioral disorders in children, and at higher levels it triggers mood swings, irritability, and loss of coordination. Historians think that the widespread use of lead in piping and drinking vessels caused the Romans to suffer a relatively high incidence of lead poisoning, which may even have contributed to the downfall of their empire. Lead-based paints may have caused a similarly high incidence among Europeans, and it is known that present-day American children living in low-income housing are at high risk because of the presence of lead-based paints that have not been removed.

Because of these damaging effects, most uses of lead that involve direct exposure for humans are being phased out. Unleaded gasoline and lead-free paints represent replacement products for two former major commercial uses of lead. Lead has proved to be indispensable, however, in the lead storage battery, which now provides the major use of this metal. Although leakage from damaged batteries is still a potential hazard, contemporary batteries are manufactured in such a way that human exposure to battery contents is minimized.

Bronze.

Pewter.

Section Exercises

20.3.1 Draw molecular pictures similar to those in Figure 20-7 that illustrate the electron pair donation that occurs when acetyl chloride (CH₃COCl, formed from acetic acid by replacing —OH with —Cl), forms an adduct with aluminum chloride.

20.3.2 Indium is a relatively soft Lewis acid. Use this fact to predict properties of indium compared with its horizontal neighbor (tin) and its diagonal neighbor (lead) in the periodic table.

20.3.3 Lead poisoning can be treated using the dianion of 2,3-dimercaptopropanol. Two ions bind one Pb^{2+} ion in a soluble tetrahedral complex that can be excreted from the body. Using hard–soft acid-base concepts, draw the expected structure for this complex.

2,3-Dimercaptopropanol dianion

SECTION 9.6 →
See Section 9.6 for a description of the electronic properties of semiconductors.

20.4 THE METALLOIDS

Six elements are classified as **metalloids:** B, Si, Ge, As, Sb, and Te. Of these, silicon is by far the most abundant, making up over 27% of the Earth's crust, more than any other element except oxygen. In fact, SiO_2 and silicate minerals account for 80% of the atoms near the Earth's surface. Despite its great abundance, silicon was not discovered until 1824, probably because the strong bonds it forms with oxygen makes silicon difficult to isolate. Two much rarer metalloids, antimony (known to the ancients) and arsenic (discovered ca. 1250 AD) were isolated and identified long before silicon.

Until quite recently, chemical interest in the metalloids consisted mainly of isolated curiosities, such as the poisonous nature of arsenic and the mildly therapeutic value of borax. With the development of metalloid semiconductors, however, these elements have become among the most intensely studied.

Boron

As is typical of second-row elements, boron has properties that distinguish it from its neighbors in Group 13 as well as from the rest of the metalloids. The unique features of boron chemistry can be attributed to characteristics of its electron configuration. As a second-row element, boron has no valence *d* orbitals that can participate in bonding. Its ionization energies are considerably higher than those of aluminum and the other Group 13 elements, all of which are metals. However, boron's three valence electrons are insufficient to form the extensive bonding networks that are characteristic of carbon, its second-row neighbor.

Boric acid (H_3BO_3) and boron trifluoride (BF_3) exemplify the bonding patterns of boron compounds. As Figure 20-8 shows, both these compounds contain delocalized π bonds. Each acts as a Lewis acid, readily adding an additional anion to form $[B(OH)_4]^-$ and $[BF_4]^-$ adducts. Both adducts are tetrahedral species in which the bonding can be described using four σ bonds formed from sp^3 hybrid orbitals on the boron atom. The changes in bonding that accompany adduct formation are reflected in differences in bond lengths. The B—F bond lengths are 131 pm in BF_3 and 139 pm in $[BF_4]^-$; B—O bond lengths are 137 pm in H_3BO_3 and 147 pm in $[B(OH)_4]^-$. The shorter bonds in the neutral species provide clear evidence for mul-

Top view   Side view   Perspective view

Figure 20-8
The Lewis structures of BF_3 and H_3BO_3 indicate that these species contain delocalized π bonds, as illustrated for BF_3. The adducts $[BF_4]^-$ and $[B(OH)_4]^-$, in contrast, contain only single σ bonds.

Lewis structures for BF_3, H_3BO_3, $[BF_4]^-$, and $[B(OH)_4]^-$:

tiple bonding. The unusual strength of the B—F bond in BF_3 also indicates the presence of π bonding: 645 kJ/mol is an energy more consistent with double bonds than single bonds.

Because of their low polarizability, BF_3 and BCl_3 are gases. The larger, more polarizable electron clouds on bromine and iodine make BBr_3 a volatile liquid and BI_3 a solid at room temperature.

Boron trihalides are strong Lewis acids that react with a wide collection of Lewis bases. Many adducts form with donor atoms from Group 15 (N, P, As) or Group 16 (O, S). Metal fluorides transfer F^- ion to BF_3 to give tetrafluoroborate salts:

$$LiF + BF_3 \longrightarrow Li^+ + [BF_4]^-$$

Tetrafluoroborate anion is an important derivative of BF_3 because it is nonreactive. With four σ bonds, $[BF_4]^-$ anion has no tendency to coordinate further ligands. Tetrafluoroborate salts are used in synthesis when a bulky inert anion is necessary.

The observed order of reactivity for the boron halides is:

$$BF_3 < BCl_3 < BBr_3 < BI_3$$

From electronegativity considerations, we might expect the opposite trend. The electronegativity difference between boron and fluorine is about 2, whereas boron and iodine differ by only 0.5. Thus fluorine atoms withdraw more electron density from boron than iodine atoms do, resulting in a more positive boron atom that should be a stronger Lewis acid. The observed reactivity trend indicates that some mechanism returns electron density to boron in the lighter boron halides. Such a mechanism is provided by the delocalized π bonding resulting from the overlap between filled halogen p orbitals and the empty p orbital on the boron atom. The π bond shifts electron density from the halogen atoms into the bonding region between the atoms. Fluorine, being the smallest halogen, forms the strongest π bond with boron, since the side-by-side overlap needed for π bonding is greater when $n = 2$ than for larger, more diffuse p orbitals.

Electronegativities: B = 2.0, F = 4.0, Cl = 3.0, Br = 2.8, and I = 2.5.

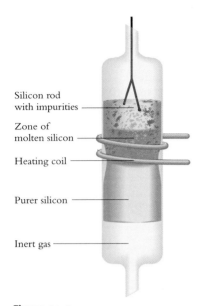

Figure 20-9
The technique of zone refining. As a heating coil passes along a rod of impure silicon, the impurities concentrate in the molten zone, leaving solid material behind that is of higher purity. Repeated passage through the coil moves the impurities to the end of the rod.

Silicon

The most abundant compounds of silicon are SiO_2 and the related silicate anions, all of which contain Si—O bonds. See Chapter 8 for a description of the structure and bonding of these compounds, which involve σ bond networks and tetrahedral geometry. As already mentioned, many minerals are combinations of hard silicate anions and hard metal cations.

Even though silicon is extremely abundant, only one silicon-containing compound appears in the list of top 50 industrial chemicals. That is sodium silicate, Na_2SiO_3, used for the manufacture of silica gel and glass. Nevertheless, with the advent of the electronic age silicon has become an extremely important substance that is the basic component of most semiconductors. Because these are microscale devices, the quantity of production of silicon remains small compared with that of fertilizers and construction materials. Although relatively small in quantity, the value of silicon products is quite high.

The main source of silicon for semiconductor chips is silicon dioxide, or silica. Silica can be reduced directly to elemental form by intense heating with coke in an electric arc furnace. At the temperature of the furnace, silicon is a gas:

$$SiO_2 + C \xrightarrow{3000\ °C} Si + 2\ CO$$

The product of this process has a purity of around 98%, which is sufficient for uses such as alloy formation but not nearly high enough for semiconductors.

To prepare pure silicon, silica is first converted to $SiCl_4$. A redox reaction between coke and chlorine gas is coupled with a metathesis reaction to give $SiCl_4$, which liquefies on cooling:

$$SiO_2 + 2\ C + 2\ Cl_2 \xrightarrow{High\ T} SiCl_4 + 2\ CO$$

Distillation yields $SiCl_4$ of very high purity, which is then reduced with magnesium:

$$SiCl_4 + 2\ Mg \longrightarrow Si + 2\ MgCl_2$$

After the removal of $MgCl_2$ by washing with water, the silicon is purified still further by the technique of zone refining, shown schematically in Figure 20-9. A rod of impure silicon is melted and resolidified many times by a heating coil that passes back and forth along the rod. During this process, impurities remain preferentially in the liquid phase, yielding solid silicon containing less than 1 part per billion of impurities.

Silicones

In addition to its uses for electronic devices, silicon is a major component of polymers called silicones, whose backbones consist of alternating silicon and oxygen atoms. The synthesis of silicone polymers begins with an organic chloride such as methyl chloride and an alloy of silicon and copper:

$$2\ CH_3Cl + Si(Cu) \longrightarrow (CH_3)_2SiCl_2 + Cu$$

Treatment of the resulting chlorosilane with water replaces the chlorine atoms with hydroxyl groups:

$$(CH_3)_2SiCl_2 + 2\ H_2O \longrightarrow (CH_3)_2Si(OH)_2 + 2\ HCl$$

Finally, the dihydroxysilane eliminates water in a condensation reaction to give a polymer with an Si—O—Si linkage:

The properties of the silicone polymer depend on the identity of the organic fragment bonded to the silicon atom. Methyl groups give a silicone polymer that is an oil with greater thermal stability and less tendency to thicken at low temperature than hydrocarbon oils. The Si—O bonds that make up the backbone of a silicone polymer give the oil greater thermal stability because they are stronger than the C—C bonds of a hydrocarbon (450 kJ/mol for Si—O vs. 345 kJ/mol for C—C).

Silicone rubbers, polymers with extensive Si—O—Si cross-linking and high molar masses, can be purchased as long single chains that cross-link at room temperature. These polymers "cure" by reaction with water in the atmosphere. Figure 20-10 shows this process. The curing process releases acetic acid, which accounts for the odor of these materials.

Approximately 70,000 tons of silicone polymers are produced each year in the United States. Silicones are used as greases, caulking, gaskets, biomedical devices, cosmetics, surfactants, antifoaming agents, hydraulic fluids, and water repellents.

Other Metalloids

The remaining metalloids—Ge, As, Sb, and Te—all are present at very low concentrations in the Earth's crust. As already mentioned, arsenic is highly poisonous. Besides being the stuff of many murder mysteries, arsenic is used as a pesticide. The major current use of antimony is in lead-acid batteries. Battery electrodes are much less likely to electrolyze water during recharging when they are made from a lead-antimony alloy containing up to 5% Sb. This virtually eliminates the danger of gas

Figure 20-10
A self-curing silicone rubber cross-links by reacting with water. A weakly bound CH$_3$CO$_2$ group reacts with water to generate acetic acid and a hydroxysilane, which then undergoes a condensation reaction to cross-link the polymer chains.

buildup and subsequent rupturing of the battery and has allowed the production of sealed batteries, which have much improved lifetimes compared with unsealed batteries that use pure lead electrodes.

A major and growing use of the minor metalloids is in semiconductor fabrication. Germanium, like silicon, exhibits semiconductor properties. Binary compounds between elements of Groups 13 and 15 also act as semiconductors. These 13-15 compounds, such as GaAs and InSb, have the same number of valence electrons as Si or Ge. By varying the composition of 13-15 compounds, the band gap between the valence and conduction bands can be varied, allowing the fine-tuning of semiconductor properties.

Section Exercises

20.4.1 The anion present in aqueous solutions of boric acid is $[B(OH)_4]^-$ rather than $H_2BO_3^-$. This is because H_3BO_3 undergoes Lewis acid-base adduct formation with one water molecule, and the adduct then transfers a proton to a second water molecule to generate $[B(OH)_4]^-$ and a hydronium ion. Write Lewis structural diagrams illustrating these two transfer reactions. Show all formal charges, and include arrows that show the movement of electrons.

20.4.2 Draw band gap diagrams (review Section 9.6 if necessary) illustrating that silicon is a semiconductor, but carbon (diamond) is not. What feature of the valence atomic orbitals accounts for this difference?

20.5 PHOSPHORUS

Of all the elements, phosphorus is the only one that was first isolated from a human source. The element was extracted from human urine in 1669 using an unsavory process: After a sample of urine was allowed to stand for several days, the putrefied liquid was boiled until only a paste remained. Further heating of the paste at high temperature produced a gas that condensed to a waxy white solid when the vapor was bubbled into water. It wasn't until 1779 that phosphorus was discovered in mineral form, as a component of phosphate minerals.

Phosphorus is the eleventh most abundant element in Earth's crustal rock. It has been estimated that world reserves of "phosphate rock" are sufficient to last for several hundred years. Virtually all phosphorus deposits contain apatite, whose general formula is $Ca_5(PO_4)_3X$, where X = F, OH, or Cl. Fluoroapatite is the least soluble, hence most abundant of the three apatite minerals. Phosphorus is found in aqueous systems as HPO_4^{2-} and $H_2PO_4^-$ ions. In biological organisms, phosphorus is a component of nucleic acids and energy-shuttling molecules such as ATP.

Elemental Phosphorus

Modern production of elemental phosphorus relies on the common practice of using coke as a reducing agent. Apatite is mixed with silica and coke and then heated strongly in the absence of oxygen. Under these conditions, coke reduces phosphate to elemental phosphorus, the silica forms liquid calcium silicate, and the fluoride ions in apatite dissolve in the liquid calcium silicate. The reactions are not fully understood, but the stoichiometry for the calcium phosphate part of apatite is as follows:

CHAPTER 19 →
The use of coke and silica in metallurgy is described in Chapter 19.

$$2\ Ca_3(PO_4)_2 + 6\ SiO_2 + 10\ C \xrightarrow{1450\ °C} P_4 + 6\ CaSiO_3 + 10\ CO$$

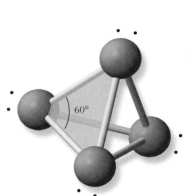

Figure 20-11
White phosphorus consists of individual P$_4$ molecules, with one P atom at each corner of a tetrahedron. The strained 60° bond angles make white phosphorus highly reactive.

Animation

Elemental phosphorus is a gas at the temperature of the reaction, so the product distills off in pure form. When cooled to room temperature the product is a waxy white solid.

White phosphorus consists of individual P$_4$ molecules with the four atoms at the corners of a tetrahedron. Each atom bonds to three others and has one lone pair of electrons, for a steric number of 4. However, the triangular geometry of the faces of the tetrahedron constrains the bond angles in the P$_4$ tetrahedron to 60°, far from the optimal four-coordinate geometry of 109.5°. As a result, P$_4$ is highly reactive. As shown in Figure 20-11, samples of white phosphorus are stored under water because P$_4$ burns spontaneously in the presence of oxygen. This form of phosphorus is very toxic: As little as 50 mg can cause death. As described later, most of the white phosphorus produced by the chemical industry is used to manufacture phosphoric acid.

If white phosphorus is heated in the absence of oxygen, the discrete P$_4$ units link together to form a chemically distinct elemental form, red phosphorus. As Figure 20-12 shows, one P—P bond of each tetrahedron breaks to allow formation of the bonds that link the P$_4$ fragments. The bond angles that involve the links are much closer to 109.5°, making red phosphorus less strained and less reactive than white phosphorus. Red phosphorus undergoes the same chemical reactions as the

The glow that emanates from P$_4$ as it burns in the dark led to its name from the Greek: *phos* ("light") and *phoros* ("bringing").

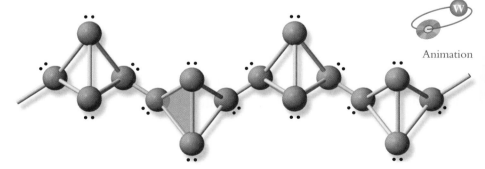

Animation

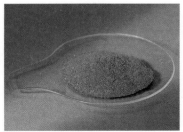

Figure 20-12
Red phosphorus consists of chains of P$_4$ tetrahedra, linked through P—P bonds.

white form, although higher temperatures are required. In addition, red phosphorus is essentially nontoxic, and therefore it is easier and safer to handle.

The most thermodynamically stable form of the element is black phosphorus, which can be prepared by heating red phosphorus under high pressure. The black form contains chains of P_4 units cross-linked by P—P bonds, making this form even more polymerized and less strained than red phosphorus. Example 20-3 explores another difference between the elemental forms of phosphorus.

Example 20-3	Melting Points of Phosphorus

The melting point of white phosphorus is 44.1 °C. In contrast, red phosphorus remains a solid up to 600 °C. Account for this very large difference in melting point.

Strategy: As described in Chapter 10, melting points of solids may depend on both covalent bonds and intermolecular forces, so we must explain the melting point difference with reference to the bonding differences between the two forms.

Solution: White phosphorus consists of individual P_4 molecules. Because there are no polar bonds, the molecules are held in the solid state only by dispersion forces. This is a molecular solid, and such solids have relatively low melting temperatures. Red phosphorus is made of long chains of P_4 groups, each of which is held together by covalent P—P bonds, as shown in Figure 20-12. Thus red phosphorus consists of macromolecules held together by dispersion forces, much like polyethylene, as described in Chapter 11. Substantially more energy is required to move the huge molecules of red phosphorus, giving it the higher melting temperature.

Phosphoric Acid

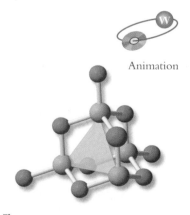

Animation

The most important commercial product of phosphorus is phosphoric acid, H_3PO_4. Phosphoric acid consistently ranks among the top ten industrial chemicals in the United States, with a production of nearly 14,000 tons in 1999.

Almost all phosphoric acid is produced directly from apatite. The ore is partially purified, crushed, and then slurried with aqueous sulfuric acid:

$$Ca_5(PO_4)_3F(s) + 5\ H_2SO_4(aq) \longrightarrow 3\ H_3PO_4(aq) + 5\ CaSO_4(s) + HF(aq)$$

The dilute phosphoric acid obtained from this process is concentrated by evaporation. It is usually dark green or brown because of the presence of many metal ion impurities in the phosphate rock. However, this impure acid is suitable for the manufacture of phosphate fertilizers, which consumes almost 90% of phosphoric acid production.

When high-purity H_3PO_4 is required, a more expensive redox process that starts from the pure element is used. Controlled combustion of white phosphorus gives phosphorus(V) oxide, P_4O_{10}, whose structure is shown in Figure 20-13:

$$P_4 + 5\ O_2 \longrightarrow P_4O_{10}$$

Addition of water to P_4O_{10} generates highly pure phosphoric acid:

$$P_4O_{10} + 6\ H_2O \longrightarrow 4\ H_3PO_4(aq)$$

More than 80% of the elemental phosphorus produced is converted to phosphoric acid. This pure product, which constitutes about 10% of the total industrial output of phosphoric acid, is the starting material for making food additives, pharmaceuticals, and detergents.

Figure 20-13
White phosphorus reacts with molecular oxygen to give phosphorus(V) oxide, P_4O_{10}. The triangular arrangement of phosphorus atoms is retained, but an oxygen atom is inserted into each P—P bond. An additional terminal O atom is double-bonded to each P atom.

The elemental phosphorus that is not converted into phosphoric acid is used mainly to produce phosphorus chlorides (PCl_3 and PCl_5) and phosphorus oxychloride ($POCl_3$). These are important reagents for production of agrochemicals, drugs, and other specialty products.

Phosphorus Fertilizers

Plants typically contain about 0.2% phosphorus by weight, but the element is easily depleted from soils. For this reason, the #1 commercial application of phosphorus is in fertilizers. There are several common phosphorus fertilizers, but the single most important one is ammonium hydrogen phosphate, $(NH_4)_2HPO_4$. This phosphate compound is particularly valuable because it is highly soluble and provides both phosphorus and nitrogen. The compound can be made by treating phosphoric acid with ammonia:

$$H_3PO_4 + 2\,NH_3 \longrightarrow (NH_4)_2HPO_4$$

This is the predominant industrial route to $(NH_4)_2HPO_4$, accounting for the single largest use of phosphoric acid.

Phosphate Condensations

As described in Chapter 11, many species that contain O—H bonds can undergo condensation reactions. Phosphoric acid can undergo phosphate condensation:

The product, $H_4P_2O_7$, is pyrophosphoric acid. A second condensation leads to triphosphoric acid:

$$H_4P_2O_7 + H_3PO_4 \longrightarrow H_5P_3O_{10} + H_2O$$

The sodium salts of these acids, sodium pyrophosphate ($Na_4P_2O_7$) and sodium triphosphate ($Na_5P_3O_{10}$), have been used widely in detergents. The polyphosphate anions are good additives for cleansing agents because they form complexes with metal ions, including those that make water "hard" (Ca^{2+}, Mg^{2+}) and those that cause color stains (Fe^{3+}, Mn^{2+}). Moreover, polyphosphates are nontoxic, nonflammable, and noncorrosive; they do not attack dyes or fabrics, and they are readily decomposed during wastewater treatment.

These advantages appear to make polyphosphates ideal for use in cleaning agents. Unfortunately, adding phosphates to water leads to an imbalance in aquatic biosystems, particularly in lakes. Photosynthetic algae require phosphate for growth. Under normal conditions, restricted phosphate concentration limits algae growth, and a lake reaches a balance in which many life forms flourish. Too much phosphate leads to runaway algae growth, which can overwhelm the lake with decaying organic matter. The process that supplies organic nutrients is *eutrophication*. Unfortunately, runaway eutrophication can be catastrophic. Organic decay consumes oxygen, depleting the lake of this life-sustaining substance and leading to the death of fish and other aquatic life forms. This contributes more organic decay and still more oxygen depletion, until eventually the lake supports no animal life.

Eutrophication is not a new phenomenon. Early in the twentieth century, Lake Zurich in Switzerland suffered from this condition as a result of the dumping of

Excess phosphate causes eutrophication of lakes, but improved water treatment and reduction of phosphate uses allow lakes to recover.

human sewage. Several North American lakes (most notably, Lake Erie) have more recently suffered eutrophication from excess phosphate. The Swiss restored Lake Zurich by treatment with $FeCl_3$, which removes excess phosphate through precipitation:

$$Fe^{3+}(aq) + PO_4{}^{3-}(aq) \longrightarrow FePO_4(s)$$

Although effective, this treatment is expensive because of the large quantities of metal salt required. Consequently, the use of phosphate detergents has been discontinued in many parts of both North America and Europe. Nevertheless, studies on the recovery of phosphates from recycled wastewater eventually may lead to a comeback for phosphate detergents.

Phosphate condensation reactions play an essential role in metabolism. Recall from Section 13.6 that the conversion of adenosine diphosphate (ADP) to adenosine triphosphate (ATP) requires an input of free energy:

$$ADP + H_3PO_4 \longrightarrow ATP + H_2O \qquad \Delta G° = +30.6 \text{ kJ}$$

As also described in that section, ATP serves as a major biochemical energy source, releasing energy in the reverse, hydrolysis, reaction. The ease of interchanging O—H and O—P bonds probably accounts for nature's choice of a phosphate condensation/hydrolysis reaction for energy storage and transport.

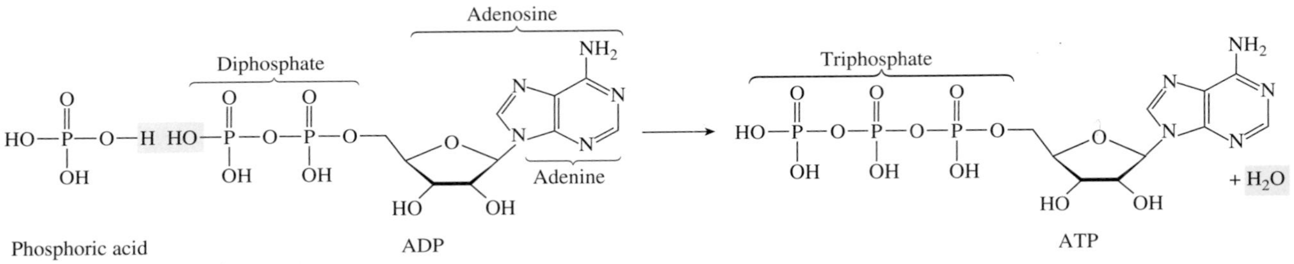

In ADP and ATP, one end of the polyphosphate chain links to adenosine through an O—C bond. These bonds form in condensation reactions between hydrogen phosphate and alcohols. Other important biochemical substances contain these linkages. Lecithins, found in brain and nervous tissue, and cephalins, involved in blood clotting, share a common structure that incorporates a phosphate group bonded through P—O—C links to two organic structures:

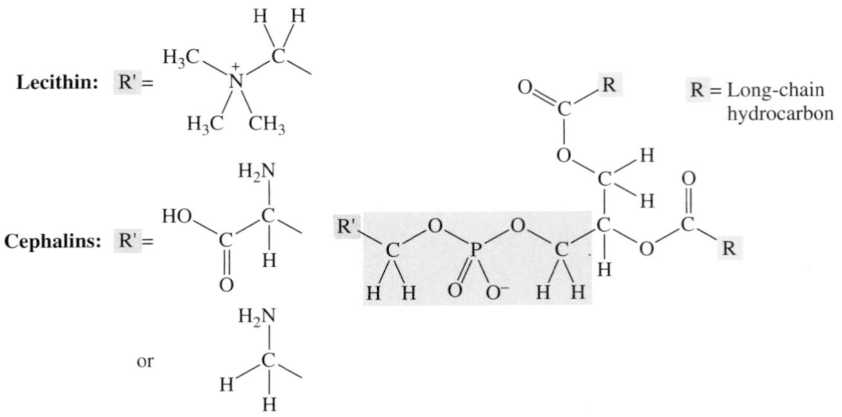

Notice that the phosphate portions of these molecules contain, in addition to the hydrophilic phosphate, polar nitrogen or oxygen atoms. These are dual-nature molecules, with phosphate-containing head groups that are water-soluble and long hydrocarbon chains that are compatible with cell membranes.

Commercial food additives make use of dual-nature phosphates. For instance, synthetic lecithins find major uses as emulsifiers in margarines. Another example is the thickening agent used in instant puddings and pie fillings, which is produced by the reaction of starch with sodium dihydrogen phosphate:

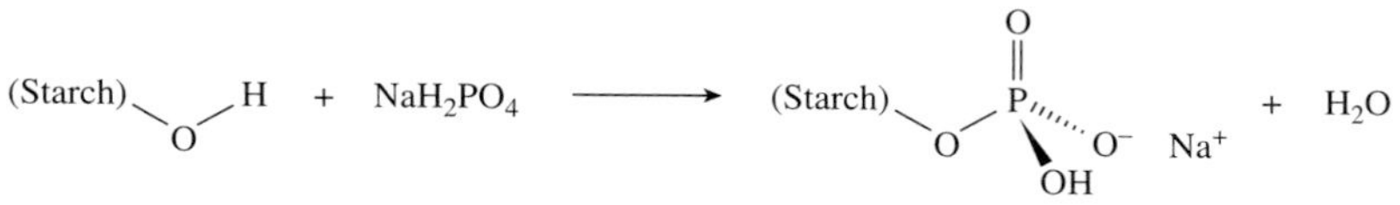

← CHAPTERS 10 & 11
Dual-nature molecules are discussed in Chapter 10, and the structure of starch is described in Chapter 11.

$$(\text{Starch})\diagdown \overset{\textstyle H}{\underset{\textstyle O}{}} \quad + \quad NaH_2PO_4 \quad \longrightarrow \quad (\text{Starch})\diagdown O \diagdown P \diagdown O^- \ Na^+ \quad + \quad H_2O$$

Starch molecules have many exposed O—H bonds, so this phosphorylation reaction results in multiple phosphate groups attached to each starch molecule. The remaining —OH group on each phosphate can condense with an O—H bond on another starch molecule. This cross-linking of starch chains gives the desired thick consistency of puddings and pies.

Organophosphorus Compounds

Strictly speaking, the biochemical substances just mentioned fall into the class of organophosphorus compounds, which are substances containing both phosphorus and carbon. However, the term is usually used more specifically to describe smaller molecules of this type. Organophosphorus compounds have varied uses, ranging from gasoline additives to herbicides.

Triphenylphosphate can be manufactured by condensing phenol with phosphoric acid in 3 : 1 ratio:

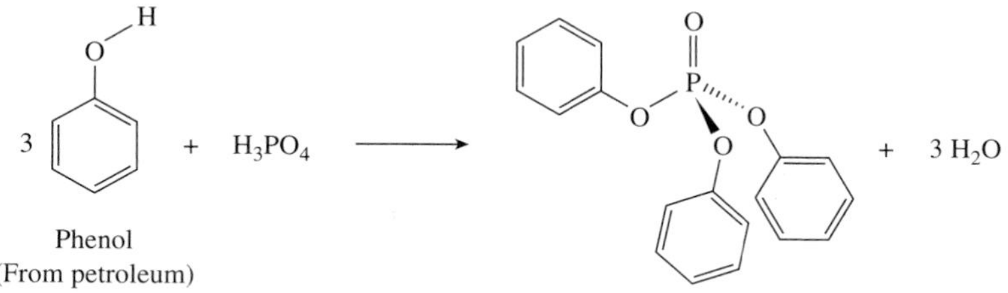

$$3 \ \underset{\substack{\text{Phenol}\\\text{(From petroleum)}}}{\text{[phenol]}} \quad + \quad H_3PO_4 \quad \longrightarrow \quad \underset{\text{Triphenyl phosphate}}{\text{[triphenyl phosphate]}} \quad + \quad 3 \ H_2O$$

This compound is added to plastics to serve as a flame retardant. If ethanol is used instead of phenol, the result is triethylphosphate, a liquid used as a specialty solvent. Similarly, 1-butanol ($CH_3CH_2CH_2CH_2OH$) condenses with phosphoric acid to form tributylphosphate, which is an excellent ligand for heavy metal cations such as thorium and the rare earth metals.

The phosphate structure plays a central role in many familiar insecticides or herbicides. These compounds often have other elements substituted in place of one or more O atoms in the phosphate group, giving structures that can penetrate living cells and disrupt their normal activities. Figure 20-14 shows the line structures of a representative set of these species. Organophosphorus compounds such as Parathion and Malathion are potent insecticides. They act by phosphorylating an essential enzyme. Herbicides are represented by amiprophos, a compound that has a highly modified phosphate group. One oxygen atom is replaced with sulfur and another with nitrogen.

Not all toxic organophosphorus compounds have uses beneficial to humans. Sarin is an extremely toxic nerve gas that is lethal to humans. In March 1995 this substance was released in a terrorist attack on a Japanese subway, resulting in several deaths and many serious injuries. Sarin and related nerve gases bind an amino acid

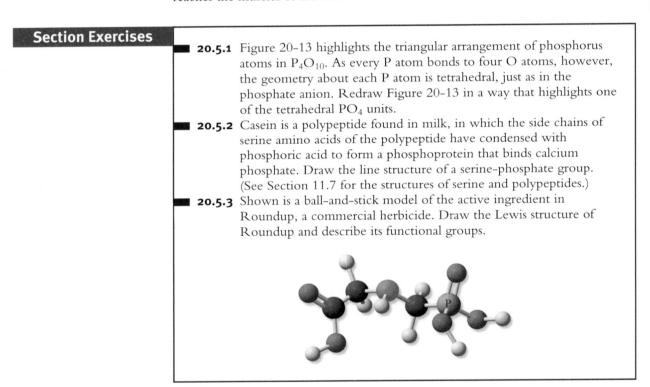

Parathion
$C_{10}H_{14}NO_5PS$

Malathion
$C_{10}H_{19}O_6PS_2$

Amiprophos
$C_{12}H_{19}N_2O_4PS$

Sarin
$C_4H_{10}FO_2P$

Figure 20-14
Line structures of four toxic organophosphorus compounds. Notice that each has a phosphate group that has been modified by replacing one or more O atoms with atoms of other elements.

in the enzyme responsible for muscle action. When this enzyme is deactivated, muscles contract but cannot relax. Even a small dose can be lethal if the nerve gas reaches the muscles of the heart.

Section Exercises

20.5.1 Figure 20-13 highlights the triangular arrangement of phosphorus atoms in P_4O_{10}. As every P atom bonds to four O atoms, however, the geometry about each P atom is tetrahedral, just as in the phosphate anion. Redraw Figure 20-13 in a way that highlights one of the tetrahedral PO_4 units.

20.5.2 Casein is a polypeptide found in milk, in which the side chains of serine amino acids of the polypeptide have condensed with phosphoric acid to form a phosphoprotein that binds calcium phosphate. Draw the line structure of a serine–phosphate group. (See Section 11.7 for the structures of serine and polypeptides.)

20.5.3 Shown is a ball-and-stick model of the active ingredient in Roundup, a commercial herbicide. Draw the Lewis structure of Roundup and describe its functional groups.

20.6 OTHER NONMETALS

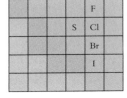

Among the nonmetals, the second-row elements carbon, nitrogen, and oxygen display a rich variety of chemical properties that provide many of the examples used throughout this textbook. Because the structure and chemistry of compounds of these elements are described in other chapters, we do not present further details here. The remaining nonmetals, all members of the *p* block, are sulfur from the third row of the periodic table and the halogens of Group 17 (F, Cl, Br, and I). Two of these elements are components of leading industrial chemicals: Sulfuric acid perennially ranks #1 by a wide margin, and molecular chlorine ranked in the top ten until very recently. Our survey of sulfur and the halogens emphasizes commercially important uses.

Sulfur

Sulfur displays rich and varied chemical behavior, but from the commercial standpoint, sulfuric acid dominates the chemistry of this element. Sulfuric acid is used in every major chemical-related industry: fertilizers (60% of annual production), chemical manufacture (6%), petroleum refining (5%), metallurgy (5%), detergents, plastics, fibers, paints and pigments, and paper-making. In many of these cases, sulfur itself is not of interest and is eventually discarded as sulfate waste. Instead, H_2SO_4 is exploited for its Brønsted acidity, its oxidizing ability, its affinity for water, and its ability to form sulfate precipitates.

Sulfur is encountered most often in sulfide minerals such as pyrite (FeS_2), molybdenite (MoS_2), chalcocite (Cu_2S), cinnabar (HgS), and galena (PbS). It is also present in huge amounts as H_2S in natural gas and in sulfur-containing organic compounds in crude oil and coal. Moreover, sulfur is found in abundance as the pure element, particularly around hot springs and volcanoes and in capping layers over natural salt deposits. References to sulfur occur throughout recorded history, dating back as far as the sixteenth century BC. As shown in Figure 20-15, elemental sulfur is a yellow crystalline solid that consists of individual S_8 molecules. The eight atoms in each molecule form a ring that puckers in such a way that four atoms lie in one plane and the other four atoms lie in a second plane.

Elemental sulfur can be obtained from underground deposits. A well is drilled into the deposit, and superheated pressurized water (165 °C) is forced into the hole. The hot water melts the sulfur (mp = 119 °C), and the liquid sulfur is forced to the surface when pressurized air is injected into the well. This method supplies about one third of the world's sulfur. However, the major present-day source of the element is from hydrogen sulfide produced as a byproduct of oil and gas refining. Many petroleum and natural gas supplies contain some sulfur—up to 25% in some cases. Besides being undesirable in the final products, sulfur poisons many of the catalysts used in oil refining; hence it must be removed from crude petroleum as a first step in refining. Sulfur is produced in a gaseous redox reaction between hydrogen sulfide and sulfur dioxide:

$$16\ H_2S + 8\ SO_2 \xrightarrow{\ Fe_2O_3\ } 3\ S_8 + 16\ H_2O$$

If a supply of SO_2 is not already available, this gas is generated by treating H_2S with molecular oxygen:

$$2\ H_2S + 3\ O_2 \longrightarrow 2\ SO_2 + 2\ H_2O$$

This process and the subsequent oxidation of sulfur to sulfuric acid are relatively inexpensive because the reactions are exothermic. Thus rather than requiring energy expenditure, these reactions produce energy.

Sulfuric acid is manufactured from elemental sulfur by the process described in our Chemistry and Technology Box in Chapter 15. About 90% of world output of sulfur is converted to sulfuric acid, and more than 60% of that sulfuric acid is used to extract phosphoric acid from phosphate minerals. This major use exploits the fact that sulfuric acid is the least expensive Brønsted acid. In addition to protonating the phosphate anions, the sulfate group of sulfuric acid sequesters Ca^{2+} as $CaSO_4$, a waste byproduct of the process. Another industrial use of sulfuric acid also exploits its acidity. The treatment of calcium fluoride (CaF_2) with sulfuric acid produces hydrogen fluoride (HF):

$$CaF_2(s) + H_2SO_4(l) \longrightarrow 2\ HF(g) + CaSO_4(s)$$

This source of hydrogen fluoride provides about 70% of the fluorine used industrially.

Sulfur is produced on an immense scale for the manufacture of sulfuric acid.

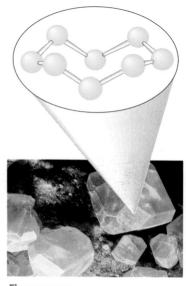

Figure 20-15
Under normal conditions, sulfur forms yellow crystals. The crystals consist of individual S_8 molecules, with the eight sulfur atoms of each molecule arranged in a puckered ring.

Sulfate solubility properties are the basis for other industrial uses of sulfuric acid. One example is in the metallurgy of titanium. One of the major ores of titanium is $FeTiO_3$, which is treated with sulfuric acid to separate titanium from iron:

$$FeTiO_3(s) + 2\ H_2SO_4(aq) + 5\ H_2O(l) \xrightarrow{150-180\ °C} TiOSO_4(aq) + FeSO_4 \cdot 7\ H_2O(s)$$

$$TiOSO_4(aq) + 2\ H_2O(l) \xrightarrow{90\ °C} TiO(OH)_2(s) + H_2SO_4(aq)$$

The $TiO(OH)_2$ precipitate is washed with dilute sulfuric acid to remove impurities and then converted to TiO_2 by heating to 1000 °C. As described in Section 19.4, TiO_2 is subsequently converted to $TiCl_4$, purified by distillation, and then reduced to titanium metal.

Another use for sulfuric acid is in water treatment. Aluminum sulfate, which is produced by treating sulfuric acid with aluminum oxide, is a top-50 industrial chemical because of its widespread use as a coagulant. In wastewater treatment and also in the paper industry, addition of $Al_2(SO_4)_3$ and $Ca(OH)_2$ generates a gelatinous precipitate of aluminum hydroxide mixed with calcium sulfate:

$$Al_2(SO_4)_3(aq) + Ca(OH)_2(aq) \longrightarrow Al(OH)_3(s) + CaSO_4(s)$$

This material traps finely suspended particulate material as it settles slowly out of the solution. This represents the primary stage of wastewater treatment, removal of materials that make the water cloudy.

Sulfuric acid plays several roles in the manufacture of synthetic dyes. Indeed, the birth of the dye industry was the accidental discovery of purple mauveine by William Perkin during an attempted oxidation of organic materials using potassium dichromate in sulfuric acid. Many modern dyes are based on either aniline or β-naphthol, both of which are produced in reactions that involve sulfuric acid:

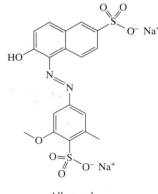

Allura red

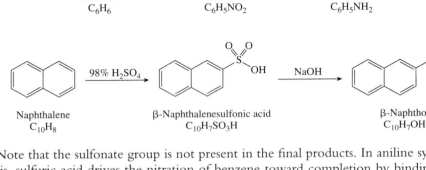

Benzene
C_6H_6

Nitrobenzene
$C_6H_5NO_2$

Aniline
$C_6H_5NH_2$

Naphthalene
$C_{10}H_8$

β-Naphthalenesulfonic acid
$C_{10}H_7SO_3H$

β-Naphthol
$C_{10}H_7OH$

Note that the sulfonate group is not present in the final products. In aniline synthesis, sulfuric acid drives the nitration of benzene toward completion by binding the water molecule produced in the reaction. In the synthesis of β-naphthol, the sulfonate group is introduced only because it is easy to replace with the hydroxyl group. On the other hand, dyes such as allura red, a food color additive, contain sulfonate groups which result from reaction with 98% sulfuric acid.

Although not as important industrially as processes involving sulfate, sulfur displays a rich chemistry with the halogens. For example, sulfur forms seven different binary fluorides. One of these, SF_4, has significant Lewis acid-base properties.

Sulfur tetrafluoride can act as either an electron pair donor or an electron pair acceptor. With a steric number of 5, SF_4 adopts a trigonal bipyramidal arrangement of valence orbitals with the lone pair of electrons in one of the equatorial positions. The molecule acts as a Lewis base but not as a sulfur lone-pair donor. Instead, SF_4 is a fluoride donor, as shown in the following reactions. The donor atom in SF_4 is fluoride, which is a harder base than sulfur. After forming an adduct with the Lewis acid, the S—F bond breaks to give the ionic products:

$$SF_4 + BF_3 \longrightarrow [SF_3]^+[BF_4]^-$$

$$SF_4 + PF_5 \longrightarrow [SF_3]^+[PF_6]^-$$

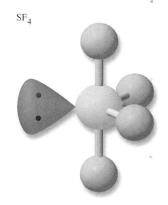

SF_4

Sulfur tetrafluoride is an important industrial fluorinating agent for both organic and inorganic compounds.

With its empty $3d$ orbitals, SF_4 also acts as a Lewis *acid*. For example, the molecule forms adducts with pyridine and with fluoride ion. Each adduct has square pyramidal geometry, with a lone pair of electrons completing the octahedral arrangement of valence orbitals:

$$SF_4 + Pyr \longrightarrow SF_4 \bullet Pyr$$

$$SF_4 + CsF \longrightarrow Cs^+ + [SF_5]^-$$

SF_5^-

Lewis acid-base aspects of sulfur chemistry account for the high toxicity of soft metal cations such as Hg^{2+} and Pb^{2+}. Anything that alters the three-dimensional structure of an enzyme may also change the shape of its active site and destroy its catalytic activity. Recall from Chapter 11 that covalent sulfur-sulfur single bonds are major determinants of protein structure. Because sulfur is a soft Lewis base, it binds preferentially to soft metal ions. Soft metal ions alter an enzyme's shape by cleaving natural sulfide bridges that link sections of the protein together and by creating new sulfide bridges that do not belong in the natural molecule.

As mentioned earlier, lead is one of the most notorious soft metal toxins. Lead ions destroy two enzymes that are essential for biosynthesis of heme. Mercury is another well-known soft metal toxin. Inorganic mercury salts are extremely poisonous, as is dimethylmercury, $(CH_3)_2Hg$.

In the nineteenth century, $Hg(NO_3)_2$ was used to stiffen the felt used in making hats. The phrase "mad as a hatter" refers to long-term effects of ingested mercury.

Another key feature of sulfur chemistry is the Lewis acidity of sulfur dioxide. Sulfur dioxide is a common atmospheric pollutant that results from burning coal to produce electricity. Most coal reserves in North America include significant amounts of sulfur-containing impurities. When coal is burned, sulfur combines with O_2 to form SO_2, a hard Lewis acid. In the atmosphere, SO_2 reacts with water to give sulfurous acid, resulting in acid rain.

In SO_2, the two S=O bonds are polarized toward the electronegative oxygen atoms, creating a partial positive charge on sulfur, as shown in Figure 20-16. The sulfur atom can use its empty d orbitals to accept a pair of electrons from a water molecule to form an adduct. A

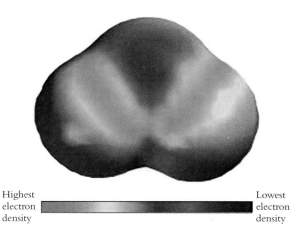

Highest electron density — Lowest electron density

Figure 20-16
An electron density model of sulfur dioxide shows that the region of lowest electron density (*blue*) is concentrated around the sulfur atom. The region of highest electron density (*red*) is concentrated on the oxygen atoms. Thus, the sulfur atom is the site of Lewis acidity.

second water molecule subsequently acts as a Brønsted base and transfers a proton from one oxygen atom to another to give H_2SO_3.

Oxidation by molecular oxygen or NO_2 converts this weak acid into H_2SO_4, a major contributor to acid rain.

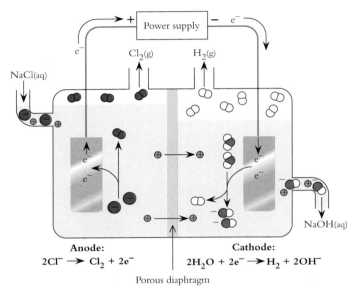

Anode:
$2Cl^- \longrightarrow Cl_2 + 2e^-$

Cathode:
$2H_2O + 2e^- \longrightarrow H_2 + 2OH^-$

Porous diaphragm

Figure 20-17
Schematic view of the electrolytic chlor-alkali process showing the molecular species involved in the redox reactions. Aqueous NaCl is fed into the cell, and the products are Cl$_2$ gas, H$_2$ gas, and aqueous NaOH.

Chlorine

As mentioned earlier, molecular chlorine has long been one of the leading industrial chemicals. Table 20-1 provides a summary of the industrial importance of chlorine.

The starting material for all industrial chlorine chemistry is sodium chloride, obtained primarily by evaporation of seawater. The chloride ion is highly stable and must be oxidized electrolytically to produce chlorine gas. This is carried out on an industrial scale using the chlor-alkali process, which is shown schematically in Figure 20-17. The electrochemistry involved in the chlor-alkali process is discussed in Section 18.7. As with all electrolytic processes, the energy costs are very high, but the process is economically feasible because it generates three commercially valuable products: H_2 gas, aqueous NaOH, and Cl_2 gas.

Although chlorine displays a range of oxidation states from -1 to $+7$, the most stable by far is Cl^-, as the positive standard reduction potentials in Table 20-2 indicate. Thus Cl_2 and all oxychloro anions are good

Table 20-1
Major Uses of Chlorine

Reactant	Intermediate	Final Products
Organic Reagents		
Alkenes + SO$_2$	Sulfonated alkenes	Surfactants, detergents
Benzene	Chlorobenzenes	Plastics, dyestuffs
Butadiene	Chloroprene	Neoprene (chlorinated rubber)
Ethylene	Chloroethenes	Plastics (PVC), solvents
Methane	Chloromethanes	Silicones
Inorganic reagents		
CO	COCl$_2$	Plastics (polycarbonate)
NaOH	NaOCl	Bleaches, disinfectants
Phosphorus	PCl$_3$, POCl$_3$	Pesticides, flame retardants
Rutile (TiO$_2$ ore)	TiCl$_4$	Paint pigment (pure TiO$_2$)

Table 20-2
Standard Reduction Potentials for Cl Species

$$Cl_2 + 2\,e^- \rightleftharpoons 2\,Cl^- \qquad +1.35827$$
$$HClO + H_3O^+ + 2\,e^- \rightleftharpoons Cl^- + 2\,H_2O \qquad +1.482$$
$$HClO_2 + 3\,H_3O^+ + 4\,e^- \rightleftharpoons Cl^- + 5\,H_2O \qquad +1.570$$
$$ClO_3^- + 6\,H_3O^+ + 6\,e^- \rightleftharpoons Cl^- + 9\,H_2O \qquad +1.451$$
$$ClO_4^- + 8\,H_3O^+ + 8\,e^- \rightleftharpoons Cl^- + 12\,H_2O \qquad +1.389$$

oxidizing agents. This property accounts for two of the major uses of chlorine compounds. Molecular chlorine is used extensively as a purifying agent for water supplies because it destroys harmful bacteria, producing harmless Cl^- in the process. Sodium hypochlorite, $NaOCl$, is the active ingredient in most bleaches. Hypochlorite ions oxidize many organic materials that are colored, breaking them into smaller, colorless substances that are easily removed by detergents.

A second major use for chlorine derives from its reactivity with organic materials, in particular with hydrocarbons. Two chlorinated hydrocarbons, ethylene dichloride (IUPAC name: 1,2-dichloroethane) and vinyl chloride (IUPAC name: chloroethene), rank among the second 10 industrial chemicals. Both substances are manufactured by reactions of Cl_2 with ethylene.

The first step in this process is a direct reaction carried out in liquid 1,2-dichloroethane with a metal chloride catalyst:

Thermal decomposition of this product yields vinyl chloride and HCl:

The HCl produced in this reaction is used in an oxychlorination reaction to chlorinate additional ethylene:

More than 95% of vinyl chloride is converted to PVC polymer.

The reaction of Cl_2 with ethylene to form vinyl chloride is just one example of a general process for the chlorination of hydrocarbons. One chlorine atom reacts with a hydrogen atom from the hydrocarbon to form HCl while a second chlorine atom replaces the hydrogen atom in the hydrocarbon. The generalized reaction is:

$$RH + Cl_2 \longrightarrow RCl + HCl$$

Organic chlorides produced in this way are key starting materials for the synthesis of more complex organic compounds. The HCl byproduct in these reactions may be exploited in an oxychlorination reaction as previously described, or it may be

PVC is used for a variety of common products.

collected in water and marketed as hydrochloric acid. Sufficient HCl is produced through various industrial chlorination reactions to account for over 90% of the production of hydrochloric acid.

In addition to making organic chlorine compounds, a significant fraction of Cl_2 production is used to make inorganic halides. One important use, described in Chapter 19, is in the metallurgy of titanium, in which molecular chlorine is used to convert TiO_2 into more readily purified $TiCl_4$:

$$TiO_2 + C + 2\,Cl_2 \longrightarrow TiCl_4 + CO_2$$

In similar fashion that is described earlier in this chapter, pure silicon is produced from molecular chlorine and SiO_2 in the presence of coke:

$$SiO_2 + 2\,C + 2\,Cl_2 \longrightarrow SiCl_4 + 2\,CO$$

Another use is the halogenation of phosphorus *en route* to the synthesis of organophosphorus compounds:

$$P_4 + 6\,Cl_2 \longrightarrow 4\,PCl_3$$

As Table 20-1 suggests, molecular chlorine is a tremendously versatile industrial chemical. Chlorine's position as a leading industrial chemical is a result of its versatility rather than of any single application, although polymers account for about one third of its uses. In recent years, however, the industrial use of chlorine has come under strong attack from many environmentally conscious groups. One major reason is that dioxins, one class of byproducts of chlorine reactions, have a very detrimental effect on biosystems. The controversy over industrial chlorine is described in our Chemistry and the Environment Box.

Other Halogens

No compound containing a halogen other than chlorine appears among the top 50 industrial chemicals, but fluorine nevertheless has considerable commercial value. Fluorine occurs as the mineral fluorite (CaF_2) and is prevalent in phosphate-bearing rock. As already mentioned, HF produced from sulfuric acid treatment of fluorite supplies some 70% of industrial HF; the remainder comes as a byproduct of phosphoric acid production.

One major use of HF is in the manufacture of fluorinated hydrocarbons. Fluorinated ethylene is used for several specialty polymers. For example, Teflon nonstick cookware is made from polytetrafluoroethylene (PTFE). This polymer is made from tetrafluoroethylene.

The starting material here is chloroform, $CHCl_3$, which is treated with HF to form the chlorofluorocarbon $CHClF_2$:

A coating of polytetrafluoroethylene gives Teflon cookware its nonstick surface.

$$\text{H}-\overset{\overset{\displaystyle Cl}{|}}{\underset{\underset{\displaystyle Cl}{|}}{\text{C}}}-\text{Cl} + 2\,\text{HF} \longrightarrow \text{H}-\overset{\overset{\displaystyle Cl}{|}}{\underset{\underset{\displaystyle F}{|}}{\text{C}}}-\text{F} + 2\,\text{HCl}$$

Subsequent heating in the presence of a platinum catalyst results in partial conversion to tetrafluoroethylene:

$$2\,\text{H}-\overset{\overset{\displaystyle Cl}{|}}{\underset{\underset{\displaystyle F}{|}}{\text{C}}}-\text{F} \xrightarrow[700\,°C]{Pt} \overset{F}{\underset{F}{}}\text{C}=\text{C}\overset{F}{\underset{F}{}} + 2\,\text{HCl}$$

| Box 20-2 | Chemistry and the Environment: The Case Against the Industrial Use of Chlorine |

Molecular chlorine has been a leading industrial chemical because it is the precursor for an immense variety of useful products—some 15,000 in all—that contain chlorine. Even a short list of the uses of chlorine indicates its versatility and commercial value: sugar refining, flame retardants, photography, deodorants, leather finishing, automobile bumpers, magnetic tape, and cosmetics. Nevertheless, groups ranging from environmental activists to government agencies have in recent years called for the reduction or even total elimination of chlorine as an industrial chemical.

The chloride anion, Cl^-, is a major species in the oceans and plays an essential role in biochemistry. In contrast, compounds containing carbon-chlorine bonds rarely occur in nature. Only marine algae, functioning in a chloride-rich environment, generate small amounts of chloromethane, CH_3Cl. Thus organic chlorine compounds are not a natural part of the biosphere, and consequently there are no natural mechanisms that degrade organic chlorine compounds.

The ecological problems caused by chlorine-containing compounds were first recognized years ago and called to public attention by Rachel Carson's best-selling book, *Silent Spring*. The chlorinated insecticide DDT accumulates in the tissues of animals. Birds of prey such as the peregrine falcon produced fragile, thin-shelled eggs that broke during incubation, and these species became endangered. Fortunately, the cause was determined and the use of DDT discontinued in time to allow gradual recovery of bird populations.

DDT

More recently, another class of organic chlorine compounds has emerged as an environmental hazard. These are the dioxins, which, like DDT, contain ring compounds with chlorine substituents. A relatively simple example is 2,3,7,8-tetrachlorodibenzo-*p*-dioxin:

Dioxin

This structure looks nothing like the structures of chlorine-containing compounds used in industrial processes. In fact, no dioxin is deliberately manufactured anywhere in the diverse chlorine industry. Nevertheless, dioxins are of concern for two reasons: First, dioxins appear to be inevitable trace byproducts of some reactions involving chlorine, particularly combustion; and second, dioxins accumulate in the biosphere, where they have highly deleterious effects.

Dioxins have effects similar to and potentially even more far-reaching than those of DDT, because they apparently affect a wide variety of species. Predatory birds are especially susceptible, and there is growing evidence that humans may be at risk. Tests have shown that when the concentration of dioxins in the blood of laboratory animals reaches a critical level, reproductive and immune system defects result. Moreover, recent data indicate that the concentration of dioxins in the blood of the average U.S. resident has nearly reached that level. A major reason is that dioxins are hydrophobic, so they accumulate in fatty tissue rather than being readily processed and excreted from the body.

Unfortunately, there is no easy way to eliminate production of dioxins without severely curtailing the use of all chlorinated compounds. This is because dioxins are formed when useful chlorine-containing compounds degrade, as for example when industrial wastes are incinerated. Consequently, several groups, including the American Public Health Association and the International Joint Commission overseeing the quality of the Great Lakes as well as environmental groups such as the Natural Resources Defense Council and the Sierra Club, have issued calls for the complete phase-out of the use of industrial chlorine.

Complete phase-out of chlorinated compounds is being resisted not only by chlorine producers, but also by the many industries that use chlorine compounds in the manufacture of products from paper to pharmaceuticals. Meanwhile, chemists seek ways to degrade dioxins to nontoxic substances.

CHAPTER 14 →
The effect of CFCs on the ozone layer is discussed in Chapter 14.

Until recently, chlorofluorocarbons (CFCs) for refrigeration were major end products of HF chemistry, but these compounds are being phased out in accord with the Montreal Protocols because of their effect on the ozone layer.

Some industrial uses of fluorine require molecular fluorine, F_2, which is produced by electrolysis of HF. As shown in Figure 20-18, the cell uses liquid HF to which KF is added as an electrolyte. The redox chemistry is straightforward:

Anode: $\qquad\qquad 2\,F^- \longrightarrow F_2 + 2\,e^-$

Cathode: $\qquad 2\,HF + 2\,e^- \longrightarrow H_2 + 2\,F^-$

Net: $\qquad\qquad\quad 2\,HF \longrightarrow F_2 + H_2$

Molecular fluorine reacts readily with a variety of organic and inorganic substances. It oxidizes metals to give metal fluorides. One useful metal fluoride is NaF, added to toothpaste to inhibit tooth decay. Uranium hexafluoride, UF_6, is used to make uranium fuel rods for nuclear power plants. We describe the production and use of UF_6 in Chapter 21. Molecular fluorine also combines with nonmetals to give binary fluorides. The most important of these is sulfur hexafluoride, an inert and nontoxic gas that is used in electrical transformers:

$$S_8 + 24\,F_2 \longrightarrow 8\,SF_6$$

In addition, F_2 reacts with the other halogens to form interhalogen compounds in which an atom of the second halogen is bonded to an odd number of F atoms. Examples are ClF_3, BrF_5, and IF_7.

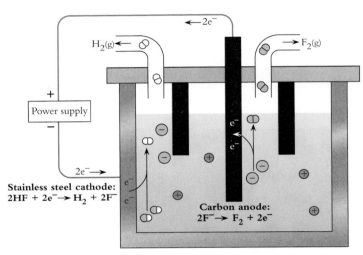

Figure 20-18
Schematic view of an electrolytic cell used for the production of molecular fluorine, showing the molecular species involved in the redox reactions.

The remaining halogens, bromine and iodine, are isolated from brines by treatment with Cl_2, which oxidizes the halide ions to their elemental forms:

$$2\,Br^-(aq) + Cl_2(g) \longrightarrow 2\,Cl^-(aq) + Br_2(l)$$

$$2\,I^-(aq) + Cl_2(g) \longrightarrow 2\,Cl^-(aq) + I_2(s)$$

Iodine is an essential element for humans because it is present in thyroxine, a hormone that regulates the rate of cellular use of oxygen. Peoples whose diets are sparse in seafoods are susceptible to iodine-deficiency diseases. Such diseases are easily prevented by providing iodine in the diet. Iodized salt, which contains 0.1% KI, performs this function.

Molecular bromine is highly toxic, as is methyl bromide (CH_3Br), a dense gas used as an insecticide. Methyl bromide is produced by bromination of methane:

$$CH_4 + Br_2 \longrightarrow CH_3Br + HBr$$

Molecular bromine is also used in the synthesis of dyes and pharmaceuticals. The most valuable bromine compound probably is AgBr, which is the light-absorbing species on which most film photography is based. Bromine compounds are also used extensively as fire retardants, particularly in carpets, rugs, and clothing.

Section Exercises

20.6.1 Determine the changes in oxidation states for the reactions that convert H_2S into elemental sulfur:

$$16\ H_2S + 8\ SO_2 \xrightarrow{Fe_2O_3} 3\ S_8 + 16\ H_2O$$

$$2\ H_2S + 3\ O_2 \longrightarrow 2\ SO_2 + 2\ H_2O$$

20.6.2 Determine the Lewis structures and describe the bonding and geometry of the following compounds that appear in this section: (a) SF_4, (b) BrF_5, and (c) $F_2C{=}CF_2$.

20.6.3 Although Cl_2 can be manufactured by electrolysis of aqueous brine, the analogous reaction cannot be used to make F_2. Explain why anhydrous HF must be used for production of F_2.

CHAPTER REVIEW

Summary and Key Terms

1. In a Lewis acid-base reaction, a bond forms between a **Lewis base,** which donates a pair of electrons, and a **Lewis acid,** which accepts this pair of electrons. Thus Lewis bases are electron pair donors, and Lewis acids are electron pair acceptors. The simplest such reactions result in formation of **adducts.**

2. Lewis acids and bases can be categorized as hard or soft according to whether their electron clouds are weakly or strongly polarizable. **Hard Lewis bases** tend to combine with **hard Lewis acids,** and **soft Lewis bases** tend to combine with **soft Lewis acids**. The exchange of bonding partners in a **metathesis reaction** occurs when the exchange matches hard and soft partners.

3. Metals in the p block of the periodic table are **main group metals.** Aluminum must be obtained from its ores by electrolysis using the **Hall-Héroult process,** and Al metal would be easily oxidized except that the oxidation product, Al_2O_3, forms an impervious film on the surface of the metal. Other main group metals (Sn and Pb) are much less reactive, but their uses are limited because of their toxicity.

4. The diagonal region of the p block contains the **metalloids.** Among metalloids, boron is unique because it has only three valence electrons and no valence d orbitals, making this element a good Lewis acid. Silicon and other metalloids have become increasingly important with the development of the semiconductor industry. Silicon also is widely used in glass-making and in silicone polymers.

5. The chemistry of phosphorus is dominated by the phosphate group, which is an essential fertilizer and a structural feature of several biochemical molecules.

6. Sulfur and the halogens are important nonmetallic elements. Sulfuric acid is the world's leading industrial chemical because it is used extensively in fertilizer production as well as for a variety of other uses. Most of the industrial chemistry of sulfuric acid exploits its acid-base and other properties rather than its sulfur content. Chlorine plays a role in a diverse collection of industrial processes, some of which make use of the oxidizing ability of Cl_2 and its derivatives, whereas others result in products containing C—Cl bonds.

Skills to Master

▶ Identifying Lewis acids and bases

▶ Depicting Lewis acid-base reactions

▶ Identifying hard and soft Lewis acids and bases

▶ Recognizing Lewis acid-base aspects of main group chemistry

Learning Exercises

20.1 There are no new "memory bank" equations in this chapter, but several concepts from earlier chapters play major roles in the chemistry of main group elements. List the key concepts from earlier chapters that you are asked to apply in this chapter.

20.2 Draw molecular pictures showing a typical Lewis acid-base reaction and a typical Brønsted acid-base reaction. Describe in words the differences and similarities of these two reactions.

20.3 State the hard-soft acid-base (HSAB) principle. Define and give examples of hard and soft acids and bases.

20.4 List the major industrial compounds containing main group elements and write a one-sentence description of the industrial importance of each.

20.5 Write the structural formula for an example of each of the following: (a) Lewis acid-base adduct; (b) polyphosphate; (c) fluorine-containing polymer; and (d) alkyllithium reagent.

Problems

Lewis Acids and Bases

20.1 In each of the following reactions, identify the Lewis acid and the Lewis base:

(a) $Ni + 4\,CO \rightarrow [Ni(CO)_4]$

(b) $SbCl_3 + 2\,Cl^- \rightarrow [SbCl_5]^{2-}$

(c) $(CH_3)_3P + AlBr_3 \rightarrow (CH_3)_3P—AlBr_3$

(d) $BF_3 + ClF_3 \rightarrow [ClF_2]^+ + [BF_4]^-$

20.2 In each of the following reactions, identify the Lewis acid and the Lewis base:

(a) $AlCl_3 + CH_3Cl \rightarrow CH_3^+ + [AlCl_4]^-$

(b) $Zn^{2+} + 4\,CN^- \rightarrow [Zn(CN)_4]^{2-}$

(c) $I_2 + I^- \rightarrow I_3^-$

(d) $CaO + CO_2 \rightarrow CaCO_3$

20.3 Make a sketch that shows the three-dimensional structure of each of the following Lewis acids. Identify the specific orbital that acts as the acceptor during adduct formation: (a) AlF_3; (b) SbF_5; and (c) SO_2.

20.4 Make a sketch that shows the three-dimensional structure of each of the following Lewis acids. Identify the specific orbital that acts as the acceptor during adduct formation: (a) $SnCl_4$; (b) PF_5; and (c) CO_2.

20.5 The reaction between CO_2 and H_2O to form carbonic acid (H_2CO_3) can be described in two steps: Lewis acid-base adduct formation followed by Brønsted proton transfer. Draw Lewis structures illustrating these two steps, showing electron and proton movement by curved arrows.

20.6 The reaction between SO_3 and H_2O to form sulfuric acid (H_2SO_4) can be described in two steps: Lewis acid-base adduct formation followed by Brønsted proton transfer. Draw Lewis structures illustrating these two steps, showing electron and proton movement by curved arrows.

Hard and Soft Lewis Acids and Bases

20.7 Rank the following ions in order of increasing polarizability, and explain your reasoning: Fe^{3+}, Fe^{2+}, Pb^{2+}, and V^{3+}.

20.8 Rank the following ions in order of increasing polarizability, and explain your reasoning: SO_3^{2-}, NO_3^-, CO_3^{2-}, and ClO_3^-.

20.9 Rank the following Lewis acids from hardest to softest and explain your reasoning: (a) BCl_3, BF_3, and $AlCl_3$; (b) Al^{3+}, Tl^{3+}, and Tl^+; and (c) $AlCl_3$, AlI_3, and $AlBr_3$.

20.10 Rank the following Lewis bases from hardest to softest and explain your rankings: (a) NH_3, SbH_3, and PH_3; (b) PO_4^{3-}, ClO_4^-, and SO_4^{2-}; and (c) O^{2-}, Se^{2-}, and S^{2-}.

20.11 Both sulfur trioxide and sulfur dioxide are Lewis acids, but SO_3 is harder than SO_2. Suggest an explanation.

20.12 Explain why iodide is a soft base but chloride is a hard base.

20.13 For each of the following pairs of substances, determine whether metathesis will occur and identify the products: (a) $AlI_3 + NaCl$; (b) $TiCl_4 + TiI_4$; (c) $CaO + H_2S$; (d) $CH_3Li + PCl_3$; and (e) $AgI + SiCl_4$.

20.14 For each of the following, state whether a reaction occurs, and write a balanced equation for any reaction:

(a) $NBr_3 + GaCl_3 \rightarrow ?$

(b) $Al(CH_3)_3 + LiCH_3 \rightarrow ?$

(c) $SiF_4 + LiF \rightarrow ?$

(d) $LiCH_2CH_2CH_2CH_3 + SnCl_4 \rightarrow ?$

The Main Group Metals

20.15 Describe in detail the bonding in Al_2Cl_6. Explain the formation of the molecule in terms of Lewis acid-base chemistry.

20.16 Write Lewis structures and describe the bonding in these three species found in the solution that is electrolyzed to form aluminum metal: $[AlF_4]^-$, $[AlF_6]^{3-}$, and $[OAlF_3]^{2-}$.

20.17 From its position in the periodic table, predict the properties of thallium (element 81).

20.18 From its position in the periodic table, predict the properties of gallium (element 31).

20.19 Determine the Lewis structure of $SnCl_4$ and explain how it functions as a Lewis acid.

20.20 Determine the Lewis structure of SbF_5 and explain how it functions as a Lewis acid.

The Metalloids

20.21 Borazine ($B_3N_3H_6$) is a planar molecule analogous to benzene (C_6H_6). Write the Lewis structure and describe the bonding of borazine.

20.22 Boron nitride (BN) is a planar covalent solid analogous to graphite. Write a portion of the Lewis structure and describe the bonding of boron nitride, which has alternating B and N atoms.

20.23 Reasoning from periodic trends, determine whether Ge or Si has a larger band gap. Use orbital overlap arguments to support your choice.

20.24 Explain why, among the 13−15 semiconductors, Ga pairs with As whereas In pairs with Sb and predict which of these has the smaller band gap.

20.25 Write the reactions that generate a silicone polymer starting from ethyl chloride (C_2H_5Cl), and draw a portion of the structure of the resulting polymer. Show at least four repeat units of the polymer.

20.26 Write the reactions that generate a silicone polymer starting from a 1:1 mixture of $(F_3CCH_2CH_2)(CH_3)Si(OH)_2$ and $(CH_3)_2Si(OH)_2$, and draw a portion of the structure of the resulting polymer. Show at least four repeat units of the polymer.

Phosphorus

20.27 Pyrophosphate ($P_2O_7^{4-}$) and triphosphate ($P_3O_{10}^{5-}$) are the first two polyphosphate anions. What is the chemical formula of the next largest polyphosphate? Draw a ball-and-stick model of this anion.

20.28 Three phosphate anions can condense to form a ring whose chemical formula is $P_3O_9^{3-}$. Determine the Lewis structure and draw a ball-and-stick model of this anion.

20.29 Describe the reactions by which apatite is converted to phosphoric acid. Identify Brønsted acid-base reactions and redox reactions, if any.

20.30 Describe the reactions by which calcium phosphate is converted to phosphoric acid. Identify Brønsted acid-base reactions and redox reactions, if any.

20.31 Write structural formulas showing the reaction of ethanol and phosphoric acid to form triethylphosphate.

20.32 Write structural formulas showing the reaction of ATP with water to form ADP. What is the other product?

Other Nonmetals

20.33 Prepare a list of the industrial reactions described in Section 20.6 that exploit the Brønsted acid character of sulfuric acid.

20.34 Prepare a list of the industrial reactions described in Section 20.6 that use sulfuric acid because of sulfate solubility characteristics.

20.35 Draw a portion of the repeating structure of polyvinylchloride.

20.36 Draw a portion of the repeating structure of polytetrafluoroethylene.

20.37 Identify the oxidizing agent, reducing agent, and changes of oxidation state that occur in the reaction forming $TiCl_4$ from TiO_2.

20.38 Identify the oxidizing agent, reducing agent, and changes of oxidation state that occur in the reaction forming $SiCl_4$ from SiO_2.

Additional Paired Problems

20.39 What mass of bauxite rock, $Al(O)OH$, must be processed to produce 2500 kg of pure aluminum if the bauxite rock is 85% pure and the processing steps have a net yield of 75%?

20.40 What mass of calcium phosphate rock, $Ca_3(PO_4)_2$, must be processed to produce 1500 kg of pure P_4 if the phosphate rock is 87% pure and the processing steps have a net yield of 68%?

20.41 Explain why SF_6 forms and is quite stable, whereas neither OF_6 nor SBr_6 is known.

20.42 Explain why the known forms of elemental carbon include forms with single, double, and triple bonds, whereas the known forms of elemental phosphorus all have single bonds.

20.43 Arsenic trichloride can act as a Lewis acid and a Lewis base. Explain why this is so and write a balanced equation for each using BF_3 and Cl^- as reaction partners.

20.44 Boron trifluoride is a very strong Lewis acid, but trimethylboron, $B(CH_3)_3$, is a mild Lewis acid. Given the π bonding character in BF_3, does this order of reactivity surprise you? Why or why not? Use bonding arguments to explain the trend in reactivity.

20.45 In basic aqueous solution, Al acts as a strong reducing agent, being oxidized to AlO_2^-. Balance this half-reaction, and determine balanced net reactions for Al reduction of the following: (a) NO_3^- to NH_3; (b) H_2O to H_2; and (c) SnO_3^{2-} to Sn.

20.46 In acidic aqueous solution, $SnCl_2$ acts as a mild reducing agent, tin being oxidized to Sn^{4+} in the process. Balance this half-reaction, and determine balanced net reactions for this reagent reducing the following: (a) MnO_4^- to Mn^{2+}; (b) $Cr_2O_7^{2-}$ to Cr^{3+}; and (c) Hg_2^{2+} to Hg.

20.47 Briefly describe the changes in chemical properties that take place from top to bottom of Group 15 of the periodic table.

20.48 Briefly describe the changes in chemical properties that take place from top to bottom of Group 13 of the periodic table.

20.49 Describe the chemical reactions by which Al metal is obtained from its ore.

20.50 Describe the chemical reactions by which Pb metal is obtained from its ore.

20.51 Describe the industrial preparation of pure Si, starting from impure SiO_2. Include balanced chemical reactions.

20.52 Describe the industrial preparation of pure H_3PO_4 starting from phosphate rock. Include balanced chemical reactions.

More Challenging Problems

20.53 Calcium dihydrogen phosphate is a common phosphorus fertilizer that is made by treating fluoroapatite with phosphoric acid. Hydrogen fluoride is a byproduct of the synthesis. Write a balanced equation for the production of this fertilizer and calculate the mass percent of phosphorus in the fertilizer.

20.54 The black tarnish that forms on pure silver metal is the sulfide Ag_2S, formed by reaction with H_2S in the atmosphere:

$$4\,Ag + 2\,H_2S + O_2 \longrightarrow 2\,Ag_2S + 2\,H_2O$$

The sulfide forms in preference to Ag_2O, even though the atmosphere is 20% O_2 with just a slight trace of H_2S. Use Lewis acid-base arguments to explain this behavior.

20.55 The pressure of gaseous Al_2Cl_6 increases more rapidly with temperature than would be predicted by the ideal gas equation. Explain this behavior.

20.56 Water in thermal hot springs often is unpalatable due to dissolved H_2S. Treatment with Cl_2 oxidizes the sulfur to S_8, which precipitates (the other product is HCl). Balance this reaction and calculate the mass of Cl_2 required to purify 6.0×10^3 L of water (the average amount used daily by one person in the United States) containing 25 ppm (by mass) dissolved H_2S.

20.57 Explain why soft metal ions such as lead and mercury are so toxic.

20.58 Thionyl chloride, $SOCl_2$, is used to remove water of hydration from metal halide hydrates. Besides the anhydrous metal halide (many of which are useful catalysts), the products are SO_2 and HCl. (a) Draw the Lewis structure of $SOCl_2$. (b) Balance the reaction of iron(III) chloride hexahydrate with $SOCl_2$.

20.59 The only important ore of mercury is cinnabar, HgS. In contrast, zinc is found in several ores, including sulfides, carbonates, silicates, and oxides. Explain these observations in terms of hard and soft acids and bases.

20.60 Ammonium dihydrogen phosphate and ammonium hydrogen phosphate are common fertilizers that provide both nitrogen and phosphorus to growing plants. In contrast, ammonium phosphate is rarely used as a fertilizer because this compound has a high vapor pressure of toxic ammonia gas. Write a balanced equation that shows how ammonium phosphate generates ammonia gas.

20.61 Boron trichloride is a gas, boron tribromide is a liquid, and boron triiodide is a solid. Explain this trend in terms of intermolecular forces and polarizability.

20.62 When PCl_5 condenses from a gas to a solid, it changes from trigonal bipyramidal molecules to an ionic crystal of the formula $[PCl_4]^+[PCl_6]^-$. Explain the reaction in terms of Lewis acids and bases:

$$2\ PCl_5(g) \longrightarrow [PCl_4]^+[PCl_6]^-(s)$$

20.63 White phosphorus and red phosphorus have strikingly different toxicities. One is quite poisonous, but the other is relatively unreactive. Use the structures of these two elemental forms to determine which form is poisonous and explain why the other form is unreactive.

20.64 The fluorides BF_3, AlF_3, SiF_4, and PF_5 are Lewis acids. They all form very stable fluoroanions when treated with lithium fluoride. In contrast, three other fluorides, CF_4, NF_3, and SF_6, do not react with lithium fluoride. Use Lewis acid-base concepts to explain this behavior.

20.65 The first commercially successful method for the production of aluminum metal was developed in 1854 by H. Deville. The process relied on earlier work by the Danish scientist H. Oersted, who discovered that aluminum chloride is produced when chlorine gas is passed over hot aluminum oxide. Deville found that aluminum chloride reacts with sodium metal to give aluminum metal. Write balanced equations for these two reactions.

20.66 Metal oxides and sulfide ores are usually contaminated with silica, SiO_2. This impurity must be removed when the ore is reduced to the pure element. Silica can be removed by adding calcium oxide to the reactor. Silica reacts with CaO to give $CaSiO_3$. Write a bal-

anced equation for this reaction, and describe the reaction in terms of Lewis acids and bases.

20.67 Complete the following reactions:
(a) $AlCl_3 + LiCH_3 \rightarrow$?
(b) $SO_3 + Excess\ H_2O \rightarrow$?
(c) $SbF_5 + LiF \rightarrow$?
(d) $SF_4 + AsCl_5 \rightarrow$?

20.68 Some pure liquid interhalogen compounds are good electrical conductors, indicating that they contain cations and anions. Show a Lewis acid-base reaction between two bromine trifluoride molecules that would generate ionic species.

20.69 Phosphorus(V) oxide has a very strong affinity for water; hence it is often used as a drying agent in laboratory desiccators. One mole of P_4O_{10} reacts with six moles of water. Based on this stoichiometry, identify the product of the reaction and balance the equation.

20.70 A company that manufactures photographic film generates 2550 L/day of aqueous waste containing 0.125 g/L of Br^- ions. To recover the bromine in the form of Br_2, the company bubbles Cl_2 gas through this waste. Calculate the volume of gas that is consumed daily if the gas is delivered at 1.05 atm and 21 °C.

20.71 Aluminum refining requires large amounts of electricity. Calculate the masses of Al and Na that are produced per mole of charge by electrolytic refining of Al_2O_3 and NaCl.

20.72 Trisodium phosphate forms strongly basic solutions that are used as cleansers. Write balanced equations that show why Na_3PO_4 solutions are strongly basic. Include pK_a values to support your equations.

Group Study Problems

20.73 Summarize the arguments for and against using phosphate-based detergents. Do you think these detergents should be used? Explain your position.

20.74 Construct a table of bond lengths that supports the existence of π bonding in boron trifluoride and boric acid. The relevant data can be found in the text. Label your table thoroughly.

20.75 Describe the ways in which the structures of DDT and dioxin (see Chemistry and the Environment Box) are similar. Describe the ways in which these structures differ.

20.76 One of the factors that controls the interaction of a Lewis base with a Lewis acid is the size of the molecules involved. For example, boron trifluoride forms a stronger adduct with tetrahydrofuran than with dimethyltetrahydrofuran. Use this example to

explain size effects in adduct formation. (Hint: Think about electron-electron repulsion.)

Tetrahydrofuran
C_4H_8O

Dimethyltetrahydrofuran
$C_6H_{12}O$

20.77 Summarize the arguments for and against using industrial chlorine. Do you think the industrial use of chlorine should be phased out? Explain your position.

Answers to Section Exercises

20.1.1

20.1.2 (a) Lewis acid: SO_3; Lewis base: OH^-. (b) Lewis acid: $SnCl_2$; Lewis base: Cl^-. (c) Lewis acid: SbF_5; Lewis base: AsF_3

20.1.3

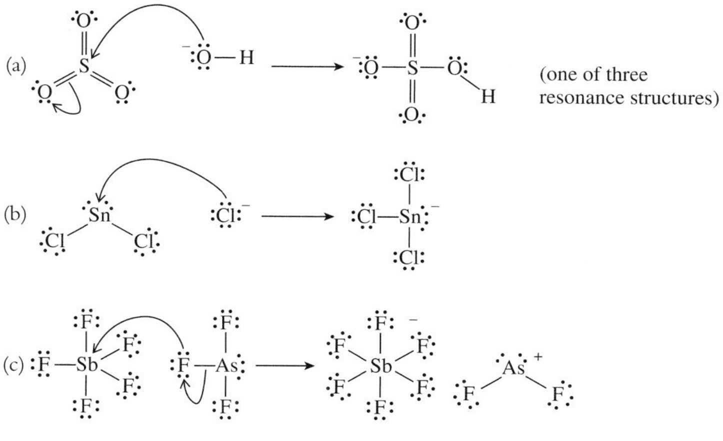

(a)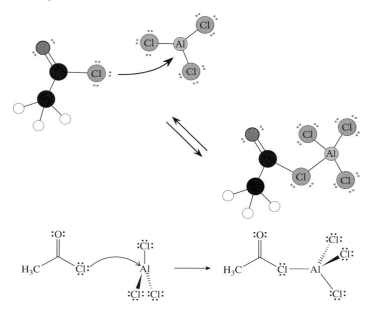

(one of three resonance structures)

(b)

(c)

20.2.1 (a) NH_3, NCl_3, NF_3; (b) Pb^{2+}, Pb^{4+}, Zn^{2+}; (c) ClO_4^-, ClO_3^-, ClO_2^-; and (d) $SbCl_3$, PCl_3, PF_3.

20.2.2 Only the softest metals, those clustered around gold in the periodic table, exist in nature in elemental form. Ru and Os are in this category. Iron forms harder cations with an affinity for the hard anions abundant in the environment, including carbonate, oxide, and the silicates.

20.2.3 (a) $4 BCl_3 + 3 SF_4 \longrightarrow 4 BF_3 + 3 SCl_2 + 3 Cl_2$

(b) Chlorine is oxidized, and sulfur is reduced.

(c) Sulfur is softer than boron, and chlorine is softer than fluorine. According to the HSAB principle, sulfur prefers to bond with the softer chlorine, and boron prefers to bond with the harder fluorine.

20.3.1

20.3.2 Indium, being relatively soft, has properties more like lead, which is softer than tin. Thus indium is toxic, occurs naturally as a sulfide, and so on.

20.3.3 Bonding is between the relatively soft Lewis base atoms, S, of 2,3-dimercaptopropanol and the Pb^{2+} cation:

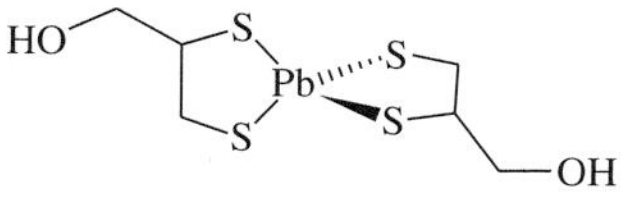

20.4.1

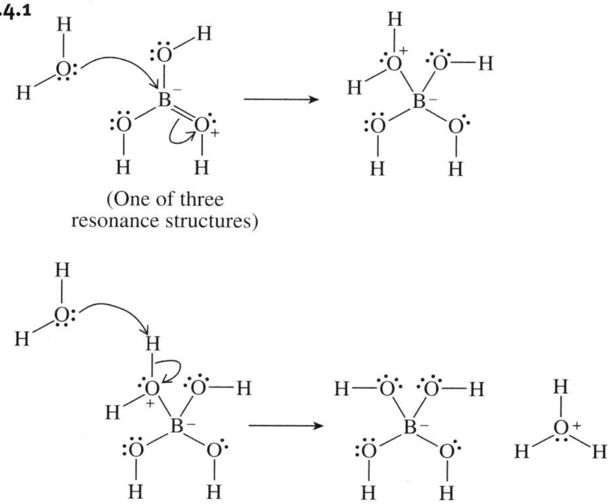

(One of three resonance structures)

20.4.2 The more diffuse $n = 3$ valence orbitals of Si do not overlap with one another as effectively as the compact $n = 2$ valence orbitals of C, so the difference in energy between bonding and antibonding orbitals (valence and conduction bands) is smaller for Si than for C.

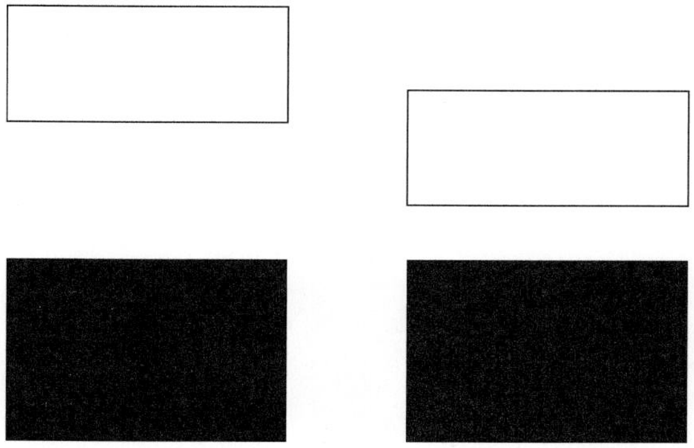

Carbon: Large band gap   Silicon: Small band gap

20.5.1

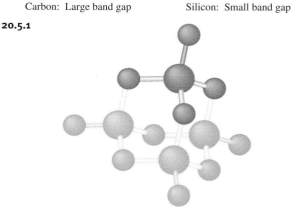

20.5.2

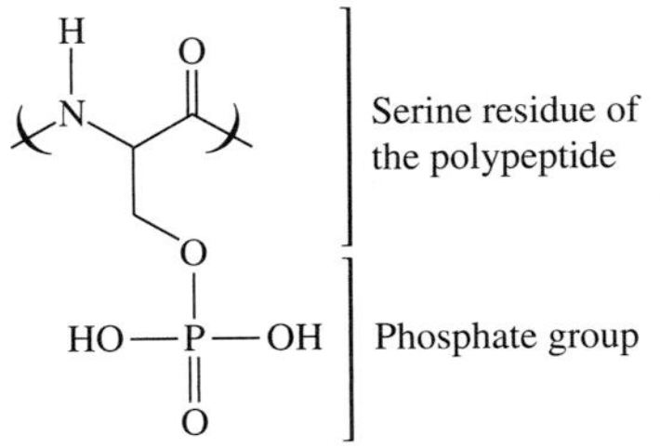

Serine residue of the polypeptide

Phosphate group

20.5.3 The structure includes a modified phosphate (P—C bond in place of one P—O bond), a carboxylic acid, and an amine.

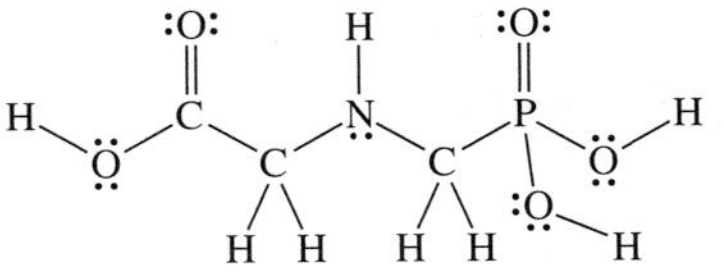

20.6.1 Oxidation states are shown beneath the atoms:

$$16 \, H_2S + 8 \, SO_2 \longrightarrow 3 \, S_8 + 16 \, H_2O$$
$$\quad -2 \qquad +4 \qquad\qquad 0$$

The sulfur in H_2S is oxidized (loses electrons), and the sulfur in SO_2 is reduced (gains electrons):

$$2 \, H_2S + 3 \, O_2 \longrightarrow 2 \, SO_2 + 2 \, H_2O$$
$$\quad -2 \qquad 0 \qquad +4 \qquad -2$$

Sulfur is oxidized and oxygen is reduced.

20.6.2 (a)

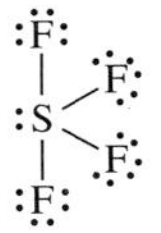

The S atom has SN = 5 and trigonal bipyramidal electron pair geometry. The lone pair is equatorial, giving a see-saw shape. Bonds can be described as formed from $2p$ atomic orbitals on F overlapping with sp^3d hybrids on S.

(b)

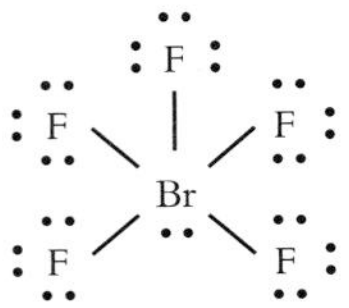

The Br atom has SN = 6 and octahedral electron geometry. The one lone pair results in a molecular shape that is square pyramidal. Bonds can be described as formed from $2p$ atomic orbitals on F overlapping with sp^3d^2 hybrids on Br.

(c)

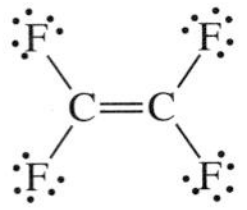

Each C atom has SN = 3, giving trigonal planar geometry and a planar molecule. There is a σ bond network that can be described using sp^2 hybrids from C and $2p$ atomic orbitals from F. Side-by-side overlap of $2p$ orbitals from C gives a π bond.

20.6.3 The oxidation reactions and their standard potentials are as follows:

$$2 \, F^- \longrightarrow F_2 + 2e^-, \; -2.866 \text{ V}$$

$$6 \, H_2O \longrightarrow O_2 + 4 \, H_3O^+(aq) + 4e^-, \; -1.229 \text{ V}$$

$$2 \, Cl^- \longrightarrow Cl_2 + 2e^-, \; -1.358 \text{ V}$$

As described in Chapter 18, even though Cl^- is thermodynamically harder to oxidize than H_2O, the reaction kinetically is much faster, so Cl^- is preferentially oxidized in aqueous solution containing Cl^-. Oxidation of F^- is thermodynamically so much more difficult that water is preferentially oxidized in aqueous solutions containing F^-.

21

Nuclear Chemistry and Radiochemistry

INTRODUCTION: THE NUCLEAR DILEMMA

L ife flourishes on Earth because of energy radiated by the sun. That energy comes from transformations of atomic nuclei. At the extreme temperature of the sun, nuclei fuse together to form larger nuclei. The vast amounts of energy released by these solar fusion reactions churn the sun's matter continuously, generating the hot spots and flares that appear in the X-ray image in our opening photo.

The sun's source of energy was a mystery until scientists studied the nucleus and learned that nuclear reactions, such as fusion, involve immense amounts of energy. Paradoxically, it was necessary to study matter at the subatomic level to unlock the energy secrets of the sun. Our inset figure shows results from an experiment carried out in an instrument called a bubble chamber. The photo shows the tracks generated by the high-energy particles formed in nuclear reactions. In passing through a liquid that is on the verge of boiling, each high-energy particle creates ions in its wake. Gas bubbles form around these ions, creating a track of bubbles along the path of the particle. Under the conditions of the experiment, the particles move in curved paths whose shapes disclose information about the structure of the nucleus. Such research on high-energy nuclear processes has helped us understand what fuels the stars.

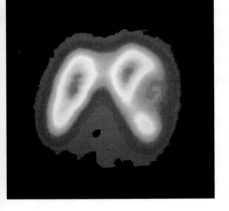

Unlocking the secrets of the nucleus was a mixed blessing, for in addition to our understanding of the sun, we also acquired nuclear weapons of immense destructive potential. The bombing of Hiroshima and Nagasaki with nuclear weapons was one of the last acts of the Second World War but the beginning of the nuclear dilemma: More than 50 years later, controversies still rage over how society should use the fruits of nuclear science.

The nuclear dilemma results partly from the devastating power of nuclear weapons. Equally troublesome are the health hazards associated with the radioactive products of nuclear reactions. These hazards include genetic effects, cancer, and other illnesses that can be fatal. Unfortunately, there is no known way to make radioactive elements nonlethal. Additionally, once radioactive substances are created, some of them will last for tens of thousands of years.

The benefits of nuclear reactions lie primarily in two areas, power generation and medical diagnosis and treatment. At present, society is divided as to whether or not these benefits outweigh their accompanying hazards. Nuclear medicine is firmly established. Many hospitals use radioactive materials for diagnosis and treatment, for example to generate the images of diseased thyroid glands shown in our margin photo. Still, society has not solved the problem of what to do with the resulting radioactive waste.

One hundred years after the discovery of radioactivity and fifty years after the dawn of the "nuclear age," society continues to debate the benefits and costs of nuclear technology. People in the United States have judged the risks to outweigh the benefits, so nuclear power plants have been out of favor. Other nations, notably France and Japan, have decided in favor of heavy reliance on nuclear power. Understanding nuclear transformations and the properties of radioactivity are prerequisites to intelligent discussions of the nuclear dilemma. In this chapter, we explore the nucleus and the nuclear processes that it undergoes. First we describe the factors that make nuclei stable or unstable. Then we describe the various types of nuclear reactions that can occur, and we conclude with the effects and applications of radioactivity.

21.1 NUCLEAR STABILITY

In earlier chapters we treated the nucleus as a structure that never changes. Although this is true for normal chemical processes, under the right conditions nuclei undergo transformations that change their structures. These processes depend strongly on energy as well as nuclear structure. Our discussion of nuclear chemistry begins with the structure and energetics of the nucleus.

Nuclear Composition

As described in Section 2.2, nuclei are unimaginably small. The radius of a nucleus is about 10^{-14} m, ten thousand times smaller than the radius of an atom. An atom the size of a football stadium would have a nucleus the size of a pea, but the density of nuclei is so great that that pea would have a mass of more than 250 million tons.

Nuclei are composed of two different fundamental particles, protons and neutrons. These nuclear particles are called **nucleons.** A proton has a charge of $+1.60218 \times 10^{-19}$ C and a mass of 1.672622×10^{-27} kg. A neutron has almost the same mass, 1.674927×10^{-27} kg, but is electrically neutral. Ionized H atoms, which form from H_2 in electrical discharges, are free protons. A free proton does not last long, because it quickly captures an electron to become a hydrogen atom. Free neutrons are often generated in the course of nuclear reactions, but they do not last long either, because they are easily captured again when they collide with nuclei. Table 21-1 summarizes the properties of these nuclear building blocks. The other fundamental atomic particle, the electron, is included for comparison.

Each particular type of nucleus is called a **nuclide.** Nuclides are characterized by the number of protons (Z) and neutrons (N) that they possess. The number of protons in a nuclide is always the same as the atomic number of the element, Z, but recall from Chapter 2 that the number of neutrons can vary. Two nuclides of an element with different numbers of neutrons are **isotopes.** Copper, for example, has two stable isotopes, one with 34 and one with 36 neutrons.

The mass number of a nuclide, A, is its total number of protons and neutrons: $A = Z + N$. Because protons and neutrons each have molar mass near 1 g/mol, A is always close to the numerical value of the molar mass of that isotope. For example, fluorine has a molar mass of 18.998 grams per mole and an A-value of 19. A particular nuclide can be described by its elemental symbol, X, preceded by the value of A as a superscript and that of Z as a subscript:

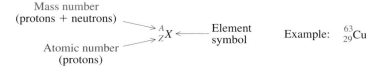

$$\underset{\text{(protons)}}{\underset{\text{Atomic number}}{\nearrow}} \underset{Z}{\overset{A}{X}} \overset{\nwarrow}{\underset{\text{Mass number}}{\text{(protons + neutrons)}}} \longleftarrow \underset{\text{symbol}}{\text{Element}} \qquad \text{Example:} \quad {}^{63}_{29}\text{Cu}$$

Examples include ${}^{63}_{29}\text{Cu}$, ${}^{4}_{2}\text{He}$, and ${}^{12}_{6}\text{C}$, whose nuclei are shown schematically in Figure 21-1. The elemental symbol identifies Z, so the subscript is often omitted

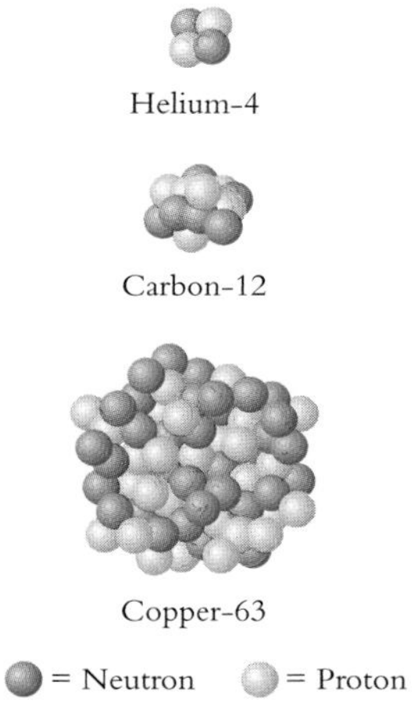

Helium-4

Carbon-12

Copper-63

● = Neutron   ○ = Proton

Figure 21-1
Each nucleus contains Z protons and N neutrons. Shown here in schematic fashion are the nuclei of ${}^{4}_{2}\text{He}$, ${}^{12}_{6}\text{C}$, and ${}^{63}_{29}\text{Cu}$.

Table 21-1
Properties of Fundamental Particles

Particle	Symbol	Charge (10^{-19} C)	Mass (10^{-27} kg)	Molar Mass (g/mol)
Proton	p	+1.60218	1.672622	1.007276
Neutron	n	0	1.674927	1.008665
Electron	e	−1.60218	0.000911	5.486×10^{-4}

(for example, ^{63}Cu). Alternatively, the name of the element is followed by its mass number, as in copper-63. Example 21-1 provides some practice in writing the symbols of nuclides.

| Example 21-1 | Nuclide Symbols |

Write the nuclear symbol for the following nuclides: (a) The one that contains 92 protons and 143 neutrons; and (b) the carbon isotope that has 8 neutrons.

Strategy: When determining symbols for nuclides, the key is to remember that the atomic number and number of protons are the same and that the mass number is the sum of the number of protons plus the number of neutrons.

Solution:

(a) The nuclide with 92 protons and 143 neutrons has an atomic number of 92 and a mass number that is the sum of the number of protons and neutrons: $A = 92 + 143 = 235$. Atomic number 92 corresponds to uranium, so the symbol for this nuclide is $^{235}_{92}$U.

(b) All carbon isotopes have the same atomic number, $Z = 6$. From the definition of A, $A = Z + N = 8 + 6 = 14$. The symbol for this nuclide is $^{14}_{6}$C.

Nuclear Binding Energy

Two positively charged particles repel each other, yet many nuclides with more than one proton are extremely stable. This stability arises from a nucleon–nucleon attraction called the *strong nuclear force*. This force is about 100 times stronger than proton–proton repulsion, but it operates only at the very small distances inside the nucleus. Because of this strong force, neutrons and protons are bound tightly in the nucleus. The energy required to remove a neutron or proton from a nucleus is called the **nuclear binding energy.**

Every nucleus contains Z protons and $(A - Z)$ neutrons, so we can visualize a nuclear "formation reaction" for any nuclide, in which protons and neutrons combine to form the product nucleus:

$$(Z)\text{p} + (A - Z)\text{n} \longrightarrow {}^{A}_{Z}X$$

> Protons and neutrons are symbolized formally as $^{1}_{1}$p and $^{1}_{0}$n, but they are often simplified to p and n.

Figure 21-2 illustrates this process schematically for fluorine:

$$9\text{p} + 10\text{n} \longrightarrow {}^{19}_{9}\text{F}$$

Because any stable nucleus is more stable than its separated nucleons, nuclear formation reactions of all stable nuclides are exothermic.

Figure 21-2
The nuclear formation reaction for $^{19}_{9}$F : $Z = 9, A - Z = 10.$

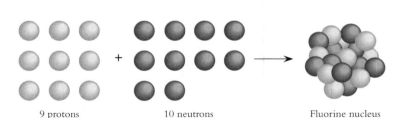

9 protons   10 neutrons   Fluorine nucleus

Formation reactions for nuclides are easy to visualize but impossible to carry out, so scientists must measure energies of nuclide formation indirectly. Accurate mass measurements are the best way to do this. Einstein recognized that mass is a form of stored energy. His famous equation links energy to mass and the speed of light, c:

$$E = mc^2 \qquad (21\text{-}1)$$

According to this equation, formation of a nucleus releases an amount of energy that is related to the change in mass accompanying the formation reaction:

$$\Delta E = (\Delta m)c^2 \qquad (21\text{-}2)$$

Equation 21-2 is the fundamental equation of nuclear energetics. It allows us to calculate the change in energy that accompanies any nuclear reaction from the masses of the particles before and after the transformation. When mass decreases, the energy change is negative. Recall that a negative energy change for a system means that energy is released to the surroundings. In other words, nuclear reactions in which mass decreases are exothermic.

Nuclide masses are ordinarily tabulated using units of grams per mole. Therefore to compute ΔE, it is useful to rewrite Equation 21-2 with energy in units of kilojoules per mole and the change in mass in grams per mole:

$$\Delta E = (\Delta m)c^2 = \Delta m (2.998 \times 10^8 \text{ m/s})^2 (10^{-3} \text{ kg/g})(10^{-3} \text{ kJ/J})$$

$$\Delta E = (\Delta m)(8.988 \times 10^{10}) \left(\frac{\text{kg m}^2 \text{ kJ}}{\text{s}^2 \text{ J g}} \right)$$

$$\Delta E = (\Delta m)(8.988 \times 10^{10} \text{ kJ/g}) \qquad (21\text{-}3)$$

$c = 2.998 \times 10^8$ m/s
$1\text{ J} = 1 \text{ kg m}^2/\text{s}^2$

The large positive exponent in Equation 21-3 indicates that a tiny change in mass results in a huge change in energy.

The change in mass that accompanies the formation of a nucleus can be found from the difference between the mass of the product nucleus and the masses of its component nucleons:

$$\Delta m = m_{\text{nucleus}} - [Z\, m_{\text{proton}} + (Z - A)m_{\text{neutron}}]$$

Once this change in mass is known, Equation 21-3 can be used to calculate the binding energy of any particular nuclide.

The tabulated molar mass of an element divided by the Avogadro constant is the *average* mass per atom of that element, but it is not the *exact* mass of an individual nucleus. There are two reasons for this. First, molar masses refer to neutral atoms. The tabulated molar mass of an element includes the mass of its electrons in addition to the mass of its nucleus. Consequently, the mass of Z electrons must be subtracted from the isotopic molar mass in computing the energy of formation of a nuclide. Second, molar masses of the elements are weighted averages of all naturally occurring isotopes of that element. As an example, the most abundant isotope of hydrogen, ^{1}H, has an isotopic molar mass of 1.007825 g/mol, but the presence of small amounts of the isotope ^{2}H makes the elemental molar mass of naturally occurring hydrogen slightly larger, 1.00794 g/mol. In making mass-energy conversions, we must work with isotopic molar masses rather than elemental molar masses.

SECTION 2.3 →
The use of mass spectrometry to measure isotopic masses with high accuracy is outlined in Section 2.3.

The isotopic molar masses of all stable and many unstable isotopes have been determined using mass spectrometry and can be found in standard data tables. We provide values as needed for these calculations. Example 21-2 illustrates the calculation of nuclear binding energies from isotopic molar masses.

| **Example 21-2** | **Nuclear Binding Energies** |

The most abundant isotope of helium has two neutrons and an isotopic molar mass of 4.00260 g/mol. The most abundant isotope of uranium is ^{238}U, with an isotopic molar mass of 238.0508 g/mol. Compute the nuclear binding energies of these nuclides.

Strategy: A particular nuclide is made from the combination of Z protons and $(A - Z)$ neutrons. Thus, a neutral atom of a specific isotope contains Z protons, Z electrons, and $(A - Z)$ neutrons. When these particles are brought together, a small amount of mass is converted to energy. To calculate that energy, first count protons, neutrons, and electrons, and then do a mass–energy calculation using Equation 21-3.

Solution:

The helium isotope has two neutrons, two electrons, and two protons:.

$$\text{Protons:} \quad 2(1.007276 \text{ g mol}^{-1}) = 2.014552 \text{ g mol}^{-1}$$

$$\text{Neutrons:} \quad 2(1.008665 \text{ g mol}^{-1}) = 2.017330 \text{ g mol}^{-1}$$

$$\text{Electrons:} \quad 2(0.0005486 \text{ g mol}^{-1}) = 0.0010972 \text{ g mol}^{-1}$$

$$\text{Total molar mass of components} = 4.032979 \text{ g mol}^{-1}$$

The actual mass of the helium isotope is slightly smaller than the mass of the component particles. The difference represents the amount of mass that is converted to energy when the nuclide forms:

$$\Delta m = (4.00260 \text{ g mol}^{-1}) - (4.032979 \text{ g mol}^{-1}) = -0.03038 \text{ g mol}^{-1}$$

Equation 21-3 lets us calculate an energy change from a mass change:

$$\Delta E = (-0.03038 \text{ g/mol}) (8.988 \times 10^{10} \text{ kJ/g}) = -2.731 \times 10^9 \text{ kJ/mol}$$

Notice that we carry all the significant figures for the masses. It is essential to do this, because a very small mass difference translates into an extremely large amount of energy. Even though the mass difference in this example is less than 1%, the energy difference is more than 10^9 kJ/mol.

The uranium nuclide has $A = 238$ and $Z = 92$. You should be able to verify that the molar mass of the component particles is 239.98496 g/mol, hence:

$$\Delta m = (238.0508 \text{ g/mol}) - (239.98496 \text{ g/mol}) = -1.9342 \text{ g/mol}$$

$$\Delta E = (-1.9342 \text{ g/mol})(8.988 \times 10^{10} \text{ kJ/g}) = -1.738 \times 10^{11} \text{ kJ/mol}$$

Both energy changes are negative, indicating that each nuclide is more stable than its separate component particles.

The calculations in Example 21-2 show that the binding energy of uranium is much larger than the binding energy of helium. As the mass number increases, so does the total binding energy of the nuclide. More important for the overall stability of the nucleus, however, is the binding energy *per nucleon*. This quantity is the total binding energy divided by the mass number A. It describes how tightly each nucleon is bound to the nucleus. As the binding energy per nucleon becomes more negative, nuclides become more stable. Figure 21-3 shows the binding energy per nucleon plotted as a function of mass number. Notice the broad minimum around $A = 60$, $Z = 26$, at a binding energy of -8.3×10^8 kJ/mol of nucleons. The most stable nuclide of all is $^{56}_{26}$Fe.

The variations in binding energy shown in Figure 21-3 indicate that there are two types of nuclear reactions that release energy. When heavy nuclides fragment in a process called **fission**, energy is released. When light nuclides combine in a process called **fusion**, energy is also released. We describe these two energy-releasing processes in Sections 21.4 and 21.5.

When the energy change accompanying a process is known, Equation 21-3 can be used to calculate how much the mass changes during the process. Example 21-3 shows how to do this for a chemical reaction.

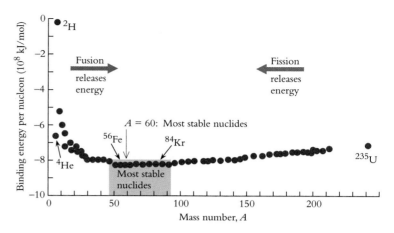

Figure 21-3
Plot of the binding energy per nucleon vs. mass number A. The most stable nuclides lie in the region around $A = 60$.

Mass Equivalence of Chemical Energy	**Example 21-3**

Hydrogen, a fuel that releases a large amount of chemical energy when it burns, is used as an energy source on the space shuttle. Use standard enthalpies of formation to calculate the change in mass that occurs when 1.00 mol of H_2 is burned.

Strategy: Whereas Example 21-2 addresses the energy change that accompanies a specific change in mass, Example 21-3 asks about the change in mass when a specific amount of energy is released in an exothermic chemical reaction. First determine ΔE for the chemical reaction and then use Equation 21-3.

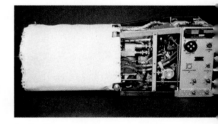

Fuel cell used in the space shuttle.

Solution: The combustion reaction per mole of hydrogen is as follows:

$$H_2(g) + \tfrac{1}{2} O_2(g) \longrightarrow H_2O(g)$$

This reaction also is the formation reaction of gaseous H_2O. Its ΔH_f° is listed in Appendix D. Enthalpy changes and energy changes are nearly the same, so $\Delta E \cong \Delta H$:

$$\Delta E \cong \Delta H = \Delta H_f^\circ = -242 \text{ kJ/mol}$$

To obtain the mass change, substitute this value into Equation 21-3:

$$\Delta E = (8.988 \times 10^{10} \text{ kJ/g}) \, \Delta m$$

$$-242 \text{ kJ/mol} = (8.988 \times 10^{10} \text{ kJ g}^{-1}) \, \Delta m$$

$$\Delta m = -2.69 \times 10^{-9} \text{ g/mol}$$

When 1.00 mole of hydrogen reacts with oxygen, a few nanograms are converted to energy. This amount, which is typical of the mass consumed in conventional chemical reactions, is too small to detect.

The binding energy of a typical nuclide is about 8×10^8 kJ/mol of nucleons. To obtain a better feel for just how immense this quantity is, compare the energy released in a nuclear reaction with the energy released in a chemical reaction. As noted in Example 21-3, the chemical reaction of H_2 with O_2 is highly exothermic. When 1.00 mol of H_2 reacts chemically with O_2, 242 kJ of energy is released. If that same 1.00 mole of H_2 could be fused with neutrons to form He nuclei, the calculation of Example 21-2 indicates that 2.73×10^9 kJ of energy would be released. There is a difference of seven orders of magnitude between these two quantities; in other words, *fusion* of hydrogen nuclei releases 10,000,000 times as much energy as *combustion* of molecular hydrogen.

Energy Barriers

If fission of large nuclides and fusion of small nuclides are vastly exothermic, why have these processes not occurred over time, converting all elements into the most stable one, iron-56? The reason is that nuclear reactions have huge energy barriers. These barriers are analogous to the activation energy barriers for conventional combustion reactions. The combustion of gasoline, for example, is highly exothermic ($\Delta H > -5000$ kJ/mol), but the rate of the reaction is negligible at room temperature because an activation energy barrier prevents combustion from occurring without an external boost.

Massive nuclides do not fragment into lighter, more stable nuclides because an immense amount of energy must be supplied to overcome strong attractions among nucleons and pull a nucleus apart. Light nuclides do not fuse to give more massive, more stable nuclides because an immense amount of energy must be supplied to overcome coulombic repulsions and bring nuclei together. The following analysis shows the magnitudes of typical activation energies for nuclear reactions.

Nuclei cannot fuse without overcoming the repulsive electrical forces between them. Recall that this repulsion prevents the nuclei in molecules from approaching closer than the lengths of chemical bonds, which are about 100 pm. For nuclei to fuse, however, they must be brought within one nuclear radius of each other, which is about 10^{-3} pm. Coulomb's law allows calculation of the magnitude of this barrier. Because isotopes of hydrogen have the smallest possible nuclear charge ($+1$), they also have the minimal energy barrier to fusion. Consider, for example, the fusion of two deuterium nuclei:

$$^2_1H + {}^2_1H \longrightarrow {}^4_2He$$

Equation 7-1 describes coulombic energy. For a pair of nuclei, the charges q_1 and q_2 are equal to the nuclear charges Z_1 and Z_2:

$$E_{coulomb} = \frac{(1.389 \times 10^5 \text{ kJ pm/mol})(Z_1)(Z_2)}{d} \tag{7-1}$$

For fusion of deuterium nuclei, Z_1 and Z_2 are each $+1$, and d is the sum of the two nuclear radii, 2.8×10^{-3} pm. Substituting these values into Equation 7-1 gives $E = 5.0 \times 10^7$ kJ/mol. This is small relative to the energy released by fusion but very large relative to typical chemical activation energies of about 100 kJ/mol. The graph in Figure 21-4 is a schematic representation of the energy barriers for fusion of deuterium.

Activation energy barriers for nuclear reactions are more than a million times larger than the activation energies of conventional chemical reactions, so nuclear processes require immense energy inputs before they can occur. As one example, deuterium nuclei cannot fuse unless they first acquire 5×10^7 kJ/mol of energy. At room temperature, the average kinetic energy of deuterium nuclei is only 3.7 kJ/mol. A sample of deuterium needs only the energy boost from a spark or flame to react chemically with oxygen, but the temperature must be raised to 5×10^6 K before deuterium nuclei acquire enough kinetic energy to overcome the barrier to fusion.

The energy profile in Figure 21-4 also indicates that nuclei cannot eject nucleons without overcoming the forces responsible for binding nucleons together. Figure 21-3 shows that, for heavier nuclei, these forces generate binding energies on the order of 10^8 kJ/mol. For example, energy greater than 7×10^8 kJ/mol must be provided to remove a neutron from a lead nucleus. The combination of strong binding forces holding nucleons in a nucleus and high coulombic repulsive forces keeping nuclei apart accounts for the stability of all the elements.

The only nuclear reaction that has no energy barrier is the capture of a free neutron by a nucleus. Free neutrons, being uncharged, are not repelled by the positive charge of a nucleus. Thus they can penetrate close enough to the nucleus to be captured by the strong nuclear attractive forces. Neutron capture is nevertheless a low-probability event for three reasons. First, free neutrons are not normally available; they must be generated by nuclear reactions, as we describe in Sections 21.3 and 21.4. Second, nuclei are so tiny that they present a very small cross-sectional area to an incoming neutron; most neutrons pass through a sample of matter without encountering a nucleus. Third, capture of a neutron releases a large amount of energy that must be transferred out of the nucleus for capture to occur; most neutrons that penetrate a nucleus escape again before this energy loss can occur.

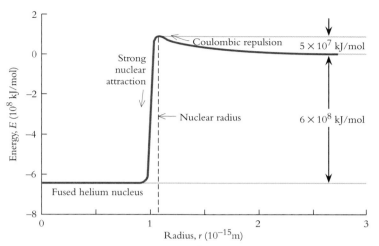

Figure 21-4
The energy profile for nuclear fusion between two deuterium nuclides.

Stable Nuclides

Figure 21-3 indicates that only those nuclides with mass numbers around 60 are stable relative to either lighter or heavier nuclides. Certain combinations of protons and neutrons are stable for indefinite times, but others undergo spontaneous reactions. For example, all hydrogen nuclides with $A > 2$ are so unstable that only one of them—tritium, ^3_1H—exists even briefly. If other nuclides of hydrogen could be made, they would decompose rapidly by expelling neutrons. Fluorine has just one stable nuclide, $^{19}_9\text{F}$, but tin has ten, with mass numbers 112, 114, 115, 116, 117, 118, 119, 120, 122, and 124. All stable nuclides, as well as a few that are unstable, are found in the Earth's crust and atmosphere.

Figure 21-5 shows all stable nuclides on a plot of the number of neutrons (N) vs. the number of protons (Z). These data reveal a striking pattern: All stable nuclides fall within a "belt of stability." Lighter nuclides lie along the $N = Z$ line, but as the mass of the nuclide increases, the $N : Z$ ratio rises slowly until it reaches 1.54. The trend is illustrated by the $N : Z$ ratios of the following four nuclides: $^{19}_9\text{F}$, 1.11; $^{93}_{41}\text{Nb}$, 1.27; $^{159}_{65}\text{Tb}$, 1.45; $^{209}_{83}\text{Bi}$, 1.54. Any nuclide whose ratio of neutrons to protons falls outside the belt of stability is unstable and decomposes spontaneously.

Unstable nuclides occur naturally either because their natural lifetimes are longer than the age of the Earth or because they are produced in naturally occurring nuclear reactions.

Figure 21-5
Plot of the *Z* and *N* values of stable nuclides. The stable nuclides fall along a belt that lies between *N* = *Z* and *N* = 1.54 *Z*.

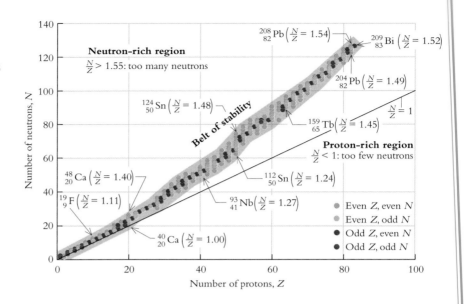

The relationships that generate this delicate balance of nuclear stability are quite complex, but they can be summarized qualitatively. Compressing more than one positively charged proton into a volume as small as the nucleus leads to strongly repulsive electrical forces that exceed the strong nuclear force of attraction, making the nucleus unstable unless neutrons are also present. Adding neutrons provides additional strong nuclear binding forces without adding electrical repulsive forces. When sufficient numbers of neutrons are present, there is enough extra binding force to hold the protons within the nuclear volume. By themselves, however, neutrons are not stable: A free neutron transforms into a proton and an electron. This fundamental instability of neutrons destabilizes a nuclide that contains too many neutrons.

Additional patterns can be observed within the region of nuclear stability. Table 21-2 shows, for example, that nuclides with even numbers of protons and neutrons are more prevalent than those with odd numbers of protons or neutrons. Almost 60% of all stable nuclides have even numbers of protons and neutrons, whereas less than 2% have odd numbers of both. Moreover, of the five stable odd-odd nuclides, four are the lightest odd-*Z* elements: $^2_1$H, $^6_3$Li, $^{10}_5$B, $^{14}_7$N. Above *Z* = 7, there are 152 stable even–even nuclides and just one, $^{158}_{57}$La, that is odd-odd.

Notice in Figure 21-5 that no nuclides above bismuth, *Z* = 83, are stable. Some elements with higher *Z* are found on the Earth, notably radium (*Z* = 88),

Table 21-2
Distribution of Stable Nuclides

Protons	Neutrons	Stable Nuclides	%
Even	Even	154	58.8
Even	Odd	53	20.2
Odd	Even	50	19.1
Odd	Odd	5	1.9

thorium ($Z = 90$), and uranium ($Z = 92$), but all such elements are unstable and eventually disintegrate into nuclides with $Z < 83$. The set of stable nuclides, those that make up the world of "normal" chemistry and provide the material for all terrestrial chemical reactions, is a small subset of all possible nuclides. As described in Section 21.5, nuclear stability in the superheated interior of a star is quite different than on Earth, and many nuclides that are stable at terrestrial temperature undergo nuclear reactions at extremely high temperatures.

Section Exercises

21.1.1 Write the symbols and determine Z, N, and A for the following nuclides: (a) a neon nucleus with the same number of neutrons and protons; (b) element 43 with 55 neutrons; and (c) the unstable nuclide of hydrogen with the lowest mass.

21.1.2 Predict whether each of the following is stable or unstable. If you predict that it is unstable, give your reason: (a) the nuclide with 94 protons and 150 neutrons; (b) the iodine nuclide with 73 neutrons; (c) $^{154}_{64}$Gd; and (d) the oxygen nuclide with 6 neutrons.

21.1.3 Fluorine has only one stable isotope, $^{19}_{9}$F Compute the total binding energy and binding energy per nucleon for this nuclide.

21.2 NUCLEAR DECAY

Unstable nuclides decompose spontaneously into other, more stable nuclides. These decompositions are called **nuclear decay,** and unstable nuclides are called **radioactive.** Three features characterize nuclear decays: the products, the energy released, and the rate of decay. The products of nuclear decomposition include electrons, helium nuclei, and high-energy photons. Nuclear decay releases energy because the decay products are more stable than the starting materials. Furthermore, each unstable nuclide decomposes at some particular rate, which can take anywhere from less than a second to many millions of years.

Nuclear transformations always obey two fundamental conservation laws:

/// *Mass number is conserved.*
/// *Electrical charge is conserved.*

That is, the sum of the number of protons and neutrons is the same after transformation as before, and the sum of the charges also is the same after transformation as before. Any description of a nuclear reaction must take these conservation requirements into account.

As an example, consider the decay of free neutrons. A neutron has $A = 1$, so its decay products must also have $A = 1$. The only other particle with $A = 1$ is a proton, so the decay must produce a proton. The neutron has zero charge, so the sum of the charges of its decay products must also be zero. Because the proton carries a $+1$ charge, another particle with a -1 charge is required. This particle must have $A = 0$ to ensure that the mass number is conserved. The only particle with these properties is the electron. Thus there is only one possible decay process for neutrons:

$$^{1}_{0}\text{n} \longrightarrow {}^{1}_{1}\text{p} + {}^{0}_{-1}\text{e}$$

Although mass *number* is conserved, total mass is *not* conserved in nuclear reactions. Instead, some mass is converted into energy, or some energy is converted into mass. The notion of conservation of mass, introduced in Chapter 2, is valid within experimental error for all chemical transformations, as Example 21-3 illustrates. For nuclear processes, however, the amounts of energy produced or consumed are large enough to generate measurable mass changes. In nuclear decay, some mass is always converted into energy. We can use Equation 21-3, the mass of the decaying nucleus, and the masses of its products to calculate how much energy is released.

This reasoning is easily applied to the decay of a free neutron. The masses of the three participants are given in Table 21-1:

Neutron: 1.008665 g/mol   Proton: 1.007276 g/mol
Electron: 0.0005486 g/mol

$$\Delta m = 1.007276 + 0.0005486 - 1.008665 = -0.000840 \text{ g/mol}$$

$$\Delta E = (-0.000840 \text{ g/mol})(8.988 \times 10^{10} \text{ kJ/g}) = -7.55 \times 10^{10} \text{ kJ/mol}$$

To summarize, the equation for a nuclear reaction is balanced when the total *charge* and total *mass number* of the products equals the total charge and total mass number of the reactants. This conservation requirement is one reason why the symbol for any nuclide includes its charge number (Z) as a subscript and its mass number as a superscript. These features provide a convenient way to keep track of charge and mass balances. Notice that in the equation for neutron decay, the sum of the subscripts for reactants equals the sum of the subscripts for products. Likewise, the sum of the superscripts for reactants equals the sum of the superscripts for products. We demonstrate how to balance equations for other reactions as they are introduced.

Decay Processes

There are five fundamental types of nuclear decay process, as listed in Table 21-3. The particular decay process for an unstable nuclide depends on the reason for its instability. Figure 21-6, an expanded region of the belt of stability from Figure 21-5, diagrams how nuclear decays affect N and Z.

Nuclides that are too massive to be stable lose mass by emitting energetic helium nuclei, reducing A by 4 units and Z by 2 units. Energetic helium nuclei are called α-*particles,* so these nuclides undergo **α-emission.** Here are some examples:

$$^{226}_{88}\text{Ra} \longrightarrow {}^{222}_{86}\text{Rn} + {}^{4}_{2}\alpha$$

Ernest Rutherford observed that the paths taken by energetic particles emitted by radioactive uranium and thorium responded in three ways to magnetic fields: slightly bent, strongly bent, and unaffected. He gave them the designations α, β, and γ. Even though scientists soon identified the particles, they still use these names to emphasize that they are nuclear decay products.

Table 21-3
Types of Nuclear Decay

Type of Decay	Type of Emitter	Emission	ΔZ of Emitter	ΔA of Emitter
α-emission	$Z > 83$	$\alpha = {}^{4}_{2}\text{He}$	-2	-4
β-emission	N/Z too large	$\beta = {}^{0}_{-1}\text{e}$	$+1$	0
β^{+} emission	N/Z too small	$\beta^{+} = {}^{0}_{+1}\text{e}$	-1	0
Electron capture	N/Z too small	X-ray photon	-1	0
γ-emission	Excited nucleus	γ-ray photon	0	0

$$\,^{222}_{86}\text{Rn} \longrightarrow \,^{218}_{84}\text{Po} + \,^{4}_{2}\alpha$$

$$\,^{218}_{84}\text{Po} \longrightarrow \,^{214}_{82}\text{Pb} + \,^{4}_{2}\alpha$$

Nuclides that lie *above* the belt of stability have ratios of neutrons to protons that are too high. To become stable, these nuclides need to increase their nuclear charges. Such a nuclide converts a neutron into a proton plus an energetic electron. The nucleus expels the electron, thereby increasing its charge by 1 unit while leaving its mass number unchanged. Energetic electrons are called *β-particles,* so these nuclides undergo **β-emission.** Here are some examples:

$$\,^{3}_{1}\text{H} \longrightarrow \,^{3}_{2}\text{He} + \,^{0}_{-1}\beta$$

$$\,^{59}_{26}\text{Fe} \longrightarrow \,^{59}_{27}\text{Co} + \,^{0}_{-1}\beta$$

$$\,^{208}_{79}\text{Au} \longrightarrow \,^{208}_{80}\text{Hg} + \,^{0}_{-1}\beta$$

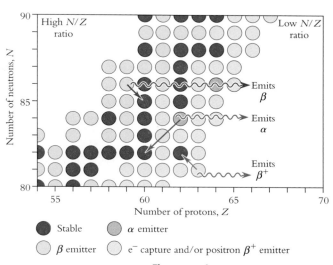

Stable ● α emitter ○

β emitter ○ e⁻ capture and/or positron β⁺ emitter ○

Figure 21-6
A detailed view of one portion of the *N* vs. *Z* plot of nuclides, illustrating the modes of nuclear decay for nuclides on either side of the belt of stability.

In β-emission, a neutron converts to a proton, but the total number of nucleons remains the same. Because the product nucleus has one new proton, its positive charge increases by +1, but the overall charge remains balanced by the −1 charge carried away by the β-particle.

Nuclides that lie *below* the belt of stability have low neutron–proton ratios and must reduce their nuclear charges to become stable. These nuclides can convert protons into neutrons by **positron emission.** Positrons (symbolized β^+) are particles with the same mass as electrons but with a charge of +1 instead of −1. Here are two examples:

$$\,^{18}_{9}\text{F} \longrightarrow \,^{18}_{8}\text{O} + \,^{0}_{+1}\beta^+$$

$$\,^{52}_{26}\text{Fe} \longrightarrow \,^{52}_{25}\text{Mn} + \,^{0}_{+1}\beta^+$$

Positrons cannot be observed directly because, as Figure 21-7 illustrates, when a positron encounters an electron, the two particles annihilate each other, converting their entire mass into a pair of photons. The occurrence of positron emission can be inferred from the observation of such a pair of photons. Each photon produced in this process has a specific energy: $E_{\text{photon}} = 9.87 \times 10^7$ kJ/mol. Photons with such high energy are called **γ-rays.**

A nuclide with a low neutron–proton ratio can also reduce its nuclear charge by capturing one of its 1*s* orbital electrons in a process called **electron capture.** The captured electron combines with a proton to give a neutron, so *Z* drops by one unit, while *A* remains fixed. For example:

$$\,^{26}_{13}\text{Al} + \,^{0}_{-1}\text{e} \longrightarrow \,^{26}_{12}\text{Mg}$$

Like positron emission, electron capture is never observed directly. However, after electron capture, the product atom is missing one of its 1*s* electrons, as shown

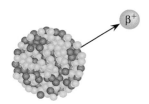

Step 1: Positron emission

Step 2: Annihilation

Figure 21-7
Positron emission can be observed only indirectly. The "signature" of positron emission is two γ-rays generated by the annihilation of a positron and an electron.

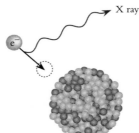

X ray

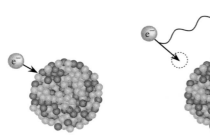

Step 1: Electron capture from 1s orbital

Step 2: Electronic transition from valence orbital to 1s orbital

Figure 21-8
Electron capture can be observed only indirectly. The "signature" of electron capture is an X ray emitted when an electron undergoes a transition from an outer to an inner orbital.

The electron captured by a nucleus during electron capture is symbolized e, not β, because β is reserved for electrons ejected during nuclear decay.

schematically in Figure 21-8. When an electron from an outer orbital occupies this vacancy in the 1s orbital, a photon is emitted whose energy falls in the X-ray region of the spectrum ($E \cong 10^6$ kJ/mol).

Nuclides that are unstable because they have odd–odd composition can be converted to stable nuclides with even–even composition in any of three processes: electron emission, positron emission, or electron capture. Each process changes Z and N by one unit. The $^{64}_{29}Cu$ nuclide provides a convenient illustration of all three processes because it decays by all three modes:

$$^{64}_{29}Cu \longrightarrow {}^{64}_{30}Zn + {}^{0}_{-1}\beta$$

$$^{64}_{29}Cu \longrightarrow {}^{64}_{28}Ni + {}^{0}_{+1}\beta^+$$

$$^{64}_{29}Cu + {}^{0}_{-1}e \longrightarrow {}^{64}_{28}Ni$$

It is unusual for a nuclide to decay by three different paths. Generally, an odd–odd nuclide decays preferentially by only one mode. However, predicting which mode of decay predominates is beyond the scope of this book.

In addition to the γ-rays that result from the annihilation of a positron and an electron, γ-rays are also emitted directly by many nuclides. The emission of a γ-ray from an unstable nuclide changes neither Z nor A, because γ-rays are photons with zero mass and zero charge. Many nuclear reactions generate γ-rays because of the release of energy that accompanies these reactions. Although most of this energy appears as kinetic energy of the products, some energy may be retained by the product nuclide as excitation energy. The excited nuclide can "shed" this excess energy by emitting a γ-ray, just as an excited atom can lose excess energy by emitting a photon. Excited nuclides (also called metastable nuclides) are identified with a superscript ^{m} after the value of A. For example, α-emission from radium-226 gives an excited radon nuclide, ^{222m}Rn, which loses its excess energy by emitting a γ-ray:

$$^{226}Ra \longrightarrow {}^{222m}Rn + \alpha$$

$$^{222m}Rn \longrightarrow {}^{222}Rn + \gamma$$

This decay sequence illustrates a common feature of nuclear decays of very heavy elements: several decays often occur in sequence, because the product of the initial decay is also unstable. As another example, notice that for the examples of α-decay given earlier, the product of the first reaction is the starting material for the second, and the product of the second is the starting material for the third. Example 21-4 traces another example of nuclear decay that occurs through a sequence of reactions.

| Example 21-4 | **Decay Sequences** |

Radon-222 is an unstable nuclide that has been detected in the air of some homes. Its presence is a concern because of high health hazards associated with exposure to its radioactivity. Radon-222 transmutes to a stable nuclide by emitting α- and β-particles in the following sequence: $\alpha,\alpha,\beta,\beta,\alpha,\beta,\beta,\alpha$. Write the sequence of nuclear reactions and identify the final product.

Decay Sequences *(continued)*	**Example 21-4**

Strategy: Charge number and mass number must be conserved in each reaction. Thus each α-particle decreases the nuclear charge by two units and the mass number by four units. Similarly, each β-emission increases the nuclear charge by one unit but leaves the mass number unchanged. Consult a periodic table to identify the elemental symbol of each product nuclide.

Solution:

The starting nuclide is ^{222}Rn, which, according to the decay sequence, emits an α-particle:

$$^{222}_{86}\text{Rn} \longrightarrow {}^{A}_{Z}\text{X} + {}^{4}_{2}\alpha$$

The product has $A = 222 - 4 = 218$ and $Z = 86 - 2 = 84$. Element 84 is Po, so radon–222 decomposes to give an α-particle and polonium–218:

$$^{222}_{86}\text{Rn} \longrightarrow {}^{4}_{2}\alpha + {}^{218}_{84}\text{Po}$$

The polonium nucleus is unstable, and the sequence indicates that it decays via α-emission:

$$^{218}_{84}\text{Po} \longrightarrow {}^{4}_{2}\alpha + {}^{214}_{82}\text{Pb}$$

The decay sequence indicates that lead–214 is unstable as well. This nuclide converts a neutron to a proton by β-emission:

$$^{214}_{82}\text{Pb} \longrightarrow {}^{A}_{Z}\text{X} + {}^{0}_{-1}\beta$$

For this product, $A = 214 - 0 = 214$ and $Z = 82 - (-1) = 83$. The product nucleus is element 83, bismuth:

$$^{214}_{82}\text{Pb} \longrightarrow {}^{214}_{83}\text{Bi} + {}^{0}_{-1}\beta$$

The fourth decay in the sequence is also β-emission, which produces another isotope of element 84, polonium–214:

$$^{214}_{83}\text{Bi} \longrightarrow {}^{214}_{84}\text{Po} + {}^{0}_{-1}\beta$$

The decay sequence continues with another α, two more βs, and one last α. We leave it to you to determine the correct equations for these reactions. Remember that the product of each reaction is the nuclear reactant for the succeeding reaction. The final, stable product is $^{206}_{82}$Pb.

Rates of Nuclear Decay

A sample of any unstable nuclide undergoes nuclear decay continuously as its individual nuclei undergo reaction. All nuclear decays obey the first-order rate law: Rate $= kc$. This rate law can be treated mathematically to give Equation 14–3, which relates concentration, c, to time, t, for a first-order process (c_0 is the concentration present at $t = 0$):

$$\ln\left(\frac{c_0}{c}\right) = kt \tag{14-3}$$

Table 21-4
Half-Lives of Representative Nuclides

Nuclide	Mode	Half-Life
^{214}Po	α	1.6×10^{-4} s
^{210}Tl	β	1.32 min
^{239}U	β	24 min
^{60}Co	β, γ	5.26 yr
^{3}H	β	12.3 yr
^{90}Sr	β	28.1 yr
^{14}C	β	5.73×10^3 yr
^{235}U	α	7.0×10^8 yr
^{40}K	β	1.28×10^9 yr
^{238}U	α	4.51×10^9 yr

Recall from our discussion of kinetics in Chapter 14 that elementary unimolecular reactions are first-order. The observed first-order kinetics of nuclear decays indicates that each nuclear decay mode is elementary and "uninuclear."

Recall also from Chapter 14 that for first-order reactions, the time required for exactly half of the substance to react is independent of how much material is present. This constant time interval is the half-life, $t_{1/2}$. Equation 14-4 relates the half-life to the reaction rate constant:

$$t_{1/2} = \frac{\ln 2}{k} \tag{14-4}$$

By convention, nuclear decay rates are expressed using half-lives rather than rate constants. Every unstable nuclide has its own characteristic half-life. Nuclide half-lives range from shorter than a second to longer than a billion years. The half-lives of some representative nuclides appear in Table 21-4.

The particles emitted during nuclear decay are so energetic that it is possible to count them individually. For this reason, the rate equations for nuclear decay are often expressed using the number of nuclei present, N, rather than molar concentrations:

> Equations 21-4 and 21-5 can be derived from the first-order kinetic equations presented in Chapter 14 by making appropriate algebraic substitutions.

$$\text{Rate} = \frac{\Delta N}{\Delta t} = \frac{-N \ln 2}{t_{1/2}} \tag{21-4}$$

$$\ln\left(\frac{N_0}{N}\right) = \frac{t \ln 2}{t_{1/2}} \tag{21-5}$$

In these equations, N is the number of nuclei present at time t, and N_0 is the number of nuclei present at $t = 0$.

As shown in Example 21-5, Equation 21-4 is used to find a nuclear half-life from measurements of nuclear decays. Equation 21-5 is used to find how much of a radioactive substance will remain after a certain time, or how long it will take for the amount of substance to fall by a given amount. Example 21-6 provides an illustration of this type of calculation. In Section 21.7, we show that these equations also provide a way to determine the age of a material that contains radioactive nuclides.

| **Nuclear Half-Life** | **Example 21-5** |

Plutonium is a synthetic element used in nuclear weapons and proposed for use as a nuclear fuel. A sample of ^{239}Pu whose mass is 1.00 mg decays at a rate of 2.3×10^6 counts/s. (Each count corresponds to the decay of one nucleus.) What is the half-life of this isotope?

Strategy: This is a quantitative problem, so our standard seven-step approach is appropriate.

Solving Quantitative Problems

Solution:

1. We are asked to determine the half-life of an isotope from information about its decay rate.

2. and **3.** A simple block diagram helps visualize the process and summarize the data:

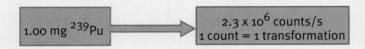

4. and **5.** The emission rate is $-\Delta N/\Delta t$, so rearrange Equation 21-4 to calculate $t_{1/2}$:

$$\frac{\Delta N}{\Delta t} = \frac{-N \ln 2}{t_{1/2}} \qquad or \qquad t_{1/2} = \frac{N \ln 2}{(-\Delta N/\Delta t)}$$

The number of nuclei can be calculated from the mass of the sample and the molar mass of the isotope. Because the molar mass of an isotope is nearly equal to its mass number, we can use the mass number without introducing significant error:

$$N = \text{Number of nuclei} = \frac{(\text{Mass})(N_A)}{(\text{Isotopic molar mass})}$$

$$N = \frac{(1.00 \times 10^{-3} \text{ g})(6.022 \times 10^{23} \text{ nuclei/mol})}{(239 \text{ g/mol})} = 2.52 \times 10^{18} \text{ nuclei}$$

6. Now we can substitute to determine the half-life:

$$t_{1/2} = \frac{(2.52 \times 10^{18} \text{ nuclei})(0.693)}{(2.3 \times 10^6 \text{ nuclei/s})} = 7.6 \times 10^{11} \text{ s}$$

This half-life is easier to interpret if expressed in years rather than seconds:

$$t_{1/2} = (7.6 \times 10^{11} \text{ s}) \left(\frac{1 \text{ min}}{60 \text{ s}}\right)\left(\frac{1 \text{ hr}}{60 \text{ min}}\right)\left(\frac{1 \text{ day}}{24 \text{ hr}}\right)\left(\frac{1 \text{ year}}{365 \text{ day}}\right)$$

$$= 2.4 \times 10^4 \text{ years.}$$

7. A half-life of many years is reasonable, given that isotopic half-lives vary from seconds or less to millions of years. It takes 24,000 years for half of the plutonium nuclei in any sample of ^{239}Pu to decay.

| Example 21-6 | Radioactive Decay |

One of the problems with radioactive nuclides such as ^{239}Pu is that their decay cannot be stopped, so any sample of the nuclide continues to emit dangerous amounts of radiation until most of it has decayed. Use the result of Example 21-5 and assume that the radiation level from ^{239}Pu is no longer a hazard when 99% of it has decayed. How long will this take?

Strategy: Again, it is appropriate to apply the seven-step strategy, but this time we do not number the steps explicitly.

Solution: The question asks for the time it takes for 99% of a sample of plutonium to decay. The half-life is known from the previous Example. Equation 21-5 relates the ratio N_0/N to time and the half-life for decay. This equation can be solved for t, the time at which the ratio reaches the desired value. First, Equation 21-5 must be rearranged to give an equation for t:

$$\ln\left(\frac{N_0}{N}\right) = \frac{t \ln 2}{t_{1/2}} \quad or \quad t = \frac{t_{1/2}\ln(N_0/N)}{\ln 2}$$

When 99% of the Pu has decayed, 1% of the original amount, in other words $0.01N_0$, remains. Use this information to evaluate the ratio in the ln term:

$$\left(\frac{N_0}{N}\right) = \frac{N_0}{0.01N_0} = 100$$

Now substitute values and solve for t:

$$t = \frac{(2.4 \times 10^4 \text{ years})(\ln 100)}{(\ln 2)} = 1.6 \times 10^5 \text{ years}$$

It will be a very long time, 160,000 years, before the plutonium that is stockpiled for use as a nuclear fuel decays sufficiently that it no longer is a potential hazard.

Section Exercises

21.2.1 Write the modes of decay that describe the following reactions: (a) ^{213}Bi to ^{213}Po; (b) ^{213}Bi to ^{209}Tl; (c) ^{213m}Bi to ^{213}Bi; and (d) ^{207}Bi to ^{207}Pb.

21.2.2 Identify the nuclides that decay in the following manner:
(a) A nuclide undergoes β- and γ- decay to give $Z = 58$ and $A = 140$.
(b) A nuclide undergoes α-decay to give polonium-218.
(c) A nuclide captures an orbital electron to give tellurium with 73 neutrons.

21.2.3 Radioisotopes are used in many research applications. Because they decay continuously, the shelf life of a radioisotope is limited. An isotope used for bone marrow scanning is ^{111}In, $t_{1/2} = 2.8$ days. Within what time must this isotope be used if it is effective down to 5% of its initial activity?

21.3 INDUCED NUCLEAR REACTIONS

Stable nuclides remain the same indefinitely, whereas unstable nuclides disintegrate continuously. In time, therefore, every element should be composed entirely of stable isotopes. On Earth, most elements have no naturally occurring unstable isotopes. With two exceptions, the unstable nuclides that are present either have half-lives longer than the age of the Earth (U, Ra, Th) or are products of the decays of these long-lived nuclides (Rn, Po). The exceptions are ^{3}H ($t_{1/2} = 12.3$ years) and ^{14}C ($t_{1/2} = 5730$ years). These half-lives are very short compared with the age of the Earth, so any ^{3}H and ^{14}C present during the Earth's formation decayed long ago. Thus the ^{3}H and ^{14}C on Earth today must come from ongoing nuclear reactions. These nuclides are not formed in any nuclear-decay schemes. Instead, each forms through a binuclear reaction in which a nuclear projectile collides with another nucleus, inducing a reaction. Such reactions, which are called **induced nuclear reactions,** are categorized according to the nature of the nuclear projectile that induces the reaction.

Neutron-Capture Reactions

One way to create unstable nuclides is by neutron capture. The Earth's atmosphere is exposed continuously to radiation from the sun. One component of solar radiation is neutrons. The most abundant nuclide in the atmosphere, nitrogen-14, can capture a neutron to form the unstable nuclide ^{15m}N. This nucleus rapidly ejects a proton, producing carbon-14:

$$^{14}_{7}\text{N} + ^{1}_{0}\text{n} \longrightarrow (^{15m}_{7}\text{N}) \longrightarrow ^{14}_{6}\text{C} + ^{1}_{1}\text{p}$$

Although carbon-14 decays via β-emission with $t_{1/2} = 5730$ years, it is replenished continuously by this neutron-capture reaction. We show in Section 21.7 how this isotope is used to estimate the age of carbon-containing artifacts.

The other short-lived nuclide in the atmosphere, ^{3}H, is also produced by disintegration of metastable ^{15}N nuclei. When nitrogen-14 captures a very energetic neutron, the metastable nucleus has sufficient excess energy that it fragments into ^{12}C and ^{3}H rather than ^{14}C and a proton:

$$^{14}_{7}\text{N} + ^{1}_{0}\text{n} \longrightarrow (^{15m}_{7}\text{N}) \longrightarrow ^{12}_{6}\text{C} + ^{3}_{1}\text{H}$$

Neutrons are electrically neutral, so there is no force of electrical repulsion hindering them from penetrating a nucleus. Consequently, almost every nuclide undergoes neutron capture if a source of neutrons is available. Unstable nuclides used in radiochemical applications are manufactured by neutron bombardment. A sample containing a suitable target nucleus is exposed to neutrons coming from a nuclear reactor. When a target nucleus captures a neutron, its mass number increases by one:

SECTION 21.4
Nuclear reactors are described in Section 21.4.

$$^{A}_{Z}\text{X} + ^{1}_{0}\text{n} \longrightarrow ^{(A+1)m}_{Z}\text{X}$$

Neutron capture always is exothermic, because the neutron is attracted to the nucleus by the strong nuclear force. Consequently, neutron capture generates a product nuclide in a metastable, excited state. These excited nuclei typically lose energy by emitting either γ-rays or protons:

$$^{(A+1)m}_{Z}\text{X} \longrightarrow ^{(A+1)}_{Z}\text{X} + \gamma$$

$$^{(A+1)m}_{Z}\text{X} \longrightarrow ^{A}_{(Z-1)}\text{Y} + ^{1}_{1}\text{p}$$

The following are specific examples of each type of decay:

$$^{98}_{42}\text{Mo} + ^{1}_{0}\text{n} \longrightarrow ^{99}_{42}\text{Mo} + \gamma$$

$$^{207}_{79}\text{Au} + ^{1}_{0}\text{n} \longrightarrow ^{208}_{79}\text{Au} + \gamma$$

$$^{14}_{7}\text{N} + ^{1}_{0}\text{n} \longrightarrow ^{14}_{6}\text{C} + ^{1}_{1}\text{p}$$

$$^{3}_{2}\text{He} + ^{1}_{0}\text{n} \longrightarrow ^{3}_{1}\text{H} + ^{1}_{1}\text{p}$$

One of the most important isotopes in nuclear medicine, $^{99\text{m}}\text{Tc}$, is produced by bombarding molybdenum with neutrons. The initial product, ^{99}Mo, has a high ratio of neutrons to protons, so it decomposes by releasing a β-particle. The product is an excited technetium nuclide that emits a γ-ray:

$$^{98}_{42}\text{Mo} + ^{1}_{0}\text{n} \longrightarrow ^{99}_{42}\text{Mo} + \gamma$$

$$^{99}_{42}\text{Mo} \longrightarrow ^{99\text{m}}_{43}\text{Tc} + ^{0}_{-1}\beta$$

$$^{99\text{m}}_{43}\text{Tc} \longrightarrow ^{99}_{43}\text{Tc} + \gamma$$

The product of this three-step nuclear process, ^{99}Tc, is also unstable but lasts much longer than its predecessor. It decays by β-emission with $t_{1/2} = 2.12 \times 10^{5}$ years.

Other Binuclear Reactions

Neutrons readily induce nuclear reactions, but they always produce nuclides on the *high* neutron–proton side of the belt of stability. To generate an unstable nuclide with a *low* neutron–proton ratio, protons must be added to the nucleus. Because protons have positive charges, this means that the bombarding particle must have a positive charge. Nuclear reactions with positively charged particles require projectile particles that possess enough kinetic energy to overcome the electrical repulsion between two positive particles.

Ernest Rutherford was the first person to observe a binuclear reaction. In 1919, he exposed a sample of nitrogen to α-particles from a naturally radioactive source. He observed the production of protons and deduced from the requirements of charge and mass balance that the other product was oxygen-17:

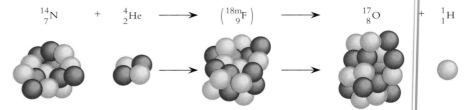

$$^{14}_{7}\text{N} \quad + \quad ^{4}_{2}\text{He} \quad \longrightarrow \quad \left(^{18\text{m}}_{9}\text{F} \right) \quad \longrightarrow \quad ^{17}_{8}\text{O} \quad + \quad ^{1}_{1}\text{H}$$

Compound Nucleus

The immediate product of a reaction between two nuclei is a **compound nucleus.** It has a charge equal to the sum of the charges of the reactants and a mass number equal to the sum of the mass numbers of the reactants. Every compound nucleus has excess energy that must be released after the two reactants bind together. Compound nuclei lose this excess energy by emitting one or more neutrons, protons, α-particles, or deuterons (^{2}H nuclei). As Example 21-7 shows, conservation of charge and mass number permit identification of the participants in binuclear reactions.

| **Balancing Binuclear Reactions** | **Example 21-7** |

Identify X and Y in each of the following nuclear reactions:
(a) $^3\text{He} + \text{n} \rightarrow (^m X) \rightarrow Y + \text{p}$
(b) $^{112}\text{Sn} + X \rightarrow (^{113m}\text{Sn}) \rightarrow {}^{113}\text{Sn} + Y$
(c) $X + \alpha \rightarrow (^{113m}\text{In}) \rightarrow Y + 2\,\text{n}$

Strategy: Charge number and mass number are conserved in nuclear reactions, so the missing components can be identified from the atomic numbers of the elements and the charge and mass numbers of elementary particles.

Solution:
(a) Both reactants are given: ^3He and n. Remember that a neutron has $Z = 0$ and $A = 1$. Therefore the compound nucleus must have $A = 4$ and $Z = 2$ (He). It then loses a proton, leaving $A = 3$ and $Z = 1$ (H). To emphasize the conservation of charge and mass number, we write the reaction showing all charge and mass numbers:

$$\tfrac{3}{2}\text{He} + \tfrac{1}{0}\text{n} \longrightarrow (^{4m}_{\ 2}\text{He}) \longrightarrow \tfrac{3}{1}\text{H} + \tfrac{1}{1}\text{p}$$

(b) Because all the nuclides are isotopes of tin, $Z = 50$ for all of them. Therefore X and Y each must have a charge number of zero. The possibilities are a neutron and a γ-ray. To balance mass numbers, X must have a mass number of one (neutron), and Y must have a mass number of zero (γ-ray):

$$^{112}_{50}\text{Sn} + \tfrac{1}{0}\text{n} \longrightarrow (^{113m}_{\ \ 50}\text{Sn}) \longrightarrow {}^{113}_{50}\text{Sn} + \gamma$$

(c) The compound nucleus is ^{113m}In, so X must have this configuration minus an α-particle, $\tfrac{4}{2}\alpha$. Thus X must have $A = 113 - 4 = 109$ and $Z = 49 - 2 = 47$, and X is $^{109}_{47}\text{Ag}$. Similarly, Y must have the configuration $^{113m}_{\ \ 49}\text{In}$ minus two neutrons. Therefore Z does not change, but $A = 113 - 2 = 111$, so Y is $^{111}_{49}\text{In}$:

$$^{109}_{47}\text{Ag} + \tfrac{4}{2}\alpha \longrightarrow (^{113m}_{\ \ 49}\text{In}) \longrightarrow {}^{111}_{49}\text{In} + 2\,\tfrac{1}{0}\text{n}$$

Making Synthetic Elements

Elements 43 (technetium), 61 (promethium), 85 (astatine), and all elements with $Z > 92$ do not exist naturally on the Earth, because no isotopes of these elements are stable. After the discovery of nuclear reactions early in the twentieth century, scientists set out to make these "missing" elements. Between 1937 and 1945, the gaps were filled and three actinides, neptunium ($Z = 93$), plutonium ($Z = 94$), and americium ($Z = 95$) also were made.

The production of synthetic elements requires binuclear reactions between two positive nuclei that must be forced together against the force of electrical repulsion. This necessitates the use of nuclear accelerators to give extremely high kinetic energies to positive projectile nuclei. The first instrument applied to this task was the **cyclotron,** a particle accelerator developed by E. O. Lawrence at the University of California, Berkeley. The operation of the cyclotron is shown

Figure 21-9
Particle accelerators can employ spiral particle paths, as in a cyclotron (*a*), or they can employ linear particle paths, as in a linear accelerator (*b*). In either case, the region where the particles are accelerated must be maintained under high vacuum to prevent trajectory-modifying collisions with gas molecules.

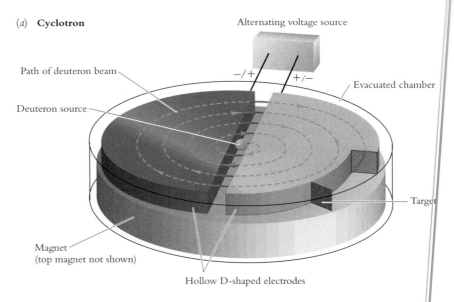

(*a*) **Cyclotron**

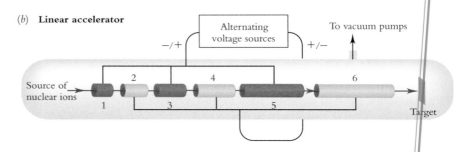

(*b*) **Linear accelerator**

schematically in Figure 21-9*a*. A combination of pulsed electric and magnetic fields is used to accelerate a beam of particles to high kinetic energy. Lawrence used high-energy deuterons to increase the atomic number of target nuclei by one unit:

$$^{98}_{42}\text{Mo} + {}^{2}_{1}\text{H} \longrightarrow {}^{99}_{43}\text{Tc} + {}^{1}_{0}\text{n}$$

$$^{238}_{92}\text{U} + {}^{2}_{1}\text{H} \longrightarrow {}^{238}_{93}\text{Np} + 2\,{}^{1}_{0}\text{n}$$

The most efficient way to make elements 93 and 94 uses neutrons produced during fission in nuclear reactors instead of accelerated positive nuclei. Neutron capture by ^{238}U followed by β-emission give isotopes with mass number of 239:

$$^{238}_{92}\text{U} + \text{n} \longrightarrow {}^{239}_{92}\text{U} + \gamma$$

$$^{239}_{92}\text{U} \longrightarrow {}^{239}_{93}\text{Np} + \beta$$

$$^{239}_{93}\text{Np} \longrightarrow {}^{239}_{94}\text{Pu} + \beta$$

As Z increases, the efficiency of these reactions falls sharply because many steps are required. Plutonium (element 94) has been produced in ton quantities by neutron bombardment of uranium-238. Up to curium (element 96), production in kilogram quantities is possible, but the yields fall by about one order of magnitude for each successive element beyond $Z = 96$.

Beyond $Z = 100$, synthesis by neutron bombardment of uranium is no longer effective. Instead, nuclides in the $Z = 95$ to 99 range are bombarded with beams of light nuclei. For example, mendelevium ($Z = 101$) was first made in 1955 by a team of Berkeley chemists led by Glenn Seaborg. They bombarded element 99 with helium nuclei that had been accelerated in the cyclotron:

$$^{253}_{99}Es + \alpha \longrightarrow {}^{256}_{101}Md + n$$

Beyond element 101, increasingly heavier nuclei must be used as the projectiles, and this requires a different type of accelerator that is linear rather than circular. The schematic diagram in Figure 21-9*b* shows how a heavy-ion **linear accelerator** works. A packet of ions is accelerated down the center of the accelerator by a series of electrically charged cylindrical tubes. The electrical potentials are varied so that the tube just ahead of the ion packet is always negatively charged, and this pulls the packet ahead at ever-increasing kinetic energy. As the photo in Figure 21-10 shows, a linear accelerator is complex and expensive to operate. Examples of reactions of this type are the following:

$$^{246}_{96}Cm + {}^{12}_{6}C \longrightarrow {}^{(258-x)}_{102}No + x\,n$$

$$^{252}_{98}Cf + {}^{11}_{5}B \longrightarrow {}^{(263-x)}_{103}Lw + x\,n$$

$$^{249}_{98}Cf + {}^{16}_{8}O \longrightarrow {}^{263}_{106}Sg + 2\,n$$

As the charge of the projectile increases, it takes ever greater acceleration for the projectile to penetrate the target nucleus. Even when successful, bombardment generates only a few nuclei of a new element, and these are so unstable that they decay in a matter of seconds or less. Nevertheless, nuclear scientists continue to work at extending the list of known elements, in part because theory predicts that the element with $Z = 114$ will be more stable than its lighter neighbors. Recently, nuclides with $Z = 118$, 116, and 114 have been reported, and their lifetimes indicate that the theoretical predictions are correct.

Figure 21-10
A view of the interior of the Fermilab linear accelerator.

Section Exercises

■ **21.3.1** The only stable isotope of cobalt is ^{59}Co. What induced reaction would generate each of the following nuclides from ^{59}Co? Identify the compound nucleus in each case: (a) ^{59}Fe; (b) ^{60}Co; (c) ^{62}Ni; and (d) ^{58}Ni.

■ **21.3.2** The two best-characterized isotopes of promethium ($Z = 61$) are ^{145}Pm and ^{147}Pm. The two elements next to Pm in the periodic table have many naturally occurring isotopes. Neodymium ($Z = 60$) has $A = 142$, 143, 144, 145, 146, 148, and 150; and samarium ($Z = 62$) has $A = 144$, 147, 148, 149, 150, 152, and 154. Write nuclear reactions describing neutron capture by naturally occurring isotopes of Nd and Sm, followed by decay to produce ^{145}Pm or ^{147}Pm.

■ **21.3.3** The following partial nuclear reactions show one reactant, the compound nucleus, and one product. Identify the other reactant and any additional products:

(a) $^{112}In \longrightarrow ({}^{113m}In) \longrightarrow {}^{113}In$
(b) $\alpha \longrightarrow ({}^{58m}Ni) \longrightarrow n$
(c) $^{60}Ni \longrightarrow ({}^{61m}Cu) \longrightarrow {}^{57}Co$

Figure 21-11
Schematic view of fission. Neutron capture produces a highly unstable nucleus that distorts and then splits into smaller nuclei and a few free neutrons. The sizes of the fragments and the number of free neutrons vary considerably.

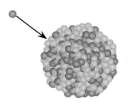

1. Neutron capture destabilizes the nucleus

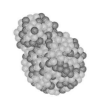

2. The unstable nucleus distorts

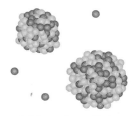

3. The nucleus splits into smaller fragment nuclei and free neutrons

21.4 NUCLEAR FISSION

The graph of binding energy in Figure 21-3 shows that large amounts of energy are released when heavy nuclei split into lighter ones **(fission)** and when light nuclei combine into heavier ones **(fusion).** Thus both fission reactions and fusion reactions can serve as sources of energy. Fission is the subject of this section, and fusion is the subject of Section 21.5.

Characteristics of Fission

Fission, shown schematically in Figure 21-11, splits a nucleus into two fragments, each with a much lower Z value. Several free neutrons are released during each fission event.

Uranium-235 is one of the most familiar examples of a nuclide that undergoes fission after it captures a neutron:

$$^{235}_{92}U + {}^1_0n \longrightarrow (^{236m}_{92}U) \begin{cases} {}^{81}_{32}Ge + {}^{152}_{60}Nd + 3\,{}^1_0n \\ \\ {}^{103}_{42}Mo + {}^{131}_{50}Sn + 2\,{}^1_0n \end{cases}$$

The compound nucleus formed on neutron capture, ^{236m}U, is highly unstable, so it quickly splits into fragment nuclei plus several neutrons. Fission results in a wide range of product nuclides, of which we show two examples. There are many modes of fragmentation, but free neutrons are always generated, and charge and mass number are always conserved.

Among naturally occurring nuclides, only ^{235}U undergoes fission, but neutron capture followed by β-decay convert two other naturally occurring nuclides, ^{238}U and ^{232}Th, into nuclides, ^{239}Pu and ^{233}U, that undergo fission.

The characteristics of nuclear fission can be summarized as follows:

1. Fission follows neutron capture by a small number of the heaviest nuclides, notably ^{235}U, ^{239}Pu, and ^{233}U.

2. Fission gives a range of product nuclides. Neutron-induced fission of ^{235}U yields the distribution shown in Figure 21-12. This distribution includes nuclides from $A = 77$ to $A = 157$. The most likely products are $A = 95$ and $A = 138$, but no single nuclide makes up more than 7% of the product fragments.

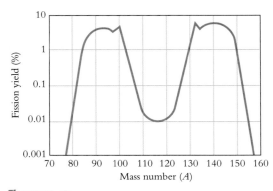

Figure 21-12
The isotopic "signature" of the nuclear fission of ^{235}U. Different mass numbers are produced in widely different percentages.

Fission yield (%)

Mass number (A)

3. Each fission reaction generates from one to four free neutrons, with two or three being most common.

4. The process of fission releases large amounts of energy. The energy released for one set of fission fragments is computed in Example 21-8. Although each set of fission fragments has a slightly different total mass, the average mass loss is about 0.2 g/mol. This translates into an average energy released in ^{235}U fission of 1.8×10^{10} kJ/mol.

5. Fission products often are radioactive. This is because the fissioning nucleus has a $N:Z$ ratio of 1.54, so its products have a similar $N:Z$ ratio. In contrast, stable nuclides in the $A = 77$ to 157 range have ratios of around 1.3, so the products of fission have excess neutrons, making them unstable.

| **Fission Energy** | **Example 21-8** |

Use isotopic molar masses (given below) to compute the energy released per mole, per nucleus, and per gram when ^{235}U undergoes fission in the following manner: $^{235}U + n \longrightarrow \, ^{138}Xe + \, ^{95}Sr + 3\,n$

Nuclide	n	^{235}U	^{138}Xe	^{95}Sr
MM, g/mol	1.0087	235.0439	137.908	94.913

Strategy: As this is a quantitative problem, the seven-step strategy is appropriate.

Solving Quantitative Problems

Solution: The problem asks for energy released during fission and provides molar masses. The nuclear reaction, which is provided, lets us visualize what is occurring. Equation 21-3 relates the energy released in a nuclear transformation to the "mass defect," Δm, which is the loss of mass per mole of reaction. The usual mole–mass conversion factors give energy released per nucleus and per gram.

The mass defect is the difference between the total mass of all products and the total mass of all reactants:

$$\Delta m = [(1 \, ^{138}Xe)(137.908 \text{ g/mol}) + (1 \, ^{95}Sr)(94.913 \text{ g/mol}) + (3\,n)(1.0087 \text{ g/mol})]$$
$$- [(1 \, ^{235}U)(235.0439 \text{ g/mol}) + (1\,n)(1.0087 \text{ g/mol})] = -0.2055 \text{ g/mol}$$

Substitute this mass change into Equation 21-3 to find the energy released:

$$\Delta E = (\Delta m)(8.988 \times 10^{10} \text{ kJ/g})$$

$$\Delta E = (8.988 \times 10^{10} \text{ kJ/g})(-0.2055 \text{ g/mol}) = -1.847 \times 10^{10} \text{ kJ/mol}$$

Finally, use the Avogadro constant and the molar mass to convert to energy per nucleus and energy per gram:

$$\Delta E_{\text{per nucleus}} = \frac{-1.847 \times 10^{10} \text{ kJ/mol}}{6.022 \times 10^{23} \text{ nuclei/mol}} = -3.07 \times 10^{-14} \text{ kJ/nucleus}$$

$$\Delta E_{\text{per gram}} = \frac{-1.847 \times 10^{10} \text{ kJ/mol}}{235.0439 \text{ g/mol}} = -7.86 \times 10^{7} \text{ kJ/g}$$

Two of the isotopic molar masses are known to just three decimal places, so the mass defect is precise to three decimal places, and the results are precise to three significant figures.

Fission of one gram of ^{235}U releases enough energy to raise the temperature of about 250 million liters of water (66 million gallons) from 25 to 100 °C. For comparison, about 1.65 million grams of octane must be burned to release the same amount of energy.

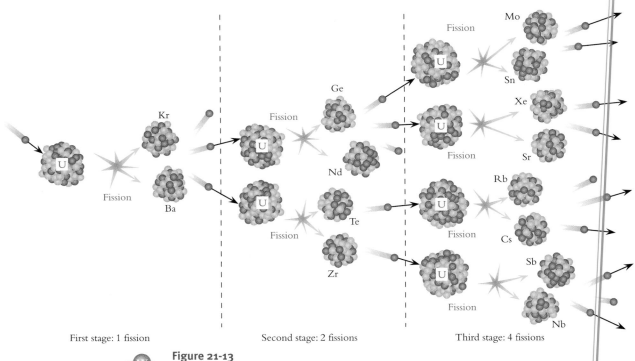

First stage: 1 fission

Second stage: 2 fissions

Third stage: 4 fissions

Figure 21-13
Schematic view of the start of a fission chain reaction. The first neutron causes fission, which generates additional neutrons. They cause more fission, and the chain continues to grow if more than one neutron, on average, is captured for every fission event.

Animation

Every fission reaction releases some neutrons, and these neutrons can be recaptured by other nuclei, causing more fission reactions. When the amount of fissionable material is small, most neutrons escape from the sample, and only a few neutrons are recaptured. Increasing the amount of material increases the likelihood that neutrons will be recaptured and cause additional fission reactions. The **critical mass** is defined as the amount of material that is just large enough to recapture one neutron, on average, for every fission reaction.

As long as the amount of fissionable material is less than the critical mass, the rate of fission events does not grow, and the rate of energy release remains low. In contrast, a sample behaves quite differently when the amount of fissionable material is larger than the critical mass. Above the critical mass, more than one neutron, on average, is recaptured for every fission that occurs. Now the number of fission reactions grows rapidly. As an illustration, consider what happens when two neutrons are recaptured from each fission reaction. As shown in Figure 21-13, the neutrons produced by the first fission reaction trigger fragmentation of two more nuclei. Neutrons from these two fission events are recaptured by four additional nuclei, and this fission cascade goes on, doubling in each successive round. The result is a "chain" reaction that grows quickly to explosive proportions.

The recapture ratio does not have to be two for this effect to occur. Any recapture value larger than 1.0 results in explosive growth of the fission chain. The critical mass is called "critical" because any mass greater than this value sustains a chain reaction and may explode.

Nuclear Weapons

The potential of nuclear fission was first realized in the atomic bomb. In 1945, the United States dropped two bombs of unprecedented power, one on Hiroshima and the other on Nagasaki, Japan. Both were fission weapons. The Hiroshima bomb contained ^{235}U, and the Nagasaki bomb contained ^{239}Pu. These tremendously destructive weapons had been developed under total secrecy in a wartime project that involved an international team that included many outstanding physicists and chemists.

The central feature of a fission explosion is a growing chain of fission reactions. There are three requirements:

1. The fissionable nuclide must be concentrated enough to become critical.
2. Subcritical portions of this fissionable nuclide must be combined into a critical mass.
3. The critical mass must be held together long enough for the chain to multiply to immense size.

The second and third requirements posed engineering problems that were met by a carefully designed detonation of a chemical explosive. This explosion propelled subcritical masses of fissionable material together and confined them while the fission chain multiplied.

The first requirement, on the other hand, posed formidable chemical problems. Of the three nuclides that can serve as nuclear fuels, ^{239}Pu, ^{233}U, and ^{235}U, only the last exists in nature. However, naturally occurring uranium contains only 0.72% ^{235}U, which is much too dilute to use in a bomb. At this concentration, nearly all the neutrons produced by fission of ^{235}U are captured by ^{238}U, which does not undergo fission. Thus no mass of natural uranium is enough to be critical. Therefore, before a bomb could be constructed from ^{235}U, a way had to be found to increase the percentage of ^{235}U in natural uranium. Neither plutonium-239 nor uranium-233 is a naturally occurring nuclide, so a bomb could not be made from these materials without first finding a means of synthesizing these nuclides.

How can ^{235}U be isolated from an isotopic mixture of ^{235}U and ^{238}U? Because isotopes react nearly identically, they cannot be separated by chemical reactions. Thus it is necessary to use a physical technique based on the small mass difference between the two isotopes. One way to accomplish this is to allow a gas to diffuse through a porous filter. Because light molecules move faster than heavy molecules, the gas that passes through the filter at the beginning of the separation process will be slightly enriched in the component with the lower molar mass.

Like other metals, uranium is not a gas except at very high temperatures. To concentrate ^{235}U by gaseous diffusion, it is necessary to prepare a uranium compound that can be vaporized at a reasonable temperature. One of the few uranium compounds that vaporizes at moderate temperature is UF_6. Uranium hexafluoride can be made by reacting elemental uranium with fluorine to make UF_4 and then treating this product with ClF_3:

$$U(s) + 2\,F_2(g) \longrightarrow UF_4(g)$$

$$UF_4(g) + ClF_3(g) \longrightarrow UF_6(g) + ClF(g)$$

When UF_6 gas diffuses through porous filters, the first emerging fraction is slightly enriched in $^{235}UF_6$. The molar mass ratio is only 1.0086 for $^{238}UF_6$ and $^{235}UF_6$, so a sample of UF_6 must pass through thousands of filters before the isotopic composition of the gas is sufficiently enriched in $^{235}UF_6$.

After purification, enriched UF_6 can be reduced to pure uranium metal in a two-step process at temperatures greater than 900 K:

$$UF_6 + H_2 + 2\,H_2O \longrightarrow UO_2 + 6\,HF$$

$$UO_2 + 2\,Mg \longrightarrow 2\,MgO + U$$

These reactions look simple, but these chemicals are highly corrosive at the elevated temperatures required for the process. As a result, special corrosion–resistant materials had to be developed and tested before isotopic enrichment could succeed.

The production of ^{239}Pu presented different challenges, because ^{239}Pu is a synthetic isotope, produced by the neutron bombardment of ^{238}U. The first nuclear reactor, constructed at the University of Chicago in 1942, was designed and used for this purpose. Significant amounts of plutonium form when natural uranium is placed in a reactor and bombarded with neutrons. The resulting mixture can be treated by appropriate chemical methods to recover the plutonium in pure form.

Nuclear Reactors

A nuclear bomb is a terrifying example of the enormous amount of energy released by nuclear fission. A bomb, however, is not the only way to extract the energy produced by nuclear fission. Instead, nuclear fission can be used to generate electrical power if the rate of fission is controlled by adjusting the number of recaptured neutrons. The fission of just one gram of ^{235}U releases as much energy as the combustion of 600 gallons of gasoline or 6 tons of coal. Consequently, nuclear power is attractive, especially for countries that lack supplies of petroleum or coal.

Nuclear power plants, such as the one diagrammed in Figure 21-14, use the energy released in fission to generate electricity. The fission reaction takes place in a core that is heated by the released energy. A circulating fluid transfers this thermal energy to a heat exchanger, where the energy is used to convert water into steam. The steam drives a turbine connected to an electrical generator. Cooling water must be supplied to recondense the steam.

Figure 21-14
Schematic view of a nuclear power plant. The energy source is the core, in which a fission reaction occurs. The rest of the plant is designed to transfer the energy released during fission and convert it into electricity.

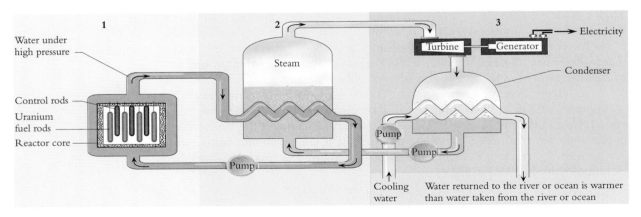

1. **Nuclear reactor:** Water under high pressure carries heat generated in the nuclear reactor core to the steam generator.

2. **Steam generator:** Heat from the reactor vaporizes water in the steam generator, creating steam.

3. **Turbine and condenser:** Steam from the steam generator powers a turbine, producing useable electricity. The condenser uses cooling water from a river or ocean to recondense the steam from the turbine.

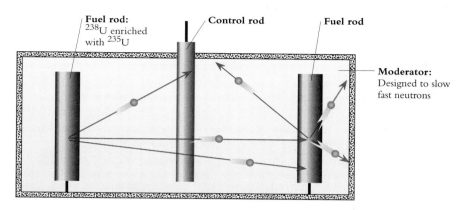

1. Fission of a ^{235}U in the fuel rod emits three fast neutrons

2. The control rod captures one of the neutrons

3. A ^{238}U in the fuel rod captures a second neutron; a ^{235}U captures the third neutron, leading to another fission event

Figure 21-15
The neutron processes occurring in the core of a fission reactor. Fission of ^{235}U gives from one to four fast neutrons, which are slowed down by the moderator. Each neutron is captured by another nuclide. In this sketch, a fission event produces three neutrons. One is captured by the control rod, one is captured by ^{238}U in the fuel rod, and one causes another fission event.

The turbine and generator components of a nuclear power plant have exact counterparts in power plants fueled by fossil fuels. The uniqueness of the nuclear power plant lies in its core. The core is a nuclear reactor where fission takes place under conditions that keep the reactor operating just below the critical level. The core contains three parts: fuel rods, moderators, and control rods. These components act on the flow of neutrons within the core, as shown in Figure 21-15. The fate of neutrons must be controlled carefully. Fission must be sustained at a steady rate that produces sufficient energy to run a generator, but the rate must not be allowed to increase and destroy the reactor.

A nuclear reactor runs on its nuclear fuel rods, which contain fissionable material such as uranium that has been enriched in ^{235}U. The fuel rods contain more than the critical mass of fuel, but the rate of fission is kept under control by movable control rods. These rods contain ^{112}Cd, which has a very high affinity for neutrons but does not undergo fission. As a result, the control rods act as a neutron "sponge," capturing neutrons that would otherwise trigger additional fission reactions. When pushed into the reactor, the control rods capture more neutrons, and the rate of fission decreases. When pulled out, the control rods capture fewer neutrons, and the rate of fission increases.

For the start up of a reactor, its core is allowed to heat up by withdrawal of the control rods until the recapture ratio is slightly greater than 1.0. When the optimum operating temperature is reached, the control rods are inserted until the capture ratio is exactly 1, and the reaction proceeds at a steady rate. For the shutdown of a reactor, its control rods are fully inserted, reducing the recapture ratio to nearly zero.

The moderator component of a reactor slows neutrons without capturing them. Moderators are used because the neutrons released in fission have such high kinetic energies that they are difficult to capture. The critical mass of a nuclear fuel is much smaller for slow neutrons than for fast neutrons, so considerably less fuel is needed in a moderated reactor. Graphite (^{12}C) or "heavy" water (D$_2$O) are good moderators because their nuclides slow neutrons without a high probability of capturing them, but ordinary water is a reasonable and inexpensive moderator that is used in many nuclear power plants.

In a nuclear power plant, heat must be transferred from the core to the turbines without any transfer of matter. This is because fission and neutron capture generate lethal radioactive products that cannot be allowed to escape from the core. A heat-transfer fluid such as liquid sodium metal flows around the core, absorbing the heat produced by nuclear fission. This hot fluid then flows through a steam generator, where its heat energy is used to vaporize water. After steam has been produced, its energy is used to produce electricity in a conventional steam turbine that has the same design as the steam turbines in fossil fuel power plants.

The radioactive materials that are an unavoidable byproduct of nuclear fission are life-threatening. To operate safely, therefore, a nuclear power plant must confine all its radioactive products until they can be disposed of safely.

The main danger in the operation of a nuclear power plant is potential loss of control over the nuclear reaction. If the core overheats, it may either explode or "melt down." In either event, radioactive materials escape from the reactor to contaminate the environment. Designers attempt to make nuclear reactors "fail-safe" by providing mechanisms that automatically shut the core down on overheating. One way this has been done is to design the control rods to fall into the core if their control mechanism fails.

Despite such safeguards, nuclear accidents have occurred. The worst occurred in 1986 at Chernobyl in Ukraine, shown in Figure 21-16a. Engineers who were not well trained in nuclear power turned off many safety systems during a test of the reactor. When the reactor cooled more than anticipated, most of the control rods were removed. In the absence of emergency cooling systems, the reactor surged out of control. The fuel rods melted and mixed with superheated water, which rapidly boiled, building up a high pressure of steam. This blew off the roof of the reactor facility, spewing substantial amounts of radioactive material into the atmosphere. In addition, the steam reacted with zirconium and graphite in the reactor, producing hydrogen gas and starting a fire that released more radioactive material. Contamination was most severe within 30 km of the plant, where radioactivity levels became so high that the entire population had to be relocated. Significant contamination was also detected 1000 km away in Germany, and increased levels of radioactivity attributable to Chernobyl appeared throughout the Northern Hemisphere.

Nuclear power plants in the United States are supposed to be designed well enough to prevent accidents as serious as the one at Chernobyl. Nevertheless, the Three Mile Island plant in Pennsylvania, an aerial view of which is shown in Figure 21-16b, experienced a partial meltdown in 1979. This accident was caused by a malfunctioning coolant system. A small amount of radioactivity was released into the environment, but because there was no explosion, the extent of contamination was minimal.

SECTION 21.6 →
We describe the lethal characteristics of nuclear radiation in Section 21.6.

Figure 21-16
Aerial views of three nuclear power plants. (a) The Chernobyl nuclear power plant, site of a major nuclear accident in 1986. (b) The Three Mile Island power plant, site of a minor nuclear accident in 1979. (c) A plant in France, which has operated nuclear power plants safely for more than 25 years.

(a)   (b)   (c)

A few nations rely heavily on nuclear power despite the possibility of accidents. In France and Japan, fission power from nuclear reactors provides two thirds or more of overall energy needs. A French plant appears in Figure 21-16c.

As a result of the accidents at Three Mile Island, Chernobyl, and recently in Japan, many people question whether safe containment and disposal can be guaranteed. Even though radioactivity is fully controlled under normal operating conditions, how to prevent accidents and how to dispose of the radioactive waste remain the subjects of intense debate.

Even when operated safely, nuclear power plants produce long-lived radioactive wastes, which must be sequestered from the biosphere until their radioactivity diminishes to acceptable levels. Plutonium-239, formed in nuclear fuel rods when ^{238}U captures neutrons and decays, is a particularly dangerous nuclear waste that is extremely toxic and has a half-life of nearly 25,000 years. In Europe, nuclear waste is placed far underground in abandoned salt mines. Deep-sea burial and ejection into outer space have also been proposed. Antinuclear activists contend that no current technology is acceptable, because there is no proof that radioactive material can be contained for the thousands of years necessary for these dangerous materials to decompose.

Modern nuclear reactors are highly technological, carefully engineered creations of advanced human societies, so it may seem impossible that a nuclear reactor could result from natural conditions. Our Chemistry and the Environment Box describes evidence indicating that such a natural nuclear reactor did exist.

Section Exercises

21.4.1 Generally, nuclides that are fission products have molar masses that average 0.09 g/mol less than their mass number. ^{106}Ru, for example, has a molar mass of 105.91 g/mol. Using this general value, calculate the mass losses and energy releases for ^{235}U fissions that release one neutron, two neutrons, and three neutrons.

21.4.2 Calculate the isotopic abundances of naturally occurring uranium 10^8 years ago. (Hint: Start with a convenient amount of uranium of present-day abundances, and use half-lives to calculate the amounts that were present 10^8 years ago.)

21.5 NUCLEAR FUSION

Fission occurs for only a few rare, extremely heavy nuclides. Fusion, in contrast, is possible for abundant light nuclides such as ^{1}H. Moreover, some fusion reactions release more energy per unit mass than fission reactions do. For example, the fusion of two hydrogen isotopes, deuterium and tritium, releases 3.4×10^8 kJ/g, compared to 7.9×10^7 kJ/g released in the fission of ^{235}U. Another attractive feature of fusion reactions is that the product nuclides are usually stable, so smaller amounts of radioactive byproducts result from fusion than from fission.

The fusion reaction of tritium and deuterium is as follows:
$$^2_1H + ^3_1H \longrightarrow ^4_2He + ^1_0n$$

The Threshold for Fusion

The major impediment to fusion reactions is that the reacting nuclei must have very high kinetic energies to overcome the electrical repulsion between positive

Box 21-1   Chemistry and the Environment: A Natural Nuclear Reactor?

Naturally occurring uranium mined today contains 99.28% ^{238}U and 0.72% ^{235}U. This is too low a percentage to sustain a fission reaction. In the past, however, the percentage of ^{235}U was higher than it is today, because ^{235}U has a shorter half-life than ^{238}U (7.0 × 10^8 yr compared with 4.5 × 10^9 yr). Calculations using these half-lives show that 1.8 billion years ago, ^{235}U made up 3% of natural uranium. This is comparable to the enrichment levels used in nuclear fuels. Long ago, then, the composition of uranium ores was such that a self-sustaining nuclear reaction could have occurred.

But did this ever happen? A sustained reaction requires a neutron recapture ratio of at least one. Besides enrichment in ^{235}U, there must be a high degree of purity that minimizes the fraction of neutrons captured by other nuclides. A sustained reaction also requires a moderator to slow the neutrons before they escape from the uranium mass. Because water is a relatively good moderator, the need for a moderator could be met by groundwater seeping through the deposit. These unique conditions are illustrated in our molecular view.

It may seem unlikely that all these conditions could have been met, but at least one deposit of uranium ore has characteristics indicating that, long ago, it operated as a natural nuclear reactor. At Oklo in the Gabon Republic near the western coast of equatorial Africa (see photo), there are uranium deposits of high purity that are about 1.8 billion years old. Thus this uranium originally contained the 3% abundance of ^{235}U needed for a sustained reaction. Currently, the deposits

in this mine differ from other uranium deposits in two ways that indicate that Oklo experienced sustained fission at one time. First, the ^{235}U content is slightly depleted, and second, there are unusual amounts of elements with mass numbers between 80 and 150.

The usual percentage of ^{235}U is 0.7207%, but the Oklo deposits contain 0.7071% of this isotope. This difference is small, but it is significant because it indicates that, at some time in the past, ^{235}U was consumed faster than its spontaneous decay rate. The question is whether or not that faster rate was caused by sustained fission.

What convinces scientists that sustained fission once occurred at Oklo is the presence of characteristic fission products in the ore. Elements of mass numbers between 75 and 160 occur in the ore in larger amounts than elsewhere. Furthermore, mass analysis of the elements in Oklo ore shows that they are distributed in the characteristic pattern shown in Figure 21-12. This isotopic "signature," which is not found in any other naturally occurring materials, is so characteristic that it has convinced most scientists that the ore deposits at Oklo once formed a huge nuclear reactor.

From the size of the ore deposit and the extent of ^{235}U depletion, scientists estimate that Oklo functioned as a natural nuclear reactor for about 100,000 years, generating 5 × 10^{14} kJ of energy. This is an average power output of about 150 kilowatts per year, which would be about enough to meet the needs of ten present-day Americans.

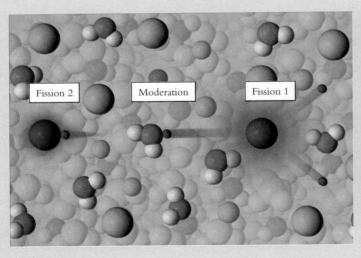

^{235}U Fast neutron

^{238}U Slow neutron

particles. The fusion of two hydrogen nuclei has the lowest possible repulsion barrier because it involves two $Z = 1$ nuclei. Even so, this reaction requires kinetic energies equivalent to temperatures of 10^7 K or greater. As Z increases, so does this energy requirement: The fusion of two ^{12}C nuclei requires kinetic energies equivalent to a temperature of 10^9 K.

There are several ways to produce nuclei with enough kinetic energy to fuse. One method uses particle accelerators to generate small quantities of fast-moving nuclei. This is very useful for studying fusion reactions, but the scale of fusion events in these experiments is too small to release useful amounts of energy. A second way to induce fusion uses the energy released in gravitational attraction to generate a temperature hot enough to start the reaction. As described later in this section, stars—including the sun—operate in this way. A third way is the fusion bomb, which uses a fission bomb to generate a temperature high enough for fusion. None of these provides a method for harnessing fusion as a practical source of power on Earth. However, nuclear scientists have proposed that radiation might be able to initiate a useful fusion reaction by heating a gaseous plasma confined in a magnetic field. If successful, this method may harness fusion as a source of power.

Fusion Bombs

A "hydrogen bomb," which uses nuclear fusion for its destructive power, is three bombs in one. A conventional explosive charge triggers a fission bomb, which in turn triggers a fusion reaction. Such bombs can be considerably more powerful than fission bombs because they can incorporate larger masses of nuclear fuel. In a fission bomb, no component of fissionable material can exceed the critical mass. In fusion, there is no critical mass because fusion begins at a threshold *temperature* and is independent of the *amount* of nuclear fuel present. Thus there is no theoretical limit on how much nuclear fuel can be squeezed into a fusion bomb.

Hydrogen bombs contain ^{2}H, ^{3}H, and ^{6}Li. The energy released in the fission explosion heats the two hydrogen isotopes hot enough to fuse:

$$^2_1H + {}^3_1H \longrightarrow ({}^5_2He) \longrightarrow {}^4_2He + {}^1_0n \qquad \Delta E = -1.7 \times 10^9 \text{ kJ/mol}$$

Lithium captures neutrons from this reaction in another energy-releasing process:

$$^6_3Li + {}^1_0n \longrightarrow ({}^7_3Li) \longrightarrow {}^4_2He + {}^3_1H \qquad \Delta E = -4.6 \times 10^8 \text{ kJ/mol}$$

Tritium nuclei produced in this reaction can fuse with additional deuterons, thus beginning the process again. All these processes occur in an extremely short time to release such an immense amount of energy that the bomb is blown apart.

Controlled Fusion

A long-standing goal of nuclear science has been to harness the energy of nuclear fusion in a sustained process. The technological problems are immense, however, because the fusing material must be confined to a volume small enough to generate high nuclear densities and many nuclear collisions. Because all materials vaporize at temperatures well below that required to sustain fusion reactions, containers that would withstand sustained fusion cannot be built.

The approach most often taken to address these problems is to use a hot, ionized gaseous plasma that contains cations of ^{2}H, ^{3}H, and ^{6}Li, along with enough electrons to maintain charge neutrality. This plasma is heated intensely, usually by powerful laser beams. To keep the plasma from flying apart, it is confined by a

A plasma is a high-temperature ionized gas consisting of free electrons and nuclei.

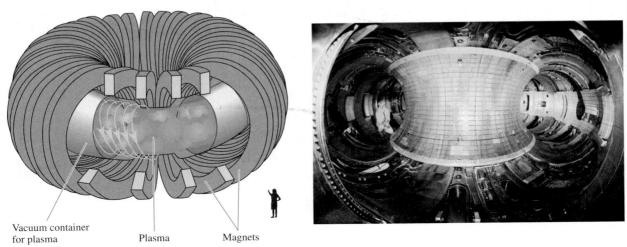

Figure 21-17
A version of a fusion reactor that has been the subject of intense development efforts is the tokamak, whose design and an experimental prototype are shown here.

Vacuum container for plasma

Plasma

Magnets

donut-shaped magnetic field. Figure 21-17 shows a diagram of such a magnetic field and a photo of a prototype that is in operation in Princeton, NJ.

Fusion cannot become a practical source of energy unless the energy released by fusion exceeds the energy used to heat and confine the plasma. Unfortunately, the energy input requirement is enormous, even in small-scale experimental studies. The best experiments conducted so far have managed to achieve fusion but not sustain it long enough to achieve any net production of energy.

Assuming that a sustained fusion reaction is achieved, some way must be found to harness its energy output. Because the field confining the plasma prevents the escape of charged particles, most of the energy output of the reactor is expected to be photons, ranging in energy from infrared rays to γ-rays. These photons will have to be absorbed by a suitable material that converts their energy, directly or indirectly, into electricity.

Stellar Nuclear Reactions

The sun and all other stars produce energy at a huge rate from sustained nuclear fusion. Over time, stars evolve through several stages, including stellar explosions. The products of a stellar explosion can form stars of more complex composition. Three distinct generations of stars have been identified, each fueled by a different set of fusion reactions.

The formation of a star begins with a cloud of matter. Nuclear reactions do not occur below a threshold temperature, so some other process must heat a cloud before it can become a star. Gravitational attraction provides the force that leads to such heating. In outer space, any cloud of material collapses on itself under the force of gravity. During collapse, gravitational potential energy is converted into kinetic energy of motion: Atoms move faster and faster, and the temperature rises. The total gravitational force increases with the mass of the cloud, so the temperature attained in the collapsing cloud increases with the total mass of the cloud. A cloud whose mass is larger than about 2×10^{29} kg eventually reaches a temperature sufficient to initiate the fusion of hydrogen nuclei.

Jupiter, the largest planet in our solar system, has a mass just below the threshold for star formation.

First-Generation Stars

Astrophysicists believe that the early universe was composed mostly of hydrogen. As a cloud of hydrogen collapses, heating breaks its hydrogen atoms into a plasma of protons and electrons. A **first-generation star** forms when the interior temperature of this plasma reaches 4×10^7 K. Above this temperature, protons combine in a reaction sequence that yields helium nuclei:

$$^1_1\text{H} + ^1_1\text{H} \longrightarrow (^2_2\text{He}) \longrightarrow ^2_1\text{H} + ^0_1\beta^+$$

$$^2_1\text{H} + ^1_1\text{H} \longrightarrow ^3_2\text{He} + \gamma$$

$$^0_1\beta^+ + ^0_{-1}\text{e} \longrightarrow 2\,\gamma$$

$$^3_2\text{He} + ^3_2\text{He} \longrightarrow (^6_4\text{Be}) \longrightarrow ^4_2\text{He} + 2\,^1_1\text{H}$$

The net reaction converts protons and electrons into helium nuclei and radiation, releasing about 2.5×10^9 kJ/mol of energy:

$$4\,^1_1\text{H} + 2\,^0_{-1}\text{e} \longrightarrow ^4_2\text{He} + 6\,\gamma \qquad \Delta E \cong -2.5 \times 10^9 \text{ kJ/mol}$$

The pressure exerted by the radiation escaping from inner portions of the star counteracts the force of gravity, and the balance of the two opposing forces keeps the volume of the star constant as long as hydrogen fusion continues.

In time (about 10^{10} years for a star the size of our sun), a star consumes most of its protons. Then the preceding sequence of nuclear reactions ceases, gravitational forces take over once more, and the star collapses until its temperature reaches about 10^8 K. At this temperature, a new fusion reaction begins in which helium nuclei fuse to form beryllium-8:

$$^4_2\text{He} + ^4_2\text{He} \rightleftharpoons (^8_4\text{Be})$$

The ^8Be nuclide decomposes back to two helium nuclei unless it collides with a third helium nucleus. Then fusion occurs to form carbon-12, a process that is exothermic by about 7.7×10^8 kJ/mol.

$$(^8_4\text{Be}) + ^4_2\text{He} \longrightarrow ^{12}_6\text{C} + \gamma \qquad \Delta E \cong -7.7 \times 10^8 \text{ kJ/mol}$$

The ^{12}C that is produced also reacts with ^4He:

$$^{12}_6\text{C} + ^4_2\text{He} \longrightarrow ^{16}_8\text{O} + \gamma$$

About 10^8 years after fusion of helium begins, a star runs out of ^4He fuel. When this happens, the star enters a new stage of gravitational collapse until the temperature increases to about 10^9 K. This triggers an entirely new set of nuclear reactions between nuclides of carbon and oxygen. The following are some examples:

$$^{12}_6\text{C} + ^{12}_6\text{C} \longrightarrow ^{20}_{10}\text{Ne} + ^4_2\text{He}$$

$$^{12}_6\text{C} + ^{16}_8\text{O} \longrightarrow ^{24}_{12}\text{Mg} + ^4_2\text{He}$$

$$^{16}_8\text{O} + ^{16}_8\text{O} \longrightarrow ^{31}_{15}\text{P} + ^1_1\text{H}$$

As a result of these and other fusion reactions, the star eventually contains nuclides with Z and A values all the way up to iron-56, which is the most stable of all nuclides.

At this stage of their evolution, some stars generate energy more rapidly than it can be dissipated. These stars explode like giant hydrogen bombs, ejecting their various nuclides, ranging from hydrogen to iron, into space. Such explosions, called **supernovae**, have been observed by astronomers over the centuries, and the

The color of a hydrogen-burning star depends on its mass; the higher the mass, the higher the temperature of the star and the more blue it appears.

A star undergoing helium fusion has a dense, hot core and a large, cooler outer mantle. These stars appear large and red, so they are called red giants.

Figure 21-18
In 1987 a supernova appeared just 160,000 light years from Earth. The two photographs are the same region of space. At top, the arrow points to the star that exploded. It had four times the mass of the sun. The picture on the bottom shows the aftermath of the explosion.

debris from supernovae explosions has been detected at numerous locations in the heavens. Johannes Kepler and Tycho Brahe, two of the most famous astronomers of the late sixteenth century, each observed a supernova, Brahe in 1572 and Kepler in 1604. Supernovae are still being observed. Figure 21-18 is a photograph of a supernova that appeared in 1987.

A supernova explosion blows much of the matter of the star out into space. The remaining core collapses on itself. Under the intense pressure that is thought to develop in this core, protons combine with electrons to form neutrons, and the remnant is an extremely dense, compact neutron star.

The drawing in Figure 21-19 shows the evolutionary stages of a first-generation star.

Second-Generation Stars

Explosions of first-generation stars spew nuclides from $Z = 1$ to $Z = 26$ into interstellar space. There, this matter mixes with interstellar hydrogen, and eventually enough matter clumps together under the force of gravity to collapse and form a **second-generation star.** Young second-generation stars, like young first-generation stars, contain large amounts of hydrogen. Unlike first-generation stars,

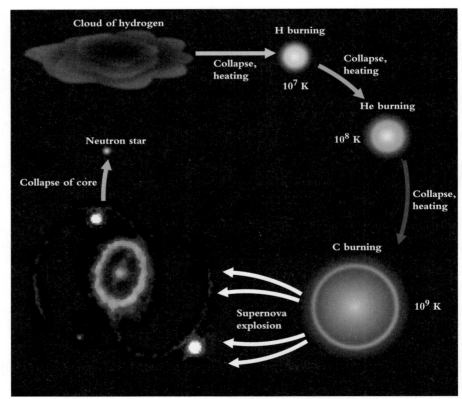

Figure 21-19
Schematic depiction of the evolutionary stages of a first-generation star. Gravitational collapse is required to heat stellar material enough for fusion to begin. The energy released during fusion then keeps the star from collapsing further until its nuclear fuel is consumed. There are three different stages of fusion.

however, they also contain higher-Z nuclides. One of these, ^{12}C, catalyzes the fusion of H to form He. As shown schematically in Figure 21-20, protons add to ^{12}C in a sequence that ultimately produces ^{4}He. Here are the reactions:

$$^{1}_{1}H + {}^{12}_{6}C \longrightarrow \left({}^{13}_{7}N\right) \longrightarrow {}^{13}_{6}C + {}^{0}_{1}\beta^{+}$$

$$^{1}_{1}H + {}^{13}_{6}C \longrightarrow {}^{14}_{7}N$$

$$^{1}_{1}H + {}^{14}_{7}N \longrightarrow \left({}^{15}_{8}O\right) \longrightarrow {}^{15}_{7}N + {}^{0}_{1}\beta^{+}$$

$$^{1}_{1}H + {}^{15}_{7}N \longrightarrow {}^{12}_{6}C + {}^{4}_{2}He$$

$$2\left[{}^{0}_{1}\beta^{+} + {}^{0}_{-1}e \longrightarrow 2\,\gamma\right]$$

$$\textit{Net:}\quad 4\,{}^{1}_{1}H + 2\,{}^{0}_{-1}e \longrightarrow {}^{4}_{2}He + 4\,\gamma \qquad \Delta E \cong -2.5 \times 10^{9}\ \text{kJ/mol}$$

When most of the hydrogen has been consumed, a second-generation star collapses until it is hot enough for helium fusion to occur. Now a larger range of reactions takes place because nuclides such as ^{13}C generate neutrons when they fuse with ^{4}He:

$$^{13}_{6}C + {}^{4}_{2}He \longrightarrow {}^{16}_{8}O + {}^{1}_{0}n$$

These neutrons are captured by iron to yield Co after β-emission:

$$^{56}_{26}Fe + {}^{1}_{0}n \longrightarrow {}^{57}_{27}Co + {}^{0}_{-1}\beta$$

Even though nuclear stability decreases beyond $Z = 26$, neutron capture (with or without β-emission) is always exothermic for nuclides lying within the belt of stability. Thus neutron capture/β-emission by Co ($Z = 27$) produces Ni ($Z = 28$), neutron capture/β-emission by Ni produces Cu ($Z = 29$), and so on up the atomic number ladder: Neutron capture and β-emission form all possible stable nuclides during the lifetime of a second-generation star.

Like first-generation stars, second-generation stars often become unstable when they reach the carbon-burning stage. If they explode in supernovae, their nuclear debris includes their content of heavier nuclides. In addition, supernova explosions generate large numbers of neutrons that can be captured by lighter nuclides in the exploding star to form still more heavy nuclides.

Earth contains significant amounts of elements all the way up to $Z = 92$. This indicates that our solar system resulted from the gravitational collapse of a cloud of matter that included debris from second-generation stellar supernovae. Thus our sun most likely is a **third-generation star.** The composition of a third-generation star includes high-Z nuclides, but the nuclear reactions are the same as those in a second-generation star.

The interiors of stars provide rich environments for fusion reactions, of which the reactions described here are representative. In addition, the synthesis of nuclides may occur in space when cosmic radiation encounters the debris from supernovae. Although the composition of various stars and the energetics of nuclear reactions are established, much about the composition of the universe remains to be discovered.

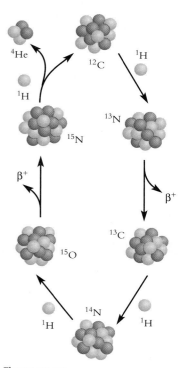

Figure 21-20
In a second-generation star, fusion of protons to produce helium occurs in a catalytic cycle by the sequential addition of protons to ^{12}C.

Section Exercises

21.5.1 One proposal for controlled fusion involves using an accelerator to propel deuterons into a lithium target, inducing the following reactions:

$$^{7}_{3}Li + {}^{2}_{1}H \longrightarrow {}^{8}_{3}Li + {}^{1}_{1}H \qquad {}^{8}_{3}Li \longrightarrow {}^{0}_{-1}\beta + 2\,{}^{4}_{2}He$$

(a) The first reaction can be viewed as the transfer of a neutron. Draw a nuclear picture illustrating this reaction.

(b) The reaction has a barrier that is 50% of the calculated coulombic barrier. What minimum kinetic energy must the deuterons have, if the radius of ^2H is 1.4×10^{-3} pm and that of ^7Li is 2.4×10^{-3} pm? (Refer to Section 21.1 to calculate the coulombic barrier.)

(c) Use the following isotopic masses, all in grams per mole, to calculate the energy change for each step of this fusion reaction:

^7Li	^8Li	^1H	^2H	^4He
7.016005	8.022488	1.007822	2.0141022	4.0026036

(d) If 5.0 kg of ^2H is completely consumed in the reaction, how much energy can be produced?

21.5.2 To show how zinc could form in a second-generation star, write a sequence of nuclear reactions that starts at ^{56}Fe, ends at ^{66}Zn, and is composed entirely of neutron capture and β-emission. Your sequence should stay within the belt of stability depicted in Figure 21-5.

21.5.3 Describe how the compositions of planets around a second-generation star differ from that of the Earth. Could life as we know it exist on such a planet? Why or why not?

21.6 EFFECTS OF RADIATION

The nuclear explosions that devastated Hiroshima and Nagasaki killed 100,000 to 200,000 people instantaneously. Probably an equal number died later, victims of the radiation released in those explosions. Millions of people were exposed to the radioactivity released by the accident at the Chernobyl nuclear power plant. The full health effects of that accident may never be known, but 31 people died of radiation sickness within a few weeks of the accident, and many more will develop cancers caused by their exposure to radiation. Even low levels of radiation can cause health problems. For this reason, workers in facilities that use radioisotopes monitor their exposure to radiation continually, and they must be rotated to other duties if their total exposure exceeds prescribed levels.

Radiation Damage

Nuclear radiation causes damage because of its high energy content. Radiation passing through matter transfers energy to atoms and molecules in its path. The major result of this energy transfer is ionization. Electrons are torn from molecules, creating positive ions. A single nuclear emission has enough energy to generate many positive ions and free electrons. A typical α-particle, for example, carries an energy of about 10^{-12} J, whereas ionization energies are typically 2×10^{-18} J. Only about one third of the energy of an α-particle goes into generating ions, the rest being converted into heat. Still, one α-particle generates around 150,000 cations. Beta and gamma rays have similar energies, and they also generate ions as they pass through matter.

As a high-energy particle passes through matter, it creates an ionization "track" that contains positive ions. These ions are chemically reactive because their bonds are weakened by the loss of bonding electrons. Even though each cation eventually recaptures an electron to return to electrical neutrality, many ions first undergo chemical reactions that are the source of the damage done by nuclear radiation.

Immediate Health Effects

Living cells are delicately balanced chemical machines. The ionization track generated by a nuclear particle upsets this balance, almost always destroying the cell in the process. Although the body has a remarkable ability to repair and replace damaged cells, exposure to radiation can overload these control mechanisms, causing weakness, illness, and even death.

Because nuclear radiation varies considerably in energy, the potential to cause damage cannot be assessed simply by counting the number of emissions. The energy of emissions must also be taken into account. Furthermore, the three different types of nuclear radiation affect human cells to different extents. When the amount, energy content, and type of radiation are taken into account, the result is a measure of the effect of radiation on the human body. This is expressed using a unit called the **rem.**

Different types of body cells show different sensitivities to nuclear radiation. Cells that divide most rapidly tend to be most easily damaged. These include bone marrow, white blood cells, blood platelets, the lining of the gastrointestinal tract, and cells in the gonads. Consequently, the symptoms of radiation sickness include loss of blood functions and gastrointestinal distress.

Individual ability to tolerate radiation damage varies, so a statistical variation exists in the relationship between dose level and health effects. Also, there are effective treatments, such as blood transfusions, for some radiation effects. The statistical patterns of human response to radiation are summarized in Table 21-5. Doses of over 600 rem are almost always fatal.

Long-Term Effects

Exposure to low doses of radiation causes no short-term damage but makes the body more susceptible to cancers. In particular, people who have been exposed to increased radiation levels have a much higher incidence of leukemia than the general population. Marie Curie, the discoverer of radium, eventually died of leukemia brought on by exposure to radiation in the course of her experiments. Medical researchers estimate that about 10% of all cancers are caused by exposure to high-energy radiation.

Table 21-5
Health Effects of Radiation Doses

Dose (rem)	Effect
0–25	Increased susceptibility to cancer; possible DNA damage
25–50	Reduced amounts of white blood cells
50–100	Fatigue and nausea in half of persons exposed
100–200	Nausea and vomiting; hair loss
200–400	Damage to bone marrow and spleen; 50% fatality rate
> 600	Usually death, even with treatment

Another cumulative effect of radiation can be an irreversible alteration of DNA sequences. If part of a DNA molecule is ionized, its molecular chain may be broken. Chain breaks are repaired in the body, but after a serious rupture, the repaired unit may have a different sequence. This type of changed sequence is a genetic mutation. Altered DNA sequences are reproduced and transmitted faithfully, thus passing on the genetic mutations to future generations. Because these effects are cumulative, individuals of childbearing age need to be especially careful about radiation exposure.

All humans are exposed to some level of radiation. Cosmic radiation continually bathes the Earth. This radiation and the naturally occurring radioactivity from nuclides such as ^{14}C, ^{40}K, and ^{222}Rn expose the average individual to about 0.1 rem per year. This exposure is increased by radiation from human activities. Medical procedures, notably the use of X rays for imaging teeth and bones, contribute the largest amount. A typical X-ray dose exposes the patient to 0.2 rem. Although this is well below the level at which immediate effects occur, the possibility of cumulative effects makes it prudent to have X-ray films taken no more often than necessary for maintaining good health. Total exposures vary considerably with human activities as well. "Frequent flyers," for example, receive higher doses of radiation because the intensity of cosmic radiation is significantly greater at high altitude than it is at ground level.

> Radon, $Z = 86$, has no stable isotopes, but it is found on Earth as a product of the decay sequence of ^{238}U. High radon levels are especially prevalent in places where there are uranium deposits, such as Montana and Idaho.

Radiation Shielding

Radiation exposure can be reduced by placing the radiation source or the potential target behind a shield that captures the radiation. A lead-lined pad is worn as a shield during dental X-ray examinations, for example, because X rays are absorbed more effectively by lead than by any other material. A lead shield a few millimeters thick is sufficient to stop X-rays.

Radiation of different types penetrates matter to different extents, as Figure 21-21 illustrates. Even though α-particles are the most damaging radiation, they are the most easily stopped by shielding. In fact, α-particles travel less than 1 mm in solid materials before losing all their excess energy. Beta particles travel much farther, requiring a shield 10 to 100 mm thick. Plastic shields several centimeters thick provide excellent shielding from both types of radiation.

Figure 21-21
Different types of radiation penetrate matter to different degrees. Although α-particles are the most damaging radiation, they lose their energy after traveling a very short distance. Gamma rays, on the other hand, are very dangerous because they travel long distances before shedding all their energy.

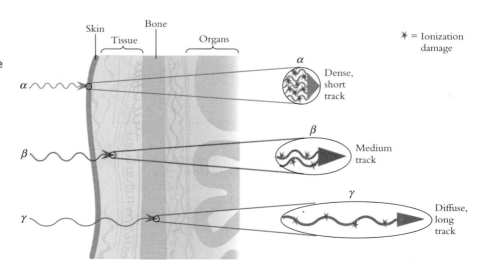

The γ-rays that accompany many nuclear decay processes can be more dangerous than α- or β-particles, because γ-rays travel long distances before losing their energy. The radiation from ^{222}Rn, which is responsible for as much as 40% of the background radiation to which humans are exposed, is typical. This nuclide decays by α-emission accompanied by γ-rays. The α-particle is stopped within 10 cm in air or about 1 mm in a solid. The accompanying γ-ray, on the other hand, may travel many meters and penetrate walls before finally losing its destructive power. Consequently, radioactive materials that emit γ-rays must be shielded using lead blocks that are many centimeters thick. The photo in Figure 21-22 illustrates γ-ray shielding.

Figure 21-22
For radioactive materials that emit γ-rays, thick shields of lead bricks are used to provide adequate shielding.

21.6.1 Potassium-40 is a naturally occurring radioactive nuclide that emits β-particles of $E = 1.32 \times 10^8$ kJ/mol and $t_{1/2} = 1.28 \times 10^9$ years. Potassium-40 is always present in our bodies, because we require potassium in our body fluids. Our bodies contain about 2.4 mg of potassium per kilogram of body weight, and ^{40}K makes up 0.0118% of the element. To how many β-particles is a person weighing 70 kg exposed daily from this nuclide, and how much energy does this decay deposit in the body?

21.6.2 Although X-ray exposures pose little threat to patients, medical technicians must be careful to avoid exposure to the beam. Suppose that an X-ray machine is "leaking" radiation that exposes the technician to 1% of the dose received by a patient. After how many exposures would the technician begin to show reduced numbers of white blood cells? If the technician administers 40 X-ray studies daily and works 250 days per year, would this damage show up?

21.7 APPLICATIONS OF RADIOACTIVITY

Considering the frighteningly destructive potential of nuclear weapons and the health risks posed by exposure to too much nuclear radiation, radioactivity may seem like a curse. Nevertheless, scientists and medical practitioners use radioactivity routinely in many beneficial ways, including dating techniques, tracing techniques, imaging, and therapy.

Dating techniques, which rely on the constant half-life of radioactive decay, allow scientists to estimate the ages of human artifacts and of the Earth itself.

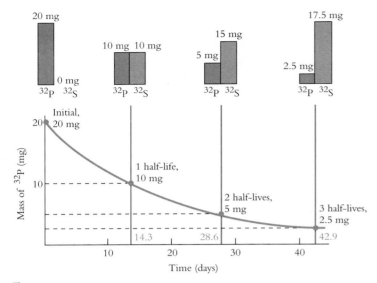

Figure 21-23
Decay of ^{32}P, a radioactive nuclide with a half-life of 14.3 days. The amount of the nuclide present is cut in half every 14.3 days.

Tracing techniques exploit the high sensitivity of radiation detectors to study complicated processes such as the mechanisms of chemical reactions and the percolation of water through geological formations. Imaging takes pictures of visually inaccessible objects, exploiting the ability of some radioactive emissions to pass through matter. Even the lethal properties of radioactivity can be used beneficially to treat some forms of cancer.

Dating Using Radioactivity

Each radioactive nuclide has a characteristic, constant half-life. This means that it acts as a clock, "ticking" (decaying) at a constant rate. Suppose that a nuclear reactor generates 20 mg of ^{32}P, an isotope with a half-life of 14.3 days. As Figure 21-23 illustrates, after 14.3 days, 10 mg will remain, and there will be 10 mg of the decay product, ^{32}S. After 28.6 days, 5 mg of ^{32}P will remain, and there will be 15 mg of ^{32}S. A chemist who analyzes a sample of ^{32}P and finds that it contains 15 mg of ^{32}S and only 5 mg of ^{32}P can calculate that the sample was generated 28.6 days ago.

Calculating the age of a sample from the amounts of its radioactive nuclides and their decay products forms the basis for radioactive dating techniques. Equation 21-5 is used for such calculations:

$$\ln\left(\frac{N_0}{N}\right) = \frac{t \ln 2}{t_{1/2}} \qquad \textbf{(21–5)}$$

Short-lived isotopes such as ^{32}P are not used for dating because other records provide the ages of recent materials. Radioactive dating proves to be especially valuable in estimating the age of materials that predate human records, such as the Earth itself.

The Earth's age can be estimated by using the half-lives of unstable nuclides that are still present and the half-lives of those that are missing. According to Table 21-4, the shortest-lived naturally occurring nuclide is ^{235}U ($t_{1/2} = 7.0 \times 10^8$ yr). Nuclides with half-lives shorter than 10^8 years, such as ^{146}Sm ($t_{1/2} = 7 \times 10^7$ yr), ^{205}Pb ($t_{1/2} = 3 \times 10^7$ yr), and ^{129}I ($t_{1/2} = 1.7 \times 10^7$ yr) are missing. If the Earth were less than 10^8 years old, its crust would contain some of these nuclides. If it were older than 10^{10} years, ^{235}U would no longer be present. Thus isotopic abundances indicate that the Earth is older than 10^8 years but younger than 10^{10} years.

More precise estimates come from accurate measurements of isotope ratios. Three pairs of radioactive isotopes and their products are abundant enough for such ratios to be measured:

$$^{238}\text{U} \longrightarrow \text{(Decay chain)} \longrightarrow ^{206}\text{Pb} \qquad t_{1/2} = 4.5 \times 10^9 \text{ yr}$$

$$^{40}\text{K} \longrightarrow ^{40}\text{Ar} + \beta \qquad t_{1/2} = 1.28 \times 10^9 \text{ yr}$$

$$^{87}\text{Rb} \longrightarrow ^{87}\text{Sr} + \beta \qquad t_{1/2} = 4.9 \times 10^{11} \text{ yr}$$

If the amount of a radioactive nuclide in a rock sample is N, the sum of this amount plus the amount of its product nuclide is N_0. For argon dating, N_0 is the sum of potassium-40 and argon-40 present in a sample of rock. Assuming that Ar gas escapes from molten rock but is trapped when the rock cools and solidifies, the lifetime obtained by substituting these values into Equation 21-5 is the time since the rock solidified. Such analyses show that the oldest rock samples on Earth are 3.8×10^9 years old.

Analyses of this type are correct only if all of the product nuclide comes from radioactive decay. This is not known with certainty, but when age estimates using different pairs of nuclides give the same age and samples from different locations also agree, the age estimate is likely to be accurate. Note also that 3.8×10^9 years agrees with the qualitative limits derived from naturally occurring radioactive nuclides.

Samples that are 4.6×10^9 years old have been found in meteorites. This is the best present estimate for the age of the solar system. Example 21-9 illustrates this type of calculation for rock from the Earth's moon.

Argon Dating   **Example 21-9**

When a sample of moon rock was analyzed by mass spectroscopy, the ratio of ^{40}K to ^{40}Ar was found to be 0.1295. Based on this ratio, how old is the moon?

Strategy: This is a question about radioactive dating. Convert the isotopic ratio of the radioactive nuclide and its product into an amount ratio, and then use Equation 21-5.

Solution: Equation 21-5 requires a value for N_0/N, the ratio of potassium-40 present when the moon was formed to the amount present today. We can determine N_0/N from the isotopic ratio N_K/N_{Ar} given in the problem. First note that $N = N_K$. Then relate N_0 to N_K/N_{Ar}. Each ^{40}K nuclide that decays generates a ^{40}Ar nuclide, so the initial amount, N_0, is the sum of the amounts currently present:

$$N_0 = N_K + N_{Ar}$$

Divide both sides of this equality by N_K:

$$\frac{N_K}{N_{Ar}} = 0.1295 \quad so \quad \frac{N_{Ar}}{N_K} = \frac{1}{0.1295} = 7.722$$

$$\frac{N_0}{N_K} = 1 + 7.722 = 8.722$$

Now rearrange Equation 21-5 to solve for t:

$$t = \frac{t_{1/2}}{\ln 2} \ln\left(\frac{N_0}{N_K}\right)$$

Find the half-life of ^{40}K from Table 21-4: $t_{1/2} = 1.28 \times 10^9$ yr:

$$t = \frac{(1.28 \times 10^9 \text{ yr})}{(0.693)} \ln(8.722) = 4.00 \times 10^9 \text{ yr}$$

According to this analysis, the moon solidified about 4 billion years ago.

A sample of rock from the moon.

Radioactivity serves as a useful clock only for times that are the same order of magnitude as the decay half-life. At times much longer than $t_{1/2}$, the amount of radioactive nuclide is too small to measure accurately. At times much shorter than $t_{1/2}$, the amount of the product nuclide is too small to measure. By an accident of nature, one naturally occurring radioisotope, ^{14}C, has a half-life that is close to the age of human civilization. As a result, this isotope is used to determine the age of human artifacts.

The logic underlying carbon-14 dating differs from that for potassium–argon dating. Recall that bombardment of the upper atmosphere by cosmic rays generates small but measurable levels of ^{14}C. This rate of production of ^{14}C by cosmic rays equals its rate of decay by β-emission, so the percentage of ^{14}C in the atmosphere remains nearly constant. Living plants incorporate CO_2 from the atmosphere into carbohydrates, and animals use these carbohydrates as food. As a result, the percentage of carbon-14 in the tissues of *living* organisms is the same as that in the atmosphere. On death, however, the uptake of carbon stops. The carbon-14 in an object that was once alive slowly disappears with $t_{1/2} = 5.73 \times 10^3$ years. This makes it possible to date objects such as bone, wood, and cloth by assaying their levels of ^{14}C.

The product of ^{14}C decay is ^{14}N, but there is so much ^{14}N already present in the atmosphere that this isotope is not an accurate measure of how much ^{14}C has decayed. Instead, the ratio of ^{14}C to ^{12}C is used for radiocarbon dating. In the atmosphere and in living objects, the fraction of ^{14}C is 1.33×10^{-12}. That is, there is 1 atom of ^{14}C for every 7.54×10^{11} atoms of ^{12}C. Although this is a tiny amount, it nevertheless results in 15.3 decays of ^{14}C per gram of carbon per minute. Using modern mass spectrometers, the fractional abundance of ^{14}C can also be measured directly.

After a living organism dies, its ^{14}C content decreases according to the first-order decay law, whereas its ^{12}C content remains fixed. After 5.73×10^3 years, there is half as much ^{14}C present, for a fractional content of 6.63×10^{-13} or one atom for every 1.508×10^{12} atoms of ^{12}C. The decay rate from a sample that is 5700 years old will be 7.65 decays g^{-1} min^{-1}.

The usual procedure for radiocarbon dating is to burn tiny samples of objects to be dated, collect the CO_2 that is produced, and compare its rate of radioactive decay with that of a fresh CO_2 sample. The ratio of counts gives N_0/N, which can then be substituted into Equation 21-5 to calculate t. Mass spectroscopic isotope analysis can also be used to obtain the N_0/N value, as Example 21-10 illustrates.

Example 21-10	Isotopic Carbon Dating

In 1988, the Shroud of Turin, claimed by some to have been used to bury Christ, was age-dated by isotopic analysis of its carbon content. The fractional content of ^{14}C from a sample from the shroud was 1.22×10^{-12}. How old is the shroud? If it were 2000 years old, what fractional content would it have?

Strategy: Calculation of an age requires the value of N_0/N, which is used in Equation 21-5 to calculate age, t. The ratio can be found from the isotopic analyses of a fresh carbon sample and the old sample.

Solution: The data include the fractional content of ^{14}C in the sample from the shroud, 1.22×10^{-12}. The text states that the fractional content for a fresh

Isotopic Carbon Dating (*continued*)   **Example 21-10**

sample is 1.33×10^{-12}. To obtain N_0/N, divide the ratio for a fresh sample by the ratio for the old sample:

$$\left(\frac{N_0}{N}\right) = \frac{1.33 \times 10^{-12}}{1.22 \times 10^{-12}} = 1.09$$

$$t = \frac{t_{1/2}}{\ln 2} \ln\left(\frac{N_0}{N}\right)$$

$$t = \frac{5730 \text{ yr}}{0.693} \ln(1.09) = 714 \text{ yr}$$

Thus the cotton or flax from which the shroud was woven was harvested around 1300 AD. The shroud might be younger than this, but it cannot be older.

Use the same equation to compute the isotopic ratio expected for an artifact that is 2000 years old:

$$\ln\left(\frac{N_0}{N}\right) = \frac{(2000 \text{ yr})(0.693)}{(5730 \text{ yr})} = 0.242$$

$$\frac{N_0}{N} = e^{0.242} = 1.274$$

$$N = \frac{N_0}{1.274} = \frac{(1.33 \times 10^{-12})}{(1.274)} = 1.04 \times 10^{-12}$$

This fraction is different enough from 1.22×10^{-12} to demonstrate conclusively that the shroud does not date from the time of Christ.

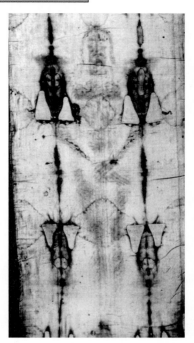

The Shroud of Turin is imprinted with an image that resembles a male human, thought by some to be a representation of Christ.

Calculations of ages, as illustrated in Example 21-10, make the assumption that the level of ^{14}C in the atmosphere at the time of death was the same as it is today. To check the validity of ^{14}C dating, its age estimates have been compared with those obtained by other methods, such as tree-ring dating. Wood from a Roman shipwreck, for example, was dated at 40 BC ± 3 years using tree-ring dating and 80 BC ± 200 years using radiocarbon dating. A large number of such comparisons show small but consistent differences, indicating that the fractional amount of ^{14}C in the atmosphere has changed slowly but steadily with time. Archaeologists use a small multiplying factor to correct for this small drift.

Tree-ring dating is more accurate than radiocarbon dating, but radiocarbon dating can be used on objects that do not retain tree rings, such as charcoal and cloth.

Radioactive Tracers

Radioactive isotopes are chemically identical to their natural, nonradioactive counterparts, but their high-energy decays allow them to be detected even though they may compose only a tiny fraction of the overall isotopic composition. As already described, radiocarbon dating detects ^{14}C at a concentration of one part in 10^{11} in carbon-containing samples. If an isotopically enriched sample is introduced into some dynamic process, the course of that process can be followed by tracing the whereabouts of the radioactive isotope.

Tracer techniques are applied in fields that range from geology to medicine. For example, a holding tank at an industrial plant might be suspected of leaking

Figure 21-24
Radioactive tracers can show flow pathways that cannot be detected using other methods of detection. Tritiated water added to a holding tank results in radioactivity in the water from wells that draw from the ground water supply into which the holding tank drains.

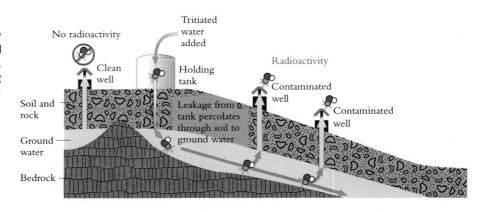

contaminated water into wells in its vicinity, as shown diagrammatically in Figure 21-24. The path of water draining from the holding tank could be traced by adding water enriched in radioactive tritium to the holding tank. The water from the wells could then be sampled for radioactivity at regular intervals. If, after a few days, water in some wells showed significant radioactivity while the remaining wells were free of contamination even after several months, that would show which of the wells were receiving water from the holding tank.

Tracer techniques are used widely in biology. Botanists, for example, work to develop new plant hybrids that grow more rapidly. One common way to determine how fast plants grow is to measure how quickly they take up elemental phosphorus from the soil. New hybrids can be planted in a plot and fertilized with phosphorus enriched in radioactive ^{32}P. When leaves from the plants are assayed for ^{32}P at regular intervals, the rate of increase in radioactivity measures the rate of phosphorus uptake.

One of the first chemical applications of radioactive tracers was a set of elegant experiments on photosynthesis performed in the 1950s by Melvin Calvin. His goal was to determine the set of reactions used by plants to transform atmospheric CO_2 into carbohydrates. Calvin supplied growing plant cells with CO_2 enriched with ^{14}C. By harvesting cells and using radioactive counting to determine which molecules gained enhanced carbon-14 radioactivity and in what sequence, Calvin was able to unravel the elaborate chain of reactions that is now known as the Calvin cycle.

Melvin Calvin received the Nobel Prize in Chemistry in 1961 for his work using ^{14}C as a tracer.

Medical Imaging

Physicians need ways to examine internal organs without disturbing those organs. X-ray films work well for teeth, bones, and lungs but do not provide clear images of other vital organs. Radioactive substances that bind to specific organs in the body make it possible to obtain an image of one particular organ. Like an X-ray film, this image reveals abnormalities, allowing physicians to diagnose ailments.

Because exposure to any radiation results in health risks, the administration of radioactive isotopes must be carefully monitored and controlled. Only γ-emitters and positron emitters are used, because they are the easiest to detect and because they cause minimal tissue damage. Specificity for a single target organ is essential so that the amount of radioactive material can be kept as small as possible. Short half-lives are advantageous, but a half-life shorter than a few hours is impractical because it does not provide time to synthesize, isolate, and administer the radioactive compound.

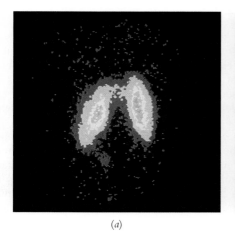
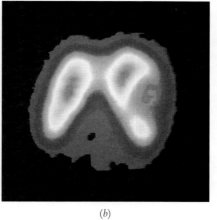

(a)   (b)

Figure 21-25
Radiation images identify abnormalities in organs. (a) The uniform distribution of ^{99m}Tc in a normal thyroid gland. (b) The asymmetrical distribution shown in this image indicates a cancerous thyroid gland.

Nuclear medicine most frequently uses a single radioactive nuclide of technetium, ^{99m}Tc. This nuclide has near-perfect properties for medical imaging. It has a half-life of 6 hours, which is long enough to deliver it into the body and generate an image but short enough to minimize long-term effects. It emits γ-rays of moderate energy (1.35×10^7 kJ/mol) that are easy to detect but are not highly damaging to living tissue. Technetium is not normally found in the body, but the element can be bound to many chemicals that the body recognizes and processes. The radioactive isotope is produced by neutron bombardment of natural molybdenum, from which ^{99m}Tc can be extracted readily in pure form.

The thyroid gland, brain, lungs, heart, liver, stomach, kidney, and bones all can be imaged by using compounds containing ^{99m}Tc, making it a very versatile imaging agent. Figure 21-25 shows how static imaging can be used to detect thyroid tumors.

Positron emission tomography (PET) is an imaging technique that makes use of positron emitters. Neutron-poor nuclides such as ^{11}C, ^{13}N, and ^{15}O decay by positron emission. Recall that positrons quickly annihilate with electrons, generating γ-rays. Because γ-rays travel long distances before being absorbed, they escape from the body and can be detected by external detectors, as shown in Figure 21-26a. One use of PET is to detect abnormal brain activity. The patient is given an injection of glucose containing a small amount of a positron emitter. Brain activity involves glucose metabolism, so a PET scan reveals brain activity patterns, as shown in Figure 21-26b. Although PET is a very powerful technique, it is also

In addition to Tc, ^{52}Fe is useful for bone marrow scans, ^{133}Xe for studying lung functions, and ^{131}I for studying the thyroid gland.

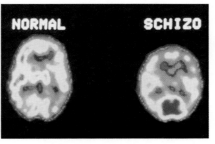

(a)   (b)

Figure 21-26
PET scans require a large array of gamma-ray detectors (a). They reveal how brain activity differs for normal and abnormal patients (b).

Box 21-2	Chemistry and Life: Preserving Food with Radiation

In a society where canned and frozen foods are so prevalent, we tend to take food preservation for granted. It is easy to forget that prevention of spoilage is a major undertaking when food is harvested at one time and place but eaten at another.

Spoilage is a biological process. Molds, bacteria, and vermin eat foodstuffs, rendering them unfit for human consumption. To stop spoilage, processors treat food to kill microorganisms, chill it to slow the metabolism of destruction, and keep it sealed to ward off pests. Even in fully developed societies, these procedures are only partially successful, and in economically developing countries, up to 50% of crops may be lost to spoilage.

Recently, the lethal properties of gamma radiation have been shown to preserve foods. After packaging, the food is exposed to enough γ-radiation to kill microorganisms. Moderate doses retard spoilage, whereas high doses can prevent spoilage completely for as long as the packaging remains intact.

Irradiation of foods can be used for a variety of purposes. Low doses help preserve potatoes, onions, and garlic by inhibiting sprouting. Such doses also kill the parasite in pork that results in trichinosis, and they inactivate insects that feed on stored grain. Medium doses inhibit the growth of mold on fruits such as strawberries. Medium doses also help to preserve meat, poultry, and fish by killing microorganisms that cause spoilage and reducing the levels of pathogens such as salmonella. High doses of radiation are sufficiently lethal to microorganisms to provide long-term protection for meat as well as spices and seasonings.

Although γ-irradiation of food is used widely in Europe, usage has been restricted in the United States. No more than 10% of packaged herbs and spices have been treated. The U.S. Food and Drug Administration has been slow to approve the process because of concern that irradiation may generate unhealthy and perhaps even carcinogenic byproducts.

Opposition to irradiated foods arises from three possible harmful effects. Some fear that irradiation will make the food itself radioactive, although there is no evidence that this occurs. A second worry is that even though irradiation kills microorganisms, it may also destroy important nutrients such as vitamins. Thirdly, in addition to destroying beneficial compounds, irradiation produces ions and free radicals such as OH that may react with foodstuffs to generate harmful compounds such as carcinogens.

It is important to keep in mind, however, that conventional methods of preservation also carry risks. For example, ethylene dibromide, which was used for years to kill pests in fruit, has been banned because it is implicated as a carcinogen. Nitrites, which are used widely in preserving bacon, bologna, and other processed meats, are now known to be unhealthy at high exposures. European regulators have taken the position that the as yet uncharted side effects of irradiation are less injurious than those associated with chemical preservatives. Regulators in the United States, on the other hand, believe that irradiation should not be permitted until its side effects are better documented. Unfortunately, no preservative may be completely free of deleterious effects, in which case societies will have to choose among high spoilage rates, the health hazards of chemical preservatives, and the possible hazards generated by the byproducts of irradiation.

extremely expensive. It requires an elaborate set of detectors and a computer to analyze the detection pattern. Moreover, positron emitters must be produced using a cyclotron, which is quite expensive to operate.

Radiation Therapy

Radioactive nuclides can be used to treat certain diseases. Some cancers respond particularly well to radiation therapy. At first this may seem paradoxical, because exposure to radiation damages cells and eventually causes cancer. Thus we might expect irradiation to worsen rather than retard this disease. The key to radiation therapy is that cancerous cells reproduce more rapidly than normal cells, and rapidly reproducing cells are more sensitive to radiation. If concentrated doses of radiation are focused on the malignant cells, a cancer may be destroyed with minimal damage to healthy tissue. Nevertheless, radiation therapy always has the unpleasant side effects associated with radiation sickness, such as nausea and hair loss.

Thyroid cancers are often treated with radioactive iodine because the thyroid gland preferentially absorbs iodine. If a patient is treated with iodine that contains radioactive ^{131}I, its radioactivity is concentrated in that gland. Inside the thyroid gland the radiation destroys cancerous cells more rapidly than normal cells. At the correct dosage, the cancer may be eliminated without destroying the healthy part of the thyroid gland.

The radioactive source need not always be introduced into the body. Inoperable brain tumors can be treated with γ-rays from an external source such as a sample of cobalt-60. The patient is placed in a position where the γ-ray beam passes through the tumor. Then the patient is moved so that the γ-rays travel different paths but always go through the tumor. In this manner the tumor receives a much higher dose than any surrounding tissues.

Our Chemistry and Life Box discusses another use for radiation, the controversial area of food preservation.

Section Exercises

21.7.1 An archaeologist discovers a new site where small charcoal bits are mixed with human remains. Burning a small sample of this charcoal gives gaseous CO_2 that, when placed in a counter, registers 1.75 decays per second. When an equal mass of CO_2 from fresh charcoal is placed in this counter, the count rate is 3.85 decays per second. How old is the archaeological site?

21.7.2 Classify each of the following radioactive nuclides as a good or a poor choice for medical imaging. State your reason in each case: (a) ^{123}I, a γ-emitter with a 13.2-hour half-life; (b) ^{131}I, a β-emitter with an 8.1-day half-life; (c) ^{60}Co, a γ-emitter with a 5.26-year half-life; and (d) ^{56}Co, a positron emitter with a 77-day half-life.

21.7.3 Describe how radioactive iodine (^{131}I) could be used to determine how rapidly each of the following proposed chemical exchange reactions occurs: (a) $I_2(s)$ exchanging with I^- in aqueous solution; (b) $I_2(s)$ exchanging with $I_2(g)$ in a closed container.

▰▰▰ CHAPTER REVIEW

Summary and Key Terms

1. Nuclei are made up of **nucleons**—protons and neutrons. Each different composition is a **nuclide**, and nuclides of an element with different numbers of neutrons are **isotopes**. The energy needed to remove a nucleon from a nucleus is its **nuclear binding energy**. When protons and neutrons bind together to form a nucleus, there is a loss of mass that appears as energy, $\Delta E = (\Delta m)c^2$. The binding energy per nucleon has a maximum value at about $Z = 26$ and has smaller values for lighter and heavier nuclides. Light nuclides can undergo **fusion**, and heavy nuclides can undergo **fission**, both of which convert mass into energy. Coulombic barriers prevent fusion under normal conditions. Stable nuclei fall within a belt of stability in which the ratio of neutrons to protons is between 1.0 and 1.54. Nuclei that are proton-rich, are neutron-rich, or have $Z > 83$ are unstable.

2. Unstable nuclides are **radioactive** and undergo **nuclear decay** by emitting energetic fragments. In such processes, mass-energy, charge number, and mass number all are conserved. The modes of decay are α**-emission** (very heavy nuclides),

β-emission (high $N:Z$ ratio), **positron emission** or **electron capture** (low $N:Z$ ratio), and γ-**ray** emission (excited nuclides). All nuclear decay processes obey first-order kinetics and have constant half-lives.

3. Induced nuclear reactions are binuclear processes such as neutron capture, proton capture, and the capture of helium nuclei. These reactions proceed through formation of a **compound nucleus** and lead to the formation of many synthetic nuclides that are unstable. The high-speed particles required for many of these reactions are produced using a **cyclotron** or a **linear accelerator.**

4. Fission of U or Pu produces two smaller nuclides, several neutrons, and large amounts of energy. Fission chain reactions, which can lead to an explosion if the mass of the nuclide exceeds a **critical mass,** are the energy sources for nuclear bombs and nuclear power plants.

5. Fusion of light nuclides releases large amounts of energy, but fusion cannot occur unless the nuclei possess the immense energy needed to overcome the coulombic barrier to their close approach. Fusion bombs have been made, but controlled fusion remains a dream. Fusion reactions produce the energy that powers the stars, including the sun. **First-generation stars** first use hydrogen fusion, then helium fusion, and finally carbon fusion as energy sources. **Second-generation stars** use these same fusion processes, catalyzed by carbon nuclides. **Supernovae** spew out nuclear debris, including nuclides heavier than iron, which subsequently becomes incorporated into **third-generation stars** such as the sun.

6. Nuclear radiation, whose effect is measured in **rem**, is energetic enough to cause substantial damage to matter through which it passes, and at sufficiently high doses, radiation causes illness and death to living creatures. People who work with radioactive materials use radiation shielding as protection.

7. Objects can be dated, processes can be mapped, organs can be imaged, and cancers can be treated using radioactivity.

Skills to Master

▶ Writing and interpreting symbols for nuclides

▶ Predicting whether or not a nuclide is stable

▶ Calculating energies of nuclear reactions

▶ Calculating coulombic barriers for nuclear reactions

▶ Describing nuclear decay reactions and binuclear reactions

▶ Working with nuclear half-lives

▶ Describing the characteristics of fission and fusion

▶ Tracing the course of stellar evolution and nuclear synthesis

▶ Dating ancient objects

Learning Exercises

21.1 Write a paragraph summarizing the important features of each of the following topics: (a) nuclear stability; (b) nuclear decay; (c) fission; (d) fusion; and (e) binding energy.

21.2 Outline the requirements for a balanced nuclear equation, and explain how they differ from those for a balanced chemical equation.

21.3 Prepare a list of the beneficial and detrimental effects that have resulted from the discovery of nuclear reactions.

21.4 Make a list of all terms introduced in Chapter 21 that are new to you. Write a one-sentence definition for each, using your own words. If you need help, consult the Glossary.

21.5 Update your list of memory bank equations to include those that apply to nuclear chemistry and radiochemistry.

Problems

Nuclear Stability

21.1 Determine Z, A, and N for each of the following nuclides: (a) ^{6}Li; (b) ^{43}Ca; (c) ^{238}U; (d) ^{130}Te; (e) the nuclide of neon that contains the same number of protons and neutrons; and (f) the nuclide of lead that contains 1.5 times as many neutrons as protons.

21.2 Determine Z, A, and N for each of the following nuclides: (a) ^{22}Ne; (b) ^{202}Pb; (c) ^{41}K; (d) ^{109}Ag; (e) the helium nuclide with one less neutron than proton; and (f) the nuclide of barium whose neutron:proton ratio is 1.25.

21.3 Write the correct elemental symbols for each of these nuclides (include the atomic number subscript): (a) helium with the same number of neutrons and protons; (b) tungsten with 110 neutrons; (c) the nuclide with $Z = 28$ and $N = 32$; and (d) the nuclide with 12 protons and 14 neutrons.

21.4 Write the correct elemental symbols for each of these nuclides (include the atomic number subscript): (a) the nuclide with $Z = 50$ and $N = 64$; (b) atomic number 86 possessing 133 neutrons; (c) sulfur with two more neutrons than protons; and (d) antimony with 70 neutrons.

21.5 Predict whether each of the following nuclides is stable or unstable and give reasons for your prediction: (a) carbon-10; (b) neptunium-239; (c) cadmium with 60 neutrons; and (d) element number 43 with 55 neutrons.

21.6 Predict whether each of the following nuclides is stable or unstable, and give reasons for your prediction: (a) carbon-15; (b) ruthenium with 58 neutrons; (c) fluorine-18; and (d) ^{211}At.

21.7 Compute the energy released in kilojoules when the sun converts 1.00 metric ton of matter into energy.

21.8 Compute the energy released in joules per event and in kilojoules per mole when antiprotons (the antimatter corresponding to protons) annihilate with protons.

21.9 Use atomic masses to compute the total binding energy and the binding energy per nucleon for elemental cesium, which has just one stable nuclide.

21.10 Use atomic masses to compute the total binding energy and the binding energy per nucleon for elemental cobalt, which has just one stable nuclide.

21.11 Is the coulombic barrier for fusion of ^{1}H and ^{6}Li larger, smaller, or about the same as that for the fusion of two ^{4}He nuclei? Which gives a more stable product? Explain.

21.12 Is the coulombic barrier for fusion of two protons larger, smaller, or about the same as that for the fusion of two deuterium nuclei? Which of the resulting helium nuclei is less stable? Explain.

Nuclear Decay

21.13 Write the correct symbol and give the name for each of the following products of nuclear decay: (a) a photon; (b) a positive particle with mass number four; and (c) a positive particle with the same mass as an electron.

21.14 What are the charge number, mass number, and symbol for each of the following: (a) an alpha particle; (b) a beta particle; and (c) a neutron?

21.15 Identify the product of each of the following decay processes: (a) Tellurium with 73 neutrons emits a γ-ray. (b) Tellurium with 71 neutrons captures an electron. (c) Tellurium with 75 neutrons undergoes β- and γ-decay.

21.16 Identify the product of each of the following decay processes: (a) ^{52}Fe emits a positron and a γ-ray. (b) ^{55}Fe captures an electron. (c) ^{59}Fe undergoes β- and γ-decay.

21.17 The iron isotope with 33 neutrons is used in medical applications. It is a β-emitter. A sample of iron containing 1.33 picograms (pico = 10^{-12}) of this isotope registers 242 decays per second. Calculate the half-life of the iron isotope.

21.18 A radioactive nuclide of mass number 94 has been prepared by neutron bombardment. If 4.7 μg of this nuclide registers 20 counts per minute on a radioactivity counter, what is the half-life of this nuclide?

21.19 Thorium-232 decays by the following sequence of emissions: α, β, β, α, α, α, β, α, β, α. Identify the final product and all the unstable intermediates.

21.20 Radon-222 decays by the following sequence of emissions: α, α, β, β, α, β, β, α. Identify the final product and all the unstable intermediates.

Induced Nuclear Reactions

21.21 Identify the compound nucleus and final product resulting from each of the following nuclear reactions: (a) Carbon-12 captures a neutron and then emits a proton. (b) The nuclide with eight protons and eight neutrons captures an α-particle and emits a γ-ray. (c) Curium-247 is bombarded with boron-11, and the product loses three neutrons.

21.22 Identify the compound nucleus and final product resulting from each of the following nuclear reactions: (a) Lithium-7 captures a proton, and the product emits a neutron. (b) The nuclide with six protons and six neutrons captures an α-particle and emits a γ-ray. (c) Plutonium-244 is bombarded with carbon-12, and the product loses two neutrons.

21.23 Draw a nuclear picture (see Figure 21-1 for pictures of nuclei) that illustrates the capture of an α-particle by ^{14}N, followed by the emission of a positron.

21.24 Draw a nuclear picture (see Figure 21-1 for pictures of nuclei) that illustrates the nuclear reaction responsible for the production of carbon-14. Include a picture of the compound nucleus.

$$^{14}\text{N} + \text{n} \longrightarrow {}^{14}\text{C} + \text{p}$$

Nuclear Fission

21.25 When uranium-235 undergoes fission, the product with higher mass is most likely to have a mass number between 132 and 142. Use Figure 21-5 to identify which elements are most likely to result.

21.26 When uranium-235 undergoes fission, the product with lower mass is most likely to have a mass number between 88 and 100. Use Figure 21-5 to identify which elements are most likely to result.

21.27 Compute the energy released in the following fission reaction:

$$^{235}\text{U} + \text{n} \longrightarrow {}^{152}\text{Nd} + {}^{81}\text{Ge} + 3\text{ n}$$

The nuclide masses in grams per mole are U, 235.0439; Nd, 151.9233; and Ge, 80.9199. Compare your result with the calculation in Section Exercise 21.4.1.

21.28 Compute the energy released in the following fission reaction:

$$^{235}\text{U} + \text{n} \longrightarrow {}^{100}\text{Mo} + {}^{134}\text{Sn} + 2\text{ n}$$

The nuclide masses in grams per mole are U, 235.0439; Mo, 99.9076; and Sn, 133.9125. Compare your result with the calculation in Section Exercise 21.4.1.

21.29 A coal-fired power plant requires just one heat exchanger, in which water is converted to the steam needed to drive the generator. A nuclear power plant requires two. A circulating fluid such as liquid sodium metal absorbs heat from the nuclear core and then transfers this heat to convert water into steam. Describe the features of nuclear reactions that require this second heat exchanger.

21.30 Describe the features of radioactivity that make an accident in a nuclear power plant more devastating than an accident in a coal-burning power plant.

Nuclear Fusion

21.31 How much energy is released if 2.50 g of deuterons fuse with tritium and 50% of the neutrons are captured by ^{6}Li? (Consult reactions and energies given in the text.)

21.32 How much energy will be released if 1.50 g of deuterons and 1.50 g of lithium-6 undergo fusion in a hydrogen bomb? (Consult reactions and energies given in the text.)

21.33 Compute the speed of a tritium nucleus with enough kinetic energy to fuse with a deuteron. Calculate nuclear radii using $r = 1.2A^{1/3}$ fm.

21.34 Compute the speed of an alpha particle with enough kinetic energy to fuse with lithium-6. Calculate nuclear radii using $r = 1.2A^{1/3}$ fm.

21.35 Describe the features of fusion that have prevented fusion power from being commercialized as fission power has been.

21.36 Describe the features of fusion that make it more attractive as a power source than fission.

21.37 Tabulate the differences in temperature and composition of a first-generation star at the hydrogen-, helium-, and carbon-burning stages of its evolution.

21.38 Repeat the tabulation in Problem 21.37 for a second-generation star. List the differences between the two tabulations.

21.39 Using the characteristics of nuclear reactions, explain why no elements with $Z > 26$ form during the evolution and eventual explosion of first-generation stars.

21.40 Second- and third-generation stars may contain significant amounts of carbon; yet carbon-burning does not begin until the third stage of their evolution. Explain this.

Effects of Radiation

21.41 The isotope ^{131}I has a half-life of 8.07 days and undergoes β-decay with an energy of 9.36×10^7 kJ/mol. If a patient with a thyroid disorder is given this isotope and 7.45 pg is incorporated into the thyroid gland, how much energy will the gland receive in one day?

21.42 A patient weighing 65 kg is given an intravenous dose of ^{99m}Tc, a radioactive isotope that decays by γ-emission with energy of 1.35×10^7 kJ/mol and with a 6.0-hour half-life. If 15.0 mL of a 2.50-nanomolar solution is administered and all of the radioactive nuclide decays within the patient's body, what total energy does the patient receive?

21.43 How many β-particles are emitted per second by the dose of ^{131}I described in Problem 21.41?

21.44 How many γ-rays are emitted per second by the dose of ^{99m}Tc described in Problem 21.42?

21.45 Why does exposure to radiation in cancer therapy result in nausea and reduced resistance to infection?

21.46 Why is it more important to protect the reproductive organs from radiation than other organs?

Applications of Radioactivity

21.47 Extremely old rock samples can be dated using the ratio of ^{87}Sr to ^{87}Rb because $t_{1/2}$ for ^{87}Rb is 4.9×10^{11} yr. What is the approximate age of a meteorite that has a ^{87}Sr to ^{87}Rb mass ratio of 0.0050?

21.48 Uranium deposits are dated by determining the ratio of ^{238}U to its final decay product, ^{206}Pb. The half-life of ^{238}U is 4.5×10^9 yr. If the ratio of ^{238}U to ^{206}Pb in an ore sample is 1.21, what is the age of the ore?

21.49 The characteristics of several isotopes of iodine follow. Which would be best suited for medical imaging? Explain your choice.

Mass no.	123	125	127	131
Half-life	13.3 hr	60.2 days	Stable	8.07 days
Decay mode	EC, γ	EC, γ	None	β, γ

21.50 The characteristics of several isotopes of iron follow. Which would be best suited for medical imaging? Explain your choice.

Mass no.	52	55	56	59
Half-life	8.2 hr	2.6 yr	Stable	45.1 days
Decay mode	β^+	EC	None	β, γ

Additional Paired Problems

21.51 Complete the following nuclear reactions: (a) alpha emission from ^{238}U (b) n + ^{60}Ni $\rightarrow$? + p; (c) ^{239}Np + ^{12}C $\rightarrow$? + 3 n; (d) p + ^{35}Cl $\rightarrow \alpha$ + ?; and (e) β-emission from ^{60}Co.

21.52 Complete the following nuclear reactions: (a) positron emission from ^{26}Si; (b) electron capture by ^{82}Sr; (c) ^{210}Po $\rightarrow \alpha$ + ?; (d) α + ^{9}Be $\rightarrow$ n + ?; and (e) ^{99m}Tc $\rightarrow \gamma$ + ?

21.53 Naturally occurring bismuth contains only one isotope, ^{209}Bi. Compute the total molar binding energy and molar binding energy per nucleon of this element.

21.54 Naturally occurring gold contains only one isotope, ^{197}Au. Compute the total molar binding energy and molar binding energy per nucleon of this element.

21.55 Polonium-210 has a half-life of 138.4 days. If a sample contains 5.0 mg of ^{210}Po, how many milligrams will remain after 365 days? How many emissions per second will this residue emit?

21.56 What fraction of ^{209}Po, which has a half-life of 103 years, will remain intact after 365 days? If a sample of polonium contains equimolar amounts of ^{210}Po and ^{209}Po, what will the isotope ratio be after 10 years (see Problem 21.55)?

21.57 Neutrons decay into protons. What is the other product of this decay? If all of the decay energy is converted into kinetic energy of this other product, how much kinetic energy does it have?

21.58 A positron has the same mass as an electron. When a positron and an electron annihilate, both masses are converted entirely into the energy of a pair of γ-rays. Calculate the energy per γ-ray and the energy of 1 mol of γ-rays.

21.59 The long-lived isotope of radium, ^{226}Ra, decays by α-emission with a half-life of 1622 years. Calculate how long it will take for 1% of this nuclide to disappear and how long until 1% of it remains.

21.60 Phosphorus-30, which has a half-life of 150 seconds, decays by positron emission. How long will it take for (a) 2% and (b) 99.5% of this nuclide to decay?

21.61 Two isotopes used in positron emission imaging are ^{11}C and ^{15}O. On which side of the belt of stability are these nuclides located? Write the nuclear reactions for their disintegrations.

21.62 Neutron bombardment is an elegant way of "doping" silicon with phosphorus to convert it into an n-type semiconductor. Bombardment generates stable ^{31}P from ^{30}Si. Write the nuclear reactions for this transformation. (Doped semiconductors are described in Chapter 9.)

21.63 Sodium-24, $t_{1/2}$ = 15.0 hr, can be used to study the sodium balance in animals. If a saline solution containing 25 μg of ^{24}Na is injected into an animal, how long can the study continue before only 1 μg is left?

21.64 What is the shelf life of a radioactive nuclide with a half-life of 14.6 days if it loses its usefulness when less than 15% of the radioactivity remains?

21.65 Free neutrons are unstable. Since they cannot be collected and weighed, it is difficult to measure their half-life accurately. Early estimates gave $t_{1/2}$ = 1100 s, but more refined experiments give $t_{1/2}$ = 876 ± 21 s. Suppose a neutron source generates 10^5 neutrons per second for 30 seconds. Assuming no neutrons are captured by nuclei, how many will be left after 5 hours, according to each of these half-lives?

21.66 Biochemical studies of DNA can be carried out using the unstable isotope phosphorus-32 ($t_{1/2}$ = 14.3 days). If 125 mg of $K_3(^{32}PO_4)$ is used in one particular study, how many milligrams will remain after 4 days?

21.67 The heaviest transuranium elements are formed by bombardment with relatively heavy nuclides such as ^{48}Ti. What nuclide could be formed by bombarding lead-208 with this nuclide?

21.68 The heaviest transuranium elements are formed by bombardment with relatively heavy nuclides such as ^{58}Fe. What nuclide could be formed by bombarding the stable bismuth nuclide with this nuclide?

21.69 Gd-152 ($MM = 151.9205$ g/mol) emits an α-particle, energy $= 3.59 \times 10^{-13}$ J to form ^{148}Sm ($MM = 147.9146$ g/mol). How much energy (in J/nucleus) is released in this transformation? What fraction of the energy is carried off by the α-particle?

21.70 Samarium-146 ($MM = 145.9129$ g/mol) emits an α-particle, $E = 3.94 \times 10^{-13}$ J and becomes ^{142}Nd ($MM = 141.9075$ g/mol). How much energy (in J/nucleus) is released in this transformation? What fraction of the energy is carried off by the α-particle?

21.71 State why each of the following nuclides is unstable: (a) carbon-14; (b) plutonium-244; (c) ^{56}Mn; and (d) a lithium nucleus with five neutrons.

21.72 State why each of the following nuclides is unstable: (a) tritium (one proton, two neutrons); (b) uranium-238; (c) ^{40}K; and (d) a beryllium nucleus with four neutrons.

21.73 Describe the roles of the control rods and the moderator in the operation of a nuclear power plant.

21.74 Describe how irradiation protects foods against spoilage, and describe how irradiation of foods might also produce undesirable byproducts.

More Challenging Problems

21.75 A small amount of NaBr containing the radioactive isotope sodium-24 is dissolved in a hot solution of sodium nitrate containing the naturally occurring nonradioactive isotope sodium-23. The solution is cooled, and sodium nitrate precipitates from the solution. Will the precipitate be radioactive? Explain your answer.

21.76 Radioactive ^{64}Cu ($MM = 63.92976$ g/mol) decays either by emission of β-particles with 9.3×10^{-14} J of energy or by emission of positrons with 1.04×10^{-13} J of energy. (a) Write the two decay reactions. (b) Calculate the molar masses of the two elemental products using mass-energy equivalence.

21.77 Nuclear power plants often use boron control rods. Boron-10 absorbs neutrons efficiently, emitting α-particles as a consequence. Write the nuclear reaction. Does this reaction pose a significant health hazard? Explain your reasoning.

21.78 The radius of a ^{12}C nucleus is 3.0×10^{-15} m. Compute the energy barrier in J and in kJ/mol for the fusion of two ^{12}C nuclei.

21.79 A 250-mg sample of CO_2 collected from a small piece of wood at an archaeological site gave 1020 counts over a 24-hour period. In the same counting apparatus, 1.00 g of CO_2 from freshly cut wood gave 18,400 counts in 20 hours. What age does this give for the site?

21.80 One isotope of nitrogen and one isotope of fluorine are positron emitters with relatively long half-lives. Identify them and write their decay reactions.

21.81 The amount of radioactive carbon in any once-living sample eventually drops too low for accurate dating. This detection limit is about 0.03/g min, whereas fresh samples exhibit a count rate of 15.3/g min. What is the upper limit for age determinations using carbon dating?

21.82 Carbon-14 dating gives 3250 years as the age of a charcoal sample, assuming a constant level of cosmic radiation. If the cosmic ray level in the atmosphere was 20% higher at the time the tree grew, what is the correct age of the sample?

21.83 Strontium-90 is a dangerous fission product because it passes into human bodies and lodges in bones. It decays in two steps to give ^{90}Zr. The molar masses of these two nuclides are 89.9073 and 89.9043 g/mol. (a) Write the decay reactions that lead from ^{90}Sr to ^{90}Zr. (b) Which nuclide has the higher molar mass? (c) Calculate the total energy released in the decay scheme. (Reminder: Include the masses of the other decay products.)

21.84 Suppose that fusion of tritium and deuterium involves two nuclei with equal kinetic energies totaling 75% of the coulombic barrier. Calculate their speeds.

21.85 Iodine-123, $t_{1/2} = 13.2$ hr, is used in medical imaging of the thyroid gland. If 0.5 mg of this isotope is injected into a patient's bloodstream and 45% of it binds to the thyroid gland, how long will it take for the amount of iodine in the thyroid gland to fall to less than 0.1 μg?

21.86 In 1934, Irene Curie and Frederic Joliot produced the first artificial radioisotope by bombarding ^{27}Al with α-particles. The resulting compound nucleus decayed by neutron emission. Write the reaction and predict whether or not the final product is stable. If it is not stable, predict its mode of decay.

21.87 Technetium has no stable isotopes. Which nuclide of this element would you predict to be nearly stable? (Make use of Figure 21-5.)

21.88 Calculate the mass of ^{235}U that reacts in a 20-kiloton bomb, given that one nucleus releases 2.9×10^{-11} J. The designation "20-kiloton" means the same energy as 20×10^3 metric tons of TNT, which releases 2500 kJ of energy per kilogram.

21.89 As discussed in Chapter 11, a condensation reaction between a carboxylic acid and an alcohol provides an ester with the elimination of water:

Explain how the unstable isotope ^{18}O can be used to show whether the oxygen atom in the water molecule comes from the carboxylic acid or from the alcohol.

21.90 A radioactivity counter gave a reading of 350 min^{-1} for a 12.5-mg sample of cobalt(II) chloride partially enriched with cobalt-60 ($t_{1/2} = 5.2$ yr). What percent of the cobalt atoms are cobalt-60?

Group Study Problems

21.91 Uranium-238 undergoes eight consecutive α-emissions to give stable lead. In a uranium ore sample, all the α-particles are quickly stopped, becoming trapped ^{4}He atoms. If analysis of a rock sample shows that it contains 6.0×10^{-5} cm^3 of helium-4 (1 atm, 298 K) and 1.3×10^{-7} g of ^{238}U per gram of rock, estimate the age of the rock.

21.92 From the mass number distribution given in Figure 21-12 and the belt of stability shown in Figure 21-4, predict which elements should be found in greatest abundance in the remains of a fission event like that of the Oklo reactor.

21.93 The Earth captures 3.4×10^{17} J/s of radiant energy, which is one part in 4.5×10^{-10} of the sun's total energy output. (a) By how much does the mass of the sun change every second? (b) The sun is in the hydrogen-burning stage of its evolution. Calculate how many moles of ^1H are converted into ^4He per second.

21.94 For how many years could ^{235}U supply the world's energy needs, if uranium contains 0.72% of ^{235}U, the world reserves of uranium are 1.0×10^7 metric tons, world energy consumption is 2.0×10^{17} kJ/yr, and each ^{235}U nucleus releases 2.9×10^{-11} J?

21.95 Smoke detectors contain small quantities of ^{241}Am, which decays by emitting an α-particle with 5.44×10^8 kJ/mol energy and a 458-year half-life. It also emits γ-rays. (a) What is the neutron : proton ratio for ^{241}Am? (b) If a smoke detector operates at 5.0 decays per second, what mass of ^{241}Am does it contain? (c) If this smoke detector malfunctions when its decay rate falls below 3.5 decays per second, for how many years could it be used? (d) If your bed were directly under this smoke detector, would you be exposed to its nuclear radiation? Explain.

21.96 Unlike most other elements, different samples of lead have different molar masses. This is because lead is the stable end product of two different radioactive decay schemes, one starting from ^{238}U and the other from ^{232}Th. Both decay schemes consist entirely of α-, β-, and γ-emissions. The most abundant stable lead isotopes have mass numbers 206, 207, and 208. Will a sample of lead, some of which came from ^{238}U, have a higher or lower molar mass than another sample of lead, some of which came from ^{232}Th? Draw a decay scheme that illustrates your reasoning.

21.97 List the advantages and disadvantages of fission power and fusion power. Based on your list, do you think that the United States should continue to develop fission power plants? What about fusion power plants?

Answers to Section Exercises

21.1.1 (a) $^{20}_{10}Ne$, $Z = 10$, $N = 10$, $A = 20$; (b) $^{98}_{43}Tc$, $Z = 43$, $N = 55$, $A = 98$; and (c) ^3_1H, $Z = 1$, $N = 2$, $A = 3$.

21.1.2 (a) Unstable, $Z > 83$; (b) unstable, odd-odd; (c) stable; (d) unstable, $N < Z$

21.1.3 Total binding energy is -1.39×10^{10} kJ/mol; binding energy per nucleon is -7.29×10^8 kJ/mol.

21.2.1 (a) β-decay; (b) α-decay; (c) γ-emission; and (d) positron emission or electron capture

21.2.2 (a) $^{140}_{57}La$; (b) $^{222}_{86}Rn$; and (c) $^{125}_{53}I$

21.2.3 12.1 days

21.3.1 (a) $^{59}_{27}Co + ^1_0n \longrightarrow (^{60m}_{27}Co) \longrightarrow ^{59}_{26}Fe + ^1_1p$
(b) $^{59}_{27}Co + ^1_0n \longrightarrow (^{60m}_{27}Co) \longrightarrow ^{60}_{27}Co + \gamma$
(c) $^{59}_{27}Co + ^4_2\alpha \longrightarrow (^{63m}_{29}Cu) \longrightarrow ^{62}_{28}Ni + ^1_1p$
(c) $^{59}_{27}Co + ^1_1p \longrightarrow (^{60m}_{28}Ni) \longrightarrow ^{58}_{28}Ni + 2 ^1_0n$

21.3.2 There are several possibilities. Here are two:
(a) $^{144}_{60}Nd + ^1_0n \longrightarrow (^{145m}_{60}Nd) \longrightarrow ^{145}_{61}Pm + ^{\;\;0}_{-1}\beta$
(b) $^{147}_{62}Sm + ^1_0n \longrightarrow (^{148m}_{62}Sm) \longrightarrow ^{147}_{61}Pm + ^1_1p$

21.3.3 (Answers are other reactant, then other product) (a) neutron, γ-ray; (b) ^{54}Fe, ^{57}Ni; and (c) proton, α-particle.

21.4.1 One neutron: -0.2151 g/mol, -1.93×10^{10} kJ/mol; Two neutrons: -0.2065 g/mol, -1.86×10^{10} kJ/mol; Three neutrons: -0.1978 g/mol, -1.78×10^{10} kJ/mol

21.4.2 99.22% ^{238}U and 0.78% ^{235}U

21.5.1 (a)

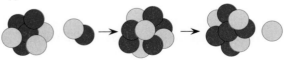

(b) 5.5×10^7 kJ/mol; (c) the first step is endothermic by 1.82×10^7 kJ/mol (but this energy could be provided by the kinetic energy of the deuterons); the second step is exothermic by -1.50×10^9 kJ/mol; and (d) 3×10^{12} kJ.

21.5.2 More than one sequence is possible. Here is one:

$$^{56}Fe \xrightarrow{+n} ^{57}Fe \xrightarrow{+n} ^{58}Fe \xrightarrow{+n} ^{59}Fe \xrightarrow{-\beta} ^{59}Co$$

$$^{59}Co \xrightarrow{+n} ^{60}Co \xrightarrow{-\beta} ^{60}Ni$$

$$^{60}Ni \xrightarrow{+n} ^{61}Ni \xrightarrow{+n} ^{62}Ni \xrightarrow{+n} ^{63}Ni \xrightarrow{-\beta} ^{63}Cu$$

$$^{63}Cu \xrightarrow{+n} ^{64}Cu \xrightarrow{+n} ^{65}Cu \xrightarrow{+n} ^{66}Cu \xrightarrow{-\beta} ^{66}Zn$$

21.5.3 Planets around a second-generation star contain no elements with $Z > 26$. Simple life as we know it requires no elements with $Z > 26$, but mammalian life depends on enzymes containing copper, zinc, and cobalt.

21.6.1 4.5×10^5 β-particles, which deposit 1.0×10^{-7} J of energy.

21.6.2 12,500 X rays; the damage would not show up in a year.

21.7.1 6.5×10^3 years

21.7.2 (a) Good choice: it is a γ-emitter with a short half-life; (b) poor choice: it is a β-emitter with a relatively long half-life; (c) poor choice because of its long half-life; and (d) good choice, even though its half-life is long, because positron emitters do little damage.

21.7.3 Prepare a sample of solid I_2 containing some of the radioactive isotope. (a) Put some of this solid in an aqueous solution containing a salt such as KI. Withdraw samples at various times, precipitate the I^- as AgI, and measure the radioactivity of the solid. (b) Put some of the solid in a closed container equipped with a gas sampling device. Collect gas samples at various times and measure the radioactivity of the gas.

Appendix A:
Scientific Notation

Special notations are useful for expressing the precisions of very large and very small quantities. The diameter of a helium atom, for instance, is 0.00000000024 meters, and the average distance from the Earth to its moon is 384,000,000 meters. Writing numbers with this many zeros is cumbersome, and it is easy to make a mistake. To shorten the writing of small and large numbers, scientists commonly use **scientific notation.**

Scientific notation is based on the fact that any number can be expressed as a number between 1 and 10 multiplied or divided by ten an appropriate number of times. For example, 384,000,000 meters can be written as:

$$(3.84)(10)(10)(10)(10)(10)(10)(10)(10) = 3.84 \times 10^8 \text{ meters.}$$

Similarly, 0.00000000024 meters is:

$$2.4 \div (10)(10)(10)(10)(10)(10)(10)(10)(10)(10) = 2.4 \times 10^{-10} \text{ meters.}$$

The numeral "8" in 10^8 and the numeral "-10" in 10^{-10} are examples of **exponents** or **powers.** The scientific notation, 3.84×10^8 m, contains an exponent (or power of ten) of eight.

▬▬ CONVERTING TO SCIENTIFIC NOTATION

When a number is *divided* by ten, the decimal point moves one place to the *left;* when a number is *multiplied* by ten, the decimal point moves one place to the *right:*

$$\frac{384}{10} = 38.4 \quad and \quad (0.0024)(10) = 0.024$$

Multiplying *or* dividing a number by ten changes its value, but multiplying *and* dividing a number by ten leaves its value unchanged. Thus to convert a number larger than ten to scientific notation, first divide by ten the number of times that give a number between one and ten, then multiply by ten that same number of times. Here, for example, is the conversion of the number of seconds in a day into scientific notation:

$$\frac{86,400 \text{ s/day}}{(10)(10)(10)(10)} \times (10)(10)(10)(10)$$

$$\downarrow \qquad\qquad\qquad \downarrow$$

$$8.6400 \qquad\qquad 10^4 \qquad = 8.6400 \times 10^4 \text{ s/day}$$

The four divisions by ten move the decimal point four places to the left, and the four multiplications by ten are written as 10^4.

A number larger than ten can be quickly converted into scientific notation by moving the decimal point to the left enough places to give a number between one and ten, and multiplying by ten raised to the positive power that equals the number of places moved. For example:

$$96{,}485 = 9.6485 \times 10^4$$
$$\text{4 places}$$

To convert a number smaller than one into scientific notation, first multiply by ten as many times as needed to give a number between one and ten, then divide by ten that same number of times. Here, for example, is the conversion into scientific notation of the fraction of a day represented by one second:

$$0.000011574 \text{ day/s } (10)(10)(10)(10)(10) \times \frac{1}{(10)(10)(10)(10)(10)}$$

$$\downarrow \qquad\qquad\qquad \downarrow$$

$$1.1574 \qquad\qquad\qquad 10^{-5} = 1.1574 \times 10^{-5} \text{ day/s}$$

To convert a number smaller than ten into scientific notation, move the decimal point to the right enough places to give a number between one and ten and multiply by ten raised to the negative power equal to the number of places moved.

With practice, these conversions can be carried out quite quickly. The key is to count the number of places the decimal point must be moved. Here are some additional examples:

$$5260 \text{ ft/mi} = 5.260 \times 10^3 \text{ ft/mi}$$
$$\text{Move left three places}$$

$$0.08206 \text{ L atm/mol K} = 8.206 \times 10^{-2} \text{ L atm/mol K}$$
$$\text{Move right two places}$$

$$384{,}000{,}000 \text{ mi} = 3.84 \times 10^8 \text{ mi}$$
$$\text{Move left eight places}$$

$$0.00000000024 \text{ m} = 2.4 \times 10^{-10} \text{ m}$$
$$\text{Move right ten places}$$

■ WORKING WITH POWERS OF TEN

Chemistry students must be able to add, subtract, multiply, and divide numbers that are expressed in scientific notation. This is easily accomplished on a calculator by entering the numbers using power-of-ten notation (10^x function key). It is also useful to understand how these operations are carried out. Multiplication and division are handled in one fashion, but addition and subtraction are handled in another fashion.

To *multiply* two numbers expressed in scientific notation, multiply the values and *add* the exponents:

$$(2.450 \times 10^2)(1.680 \times 10^3) = (2.450)(1.680) \times 10^{(2 + 3)} = 4.116 \times 10^5$$

Retain signs when adding exponents:

$$(2.450 \times 10^2)(1.680 \times 10^{-3}) = (2.450)(1.680) \times 10^{(2 - 3)} = 4.116 \times 10^{-1}$$

To *divide* numbers expressed in power of ten notation, divide the values and *subtract* the exponent of the divisor from that of the number divided:

$$\frac{2.450 \times 10^2}{1.680 \times 10^3} = \frac{2.450}{1.680} \times 10^{(2 - 3)} = 1.458 \times 10^{-1}$$

As with multiplication, retain signs when subtracting exponents:

$$\frac{2.450 \times 10^2}{1.680 \times 10^{-3}} = \frac{2.450}{1.680} \times 10^{[2 - (-3)]} = 1.458 \times 10^5$$

Addition and subtraction require a different approach. Consider, for example, the addition of these two numbers:

$$2.450 \times 10^2 + 1.680 \times 10^3 = ?$$

Using standard notation, this addition is as follows:

$$
\begin{array}{r}
1680 \\
\underline{245.0} \\
1925
\end{array}
$$

To obtain this result in scientific notation, express both numbers using the *same* power of ten, most conveniently the largest such power:

$$2.450 \times 10^2 = 0.2450 \times 10^3$$

$$0.2450 \times 10^3 + 1.680 \times 10^3 = 1.925 \times 10^3$$

To add or subtract numbers expressed in power of ten notation, first express all numbers using the *same power of ten*. Then add or subtract the values and retain the same power of ten.

Subtraction may give a result that is smaller than one. When this happens, the convention is to change the exponent to make the numerical value between one and ten. Here, for example, is the difference between the atomic radii of oxygen atoms and fluorine atoms:

O atom		F atom		Difference		Proper notation
1.40×10^{-10} m	$-$	1.35×10^{-10} m	$=$	0.05×10^{-10} m	$=$	5×10^{-12} m

To express this result using a number between one and ten, move the decimal point two places to the right and increase the negative power of ten by two.

This last example helps to show how the use of scientific notation simplifies the operations with the very small and very large numbers encountered in chemistry. Carried out in standard notation, this subtraction requires a whole lot of zeros:

$$
\begin{array}{r}
0.000000000140 \text{ m} \\
- \underline{0.000000000135 \text{ m}} \\
0.000000000005 \text{ m}
\end{array}
$$

Appendix B: Quantitative Observations

Chemists usually want to know not only *what* is happening but also to what extent (*how much*). Quantitative measurements have three equally important parts: a *numerical value*, appropriate *units*, and a *precision*. Any experimentally measured quantity *always includes all these parts.* Numerical value refers to the numbers. Units are what allow us to scale a numerical value appropriately. For example, 20 *miles* is a very different length than 20 *centimeters.* Precision requires further description.

PRECISION AND SIGNIFICANT FIGURES

The precision of a quantitative value is the degree of certainty with which it is known. For example, "about 20 miles" is a less precise statement than "21.5 miles." The basic rule for precision is that the number of digits in the numerical value expresses the precision of the measurement.

As an example of this rule, consider measuring the length of a table. After a quick measurement, you might estimate, "This table is 1.8 meters long." You have expressed the length as a two-digit number, meaning that you know the table to be longer than 1.7 but shorter than 1.9 meters. A more careful measurement with a tape measure might yield a result containing four digits, 1.826 meters. This means that the table is longer than 1.825 meters but shorter than 1.827 meters.

The number of digits in a numerical result is called its number of **significant figures.** Unless otherwise stated, the measurement is precise to within one unit in the last significant figure. Thus, a result reported as 1.826 meters has four significant figures. Unless additional information is given, the last digit (6) is uncertain by one unit: the length is $1.826 \pm .001$ meters. That is, the length falls somewhere between 1.825 meters and 1.827 meters.

When quantities are expressed in scientific notation, extra zeros indicate extra precision. For instance, 3.840×10^8 meters means "not less than 3.839×10^8, nor more than 3.841×10^8 meters," whereas 3.84×10^8 meters means "not less than 3.83×10^8, nor more than 3.85×10^8 meters."

The precision of a measurement depends on the quality of the measuring device used to obtain the measurement. Whereas very sensitive instruments yield measurements of high precision, less sensitive instruments yield results of lower precision. An electronic balance can determine the mass of 24 pennies to the nearest 0.0001 g, giving a measurement with six significant figures: 63.5465 g. A pan balance, on the other hand, can determine this same mass only to the nearest 0.01 g, giving a measurement with four significant figures: 63.55 g.

When a quantity fluctuates in value, the precision in its measurement is also limited by these fluctuations. For example, the distance between the Earth and its moon fluctuates with time, because the moon's orbit is elliptical rather than perfectly circular. The average distance is 3.844×10^5 km, but the actual distance varies between 3.564 and 4.067×10^5 km. Thus, the instantaneous distance between the Earth and its moon varies by 4.97×10^4 kilometers in the course of a month.

RELATIVE AND ABSOLUTE PRECISION

The degree of precision in a given experimental value can be expressed either as an **absolute** precision or as a **relative** precision. Absolute precision is the numerical uncertainty in the experimental value, and relative precision is the absolute precision divided by the experimental value. Absolute precisions have units, but relative precisions are always dimensionless ratios. Each of these ways of looking at precision has useful applications.

To illustrate these concepts, return to the example of table length. If a table has been measured to be 1.826 meters long, the absolute precision of the measurement is ±0.001 meters. The relative precision is the absolute precision divided by the length, $(0.001 \text{ m})/(1.826 \text{ m})$, or 5×10^{-4}. This relative precision can be expressed as 1/1826, or 5×10^{-4}, or 0.05%, or 5 parts per 10,000. Each of these expressions is a ratio of two values that have the same units and is therefore dimensionless. Notice that because the absolute precision contains only one significant figure, this relative precision also contains only one significant figure.

When measurements are added or subtracted to compute a result, the *absolute* precision is more useful. When measurements are multiplied or divided to compute a result, the *relative* precision is more useful. As illustrations, return to the table example and consider the area and perimeter of the table top. Suppose that measurements show its width to be 3.20×10^{-1} meters. The absolute precision of this measurement is 0.001 meters. The relative precision of the width measurement is $(0.001 \text{ m})/(3.20 \times 10^{-1} \text{ m})$, which is 1/320, or 3×10^{-3}.

Now combine these measurements to determine the perimeter P and the area A. Perimeter is $P = L + L + W + W = 2(1.826) + 2(3.20 \times 10^{-1}) = 4.292$ m, and area is $A = (L)(W) = (1.826)(3.20 \times 10^{-1}) = 5.84 \times 10^{-1}$ m². How precisely is each of these computed values known? We find the precision of a computed result from the precisions of individual measurements by considering the largest possible variation in each individual measurement.

First, consider the table's perimeter. According to the precision of the length measurement, the length might be as large as 1.827 m or as small as 1.825 m. The width might be as large as 0.321 m or as small as 0.319 m. Thus the perimeter could be as large as $2(1.827) + 2(.321) = 4.296$ m, and it might be as small as $2(1.825) + 2(.319) = 4.288$ m. The precision in the perimeter is ±0.004 m. This is the *sum of the absolute precisions* of the individual values: $0.001 + 0.001 + 0.001 + 0.001 = 0.004$.

Applying the same logic to the area reveals that it might be as large as $1.827 \times 0.321 = 5.87 \times 10^{-1}$ m² or as small as $1.825 \times 0.319 = 5.82 \times 10^{-1}$ m². The area is $5.84 \times 10^{-1} \pm 0.025 \times 10^{-1}$ m², meaning that there is an imprecision of $(0.025 \times 10^{-1})/(5.84 \times 10^{-1}) = 4 \times 10^{-3}$. This is the *sum of the relative precisions* of length (5×10^{-4}) and width (3×10^{-3}).

▬▬ RULES FOR DETERMINING PRECISION

The precision of a composite result can always be determined by the type of analysis described above, but this procedure is tedious. Fortunately the outcome is always the same. For *addition and subtraction,* the *absolute* precision of the result is the sum of the *absolute* precisions of the individual values. For *multiplication and division,* the *relative* precision of the result is the sum of the *relative* precisions of the individual values.

Another example illustrates the application of these rules. To determine the density of a liquid, a student filled a 25-mL graduated cylinder until it contained 25.0 mL of liquid. The mass of the full cylinder was 47.5764 g. The student removed liquid until the cylinder contained 20.0 mL of liquid, whereupon its mass was 43.0464 g. Table B-1 summarizes the measurements, computations, and precisions involved in this example.

Table B-1
Measured and Derived Quantities and Precisions

Quantity	Value	Absolute Precision	Relative Precision
Measured Quantities			
Initial cylinder volume	25.0 mL	0.1 mL	1/250
Initial cylinder mass	47.5764 g	0.0001 g	1/475,000
Final cylinder volume	20.0 mL	0.1 mL	1/200
Final cylinder mass	43.0464 g	0.0001 g	1/430,000
Computed Quantities			
Volume transferred	5.0 mL	**0.2 mL**	2/50 = 1/25
Mass transferred	4.5300 g	**0.0002 g**	1/20,000
Liquid density	0.9060 g/mL	0.04	**1/25**

The absolute precision of each *measured quantity* is found directly from the measurement. The absolute precision of the mass of the cylinder is the precision of the balance, 0.0001 g. Each relative precision is the absolute precision divided by the measured value; thus, the relative precision of the mass of the cylinder is 1/475,000. The precision of each *computed quantity* is determined by adding the precisions of quantities used in the calculation. The values in bold face are found directly. The volume transferred (5.0 mL) is obtained by *subtracting* two volumes, so its *absolute* precision is the sum of the *absolute* precisions of the volumes (0.1 + 0.1 = 0.2 mL). Similar reasoning applies to the mass transferred. The density (0.9060 g/mL) is obtained by *dividing* mass by volume, so its *relative* precision is the sum of the *relative* precisions of transferred mass and transferred volume (1/25 + 1/20,000 = 1/25). Once the relative precision of the density has been calculated, it can be used to compute the absolute precision. The absolute precision of the density is (1/25)(0.9060 g/mL) = ± 0.04 g/mL.

ROUNDING OFF

In the above example, the density might be as large as 0.94 g/mL or as small as 0.87 g/mL. It would be incorrect to write it as 0.9060 g/mL, because the use of four significant figures implies that the value is known to ±0.0001 g/mL. All non-significant figures must be eliminated, a process that is called **rounding off.** Because the uncertainty in the density is in the *second* decimal place, this result is rounded off to two decimals: 0.91 g/mL. This still overstates the precision, because 0.91 g/mL implies a precision of 0.01 g/mL, but this result is actually known only to ±0.04 g/mL. However, rounding off one more place to give 0.9 g/mL would imply an uncertainty of 0.1 g/mL, which is larger than the actual uncertainty. The convention is to round off until dropping one more digit would result in an uncertainty larger than the actual uncertainty.

When the digit following the last significant digit is 5 or greater, the remaining digit is increased by 1 unit. For example, 0.9060 becomes 0.91. If the digit following the last significant digit is less than 5, the remaining digit remains unchanged. For example, 0.9045 rounded to 2 significant figures is 0.90.

SHORTCUTS TO PRECISION

You can spend much effort figuring out the appropriate number of significant figures in a result. Every time a scientist makes a measurement, precision and significant figures are a concern, but scientists are more interested in what experiments reveal than they are in significant figures. Precision may be extremely important, for example when determining the level of a particular carcinogen in a sample of ground water. Because the amount of carcinogen may be critical to human health, the precision of this measurement would have to be carefully stated. Precision is not so important to a chemist whose goal is to prepare new compounds. The chemist wishes to achieve a high yield but is not concerned about whether a yield is 90% or 90.05%.

The following simple guidelines are sufficient to determine the appropriate number of significant figures in General Chemistry:

1. To determine the number of significant figures in an individual measurement, count the digits, left-to-right, beginning with the first one that is non-zero.

305	3 sig. figs.
1.00	3 sig. figs.
0.020	2 sig. figs.

2. When adding or subtracting, set the number of *decimal places* in the answer equal to the number of *decimal places* in the number with the fewest places. The number of significant figures is irrelevant:

	0.12	2 decimal places
	1.6	**1 decimal place**
	11.490	3 decimal places
Sum:	13.2	**1 decimal place**

3. When multiplying or dividing, set the number of *significant figures* in the answer equal to that of the quantity with the fewest *significant figures*. The number of decimal places is irrelevant:

	1.365	$\times$	2.63	$\times$	0.33	=	1.2
Sig. figs.:	**4**		3		**2**		**2**

The above example is one where the simplified procedure gives a different result than the more elaborate one. One number (0.33) has a relative precision of 1/33, so the result, 1.1847, could be rounded to 1.18 (1 part in 118) rather than to 1.2 (1 part in 12). If the importance of the result is its *quantitative value*, 1.18 would be the more appropriate number to report; if its importance is in its *qualitative* significance, report 1.2. (But don't spend a lot of time worrying about it: either way of reporting is legitimate! What is *not* legitimate is to report this result as **1.1847.**)

4. Calculators do not necessarily give results with the correct number of significant figures. They automatically drop trailing zeroes even when they are significant (try multiplying or adding 3.00 and 5.00 on your calculator) and they carry extra decimal places even when they are insignificant (try dividing 5.00 by 3.00 on your calculator). *Never believe the number of significant figures on your calculator!*

Appendix C: Ionization Energies and Electron Affinities of the First 36 Elements

Z	Symbol	EA	IE_1	IE_2	IE_3	$n^\dagger$
1	H	−72.8	1312	—	—	
2	He	>0*	2372	5250	—	1
3	Li	−59.7	520.2	7298	11,815	
4	Be	>0	899.4	1757	14,848	
5	B	−26.8	800.6	2427	3660	
6	C	−121.9	1086	2353	4620	
7	N	>0	1402	2856	4578	
8	O	−141.1	1314	3388	5300	
9	F	−328.0	1681	3374	6050	2
10	Ne	>0	2081	3952	6122	
11	Na	−52.9	495.6	4562	6912	
12	Mg	>0	737.7	1451	7733	
13	Al	−42.7	577.6	1817	2745	
14	Si	−133.6	786.4	1577	3232	
15	P	−72.0	1012	1908	2912	
16	S	−200.4	999.6	2251	3357	3
17	Cl	−348.8	1251	2297	3822	
18	Ar	>0	1520	2666	3931	
19	K	−48.4	418.8	3051	4420	
20	Ca	−2.4	589.8	1145	4912	
21	Sc	−18.2	633	1235	2389	
22	Ti	−7.7	658	1310	2653	
23	V	−50.8	650	1414	2828	
24	Cr	−64.4	652.8	1591	2987	
25	Mn	>0	717.4	1509	3248	
26	Fe	−14.6	763	1561	2957	
27	Co	−63.9	758	1646	3232	
28	Ni	−111.6	736.7	1753	3396	4
29	Cu	−119.2	745.4	1958	3554	
30	Zn	>0	906.4	1733	3833	
31	Ga	−28.9	578.8	1979	2963	
32	Ge	−119	762.1	1537	3302	
33	As	−78	947	1798	2736	
34	Se	−195.0	940.9	2045	2974	
35	Br	−324.6	1140	2103	3500	
36	Kr	>0	1351	2350	3565	

All values are in kJ/mol.

* A value >0 means that the anion is unstable, so its electron affinity cannot be experimentally determined.

† n is the principal quantum number of the electron whose ionization energy is listed.

Source: *Handbook of Chemistry and Physics, 82nd Edition,* David R. Lide, editor-in-chief, CRC Press LLC, Boca Raton, FL (2001), pp. **10**−47, 48, 175.

Appendix D:
Standard Thermodynamic Functions

$T = 298.15$ K and $P = 1.000$ atm;
Aqueous species at 1.000 M

Substance	ΔH_f° (kJ/mol)	ΔG_f° (kJ/mol)	S° (J/mol K)	Substance	ΔH_f° (kJ/mol)	ΔG_f° (kJ/mol)	S° (J/mol K)
Aluminum				**Carbon**			
Al(s)	0	0	28.3	C(s, graphite)	0	0	5.7
Al_2O_3(s)	−1675.7	−1582.3	50.9	C(s, diamond)	1.9	2.9	2.4
$AlCl_3$(s)	−704.2	−628.8	109.3	CH_4(g)	−74.6	−50.5	186.3
Antimony				C_2H_2(g)	227.4	209.9	200.9
Sb(s)	0	0	45.7	C_2H_4(g)	52.4	68.4	219.3
Sb_4O_6(s)	−1417.1	−1253.0	246.0	C_2H_6(g)	−84.0	−32.0	229.2
Argon				C_3H_6(g)	20.0	74.62	226.9
Ar(g)	0	0	154.843	CO(g)	−110.5	−137.2	197.7
Arsenic				CO_2(g)	−393.5	−394.4	213.8
As(s)	0	0	35.1	HCN(g)	135.1	124.7	201.8
As_2O_5(s)	−924.9	−782.3	105.4	CS_2(l)	89.0	64.6	151.3
$AsCl_3$(l)	−305.0	−259.4	216.3	CCl_4(l)	−128.2	−65.21	216.40
Barium				CH_3CHO(g)	−166.2	−127.6	263.8
Ba(s)	0	0	62.5	CH_3CO_2H(l)	−484.3	−389.9	159.8
BaO(s)	−548.0	−520.3	72.1	CH_3OH(l)	−239.2	−166.6	126.8
$BaCO_3$(s)	−1213.0	−1134.4	112.1	CH_3CH_2OH(l)	−277.6	−174.8	160.7
$BaCl_2$(s)	−855.0	−806.7	123.7	C_3H_5N(l)	172.9	208.6	188
$BaSO_4$(s)	−1473.2	−1362.2	132.2	**Chlorine**			
Boron				Cl_2(g)	0	0	223.1
B_2O_3(s)	−1273.5	−1194.3	54.0	Cl^-(aq)	−167.1	−131.0	56.5
H_3BO_3(s)	−1094.3	−968.9	90.0	HCl(g)	−92.3	−95.3	186.9
Bromine				ClO^-(aq)	−107.1	−36.8	42
Br_2(l)	0	0	152.2	ClO_2^-(aq)	−67	17	101
Br_2(g)	30.9	3.1	245.5	ClO_3^-(aq)	−104	−3	162
Br^-(aq)	−121.4	−104.0	82.4	ClO_4^-(aq)	−128.1	−8.52	184.0
BrO^-(aq)	−94.1	−33.4	42	ClO(g)	101.8	98.1	226.6
HBr(g)	−36.3	−53.4	198.7	ClO_2(g)	102.5	120.5	256.8
Cadmium				**Chromium**			
Cd(s)	0	0	51.8	Cr(s)	0	0	23.8
CdO(s)	−258.4	−228.7	54.8	Cr_2O_3(s)	−1139.7	−1058.1	81.2
$CdCl_2$(s)	−391.5	−343.9	115.3	$CrCl_3$(s)	−556.5	−486.1	123.0
Calcium				**Cobalt**			
Ca(s)	0	0	41.6	Co(s)	0	0	30.0
Ca^{2+}(aq)	−543.0	−553.6	−56.2	CoO(s)	−237.9	−214.2	53.0
CaO(s)	−634.9	−603.3	38.1	$CoCl_2$(s)	−312.5	−269.8	109.2
$Ca(OH)_2$(s)	−985.2	−897.5	83.4	**Copper**			
$CaCO_3$(s)	−1207.6	−1129.1	91.7	Cu(s)	0	0	33.2
$CaSO_4$(s)	−1434.5	−1322.0	106.5	CuO(s)	−157.3	−129.7	42.6
$CaCl_2$(s)	−795.4	−748.8	108.4	CuS(s)	−53.1	−53.6	66.5

Substance	ΔH_f° (kJ/mol)	ΔG_f° (kJ/mol)	S° (J/mol K)	Substance	ΔH_f° (kJ/mol)	ΔG_f° (kJ/mol)	S° (J/mol K)
Copper (cont.)				**Magnesium (cont.)**			
$CuCl_2(s)$	−220.1	−175.7	108.1	$MgSO_4(s)$	−1284.9	−1170.6	91.6
$CuCl(s)$	−137.2	−119.9	86.2	**Manganese**			
$CuBr(s)$	−104.6	−100.8	96.1	$Mn(s)$	0	0	32.0
$CuI(s)$	−67.8	−69.5	96.7	$MnO(s)$	−385.2	−362.9	59.7
$CuSO_4(s)$	−771.4	−662.2	109.2	$MnO_2(s)$	−520.0	−465.1	53.1
Fluorine				$MnCl_2(s)$	−481.3	−440.5	118.2
$F_2(s)$	0	0	202.8	$MnSO_4(s)$	−1065.25	−957.36	112.1
$F^-(aq)$	−335.4	−278.79	−13.8	**Mercury**			
$HF(g)$	−273.3	−275.4	173.8	$Hg(l)$	0	0	75.9
Germanium				$Hg(g)$	61.4	31.8	175.0
$Ge(s)$	0	0	31.1	$HgO(s)$	−90.8	−58.5	70.3
$GeCl_4(g)$	−495.8	−457.3	347.7	$HgCl_2(s)$	−224.3	−178.6	146.0
$GeO_2(s)$	−580.0	−521.4	39.7	$Hg_2Cl_2(s)$	−265.4	−210.7	191.6
Gold				**Neon**			
$Au(s)$	0	0	47.4	$Ne(g)$	0	0	146.328
Helium				**Nickel**			
$He(g)$	0	0	126.153	$Ni(s)$	0	0	29.9
Hydrogen				$NiO(s)$	−239.7	−211.7	37.99
$H_2(g)$	0	0	130.680	$NiCl_2(s)$	−305.3	−259.0	97.7
$H^+(aq)$	0	0	0	**Nitrogen**			
$H_3O^+(aq)$	−285.83	−237.1	69.95	$N_2(g)$	0	0	191.61
Iodine				$NH_3(g)$	−45.9	−16.4	192.8
$I_2(s)$	0	0	116.1	$NH_3(aq)$	−80.29	−26.50	111.3
$I_2(g)$	62.4	19.3	260.7	$NH_4^+(aq)$	−133.3	−79.31	111.2
$I^-(aq)$	−56.78	−51.57	106.5	$N_2H_4(l)$	50.6	149.3	121.2
$HI(g)$	26.5	1.7	206.6	$NO(g)$	91.3	87.6	210.8
Iron				$NO_2(g)$	33.2	51.3	240.1
$Fe(s)$	0	0	27.3	$N_2O(g)$	81.6	103.7	220.0
$Fe_2O_3(s)$	−824.2	−742.2	87.4	$N_2O_4(g)$	11.1	99.8	304.4
$Fe_3O_4(s)$	−1118.4	−1015.4	146.4	$N_2O_5(g)$	13.3	117.1	355.7
$Fe(OH)_3(s)$	−823.0	−696.5	106.7	$NH_4NO_3(s)$	−365.6	−183.9	151.1
$FeCl_2(s)$	−341.8	−302.3	118.0	$(NH_2)_2CO(s)$	−333.1	−198	105
$FeCl_3(s)$	−399.5	−334.0	142.3	$HNO_3(g)$	−133.9	−73.5	266.9
$FeSO_4(s)$	−928.4	−820.8	107.5	**Oxygen**			
$FeS(s)$	−100.0	−100.4	60.3	$O_2(g)$	0	0	205.152
Krypton				$O_3(g)$	142.7	163.2	238.9
$Kr(g)$	0	0	164.085	$OH^-(aq)$	−230.0	−157.244	−10.9
Lead				$H_2O(l)$	−285.83	−237.1	69.95
$Pb(s)$	0	0	64.8	$H_2O(g)$	−241.83	−228.72	188.835
$PbO(s)$	−217.3	−187.9	68.7	$H_2O_2(l)$	−187.8	−120.4	109.6
$PbO_2(s)$	−277.4	−217.3	68.6	**Phosphorus**			
$PbS(s)$	−100.4	−98.7	91.2	$P(s, white)$	0	0	41.1
Lithium				$P(s, red)$	−17.6	−12.1	22.8
$Li(s)$	0	0	29.1	$P_4(g)$	58.9	24.4	280.0
$Li^+(aq)$	−278.47	−293.31	12.2	$PH_3(g)$	5.4	13.4	210.2
$Li_2O(s)$	−597.9	−561.2	37.6	$P_4O_{10}(s)$	−2984.0	−2697.7	228.86
$LiCl(s)$	−408.6	−384.4	59.3	$PCl_5(g)$	−374.9	−305.0	364.6
Magnesium				$PCl_3(g)$	−287.0	−267.8	311.8
$Mg(s)$	0	0	32.67	**Potassium**			
$Mg^{2+}(aq)$	−467.0	−454.8	−137	$K(s)$	0	0	64.7
$MgO(s)$	−601.6	−569.3	27.0	$K^+(aq)$	−252.1	−283.27	101.2
$Mg(OH)_2(s)$	−924.5	−833.5	63.2	$KO_2(s)$	−284.9	−239.4	116.7
$MgCl_2(s)$	−641.3	−591.8	89.6	$KOH(s)$	−424.6	−378.9	78.9
$MgF_2(s)$	−1124.2	−1071.1	57.2	$KCl(s)$	−436.5	−408.5	82.6

Substance	ΔH_f° (kJ/mol)	ΔG_f° (kJ/mol)	S° (J/mol K)	Substance	ΔH_f° (kJ/mol)	ΔG_f° (kJ/mol)	S° (J/mol K)
Selenium				**Sulfur**			
Se(s)	0	0	42.4	S_8(s)	0	0	256.8
H_2Se(g)	29.7	15.9	219.0	S(g)	102.30	49.63	430.98
Silicon				H_2S(g)	−20.6	−33.4	205.8
Si(s)	0	0	18.8	SO_2(g)	−296.8	−300.1	248.2
SiH_4(g)	34.3	56.9	204.6	SO_3(g)	−395.7	−371.1	256.8
SiO_2(s)	−910.7	−856.3	41.5	H_2SO_4(l)	−814	−690.0	156.9
$SiCl_4$(l)	−687.0	−619.8	239.7	SO_4^{2-}(aq)	−909.3	−774.53	18.5
SiC(s)	−65.3	−62.8	16.6	SF_6(g)	−1220.5	−1116.5	291.5
Silver				**Tellurium**			
Ag(s)	0	0	42.6	Te(s)	0	0	49.7
Ag^+(aq)	105.8	77.107	73.4	TeO_2(s)	−322.6	−270.3	79.5
Ag_2O(s)	−31.1	−11.2	121.3	**Tin**			
AgCl(s)	−127.0	−109.8	96.3	Sn(s, white)	0	0	51.2
AgBr(s)	−100.4	−96.9	107.1	Sn(s, gray)	−2.1	0.13	44.1
Sodium				SnO(s)	−280.7	−251.9	57.2
Na(s)	0	0	51.3	SnO_2(s)	−577.6	−515.8	49.0
Na^+(aq)	−240.3	−261.905	58.5	$SnCl_4$(s)	−511.3	−440.1	258.6
Na_2O(s)	−414.2	−375.5	75.1	**Uranium**			
NaOH(s)	−425.6	−379.5	64.5	U(s)	0	0	50.2
NaF(s)	−576.6	−546.3	51.1	UO_2(s)	−1085.0	−1031.8	77.0
NaCl(s)	−411.2	−384.1	72.1	UF_6(g)	−2147.4	−2063.7	377.9
NaBr(s)	−361.1	−349.0	86.8	**Xenon**			
NaI(s)	−287.8	−286.1	98.5	Xe(g)	0	0	169.685
Strontium				XeF_4(s)	−261.5	−138	316
Sr(s)	0	0	55.0	**Zinc**			
SrO(s)	−592.0	−561.9	54.4	Zn(s)	0	0	41.63
$SrCl_2$(s)	−828.9	−781.1	114.9	ZnO(s)	−350.5	−320.5	43.7
				$ZnCl_2$(s)	−415.1	−369.4	111.5

Main Source: *Handbook of Chemistry and Physics, 82nd Edition,* David R. Lide, editor-in-chief, CRC Press LLC, Boca Raton, FL (2001), pp. **5**-1 through **5**-60.

Appendix E: Equilibrium Constants

K_a VALUES

Name	Formula	pK_a	K_a	Name	Formula	pK_a	K_a
Oxyacids				**Oxyacids (cont.)**			
Arsenic	H_3AsO_4	2.26	5.5×10^{-3}	Nitrous	HNO_2	3.25	5.6×10^{-4}
Boric	H_3BO_3	9.27	5.4×10^{-10}	Paraperiodic	H_5IO_6	1.55	2.8×10^{-2}
Carbonic	H_2CO_3	6.35	4.5×10^{-7}	Periodic	HIO_4	1.64	7.3×10^{-2}
K_{a2}	HCO_3^-	10.33	4.7×10^{-11}	Phosphoric	H_3PO_4	2.16	6.9×10^{-3}
Chlorous	$HClO_2$	1.94	1.1×10^{-2}	K_{a2}	$H_2PO_4^-$	7.21	6.2×10^{-8}
Chromic	H_2CrO_4	0.74	1.8×10^{-1}	K_{a3}	HPO_4^{2-}	12.32	4.8×10^{-13}
K_{a2}	$HCrO_4^-$	6.49	3.2×10^{-7}	Phosphorous	H_3PO_3	1.3	5.0×10^{-2}
Hypobromous	HBrO	8.55	2.8×10^{-9}	K_{a2}	$H_2PO_3^-$	6.70	2.0×10^{-7}
Hypochlorous	HClO	7.40	4.0×10^{-8}	Sulfuric	H_2SO_4	strong	
Hypoiodous	HIO	10.5	3.2×10^{-11}	K_{a2}	HSO_4^-	1.99	1.0×10^{-2}
Iodic	HIO_3	0.78	1.7×10^{-1}	Sulfurous	H_2SO_3	1.85	1.4×10^{-2}
				K_{a2}	HSO_3^-	7.20	6.3×10^{-8}

Name	*Formula*	pK_a	K_a
Carboxylic acids			
Acetic	CH_3CO_2H	4.75	1.8×10^{-5}
Benzoic	$C_6H_5CO_2H$	4.20	6.3×10^{-5}
Chloroacetic	$ClCH_2CO_2H$	2.87	1.4×10^{-3}
Formic	HCO_2H	3.75	1.8×10^{-4}
Oxalic	$(CO_2H)_2$	1.25	5.6×10^{-2}
K_{a2}	$HO_2CCO_2^-$	3.81	1.5×10^{-4}
Propanoic	$C_2H_5CO_2H$	4.87	1.3×10^{-5}
Trichloroacetic	Cl_3CCO_2H	0.66	2.2×10^{-1}
Other acids			
Cyanic	$HCNO$	3.46	3.5×10^{-4}
Hydrazoic	HN_3	4.6	2.5×10^{-5}
Hydrocyanic	HCN	9.21	6.2×10^{-10}
Hydrofluoric	HF	3.20	6.3×10^{-4}
Hydrogen peroxide	H_2O_2	11.62	2.4×10^{-12}
Hydrogen sulfide	H_2S	7.05	8.9×10^{-8}
K_{a2}	HS^-	19	1×10^{-19}
Phenol	C_6H_5OH	9.99	1.0×10^{-10}

K_b VALUES

Name	*Formula*	pK_b	K_b
Ammonia	NH_3	4.75	1.8×10^{-5}
Aniline	$C_6H_5NH_2$	9.13	7.4×10^{-10}
Diethylamine	$(C_2H_5)_2NH$	3.16	6.9×10^{-4}
Dimethylamine	$(CH_3)_2NH$	3.27	5.4×10^{-4}
Ethylamine	$C_2H_5NH_2$	3.35	4.5×10^{-4}
Ethylenediamine	$(CH_2NH_2)_2$	4.08	8.3×10^{-5}
Hydrazine	H_2NNH_2	5.89	1.3×10^{-6}
Hydroxylamine	$HONH_2$	8.06	8.7×10^{-9}
Methylamine	CH_3NH_2	3.34	4.6×10^{-4}
Pyridine	C_5H_5N	8.77	1.7×10^{-9}
Triethylamine	$(C_2H_5)_3N$	3.25	5.6×10^{-4}
Trimethylamine	$(CH_3)_3N$	4.19	6.5×10^{-5}
Urea	H_2NCONH_2	13.82	1.5×10^{-14}

K_{sp} VALUES, SALTS AT 25 °C

Formula	pK_{sp}	K_{sp}
Bromides		
AgBr	12.27	5.35×10^{-13}
CuBr	8.20	6.3×10^{-9}
Chlorides		
AgCl	9.74	1.8×10^{-10}
CuCl	6.76	1.7×10^{-7}
$PbCl_2$	4.77	1.7×10^{-5}
Carbonates		
Ag_2CO_3	11.07	8.46×10^{-12}
$BaCO_3$	8.59	2.6×10^{-9}
$CaCO_3$	8.47	3.36×10^{-9}
$CdCO_3$	12.00	1.0×10^{-12}

Formula	pK_{sp}	K_{sp}
Carbonates (cont.)		
$CoCO_3$	12.85	1.4×10^{-13}
$CuCO_3$	9.85	1.4×10^{-10}
$FeCO_3$	10.51	3.1×10^{-11}
$MgCO_3$	5.17	6.8×10^{-6}
$MnCO_3$	10.66	2.2×10^{-11}
$NiCO_3$	6.84	1.4×10^{-7}
$PbCO_3$	13.13	7.4×10^{-14}
$SrCO_3$	9.25	5.6×10^{-10}
$ZnCO_3$	9.82	1.5×10^{-10}
Fluorides		
BaF_2	6.74	1.8×10^{-7}
CaF_2	10.46	3.45×10^{-11}
PbF_2	7.48	3.3×10^{-8}
Hydroxides		
$Al(OH)_3$	32.34	4.6×10^{-33}
$Ba(OH)_2$	3.59	2.55×10^{-4}
$Ca(OH)_2$	5.30	5.0×10^{-6}
$Cd(OH)_2$	14.14	7.2×10^{-15}
$Co(OH)_2$	14.23	5.9×10^{-15}
$Co(OH)_3$	43.8	1.6×10^{-44}
$Cu(OH)_2$	14.96	1.1×10^{-15}
$Fe(OH)_2$	16.32	4.8×10^{-17}
$Fe(OH)_3$	38.55	2.8×10^{-39}
$Mg(OH)_2$	11.25	5.6×10^{-12}
$Ni(OH)_2$	15.26	5.5×10^{-16}
$Pb(OH)_2$	19.85	1.4×10^{-20}
$Sn(OH)_2$	26.26	5.5×10^{-27}
$Zn(OH)_2$	16.52	3×10^{-17}
Phosphates		
$Ba_3(PO_4)_2$	22.47	3.4×10^{-23}
$Ca_3(PO_4)_2$	32.70	2.0×10^{-33}
Sulfates		
Ag_2SO_4	4.92	1.2×10^{-5}
$BaSO_4$	9.96	1.1×10^{-10}
$CaSO_4$	4.31	4.9×10^{-5}
$PbSO_4$	7.60	2.5×10^{-8}
Sulfides		
Ag_2S	49.20	6.3×10^{-50}
CoS	20.40	4.0×10^{-21}
Cu_2S	47.60	2.5×10^{-48}
CuS	25.20	6.3×10^{-26}
FeS	18.80	1.6×10^{-19}
Hg_2S	47.00	1.0×10^{-47}
HgS	52.40	4.0×10^{-53}
MnS	13.34	4.6×10^{-14}
NiS	20.97	1.1×10^{-21}
PbS	28.05	8.9×10^{-29}
SnS	27.49	3.2×10^{-28}
ZnS	24.53	3.0×10^{-25}

Main Source: *Handbook of Chemistry and Physics, 82nd Edition,* David R. Lide, editor-in-chief, CRC Press LLC, Boca Raton, FL (2001), pp. **8**-44 through **8**-56 and **8**-117 through **8**-120.

Appendix F:
Standard Reduction Potentials, $E°$

$T = 298.15$ K, and $P = 1.00$ atm.
Listed alphabetically by element; all values are in volts.

All ionic species are aqueous. H_2O is liquid.
All other neutrals are solids unless otherwise specified.

Aluminum
$Al^{3+} + 3\,e^- \rightleftharpoons Al$	-1.662

Arsenic
$HAsO_2 + 3\,H_3O^+ + 3\,e^- \rightleftharpoons As + 5\,H_2O$	$+0.248$

Barium
$Ba^{2+} + 2\,e^- \rightleftharpoons Ba$	-2.912
$Ba(OH)_2 + 2\,e^- \rightleftharpoons Ba + 2\,OH^-$	-2.99

Boron
$H_3BO_3 + 3\,H_3O^+(aq) + 3\,e^- \rightleftharpoons B + 6\,H_2O$	-0.8698

Bromine
$Br_2(aq) + 2\,e^- \rightleftharpoons 2\,Br^-$	$+1.0873$
$Br_2(l) + 2\,e^- \rightleftharpoons 2\,Br^-$	$+1.066$
$HBrO + H_3O^+(aq) + 2\,e^- \rightleftharpoons Br^- + 2\,H_2O$	$+1.331$

Cadmium
$Cd(OH)_2 + 2\,e^- \rightleftharpoons Cd^{2+} + 2\,OH^-$	-0.809

Calcium
$Ca^{2+} + 2\,e^- \rightleftharpoons Ca$	-2.868
$Ca(OH)_2 + 2\,e^- \rightleftharpoons Ca + 2\,OH^-$	-3.02

Carbon
$CO_2(g) + 2\,H_3O^+(aq) + 2\,e^- \rightleftharpoons HCO_2H + 2\,H_2O$	-0.199
$2\,CO_2(g) + 2\,H_3O^+(aq) + 2\,e^- \rightleftharpoons$	
$\qquad\qquad H_2C_2O_4 + 2\,H_2O$	-0.49
$CNO^- + H_2O + 2\,e^- \rightleftharpoons CN^- + 2\,OH^-$	-0.970
$p\text{-Benzoquinone} + H_3O^+(aq) + 2\,e^- \rightleftharpoons$	
$\qquad\qquad \text{Hydroquinone} + H_2O$	$+0.6992$

Cesium
$Cs^+ + e^- \rightleftharpoons Cs$	-3.026

Chlorine
$Cl_2(g) + 2\,e^- \rightleftharpoons 2\,Cl^-$	$+1.35827$
$ClO_3^- + 6\,H_3O^+(aq) + 6\,e^- \rightleftharpoons Cl^- + 9\,H_2O$	$+1.451$
$ClO_4^- + 2\,H_3O^+(aq) + 2\,e^- \rightleftharpoons ClO_3^- + 3\,H_2O$	$+1.189$
$2\,ClO_4^- + 16\,H_3O^+(aq) + 14\,e^- \rightleftharpoons$	
$\qquad\qquad Cl_2(g) + 24\,H_2O$	$+1.39$
$ClO_4^- + 6\,H_3O^+ + 6\,e^- \rightleftharpoons ClO^- + 9\,H_2O$	$+1.36$
$ClO_4^- + 8\,H_3O^+(aq) + 8\,e^- \rightleftharpoons Cl^- + 12\,H_2O$	$+1.389$

Chromium
$Cr^{3+} + 3\,e^- \rightleftharpoons Cr$	-0.744
$Cr^{2+} + 2\,e^- \rightleftharpoons Cr$	-0.913
$Cr_2O_7^{2-} + 14\,H_3O^+(aq) + 6\,e^- \rightleftharpoons$	
$\qquad\qquad 2\,Cr^{3+} + 21\,H_2O$	$+1.232$
$Cr(OH)_3 + 3\,e^- \rightleftharpoons Cr + 3\,OH^-$	-1.48

Cobalt
$Co^{3+} + e^- \rightleftharpoons Co^{2+}$	$+1.92$
$Co^{2+} + 2\,e^- \rightleftharpoons Co$	-0.28

Copper
$Cu^{2+} + 2\,e^- \rightleftharpoons Cu$	$+0.3419$
$Cu^{2+} + e^- \rightleftharpoons Cu^+$	$+0.153$
$Cu^+ + e^- \rightleftharpoons Cu$	$+0.521$
$Cu(OH)_2 + 2\,e^- \rightleftharpoons Cu + 2\,OH^-$	-0.222

Fluorine
$F_2(g) + 2\,e^- \rightleftharpoons 2\,F^-$	$+2.866$

Gold
$Au^{3+} + 3\,e^- \rightleftharpoons Au$	$+1.498$
$Au(CN)_2^- + e^- \rightleftharpoons Au + 2\,CN^-$	-0.60

Hydrogen
$2\,H_3O^+(aq) + 2\,e^- \rightleftharpoons H_2(g) + 2\,H_2O$	0 (by definition)
$2\,H_2O + 2\,e^- \rightleftharpoons H_2(g) + 2\,OH^-$	-0.828

Iodine
$I_2 + 2\,e^- \rightleftharpoons 2\,I^-$	$+0.5355$
$I_3^- + 2\,e^- \rightleftharpoons 3\,I^-$	$+0.536$
$2\,IO_3^- + 12\,H_3O^+ + 10\,e^- \rightleftharpoons I_2 + 18\,H_2O$	$+1.195$
$H_5IO_6 + H_3O^+ + 2\,e^- \rightleftharpoons IO_3^- + 4\,H_2O$	$+1.601$

Iron
$Fe^{3+} + 3\,e^- \rightleftharpoons Fe$	-0.037
$Fe^{2+} + 2\,e^- \rightleftharpoons Fe$	-0.447
$Fe^{3+} + e^- \rightleftharpoons Fe^{2+}$	$+0.771$

Lead
$Pb^{2+} + 2\,e^- \rightleftharpoons Pb$	-0.1262
$PbO_2 + SO_4^{2-} + 4\,H_3O^+(aq) + 2\,e^- \rightleftharpoons$	
$\qquad\qquad PbSO_4 + 6\,H_2O$	$+1.6913$
$PbO_2 + 4\,H_3O^+(aq) + 2\,e^- \rightleftharpoons Pb^{2+} + 6\,H_2O$	$+1.455$
$PbSO_4 + 2\,e^- \rightleftharpoons Pb + SO_4^{2-}$	-0.3588

Lithium
$Li^+ + e^- \rightleftharpoons Li$	-3.0401

Magnesium
$Mg^{2+} + 2\,e^- \rightleftharpoons Mg$	-2.37
$Mg(OH)_2 + 2\,e^- \rightleftharpoons Mg + 2\,OH^-$	-2.69

Manganese
$Mn^{2+} + 2\,e^- \rightleftharpoons Mn$	-1.185
$MnO_4^- + 4\,H_3O^+(aq) + 3\,e^- \rightleftharpoons$	
$\qquad\qquad MnO_2 + 6\,H_2O$	$+1.679$
$MnO_4^- + 2\,H_2O + 3\,e^- \rightleftharpoons MnO_2 + 4\,OH^-$	$+0.595$
$MnO_4^- + 8\,H_3O^+(aq) + 5\,e^- \rightleftharpoons$	
$\qquad\qquad Mn^{2+} + 12\,H_2O$	$+1.507$
$MnO_2 + 4\,H_3O^+(aq) + 2\,e^- \rightleftharpoons Mn^{2+} + 6\,H_2O$	$+1.224$

Mercury
$Hg^{2+} + 2\,e^- \rightleftharpoons Hg(l)$	$+0.851$
$Hg_2Cl_2 + 2\,e^- \rightleftharpoons 2\,Hg(l) + 2\,Cl^-$	$+0.26808$

Nickel

$Ni^{2+} + 2\,e^- \rightleftharpoons Ni$ -0.257

$Ni(OH)_2 + 2\,e^- \rightleftharpoons Ni + 2\,OH^-$ -0.72

$NiO_2 + 4\,H_3O^+(aq) + 2\,e^- \rightleftharpoons Ni^{2+} + 6\,H_2O$ $+1.678$

Nitrogen

$N_2(g) + 8\,H_3O^+(aq) + 6\,e^- \rightleftharpoons 2\,NH_4^+ + 8\,H_2O$ $+0.092$

$NO_2(g) + 2\,H_3O^+(aq) + 2\,e^- \rightleftharpoons NO(g) + 3\,H_2O$ $+1.03$

$NO_3^- + 4\,H_3O^+(aq) + 3\,e^- \rightleftharpoons NO(g) + 6\,H_2O$ $+0.957$

$2\,NO(g) + 2\,H_3O^+ + 2\,e^- \rightleftharpoons N_2O(g) + 3\,H_2O$ $+1.591$

$NO_3^- + 2\,H_2O + 3\,e^- \rightleftharpoons NO(g) + 4\,OH^-$ $+0.109$

Oxygen

$O_2(g) + 2\,H_3O^+(aq) + 2\,e^- \rightleftharpoons H_2O_2(l) + 2\,H_2O$ $+0.695$

$O_2(g) + 2\,H_2O + 2\,e^- \rightleftharpoons H_2O_2(l) + 2\,OH^-$ -0.146

$O_2(g) + 4\,H_3O^+(aq) + 4\,e^- \rightleftharpoons 6\,H_2O$ $+1.229$

$O_2(g) + 2\,H_2O + 4\,e^- \rightleftharpoons 4\,OH^-$ $+0.401$

$O_3(g) + 2\,H_3O^+(aq) + 2\,e^- \rightleftharpoons O_2(g) + 3\,H_2O$ $+2.076$

Platinum

$Pt^{2+} + 2\,e^- \rightleftharpoons Pt$ $+1.18$

Potassium

$K^+ + e^- \rightleftharpoons K$ -2.931

Selenium

$SeO_3^{2-} + 3\,H_2O + 4\,e^- \rightleftharpoons Se + 6\,OH^-$ -0.366

Silver

$Ag^+ + e^- \rightleftharpoons Ag$ $+0.7996$

$AgCl + e^- \rightleftharpoons Ag + Cl^-$ $+0.22233$

$AgBr + e^- \rightleftharpoons Ag + Br^-$ $+0.07133$

$AgI + e^- \rightleftharpoons Ag + I^-$ -0.15224

Sodium

$Na^+ + e^- \rightleftharpoons Na$ -2.71

Sulfur

$S + 2\,H_3O^+(aq) + 2\,e^- \rightleftharpoons H_2S + 2\,H_2O$ $+0.14$

$S_4O_6^{2-} + 2\,e^- \rightleftharpoons 2\,S_2O_3^{2-}$ $+0.08$

Tin

$Sn^{2+} + 2\,e^- \rightleftharpoons Sn$ -0.137

$Sn^{4+} + 2\,e^- \rightleftharpoons Sn^{2+}$ $+0.151$

Titanium

$Ti^{2+} + 2\,e^- \rightleftharpoons Ti$ -1.630

$Ti^{3+} + e^- \rightleftharpoons Ti^{2+}$ -0.85

Zinc

$Zn^{2+} + 2\,e^- \rightleftharpoons Zn$ -0.7618

$[Zn(NH_3)_4]^{2+} + 2\,e^- \rightleftharpoons Zn + 4\,NH_3(aq)$ -1.04

Main Source: *Handbook of Chemistry and Physics, 82nd Edition,* David R. Lide, editor-in-chief, CRC Press LLC, Boca Raton, FL (2001), pp. **8**-21 through **8**-26.

Solutions to Odd-Numbered Problems

1.1 Examples include the following: What are appropriate standards for "clean" air and water? What should be the regulations for testing new drugs before they are approved for human use? How can toxic waste sites best be cleaned up? Should chlorofluorocarbons be banned because of potential damage to the ozone layer?

1.3 Examples include the following: weighing, volume measurements and unit conversions; knowing chemical compatibility of different drugs; identifying similar or chemically equivalent drugs; and protecting drugs from degradation because of exposure to adverse conditions.

1.5 The chemists should redo the experiment, adjusting conditions if necessary to determine whether or not the results are correct. If repeated measurements show that the results consistently differ from what theory predicts, then the chemist should examine how to revise the theory to accommodate the results.

1.7 (a) H; (b) He; (c) Hf; (d) N; (e) Ne; and (f) Nb

1.9 (a) arsenic; (b) argon; (c) aluminum; (d) americium; (e) silver; (f) gold; (g) astatine; and (h) actinium

1.11 (a) Br_2; (b) HCl; (c) C_2H_5I; (d) C_3H_6O; (e) $C_3H_6O_2$; and (f) $C_7H_5N_3O_6$

1.13 (a) CCl_4; (b) H_2O_2; (c) P_4O_{10}; and (d) Fe_2S_3

1.15 Cesium

1.17 Vertical neighbors are oxygen (O) and selenium (Se); horizontal neighbors are phosphorus (P) and chlorine (Cl). Oxygen and selenium have similar chemical properties.

1.19 Metals from Group 1: Li, Na, K, Rb, and Cs; other metals: Cu, Ag, and Au

1.21 Lithium, Li; beryllium, Be; boron, B; carbon, C; oxygen, O; fluorine, F; and neon, Ne

1.23 (a) Pure substance; (b) solution; (c) heterogeneous mixture; (d) solution; (e) heterogeneous mixture; and (f) heterogeneous mixture

1.25 (a) liquid; (b) solid; (c) solid; and (d) gas

1.27 (a) physical; (b) physical; and (c) chemical

1.29 (a) mixture; (b) compound; (c) mixture; (d) element; (e) mixture; and (f) mixture

1.31 (a) 1.00000×10^5; (b) 1.0×10^4; (c) 4.00×10^{-4}; (d) 3×10^{-4}; and (e) 2.753×10^2

1.33 (a) 4.32×10^5 g; (b) 6.24×10^{-10}; (c) 1.024×10^{-6} g; (d) 9.3000×10^7 m; (e) 8.6400×10^4 s; and (f) 1.08×10^{-3} m

1.35 7.10×10^{-3} kg

1.37 9.46×10^2 g

1.39 0.79 g/cm^3

1.41 5.70 cm^3

1.43 9.98×10^2 kg/m^3

1.45 lead cube

1.47 11.7 mL

1.49 8: Ti (titanium), Tc (technetium), Te (tellurium), Ta (tantalum), Tl (thallium), Tb (terbium), Tm (thulium), and Th (thorium)

1.51

1.53 Lead (Pb)

1.55 11.12 s

1.57 Silver and copper

1.59

1.61 B matches D; C matches E; A matches F

1.63 261.7 K

1.65 (a) 4.52×10^{34}; and (b) 1×10^{-4}

1.67 Intensive: (a), (c), and (e); extensive: (b) and (d)

1.69 (a) $CHClF_2$; (b) CH_2O_2; (c) BrF_3; and (d) C_4H_{10}

1.71 (a) 7.44×10^{-3} m^3; (b) 16 m s^{-1}; (c) 2.1 m; and (d) 1.03×10^2 kg

1.73 (a) Sr; (b) In; (c) Br and O; (d) Co; (e) Ne; and (f) Pu

1.75 HCN   H$_2$O   CO   N$_2$O

1.77 393 minutes

1.79 Physical properties: appearance, melting point, softness, density; chemical properties: reaction with chlorine and reaction with water

1.81 (a) fluorine (F), chlorine (Cl), bromine (Br), iodine (I); (b) beryllium (Be), magnesium (Mg), calcium (Ca), strontium (Sr), barium (Ba); (c) examples include actinium (Ac), uranium (U), and berkelium (Bk); and (d) examples include helium (He), neon (Ne), argon (Ar), and krypton (Kr)

1.83 9.4604×10^{12} km

1.85 1.3×10^8 atoms

1.87 3.6×10^8 ft; 1.1×10^5 km

1.89 3.1557×10^9 s

1.91 Your list might include questions about which you have a special interest. Our list includes pollution questions such as how to reverse the processes that form the ozone hole; health questions such as how to design new drugs to combat bacterial infections and how important for our health are trace elements in our diets; biochemical questions such as what molecular processes take place when we smell an odor; and technological questions such as how to design materials that are high-temperature superconductors.

1.93 (a) oil; (b) table salt; and (c) iron

2.1 (a) gaseous helium (b) solid tungsten and (c) Liquid gallium

2.3 (1) All starting materials and products are made from atoms of carbon and oxygen. (2) All oxygen atoms are in diatomic molecules in the starting materials and CO molecules in the products; likewise, C atoms behave all in the same way. (3) Carbon atoms and oxygen atoms do different things in the reaction. (4) C and O combine in 1 : 1 atomic ratio to give CO molecules. (5) Atoms of each type are conserved: 4 O atoms and 13 C atoms.

2.5 When magnesium burns in air, magnesium atoms combine with oxygen from the air to form magnesium oxide. The mass of the solid increases, but the mass of solid plus gas remains constant.

2.7

Solid bromine (temp $<$ 266K)	liquid bromine (266K $<$ temp $<$ 332K)	gaseous bromine (temp $>$ 332K)

2.9 The molecules that give roses their aroma evaporate from the surface of the flower. Once in the gas phase, they collide countless times with other gas molecules, moving slowly away from the rose until, when they reach a nose, they are sensed by the olfactory sensors.

2.11 Iodine molecules sublime from the crystals at the bottom of the flask into the gas phase, where their presence imparts a pale violet color to the gas. Some of these molecules then condense on the surfaces of the flask, forming crystals. At any time, there is a constant number of molecules in the gas phase, but some molecules are subliming into the gas from the crystals while equal numbers of molecules are condensing onto the crystals.

2.13 (a) 6.24×10^{12} electrons; and (b) 5.68×10^{-18} kg

2.15 Negatively-charged particles would be attracted downward, but now positively-charged particles would be attracted upward. Under these conditions, the proper electrical field would cause some of the positively-charged particles to be suspended in space. Because each of these particles had lost one or more electrons, the charge on the electron could be determined from these observations.

2.17 9.0×10^{23} protons; 1.4×10^5 C

2.19 (a) 2; and (b) 2.7205×10^{-4} or 2.741×10^{-4} (See Solutions Manual for details.)

2.21 (a) 8p, 8n, 10e; (b) 5p, 6n, 5e; (c) 25p, 30n, 22e; (d) 17p, 18n, 18e; and (e) 17p, 20n, 16e

2.23 (a) $^{56}_{26}Fe$; (b) $^{236}_{92}U$; (c) $^{38}_{18}Ar$; and (d) $^{19}_{9}F$.

2.25

2.27

2.29 cations: Cl_2^+, CO^+, Cr^{3+}; anions: Cl^-, $Cr_2O_7^{2-}$; and neutrals: C, CCl_4, CO_2

2.31 (a) H_2O; (b) Na^+; (c) Cl^-; and (d) O^{2-}

2.33 (a)

(b)

2.35 (a) Rb^+; (b) F^-; and (c) Ba^{2+}

2.37 RbF and BaF_2

2.39

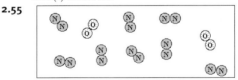

2.41 Al_2O_3; AlF_3

2.43 (a) On the tree, an apple possesses some gravitational potential energy. As an apple falls, that potential energy is converted into kinetic energy of motion. (b) When an apple hits the ground, the impact transfers energy to molecules in the earth and in the apple. As a result, there is a slight increase in temperature; the kinetic energy of the apple has been converted into thermal energy.

2.45 9.43×10^{-20} J

2.47 1.73×10^{-16} J

2.49 (a) Radiant energy consumed; thermal energy produced; (b) gravitational potential energy consumed; kinetic energy produced; and (c) chemical potential energy consumed; thermal energy produced

2.51 2.12×10^2 m/s

2.53 (a) 6 electrons added; (b) 2 electrons removed; and (c) conserved

2.55

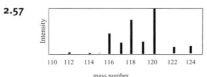

2.57

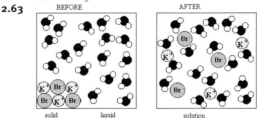

2.59 (a) 43 p, 56 n, 43 e; (b) 26 p, 26 n, 26 e; (c) 54 p, 79 n, 54 e; and (d) 53 p, 78 n, 53e

2.61 3.37×10^5 J

2.63

2.65 $^{40}_{17}Cl$, $^{40}_{18}Ar$, $^{40}_{19}K$, $^{40}_{20}Ca$, $^{40}_{21}Sc$

2.67

2.69 (a) 11 p, 10 e; (b) 7 p, 10 e; (c) 22 p, 18 e; and (d) 53 p, 54 e

2.71 (a) Ca^{2+}, Cl^-; (b) $CaCl_2$; and

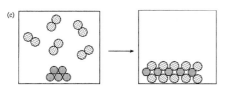

2.73

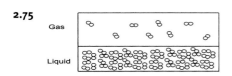

Solid   Liquid

2.75

Gas

Liquid

2.77 $^{45}_{20}Ca$, $^{54}_{24}Cr$, $^{63}_{28}Ni$, $^{72}_{32}Ge$, $^{81}_{36}Kr$, and $^{90}_{40}Zr$

2.79 not correct;

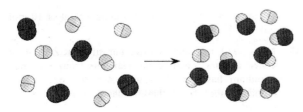

2.81 3 peaks. Mass 70 is most intense and mass 74 is least intense.

3.1 (a) CH_4; (b) C_2H_4; (c) C_2H_6O; (d) HBr; (e) PCl_3; (f) CH_4N_2O; and (g) C_2H_5I

3.3

(a) H—C—H

(e) P with Cl, Cl, Cl

(b) C=C with H

(f) urea structure

(c) H₃C—O—CH₃

(d) H—Br

(g) H—C—C—H with I

3.5 (a), (b), (c), (d), (e), (f)

3.7 (a), (b), (c), (d), (e), (f), (g), (h), (i)

3.9 (a) CO_2, carbon dioxide; (b) HCl, hydrogen chloride; and (c) CCl_4, carbon tetrachloride

3.11 (a) H_2S, hydrogen sulfide; (b) SF_4, sulfur tetrafluoride; and (c) HF, hydrogen fluoride

3.13 (a) CH_4; (b) HF; (c) CaH_2; (d) PCl_3; (e) N_2O_5; (f) SF_6; and (g) BF_3

3.15 (a) disulfur dichloride; (b) iodine heptafluoride; (c) hydrogen bromide; (d) dinitrogen trioxide; (e) silicon carbide; and (f) methanol

3.17 (a) not ionic, HF; (b) ionic, CaF_2; (c) ionic, $Al_2(SO_4)_3$; (d) ionic, $(NH_4)_2S$; (e) not ionic, SO_2; and (f) not ionic, CCl_4

3.19 (a) not ionic, dichloromethane; (b) not ionic, carbon dioxide; (c) ionic, calcium oxide; (d) ionic, potassium carbonate; (e) not ionic, phosphorus tribromide; (f) not ionic, hydrogen fluoride; and (g) ionic, sodium hydrogen phosphate

3.21 (a) Na_2SO_4; (b) K_2S; (c) KH_2PO_4; (d) $CoF_2 \cdot 4H_2O$; (e) PbO_2; (f) $NaHCO_3$; and (g) $LiBrO_4$

3.23 (a) calcium chloride hexahydrate; (b) iron(II) ammonium sulfate; (c) potassium carbonate; (d) tin(II) chloride dihydrate; (e) sodium hypochlorite; (f) silver sulfate; (g) copper(II) sulfate; (h) potassium dihydrogen phosphate; (i) sodium nitrate; (j) calcium sulfite; and (k) potassium permanganate

3.25 (a) 0.141 mol; (b) 5.45×10^{-6} mol; (c) 1.67×10^{-4} mol; and (d) 5.41×10^4 mol

3.27 39.948 g/mol

3.29 (a) 153.8 g/mol; (b) 110.27 g/mol; (c) 48.00 g/mol; (d) 86.84 g/mol; (e) 144.64 g/mol; and (f) 169.88 g/mol

3.31 (a) $C_9H_{11}NO_3$, 181.19 g/mol; (b) $C_{11}H_{12}N_2O_2$, 204.23 g/mol; (c) $C_5H_9NO_4$, 147.13 g/mol; and (d) $C_6H_{14}N_2O_2$, 146.19 g/mol

3.33 (a) 3.92×10^{20} atoms; (b) 1.14×10^{20} atoms; (c) 3.87×10^{19} atoms; and (d) 1.48×10^{19} atoms

3.35 (a) 9.99×10^{-18} g; (b) 7.6×10^{-13} g; and (c) 1.484×10^{-21} g

3.37 2.6×10^{-6} mol; 1.6×10^{18} molecules;
$\#_H = 4.8 \times 10^{19}$ atoms; 8.0×10^{-2} mg

3.39 CaO: O, 28.53%, Ca, 71.47%; SiO_2, O, 53.25%,
Si, 46.75%; Al_2O_3, O, 47.08%, Al, 52.92%; Fe_2O_3,
O 30.06%, Fe, 69.94%

3.41 C_5H_7N; $C_{10}H_{14}N_2$

3.43 % C = 64.4%; % H = 5.41%; % Fe = 30.2%; $C_{10}H_{10}Fe$

3.45 $C_6H_{10}OS_2$

3.47 (a) $M_{Mg^{2+}} = 0.328$ M; $M_{Cl^-} = 0.656$ M; and (b)

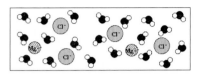

3.49 (a) $[K^+] = [OH^-] = 0.308$ M;
(b) $[K^+] = [OH^-] = 0.077$ M; and

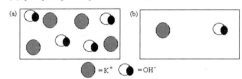

= K^+ = OH^-

3.51 5.17 mL

3.53 (a) $[CO_3^{2-}] = 0.175$ M; $[Na^+] = 0.350$ M; (b) $[NH_4^+] =$
$[Cl^-] = 0.0331$ M; and (c) $[SO_4^{2-}] = 1.84 \times 10^{-3}$ M;
$[K^+] = 3.69 \times 10^{-3}$ M

3.55 1.49 g

3.57 $C_4H_7Cl_3$

3.59 HgS

3.61 (a) $C_{15}H_{20}O_4$, $MM = 264.31$ g/mol; (b) $C_{10}H_9NO_2$,
$MM = 175.18$ g/mol; and (c) $C_{10}H_{13}N_5O$,
$MM = 219.25$ g/mol

3.63 (a) $NaNO_2$ and $NaNO_3$; (b) K_2CO_3 and $KHCO_3$;
(c) FeO and Fe_2O_3; and (d) I_2 and I^-

3.65 (a) 342.17 g/mol; (b) 7.31×10^{-2} mol; (c) Al: 15.77%,
S: 28.12%, O: 56.11%; and (d) 28.5 g

3.67 (a) 0.49 mol; (b) 0.72 g; (c) 3.7×10^{-5} mol;
2.2×10^{19} molecules; and (d) 7.4×10^{-6} mol

3.69 7.696×10^{13} m

3.71 (a) carbon dioxide; (b) potassium nitrate; (c) sodium chloride; (d) sodium hydrogen carbonate; (e) sodium carbonate;
(f) sodium hydroxide; (g) calcium oxide; and (h) magnesium hydroxide

3.73 (a) Fe = 54.81%; Si = 13.78%; O = 31.40%; (b) Na =
8.77%; Al = 10.29%; Si = 32.13%; O = 24.41%; (c) Al =
20.90%; Si = 21.76%; O = 55.78%; H = 1.56%; and
(d) Mg = 5.62%; Si = 25.97%; O = 66.55%; H = 1.86%

3.75 (a) H_2O, NH_4^+, SO_4^{2-}; (b) H_2O, CO_2; (c) H_2O, Na^+, F^-;
(d) H_2O, K^+, CO_3^{2-}; (e) H_2O, Na^+, HSO_4^-; and (f) H_2O,
Cl_2

3.77 2.86 kg; 4.72 kg; 2.14 kg; 4.71 kg

3.79 (a) ammonium chloride; (b) xenon tetrafluoride; (c) iron(III) oxide; (d) sulfur dioxide; and (e) potassium perchlorate

3.81 (a) 454.59 g/mol; (b) 2.640×10^{-4} mol; and
(c) 3.180×10^{20} atoms N

3.83 $C_{20}H_{24}O_2N_2$

3.85 1.36×10^3 g/mol

3.87 0.205 g

3.89 1.818×10^8 mol; 5.6×10^6 kg

3.91 To determine a molecular formula, mass percent composition and an approximate molar mass must be known. The mass of compound that was burned must be known. Unless the percentages of C and H together total 100%, further information about the other elements present in the compound is also needed.

3.93 56

3.95 23.682% C; 3.180% H; 13.809% N; 41.008% O; 18.321% P

3.97 (a) 1.44 g Al; (b) 2.96×10^{19} phosphate ions; and (c) +2

3.99 (a) 54.8 g; (b) 8.3×10^{22} molecules; (c) 1.25 mol;
and (d) 0.486 M

3.101 (a) 14 M; and (b) 1.1×10^2 mL

4.1 (a) $NH_4NO_3 \rightarrow N_2O + 2\ H_2O$;
(b) $P_4O_{10} + 6\ H_2O \rightarrow 4\ H_3PO_4$;
(c) $2\ HIO_3 \rightarrow I_2O_5 + H_2O$; and
(d) $2\ As + 5\ Cl_2 \rightarrow 2\ AsCl_5$

4.3

4.5 (a) $2\ H_2 + CO \rightarrow CH_3OH$;
(b) $CaO + 3\ C \rightarrow CO + CaC_2$; and
(c) $2\ C_2H_4 + O_2 + 4\ HCl \rightarrow 2\ C_2H_4Cl_2 + 2\ H_2O$

4.7

4.9 (a) $3\ Ca(OH)_2 + 2\ H_3PO_4 \rightarrow 6\ H_2O + Ca_3(PO_4)_2$;
(b) $Na_2O_2 + 2\ H_2O \rightarrow 2\ NaOH + H_2O_2$;
(c) $BF_3 + 3\ H_2O \rightarrow 3\ HF + H_3BO_3$; and
(d) $2\ NH_3 + 3\ CuO \rightarrow 3\ Cu + N_2 + 3\ H_2O$

4.11 (a) 34.7 g CO; (b) 3.21 g C; and (c) 2.85 g O_2

4.13 1.77×10^3 g NaI

4.15 512 g ethyl alcohol

4.17 45.5 kg HF; 138 kg CCl_2F_2; and 83.0 kg HCl

4.19 68.6%

4.21 28 kg

4.23 76.3%; 258 kg CCl_4; and 67.2 kg HF

4.25 (a) Limiting reactant = CO, 1.14 metric ton; (b) Limiting reactant = CaO, 0.499 metric ton CO and 1.14 metric ton CaC_2; and (c) Limiting reactant = HCl, 1.36 metric ton $C_2H_4Cl_2$ and 0.247 metric ton H_2O

4.27 91.3 kg

4.29 lettuce; 66

4.31 8.60 g; 1.71g O_2

4.33 (a) NH_4^+, Cl^-, H_2O; (b) Fe^{2+}, ClO_4^-, H_2O; (c) Na^+,
SO_4^{2-}, H_2O; (d) Br_2, H_2O; and (e) K^+, Br^-, H_2O

4.35 (a) K^+, HPO_4^-, H_2O; (b) CH_3CO_2H, H_2O; (c) Na^+,
$CH_3CO_2^-$, H_2O; (d) NH_3, H_2O; and (e) NH_4^+, Cl^-, H_2O

4.37 (a) AgCl precipitates with a; No precipitate with b, c, or d;
AgBr precipitates with e; (b) No precipitate with a, c, d, or
e; $FeCO_3$ precipitates with b; and (c) No precipitate with a,
d, or e; $Fe(OH)_2$ precipitates with b; $BaSO_4$ precipitates
with c.

4.39 (a) $Ag^+(aq) + OH^-(aq) \rightarrow AgOH(s)$, spectator ions:
NO_3^-, K^+; (b) $2\ Fe^{3+}(aq) + 3\ C_2O_4^{2-}(aq) \rightarrow Fe_2(C_2O_4)_3(s)$,
spectator ions: ClO_4^-, NH_4^+; (c) $Pb^{2+}(aq) + 2\ Br^-(aq) \rightarrow$
$PbBr_2(s)$, spectator ions: NO_3^-, Na^+; and (d) $Ni^{2+}(aq) + 2$
$OH^-(aq) \rightarrow Ni(OH)_2(s)$, spectator ions: K^+, SO_4^{2-}

4.41 0.381 g; CO_3^{2-}, NO_3^-, and K^+

4.43 (a) $H_3O^+(aq) + OH^-(aq) \rightarrow 2H_2O(l)$, spectator ions: Ca^{2+}, Cl^-; (b) $H_3PO_4(aq)+3OH^-(aq) \rightarrow 3H_2O(l) + PO_4^{3-}(aq)$, spectator ions: Li^+; (c) $NH_3(aq) + H_3O^+(aq) \rightarrow H_2O(l) + NH_4^+(aq)$, spectator ions: NO_3^-; and (d) $CH_3CO_2H(aq) + OH^-(aq) \rightarrow H_2O(l) + CH_3CO_2^-(aq)$, spectator ions: K^+

4.45 0.705 g

4.47 $[H_3O^+] = 0$; $[OH^-] = 4.00 \times 10^{-3}$ M; $[Ba^{2+}] = 1.20 \times 10^{-2}$ M; $[Cl^-] = 2.00 \times 10^{-2}$ M

4.49 0.660 M

4.51 8.886×10^{-2} M

4.53 (a) and (b) no rxn; (c) $Cu(s) + 2 Ag^+(aq) \rightarrow Cu^{2+}(aq) + 2 Ag(s)$; and (d) $2 K(s) + 2 H_2O(l) \rightarrow 2 K^+(aq) + H_2(g) + 2 OH^-(aq)$

4.55 (a) $2 Sr + O_2 \rightarrow 2 SrO$; (b) $4 Cr + 3 O_2 \rightarrow 2 Cr_2O_3$; and (c) $Sn + O_2 \rightarrow SnO_2$.

4.57 3.98×10^{-2} g

4.59 (a) $Na^+ + HCO_3^- \rightarrow NaHCO_3(s)$; (b) NH_4^+ and Cl^- are spectator ions; and (c) 55.6%; (d) $[NH_4^+] = 0.750$ M; $[Cl^-] = 3.00$ M; $[Na^+] = 2.58$ M; $[HCO_3^-] = 0.333$ M.

4.61 $4 Ru + 3 O_2 \rightarrow 2 Ru_2O_3$, 3 electrons; $Ru + O_2 \rightarrow RuO_2$, 4 electrons; $2 Ru + 3 O_2 \rightarrow 2 RuO_3$, 6 electrons; $Ru + 2 O_2 \rightarrow RuO_4$. 8 electrons

4.63 54 metric ton

4.65 (a) $2 Al + 3 O_2 \rightarrow 2 Al_2O_3$ (redox reaction); (b) $C_3H_8(g) + 5 O_2(g) \rightarrow 3 CO_2(g) + 4 H_2O(g)$ (redox reaction); (c) $Mg(s) + 2 H_3O^+(aq) \rightarrow Mg^{2+}(aq) + H_2(g) + 2 H_2O(l)$ (redox reaction); (d) $OH^-(aq) + H_3O^+(aq) \rightarrow 2 H_2O(l)$ (acid-base reaction); (e) $Pb^{2+}(aq) + CO_3^{2-}(aq) \rightarrow PbCO_3(s)$ (precipitation reaction); and (f) $OH^-(aq) + H_3O^+(aq) \rightarrow 2 H_2O(aq)$ (acid-base reaction)

4.67 50.6%; 2.47 g

4.69 $Y + 2 X \rightarrow YX_2$

4.71 (a) $5 H_2 + 2 NO \rightarrow 2 NH_3 + 2 H_2O$; (b) $2 CO + 2 NO \rightarrow N_2 + 2 CO_2$; (c) $2 NH_3 + 2 O_2 \rightarrow N_2O + 3 H_2O$; (d) $6 NO + 4 NH_3 \rightarrow 5 N_2 + 6 H_2O$; (e) $6 H_2O + 4 NO \rightarrow 5 O_2 + 4 NH_3$; and (f) $2 H_2 + O_2 \rightarrow 2 H_2O$

4.73 (a) Mix solutions of Na_3PO_4 and $FeCl_3$. The net reaction is $Fe^{3+}(aq) + PO_4^{3-}(aq) \rightarrow FePO_4(aq)$; 2.69×10^3 g $FeCl_3$ and 2.72×10^3 g Na_3PO_4; (b) Mix solutions of NaOH and $ZnCl_2$. The net reaction is $Zn^{2+}(aq) + 2 OH^-(aq) \rightarrow Zn(OH)_2(s)$; 3.42×10^3 g of $ZnCl_2$ and 2.01×10^3 g NaOH; and (c) Mix solutions of Na_2CO_3 and $NiCl_2$. The net reaction is $Ni^{2+}(aq) + CO_3^{2-}(aq) \rightarrow NiCO_3(s)$; 2.73×10^3 g $NiCl_2$ and 2.24×10^3 g Na_2CO_3.

4.75 (a) 0.30 mol; (b) 1.8×10^{23} molecules; (c) and 5.4 g

4.77 (a) H_2; (b)

and (c) 170.3 g NH_3 produced and 28.02 g N_2 remaining

4.79 (a) species present are NH_3, H_2O, H_3O^+ and Cl^-; $NH_3 + H_3O^+ \rightarrow NH_4^+ + H_2O$; (b) H_2O, Ca^{2+}, Cl^-, Na^+, and SO_4^{2-}; $Ca^{2+} + SO_4^{2-} \rightarrow CaSO_4(s)$; (c) H_2O, K^+, OH^-, H_3O^+, and Br^-; $H_3O^+ + OH^- \rightarrow 2 H_2O$; and (d) HNO_2, H_2O, K^+, and OH^-; $HNO_2 + OH^- \rightarrow NO_2^- + H_2O$

4.81 $H_3O^+ + OH^- \rightarrow 2H_2O$

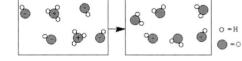

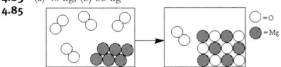

4.83 (a) 48 kg; (b) 35 kg

4.85

4.87 (a) $CaCO_3 + 2 H_3O^+ \rightarrow Ca^{2+} + CO_2 + 3 H_2O$; (b) 1.1g; and (c) $[Cl^-] = 0.10$ M and $[Ca^{2+}] = 0.050$ M.

4.89 (a) $Ca + 2 H_3O^+ \rightarrow Ca^{2+} + H_2 + 2 H_2O$ (redox); (b) $4 Li + O_2 \rightarrow 2 Li_2O$ (redox); (c) $NH_3 + H_3O^+ \rightarrow NH_4^+ + 2 H_2O$ (acid-base); and (d) $C_3H_8 + 5 O_2 \rightarrow 3 CO_2 + 4 H_2O$ (redox)

4.91 $4 C_3H_5N_3O_9(l) \rightarrow 6 N_2(g) + 12 CO_2(g) + 10 H_2O(g) + O_2(g)$

4.93 Theoretical yield = 102 kg; % yield = 67%; 28 kg N_2 and 12 kg H_2

4.95 1:3.03

4.97 10.0% impurities

4.99 (a) $2 H_2SO_4 + Ca_3(PO_4)_2 \rightarrow 2 CaSO_4 + Ca(H_2PO_4)_2$; (b) $Ca(H_2PO_4)_2$: 30.6 g; H_2SO_4: 19.4 kg; and (c) 198 mol

4.101 Cu: 13.0%; Ag: 87.0%

4.103 $2 C_5H_{12}O + 15 O_2 \rightarrow 12 H_2O + 10 CO_2$; 0.132 mol

4.105 76 kg

5.1

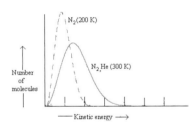

5.3

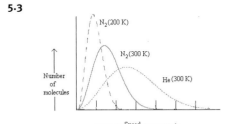

5.5

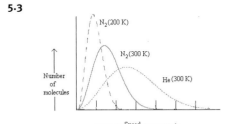

5.7 (a) 1.93×10^3 m/s; 11.2 kJ/mol; (b) 394m/s; 3.74 kJ/mol; and (c) 319m/s; 11.2 kJ/mol

5.9 (a) Xenon atom (greater mass); (b) methane at 700 K (greater temperature); and (c) F_2 (greater mass)

5.11 (a) At very high pressure, molecules are very close together, so their volumes are significant compared to the volume of their container; the first condition is not met and the gas is not ideal. (b) At very low temperature, molecules move very slowly, so the forces between molecules, even though small, are sufficient to influence molecular motion; the second condition is not met and the gas is not ideal.

5.13 1 atm

5.15 A pinhole in the top of the tube would let air leak into the space until the internal pressure matched the external, atmospheric pressure. The barometer would then indicate zero pressure.

5.17 (a) 6.07×10^4 Pa; (b) 2.48×10^5 Pa; (c) 61 Pa; and (d) 1.35×10^2 Pa

5.19 4.09 mol; 1.00×10^2 L

5.21 (a) $P_i/T_i = P_f/T_f$; (b) $V = nRT/P$; and (c) $P_iV_i = P_fV_f$

5.23 0.221 L

5.25 As a gas cools, its molecules move more slowly, so they impart smaller impulses on the walls of their container. This reduces the internal pressure, so the balloon collapses until the increase in gas density inside the balloon brings the internal pressure back up to 1.000 atm.

5.27 (a) equation is valid; (b), (c), and (d) not valid

5.29 121 g/mol; CF_2Cl_2

5.31 7.8×10^{-7} atm

5.33 $p(N_2) = 593.4$ torr; $p(O_2) = 159.2$ torr; $p(Ar) = 7.10$ torr; $p(CO_2) = 2.47 \times 10^{-1}$ torr

5.35 (a) He; (b) He; and (c) 0.67

5.37 $p(CH_4) = 2.05$ atm; $p(C_2H_6) = 0.28$ atm; $p(C_3H_8) = 9.5 \times 10^{-3}$ atm

5.39 3.94 L

5.41 0.438 atm

5.43 2.28×10^5 g

5.45 3.0×10^4 g

5.47 $+0.0048$

5.49 20.8 °C

5.51 6.7×10^2 L

5.53 2×10^{12} molecules/L

5.55 0.911 L

5.57 10^{-9} atm; 3×10^{10} molecules/cm^3

5.59 6.18 g; 4.44×10^{22} atoms

5.61 (a) chamber B; (b) 0.50 atm; (c) 4.5 atm; and (d) 0.38 atm

5.63 359 m/s

5.65 5.89 g/L

5.67 (a) 30.9 °C; (b) 4.6 °C; and (c) 6.1 °C

5.69 Pump out a bulb of known volume, weigh it, fill with oxygen at a measured pressure, and weigh again. If gaseous oxygen were monatomic, this experiment would give $MM = 16.00$ g/mol, while the diatomic gas gives $MM = 32.00$ g/mol.

5.71 $m(C_3H_6) = 0.692$ g; $m(O_2) = 2.10$ g

5.73 (a) H_2; (b) O_2; (c) O_2; and (d) H_2

5.75 (a) 4×10^{-21} J; and (b) 4×10^2 m/s

5.77 Sulfur in coal combines with oxygen from the air to form SO_2, and SO_2 reacts with water to form sulfuric acid: $S + O_2 \rightarrow SO_2$; $2\ SO_2 + O_2 \rightarrow 2\ SO_3$; $SO_3 + H_2O \rightarrow H_2SO_4$.

5.79 (a) 2.09 L; (b) 788 torr; and (c) 621 torr

5.81 (a) Your drawing should show the same atomic density as Figure 5-8 but with longer "tails" on the atoms, indicating that they are moving at higher speeds. (b) Your drawing should show the same lengths of "tails" as Figure 5-8, but there should be 1/3 fewer atoms. (c) Your drawing should be the same as Figure 5-8 except that half the atoms should be replaced with diatomic molecules.

5.83 (a) P is proportional to number of moles of gas; (b) true; and (c) P/T is constant

5.85 $p(H_2O) = 1.6 \times 10^{-2}$ atm; $p(\text{dry air}) = 0.991$ atm; $p(N_2) = 0.774$ atm; $p(O_2) = 0.208$ atm; $p(Ar) = 9.26 \times 10^{-3}$ atm; $p(CO_2) = 3.22 \times 10^{-4}$ atm; $p(Ne) = 1.80 \times 10^{-5}$ atm; $p(He) = 5.19 \times 10^{-6}$ atm; $p(CH_4) = 1.4 \times 10^{-6}$ atm.

5.87 $C_6H_{15}N$

5.89 10.2 atm

5.91 (a) 1.18 g/L; (b) 1.17 g/L

5.93 4.21×10^6 L

5.95 69.9%

5.97 $Ni(CO)_4$

5.99

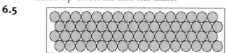

5.101 At higher elevations, Mount Everest, the atmospheric pressure, 250 mm, is much lower than at ground level, 760 mm. A lower pressure corresponds to a smaller molecular density. "The air is thin" refers to the fact that there is less air at higher elevations.

6.1 (a) $V_{Ag} = 1.706 \times 10^{-29}$ m^3/atom, $V_{Pb} = 3.034 \times 10^{-29}$ m^3/atom; (b) $d_{Ag} = 2.57 \times 10^{-10}$ m, $d_{Pb} = 3.12 \times 10^{-10}$ m; and (c) thickness$_{Ag} = 1.7 \times 10^{-3}$ m, thickness$_{Pb} = 2.0 \times 10^{-3}$ m

6.3 Gas pressure results from transfer of momentum between atoms and container walls, indicating that atoms have momentum and mass; mass spectrometers sort atomic and molecular ions according to their masses; all matter is made up of atoms and has mass.

6.5

6.7 (a) 6.92×10^{16} Hz; (b) 1.28×10^{18} Hz; (c) 4.08×10^8 Hz; and (d) 6.56×10^{13} Hz

6.9 (a) 6.29×10^{-2} m; (b) 1.04×10^6 cm; (c) 5×10^9 mm; and (d) 1.04×10^8 μm

6.11 (a) 2.438×10^2 kJ/mol; (b) 4.69×10^3 kJ/mol; and (c) 1.0150×10^{-2} kJ/mol

6.13 2×10^{16} photons

6.15 (a) 161 nm, 1.86×10^{15} Hz; and (b) 560 nm, 5.35×10^{14} Hz

6.17 (a) 231 nm; (b) 3.4×10^{-19} J; and (c) 580 nm

6.19

(Energy level diagram with vertical axis labeled "Energy"; two sets of horizontal levels labeled bE, one under "Cesium" and one under "Chromium.")

6.21 (a) 1 to 1000 m; (b) yellow; and (c) TV, FM region

6.23 745 kJ/mol

6.25 $\lambda_{8-1} = 92.57$ nm, $\lambda_{9-1} = 92.27$ nm; both are in the ultraviolet region

6.27 Under normal conditions, absorptions occur between the lowest state and any higher-energy state. Emission originates from any higher state, not only the lowest state.

6.29 5.4857990×10^{-4} g/mol

6.31 (a) 1.45 nm; (b) 6.38 nm; and (c) 4.41 μm

6.33 (a) 1.71×10^{-20} J; (b) 1.11×10^{-38} J; and (c) 3.08×10^{-33} J

6.35

n	l	m_l	m_s
6	1	+1	+1/2
6	1	+1	−1/2
6	1	0	+1/2
6	1	0	−1/2
6	1	−1	+1/2
6	1	−1	−1/2

6.37 For $n = 3$, l can be 0 with $m_l = 0$, l can be 1 with $m_l = +1, 0,$ or -1; and l can be 2 with $m_l = +2, +1, 0, -1,$ or -2; $m_s = +1/2$ or $-1/2$ in all cases.

6.39 (a) non-existent: m_s must be $+1/2$ or $-1/2$; (b) actual; (c) non-existent: l must be less than n; and (d) actual

6.41 3, 2, 2, +1/2

6.43

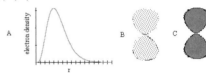

6.45

6.47 (a) = 2p, $n = 2$, $l = 1$; (b), and (c) = 3d, $n = 3$, $l = 2$

6.49 Electron density plots accurately show how the wave function varies along one axis, but they fail to show whether or not an orbital has spherical symmetry, and they do not directly show electron densities. Thus, they fail to give a good picture of how an orbital appears in three dimensions.

6.51

6.53

6.55 $E = 352$ kJ/mol; $E_{\text{kinetic}} = 2.10 \times 10^{-19}$ J

6.57 (a) O_2, N_2, and O_3 (thermosphere); and (b) O_3 (stratosphere)

6.59 (a) ~50km; (b) upper stratosphere; and (c) formation of ozone

6.61 (a) 6.7×10^{-6} m; (b) 3.0×10^{-20} J; and (c) 1.54 s

6.63 492 nm; yes

6.65 six;

n	l	m_l	m_s
4	1	1	1/2
4	1	1	−1/2
4	1	0	1/2
4	1	0	−1/2
4	1	−1	1/2
4	1	−1	−1/2

6.67

6.69 552.7 nm

6.71 (a) 3.30×10^{-19} J; (b) 3.31×10^{-7} m; and (c) 9.45×10^{-10} m

6.73 11.0 m; 1.81×10^{-26} J

6.75 (a) 2×10^{-4} atm; (b) N_2, O_2, O_3; (c) ~35 km; and (d) stratosphere

6.77 6.16×10^{14} s^{-1}, 246 kJ/mol; 5.83×10^{14} s^{-1}, 233 kJ/mol; 5.52×10^{14} s^{-1}, 220 kJ/mol; 5.42×10^{14} s^{-1}, 216 kJ/mol; 5.19×10^{14} s^{-1}, 207 kJ/mol

6.79 95.0 nm; UV

6.81 no

6.83 $V_{\text{atom}} = 4 \times 10^{-30}$ m^3; $V_{\text{nucleus}} = 4 \times 10^{-45}$ m^3; 10^{-15}

6.85 6 photons

6.87 electron: 4.853×10^{-11} m; proton: 2.642×10^{-14} m

6.89 Photon$_1$ = 565 nm (*d to c*); Photon$_2$ = 121 nm (*c to b*); Photon$_3$ = 152 nm (*b to a*). The energies relative to *a* are: *b*, 1.31×10^{-18} J; *c*, 2.95×10^{-18} J; *d*, 3.30×10^{-18} J.

6.91 (a) $E_{488} = 4.07 \times 10^{-19}$ J; $E_{514} = 3.86 \times 10^{-19}$ J; (b)

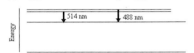

and (c) 4.17×10^{15} s^{-1} and 71.9 nm

6.93 (a) $E_{589.6} = 202.8$ kJ/mol; $E_{590.0} = 202.7$ kJ/mol; (b)

and (c) 423 nm

6.95 $E_A = 2.7 \times 10^{-19}$ J, $E_B = 4.3 \times 10^{-19}$ J; Metal B has the higher binding energy. (b) 1.59×10^{-18} J; and (c) 460 nm to 750 nm

7.1 (a) He 1s is more stable than He 2s because of its lower value of n; (b) Kr 5s is more stable than Kr 5p because of its lower value of l; and (c) He$^+$ 2s is more stable than He 2s because it is less screened by 1s electrons.

7.3 A hydrogen atom contains just one electron, so there is no screening effect. In the absence of screening, orbital energy depends only on n and Z, so all $n = 3$ orbitals have identical energy. In a helium atom, on the other hand, an electron in an $n = 3$ orbital is screened from the nucleus by the second electron, and the amount of screening decreases as l increases.

7.5 (a) The ionization energy (*IE*) of the He 2*p* orbital $(0.585 \times 10^{-18}$ J) is not much larger than that of the H 2*p* orbital $(0.545 \times 10^{-18}$ J). (b) The *IE* of the He^+ 2*p* orbital $(2.18 \times 10^{-18}$ J) is four times larger than that of the H 2*p* orbital $(0.545 \times 10^{-18}$ J).

7.7

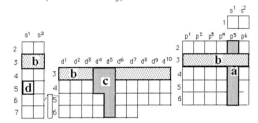

7.9 Atomic number = 114; valence orbital = 7*p*

7.11 Row = 7; column = Group 11 (directly below gold)

7.13 O has 6 valence electrons; V, 5; Rb, 1; Sn, 4; Cd, 2

7.15 Be, 4 electrons:

n	*l*	m_l	m_s
1	0	0	+1/2
1	0	0	−1/2
2	0	0	+1/2
2	0	0	−1/2

O, 8 electrons:

n	*l*	m_l	m_s
1	0	0	+1/2
1	0	0	−1/2
2	0	0	+1/2
2	0	0	−1/2
2	1	1	+1/2
2	1	0	+1/2
2	1	−1	+1/2
2	1	1	−1/2

Ne, 10 electrons:

n	*l*	m_l	m_s
1	0	0	+1/2
1	0	0	−1/2
2	0	0	+1/2
2	0	0	−1/2
2	1	1	+1/2
2	1	1	−1/2
2	1	0	+1/2
2	1	0	−1/2
2	1	−1	+1/2
2	1	−1	−1/2

P, 15 electrons: first 10 electrons are the same as for Ne; remaining 5 electrons are:

n	*l*	m_l	m_s
3	0	0	+1/2
3	0	0	−1/2
3	1	1	+1/2
3	1	0	+1/2
3	1	−1	+1/2

7.17

O P

2*p* ⇅ ↿ ↿   3*p* ↿ ↿ ↿

7.19 (a) Pauli-forbidden; (b) Pauli-forbidden; (c) excited-state configuration; (d) excited-state configuration; (e) ground-state configuration; (f) non-existent orbital; and (g) non-existent orbital

7.21

Mo   Tc

5*s* ↿   4*d* ↿↿↿↿↿   5*s* ⇅   4*d* ↿↿↿↿↿

7.23 (a) Cr, $4s^1 3d^5$; Cu, $4s^1 3d^{10}$; (b) In column 6, Cr and Mo have $s^1 d^5$ configurations, while the second two elements in this column, W and Sg, have $s^2 d^4$ configurations; and (c) Pa, U, Np, and Cm

7.25 6: $1s^1 2s^2 2p^4$; $2s^2 2p^5$; $1s^1 2s^1 2p^5$; $1s^2 2s^1 2p^4$; $1s^2 2p^5$; $2s^1 2p^6$; $1s^1 2p^6$

7.27 Ar > Cl > K > Cs

7.29 Column 1 (Cs)

7.31 For N, the added electron adds to an already-occupied orbital; electron-electron repulsion makes this a disfavored process. For Mg, the added electron adds to the next higher orbital, which is significantly higher in energy. For Zn, the added electron adds to the next higher orbital, which is significantly higher in energy.

7.33 $Y^{3+} < Sr^{2+} < Rb^+ < Br^- < Se^{2-} < As^{3-}$

7.35 F, Cl, Br, I, (At)

7.37 −270 kJ/mol

7.39 Stable anion: Cl; stable cations: Ca, Cu, Cs, Cr

7.41 −607 kJ/mol

7.43

Electron affinity - 328.0 kJ/mol
1/2 Bond energy 77.5 kJ/mol
Ionization energy 520.7 kJ/mol
Vaporization 159 kJ/mol
Lattice energy - 1036 kJ/mol
Calculated overall energy change - 607.3 kJ/mol

7.45 Metal: Po; nonmetal: O, S, Se; metalloid: Te

7.47 Metal: Ca, Cu, Cs, Cr; nonmetal: C, Cl

7.49 Al, Ga, Sn, Bi

7.51 C: $1s^2 2s^2 2p^2$; Cr: $1s^2 2s^2 2p^6 3s^2 3p^6 4s^1 3d^5$; Sb: $1s^2 2s^2 2p^6 3s^2 3p^6 4s^2 3d^{10} 4p^6 5s^2 4d^{10} 5p^3$; Br: $1s^2 2s^2 2p^6 3s^2 3p^6 4s^2 3d^{10} 4p^5$

7.53 Mn^{2+}: $1s^2 2s^2 2p^6 3s^2 3p^6 3d^5$;

n	*l*	m_l	m_s
3	2	2	+1/2
3	2	1	+1/2
3	2	0	+1/2
3	2	−1	+1/2
3	2	−2	+1/2

7.55 P: spin = 3/2; Br^-: spin = 0; Cu^+: spin = 0

7.57 Z = 9 is F: [He] $2s^2 2p^5$; Z = 20 is Ca: [Ar] $4s^2$; and Z = 33 is As: [Ar] $4s^2 3d^{10} 4p^3$

7.59 Na < O < N < Ne < Na^+

7.61 $Br^- > Cl^- > Cl > K^+$

7.63 (a) In all one-electron atoms, orbital energy depends only on *n* and Z, both of which are the same for the hydrogen atom 2*s* and 2*p* orbitals. and (b) In multi-electron atoms, *n* orbitals with different *l* values are screened to different extents. The orbital with the lower *l* value is less screened, hence more stable.

7.65 (a) From element 80 (Hg) to 81 (Tl); (b) Cs, element 55; (c) Z = 56 to 71, Z = 39 to 46, and Z = 20 to 29; and (d) Elements 15 (P), 16 (S), 33 (As), 34 (Se), 53 (I), 80 (Hg), 85 (At), and 86 (Rn)

7.67

Cu^+ is $4s^0 3d^{10}$

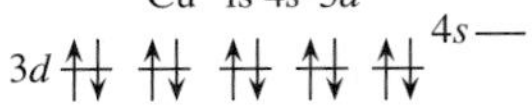

$3d\ \uparrow\downarrow\ \uparrow\downarrow\ \uparrow\downarrow\ \uparrow\downarrow\ \uparrow\downarrow\quad 4s\ —$

Mn^{2+} is $4s^0 3d^5$

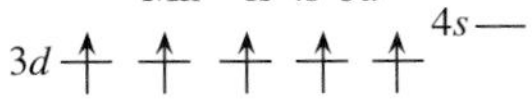

$3d\ \uparrow\ \uparrow\ \uparrow\ \uparrow\ \uparrow\quad 4s\ —$

Au^{3+} is $6s^0 5d^8$

$5d\ \uparrow\downarrow\ \uparrow\downarrow\ \uparrow\downarrow\ \uparrow\ \uparrow\quad 6s\ —$

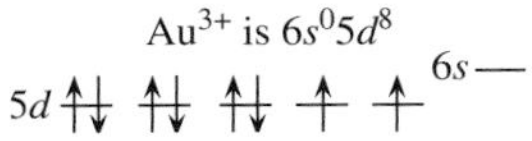

7.69 S^+

7.71 (a) B (most stable), A, C; (b) $n = 3$, $l = 0$, $m_l = 0$, $m_s = +1/2$; $n = 3$, $l = 0$, $m_l = 0$, $m_s = -1/2$; (c) any from $Z = 12$ to $Z = 20$; (d) Fe^{3+}; (e) 8; and (f) smaller.

7.73 $8s$, then $5g$ (or perhaps $6f$)

7.75 The data in Appendix C indicate that noble gas elements have unfavorable electron affinities (<0) and relatively large first ionization energies (>1350 kJ/mol). Consequently, they are unlikely either to gain or lose electrons to form stable chemical compounds. Nevertheless, Xe and, to a lesser extent Kr, do form a few compounds.

7.77 Be: [He] $2s^1 2p^1$; O^{2-}: [He] $2s^2 2p^5 3s^1$; Br^-: [Ar] $4s^2 3d^{10} 4p^5 5s^1$; Ca^{2+}: [Ne] $3s^2 3p^5 4s^1$; Sb^{3+}: [Kr] $4d^{10} 5s^1 5p^1$

7.79 The first ionization energies involve the $2s$ orbital for both atoms, so Be, with larger Z, has a larger IE. The second electron removed from Li is a core electron, so the IE is much greater than the second IE for Be.

7.81
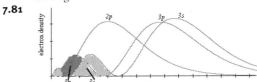

8.1 $1s^2\ 2s^2$; Its two $n = 2$ electrons

8.3 Two sodium atoms form a bond by overlap of their $3s$ orbitals to produce an orbital with high electron density between the nuclei:

8.5 Al, $3s$ and $3p$; As, $4s$ and $4p$; F, $2s$ and $2p$; Sn, $5s$ and $5p$

8.7 The $2s$ orbital of a lithium atom can overlap with the $1s$ orbital of a hydrogen atom to produce an orbital with high electron density between the nuclei:

8.9 Cr (1.6) and Mn (1.5); Cu (1.9) and Zn (1.6); Pd (2.2), Ag (1.9), and Cd (1.7); Au (2.4) and Hg (1.9); Cr (1.6) and Mo (1.8); Mn (1.5) and Tc (1.9)

8.11 Non-polar: F_2; ionic: NaF and CaO; polar: HF and NaH

8.13 (a) N; (b) S; (c) I; and (d) S

8.15 $PH_3 < H_2S < NH_3 < H_2O$

8.17

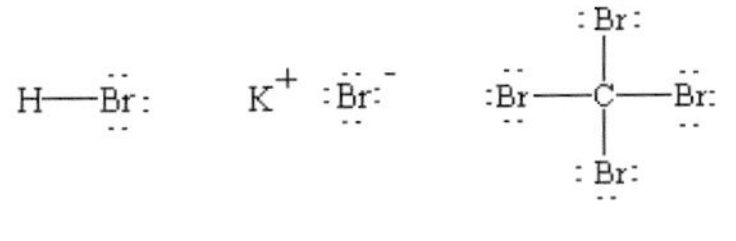

H—Br: K⁺ :Br:⁻   :Br—C—Br:
                              |
                            :Br:

8.19

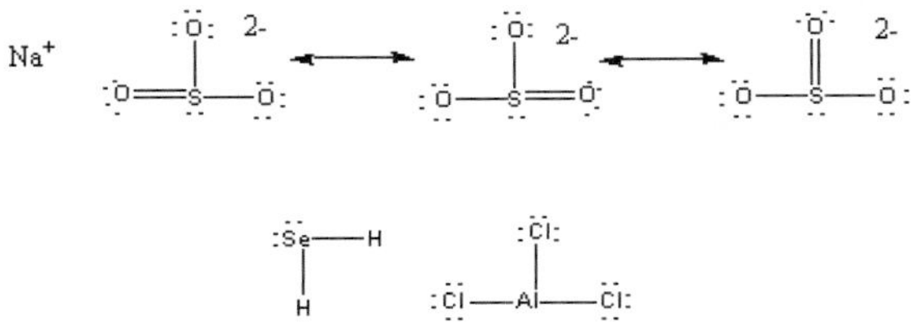

Na⁺

[O=S—O ⟷ O—S=O ⟷ O—S—O]²⁻

Se—H
|
H

Cl—Al—Cl
      |
     Cl

8.21

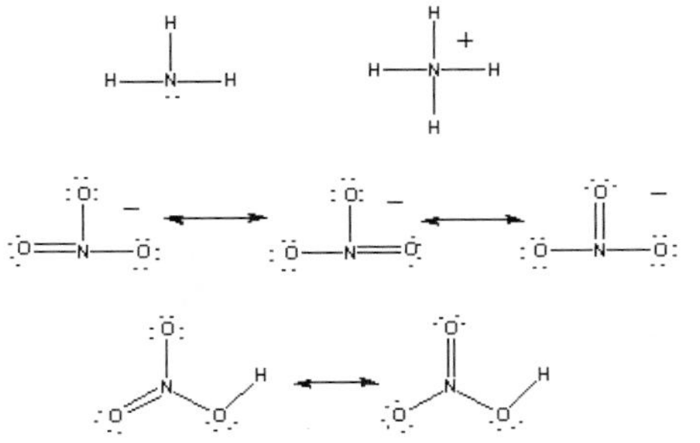

H—N—H
    |
    H

[H—N—H]⁺
     |
     H

[O=N—O ⟷ O—N=O ⟷ O—N—O]⁻

[O₂N—O—H ⟷ O₂N—O—H]

8.23

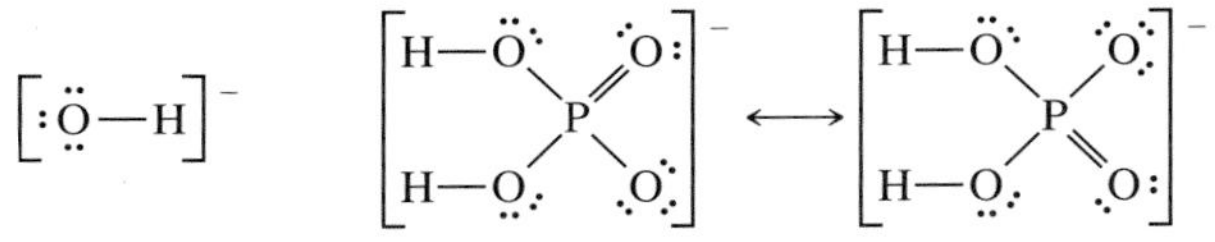

[:O—H]⁻

[H—O—P(=O)—O—H with O] ⟷ [H—O—P(=O)—O—H with O]⁻

8.25

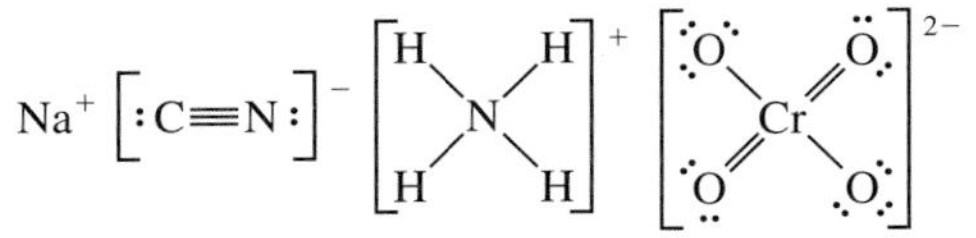

Na⁺ [:C≡N:]⁻

[H—N(H)(H)—H]⁺

[O—Cr(=O)(=O)—O]²⁻

8.27

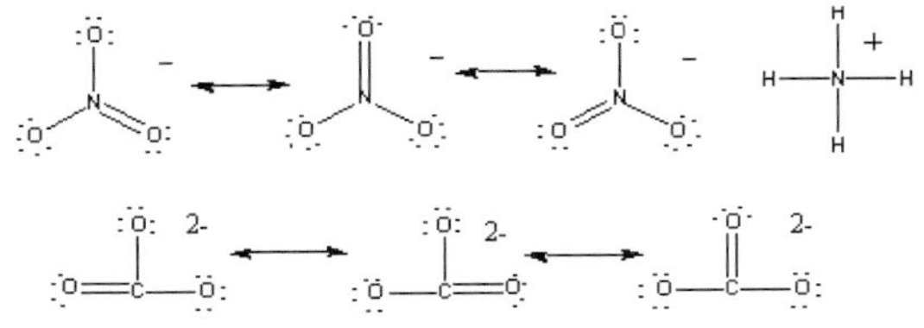

[O=N—O ⟷ O—N=O ⟷ O—N—O]⁻

[H—N(H)(H)—H]⁺

[O=C—O ⟷ O—C=O ⟷ O—C—O]²⁻

8.29

Cl—C(—H)(—H)—Cl
(Cl₂CH₂)

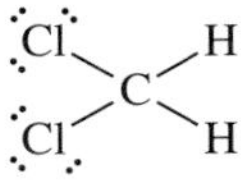

The steric number of the carbon atom is 4. The bonding about it can be described using sp^3 hybrid orbitals overlapping with $1s$ orbitals of H and $3p$ orbitals of Cl.

The geometry about the carbon atom is tetrahedral.

8.31

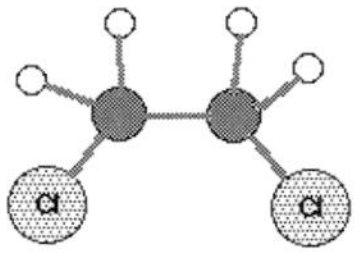

8.33 "Straight chain": C—C—C—C—C—C

C—C—C—C—C   C—C—C—C—C
    | |
    C   C

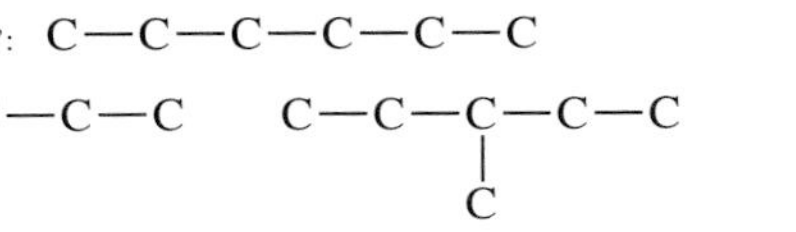

Putting the branched carbon in position 4 gives the same isomer as putting it in position 2

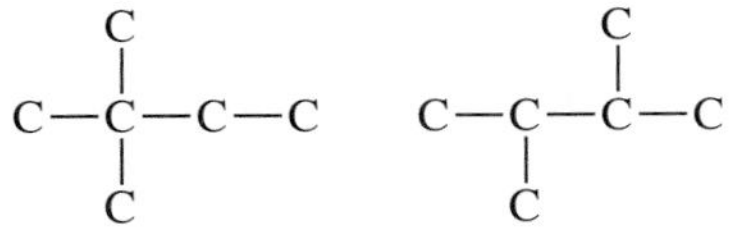

C—C—C—C   C—C—C—C
    | |
    C   C

8.35

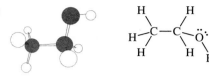

Methane molecule with left hydrogen replaced with methyl, top hydrogen replaced with —OH

8.37

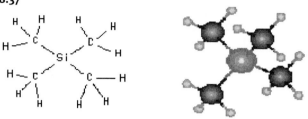

Hybridization = sp^3; geometry = tetrahedral

8.39

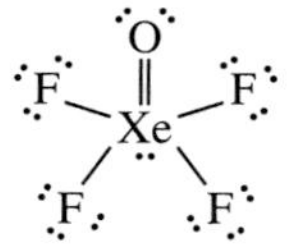

Hybridization: d^2sp^3; geometry: square pyramid with bond angles near 90°

8.41 GeF$_4$, tetrahedron, sp^3; SeF$_4$, seesaw, dsp^3; XeF$_4$, square plane, d^2sp^3

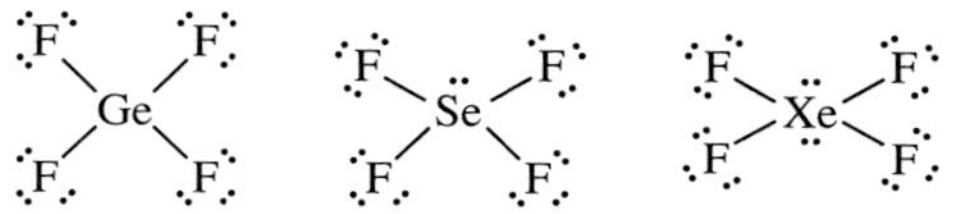

8.43 H$_2$S and NF$_3$ have dipole moments.

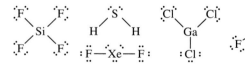

8.45 (a) SiF$_4$, 109.5°; (b) H$_2$S, < 109.5°; (c) XeF$_2$, 180°; (d) GaCl$_3$, 120°; and (e) NF$_3$, < 109.5°

8.47 (a)

(b)

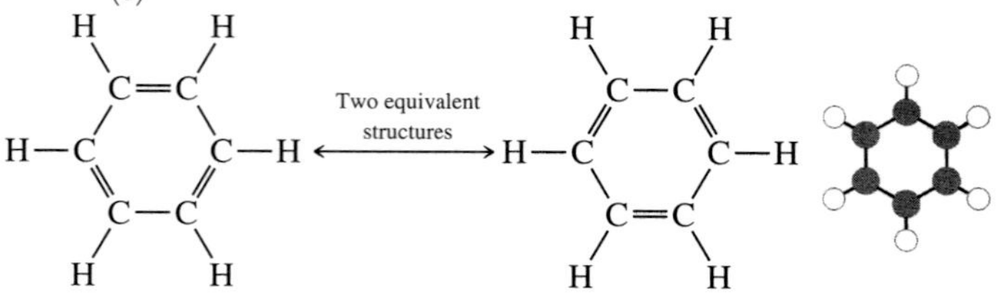

Two equivalent structures

(c)

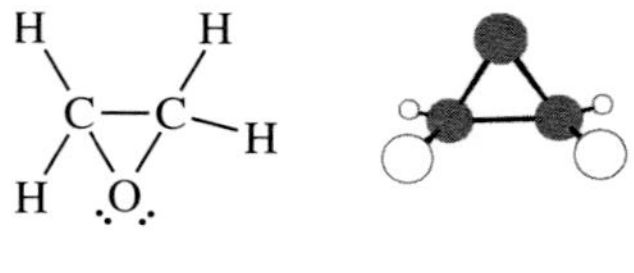

8.49 SiCl$_4$ and CI$_4$, tetrahedral; SeF$_4$, seesaw;

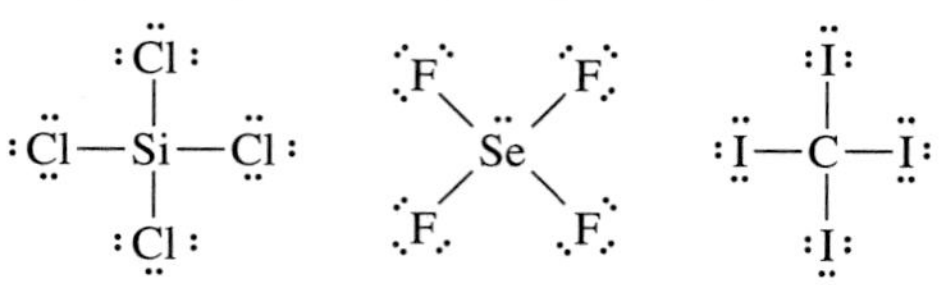

8.51 Two isomers

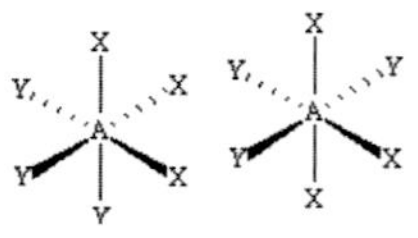

8.53 The molecule has trigonal planar structure with Al—Cl bonds that can be represented using sp^2 hybrids on Al and $3p$ orbitals on Cl.

8.55 The empirical chemical formula of the silicon-oxygen network of zircon, an orthosilicate, is SiO$_4^{4-}$. The Si—O bonds are formed by the overlap of the oxygens $2p$ orbital with one of the silicon's sp^3 orbitals. The molecular geometry of the SiO$_4^{4-}$ anions are tetrahedral. Orthosilicates networks are ionic, containing discrete (not connected) SiO$_4^{4-}$ anions and metal cations.

8.57 (a) FC$_{Br}$ = 0

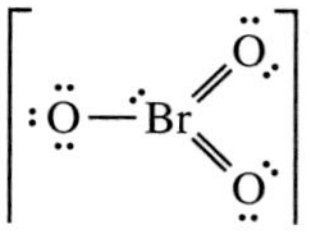

There are two additional resonance structures

(b) FC$_N$ = 0

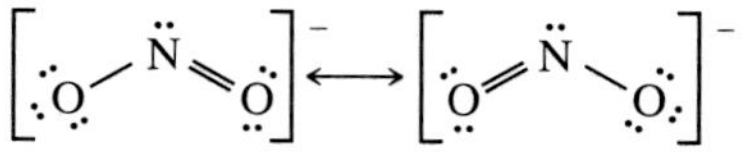

(c) FC$_P$ = 0

There are two additional resonance structures

(d) FC$_C$ = 0; FC$_O$ = 0

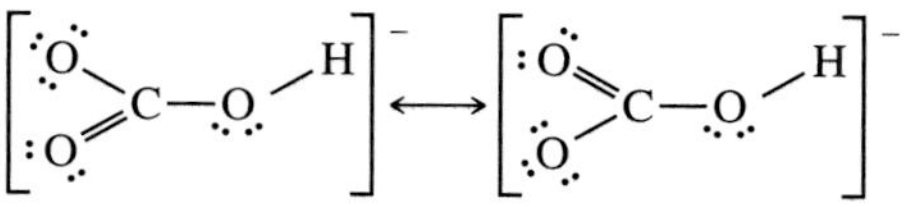

8.59 The bond angles about carbon are 109° and those about nitrogen are < 109°;

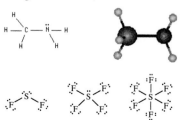

8.61

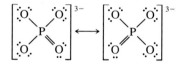

8.63 SF$_2$, bent, sp^3; SF$_4$, seesaw, dsp^3; SF$_6$, octahedral, d^2sp^3

8.65 The central carbon has SN = 3, trigonal planar geometry, and its bonding orbitals can be described using sp^2 hybrid orbitals. The methyl carbons have SN = 4, tetrahedral geometry, and their bonding orbitals can be described using sp^3 hybrid orbitals.

8.67 A bond angle of 92.2° indicates that the bonding can be well described using p orbitals rather than hybrid orbitals, whereas an angle of 104.5° indicates significant distortion from pure p orbital bonding. Space-filling models of the two molecules shows that the smaller oxygen atom cannot accommodate two H atoms at right angles, but the larger S atom can.

8.69

8.71 Si—H < C—H < N—H < O—H < F—H

8.73 Same compound, because they can be superimposed after rotating one by 109.5°

8.75 Compounds with formula XF_5 have five electron pairs associated with the inner atom. This is possible for phosphorus, a third row element that has d orbitals available for bonding. It is not possible for nitrogen, a second row element which lacks valence d orbitals.

8.77 (a)

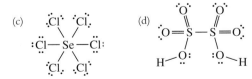

(b) When like atoms are opposite each other, there is no net dipole moment. When like atoms occupy adjacent positions the molecule has a net dipole moment.

8.79 (a) TeF_2; (b) TeF_3^-; (c) TeF_5^-; (d) TeF_5^+; (e) TeF_6; (f) TeF_4

8.81 IF_7

8.83 (a)

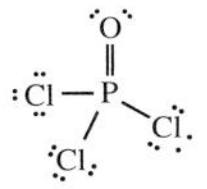

(b) These are two equivalent structures

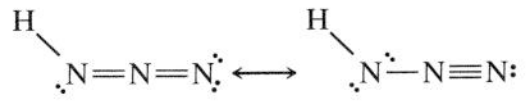

(c)   (d)

8.85 Each C atom has four bonds, SN = 4, uses sp^3 hybrids, and should have tetrahedral geometry. The four-carbon rings are distorted squares, however, so the two rings are strained with bond angles around 90° rather than the tetrahedral 109.5°

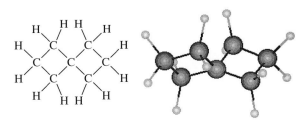

9.1 Acetone has three inner C atoms. Two have only single bonds (one C—C and three C—H); these have SN = 4. The atom bonded to O has one π bond and SN = 3.

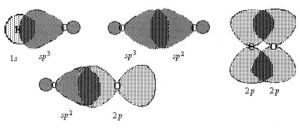

9.3 Some of the inner atoms have all single bonds, SN = 4, sp^3 hybridization, and tetrahedral geometry; others have one double bond, SN = 3, sp^2 hybridization, and trigonal planar geometry. The SN values for the 20 carbon atoms are shown in the line structure below.

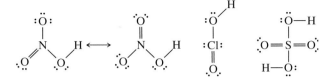

9.5

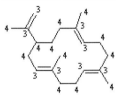

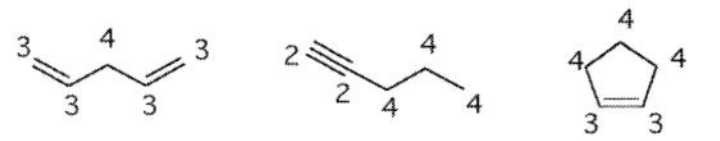

1,4-Pentadiene   1-Pentyne   Cyclopentene

9.7 H—N < N≡N < C=O < N—N < Cl—N

9.9 C—C < H—N < C=C < C=O < N≡N

9.11 165 kJ

9.13 −475 kJ; −90 kJ; −205 kJ

9.15 The orbital energy ladder should show that Na_2 has two bonding and no antibonding electrons, indicating a stable molecule with a single bond.

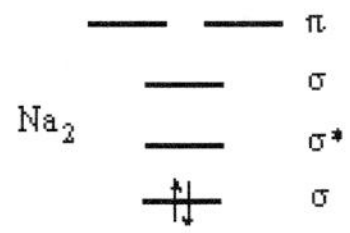

9.17 (a) CO; (b) N_2; (c) CN^-

9.19 There are eight bonding electrons and six antibonding electrons, for a net of two bonding electrons and a single bond for this ionic species.

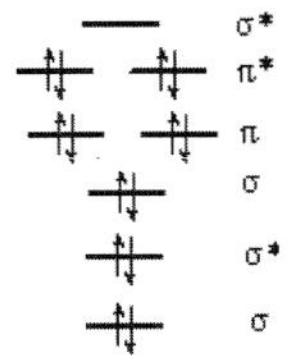

9.21

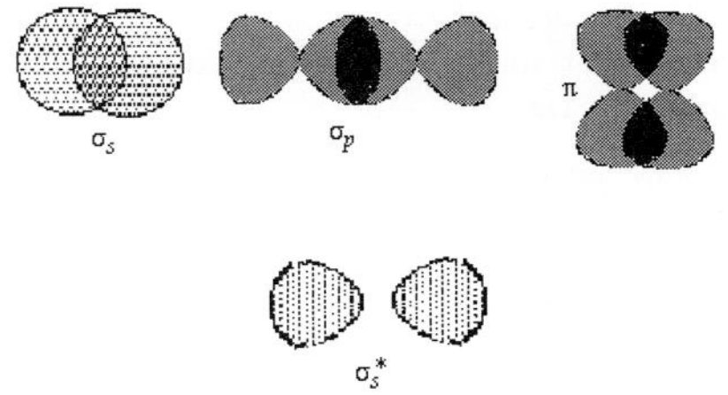

9.23 The bonding pattern for CS_2 is like that for CO_2, except that the S atoms have $n = 3$ valence orbitals rather than $n = 2$ orbitals. The inner atom can be described using sp hybrids which overlap with $3p$ orbitals from the S atoms to form two σ bonds. There are two delocalized π systems at right angles to each other, each made up of a $2p$ orbital on the C atoms overlapping side-by-side with a $3p$ orbital on each S atom. As in CO_2, eight electrons occupy the delocalized π orbitals.

9.25 NO_2^+ is isoelectronic with CO_2. It has no lone pair on the inner N atom, giving SN = 2 and sp hybridization. NO_2^- is isoelectronic with O_3. Its 18 valence electrons include a lone pair on the inner atom, SN = 3 and sp^2 hybridization to accommodate three electron pairs:

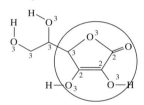

9.27 N bonded to H: bent, sp^2 hybridized, angle between $109-120°$; outer N is not hybridized, middle N atom is linear, sp hybridized.

9.29 Xanthin has a delocalized π system in which all 13 π bonds contribute.

9.31 (a) Hybridization is as shown on the line drawing ($3 = sp^3$, $2 = sp^2$).

(b–c) The Lewis structure shows 2 π bonds, on adjacent atoms, so as in butadiene there are four electrons in the extended π system; and (d) The eight atoms circled in the line drawing lie in the same plane.

9.33 Lewis structure:

$$:\!O\!=\!C\!=\!C\!=\!C\!=\!O\!:$$

The molecule is linear and the σ bond network can be described using sp hybrids. There are two $sp-sp$ bonds between carbon atoms and two $sp-2p$ bonds between carbon and oxygen atoms:

If the molecular axis is the z-axis, a p_x and a p_y orbital on each atom overlap side-by-side, giving two sets of delocalized π orbitals, each extending over all five atoms:

9.35 The molecule has tetrahedral geometry and the σ bond framework can be described using sp^3 hybrids on Cl overlapping with $2p$ orbitals from O. The resonance structures signal that the π system is extended over all five atoms incorporating the d orbitals from the inner Cl atom, and there are three bonding π orbitals occupied by six electrons.

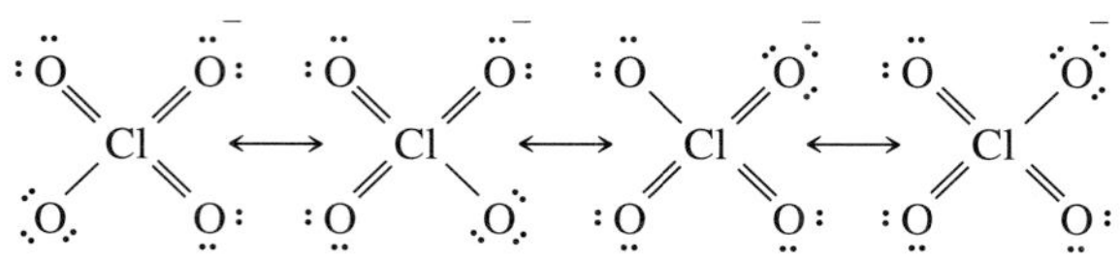

9.37 (a) undoped; (b) n-type; and (c) undoped

9.39 Whereas each potassium atom contributes just one valence electron to the valence (bonding) band of the solid, each iron atom contributes eight electrons. Thus iron has much greater bonding, making it harder and giving it a higher melting point than potassium.

9.41 (a) C—N is shorter but weaker than C—C; (b) C—N is also longer but stronger than N—N.

9.43 ethanol is more stable

9.45 (a) n-type; (b) p-type; and (c) p-type

9.47 shortest: O—H; longest: C=O

9.49 H—C < H—O < C=O < C≡N

9.51 The sulfur atom in each compound has SN = 3, so sp^2 hybrids overlapping with oxygen $2p$ orbitals describe the σ bonds: two in SO_2 and three in SO_3. SO_2 is bent, like O_3, and SO_3 is trigonal planar, like NO_3^-. The Lewis structures indicate the presence of two π bonds in SO_2 and three π bonds in SO_3. All the π orbitals extend over the entire molecule. Because sulfur is a third-row element, its $3d$ orbitals contribute to the extended π bonding orbitals, which form from side-by-side overlap of oxygen $2p$ orbitals with sulfur $3p$ and $3d$ orbitals.

9.53 (a)

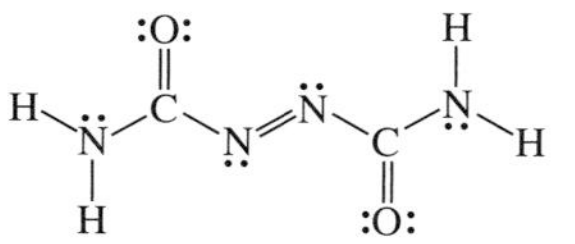

(b) The end N atoms have sp^3 hybridization, and tetrahedral geometry. The C atoms and the center N atoms have sp^2 hybridization, and trigonal planar geometry. and (c) all but the end NH_2 groups form a π system containing 6 electrons.

9.55 $\Delta E_{ethane} = -46.3$ kJ/g; $\Delta E_{ethylene} = -45.6$ kJ/g; $\Delta E_{acetylene} = -46.7$ kJ/g; Estimates indicate that acetylene releases slightly more energy per unit mass than ethane or ethylene.

9.57

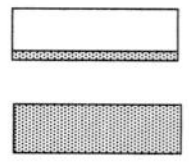

9.59 -45 kJ for HCl reaction; -150 kJ for Cl_2 reaction

9.61 All outer atoms (including H) always use atomic orbitals.

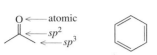

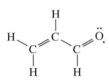

All C's use sp^2 C and N use sp^3

9.63 Each carbon atom has trigonal planar geometry, and can be described using sp^2 hybridization. This leaves one p orbital on each atom, plus one from the oxygen atom, to form a delocalized π network extending over four atoms, as in butadiene.

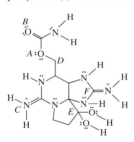

9.65 (a) 3; (b) A, and D use sp^3, B uses atomic $2p$ orbitals, C uses sp^2 hybrids; (c) D and E, 109.5°, F 120°; and (d)

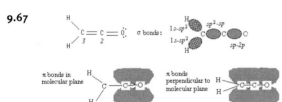

9.67

9.69 N_2 has two antibonding electrons, for a net bond order of three; O_2 has four antibonding electrons, for a net bond order of two; F_2 has six antibonding electrons, for a net bond order of one. The bond lengths, which increase in the order $N_2 < O_2 < F_2$, reflect this trend in bond order.

9.71

9.73 -430 kJ; -130 kJ

9.75 O_2: $(\sigma_s)^2(\sigma_s^*)^2(\sigma_p)^2(\pi_x)^2(\pi_y)^2(\pi_x^*)^1(\pi_y^*)^1$ Bond order = 2
O_2^-: $(\sigma_s)^2(\sigma_s^*)^2(\sigma_p)^2(\pi_x)^2(\pi_y)^2(\pi_x^*)^2(\pi_y^*)^1$ Bond order = 1.5
O_2^{2-}: $(\sigma_s)^2(\sigma_s^*)^2(\sigma_p)^2(\pi_x)^2(\pi_y)^2(\pi_x^*)^2(\pi_y^*)^2$ Bond order = 1
O_2 has the strongest, shortest bond and O_2^{2-} the longest, weakest bond. O_2 and O_2^- are both magnetic, with O_2 showing the largest magnetism.

9.77 The strength of a bond increases with the number of electrons shared by the two atoms, so a triple bond, with three electron pairs, is stronger than a double bond, with two electron pairs. However, the energy *per electron pair* is less for a triple bond than for a double bond.

9.79 The σ bond framework of both ions can be described using sp^3 hybrids on Cl overlapping with $2p$ orbitals from O. The π system of the chlorate ion is extended over all four atoms, and there are two bonding π orbitals occupied by four electrons. The π system of the chlorite ion is extended over all three atoms, with one bonding π orbital occupied by two electrons.

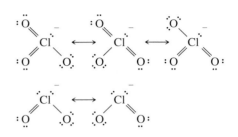

9.81 Higher bond order should be accompanied by increased bond strength and decreased bond length. Bond length and bond strength can be measured.

9.83

9.85 The differences in behavior between $n = 2$ elements and their $n = 3$ counterparts from the same column of the periodic table can be explained by the difference in side-by-side π overlap for $n = 2$ and $n = 3$. Multiple bonds involving π overlap are strong for N and O, but π overlap is very slight for P and S.

9.87 (1) N≡N (420 kJ) stronger than N—N (160 kJ)
(2) N—F (285 kJ) stronger than N—N (160 kJ)
(3) N—F (285 kJ) stronger than N—Cl (200 kJ)

9.89 (a) The carbon atoms with SN = 3 have σ bonding that can be described using sp^2 hybrid orbitals, and the σ bonding of the carbon atom with SN = 2 can be described using sp hybrids.

(b)

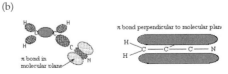

and (c) There is an extended π network that includes the three C atoms and the N atom.

9.91 There are 7 bonding electrons and 2 antibonding electrons, for a net of 5, or bond order 2.5. The least stable occupied orbital is a σ_p orbital:

σ_p

9.93 Its geometry is tetrahedral, the σ bonds can be described using sp^3 hybrids from Cl, and the molecule has a bond angle near 109°. There is an extended π system formed from d orbitals on Cl overlapping side-by-side with $2p$ orbitals from the two O atoms. It is unusual because it has an odd number of electrons.

10.1

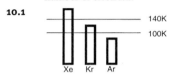

140K
100K

Xe   Kr   Ar

10.3 (a) less significant; (b) more significant

10.5

Ag$_{(s)}$ Ar$_{(g)}$ Hg$_{(l)}$

10.7 −7.9%

10.9 21.5 atm; −25% deviation from ideal gas behavior

10.11 CH_4 (hardest to liquefy) $< CF_4 < CCl_4$ (easiest to liquefy)

10.13

CH$_4$ CCl$_4$

10.15 Propane (lowest bp) $< n$-pentane $<$ ethanol (highest bp)

10.17 (a) and (c) no hydrogen bonding; (b) and (d) hydrogen bonding

b)

d)

10.19

(a) :N⟨H H---:N⟨H H

(b) :Ö—H---:N⟨H H   :N⟨H H---:Ö—H

10.21 Sn: metallic solid; S$_8$: molecular solid; Se: network solid; SiO_2: network solid; Na_2SO_4: ionic solid

10.23 (a) The bonding in metals comes from extended networks of delocalized electrons, while the bonding in network solids includes many individual covalent bonds; (b) Metals conduct electricity, are malleable and ductile, and are shiny in appearance. Network solids are non-conductors or semiconductors, are brittle, and often have dull appearances.

10.25 (a) molecular solid; (b) ionic solid; (c) metallic solid; (d) network solid; and (e) molecular solid

10.27 The density of a solid depends not only on the nature of the elements it contains but also on the tightness of bonding. Graphite has a more open bonding pattern than diamond.

10.29 $CaTiO_3$

10.31

C   C   C   C
 \ /   \ /
  C—Si   Si—C
 /   \
C   C
 \   /
  Si   C Si   C
 / \ / \/ \ /
C   C C C   C

10.33 Pentane $(C_5) <$ gasoline $(C_8) <$ fuel oil (C_{12})

10.35 The hydrocarbon molecules in salad oil have dispersion-type intermolecular forces but low polarity and no hydrogen-bonding capability. A solution of acetic acid in water has high polarity and a large hydrogen-bonding capability. Hydrogen bonds would have to be broken for the two liquids to mix, making mixing energetically unfavorable.

10.37 Ionic salts and polar organic materials such as alcohols

10.39 6 hydrogen bonds to water molecules:

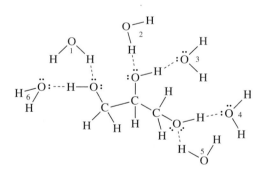

10.41 (a) Heptane (left peak), methylpentyl ether (middle peak), 1-hexanol (right peak); (b) largest amount: 1-hexanol; smallest amount: methylpentyl ether

10.43 −5.5 °C

10.45 Ethanol is volatile, so the equation does not apply and the boiling point of the solution cannot be calculated from the information provided.

10.47 (a) 287 g/mol; (b) 132 g/mol

10.49 Best surfactant: lauryl sulfate; worst surfactant: propanoic acid.

10.51 Non-polar, because the non-polar "tails" of the surfactant are attracted by the liquid

10.53 Gas chromatography

10.55 11 mol/kg

10.57 Benzene (lowest solubility) < pentanol < erythritol (highest solubility)

10.59 Hydrophobic; it will concentrate in fatty tissues; the effects of DDT are cumulative, especially for organisms high on the "food chain" such as predator birds.

10.61 Iodine is a non-polar solid that will differentially dissolve in a non-polar liquid. Water is highly polar and CCl_4 is non-polar, so these two liquids are immiscible and I_2 preferentially dissolves in CCl_4.

10.63 (a) dispersion interactions and hydrogen bonds; (b) dispersion interactions and polarity; (c) dispersion interactions; and (d) dispersion interactions

10.65 (a) CH_3OH (hydrogen bond); (b) SiO_2 (covalent bonds); (c) HF (hydrogen bond); and (d) I_2 (larger dispersion forces)

10.67 *trans*, no dipole moment, 47 °C; *cis*, dipole moment, 60 °C

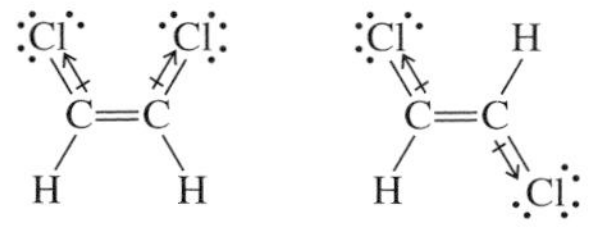

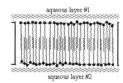

cis
dipole moment
bp = 60 °C

trans
no dipole moment
bp = 47 °C

10.69 nonpolar substances (oils and greases) and network solids (metal oxides)

10.71 Cl_2

10.73 Yes

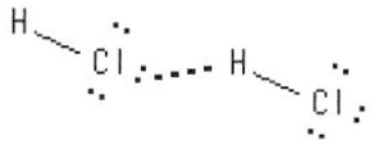

10.75 Linear water would be nonpolar like CO_2, so it would not be a good solvent of salts.

10.77 The magnitudes of dispersion forces depend on how extended the valence electrons are. In a molecule, valence electrons are spread over a larger volume because they are shared between atoms.

10.79 Pentane boils at a higher temperature than dimethylpropane, indicating that dispersion forces are greater for the linear chain than the branched chain. Both compounds have the same number of atoms and electrons, so the effect must be due to extension of the electron clouds.

10.81 29 atm

10.83

aqueous layer #1

aqueous layer #2

10.85 The molecule in figure (a) will strike the wall with greater force than the molecule in figure (b).

10.87 2,2-dimethylpropane (smallest viscosity) < *n*-pentane < butanol < propane-1,3-diol

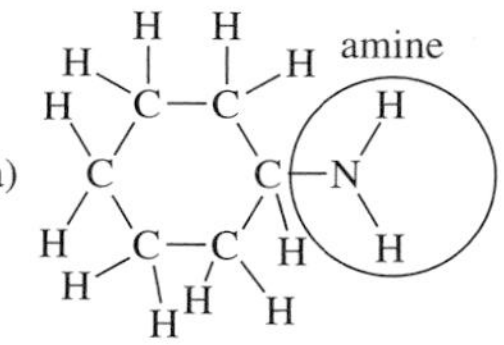

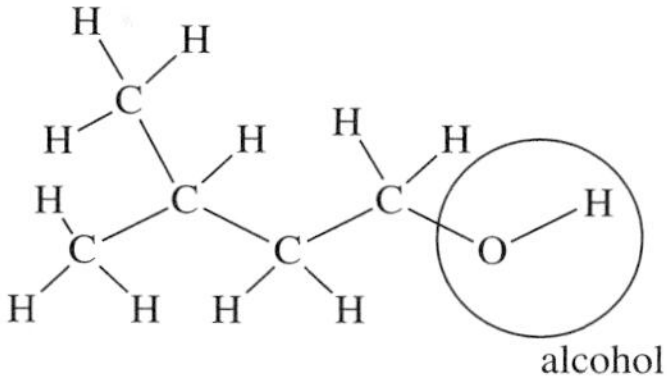

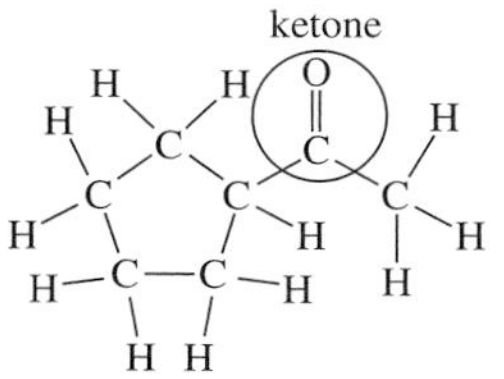

11.1

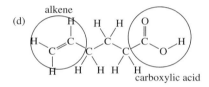

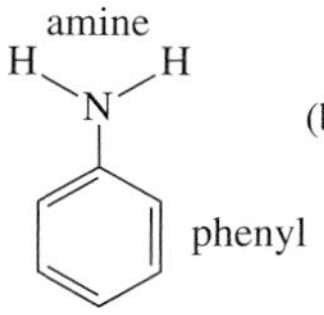

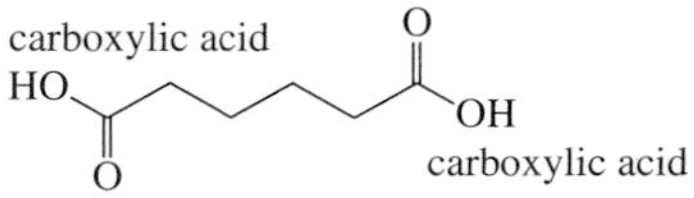

11.3

11.5

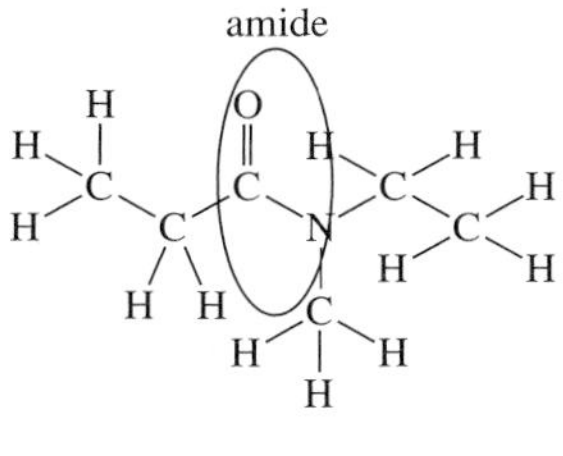

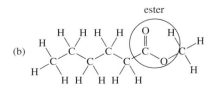

(c) thiol

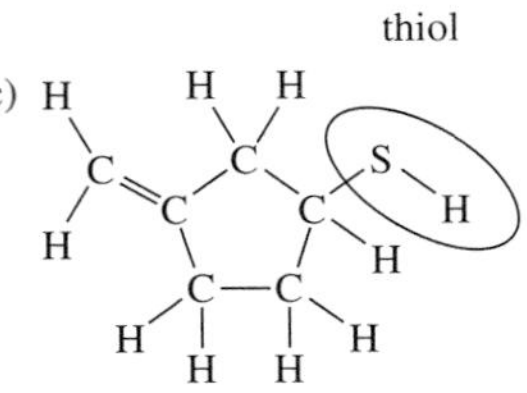

(d) ether

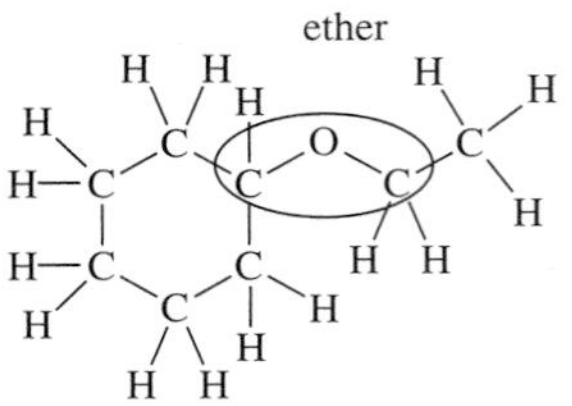

11.7

(a)

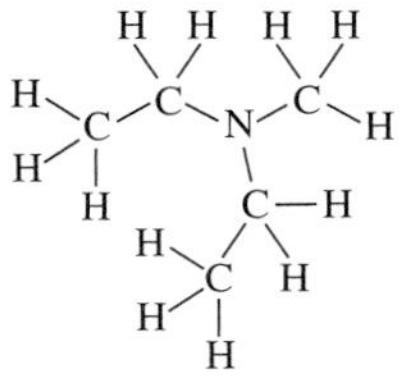

Other isomers are also possible.

(b)

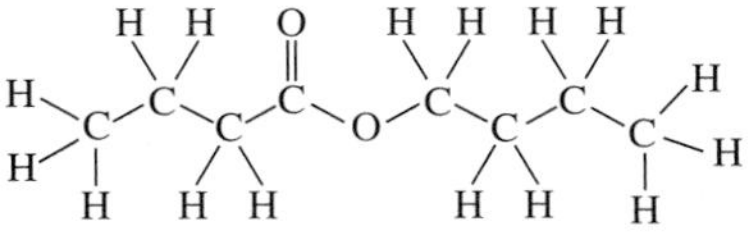

Other isomers are also possible.

(c)

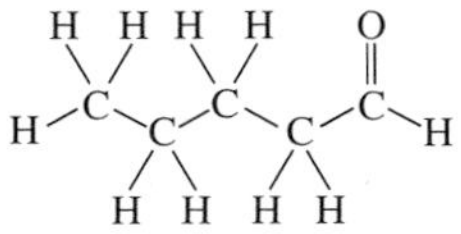

Others are also possible.

(d)

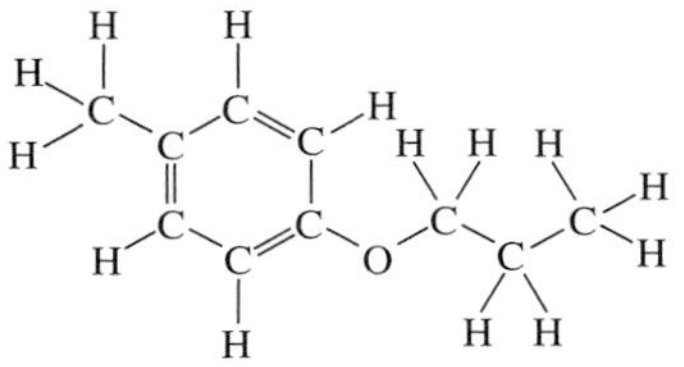

11.9

(a)

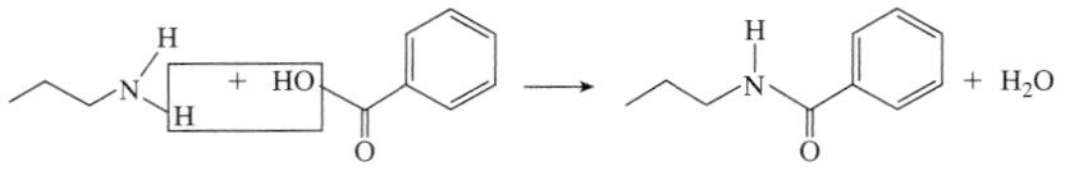

+ H_2O

(b)

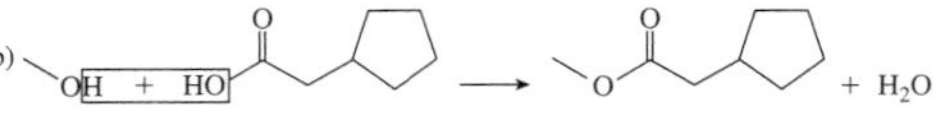

+ H_2O

(c)

+ H_2O

11.11

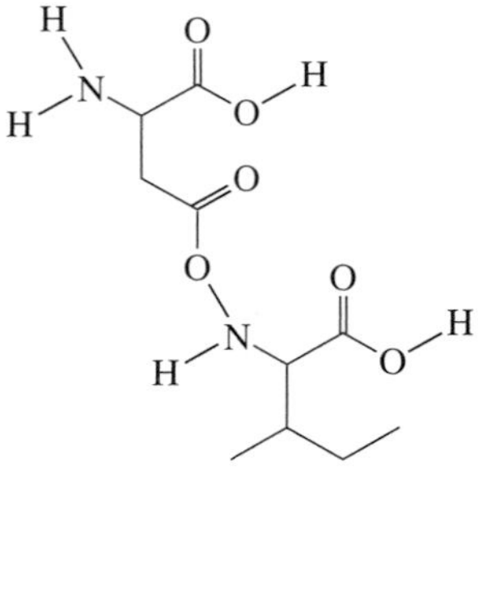

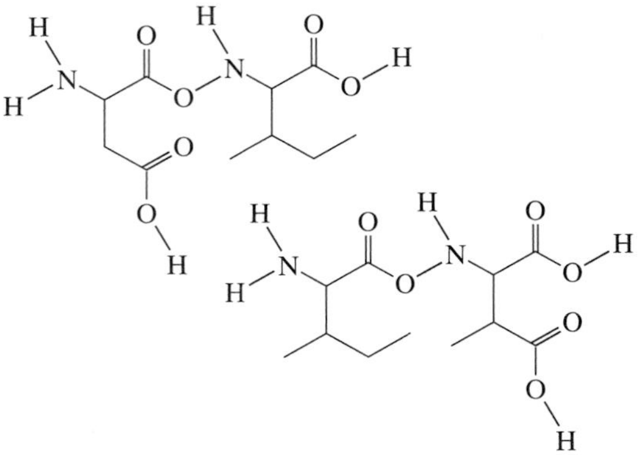

11.13

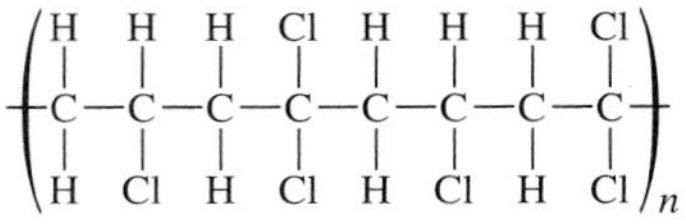

11.15

(a)

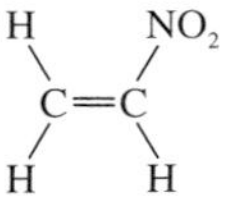

Nitroethene or nitroethylene

(b)

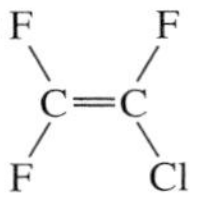

Chlorotrifluoroethene or chlorotrifluoroethylene

11.17

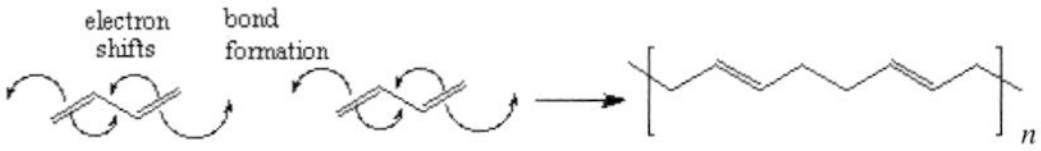

11.19

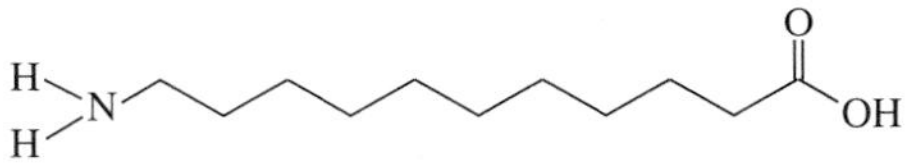

11.21

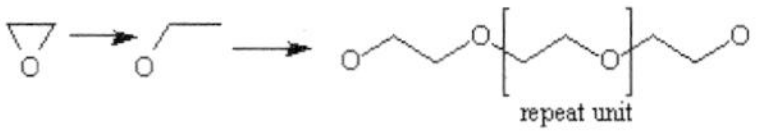

11.23 automobile tire

11.25 (a) elastomers; (b) fibers; and (c) plastics

11.27 Dioctyl phthalate reduces the amount of cross-linking and adds a fluid component to the polymer, so the polymer becomes more flexible.

11.29

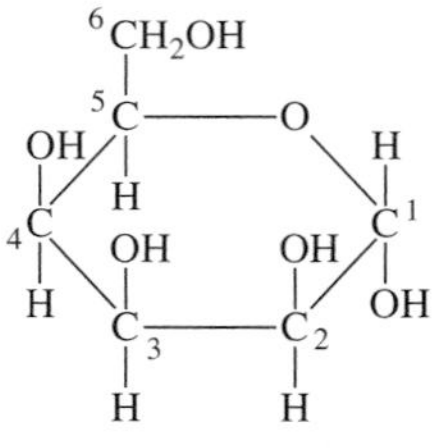

α-talose

11.31

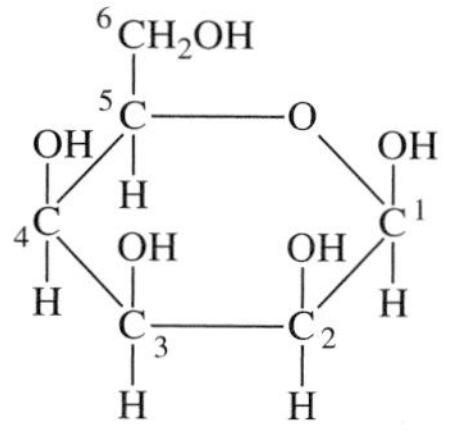

β-talose

11.33 Polymers of β-glucose (e.g., glycogen) coil on themselves; polymers of α-glucose (e.g., cellulose) form planar sheets.

11.35 T-T-A-C-G-T-G-A-C

11.37

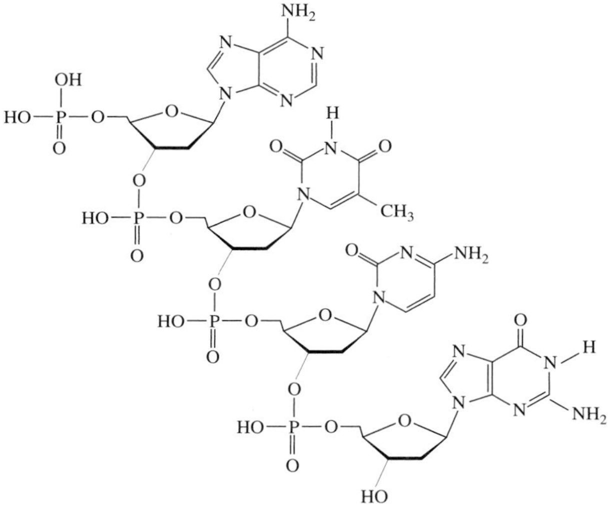

11.39 The complementary sequence is T-A-G

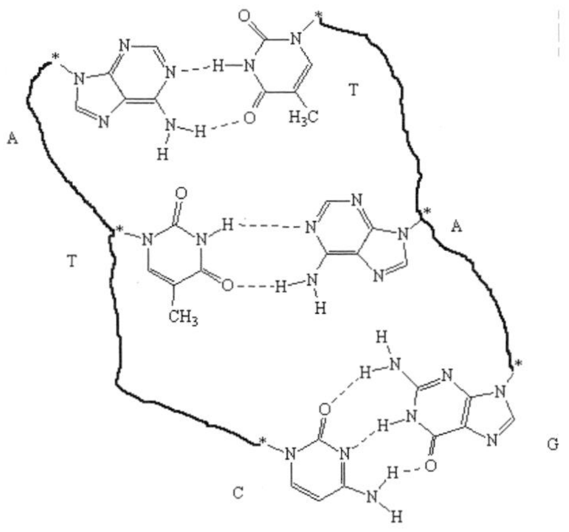

11.41

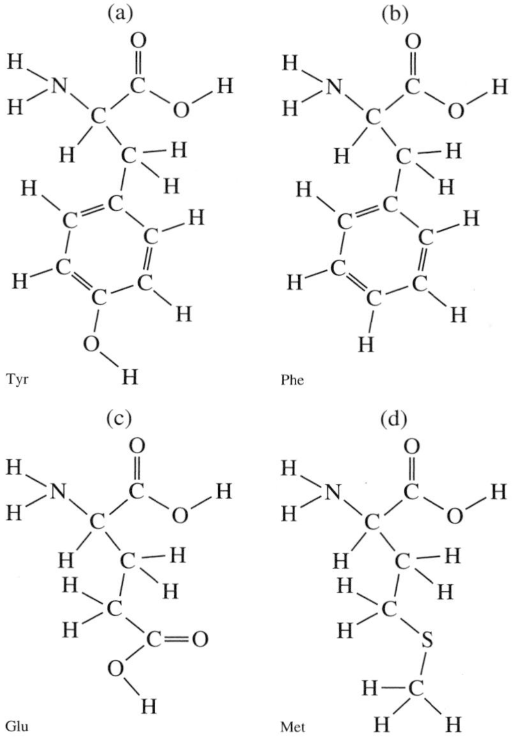

(a)

Tyr

(b)

Phe

(c)

Glu

(d)

Met

11.43 Hydrophilic side chains: Tyr and Glu; hydrophobic side chains: Met and Phe

11.45 Glycine, hydrophobic side chain; (b) serine, hydrophilic; and (c) cysteine, hydrophilic

11.47

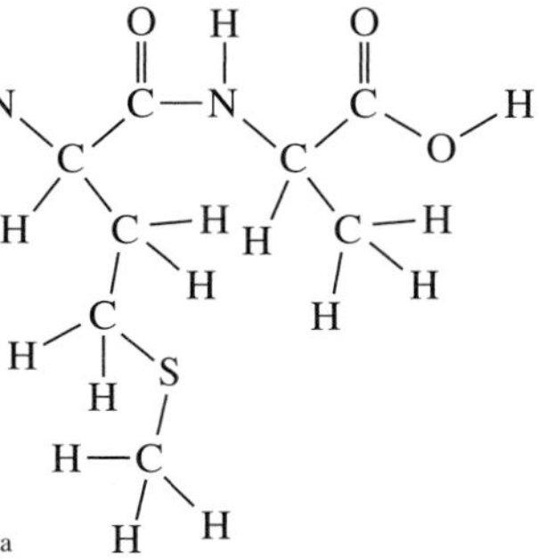

Met-Ala

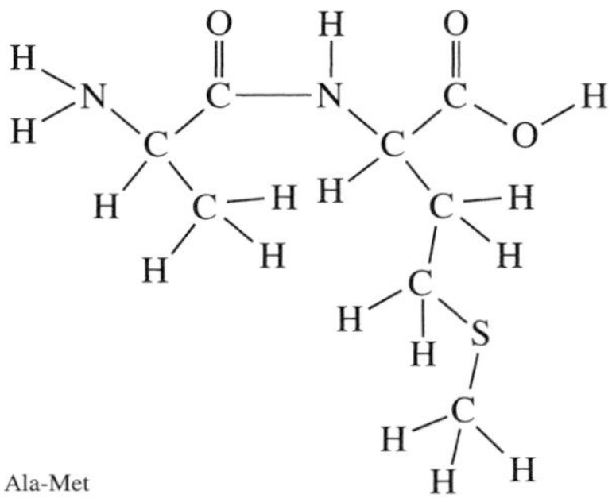

Ala-Met

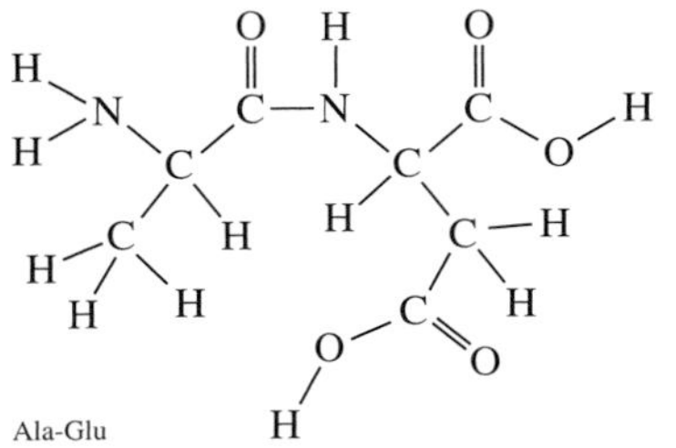

Ala-Glu

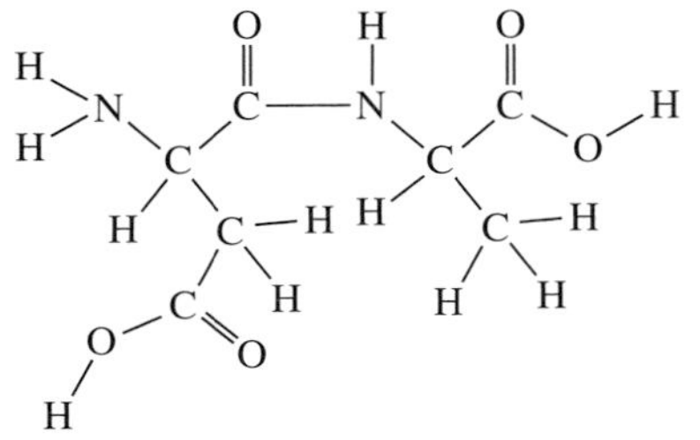

Glu-Ala

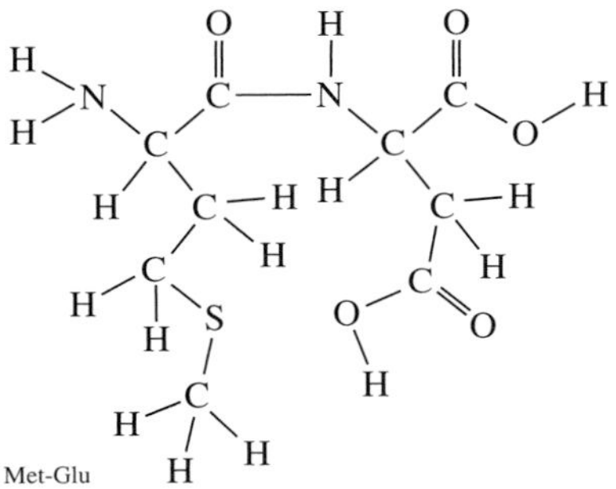

Met-Glu

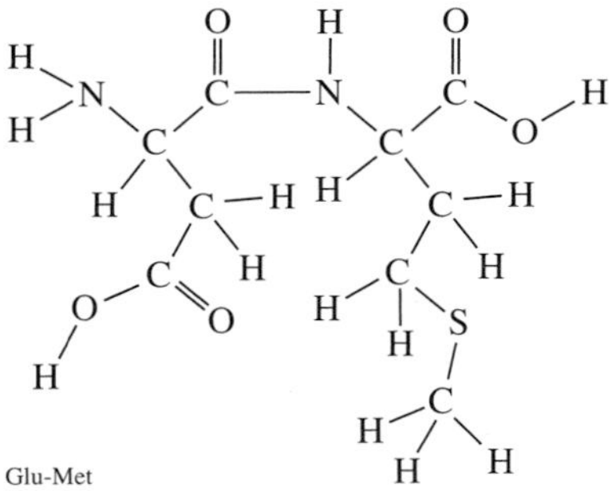

Glu-Met

11.49 2.09×10^4 g/mol

11.51 16,777,216

11.53 (a) random; (b)

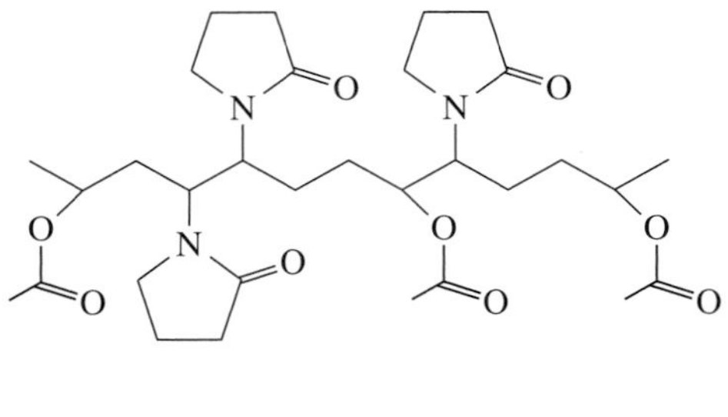

and (c) The backbone of this polymer is hydrophobic, but its side chains contain highly polar $C=O$ groups that interact with one another and with polar and H-bonding groups on hair.

11.55 G–C

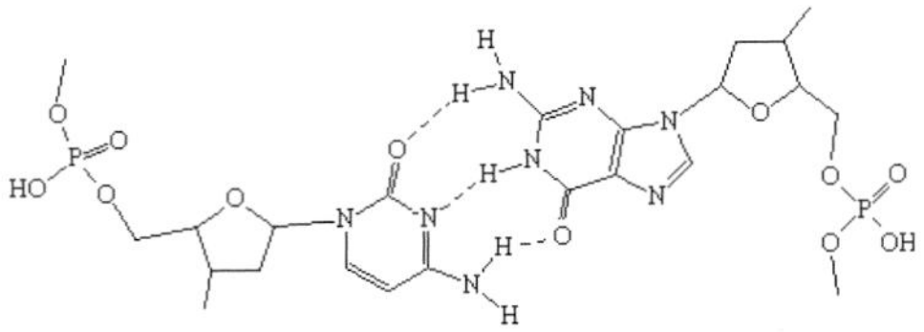

11.57 Initiation

Propagation

Termination:

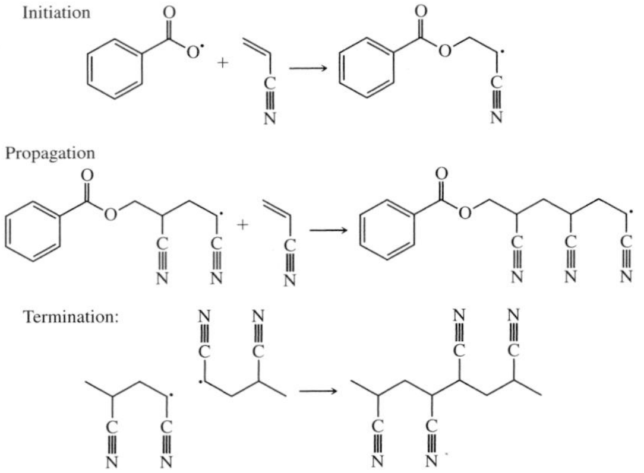

11.59

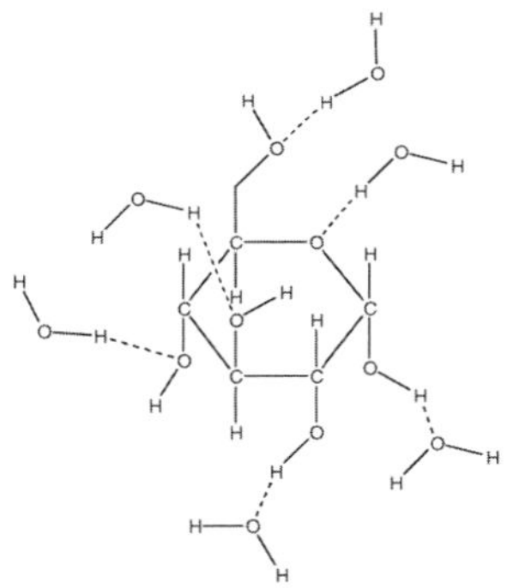

11.61 Nylons and proteins both form by condensation reactions between carboxylic acids and amines, so their linkages are amide groups. Nylons contain one or at most two different monomers, each of which typically contains several carbon atoms that form part of the backbone of the polymer. Proteins contain backbones that are absolutely regular repetitions of amide—C—amide bonding, but the carbon atoms in the backbone have a variety of substituent groups attached to them.

11.63 (a) terephthalic acid and phenylenediamine; (b) PET is made from ethylene glycol and terephthalic acid; and (c) styrene

11.65 High-density polyethylene is all straight chains that nest together readily; low-density polyethylene has many side chains that cannot nest easily, so low-density polyethylene has more open space.

11.67 1 = cytosine; 2 = uracil; 3 = adenine; 4 = guanine

11.69 The Watson-Crick model of DNA requires that there be a complementary base for each base in a given strand. This requires that the molar ratios of A to T and G to C be 1.0. Chargoff's observations indicate that this relationship holds even though DNA from different sources has different sequences, some A, T-rich and others G, C-rich.

11.71 When a protein is in contact with a hydrophilic medium such as aqueous solution, its hydrophilic amino acids are most stable when facing outward, in contact with the solvent, while its hydrophobic amino acids are most stable when facing inward, in contact with the protein backbone. The opposite is the case when a protein is immersed in a hydrophobic medium. The tertiary structure of a protein is the manner in which the individual amino acids orient themselves, so solvent interactions are the primary determinants of this tertiary structure.

11.73

11.75

11.77 4

11.79

glucose   diphosphate   uridine

11.81 sp^2

11.83

gentobiose

12.1 (a) open; (b) isolated; (c) open; and (d) closed

12.3 state variables are (b) and (c)

12.5 automobile and driver, automobile, driver, cooling system, engine, etc

12.7 (a) 17.8 °C; (b) 296 K; (c) 299 K; and (d) 22.2 °C

12.9 kettle: 4.37×10^4 J; water: 8.29×10^5 J

12.11 23.6 °C

12.13 131 kJ

12.15 -2.5×10^2 J

12.17 1.4 km

12.19 14.4 kJ/°C

12.21 (a) $2\ C_7H_6O_2 + 15\ O_2 \rightarrow 14\ CO_2 + 6\ H_2O$; (b) -3.221×10^3 kJ/mol; and (c) -4.295×10^2 kJ/mol

12.23 -4.96×10^3 kJ/mol

12.25 (a) -1411.1 kJ; (b) 91.8 kJ; (c) -1597.0 kJ; and (d) -21.2 kJ

12.27 (a) -1406.1 kJ; (b) 86.8 kJ; (c) -1597.0 kJ; and (d) -11.3 kJ

12.29 (a) $3\ K(s) + P(s) + 2\ O_2(g) \rightarrow K_3PO_4(s)$; (b) $2\ C(graphite) + 2\ H_2(g) + O_2(g) \rightarrow CH_3CO_2H(l)$; (c) $3\ C(graphite) + \frac{9}{2}\ H_2(g) + \frac{1}{2}\ N_2(g) \rightarrow (CH_3)_3N(g)$; and (d) $2\ Al(s) + \frac{3}{2}\ O_2(g) \rightarrow Al_2O_3(s)$

12.31 (a) -1166.2 kJ; and (b) -1531.4 kJ

12.33 -8.3×10^4 J/mol

12.35 (a) absorbed; (b) 206 kJ; and (c) 0.924 K

12.37 Energy is released because of the coulombic attraction between cations and anions, but energy is absorbed to overcome the ion-dipole attractions between ions and water molecules. Solid formation is endothermic when the sum of all ion-dipole attractions in solution is greater than the ion-ion interactions in the solid.

12.39 (a) Methane is a smaller molecule, thus it has smaller dispersion forces and a lower heat of vaporization. (b) Ethanol has a significantly higher heat of vaporization than diethyl ether because of strong hydrogen bonding, and (c) Argon has a higher heat of fusion than methane because it has a higher polarizability due to its larger number of electrons.

12.41 while idling

12.43 2.93 g

12.45 (a) $C_6H_{12}O_6 + 6\ O_2 \longrightarrow 6\ CO_2 + 6\ H_2O$; (b) -2.83×10^3 kJ/mol; and (c) -1.25×10^3 kJ/mol

12.47 It will die.

12.49 20.1 J/mol K

12.51 $q = -4.05 \times 10^3$ J; $w = 4.05 \times 10^3$ J; $\Delta E = 0$

12.53 (a) -197.8 kJ; (b) -55.3 kJ; and (c) -851.5 kJ

12.55 -197 kJ

12.57 $C = 0.382$ J/g K; Cu

12.59 ΔH_{sys} negative; ΔE_{surr} positive; $\Delta E_{universe} = 0$

12.61 44.00 kJ; ΔH_{vap} is at the boiling pt, 373 K. $\Delta H_{reaction}$ is at standard temperature, 298 K.

12.63 (a) −539.8 kJ/mol; and (b) −534.3 kJ/mol

12.65 (a) 0 J; (b) 4.96 × 10³ J; and (c) −4.96 × 10³ J

12.67 (a) $I_2(s) \rightarrow I_2(g)$; (b) $\frac{1}{2} I_2(s) \rightarrow I(g)$;
(c) $2\ C(graphite) + \frac{3}{2} H_2(g) + \frac{1}{2} Cl_2(g) \rightarrow C_2H_3Cl(g)$;
and (d) $Na_2SO_4(s) \rightarrow 2\ Na^+(aq) + SO_4^{2-}(aq)$.

12.69 (a) 653 J; and (b) 6.31 × 10³ J

12.71 638 kJ

12.73 (a) 6.48 × 10⁹ J; and (b) 1.6 × 10⁵ g

12.75 $T = 348$ K; more effective

12.77 15 g

12.79 (a) negative; (b) 0; and (c) positive

12.81 1.1 × 10³ g

12.83 5.85 °C

12.85 $q = \Delta H = 304$ kJ; $w = -26.1$ kJ; $\Delta E = 2.8 \times 10^2$ kJ

13.1 (a) A sand castle represents an ordered structure constructed by a person. Waves destroy that structure, returning the sand grains to a disordered arrangement; (b) Two separate liquids represent an ordered arrangement (all molecules of one kind in one container). Upon mixing, the molecules in the liquids are distributed randomly throughout the container, an increase in disorder; (c) Sticks in a bundle are ordered (all aligned in the same direction). When dropped, the sticks lose their alignment and become more disordered; (d) Water in a puddle is relatively ordered, as it is confined to a small volume. When the water evaporates, the molecules spread over a much larger volume and become more disordered.

13.3 Initially the ink is organized in a droplet, but collisions with water molecules generate disorder as the ink molecules become scattered throughout the liquid.

13.5 (a) The air molecules in a tire are relatively ordered, because they are confined to a specific, small volume. A puncture allows gas molecules to escape from the tire and fill a much larger volume, becoming less ordered in the process; (b) The fragrant molecules in a perfume bottle are relatively ordered because they are confined to a specific, small volume. When the bottle is open, molecules escape from the confined volume into a much larger space of the room, becoming more disordered in the process.

13.7 72

13.9 (a) 16.8 J/K; (b) −15.3 J/K; and (c) 1.5 J/K

13.11 (a) ΔS_{sys} is positive., ΔS_{surr} is negative; (b) ΔS_{sys} is positive, ΔS_{surr} is negative; and (c) ΔS_{sys} is positive, ΔS_{surr} is negative

13.13 −94.0 J/K; $\Delta S_{surr} >$ 94.0 J/K

13.15 (a) 16 J/mol K; (b) 9.3 J/mol K; (c) 48.8 J/mol K; and (d) 1.00 × 10² J/mol K

13.17 (a) $MgCl_2$ has the larger molar entropy because $MgCl_2$ produces more moles of ions per mole of substance; (b) HgS has the larger molar entropy because HgS has a higher molar mass; and (c) $Br_2(l)$ has the larger molar entropy because Br_2 is a liquid.

13.19 Ozone has three atoms per molecule, whereas O_2 has only two. Methane has fewer distinguishable orientations in space.

13.21 He: $S° = 126.153$ J/mol K; H_2: $S° = 65.340$ J/mol K; CH_4: $S° = 37.26$ J/mol K; C_3H_6: $S° = 25.21$ J/mol K. Entropy per mole of atoms decreases as the number of atoms in a molecule increase, because tying together atoms into a molecule increases the amount of order among those atoms.

13.23 (a) −198.05 J/K; (b) −137.7 J/K; (c) 12.3 J/K; and (d) −267.3 J/K.

13.25 (a) There is a relatively large negative value because of the reduction in number of moles of gaseous substances; (b) There is a relatively large negative value because of the reduction in number of moles of gaseous substances; (c) This reaction has a near-zero entropy change because all reagents are relatively ordered solid substances; and (d) There is a relatively large negative value because of the reduction in number of moles of gaseous substances.

13.27 (a) 416 J/K; (b) 1.7 × 10² J/K; and (c) 9.0 × 10¹ J/K

13.29 (a) $\Delta G_{system} <$ 0 for any spontaneous process at constant T and P. (b) The free energy of a system decreases in any process at constant T and P. and (c) $\Delta G = \Delta H - \Delta(TS)$

13.31 (a) −32.8 kJ; (b) 326.4 kJ; (c) −206.1 kJ; and (d) −1331.4 kJ

13.33 (a) −7.6 kJ; (b) 343.9 kJ; (c) −207.2 kJ; and (d) −1297 kJ

13.35 $\Delta H°_{reaction} = -559.6$ kJ; $\Delta S°_{reaction} = -574.9$ J/K; $\Delta G°_{reaction} = -388.2$ kJ

13.37

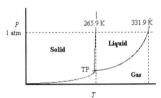

13.39 (a) At $T = 400$ K, $P = 1.00$ atm, Br_2 is a gas. As it cools under a pressure of 1.00 atm, it liquefies at 331.9 K and solidifies at 265.9 K. (b) At $T = 265.8$ K, $P = 1.00 \times 10^{-3}$ atm, Br_2 is a gas. Compressing it at this temperature causes it to liquefy at about $P = 6 \times 10^{-2}$ atm and solidify at about 0.5 atm; and (c) $P = 2.00 \times 10^{-2}$ atm is below the triple point of Br_2. Heating a solid at this pressure from 250 to 400 K causes sublimation to the vapor at about 265 K.

13.41 The first "reaction" would be child 1 starting on the ground and rising into the air. The second "reaction" would be child 2 starting in the air and falling to the ground. In this case child 2 would be the spontaneous reaction, child 1 would be non-spontaneous, so as child 2 is falling he will cause the first child to raise. When they are at equal heights they will switch spontaneity.

13.43 $ATP + C_6H_{12}O_6 (fructose) + C_6H_{12}O_6 (glucose) \longrightarrow ADP + C_{12}H_{22}O_{11;}$ $\Delta G° = -7.6$ kJ

13.45 (a) $\Delta E_{universe} = 0$; (b) $\Delta E_{teaspoon} > 0$; (c) $\Delta S_{universe} > 0$; (d) $\Delta S_{water} < 0$; and (e) $q_{teaspoon} > 0$

13.47 (0.5 mol, l, 298 K) < (1 mol, l, 298 K) < (1 mol, g, 1 atm, 373 K) < (1 mol, g, 0.1 atm, 373 K)

13.49 (a) Not spontaneous; (b) absorbs heat; and (c) more ordered

13.51 (a) −83 J/K; (b) 328.7 J/K; and (c) 365.3 J/K

13.53 (a) −40 kJ; (b) −148.4 kJ; and (c) −1997.7 kJ

13.55 (a) −22 kJ; (b) −257 kJ; and (c) −1662.1 kJ

13.57 (a) $w_{sys} = 0$ (no volume change); (b) $q_{sys} > 0$; (energy must be supplied to break chemical bonds); and (c) $\Delta S_{surr} < 0$ ($q_{sys} > 0$, $q_{surr} < 0$)

13.59 (a) not thermodynamically feasible; and (b) thermodynamically feasible

13.61 (a) −189 J/K; (b) 15 J/K; and (c) yes

13.63 (a) spontaneous; (b) absorbs heat; and (c) decreases

13.65 When a cell needs energy, it "spends" ATP; when fat or carbohydrates ("capital") are consumed, the energy is stored by converting ADP to ATP ("buying" ATP).

13.67 (a) N_2 becomes a gas; (b) N_2 remains a gas; and (c) N_2 is solid.

13.69 Ammonia is a toxic gas, while urea and ammonium nitrate both are relatively non-toxic solids. Thus the transport and application of ammonia entail significant risks to humans. Even in aqueous solution, ammonia is highly irritating, as anyone knows who has used strong ammonia as a cleanser.

13.71 palmitic acid

13.73 (a)

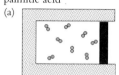

(b) $\Delta S_{universe} = 0$

13.75 attractive forces are larger for CaO

13.77 (a) $q_v = \Delta E$; (b) $q_p = \Delta H$; and (c) $q_T = T\Delta S$

13.79 (a) ΔS is negative; (b) ΔS is negative; and (c) ΔS is positive

13.81 (a) ΔH is negative, ΔS is negative; (b) ΔH is positive, ΔS is positive; and (c) ΔH is expected to be negative, ΔS is expected to be small

13.83 Standard conditions refers to 1 atm, 298 K, all solutes at 1 M. In a cell, temperature is 37 °C (310 K) and no solutes are present at 1 M. In particular, hydronium ion concentration is around 10^{-7} M, far from standard conditions.

13.85 40.6 %; 7.8 g

13.87 both must have $\Delta G < 0$

13.89

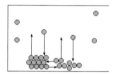

14.1 (a) Pouring the coffee from the urn into the cup; (b) Entering the items on the cash register (if the market has a good laser scanner, paying and receiving change may be rate-determining); and (c) Preparing for the jump and passing through the door.

14.3 (a) $I_2 \longrightarrow I + I$; (b) $H_2 + I_2 \longrightarrow H_2I_2$; and (c) $H_2 + I_2 \longrightarrow H + HI_2$

14.5 (a)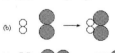

(b)

(c)

14.7 (a) $I_2 \longrightarrow I + I$
$I + H_2 \longrightarrow HI + H$
$H + I \longrightarrow HI$
(b) $H_2 + I_2 \longrightarrow H_2I_2$
$H_2I_2 \longrightarrow HI + HI$ and
(c) $H_2 + I_2 \longrightarrow H + HI_2$
$H + HI_2 \longrightarrow HI + HI$

14.9 (a) Rate $= -\dfrac{\Delta[Cl_2]}{\Delta t}$

(b) $-\dfrac{\Delta[Cl_2]}{\Delta t} = \dfrac{1}{2}\dfrac{\Delta[NOCl]}{\Delta t}$; and

(c) NOCl appears at a rate of 94 M s^{-1}

14.11

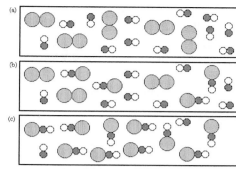

14.13 (a) 2.2×10^3 mol/min; and (b) 1.1×10^{-1} mol

14.15 (a)

(b) After 20 minutes, the amount reacted is
(20 min)(0.25 molecules/min) = 5 molecules

14.17 (a) 60; (b) $12/N_A$; and (c) in first-order reactions, the fraction reacting is independent of concentration.

14.19 (a) Rate $= k[C][AB]$; (b) units $= (conc.)^{-1} (time)^{-1}$; and
(c) $C + AB \longrightarrow BC + A$
$A + AB \longrightarrow B + A_2$
$B + C \longrightarrow BC$
Net: $2\,C + 2\,AB \longrightarrow 2\,BC + A_2$

14.21 (a) Rate $= k[C]^2[AB]$; (b) units $= (conc.)^{-2} (time)^{-1}$; and
(c) $2\,C + AB \longrightarrow BC + AC$
$AC + AB \longrightarrow BC + A_2$
Net: $2\,C + 2\,AB \longrightarrow 2\,BC + A_2$

14.23 B will react slower by a factor of 0.56.

14.25 (a) This reaction is first-order; (b) $k = 6.14 \times 10^{-4}$ s^{-1};
(c) 0.936 atm; and (d) 2.62×10^3 s

14.27 (a) 2.4 min; and (b) 3.2×10^{-4} M

14.29 (a) 12.0 hr; and (b) 0.0324 M

14.31 This reaction is first-order. $k = 2.55 \times 10^{-3}$ s^{-1}

14.33 Rate $= k_2 \left(\dfrac{k_1}{k_{-1}}\right)^2 \dfrac{[C]^2\,[AB]^2}{[BC]}$

14.35 (a) $O_3 + O \rightarrow 2\,O_2$; (b) Rate $= k_2 \dfrac{k_1}{k_{-1}}[O_3][O_2]$; and

(c) it is improbable to have two simultaneous fragmentations of O_5

14.37 (a) Rate $= k_2\left(\dfrac{k_1}{k_{-1}}\right)[NO]^2[O_2]$; (b) yes; and

(c) the intermediate with an N—N bond is the most likely

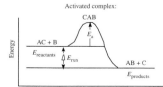

14.39 When $E_a = 0$, k is independent of temperature. This happens when a reaction can occur without first breaking any chemical bonds.

14.41

14.43

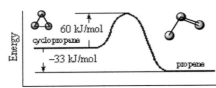

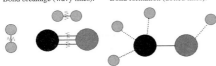

14.45 24.9 kj/mol

14.47 Bond breakage (wavy lines):   Bond formation (dotted lines):

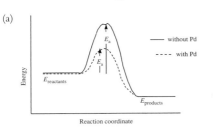

14.49 (a)

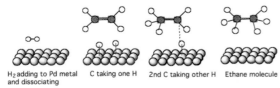

(b) catalyst: Pd metal; intermediate: H atoms on the Pd metal surface; and (c)

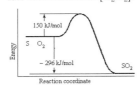

H₂ adding to Pd metal and dissociating   C taking one H   2nd C taking other H   Ethane molecule

14.51 (a) nine-fold increase; (b) no change; and (c) rate increases by 5.2 times.

14.53 $O + NO \rightarrow NO_2$; (b)

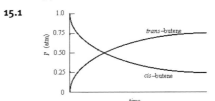

14.55 (a) Rate $\dfrac{\Delta[C_6H_6]}{\Delta t} = -\dfrac{1}{3}\left(\dfrac{\Delta[C_2H_2]}{\Delta t}\right)$; and

(b) Experiments would have to be carried out measuring the rate as a function of $[C_2H_2]$.

14.57

14.59 The rate for B will be twice that for A, because the concentration of X_2 is twice as great.

14.61 10%: 1.9×10^2 s; 50%: 1.3×10^3 s; 99.9%: 1.3×10^4 s.

14.63 (a) False; (b) False; (c) True; and (d) True

14.65 (a) first-order in N_2O_5 and first-order overall; (b) second-order in NO, first-order in H_2, and third-order overall; and (c) first-order in enzyme and first-order overall.

14.67 (a) $2\,NO + 2\,H_2 \longrightarrow N_2 + 2\,H_2O$;

(b) Rate $= k_2\dfrac{k_1}{k_{-1}}\,[NO]^2[H_2]$

14.69 $CF_2Cl_2 \xrightarrow{h\upsilon} CF_2Cl + Cl$

$O_3 \xrightarrow{h\upsilon} O_2 + O$

$Cl + O_3 \longrightarrow ClO + O_2$

$ClO + O \longrightarrow Cl + O_2$

14.71 The speed of a chemical reaction refers to how fast it proceeds. The spontaneity of a chemical reaction refers to whether or not the reaction can go in the direction written without outside intervention.

14.73 (a) At the molecular level, a catalyst binds to one or more of the reactants in a way that weakens chemical bonds and makes it easier for bonds to rearrange to form the products; (b) When temperature increases, the average kinetic energies of the molecules increases, with the result that enough energy is present for reaction to occur in a larger fraction of molecular collisions; and (c) When concentration increases, the molecular density increases. There are more molecules to react. In addition, there is a higher rate of molecular collisions. Both factors contribute to a greater rate of reaction.

14.75 Rate $= k[A]^2$

14.77 1.1×10^{14}

14.79 7.0 min

14.81 d and f

14.83 (a) Rate $= k[H_2][X_2]$; (b) Rate $= k[X_2]$; and

(c) $k_1[X_2] = $ Rate $= k_2\left(\dfrac{k_1}{k_{-1}}\right)^{1/2}[H_2][X_2]^{1/2}$.

14.85 (a) $Cl_2 + CHCl_3 \rightarrow HCl + CCl_4$; (b) Cl and CCl_3; and

(c) Rate $= k_2\left(\dfrac{k_1}{k_{-1}}\right)^{1/2}[CHCl_3][Cl_2]^{1/2}$

14.87 Neither intermediates nor catalysts appear in the overall stoichiometry of the reaction, so any species that appears in the mechanism but not in the overall stoichiometry is either an intermediate or a catalyst. Catalysts are consumed in early steps and regenerated in later steps, while intermediates are produced in early steps and consumed in later steps.

14.89 (a) Rate $= k[CH_3CHO]^2$; (b) 4.5×10^{-3} M^{-1} s^{-1}; and (c) 2.7×10^3 s

15.1

15.3 (a) $Cl^-(aq) + ClO_2^-(aq) \xrightarrow{k_{-1}} ClO^-(aq) + ClO^-(aq)$

$Cl^-(aq) + ClO_3^-(aq) \xrightarrow{k_{-2}} ClO^-(aq) + ClO_2^-(aq)$

(b) $K_{eq} = \dfrac{[ClO_3^-]_{eq}[Cl^-]_{eq}^2}{[ClO^-]_{eq}^3}$; and (c) $K_{eq} = \dfrac{k_1k_2}{k_{-1}k_{-2}}$

15.5 Your molecular pictures of elementary reactions should show the reactants, the products, and (if applicable) the intermediate collision complex

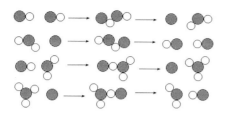

15.7 To test the reversibility of a reaction, set up a system containing the products and observe whether or not reactants form. Here, a solution containing Cl^- and ClO_3^- ions should react to form some ClO^- ions.

15.9 (a) $K_{eq} = \dfrac{(p_{IF_5})_{eq}^2}{(p_{F_2})_{eq}^5}$; (b) $K_{eq} = \dfrac{1}{(p_{O_2})_{eq}^5}$;

(c) $K_{eq} = (p_{CO})_{eq}^2$; (d) $K_{eq} = \dfrac{1}{(p_{CO})_{eq}(p_{H_2})_{eq}^2}$;

(e) $K_{eq} = \dfrac{[H_3O^+]_{eq}^3[PO_4^{3-}]_{eq}}{[H_3PO_4]_{eq}}$

15.11 (a) $2IF_5(g) \rightleftharpoons I_2(s) + 5 F_2(g)$; $K_{eq} = \dfrac{[pF_2]_{eq}^5}{[pIF_5]_{eq}^2}$;

(b) $P_4O_{10}(s) \rightleftharpoons P_4(s) + 5 O_2(g)$; $K_{eq} = (p_{O_2})_{eq}^5$;

(c) $BaO(s) + 2 CO(g) \rightleftharpoons BaCO_3(s) + C(s)$;

$K_{eq} = \dfrac{1}{(p_{CO})_{eq}^2}$;

(d) $CH_3OH(l) \rightleftharpoons CO(g) + 2 H_2(g)$; $K_{eq} = (p_{CO_2})_{eq}(p_{H_2})_{eq}^2$;

(e) $PO_4^{3-}(aq) + 3 H_3O^+(aq) \rightleftharpoons H_3PO_4(aq) + 3 H_2O(l)$

$K_{eq} = \dfrac{[H_3PO_4]_{eq}}{[H_3O^+]_{eq}^3[PO_4^{3-}]_{eq}}$

15.13 (a) $p = 1$ atm for F_2 and IF_5, $X = 1$ for I_2; (b) $p = 1$ atm for O_2, $X = 1$ for P_4 and P_4O_{10}; (c) $p = 1$ atm for CO, $X = 1$ for others; (d) $p = 1$ atm for CO and H_2, $X = 1$ for CH_3OH; and (e) $c = 1$ M for H_3PO_4, H_3O^+, and PO_4^{3-}, $X = 1$ for H_2O

15.15 A molecular picture of an equilibrium between phases shows the concentrations of the species and arrows indicating transfers between phases:

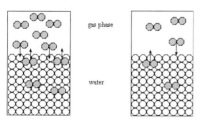

15.17 47%

15.19 (a) 3.3×10^3; (b) 1×10^{90}; (c) 3×10^{-60}; and (d) 2.9×10^{39}

15.21 2×10^{109}; 3×10^{-75}

15.23 1.8×10^3; 9.2×10^{18}

15.25 (a), (b), and (d), adding CO causes the reaction to go to the right, forming products; (c), adding CO causes the reaction to proceed to the left, forming reactants.

15.27 (a) no effect; (b) more solid will dissolve; (c) some solid will precipitate; and (d) no effect

15.29 This reaction can be driven to the left by removing SO_2, removing Cl_2, adding SO_2Cl_2, or increasing the temperature.

15.31 0.92

15.33 1.3×10^{-5}

15.35 $(p_{H_2})_{eq} = (p_{Br_2})_{eq} = 2.5 \times 10^{-2}$ atm; $(p_{HBr})_{eq} = 10.0$ atm

15.37 $(p_{CO_2})_{eq} = 1.4$ atm and $(p_{CO})_{eq} = 3.6$ atm

15.39 (a) H_2O and CH_3CO_2H; (b) H_2O, NH_4^+ and Cl^-; (c) H_2O, K^+ and Cl^-; (d) H_2O, Na^+ and $CH_3CO_2^-$; and (e) H_2O, Na^+ and OH^-

15.41 (a) $CH_3CO_2H(aq) + H_2O(l) \rightleftharpoons CH_3CO^-(aq) + H_3O^+(aq)$; (b) $NH_4^+(aq) + H_2O(l) \rightleftharpoons NH_3(aq) + H_3O^+(aq)$; (c) There are no equilibria other than the water equilibrium; (d) $CH_3CO_2^-(aq) + H_2O(l) \rightleftharpoons CH_3CO_2H(aq) + OH^-(aq)$; and (e) There are no equilibria other than the water equilibrium.

15.43 (a) $(CH_3)_2CO$ (acetone) and H_2O; (b) H_2O, K^+, and Br^-; (c) H_2O, Li^+, and OH^-; and (d) H_2O, H_3O^+, and HSO_4^-

15.45 (a) $K = \dfrac{[ClO_2^-][H_3O^+]}{[HClO_2]} = K_a$;

(b) $K = \dfrac{1}{[Fe^{3+}][OH^-]^3} = \dfrac{1}{K_{sp}}$; and

(c) $K = \dfrac{[HCN]}{[CN^-][H_3O^+]} = \dfrac{1}{K_a}$

15.47 (a) Na^+; (b) Cl^- and K^+; and (c) K^+ and NO_3^-

15.49 (a) $K_{eq} = (p_{CO_2})_{eq}(p_{H_2O})_{eq}$; (b) $K_{eq} = \dfrac{(p_{NH_3})_{eq}^4(p_{O_2})_{eq}^3}{(p_{N_2})_{eq}^2}$;

(c) $K_{eq} = \dfrac{(p_{CH_3CHO})_{eq}^2}{(p_{C_2H_4})_{eq}^2(p_{O_2})_{eq}}$; (d) $K_{eq} = [Ag^+]_{eq}^2[SO_4^{2-}]_{eq}$; and (e) $K_{eq} = (p_{H_2S})_{eq}(p_{NH_3})_{eq}$

15.51 $(p_{CO})_{eq} = 7.2 \times 10^{-2}$ atm; $(p_{COCl_2})_{eq} = 6.2$ atm

15.53 A test tube has a smaller surface area than a petri dish, so fewer molecules escape per unit time from the test tube. However, the rate of escape per unit surface area is the same for both samples, so equilibrium is established at the same pressure.

15.55 The molecular picture should show 2 each of reactants and 10 each of products:

15.57 7.34 atm

15.59 3 at 298 K; 2.1×10^{-4} at 525 K

15.61 (a) $(CH_3)_3N + H_2O \rightleftharpoons (CH_3)_3NH^+ + OH^-$,

$K_b = \dfrac{[(CH_3)_3NH^+]_{eq}[OH^-]_{eq}}{[(CH_3)_3N]_{eq}}$;

(b) $HF + H_2O \rightleftharpoons F^- + H_3O^+$, $K_a = \dfrac{[F^-]_{eq}[H_3O^+]_{eq}}{[HF]_{eq}}$;

(c) $CaSO_4(s) \rightleftharpoons Ca^{2+}(aq) + SO_4^{2-}(aq)$, $K_{sp} = [Ca^{2+}]_{eq}[SO_4^{2-}]_{eq}$

15.63 (a) right; (b) left; and (c) no effect

15.65 4.28×10^{-29} atm; 9.49×10^5 L

15.67 0.75

15.69 (a) $K_{eq} = \dfrac{[CO_3^{2-}]_{eq}}{(p_{CO_2})_{eq}[OH^-]_{eq}^2}$

(b) increases; (c) increases

15.71 (a) 1.6×10^{11}; (b) 792 K; and (c) 2.2×10^{-2}

15.73 1.20×10^2 atm

15.75 (a)

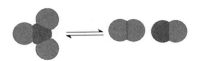

(b) 48

16.1

16.3 $[H_3O^+] = 1.25 \times 10^{-3}$ M; $[OH^-] = 8.00 \times 10^{-12}$ M
16.5 $[H_3O^+] = [Cl^-] = 0.121$ M; $[OH^-] = 8.26 \times 10^{-14}$ M
16.7 $[H_3O^+] = 4.12 \times 10^{-2}$ M; $[OH^-] = 2.43 \times 10^{-13}$ M
16.9 (a) -0.60; (b) 5.426; (c) 2.32; and (d) 3.593
16.11 (a) 14.60; (b) 8.574; (c) 11.68; and (d) 10.407
16.13 (a) 0.22 M; (b) 1.4×10^{-8} M; (c) 2.1×10^{-4} M; and
(d) 4.7×10^{-15} M
16.15 (a) 4.6×10^{-14} M; (b) 7.1×10^{-7} M; (c) 4.8×10^{-11} M;
and (d) 2.1 M
16.17 (a) 14.18; (b) 9.70; (c) 10.04; and (d) 1.80
16.19

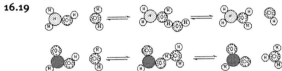

16.21 (a) Major species: HN_3 and H_2O, Minor species: N_3^- and
H_3O^+; (b) $[H_3O^+] = [N_3^-] = 6.1 \times 10^{-3}$ M, and $[HN_3] =$
1.50 M; (c) 2.21; and (d)

16.23 (a) Major species: $N(CH_3)_3$, H_2O, Minor species:
$HN(CH_3)_3^+$, OH^-; (b) $[OH^-] = [HN(CH_3)_3^+] =$
4.8×10^{-3} M, $[N(CH_3)_3] = 0.345$ M; (c) 11.68; and (d)

16.25 (a) weak base; (b) weak acid, (c) weak acid; and (d) strong
base
16.27 C_5H_5N conjugate acid is $C_5H_5NH^+$; $HONH_2$ conjugate
acid is $HONH_3^+$; HCO_2H conjugate base is HCO_2^-; NH_3
conjugate acid is NH_4^+; HCNO conjugate base is CNO^-;
HClO conjugate base is ClO^-
16.29 (a)

base   acid
$$NH_3 + H_2O \rightleftharpoons NH_4^+ + OH^-$$
acid   base

(b)

base   acid
$$H_2O + HCNO \rightleftharpoons H_3O^+ + CNO^-$$
acid   base

(c)

base   acid
$$H_2O + HClO \rightleftharpoons H_3O^+ + ClO^-$$
acid   base

(d)

base   acid
$$OH^- + H_3O^+ \rightleftharpoons H_2O + H_2O$$
acid   base

16.31 (a) Major species are Na^+, SO_3^{2-}, and H_2O;
(b) $H_2O + SO_3^{2-} \rightleftharpoons HSO_3^- + OH^-$; and (c) 10.43
16.33 (a) Major species are NH_4^+, NO_3^-, and H_2O;
(b) $H_2O + NH_4^+ \rightleftharpoons NH_3 + H_3O^+$; and (c) 5.63
16.35 $H_2O + SO_2 \rightleftharpoons H_2SO_3$
$H_2SO_3 + H_2O \rightleftharpoons HSO_3^- + H_3O^+$
$HSO_3^- + H_2O \rightleftharpoons SO_3^{2-} + H_3O^+$

16.37 $FeO(s) + 2 H_3O^+(aq) \rightleftharpoons Fe^{2+}(aq) + 3 H_2O(l)$
16.39 (a) H_2SO_4 (anions are poorer proton donors than neutrals);
(b) HClO (Cl is more electronegative); and (c) $HClO_2$.
(more O atoms)
16.41

16.43 $[Na^+] = [C_2H_3O_2^-] = 0.250$ M; $[OH^-] = 1.2 \times 10^{-5}$ M;
$[H_3O^+] = 8.3 \times 10^{-10}$ M
16.45 $[H_3O^+] = [HCO_3^-] = 8.4 \times 10^{-5}$ M;
$[CO_3^{2-}] = 4.7 \times 10^{-11}$ M; $[OH^-] = 1.2 \times 10^{-10}$ M
16.47

16.49 pH = 8.30

16.51 $[H_3O^+] = 2.01$ M; $[SO_4^{2-}] = 1.0 \times 10^{-2}$ M; $[HSO_4^-] = 1.99$ M

16.53 (a) Major species: H_2O, HSO_4^-, and H_3O^+;
$H_2O + HSO_4^- \rightleftarrows SO_4^{2-} + H_3O^+$;
(b) Major species: H_2O, SO_4^{2-}, and Na^+;
$SO_4^{2-} + H_2O \rightleftarrows HSO_4^- + OH^-$;
(c) Major species: CO_2, H_2CO_3, and H_2O;
$H_2O + H_2CO_3 \rightleftarrows HCO_3^- + H_3O^+$; and
(d) Major species: H_2O, Cl^-, and NH_4^+;
$H_2O + NH_4^+ \rightleftarrows NH_3 + H_3O^+$

16.55 $K_{eq} = 48$

16.57 (a)

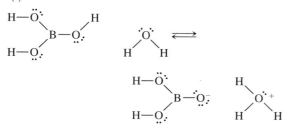

(b) 5.28

16.59 10.81

16.61 benzoic acid

16.63 (a)

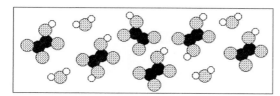

(b)

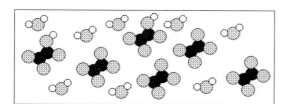

(c)

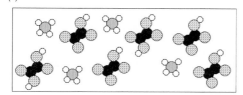

16.65 (a) The major species present are H_2O, and H_3PO_4;
(b) The minor species are $H_2PO_4^-$, H_3O^+, HPO_4^{2-}, OH^-, and PO_4^{3-}; and (c) pH = 1.92

16.67

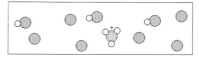

16.69 (a) $OH^- + C_6H_5CO_2H \rightleftarrows H_2O + C_6H_5CO_2^-$;
(b) $H_3O^+ + (CH_3)_3N \rightleftarrows H_2O + (CH_3)_3NH^+$;
(c) $SO_4^{2-} + CH_3CO_2H \rightleftarrows HSO_4^- + CH_3CO_2^-$;
(d) $OH^- + NH_4^+ \rightleftarrows H_2O + NH_3$; and
(e) $HPO_4^{2-} + NH_3 \rightleftarrows PO_4^{3-} + NH_4^+$

17.1 (a) no buffer properties; (b) buffer; 9.25; (c) no buffer properties; and (d) buffer; 9.25

17.3 8.88

17.5 1.4×10^{-3} mol

17.7 For pH = 3.50, the HCO_2H/HCO_2^- system should be used. Sodium formate, could be used along with HCl solution; For pH = 12.60, the HPO_4^{2-}/PO_4^{3-} system should be used. Potassium phosphate, could be used along with HCl solution.

17.9 37 g

17.11 55.0 mL

17.13 (a) $NH_4^+(aq) + H_2O(l) \rightleftarrows NH_3(aq) + H_3O^+(aq)$; pH < 7;
(b) $H_2O(aq) + H_2O(l) \rightleftarrows OH^-(aq) + H_3O^+(aq)$; pH = 7;
and (c) $CH_3CO_2H(aq) + H_2O(l) \rightleftarrows CH_3CO_2^-(aq) + H_3O^+(aq)$; pH < 7

17.15 (a) 2.8; (b) 7.8; and (c) 3.5

17.17 phenol red

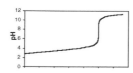

17.19 8.79; thymol blue

17.21 (a) $AgCl(s) \rightleftarrows Ag^+(aq) + Cl^-(aq)$; $K_{sp} = [Ag^+]_{eq}[Cl^-]_{eq}$;
(b) $BaSO_4(s) \rightleftarrows Ba^{2+}(aq) + SO_4^{2-}(aq)$; $K_{sp} = [Ba^{2+}]_{eq}[SO_4^{2-}]_{eq}$; (c) $Fe(OH)_2(s) \rightleftarrows Fe^{2+}(aq) + 2\,OH^-(aq)$;
$K_{sp} = [Fe^{2+}]_{eq}[OH^-]_{eq}^2$; and (d) $Ca_3(PO_4)_2(s) \rightleftarrows 3\,Ca^{2+}(aq) + 2\,PO_4^{3-}(aq)$; $K_{sp} = [Ca^{2+}]_{eq}^3[PO_4^{3-}]_{eq}^2$

17.23 (a) 9.1×10^{-4} g; (b) 1.2×10^{-3} g; (c) 9.8×10^{-5} g; and (d) 1.6×10^{-5} g

17.25 2.3×10^{-9}

17.27 0.53 g

17.29 1.6×10^3

17.31 0.125 M

17.33 (a) $[Fe(CN)_6]^{3-}$; (b) $[Zn(NH_3)_4]^{2+}$; and (c) $[V(en)_3]^{3+}$

17.35 $[Cl^-] = 0.14$ M; $[Pb^{2+}] = 8.5 \times 10^{-15}$ M; $[Na^+] = 0.17$ M; $[PbCl_4^{2-}] = 0.0075$ M

17.37 $Mn^{2+} + en \rightleftarrows [Mn(en)]^{2+}$
$[Mn(en)]^{2+} + en \rightleftarrows [Mn(en)_2]^{2+}$
$[Mn(en)_2]^{2+} + en \rightleftarrows [Mn(en)_3]^{2+}$

17.39 3.8 g

17.41 2.0;

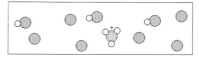

17.43 all the salt dissolves

17.45 (a) 6.81; (b) 0.06; and (c) 7.5×10^{-3} mol

17.47 precipitate forms

17.49 (a) yellow; (b) purple; (c) purple; and (d) purple

17.51 4.4×10^{-3}

17.53 3.2×10^{-8}

17.55 (a) 9.43; (b) 0.050 mol; and (c) 9.32

17.57 2.0×10^{-33}

17.59 (a) 8.52; (b) 3.74; (c) 2.23; and (d) thymol blue

17.61 2.16 g; $[Mn^{2+}] = 3.5 \times 10^{-10}$ M

17.63 2.0×10^{-4} g

17.65 Mix together 15 g sodium acetate, 0.17 L of 1.0 M acetic acid, and enough water to make 1.0 L.

17.67 (a) $MgF_2(s) \rightleftharpoons Mg^{2+}(aq) + 2 F^-(aq)$, $K_{sp} = [Mg^{2+}]_{eq}[F^-]_{eq}^2$; (b) 5.93×10^{-9}; and (c) 2.1×10^{-10}

17.69 2.6×10^2 L

17.71 (a) 6.2×10^{-13} M; and (b) 1.0×10^3 g

17.73 at 25 °C, $K_{sp} = 1.29 \times 10^{-9}$; $\Delta G^\circ = 50.7$ kJ/mol; at 100 °C $K_{sp} = 2.5 \times 10^{-5}$; $\Delta G^\circ = 33$ kJ/mol

17.75 1.1×10^{-4} M

17.77 more soluble in acidic solution: Ag_2CO_3, Ag_2SO_4, and Ag_2S. Independent of pH: AgBr and AgCl

17.79 7.79; 7.65

17.81 Point *A*: Major species are H_2O and HCO_2H; point *B*: Major species are H_2O, HCO_2H, and HCO_2^-; point *C*: Major species are H_2O and HCO_2^-; point *D*: Major species are H_2O, OH^-, and HCO_2^-; the cation of the strong base will also be present as a major species at points *B*, *C*, and *D*.

17.83 (a) 4.27; (b)

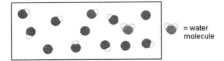

and (c)

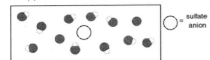

18.1 (a) O is -2, H is $+1$, and Fe is $+3$; (b) F is -1, N $+3$; (c) O is -2, H is $+1$, C is -2; (d) $K^+ = +1$, O is -2, and C is $+4$; (e) in NH_4^+, H is $+1$ and N is -3; in NO_3^-, O is -2 and N is $+5$; (f) Cl is -1, Ti is $+4$; (g) Pb is $+2$; in sulfate, O is -2, S is $+6$; and (h) P is 0

18.3 (a) not redox; (b) redox; (c) not redox; (d) redox; and (e) redox

18.5 (a) $+5$; (b) -1; (c) $+7$; (d) 0; (e) $+3$; and (f) $+1$

18.7 (a) $Na \rightarrow Na^+ + e^-$ and $2 H_3O^+ + 2 e^- \rightarrow H_2 + 2H_2O$; (b) $Au + 4 Cl^- \rightarrow [AuCl_4]^- + 3 e^-$ and $NO_3^- + 4 H_3O^+ + 3 e^- \rightarrow NO + 6 H_2O$; (c) $MnO_4^- + 8 H_3O^+ + 5 e^- \rightarrow Mn^{2+} + 12 H_2O$ and $C_2O_4^{2-} \rightarrow 2 CO_2 + 2 e^-$; and (d) $C + 3 H_2O \rightarrow CO + 2 H_3O^+ + 2e^-$ and $2 H_3O^+ + 2e^- \rightarrow H_2 + 2 H_2O$

18.9 (a) $Cu^+ + 3 H_2O \rightarrow CuO + 2 H_3O^+ + e^-$; (b) $S + 2 H_3O^+ + 2 e^- \rightarrow H_2S + 2 H_2O$; (c) $Ag_2O + H_2O + 2 e^- \rightarrow 2 Ag + 2 OH^-$; (d) $I^- + 6 OH^- \rightarrow IO_3^- + 3 H_2O + 6 e^-$; (e) $IO_3^- + 2 H_2O + 4 e^- \rightarrow IO^- + 4 OH^-$; and (f) $H_2CO + 5 H_2O \rightarrow CO_2 + 4 H_3O^+ + 4 e^-$

18.11 (a) $2 Cu^+ + 4 H_2O + S \rightarrow 2 CuO + 2 H_3O^+ + H_2S$; (b) $3 Ag_2O + I^- \rightarrow 6 Ag + IO_3^-$; (c) $2I^- + IO_3^- \rightarrow 3 IO^-$; and (d) $2S + H_2CO + H_2O \rightarrow 2 H_2S + CO_2$

18.13 (a) Cl^-; (b) MnO_4^-; (c) MnO_4^-; (d) Cl^-; (e) MnO_4^-; and (f) Cl^-

18.15 (a) $3 CN^- + 2 MnO_4^- + H_2O \rightarrow 3 CNO^- + 2 MnO_2 + 2 OH^-$; (b) $4 As + 3 O_2 + 2 H_2O \rightarrow 4 HAsO_2$; (c) $Br^- + 2 MnO_4^- + H_2O \rightarrow BrO_3^- + 2 MnO_2 + 2 OH^-$;

(d) $3 NO_2 + 3 H_2O \rightarrow 2 NO_3^- + NO + 2 H_3O^+$; (e) $ClO_4^- + 6 Cl^- + 6 H_3O^+ \rightarrow ClO^- + 9 H_2O + 3 Cl_2$; and (f) $AlH_4^- + 4 H_2CO + 4 H_2O \rightarrow Al^{3+} + 4 CH_3OH + 4 OH^-$

18.17 (a), (b), (c) and (d) spontaneous

18.19

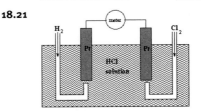

18.21

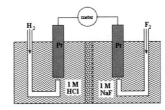

18.23 The silver-silver chloride electrode (18.19) is active, whereas the platinum electrodes (18.21) are passive.

18.25 (a) 1.565 V; (b) 0.981 V; and (c) 0.002 V

18.27 $8 NO + 3 H_2O \longrightarrow 3 N_2O + 2 NO_3^- + 2 H_3O^+$; $E^\circ_{cell} = 0.634$ V

18.29 Set up a standard hydrogen electrode on one side and a Pt electrode immersed in 1 M solution of NaF with F_2 bubbling over the electrode. The H_2/H_3O^+ electrode will be the anode.

18.31 Ba, Ca, Cs, Li, Mg, K, and Na. All these metals lie in the *s* block of the periodic table and easily lose 1 or 2 electrons.

18.33 (a) -906.0 kJ; (b) -1.14×10^3 kJ; and (c) -1 kJ

18.35 $E = 1.35$ V

18.37 $m_{Pb} = 9.5 \times 10^{-2}$ g; $m_{PbO_2} = 0.11$ g

18.39 1.35×10^3 s

18.41 1.98 g

18.43 $Q = \dfrac{[OH^-]_{cathode}}{[OH^-]_{anode}}$; as the battery operates, the anode concentration decreases while the cathode concentration increases, $Q > 1$ and ln $Q > 0$, and the potential of the battery decreases with use.

18.45 Cr corrodes.

18.47 Mercury batteries are characterized by a stable potential and compact size but limited current capacity, making them well suited for use where large currents are not needed, such as pacemakers and cameras. They cannot supply the large current needed to start an automobile engine; moreover, they are irreversible so would have to be replaced very frequently.

18.49 19.9 g

18.51 The oxidation reaction is $Cu \rightarrow Cu^{2+} + 2\ e^-$. The reduction of H_3O^+ occurs:
$2\ H_3O^+(aq) + 2\ e^- \rightarrow 2H_2O + H_2\ (E^\circ = 0\ V)$

18.53 (a) Attach the negative wire to the Cd eletrode; $Cd(s) + 2\ OH^-(aq) \rightleftarrows Cd(OH)_2(s) + 2\ e^-$; (b) 4.56 hr

18.55 1.90×10^3 C; 2.10 A

18.57

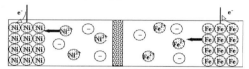

18.59 $K_{eq} = 10^{-14}$

18.61 3.0×10^9

18.63 (a) $3\ H_2 + 2\ Cr(OH)_3 \rightarrow 6\ H_2O + 2\ Cr$; (b) -0.65 V; and (c) 3.8×10^2 kJ

18.65 (a) $Cr_2O_7^{2-} + 14\ H_3O^+(aq) + 6\ e^- \rightarrow 2\ Cr^{3+} + 21\ H_2O$; and (b) 16.2 g

18.67 (a) On the left, $Ni^{2+} + 2e^- \rightarrow Ni$, $E^\circ = -0.257$ V; on the right, $Fe^{2+} + 2e^- \rightarrow Fe$, $E^\circ = -0.447$ V; (b) $E^\circ = 0.190$ V; (c) Fe electrode is the anode and Ni is the cathode; and (d)

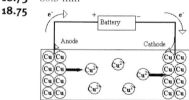

18.69 $[Ni^{2+}]$ must be reduced to 3.8×10^{-7} M to reach a potential of 0 V.

18.71 (a) $2\ MnO_4^- + 5\ H_2SO_3 + H_3O^+(aq) \rightarrow 2\ Mn^{2+} + 5\ HSO_4^- + 4\ H_2O$; (b) $2\ MnO_4^- + 5\ SO_2 + H_2O + H_3O^+(aq) \rightarrow 2\ Mn^{2+} + 5\ HSO_4^-$; (c) $8\ MnO_4^- + 5\ H_2S + 19\ H_3O^+(aq) \rightarrow 8\ Mn^{2+} + 5\ HSO_4^- + 31\ H_2O$; and (d) $8\ MnO_4^- + 5\ H_2S_2O_3 + 14\ H_3O^+(aq) \rightarrow 8\ Mn^{2+} + 10\ HSO_4^- + 21\ H_2O$

18.73 80.3 min

18.75

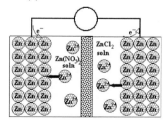

18.77 (a) MnO_4^-; (b) O_2; and (c) Sn^{2+}

18.79 O_2 can oxidize Cu, Ag, Fe^{2+}, H_2, and I^-

18.81 (a) $Zn(NO_3)_2$ soln (cathode): $Zn^{2+} + 2\ e^- \rightarrow Zn$; $ZnCl_2$ soln (anode): $Zn \rightarrow Zn^{2+} + 2\ e^-$; (c) 0.0207 V; and (b)

18.83 2.3×10^2 hr

18.85 5.6×10^{-8} M

18.87 (a) $Na^+ + e^- \rightarrow Na$ and $2\ Cl^- \rightarrow Cl_2 + 2\ e^-$; (b) The cathode is the Pt electrode connected to the negative pole, and the anode is the Pt electrode connected to the positive pole of the battery; and

(c)
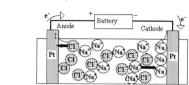

18.89

18.91 0.368 V

19.1 (a) Mn is +2; (b) Mo is +5; (c) V is +5; (d) Au is +3; and (e) Fe is +3

19.3 (a) chromium, Cr; (b) cadmium, Cd; and (c) copper, Cu

19.5 (a) Pd; (b) Au; and (c) Co

19.7 (a) Ru has oxidation state $+2$; d^6; (b) Cr has oxidation state $+3$; d^3; (c) Pd has oxidation state $+2$; d^8; (d) Ir has oxidation state $+3$; d^6; and (e) Ni has oxidation state 0; $s^2 d^8$

19.9 (a) hexaammineruthenium(II) chloride; (b) *trans*-(ethylenediamine)diiodochromium(III) iodide; (c) *cis*-dichlorobis(trimethylphosphine)palladium(II); (d) *fac*-triamminetrichloroiridium(III); and (e) tetracarbonylnickel(0)

19.11 (a) Six NH_3 in an octahedron around the central Ru

(b) Two I at opposite ends of one axis

(c) Square planar, two Cl adjacent

(d) Three Cl form a triangular face

$$H_3N-Ir-Cl$$ (with Cl, Cl on top, Cl right, H_3N below, N at bottom with H H)

(e) Tetrahedral

Ni(CO)$_4$ structure: O≡C, O≡C—Ni—C≡O, C≡O

19.13 (a) cis-[Co(NH$_3$)$_4$ClNO$_2$]$^+$; (b) [PtNH$_3$Cl$_3$]$^-$;
(c) trans-[Cu(en)$_2$(H$_2$O)$_2$]$^{2+}$; and (d) [FeCl$_4$]$^-$

19.15 (a) [O$_2$N, NH$_3$, NH$_3$, Cl, NH$_3$, NH$_3$ around Co]$^+$ (b) [Cl, Cl, Cl, NH$_3$ around Pt]$^-$ (c) [H$_2$O, N, N, N, N, H$_2$O around Cu]$^{2+}$ (d) [Cl, Cl, Cl, Cl around Fe]$^-$

19.17 (a) d^2

(b) d^3

(c and d) d^5 low field / high field

19.19 (a) diamagnetic; (b) paramagnetic with four unpaired electrons; (c) paramagnetic with two unpaired electrons; and (d) diamagnetic.

19.21 The colors of transition metal complexes are generally determined by d-d transitions, but Zr^{4+} has no valence electron that can undergo a transition that would absorb visible light.

19.23 (a) CuFeS$_2$ + 2 O$_2$ → CuO + FeO + 2 SO$_2$; (b) Si + O$_2$ + CaO → CaSiO$_3$; and (c) TiCl$_4$ + 4 Na → 4 NaCl + Ti

19.25 1.06 × 10^6 g; 2.05 × 10^5 L

19.27 183.3 kJ; 63.3 kJ

19.29 CaWO$_4$ + 2 H$_3$O$^+$ ⟶ Ca^{2+} + H$_2$WO$_4$ + 2 H$_2$O; this reaction dissolves the ore;
H$_2$WO$_4$+ 2 NH$_3$ ⟶ (NH$_4$)$_2$WO$_4$ $\xrightarrow{heat}$ WO$_3$ + 2 NH$_3$ + H$_2$O; this reaction removes all elements except W and O; WO$_3$ + 3 H$_2$ → W + 3 H$_2$O; this reaction reduces the oxide to the pure metal

19.31 The coinage metals are copper, silver, and gold. They are used for money, for electrical wire, for jewelry, and for other decorative objects.

19.33 Titanium is used as an engineering metal because of its relatively low density, high bond strength, resistance to corrosion, and ability to withstand high temperatures, all of which make it a favored structural material.

19.35 Hemoglobin has four subunits, whereas myoglobin has just one. As a consequence, hemoglobin has a more complex cooperative chemical behavior than myoglobin.

19.37 remove 1 electron

Fe^{2+} → Fe^{3+}

19.39

(a) mer and fac isomers of Ir with Cl and NH$_3$

(b) Cl—Pd—Cl with P(CH$_3$)$_3$ trans

(c) OC, Cr, CO, Br trans and cis isomers

(H$_3$C)$_3$P—Pd—Cl with P(CH$_3$)$_3$, Cl cis

(d) Cr complexes with N, I, NH$_3$ (three isomers)

19.41 Bidentate ligands form complexes with two links. Each Fe ion forms an octahedral complex with three oxalate anions:

[Fe(oxalate)$_3$ structure]

19.43 (a) Cr is [Ar]$4s^1$ $3d^5$; Cr^{2+} is [Ar] $3d^4$; Cr^{3+} is [Ar] $3d^3$;
(b) V is [Ar]$4s^2$ $3d^3$; V$^-$ is [Ar]$4s^2$ $3d^4$; V$^+$ is [Ar]$4s^1$ $3d^3$;

V^{2+} is [Ar] $3d^3$; V^{3+} is [Ar] $3d^2$; V^{4+} is [Ar] $3d^1$; V^{5+} is [Ar]; and (c) Ti is [Ar]$4s^2 3d^2$; Ti^{2+} is [Ar] $3d^2$; Ti^{4+} is [Ar]

19.45 (a) *cis*-tetraaquadichlorochromium(III) chloride;
(b) bromopentacarbonylmanganese(I); and
(c) *cis*-diamminedichloroplatinum(II)

19.47 $2 O_2^- + 2 H_3O^+ \xrightarrow{\text{SOD}} O_2 + H_2O_2 + 2 H_2O$

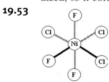

19.49 Cu(II) in water forms an aqua complex. Addition of fluoride produces the insoluble green salt, CuF_2, while addition of chloride produces the tetrachlorocopper(II) complex, $[CuCl_4]^{2-}$:
$Cu^{2+}(aq) + 2 F^-(aq) \longrightarrow CuF_2(s)$
$Cu^{2+}(aq) + 4 Cl^-(aq) \longrightarrow [CuCl_4]^{2-}(aq)$

19.51 Silver is difficult to oxidize, so it has good resistance to corrosion and is suitable for jewelry. Vanadium is readily oxidized, so it corrodes rapidly and is unsuited to jewelry.

19.53

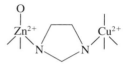

19.55 silver sulfide;
$4 Ag(s) + 2 H_2S(g) + O_2(g) \rightarrow 2 Ag_2S(s) + 2 H_2O(l)$

19.57 $[Ni(CO)_4]$ has tetrahedral geometry, colorless; $[Zn(CN)_4]^{2-}$ has tetrahedral geometry, colorless

19.59 Brass is an alloy of zinc and copper, and superoxide dismutase contains zinc and copper ions in its reaction center.

19.61 $[Ni(CN)_4]^{2-}$ is square planar and $[NiCl_4]^{2-}$ is tetrahedral

19.63 When ferritin is neither empty nor filled to capacity, it can provide iron as needed for hemoglobin synthesis, or store iron if an excess is absorbed by the body.

19.65 Because Zn^{2+} has d^{10} configuration, its d orbitals are completely filled and the lowest unoccupied orbital is quite high in energy. Co^{2+} has d^7 configuration, giving this cation unfilled d orbitals. Consequently, metalloproteins that contain Co^{2+} absorb visible light, making it possible to study them with uv-visible spectroscopy.

19.67 The redox reaction for Zn has a lower reduction potential than less active metals. Therefore it will preferentially be oxidized to its cation while the less active metals ions will precipitate out.

20.1 (a) Lewis acid is Ni, Lewis base is CO; (b) Lewis acid is $SbCl_3$, Lewis base is Cl^-; (c) Lewis acid is $AlBr_3$, Lewis base is $P(CH_3)_3$; and (d) Lewis acid is BF_3, Lewis base is ClF_3

20.3 (a) acceptor: $3p$ orbital

(b) acceptor: $3d$ orbital

(c) acceptor: delocalized π orbital

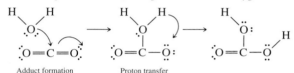

20.5 In the first step, an electron pair from the O atom of H_2O displaces a π bond in Lewis acid-base adduct formation. Then a proton from H_2O migrates to a C—O oxygen atom:

20.7 V^{3+} (lowest Z, high charge) < Fe^{3+} (high charge) < Fe^{2+} < Pb^{2+} (large **n**).

20.9 (a) the hardest acid is BF_3 (both elements from row 2), then BCl_3, and $AlCl_3$ is softest (both elements from row 3); (b) the hardest acid is Al^{3+} (row 3), then Tl^{3+} (row 6), and Tl^+ (low charge) is softest; and (c) Polarizability increases with **n**, so hardest is $AlCl_3$, then $AlBr_3$, and AlI_3 is softest.

20.11 When an electronegative O atom bonds to a less negative S atom, it withdraws electron density from S, decreasing the polarizability about S and increasing the hardness of the base.

20.13 (a) Metathesis occurs, giving $AlCl_3$ and NaI; (b) no reaction occurs; (c) metathesis occurs, giving H_2O and CaS; (d) metathesis occurs, giving $(CH_3)_3P$ and LiCl; and (e) no reaction occurs.

20.15 Al_2Cl_6 contains two "bridging" chlorine atoms. Standard procedures would predict tetrahedral geometry about all inner atoms, but the Al—Cl—Al bond angles of 91° indicate that Cl uses p orbitals. Each Al atom can be described as using sp^3 hybrids to form four σ bonds to four different Cl atoms. In Lewis acid-base terms, the bridged molecule forms from two $AlCl_3$ units linking together in double adduct formation between the Al Lewis acid atoms and two Cl Lewis base atoms.

20.17 Thallium lies below indium and gallium, so its properties should be similar to those metals: valence of 3, soft Lewis acid. Like its neighbor, Pb, it is toxic.

20.19 $SnCl_4$ can function as a Lewis acid because Sn has empty d orbitals from which it can accept electrons to form more bonds.

20.21 Lewis structure:

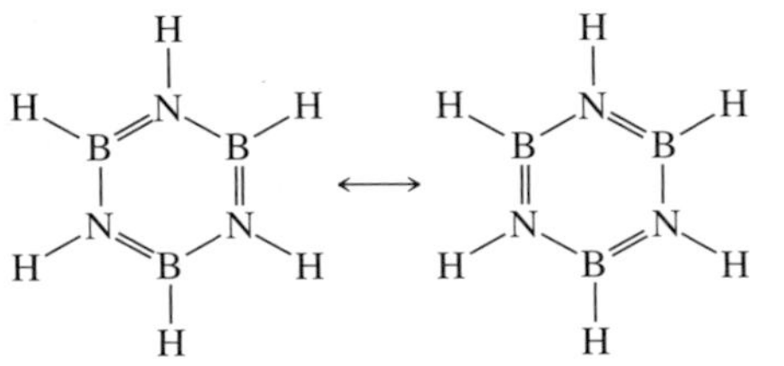

$$
\text{H}_2\text{B}_3\text{N}_3\text{H}_3 \text{ (borazine ring structures shown as two resonance forms)}
$$

The B and N atoms have bonding and geometry that can be described using sp^2 hybrid orbitals, resulting in 3 σ bonds around each ring atom. In addition, there is a delocalized π bonding network encompassing all six ring atoms and containing 6 electrons.

20.23 Si has a larger band gap

20.25 $2 C_2H_5Cl + Si(Cu) \longrightarrow (C_2H_5)_2SiCl_2 + Cu$
$(C_2H_5)_2SiCl_2 + 2 H_2O \longrightarrow (C_2H_5)_2Si(OH)_2 + 2 HCl$
condensation will eliminate water to give the polymer:

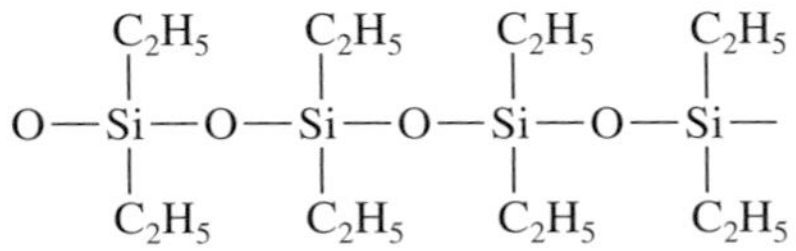

$$
\text{O—Si—O—Si—O—Si—O—Si—} \quad (\text{with } C_2H_5 \text{ groups above and below each Si})
$$

20.27 $P_4O_{13}{}^{6-}$:

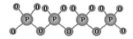

20.29 Brønsted acid-base reaction:
$Ca_5(PO_4)_3F(s) + 5 H_2SO_4(aq) \longrightarrow$
$3 H_3PO_4(aq) + 5 CaSO_4(s) + HF(aq)$

20.31

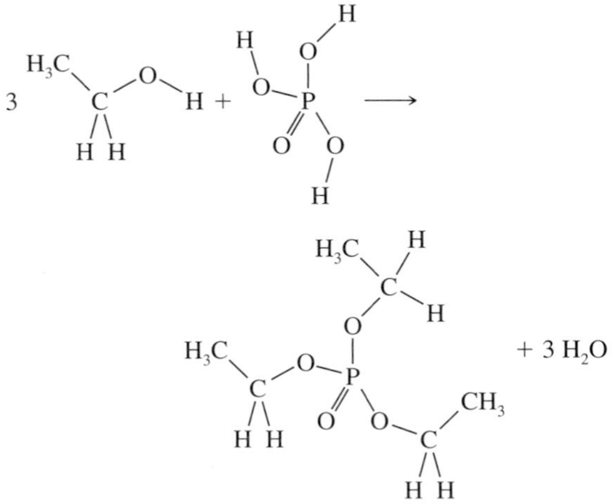

$$
3 \; \text{H}_3\text{C—CH(OH)}_2\text{(CH}_3\text{)} + \text{H}_3\text{PO}_4 \longrightarrow \text{(triethyl phosphate structure)} + 3 H_2O
$$

20.33 There are two industrial reactions in Section 20.6 in which sulfuric acid acts as a Brønsted acid:
$CaF_2(s) + H_2SO_4(l) \longrightarrow 2 HF(g) + CaSO_4(s)$
$Ca_5(PO_4)_3F(s) + 5H_2SO_4(aq) \longrightarrow 3H_3PO_4(aq) + 5CaSO_4(s) +$
$\qquad\qquad\qquad\qquad\qquad\qquad\qquad\qquad\qquad HF(aq)$

20.35

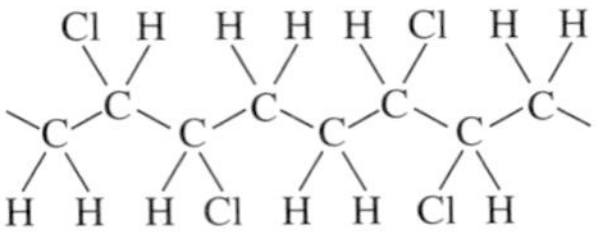

$$
\text{Cl H H H H Cl H H} \\
\text{(chlorinated hydrocarbon chain of carbon atoms)} \\
\text{H H H Cl H H Cl H}
$$

20.37 $TiO_2(s) + C(s) + 2 Cl_2(g) \longrightarrow TiCl_4(l) + CO_2(g)$
C goes from zero to $+4$ oxidation state, so it is oxidized and serves as the reducing agent.
Cl goes from zero to -1 oxidation state, so it is reduced and serves as the oxidizing agent.

20.39 8.7×10^3 kg bauxite rock

20.41 Sulfur has d orbitals available for bond formation, allowing the formation of SF_6, in which the bonding can be described using d^2sp^3 hybrid orbitals on the S atom. Oxygen has no valence d orbitals available. In principle, SBr_6 could also form, but the Br atom is too large for six Br atoms to be accommodated around a central S atom.

20.43 The As atom in $AsCl_3$ has a lone pair that it donates, giving this compound Lewis base character. The As atom also can accommodate additional electron pairs by using valence d orbitals, giving this compound Lewis acid character.
$AsCl_3 + BF_3 \longrightarrow Cl_3As—BF_3$ (As acts as Lewis base)
$AsCl_3 + Cl^- \longrightarrow AsCl_4{}^-$ (As acts as Lewis acid).

20.45 $Al + 2 H_2O \rightarrow AlO_2{}^- + 4 H^+ + 3 e^-$; (a) $8 Al + 5 OH^- + 3 NO_3{}^- + 2 H_2O \rightarrow 3 NH_3 + 8 AlO_2{}^-$;
(b) $2 Al + 2 OH^- + 2 H_2O \rightarrow 3 H_2 + 2 AlO_2{}^-$;
(c) $4 Al + 3 SnO_3{}^{2-} + H_2O \rightarrow 3 Sn + 2 OH^- + 4 AlO_2{}^-$.

20.47 Nitrogen, at the top of Group 15, is a non-metal showing a valence of three that forms polar bonds most readily with other non-metals (B, C, O, the halogens). Phosphorus has a similar pattern of reactivity but can also involve d orbitals in its bonding. Arsenic and antimony are metalloids with useful semiconductor properties, and Bismuth is metallic.

20.49 $Al(O)OH(s) + OH^-(aq) + H_2O \longrightarrow [Al(OH)_4]^-(aq)$
$[Al(OH)_4]^-(aq) \longrightarrow Al(OH)_3(s) + OH^-(aq)$
$2 Al(OH)_3(s) \xrightarrow{1250\,°C} Al_2O_3(s) + 3 H_2O(g)$
$2Al_2O_3(s) + 3 C(s) \longrightarrow 3 CO_2(g) + 4 Al(l)$

20.51 Silicon dioxide is first converted into silicon tetrachloride by reaction with molecular chlorine:
$SiO_2(s) + 2 C(s) + 2 Cl_2(g) \xrightarrow{\text{high } T} SiCl_4(g) + 2 CO(g)$
The $SiCl_4$ is purified by distillation and then reduced by reaction with Mg metal:
$SiCl_4 + 2 Mg \longrightarrow Si + 2 MgCl_2$

20.53 $Ca_5(PO_4)_3F(s) + 7 H_3PO_4(aq) \rightarrow + 5 Ca(H_2PO_4)_2(s) + HF(aq)$; 26.47%

20.55 Gaseous Al_2Cl_6 is in equilibrium with gaseous $AlCl_3$:
$Al_2Cl_6 \rightleftharpoons 2 AlCl_3$.
Le Chatelier's principle predicts that because the forward reaction is endothermic (bonds must be broken), an increase in temperature shifts the position of the equilibrium to the right. Thus as temperature increases, the number of moles of gaseous substance increases, so the pressure increases faster than would be predicted by the ideal gas equation.

20.57 Soft metal ions are toxic because they are very soft Lewis acids which react readily with soft Lewis bases such as the sulfur atoms that hold essential enzymes in the conformations that those enzymes require to function properly. When a soft metal ion reacts with an S—S link in an enzyme, the conformation of the enzyme becomes distorted or destroyed and the enzyme can no longer function.

20.59 Mercury is a very soft Lewis acid, while zinc is intermediate in hard/soft character. Thus mercury preferentially associates with sulfur, a soft Lewis base, while zinc forms stable bonds with both soft (sulfur-containing) and hard (oxygen-containing) Lewis bases.

20.61 As the size of an atom increases, so does its polarizability, and greater polarizability leads to larger dispersion forces. Thus the larger the atoms in a molecule, the larger the intermolecular forces and the easier it is to condense the substance. In the sequence BCl_3, BBr_3, BI_3, the halogens are increasing in size, accounting for the different stable phases at room temperature.

20.63 White phosphorus is toxic

20.65 $2\ Al_2O_3 + 6\ Cl_2 \rightarrow 4\ AlCl_3 + 3\ O_2$;
$AlCl_3 + 3\ Na \rightarrow Al + 3\ NaCl$.

20.67 (a) $AlCl_3 + 3\ LiCH_3 \rightarrow Al(CH_3)_3 + 3\ LiCl$;
(b) $SO_3 + H_2O \rightarrow H_2SO_4(aq)$; (c) $SbF_5 + LiF \rightarrow Li(SbF_6)$;
and (d) no reaction.

20.69 $P_4O_{10} + 6\ H_2O \rightarrow 4\ H_3PO_4$

20.71 8.99 g Al and 23.0 g Na

21.1

Part:	(a)	(b)	(c)	(d)	(e)	(f)
Z	3	20	92	52	10	82
A	6	43	238	130	20	205
N	3	23	146	78	10	123

21.3 (a) ^4_2He; (b) $^{184}_{74}W$; (c) $^{60}_{28}Ni$; and (d) $^{26}_{12}Mg$

21.5 (a) unstable, too few neutrons; (b) unstable, $Z > 83$; (c) stable; and (d) unstable, odd-odd.

21.7 $\Delta E = 8.99 \times 10^{16}$ kJ

21.9 $\Delta E = -1.079 \times 10^{11}$ kJ/mol; ΔE (per nucleon) $= -8.11 \times 10^8$ kJ/mol nucleon

21.11 The coulombic barrier is greater for $^1H + {}^6Li \rightarrow {}^7Be$ than for $^4He + {}^4He \rightarrow {}^8Be$; ^7Be is more stable

21.13

description	symbol	name
(a) high energy photon	γ	gamma ray
(b) mass number 4	α	alpha particle
(c) positive particle with m_e	β^+	positron.

21.15 (a) $^{125}_{52}Te$; (b) $^{123}_{51}Sb$; and (c) $^{127}_{53}I$

21.17 $t_{1/2} = 1.2$ yr

21.19 $^{232}_{90}Th \rightarrow \alpha + {}^{228}_{88}Ra \rightarrow \beta + {}^{228}_{89}Ac \rightarrow \beta + {}^{228}_{90}Th \rightarrow \alpha + {}^{224}_{88}Ra \rightarrow \alpha + {}^{220}_{86}Rn \rightarrow \alpha + {}^{216}_{84}Po$;
$^{216}_{84}Po \rightarrow \beta + {}^{216}_{85}At \rightarrow \alpha + {}^{212}_{83}Bi \rightarrow \beta + {}^{212}_{84}Po \rightarrow \alpha + {}^{208}_{82}Pb$.

21.21 (a) $^{12}_6C + n \rightarrow [{}^{13}_6C] \rightarrow {}^{12}_5B + p$;
(b) $^{16}_8O + \alpha \rightarrow [{}^{20m}_{10}Ne] \rightarrow {}^{20}_{10}Ne + \gamma$; and
(c) $^{247}_{96}Cm + {}^{11}_5B \rightarrow [{}^{258}_{101}Md] \rightarrow {}^{255}_{101}Md + 3\ n$.

21.23 The reaction in this problem is:
$^{14}_7N + \alpha \rightarrow [{}^{18}_9F] \rightarrow {}^{18}_8O + \beta^+$

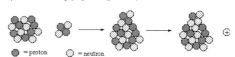

● = proton   ◎ = neutron

21.25 Cs, Ba, La, Ce

21.27 $\Delta E = -1.648 \times 10^{10}$ kJ/mol; this result is somewhat less than the result of the general calculation of Section Exercise 21.4.1 for a net change of 2 neutrons.

21.29 The core of a nuclear reactor generates radiation that converts any material in its vicinity into radioactive substances. Thus the heat exchanger in immediate contact

with the core becomes radioactive and must be separated from the turbine that generates electricity. The primary heat exchanger transfers energy to a secondary heat exchanger, which does not become radioactive and can safely drive the turbine.

21.31 $\Delta E_{total} = -2.4 \times 10^9$ kJ

21.33 $u = 5.3 \times 10^6$ m/s

21.35 Your description should include the extremely high energies required to initiate fusion and the difficulties in containing the fusion components at the temperature required for nuclei to have these high energies.

21.37

Stage	Temperature	Composition
H-burning	4×10^7 K	C, N, H, He, e^-
He-burning	10^8 K	He, Be, C, and O
C-burning	10^9 K	all nuclides from $Z = 6$ (C) up to $Z = 26$ (Fe)

21.39 Your description should include the fact that elements beyond $Z = 26$ are less stable than Fe, so they cannot be generated by fusion of lighter elements.

21.41 $E = 4.4 \times 10^{-4}$ J

21.43 There are 3.41×10^4 decays/s

21.45 Exposure to radiation results first in damage to those cells that reproduce most quickly, including the white blood cells that are responsible for fighting infection and the mucous membrane lining of the intestinal tract. Thus the early symptoms of radiation exposure include reduced resistance to infection and nausea due to disruption of the digestive tract.

21.47 3.5×10^9 yr

21.49 Among the isotopes of iodine listed, $A = 123$ would be best because of its 13.3-hr half-life and EC/γ mode of decay.

21.51 (a) $^{234}_{90}Th$; (b) $^{60}_{27}Co$; (c) $^{248}_{99}Es$; (d) $^{32}_{16}S$; and (e) $^{60}_{28}Ni$

21.53 $\Delta E = -1.583 \times 10^{11}$ kJ/mol; ΔE (per nucleon) $= -7.574 \times 10^8$ kJ/mol nucleon

21.55 $N = 0.80$ mg; $\dfrac{\Delta N}{\Delta t} = 1.3 \times 10^{11}$/s

21.57 The other product of neutron decay must be a particle with -1 charge and 0 mass, which is an electron: $n \rightarrow p + e$; $E_{kinetic}$ (electron) $= 1.25 \times 10^{-13}$ J.

21.59 When 1% has decayed, $t = 24$ yr; when 1% remains, $t = 1.1 \times 10^4$ yr

21.61 Both isotopes are neutron-deficient and lie below (to the right of) the belt of stability: $^{11}C \rightarrow \beta^+ + {}^{11}B$ and $^{15}O \rightarrow \beta^+ + {}^{15}N$

21.63 70 hr

21.65 If $t_{1/2} = 1100$ s, $N = 36$; If $t_{1/2} = 876$ s, $N = 2$

21.67 $^{256}_{104}Rf$

21.69 ΔE (per event) $= -4.9 \times 10^{-13}$ J; Fraction carried off by the α-particle $= 0.73$

21.71 (a) too many neutrons; (b) $Z > 83$; (c) odd-odd; and (d) too many neutrons.

21.73 In a nuclear reactor, the moderator serves to slow down fast neutrons so they are more efficiently captured by the nuclear fuel. This reduces the amount of fuel required to sustain the reaction. The control rods serve to absorb some of the neutrons, allowing the reactor to operate just below its critical point and generate large quantities of heat without heating up beyond control.

21.75 The precipitate will contain radioactive Na.

21.77 $^{10}_{5}B + n \rightarrow ^{11}_{5}B \rightarrow ^{7}_{3}Li + \alpha$; The reaction does not pose a significant health hazard

21.79 1.4×10^4 yr

21.81 5.2×10^4 yr

21.83 (a) $^{90}_{38}Sr \rightarrow ^{90}_{39}Y + \beta$; $^{90}_{39}Y \rightarrow ^{90}_{40}Zr + \beta$; (b) Sr has the larger mass; and (c) $\Delta E = -1.7 \times 10^8$ kJ/mol

21.85 1.5×10^2 hr

21.87 $^{99}_{43}Tc$

21.89 To determine where the oxygen atom in the water molecule comes from, prepare a sample of the alcohol that is enriched in radioactive ^{18}O and run the reaction using this sample. Separate the products and measure the radioactivity of the ester and the water. If the C—OH bond in the alcohol breaks during the condensation, the ^{18}O will appear in the water, while if the C—OH bond in the carboxylic acid breaks during the condensation, the ^{18}O will appear in the ester.

Glossary

Absolute entropy. (13.3) The amount of disorder contained in a chemical substance.

Absolute zero. (5.4) The temperature at which the molecular energy of motion is at a minimum.

Absorption spectrum. (6.3) The distribution of wavelengths of light absorbed by a species.

Accuracy. (1.5) How close a measurement is to the true value.

Acid. (Brønsted definition) (4.6) A substance that acts as a proton donor.

Acid ionization constant (K_a). (15.6) The equilibrium constant for proton transfer between an acid and water

$$K_a = \frac{[H_3O^+]_{eq}\,[A^-]_{eq}}{[HA]_{eq}}.$$

Acid rain. (5.7) Rain whose pH is lower than 5.6, which is the pH of water saturated with carbon dioxide.

Actinide element. (1.3) Any of the elements in the $5f$ block of the periodic table, between $Z = 89$ and $Z = 102$.

Activated complex. (14.6) The unstable molecular species through which an elementary reaction proceeds as it evolves from reactant(s) to product(s).

Activation energy (E_a). (14.6) The minimum amount of energy reactants must possess to react to form products.

Activation energy diagram. (14.6) A graph showing how the energy of a reacting set of molecules varies with the course of the reaction.

Active electrode. (18.3) An electrode whose chemical constituents take part in the redox reaction that occurs at its surface.

Active site. (14.7) The specific location in an enzyme where catalytic activity takes place.

Activity series. (4.7) A list of elements in order of how easily they are oxidized in aqueous solution.

Actual yield. (4.3) The amount of chemical product formed in a chemical reaction.

Adduct. (20.1) A compound formed by bond formation between a Lewis base and a Lewis acid.

Adenosine triphosphate (ATP). (13.6) A biochemical molecule containing adenine and three phosphate groups instrumental in biochemical energetics.

Adhesive force. (10.5) A force of attraction between molecules in one phase and different molecules in another phase.

Adsorption. (14.7) The physical attachment of molecules to a surface.

Alcohol. (11.1) An organic compound that contains the hydroxyl group, —OH.

Aldehyde. (11.1) An organic compound that contains the carbonyl group (C=O) attached to a hydrogen atom, —CHO.

Alkali metal. (1.3) Any of the Group 1 elements, all of which are reactive metals with s^1 valence configurations.

Alkaline earth metal. (1.3) Any of the Group 2 elements, all of which are reactive metals with s^2 valence configurations.

Alkane. (3.2) A compound containing only carbon and hydrogen, with general formula $C_nH_{(2n+2)}$, in which all carbon atoms possess four σ bonds.

Alkene. (3.2) A compound containing only carbon and hydrogen in which there is at least one C=C bond.

Alkyne. (3.2) A compound containing only carbon and hydrogen in which there is at least one C≡C bond.

Alloy. (10.5) A metallic solution or mixture of two or more metals.

Alpha emission. (21.2) Nuclear decay in which a helium nucleus is emitted by the decaying nucleus.

Alpha particle (α). (2.2) An energetic helium nucleus emitted by a radioactive nuclide.

Amalgam. (10.5) A solution of a metal in mercury.

Amide. (11.1) An organic compound that contains the $-\!\!\overset{\displaystyle O}{\underset{}{\overset{\|}{C}}}\!-\!\overset{\displaystyle H}{\underset{}{\overset{|}{N}}}\!-$ linkage.

Amine. (11.1) A compound whose molecules can be viewed as ammonia with one or more N—H bonds replaced by N—C bonds.

Amino acid. (11.7) A compound that contains both an amine ($-NH_2$) and a carboxylic acid ($-CO_2H$) group.

Amorphous. (10.4) Without any organized regular repeating pattern.

Ampere (A). (1.5) The base unit of electric current in the SI system.

Amphiprotic. (16.1) Able to act as both an acid and a base.

Amplitude. (6.2) The height of a wave.

Analytical balance. (1.5) An accurate mass-measuring instrument.

Anion. (2.4) An ion that possesses negative charge.

Anode. (18.3) The electrode at which oxidation occurs.

Antibonding orbital. (9.3) A molecular orbital that has electron density concentrated outside the bonding region, making it less stable than the atomic orbitals from which it forms.

Aqueous. (3.7) A solution with water as the solvent.

Area. (1.5) The total surface of something.

Arrhenius equation. (14.6) The equation describing how rate constants depend on temperature and activation energy, $k = Ae^{-E_a/RT}$.

Atmosphere (atm). (5.3) A unit of pressure based on the normal pressure exerted by the Earth's atmosphere at sea level. 1 atm = 101.325 kPa.

Atom. (1.2) The basic unit of chemical matter, consisting of a nucleus and enough electrons to convey electrical neutrality.

Atomic number (Z). (2.3) The number of protons in an atomic nucleus.

Atomic orbital. (6.5) A description of an atomic electron that provides the distribution of electron density about the nucleus.

Atomic radius. (7.4) The distance from the nucleus of an atom at which electron-electron repulsion prevents closer approach of another atom.

Atomic symbol. (2.3) The letter designation for an element.

Atomic theory. (2.1) The description of matter as composed of atoms that retain their identities during all physical and chemical processes.

Atomic view. (1.2) How matter looks when viewed at the level of atoms.

Aufbau principle. (7.2) The statement that the most stable arrangement of electrons in an atom results from placing each successive electron in the most stable available atomic orbital.

Avogadro constant (N_A). (3.4) The number of particles contained in 1 mol, 6.022142×10^{23} particles/mol.

Axial position. (8.6) A position along the z axis in a trigonal bipyramid.

Azimuthal quantum number (l). (6.5) The quantum number, restricted to integers from 0 to $n - 1$, that indexes the shape of an atomic orbital.

Balanced chemical equation. (4.1) A description of a chemical reaction using chemical formulas in which the coefficients describe the ratios of molecules of each species that react.

Ball-and-stick model. (3.1) A representation of a molecule that shows the atoms as small balls and the chemical bonds as sticks.

Band gap. (9.6) The difference in energy between the highest filled orbital and the lowest vacant orbital in a solid.

Band theory. (9.6) The description of bonding in solids using delocalized orbitals.

Barometer. (5.3) An instrument for measuring atmospheric pressure.

Base (Brønsted definition). (4.6) A substance that acts as a proton acceptor.

Base ionization constant (K_b). (15.6) The equilibrium constant for the proton transfer from water to a base,

$$K_b = \frac{[BH^+]_{eq}[OH^-]_{eq}}{[B]_{eq}}.$$

Base pairing. (11.6) The formation of hydrogen bonds between strands of DNA or RNA.

Battery. (18.6) One or more galvanic cells that serve as a source of electric current or voltage.

Bent shape. (8.5) The arrangement of a triatomic molecule having a bond angle less than 180°.

Beta emission. (21.2) Nuclear decay in which an energetic electron is emitted by the decaying nucleus.

Beta particle (β). (21.2) An energetic electron emitted by a radioactive nuclide.

Bilayer. (10.7) A double layer of atoms or molecules.

Bimolecular reaction. (14.1) An elementary step in which two molecules collide and undergo a chemical reaction.

Binary compound. (3.2) A chemical substance that contains only two different elements.

Body centered cubic (BCC). (10.4) The crystal structure whose unit cell is a cube with identical atoms at its corners and at its center.

Boiling point. (10.1) The temperature at which the vapor pressure of a liquid matches the external pressure.

Boiling point elevation. (10.6) The increase in boiling point of a liquid caused by the presence of nonvolatile solutes in solution.

Boltzmann distribution. (14.6) The way that molecular kinetic energies are distributed among a collection of molecules.

Bond. (3.1) A strong attractive force generated by sharing of electrons between atoms or coulombic attraction between ions in a substance.

Bond angle. (8.4) The angle between two atoms that are bonded to a third atom.

Bond energy. (8.1) The energy (usually expressed per mole) required to break one particular chemical bond in a gaseous substance.

Bonding orbital. (8.1) An orbital with high electron density between atoms, formed by constructive interaction between atomic orbitals.

Bond length. (8.1) The average distance between the nuclei of two bonded atoms in a molecule.

Bond order. (9.3) The net number of pairs of bonding electrons in a chemical bond.

Bonding molecular orbital. (9.3) An orbital formed from atomic orbitals that conveys bonding by placing electron density between the bonded atoms.

Brønsted-Lowry. (16.1) The proton-transfer definition of acids and bases.

Buffer capacity. (17.2) The amount of added hydronium or hydroxide ions that can be added to a buffer solution without exceeding a specified pH range.

Buffer equation. (17.1) The equation linking the pH of a buffer solution with the pK_a and concentrations of the acid-base pair.

Buffer solution. (17.1) A solution containing both a weak acid and its conjugate base whose pH changes slowly on addition of small amounts of an acid or base.

Buret. (4.6) A volume-delivering device that is calibrated to allow measurement of the amount of fluid delivered.

Calorimeter. (12.3) An apparatus for measuring the amount of heat absorbed or emitted during some process.

Capillary action. (10.5) The upward movement of a liquid in a narrow tube against the force of gravity.

Carbohydrate. (11.5) A biochemical substance whose empirical formula is $(CH_2O)_n$.

Carbonyl group. (11.1) The $C{=}O$ functional group.

Carboxylic acid. (11.1) An acid that contains the $-CO_2H$ functional group.

Catalyst. (14.7) A substance that increases the rate of a chemical reaction without being consumed or produced in the reaction.

Cathode. (18.3) The electrode at which reduction occurs.

Cation. (2.4) A positively charged ion.

Cell potential. (18.4) The difference in electrical potential (voltage) between the two electrodes in a galvanic cell.

Celsius scale. (1.5) The temperature scale defined by 0 °C as the normal freezing point and 100 °C as the normal boiling point of water.

Change of state. (12.1) Any process in which there is a change in one or more of the variables (P, V, T, n) describing a system.

Chelate. (19.2) A complex formed by polydentate ligands.

Chelate effect. (17.5) The stabilization of a metal complex by a multidentate ligand.

Chemical energy. (2.5) Energy stored in bonds between atoms or ions.

Chemical equation. (4.1) An equation using chemical formulas that describes the identities and relative amounts of reactants and products in a chemical reaction.

Chemical equilibrium. (15.0) The condition of a system in which no net chemical change occurs, because forward and reverse processes proceed at equal rates.

Chemical formula. (1.2) An expression using chemical symbols that describes the identities and relative amounts of elements in a substance.

Chemical kinetics. (14.2) The study of how fast chemical reactions occur.

Chemical nomenclature. (3.2) The organized pattern for naming chemical substances.

Chemical transformation. (1.4) A process in which chemical substances react to form other chemical substances.

Chemistry. (1.1) The study of matter and its interactions.

Chromatography. (10.8) A class of separation techniques that make use of differing rates of movement of substances moving with a mobile phase over a stationary phase.

Close packed. (10.4) The most efficient arrangement for packing atoms, molecules, or ions in a regular crystal.

Closed system. (12.1) A system that can exchange energy but not matter with its surroundings.

Cohesive force. (10.5) A force of attraction between like molecules in the same phase.

Colligative property. (10.6) A property of a solution that is proportional to the concentration of solute species.

Combustion. (3.6) A reaction of a substance with oxygen that releases chemical energy.

Common ion effect. (17.4) Reduction in solubility caused by the presence of one of the ions contained in a salt.

Common name. (3.2) A name for a chemical substance used in ordinary conversation.

Complementary base pair. (11.6) Two nitrogen-containing organic bases that hydrogen bonds in DNA or RNA.

Complementary colors. (19.3) Colors that are related in that if light of one color is absorbed by a substance, the substance displays the complementary color.

Complex ion. (15.6) A species formed by bonding of species containing lone pairs of electrons to a metal ion.

Compound. (1.4) A substance formed of atoms of two or more elements chemically bonded in fixed proportions.

Compound nucleus. (21.3) The transient nucleus that forms from the collision of a nuclear projectile with another nucleus.

Concentration. (3.7) The amount of a solute contained in a standard amount of a solution.

Concentration table. (15.5) A table of amounts for a chemical reaction.

Condensation reaction. (11.1) A reaction that joins two molecules, accompanied by the elimination of a small molecule such as water.

Conduction band. (9.6) The set of delocalized orbitals in a solid that, when partially occupied, contains mobile electrons and conveys high electrical conductivity.

Conductor. (9.6) A substance that is a good conductor of electrical current.

Conjugate acid–base pair. (16.4) A Brønsted acid and the base that results when the acid loses one proton.

Conservation law. (2.1) A law stating that a quantity is conserved, in other words has a constant amount.

Conserved. (2.1) Unchanging in amount.

Conversion ratio. (1.5) A ratio equal to 1 that converts a quantity from one unit to another.

Coordination complex. (19.2) A combination of a metal atom or cation with two or more species (ligands) that bind covalently to the metal.

Coordination number. (8.5) The number of atoms bonded to an inner atom in a molecule or metal complex.

Copolymerization. (11.2) The formation of a polymer from two or more different monomers.

Core electrons. (7.2) The inner atomic electrons with principal quantum number less than that of the valence electrons.

Corrosion. (18.6) The oxidation of a metal in contact with its environment.

Coulomb's law. (7.5) The law describing the interaction between two electrically charged objects, $F_{el} = k \dfrac{q_1 q_2}{r^2}$.

Coupled reactions. (13.5) A pair of reactions, one of which is driven in what is otherwise its nonspontaneous direction by the influence of the second, highly spontaneous reaction.

Covalent bond. (8.1) A bond resulting from the sharing of electrons between atoms.

Critical mass. (21.4) The minimum mass of fissionable material required to generate a self-sustaining nuclear fission reaction.

Cross-linking. (11.2) The formation of additional chemical bonds between the chains of a polymer.

Crystal field splitting energy. (19.3) The difference in energy atomic d orbitals created by the effects of neighboring ions in a crystal.

Crystal field theory. (19.3) A model describing the electronic structure of transition metal complexes in terms of the interactions between metal d orbitals and the electric field of the ligands.

Crystalline. (10.4) Containing a regular array of atoms, molecules, or ions.

Cubic close packing. (10.4) The close packing arrangement in which hexagonal layers are stacked in an ABCABC . . . pattern.

Cyclotron. (21.3) A particle accelerator used to generate high energy beams of nuclear particles.

Dalton's law of partial pressures. (5.5) The law stating that in a gas mixture, each gas exerts a pressure equal to the pressure that it would exert if present by itself under otherwise identical conditions.

de Broglie equation. (6.4) The equation describing the wave nature of particles, $\lambda = h/mu$.

Degenerate orbitals. (7.2) Those orbitals in a species that have identical energies.

Delocalized. (6.4) Spread out over space.

Delocalized orbital. (9.4) An orbital in which electron density is distributed over more than two atoms.

Density (ρ). (1.6) The ratio of an object's mass to its volume, $\rho = m/V$.

Deoxyribonucleic acid (DNA). (11.6) A polymer of nucleotide units that stores genetic information in the chromosomes.

Desorption. (14.7) Detachment of an adsorbed molecule from a surface.

Deuterium (D). (2.3) An atom (an isotope of hydrogen) containing one proton and one neutron in its nucleus.

Diamagnetic. (19.3) The quality of being repelled by a magnetic field. Substances with no unpaired electrons are diamagnetic.

Diatomic molecule. (1.2) A molecule that contains exactly two atoms.

Diffusion. (2.1) The gradual mixing of a solute in a solution by molecular motion.

Dilution. (3.7) Production of a less-concentrated solution by addition of solvent.

Dipolar force. (10.2) The attractive force between polar molecules that results from the negative end of one molecule aligning with the positive end of its neighbor.

Dipole moment. (8.7) The net electrical character arising from an asymmetric charge distribution.

Diprotic acid. (16.4) An acid each of whose molecules contains two acidic hydrogen atoms.

Disorder. (13.0) The absence of order.

Dispersion force. (10.2) The net attractive force between molecules generated by temporary dipole moments arising from polarization of their electron clouds.

Disulfide bridge. (11.7) An S—S bond linking two portions of a protein molecule.

Dominant equilibrium. (15.6) The equilibrium that is most important in determining equilibrium concentrations of major species in a solution.

Donor atom. (19.2) An atom that possesses one or more lone pairs of electrons that it can contribute to form a covalent bond.

Doped semiconductor. (9.6) A metalloid to which a small amount of another element has been added, usually to convey higher conductivity.

Double bond. (3.1) A chemical bond containing two pairs of bonding electrons.

Double helix. (11.6) The pair of intertwined spirals that is the secondary structure of DNA.

Ductile. (1.3) Able to be drawn into tubes or wires.

Dynamic. (2.1) Containing objects that are in continual motion.

Dynamic equilibrium. (2.1) The condition in which a forward and a reverse process occur at equal rates, so the system undergoes no net change.

Effective collision. (14.1) A molecular collision that leads to chemical reaction.

Effective nuclear charge. (7.1) The net positive charge, equal to the nuclear charge minus the effects of screening, that an electron in an atomic orbital experiences.

Elastomer. (11.4) A flexible polymer.

Electrical Energy. (2.5) Energy of an object or system created by the action of electric force.

Electrical Force. (2.2) The attraction or repulsion between objects caused by their electric charge.

Electrochemical reaction. (18.3) A redox reaction that occurs in a manner that generates the passage of electrical current through an external circuit.

Electrode. (18.3) An electrical conductor that establishes an interface between an external circuit and an electrochemical cell.

Electrolysis. (18.7) The use of electrical energy to drive a nonspontaneous chemical reaction.

Electrolytic cell. (18.7) An electrochemical cell whose redox reaction is driven by an externally applied electrical potential.

Electromagnetic radiation. (6.2) Wave phenomena having both electrical and magnetic components; visible light is one type.

Electron. (2.2) The smallest unit of negative electrical charge, whose mass is 9.11×10^{-31} kg and whose charge is -1.602×10^{-19} C.

Electron affinity. (7.4) The energy change accompanying the attachment of an electron to an atom or anion.

Electron capture. (21.2) The mode of nuclear decay in which an unstable nucleus captures a core electron and converts a proton into a neutron.

Electron configuration. (7.3) The distribution of electrons among the various orbitals of an atom, molecule, or ion.

Electron contour drawing. (6.6) A depiction of an orbital that shows a contour surface within which most of the electron density is located.

Electron deficient. (20.1) Lacking enough electrons to fill the available valence orbitals.

Electron density. (6.6) The distribution of electron probability in space around an atom or molecule.

Electronegativity. (8.2) A measure of the ability of an atom in a molecule to attract the shared electrons in a chemical bond.

Electroplating. (18.7) The electrolytic deposition of a thin metal film on the surface of an object.

Element. (1.2) A substance composed of atoms all of whose nuclei have the same amount of positive charge.

Elemental analysis. (3.6) Determination of the percent by mass of each element in a compound.

Elementary reaction. (14.1) A chemical reaction that describes a process as it occurs at the molecular level.

Emission spectrum. (6.3) The distribution of wavelengths of light given off by a species in an excited state.

Empirical formula. (3.6) The chemical formula of a substance expressed using the smallest possible integers.

Endothermic. (12.3) Accompanied by the absorption of energy.

Energy. (2.5) The capacity to do work or move matter.

Energy level. (6.3) A specific energy value for a species.

Energy level diagram. (6.2) A diagram of the specific energy levels of a species, showing increasing energy on the y axis.

Enthalpy (H). (12.4) The energy-related thermodynamic state function defined by the equation $H = E + PV$.

Entropy (S). (13.2) The thermodynamic state function that measures the amount of disorder.

Enzyme. (11.7) A biological catalyst.

Equatorial position. (8.6) A position within the trigonal plane of a trigonal bipyramid.

Equilibrium. (2.1) The state in which a system shows no change in properties over time.

Equilibrium constant (K). (15.1) The value of the ratio of equilibrium concentrations of products to equililbrium concentrations of reactants, each raised to the power equal to its stoichiometric coefficient.

Ester. (11.1) An organic compound that contains a $-\overset{\overset{\textstyle O}{\|}}{C}-O-$ linkage.

Ether. (11.1) An organic compound that contains a C—O—C linkage.

Evaporation. (5.7) The escape of molecules from the liquid to the gas phase.

Excited state. (6.2) An orbital description of a species that is not the most stable description because one or more electrons possesses more energy than in the most stable description.

Exothermic. (12.3) Accompanied by the release of energy.

Extensive property. (1.6) A property that increases in proportion to the amount of material in the sample.

Faraday (F). (18.5) The electrical charge of 1 mol of electrons, 96,485 C.

Fiber. (11.4) A long thin strand of polymeric material.

Fibrous protein. (11.7) A protein that forms long linear strands or coils that are water insoluble.

First law of thermodynamics. (12.2) The statement that energy is conserved, often expressed as an equation for change in energy, $\Delta E = q + w$.

First-generation star. (21.5). A star formed from a collapsing cloud that contains only hydrogen and electrons.

First-order. (14.3) Dependent on concentration to the first power.

Fission. (21.1) A nuclear reaction in which one nucleus fragments into two smaller ones.

Formal charge. (8.3) The charge that an atom in a molecule would have if each of its bonding electrons were equally shared with its bonding partner.

Formation constant. (15.6) The equilibrium constant for the formation of a complex ion.

Formation reaction. (12.4) A reaction in which 1 mol of a compound forms from chemical elements in their standard states.

Free energy (G). (13.4) The energy-related thermodynamic state function defined by the equation, $G = H - TS$.

Free radical. (11.2) A highly reactive chemical species containing an unpaired lone electron.

Free radical polymerization. (11.2) Polymer formation that proceeds by a mechanism that involves free radicals.

Freezing point. (10.6) The temperature at which a liquid crystallizes to form a solid.

Freezing point depression. (10.6) The reduction in freezing point caused by the presence of solutes in a solution.

Frequency (v). (6.2) The number of wave crests of a wave that pass a fixed point in unit time.

Functional group. (3.2) An atom or small group of atoms in a molecule that gives the molecule characteristic chemical properties.

Fusion. (21.1) A nuclear reaction in which two nuclei combine (fuse) into one more massive nucleus.

Galvanic cell. (18.3) An electrochemical cell operating spontaneously to produce a voltage from a chemical reaction.

Gamma ray (γ). (21.2) A high-energy photon emitted in the course of nuclear decay.

Gas. (1.4) A fluid in which intermolecular forces are so small that the substance expands, contracts, and flows freely to take on the size and shape of its container.

Gas constant (R). (5.1) The fundamental constant linking temperature with molar energy units. $R = 8.314$ J/mol K $= 0.08206$ L atm/mol K.

Geometric isomer. (19.2) One of two or more isomers that have identical chemical bonds distributed differently in space.

Glass. (10.4) An amorphous solid that has cooled to a solid state without crystallizing.

Globular protein. (11.7) A protein with a compact, roughly spherical tertiary structure.

Gravitational force. (2.2) The attraction between objects caused by their mass.

Greenhouse effect. (6.7) The heating of the Earth's surface caused by gases in the atmosphere that absorb outgoing infrared radiation.

Ground state. (6.2) An orbital description of a species in which the electrons are in the most stable description possible.

Group. (1.3) A column of the periodic table.

Haber process. (4.0) The industrial process for preparing ammonia from molecular nitrogen and hydrogen.

Half-life ($t_{1/2}$). (14.4) The time required to consume half the initial concentration of a reactant, $t_{1/2} = \dfrac{\ln 2}{k}$.

Half-reaction. (4.7) A reaction that isolates a reduction or oxidation and explicitly shows the electrons involved.

Hall–Héroult process. (20.3) The industrial process for producing aluminum electrolytically.

Halogen. (1.3) Any of the elements in Group 17 of the periodic table.

Hard acid. (20.2) A Lewis acid whose acceptor atom has a low polarizability.

Hard base. (20.2) A Lewis base whose donor electron pairs are tightly bound, resulting in low polarizability.

Heat (q). (12.2) A transfer of thermal energy between a system and its surroundings.

Heat capacity. (12.2) The amount of energy required to raise the temperature of a given quantity of a substance by 1 K (or 1 °C).

Heat of dilution. (12.5) The enthalpy change accompanying a dilution process.

Heat of solution. (12.5) The enthalpy change accompanying the dissolving of a solute.

Heat of vaporization. (12.5) The enthalpy change accompanying a vaporization process.

Henry's law. (15.2) The statement that the concentration of a gas in solution is directly proportional to the partial pressure of that gas above the solution.

Henry's law constant. (15.2) The proportionality constant relating the solubility of a gas to its partial pressure.

Hess' law. (12.4) The statement that the enthalpy change for any process is equal to the sum of the enthalpy changes for any set of steps that leads from the initial to final conditions.

Heterogeneous. (1.4) Containing more than one physically distinct component and therefore not uniform in composition.

Hexagonal close packing. (10.4) The close packing arrangement in which hexagonal layers are stacked in an ABAB . . . pattern.

High-spin complex. (19.3) A complex in which the maximum number of electrons is unpaired.

Homogeneous. (1.4) Uniform in composition.

Hund's rule. (7.3) The observation that the most stable arrangement of electrons among degenerate orbitals is the one that maximizes the number of unpaired electron spins.

Hybrid orbital. (8.4) An atomic orbital obtained by combining two or more valence orbitals on the same atom.

Hybridization. (8.4) The formation of a set of hybrid orbitals with favorable directional characteristics by mixing together two or more valence orbitals of the same atom.

Hydrate. (3.3) A solid compound or complex ion that has water molecules incorporated into its structure.

Hydrocarbon. (8.4) A compound that contains only hydrogen and carbon.

Hydrogen bond. (10.2) A moderately strong intermolecular attraction caused by the partial sharing of electrons between a highly electronegative atom of F, O, or N and the polar hydrogen atom in a F—H, O—H, or N—H bond.

Hydrophilic. (10.7) Compatible with water.

Hydrophobic. (10.7) Incompatible with water.

Hydroxyl group. (11.1) The —OH group.

Hypothesis. (1.1) An explanatory proposition that is not yet established.

Ideal gas. (5.2) A gas in which molecular volumes and intermolecular forces both are negligible.

Ideal gas equation. (5.2) The equation describing the behavior of an ideal gas, $PV = nRT$.

Indicator. (17.3) A substance that can be used to identify the stoichiometric point of an acid-base titration because its color is sensitive to pH changes.

Induced fit. (14.7) The process by which an enzyme and the reactant binding to it accommodate their shapes to each other.

Induced nuclear reaction. (21.3) A reaction that occurs as a result of a nuclear projectile colliding with an atomic nucleus.

Ineffective collision. (14.1) A molecular collision that does not lead to chemical reaction.

Initial concentration. (15.5) The concentration of a reagent immediately after a system is prepared but before any chemical reaction takes place.

Initiation. (11.2) A chemical reaction that begins the process of polymerization.

Initiator. (11.2) A substance that starts a polymerization chain reaction.

Inner atom. (8.3) Any atom in a molecule that is bonded to more than one other atom.

Insoluble. (17.4) Unable to dissolve in a liquid. Salts with $K_{sp} \ll 1$ are classified as insoluble in water.

Integrated rate law. (14.4) A rate statement giving the concentration as a function of time.

Intensity. (6.2) The brightness (number of photons) of light.

Intensive property. (1.6) A property whose magnitude is independent of the amount of the substance present.

Intermediate. (14.1) A species that is not part of the overall stoichiometry but is produced in an early step of a reaction mechanism and consumed in a later step.

Intermolecular forces. (10.1) Forces that exist between molecules.

International System of Units (SI). (1.5) The set of metric units that have been chosen by international agreement for expressing scientific quantities.

Intramolecular forces. (10.1) Forces that exist within a molecule.

Ion. (2.4) A charged species resulting from the gain or loss of electrons from a neutral atom or molecule.

Ion exchange. (10.8) A process for exchanging one ionic species in solution for another by passing the solution through a column containing a resin charged with ions.

Ionic compound. (2.4) A neutral compound composed of cations and anions.

Ionic solid. (10.3) A solid containing cations and anions that are attracted to each other by coulombic interactions rather than covalent bonds.

Ionization energy. (7.1) The energy required to remove an electron from an isolated species.

Isoelectronic. (7.3) Having the same number of electrons.

Isolated system. (12.1) A system that can exchange neither energy nor matter with its surroundings.

Isolation method. (14.4) An experimental technique in chemical kinetics that sets the initial concentration of one reactant substantially lower than those of all others.

Isomers. (3.1) Molecules that have the same chemical formulas but different molecular structures.

Isotonic. (10.6) Having identical total molarity of solutes.

Isotopes. (2.3) Atoms of the same element whose nuclei contain different numbers of neutrons.

Joule (J). (2.5) The SI energy unit, defined to be $1 \text{ J} = 1 \text{ kg m}^2 \text{ s}^{-2}$.

Kelvin (K). (1.5) The SI temperature unit, based on absolute zero with a unit size equal to 1/100 the difference between the normal freezing point and normal boiling point of water.

Ketone. (11.1) An organic compound that contains the carbonyl group (C=O) linked to two carbon atoms.

Kilogram (kg). (1.5) The SI base unit for mass.

Kinetic energy. (2.5) The energy of motion of an object, $E_{kinetic} = \frac{1}{2}mu^2$.

Kinetics. (14.2) The study of chemical reaction rates.

Lanthanide element. (1.3) Any of the elements in the $4f$ block of the periodic table, between $Z = 57$ and $Z = 70$.

Lattice. (7.5) An organization of ions in a regular, alternating array.

Lattice energy. (7.5) The energy released when individual gaseous cations and anions condense to form a regular three-dimensional array.

Law of conservation of energy. (2.5) The statement that energy cannot be created or destroyed.

Law of conservation of mass. (2.1) The statement that mass cannot be created or destroyed.

Lead battery. (18.6) A voltaic cell, widely used for automobile storage batteries, in which both half-reactions involve reactions of lead.

Le Châtelier's principle. (15.4) The observation that when a change of conditions is applied to a system at equilibrium, the system responds in the manner that reduces the amount of change.

Length. (1.5) The distance that something extends.

Lewis acid. (20.1) An electron-pair acceptor.

Lewis base. (20.1) An electron-pair donor.

Lewis structure. (8.3) A representation of covalent bonding that uses symbols for the elements, dots for nonbonding valence electrons, and lines for pairs of bonding valence electrons.

Ligand. (19.2) A molecule or anion that acts as a Lewis base, forming a bond to a metal atom or cation.

Light. (6.2) Electromagnetic radiation in the visible portion of the spectrum, between $\lambda = 400$ nm and $\lambda = 700$ nm.

Limiting reactant. (4.4) The reactant whose amount falls to zero in a chemical reaction.

Linear. (8.6) Lying along a straight line.

Linear accelerator. (21.3) An apparatus that accelerates atomic nuclei to high energy along a linear trajectory.

Line structure. (3.1) A compact representation of a carbon-containing molecule that shows the molecular structure in a simplified fashion.

Liquid. (1.4) The phase of matter in which intermolecular forces are large enough that molecules cannot escape into space but small enough that molecules move freely past one another. A liquid has a distinct volume but no distinct shape.

Liter (L). (1.5) A unit of volume equal to 1000 cm^3 (10^{-3} m^3).

Lone pair (nonbonding pair). (8.3) A pair of valence electrons that is localized on an atom rather than involved in bonding.

Low-spin complex. (19.3) A complex in which the maximum number of electrons is paired.

Macromolecule. (11.1) A molecule that contains a large number of atoms (typically, more than several hundred).

Macroscopic. (1.2) Visible to the naked eye.

Magnet. (2.2) An object that attracts or repels a moving electrical charge but carries no net electric charge.

Magnetic Force. (2.2) The attraction or repulsion between objects caused by their magnetism.

Magnetic quantum number (m_l). (6.5) The quantum number, restricted to integers between $+l$ and $-l$, that indexes the orientation in space of an atomic orbital.

Magnitude. (1.5) The size of a number.

Main group element. (1.3) Any of the elements in the p block of the periodic table.

Main group metal. (20.3) Any element displaying metallic properties and located in the p block of the periodic table.

Major species. (15.6) The species (cations, anions, molecules) present in relatively high concentration in aqueous solution.

Malleable. (1.3) Able to be formed into various shapes, including thin sheets.

Manometer. (5.3) A tube, containing a liquid and open at both ends, used to measure differences in pressure.

Mass. (1.5) The quantity of matter in a substance.

Mass number (A). (2.3) The total number of protons and neutrons in a nucleus.

Mass percent composition. (3.6) The makeup of a chemical substance expressed in parts per hundred by mass of each component.

Mass spectrometer. (2.3) An instrument that separates the atomic/molecular components of a sample according to their masses.

Matter. (1.4) Anything that occupies space and possesses mass.

Mechanism. (14.1) The detailed molecular processes by which a chemical reaction proceeds.

Meniscus. (10.5) The curved surface of a liquid contained in a narrow tube.

Mesosphere. (6.7) The region of the Earth's atmosphere between 50 km and 85 km above its surface.

Metabolism. (13.6) The process of breaking down organic molecules in biochemical cells.

Metal. (1.3) An element that is lustrous, conducts heat and electricity well, and tends to lose electrons to form cations.

Metalloid. (1.3) An element with properties intermediate between those of metals and nonmetals.

Metallurgy. (19.4) The science of extracting, purifying, and forming useful objects from metals.

Metathesis reaction. (20.2) A chemical reaction in which there is an exchange of bonding partners between two atoms or ions.

Meter (m). (1.5) The SI unit of length.

Micelle. (10.7) A cluster of molecules in aqueous solution organized with their hydrophobic ends pointing inward and their hydrophilic ends pointing outward.

Microscopic. (1.2) Requiring magnification to be seen.

Midpoint. (17.3) The point in a titration where the amount of added titrant is exactly half the amount required for complete reaction.

Minor species. (16.1) The species (ions, molecules) present in relatively low concentration in aqueous solution.

Miscible. (10.5) Soluble in all proportions.

Mixture. (1.4) A material containing two or more substances.

Moderator. (21.4) A substance used to slow down neutrons in a nuclear reactor.

Molar heat of solution. (12.5) The net energy change that occurs as a substance dissolves.

Molar mass (*MM*). (3.4) The mass in grams of 1 mol of atoms or molecules.

Molarity (*M*). (3.7) The amount of moles of solute dissolved in 1 L of solution.

Mole. (3.4) The amount of a substance that contains the same number of units as the number of atoms in exactly 12 g of carbon-12, 6.022×10^{23}.

Mole fraction. (5.5) The ratio of the number of moles of one component to the total number of moles of all components.

Molecular density. (5.1) The number of molecules per unit volume.

Molecular geometry. (8.4) The overall shape of a molecule, arising from the relative positions of its atomic nuclei.

Molecular view. (1.2) How matter looks when viewed at the level of molecules.

Molecule. (1.2) A group of atoms linked together by chemical bonds.

Monodentate. (19.2) A ligand with one donor atom.

Monolayer. (10.7) A single two-dimensional layer of atoms or molecules.

Monomer. (11.1) A molecule from which a polymer is synthesized.

Near-degenerate orbitals. (7.3) Those orbitals in a species that have nearly identical energies.

Nernst equation. (18.5) The equation that relates the potential of an electrochemical cell to standard potentials and concentrations, $E = E^\circ - \dfrac{RT}{nF} \ln Q$.

Net ionic equation. (4.5) A chemical equation showing the actual participants in a reaction of ions in solution.

Network solid. (10.3) A solid made up of atoms held together in a crystalline array by covalent bonds.

Neutral atom. (2.3) An atom containing the same number of electrons as it has protons in its nucleus.

Neutralization reaction. (4.6) Proton transfer between an acid and a base to generate a pair of neutral molecules.

Neutron. (2.2) A fundamental nuclear particle that possesses mass but zero electric charge.

Newton. (5.3) The SI unit for force.

Nitrogen fixation. (13.5) The conversion of molecular nitrogen into nitrogen-containing species such as ammonia or nitrate.

Noble gas. (1.3) Any of the elements of Group 18 of the periodic table.

Node. (6.6) A point, line, or surface where the electron density of an orbital is zero.

Nomenclature. (3.2) A systematic procedure for naming chemical compounds.

Nonbonding electron. (8.1) A valence electron that does not participate in bond formation.

Nonmetal. (1.3) An element that lacks the properties of metals, in particular the tendency to form cations.

Normal boiling point. (10.1) The boiling point of a substance under one atmosphere pressure.

n-type semiconductor. (9.6) A metalloid that contains a dopant that gives it excess valence electrons.

Nuclear binding energy. (21.1) The amount of energy per nucleon that binds an atomic nucleus together.

Nuclear decay. (21.2) The spontaneous decomposition of an unstable nucleus.

Nucleic acid. (11.6) A biochemical macromolecule containing nucleotide units.

Nucleon. (21.1) One of the protons or neutrons in a nucleus.

Nucleotide. (11.6) The repeating unit in DNA and RNA, containing a base, a five-carbon sugar, and a phosphate group.

Nucleus. (2.2) The central core of an atom, where nearly all the mass is concentrated.

Nuclide. (21.1) One particular nucleus, characterized by its charge number (Z) and mass number (A).

Octahedral. (8.6) Having the shape of a regular octahedron, with eight triangular faces and six vertices.

Open system. (12.1) A system that can exchange both mass and energy with its surroundings.

Orbital. (6.5) A three-dimensional wave describing a bound electron.

Orbital density picture. (6.6) A two-dimensional dot drawing representing the distribution of electron density in an orbital.

Orbital overlap. (8.1) The extent to which two orbitals on different atoms interact.

Order. (13.0) The quality of being regularly arranged.

Order of reaction. (14.3) The exponent to which a concentration is raised in a rate law.

Organic chemistry. (3.2) The chemistry of carbon and its compounds.

Osmosis. (10.6) The net movement of solvent molecules through a semipermeable membrane.

Osmotic pressure (Π). (10.6) The pressure difference that must be applied to a solution to prevent osmosis from pure solvent, $\Pi = MRT$.

Outer atom. (8.3) Any atom in a molecule that bonds to only one other atom.

Overall order. (14.3) The sum of the reaction orders of all species in a rate law.

Oxidation. (4.7) Loss of electrons by a chemical species.

Oxidation number. (18.1) The charge that an atom would have if each of its bonding electrons were assigned to the more electronegative atom involved in the bond.

Oxidation–reduction reactions. (4.7) A class of chemical reactions that involves transfers of electrons between chemical species.

Oxidizing agent. (4.7) A chemical species that can gain electrons from another substance.

Oxyacid. (16.4) An acid containing an inner atom bonded to OH groups and O atoms.

Oxyanion. (3.3) An anion of general formula XO_m, containing a central atom bonded to two or more oxygen atoms.

Ozone layer. (6.7) The region of the Earth's atmosphere that contains ozone (O_3) and absorbs potentially lethal ultraviolet radiation.

Pairing energy. (19.3) The electron-electron repulsion energy arising from placement of two electrons in the same orbital.

Paramagnetic. (7.3) Attracted by a magnetic field, as a consequence of having unpaired electron spins.

Partial pressure. (5.5) The pressure exerted by one component of a gaseous mixture.

Parts per billion. (5.5) How many of one particular component are present in one billion objects.

Parts per million. (5.5) How many of one particular component are present in one million objects.

Pascal (Pa). (5.3) The SI unit of pressure, one newton per square meter.

Passive electrode. (18.3) An electrode that serves only to conduct electrons between a wire and a solution; its chemical constituents do not take part in the redox reaction that occurs at its surface.

Path function. (12.1) A quantity whose change depends on the path along which a change takes place.

Pauli exclusion principle. (7.2) The requirement that no two electrons in a chemical species can be described by the same wave function.

Peptide. (11.7) A small polymer of amino acids.

Percent yield. (4.3) The actual yield of a reaction divided by its theoretical yield and multiplied by 100%.

Periodic table. (1.3) The table of the chemical elements arranged in rows of increasing atomic number and columns of similar chemical behavior.

pH. (16.2) The negative logarithm of the hydronium ion concentration.

pH meter. (18.5) A device for measuring solution pH.

Phase. (1.4) Any of the three states of matter: gas, liquid, or solid.

Phase change. (1.4) Transformation from one state of matter to another.

Phase diagram. (13.5) A pressure-temperature graph showing the conditions under which a substance exists as solid, liquid, and gas.

Phospholipid. (10.7) A biochemical surfactant molecule that is one component of cell membranes.

Photoelectric effect. (6.2) The ejection of electrons from a metal surface by light.

Photon. (6.2) A particle of light, characterized by energy $E = h\nu$.

Pi (π) bond. (9.1) A chemical bond formed by side-by-side orbital overlap so that electron density is concentrated above and below the bond axis.

Physical transformation. (1.4) A change in properties that is not accompanied by chemical rearrangements.

Physical property. (1.5) A property that can be observed without causing any chemical rearrangements.

Pipet. (3.7) A volumetric device designed to deliver a measured quantity of a liquid.

Planck's constant (h). (6.2) The physical constant, 6.63×10^{-34} J s, that relates the energy of a photon to its frequency.

Plastic. (11.4) A polymer that exists as blocks or sheets.

Plasticizer. (11.4) A substance added to a plastic to make it more flexible.

Pleated sheet. (11.7) The protein secondary structure formed like a corrugated plane.

Polar bond. (8.2) A bond that possesses an asymmetric distribution of electrons.

Polarizability. (10.2) The ease with which the electron density about an atom or molecule can be distorted.

Polyamide. (11.3) A polymer containing the amide linkage group.

Polyatomic. (3.2) Containing many bonded atoms.

Polyester. (11.3) A polymer containing the ester linkage group.

Polymer. (11.1) A molecule that contains a large number of identical individual units (monomers) linked together.

Polypeptide. (11.7) A protein molecule.

Polyprotic acid. (16.4) A molecule that contains more than one acidic hydrogen atom.

Positron. (21.2) A subatomic particle with the mass of an electron but a positive unit charge.

Positron emission. (21.2) Nuclear decay by emission of a positron, which decreases the atomic number by one unit without changing the mass number.

Potential energy. (2.5) The energy an object has by virtue of some force (for example, gravitational or electrical) acting on it.

Precipitate. (3.7) An insoluble solid that separates from a solution.

Precipitation. (3.7) Formation of an insoluble solid that separates from a solution.

Precision. (1.5) How reproducible an experimental quantity is.

Pressure. (5.3) Force per unit area.

Primary structure. (11.7) The sequence of amino acids making up a protein.

Principal quantum number (n). (6.5) The quantum number, restricted to positive integers, that indexes the energy and size of an atomic orbital.

Product. (4.1) A substance that is formed in a chemical reaction.

Propagation. (11.2) The continuation of a polymerization chain reaction.

Protein. (11.7) A biochemical polymer composed of amino acids.

Proton. (2.2) The subatomic constituent of nuclei that possesses unit positive charge, 1.602×10^{-19} C, and a mass of 1.673×10^{-27} kg.

Proton transfer. (4.6) The transfer of H^+ from one chemical species to another.

p-type semiconductor. (9.6) A metalloid containing a dopant that gives it a deficiency of valence electrons.

Quantized. (6.3) Having discrete allowed values.

Quantum number. (6.5) An integer or half-integer describing the allowed values of some quantized property.

Quantum theory. (6.5) The theory of atomic and molecular structure that includes wave behavior.

Radiant energy. (2.5) The energy possessed by electromagnetic radiation (photons).

Radiation therapy. (21.7) The use of radioactive nuclides to treat cancers.

Radioactivity. (2.2) The spontaneous breakdown of a nucleus by giving off energetic particles.

Rate. (14.2) Change per unit time.

Rate constant (k). (14.3) The constant of proportionality linking a reaction rate with concentrations of reagents.

Rate law. (14.3) An expression relating the rate of a reaction to the concentrations of reagents.

Rate-determining step. (14.1) The slowest step in a reaction mechanism.

Reactant. (4.1) A species that is consumed in a chemical reaction.

Reaction coordinate. (14.6) The course of a reaction as reactants are converted to products.

Reaction quotient (Q). (13.4) The ratio of concentrations of products to concentrations of reactants, each raised to its stoichiometric coefficient.

Redox reaction. (4.7) A reaction in which electrons are transferred between species.

Reducing agent. (4.7) A substance that loses electrons during a reaction.

Reduction. (4.7) Gain of electrons.

Rem. (21.6) The unit of measure for the effect on humans of the energetic emissions from radioactive elements.

Resonance. (8.3) The use of two or more equivalent Lewis structures to describe a substance that contains delocalized electrons.

Resonance structure. (8.3) One of two or more Lewis structures that are equivalent to one another.

Reversible reaction. (14.5) A molecular process that can proceed readily in both directions.

Reversibility. (15.2) The ability of any elementary chemical reaction to proceed in either direction.

Ribonucleic acid (RNA). (11.6) A polymer of nucleotide units that transmits genetic information in the cell.

Root-mean-square. (5.1) Average obtained by taking the square root of the mean value of the squares of the individual values.

Rounding. (1.6) The procedure for expressing a numerical result with the correct precision (number of significant figures).

Saturated. (10.5) Containing the maximum possible concentration of a solute.

Schrödinger equation. (6.5) The equation describing the behavior of electrons in atoms and molecules.

Scientific notation. (1.5) Expression of a number as a value between 1 and 10

multiplied by the appropriate power of ten.

Screening. (7.1) The reduction in effective nuclear charge caused by electrons in orbitals.

Second (s). (1.5) The SI base unit of time.

Second law of thermodynamics. (13.1) The assertion that entropy (disorder) always increases.

Second-generation star. (21.5) A star formed from the collapse of interstellar matter, including (besides hydrogen and electrons) elements with $Z \leq 26$ that are the debris from supernovae explosions of first-generation stars.

Second-order reaction. (14.3) A reaction whose rate law has an overall order equal to two.

Secondary structure. (11.7) The structural arrangement of a string of amino acids.

Seesaw shape. (8.6) The molecular shape that resembles a seesaw.

Semiconductor. (1.3) A substance that is intermediate in electrical conductivity between metals (good conductors) and nonmetals (poor conductors).

Semipermeable membrane. (10.6) A thin sheet that allows the passage of some types of molecules (typically, solvent) but prevents the passage of others (typically, solutes).

SI (Système International). (1.5) The system of units that has been adopted by scientists for general use.

Sigma (σ) bond. (9.1) A bond formed by end-on overlap of atomic orbitals, giving electron density that is concentrated along the bond axis.

Significant figure. (1.5) A digit in a numerical value that is known with certainty or has an uncertainty of one unit.

Silane. (8.5) A binary compound of silicon and hydrogen.

Silicate. (8.5) A compound containing one or more metal cations and a network of Si—O bonds.

Single bond. (3.1) A chemical bond formed by one pair of electrons shared between two atoms.

Slightly soluble. (17.4) Able to dissolve to a modest extent. Salts whose K_{sp} lies between 10^{-2} and 10^{-5} are classified as slightly soluble.

Soft acid. (20.2) A Lewis acid whose acceptor atom has a high polarizability.

Soft base. (20.2) A Lewis base whose donor electron pairs are loosely bound, resulting in high polarizability.

Solid. (1.4) The state of matter characterized by a defined volume and shape.

Solubility. (4.5) The amount of a solute that will dissolve in a given amount of solution.

Solubility product (K_{sp}). (15.6) The equilibrium constant for the solubility equilibrium of an ionic compound in water.

Soluble. (17.4) Able to dissolve in a liquid. Salts with $K_{sp} > 10^{-2}$ are classified as soluble in water.

Solute. (3.7) A substance that dissolves in a solvent to form a solution.

Solution. (1.4) A homogeneous mixture of two or more substances.

Solvent. (3.7) The component of a solution that defines its phase. Generally, the solvent is the component present in largest amount.

Space-filling model. (3.1) A representation of a molecule that shows the space occupied by its electron cloud.

Spectator ions. (15.6) Ions that are present in a solution but do not participate in a chemical reaction.

Spectrochemical series. (19.3) The listing of ligands in order of increasing energy-level splitting.

Spectrum. (6.3) A graph of the intensity of light as a function of either frequency or wavelength.

Spin orientation quantum number (m_s). (6.5) The quantum number, restricted to either $+\frac{1}{2}$ or $-\frac{1}{2}$, that indexes the orientation of electron spin.

Splitting energy. (19.3) The difference in energy between d orbitals in the central metal species of a coordination complex.

Spontaneous. (13.1) Able to occur without outside intervention.

Square planar. (8.6) Having the shape of a square.

Square pyramid. (8.6) A pyramid with a square base.

Standard conditions. (12.4) Unit concentrations (1 M for solutes, 1 atm for gases). Unless otherwise specified, standard conditions also means 298 K.

Standard enthalpy of formation (ΔH_f°). (12.4) The enthalpy change accompanying a formation reaction.

Standard hydrogen electrode (SHE). (18.4) The reference standard for standard reduction potentials, with a defined value of exactly 0 V. The electrode is a platinum wire immersed in an acid solution that is 1.00 M in hydronium ion, over which hydrogen gas passes at a pressure of 1.00 atm.

Standard reduction potential (E°). (18.4) The electrode potential for reduction under standard conditions.

Standard solution. (4.6) A solution whose concentration is accurately known.

Standard state. (12.4) The most stable phase of a substance under standard conditions.

Standardization. (4.6) Accurate determination of the concentration of a solution.

Starch. (11.5) The carbohydrate that plants use to store chemical energy.

State function. (12.1) A property that depends only on the present state but not on the previous history of the system.

State of a system. (12.1) A complete description of a system.

Steric number. (8.5) The sum of the coordination number and lone pairs for an inner atom.

Stoichiometric coefficient. (4.1) An integer giving the relative number of molecules of a species that react in a chemical reaction.

Stoichiometric point. (4.6) The point in a titration at which the amount of added titrant is exactly enough to react completely with the species being titrated.

Stoichiometry. (4.2) The amount relationships among chemical substances undergoing reactions.

Stratosphere. (6.7) The region of the Earth's atmosphere between 10 km and 50 km above its surface.

Strong acid. (4.6) An acid that generates virtually stoichiometric amounts of hydronium ions in water.

Strong base. (4.6) A base that generates virtually stoichiometric amounts of hydroxide ions in water.

Structural formula. (3.1) A molecular formula that shows how atoms are bonded together.

Structural isomers. (8.4) Compounds that have identical chemical formulas but different molecular structures.

Sublimation. (12.5) The phase change between solid and vapor.

Supernova. (21.5) An exploding star, which produces unusually bright emission.

Surface tension. (10.5) The resistance of a liquid to an increase in its surface area.

Surfactant. (10.7) A molecule containing hydrophilic and hydrophobic parts that is used to modify the behavior of aqueous solutions.

Surroundings. (12.1) All of the universe outside of a system.

System. (12.1) Any specific, well-defined part of the universe that is of interest.

Temperature. (1.5) The property of an object that measures the amount of random energy of motion of its molecules and determines the direction of spontaneous heat flow.

Termination. (11.2) The completion of a polymerization chain reaction.

Termolecular reaction. (14.1) An elementary reaction in which three molecular species collide and react.

Tertiary structure. (11.7) The overall shape of a protein molecule.

Tetrahedron. (8.4) Pyramid with four identical faces that are equilateral triangles.

Theoretical yield. (4.3) The amount of a product that would be formed if a reaction proceeded to completion without any losses.

Theory. (1.1) A unifying principle that explains a collection of facts.

Thermal energy. (2.5) Energy associated with the random motion of atoms and molecules.

Thermal pollution. (13.5) Heating of the environment as a byproduct of industrial operations.

Thermodynamics. (12.1) The scientific study of the relationships among heat and other forms of energy.

Thermosphere. (6.7) The region of the Earth's atmosphere more than 90 km above its surface.

Thiol. (11.1) An organic compound that contains the —SH group.

Third law of thermodynamics. (13.3) The statement that the entropy of any pure, perfect crystalline substance is zero at 0 K.

Third-generation star. (21.5) A star formed from the collapse of interstellar matter, including (besides hydrogen and electrons) elements with all Z values that are the debris from supernovae explosions of second-generation stars.

Titrant. (4.6) The liquid solution added during a titration.

Titration. (4.6) The gradual addition of measured amounts of one solution to another until a chemical reaction between them is complete.

Transition metal. (1.3) Any of the elements in the d block of the periodic table.

Trigonal bipyramid. (8.6) A double pyramid with a triangular base and two apices along a linear axis perpendicular to the base.

Trigonal planar. (8.6) The molecular shape in which a central atom is bonded to three other atoms lying in a plane at 120° angles to one another.

Trigonal plane. (8.6) The arrangement of three atoms bonded to a central atom in which all the atoms lie in a plane.

Trigonal pyramid. (8.5) A pyramid with a base that is an equilateral triangle and sides that are isosceles triangles.

Triple bond. (3.1) A bond between two atoms consisting of three pairs of bonding electrons.

Triple point. (13.5) The temperature and pressure at which solid, liquid, and vapor can coexist at equilibrium.

Troposphere. (5.7) The region of the Earth's atmosphere between its surface and an altitude of 10 km.

T-shaped. (8.6) The molecular shape that resembles the letter T.

Uncertainty principle. (6.4) The assertion that position and momentum cannot both be exactly known.

Unimolecular reaction. (14.1) An elementary reaction in which there is only one reactant molecule.

Unit. (1.5) A standard reference value for a quantity.

Unit cell. (10.4) The simplest repeating unit of a regular pattern, such as an atomic or molecular crystal.

Valence electrons. (7.2) The electrons of an atom that occupy orbitals of highest principal quantum number and incompletely filled orbitals.

Valence shell electron pair repulsion (VSEPR). (8.4) The principle of minimizing electron-electron repulsion by placing electron pairs as far apart as possible.

Vapor pressure. (5.7) The partial pressure of a vapor at equilibrium with a condensed phase.

Vesicle. (10.7) An enclosed bilayer made up of surfactant molecules.

Viscosity. (10.5) The resistance to flow of a fluid.

Visible light. (6.2) Photons in the wavelength range between 400 and 700 nm, to which the human eye is sensitive.

Volt (V). (18.4) The SI unit for electrical potential.

Volume. (1.5) Three-dimensional space occupied by something.

Volumetric flask. (3.7) A vessel calibrated to hold a specified volume of liquid.

Water equilibrium constant (K_w). (16.1) The equilibrium constant for proton transfer between two water molecules, $K_w = [H_3O^+]_{eq} [OH^-]_{eq}$.

Wave. (6.2) A periodic variation that can be described by amplitude, wavelength, and frequency.

Wavelength. (6.2) The distance between successive crests in a wave.

Weak acid. (4.6) An acid that undergoes incomplete proton-transfer in water.

Weak base. (4.6) A base that undergoes incomplete proton-transfer in water.

Work (w). (12.2) Energy transfer that is described by the product of a force times a displacement, $w = fd$.

Photo Credits

Chapter 1

Pages 1 & 4 *(top and bottom left):* Courtesy NASA. Page 4 *(bottom right):* Courtesy Dudley Foster, Woods Hole Oceanographic Institution. Page 6 *(left):* Arthur S. Aubry/PhotoDisc. Page 6 *(center):* Courtesy The Hendrix Group, Inc., Houston, Texas. Page 8 *(top inset):* Michael Dalton/Fundamental Photographs. Page 8 *(left):* Charles Lenars/Corbis Images. Page 10 *(left and center):* Courtesy IBM Almaden Research Center. Page 10 *(right and center):* Courtesy Digital Instruments. Page 10 *(bottom):* Courtesy IBM Almaden Research Center. Page 12 *(bottom):* Stephen Frisch. Page 13 *(top left):* Andy Washnik. Page 13 *(top right):* Richard Megna/Fundamental Photographs. Page 13 *(center):* Ken Karp. Page 14 *(left):* Peter Lamberti/Stone. Page 14 *(center):* Don Mason/Corbis Stock Market. Page 14 *(right):* Manfred Kage/Peter Arnold, Inc. Page 15: Layne Kennedy/Corbis Images. Page 16: Pat O'Hara/Corbis Images. Page 18 *(top):* Michael Watson. Page 18 *(bottom left):* Spencer Grant/PhotoEdit. Page 18 *(bottom right):* Yoav Levy/Phototake. Page 21 *(top):* Don Mason/Corbis Stock Market. Page 21 *(bottom):* Tom McHugh/Corbis Stock Market. Page 25 *(center):* Andy Washnik. Page 25 *(bottom):* Steve Taylor/Stone. Page 27: Courtesy Casio. Page 30: Courtesy International Silver Plating, Inc. Page 34 *(left):* OPC, Inc. Page 34 *(center):* Peter Lerman. Pages 34 *(right)* and 35: Stephen Frisch.

Chapter 2

Page 37: Courtesy International Business Machines Corporation. Page 37 *(inset):* IBM/Phototake. Page 39 *(top):* Richard Megna/Fundamental Photographs. Page 39 *(bottom):* Archive Photos. Pages 40 & 41: Patrick Watson. Page 43 *(top right):* Courtesy IBM Almaden Research Center. Page 43 *(bottom left):* From Jie Han, Al Globus, Richard Jaffe and Glen Deardorff, Numerical Aerospace Simulation (NAS) Systems Division at NASA Ames Research Center, "Molecular Dynamics Simulation of Carbon Nanotube Based Gears," *Nanotechnology*, vol. 8, #3, Sept. 3, 1997, pages 95–102. Pages 45 & 47: Patrick Watson. Page 46: Jump Run Productions/The Image Bank. Page 52: M. Freeman/PhotoLink/PhotoDisc. Page 52 *(inset):* Courtesy Sachtleben Chemie GmbH. Page 58 *(top left):* Michael Watson. Page 58 *(bottom left):* Ken Karp. Page 58 *(bottom and center):* A. Fenn/ Time Life Books. Page 58 *(bottom right):* Charles Falco/Photo Researchers. Page 59: Ken Karp. Page 60: Ross Harrison Koty/Stone. Page 65 *(top):* Peter Lerman. Page 65 *(bottom):* John Olmsted. Page 67: Michael Scott/Stone. Page 68: Richard Megna/Fundamental Photographs.

Chapter 3

Page 70: Dr. Jeremy Burgess/Science Photo Library/Photo Researchers. Page 71: Andy Washnik. Page 79: Lee Snyder/The Image Works. Page 88: John Olmsted. Page 89: L.S. Stepanowicz/Visuals Unlimited. Page 92: Andy Washnik. Page 93: Courtesy Edgar Fahs Smith Collection. Page 100: Courtesy Michael P. Doukas, USGS. Page 104: Courtesy Key Technology, Inc. Page 108: Richard Megna/Fundamental Photographs. Page 115: Stephen Frisch. Page 118: John Olmsted. Pages 119 & 120: Tony Freeman/PhotoEdit. Page 121: Richard Megna/Fundamental Photographs.

Chapter 4

Page 133: Corbis Images. Page 134 *(top):* Robert Frerck/Stone. Page 134 *(center):* Santokh Kochar/PhotoDisc. Page 143: Courtesy Cargill, Inc. Page 145: Richard Megna/Fundamental Photographs. Pages 154 & 155: Stephen Frisch. Page 156: Michael Dalton/Fundamental Photographs. Page 158: Patrick Watson. Page 159: Courtesy Eastman Kodak Company. Page 161 *(top):* CNRI/Science Photo Library/Photo Researchers. Page 161 *(bottom):* Courtesy Newport Corporation. Page 168: Stephen Frisch. Page 171: Richard Megna/Fundamental Photographs. Page 173 *(top left):* Royal Ontario Museum/Corbis Images. Page 173 *(bottom left):* Ken Whitmore/Stone. Page 174: Patrick Watson. Page 177: Stephen Frisch. Page 178: Kevin R. Morris/Corbis Images. Page 181: John Olmsted. Page 183 *(center):* Patrick Watson. Page 183 *(bottom):* Richard Megna/Fundamental Photographs.

Chapter 5

Page 189: Ron Watts/Corbis Images. Page 190 *(top):* Danny Lehman/Corbis Images. Page 190 *(bottom):* Ted Spiegel/Corbis Images. Page 196: Patrick Watson. Page 197: Felicia Martinez/PhotoEdit. Page 202: Courtesy Thermionics Laboratory, Inc., San Leandro, California. Page 203: Courtesy Stanford Linear Accelerator Center, US Department of Energy. Page 212: Joseph Nettis/Photo Researchers. Page 215: Jonathan Blair/

Corbis Images. Page 216: U. Wallin/The Image Bank. Page 219: Courtesy Armfield Limited. Page 221: Richard Megna/Fundamental Photographs. Page 223: Courtesy NASA. Page 225 (top): Hans Pfletschinger/Peter Arnold, Inc. Page 225 (bottom): Corbis Images. Page 227 (top): Jim Zuckerman/Corbis Images. Page 227 (bottom): Bernard Edmaier/Photo Researchers. Page 229: Wolfgang Kaehler/Corbis Images.

Chapter 6

Page 237: Courtesy Patrick Murphy, Pangolin Laser Systems, Inc. Page 237 (inset): Courtesy Electron Physics Group, National Institute of Standards and Technology. Page 238 (center): Roger Ressmeyer/Corbis Images. Page 238 (bottom): Yoav Levy/Phototake. Page 239: John Olmsted. Page 240: Pal Hermansen/Stone. Page 243: Courtesy Bausch & Lomb. Page 244: Albert Einstein™ is licensed by the Hebrew University of Jerusalem. Represented by The Roger Richman Agency, Inc., www.albert-einstein.net. Photo provided courtesy of the Archives of the California Institute of Technology. Page 247: Michael Dalton/Fundamental Photographs. Page 249: Courtesy Bausch & Lomb. Page 254: Tom Tracy/Corbis Stock Market. Page 256 (center): William Sterne, Jr./Sterne Photography. Page 256 (bottom): Andy Washnik. Page 257 (top): E.R. Degginger/Color-Pic, Inc. Page 257 (center): VU-NIH/Visuals Unlimited. Page 257 (bottom): Courtesy IBM Research Division. Page 262: Andy Washnik. Page 269: Sanford/Agliolo/Corbis Stock Market. Page 271: Paul Silverman/Fundamental Photographs.

Chapter 7

Page 281: Richard Megna/Fundamental Photographs. Page 290: Science Photo Library/Photo Researchers. Page 295 (left): Courtesy Bente Lebech, Materials Research Department, Risoe National Laboratory. Page 295 (right): ©Index Stock. Page 304: Paul Silverman/Fundamental Photographs. Page 313: Michael Watson. Page 316: John Olmsted. Page 318: C. Van Der Lende/The Image Bank. Page 319: Walt Anderson/Visuals Unlimited.

Chapter 8

Page 326: Courtesy James Gimzewski, IBM Research, Zurich Research Laboratory. Page 327: Courtesy Stephen R. Wilson and Austin N. Kirschner, New York University. Page 334: Roger Ressmeyer/Corbis Images. Page 350: Courtesy Bancroft Library, University of California/LBNL Image Library. Page 355: Rich LaSalle/Stone. Page 359: Geoff Topkinson/Science Photo Library/Photo Researchers. Page 361 (left): Paul Silverman/Fundamental Photographs. Page 361 (center): Charles D. Winters/Photo Researchers. Page 361 (right): Gary Retherford/Photo Researchers. Page 361 (far right): Mark A. Schneider/Photo Researchers. Page 370: Photo provided courtesy of Richard L. Battaglia, Department of Microelectronic Engineering, Rochester Institute of Technology, Rochester, NY. Page 373: Stephen Frisch.

Chapter 9

Page 381: John Kieffer/Peter Arnold, Inc. Page 382: Newcomb & Wergin/Stone. Page 385 (left): Joseph Van Os/The Image Bank. Page 385 (right): Tom Mareschal/The Image Bank. Page 385 (right inset): Michael Newman/PhotoEdit. Page 397: Courtesy NASA. Page 405: Courtesy Bassam Z. Shakhashiri. Page 411: Richard T. Nowitz/Corbis Images. Page 414: Dave. G. Houser/Corbis Images. Page 416 (top): John Olmsted. Page 416 (center): Andy Washnik. Page 417: John Olmsted. Page 421: AP Wide World Photos.

Chapter 10

Page 428: ©SUPERSTOCK. Page 429: Andy Washnik. Page 430 (top): Yoav Levy/Phototake. Page 430 (center): John Olmsted. Page 430 (bottom): Richard Megna/Fundamental Photographs. Page 431: Lester V. Bergman/Corbis Images. Page 433: ©SUPERSTOCK. Page 444: Stephen Frisch. Page 446 (top): Richard Megna/Fundamental Photographs. Page 446 (bottom): Adam Woolfitt/Corbis Images. Page 448: Courtesy American Superconductor. Reproduced with permission. Page 449: John Olmsted. Page 450 (left): Sinclair Stammers/Science Photo Library/Photo Researchers. Page 450 (center): Geoff Tompkinson/Science Photo Library/Photo Researchers. Page 450 (top right): Mark A. Schneider/Photo Researchers. Page 450 (bottom right): Jose Manuel Sanchis Calvete/Corbis Images. Page 453: Art by M.C. Escher. © 2001 Cordon Art-Baarn-Holland. All rights reserved. Page 454 (top): Andy Washnik. Page 454 (bottom left): Richard Hutchings/PhotoEdit. Page 454 (bottom right): Richard Megna/Fundamental Photographs. Page 455: Vaughan Fleming/Science Photo Library/Photo Researchers. Page 457 (top): Alfred Pasieka/Peter Arnold, Inc. Page 457 (center): Richard Megna/Fundamental Photographs. Page 458: Lester V. Bergman/Corbis Images. Page 460: L.S. Stepanowicz/Visuals Unlimited. Page 464: Frank Pedrick/The Image Works. Page 467: Diane Hirsch/Fundamental Photographs. Page 470: David M. Phillips/Visuals Unlimited. Page 475: Dr. Patricia J. Shulz/Peter Arnold, Inc.

Chapter 11

Page 483: John Olmsted. Page 484 (top): Hans Pfletschinger/Peter Arnold, Inc. Page 484 (bottom): ©Nik Wheeler. Page 486: Tom J. Ulrich/Visuals Unlimited. Page 493: Felicia Martinez/PhotoEdit. Page 494 (left): Richard T. Nowitz/Corbis Images. Page 494 (right): Laurence Fordyce; Eye Ubiquitous/Corbis Images. Page 498 (top): Andy Washnik. Page 498 (bottom): ©Telegraph

Colour Library/FPG International. Page 499: Tom Pantages. Page 500: ©VU/SIU/Visuals Unlimited. Page 503 (left): ©SUPERSTOCK. Page 503 (right): ©Science Vu/EP-AS/Visuals Unlimited. Page 504 (top): ©SUPERSTOCK. Page 504 (bottom): M. Greenlar/The Image Works. Page 505: Leonard Lessin/Peter Arnold, Inc. Page 506 (top): Arthur Tilley/Stone. Page 506 (bottom): Richard Megna/Fundamental Photographs. Page 507: Diane Hirsch/Fundamental Photographs. Page 512: Alvin E. Staffan/Photo Researchers. Page 513: Andrew Syred/Photo Researchers. Page 514: James Bell/Photo Researchers. Page 515: Michael J. Doolittle/The Image Works. Page 519: Courtesy Oesper Collection in the History of Chemistry, University of Cincinnati. Page 520: Ken Eward/Biografix/Photo Researchers. Page 528: Patrick Watson.

Chapter 12

Page 540: Audobon Productions/The Image Bank. Page 541: Peter Turnley/Corbis Images. Page 542 (center): Chris Sorenson/Corbis Stock Market. Page 542 (bottom, left): Gary Walts/The Image Works. Page 542 (bottom, center): Courtesy Mazda North American Operations. Page 542 (bottom, right): Courtesy Saturn Corporation. Page 543: Paul Silverman/Fundamental Photographs. Page 552 (left): Danny Lehman/Corbis Images. Page 552 (right): Monika Graff/The Image Works. Page 553 (top): Jeff Greenberg/Photo Researchers. Page 553 (bottom): Picture Press/Corbis Images. Page 560: Patrick Watson. Page 561: Courtesy Parr Instrument Company. Page 562: Culver Pictures, Inc. Page 569: Stephen Frisch. Page 576: Richard Megna/Fundamental Photographs. Page 578: Stephen Frisch. Page 580: Photex/Corbis Images. Page 581 (left): Patrick Watson. Page 581 (right): Don Hammond/Corbis Images.

Chapter 13

Page 590: Robert Maass/Corbis Images. Page 590 (inset): From Jean-Marie Lehn, Chem. Eur. J., Vol. 3, issue 1 (January 1997). Reproduced with permission. Page 591 (top): M. Antman/The Image Works. Page 591 (bottom): Frozen Images/The Image Works. Page 593: Patrick Watson. Page 594: Dennis M. Gottlieb/Corbis Stock Market. Page 602 (top): Courtesy The Sherwin-Williams Company. Page 602 (bottom): John Olmsted. Page 614 (left): Stephen Frisch. Page 614 (center): Courtesy John Murdzek. Page 614 (right): Martin Bond/Science Photo Library/Photo Researchers. Page 606: Courtesy J.F. Allen, St. Andrews University. Page 610: Peter Tenzer/International Stock Photo. Page 621: John Olmsted. Page 624: Jim Sugar Photography/Corbis Images. Page 626: Pat O'Hara/Corbis Images. Page 632: Dr. R. Clark & M. Goff/Science Photo Library/Photo Researchers.

Chapter 14

Page 643: Courtesy NASA Goddard Space Fight Center. Page 645: Richard Megna/Fundamental Photographs. Page 649 (top): Kevin Fleming/Corbis Images. Page 649 (bottom): ©The Image Bank. Page 663: Stuart Cohen/The Image Works. Page 666: Layne Kennedy/Corbis Images. Page 686: Stephen Frisch. Page 690: Courtesy Walker Manufacturing. Page 691 (top): Courtesy Englehard Corporation. Page 691 (center): Telegraph Colour Library/FPG International. Page 691 (bottom): Peticolas/Megna/Fundamental Photographs. Page 694: Courtesy Thomas Steitz, Yale University.

Chapter 15

Page 706: Nigel Cattlin/Holt Studios International/Photo Researchers. Page 707 (top): Craig Aurness/Corbis Images. Page 707 (bottom): Courtesy Cargill, Inc. Page 717 (top): Richard Megna/Fundamental Photographs.

Page 717 (bottom): ©John S. Reid Photography. Page 718: Chris Sorenson/Corbis Stock Market. Page 720 (left): Jon Feingersh/Tom Stack & Associates. Page 720 (right): David Hall/The Image Works. Page 742: John Olmsted. Page 748: CNRI/Science Photo Library/Photo Researchers. Page 753: John Olmsted. Page 754: Lawrence Migdale/Science Source/Photo Researchers.

Chapter 16

Page 763: Foodpix. Page 764 (top): Karen Tweedy-Holmes/Corbis Images. Page 764 (bottom): Carlos Spaventa/FPG International. Page 770 (top): Andrew McClenaghan/Science Photo Library/Photo Researchers. Page 770 (bottom): Yoav Levy/Phototake. Page 771: John Olmsted. Page 773: Richard Megna/Fundamental Photographs. Page 774: ©1970 FP/Fundamental Photographs. Page 787: OPC, Inc. Page 796: Andy Washnik.

Chapter 17

Page 804: David Muench/Corbis Images. Page 805 (top): Luis Veiga/The Image Bank. Page 805 (top inset and bottom): Dave Bunnell. Page 814: John Olmsted. Page 818: Stephen Frisch. Page 820 (top right): Moredun Animal Health LTD/Science Photo Library/Photo Researchers. Page 820 (bottom right): Jean-Marc Favre, Agence Vandystadt/Photo Researchers. Pages 825 & 827: Patrick Watson. Page 829: Stephen Frisch. Page 830 (top): Paul Chesley/Stone. Page 830 (center): Ralph White/Corbis Images. Page 830 (bottom): Courtesy Betz Company. Page 831: Patrick Watson. Page 844: Paul Silverman/Fundamental Photographs. Page 845: Charles D. Winters/Photo Researchers.

Chapter 18

Page 853: Kevin R. Morris/Corbis Images. Page 854 (top): Roger Ressmeyer/Corbis Images. Page 854 bottom): Fritz Hoffman/The Image Works. Page 855: Richard Megna/Fundamental Photographs. Page 868 (left): Stephen Frisch. Page 868 (right): Tom Pantages. Pages 869, 871 & 874: Richard Megna/Fundamental Photographs. Page 875: Thomas Del Brase/Stone. Page 877: Andy Washnik. Page 882: Tom Pantages. Page 884: Alexander Blaikley, *Michael Faraday Lecturing in the Theatre of the Royal Institution,* c. 1856. Colored lithograph. The Royal Institution, UK/Bridgeman Art Library/NY. Page 889: Charles D. Winters/Photo Researchers. Page 891 (top): Patrick Watson. Page 891 (bottom): Martin Bond/Photo Researchers. Page 893: Tony Freeman/PhotoEdit. Page 896: Courtesy DaimlerChrysler Corporation. Page 898: Stephen Frisch. Page 901: Courtesy International Silver Plating, Inc. Page 902: Patrick Ward/Corbis Images.

Chapter 19

Page 909: Holt Studios/Nigel Cattlin/Photo Researchers. Pages 912 & 923: John Olmsted. Pages 916 & 930: Stephen Frisch. Pages 931 & 932: Patrick Watson. Page 935: Albert Copley/Visuals Unlimited. Page 937: Phil Degginger/Stone. Page 940 (top): M. Antman/The Image Works. Page 940 (center): James L. Amos/Corbis Images. Page 940 (bottom): Charles D. Winters/Photo Researchers. Page 941 (center): Stewart

Cohen/Stone. Page 941 (bottom): ©FPG International. Page 942 (top): Ernst Haas/Stone. Page 942 (center): Bill Bachmann/PhotoEdit. Page 942 (bottom): Stephen Frisch. Page 943 (top): George Holton/Photo Researchers. Page 943 (bottom): Andy Levin/Photo Researchers. Page 945 (top): Peter Cade/Stone. Page 945 (bottom): James L. Amos/Corbis Images. Page 952: ©Photo Researchers.

Chapter 20

Page 958: Alfred Pasieka/Photo Researchers. Page 970 (top right): Courtesy Alcoa. Page 970 (center): James L. Amos/Corbis Images. Page 970 (bottom): Tony Freeman/PhotoEdit. Page 971: Charles D. Winters/Photo Researchers. Page 973 (top): Gianni Dagli Orti/Corbis Images. Page 973 (bottom): David J. & Janice L. Frent Collection/Corbis Images. Page 977: Leonard Lessin/Peter Arnold, Inc. Page 979 (top): Richard Megna/Fundamental Photographs. Page 979 (center): Andy Washnik. Page 979 (bottom): Stephen Frisch. Page 981: Courtesy BioWorld USA.com, Visalia, California. Photo by Dale P. Barnes. Page 985 (top): Edy Pumono/The Image Works. Page 985 (bottom): Gary Retherford/Photo Researchers. Page 986: Paul Silverman/Fundamental Photographs. Page 989: Charles D. Winters/Photo Researchers. Page 990: Courtesy Regal Ware, Inc. Page

991: Thomas D. Mangelsen/Peter Arnold, Inc.

Chapter 21

Page 999: Courtesy Astronomical Society of the Pacific. Page 999 (inset): Science Photo Library/Photo Researchers. Page 1000: Dept. of Nuclear Medicine, Charing Cross Hospital/Photo Researchers. Page 1005: Courtesy NASA. Page 1021: Courtesy Fermilab. Page 1025: Courtesy U.S. Department of Defense. Page 1028 (left): Agence France Presse/Corbis Images. Page 1028 (center): George Lepp/Corbis Images. Page 1028 (right): AP/Wide World Photos. Page 1030: Courtesy Hirsohi Hidaka, Hiroshima University. Page 1032: Courtesy Princeton University Plasma Physics Lab. Page 1033: Courtesy NASA. Page 1034 (top): Courtesy David Malin, Anglo Australian Telescope Board. Page 1034 (bottom): Courtesy NASA and Space Telescope Science Institute. Page 1039: Courtesy Max-Plack-Institut für Physik. Page 1041: Courtesy NASA. Page 1043: Agence France Presse/Corbis Images. Page 1045 (top left): Simon Fraser/Photo Researchers. Page 1045 (top right): Dept. of Nuclear Medicine, Charing Cross Hospital/Photo Researchers. Page 1045 (bottom left): Hank Morgan/Photo Researchers. Page 1045 (bottom right): NIH/Photo Researchers. Page 1046: Courtesy MDS Nordion.

Index

Note: The following codes are used after page numbers in this index: A b denotes a Box; f denotes a figure; n denotes a marginal note; t denotes a table; tf denotes a table footnote. An A preceding a page number denotes an Appendix page.

Electron configurations—cont'd
  in coordination complexes, 924–926, 927–928, 927f
  noble gas configuration, 298–299
  of ions, 301–302
  shorthand notation, 297–299
  ways to represent them, 297–299, 297f, 298f, 299f
Electron contour drawings, 265–266, 265f
Electron density, 264–265
  and bond polarity, 333
  and bond properties, 390
  and bond strength, 395–396
  and electronegativity, 333
  and molecular orbitals, 403, 403f
  and orbital overlap, 395–396
  and screening, 285, 287
  and symmetry, 332–333, 372
  depictions, 265–266, 265f
  in bonds, 329, 352
    effects on oxyacids, 791, 791f
  node of, 265f, 266
  of lone pairs vs. bonding orbitals, 371
  of sigma (σ) and pi (π) bonds, 383–384, 384f, 386f
Electron-electron repulsion, 239, 299–300, 328, 328f
  and bond angles, 371, 372
  and orbital stability, 311
  and screening, 285, 285f
  and shape of molecule, 352, 352f, 371
  and size of atoms vs. ions, 312
  in coordination complexes, 924, 925f
  of bonding vs. nonbonding orbitals, 371
  pairing energy, 926, 927f
Electron pairs, See Paired electrons
Electron spin resonance spectroscopy, 297n
Electron transfer, 853–908 (See also Redox reactions)
  and band gap, 419
  and electrical conduction in metals, 416, 417, 417f
  and electrical conduction in semiconductors, 418–419, 419f
  and first ionization energy, 868–869
  and photography, 944b
  between energy bands, 418–419, 419f
  between species, 171, 172f, 174f
  corrosion as, 854
  direct, 854, 869–870, 869f
  everyday examples, 854
  for purifying groundwater, 860b
  forest fire as, 854
  in iron-sulfur proteins, 910
  in photosynthesis, 382
  in redox reactions, 855–856, 856f
  indirect, 854, 870, 870f
  recognizing it, 856
Electron transfer proteins, 950–952
Electronegativity, 333–335, 333f, 334f, 335t (See also Polar entries)
  and acid strength, 791, 791f
  and acids, 777
  and bond length, 392
  and bond strength, 395
  and hydrogen bonding, 439, 439n, 441, 463
  and Lewis structures, 340, 341
  and oxidation number, 856–857
  and polarity, 333, 334–335, 335t

Electronegativity—cont'd
  and polarizability, 965
  of elements, 856f
  Pauling values, 333, 334f
  periodicity, 333–335, 334f
  vs. electron affinity vs. ionization energy, 333, 333f
Electrons, 46–48, 50, 50f, 50t, 239, 1001t (See also Charge; Ions; Orbitals)
  addition to atoms, 51, 57, 57f, 309
  and atomic number, 240
  and charged particles, 47, 55, 57, 57f
  and light, 243
  and neutron decay, 1009
  and positron emission, 1011, 1011f
  and structure of periodic table, 291
  antibonding, 404, 447
  at anodes and cathodes, 871, 874, 880
  attraction and repulsion, 240, 285, 285f
    and covalent bond formation, 328
    and screening, 285, 285f
    and shape of molecule, 326, 352, 352f
    in bonding, 328, 332–333
  attraction to nuclei, 328, 328f, 329–330, 330f
  bonding, 332–333
    vs. nonbonding, 336
    effects of antibonding electrons, 404
    in Lewis structures, 336
    in transition metals, 447
  bound
    and Pauli exclusion principle, 290
    properties, 260, 264
    stability of, 309–310
    what they are, 260
  charge on, 48, 48f, 50t, 239, 883–884
  charge per mole, 884, 889
  charge/mass ratio, 47
  conservation of, 57
    and redox reactions, 172, 174, 855
  core electrons, 294, 296, 305
    and bonding, 336
    and Lewis structures, 336
    and screening, 307
    removal, 309, 316
  delocalization, 259, 265
  delocalized, 259, 265, 382, 407n
  distance from nucleus
    and screening, 286–287, 286f, 287f
    and stability, 305
  electron-electron interactions, 328, 328f, 329–330, 330f
  electronegativity, 333–335
  energy
    and orbital size, 261
    and principal quantum number, 261–262
    binding energy, 244–245, 244f
    changes in, 250–251, 252f
    kinetic energy, 243, 243f, 244, 244f, 245, 258–259, 260
    quantization, 250, 252f, 260–261
  equal sharing, 332–333
  flow of
    and electrical circuit, 854, 871, 873, 873f
    and electrical current, 874
    and electrical potential differences, 876
    and electrochemistry, 872
    and free energy change, 883
    "downhill," 875, 875f, 877, 877f
    from anode to cathode, 877f, 880
    sustaining it, 870–871, 871f

Electrons—cont'd
  free
    and electrical conduction, 416
    properties, 258t
    speed of
      equation for, 258t
      wavelength, 258–259
        equation for, 258t
    in electrical and heat conduction, 912
  in redox half-reactions, 174, 179, 861–862, 864, 867, 879, 880f, 881
  in redox reactions, 171, 172–179, 172f, 174f, 176t
    gains and losses, 172, 172f, 174f, 177, 178, 855, 861
    keeping track of them, 174
    number transferred, 884, 889, 889n
  lone pairs (See Lone pairs of electrons)
  loss (removal, emission) from atoms, 51, 55, 57, 57f, 250, 309, 316, 318
    and ionization energy, 308
    and photoelectron spectroscopy, 288b, 288f
    energy required, 62
  magnetic properties, 255, 257, 263, 302
  mass of, 48, 50, 50t, 51
  maximum number per orbital, 291
  nonbonding electrons, 332, 336. (See also Lone pairs of electrons)
    and Lewis acids and bases, 960, 960f, 961
    in Lewis structures, 340, 341, 341f
      assigning of leftovers, 341, 341f
  octets, 340, 341, 342, 348
    and oxidation number, 856
    in ethylene, 383
  orbital distribution (See Electron configurations)
  paired electrons, 297
    and diamagnetism, 302
    and reactive functional groups, 485
    in polymerization, 490, 491
    pairing energy, 926, 927f
    spin of, 297, 302, 927
  pi (π) electrons
    and reactive functional groups, 485, 485f
    in polymerization, 490, 491
    in ring structure, 412
  properties, 255–260
  quantum number descriptions, 290–291
  quantum numbers, 261–264 (See also Quantum numbers)
  repulsions between (See Electron-electron repulsion)
  sharing, 74, 74f, 317, 318
    and bond strength, 395
    between nuclei, 328, 333, 336
    unequal, 332–335
  spatial distribution, 261
    depictions of, 265–266, 265f
  spin, 255, 257, 263, 263f
    and atom stability, 299–300
    and chemical bonds, 303
    and electron configuration, 297–298, 297f, 298f, 302–303
    and electron spin resonance spectroscopy, 297n
    and magnetic fields, 297n
    and magnetic properties, 255, 257, 302–303, 927–928
    high vs. low, 926, 927–928
    and crystal field splitting energy, 928

Index of Equations

Bold entries in parentheses are equation numbers used in the textbook.

List of Boxes